IEEE
Standard Dictionary
of
Electrical and
Electronics
Terms

Second Edition

IEEE
Standard Dictionary
of
Electrical and
Electronics
Terms

Second Edition

Frank Jay
Editor in Chief

Published by
The Institute of Electrical and Electronics Engineers, Inc.
New York, NY

Distributed in cooperation with
Wiley-Interscience, a division of John Wiley & Sons, Inc.

December 1, 1977

SH07088

Introduction

From their earliest years, both the American Institute of Electrical Engineering (AIEE) (1884) and the Institute of Radio Engineers (IRE) (1912) published standards defining technical terms. They have maintained this practice since they were combined in 1963 to become the IEEE (Institute of Electrical and Electronics Engineers).

In 1928 the AIEE organized Sectional Committee C42 on Definitions of Electrical Terms under the procedures of the American Standards Association, now the American National Standards Institute. In 1941 AIEE published its first edition of *American Standard Definitions of Electrical Terms* in a single volume. However, by the time a second edition was ready, the highly accelerated development of new terms made it impracticable to publish in a single volume, and 17 separate documents, each limited to a specific field, were published from 1956 to 1959.

Over the years, IRE published a large number of standards that either included definitions or were devoted entirely to definitions. In 1961 it published all of its then-approved definitions in an alphabetically arranged single volume.

The 1972 edition of the IEEE Standard Dictionary of Electrical and Electronics Terms included all terms and definitions that had been standardized previously by IEEE, as well as many American National Standards and IEC terms. The current edition derives the bulk of its new definitions from standards published between 1968 and 1977. Most of the definitions that have been continued from the 1972 edition have been reaffirmed because of their inherent usefulness. The committees that generated the definitions originally, still consider them to be equally appropriate at the present time.

How To Use This Dictionary

The terms defined in this dictionary are listed in alphabetical order. Terms made up of more than two words appear in the order most familiar to the people who use them. In some cases cross-references are given.

Some terms take on different meanings in different fields. When this happens the different definitions are numbered, identified as to area of origin, coded, and listed under the main entry.

If a reader wants to know the source of a definition he need only look up the code number following the definition in the SOURCES section that appears at the back of the book between pages 792 and 795.

Foreword

A definition is an attempt to approximate, precisely and economically in language, an aspect of reality perceived and understood by the senses or the intellect. Beginning with the broadest category to which the entity under consideration belongs, the process of definition—a function of human reason—narrows, refines, and distills until that entity emerges as individuated and unique, while its relationships to the broader categories from which it was derived remain clearly apparent.

In every era of world history definitions have been necessary and have been attempted within the body of knowledge available to man at the time.

Lightning, for the ancient but immensely reasonable Hellenic mythologists, was the awesome fiery weapon of Zeus. Tide rise and fall reflected the breathing of the blue-scaled Poseidon, god of the seas, the earth-shaker. The Greeks were trying to explain the realities of their world. Their definitions weren't bad, if we consider what they had to work with.

The body of human knowledge that had been accumulating on a gradually rising plateau from the beginnings of written language in the Western world, suddenly exploded upward in the Renaissance. With a new focus on the physical world and time, rather than on heaven and eternity, the necessity to identify and define the newly perceived physical realities became imperative. Printing, recently discovered, helped preserve and record these definitions which could then survive their definers and lie in archives for centuries or until scholars could synthesize them into still newer applications.

Knowledge generates knowledge. In our time with the help of electronics—memory banks, instant selective retrieval—the sum total of information available for the generation of world knowledge is proliferating faster and faster. The Renaissance man, who believed that in his own lifetime he could learn everything the world had to teach, has been replaced. Always in the forefront of an even more gigantic upsurge of new information than that of the Renaissance, the serious student of the world now—the physicist, the biologist, the engineer in every field—must specialize deeply and intensively for his entire career. His special efforts, managed and coordinated with the work of others by systems science, result in substantive, profound, and incredibly fast changes in the state-of-the-art. Every change results in a new detail of a complex new reality. Every new detail needs a name, a definition.

The scientists who are generating new concepts in the many related fields of electricity and electronics are working so fast and producing such radical and rapid change in their competitive research, that language, attempting to follow, can only stagger, always, behind.

The work of the definitions makers then, in the face of such dynamic change can never exactly reflect all the most immediate current sophistications. A dictionary of these definitions is therefore prepared with the knowledge that, by the time the book is published, although some of its terms will have been superseded, the bulk of the work will remain current for future reference.

List of Contributing Engineers

The following list is made up of the names of engineers who, over the years, have served as chairmen of committees that have generated standards. To a large extent the definitions in this compilation come from those standards.

IEEE has profound appreciation for the work that these men so willingly undertook and so carefully fulfilled.

We regret that the limitations of space do not allow us to list the names of the thousands of committee-and subcommittee members without whose painstaking efforts this dictionary would never have been possible.

Abrahams, I. C.
Andrews, F. T.
Angelo, S. J.
Angus, A. C.
Armstrong, J. H.
Arthur, M. G.
Avins, J.
Axelby, G. S.

Baker, J. M.
Baldwin, M. S.
Bangert, J. T.
Bargellini, P. L.
Barrow, B. B.
Barstow, J. M.
Bartheld, R. G.
Bauer, J. T.
Baum, J. F.
Bellack, J. H.
Bixby, W. E.
Blachman, N. M.
Blodgett, E. D.
Bloomquist, W. C.
Bobo, P. O.
Bochnak, P. M.
Boice, Jr., C. W.
Borst, D. W.
Bowers, G. H.
Brainerd, J. G.
Brereton, D. S.
Brociner, V.
Brockwell, K. C.
Broome, W. M.
Brown, R. D.
Buhl, H. A.

Cameron, A. W. W.
Caslake, S. G.
Chappell, J. F.
Chase, A. A.
Chiappetta, C. M.
Clark, R. A.
Clevenger, C. M.
Cohn, S. I.
Cook, W. H.
Copel, M.
Costrell, L.
Cottony, H. V.
Cox, V. L.
Curdts, E. B.

Dallas, J. P.
Davidoff, F.
Denkowski, W. J.
Desch, R. F.
Deschamps, G. A.
Dietrich, R. E.
Dietsch, C. G.
Doba, Jr., S.
Doble, F. C.
Donahoe, F. J.
Duncan, R. O.

Early, J. M.
Easley, G. J.
Eiselein, J. E.
Elias, P.
Eliasson, I. E.
Ellerbruch, D. A.
Espersen, G. A.

Evans, C. T.
Evendorff, S.
Ewing, J. S.

Farr, N. C.
Fields, C. V.
Fink, L.
Forster, J.
Fox, A. G.
Fredendall, G. L.
Fricke, C. A.
Frihart, H. N.

Garschick, A.
Gerber, E. A.
Giles, W. F.
Graham, J. D.
Gressitt, T. J.
Griffith, M. S.

Hackley, R. A.
Hall, J. R.
Hannan, P. W.
Hansell, C. D.
Hansen, R. C.
Hanver, G. N.
Harper, W. E.
Harvey, F. K.
Haymes, T. W.
Hedrick, D. L.
Hefele, J. R.
Hendrix, K. D.
Hilibrand, J.
Hillen, R. J.

Hirtler, R. W.
Hissey, T. W.
Holland, M. G.
Hovey, L. M.
Hubbs, J. C.
Huber, R. F.
Hvizd, A.

Jacobs, I. M.
Jasik, H.
Jesch, R. L.
Johnson, I. B.
Johnson, W. J.
Johnston, F. C.
Jones, J. L.

Kaenel, R. A.
Kaufmann, R. H.
Keezer, D. C.
Kerny, I.
Kerwein, A. E.
Klein, P. H.
Knowles, C. H.
Koen, Jr., H. R.
Kolb, Jr., F. J.
Kolcio, N.
Kotter, F. R.
Kreer, Jr., J. G.
Kurth, C. F.
Kurtz, S. K.

Lang, W. W.
Lee, R.
Lokken, G.
Louden, V. J.
Lougher, E. H.
Luehring, E. L.
Lynch, R. D.
Lynch, W. A.
Lynn, G. E.

McFarlin, V. F.
McGee, A. A.
McGrath, J. N.
McGrath, T. J.
McKean, A. L.
McKiernan, J.
McKnight, J. G.
McWilliams, D. W.
Madison, L. C.
Marieni, G. J.
Martin, T. J.
Mattingly, R. L.
Mayer, R. P.
Meindl, J. D.
Mertz, P.
Michael, D. T.
Miles, H. C.
Miller, G. L.

Mitsanas, H.
Moran, R. J.
Morris, C. R.
Morrison, G. E.
Morrison, S. M.
Mortenson, K.
Morton, G. A.
Moses, G. L.
Mulhern, W. J.
Muller, C. R.

Nalley, L. J.
Neiswender, W. J.
Neubauer, J. R.

Oliner, A. A.

Page, C. H.
Palmer, W.
Parker, J. C.
Pelc, T.
Penn, W. B.
Phillips, V. E.
Piccione, N.
Powers, K. H.
Pritchard, R. L.

Ray, K. A.
Redhead, P. A.
Rediker, R. H.
Reynolds, J. N.
Rice, B. M.
Rietz, E. B.
Roberts, W. K.
Rohlfs, A. F.
Rook, L. E.
Rose, 2nd, R. H.
Rothauser, E. H.
Rubin, S.
Ruete, R. C.
Russ, J. C.

Samuel, C. H.
Schaufelberger, F.
Schmidt, P. L.
Schwalbe, C. A.
Schwartz, J. D.
Schwertz, F. A.
Scoville, M. E.
Shackman, N.
Sharrow, R. F.
Shea, R. F.
Sheckler, A. C.
Shields, F. J.
Shipman, W. A.
Shores, R. B.
Showers, R. M.
Sidway, C. L.
Silbiger, H. R.

Simmons, A. J.
Simpson, H. A.
Singer, G. A.
Skolnik, M. I.
Smith, J. H.
Smith, P. H.
Sorensen, D. K.
Sorger, G. U.
Spitzer, C. F.
Spurgin, A. J.
Stadtfeld, Jr., N.
Stewart, J. A.
Stuckert, P. E.
Sullivan, J. B.
Sullivan, R. J.

Talaat, M. E.
Tebo, J. D.
Test, L. D.
Thurell, J. R.
Tillinger, H. I.
Tilston, W. V.
Tjepkema, S.
Toman, K.
Toppeto, A. A.
Tuller, W. G.

Unnewehr, L. E.
Ure, R. W.

Vadersen, C. W.
Vlahos, P.
von Recklinghausen, D. R.
von Roeschlaub, F.

Wagner, C. L.
Wagner, S.
Wahlgren, W. W.
Weber, E.
Weddendorf, W. A.
Weinberg, L.
Weinschel, B. O.
Weitzel, H. B.
White, H. F.
White, J. C.
Wickham, W. H.
Williamson, R. A.
Wintringham, W. T.
Wolfe, P. N.
Woods, D. E.
Wroblewski, J. J.

Yasuda, E. J.
Yates, E. S.
Younkin, G. W.

Zucker, M.

A

AA (transformer classification). *See:* **transformer, oil-immersed.**

aa auxiliary switch. *See:* **auxiliary switch; aa contact.**

aa contact. A contact that is open when the operating mechanism of the main device is in the standard reference position and that is closed when the operating mechanism is in the opposite position. *See:* **standard reference position.** 103

A and R display (radar). An *A* display, any portion of which may be expanded. *See also:* **navigation.** 13

a auxiliary switch. *See:* **a contact; auxiliary switch.**

abampere. The unit of current in the centimeter-gram-second (cgs) electromagnetic system. The abampere is 10 amperes. 172

abandoned call (telephone switching systems). A call during which the calling station goes on-hook prior to its being answered. 55

A battery. A battery designed or employed to furnish current to heat the filaments of the tubes in a vacuum-tube circuit. *See also:* **battery (primary or secondary).** 328

abbreviated dialing (telephone switching systems). A feature permitting the establishment of a call with an input of fewer digits than required under the numbering plan. 55

abbreviation. A shortened form of a word or expression. *See:* **functional designation; graphic symbol; letter combination; mathematical symbol; reference designation; symbol for a quantity; symbol for a unit.** 173

abnormal decay (charge-storage tubes). The dynamic decay of multiply-written, superimposed (integrated) signals whose total output amplitude changes at a rate distinctly different from that of an equivalent singly-written signal. *Note:* Abnormal decay is usually very much slower than normal decay and is observed in bombardment-induced conductivity type of tubes. *See also:* **charge-storage tube.** 174

abnormal glow discharge (gas tube). The glow discharge characterized by the fact that the working voltage increases as the current increases. *See also:* **discharge.** 175

absolute accuracy. Accuracy as measured from a reference that must be specified. 224, 207

absolute address (computing machines). (1) An address that is assigned by the machine designer to a physical storage location. (2) A pattern of characters that identifies a unique storage location without further modification. *See also:* **machine address.** 255, 77

absolute altimeter (electronic navigation). A device that measures altitude above local terrain. 13, 187

absolute block (automatic train control). A block governed by the principle that no train shall be permitted to enter the block while it is occupied by another train. 328

absolute capacitivity (absolute dielectric constant) (permittivity). Of a homogeneous, isotropic, insulating material or medium, in any system of units, the product of its relative capacitivity and the electric constant appropriate to that system of units. *See also:* **electric constant.** 210

absolute delay (loran). The interval of time between the transmission of a signal from the master station and transmission of the next signal from the slave station. *See also:* **navigation.** 13, 187

absolute dielectric constant. *See:* **absolute capacitivity.**

absolute dimension. A dimension expressed with respect to the initial zero point of a coordinate axis. *See:* **coordinate dimension word.** 224, 207

absolute error. (1) The amount of error expressed in the same units as the quantity containing the error. (2) Loosely, the absolute value of the error, that is, the magnitude of the error without regard to algebraic sign. 255, 77, 54

absolute measurement (system of units). Measurement in which the comparison is directly with quantities whose units are basic units of the system. *Notes:* (1) For example, the measurement of speed by measurements of distance and time is an absolute measurement, but the measurement of speed by a speedometer is not an absolute measurement. (2) The word absolute implies nothing about precision or accuracy. 210

absolute permissive block (automatic train control). A term used for an automatic block signal system on a track signaled in both directions. For opposing movements the block is from siding to siding and the signals governing entrance to this block indicate STOP. For following movements the section between sidings is divided into two or more blocks and train movements into these blocks, except the first one, are governed by intermediate signals usually displaying STOP; then proceed at restricted speed, as their most restrictive indication. 328

absolute refractory state (medical electronics). The portion of the electrical recovery cycle during which a biological system will not respond to an electric stimulus. 192

absolute steady-state deviation (control). The numerical difference between the ideal value and the final value of the directly controlled variable (or another variable if specified). *See:* **deviation (control); percent steady-state deviation.** 219, 206

absolute system deviation (control). At any given point on the time response, the numerical difference between the ideal value and the instantaneous value of the directly controlled variable (or another variable if specified). *See:* **deviation.** 219, 206

absolute threshold. The luminance threshold or minimum perceptible luminance (photometric brightness) when the eye is completely dark adapted. *See also:* **visual field.** 167

absolute transient deviation (control). The numerical difference between the instantaneous value and the final value of the directly controlled variable (or another variable if specified). *See:* **deviation; percent transient deviation.** 219, 206

absolute value (number). The absolute value of a number u (real or complex) is that positive real number $|u|$ given by

$$|u| = + (u_1{}^2 + u_2{}^2)^{1/2}$$

where u_1 and u_2 are respectively the real and imaginary parts of u in the equation

$$u = u_1 + ju_2.$$

If u is a real number, $u_2 = 0$. 210

absolute-value device. A transducer that produces an output signal equal in magnitude to the input signal but always of one polarity. *See also:* **electronic analog computer.** 9, 77

absorptance (illuminating engineering). The ratio of the absorbed flux to the incident flux. *Note:* The sum of the hemispherical reflectance, the hemispherical transmittance, and the absorptance is one. *See also:* **lamp.** 167

absorption (1) (transmission of waves). The loss of energy over radio or wire paths due to conversion into heat or other forms of energy. In wire transmission, the term is usually applied only to a loss of energy in extraneous media. 328
(2) (radio wave propagation). The irreversible conversion of the energy of an electromagnetic wave into another form of energy as a result of its interaction with matter. *See also:* **radio wave propagation.** 146, 180, 199, 59
(3) (illuminating engineering). A general term for the process by which incident flux is dissipated. *Note:* All of the incident flux is accounted for by the processes of reflection, transmission, and absorption. *See also:* **lamp.** 167

absorption (laser-maser). The transfer of energy from a radiation field to matter. 363

absorption current (rotating machinery). A reversible component of the measured current, which changes with time of voltage application, resulting from the phenomenon of dielectric absorption within the insulation when stressed by direct voltage. 63, 6

absorption frequency meter (reaction frequency meter) (waveguide). A one-port cavity frequency meter that, when tuned, absorbs electromagnetic energy from a waveguide. *See:* **waveguide.** 179

absorption loss (data transmission). The loss of power in a transmission circuit that results from coupling to a neighboring circuit or conductor. 59

absorption modulation. A method for producing amplitude modulation of the output of a radio transmitter by means of a variable-impedance (principally resistive) device inserted in or coupled to the output circuit. 145, 211

absorptive attenuator (waveguide). *See:* **resistive attenuator.**

abstract quantity. *See:* **mathematico-physical quantity.**

ac. *See:* **alternating current.**

ACA. *See:* **adjacent channel attenuation.**

accelerated life test (1) (cable). A test in which certain factors such as voltage, temperature, etcetera, to which a cable is subjected are increased in magnitude above normal operating values to obtain observable deterioration in a reasonable period of time, and thereby afford some measure of the probable cable life under operating voltage, temperature, etcetera. 64
(2) (test, measurement and diagnostic equipment). A test in which certain factors, such as voltage, temperature, and so forth, are increased or decreased beyond normal operating values to obtain observable deterioration in a reasonable period of time, and thereby afford some measure of the probable life under normal operating conditions or some measure of the durability of the equipment when exposed to the factors being aggravated. *See also:* **power distribution, underground construction.** 54

accelerated test (reliability). A test in which the applied-stress level is chosen to exceed that stated in the reference conditions, in order to shorten the time required to observe the stress responses of the item, or magnify the response in a given time. To be valid, an accelerated test shall not alter the basic modes and/or mechanisms of failure, or their relative prevalence. *See also:* **reliability.** 182, 164

accelerating (rotating machinery). The process of running a motor up to speed after breakaway. *See also:* **asynchronous machine; direct-current commutating machine.** 63

accelerating electrode. An electrode to which a potential is applied to increase the velocity of the electrons or ions in the beam. 190, 117, 125

accelerating grid (electron tubes). *See:* **accelerating electrode.**

accelerating relay (rotating machinery). A programming relay whose function is to control the acceleration. 103, 202, 60, 127

accelerating time (control) (industrial control). The time in seconds for a change of speed from one specified speed to a higher specified speed while accelerating under specified conditions. *See:* **electric drive.** 219, 206

accelerating torque (rotating machinery). Difference between the input torque to the rotor (electromagnetic for a motor or mechanical for a generator) and the sum of the load and loss torques; the net torque available for accelerating the rotating parts. *See:* **rotor.** 63

accelerating voltage (oscilloscope). The cathode-to-viewing-area voltage applied to a cathode-ray tube for the purpose of accelerating the electron beam. *See:* **oscillograph.** 185

acceleration (electric drive). Operation of raising the motor speed from zero or a low level to a higher level. *See also:* **electric drive.** 206

acceleration factor (reliability). The ratio between the times necessary to obtain the same stated proportion of failures in two equal samples under two different sets of stress conditions involving the same failure modes and mechanisms. 164

acceleration-insensitive drift rate (gyro). That component of systematic drift rate which has no correlation with acceleration. It is usually due to torques such as flexlead torques, magnetic torques and certain fluid torques. 46

acceleration, programmed. A controlled velocity increase to the programmed rate. 224, 207

acceleration-sensitive drift rate (gyro). Those components of systematic drift rates that are correlated with the first power of linear acceleration applied to the gyro case. The relationship of these components of drift rate to acceleration can be stated by means of coefficients having dimensions of angular displacement per unit time per unit acceleration for accelerations along each of the principal axes of the gyro, for example, drift rate caused by mass unbalance. 46

acceleration space (velocity-modulated tube). The part of the tube following the electron run in which the emitted electrons are accelerated to reach a determined velocity. *See also:* **velocity-modulated tube.** 244, 190

acceleration-squared-sensitive drift rate (gyro). Those components of systematic drift rates that are correlated with the second power or product of linear acceleration applied to the gyro case. The relationship of these components of drift rate to acceleration squared can be stated by means of coefficients having dimensions of angular displacement per unit time per unit acceleration squared for accelerations along each of the principal axes of the gyro and angular displacement per unit time per the product of accelerations along combinations of two principal axes of the gyro for example, drift rate caused by anisoelasticity. 46

acceleration, timed (industrial control). A control function that accelerates the drive by automatically controlling the speed change as a function of time. *See:* **electric drive.**
219, 206

accelerator, electron, linear. *See:* **linear electron accelerator.**

accelerator, particle. *See:* **particle accelerator.**

accelerometer (1) (electronic navigation). A device that senses inertial reaction to measure linear or angular acceleration. *Note:* In its simplest form, an accelerometer consists of a case-mounted spring and mass arrangement where displacement of the mass from its rest position relative to the case is proportional to the total nongravitational acceleration experienced along the instrument's sensitive axes. *See also:* **navigation.** 13, 187
(2) (gyro). A device that senses the inertial reaction of a proof mass for the purpose of measuring linear or angular acceleration. 46

accent lighting. Directional lighting to emphasize a particular object or draw attention to a part of the field of view. *See also:* **general lighting.** 167

acceptable. Demonstrated to be adequate by the safety analysis of the station. 31, 102

acceptance angle (1) (phototube housing). The solid angle within which all received light reaches the phototube cathode. *See also:* **electronic controller.** 206
(2) (acousto-optic device). The angular spread of the incoming light at which the diffraction efficiency will fall to 50 percent of the value it obtains when the input light is collimated. 82
(3) (acoustically tunable optical filter). The full angle of a cone of light which, when incident on the device, will result in a negligible (typically 5 percent) decrease in the dynamic transmission and a negligible broadening (typically 5 percent) in the optical bandwidth as compared to collimated input light. 72

acceptance proof test (rotating machinery). A test applied to new insulated winding before commercial use. It may be performed at the factory or after installation, or both.
6

acceptance test (1) (general) (A). A test to demonstrate the degree of compliance of a device with purchaser's requirements. (B) A **conformance test** demonstrates the quality of the units of a consignment, without implication of contractual relations between buyer and seller. *See:* **routine test; test (instrument or meter); conformance test.**
91, 210, 212, 203, 103, 63, 202
(2) *See:* **conformance tests.** [*Note:* American National Standards should use the term *conformance test,* as directed by the Standards Council of ANSI, rather than the term *acceptance test.*] Use of the term *conformance test* avoids the implication of contractual relations between buyer and seller. 103
(3) (nuclear power generating stations) (lead storage bat-

teries**).** A capacity test made on a new battery to determine that it meets specifications or manufacturer's ratings, or both. 31, 38

acceptor (semiconductor). *See:* **impurity, acceptor.** *See also:* **semiconductor.**

access. *See:* **random access; serial access.**

access code (telephone switching systems). One or more digits required in certain situations in lieu of or preceding an area or office code. 55

access fitting. A fitting permitting access to the conductors in a raceway at locations other than at a box. *See also:* **raceway.** 328

accessibility (telephone switching systems). The ability of a given inlet to reach the available outlets. 55

accessible (1) (equipment). Admitting close approach because not guarded by locked doors, elevation, or other effective means. 328, 53
(2) (wiring methods). Not permanently closed in by the structure or finish of the building; capable of being removed without disturbing the building structure or finish.
328
(3) (transformer). Admitting close approach to contact by persons due to lack of locked doors, elevation or other effective safeguards. 53

accessible, readily. Capable of being reached quickly without requiring those to whom ready access is requisite to climb over or remove obstacles or to resort to portable ladders, chairs, etc. 53

accessible terminal (network). A network node that is available for external connections. *See also:* **network analysis.** 210

accessories. Devices that perform a secondary or minor duty as an adjunct or refinement to the primary or major duty of a unit of equipment. 103, 202, 53

accessory (1) (test, measurement and diagnostic equipment). An assembly of a group of parts or a unit which is not always required for the operation of a test set or unit as originally designed but serves to extend the functions or capabilities of the test set; similarly as headphones for a radio set supplied with a loudspeaker; a vibrator power unit for use with a set having a built-in power supply, or a remote control unit for use with a set having integral controls. 54
(2) (electric and electronics parts and equipment). A basic part, subassembly, or assembly designed for use in conjunction with or to supplement another assembly, unit, or set, contributing to the effectiveness thereof without extending or varying the basic function of the assembly or set. An accessory may be used for testing, adjusting, or calibrating purposes. Typical examples: test instrument, recording camera for radar set, headphones, emergency power supply. 17

accessory equipment (Class 1E motor) (nuclear power generating station). Devices other than the principal motor components that are furnished with or built as a part of the motor structure and are necessary for the operation of the motor. 104

access time (1). A time interval that is characteristic of a storage device, and is essentially a measure of the time required to communicate with that device. *Note:* Many definitions of the beginning and ending of this interval are in common use. 235, 210

(2) (A) The time interval between the instant at which data are called for from a storage device and the instant delivery is completed, that is, the read time. (B) The time interval between the instant at which data are requested to be stored and the instant at which storage is completed, that is, the write time. 255, 77

(3) (acousto-optic deflector). The minimum time to randomly deflect the light beam from one spot position to another. It is given by the time it takes the acoustic beam to cross the optical beam, viz, $\tau = S/V$, with τ the access time, S the optical beam dimension, and V the acoustic velocity. 72

(4) (acoustically tunable optical filter). The minimum time to randomly tune the filter from one wavelength to another. It is given by the time it takes the acoustic beam to cross the optical beam, namely: $\tau = S/V$, with S the length of the optical beam along the acoustic beam direction and V the acoustic velocity. 72

access tools (tamper-resistant switchgear assembly) (relaying). Keys or other special accessories with unique characteristics that make them suitable for gaining access to the tamper-resistant switchgear assembly. 79

accommodation (1) (general). The process by which the eye changes focus from one distance to another. *See also:* **visual field.** 167

(2) (laser-maser). The ability of the eye to change its power and thus focus for different object distances. 363

accommodation, electrical (biology) (electrobiology). A rise in the stimulation threshold of excitable tissue due to its electrical environment, often observed following a previous stimulation cycle. *See also:* **excitability.** 192

accumulating stimulus (electrotherapy). A current that increases so gradually in intensity as to be less effective than it would have been if the final intensity had been abruptly attained. *See also:* **electrotherapy.** 192

accumulator. (1) A device that retains a number (the augend) adds to it another number (the addend) and replaces the augend with the sum. (2) Sometimes only the part of (1) that retains the sum. *Note:* The term is also applied to devices that function as described but that also have other properties. 235, 210

accuracy (1) (instrumentation and measurement). The quality of freedom from mistake or error, that is, of conformity to truth or to a rule. *Notes:* (A) Accuracy is distinguished from precision as in the following example: A six-place table is more precise than a four-place table. However, if there are errors in the six-place table, it may be more or less accurate than the four-place table. (B) The accuracy of an indicated or recorded value is expressed by the ratio of the error of the indicated value to the true value. It is usually expressed in percent. Since the true value cannot be determined exactly, the measured or calculated value of highest available accuracy is taken to be the true value or reference value. Hence, when a meter is calibrated in a given echelon, the measurement made on a meter of a higher-accuracy echelon usually will be used as the reference value. Comparison of results obtained by different measurement procedures is often useful in establishing the true value. *See also:* **dynamic accuracy; electronic analog computer; measurement system; static accuracy.** 213, 183

(2) (analog computers). (A) Conformity of a measured value to an accepted standard value. (B) A measure of the degree by which the actual output of a device approximates the output of an ideal device nominally performing the same function. *See also:* **electronic analog computer.** 10

(3) (power supply). Used as a specification for the output voltage of power supplies, accuracy refers to the absolute voltage tolerance with respect to the stated nominal output. *See also:* **power supply.** 228, 186

(4) (numerically controlled machine). Conformity of an indicated value to the true value, that is, an actual or an accepted standard value. *Note:* Quantitatively, it should be expressed as an error or an uncertainty. The property is the joint effect of method, observer, apparatus, and environment. Accuracy is impaired by mistakes, by systematic bias such as abnormal ambient temperature, or by random errors (imprecision). The accuracy of a control system is expressed as the system deviation (the difference between the ultimately controlled variable and its ideal value), usually in the steady state or at sampled instants. *See:* **precision and reproducibility.** 56, 207

(5) (electronic navigation). Generally, the quality of freedom from mistake or error; that is, of conformity to truth or a rule. Specifically, the difference between the mean value of a number of observations and the true value. *Note:* Often refers to a composite character including both accuracy and precision. *See also:* **navigation; precision.** 13, 187

(6) (signal-transmission system). Conformity of an indicated value to an accepted standard value, or true value. *Note:* Quantitatively, it should be expressed as an error or uncertainty. The accuracy of a determination is affected by the method, observer, environment, and apparatus, including the working standard used for the determination. *See also:* **signal.** 56, 188

(7) (measurement) (control equipment). The degree of correctness with which a measurement device yields the true value of measured quantity. 94

(8) (indicated or recorded value). The accuracy of an indicated or recorded value is expressed by the ratio of the error of the indicated value to the true value. It is usually expressed in percent. *See also:* **accuracy rating of an instrument.** 328

(9) (instrument transformer). The extent to which the current or voltage in the secondary circuit reproduces the current or voltage of the primary circuit in the proportion stated by the marked ratio, and represents the phase relationship of the primary current or voltage. 53

(10) (test, measurement and diagnostic equipment). The degree of correctness with which a measured value agrees with the true value. *See:* **precision.** 54

(11) (electrothermic power meters). The degree of correctness with which a measurement device yields the true value of a measured quantity; quantitatively expressed by uncertainty. *See also:* **uncertainty.** 47

(12) (nuclear power generating stations). The quality of freedom from mistake or error. 31, 41

(13) (pulse measurement). The degree of agreement between the result of the application of a pulse measurement process and the true magnitude of the pulse characteristic, property, or attribute being measured. 15

(14) (excitation control system). The degree of correspondence between the controlled variable and the desired value under specified conditions such as load changes, ambient temperature, humidity, frequency, and supply voltage variations. Quantitatively, it is expressed as the ratio of difference between the controlled variable and the desired value to the desired value.   105

accuracy burden rating. A burden that can be carried at a specified accuracy for an unlimited period without causing the established limitations to be exceeded. *See:* **instrument transformer.**   203

accuracy class (instrument transformer). The highest of the standard accuracy classes, the requirements of which are fulfilled by the values of the instrument-transformer correction factor under specified standard conditions. *See:* **instrument transformer.**   212, 328

accuracy classes for metering (instrument transformer). Limits of a transformer correction factor, in terms of percent error, that have been established to cover specific performance ranges for line power-factor conditions between 1.0 and 0.6 lag.   303, 203

accuracy classes for relaying (instrument transformer). Limits, in terms of percent ratio error, that have been established.   303, 203

accuracy, dynamic. *See:* **dynamic accuracy.**

accuracy rating (class) (1) (general) (electric instrument). The accuracy classification of the instrument. It is given as the limit, usually expressed as a percentage of full-scale value, that errors will not exceed when the instrument is used under reference conditions. *Notes:* (1) The accuracy rating is intended to represent the tolerance applicable to an instrument in an "as-received condition." Additional tolerances for the various influences are permitted when applicable. It is required that the accuracy, as received, be directly in terms of the indications on the scale and without the application of corrections from a curve, chart, or tabulation. Over that portion of the scale where the accuracy tolerance applies, all marked division points shall conform to the stated accuracy class. (2) Generally the accuracy of electrical indicating instruments is stated in terms of the electrical quantities to which the instrument responds. In instruments with the zero at a point other than one end of the scale, the arithmetic sum of the end-scale readings to the right and to the left of the zero point shall be used as the full-scale value. Exceptions: (A) The accuracy of frequency meters shall be expressed on the basis of the percentage of actual scale range. Thus, an instrument having a scale range of 55 to 65 hertz would have its error expressed as a percentage of 10 hertz. (B) The accuracy of a power-factor meter shall be expressed as a percentage of scale length. (C) The accuracy of instruments that indicate derived quantities, such as series type ohmmeters, shall be expressed as a percentage of scale length. (3) In the case of instruments having nonlinear scales, the stated accuracy only applies to those portions of the scale where the divisions are equal to or greater than two-thirds the width they would be if the scale were even divided. The limit of the range at which this accuracy applies may be marked with a small isosceles triangle whose base marks the limit and whose point is directed toward the portion of the scale having the specified accuracy. (4) Instruments having an accuracy rating of 0.1 percent are frequently referred to as laboratory standards.

Portable instruments having an accuracy rating of 0.25 percent are frequently referred to as portable standards.   280

(2) (automatic control system). The limit which the system deviation will not exceed under specified operating conditions.   56

accuracy ratings for metering. The accuracy class followed by a standard burden for which the accuracy class applies. Accuracy rating applies only over the specified current or voltage range and at the stated frequency.   280, 203

accuracy ratings for relaying. The relay accuracy class is described by a letter denoting whether the accuracy can be obtained by calculation or must be obtained by test, followed by the maximum secondary terminal voltage that the transformer will produce at 20 times secondary current with one of the standard burdens, without exceeding the relay accuracy class limit.   203

accuracy ratings of instrument transformers. Means of classifying transformers in terms of percent error limits under specified conditions of operation.   305, 203

accuracy, static. *See:* **static accuracy.**

accuracy, synchronous-machine regulating system. The degree of correspondence (or ratio) between the actual and the ideal values of a controlled variable of the synchronous-machine regulating system under specified conditions, such as load changes, drift, ambient temperature, humidity, frequency, and supply voltage.   63

accuracy test (instrument transformer). A test to determine the degree to which the value of the quantity obtained from the secondary reflects the value of the quantity applied to the primary.   305, 203

acetate disks. Mechanical recording disks, either solid or laminated, that are made of various acetate compounds.   176

achromatic locus (achromatic region). A region including those points in a chromaticity diagram that represent, by common acceptance, arbitrarily chosen white points (white references). *Note:* The boundaries of the achromatic locus are indefinite, depending on the tolerances in any specific application. Acceptable reference standards of illumination (commonly referred to as white light) are usually represented by points close to the locus of Planckian radiators having temperatures higher than about 2000 kelvins*. While any point in the achromatic locus may be chosen as the reference point for the determination of dominant wavelength, complementary wavelength and purity for specification of object colors, it is usually advisable to adopt the point representing the chromaticity of the luminator. Mixed qualities of illumination, and luminators with chromaticities represented very far from the Planckian locus, require special consideration. Having selected a suitable reference point, dominant wavelength may be determined by noting the wavelength corresponding to the intersection of the spectrum locus with the straight line drawn from the reference point through the point representing the sample. When the reference point lies between the sample point and the intersection, the intersection indicates the complementary wavelength. Any point within the achromatic locus, chosen as a reference point, may be called an achromatic point. Such points have also been called white points. *See:* **kelvin.**   18

acid-resistant (industrial control). So constructed that it will not be injured readily by exposure to acid fumes.
225, 206

acknowledger (forestaller). A manually operated electric switch or pneumatic valve by means of which, on a locomotive equipped with an automatic train stop or train control device, an automatic brake application can be forestalled, or by means of which, on a locomotive equipped with an automatic cab signal device, the sounding of the cab indicator can be silenced. 328

acknowledging (forestalling).· The operating by the engineman of the acknowledger associated with the vehicle-carried equipment of an automatic speed control or cab signal system to recognize the change of the aspect of the vehicle-carried signal to a more restrictive indication. The operation stops the sounding of the warning whistle, and in a locomotive equipped with speed control it also forestalls a brake application. 328

acknowledging device. *See:* **acknowledger (forestaller).**

acknowledging switch. *See:* **acknowledger (forestaller).**

acknowledging whistle. An air-operated whistle that is sounded when the acknowledging switch is operated. Its purpose is to inform the fireman that the engineman has recognized a more restrictive signal indication. 328

a contact (front contact). A contact that is open when the main device is in the standard reference position, and that is closed when the device is in the opposite position. *Notes:* (1) *a* contact has general application. However, this meaning for front contact is restricted to relay parlance. (2) For indication of the specific point of travel at which the contact changes position, an additional letter or percentage figure may be added to *a* as detailed in Sections 9.4.4.1. and 9.4.4.2 of American National Standard C37.2-1962. *See also:* **standard reference position.**
103, 202

acoustic, acoustical. Used as qualifying adjectives to mean containing, producing, arising from, actuated by, related to, or associated with sound. Acoustic is used when the term being qualified designates something that has the properties, dimensions, or physical characteristics associated with sound waves; acoustical is used when the term being qualified does not designate explicitly something that has such properties, dimensions, or physical characteristics. *Notes:* (1) The following examples qualify as having the properties of physical characteristics associated with sound waves and hence would take acoustic; impedance, inertance, load (radiation field), output (sound power), energy, wave, medium, signal, conduit, absorptivity, transducer. (2) The following examples do not have the requisite physical characteristics and therefore take acoustical: society, method, engineer, school, glossary, symbol, problem, measurement, point of view, end-use, device. (3) As illustrated in the preceding notes, usually the generic term is modified by acoustical, whereas the specific technical implication calls for acoustic. 176

acoustical. *See:* **acoustic.**

acoustically tunable optical filter. An optical filter which is driven by an acoustic wave and which is tunable by varying the acoustic frequency. 72

acoustical reciprocity theorem. In an acoustic system comprising a fluid medium having bounding surfaces S_1, $S_2, S_3, \ldots$, and subject to no impressed body forces, if two distributions of normal velocities v_n' and v_n'' of the bounding surfaces produce pressure fields p' and p'', respectively, throughout the region, then the surface integral of $(p''v_n' - p'v_n'')$ over all the bounding surfaces $S_1, S_2, S_3, \ldots$ vanishes. *Note:* If the region contains only one simple source, the theorem reduces to the form ascribed to Helmholtz, namely, in a region as described, a simple source at A produces the same sound pressure at another point B as would have been produced at A had the source been located at B. 176

acoustical units. In different sections of the field of acoustics at least three systems of units are in common use: the meter-kilogram-second (mks), the centimeter-gram-second (cgs), and the British. The following table (pp. 8–9) facilitates conversion from one system of units to another. 176

acoustic center, effective (acoustic generator). The point from which the spherically divergent sound waves appear to diverge when observed at remote points. *See also:* **loudspeaker.** 176

acoustic coupler (data communication). A type of data communication equipment which has sound transducers that permit the use of a telephone handset as a connection to a voice communication system for the purpose of data transmission. 12

acoustic delay line. A delay line whose operation is based on the time of propagation of sound waves. *See:* **sonic delay line.** 255, 77

acoustic horn. A tube of varying cross section having different terminal areas that provides a change of acoustic impedance and control of the direction pattern. *Syn.* **horn.** *See also:* **loudspeaker.** 176

acoustic impedance. The acoustic impedance of a sound medium on a given surface lying in a wave front is the complex quotient of the sound pressure (force per unit area) on that surface by the flux (volume velocity, or linear velocity multiplied by the area), through the surface. When concentrated, rather than distributed, impedances are considered, the impedance of a portion of the medium is defined by the complex quotient of the pressure difference effective in driving that portion, by the flux (volume velocity). The acoustic impedance may be expressed in terms of mechanical impedance, acoustic impedance being equal to the mechanical impedance divided by the square of the area of the surface considered. The commonly used unit is the acoustical ohm. *Notes:* (1) Velocities in the direction along which the impedance is to be specified are considered positive. (2) The terms and definitions to which this note is appended pertain to single-frequency quantities in the steady state, and to systems whose properties are independent of the magnitudes of these quantities. 334

acoustic interferometer. An instrument for the measurement of wavelength and attenuation of sound. Its operation depends upon the interference between reflected and direct sound at the transducer in a standing-wave column. *See also:* **instrument.** 328

acoustic output (telephone set). The sound pressure level developed in an artificial ear, measured in decibels referred to 0.00002 newton per square meter. *See also:* **telephone station.** 122, 196

acoustic pickup (sound box) (phonograph). A device that transforms groove modulations directly into acoustic vi-

brations. *See also:* **loudspeaker; phonograph pickup.**
176

acoustic properties of water, representative. Numerical values for some important acoustic properties of water at representative temperatures and salinities are listed in the following table (p. 9).
176

acoustic radiating element. A vibrating surface in a transducer that can cause or be actuated by sound waves.
176

acoustic radiation pressure. A unidirectional steady-state pressure component exerted upon a surface by an acoustic wave.
176

acoustic radiator. A means for radiating acoustic waves. *See also:* **loudspeaker.**
328

acoustic radiometer. An instrument for measuring acoustic radiation pressure. *See also:* **instrument.**
176

acoustic refraction. The process by which the direction of sound propagation is changed due to spatial variation in the speed of sound in the medium.
176

acoustics. The science of sound or the application thereof.
176

acoustic scattering. The irregular reflection, refraction, or diffraction of a sound in many directions.
176

acoustic transmission system. An assembly of elements for the transmission of sound.
176

acoustic wave filter. A filter designed to separate acoustic waves of different frequencies. *Note:* Through electroacoustic transducers such a filter may be associated with electric circuits. *See also:* **filter.**
328

acousto-optic device. A device which is used to modulate light in amplitude, frequency, phase, polarization, or spatial position by virtue of optical diffraction from an acoustically generated diffraction grating.
82

acquisition (radar). The process of establishing a stable track on a target which is designated in one or more coordinates. A search of a given volume of coordinate space is usually required, because generally the designation data are neither complete nor accurate and consist of two or three coordinates with some finite distribution of errors, describable on a statistical basis.
13

acquisition probability (radar). The summation, over all resolution elements of a designated volume of radar coordinate space, of the detection probability for each resolution element multiplied by the probability that the target actually lies in that element.
13

across-the-line starter (rotating machinery). A device that connects the motor to the supply without the use of a resistance or autotransformer to reduce the voltage. *Note:* It may consist of a manually operated switch or a master switch that energizes an electromagnetically operated contactor. *See also:* **starter.**
3

across-the-line starting (rotating machinery). The process of starting a motor by connecting it directly to the supply at rated voltage.
63

ACSR. *See:* **aluminum cable steel reinforced.**

action potential (action current) (medical electronics). The instantaneous value of the potential observed between excited and resting portions of a cell or excitable living structure. *Note:* It may be measured direct or through a volume conductor.
192

action spike (medical electronics). The greatest in magnitude and briefest in duration of the characteristic negative

waves seen during the observation of action potentials.
192

activation (cathode) (thermionics). The treatment applied to a cathode in order to create or increase its emission. *See also:* **electron emission.**
244, 190

activation polarization. The difference between the total polarization and the concentration polarization. *See also:* **electrochemistry.**
328

active (corrosion). A state wherein passivity is not evident.
221, 205

active area (1) (semiconductor rectifier cell). The portion of the rectifier junction that effectively carries forward current. *See also:* **semiconductor.**
234, 66
(2) (solar cell). The illuminated area normal to light incidence, usually the face area less the contact area. *Note:* For the purpose of determining efficiency, the area covered by collector grids is considered a part of the active area. *See also:* **semiconductor.**
186, 113

active current (rotating machinery). The component of the alternating current that is in phase with the voltage. *See also:* **asynchronous machine.**
63

active-current compensator (rotating machinery). A compensator that acts to modify the functioning of a voltage regulator in accordance with active current.
63

active electric network. An electric network containing one or more sources of power. *See also:* **network analysis.**
328

active electrode (electrobiology). (1) A pickup electrode that, because of its relation to the flow pattern of bioelectric currents, shows a potential difference with respect to ground or to a defined zero, or to another (reference) electrode on related tissue. (2) Any electrode, in a system of stimulating electrodes, at which excitation is produced. (3) A stimulating electrode (different electrode) applied to tissue for stimulation and distinguished from another (inactive, dispersive, diffuse, or indifferent) electrode by having a smaller area of contact thus affording a higher current density. *See also:* **electrobiology.**
192

active maintenance time. The time during which maintenance actions are performed on an item either manually or automatically. *Notes:* (1) Delays inherent in the maintenance operation, for example those due to design or to prescribed maintenance procedures, shall be included. (2) Active maintenance may be carried out while the item is performing its intended function.
164

active materials (storage battery). The materials of the plates that react chemically to produce electric energy when the cell discharges and that are restored to their original composition, in the charged condition, by oxidation and reduction processes produced by the charging current. *See also:* **battery.**
328

active power (1) (rotating machinery). A term used for power when it is necessary to distinguish among apparent power, complex power and its components, active and reactive power. *See also:* **asynchronous machine.**
63
(2) (general). *See:* **power, active.**

active-power relay. A power relay that responds to active power. *See also:* **relay; power relay; reactive power relay.**
60, 103, 127

active preventive maintenance time. That part of the active maintenance time in which preventive maintenance is carried out. *Notes:* (1) Delays inherent in the preventive

Relations among various acoustical units

Quantity	Dimension	centimeter-gram-second (cgs) Unit	meter-kilogram-second (mks) Unit	Conversion Factor*	British Unit	Conversion Factor†
Mass	M	gram	kilogram	10^{-3}	slug	6.854×10^{-5}
Velocity (linear)	LT^{-1}	centimeter per second	meter per second	10^{-2}	foot per second	3.281×10^{-2}
Force	MLT^{-2}	dyne	newton	10^{-5}	pound-force	2.248×10^{-6}
Sound pressure	$ML^{-1}T^{-2}$	dyne per square centimeter [microbar]	newton per square meter	10^{-1}	pound per square foot	2.089×10^{-3}
Volume velocity	$L^{3}T^{-1}$	cubic centimeter per second	cubic meter per second	10^{-6}	cubic foot per second	3.531×10^{-5}
Sound energy	$ML^{2}T^{-2}$	erg	joule	10^{-7}	foot-pound	7.376×10^{-7}
Sound-energy density	$ML^{-1}T^{-2}$	erg per cubic centimeter	joule per cubic meter	10^{-1}	foot-pound per cubic foot	2.089×10^{-3}
Sound-energy flux [sound power of source]	$ML^{2}T^{-3}$	erg per second	watt	10^{-7}	foot-pound per second	7.376×10
Sound-energy-flux density [sound intensity]	MT^{-3}	erg per second per square centimeter	watt per square meter	10^{-3}	(foot-pound per second) per square foot	6.847×10^{-5}
Mechanical impedance	MT^{-1}	mechanical ohm [dyne second per centimeter]	mks mechanical ohm [newton second per meter]	10^{-3}	pound-second per foot	6.854×10^{-5}
Acoustic impedance [resistance, reactance]	$ML^{-4}T^{-1}$	acoustical ohm	mks acoustical ohm	10^{5}	(pound per square foot) per (cubic foot per second)	59.16
Specific acoustic impedance	$ML^{-2}T^{-1}$	rayl [acoustical ohm × square centimeter]	mks rayl [mks acoustical ohm × square meter]	10	(pound per square foot) per (foot per second)	6.366×10^{-3}
Acoustic inertance	ML^{-4}	gram per centimeter to the fourth power	kilogram per meter to the fourth power	10^{5}	slug per (foot to the fourth power)	59.16

Quantity	Dimensions	cgs unit	cgs→mks factor*	mks unit	British unit	cgs→British factor†
Acoustic stiffness	$ML^{-4}T^{-2}$	(gram per centimeter to the fourth power) per square second	10^5	(kilogram per meter to the fourth power) per square second	(slug per foot to the fourth power) per square second	59.16
Acoustic compliance	$M^{-1}L^4T^2$	(centimeter to the fifth power) per dyne	10^5	(meter to the fifth power) per newton	(foot to the fifth power) per pound	1.690×10^{-2}

* Multiply a magnitude expressed on centimeter-gram-second (cgs) units by the tabulated conversion factor to obtain magnitude in meter-kilogram-second (mks) units.
† Multiply a magnitude expressed in centimeter-gram-second (cgs) units by the tabulated conversion factor to obtain magnitude in British units. These conversion factors were calculated on the basis of standard acceleration due to gravity.

NOTE: M, L, T represent mass, length, and time, respectively, in the sense of the theory of dimensions. For meter-kilogram-second (mks) mechanical ohm and meter-kilogram-second (mks) acoustical ohm, rayl and meter-kilogram-second (mks) rayl are proposed terms. Alternative terms and units are in square brackets.

Speed of sound, density, and characteristic impedance of water at atmospheric pressure for various temperatures and salinities

	Fresh Water				Sea Water							
Salinity in per mil (0/00)	0				30				35			
Temperature in degrees Celsius	0	4	15	20	0	4	15	20	0	4	15	20
Speed of sound in meters per second	1403	1422	1466	1483	1443	1461	1501	1516	1449	1467	1507	1522
Density in kilograms per cubic meter	0999.8	1000.0	0999.1	0998.2	1024.1	1023.8	1022.2	1021.0	1028.1	1027.8	1026.0	1024.8
Characteristic impedance $\times$ 10^{-6} meter-kilogram-second units (rayls)	1.402	1.422	1.465	1.480	1.478	1.496	1.534	1.548	1.490	1.508	1.546	1.560

maintenance operation, for example those due to design or prescribed maintenance procedures, shall be included. (2) Active preventive maintenance time does not include any time taken to maintain an item which has been replaced. 164

active redundancy. *See:* **redundancy, active.**

active repair time. The time during which corrective maintenance actions are performed on an item either manually or automatically. *Notes:* (1) Delays inherent in the repair operation, for example those due to design or to prescribed maintenance procedures, shall be included. (2) Active repair time does not include any time taken to repair an item which has been replaced as part of the corrective maintenance action under consideration. 164

active testing (test, measurement and diagnostic equipment). The process of determining equipment static and dynamic characteristics by performing a series of measurements during a series of known operating conditions. Active testing may require an interruption of normal equipment operations and it involves measurements made over the range of equipment operation. *See:* **interference testing.** 54

active transducer. *See:* **transducer, active.**

actual transient recovery voltage. The transient recovery voltage that actually occurs across the terminals of a pole of a circuit-interrupting device on a particular interruption. *Note:* This is the modified circuit transient recovery voltage with whatever distortion may be introduced by the circuit-interrupting device. 103, 202

actual zero. Actual zero is the instant at which the switching impulse voltage first departs from zero value. 108

actuated equipment (nuclear power generating stations). (1) A component or assembly of components that performs a protective function such as reactor trip, containment isolation, or emergency coolant injection. *Note:* The following are examples of actuated equipment: an entire control rod and its release mechanism, a containment isolation valve and its operator, and a safety injection pump and its prime mover. 31, 109 (2) A component or assembly of components that performs, or directly contributes to the performance of, a protective function such as reactor trip, containment isolation, or emergency coolant injection. The following are examples of actuated equipment: an entire control rod with its release or drive mechanism, a containment isolation valve with its operator, and a safety injection pump with its prime mover. 355 (3) The assembly of prime movers and driven equipment used to accomplish a protective action. *Note:* Examples of prime movers are: turbines, motors, and solenoids. Examples of driven equipment are: control rods, pumps, and valves. 356

actuating current (automatic line sectionalizer). The root-mean-square current that actuates a counting operation or an automatic operation. 103, 202

actuating device (protective signaling). A manually or automatically operated mechanical or electric device that operates electric contacts to effect signal transmission. *See also:* **protective signaling.** 328

actuating signal (industrial control). The reference input signal minus the feedback signal. *See also:* **control system, feedback.** 206, 56, 219

actuation device (actuator) (nuclear power generating stations). (1) A component or assembly of components that directly controls the motive power (electricity, compressed air, etcetera) for actuated equipment. *Note:* The following are examples of an actuation device: a circuit breaker, a relay, and a pilot valve used to control compressed air to the operator of a containment isolation valve. 31, 109 (2) A component or assembly of components (or module) that directly controls the motive power (electricity, compressed air, etcetera) for actuated equipment. The following are examples of an actuation device: a circuit breaker, a relay, a valve (with its operator) used to control compressed air to the operator of a containment isolation valve, (and a module containing such equipment). 355 (3) A component or assembly of components that directly controls the motive power (electricity, compressed air, hydraulic fluid, etcetera) for actuated equipment. *Note:* Examples of actuation devices are: circuit breakers, relays, and pilot valves. 356

actuation time, relay. *See:* **relay actuation time.**

actuator (automatic train control). A mechanical or electric device used for automatic operation of a brake valve. 328

actuator, centrifugal (rotating machinery). Rotor-mounted element of a centrifugal starting switch. *See also:* **centrifugal starting switch.** 63

actuator, relay. *See:* **relay actuator.**

actuator valve. An electropneumatic valve used to control the operation of a brake valve actuator. 328

acyclic machine (homopolar machine*) (unipolar machine*) (rotating machinery). A direct-current machine in which the voltage generated in the active conductors maintains the same direction with respect to those conductors. *See also:* **direct-current commutating machine.** 63

* Deprecated

adaptation. The process by which the retina becomes accustomed to more or less light than it was exposed to during an immediately preceding period. It results in a change in the sensitivity of the photoreceptors to light. *Note:* Adaptation is also used to refer to the final state of the process, as reaching a condition of dark adaptation or light adaptation. *See:* **scotopic vision; photopic vision.** *See also:* **visual field.** 167

adapter (1) (general). A device for connecting parts that will not mate. An accessory to convert a device to a new or modified use. 185

(2) (test, measurement and diagnostic equipment). A device or series of devices designed to provide a compatible connection between the unit under test and the test equipment. May include proper stimuli or loads not contained in the test equipment. *See:* **interface.** 54

adapter kit (test, measurement and diagnostic equipment). A kit containing an assortment of cables and adapters for use with test or support equipment. 54

adapter, standard. A two-port device having standard connectors for joining together two waveguides or transmission lines with nonmating standard connectors. 110

adapter, waveguide (1) (general). A two-port device for joining two waveguides having nonmating connectors. *See also:* **waveguide.** 185

(2) (waveguide components). A structure used to interconnect two waveguides which differ in size or type. If the modes of propagation also differ, the adapter functions as a mode transducer.   166

adapting. *See:* **self-adapting.**

adaptive antenna system. An antenna system having circuit elements associated with its radiating elements such that some of the antenna properties are controlled by the received signal. *See also:* **antenna.**   179, 111

adaptive control system. *See:* **control system, adaptive.**

adaptive equalization. *See:* **adaptive systems.**

adaptive system. A system that has a means of monitoring its own performance and a means of varying its own parameters by closed-loop action to improve its performance. *See also:* **system science.**   209

Adcock antenna. A pair of vertical antennas separated by a distance of one-half wavelength or less, and connected in phase opposition to produce a radiation pattern having the shape of the figure eight in all planes containing the centers of the two antennas. *See also:* **antenna.**   111

add and subtract relay. A stepping relay that can be pulsed to rotate the movable contact arm in either direction. *See also:* **relay.**   259

adder. A device whose output is a representation of the sum of the two or more quantities represented by the inputs. *See:* **half-adder.** *See also:* **electronic analog computer.**   235, 210, 54, 77

addition agent (electroplating). A substance that, when added to an electrolyte, produces a desired change in the structure or properties of an electrodeposit, without producing any appreciable change in the conductivity of the electrolytes, or in the activity of the metal ions or hydrogen ions. *See also:* **electroplating.**   328

address (electronic computations and data processing). (1) An identification, as represented by a name, label, or number, for a register, location in storage, or any other data source or destination such as the location of a station in a communication network. (2) Loosely, any part of an instruction that specifies the location of an operand for the instruction. (3) (electronic machine-control system). A means of identifying information or a location in a control system. *Example:* The x in the command x 12345 is an address identifying the numbers 12345 as referring to a position on the x axis.   224, 207, 255, 77

address, effective (computing systems). The address that is derived by applying any specified rules (such as rules relating to an index register or indirect address) to the specified address and that is actually used to identify the current operand.   77

address format (computing machines). The arrangement of the address parts of an instruction. *Note:* The expression plus-one is frequently used to indicate that one of the addresses specifies the location of the next instruction to be executed, such as one-plus-one, two-plus-one, three-plus-one, four-plus-one.   255, 77

address part. A part of an instruction that usually is an address, but that may be used in some instructions for another purpose. *See also:* **instruction code.**   235

address register (computing machines). A register in which an address is stored.   255, 77

address, tag. *See:* **symbolic address.**

ADF. *See:* **automatic direction finder.**

***A* display (radar).** A display in which targets appear as vertical deflections from a line representing a time base. Target distance is indicated by the horizontal position of the deflection from one end of the time base. The amplitude of the vertical deflection is a function of the signal intensity. See accompanying figure. *See also:* **navigation.**
13, 187

A display.

adjacent channel (data transmission). The channel whose frequency is adjacent to that of the reference channel.
59

adjacent-channel attenuation (receivers). *See:* **selectance.**

adjacent-channel interference. Interference in which the extraneous power originates from a signal of assigned (authorized) type in an adjacent channel. *See also:* **interference; radio transmission.**   339

adjacent-channel selectivity and **desensitization (receiver performance) (receiver).** A measure of the ability to discriminate against a signal at the frequency of the adjacent channel. Desensitization occurs when the level of any off-frequency signal is great enough to alter the useable sensitivity. *See also:* **receiver performance.**   181

adjoint system. (1) A method of computation based on the reciprocal relation between a system of ordinary linear differential equation and its adjoint. *Note:* By solution of the adjoint system it is possible to obtain the weighting function (response to a unit impulse) $W(T, t)$ of the original system for fixed T (the time of observation) as a function of t (the time of application of the impulse). Thus, this method has particular application to the study of systems with time-varying coefficients. The weighting function then may be used in convolution to give the response of the original system to an arbitrary input. *See also:* **electronic analog computer.**   9, 77
(2) For a system whose state equations are $dx(t)/dt = f(x(t),u(t),t)$, the adjoint system is defined as that system whose state equations are $dy(t)/dt = -y(t)$, where $A*$ is the conjugate transpose of the matrix whose i,j element is $\partial f_i/\partial x_j$. *See also:* **control system.**   198

adjust (instrument). Change the value of some element of the mechanism, or the circuit of the instrument or of an auxiliary device, to bring the indication to a desired value, within a specified tolerance for a particular value of the quantity measured. *See also:* **instrument.** 328

adjustable capacitor. A capacitor, the capacitance of which can be readily changed. 210

adjustable constant-speed motor. A motor, the speed of which can be adjusted to any value in the specified range, but when once adjusted the variation of speed with load is a small percentage of that speed. For example, a direct-current shunt motor with field-resistance control designed for a specified range of speed adjustment. *See also:* **asynchronous machine; direct-current commutating machine.** 63

adjustable impedance-type ballast (illuminating engineering). A reference ballast consisting of an adjustable inductive reactor and a suitable adjustable resistor in series. These two components are usually designed so that the resulting combination has sufficient current-carrying capacity and range of impedance to be used with a number of different sizes of lamps. The impedance and power factor of the reactor-resistor combination are adjusted and checked each time the unit is used. 271

adjustable inductor. An inductor in which the self or mutual inductance can be readily changed. 210

adjustable resistor. A resistor so constructed that its resistance can be changed by moving and setting one or more movable contacting elements. 210

adjustable-speed drive (industrial control). An electric drive designed to provide easily operable means for speed adjustment of the motor, within a specified speed range. *See also:* **electric drive.** 206

adjustable-speed motor. A motor the speed of which can be varied gradually over a considerable range, but when once adjusted remains practically unaffected by the load; such as a direct-current shunt-wound motor with field resistance control designed for a considerable range of speed adjustment. *See also:* **asynchronous machine; direct-current commutating machine.** 328

adjustable varying-speed motor. A motor the speed of which can be adjusted gradually, but when once adjusted for a given load will vary in considerable degree with change in load; such as a direct-current compound-wound motor adjusted by field control or a wound-rotor induction motor with rheostatic speed control. *See also:* **asynchronous machine; direct-current commutating machine.** 328

adjustable varying-voltage control (industrial control). A form of armature-voltage control obtained by impressing on the armature of the motor a voltage that may be changed by small increments, but when once adjusted for a given load will vary considerably with change in load with a consequent change in speed, such as may be obtained from a differentially compound-wound generator with adjustable field current or by means of an adjustable resistance in the armature circuit. *See also:* **control.** 206

adjustable voltage control. A form of armature-voltage control obtained by impressing on the armature of the motor a voltage that may be changed in small increments; but when once adjusted, it, and consequently the speed of the motor, are practically unaffected by a change in load. *Note:* Such a voltage may be obtained from an individual shunt-wound generator with adjustable field current, for each motor. *See also:* **control.** 206

adjusted speed (industrial control). The speed obtained intentionally through the operation of a control element in the apparatus or system governing the performance of the motor. *Note:* The adjusted speed is customarily expressed in percent (or per unit) of base speed (for direct-current shunt motors) or of rated full-load speed (for all other motors). *See also:* **electric drive.** 206

adjuster (rotating machinery). An element or group of elements associated with a feedback control system by which manual adjustment of the level of a controlled variable can be made. 63, 105

adjuster, synchronous-machine voltage-regulator. An adjuster associated with a synchronous-machine voltage regulator by which manual adjustment of the synchronous-machine voltage can be made. 63

adjustment (test, measurement and diagnostic equipment). The act of manipulating the equipment's controls to achieve a specified condition. 54

adjustment accuracy of instrument shunts (electric power systems). The limit of error, expressed as a percentage of the rated voltage drop, of the initial adjustment of the shunt by resistance or low-current methods. 201

adjustment, relay. *See:* **relay adjustment.**

Adler tube*. *See:* **beam parametric-amplifier.**

* Deprecated

administrative authority (transmission and distribution). An organization exercising jurisdiction over application of the National Electrical Safety Code. 262

administrative control (nuclear power generating stations). Rules, orders instructions, procedures, policies, practices, and designations of authority and responsibility written by management to obtain assurance of safety and high quality operation and maintenance of a nuclear power plant. 20

admissible control input set (control system). A set of control inputs that satisfy the control constraints. *See also:* **control system.** 198

admittance (1) (linear constant-parameter system). (1) The corresponding admittance function with p replaced by $j\omega$ in which ω is real. (2) The ratio of the phasor equivalent of a steady-state sine-wave current or current-like quantity (response) to the phasor equivalent of the corresponding voltage or voltage-like quantity (driving force). The real part is the conductance and the imaginary part is the susceptance. *Note:* Definitions (1) and (2) are equivalent. *Editor's Note:* The ratio Y is commonly expressed in terms of its orthogonal components, thus:

$$Y = G + jB$$

where Y, G, and B are respectively termed the admittance, conductance, and susceptance, all being measured in mhos (reciprocal ohms). In a simple circuit consisting of R, L, and C all in parallel, Y becomes

$$Y = G + j\left(\omega C - \frac{1}{\omega L}\right),$$

where $\omega = 2\pi f$ and f is the frequency. In this special case, $G = 1/R$. Historically, some authors have preferred an opposite sign convention for susceptance, thus: $Y = G - jB$ (now deprecated). Thus, in the simple parallel circuit

$$Y = G - j \left(\frac{1}{\omega L} - \omega C \right).$$

The reader will note that according to either convention, the end result is that any predominantly capacitive susceptance will advance the phase of the response, relative to the driving force. However, the $Y = G - jB$ convention* requires that a predominantly capacitive susceptance B, like a predominantly capacitive reactance X, shall be denoted as negative. The reader is therefore advised to become aware of his author's preference. *See:* **conductance; susceptance.** 210,150

* Deprecated

(2) (electric machine). A linear operator expressing the relation between current (incrementals) and voltage (incrementals). Its inverse is called the impedance of an electric machine. *Notes:* (1) If a matrix has admittances as its elements, it is usually referred to as admittance matrix. Frequently the admittance matrix is called admittance for short. (2) Most admittances are usually defined with the mechanical angular velocity of the machine at steady state. *See also:* **asynchronous machine; direct-current commutating machine.** 63

admittance, effective input (electron tube or valve). The quotient of the sinusoidal component of the control-grid current by the corresponding component of the control voltage, taking into account the action of the anode voltage on the grid current; it is a function of the admittance of the output circuit and the interelectrode capacitance. *Note:* It is the reciprocal of the effective input impedance. *See:* **electron-tube admittances.** 244,189

admittance, effective output (electron tube or valve). The quotient of the sinusoidal component of the anode current by the corresponding component of the anode voltage, taking into account the output admittance and the interelectrode capacitance. *Note:* It is the reciprocal of the effective output impedance. *See* **electron-tube admittances.** 244, 189

admittance, electrode (jth electrode of the n-electrode electron tube). *See:* **electrode admittance.**

admittance function (defined for linear constant-parameter systems or parts of such systems). That mathematical function of p that is the ratio of a current or current-like quantity (response) to the corresponding voltage-like quantity (driving force) in the hypothetical case in which the latter is e^{pt} (where e is the base of the natural logarithms, p is arbitrary but independent of t, and t is an independent variable that physically is usually time), and the former is a steady-state response in the form of

$$Y(p)e^{pt}.$$

Note: In electric circuits voltage is always the driving force and current is the response even though as in nodal analysis the current may be the independent variable; in electromagnetic radiation electric field strength is always considered the driving force and magnetic field strength the response, and in mechanical systems mechanical force is always considered as a driving force and velocity as a response. In a general sense the dimension (and unit) of admittance in a given application may be whatever results from the ratio of the dimensions of the quantity chosen as the response to the dimensions of the quantity chosen as the driving force. However, in the types of systems cited above any deviation from the usual convention should be noted. *See also:* **network analysis.** 210

admittance matrix, short-circuit (multiport network) (circuits and systems). A matrix whose elements have the dimension of admittance and, when multiplied into the vector of port voltages, gives the vector of port currents. 67

admittance, short-circuit (circuits and systems). (1) (general) An admittance of a network that has a specified pair or group of terminals short-circuited. (2) (four-terminal network or line) The input, output or transfer admittance parameters y_{11}, y_{22}, and y_{12} of a four-terminal network when the far-end is short circuited. 67

admittance, short-circuit driving-point (jth terminal of an n-terminal network). The driving-point admittance between that terminal and the reference terminal when all other terminals have zero alternating components of voltage with respect to the reference point. *See also:* **electron-tube admittances.** 125

admittance, short-circuit feedback (electron-device transducer). The short-circuit transfer admittance from the physically available output terminals to the physically available input terminals of a specified socket, associated filters, and electron device. *See also:* **electron-tube admittances.** 125

admittance, short-circuit forward (electron-device transducer). The short-circuit transfer admittance from the physically available input terminals to the physically available output terminals of a specified socket, associated filters, and electron device. *See also:* **electron-tube admittances.** 125

admittance, short-circuit input (electron-device transducer). The driving-point admittance at the physically available input terminals of a specified socket, associated filters, and tube. All other physically available terminals are short-circuited. *See also:* **electron-tube admittances.** 125, 190

admittance, short-circuit output (electron-device transducer). The driving-point admittance at the physically available output terminals of a specified socket, associated filters, and tube. All other physically available terminals are short-circuited. *See also:* **electron-tube admittances.** 125, 190

admittance, short-circuit transfer (from the jth terminal to the lth terminal of an n-terminal network). The transfer admittance from terminal j to terminal l when all terminals except j have zero complex alternating components of voltage with respect to the reference point. *See also:* **electron-tube admittances.** 190, 125

advance ball (mechanical recording). A rounded support (often sapphire) attached to a cutter that rides on the surface of the recording medium so as to maintain a uniform depth of cut and correct for small irregularities of the disk surface. 176

adverse weather (electric power systems). Weather conditions that cause an abnormally high rate of forced outages for exposed components during the periods such conditions persist. *Note:* Adverse weather conditions can be defined for a particular system by selecting the proper values and combinations of conditions reported by the

Weather Bureau: thunderstorms, tornadoes, wind velocities, precipitation, temperature, etcetera. *See also:* **outage.**
200

aeolight (optical sound recording). A glow lamp employing
a cold cathode and a mixture of permanent gases in which
the intensity of illumination varies with the applied signal
voltage. 176

aeration cell. *See:* **differential aeration cell.**

aerial cable. An assembly of insulated conductors installed
on a pole line or similar overhead structures; it may be
self-supporting or installed on a supporting messenger
cable. *See also:* **cable.** 64

aerial lug. *See* **external connector (pothead).**

aerodrome beacon. An aeronautical beacon used to indicate the location of an aerodrome. *Note:* An aerodrome
is any defined area on land or water, including any
buildings, installations, and equipment intended to be used
either wholly or in part for the arrival, departure, and
movement of aircraft. 167

aeronautical beacon. An aeronautical ground light visible
at all azimuths, either continuously or intermittently, to
designate a particular location on the surface of the earth.
167

aeronautical ground light. Any light specially provided as
an aid to air navigation, other than a light displayed on an
aircraft. 167

aeronautical light. Any luminous sign or signal that is
specially provided as an aid to air navigation. 167

aerophare (air operations). A name for radio beacon. *See
also:* **navigation.** 13, 187

aerosol (laser-maser). A suspension of small solid or liquid
particles in a gaseous medium. Typically, the particle sizes
may range from 100 μm to 0.01 μm or less. 363

aerosol development (electrostatography). Development
in which the image-forming material is carried to the field
of the electrostatic image by means of a suspending gas.
See also: **electrostatography.** 236, 191

**aerospace support equipment (test, measurement and diag-
nostic equipment).** All equipment [implements, tools, test
equipment, devices (mobile or fixed), and so forth], both
airborne and ground, required to make an aerospace system (aircraft, missile, and so forth) operational in its intended environment. Aerospace support equipment includes ground support equipment. 54

AEW. Airborne early warning radar.

AF. *See:* **analog-to-frequency converter.**

AFC. *See:* **automatic frequency control.**

afterimage. A visual response that occurs after the stimulus
causing it has ceased. *See also:* **visual field.** 167

afterpulse (photo multipliers). A spurious pulse induced in
a photomultiplier by a previous pulse. *See also:* **phototube.**
117

AGC. *See:* **automatic gain control.**

aggressive carbon dioxide (corrosion). Free carbon dioxide
in excess of the amount necessary to prevent precipitation
of calcium as calcium carbonate. 221, 205

aging (metallic rectifier) (semiconductor rectifier cell). Any
persisting change (except failure) that takes place for any
reason in either the forward or reverse resistance characteristic. *See also:* **rectification.** 237, 66

agitator (hydrometallurgy) (electrowinning). A receptacle
in which ore is kept in suspension in a leaching solution.

See also: **electrowinning.** 328

aided tracking (radar). A tracking system in which the
manual correction of the tracking error automatically
corrects the rate of motion of the tracking mechanism.
13

air (1) (industrial control) (prefix). Applied to a device that
interrupts an electric circuit, indicates that the interruption
occurs in air. 225, 206
(2) (rotating machinery). *Note:* When a definition
mentions the term air, this term can be replaced, where
appropriate, by the name of another gas (for example,
hydrogen). Similarly the term water can be replaced by
the name of another liquid. *See also:* **asynchronous machine; direct-current commutating machines.** 63

AI radar (airborne intercept radar). A fire control radar
for use in interceptor aircraft. 13

air baffle (rotating machinery). *See:* **air guide.**

air-blast circuit breaker. *See:* Note under **circuit breaker.**

air cell. A gas cell in which depolarization is accomplished
by atmospheric oxygen. *See also:* **electrochemistry.**
328

air circuit breaker. *See:* Note under **circuit breaker.**

air conduction (hearing). The process by which sound is
conducted to the inner ear through the air in the outer ear
canal as part of the pathway. 176

air-cooled (rotating machinery). Cooled by air at atmospheric pressure. *See also:* **asynchronous machine; direct-current commutating machine.** 63

air cooler (rotating machinery). A cooler using air as one
of the fluids. *See also:* **fan (rotating machinery).** 63

air-core inductance (winding inductance). The effective
self-inductance of a winding when no ferromagnetic materials are present. *Note:* The winding inductance is not
changed when ferromagnetic materials are present.
197

aircraft aeronautical light. Any aeronautical light specially
provided on an aircraft. 167

aircraft bonding. The process of electrically interconnecting all parts of the metal structure of the aircraft as
a safety precaution against the buildup of isolated static
charges and as a means of reducing radio interference.
328

aircraft induction motor (rotating machinery). A motor
designed for operation in the environment seen by aircraft
yet having minimum weight and utmost reliability for a
limited life. *See also:* **asynchronous machine.** 63

air-derived navigation data. Data obtained from measurements made at an airborne vehicle. *See also:* **navigation.** 187, 13

air duct (air guide) (ventilating duct) (rotating machinery).
Any passage designed to guide ventilating air. 63

air ducting (rotating machinery). Any separate structure
mounted on a machine to guide ventilating air to or from
a heat exchanger, filter, fan or other device mounted on
the machine. 63

air filter (rotating machinery). Any device used to remove
suspended particles from air. *See also:* **fan (rotating machinery).** 63

air gap (gap) (rotating machinery). A separating space
between two parts of magnetic material, the combination
serving as a path for magnetic flux. *Note:* This space is

normally filled with air or hydrogen and represents clearance between rotor and stator of an electric machine. *See also:* **direct-current commutating machine.** 63

air-gap factor (fringing coefficient) (rotating machinery). A factor used in machine design calculations to determine the effective length of the air gap from the actual separation of rotor and stator. *Note:* The use of the fringing coefficient is necessary to account for several geometric effects such as the presence of slots and cooling ducts in the rotor and stator. *See also:* **rotor (rotating machinery); stator.** 63

air-gap line. The extended straight line part of the no-load saturation curve. 63, 105

air gap, relay. *See:* **relay air gap.**

air guide (air duct) (rotating machinery). Any structure designed to direct the flow of ventilating air. 63

air horn. A horn having a diaphragm that is vibrated by the passage of compressed air. *See also:* **protective signaling.** 328

air-insulated terminal box. A terminal box so designed that the protection of phase conductors against electrical failure within the terminal box is [achieved] by adequately spacing bare conductors with appropriate insulation supports. 63

air mass. The mass of air between a surface and the sun that affects the spectral distribution and intensity of sunlight. *See also:* **air mass one; air mass zero.** 113

air mass one. A term that specifies the spectral distribution and intensity of sunlight on earth at sea level with the sun directly overhead and passing through a standard atmosphere. *See also:* **air mass; air mass zero.** 113

air mass zero. A term that specifies the spectral distribution and intensity of sunlight in near-earth space without atmospheric attenuation. *Note:* The air mass must be specified when reporting the efficiency of solar cells; for example, 10 percent efficient at air mass zero, 60 degrees Celsius. *See also:* **air mass; air mass one.** 113

air opening (rotating machinery). A port for the passage of ventilation air. 63

air pipe (rotating machinery). Any separate structure designed for attaching to a machine to guide the inlet ventilating air to the machine or the exhaust air away from the machine. 63

airport beacon. *See:* aerodrome beacon.

airport surface detection equipment (ASDE). A ground based radar for observation of the positions of aircraft on the surface of an airport. 13

airport surveillance radar (ASR). A medium-range (~60 nautical miles) Surveillance Radar used to control aircraft in the vicinity of an airport. 13

air-position indicator (API) (navigation). An airborne computing system which presents a continuous indication of the aircraft position on the basis of aircraft heading, airspeed and elapsed time. 13

air-route surveillance radar (ARSR). A long-range (~200 nautical miles) Surveillance Radar used to control aircraft on airways beyond the coverage of airport surveillance radar **(ASR).** 13

air shield (rotating machinery). An air guide used to prevent ventilating air from returning to the blower inlet before passing through the ventilating circuit. 63

air speed. The rate of motion of a vehicle relative to the air mass. *See also:* **navigation.** 13, 187

air switch. A switching device designed to close and open one or more electric circuits by means of guided separable contacts that separate in air. 79

air terminal (lightning protection). The combination of elevation rod and brace, or footing placed on upper portions of structures, together with tip or point if used. *See also:* **lightning protection and equipment.** 328

air-to-air characteristic (telephony). The acoustic output level of a telephone set as a function of the acoustic input level of another telephone set to which it is connected. The output is measured in an Artificial Ear, and the input is measured in free field at a specified location relative to the reference point of an Artificial Mouth. 114

airway beacon. An aeronautical beacon used to indicate a point on the airway. 167

alarm checking (telephone switching system). The identification of an alarm from a remote location by communicating with its point of origin. 55

alarm point (power-system communication). A supervisory control status point considered to be an alarm. *See also:* **supervisory control system.** 59

alarm relay. A monitoring relay whose function is to operate an audible or visual signal to announce the occurrence of an operation or a condition needing personal attention, and usually provided with a signaling cancellation device. *See also:* **relay.** 60, 127, 103

alarm sending (telephone switching system). The extension of alarms from an office to another location. 55

alarm signal. A signal for attracting attention to some abnormal condition. 193

alarm summary printout (sequential events recording systems). The recording of all inputs currently in the alarm state. 48, 58

alarm switch (switching device). An auxiliary switch that actuates a signaling device upon the automatic opening of the switching device with which it is associated. 103, 202

alarm system (protective signaling). An assembly of equipment and devices arranged to signal the presence of a hazard requiring urgent attention. *See also:* **protective signaling.** 328

albedo (photovoltaic power system). The reflecting power expressed as the ratio of light reflected from an object to the total amount falling on it. *See also:* **photovoltaic power system; solar cells (photovoltaic power system).** 186

ALC. *See:* **automatic load (level) control.**

Alford loop antenna. A multielement antenna having approximately equal amplitude currents which are in phase and uniformly distributed along each of its peripheral elements producing a substantially circular radiation pattern in the plane of polarization (originally developed as a four-element horizontally polarized UHF loop antenna). 111

algorithm. A prescribed set of well-defined rules or processes for the solution of a problem in a finite number of steps, for example, a full statement of an arithmetic procedure for evaluating sin x to a stated precision. *See also:* **heuristic.** 255, 77, 54

algorithmic language (test, measurement and diagnostic equipment). A language designed for expressing algorithms. 54

align (test, measurement and diagnostic equipment). To adjust a circuit, equipment or system so that their functions are properly synchronized or their relative positions properly oriented. For example, trimmers, padders, or variable inductances in tuned circuits are adjusted to give a desired response for fixed tuned equipment or to provide tracking for tuneable equipment. 54

aligned-grid tube (or valve). A vacuum multigrid tube or valve in which at least two of the grids are aligned the one behind the other so as to obtain a particular effect (canalizing an electron beam, suppressing noise, etcetera). *See also:* **electron tube.** 244, 190

alignment (1) (communication practice). The process of adjusting a plurality of components of a system for proper interrelationship. *Note:* The term is applied especially to (A) the adjustment of the tuned circuits of an amplifier for desired frequency response, and (B) the synchronization of components of a system. *See also:* **radio transmission.** 59
(2) (inertial navigation equipment). The orientation of the measuring axes of the inertial components with respect to the coordinate system in which the equipment is used. *Note:* Initial alignment refers to the result of the process of bringing the measuring axes into a desired orientation with respect to the coordinate system in which the equipment is used, prior to departure. The initial alignment can be refined by the use of non-inertial sensors while in operational use. *See also:* **navigation.** 187, 13

alignment kit (test, measurement and diagnostic equipment). A kit containing all the instruments or tools necessary for the alignment of electrical or mechanical components. 54

alignment tool (test, measurement and diagnostic equipment). A small screwdriver, socket wrench, or special tool used for adjusting electronic, mechanical, or optical units, usually constructed of non-magnetic materials. 54

alive (1) (electric system). Electrically connected to a source of potential difference, or electrically charged so as to have a potential different from that of the ground. *Note:* The term **alive** is sometimes used in place of the term **current-carrying**, where the intent is clear, to avoid repetitions of the longer term. *Syn.* **live.** *See also:* **insulated.** 262
(2) (transmission and distribution) (live). *See:* **energized.** 262

alkaline cleaning (electroplating). Cleaning by means of alkaline solutions. *See also:* **electroplating.** 328

alkaline storage battery. A storage battery in which the electrolyte consists of an alkaline solution, usually potassium hydroxide. *See also:* **battery (primary or secondary).** 328

allocation (computing machine). *See:* **storage allocation.** *See also:* **electronic digital computer.**

allotting (telephone switching system). The preselecting by a common control of an idle circuit. 55

alloy junction (semiconductor). A junction formed by recrystallization on a base crystal from a liquid phase of one or more components and the semiconductor. *See also:* **semiconductor.** 237, 66

alloy plate. An electrodeposit that contains two or more metals codeposited in combined form or in intimate mixtures. *See also:* **electroplating.** 328

all-pass function (linear passive network). A transmittance

that provides only phase shift, its magnitude characteristic being constant. *Notes:* (1) For lumped-parameter networks, this is equivalent to specifying that the zeros of the function are the negatives of the poles. (2) A realizable all-pass function exhibits non-decreasing phase lag with increasing frequency. (3) A trivial all-pass function has zero phase at all frequencies. *See also:* **linear passive networks.** 238

all-pass network (all-pass transducer). A network designed to introduce phase shift or delay without introducing appreciable attenuation at any frequency. *See also:* **network analysis.** 328

all-pass transducer. *See:* **all-pass network.**

all-relay system. An automatic telephone switching system in which all switching functions are accomplished by relays. 328

alphabet. A character set arranged in certain order. *Note:* Character sets are finite quantities of letters of the normal alphabet, digits, punctuation marks, control signals, such as carriage return and other ideographs. Characters are usually represented by letters (graphics) or technically realized in the form of combinations of punched holes, sequences of electric pulses, etcetera. 194

alphameric. *See:* **alphanumeric.**

alphanumeric. Pertaining to a character set that contains both letters and digits, but usually some other characters such as punctuation symbols. *Syn.* **alphameric.** 77

alteration (elevator, dumbwaiter, or escalator). Any change or addition to the equipment other than ordinary repairs or replacements. *See also:* **elevator.** 328

alternate-channel interference (second-channel interference). Interference caused in one communication channel by a transmitter operating in a channel next beyond an adjacent channel. *See also:* **radio transmission.** 328

alternate display (oscillography). A means of displaying output signals of two or more channels by switching the channels in sequence. *See:* **oscillograph.** 185

alternate route (data transmission). A secondary communications path used to reach a destination if the primary path is unavailable. 59

alternate-route trunk group (telephone switching system). A trunk group which accepts alternate-routed traffic. 55

alternate routing (telephone switching system). A means of selectively distributing traffic over a number of routes ultimately leading to the same destination. 55

alternating charge characteristic (nonlinear capacitor). The function relating the instantaneous values of the alternating component of transferred charge, in a steady state, to the corresponding instantaneous values of a specified applied periodic capacitor-voltage. *Note:* The nature of this characteristic may depend upon the nature of the applied voltage. *See also:* **non-linear capacitor.** 191

alternating current. A periodic current the average value of which over a period is zero. *Note:* Unless distinctly specified otherwise, the term **alternating current** refers to a current which reverses at regularly recurring intervals of time and which has alternately positive and negative values. 3, 53

alternating-current analog computer. *See:* **analog computer, alternating current.**

alternating-current and direct-current (ac-dc) ringing

(telephone switching systems). Ringing in which alternating current activates the ringer and direct current controls the removal of ringing upon answer.   55

alternating-current circuit. A circuit that includes two or more interrelated conductors intended to be energized by alternating current.   210

alternating-current commutator motor. An alternating-current motor having an armature connected to a commutator and included in an alternating-current circuit. *See also:* **asynchronous machine.**   63

alternating-current component. The current remaining when the average value has been subtracted from an alternating current. *See:* **symmetrical component (total current).**   210, 103

alternating-current–direct-current general-use snap-switch. A form of general-use snap-switch suitable for use on either direct- or alternating-current circuits for controlling the following: (1) Resistive loads not exceeding the ampere rating at the voltage involved. (2) Inductive loads not exceeding one-half the ampere rating at the voltage involved, except that switches having a marked horsepower rating are suitable for controlling motors not exceeding the horsepower rating of the switch at the voltage involved. (3) Tungsten filament lamp loads not exceeding the ampere rating at 125 volts, when marked with the letter "T" Alternating-current–direct-current general-use snap-switches are not generally marked alternating-current–direct-current, but are always marked with their electrical rating. *See also:* **switch.**   256

alternating-current–direct-current ringing. Ringing in which a combination of alternating and direct currents is utilized, the direct current being provided to facilitate the functioning of the relay that stops the ringing. *See also:* **telephone switching system.**   328

alternating-current distribution. The supply to points of utilization of electric energy by alternating current from its source to one or more main receiving stations. *Note:* Generally a voltage is employed that is not higher than that which could be delivered or utilized by rotating electric machinery. Step-down transformers of a capacity much smaller than that of the line are usually employed as links between the moderate voltage of distribution and the lower voltage of the consumer's apparatus.   64

alternating-current electric locomotive. An electric locomotive that collects propulsion power from an alternating-current distribution system. *See also:* **electric locomotive.**   328

alternating-current erasing head (magnetic recording). A head that uses alternating current to produce the magnetic field necessary for erasing. *Note:* Alternating-current erasing is achieved by subjecting the medium to a number of cycles of a magnetic field of a decreasing magnitude. The medium is, therefore, essentially magnetically neutralized.   176

alternating-current floating storage-battery system. A combination of alternating-current power supply, storage battery, and rectifying devices connected so as to charge the storage battery continuously and at the same time to furnish power for the operation of signal devices.   328

alternating-current general-use snap-switch. A form of general-use snap-switch suitable only for use on alternating-current circuits for controlling the following: (1)

Resistive and inductive loads (including electric discharge lamps) not exceeding the ampere rating at the voltage involved. (2) Tungsten filament lamp loads not exceeding the ampere rating at 120 volts. (3) Motor loads not exceeding 80 percent of the ampere rating of the switches at the rated voltage. *Note:* All alternating-current general-use snap-switches are marked ac in addition to their electrical rating. *See also:* **switch.**   256

alternating-current generator. A generator for the production of alternating-current power.   63

alternating-current magnetic biasing (magnetic recording). Magnetic biasing accomplished by the use of an alternating current, usually well above the signal-frequency range. *Note:* The high-frequency linearizing (biasing) field usually has a magnitude approximately equal to the coercive force of the medium.   176

alternating-current motor. An electric motor for operation by alternating current.   63

alternating-current pulse. An alternating-current wave of brief duration. *See also:* **pulse.**   328

alternating-current relay. *See:* **relay, alternating-current.**

alternating current root-mean-square voltage rating (semiconductor rectifiers). The maximum root-mean-square value of applied sinusoidal voltage permitted by the manufacturer under stated conditions. *See:* **semiconductor rectifier stack.**   208

alternating-current saturable reactor. A reactor whose impedance varies cyclically with the alternating current (or voltage).   53

alternating-current transmission. The transfer of electric energy by alternating current from its source to one or more main receiving stations for subsequent distribution. *Note:* Generally a voltage is employed that is higher than that which would be delivered or utilized by electric machinery. Transformers of a capacity comparable to that of the line are usually employed as links between the high voltage of transmission and the lower voltage used for distribution or utilization. *See also:* **alternating-current distribution.**   64

alternating-current transmission (television). That form of transmission in which a fixed setting of the controls makes any instantaneous value of signal correspond to the same value of brightness only for a short time. *Note:* Usually this time is not longer than one field period and may be as short as one line period. *See also:* **television.**   328

alternating-current winding (rectifier transformer). The primary winding that is connected to the alternating-current circuit and usually has no conductive connection with the main electrodes of the rectifier. *See also:* **rectifier transformer.**   203, 53

alternating function. A periodic function whose average value over a period is zero. For instance, $f(t) = B \sin \omega t$ is an alternating function (ω, B assumed constants).   210

alternating sparkover voltage (arrester). *See:* **arrester, alternating sparkover voltage.**

alternating voltage. *See:* **alternating current.**

alternative (electric power system). A qualifying word identifying a power circuit, equipment, device, or component available to be connected or switched into the circuit to perform a function when the preferred component has failed or is inoperative. *See also: reserve.* 58

alternator transmitter. A radio transmitter that utilizes power generated by a radio-frequency alternator. *See also: radio transmitter.* 111, 240

altitude (1) (astronomy). The angular distance of a heavenly body measured on that great circle that passes perpendicular to the plane of the horizon through the body and through the zenith. *Note:* It is measured positively from the horizon to the zenith, from 0 to 90 degrees. *See also: light; sunlight.* 167

(2) (series capacitor). The elevation of the series capacitor above mean sea level. 86

altitude-treated current-carrying brush. A brush specially fabricated or treated to improve its wearing characteristics at high altitudes (over 6000 meters). *See also: air-transportation electric equipment.* 328

aluminum cable steel reinforced (ACSR). A composite conductor made up of a combination of aluminum and steel wires. In the usual construction the aluminum wires surround the steel. *See also: conductor.* 64

aluminum conductor. A conductor made wholly of aluminum. *See also: conductor.* 64

aluminum-covered steel wire (power distribution underground cables). A wire having a steel core to which is bonded a continuous outer layer of aluminum. *See also: power distribution, underground construction.* 57

AMA. *See: automatic message accounting system.*

amalgam (electrolytic cells). The product formed by mercury and another metal in an electrolytic cell. 328

ambient air temperature (metal enclosed bus) (relaying). The temperature of the surrounding air which comes in contact with equipment. *Note:* Ambient air temperature, as applied to enclosed bus or switchgear assemblies, is the average temperature of the surrounding air that comes in contact with the enclosure. 78, 79

ambient conditions. Characteristics of the environment, for example, temperature, humidity, pressure. *See also: measurement system.* 185, 54

ambient level (electromagnetic compatibility). The values of radiated and conducted signal and noise existing at a specified test location and time when the test sample is not activated. *See also: electromagnetic compatibility.* 197

ambient noise (1) (room noise). Acoustic noise existing in a room or other location. Magnitudes of ambient noise are usually measured with a sound level meter. *Note:* The term room noise is commonly used to designate ambient noise at a telephone station. *See also: circuit noise; circuit noise level; line noise.* 59

(2) (mobile communication). The average radio noise power in a given location that is the integrated sum of atmospheric, galactic, and man-made noise. *See also: telephone station.* 181

ambient operating-temperature range (power supply). The range of environmental temperatures in which a power supply can be safely operated. For units with forced-air cooling, the temperature is measured at the air intake. *See also: power supply.* 228, 186

ambient radio noise. *See: ambient level.*

ambient temperature (1) (general). The temperature of the medium such as air, water, or earth into which the heat of the equipment is dissipated. *Notes:* (A) For self-ventilated equipment, the ambient temperature is the average temperature of the air in the immediate neighborhood of the equipment. (B) For air- or gas-cooled equipment with forced ventilation or secondary water cooling, the ambient temperature is taken as that of the ingoing air or cooling gas. (C) For self-ventilated enclosed (including oil-immersed) equipment considered as a complete unit, the ambient temperature is the average temperature of the air outside of the enclosure in the immediate neighborhood of the equipment. 225, 206, 53, 124, 27, 91

(2) (power switchgear). The temperature of the surrounding medium that comes in contact with the device or equipment. 103

(3) (outdoor apparatus bushing). The temperature of the surrounding air that comes in contact with the device or equipment in which the bushing is mounted. 168

(4) (light emitting diode) (free air temperature) (T_A). The air temperature measured below a device, in an environment of substantially uniform temperature, cooled only by natural air convection and not materially affected by reflective and radiant surfaces. 162

ambient temperature rating at quarter-rated thermal burden. Maximum ambient temperature at which a potential transformer can be operated without exceeding the specified temperature limitations, when operated at rated voltage and frequency while supplying 25 percent of the rated thermal burden. 203

ambient temperature time constant. At a constant operating resistance, the time required for the change in (bolometer unit) bias power to reach 63 percent of the total change in bias power after an abrupt change in ambient temperature. 115

ambiguity (navigation). The condition obtained when navigation or radar coordinates define more than one point, direction, line of position or surface of position. 13

ambiguity function (radar). The squared magnitude

$$|\chi(\tau, f_d)|^2$$

of the function which describes the response of a radar receiver to targets displaced in range delay τ and Doppler frequency f_d from a reference position, where $|\chi(0, 0)|$ is normalized to unity. Mathematically,

$$\chi(\tau, f_d) = \int u(t)u^* (t + \tau) \exp (2\pi j f_d t) \, dt$$

where u(t) is the transmitted envelope waveform, suitably normalized; $u^*(t + \tau)$ is the complex conjugate of the same waveform with argument $(t + \tau)$; positive τ indicates a target beyond the reference delay; and positive f_d indicates an incoming target. *Note:* Used to examine the suitability of radar waveforms for achieving accuracy, resolution, freedom from ambiguities, and reduction of unwanted clutter. 13

American Morse Code. *See: Morse Code.*

α_{min} (laser-maser). The limiting angular subtense. 363

ammeter (1) (general). An instrument for measuring the magnitude of an electric current. *Note:* It is provided with a scale, usually graduated in either amperes, milliamperes, microamperes, or kiloamperes. If the scale is graduated in milliamperes, microamperes, or kiloamperes, the instrument is usually designated as a milliammeter, a mi-

croammeter, or a kiloammeter. *See also:* **instrument.**
328

(2) (circuits and systems). An instrument for measuring electric current in amperes. 67

amortisseur. A permanently short-circuited winding consisting of conductors embedded in the pole shoes of a synchronous machine and connected together at the ends of the poles, but not necessarily connected between poles. *Note:* This winding when used in salient-pole machines sometimes includes bars that do not pass through the pole shoes, but are supported in the interpolar spaces between the pole tips. 3

amortisseur bar (damper bar) (rotating machinery). A single conductor that is a part of an amortisseur winding or starting winding. *See also:* **rotor (rotating machinery); stator.** 63

amortisseur winding. *See:* **damper winding; damping winding.**

ampacity. Current-carrying capacity, expressed in amperes, of a wire or cable under stated thermal conditions. 256, 57, 53, 124

ampere (1) (general). That constant current that, if maintained in two straight parallel conductors of infinite length, of negligible circular cross section, and placed 1 meter apart in vacuum, would produce between these conductors a force equal to 2×10^{-7} newton per meter of length. *See also:* **abampere.** 233

(2) (circuits and systems). A unit of electric current flow equivalent to the motion of 1 coulomb of charge or 6.24×10^{18} electrons past any cross section in 1 second. 67

ampere-conductors (distributed winding) (rotating machinery). The product of the number of effective conductors round the periphery of the winding and the current (in amperes, root-mean-square for alternating-current and average for direct-current) circulating in these conductors. *See:* **ampere-turns.** 63

ampere-hour capacity (storage battery). The number of ampere-hours that can be delivered under specified conditions as to temperature, rate of discharge, and final voltage. *See also:* **battery (primary or secondary).** 328

ampere-hour efficiency (storage cell) (storage battery). The electrochemical efficiency expressed as the ratio of the ampere-hours output to the ampere-hours input required for the recharge. *See also:* **charge.** 328

ampere-hour meter. An electricity meter that measures and registers the integral, with respect to time, of the current of the circuit in which it is connected. *Note:* The unit in which this integral is measured is usually the ampere-hour. *See also:* **electricity meter (meter).** 328

Ampere's law. *See:* **magnetic field strength produced by an electric current.**

ampere-turn per meter. The unit of magnetic field strength in SI units (International System of Units). The ampere-turn per meter is the magnetic field strength in the interior of an elongated uniformly wound solenoid that is excited with a linear current density in its winding of one ampere per meter of axial distance. 210

ampere-turns (rotating machinery). The product of the number of turns of a coil or a winding (distributed or concentrated) and the net current in amperes per turn. *See also:* **ampere-conductors; asynchronous machine; direct-**

current commutating machine. 63

amplification (1) (signal-transmission system). (A) The ratio of output magnitude to input magnitude in a device which is intended to produce an output that is an enlarged reproduction of its input. *Note:* It may be expressed as a ratio or, by extension of the term, in decibels. (B) The process causing this increase. *See also:* **signal.**
239, 59, 188

(2) (automatic control). The ratio of output to input, in a device intended to increase this ratio. *Note:* ASA C85 prefers sensitivity for devices intended to indicate or record a variable; gain for sinusoidal signals. 56

amplification, current. *See:* **current amplification.**

amplification factor. The μ factor for a specified electrode and the control grid of an electron tube under the condition that the anode current is held constant. *Notes:* (1) In a triode this becomes the μ factor for the anode and control-grid electrodes. (2) In multielectrode tubes connected as triodes the term anode applies to the combination of electrodes used as the anode. (3) The voltage gain of an amplifier with the output unloaded. *See also:* **electron-tube admittances.** 190, 67, 125

amplification factor, gas (gas phototube). *See:* **gas amplification factor (gas phototube).**

amplification, voltage. *See:* **voltage amplification.**

amplified spontaneous emission (laser-maser). The radiation resulting from amplification of **spontaneous emission.**
363

amplifier (1) (general). A device that enables an input signal to control power from a source independent of the signal and thus be capable of delivering an output that bears some relationship to, and is generally greater than, the input signal. 111, 239, 9, 77

(2) (photomultipliers for scintillation counting). A device whose output is an enlarged reproduction of the essential features of an input signal and which draws power from a source other than the input signal. 117

(3) (analog computer). A device that enables an input signal to control a source of power and thus is capable of delivering at its output a reproduction or analytic modification of the essential characteristics of the signal. 9

amplifier, balanced (push-pull amplifier). An amplifier in which there are two identical signal branches connected so as to operate in phase opposition and with input and output connections each balanced to ground.
111, 239, 240, 59

amplifier, bridging. An amplifier with an input impedance sufficiently high so that its input may be bridged across a circuit without substantially affecting the signal level of the circuit across which it is bridged. *See also:* **amplifier.**
239

amplifier, buffer (signal-transmission system). *See:* **amplifier, isolating.**

amplifier, carrier (signal-transmission system). An alternating-current amplifier capable of amplifying a prescribed carrier frequency and information side-bands relatively close to the carrier frequency. *See also:* **signal.**
188

amplifier, chopper (signal-transmission system). A modulated amplifier in which the modulation is achieved by an electronic or electromechanical chopper, the resultant wave being substantially square. *See also:* **signal.** 188

amplifier class ratings (electron tube).

(1) class-A amplifier. An amplifier in which the grid bias and alternating grid voltages are such that anode current in a specific tube flows at all times. *Note:* The suffix 1 is added to the letter or letters of the class identification to denote that grid current does not flow during any part of the input cycle. The suffix 2 is used to denote that current flows during some part of the cycle. *See also:* **amplifier.**
190, 111, 125

(2) class-AB amplifier. An amplifier in which the grid bias and alternating grid voltages are such that anode current in a specific tube flows for appreciably more than half but less than the entire electrical cycle. *Note:* The suffix 1 is added to the letter or letters of the class identification to denote that grid current does not flow during any part of the input cycle. The suffix 2 is used to denote that current flows during some part of the cycle. *See also:* **amplifier.**
190, 111, 125

(3) class-B amplifier. An amplifier in which the grid bias is approximately equal to the cutoff value so that the anode current is approximately zero when no exciting grid voltage is applied, and so that anode current in a specific tube flows for approximately one half of each cycle when an alternating grid voltage is applied. *Note:* The suffix 1 is added to the letter or letters of the class identification to denote that grid current does not flow during any part of the input cycle. The suffix 2 is used to denote that current flows during some part of the cycle. *See also:* **amplifier.**
190, 111, 125

(4) class-C amplifier. An amplifier in which the grid bias is appreciably greater than the cutoff value so that the anode current in each tube is zero when no alternating grid voltage is applied, and so that anode current in a specific tube flows for appreciably less than one half of each cycle when an alternating grid voltage is applied. *Note:* The suffix 1 is added to the letter or letters of the class identification to denote that grid current does not flow during any part of the input cycle. The suffix 2 is used to denote that current flows during some part of the cycle. *See also:* **amplifier.**
190, 111, 125

amplifier, clipper. An amplifier designed to limit the instantaneous value of its output to a predetermined maximum. *See also:* **amplifier.**
239

amplifier, difference. *See:* **differential amplifier.**

amplifier, differential. *See:* **differential amplifier.**

amplifier, distribution. A power amplifier designed to energize a speech or music distribution system and having sufficiently low output impedance so that changes in load do not appreciably affect the output voltage. *See also:* **amplifier.**
239

amplifier ground (signal-transmission system). *See:* **receiver ground.**

amplifier, horizontal. *See:* **horizontal amplifier.**

amplifier, integrating. *See:* **integrating amplifier.**

amplifier, intensity. *See:* **intensity amplifier.**

amplifier, inverting. *See:* **inverting amplifier.**

amplifier, isolating (signal-transmission system). An amplifier employed to minimize the effects of a following circuit on the preceding circuit. *Example:* An amplifier having effective direct-current resistance and/or alternating-current impedance between any part of its input circuit and any other of its circuits that is high compared to some critical resistance or impedance value in the input circuit. *See also:* **signal.**
239, 188

amplifier, isolation (buffer). An amplifier employed to minimize the effects of a following circuit on the preceding circuit. *See also:* **amplifier.**
239, 178

amplifier, line. An amplifier that supplies a transmission line or system with a signal at a stipulated level. *See also:* **amplifier.**
239, 178

amplifier, modulated (signal-transmission signal). *See:* **modulated amplifier.**

amplifier, monitoring (electroacoustics). An amplifier used primarily for evaluation and supervision of a program. *See also:* **amplifier.**
239, 178

amplifier, peak limiting. *See:* **peak limiter.**

amplifier, power. An amplifier that drives a utilization device such as a loudspeaker. *See also:* **amplifier.**
239

amplifier, program. *See:* **amplifier, line.**

amplifier, relay. *See:* **relay amplifier.**

amplifier, servo. *See:* **servo amplifier.**

amplifier, summing. *See:* **summing amplifier.**

amplifier, vertical. *See:* **vertical amplifier.**

amplifier, X-axis. *See:* **horizontal amplifier.**

amplifier, Y-axis. *See:* **vertical amplifier.**

amplifier, Z-axis. *See:* **Z-axis amplifier; intensity amplifier.**

amplitude (sine wave). A in $A \sin(\omega t + \theta)$ where A, ω, θ are not necessarily constants, but are specified functions of t. In amplitude modulation, for example, the amplitude A is a function of time. In electrical engineering, the term **amplitude** is often used for the modulus of a complex quantity. Amplitude with a modifier, such as peak or maximum, minimum, root-mean-square, average, etcetera, denotes values of the quantity under discussion that are either specified by the meanings of the modifiers or otherwise understood. *See:* **amplitude (simple sine wave).**
210

amplitude (simple sine wave). The positive real A in $A \sin(\omega t + \theta)$, where A, ω, θ are constants. In this case, amplitude is synonymous with maximum or peak value. *See:* **amplitude (sine wave).**
210

amplitude balance control (electronic navigation). The portion of a system that may be varied to adjust the relative output levels of two related signals. Originally used in instrument landing systems and later in loran. *See also:* **navigation.**
13, 187

amplitude characteristic. *See:* **amplitude-frequency characteristic.**

amplitude-comparison monopulse (radar). A variant of monopulse for determining the angle of arrival of a source of radiation by simultaneously comparing the relative amplitudes from two or more antenna beams whose patterns are squinted off a common bore-sight, for example, by lateral displacement of the feeds from the antenna focus. *Note:* If the beams have a common phase center, the monopulse is pure amplitude-comparison; otherwise, it is a combination of amplitude-comparison and phase-comparison. *See also:* **monopulse** and **phase-comparison monopulse.**
13

amplitude discriminator (radar). A circuit whose output is a function of the relative magnitudes of two signals. *See also:* **navigation.**
13, 187

amplitude distortion (nonlinear distortion) (control systems). That form of distortion of output wave shape, relative to the fundamental sinusoid of the input, which occurs under steady-state conditions when the output amplitude is not a linear function of the input amplitude. 56

amplitude factor (restriking voltage) (transient recovery voltage) (1) (surge arrester). The ratio between the peak restriking voltage and the peak value ($\sqrt{2}$ times the root-mean-square value) of the recovery voltage. *See also:* **surge arrester (surge diverter).** 308, 62
(2) (power switchgear). The ratio of the highest peak of the transient recovery voltage to the peak value of the normal-frequency recovery voltage. *Note:* In tests made under one condition to simulate duty under another, as in single-phase tests made to simulate duty on three-phase ungrounded faults, the amplitude factor is expressed in terms of the duty being simulated. 103, 202

amplitude-frequency characteristic (amplitude characteristic). The variation with frequency of the amplitude (that is, modulus or absolute magnitude) of a phasor quantity. *Note:* The magnitudes of transfer admittances, transfer ratios, amplification, etcetera, plotted against frequency are a few examples; measures in decibels or other units of these same quantities plotted against frequency are included. 210

amplitude-frequency distortion. Distortion due to an undesired amplitude-frequency characteristic. *Notes:* (1) The usual desired characteristic is flat over the frequency range of interest. (2) Also sometimes called amplitude distortion or frequency distortion. *See also:* **distortion.** 241

amplitude-frequency response. The variation of gain, loss, amplification, or attenuation as a function of frequency. *Note:* This response is usually measured in the region of operation in which the transfer characteristic of the system or transducer is essentially linear. *See also:* **transmission characteristics.** 239, 178, 59

amplitude gate. *See:* **slicer.**

amplitude locus (control system, feedback) (for a nonlinear system or element whose gain is amplitude dependent). A plot of the describing function, in any convenient coordinate system. *See also:* **control system, feedback.** 56

amplitude-modulated transmitter. A transmitter that transmits an amplitude-modulated wave. *Note:* In most amplitude-modulated transmitters, the frequency is stabilized. *See also:* **radio transmitter.** 111, 240

amplitude modulation (signal-transmission system). (1) The process, or the result of the process, whereby the amplitude of one electrical quantity is varied in accordance with some selected characteristic of a second quantity, which need not be electrical in nature. *See also:* **modulating systems; signal.** 188
(2) Modulation in which the amplitude of a wave is the characteristic varied. 111, 242, 194, 59

amplitude-modulation noise. The noise produced by undesired amplitude variations of a radio-frequency signal. *See also:* **radio transmission.** 240

amplitude-modulation noise level. The noise level produced by undesired amplitude variations of a radio-frequency signal in the absence of any intended modulation. 111

amplitude noise (radar). Used variously to describe **target fluctuation** and **scintillation error.** Use of one of these specific terms is recommended to avoid ambiguity. 13

amplitude permeability (magnetic core testing). The value of permeability at a stated value of field strength (or induction), the field strength varying periodically with time and with no static magnetic field being present.

$$\mu_a = \frac{1}{\mu_0}\frac{B}{H}$$

μ_a = relative amplitude permeability. Maximum permeability is the maximum value of the amplitude permeability as a function of the field strength (or of the induction). 165

amplitude, pulse. A general term indicating the magnitude of a pulse. *Note:* Pulse amplitude is measured with respect to the nominally constant baseline, unless otherwise stated. For specific designation, adjectives such as average, instantaneous, peak, root-mean-square, etcetera, should be used to indicate the particular meaning intended. *See also:* **pulse.** 185

amplitude range (electroacoustics). The ratio, usually expressed in decibels, of the upper and lower limits of program amplitudes that contain all significant energy contributions. 239

amplitude ratio. *See:* **gain; subsidence ratio.**

amplitude reference level (pulse techniques). The arbitrary reference level from which all amplitude measurements are made. *Note:* The arbitrary reference level normally is considered to be at an absolute amplitude of zero but may, in fact, have any magnitude of either polarity. If this arbitrary reference level is other than zero, its value and polarity must be stated. *See also:* **pulse.** 185

amplitude resonance. Resonance in which amplitude is stationary with respect to frequency. 210

amplitude response (camera tubes). The ratio of (1) the peak-to-peak output from the tube resulting from a spatially periodic test pattern, to (2) the difference in output corresponding to large-area blacks and large-area whites, having the same illuminations as the test pattern minima and maxima, respectively. *Note:* The amplitude response is referred to as modulation transfer (sine-wave response) when a sinusoidal test pattern is used and as square-wave response when the pattern consists of alternate black and white bars of equal width. *See also:* **camera tube.** 190

amplitude response characteristic (camera tubes). The relation between (1) amplitude response and (2) television line number (camera tubes) or (image tubes) test-pattern spatial frequency, usually in line pairs per millimeter. *See also:* **camera tube.** 190

amplitude selection. A summation of one or more variables and a constant resulting in a sudden change in rate or level at the output of a computing element as the sum changes sign. *See also:* **electronic analog computer.** 9, 77

amplitude suppression ratio (frequency modulation). The ratio of the undesired output to the desired output of a frequency-modulation receiver when the applied signal has simultaneous amplitude modulation and frequency modulation. *Note:* This ratio is generally measured with an applied signal that is amplitude modulated 30 percent at a 400-hertz rate and is frequency modulated 30 percent of maximum system deviation at a 1000-hertz rate. *See also:* **frequency modulation.** 328

AMTI (radar). Airborne moving target indication. *See:* **moving target indication (MTI).** 13

AM to FS converter. *See:* **transmitting converter, facsimile.** *See also:* **facimile (electrical communication).**

anaerobic. Free of uncombined oxygen. 221, 205

analog (1) (data transmission). Pertaining to data in the form of continuously variable physical quantities. *See:* **digital.** 255, 77

(2) (adjective). Pertaining to representation by means of continuously variable physical quantities, for example, to describe a physical quantity, such as voltage or shaft position, that normally varies in a continuous manner, or devices such as potentiometers and synchros that operate with such quantities. *See also:* **electronic analog computer.** 9

(3) (industrial control). Pertains to information content that is expressed by signals dependent upon magnitude. *See also:* **control system, feedback.** 219, 206

(4) (electronic computers). A physical system on which the performance of measurements yields information concerning a class of mathematical problems. 210

analog and digital data. Analog data implies continuity as contrasted to digital data that is concerned with discrete states. *Note:* Many signals can be used in either the analog or digital sense, the means of carrying the information being the distinguishing feature. The information content of an analog signal is conveyed by the value or magnitude of some characteristics of the signal such as the amplitude, phase, or frequency of a voltage, the amplitude or duration of a pulse, the angular position of a shaft, or the pressure of the fluid. To extract the information, it is necessary to compare the value or magnitude of the signal to a standard. The information content of the digital signal is concerned with discrete states of the signal, such as the presence or absence of a voltage, a contact in the open or closed position, or a hole or no hole in certain positions on a card. The signal is given meaning by assigning numerical values or other information to the various possible combinations of the discrete states of the signal. *See also:* **analog data; digital data.** 224, 207

analog computer (1) (general). An automatic computing device that operates in terms of continuous variation of some physical quantities, such as electrical voltages and currents, mechanical shaft rotations, or displacements, and which is used primarily to solve differential equations. The equations governing the variation of the physical quantities have the same or very nearly the same form as the mathematical equations under investigation and therefore yield a solution analogous to the desired solution of the problem. Results are measured on meters, dials, oscillograph recorders, or oscilloscopes. *See also:* **simulator.** 9

(2) (direct-current). An analog computer in which computer variables are represented by the instantaneous values of voltages. 9

(3) (alternating-current). An analog computer in which electrical signals are of the form of amplitude modulated suppressed carrier signals where the absolute value of a computer variable is represented by the amplitude of the car(0 or 180 degrees) of the carrier relative to the reference alternating-current signal. 9

analog device (control equipment). A device that operates with variables represented by continuously measured quantities such as voltages, resistances, rotations, pressures, etcetera. 94

analog output. One type of continuously variable quantity used to represent another; for example, in temperature measurement, an electric voltage or current output represents temperature input. *See also:* **signal.** 188

analog signal (control) (industrial control). A signal that is solely dependent upon magnitude to express information content. *See also:* **control system, feedback.** 219, 206

analog switching (telephone switching systems). Switching of continuously-varying-level information signals. 55

analog telemetering. Telemetering in which some characteristic of the transmitter signal is proportional to the quantity being measured. 103, 202

analog-to-digital converter (1) (data processing). A device that converts a signal that is a function of a continuous variable into a representative number sequence. 198, 54

(2) (analog-to-digital). (A–D). A circuit whose input is information in analog form and whose output is the same information in digital form. *See also:* **analog; digital.** 59

(3) (digitizer). A device or a group of devices that converts an analog quantity or analog position input signal into some type of numerical output signal or code. *Note:* The input signal is either the measurand or a signal derived from it. 103, 202

(4) (hybrid computer linkage components) (analog-to-digital converter). (ADC). Provides the means of obtaining a digital number representation of a specific analog voltage value. 10

analog-to-frequency (A–F) converter. A circuit whose input is information in an analog form other than frequency and whose output is the same information as a frequency proportional to the magnitude of the information. *See also:* **analog.** 59

analysis (nuclear power generating systems). A process of mathematical or other logical reasoning that leads from stated premises to the conclusion concerning specific capabilities of equipment and its adequacy for a particular application. *See also:* **numerical analysis.** 31, 120

analytic inertial-navigation equipment. The class of inertial-navigation equipment in which geographic navigational quantities are obtained by means of computations (generally automatic) based upon the outputs of accelerometers whose orientations are maintained fixed with respect to inertial space. *See also:* **navigation.** 187

analytic signal (pre-envelope). A complex function of a real variable (time, for example) whose imaginary part is the Hilbert transform of its real part. *Note:* The analytic signal associated with a real signal is that one whose real part is the given real signal. *See also:* **Hilbert transform; network analysis.** 61

analyzer. *See:* **differential analyzer; digital differential analyzer; network analyzer.**

anchor (conductor stringing equipment). A device that serves as a reliable support to hold an object firmly in place. The term "anchor" is normally associated with cone, plate, screw or concrete anchors, but the terms "snub," "deadman" and "anchor log" are usually associated with pole stubs or logs set or buried in the ground to serve as temporary anchors. The latter are often used at pull and tension sites. *See:* **anchor log.** 45

anchor guy guard. A protective cover over the guy, often a length of sheet metal or plastic shaped to a semicircular or tubular section and equipped with means of attachment to the guy. It may also be of wood. *See also:* **tower.** 64

anchor light. (1) A lantern hung in the rigging at a prescribed height to indicate to navigators of nearby vessels that a ship is at anchor. 328

(2) An aircraft light designed for use on a seaplane or amphibian to indicate its position when at anchor or moored. 167

anchor log (dead man). A piece of rigid material such as timber, metal, or concrete, usually several feet in length, buried in earth in a horizontal position and at right angles to anchor rod attachment. *See also:* **tower.** 64

anchor rod. A steel or other metal rod designed for convenient attachment to a buried anchor and also to provide for one or more guy attachments above ground. *See also:* **tower.** 64

ancillary equipment (test, measurement and diagnostic equipment). Equipment which is auxiliary or supplementary to an automatic test equipment installation. Ancillary equipment usually consists of standard off-the-shelf items such as an oscilloscope and distortion analyzer. 54

AND. A logic operator having the property that if P is a statement, Q is a statement, R is a statement, . . . , then the AND of $P, Q, R, . . .$ is true if all statements are true, false if any statement is false. P AND Q is often represented by $P \cdot Q, PQ, P \wedge Q$. 79, 255

AND-circuit. *See:* AND-**gate.**

Anderson bridge. A 6-branch network in which an outer loop of 4 arms is formed by three nonreactive resistors and the unknown inductor, and an inner loop of 3 arms is formed by a capacitor and a fourth resistor in series with each other and in parallel with the arm that is opposite the unknown inductor, the detector being connected between the junction of the capacitor and the fourth resistor and that end of the unknown inductor that is separated from a terminal of the capacitor by only one resistor, while the source is connected to the other end of the unknown inductor and to the junction of the capacitor with two resistors of the outer loop. *Note:* Normally used for the comparison of self-inductance with capacitance. The balance is independent of frequency. *See also:* **bridge.** 328

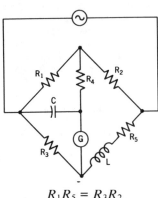

$$R_1 R_5 = R_3 R_2$$

$$L = CR_3 \left[R_4 \left(1 + \frac{R_2}{R_1} \right) + R_2 \right]$$

Anderson bridge.

AND gate (1) (general). A combinational logic element such that the output channel is in its ONE state if and only if each input channel is in its ONE state. 235,

(2) A gate that implements the logic AND operator. 255, 77

(3) (**AND circuit**). A gate whose output is energized when and only when every input is in its prescribed state. An AND gate performs the function of the logical AND. 210

anechoic chamber. An enclosure especially designed with boundaries that absorb sufficiently well the sound incident thereon to create an essentially free-field condition in the frequency range of interest. 176

anechoic enclosure (radio frequency). An enclosure whose internal walls have low reflection characteristics. *See also:* **electromagnetic compatibility.** 199

anelectrotonus (electrobiology). Electrotonus produced in the region of the anode. *See also:* **excitability.** 192

angel echoes. Radar returns which are not desired in a particular situation and which are caused by atmospheric inhomogeneities, refractive index discontinuities, insects, birds, or unknown phenomena. *Note:* Originally when some physical target could not be identified through direct visual observation, all echoes from such unknown causes were designated as "angels." 13

angle, bunching (electron stream). *See:* **bunching angle.**

angle, effective bunching (reflex klystrons). *See:* **bunching angle, effective.**

angle, flow (gas tubes). That portion, expressed as an angle, of the cycle of an alternating voltage during which current flows. *See also:* **gas tubes.** 190

angle, maximum-deflection. The maximum plane angle subtended at the deflection center by the usable screen area. *Note:* In this term the hyphen is frequently omitted. 178, 190, 125

angle modulation. Modulation in which the angle of a sine-wave carrier is the characteristic varied from its reference value. *Notes:* (1) Frequency modulation and phase modulation are particular forms of angle modulation; however, the term frequency modulation is often used to designate various forms of angle modulation. (2) The reference value is usually taken to be the angle of the unmodulated wave. *See also:* **modulation index.** 111, 242

angle noise (radar). The noise-like variation in the apparent angle of arrival of a signal received from a target, caused by changes in phase and amplitude of target scattering sources and including angular components of both **glint** and **scintillation error.** 13

angle of advance (1) (power inverter). The time interval in electrical degrees by which the beginning of anode conduction leads the moment at which the anode voltage would attain a negative value equal to that of the succeeding anode in the commutating group. *See also:* **rectification.** 291

(2) (**semiconductor rectifiers**) (**semiconductor power converter**). The angle by which forward conduction is advanced by the control means only, in the incoming circuit element, ahead of the instant in the cycle at which the incoming commutating voltage passes through zero in the direction to produce forward conduction in the outgoing circuit element. *See also:* **rectification; semiconductor rectifier stack.** 208

angle-of-approach lights. Aeronautical ground lights arranged so as to indicate a desired angle of descent during an approach to an aerodrome runway. Also called **optical glide-path lights.** 167

angle of cut (navigation). The angle at which two lines of position intersect. *See also:* **navigation.** 187

angle of extinction (industrial control). The phase angle of the stopping (extinction) instant of anode-current flow in a glass tube with respect to the starting instant of the corresponding positive half cycle of the anode voltage of the tube. *See also:* **electronic controller.** 206

angle of ignition (industrial control). The phase angle of the starting instant of anode-current flow in a gas tube with respect to the starting instant of the corresponding positive half cycle of the anode voltage of the tube. *See also:* **electronic controller.** 206

angle of incidence (acoustic-optic device). The angle in air between the acoustic wavefront and the normal to the optical wavefront. For operation in the Bragg region, maximum diffraction into the first order occurs when the angle of incidence is equal to the Bragg angle, θ_B, which is given by the equation $\sin \theta_B = \lambda_0/2\Lambda$. 82

angle of lag. The angle of lag of one simple sine wave with respect to a second having the same period is the angle by which the second must be assumed to be shifted backward to make it coincide in position with the first. 210

angle of lead. The angle of lead of one simple sine wave with respect to a second having the same period is the angle by which the second must be assumed to be shifted forward to make it coincide in position with the first. 210

angle of protection (lightning). The angle between the vertical plane and a plane through the ground wire, within which the line conductors must lie in order to ensure a predetermined degree of protection against direct lightning strokes. *See also:* **surge arrester (surge diverter).**
244, 62

angle of retard (1) (semiconductor rectifiers) (semiconductor rectifier operating with phase control). The angle by which forward conduction is delayed by the control means only, beyond the instant in the cycle at which the incoming commutating voltage passes through zero in the direction to produce forward conduction in the incoming circuit element. *See also:* **semiconductor rectifier stack.** 208
(2) (power rectifier). The angle by which the beginning of anode conduction leads the point at which the incoming and outgoing rectifying elements of a commutating group have equal negative fundamental components of alternating voltage. *See also:* **rectification.** 328

angle optimum bunching. *See:* **optimum bunching.**

angle or phase (sine wave). The measure of the progression of the wave in time or space from a chosen instant or position or both. *Notes:* (1) In the expression for a sine wave, the angle or phase is the value of the entire argument of the sine function. (2) In the representation of a sine wave by a phasor or rotating vector, the angle or phase is the angle through which the vector has progressed. *See also:* **modulating systems; wavefront.** 242, 111

angle, overlap. *See:* **overlap angle.**

angle, roll over (conductor stringing equipment). For tangent stringing, the sum of the vertical angles between the conductor and the horizontal on both sides of the traveler. Resultants of these angles must be considered when stringing through line angles. Under some stringing conditions, such as stringing large diameter conductors, excessive roll over angles can cause premature failure of a conductor splice if it is allowed to pass over the travelers.
45

angle tower. A tower located where the line changes horizontal direction sufficiently to require special design of the tower to withstand the resultant pull of the wires and to provide adequate clearance. *See also:* **tower.** 64

angle tracking. *See:* **tracking.**

angle, transit. *See:* **transit angle.**

angular acceleration sensitivity (accelerometer). The output (divided by the scale factor) of a linear accelerometer that is produced per unit of angular acceleration input about a specified axis. 46

angular accelerometer. A device that senses angular acceleration about an input axis. An output signal is produced by the reaction of the moment of inertia of a proof mass to an angular acceleration input. The output is usually an electrical signal proportional to applied angular acceleration. 46

angular accuracy (radar). The degree to which the measurement of the angular location of a target with respect to a given reference represents the true angular location of the target with respect to this reference. 13

angular case motion sensitivity (tuned rotor gyro). The drift rate resulting from an oscillatory angular input about an axis normal to the spin axis at twice the rotor spin frequency. This effect is due to the single degree of freedom of the gimbal relative to the support shaft and is proportional to the input amplitude and phase relative to the flexure axes. 46

angular deviation loss (acoustic transducer). The ratio of the response in a specific direction to the response on the principal axis, usually expressed in decibels. *See also:* **loudspeaker.** 176

angular deviation sensitivity (electronic navigation). The ratio of change of course indication to the change of angular displacement from the course line. *See also:* **navigation.** 13, 187

angular displacement (1) (polyphase regulator). The phase angle expressed in degrees between the line-to-neutral voltage of the reference identified source voltage terminal and the line-to-neutral voltage of the corresponding identified load voltage terminal. *Note:* The connection and arrangement of terminal markings for three-phase regulators is a wye-wye connection having an angular displacement of zero degrees. *See also:* **voltage regulator.**
257

(2) (polyphase transformer). The phase angle expressed in degrees between the line-to-neutral voltage of the reference identified high-voltage terminal and the line-to-neutral voltage of the corresponding identified low-voltage of terminal. *Note:* The preferred connection and arrangement of terminal markings for polyphase transformers are those which have the smallest possible phase-angle displacements and are measured in a clockwise direction from the line-to-neutral voltage of the reference identified high-voltage terminal. Thus, standard three-phase transformers have angular displacements of either zero or 30 degrees. *See:* **routine test.** 203, 53

(3) (polyphase rectifier transformer). The time angle expressed in degrees between the line-to-neutral voltage of the reference identified alternating-current winding terminal and the line-to-neutral voltage of the corresponding identified direct-current winding terminal. *See also:* **rectifier transformer.** 258

angular frequency (periodic function). 2π times the frequency. *Note:* This definition applies to any definition of frequency. *See also:* **network analysis.**

210, 180, 61, 146

angular resolution (radar). The ability to distinguish between two targets solely by the measurement of angles; generally expressed in terms of the minimum angle by which two targets of equal strength within the same range resolution cell must be spaced to be separately distinguishable. 13

angular swing (acousto-optic deflector). The center-to-center angular separation between the deflected light beams obtained upon application of the maximum and minimum acoustic frequency of the frequency range.

72

angular variation (1) (alternating-current circuits). The maximum angular displacement, expressed in electrical degrees, of corresponding ordinates of the voltage wave and of a wave of constant frequency, equal to the average frequency of the alternating-current circuit in question.

328

(2) (synchronous generator). The maximum angular displacement, expressed in electrical degrees, of corresponding ordinates of the voltage wave and of a wave of absolutely constant frequency, equal to the average frequency of the synchronous generator in question. 63

angular velocity sensitivity (accelerometer). The output (divided by the scale factor) of a linear accelerometer that is produced per unit of angular velocity input about a specific axis. 46

angular width (electronic navigation). *See:* **course.** *See also:* **width; navigation.**

anion. A negatively charged ion or radical that migrates toward the anode under the influence of a potential gradient. *See also:* **ion.** 221, 205

anisoelasticity (gyro). The inequality of compliance of a structure in different directions. *See:* **acceleration squared sensitive drift rate.** 46

anisoinertia (gyro). The inequality of the moments of inertia about the gimbal principal axes. When the gyro is subjected to angular rates about the input and spin axes and the moments of inertia about these axes are unequal, a torque is developed about the output axis which is proportional to the difference of the inertias about the input and spin axes multiplied by the product of the rates about these two axes. This torque results in a drift rate (anisoinertia drift) proportional to the product of the two input rates. In addition, when the gimbal is displaced about the output axis, a torque is developed which results in a drift rate (anisoelastic coupling drift) proportional to the product of this (gimbal) angle, the difference in inertias and the difference of the squares of the two input rates.

46

annotation. An added descriptive comment or explanatory note. 255, 77

announcement system. A general arrangement for supplying information by means of periodic announcements distributed to the various central offices over one-

way distribution circuits. *Note:* It is used for time of day, weather, etcetera. 328

annular slot antenna. A slot antenna with the radiating slot having the shape of an annulus (ring). 111

annunciator. A visual signal device consisting of a number of pilot lights or drops, each one indicating the condition that exists or has existed in an associated circuit, and being labeled accordingly. 328

annunciator relay. A relay that indicates visually whether a current is flowing or has flowed in one or more circuits. *See also:* **relay.** 259

anode (1). An electrode through which current enters any conductor of the nonmetallic class. Specifically, an electrolytic anode is an electrode at which negative ions are discharged, or positive ions are formed, or at which other oxidizing reactions occur. 328
(2). An electrode or portion of an electrode at which a net oxidation-reaction occurs. *See also:* **electrochemical cell.**

223, 186, 205

(3) (thyristor). The electrode by which current enters the thyristor, when the thyristor is in the ON state with the gate open-circuited. *Note:* This term does not apply to bidirectional thyristors. 243, 66, 208, 191
(4) (electron tube or valve). An electrode through which a principal stream of electrons leaves the interelectrode space. *See also:* **electrode (electron tube).**

190, 117, 125

(5) (semiconductor rectifier diode). The electrode from which the forward current flows within the cell. *See also:* **semiconductor.** 237, 66
(6) (X-ray tube). *See:* **target (X-ray tube).**
(7) (light emitting diodes). The electrode from which the forward current is directed within the device. 162

anode breakdown voltage (glow-discharge cold-cathode tube). The anode voltage required to cause conduction across the main gap with the starter gap not conducting and with all other tube elements held at cathode potential before breakdown. *See also:* **electrode voltage (electron tube).** 190

anode butt. A partially consumed anode. *See also:* **fused electrolyte.** 328

anode characteristic. *See:* **anode-to-cathode voltage-current characteristic.**

anode circuit (industrial control). A circuit that includes the anode-cathode path of an electron tube in series connection with other elements. *See also:* **electronic controller.**

206

anode circuit breaker. A low-voltage power circuit breaker: (1) designed for connection in an anode of a mercury-arc power rectifier unit, (2) that trips automatically only on reverse current and starts reduction of current in a specified time when the arc-back occurs at the end of the forward current conduction, and (3) substantially interrupts the arc-back current within one cycle of the fundamental frequency after the beginning of the arc-back. *Note:* The specified time in present practice is 0.008 second or less (at an ac frequency of 60 Hz). 103

anode cleaning (reverse-current cleaning) (electroplating). Electrolytic cleaning in which the metal to be cleaned is made the anode. *See also:* **battery (primary or secondary).**

328

anode corrosion efficiency. The ratio of the actual corrosion of an anode to the theoretical corrosion calculated

from the quantity of electricity that has passed.
221, 205

anode current (electron tubes). *See:* **electrode current; electronic controller.**

anode dark space (gas tubes) (gas). A narrow dark zone next to the surface of the anode. *See also:* **discharge (gas).**
244, 190

anode differential resistance. *See:* **anode resistance.**

anode effect. A phenomenon occurring at the anode, characterized by failure of the electrolyte to wet the anode and resulting in the formation of a more or less continuous gas film separating the electrolyte and anode and increasing the potential difference between them. *See also:* **fused electrolyte.**
328

anode efficiency. The current efficiency of a specified anodic process. *See also:* **electrochemistry.**
328

anode, excitation (pool-cathode tube). *See:* **excitation anode.**

anode (potential) fall (gas). The fall of potential due to the space charge near the anode. *See also:* **discharge (gas).**
244, 190

anode firing (industrial control). The method of initiating conduction of an ignitron by connecting the ignitor through a rectifying element to the anode of the ignitron to obtain power for the firing current pulse. *See also:* **electronic controller.**
206

anode glow (gas tubes) (gas). A very bright narrow zone situated at the near end of the positive column with respect to the anode. *See also:* **discharge (gas).**
244, 190

anode layer. A molten metal or alloy, serving as the anode in an electrolytic cell, that floats on the fused electrolyte or upon which the fused electrolyte floats. *See also:* **fused electrolyte.**
328

anode, main (pool-cathode tube). *See:* **main anode.**

anode mud. *See:* **slime, anode.**

anode paralleling reactor. A reactor with a set of mutually coupled windings connected to anodes operating in parallel from the same transformer terminal. *See:* **alternating-current saturable reactor; filter reactor; shunt reactor; current-limiting reactor; bus reactor; feeder reactor; neutral grounding reactor; starting reactor; synchronizing reactor; paralleling reactor.**
203, 53

anode power supply (electron tube) (plate power supply). The means for supplying power to the plate at a voltage that is usually positive with respect to the cathode. *See also:* **power pack.**
328

anode region (gas tubes) (gas). The group of regions comprising the positive column, anode glow, and anode dark space. *See also:* **discharge (gas).**
244, 190

anode-reflected-pulse rise time. The rise time of a pulse reflected from the anode. *Note:* This time can be measured with a time-domain reflectometer.
117

anode relieving (pool-cathode tube) (gas tube). An anode that provides an alternative conducting path to reduce the current to another electrode. *See also:* **electrode (electron tube); gas filled rectifier.**
244, 190

anode resistance (anode differential resistance) (electron tube). The quotient of a small change in anode voltage by a corresponding small change of the anode current, all the other electrode voltages being maintained constant. It is equal to the reciprocal of the anode conductance. *See also:* **ON period (electron tubes).**
244, 190

anode scrap. That portion of the anode remaining after the

schedule period for the electrolytic refining of the bulk of its metal content has been completed. *See also:* **electrorefining.**
328

anode slime. *See:* **slime, anode.**

anode strap (magnetron). A metallic connector between selected anode segments of a multicavity magnetron, principally for the purpose of mode separation. *See also:* **magnetron.**
190, 125

anode supply voltage (industrial control). The voltage at the terminals of a source of electric power connected in series in the anode circuit. *See also:* **electronic controller.**
206

anode terminal (1) (semiconductor device). The terminal by which current enters the device. *See also:* **semiconductor; semiconductor device.**
245, 66

(2) (semiconductor diode). The terminal that is positive with respect to the other terminal when the diode is biased in the forward direction. *See also:* **semiconductor device.**
210

(3) (semiconductor rectifier diode or rectifier stack). The terminal to which forward current flows from the external circuit. *Note:* In the semiconductor rectifier components field, the anode terminal is normally marked negative. *See also:* **semiconductor device; semiconductor rectifier cell.**
237, 66, 208

(4) (thyristor). The terminal that is connected to the anode. *Note:* This term does not apply to bidirectional thyristors. *See also:* **anode.**
243, 66, 191

anode-to-cathode voltage (anode voltage) (thyristor). The voltage between the anode terminal and the cathode terminal. *Note:* It is called positive when the anode potential is higher than the cathode potential and called negative when the anode potential is lower than the cathode potential. *See also:* **electronic controller; principal voltage-current characteristic (principal characteristic).**
243, 66, 208, 191

anode-to-cathode voltage-current characteristic (thyristor). A function, usually represented graphically, relating the anode-to-cathode voltage to the principal current with gate current, where applicable, as a parameter. *Note:* This term does not apply to bidirectional thyristors. *Syn.* **anode characteristic.** *See also:* **principal voltage-current characteristic (principal characteristic).**
243, 66, 191

anode voltage. *See:* **anode-to-cathode voltage.**

anode voltage (electron tubes). *See:* **electrode voltage; electronic controller.**

anode voltage drop (glow-discharge cold-cathode tube). The main gap voltage drop after conduction is established in the main gap.
190

anode voltage, forward, peak. *See:* **peak forward anode voltage.**

anode voltage, inverse, peak. *See:* **peak inverse anode voltage.**

anodic polarization. Polarization of an anode. *See also:* **electrochemistry.**
328

anolyte. The portion of an electrolyte in an electrolytic cell adjacent to an anode. If a diaphragm is present, it is the portion of electrolyte on the anode side of the diaphragm. *See also:* **electrolytic cell.**
205

A–N radio range (navigation). A radio range providing four radial lines of position identified aurally as a continuous tone resulting from the interleaving of equal amplitude A and N International Morse Code letters. The sense of deviation from these lines is indicated by deterioration of the steady tone into audible A or N code signals. *See also:* **navigation; radio navigation.** 13

answered call (telephone switching systems). A call on which an answer signal occurred. 55

answering plug and cord. A plug and cord used to answer a calling line. 328

answer signal (telephone switching systems). A signal that indicates that the call has been answered. 55

antenna (aerial). A means for radiating or receiving radio waves. 111

antenna effect (1) (radio direction-finding). The presence of output signals having no directional information and caused by the directional array acting as a simple nondirectional antenna; the effect is manifested by (A) angular displacement of the nulls, or (B) a broadening of the nulls. *See also:* **antenna; navigation.** 187

(2) (loop antenna) (old usage). Any spurious effect resulting from the capacitance of the loop to ground. *See also:* **antenna.** 111, 179

antenna, effective area (in a given direction). The ratio of the power available at the terminals of an antenna to the incident power density of a plane wave from that direction polarized coincident with the polarization that the antenna would radiate. *See also:* **antenna.** 179

antenna, effective height. (1) The height of its center of radiation above the effective ground level. (2) In low-frequency applications the term effective height is applied to loaded or nonloaded vertical antennas and is equal to the moment of the current distribution in the vertical section, divided by the input current. *Note:* For an antenna with symmetrical current distribution the center of radiation is the center of distribution. For an antenna with asymmetrical current distribution the center of radiation is the center of current moments when viewed from directions near the direction of maximum radiation. *See also:* **antenna.** 179, 59

(3) The height of the center of a vertical antenna (of at least ¼ wavelength) above the effective ground plane of the vehicle on which the antenna is mounted. *See also:* **antenna; mobile communication system.** 181

antenna, effective height base station (mobile communication). The height of the physical center of the antenna above the effective ground plane. *See also:* **mobile communication system.** 181

antenna, effective length (effective height, low-frequency usage). (1) general. (A) For an antenna radiating linearly polarized waves, the length of a thin straight conductor oriented perpendicular to the direction of maximum radiation, having a uniform current equal to that at the antenna terminals and producing the same far field strength as the antenna. (B) Alternatively, for the same antenna receiving linearly polarized waves from the same direction, the ratio of the open-circuit voltage developed at the terminals of the antenna to the component of the electric field strength in the direction of antenna polarization. *Notes:* (1) The two definitions yield equal effective lengths. (2) In low-frequency usage the effective length of a ground-based antenna is taken in the vertical direction and is frequently referred to as effective height. Such usage should not be confused with effective height (of an antenna, high-frequency usage). *See also:* **antenna.** 179

(2) (electromagnetic compatibility). The ratio of the antenna open-circuit voltage to the strength of the field component being measured. *See also:* **electromagnetic compatibility.** 199

antenna efficiency (aperture-type antenna). For an antenna with a specified planar aperture, the ratio of the maximum effective area of the antenna to the aperture area. 111

antenna factor (field-strength meter). That factor that, when properly applied to the meter reading of the measuring instrument, yields the electric field strength in volts/meter or the magnetic field strength in amperes/meter. *Notes:* (1) This factor includes the effects of antenna effective length and mismatch and transmission line losses. (2) The factor for electric field strength is not necessarily the same as the factor for the magnetic field strength. *See also:* **electromagnetic compatibility.** 199

antenna figure of merit (communication satellite). An antenna performance parameter equalling the antenna gain G divided by the antenna noise temperature T, measured at the antenna terminals. It can be expressed as a ratio, $M = G/T$ or logarithmically

$$M(dB) = 10 \log_{10} G - 10 \log_{10} T$$

See also: **noise temperature, average operating.** 85

antenna pattern. *See:* **radiation pattern.**

antenna resistance (1) (general). The real part of the input impedance of an antenna. 111

(2) (test procedures for antennas). The ratio of the power accepted by the entire antenna circuit to the mean-square antenna current referred to a specified point. *Note:* Antenna resistance is made up of such components as radiation resistance, ground resistance, radio-frequency resistance of conductors in the antenna circuit, and equivalent resistance due to corona, eddy currents, insulator leakage, and dielectric power loss. *See also:* **antenna.** 246, 179

antenna terminal conducted interference (electromagnetic compatibility). Any undesired voltage or current generated within a receiver, transmitter, or their associated equipment appearing at the antenna terminals. *See also:* **electromagnetic compatibility.** 199

anticathode (X-ray tube)*. *See:* **anode.**

* Deprecated

anticlutter circuits (radar). Circuits which attenuate undesired reflections to permit detection of targets otherwise obscured by such reflections. *See also:* **navigation.** 13, 187

anticlutter gain control (radar). A device that automatically and smoothly increases the gain of a radar receiver from a low level to the maximum, within a specified period after each transmitter pulse, so that short-range echoes producing clutter are amplified less than long-range echoes. *See also:* **radar.** 328

anticoincidence (radiation counter). The occurrence of a count in a specified detector unaccompanied simultaneously or within an assignable time interval by a count in one or more other specified detectors. 190

anticoincidence circuit (pulse techniques). A circuit that produces a specified output pulse when one (frequently predesignated) of two inputs receives a pulse and the other receives no pulse within an assigned time interval. *See also:* **pulse, anticoincidence (radiation counter).** 117

anticollision light. A flashing aircraft aeronautical light or system of lights designed to provide a red signal throughout 360 degrees of azimuth for the purpose of giving long-range indication of an aircraft's location to pilots of other aircraft. 167

antiferroelectric material. A material that exhibits structural phase changes and anomalies in the dielectric permittivity as do ferroelectrics, but has zero net spontaneous polarization, and hence exhibits no hysteresis phenomena. *Note:* In some cases it is possible to apply electric fields sufficiently high to produce a structural transition to a ferroelectric phase as evidenced by appearance of a double hysteresis loop. *See also:* **ferroelectric material, ferroelectric domain, paraelectric.** 80

antifouling. Pertaining to the prevention of marine organism attachment and growth on a submerged metal surface (through the effects of chemical action). 221, 205

antifreeze pin, relay. *See:* **relay antifreeze pin.**

antifriction bearing (rotating machinery). A bearing incorporating a peripheral assembly of separate parts that will support the shaft and reduce sliding friction by rolling as the shaft rotates. *See also:* **bearing.** 63

antinode (standing wave). *See:* **loop (standing wave).**

antinoise microphone. A microphone with characteristics that discriminate against acoustic noise. *See also:* **microphone.** 328

anti-overshoot (industrial control). The effect of a control function or a device that causes a reduction in the transient overshoot. *Note:* Anti-overshoot may apply to armature current, armature voltage, field current, etcetera. *See also:* **control system, feedback.** 225, 206

antiplugging protection (industrial control). The effect of a control function or a device that operates to prevent application of counter torque by the motor until the motor speed has been reduced to an acceptable value. *See also:* **control system, feedback.** 225, 206

antipump device (pump-free device). A device that prevents reclosing after an opening operation as long as the device initiating closing is maintained in the position for closing. 103, 202

antiresonant frequency (circuits and systems). Usually in reference to a crystal unit or the parallel combination of a capacitor and inductor. The frequency at which, neglecting dissipation, the impedance of the object under consideration is infinite. 67

antisidetone induction coil. An induction coil designed for use in an antisidetone telephone set. *See also:* **telephone station.** 328

antisidetone telephone set. A telephone set that includes a balancing network for the purpose of reducing sidetone. *See also:* **telephone station; sidetone.** 328

anti-single-phase tripping device. A device that operates to open all phases of a circuit by means of a polyphase switching device, in response to the interruption of the current in one phase. *Notes:* (1) This device prevents single phasing of connected equipment resulting from the interruption of any one phase of the circuit. (2) This device may sense operation of a specific single-phase interrupting device or may sense loss of single phase potential. 103

anti-transmit-receive box (ATR box) (electronic navigation). *See:* anti-transmit-receive switch.

anti-transmit-receive switch (ATR switch) (radar). An RF switch that automatically decouples the transmitter from

the antenna during the receiving period; it is employed when a common transmitting and receiving antenna is used. *See also:* **radar.** 13, 187

anti-transmit-receive tube (ATR tube). A gas-filled radio-frequency switching tube used to isolate the transmitter during the interval for pulse reception. *See also:* **gas tubes.** 190, 125

A **operator.** An operator assigned to an *A* switchboard. 328

APC. *See:* **automatic phase control.**

aperiodically sampled equivalent time format (pulse measurement). A format which is identical to the aperiodically sampled real time format except that the time coordinate is equivalent to and convertible to real time. Typically, each datum point is derived from a different measurement on a different wave in a sequence of waves. *See also:* **sampled format.** 15

aperiodically sampled real time format (pulse measurement). A format which is identical to the periodically sampled real time format except that the sampling in real time is not periodic and wherein the data exists as coordinate point pairs $t_1, m_1; t_2, m_2; \cdots; t_n, m_n$. *See also:* **sampled format.** 15

aperiodic circuit. A circuit in which it is not possible to produce free oscillations. *See also:* **oscillatory circuit.** 244, 59

aperiodic component of short-circuit current (rotating machinery). The component of current in the primary winding immediately after it has been suddenly short-circuited when all components of fundamental and higher frequencies have been subtracted. *See also:* **asynchronous machine.** 63

aperiodic damping*. *See:* **overdamping.**

* Deprecated

aperiodic function. Any function that is not periodic. 210

aperiodic time constant (rotating machinery). The time constant of the aperiodic component when it is essentially exponential, or of the exponential which can most nearly be fitted. 63

See also: **asynchronous machine; direct current comutating machine.** 63

aperture (antenna). A surface, near or on an antenna, on which it is convenient to make assumptions regarding the field values for the purpose of computing fields at external points. *Notes:* (1) In some cases the aperture may be considered as a line. (2) In the case of a unidirectional antenna the aperture is often taken as that portion of a plane surface near the antenna, perpendicular to the direction of maximum radiation, through which the major part of the radiation passes. *See also:* **antenna.** 246, 179, 59, 111

aperture blockage (antenna). A blocking of or interfering with the radiation from the feed or secondary radiator by obstacles such as the feed itself or support struts. 111

aperture compensation (television). *See:* **aperture equalization.**

aperture correction (television). *See:* **aperture equalization.**

aperture efficiency (antenna) (for an antenna aperture). The ratio of its directivity to the directivity obtained when the

aperture illumination is uniform. *See also:* **antenna.**
179, 111

aperture equalization (1) (television). Electrical compensation for the distortion introduced by the size of a scanning aperture. *See also:* **television.**   178
(2) (aperture correction, aperture compensation). Compensation for the distortion introduced by the finite size of the aperture of a scanning beam by means of circuits inserted in the signal channel.   328

aperture illumination (excitation). The field over the aperture as described by amplitude, phase, and polarization distributions.   111

aperture seal (nuclear power generating stations). A seal between the containment aperture and the electrical penetration assembly.   31, 26

API. *See:* **air-position indicator.**

APL. *See:* **average picture level.**

apoapsis (communication satellite). The most distant point from the center of a primary body (or planet) to an orbit around it.   74

apogee (communication satellite). The most distant point from the center of the earth to an orbit around it.   74

apparatus (1) (electric). The terms appliance and device generally refer to small items of electric equipment, usually located at or near the point of utilization, but with no established criterion for the meaning of small. The term apparatus then designates the large items of electric equipment, such as generators, motors, transformers, or circuit breakers. *See also:* **generating station.**   260
(2) (transformer). A general designation for large electrical equipment such as generators, motors, tramsformers, circuit breakers, etcetera.   53

apparatus insulator (cap and pin, post). An assembly of one or more apparatus-insulator units, having means for rigidly supporting electric equipment. *See also:* **insulator.**
261

apparatus insulator unit. The assembly of one or more elements with attached metal parts, the function of which is to support rigidly a conductor, bus, or other conducting elements on a structure or base member. *See also:* **tower.**
64

apparatus termination (high voltage AC cable terminations). A termination intended for use in apparatus where the ambient temperature of the medium immediately surrounding the termination may reach 55 °C.   4

apparent bearing (direction finding). A bearing from a direction-finder site to a target transmitter determined by averaging the readings made on a calibrated direction-finder test standard; the apparent bearing is then used in the calibration and adjustment of other direction-finders at the same site. *See also:* **navigation.**   187

apparent candlepower (extended source). At a specified distance, the candlepower of a point source that would produce the same illumination at that distance.   167

apparent dead time. *See:* **dead time.**

apparent inductance. The reactance between two terminals of a device or circuit divided by the angular frequency at which the reactance was determined. This quantity is defined only for frequencies at which the reactance is positive. *Note:* Apparent inductance includes the effects of the real and parasitic elements that comprise the device or circuit and is therefore a function of frequency and other

operating conditions.   197

apparent power (1) (rotating machinery). The product of the root-mean-square current and the root-mean-square voltage. *Note:* It is a scalar quantity equal to the magnitude of the phasor power. *See also:* **asynchronous machine.**
63

(2) (general). *See:* **power, apparent.**

apparent-power loss (volt-ampere loss) (electric instrument). Of the circuit for voltage-measuring instruments, the product of end-scale voltage and the resulting current; and for current-measuring instruments, the product of the end-scale current and the resulting voltage. *Notes:* (1) For other than current-measuring or voltage-measuring instruments, for example, wattmeters, the apparent power loss of any circuit is expressed at a stated value of current or of voltage. (2) Computation of loss: for the purpose of computing the loss of alternating-current instruments having current circuits at some selected value other than that for which it is rated, the actual loss at the rated current is multiplied by the square of the ratio of the selected current to the rated current. *Example:* A current transformer with a ratio of 500:5 amperes is used with an instrument having a scale of 0–300 amperes and, therefore, a 3-ampere field coil, and the allowable loss at end scale is as stated on the Detailed Requirement Sheet. The allowable loss of the instrument referred to a 5-amperes basis is as follows: Allowable loss in volt-amperes equals (allowable loss end-scale volt-amperes) $(5/3)^2$. *See also:* **accuracy rating (instrument).**   280

apparent sag (wire in a span). (1) The maximum departure in the vertical plane of the wire in a given span from the straight line between the two points of support of the span, at 60 degrees Fahrenheit, with no wind loading. *Note:* Where the two supports are at the same level this will be the sag. *See also:* **tower.**   64, 262
(2) The departure in the vertical plane of the wire at the particular point in the span from the straight line between the two points of support.   64

apparent sag at any point in the span (transmission and distribution). The departure of the wire at the particular point in the span from the straight line between the two points of support of the span.   262

apparent sag of a span (transmission and distribution). The maximum departure of the wire in a given span from the straight line between the two points of support of the span.
262

apparent time constant (thermal converter) (63-percent response time) (characteristic time). The time required for 63 percent of the change in output electromotive force to occur after an abrupt change in the input quantity to a new constant value. *See:* Note 1 of **Response time of a thermal converter.** *See also:* **thermal converter.**   280

apparent vertical (electronic navigation). The direction of the vector sum of the gravitational and all other accelerations. *See also:* **navigation.**   187

apparent visual angle (laser-maser). The angular subtense of the source as calculated from the source size and distance from the eye. It is not the beam divergence of two sources.   363

appliance (1) (electric). A utilization item of electric equipment, usually complete in itself, generally other than industrial, normally built in standardized sizes or types

that transforms electric energy into another form, usually heat or mechanical motion, at the point of utilization. For example, a toaster, flatiron, washing machine, dryer, hand drill, food mixer, air conditioner.   256, 260, 210
(2) (transmission and distribution). Current-consuming equipment, fixed or portable; for example, heating, cooking, and small motor-operated equipment.   262
appliance branch circuit. A circuit supplying energy to one or more outlets to which appliances are to be connected; such circuits to have no permanently connected lighting fixtures not a part of an appliance.   328
appliance, fixed (electric system). An appliance that is fastened or otherwise secured at a specific location. *See also:* **appliances.**   210, 256
appliance outlet (household electric ranges). An outlet mounted on the range and to which a portable appliance may be connected by means of an attachment plug cap.   263
appliance, portable (electric system). An appliance that is actually moved or can easily be moved from one place to another in normal use. *See also:* **appliances.**   210, 256
appliance, stationary (electric system). An appliance that is not easily moved from one place to another in normal use. *See also:* **appliances.**   210, 256
application valve (brake application valve). An air valve through the medium of which brakes are automatically applied.   328
applicator (electrodes) (dielectric heating). Appropriately shaped conducting surfaces between which is established an alternating electric field for the purpose of producing dielectric heating. *See also:* **dielectric heating.** 14, 114
applied fault protection. A protective method in which, as a result of relay action, a fault is intentionally applied at one point in an electric system in order to cause fuse blowing or further relay action at another point in the system.   103, 202, 60, 127
applied-potential tests (electric power). Dielectric tests in which the test voltages are low-frequency alternating voltages from an external source applied between conducting parts, and between conducting parts and ground.   91
applied voltage tests. Dielectric tests in which the test voltages are low-frequency alternating voltages from an external source applied between conducting parts and ground without exciting the core of the transformer being tested.   53
approach circuit. A circuit used to announce the approach of trains at block or interlocking stations.   328
approach indicator. A device used to indicate the approach of a train.   328
approach-light beacon. An aeronautical ground light placed on the extended centerline of the runway at a fixed distance from the runway threshold to provide an early indication of position during the approach to a runway. *Note:* The runway threshold is the beginning of the runway usable for landing.   167
approach lighting. An arrangement of circuits so that the signal lights are automatically energized by the approach of a train.   328
approach-lighting relay. A relay used to close the lighting circuit for signals upon the approach of a train.   328
approach lights. A configuration of aeronautical ground

lights located in extension of a runway or channel before the threshold to provide visual approach and landing guidance to pilots.   167
approach locking (electric approach locking). Electric locking effective while a train is approaching, within a specified distance, a signal displaying an aspect to proceed, and that prevents, until after the expiration of a predetermined time interval after such signal has been caused to display its most restrictive aspect, the movement of any interlocked or electrically locked switch, movable-point frog, or derail in the route governed by the signal, and that prevents an aspect to proceed from being displayed for any conflicting route. *See also:* **interlocking.**   328
approach navigation. Navigation during the time that the approach to a dock, runway, or other terminal facility is of immediate importance. *See also:* **navigation.** 13, 187
approach path (electronic navigation). The portion of the flight path between the point at which the descent for landing is normally started and the point at which the aircraft touches down on the runway. *See also:* **navigation; radio navigation.**   187
approach signal. A fixed signal used to govern the approach to one or more other signals.   328
approval plate (mining). A label that the United States Bureau of Mines requires manufacturers to attach to every completely assembled machine or device sold as permissible mine equipment. *Note:* By this means, the manufacturer certifies to the permissible nature of the machine or device.   328
approval test (acceptance test). The testing of one or more meters or other items under various controlled conditions to ascertain the performance characteristics of the type of which they are a sample. *See also:* **service test (field test).**   212
approved. Approved by the enforcing authority.   328
APT (numerically controlled machines). *See:* **automatic programmed tools.**
arc. (1) A discharge of electricity through a gas, normally characterized by a voltage drop in the immediate vicinity of the cathode approximately equal to the ionization potential of the gas. *See:* **gas tube.**   125
(2) A continuous luminous discharge of electricity across an insulating medium, usually accompanied by the partial volatilization of the electrodes.   210
arc-back (gas tube). A failure of the rectifying action that results in the flow of a principal electron stream in the reverse direction, due to the formation of a cathode spot on an anode. *See also:* **gas tubes; rectification.**   125
arc cathode (gas tube). A cathode the electron emission of which is self-sustaining with a small voltage drop approximately equal to the ionization potential of the gas. *See also:* **gas-filled rectifier.**   244
arc chute (switching device). A structure affording a confined space or passageway, usually lined with arc-resisting material, into or through which an arc is directed to extinction. *See also:* **contactor.**   103, 202, 206
arc, clockwise (numerically controlled machines). An arc generated by the coordinated motion of two axes in which curvature of the path of the tool with respect to the workpiece is clockwise, when viewing the plane of motion in the negative direction of the perpendicular axis.   224, 207

arc converter. A form of negative-resistance oscillator utilizing an electric arc as the negative resistance. *See also:* **radio transmission.** 240

arc, counterclockwise (numerically controlled machines). An arc generated by the coordinated motion of two axes in which curvature of the path of the tool with respect to the workpiece is counterclockwise, when viewing the plane of motion in the negative direction of the perpendicular axis. 224, 207

arc discharge. An electric discharge characterized by high cathode current densities and a low voltage drop at the cathode. *See also:* **lamp.** 167

arc-discharge tube (valve). A gas-filled tube or valve in which the required current is that of an arc discharge. 190

arc-drop loss (gas tube). The product of the instantaneous values of the arc-drop voltage and current averaged over a complete cycle of operation. *See:* **gas tube.** 125

arc-drop voltage (gas tube). The voltage drop between the anode and cathode of a rectifying device during conduction. *See also:* **electrode voltage (electron tube); tube voltage drop.** 190

arc-extinguishing medium (fuse filler) (fuse). Material included in the fuse to facilitate current interruption. 103, 202

arc furnace. An electrothermic apparatus the heat energy for which is generated by the flow of electric current through one or more arcs internal to the furnace. *See also:* **electrothermics.** 328

arcing chamber (expulsion-type arrester). The part of an expulsion-type arrester that permits the flow of discharge current to the ground and interrupts the follow current. *See:* **surge arrester (surge diverter).** 62, 308

arcing contacts (switching device). The contacts on which the arc is drawn after the main (and intermediate, where used) contacts have parted. 103, 202, 27

arcing horn. One of a pair of diverging electrodes on which an arc is extended to the point of extinction after the main contacts of the switching device have parted. *Note:* Arcing horns are sometimes referred to as arcing runners. 103, 202, 27

arcing time (1) (fuse). The time elapsing from the severance of the current-responsive element to the final interruption of the circuit. 103, 202
(2) (mechanical switching device). The interval of time between the instant of the first initiation of the arc and the instant of final arc extinction in all poles. *Note:* For switching devices that embody switching resistors, a distinction should be made between the arcing time up to the instant of the extinction of the main arc, and the arcing time up to the instant of the breaking of the resistance current. 103, 202

arc loss (switching tube). The decrease in radio-frequency power measured in a matched termination when a fired tube, mounted in a series or shunt junction with a waveguide, is inserted between a matched generator and the termination. *Note:* In the case of a pretransmit-receive tube, a matched output termination is also required for the tube. *See also:* **gas tubes.** 190, 125

arc-shunting-resistor-current arcing time. The interval between the parting of the secondary arcing contacts and the extinction of the arc-shunting-resistor current. 103, 202

arc suppression (rectifier). The prevention of the recurrence of conduction, by means of grid or ignitor action, or both, during the idle period, following a current pulse. *See also:* **rectification.** 291

arc-through (gas tube). A loss of control resulting in the flow of a principal electron stream through the rectifying element in the normal direction during a scheduled nonconducting period. *See also:* **rectification.** 190, 125

arc-tube relaxation oscillator. *See:* **gas-tube relaxation oscillator.**

arc-welding engine-generator. A device consisting of an engine mechanically conected to and mounted with one or more arc-welding generators. 264

arc-welding motor-generator. A device consisting of a motor mechanically connected to and mounted with one or more arc-welding generators. 264

area. *See:* **effective area of an antenna; equivalent flat plate area of a scattering object.** 111

area assist action (electric power system). The component of area supplementary control that involves the temporary assignment of generation changes to minimize the area control error prior to the assignment of generation changes on an economic dispatch basis. *See also:* **speed-governing system.** 94

area code (telephone switching system). A one, two-, or three-digit number that, for the purpose of distance dialing, designates a geographical subdivision of the territory covered by a separate national or integrated numbering plan. 55

area control error (isolated-power system consisting of a single control area). The frequency deviation (of a control area on an interconnected system) is the net interchange minus the biased scheduled-net interchange. *Note:* The above polarity is that which has been generally accepted by electric power systems and is in wide use. It is recognized that this is the reverse of the sign of control error generally used in servomechanism and control literature, which defines control error as the reference quantity minus the controlled quantity. 200, 94

area frequency-response characteristic (control area). The sum of the change in total area generation caused by governor action and the change in total area load, both of which result from a sudden change in system frequency, in the absence of automatic control action. 94

area load-frequency characteristic (control area). The change in total area load that results from a change in system frequency. 94

Area MTI (radar). A moving target indicator (MTI) system in which the moving target is selected on the basis of its change in position between looks, rather than its Doppler frequency. 13

area supplementary control (electric power system). The control action applied, manually or automatically, to area generator speed governors in response to changes in system frequency, tie-line loading, or the relation of these to each other, so as to maintain the scheduled system frequency and/or the established net interchange with other control areas within predetermined limits. 94

area tie line (electric power system). A transmission line connecting two control areas. *Note:* Similar to interconnection tie. *See also:* **transmission line.** 94

argument (information processing). An independent variable, for example, in looking up a quantity in a table, the number, or any of the numbers, that identifies the location of the desired value. 77, 255

arithmetic apparent power (polyphase circuit). At the terminals of entry of a polyphase circuit, a scalar quantity equal to the arithmetic sum of the apparent powers for the individual terminals of entry. *Notes:* (1) The apparent power for each terminal of entry is determined by considering each phase conductor and the common reference terminal as a single-phase circuit, as described for distortion power. The common reference terminal shall be taken at the neutral terminal of entry if one exists, otherwise as the true neutral point. (2) If the ratios of the components of the vector powers, for each of the terminals of entry, to the corresponding apparent power are the same for every terminal of entry, the arithmetic apparent power is equal to the apparent power for the polyphase circuit; otherwise the arithmetic apparent power is greater than the apparent power. (3) If the voltages have the same wave form as the corresponding currents, the arithmetic apparent power is equal to the amplitude of the phasor power. (4) Arithmetic apparent power is expressed in voltamperes when the voltages are in volts and the currents in amperes. 210

arithmetic element. The part of a computer that performs arithmetic operations. *See also:* **arithmetic unit.** 210

arithmetic operations (test, measurement and diagnostic equipment). Operations in which numerical quantities form the elements of the calculation. 54

arithmetic reactive factor. The ratio of the reactive power to the arithmetic apparent power. 210

arithmetic shift. (1) A shift that does not affect the sign position. (2) A shift that is equivalent to the multiplication of a number by a positive or negative integral power of the radix. 77, 255, 54

arithmetic unit. The unit of a computing system that contains the circuits that perform arithmetic operations. *See also:* **arithmetic element.** 77, 255, 54

arm. *See:* **branch.** *See also:* **network analysis.**

armature (1) (rotating machinery). The member of an electric machine in which an alternating voltage is generated by virtue of relative motion with respect to a magnetic flux field. In direct-current universal, alternating-current series, and repulsion-type machines, the term is commonly applied to the entire rotor. 63

(2) (relay) (electromechanical relay). The moving element that contributes to the designed response of the relay and that usually has associated with it a part of the relay contact assembly. *See:* **relay armature.** *See also:* **relay.** 103, 60

(3) (magnet). A piece or an assembly of pieces of ferromagnetic material associated with the pole pieces of a magnet in such a manner that it and the pole pieces can move in relation to each other. 210

armature band (rotating machinery). A thin circumferential structural member applied to the winding of a rotating armature to restrain and hold the coils so as to counteract the effect of centrifugal force during rotation. *Note:* Armature bands may serve the further purpose of arch-binding the coils. They may be on the end windings only

or may be over the coils within the core. 63

armature band insulation (rotating machinery). An insulation member placed between a rotating armature winding and an armature band. *See also:* **armature.** 63

armature bar (half coil) (rotating machinery). Either of two similar parts of an armature coil, comprising an embedded coil side and two end sections, that when connected together form a complete coil. *See also:* **armature.** 63

armature coil (rotating machinery). A unit of the armature winding composed of one or more insulated conductors. *See:* **armature; asynchronous machine; direct-current commutating machine.** 328

armature core (rotating machinery). A core on or around which armature windings are placed. *See also:* **armature.** 63

armature I^2R loss (synchronous machine). The sum of the I^2R losses in all of the armature current paths. *Note:* The I^2R loss in each current path shall be the product of its resistance in ohms, as measured with direct current and corrected to a specified temperature, and the square of its current in amperes. 63, 298

armature quill. *See:* **armature spider.**

armature reaction (rotating machinery). The magnetomotive force due to armature-winding current. 63

armature-reaction excited machine. A machine having a rotatable armature, provided with windings and a commutator, whose load-circuit voltage is generated by flux that is produced primarily by the magnetomotive force of currents in the armature winding. *Notes:* (1) By providing the stationary member of the machine with various types of windings different characteristics may be obtained, such as a constant-current characteristic or a constant-voltage characteristic. (2) The machine is normally provided with two sets of brushes, displaced around the commutator from one another, so as to provide primary and secondary circuits through the armature. (3) The primary circuit carrying the excitation armature current may be completed externally by a short-circuit connection, or through some other external circuit, such as a field winding or a source of power supply; and the secondary circuit is adapted for connection to an external load. *See also:* **direct-current commutating machine.** 328

armature sleeve. *See:* **armature spider.**

armature spider (armature sleeve) (armature quill). A support upon which the armature laminations are mounted and which in turn is mounted on the shaft. *See also:* **armature (rotating machinery).** 328

armature terminal (rotating machinery). A terminal connected to the armature winding. *See also:* **armature.** 63

armature-voltage control. A method of controlling the speed of a motor by means of a change in the magnitude of the voltage impressed on its armature winding. *See:* **control.** 206

armature winding (rotating machinery). The winding in which alternating voltage is generated by virtue of relative motion with respect to a magnetic flux field. *See:* **asynchronous machine; direct-current commutating machine.** 63

armature winding cross connection. *See:* **armature winding equalizer.**

armature winding equalizer (armature winding cross connection) (rotating machinery). An electric connection to normally equal-potential points in an armature circuit having more than two parallel circuits. *See also:* **armature (rotating machinery).** 63

armed sweep. *See:* **single sweep mode.**

armor clamp (wiring methods). A fitting for gripping the armor of a cable at the point where the armor terminates or where the cable enters a junction box or other piece of apparatus. 64

armored cable (interior wiring). A fabricated assembly of insulated conductors and a flexible metallic covering. *Note:* Armored cable for interior wiring has its flexible outer sheath or armor formed of metal strip, helically wound and with interlocking edges. Armored cable is usually circular in cross section but may be oval or flat and may have a thin lead sheath between the armor and the conductors to exclude moisture, oil, etcetera, where such protection is needed. *See:* **non-metallic sheathed cable.** 328

arm, thermoelectric. The part of a thermoelectric device in which the electric-current density and temperature gradient are approximately parallel or antiparallel and that is electrically connected only at its extremities to a part having the opposite relation between the direction of the temperature gradient and the electric-current density. *Note:* The term thermoelement is ambiguously used to refer to either a thermoelectric arm or to a thermoelectric couple, and its use is therefore not recommended. *See also:* **thermoelectric device.** 248, 191

arm, thermoelectric, graded. A thermoelectric arm whose composition changes continuously along the direction of the current density. *See also:* **thermoelectric device.** 248, 191

arm, thermoelectric, segmented. A thermoelectric arm composed of two or more materials having different compositions. *See also:* **thermoelectric device.** 248, 191

array (1) (photovoltaic converter). A combination of panels coordinated in structure and function. *See also:* **semiconductor.** 186

(2) (solar cell). A combination of solar-cell panels or paddles coordinated in structure and function. 113

array antenna. An antenna comprising a number of radiating elements, generally similar, that are arranged and excited to obtain directional radiation patterns. *See also:* **antenna.** 179

array element (array antenna) (antenna). A single radiating element or a convenient grouping of radiating elements that have a fixed relative excitation. *See also:* **antenna.** 179

array factor. The radiation pattern of an array antenna when the pattern of each array element is considered to be isotropic. 111

arrester alternating sparkover voltage. The root-mean-square value of the minimum 60-hertz sine-wave voltage that will cause sparkover when applied between its line and ground terminals. *See also:* **current rating, 60-hertz (arrester); surge arrester (surge diverter).** 62

arrester discharge capacity. The crest value of the maximum discharge current of specified wave shape that the arrester can withstand without damage to any of its parts. *See:* **surge arrester (surge diverter).** 62

arrester discharge voltage-current characteristic. The variation of the crest values of discharge voltage with re-

spect to discharge current. *Note:* This characteristic is normally shown as a graph based on three or more current surge measurements of the same wave shape but of different crest values. *See also:* **lightning; surge arrester (surge diverter); current rating, 60-hertz (arrester).** 2, 62

arrester discharge voltage-time curve. A graph of the discharge voltage as a function of time while discharging a current surge of given wave shape and magnitude. *See:* **surge arrester (surge diverter).** 62

arrester disconnector (surge arrester). A means for disconnecting an arrester in anticipation of, or after, a failure in order to prevent a permanent fault on the circuit and to give indication of a failed arrester. *Note:* Clearing of the power current through the arrester during disconnection generally is a function of the nearest source-side overcurrent-protective device. 2

arrester, expulsion-type. An arrester having an arcing chamber in which the follow-current arc is confined and brought into contact with gas-evolving or other arc-extinguishing material in a manner that results in the limitation of the voltage at the line terminal and the interruption of the follow current. *Note:* The term **expulsion arrester** includes any external series-gap or current-limiting resistor if either or both are used as a part of the complete device as installed for service. 62, 308

arrester ground. An intentional electric connection of the arrester ground terminal to the ground. *See:* **surge arrester (surge diverter).** 62

arresters, classification of. Arrester classification is determined by prescribed test requirements. These classifications are: station valve arrester; intermediate valve arrester; distribution valve arrester; distribution expulsion arrester; secondary valve arrester; protector tube. 2

arrester, valve-type. An arrester having a characteristic element consisting of a resistor with a nonlinear volt-ampere characteristic that limits the follow current to a value that the series gap can interrupt. *Note:* If the arrester has no series gap the characteristic element limits the follow current to a magnitude that does not interfere with the operation of the system. *See:* **nonlinear-resistor type arrester (valve type).** *See also:* **surge arrester (surge diverter).** 62

ARSR (air route surveillance radar). *See:* **surveillance radar.**

articulated unit substation. A unit substation in which the incoming, transforming and outgoing sections are manufactured as one or more subassemblies intended for connection in the field. 103, 202, 53

(1) radial type. One which has a single stepdown transformer and which has an outgoing section for the connection of one or more outgoing radial (stub end) feeders.

(2) distributed-network type. One which has a single stepdown transformer having its outgoing side connected to a bus through a circuit breaker equipped with relays which are arranged to trip the circuit breaker on reverse power flow to the transformer and to reclose the circuit breaker upon the restoration of the correct voltage, phase angle and phase sequence at the transformer secondary. The bus has one or more outgoing radial (stub end) feeders and one or more tie connections to a similar unit substation.

(3) spot-network type. One which has two stepdown transformers, each connected to an incoming high-voltage circuit. The outgoing side of each transformer is connected to a common bus through circuit breakers equipped with relays which are arranged to trip the circuit breaker on reverse power flow to the transformer and to reclose the circuit breaker upon the restoration of the correct voltage, phase angle and phase sequence at the transformer secondary. The bus has one or more outgoing radial (stub end) feeders.
(4) secondary-selective type (low-voltage-selective type). One which has two stepdown transformers, each connected to an incoming high-voltage circuit. The outgoing side of each transformer is connected to a separate bus through a suitable switching and protective device. The two sections of bus are connected by a normally open switching and protective device. Each bus has one or more outgoing radial (stub end) feders.
(5) duplex type (breaker-and-a-half arrangement). One which has two stepdown transformers, each connected to an incoming high-voltage circuit. The outgoing side of each transformer is connected to a radial (stub end) feeder. These feeders are joined on the feeder side of the power circuit breakers by a normally-open-tie circuit breaker.
53
articulation (percent articulation) and intelligibility (percent intelligibility). The percentage of the speech units spoken by a talker or talkers that is correctly repeated, written down, or checked by a listener or listeners. *Notes:* (1) The word articulation is used when the units of speech material are meaningless syllables or fragments; the word intelligibility is used when the units of speech material are complete, meaningful words, phrases, or sentences. (2) It is important to specify the type of speech material and the units into which it is analyzed for the purpose of computing the percentage. The units may be fundamental speech sounds, syllables, words, sentences, etcetera. (3) The percent articulation or percent intelligibility is a property of the entire communication system; talker, transmission equipment or medium, and listener. Even when attention is focused upon one component of the system (for example, a talker, a radio receiver), the other components of the system should be specified. (4) The kind of speech material used is identified by an appropriate adjective in phrases such as syllable articulation, individual sound articulation, vowel (or consonant) articulation, monosyllabic word intelligibility, discrete word intelligibility, discrete sentence intelligibility. *See also:* **volume equivalent.**   176
articulation equivalent (complete telephone connection). A measure of the articulation of speech reproduced over it. The articulation equivalent of a complete telephone connection is expressed numerically in terms of the trunk loss of a working reference system when the latter is adjusted to give equal articulation. *Note:* For engineering purposes, the articulation equivalent is divided into articulation losses assignable to (1) the station set, subscriber line, and battery supply circuit that are on the transmitting end, (2) the station set, subscriber line, and battery supply circuit that are on the receiving end, (3) the trunk, and (4) interaction effects arising at the trunk terminals. *See also:* **volume equivalent.**   328
artificial antenna (dummy antenna). A device that has the

necessary impedance characteristics of an antenna and the necessary power-handling capabilities, but which does not radiate or receive radio waves. *See also:* **antenna.**
111, 179
artificial dielectric (antennas). A medium containing metallic conductors in an arrangement which appears as a dielectric to electromagnetic waves. *Note:* The metallic conductors are usually small compared to a wavelength and embedded in a dielectric material of low effective permittivity and density.   111
artificial ear (1) (general). A device for the measurement of the acoustic output of earphones in which the artificial ear presents to the earphone an acoustic impedance approximating the impedance presented by the average human ear and is equipped with a microphone for measurement of the sound pressures developed by the earphone.   122, 196, 176
(2) (telephony). A device for the measurement of the acoustic output of telephone receivers that presents to the receiver an acoustic impedance approximating the impedance presented by the average human ear.   1 22
artificial hand (electromagnetic compatibility). A device simulating the impedance between an electric appliance and the local earth when the appliance is grasped by the hand. *See also:* **electromagnetic compatibility.** 220, 199
artificial language. A language based on a set of prescribed rules that are established prior to its usage. *See:* **natural language.**   77, 255
artificial line. An electric network that simulates the electrical characteristics of a line over a desired frequency range. *Note:* Although the term basically is applied to the case of simulation of an actual line, by extension it is used to refer to all periodic lines that may be used for laboratory purposes in place of actual lines, but that may represent no physically realizable line. For example, an artificial line may be composed of pure resistances. *See also:* **network analysis.**   111, 210, 59
artificial load. A dissipative but essentially nonradiating device having the impedance characteristics of an antenna, transmission line, or other practical utilization circuit.
111, 210, 59
artificial mains network (electromagnetic compatibility). A network inserted in the supply mains lead of the apparatus to be tested that provides a specified measuring impedance for interference voltage measurements and isolates the apparatus from the supply mains at radio frequencies. *See also:* **electromagnetic compatibility.**   220, 199
artificial mouth (telephony). An electroacoustic transducer that produces a sound field, with respect to a reference point, simulating that of a typical human talker. *Note:* The reference point corresponds to the center of the lips of the talker. *See also:* **telephone station.**   122, 196
artificial pupil. A diaphragm or other limitation that confines the light entering the eye to a smaller aperture than does the normal pupil. *See also:* **visual field.**   167
artificial voice (close-talking pressure-type microphone). A sound source for microphone measurements consisting of a small loudspeaker mounted in a shaped baffle proportioned to simulate the acoustic constants of the human head. *See also:* **close-talking pressure-type microphone; loudspeaker.**   249
A scan (electronic navigation). *See:* **A and R display.**

ascending node (communication satellite). The point on the line of nodes that the satellite passes through as the satellite travels from below to above the equatorial plane.
74

A scope. A cathode-ray oscilloscope arranged to present an *A* display. *See: A* **display.** *See also:* **radar.**   13

ASDE. *See:* **airport surface detection equipment.**

askarel. A generic term for a group of synthetic, fire-resistant, chlorinated aromatic hydrocarbons used as electrical insulating liquids. They have a property under arcing conditions such that any gases produced will consist predominantly of noncombustible hydrogen chloride with lesser amounts of combustible gases.   53

aspect ratio (television). The ratio of the frame width to the frame height as defined by the active picture. 163, 178

asphalt (rotating machinery). A dark brown to black cementitious material, solid or semisolid in consistency, in which the predominating constituents are bitumens that occur in nature as such or are obtained as residue in refining of petroleum.   214, 63

ASR. *See:* **airport surveillance radar.**

assemble (computing machine). To prepare a machine language program from a symbolic language program by substituting absolute operation codes for symbolic operation codes and absolute or relocatable addresses for symbolic addresses.   77, 255

assembler (1) (computing machine). A program that assembles.   77, 255
(2) (test, measurement and diagnostic equipment). A computer program that is one step more automatic than a translator; it translates not only operations but also data and input-output quantities from symbolic to machine language form in a one to one ratio. An assembler program may have the capability to assign locations within a storage device.   54

assembly (1) (industrial control). A combination of all or of a portion of component parts included in an electric apparatus, mounted on a supporting frame or panel, and properly interwired.   206
(2) (electric and electronics parts and equipments). A number of basic parts or subassemblies, or any combination thereof, joined together to perform a specific function. The application, size, and construction of an item may be factors in determining whether an item is regarded as a unit, an assembly, a subassembly, or a basic part. A small electric motor might be considered as a part if it is not normally subject to disassembly. The distinction between an assembly and a subassembly is not always exact; an assembly in one instance may be a subassembly in another where it forms a portion of an assembly. Typical examples are: electric generator, audio-frequency amplifier, power supply.   17
(3) (nuclear power generating stations) (seismic qualification of class 1E equipment). Two or more devices sharing a common mounting or supporting structure. *Note:* Examples are control panels and diesel generators.   31, 28

assembly language (test, measurement and diagnostic equipment). A language in which machine operations and locations are represented by mnemonic symbols.   54

assembly, microelectronic device (electric and electronics parts and equipments). An assembly of inseparable parts, circuits, or combination thereof. Typical examples are:

microcircuit, integrated-circuit package, micromodule.
17

assessed mean active maintenance time. The active maintenance time determined as the limit or the limits of the confidence interval associated with a stated confidence level, and based on the same data as the observed mean active maintenance time or nominally identical items. *Notes:* (1) The source of the data shall be stated. (2) Results can be accumulated (combined) only when all conditions are similar. (3) It should be stated whether a one-sided or two-sided interval is being used. (4) The assumed underlying distribution of mean active maintenance times shall be stated with the reason for the assumption. (5) When one value is given this is usually the upper limit.
164

assessed reliability. The reliability of an item determined by a limiting value or values of the confidence interval associated with a stated confidence level, based on the same data as the observed reliability of nominally identical items. *Notes:* (1) The source of the data shall be stated. (2) Results can be accumulated (combined) only when all conditions are similar. (3) The assumed underlying distribution of failures against time shall be stated. (4) It should be stated whether a one-sided or a two-sided interval is being used. (5) Where one limiting value is given this is usually the lower limit.   164

assigned value. The best estimate of the value of a quantity. The assigned value may be from an instrument reading, a calibration result, a calculation, or other.   115, 47

assistance call (telephone switching system). A call to an operator for help in making a call.   55

associated circuits (nuclear power generating stations). Non-Class 1E circuits that share power supplies, enclosures, or raceways with Class 1E circuits or are not physically separated from Class 1E circuits by acceptable separation distance or barriers.   31

associative storage. A storage device in which storage locations may be identified by specifying part or all of their contents. *Note:* Also called parallel-search storage or content-addressed storage.   235, 77

aster rectifier circuit. A circuit that employs twelve or more rectifying elements with a conducting period of 30 electrical degrees plus the commutating angle. *See also:* **rectification.**   328

astigmatism (electron optical). In an electron-beam tube, a focus defect in which electrons in different axial planes come to focus at different points. *See also:* **oscillograph.**
190, 125

Aston dark space (gas). The dark space in the immediate neighborhood of the cathode, in which the emitted electrons have a velocity insufficient to excite the gas. *See also:* **discharge (gas).**   244, 190

astrodynamics (communication satellite). Engineering application of celestial mechanics.   74

astro-inertial navigation equipment. *See:* **celestial inertial navigation equipment.**

astronomical unit (communication satellite). Abbreviated AU; the mean distance between the centers of the sun and the earth, 149.6×10^6 kilometers, 92.98×10^6 miles or 80.78×10^6 nautical miles.   74

astronomical unit of distance. The length of the radius of the unperturbed circular orbit of a body of negligible mass moving around the sun with a sidereal angular velocity of 0.017 202 098 950 radian per day of 86 400 ephemeris seconds. In the system of astronomical constants of the International Astronomical Union the value adopted for it is 1 AU = 149 600 $\times$ 10^6 m. 21

A switchboard (telephone switching systems). A telecommunications switchboard in a local central office, used primarily to extend calls received from local stations. 55

asymmetrical cell. A cell in which the impedance to the flow of current in one direction is greater than in the other direction. *See also:* **electrolytic capacitor.** 328

asymmetrical current. *See:* **total asymmetrical current.**

asymmetric terminal voltage (electromagnetic compatibility). Terminal voltage measured with a delta network between the midpoint of the resistors across the mains lead and ground. *See also:* **electromagnetic compatibility.** 220, 199

asynchronous computer. A computer in which each event or the performance of each operation starts as a result of a signal generated by the completion of the previous event or operation, or by the availability of the parts of the computer required for the next event or operation. 77, 255

asynchronous impedance (rotating machinery). The quotient of the voltage, assumed to be sinusoidal and balanced, supplied to a rotating machine out of synchronism, and the same frequency component of the current. *Note:* The value of this impedance depends on the slip. *See also:* **asynchronous machine.** 63

asynchronous machine (1) (rotating machinery). An alternating-current machine in which the rotor does not turn at synchronous speed. 63

(2) (electric installations on shipboard). A machine in which the speed of operation is not proportional to the frequency of the system to which it is connected. 3

asynchronous operation (rotating machinery). Operation of a machine where the speed of the rotor is other than synchronous speed. *See also:* **asynchronous machine.** 63

asynchronous reactance (rotating machinery). The quotient of the reactive component of the average voltage at rated frequency, assumed to be sinusoidal and balanced, applied to the primary winding of a machine rotating out of synchronism, and the average current component at the same frequency. 63

asynchronous resistance (rotating machinery). The quotient of (a) the active component of the average voltage at rated frequency assumed to be sinusoidal and balanced, applied to the primary winding of a machine rotating out of synchronism, and (b) the average current component at the same frequency. 63

asynchronous transmission (data communication). A mode of data transmission such that the time occurrence of the bits within each character, or block of characters, relates to a fixed time frame, but the start of each character, or block of characters, is not related to this fixed time frame. *See also:* **nonsynchronous transmission.** 12, 59

ATC (radar). air traffic control. 13

ATE (automatic test equipment) control software (test, measurement and diagnostic equipment). Software used during execution of a test program which controls the nontesting operations of the ATE. This software is used to execute a test procedure but does not contain any of the stimuli or measurement parameters used in testing the Unit Under Test (UUT). Where test software and control software are combined in one inseparable program, that program will be treated as test software not control software. 54

ATE (automatic test equipment) oriented language (test, measurement and diagnostic equipment). A computer language used to program an automatic test equipment to test units under test (UUT's), whose characteristics imply the use of a specific ATE system or family of ATE systems. 54

ATE (automatic test equipment) support software (test, measurement and diagnostic equipment). Computer programs which aid in preparing, analyzing, and maintaining test software. Examples are: ATE compilers, translation/analysis programs, and punch/print programs. 54

atmospheric absorption (1) (general). The loss of energy in transmission of radio waves, due to dissipation in the atmosphere. *See also:* **radiation.** 328

(2) (communication satellite). Absorption, by the atmosphere, of electromagnetic energy traversing it. 85

atmospheric duct (radio wave propagation). A stratum of the troposphere within which the variation of refractive index is such as to confine within the limits of the stratum the propagation of an abnormally large proportion of any radiation of sufficiently high frequency. *Note:* The duct extends from the level of a local minimum of the modified index of refraction as a function of height down to the level where the minimum value is again encountered, or down to the surface bounding the atmosphere if the minimum value is not again encountered. *See also:* **radiation; radio wave propagation.** 180, 146

atmospheric noise (communication satellite). Noise radiated by the atmosphere into a space communications receiver antenna. 85

atmospheric paths (atmospheric correction factors to dielectric tests). Paths entirely through atmospheric air, such as along the porcelain surface of an outdoor bushing. 50

atmospheric radio noise (electromagnetic compatibility). Noise having its source in natural atmospheric phenomena. *See also:* **electromagnetic compatibility.** 199

atmospheric radio wave. A radio wave that is propagated by reflection in the atmosphere. *Note:* It may include either the ionospheric wave or the tropospheric wave, or both. *See also:* **radiation.** 328

atmospherics. *See:* **static (atmospherics).**

atmospheric transmissivity. The ratio of the directly transmitted flux incident on a surface after passing through unit thickness of the atmosphere to the flux which would be incident on the same surface if the flux had passed through a vacuum. 167

A-trace (loran). The first (upper) trace on the scope display. *See also:* **navigation.** 187

ATR box. *See:* **anti-transmit-receive box.**

ATR switch. *See:* **anti-transmit-receive switch.**

ATR tube. *See:* **anti-transmit-receive tube.**

attachment (electric and electronics parts and equipments). A basic part, subassembly, or assembly designed for use in conjunction with another assembly, unit, or set, contributing to the effectiveness thereof by extending or varying the basic function of the assembly, unit, or set. A typical example is: ultra-high-frequency (UHF) converter for very-high-frequency (VHF) receiver. 17

attachment plug (plug cap*) (cap*) (interior wiring). A device, that by insertion in a receptacle, establishes connection between the conductors of the attached flexible cord and the conductors connected permanently to the receptacle. 3, 328

* Deprecated

attachments. Accessories to be attached to switchgear apparatus, as distinguished from auxiliaries.
103, 202, 27

attack time (electroacoustics). The interval required, after a sudden increase in input signal amplitude to a system or transducer, to attain a stated percentage (usually 63 percent) of the ultimate change in amplification or attenuation due to this increase. 239

attendant (telephone switching systems). A private branch exchange or centrex operator. 55

attenuating pad. *See:* **pad.**

attenuation. (1) A general term used to denote a decrease in magnitude in transmission from one point to another. *Note:* It may be expressed as a ratio or, by extension of the term, in decibels. 199, 59
(2) **(quantity associated with a traveling wave in a homogeneous medium).** The decrease of its amplitude with increasing distance from the source, excluding the decrease due to spreading. *See also:* **radio wave propagation.**
180, 146
(3) **(signal-transmission system).** A decrease in signal magnitude. *Note:* Attenuation may be expressed as the scalar ratio of the input magnitude to the output magnitude or in decibels as 20 times the logarithm of that ratio.
188
(4) **(per unit length) (coaxial transmission line).** The decrease with distance in the direction of propagation of a quantity associated with a traveling transverse electromagnetic wave. *See also:* **wave front.** 265
(5) **(waveguide) (quantity associated with a traveling waveguide or transmission-line wave).** The decrease with distance in the direction of propagation. *Note:* Attenuation of power is usually measured in terms of decibels or decibels per unit length. 250, 180
(6) **(control systems).** (A) A decrease in signal magnitude between two points, or between two frequencies. (B) The reciprocal of gain. *Note:* It may be expressed as a scalar ratio or in decibels as 20 times the log of that ratio. A decrease with time is usually called damping or "subsidence." *See:* **subsidence ratio.** 56
(7) **(laser-maser).** The decrease in the radiant flux as it passes through an absorbing or scattering medium. 363

attenuation band (uniconductor waveguide). Rejection band. *See also:* **waveguide.** 250, 179

attenuation constant. The real part of the propagation constant. *Note:* Unit: Neper per unit length. (1 neper equals 8.686 decibels). *See also:* **radio transmission.**
210, 180, 185, 146

attenuation, current. Either (1) a decrease in signal current magnitude, in transmission from one point to another, or the process thereof, or (2) of a transducer, the scalar ratio of the signal input current to the signal output current.

Note: By incorrect extension of the term decibel, this ratio is sometimes expressed in decibels by multiplying its common logarithm by 20. It may be correctly expressed in decilogs. *See:* **decibel.** *See also:* **attenuation.** 239

attenuation distortion (frequency distortion*). Attenuation distortion is either (1) departure, in a circuit or system, from uniform amplification or attenuation over the frequency range required for transmission, or (2) the effect of such departure on a transmitted signal. 59

* Deprecated

attenuation equalizer. A corrective network that is designed to make the absolute value of the transfer impedance, with respect to two chosen pairs of terminals, substantially constant for all frequencies within a desired range. *See also:* **network analysis.** 59

attenuation loss. The reduction in radiant power surface density, excluding the reduction due to spreading. 210

attenuation network. A network providing relatively little phase shift and substantially constant attenuation over a range of frequency. 210

attenuation ratio. The magnitude of the propagation ratio. *See also:* **radiation; radio wave propagation.** 146, 180

attenuation vector (field quantity) (radio wave propagation). The vector pointing in the direction of maximum decrease of amplitude, whose magnitude is the attenuation constant. *See also:* **radio wave propagation.** 180, 146

attenuation vector in physical media (antennas). The real part of the propagation vector. 111

attenuation, voltage. Either (1) a decrease in signal voltage magnitude in transmission from one point to another, or the process thereof, or (2) a transducer, the scalar ratio of the signal input voltage to the signal output voltage. *Notes:* (1) By incorrect extension of the term decibel, this ratio is sometimes expressed in decibels by multiplying its common logarithm by 20. It may be correctly expressed in decilogs. (2) If the input and/or output power consist of more than one component, such as multifrequency signal or noise, then the particular components used and their weighting must be specified. *See:* **decibel.** *See also:* **attenuation; signal; transducer.** 239, 210

attenuator (1) (general). An adjustable transducer for reducing the amplitude of a wave without introducing appreciable distortion. 59
(2) An adjustable passive network that reduces the power level of a signal without introducing appreciable distortion. *See also:* **transducer.** 239
(3) **(analog computer).** A device for reducing the amplitude of a signal without introducing appreciable distortion.
9

attenuator tube (electron tubes). A gas-filled radio-frequency switching tube in which a gas discharge, initiated and regulated independently of radio-frequency power, is used to control this power by reflection or absorption. *See also:* **gas tubes.** 190, 125

attenuator, waveguide (waveguide components). A waveguide component that reduces the output power relative to the input, by any means, including absorption and reflection. *See:* **waveguide.** 166, 179

attitude (communication satellite). Orientation of a satellite vehicle with respect to a reference coordinate system. Deviations of the satellite axes from the reference system are called roll, pitch and yaw. The reference system is generally an orbital reference system with the x-axis (roll axis) in the orbital plane in direction of the satellite motion, the y-axis (pitch axis) normal to the orbital plane and the z-axis (yaw axis) in the orbital plane in direction of the center of the earth. 74

attitude-effect error (electronic navigation). A manifestation of polarization error; an error in indicated bearing that is dependent üpon the attitude of the vehicle with respect to the direction of signal propagation. *See:* **heading-effect error.** *See also:* **navigation.** 187

attitude gyro-electric indicator. An electrically driven device that provides a visual indication of an aircraft's roll and pitch attitude with respect to the earth. *Note:* It is used in highly maneuverable aircraft and differs from the gyro-horizon electric indicator in that the gyro is not limited by stops and has complete freedom about the roll and pitch axes. 328

attitude stabilized satellite (communication satellite). A satellite with at least one axis maintained in a specified direction, namely toward the center of the earth, the sun or a specified point in space. 74

attitude storage (gyro). The transient deviation of the output of a rate integrating gyro from that of an ideal integrator when the gyro is subjected to an input rate. It is a function of the gyro characteristic time. *See:* **float storage, torque command storage.** 46

audible busy signal (busy tone). An audible signal connected to the calling line to indicate that the called line is in use. 328

audible cab indicator. A device (usually an air whistle, bell, or buzzer) located in the cab of a vehicle equipped with cab signals or continuous train control designed to sound when the cab signal changes to a more restrictive indication. 328

audible signal device (protective signaling). A general term for bells, buzzers, horns, whistles, sirens, or other devices that produce audible signals. *See also:* **protective signaling.** 328

audio. Pertaining to frequencies corresponding to a normally audible sound wave. *Note:* These frequencies range roughly from 15 hertz to 20 000 hertz. *See also:* **signal wave.** 59

audio frequency (1) (general). Any frequency corresponding to a normally audible sound wave. *Notes:* (A) Audio frequencies range roughly from 15 hertz to 20 000 hertz. (B) This term is frequently shortened to audio and used as a modifier to indicate a device or system intended to operate at audio frequencies, for example, audio amplifier. 111, 239, 176, 197

(2) (interference terminology). Components of noise having frequencies in the audio range. *See also:* **signal.** 188

audio-frequency distortion. The form of wave distortion in which the relative magnitudes of the different frequency components of the wave are changed on either a phase or amplitude basis. 181

audio-frequency harmonic distortion. The generation in a system of integral multiples of a single audio-frequency input signal. *See also:* **modulation.** 111

audio-frequency noise. *See:* **noise, audio frequency.**

audio-frequency oscillator (audio oscillator). A nonrotating

device for producing an audio-frequency sinusoidal electric wave, whose frequency is determined by the characteristics of the device. *See:* **oscillatory circuit.** 239

audio-frequency peak limiter. A circuit used in an audio-frequency system to cut off peaks that exceed a predetermined value. 111

audio-frequency response (receiver performance). The measure of the relative departure of all audio-frequency signal levels within a specified bandwidth, from a specified reference frequency signal power level. 123

audio-frequency spectrum (audio spectrum). The continuous range of frequencies extending from the lowest to the highest audio frequency. 239

audio-frequency transformer. A transformer for use with audio-frequency currents. 197

audiogram (threshold audiogram). A graph showing hearing level as a function of frequency. 176

audio input power (transmitter performance). The input power level to the modulator, expressed in decibels referred to a 1 milliwatt power level. *See also:* **audio-frequency distortion.** 181

audio input signal (transmitter performance). That composite input to the transmitter modulator that consists of frequency components normally audible to the human ear. *See also:* **audio-frequency distortion.** 181

audiometer. An instrument for measuring hearing level. *See also:* **instrument.** 176

audio oscillator. *See:* **audio-frequency oscillator.**

audio output power (receiver performance) (receiver). The audio-frequency power dissipated in a load across the output terminals. *See also:* **receiver performance.** 181

audio power output (receiver performance). The measure of the audio-frequency energy dissipated in a specified output load. 123

audio-tone channel. *See:* **voice-frequency carrier-telegraph.**

auditable data (nuclear power generating stations). Technical information which is documented and organized in a readily understandable and traceable manner that permits independent auditing of the inferences or conclusions based on the information. 31, 120

auditory sensation area. (1) The region enclosed by the curves defining the threshold of feeling and the threshold of audibility each expressed as a function of frequency. (2) The part of the brain (temporal lobe of the cortex) that is responsive to auditory stimuli. 176

A unit. A motive power unit so designed that it may be used as the leading unit of a locomotive, with adequate visibility in a forward direction, and which includes a cab and equipment for full control and observation of the propulsion power and brake applications for the locomotive and train. *See also:* **electric locomotive.**

aural harmonic. A harmonic generated in the auditory mechanism. 176

aural radio range. *See:* **A-N radio range.**

aural transmitter. The radio equipment used for the transmission of the aural (sound) signals from a television broadcast station. *See also:* **television.** 211, 111

austenitic. The face-centered cubic crystal structure of ferrous metals. 221, 205

authorities (nuclear power generating stations). Any governmental agencies or recognized scientific bodies which by their charter define regulations or standards dealing with radiation protection. 31

authorized bandwidth (mobile communication). The frequency band containing those frequencies upon which a total of 99 percent of the radiated power appears. *See also:* **mobile communication system.** 181

auto alarm. A radio receiver that automatically produces an audible alarm when a prescribed radio signal is received. 328

autocondensation* (electrotherapy). A method of applying alternating currents of frequencies exceeding 100 kilohertz to limited areas near the surface of the human body through the use of one very large and one small electrode, the patient becoming part of the capacitor. *See also:* **electrotherapy.** 192

* Deprecated

autoconduction* (electrotherapy). A method of applying alternating currents, of frequencies exceeding 100 kilohertz for therapeutic purposes, by electromagnetic induction, the patient being placed inside a large solenoid. *See also:* **electro-therapy.** 192

* Deprecated

autodyne reception. A system of heterodyne reception through the use of a device that is both an oscillator and a detector. 328

autoerection (gyro). The process by which gimbal axis friction causes the spin axis of a free gyro to tend to align with the axis about which the case is rotated. The resulting drift rate is a function of the angular displacement between the spin axis and the rotation axis. 46

automatic (1) (industrial control). Self-acting, operating by its own mechanism when actuated by some impersonal influence; as, for example, a change in current strength, pressure, temperature, or mechanical configuration. 206

(2) (transmission and distribution). Self-acting, operating by its own mechanism when actuated by some impersonal influence as, for example, a change in current strength; not manual, without personal intervention. Remote control that requires personal intervention is not automatic, but manual. 262

automatic acceleration (1) (automatic train control). Acceleration under the control of devices that function automatically to maintain, within relatively close predetermined values or schedules, current passing to the traction motors, the tractive force developed by them, the rate of vehicle acceleration, or similar factors affecting acceleration. *See also:* **electric drive; multiple-unit control.** 328

(2) (industrial control). Acceleration under the control of devices that function automatically to raise the motor speed. *See also:* **electric drive; multiple unit control.** 206

automatically regulated (rotating machinery). Applied to a machine that can regulate its own characteristics when associated with other apparatus in a suitable closed-loop circuit. *See also:* **direct-current commutating machine; asynchronous machine.** 63

automatically reset relay. *See:* **self-reset relay.**

automatic approach control. A system that integrates signals, received by localizer and glide path receivers, into the automatic pilot system, and guides the airplane down the localizer and glide path beam intersection. *See also:* **air transportation electronic equipment.** 328

automatic block signal system. A series of consecutive blocks governed by block signals, cab signals, or both, operated by electric, pneumatic, or other agency actuated by a train or by certain conditions affecting the use of a block. *See also:* **block signal system.** 328

automatic cab signal system. A system that provides for the automatic operation of cab signals. *See also:* **automatic train control.** 328

automatic call distributor (ACD) (telephone switching systems). The facility for allotting incoming traffic to idle operators or attendants. 55

automatic capacitor control equipment. An equipment that provides automatic control for functions related to capacitors, such as their connection to and disconnection from a circuit in response to predetermined conditions such as voltage, load, or time. 103

automatic carriage. A control mechanism for a typewriter or other listing device that can automatically control the feeding, spacing, skipping, and ejecting of paper or preprinted forms. 77, 255

automatic chart-line follower (electronic navigation). A device that automatically derives error signals proportional to the deviation of the position of a vehicle from a predetermined course line drawn on a chart. *See also:* **navigation.** 13, 187

automatic check. *See:* **check, automatic.**

automatic circuit recloser. A self-controlled device for automatically interrupting and reclosing an alternating-current circuit, with a predetermined sequence of opening and reclosing followed by resetting or Lockout. *Note:* When applicable it includes an assembly of control elements required to detect overcurrents and control the recloser operation. 79

automatic combustion control. A method of combustion control that is effected automatically by mechanical or electric devices. 64

automatic computer. A computer that can perform a sequence of operations without intervention by a human operator. 77, 255

automatic component interconnection matrix (analog computer). A hardware system for connecting inputs and outputs of parallel computing components according to a predetermined program. This system, which may consist of a matrix of mechanical and/or electronic switches, replaces the manual program patch panel and patch cords on analog computers. *See:* **problem board.** 9

automatic control (1) (electrical controls). An arrangement that provides for switching or otherwise controlling, or both, in an automatic sequence and under predetermined conditions, the necessary devices comprising an equipment. *Note:* These devices thereupon maintain the required character of service and provide adequate protection against all usual operating emergencies. 103, 202

(2) (analog computer). A method of computer operation using auxiliary automatic equipment to perform computer Control State selections, switching operations, or component adjustments in accordance with previously selected criteria. Such auxiliary automatic equipment usually consists of programmable digital logic which is part of the analog, a separate digital computer, or both. The case of the digital computer controlling the analog computer is an example of a hybrid computer. 9

automatic control equipment (power switchgear). Equipment that provides automatic control for a specified type of power circuit or apparatus. 103

automatic controller. *See:* controller, automatic.

automatic direct-control telecommunications system (telephone switching systems). A system in which the connections are set directly in response to pulsing from the originating calling device. 55

automatic direction-finder (electronic navigation). A direction-finder that automatically and continuously provides a measure of the direction of arrival of the received signal. Data are usually displayed visually. *See also:* **navigation.** 13, 187

automatic dispatching system (electric power systems). A controlling means for maintaining the area control error or station control error at zero by automatically loading generating sources, and it also may include facilities to load the sources in accordance with a predetermined loading criterion. 94

automatic equipment (for a specified type of power circuit or apparatus). Equipment that provides automatic control. 103, 202

automatic fire-alarm system. A fire-alarm system for automatically detecting the presence of fire and initiating signal transmission without human intervention. *See also:* **protective signaling.** 328

automatic fire detector (fire protection devices). A device designed to detect the presence of fire and initiate action. 71

automatic frequency control. An arrangement whereby the frequency of an oscillator or the tuning of a circuit is automatically maintained within specified limits with respect to a reference frequency. *See also:* **radio transmitter; receiver performance.** 240

automatic-frequency-control synchronization. A process for locking the frequency (phase) of a local oscillator to that of an incoming synchronizing signal by the use of a comparison device whose output continuously corrects the local-oscillator frequency (phase). 328

automatic gain control (AGC). (1) A process or means by which gain is automatically adjusted in a specified manner as a function of input or other specified parameters. 239, 178
(2) A method of automatically obtaining a substantially constant output of some amplitude characteristic of the signal over a range of variation of that characteristic at the input. The term is also applied to a device for accomplishing this result. 59

automatic generation control. The regulation of the power output of electric generators within a prescribed area in response to change in system frequency, tie-line loading, or the relation of these to each other, so as to maintain the scheduled system frequency or the established interchange with other areas within predetermined limits or both. 200, 94

automatic grid bias. Grid-bias voltage provided by the difference of potential across resistance(s) in the grid or cathode circuit due to grid or cathode current or both. *See also:* **radio receiver.** 111

automatic hold (analog computer). Attainment of the hold condition automatically through amplitude comparison of a problem variable or through an overload condition.

See also: **electronic analog computer.** 9, 77

automatic holdup alarm system. An alarm system in which the signal transmission is initiated by the action of the robber. *See also:* **protective signaling.** 328

automatic-identified outward dialing (AIOD) (telephone switching system). A method of automatically obtaining the identity of a calling station from a private branch exchange over a separate data link for use in automatic message accounting. 55

automatic indirect-control telecommunications system (telephone switching system). A system in which the pulsing from the originating calling device are stored in a register temporarily associated with the call, for the subsequent establishing of connections. 55

automatic interlocking. An arrangement of signals, with or without other signal appliances, that functions through the exercise of inherent powers as distinguished from those whose functions are controlled manually, and that are so interconnected by means of electric circuits that their movements must succeed one another in proper sequence. *See also:* **interlocking (interlocking plant).** 328

automatic keying device. A device that, after manual initiation, controls automatically the sending of a radio signal that actuates the auto alarm. *Note:* The prescribed signal is a series of twelve dashes, each of four seconds duration, with one-second interval between dashes, transmitted on the radiotelegraph distress frequency in the medium-frequency band. This signal is used only to proceed distress calls or urgent warnings. 328

automatic line sectionalizer. A self-controlled circuit-opening device that opens the main electric circuit through it while disconnected from the source of power after the device has sensed and responded to a predetermined number of successive main-current impulses of predetermined or greater magnitude. *Note:* It may have provision to be manually operated to interrupt loads up to its rated continuous current. 103, 202

automatic load (armature current division) (industrial control). The effect of a control function or a device to automatically divide armature currents in a prescribed manner between two or more motors or two or more generators connected to the same load. *See:* **control system, feedback.** 225, 206

automatic load (level) control (ALC) (power-system communication). A method of automatically maintaining the peak power of a single-sideband suppressed-carrier transmitter at a constant level. *See also:* **radio transmitter.** 59

automatic machine control equipment (power switchgear). An equipment that provides automatic control for functions related to rotating machines or power rectifiers. 103

automatic message accounting (AMA) (telephone switching system). An arrangement for automatically collecting, recording, and processing information relating to calls for billing purposes. 55

automatic message accounting system (AMA). An arrangement of apparatus for recording and processing on continuous tapes the data required for computing telephone charges on certain classes of calls. *Note:* This system may include provision for compiling all charges and credits that affect the customer's bill and for automatic printing of the bill. 328

automatic number identification (ANI) (telephone switching system). The automatic obtaining of a calling station directory or equipment number for use in automatic message accounting. 55

automatic opening (switching device) (automatic tripping). The opening under predetermined conditions without the intervention of an attendant. 103, 202

automatic operation (1) (elevator). Operation wherein the starting of the elevator car is effected in response to the momentary actuation of operating devices at the landing, and/or of operating devices in the car identified with the landings, and/or in response to an automatic starting mechanism, and wherein the car is stopped automatically at the landings. *See:* **control.** 328
(2) (switching device). The ability to complete an assigned sequence of operations by automatic control without the assistance of an attendant. 103, 202

automatic phase control (television). A process or means by which the phase of an oscillator signal is automatically maintained within specified limits by comparing its phase to the phase of an external reference signal and thereby supplying correcting information to the controlled source or a device for accomplishing this result. *Note:* Automatic phase control is sometimes used for accurate frequency control and under these conditions is often called automatic frequency control. *See:* **television.** 328

automatic pilot (electronic navigation). Equipment that automatically controls the attitude of a vehicle about one or more of its rotational axes (pitch, roll, and yaw), and may be used to respond to manual or electronic commands. *See also:* **navigation.** 187

automatic-pilot servo motor. A device that converts electric signals to mechanical rotation so as to move the control surfaces of an aircraft. 328

automatic programmed tools (APT) (numerically controlled machines). A computer-based numerical control programming system that uses English-like symbolic descriptions of part and tool geometry and tool motion. 224, 203

automatic programming (analog computer). A method of computer operation using auxiliary automatic equipment to perform computer control state selections, switching operations, or component adjustments in accordance with previously selected criteria. *See also:* **electronic analog computer.** 9, 77

automatic reclosing equipment. An automatic equipment that provides for reclosing a switching device as desired after it has opened automatically under abnormal conditions. *Note:* Automatic reclosing equipment may be actuated by conditions sensed on either or both sides of the switching device as designed. 103, 202

automatic-reset manual release of brakes (control) (industrial control). A manual release that, when operated, will maintain the braking surfaces in disengagement but will automatically restore the braking surfaces to their normal relation as soon as electric power is again applied. *See:* **control system, feedback.** 225, 206

automatic-reset relay. *See:* **relay, automatic-reset.**

automatic-reset thermal protector (rotating machinery). A thermal protector designed to perform its function by opening the circuit to or within the protected machine and then automatically closing the circuit after the machine cools to a satisfactory operating temperature. *See also:* **starting switch assembly.** 63

automatic reversing (industrial control). Reversing of an electric drive, initiated by automatic means. *See also:* **electric drive.** 206

automatic signal. A signal controlled automatically by the occupancy or certain other conditions of the track area that it protects. 328

automatic smoke alarm system. An alarm system designed to detect the presence of smoke and to transmit an alarm automatically. *See also:* **protective signaling.** 328

automatic speed adjustment (industrial control). Speed adjustment accomplished automatically. *See:* **automatic.** *See also:* **electric drive.** 206

automatic starter (industrial control). A starter in which the influence directing its performance is automatic. *See also:* **starter.** 206, 3

automatic station (generating stations and substations). A station (usually unattended) that under predetermined conditions goes into operation by an automatic sequence; that thereupon by automatic means maintains the required character of service within its capability; that goes out of operation by automatic sequence under other predetermined conditions, and includes protection against the usual operating emergencies. *Note:* An automatic station may go in and out of operation in response to predetermined voltage, load, time, or other conditions, or in response to supervisory control or to a remote or local manually operated control device. 103, 202

automatic switchboard. A switchboard in which the connections are made by apparatus controlled from remote calling devices. 193, 101

automatic switching system (1) (machine switching system). A system in which connections between customers station equipment are ordinarily established as a result of signals produced by a calling device. 193
(2) (telephone switching systems). The switching entity for an automatic telecommunication system. 55

automatic system. A system in which the operations are performed by electrically controlled devices without the intervention of operators. 194

automatic telecommunications exchange (telephone switching system). A telecommunications exchange in which connections between stations are automatically established as a result of signals produced by calling devices. 55

automatic telecommunications system (telephone switching system). A system in which connections between stations are automatically established as a results of signals produced by calling devices. 55

automatic telegraphy. That form of telegraphy in which transmission or reception of signals, or both, are accomplished automatically. *See also:* **telegraphy.** 194

automatic test equipment (ATE) (test, measurement and diagnostic equipment). Equipment that is designed to conduct analysis of functional or static parameters to evaluate the degree of performance degradation and may be designed to perform fault isolation of unit malfunctions. The decision making, control, or evaluative functions are conducted with minimum reliance on human intervention. 54

automatic throw-over equipment. *See:* **automatic transfer equipment.**

automatic throw-over equipment of the fixed preferential type. *See:* **automatic transfer equipment of the fixed preferential type.**

automatic throw-over equipment of the nonpreferential type. *See:* **automatic transfer equipment of the nonpreferential type.**

automatic throw-over equipment of the selective-preferential type. *See:* **automatic transfer equipment of the selective-preferential type.**

automatic ticketing (telephone switching system). An arrangement for automatically recording information relating to calls, for billing purposes. 55

automatic TMDE (test, measurement and diagnostic equipment). *See:* **automatic test equipment.**

automatic track-follower (electronic navigation). *See:* **automatic chart-line follower.**

automatic tracking (1) (radar). Tracking in which a system employs some mechanism, for example, servo, or computer, to automatically follow some characteristic of the signal. 13

(2) (electronic navigation). Tracking in which a servomechanism automatically follows some characteristic of the signal. *See also:* **radar.** 13, 187

automatic train control (automatic speed control) (train control). A system or an installation so arranged that its operation on failure to forestall or acknowledge will automatically result in either one or the other or both of the following conditions: (1) Automatic train stop: The application of the brakes until the train has been brought to a stop; (2) Automatic speed control: The application of the brakes when the speed of the train exceeds a prescribed rate and continued until the speed has been reduced to a predetermined and prescribed rate. 328

automatic train control application. An application of the brake by the automatic train control device. 328

automatic transfer equipment (automatic throw-over equipment). An equipment that automatically transfers a load to another source of power when the original source to which it has been connected fails, and that will automatically retransfer the load to the original source under desired conditions. *Notes:* (1) It may be of the nonpreferential, fixed-preferential, or selective-preferential type. (2) Compare with transfer switch where transfer is accomplished without current interruption. 103, 202

automatic transfer equipment of the fixed-preferential type (automatic throw-over equipment of the fixed preferential type). Equipment in which the original source always serves as the preferred source and the other source as the emergency source. The automatic transfer equipment will retransfer the load to the preferred source upon its reenergization. *Note:* The restoration of the load to the preferred source from the emergency source upon the reenergization of the preferred source after an outage may be of the continuous-circuit restoration type or the interrupted-circuit restoration type. 103, 202

automatic transfer equipment of the nonpreferential type (automatic throw-over equipment of the nonpreferential type). Equipment that automatically retransfers the load to the original source only when the other source, to which it has been connected, fails. 103, 202

automatic transfer equipment of the selective-preferential type (automatic throw-over equipment of the selective-preferential type). Equipment in which either source may serve as the preferred or the emergency source of preselection as desired, and which will retransfer the load to the preferred source upon its reenergization. *Note:* The restoration of the load to the preferred source from the emergency source upon the reenergization of the preferred source after an outage may be of the continuous-circuit restoration type or the interrupted-circuit restoration type. 103, 202

automatic transfer switch (1) general. An automatic transfer switch is self-acting equipment for transferring one or more load conductor connections from one power source to another. *Note:* (A) It may be of the nonpreferential, fixed-preferential, or selective-preferential type. (B) Compare with transfer switch where transfer is accomplished without current interruption. 70

(2) fixed-preferential type. Equipment in which the original source always serves as the preferred source and the other source as the emergency source. The automatic transfer switch will retransfer the load to the preferred source upon its reenergization. *Note:* The restoration of the load to the preferred source from the emergency source upon the reenergization of the preferred source after an outage may be of the continuous- circuit restoration type or the interrupted-circuit restoration type. 70

(3) nonpreferential type. Equipment that automatically retransfers the load to the original source only when the other source, to which it has been connected, fails. 70

(4) selective-preferential type. Equipment in which either source may serve as the preferred or the emergency source of preselection as desired, and that will retransfer the load to the preferred source upon its reenergization. *Note:* The restoration of the load to the preferred source from the emergency source upon the reenergization of the preferred source after an outage may be of the continuous-circuit restoration type or the interrupted-circuit restoration type. 70

(5) (emergency and standby power systems). Self-acting equipment for transferring one or more load connections from one power source to another. 73

automatic transformer control equipment (power switchgear). An equipment that provides automatic control for functions relating to transformers, such as their connection, disconnection or regulation in response to predetermined such as system load, voltage or phase angle. 103

automatic triggering (oscilloscope). A mode of triggering in which one or more of the triggering-circuit controls are preset to conditions suitable for automatically displaying repetitive waveforms. *Note:* The automatic mode may also provide a recurrent trigger of recurrent sweep in the absence of triggering signals. *See:* **oscillograph.** 185

automatic tripping. *See:* **automatic opening.**

automatic volume control (AVC). A method of automatically maintaining a substantially constant audio output volume over a range of variation of input volume. *Note:* The term is also applied to a device for accomplishing this result in a system of transducer. 239, 59

automation. (1) The implementation of processes by au-

tomatic means. (2) The theory, art, or technique of making a process more automatic. (3) The investigation, design, development, and application of methods of rendering processes automatic, self-moving, or self-controlling.
77, 255

autonavigator. Navigation equipment that includes means for coupling the output navigational data derived from the navigation sensors to the control system of the vehicle. *See also:* **navigation.** 187

autopatch. *See:* **automatic component interconnection matrix.**

autopilot. *See:* **automatic pilot.**

autopilot coupler (electronic navigation). The means used to link the navigation system receiver output to the automatic pilot. *See also:* **navigation.** 13, 187

autoradar plot (electronic navigation). A particular chart comparison unit using a radar presentation of position. *See also:* **navigation.** 187

autoregulation induction heater. An induction heater in which a desired control is effected by the change in characteristics of a magnetic charge as it is heated at or near its Curie point. *See also:* **dielectric heating; induction heating; coupling.** 14, 114

autotrack (communication satellite). The capability of a space communications receiver antenna to automatically track an orbiting satellite vehicle, for example, by using a monopulse system. 84

autotransformer. A transformer in which at least two windings have a common section. 53

autotransformer, individual-lamp. A series autotransformer that transforms the primary current to a higher or lower current as required for the operation of an individual street light. *See:* **transformer, speciality.** 203

autotransformer starter (industrial control). A starter that includes an autotransformer to furnish reduced voltage for starting of an alternating-current motor. It includes the necessary switching mechanism and it is frequently called a compensator or autostarter. *See also:* **starter.**
206

autotransformer starting (rotating machinery). The process of starting a motor at reduced voltage by connecting the primary winding to the supply initially through an auto-transformer and reconnecting the winding directly to the supply at rated voltage for the running conditions. *See:* **asynchronous machine.** 63

auxiliaries (1) (switchgear). Accessories to be used with switchgear apparatus but not attached to it, as distinguished from attachments. 103, 202, 27
(2) (power generation) (collective). For more than one auxiliary; that is, auxiliaries bus, auxiliaries power transformer, etc. 58

auxiliary. Any item not directly a part of a specific component or system but required for its functional operation.
58

auxiliary anode (industrial control). An anode located adjacent to the pool cathode in an ignitron to facilitate the maintenance of a cathode spot under conditions adverse to its maintenance by the main anode circuit. *See also:* **electronic controller.** 206

auxiliary burden (capacitance potential device). A variable burden furnished, when required, for adjustment purposes. *See:* **outdoor coupling capacitor.** 341

auxiliary device. Any electrical device other than motors and motor starters necessary to fully operate the machine or equipment. 124

auxiliary device to an instrument. A separate piece of equipment used with an instrument to extend its range, increase its accuracy, or otherwise assist in making a measurement or to perform a function additional to the primary function of measurement. 328

auxiliary equipment (1) (nuclear power generating stations) (class 1E motor). Equipment that is not part of the motor but is necessary for the operation of the motor and will be installed within the containment. 31
(2) (test, measurement and diagnostic equipment). *See:* **ancillary equipment.**

auxiliary function (numerically controlled machine). A function of a machine other than the control of the coordinates of a workpiece or tool. Includes functions such as miscellaneous, feed, speed, tool selection, etcetera. *Note:* Not a preparatory function. 224, 207

auxiliary generator. A generator, commonly used on electric motive power units, for serving the auxiliary electric power requirements of the unit. *See also:* **traction motor.** 328

auxiliary generator set. A device usually consisting of a commonly mounted electric generator and a gasoline engine or gas turbine prime mover designed to convert liquid fuel into electric power. *Note:* It provides the aircraft with an electric power supply independent of the aircraft propulsion engines. 328

auxiliary lead (rotating machinery). A conductor joining an auxiliary terminal to the auxiliary device. 63

auxiliary means. A system element or group of elements that changes the magnitude but not the nature of the quantity being measured to make it more suitable for the primary detector. In a sequence of measurement operations it is usually placed ahead of the primary detector. *See also:* **measurement system.** 328

auxiliary operation. An operation performed by equipment not under continuous control of the central processing unit.
77, 255

auxiliary or shunt capacitance (capacitance potential device). The capacitance between the network connection and ground, if present. *See also:* **outdoor coupling capacitor.**
341

auxiliary power supply (industrial control). A power source supplying power other than load power as required for the proper functioning of a device. *See also:* **electronic controller.** 206

auxiliary power transformer. A transformer having a fixed phase position used for supplying excitation for the rectifier station and essential power for the operation of rectifier equipment auxiliaries. *See also:* **transformer.** 258

auxiliary relay. A relay whose function is to assist another relay or control device in performing a general function by supplying supplementary actions. *Notes:* (1) Some of the specific functions of an auxiliary relay are:

(A) Reinforcing contact current-carrying capacity of another relay or device.

(B) Providing circuit seal-in functions.

(C) Increasing available number of independent contacts.

(D) Providing circuit-opening instead of circuit-closing

contacts or vice-versa.

(E) Providing time delay in the completion of a function.

(F) Providing simple functions for interlocking or programming.

(G) Controls signals, lights, or other secondary circuit functions.

(2) The operating coil or the contacts of an auxiliary relay may be used in the control circuit of another relay or other control device. *Example:* An auxiliary relay may be applied to the auxiliary contact circuits of a circuit breaker in order to coordinate closing and tripping control sequences. (3) A relay that is auxiliary in its functions even though it may derive its driving energy from the power system current or voltage is a form of auxiliary relay. *Example:* A timing relay operating from current or potential transformers. (4) Relays that, by direct response to power system input quantities, assist other relays to respond to such quantities with greater discrimination are *not* auxiliary relays. *Example:* Fault-detector relay. (5) Relays that are limited in function by a control circuit, but are actuated primarily by system input quantities, are *not* auxiliary relays. *Example:* Torque-controlled relays. *See also:* **relay.**   103, 202, 60, 127, 259

auxiliary relay driver (relaying). A circuit which supplies an input to an auxiliary relay.   79

auxiliary rope-fastening device (elevator). A device attached to the car or counterweight or to the overhead dead-end rope-hitch support that will function automatically to support the car or counterweight in case the regular wire-rope fastening fails at the point of connection to the car or counterweight or at the overhead dead-end hitch. *See also:* **elevator.**   328

auxiliary secondary terminals. The auxiliary secondary terminals provide the connections to the auxiliary secondary winding, when furnished. *See:* **auxiliary secondary winding.**   341

auxiliary secondary winding (capacitance potential device). The auxiliary secondary winding is an additional winding that may be provided in the capacitance potential device when practical considerations permit. *Note:* It is a separate winding that provides a potential that is substantially in phase with the potential of the main winding. The primary purpose of this winding is to provide zero sequence voltage by means of a broken delta connection of three single-phase devices. *See:* **auxiliary secondary terminals.** *See also:* **outdoor coupling capacitor.**   341

auxiliary storage (computing machine). A storage that supplements another storage.   77, 255

auxilary supporting features (nuclear power generating station). (1) Systems or components which provide services (such as cooling, lubrication and energy supply) which are required for the safety system to accomplish its protective functions.   356
(2) Installed systems or components which provide services such as cooling, illumination and energy supply and which are required by the Post Accident Monitoring Instrumentation to perform its functions.   361

auxiliary switch. A switch mechanically operated by the main device for signaling, interlocking, or other purposes. *Note:* Auxiliary switch contacts are classed as *a, b, aa, bb, LC,* etcetera, for the purpose of specifying definite contact

positions with respect to the main device.   103, 202

auxiliary terminal (rotating machinery). A termination for parts other than the armature or field windings.   63

auxiliary winding (single-phase induction motor). A winding that produces poles of a magnetic flux field that are displaced from those of the main winding, that serves as a means for developing torque during starting operation, and that, in some types of design, also serves as a means for improvement of performance during running operation. An auxiliary winding may have a resistor or capacitor in series with it and may be connected to the supply line or across a portion of the main winding. *See also:* **asynchronous machine.**   63

available accuracy (noise temperature of noise generators). An accuracy that is readily available to the public at large, such as may be announced in calibration service bulletins or instrument catalogs. This term shall not include accuracies that may be obtainable at any echelon by employing special efforts and expenditures over and above those invested in producing the advertised or announced accuracies, nor shall it include accuracies of calibration or measurement services that are not readily available to any and all customers and clients.   155

availability (1) (general). The fraction of time that the system is actually capable of performing its mission. *See also:* **system.**   209, 89
(2) (nuclear power generating station). (A) The characteristic of an item expressed by the probability that it will be operational at a randomly selected future instant in time.   29, 31, 357
(B) Relates to the accessibility of information to the operator on a "continuous", "sequence" or "as called for" basis.   358
(3) (telephone switching system). The number of outlets of a group that can be reached from a given inlet in a switching stage or network.   55
(4) (reliability). The ability of an item—under combined aspects of its reliability, maintainability and maintenance support—to perform its required function at a stated instant of time or over a stated period of time. *Note:* The term availability is also used as an availability characteristic denoting either the probability of performing at a stated instant of time or the probability related to an interval of time.   164

availability factor. The ratio of the time a generating unit or piece of equipment is ready for or in service to the total time interval under consideration. *See also:* **generating station.**   64

available. *See also:* **generic term being modified.**

available conversion power gain (conversion transducer). The ratio of the available output-frequency power from the output terminals of the transducer to the available input-frequency power from the driving generator with terminating conditions specified for all frequencies that may affect the result. *Notes:* (1) This applies to outputs of such magnitude that the conversion transducer is operating in a substantially linear condition. (2) The maximum available conversion power gain of a conversion transducer is obtained when the input termination admittance, at input frequency, is the conjugate of the input-frequency driving-point admittance of the conversion transducer. *See also:* **transducer.**   252, 125

available current (prospective current) (circuit) (1) (industrial control). The current that would flow if each pole of the breaking device under consideration were replaced by a link of negligible impedance without any change of the circuit or the supply. *See:* **contactor; prospective current.**
244, 206

(2) (of a circuit with respect to a switching device situated therein). The current that would flow in that circuit if each pole of the switching device were to be replaced by a link of negligible impedance without any other change in the circuit or the supply. 103

available line. The portion of the scanning line which can be used specifically for picture signals. 12

available power (at a port) (hydraulic turbines). The maximum power which can be transferred from the port to a load. *Note:* At a specified frequency, maximum power transfer will take place when the impedance of the load is the conjugate of that of the source. The source impedance must have a positive real part. 125

available power efficiency (electroacoustics). Of an electroacoustic transducer used for sound reception, the ratio of the electric power available at the electric terminals of the transducer to the acoustic power available to the transducer. *Notes:* (1) For an electroacoustic transducer that obeys the reciprocity principle, the available power efficiency in sound reception is equal to the transmitting efficiency. (2) In a given narrow frequency band the available power efficiency is numerically equal to the fraction of the open-circuit mean-square thermal noise voltage present at the electric terminals that contributed by thermal noise in the acoustic medium. *See also:* **microphone.** 176

available power gain (1) (linear transducer). The ratio of the available power from the output terminals of the transducer, under specified input termination conditions, to the available power from the driving generator. *Note:* The maximum available power gain of an electric transducer is obtained when the input termination admittance is the conjugate of the driving-point admittance at the input terminals of the transducer. It is sometimes called completely matched power gain. *See also:* **transducer.**
252, 210

(2) (two-port linear transducer). At a specified frequency, the ratio of (1) the available signal power from the output port of the transducer, to (2) the available signal power from the input source. *Note:* The available signal power at the output port is a function of the match between the source impedance and the impedance of the input port. *See also:* **network analysis.** 125

(3) (circuits and systems). The maximum power gain that can be obtained from a signal source. For a source of internal impedance $Z_s = R_s + jX_s$ the maximum power gain is obtained when the source is connected to a conjugate matched load; i.e. if $Z_2 = R_s - jX_s$. It is sometimes called completely matched power gain or available gain. *See also:* **network analysis; transducer.** 67

available power response (electroacoustics) (electroacoustic transducer used for sound emission). The ratio of the mean-square sound pressure apparent at a distance of 1 meter in a specified direction from the effective acoustic center of the transducer to the available electric power from the source. *Notes:* (1) The sound pressure apparent at a distance of 1 meter can be found by multiplying the sound pressure observed at a remote point where the sound field is spherically divergent by the number of meters from the effective acoustic center to that point. (2) The available power response is a function not only of the transducer but also of some source impedances, either actual or nominal, the value of which must be specified. *See also:* **loudspeaker.** 176

available short-circuit current (at a given point in a circuit) (prospective short-circuit current). The maximum current that the power system can deliver through a given circuit point to any negligible-impedance short-circuit applied at the given point, or at any other point that will cause the highest current to flow through the given point. *Notes:* (1) This value can be in terms of either symmetrical or asymmetrical; peak or root-mean-square current, as specified. (2) In some resonant circuits the maximum available short-circuit current may occur when the short circuit is placed at some other point than the given one where the available current is measured. 103, 202

available short-circuit test current (at the point of test) (prospective short-circuit test current). The maximum short-circuit current for any given setting of a testing circuit that the test power source can deliver at the point of test, with the test circuit short-circuited by a link of negligible impedance at the line terminals of the device to be tested. *Note:* This value can be in terms of either symmetrical or asymmetrical, peak or root-mean-square current, as specified. 103, 202

available signal-to-noise ratio (at a point in a circuit). The ratio of the available signal power at that point to the available random noise power. *See also:* **signal-to-noise ratio.** 328

available time (electric drive) (industrial control). The period during which a system has the power turned on, is not under maintenance, and is known or believed to be operating correctly or capable of operating correctly. *See:* **electric drive.** 219, 206

avalanche (gas-filled radiation counter tube). The cumulative process in which charged particles accelerated by an electric field produce additional charged particles through collision with neutral gas molecules or atoms. It is therefore a cascade multiplication of ions. *See:* **amplifier.**
190, 96, 125

avalanche breakdown (semiconductor device). A breakdown that is caused by the cumulative multiplication of charge carriers through field-induced impact ionization. *See also:* **semiconductor; semiconductor device.**
245, 210, 66, 118, 119

avalanche impedance* (semiconductor). *See:* **breakdown impedance.** *See also:* **semiconductor.**

* Deprecated

avalanche luminescent diode (light emitting diodes). A semiconductor device containing a semiconductor junction in which visible light is non-thermally produced when a controlled reverse current in the breakdown region flows as a result of an applied voltage. 162

average absolute burst magnitude (audio and electroacoustics). The average of the instantaneous burst magnitude taken over the burst duration. *See:* figure under **burst duration.** *See also:* **burst (audio and electroacoustics).**
253, 176

average absolute pulse amplitude. The average of the absolute value of the instantaneous amplitude taken over the pulse duration. 254

average absolute value (periodic quantity). The average of the instantaneous absolute value $|x|$ of the quantity over one period:

$$\frac{1}{T} \int_t^{t+T} |x|\, \mathrm{d}t$$

where the lower limit of the integral may be any value of t, and the upper limit T greater. *Note:* Average value of a periodic quantity and average absolute value of a periodic quantity are not equivalent, and care must be taken to specify the definition being used. 210

average active power (single-phase, two-wire or poly-phase circuit) (average power). The time average of the values of active power when the active power varies slowly over a specified period of time (long in comparison with the period of a single cycle). This situation is normally encountered because electric system voltages or currents or both are regularly quasi-periodic. The average active power is readily obtained by dividing the energy flow during the specified period of time, by the time. *See:* **power, active; power, instantaneous.** 210

average current (periodic current). The value of the current averaged over a full cycle unless otherwise specified. *See also:* **rectification.** 237, 66

average detector. A detector, the output voltage of which approximates to the average value of the envelope of an applied signal. *See also:* **electromagnetic compatibility.** 220, 199

average electrode current (electron tube). The value obtained by integrating the instantaneous electrode current over an averaging time and dividing by the averaging time. *See also:* **electrode current (electron tube).** 190, 125

average forward-current rating (rectifier circuit element). The maximum average value of forward current averaged over a full cycle, permitted by the manufacturer under stated conditions. 237, 66, 208

average information content (per symbol) (information rate from a source, per symbol). The average of the information content per symbol emitted from a source. *Note:* The terms entropy and negentropy are sometimes used to designate average information content. *See also:* **information theory.** 160

average inside air temperature (enclosed switchgear). The average temperature of the surrounding cooling air that comes in contact with the heated parts of the apparatus within the enclosure. 103, 202

average magner. *See:* **average reactive power (single-phase, two-wire, or polyphase circuit).**

average noise factor. *See:* **average noise figure.**

average noise figure (transducer) (average noise factor). The ratio of total output noise power to the portion thereof attributable to thermal noise in the input termination, the total noise being summed over frequencies from zero to infinity, and the noise temperature of the input termination being standard (290 kelvins). *See:* **noise figure.** *See also:* **signal-to-noise ratio.** 328

average noise temperature (antenna). The noise temperature of an antenna averaged over a specified frequency band. 111

average picture level (APL) (television). The average signal level, with respect to the blanking level, during the active picture scanning time (averaged over a frame period, excluding blanking intervals) expressed as a percentage of the difference between the blanking and reference white levels. *See:* **television.** 178

average power. *See:* **average active power (single-phase, two-wire or polyphase circuit).**

average power factor (single-phase, two-wire, or a polyphase circuit, when the voltages and currents are sinusoidal). The ratio of the average active power to the amplitude of the average phasor power, for a specified period of time. 210

average power output (amplitude-modulated transmitter). The radio-frequency power delivered to the transmitter output terminals averaged over a modulation cycle. *See also:* **radio transmitter.** 111

average pulse amplitude. The average of the instantaneous amplitude taken over the pulse duration. 254, 210

average reactive factor (single-phase, two-wire, or a polyphase circuit, when the voltages and currents are sinusoidal). The ratio of the average reactive power to the amplitude of the average phasor power, for a specified period of time. 210

average reactive power (single-phase, two-wire, or polyphase circuit) (average magner). The time average of the values of reactive power when the reactive power varies slowly over a specified period of time (long in comparison with the period of a single cycle). This situation is normally encountered because electric power system voltages or currents, or both, are regularly quasi-periodic. *Note:* The average reactive power is readily obtained by dividing the quadergy flow during the specified period, by the time. *See:* **power, instantaneous (two-wire circuit).** 210

average transinformation (output symbols and input symbols). Transinformation averaged over the ensemble of pairs of transmitted and received symbols. *See also:* **information theory.** 160

average value of a function. The average value of a function, over an interval $a \leq x \leq b$, is

$$\frac{1}{b-a} \int_a^b f(x)\,\mathrm{d}x \qquad 210$$

average value of a periodic quantity. The average of the instantaneous value x of the quantity over one period:

$$\frac{1}{T} \int_t^{t+T} x\,\mathrm{d}t$$

where the lower limit of the integral may be any value of t, and the upper limit T greater. *See also:* **average absolute value (periodic quantity).** 210

average voltage (1) (periodic voltage). The value of the voltage averaged over a full cycle unless otherwise specified. *See also:* **battery; rectification.** 237, 66
(2) (storage battery) (storage cell). The average value of the voltage during the period of charge or discharge. It is conveniently obtained from the time integral of the voltage curve. *See also:* **battery.** 328

averaging time, electrode current. *See:* **electrode current averaging time.**

A-weighted sound level (speech quality measurements). A weighted sound pressure level obtained by the use of a metering characteristic and the weighting *A* specified in USAS S1.4-1961 (General Purpose Sound Level Meters). 126

axially extended interaction tube (microwave tubes). A klystron tube utilizing an output circuit having more than one gap. *See also:* **microwave tube or valve.** 190

axial magnetic centering force (rotating machinery). The axial force acting between rotor and stator resulting from the axial displacement of the rotor from magnetic center. *Note:* Unless other conditions are specified, the value of axial magnetic centering force will be for no load and rated voltage, and for rated no load field current and rated frequency as applicable. 63

axial mode (laser-maser). The mode in a beamguide or beam resonator which has one or more maxima for the transverse field intensities over the cross-section of the beam. 363

axial ratio (1) (antennas and waveguides). The ratio of the major axis to the minor axis of the polarization ellipse. *Note:* This is preferred to ellipticity because mathematically ellipticity is 1 minus the reciprocal of the axial ratio. *See also:* **radio transmission.** 267, 179

(2) (antennas). The ratio of the axes of the polarization ellipse. *See:* Note (3) under **polarization of a field vector.** 111

axis *See:* **magnetic axis; direct axis; quadrature axis.**

axis-of-freedom (gyro). The axis about which a gimbal provides a degree-of-freedom. 46

axle bearing. A bearing that supports a portion of the weight of a motor or a generator on the axle of a vehicle. *See:* **bearing; traction motor.** 328

axle-bearing cap. The member bolted to the motor frame supporting the bottom half of the axle bearing. *See:* **bearing.** 328

axle-bearing-cap cover. A hinged or otherwise applied cover for the waste and oil chamber of the axle bearing. *See:* **bearing.** 328

axle circuit. The circuit through which current flows along one of the track rails to the train, through the wheels and axles of the train, and returns to the source along the other track rail. 328

axle current. The electric current in an axle circuit. 328

axle generator. An electric generator designed to be driven mechanically from an axle of a vehicle. *See also:* **axle-generator system.** 328

axle-generator pole changer. A mechanically or electrically actuated changeover switch for maintaining constant polarity at the terminals of an axle generator when the direction of the rotation of the armature is reversed due to a change in direction of movement of a vehicle on which the generator is mounted. *See also:* **axle-generator system.** 328

axle-generator regulator. A control device for automatically controlling the voltage and current of a variable-speed axle generator. *See also:* **axle-generator system.** 328

axle-generator system. A system in which electric power for the requirements of a vehicle is supplied from an axle generator carried on the vehicle, supplemented by a storage battery. 328

axle-hung motor (or generator). A traction motor (or generator), a portion of the weight of which is carried directly on the axle of a vehicle by means of axle bearings. *See also:* **traction motor.** 328

azimuth (radar). (1) The direction of a celestial point from a terrestrial point, expressed as the angle in the horizontal plane between a reference line and the horizontal projection of the line joining the two points. *Note:* True north is usually but not always implied when no reference direction is stated. (2) In directional antenna systems, the pointing direction of an antenna, expressed as the angle in the horizontal plane between a reference direction and the horizontal component of the pointing direction of the boresight of the antenna. (3) Bearing. *See also:* **navigation; sunlight.** 13, 187

azimuth discrimination* (electronic navigation). *See:* **angular resolution.**
* Deprecated

azimuth marker (radar). *See:* **calibration mark.**

azimuth-stabilized plan-position indicator (electronic navigation). A plan-position indicator on which the reference bearing remains fixed with respect to the indicator, regardless of the vehicle orientation. *See also:* **radar.** 13, 187

B

babble. The aggregate crosstalk from a large number of interfering channels. *See also:* **signal-to-noise ratio.** 239

back (motor or generator) (turbine or drive end). The end that carries the largest coupling or driving pulley. *See:* **armature.** 63

back connected switch. A switch in which the current-carrying conductors are connected to the studs in back of the mounting base. 27

back contact (1) (electric power apparatus relaying). A contact which is closed when the relay is reset. *Syn.* **"b" contact.** 127

(2) (utility-consumer interconnections relaying). A contact that is closed when the relay is deenergized. 128

back course (instrument landing systems). The course that is located on the opposite side of the localizer from the runway. *See also:* **navigation.** 187

backed stamper (phonograph techniques) (mechanical recording). A thin metal stamper that is attached to a backing material, generally a metal disk of desired thickness. *See also:* **phonograph pickup.** 176, 256

backfire antenna. An antenna consisting of a radiating feed, a reflector element, and a reflecting surface such that the antenna functions as an open resonator, with radiation from the open end of the resonator. 111

back flashover (lightning). A flashover of insulation resulting from a lightning stroke to part of a network or electric installation that is normally at ground potential. *See also:* **direct-stroke protection (lightning).** 64

back focal length (laser-maser). The distance from the last optical surface of a lens to the focal point. 363

background (test, measurement and diagnostic equipment). Those effects present in physical apparatus or surrounding environment which limit the measurement or observation of low level signals or phenomenon; commonly referred to as noise (background acoustical noise, background electromagnetic radiation, background ionizing radiation). 54

background counts (radiation counters). Counts caused by ionizing radiation coming from sources other than that to be measured; and by any electric disturbance in the circuitry that is used to record the counts. *See also:* **ionizing radiation.** 190, 117, 96, 125

background ionization voltage (surge arrester). A high-frequency voltage appearing at the terminals of the apparatus to be tested that is generated by ionization extraneous to the apparatus. *Note:* While this voltage does not add arithmetically to the radio influence or internal ionization voltage, it affects the sensitivity of the test. *See:* **surge arrester (surge diverter).** 229, 62

background level (sound measurement). Any sound at the points of measurement other than that of the machine being tested. It also includes the sound of any test support equipment. 129

background noise (1) (general). Noise due to audible disturbances of periodic and/or random occurrence. *See also:* **modulation.** 111

(2) (electroacoustics) (recording and reproducing). The total system noise in the absence of a signal. *See also:* **phonograph pickup.** 176

(3) (receivers). The noise in the absence of signal modulation on the carrier. 339

(4) (telephone practice). The total system noise independent of the presence or absence of a signal. 24

(5) (radio noise from overhead power lines). The total system noise independent of the presence or absence of radio noise from the power line. *Note:* Background noise is not to be included as part of the radio noise measured from the power line. 36

(6) (communication satellite). That part of the receiving system noise power produced by noise sources in the celestial background of the radiation pattern of the receiving antenna. Typical sources are the galaxy (galactic noise) the sun and radio stars. 85

background response (radiation detectors). Response caused by ionizing radiation coming from sources other than that to be measured. *See also:* **ionizing radiation.** 117

background returns (radar). *See:* **clutter.**

backing (planar structure) (rotating machinery). A fabric, mat, film, or other material used in intimate conjunction with a prime material and forming a part of the composite for mechanical support or to sustain or improve its properties. 63

backing lamp (backing light) (backup light). *See:* **backup lamp.** 328

backlash (1) (general). A relative movement between interacting mechanical parts, resulting from looseness. *See*

also: **control system, feedback; industrial control.** 266, 207, 219, 206

(2) (signal generator). The difference in actual value of a parameter when the parameter is set to an indicated value by a clockwise rotation of the indicator, and when it is set by a counterclockwise rotation. *See also:* **signal generator.** 185

(3) (tunable microwave tube). The amount of motion of the tuner control mechanism (in a mechanically tuned oscillator) that produces no frequency change upon reversal of the motion. *See also:* **tunable microwave oscillators.** 174, 190

back light (television). Illumination from behind the subject in a direction substantially parallel to the plane through the optical axis of the camera. *See also:* **television.** 167

back pitch (rotating machinery). The coil pitch at the non-connection end of a winding (usually in reference to a wave winding). 63

back plate (signal plate) (camera tubes). The electrode in an iconoscope or orthicon camera tube to which the stored charge image is capacitively coupled. *See:* **television.** 178

back porch (1) (monochrome composite picture signal). The portion that lies between the trailing edge of a horizontal synchronizing pulse and the trailing edge of the corresponding blanking pulse.

(2) (National Television System Committee composite color-picture signal). The portion that lies between the color burst and the trailing edge of the corresponding blanking pulse. *See also:* **television.** 178

back scatter (radar). Energy reflected with a direction component toward the transmitter. *See also:* **navigation.** 187, 13

backscatter coefficient. A measure of a distributed radar target's ability to reflect energy to a radar receiving antenna colocated with the transmitting antenna. The measure is defined as 4π times the ratio of the reflected power per unit solid angle in the direction of the source to the power per unit area in the incident wave at the target, divided by the area of a radar resolution element. 13

back scattering coefficient (echoing area) (data transmission). The back scattering coefficient (B) of an object for an incident plane wave is 4π times the ratio of the reflected power per unit solid angle ϕ_r in the direction of the source to the power per unit area (W_1) in the incident wave:

$$B = 4\pi\,\frac{\phi_r}{W_1} = 4\pi r^2\,\frac{W_r}{W_1}$$

where W_r is the power per unit area at distance r. *Note:* For large objects, the back scattering coefficient of an object is approximately the product of its interception area and its scattering gain in the direction of the source, where the interception area is the projected geometrical area and the scattering gain is the reradiated power gain relative to an isotropic radiator. 59

back-scattering cross section. The scattering cross-section in the direction towards the source. *See:* **back-scattering coefficient; monostatic cross section; radar cross section; target echoing area.** *See also:* **antenna.** 179, 111

back-shunt keying. A method of keying a transmitter in which the radio-frequency energy is fed to the antenna when the telegraph key is closed and to an artificial load

when the key is open. *See also:* **radio transmission.**
111, 211

backstop, relay. *See:* **relay backstop.**

backswing (1) (circuits and systems) (distortion pulse). A contribution to pulse distortion resulting from overshoot of the baseline during pulse turnoff. 67

(2) (low power pulse transformers) (high power pulse transformers) (last transition overshoot) (A_{BS}). The maximum amount by which the instantaneous pulse value is below the zero axis in the region following the fall time. It is expressed in amplitude units or in percent of A_M. *See also:* **input pulse shape.** 32, 33

backup current-limiting fuse. *See* **function class "a" (backup) current-limiting fuse.**

backup gap (series capacitor). A supplementary gap which may be used in parallel with the protective power gap. 86

backup lamp. A lighting device mounted on the rear of a vehicle for illuminating the region near the back of the vehicle while moving in reverse. *Note:* It normally can be used only while backing up. *See also:* **headlamp.** 167

backup overcurrent protective device or apparatus (nuclear power generating stations). A device or apparatus which performs the circuit interrupting function in the event the primary protective device or apparatus fails or is out of service. 26

backup protection (relay system). A form of protection that operates independently of specified components in the primary protective system and that is intended to operate if the primary protection fails or is temporarily out of service. 103, 202, 60, 127

backward-acting regulator. A transmission regulator in which the adjustment made by the regulator affects the quantity that caused the adjustment. *See also:* **transmission regulator.** 328

backward wave (traveling-wave tubes). A wave whose group velocity is opposite to the direction of electron-stream motion. *See:* **amplifier.** *See also:* **electron devices, miscellaneous.** 125, 190

backward-wave oscillator. *See:* **carcinotron.**

backward-wave (BW) structure (microwave tubes). A slow-wave structure whose propagation is characterized on an ω/β diagram (sometimes called a Brillouin diagram) by a negative slope in the region $0 < \beta < \pi$ (in which the phase velocity is therefore of opposite sign to the group velocity). *See also:* **microwave tube (or valve).** 190

back wave. A signal emitted from a radio telegraph transmitter during spacing portions of the code characters and between the code characters. *See also:* **radio transmission.** 211

bactericidal (germicidal) effectiveness (illuminating engineering). The capacity of various portions of the ultraviolet spectrum to destroy bacteria, fungi, and viruses. *See also:* ultraviolet radiation. 167

bactericidal (germicidal) efficiency (radiant flux). For a particular wavelength, the ratio of the bactericidal effectiveness of that wavelength to that of wavelength 265.0 nanometers, which is rated as unity. *Note:* Tentative bactericidal efficiency of various wavelengths of radiant flux is given in the following table.

Erythemal and bactericidal efficiency of ultraviolet radiation

Wavelength (nanometers)	Erythemal Efficiency	Tentative Bactericidal Efficiency
*235.3	——	0.35
240.0	0.56	——
*244.6	0.57	0.58
*248.2	0.57	0.70
250.0	0.57	——
*253.7	0.55	0.85
*257.6	0.49	0.94
260.0	0.42	——
265.0	——	1.00
*265.4	0.25	0.99
*267.5	0.20	0.98
*270.0	0.14	0.95
*275.3	0.07	0.81
*280.4	0.06	0.68
285.0	0.09	——
*285.7	0.10	0.55
*289.4	0.25	0.46
290.0	0.31	——
*292.5	0.70	0.38
295.0	0.98	——
*296.7	1.00	0.27
300.0	0.83	——
*302.2	0.55	0.13
305.0	0.33	——
310.0	0.11	——
*313.0	0.03	0.01
315.0	0.01	——
320.0	0.005	——
325.0	0.003	——
330.0	0.000	——

* Emission lines in the mercury spectrum; other values interpolated.

See also: **ultraviolet radiation.** 167

bactericidal (germicidal) exposure. The product of bactericidal flux density on a surface and time. It usually is measured in bactericidal microwatt minutes per square centimeter or bactericidal watt-minutes per square foot. *See also:* **ultraviolet radiation.** 167

bactericidal (germicidal) flux. Radiant flux evaluated according to its capacity to produce bactericidal effects. It usually is measured in microwatts of ultraviolet radiation weighted in accordance with its bactericidal efficiency. Such quantities of bactericidal flux would be in bactericidal microwatts. *Note:* Ultraviolet radiation of wavelength 253.7 nanometers usually is referred to as "ultraviolet microwatts" or "UV watts." These terms should not be confused with "bactericidal microwatts." *See also:* **ultraviolet radiation.** 167

bactericidal (germicidal) flux density. The bactericidal flux per unit area of the surface being irradiated. *Note:* It is equal to the quotient of the incident bactericidal flux divided by the area of the surface when the flux is uniformly distributed. It usually is measured in microwatts per square centimeter or watts per square foot of bactericidally weighted ultraviolet radiation (bactericidal microwatts per square centimeter or bactericidal watts per square foot). *See also:* **ultraviolet radiation.** 167

baffle (1) (audio and electroacoustics). A shielding structure or partition used to increase the effective length of the transmission path between two points in an acoustic system as for example, between the front and back of an electroacoustic transducer. *Note:* In the case of a loudspeaker, a baffle is often used to increase the acoustic loading of the diaphragm. *See also:* **loudspeaker.** 176
(2) (light sources). A single opaque or translucent element to shield a source from direct view at certain angles or to absorb unwanted light. *See also:* **bare (exposed) lamp.** 167
(3) (gas tube). An auxiliary member, placed in the arc path and having no separate external connection. *Note:* A baffle may be used for: (A) controlling the flow of mercury vapor or mercury particles; (B) controlling the flow of radiant energy; (C) forcing a distribution of current in the arc path; or (D) deionizing the mercury vapor following conduction. It may be of either conducting or insulating material. *See also:* **electrode (electron tube).** 190

bag-type construction (dry cell) (primary cell). A type of construction in which a layer of paste forms the principal medium between the depolarizing mix, contained within a cloth wrapper, and the negative electrode. *See also:* **electrolytic cell.** 328

balance beam (relay). A lever form of relay armature, one end of which is acted upon by one input and the other end restrained by a second input. 127, 103, 202, 60

balance check (analog computer). The computer control state in which all amplifier summing junctions are connected to the computer zero reference level (usually signal ground) to permit zero balance of the operational amplifier. *Note:* Integrator capacitors may be shunted by a resistor to permit the zero balance of an integrator. This control state may not be found in some analog computers. 9

balanced (1) (general). Used to signify proper relationship between two or more things, such as stereophonic channels.
(2) (communication practice). Usually signifies (A) electrically alike and symmetrical with respect to a common reference point, usually ground, or (B) arranged to provide conjugacy between certain sets of terminals. 239, 59

balanced amplifier (push-pull amplifier). *See:* **amplifier, balanced.**

balanced capacitance (between two conductors) (mutual capacitance between two conductors*). The capacitance between two conductors when the changes in the charges on the two are equal in magnitude but opposite in sign and the potentials of the other $n - 2$ conductors are held constant. *See:* **direct capacitances (system of conductors).** 210, 185

* Deprecated

balanced circuit (1) (signal-transmission system). A circuit, in which two branches are electrically alike and symmetrical with respect to a common reference point, usually ground. *Note:* For an applied signal difference at the input, the signal relative to the reference at equivalent points in the two branches must be opposite in polarity and equal in amplitude. 185
(2) (electric power system). A circuit in which there are substantially equal currents, either alternating or direct, in all main wires and substantially equal voltages between main wires and between each main wire and neutral (if one exists). *See also:* **center of distribution.** 64

balanced conditions (1) (time domain) (rotating machinery). A set of polyphase quantities (phase currents, phase voltages, etcetera) that are sinusoidal in time, that have identical amplitudes, and that are shifted in time with respect to each other by identical phase angles.
(2) (space domain). In space, a set of coils (for example, of a rotating machine) each having the same number of effective turns, with their magnetic axes shifted by identical angular displacements with respect to each other. *Notes:* (A) The impedance (matrix) of a balanced machine is balanced. A balanced set of currents will produce a balanced set of voltage drops across a balanced set of impedances. (B) If all sets of windings of a machine are balanced and if the magnetic structure is balanced, the machine is balanced. *See also:* **asynchronous machine; direct-current commutating machine.** 63

balanced currents (on a balanced line). Currents flowing in the two conductors of a balanced line that, at every point along the line, are equal in magnitude and opposite in direction. *See also:* **transmission line.** 267, 179

balanced line. A transmission line consisting of two single or two interconnected groups of conductors capable of being operated in such a way that when the voltages of the two groups of conductors at any transverse plane are equal in magnitude and opposite in polarity with respect to ground, the total currents along the two groups of conductors are equal in magnitude and opposite in direction; for example, a line in which the two conductors are identical in cross section and symmetrically positioned with respect to the electric ground. *Note:* A balanced line may be operated under unbalanced conditions and the aggregate then does not form a balanced line system. *See also:* **transmission line.** 267, 179, 185

balanced line system. A system consisting of generator, balanced line, and load adjusted so that the voltages of the two conductors at all transverse planes are equal in magnitude and opposite in polarity with respect to ground. *Note:* Balanced line system is frequently shortened to balanced line. Care should be taken not to confuse this abbreviated terminology with the standard definition of balanced line. *See also:* **transmission line.** 267, 179

balanced mixer (1) (circuits and systems) (single, double). A type of mixer that forms from two signals A & B a third signal C having the form $C = (a + A)(b + B)$. Single balanced implies $a = 0$, $b \neq 0$; double balanced implies $a = b = 0$. Note: Such mixers can suppress a RF carrier and/or a local oscillator in their output spectrum. *Syn.* **balanced modulator.** 67
(2) A hybrid junction with crystal receivers in one pair of uncoupled arms, the arms of the remaining pair being fed from a signal source and a local oscillator. *Note:* The resulting intermediate-frequency signals from the crystals are added in such a manner that the effect of local-oscillator noise is minimized. *See:* **converter; hybrid junction;**

radio receiver; waveguide.   244, 179

balanced modulator (signal-transmission system). A modulator, specifically a push-pull circuit, in which the carrier and modulating signal are so introduced that after modulation takes place the output contains the two sidebands without the carrier. *See also:* **modulation; modulator, symmetrical.**   111, 178

balanced oscillator. An oscillator in which at the oscillator frequency the impedance centers of the tank circuit are at ground potential and the voltages between either end and their centers are equal in magnitude and opposite in phase. *See also:* **oscillatory circuit.**   111, 211

balanced polyphase load. A load to which symmetrical currents are supplied when it is connected to a system having symmetrical voltages. *Note:* The term balanced polyphase load is applied also to a load to which are supplied two currents having the same wave form and root-mean-square value and differing in phase by 90 electrical degrees when it is connected to a quarter-phase (or two-phase) system having voltages of the same wave form and root-mean-square value. *See also:* **generating station.**   64

balanced polyphase system. A polyphase system in which both the currents and voltages are symmetrical. *See also:* **alternating-current distribution.**   64

balanced telephone-influence factor (three-phase synchronous machine). The ratio of the square root of the sum of the squares of the weighted root-mean-square values of the fundamental and the non-triple series of harmonics to the root-mean-square value of the normal no-load voltage wave.   63

balanced termination (system or network having two output terminals). A load presenting the same impedance to ground for each of the output terminals. *See also:* **network analysis.**   267, 179, 185

balanced three-wire system. A three-wire system in which no current flows in the conductor connected to the neutral point of the supply. *See:* **three-wire system.** *See also:* **alternating-current distribution.**   64

balanced voltages (1) (on a balanced line). Voltages (relative to ground) on the two conductors of a balanced line that, at every point along the line, are equal in magnitude and opposite in polarity. *See also:* **transmission line.**   267, 179

(2) (signal-transmission system). The voltages between corresponding points of a balanced circuit (voltages at a transverse plane) and the reference plane relative to which the circuit is balanced. *See also:* **signal.**   188

balanced wire circuit. A circuit whose two sides are electrically alike and symmetrical with respect to ground and other conductors. *Note:* The term is commonly used to indicate a circuit whose two sides differ only by chance. *See also:* **transmission line.**   59

balancer. That portion of a direction-finder that is used for the purpose of improving the sharpness of the direction indication. *See also:* **radio receiver.**   328

balance relay. A relay that operates by comparing the magnitudes of two similar input quantities. *Note:* The balance may be effected by counteracting electromagnetic forces on a common armature, or by counteracting magnetomotive forces in a common magnetic circuit, or by similar means, such as springs, levers, etcetera.   127, 103, 202, 60

balance test (rotating machinery). A test taken to enable a rotor to be balanced within specified limits. *See also:* **rotor (rotating machinery).**   63

balancing network. An electric network designed for use in a circuit in such a way that two branches of the circuit are made substantially conjugate, that is, such that an electromotive force inserted in one branch produces no current in the other branch. *See also:* **network analysis.**   59

balancing of an operational amplifier (analog computer). The act of adjusting the output level of an operational amplifier to coincide with its input reference level, usually ground or zero voltage, in the balance check computer control state. This operation may not be required in some amplifiers and there may be no provision for performing it.   9

ballast (1) (general). An impedor connected in series with a circuit or device, such as an arc, that normally is unstable when connected across a constant-voltage supply, for the purpose of stabilizing the current.   210

(2) (fluorescent lamps or mercury lamps). Devices that by means of inductance, capacitance, or resistance, singly or in combination, limit the lamp current of fluorescent or mercury lamps, to the required value for proper operation, and also, where necessary, provide the required starting voltage and current and, in the case of ballasts for rapid-start lamps, provide for low-voltage cathode heating. *Note:* Capacitors for power-factor correction and capacitor-discharge resistors may form part of such a ballast.   268, 269

(3) (fixed-impedance type) (reference ballast). Designed for use with one specific type of lamp that, after adjustment during the original calibration, is expected to hold its established impedance through normal use.   270, 271

(4) (variable-impedance type). An adjustable inductive reactor and a suitable adjustable resistor in series. *Note:* These two components are usually designed so that the resulting combination has sufficient current-carrying capacity and range of impedance to be used with a number of different sizes of lamps. The impedance and power factor of the reactor-resistor combination are adjusted, or rechecked, each time the unit is used.   270, 271

ballast leakage. The leakage of current from one rail of a track circuit to another through the ballast, ties, earth, etcetera.   328·

ballast resistance. The resistance offered by the ballast, ties, earth, etcetera, to the flow of leakage current from one rail of a track circuit to another.   328

ballast tube (ballast lamp). A current-controlling resistance device designed to maintain substantially constant current over a specified range of variation in the applied voltage or the resistance of a series circuit.   328

ball bearing (rotating machinery). A bearing incorporating a peripheral assembly of balls. *See also:* **bearing.**   63

ball burnishing. Burnishing by means of metal balls. *See also:* **electroplating.**   328

ballistic focusing (microwave tube). A focusing system in which static electric fields cause an initial convergence of the beam and the electron trajectories are thereafter determined by momentum and space charge forces only. *See* *also:* **microwave tube or valve.** 189

ball lightning. A type of lightning discharge reported from visual observations to consist of luminous, ball-shaped regions of ionized gases. *Note:* In reality ball lightning may or may not exist. *See also:* **direct-stroke protection (lightning).** 64

balun. (1) A passive device having distributed electrical constants used to couple a balanced system or device to an unbalanced system or device. *Note:* The term is derived from **balance to unbalance transformer.** 197
(2) A network for the transformation from an unbalanced transmission line or system to a balanced line or system, or vice versa. 185

banana plug. A single-conductor plug with a spring metal tip that somewhat resembles a banana in shape. 329

band (1) (electronic computers). A group of circular recording tracks, on a moving storage device such as a drum or disc. *See also:* **channel.** 255, 77, 54
(2) (data transmission). Range of frequency between two defined limits. 59

band, effective (facsimile). The frequency band of a facsimile signal wave equal in width to that between zero frequency and maximum keying frequency. *Note:* The frequency band occupied in the transmission medium will in general be greater than the effective band. *See also:* **facsimile signal (picture signal).** 12

band-elimination filter. *See:* **filter, band-elimination.**

band gap. The difference in energy between the energy level of the bottom of the conduction band and the energy level of the top of the valence band. 118, 119

banding insulation (rotating machinery). Insulation between the winding overhang and the binding bands. 63

band of regulated voltage. The band or zone, expressed in percent of the rated value of the regulated voltage, within which the excitation system will hold the regulated voltage of an electric machine during steady or gradually changing conditions over a specified range of load. *See:* **direct-current commutating machine.** 328

band-pass filter. *See:* **filter, band-pass.**

bandpass tube (microwave gas tubes). *See:* **broad-band tube.**

band spreading. (1) The spreading of tuning indications over a wide scale range to facilitate tuning in a crowded band of frequencies. (2) The method of double-sideband transmission in which the frequency band of the modulating wave is shifted upward in frequency so that the sidebands produced by modulation are separated in frequency from the carrier by an amount at least equal to the bandwidth of the original modulating wave, and second-order distortion products may be filtered from the demodulator output. *See also:* **radio receiver.** 328

band switch. A switch used to select any one of the frequency bands in which an electric transmission apparatus may operate. 328

bandwidth (1) (continuous frequency band). The difference between the limiting frequencies. 210
(2) (device). The range of frequencies within which performance, with respect to some characteristic, falls within specific limits. *See also:* **radio receiver.**
(3) (wave). The least frequency interval outside of which the power spectrum of a time-varying quantity is everywhere less than some specified fraction of its value at a reference frequency. *Warning:* This definition permits the spectrum to be less than the specified fraction within the interval. *Note:* Unless otherwise stated, the reference frequency is that at which the spectrum has its maximum value. 339
(4) (burst) (burst measurements). The smallest frequency interval outside of which the integral of the energy spectrum is less than some designated fraction of the total energy of the burst. *See also:* **burst.** 272
(5) (antenna). The range of frequencies within which its performance, in respect to some characteristics, conforms to a specified standard. *See also:* **antenna.** 111, 179
(6) (facsimile). The difference in hertz between the highest and the lowest frequency components required for adequate transmission of the facsimile signals. *See also:* **facsimile (electrical communication).** 12
(7) (industrial control) (excitation control systems). The interval separating two frequencies between which both the gain and the phase difference (of sinusoidal output referred to sinusoidal input) remain within specified limits. *Note:* For control systems and many of their components, the lower frequency often approaches zero. *See also:* **control system, feedback.** 266, 219, 206, 329, 353
(8) (pulse terms). The two portions of a pulse waveform which represents the first nominal state from which a pulse departs and to which it ultimately returns.

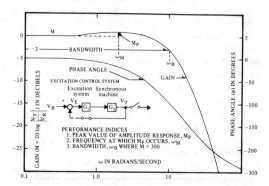

Typical closed-loop frequency response of an excitation control system with the synchronous machine open circuited.

(9) (signal-transmission system). The range of frequencies within which performance, with respect to some characteristic, falls within specific limits. *Note:* For systems capable of transmitting at zero frequency the frequency at which the system response is less than that at zero frequency by a specified ratio. For carrier-frequency systems: the difference in the frequencies at which the system response is less than that at the frequency of reference response by a specified ratio. For both types of systems, bandwidth is commonly defined at the points where the response is 3 decibels less than the reference value (0.707 root-mean-square voltage ratio). *See:* **equivalent noise bandwidth.** 188
(10) (oscilloscope). The difference between the upper and lower frequency at which the response is 0.707 (−3 decibels) of the response at the reference frequency. Usually

both upper and lower limit frequencies are specified rather than the difference between them. When only one number appears, it is taken as the upper limit. *Notes:* (A) The reference frequency shall be at least 20 times greater for the lower bandwidth limit and at least 20 times less for the upper bandwidth limit than the limit frequency. The upper and lower reference frequencies are not required to be the same. In cases where exceptions must be made, they shall be noted. (B) This definition assumes the amplitude response to be essentially free of departures from a smooth roll-off characteristic. (C) If the lower bandwidth limit extends to zero frequency, the response at zero frequency shall be equal to the response at the reference frequency, not −3 decibels from it. 185

(11) (dispersive and nondispersive delay lines). A specified frequency range over which the amplitude response does not vary more than a defined amount. *Note:* Typically amplitude range is 1 dB bandwidth, 3 dB bandwidth. 81

(12) (analog computer). (A) Of a signal, the difference between the limiting frequencies encountered in the signal. (B) Of a device, the range of frequencies within which performance in respect to some characteristic falls within specific limits. 9

bandwidth, effective (bandpass filter in a signal transmission system). The width of an assumed rectangular bandpass filter having the same transfer ratio at a reference frequency and passing the same mean square of a hypothetical current and voltage having even distribution of energy over all frequencies. *Note:* For a nonlinear system, the bandwidth at a specified input level. *See also:* **network analysis; signal.** 111, 188

bandwidth, root-mean-square. The root mean squared (rms) deviation of the power spectrum of the received signal relative to zero frequency or the spectral center, in units of radians per second or hertz. This bandwidth, β, may be defined as

$$\beta^2 = \frac{\int_{-\infty}^{\infty} [2\pi(f - f_0)]^2 \, |S(f)|^2 \, df}{\int_{-\infty}^{\infty} |S(f)|^2 \, df}$$

where S(f) is the Fourier transform of the signal $s(t - \tau_0)$ with true time delay τ_0 and f_0 is the center frequency of the spectrum. *Note:* β^2 is the normalized second moment of the spectrum $|S(f)|^2$ about the mean. *See also:* **bandwidth.**

bank. An aggregation of similar devices (for example, transformers, lamps, etcetera) connected together and used in cooperation. *Note:* In automatic switching, a bank is an assemblage of fixed contacts over which one or more wipers or brushes move in order to establish electric connections. *See:* **relay level.** 328

bank-and-wiper switch (telephone switching system). A switch in which an electromagnetic ratchet or other mechanisms are used, first, to move the wipers to a desired group of terminals, and second, to move the wipers over the terminals of this group to the desired bank contacts. 328

banked winding. *See:* **bank winding.**

bank winding (banked winding). A compact multilayer form of coil winding, for the purpose of reducing distributed capacitance, in which single turns are wound successively in each of two or more layers, the entire winding proceeding from one end of the coil to the other, without return. 329

bar (lights). A group of three or more aeronautical ground lights placed in a line transverse to the axis, or extended axis, of the runway. 167

bare conductor. A conductor not covered with insulating material. 64

bare (exposed) lamp. A light source with no shielding. *See also:* **light.** 167

barette. A short bar in which the lights are closely spaced so that from a distance they appear to be a linear light. *Note:* Barettes are usually less than 15 feet in length. *See:* **bar (lights).** 167

bar generator (television). A generator of pulses that are uniformly spaced in time and are synchronized to produce a stationary bar pattern on a television screen. *See also:* **television.** 339

Barkhausen-Kurz oscillator. An oscillator of the retarding-field type in which the frequency of oscillation depends solely upon the electron transit-time within the tube. *See also:* **oscillatory circuit.** 111

Barkhausen tube. *See:* **positive-grid oscillator tube.**

bar pattern (television). A pattern of repeating lines or bars on a television screen. When such a pattern is produced by pulses that are equally separated in time, the spacing between the bars on the television screen can be used to measure the linearity of the horizontal or vertical scanning systems. *See also:* **television.** 328

barrel plating. Mechanical plating in which the cathodes are kept loosely in a container that rotates. *See also:* **electroplating.** 328

barretter (waveguide components). A form of bolometer element having a positive temperature coefficient of resistivity which typically employs a power-absorbing wire or thin metal film. 166

barrier (1). A partition for the insulation of isolation of electric circuits or electric arcs. 27, 103, 202
(2) (in a semiconductor) (obsolete). *See:* **depletion layer.**
(3) (nuclear power generating stations) (class 1E systems and equipment). A device or structure interposed between Class 1E equipment or circuits and a potential source of damage to limit damage to Class 1E systems to an acceptable level. 31, 131

barrier grid (charge-storage tubes). A grid, close to or in contact with a storage surface, which grid establishes an equilibrium voltage for secondary-emission charging and serves to minimize redistribution. *See also:* **charge-storage tube.** 174, 190

barrier wiring techniques (coupling in control systems). Those wiring techniques which obstruct electric or magnetic fields, excluding or partially excluding the fields from a given circuit. Barrier techniques are often effective against electromagnetic radiation also. In general, these techniques change the coupling coefficients between wires connected to a noise source and the signal circuit. Example: placement of signal lines within steel conduit to isolate them from an existing magnetic field. *See also:* **compen-**

satory wiring techniques; suppressive wiring techniques. 43

barring hole (rotating machinery). A hole in the rotor to permit insertion of a pry bar for the purpose of turning the rotor slowly or through a limited angle. *See also:* **rotor (rotating machine).** 63

bar, rotor (rotating machinery). *See:* **rotor bar.**

base (1) (number system). An integer whose successive powers are multiplied by coefficients in a positional notation system. *See:* **positional notation; radix.** 235
(2) (rotating machinery). A structure, normally mounted on the foundation, that supports a machine or a set of machines. In single-phase machines rated up through several horsepower, the base is normally a part of the machine and supports it through a resilient or rigid mounting to the end shields. 63
(3) (electron tube or valve). The part attached to the envelope, carrying the pins or contacts used to connect the electrodes to the external circuit and that plugs in to the holder. *See also:* **electron tube.** 244, 190
(4) (basis or base metals) (electroplating). The object upon which the metal is electroplated. *See also:* **electroplating.** 328
(5) (transistor). A region that lies between an emitter and a collector of a transistor and into which minority carriers are injected. *See also:* **transistor.** 210, 245, 66
(6) (high-voltage fuse). The supporting member to which the insulator unit or units are attached. 79
(7) (pulse terms). The two portions of a pulse waveform which represents the first nominal state from which a pulse departs and to which it ultimately returns. 254

base active power (1) (synchronous generators and motors). The total (generator) output or (motor) input power at base voltage and base current with a power factor of unity.
(2) (induction motors). The rated output power. *Note:* Base active power is usually expressed in watts, but any consistent set of units may be used. 5

base address. A given address from which an absolute address is derived by combination with a relative address. 255, 77

base ambient temperature (power distribution underground cables) (cable or duct). The no-load temperature in a group with no load on any cable or duct in the group. *See also:* **power distribution, underground construction.** 57

base apparent power (synchronous generators and motors). The total rated apparent power at rated voltage and rated current. *Note:* Base apparent power is usually expressed in volt-amperes, but any consistent set of units may be used. 5

baseband (carrier or subcarrier wire or radio transmission system). The band of frequencies occupied by the signal before it modulates the carrier (or subcarrier) frequency to form the transmitted line or radio signal. *Note:* The signal in the baseband is usually distinguished from the line or radio signal by ranging over distinctly lower frequencies, which at the lower end relatively approach or may include direct current (zero frequency). In the case of a facsimile signal before modulation on a subcarrier, the baseband includes direct current. *See also:* **facsimile transmission.** 12, 178

baseband-multiplexer (data transmission). The frequency band occupied by the aggregate of the transmitted signals applied to the facility interconnecting the multiplexing and line equipment. The multiplex baseband is also defined as the frequency band occupied by the aggregate of the received signals obtained from the facility interconnecting the line and the multiplex equipment. 59

base current (1) (synchronous generators and motors). The rated phase current. The value of the base current is the value of the rated line current for a 3-phase wye-connected machine, and is the value of the rated line current divided by $\sqrt{3}$ for a delta-connected machine. 5
(2) (induction motors). The phase current corresponding to the base power and base voltage with a power factor of unity. The value of the base current is the value of the base power per phase, divided by the value of the base voltage. *Note:* Base current is usually expressed in amperes, but any consistent set of units may be used. 5

base electrode (transistor). An ohmic or majority-carrier contact to the base region. *See also:* **transistor base region; base.** 66

base impedance (ac rotating machinery). The value of the base impedance is the value of the base voltage divided by the value of the base current. *Note:* Base impedance is usually expressed in ohms, but any consistent set of units may be used. 5

base light (television). Uniform, diffuse illumination approaching a shadowless condition, that is sufficient for a television picture of technical acceptibility, and that may be supplemented by other lighting. *See also:* **television.** 167

baseline (1) (electronic navigation). The line joining the two points between which electrical phase or time is compared in determining navigation coordinates; for two ground stations, this is the great circle joining the two stations, and, in the case of a rotating collector system, it is the line joining the two sides of the collector. *See also:* **navigation.** 13, 187
(2) (pulse techniques). That amplitude level from which the pulse waveform appears to originate. *See also:* **pulse.** 185
(3) (at pulse peak). The instantaneous value that the voltage would have had at the time of the pulse peak in the absence of that pulse. 118, 119, 23

baseline delay (loran). The time interval needed for a signal from a loran master station to travel to the slave station. *See also:* **navigation.** 187

baseline offset (pulse techniques). The algebraic difference between the amplitude of the baseline and the amplitude reference level. *See also:* **pulse.** 185

baseline overshoot (pulse techniques). *See:* **distortion, pulse.**

base load (electric power utilization). The minimum load over a given period of time. *See also:* **generating stations.** 64, 200

base load control (electric generating unit or station). A mode of operation in which the unit or station generation is held constant. *See also:* **speed-governing system.** 94

base magnitude (pulse terms). The magnitude of the base as obtained by a specified procedure or algorithm. Unless otherwise specified, both portions of the base are included in the procedure or algorithm. *See:* The single pulse diagram below the **waveform epoch** entry. *See also:* IEEE

Std 181-1975, Pulse Measurement and Analysis by Objective Techniques, Section 4.3, for suitable algorithms. 254

basement. The rock region underlying the overburden largely comprising aged rock types, often crystalline and of low conductivity. 132

base-minus-ones complement. A number representation that can be derived from another by subtracting each digit from one less than the base. Nines complements and ones complements are base-minus-ones complements. 235

base-mounted electric hoist. A hoist similar to an overhead electric hoist except that it has a base or feet and may be mounted overhead, on a vertical plane, or in any position for which it is designed. *See also:* **hoist.** 328

base power (ac rotating machinery). Either base apparent power or base active power. *Note:* The same numerical value must be used for base apparent power and base active power. 5

base rate (telephone switching systems). A fixed amount charged each month for any one of the classes-of-service that is provided to a customer. 55

base-rate area (telephone switching systems). The territory in which the tariff applies. 55

base region (transistor). The interelectrode region of a transistor into which minority carriers are injected. *See also:* **transistor.** 328

base resistivity. The electrical resistivity of the material composing the base of a semiconductor device. 113

base speed (1) (adjustable-speed motor). The lowest rated speed obtained at rated load and rated voltage at the temperature rise specified in the rating. *See:* **asynchronous machine; direct-current commutating machine.** 3
(2) (ac rotating machinery). The rated synchronous speed. *Note:* Base speed is usually expressed in revolutions per minute, but any consistent set of units may be used. 5

base station (mobile communication). A land station in the land-mobile service carrying on a radio communication service with mobile and fixed radio stations. *See also:* **mobile communication system.** 181

base torque (ac rotating machinery). The torque at synchronous speed corresponding to the base active power. The value of the base torque (in pound-force feet) is 7.04 times the value of the base active power (in watts) divided by the value of the base speed (in revolutions per minute). *Note:* Base torque is usually expressed in pound-force feet or newton meters, but any consistent set of units may be used. 5

base value (rotating machinery). A normal or nominal or reference value in terms of which a quantity is expressed in per unit or percent. *See also:* **asynchronous machine; direct-current commutating machines.** 63

base voltage (AC rotating machinery). The rated phase voltage. The value of the base voltage is the value of the rated line voltage for a delta-connected machine, and is the value of the rated line voltage divided by $\sqrt{3}$ for a wye-connected machine. *Note:* Base voltage is usually expressed in volts, but any consistent set of units may be used. 5

basic alternating voltage (power rectifier). The sustained sinusoidal voltage that must be impressed on the terminal of the alternating-current winding of the rectifier transformer, when set on the rated voltage tap, to give rated output voltage at rated load with no phase control. *See also:* **rectification.** 328

basic device (supervisory system). *See:* **common device.**

basic element (measurement system). A measurement component or group of components that performs one necessary and distinct function in a sequence of measurement operations. *Note:* Basic elements are single-purpose units and provide the smallest steps into which the measurement sequence can be classified conveniently. Typical examples of basic elements are: a permanent magnet, a control spring, a coil, and a pointer and scale. *See:* **measurement system.** 328

basic frequency. Of an oscillatory quantity having sinusoidal components with different frequencies, the frequency of the component considered to be the most important. *Note:* In a driven system, the basic frequency would, in general, be the driving frequency, and in a periodic oscillatory system, it would be the fundamental frequency. 176

basic functions (industrial control) (controller). The functions of those of its elements that govern the application of electric power to the connected apparatus. *See also:* **electric controller.** 206

basic impulse insulation levels (1) (general). Reference insulation levels expressed as the impulse crest value of withstand voltage of a specified full impulse voltage wave. Nominal 1.2×50-microsecond wave. *Notes:* (Λ) Impulse waves are defined by a combination of two numbers. The first number is the time from the start of the wave to the instant of crest value, and the second number if the time from the start to the instant of half-crest value on the tail of the wave. (B) In practice it is necessary to determine the starting point of the wave in a prescribed manner, set forth in the test code. The starting point so determined is called the virtual time zero. (C) Various specifications for test-voltage-waves include: 1.2×50 microseconds and 1.5×40 microseconds. 203
(2) (electric power). Basic impulse insulation levels are reference levels expressed in impulse crest voltage with a standard wave not longer than 1.5×40-microsecond wave. *See also:* **insulation.** 91, 2, 103, 202, 273, 62, 274, 275, 276
(3) (surge arrester). A reference impulse insulation strength expressed in terms of the crest value of withstand voltage of a standard full impulse voltage wave. *Note:* See American National Standard Voltage Values for Preferred Basic Impulse Insulation Levels. C92.1-1971, for standard wave shape. 2
(4) (outdoor apparatus bushings). A reference insulation level expressed as the impulse crest voltage of the 1.2×50 microsecond wave which the bushing will withstand when tested in accordance with specified conditions. 168

basic lightning impulse insulation level (BIL). A specific insulation level expressed in kilovolts of the crest value of a standard lightning impulse. 53

basic metallic rectifier. One in which each rectifying element consists of a single metallic rectifying cell. *See also:* **rectification.** 328

basic network. An electric network designed to simulate the iterative impedance, neglecting dissipation, of a line at a particular termination. *See also:* **network analysis.**

basic numbering plan USA (telephony). The plan whereby every telephone station is identified for nationwide dialing by a code for routing and a number of digits. 193

basic part (electric and electronics parts and equipments). One piece, or two or more pieces joined together, which are not normally subject to disassembly without destruction of designed use. The application, size, and construction of an item may be factors in determining whether an item is regarded as a unit, an assembly, a subassembly, or a basic part. A small electric motor might be considered as a part if it is not normally subject to disassembly. Typical examples: electron tube, resistor, relay, power transformer, microelectronic device. 17

basic reference designation (electric and electronics parts and equipments). The simplest form of a reference designation, consisting only of a class letter portion and a number (namely, without mention of the item within which the reference-designated item is located). The reference designation for a unit consists of only a number. 17

basic reference standards (laboratory). Those standards with which the values of the electrical units are maintained in the laboratory, and that serve as the starting point of the chain of sequential measurements carried out in the laboratory. *See also:* **measurement system.** 212

basic repetition frequency (loran). The lowest pulse repetition frequency of each of the several sets of closely spaced repetition frequencies employed. *See also:* **navigation.** 187

basic repetition rate. *See:* **basic repetition frequency.**

basic switching impulse insulation level (BSL). A specific insulation level expressed in kilovolts of the crest value of a standard switching impulse. 53

basic units (system of units). Those particular units arbitrarily selected to serve as a basis, in terms of which the other units of the system may be conveniently derived. *Note:* A basic unit is usually determined by a specified relationship to a single prototype standard. 210

bass boost. An adjustment of the amplitude-frequency response of a system or transducer to accentuate the lower audio frequencies. 328

bath voltage. The total voltage between the anode and cathode of an electrolytic cell during electrolysis. It is equal to the sum of (1) equilibrium reaction potential, (2) *IR* drop, (3) anode polarization, and (4) cathode polarization. *See:* **tank voltage.** *See also:* **electrolytic cell.** 328

battery (primary or secondary). Two or more cells electrically connected for producing electric energy. (Common usage permits this designation to be applied also to a single cell used independently. In this Dictionary, unless otherwise specified, the term battery will be used in this dual sense.) 328

battery-and-ground pulsing (telephone switching systems). Dial pulsing using battery-and-ground signaling. 55

battery-and-ground signaling (telephone switching systems). A method of loop signaling, used to increase the range, in which battery and ground at both ends of the loop are poled oppositely. 55

battery chute. A small cylindrical receptacle for housing track batteries and so set in the ground that the batteries will be below the frost line. 328

battery-current regulation (generator). That type of au-

tomatic regulation in which the generator regulator controls only the current used for battery charging purposes. *See also:* **axle generator system.** 328

battery, electric. A device that transforms chemical energy into electric energy. *See also:* **battery.** 59

battery eliminator. A device that provides direct-current energy from an alternating-current source in place of a battery. *See also:* **battery.** 59

battery, power station (1) (communications). A battery that is a separate source of energy for communication equipment in power stations.
(2) (control). A battery that is a separate source of energy for the control of power apparatus in a power station. *See also:* **battery.** 59

battery rack (nuclear power generating stations) (lead storage batteries). A rigid structure used to accommodate a group of cells. 31, 38

baud (1) (general). A unit of signalling speed equal to the number of discrete conditions or signal events per second. For example, one baud equals one half dot cycle per second in Morse code, one bit per second in a train of binary signals, and one 3-bit value per second in a train of signals each of which can assume one of 8 different states. *See also:* **telegraphy.** 255, 77

(2) (telegraphy). The unit of telegraph signaling speed, derived from the duration of the shortest signaling pulse. A telegraphic speed of one baud is one pulse per second. *Note:* The term unit pulse is often used for the same meaning as the baud. A related term, the dot cycle, refers to an ON-OFF or MARK-SPACE cycle in which both mark and space intervals have the same length as the unit pulse. 111

b auxiliary switch. *See:* **auxiliary switch; b contact.**

bay (computing system). *See:* **patch bay.** *See also:* **electronic analog computer.**

B battery. A battery designed or employed to furnish the plate current in a vacuum-tube circuit. *See also:* **battery (primary or secondary).** 328

bb auxiliary switch. *See:* **auxiliary switch; bb contact.**

bb contact. A contact that is closed when the operating mechanism of the main device is in the standard reference position and that is open when the operating mechanism is in the opposite position. *See:* **standard reference position.** 103, 202

b contact (back contact) (relay). A contact that is closed when the main device is in the standard reference position and that is open when the device is in the opposite position. *Notes:* (1) *b* contact has general application. However, this meaning for **back contact** is restricted to relay parlance. (2) For indication of the specific point of travel at which the contact changes position, an additional letter or percentage figure may be added to *b* as detailed in ANSI C37.2-1970, or revisions thereof in which see **standard reference position.** 103, 202

B-display (radar). An intensity modulated rectangular display in which each target appears as a blip, with bearing indicated by the horizontal coordinate and distance (range) by the vertical coordinate. 13

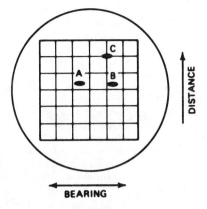

B display.

beacon. A light (or mark) used to indicate a geographic location. 167

beacon equation. An equation which gives maximum detection range or signal-to-noise ratio (SNR) of a transponder (secondary radar) as a function of system parameters for a given set of conditions. It is the counterpart of the radar equation, for one-way transmission. 13

beacon receiver. A radio receiver for converting waves, emanating from a radio beacon, into perceptible signals. *See also:* **radio receiver; radio beacon.** 328

beam (1) (antenna). The major lobe of the radiation pattern. *See also:* **radiation.** 179, 111
(2) (laser-maser). A collection of rays which may be parallel, divergent, or convergent. 363

beam alignment (camera tubes). An adjustment of the electron beam, performed on tubes employing low-velocity scanning, to cause the beam to be perpendicular to the target at the target surface. 125, 190

beam angle. *See* **scan angle.** 111

beam bending (camera tubes). Deflection of the scanning beam by the electrostatic field of the charges stored on the target. 125, 190, 178

beam current (storage tubes). The current emerging from the final aperture of the electron gun. *See also:* **storage tube.** 174, 190

beam-deflection tube. An electron-beam tube in which current to an output electrode is controlled by the transverse movement of an electron beam. 125, 190

beam diameter (laser-maser). The distance between diametrically opposed points in that cross section of a beam where the power per unit area is $1/e$ times that of the peak power per unit area. 363

beam divergence, ϕ (laser-maser). The full angle of the beam spread between diametrically opposed $1/e$-irradiance points; usually measured in milliradians (one milliradian = 3.4 minutes of arc). 363

beam error (navigational systems using directionally propagated signals). The lateral or angular difference between the mean position of the actual course and the desired course position. *Note:* Sometimes called course error. *See also:* **navigation.** 190, 13

beam expander (laser-maser). A combination of optical elements which will increase the diameter of a laser beam. 363

beam finder (oscilloscope). A provision for locating the spot when it is not visible. 106

beamguide (laser-maser). A set of beam-forming elements spaced in such a way as to conduct a well-defined beam of radiation. Analogs are waveguides and fiber optic filaments. 363

beam-indexing color tube. A color-picture tube in which a signal, generated by an electron beam after deflection, is fed back to a control device or element in such a way as to provide an image in color. 125, 178, 190

beam landing error (camera tube). A signal non-uniformity resulting from beam electrons arriving at the target with a spatially varying component of velocity parallel to the target. *See also:* **camera tube.** 190

beam locator. *See:* **beam finder.**

beam modulation, percentage (image orthicons). One hundred times the ratio of (1) the signal output current for highlight illumination on the tube to (2) the dark current. 125, 178

beam noise (navigational systems using directionally propagated signals). Extraneous disturbances tending to interfere with ideal system performance. *Note:* Beam noise is the aggregate effect of bends, scalloping, roughness, etcetera. *See also:* **navigation.** 190

beam parametric amplifier. A parametric amplifier that uses a modulated electron beam to provide a variable reactance. *See also:* **parametric device.** 277, 191

beam pattern. *See:* **directional response pattern.**

beam pointing (communication satellite). The ability to orient the beam of a high gain antenna into a specific direction in a coordinate system. 84

beam power tube. An electron-beam tube in which use is made of directed electron beams to contribute substantially to its power-handling capability, and in which the control grid and the screen grid are essentially aligned. 125, 190

beam resonator (laser-maser). A resonator which serves to confine a beam of radiation to a given region of space without continuous guidance along the beam. 363

beam rider guidance. That form of missile guidance wherein a missile, through a self-contained mechanism, automatically guides itself along a beam. *See also:* **guided missile.** 328

beamshape loss (radar). A loss included in the radar equation to account for the use of the peak antenna gain in the radar equation instead of the actual gain that results when the received train of pulses is modulated by the two-way pattern of the scanning antenna. *Syn.* **antenna-pattern loss.** 13

beam shaping (communication satellite). Controlling the shape of an antenna beam, by design of the surfaces of the antenna or by controlling the phasing of the signals radiated from the antenna. 85

beam splitter (laser-maser). An optical device which uses controlled reflection to produce two beams from a single incident beam. 363

beam spread (any plane) (1) (illuminating engineering). The angle between the two directions in the plane in which the candlepower is equal to a stated percent (usually 10 percent) of the maximum candlepower in the beam. *See also:* **lamp.** 167
(2) (light emitting diodes). (source of light, φy, where y is the

stated percent). *See:* **(1) (illuminating engineering)** above. 162

beam steering (antenna). Changing the direction of the major lobe of a radiation pattern. *See also:* **radiation.** 179

beam waveguide. A structure consisting of a sequence of lenses or mirrors that can guide an electromagnetic wave. *See:* **waveguide.** 179

beamwidth. *See* **half-power beamwidth.**

bearing (1) (electronic navigation). (A) The horizontal direction of one terrestrial point from another, expressed as the angle in the horizontal plane between a reference line and the horizontal projection of the line joining the two points. (B) Azimuth. *See also:* **navigation.** 13, 187
(2) (rotating machinery). (A) A stationary member or assembly of stationary members in which a shaft is supported and may rotate. (B) In a ball or roller bearing, a combination (frequently preassembled) of stationary and rotating members containing a peripheral assembly of balls or rollers, in which a shaft is supported and may rotate. 63

bearing accuracy, instrumental (direction-finding). (1) The difference between the indicated and the apparent bearings in a measurement of the same signal source. (2) As a statement of overall system performance, a difference between indicated and correct bearings whose probability of being exceeded in any measurement made on the system is less than some stated value. *See also:* **navigation.** 187

bearing bracket (rotating machinery). A bracket which supports a bearing, but including no part thereof. A bearing bracket is not specifically constructed to provide protection for the windings or rotating parts. *See:* **bracket-end shield.** 63

bearing cap (bearing bracket cap) (rotating machinery). A cover for the bearing enclosure of a bearing bracket type machine or the removable upper half of the enclosure for a bearing. *See:* **bearing.** 63

bearing cartridge (rotating machinery). A complete enclosure for a ball or roller bearing, separate from the bearing bracket or end shield. *See:* **bearing.** 63

bearing clearance (rotating machinery). (1) The difference between the bearing inner diameter and the journal diameter. (2) The total distance for axial movement permitted by a double-acting thrust bearing. *See:* **bearing.** 63

bearing dust-cap (rotating machinery). A removable cover to prevent the entry of foreign material into the bearing. *See also:* **bearing.** 63

bearing error curve (direction-finding). (1) A plot of the combined instrumental bearing error (of the equipment) and site error versus indicated bearings. (2) A plot of the instrumental bearing errors versus either indicated or correct bearings. *See also:* **navigation.** 278, 187, 13

bearing housing (rotating machinery). A structure supporting the actual bearing liner or ball or roller bearing in a bearing assembly. *See also:* **bearing.** 63

bearing insulation (rotating machinery). Insulation that prevents the circulation of stray currents by electrically insulating the bearing from its support. *See also:* **bearing.** 63

bearing liner (rotating machinery). The assembly of a bearing shell together with its lining. *See also:* **bearing.** 63

bearing lining (rotating machinery). The element of the journal bearing assembly in which the journal rotates. *See also:* **bearing.** 63

bearing locknut (rotating machinery). A nut that holds a ball or roller bearing in place on the shaft. *See also:* **bearing.** 63

bearing lock washer (rotating machinery). A washer between the bearing locknut and the bearing that prevents the locknut from turning. *See also:* **bearing.** 63

bearing offset, indicated (direction-finding) (electronic navigation). The mean difference between the indicated and apparent bearings of a number of signal sources, the sources being substantially uniformly distributed in azimuth. *See also:* **navigation.** 278, 187, 172

bearing oil seal (bearing seal) (rotating machinery). *See:* **oil seal.**

bearing oil system (oil-circulating system) (rotating machinery). All parts that are provided for the flow, treatment, and storage of the bearing oil. *See also:* **oil cup (rotating machinery).** 63

bearing pedestal (rotating machinery). A structure mounted from the bedplate or foundation of the machine to support a bearing, but not including the bearing. *See also:* **bearing.** 63

bearing-pedestal cap (rotating machinery). The top part of a bearing pedestal. *See also:* **bearing.** 63

bearing, reciprocal (electronic navigation). *See:* **reciprocal bearing.**

bearing reservoir (oil tank) (oil well) (rotating machinery). A container for the oil supply for the bearing. It may be a sump within the bearing housing. *See also:* **oil cup (rotating machinery).** 63

bearing seal (bearing oil seal) (rotating machinery). *See:* **oil seal.**

bearing seat (rotating machinery). The surface of the supporting structure for the bearing shell. *See also:* **bearing.** 63

bearing sensitivity (electronic navigation). The minimum field strength input to a direction-finder system to obtain repeatable bearings within the bearing accuracy of the system. *See also:* **navigation.** 187, 172

bearing shell (rotating machinery). The element of the journal bearing assembly that supports the bearing lining. *See also:* **bearing.** 63

bearing shoe (rotating machinery). *See:* **segment shoe.**

bearing-temperature detector (rotating machinery). A temperature detector whose sensing element is mounted at or near the bearing surface. *See also:* **bearing.** 63

bearing-temperature relay (bearing thermostat) (rotating machinery). A relay whose temperature sensing element is mounted at or near the bearing surface. *See also:* **bearing.** 63

bearing thermometer (rotating machinery). A thermometer whose temperature sensing element is mounted at or near the bearing surface. *See also:* **bearing.** 63

bearing thermostat (rotating machinery). *See:* **bearing-temperature relay.**

beating. A phenomenon in which two or more periodic quantities of different frequencies produce a resultant having pulsations of amplitude. *See also:* **beats; signal wave.**   339

beat note. The wave of difference frequency created when two sinusoidal waves of different frequencies are supplied to a nonlinear device. *See also:* **radio receiver.**   339

beat reception. *See:* **heterodyne reception.**

beats (1) (general). Periodic variations that result from the superposition of waves having different frequencies. *Note:* The term is applied both to the linear addition of two waves, resulting in a periodic variation of amplitude, and to the nonlinear addition of two waves, resulting in new frequencies, of which the most important usually are the sum and difference of the original frequencies. *See also:* **signal wave.**   244, 55

(2) (automatic control). Periodic variations that result from the superposition of periodic signals having different frequencies.   56

bel. The fundamental division of a logarithmic scale for expressing the ratio of two amounts of power, the number of bels denoting such a ratio being the logarithm to the base 10 of this ratio. *Note:* With P_1 and P_2 designating two amounts of power and N the number of bels denoting their ratio, $N = \log_{10}(P_1/P_2)$ bels.   111

bell box (ringer box). An assemblage of apparatus, associated with a desk stand or hand telephone set, comprising a housing (usually arranged for wall mounting) within which are those components of the telephone set not contained in the desk stand or hand telephone set. These components are usually one or more of the following: induction coil, capacitor assembly, signaling equipment, and necessary terminal blocks. In a magneto set a magneto and local battery may also be included. *See also:* **telephone station.**   328

bell crank. A lever with two arms placed at an angle diverging from a given point, thus changing the direction of motion of a mechanism.   79, 27

bell crank hanger. A support for a bell crank.   27

belt (rotating machinery). A continuous flexible band of material used to transmit power between pulleys by motion.   63

belt-drive machine (elevators). An indirect-drive machine having a single belt or multiple belts as the connecting means. *See also:* **driving machine (elevators).**   328

belted-type cable. A multiple-conductor cable having a layer of insulation over the assembled insulated conductors. *See also:* **power distribution, underground construction.**   64

belt insulation (rotating machinery). A form of overhang packing inserted circumferentially between adjacent layers in the winding overhang. *See also:* **rotor (rotating machinery); stator.**   63

belt leakage flux (rotating machinery). The low-order harmonic airgap flux attributable to the phase belts of a winding. The magnitude of this leakage flux varies with winding pitch. *See also:* **rotor (rotating machinery); stator.**   63

benchboard. a combination of a control desk and a vertical switchboard in a common assembly.   103, 202

benchmark problem (computers). A problem used to evaluate the performance of computers relative to each other.   255, 77

bend (electronic navigation). A departure of the course line from the desired direction at such a rate that it can be followed by the vehicle. *Note:* Sometimes spuriously produced by defects in the design of the receiver. *See also:* **navigation.**   187

bend amplitude (navigation) (electronic navigation). The measured maximum amount of course deviation d ie to bend; measurement is made from the nominal or bend-free position of the course. *See also:* **bend; bend frequency; navigation.**   187

bend frequency (electronic navigation). The frequency at which the course indicator oscillates when the vehicle track is straight and the course contains bends; bend frequency is a function of vehicle velocity. *See also:* **navigation.**   187

bend ratio (cable plowing). The radius of a bend (segment of a circle) divided by the outside diameter of a cable, pipe, etcetera.   52

bend reduction factor (electronic navigation). The ratio of bend amplitude existing before the introduction of bend-reducing features to that existing afterward. *See also:* **navigation.**   187

bend, waveguide (waveguide components). A section of waveguide or transmission line in which the direction of the longitudinal axis is changed. In common usage the waveguide corner formed by an abrupt change in direction is considered to be a bend.   166, 179

beta (circuits and systems). The ratio of the collector current to the base current of a bipolar transistor, commonly referred to as either the common-emitter current gain or the current amplification factor.   67

beta (β) circuit (feedback amplifier). That circuit that transmits a portion of the amplifier output back to the input. *See also:* **feedback.**   328

betatron. An electric device in which electrons revolve in a vacuum enclosure in a circular or a spiral orbit normal to a magnetic field and have their energies continuously increased by the electric force resulting from the variation with time of the magnetic flux enclosed by their orbits. *See also:* **electron devices, miscellaneous.**   190

bevatron. A synchrotron designed to produce ions of a billion (10^9) electron-volts energy or more. *See also:* **electron devices, miscellaneous.**   190

beveled brush corners (electric machines). Where material has been removed from a corner, leaving a triangular surface. *See also:* **brush.**   279

beveled brush edges (electric machines). The removal of an edge to provide a slanting surface from which a shunt connection can be made or for clearance of pressure fingers or for any other purpose. *See also:* **brush.**   279

beveled brush ends and toes (electric machines). The angle included between the beveled surface and a plane at right angles to the length. The toe is the uncut or flat portion on the beveled end. When a brush has one or both ends beveled, the front of the brush is the short side of the side exposing the face level. *See also:* **brush.**   279

Beverage antenna. A directional antenna composed of a system of parallel, horizontal conductors from one-half to several wavelengths long, and terminated to ground at the far end in its characteristic impedance. *Syn.* **wave antenna.** *See also:* **antenna.**   111, 179

bezel (cathode-ray oscilloscope). The flange or cover used for holding an external graticule or cathode-ray tube cover in front of the cathode-ray tube. It may also be used for mounting a trace recording camera or other accessory item.   106

BF. *See:* **ballistic focusing.**

bias (1) (light emitting diodes). To influence or dispose to one direction, as, for example, with a direct voltage or with a spring.   162

(2) (telegraph transmission). A uniform displacement of like signal transitions resulting in a uniform lengthening or shortening of all marking signal intervals. *See also:* **telegraphy.**   194

(3) (semiconductor radiation detector). The voltage applied to a detector to produce the electric field to sweep out the signal charge.   118, 119, 23

(4) (accelerometer). An accelerometer output when no acceleration is applied.   46

bias current or power. The direct and/or alternating current or power required to operate a bolometer at a specified resistance under specified ambient conditions.   115

bias distortion (data transmission). A measure in the difference in the pulse width of the positive and negative pulses of a dotting signal. Usually expressed in percent of a full signal.   59

biased amplifier (semiconductor radiation detectors). An amplifier giving essentially zero output for all inputs below a threshold and having constant incremental gain for all inputs above the threshold up to a specified maximum amplitude.   23, 118, 119

biased induction. At a point in a magnetic material that is subjected simultaneously to a periodically varying magnetizing force and a biasing magnetizing force, the mean value of the magnetic induction at the point.   210

biased scheduled net interchange of a control area (electric power systems). The scheduled net interchange plus the frequency and/or other bias.   94, 200

biased telephone ringer. A telephone ringer whose clapper-driving element is normally held toward one side by mechanical forces or by magnetic means, so that the ringer will operate on alternating current or on electric pulses in one direction, but not on pulses in the other direction. *See also:* **telephone station.**

bias error (radar). A systematic error, whether due to equipment or propagation conditions. *See also:* **random errors.**   13

bias, grid, direct. *See:* **direct grid bias.**

biasing (laser gyro). The action of intentionally imposing a real or artificial rate into the laser gyro to avoid the region in which lock-in occurs.   46

biasing magnetizing force. At a point in a magnetic material that is subjected simultaneously to a periodically varying magnetizing force and a constant magnetizing force, the mean value of the combined magnetizing forces.   210

bias resistor (semiconductor radiation detector). The resistor through which bias voltage is applied to a detector.   23

bias spectrum. At reference ambient a specification of the fractions of total bias power in the dc and ac components and the frequency of the ac component. *Note:* The polarity of the dc component should also be given.   115

bias telegraph distortion. Distortion in which all mark pulses are lengthened (positive bias) or shortened (negative

bias). It may be measured with a steady stream of unbiased reversals, square waves having equal-length mark and space pulses. The average lengthening or shortening gives true bias distortion only if other types of distortion are negligible. *See also:* **modulation.**   111

bias winding (1) (saturable reactor). A control winding through which a biasing magnetomotive force is applied.   328

(2) (relay). *See:* **relay bias winding.**

biconical antenna. An antenna consisting of two conical conductors having a common axis and vertex.   111
See also: **antenna.**   179

bidirectional antenna. An antenna having two directions of maximum response. *See also:* **antenna.**   179

bidirectional bus (programmable instrumentation). A bus used by any individual device for two-way transmission of messages, that is, both input and output.   40

bidirectional diode-thyristor. A two-terminal thyristor having substantially the same switching behavior in the first and third quadrants of the principal voltage-current characteristic. *See also:* **thyristor.**   243, 66, 191

bidirectional pulses. Pulses, some of which rise in one direction and the remainder in the other direction. *See also:* **pulse.**   111

bidirectional relay or add-and-subtract relay. A stepping relay in which the rotating wiper contacts may move in either direction.   329

bidirectional transducer (bilateral transducer). A transducer that is not a unidirectional transducer. *See also:* **transducer.**   210

bidirectional triode-thyristor. A three-terminal thyristor having substantially the same switching behavior in the first and third quadrants of the principal voltage-current characteristic. *See also:* **thyristor.**   243, 66, 191

bifilar suspension. A suspension employing two parallel ligaments, usually of conducting material, at each end of the moving element.   280

bifurcated feeder. A stub feeder that connects two loads in parallel to their only power source.   103

BIL (1). *See:* **basic impulse insulation level.**

(2) (insulation strength). *See:* **preferred basic impulse insulation level.**

bilateral-area track (electroacoustics). A photographic sound track having the two edges of the central area modulated according to the signal. *See also:* **phonograph pickup.**   176

bilateral network. A network that is not a unilateral network. *See also:* **network analysis.**   210

bilateral transducer. A transducer capable of transmission simultaneously in both directions between at least two terminations.   252

bilevel operation. Operation of a storage tube in such a way that the output is restricted to one or the other of two permissible levels. *See also:* **storage tube.**   174, 190

billing demand (electric power utilization). The demand that is used to determine the demand charges in accordance with the provisions of a rate schedule or contract. *See also:* **alternating-current distribution.**   200

bimetallic element. An actuating element consisting of two strips of metal with different coefficients of thermal expansion bound together in such a way that the internal strains caused by temperature changes bend the compound strip. *See also:* **relay.** 259

bimetallic thermometer. A temperature-measuring instrument comprising an indicating pointer and appropriate scale in a protective case and a bulb having a temperature-sensitive bimetallic element. The bimetallic element is composed of two or more metals mechanically associated in such a way that relative expansion of the metals due to temperature change produces motion. 7

binary (computers). (1) Pertaining to a characteristic or property involving a selection, choice, or condition in which there are two possibilities. (2) Pertaining to the numeration system with a radix of two. *See also:* **column binary; positional notation; row binary.** 255, 77

binary cell (computers). An elementary unit of storage that can be placed in either of two stable states. *Note:* It is therefore a storage cell of one binary digit capacity, for example, a single-bit register. 210, 255, 77

binary code. (1) A code in which each code element may be either of two distinct kinds or values, for example, the presence or absence of a pulse. 59, 194
(2) A code that makes use of members of an alphabet containing exactly two characters, usually 0 and 1. The binary number system is one of many binary codes. *See also:* **information theory; pulse; reflected binary code.**
77, 224, 207

binary-coded-decimal. Pertaining to a number-representation system in which each decimal digit is represented by a unique arrangement of binary digits (usually four), for example, in the 8-4-2-1 binary-coded-decimal notation, the number 23 is represented as 0010 0011 whereas in binary notation, 23 is represented as 10111.
235, 255, 77

binary-coded-decimal number (BCD). The representation of the cardinal numbers 0 through 9 by 10 binary codes of any length. Note that the minimum length is four and that there are over 29×10^9 possible four-bit binary-coded-decimal codes.
Note: An example of 8-4-2-1 binary-coded decimal code follows for numbers 0 through 9.

Number	$r4$	$r3$	$r2$	$r1$
0	0	0	0	0
1	0	0	0	1
2	0	0	1	0
3	0	0	1	1
4	0	1	0	0
5	0	1	0	1
6	0	1	1	0
7	0	1	1	1
8	1	0	0	0
9	1	0	0	1

Where $r1$ is termed the **least significant binary digit (bit).** *See also:* **digital.** 59

binary digit. A character used to represent one of the two digits in the numeration system with a radix of two. Abbreviated **bit.** *See:* **equivalent binary digits.** 255, 77

binary number. Loosely, a binary numeral. 255, 77

binary number system. *See:* **positional notation.**

binary numeral. The binary representation of a number, for example, 101 is the binary numeral and V is the Roman numeral of the number of fingers on one hand. 255, 77

binary point. *See:* **point.**

binary search. A search in which a set of items is divided into two parts, one part is rejected, and the process is repeated on the accepted part until those items with the desired property are found. *See:* **dichotomizing search.**
255, 77

binary word (power-system communication). A binary code of stated length given a specific meaning. *See also:* **code character; digital.** 59

binder (bond) (1) (rotating machinery). A solid, liquid, or semiliquid composition that exhibits marked ability to act as an adhesive, and that, when applied to wires, insulation components, or other parts, will solidify, hold them in position, and strengthen the structure. 63
(2) (electroacoustics). A resinous material that causes the various materials of a record compound to adhere to one another. *See also:* **phonograph pickup.** 176

binder, load. *See:* **load binder.**

binding band (rotating machinery). A band of material, encircling stator or rotor windings to restrain them against radial movement. *See:* **rotor (rotating machinery).** 63

binding post. *See:* **binding screw.**

binding screw (binding post) (terminal screw) (clamping screw). A screw for holding a conductor to the terminal of a device or equipment. 328

binocular visual field. That portion of the visual field where the fields of the two eyes overlap. *Note:* It has a half angle of roughly 60 degrees. *See also:* **visual field.** 167

binomial array. A linear array in which the currents in successive elements are made proportional to the binomial coefficients of $(x + y)^{n-1}$ for the purpose of reducing minor lobes. *See also:* **antenna.** 179

bioelectric null (zero lead) (medical electronics). A region of tissue or other area in the system, which has such electric symmetry that its potential referred to infinity does not significantly change. *Note:* This may or may not be ground potential. 192

Biot-Savart law. *See:* **magnetic field strength produced by an electric current.**

biparting door (elevator). A vertically sliding or a horizontally-sliding door, consisting of two or more sections so arranged that the sections or groups of sections open away from each other and so interconnected that all sections operate simultaneously. *See also:* **hoistway (elevator or dumbwaiter).** 328

bipolar (power supplies). Having two poles, polarities, or directions. *Note:* Applied to amplifiers or power supplies, it means that the output may vary in either polarity from zero; as a symmetrical program it need not contain a direct-current component. *See:* **unipolar.** *See also:* **power supply.** 228

bipolar device (circuits and systems). An electronic device whose operation depends on the transport of both holes and electrons. 67

bipolar electrode. An electrode, without metallic connection with the current supply, one face of which acts as an anode surface and the opposite face as a cathode surface when an electric current is passed through the cell. *See also:* **electrolytic cell.** 328

bipolar electrode system (electrobiology). Either a pickup or stimulating system consisting of two electrodes whose relation to the tissue currents is roughly symmetrical. *See also:* **electrobiology.** 192

bipolar pulse (pulse terms). Two pulse waveforms of opposite polarity which are adjacent in time and which are considered or treated as a single feature. 254

bipolar video (radar). A radar video signal derived from a synchronous or non-synchronous phase detection process. Coherent detection produces one type of bipolar video. 13

biquinary. Pertaining to the number representation system in which each decimal digit N is represented by the digit pair AB, where $N = 5A + B$, and where $A = 0$ or 1 and $B = 0,1,2,3,$ or 4; for example, decimal 7 is represented by biquinary 12. This system is sometimes called a mixed-radix system having the radices 2 and 5. 255, 77

bistable (1) (general). The ability of a device to assume either of two stable states. 219, 206
(2) (industrial control). Pertaining to a device capable of assuming either one or two stable states. *See:* **control system, feedback.** 255, 77

bistable amplifier (industrial control). An amplifier with an output that can exist in either of two stable states without a sustained input signal and can be switched abruptly from one state to the other by specified inputs. *See:* **control system, feedback; rating and testing magnetic amplifiers.** 219, 206

bistable operation. Operation of a charge-storage tube in such a way that each storage element is inherently held at either of two discrete equilibrium potentials. *Note:* Ordinarily this is accomplished by electron bombardment. *See also:* **charge-storage tube.** 174, 190

bistatic radar. A radar using antennas at different locations for transmission and reception. 13

bit (1) (electronic computers). (A) An abbreviation of binary digit. (B) A single occurrence of a character in a language employing exactly two distinct kinds of characters. (C) A unit of storage capacity. The capacity, in bits, of a storage device is the logarithm to the base two of the number of possible states of the device. *See also:* **storage capacity.** 54, 235, 210
(2) (information theory). A unit of information content equal to the information content of a message the *a priori* probability of which is one-half. *Note:* If, in the definition of information content, the logarithm is taken to the base two, the result will be expressed in bits. One bit equals $\log_{10} 2 \times$ hartley.
See also: **information theory; check bit; parity bit.** 160, 194

bit error. *See:* **error rate.**

bit-parallel (programmable instrumentation). Refers to a set of concurrent data bits present on a like number of signal lines used to carry information. Bit-parallel data bits may be acted upon concurrently as a group (byte) or independently as individual data bits. 40

bit rate (data transmission). The speed at which bits are transmitted, usually expressed in bits per second. 59

bits per unit time (test, measurement and diagnostic equipment). Operating number of bits, handled by a device in a given unit of time, under specified conditions. 54

black and white. *See:* **monochrome.**

blackbody. A temperature radiator of uniform temperature whose radiant existence in all parts of the spectrum is the maximum obtainable from any temperature radiator at the same temperature. *Notes:* (1) Such a radiator is called a blackbody because it will absorb all the radiant energy that falls upon it. All other temperature radiators may be classed as nonblackbodies. They radiate less in some or all wavelength intervals than a blackbody of the same size and the same temperature. (2) The blackbody is practically realized in the form of a cavity with opaque walls at a uniform temperature and with a small opening for observation purposes. It is variously called a standard radiator or an ideal radiator. *See also:* **radiant energy.** 167

blackbody (Planckian) locus. The locus of points on a chromaticity diagram representing the chromaticities of blackbodies having various color temperatures. *See also:* **color.** 167

black compression (black saturation) (television). The reduction in gain applied to a picture signal at those levels corresponding to dark areas in a picture with respect to the gain at that level corresponding to the mid-range light value in the picture. *Notes:* (1) The gain referred to in the definition is for a signal amplitude small in comparison with the total peak-to-peak picture signal involved. A quantitative evaluation of this effect can be obtained by a measurement of differential gain. (2) The over-all effect of black compression is to reduce contrast in the low lights of the picture as seen on a monitor. *See also:* **television.** 178

black level (television). The level of the picture signal corresponding to the maximum limit of black peaks. *See also:* **television.** 178

black light. The popular term for ultraviolet energy near the visible spectrum. *Note:* For engineering purposes the wavelength range 320 to 400 nanometers has been found useful for rating lamps and their effectiveness upon fluorescent materials (excluding phosphors used in fluorescent lamps). By confining black-light applications to this region, germicidal and erythemal effects are, for practical purposes, eliminated. *See also:* **ultraviolet radiation.** 167

black-light flux. Radiant flux within the wavelength range 320 to 400 nanometers. It is usually measured in milliwatts. *Note:* The fluoren is used as a unit of blacklight flux and is equal to one milliwatt of radiant flux in the wavelength range 320 to 400 nanometers. Because of the variability of the spectral sensitivity of materials irradiated by black-light in practice, no attempt is made to evaluate black-light flux according to its capacity to produce effects. *See also:* **ultraviolet radiation.** 167

black-light flux density. Black-light flux per unit area of the surface being irradiated. It is equal to the incident black-light flux divided by the area of the surface when the flux is uniformly distributed. It usually is measured in milliwatts per square foot of black-light flux. *See also:* **ultraviolet radiation.** 167

black peak (television). A peak excursion of the picture signal in the black direction. 178

black recording (1) (amplitude-modulation facsimile system). The form of recording in which the maximum received power corresponds to the maximum density of the record medium.

(2) (frequency-shift, facsimile system). The form of recording in which the lowest received frequency corresponds to the maximum density of the record medium. *See also:* **recording (facsimile).**   12

black saturation (television). *See:* **black compression.**

black signal (at any point in a facsimile system). The signal produced by the scanning of a maximum-density area of the subject copy. *See also:* **facsimile signal (picture signal).**   12

black transmission (1) (amplitude-modulation facsimile system). The form of transmission in which the maximum transmitted power corresponds to the maximum density of the subject copy.

(2) (frequency-modulation facsimile system). The form of transmission in which the lowest transmitted frequency corresponds to the maximum density of the subject copy. *See also:* **facsimile transmission.**   12

blade (disconnecting blade) (switching device). The moving contact member that enters or embraces the contact clips. *Note:* In cutouts the blade may be a fuse carrier or fuseholder on which a nonfusible member has been mounted in place of a fuse link. When so used the nonfusible member alone is also called a blade in fuse parlance. *See also:* **relay blades.**   103, 202, 27

blade antenna. A form of monopole antenna which is blade shaped for strength and low aerodynamic drag.   111

blade control deadband (speed governing systems, hydraulic turbines). The magnitude of the change in the blade control cam follower position required to reverse the travel of the blade control servomotor. The deadband is expressed in percent of the change in cam follower position required to move the blades from extreme "flat" to extreme "steep".   8, 58

blade guide. An attachment to insure proper alignment of the blade and contact clip when closing the switch.   27, 79

blade latch (switch). A latch used on a switch to hold the switch blade in a fixed position.   79

blank (test, measurement and diagnostic equipment). (1) A place of storage where data may be stored (synonymous with space); (2) A character, used to indicate an output space on a printer in which nothing is printed; and (3) A condition of no information at all in a given column of a punched card or in a given location on perforated tape.   54

blank character. A character used to produce a character space on an output medium.   255, 77

blanked picture signal (television). The signal resulting from blanking a picture signal. *Notes:* (1) Adding synchronizing signal to the blanked picture signal forms the composite picture signal. (2) This signal may or may not contain setup. A blanked picture signal with setup is commonly called a non-composite signal. *See also:* **television.**   178, 87

blanketing. The action of a powerful radio signal or interference in rendering a receiving set unable to receive desired signals. *See also:* **radiation.**   328

blanking (1) (general). The process of making a channel or device noneffective for a desired interval.

(2) (television). Blanking is the substitution for the picture signal, during prescribed intervals, of a signal whose instantaneous amplitude is such as to make the return trace invisible. *See also:* **television.**   328

(3) (oscilloscopes). Extinguishing of the spot. Retrace blanking is the extinction of the spot during the retrace portion of the sweep waveform. The term does not necessarily imply blanking during the holdoff interval or while waiting for a trigger in a triggered sweep system.   185

blanking, chopped. *See:* **blanking; chopping transient.**

blanking, chopping transient. The process of blanking the indicating spot during the switching periods in chopped display operation.   106

blanking, deflection (oscilloscope). Blanking by means of a deflection structure, in the cathode-ray tube electron gun which traps the electron beam inside the gun, to extinguish the spot, permitting blanking during retrace and between sweeps regardless of intensity setting.   106

blanking level (television). That level of a composite picture signal which separates the range containing picture information from the range containing synchronizing information. *Note:* This term should be used for controls performing this function.   87

blanking signal (television). A wave constituted of recurrent pulses, related in time to the scanning process, used to effect blanking. *Note:* In television, this signal is composed of pulses at line and field frequencies, which usually originate in a central synchronizing generator and are combined with the picture signal at the pickup equipment in order to form the blanked picture signal. The addition of synchronizing signal completes the composite picture signal. The blanking portion of the composite picture signal is intended primarily to make the return trace on a picture tube invisible. The same blanking pulses or others of somewhat shorter duration are usually used to blank the pickup device also. *See also:* **television.**   337

blanking, transient. *See:* **blanking, chopping transient.**

blaster. *See:* **blasting unit.**

blasting circuit. A shot-firing cord together with connecting wires and electric blasting caps used in preparation for the firing of a blast in mines, quarries, and tunnels. *See also:* **blasting unit.**   328

blasting switch. A switch used to connect a power source to a blasting circuit. *Note:* A blasting switch is sometimes used to short-circuit the leading wires as a safeguard against premature blasts. *See also:* **blasting unit.**   328

blasting unit (blaster) (exploder) (shot-firing unit). A portable device including a battery or a hand-operated generator designed to supply electric energy for firing explosive charges in mines, quarries, and tunnels.   328

bleaching (laser-maser). The decrease of optical **absorption** produced in a medium by radiation or by external forces.   363

bleeder. A resistor connected across a power source to improve voltage regulation, to drain off the charge remaining in capacitors when the power is turned off, or to protect equipment from excessive voltages if the load is removed or substantially reduced.   328

blemish (television). A small area brightness gradient in the reproduced picture, not present in the original scene.   87

blemish charge (storage tubes). A localized imperfection of the storage assembly that produces a spurious output.

See also: **storage tube.** 174, 190, 125

blind speed (radar systems using moving target indicators). The radial velocity of a target with respect to the radar for which the response is approximately zero. *Note:* In a coherent moving target indicator system using a uniform repetition rate, the blind speed is the radial velocity at which the moving target changes distance by one-half wavelength (or multiples thereof) during each pulse period. *See also:* **navigation.** 187, 13

blinding glare. Glare that is so intense that for an appreciable length of time no object can be seen. *See also:* **visual field.** 167

blinker signal. *See:* **Morse signal light.**

blinking (pulse navigation systems). A method of providing information by modifying the signal at its source so that the signal presentation on the display at the receiver alternatively appears and disappears; for example, in loran, blinking is used to indicate that a station is malfunctioning. *See also:* **navigation.** 13, 187

blip (radar display). A deflection, or a spot of contrasting luminescence, caused by the presence of a target. *See also:* **navigation.** 13, 187

blip/scan ratio (radar). The fraction of scans on which a blip is observed at a given range. *Note:* Corresponds to probability of detection when observer's integration time is less than scan period. 13

block (1) (data processing and computation). (A) A set of things, such as words, characters, or digits, handled as a unit. (B) A collection of contiguous records recorded as a unit. (C) In data communication, a group of contiguous characters formed for transmission purposes. *Notes:* (1) Blocks are separated by interblock gaps and each block may contain one or more records. (2) The groups are separated by interblock characters. 77

(2) (railway practice). A length of track of defined limits on which the movement of trains is governed by block signals, cab signals, or both. *See also:* **absolute block.** 328

(3) (conductor stringing equipment). A device designed with one or more singles sheaves, a wood or steel shell, and an attachment hook or shackle. When rope is reeved through two of these devices, the assembly is commonly referred to as a "block and tackle." A "Set of 4's" refers to a "block and tackle" arrangement utilizing two 4-inch double sheave blocks to obtain four load bearing lines. Similarly, a "Set of 5's or a "Set of 6's" refers to the same number of load bearing lines obtained using two 5-inch or two 6-inch double sheave blocks respectively. 45

(4) (relaying). An output signal of constant amplitude and specified polarity derived from an alternating input and with the duration controlled by the polarity of the input quantity. 79

block-block element (relaying). A signal element in which two blocks are compared as to coincidence or sequence. 79

block cable (communication practice). A distribution cable installed on poles or outside building walls, in the interior of a block, including cable run within buildings from the point of entrance to a cross-connecting box, terminal frame, or point of connection to house cable. *See also:* **cable.** 328

block count readout. Display of the number of blocks that have been read from the tape derived by counting each block as it is read. *See:* **sequence number readout.** 224, 207

block diagram. A diagram of a system, instrument, computer, or program in which selected portions are represented by annotated boxes and interconnecting lines. 255, 77

block error (data transmission). Discrepancy of information in a block as detected by a checking code or technique. 59

blocked impedance (transducer). The input impedance of the transducer when its output is connected to a load of infinite impedance. *Note:* For example, in the case of an electromechanical transducer, the blocked electric impedance is the impedance measured at the electric terminals when the mechanical system is blocked or clamped; the blocked mechanical impedance is measured at the mechanical side when the electric circuit is open-circuited. *See also:* **self-impedance.** 176

block, hold down (conductor stringing equipment). A device designed with one or two single groove sheaves and latch release to be placed on the conductor and used as a means of holding it down. This device functions essentially as a traveler used in an inverted position. It is normally used in mid-span to control conductor uplift caused by stringing tensions, or at splicing locations to control the conductor as it is allowed to rise after splicing is completed. *Syn.* **hold down roller; hold down traveler; splice release block.** 45

block indicator. A device used to indicate the presence of a train in a block. 328

blocking (1) (tube rectifier). The prevention of conduction by means of grid or ignitor action, or both, when forward voltage is applied across a tube. 204

(2) (semiconductor rectifier). The action of a semiconductor rectifier cell that essentially prevents the flow of current. *See also:* **rectification.** 237, 66

(3) (relay system). A relaying function which prevents action that would otherwise be initiated by the relay system. *See also:* **relay.** 281, 60, 103, 127

(4) (rotating machinery). A structure or combination of parts, usually of insulating material, formed by hold coils in relative position for mechanical support. *Note:* Usually inserted in the end turns to resist forces during running and abnormal conditions. *See also:* **stator.** 63

(5) (telephone switching system). The inability of a telecommunication system to establish a connection due to the unavailability of paths. 55

blocking capacitor (1) (blocking condenser*). A capacitor that introduces a comparatively high series impedance for limiting the current flow of low-frequency alternating current or direct current without materially affecting the flow of high-frequency alternating current. *See also:* **electrolytic capacitor.** 329

*Deprecated

(2) (check valve) An asymmetrical cell used to prevent flow of current in a specified direction. *See also:* **electrolytic capacitor.** 328

blocking condenser. *See:* **blocking capacitor.**

blocking oscillator (1). A relaxation oscillator consisting of an amplifier (usually single-stage) with its output coupled back to its input by means that include capacitance, resistance, and mutual inductance. *See also:* **os-cillatory circuit.** 328

(2) (squegging oscillator). An electron-tube oscillator operating intermittently with grid bias increasing during oscillation to a point where oscillations stop, then decreasing until oscillation is resumed. *Note:* Squegge rhymes with wedge. *See also:* **oscillatory circuit.** 111

blocking period (1) (rectifier-circuit element). The part of an alternating-voltage cycle during which reverse voltage appears across the rectifier-circuit element. *Note:* The blocking period is not necessarily the same as the reverse period because of the effect of circuit parameters and semiconductor rectifier cell characteristics. *See also:* **rectifier circuit element.** 237, 66

(2) (gas tube). The part of the idle period corresponding to the commutation delay due to the action of the control grid. *See also:* **gas-filled rectifier.** 244, 190

blocking relay. An auxiliary relay whose function is to render another relay or device ineffective under specified conditions. 103, 60, 202, 127

blocking switching network (telephone switching system). A switching network in which a given outlet cannot be reached from any given inlet under certain traffic conditions. 55

block, input (test, measurement and diagnostic equipment). (1) A section of internal storage of a computer, reserved for the receiving and processing of input information (synonymous with input area); (2) A block used as an input buffer; and (3) A block of machine words, considered as a unit and intended to be transferred from an external source or storage or storage medium to the internal storage of the computer. 54

block-interval demand meter. *See:* **integrated-demand meter.** *See also;* **demand meter.**

block signal. A fixed signal installed at the entrance of a block to govern trains entering and using that block. 328

block-signal system. A method of governing the movement of trains into or within one or more blocks by block signals or cab signals. 328

block, snatch (conductor stringing equipment). A device normally designed with a single sheave, wood or steel shell and hook. One side of the shell usually opens to eliminate "threading" of the line. Commonly used for lifting loads on a single line, or as a device to control the position and/or direction of a "fall" line or pulling line. *Syn.* **skookum; washington; western.** 45

block-spike element (relaying). A signal element in which a block and a spike are compared as to coincidence. 79

block station. A place at which manual block signals are displayed. 328

Blondel diagram (rotating machinery). A phasor diagram intended to illustrate the currents and flux linkages of the primary and secondary windings of a transformer, and the components of flux due to primary and secondary winding currents acting alone. *Note:* This diagram is also useful as an aid in visualizing the fluxes in an induction motor. *See also:* **asynchronous machine.** 63

blooming (1) (radar). An increase in the blip size on the display as a result of an increase in signal intensity or its duration; may be employed in navigational systems with intensity modulation displays for the purpose of conveying information. *See also:* **navigation.** 187, 13

(2) (television picture tube). Excessive luminosity of the spot due to an excessive beam current. *See also:* **cathode-ray tubes.** 244, 190

blower blade (rotating machinery). An active element of a fan or blower. *See also:* **fan (rotating machinery).** 63

blower housing. *See:* **fan housing.**

blowoff valve (gas turbines). A device by means of which a part of the air flow bypasses the turbine(s) and/or the regenerator to reduce the rate of energy input to the turbine(s). *Note:* It may be used in the speed governing system to control the speed of the turbine(s) at rated speed when fuel flow permitted by the minimum fuel limiter would otherwise cause the turbine to operate at a higher speed. *See also:* **asynchronous machine; direct-current commutating machine.** 98, 58

blowout coil. An electromagnetic device that establishes a magnetic field in the space where an electric circuit is broken and helps to extinguish the arc by displacing it, for example, into an arc chute. *See also:* **contactor; relay.** 259, 244, 206

blowout magnet. A permanent-magnet device that establishes a magnetic field in the space where an electric circuit is broken and helps to extinguish the arc by displacing it. *See also:* **relay.** 259

blue dip (electroplating). A solution containing a mercury compound, and used to deposit mercury upon an immersed metal, usually prior to silver plating. *See also:* **electroplating.** 328

blur (null-type direction-finding system). The output (including noise) at the bearing of minimum response expressed as a percentage of the output at the bearing of maximum response. *See also:* **navigation.** 278, 187, 13

board (computing system). *See:* **problem board.** *See also:* **electronic analog computer.**

board, running. *See:* **running board.**

boatswain chair (conductor stringing equipment). A seat designed to be suspended on a line reeved through a block and attached to a pulling device to hoist a workman to an elevated position. *Syn.* **bosun chair.** 45

bobbin (1) (primary cell). A body in a dry cell consisting of a depolarizing mix molded around a central rod of carbon and constituting the positive electrode in the assembled cell. *See also:* **electrolytic cell.** 328

(2) (rotating machinery). Spool-shaped ground insulation fitting tightly on a pole piece, into which field coil is wound or placed. *See also:* **rotor (rotating machinery); stator.** 63

bobbin core. A tape-wound core in which the ferromagnetic tape has been wrapped on a form or bobbin that supplies mechanical support to the tape. *Note:* The dimensions of a bobbin are illustrated in the accompanying figure. Bobbin I.D. is the center-hole diameter (D) of the bobbin. Bobbin O.D. is the over-all diameter (E) of the bobbin. The bobbin height is the over-all axial dimension (F) of the bobbin. Groove diameter is the diameter (G) of the center portion of the bobbin on which the first tape wrap is placed. The groove width is the axial dimension (H) of the bobbin measured inside the groove at the groove diameter. 331

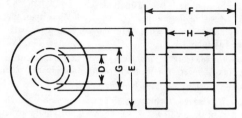

Dimensions of a bobbin.

bobbin height. *See:* **bobbin core; tape-wound core.**
bobbin I.D. *See:* **bobbin core; tape-wound core.**
bobbin O.D. *See:* **bobbin core; tape-wound core.**
Bode diagram (automatic control). A plot of log-gain and phase-angle values on a log-frequency base, for an element transfer function $G(j\omega)$, a loop transfer function $GH(j\omega)$, or an output transfer function $G(j\omega)/[1 + GH(j\omega)]$. the generalized Bode diagram comprises similar plots of functions of the complex variable $s = \sigma + j\omega$. *Note:* Except for functions containing lightly damped quadratic factors, the gain characteristic may be approximated by asymptotic straight-line segments that terminate at corner frequencies. The ordinate may be expressed as a gain, a log-gain, or in decibels as 20 times log-gain; the abscissa as cycles per unit time, radians per unit time, or as the ratio •of frequency to an arbitrary reference frequency. *See also:* **control system, feedback.**   56, 329
body capacitance. Capacitance introduced into an electric circuit by the proximity of the human body.   328
body-capacitance alarm system. A burglar alarm system for detecting the presence of an intruder through his body capacitance. *See also:* **protective signaling.**   328
body generator suspension. A design of support for an axle generator in which the generator is supported by the vehicle body. *See also:* **axle generator system.**   328
bolometer (1) (waveguide components). A term commonly used to denote the combination of a bolometer element and a bolometer mount; sometimes used imprecisely to refer to a bolometer element. *Syn.* **bolometer unit; bolometric instrument.**   166
(2) (laser-maser). A radiation detector of the thermal type in which absorbed radiation produces a measurable change in the physical property of the sensing element. The change in state is usually that of electrical resistance.   363
bolometer bridge. A bridge circuit with provisions for connecting a bolometer in one arm and for converting bolometer-resistance changes to indications of power. *See also:* **bolometric power meter.**   185
bolometer bridge, balanced. A bridge in which the bolometer is maintained at a prescribed value of resistance before and after radio-frequency power is applied, or after a change in radio-frequency power, by keeping the bridge in a state of balance. *Note:* The state of balance can be achieved automatically or manually by decreasing the bias power when the radio-frequency power is applied or increased and by increasing the bias power when the radio-frequency power is turned off, or decreased. The change in the bias power is a measure of the applied radio-frequency power. *See also:* **bolometric power meter.**   115, 185

bolometer bridge, unbalanced. A bridge in which the resistance of the bolometer changes after the radio-frequency power is applied and unbalances the bridge. The degree of bridge unbalance is a measure of the radio-frequency power dissipated in the bolometer. *See also:* **bolometric power meter.**   185
bolometer-coupler unit. A directional coupler with a bolometer unit attached to either the side arm or the main arm, normally used as a feed-through power-measuring system. *Note:* Typically, a bolometer unit is attached to the side arm of the coupler so that the radio-frequency power at the output port can be determined from a measurement of the substitution power in the side arm. This system can be used as a terminating power meter by terminating the output port of the directional coupler. *See also:* **bolometric power meter.**   185, 115
bolometer element (waveguide components). A power-absorbing element which uses the resistance change related to the temperature coefficient of resistivity (either positive or negative) as a means of measuring or detecting the power absorbed by the element. *See also:* **barretter, thermistor.** *Syn.* **bolometric detector.**   166
bolometer mount: (1) (general). A waveguide or transmission-line termination that houses a bolometer element(s). *Note:* It normally contains internal matching devices or other reactive elements to obtain specified impedance conditions when a bolometer element is inserted and appropriate bias power is applied. Bolometer mounts may be subdivided into tunable, fixed-tuned, and broad-band untuned types. *See also:* **bolometric power meter.**   185, 115
(2) (waveguide components). A waveguide or transmission line termination that can house a bolometer element.   166
bolometer unit. An assembly consisting of a bolometer element or elements and bolometer mount in which they are supported. *See also:* **bolometric power meter.**   185, 115
bolometer unit, dual element. An assembly consisting of two bolometer elements and a bolometer mount in which they are supported. *Note:* The bolometer elements are effectively in series to the bias power and in parallel to the radio frequency power.   185
bolometric detector (bolometer). The primary detector in a bolometric instrument for measuring power or current and consisting of a small resistor, the resistance of which is strongly dependent on its temperature. *Notes:* (1) Two forms of bolometric detector are commonly used for power or current measurement: (A) The barretter that consists of a fine wire or metal film; and (B) the thermistor that consists of a very small bead of semiconducting material having a negative temperature-coefficient of resistance; either is usually mounted in a waveguide or coaxial structure and connected so that its temperature can be adjusted and its resistance measured. (2) Bolometers for measuring radiant energy usually consist of blackened metal-strip temperature-sensitive elements arranged in a bridge circuit including a compensating arm for ambient temperature compensation. *See also:* **instrument; bolometric instrument.**   185
bolometric instrument (bolometer). An electrothermic instrument in which the primary detector is a resistor, the

resistance of which is temperature sensitive, and that depends for its operation on the temperature difference maintained between the primary detector and its surroundings. Bolometric instruments may be used to measure nonelectrical quantities, such as gas pressure or concentration, as well as current and radiant power. *See also:* **instrument.** 328

bolometric power meter. A device consisting of a bolometer unit and associated bolometer-bridge circuit(s).
185, 115

bolometric technique (power measurement). A technique wherein the heating effect of an unknown amount of radio-frequency power is compared with that of a measured amount of direct-current or audio-frequency power dissipated within a temperature sensitive resistance element (bolometer). *Note:* The bolometer is generally incorporated into a bridge network, so that a small change in its resistance can be sensed. This technique is applicable to the measurement of low levels of radio-frequency power, that is, below 100 milliwatts. 185

bombardment-induced conductivity (storage tubes). An increase in the number of charge carriers in semiconductors or insulators caused by bombardment with ionizing particles. *See also:* **storage tube.** 174, 190

bomb-control switch. A switch that closes an electric circuit, thereby tripping the bomb-release mechanism of an aircraft, usually by means of a solenoid. *See also:* **air transportation wiring and associated equipment.** 328

bond (electrolytic cell line). A reliable connection to assure the required electrical conductivity between conductive parts required to be electrically connected. *See also:* **binder.** 133

bonded (conductor stringing equipment). The mechanical interconnection of conductive parts to maintain a common electrical potential. *See also:* **bonding.** *Syn.* **connected.** 45

bonded motor (rotating machinery). A complete motor in which the stator and end shields are held together by a cement, or by welding or brazing. 63

bonding (1) (electric cables). The electric interconnecting of cable sheaths or armor to sheaths or armor of adjacent conductors. *See:* **cable bond cross-cable bond; continuity-cable bond.** 64
(2) (transmission and distribution). The electrical interconnecting of conductive parts, designed to maintain a common electrical potential 262

bone conduction (hearing). The process by which sound is conducted to the inner ear through the cranial bones. 176

Boolean. (1) Pertaining to the processes used in the algebra formulated by George Boole. (2) Pertaining to the operations of formal logic. 255, 77

boost. The act of increasing the power output capability of an operational amplifier by circuit modification in the output stage. *See also:* **electronic analog computer.** 9

boost charge (quick charge) (storage battery). A partial charge, usually at a high rate for a short period. *See also:* **charge.** 328

booster. An electric generator inserted in series in a circuit so that it either adds to or subtracts from the voltage furnished by another source. *See:* **direct-current commutating machine.** 328

booster coil. An induction coil utilizing the aircraft direct-current supply to provide energy to the spark plugs of an aircraft engine during its starting period. *See also:* **air transportation electric equipment.** 328

booster dynamotor. A dynamotor having a generator mounted on the same shaft and connected in series for the purpose of adjusting the output voltage. *See:* **converter.** 328

bootleg (railway techniques). A protection for track wires when the wires leave the conduit or ground near the rail. 328

bootstrap (computers). A technique or device designed to bring itself into a desired state by means of its own action, for example, a machine routine whose first few instructions are sufficient to bring the rest of itself into the computer from an input device. 255, 77

bootstrap circuit (1) (general). A single-stage electron-tube amplifier circuit in which the output load is connected between cathode and ground or other common return, the signal voltage being applied between the grid and the cathode. *Note:* The name bootstrap arises from the fact that a change in grid voltage changes the potential of the input source with respect to ground by an amount equal to the output signal. 211
(2) (circuits and systems). A circuit in which an increment of the applied input signal is partially fed back across the input impedance resulting in a higher effective input impedance. 67

bootstrap loader (test, measurement and diagnostic equipment). A routine which must be entered initially into a computer to allow the automatic loading of all subsequent instructions. 54

B operator. An operator assigned to a B switchboard. *See also:* **telephone system.** 328

bore (rotating machinery). The surface of a cylindrical hole (for example, stator bore). *See also:* **stator.** 63

bore-hole lead insulation (rotating machinery). Special insulation surrounding connections that pass through a hollow shaft. *See:* **rotor (rotating machinery).** 63

borehole cable (mining). A cable designed for vertical suspension in a borehole or shaft and used for power circuits in mines. *See also:* **mine feeder circuit.** 328

boresight. *See:* **electrical boresight; reference boresight.** 111

boresight error (antenna). The angular deviation of the electrical boresight of an antenna from its reference boresight. *See also:* **antenna.** 111, 246, 179

boresighting (radar). The process of aligning the electrical and mechanical axes of a directional antenna system, usually by an optical procedure. 13

borrow. In direct subtraction, a carry that arises when the result of the subtraction in a given digit place is less than zero. 235

bottom-car clearance (elevator). The clear vertical distance from the pit floor to the lowest structural or mechanical part, equipment, or device installed beneath the car platform, except guide shoes or rollers, safety jaw assemblies, and platform aprons or guards, when the car rests on its fully compressed buffers. *See also:* **hoistway (elevator or dumbwaiter).** 328

bottom-coil slot (radially outer-coil side) (rotating machinery). The coil side of a stator slot farthest from the

bore of the stator or from the slot wedge. *See also:* **stator.**
63

bottom-half bearing (rotating machinery). The bottom half of a split-sleeve bearing. *See also:* **bearing.** 63

bottom-terminal landing (elevators). The lowest landing served by the elevator that is equipped with a hoistway door and hoistway-door locking device that permits egress from the hoistway side. *See also:* **elevator landing.** 328

bounce (television). A transient disturbance affecting one or more parameters of the display and having duration much greater than the period of one frame. *Note:* The term is usually applied to changes in vertical position or in brightness. *See:* **television.** 178

boundary (region). Comprises all the boundary points of that region. 210

boundary lights. Aeronautical ground-lights delimiting the boundary of a land aerodrome without runways.
167

boundary marker (instrument landing systems). A radio transmitting station, near the approach end of the landing runway, that provides a fix on the localizer course. *See also:* **radio navigation.** 187, 13

boundary, *p-n* (semiconductor). A surface in the transition region between *p*-type and *n*-type material at which the donor and acceptor concentrations are equal. *See also:* **semiconductor; transistor.** 210, 245, 66

boundary point (region). A point any neighborhood of which contains points that belong and points that do not belong to the region. 210

boundary potential. The potential difference, of whatever origin, across any chemical or physical discontinuity or gradient. *See also:* **electrobiology.** 192

bounded function. A function whose absolute value never exceeds some finite number M. 210

Bourdon. A closed and flattened tube formed in a spiral, helix, or arc, which changes in shape when internal pressure changes are applied. *Note:* Bourdon tube, or simply Bourdon, has at times been used more restrictively to mean only the C-shaped member invented by Bourdon. 7

bowl (illumination techniques). An open-top diffusing glass or plastic enclosure used to shield a light source from direct view and to redirect or scatter the light. *See also:* **bare (exposed) lamp.** 167

box frame (rotating machinery). A stator frame in the form of a box with ends and sides and that encloses the stator core. *See also:* **rotor (rotating machinery).** 63

bracket (rotating machinery). A solid or skeletal structure usually consisting of a central hub and a plurality of arms extending (often radially) outward from the hub to a supporting structure. The supporting structure usually is the stator frame when the axis of the shaft is horizontal. When the axis of the shaft is vertical, the stator usually supports the upper bracket and the foundation supports the lower bracket. *See also:* **bearing bracket.** 63

bracket arm (rotating machinery). One of several structural members (beams) extending from the hub portion of a bracket to the supporting structure. The arms may be individual or parallel pairs extending radially or near-radially from the hub. 63

bracket function [x]. Equal to x if x is an integer and is the lesser of the two integers between which x falls if x is not an integer. *Note:* Thus, if $x = 2$ then $[x] = 2$, if $x = 1.5$

then $[x] = 1$, if $x = -3.7$ then $[x] = -4$ (noting that lesser means less positive). 210

bracket mast arm (illumination techniques). An attachment to a lamp post or pole from which a luminaire is suspended. *See also:* **street-lighting luminaire.** 167

bracket-type handset telephone (suspended-type handset telephone). *See:* **hang-up hand telephone set.**

Bragg region (acousto-optic device). The region that occurs when the length of the acoustic column in the direction of light propagation, L, satisfies the inequality $L > n \Lambda^2/\lambda_0$, with n the index of refraction at wavelength λ_0 and Λ the acoustic wavelength. 82

brake assembly (rotating machinery). All parts that are provided to apply braking to the rotor. *See also:* **rotor (rotating machinery).** 63

brake control (industrial control). The provision for controlling the operation of an electrically actuated brake. *Note:* Electrical energizing of the brake may either release or set the brake, depending upon its design. *See:* **electric controller.** 225, 206

brake drum. *See:* **brake ring.**

brake ring (brake drum) (rotating machinery). A rotating ring mounted on the rotor that provides a bearing surface for the brake shoes. *See also:* **rotor (rotating machinery).** 63

braking (industrial control). The control function of retardation by dissipating the kinetic energy of the drive motor and the driven machinery. *See also:* **electric drive.** 206

braking magnet. *See:* **retarding magnet.**

braking resistor. A resistor commonly used in some types of dynamic braking systems, the prime purpose of which is to convert the electric energy developed during dynamic braking into heat and to dissipate this energy to the atmosphere. *See also:* **dynamic braking.** 328

braking test (rotating machinery). (1) A test in which the mechanical power output of a machine acting as a motor is determined by the measurement of the shaft torque, by means of a brake, dynamometer, or similar device, together with the rotational speed. (2) A test performed on a machine acting as a generator, by means of a dynamometer or similar device, to determine the mechanical power input. *See also:* **asynchronous machine; direct-current commutating machine.** 63

braking torque (synchronous motor). Any torque exerted by the motor in the same direction as the load torque so as to reduce its speed. 63

branch (1) (arm) (network). A portion of a network consisting of one or more two-terminal elements in series, comprising a section between two adjacent branch-points.
210

(2) (network analysis). A line segment joining two nodes, or joining one node to itself. *See also:* **directed branch; network analysis; linear signal flow graphs.** 282

(3) (electronic computer). (A) A set of instructions that are executed between two successive decision instructions. (B) To select a branch as in (A). (C) Loosely, a conditional jump. *See:* **conditional jump.** 255

(4) (circuits and systems). A portion of a network consisting of one or more two-terminal elements, comprising a section between two adjacent branch-points. 67

branch circuit. That portion of a wiring system extending beyond the final overcurrent device protecting the circuit. *See:* **appliance branch circuit; combination lighting and appliance; lighting branch circuit; motor branch circuit; receptacle circuit.** 328

branch-circuit distribution center. A distribution center at which branch circuits are supplied. *See also:* **distribution center.** 328

branch circuit, general purpose. A branch circuit that supplies a number of outlets for lighting and appliances. 256

branch circuit, individual. A branch circuit that supplies only one utilization equipment. 256

branch circuit, multiwire. A circuit consisting of two or more ungrounded conductors having a potential difference between them, and an identified grounded conductor having equal potential difference between it and each ungrounded conductor of the circuit and that is connected to the neutral conductor of the system. 256

branch conductor (lightning protection). A conductor that branches off at an angle from a continuous run of conductor. *See also:* **lightning protection and equipment.** 328

branch current. The current in a branch of a network. 210

branch impendance. The impedance of a passive branch when no other branch is electrically connected to the branch under consideration. 210

branch input signal (network analysis). The signal *xj* at the input end of branch *jk*. *See also:* **linear signal flow graphs.** 282

branch instruction (test, measurement and diagnostic equipment). An instruction in the program that provides a choice between alternative subprograms in accordance with the test logic. 54

branch joint (1) (general). A joint used for connecting a branch conductor or cable to a main conductor or cable, where the latter continues beyond the branch. *Note:* A branch joint may be further designated by naming the cables between which it is made, for example, single-conductor cables; three-conductor main cable to single-conductor branch cable, etcetera. With the term **multiple joint** it is customary to designate the various kinds as 1-way, 2-way, 3-way, 4-way, etcetera, multiple joint. *See:* **cable joint; straight joint; reducing joint.** *See also:* **power distribution, underground construction.** 64

(2) (power cable joints). A cable joint used for connecting one or more cables to a main cable. *Note:* A branch joint may be further designated by naming the cables between which it is made, for example, single conductor cable, three conductor cable, three conductor main cable to single conductor branch, etcetera. It is customary to designate the various kinds as Y joint, T joint, H joint, cross joint, etcetera. 34

branch output signal (branch *jk*) (network analysis). The component of signal *x_k* contributed to node *k* via branch *jk*. *See also:* **linear signal flow graphs.** 282

branch point (1) (electric network). A junction where more than two conductors meet. *See also:* **network analysis; node.** 328

(2) (computers). A place in a routine where a branch is selected. *See also:* **network analysis.** 255

branch, thermoelectric. Alternative term for thermoelectric

arm. *See also:* **thermoelectric device.** 248, 191

branch transmittance (network analysis). The ratio of branch output signal to branch input signal. *See also:* **linear signal flow graphs.** 282

branch voltage. The voltage between the two nodes terminating the branch, the path of integration coinciding with the branch. 210

breadboard construction (communication practice). An arrangement in which components are fastened temporarily to a board for experimental work. 328

break (1) (circuit-opening device). The minimum distance between the stationary and movable contacts when these contacts are in the open position. (A) The length of a single break is as defined above. (B) The length of a multiple break (breaks in series) is the sum to two or more breaks. *See:* **contactor.** 225, 206

(2) (communication circuit). For the receiving operator or listening subscriber to interrupt the sending operator or talking subscriber and take control of the circuit. *Note:* The term is used especially in connection with half-duplex telegraph circuits and two-way telephone circuits equipped with voice-operated devices. *See also:* **telegraphy.** 328

breakaway. The condition of a motor at the instant of change from rest to rotation. *See also:* **asynchronous machine; direct-current commutating machine.** 63

breakaway starting current (alternating-current motor) (rotating machinery). The highest root mean square current absorbed by the motor when at rest, and when it is supplied at the rated voltage and frequency. *Note:* This is a design value and transient phenomena are ignored. 63

breakaway torque (rotating machinery). The torque that a motor is required to develop to break away its load from rest to rotation. *See:* **asynchronous machine; direct-current commutating machine.** 63

break distance (switching device). The minimum open-gap distance between the main-circuit contacts, or live parts connected thereto, when the contacts are in the open position. *Note:* In a multiple-break device, it is the sum of the breaks in series. 27, 103

breakdown (1) (gas) (electron device). The abrupt transition of the gap resistance from a practically infinite value to a relatively low value. *See:* **gas tubes.** 190, 125

(2) (semiconductor diode) (light emitting diode). A phenomenon occurring in a reverse-biased semiconductor diode (junction) the initiation of which is observed as a transition from a region of high dynamic resistance to a region of substantially lower dynamic resistance for increasing magnitude of reverse current. *See also:* **rectification; semiconductor.** 245, 66, 210, 118, 162, 119

(3) (rotating machinery). The condition of operation when a motor is developing breakdown torque. *See also:* **asynchronous machine; direct-current commutating machine.** 63

(4) (puncture) (lightning). A disruptive discharge through insulation or air. *See also:* **surge arrester (surge diverter).** 64, 62

(5) (thyristor converter). A failure that permanently deprives a rectifier diode or a thyristor of its property to block voltage in the reverse direction (reverse breakdown) or a thyristor in the forward direction (forward breakdown). 121

breakdown current (semiconductor). The current at which the breakdown voltage is measured. 66

breakdown impedance (semiconductor diode). The small-signal impedance at a specified direct current in the breakdown region. *See also:* **semiconductor.**
245, 66, 210

breakdown region (1) (semiconductor diode characteristic). The entire region of the volt-ampere characteristic beyond the initiation of breakdown for increasing magnitude of reverse current. *See also:* **rectification; semiconductor.**
237, 66, 245, 210, 119, 118

(2) (light emitting diodes). A region of the volt-ampere characteristic beyond the initiation of breakdown for an increasing magnitude of reverse current. 162

breakdown strength. *See:* **dielectric strength.**

breakdown torque (1) (rotating machinery). The maximum shaft-output torque that an induction motor (or a synchronous motor operating as an induction motor) develops when the primary winding is connected for running operation, at normal operating temperature, with rated voltage applied at rated frequency. *Note:* A motor with a continually increasing torque as the speed decreases to standstill, is not considered to have a breakdown torque.
63

(2) (electric installations on shipboard). The maximum torque of a motor which it will develop with rated voltage applied at rated frequency, without an abrupt drop in speed. 3

breakdown-torque speed (rotating machinery). The speed of rotation at which a motor develops breakdown torque. *See also:* **asynchronous machine; direct-current commutating machine.** 63

breakdown transfer characteristic (gas tube). A relation between the breakdown voltage of an electrode and the current to another electrode. *See:* **gas tubes.** 190, 125

breakdown voltage (1) (insulation). The voltage at which a disruptive discharge takes place through or over the surface of the insulation. *See also:* **insulation testing (large alternating-current rotating machinery).** 6

(2) (gas) (electron device). The voltage necessary to produce a breakdown. *See:* **gas tues.** 190

(3) (electrode of a gas tube). The voltage at which breakdown occurs to that electrode. *Notes:* (A) The breakdown voltage is a function of the other electrode voltages or currents and of the environment. (B) In special cases where the breakdown voltage of an electrode is referred to an electrode other than the cathode, this reference electrode shall be indicated. (C) This term should be used in preference to pickup voltage, firing voltage, starting voltage, etcetera, which are frequently used for specific types of gas tubes under specific conditions. *See:* **critical grid voltage (multielectrode gas tubes).** *See also;* **gas tubes.**
190, 125

breaking capacity (interrupting capacity) (industrial control). The current that the device is capable of breaking at a stated recovery voltage under prescribed conditions of use and behavior. *See:* **control.** 244, 206

breaking current (pole of a breaking device) (industrial control). The current in that pole at the instant of contact separation, expressed as a root-mean-square value. *See:* **contactor.** 244, 206

breaking point (transmission system or element thereof). A

level at which there occurs an abrupt change in distortion or noise that renders operation unsatisfactory. *See also:* **level.** 328

break-in keying. A method of operating a continuous-wave radio telegraph communication system in which the receiver is capable of receiving signals during transmission of spacing intervals. *See also:* **modulation; radio transmission.** 328

break-in period (nuclear power generating stations). That early period, beginning at some stated time and during which the failure rate of some items is decreasing rapidly. *Syn.* **early failure period.** 29, 31

breakover current (thyristor). The principal current at the breakover point. *See also:* **principal current.**
243, 191, 66, 208

breakover point (thyristor). Any point on the principal voltage-current characteristic for which the differential resistance is zero and where the principal voltage reaches a maximum value. *See also:* **principal voltage-current characteristic (principal characteristic).** 243, 191, 66

breakover voltage (thyristor). The principal voltage at the breakover point. *See also:* **principal voltage-current characteristic (principal characteristic).**
243, 191, 66, 208

breakpoint (1) (plotted curve) (automatic control). The junction of two confluent straight-line segments of a plotted curve. *Note:* In the asymptotic approximation of a log-gain versus log-frequency relation in a Bode diagram, the value of the abscissa is called the corner frequency. *See also:* **control system, feedback.** 56, 329

(2) (computer routine). (A) Pertaining to a type of instruction, instruction digit, or other condition used to interrupt or stop a computer at a particular place in a routine when manually requested. (B) A place in a routine where such an interruption occurs or can be made to occur.
235, 255, 77

breakthrough (thyristor converter). The failure of the forward-blocking action of an arm of a thyristor connection during a normal off-state period with the result that it allows on-state current to pass during a part of this period. *Note:* Breakthrough can occur in rectifier operation as well as inverter operation and for various reasons, for example, excessive virtual junction temperature, voltage surges in excess of rated peak off-state voltage, excessive rate of rise of off-state voltage, advance gating, or forward breakdown. 121

breather. A device fitted in the wall of an explosion-proof compartment, or connected by piping thereto, that permits relatively free passage of air through it, but that will not permit the passage of incendiary sparks or flames in the event of gas ignition inside the compartment. 328

breathing (carbon microphones). The phenomenon manifested by a slow cyclic fluctuation of the electric output due to changes in resistance resulting from thermal expansion and contraction of the carbon chamber. *See also:* **close-talking pressure-type microphone.** 249

breezeway (television synchronizing waveform for color transmission). The time interval between the trailing edge of the horizontal synchronizing pulse and the start of the color burst. 178

bridge (bridge network). (1) A network with a minimum of two ports or terminal pairs capable of being operated

in such a manner that when power is fed into one port, by suitable adjustment of the elements in the network or of elements connected to one or more other ports, zero output can be obtained at another port. Under these conditions, the bridge is balanced. *Note:* The described networks include hybrids and two-channel nulling circuits.   185
(2) An instrument or an intermediate means in a measurement system that embodies part or all of a bridge circuit, and by means of which one or more of the electrical constants of a bridge circuit may be measured. *Note:* The operation of a bridge consists of the insertion of a suitable electromotive force and a suitable detecting device in branches that can be made conjugate and that do not include the branch whose constants are to be measured, followed by the adjustment of one or more of the branches until the response of the detecting device becomes zero or an amount measurable by the detector for the purpose of interpolation.   328

bridge control.   Apparatus and arrangement providing for direct control from the bridge or wheelhouse of the speed and direction of a vessel.   328

bridge current (power supply).   The circulating control current in the comparison bridge. *Note:* Bridge current equals the reference voltage divided by the reference resistor. Typical values are 1 milliampere and 10 milliamperes, corresponding to control ratios of 1000 ohms per volt and 100 ohms per volt, respectively. *See also:* **power supply.**   228, 186

bridged-T network.   A T network with a fourth branch connected across the two series arms of the T, between an input terminal and an output terminal. *See also:* **network analysis.**   210

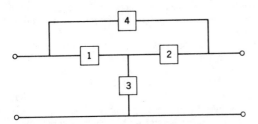

Bridged-T network. Branches 1, 2, and 3 comprise the T network and branch 4 is the fourth branch.

bridge duplex system.   A duplex system based on the Wheatstone bridge principle in which a substantial neutrality of the receiving apparatus to the sent currents is obtained by an impedance balance. *Note:* Received currents pass through the receiving relay that is bridged between the points that are equipotential for the sent currents. *See also:* **telegraphy.**   328

bridge limiter.   *See:* **limiter circuit.** *See also:* **electronic analog computer.**

bridge network.   Any network (of relatively few branches), containing only two-terminal elements, that is not a series-parallel network. *Note:* Elementary bridges such as the Wheatstone bridge are special cases of a bridge network. *See also:* **network analysis.**   210

bridge rectifier (power semiconductor).   A rectifier unit which makes use of a bridge-rectifier circuit.   66

bridge rectifier circuit.   A full-wave rectifier with four rectifying elements connected as the arms of a bridge circuit. *See also:* **rectifier; single-way rectifier circuit; double-way rectifier circuit.**   111

bridge transition.   A method of changing the connection of motors from series to parallel in which all of the motors carry like currents throughout the transfer due to the Wheatstone bridge connection of motors and resistors. *See also:* **multiple-unit control.**   328

bridging (1) (signal circuits).   The shunting of one signal circuit by one or more circuits usually for the purpose of deriving one or more circuit branches. *Note:* A bridging circuit often has an input impedance of such a high value that it does not substantially affect the circuit bridged.   239

(2) (soldered connections).   Solder that forms an unwanted conductive path. *See also:* **soldered connections (electronic and electrical applications).**   284

(3) (relays).   *See:* **relay bridging.**

bridging amplifier.   *See:* **amplifier, bridging.**

bridging connection.   A parallel connection by means of which some of the signal energy in a circuit may be withdrawn, frequently with imperceptible effect on the normal operation of the circuit.   328

bridging gain.   The ratio of the signal power a transducer delivers to its load (Z_B) to the signal power dissipated in the main circuit load (Z_M) across which the imput of the transducer is bridged. *Note:* Bridging gain is usually expressed in decibels. See accompanying figure.   239

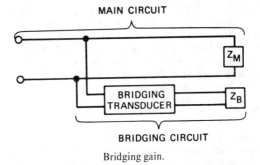

Bridging gain.

bridging loss.   (1) The ratio of the signal power delivered to that part of the system following the bridging point before the insertion of the bridging element to the signal power delivered to the same part after the bridging. (2) The ratio of the power dissipated in a load B across which the input of a transducer is bridged, to the power the transducer delivers to its load A. *Notes:* (1) The inverse of bridging gain. Bridging loss is usually expressed in decibels. (2) It may be considered a special case of insertion loss. *See also:* **transmission loss.**   239, 59

bridle wire.   Insulated wire for connecting conductors of an open wire line to associated pole-mounted apparatus. *See also:* **open wire.**   328

bright dip (electroplating).   A dip used to produce a bright surface on a metal. *See also:* **electroplating.**   328

brightener (electroplating).   An addition agent used for the purpose of producing bright deposits. *See also:* **electroplating.**   328

brightness. The attribute of visual perception in accordance with which an area appears to emit more or less light. *Note:* Luminance is recommended for the photometric quantity that has been called brightness. Luminance is a purely photometric quantity. Use of this name permits brightness to be used entirely with reference to the sensory response. The photometric quantity has been often confused with the sensation merely because of the use of one name for two distinct ideas. Brightness will continue to be used, properly, in nonquantitative statements, especially with reference to sensations and perceptions of light. Thus, it is correct to refer to a brightness match, even in the field of a photometer, because the sensations are matched and only by inference are the photometric quantities (luminances) equal. Likewise, a photometer in which such matches are made will continue to be called an equality-of-brightness photometer. A photoelectric instrument, calibrated in foot-lamberts, should not be called a brightness meter. If correctly calibrated, it is a luminance meter. A troublesome paradox is eliminated by the proposed distinction of nomenclature. The luminance of a surface may be doubled, yet it will be permissible to say that the brightness is not doubled, since the sensation that is called brightness is generally judged to be not doubled. 18, 177, 185

brightness (perceived light source color). The attribute in accordance with which the source seems to emit more or less luminous flux per unit area. *See also:* **color.** 167

brightness channel* (television). *See:* **luminance channel.** *See also:* **television.**

* Deprecated

brightness control (television). A control, associated with a picture display device, for adjusting the average luminance of the reproduced picture. *Note:* In a cathode-ray tube the adjustment is accomplished by shifting bias. This affects both the average luminance and the contrast ratio of the picture. In a color-television system, saturation and hue are also affected. 328

brightness signal.* *See:* **luminance signal.**

* Deprecated

brine. A salt solution. 328

broadband interference (measurement) (electromagnetic compatibility). A disturbance that has a spectral energy distribution sufficiently broad, so that the response of the measuring receiver in use does not vary significantly when tuned over a specified number of receiver bandwidths. *See also:* **electromagnetic compatibility.** 199

broadband radio noise. Radio noise having a spectrum broad in width as compared to the nominal bandwidth of the radio-noise meter, and whose spectral components are sufficiently close together and uniform so that the receiver cannot resolve them. *See also:* **radio-noise field strength.** 285

broadband response spectrum (nuclear power generating stations) (seismic qualification of class 1E equipment). A response spectrum that describes the motion indicating that multiple frequency excitation predominates. 28, 13

broadband tube (microwave gas tube). A gas-filled fixed-tuned tube incorporating a bandpass filter of geometry suitable for radio-frequency switching. *See:* **gas tube; pretransmit-receive tube; transmit-receive tube.** 125, 190

broadside array. A linear or planar array antenna whose direction of maximum radiation is perpendicular to the line or plane, respectively, of the array. *See also:* **antenna.** 179, 111

bronze conductor. A conductor made wholly of an alloy of copper with other than pure zinc. *Note:* The copper may be alloyed with tin, silicon, cadmium, manganese, or phosphorus, for instance, or several of these in combination. *See also:* **conductor.** 64

bronze leaf brush (rotating machinery). A brush made up of thin bronze laminations. *See also:* **brush.** 63

brush (1) (electric machines). A conductor, usually composed in part of some form of the element carbon, serving to maintain an electric connection between stationary and moving parts of a machine or apparatus. *Note:* Brushes are classified according to the types of material used, as follows: carbon, carbon-graphite, electrographitic, graphite, and metal-graphite. 279, 63
(2) (relay). *See:* **relay wiper.**

brush box (rotating machinery). The part of a brush holder that contains a brush. *See also:* **brush.** 63

brush chamfers (electric machines). The slight removal of a sharp edge. *See also:* **brush.** 279

brush contact loss (rotating machinery). The I^2R loss in brushes and contacts of the field collector ring or the direct-current armature commutator. *See also:* **brush.** 63

brush convex and concave ends (electric machines). Partially cylindrical surfaces of a given radius. *Note:* When concave bottoms are applied to bevels, both bevel angle and radius shall be given. *See also:* **brush.** 279

brush corners (electric machines). The point of intersection of any three surfaces. *Note:* They are designated as top or face corners. *See also:* **brush.** 279

brush diameter (electric machines). The dimension of the round portion that is at right angles to the length. *See also:* **brush.** 279

brush discharge (gas). An intermittent discharge of electricity having the form of a mobile brush, that starts from a conductor when its potential exceeds a certain value, but remains too low for the formation of an actual spark. *Note:* It is generally accompanied by a whistling or crackling noise. *See also:* **discharge (gas).** 210

brush edges (electric machines). The intersection of any two brush surfaces. *See also:* **brush.** 279

brush ends (electric machines). The surfaces defined by the width and thickness of the brush. *Note:* They are designated as top or holder end and bottom or commutator end. The end that is in contact with the commutator or ring is also known as the brush face. *See also:* **brush.** 279

brush friction loss (rotating machinery). The mechanical loss due to friction of the brushes normally included as part of the friction and windage loss. *See also:* **brush.** 63

brush hammer, lifting, or guide clips (electric machines). Metal parts attached to the brush that serve to accommodate the spring finger or hammer or to act as guides. *Note:* Where these serve to prevent the wear of the carbon due to the pressure finger, they are called hammer or finger clips. Rotary converter brushes may have clips that serve the dual purpose of lifting the brushes and of preventing wear from the spring finger. These are generally called lifting clips. *See also:* **brush.** 279

brush holder (rotating machinery). A structure that supports a brush and that enables it to be maintained in contact with the sliding surface. *See also:* **brush.** 63

brush-holder bolt insulation (rotating machinery). A combination of members of insulating materials that insulate the brush yoke mounting bolts, brush yoke, and brush holders. *See also:* **brush.** 63

brush-holder insulating barriers (rotating machinery). Pieces of sheet insulation installed in the brush yoke assembly to provide longer leakage paths between live parts and ground, or between live parts of different polarities. *See also:* **brush.** 63

brush-holder spindle insulation (rotating machinery). Insulation members that (when required by design) insulate the spindle on which the brush spring is mounted from the brush yoke and the brush holder. *See also:* **brush.** 63

brush-holder spring. That part of the brush holder that provides pressure to hold the brush against the collector ring or commutator. *See also:* **brush.** 63

brush holder stud (rotating machinery). An intermediate member between the brush holder and the supporting structure. *See also:* **brush.** 63

brush-holder-stud insulation (rotating machinery). An assembly of insulating material that insulates the brush holder or stud from the supporting structure. *See also:* **brush.** 63

brush-holder support (rotating machinery). The intermediate member between the brush holder or holders and the supporting structure. *Note:* This may be in the form of plates, spindles, studs, or arms. *See also:* **brush.** 63

brush-holder yoke. A rocker arm, ring, quadrant, or other support for maintaining the brush holders or brush-holder studs in their relative positions. *See also:* **brush.**

brush length (electric machines). The maximum overall dimension of the carbon only, measured in the direction in which the brush feeds to the commutator or collector ring. *See also:* **brush.** 279

brushless (rotating machinery). Applied to machines with primary and secondary or field windings that are constructed such that all windings are stationary, or in which the conventional brush gear is eliminated by the use of transformers having both moving and stationary windings, or by the use of rotating rectifiers. *See also:* **brush.** 63

brushless exciter. An alternating-current (rotating armature type) exciter whose output is rectified by a semiconductor device to provide excitation to an electric machine. The semiconductor device would be mounted on and rotate with the alternating-current exciter armature. *See:* **asynchronous machine.** 3

brushless synchronous machine. A synchronous machine having a brushless exciter with its rotating armature and semiconductor devices on a common shaft with the field of the main machine. *Note:* This type of machine with its exciter has no collector, commutator, or brushes. 3

brush or sponge plating (electroplating). A method of plating in which the anode is surrounded by a brush or sponge or other absorbent to hold electrolyte while it is moved over the surface of the cathode during the plating operation. *See also:* **electroplating.** 328

brush rigging (rotating machinery). The complete assembly of parts whose main function is to position and support all of the brushes for a commutator or collector. *See also:* **brush.** 63

brush rocker (brush yoke) (rotating machinery). A structure from which the brush holders are supported and fixed relative to each other and so arranged that the whole assembly may be moved circumferentially. *See also:* **brush.** 63

brush-rocker gear (brush-yoke gear) (rotating machinery). The worm wheel or other gear by means of which the position of the brush rocker may be adjusted. *See also:* **brush.** 63

brush shoulders (electric machines). When the top of the brush has a portion cut away by two planes at right angles to each other, this is designated as a shoulder. *See also:* **brush.** 279

brush shunt (rotating machinery). A stranded cable or other flexible conductor attached to a brush to connect it electrically to the machine or apparatus. *Note:* Its purpose is to conduct the current that would otherwise flow from the brush to the brush holder or brushholder finger. *See:* **brush.** 63

brush shunt length (electric machines). The distance from the extreme top of the brush to the center of the hole or slot in the terminal, or the center of the inserted portion of a plug terminal or, if there is no terminal, to the end of the shunt. *See also:* **brush.** 279

brush sides (electric machines). (1) Front and back (bounded by width and length). *Note:* (1) If the brush has one or both ends beveled, the short side of the brush is the front. (2) If there are no top or bottom bevels and width is greater than thickness and there is a top clip, the side to which the clip is attached is the back, except in the case of angular clips where the front or short side is determined by the slope of the clip and not by the side to which it is attached. (3) Left side and right side (bounded by thickness and length). *See also:* **brush.** 279

brush slots, grooves, and notches (electric machines). Hollows in the brush. *See also:* **brush.** 279

brush spring (rotating machinery). The portion of a brush holder that exerts pressure on the brush to hold it in contact with the sliding surface. 63

brush thickness (electric machines). The dimension at right angles to the length in the direction of rotation. *See also:* **brush.** 279

brush width (electric machines). The dimension at right angles to the length and to the direction of rotation. *See also:* **brush.** 279

brush yoke. *See:* **brush rocker.**

B scan (electronic navigation). *See:* **B display.**

B scope (class-*B* oscilloscope) (electronic navigation) (radar). A cathode-ray oscilloscope arranged to present a type *B* display. *See also:* **B display.** 13

B stage. An intermediate stage in the reaction of certain thermosetting resin in which the material swells when in contact with certain liquids and softens when heated, but may not entirely dissolve or fuse. *Note:* The resin in an uncured thermosetting moulding compound is usually in this stage. 63

B-station* (electronic navigation). *See:* **B-trace.**

* Deprecated

B switchboard (telephone switching systems). A telecommunications switchboard in a local central office, used primarily to complete calls received from other central offices. 55

B-trace (loran). The second (lower) trace on the scope display. *See also:* **navigation.** 187

Buchmann-Meyer pattern (mechanical recording). *See:* **light pattern.**

buck arm. A crossarm placed approximately at right angles to the line crossarm and used for supporting branch or lateral conductors or turning large angles in line conductors. *See also:* **tower.** 64

bucket (conductor stringing equipment). A device designed to be attached to the boom tip of a line truck, crane or aerial lift, and support a workman in an elevated working position. It is normally constructed of fiberglass to reduce its physical weight, maintain strength and obtain good dielectric characteristics. *Syn.* **basket.** 45

buffer (1) (circuit design). An isolating circuit used to avoid reaction of a driven circuit upon the corresponding driving circuit.

(2) (data processing and computation). A storage device used to compensate for a difference in rate of flow of information or time of occurrence of events when transmitting information from one device to another.
235, 210, 255, 77

(3) (elevator design). A device designed to stop a descending car or counterweight beyond its normal limit of travel by storing or by absorbing and dissipating the kinetic energy of the car or counterweight. *See also:* **elevator.** 328

(4) (relay). *See:* **relay spring stud.**

buffer amplifier (1) (general). An amplifier in which the reaction of output-load-impedance variation on the input circuit is reduced to a minimum for isolation purposes. *See also:* **amplifier.** 111, 77

(2) (analog computer). An amplifier in which the reaction of the output-load-impedance variation on the input circuit is reduced to a constant for isolation purposes on the input circuit. *See:* **unloading amplifier.** 9

buffer memory (sequential events recording systems). The memory used to compensate for the difference in rate of flow of information or time of occurrence of events when transmitting information from one device to another. *See also:* **buffer; event; storage.** 48

buffer salts. *See:* **buffers.**

buffer storage. An intermediate storage medium between data input and active storage. 224, 207

buffers (buffer salts). Salts or other compounds that reduce the changes in the pH of a solution upon the addition of an acid or alkali. *See also:* **ion.** 328

buffing (electroplating). The smoothing of a metal surface by means of flexible wheels, to the surface of which fine abrasive particles are applied, usually in the form of a plastic composition or paste. *See also:* **electroplating.** 328

bug. A semiautomatic telegraph key in which movement of a lever to one side produces a series of correctly spaced dots and movement to the other side produces a single dash. *See also:* **telegraphy.** 328

bugduster. An attachment used on shortwall mining machines to remove cuttings (bugdust) from back of the cutter and to pile them at a point that will not interfere with operation. 328

building. A structure that stands alone or that is cut off from adjoining structures by fire walls with all openings therein protected by approved fire doors. 256

building block (test, measurement and diagnostic equipment). Any programmable measurement or stimulus device, such as multimeter, power supply, switching unit, frequency meter, installed as an integral part of the automatic test equipment. 54

building bolt (rotating machinery). A bolt used to insure alignment and clamping of parts. 63

building out (communication practice). The addition to an electric structure of an element or elements electrically similar to an element or elements of the structure, in order to bring a certain property of characteristic to a desired value. *Note:* Examples are building-out capacitors, building-out sections of line, etcetera. 328

building-out capacitor (building-out condenser*). A capacitor employed to increase the capacitance of an electric structure to a desired value. 329

* Deprecated

building-out network. An electric network designed to be connected to a basic network so that the combination will simulate the sending-end impedance, neglecting dissipation, of a line having a termination other than that for which the basic network was designed. *See also:* **network analysis.** 328

building pin (rotating machinery). A dowel used to insure alignment of parts. 63

building up (electroplating). Electroplating for the purpose of increasing the dimensions of an article. *See also:* **electroplating.** 328

built-in ballast (mercury lamp). A ballast specifically designed to be built into a lighting fixture. 269

built-in check (computers). *See:* **automatic check.**

built-in-test (BIT) (test, measurement and diagnostic equipment). A test approach using **BITE** or self test hardware or software to test all or part of the unit under test. 54

built-in-test equipment (BITE) (test, measurement and diagnostic equipment). Any device which is part of an equipment or system and is used for the express purpose of testing the equipment or system. BITE is an identifiable unit of the equipment or system. *See:* **self-test.** 54

built-in transformer. A transformer specifically designed to be built into a luminaire. 274

built-up connection. A toll call that has been relayed through one or more switching points between the originating operator and the receiving exchange. *See also:* **telephone system.** 328

bulb (electron tubes and electric lamp). (1) The glass envelope used in the assembly of an electron tube or an electric lamp. (2) The glass component part used in a bulb assembly. 286

bulk storage (test, measurement and diagnostic equipment). A supplementary large volume memory or storage device. 54

bull ring. A metal ring used in overhead construction at the junction point of three or more guy wires. *See also:* **tower.** 64

bullwheel (conductor stringing equipment). A wheel incorporated as an integral part of a "bullwheel puller or tensioner" to generate pulling or braking tension on conductors and/or pulling lines through friction. A puller or tensioner normally has one or more pairs arranged in tandem incorporated in its design. The physical size of the wheels will vary for different designs, but 17 inch face widths and diameters of 5 feet are common. The wheels are power driven or retarded and lined with single or multiple groove neoprene or urethane linings. Friction is accomplished by reeving the pulling line or conductor around the groove of each pair.　　　45

bump. *See:* **distortion pulse.**

bumper (elevators). A device other than an oil or spring buffer designed to stop a descending car or counterweight beyond its normal limit of travel by absorbing the impact. *See also:* **elevator.**　　　328

buncher space (velocity-modulated tube). The part of the tube following the acceleration space where there is a high-frequency field, due to the input signal, in which the velocity modulation of the electron beam occurs. *Note:* It is the space between the input resonator grids. *See also:* **velocity-modulated tube.**　　　244, 190

bunching. The action in a velocity-modulated electron stream that produces an alternating convection-current component as a direct result of the differences of electron transit time produced by the velocity modulation. *See:* **optimum bunching; overbunching; reflex bunching; space-charge debunching; underbunching.** *See also:* **electron device.**　　　125, 190

bunching angle (electron steam) (given drift space). The average transit angle between the processes of velocity modulation and energy extraction at the same or different gaps. *See also:* **bunching angle, effective (reflex klystrons); electron device.**　　　125, 190

bunching angle, effective (reflex klystrons) (given drift space). The transit angle that would be required in a hypothetical drift space in which the potentials vary linearly over the same range as in the given space and in which the bunching action is the same as in the given space. *See:* **magnetrons.** *See also:* **electron device.**　　　190

bunching, optimum. *See:* **optimum bunching.**

bunching parameter. One-half the product of: (1) the bunching angle in the absence of velocity modulation and (2) the depth of velocity modulation. *Note:* In a reflex klystron the effective bunching angle must be used. *See also:* **electron device.**　　　125, 190

bunching time, relay. *See:* **relay bunching time.**

bundle (two conductor; three conductor; four conductor) (conductor stringing equipment). A circuit phase consisting of more than one conductor. Each conductor of the phase is referred to as a "subconductor." A "two conductor bundle" has two "subconductors" per phase. These may be arranged in a vertical or horizontal configuration. Similarly, a "three conductor bundle" has three "subconductors" per phase. These are usually arranged in a triangular configuration with the vertex of the triangle up or down. A "four conductor bundle" has four "subconductors" per phase. These are normally arranged in a square configuration. Although other configurations are possible, those listed are the most common. *Syn.* **twin-bundle; tri-bundle; quad-bundle.**　　　45

bundled conductor (transmission and distribution). An assembly of two or more conductors used as a single conductor and employing spacers to maintain a predetermined configuration. The individual conductors of this assembly are called subconductors.　　　262

B unit. A motive power unit designed primarily for use in multiple with an *A* unit for the purpose of increasing locomotive power, but not equipped for use as the leading unit of a locomotive or for full observation of the propulsion power and brake applications for a train. *Note: B* units are normally equipped with a single control station for the purpose of independent movement of the unit only, but are not usually provided with adequate instruments for full observation of power and brake applications. *See also:* **electric locomotive.**　　　328

burden (1) (instrument transformer). That property of the circuit connected to the secondary winding that determines the active and reactive power at the secondary terminals. *Note:* it is expressed either as total ohms impedance together with the effective resistance and reactance components of the impedance or as the total volt-amperes and power factor of the secondary devices and leads at the specified values of frequency and current or voltage. *See:* **instrument transformer.**　　　212, 203

(2) (power system relaying). Load imposed by a relay on an input circuit, expressed in ohms or volt-amperes. *See also:* **relay.**　　　60

(3) (relay). Load impedance imposed by a relay on an input circuit expressed in ohms and phase angle at specified conditions. *Note:* If burden is expressed in other terms such as voltamperes, additional parameters such as voltage, current, and phase angle must be specified.　　　103

burden regulation (capacitance potential device). Refers to the variation in voltage ratio and phase angle of the secondary voltage of the capacitance potential device as a function of burden variation over a specified range, when energized with constant, applied primary line-to-ground voltage. *See also:* **outdoor coupling capacitor.**　　　341

burglar-alarm system. An alarm system signaling an entry or attempted entry into the area protected by the system. *See also:* **protective signaling.**　　　328

burial depth (cable plowing). The depth of soil cover over buried cable, pipe, etcetera measured on level ground.　　　52

buried cable. A cable installed under the surface of the ground in such a manner that it cannot be removed without disturbing the soil. *See also:* **cable; power distribution, underground construction.**　　　328

burn-in (reliability). The operation of items prior to their ultimate application intended to stabilize their characteristics and to identify early failures. *See also:* **reliability.**　　　182

burnishing. The smoothing of metal surfaces by means of a hard tool or other article. *See also:* **electroplating.**　　　328

burnishing surface (mechanical recording). The portion of the cutting stylus directly behind the cutting edge, that smooths the groove. *See also:* **phonograph pickup.**　　　176

burnt deposit. A rough or noncoherent electrodeposit produced by the application of an excessive current density. *See also:* **electroplating.**　　　328

burnup, nuclear (electric power supply). (1) A measure of nuclear reactor fuel consumption, usually expressed as energy produced per unit weight of fuel exposed (megawatt-days per metric ton of fuel). (2) Percentage of fuel atoms that have undergone fission (atom percent burnup).　　　112

burst (1) (pulse techniques). A wave or waveform composed of a pulse train or repetitive waveform that starts as a prescribed time and/or amplitude, continues for a relatively short duration and/or number of cycles, and upon completion returns to the starting amplitude. *See also:* **pulse.** 185

(2) (audio and electroacoustics). An excursion of a quantity (voltage, current, or power) in an electric system that exceeds a selected multiple of the long-time average magnitude of this quantity taken over a period of time sufficiently long that increasing the length of this period will not change the result appreciably. This multiple is called the upper burst reference. *Notes:* (1) If measurements are made at different points in a system, or at different times, the same quantity must be measured consistently. (2) The excursion may be an electrical representation of a change of some other physical variable such as pressure, velocity, displacement, etcetera. 253, 176

burst build-up interval. The time interval between the burst leading-edge time and the instant at which the upper burst reference is first equaled. *See:* The figure attached to the definition of **burst duration.** *See also:* **burst.** 253, 176

burst decay interval (audio and electroacoustics). The time interval between the instant at which the peak burst magnitude occurs and the burst trailing-edge time. *See:* The figure attached to the definition of **burst duration.** *See also:* **burst.** 253, 176

burst duration (audio and electroacoustics). The time interval during which the instantaneous magnitude of the quantity exceeds the lower burst reference, disregarding brief excursions below the reference, provided the duration of any individual excursion is less than a burst safeguard interval of selected length. *Notes:* (1) If the duration of an excursion is equal to or greater than the burst safeguard interval, the burst has ended. (2) These terms, as well as those defined below, are illustrated in the accompanying figure.
(A) A burst is found with the aid of a "window" that is slid horizontally to the right with its base resting on the lower burst reference. The width of the window equals the burst safeguard interval and the height of the window equals the difference between the upper and lower burst references. The window is slid to the right until the trace crosses the top of the window. The upper burst reference has then been reached and a burst has occurred. (B) The burst leading-edge time is found by sliding the window to the left until the trace disappears from the window. The right-hand side of the window marks the burst leading-edge time. (C) The burst trailing edge time is found by a similar procedure. The window is slid to the right past its position in (A) until the trace disappears from the window. The left-hand side of the window marks the burst trailing-edge time. (D) Terms used in defining a burst: **burst leading-edge time,** t_1; **burst build-up interval,** $t_2 - t_1$; **burst rise interval,** $t_3 - t_1$; **burst trailing-edge time,** t_5; **burst decay interval,** $t_5 - t_3$; **burst fall-off interval,** $t_5 - t_4$; **burst duration,** $t_5 - t_1$; **upper burst reference,** U; **lower burst reference,** L; **long-time average power,** P. *See also:* **burst.** 253, 176

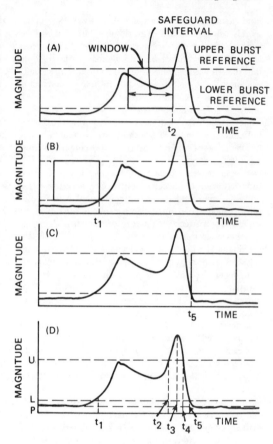

Plot of instantaneous magnitude versus time to illustrate terms used in defining a burst.

burst duty factor (audio and electroacoustics). The ratio of the average burst duration to the average spacing. *Note:* This is equivalent to the product of the average burst duration and the burst repetition rate. *See:* The figure attached to the definition of **burst duration.** *See also:* **burst.** 253, 176

burst fall-off interval (audio and electroacoustics). The time interval between the instant at which the upper burst reference is last equaled and the burst trailing edge time. *See:* The figure attached to the definition of **burst duration.** *See also:* **burst.** 253, 176

burst flag (television). A keying or gating signal used in forming the color burst from a chrominance subcarrier source. *See:* **television.** 178

burst gate (television). A keying or gating device or signal used to extract the color burst from a color picture signal. *See:* **television.** 178

burst keying signal (television). *See:* **burst flag; television.**

burst leading-edge time (audio and electroacoustics). The instant at which the instantaneous burst magnitude first equals the lower burst reference. *See:* The figure attached to the definition of **burst duration.** *See also:* **burst (audio and electroacoustics).** 253, 176

burst magnitude, instantaneous. *See:* **instantaneous burst magnitude.**

burst measurements. *See:* **energy density spectrum.**

burst quiet interval (audio and electroacoustics). The time interval between successive bursts during which the instantaneous magnitude does not equal the upper burst reference. *See:* The figure attached to the definition of **burst duration.** *See also:* **burst (audio and electroacoustics).**
253, 176

burst repetition rate (audio and electroacoustics). The average number of bursts per unit of time. *See:* The figure attached to the definition of **burst duration.** *See also:* **burst (audio and electroacoustics).**
253, 176

burst rise interval (audio and electroacoustics). The time interval between the burst leading-edge time and the instant at which the peak burst magnitude occurs. *See:* The figure attached to the definition of **burst duration.** *See also:* **burst (audio and electroacoustics).**
253, 176

burst safeguard interval (audio and electroacoustics). A time interval of selected length during which excursions below the lower burst reference are neglected; it is used in determining those instants at which the lower burst reference is first and last equaled during a burst. *See:* The figure attached to the definition of **burst duration.** *See also:* **burst (audio and electroacoustics).**
253, 176

burst spacing (audio and electroacoustics). The time interval between the burst leading-edge times of two consecutive bursts. *See:* The figure attached to the definition of **burst duration.** *See also:* **burst (audio and electroacoustics).**
253, 176

burst trailing-edge time (audio and electroacoustics). The instant at which the instantaneous burst magnitude last equals the lower burst reference. *See:* The figure attached to the definition of **burst duration.** *See also:* **burst (audio and electroacoustics).**
253, 176

burst train (audio and electroacoustics). A succession of similar bursts having comparable adjacent burst quiet intervals. *See:* The figure attached to the definition of **burst duration.** *See also:* **burst (audio and electroacoustics).**
253, 176

bus (1) (electric power). A conductor, or group of conductors, that serve as a common connection for two or more circuits.
103, 202
(2) (electronic computers). One or more conductors used for transmitting signals or power from one or more sources to one or more destinations.
235

bushing (1) (general). An insulating structure including a through conductor, or providing a passageway for such a conductor, with provision for mounting on a barrier, conducting or otherwise, for the purpose of insulating the conductor from the barrier and conducting current from one side of the barrier to the other. *See:* **floor bushing; indoor wall bushing; outdoor wall bushing; roof bushing; wall bushing.**
202, 53
(2) (rotating machinery) (electrical). Insulator to permit passage of a lead through a frame or housing.
63
(3) (relay). *See:* **relay spring stud.**

bushing potential tap (outdoor apparatus bushings). An insulated connection to one of the conducting layers of a bushing providing a capacitance voltage divider to indicate the voltage applied to the bushing.
168

bushing, rotor. *See:* **rotor bushing.**

bushing test tap (outdoor apparatus bushings). An insulated connection to one of the conducting layers of a bushing for the purpose of making insulation power factor tests.
168

bushing-type current transformer. One which has an annular core and a secondary winding insulated from and permanently assembled on the core but has no primary winding nor insulation for a primary winding. This type of current transformer is for use with a fully insulated conductor as the primary winding. A bushing-type current transformer usually is used in equipment where the primary conductor is a component part of other apparatus.
53

bushing well. An apparatus bushing having a cavity for insertion of a connector element.
134

bus line (railway terminology). A continuous electric circuit other than the electric train line, extending through two or more vehicles of a train, for the distribution of electric energy. *See also:* **multiple-unit control.**
328

bus reactor. A current-limiting reactor for connection between two different buses or two sections of the same bus for the purpose of limiting and localizing the disturbance due to a fault in either bus.
53

bus structure. An assembly of bus conductors, with associated connection joints and insulating supports.
103, 202

bus support. An insulating support for a bus. *Note:* It includes one or more insulator units with fittings for fastening to the mounting structure and for receiving the bus.
103, 202, 27

bus-type shunts (electric power systems). Instrument shunts intended for switchboard use that are designed to be installed in the bus or connection bar structure of the circuit whose current is to be measured.
201

busy hour (1) (telephone switching systems). That uninterrupted period of sixty minutes during the day when the traffic offered is a maximum.
55
(2) (data transmission). The peak 60-minute period during a 24 hour period when the largest volume of communications traffic is handled.
59

busy test (telephone switching systems). A test made to determine if certain facilities, such as a line, link, junctor, trunk, or other servers are available for use.
55, 193

busy tone. *See:* **audible busy signal.**

busy verification (telephone switching systems). A procedure for checking whether or not a called station is in use or out-of-order.
55

butt contacts (industrial control). An arrangement in which relative movement of the cooperating members is substantially in a direction perpendicular to the surface of contact. *See:* **contactor.**
244, 206

butt joint (waveguides). A connection between two waveguides or transmission lines that provides physical contact between the ends of the waveguides in order to maintain electric continuity. *See:* **waveguides.**
166

buzz (electromagnetic compatibility). A disturbance of relatively short duration, but longer than a specified value as measured under specified conditions. *Note:* For the specified values and conditions, guidance should be found in documents of the International Special Committee on Radio Interference. *See also:* **electromagnetic compatibility.**
226, 199

buzz stick. A device for testing suspension insulator units for fault when the units are in position on an energized line. *Note:* It consists of an insulating stick, on one end of which are metal prongs of the proper dimensions for spanning and short-circuiting the porcelain of one insulator unit at a time, and thereby checking conformity to normal voltage gradient. *See also:* **tower.** 64

buzzer. A signaling device for producing a buzzing sound by the vibration of an armature. 328

BW. *See:* **backward wave structure.**

by-link (telephone switching systems). A temporary connection between trunks and registers set up before the normal connection between them can be established. 55

bypass capacitor (bypass condenser*). A capacitor for providing an alternating-current path of comparatively low impedance around some circuit element. 328

* Deprecated

bypass current (series capacitor). The current flowing through the bypass device, either a gap or switch, in parallel with the series capacitor. When the bypass closes on an energized capacitor, the bypass current includes the capacitor discharge current in addition to the system rms current. 86

byte (1) (general). A sequence of adjacent binary digits operated upon as a unit and usually shorter than a word. 255, 77

(2) (programmable instrumentation). A group of adjacent binary digits operated on as a unit and usually shorter than a computer word (frequently-connotes a group of eight bits). 40

byte-serial (programmable instrumentation). A sequence of bit-parallel data bytes used to carry information over a common bus. 40

C

cab signal. A signal located in the engineman's compartment or cab indicating a condition affecting the movement of a train or engine and used in conjunction with interlocking signals and in conjunction with or in lieu of block signals. *See also:* **automatic train control.** 328

cabinet. An enclosure designed either for surface or flush mounting, and provided with a frame, mat, or trim in which swinging doors are hung. 328

cabinet for safe (burglar-alarm system). Usually a wood enclosure, having protective linings on all inside surfaces and traps on the doors, built to surround a safe and designed to produce an alarm condition in a protection circuit if any attempt is made to attack the safe. *See also:* **protective signaling.** 328

cable (1) (electric power). Either a stranded conductor (single-conductor cable), or a combination of conductors insulated from one another (multiple-conductor cable). *Note:* The first kind of cable is a single conductor, while the second kind is a group of several conductors. The component conductors of the second kind of cable may be either solid or stranded, and this kind of cable may or may not have a common insulating covering. The term cable is applied by some manufacturers to a solid wire heavily insulated and lead covered; this usage arises from the manner of the insulation, but such a conductor is not included under this definition of cable. The term cable is a general one, and in practice, it is usually applied only to the larger sizes. A small cable is called a stranded wire or a cord. Cables may be bare or insulated, and the latter may be sheathed with lead, or armored with wires or bands. 345

(2) (signal-transmission system). A transmission line or group of transmission lines mechanically assembled into a complex flexible form. *Note:* The conductors are insulated and are closely spaced and usually have a common outer cover which may be an electric portion of the cable. This definition also includes a twisted pair. 267, 188

(3) (transmission and distribution). A conductor with insulation, or a stranded conductor with or without insulation and other coverings (single-conductor cable) or a combination of conductors insulated from one another (multiple-conductor cable). *See also:* **spacer cable.** 262

cable bedding (power distribution, underground cables). A relatively thick layer of material, such as a jute serving, between two elements of a cable to provide a cushion effect, or gripping action, as between the lead sheath and wire armor of a submarine cable. *See also:* **power distribution, underground construction.** 57

cable bond. An electric connection across a joint in the armor or lead sheath of a cable, or between the armor or lead sheath and the earth, or between the armor or sheath of adjacent cables. *See:* **continuity cable bond; and cross cable bond.** *See also:* **power distribution, underground construction.** 328

cable car (conductor stringing equipment). A seat or basket shaped device designed to be suspended by a frame-work and two sheaves arranged in tandem to enable a workman to ride a single conductor, wire or cable. *Syn.* **cable trolley.** 45

cable charging current. Current supplied to an unloaded cable. *Note:* Current is expressed in rms amperes. 130, 103

cable complement (communication practice). A group of pairs in a cable having some common distinguishing characteristic. *See also:* **cable.** 328

cable core (cable). The portion lying under other elements of a cable. *See also:* **power distribution, underground construction.** 64

cable core binder. A wrapping of tapes or cords around the several conductors of a multiple-conductor cable used to hold them together. *Note:* Cable core binder is usually supplemented by an outer covering of braid, jacket, or sheath. *See also:* **power distribution, underground con-**

struction. 64

cable coupler (rotating machinery). A form of termination in which the ends of the machine winding are connected to the supply leads by means of a plug-and-socket device. 63

cable entrance fitting (pothead). A fitting used to seal or attach the cable sheath or armor to the pothead. *Note:* A cable entrance fitting is also used to attach and support the cable sheath or armor where a cable passes into a transformer removable cable terminating box without the use of potheads. *See also:* **transformer.** 288, 289

cable fill. The ratio of the number of pairs in use to the total number of pairs in a cable. *Note:* The maximum cable fill is the percentage of pairs in a cable that may be used safely and economically without serious interference with the availability and continuity of service. *See also:* **cable.** 328

cable filler. The material used in multiple-conductor cables to occupy the interstices formed by the assembly of the insulated conductors, thus forming a cable core of the desired shape (usually circular). 64

cableheads. *See:* **submersible entrance terminals (distribution oil cutouts).**

cable jacket (transmission and distribution). A protective covering over the insulation, core, or sheath of a cable. 262

cable joint (cable splice) (1) (transmission and distribution). A connection between two or more separate lengths of cable with the conductors in one length connected individually to conductors in other lengths and with the protecting sheaths so connected as to extend protection over the joint. *Note:* Cable joints are designated by naming the conductors between which the joint is made, for example, 1 single-conductor to 2 single-conductor cables; 1 single-conductor to 3 single-conductor cables; 1 concentric to 2 concentric cables; 1 concentric to 1 single-conductor cable; 1 concentric to 2 single-conductor cables; 1 concentric to 4 single-conductor cables; 1 three-conductor to 3 single-conductor cables. *See:* **branch joint; regular straight joint; reducing joint.** 64

(2) (power cable joints). A complete insulated splice or group of insulated splices contained within a single protective covering or housing. In some designs, the insulating material may also serve as the protective covering. Insulated end caps are considered joints in this context. *See also:* **straight joint; branch joint; insulating (isolating) joint; transition joint.** 34

cable Morse code. A three-element code, used mainly in submarine cable telegraphy, in which dots and dashes are represented by positive and negative current impulses of equal length, and a space by absence of current. *See also:* **telegraphy.** 328

cable rack (hanger) (shelf*). A device usually secured to the wall of a manhole, cable raceway, or building to provide support for cables. 64

* Deprecated

cable reel (mining). A drum on which conductor cable is wound, including one or more collector rings and associated brushes, by means of which the electric circuit is made between the stationary winding on the locomotive or other mining device and the trailing cable that is wound on the drum. *Note:* The drum may be driven by an electric motor, a hydraulic motor, or mechanically from an axle on the machine. *See also:* **mine feeder circuit.** 328

cable separator (power distribution, underground cables). A serving of threads, tapes, or films to separate two elements of the cable, usually to prevent contamination or adhesion. 57

cable sheath (1) (insulated conductor). A tubular impervious metallic protective covering applied directly over the cable core. 64, 57

(2) (transmission and distribution). A conductive protective covering applied to cables. *Note:* A cable sheath may consist of multiple layers of which one or more is conductive. 262

cable sheath insulator (pothead). An insulator used to insulate an electrically conductive cable sheath or armor from the metallic parts of the pothead or transformer removable cable terminating box in contact with the supporting structure for the purpose of controlling cable sheath currents. *See also:* **transformer; transformer removable cable terminating box.** 288, 289

cable shield. *See:* **duct edge shield.**

cable splicer (mining). A short piece of tubing or a specially formed band of metal generally used without solder in joining ends of portable cables for mining equipment. *See also:* **mine feeder circuit.** 328

cable spreading room (cable systems). The cable spreading room is normally the area adjacent to the control room where cables leaving the panels are dispersed into various cable trays for routing to all parts of the plant. 35

cable terminal. A structure adapted to be associated with a cable, by means of which electric connection is made available for any predetermined group of cable conductors in such a manner as to permit of their individual selection and extension by conductors outside of the cable. *See also:* **cable.** 328

cable terminal (pothead) (end bell) (power work). A device that seals the end of a cable and provides insulated egress for the conductors. 64

cable type (nuclear power generating stations). A cable type for purposes of qualification testing shall be representative of those cables having the same materials, similar construction, and service rating, as manufactured by a given manufacturer. 31

cable vault. *See also:* **manhole; power distribution, underground construction; splicing chamber.**

CADF (electronic navigation). *See:* **commutated-antenna direction-finder.**

cage (Faraday cage). A system of conductors forming an essentially continuous conducting mesh or network over the object protected and including any conductors necessary for interconnection to the object protected and an adequate ground. *See also:* **lightning protection and equipment.** 328

cage synchronous motor (rotating machinery). A salient pole synchronous motor having an amortisseur (damper) winding embedded in the pole shoes, the primary purpose of this winding being to start the motor. 63

cage winding. *See:* **squirrel-cage winding.**

caging (gyro). The process of orienting and mechanically locking one or more gyro axes or gimbal axes to a reference position. 46

calculations (International System of Units) (SI). Errors in calculations can be minimized if the base and the coherent derived SI units are used and the resulting numerical values are expressed in powers-of-ten notation instead of using prefixes. *See also:* **units and letter symbols; prefixes and symbols.**   21

calculator. (1) A device capable of performing arithmetic. (2) A calculator as in (1) that requires frequent manual intervention. (3) Generally and historically, a device for carrying out logic and arithmetic digital operations of any kind.   255

calibrate (1) (instrument). To ascertain, usually by comparison with a standard, the locations at which scale graduations should be placed to correspond to a series of values of the quantity which the instrument is to measure. *Note:* This definition of **calibrate** is to be considered as reflecting usage only in the electric instrument and meter industry and in electric metering practice.   328

(2) (nuclear power generating stations). Adjustment of the system and the determination of system accuracy using one or more sources traceable to the NBS (National Bureau of Standards).   31

calibrated (industrial control). Checked for proper operation at selected points on the operating characteristic.   206

calibrated-driving-machine test (rotating machinery). A test in which the mechanical input or output of an electric machine is calculated from the electric input or output of a calibrated machine mechanically coupled to the machine on test. *See also:* **asynchronous machine; direct-current commutating machine.**   63

calibration (1) (device). The adjustment to have the designed operating characteristics and the subsequent marking of the positions of the adjusting means, or the making of adjustments necessary to bring operating characteristics into substantial agreement with standardized scales or marking.   103, 202, 60, 127

(2) (nuclear power generating stations) (measuring and test equipment). Comparison of an item of measuring and test equipment with a reference standard or with an item of measuring and test equipment of equal or closer tolerance to detect and quantify inaccuracies and to report or eliminate those inaccuracies.   31, 41

calibration error. In operation the departure under specified conditions, of actual performance from performance indicated by scales, dials, or other markings on the device. *Note:* The indicated performance may be by calibration markings in terms of input or performance quantities (amperes, ohms, seconds, etcetera) or by reference to specific performance data recorded elsewhere. *See also:* **relay; setting error.**   281, 60, 79

calibration factor (1) (bolometer-coupler unit). The ratio of the substitution power in the bolometer attached to the side arm of the directional coupler to the microwave power incident on a nonreflecting load connected to the output port of the main arm of the directional coupler. *Note:* If the bolometer unit is attached to the main arm of the directional coupler, the calibration factor is the ratio of the substitution power in the bolometer unit attached to the main arm of the directional coupler to the microwave power incident upon a nonreflecting load connected to the

output port of the side arm of the directional coupler.   185, 115

(2) (bolometer unit). The ratio of the substitution power to the radio-frequency power incident upon the bolometer unit. The ratio of the bolometer-unit calibration factor to the effective efficiency is determined by the reflection coefficient of the bolometer unit. The two terms are related as follows:

$$K_b/\eta_e = 1 - |\Gamma|^2$$

where K_b, η_e, and Γ are the calibration factor, effective efficiency, and reflection coefficient of the bolometer unit, respectively.   185

(3) (electrothermic unit). The ratio of the substituted reference power (dc, audio, or rf) in the electrothermic unit to the power incident upon the electrothermic unit for the same dc output voltage from the electrothermic unit at a prescribed temperature. *Notes:* (A) Calibration factor and effective efficiency are related as follows:

$$\frac{K_b}{\eta_c} = 1 - |\Gamma|^2$$

where K_b, η_c and Γ are the calibration factor, effective efficiency and reflection coefficient of the electrothermic unit, respectively. (B) The reference frequency is to be supplied with the calibration factor.   47

(4) (electrothermic-coupler unit). The ratio of the substituted reference power (dc, audio, or rf) in the electrothermic unit attached to the side arm of the directional coupler to the power incident upon a nonreflecting load connected to the output port of the main arm of the directional coupler for the same dc output voltage from the electrothermic unit at a prescribed temperature. If the electrothermic unit is attached to the main arm of the directional coupler, the calibration factor is the ratio of the substituted reference (dc, audio, or rf) power in the electrothermic unit attached to the main arm of the directional coupler to the power incident upon a nonreflecting load connected to the output port of the side arm of the directional coupler for the same dc output voltage from the electrothermic unit at a prescribed temperature. *Note:* The reference frequency is to be supplied with the calibration factor.   47

calibration interval, period or cycle (test, measurement and diagnostic equipment). The maximum length of time between calibration services during which each standard and test and measuring equipment is expected to remain within specific performance levels under normal conditions of handling and use.   54

calibration level (signal generator). The level at which the signal generator output is calibrated against a standard. *See also:* **signal generator.**   185

calibration marks (radar). Indications superimposed on a display to provide a numerical scale of the parameters displayed. *See also:* **navigation.**   187, 13

calibration or conversion factor (calibration) (loosely called antenna factor). The factor or set of factors that, at given frequency, expresses the relationship between the field strength of an electromagnetic wave impinging upon the antenna of a field-strength meter and the indication of the field-strength meter. *Note:* The composite of antenna characteristic, balun and transmission line effects, receiver sensitivity and linearity, etcetera. *See also:* **measurement system.**   213

calibration procedure (test, measurement and diagnostic equipment). A document which outlines the steps and operations to be followed by standards and calibration laboratory and field calibration activity personnel in the performance of an instrument calibration.   54

calibration programming (power supplies). Calibration with reference to power-supply programming describes the adjustment of the control-bridges current to calibrate the programming ratio in ohms per volt. *Note:* Many programmable supplies incorporate a calibrate control as part of the reference resistor that performs this adjustment. *See also:* **power supply.**   228, 186

calibration scale. A set of graduations marked to indicate values of quantities, such as current, voltage, or time at which an automatic device can be set to operate.   103, 202, 60

calibration voltage. The voltage applied during the adjustment of a meter. *See also:* **test (instrument or meter).**   212

calibrator (oscilloscopes). A signal generator whose output is used for purposes of calibration, normally either amplitude or time or both.   106

caliche (cable plowing). Common sedimentary rock normally formed from ancient marine life.   52

call (1) (computers). To transfer control to a specified closed subroutine.
    (2) (communications). The action performed by the calling party, or the operations necessary in making a call, or the effective use made of a connection between two stations.   255, 101
    (3) (telephone switching systems). A demand to set up a connection.   55

call announcer (automatic telephone office). A device for receiving pulses and audibly reproducing the corresponding number in words so that it may be heard by a manual operator.   328

call circuit (manual switching). A communication circuit between switching points used by the traffic forces for the transmission of switching instructions.   328

called-line release (telephone switching systems). Release under the control of the line to which the call was directed.   55

called-line timed release (telephone switching systems). Timed release initiated by the called line.   55

call forwarding (telephone switching systems). A feature that permits a customer to instruct the switching equipment to transfer calls intended for his station to another station.   55

call indicator. A device for receiving pulses from an automatic switching system and displaying the corresponding called number before an operator at a manual switchboard.   328

calling device (telephone switching systems). An apparatus that generates the signals required for establishing connections in an automatic switching system.   193, 55

calling line identification (telephone switching systems). Means for automatically identifying the source of calls.   55

calling-line release (telephone switching systems). Release under the control of the line from which the call originated.   55

calling-line timed release (telephone switching systems). Timed release initiated by the calling line.   55

calling plug and cord. A plug and cord that are used to connect to a called line.   328

calling sequence (computers). A specified arrangement of instructions and data necessary to set up and call a given subroutine.   255, 77

call packing (telephone switching systems). A method of selecting paths in a switching network according to a fixed hunting sequence.   55

call rate (telephone switching systems). The number of calls per unit of time.   55

call splitting (telephone switching systems). Opening the transmission path between the parties of a call.   55

call tracing (telephone switching systems). A means for manually identifying the source of calls.   55

call transfer (telephone switching systems). A feature that allows a customer to instruct the switching equipment or operator to transfer his call to another station.   55

call waiting (telephone switching systems). A feature providing a signal to a busy called line to indicate that another call is waiting.   55

call-waiting tone (telephone switching systems). A tone used in the call-waiting feature.   55

calomel electrode. *See:* **calomel half-cell.**

calomel half-cell (calomel electrode). A half-cell containing a mercury electrode in contact with a solution of potassium chloride of specified concentration that is saturated with mercurous chloride of which an excess is present. *See also:* **electrochemistry.**   328

calorimeter (laser-maser). A device for measuring the total amount of energy absorbed from a source of electromagnetic radiation.   363

calorimetric test (rotating machinery). A test in which the losses in a machine are deduced from the heat produced by them. The losses are calculated from the temperature rises produced by this heat in the coolant or in the surrounding media. *See also:* **asynchronous machine; direct-current commutating machine.**   63

CAMAC (Computer Automated Measurement and Control). CAMAC is a standard modular instrumentation and digital interface system.   51

CAMAC branch driver. *See:* **CAMAC parallel highway driver.**   51

CAMAC branch highway. *See:* **CAMAC parallel highway.**   51

CAMAC compatible crate. A mounting unit for CAMAC plug-in units that does not conform to the full requirements for a CAMAC crate but in which CAMAC modules can be mounted and operated in accordance with the dataway requirements of IEEE Standard 583.   51

CAMAC crate. A mounting unit for CAMAC plug-in units that includes a CAMAC dataway and conforms to the mandatory requirements for a CAMAC crate as specified in IEEE Standard 583.   51

CAMAC crate assembly. An assembly of a CAMAC crate controller and one or more CAMAC modules mounted in a CAMAC crate (or CAMAC compatible crate), and operable in conformity with the dataway requirements of

IEEE Standard 583. 51

CAMAC crate controller. A functional unit that when mounted in the control station and one or more normal stations of a CAMAC crate (or CAMAC compatible crate) communicates with the dataway in accordance with IEEE Standard 583. 51

CAMAC dataway. An interconnection between CAMAC plug-in units which conforms to the mandatory requirements for a CAMAC dataway as specified in IEEE Standard 583. 51

CAMAC module. A CAMAC plug-in unit that when mounted in one or more normal stations of a CAMAC crate is compatible with IEEE Standard 583. 51

CAMAC parallel highway. A standard highway (for a CAMAC system) in which the data is transferred in parallel and which conforms to the requirements of IEEE Standard 596. *Syn.* **CAMAC branch highway.** 51

CAMAC parallel highway driver. A unit that communicates via the CAMAC parallel highway with up to seven CAMAC crates and conforms to the requirements as specified in IEEE Standard 596. *Syn.* **CAMAC branch driver.** 51

CAMAC plug-in unit. A functional unit that conforms to the mandatory requirements for a plug-in unit as specified in IEEE Standard 583. 51

CAMAC serial highway. A standard highway (for a CAMAC system) in which the data is transferred in bit or byte serial and which conforms to the requirements of IEEE Standard 595. 51

CAMAC system. A system including at least one CAMAC crate assembly. 51

cam contactor (cam switch). A contactor or switch actuated by a cam. *See also:* **control switch.** 1

camera storage tube. A storage tube into which the information is introduced by means of electromagnetic radiation, usually light, and read at a later time as an electric signal. *See also:* **storage tube.** 174, 190

camera tube (television). A tube for conversion of an optical image into an electric signal. 163

cam-operated switch (industrial control). A switch consisting of fixed contact elements and movable contact elements operated in sequence by a camshaft. *See:* **switch.** 244, 206

camp-on busy (telephone switching systems). A feature whereby a call encountering a busy condition can be held and subsequently connected automatically when the busy condition is removed. 55

cam-programmed (test, measurement and diagnostic equipment). (1) A programming technique that uses a rotating shaft, having specifically oriented, eccentric projections which control a series of switches that set up the proper circuits for a test; and (2) A cam-follower system used to set positions or values of a shafted instrument for programming instructions to the test system. 54

cam-shaft position (electric power system). The angular position of the main shaft directly operating the governor-controlled valves. *See also:* **speed-governing system.** 94

can (dry cell). A metal container, usually zinc, in which the cell is assembled and that serves as its negative electrode. *See also:* **electrolytic cell.** 328

cancel (numerically controlled machines). A command that will discontinue any fixed cycles or sequence commands. 224, 207

cancellation ratio (radar moving-target indicator). The ratio of (1) canceller-voltage amplification for fixed-target echoes received with a fixed antenna to (2) the gain for a single pulse passing through the undelayed channel of the canceller. *See also:* **navigation.** 187

cancelled video (radar moving-target indicator). The video output remaining after the cancellation process. *See also:* **navigation.** 13, 187

canceller (radar). That portion of the system in which cancellation processing is performed to suppress unwanted signals such as clutter, sidelobe interference, or fixed targets. *See:* **moving target indication; sidelobe canceller.** 187, 13

candela (cd). The luminous intensity, in the perpendicular direction, of a surface of 1/600 000 square meter of a blackbody at the temperature of freezing platinum under a pressure of 101 325 newtons per square meter. *Notes:* (1) Values for standards having other spectral distributions are derived by the use of accepted spectral luminous efficiency data for photopic vision. (2) From 1909 until the introduction of the present photometric system on January 1, 1948, the unit of luminous intensity in the United States, as well as in France and Great Britain, was the **international candle** which was maintained by a group of carbon-filament vacuum lamps. For the present unit as defined above, the internationally accepted term is **candela.** The difference between the candela and the old **international candle** is so small that only measurements of high precision are affected. *See:* **spectral luminous efficiency of radiant flux; spectral luminous efficiency for photopic vision.** *See also:* **light.** 167

C and I (C & I). *See:* **supervisory control point, control and indication.**

candle. *See:* **candela.**

candlepower.* Luminous intensity as measured in candelas. *See:* **candela.** 162

* Deprecated

candlepower distribution curve. A curve, generally polar, that represents the variation of luminous intensity of a lamp or luminaire in a plane through the light center. *Note:* A vertical candlepower distribution curve is obtained by taking measurements at various angles of elevation in a vertical plane through the light center; unless the plane is specified, the vertical curve is assumed to represent an average such as would be obtained by rotating the lamp or luminaire about its vertical axis. A horizontal candlepower distribution curve represents measurements made at various angles of azimuth in a horizontal plane through the light center. *See also:* **colorimetry.** 167

can loss (rotating machinery). Electric losses in a can used to protect electric components from the environment. *See also:* **asynchronous machine; direct-current commutating machine.** 63

canned (rotating machinery). Completely enclosed and sealed by a metal sheath. 63

canned cycle (numerically controlled machines). *See:* **fixed cycle.**

capability (electric power supply). The maximum load-carrying ability expressed in kilovolt-amperes or kilowatts of generating equipment or other electrical apparatus under specified conditions for a given time interval. *See:* **emergency transfer capability; extended capability; hydro capability; net capability; net assured capability; net dependable capability; normal transfer capability; pumped-storage hydro capability; short time capability; steam capability; total for load capability.** 112

capability margin (electric power supply). The difference between (1) total capability for load and (2) system load responsibility. It is the margin of capability available to provide for scheduled maintenance, emergency outages, adverse system operating requirements, and unforeseen loads. *See also:* **generating station.** 200

capacitance (capacity*) (1) (general). The property of a system of conductors and dielectrics that permits the storage of electrically separated charges when potential differences exist between the conductors. *Notes:* (A) Its value is expressed as the ratio of a quantity of electricity to a potential difference. (B) As defined elsewhere, self-, direct, balanced, and plenary are used to indicate the capacitances corresponding to different groupings of the n conductors of a multiple-conductor system; and the names are used even if, in addition to the n conductors constituting the working system, additional conductors are present that, either intentionally or because they are inaccessible, are not explicitly considered either experimentally or theoretically. Two or more conductors when connected conductively together are regarded as one conductor. Ground is regarded as one of the n conductors of the system. *See also:* **electron-tube admittances.** 3, 210

* Deprecated

(2) (semiconductor diode) (semiconductor radiation conductor). The small-signal capacitance measured between the terminals of the diode or detector under specified conditions of bias and frequency. *See also:* **rectification; semiconductor; semiconductor device.**
237, 245, 66, 210, 23, 119, 118

(3) (control systems). A property expressible by the ratio of the time integral of the flow rate of a quantity, such as heat, or electric charge to or from a storage, divided by the related potential change. *Note:* Typical units are microfarads, Btu/deg F, lb/psi, gal/ft. 56

(4) (outdoor apparatus bushings). (A) The main capacitance, Cl, of a condenser bushing is the value in picofarads between the high-voltage conductor and the potential tap or the test tap. (B) The tap capacitance, C2, of a condenser bushing is the value in picofarads between the potential tap and mounting flange (ground). (C) The capacitance, C, of a bushing without a potential or test tap is the value in picofarads between the high-voltage conductor and the mounting flange (ground).

(5) (VLF insulation testing). Capacitance, as used here, and distinguished from power-frequency capacitance, is that value which would result from a measurement at VLF, that is, 0.1 Hz ± 25 percent. In magnitude, it would tend to be greater than the power-frequency capacitance, to the extent of increased contributions made by dipole and interfacial polarizations. 135

capacitance between two conductors. The capacitance between two conductors, although it is a function only of the total geometry and the absolute capacitivity of the

medium, is defined most conveniently as the ratio of the change in charge on one conductor to the corresponding change in potential difference between it and the second conductor with the charges or the potentials of the other conductors of the system specified. *Notes:* (1) In the ideal case where one, and only one, conductor is completely surrounded by another, the capacitance between them is not affected by the presence of the other $n - 2$ conductors. (2) The capacitance between two conductors may be made independent of all except a third if the third conductor completely surrounds the two under discussion. 210

capacitance between two conductors, balanced. *See:* **balanced capacitance.**

capacitance coupling (signal-transmission system). *See:* **coupling capacitance.**

capacitance current (or component). A reversible component of the measured current on charge or discharge of the winding and is due to the geometrical capacitance, that is, the capacitance as measured with alternating current of power or higher frequencies. With high direct voltage this current has a very short time constant and so does not affect the usual measurements. *See also:* **insulation testing (large alternating-current rotating machinery).** 6

capacitance, discontinuity (waveguide or transmission line). The shunt capacitance that, when inserted in a uniform waveguide or transmission line, would cause reflected waves of the dominant mode equal to those resulting from the given discontinuity. *See also:* **waveguide.** 194

capacitance, effective. The imaginary part of a capacitive admittance divided by the angular frequency. 194

capacitance, input (n-terminal electron tubes). The short-circuit transfer capacitance between the input terminal and all other terminals, except the output terminal, connected together. *Note:* This quantity is equivalent to the sum of the interelectrode capacitances between the input electrode and all other electrodes except the output electrode. *See also:* **electron-tube admittances.** 125, 190

capacitance meter. An instrument for measuring capacitance. *Note:* If the scale is graduated in microfarads the instrument is usually designated as a microfaradmeter. *See also:* **instrument.** 328

capacitance, output (n-terminal electron tube). The short-circuit transfer capacitance between the output terminal and all other terminals, except the input terminal, connected together. *See also:* **electron-tube admittances.** 190

capacitance potential device. A voltage-transforming equipment or network connected to one conductor of a circuit through a capacitance, such as a coupling capacitor or suitable high-voltage bushing, to provide a low voltage such as required for the operation of instruments and relays. *Notes:* (1) The term **potential device** applies only to the network and is exclusive of the coupling capacitor or high-voltage bushing. (2) The term **coupling-capacitor potential device** indicates use with coupling capacitators. (3) The term **bushing potential device** indicates use with bushings. (4) The term **capacitance potential device** indicates use with any type of capacitance coupling. (5) Capacitance potential devices and their associated coupling capacitors or bushings are designed for line-to-ground connection, and not line-to-line connection. The potential device is a single-phase device, and, in combination with its coupling capacitor or bushing, is connected line-to-ground. The low voltage thus provided is a function of the line-to-ground voltage and the constants of the capacitance potential device. Two or more capacitance potential devices, in combination with their coupling capacitors or bushings, may be connected line-to-ground on different high-voltage phases to provide low voltages of other desired phase relationships. (6) Zero-sequence voltage may be obtained from the broken-delta connection of the auxiliary windings or by the use of one device with three coupling capacitors or bushings. In the latter case, the three operating-tap connection-points are joined together and one device connected between this common point and ground. Although used in combination with three coupling capacitors or bushings, the device output and accuracy rating standards are based on the single-phase conditions. *See also:* **outdoor coupling capacitor.** 341

capacitance ratio (nonlinear capacitor). The ratio of maximum to minimum capacitance over a specified voltage range, as determined from a capacitance characteristic, such as a differential capacitance characteristic, or a reversible capacitance characteristic. *See also:* **nonlinear capacitor.** 191

capacitance, short-circuit input (*n*-terminal electron device). The effective capacitance determined from the short-circuit input admittance. *See also:* **electron-tube admittances.** 125, 190

capacitance, short-circuit output (*n*-terminal electron device). The effective capacitance determined from the short-circuit output admittance. *See also:* **electron-tube admittances.** 125, 190

capacitance, short-circuit transfer (electron tubes). The effective capacitance determined from the short-circuit transfer admittance. *See also:* **electron-tube admittances.** 125, 190

capacitance, signal electrode. *See:* **electrode capacitance.**

capacitance, stray (electric circuits). Capacitance arising from proximity of component parts, wires, and ground. *Note:* It is undesirable in most circuits, although in some high-frequency applications it is used as the tuning capacitance. In bridges and other measuring equipment, its effect must be eliminated by preliminary balancing out, or known and included in the results of any measurement performed. *See also:* **measurement system.** 185

capacitance-switching transient overvoltage ratio (1) (general). The transient overvoltage ratio for a capacitance load. 103

(2) (high voltage air switches, insulators, and bus supports). The ratio of the peak value of voltage above ground, during the transient conditions resulting from the operation of the switch, to the peak value of the steady-state line-to-neutral voltage. *Note:* It is measured at either terminal of the switch, which ever is higher, and is expressed in multiples of the peak values of the operating line-to-ground voltage at the switch with the capacitance connected. 27

capacitance, target (camera tubes). *See:* **target capacitance (camera tubes).**

capacitance unbalance protection device (series capacitor). A device to detect objectionable unbalance in capacitance between capacitor groups within a phase, such as that caused by blown capacitor fuses or faulted capacitor, and

to initiate an alarm or the closing of the capacitor bypass switch. 86

capacitive current (component) (rotating machinery). A reversible component of the measured current on charge or discharge of the winding which is due to the geometrical capacitance, that is, the capacitance as measured with alternating current of power or higher frequencies. With high direct voltage this current has a very short time constant and so does not affect the usual measurements. 6

capacitivity (free space). *See:* **electric constant.** *See also:* **absolute capacitivity.**

capacitor (condenser*). A device, the primary purpose of which is to introduce capacitance into an electric circuit. Capacitors are usually classified, according to their dielectrics, as air capacitors, mica capacitors, paper capacitors, etc. 3
* Deprecated

capacitor antenna (condenser antenna*). An antenna consisting of two conductors or systems of conductors, the essential characteristic of which is its capacitance. *See also:* **antenna.** 197
* Deprecated

capacitor bank (shunt power capacitors). An assembly at one location of capacitors and all necessary accessories, such as switching equipment, protective equipment, controls, etcetera required for a complete operating installation. It may be a collection of components assembled at the operating site or may include one or more factory-assembled equipments. 138

capacitator braking (rotating machinery). A form of dynamic braking for induction motors in which a capacitator is used to magnetize the motor. *See:* **asynchronous machine.** 63

capacitor bus (series capacitor). The main conductors which serve to connect the capacitor assemblies in series relation to the line. 86

capacitor bushing (condenser bushing)* (outdoor apparatus bushings). A bushing in which cylindrical conducting layers are arranged coaxially with the conductor within the insulating material for the purpose of controlling the electric field of the bushing. 168

* Deprecated
capacitor bypass switch (series capacitor). A switch device with moving and stationary contacts that functions as a means of bypassing the capacitor. This switch may also have the capability of inserting the capacitor against a specified level of current. 86

capacitor-bypass-switch interlocking devices (series capacitor). Devices that perform the function of having all three integral bypass switches of a capacitor step take the same open or close position. 86

capacitor element (series capacitor). An individual part of a capacitor unit consisting of coiled conductors separated by dielectric material. 86

capacitor group (series capacitor). An assembly of more than one capacitor unit solidly connected in parallel between two buses or terminals. 86

capacitor enclosure. The case in which the capacitor is mounted. 328

capacitor, ideal. *See:* **ideal capacitor.**

capacitor indicating fuse (series capacitor). A capacitor-unit fuse which provides an externally visible indication when the fuse blows. 86

capacitor loudspeaker. *See:* **electrostatic loudspeaker.**

capacitor microphone. *See:* **electrostatic microphone.**

capacitor motor. A single-phase induction motor with a main winding arranged for direct connection to a source of power and an auxiliary winding connected in series with a capacitor. The capacitor may be directly in the auxiliary circuit or connected into it through a transformer. *See also:* **asynchronous machine; capacitor-start motor; permanent-split capacitor motor; two-value capacitor motors.** 63

capacitor mounting strap. A device by means of which the capacitor is affixed to the motor. 328

capacitor pickup. A phonograph pickup that depends for its operation upon the variation of its electric capacitance. *See also:* **phonograph pickup.** 290

capacitor platform (series capacitor). A structure which supports the capacitor-rack assemblies and all associated equipment, and protective device, and is supported on insulators compatible with line-to-ground insulation requirements. 86

capacitor rack (series capacitor). A frame which supports the capacitor units in multiples. 86

capacitor segment (series capacitor). An assembly of groups of capacitors, voltage-limiting device, switch, and relays to protect the capacitors from overvoltage and overloads and to provide bypass and insertion switching functions. Ground-level control means is also included. 86

capacitor start-and-run motor. A capacitor motor in which the auxiliary primary winding and series-connected capacitors remain in circuit for both starting and running. *See:* **permanent-split capacitor motor.** *See also:* **asynchronous machine.** 63

capacitor-start motor. A capacitor motor in which the auxiliary winding is energized only during the starting operation. *Note:* The auxiliary-winding circuit is open-circuited during running operation. *See also:* **asynchronous machine.** 63

capacitor switch (power switchgear). A switch capable of making and breaking capacitive currents of capacitor banks. 103

capacitor switching step (series capacitor). A function which consists of three capacitor segments, one per phase with ground-level control devices, and platform-to-ground signaling for interlocked operation when bypassing or inserting the capacitor segments. 86

capacitor unit (1) (general). A single assembly of dielectric and electrodes in a container with terminals brought out. *See also:* **alternating-current distribution; indoor capacitor unit; outdoor capacitor unit.** 138, 210

(2) (series capacitor). An assembly of one or more capacitor elements in a single container, with one or more insulated terminals brought out. 86

capacitor voltage. The voltage across two terminals of a capacitor. 191

capacity (1) (electric-power devices). (A) The capacity of a machine, apparatus, or appliance is the rated load (rated capacity). (B) The capacity of a machine, apparatus, or appliance is the maximum load of which it is capable under existing service conditions (maximum capacity). 210

(2) (data processing and computation). *See:* **storage capacity.** 235, 77

* Deprecated for (1)

(3) (data transmission). (A) Number of digita or characters in a machine word regularly handled in a computer. (B) Upper and lower limits of the numbers which may be regularly handled in a computer. (C) Maximum number of binary digits which can be transmitted by a communications channel in one second. 59

(4) (automatic control). (A) Capability of storing matter, energy, or information. (B) Capability of transmitting matter, energy, electric current, or information. (C) Quantity of volume, mass, electric charge, heat, or information which an element can store, or quantity per unit time at which it can transmit. 56

capacity factor. The ratio of the average load on a machine or equipment for the period of time considered to the capacity of the machine or equipment. *See also:* **generating station.** 64, 112

capacity, firm, purchases or sales (electric power supply). Firm capacity which is purchased, or sold, in transactions with other systems and which is not from designated units, but is from the overall system of the seller. *Note:* It is understood that the seller treats this type of transaction as a load obligation. 112

capacity, specific unit, purchases or sales (electric power supply). Capacity which is purchased, or sold, in transactions with other systems and which is from a designated unit on the system of the seller. 112

cap-and-pin insulator. An assembly of one or more shells with metallic cap and pin, having means for direct and rigid mounting. *See also:* **insulator.** 261

capture effect (1) (modulation systems). The effect occuring in a transducer (usually a demodulator) whereby the input wave having the largest magnitude controls the output. 242

(2) (radar). The tendency of a receiver to suppress the weaker of two signals within its passband. 13

capturing (gyro; accelerometer). The use of a torquer (forcer) in a servo loop to restrain a gyro gimbal or accelerometer proof mass to a specified reference position. 46

car annunciator. An electric device in the car that indicates visually the landings at which an elevator-landing signal-registering device has been actuated. *See also:* **elevator.** 328

carbon-arc lamp. An electric-discharge lamp employing an arc discharge between carbon electrodes. *Note:* One or more of these electrodes may have cores of special chemicals which contribute importantly to the radiation. *See also:* **lamp.** 167

carbon brush (motors and generators). (1) A specific type of brush composed principally of amorphous carbon. *Note:* This type of brush is usually hard and is adapted to low speeds and moderate currents. (2) A broader classification of brush, containing carbon in appreciable amount. *See:* **brush (rotating machinery).** 63

carbon-consuming cell (carbon-combustion cell). A cell for the production of electric energy by galvanic oxidation of carbon. *See also:* **electrochemistry.** 328

carbon-contact pickup. A phonograph pickup that depends for its operation upon the variation in resistance of carbon contacts. *See also:* **phonograph pickup.** 328

carbon-dioxide system (rotating machinery). A fire-protection system using carbon-dioxide gas as the extinguisher. 63

carbon-graphite brush (electric machines). A carbon brush to which graphite is added. This type of brush can vary from medium hardness to very hard. It can carry only moderate currents and is adapted to moderate speeds. *See:* **brush (rotating machinery).** 279

carbon noise (carbon microphones). The inherent noise voltage of the carbon element. *See also:* **close-talking pressure-type microphone.** 249

carbon-pressure recording facsimile. That type of electromechanical recording in which a pressure device acts upon carbon paper to register upon the record sheet. *See also:* **recording (facsimile).** 12

carbon telephone transmitter. A telephone transmitter that depends for its operation upon the variation in resistance of carbon contacts. *See also:* **telephone station.** 328

car, cable. *See:* **cable car.**

carcinotron (*M*-type backward-wave oscillator) (microwave tubes). A crossed-field oscillator tube in which an electron stream interacts with a backward wave on the nonreentrant circuit. The oscillation frequency is a function of anode-to-sole voltage. *See also:* **microwave tube.** 190

car, conductor. *See:* **conductor car.**

card (computers). *See:* **magnetic card; punched card; tape to card.**

card extender (relaying). A device for testing static relay (circuit) cards which provides access to components on the card while maintaining all the electrical connections to the card. 79

card feed (test, measurement and diagnostic equipment). The mechanism which moves cards serially into a machine. 54

card field (test, measurement and diagnostic equipment). An area (one or more columns) of a card which is regularly assigned for the same information item. 54

card hopper (computers). A device that holds cards and makes them available to a card-feed mechanism. *See:* **input magazine; card stacker.** 255, 77

card image (computers). A one-to-one representation of the contents of a punched card, for example, a matrix in which a 1 represents a punch and a 0 represents the absence of a punch. 255, 77

cardinal plane (antenna). Any plane normal to the surface of a planar array and passing through the nearest adjacent array elements. *Note:* This term is used to relate the regular geometry of antenna array elements to the radiation pattern. 111

cardiogram. *See:* **electrocardiogram.**

car-door or gate electric contact. An electric device, the function of which is to prevent operation of the driving machine by the normal operating device unless the car door or gate is in the closed position. *See also:* **elevator.** 328

car-door or gate power closer. A device or assembly of devices that closes a manually opened car door or gate by power other than by hand, gravity, springs, or the movement of the car. *See also:* **elevator.** 328

card-programmed (test, measurement and diagnostic equipment). The capability of performing a sequence of tests according to instructions contained in one or a deck of punched cards. 54

card reader (test, measurement and diagnostic equipment). A mechanism that senses and obtains information from punched cards. 54

card, relay. *See:* **relay armature card.**

card stacker (computers). An output device that accumulates punched cards in a deck. *See:* **card hopper.** 255, 77, 54

card tester (test, measurement and diagnostic equipment). An instrument for testing and diagnosing printed circuit cards. 54

car enclosure (elevator). Consists of the top and the walls resting on and attached to the car platform. *See also:* **elevator.** 328

car-frame sling. The supporting frame to which the car platform, upper and lower sets of guide shoes, car safety, and the hoisting ropes or hoisting-rope sheaves, or the plunger of a direct-plunger elevator are attached. *See also:* **elevator.** 328

car or counterweight safety. A mechanical device attached to the car frame or to an auxiliary frame, or to the counterweight frame, to stop and hold the car or counterweight in case of predetermined overspeed or free fall, or if the hoisting ropes slacken. *See also:* **elevator.** 328

car or hoistway door or gate (elevators). The sliding portion of the car or the hinged or sliding portion in the hoistway enclosure that closes the opening giving access to the car or to the landing. *See also:* **hoistway (elevator or dumbwaiter).** 328

car platform (elevators). The structure that forms the floor of the car and that directly supports the load. *See also:* **hoistway (elevator or dumbwaiter).** 328

car retarder. A braking device, usually power operated, built into a railway track and used to reduce the speed of cars by means of brake shoes that when set in braking position press against the sides of the lower portions of the wheels. *See:* **control machine; master controller (pressure regulator); switch machine; trimmer signal.** 328

carriage (typewriter). *See:* **automatic carriage.**

carriage return (typewriter). The operation that causes the next character to be printed at the left margin. 255, 77

carried traffic (telephone switching systems). A measure of the calls served during a given period of time. 55

carrier (1) (signal transmission system). (A) A wave having at least one characteristic that may be varied from a known reference value by modulation. (B) That part of the modulated wave that corresponds in a specified manner to the unmodulated wave, having, for example, the carrier-frequency spectral components. *Note:* Examples of carriers are a sine wave and a recurring series of pulses. 111, 242

(2) (semiconductor). A mobile conduction electron or hole. *See also:* **semiconductor device.** 54

(3) (electrostatography). The substance in a developer that conveys a toner, but does not itself become a part of the viewable record. *See also:* **electrostatography.** 236, 191

carrier-amplitude regulation. The change in amplitude of the carrier wave in an amplitude-modulated transmitter when modulation is applied under conditions of symmetrical modulation. *Note:* The term **carrier shift,** often applied to this effect, is deprecated. 111

carrier beat (facsimile). The undesirable heterodyne of signals each synchronous with a different stable reference oscillator causing a pattern in received copy. *Note:* Where one or more of the oscillators is fork controlled, this is called fork beat. *See also:* **facsimile transmission.** 12

carrier bypass (power-system communication). A path for carrier current of comparatively low impedance around some circuit element. *See also:* **power-line carrier.** 59

carrier chrominance signal. *See:* **chrominance signal.**

carrier, communications common (data transmission). A company recognized by an appropriate regulatory agency as having a vested interest in furnishing communications services. 59

carrier-controlled approach system (CCA) (electronic navigation). An aircraft-carrier radar system providing information by which aircraft approaches may be directed via radio communication. *See also:* **navigation.** 13, 187

carrier current. The current associated with a carrier wave. *See also:* **carrier.** 59

carrier-current choke coil (capacitance potential device). A reactor or choke coil connected in series between the potential tap of the coupling capacitor and the potential device transformer unit, to present a low impedance to the flow of power current and a high impedance to the flow of carrier-frequency current. Its purpose is to limit the loss of carrier-frequency current through the potential-device circuit. *See also:* **outdoor-coupling capacitor.** 341

carrier-current coupling capacitor (power-system communication). An assembly of capacitor units for coupling carrier-current receiving and transmitting equipment to a power line. *See also:* **power-line carrier.** 59

carrier-current drain coil (capacitance potential device). A reactor or choke coil connected between the carrier-current lead and ground, to present a low impedance to the flow of power current and a high impedance to the flow of carrier-frequency current. *Note:* Its purpose is to prevent high voltage at power frequency from being impressed on the carrier-current lead and to limit the loss of carrier-frequency energy to ground. *See also:* **outdoor coupling capacitor.** 59

carrier-current grounding-switch and gap. Consists of a protective gap for limiting the voltage impressed on the carrier-current equipment and the line tuning unit (if used); and a switch that, when closed, solidly grounds the carrier equipment for maintenance or adjustment without interrupting either high-voltage line or potential-device operation. *See also:* **outdoor coupling capacitor.** 341

carrier-current lead (power-system communication). A cable for interconnecting carrier-current coupling capacitor, tuning units, carrier-current transmitter, and carrier-current receiver equipment. *See also:* **outdoor coupling capacitor; power-line carrier.** 59

carrier-current line trap (power-system communication). A network of inductance and capacitance inserted into a power line that offers a high impedance to one or more carrier-current frequencies and a low impedance to power-frequency current. *See also:* **power-line carrier.** 59

carrier-current line trap, single-frequency (power-system communication). A carrier-current line trap that offers high impedance to only one carrier-current frequency. *See also:* **power-line carrier.** 59

carrier-current line trap, two-frequency (power-system communication). A carrier-current line trap that offers high impedance to two separate carrier-current frequencies. *See also:* **power-line carrier.** 59

carrier frequency (1) (in a periodic carrier). The reciprocal of its period. *Note:* The frequency of a periodic pulse carrier often is called the pulse-repetition frequency in a signal-transmission system.

(2) (modulated amplifier). The frequency that is used to modulate the input signal for amplification. *See also:* **carrier.** 111, 242, 188, 59

carrier-frequency choke coil, power line. A reactor that is connected between a carrier-current coupling capacitor and a grounded shunt capacitor that may be used for power-frequency voltage measurement. *Note:* The coil is a low impedance to the flow of power-frequency current and a high impedance to the flow of carrier current. *See also:* **power-line carrier.** 59

carrier-frequency pulse. A carrier that is amplitude modulated by a pulse. *Notes:* (1) The amplitude of the modulated carrier is zero before and after the pulse. (2) Coherence of the carrier (with itself) is not implied. 254

carrier frequency range (transmitter). The continuous range of frequencies within which the transmitter may be adjusted for normal operation. A transmitter may have more than one carrier-frequency range. *See also:* **radio transmitter.** 111, 240, 59

carrier frequency stability (radio transmitter) (transmitter performance). The measure of the ability to remain on its assigned channel as determined on both a short term (1-second) and a long term (24-hour) basis. 181, 59

carrier, fuse. *See:* **fuse carrier.**

carrier group (base group) (data transmission). Common practice is to assemble twelve voice channels by modulation processes in the basic group frequency band, 60 hertz to 108 hertz, where they appear as lower sidebands of twelve-carrier frequencies. This carrier group is the basic building block of the larger system. These groups may be assembled by further modulation processes to adjacent positions in the basic super-group-frequency. 59

carrier isolating choke coil. An inductor inserted, in series with a line on which carrier energy is applied, to impede the flow of carrier energy beyond that point. 59, 329

carrier modulation (data transmission). A process where a high frequency carrier wave is altered by a signal containing the information to be transmitted. Most communication engineers are familiar with carrier systems where a frequency spectrum is channelized into voice bandwidth carrier channels. Each channel has a carrier frequency which is modulated by a voice or data signal. 59

carrier noise level (residual modulation). The noise produced by undesired variations of radio-frequency signal in the absence of any intended modulation. *See also:* **radio transmission; modulation.** 111

carrier-pilot protection (relays) (power switchgear). A form of pilot protection in which the communication means between relays is a carrier current channel. 127, 103

carrier power output (transmitter performance). The radio-frequency power available at the antenna terminal when no modulating signal is present. *See also:* **audio-frequency distortion.** 181

carrier reinsertion (power-system communication). The process of mixing a locally generated carrier frequency with a received single-sideband suppressed-carrier signal to reestablish a desired amplitude-modulated signal that can be demodulated. *See also:* **power-line carrier.** 59

carrier relaying protection. A form of pilot protection in which high-frequency current is used over a metallic circuit (usually the line protected) for the communicating means between the relays at the circuit terminals.
103, 202

carrier repeater. A repeater for use in carrier transmission.
59

carrier shift (frequency departure or frequency deviation) (data transmission). (1) Difference between the steady state, mark and space frequencies in a system utilizing frequency shift modulation. (2) The change in frequency between transmitting and receiving terminals in nonsynchronized carrier systems. 59

carrier start time (power-system communication). The interval between the instant that a keyer causes a carrier transmitter to increase its output and the instant the output reaches a specified upper amplitude, namely 90 percent of the peak amplitude. *See also:* **power-line carrier.** 59

carrier stop time (power-system communication). The interval between the instant that a keyer causes a carrier transmitter to decrease its output and the instant the output reaches a specified lower amplitude, namely 10 percent of the peak amplitude. *See also:* **power-line carrier.**
59

carrier suppression (radio communication). The method of operation in which the carrier wave is not transmitted. *See also:* **modulation.** 111

carrier system (data transmission). A communication using frequency multiplexing to a number of channels over a single path by modulating each channel upon a different carrier frequency and demodulating at the receiving point to restore the signals to their original form. 59

carrier tap choke coil. A carrier-isolating choke coil inserted in series with a line tap. 329

carrier telegraphy. The form of telegraphy in which, in order to form the transmitted signals, alternating-current is supplied to the line after being modulated under the control of the transmitting apparatus. *See also:* **telegraphy.**
59, 194

carrier telephone channel. A telephone channel employing carrier transmission. *See also:* **channel.** 328

carrier telephony. The form of telephony in which carrier transmission is used, the modulating wave being a voice-frequency wave. *Note:* This term is ordinarily applied only to wire telephony. 59

carrier terminal. The assemblage of apparatus at one end of a carrier transmission system, whereby the processes of modulation, demodulation, filtering, amplification, and associated functions are effected. 59

carrier terminal grounding switch (power-system communication). A switch provided with power-line carrier coupling capacitors to protect personnel during equipment adjustment. *Note:* The switch closes a circuit across the power-line carrier terminal between the power-line coupling capacitor and ground. *See also:* **power-line carrier.**
59

carrier terminal protective gap (power-system communi-

cation). A gap used to limit voltage impressed on power-line carrier terminal equipment by electric disturbances on the power line. *See also:* **power-line carrier.**
59

carrier test switch (CTS) (power-system communication). A switch on power-line carrier transmitters for applying carrier energy to the power line for test purposes. *See also:* **power-line carrier.** 59

carrier-to-noise ratio. The ratio of specified measures of the carrier and the noise after specified band limiting and before any nonlinear process such as amplitude limiting and detection. *Note:* This ratio is expressed in many different ways, for example, in terms of peak values in the case of impulse noise and in terms of mean-square or root-mean-square values for other types of noise. *See also:* **amplitude modulation.** 242, 59

carrier transmission. That form of electric transmission in which the transmitted electric wave is a wave resulting from the modulation of a single-frequency wave by a modulating wave. *See also:* **carrier.** 328

carrier wave. *See:* **carrier.**

carry. (1) A character or characters, produced in connection with an arithmetic operation on one digit place of two or more number representations in positional notation, and forwarded to another digit place for processing there. (2) The number represented by the character or characters in (1). (3) Usually, a signal or expression as defined in (1) which arises in adding, when the sum of two digits in the same digit place equals or exceeds the base of the number system in use. *Note:* If a carry into a digit place will result in a carry out of the same digit place, and if the normal adding circuit is bypassed when generating this new carry, it is called a high-speed carry, or standing-on-nines carry. If the normal adding circuit is used in such a case, the carry is called a cascaded carry. If a carry resulting from the addition of carries is not allowed to propagate (for example, when forming the partial product in one step of a multiplication process), the process is called a partial carry. If it is allowed to propagate, the process is called a complete carry. If a carry generated in the most-significant-digit place is sent directly to the least-significant place (for example, when adding two negative numbers using nines complements) that carry is called an end-around carry. (4) A carry, in direct subtraction, is a signal or expression as defined in (1) that arises when the difference between the digits is less than zero. Such a carry is frequently called a borrow. (5) To carry is the action of forwarding a carry. (6) A carry is the command directing a carry to be forwarded. *See:* **cascaded carry; complete carry; end-around carry; high-speed carry; partial carry; standing-on-nines carry.** 235, 210

car-switch automatic floor-stop operation (elevators). Operation in which the stop is initiated by the operator from within the car with a definite reference to the landing at which it is desired to stop, after which the slowing down and stopping of the elevator is effected automatically. *See:* **control.** 328

car-switch operation (elevators). Operation wherein the movement and direction of travel of the car are directly and solely under the control of the operator by means of a manually operated car switch of continuous-pressure buttons in the car. *See:* **control.** 328

cartridge fuse. A low-voltage fuse consisting of a current-responsive element inside a fuse tube with terminals on both ends.   103, 202

cartridge size (cartridge fuse). The range of voltage and ampere ratings assigned to a fuse cartridge with specific dimensions and shape.   103, 202

cartridge-type bearing (rotating machinery). A complete ball or roller bearing assembly consisting of a ball or roller bearing and bearing housing that is intended to be inserted into a machine endshield. *See also:* **bearing.**   63

cart, splicing. *See:* **splicing cart.**

car-wiring apparatus. *See:* **electric train-line; train-line coupler; wire.** *See also:* **multiple-unit control.**

cascade. *See:* **tandem.**

cascade (electrolyte cells). A series of two or more electrolytic cells or tanks so placed that electrolyte from one flows into the next lower in the series, the flow being favored by differences in elevation of the cells, producing a cascade at each point where electrolyte drops from one cell to the next. *See also:* **electrowinning.**   328

cascade connection (cascade). A tandem arrangement of two or more similar component devices in which the output of one is connected to the input of the next. *See:* **tandem.**
                                                                      328

cascade control (1) (street lighting system). A method of turning street lights on and off in sections, each section being controlled by the energizing and de-energizing of the preceding section. *See also:* **alternating-current distribution; direct-current distribution.**   64

(2) (automatic control). *See:* **control system, cascade.**
                                                                       56

cascaded carry (parallel addition). A carry process in which the addition of two numerals results in a partial-sum numeral and a carry numeral that are in turn added together, this process being repeated until no new carries are generated. *See:* **carry; high-speed carry.**   255, 77

cascade development (electrostatography). Development in which the image-forming material is carried to the field of the electrostatic image by means of gravitational forces, usually in combination with a granular carrier. *See also:* **electrostatography.**   236, 191

cascaded thermoelectric device. A thermoelectric device having two or more stages arranged thermally in series. *See also:* **thermoelectric device.**   248, 191

cascade-type voltage transformer. An insulated-neutral terminal type voltage transformer with the primary winding distributed on several cores with the cores electromagnetically coupled by coupling windings. The secondary winding is on the core at the neutral end of the high voltage winding. Each core of this type of transformer is insulated from the other cores and is maintained at a fixed voltage with respect to ground and the line to ground voltage.   53

cascade node (branch) (network analysis). A node (branch) not contained in a loop. *See also:* **linear signal flow graphs.**
                                                                      282

cascade rectifier (cascade rectifier circuit). A rectifier in which two or more similar rectifiers are connected in such a way that their direct voltages add, but their commutations do not coincide. *Note:* When two or more rectifiers operate so that their commutations coincide, they are said to be in parallel if the direct currents add, and in series if

the direct voltages add. *See also:* **power rectifier; rectification; rectifier circuit element.**   237, 291, 208, 66

cascading (switching devices). The application in which the devices nearest the source of power have interrupting ratings equal to, or in excess of, the available short-circuit current, while devices in succeeding steps farther from the source, have successively lower interrupting ratings.
                                                                      103, 202

case (1) (storage battery) (storage cell). A multiple compartment container for the elements and electrolyte of two or more storage cells. Specifically wood cases are containers for cells in individual jars. *See also:* **battery (primary or secondary).**   328

(2) (electrotyping). A metal plate to which is attached a layer of wax to serve as a matrix. *See also:* **electroforming.**
                                                                      328

(3) (semiconductor device). The housing of a semiconductor device.   66

(4) (gyro; accelerometer). The structure which provides the mounting surfaces and establishes the reference axes.
                                                                       46

case capacitance. The difference between the total varactor capacitance and the nonlinear element capacitance.
                                                                      136

case (frame) ground protection (relays) (power switchgear). Overcurrent relay protection used to detect current flow in the ground or earth connection of the equipment or machine.   127, 103

case shift (telegraphy). The change-over of the translating mechanism of a telegraph receiving machine from letters-case to figures-case or vice versa. *See also:* **telegraphy.**
                                                                      194

case temperature (1) (semiconductor device). The temperature of a semiconductor device measured at a standard or specified location on or near the surface of the device case.   66

(2) (light emitting diodes). (T_C) The temperature measured at a specified location on the case of a device.
                                                                      162

cassegrainian feed (communication satellite). A feed system used for parabolic reflector antennas, where a small hyperbolic subreflector is placed near the focus of the paraboloid. The cassegrainian feed system prevents spillover to the back of the reflector, thus a better noise performance is achieved.   83

Cassegrain reflector antenna. A double-reflector antenna with a subreflector (usually hyperboloidal) located between the focal point and the vertex of the main reflector.
                                                                      111

casting (electrotyping). The pouring of molten electrotype metal upon tinned shells. *See also:* **electroforming.**
                                                                      328

catastrophic failure (reliability). *See:* **failure, catastrophic.**

catcher (electron tubes). *See:* **output resonator.**

catcher space (velocity-modulated tube). The part of the tube following the drift space, and where the density modulated-electron beam excites oscillations in the output resonator. It is the space between the output-resonator grids. *See also:* **velocity-modulated tube.**   244, 190

catelectrotonus (electrobiology). Electrotonus produced in the region of the cathode. *See also:* **excitability, elec-**

trotonus (electrobiology). 192

cathode (1) (electron tube or valve). An electrode through which a primary stream of electrons enters the interelectrode space. *See:* **electrode (electron tube) anode.**

125, 190

(2) (semiconductor rectifier diode). The electrode to which the forward current flows within the cell. *See also:* **semiconductor.** 237, 66

(3) (electrolytic). An electrode through which current leaves any conductor of the nonmetallic class. Specifically, an electrolytic cathode is an electrode at which positive ions are discharged, or negative ions are formed, or at which other reducing reactions occur. *See also:* **electrolytic cell.** 328

(4) (thyristor). The electrode by which current leaves the thyristor when the thyristor is in the ON state with the gate open-circuited. *Note:* This term does not apply to bidirectional thyristors. 243, 191, 66, 208

(5) (light emitting diodes). The electrode to which the forward current is directed within a device. 162

cathode border (gas) (gas tube). The distinct surface of separation between the cathode dark space and the negative glow. *See also:* **discharge (gas).** 244

cathode cleaning (electroplating). Electrolytic cleaning in which the metal to be cleaned is the cathode. *See also:* **battery (primary or secondary).** 328

cathode coating impedance (electron tube). The impedance, excluding the cathode interface (layer) impedance, between the base metal and the emitting surface of a coated cathode. 125, 190

cathode, cold. *See:* **cold cathode.**

cathode current. *See:* **electrode current (electron tube); electronic controller.**

cathode current, peak (1) (fault). The highest instantaneous value of a nonrecurrent pulse of cathode current occurring under fault conditions. *See also:* **electrode current (electron tube).** 190

(2) (steady state). The maximum instantaneous value of a periodically recurring cathode current. *See also:* **electrode current (electron tube).** 190

(3) (surge). The highest instantaneous value of a randomly recurring pulse of cathode current. *See also:* **electrode current (electron tube).** 190

cathode dark space (Crookes dark space) (gas tube). The relatively nonluminous region in a glow-discharge cold-cathode tube between the cathode glow and the negative glow. *See:* **gas tube.** 190

cathode efficiency. The current efficiency of a specified cathodic process. *See also:* **electrochemistry.** 328

cathode (potential) fall (gas). The difference of potential due to the space charge near the cathode. *See also:* **discharge (gas).** 244, 190

cathode follower. A circuit in which the output load is connected in the cathode circuit of an electron tube and the input is applied between the control grid and the remote end of the cathode load, which may be at ground potential. *Note:* The circuit is characterized by low output impedance, high input impedance, gain less than unity, and negative feedback. 111, 211

cathode glow (gas tube). The luminous glow that covers all, or part, of the surface of the cathode in a glow-discharge cold-cathode tube, between the cathode and the cathode

dark space. *See:* **gas tube.** 125, 190

cathode heating time (vacuum tube). The time required for the cathode to attain a specified condition, for example: (1) a specified value of emission or (2) a specified rate of change of emission. *Note:* All electrode voltages are to remain constant during measurement. The tube elements must all be at room temperature at the start of the test. *See:* **operation time.** 125, 190

cathode interface (layer) capacitance (electron tube). A capacitance that, in parallel with a suitable resistance, forms an impedance approximating the cathode interface impedance. *Note:* Because the cathode interface impedance cannot be represented accurately by the two-element resistance-capacitance circuit, this value of capacitance is not unique. 125, 190

cathode interface (layer) impedance (electron tube). An impedance between the cathode base and coating. *Note:* This impedance may be the result of a layer of high resistivity or a poor mechanical bond between the cathode base and coating. 125, 190

cathode interface (layer) resistance (electron tube). The low-frequency limit of cathode interface impedance. 125, 190

cathode, ionic-heated. A hot cathode that is heated primarily by ionic bombardment of the emitting surface. 190

cathode layer. A molten metal or alloy forming the cathode of an electrolytic cell and which floats on the fused electrolyte, or upon which fused electrolyte floats. *See also:* **fused electrolyte.** 328

cathode luminous sensitivity (multiplier phototube). *See:* **sensitivity, cathode luminous.**

cathode modulation. Modulation produced by application of the modulating voltage to the cathode of any electron tube in which the carrier is present. *Note:* Modulation in which the cathode voltage contains externally generated pulses is called cathode pulse modulation. 328

cathode (or anode) sputtering (gas). The emission of fine particles from the cathode (or anode) produced by positive ion (or electron) bombardment. *See also:* **discharge (gas).** 244, 190

cathode, pool. A cathode at which the principal source of electron emission is a cathode spot on a metallic pool electrode. 190

cathode preheating time (electron tube). The minimum period of time during which the heater voltage should be applied before the application of other electrode voltages. *See:* **heater current (electron device).** 190

cathode pulse modulation. Modulation produced in an amplifier or oscillator by application of externally generated pulses to the cathode circuit. *See also:* **modulating function.** 111

cathode-ray charge-storage tube. A charge-storage tube in which the information is written by means of a cathode-ray beam. *Note:* Dark-trace tubes and cathode-ray tubes with a long persistence are examples of cathode-ray storage tubes that are not charge-storage tubes. Most television camera tubes are examples of charge-storage tubes that are not cathode-ray storage tubes. *See also:* **charge-storage tube.** 174, 190

cathode-ray instrument. *See:* **electron beam instrument.**

cathode-ray oscillograph. An oscillograph in which a

photographic or other record is produced by means of the electron beam of a cathode-ray tube. *Note:* The term cathode-ray oscillograph has frequently been applied to a cathode-ray oscilloscope but this usage is deprecated. *See also:* oscillograph. 328

cathode-ray oscilloscope. An oscilloscope that employs a cathode-ray tube as the indicating device. *See also:* oscillograph. 328

cathode-ray storage tube. A storage tube in which the information is written by means of a cathode-ray beam. *See also:* storage tube. 174, 190

cathode-ray tube. An electron-beam tube in which the beam can be focused to a small cross section on a luminescent screen and varied in position and intensity to produce a visible pattern. 328

cathode-ray-tube display area. *See:* graticule area.

cathode region (gas) (gas tube). The group of regions that extends from the cathode to the Faraday dark space inclusively. *See also:* discharge (gas). 244, 290

cathode spot (arc). An area on the cathode of an arc from which electron emission takes place at a current density of thousands of amperes per square centimeter and where the temperature of the electrode is too low to account for such currents by thermionic emission. *See:* gas tube. 290

cathode sputtering. *See:* sputtering (electroacoustics).

cathode terminal (1) (semiconductor device). The terminal from which forward current flows to the external circuit. *Note:* In the semiconductor rectifier components field, the cathode terminal is normally marked positive. *See also:* semiconductor; semiconductor rectifier cell. 237, 66, 208

(2) (thyristor). The terminal that is connected to the cathode. *Note:* This term does not apply to bidirectional thyristors. *See also:* anode. 243, 66, 191

cathodic corrosion. An increase in corrosion of a metal by making it cathodic. *See also:* stray current corrosion. 221, 205

cathodic polarization. Polarization of a cathode. *See also:* electrochemistry. 221, 205

cathodic protection. Reduction or prevention of corrosion by making a metal the cathode in a conducting medium by means of a direct electric current (which is either impressed or galvanic). 221, 205

catholyte. The portion of an electrolyte in an electrolytic cell adjacent to a cathode. If a diaphragm is present, it is the portion of electrolyte on the cathode side of the diaphragm. *See also:* electrolytic cell. 328

cation. A positively charged ion or radical that migrates toward the cathode under the influence of a potential gradient. *See also:* ion. 221, 205

catwhisker. A small, sharp-pointed wire used to make contact with a sensitive point on the surface of a semiconductor. 328

caustic embrittlement. Stress-corrosion cracking in alkaline solutions. 221, 205

caustic soda cell. A cell in which the electrolyte consists primarily of a solution of sodium hydroxide. *See also:* electrochemistry. 328

cavitation (liquid). Formation, growth, and collapse of gaseous and vapor bubbles due to the reduction of pressure of the cavitation point below the vapor pressure of the fluid at the working temperature. 176

cavitation damage. Deterioration caused by formation and collapse of cavities in a liquid. 221, 205

cavity. *See:* tuned grid; unloaded applicator impedance.

cavity dumpers (acousto-optic device). Generally, a fast rise time pulse modulator used intracavity. 72

cavity magnetron*. *See:* magnetron, cavity resonator.

* Deprecated

cavity resonator (1) (antenna). A space normally bounded by an electrically conducting surface in which oscillating electromagnetic energy is stored, and whose resonant frequency is determined by the geometry of the enclosure. *See also:* waveguide. 111

(2) (waveguide components). A resonator formed by a volume of propagating medium bounded by reflecting surfaces. *See also:* resonator, waveguide. 166

cavity-resonator frequency meter (electromagnetic wave). A cavity resonator used to determine frequency. *See also:* waveguide. 166, 179

CAX. *See:* unattended automatic exchange.

C-band radar. A radar operating at frequencies between 4 and 8 gigahertz usually in the International Telecommunications Union assigned band 5.2 to 5.9 gigahertz. 13

C **battery.** A battery designed or employed to furnish voltage used as a grid bias in a vacuum-tube circuit. *See also:* battery (primary or secondary). 328

CCA (electronic navigation). *See:* carrier-controlled approach system.

CCITT (data transmission). The International Telegraph and Telephone Consultative Committee (CCITT) is an advisory committee established under the United Nations in accordance with Article 13 of the International Tele-Communications Convention (Geneva 1959) to study and recommend solutions for questions on technical operation and tariffs. The organization is attempting to establish standards for inter-country operation on a world-wide basis. 59

C **display (radar).** A rectangular display in which each target appears as a blip with bearing indicated by the horizontal coordinate and angle of elevation by the vertical coordinate. *See also:* navigation. 13

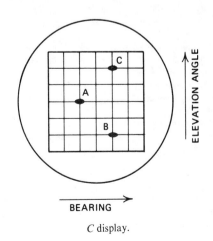

C display.

CDO. *See:* **unattended automatic exchange.**

ceiling area lighting. A general lighting system in which the entire ceiling is, in effect, one large luminaire. *Note:* Ceiling area lighting includes luminous ceilings and louvered ceilings. *See also:* **general lighting.** 167

ceiling direct voltage (direct potential rectifier unit). The average direct voltage at rated direct current with rated sinusoidal voltage applied to the alternating-current line terminals, with the rectifier transformer set on rated voltage tap and with voltage regulating means set for maximum output. *See also:* **power rectifier; rectification.** 291, 208

ceiling ratio (illuminating engineering). The ratio of the luminous flux that reaches the ceiling directly to the upward component from the luminaire. *See also:* **inverse-square law.** 167

ceiling voltage (synchronous-machine excitation system). The maximum output voltage that may be attained by an excitation system under specified conditions. 63, 105

ceiling voltage, exciter nominal. The ceiling voltage of an exciter loaded with a resistor having an ohmic value equal to the resistance of the field winding to be excited and with this field winding at a temperature of (1) 75°C for field windings designed to operate at rating with a temperature rise of 60°C or less; or (2) 100°C for field windings designed to operate at rating with a temperature rise greater than 60°C. 105

celestial-inertial navigation equipment. An equipment employing both celestial and inertial sensors. *See also:* **navigation.** 187, 13

celestial mechanics (communication satellite). The mechanics of motion of celestial bodies, including satellites. 74

cell (information storage). An elementary unit of storage, for example, binary cell, decimal cell. 235, 210

cell cavity (electrolysis). The container formed by the cell lining for holding the fused electrolyte. *See also:* **fused electrolyte.** 328

cell connector (storage cell). An electric conductor used for carrying current between adjacent storage cells. *See also:* **battery (primary or secondary).** 328

cell constant (electrolytic cell). The resistance in ohms of that cell when filled with a liquid of unit resistivity. 328

cell cover. The transparent medium (glass, quartz, etc.) that protects the solar cells from space particulate radiation. 113

cell line (electrolytic cell line). An assembly of one or more electrically interconnected electrolytic cells supplied from a source of dc power. 133

cell line potential (electrolytic cell line). The dc voltage applied to the positive and negative buses supplying power to a cell line. 133

cell line working zone (electrolytic cell line). That space within which normal operation and maintenance of electrolytic cells and their attachments may be performed. 133

cell potential (electrolytic cell). The dc voltage between the positive and negative terminals of one electrolytic cell. 133

cell-type tube (microwave gas). A gas-filled radio-frequency switching tube that operates in an external resonant circuit. *Note:* A tuning mechanism may be incorporated in either the external resonant circuit or the tube. *See:* **gas tubes.** 125, 190

cellular metal floor raceway. A hollow space of a cellular metal floor suitable for use as a raceway. *See also:* **raceway.** 328

cent (acoustics). The interval between two sounds whose basic frequency ratio is the twelve-hundredth root of 2. *Note:* The interval, in cents, between any two frequencies is 1200 times the logarithm to the base 2 of the frequency ratio. Thus 1200 cents equal 12 equally tempered semitones equal 1 octave. 176, 334

center-break switching device. A mechanical switching device in which both contacts are movable and engage at a point substantially midway between their supports. 103, 202

center frequency (1) (frequency modulation). The average frequency of the emitted wave when modulated by a symmetrical signal. *See also:* **frequency modulation.** 111, 59

(2) (burst measurements). The arithmetic mean of the two frequencies that define the bandwidth of a filter. *See also:* **burst.** 292

(3) (non-real time spectrum analyzer). That frequency which corresponds to the center of a frequency span, (hertz). 68

center frequency delay (dispersive and nondispersive delay line). The frequency delay of the device at the center frequency, F_0, generally expressed in microseconds. 81

center of distribution (primary distribution). The point from which the electric energy must be supplied if the minimum weight of conducting material is to be used. *Note:* The center of distribution is commonly considered to be that fixed point that, in practice, most nearly meets the ideal conditions stated above. 64

centimeter-gram-second (system of units). *See:* **cgs.**

central control room (nuclear power generating station). A continuously manned, protected enclosure from which actions are normally taken to operate the nuclear generating station under normal and abnormal conditions. 358

centralized accounting, automatic message (CAMA) (telephone switching systems). An arrangement at an intermediate office for collecting automatic message accounting information. 55

centralized computer network (data communication). A computer network configuration in which a central node provides computing power, control, or other services. *See:* **decentralized computer network.** 12

centralized test system (test, measurement and diagnostic equipment). A test system which processes, records or displays at a central location, information gathered by test point data sensors at more than one remotely located equipment or system under test. 54

centralized traffic-control machine (railway practice). A control machine for operation of a specific type of traffic control system of signals and switches. *See also:* **centralized traffic-control system.** 328

centralized traffic-control system (railway practice). A specific type of traffic control system in which the signals and switches for a designated section of track are controlled from a remotely located centralized traffic control machine. *See also:* **block signal system; centralized traffic-control machine; control machine; electropneumatic interlocking machine.** 328

central office (1) (data transmission). The place where communications common carriers terminate customer lines and locate the equipment which interconnects those lines. Usually the junction point between metallic pair and carrier system. 59

(2) (telephone switching systems). A switching entity that has one or more office codes and a system control serving a telecommunication exchange. 55

central office diagram (telephone switching systems). A simplified switching network plan for a given installation, specifying types and quantities of equipment and trunk groups and other parameters. 55

central office exchange (data transmission). The place where a communication common carrier locates the equipment which interconnects incoming subscribers and circuits. 59

central processing unit (computing system). The unit of a computing system that includes the circuits controlling the interpretation of instructions and their execution. 255, 77

central station (protective signaling). An office to which remote alarm and supervisory signaling devices are connected, where operators supervise the circuits, and where guards are maintained continuously to investigate signals. *Note:* Facilities may be provided for transmission of alarms to police and fire departments or other outside agencies. *See also:* **protective signaling.** 328

central station equipment (protective signaling). The signal receiving, recording, or retransmission equipment installed in the central station. *See also:* **protective signaling.** 328

central station switchboard (protective signaling). That portion of the central station equipment on or in which are mounted the essential control elements of the system. *See also:* **protective signaling.** 328

central station system (protective signaling) (central office system). A system in which the operations of electric protection circuits and devices are signaled automatically to, recorded in, maintained, and supervised from a central station having trained operators and guards in attendance at all times. *See also:* **protective signaling.** 328

central (foveal) vision. The seeing of objects in the central or foveal part of the visual field, approximately two degrees in diameter. *Note:* It permits seeing much finer detail than does peripheral vision. *See also:* **visual field.** 167

central visual field. That region of the visual field that corresponds to the foveal portion of the retina. *See also:* **visual field.**

centrex CO (company) (telephone switching systems). The provision of centrex service by switching equipment located on telephone company owned or leased premises; the station equipment and attendant facilities are located on the premises of the customer. 55

centrex CU (customer) (telephone switching systems). The provision of centrex service by switching, station equipment, and attendant facilities located on the premises of the customer. 55

centrex service (telephone switching systems). A service

that provides direct inward dialing and identified outward dailing in accordance with the national numbering plan for stations served as they would be by a private branch exchange. 55

centrifugal actuator. *See:* **actuator, centrifugal.**

centrifugal-mechanism pin (governor pin). A component of the linkage between the centrifugal mechanism weights and the short-circuiting device. *See:* **centrifugal starting switch.** 328

centrifugal-mechanism spring (governor spring). A spring that opposes the centrifugal action of the centrifugal-mechanism weights in determining the motor speed at which the switch or short-circuiting device is actuated. *See:* **centrifugal starting switch.** 328

centrifugal-mechanism weights (governor weights). Moving parts of the centrifugal-mechanism assembly that are acted upon by centrifugal force. *See:* **centrifugal starting switch.** 328

centrifugal relay. An alternating-current frequency-selective relay in which the contacts are operated by a flyball governor or centrifuge driven by an induction motor. 328

centrifugal starting-switch (rotating machinery). A centrifugally operated automatic mechanism used to perform a circuit-changing function in the primary winding of a single-phase induction motor after the rotor has attained a predetermined speed, and to perform the reverse circuit-changing operation prior to the time the rotor comes to rest. *Notes:* (1) One of the circuit changes that is usually performed is to open or disconnect the auxiliary winding circuit. (2) In the usual form of this device, the part that is mounted to the stator frame or end shield is the starting switch, and the part that is mounted on the rotor is the centrifugal actuator. 63

CEP (electronic navigation). *See:* **circular probable error.**

certification (test, measurement and diagnostic equipment). Attestation that a support test system is capable, at the time of certification demonstration, of correctly assessing the quality of the items to be tested. This attestation is based on an evaluation of all support test system elements and establishment of acceptable correlation among similar test systems. 54

certified unit (test, measurement and diagnostic equipment). A unit whose demonstrated ability to perform in accordance with preestablished criteria has been attested. 54

CFAR. *See:* **constant false alarm rate.**

cgs (centimeter-gram-second) electromagnetic system of units. A system in which the basic units are the centimeter, gram, second, and abampere. *Notes:* (1) The abampere is a derived unit defined by assigning the magnitude 1 to the unrationalized magnetic constant (sometimes called the permeability of space). (2) Most electrical units of this system are designated by prefixing the syllable "ab-" to the name of the corresponding unit in the mksa system. Exceptions are the maxwell, gauss, oersted, and gilbert. 210

cgs electrostatic system of units. The system in which the basic units are the centimeter, gram, second, and statcoulomb. *Notes:* (1) The statcoulomb is a derived unit defined by assigning the magnitude 1 to the unrationalized electric constant (sometimes called the permittivity of space). (2) Each electrical unit of this system is commonly designated by prefixing the syllable "stat-" to the name of the corresponding unit in the International System of Units. 210

cgs system of units. A system in which the basic units are the centimeter, gram, and second. 210

chad. The piece of material removed when forming a hole or notch in a storage medium such as punched tape or punched cards. 255, 77, 54

chadded. Pertaining to the punching of tape in which chad results. 255, 77

chadless. Pertaining to the punching of tape in which chad does not result. 255, 77, 54

chadless tape. A punched tape wherein only partial perforation is completed and the chad remains attached to the tape. 224, 207

chaff (radar). An airborne cloud of lightweight reflecting objects typically consisting of strips of aluminum foil or metallic coated fibers which produce clutter echoes in a region of space. 13

chafing strip. *See:* **drive strip.**

chain (electronic navigation). A network of similar stations operating as a group for determination of position or for furnishing navigational information. 13, 187

chain code (computing system). An arrangement in a cyclic sequence of some or all of the possible different *n*-bit words, in which adjacent words are related such that each word is derivable from its neighbor by displacing the bits one digit position to the left, or right, dropping the leading bit, and inserting a bit at the end. The value of the inserted bit needs only to meet the requirement that a word must not recur before the cycle is complete, for example, 000 001 010 101 011 111 110 100 000 . . . 255, 77

chain-drive machine (elevators). An indirect-drive machine having a chain as the connecting means. *See also:* **driving machine (elevators).** 328

chain matrix (circuits and systems). The 2×2 matrix relating voltage and current at one port of a two port network to voltage and current at the other port, as indicated below:

$$\begin{pmatrix} v_1 \\ i_1 \end{pmatrix} = \begin{pmatrix} A & B \\ C & D \end{pmatrix} \begin{pmatrix} v_2 \\ i_2 \end{pmatrix}$$

Syn. **transmission or cascade matrix.** 67

chair, boatswain. *See:* **boatswain chair.**

chalking (corrosion). The development of loose removable powder at or just beneath a coating surface. 221, 205

challenge (radar). *See:* **interrogation.**

challenger (radar). *See:* **interrogator.**

changeover switch. A switching device for changing electric circuits from one combination to another. *Note:* It is usual to qualify the term changeover switch by stating the purpose for which it is used, such as a series-parallel changeover switch, trolley-shoe changeover switch, etcetera. *See also:* **multiple-unit control.** 328

CHANHI. Abbreviation for upper channel corresponding to the half-amplitude point of a distribution. 117

CHANLO. Abbreviation for lower channel corresponding to the half-amplitude point of a distribution. 117

channel (1) (electric communication). (A) A single path for transmitting electric signals, usually in distinction from other parallel paths. (B) A band of frequencies. *Note:* The word path is to be interpreted in a broad sense to include separation by frequency division or time division. The term channel may signify either a one-way path, providing transmission in one direction only, or a two-way path, providing transmission in two directions. 328

(2) (electronic computers). (A) A path along which signals can be sent, for example, data channel, output channel. (B) The portion of a storage medium that is accessible to a given reading station. *See also:* **track.** 235, 54

(3) (information theory). A combination of transmission media and equipment capable of receiving signals at one point and delivering related signals at another point. *See also:* **information theory.** 160

(4) (illuminating engineering). An enclosure containing the ballast, starter, lamp holders, and wiring for a fluorescent lamp, or a similar enclosure on which filament lamps (usually tubular) are mounted. *See also:* **luminaire.** 167

(5) (nuclear power generating stations). An arrangement of components and modules as required to generate a single protective action signal when required by a generating station condition. A channel loses its identity where single action signals are combined. 20, 31, 356

channel, analog (data transmission). A channel on which the information transmitted can take any value between the limits defined by the channel. Voice channels are analog channels. 59

channel-busy tone (telephone switching system). A tone that indicates that a server other than a destination outlet is either busy or not accessible. 55

channel capacity (information theory). The maximum possible information rate through a channel subject to the constraints of that channel. *Note:* channel capacity may be either per second or per symbol. *See also:* **information theory.** 160

channel failure alarm (power-system communication). A circuit to give an alarm if a communication channel should fail. *See also:* **power-line carrier.** 59

channel group (group). A number of channels regarded as a unit. *Note:* The term is especially used to designate part of a larger number of channels. *See also:* **channel.** 59

channeling, lattice (semiconductor radiation detector). A phenomenon that results in a crystallographic directional dependence of the rate of energy loss of ionizing particles. 118, 119

channel lights. Aeronautical ground lights arranged along the sides of a channel of a water aerodrome. *See also:* **signal lighting.** 167

channel, melting. *See:* **melting channel.**

channel multiplier. A tubular electron-multiplier with a continuous interior surface of secondary-electron emissive material. *See also:* **amplifier; camera tube.** 190

channel, radio. *See:* **radio channel.**

channel spacing (radio communication). The frequency increment between the assigned frequency of two adjacent radio-frequency channels. *See also:* **dispatch operation; radio transmission; single-frequency simplex operation; two-frequency simplex operation.** 181

channel supergroup (supergroup). A number of channel

groups regarded as a unit. *Note:* The term is especially used to designate part of a larger number of channels. *See also:* **channel.**   59

channel, surface (semiconductor radiation detector). A thin region at a semiconductor surface of p- or n-type conductivity created by the action of an electric field; for example, that due to trapped surface charge.   118, 119

channel utilization index (information theory). The ratio of the information rate (per second) through a channel to the channel capacity (per second). *See also:* **information theory.**   160

channel, voice grade (data transmission). A channel which permits transmission of frequencies within the voice band.   59

character (1) (electronic computers). (A) An elementary mark or event that may be combined with others, usually in the form of a linear string, to form data or represent information. If necessary to distinguish from (B) below, such a mark may be called a character event. (B) A class of equivalent elementary marks or events as in (A) having properties in common, such as shape or amplitude. If necessary to distinguish from (A) above, such a class may be called a character design. There are usually only a finite set of character designs in a given language. *Notes:* (1) In "bookkeeper" there are six character designs and ten character events, while in "1010010" there are two character designs and seven character events. (2) A group of characters, in one context, may be considered as a single character in another, as in the binary-coded-decimal system. *See:* **blank character; check character; control character; escape character; numerical control; special character.**   235, 54

(2) (data transmission). One of a set of elementary symbols which normally include both alpha and numeric codes plus punctuation marks and any other symbol which may be read, stored or written.   59

character density (test, measurement and diagnostic equipment). The number of characters that can be stored per unit area or length.   54

character distortion (data transmission). The normal and predictable distortion of data bit produced by characteristics of a given circuit at a particular transmission speed.   59

character-indicator tube (electron device). A glow-discharge tube in which the cathode glow displays the shape of a character, for example, letter, number, or symbol.   190

character interval (data transmission). In start-stop operation the duration of a character expressed as the total number of unit intervals (including information, error checking and control bits and the start and stop elements) required to transmit any given character in any given communications system.   59

characteristic (semiconductor device). An inherent and measurable property of a device. Such a property may be electrical, mechanical, thermal, hydraulic, electro-magnetic or nuclear and can be expressed as a value for stated or recognized conditions. A characteristic may also be a set of related values, usually shown in graphical form.   66

characteristic angular phase difference. The characteristic angular phase difference of a set of polyphase voltages or currents of m phases is the minimum phase angle difference by which each member of a polyphase symmetrical set of polyphase voltages or currents may lag the preceding member of the set, when the members are arranged in the normal sequence. The characteristic angular phase difference is $2\pi/m$ radians, where m is the number of phases.   210

characteristic curve (1) (illuminating engineering). A curve that expresses the relationship between two variable properties of a light source, such as candle-power and voltage, flux and voltage, etcetera. *See also:* **colorimetry.**   167

(2) (Hall generator). A plot of Hall output voltage versus control current, magnetic flux density, or the product of magnetic flux density and control current.   107

characteristic curves (rotating machinery). The graphical representation of the relationships between certain quantities used in the study of electric machines. *See also:* **asynchronous machine; direct-current commutating machine.**   63

characteristic distortion (telegraphy). A displacement of signal transitions resulting from the persistence of transients caused by preceding transitions. *See also:* **telegraphy.**   194

characteristic element (surge arresters). The element that in a valve-type arrester determines the discharge voltage and the follow current, and in an expulsion-type arrester determines the discharge voltage and interrupts the follow current. *See:* **surge arrester (surge diverter).**   62

characteristic equation (feedback control system). The relation formed by equating to zero the denominator of a transfer function of a closed loop. *See also:* **control system, feedback.**   329

characteristic impedance (1) (circular waveguide.) For a traveling wave in the dominant ($TE_{1,1}$) mode of a lossless circular waveguide at a specified frequency above the cutoff frequency, (A) the ratio of the square of the root-mean-square voltage along the diameter where the electric vector is a maximum to the total power flowing when the guide is match terminated, (B) the ratio of the total power flowing to the square of the total root-mean-square longitudinal current flowing in one direction when the guide is match terminated, (C) the ratio of the root-mean-square voltage along the diameter where the electric vector is a maximum to the total root-mean-square longitudinal current flowing along the half surface bisected by this diameter when the guide is match terminated. *Note:* Under definition (A) the power $W = V^2/Z_{(W,V)}$ where V is the voltage and $Z_{(W,V)}$ is the characteristic impedance defined in (A). Under definition (B) the power $W = I^2Z_{(W,I)}$ where I is the current and $Z_{(W,I)}$ is the characteristic impedance defined in (B). The characteristic impedance $Z_{(V,I)}$ as defined in (C) is the geometric mean of the values given by (A) and (B). Definition (C) can be used also below the cutoff frequency. *See also:* **self-impedance; waveguide.**

(2) (rectangular waveguide). For a traveling wave in the dominant ($TE_{1,0}$) mode of a lossless rectangular waveguide at a specified frequency above the cutoff frequency, (A) the ratio of the square of the root-mean-square voltage between midpoints of the two conductor faces normal to the electric vector, to the total power flowing when the guide is match terminated, (B) the ratio of the total power flowing to the square of the root-mean-square longitudinal current, flowing on one face normal to the electric vector when the guide is match terminated, (C) the ratio of the root-mean-square voltage, between midpoints of the two conductor faces normal to the electric vector, to the total root-mean-square longitudinal current, flowing on one face when the guide is match terminated. *Note:* Under definition (A) the power $W = V^2/Z_{(W,V)}$ where V is the voltage, and $Z_{(W,V)}$ the characteristic impedance defined in (A). Under definition (B) the power $W = I^2 Z_{(W,I)}$ where I is the current and $Z_{(W,I)}$ the characteristic impedance defined in (B). The characteristic impedance $Z_{(V,I)}$ as defined in (C) is the geometric mean of the values given by (A) and (B). Definition (C) can be used also below the cutoff frequency. *See:* **waveguide.** *See also:* **self-impedance.**

(3) (two-conductor transmission line) (for a traveling transverse electromagnetic wave). The ratio of the complex voltage between the conductors to the complex current on the conductors in the same transverse plane with the sign so chosen that the real part is positive. *See also:* **self-impedance; transmission line; waveguide.**

(4) (coaxial transmission line). The driving impedance of the forward-traveling transverse electromagnetic wave. *See also:* **self-impedance; transmission line.**

267, 265, 179

(5) (surge impedance) (surge arrester). The driving-point impedance that the line would have if it were of infinite length. *Note:* It is recommended that this term be applied only to lines having approximate electric uniformity. For other lines or structures the corresponding term is iterative impedance. *See also:* **self-impedance.**

210, 62, 179, 185, 59

characteristic insertion loss (1) (waveguide and transmission line). The insertion loss in a transmission system that is reflectionless looking toward both the source and the load from the inserted transducer. *Notes:* (A) This loss is a unique property of the inserted transducer. (B) The frequency, internal impedance, and available power of the source and the impedance of the load have the same value before and after the transducer is inserted. *See also:* **waveguide.** 185

(2) (fixed and variable attenuators).

P_{INPUT} = Incident power from Z_0 source
P_{OUTPUT} = Net power into Z_0 load

$$\text{Characteristic insertion loss} = 10 \log_{10} \frac{P_{INPUT}}{P_{OUTPUT}} \text{ (dB)}$$

110

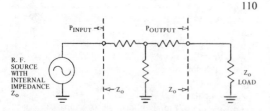

characteristic insertion loss, incremental. The change in the characteristic insertion loss of an adjustable device between two settings. *See also:* **waveguide.** 185

characteristic insertion loss, residual. The characteristic

insertion loss of an adjustable device at an indicated minimum position. *See also:* **waveguide.** 185

characteristic phase shift. For a 2-port device inserted into a stable, nonreflecting system between the generator and its load, the magnitude of the phase change of the voltage wave incident upon the load before and after insertion of the device, or change of the device from initial to final condition. *Note:* The following conditions apply: (1) The frequency, the load impedance, and the generator characteristics, internal impedance and available power, initially have the same values as after the device is inserted; (2) the joining devices, connectors or adapters belonging to the system conform to some set of standard specifications, the same specifications to be used by different laboratories, if measurements are to agree precisely; (3) the nonreflecting conditions are to be obtained in uniform, standard sections of waveguide on the system sides of the connectors at the place of insertion. *See also:* **measurement system.** 293, 183

characteristic telegraph distortion. Distortion that does not affect all signal pulses alike, the effect on each transition depending upon the signal previously sent, due to remnants of previous transitions or transients that persist for one or more pulse lengths. *Note:* Lengthening of the mark pulse is positive, and shortening, negative. Characteristic distortion is measured by transmitting biased reversals, square waves having unequal mark and space pulses. The average lengthening or shortening of mark pulses, expressed in percent of unit pulse length, gives a true measure of characteristic distortion only if other types of distortion are negligible. *See also:* **modulation.** 111

characteristic time (1) (gyro). The time required for the output to reach 63% of its final value for a step input. For a single degree of freedom rate integrating gyro, it is numerically equal to the ratio of the float moment of inertia to the damping coefficient about the output axis. For certain fluid filled gyros, the float moment of inertia may include other effects, such as that of transported fluid.

46

(2) (thermal converter). *See:* **apparent time constant.**

characteristic wave impedance. *See:* **wave impedance, characteristic.**

character recognition. The identification of graphic, phonic, or other characters by automatic means. *See:* **magnetic-ink character recognition; optical character recognition.** 255, 77

charge (1) (storage battery) (storage cell). The conversion of electric energy into chemical energy within the cell or battery. *Note:* This restoration of the active materials is accomplished by maintaining a unidirectional current in the cell or battery in the opposite direction to that during discharge; a cell or battery that is said to be charged is understood to be fully charged. 328

(2) (induction and dielectric-heating usage). *See:* **load.**

(3) (electric power supply). The amount paid for a service rendered or facilities used or made available for use.

112

charge, apparent (dielectric tests). That charge of a partial discharge which, if injected instantaneously between the terminals of the test object, would momentarily change the voltage between its terminals by the same amount as the partial discharge itself. The apparent charge should not be confused with the charge transferred across the discharging cavity in the dielectric. Apparent charge is expressed in coulombs. *Syn: terminal charge.*   139

charge carrier (semiconductor). A mobile conduction electron or mobile hole. *See also: semiconductor.*
23, 245, 66, 210, 118

charge collection time (semiconductor radiation detector). The time interval, after the passage of an ionizing particle, for the integrated current flowing between the terminals of the detector to increase from 10 to 90 percent of its final value.   23, 118, 119

charge, connection (electric power supply). The amount paid by a customer for connecting the customer's facilities to the supplier's facilities.   112

charge, customer (electric power utilization). The amount paid periodically by a customer without regard to demand or energy consumption.   112

charge-delay interval (telephone switching systems). The recognition time for a valid answer signal in message charging.   55

charge, demand (electric power utilization). That portion of the charge for electric service based upon a customer's demand.   112

charge, energy (electric power utilization). That portion of the charge for electric service based upon the electric energy consumed or billed.   112

charge, facilities (electric power utilization). The amount paid by the customer as lump sum, or periodically, as reimbursement for facilities furnished. The charge may include operation and maintenance as well as fixed costs.
112

charge-resistance furnace. A resistance furnace in which the heat is developed within the charge acting as the resistor. *See also: electrothermics.*

charge, space. *See: space charge.*

charge-storage tube (electrostatic memory tube). A storage tube in which the information is retained on the storage surface in the form of a pattern of electric charges.
174, 190

charge, termination (electric power utilization). The amount paid by a customer when service is terminated at the customer's request.   112

charge-to-third-number call (telephone switching systems). A call for which the charges are billed to a number other than that of the calling or called number.   55

charge transit time. *See: transit time.*

charge, wheeling (electric power supply). The amount paid to an intervening system for the use of its transmission facilities.   112

charging (electrostatography). *See: sensitizing. See also: electrostatography.*

charging circuit (surge generator) (surge arresters). The portion of the surge generator connections through which electric energy is stored up prior to the production of a surge. *See: surge arrester (surge diverter).*   62, 64

charging current (transmission line). The current that flows in the capacitance of a transmission line when voltage is applied at its terminals. *See also: transmission line.* 64

charging rack (mining). A device used for holding batteries for mining lamps and for connecting them to a power supply while the batteries are being recharged. *See also: mine feeder circuit.*   328

charging rate (storage battery) (storage cell). The current expressed in amperes at which a battery is charged. *See also: charge.*   328

charles or kino gun. *See: end injection.*

chart (recording instrument). The paper or other material upon which the graphic record is made. *See also: moving element (instrument).*   328

chart comparison unit (electronic navigation). A device for the simultaneous viewing of a navigational position presentation and a navigational chart in such a manner that one appears superimposed upon the other. *See also: navigation.*   13, 187

chart mechanism (recording instrument). The parts necessary to carry the chart. *See also: moving element (instrument).*   294

chart scale (recording instrument). The scale of the quantity being recorded, as marked on the chart. *Note:* Independent of and generally in quadrature with the chart scale is the time scale which is graduated and marked to correspond to the principal rate at which the chart is advanced in making the recording. This quadrature scale may also be used for quantities other than time. *See also: moving element (instrument).*   328

chart scale length (recording instrument). The shortest distance between the two ends of the chart scale. *See also: instrument.*   328

chassis (frame connection; equivalent chassis connection) (printed-wiring boards). A conducting connection to a chassis or frame, or equivalent chassis connection of a printed-wiring board. The chassis or frame (or equivalent chassis connection of a printed-wiring board) may be at substantial potential with respect to the earth or structure in which this chassis or frame (or printed-wiring board) is mounted.   25

chatter, relay. *See: relay chatter time; relay contact chatter.*

check (1) (standardize) (instrument or meter). Ascertain the error of its indication, recorded value, or registration. *Note:* The use of the word **standardize** in place of **adjust** to designate the operation of adjusting the current in the potentiometer circuit to balance the standard cell is deprecated. *See also: test (instrument or meter).*   328

(2) (computer-controlled machines). A process of partial or complete testing of (A) the correctness of machine operations, or (B) the existence of certain prescribed conditions within the computer. A check of any of these conditions may be made automatically by the equipment or may be programmed.   235, 210

(3) (nuclear power generating stations). The use of a source to determine if the detector and all electronic components of the system are operating correctly.   31

check, automatic (electronic computation). A check performed by equipment built into the computer specifically for that purpose and automatically accomplished each time the pertinent operation is performed. *Note:* Sometimes referred to as a **built-in check.** Machine check can refer to an automatic check or to a programmed check of machine functions.   210

check back. The retransmission from the receiving end to the initiating end of a coded signal or message to verify, at the initiating end, the initial message before proceeding with the transmitting of data or a command.   103

check before operate (data transmission). A message and control technique providing for confirmation of control request before operation. *Syn.* **check back.** 59

check bit. A binary check digit, for example, a parity bit. 77

check bits (data transmission). Associated with a code character or block for the purpose of checking the absence of error within the code character or block. *See also:* **data processing.** 194

check character. A character used for the purpose of performing a check, but often otherwise redundant. 77

check digit. A digit used for the purpose of performing a check, but often otherwise redundant. *See:* **check, forbidden-combination.** 77

check, forbidden-combination (electronic computation). A check (usually an automatic check) that tests for the occurrence of a nonpermissible code expression. *Notes:* (1) A self-checking code (or error-detecting code) uses a code expression such that one (or more) error(s) in a code expression produces a forbidden combination. (2) A parity check makes use of a self-checking code employing binary digits in which the total number of 1's (or 0's) in each permissible code expression is always even or always odd. A check may be made for either even parity or odd parity. (3) A redundancy check employs a self-checking code that makes use of redundant digits called check digits. 210

checkout (test, measurement and diagnostic equipment). A sequence of tests for determining whether or not a device or system is capable of, or is actually performing, a required operation or function. 54

checkout equipment (test, measurement and diagnostic equipment). Electric, electronic, optical, mechanical, hydraulic, or pneumatic equipment, either automatic, semiautomatic, or any combination thereof, which is required to perform the checkout function. 54

checkout time (test, measurement and diagnostic equipment). Time required to determine whether designated characteristics of a system are within specified values. 54

checkpoint (1) (electronic computation). A place in a routine where a check, or a recording of data for restart purposes, is performed.

(2) (electronic navigation). *See:* **way point.** 255, 77, 13

check problem (electronic computation). A routine or problem that is designed primarily to indicate whether a fault exists in the computer, without giving detailed information on the location of the fault. Also called check routine. *See:* **diagnostic; test.** *See also:* **programmed check.** 235

check, programmed (electronic computation). *See:* **programmed check.**

check routine. Same as check problem. 54

check, redundant (checking code) (data transmission). A check which uses extra digits (check bits) short of complete duplication, to help detect malfunctions and mistakes. 59

check, selection (electronic computation). *See:* **selection check.**

check solution. A solution to a problem obtained by independent means to verify a computer solution. *See also:* **electronic analog computer.** 9, 77, 10

check, transfer (electronic computation). *See:* **transfer check.**

check valve. *See:* **blocking capacitor.**

cheek, field-coil flange (washer). *See:* **collar.**

cheese antenna. A reflector antenna having a cylindrical reflector enclosed by two parallel conducting plates perpendicular to the cylinder, spaced more than one wavelength apart. *Syn:* **pillbox antenna.** 111

chemical conversion coating. A protective or decorative coating produced in situ by chemical reaction of a metal with a chosen environment. 221, 205

Child-Langmuir equation (thermionics). An equation representing the cathode current of a thermionic diode in a space-charge-limited-current state.

$$I = GV^{3/2}$$

where I is the cathode current, V is the anode voltage of a diode or the equivalent diode of a triode or of a multielectrode value or tube, and G is a constant (perveance) depending on the geometry of the diode or equivalent diode. *See also:* **electron emission.** 244, 190

chip (mechanical recording). The material removed from the recording medium by the recording stylus while cutting the groove. *See also:* **phonograph pickup.** 176

chirp (radar). A technique for pulse compression which uses linear frequency modulation during the pulse. 13

choice (telephone switching systems). The position of an outlet in a group with respect to the order of selection. 55

choke (waveguide). A device for preventing energy within a waveguide in a given frequency range from taking an undesired path. *See:* **waveguide.** 179

choke coil. An inductor used in a special application to impede the current in a circuit over a specified frequency range while allowing relatively free passage of the current at lower frequencies. 197

choke flange (microwave technique). A flange in whose surface is cut a groove so dimensioned that the flange may form part of a choke joint. *See also:* **waveguide.** 179

choke joint (1) (waveguide components). A connection designed for essentially complete transfer of power between two waveguides without metallic contact between the inner walls of the waveguides. It typically consists of one cover flange and one choke flange. *Syn.* **choke coupling.** 166

(2) (microwave transmission lines) (microwave technique). A connector between two sections of transmission line in which the gap between sections to be connected is built out to form a series-branching transmission line carrying a standing wave in which actual contact falls at or near a current minimum. *Note:* The series-branching line is typically one-half wave in length. The connection then occurs at a quarter-wave point, and the closed end of the line is contained wholly within one of the sections. Such joints are used to prevent radio-frequency leakage and high ohmic losses, and for other purposes. 328

choke piston (choke plunger) (noncontact plunger) (waveguide). A piston in which there is no metallic contact with the walls of the waveguide at the edges of the reflecting surface; the short-circuit to high-frequency currents is achieved by a choke system. *Note:* This definition covers a number of configurations: dumbbell; Z-slot; inverted bucket; etcetera. *See also:* **waveguide.** 179

choke plunger. *See:* **choke piston.**

chopped display (oscilloscopes). a time-sharing method of

displaying output signals of two or more channels with a single cathode-ray-tube gun, at a rate that is higher than and not referenced to the sweep rate. *See:* **oscillograph.**
185

chopped impulse voltage. A transient voltage derived from a full impulse voltage that is interrupted by the disruptive discharge of an external gap or the external portion of the test specimen causing a sudden collapse in the voltage, practically to zero value. *Note:* The collapse can occur on the front, at the peak, or on the tail. *See also:* **test voltage and current.**

chopped impulse wave (surge arresters). An impulse wave that has been caused to collapse suddenly by a flash-over.
244, 62

chopped wave. A voltage impulse that is terminated intentionally by sparkover of a gap.
64, 91

chopped-wave lightning impulse test (transformer). A voltage impulse that is terminated intentionally by sparkover of a gap, which occurs subsequent to the maximum crest of the impulse wave voltage, with a specified minimum crest voltage, and a specified minimum time to flashover.
53

chopper (1) (communications). A device for interrupting a current or a light beam at regular intervals. Choppers are frequently used to facilitate amplification.
328

(2) (capacitance devices). A special form of pulsing relay having contacts arranged to rapidly interrupt, or alternately reverse, the direct-current polarity input to an associated circuit.
351

(3) (analog computer). A mechanical, electrical, or electromechanical device that converts dc into a square wave. *Note:* As applied to a direct coupled operational amplifier, it is a modulator used to convert the direct current at the summing junction to alternating current for amplification and reinsertion as a correcting voltage to reduce offset.
9

chopping frequency. *See:* **chopping rate.**

chopping rate (oscilloscopes). The rate at which channel switching occurs in chopped-mode operation. *See:* **oscillograph.**
185

chroma. *See:* **Munsell chroma.**

chromaticity (1) (general). The color quality of light definable by its chromaticity coordinates, or by its dominant (or complementary) wavelength and its purity, taken together. *See also:* **color.**
244, 178

(2) (television). That color attribute of light definable by its chromaticity coordinates. *Note:* When a specific white, the value of dominant (or complementary) wavelength, and saturation are given, there will be a corresponding set of unique chromaticity coordinates.
18

chromaticity coordinate (light). The ratio of any one of the tristimulus values of a sample to the sum of the three tristimulus values. *Note:* In the standard colorimetric system CIE (1931) the symbols x, y, z are recommended for the chromaticity coordinates. *See also:* **CIE.**
18, 244, 178

chromaticity diagram. A plane diagram formed by plotting one chromaticity coordinate against another. *Notes:* (1) A commonly used chromaticity diagram is in 1931 CIE (x,y) diagram. (2) Another chromaticity diagram coming into use is defined in the 1960 CIE (u,v) Uniform Chromaticity System (UCS). In contrast with the CIE (x,y)

diagram, chromaticities which have a just noticeable difference (jnd) are spaced by essentially equal distances over the entire diagram. Coordinate values in the two systems are related by the transformations:

$$u = \frac{4x}{-2x + 12y + 3}$$

$$v = \frac{6y}{-2x + 12y + 3}$$

The (x,y) diagram has been plotted from values of x and y for a 2° field of view, given to four decimal places in Table 2(a) on page 7 of Monograph 104 entitled "Colorimetry" issued by the National Bureau of Standards. Values of u and v were calculated by means of the transformations given above. Values of x,y and u,v for the source temperatures on the Planckian Locus are given in Table 12, page 37, of the same document.
8

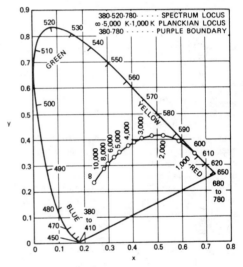

1931 CIE (x,y) Chromaticity Diagram

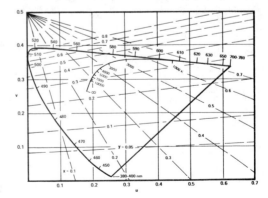

1960 CIE-UCS (u,v) Chromaticity Diagram

chromaticity flicker. The flicker that results from fluctuation of chromaticity only.
18, 178

chrominance. The colorimetric difference between any color and a reference color of equal luminance, the reference color having a specified chromaticity. *Notes:* (1) in three-dimensional color space, chrominance is a vector that lies in a plane of constant luminance. In that plane it may be resolved into components, called chrominance components. (2) In color television transmission, for example, the chromaticity of the reference color may be that of a specified white.   18, 178

chrominance channel (color television system) (television). Any path that is intended to carry the chrominance signal. *See also:* **television.**   63, 178

chrominance channel bandwidth (television). The bandwidth of the path intended to carry the chrominance signal. *See also:* **television.**   163, 178

chrominance components. *See:* **chrominance.**

chrominance demodulator (color television reception). A demodulator used for deriving video-frequency chrominance components from the chrominance signal and a sine wave of chrominance subcarrier frequency.   18, 178

chrominance modulator (color television transmission). A modulator used for generating the chrominance signal from the video-frequency chrominance components and the chrominance subcarrier.   18, 178

chrominance primary (color television). A transmission primary that is one of two whose amounts determine the chrominance of a color. *Note:* Chrominance primaries have zero luminance and are nonphysical. The term is useful only in a linear system.   18

chrominance signal (color television). The sidebands of the modulated chrominance subcarrier that are added to the monochrome signal to convey color information.   18

chrominance signal component (television). A signal resulting from suppressed-carrier modulation of a chrominance subcarrier voltage at a specified phase, by a chrominance primary signal such as the I Video Signal or the Q Video Signal.   163

chrominance signal (television). *See:* **television.**

chrominance subcarrier (color television). The carrier whose modulation sidebands are added to the luminance signal to convey color information.   18

chronaxie (medical electronics). The minimum duration of time required to stimulate with a current of twice the rheobase.   192

chute.* *See:* **feed tube.**

* Deprecated

CIE. Abbreviation for Commission Internationale de l'Eclairage. *Note:* These are the initials of the official French name of the International Commission on Illumination. This translated name is approved for usage in English-speaking countries, but at its 1951 meeting the Commission recommended that only the initials of the French name be used. The initials ICI, which have been used commonly in this country, are deprecated because they conflict with an important trademark registered in England and because the initials of the name translated into other languages are different.   18, 178

CIE standard chromaticity diagram. One in which the x and y chromaticity coordinates are plotted in rectangular coordinates. *See also:* **color.**   167

CIE (1931) standard colorimetric observer. Receptor of

radiation whose colorimetric characteristics correspond to the distribution coefficients $\bar{x}_\lambda, \bar{y}_\lambda, \bar{z}_\lambda$ adopted by the International Commission on Illumination in 1931. *See also:* **color.**   244, 178

circle diagram (rotating machinery). (1) Circular locus describing performance characteristics (current, impedance, etcetera) of a machine or system. In case of rotating machinery, the term **circle diagram** has, in addition, some specific usages: The locus of the armature current phasor of an induction machine, or of some other type of asynchronous machine, displayed on the complex plane, with the shaft speed as the variable (parameter), when the machine operates at a constant voltage and at a constant frequency. (2) The locus of the current vector(s) of a nonsalient-pole synchronous machine, displayed in a synchronously rotating reference frame (Park transform, d-q coordinates), with the active component of the load, hence with the rotor displacement angle, as the variable (parameter), when the machine operates at a constant field current. (3) The locus of the current phasor(s) of (2). *see:* **asynchronous machine.**   63

circling guidance lights. Aeronautical ground lights provided to supply additional guidance during a circling approach when the circling guidance furnished by the approach and runway lights is inadequate. *See also:* **atmospheric transmissivity.**   167

circuit (1). A conductor or system of conductors through which an electric current is intended to flow. *See also:* **center of distribution; service.**   178, 64, 262
(2). A network providing one or more closed paths. *See also:* **network analysis.**   210
(3) (machine winding). The element of a winding that comprises a group of series-connected coils. A single-phase winding or one phase of a polyphase winding may comprise one circuit or several circuits connected in parallel.   63
(4) (circuits and systems). An interconnection of electrical elements. *See also:* **network.**   67

circuit analyzer (multimeter). The combination in a single enclosure of a plurality of instruments or instrument circuits for use in measuring two or more electrical quantities in a circuit. *See also:* **instrument.**   328

circuitation. *See:* **circulation.**

circuit, balanced. *See:* **balanced circuit.**

circuit breaker (1) (general) (A). A device designed to open and close a circuit by nonautomatic means, and to open the circuit automatically on a predetermined overload of current, without injury to itself when properly applied within its rating.   256, 124
(B) A mechanical switching device capable of making, carrying, and breaking currents under normal circuit conditions and also, making, carrying for a specified time, and breaking currents under specified abnormal circuit conditions such as those of short-circuit.
Notes: (1) A circuit breaker is usually intended to operate infrequently, although some types are suitable for frequent operation. (2) The medium in which circuit interruption is performed may be designated by suitable prefix, such as, air-blast circuit breaker, air circuit breaker, compressed-air circuit breaker, gas circuit breaker, oil circuit breaker, vacuum circuit breaker, etcetera. (3) Circuit breakers are classified according to their application or characteristics and these classifications are designated by

the following modifying words or clauses delineating the several fields of application, or pertinent characteristics:

High-voltage power—Rated 1000 volts alternating current or above.

Molded-case—See separate definition.

Low-voltage power—Rated below 1000 volts alternating current or 3000 volts direct current and below, but not including molded-case circuit breakers.

Direct-current low-voltage power circuit breakers are subdivided according to their specified ability to limit fault-current magnitude by being called general purpose, high-speed, semi-high-speed, or anode. For specifications of these restrictions see the latest revision of the applicable American National Standard. *See also:* **alternating-current distribution; switch.**   103, 202

(2) (transmission and distribution). A switching device capable of making, carrying and breaking currents under normal circuit conditions and also making, carrying for a specified time, and breaking currents under specified abnormal conditions such as those of short circuit.   262

circuit breaker, field discharge (enclosed field discharge circuit breakers for rotating electric machinery). A circuit breaker having main contacts for energizing and de-energizing the field of a generator, motor, synchronous condenser, or rotating exciter and having discharge contacts for short-circuiting the field through the discharge resistor at the instant preceding the opening of the circuit breaker main contacts. The discharge contacts also disconnect the field from the discharge resistor at the instant following the closing of the main contacts. For direct-current generator operation, the discharge contacts may open before the main contacts close.

Note: When used in the main field circuit of an alternating or direct-current generator, motor, or synchronous condenser, the circuit breaker is designated as a main field discharge circuit breaker. When used in the field circuit of the rotating exciter of the main machine, the circuit breaker is designated as an exciter field discharge circuit breaker.   359

circuit breaker general purpose low-voltage dc power (low voltage dc power circuit breaker). A circuit breaker, which during interruption does not usually prevent the fault current from rising to its sustained value.   360

circuit breaker, high-speed low-voltage dc power (low voltage dc power circuit breaker). A circuit breaker, which, when applied in a circuit with the parameter values specified in ANSI Standard C37.16, Table 12, tests "b" (5 Amperes per microsecond initial rate-of-rise current), forces a current crest during interruption within 0.01 seconds after the current reaches the pick-up setting of the instantaneous trip device. *Note:* For total performance characteristics at other than test circuit parameter values, consult the manufacturer.   360

circuit breaker interrupting rating (rated interrupting current) (rated interrupting capacity). The highest rms current at a specified operating voltage which a circuit breaker is required to interrupt under the operating duty specified and with a normal frequency recovery voltage equal to the specified operating voltage. *Note:* (1) The current is the rms value, including the dc component, at the instant of contact separation as determined from the

envelope of the current wave. Where limited by testing equipment, the maximum tolerance for normal frequency recovery voltage is 15 percent of the specified operating voltage.

(2) For dc breakers the rated interrupting current is the maximum value of direct current.   3

circuit breaker, semi-high-speed low voltage dc power (low voltage dc power circuit breaker). A circuit breaker which, when applied in a circuit with the parameter values specified in ANSI Standard C37.16, Table 11, tests "b" (1.7 amperes per microsecond initial rate-of-rise of current), forces a current crest during interruption within 0.030 seconds after the current reaches the pick-up setting of the instantaneous trip device. *Note:* For total performance at other than test circuit parameter values, consult the manufacturer.   360

circuit-commutated turn-off time (thyristor). The time interval between the instant when the principal current has decreased to zero after external switching of the principal voltage circuit and the instant when the thyristor is capable of supporting a specified principal voltage without turning on. *See also:* **principal voltage-current characteristic (principal characteristic).**   243, 66, 208, 191

circuit controller. A device for closing and opening electric circuits.   328

circuit efficiency (output circuit of electron tubes). The ratio of (1) the power at the desired frequency delivered to a load at the output terminals of the output circuit of an oscillator or amplifier to (2) the power at the desired frequency delivered by the electron stream to the output circuit. *See also:* **network analysis.**   125, 190

circuit element. A basic constituent part of a circuit, exclusive of interconnections.   328

circuit interrupter. A manually operated device designed to open under abnormal conditions a current-carrying circuit without damage to itself.   124

circuit malfunction analysis (test, measurement and diagnostic equipment). The logical, systematic examination of circuits and their diagrams to (1) identify and analyze the probability and consequence of potential malfunctions and (2) for determining related maintenance and support requirements to investigate effects of failures.   54

circuit, multipoint. *See:* **multipoint circuit.**

circuit noise (telephone practice). Noise that is brought to the receiver electrically from a telephone system, excluding noise picked up acoustically by the telephone transmitters. *See also:* **signal-to-noise ratio.**   59

circuit noise level (at any point in a transmission system) (telecommunication). The ratio of the circuit noise at that point to some arbitrary amount of circuit noise chosen as a reference. *Note:* This ratio is usually expressed in decibels above reference noise, abbreviated "dBrn," signifying the reading of a circuit noise meter, or in adjusted decibels, abbreviated "dBa," signifying circuit noise meter reading adjusted to represent interfering effect under specified conditions. *See also:* **signal-to-noise ratio.**   59

circuit noise meter (noise measuring set). An instrument for measuring circuit noise level. Through the use of a suitable frequency-weighting network and other characteristics, the instrument gives equal readings for noises that are approximately equally interfering. The readings are expressed as circuit noise levels in decibels above reference

noise. *See:* **circuit noise level.** *See also:* **instrument.**
328

circuit parameters. The values of physical quantities associated with circuit elements. *Note:* For example, the resistance (parameter) of a resistor (element), the amplification factor and plate resistance (parameters) of a tube (element), the inductance per unit length (parameter) of a transmission line (element), etcetera. *See also:* **network analysis.**
210

circuit switch (data transmission). A communications switching system which completes a circuit from sender to receiver at the time of transmission (as opposed to a message switch).
59

circuit switching (data communication). A method of communications where an electrical connection between calling and called stations is established on demand for exclusive use of the circuit until the connection is released. *See also:* **packet switching, store-and-forward switching, message switching.**
12

circuit switching element (inverters). A group of one or more simultaneously conducting thyristors, connected in series or parallel or any combination of both, bounded by no more than two main terminals and conducting principal current between these main terminals. *See also:* **self-commutated inverters.**
208

circuit switching system (telephone switching systems). A switching system providing through connections for the exchange of messages.
55

circuit transient recovery voltage. The transient recovery voltage characterizing the circuit and obtained with 100-percent normal-frequency recovery voltage, symmetrical current, and no modifying effect of the interrupting device. *Note:* This voltage indicates the inherent severity of the circuit with respect to recovery voltage phenomena.
103, 202

circuit voltage class (electric power system). A phase-to-phase reference voltage that is used in the selection of insulation class designations for neutral grounding devices.
91

circular array (antennas). An arrangement of elements with centers on a circle or on coplanar concentric circles.
111

circular electric wave. A transverse electric wave for which the lines of electric force form concentric circles. *See also:* **waveguide.**
267, 179

circular interpolation (numerically controlled machines). A mode of contouring control that uses the information contained in a single block to produce an arc of a circle. *Note:* The velocities of the axes used to generate this arc are varied by the control.
224, 207

circularly polarized field vector (antennas). A field vector for which the polarization ellipse is a circle.
111

circularly polarized plane wave (antennas). A plane wave in which the electric field is circularly polarized.
111

circularly polarized wave (1) (general). An elliptically polarized wave in which the ellipse is a circle in a plane perpendicular to the direction of propagation. *See also:* **radiation.**
328

(2) (radio wave propagation). An electromagnetic wave for which either the electric or the magnetic field vector at a fixed point describes a circle. *Note:* This term is usually applied to transverse waves. *See also:* **radiation; radio**

wave propagation.
180, 146

circular magnetic wave. A transverse magnetic wave for which the lines of magnetic force form concentric circles. *See also:* **waveguide.**
267, 179

circular mil. A unit of area equal to $\pi/4$ of a square mil (= 0.7854 square mil). The cross-sectional area of a circle in circular mils is therefore equal to the square of its diameter in mils. A circular inch is equal to one million circular mils. *Note:* A mil is one-thousandth part of an inch. There are 1974 circular mils in a square millimeter.
341

circular orbit (communication satellite). An orbit of a satellite in which the distance between the centers of mass of the satellite and of the primary body is constant.
74

circular probable error (CPE or CEP) (two-dimensional error distribution). The radius of a circle encompassing half of all errors. *See also:* **navigation.**
187

circular scanning (radio). Scanning in which the direction of maximum response generates a plane or a right circular cone whose vertex angle is close to 180 degrees.
179

circulating memory. *See:* **circulating register.**

circulating register (1) (data processing and computation). A register that retains data by inserting it into a delaying means, and regenerating and reinserting the data into the register.
235, 210

(2) Shift register in which data moved out of one end of the register are reentered into the other end as in a closed loop. *See:* **cyclic shift.**
255, 77

circulation (circuitation). The circulation of a vector field is its line integral over a closed curve *C*.
210

circulation of electrolyte. A constant flow of electrolyte through a cell to facilitate the maintenance of uniform conditions of electrolysis. *See also:* **electrorefining.**
328

circulator (waveguide system). A passive waveguide junction of three or more arms in which the arms can be listed in such an order that when power is fed into any arm it is transferred to the next arm on the list, the first arm being counted as following the last in order. *See also:* **transducer; waveguide.**
244, 179

CIRGE. International Conference on Large High Voltage Electric Systems.
59

CISPR. International Special Committee on Radio Interference.

clamp. *See:* **clamping circuit.**

clamp, cable (conductor stringing equipment). A device designed to clamp cables together. It consists of a "u" bolt threaded on both ends, two nuts and a base, and is commonly used to make temporary "bend back" eyes on wire rope. *Syn.* **clip; crosby; crosby clip.**
45

clamper (data transmission). When used in broadband transmissions reinserts low frequency signal components which were not faithfully transmitted.
59

clamping (1) (control) (industrial control). A function by which the extreme amplitude of a waveform is maintained at a given level. *See:* **control system, feedback.**
219, 206

(2) (pulse terms) (operations on a pulse). A process in which a specified instantaneous magnitude of a pulse is fixed at a specified magnitude. Typically, after clamping, all instantaneous magnitudes of the pulse are offset, the pulse shape remaining unaltered.

clamping circuit (clamper) (clamp) (1) (electronic circuits). A circuit that adds a fixed bias to a wave at each occurrence of some predetermined feature of the wave so that the voltage or current of the feature is held at or "clamped" to some specified level. The level may be fixed or adjustable. 328

(2) (analog computers). A circuit used to provide automatic hold and reset action electronically for the purposes of switching or supplying repetitive operation in an analog computer. *See also:* **electronic analog computer.** 9

clamping screw. *See:* **binding screw.**

clamp, strand restraining (conductor stringing equipment). An adjustable circular clamp commonly used to keep the individual strands of a conductor in place and prevent them from spreading when the conductor is cut. It is used as required during conductor splicing. *Syn.* **cable binding block; hose clamp; vise grip plier clamp (homemade).**

clapper. An armature that is hinged or pivoted. 259

class (electric instrument). *See also:* **accuracy rating (electric instrument).**

class-A amplifier (electron tubes). *See:* **amplifier ratings.**

class-AB amplifier (electron tubes). *See:* **amplifier ratings.**

class-AB operation. *See:* **amplifier ratings.**

class-A insulation. *See:* **insulation ratings.**

class-A modulator. A class-A amplifier that is used specifically for the purpose of supplying the necessary signal power to modulate a carrier. *See also:* **modulation.** 111, 240

class-A operation. *See:* **amplifier class ratings.**

class-A push-pull sound track. A class-A push-pull photographic sound track consists of two single tracks side by side, the transmission of one being 180 degrees out of phase with the transmission of the other. Both positive and negative halves of the sound wave are linearly recorded on each of the two tracks. *See also:* **phonograph pickup.** 176

class-B amplifier (electron tubes). *See:* **amplifier class ratings.**

class-B insulation. *See:* **insulation class ratings.**

class-B modulator. A class-B amplifier that is used specifically for the purpose of supplying the necessary signal power to modulate a carrier. *Note:* In such a modulator the class-B amplifier is normally connected in push-pull. *See also:* **modulation.** 111, 240

class-B operation. *See:* **amplifier class ratings.**

class-B push-pull sound track. A class-B push-pull photographic sound track consists of two tracks side by side, one of which carries the positive half of the signal only, and the other the negative half. *Note:* During the inoperative half-cycle, each track transmits little or no light. *See also:* **phonograph pickup.** 176

class-C amplifier (electron tubes). *See:* **amplifier class ratings.**

class-C insulation. *See:* **insulation class ratings.**

class-C operation. *See:* **amplifier class ratings.**

class designation of a watthour meter. Denotes the maximum of the load range in amperes. *See also:* **watthour meter.** 212

class-F insulation. *See:* **insulation class ratings.**

class-H insulation. *See:* **insulation class ratings.**

class-O insulation. *See:* **insulation class ratings.**

class 1 electric equipment. *See:* **nuclear power generating stations, class ratings.**

class 1 structures and equipment. *See:* **nuclear power generating stations, class ratings.**

class 1E (nuclear power generating station). The safety classification of the electric equipment and systems that are essential to emergency reactor shutdown, containment isolation, reactor core cooling, and containment and reactor heat removal, or are otherwise essential in preventing significant release of radioactive material to the environment. *See:* **nuclear power generating stations, class ratings.** 356, 361

class 1E control switchboard. *See:* **nuclear power generating stations, class ratings.**

class 1E electric systems. *See:* **nuclear power generating stations, class ratings.**

class 2 structures and equipment. *See:* **nuclear power generating stations, class ratings.**

class 3 structures and equipment. *See:* **nuclear power generating stations, class ratings.**

class-90 insulation. *See:* **insulation, class ratings.**

class-105 insulation. *See:* **insulation, class ratings.**

class-130 insulation. *See:* **insulation, class ratings.**

class-155 insulation. *See:* **insulation, class ratings.**

class-180 insulation. *See:* **insulation, class ratings.**

class-220 insulation. *See:* **insulation, class ratings.**

class-over-220 insulation. *See:* **insulation, class ratings.**

classification lamp (classification light). A signal lamp placed at the side of the front end of a train or vehicle, displaying light of a particular color to identify the class of service in which the train or vehicle is operating. 328

classification light. *See:* **classification lamp.**

classification of arresters. Arrester classification is determined by prescribed test requirements. These classifications are: station valve arrester, intermediate valve arrester, distribution valve arrester, distribution expulsion arrester, secondary valve arrester, protector tube. 2

class-of-service indication (telephone switching systems). An indication of the features assigned to a switching network termination. 55

class-of-service tone (telephone switching systems). A tone that indicates to an operator that a certain class-of-service is appropriate to a call. 55

"class two" transformer. A step-down transformer of the low-secondary-voltage type, suitable for use in class 2 remote-control low-energy circuits. It shall be of the energy-limiting type, or of a non-energy-limiting type equipped with an overcurrent device. *Note:* Low-secondary-voltage, as used here, has a value of approximately 24 volts. 53

cleaner (electroplating). A compound or mixture used in degreasing, which is usually alkaline. 328

cleaning (electroplating). The removal of grease or other foreign material from a metal surface, chiefly by physical means. *See also:* **electroplating.** 328

clear (electronic computers). (1) To preset a storage or memory device to a prescribed state, usually that denoting zero. (2) To place a binary cell in the zero state. *See:* **reset; set.** 235, 210, 255, 77

clearance (1) (electronic navigation). (A) In instrument landing systems (ILS), a difference in depth of modulation (DDM) in excess of that required to produce full-scale deflection of the course deviation indicator in flight areas outside the on-course sector; when the difference in depth of modulation is too low the indicator falls below full-scale deflection and the condition of low clearance exists. (B) In air-traffic control, permission by a control facility to the pilot to proceed in a mutually understood manner. *See also:* **navigation.** 187

(2) (transmission and distribution). The minimum separation between two conductors, between conductors and supports or other objects, or between conductors and ground. *See also:* **tower.** 64

(3) (conductor stringing equipment). (A) The de-energizing of a circuit to enable work to be performed safely. In the event of an accidental electrical interruption of a circuit where repairs will be required or where a potential hazard exists from contact with an energized circuit, a "clearance" is normally requested prior to starting work. *Syn.* **outage; permit; restriction.** 45
(B) The minimum separation between two conductors, between conductors and supports or other objects, or between conductors and ground, or the clear space between any objects. 45

(4) (power switchgear). *See:* **minimum clearance.** 103

clearance antenna array (directional localizer). The antenna array that radiates a localizer signal on a separate frequency within the pass band of the receiver and provides the required signals in the clearance sectors as well as a back course. *See also:* **navigation.** 187

clearance lamp. A lighting device for the purpose of indicating the width of a vehicle. *See also:* **headlamp.** 167

clearance point. The location on a turnout at which the carrier's specified clearance is provided between tracks. 328

clearance sector (instrument landing systems). The sector extending around either side of the localizer from the course sector to the back course sector, and within which the deviation indicator provides the required off-course indication. *See also:* **navigation.** 187

clearing circuit. A circuit used for the operation of a signal in advance of an approaching train. 328

clearing time (1) (fuse) (total clearing time). The time elapsing from the beginning of an overcurrent to the final circuit interruption. *Note:* The clearing time is equal to the sum of melting time and arcing time. 103, 202
(2) (mechanical switching device). The interval between the time the actuating quantity in the main circuit reaches the value causing actuation of the release and the instant of final arc extinction on all poles of the primary arcing contacts. *Note:* Clearing time is numerically equal to the sum of contact parting time and arcing time. 103

clearing-out drop (cord circuit or trunk circuit). A drop signal that is operated by ringing current to attract the attention of the operator. 328

clear sky. A sky that has less than 30-percent cloud cover. *See also:* **sunlight.** 167

cleat. An assembly of two pieces of insulating material provided with grooves for holding one or more conductors at a definite spacing from the surface wired over and from each other, and with screw holes for fastening in position. *See also:* **raceway.** 328

clerestory. That part of a building that rises clear of the roofs or other parts and whose walls contain windows for lighting the interior. *See also:* **sunlight.** 167

click. A disturbance of a duration less than a specified value as measured under specified conditions. *Note:* For the specified values and conditions, guidance should be found in International Special Committee on Radio Interference (CISPR) publications. *See also:* **electromagnetic compatibility.** 226, 199

climbing space (1) (wiring system). The vertical space reserved along the side of a pole or tower to permit ready access for linemen to equipment and conductors located thereon. 64, 252
(2) (transmission and distribution). The vertical space reserved along the side of a pole or structure to permit ready access for lineman to equipment and conductors located on the pole structure. 262

clip (clipping) (radiation detectors). A limiting operation such as (1) use of a high-pass filter (*See:* **differentiated**), or (2) a nonlinear operation such as diode limiting of pulse amplitude. 23, 118, 119

clipper (peak chopper). A device that automatically limits the instantaneous value of the output to a predetermined maximum value. *Note:* The term is usually applied to devices that transmit only portions of an input wave lying on one side of an amplitude boundary. 328

clipper amplifier. *See:* **amplifier, clipper.**

clipper limiter. A transducer that gives output only when the input lies above a critical value and a constant output for all inputs above a second higher critical value. *Note:* This is sometimes called an amplitude gate, or slicer. *See also:* **transducer.** 111

clipping (voice-operated telephone circuit). The loss of initial or final parts of words or syllables due to nonideal operation of the voice-operated devices. 328

clipping-in (conductor stringing equipment). The transferring of sagged conductors from the travelers to their permanent suspension positions and the installing of the permanent suspension clamps. *Syn.* **clamping-in; clipping.** 45

clipping offset (conductor stringing equipment). A calculated distance, measured along the conductor immediately after initial sag while the conductor is in the travelers, between a point on the conductor which lies vertically below the insulator point of suspension on the structure and a point on the conductor at which the center of the suspension clamp is to be placed. The application of clipping offsets will balance the horizontal forces on each structure. 45

clip. *See:* **contact clip; fuse clip.**

clock. (1) A device that generates periodic signals used for synchronization. (2) A device that measures and indicates time. (3) A register whose content changes at regular intervals in such a way as to measure time. 77

clocked logic (power-system communication). The technique whereby all the memory cells (flip-flops) of a logic network are caused to change in accordance with logic input levels but at a discrete time. *See also:* **digital.** 59

clocking (data transmission). The generation of periodic signals used for synchronization. *See also:* **data processing.** 194

close coupling (tight coupling). Any degree of coupling greater than the critical coupling. *See also:* **coupling; critical coupling.** 328

closed air circuit (rotating machinery). A term referring to duct-ventilated apparatus used in conjunction with external components so constructed that while it is not necessarily airtight, the enclosed air has no deliberate connection with the external air. *Note:* The term must be qualified to describe the means used to circulate the cooling air and to remove the heat produced in the apparatus. 63

closed amortisseur. An amortisseur that has the end connections connected together between poles by bolted or otherwise separable connections.
328

closed circuit cooling (rotating machinery). A method of cooling in which a primary coolant is circulated in a closed circuit through the machine and if necessary a heat exchanger. Heat is transferred from the primary coolant to the secondary coolant through the structural parts or in the heat exchanger. 63

closed-circuit principle. The principle of circuit design in which a normally energized electric circuit, on being interrupted or de-energized, will cause the controlled function to assume its most restrictive condition. 328

closed-circuit signaling (telecommunication). That type of signaling in which current flows in the idle condition, and a signal is initiated by increasing or decreasing the current.
59

closed-circuit transition (industrial control). As applied to reduced-voltage controllers, including star-delta controllers, a method of starting in which the power to the motor is not interrupted during the starting sequence. *See:* **electric controller.** 225, 206

closed circuit transition auto-transformer starting (rotating machinery). The process of auto-transformer starting whereby the motor remains connected to the supply during the transition from reduced to rated voltage. 63

closed-circuit voltage (working voltage) (battery). The voltage at its terminals when a specified current is flowing. *See also:* **battery (primary or secondary); working voltage.**
328

closed loop (feedback loop) (automatic control) (industrial control). A signal path that includes a forward path, a feedback path, and a summing point and that forms a closed circuit. *See:* **control system, feedback.** 219, 206

closed-loop control system (control system, feedback). A control system in which the controlled quantity is measured and compared with a standard representing the desired performance. *Note:* Any deviation from the standard is fed back into the control system in such a sense that it will reduce the deviation of the controlled quantity from the standard. *See:* **control; network analysis.**
151, 277, 206, 94

closed-loop gain (operational gain) (power supplies). The gain, measured with feedback, is the ratio of the voltage appearing across the output terminal pair to the causative voltage required at the input resistor. If the open-loop gain is sufficiently large, the closed-loop gain can be satisfactorily approximated by the ratio of the feedback resistor to the input resistor. *See:* **open -loop gain.** *See also:* **power supply.** 228, 176

closed-loop series street lighting system. Street lighting system that employs two-wire series circuits in which the return wire is always adjacent. *See also:* **alternating-current distribution; direct-current distribution.** 328

closed loop testing (test, measurement and diagnostic equipment). Testing in which the input stimulus is controlled by the equipment output monitor. 54

closed-numbering plan (telephone switching systems). A numbering plan in which a fixed number of digits is always dialed. 55

closed region. An open region and its boundary. 210

closed subroutine (computing system). A subroutine that can be stored at one place and can be connected to a routine by linkages at one or more locations. *See:* **open subroutine; subroutine, closed.** 255, 77

close-open operation (switching device). A close operation followed immediately by an open operation without purposely delayed action. *Note:* The letters CO signify this operation: close-open. 103, 202

close operation (switching device). The movement of the contacts from the normally open to the normally closed position. *Note:* The letter C signifies this operation: close. 103, 202

close-talking microphone. A microphone designed particularly for use close to the mouth of the speaker. *See also:* **microphone.** 328

close-talking pressure-type microphones. An acoustic transducer that is intended for use in close proximity to the lips of the talker and is either hand-held or boom-mounted. *Notes:* (1) Various types of microphones are currently used for close-talking applications. These include carbon, dynamic, magnetic, piezoelectric, electrostrictive, and capacitor types. Each of these microphones has only one side of its diaphragm exposed to sound waves, and its electric output substantially corresponds to the instantaneous sound pressure of the impressed sound wave. (2) Since a close-talking microphone is used in the near sound field produced by a person's mouth, it is necessary when measuring the performance of such microphones to utilize a sound source that approximates the characteristics of the human sound generator. 249

close-time delay-open operation (switching device). A close operation followed by an open operation after a purposely delayed action. *Note:* The letters CTO signify this operation: close-time delay-open. 103, 202

closing coil (switching device). A coil used in the electromagnet that supplies power for closing the device. *Note:* In an air-operated, or other stored-energy-operated device, the closing coil may be the coil used to release the air or other stored energy that in turn closes the device.
103, 202

closing operating time (switch). The interval during which the switch is being operated to move from the fully open position to the fully closed position. 118, 103

closing relay (electrically operated device). A form of auxiliary relay used to control the closing and opening of the closing circuit of the device so that the main closing current does not pass through the control switch or other initiating device. 103, 202, 60, 127

closing time (mechanical switching device). The interval of time between the initiation of the closing operation and the instant when metallic continuity is established in all poles. *Notes:* (1) It includes the operating time of any auxiliary equipment necessary to close the switching device, and that forms an integral part of the switching device. (2) For switching devices that embody switching resistors, a distinction should be made between the closing time up to the instant of establishing a circuit at the secondary arcing contacts, and the closing time up to the establishment of a circuit at the main or primary arcing contacts, or both.   103, 202, 27

cloud chamber smoke detector (fire protection devices). A device which is a form of sampling detector. The air pump draws a sample of air into a high humidity chamber within the detector. After the air is in the humidity chamber, the pressure is lowered slightly. If smoke particles are present, the moisture in the air condenses on them forming a cloud in the chamber. The density of this cloud is then measured by the photoelectric principle. When the density is greater than a predetermined level, the detector responds to the smoke.   71

cloud pulse (charge-storage tubes). The output resulting from space-charge effects produced by the turning on or off of the electron beam. *See also:* **charge-storage tube.**
174, 125, 190

cloudy sky. A sky that has more than 70-percent cloud cover. *See also:* **sunlight.**   167

CLR. *See:* **recording-completing trunk (combined line and recording trunk).**

clutter (radar). Unwanted echoes, typically from the ground, sea, rain and other hydrometeors, chaff, birds, insects and aurora.   13

clutter attenuation (radar moving-target indicator). The ratio of clutter power at the canceller input to clutter residue at the output, normalized to the attenuation for a single pulse passing through the unprocessed channel of the canceller. *Syn.* **clutter cancellation** or **suppression.** *See:* **moving-target indicator improvement factor.**   13

clutter improvement factor (radar moving-target indicator). *See:* **moving-target indicator improvement factor.**   13

clutter reflectivity (radar). The backscattering coefficient of a unit volume, or of a unit area, containing clutter sources. *See also:* **navigation.**   187

clutter residue (radar moving-target indicator). The clutter power remaining at the output of a moving-target indicator system. *Note:* It is the sum of several (generally uncorrelated) components resulting from radar instabilities, antenna scanning, relative motion of the radar with respect to the sources of clutter and fluctuations of the clutter reflectivity. *See also:* **navigation.**   187

clutter visibility factor (radar). The predetection signal-to-clutter ratio that provides stated probabilities of detection and false alarm on a display; in moving-target indicator systems, it is the ratio after cancellation or Doppler filtering. When the clutter residue has the spectrum of thermal noise, this factor is the same as the visibility factor.   13, 187

CMRR. *See:* **common-mode rejection ratio.**

C **network.** A network composed of three impedance branches in series, the free ends being connected to one pair of terminals, and the junction points being connected to another pair of terminals. *See also:* **network analysis.**
328

CO. *See:* **close-open operation (switching device).**

coagulating current. *See:* **Tesla current (electrotherapy).**

coal cleaning equipment. Equipment generally electrically driven, to remove impurities from the coal as mined, such

as slate, sulphur, pyrite, shale, fire clay, gravel, and bone.
328

coarse chrominance primary (National Television System Committee color television).* *See:* **Q chrominance signal.**
18

* Deprecated

coated fabric (coated mat) (rotating machinery). A fabric or mat in which the elements and interstices may or may not in themselves be coated or filled but that has a relatively uniform compound or varnish finish on either one or both surfaces. *See also:* **rotor (rotating machinery); stator.**   63

coated magnetic tape (magnetic powder-coated tape). A tape consisting of a coating of uniformly dispersed, powdered ferromagnetic material (usually ferromagnetic oxides) on a nonmagnetic base. *See also:* **magnetic tape; phonograph pickup.**   176

coated mat. *See:* **coated fabric.**

coating (electroplating). The layer deposited by electroplating. *See also:* **electroplating.**   328

coaxial (coaxial line) (concentric line) (coaxial pair*). A transmission line formed by two coaxial conductors. *See also:* **cable; transmission line.**   210

* Deprecated

coaxial antenna. An antenna comprised of a quarter-wavelength extension to the inner conductor of a coaxial line and a radiating sleeve which in effect is formed by folding back the outer conductor of the coaxial line for approximately one-quarter wavelength. *See also:* **antenna.**   111, 179

coaxial cable. A cable containing one or more coaxial lines. *See also:* **cable.**   59

coaxial conductor. An electric conductor comprising outgoing and return current paths having a common axis, one of the paths completely surrounding the other throughout its length.   14

coaxial line. *See:* **coaxial.**

coaxial pair*. *See:* **coaxial.**

* Deprecated

coaxial relay. A relay that opens and closes an electric contact switching high-frequency current as required to maintain minimum losses. *See also:* **relay.**   341

coaxial stop filter (electromagnetic compatibility). A tuned movable filter set round a conductor in order to limit the radiating length of the conductor for a given frequency. *See also:* **electromagnetic compatibility.**   220, 199

coaxial stub. A short length of coaxial that is joined as a branch to another coaxial. *Note:* Frequently a coaxial stub is short-circuited at the outer end and its length is so chosen that a high or low impedance is presented to the main coaxial in a certain frequency range. *See also:* **waveguide.**   328

coaxial switch. A switch used with and designed to simulate the critical electric properties of coaxial conductors.   346

coaxial transmission line. A transmission line consisting of two coaxial cylindrical conductors. *See also:* **radio receiver; transmission line.**   267, 179

co-channel interference. Interference caused in one communication channel by a transmitter operating in the same channel. *See also:* **radio transmission.**   178

code (1) (general). A plan for representing each of a finite number of values or symbols as a particular arrangement or sequence of discrete conditions or events.   328

(2) (electronic computers). (A) The characters or expressions of an originating or source language, each correlated with its equivalent expression in an intermediate or target language, for example, alphanumeric characters correlated with their equivalent 6-bit expressions in a binary machine language. *Note:* For punched or magnetic tape; a predetermined arrangement of possible locations of holes or magnetized areas and rules for interpreting the various possible patterns. (B) Frequently, an expression in the target language. (C) Frequently, the set of expressions in the target language that represent the set of characters of the source language. (D) To encode is to express given information by means of a code. (E) To translate the program for the solution of a problem on a given computer into a sequence of machine-language or pseudo instructions acceptable to that computer.

235, 224, 207

code character. A particular arrangement of code elements representing a specific symbol or value.   194, 59

code conversion (telephone switching systems). The substitution of a routing code for a destination code.   55

coded-decimal code. The decimal number system with each decimal digit expressed by a code.   224, 207

coded fire-alarm system. A local fire-alarm system in which the alarm signal is sounded in a predetermined coded sequence. *See also:* **protective signaling.**   328

coded pulse (radar). A pulse with internal (intra-pulse) amplitude, frequency, or phase modulation used for identification and signal processing purposes.   13

coded track circuit. A track circuit in which the energy is varied or interrupted periodically.   328

code element. One of the discrete conditions or events in a code, for example, the presence or absence of a pulse. *See also:* **data processing; information theory.**   194

code letter (locked-rotor kilovolt-amperes). A letter designation under the caption "code" on the nameplate of alternating-current motors (except wound-rotor motors) rated $\frac{1}{20}$ horsepower and larger to designate the locked-rotor kilovolt-amperes per horsepower as measured at rated voltage and frequency.   63

coder (1) (general). A device that sets up a series of signals in code form.   328

(2) (code transmitter). A device used to interrupt or modulate the track or line current periodically in various ways in order to establish corresponding controls in other apparatus.   328

code ringing (telephone switching systems). Ringing wherein the number of rings or the duration, or both, indicate which station on a party line is being called.   55

code system. A system of control of wayside signals, cab signals, train stop or continuous train control in which electric currents of suitable character are supplied to control apparatus, each function being controlled by its own distinctive code. *See also:* **block signal system.**

328

code translator. *See:* **digital converter.**

coding (1) (communication). The process of transforming messages or signals in accordance with a definite set of rules.   242

(2) (computing systems). Loosely, a routine. *See:* **relative coding; straight-line coding; symbolic coding.**   77

(3) (test, measurement and diagnostic equipment). A part of the programming process in which a completely defined, detailed sequence of operations is translated into computer-entry language.   54

coding delay (loran). An arbitrary time delay in the transmission of pulse signals from the slave station to permit the resolution of ambiguities. *Note:* The term **suppressed time delay** more accurately represents what is being accomplished and should be used instead of **coding delay.** *See also:* **navigation.**   13

coding fan. *See:* **electrode radiator.**

coding siren. A siren having an auxiliary mechanism to interrupt the flow of air through the device, thereby enabling it to produce a series of sharp blasts as required in code signaling. *See also:* **protective signaling.**   328

coefficient of attenuation (illuminating engineering). The decrement in flux per unit distance in a given direction within a medium and is defined by the relation: $\Phi_x = \Phi_0 e^{-\mu} x$ where Φ_x is the flux at any distance x from a reference point having flux Φ_0. *See also:* **lamp.**   167

coefficient of beam utilization. The ratio of (1) the luminous flux (lumens) reaching a specified area directly from a floodlight or projector to (2) total beam luminous flux. *See also:* **inverse-square law.**   167

coefficient of capacitance (system of conductors) (coefficients of capacitance and induction—Maxwell). The coefficients in the array of linear equations that express the charges on the conductors in terms of their potentials. For a set of n conductors, $n - 1$ of which are insulated from each other and mounted on insulating supports within a closed conducting shell or on one side of a conducting plane of infinite extent, the shell or plane as a common conductor (or ground) is taken to be at zero potential and the charge on each insulated conductor is regarded as having been transferred from the common conductor. Hence the algebraic sum of the charges on the $n - 1$ insulated conductors is equal and opposite in sign to the charge on the common conductor. The charges on the insulated conductors are given by the equations

$$Q_1 = c_{11}V_1 + c_{12}V_2 + c_{13}V_3 \cdots$$
$$+ c_{1(n-1)}V_{n-1}$$
$$Q_2 = c_{21}V_1 + c_{22}V_2 + c_{23}V_3 \cdots$$
$$+ c_{2(n-1)}V_{n-1}$$
$$Q_{n-1} = c_{(n-1)1}V_1 + c_{(n-1)2}V_2$$
$$+ c_{(n-1)3}V_3 + \cdots + c_{(n-1)(n-1)}V_{n-1}$$

The coefficients c_{rr} are the self-capacitances and the coefficients c_{rp} are the negatives of the direct (or mutual) capacitances of the system. *Note:* Under the conventions stated both self and direct capacitances have positive signs; also $c_{rp} = c_{pr}$. If the mth conductor completely surrounds the kth conductor and no other, then $c_{kp} = 0$ for $p \neq k$ or m; and $c_{kk} = c_{km}$. *See also:* **network analysis.**   210

coefficient of coupling. *See* **coupling coefficient.**

coefficient of grounding (surge arrester). The ratio (Elg/Ell) expressed as a percentage, of the highest root-mean-square line-to-ground power-frequency voltage (Elg) on a sound phase, at a selected location, during a fault to earth (ground) affecting one or more phases to the line-to-line power-frequency voltage (Ell) that would be obtained, at the selected location, with the fault removed. *Notes:* (1) Coefficients of grounding for three-phase systems are calculated from the phase-sequence impedance components as viewed from the selected location. For machines use the subtransient reactance. (2) The coefficient of grounding is useful in the determination of an arrester rating for a selected location. (3) A value not exceeding 80 percent is obtained approximately when for all system conditions the ratio of zero-sequence reactance to positive-sequence reactance is positive and less than three and the ratio of zero-sequence resistance to positive-sequence reactance is positive and less than one.   2, 53

coefficient of performance (1) (thermoelectric cooling couple). The quotient of (A) the net rate of heat removal from the cold junction by the thermoelectric couple by (B) the electric power input to the thermoelectric couple. *Note:* This is an idealized coefficient of performance assuming perfect thermal insulation of the thermoelectric arms. *See also:* **thermoelectric device.**   248, 191

(2) (thermoelectric cooling device). The quotient of (A) the rate of heat removal from the cooled body by (B) the electric power input to the device. *See also:* **thermoelectric device.**   248, 191

(3) (thermoelectric heating device). The quotient of (A) the rate of heat addition to the heated body by (B) the electric power input to the device. *See also:* **thermoelectric device.**   248, 191

(4) (thermoelectric heating couple). The quotient of (A) the rate of heat addition to the hot junction by the thermoelectric couple by (B) the electric power input to the thermoelectric couple. *Note:* This is an idealized coefficient of performance assuming perfect thermal insulation of the thermoelectric arms. *See also:* **thermoelectric device.**   248, 191

coefficient of performance, reduced (thermoelectric device). The ratio of (1) a specified coefficient of performance to (2) the corresponding coefficient of performance of a Carnot cycle. *See also:* **thermoelectric device.**   248, 191

coefficient of trip point repeatability. *See:* **trip-point repeatability coefficient.**

coefficient of utilization (illuminating engineering). The ratio of (1) luminous flux (lumens) calculated as received on the work plane to (2) the rated lumens emitted by the lamps alone. *See also:* **inverse-square law.**   167

coefficient of zero error. *See:* **environmental coefficient.**

coefficient potentiometer. *See also:* **electronic analog computer; parameter potentiometer.**

coefficient, sensitivity. *See:* **sensitivity coefficient.**

coefficients of capacitance and induction—Maxwell. *See:* **coefficients of capacitance (system of conductors).**   210

coefficients of elastance (system of conductors) (coefficients of potential—Maxwell). The coefficients in the array of linear equations that express the potentials of the conductors in terms of their charges. For the system of n conductors (including ground) described in **coefficients of capacitance (system of conductors)** the equations for the potentials of the insulated conductors are

$$V_1 = s_{11}Q_1 + s_{12}Q_2 + s_{13}Q_3$$
$$+ \cdots + s_{1(n-1)}Q_{n-1}$$

$$V_2 = s_{21}Q_1 + s_{22}Q_2 + s_{23}Q_3$$
$$+ \cdots + s_{s2(n-1)}Q_{n-1}$$

$$\vdots$$

$$V_{n-1} = s_{(n-1)1}Q_1 + s(n-1)_2Q_2 + s_{(n-1)3}Q_3$$
$$+ \cdots + s_{(n-1)(n-1)}Q_{n-1}$$

The coefficients s_{rr} are the self-elastances and the coefficients s_{rp} are the negatives of the mutual elastances of the system. *Note:* The matrix of the coefficients of elastance is the reciprocal of the matrix of the coefficients of capacitance. *See also:* **network analysis.**   210

coefficients of potential—Maxwell. *See:* **coefficients of elastance (system of conductors); elastances (system of conductors).**   210

coercive electric field (ferroelectric material). For a given hysteresis loop, the value of the applied electric field at which the polarization equals zero. *Note:* The value of the coercive field E_C (see accompanying figure) at a given temperature is dependent on the frequency, amplitude and wave shape of the applied alternating electric field. To facilitate comparison of data, the coercive field, unless otherwise specified, should be determined with a 60-hertz sinusoidal electric field with an amplitude at least three times the coercive field to insure saturation. The coercive field in the meter-kilogram-second-ampere (MKSA) system is expressed in volts per meter. *See also:* **remanent polarization: poling.**   80

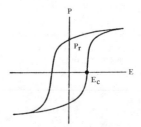

Ferroelectric hysteresis loop

coercive force (1) (general). The magnetizing force at which the magnetic flux density is zero when the material is in a symmetrically cyclically magnetized condition. *Note:* Coercive force is not a unique property of a magnetic material but is dependent upon the conditions of measurement. *See also:* **coercivity.**   210

(2) (magnetic core testing) (H_C). The magnetic field strength at which the magnetic induction is zero, when the core material is in a symmetrically cyclically magnetized condition, with a specified maximum value of field strength, (that is, loci of points on the hysteresis curve when B = 0).   165

coercive voltage (ferroelectric device). The voltage at which the charge on a continuously cycled hysteresis loop is zero. *Note:* Coercive voltage (V_c) is dependent on the frequency, amplitude, and wave shape of the alternating signal. In order to have a basis of comparison, this quantity should be determined from a 60-hertz sinusoidal hysteresis loop with the applied signal of sufficient amplitude to insure saturation. The peak applied voltage and the sample thickness should accompany the data. *See:* figure under **total charge.**   247

coercivity. The property of a magnetic material measured by the coercive force corresponding to the saturation induction for the material. *Note:* This is a quasi-static property only.   210

cofactor (or path cofactor) (network analysis). *See:* **path (loop) factor.**

coffer (illuminating engineering). A recessed panel or dome in the ceiling. *See also:* **luminaire.**   167

cogging (rotating machinery). Variations in motor torque at very low speeds caused by variations in magnetic flux due to the alignment of the rotor and stator teeth at various positions of the rotor. *See also:* **rotor (rotating machinery); stator.**   63

cohered video (in radar moving-target indicator). Video-frequency signal output employed in a coherent system. *See also:* **navigation.**   13, 187

coherence (laser-maser). The correlation between electromagnetic fields at points which are separated in space and in time or both.   363

coherence area (laser-maser). A quantitative measure of **spatial coherence.** The largest cross-sectional area of a light beam, such that light from this area (passing through any two pin holes placed in this area) will produce interference fringes.   363

coherence time (laser-maser). A quantitative measure of **temporal coherence.** The maximum delay time which can be introduced between the two beams in a Michelson interferometer before the interference fringes disappear.
                                                          363

coherent (laser-maser). A light beam is said to be coherent when the electric vector at any point in it is related to that at any other point by a definite, continuous sinusoidal function.   363

coherent interrupted waves. Interrupted continuous waves occurring in wave trains in which the phase of the waves is maintained through successive wave trains. *See also:* **wave front.**   328

coherent moving-target indicator (radar moving-target indicator). A system in which the target echo is selected on the basis of its Doppler frequency when compared to a local reference frequency maintained by a coherent oscillator. *See also:* **navigation.**   187

coherent oscillator (coho) (radar moving-target indicator). An oscillator that provides a reference phase by which changes in the radio-frequency phase of successively received pulses may be recognized. *See also:* **navigation.**
                                                       13, 187

coherent pulse operation. The method of pulse operation in which a fixed phase relationship is maintained from one pulse to the next. *See also:* **pulse.**   328

coherent signal processing (radar). Echo integration, filtering or detection using amplitude and phase of the signal referred to a coherent oscillator, usually by splitting up into an inphase and a quadrature component of the signal.
                                                           13

coherent video (radar). Bipolar video obtained from a synchronous (coherent) detector.   13

coho. Abbreviation for coherent oscillator.   13, 187

coil (1) (general). An assemblage of successive convolutions of a conductor.

(2) (rotating machinery). A unit of a winding consisting of one or more insulated conductors connected in series and surrounded by common insulation, and arranged to link or produce magnetic flux. *See also:* **rotor (rotating machinery); stator.**   210, 63

coil brace (1) (coil support). A structure for the support or restraint for one or more coils.

(2) (V wedge, salient-pole construction). A trapezoidal insulated insert clamped between field poles, to provide radial restraint for the field coil turns against centrifugal force and to brace the coils tangentially. *See also:* **stator.**
                                                           63

coil end-bracing (rotating machinery). *See:* **end-winding support.** *See also:* **rotor (rotating machinery); stator.**

coil insulation (rotating machinery). The main insulation to ground or between phases surrounding a coil, additional to any conductor or turn insulation. *See also:* **rotor (rotating machinery); stator.**   63

coil lashing (rotating machinery). The binding used to attach a coil end to the supporting structure. *See also:* **rotor (rotating machinery); stator.**   63

coil loading. Loading in which inductors, commonly called loading coils, are inserted in a line at intervals. *Note:* The loading coils may be inserted either in series or in shunt. As commonly understood, coil loading is a series loading in which the loading coils are inserted at uniformly spaced recurring intervals. *See also:* **loading.**   328

coil pitch (rotating machinery). The distance between the two active conductors (coil sides) of a coil, usually expressed as a percentage of the pole pitch. *See:* **armature.**
                                                           63

coil Q (dielectric heating usage). Ratio of reactance to resistance measured at the operating frequency. *Note:* The loaded-coil Q is that of a heater coil with the charge in position to be heated. Correspondingly, the unloaded-coil Q is that of a heater coil with the charge removed from the coil. *See also:* **dielectric heating.**   14

coil section (rotating machinery). The basic electrical element of a winding comprising an assembly of one or more turns insulated from one another.   63

coil shape factor (dielectric heating usage). A correction factor for the calculation of the inductance of a coil based on its diameter and length. *See also:* **dielectric heating.**
                                                           14

coil side (rotating machinery). Either of the two normally straight parts of a coil that lie in the direction of the axial length of the machine. *See also:* **rotor (rotating machinery); stator.**   63

coil-side separator (rotating machinery). Additional insulation used to separate embedded coil sides. *See also:* **rotor (rotating machinery); stator.**   63

coil space factor. The ratio of the cross-sectional area of the conductor metal in a coil to the total cross-sectional area of the coil. *Note:* If the overall insulation, such as spool bodies or stop linings, is omitted from consideration when the space factor is calculated, the omission should be specifically stated. *See:* **asynchronous machine; direct-current commutating machine.**   328

coil span (rotating machinery). *See:* **coil pitch.** *See also:*

rotor (rotating machinery); stator.

coil support bracket (rotating machinery). A bracket used to mount a coil support ring or binding band. *See also:* rotor (rotating machinery); stator. 63

coin box. A telephone set equipped with a device for collecting coins in payment for telephone messages. *See also:* telephone station. 328

coin call (telephone switching systems). A call in which a coil collection device is used. 55

coincidence (radiation counters). The practically simultaneous production of signals from two or more counter tubes. *Note:* A genuine or true coincidence is due to signals from related events (passage of one particle or of two or more related particles through the counter tubes); an accidental, spurious, or chance coincidence is due to unrelated signals that coincide accidentally. *See:* **anticoincidence (radiation counters).** 190

coincidence circuit. A circuit that produces a specified output pulse when and only when a specified number (two or more) or a specified combination of input terminals receives pulses within an assigned time interval. *See also:* anticoincidence (radiation counters); pulse. 117

coincidence factor (electric power utilization). The ratio of the maximum coincident total demand of a group of consumers to the sum of the maximum power demands of individual consumers comprising the group both taken at the same point of supply for the same time. *See also:* generating station. 64

coincident-current selection. The selection of a magnetic cell for reading or writing, by the simultaneous application of two or more currents. 77

coincident demand (electric power utilization). Any demand that occurs simultaneously with any other demand, also the sum of any set of coincident demands. *See also:* alternating-current distribution. 200

coin-control signal (telephone switching systems). On a coin call, one of the signals used for collecting or returning coins. 55

coin-denomination tone (telephone switching systems). The tone that indicates the value of coins when they are deposited in a coin telephone. 55

coin tone (telephone switching systems). A class-of-service tone that indicates to an operator that the call has originated from a coin telephone. 55

cold cathode. A cathode that functions without the application of heat. *See also:* electrode (electron tube). 328

cold-cathode glow-discharge tube (glow tube). A gas tube that depends for its operation on the properties of a glow discharge. 190

cold-cathode lamp. (1) An electric-discharge lamp whose mode of operation is that of a glow discharge, and that has electrodes so spaced that most of the light comes from the positive column between them. (2) An electric-discharge lamp in which the electrodes, operating at less than incandescent temperatures, furnish an electron current by field emission, and in which the cathode drop is relatively high (75–150 volts). *Note:* The current density at the cathodes is relatively low, and cathodes become impracticably large for currents greater than a few hundred milliamperes. *See also:* lamp. 167, 268

cold-cathode stepping tube (electron device). A glow dis-

charge tube having several main gaps with or without associated auxiliary gaps, and in which the main discharge has two or more stable positions and can be made to step in sequence, when a suitable shaped signal is applied to an input electrode, or a group of input electrodes. 190

cold-cathode tube. An electron tube containing a cold cathode. 125, 190

cold reserve. Thermal generating capacity available for service but not maintained at operating temperature. 64

cold test (test, measurement and diagnostic equipment). *See:* passive test.

collapsing loss (radar). Loss of information occurring when envelope-detected noise from resolution elements not containing the signal is added to the signal during processing, for example, occurs when radar return containing range, azimuth, and elevation information is constrained to a two dimensional display. 13

collapsing ratio (radar). The total number of envelope detected noise samples added to the signal divided by the number which originated in the resolution cell containing the signal. 13

collar (cheek, field-coil flange) (washer) (rotating machinery). Insulation between the field coil and the pole shoe (top collar) and between the field coil and the member carrying the pole body (bottom collar). *See also:* rotor (rotating machinery). 63

collate. To compare and merge two or more similarly ordered sets of items into one ordered set. 255, 77

collating sequence. An ordering assigned to a set of items, such that any two sets in that assigned order can be collated. 255, 77

collator. A device to collate sets of punched cards or other documents into a sequence. 255, 77

collect call (telephone switching systems). A call for which the called customer agrees to pay. 55

collection efficiency (quantum yield). The number of carriers crossing the p-n junction per incident photon. 113

collector (1) (rotating machinery). An assembly of collector rings, individually insulated, on a supporting structure. *See:* asynchronous machine. 63

(2) (electron tube). An electrode that collects electrons or ions that have completed their functions within the tube. *See also:* electrode (electron tube). 190, 125

(3) (transistor). A region through which primary flow of charge carriers leaves the base. 245, 66, 210

collector grid (1) (photoelectric converter). A specially designed metallic conductor electrically bonded to the *p* or *n* layer of a photoelectric converter to reduce the spreading resistance of the device by reducing the mean path of the current carriers within the semiconductor. *See also:* semiconductor. 186

(2) (solar cells). A pattern of conducting material making ohmic contact to the active surface of a solar cell to reduce the series resistance of the device by reducing the mean path of the current carriers within the semiconductor. 113

collector junction. *See:* junction, collector.

collector plates. Metal inserts embedded in the cell lining to minimize the electric resistance between the cell lining and the current leads. *See also:* fused electrolyte. 328

collector ring (slip ring). A metal ring suitably mounted on an electric machine that (through stationary brushes bearing thereon) conducts current into or out of the rotating member. *See:* **asynchronous machine.** 328

collector-ring (slip-ring) lead insulation (rotating machinery). Additional insulation, applied to the leads that connect the collector rings to the windings of the rotating member, to prevent grounding to the metallic parts of the rotating members, and to provide electrical separation between leads. *See also:* **rotor (rotating machinery).** 63

collector-ring (slip-ring) shaft insulation (rotating machinery). The combination of insulating members that insulate the collector rings from the parts of the structure that are mounted on the shaft. *See also:* **rotor (rotating machinery).** 63

collimate (storage tubes). To modify the paths of electrons in a flooding beam or of various rays of a scanning beam in order to cause them to become more nearly parallel as they approach the storage assembly. *See also:* **storage tube.** 174, 190

collimated beam (laser-maser). Effectively, a parallel beam of light with very low divergence or convergence. 363

collimating lens (storage tubes). An electron lens that collimates an electron beam. *See also:* **storage tube.** 174, 190

collimator (laser-maser). An optical device for converting a diverging or converging beam of light into a collimated or parallel one. 363

collinear array (antenna). A linear array of radiating elements, usually dipoles, with their axes lying in a straight line. *See also:* **antenna.** 179, 111

collision frequency (plasma) (radio wave propagation). The average number of collisions per second of a charged particle of a given species with particles of another or the same species. *See also:* **radio wave propagation.** 180, 146

colloidal ions. Ions suspended in a medium, that are larger than atomic or molecular dimensions but sufficiently small to exhibit Brownian movement. 328

color. That characteristic of visual sensation in the photoptic range that depends on the spectral composition of light entering the eye. 18

color breakup (color television). Any transient or dynamic distortion of the color in a television picture. *Note:* This effect may originate in video-tape equipment, in a television camera, or in a receiver. (1) In video-tape recording or playback, it occurs as intermittent misphasing or loss of the chrominance signal. (2) In a field-sequential system, it may be caused at the camera by rapid motion of the image of the camera sensor, through motion of either the camera or subject. It may be caused at the receiver by rapid changes in viewing conditions such as blinking or motion of the eyes. 18

color burst (color television). The portion of the composite or noncomposite color-picture signal, comprising a few cycles of a sine wave of chrominance subcarrier frequency, that is used to establish a reference for demodulating the chrominance signal. 18

color-burst flag (color-burst keying signal) (television). A keying signal used to form the color burst from a color-subcarrier signal source. *See:* **burst flag.** 328

color-burst gate (television). A keying or gating signal used to extract the color burst from a color-television signal. *See:* **burst gate; television.** 328

color-burst keying signal. *See:* **color-burst flag.**

color carrier (color television). *See:* **chrominance subcarrier.**

color cell (repeating pattern of phosphors on the screen of a color-picture tube). The smallest area containing a complete set of all the primary colors contained in the pattern. *Note:* If the cells are described by only one dimension as in the line type of screen, the other dimension is determined by the resolution capabilities of the tube. 125, 190

color center (color-picture tubes). A point or region (defined by a particular color-selecting electrode and screen configuration) through which an electron beam must pass in order to strike the phosphor array of one primary color. *Note:* This term is not to be used to define the color-triad center of a color-picture tube screen. 125, 190

color code (electrical). A system of standard colors adopted for identification of conductors for polarity, etcetera, and for identification of external terminals of motors and starters to facilitate making power connections between them. *See also:* **mine feeder circuit.** 328

color coder (color television transmission). *See:* **color encoder,** the preferred term in the United States. In Europe, the term **color coder** is commonly used. 18

color contamination (color television). An error of color rendition caused by incomplete separation of paths carrying different color components of the picture. *Note:* Such errors can arise in the optical, electronic, or mechanical portions of a color-television system as well as in the electric portions. 18, 178

color coordinate transformation (color television). Computation of the tristimulus values of colors in terms of one set of primaries from the tristimulus values of the same colors in another set of primaries. *Note:* This computation may be performed electrically in a color-television system. 18, 178

color decoder (color television). An apparatus for deriving the signals for the color-display device from the color-picture signal and the color-burst. 18, 178

color-difference signal (color television). An electric signal that when added to the luminance signal produces a signal representative of one of the tristimulus values (with respect to a stated set of primaries) of the transmitted color. 18, 178

color discrimination. The perception of differences between two or more colors. *See also:* **visual field.** 167

color encoder (National Television System Committee color television). An apparatus for generating the color picture signal and the color burst from camera signals (or equivalents) and the chrominance subcarrier. 18

color-field corrector (electron tubes). A device located external to the tube producing an electric or magnetic field that affects the beam after deflection as an aid in the production of uniform color fields. 125, 190

color flicker. The flicker that results from fluctuation of both chromaticity and luminance. 18, 178

color fringing (color television). Spurious chromaticity at boundaries of objects in the picture. *Note:* Color fringing can be caused by a change in relative position of the televised object from field to field (in a field-sequential system), or by misregistration in either camera or receiver; in the case of small objects, it may cause them to appear separated into different colors. 18

colorimetric purity (light). The ratio L_1/L_2 where L_1 is the luminance (photometric brightness) of the single frequency component that must be mixed with a reference standard to match the color of the light and L_2 is the luminance (photometric brightness) of the light. *See also: color.* 167

colorimetry. The techniques for the measurement of color and for the interpretation of the results of such measurements. *Note:* The measurement of color is made possible by the properties of the eye and is based on a set of conventions. 18, 244, 178

coloring (electroplating) (1) (chemical). The production of desired colors on metal surfaces by appropriate chemical action. *See also: electroplating.* 328
(2) (buffing). Light buffing of metal surfaces, for the purpose of producing a high luster. *See also: electroplating.* 328

color light signal. A fixed signal in which the indications are given by the color of a light only. 328

color match. The condition in which the two halves of a structureless photometric field are judged by the observer to have exactly the same appearance. *Note:* A color match for the standard observer may be calculated. 18, 178

color mixture. Color produced by the combination of light of different colors. *Notes:* (1) The combination may be accomplished by successive presentation of the components, provided the rate of alternation is sufficiently high, or the combination may be accomplished by simultaneous presentation, either in the same area or on adjacent areas, provided they are small enough and close enough together to eliminate pattern effects. (2) A color mixture as here defined is sometimes denoted as an additive color mixture, to distinguish it from combinations of dyes, pigments, and other absorbing substances. Such mixtures of substances are sometimes called subtractive color mixtures, but might more appropriately be called colorant mixtures.
18, 178

color-mixture data. *See: tristimulus values.*

color-picture tube. An electron tube used to provide an image in color by the scanning of a raster and by varying the intensity of excitation of phosphors to produce light of the chosen primary colors. 125, 190

color plane (multibeam color-picture tubes). A surface approximating a plane containing the color centers.
125, 190

color-position light signal. A fixed signal in which the indications are given by the color and the position of two or more lights. 328

color-purity magnet. A magnet in the neck region of a color-picture tube to alter the electron beam path for the purpose of improving color purity. 125, 190

color-selecting-electrode system. A structure containing a plurality of openings mounted in the vicinity of the screen of a color-picture tube (electron tubes), the function of this structure being to cause electron impingement on the proper screen area by using either masking, focusing, deflection, reflection, or a combination of these effects. *Note:* For examples see **shadow mask.** 125, 190

color-selecting-electrode system transmission (electron

tubes). The fraction of incident primary electron current that passes through the color-selecting-electrode system.
125, 190

color signal (color-television system). Any signal at any point for wholly or partially controlling the chromaticity values of a color-television picture. *Note:* This is a general term that encompasses many specific connotations, such as are conveyed by the words, color-picture signal, chrominance signal, carrier color signal, luminance signal (in color television), etcetera. 18, 178

color sync signal (television). A signal used to establish and to maintain the same color relationships that are transmitted. *Note:* In Rules governing Radio Broadcast Services, Part 3, of the Federal Communications Commission, the color sync signal consists of a sequence of color bursts that recur every line except for a specified time interval during the vertical interval, each burst occurring on the back porch. 163, 178

color temperature (1) (light source). The absolute temperature at which a blackbody radiator must be operated to have a chromaticity equal to that of the light source. *See also: color.* 167
(2) (television). The absolute temperature of the full (black-body) radiator for which the ordinates of the spectral distribution curve of emission are proportional (or approximately so) in the visible region, to those of the distribution curve of the radiation considered, so that both radiations have the same chromaticity. *Note:* In certain countries, by extension, the term color temperature is used in the case of a selective radiator when, for the colorimetric standard observer, this radiator has the same color (or at least approximately the same color) as a full radiator at a certain temperature; this temperature is then called the color temperature of the selective radiator. 18

color tracking (television). (1) The degree to which color balance is maintained over the complete range of the achromatic (neutral gray) scale. 328
(2) A qualitative term indicating the degree to which constant chromaticity within the achromatic region in the chromaticity diagram is achieved on a color-display device over the range of luminances produced from a monochrome signal. *See: television.* 178

color transmission (color television). The transmission of a signal wave for controlling both the luminance values and the chromaticity values in a picture. *See also: television.* 18, 178

color triad (phosphor-dot screen). A color cell of a three-color phosphor-dot screen. 125, 190

color triangle. A triangle drawn on a chromaticity diagram, representing the entire range of chromaticities obtainable as additive mixtures of three prescribed primaries, represented by the corners of the triangle.
18, 178

Colpitts oscillator. An electron tube or solid state circuit in which the parallel-tuned tank circuit is connected between grid and plate, the capacitive portion of the tank circuit being comprised of two series elements, the connection between the two being at cathode potential with the feedback voltage obtained across the grid-cathode portion of the capacitor. *See also: radio-frequency generator (electron tube type).* 14, 211

column (positional notation). (1) A vertical arrangement of characters or other expressions. (2) Loosely, a digit place. *See:* **place.** 235, 255, 77

column binary. Pertaining to the binary representation of data on punched cards in which adjacent positions in a column correspond to adjacent bits of data, for example, each column in a 12-row card may be used to represent 12 consecutive bits of a 36-bit word. 255, 77

column, positive. *See:* **positive column.**

combinational logic element. (1) A device having zero or more input channels and one output channel, each of which is always in one of exactly two possible physical states, except during switching transients. *Note:* On each of the input channels and the output channel, a single state is designated arbitrarily as the "one" state, for that input channel or output channel, as the case may be. For each input channel and output channel, the other state may be referred to as the "zero" state. The device has the property that the output channel state is determined completely by the contemporaneous input-channel-state combination, to within switching transients. (2) By extension, a device similar to (1) except that one or more of the input channels or the output channel, or both, have a finite number, but more than two, possible physical states each of which is designated as a distinct logic state. The output channel state is determined completely by the contemporaneous input-channel-state combination, to within switching transients. (3) A device similar to (1) or (2) except that it has more than one output channel. *See:* AND gate; OR gate. 235

combinational logic function. A logic function wherein for each combination of states of the input or inputs, there corresponds one and only one state of the output or outputs. *Note:* The terms combinative and combinatorial have also been used to mean combinational. 88

combination controller (industrial control). A full magnetic or semimagnetic controller with additional externally operable disconnecting means contained in a common enclosure. The disconnecting means may be a circuit breaker or a disconnect switch. *See:* **electric controller.** 225, 206

combination current and voltage regulation. That type of automatic regulation in which the generator regulator controls both the voltage and current output of the generator. *Note:* This type of control is designed primarily for the purpose of insuring proper charging of storage batteries on cars or locomotives. *See also:* **axle generator system.** 328

combination detector (fire protection devices). A device that either responds to more than one of the fire phenomena (heat, smoke, or flame) or employs more than one operating principle to sense one of these phenomena. 71

combination electric locomotive. An electric locomotive, the propulsion power for which may be drawn from two or more sources, either located on the locomotive or elsewhere. *Note:* The prefix combination may be applied to cars, buses, etcetera, of this type. *See also:* **electric locomotive.** 328

combination lighting and appliance branch circuit. A circuit supplying energy to one or more lighting outlets and to one or more appliance outlets. *See also:* **branch circuit.** 328

combination microphone. A microphone consisting of a combination of two or more similar or dissimilar microphones. *Examples:* Two oppositely phased pressure microphones acting as a gradient microphone; a pressure microphone and velocity microphone acting as a unidirectional microphone. *See also:* **microphone.** 328

combination rubber tape. The assembly of both rubber and friction tape into one tape that provides both insulation and mechanical protection for joints. 328

combinations of pulses and waveforms (pulse terms). *See:* **double pulse; bipolar pulse; staircase.**

combination starter. A starter having manually operated disconnecting means built into the same enclosure with the magnetic contactor. 124

combination thermoplastic tape. An adhesive tape composed of a thermoplastic compound that provides both insulation and mechanical protection for joints. 328

combination watch-report and fire-alarm system. A coded manual fire-alarm system, the stations of which are equipped to transmit a single watch-report signal or repeated fire-alarm signals. *See also:* **protective signaling.** 328

combined-line-recording trunk (CLR) (telephone switching systems). A one-way trunk for operator recording and extending of toll calls. 55

combined mechanical and electrical strength (insulator). The loading in pounds at which the insulator fails to perform its function either electrically or mechanically, voltage and mechanical stress being applied simultaneously. *Note:* The value will depend upon the conditions under which the test is made. *See also:* **insulator; tower.** 64

combined telephone set. A telephone set including in a single housing all the components required for a complete telephone set except the handset which it is arranged to support. *Note:* Wall hand telephone sets are of this type, but the term is usually reserved for a self-contained desk telephone set to distinguish it from desk telephone sets requiring an associated bell box. A desk local-battery telephone set may be referred to as a combined set if it includes in its mounting all components except its associated local batteries. *See also:* **telephone station.** 328

combined voltage and current influence (wattmeter). The percentage change (of full-scale value) in the indication of an instrument that is caused solely by a voltage and current departure from specified references while constant power at the selected scale point is maintained. *See also:* **accuracy rating (instrument).** 280

combustible materials (transformer). Materials which are external to the apparatus and made of or surfaced with wood, compressed paper, plant fibers, or other materials that will ignite and support flame. 53

combustion control. The regulation of the rate of combination of fuel with air in a furnace. 64

command (1) (electronic computation). (A) One of a set of several signals (or groups of signals) that occurs as a result of interpreting an instruction; the commands initiate the individual steps that form the process of executing the instruction's operation. (B) Loosely: an instruction in machine language. (C) Loosely: a mathematical or logic operator. (D) Loosely: an operation. 235, 210, 255, 77, 54

(2) (industrial control). An input variable established by means external to, and independent of, the feedback (automatic) control system. It sets, is equivalent to, and is expressed in the same units as the ideal value of the ultimately controlled variable. *See:* **control system, feedback; set point.** 56, 219, 206

command control (electric power systems). A control mode in which each generating unit is controlled to reduce unit control error. *See also:* **speed-governing system.** 94

command guidance. That form of missile guidance wherein control signals transmitted to the missile from an outside agency cause it to traverse a directed path through space. *See also:* **guided missile.** 328

command link (communication satellite). A data transmission link (generally earth to spacecraft or satellite) used to command a satellite or spacecraft in space. 83

command rate (gyro). The input rate equivalent of a torquer command signal. 46

command reference (servo or control system) (power supplies). The voltage or current to which the feedback signal is compared. As an independent variable, the command reference exercises complete control over the system output. *See:* **operational programming.** *See also:* **power supply.** 228, 186

commercial power. Power furnished by an electric power utility company; when available, it is usually the prime power source. 89

commercial tank (electrorefining). An electrolytic cell in which the cathode deposit is the ultimate electrolytically refined product. *See also:* **electrorefining.** 328

commissioning tests (rotating machinery). Tests applied to a machine at site under normal service conditions to show that the machine has been erected and connected in a correct manner and is able to work satisfactorily. *See also:* **asynchronous machine; direct-current commutating machine.** 63

common-battery central office. *See:* **common-battery office.**

common-battery office (telephone switching systems). A central office that supplies transmitter and signaling currents for its associated stations and current for the central office equipment from a power source located in the central office. 55

common-battery switchboard. A telephone switchboard for serving common-battery telephone sets. 193

common carrier (data communication). In telecommunications, a public utility company that is recognized by an appropriate regulatory agency as having a vested interest and responsibility in furnishing communication services to the general public. *See also:* **specialized common carrier, value added service.** 12

common-channel interoffice (CCIS) signaling (telephone switching systems). The use of separate paths between switching entities to carry the signaling associated with a group of communication paths. 55

common control (telephone switching systems). An automatic switching arrangement in which the control equipment necessary for the establishment of connections is shared, being associated with a given call only during the period required to accomplish the control function. 55

common device (supervisory system) (basic device). A device in either the master or remote station that is required for

the basic operation and is not part of the equipment for the individual points. 103, 202

common failure mode (nuclear power generating stations). A mechanism by which a single design basis event can cause redundant equipment to be inoperable. 31, 102

common-mode conversion (interference terminology). The process by which differential-mode interference is produced in a signal circuit by a common-mode interference applied to the circuit. *Note:* See the accompanying figure.

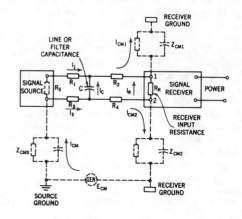

Common-mode conversion.

Common-mode currents are converted to differential-mode voltages by impedances R_1, R_2, R_3, R_4, R_S, R_R, and c. The differential-mode voltage at the receiver resulting from the conversion is the algebraic summation of the voltage drops produced by the various currents in these impedances. Various of the impedances may be neglected at particular frequencies. At direct current,

$$V_{CM} = I_r R_r \approx I_{CM1}(R_s + R_1 + R_2)$$
$$- I_{CM2}(R_3 + R_4)$$

At

$$f > \frac{I}{c(R_1 + R_3 + R_s)},$$
$$V_{CM} \approx I_c X_c \frac{R_R}{R_2 + R_4 + R_R}$$

See also: **interference.** 188

common-mode failure (nuclear power generating stations). Multiple failures attributable to a common cause. 29, 31, 356

common-mode interference (automatic null-balanced electrical instruments). (1) Interference that appears between both signal leads and a common reference plane (ground) and causes the potential of both sides of the transmission path to be changed simultaneously and by the same amount relative to the common reference plane (ground). *See:* **interference.** 188
(2). A form of interference that appears between any measuring circuit terminal and ground. *See also:* **accuracy rating (instrument).** 295

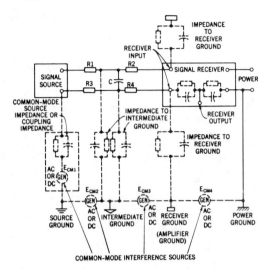

COMMON-MODE INTERFERENCE SOURCES

Common-mode interference—sources and current paths. The common-mode voltage V_{CM} in any path is equal to the sum of the common-mode generator voltages in that path; for example, in the source-receiver path,

$$V_{CM} = E_{CM1} + E_{CM2} + E_{CM3}.$$

common-mode noise (longitudinal) (cable systems). The noise voltage which appears equally and in phase from each signal conductor to ground. Common mode noise may be caused by one or more of the following: (1) Electrostatic induction. With equal capacitance between the signal wires and the surroundings, the noise voltage developed will be the same on both signal wires. (2) Electromagnetic induction. With the magnetic field linking the signal wires equally, the noise voltage developed will be the same on both signal wires. 35

common-mode rejection (in-phase rejection). The ability of certain amplifiers to cancel a common-mode signal while responding to an out-of-phase signal. *See:* **degeneration negative feedback.** 192

common-mode rejection quotient (in-phase rejection quotient). The quotient obtained by dividing the response to a signal applied differentially by the response to the same signal applied in common mode, or the relative magnitude of a common-mode signal that produces the same differential response as a standard differential input signal. 192

common-mode rejection ratio (CMRR). (1) (signal transmission system). The ratio of the common-mode interference voltage at the input terminals of the system to the effect produced by the common-mode interference, referred to the input terminals for an amplifier. For example,

$$\text{CMRR} = \frac{V_{CM}(\text{root-mean-square}) \text{ at input}}{\text{effect at output/amplifier gain}}$$

See also: **interference.** 188
(2) (oscilloscopes). The ratio of the deflection factor for a common-mode signal to the deflection factor for a differential signal applied to a balanced-circuit input. *See:* **oscillograph.**

common-mode signal (1) (general). The instantaneous algebraic average of two signals applied to a balanced circuit, both signals referred to a common reference. *See:* **oscillograph.** 185
(2) (in-phase signal) (medical electronics). A signal applied equally and in phase to the inputs of a balanced amplifier or other differential device. 192
common-mode signal maximum. The largest common-mode signal at which the specified common-mode rejection ratio is valid. 106
common mode to normal mode conversion (cable systems). In addition to the common mode voltages which are developed in the signal conductors by the general environmental sources of electrostatic and electromagnetic radiation, differences in voltage exist between different ground points in a facility due to the flow of ground currents. These voltage differences are considered common mode when connection is made to them either intentionally or accidentally and the currents they produce are common mode. These common mode currents can develop normal mode noise voltage across circuit impedances. 35
common return. A return conductor common to several circuits. *See also:* **center of distribution.** 64
common spectrum multiple access (CSMA) (communication satellite). A method of providing multiple access to a communication satellite in which all of the participating earth stations use a common time-frequency domain. Signal processing is employed to detect a wanted signal in the presence of others. Three typical approaches utilizing these techniques are spread spectrum, frequency-time matrix, and frequency-hopping. 84
common trunk (telephone switching systems). A trunk, link, or junctor accessible from all input groups of a grading. 55
common use (wiring system). Simultaneous use by two or more utilities of the same kind. 262
common winding (autotransformer). That part of the autotransformer winding which is common to both the primary and the secondary circuits. *See:* **autotransformer.** 203, 257, 53
communication (telecommunication) (electric system). The transmission of information from one point to another by means of electromagnetic waves. *Note:* While there is no sharp line of distinction between the terms "communication" and "telecommunication," the latter is usually applied where substantial distances are involved. 59
communication band. *See:* **frequency band of emission.**
communication circuits. Circuits used for audible and visual signals and communication of information from one place to another, within or on the vessel. 3
communication conductor (telephone equipment). A conductor used in a communication network. *See:* **conductor.** 39
communication control character (data communication). A functional character intended to control or facilitate transmission over data networks. Control characters form the basis for character-oriented communications control procedures. 12
communication lines (transmission and distribution). The conductors and their supporting or containing structures which are used for public or private signal or communication service, and which operate at potentials not ex-

ceeding 400 volts to ground or 750 volts between any two points of the circuit, and the transmitted power of which does not exceed 150 watts. When operating at less than 150 volts, no limit is placed on the transmitted power of the system. Under specified conditions, communication cables may include communication circuits exceeding the preceding limitation where such circuits are also used to supply power solely to communication equipment. *Notes:* (1)Telephone, telegraph, railroad-signal, data, clock, fire, police-alarm, cable televison and other systems conforming with the above are included. (2) Lines used for signaling purposes, but not included under the above definition, are considered as supply lines of the same voltage and are to be so installed. 262

communication reliability (mobile communication). A specific criterion of system performance related to the percentage of times a specified signal can be received in a defined area during a given interval of time. *See also:* **mobile communication system.** 181

communication satellite. A satellite used for communication between two or more ground points by transmitting the messages to the satellite and retransmitting them to the participating ground stations. 83

communications common carrier. *See:* **carrier, communications common.**

communications computer (data communication). A computer that acts as the interface between another computer or terminal and a network, or a computer controlling data flow in a network. *See also:* **front end computer, concentrator.** 12

communications interface equipment. A portion of a relay system which transmits information from the relay logic to a communications link, or conversely to logic, for example, audio tone equipment, a carrier transmitter-receiver when an integral part of the relay system. 90

communications link. Any of the communications media, for example, microwave, power line carrier, wire line. 90

community dial office (telephone switching systems). A small automatic central office that serves a separate exchange area which ordinarily has no permanently assigned central office operating or maintenance forces. 55

community-of-interest (telephone switching systems). A characteristic of traffic resulting from the calling habits of the customers. 55

communication theory (information theory). The mathematical theory underlying the communication of messages from one point to another. 59

commutated-antenna direction-finder (CADF). A system using a multiplicity of antennas in a circular array and a receiver that is connected to the antennas in sequence through a commutating device for finding the direction of arrival of radio waves; the directional sensing is related to phase shift that occurs as a result of the commutation. *See also:* **navigation.** 187, 13

commutating angle (1) (rectifier circuits). The time, expressed in degrees, during which the current is commutated between two rectifying elements. *See also:* **rectification; rectifier circuit element.** 237, 66, 208
(2) (thyristor converter circuit). (μ) The time, expressed in degrees (1 cycle of the ac wave form = 360°), during which the current is commutated between two thyristor

converter circuit elements. 121

commutating-field winding. An assembly of field coils located on the commutating poles, that produces a field strength approximately proportional to the load current. The commutating field is connected in direction and adjusted in strength to assist the reversal of current in the armature coils for successful commutation. This field winding is used alone, or supplemented by a compensating winding. *See:* **asynchronous machine; direct-current commutating machine.** 328

commutating group (rectifier circuit). A group of rectifier-circuit elements and the alternating-voltage supply elements conductively connected to them in which the direct current of the group is commutated between individual elements that conduct in succession. *See also:* **circuit element; rectification; rectifier.** 237, 66, 208

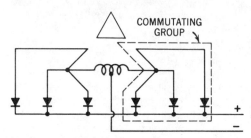

Commutating group in a single-way rectifier circuit.

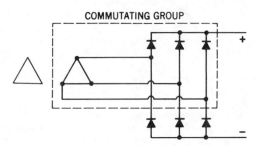

Commutating group in a double-way rectifier circuit.

commutating impedance (rectifier transformer). The impedance that opposes the transfer of current between two direct-current winding terminals of a commutating group, or a set of commutating groups. *See also:* **rectifier transformer.** 258

commutating period (inverters). The time during which the current is commutated. *See also:* **self-commutated inverters.** 208

commutating pole (interpole). An auxiliary pole placed between the main poles of a commutating machine. Its exciting winding carries a current proportional to the load current and produces a flux in such a direction and phase as to assist the reversal of the current in the short-circuited coil. 328

commutating reactance (1) (rectifier transformer). Reactive component of the commutating impedance. *Note:* This value is defined as one-half the total reactance in the commutating circuit expressed in ohms referred to the total direct-current winding. For wye, star, and multiple-wye circuits, this is the same value as derived in ohms on a phase-to-neutral voltage base; while with diametric and zig-zag circuits it must be expressed as one-half the total due to both halves being mutually coupled on the same core leg or phase. *See also:* **commutating impedance; rectification; rectifier transformer.** 258

(2) **(thyristor converter).** The reactance that effectively opposes the transfer of current between thyristor converter circuit elements of a commutating group or set of commutating groups. *Note:* For convenience, the reactance from phase to neutral, or one half the total reactance in the commutating circuit, is the value usually employed in computations, and is designated as the commutating reactance. 121

commutating reactance factor (rectifier circuit). The line-to-neutral commutating reactance in ohms multiplied by the direct current commutated and divided by the effective (root-mean-square) value of the line-to-neutral voltage of the transformer direct-current winding. *See also:* **circuit element; rectification; rectifier circuit.** 237, 208

commutating reactance transformation constant. A constant used in transforming line-to-neutral commutating reactance in ohms on the direct-current winding to equivalent line-to-neutral reactance in ohms referred to the alternating-current winding. *See also:* **rectification.** 208

commutating reactor. A reactor used primarily to modify the rate of current transfer between rectifying elements. *See:* **reactor; alternating-current saturable reactor; filter reactor; shunt reactor; current-limiting reactor; bus reactor; feeder reactor; neutral grounding reactor; starting reactor; synchronizing reactor; paralleling reactor.** 203, 53

commutating resistance (rectifier transformer). The resistance component of the commutating impedance. *See also:* **rectifier transformer.** 258

commutating voltage (rectifier circuit). The phase-to-phase alternating-current voltage of a commutating group which causes the current to commutate. *See also:* **rectifier circuit element.** 237, 66

commutation (rectifier) (rectifier circuit) (thyristor converter). The transfer of unidirectional current between rectifier circuit elements or thyristor converter circuit elements that conduct in succession. *See also:* **rectification; rectifier circuit element.** 237, 66, 208, 291, 121

commutation elements (semiconductor rectifiers). The circuit elements used to provide circuit-commutated turnoff time. *See also:* **semiconductor rectifier stack.** 208

commutation factor (1) (rectifier circuits). The product of the rate of current decay at the end of conduction, in amperes per microsecond, and the initial reverse voltage, in kilovolts. *See also:* **element; rectification; rectifier circuit.** 66, 208, 291

(2) **(gas tubes).** The product of the rate of current decay and the rate of the inverse voltage rise immediately following such current decay. *Note:* The rates are commonly stated in amperes per microsecond and volts per microsecond. *See also:* **heterodyne conversion transducer (converter); gas tubes; rectification.** 125, 190

commutator (rotating machinery). An assembly of conducting members insulated from one another, in the radial-axial plane, against which brushes bear, used to enable current to flow from one part of the circuit to another by sliding contact. 63

commutator bars. *See:* **commutator segments.**

commutator bore. Diameter of the finished hole in the core that accommodates the armature shaft. *See:* **commutator.** 328

commutator brush track diameter. That diameter of the commutator segment assembly that after finishing on the armature is in contact with the brushes. *See:* **commutator.** 328

commutator core. The complete assembly of all of the retaining members of a commutator. *See:* **commutator.** 328

commutator-core extension. That portion of the core that extends beyond the commutator segment assembly. *See:* **commutator.** 328

commutator inspection cover. A hinged or otherwise attached part that can be moved to provide access to commutator and brush rigging for inspection and adjustment. *See:* **commutator.** 328

commutator insulating segments (rotating machinery). The insulation between commutator segments. 63

commutator insulating tube (rotating machinery). The insulation between the underside of the commutator segment assembly and the core. *See also:* **commutator.** 63

commutator motor meter. A motor type of meter in which the rotor moves as a result of the magnetic reaction between two windings, one of which is stationary and the other assembled on the rotor and energized through a commutator and brushes. *See also:* **electricity meter (meter).** 328

commutator nut. The retaining member that is used in combination with a vee ring and threaded shell to clamp the segment assembly. *See:* **commutator.** 328

commutator riser (rotating machinery). A conducting element for connecting a commutator segment to a coil. *See:* **commutator.** 63

commutator-segment assembly. A cylindrical ring or disc assembly of commutator segments and insulating segments that are bound and ready for installation. *Note:* The binding used may consist of wire, temporary assembly rings, shrink rings, or other means. *See:* **commutator.** 328

commutator segments (commutator bars). Metal current-carrying members that are insulated from one another by insulating segments and that make contact with the brushes. *See:* **commutator.** 328

commutator shell. The support on which the component parts of the commutator are mounted. *Note:* The commutator may be mounted on the shaft, on a commutator spider, or it may be integral with a commutator spider. *See:* **commutator.** 328

commutator-shell insulation (rotating machinery). The insulation between the under (or in the case of a disc commutator, the back) side of the commutator assembled segments and the commutator shell. *See also:* **commutator.** 63

commutator shrink ring. A member that holds the commutator-segment assembly together and in place by being shrunk on an outer diameter of and insulated from the commutator-segment assembly. *See:* **commutator.** 328

commutator vee ring. The retaining member that, in combination with a commutator shell, clamps or binds the commutator segments together. *See:* **commutator.**

328

commutator vee ring insulation (rotating machinery). The insulation between the V-ring and the commutator segments.

63

commutator vee-ring insulation extension (rotating machinery). The portion of the vee-ring insulation that extends beyond the commutator segment assembly. *See:* **commutator.**

63

companding (modulation systems). A process in which compression is followed by expansion. *Note:* Companding is often used for noise reduction, in which case the compression is applied before the noise exposure and the expansion after the exposure.

242, 59

companding, instantaneous. *See:* **instantaneous companding.**

compandor. A combination of a compressor at one point in a communication path for reducing the amplitude range of signals, followed by an expander at another point for a complementary increase in the amplitude range. *Note:* The purpose of the compandor is to improve the ratio of the signal to the interference entering in the path between the compressor and expander. *Editorial Note:* **Compandor** is the preferred spelling rather than **compander.**

239, 59

comparative tests (test, measurement and diagnostic equipment). Comparative tests compare end item signal or characteristic values with a specified tolerance band and present the operator with a go/no-go readout; a go for signals within tolerances, and a no-go for signals out of tolerance.

54

comparator (1). A circuit for performing amplitude selection between either two variables or between a variable and a constant.

77

(2) (test, measurement and diagnostic equipment). A device capable of comparing a measured value with predetermined limits to determine if the value is within these limits.

54

(3) (analog computer). A circuit, having only two logic output states, for comparing the relative amplitudes of two analog variables, or of a variable and a constant, such that the logic signal output of the comparator uniquely determines which variable is the larger at all times.

9

comparer (relaying). A signal element which performs an "AND" logic function.

79

comparison amplifier (power supplies). A high-gain non-inverting direct-current amplifier that, in a bridge-regulated power supply, has as its input the voltage between the null junction and the common terminal. The output of the comparison amplifier drives the series pass elements. *See also:* **power supply.**

228, 186

comparison bridge (power supplies). A type of voltage-comparison-circuit whose configuration and principle of operation resemble a four-arm electric bridge. See the accompanying figure.

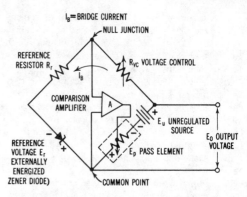

Comparison bridge connected as a voltage regulator. U.S. Patent No. 3,028,538—Foxbro Design Corp.

Note: The elements are so arranged that, assuming a balance exists in the circuit, a virtual zero error signal is derived. Any tendency for the output voltage to change, in relation to the reference voltage, creates a corresponding error signal, that by means of negative feedback, is used to correct the output in the direction toward restoring bridge balance. *See:* **error signal.** *See also:* **power supply.**

228, 186

comparison lamp. A light source having a constant, but not necessarily known, luminous intensity with which standard and test lamps are successively compared. *See also:* **primary standards.**

167

compartment. A space within the base, frame, or column of the industrial equipment.

124

compartment-cover mounting plate (pothead). A metal plate used for covering the flange opening. It may be altered in the field to accommodate various cable entrance fittings. It may also be used to attach single pothead mounting plates to the compartment flange.

288

compass bearing. Bearing relative to compass north.

13

compass-controlled directional gyro. A device that uses the earth's magnetic field as a reference to correct a directional gyro. *Note:* The direction of the earth's field is sensed by a remotely located compass that is connected electrically to the gyro.

328

compass course. Course relative to compass north. 13

compass declinometer. *See:* **declinometer.**

compass deviation. *See:* **magnetic deviation.**

compass heading. Heading relative to compass north.

13

compass locator (electronic navigation). *See:* **non-directional beacon.**

compass repeater. An indicator, located remotely from a master compass, so controlled that it continuously indicates (repeats) the bearing shown by the master compass.

328

compatibility (1) (color television). The property of a color television system that permits substantially normal monochrome reception of the transmitted signal by typical unaltered monochrome receivers.

18, 178

(2) (programmable instrumentation). The degree to which devices may be interconnected and used, without modification, when designed as defined according to functional, electrical, and mechanical specifications.

40

compensated-loop direction-finder. A direction-finder employing a loop antenna and a second antenna system to compensate polarization error. *See also:* **radio receiver.** 328

compensated repulsion motor. A repulsion motor in which the primary winding on the stator is connected in series with the rotor winding via a second set of brushes on the commutator in order to improve the power factor and commutation. 63

compensated series-wound motor. A series-wound motor with a compensating-field winding. The compensating-field winding and the series-field winding may be combined into one field winding. *See also:* **asynchronous machine; direct-current commutating machine.** 236, 63

compensating-field winding (rotating machinery). Conductors embedded in the pole shoes and their end connections. It is connected in series with the commutating-field winding and the armature circuit. *Note:* A compensating-field winding supplements the commutating-field winding, and together they function to assist the reversal of current in the armature coils for successful commutation. *See also:* **asynchronous machine; direct-current commutating machine.** 328

compensating-rope sheave switch. A device that automatically causes the electric power to be removed from the elevator driving-machine motor and brake when the compensating sheave approaches its upper or lower limit of travel. *See also:* **hoistway (elevator or dumbwaiter).** 328

compensation (control system, feedback). A modifying or supplementary action (also, the effect of such action) intended to improve performance with respect to some specified characteristic. *Note:* In control usage, this characteristic is usually the system deviation. Compensation is frequently qualified as series, parallel, feedback, etcetera, to indicate the relative position of the compensating element. *See also:* **control system, feedback; equalization.** 56, 105

compensation theorem. States that if an impedance ΔZ is inserted in a branch of a network, the resulting current increment produced in any branch in the network is equal to the current that would be produced at that point by a compensating voltage, acting in series with the modified branch, whose value is $-I\Delta Z$, where I is the original current that flowed where the impedance was inserted before the insertion was made. 328

compensator (1) (rotating machinery). An element or group of elements that acts to modify the functioning of a device in accordance with one or more variables. *See also:* **asynchronous machine.** 63

(2) (radio direction-finders). That portion of a direction-finder that automatically applies to the direction indication all or a part of the correction for the deviation. *See also:* **radio receiver.** 328

(3) (excitation systems). A feedback element of the regulator that acts to compensate for the effect of a variable by modifying the function of the primary detecting element. *Notes:* (A) Examples are reactive current compensator and active current compensator. A reactive current compensator is a compensator that acts to modify the function of a voltage regulator in accordance with reactive current. An active current compensator is a

compensator that acts to modify the function of a voltage regulator in accordance with active current. (B) Historically, terms such as equalizing reactor and cross-current compensator have been used to describe the function of a reactive compensator. These terms are deprecated. (C) Reactive compensators are generally applied with generator voltage regulators to obtain reactive current sharing among generators operating in parallel. They function in the following two ways. (1) Reactive droop compensation is the more common method. It creates a droop in generator voltage proportional to reactive current and equivalent to that which would be produced by the insertion of a reactor between the generator terminals and the paralleling point. (2) Reactive differential compensation is used where droop in generator voltage is not wanted. It is obtained by a series differential connection of the various generator current transformer secondaries and reactive compensators. The difference current for any generator from the common series current creates a compensating voltage in the input to the particular generator voltage regulator which acts to modify the generator excitation to reduce to minimum (zero) its differential reactive current. (D) Line drop compensators modify generator voltage by regulator action to compensate for the impedance drop from the machine terminals to a fixed point. Action is accomplished by insertion within the regulator input circuit of a voltage equivalent to the impedance drop. The voltage drops of the resistance and reactance portions of the impedance are obtained, respectively, in per unit quantities by an active compensator and a reactive compensator. 105

compensatory leads. Connections between an instrument and the point of observation so contrived that variations in the properties of leads, such as variations of resistance with temperature, are so compensated that they do not affect the accuracy of the instrument readings. *See also:* **auxiliary device to an instrument.** 328

compensatory wiring techniques (coupling in control systems). Those wiring techniques which result in a substantial cancellation or counteracting of the effects of rates of change of electric or magnetic fields, without actually obstructing or altering the intensity of the fields. If the signal wires are considered to be part of the control circuit, these techniques change the susceptibility of the circuit. Example: twisting of signal and return wires associated with a susceptable instrument so as to cancel the voltage difference between wires caused by an existing, varying magnetic field. *See also:* **barrier wiring techniques; suppressive wiring techniques.** 43

compile (computing system). To prepare a machine language program from a computer program, written in another programming language, by making use of the overall logic structure of the program, or generating more than one machine instruction for each symbolic statement, or both, as well as performing the function of an assembler. 255, 77

compiler (test, measurement and diagnostic equipment). A software system used as an automatic means of translating statements from problem oriented language to machine oriented language. 54

complement. (1) A number whose representation is derived from the finite positional notation of another by one of the following rules: (A) true complement—subtract each digit from one less than the base; then add one to the least-significant digit, executing all carries required. (B) base minus ones complement—subtract each digit from one less than the base (for example, "nines complement" in the base 10, "ones complement" in the base 2, etcetera). (2) To complement is to form the complement of a number. *Note:* In many machines, a negative number is represented as a complement of the corresponding positive number. *See:* **nines complement; ones complement; radix complement; radix-minus-ones complement; tens complement; true complement; twos complement.** 210

complementary function (automatic control). The solution of a homogeneous differential equation, representing a system or element, which describes a free motion. 56

complementary functions. Two driving-point functions whose sum is a positive constant. *See also:* **linear passive networks.** 238

complementary tracking (power supplies). A system of interconnection of two regulated supplies in which one (the master) is operated to control the other (the slave). The slave supply voltage is made equal (or proportional) to the master supply voltage and of opposite polarity with respect to a common point. See the accompanying figure. *See also:* **power supply.** 228, 186

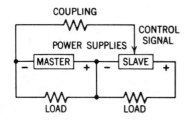

Complementary tracking.

complementary wavelength (color). The wavelength of a spectrum light that, when combined in suitable proportions with the light considered, yields a match with the specified achromatic light. *See:* **dominant wavelength.** 18

complete carry. A carry process in which a carry resulting from addition of carries is allowed to propagate. Contrasted with partial carry. *See also:* **carry.** 235

completed call (telephone switching systems). An answered call that has been released. 55

complete diffusion (illuminating engineering). Diffusion in which the diffusing medium redirects the flux incident upon it so that none is in an image-forming state. *See also:* **lamp.** 167

complete failure. *See:* **failure, complete.**

completely mesh-connected circuit. A mesh-connected circuit in which a current path extends directly from the terminal of entry of each phase conductor to the terminal of entry of every other phase conductor. *See also:* **network analysis.** 210

complete operating test equipment (test, measurement and diagnostic equipment). Equipment together with the necessary detail parts, accessories, and components, or any combination thereof, required for the testing of a specified operational function. 54

complete reference designation (electric and electronics parts

and equipments). A reference designation that consists of a basic reference designation and, as prefixes, all the reference designations that apply to the subassemblies or assemblies within which the item is located, including those of the highest level needed to designate the item uniquely. The reference designation for a unit consists of only a number. 17

complex apparent permeability. The complex (phasor) ratio of induction to applied magnetizing force. *Notes:* (1) In ordinary ferromagnetic materials, there is a phase shift between the applied magnetizing force and the resulting total magnetizing force, caused by eddy currents. (2) In anisotropic media, complex apparent permeability becomes a matrix. 210

complex capacitivity. *See:* **relative complex dielectric constant.**

complex conductivity (physical media). For isotropic media the ratio of the current density to the electric field, in which the current density comprises all currents at the point under consideration. *Note:* This term is used to describe both the conductive and dielectric properties of a medium. 111

complex dielectric constant. *See:* **relative permittivity in physical media.**

complex dielectric constant, relative. *See:* **relative complex dielectric constant.**

complex permeability (1) (general). The complex (phasor) ratio of induction to magnetizing force. *Notes:* (A) This is related to a phenomenon wherein the induction is not in phase with the total magnetizing force. (B) In anisotropic media, complex permeability becomes a matrix. 210

(2) (magnetic core testing). Under stated conditions, the complex quotient of vectors representing induction and field strength inside the core material. One of the vectors is made to vary sinusoidally and the other referenced to it.

Series representation:

$$\bar{\mu} = \mu'_s - j\mu''_s = \frac{1}{\mu_0}\frac{\bar{B}}{H}$$

Parallel representation:

$$\frac{1}{\bar{\mu}} = \frac{1}{\mu'_p} - \frac{1}{j\mu''_p} = \mu_0\frac{\overline{H}}{\overline{B}}$$

165

complex permittivity. *See:* **relative complex dielectric constant.**

complex permittivity in physical media (antenna). For isotropic media the ratio of the electric flux density to the electric field, in which the displacement current density represents the total current density. *Note:* This term is used to describe both the conductive and dielectric properties of a medium. 111

complex plane (automatic control). A plane defined by two perpendicular reference axes, used for plotting a complex variable or functions of this variable, such as a transfer function. 56

complex power (rotating machinery). *See:* **power, phasor; phasor power.**

complex target (radar). A target composed of a number of scatterers where the target extent is smaller in all dimensions than the radar resolution cell. 13

complex tone. (1) A sound containing simple sinusoidal components of different frequencies. (2) A sound sensation characterized by more than one pitch. 176

complex variable (automatic control). A convenient mathematical concept having a complex value, that is having a real part and an imaginary part. *Note:* In control systems, the pertinent independent variable is a generalized frequency $s = \sigma + j\omega$ used in the Laplace transform. 56

complex waveforms (pulse terms). *See:* **combinations of pulses and transitions; waveforms produced by magnitude superposition; waveforms produced by continuous time superposition of simpler waveforms; waveforms produced by noncontinuous time superposition of simpler waveforms; waveforms produced by operations on waveforms.**

compliance (industrial control). A property reciprocal to stiffness. *See:* **control system, feedback.** 219, 206

compliance extension (power supply). A form of master/slave interconnection of two or more current-regulated power supplies to increase their compliance voltage range through series connection. *See also:* **power supply; compliance voltage.** 228, 186

compliance voltage (power supplies). The output voltage of a direct-current power supply operating in constant-current mode. *Note:* The compliance range is the range of voltages needed to sustain a given value of constant current throughout a range of load resistances. *See also:* **power supply.** 228, 186

component (1). A piece of equipment, a line, a section of line, or a group of items that is viewed as an entity for purposes of reporting, analyzing, and predicting outages. *See also:* **computer component; solid-state component.** 200

(2) (power-system communication). An elementary device. 59

components (1) (vector). The components of a vector are any vectors the sum of which gives the original vector. The components of a vector **V** in a specified coordinate system are the directed projections of **V** on the system. Thus if V_x, V_y, and V_z are the scalar values of the projections of **V** on the x, y, and z axes, respectively and **i, j, k,** are unit vectors along the axes, then iV_x, jV_y, and kV_z are the components of V. Moreover

$$V = iV_x + jV_y + kV_z.$$

Again if V is the magnitude of the vector **V**, and θ its polar angle with the x axis and ϕ its azimuth with the x-y plane, then

$$V_x = V\cos\theta$$
$$V_y = V\sin\theta\cos\phi$$
$$V_z = V\sin\theta\sin\phi$$
$$\text{and } V = iV_x + jV_y + kV_z$$

Note: A right-handed system of coordinates for a rectangular system is such that if **i, j, k** are unit vectors measured along the x, y, and z axes, respectively, then the 90-degree rotation from **i** to **j** of a right-handed screw causes that screw to progress along the direction of **k**. 210

(2) (nuclear power generating stations). (A) Items from which the system is assembled (for example, resistors, capacitors, wires, connectors, transistors, tubes, switches, springs, etcetera). 20, 31, 120
(B) Discrete items from which a system is assembled (for example, resistors, capacitors, wires, connectors, transistors, integrated circuits, switches, motors, relays or solenoids).

composite bushing (1) (electric apparatus). A bushing in which the insulation consists of several coaxial layers of different insulating materials. 287
(2) (outdoor apparatus bushing). A bushing in which the major insulation consists of several coaxial layers of different insulation materials. 168

composite cable (communication practice). A cable in which conductors of different gauges or types are combined under one sheath. *Note:* Differences in length of twist are not considered here as constituting different types. *See also:* **cable.** 328

composite color-picture signal (National Television System Committee color television). The electric signal that represents complete color-picture information and all synchronizing signals. 18

composite color signal (color television). The color-picture signal plus blanking and all synchronizing signals. 18

composite color synchronizing (color television). The signal comprising all the synchronizing signals necessary for proper operation of a color receiver. *Note:* This includes the deflection synchronizing signals to which the color synchronizing signal is added in the proper time relationship. 18, 178

composite conductor. A composite conductor consists of two or more strands consisting of two or more materials. *See also:* **conductor.** 64

composite controlling voltage (electron tube). The voltage of the anode of an equivalent diode combining the effects of all individual electrode voltages in establishing the space-charge-limited current. *See:* **excitation (drive).** 125, 190

composited circuit. A circuit that can be used simultaneously for telephony and direct-current telegraphy or signaling, separation between the two being accomplished by frequency discrimination. *See also:* **transmission line.** 328

composite error (gyro; accelerometer). The maximum deviation of the output data from a specified output function. Composite error is due to the composite effects of hysteresis, resolution, non-linearity, non-repeatability, and other uncertainties in the output data. It is generally expressed as a percentage of half the output span. *See:* **input-output characteristics.** 46

composite picture signal (television). The signal that results from combining a blanked picture signal with the synchronizing signal. *See also:* **television.** 178

composite plate (electroplating). An electrodeposit consisting of two or more layers of metals deposited separately. *See also:* **electroplating.** 328

composite pulse (pulse navigation systems). A pulse composed of a series of overlapping pulses received from the same signal source but via different paths. *See also:* **navigation.** 13, 187

composite set. An assembly of apparatus designed to

provide one end of a composited circuit. 328

composite signaling (CX) (telephone switching systems). A form of polar-duplex signaling capable of simultaneously serving a number of circuits using low-pass filters to separate the signaling currents from the voice currents. 55

composite supervision. The use of a composite signaling channel for transmitting supervisory signals between two points in a connection. 328

composite waveform (pulse terms). A waveform which is, or which for analytical or descriptive purposes is treated as, the algebraic summation of two or more waveforms. *See:* composite waveform diagram below.

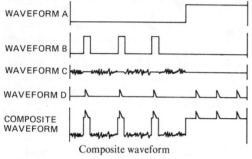

Composite waveform

compound (rotating machinery). (1) A definite substance resulting from the combination of specific elements or radicals in fixed proportions: distinguished from mixture. (2) The intimate admixture of resin with ingredients such as fillers, softeners, plasticizers, catalysts, pigments, or dyes. *See also:* **rotor (rotating machinery); stator.**
217, 63

compound-filled (reactor, transformer) (grounding device). Having the coils/windings encased in an insulating fluid that becomes solid or remains slightly plastic at normal operating temperatures. *See also:* **instrument transformer; reactor.** 257, 258, 203, 91

compound-filled bushing (outdoor electric apparatus). A bushing in which the space between the inside surface of the porcelain/weather casing and the major insulation (or conductor where no major insulating is used) is filled with compound. 287, 168

compound filled joints (power cable joints). Joints in which the joint housing is filled with an insulating compound that is non-fluid at normal operating temperatures. 34

compound-filled transformer. A transformer in which the windings are enclosed with an insulating fluid which becomes solid, or remains slightly plastic, at normal operating temperatures. *Note:* The shape of the compound-filled transformer is determined in large measure by the shape of the container or mold used to contain the fluid before solidification. 53

compounding curve (direct-current generator). A regulation curve of a compound-wound direct-current generator. *Note:* The shunt field may be either self or separately excited. *See:* **direct-current commutating machine.** 328

compound interferometer system. An antenna system consisting of two or more interferometer antennas whose outputs are combined using nonlinear circuit elements such that grating lobe effects are reduced. 111

compound target* (radar). *See:* distributed target. 13

* Deprecated

compound-wound. A qualifying term applied to a direct-current machine to denote that the excitation is supplied by two types of windings, shunt and series. *Note:* When the electromagnetic effects of the two windings are in the same direction, it is termed cumulative compound wound; when opposed, differential compound wound. *See:* **direct-current commutating machine.** 328

compound-wound generator. A direct-current generator that has two separate field windings—one, supplying the predominating excitation, is connected in parallel with the armature circuit and the other, supplying only partial excitation, is connected in series with the armature circuit and of such proportion as to require an equalizer connection for satisfactory parallel operation. *See also:* **direct-current commutating machine.** 3

compound-wound motor. A direct-current motor which has two separate field windings: one, usually, the predominating field, connected in parallel with the armature circuit, and the other connected in series with the armature circuit. *Note:* The characteristics as regards speed and torque are intermediate between those of shunt and series motors. 3

compressed-air circuit breaker. *See:* **circuit breaker.**

compression (1) (modulation systems). A process in which the effective gain applied to a signal is varied as a function of the signal magnitude, the effective gain being greater for small than for large signals. 242, 59
(2) (television). The reduction in gain at one level of a picture signal with respect to the gain at another level of the same signal. *Note:* The gain referred to in the definition is for a signal amplitude small in comparison with the total peak-to-peak picture signal involved. A quantitative evaluation of this effect can be obtained by a measurement of differential gain. *See:* **black compression; white compression.** *See also:* **television.** 178
(3) (oscillography). An increase in the deflection factor usually as the limits of the quality area are exceeded. *See:* **oscillograph.** 185

compression ratio (gain or amplification). The ratio of (1) the magnitude of the gain (or amplification) at a reference signal level to (2) its magnitude at a higher stated signal level. *See:* **amplifier.** 125, 190

compressional wave. A wave in an elastic medium that is propagated by fluctuation in elemental volume, accompanied by velocity components along the direction of propagation only. *Note:* A compressional plane wave is a longitudinal wave. 176

compressor (1) (audio and electroacoustics). A transducer that for a given input amplitude range produces a smaller output range. *See also:* **attenuation.** 239
(2) (data transmission). A transducer which for a given amplitude range of input voltages produces a smaller range of output voltages. One important type of compressor employs the envelope of speech signals to reduce their volume range by amplifying weak signals and attenuating strong signals. 59

compressor stator-blade-control system (gas turbines). A means by which the turbine compressor stator blades are adjusted to vary the operating characteristics of the compressor. *See also:* **speed-governing system (gas turbines).** 98, 58

computation. *See:* **implicit computation.**

computer. (1) A machine for carrying out calculations. (2) By extension, a machine for carrying out specified transformations on information. (3) A stored-program data-processing system.   235, 9, 210

computer-assisted tester (test, measurement and diagnostic equipment). A tester not directly programmed by a computer but which operates in association with a computer by using some arithmetic functions of the computer.   54

computer code. A machine code for a specific computer.   255, 77

computer component. Any part, assembly, or subdivision of computer, such as resistor, amplifier, power supply, or rack.   9, 77, 10

computer control (physical process) (electric power systems). A mode of control wherein a computer, using as input the process variables, produces outputs that control the process. *See also:* **power system.**   200, 94

computer control state (analog computer). One of several distinct and selectable conditions of the computer control circuits. *See:* **balance check; hold; operate; reset potentiometer reset; static test.**   77, 9

computer diagram (analog computer). A functional drawing showing interconnections between computing elements, such interconnections being specified for the solution of a particular set of equations. *See:* **computer program; problem board.**   9

computer equation (machine equation) (analog computer). An equation derived from a mathematical model for use on a computer which is equivalent or proportional to the original equation. *See also:* **scale factor.**   9

computer instruction. A machine instruction for a specific computer.   255, 77

computer interface equipment (surge withstand capability). A device which interconnects a protective relay system to an independent computer, for example, an analog to digital converter, a scanner, a buffer amplifier.   90

computer network (1) (general). A complex consisting of two or more interconnected computing units.   255, 77
(2) (data communication). An interconnection of assemblies of computer systems, terminals and communications facilities.   12

computer program (1) (general). A plan or routine for solving a problem on a computer, as contrasted with such terms as fiscal program, military program, and development program.   255, 77, 54
(2) (analog computer). That combination of computer diagram, potentiometer list, amplifier list, trunk list, switch list, scaled equations, and any other documentation that defines the analog configuration for the particular problem to be solved. This term sometimes is used to include the problem patch board as well, and, in some loose usage, the computer program may be (incorrectly) used to refer solely to the program patch panel.   9

computer time. *See:* **time.**

computer variable (1). A dependent variable as represented on the computer. *See also:* **time.**   9

(2) (machine variable). *See:* **scale factor.**

computer word. A sequence of bits or characters treated as a unit and capable of being stored in one computer location. *See:* **machine word.**   255, 77

computing element (analog computer). A computer component that performs a mathematical operation required for problem solution. It is shown explicitly in computer diagrams, or computer programs.   9

concealed (wiring methods). Rendered inaccessible by the structure or finish of the building. Wires in concealed raceways are considered concealed, even though they may become accessible by withdrawing them.   256

concentrate (metallurgy). The product obtained by concentrating disseminated or lean ores by mechanical or other processes thereby eliminating undesired minerals or constituents. *See also:* **electrowinning.**   328

concentrated winding (rotating machinery). A winding, the coils of which occupy one slot pole; or a field winding mounted on salient poles. *See also:* **asynchronous machine; direct-current commutating machine.**   63

concentration cell. (1) An electrolyte cell, the electromotive force of which is due to differences in composition of the electrolyte at anode and cathode areas.   221, 205
(2) A cell of the two-fluid type in which the same dissolved substance is present in differing concentrations at the two electrodes. *See also:* **electrochemistry.**   328

concentration polarization. (1) That part of the total polarization that is caused by changes in the activity of the potential-determining components of the electrolyte. *See also:* **electrochemistry.**   328
(2) That portion of the polarization of an electrode produced by concentration changes at the metal-environment interface.   221, 205

concentrator (1) (telephone switching systems). A switching entity for connecting a number of inlets to a smaller number of outlets.   55
(2) (data transmission). A communication technique wherein several subscribers or data terminals simultaneously share a single communication circuit by using time division or other techniques.   59
(3) (data communications). A device that provides communications capability between many low-speed, usually asynchronous channels and fewer high-speed, usually synchronous channels the sum of whose data rates is (usually) less than the sum of the data rates of the low-speed channels.   12

concentric electrode system (electrobiology) (coaxial electrode system). An electrode system that is geometrically coaxial but electrically unsymmetrical. *Example:* One electrode may have the form of a cylindrical shell about the other so as to afford electrical shielding. *See also:* **electrobiology.**   192

concentric groove (disk recording). *See:* **locked groove.**

concentricity of coaxial connectors (fixed and variable attenuator). Total indicator runout between the diameter of outer conductor and that diameter of that portion of inner conductor which engages with the corresponding diameters of mating connector. *Note:* This does not apply to precision connectors with only butt contacts. .   110

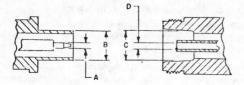

$A \equiv$ Minimum diameter of circle capable to enclose maximum diameter of shank

$B \equiv$ Minimum diameter of circle capable of enclosing maximum diameter of outer conductor

$C \equiv$ Maximum diameter of circle fitting within minimum diameter of outer contact

$D \equiv$ If contact is fully opened by insertion of nominal-size male shank, diameter of smallest circle enclosing inserted shank at end of female inner contact

Definitions of concentricity:

Male connectors: Total indicator runout between circles A and B

Female connectors: Total indicator runout between circles C and D

concentric-lay cable. (1) A concentric-lay conductor as defined below or (2) a multiple-conductor cable composed of a central core surrounded by one or more layers of helically laid insulated conductors. *See also:* **power distribution, underground construction.** 64

concentric-lay conductor. A conductor composed of a central core surrounded by one or more layers of helically laid wires. *Note:* In the most common type of concentric-lay conductor, all wires are of the same size and the central core is a single wire. *See also:* **conductor.** 64

concentric line. *See:* **coaxial.**

concentric resonator (laser-maser). A beam resonator comprising a pair of spherical mirrors having the same axis of rotational symmetry and positioned so that their centers of curvature coincide on this axis. 363

concentric winding (rotating machinery). A winding in which the two coil sides of each coil of a phase belt, or of a pole of a field winding, are symmetrically located so as to be equidistant from a common axis. *See also:* **asynchronous machine.** 63

concentric windings (transformer). An arrangement of transformer windings where the primary and secondary windings, and the tertiary winding, if any, are located in radial progression about a common core. 53

concrete quantity. *See:* **physical quantity.**

concrete-tight fitting (for conduit). A fitting so constructed that embedment in freshly mixed concrete will not result in the entrance of cement into the fitting. 296

condensed-mercury temperature (mercury-vapor tube). The temperature measured on the outside of the tube envelope in the region where the mercury is condensing in a glass tube or at a designated point on a metal tube. *See:* **gas tube.** 125, 190

condenser* (1). *See:* **capacitor.**

* Deprecated

(2) (fuse). *See:* **fuse condenser.**

condenser antenna*. *See:* **capacitor antenna.**

* Deprecated

condenser box*. *See:* **subdivided capacitor.**

* Deprecated

condenser bushing*. *See:* **capacitor bushing.**

* Deprecated

condenser loudspeaker* (capacitor loudspeaker). *See:* **electrostatic loudspeaker.**

* Deprecated

condenser microphone* (capacitor microphone). *See:* **electrostatic microphone.**

* Deprecated

condition (computing system). *See:* **initial condition.**

conditional information content (first symbol given a second symbol). The negative of the logarithm of the conditional probability of the first symbol, given the second symbol. *Notes:* (1) The choice of logarithmic base determines the unit of information content. *See:* **bit** and **hartley.** (2) The conditional information content of an input symbol given an output symbol, averaged over all input-output pairs, is the equivocation. (3) The conditional information content of output symbols relative to input symbols, averaged over all input-output pairs, has been called spread, prevarication, irrelevance, etcetera. *See also:* **information theory.** 160

conditional jump. To cause, or an instruction that causes, the proper one of two (or more) addresses to be used in obtaining the next instruction, depending upon some property of one or more numerical expressions or other conditions. Sometimes called a branch. *See also:* **jump.** 235, 210

conditional transfer of control. Same as **conditional jump.** 235, 210

conditioning (data communications). The addition of equipment to or selection of communication facilities to provide the performance characteristics required for certain types of data transmission. 12

conditioning stimulus (medical electronics). A stimulus of given configuration applied to a tissue before a test stimulus. 192

conductance. (1) That physical property of an element, device, branch, network or system, that is the factor by which the mean square voltage must be multiplied to give the corresponding power lost by dissipation as heat or as other permanent radiation or loss of electromagnetic energy from the circuit. (2) The real part of admittance. *Note:* Definitions (1) and (2) are not equivalent but are supplementary. In any case where confusion may arise, specify the definition being used. 210, 185

conductance, electrode. *See:* **electrode conductance.**

conductance for rectification (electron tube). The quotient of (1) the electrode alternating current of low frequency by (2) the in-phase component of the electrode alternating voltage of low frequency, a high frequency sinusoidal voltage being applied to the same or another electrode and all other electrode voltages being maintained constant. *See also:* **rectification factor.** 190

conductance relay. A mho* relay for which the center of the operating characteristic on the $R\text{-}X$ diagram is on the R axis. *Note:* The equation that describes such a characteristic is $Z = K \cos \theta$ where K is a constant and θ is the phase angle by which the input voltage leads the input current. 103, 202, 60, 127

* Colloquial

conducted heat. The thermal energy transported by thermal conduction. *See also:* **thermoelectric device.**
248, 191

conducted interference (electromagnetic compatibility). Interference resulting from conducted radio noise or unwanted signals entering a transducer (receiver) by direct coupling. *See also:* **electromagnetic compatibility.**
199

conducted radio noise (electromagnetic compatibility). Radio noise propagated along circuit conductors. *Note:* It may enter a transducer (receiver) by direct coupling or by an antenna as by subsequent radiation from some circuit element. *See also:* **electromagnetic compatibility.**
199

conducting element (fuse) (fuse link). The conducting means, including the current-response element, for completing the electric circuit between the terminals of a fuseholder or fuse unit. 103, 202

conducting material. A conducting medium in which the conduction is by electrons, and whose temperature coefficient of resistivity is, except for certain alloys, nonnegative at all temperatures below the melting point.
210

conducting mechanical joint. The juncture of two or more conducting surfaces held together by mechanical means. *Note:* Parts jointed by fusion processes, such as welding, brazing, or soldering, are excluded from this definition.
103, 202, 27

conducting paint (rotating machinery). A paint in which the pigment or a portion of pigment is a conductor of electricity and the composition is such that when it is converted to a solid film, the electric conductivity of the film approaches that of metallic substances. 63

conducting parts (industrial control). The parts that are designed to carry current or that are conductively connected therewith. 225, 206

conducting (conduction) period (1) (rectifier circuit element) (semiconductor). That part of an alternating voltage cycle during which the current flows in the forward direction. *Note:* The forward period is not necessarily the same as the conducting period because of circuit parameters and semiconductor rectifier diode characteristics. 66
(2) (gas tube). That part of an alternating-voltage cycle during which a certain arc path is carrying current. *See also:* **gas-filled rectifier.** 244, 190
(3) (rectifying element). That part of an alternating-voltage cycle during which current flows from the anode to the cathode. *See also:* **rectification.** 328

conducting salts. Salts that, when added to a plating solution, materially increase its conductivity. *See also:* **electroplating.** 328

conduction band (semiconductor). A range of states in the energy spectrum of a solid in which electrons can move freely. *See also:* **semiconductor.** 245, 66, 210, 186

conduction current. Through any surface, the integral of the normal component of the conduction current density over that surface. *Notes:* (1) Conduction current is a scalar and hence has no direction. (2) Current does not flow. In speaking of conduction current through a given surface it is appropriate to refer to the charge flowing or moving through that surface, but not to the current "flow". (3) For the use on circuit diagrams of arrows or other graphic symbols in connection with currents, *see* **reference direction (current).** 210

conduction electron. *See:* **electrons, conduction.**

conductive coating (rotating machinery). Conducting paint applied to the slot portion of a coil-side, to carry capacitive and leakage currents harmlessly between insulation and grounded iron. 63

conduction-through (thyristor converter). The failure to achieve forward blocking, during inverter operation, of an arm of a thyristor connection at the end of the normally conducting period, thus enabling the direct current to continue to pass during the period when the thyristor is normally in the off state. *Note:* A conduction-through occurs, for example, when the margin angle is too small or because of a misgating in the succeeding arm. *Syn.* **shoot through.** 121

conductive coupling (interference terminology). *See:* **coupling, conductance.**

conductivity (material). A factor such that the conduction-current density is equal to the electric-field intensity in the material multiplied by the conductivity. *Note:* In the general case it is a complex tensor quantity. *See also:* **transmission line.** 210, 185

conductivity in physical media (antenna). The real part of the complex conductivity. *See also:* **complex permittivity.**
111

conductivity modulation (semiconductor). The variation of the conductivity of a semiconductor by variation of the charge-carrier density. *See also:* **semiconductor; semiconductor device.** 245, 66, 210

conductivity, *n*-type (semiconductor). The conductivity associated with conduction electrons in a semiconductor. *See also:* **semiconductor.** 245, 66, 210, 186

conductivity, *p*-type (semiconductor). The conductivity associated with holes in a semiconductor. *See also:* **semiconductor.** 245, 66, 210, 186

conductor (1) (general). (A) A substance or body that allows a current of electricity to pass continuously along it. *See:* **conducting material.** 210, 244, 63
(B) A wire or combination of wires not insulated from one another, suitable for carrying an electric current. It may be, however, bare or insulated. 345
(C) The portion of a lightning-protection system designed to carry the lightning discharge between air terminal and ground. 297
(2) (conductor stringing equipment). A wire or combination of wires not insulated from one another, suitable for carrying an electric current. It may be, however, bare or insulated. *Syn.* **cable; wire.** 45
(3) (transmission and distribution). A material, usually in the form of a wire, cable, or bus bar, suitable for carrying an electric current. *See also:* **bundled conductor; covered conductor; grounded conductor; grounding conductor; insulated conductor; lateral conductor; line conductor; open conductor (open wire).** 262

conductor, bare. One having no covering or insulation whatsoever. 256

conductor car (conductor stringing equipment). A device designed to carry workmen and ride on "sagged bundle conductors," thus enabling them to inspect the conductors for damage and install spacers and dampers where required. These devices may be manual or powered. *Syn.* **cable buggy; cable car; spacer buggy; spacer cart.** 45

conductor, coaxial. *See:* **coaxial conductor.** 54

conductor clearance. *See:* ANSI C2, current edition.

conductor-cooled (rotating machinery). A term referring to windings in which coolant flows in close contact with the conductors so that the heat generated within the principal portion of the windings reaches the cooling medium without flowing through the major ground insulation. 244

conductor, covered (1) (general). A conductor having one or more layers of nonconducting materials that are not recognized as insulation under the electric code. 256 **(2) (transmission and distribution).** A conductor covered with a dielectric having no rated insulating strength or having a rated insulating strength less than the voltage of the circuit in which the conductor is used. 64

conductor insulation (rotating machinery). The insulation on a conductor or between adjacent conductors. 63

conductor loading (mechanical). The combined load per unit length of a conductor due to the weight of the wire plus the wind and ice loads. *See also:* **tower.** 64

conductor-loop resistance (telephone switching systems). The series resistance of the conductors of a line or trunk loop, excluding terminal equipment or apparatus. 55

conductor shielding (1) (power distribution underground cables). A conducting or semiconducting element in direct contact with the conductor and in intimate contact with the inner surface of the insulation so that the potential of this element is the same as the conductor. Its function is to eliminate ionizable voids at the conductor and provide uniform voltage stress at the inner surface of the insulating wall. 57 **(2) (cable systems).** A conducting material applied in manufacture directly over the surface of the conductor and firmly bonded to the inner surface of the insulation. 35

conductor support box. A box that is inserted in a vertical run of raceway to give access to the conductors for the purpose of providing supports for them. *See also:* **cabinet.** 328

conduit (1) (electric power). A structure containing one or more ducts. *Note:* Conduit may be designated as iron pipe conduit, tile conduit, etcetera. If it contains one duct only it is called "single-duct conduit," if it contains more than one duct it is called "multiple-duct conduit," usually with the number of ducts as a prefix, namely, two-duct multiple conduit. *See also:* **cable.** 64, 262 **(2) (aircraft).** An enclosure used for the radio shielding or the mechanical protection of electric wiring in an aircraft. *Note:* It may consist of either rigid or flexible, metallic or nonmetallic tubing. Conduit differs from pipe and metallic tubing in that it is not normally used to conduct liquids or gases. *See also:* **air transportation wiring and associated equipment; flexible metal conduit; rigid metal conduit.** 328 **(3) (packaging machinery).** A tubular raceway for holding wires or cables, which is designed expressly for, and used

solely for, this purpose. 124

conduit fitting. An accessory that serves to complete a conduit system, such as bushings and access fittings. *See also:* **raceways.** 296

conduit knockout. *See:* **knockout.**

conduit run. *See:* **duct bank.**

cone (1) (cathode-ray tube). The divergent part of the envelope of the tube. *See also:* **cathode-ray tubes.** 244, 190 **(2) (vision).** Retinal elements that are primarily concerned with the perception of detail and color by the light-adapted eye. *See also:* **retina.** 167

cone, leader. *See:* **leader cone.**

cone of ambiguity (electronic navigation). A generally conical volume of airspace above a navigation aid within which navigational information from that facility is unreliable. *See also:* **navigation.** 187

cone of nulls. A conical surface formed by directions of negligible radiation. *See also:* **antenna.** 179

cone of protection (lightning). The space enclosed by a cone formed with its apex at the highest point of a lightning rod or protecting tower, the diameter of the base of the cone having a definite relation to the height of the rod or tower. *Note:* This relation depends on the height of the rod and the height of the cloud above the earth. The higher the cloud, the larger the radius of the base of the protecting cone. The ratio of radius of base to height varies approximately from one to two. When overhead ground wires are used, the space protected is called a zone of protection or protected zone. *See also:* **lightning protection and equipment.** 297, 64

cone of silence (electronic navigation). A conically shaped region above an antenna where the field strength is relatively weak because of the configuration of the antenna system. *See also:* **navigation.** 13, 187

conference call (telephone switching systems). A call in which communication is provided among more than two main stations. 55

conference connection. A special connection for a telephone conversation among more than two stations. 328

confidence test (test, measurement and diagnostic equipment). A test primarily performed to provide a high degree of certainty that the unit under test is operating acceptably. 54

confidence tester (test, measurement and diagnostic equipment). Any test equipment, either automatic, semiautomatic, or manual, which is used expressly for performing a test or series of tests to increase the degree of certainty that the unit under test is operating acceptably. 54

configuration (group of electric conductors). Their geometrical arrangement, including the size of the wires and their relative positions with respect to other conductors and the earth. *See also:* **inductive coordination.** 328

conflict, antenna. *See:* **antenna conflict.**

confocal resonator (laser-beam). A beam resonator comprising a pair of spherical mirrors having the same axis of rotational symmetry and positioned so that their focal points coincide on this axis. 363

conformance tests (acceptance tests) (1) (surge arrester). Tests made, when required, to demonstrate selected performance characteristics of a product or representative samples thereof.   2, 62

(2) (general). Tests that are specifically made to demonstrate conformity with applicable standards.

103, 202, 53

(3) (mechanical switching device). Those tests that are specifically made to demonstrate the conformity of switchgear or its component parts with applicable standards.   92, 300

(4) (transformer). Tests which are made by agreement between the manufacturer and the purchaser at the time the order is placed. In some cases, by mutual agreement, certain Design Tests may be made as Conformance Tests.

53

conformity (1) (potentiometer). The accuracy of its output; used especially in reference to a function potentiometer.

9, 77, 10

(2) (curve) (automatic control). The closeness with which it approximates the specified functional curve (for example logarithmic, parabolic, cubic, etcetera). *Note:* It is usually expressed in terms of a nonconformity, for example the maximum deviation. For "independent conformity," any shift or rotation is permissible to reduce this deviation. For "terminal conformity," the specified functional curve must be drawn to give zero output at zero input and maximum output at maximum input, but the actual deviation at these points is not necessarily zero.   56

conical horn. A horn whose cross-sectional area increases as the square of the axial length. *See also:* **loudspeaker.**

176

conical scanning (1) (antenna). A form of sequential lobing in which the direction of maximum radiation generates a cone where the vertex angle is of the order of the antenna half-power beamwidth. *Note:* Such scanning may be either rotating or nutating according to whether the direction of polarization rotates or remains unchanged. *See also:* **antenna.**   246, 179

(2) (radar). A form of angular tracking in which the direction of maximum radiation is offset from the axis of the antenna. Rotation of the beam about the axis generates a cone whose vertex angle is of the order of the beamwidth; such scanning may be either rotating or nutating, according to whether the direction of polarization rotates or remains unchanged.   13

conical wave (radio wave propagation). A wave whose equiphase surfaces asymptotically form a family of coaxial circular cones. *See also:* **radio wave propagation.**   180

coning effect (gyro). The apparent drift rate caused by motion of an input axis in a manner which generally describes a cone. This usually results from a combination of oscillatory motions about the principal axes of single-degree-of-freedom gyros. The apparent drift rate is a function of the amplitudes and frequencies of oscillations present and the phase angles between them.   46

conjugate branches. Any two branches of a network such that a driving force impressed in either branch produces no response in the other. *See also:* **network analysis.**

210

conjugate bridge. The detector circuit and the supply circuit are interchanged as compared with a normal bridge of the given type. *See also:* **bridge.**   328

connected (1) (network). A network is connected if there exists at least one path, composed of branches of the network, between every pair of nodes of the network. *See also:* **network analysis.**   332

(2) (circuits and systems) (graph). A graph is connected if there exists at least one path between every pair of its vertices.   67

connected load. The sum of the continuous ratings of the load-consuming apparatus connected to the system or any part thereof. *See also:* **generating station.**   64

connected position(switchgear-assembly removable element). That position of the removable element in which both primary and secondary disconnecting devices are in full contact.   103, 202

connecting rod or shaft (high voltage air switches, insulators, and bus supports). A component of a switch operating mechanism designed to transmit motion from an offset bearing or bell crank to a switch pole unit.   27

connecting wire (mining). A wire generally of smaller gauge than the shot-firing cord and used for connecting the electric blasting-cap wires from one drill hole to those of an adjoining one in mines, quarries, and tunnels. *See also:* **mine feeder circuit.**   328

connection (1) (rotating machinery). Any low-impedance tie between electrically conducting components.   63

(2) (nuclear power generating stations). A cable terminal, splice, or hostile environment boundary seal at the interface of cable and equipment.   31, 141

connection box (mine type). A piece of apparatus with enclosure within which electric connections between sections of cable can be made. *See also:* **mine feeder circuit.**

124

connection diagram. A diagram that shows the connection of an installation or its component devices, controllers, and equipment. *Notes:* (1) It may cover internal or external connections, or both, and shall contain such detail as is needed to make or trace connections that are involved. It usually shows the general physical arrangement of devices and device elements and also accessory items such as terminal blocks, resistors, etcetera. (2) A connection diagram excludes mechanical drawings, commonly referred to as wiring templates, wiring assemblies, cable assemblies, etcetera.   210, 225, 206, 301

connection insulation (joint insulation) (rotating machinery). The insulation at an electric connection such as between turns or coils or at a bushing connection. *See also:* **stator.**

63

connections of polyphase circuits (rotating machinery). *See:* **mesh-connected circuit; star-connected circuit; zig-zag connection of polyphase circuits.**

connector (1). A coupling device employed to connect conductors of one circuit or transmission element with those of another circuit or transmission element. *See also:* **auxiliary device to an instrument.**   185

(2) (wires). A device attached to two or more wires or cables for the purpose of connecting electric circuits without the use of permanent splices.   1

(3) (splicing sleeve). A metal sleeve, that is slipped over and secured to the butted ends of the conductors in making up a joint.   64

(4) (waveguides). A mechanical device, excluding an adapter, for electrically joining separable parts of a waveguide or transmission-line system.   179, 185

(5) (power cable joints). A metallic device of suitable

electric conductance and mechanical strength, used to splice the ends of two or more cable conductors, or as a terminal connector on a single conductor. Connectors usually fall into one of the following types: solder, welded, mechanical, and compression or indent. Conductors are sometimes spliced without connectors, by soldering, brazing or welding. 34

connector base (motor plug*) (motor attachment plug cap*). A device, intended for flush or surface mounting on an appliance, that serves to connect the appliance to a cord connector. 328

* Deprecated

connector, precision (waveguide or transmission line). A connector that has the property of making connections with a high degree of repeatability without introducing significant reflections, loss or leakage. *See also:* **auxiliary device to an instrument.** 185

connector switch (connector). A remotely controlled switch for connecting a trunk to the called line. 328

connector, waveguide (fixed and variable attenuator). A mechanical device, excluding an adapter, for electrically joining separable parts of a waveguide or transmission-line system. 110

conservator (expansion tank) system (transformer or regulator). A method of oil preservation in which the oil in the main tank is sealed from the atmosphere, over the temperature range specified, by means of an auxiliary tank partly filled with oil and connected to the completely filled main tank. *See:* **transformer, oil-immersed.**
203, 257, 91, 53

conservator/diaphragm system. A system in which the oil in the main tank is completely sealed from the outside atmosphere, and is connected to an elastic diaphragm tank contained inside a tank mounted at the top of the transformer. As oil expands and contracts within a specified temperature range the system remains completely sealed with an approximately constant pressure. 53

consistency. *See:* **precision.**

consol (electronic navigation). A keyed continuous-wave short-baseline radio navigation system operating in the low- and medium-frequency bands, generally useful to about 1500 nautical miles (2800 kilometers), and using three radiators to provide a multiplicity of overlapping lobes of dot-and-dash patterns that form equisignal hyperbolic lines of position. *Note:* These lines of position are moved slowly in azimuth by changing radio-frequency phase, thus allowing a simple listening and counting or timing operation to be used to determine a line of position within the sector bounded by any pair of equisignal lines. *See also:* **navigation.** 187, 13

consolan (electronic navigation). A form of consol using two radiators instead of three. *See also:* **navigation.**
187, 13

console (telephony) (1) (switchgear). A control cabinet located apart from the associated switching equipment arranged to control those functions for which an attendant or an operator is required. 193

(2) (computing system). The part of a computer used for communication between the operator or maintenance engineer and the computer. 255

(3) (telephone switching systems). A desk or desk-top cordless switchboard which may include display elements in addition to those required for supervisory purposes is required. 55

consonant articulation (percent consonant articulation). The percent articulation obtained when the speech units considered are consonants (usually combined with vowels into meaningless syllables). *See also:* **volume equivalent.**
328

conspicuity. The capacity of a signal to stand out in relation to its background so as to be readily discovered by the eye. *See also:* **atmospheric transmissivity.** 167

constancy (probe coupling). *See:* **residual probe pickup.**

constant. *See:* **time constant of integrator.**

constant-amplitude recording (mechanical recording). A characteristic wherein, for a fixed amplitude of a sinusoidal signal, the resulting recorded amplitude is independent of frequency. *See also:* **phonograph pickup.**
176

constant-available-power source (telephony). A signal source with a purely resistive internal impedance and a constant open-circuit terminal voltage, independent of frequency. *See also:* **telephone station.** 122, 196

constant-current arc-welding power supply. A power supply that has characteristically drooping volt-ampere curves producing relatively constant current with a limited change in load voltage. *Note:* This type of supply is conventionally used in connection with manual-stick-electrode or tungsten-inert-gas arc welding. 264

constant-current characteristic (electron tubes). The relation, usually represented by a graph, between the voltages of two electrodes, with the current to one of them as well as all other voltages maintained constant.
190, 125

constant-current charge (storage battery) (storage cell). A charge in which the current is maintained at a constant value. *Note:* For some types of lead-acid batteries this may involve two rates called the starting and finishing rates. *See also:* **charge.** 328

constant-current (Heising) modulation. A system of amplitude modulation wherein the output circuits of the signal amplifier and the carrier-wave generator or amplifier are directly and conductively coupled by means of a common inductor that has ideally infinite impedance to the signal frequencies and that therefore maintains the common plate-supply current of the two devices constant. *Note:* The signal-frequency voltage thus appearing across the common inductor appears also as modulation of the plate supply to the carrier generator or amplifier with corresponding modulation of the carrier output. 111

constant-current power supply. A power supply that is capable of maintaining a preset current through a variable load resistance. *Note:* This is achieved by automatically varying the load voltage in order to maintain the ratio V_{load}/R_{load} constant. *See also:* **power supply.** 228, 186

constant-current regulation (generator). That type of automatic regulation in which the regulator maintains a constant-current output from the generator. *See also:* **axle generator system.** 328

constant-current (series) incandescent filament lamp transformer. *See:* **incandescent filament lamp transformer (series type).**

constant-current street-lighting system (series street-lighting

system). A street-lighting system employing a series circuit in which the current is maintained substantially constant. *Note:* Special generators or rectifiers are used for direct current while suitable regulators or transformers are used for alternating current. *See also:* **alternating-current distribution; direct-current distribution.**   64

constant-current (series) mercury-lamp transformer. A transformer that receives power from a current-regulated series circuit and transforms the power to another circuit at the same or different current from that in the primary circuit. *Note:* It also provides the required starting and operating voltage and current for the specified lamp. Further, it provides protection to the secondary circuit, casing, lamp, and associated luminaire from the high voltage of the primary circuit.   274

constant-current transformer. A transformer that automatically maintains an approximately constant current in its secondary circuit under varying conditions of load impedance when supplied from an approximately constant-potential source.   53

constant cutting speed (numerically controlled machine). The condition achieved by varying the speed of rotation of the workpiece relative to the tool inversely proportional to the distance of the tool from the center of rotation.   224

constant-delay discriminator (electronic navigation). *See:* **pulse decoder.**   182

constant failure rate period (reliability). That possible period during which the failures occur at an approximately uniform rate. *Note:* The curve below shows the failure rate pattern when the terms of minor failure, early failure period, and constant failure rate period all apply to the item.   164

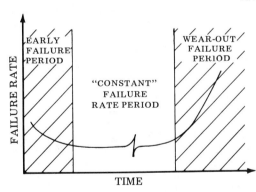

constant false alarm rate (radar). The property of threshold or gain control devices specially designed to suppress false alarms caused by noise, clutter or ECM (electronic counter measures) of varying level.   13

constant fraction discriminator. A pulse amplitude discriminator in which the threshold for acceptance of a pulse is a constant fraction of the peak amplitude of the pulse.   118

constant-frequency control (power system). A mode of operation under load-frequency control in which the area control error is the frequency deviation. *See also:* **speed-governing system.**   94

constant-horsepower motor. *See:* **constant-power motor.**

constant-horsepower range (electric drive). The portion of its speed range within which the drive is capable of maintaining essentially constant horsepower. *See also:* **electric drive.**   206

constant-*K* network. A ladder network for which the product of series and shunt impedances is independent of frequency within the range of interest. *See also:* **network analysis.**   210

constant-luminance transmission (color television). A type of transmission in which the sole control of luminance is provided by the luminance signal, and no control of luminance is provided by the chrominance signal. *Notes:* (1) In such a system, noise signals falling within the bandwidth of the chrominance channel produce only chromaticity variations at the outputs of the chrominance demodulators. Coarse-structured chromaticity variations thus produced are subjectively less objectionable than correspondingly coarse-structured luminance variations. (2) Because of the use of gamma correction in the camera, these ideal conditions are not completely realized, especially for colors of high saturation.   18, 178

constant-net-interchange control (power system). A mode of operation under load-frequency control in which the area control error is determined by the net interchange deviation. *See also:* **speed-governing system.**   94

constant-power motor (constant-horsepower motor). A multispeed motor that develops the same related power output at all operating speeds. The torque then is inversely proportional to the speed. *See also:* **asynchronous machine.**   63

constant-resistance (conductance) network. A network having at least one driving-point impedance (admittance) that is a positive constant. *See also:* **linear passive networks.**   238

constant-speed motor. A motor in which the speed is constant or substantially constant over the normal range of loads. *Note:* For example, a synchronous motor, an induction motor with small slip, or a direct-current shunt-wound motor with constant excitation. *See also:* **asynchronous machine; direct-current commutating machine.**   63, 3

constant-torque motor. Multispeed motor that is capable of developing the same torque for all design speeds. The rated power output varies directly with the speed. *See also:* **asynchronous machine.**   63

constant-torque range (electric drive). The portion of its speed range within which the drive is capable of maintaining essentially constant torque. *See also:* **electric drive.**   206

constant-torque resistor (motors). A resistor for use in the armature or rotor circuit of a motor in which the current remains practically constant throughout the entire speed range. *See also:* **asynchronous machine; direct-current commutating machine; electric drive.**   3, 225, 206

constant-torque speed range (industrial control). The portion of the speed range of a drive within which the drive is capable of maintaining essentially constant torque.   219, 206

constant-velocity recording (mechanical recording). A characteristic wherein for a fixed amplitude of a sinusoidal signal, the resulting recorded amplitude is inversely proportional to the frequency. *See also:* **phonograph pickup.**   176

constant-voltage arc-welding power supply. Power supply (arc welder) that has characteristically flat volt-ampere curves producing relatively constant voltage with a change in load current. This type of power supply is conventionally used in connection with welding processes involving consumable electrodes fed at a constant rate. 264

constant-voltage charge (storage battery) (storage cell). A charge in which the voltage at the terminals of the battery is held at a constant value. *See also:* **charge.** 328

constant-voltage power supply (voltage regulator). A power supply that is capable of maintaining a preset voltage across a variable load-resistance. This is achieved by automatically varying the output current in order to maintain the product of load current times load resistance constant. *See also:* **power supply.** 228

constant-voltage regulation (generator). That type of automatic regulation in which the regulator maintains constant voltage of the generator. *See also:* **axle generator system.** 328

constant-voltage transformer. A transformer that maintains an approximately constant voltage ratio. 53

constraints (1). Limits on the ranges of variables or system parameters because of physical or system requirements. *See also:* **system.** 209
(2) (control system). A restriction placed on the control signal, control law, or state variables. *See also:* **control system.** 329

construction diagram (industrial control). A diagram that shows the physical arrangement of parts, such as wiring, buses, resistor units, etc. *Example:* A diagram showing the arrangement of grids and terminals in a grid-type resistor. 210, 206

construction tests (nuclear power generating station). A construction test is defined as a test to verify proper installation and operation of individual components in a system prior to operation of the system as an entity. 354

contact. A conducting part that co-acts with another conducting part to make or break a circuit. 103, 202, 127, 27

contact area (1) (photoelectric converter). The area of ohmic contact provided on either the *p* or *n* faces of a photoelectric converter for electrical circuit connections. *See also:* **semiconductor.** 186
(2) (solar cells). That area of ohmic contact provided on either the *p* or *n* surface of a solar cell for electric circuit connections. 113

contact bounce (sequential events recording systems). The undesired intermittent closure of open contacts or opening of closed contacts. *Note:* It may occur either when the contact device is operated or released, and is solely a characteristic of the contact device. 48

contact chatter, relay. *See:* **relay contact chatter.**

contact clip (mechanical switching device). The clip that the switchblade enters or embraces. 103, 202, 27

contact conductor (1) (electric traction). The part of the distribution system other than the track rails, that is in immediate electric contact with current collectors of the cars or locomotives. *See:* **contact wire (trolley wire); trolley; underground collector or plow.** *See also:* **multiple-unit control.** 1
(2) (contact electrode) (electrochemistry). A device to lead electric current into or out of a molten or solid metal or alloy that itself serves as the active electrode in the cell. *See also:* **fused electrolyte.** 328

contact converter (relaying). A buffer element used to produce a prescribed output as the result of the opening or closing of a contact. 79

contact corrosion. *See:* **crevice corrosion.**

contact current-carrying rating (relay). The current that can be carried continuously, or for stated periodic intervals, without impairment of the contact structure or interrupting capability. *See also:* **relay.** 60, 127

contact current-closing rating (relay). The current that the device can close successfully with prescribed operating duty and circuit conditions without significant impairment of the contact structure. *See also:* **relay.** 60

contact follow-up (relays, switchgear, and industrial control). The distance between the position one contact face would assume, were it not blocked by the second (mating) contact, and the position the second contact face would assume were the first contact removed, when the actuating member is fixed in its final contact-closed position. *See:* **electric controller.** *See also:* **initial contact pressure.** 302, 225, 206

contact gap (break) (industrial control). The final length of the isolating distance of a contact in the open position. *See:* **contactor.** 244, 206

contact, high recombination rate (semiconductor). A semiconductor-semiconductor or metal-semiconductor contact at which thermal equilibrium charge-carrier concentrations are maintained substantially independent of current density. *See also:* **semiconductor; semiconductor device.** 210, 245, 66

contact interrupting rating (relay). The current that the device can interrupt successfully, with prescribed operating duty and circuit conditions without significant impairment of the contact structure. *See also:* **relay.** 60, 127

contact joint (contact coupling) (waveguide components). A connection designed for essentially complete transfer of power between two waveguides by means of metallic contact between the inner walls of the waveguides. *Note:* It typically consists of two contact flanges. 166

contactless vibrating bell. A vibrating bell whose continuous operation depends upon application of alternating-current power without circuit-interrupting contacts. *See also:* **protective signaling.** 328

contact-making clock (demand meter). A device designed to close momentarily an electric circuit to a demand meter at definite and consecutive intervals. *See also:* **demand meter.** 212

contact mechanism (demand meter). A device for attachment to an electricity meter or to a demand-totalizing relay for the purpose of providing electric impulses for transmission to a demand meter or relay. *See also:* **demand meter.** 328

contact nomenclature. *See:* **relay terms.**

contact opening time (relay). The time a contact remains closed, while in process of opening, following a specified change of input. *Note:* This term is properly restricted to the inherent delay while the contact is actually being operated and is not to include delay that may occur before contact operation begins. *For example:* contact-opening time while opening results from the flexing and wiping together of contacts when they are closed. *See also:* **drop-out time.** 103, 202, 60

contactor (1) (industrial control) (transformer). A device for repeatedly establishing and interrupting an electric power circuit. 206, 53

(2) (land transportation). A device which upon receipt of an electrical signal establishes or opens repeatedly an electrical circuit with a nominal current rating of 5 amperes minimum for its main contacts. 1

contactor, load. *See:* **load switch (load contactor).**

contactor of switch, load. *See:* **load switch of contactor.** 14

contactor or unit switch. A device operated other than by hand for repeatedly establishing and interrupting an electric power circuit under normal conditions. *See also:* **control switch.** 1

contact parting time (mechanical switching device). The interval between the time when the actuating quantity in the main circuit reaches the value causing actuation of the release and the instant when the primary arcing contacts have parted in all poles. *Note:* Contact parting time is the numerical sum of release delay and opening time. 103, 202

contact piston (contact plunger) (waveguide). A piston with sliding metallic contact with the walls of a waveguide. *See:* **waveguide.** 179

contact plating. The deposition, without the application of an external electromotive force, of a metal coating upon a base metal, by immersing the latter in contact with another metal in a solution containing a compound of the metal to be deposited. *See also:* **electroplating.** 328

contact plunger. *See:* **contact piston.**

contact potential. The difference in potential existing at the contact of two media or phases. *See also:* **biological contact potential; electrolytic cell.** *See:* **depolarization (biological); depolarization front; injury potential (electrobiology); negative after-potential (electrobiology); positive after-potential (electrobiology).** 328

contact-potential difference. The difference between the work functions of two materials divided by the electronic charge. 190, 125

contact pressure, final (industrial control). The force exerted by one contact against the mating contact when the actuating member is in the final contact-closed position. *Note:* Final contact pressure is usually measured and expressed in terms of the force that must be exerted on the yielding contact while the actuating member is held in the final contact-closed position, and with the mating contact fixed in position, in order to separate the mating contact surfaces. *See:* **contactor.** 225, 206

contact pressure, initial (industrial control). The force exerted by one contact against the mating contact when the actuating member is in the initial contact-touch position. *Note:* The initial contact pressure is usually measured and expressed in terms of the force that must be exerted on the yielding contact while the actuating member is held in the initial contact-touch position in order to separate the mating contact surface against the action of the spring or other contact pressure device. *See:* **electric controller.** 225, 206

contact race (power systems relaying). A circuit design

condition wherein two or more independently operated contacts compete for the control of a circuit which they will open and close. 60

contact rectifier. A rectifier consisting of two different solids in contact, in which rectification is due to greater conductivity across the contact in one direction than in the other. *See also:* **rectifier.** 111

contacts (1). Conducting parts which co-act to complete or to interrupt a circuit. 328

(2) (nonoverlapping) (industrial control). Combinations of two sets of contacts, actuated by a common means, each set closing in one of two positions, and so arranged that the contacts of one set open before the contacts of the other set close. *See:* **electric controller.** 225, 206

(3) (auxiliary) (industrial control) (switching device). Contacts in addition to the main circuit contacts that function with the movement of the latter. *See also:* **contactor.** 206

(4) (overlapping, industrial control). Combinations of two sets of contacts, actuated by a common means, each set closing in one of two positions, and so arranged that the contacts of one set open after the contacts of the other set have been closed. *See:* **electric controller.** 225, 206

contact surface. That surface of a contact through which current is transferred to the co-acting contact. 103, 202, 27

contact voltage (human safety). A voltage accidentally appearing between two points with which a person can simultaneously make contact. 244, 62

contact-wear allowance (industrial control). The total thickness of material that may be worn away before the co-acting contacts cease to perform adequately. *See also:* **contactor.** 103, 202, 206, 27

contact wire (trolley wire). A flexible contact conductor, customarily supported above or to one side of the vehicle. *See also:* **contact conductor.** 1

containment (nuclear power generating stations). That portion of the engineered safety features designed to act as the principal barrier, after the reactor system pressure boundary, to prevent the release, even under conditions of a reactor accident, of unacceptable quantities of radioactive material beyond a controlled zone. 31, 104, 120, 141

contamination (rotating machinery). This deteriorates electrical insulation by actually conducting current over insulated surfaces, or by attacking the material reducing its electrical insulating quality or its physical strength, or by thermally insulating the material forcing it to operate at higher than normal temperatures. *Note:* Included here are: wetness or extreme humidity, oil or grease, conducting dusts and particles, non-conducting dusts and particles, and chemicals of industry. 37

content addressed storage (computing system). *See:* **associative storage.**

content, average information. *See:* **average information content.** *See also:* **information theory.**

content, conditional information. *See:* **conditional information content.** *See also:* **information theory.**

contention (1) (data transmission). A condition on a multipoint communication channel when two or more locations try to transmit at the same time. 59

(2) (data communication). A condition on a communications channel when two or more stations may try to seize the channel at the same time. 12

continuing current (lightning). The low-magnitude current that may continue to flow between components of a multiple stroke. *See also:* **direct-stroke protection (lightning).** 64

continuity cable bond. A cable bond used for bonding of cable sheaths and armor across joints between continuous lengths of cable. *See:* **cable bond; cross cable bond.** *See also:* **power distribution, underground construction.** 64

continuity test (test, measurement and diagnostic equipment). A test for the purpose of detecting broken or open connections and ground circuits in a network or device. 54

continuity tester (test, measurement and diagnostic equipment). An electrical tester used to determine the presence and location of broken or open connections and grounded circuits. 54

continuous-current tests. Tests made at a rated current, until temperature rise ceases, to determine that the device or equipment can carry its rated continuous current without exceeding its allowable temperature rise. 103, 202

continuous data. Data of which the information content can be ascertained continuously in time. 224, 207

continuous duty (rating of electric equipment). A duty that demands operation at a substantially constant load for an indefinitely long time. *See also:* **asynchronous machine; direct-current commutating machine; transformer.** 3, 210, 257, 203, 225, 206, 53

continuous-duty rating. The rating applying to operation for an indefinitely long time. 111

continuous electrode. A furnace electrode that receives successive additions in length at the end remote from the active zone of the furnace to compensate for the length consumed therein. *See also:* **electrothermics.** 328

continuous inductive train control. *See:* **continuous train control.**

continuous lighting (railway practice). An arrangement of circuits so that the signal lights are continuously energized. 328

continuous load. A load where the maximum current is expected to continue for three hours or more. 256

continuous load rating (power inverter unit). Defines the maximum load that can be carried continuously without exceeding established limitations under prescribed conditions of test, and within the limitations of established standards. *See also:* **self-commutated inverters.** 208

continuously adjustable capacitor (continuously variable capacitor*). An adjustable capacitor in which the capacitance can have every possible value within its range. 210

* Deprecated

continuously adjustable inductor (continuously variable inductor*) (variable inductor*). An adjustable inductor in which the inductance can have every possible value within its range. 210

* Deprecated

continuously adjustable resistor (continuously variable resistor*). An adjustable resistor in which the resistance can have every possible value within its range. 210

* Deprecated

continuous noise (electromagnetic compatibility). Noise, the effect of which is not resolvable into a succession of discrete impulses. *See also:* **electromagnetic compatibility.** 199

continuous periodic rating (industrial control). The load that can be carried for the alternate periods of load and rest specified in the rating and repeated continuously without exceeding the specified limitation. 206

continuous-pressure operation (elevators). Operation by means of buttons or switches in the car and at the landings, any one of which may be used to control the movement of the car as long as the button or switch is manually maintained in the actuating position. *See:* **control.** 328

continuous rating (1) (electric equipment). The maximum constant load that can be carried continuously without exceeding established temperature-rise limitations under prescribed conditions of test and within the limitations of established standards. *See also:* **duty; rectification.** 203, 208, 303, 257, 258

(2) (diesel-generator unit, nuclear power generating stations). The electric power output capability which the diesel-generator unit can maintain in the service environment for 8760 hours of operation per (common) year with only scheduled outages for maintenance. 31

(3) transformer. The maximum constant load that can be carried continuously without exceeding established temperature-rise limitations under prescribed conditions. 53

continuous-scan system, supervisory control. *See:* **supervisory control system, continuous-scan.**

continuous-speed adjustment (industrial control). Refers to an adjustable-speed drive capable of being adjusted with small increments, or continuously, between minimum and maximum speed. *See also:* **electric drive.** 206

continuous test (battery). A service test in which the battery is subjected to an uninterrupted discharge until the cutoff voltage is reached. *See also:* **battery (primary or secondary); cutoff voltage.** 328

continuous-thermal-current rating factor (1) (current transformer). The factor by which the rated primary current is multiplied to obtain the maximum allowable primary current based on the limiting temperature rise on a continuous basis. *Note:* The rated primary current is established in many instances as a basis for accuracy classification and to place the ratio on the basis of a 5-ampere secondary. In many designs the rated primary current will produce the limiting temperature rise; in such cases, the factor is unity. Some transformers have an additional continuous thermal-current capacity in excess of the rated primary current; in such cases the factor will be greater than unity. 212

(2) (instrument transformer). The specified factor by which the rated primary current of a current transformer can be multiplied to obtain the maximum primary current that can be carried continuously without exceeding the limiting temperature rise from 30°C ambient air temperature. *Note:* When current transformers are incorporated internally as parts of larger transformers or power circuit breakers, they shall meet allowable average winding and hot spot temperatures under the specific conditions and requirements of the larger apparatus. 53

continuous train control (continuous inductive train control). A type of train control in which the locomotive apparatus is constantly in operative relation with the track circuit and is immediately responsive to a change in the character of the current flowing in the track circuit of the track on which the locomotive is traveling. *See also:* **automatic train control.** 328

continuous-type control (electric power systems). A control mode that provides a continuous relation between the deviation of the controlled variable and the position of the final controlling element. *See also:* **speed-governing system.** 94

continuous-voltage-rise test (rotating machinery). A controlled overvoltage test in which voltage is increased in continuous function of time, linear or otherwise. *See also:* **asynchronous machine; direct-current commutating machines.** 63

continuous wave (cw) (laser-maser). The output of a laser which is operated in a continuous rather than pulsed mode. In this standard, a laser operating with a continuous output for a period greater than 0.25 second is regarded as a cw laser. 363

continuous-wave modulation (data transmission). In continuous-wave modulation, the amplitude (amplitude modulation), frequency (frequency modulation) or phase (phase modulation) of a sine wave (the carrier) is altered in accordance with the information to be transmitted. *Note:* The information to be transmitted is, of course, the alarm, supervisory control and telemetering data. 59

continuous waves (CW). Waves, the successive oscillations of which are identical under steady-state conditions. *See also:* **radio transmission; wave front.** 111, 59

contouring control system (numerically controlled machines). A system in which the controlled path can result from the coordinated, simultaneous motion of two or more axes. 224

contract demand (electric-power utilization). The demand that the supplier of electric service agrees to have available for delivery. *See also:* **alternating-current distribution.** 200

contrast (display presentation). The subjective assessment of the difference in appearance of two parts of a field of view seen simultaneously or successively. (Hence: luminosity contrast, lightness contrast, color contrast, simultaneous contrast, successive contrast). *See also:* **photometry; television.** 244, 178

contrast control. A control, associated with a picture-display device, for adjusting the contrast ratio of the reproduced picture. *Note:* The contrast control is normally an amplitude control for the picture signal. In a monochrome-television system, both average luminance and the contrast ratio are affected. In a color-television system, saturation and hue also may be affected. *See also:* **television.** 328

contrast ratio (1) (television). The ratio of the maximum to the minimum luminance values in a television picture or a portion thereof. *Note:* Generally the entire area of the picture is implied but smaller areas may be specified as in detail contrast. *See also:* **television.** 163, 178

(2) (amplitude, frequency, and pulse modulation). For any diffraction order, the ratio of the maximum light intensity to the minimum light intensity in the order, so that $C = I_{max}/I_{min}$, where C is the contrast ratio. *Note:* In the limiting case when the depth of modulation is equal to 1, the minimum light intensity is due to background light, so that $C = I_{max}/I_b$. In the other extreme when m = 0, the contrast ratio is equal to 1. 72

(3) (acoustically tunable optical filter). The ratio of the dynamic transmission at a given acoustic frequency and power level to the dynamic transmission with no applied acoustic power. *Note:* The contrast ratio is a measure of light leakage through the device. It should be specified for either a monochromatic or white light source input, and the angular spread of the input light. 72

(4) (television). The ratio of the maximum to the minimum luminance values in a television picture or a portion thereof. *Note:* Generally the entire area of the picture is implied, but smaller areas may be specified as in detail contrast. 163

contrast sensitivity. The ability to detect the presence of luminance (photometric brightness) differences. *Note:* Quantitatively, it is equal to the reciprocal of the contrast threshold. *See also:* **visual field.** 167

contrast threshold. (1) The minimal perceptible contrast for a given state of adaption of the eye. (2) The luminance contrast that can be detected during some specific fraction of the times it is presented to an observer, usually 50 percent. *See also:* **visual field.** 167

control (1) (industrial). (A) Broadly, the methods and means of governing the performance of any electric apparatus, machine, or system. 206, 124

(B) The system governing the starting, stopping, direction of motion, acceleration, speed, and retardation of the moving member. 328

(C) A designation of how the equipment is governed, that is, by an attendant, by automatic means, or partially by automatic means and partially by an attendant. *Note:* The word **control** is often used in a broad sense to include **indication** also. 103, 202

(D) Frequently, one or more of the components in any mechanism responsible for interpreting and carrying out manually-initiated directions. *Note:* Sometimes called manual control.

(E) (verb). To execute the function of control as defined above. 206

(2) (electronic computation). (A) Usually, those parts of a digital computer that effect the carrying out of instructions in proper sequence, the interpretation of each instruction, and the application of the proper signals to the arithmetic unit and other parts in accordance with this interpretation. (B) In some business applications of mathematics, a mathematical check. *See:* **computer control state.** 235, 210

(3) (cryotron). an input element of a cryotron. *See also:* **super conductivity.** 191

control accuracy (1) (industrial control). The degree of correspondence between the final value and the ideal value of the directly controlled variable. *See:* **control system, feedback.** 219, 206

(2) (automatic control). The degree of correspondence between the ultimately controlled variable and the ideal value. 56

control action (automatic control). Of a control element or a controlling system, the nature of the change of the output effected by the input. *Note:* The output may be a signal or the value of a manipulated variable. The input may be the control loop feedback signal when the command is constant, an actuating signal, or the output of another control element. One use of control action is to effect compensation. *See:* **compensation.** 56

control action, derivative (1) (industrial control). The component of control action for which the output is proportional to the rate of change of the input. *See:* **control system, feedback.** 219, 206, 94

(2) (automatic control). Action in which the output of the controller is proportional to the first time derivative of the input. 94

(3) (control systems). That component of control action for which the output is proportional to the rate of change of the input. *See:* **control action, proportional plus derivative, and control action, proportional plus integral plus derivative.** 56

control action, integral (1) (industrial control). Control action in which the output is proportional to the time integral of the input. *See also:* **control system, feedback.** 219, 206, 94

(2) (control system). Control action in which the output is proportional to the time integral of the input, that is the rate of change of output is proportional to the input. *Note:* In the practical embodiment of integral control action the relation between output and input, neglecting high frequency terms, is given by

$$\frac{Y}{X} = \pm \frac{I/s}{\frac{bI}{s} + 1} \quad 0 \leqq b \ll 1$$

where b = reciprocal of static gain

$1/2\,\pi$ = gain cross-over frequency in cycles per unit time

s = complex variable

X = input transform

Y = output transform

control action, lead. *See:* **lead, first order.**

control action, proportional (1) (industrial control). Control action in which there is a continuous linear relation between the output and the input. *Note:* This condition applies when both the output and input are within their normal operating ranges. 219, 206

(2) (control systems). Control action in which there is a continuous linear relation between the output and the input. *Note:* This condition applies when both the output and input are within their normal operating ranges and when operation is at a frequency below a limiting value. *See:* **gain, proportional.** 56

(3) (automatic control). Action in which there is a linear relation between the output and the input of the controller. *Note:* The ratio of the change in output produced by the proportional control action to the change in input is defined as the proportional gain. 94

control action, proportional plus derivative (control systems). Control action in which the output is proportional to a linear combination of the input and the time rate-of-change of input. *Syn.* **P.D.** *Note:* In the practical embodiment of proportional plus derivative control action the relationship between output and input, neglecting high

frequency terms, is

$$\frac{Y}{X} = \pm P \frac{1 + sD}{1 + sD/a} \quad a > 1$$

where a = derivative action gain

D = derivative action time constant

P = proportional gain

s = complex variable

X = input transform

Y = output transform 56

control action, proportional plus integral (control systems). Control action in which the output is proportional to a linear combination of the input and the time integral of the input. *Syn.* **P.I.** *Note:* In the practical embodiment of proportional plus integral action the relation between output and input, neglecting high frequency terms, is

$$\frac{Y}{X} = \pm P \frac{\dfrac{I}{s} + 1}{\dfrac{bI}{s} + 1} \quad 0 \leqq b \ll 1$$

where b = proportional gain/static gain

I = integral action rate

P = proportional gain

s = complex variable

X = input transform

Y = output transform 56

control action, proportional plus integral plus derivative (control systems). Control action in which the output is proportional to a linear combination of the input, the time integral of input and the time rate-of-change of input. *Syn.* **P.I.D.** *Note:* In the practical embodiment of proportional plus integral plus derivative control action the relationship of output and input, neglecting high frequency terms, is

$$\frac{Y}{X} = \pm P \frac{\dfrac{I}{s} + 1 + Ds}{\dfrac{bI}{s} + 1 + \dfrac{Ds}{a}} \quad \begin{array}{l} a > 1 \\[4pt] 0 \leqq b < 1 \end{array}$$

where a = derivative action gain

b = proportional gain/static gain

D = derivative action time constant

I = integral action rate

P = proportional gain

s = complex variable

X = input transform

Y = output transform 56

control and indication point, supervisory control. *See:* **supervisory control point, control and indication.**

control apparatus. A set of control devices used to accomplish the intended control functions. *See:* **control.** 206

control area (electric power). A part of a power system or a combination of systems to which a common generation control scheme is applied. 94

control battery (industrial control). A battery used as a source of energy for the control of an electrically operated device. 206

control bus. A bus to distribute power for operating electrically controlled devices. 103, 202

control cable (cable systems). Applied at relatively low current levels or used for intermittent operation to change the operating status of a utilization device of the plant auxiliary system. 35

control center. An assembly of devices for the purpose of switching and protecting a number of load circuits. The control center may contain transformers, contactors, circuit breakers, protective and other devices intended primarily for energizing and de-energizing load circuits. 58

control character. A character whose occurrence in a particular context initiates, modifies, or stops a control operation, for example, a character to control carriage return. 255, 77

control characteristic (gas tube). A relation, usually shown by a graph, between critical grid voltage and anode voltage. *See:* **gas tube.** 125, 190

control circuit (industrial control). The circuit that carries the electric signals directing the performance of the controller but does not carry the main power circuit. *See:* **control.** 3, 206, 124

control-circuit limit switch. A limit switch the contacts of which are connected only into the control circuit. *See:* **control; switch.** 206

control-circuit transformer. A voltage transformer utilized to supply a voltage suitable for the operation of control devices. *See:* **control.** 206, 124

control circuit voltage. The voltage provided for the operation of shunt coil magnetic devices. 124

control compartment (packaging machinery). A space within the base, frame, or column of the machine, used for mounting the control panel. 124

control current (Hall-effect devices). The current through the Hall plate that by its interaction with a magnetic flux density generates the Hall voltage. 107

control current sensitivity (Hall-effect devices). The ratio of the voltage across the Hall terminals to the control current for a given magnitude of magnetic flux density. 107

control current terminals (Hall-effect devices). The terminals through which the control current flows. 107

control cut-out switch (land transportation vehicles). An isolating switch that isolates the control circuits of a motor controller from the source of energy. 1

control designation symbol. A symbol that identifies the particular manner, permissible or required, in which an input variable (possibly in combination with other variables) causes the logic element to perform according to its defined function. 88

control desk (1) (industrial control). An assembly of master switches, indicators, instruments, and the like on a structure in the form of a desk for convenience in manipulation and control supervision by the operator. *See:* **control.** 206, 103

(2) (switchgear assemblies). A (control) switchboard consisting of one or more relatively short horizontal or inclined panels mounted on an assembly of such a height that the panel-mounted devices are within convenient reach of an attendant. 93, 103

control device. An individual device used to execute a control function. *See:* **control.** 206

control electrode (electron tubes). An electrode used to initiate or vary the current between two or more electrodes. *See also:* **electrode (electron tube).** 125, 190

control-electrode discharge recovery time (attenuator tubes). The time required for the control-electrode discharge to deionize to a level such that a specified fraction of the critical high-power level is required to ionize the tube. *See:* **gas tube.** 125, 190

control enclosure. The metal housing for the control panel, whether mounted on the industrial equipment, or separately mounted. 124

control exciter (rotating machinery). An exciter that acts as a rotary amplifier in a closed-loop circuit. *See also:* **asynchronous machines.** 63

control generator. A generator, commonly used on electric motive power units for the generation of electric energy in proportion to vehicle speed, prime mover speed, or some similar function, thereby serving as a guide for initiating appropriate control functions. *See also:* **traction motor.** 328

control grid (electron tube). A grid, ordinarily placed between the cathode and an anode, for use as a control electrode. *See also:* **electrode (electron tube); grid.** 125, 190

control initiation. The function introduced into a measurement sequence for the purpose of regulating any subsequent control operations in relation to the quantity measured. *Note:* The system element comprising the control initiator is usually included in the end device but may be associated with the primary detector or the intermediate means. *See also:* **measurement system.** 328

control interaction factors (control systems). In a proportional plus integral plus derivative control action unit, the ratio of the effective values to the values that would be measured when the product (integral action rate) (derivative action time constant) is zero. Example: Assume a control unit composed of elements whose ratios of output to input are $1 + D's$ and $P'(l'/s + 1)$ connected so that the output of one is the input of the other. The ratio of output to input of the combination is

$$\frac{Y}{X} = P'(1 + l'D') \left[\frac{l'/s}{1 + l'D'} + 1 + \frac{D's}{1 + l'D'} \right]$$

By comparison with the equation

$$\frac{Y}{X} = P \left[\frac{l}{s} + 1 + Ds \right]$$

it is seen that the effective values are
$P = P'(1 + l'D') =$ proportional gain
$l = l'/(1 + l'D') =$ integral action rate
$D = D'/(1 + l'D') =$ derivative action time constant
When either l' or D' is set equal to zero the factor $1 + l'D'$ equals unity and the measured values are P', l' and D'. Consequently, $1 + l'D'$ is the "proportional interaction factor" and $1/(1 + l'D')$ is both the "integral action rate interaction factor" and the "derivative action time interaction factor." 56

controllability (control systems). In comparison of processes, a qualitative term indicating the relative ease with which they can be controlled. *Note:* The type of disturbance for which the comparison is made should be specified. *See also:* **inherent regulation.** 56

controllable. A property of a component of a state whereby, given an initial value of the component at a given time, there exists a control input that can change this value to any other value at a later time. *See also:* **control system.**
329

controllable, completely. The property of a plant whereby all components of the state are controllable within a given time interval. *See also:* **control system.** 329

control law. A function of the state of a plant and possibly of time, generated by a controller to be applied as the control input to a plant. *See also:* **control system.** 329

control law, closed-loop. A control law specified in terms of some function of the observed state. *See also:* **control system.**
329

control law, open-loop. A control law specified in terms of the initial state only and possibly of time. *See also:* **control system.**
329

controlled access (communication satellite). A mode of operation of a communication satellite in which an earth station desiring access to the system must request and obtain access to the system via a network management facility. 84

controlled area (laser-maser). An area where the occupancy and activity of those within is subject to control and supervision for the purpose of protection from radiation hazards. 363

controlled-avalanche rectifier diode (semiconductor). A rectifier diode that has specified maximum and minimum breakdown-voltage parameters and is specified to operate under steady-state conditions in the breakdown region of its reverse characteristic. *See:* **breakdown.** 66

controlled carrier (floating carrier) (variable carrier). A system of compound modulation wherein the carrier is amplitude modulated by the signal frequencies in any conventional manner, and, in addition, the carrier is simultaneously amplitude modulated in accordance with the envelope of the signal so that the percentage of modulation, or modulation factor, remains approximately constant regardless of the amplitude of the signal.
111, 211

controlled manual block signal system. A series of consecutive blocks governed by block signals, controlled by continuous track circuits, operated manually upon information by telegraph, telephone, or other means of communication, and so constructed as to require the cooperation of the signalmen at both ends of the block to display a clear or permissive block signal. *See also:* **block signal system.** 328

controlled overvoltage test (1) (rotating machinery). A test of an insulation system, in which a voltage exceeding the normal operating value is applied and increased by manual or automatic control, according to a prearranged time schedule or function, under designated conditions of temperature, humidity, and frequency. *Note:* Usually a direct-current test. *See also:* **asynchronous machine; direct-current commutating machine.** 63
(2) (direct-current leakage, measured current or step voltage test). A test in which the increase of applied direct voltage is controlled, and measured currents continuously observed for abnormalities, with the intention of stopping the test before breakdown occurs. 6, 63

controlled rectifier. A rectifier in which means for controlling the current flow through the rectifying devices is provided. *See also:* **electronic controller; rectification.**
328

controlled-speed axle generator. An axle generator in which the speed of the generator is maintained approximately constant at all vehicle speeds above a predetermined minimum. *See also:* **axle generator system.** 328

controlled system (automatic control). The apparatus, equipment, or machine used to effect changes in the value of the ultimately controlled variable. *See also:* **control system.** 56

controller. The part of a control system that implements the control law. *See also:* **control system; electric controller.** 329

controller, automatic (process control). A device that operates automatically to regulate a controlled variable in response to a command and a feedback signal. *Note:* The term originated in process control usage. Feedback elements and final control elements may also be part of the device. *See also:* **control system, feedback.** 56

controller (CAMAC). *See:* **CAMAC crate.**

controller diagram (electric-power devices). A diagram that shows the electric connections between the parts comprising the controller and that shows the external connections. 210, 206

controller, self-operated (automatic control). A control device in which all the energy to operate the final controlling element is derived from the controlled system through the primary detecting element. 56

controllers for steel-mill accessory machines. Controllers for machines that are not used directly in the processing of steel, such as pumps, machine tools, etcetera. *See also:* **electric controller.** 225, 206

controllers for steel-mill auxiliaries. Controllers for machines that are used directly in the processing of steel, such as screwdowns and manipulators but not cranes and main rolling drives. *See also:* **electric controller.** 225, 206

controller, time schedule (process control). A controller in which the command (or reference input signal) automatically adheres to a pre-determined time schedule. *Note:* The time schedule mechanism may be programmed to switch motors or other devices. 56

controlling elements (control system, feedback). The functional components of a controlling system. *See also:* **control system, feedback.** 56, 329

controlling element, final (control systems). That forward controlling element which directly changes the value of the manipulated variable. 56

controlling elements, forward (control system, feedback). The elements in the controlling system that change a variable in response to the actuating signal. *See also:* **control system, feedback.** 56, 329

controlling means (of an automatic control system). Consists of those elements that are involved in producing a corrective action. 94

controlling section. A length of track consisting of one or more track circuit sections, by means of which the roadway elements or the device that governs approach to or movement within a block are controlled. 328

controlling system (1) (automatic control system without feedback). That portion of the control system that manipulates the controlled system. 56, 329

(2) (control system, feedback). The portion that compares functions of a directly controlled variable and a command and adjusts a manipulated variable as a function of the difference. *Note:* It includes the reference input elements; summing point; forward and final controlling elements; and feedback elements. *See also:* **control system, feedback.** 56, 329

controlling voltage, composite. *See:* **composite controlling voltage.**

control machine (railroad practice). (1) An assemblage of manually operated levers or other devices for the control of signals, switches, or other units, without mechanical interlocking, usually including a track diagram with indication lights. (2) A group of levers or equivalent devices used to operate the various mechanisms and signals that constitute the car retarder installation. *See also:* **car retarder; centralized traffic control system.** 328

control, manual. Those elements in the excitation control system which provide for manual adjustment of the synchronous machine terminal voltage by open-loop control. 105

control metering point (tie line) (electric power systems). The location of the metering equipment that is used to measure power on the tie line for the purpose of control. *See also:* **center of distribution; power system.** 94, 200

control module (rotating machinery). Control-circuit subassembly. *See also:* **starting switch assembly.** 63

control panel. A part of a computer console that contains manual controls. *See:* **plugboard.** *See also:* **electronic analog computer.** 225, 77

control point selector (test, measurement and diagnostic equipment). A device capable of selecting and controlling the proper stimuli, power of loads, and applying it to the unit under test, in accordance with instructions from the programming device. 54

control position electric indicator. A deviced that provides an indication of the movement and position of the various control surfaces or structural parts of an aircraft. It may be used for wing flaps, cowl flaps, trim tabs, oil-cooler shutters, landing gears, etcetera. 328

control positioning accuracy, precision, or reproducibility (numerically controlled machines). Accuracy, precision, or reproducibility of position sensor or transducer and interpreting system and including the machine positioning servo. *Note:* May be the same as machine positioning accuracy, precision, or reproducibility in some systems. 224, 207

control-power winding (or transformer) (for transformers only). The winding (or transformer) which supplies power to motors, relays, and other devices used for control purposes. 53

control precision (control systems). Precision evidenced by either the directly or the indirectly controlled variable, as specified. 56

control procedure (data communications). The means used to control the orderly communication of information between stations on a data link. 12

control ratio (1) (gas tube). The ratio of the change in anode voltage to the corresponding change in critical grid voltage, with all other operating conditions maintained constant. *See:* **gas tubes.** 125, 190

(2) (power supplies). The required change in control re-

sistance to produce a one-volt change in the output voltage. The control ratio is expressed in ohms per volt and is reciprocal of the bridge current. *See also:* **power supply.** 228, 186

control relay. An auxiliary relay whose function is to initiate or permit the next desired operation in a control sequence. 103, 202, 60, 127

control ring. *See:* **grading ring.**

control sequence table (electric-power devices). A tabulation of the connections that are made for each successive position of the controller. 210, 206

control station (mobile communication). A base station, the transmission of which is used to control automatically the emission or operation of another radio station. *See also:* **mobile communication system.** 181

control switch. A manually operated switching device for controlling power-operated devices. *Notes:* (1) It may include signaling, interlocking, etcetera, as dependent functions. (2) Control switches derive their names from the function of the apparatus that they control, such as master control switches, reset switches, etcetera. 1, 103, 202

control switchboard. A type of switchboard including control, instrumentation, metering, protective (relays), or regulating equipment for remotely controlling other equipment. *Notes:* (1) Control switchboards do not include the primary power circuit-switching devices or their connections. (2) A power plant auxiliary control switchboard is a switchboard which includes some or all of the components of an electrical control switchboard plus hydraulic or pneumatic control, instrumentation, or metering devices and their interconnecting piping such as commonly used for control of boilers and auxiliaries in power plants. The electrical devices and wiring on such switchboards are subject to the same standards as used for similar electrical devices and wiring on purely electrical switchboards. (3) Control switchboards may be further classified by type of construction. 103, 202, 31, 93

control-switching point (CSP) (telephone switching systems). A switching entity arranged for routing and control in the distance dialing network, at which intertoll trunks are interconnected. 55

control switch point (telephone networks). A switching point (arranged for routing and control) in the direct distance network at which intertoll trunks are connected to other intertoll trunks. 193

control system (1) (broadly). An assemblage of control apparatus coordinated to execute a planned set of controls. *See:* **control.** 206

(2) A system in which a desired effect is achieved by operating on the various inputs to the system until the output, which is a measure of the desired effect, falls within an acceptable range of values. *See:* **closed-loop control system (control system, feedback); control; network analysis; open-loop control system; transfer function** 151, 304

(3) (automatic control). A system in which deliberate guidance or manipulation is used to achieve a prescribed value of a variable. *Note:* It is subdivided into a controlling system and a controlled system. 56

control system, adaptive (industrial control). A control system within which automatic means are used to change the system parameters in a way intended to improve the performance of the control system. *See* **control system, feedback.**   219, 206, 56, 329

control system, automatic. A control system that operates without human intervention. *See also:* **control system, feedback.**   56, 329

control system, automatic feedback. A feedback control system that operates without human intervention. *See also:* **control system, feedback.**   56, 329

control system, cascade. A control system in which the output of one subsystem is the input for another subsystem. *See also:* **control system, feedback.**   56, 329

control system, closed-loop. A control system in which the controlled quantity is measured and compared with a standard representing the desired performance. Any deviation from the standard is fed back into the control system in such a sense that it will reduce the deviation of the controlled quantity from the standard. *Note:* In automatic generation control, the controlled quantities are frequency, unit generation, and net interchange.   94

control system, coarse-fine (industrial control). A control system that uses some elements to reduce the difference between the directly controlled variable and its ideal value to a small value and that uses other elements to reduce the remaining difference to a smaller value.   219, 206, 329

control system, dual-mode. A control system in which control alternates between two predetermined modes. *Note:* The condition for change from one mode to the other is often a function of the actuating signal. One use of dual-mode action is to provide rapid recovery from large deviations without incurring large overshoot. *See also:* **control system, feedback.**   56, 329

control system, duty factor (automatic control). A control system in which the signal to the final controlling element consists of periodic pulses whose duration is varied to relate, in some prescribed manner, the time average of the signal to the actuating signal. *Note:* This mode of control differs from two-step control in that the period of the pulses in duty-factor control is predetermined.   56

control system, feedback (1) (general). A control system that operates to achieve prescribed relationships between selected system variables by comparing functions of these variables and using the comparison to effect control. See the accompanying figures.
(2) (speed governing of hydraulic turbines). A closed-loop or feedback control system is a control system in which the controlled quantity is measured and compared with a standard representing the desired value of the controlled quantity. In hydraulic governors, any deviation from the standard is fed back into the control system in such a sense that it will reduce the deviation between the controlled quantity and the standard providing negative feedback.   8

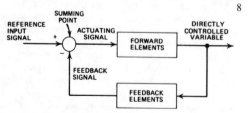

Simplified block diagram indicating essential elements of an automatic control system.

control system, floating (automatic control). A control system in which the rate of change of the manipulated variable is a continuous (or at least a piecewise continuous) function of the actuating signal. *Note:* The manipulated variable can remain at any value in its operating range when the actuating signal is zero and constant. Hence the manipulated variable is said to "float." When the forward elements in a control loop have integral control action only, the mode of control has been called "proportional-speed floating." The use of the term integral control action is recommended as a replacement for "proportional-speed floating control." *Syn.* **floating control.** *See also:* **control system, single-speed floating; control system, multiple-speed floating; control action, integral; neutral zone.**   56

control system, multiple-speed floating (automatic control). A form of floating control system in which the manipulated variable may change at two or more rates each corresponding to a definite range of values of the actuating signal.   56

control system, multi-step (automatic control). *See:* **control system, step.**

control system, on-off. A two-step control system in which a supply of energy to the controlled system is either on or off. *See also:* **control system, feedback.**   56, 329

control system, positioning (automatic control). A control system in which there is a predetermined relation between the actuating signal and the position of a final controlling element. *Note:* In a "proportional-position control system" there is a continuous linear relation between the value of the actuating signal and the position of a final controlling element.   56

control system, ratio (automatic control). A system that maintains two or more variables at a predetermined ratio. *Note:* Frequently some function of the value of an uncontrolled variable is the command to a system controlling another variable.   56

control system, sampling. Control using intermittently observed values of signals such as the feedback signal or the actuating signal. *Note:* The sampling is often done periodically. *See also:* **control system, feedback.**   56, 329

control system, single-speed floating (automatic control). A floating control system in which the manipulated variable changes at a fixed rate, increasing or decreasing depending on the sign of the actuating signal. *Note:* A neutral zone of values of the actuating signal, in which no action occurs, may be used.   56

control system, step (automatic control). A system in which the manipulated variable assumes discrete predetermined values. *Note:* The condition for change from one predetermined value to another is often a function of the value of the actuating signal. When the number of values of the manipulated variable is two, it is called a two-step control system; when more than two, a multi-step control system.   56

control system, two-step. A control system in which the manipulated variable alternates between two predetermined values. *Note:* A control system in which the manipulated variable changes to the other predetermined value whenever the actuating signal passes through zero is called a two-step single-point control system. A two-step

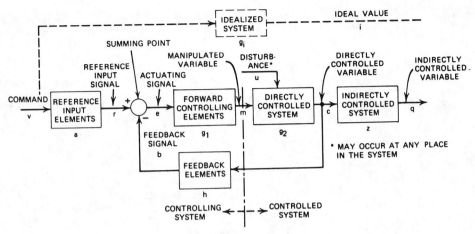

Block diagram of an automatic control system illustrating expansion of the simplified block diagram to a more complex system.

neutral-zone control system is one in which the manipulated variable changes to the other predetermined value when the actuating signal passes through a range of values known as the neutral zone. The neutral zone may be produced by a mechanical differential gap. The neutral zone is also called overlap, and two-step neutral-zone control overlap control. *See also:* **control system, feedback.**
56, 329

control system, two-step neutral zone (automatic control). *See:* **control system, two-step.**

control system, two-step single-point (automatic control). *See:* **control system, two-step.**

control terminal (base station) (mobile communication). Equipment for manually or automatically supervising a multiplicity of mobile and/or radio stations including means for calling or receiving calls from said stations. *See also:* **mobile communication system.** 181

control track (electroacoustics). A supplementary track usually placed on the same medium with the record carrying the program material. *Note:* Its purpose is to control, in some respect, the reproduction of the program, or some related phenomenon. Ordinarily, the control track contains one or more tones, each of which may be modulated either as to amplitude, frequency, or both. *See also:* **phonograph pickup.** 176

control transformers. Step-down transformers generally used in circuits which are characterized by low power levels and which contribute to a control function, such as in heating and air conditioning, printing and general industrial controls. 53

control unit (1) (digital computer). The parts that effect the retrieval of instructions in proper sequence, the interpretation of each instruction, and the application of the proper signals to the arithmetic unit and other parts in accordance with this interpretation. 255, 77

(2) (mobile station) (mobile communication). Equipment including a microphone and/or handset and loudspeaker together with such other devices as may be necessary for controlling a mobile station. *See also:* **mobile communication system.** 181

control voltage (power switchgear). The voltage applied to the operating mechanism of a device to actuate it, usually measured at the control power terminals of the mechanism. 103

control winding (1) (rotating machinery). An excitation winding that carries a current controlling the performance of a machine. *See also:* **asynchronous machines.** 63

(2) (saturable reactor). A winding by means of which a controlling magnetomotive force is applied to the core. *See also:* **magnetic amplifier.** 328

convection current. In an electron stream, the time rate at which charge is transported through a given surface. *See:* **electron emission.** 125, 190

convection-current modulation. The time variation in the magnitude of the convection current passing through a surface, or the process of directly producing such a variation. *See:* **electron emission.** 125, 190

convection heater. A heater that dissipates its heat mainly by convection and conduction. *See also:* **appliances (including portable).** 328

convective discharge (effluve*) (electric wind) (static breeze) (medical electronics). The movement of a visible or invisible stream of particles carrying away charges from a body that has been charged to a sufficiently high voltage. 192

---

* Deprecated

convenience outlet. *See:* **receptacle (electric distribution).**

conventionally cooled (rotating machinery). A term referring to windings in which the heat generated within the principal portion of the windings must flow through the major ground insulation before reaching the cooling medium. 298, 63

convergence (multibeam cathode-ray tubes). A condition in which the electron beams intersect at a specified point. 125, 190

convergence, dynamic (multibeam cathode-ray tubes). The process whereby the locus of the point of convergence of electron beams is made to fall on a specified surface during scanning. 125, 190

convergence electrode (multibeam cathode-ray tubes). An electrode whose electric field converges two or more electron beams. 125, 190

convergence magnet (multibeam cathode-ray tubes). A magnet assembly whose magnetic field converges two or more electron beams. 125, 190

convergence plane (multibeam cathode-ray tubes). A plane containing the points at which the electron beams appear to experience a deflection applied for the purpose of obtaining convergence. 125, 190

convergence surface (multibeam cathode-ray tubes). The surface generated by the point of intersection of two or more electron beams during the scanning process. 125, 190

conversion efficiency (1) (electrical conversion). In alternating-current to direct-current conversion equipment, the ratio of the product of output direct-current and voltage to input watts expressed in percent. *Note:* It reflects alternating-current power capacity required for a given voltage and current output and does not necessarily reflect watts lost.

Conversion Efficiency

$$= \frac{(E_{\text{dc}})(I_{\text{dc}})}{P} \, (100 \text{ percent})$$

See also: **electric conversion.** 186

(2) (overall) (photoelectric converter). The ratio of available power output to total incident radiant power in the active area for photovoltaic operation. *Note:* This depends on the spectral distribution of the source and junction temperature. *See also:* **semiconductor.** 186

(3) (klystron oscillator). The ratio of the high-frequency output power to the direct-current power supplied to the beam. *See also:* **velocity-modulated tube.** 244, 190

(4) (metallic rectifier). The ratio of the product of (A) the average values of the unidirectional voltage and current (B) the total power input on the alternating-voltage side. *See also:* **rectification.** 328

(5) (solar cells). The ratio of the solar cell's available power output (at a specified voltage) to the total incident radiant power. The cell active area shall be used in this calculation; that is, ohmic contact (but not grid lines) areas on the irradiated side shall be deducted from the total irradiated cell area to determine active area. The spectral distribution of the source and the junction temperature must be specified. 113

conversion rate (1) (analog-to-digital converter). The maximum rate at which the start conversion commands can be applied to the converter, to which the converter will respond by providing the desired signal at the output to within a given accuracy. 10

(2) (analog-to-digital converter with multiplexor with sample and hold). The maximum rate at which the start sample commands can be applied to the system to which the system will respond by providing the desired signal at the output to within a given accuracy. (Pre-selected channel.) 10

conversion time (1) (analog-to-digital converter). That time required from the instant at which a conversion command is received and a final digital representation is available for external output to within a given accuracy. 10

(2) (analog-to-digital converter with multiplexor with sample and hold). That time required from the time at which a sample command is received and a final digital representation is available for external output to within a given accuracy. (Pre-selected channel.) 10

conversion transconductance (heterodyne conversion transducer). The quotient of (1) the magnitude of the desired output-frequency component of current by (2) the magnitude of the input-frequency (signal) component of voltage, when the impedance of the output external termination is negligible for all of the frequencies that may affect the result. *Note:* Unless otherwise stated, the term refers to the cases in which the input-frequency voltage is of infinitesimal magnitude. All direct electrode voltages, and the magnitude of the local-oscillator voltage, must remain constant. *See:* **modulation; transducer.**
  125, 190, 252, 210

conversion transducer (1) (general). A transducer in which the signal undergoes frequency conversion. *Note:* The gain or loss of a conversion transducer is specified in terms of the useful signal. *See also:* **transducer.** 328

(2) An electric transducer in which the input and the output frequencies are different. *Note:* If the frequency-changing property of a conversion transducer depends upon a generator of frequency different from that of the input or output frequencies, the frequency and voltage or power of this generator are parameters of the conversion transducer. *See also:* **heterodyne conversion transducer (converter).** 125, 190, 252

conversion voltage gain (conversion transducer). The ratio of (1) the magnitude of the output-frequency voltage across the output termination, with the transducer inserted between the input-frequency generator and the output termination, to (2) the magnitude of the input-frequency voltage across the input termination of the transducer.
  190

convert (data processing). To change the representation of data from one form to another, for example, to change numerical data from binary to decimal or from cards to tape. 255, 77

converter (1) (general). A machine or device for changing alternating-current power to direct-current power or vice versa, or from one frequency to another. 63

(2) (frequency converter) (heterodyne reception). The portion of the receiver that converts the incoming signal to the intermediate frequency. 59

(3) (facsimile). A device that changes the type of modulation. *See also:* **facsimile (electrical communication).**
  12

(4) (industrial control). A network or device for changing the form of information or energy. 219, 206

(5) (data transmission). A device for changing one form of information language to another so as to render the language acceptable to a different machine, that is card to tape conversion. 59

(6) (test, measurement and diagnostic equipment). A device which changes the manner of representing information from one form to another. 54

converter, analog-to-digital (ADC) (hybrid computer linkage components). *See:* **analog to digital converter.**

converter, digital-to-analog (DAC) (hybrid computer linkage components). *See:* **digital-to-analog converter.**

converter, reversible power. *See:* **reversible power converter.**

converters, semiconductor. *See:* **semiconductor converters.**

converter, static solid state. *See:* **static solid state converter.**

converter tube. An electron tube that combines the mixer and local-oscillator functions of a heterodyne conversion transducer. *See also:* **heterodyne conversion transducer (converter).**   125, 190

conveyor. A mechanical contrivance, generally electrically driven, that extends from a receiving point to a discharge point and conveys, transports, or transfers material between those points. *See:* **conveyor, belt-type; conveyor, chain-type; conveyor, shaker-type; conveyor, vibrating-type.**   328

conveyor, belt-type. A conveyor consisting of an endless belt used to transport material from one place to another. *See also:* **conveyor.**   328

conveyor, chain-type. A conveyor using a driven endless chain or chains, equipped with flights that operate in a trough and move material along the trough. *See also:* **conveyor.**   328

conveyor, shaker-type. A conveyor designed to transport material along a line of troughs by means of a reciprocating or shaking motion. *See also:* **conveyor.**   328

conveyor, vibrating-type. A conveyor consisting of a movable bed mounted at an angle to the horizontal, that vibrates in such a way that the material advances. *See also:* **conveyor.**   328

convolution function (burst measurements). The integral of the function $x(\tau)$ multiplied by another function $y(-\tau)$ shifted in time by t

$$\int_{-\infty}^{\infty} x(\tau)y(t - \tau)\mathrm{d}\tau$$

See also: **burst.**   292

convolution integral (automatic control). A mathematical integral operation which is used to describe the time response of a linear element to an input function in terms of the weighting function of the element. The integral generally takes the form $\int_{o}^{t} f(x)g(t - x)dx$ where $f(x)$ is an arbitrary input, and $g(t - x)$ is a weighting function which extends backward from instant t through x as far as zero.   56

cooking unit, counter-mounted. An assembly of one or more domestic surface heating elements for cooking purposes, designed for flush mounting in, or supported by, a counter, and which assembly is complete with inherent or separately mountable controls and internal wiring. *See:* **oven, wall-mounted.**   256

coolant. *See:* **cooling medium.**

cooler (heat exchanger) (rotating machinery). A device used to transfer heat between two fluids without direct contact between them.   63

Coolidge tube. An X-ray tube in which the needed electrons are produced by a hot cathode. *See also:* **electron devices, miscellaneous.**   190

cooling (power supplies). The cooling of regulator elements refers to the method used for removing heat generated in the regulating process. *Note:* Methods include radiation, convection, and conduction or combinations thereof. *See*

also: **power supply.**   228, 186

cooling coil (rotating machinery). A tube through whose wall, heat is transferred between two fluids without direct contact between them.   63

cooling, convection (power supplies). A method of heat transfer that uses the natural upward motion of air warmed by the heat dissipators. *See also:* **power supply.**   228, 186

cooling duct. *See:* **ventilating duct (rotating machinery).**

cooling fin (electron device). A metallic part or fin extending the cooling area to facilitate the dissipation of the heat generated in the device. *See:* **electron device.**   190

cooling, lateral forced-air (power supplies). An efficient method of heat transfer by means of side-to-side circulation that employs blower movement of air through or across the heat dissipators. *See also:* **power supply.**   228, 186

cooling medium (rotating machinery) (coolant). A fluid, usually air, hydrogen, or water, used to remove heat from a machine or from certain of its components.   63

cooling system (1) (rectifier). Equipment, that is, parts and their interconnections, used for cooling a rectifier. *Note:* It includes all or some of the following: rectifier water jacket, cooling coils or fins, heat exchanger, blower, water pump, expansion tank, insulating pipes, etcetera. *See also:* **rectification.**   208

(2) (thyristor converter). Equipment, that is, parts and their interconnections, used for cooling a thyristor converter. *Note:* It includes all or some of the following: thyristor heat sink, cooling coils or fins, heat exchanger, fan or blower, water pump, expansion tank, insulating pipes, etcetera.   121

cooling system, direct raw-water (thyristor converter). A cooling system in which water, received from a constantly available supply, such as a well or water system, is passed directly over the cooling surfaces of the thyristor converter and discharged.   121

cooling system, direct raw-water, with recirculation (thyristor converter). A direct raw-water cooling system in which part of the water passing over the cooling surfaces of the thyristor converter is recirculated and raw water is added as needed to maintain the required temperature, the excess being discharged.   121

cooling system, forced-air (thyristor converter). An air cooling system in which heat is removed from the cooling surfaces of the thyristor converter by means of a flow of air produced by a fan or blower.   121

cooling system, heat-exchanger (thyristor converter). A cooling system in which the coolant, after passing over the cooling surfaces of the thyristor converter, is cooled in a heat exchanger and recirculated. *Note:* Heat may be removed from the thyristor converter cooling surfaces by liquid or air using the following types of heat exchangers: (1) water-to-water, (2) water-to-air, (3) air-to-water, (4) air-to-air, (5) refrigeration cycle. The liquid in the closed system may be other than water, and the gas in the closed system may be other than air.   121

cooling system, natural-air (thyristor converter). An air cooling system in which heat is removed from the cooling surfaces of the thyristor converter only by the natural action of the ambient air.   121

cooling-water system (rotating machinery). All parts that are provided for the flow, treatment, or storage of cooling water. 63

coordinate dimension word (numerically controlled machines). A word defining an absolute dimension. 224, 207

coordinated transpositions (electric supply or communication circuits). Transpositions that are installed for the purpose of reducing inductive coupling, and that are located effectively with respect to the discontinuities in both the electric supply and communication circuits. *See also:* **inductive coordination.** 328

coordinate system (pulse terms). Throughout the following, a rectangular Cartesian coordinate system is assumed in which, unless otherwise specified: (1) Time (t) is the independent variable taking alone the horizontal axis, increasing in the positive sense from left to right. (2) Magnitude (m) is the dependent variable taken along the vertical axis, increasing in the positive sense or polarity from bottom to top. (3) The following additional symbols are used:
(a) e—The base of natural logarithms.
(b) $a, b, c,$ etcetera—Real constants which, unless otherwise specified, may have any value and either sign.
(c) n—A positive integer

coordination of insulation (1) (lightning insulation strength). The steps taken to prevent damage to electric equipment due to overvoltages and to localize flashovers to points where they will not cause damage. *Note:* In practice, coordination consists of the process of correlating the insulating strengths of electric equipment with expected overvoltages and with the characteristics of protective devices. *See also:* **basic impulse insulation level (insulation strength).** 224, 62
(2) (transformer). The process of correlating the insulation strengths of electrical equipment with expected overvoltages and with the characteristics of surge protective devices. 53

copper brush (rotating machinery). A brush composed principally of copper. *See also:* **brush.** 63

copper-clad steel. Steel with a coating of copper welded to it, as distinguished from copper-plated or copper-sheathed material. *See also:* **lightning protection and equipment.** 328

copper-covered steel wire. A wire having a steel core to which is bonded a continuous outer layer of copper. *See also:* **conductor.** 64, 57

copper losses. *See:* **load losses.**

copy (electronic data processing). (1) To reproduce data leaving the original data unchanged. 255
(2) To produce a sequence of character events equivalent, character by character, to another sequence of character events.
(3) The sequence of character events produced in (2). *See also:* **transfer.** 235

cord. One or a group of flexible insulated conductors, enclosed in a flexible insulating covering and equipped with terminals. 328

cord adjuster. A device for altering the pendant length of the flexible cord of a pendant. *Note:* This device may be a ratchet reel, a pulley and counterweight, a tent-rope stick, etcetera. 328

cord circuit (telephone switching systems). A connecting circuit, usually terminating in a plug at one or both ends, used at switchboard positions in establishing telephone connections. 55

cord-circuit repeater. A repeater associated with a cord circuit so that it may be inserted in a circuit by an operator. *See also:* **repeater.** 328

cord grip (strain relief). A device by means of which the flexible cord entering a device or equipment is gripped in order to relieve the terminals from tension in the cord. 328

cord connector (cord connector body*) (table tap*). A plug receptacle provided with means for attachment to flexible cord. 328

* Deprecated

cordless switchboard (telephone switching systems). A telecommunications switchboard in which manually operated keys are used to make connections. 55

core (1) (magnetic core) (rotating machinery) (transformer). An element made of magnetic material, serving as part of a path for magnetic flux. *Note:* In a rotating machine, this is frequently part of the stator, a hollow cylinder of laminated magnetic steel, slotted on the inner surface for the purpose of containing the stator windings. Such a stator core frequently contains ducts through which a cooling medium may flow to remove heat generated in the core and windings. *See:* **magnetic core.** 63, 53
(2) (electronic information storage). *See:* **digital computer.**
(3) (mechanical recording). The central layer or basic support of certain types of laminated media. 176
(4) (electromagnet). The part of the magnetic structure around which the magnetizing winding is placed. 210

core duct (rotating machinery). The space between or through core laminations provided to permit the radial or axial flow of coolant gas. *See also:* **rotor (rotating machinery).** 63

core end plate (end plate) (flange) (rotating machinery). A plate or structure at the end of a laminated core to maintain axial pressure on the laminations. 63

core-form transformer. A transformer in which those parts of the magnetic-circuit surrounded by the windings have the form of legs with two common yokes. 53

core length (rotating machinery). The dimension of the stator, or rotor, core measured in the axial direction. *See also:* **rotor (rotating machinery); stator.** 63

coreless-type induction heater or furnace. A device in which a charge is heated by induction and no magnetic core material links the charge. *Note:* Magnetic material may be used elsewhere in the assembly for flux guiding purposes. *See also:* **dielectric heating; induction heating.** 14, 114

core loss (1). The power dissipated in a magnetic core subjected to a time-varying magnetizing force. *Note:* Core loss includes hysteresis and eddy-current losses of the core. 197, 53
(2) (synchronous machine). The difference in power required to drive the machine at normal speed, when excited to produce a voltage at the terminals on open circuit corresponding to the calculated internal voltage, and the power required to drive the unexcited machine at the same speed. *Note:* The internal voltage shall be determined by

correcting the rated terminal voltage for the resistance drop only.   244, 63

(3) (electronics power transformer). The measured power loss, expressed in watts, attributable to the material in the core and associated clamping structure, of a transformer that is excited, with no connected load, at a core flux density and frequency equal to that in the core when rated voltage and frequency is applied and rated load current is supplied.   95

core-loss current. The in-phase component (with respect to the induced voltage) of the exciting current supplied to a coil. *Note:* It may be regarded as a hypothetical current, assumed to flow through the equivalent core-loss resistance.   210, 197

core loss, open-circuit (rotating machinery). The difference in power required to drive a machine at normal speed, when excited to produce a specified voltage at the open-circuited armature terminals, and the power required to drive the unexcited machine at the same speed.   244

core-loss test (rotating machinery). A test taken on a built-up (usually unwound) core of a machine to determine its loss characteristic. *See also:* **stator.**   63

core, relay. *See:* **relay core.**

core test (rotating machinery). A test taken on a built-up (usually unwound) core of a machine to determine its loss characteristics or its magnetomotive force characteristics, or to locate short-circuited laminations. *See also:* **rotor (rotating machinery); stator.**   63

core-type induction heater or furnace. A device in which a charge is heated by induction and a magnetic core links the inducing winding with the charge. *See also:* **dielectric heating; induction heating.**   14, 114

core-type transformer. A transformer in which those parts of the magnetic-circuit surrounded by the windings have the form of legs with two common yokes.   53

Coriolis correction (navigation). A correction, that must be applied to measurements made with respect to a coordinate system in translation, to compensate for the effect of any angular motion of the coordinate system with respect to inertial space. *See also:* **navigation.**   187, 13

cornea (laser-maser). The transparent outer coat of the human eye which covers the iris and the crystalline lens. It is the main refracting element of the eye.   363

corner (waveguide technique). An abrupt change in the direction of the axis of a waveguide. *Note:* Also termed **elbow.** *See also:* **waveguide.**   244, 179

corner reflector (1) (antenna). A reflecting object consisting of two or three mutually intersecting conducting flat surfaces. *Note:* Dihedral forms of corner reflectors are frequently used in antennas; trihedral forms are more often used as radar targets. *See also:* **antenna; radar.**
                                                        179, 111

(2) (radar). Two (dihedral) or three (trihedral) conducting surfaces, mutually intersecting at right angles, designed to return electromagnetic radiations toward their sources and used to render a target more conspicuous to radar observations.   13

corner-reflector antenna. An antenna consisting of a feed and a corner reflector. *See also:* **antenna.**   179, 111

corner, waveguide. *See:* **bend, waveguide.**

cornice lighting. Light sources shielded by a panel parallel to the wall and attached to the ceiling and distributing

light over the wall. *See also:* **general lighting.**   167

corona (1) (air). A luminous discharge due to ionization of the air surrounding a conductor caused by a voltage gradient exceeding a certain critical value. *See also:* **tower.**
                                                           64

(2) (gas). A discharge with slight luminosity produced in the neighborhood of a conductor, without greatly heating it, and limited to the region surrounding the conductor in which the electric field exceeds a certain value. *See also:* **discharge (gas).**   244, 190

(3) (transformer). *See:* **partial discharge.**

corona charging (electrostatography). Sensitizing by means of gaseous ions of a corona. *See also:* **electrostatography.**
                                                      236, 191

corona-discharge tube. A low-current gas-filled tube utilizing the corona-discharge properties.   190

corona effect. The particular form of the glow discharge that occurs in the neighborhood of electric conductors where the insulation is subject to high electric stress.
                                                          210

corona inception test. *See:* **discharge inception test.**

corona level (power distribution, underground cables). *See:* **ionization extinction voltage.**

corona shielding (corona grading) (rotating machinery). A means adapted to reduce potential gradients along the surface of coils. *See also:* **asynchronous machine; direct-current commutating machine.**   63

corona voltage level. The voltage at which corona discharge does not exceed a specified level following the application of a specified higher voltage.   134

corona voltmeter. A voltmeter in which the crest value of voltage is indicated by the inception of corona. *See also:* **instrument.**   328

corrected compass course. *See:* **magnetic course.**

corrected compass heading. *See:* **magnetic heading.**

correcting signal. *See:* **synchronizing signal.**

correction (1) (digital computer). A quantity (equal in absolute value to the error) added to a calculated or observed value to obtain the true value. *See also:* **accuracy rating (instrument); error.**   235

(2) (automatic control). (A) The quantity added to a calculated or observed value to obtain the correct value. (B) The act of such addition. (C) The changing value of a manipulated variable intended to reduce a system deviation.   56

correction rate (1) (industrial control). The velocity at which the control system functions to correct error in register.
                                                          206

(2) (automatic control). The instantaneous rate at which the transient deviation of the controlled variable is reduced; the slope of the time-response curve.   56

correction time. *See:* **time, settling.**

corrective maintenance (1) (availability, reliability, and maintainability). The maintenance carried out after a failure has occurred and intended to restore an item to a state in which it can perform its required function.
                                                          164

(2) (test, measurement and diagnostic equipment). Actions performed to restore a failed or degraded equipment. It includes fault isolation, repair or replacement of defective units, alignment and checkout.   54

corrective network. An electric network designed to be

inserted in a circuit to improve its transmission properties, its impedance properties, or both. *See also:* **network analysis.** 328

correct relaying-system performance. The satisfactory operation of all equipment associated with the protective-relaying function in a protective-relaying system. *Note:* It includes the satisfactory presentation of system input quantities to the relaying equipment, the correct operation of the relays in response to these input quantities, and the successful operation of the assigned switching device or devices. 103, 202, 60, 127

correct relay operation. An output response by the relay which agrees with the operating characteristic for the input quantities applied to the relay. *See also:* **correct relaying-system performance.** 103, 202, 60, 127

correlated color temperature (light source). The absolute temperature of a blackbody whose chromaticity most nearly resembles that of the light source. *See also:* **color.** 167

correlation (test, measurement and diagnostic equipment). That portion of certification which establishes the mutual relationships between similar or identical support test systems by comparing test data collected on specimen hardware or simulators. 54

correlation detection (modulation systems). Detection based on the averaged product of the received signal and a locally generated function possessing some known characteristic of the transmitted wave. *Notes:* (1) The averaged product can be formed, for example, by multiplying and integrating, or by the use of a matched filter whose impulse response, when reversed in time, is the locally generated function. (2) Strictly, the above definition applies to detection based on cross correlation. The term correlation detection may also apply to detection involving autocorrelation, in which case the locally generated function is merely a delayed form of the received signal. 242

corrosion. The deterioration of a substance (usually a metal) because of a reaction with its environment. 221, 205

corrosion fatigue. Reduction in fatigue life in a corrosive environment. 221, 205

corrosion fatigue limit. The maximum repeated stress endured by a metal without failure in a stated number of stress applications under defined conditions of corrosion and stressing. 221, 205

corrosion rate. The rate at which corrosion proceeds. 221, 205

corrosion-resistant (industrial control). So constructed, protected or treated that corrosion will not exceed specified limits under prescribed test conditions. 70, 53

corrosion-resistant parts. *Notes:* (1) Where essential to minimize deterioration due to marine atmospheric corrosion, corrosion-resisting materials or other materials treated in a satisfactory manner to render them adequately resistant to corrosion should be used. (2) Corrosion-resisting materials: Silver, corrosion-resisting steel, copper, brass, bronze, copper-nickel, certain nickel-copper alloys, and certain aluminum alloys are considered satisfactory corrosion-resisting materials within the intent of the foregoing. (3) Corrosion-resistant treatments: The following treatments when properly done and of a sufficiently heavy coating are considered satisfactory corrosion-re-

sistant treatments within the intent of the foregoing. Electroplating of: cadmium, chromium, copper, nickel, silver, and zinc; sherardizing, galvanizing, dipping and painting with phosphate or suitable cleaning, followed by the application of zinc chromate primer or equivalent. (4) These provisions should apply to the following: Parts— Interior small parts that are normally expected to be removed in service, such as bolts, nuts, pins, screws, cap screws, terminals, brush-holder studs, springs, etcetera. Assemblies, subassemblies and other units—Where necessary due to the unit function, or for interior protection, such as shafts within a motor or generator enclosure, and surface of stator and rotor. Enclosures and their fastenings and fittings—Such as enclosing cases for control apparatus and outer cases for signal and communication systems (both outside and inside), and all their fastenings and fittings that would be seriously damaged or rendered ineffective by corrosion. 3

cosecant-squared antenna. A shaped-beam antenna in which the radiation intensity over a part of its pattern in some specified plane (usually the vertical) is proportional to the square of the cosecant of the angle measured from a specified direction in that plane (usually the horizontal). *Note:* Its purpose is to lay down a uniform field along a line that is parallel to the specified direction but that does not pass through the antenna. *See also:* **antenna.** 179

cosecant-squared pattern (radar). An antenna field in which the signal-power pattern in the vertical plane varies as the square of the cosecant of the elevation angle. *Note:* Similar objects at the same altitude give equal response regardless of distance. *See also:* **navigation.** 13, 187

cosine-cubed law (illuminating engineering). An extension of the cosine law in which the distance d between the source and surface is replaced by $h/\cos\theta$, where h is the perpendicular distance of the source from the plane in which the point is located. It is expressed by $E = (I\cos^3\theta)/h^2$. *See also:* **inverse-square law (illuminating engineering).** 167

cosine function. A periodic function of the form $A\cos x$, or $A\cos wt$. The amplitude of the cosine function is A and its argument is x or ωt. 210

cosine law (illuminating engineering). A law stating that the illumination on any surface varies as the cosine of the angle of incidence. The angle of incidence θ is the angle between the normal to the surface and the direction of the incident light. The inverse-square law and the cosine law can be combined as $E = (I\cos\theta)/d^2$. *See also:* **inverse-square law (illuminating engineering).** 167

costate (control system). The state of the adjoint system. *See also:* **control system.** 329

cost of incremental fuel (electric power systems) (usually expressed in cents per million British thermal units). The ultimate replacement cost of the fuel that would be consumed to supply an additional increment of generation. *See also:* **power system, low-frequency and surge testing.** 94

costs (transmission and distribution). (1) General. Monies associated with investment or use of electrical plant. (2) Fixed investment. Monies associated with investment in plant. (3) Fixed operation. Monies other than those associated with investment in plant, which do not vary or fluctuate with changes in operation or utilization of plant.

(4) Variable operating. Monies that vary or fluctuate with operation or utilization of plant. 112

coulomb. The unit of electric charge in SI units (International System of Units). The coulomb is the quantity of electric charge that passes any cross section of a conductor in one second when the current is maintained constant at one ampere. 210

Coulomb friction (1) (general). A constant velocity-independent force that opposes the relative motion of two surfaces in contact. *See also:* **control system, feedback.** 329

(2) (control systems). That occurring between dry (that is unlubricated) surfaces. *Note:* It is usually idealized as a constant force whose sign is such as to cause the force to oppose the relative motion of the rubbing elements. *Syn:* **dry friction.** *See also:* **stiction.** 56

Coulomb's law (electrostatic attraction). The force of repulsion between two like charges of electricity concentrated at two points in an isotropic medium is proportional to the product of their magnitudes and inversely proportional to the square of the distance between them and to the dielectric constant of the medium. *Note:* The force between unlike charges is an attraction. 210

coulometer (voltameter). An electrolytic cell arranged for the measurement of a quantity of electricity by the chemical action produced. *See also:* **electricity meter (meter).** 328

count (radiation counters). A single response of the counting system. *See also:* **gas-filled radiation-counter tubes; scintillation counter; tube count.** 125, 190, 117, 96

count-down (transponder). The ratio of the number of interrogation pulses not answered to the total number of interrogation pulses received. *See also:* **navigation.** 13, 187

counter (1) (electronic computation). (A) A device, capable of changing from one to the next of a sequence of distinguishable states upon each receipt of an input signal. *Note:* One specific type is a circuit that produces one output pulse each time it receives some predetermined number of input pulses. The same term may also be applied to several such circuits connected in cascade to provide digital counting. (B) Less frequently, a counter is an accumulator. (C) (ring). A loop of interconnected bistable elements such that one and only one is in a specified state at any given time and such that, as input signals are counted, the position of the one specified state moves in an ordered sequence around the loop. *See also:* **trigger circuit.** 210

(2) (test, measurement and diagnostic equipment). (A) A device such as a register or storage location used to represent the number of occurrences of an event; and (B) An instrument for storing integers, permitting these integers to be increased or decreased sequentially by unity or by an arbitrary integer, and capable of being reset to zero or to an arbitrary integer. 54

counter cells. *See:* **counter-electromotive-force cells.**

counterclockwise polarized wave (radio wave propagation). *See:* **left-handed polarized wave.**

counter electromotive force (any system). The effective electromotive force within the system that opposes the passage of current in a specified direction. 328

counter-electromotive-force cells (counter cells). Cells of practically no ampere-hour capacity used to oppose the battery voltage. *See also:* **battery (primary or secondary).** 328

counterpoise (1) (antenna). A system of conductors, elevated above and insulated from the ground, forming a lower system of conductors of an antenna. *Note:* The purpose of a counterpoise is to provide a relatively high capacitance and thus a relatively low impedance path to earth. The counterpoise is sometimes used in medium- and low-frequency applications where it would be more difficult to provide an effective ground connection. 111

(2) (overhead line) (lightning protection). A conductor or system of conductors, arranged beneath the line, located on, above, or most frequently below the surface of the earth, and connected to the footings of the towers or poles supporting the line. *See also:* **antenna; ground.** 64

counter, radiation. *See:* **radiation counter.**

counter tube (radiation counters). A device that reacts to individual ionizing events, thus enabling them to be counted. (1) (externally quenched). A radiation-counter tube that requires the use of an external quenching circuit to inhibit reignition. (2) (gas-filled, radiation). A gas tube used for detection of radiation by means of gas ionization. (3) (gas-flow). A radiation-counter tube in which an appropriate atmosphere is maintained by a flow of gas through the tube. (4) (Geiger-Mueller). A radiation-counter tube operated in the Geiger-Mueller region. (5) (proportional). A radiation-counter tube operated in the proportional region. (6) (self-quenched). A radiation-counter tube in which reignition of the discharge is inhibited by internal processes. *See:* **anticoincidence (radiation counters).** *See also:* **gas-filled radiation-counter tube.** 96, 125, 190

counting efficiency (1) (radiation-counter tubes). The average fraction of the number of ionizing particles or quanta incident on the sensitive area that produce tube counts. *Note:* The operating conditions of the counter and the condition of irradiation must be specified. *See:* **gas-filled radiation-counter tubes.** 125, 190

(2) (scintillation counters). The ratio of (A) the average number of photons or particles of ionizing radiation that produce counts to (B) the average number incident on the sensitive area. *Note:* The operating conditions of the counter and the conditions of irradiation must be specified. *See also:* **scintillation counter.** 117

counting mechanism (automatic line sectionalizer). A device that counts the number of electric impulses and, following a predetermined number of successive electric impulses, actuates a releasing mechanism. *Note:* It resets if the total predetermined number of successive impulses do not occur in a predetermined time. 103, 202

counting operation (automatic line sectionalizer) (automatic circuit recloser). Each advance of the counting mechanism towards an opening operation. 103, 202

counting-operation time (automatic line sectionalizer) (automatic circuit recloser). The time between the cessation of a current above the minimum actuating current value and the completion of a counting operation. 103, 202

counting rate. Number of counts per unit time. *See:* **anticoincidence (radiation counters).** 190

counting-rate meter (pulse techniques). A device that indicates the time rate of occurrence of input pulses averaged over a time interval. *See also:* **scintillation counter.**

117

counting rate versus voltage characteristic (gas-filled radiation-counter tube). The counting rate as a function of applied voltage for a given constant average intensity of radiation. *See:* **gas-filled radiation-counter tubes.**

125, 190

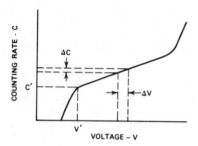

Counting rate-voltage characteristic in which

$$\text{Relative plateau slope} = 100 \frac{\Delta C/C}{\Delta V}$$

$$\text{Normalized plateau slope} = \frac{\Delta C/\Delta V}{C'/V'} = \frac{\Delta C/C'}{\Delta V/V'}$$

country code (telephone switching systems). The one-, two-, or three-digit number that, in the world numbering plan, identifies each country or integrated numbering plan area in the world. The initial digit is always the world-zone area normally identifying a specific country. On an international call, this code is dialed ahead of the national number.

55

counts background. *See:* **background counts.**

counts, tube, multiple (radiation-counter tubes). *See:* **multiple tube counts.**

counts, tube, spurious (radiation-counter tubes). *See:* **spurious tube counts.**

couple (1) (storage cell). An element of a storage cell consisting of two plates, one positive and one negative. *Note:* The term couple is also applied to a positive and a negative plate connected together as one unit for installation in adjacent cells. *See also:* **battery (primary or secondary); galvanic cell.**

328

(2) (thermoelectric). A thermoelectric device having two arms of dissimilar composition. *Note:* The term thermoelement is ambiguously used to refer to either a thermoelectric arm or to a thermoelectric couple, and its use is therefore not recommended. *See also:* **thermoelectric device.**

248, 191

coupled circuit. A network containing at least one element common to two meshes. *See also:* **network analysis.**

210

coupler (navigation). The portion of a navigational system that receives signals of one type from a sensor and transmits signals of a different type to an actuator. *See:* **autopilot coupler.** *See also:* **navigation.**

13, 187

coupler, 3-decibel. *See:* **hybrid control.**

coupling (1) (electric circuits). The circuit element or elements, or the network, that may be considered common to the input mesh and the output mesh and through which

energy may be transferred from one to the other.

(2) The association of two or more circuits or systems in such a way that power or signal information may be transferred from one to another. *Note:* Coupling is described as close or loose. A close-coupled process has elements with small phase shift between specified variables; close-coupled systems have large mutual effect shown mathematically by cross-products in the system matrix. *See also:* **network analysis.** 210, 206, 59, 197

(3) (interference terminology) (electric circuits). The effect of one system or subsystem upon another. (A) For interference, the effect of an interfering source on a signal transmission system. (B) The mechanism by which an interference source produces interference in a signal circuit. *See also:* **interference.** 56, 188

(4) (induction heating). The percentage of the total magnetic flux, produced by an inductor, that is effective in heating a load or charge. *See also:* **dielectric heating; induction heating.** 14, 114

(5) (metal raceway). A fitting intended to connect two lengths of metal raceway or perform a similar function. 328

(6) (rotating machinery). A part or combination of parts that connects two shafts for the purpose of transmitting torque or maintaining alignment of the two shafts. 63

(7) (automatic control). The effect of one system or subsystem upon another. *Note:* Coupling is described as either close or loose. A close-coupled process has elements with small phase shift between specified variables; close-coupled systems have a large mutual effect shown mathematically by cross-products in the system matrix. 56

(8) (power-system communication) (A) (phase to ground). A coupling circuit employing a coupling capacitor connected to one phase of a power line (or power lines) with the return path completed through ground or earth. *See also:* **power-line carrier.** 59

(B) (phase to phase). A coupling circuit employing coupling capacitors connected to two phases of a power line (or power lines). *Note:* It is a metallic circuit, that is, one in which ground or earth forms no intentional part. *See also:* **power-line carrier.** 59

coupling aperture (waveguide components). An aperture in the bounding surface of a cavity resonator, waveguide, transmission line, or waveguide component, which permits the flow of energy to or from an external circuit. *Syn:* **coupling hole; coupling slot.** 166, 179

coupling, broadband, aperiodic (power-system communication). Coupling that is equally responsive to all frequencies (over a stated range). As applied to power-line-carrier coupling devices, a broadband or an aperiodic coupler does not require tuning to the particular carrier frequencies being used. *See also:* **power-line carrier.**

59

coupling capacitance (1) (general). The association of two or more circuits with one another by means of capacitance mutual to the circuits. *See also:* **coupling.** 59

(2) (interference terminology). The type of coupling in which the mechanism is capacitance between the interference source and the signal system, that is, the interference is induced in the signal system by an electric field produced by the interference source. *See also:* **interference.** 188

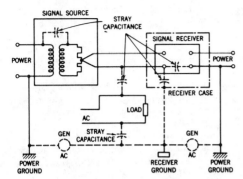

Capacitance coupling (interference).

coupling capacitor (power-system communication). A capacitor employed to connect the carrier lead-in conductor to the high-voltage power-transmission line. *See also:* **power-line carrier.** 59

coupling coefficient (coefficient of coupling). The ratio of impedance of the coupling to the square root of the product of the total impedances of similar elements in the two meshes. *Notes:* (1) Used only in the case of resistance, capacitance, self-inductance, and inductance coupling. (2) Unless otherwise specified, coefficient of coupling refers to inductance coupling, in which case it is equal to $M/(L_1L_2)^{1/2}$, where M is the mutual inductance, L_1 the total inductance of one mesh, and L_2 the total inductance of the other. *See also:* **network analysis.** 210, 185

coupling coefficient, small-signal (electron stream). The ratio of (1) the maximum change in energy of an electron traversing the interaction space to (2) the product of the peak alternating gap voltage by the electronic charge. *See also:* **coupling; coupling coefficient; electron emission.**
125, 190, 185

coupling, common (joint) (power-system communication). Line coupling employing tuning circuits that accommodate more than one carrier channel through a common set of coupling capacitors. *See also:* **power-line carrier.**
59

coupling, conductance (interference terminology). The type of coupling in which the mechanism is conductance between the interference source and the signal system. *See also:* **interference; raceway.** 188

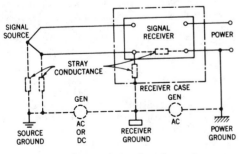

Conductance coupling (interference).

coupling, electric (rotating machinery). (1) A device for transmitting torque by means of electromagnetic force in

which there is no mechanical torque contact between the driving and driven members. *Note:* The slip-type electric coupling has poles excited by direct current on one rotating member, and an armature winding, usually of the double-squirrel-cage type, on the other rotating member. (2) A rotating machine that transmits torque by electric or magnetic means or in which the torque is controlled by electric or magnetic means. 3, 63

coupling factor (1) (lightning). The ratio of the induced voltage to the inducing voltage on parallel conductors. *See also:* **direct-stroke protection (lightning).** 64

(2) (directional coupler). The ratio of the incident power fed into the main port, and propagating in the preferred direction, to the power output at an auxiliary port, all ports being terminated by reflectionless terminations. *See also:* **waveguide.** 185

coupling flange (flange) (rotating machinery). The disc-shaped element of a half coupling that permits attachment to a mating half coupling. *See also:* **rotor (rotating machinery).** 63

coupling hole. *See:* **coupling aperture.**

coupling, hysteresis. An electric coupling in which torque is transmitted from the driving to the driven member by magnetic forces arising from the resistance to reorientation of established magnetic flux fields within ferromagnetic material usually of high coercivity. *Note:* The magnetic flux field is normally produced by current in the excitation winding, provided by an external source. 63

coupling, inductance (interference terminology). The type of coupling in which the mechanism is mutual inductance between the interference source and the signal system, that is, the interference is induced in the signal system by a magnetic field produced by the interference source. See the following figure. *See also:* **interference.** 188

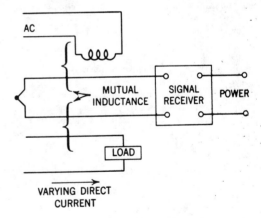

Inductance coupling (interference).

coupling, induction. An electric coupling in which torque is transmitted by the interaction of the magnetic field produced by magnetic poles on one rotating member and due to an induced voltage in the other rotating member. *Note:* The magnetic poles may be produced by direct-current excitation, permanent-magnet excitation, or alternating-current excitation. Currents due to the induced voltages may be carried in a wound armature, cylindrical cage, or may be present as eddy currents in an electrically conductive disc or cylinder. Couplings utilizing a wound armature or a cylindrical cage are known as slip or magnetic couplings. Couplings utilizing eddy-current effects are known as eddy-current couplings.   63

coupling loop (waveguide components). A conducting loop which permits the flow of energy between a cavity resonator, waveguide, transmission line, or waveguide component and an external circuit.   166, 179

coupling, magnetic friction. An electric coupling in which torque is transmitted by means of mechanical friction. Pressure normal to the rubbing surfaces is controlled by means of an electromagnet and a return spring. *Note:* Couplings may be either magnetically engaged or magnetically released depending upon application.   63

coupling, magnetic-particle. A type of electric coupling in which torque is transmitted by means of a fluid whose viscosity is adjustable by virtue of suspended magnetic particles. *Note:* The coupling fluid is incorporated in a magnetic circuit in which the flux path includes the two rotating members, the fluid, and a magnetic yoke. Flux density, and hence the fluid viscosity, are controlled through adjustment of current in a magnetic coil linking the flux path.   63

coupling probe (waveguide components). A probe which permits the flow of energy between a cavity resonator, waveguide, transmission line, or waveguide component and an external circuit.   166

coupling, radiation (interference terminology). The type of coupling in which the interference is induced in the signal system by electromagnetic radiation produced by the interference source. *See also:* **interference.**   188

couplings (pothead). Entrance fittings which may be provided with a rubber gland to provide a hermetic seal at the point where the cable enters the box and may have, in addition, a threaded portion to accommodate the conduit used with the cable or have an armor clamp to clamp and ground the armored sheath on armor-covered cable.   289

coupling slat. *See:* **coupling aperture.**

coupling, synchronous (rotating machinery). A type of electric coupling in which torque is transmitted at zero slip, either between two electromagnetic members or like number of poles, or between one electromagnetic member and a reluctance member containing a number of saliencies equal to the number of poles. *Note:* Synchronous couplings may have induction members or other means for providing torque during nonsynchronous operation such as starting. *See also:* **electric coupling.**   63

course-deviation indicator. *See:* **course-line deviation indicator.**

course (navigation). (1) The intended direction of travel, expressed as an angle in the horizontal plane between a reference line and the course line, usually measured clockwise from the reference line. (2) The intended direction of travel as defined by a navigational facility. (3) Common usage for course line. *See also:* **navigation.**   13, 187

course line (electronic navigation). The projection in the horizontal plane of a flight path (proposed path of travel). *See also:* **navigation.**   13, 187

course-line computer (electronic navigation). A device,

usually carried aboard a vehicle, to convert navigational signals such as those from very-high-frequency omnidirectional range and distance-measuring equipment into courses extending between any desired points regardless of their orientation with respect to the source of the signals. *See also:* **navigation.**   187

course-line deviation (electronic navigation). The amount by which the track of a vehicle differs from its course line, expressed in terms of either an angular or linear measurement. *See also:* **navigation.**   187

course-line deviation indicator (course-deviation indicator) (electronic navigation). A device providing a visual display of the direction and amount of deviation from the course. Also called flight-path deviation indicator. *See also:* **navigation.**   187, 13

course linearity (instrument landing systems). A term used to describe the change in the difference in depth of modulation of the two modulation signals with respect to displacement of the measuring position from the course line but within the course sector. *See also:* **navigation.**   187, 13

course made good (navigation). The direction from the point of departure to the position of the vehicle. *See also:* **navigation.**   187, 13

course push (or pull) (electronic navigation). An erroneous deflection of the indicator of a navigational aid, produced by altering the attitude of the receiving anténna. *Note:* This effect is a manifestation of polarization error and results in an apparent displacement of the course line. *See also:* **navigation.**   13, 187

course roughness (electronic navigation). A term used to describe the imperfections in a visually indicated course when such imperfections cause the course indicator to make rapid erratic movements. *See:* **scalloping.** *See also:* **navigation.**   187, 13

course scalloping (electronic navigation). *See:* **scalloping.**

course sector (instrument landing systems). A wedge-shaped section of airspace containing the course line and spreading with distance from the ground station; it is bounded on both sides by the loci of points at which the difference in depth of modulation is a specified amount, usually of the difference in depth of modulation giving full-scale deflection of the course-deviation indicator. *See also:* **course linearity; difference in depth of modulation; navigation.**   187, 13

course sector width (instrument landing systems). The transverse dimension at a specified distance, or the angle in degrees between the sides of the course sector. *See also:* **navigation.**   187, 13

course sensitivity (electronic navigation systems). The relative response of a course-line deviation indicator to the actual or simulated departure of the vehicle from the course line. *Note:* In very-high-frequency omnidirectional ranges, tacan, or similar omnirange systems, course sensitivity is often taken as the number of degrees through deflection of the course-line deviation indicator from full scale on one side to full scale on the other, while the receiver omnibearing input signal is held constant. *See also:* **navigation.**   187, 13

course softening (electronic navigation). The intentional decrease in course sensitivity upon approaching a navigational aid such that the ratio of indicator deflection to linear displacement from the course line tends to remain constant. *See also:* **navigation.**   13, 187

course width (electronic navigation). Twice the displacement (of the vehicle), in degrees, to either side of a course line, that produces a specified amount of indication on the course-deviation indicator. *Note:* Usually the specified amount is full scale. *See also:* **navigation.**   187, 13

cove lighting. Light sources shielded by a ledge or horizontal recess, and distributing light over the ceiling and upper wall. *See also:* **general lighting.**   167

cover (electric machine). A protective covering used to enclose, or to enclose partially, parts that are external to the stator frame and end shields. *Note:* In general, the word cover will be preceded by the name of the part that is covered.   63

coverage area (mobile communication). The area surrounding the base station that is within the signal-strength contour that provides a reliable communication service 90 percent of the time. *See also:* **mobile communication system.**   181

covered conductor (transmission and distribution). A conductor covered with a dielectric having no rated insulating strength or having a rated insulating strength less than the voltage of the circuit in which the conductor is used.
                    262

covered plate (storage cell). A plate bearing a layer of oxide between perforated sheets. *See also:* **battery (primary or secondary).**   328

cover plate (rotating machinery). An essentially flat removable part used as a closure for an opening.   63

CPE. *See:* **circular probable error.**

crab angle* (electronic navigation). *See:* **drift correction angle.**

* Deprecated

cradle base (rotating machinery). A device that supports the machine at the bearing housings.   63

crane. A machine for lifting or lowering a load and moving it horizontally, in which the hoisting mechanism is an integral part of the machine. *Note:* It may be driven manually or by power and may be a fixed or a mobile machine. *See also:* **elevators.**   328

crate (CAMAC). *See:* **CAMAC crate.**

crawling (rotating machinery). The stable but abnormal running of a synchronous or asynchronous machine at a speed near to a submultiple of the synchronous speed. *See also:* **asynchronous machine.**   63

credit-card call (telephone switching systems). A call in which a credit-card identity is used for billing purposes.
                    55

creep (watthour meter). A meter creeps if, with the load wires removed, and with test voltage applied to the voltage circuits of the meter, the rotor moves continuously. *Note:* For the practical recognition of creep, a meter in service is considered to creep when, with all load wires disconnected, the rotor makes one complete revolution in ten minutes or less. *See also:* **accuracy rating (instrument).**
                    212

creepage. The travel of electrolyte up the surface of electrodes or other parts of the cell above the level of the main body of electrolyte. *See also:* **electrolytic cell.**   328

creepage distance (switchgear) (transformer). The shortest distance between two conducting parts measured along the surface or joints of the insulating material between them.   103, 202, 149, 53

creepage surface (rotating machinery). An insulating-material surface extending across the separating space between components at different electric potential, where the physical separation provides the electrical insulation. *See also:* **asynchronous machine; direct-current commutating machine.**   63

creep distance (outdoor apparatus bushings). The distance measured along the external contour of the weather casing separating the metal parts which have the operating line-to-ground voltage between them.   168

creeping stimulus. *See:* **accumulating stimulus.**

crest (wave, surge, or impulse) (surge arresters). The maximum value that it attains. *Syn.* **peak.** *See:* **lightning.**
                    2, 299, 62

crest factor (1) (periodic function). The ratio of its crest (peak, maximum) value to its root-mean-square (rms) value.   210, 53

(2) (pulse carrier). The ratio of the peak pulse amplitude to the root-mean-square amplitude. *See also:* **carrier.**
                    111, 254

(3) (semiconductor radiation detectors)(average reading or root-mean-square voltmeter). The ratio of the peak voltage value that an average reading or root-mean-square voltmeter will accept without overloading to the full scale value of the range being used for measurement.
                    23, 118, 119

crest value (peak value) (1) (transformer). The maximum absolute value of a function when such a maximum exists.
                    53

(2) (surge arrester) (wave, surge, or impulse). The maximum value that it attains.   2

crest voltage. The maximum instantaneous value that an alternating voltage attains during a cycle.   301

crest voltmeter. A voltmeter depending for its indications upon the crest or maximum value of the voltage applied to its terminals. *Note:* Crest voltmeters should have clearly marked on the instrument whether readings are in equivalent root-mean-square values or in true crest volts. It is preferred that the marking should be root-mean-square values of the sinusoidal wave having the same crest value as that of the wave measured. *See also:* **instrument.**
                    328

crest working voltage (between two points) (semiconductor rectifiers). The maximum instantaneous difference of voltage, excluding oscillatory and transient overvoltages, that exists during normal operation. *See also:* **crest; rectification; semiconductor rectifier stack.**   291, 208

crevice corrosion. Localized corrosion as a result of the formation of a crevice between a metal and a nonmetal, or between two metal surfaces.   221, 205

critical anode voltage (multielectrode gas tubes). Synonymous with anode breakdown voltage (gas tubes). *See:* **gas tube.**   125, 190

critical build-up resistance (rotating machinery). The highest resistance of the shunt winding circuit supplied from the primary winding for which the machine voltage builds up under specified conditions.   63

critical build-up speed (rotating machinery). The limiting speed below which the machine voltage will not build up under specified condition of field-circuit resistance. *See:* **direct-current commutating machine.** 63

critical controlling current (cryotron). The current in the control that just causes direct-current resistance to appear in the gate, in the absence of gate current and at a specified temperature. *See also:* **superconductivity.** 191

critical coupling. That degree of coupling between two circuits, independently resonant to the same frequency, that results in maximum transfer of energy at the resonance frequency. *See also:* **coupling.** 328

critical current (superconductor). The current in a superconductive material above which the material is normal and below which the material is superconducting, at a specified temperature and in the absence of external magnetic fields. *See also:* **superconductivity.** 191

critical damping. *See:* **damped harmonic system (2) (critical damping).**

critical dimension (waveguide). The dimension of the cross-section that determines the cutoff frequency. *See also:* **waveguide.** 328

critical failure. *See:* **failure, critical.**

critical field (magnetrons). The smallest theoretical value of steady magnetic flux density, at a steady anode voltage, that would prevent an electron emitted from the cathode at zero velocity from reaching the anode. *See also:* **magnetrons.** 125, 190

critical frequency (1) (radio propagation by way of the ionosphere). The limiting frequency below which a wave component is reflected by, and above which it penetrates through, an ionospheric layer at vertical incidence. *Note:* The existence of the critical frequency is the result of electron limitation, that is, the inadequacy of the existing number of free electrons to support reflection at higher frequencies. *See also:* **radiation; radio wave propagation.** 180, 59, 146

(2) (circuits and systems) (network or system). A pole or zero of a transfer or driving-point function. 67

critical grid current (multielectrode gas tubes). The grid current corresponding to the critical grid voltage, before anode breakdown. *Note:* The critical grid current is a function of the other electrode voltages or currents and of the environment. 125

critical grid voltage (multielectrode gas tubes). The grid voltage at which anode breakdown occurs. *Note:* The critical grid voltage is a function of the other electrode voltages or currents and of the environment. *See also:* **breakdown voltage (electrode of a gas tube).** 125

critical high-power level (attenuator tubes). The radio-frequency power level at which ionization is produced in the absence of a control-electrode discharge. 125, 190

critical humidity (corrosion). The relative humidity above which the atmospheric corrosion rate of a given metal increases sharply. 221, 205

critical impulse (relay). The maximum impulse in terms of duration and input magnitude that can be applied suddenly to a relay without causing pickup. 103, 202, 60, 127

critical impulse flashover voltage (insulator). The crest value of the impulse wave that, under specified conditions, causes flashover through the surrounding medium on 50 percent of the applications. *See also:* **impulse flashover voltage.** 261, 91

critical impulse time (relay). The duration of a critical impulse under specified conditions. 103, 202, 60, 127

criticality (nuclear power generation). The state of an assembly of fissionable material in which a stable, self-sustaining chain reaction exists. At this condition a nuclear reactor will produce energy at a constant rate and the effective multiplication factor, k_{eff}, is exactly equal to 1. 112

critically damped (1) (control systems). Describing a linear second-order system which is damped just enough to prevent any overshoot of the output following an abrupt stimulus. *Note:* The characteristic equation has a real root once repeated. *See also:* **damping.** 56

(2) (industrial control). Damping that is sufficient to prevent any overshoot of the output following an abrupt stimulus. *See:* **control; damping (note).** 219, 206

critical magnetic field (superconductor). The field below which a superconductive material is superconducting and above which the material is normal, at a specified temperature and in the absence of current. *See also:* **superconductivity.** 191

critical overtravel time (relay). The time following a critical impulse until movement of the responsive element ceases just short of pickup. 103, 202, 60, 127

critical point (feedback control system) (1) (Nichols chart). The bound of stability for the GH $(j\omega)$ plot; the intersection of $|\text{GH}| = 1$ with ang $GH = -180$ degrees.

(2) (Nyquist diagram). The bound of stability for the locus of the loop transfer function $GH(j\omega)$; the $(-1.j0)$ point. 56, 329

critical rate of rise of communication voltage (bidirectional thyristor). The minimum value of the rate of rise of principal voltage which will cause switching from the OFF-state to the ON-state immediately following ON-state current conduction in the opposite quadrant. 66

critical rate-of-rise of OFF-state voltage (thyristor). The minimum value of the rate-of-rise of principal voltage that will cause switching from the OFF-state to the ON-state. *See also:* **principal voltage-current characteristic (principal characteristic).** 243, 66, 208, 191

critical rate-of-rise of ON-state current (thyristor). The maximum value of the rate-of-rise of ON-state current that a thyristor can withstand without deleterious effect. *See also:* **principal current.** 243, 66, 208, 191

critical speed (rotating machinery). A speed at which the amplitude of the vibration of a rotor due to shaft transverse vibration reaches a maximum value. *See also:* **rotor (rotating machinery).** 63

critical temperature (1) (superconductor). The temperature below which a superconductive material is superconducting and above which the material is normal, in the absence of current and external magnetic fields. *See also:* **superconductivity.** 191

(2) (storage cell or battery). The temperature of the electrolyte at which an abrupt change in capacity occurs. *See also:* **initial test temperature.** 328

critical torsional speed (rotating machinery). A speed at which the amplitude of the vibration of a rotor due to shaft torsional vibration reaches a maximum value. *See also:* **rotor (rotating machinery).** 63

critical travel (relay). The amount of movement of the responsive element of a relay during a critical impulse, but not subsequent to the impulse.   103, 202, 60, 127

critical voltage (1) (magnetron). The highest theoretical value of steady anode voltage, at a given steady magnetic flux density, at which electrons emitted from the cathode at zero velocity would fail to reach the anode. *See also:* **magnetron.**   125, 190

(2) (relay). *See:* **relay critical voltage.**

critical-voltage parabola (cutoff parabola) (magnetrons). The curve representing in Cartesian coordinates the variation of the critical voltage as a function of the magnetic induction. *See:* **magnetron.**   244, 190

critical wavelength. *See:* **cutoff wavelength.**

critical withstand current (surge). The highest crest value of a surge of given waveshape and polarity that can be applied without causing disruptive discharge on the test specimen.   62

critical withstand voltage (impulse). The highest crest value of an impulse, of given waveshape and polarity, that can be applied without causing disruptive discharge on the test specimen.   62

Crookes dark space. *See:* **cathode dark space.**

cross acceleration (accelerometer). The acceleration applied in a plane normal to an accelerometer input reference axis.   46

crossarm. A horizontal member (usually wood or steel) attached to a pole, post, tower or other structure and equipped with means for supporting the conductors. *Note:* The crossarm is placed at right angles to conductors on straight line poles, but splits the angle on light corners. *See also:* **tower.**   64

crossarm guy. A tensional support for a crossarm used to offset unbalanced conductor stress.   64

cross axis sensitivity (accelerometer). The proportionality constant that relates a variation of accelerometer output to cross acceleration. This sensitivity can vary depending on the direction of cross acceleration.   46

crossbar switch. A switch having a plurality of vertical paths, a plurality of horizontal paths, and electromagnetically-operated mechanical means for interconnecting any one of the vertical paths with any one of the horizontal paths.   56

crossbar system. An automatic telephone switching system that is generally characterized by the following features: (1) The selecting mechanisms are crossbar switches. (2) Common circuits select and test the switching paths and control the operation of the selecting mechanisms. (3) The method of operation is one in which the switching information is received and stored by controlling mechanisms that determine the operations necessary in establishing a telephone connection.   328

cross cable bond. A cable bond used for bonding between the armor or lead sheath of adjacent cables. *See:* **cable bond; continuity cable bond.** *See also:* **power distribution, underground construction.**   64

crossconnection (telephone switching system). Easily changed or removed wire that is run loosely between equipment terminals to establish an electrical association.   55

cross-correlation (excitation control system). The cross-correlation of two random signals $x_1(t)$ and $x_2(t)$ is $R_{12}(t)$ defined by

$$R_{12}(t) = \int_0^t x_1(t - \tau) \, x_2(\tau) d\tau.$$

If $x_1(t)$ is a random input to a linear stationary system and $x_2(t)$ is the response, then $R_{12}(t)$ is the inverse Laplace transform of the transfer function of the system.   353

cross coupling (transmission medium). A measure of the undesired power transferred from one channel to another. *See also:* **coupling; transmission line.**   267, 199

cross coupling coefficient (accelerometer). The proportionality constant that relates a variation of accelerometer output to the product of acceleration applied normal and parallel to an input reference axis. This coefficient can vary depending on the direction of cross acceleration.   46

cross coupling errors (gyro). The errors in the gyro output resulting from gyro sensitivity to inputs about axes normal to an input reference axis.   46

crossed-field amplifier (microwave tubes). A crossed-field tube or valve, with a nonreentrant slow-wave structure, used as an amplifier. *See also:* **microwave tube or valve.**   190

crossed-field tube (microwave). A high-vacuum electron tube in which a direct, alternating, or pulsed voltage is applied to produce an electric field perpendicular both to a static magnetic field and to the direction of propagation of a radio-frequency delay line. *Note:* The electron beam interacts synchronously with a slow wave on the delay line. *See also:* **microwave tube (or valve).**   190

crossfire. Interfering current in one telegraph or signaling channel resulting from telegraph or signaling currents in another channel. *See also:* **telegraphy.**   328

crossing angle (electronic navigation). *See:* **angle of cut.**

crossing structure (conductor stringing equipment). A structure built of poles and sometimes rope nets. It is used whenever conductors are being strung over roads, power lines, communications circuits, highways or railroads and normally constructed in such a way that it will prevent the conductor from falling onto or into any of these facilities in the event of equipment failure, broken pulling lines, loss of tension, etc. *Syn.* **guard structure; h-frame; rider structure; temporary structure.**   45

cross light (television). Equal illumination in front of the subject from two directions at substantially equal and opposite angles with the optical axis of the camera and a horizontal plane. *See also:* **television lighting.**   167

cross modulation. A type of intermodulation due to modulation of the carrier of the desired signal by an undesired signal wave.   111, 59

cross neutralization. A method of neutralization used in push-pull amplifiers whereby a portion of the plate-cathode alternating voltage of each tube is applied to the grid-cathode circuit of the other tube. *See also:* **amplifier; feedback.**   111, 211

crossover (cathode-ray tube). The first focusing of the beam that takes place in the electron gun. *See also:* **cathode-ray tubes.**   244, 190

crossover, automatic voltage-current (power supplies). The characteristic of a power supply that automatically changes the method of regulation from constant voltage to constant current (or vice versa) as dictated by varying load conditions (see the following figure). *Note:* The constant-voltage and constant-current levels can be independently adjusted within the specified voltage and current limits of the power supply. The intersection of constant-voltage and constant-current lines is called the crossover point E, I and may be located anywhere within the volt-ampere range of the power supply. *See also:* **power supply.** 228, 186

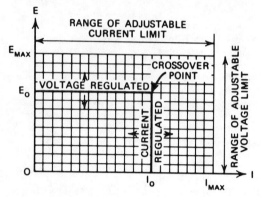

Automatic voltage-current crossover.

crossover characteristic curve (navigation systems such as very-high-frequency omnidirectional ranges and instrument landing systems). The graphical representation of the indicator current variation with change of position in the crossover region. *See also:* **navigation.** 13, 187

crossover frequency (1) (frequency-dividing networks). The frequency at which equal power is delivered to each of two adjacent channels when all channels are properly terminated. *See also:* **loudspeaker; transition frequency.** 176

(2) (automatic control). *See:* **gain crossover frequency; phase crossover frequency.**

crossover loss (radar). The reduction in signal-to-noise ratio for a target on the tracking axis relative to that for a target on the peak two-way antenna gain of the beam, for a tracker which uses an offset beam, such as a conical scan tracker. The crossover loss factor is the ratio of the signal-to-noise ratio for a target on the peak two-way antenna gain to that for a target on the tracking axis. 13

crossover network. *See:* **dividing network.**

crossover region (navigation system). A loosely defined region in space containing the course line and within which a transverse flight yields information useful in determining course sensitivity and flyability. *See also:* **navigation.** 187, 13

crossover spiral. *See:* **lead-over groove.**

crossover time (semiconductor radiation detectors). The instant at which the waveform of bipolar pulse passes through a designated level. 23, 118, 119

crossover voltage, secondary-emission (charge-storage tubes). The voltage of a secondary-emitting surface, with respect to cathode voltage, at which the secondary-emission ratio is unity. The crossovers are numbered in progression with increasing voltage. See the following figure. *Note:* The qualifying phrase **secondary-emission** is frequently dropped in general usage. *See also:* **charge-storage tube.** 174, 190

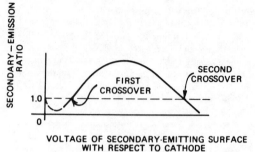

VOLTAGE OF SECONDARY-EMITTING SURFACE WITH RESPECT TO CATHODE

Typical secondary-emission curve.

crossover walk (pulse) (semiconductor radiatior detectors). The deviation of the crossover time for some variable; such as amplitude. 23, 118, 119

crosspoint (telephone switching systems). A controlled device used in extending a transmission or control path. 55

cross polarization (antenna). The polarization orthogonal to a reference polarization. *Notes:* (1) Two fields have orthogonal polarizations if their polarization ellipses have the same axial ratio, major axes at right angles, and opposite senses of rotation. (2) If the reference polarization is right-handed circular, the cross polarization is left-handed circular, and vice versa. '111

cross protection. An arrangement to prevent the improper operation of devices from the effect of a cross in electric circuits. 328

cross product. *See:* **vector product.**

cross rectifier circuit. A circuit that employs four or more rectifying elements with a conducting period of 90 electrical degrees plus the commutating angle. *See also:* **rectification.** 328

cross section (1) (radar). *See:* **radar cross section.**

(2) (antenna). *See:* **back-scattering cross section; radar cross section; scattering cross section.**

cross-sectional area (conductor) (cross section of a conductor). The sum of the cross-sectional areas of its component wires, that of each wire being measured perpendicular to its individual axis. 59

crosstalk (1). Undesired energy appearing in one signal path as a result of coupling from other signal paths. *Note:* Path implies wires, waveguides, or other localized or constrained transmission systems. *See also:* **coupling.** 239

(2) (electroacoustics). The unwanted sound reproduced by an electroacoustic receiver associated with a given transmission channel resulting from cross coupling to another transmission channel carrying sound-controlled electric waves or, by extension, the electric waves in the disturbed channel that result in such sound. *Note:* In practice, crosstalk may be measured either by the volume of the overheard sounds or by the magnitude of the coupling between the disturbed and disturbing channels. In the latter case, to specify the volume of the overheard sounds, the volume in the disturbing channel must also be given. *See also:* **coupling.** 53

(3) (cable systems). The noise or extraneous signal caused by alternating-current or pulse type signals in adjacent circuits. 35

crosstalk coupling (crosstalk loss). Cross coupling between speech communicatior channels or their component parts. *Note:* Crosstalk coupling is measured between specified points of the disturbing and disturbed circuits and is preferably expressed in decibels. *See also:* **coupling.**
328

crosstalk, electron beam (charge-storage tubes). Any spurious output signal that arises from scanning or from the input of information. *See also:* **charge-storage tube.**
174, 190

crosstalk loss. *See:* **crosstalk coupling.**

crosstalk unit*. Crosstalk coupling is sometimes expressed in crosstalk units through the relation

$$\text{Crosstalk units} = 10^{[6-(L/20)]}$$

where L = crosstalk coupling in decibels. *Note:* For two circuits of equal impedance, the number of crosstalk units expresses the current in the disturbed circuit as millionths of the current in the disturbing circuit. *See also:* **coupling.**
328

* Obsolescent

CR–RC **shaping.** The pulse shaping present in an amplifier that has a simple high-pass filter consisting of a capacitor and a resistor together with a simple low-pass filter, separated by impedance isolation. Pulse shaping in such an amplifier cuts off at 6 decibels per octave at both ends of the band.
23, 118, 119

crude metal. Metal that contains impurities in sufficient quantities to make it unsuitable for specified purposes or that contains more valuable metals in sufficient quantities to justify their recovery. *See also:* **electrorefining.** 328

crust. A layer of solidified electrolyte. *See also:* **fused electrolyte.**
328

cryogenics (1) (general). The study and use of devices utilizing properties of materials near absolute-zero temperature.
255, 77

(2) (laser-maser). The branch of physics dealing with very low temperatures.
363

cryotron. A superconductive device in which current in one or more input circuits magnetically controls the superconducting-to-normal transition in one or more output circuits, provided the current in each output circuit is less than its critical value. *See also:* **superconductivity.** 191

crystal (communication practice). (1) A piezoelectric crystal. (2) A piezoelectric crystal plate. (3) A crystal rectifier.
328

crystal-controlled oscillator. *See:* **crystal oscillator.**

crystal-controlled transmitter. A transmitter whose carrier frequency is directly controlled by a crystal oscillator. *See also:* **radio transmitter.**
59

crystal diode. A rectifying element comprising a semiconducting crystal having two terminals designed for use in circuits in a manner analogous to that of electron-tube diodes. *See also:* **rectifier.** 328

crystal filter. An electric wave filter employing piezoelectric crystals for its reactive elements.
59

crystal loudspeaker (piezoelectric loudspeaker). A loudspeaker in which the mechanical displacements are produced by piezoelectric action. *See also:* **loudspeaker.**
328

crystal microphone (piezoelectric microphone). A microphone that depends for its operation on the generation of an electric charge by the deformation of a body (usually crystalline) having piezoelectric properties. *See also:* **microphone.**
328

crystal mixer (mixer). A crystal receiver that can be fed simultaneously from a local oscillator and signal source, for the purpose of frequency changing. *See:* **waveguide.**
244, 179

crystal oscillator (crystal-controlled oscillator). An oscillator in which the principal frequency-determining factor is the mechanical resonance of a piezoelectric crystal. *See also:* **oscillatory circuit.**
211

crystal pickup (piezoelectric pickup). A phonograph pickup that depends for its operation on the generation of an electric charge by the deformation of a body (usually crystalline) having piezoelectric properties. *See also:* **phonograph pickup.**
328

crystal pulling. A method of crystal growing in which the developing crystal is gradually withdrawn from a melt. *See also:* **electron devices, miscellaneous.**
66

crystal receiver. A waveguide incorporating a crystal detector for the purpose of rectifying received electromagnetic signals. *See:* **waveguide.**
244, 179

crystal spots. Spots produced by the growth of metal sulfide crystals upon metal surfaces with a sulfide finish and lacquer coating. The appearance of crystal spots is called spotting in. *See also:* **electroplating.**
328

crystal-stabilized transmitter. A transmitter employing automatic frequency control, in which the reference frequency is that of a crystal oscillator. *See also:* **radio transmitter.**
111

crystal-video receiver. A receiver consisting of a crystal detector and a video amplifier.
328

C **scan (electronic navigation).** *See:* *C* **display.**

C **scope (1) (class *C* oscilloscope).** A cathode ray oscilloscope arranged to present a type *C*-display. *See also:* **radar.**
328

(2) (radar). *See:* *C* **display.** *See also:* **radar.**

CTS. *See:* **carrier test switch.**

cube tap (plural tap*). *See:* **multiple plug.**

* Deprecated

cubic natural spline (pulse terms). A catenated piecewise sequence of cubic polynominal functions $p(1, 2)$, $p(2, 3)$, . . ., $p(n-1, n)$ between knots $t_1 m_1$ and $t_2 m_2$, $t_2 m_2$ and $t_3 m_3$, . . ., $t_{n-1} m_{n-1}$ and $t_n m_n$, respectively, wherein:
(1) At all knots the first and second derivatives of the adjacent polynominal functions are equal, and
(2) For all values of t less than t_1 and greater than t_n the function is linear. *See:* Pulse burst envelopes diagram below the **knot** entry. *See also:* **waveforms produced by operations on waveforms.**
254

cumulative compound (rotating machinery). Applied to a compound machine to denote that the magnetomotive forces of the series and the shunt field windings are in the same direction. *See:* **direct-current commutating machine; magnetomotive force.**
63

cumulative demand meter (or register). An indicating demand meter in which the accumulated total of maximum demands during the preceding periods is indicated during the period after the meter has been reset and before it is reset again. *Note:* The maximum demand for any one period is equal or proportional to the difference between the accumulated readings before and after reset. *See also:* **electricity meter (meter).**　　212

cumulative detection probability (radar). The probability that a target is detected on at least one of N scans of a surveillance radar.　　13

cuprous chloride cell. A primary cell in which depolarization is accomplished by cuprous chloride. *See also:* **electrochemistry.**　　328

Curie point (1) (general). A temperature marking a transition between ferromagnetic (ferroelectric, antiferromagnetic) properties and nonferromagnetic (nonferroelectric, nonantiferromagnetic) properties of a material. *Notes:* (A) Ferromagnetic materials have a single Curie point, above which they are paramagnetic. (B) The susceptibility of antiferromagnetic material increases with temperature, up to the Curie point, decreasing therafter. (C) Some ferroelectric materials have both upper and lower Curie points, and some have different Curie points associated with different crystal axes. (D) Experimentally, the transition may cover an appreciable temperature range. *See also:* **dielectric heating; induction heating.**　　210

(2) (induction heating). The temperature in a ferromagnetic material above which the material becomes substantially non-magnetic. *Note:* The Curie point may be reduced slightly in some cases of high-density induction heating by stresses resulting from high temperature gradients. *See:* **recalescent point.**　　14

Curie-Weiss temperature (ferroelectric material). The intercept of the linear portion of the plot of $1/\kappa$ versus T, where κ is the small-signal relative dielectric permittivity measured at zero bias field along the polar axis and T is the absolute temperature in the region above the ferroelectric Curie point. *Note:* In many ferroelectrics κ follows the Curie-Weiss relation

$$\kappa = \epsilon/\epsilon_0 = c/(T - \theta),$$

where ϵ is the small-signal absolute permittivity, ϵ_0 is the permittivity of free space ($8.854. \times 10^{-12}$ coulomb per voltmeter), C is the Curie constant, and θ is the Curie-Weiss temperature. *See also:* **paraelectric material.**　　80

curl (vector field). A vector that has a magnitude equal to the limit of the quotient of the circulation around a surface element on which the point is located by the area of the surface, as the area approaches zero, provided the surface is oriented to give a maximum value of the circulation; the positive direction of this vector is that traveled by a right hand screw turning about an axis normal to the surface element when an integration around the element in the direction of the turning of the screw gives a positive value to the circulation. If the vector **A** of a vector field is expressed in terms of its three rectangular components A_x, A_y, and A_z, so that the values of A_x, A_y, and A_z are each given as a function of x, y, and z, the curl of the vector field (abbreviated curl **A** or $\nabla \times$ **A**) is the vector sum of the partial derivatives of each perpendicular to it, or

$$\text{curl } \mathbf{A} = \nabla \times \mathbf{A} = \begin{vmatrix} \mathbf{i} & \mathbf{j} & \mathbf{k} \\ \dfrac{\delta}{\delta x} & \dfrac{\delta}{\delta y} & \dfrac{\delta}{\delta z} \\ A_x & A_y & A_z \end{vmatrix}$$

$$= \mathbf{i}\left(\frac{\delta A_z}{\delta y} - \frac{\delta A_y}{\delta z}\right) + \mathbf{j}\left(\frac{\delta A_x}{\delta z} - \frac{\delta A_z}{\delta x}\right)$$

$$+ \mathbf{k}\left(\frac{\delta A_y}{\delta x} - \frac{\delta A_x}{\delta y}\right)$$

$$= \mathbf{i}(D_y A_z - D_z A_y)$$

$$+ \mathbf{j}(D_z A_x - D_x A_z) + \mathbf{k}(D_x A_y - D_y A_z)$$

where **i**, **j**, and **k** are unit vectors along the x, y, and z axes, respectively. Example: The curl of the linear velocity of points in a rotating body is equal to twice the angular velocity. The curl of the magnetic field strength at a point within an electric conductor is equal to k times the current density at the point where k is a constant depending on the system of units.　　210

current (electric) (1) (general). A generic term used when there is no danger of ambiguity to refer to any one or more of the currents specifically described. *Notes:* (A) For example, in the expression "the current in a simple series circuit," the word current refers to the conduction current in the wire of the inductor and the displacement current between the plates of the capacitor. (B) A direct current is a unidirectional current in which the changes in value are either zero or so small that they may be neglected. A given current would be considered a direct current in some applications, but would not necessarily be so considered in other applications.　　210

(2) (modified by an adjective). The use of certain adjectives before "current" is often convenient, as in convection current, anode current, electrode current, emission current, etcetera. The definition of conducting current usually applies in such cases and the meaning of adjectives should be defined in connection with the specific applications.　　210

(3) (cable insulation materials). Sum of the polarization and conductance currents.　　97

current amplification (1) (general). An increase in signal current magnitude in transmission from one point to another or the process thereof.

(2) (transducer). The scalar ratio of the signal output current to the signal input current. *Warning:* By incorrect extension of the term decibels, this ratio is sometimes expressed in decibels by multiplying its common logarithm by 20. It may be correctly expressed in decilogs.　　328

(3) (multiplier phototube). The ratio of the output current to the cathode current due to photoelectric emission at constant electrode voltages. *Notes:* (A) The term output current and photocathode current as here used does not include the dark current. (B) This characteristic is to be measured at levels of operation that will not cause saturation. *See also:* **phototube.**　　190, 125, 117

(4) (magnetic amplifier). The ratio of differential output current to differential control current.　　171

(5) (electron multipliers). The ratio of the signal output current to the current applied to the input. *See also:* **amplifier.**　　190

current, anode. *See:* **electrode current.**

current attenuation. *See:* **attenuation, current.**

current, average discharge (dielectric tests). The sum of the rectified charge quantities passing through the terminals of the test object due to partial discharges during a time interval, divided by this interval. The average discharge current is expressed in coulombs per second (amperes).
139

current balance ratio. The ratio of the metallic-circuit current or noise-metallic (arising as a result of the action of the longitudinal-circuit induction from an exposure on unbalances outside the exposure) to the longitudinal circuit current or noise-longitudinal in sigma at the exposure terminals. It is expressed in microamperes per milliampere or the equivalent. *See also:* **inductive coordination.**
328

current-balance relay. A relay that operates by comparing the magnitudes of two current inputs. *See also:* **relay.**
60, 127, 103

current-balancing reactor. *See:* **reactor, current-balancing.**

current capacity (electric penetration assembly). The maximum current that each conductor in the assembly is specified to carry for its duty cycle in the design service environment without causing stabilized temperatures of the conductors or the penetration nozzle—concrete interface (if applicable) to exceed their design limits.   26

current carrier. In a semiconductor, a mobile conduction electron or hole.
113

current-carrying capacity (contacts). The maximum current that a contact is able to carry continuously or for a specified period of time. *See:* **contactor.**   244, 206

current-carrying part. A conducting part intended to be connected in an electric circuit to a source of voltage. *Note:* Non-current carrying parts are those not intended to be so connected. *See also:* **center of distribution.**
262, 64, 103, 202, 27

current circuit (electric instrument). The combination of conductors and windings of the instrument that carries the current of the circuit in which a given electrical quantity is to be measured, or a definite fraction of the current, or a current dependent upon it. *See also:* **moving element of an instrument; watthour meter.**
280

current, conduction (cable insulation materials). Current in the specimen under steady-state conditions. *Notes:* (1) This is sometimes called "leakage" current. (2) Absorption and capacitive effects are assumed to have been made negligible under steady-state conditions. (3) Surface leakage current is assumed excluded from the measured current. *Syn:* **conductance current.**
97

current crest factor (mercury-lamp—ballast combination). The ratio of the peak value of lamp current to the root mean-square value of lamp current.
271

current cutoff (power supplies). An overload protective mechanism designed into certain regulated power supplies to reduce the load current automatically as the load resistance is reduced. This negative resistance characteristic reduces overload dissipation to negligible proportions and protects sensitive loads. The accompanying figure shows the E-I characteristic of a power supply equipped with a current cutoff overload protector. *See also:* **power supply.**
228, 186

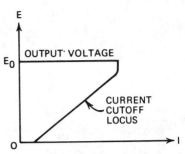

Output characteristics of a power supply equipped with a current cutoff overload protector.

current density (1). A generic term used where there is no danger of ambiguity to refer either to conduction-current density or to displacement-current density, or to both.
210

(2) (electrolytic cells). On a specified electrode in an electrolytic cell the current per unit area of that electrode. *See also:* **electrolytic cell.**
328

(3) (semiconductor rectifier). The current per unit active area. *See also:* **rectification.**
237, 66

current derived voltage (protective relaying). A voltage produced by a combination of currents. *Notes:* (1) The element used to create this voltage in a pilot system is popularly referred to as a filter. A typical example is a filter that is supplied three-phase currents and produces an output voltage proportional to the symmetrical component content of these currents. (For example, $V_F = K_1 IA_1 + K_2 IA_2 + K_0 IA_0$ where IA_1, IA_2, and IA_0 are the symmetrical components of the A phase current and the K are weighting factors.)
128

current efficiency (specified electrochemical process). The proportion of the current that is effective in carrying out that process in accordance with Faraday's law. *See also:* **electrochemistry.**
328

current generator (signal-transmission system). A two-terminal circuit element with a terminal current substantially independent of the voltage between its terminals. *Note:* An ideal current generator has zero internal admittance. *See also:* **network analysis; signal.**   125, 60

current-limit (control) (industrial control). A control function that prevents a current from exceeding its prescribed limits. *Note:* Current-limit values are usually expressed as percent of rated-load value. If the current-limit circuit permits the limit value to increase somewhat instead of being a single value, it is desirable to provide either a curve of the limit value of current as a function of some variable such as speed or to give limit values at two or more conditions of operation. *See:* **control system, feedback.**
219, 206

current-limit acceleration (electric drive) (industrial control). A system of control in which acceleration is so governed that the motor armature current does not exceed an adjustable maximum value. *See also:* **electric drive.**   206

current-limit control (electric drive). A system of control in which acceleration, or retardation, or both, are so governed that the armature current during speed changes does not exceed a predetermined value. *See also:* **electric drive.**
328

current limiting, automatic (power supplies). An overload protection mechanism that limits the maximum output current to a preset value, and automatically restores the output when the overload is removed. *See the accompanying figure. See:* **short-circuit protection.** *See also:* **power supply.**   228

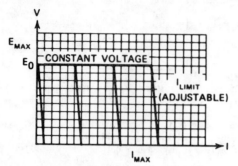

Plot of typical current-limiting curves.

current-limiting fuse. A fuse that, when it is melted by a current within its specified current-limiting range, abruptly introduces a high arc voltage to reduce the current magnitude and duration. *Note:* The values specified in standards for the threshold ratio, peak let-through current, and $I^2 t$ characteristic are used as the measures of current-limiting ability.   103, 202

current-limiting characteristic curve (current-limiting fuse) (peak let-through characteristic curve) (cutoff characteristic curve). A curve showing the relationship between the maximum peak current passed by a fuse and the correlated root-mean-square available current magnitudes under specified voltage and circuit impedance conditions. *Note:* The root-mean-square available current may be symmetrical or asymmetrical.   103, 202

current-limiting range (current-limiting fuse). That specified range of currents between the threshold current and the rated interrupting current within which current limitation occurs.   103, 202

current-limiting reactor. A reactor intended for limiting the current that can flow in a circuit under short-circuit conditions, or under other operating conditions such as starting, synchronizing, etc.   53

current-limiting resistor (industrial control). A resistor inserted in an electric circuit to limit the flow of current to some predetermined value. *See:* **control system, feedback.**   328

current loss (electric instrument) (voltage circuit current drain) (parallel loss of an electric instrument). In a voltage-measuring instrument, the value of the current when the applied voltage corresponds to nominal end-scale indication. *Note:* In other instruments it is the current in the voltage circuit at rated voltage. *See also:* **accuracy rating (instrument).**   280

current margin (neutral direct-current telegraph system). The difference between the steady-state currents flowing through a receiving instrument, corresponding, respectively, to the two positions of the telegraph transmitter. *See also:* **telegraphy.**   328

current of traffic. The movement of trains on a main track in one direction specified by the rules. *See also:* **railway signal and equipment.**   328

current, peak (low voltage dc power circuit breakers). The instantaneous value of current at the time of its maximum value.   360

current phase-balance protection. A method of protection in which an abnormal condition within the protected equipment is detected by the current unbalance between the phases of a normally balanced polyphase system.   103, 202, 60, 127

current, polarization. Time-dependent, decaying current in the specimen, following the instant that a constant voltage is applied until steady-state conditions have been obtained. *Note:* Polarization current does not include the conductance current. The sum of the polarization and conductance currents in the specimen is that which is normally observed during measurements. *Note:* Polarization current includes both polarization absorption and capacitive-charge currents.   97

current pulsation (rotating machinery). The difference between maximum and minimum amplitudes of the motor current during a single cycle corresponding to one revolution of the driven load expressed as a percentage of the average value of the current during this cycle. *See also:* **asynchronous machine.**   63

current, rated. *See:* **rated current.**

current rating (1) (rectifier transformer). The root-mean-square equivalent of a rectangular current waveshape based on direct-current rated load commutated with zero commutating angle. *See also:* **rectifier transformer.**   258

(2) (relay). The current at specified frequency that may be sustained by the relay for an unlimited period without causing any of the prescribed limitations to be exceeded.   127

(3) (separable insulated connector). (A) (continuous). The designated rms alternating or direct current which the connector can carry continuously under specified conditions. (B) (fault-closure). The designated root-mean-square fault current which a load-break connector can close under specified conditions.   134

current rating, 60-hertz (arrester). A designation of the range of the symmetrical root-mean-square fault currents of the system for which the arrester is designed to operate. *Notes:* (1) An expulsion arrester is given a maximum current rating and may also have a minimum current rating. (2) The designation of the maximum and minimum current ratings of an expulsion arrester not only specifies the useful operating range of the arrester between those extreme values for symmetrical root-mean-square short-circuit current, but indicates that at the point of application of the arrester the root-mean-square short-circuit current for the system should neither be greater than the maximum nor less than the minimum current rating.   328

current ratio (series transformer) (mercury lamp). The ratio of the (root-mean-square) primary current to the root-mean-square secondary current under specified conditions of load.   203, 274

current-recovery ratio (arc-welding apparatus). With a welding power supply delivering current through a short-circuited resistor whose resistance is equivalent to the load setting on the power supply, and with the short-circuit suddenly removed, the ratio of (1) the minimum transient value of current upon the removal of the short-circuit to (2) the final steady-state value is the current-recovery ratio. 264

current regulation (constant-current transformer). The maximum departure of the secondary current from its rated value, with rated primary voltage at rated frequency applied, and at rated secondary power factor, and with the current variation taken between the limits of a short-circuit and rated load. *Note:* This regulation may be expressed in per unit, or percent, on the basis of the rated secondary current. 203, 53

current relay (1) (general). A relay that functions at a predetermined value of current. *Note:* It may be an over-current, undercurrent, or reverse-current relay. *See also:* **relay.** 1
(2) **(power systems relaying).** A relay which responds to current. 79, 127

current-responsive element (fuse). That part with predetermined characteristics, the melting and severance or severances of which initiate the interrupting function of the fuse. *Note:* The current-responsive element may consist of one or more fusible elements combined with a strain element and/or other component(s) that affect(s) the current-responsive characteristic. 103, 202

current-sensing resistor (power supplies). A resistor placed in series with the load to develop a voltage proportional to load current. A current-regulated direct-current power supply regulates the current in the load by regulating the voltage across the sensing resistor. *See also:* **power supply.** 228, 186

current, short-circuit (transformer-rectifier system). The steady-state value of the input alternating current that flows when the output direct current terminals are short-circuited and rated line alternating voltage is applied to the line terminals. This current is normally of interest when using current limiting transformers or checking current limiting devices. 95

current tap*. *See:* **multiple lampholder; plug adapter lampholder.**

* Deprecated

current terminals of instrument shunts (electric power systems). Those terminals that are connected into the line whose current is to be measured and that will carry the current of the shunt. 201

current transformer (instrument transformer). Intended to have its primary winding connected in series with the conductor carrying the current to be measured or controlled. (In window type current transformers, the primary winding is provided by the line conductor and is not an integral part of the transformer.) 53

current-type telemeter. A telemeter that employs the magnitude of a single current as the translating means. *See also:* **telemetering.** 103, 202

curvature (coaxial transmission line). The radial departure from a straight line between any two points on the external surface of a conductor. *See also:* **waveguide.** 265

curve follower. *See:* **curve-follower function generator.** 9

curve, integrated energy (electric power utilization). A curve of demand versus energy showing the amount of energy represented under a load curve, or a load duration curve

above any point of demand. *Syn.* **peak percent curve.** *See also:* **generating station.** 112

curve, load duration (electric power utilization). A curve of loads, plotted in descending order of magnitude, against time intervals for a specified period. *See also:* **generating station.** 2

curve, monthly peak duration (electric power utilization). A curve showing the total number of days within the month during which the net 60 minute clock-hour integrated peak demand equals or exceeds the percent of monthly peak values shown. *See also:* **generating station.** 112

curve, reservoir operating rule (electric power supply). A curve, or family of curves (reservoir capability versus time), indicating how a reservoir is to be operated under specified conditions to obtain best or predetermined results. 112

cushion clamp (rotating machinery). A device for securing the cushion to the supporting member. 328

cushioning time (speed governing of hydraulic turbines). The elapsed time during which the (closing) rate of servomotor travel is retarded by the slow closure device. *See also:* **slow closure device.** 8, 58

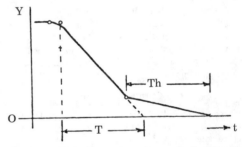

cut-in loop. A circuit on the roadway energized to automatically cut in the train control or cab signal apparatus on a passing vehicle. 328

cutoff. *See:* **cutoff frequency.**

cutoff angle (luminaire). The angle, measured up from nadir, between the vertical axis and the first line of sight at which the bare source is not visible. *See also:* **bare (exposed) lamp.** 167

cutoff attenuator. An adjustable length of waveguide used below its cutoff frequency to introduce variable non-dissipative attenuation. *See also:* **waveguide.**

cutoff characteristic (current-limiting fuse). *See:* **current-limiting characteristic curve.**

cutoff frequency (1) (general). The frequency that is identified with the transition between a pass band and an adjacent attenuation band of a system or transducer. *Note:* It may be either a theoretical cutoff frequency or an effective cutoff frequency. 239
(2) **(waveguide).** For a given transmission mode in a non-dissipative uniconductor waveguide, the frequency below which the propagation constant is real. *See also:* **waveguide.** 267, 179

cutoff frequency, effective (electric circuit). A frequency at which its insertion loss between specified terminating impedances exceeds by some specified amount the loss at some reference point in the transmission band. *Note:* The specified insertion loss is usually 3 decibels. *See also:* **cutoff frequency; network analysis.** 328

cutoff relay (telephony). A relay associated with a subscriber line, that disconnects the line relay from the line when the line is called or answered.   328

cutoff voltage (1) (final voltage) (battery). The prescribed voltage at which the discharge is considered complete. *Note:* The cutoff or final voltage is usually chosen so that the useful capacity of the battery is realized. The cutoff voltage varies with the type of battery, the rate of discharge, the temperature, and the kind of service. The term **cutoff voltage** is applied more particularly to primary batteries, and **final voltage** to storage batteries. *See also:* **battery (primary or secondary).**   328

(2) (electron tube). The electrode voltage that reduces the value of the dependent variable of an electron-tube characteristic to a specified low value. *Note:* A specific cutoff characteristic should be identified as follows: current versus grid cutoff voltage, spot brightness versus grid cutoff voltage, etcetera. *See also:* **electrode voltage (electron tube).**   190, 125

(3) (magnetrons). *See:* **critical voltage (magnetrons).**

cutoff waveguide (evanescent waveguide). A waveguide used as a frequency below its cutoff frequency. *See:* **waveguide.**
                                                                244, 179

cutoff wavelength (1) (mode in a waveguide) (critical wavelength). That wavelength, in free space or in the unbounded guide medium, as specified, above which a traveling wave in that mode cannot be maintained in the guide. *Note:* For $TE_{m,n}$ or $TM_{m,n}$ waves in hollow rectangular cylinders,

$$\lambda_c = 2/[(m/a)^2 + (n/b)^2]^{1/2}$$

where a is the width of the waveguide along the x coordinate and b is the height of the waveguide along the y coordinate.

The following table gives the ratio of cutoff wavelengths to diameters for $TM_{n,m}$ waves in hollow circular metal cylinders:

$n =$	0	1	2	3	4	5
1	1.307	0.820	0.613	0.483	0.414	0.358
2	0.569	0.448	0.373	0.322	0.284	0.2547
$m =$ 3	0.363	0.309	0.270	0.241	0.219	0.200
4	0.267	0.2375	0.2127	0.1938		
5	0.2106	0.1910	0.1750			

The following table gives the ratio of cutoff wavelengths to diameters for $TE_{m,n}$ waves in hollow circular metal cylinders:

$n =$	0	1	2	3	4
1	0.820	1.708	1.030	0.748	0.590
2	0.448	0.59	0.468	0.382	0.338
$m =$ 3	0.309	0.368	0.315	0.277	0.247
4	0.2375	0.27	0.24	0.22	0.198
5	0.1910	0.21	0.194	0.18	

See: **guided wave.** *See also:* **waveguide.**   244, 179

(2) (uniconductor waveguide). The ratio of the velocity of electromagnetic waves in free space to the cutoff frequency. *See also:* **waveguide.**   267

cutout (1) (general). An electric device used manually or automatically to interrupt the flow of current through any particular apparatus or instrument.   328

(2) (electric distribution systems). An assembly of a fuse support with either a fuseholder, fuse carrier, or disconnecting blade. *Notes:* (A) The fuseholder or fuse carrier may include a conducting element (fuse link), or may act as a disconnecting blade by the inclusion of a nonfusible member. (B) The term **cutout** as defined here is restricted in practice to equipment used on distribution systems. *See:* **distribution** and **distribution cutout.** For fuses having similar components used on power systems. *See:* **power; power fuse.**   103, 202

cutout base*. *See:* **fuse holder.**

* Deprecated

cutout box (interior wiring). An enclosure designed for surface mounting and having swinging doors or covers secured directly to and telescoping with the walls of the box proper. *See also:* **cabinet.**   328

cutout loop (railway practice). A circuit in the roadway that cooperates with vehicle-carried apparatus to cut out the vehicle train control or cab signal apparatus.   328

cut paraboloidal reflector. A reflector that is not symmetrical with respect to its axis. *See also:* **antenna.**
                                                                111

cut-section. A location within a block other than a signal location where two adjacent track circuits end.   328

cut-set (network). A set of branches of a network such that the cutting of all the branches of the set increases the number of separate parts of the network, but the cutting of all the branches except one does not. *See also:* **network analysis.**   210

cutter (audio and electroacoustics) (mechanical recording head). An electromechanical transducer that transforms an electric input into a mechanical output, that is typified by mechanical motions that may be inscribed into a recording medium by a cutting stylus. *See also:* **phonograph pickup.**   176

cutter compensation (numerically controlled machines). Displacement, normal to the cutter path, to adjust for the difference between actual and programmed cutter radii or diameters.   224, 207

cutting down (metal or electrodeposit) (electroplating). Polishing for the purpose of removing roughness or irregularities. *See also:* **electroplating.**   328

cutting stylus (electroacoustics). A recording stylus with a sharpened tip that, by removing material, cuts a groove into the recording medium. *See also:* **phonograph pickup.**
                                                                176

CW radar. A radar which transmits a continuous wave transmitted signal.   13

cybernetics. *See:* **system science.**

cycle (1). (A) An interval of space or time in which one set of events or phenomena is completed. (B) Any set of operations that is repeated regularly in the same sequence. The operations may be subject to variations on each repetition.   255, 77

(2) The complete series of values of a periodic quantity that occurs during a period. *Note:* It is one complete set of positive and negative values of an alternating current. *See:* **major cycle; minor cycle.**   3, 210, 53

(3) (pulse terms). The complete range of states or magnitudes through which a periodic wave-form or a periodic feature passes before repeating itself identically. 254

cycle of operation (storage cell or battery). The discharge and subsequent recharge of the cell or battery to restore the initial conditions. *See also:* **charge.** 328

cyclically magnetized condition. A condition of a magnetic material when, under the influence of a magnetizing force that is a cyclic (but not necessarily periodic) function of time having one maximum and one minimum per cycle, it follows identical hysteresis loops on successive cycles.   210

cyclic code error detection (power-system communication). The process of cyclically computing bits to be added at the end of a word such that an identical computation will reveal a large portion of errors that may have been introduced in transmission. *See also:* **digital.** 59

cyclic duration factor (rotating machinery). The ratio between the period of loading including starting and electric braking, and the duration of the duty cycle, expressed as a percentage. *See also:* **asynchronous machine; direct-current commutating machine.**   63

cyclic function. A function that repetitively assumes a given sequence of values at an arbitrarily varying rate. *Note:* That is, if y is a periodic function of x and x in turn is a monotonic nondecreasing function of t, then y is said to be a cyclic function of t.   210

cyclic irregularity (rotating machinery). The periodic fluctuation of speed caused by irregularity of the prime-mover torque. *See also:* **asynchronous machine; direct-current commutating machine.**   63

cyclic shift (digital computers). (1) An operation that produces a word whose characters are obtained by a cyclic permutation of the characters of a given word.   235, 210

(2) A shift in which the data moved out of one end of the storing register are reentered into the other end, as in a closed loop. *See:* **circulating register.** 255, 77

cyclometer register. A set of 4 or 5 wheels numbered from zero to 9 inclusive on their edges, and so enclosed and connected by gearing that the register reading appears as a series of adjacent digits. *See also:* **watthour meter.**   328

cyclotron. A device for accelerating positively charged particles (for example, protons, deuterons, etcetera) to high energies. The particles in an evacuated tank are guided in spiral paths by a static magnetic field while they are accelerated many times by an electric field of mixed frequency. *See also:* **electron devices, miscellaneous.**   190

cyclotron frequency (1). The frequency at which an electron traverses an orbit in a steady uniform magnetic field and zero electric field. *Note:* It is given by the product of the electron charge and the magnetic flux density, divided by 2π times the electron mass.

(2) (radio wave propagation). Gyrofrequency. *See:* **gyrofrequency; magnetrons.**   125, 190

cyclotron, frequency-modulated. *See:* **frequency-modulated cyclotron.**

cyclotron-frequency magnetron oscillations. Those oscillations whose frequency is substantially the cyclotron frequency.   125

cylindrical antenna. An antenna having the shape of a circular cylinder. *Note:* This term usually describes a linear antenna.   111

cylindrical array (antenna). An arrangement of elements whose centers are on a cylindrical surface. 111

cylindrical reflector. A reflector that is a portion of a cylinder. *Note:* This cylinder is usually parabolic, although other shapes may be used. *See also:* **antenna.** 111, 179

cylindrical-rotor generator (solid-iron cylindrical-rotor generator) (turbine generator*). An alternating-current generator driven by a high-speed turbine (usually steam) and having an exciting winding embedded in a cylindrical steel rotor. *See:* **generating station.** 63

* Deprecated. Favor steam-turbine or gas-turbine generator as distinguished from hydraulic-turbine generator.

cylindrical-rotor machine. A machine having a cylindrically-shaped rotor the periphery of which is slotted to accommodate the coil sides of a field winding. *See:* **rotor (rotating machinery).**   63

cylindrical wave (radio wave propagation). A wave whose equiphase surfaces form a family of coaxial or confocal cylinders. *See also:* **radiation; radio wave propagation.**   146, 180

D

damped harmonic system (1) (linear system with one degree of freedom). A physical system in which the internal forces, when the system is in motion, can be represented by the terms of a linear differential equation with constant coefficients, the order of the equation being higher than the first. Example: The differential equation of a damped system is often of the form

$$M \frac{d^2 x}{dt^2} + F \frac{dx}{dt} + Sx = f(t)$$

where M, F, and S are positive constants of the system; x is the dependent variable of the system (displacement in mechanics, quantity of electricity, etcetera); and $f(t)$ is the applied force. Examples: A tuning fork is a damped harmonic system in which M represents a mass; F a coefficient of damping; S a coefficient of restitution; and x a displacement. Also an electric circuit containing constant inductance, resistance, and capacitance is a damped harmonic system, in which case M represents the self-inductance of the circuit; F the resistance of the circuit; S the reciprocal of the capacitance; and x the charge that has passed through a cross section of the circuit.

(2) (critical damping). The name given to that special case of damping that is the boundary between underdamping and overdamping. A damped harmonic system is critically damped if $F^2 = 4MS$. *See also:* **network analysis.**

(3) (overdamping) (aperiodic damping*). The special case of a damping in which the free oscillation does not change sign. A damped harmonic system is overdamped if $F^2 > 4MS$. *See also:* **network analysis.**

* Deprecated

(4) (underdamping) (periodic damping*). The special case of damping in which the free oscillation changes sign at least once. A damped harmonic system is underdamped if $F^2 < 4MS$. 210

* Deprecated

(5) (underdamped). Damped insufficiently to prevent oscillation of the output following an abrupt input stimulus. *Note:* In an underdamped linear second-order system, the roots of the characteristic equation have complex values. *See:* **control; critical damping; damping.** *See also:* **control system, feedback.** 219, 206

damped oscillation (damped vibration). An oscillation in which the amplitude of the oscillating quantity decreases with time. *Note:* If the rate of decrease can be expressed by means of a mathematical function, the name of the function may be used to describe the damping. Thus if the rate of decrease is exponential, the system is said to be an exponentially damped system. 210

damped sinusoidal function. A function having the form $f_1(t) \sin(\omega t + \theta)$ where $f_1(t)$ is a monotonic decreasing function, and ω and θ are constants. 210

damped vibration. See: **damped oscillation.**

damped wave. A wave in which, at every point, the amplitude of each sinusoidal component is a decreasing function of the time. *See also:* **signal; wave front.** 210

damper bar. See: **amortisseur bar.**

damper segment (rotating machinery). One portion of a short-circuiting end ring (of an amortisseur winding) that can be separated into parts for mounting or removal without access to a shaft end. *See also:* **rotor (rotating machinery).** 63

damper winding (amortisseur winding) (rotating machinery). A winding consisting of a number of conducting bars short-circuited at the ends by conducting rings or plates and distributed on the field poles of a synchronous machine to suppress pulsating changes in magnitude or position of the magnetic field linking the poles. 63

damping (1) (noun). The temporal decay of the amplitude of a free oscillation of a system, associated with energy loss from the system.

(2) (adjective). Pertaining to or productive of damping. *Note:* The damping of many physical systems is conveniently approximated by a viscous damping coefficient in a second-order linear differential equation (or a quadratic factor in a transfer function). In this case the system is said to be critically damped when the time response to an abrupt stimulus is as fast as possible without overshoot; underdamped (oscillatory) when overshoot occurs; overdamped (aperiodic) when response is slower than critical. The roots of the quadratic are, respectively, real and equal; complex, and real and unequal. *See also:* **control system, feedback.**

(3) (relative underdamped system). A number expressing the quotient of the actual damping of a second-order linear system or element by its critical damping. *Note:* For any system whose transfer function includes a quadratic factor $s^2 + 2z\omega_n s + \omega_n^2$, relative damping is the value of z, since $z = 1$ for critical damping. Such a factor has a root $-\sigma + j\omega$ in the complex s plane, from which $z = \sigma/\omega_n = \sigma/(\sigma^2 + \omega^2)^{1/2}$. *See also:* **control system, feedback.**

(4) (Coulomb). That due to Coulomb friction. *See also:* **control system, feedback.**

(5) (instrument). Term applied to the performance of an instrument to denote the manner in which the pointer settles to its steady indication after a change in the value of the measured quantity. Two general classes of damped motion are distinguished as follows: (A) periodic (underdamped) in which the pointer oscillates about the final position before coming to rest, and (B) aperiodic (overdamped) in which the pointer comes to rest without overshooting the rest position. The point of change between periodic and aperiodic damping is called critical damping. *Note:* An instrument is considered for practical purposes to be critically damped when overshoot is present but does not exceed an amount equal to one-half the rated accuracy of the instrument. *See also:* **accuracy rating (instrument); moving element (instrument).** 210, 206, 56, 280

damping amortisseur. An amortisseur the primary function of which is to oppose rotation or pulsation of the magnetic field with respect to the pole shoes. 328

damping coefficient. *See:* **damping coefficient, viscous; damping; (4) (Coulomb); decrement, logarithmic; damping, relative.**

damping coefficient, viscous. In the characteristic equation of a second-order linear system, say $Mx'' + Bx' + Kx = 0$, the coefficient B of the first derivative of the damped variable. *See also:* **control system, feedback.** 56

damping factor (1) (free oscillation of a second-order linear system). A measure of damping, expressed (without sign) as the quotient of the greater by the lesser of a pair of consecutive swings of the variable (in opposite directions) about an ultimate steady-state value. *Note:* The ratio is a constant whose Napierian logarithm (used with a negative sign) is called the logarithmic decrement, whose square relates adjoining swings in the same direction, and whose value following a step input from zero to a value x_u is conveniently measured as the ratio of x_u to the initial overshoot. *See also:* **control system, feedback; moving element (instrument).**

(2) (electric instruments). (A) Since, in some instruments, the damping factor depends upon the magnitude of the deflection, it is measured as the ratio in angular degrees of the steady deflection to the difference between angular maximum momentary deflection and the angular steady deflection produced by a sudden application of a constant electric power. The damping factor shall be determined with sufficient constant electric power applied to carry the pointer to end-scale deflection on the first swing. The damping shall be due to the instrument and its normal accessories only. For purposes of testing damping, in instances where the damping is influenced by the circuit impedance, the impedance of the driving source shall be at least 100 times the impedance of the instrument and its normal accessories. (B) In the determination of the damping factor of instruments not having torque-spring control, such as a power-factor meter or a frequency meter, the instrument shall be energized to give center-scale deflection. The pointer shall then be moved mechanically to end scale and suddenly released. *See also:* **moving element (instrument).**

(3) (underdamped harmonic system). The quotient of the logarithmic decrement by the natural period. *Note:* The damping factor of an underdamped harmonic system if $F/2M$. *See:* **damped harmonic system.**   280, 56, 210

damping magnet. A permanent magnet so arranged in conjunction with a movable conductor such as a sector or disk as to produce a torque (or force) tending to oppose any relative motion between them. *See also:* **moving element (instrument).**   328

damping torque (synchronous machine). The torque produced, such as by action of the amortisseur winding, that opposes the relative rotation, or changes in magnitude, of the magnetic field with respect to the rotor poles.   63

damping torque coefficient (synchronous machine). A proportionality constant that, when multiplied by the angular velocity of the rotor poles with respect to the magnetic field, for specified operating conditions, results in the damping torque.   63

damp location (electric system). A location subject to a moderate degree of moisture, such as some basements, some barns, some cold-storage warehouses, and the like. *See also:* **distribution center.**   256

dark adaptation. The process by which the retina becomes adapted to a luminance less than about 0.01 footlambert (0.03 nit). *See also:* **visual field.**   167

dark current (photoelectric device) (electron device). The current flowing in the absence of irradiation. *See:* **photoelectric effect; dark current; electrode; dark-current pulses (phototubes).**   244, 190, 117, 125

dark-current pulses (phototubes). Dark-current excursions that can be resolved by the system employing the phototube. *See also:* **phototube.**   335

darkening (electroplating). The production by chemical action, usually oxidation, of a dark colored film (usually a sulfide) on a metal surface. *See also:* **electroplating.**   328

dark pulses. Pulses observed at the output electrode when the photomultiplier is operated in total darkness. These pulses are due primarily to electrons originating at the photocathode.   117

dark space, cathode. *See:* **cathode dark space.**

dark space, Crookes. *See:* **cathode dark space.**

dark-trace screen (cathode-ray tube). A screen giving a spot darker than the remainder of the surface. *See also:* **cathode-ray tubes.**   244, 190

dark-trace tube (skiatron) (1) (electronic navigation). A cathode-ray tube having a special screen that changes color but does not necessarily luminesce under electron impact, showing, for example, a dark trace on a bright background. *See also:* **cathode-ray tubes; navigation.**   244, 190

(2) (radar). A cathode-ray tube, of which the face is bright, and signals are displayed as dark traces or dark blips. *See also:* **skiatron.**   13

D'Arsonval current* (solenoid current*) (medical electronics). The current of intermittent and isolated trains of

heavily damped oscillations of high frequency, high voltage, and relatively low amperage. *See:* Note under **D'Arsonvalization.**   192

* Deprecated

D'Arsonvalization* (medical electronics). The therapeutic use of intermittent and isolated trains of heavily damped oscillations of high frequency, high voltage, and relatively low amperage. *Note:* This term is deprecated because it was initially ill-defined and because the technique is not of contemporary interest.   192

* Deprecated

data. Any representations such as characters or analog quantities to which meaning might be assigned.   255, 77, 194

data communication equipment. The equipment that provides the functions required to establish, maintain, and terminate a connection, as well as the signal conversion, and coding required for communication between data terminal equipment and data circuit.   12

data communications (data transmission). The movement of encoded information by means of communication techniques.   59

data-hold (data processing). A device that converts a sampled function into a function of a continuous variable. The output between sampling instants is determined by an extrapolation rule or formula from a set of past inputs.   198

data link (1) (data communication). An assembly of data terminals and the interconnecting circuits operating according to a particular method that permits information to be exchanged between the terminals.   12, 194

(2) (test, measurement and diagnostic equipment). Any information channel used for connecting data processing equipment to any input, output, display device, or other data processing equipment, usually at a remote location.   54

data logger (power-system communication). A system to measure a number of variables and make a written tabulation and/or record in a form suitable for computer input. *See also:* **digital.**   59

data logging (power switchgear). An arrangement for the alphanumerical representation of selected quantities on log sheets, papers, magnetic tape, or the like, by means of an electric typewriter or other suitable devices.   103

data-logging equipment. Equipment for numerical recording of selected quantities on log sheets or paper or magnetic tape or the like, by means of an electric typewriter or other suitable device.   103, 202

data processing. Pertaining to any operation or combination of operations on data.   255, 77

data processor. Any device capable of performing operations on data, for example, a desk calculator, a tape recorder, an analog computer, a digital computer.   255, 77

data reconstruction (date processing). The conversion of a signal defined on a discrete-time argument to one defined on a continuous-time argument.   198

data reduction. The transformation of raw data into a more useful form, for example, smoothing to reduce noise.   255, 77, 54

data set (data transmission). A modem serving as a conversion element and interface between a data machine and communication facilities. *See:* **modem.** 59, 12

data sink (data transmission). The equipment which accepts data signals after transmission. 59

data source (data transmission). The equipment which supplies data signals to be transmitted. 59

data stabilization (vehicle-borne navigation systems). The stabilization of the output signals with respect to a selected reference invariant with vehicle orientation. *See also:* **navigation.** 187

data terminal (data transmission). A device which modulates and/or demodulates data between one input-output device and a data transmission link. 59

data terminal equipment (data communication). The equipment comprising the data source, the data sink, or both. 12

data transmission (data link). The sending of data from one place to another or from one part of a system to another. 59

dataway (CAMAC). *See:* **CAMAC dataway.**

daylight factor. A measure of daylight illumination at a point on a given plane expressed as a ratio of the illumination on the given plane at that point to the simultaneous exterior illumination on a horizontal plane from the whole of an unobstructed sky of assumed or known luminance (photometric brightness) distribution. *Note:* Direct sunlight is excluded from both interior and exterior values of illumination. *See also:* **sunlight.** 167

dB. *See:* **decibel.**

dBm. A unit for expression of power level in decibels with reference to a power of one milliwatt (0.001 watt). *See:* **power level (dBm).** 239

dBV. *See:* **voltage level.**

dc. *See:* **direct current.**

D cable. A two-conductor cable, each conductor having the shape of the capital letter D with insulation between the conductors themselves and between conductors and sheath. *See also:* **power distribution, underground construction.** 64

DDA (computing systems). *See:* **digital differential analyzer.**

***D* dimension (motor).** The standard designation of the distance from the centerline of the shaft to the plane through the mounting surface bottom of the feet, in National Electrical Manufacturers Association approved designations. 63

***D* display (radar).** A *C* display in which the blips extend vertically to give a rough estimate of distance. *See also:* **navigation.** 13, 187

DDM (electronic navigation). *See:* **difference in depth of modulation.**

deactivation (corrosion). The process of removing active constituents from a corroding liquid (as removal of oxygen from water). 221, 205

dead (electric system). Free from any electric connection to a source of potential difference and from electric charge; not having a potential different from that of the ground. The term is used only with reference to current-carrying parts that are sometimes alive. *See also:* **insulated.** 178

dead band (1) (general). The ra6ge through which the measured signal can be varied without initiating response. *Notes:* (A) This is usually expressed in percent of span. (B) Threshold and ultimate sensitivity are defined as one-half the dead band. When the instrument is balanced at the center of the dead band, it denotes the minimum change in measured signal required to initiate response. Since this condition of balance is not normally achieved and is not readily recognizable, these terms are deprecated and the various requirements should preferably be stated in terms of dead band. Sensitivity is frequently used to denote the same quantity. However, its usage in this sense is also deprecated since it is not in accord with accepted standard definitions of the term. (C) The presence of dead band may be intentional, as in the case of a differential gap or neutral zone, or undesirable, as in the case of play in linkages or backlash in gears.

(2) (gas turbines). The total range through which an input can be varied with no resulting measurable corrective action of the fuel-control valve. *Note:* Speed deadband is expressed in percent of rated speed. *See also:* **control system, feedback; instrument.** 98, 58, 295, 219, 206, 94

(3) (nuclear power generating station). The range through which an input can be varied without initiating output response. 355

dead-band rating. The limit that the dead band will not exceed when the instrument is used under rated operating conditions. *See also:* **accuracy rating (instrument).** 295

dead-break connector (transformer). A connector designed to be separated and engaged on deenergized circuits only. 53

dead-end guy. An installation of line or anchor guys to hold the pole at the end of a line. *See:* **pole guy.** *See also:* **tower.** 64

dead-end tower. A tower designed to withstand unbalanced pull from all of the conductors in one direction together with wind and vertical loads. *See also:* **tower.** 64

dead-front (industrial control) (transformer). So constructed that there are no exposed live parts on the front of the assembly. *See:* **dead-front switchboard.** 206, 53

dead-front mounting (switching device). A method of mounting in which a protective barrier is interposed between all live parts and the operator, and all exposed operating parts are insulated or grounded. *Note:* The barrier is usually grounded metal. 103, 202

dead-front switchboard. A switchboard that has no exposed live parts on the front. *Note:* The switchboard panel is normally grounded metal and provides a barrier between the operator and all live parts. 103, 202, 79

dead layer thickness (semiconductor radiation detector). The thickness of an inactiver region (in the form of a layer) through which incident radiation must pass to reach the sensitive volume. 23, 118, 119

dead man (overhead construction). *See:* **anchor log.**

deadman's handle (industrial control). A handle of a controller or master switch that is designed to cause the controller to assume a preassigned operating condition if the force of the operator's hand on the handle is released. *See:* **electric controller.** 225, 206

deadman's release (industrial control). The effect of that feature of a semiautomatic or nonautomatic control system that acts to cause the controlled apparatus to assume a preassigned operating condition if the operator becomes incapacitated. *See:* **control.** 225, 206

dead-metal part. A part, accessible or inaccessible, which is conductively connected to the grounded circuit under conditions of normal use of the equipment. *See also:* **dead-front.** 53

dead reckoning (navigation). The determining of the position of a vehicle at one time with respect to its position at a different time by the application of vector(s) representing course(s) and distance(s). *See also:* **navigation.** 187, 13

dead room (audio and electroacoustics) (acoustics). A room that has an unusually large amount of sound absorption. 176

dead spots (mobile communication). Those locations completely within the coverage area where the signal strength is below the level needed for reliable communication. *See also:* **mobile communicating system.** 181

dead-tank switching device. A switching device in which a vessel(s) at ground potential surrounds and contains the interrupter(s) and the insulating medium. 79

dead time (1) (automatic control). The interval of time between initiation of an input change or stimulus and the start of the resulting response. *Notes:* (A) It may be qualified as effective if extended to the start of build-up time; theoretical if the dead band is negligible; apparent if it includes the time spent within an appreciable dead band. (B) The interval of inaction is a result that may arise from any of several causes; for example, the quotient distance/velocity when the speed of signal propagation is finite; the presence of a hysteretic zone, a resetting action, Coulomb friction, play in linkages, or backlash in gears; the purposeful introduction of a periodic sampling interval, or a neutral zone. *See also:* **control system, feedback.** 56, 329

(2) (navigation). The time interval in an equipment's cycle of operation during which the equipment is prevented from providing normal response. For example, in a radar display, the portion of the interpulse interval which is not displayed; or, in secondary radar, the interval immediately following the transmission of a pulse reply during which the transponder is insensitive to interrogations. 13

(3) (recovery time) (radiation counter). The time from the start of a counted pulse until an observable succeeding pulse can occur. *See also:* **gas-filled radiation-counter tubes.** 190, 96, 125

(4) (circuit breaker on a reclosing operation). The interval between interruption in all poles on the opening stroke and reestablishment of the circuit on the reclosing stroke. *Notes:* In breakers using arc-shunting resistors, the following intervals are recognized and the one referred to should be stated. (A) Dead time from interruption on the primary arcing contacts to reestablishment through the primary arcing contacts. (B) Dead time from interruption on the primary arcing contacts to reestablishment through the secondary arcing contacts. (C) Dead time from interruption on the secondary arcing contacts to reestablishment on the primary arcing contacts. (D) Dead time from interruption on the secondary arcing contacts to reestablishment on the secondary arcing contacts. (E) The dead time of an arcing fault on a reclosing operation is not

necessarily the same as the dead time of the circuit breakers involved, since the dead time of the fault is the interval during which the faulted conductor is deenergized from all terminals. 103, 202

dead-time correction (radiation counters). A correction to the observed counting rate to allow for the probability of the occurrence of events within the dead time of the system. *See:* **anticoincidence (radiation counters).** 190

dead zone (industrial control). The period(s) in the operating cycle of a machine during which corrective functions cannot be initiated. 206

debug (1). To examine or test a procedure, routine, or equipment for the purpose of detecting and correcting errors. 194

(2) (computing systems). To detect, locate, and remove mistakes from a routine or malfunctions from a computer. *See:* **troubleshoot.** 255, 77

debugging (reliability). The operation of an equipment or complex item prior to use to detect and replace parts that are defective or expected to fail, and to correct errors in fabrication or assembly. *See also:* **reliability.** 182, 54

Debye length (radio wave propagation). The distance in a plasma over which a free electron may move under its own kinetic energy before it is pulled back by the electrostatic restoring forces of the polarization cloud surrounding it. *Note:* Over this distance a net charge density can exist in an ionized gas. The Debye length is given by

$$l_D = \left(\frac{\epsilon_0 k T_e}{e^2 N_e} \right)^{1/2}$$

where ϵ_0 is the permittivity of vacuum, k is Boltzmann's constant, e is the charge of the electron, T_e is the electron temperature, and N_e is the electron number density. *See also:* **radio wave propagation.** 180, 146

decade (electric circuit theory). The frequency interval between two frequencies having a ratio of 10 to 1. *Note:* This definition does not conform with the usual dictionary definition and does not agree with other uses of decade in electrical engineering: for example, as in decade resistance box. 210

decalescent point (induction heating usage). The temperature at which there is a sudden absorption of heat when metals are raised in temperature. *See also:* **dielectric heating; induction heating; coupling; recalescent point.** 14, 114, 125

decay (storage tubes). A decrease in stored information by any cause other than erasing or writing. *Note:* Decay may be caused by an increase, a decrease, or a spreading of stored charge. *See also:* **storage tube.** 174, 190, 125

decay characteristic (luminescent screen). *See:* **persistence characteristic.**

decay, dynamic (storage tubes). Decay caused by an action such as that of the reading beam, ion currents, field emission, or holding beam. *See also:* **storage tube.** 174, 190

decay, static (charge-storage tubes). Decay that is a function only of the target properties, such as lateral and transverse leakage. *See also:* **charge-storage tube.** 174, 190

decay time (storage tubes). The time interval during which the stored information decays to a stated fraction of its initial value. *Note:* Information may not decay exponentially. *See also:* **storage tube.**   174, 190, 125

decay time constant (semiconductor radiation detector). The time for a true single-exponential waveform to decay to a value of $1/e$ of the original step height.
           118, 23, 119

Decca. A radio navigation system transmitting on several related frequencies near 100 kilohertz useful to about 200 nautical miles (370 kilometers), in which sets of hyperbolic lines of position are determined by comparison of the phase of (1) one reference continuous wave signal from a centrally located master with (2) each of several continuous wave signals from slave transmitters located in a star pattern, each about 70 nautical miles (130 kilometers) from the master. *See also:* **navigation.**   187, 13

decelerating electrode (electron-beam tubes). An electrode the potential of which provides an electric field to decrease the velocity of the beam electrons. 190, 117, 125

decelerating relay (industrial control). A relay that functions automatically to maintain the armature current or voltage within limits, when decelerating from speeds above base speed, by controlling the excitation of the motor field. *See also:* **relay.**   225, 206

decelerating time (industrial control). The time in seconds for a change of speed from one specified speed to a lower specified speed while decelerating under specified conditions. *See:* **control system, feedback.** 219, 206

deceleration (industrial control). *See:* **retardation.**

deceleration, programmed (numerically controlled machines). A controlled velocity decrease to a fixed percent of the programmed rate.   224, 207

deceleration, timed (industrial control). A control function that decelerates the drive by automatically controlling the speed change as a function of time. *See:* **control system, feedback.**   219, 206

decentralized computer network (data communication). A computer network, where some of the computing power and network control functions are distributed over several network nodes. *See also:* **centralized network.** 12

decentralized switching entity (telephone switching systems). A portion of a central office that is intended to be located away from the serving system control. 55

decibel (1) (general). One-tenth of a bel, the number of decibels denoting the ratio of the two amounts of power being ten times the logarithm to the base 10 of this ratio. *Note:* The abbreviation dB is commonly used for the term decibel. With P_1 and P_2 designating two amounts of power and n the number of decibels denoting their ratio,

$$n = 10 \log_{10}(P_1/P_2) \text{ decibel.}$$

When the conditions are such that ratios of currents or ratios of voltages (or analogous quantities in other fields) are the square roots of the corresponding power ratios, the number of decibels by which the corresponding powers differ is expressed by the following equations:

$$n = 20 \log_{10}(I_1/I_2) \text{ decibel}$$
$$n = 20 \log_{10}(V_1/V_2) \text{ decibel}$$

where I_1/I_2 and V_1/V_2 are the given current and voltage ratios, respectively. By extension, these relations between numbers of decibels and ratios of currents or voltages are sometimes applied where these ratios are not the square

roots of the corresponding power ratios; to avoid confusion, such usage should be accompanied by a specific statement of this application. Such extensions of the term described should preferably be avoided. 111, 120, 197, 59

(2) (automatic control). A logarithmic scale unit relating a variable x (e.g., angular displacement) to a specified reference level x_0; $db = 20 \log x/x_0$. *Note:* The relation is strictly applicable only where the ratio x/x_0 is the square root of the power ratio P/P_0, as is true for voltage or current ratios. The value $db = 10 \log_{10} P/P_0$ originated in telephone engineering, and is approximately equivalent to the old "transmission unit." 56

(3) (excitation control systems). In control usage, a logarithmic scale unit relating a variable x to a specified reference level x_0; $db = 20 \log x/x_0$. *Note:* The relation is strictly applicable only where the ratio x/x_0 is the square root of the power ratio P/P_0, as is true for voltage or current ratios.   353

decibel meter. An instrument for measuring electric power level in decibels above or below an arbitrary reference level. *See also:* **instrument.**   328

decilog (dg). A division of the logarithmic scale used for measuring the logarithm of the ratio of two values of any quantity. *Note:* Its value is such that the number of decilogs is equal to 10 times the logarithm to the base 10 of the ratio. One decilog therefore corresponds to a ratio of $10^{0.1}$ (that is 1.25892+).   59

decimal. (1) Pertaining to a characteristic or property involving a selection, choice, or condition in which there are ten possibilities. (2) Pertaining to the numeration system with a radix of ten. *See:* **binary-coded decimal.**
           255, 77

decimal code. A code in which each allowable position has one of 10 possible states. The conventional decimal number system is a decimal code.   244, 207

decimal number system. *See:* **positional notation.**

decimal point. *See:* **point.**

decineper. One-tenth of a neper.   111

decision gate (electronic navigation). A specified point near the lower end of an instrument-landing-system approach at which a pilot must make a decision either to complete the landing or to execute a missed-approach procedure. *See also:* **navigation.**   187, 13

decision instruction (computing systems). An instruction that effects the selection of a branch of a program, for example, a conditional jump instruction. 255, 77

decision table (computing systems). A table of all contingencies that are to be considered in the description of a problem, together with the actions to be taken. Decision tables are sometimes used in place of flow charts for problem description and documentation. 255, 77

deck (computing systems). A collection of punched cards.
           255, 77

declinometer (compass declinometer). A device for measuring the direction of a magnetic field relative to astronomical or survey coordinates. *See also:* **magnetometer.**
           328

decode. (1) To recover the original message from a coded form of the message.

(2) To apply a code so as to reverse some previous encoding.   255, 77

(3) To produce a single output signal from each combi-

nation of a group of input signals. *See also:* **translate; matrix; encode.** 235

decoder (1) (telephone practice). A device for decoding a series of coded signals. *Note:* In automatic telephone switching, a decoder is a relay-type translator that determines from the office code of each call the information required for properly recording the call through the switching train. Each decoder has means, such as a cross-connecting field, for establishing the controls desired and for readily changing them. 328

(2) (electronic computation and control). A matrix of logic elements that selects one or more output channels according to the combination of input signals present.
210, 255, 77

(3) (railway practice). A device adopted to select controls for wayside signal apparatus, train control apparatus, or cab signal apparatus in accordance with the code received from the track of the line circuit. 328

decomposition potential (decomposition voltage). The minimum potential (excluding *IR* drop) at which an electrochemical process can take place continuously at an appreciable rate. *See also:* **electrochemistry.** 328

decoupling. The reduction of coupling. *See also:* **coupling.** 328

Dectra. An experimental adaptation of the Decca low-frequency radio navigation system in which two pairs of continuous-wave transmitters are oriented so that the center lines of both pairs are along and at opposite ends of the same great-circle path, to provide course guidance along and adjacent to the great-circle path. *Note:* Distance along track may be indicated by synchronized signals from one transmitter of each pair. *See also:* **navigation.** 187, 13

along and adjacent to the great-circle path. *Note:* Distance along track may be indicated by synchronized signals from one transmitter of each pair. *See also:* **navigation.** 187, 13

de-emphasis (post emphasis) (post equalization). The use of an amplitude-frequency characteristic complementary to that used for pre-emphasis earlier in the system. *See also:* **pre-emphasis.** 328

de-emphasis network. A network inserted in a system in order to restore the pre-emphasized frequency spectrum to its original form. 111

de-energize (relay). To disconnect the relay from its power source. 259

de-energized (conductor stringing equipment). Free from any electric connection to a source of potential difference and from electric charge; not having a potential different from that of the ground. The term is used only with reference to current-carrying parts that are sometimes alive (energized). To state that a circuit has been "de-energized" means that the circuit has been disconnected from all intended electrical sources. However, it could be electrically charged through induction from energized circuits in proximity to it, particularly if the circuits are parallel. *Syn.* **dead.** 45

(2) (transmission and distribution) (dead). Free from any electrical connection to a source of potential difference and from electric charge; not having a potential different from that of the earth. *Note:* The term is used only with reference to current-carrying parts which are sometimes energized (alive). 262

deep-bar rotor. A squirrel-cage induction-motor rotor having a winding that is narrow and deep giving the effect of varying secondary resistance, large at standstill and decreasing as the speed rises. *See also:* **rotor (rotating machinery).** 63

deep space (communication satellite). Space at distances from the earth approximately equal to or greater than the distance between the earth and the moon. 74

deep space instrumentation facility (DSIF) (communication satellite). A ground network of worldwide communication stations (earth terminals) maintained for providing communications to and from lunar and inter-planetary spacecraft and deep space probes. Each earth terminal utilizes large antennas, low-noise receiving systems and high-power transmitters. 83

deepwell pump. An electrically-driven pump located at the low point in the mine to discharge the water accumulation to the surface. 328

defeater (industrial control). *See:* **interlocking deactivating means.**

defect (solar cells). A localized deviation of any type from the regular structure of the atomic lattice of a single crystal. 113

defined pulse width (semiconductor radiation detectors). The time elapsed between the first and final crossings of the defined zero level for the maximum rated output pulse amplitude. 23

defined reference pulse waveform (pulse measurement). A reference pulse waveform which is defined without reference to any practical or derived pulse waveform. Typically, a defined reference pulse waveform is an ideal pulse waveform. 15

defined zero (semiconductor radiation detectors). An arbitrarily chosen voltage level at the amplifier output resolvable from zero by the measuring apparatus. 23

definite minimum-time relay. An inverse-time relay in which the operating time becomes substantially constant at high values of input. 60, 103, 127

definite purpose circuit breakers (capacitance current switching). A definite purpose circuit breaker is one that is designed specifically for capacitance current switching. 130

definite-purpose controller (industrial control). Any controller having ratings, operating characteristics, or mechanical construction for use under service conditions other than usual or for use on a definite type of application. *See:* **electric controller.** 225, 206

definite-purpose motor. Any motor designed, listed, and offered in standard ratings with standard operating characteristics or mechanical construction for use under service conditions other than usual or for use on a particular type of application. *Note:* Examples: crane, elevator, and oil-burner motors. *See:* **asynchronous machine; direct-current commutating machine.** 328

definite time (relays). A qualifying term indicating that there is purposely introduced a delay in action, which delay remains substantially constant regardless of the magnitude of the quantity that causes the action. *See:* **relay.**
103, 202

definite-time acceleration (electric drive) (industrial control). A system of control in which acceleration proceeds on a definite-time schedule. *See also:* **electric drive.** 206

definite-time relay (1) (general). A relay in which the operating time is substantially constant regardless of the magnitude of the input quantity. *See also:* **relay.**

281, 60, 127

(2) (power switchgear). A qualifying term indicating that there is purposely introduced a delay in action, which delay remains substantially constant regardless of the magnitude of the quantity that causes the action. 103

definition (1) (general). The fidelity with which the detail of an image is reproduced. *See also:* **recording (facsimile).**

328

(2) (facsimile). Distinctness or clarity of detail or outline in a record sheet, or other reproduction. 12

deflecting electrode (electron-beam tubes). An electrode the potential of which provides an electric field to produce deflection of an electron beam. *See also:* **electrode (electron tube).** 190, 125

deflecting force (direct-acting recording instrument). At any part of the scale (particularly full scale), the force for that position, measured at the marking device, and produced by the electrical quantity to be measured, acting through the mechanism. *See also:* **accuracy rating (instrument).** 328

deflecting voltage (cathode-ray tube). Voltage applied between the deflector plates to create the deflecting electric field. *See also:* **cathode-ray tubes.** 244, 190

deflecting yoke. An assembly of one or more coils that provide a magnetic field to produce deflection of an electron beam. 190

deflection (cathode-ray tube) (electron tubes). The displacement of the beam or spot on the screen under the action of the deflecting field. *See also:* **cathode-ray tubes.** 244, 190

deflection axis, horizontal (oscilloscope). The horizontal trace obtained when there is a horizontal deflection signal but no vertical deflection signal. 106

deflection axis, vertical (oscilloscope). The vertical trace obtained when there is a vertical deflection signal and no horizontal deflection signal. 106

deflection blanking (oscilloscopes). Blanking by means of a deflection structure in the cathode-ray tube electron gun that traps the electron beam inside the gun to extinguish the spot, permitting blanking during retrace and between sweeps regardless of intensity setting. *See:* **oscillograph.** 185

deflection center (electron-beam tube). The intersection of the forward projection of the electron path prior to deflection and backward projection of the electron path in the field-free space after deflection. 190, 125

deflection coefficient. *See:* **deflection factor.**

deflection defocusing (cathode-ray tube). A fault of a cathode-ray tube characterized by the enlargement, usually nonuniform, of the deflected spot which becomes progressively greater as the deflection is increased. *See also:* **cathode-ray tubes.** 244, 190

deflection factor (1) (inverse sensitivity) (general). The reciprocal of sensitivity. *Note:* It is, for example, often used to describe the performance of a galvanometer by expressing this in microvolts per millimeter (or per division) and for a mirror galvanometer at a specified scale distance,

usually 1 meter. *See also:* **accuracy rating (instrument).** 125

(2) (oscilloscopes). The ratio of the input signal amplitude to the resultant displacement of the indicating spot, for example, volts per division. *See also:* **oscillograph.** 185, 125

(3) (non-real time spectrum analyzer). The ratio of the input signal amplitude to the resultant output indication. The ratio may be in terms of volts (root-mean-square) per division, decibels above (or below) one milliwatt per division, watts per division, or any other specified factor. 68

deflection plane (cathode-ray tubes). A plane perpendicular to the tube axis containing the deflection center. 190

deflection polarity (oscilloscopes). The relation between the polarity of the applied signal and the direction of the resultant displacement of the indicating spot. *Note:* Conventionally a positive-going voltage causes upward deflection or deflection from left to right. *See:* **oscillograph.** 185

deflection sensibility (oscilloscopes). The number of trace widths per volt that can be simultaneously resolved anywhere within the quality area. *See:* **oscillograph.** 185

deflection sensitivity (magnetic-deflection cathode-ray tube and yoke assembly). The quotient of the spot displacement by the change in deflecting-coil current. 190, 125

deflection yoke (television). An assembly of one or more coils, whose magnetic field deflects an electron beam. *See:* **deflecting yoke.** See also: **television.** 163, 178, 125

deflection-yoke pull-back (cathode-ray tubes). (1) color. The distance between the maximum possible forward position of the yoke and the position of the yoke to obtain optimum color purity. **(2) monochrome.** The maximum distance the yoke can be moved along the tube axis without producing neck shadow. 190, 125

deflector (1) (surge arrester). Means for directing the flow of the gas discharge from the vent of the arrester. 2, 299, 62

(2) (acousto-optic). A device which directs a light beam to an angular position in space upon application of an acoustic frequency. 72

deflector plates. *See:* **deflecting electrode.**

defruiter (radar). Equipment which deletes random nonsynchronous unintentional returns in a beacon system. 13

degassing (electron tube) (rectifier). The process of driving out and exhausting occluded and remanent gases within the vacuum tank or tube, anodes, cathode, etcetera, that are not removed by evacuation alone. *See also:* **rectification.** 328

degauss. To neutralize or bias a ship's magnetic polarity by electric means so as to approximate the effect of a nonmagnetic ship. *See:* **degaussing coil; degaussing generator; deperm; magnetic mine; nonmagnetic ship.** 328

degaussing coil. A single conductor or a multiple-conductor cable so disposed that passage of current through it will neutralize or bias the magnetic polarity of a ship or portion of a ship. *Note:* Continuous application of current, with adjustment to suit changes of position or heading of the ship, is required to maintain a degaussed condition. *See also:* **degauss.** 328

degaussing generator. An electric generator provided for the purpose of supplying current to a degaussing coil or coils. *See also:* **degauss.**   328

degeneracy (resonant device). The condition where two or more modes have the same resonance frequency. *See also:* **waveguide.**   179

degenerate gas. A gas formed by a system of particles whose concentration is very great, with the result that the· Maxwell-Boltzmann law does not apply. *Example:* An electronic gas made up of free electrons in the interior of the crystal lattice of a conductor. *See also:* **discharge (gas).**   244, 190

degenerate parametric amplifier. An inverting parametric device for which the two signal frequencies are identical and equal to one-half the frequency of the pump. *Note:* This exact but restrictive definition is often relaxed to include cases where the signals occupy frequency bands that overlap. *See also:* **parametric device.**   277, 191

degeneration. Same as negative feedback. *See also:* **feedback.**   111

degradation failure (reliability). *See:* **failure, degradation.**

degreasing machine. An electrically d6iven machine including high-pressure pump and special cleaning solution for removing grease and oil from underground mine machines as a prevention of mine fires.   328

degree of asymmetry (current at any time). The ratio of the direct-current component to the peak value of the symmetrical component determined from the envelope of the current wave at that time. *Note:* This value is 100 percent when the direct-current component equals the peak value of the symmetrical component.   103, 202

degree of distortion (at the digital interface for binary signals) (data transmission). A measure of the time displacement of the transitions between signal states from their ideal instants. The degree of distortion is generally expressed as a percentage of the unit interval.   59

degrees of freedom (1) (mesh basis). *See:* **nullity.**
(2) (node basis). *See:* **rank.**
(3) (gyro). An allowable mode of angular motion of the spin axis with respect to the case. The number of degrees-of-freedom is the number of orthogonal axes about which the spin axis is free to rotate. All gyros are either single-degree-of-freedom (SDF) or two-degree-of-freedom (TDF).   46

degree of gross start-stop distortion (data transmission). The degree of start-stop distortion determined using the unit interval which corresponds to the actual mean modulation rate of the signal involved.   59

degree of individual distortion (of a particular signal transition) (data transmission). The ratio to the unit interval of the displacement, expressed algebraically, of this transition from its ideal instant. *Note:* This displacement is considered positive when the transition occurs after its ideal instant (late).   59

degree of isochronous distortion (data transmission). A ratio to the unit interval of the maximum measured difference, irrespective of sign, between the actual and the theoretical intervals separating any two transitions of modulation (or of restitution), these transitions being not necessarily consecutive. The b-algebraical difference between the highest and lowest value of individual distortion

affecting the transitions of isochronous modulation. *Notes:* (1) This difference is independent of the choice of the reference ideal instant. (2) The degree of distortion (of an isochronous modulation or restitution) is usually expressed as a percentage.   59

degree of start-stop distortion (data transmission). The ratio to the unit interval of the maximum measured difference, irrespective of sign, between the actual interval and the theoretical interval (the appropriate integral multiple of unit intervals) separating any transition from the start transition preceding it.   59

deionization time (gas tube). The time required for the grid to regain control after anode-current interruption. *Note:* To be exact the deionization time of a gas tube should be presented as a family of curves relating such factors as condensed-mercury temperature, anode and grid currents, anode and grid voltages, and regulation of the grid current. *See:* **gas tubes.**   190, 125

deionizing grid (gas tubes). A grid accelerating deionization in its vicinity in a gas-filled valve or tube, and forming a screen between two regions within the envelope.   244, 190

delay. (1) The amount of time by which an event is retarded.   255, 77
(2) The amount of time by which a signal is delayed. *Note:* It may be expressed in time (milliseconds, microseconds, etcetera) or in number of characters (pulse times, word times, major cycles, minor cycles, etcetera).   235

delay, absolute (loran). *See:* **absolute delay.**

delay angle (α) (thyristor converter). The time, expressed in degrees (1 cycle of the ac waveform = 360°), by which the starting point of commutation is delayed by phase control in relation to rectifier operation without phase control, including possible inherent delay angle.   121

delay angle, inherent (αp) (thyristor converter). The delay angle which occurs in some connections (for example, 12-pulse connections) in certain operating conditions even where no phase control is applied.   121

delay circuit (pulse techniques). A circuit that produces an output signal that is delayed intentionally with respect to the input signal. *See also:* **pulse.**   117

delay coincidence circuit (pulse techniques). A coincidence circuit that is actuated by two pulses, one of which is delayed by a specified time interval with respect to the other. *See also:* **pulse.**   117

delay dispersion (dispersive delay line). The change in phase delay over a specified operating frequency range.   81

delay distortion (1) (general). Either (A) phase-delay distortion (also called phase distortion) that is either departure from the flatness in the phase delay of a circuit or system over the frequency range required for transmission or the effect of such departure on a transmitted signal, or (B) envelope-delay distortion, that is either departure from flatness in the envelope delay of a circuit or system over the frequency range required for transmission or the effect of such departure on a transmitted signal. *See:* **distortion; distortion, envelope delay.**   59

(2) (facsimile). *See:* **envelope delay distortion.**

delayed application (railway practice). The application of the brakes by the automatic train control equipment after the lapse of a predetermined interval of time following its initiation by the roadway apparatus. *See also:* **automatic**

train control.   328

delayed overcurrent trip. *See:* **delayed release (delayed trip); overcurrent release (overcurrent trip).**

delayed plan-position indicator (radar). A plan-position indicator in which the initiation of the time base is delayed. *See also:* **navigation.**   13, 187

delayed release (delayed trip). A release with intentional delay introduced between the instant when the activating quantity reaches the release setting and the instant when the release operates.   103, 202

delayed sweep (oscilloscopes). (1) A sweep that has been delayed either by a predetermined period or by a period determined by an additional independent variable. (2) A mode of operation of a sweep, as defined above. *See also:* **radar.**   185

delayed test. A service test of a battery made after a specified period of time, which is usually made for comparison with an initial test to determine shelf depreciation. *See also:* **battery (primary or secondary).**   328

delay electric blasting cap. An electric blasting cap with a delay element between the priming and detonating composition to permit firing of explosive charges in sequence with but one application of the electric current. *See also:* **blasting unit.**   328

delay, envelope. *See:* **envelope delay.**

delay equalizer (facsimile). A corrective network that is designed to make the phase delay or envelope delay of a circuit or system substantially constant over a desired frequency range. *See also:* **facsimile transmissions; network analysis.**   210, 178, 12

delaying (operations on a pulse) (pulse terms). A process in which a pulse is delayed in time by active circuitry or by propagation.   254

delaying sweep (oscilloscopes). A sweep used to delay another sweep. *See:* **delayed sweep; oscillograph.**   185

delay line (1) (general). Originally, a device utilizing wave propagation for producing a time delay of a signal.   210

(2) (scintillation counting). Commonly, any real or artificial transmission line or equivalent device designed to introduce delay.   117

(3) (digital computer). A sequential logic element or device with one input channel in which the output-channel state at a given instant t is the same as the input-channel state at the instant $t—n$, that is, the input sequence undergoes a delay of n units. There may be additional taps yielding output channels with smaller values of n. *See:* **acoustic delay line; electromagnetic delay line; magnetic delay line; sonic delay line.** *See also:* **pulse.**   235, 255, 77

(4) (sonics and ultrasonics). A device which operates over some defined range of electrical and environmental conditions as a linear passive circuit element. The transfer characteristic has a modulus and argument (phase) which can be constant or a function of frequency.   81

delay line, digital. A delay line designed specifically to accept digital (video) electrical signals. The signals are specified usually as bipolar, RZ, or NRZ. The definitions are based on the output signal being a doublet generated by an input step function.   81

delay line, dispersive. A delay line which has a transfer characteristic with a constant modulus and an argument (phase) which is a nonlinear function of frequency. The

phase characteristic of devices of common interest is a quadratic function of frequency, but in general may be represented by higher order polynominals and/or other nonlinear functions.   81

delay line, nondispersive. A delay line which nominally has constant delay over a specified frequency band. The argument (phase) of the transfer function is a linear function of frequency.   81

delay-line memory. *See:* **delay-line storage.**

delay-line storage (delay-line memory) (electronic computation). A storage or memory device consisting of a delay line and means for regenerating and reinserting information into the delay line.   210, 54

delay logic function. A sequential logic function wherein each state transition of the input signal causes a single delayed state transition at the output.   88

delay, phase. The ratio of the total phase shift (radians) experienced by a sinusoidal signal in transmission through a system or transducer, to the frequency (radians/second) of the signal. *Note:* The unit of phase delay is the second.   239

delay pickoff (oscilloscopes). A means of providing an output signal when a ramp has reached an amplitude corresponding to a certain length of time (delay interval) since the start of the ramp. The output signal may be in the form of a pulse, a gate, or simply amplification of that part of the ramp following the pickoff time.   185

delay, pulse (1) (general). The time interval by which a pulse is time retarded with respect to a reference time. *Note:* Pulse delay may be of either polarity, but **negative delay** is sometimes known as **pulse advance.** *See also:* **pulse.**   185

(2) (transducer). The interval of time between a specified point on the input pulse and a specified point on the related output pulse. *Notes:* (A) This is a general term which applies to the pulse delay in any transducer, such as receiver, transmitter, amplifier, oscillator, etcetera. (B) Specifications may require illustrations. *See also:* **transducer.**   254, 59, 210

delay pulsing (telephone switching systems). A method of pulsing control and trunk integrity check wherein the sender delays the sending of the address pulses until it receives from the far end an off-hook signal (terminating register not yet attached), followed by a steady on-hook signal (terminating register attached).   55

delay relay. A relay having an assured time interval between energization and pickup or between de-energization and dropout. *See also:* **relay.**   259, 341

delay, signal. *See:* **signal delay.**

delay slope (dispersive delay line). The ratio of the delay dispersion to the dispersive bandwidth.   81

delay switching system (telephone switching systems). A switching system in which a call is permitted to wait until a path becomes available.   55

delay time (1) (railway practice). The period or interval after the initiation of an automatic train-control application by the roadway apparatus and before the application of the brakes becomes effective. *See also:* **automatic train control.**   328

(2) (nondispersive delay line). The transit time of the envelope of an RF tone burst.   81

(3) (thyristor). *See:* **gate controlled delay time (thyristor).**

delimited region (electric-circuit purposes). A space from which, or into which, energy is transmitted by an electric circuit, so chosen that the boundary surface satisfies certain conditions. *Note:* In order that conventional measurements of electric voltage, current, and power may be correctly interpreted, it is necessary: (1) that the surface bounding the region cut each current-carrying conductor that enters the region in a cross-section that is perpendicular to the electric-field intensity within the conductor; (2) that the component of the magnetizing force normal to the boundary surface at every point be constant or negligibly small; (3) that the component of the displacement current normal to the boundary surface at every point (not included within a conductor) be negligibly small; (4) that the paths along which voltages are determined must lie in the boundary surface or be chosen so that equal voltages are obtained. Examples of regions that may be delimited for practical purposes include: a generator, a generating station, a transformer, a substation, a manufacturing establishment, a motor, a lamp, a geographical area, etcetera. *See also:* **network analysis.** 210

delimiter (computing systems). A flag that separates and organizes items of data. *See:* **separator.** 255, 77

Dellinger effect. *See:* **radio fadeout.**

delta. The difference between a partial-select output of a magnetic cell in a ONE state and a partial-select output of the same cell in a ZERO state. *See:* **coincident-current selection.** 331

delta B (ΔB). *See:* **delta induction.**

delta-connected (Δ-connected) circuit. A three-phase circuit that is (simply or completely) mesh connected. *See also:* **network analysis.** 210

delta connection. So connected that the windings of a three-phase transformer (or the windings for the same rated voltage of single-phase transformers associated in a three-phase bank) are connected in series to form a closed circuit. 53

delta-function light source (scintillation counting). A light source whose rise time, fall time, and FWHM (full width at half maximum) are no more than one third of the corresponding parameters of the output pulse of the photomultiplier. 117

delta induction (toroidal magnetic amplifier cores). The change in induction (flux density) when a core is in a cyclically magnetized condition. *Syn.* **delta flux density.** 170

delta, minimum (power supplies). A qualifier, often appended to a percentage specification to describe that specification when the parameter in question is a variable, and particularly when that variable may approach zero. The qualifier is often known as the minimum delta V, or minimum delta I, as the case may be. *See also:* **power supply.** 228, 186

delta network (1) (circuits and systems). A network or that part of a network that consists of three branches connected among three terminals. 67

(2) (electromagnetic compatibility). An artificial mains network of specified symmetric and asymmetric impedance used for two-wire mains operation and comprising resistors connected in delta formation between the two conductors, and each conductor and earth. *See also:* **electromagnetic compatibility.** 220, 197

delta tan delta (Δ tan δ). The increment in the dielectric dissipation factor (tan δ) of the insulation measured at two designated voltages. *Note:* When the values of power factors or dissipation factors are in the 0–0.10 range (see dielectric dissipation factor), the value of delta tan delta may be used as the equivalent of the power-factor tip-up value. *See also:* **power factor tip-up.** 22

demand (1) (electric power utilization). The load integrated over a specified interval of time. *See also:* **alternating-current distribution.** 200

(2) (installation or system). The load at the receiving terminals averaged over a specified interval of time. *Note:* Demand is expressed in kilowatts, kilovolt-amperes, kilovars, amperes, or other suitable units. *See also:* **alternating-current distribution.** 212, 64

demand assigned multiple access (communication satellite). A method of multiple access in which the satellite channels are variably assigned according to traffic offered. 84

demand, billing (electric power utilization). The demand which is used to determine the demand charges in accordance with the provisions of a rate schedule or contract. 112

demand charge (electric power utilization). That portion of the charge for electric service based upon a customer's demand. *See also:* **alternating-current distribution.** 200

demand clause, ratchet (electric power utilization). A clause in a rate schedule which provides that maximum past or present demands be taken into account to establish billings for previous or subsequent periods. *See also:* **alternating-current distribution.** 112

demand, coincident (electric power utilization). Any demand that occurs simultaneously with any other demand; also the sum of any set of coincident demands. *See also:* **alternating-current distribution.** 112

demand, contract (electric power utilization). The demand that the supplier of electric service agrees to have available for delivery. *See also:* **alternating-current distribution.** 112

demand deviation (demand meter or register). The difference between the indicated or recorded demand and the true demand, expressed as a percentage of the full-scale value of the demand meter or demand register. *See also:* **demand meter.** 212

demand factor. The ratio of the maximum demand of a system to the total connected load of the system. *Note:* The demand factor of a part of the system may be similarly defined as the ratio of the maximum demand of the part of the system to the total connected load of the part of the system under consideration. *Syn.* **system demand factor.** *See also:* **alternating-current distribution; direct-current distribution.** 64, 3

demand, instantaneous (electric power utilization). The load at any instant. 112

demand, integrated (electric power utilization). The demand integrated over a specified period. 112

demand interval (demand meter or register). The length of the interval of time upon which the demand measurement is based. *Note:* The demand interval of a block-interval demand meter is a specific period of time such as 15, 30, or 60 minutes during which the electric energy flow is averaged. The demand interval of a lagged-demand meter is the time required to indicate 90 percent of the full value of a constant load suddenly applied. Some meters record the highest instantaneous load. *See also:* **demand meter; electricity meter (meter).**   212, 200

demand-interval deviation (demand meter or register). The difference between the actual (measured) demand interval and the rated demand interval, expressed as a percentage of the rated demand interval. *See also:* **demand meter.**
                    212

demand meter. A device that indicates or records the demand or maximum demand. *Note:* Since demand involves both an electric factor and a time factor, mechanisms responsive to each of these factors are required as well as in indicating or recording mechanism. These mechanisms may be either separate from or structurally combined with one another. Demand meters may be classified as follows: Class 1, curve-drawing meters; Class 2, integrated-demand meters; Class 3, lagged-demand meters.

demand, native system (electric power utilization). The net 60 minute clock-hour peak integrated demand within the system less interruptible loads. *See also:* **alternating-current distribution.**   112

demand, noncoincident (electrical power utilization). The sum of the individual maximum demands regardless of time of occurrence within a specified period. *See also:* **alternating-current distribution.**   112

demand register. A mechanism, intended for mounting in an integrating electricity meter, that indicates maximum demand and that also registers electric energy (or other integrated quantity). *See also:* **demand meter.**   212

demand-totalizing relay. A device designed to receive and totalize electric pulses from two or more sources for transmission to a demand meter or to another relay. *See also:* **demand meter.**   212

demarcation potential (demarcation current*) (current of injury*) (electrobiology). *See:* **injury potential.**

* Deprecated

demarcation strip (data transmission). The terminals at which the telephone company's service ends and the customers' equipment is connected.   59

demineralization. The process of removing dissolved minerals (usually by chemical means).   221

demodulation. A modulation process wherein a wave resulting from previous modulation is employed to derive a wave having substantially the characteristics of the original modulating wave. *Note:* The term is sometimes used to describe the action of a frequency converter or mixer, but this practice is deprecated.   328

demodulator. A device to effect the process of demodulation. *See also:* **demodulation.**   242

demonstration (nuclear power generating stations). A course of reasoning showing that a certain result is a consequence of assumed premises; an explanation or illustration, as in teaching by use of examples.   31

densitometer. A photometer for measuring the optical density (common logarithm of the reciprocal of the transmittance) of materials. *See also:* **photometry.**
                    167

density (1) (facsimile). A measure of the light-transmitting or -reflecting properties of an area. *Notes:* (A) It is expressed by the common logarithm of the ratio of incident to transmitted or reflected light flux. (B) There are many types of density that will usually have different numerical values for a given material; for example, diffuse density, double diffuse density, specular density. The relevant type of density depends upon the geometry of the optical system in which the material is used. *See also:* **scanning.**   12

(2) (electron or ion beam). The density of the electron or ion current of the beam at any given point.   244, 190

(3) (computing systems). *See:* **packing density.**

density coefficient. *See:* **environmental coefficient.**

density-modulated tube (space-charge-control tube) (microwave tubes). Microwave tubes or valves characterized by the density modulation of the electron stream by a gating electrode. *Note:* The electron stream is collected on those electrodes that form a part of the microwave circuit, principally the anode. These electrodes are often small compared to operating wavelength so that for this reason space-charge-control tubes or valves are often not considered to be microwave tubes even though they are used at microwave frequencies.   190

density modulation (electron beam). The process whereby a desired time variation in density is impressed on the electrons of a beam. *See also:* **velocity-modulated tube.**
                    244, 190

density-tapered array antenna. *See:* **space-tapered array antenna.**

denuder. That portion of a mercury cell in which the metal is separated from the mercury.   328

dependability (relay or relay system). The facet of reliability that relates to the degree of certainty that a relay or relay system will operate correctly.   127, 103

dependent contact. A contacting member designed to complete any one of two or three circuits, depending on whether a two- or a three-position device is considered.
                    328

dependent manual operation. An operation solely by means of directly applied manual energy, such that the speed and force of the operation are dependent upon the action of the attendant.   103, 202

dependent node (network analysis). A node having one or more incoming branches. *See also:* **linear signal flow graphs.**   282

dependent power operation. An operation by means of energy other than manual, where the completion of the operation is dependent upon the continuity of the power supply (to solenoids, electric or pneumatic motors, etc).
                    103

deperm. To remove, as far as practicable, the permanent magnetic characteristic of a ship's hull by powerful external demagnetizing coils. *See also:* **degauss.**   328

depletion region (semiconductor) (semiconductor radiation detectors). A region in which the charge-carrier charge density is insufficient to neutralize the net fixed charge density of donors and acceptors. In a diode-type semiconductor radiation detector the depletion region is the sensitive region of the device.   23

depolarization (1) (electrochemistry). A decrease in the polarization of an electrode at a specified current density. *See also:* **electrochemistry.**   328

(2) (medical electronic biology). A reduction of the voltage between two sides of a membrane or interface below an

initial value. *See also:* **contact potential; electrochemistry.**
   192

(3) (corrosion). Reduction or elimination of polarization by physical means or by change in environment.
   221, 205

(4) (antenna). The conversion of power from a reference polarization into the cross polarization. 111

depolarization front (medical electronics). The border of a wave of electric depolarization, traversing an excitable tissue that has appreciable width and thickness as well as length. *See also:* **contact potential.** 192

depolarizer (1). A substance or a means that produces depolarization. 328

(2) (primary cell). A cathodic depolarizer that is adjacent to or a part of the positive electrode. *See also:* **electrochemistry; electrolytic cell.** 328

depolarizing mix (primary cell). A mixture containing a depolarizer and a material to improve conductivity. *See also:* **electrolytic cell.** 328

deposit attack (deposition corrosion). Pitting corrosion resulting from deposits on a metallic surface. 221, 205

deposited-carbon resistor. A resistor containing a thin coating of carbon deposited on a supporting material.
   341

deposition corrosion. *See:* **deposit attack.**

depot maintenance. *See:* **maintenance, depot.**

depth control (cable plowing). The means used to maintain a predetermined plowing depth. 52

depth of current penetration (induction-heating usage). The thickness of a layer extending inward from the surface of a conductor which has the same resistance to direct current as the conductor as a whole has to alternating current of a given frequency. *Note:* About 87 percent of the heating energy of an alternating current is dissipated in the so-called **depth of penetration.** *See also:* **dielectric heating; induction heating.** 14, 114

depth of heating (dielectric-heating usage). The depth below the surface of a material in which effective dielectric heating can be confined when the applicator electrodes are applied adjacent to one surface only. *See also:* **dielectric heating.** 14, 114

depth of modulation (amplitude, frequency, and pulse modulation). The ratio of the maximum minus minimum light intensity to the sum of the maximum and minimum light intensity, namely: $m = (I_{max} - I_{min})/(I_{max} + I_{min})$. This applies to either the diffracted or the zero order. 72

depth of penetration (induction-heating usage). *See:* **depth of current penetration.**

depth of velocity modulation (electron beams). The ratio of (1) the amplitude of a stated frequency component of the varying velocity of an electron beam, to (2) the average beam velocity. 190, 125

derating (reliability). The intentional reduction of stress/strength ratio in the application of an item, usually for the purpose of reducing the occurrence of stress-related failures. *See also:* **reliability.** 182

derivative action gain (automatic control). The ratio of maximum gain resulting from proportional plus derivative control action to the gain due to proportional action alone. *Note:* Derivative action gain has been called "rate gain."
   56

derivative control action (electric power systems). *See:* **control action, derivative.**

derivative time (control systems) (speed governing of hydraulic turbines). The derivative time, T_n, of a derivative element is also the derivative gain, G_n. The derivative gain is the ratio of the element's percent output to the time derivative of the element's percent input (input slope with respect to time). 58, 8

derived envelope (loran-C). The waveform equivalent to the summation of the video envelope of the radio-frequency received pulse and the negative of its derivative, in proper proportion. *Note:* The resulting envelope has a zero crossing at a standard point (for example, 25 microseconds) from the pulse beginning, serving an accurate reference point for envelope time-difference measurements and as a gating point in rejecting the latter part of the received pulse, which may be contaminated by sky-wave transmissions. *See also:* **navigation.** 187, 13

derived pulse (loran-C). A pulse derived by summing the received radio-frequency pulse and an oppositely phased radio-frequency pulse so that it has an envelope that is the derivative of the received radio-frequency pulse envelope. *Note:* The resultant envelope has a zero point and a radio-frequency phase reversal at a standard interval (for example, 25 microseconds) from the pulse beginning and it serves as an accurate reference for cycle time-difference measurements and as a gating point in rejecting the latter part of the received pulse, which may be contaminated by skywave transmissions. *See also:* **navigation.** 187, 13

derived reference pulse waveform (pulse measurement). A reference pulse waveform which is derived by a specified procedure or algorithm from the pulse waveform which is being analyzed in a pulse measurement process. 15

derived units. The derived units of a system of units are those that are not selected as basic but are derived from the basic units. *Note:* A derived unit is defined by specifying an operational procedure for its realization. 210

derrick. An apparatus consisting of a mast or equivalent members held at the top by guys or braces, with or without a boom, for use with a hoisting mechanism and operating ropes. *See also:* **elevators.** 328

descending node (communication satellite). The point on the line of nodes that the satellite passes through as the satellite travels from above to below the equatorial plane.
   74

descriptive adjectives (pulse terms). (1) major (minor). Having or pertaining to greater (lesser) importance, magnitude, time, extent, or the like, than another similar feature(s). 254

(2) ideal. Of or pertaining to perfection in, or existing as a perfect exemplar of, a waveform or a feature. 254

(3) reference. Of or pertaining to a time, magnitude, waveform, feature, or the like which is used for comparison with, or evaluation of, other times, magnitudes, waveforms, features, or the like. A reference entity may, or may not, be an ideal entity. 254

design. *See:* **logic design.**

design *A* motor. An integral-horsepower polyphase squirrel-cage induction motor designed for full-voltage starting with normal values of locked-rotor torque and breakdown torque, and with locked-rotor current higher than that specified for design *B, C,* and *D* motors.   63

designation (radar). Identification and location of a particular target.   13

design basis earthquake (DBE) (nuclear power generating stations). That earthquake producing the maximum vibratory ground motion that the nuclear power generating station is designed to withstand without functional impairment of those features necessary to shut down the reactor, maintain the station in a safe condition, and prevent undue risk to the health and safety of the public.   31

design basis event conditions (nuclear power generating stations). Conditions calculated to occur as a result of the design basis events.   31, 104

design basis events (DBE) (nuclear power generating stations). (1) Postulated (abnormal) events (specified by the safety analysis of the station) used in the design to establish performance requirements of structures, systems, and components.   20, 26

(2) Postulated abnormal events used in the design to establish the acceptable performance requirements of the structures and systems, and components.   356

design *B* motor. An integral-horsepower polyphase squirrel-cage induction motor, designed for full-voltage starting, with normal locked-rotor and breakdown torque and with locked-rotor current not exceeding specified values.   63

design *C* motor. An integral-horsepower polyphase squirrel-cage induction motor, designed for full-voltage starting with high locked-rotor torque and with locked-rotor current not exceeding specified values.   63

design current (glow lamp). The value of current flow through the lamp upon which rated-life values are based.   283

design *D* motor. An integral-horsepower polyphase squirrel-cage induction motor with rated-load slip of at least 5 percent and designed for full-voltage starting with locked-rotor torque at least 275 percent of rated-load torque and with locked-rotor currents not exceeding specified values.   63

designed availability (nuclear power generating stations). The probability that an item will be operable when needed as determined through the design analyses.   31, 75

design letters. Terminology established by the National Electrical Manufacturers Association to describe a standard range of characteristics. The characteristics covered are slip at rated load, locked-rotor and breakdown torque, and locked-rotor current.   63

design life (nuclear power generating stations). The time during which satisfactory performance can be expected for a specific set of service conditions.   31

design limits. Design aspects of the instrument in terms of certain limiting conditions to which the instrument may be subjected without permanent physical damage or impairment of operating characteristics. *See also:* **instrument.**   295

design *L* motor. An integral-horsepower single-phase motor, designed for full-voltage starting with locked-rotor current not exceeding specified values, which are higher than those for design *M* motors.   63

design load (nuclear power generating stations). That combination of electrical loads, having the most severe power demand characteristic, which is provided with electric energy from a diesel-generator unit for the operation of engineered safety features and other systems required during the following shutdown of the reactor.   31

design *M* motor. An integral-horsepower single-phase motor, designed for full-voltage starting with locked-rotor current not exceeding specified values, which are lower than those for design *L* motors.   63

design *N* motor. A fractional-horsepower single-phase motor, designed for full-voltage starting with locked-rotor current not exceeding specified values, which are lower than those for design *O* motors.   63

design *O* motor. A fractional-horsepower single-phase motor, designed for full-voltage starting with locked-rotor current not exceeding specified values, which are higher than those for design *N* motors.   63

design tests (1) (general). Those tests made to determine the adequacy of the design of a particular type, style or model of equipment or its component parts to meet its assigned ratings and to operate satisfactorily under normal service conditions or under special conditions if specified. *Note:* Design tests are made only on representative apparatus to substantiate the ratings assigned to all other apparatus of basically the same design. These tests are not intended to be used as a part of normal production. The applicable portion of these design tests may also be used to evaluate modifications of a previous design and to assure that performance has not been adversely affected. Test data from previous similar designs may be used for current designs, where appropriate.   103, 53

(2) **(surge arrester).** Tests made by the manufacturer on each design to establish the performance characteristics and to demonstrate compliance with the appropriate standards of the industry. Once made they need not be repeated unless the design is changed so as to modify performance.   2

(3) **(cable termination).** Tests made by the manufacturer to obtain data for design or application, or to obtain information on the performance of each type of high-voltage cable termination.   4

(4) **(metal enclosed bus).** Those tests made to determine the adequacy of a particular type, style, or model of metal-enclosed bus or its component parts to meet its assigned ratings and to operate satisfactorily under normal service conditions or under special conditions if specified. *See:* General note above.   78

(5) **(power cable joints).** Tests made on typical joint designs to obtain data to substantiate the design. These tests are of such nature that after they have once been made, they need not be repeated unless significant changes are made in the material or design which may change the performance of the joint.   34

design voltage. The voltage at which the device is designed to draw rated watts input. *See also:* **appliance outlet.**   263

desired track (electronic navigation). *See:* **course line.**

desired value. *See:* **value, ideal.**

deskstand. A movable pedestal or stand (adapted to rest on a desk or table) that serves as a mounting for the transmitter of a telephone set and that ordinarily includes a hook for supporting the associated receiver when not in use. *See also:* **telephone station.** 328

deskstand telephone set. A telephone set having a deskstand. *See also:* **telephone station.** 328

despun antenna. On a rotating vehicle, an antenna whose beam is scanned such that, with respect to a point external to the antenna, the beam is stationary. 111

destination code (telephone switching systems). A combination of digits providing a unique termination address in a communication network. 55

destination-code routing (telephone switching systems). The means of using the area and office codes to direct a call to a particular destination regardless of its point of origin. 55

destructive read (computing systems). A read process that also erases the data in the source. 255, 77

destructive reading (charge-storage tubes). Reading that partially or completely erases the information as it is being read. *See also:* **charge-storage tube.** 174, 190

destructive testing (test, measurement and diagnostic equipment). (1) Prolonged endurance testing of equipment or a specimen until it fails in order to determine service life or design weakness; and (2) Testing in which the preparation of the test specimen or the test itself may adversely affect the life expectancy of the unit under test or render the sample unfit for its intended use. 54

detail contrast (television). *See:* **resolution response.**

detailed-billed call (telephone switching systems). A call for which there is a record including the calling and called line identities which will appear in a customer's billing statement. 55

detailed-record call (telephone switching systems). A call for which there is a record including the calling and called line identities that may be used in the billing process as well as for other purposes. 55

detectability factor (1) (radar). In pulsed radar, the ratio of single-pulse signal energy to noise power per unit bandwidth that provides stated probabilities of detection and false alarm, measured in the intermediate-frequency amplifier and using an intermediate-frequency filter matched to the single pulse, followed by optimum video integration. 13

(2) (continuous-wave radar). The ratio of single-look signal energy to noise power per unit bandwidth, using a filter matched to the time on target. 187

detectable failures (nuclear power generating station). Detectable failures are those that will be identified through periodic testing or will be revealed by alarm or anomolous indication. Component failures which are detected at the channel or system level are detectable failures. 356

detecting element. *See:* **element, primary detecting.**

detecting means. The first system element or group of elements that responds quantitatively to the measured variable and performs the initial measurement operation. The detecting means performs the initial conversion or control of measurement energy. *See also:* **instrument.** 295

detection. (1) Determination of the presence of a signal. (2) Demodulation. The process by which a wave corre-

sponding to the modulating wave is obtained in response to a modulated wave. *See also:* **linear detection; power detection; square-law detection.** 111, 242, 59

detection probability (radar). The probability that the signal, when present, will be detected, when a decision is made as to whether signal plus noise was present, or noise alone. 13

detector (1) (FM broadcast receiver). (A) A device to effect the process of detection.
(B) A mixer in a superheterodyne receiver. *Note:* In definition (B), the device is often referred to as a first detector and the device is not used for detection as defined above. 339

(2) (electromagnetic energy). A device for the indication of the presence of electromagnetic fields. *Note:* In combination with an instrument, a detector may be employed for the determination of the complex field amplitudes. *See also:* **auxiliary device to an instrument.** 185

(3) (nuclear power generating stations). Any device for converting radiation flux to a signal suitable for observation and measurement. 31

detector, average. *See:* **average detector.**

detector geometry (semiconductor radiation detectors). The physical configuration of a solid-state detector. 23, 118, 119

detector, totally depleted. *See:* **totally depleted detector.**

detector, transmission. *See:* **transmission detector.**

determinant (network) (1) (branch basis). The determinant formed by the coefficients of the branch currents after the application of Kirchhoff's two laws. This determinant is not necessarily symmetrical in the case of a passive network.

(2) (mesh or loop basis). The determinant formed by the impedance coefficients of the mesh or loop currents in a complete set of mesh or loop equations.

(3) (node basis). The determinant formed by the (admittance) coefficients of the node voltages in a complete set of node equations. *See also:* **network function.** 210

(4) (circuits and systems). A square array of numbers or elements bordered on either side by a straight line. The value of the determinant is a function of its elements. 67

developer (electrostatography). A material or materials that may be used in development. *See also:* **electrostatography.** 236, 191

development (electrostatography). The act of rendering an electrostatic image viewable. *See also:* **electrostatography.** 236, 191

deviation (automatic control). Any departure from a desired or expected value or pattern. 56, 219, 206

deviation, frequency. *See:* **frequency deviation.**

deviation distortion. Distortion in a frequency-modulation receiver caused by inadequate bandwidth, inadequate amplitude-modulation rejection, or inadequate discriminator linearity. *See also:* **distortion.** 59

deviation factor (wave) (rotating machinery). The ratio of the maximum difference between corresponding ordinates of the wave and of the equivalent sine wave to the maximum ordinate of the equivalent sine wave when the waves are superposed in such a way as to make this maximum difference as small as possible. *Note:* The equivalent sine wave is defined as having the same frequency and the same

root-mean-square value as the wave being tested. *See:* **direct-axis synchronous impedance (rotating machinery).**
63

deviation integral, absolute (automatic control). The time integral of the absolute value of the system deviation following a stimulus specified as to location, magnitude, and time pattern. *Note:* The stimulus commonly employed is a step input.   56

deviation ratio (frequency-modulation system). The ratio of the maximum frequency deviation to the maximum modulating frequency of the system. *See also:* **frequency modulation.**   111, 242

deviation sensitivity (1) (frequency-modulation receivers). The least frequency deviation that produces a specified output power.   339
(2) (radar). The rate of change of course indication with respect to the change of displacement from the course line. *See also:* **navigation.**   13, 187

deviation, steady-state (control). The system deviation after transients have expired. *Note:* For the purpose of this definition, drift is not considered to be a transient. *See also:* **deviation.**   56, 219, 206, 329

deviation, system (control). The instantaneous value of the ultimately controlled variable minus the command. *Note:* The use of system error to mean a system deviation with its sign changed is deprecated. *Syn.* **system overshoot.** *See also:* **deviation.**   56, 219, 206, 105

deviation, transient (control). The instantaneous value of the ultimately controlled variable minus its steady-state value. *Syn.* **transient overshoot.** *See also:* **deviation.**
56, 219, 206, 105

device (electric system) (1). A mechanical or an electric contrivance to serve a useful purpose. *See also:* **absolute-value device; control; storage device.**   210
(2) A smallest subdivision of a system that still has a recognizable function of its own.   59
(3) A unit of an electric system that is intended to carry but not utilize electric energy.   210, 256, 178, 124
(4) (electric). An item of electric equipment that is used in connection with, or as an auxiliary to, other items of electric equipment. For example, thermostat, relay, push button or switch, or instrument transformers. *See also:* **control.**   260
(5) (nuclear power generating stations). An item of electric equipment that is used in connection with, or as an auxiliary to, other items of electric equipment.   28, 31

device rise time (DRT) (photomultipliers for scintillation counting). The mean time difference between the 10- and 90-percent amplitude points on the output waveform for full cathode illumination and delta-function excitation. DRT is measured with a repetitive delta-function light source and a sampling oscilloscope. The trigger signal for the oscilloscope may be derived from the device output pulse, so that light sources such as the scintillator light source may be employed.   117

dezincification. Parting of zinc from an alloy (parting is the preferred term). *Note:* Other terms in this category, such as denickelification, dealuminification, demolybdenization, etcetera, should be replaced by the term parting. *See:* **electrometallurgy.**   221, 205

DF. *See:* **direction-finder; radio direction-finder.**

DF noise level. In the absence of the desired signals, the average power or rms voltage at any specified point in a direction finder system circuit. *Note:* In radio-frequency and audio channels, the direction finding noise level is usually measured in terms of the power dissipated in suitable termination. In a video channel, it is customarily measured in terms of voltage across a given impedance, or of the cathode-ray deflection.   13

DF sensitivity. That field strength at the DF antenna, in microvolts per meter, which produces a ratio of signal-plus-noise to noise equal to 20 decibels in the receiver output, the direction of arrival of the signal being such as to produce maximum pickup in the direction finding antenna system.   13

dg. *See:* **decilog.**

diagnostic. Pertaining to the detection and isolation of either a malfunction or mistake. *See also:* **check problem.**
235, 255, 77

diagnostic routine (1) (computer). A routine designed to locate either a malfunction in the computer or a mistake in coding. *See also:* **programmed check.**   210
(2) (test, measurement and diagnostic equipment). A logical sequence of tests designed to locate a malfunction in the unit under test.   54

diagnostic test (test, measurement and diagnostic equipment). A test performed for the purpose of isolating a malfunction in the unit under test or confirming that there actually is a malfunction.   54

diagram. *See:* **block diagram; logic diagram; Venn diagram.**

dial (1) (industrial control). A plate or disc, suitably marked, that serves to indicate angular position, as for example the position of a handwheel.   206
(2) (automatic switching). A type of calling device used in automatic switching that, when wound up and released, generates pulses required for establishing connections.
192

dialing (telephone switching systems). The act of using a calling device.   55

dialing pattern (telephone switching systems). The implementation of a numbering plan with reference to an individual automatic exchange.   55

dial-mobile telephone system (mobile communication). A mobile communication system that can be interconnected with a telephone network by dialing, or a mobile communication system connected on a dial basis with a telephone network. *See also:* **mobile communication system.**
181

dial pulsing (telephone switching systems). A means of pulsing consisting of regular, momentary interruptions of a direct or alternating current path at the sending end in which the number of interruptions corresponds to the value of the digit or character.   55

dial telephone set. A telephone set equipped with a dial. *See also:* **telephone station.**   328

dial tone (telephone switching systems). The tone that indicates that the switching equipment is ready to receive signals from a calling device.   55

dial train (register). All the gear wheels and pinions used to interconnect the dial pointers. *See also:* **watt-hour meter.**   328

diamagnetic material. A material whose relative permeability is less than unity.   210

diametric rectifier circuit. A circuit that employs two or more rectifying elements with a conducting period of 180 electrical degrees plus the commutating angle. *See also:* **rectification.**   328

diamond winding (rotating machinery). A distributed winding in which the individual coils have the same shape and coil pitch.   63

diaphragm (electrolytic cells). A porous or permeable membrane separating anode and cathode compartments of an electrolytic cell from each other or from an intermediate compartment for the purpose of preventing admixture of anolyte and catholyte. *See also:* **electrolytic cell.**   328

diathermy (medical electronics). The therapeutic use of alternating currents to generate heat within some part of the body, the frequency being greater than the maximum frequency for neuromuscular response.   192

dibit (data transmission). Two bits; two binary digits.   59

dichotomizing search. *See:* **binary search.**

dichromate cell. A cell having an electrolyte consisting of a solution of sulphuric acid and a dichromate. *See also:* **electrochemistry.**   328

dielectric (surge arresters). A medium in which it is possible to maintain an electric field with little or no supply of energy from outside sources.   210, 62

dielectric absorption. A phenomenon that occurs in imperfect dielectrics whereby positive and negative charges are separated and then accumulated at certain regions within the volume of the dielectric. This phenomenon manifests itself usually as a gradually decreasing current with time after the application of a fixed direct voltage.   210

dielectric attenuation constant. Per unit of length, the reciprocal of the distance that a plane electromagnetic wave travels through a dielectric before its amplitude is reduced to $1/e$ of the original amplitude, where e is the base of the natural logarithms. *Note:* The dielectric attenuation constant is related to the dielectric dissipation factor by the equation

$$\alpha = \frac{2\pi}{\lambda}\left[\frac{\epsilon'\mu}{2}\left((1+\tan^2\delta)^{1/2}-1\right)\right]^{1/2}$$

$$(\text{neper}/\text{meter})$$

where ϵ' is the relative dielectric constant, μ is the relative magnetic permeability, λ is the wavelength in a vacuum, and tan δ is the dissipation factor. This expression is valid for dielectric in free space or coaxial transmission line, but must be modified if the wave is in a dielectric filled waveguide.   210

dielectric constant (1) (dielectric). That property which determines the electrostatic energy stored per unit volume for unit potential gradient. *Note:* This numerical value usually is given relative to a vacuum. *See also:* **dielectric heating.**   14

(2) (antenna). *See:* **relative dielectric constant in physical media.**

dielectric dispersion. The phenomenon in which the magnitude of the dielectric constant of a material decreases or increases with increasing frequency.   210

dielectric dissipation factor. (1) The cotangent of the di-

electric phase angle of a dielectric material or the tangent of the dielectric loss angle. *See also:* **dielectric heating.** (2) The ratio of the loss index ϵ'' to the relative dielectric constant ϵ'. *See:* **relative complex dielectric constant.**
                                                           14, 114, 210, 22

dielectric guide. A waveguide in which the waves travel through solid dielectric material. *See also:* **waveguide.**
                                                                 328

dielectric heater. A device for heating normally insulating material by applying an alternating-current field to cause internal losses in the material. *Note:* The normal frequency range is above 10 megahertz. *See also:* **interference.**
                                                                 188

dielectric heating. The heating of a nominally insulating material in an alternating electric field due to its internal dielectric losses.   14, 122

dielectric, imperfect. A dielectric in which a part of the energy required to establish an electric field in the dielectric is not returned to the electric system when the field is removed. *Note:* The energy that is not returned is converted into heat in the dielectric.   210

dielectric lens. A lens made of dielectric material and used for refraction of radio-frequency energy. *See:* **antenna; waveguide.**   244, 179

dielectric loss. The time rate at which electric energy is transformed into heat in a dielectric when it is subjected to a changing electric field. *Note:* The heat generated per unit volume of material upon the application of a sinusoidal electric field is given by the expression

$$W = \frac{5}{9}\,\epsilon''fE^2 \times 10^{-12}\ \text{watt}/\text{centimeter}^3$$

where E is the root-mean-square field strength in volts per centimeter, f is the frequency, and ϵ'' is the loss index.
                                                                 210

dielectric loss angle (1) (general). The angle whose tangent is the dissipation factor, or arc tan ϵ''/ϵ'. *See:* **dielectric dissipation factor; relative complex dielectric constant.**
                                                                 210

(2) (rotating machinery). (δ) The angle whose tangent is the dissipation factor.   22

dielectric loss factor*. *See:* **loss factor.**

* Deprecated

dielectric loss index ϵ'' **(dielectric loss factor*) (homogeneous isotropic material).** The negative of the imaginary part of the relative complex dielectric constant. *See:* **relative complex dielectric constant.**   210

* Deprecated

dielectric, perfect (ideal dielectric). A dielectric in which all of the energy required to establish an electric field in the dielectric is recoverable when the field or impressed voltage is removed. Therefore, a perfect dielectric has zero conductivity and all absorption phenomena are absent. A complete vacuum is the only known perfect dielectric.
                                                                 210

dielectric phase angle. (1) The angular difference in phase between the sinusoidal alternating voltage applied to a dielectric and the component of the resulting alternating current having the same period as the voltage. *See also:* **dielectric heating.** (2) The angle whose contangent is the dissipation factor, or arc cot ϵ''/ϵ'. *See:* **dielectric dissipation factor; relative complex dielectric constant; dielectric heating.** 14, 114, 210, 22

dielectric phase constant. Per unit of length in a dielectric, 2π divided by the wavelength of the electromagnetic wave in the dielectric. *Note:* The phase constant is related to the dielectric constant by the following equation

$$\beta = \frac{2\pi}{\lambda}\left[\frac{\epsilon'\mu}{2}\left[(1 + \tan^2\delta)^{1/2} + 1\right]\right]^{1/2}$$

where λ is the wavelength of the wave in a vacuum, ϵ' is the relative dielectric constant, μ is the relative magnetic permeability, and $\tan\delta$ is the dissipation factor. This expression is valid for dielectric in free space or a coaxial transmission line but must be modified if the wave is in a dielectric-filled waveguide. 210

dielectric power factor. The cosine of the dielectric phase angle (or the sine of the dielectric loss angle). *See also:* **dielectric heating.** 14, 114, 210, 22

dielectric rod antenna. An antenna that employs a shaped dielectric rod as the electrically significant part of a radiating element. *See:* **dielectric lens.** *See also:* **antenna.** 179, 111

dielectric strength (1) (general) (material) (electric strength) (breakdown strength). The potential gradient at which electric failure or breakdown occurs. To obtain the true dielectric strength the actual maximum gradient must be considered, or the test piece and electrodes must be designed so that uniform gradient is obtained. The value obtained for the dielectric strength in practical tests will usually depend on the thickness of the material and on the method and conditions of test. 210, 62 **(2) (dielectric heating).** The maximum potential gradient that the material can withstand without rupture. 14

dielectric tests (1) (transformer) (general). Tests which consist of the application of a voltage higher than the rated voltage, for a specified time to assure the withstand strength of insulation materials and spacing. These various types of dielectric tests have been developed to allow selectivity testing the various insulation components of a transformer, without overstressing other components; or to simulate transient voltages which transformers may encounter in service. *See also:* **applied voltage tests; induced voltage tests; impulse tests.** 53 **(2) (high voltage air switches).** Tests that consist of the application of a standard test voltage for a specified time and are designed to determine the adequacy of insulating materials and spacings. 144 **(3) (neutral grounding device).** Tests that consist of the application of a voltage, higher than the rated voltage, for a specified time to prove compliance with the required voltage class of the device. 91

dielectric waveguide. A waveguide consisting of a dielectric structure. *See also:* **waveguide.** 267, 179

dielectric withstand-voltage tests. Tests made to determine the ability of insulating materials and spacings to withstand specified overvoltages for a specified time without flashover or puncture. *Note:* The purpose of the tests is to determine the adequacy against breakdown of insulating materials and spacings under normal or transient conditions. *See also:* **routine test; service test.**

91, 212, 257, 103, 202, 203, 225, 206, 53

diesel-electric drive (oil-electric drive). A self-contained system of power generation and application in which the power generated by a diesel engine is transmitted electrically by means of a generator and a motor (or multiples of these) for propulsion purposes. *Note:* The prefix diesel-electric is applied to ships, locomotives, cars, buses, etcetera, that are equipped with this drive. *See also:* **electric locomotive.** 328

diesel-generator unit (nuclear power generating stations). The assembly or aggregate of assemblies of one or more single or multiple diesel-engine-generators, associated auxiliary systems, and control, surveillance, and protection systems which make up an individual unit of a diesel-generator standby power supply. 31

difference amplifier. *See:* **differential amplifier.**

difference channel (monopulse radar). (1) A part of a monopulse receiver dedicated to the amplification, filtering and other processing of a "difference" signal, which is generated by comparison of signals received by two (or two sets of) antenna beams, and indicating the departure of the target from the boresight axis. (2) A signal path through a monopulse receiver for processing a difference signal (which is commonly generated by comparison of two or more signals received by two antenna beams or two sets of antenna beams, that is by simultaneous lobing). 13

difference detector. A detector circuit in which the output is a function of the difference of the peak amplitudes or root-mean-square amplitudes of the input waveforms. *See also:* **navigation.** 13, 187

difference frequency (parametric device). The absolute magnitude of the difference between a harmonic nf_p of the pump frequency f_p and the signal frequency f_s, where n is a positive integer. *Note:* Usually n is equal to one. *See also:* **parametric device.** 277, 191

difference-frequency parametric amplifier*. *See:* **inverting parametric device.**

* Deprecated

difference in depth of modulation (DDM) (electronic navigation) (directional systems employing overlapping lobes with modulated signals (such as instrument landing systems)). A fraction obtained by subtracting from the percentage of modulation of the larger signal the percentage of modulation of the smaller signal and dividing by 100. *See also:* **navigation.** 13, 187

difference limen (differential threshold) (just noticeable difference). The increment in a stimulus that is just noticeable in a specified fraction of trials. *Note:* The relative difference limen is the ratio of the difference limen to the absolute magnitude of the stimulus to which it is related. *See also:* **phonograph pickup.** 176

difference pattern (monopulse radar). A description, often given graphically, of the variation in the strength of the signal in a difference channel as a function of the difference in angle between the source of the received signal (an emitter or reflector) and the boresight axis. 13

difference signal. *See:* **differential signal.**

difference slope (monopulse radar). The slope of the difference pattern voltage with respect to target angle from the boresight axis. *Note:* The reference direction is usually the direction that produces zero angle-error voltage, and the difference slope for this point on the curve is the most commonly specified value of the difference slope. 13

differential (photoelectric lighting control). The difference in foot-candles between the light levels for turn-on and turn-off operation. *See also:* **photoelectric control.**
206

differential aeration cell. An oxygen concentration cell. *See:* **electrolytic cell.** 221, 205

differential amplifier (1). An amplifier whose output signal is proportional to the algebraic difference between two input signals. *See also:* **amplifier.** 185
(2) (signal-transmission system). An amplifier that produces an output only in response to a potential difference between its input terminals (differential-mode signal) and in which outputs from common-mode interference voltages on its input terminals are suppressed. *Note:* An ideal differential amplifier produces neither a differential-mode nor a common-mode output in response to a common-mode interference input. *See also:* **amplifier; signal.**
188

differential analyzer (analog computer). A computer designed primarily for the convenient solution of differential equations. 9

differential capacitance (nonlinear capacitor). The derivative with respect to voltage of a charge characteristic, such as an alternating charge characteristic or a mean charge characteristic, at a given point on the characteristic. *See also:* **nonlinear capacitor.** 191

differential-capacitance characteristic (nonlinear capacitor). The function relating differential capacitance to voltage. *See also:* **nonlinear capacitor.** 191

differential capacitance voltage (switching). The difference in magnitudes of the rms system normal frequency line-to-neutral voltage multiplied by the square root of two, with and without the capacitance connected. *Note:* This can be calculated from the equations:

$$\Delta V = \sqrt{2}\, E_S \frac{X_L}{X_C - X_L}$$

or

$$\Delta V = \sqrt{2}\, E_S \frac{\text{kVAR}}{\text{kVA}_{SC} - \text{kVAR}}$$

where

ΔV = differential capacitance voltage in volts
E_S = system phase-to-neutral voltage in volts rms
X_L = source inductive reactance to point of application, in ohms per phase
X_C = capacitive reactance of bank being switched in ohms per phase
kVAR = size of bank being switched (three phase)
kVA$_{SC}$ = system short-circuit kVA at point of capacitor application (symmetrical three phase)

27, 103

differential compounded (rotating machinery). Applied to a compound machine to denote that the magnetomotive forces of the series field winding is opposed to that of the shunt field winding. *See:* **direct-current commutating machine.** 63

differential control. A system of load control for self-propelled electrically driven vehicles wherein the action of a differential field wound on the field poles of a main generator (or of an exciter) and connected in circuit between the main generator and the traction motors, serves to limit the power demand from the prime mover. *See also:* **multiple-unit control.** 328

differential control current (magnetic amplifier). The total absolute change in current in a specified control winding necessary to obtain differential output voltage when the control current is varied very slowly (a quasistatic characteristic). 171

differential control voltage (magnetic amplifier). The total absolute change in voltage across the specified control terminals necessary to obtain differential output voltage when the control voltage is varied very slowly (a quasistatic characteristic). 171

differential dE/dx detector (semiconductor radiation detectors). A transmission detector whose thickness is small compared to the range of the incident particle.
23, 118, 119

differential duplex system. A duplex system in which the sent currents divide through two mutually inductive sections of the receiving apparatus, connected respectively to the line and to a balancing artificial line, in opposite directions so that there is substantially no net effect on the receiving apparatus; whereas the received currents pass mainly through one section, or through the two sections in the same direction, and operate the apparatus. *See also:* **telegraphy.** 328

differential gain (video transmission system). The difference between (1) the ratio of the output amplitudes of a small high-frequency sine-wave signal at two stated levels of a low frequency signal on which it is superimposed, and (2) unity. *Notes:* (A) Differential gain may be expressed in percent by multiplying the above difference by 100. (B) Differential gain may be expressed in decibels by multiplying the common logarithm of the ratio described in (1) above by 20. (C) In this definition, level means a specified position on an amplitude scale applied to a signal waveform. (D) The low- and high-frequency signals must be specified. *See also:* **television.** 306, 178

differential-gain control (gain sensitivity control). A device for altering the gain of a radio receiver in accordance with an expected change of signal level, to reduce the amplitude differential between the signals at the output of the receiver. *See also:* **radio receiver.** 328

differential-gain-control circuit (electronic navigation). The circuit of a receiving system that adjusts the gain of a single radio receiver to obtain desired relative output levels from two alternately applied or sequentially unequal input signals. *Note:* This may be accomplished automatically or manually; if automatic, it is referred to as automatic differential-gain control. Example: Loran circuits that adjust gain between successive pulses from different ground stations. 187, 13

differential-gain-control range. The maximum ratio of signal amplitudes (usually expressed in decibels), at the input of a single receiver, over which the differential-gain-control circuit can exercise proper control and maintain the desired output levels. *See also:* **navigation.**
187, 13

differential gap. *See:* **neutral zone.**

differential-mode-interference (interference terminology). *See:* **interference, differential mode; interference, normal-mode.** *See also:* **interference; accuracy rating (instrument).**

differential nonlinearity (percent) (semiconductor radiation detectors). The percentage departure of the slope of the plot of output versus input from the slope of a reference line. 23

differential nonreversible output voltage. *See:* **differential output voltage.**

differential output current (magnetic amplifier). The ratio of differential output voltage to rated load impedance. 171

differential output voltage (magnetic amplifier) (1) (nonreversible output). The voltage equivalent to the algebraic difference between maximum test output voltage and minimum test output voltage. 171
(2) (reversible output). The voltage equivalent to the algebraic difference between positive maximum test output voltage and negative maximum test output voltage. 171

differential permeability (1) (general). The derivative of the scalar magnitude of the magnetic flux density with respect to the scalar magnitude of the magnetizing force. *Notes:* (A) This term has also been used for the slope of the normal induction curve. (B) In anisotropic media, differential permeability becomes a matrix. 210
(2) (magnetic core testing). The rate of change of the induction with respect to the magnetic field strength.

$$\mu_{dif} = \frac{1}{\mu_0} \frac{dB}{dH}$$

μ_{dif} = relative differential permeability
dH = infinitely small change in field strength
dB = corresponding change induction
165

differential phase (video transmission system). The difference in output phase of a small high-frequency sine-wave signal at the two stated levels of a low-frequency signal on which it is superimposed. *Note:* Notes C and D appended to **differential gain** apply also to **differential phase.** *See also:* **television.** 306, 178

differential phase shift. A change in phase of a field quantity at the output port of a network that is produced by an adjustment of the electrical properties, or characteristics, of the network. *Note:* Differential phase shift may also be the difference between the insertion phase shifts of two 2-port networks. *See also:* **measurement system.** 293, 183

differential-phase-shift keying (modulation systems). A form of phase-shift keying in which the reference phase for a given keying interval is the phase of the signal during the preceding keying interval. 242

differential protection. A method of apparatus protection in which an internal fault is identified by comparing electrical conditions at all terminals of the apparatus. 103, 202, 60, 127

differential relay (1). A relay that by its design or application is intended to respond to the difference between incoming and outgoing electrical quantities associated with the protected apparatus. *See also:* **relay.** 103, 202, 60, 127
(2). A relay with multiple windings that functions when the voltage, current, or power difference between windings reaches a predetermined value. *See also:* **relay.** 259
(3). A relay with multiple windings that functions when the power developed by the individual windings is such that pickup or dropout results from the algebraic summation of the fluxes produced by the effective windings. 341
(4) (industrial control). A relay that operates when the vector difference of two or more similar electrical quantities exceeds a predetermined amount. *Note:* This term includes relays heretofore known as **ratio balance relays, biased relays,** and **percentage differential relays.** 225, 206

differential resistance (semiconductor rectifiers). The differential change of forward voltage divided by a stated increment of forward current producing this change. *See also:* **semiconductor rectifier stack.** 208

differential reversible output voltage. *See:* **differential output voltage.**

differential signal. The instantaneous, algebraic difference between two signals. *See:* **oscillograph.** 185

differential threshold. See: **difference limen.**

differential trip signal (magnetic amplifier). The absolute magnitude of the difference between trip OFF and trip ON control signal. *See also:* **rating and testing magnetic amplifiers.** 171

differentiated (pulse) (pulse amplifier jargon) (semiconductor radiation detectors). A pulse is differentiated when it is passed through a high-pass network, such as a CR filter. 23, 118, 119

differentiating network. *See:* **differentiator.**

differentiator (1) (analog computer). A device producing an output proportional to the derivate of one variable with respect to another, usually time. *See also:* **electronic analog computer.** 9
(2) (electronic computers). A device, usually of the analog type, whose output is proportional to the derivative of an input signal. *See also:* **network analysis.** 210
(3) (electronic circuits). A device whose output function is reasonably proportional to the derivative of the input function with respect to one or more variables, for example, a resistance-capacitance network used to select the leading and trailing edges of a pulse signal. 255, 77
(4) (modulation circuits and industrial control) (differentiating circuit) (differentiating network). A transducer whose output waveform is substantially the time derivative of its input waveform. *Note:* Such a transducer preceding a frequency modulator makes the combination a phase modulator; or following a phase detector makes the combination a frequency detector. Its ratio of output amplitude to input amplitude is proportional to frequency and its output phase leads its input phase by 90 degrees. 210, 206, 111
(5) (relaying). A transducer whose output wave form is substantially the time derivative of its input wave form. 79

diffracted wave (1). When a wave in a medium of certain propagation characteristics is incident upon a discontinuity or a second medium, the diffracted wave is the wave component that results in the first medium in addition to the incident wave and the waves corresponding to the reflected rays of geometrical optics. *See also:* **radiation.** 328

(2) (audio and electroacoustics). A wave whose front has been changed in direction by an obstacle or other nonhomogeneity in a medium, rather than by reflection or refraction. *See also:* **radiation.** 176

diffraction (1) (general). A process that produces a diffracted wave. *See also:* **radiation.** 176

(2) (laser-maser). Deviation of part of a beam, determined by the wave nature of radiation, and occurring when the radiation passes the edge of an opaque obstacle. 363

diffraction angle (acousto-optic device). The angle between the Nth order diffracted beam and the zeroth order beam. It is given by the ratio of the optical wavelength λ_O to the acoustic wavelength, times the order of the diffracted beam $N = \pm 1, \pm 2, \pm 3, \ldots$ so that $\Theta_N = N\lambda_O/\Lambda$. 82

diffraction efficiency (acousto-optic device). For the Nth order, the percent ratio of the light intensity diffracted into the Nth order divided by the light intensity in the zeroth order with the acoustic drive power off, thus

$$\eta_N = (I_N/I_O) \times 100$$

For a device of fixed design, the diffraction efficiency will depend on the optical wavelength, beam diameter, angle of incidence, and acoustic drive power. 82

diffraction loss (laser-maser). That portion of the loss of power in a propagating wave (beam) which is due to diffraction. 363

diffused junction. *See:* **junction, diffused.**

diffused junction detector. A semiconductor detector in which the p-n or n-p junction is produced by diffusion of donor or acceptor impurities. 118, 119

diffused lighting. Lighting that provides on the work plane or on an object, light that is not incident predominantly from any particular direction. *See also:* **general lighting.** 244

diffuser. A device to redirect or scatter the light from a source, primarily by the process of diffuse transmission. 244

diffuse reflectance. The ratio of (1) the flux leaving a surface or medium by diffuse reflection to (2) the incident flux. *See also:* **lamp.** 244

diffuse reflection (1) (illumination engineering). Diffuse reflection is that process by which incident flux is redirected over a range of angles. *See also:* **lamp.** 244

(2) (laser-maser). Change of the spatial distribution of a beam of radiation when it is reflected in many directions by a surface or by a medium. 363

diffuse sound field. A sound field in which the time average of the mean-square sound pressure is everywhere the same and the flow of energy in all directions is equally probable. *See also:* **loudspeaker.** 176

diffuse transmission (illuminating engineering). That process by which the incident flux passing through a surface or medium is scattered. *See also:* **transmission.** 244

diffuse transmission density. The value of the photographic transmission density obtained when the light flux impinges normally on the sample and all the transmitted flux is collected and measured. 176

diffuse transmittance (illuminating engineering). The ratio of the diffusely transmitted flux leaving a surface or medium to the incident flux. *See also:* **transmission (illuminating engineering).** 244

diffusing panel. A translucent material covering the lamps in a luminaire in order to reduce the brightness by distributing the flux over an extended area. *See also:* **bare (exposed) lamp.** 244

diffusing surfaces and media. Those that redistribute some of the incident flux by scattering in all directions. *See also:* **lamp.** 244

diffusion (laser-maser). Change of the spatial distribution of a beam of radiation when it is deviated in many directions by a surface or by a medium. 363

diffusion capacitance (semiconductor junction). The rate of change of stored minority-carrier charge with the voltage across the junction. *See also:* **semiconductor.** 245, 66, 210

diffusion constant (charge carrier) (homogeneous semiconductor). The quotient of diffusion current density by the charge-carrier concentration gradient. It is equal to the product of the drift mobility and the average thermal energy per unit charge of carriers. *See also:* **semiconductor.** 210, 245, 66

diffusion depth (semiconductor radiation detector). *See:* **junction depth.**

diffusion length, charge-carrier (homogeneous semiconductor). The average distance to which minority carriers diffuse between generation and recombination. *Note:* The diffusion length is equal to the square root of the product of the charge-carrier diffusion constant and the volume lifetime. *See also:* **semiconductor.** 186, 245, 66, 210

digit (positional notation) (notation). (1) (A) A character that stands for an integer. (B) Loosely, the integer that the digit stands for. (C) Loosely, any character.

(2) A character used to represent one of the nonnegative integers smaller than the radix, for example, in decimal notation, one of the characters 0 to 9. 255, 77

digit absorption (telephone switching systems). The interpretation and rejection of those digits received, but not required, in the setting of automatic direct control system crosspoints. 55

digital. (1) Pertaining to data in the form of digits. *See:* **analog.** 235, 255, 77, 54

(2) Information in the form of one of a discrete number of codes. 59

digital coefficient attenuator (DCA) (hybrid computer linkage components). Essentially the same as a digital-to-analog multiplier (DAM). This term is generally reserved for those components that are used as the high speed hybrid replacement for manual and servo potentiometers. *Syn.* **digital potentiometer.** 10

digital computer (1) (information processing). A computer that operates on discrete data by performing arithmetic and logic processes on these data. Contrast with analog computer. 255, 77

(2) (test, measurement and diagnostic equipment). A computer in which discrete quantities are represented in digital form and which generally is made to solve mathematical problems by iterative use of the fundamental processes of addition, subtraction, multiplication, and division. 54

digital controller (data processing). A controller that accepts an input sequence of numbers and processes them to produce an output sequence of numbers. 198

digital converter (code translator). A device, or group of devices, that converts an input numerical signal or code of one type into an output numerical signal or code of another type. 103, 202

digital data. Pertaining to data in the form of digits, or integral quantities. Contrast with analog data. *See:* **analog and digital data.** 194

digital device (control equipment). A device that operates on the basis of discrete numerical techniques in which the variables are represented by coded pulses or states. 94

digital differential analyzer (DDA) (analog computer). A special purpose digital computer consisting of many computing elements, all operating in parallel, that performs integration by means of a suitable integration code on incremental quantities and that can be programmed for the solution of differential equations in a manner similar to an analog computer. 9

digital logic elements (analog computer). In an analog computer, a number of digital functional modules, consisting of logic gates, registers, flip-flops, timers, etc., all operating in parallel, either synchronously or asynchronously, and whose inputs and outputs are interconnected, according to a "logic program," via patch cords, on a patch board. 9

digital phase lock loop (communication satellite). A circuit for synchronizing the received waveform, by means of discrete corrections. 83

digital potentiometer. *See:* **digital coefficient attenuator.**

digital readout clock. A clock that gives (usually with visual indication) a voltage or contact-closure pattern of electric circuitry for a readout of time. *Note:* A digital readout calendar clock also includes a readout of day, month, and year, usually also with indication. 103, 202

digital switching (telephone switching systems). Switching of discrete-level information signals. 55

digital telemeter indicating receiver. A device that receives the numerical signal transmitted from a digital telemeter transmitter and gives a visual numerical display of the quantity measured. 103, 202

digital telemetering. Telemetering in which a numerical representation, as for example some form of pulse code, is generated and transmitted; the number being representative of the quantity being measured. 103, 202

digital telemeter receiver. A device that receives the numerical signal transmitted by a digital telemeter transmitter and stores it and/or converts it to a usable form, or both, for such purposes as recording, indication, or control. 103, 202

digital telemeter transmitter. A device that converts its input signal to a numerical form for transmission to a digital telemeter receiver over an interconnecting channel. 103, 202

digital-to-analog converter (DAC) (1) (general). A device, or group of devices, that converts a numerical input signal or code into an output signal some characteristic of which is proportional to the input. 103

(2) (power-system communication). A circuit or device whose input is information in digital form and whose output is the same information in an analog form. *See also:* **analog; digital.** 59

(3) (data processing). A device that converts an input number sequence into a function of a continuous variable. 198

(4) (hybrid computer linkage components) (DAC). A circuit or device whose input is information in digital form and whose output is the same information in analog form. In a hybrid computer, the input is a number sequence (or word) coming from the digital computer, while the output is an analog voltage proportional to the digital number. 10

digital-to-analog multiplier (DAM) (hybrid computer linkage components). A device which provides the means of obtaining the continuous multiplication of a specific digital value with a changing analog variable. The product is represented by a varying analog voltage. *Syn.* **multiplying digital-to-analog converter.** 10

digit deletion (telephone switching systems). In the processing of a call, the elimination of a portion of the destination code. 55

digitize. (1) To express data in a digital form. 255, 77
(2) The conversion of analog to digital data. (A/D). 224, 203, 54

dimension (physical quantity). A convenience label that is determined by the kind of quantity (for example, mass, length, voltage) and that is an element of multiplicative group. *Notes:* (1) The product of any pair of dimensions is a dimension. (2) The unit element of the group is "numeric" or "pure number" which is therefore a dimension. *See:* **incremental dimension; normal dimension; short dimension.** 210

dimensional system. A multiplicative group of dimensions characterized by the independent dimensions generating the group. *Notes:* (1) Derived dimensions are often given individual names (for example, energy and speed, when mass, length, and time are the generators). (2) The basic dimensions and basic units may be chosen independently of each other. In the special case in which the basic dimensions are the dimensions of the quantities to which the basic units are assigned, dimensional analysis indicates the manner in which the measure of a quantity would change with an assumed virtual variation in the basic units. 210

dimensions, critical mating (standard connector). Those longitudinal and transverse dimensions assuring nondestructive mating with a corresponding standard connector. 110

dimming reactor (thyristor). A reactor that may be inserted in a lamp circuit at will for reducing the luminous intensity of the lamp. *Note:* Dimming reactors are normally used to dim headlamps, but may be applied to other circuits, such as gauge lamp circuits. 328

diode (1) (electron tube). A two-electrode electron tube containing an anode and a cathode. *See also:* **equivalent diode.** 190, 125

(2) (semiconductor). A semiconductor device having two terminals and exhibiting a nonlinear voltage-current characteristic; in more-restricted usage, a semiconductor device that has the asymmetrical voltage-current characteristic exemplified by a single *p-n* junction. *See also:*

semiconductor.   245, 66, 210

diode characteristic (multielectrode tube). The composite electrode characteristic taken with all electrodes except the cathode connected together.   190, 125

diode equivalent. The imaginary diode consisting of the cathode of a triode or multigrid tube and a virtual anode to which is applied a composite controlling voltage such that the cathode current is the same as in the triode or multigrid tube.   190, 125

diode fuses (semiconductor rectifiers). Fuses of special characteristics connected in series with one or more semiconductor rectifier diodes to disconnect the semiconductor rectifier diode in case of failure and protect the other components of the rectifier. *Note:* Diode fuses may also be employed to provide coordinated protection in case of overload or short-circuit. *See also:* **semiconductor rectifier stack.**   208

dip (electroplating). A solution used for the purpose of producing a chemical reaction upon the surface of a metal. *See also:* **electroplating.**   328

diplex operation. The simultaneous transmission or reception of two signals using a specified common feature, such as a single antenna or a single carrier. *See also:* **telegraphy.**   328

diplex radio transmission. The simultaneous transmission of two signals using a common carrier wave. *See also:* **radio transmission.**   111

dip needle. A device for indicating the angle between the magnetic field and the horizontal. *See also:* **magnetometer.**   328

dipole. *See:* **dipole antenna; folded dipole antenna; electric dipole; magnetic dipole.**

dipole antenna. Any one of a class of antennas producing a radiation pattern approximating that of an elementary electric dipole. *Note:* Common usage considers the dipole antenna to be a metal radiating structure which supports a line current distribution similar to that of a thin straight wire so energized that the current has a node only at each end. *Syn:* **doublet antenna.**   111

dipole molecule. A molecule that possesses a dipole moment as a result of the permanent separation of the centroid of positive charge from the centroid of negative charge for the molecule as a whole.   210

dip plating. *See:* **immersion plating.**

dip soldering (soldered connections). The process whereby assemblies are brought in contact with the surface of molten solder for the purpose of making soldered connections.   284

direct-acting machine voltage regulator. A machine voltage regulator having a voltage-sensitive element that acts directly without interposing power-operated means to control the excitation of an electric machine.   103, 202

direct-acting overcurrent trip device. *See:* **direct release (series trip)** and **overcurrent release (trip).**

direct-acting recording instrument. A recording instrument in which the marking device is mechanically connected to, or directly operated by, the primary detector. *See also:* **instrument.**   328

direct address (computing systems). An address that specifies the location of an operand. *See:* **one-level address.**   255, 77

direct-arc furnace. An arc furnace in which the arc is

formed between the electrodes and the charge. *See also:* **electrothermics.**   328

direct axis (synchronous machine). The axis that represents the direction of the plane of symmetry of the no-load magnetic-flux density, produced by the main field winding current, normally coinciding with the radial plane of symmetry of a field pole. *See:* **direct-axis synchronous reactance.**   63

direct-axis component of armature current. That component of the armature current that produces a magnetomotive force distribution that is symmetrical about the direct axis.   328

direct-axis component of armature voltage. That component of the armature voltage of any phase that is in time phase with the direct-axis component of current in the same phase. *Note:* A direct-axis component of voltage may be produced by: (1) Rotation of the quadrature-axis component of magnetic flux, (2) Variation (if any) of the direct-axis component of magnetic flux, (3) Resistance drop caused by flow of the direct-axis component of armature current. As shown in the phasor diagram, the direct-axis component of terminal voltage, assuming no field magnetization in the quadrature-axis, is given by

$$\mathbf{E}_{ad} = -R\mathbf{I}_{ad} - jX_q\mathbf{I}_{aq} \qquad 328$$

direct-axis component of magnetomotive force (rotating machinery). The component of magnetomotive force that is directed along the direct axis. *See also:* **asynchronous machine; direct-axis synchronous impedance (rotating machinery); direct-current commutating machine.**   63

direct-axis current (rotating machinery). The current that produces direct-axis magnetomotive force. *See:* **direct-axis synchronous reactance.**   63

direct-axis magnetic-flux component (rotating machinery). The magnetic-flux component directed along the direct axis. *See:* **direct-axis synchronous reactance.**   63

direct-axis subtransient impedance (rotating machinery). The magnitude obtained by the vector addition of the value for armature resistance and the value for direct-axis subtransient reactance. *Note:* The resistance value to be applied in this case will be a function of frequency depending on rotor iron losses.   63

direct-axis subtransient open-circuit time constant. The time in seconds required for the rapidly decreasing component (negative) present during the first few cycles in the direct-axis component of symmetrical armature voltage under suddenly removed symmetrical short-circuit condition, with the machine running at rated speed to decrease to $1/e \approx 0.368$ of its initial value. *Note:* If the rotor is made of solid steel no single subtransient time constant exists but a spectrum of time constants will appear in the subtransient region. *See:* **direct-axis synchronous impedance (rotating machinery).**   63

direct-axis subtransient reactance (rotating machinery). The quotient of the initial value of a sudden change in that fundamental alternating-current component of armature voltage, which is produced by the total direct-axis primary flux, and the value of this simultaneous change in fundamental alternating-current component of direct-axis armature current, the machine running at rated speed. *Note:* The rated current value is obtained from the tests for the rated current value of direct-axis transient reactance. The rated voltage value is that obtained from a sudden short-circuit test at the terminals of the machine at rated armature voltage, no load. *See:* **direct-axis synchronous reactance (rotating machinery).**   63

direct-axis subtransient short-circuit time constant. The time required for the rapidly changing component, present during the first few cycles in the direct-axis alternating component of a short-circuit armature current, following a sudden change in operating conditions, to decrease to $1/e \approx 0.368$ of its initial value, the machine running at rated speed. *Note:* The rated current value is obtained from the test for the rated current value of the direct-axis transient reactance. The rated voltage value is obtained from the test for the rated voltage value of direct-axis transient reactance. *See:* **direct-axis synchronous reactance.**   63

direct-axis subtransient voltage (rotating machinery). The direct-axis component of the terminal voltage which appears immediately after the sudden opening of the external circuit when the machine is running at a specified load, before any flux variation in the excitation and damping circuits has taken place.   63

direct-axis synchronous impedance (synchronous machine) (rotating machinery). The magnitude obtained by the vector addition of the value for armature resistance and the value for direct-axis synchronous reactance.   63

direct-axis synchronous reactance. The quotient of a sustained value of that fundamental alternating-current component of armature voltage that is produced by the total direct-axis flux due to direct-axis armature current and the value of the fundamental alternating-current component of this current, the machine running at rated speed. Unless otherwise specified, the value of synchronous reactance will be that corresponding to rated armature current. For most machines, the armature resistance is negligibly small compared to the synchronous reactance. Hence the synchronous reactance may be taken also as the synchronous impedance.   63

direct-axis transient impedance (rotating machinery). The magnitude obtained by the vector addition of the value for armature resistance and the value for direct-axis transient reactance.   63

direct-axis transient open-circuit time constant. The time in seconds required for the root-mean-square alternating-current value of the slowly decreasing component present in the direct-axis component of symmetrical armature voltage on open-circuit to decrease to $1/e \approx 0.368$ of its initial value when the field winding is suddenly short-circuited with the machine running at rated speed. *See:* **direct-axis synchronous impedance (rotating machinery).**   328

direct-axis transient reactance (rotating machinery). The quotient of the initial value of a sudden change in that fundamental alternating-current component of armature voltage, which is produced by the total direct-axis flux, and the value of the simultaneous change in fundamental alternating-current component of direct-axis armature current, the machine running at rated speed and the high-decrement components during the first cycles being excluded. *Note:* The rated current value is that obtained from a three-phase sudden short-circuit test at the terminals of the machine at no load, operating at a voltage such as to give an initial value of the alternating compo-

nent of current, neglecting the rapidly decaying component of the first few cycles, equal to the rated current. This requirement means that the per-unit test voltage is equal to the rated current value of transient reactance (per unit). In actual practice, the test voltage will seldom result in initial transient current of exactly rated value, and it will usually be necessary to determine the reactance from a curve of reactance plotted against voltage. The rated voltage value is that obtained from a three-phase sudden short-circuit test at the terminals of the machine at rated voltage, no load. *See:* **direct-axis synchronous reactance.** *See also:* **direct-axis synchronous impedance (rotating machinery).**   63

direct-axis transient short-circuit time constant. The time in seconds required for the root-mean-square value of the slowly decreasing component present in the direct-axis component of the alternating-current component of the armature current under suddenly applied symmetrical short-circuit conditions with the machine running at rated speed, to decrease to $1/e \approx 0.368$ of its initial value.

direct-axis transient voltage (rotating machinery). The direct-axis component of the armature voltage that appears immediately after the sudden opening of the external circuit when running at a specified load, the components that decay very fast during the first few cycles, if any, being neglected. *See:* **direct-axis synchronous reactance.**   63

direct-axis voltage (rotating machinery). The component of voltage that would produce direct-axis current when resistance-limited. *See:* **direct-axis synchronous reactance.**   63

direct-buried transformer. A transformer designed to be buried in the earth with connecting cables.   53

direct capacitance (between two conductors). The ratio of the change in charge on one conductor to the corresponding change in potential of the second conductor when the second conductor is the only one whose potential is permitted to change. *See:* **direct capacitances (system of conductors).**   210

direct capacitances (system of conductors). The direct capacitances of a system of n conductors such as that considered in **coefficients of capacitance (system of conductors)** are the coefficients in the array of linear equations that express the charges on the conductors in terms of their differences in potential, instead of potentials relative to ground.

$$Q_1 = O + C_{12}(V_1 - V_2) + C_{13}(V_1 - V_3)$$
$$+ \cdots + C_{1(n-1)}(V_1 - V_{n-1}) + C_{10}V_1$$

$$Q_2 = C_{21}(V_2 - V_1) + O + C_{23}(V_2 - V_3)$$
$$+ \cdots + C_{2(n-1)}(V_2 - V_{n-1}) + C_{20}V_2$$

$$Q_{n-1} = C_{(n-1)1}(V_{n-1} - V_1, + C_{(n-1)2}(V_{n-1}$$
$$- V_2) + \cdots + O + C_{(n-1)0}V_{n-1}$$

with $C_{rp} = C_{pr}$ and C_{re} not involved but defined as zero. *Note:* The coefficients of capacitance c are related to the direct capacitances C as follows

$$c_{rp} = -C_{rp}, \text{ for } r \neq p$$

and

direct component

$$c_{rr} = \sum_{p=1}^{p=n} C_{rp}$$

Note: The relationships of the direct capacitances in a four-conductor system to the other defined capacitances are given in the following table. As in the diagram, the direct capacitances between the conductors of the system, including ground, are indicated as C_{10}, C_{12}, C_{13}, etcetera. Here the subscript 0 is used to denote the ground (nth) conductor. 210

Conductor	Self Capacitance
1	$C_{10} + C_{12} + C_{13}$
2	$C_{20} + C_{21} + C_{23}$
3	$C_{30} + C_{31} + C_{32}$

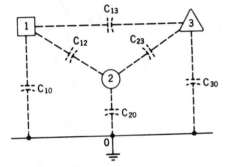

Capacitance diagram showing the equivalent direct capacitance network of a 4-conductor system.

Pair of Conductors: 1 and 2

$$\text{Plenary Capacitance} = C_{12} + \frac{C_{30}(C_{10} + C_{13})(C_{20} + C_{23}) + C_{10}C_{20}(C_{31}C_{32}) + C_{31}C_{32}(C_{10} + C_{20})}{C_{30}(C_{10} + C_{13} + C_{20} + C_{23}) + (C_{10} + C_{20})(C_{31} + C_{32})}.$$

$$\text{Balanced Capacitance} = C_{12} + \frac{(C_{13} + C_{10})(C_{23} + C_{20})}{C_{13} + C_{10} + C_{23} + C_{20}}.$$

Pair of Conductors: 2 and 3

$$\text{Plenary Capacitance} = C_{23} + \frac{C_{10}(C_{20} + C_{21})(C_{30} + C_{31}) + C_{20}C_{30}(C_{31} + C_{21}) + C_{31}C_{21}(C_{20} + C_{30})}{C_{10}(C_{20} + C_{21} + C_{30} + C_{31}) + (C_{20} + C_{30})(C_{31} + C_{21})}.$$

$$\text{Balanced Capacitance} = C_{23} + \frac{(C_{21} + C_{20})(C_{31} + C_{30})}{C_{21} + C_{20} + C_{31} + C_{30}}.$$

Pair of Conductors: 3 and 1

$$\text{Plenary Capacitance} = C_{31} + \frac{C_{20}(C_{10} + C_{12})(C_{30} + C_{32}) + C_{10}C_{30}(C_{32} + C_{12}) + C_{12}C_{32}(C_{10} + C_{30})}{C_{20}(C_{10} + C_{12} + C_{30} + C_{32}) + (C_{10} + C_{30})(C_{12} + C_{32})}.$$

$$\text{Balanced Capacitance} = C_{31} + \frac{(C_{32} + C_{30})(C_{12} + C_{10})}{C_{32} + C_{30} + C_{12} + C_{10}}$$

direct component (illuminating engineering). That portion of the light from a luminaire that arrives at the work plane without being reflected by room surfaces. *See also:* **inverse-square law (illuminating engineering).** 167

direct-connected exciter (rotating machinery). An exciter mounted on or coupled to the main machine shaft so that both machines operate at the same speed. *See also:* **asynchronous machine.** 63

direct-connected system. *See:* **headquarters system.**

direct-coupled amplifier (signal-transmission system). A direct-current amplifier in which all signal connections between active channels are conductive. *See also:* **signal.** 188

direct-coupled attenuation (transmit-receive, pretransmit-receive, and attenuator tubes). The insertion loss measured with the resonant gaps, or their functional equivalent, short-circuited. 333

direct coupling. The association of two or more circuits by means of self-inductance, capacitance, resistance, or a combination of these, that is common to the circuits. *Note:* **Resistance coupling** is the case in which the common branch contains only resistance; **capacitance coupling** that in which the branch contains only capacitance; **direct-inductance** or **self-inductance coupling** that in which the branch contains only self-inductance. *See also:* **coupling.** 210

direct-current (dc). A unidirectional current in which the changes in value are either zero or so small that they may be neglected. (As ordinarily used, the term designates a practically non-pulsating current.) 3

direct-current amplifier (1). An amplifier capable of amplifying waves of infinitesimal frequency. *See also:* **amplifier.** 111

(2) (signal-transmission system). An amplifier capable of producing a sustained single-valued, unidirectional output in response to a similar but smaller input. *Note:* It generally employs between stages either resistance coupling alone or resistance coupling combined with other forms of coupling. *See also:* **amplifier.** 59, 188

direct-current analog computer. *See:* **analog computer.**

direct-current balance (amplifiers). An adjustment to avoid a change in direct-current level when changing gain. *See:* **amplifier.** 185

direct-current balancer (direct-current compensator*). A machine that comprises two or more similar direct-current machines (usually with shunt or compound excitation) directly coupled to each other and connected in series across the outer conductors of a multiple-wire system of distribution, for the purpose of maintaining the potentials of the intermediate conductors of the system, which are connected to the junction points between the machines. *See:* **converter.** 3

* Deprecated

direct-current blocking voltage rating (rectifier circuit element). The maximum continuous direct-current reverse voltage permitted by the manufacturer under stated conditions. *See also: rectifier circuit element.*

237, 66, 208

direct-current circuit. A circuit that includes two or more interrelated conductors intended to be energized by direct current. 210

direct-current commutator machine (rotating machinery). A direct-current machine incorporating an armature winding connected to a commutator and magnetic poles which are excited from a direct-current source or which are permanent magnets. *Note:* Specific types of direct-current commutating machines are: **direct-current generators, motors, synchronous converters, boosters, balancers,** and **dynamotors.** 3

direct-current compensator. *See:* **direct-current balancer.**

direct-current component (total current). That portion of the total current that constitutes the asymmetry.

103, 202

direct-current component of a composite picture signal, blanked picture signal, or picture signal (television). The difference in level between the average value, taken over a specified time interval, and the peak value in the black direction, which is taken as zero. *Note:* The averaging period is usually one line interval or greater. *See:* **television.**

178

direct-current distribution. The supply, to points of utilization, of electric energy by direct current from its point of generation or conversion. 64

direct-current dynamic short-circuit ratio. The ratio of the maximum transient value of a current, after a suddenly applied short circuit, to the final steady-state value.

264

direct-current electric locomotive. An electric locomotive that collects propulsion power from a direct-current distribution system. *See also:* **electric locomotive.** 328

direct-current electron-stream resistance (electron tubes). The quotient of electron-stream potential and the direct-current component of stream current. *See also:* **electrode current (electron tube).** 190, 125

direct-current erasing head (magnetic recording). One that uses direct current to produce the magnetic field necessary for erasing. *Note:* Direct-current erasing is achieved by subjecting the medium to a unidirectional field. Such a medium is, therefore, in a different magnetic state than one erased by alternating current. *See also:* **photograph pickup.** 176

direct-current generator. A generator for production of direct-current power. *See:* **direct-current commutating machine.** 63

direct-current leakage. *See:* **controlled overvoltage test.**

direct-current magnetic biasing (magnetic recording). Magnetic biasing accomplished by the use of direct current. *See also:* **phonograph pickup.** 176

direct-current neutral grid. A network of neutral conductors, usually grounded, formed by connecting together within a given area of all the neutral conductors of a low-voltage direct-current supply system. *See also:* **center of distribution.** 64

direct-current offset (amplifiers). A direct-current level that may be added to the input signal, referred to the input terminals. *See:* **amplifier.** 185

direct-current quadruplex system. A direct-current tele-graph system that affords simultaneous transmission of two messages in each direction over the same line, operation being obtained by superposing neutral telegraph upon polar telegraph. *See also:* **telegraphy.** 328

direct-current rated (thyristor converter). The current in terms of which all test and service current ratings are specified (for example, the per-unit base), except in the case of high-peak loads which are specified in terms of peak load duty. 121

direct-current relay. *See:* **relay, direct-current.**

direct-current restoration (television). The re-establishment of the direct-current and the low-frequency components of a video signal that have been lost by alternating-current transmission. 163

direct-current restorer (1) (general). A means, used in a circuit incapable of transmitting slow variations but capable of transmitting components of higher frequency, by which a direct-current component is reinserted after transmission, and in some cases other low-frequency components are also reinserted. 328

(2) (television). A device for accomplishing direct-current restoration. 163

direct-current self-synchronous system. A system for transmitting angular position or motion, comprising a transmitter and one or more receivers. The transmitter is an arrangement of resistors that furnishes the receiver with two or more voltages that are functions of transmitter shaft position. The receiver has two or more stationary coils that set up a magnetic field causing a rotor to take up an angular position corresponding to the angular position of the transmitter shaft. *See:* **synchro system.** 328

direct-current telegraphy. That form of telegraphy in which, in order to form the transmitted signals, direct current is supplied to the line under the control of the transmitting apparatus. *See also:* **telegraphy.** 328

direct-current transmission (1) (electric energy). The transfer of electric energy by direct current from its source to one or more main receiving stations. *Note:* For transmitting large blocks of power, high voltage may be used such as obtained with generators in series, rectifiers, etcetera. *See also:* **direct-current distribution.** 64

(2) (television). A form of transmission in which the direct-current component of the video signal is transmitted. *Note:* In an amplitude-modulated signal with direct-current transmission, the black level is represented always by the same value of envelope. In a frequency-modulated signal with direct-current transmission, the black level is represented always by the same value of the instantaneous frequency. 163

direct-current winding (rectifier transformer). The secondary winding that is conductively connected to the main electrodes of the rectifier, and that conducts the direct current of the rectifier. *See:* **rectifier transformer.**

203, 53

direct digital control (electric power systems). A mode of control wherein digital computer outputs are used to directly control a process. *See also:* **power system.**

200, 94

direct distance dialing (DDD) (telephone switching systems). The automatic establishing of toll calls in response to signals from the calling device of a customer. 55

direct-drive machine. An electric driving machine the

motor of which is directly connected mechanically to the driving sheave, drum, or shaft without the use of belts or chains, either with or without intermediate gears.   328

directed branch (network analysis). A branch having an assigned direction. *Note:* In identifying the branch direction, the branch jk may be thought of as outgoing from node j and incoming at node k. Alternatively, branch jk may be thought of as originating or having its input at node j and terminating or having its output at node k. The assigned direction is conveniently indicated by an arrow pointing from node j toward node k. *See also:* **linear signal flow graphs.**   282

directed reference flight (electronic navigation). The type of stabilized flight that obtains control information from external signals that may be varied as necessary to direct the flight. For example, flight of a guided missile or a target aircraft. *See also:* **navigation.**   13, 187

direct feeder. A feeder that connects a generating station, substation, or other supply point to one point of utilization. *See:* **radial feeder.** *See also:* **center of distribution.**   64

direct glare. Glare resulting from high brightnesses or insufficiently shielded light sources in the field of view or from reflecting areas of high brightness. *Note:* It usually is associated with bright areas, such as luminaires, ceilings, and windows that are outside the visual task or region being viewed. *See also:* **visual field.**   167

direct grid bias. The direct component of grid voltage. *Note:* This is commonly called grid bias. *See also:* **electrode voltage (electron tube).**   190, 125

direct-indirect lighting. A variant of general diffuse lighting in which the luminaires emit little or no light at angles near the horizontal. *See also:* **general lighting.**
                                                          167

direct inductance coupling. *See:* **inductance coupling (communication circuits).**

direct interelectrode capacitance (electron tubes). The direct capacitance between any two electrodes excluding all capacitance between either electrode and any other electrode or adjacent body.   244, 190

direct inward dialing (DID) (telephone switching systems). A private automatic branch exchange or centrex service feature that permits outside calls to be dialed directly to the stations.   55

direction (navigation). The position of one point in space relative to another without reference to the distance between them. *Notes:* (1) Direction may be either three dimensional or two dimensional, and it is not an angle but is often indicated in terms of its angular difference from a reference direction. (2) Five terms used in navigation: azimuth, bearing, course, heading, and track, involve measurement of angles from reference directions. To specify the reference directions, certain modifiers are used. These are: true, magnetic, compass, relative grid, and gyro. *See also:* **navigation.**   13, 187

directional antenna. An antenna having the property of radiating or receiving radio waves more effectively in some directions than others. *See also:* **antenna.**   111, 179

directional-comparison protection. A form of pilot protection in which the relative operating conditions of the directional units at the line terminals are compared to determine whether a fault is in the protected line section. *Note:* Commonly, the directional units are relays, and the

comparison is made by noting the relative positions of the contacts of the relay directional units. *See also:* **relay.**
                                                          127, 103

directional control (protective relay or relay scheme). A qualifying term that indicates a means of controlling the operating force in a nondirectional relay so that it will not operate until the two or more phasor quantities used to actuate the controlling means (directional relay) are in a predetermined band of phase relations with a reference input.   103, 127

directional coupler (1) (transmission lines). A transmission coupling device for separately (ideally) sampling (through a known coupling loss for measuring purposes) either the forward (incident) or the backward (reflected) wave in a transmission line. *Notes:* (A) Similarly, it may be used to excite in the transmission line either a forward or backward wave. (B) A unidirectional coupler has available terminals or connections for sampling only one direction of transmission; a bidirectional coupler has available terminals for sampling both directions. *See also:* **auxiliary device to an instrument.**   328
(2) (waveguide components). A four port junction consisting of two waveguides coupled together in such a manner that a single traveling wave in either guide will induce a single traveling wave in the other, the direction of the latter wave being determined by the direction of the former.   166

directional-current tripping. *See:* **directional-overcurrent protection; directional-overcurrent relay.**

directional gain directivity index (transducer) (audio and electroacoustics). In decibels, 10 times the logarithm to the base 10 of the directivity factor.   176

directional-ground relay. A directional relay used primarily to detect single-phase-to-ground faults, but also sensitive to double-phase-to-ground faults. *Note:* This type of relay is usually operated from the zero-sequence components of voltage and current, but is sometimes operated from negative-sequence quantities.   127, 103

directional gyro. A two-degree-of-freedom gyro with provision for maintaining the spin axis approximately horizontal. In this gyro, an output signal is produced by gimbal angular displacement which corresponds to the angular displacement of the case about an axis which is nominally vertical.   46

directional gyro electric indicator. An electrically driven device for use in aircraft for measuring deviation from a fixed heading.   328

directional homing (navigation). The process of homing wherein the navigational quantity maintained constant is the bearing. *See also:* **radio navigation.**   13, 187

directional lighting. Lighting provided on the work plane or on an object light that is predominantly from a preferred direction. *See also:* **general lighting.**   167

directional localizer (instrument landing systems). A localizer in which maximum energy is directed close to the runway centerline, thus minimizing extraneous reflections. *See also:* **navigation.**   187, 13

directional microphone. A microphone the response of which varies significantly with the direction of sound incidence. *See also:* **microphone.**   328

directional-overcurrent protection. A method of protection in which an abnormal condition within the protected equipment is detected by the current being in excess of a predetermined amount and in a predetermined phase relation with a reference input.   103, 202, 127

directional-overcurrent relay. A relay consisting of an overcurrent unit and a directional unit combined to operate jointly.   103, 202, 60, 127

directional pattern (radiation pattern). The directional pattern of an antenna is a graphical representation of the radiation or reception of the antenna as a function of direction. *Note:* Cross sections in which directional patterns are frequently given are vertical planes and the horizontal planes or the principal electric and magnetic polarization planes. *See also:* **antenna.**   328

directional phase shifter (directional phase changer) (nonreciprocal phase shifter). A passive phase changer in which the phase change for transmission in one direction differs from that for transmission in the opposite direction. *See:* **transmission line.**   179

directional-power relay. A relay that operates in conformance with the direction of power flow.   103, 202, 60

directional-power tripping. *See:* **directional-power relay.**

directional relay. A relay that responds to the relative phase position of a current with respect to another current or voltage reference. *Note:* The above definition, which applies basically to a single-phase directional relay, may be extended to cover a polyphase directional relay.   103, 202, 60, 127

directional response pattern (beam pattern) (electroacoustics) (transducer used for sound emission or reception). A description, often presented graphically, of the response of the transducer as a function of the direction of the transmitted or incident sound waves in a specified plane and at a specified frequency. *Notes:* (1) A complete description of the directional response pattern of a transducer would require a three-dimensional presentation. (2) The directional response pattern is often shown as the response relative to the maximum response. *See also:* **loudspeaker.**   176

direction-finder (DF). *See:* **navigation; radio direction-finder.**

direction-finder antenna. Any antenna used for radio direction finding. *See also:* **navigation.**   187, 13

direction-finder deviation. The amount by which an observed radio bearing differs from the corrected bearing. *See also:* **navigation.**   187, 13

direction-finder noise level (in the absence of the desired signals). The average power or root-mean-square voltage at any specified point in a direction-finder system circuit. *Note:* In radio-frequency and audio channels, the direction-finder noise level is usually measured in terms of the power dissipated in suitable termination. In a video channel, it is customarily measured in terms of voltage across a given impedance or of the cathode-ray deflection. *See also:* **navigation.**   278, 187, 13

direction-finder sensitivity. The field strength at the direction-finder antenna, in microvolts per meter, that produces a ratio of signal-plus-noise to noise equal to 20 decibels in the receiver output, the direction of arrival of the signal being such as to produce maximum pickup in the direction-finder antenna system. *See also:* **navigation.**   278, 187, 13

direction finding. *See:* **radio direction-finder.**

direction finding (DF) antenna. Any antenna used for radio direction finding.   13

direction finding (DF) antenna system. One or more DF antennas, their combining circuits and feeder systems, together with the shielding and all electrical and mechanical items up to the termination at the receiver input terminals.   13

direction of energy flow (specified circuit). With reference to the boundary (of a delimited region), the direction in which electric energy is being transmitted past the boundary, into or out of the region.   210

direction of lay (cables). The lateral direction, designated as left-hand or right-hand, in which the elements of a cable run over the top of the cable as they recede from an observer looking along the axis of the cable. *See also:* **power distribution, underground construction.**   57, 64

direction of polarization (radio propagation) (1) (linearly polarized wave). The direction of the electric intensity. **(2) (elliptically polarized wave).** The direction of the major axis of the electric vector ellipse. *See also:* **radiation; radio wave propagation.**   180, 146

direction of propagation (point in a homogeneous isotropic medium) (1). The normal to an equiphase surface taken in the direction of increasing phase lag. *See also:* **radiation.**   328

(2). The direction of time-average energy flow. *Notes:* (A) In a uniform waveguide the direction of propagation is often taken along the axis. (B) In the case of a uniform lossless waveguide the direction of propagation at every point is parallel to the axis and in the direction of time-average energy flow. *See also:* **radiation; waveguide.**   267, 179, 146

direction of rotation of phasors. Phasor diagrams should be drawn so that an advance in phase of one phasor with respect to another is in the counterclockwise direction. In the following figure, vector 1 is 120 degrees in advance of vector 2, and the phase sequence is 1, 2, 3.   53

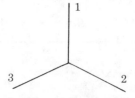

directive gain (antenna) (given direction). 4π times the ratio of the radiation intensity in that direction to the total power radiated by the antenna. *Note:* The directive gain is fully realized on reception only when the incident polarization is the same as the polarization of the antenna on transmission. *See also:* **antenna.**   179, 111

directive gain in physical media. In a given direction and at a given point in the far field, the ratio of the power flux per unit area from an antenna to the power flux per unit area from an isotropic radiator at a specified location delivering the same power from the antenna to the medium. *Notes:* (1) The isotropic radiator must be within the smallest sphere containing the antenna. Suggested locations are antenna terminals and points of symmetry if such exist. (2) See IEEE Std 270-1966. Definitions of General (Fundamental and Derived) Electrical and Electronics

Terms, and IEEE Std 211-1969, Definitions of Terms for Radio Wave Propagation, referring to the use of power flux and power flux density, respectively.   111

directivity (1) (antenna) (gain). The value of the directive gain in the direction of its maximum value. *See also:* **antenna.**   111, 246, 179, 311

(2) (directional coupler). The ratio of the power output at an auxiliary port, when power is fed into the main waveguide or transmission line in the preferred direction, to the power output at the same auxiliary port when power is fed into the main guide or line in the opposite direction, the incident power fed into the main guide or line being the same in each case, and reflectionless terminations being connected to all ports. *Note:* The ratio is usually expressed in decibels.   185

directivity factor (audio and electroacoustics) (1) (transducer used for sound emission). The ratio of the sound pressure squared, at some fixed distance and specified direction, to the mean-square sound pressure at the same distance averaged over all directions from the transducer. *Note:* The distance must be great enough so that the sound pressure appears to diverge spherically from the effective acoustic center of the transducer. Unless otherwise specified, the reference direction is understood to be that of a maximum response. The frequency must be stated. *See also:* **loudspeaker.**   176

(2) (transducer used for sound reception). The ratio of the square of the open-circuit voltage produced in response to sound waves arriving in a specified direction to the mean-square voltage that would be produced in a perfectly diffused sound field of the same frequency and mean-square sound pressure. *Notes:* (A) This definition may be extended to cover the case of finite frequency bands whose spectrum may be specified. (B) The average free-field response may be obtained in various ways, such as (1) by the use of a spherical integrator, (2) by numerical integration of a sufficient number of directivity patterns corresponding to different planes, or (3) by integration of one or two directional patterns whenever the pattern of the transducer is known to possess adequate symmetry. *See also:* **microphone.**   176

direct lighting. Lighting involving luminaires that distribute 90 to 100 percent of the emitted light in the general direction of the surface to be illuminated. *Note:* The term usually refers to light emitted in a downward direction. *See also:* **general lighting.**   167

direct liquid cooling system (semiconductor rectifiers). A cooling system in which a liquid, received from a constantly available supply, is passed directly over the cooling surfaces of the semiconductor power converter and discharged. *See also:* **semiconductor rectifier stack.**   208

direct liquid cooling system with recirculation (semiconductor rectifiers). A direct liquid cooling system in which part of the liquid passing over the cooling surfaces of the semiconductor power converter is recirculated and additional liquid is added as needed to maintain the required temperature, the excess being discharged. *See also:* **semiconductor rectifier stack.**   208

directly controlled variable (industrial control) (automatic control). The variable in a feedback control system whose value is sensed to originate the primary feedback signal. *See:* **control system, feedback.**   219, 206, 105

direct on-line starting (rotating machinery). The process of starting a motor by connecting it directly to the supply at a rated voltage. *See:* **asynchronous machine; direct-current commutating machine.**   63

direct operation. Operation by means of a mechanism connected directly to the main operating shaft or an extension of the same.   103, 202, 27

direct orbit (communication satellite). An inclined orbit with an inclination between zero and ninety degrees.   74

director element. A parasitic element located forward of the driven element of an antenna, intended to increase the directive gain of the antenna in the forward direction. *See also:* **antenna.**   179, 111

directory-assistance call (telephone switching systems). A call placed to request the directory number of a customer.   55

directory number (telephone switching systems). The full complement of digits required to designate a customer in a directory.   55

directory-numbering plan (telephone switching systems). The arrangement whereby each customer is identified by an office and main-station code.   55

direct outward dialing (DOD) (telephone switching systems). A private automatic branch exchange or centrex service feature that permits stations to dial outside numbers without intervention of an attendant.   55

direct-plunger driving machine (elevators). A machine in which the energy is applied by a plunger or piston directly attached to the car frame or platform and that operates in a cylinder under hydraulic pressure. *Note:* It includes the cylinder and plunger or piston. *See also:* **driving machine (elevators).**   328

direct-plunger elevator. A hydraulic elevator having a plunger or piston directly attached to the car frame or platform. *See also:* **elevators.**   328

direct-point repeater. A telegraph repeater in which the receiving relay, controlled by the signals received over a line, repeats corresponding signals directly into another line or lines without the interposition of any other repeating or transmitting apparatus. *See also:* **telegraphy.**   328

direct polarity indication. *See:* **mixed logic.**

direct ratio (illuminating engineering). The ratio of the luminous flux that reaches the work plane directly to the downward component from the luminaire. *See also:* **inverse-square law (illuminating engineering).**   167

direct raw-water cooling system (rectifier). A cooling system in which water, received from a constantly available supply, such as a well or water system, is passed directly over the cooling surfaces of the rectifier and discharged. *See:* **direct liquid cooling system.** *See also:* **rectification.**   208

direct raw-water cooling system with recirculation. A direct raw-water cooling system in which part of the water passing over the cooling surfaces of the rectifier is recirculated and raw water is added as needed to maintain the required temperature, the excess being discharged. *See:* **direct liquid cooling system with recirculation.** *See also:* **rectification.**   208

direct-recording (facsimile). That type of recording in which a visible record is produced, without subsequent processing, in response to the received signals. *See also:* **recording (facsimile).** 12

direct release (series trip). A release directly energized by the current in the main circuit of a switching device. 103, 202

direct stroke. A lightning stroke direct to any part of a network or electric installation. 244, 62

direct-stroke protection (lightning). Lightning protection designed to protect a network or electric installation against direct strokes. 64

direct support maintenance. *See:* **intermediate maintenance.**

direct vacuum-tube current (medical electronics). A current obtained by applying to the part to be treated an evacuated glass electrode connected to one terminal of a generator of high-frequency current (100 to 10 000 kilohertz), the other terminal being grounded. *Note:* Deprecated as confusing and as representing an ill-defined and obsolescent procedure. 192

direct-voltage high-potential test (rotating machinery). A test that consists of the application of a specified unidirectional voltage higher than the rated root-mean-square value for a specified time for the purpose of determining (1) the adequacy against breakdown of the insulation system under normal conditions, or (2) the resistance characteristic of the insulation system. *See also:* **asynchronous machines; direct-current commutating machines.** 63

direct wave (radio wave propagation). A wave propagated directly from a source to a point. *See also:* **radiation; radio wave propagation.** 180

direct-wire circuit (one-wire circuit). A supervised circuit, usually consisting of one metallic conductor and a ground return, and having signal receiving equipment responsive to either an increase or a decrease in current. *See also:* **protective signaling.** 328

disability glare. Glare that reduces visual performance and visibility and often is accompanied by discomfort. *See:* **veiling brightness.** *See also:* **visual field.** 167

disaster, major storm (transmission and distribution). Designates weather which exceeds design limits of plant and which satisfies all of the following: (1) Extensive mechanical damage to plant. (2) More than a specified percentage of customers out of service. (3) Service restoration times longer than a specified time. *Note:* It is suggested that the specified percentage of customers out of service and restoration times be 10 percent and 24 hours. Percentage of customers out of service may be related to a company operating area rather than to an entire company. Examples of major storm disasters are hurricanes and major ice storms. 112

disc (disk). *See:* **magnetic disc; disc recorder.**

disc-and-wiper-lubricated bearing (rotating machinery). A bearing in which a disc mounted on and concentric with the shaft dips into a reservoir of oil. *Note:* As the shaft rotates the oil is diverted from the surface of the disc by a scraper action into the bearing. *See also:* **bearing.** 63

discharge (1) (storage cell). The conversion of the chemical energy of the battery into electric energy. *See also:* **charge.** 328

(2) (gas). The passage of electricity through a gas. 244, 190

discharge capacity (arrester). *See:* **arrester discharge capacity.**

discharge circuit (surge generator). That portion of the surge-generator connections in which exist the current and voltage variations constituting the surge generated. 64, 62

discharge counter (surge arrester). A means for recording the number of arrester discharge operations. *See:* **lightning; current rating, 60-hertz (arrester).** 2

discharge current (surge arrester). The surge current that flows through an arrester when spark-over occurs. 2, 62

discharge-current-limiting device (series capacitor). A reactor or equivalent device to limit the magnitude and frequency of the discharge of the capacitor segment upon sparkover of the protective power gap or closing of the capacitor bypass switch. 86

discharge detector (ionization or corona detector) (rotating machinery). An instrument that can be connected in or across an energized insulation circuit to detect current or voltage pulses produced by electric discharges within the circuit. *See also:* **instrument.** 63

discharge device (capacitor). An internal or external device intentionally connected in shunt with the terminals of a capacitor for the purpose of reducing the residual voltage after the capacitor is disconnected from an energized line. 86, 138

discharge-energy test (rotating machinery). A test for determining the magnitude of the energy dissipated by a discharge or discharges within the insulation. *See also:* **asynchronous machine; direct current commutating machine.** 63

discharge extinction voltage (ionization or corona extinction voltage) (rotating machinery). The voltage at which discharge pulses that have been observed in an insulation system, using a discharge detector of specified sensitivity, cease to be detectable as the voltage applied to the system is decreased. *See also:* **asynchronous machines; direct-current commutating machine.** 63

discharge inception test (corona inception test) (rotating machinery). A test for measuring the lowest voltage at which discharges of the specified magnitude recur in successive cycles when an increasing alternating voltage is applied to insulation. *See also:* **asynchronous machine; direct-current commutating machine.** 63

discharge inception voltage (ionization or corona inception voltage) (1) (rotating machinery). The voltage at which discharge pulses in an insulation system become observable with a discharge detector of specified sensitivity, as the voltage applied to the system is raised. *See also:* **asynchronous machine; direct-current commutating machine.** 63

(2) (surge arrester). The root-mean-square value of the power-frequency voltage at which discharges start, the measurement of their intensity being made under specified conditions. 244, 62

discharge indicator (surge arrester). A means for indicating that the arrester has discharged. 2

discharge opening (rotating machinery). A port for the exit of ventilation air. 63

discharge oscillations (laser gyro). Periodic variations in voltage and current at the terminals of a direct current discharge tube which are supported by the negative resistance of the discharge tube itself.   46

discharge probe (ionization or corona probe) (rotating machinery). A portable antenna, safely insulated, and designed to be used with a discharge detector for locating sites of discharges in an energized insulation system. *See also:* **instrument.**   63

discharge resistor. A resistor that, upon interruption of excitation source current, is connected across the field windings of a generator, motor, synchronous condenser, or an exciter to limit the transient voltage in the field circuit and to hasten the decay of field current of these machines.   103

discharge tube. An evacuated enclosure containing a gas at low pressure that permits the passage of electricity through the gas upon application of sufficient voltage. *Note:* The tube is usually provided with metal electrodes, but one form permits an electrodeless discharge with induced voltage.   328

discharge voltage (surge arrester). The voltage that appears across the terminals of an arrester during passage of discharge current.   2

discharge voltage-current characteristic (surge arrester). The variation of the crest values of discharge voltage with respect to discharge current. *Note:* This characteristic is normally shown as a graph based on three or more current surge measurements of the same wave shape but of different crest values.   2

discharge voltage-time curve (arrester). *See:* **arrester discharge voltage-time curve.**

discharge withstand current rating (surge arrester). The specified magnitude and wave shape of a discharge current that can be applied to an arrester a specified number of times without causing damage to it.   2

discomfort glare. Glare that produces discomfort. It does not necessarily interfere with visual performance or visibility. *See also:* **visual field.**   167

discomfort-glare factor. The numerical assessment of the capacity of a single source of brightness, such as a luminaire, in a given visual environment for producing discomfort. *See:* **glare; discomfort glare.** *see also:* **inverse-square law (illuminating engineering).**   167

discomfort-glare rating. A numerical assessment of the capacity of a number of sources of brightness, such as luminaires, in a given visual environment for producing discomfort. *Note:* It usually is derived from the discomfort-glare factors of the individual sources. *See also:* **inverse-square law (illuminating engineering).**   167

discone antenna. A biconical antenna with one cone having a vertex angle of 180°.   111

disconnect (release) (telephony). To disengage the apparatus used in a telephone connection and to restore it to its condition when not in use.   193

disconnectable device. A grounding device that can be disconnected from ground by the operation of a disconnecting switch, circuit breaker, or other switching device.   91

disconnected position (switchgear-assembly removable element). That position in which the primary and secondary disconnecting devices of the removable element are separated by a safe distance from the stationary element contacts. *Note:* Safe distance, as used here, is a distance

at which the equipment will meet its withstand ratings, both power frequency and impulse, between line and load stationary terminals and phase-to-phase and phase-to-ground on both line and load stationary terminals with the switching device in the closed position.   103

disconnecting blade. *See:* **blade (disconnecting blade) (switching device).**

disconnecting cutout. A cutout having a disconnecting blade for use as a disconnecting or isolating switch.   103, 202

disconnecting device. A device whereby the conductors of a circuit can be disconnected from their source of supply.   124

disconnecting fuse. *See:* **fuse-disconnecting switch.**

disconnecting means. A device, group of devices, or other means whereby the conductors of a circuit can be disconnected from their source of supply. *See also:* **switch.**   256

disconnecting or isolating switch (disconnector, isolator). A mechanical switching device used for changing the connections in a circuit or for isolating a circuit or equipment from the source of power. *Note:* It is required to carry normal load current continuously and also abnormal or short-circuit currents for short intervals as specified. It is also required to open or close circuits either when negligible current is broken or made or when no significant change in the voltage across the terminals of each of the switch poles occurs.   103, 202, 27

disconnection (control) (industrial control). Connotes the opening of a sufficient number of conductors to prevent current flow.   225, 206

disconnector. A switch that is intended to open a circuit only after the load has been thrown off by some other means. *Note:* Manual switches designed for opening loaded circuits are usually installed in circuit with disconnectors, to provide a safe means for opening the circuit under load.   178

disconnect signal (telephony). A signal transmitted from one end of a subscriber line or trunk to indicate that the relevant party has released.   193, 101

disconnect-type pothead. A pothead in which the electric continuity of the circuit may be broken by physical separation of the pothead parts, part of the pothead being on each conductor end after the separation. *See also:* **pothead.**   4

discontinuity (1). An abrupt nonuniformity in a uniform waveguide or transmission line that causes reflected waves. *See:* **capacitance, discontinuity.** *See also:* **waveguide.**   185

(2) (inductive coordination). An abrupt change at a point, in the physical relations of electric supply and communication circuits or in electrical parameters of either circuit, that would materially affect the coupling. *Note:* Although technically included in the definition, transpositions are not rated as discontinuities because of their application to coordination. *See also:* **inductive coordination.**   328

disc recorder (phonograph techniques). A mechanical recorder in which the recording medium has the geometry of a disc. *See also:* **phonograph pickup.**   176

discrete sentence intelligibility. The percent intelligibility obtained when the speech units considered are sentences (usually of simple form and content). *See also:* **volume**

equivalent. 328

discrete word intelligibility. The percent intelligibility obtained when the speech units considered are words (usually presented so as to minimize the contextual relation between them). *See also:* **volume equivalent.** 328

discrimination (1) (any system or transducer). The difference between the losses at specified frequencies, with the system or transducer terminated in specified impedances. *See also:* **transmission loss.** 328

(2) (radar). Separation or identification of the differences between nonsimilar signals. 13

discrimination ratio (ferroelectric device). The ratio of signal charge to induced charge. *Note:* Discrimination ratio is dependent on the magnitude of the applied voltage, which therefore should be specified. *See also:* **ferroelectric domain.** 247

discriminator (1). A device in which amplitude variations are derived in response to frequency or phase variations. *Note:* The device is termed a frequency discriminator or phase discriminator according to whether it responds to variations of frequency or phase. 59

(2) (radar). A circuit in which the output is dependent upon how an input signal differs in some aspect from a standard or from another signal. 13

discriminator, amplitude (pulse techniques). *See:* **discriminator, pulse-height.**

discriminator, constant-fraction pulse-height. A pulse-height discriminator in which the threshold changes with input amplitude in such a way that the triggering point corresponds to a constant fraction of the input pulse height. 117

discriminator, pulse-height (pulse techniques). A circuit that produces a specified output pulse if and only if it receives an input pulse whose amplitude exceeds an assigned value. *See also:* **pulse.** 335

dish (1) (radio practice). A reflector the surface of which is concave as, for example, a part of a sphere or of a paraboloid of revolution. *See also:* **antenna.** 328

(2) (radar). A colloquial term for a parabolic microwave antenna reflecting surface. 13

disk. *See:* **disc.**

dispatching system (mining practice). A system employing radio, telephones, and/or signals (audible or light) for orderly and efficient control of the movements of trains of cars in mines. *See:* **mine fan signal system; mine radio telephone system.** 328

dispatch operation (radio-communication circuit). A method for permitting a maximum number of terminal devices to have access to the same two-way radio communication circuit. *See also:* **channel spacing.** 181

dispenser cathode (electron tubes). A cathode that is not coated but is continuously supplied with suitable emission material from a separate element associated with it. *See also:* **electron tube.** 244, 190

dispersed magnetic powder tape (impregnated tape). *See:* **magnetic powder-impregnated tape.**

dispersive bandwidth (dispersive delay line). The operating frequency range over which the delay dispersion is defined. 81

displacement current (any surface). The integral of the normal component of the displacement current density over that surface. *Note:* Displacement current is a scalar and hence has no direction. 210

displacement current density (any point in an electric field). The time rate of change in SI units (International System of Electrical Units) of the electric flux density vector at that point. 210

displacement power factor (rectifier) (phasor power factor) (rectifier unit). The displacement component of power factor; the ratio of the active power of the fundamental wave, in watts, to the apparent power of the fundamental wave, in root-mean-square voltamperes (including the exciting current of the rectifier transformer). *See also:* **power rectifier; rectification.** 204, 208

display (1) (navigation systems). The visual representation of output data. *Note:* See display by type designation, for example, **B** display, **C** display, etcetera. 187

(2) (test, measurement and diagnostic equipment). A mechanical, optical, electro-mechanical, or electronic device for presenting information to the operator or maintenance technician about the state or condition of the unit under test or the checkout equipment itself. 54

(3) (oscilloscopes). The visual presentation on the indicating device of an oscilloscope. 106

display flatness (non-real time spectrum analyzer). The peak-to-peak variation in amplitude over a specified span, (decibels). 68

display frequency (non-real time spectrum analyzer). The input frequency as indicated by the spectrum analyzer (hertz). 68

display law (non-real time spectrum analyzer). The mathematical law that defines the input-output function of an instrument. *Note:* The three methods of displaying an input signal are: (1) linear—a display in which the scale division is a linear function of the input voltage; (2) square law (power)—a display in which the scale division is a linear function of the input power; (3) logarithmic—a display in which the scale division is a logarithmic function of the input signal. 68

display primaries (receiver primaries). The colors of constant chromaticity and variable luminance produced by the receiver or any other display device that, when mixed in proper proportions, are used to produce other colors. *Note:* Usually three primaries used are: red, green and blue. 18, 178

display reference level (non-real time spectrum analyzer). A designated vertical position representing specified input levels. The level may be expressed in dBm, volts, or any other units. 68

displays (nuclear power generating station). Devices which convey information to the operator. 358

display storage tube. A storage tube into which the information is introduced as an electric signal and read at a later time as a visible output. *See also:* **storage tube.** 174, 190

display tube. A tube, usually a cathode-ray tube, used to display data. 255, 77

disruptive discharge (1) (general). The phenomena associated with the failure of insulation, under electric stress, that include a collapse of voltage and the passage of current; the term applies to electrical breakdown in solid, liquid, and gaseous dielectrics and combinations of these. *Note:* A disruptive discharge in a solid dielectric produces permanent loss of electric strength; in a liquid or gaseous dielectric the loss may be only temporary. *See:* **low-frequency and surge testing; power systems; test voltage and current.** 307, 201, 62

(2) (surge arrester). The sudden and large increase in current through an insulating medium due to the complete failure of the medium under the electrostatic stress. 2

disruptive discharge voltage (1). The value of the test voltage for which disruptive discharge takes place. *Note:* The disruptive discharge voltage is subject to statistical variation which can be expressed in different ways as, for example, by the mean, the maximum, and the minimum values of series of observations, or by the mean and standard deviation from the mean, or by a relation between voltage and probability of a disruptive discharge. *See also:* **test voltage and current.** 307, 57

(2) (50 percent) (surge arresters). The voltage that has a 50-percent probability of producing a disruptive discharge. *Note:* The term applies mostly to impulse tests and has significance only in cases when the loss of electric strength resulting from a disruptive discharge is temporary. 307, 201, 62, 308

(3) (100 percent) (surge arresters). The specified voltage that is to be applied to a test object in a 100-percent disruptive discharge test under specified conditions. *Note:* The term applies mostly to impulse tests and has significance only in cases when the loss of electric strength resulting from a disruptive discharge is temporary. During the test, in general, all voltage applications should cause disruptive discharge. 308, 62, 307, 201

dissector. *See:* **image dissector.**

dissector tube. A camera tube having a continuous photocathode on which is formed a photoelectric-emission pattern that is scanned by moving its electron optical image over an aperture. *See also:* **image dissector tube.** 328

dissipation (electrical energy). Loss of electric energy as heat. 231, 197

dissipation, electrode. *See:* **electrode dissipation.**

dissipation factor (circuits and systems). (1) The ratio of energy dissipated to the energy stored in an element for one cycle, (2) the loss tangent of an element and (3) the inverse of Q. *See:* **dielectric dissipation factor.** 67

dissipation-factor test (rotating machinery). *See:* **loss-tangent test.**

dissymmetrical transducer. Dissymmetrical with respect to a specified pair of terminations when the interchange of that pair of terminations will affect the transmission. *See also:* **transducer.** 328

distance. *See:* **Hamming distance; signal distance.**

distance dialing (telephone switching systems). The automatic establishing of toll calls by means of signals from the calling device of either a customer or an operator. 55

distance mark (range mark) (on a radar display). A calibration marker used on a cathode-ray screen in determining target distance. *See also:* **radar.** 187

distance-measuring equipment. A radio aid to navigation which provides distance information by measuring total round-trip time of transmission from an interrogator to a transponder and return. *Note:* Commonly abbreviated DME. 13, 187

distance protection. A method of line protection in which an abnormal condition within a predetermined electrical distance of a line terminal on the protected circuit is detected by measurement of system conditions at that terminal. 103, 202, 127

distance relay. A generic term covering those forms of protective relays in which the response to the input quantities is primarily a function of the electrical circuit distance between the relay location and the point of fault. *Note:* Distance relays may be single-phase devices or they may be polyphase devices. 103, 202, 60, 127

distance resolution (radar). The ability to distinguish between two targets solely by the measurement of distances; generally expressed in terms of the minimum distance by which two targets of equal strength at the same azimuth and elevation angles must be spaced to be separately distinguishable and measureable. 13

distinctive-shape logic symbol. A logic symbol whose form uniquely identifies its logic function. 88

distortion (data transmission). (1) An undesired change in waveform. *Note:* The principal sources of distortion of a waveform are: (A) nonlinear relationship between input and output, (B) nonuniform transmission at different frequencies, and (C) phase shift not proportional to frequency. *See:* **electrical noise; fortuitous telegraph distortion; frequency distortion; nonlinear distortion; total telegraph distortion.** (2) Any departure from a specified input-output relationship over a range of frequencies, amplitudes, or phase shifts, or during a time interval. (3) Any undesired change in a transmitted pattern or picture. (4) (electrical conversion). The portion of an alternating-current waveform caused by frequencies other than the fundamental. It provides an indication of the harmonic content of the alternating-current wave and is expressed as a percent of the fundamental.

Distortion Factor =

$$\left[\frac{\text{(sum of squares of amplitudes of all harmonics)}}{\text{(square of amplitude of fundamental)}}\right]^{1/2} (100\%)$$

Note: For distortion factors less than 10 percent a distortion-factor meter will provide a method of accurate measurement. 59

distortion, amplitude-frequency (electroacoustics). *See:* **amplitude-frequency distortion; distortion.**

distortion, barrel (camera tubes or image tube). A distortion that results in a progressive decrease in radial magnification in the reproduced image away from the axis of symmetry of the electron optical system. *Note:* For a camera tube, the reproducer is assumed to have no geometric distortion. 125

distortion, envelope delay (1) (general). Of a system or transducer, the difference between the envelope delay at one frequency and the envelope delay at a reference frequency. 239

(2) (facsimile). That form of distortion which occurs when the rate of change of phase shift with frequency of a circuit or system is not constant over the frequency range required for transmission. *Note:* In facsimile, envelope delay distortion is usually expressed as one-half the difference in microseconds between the maximum and the minimum envelope delays existing between the two extremes of frequency defining the channel used. *See also:* **facsimile transmission.**　　　　　　　　　　　　　　　　12

distortion factor (1) (power-system communication). The ratio of the root-mean-square value of the harmonic content to the root-mean-square value of the nonsinusoidal quantity. *See also: distortion.*　　　　　244, 59

(2) (wave) (rotating machinery). The ratio of the root-mean-square value of the residue of a voltage wave after the elimination of the fundamental to the root-mean-square value of the original wave.　　　　　63

distortion, field-time waveform (FD). The linear TV waveform distortion of time components from 64 μs to 16 ms, that is, time components of the field-time domain.　　　　　42

distortion, frequency. *See:* **amplitude-frequency distortion.**

distortion, harmonic. Nonlinear distortion of a system or transducer characterized by the appearance in the output of harmonics other than the fundamental component when the input wave is sinusoidal. *Note:* Subharmonic distortion may also occur. *See: distortion.*　　　239, 210

distortion, intermodulation. Nonlinear distortion of a system or transducer characterized by the appearance in the output of frequencies equal to the sums and differences of integral multiples of the two or more component frequencies present in the input wave. *Note:* Harmonic components also present in the output are usually not included as part of the intermodulation distortion. When harmonics are included, a statement to that effect should be made. *See also: distortion.*　　　239

distortion, keystone (camera tubes). A distortion such that the slope or the length of a horizontal line trace or scan line is linearly related to its vertical displacement. *Note:* A system having keystone distortion distorts a rectangular pattern into a trapezoidal pattern.　　　125

distortion, linear. That distortion of an electrical signal which is independent of the signal amplitude. *Note:* A small-signal nonuniform frequency response is an example of linear distortion. By contrast, nonlinear distortions of an electrical signal are those distortions that are dependent on the signal amplitude, for example, compression, expansion, and harmonic distortion, etc.　　　42

distortion, linear TV waveform. The distortion of the shape of a waveform signal where this distortion is independent of the amplitude of the signal. *Notes:* (1) A TV video signal may contain time components with durations from as long as a TV field to as short as a picture element. The shapes of all these time components are subject to distortions. For ease of measurement it is convenient to group these distortions in three separate time domains; short-time waveform distortion, line-time waveform distortion, and field-time waveform distortion. (2) The waveform distortions for times from one field to tens of seconds is not within the scope of this standard.　　　42

distortion, line-time waveform (LD). The linear TV waveform distortion of time components from 1 μs to 64 μs, that is, time components of the line-time domain.　　　42

distortion, pattern (oscilloscopes). Any deformation of the pattern from its intended form. (IEC 151-14.) *Notes:* (1) In an oscilloscope the intended form is rectilinear and rectangular. (2) An oscilloscope control that affects pattern distortion may be labeled "pattern" or "geometry."　　　106

distortion, percent harmonic (electroacoustics). A measure of the harmonic distortion in a system or transducer, numerically equal to 100 times the ratio of the square root of the sum of the squares of the root-mean-square voltages (or currents) of each of the individual harmonic frequencies, to the root-mean-square voltage (or current) of the fundamental. *Note:* It is practical to measure the ratio of the root-mean-square amplitude of the residual harmonic voltages (or currents), after the elimination of the fundamental, to the root-mean-square amplitude of the fundamental and harmonic voltages (or currents) combined. This measurement will indicate percent harmonic distortion with an error of less than 5 percent if the magnitude of the distortion does not exceed 30 percent. *See: distortion.*　　　239

distortion, phase delay (system or transducer). The difference between the phase delay at one frequency and the phase delay at a reference frequency.　　　239

distortion, phase-frequency. *See: distortion, phase delay.*

distortion, pincushion (camera tubes or image tubes). A distortion that results in a progressive increase in radial magnification in the reproduced image away from the axis of symmetry of the electron optical system. *Note:* For a camera tube, the reproducer is assumed to have no geometric distortion. *See:* **distortion, amplitude-frequency (electroacoustics); distortion factor; distortion, percent harmonic (electroacoustics); hiss (in an electron device).**　　　178, 186, 190, 125

distortion power (1) (single-phase two-wire circuit). At the two terminals of entry of a single-phase two-wire circuit into a delimited region, a scalar quantity having an amplitude equal to the square root of the difference of the squares of the apparent power and the amplitude of the phasor power. *Note:* Mathematically the amplitude of the distortion power D is given by the equation

$$D = (U^2 - S^2)^{1/2}$$
$$= (U^2 - P^2 - Q^2)^{1/2}$$
$$= \left(\sum_{r=1}^{r=\infty} \sum_{q=1}^{q=\infty} \{ E_r{}^2 I_q{}^2 \right.$$
$$\left. - E_r E_q I_r I_q \cos \left[(\alpha_r - \beta_r) - (\alpha_q - \beta_q) \right] \} \right)^{1/2}$$

where the symbols are as in **power, apparent (single-phase two-wire circuit).** If the voltage and current are quasi-periodic and the amplitudes are slowly varying, the distortion power at any instant may be taken as the value derived from the amplitude of the apparent power and phasor power at that instant. By this definition the sign of distortion power is not definitely determined, and it may be given either sign. In the absence of other definite information, it is to be taken the same as for the active power. Distortion power is expressed in volt-amperes when the voltage is in volts and the current in amperes. The distortion power is zero if the voltage and the current have the same waveform. This condition is fulfilled when the voltage and current are sinusoidal and have the same period, or when the circuit consists entirely of noninductive resistors.　　　210

distortion power (polyphase circuit). At the terminals of entry of a polyphase circuit, equal to the sum of the distortion powers for the individual terminals of entry. *Notes:* (1) The distortion power for each terminal of entry is determined by considering each phase conductor, in turn, with the common reference point as a single-phase, two-wire circuit and finding the distortion power for each in accordance with the definition of **distortion power (single-phase two-wire circuit).** The common reference terminal shall be taken as the neutral terminal of entry, if one exists, otherwise as the true neutral point. The sign given to the distortion power for each single-phase current, and therefore to the total for the polyphase circuit, shall be the same as that of the total active power. (2) Distortion power is expressed in volt-amperes when the voltages are in volts and the current in amperes. (3) The distortion power is zero if each voltage has the same wave form as the corresponding current. This condition is fulfilled, of course, when all the currents and voltages are sinusoidal.   210

distortion, pulse (pulse techniques). The unwanted deviation of a pulse waveform from a reference waveform. *Note:* Some specific forms of pulse distortion have specific names. They include, but are not exclusive to, the following: **overshoot, ringing, preshoot, tilt (droop*), rounding (undershoot and dribble-up*), glitch, bump, spike,** and **backswing.** For further explanation of the forms of pulse distortion, see the following illustrations and IEEE Standard 194 (1976). *See also:* **pulse.**   185

* Deprecated

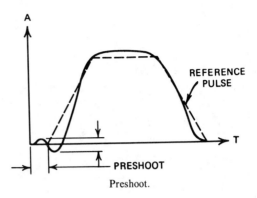
Preshoot.

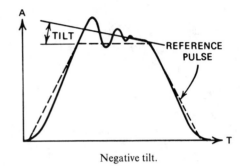
Negative tilt.

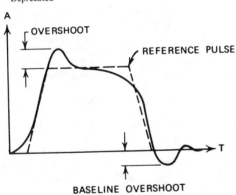
BASELINE OVERSHOOT
Overshoot.

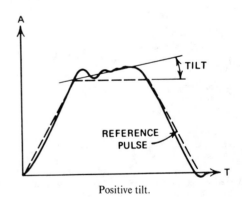
Positive tilt.

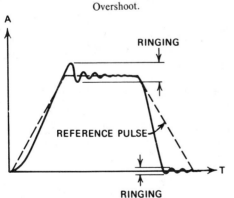
Ringing.

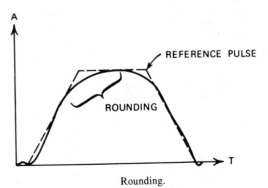
Rounding.

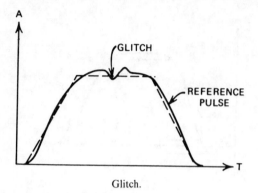

Glitch.

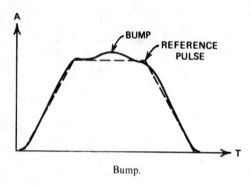

Bump.

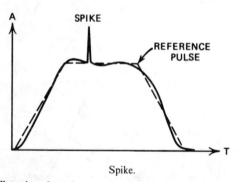

Spike.

distortion, short-time waveform (SD) (TV). The linear TV waveform distortion of time components from 125 ns to 1 μs, that is, time components of the short-time domain.

42

distortion, spiral (camera tubes or image tubes using magnetic focusing). A distortion in which image rotation varies with distance from the axis of symmetry of the electron optical system.   178, 190, 125

distortion tolerance (telegraph receiver). The maximum signal distortion that can be tolerated without error in reception. *See also:* **telegraphy.**   194

distortion, waveform (oscilloscopes). A displayed deviation from the correct representation of the input reference signal. *See also:* **oscillograph.**   106

distributed. Spread out over an electrically significant length or area.   328

distributed constant (waveguide). A circuit parameter that exists along the length of a waveguide or transmission line.

Note: For a transverse electromagnetic wave on a two-conductor transmission line, the distributed constants are series resistance, series inductance, shunt conductance, and shunt capacitance per unit length of line. *See also:* **waveguide.**   267, 179

distributed element circuit (microwave tubes). A circuit whose inductance and capacitance are distributed over a physical distance that is comparable to a wavelength. *See also:* **microwave tube or valve.**   190

distributed target (radar). A target composed of a number of scatterers, where the target extent in any dimension is greater than the radar resolution in that dimension. *See also:* **complex target.**   13

distributed winding (rotating machinery). A winding, the coils of which occupy several slots per pole. *See also:* **armature; rotor (rotating machinery); stator.**   63

distributing cable. *See:* **distribution cable.**

distributing frame. A structure for terminating permanent wires of a central office, private branch exchange, or private exchange and for permitting the easy change of connections between them by means of cross-connecting wires.   193

distributing valve (speed governing systems, hydraulic turbines). The element of the governor-control actuator which controls the flow of hydraulic fluid to the turbine-control servomotor(s).   8, 58

distribution (adjective). A general term used, by reason of specific physical or electrical characteristics, to denote application or restriction of the modified term, or both, to that part of an electric system used for conveying energy to the point of utilization from a source or from one or more main receiving stations. *Notes:* (1) From the standpoint of a utility system, the area described is between the generating source, or intervening substations, and the customer's entrance equipment. (2) From the standpoint of a customer's internal system, the area described is between a source or receiving station within the customer's plant and the points of utilization.

103, 202

distribution amplifier. *See:* **amplifier, distribution.**

distribution box (mine type). A portable piece of apparatus with enclosure by means of which an electric circuit is carried to one or more machine trailing cables from a single incoming feed line, each trailing cable circuit being connected through individual overcurrent protective devices. *See also:* **distributor box; mine feeder circuit.**

328

distribution cable (distributing cable) (communication practice). A cable extending from a feeder cable into a specific area for the purpose of providing service to that area. *See also:* **cable.**   328

distribution center (secondary distribution). A point at which is located equipment consisting generally of automatic overload protective devices connected to buses, the principal functions of which are subdivision of supply and the control and protection of feeders, subfeeders, or branch circuits, or any combination of feeders, subfeeders, or branch circuits.   3

distribution coefficients (color). The tristimulus values of monochromatic radiations of equal power. *Note:* Generally represented by overscored, lower-case letters, such as $\bar{x}, \bar{y}, \bar{z}$ in the CIE (International Commission on Illumination) system.   18, 178

distribution cutout. A fuse or disconnecting device consisting of any one of the following assemblies: (1) A fuse support and fuseholder which may or may not include the conducting element (or fuse link). (2) A fuse support and disconnecting blade. (3) A fuse support and fuse carrier which may or may not include the conducting element (fuse link) or disconnecting blade. *Note:* In addition the distribution cutout is identified by the following characteristics: (A) dielectric withstand (basic impulse insulation level) strengths at distribution levels; (B) application primarily on distribution feeders and circuits; (C) mechanical construction basically adapted to pole or crossarm mounting except for the distribution oil cutout; (D) operating voltage limits corresponding to distribution systems voltage.   103, 202

distribution disconnecting cutout. *See:* **distribution cutout; disconnecting cutout.**

distribution enclosed single-pole air switch (distribution enclosed air switches). A single-pole disconnecting switch in which the contacts and blade are mounted completely within an insulated enclosure. (Cannot be converted into a distribution cutout or disconnecting fuse). *Note:* The distribution enclosed air switch is identified by the following characteristics: (1) dielectric withstand (basic impulse insulation level) strengths at distribution level; (2) application primarily on distribution feeders and circuits; (3) mechanical construction basically adapted to cross-arm mounting; (4) operating voltage limits correspond to distribution voltage. (5) Unless incorporating load-break means, it has no interrupting (load-break current) rating. (Some load-break ability is inherent in the device. This ability can best be evaluated by the user, based on experience under operating conditions).

103, 202, 79

distribution factor (rotating machinery). A factor related to a distributed winding, taking into account the spatial distribution of the slots in which the winding considered is laid, that is the decrease in the generated voltage, as a result of a geometrical addition of the corresponding representative vectors.   63

distribution feeder. *See:* **primary distribution feeder; secondary distribution feeder.** *See also:* **distribution center.**

distribution fuse cutout. *See:* **distribution cutout; fuse cutout.**

distribution main. *See:* **primary distribution mains; secondary distribution mains.** *See also:* **center of distribution.**

distribution network. *See:* **primary distribution network; secondary distribution network.** *See also:* **center of distribution.**

distribution oil cutout. *See:* **distribution; oil cutout.**

distribution open ctout. *See:* **distribution; open cutout.**

distribution open-link cutout. *See:* **distribution; open-link cutout.**

distribution switchboard. A power switchboard used for the distribution of electric energy at the voltages common for such distribution within a building. *Note:* Knife switches, air circuit breakers, and fuses are generally used for circuit interruption on distribution switchboards, and voltages seldom exceed 600. However, such switchboards often include switchboard equipment for a high-tension

incoming supply circuit and a step-down transformer.

103, 202

distribution system. That portion of an electric system which delivers electric energy from transformation points on the transmission, or bulk power system to the consumers.   112

distribution temperature (light source). The absolute temperature of a blackbody whose relative spectral distribution is the same (or nearly so) in the visible region of the spectrum as that of the light source. *See also:* **color.**

167

distribution transformer. A transformer for transferring electrical energy from a primary distribution circuit to a secondary distribution circuit or consumer's service circuit. *Note:* Distribution transformers are usually rated in the order of 5–500 kVA.   53

distribution trunk line. *See:* **primary distribution trunk line.** *See also:* **center of distribution.**

distributor box. A box or pit through which cables are inserted or removed in a draw-in system of mains. It contains no links, fuses, or switches and its usual function is to facilitate tapping into a consumer's premises. *See also:* **distribution box; tower.**   64

distributor duct. A duct installed for occupancy of distribution mains. *See:* **service pipe.**   64

distributor suppressor (internal-combustion engine terminology). A suppressor designed for direct connection to the high-voltage terminals of a distributor cap. *See also:* **electromagnetic compatibility.**   220, 199

disturbance (1) (general). An undesired variable applied to a system that tends to affect adversely the value of a controlled variable.   56, 105

(2) (communication practice). Any irregular phenomenon associated with transmission that tends to limit or interfere with the interchange of intelligence.   328

(3) (interference terminology). *See:* **interference.**

(4) (industrial control). An undesired input variable that may occur at any point within a feedback control system. *See also:* **control system, feedback.**   219, 206

(5) (storage tubes). That type of spurious signal generated within a tube that appears as abrupt variations in the amplitude of the output signal. *Notes:* (A) These variations are spatially fixed with reference to the target area. (B) The distinction between this and shading. (C) A blemish, a mesh pattern, and moire present in the output are forms of disturbance. Random noise is not a form of disturbance. *See also:* **storage tube.**   174, 190

disturbed-ONE output (magnetic cell). A ONE output to which partial-read pulses have been applied since that cell was last selected for writing. *See:* **coincident-current selection.** *See also:* **one output.**   331

disturbed-ZERO output (magnetic cell). A ZERO output to which partial-write pulses have been applied since that cell was last selected for reading. *See:* **coincident-current selection.**   331

dither (control circuits). A useful oscillation of small amplitude, introduced to overcome the effects of friction, hysteresis, or clogging. *See:* **control system, feedback.**

56, 219, 206

divergence (vector field at a point). A scalar equal to the limit of the quotient of the outward flux through a closed surface that surrounds the point by the volume within the surface, as the volume approaches zero. *Note:* If the vector **A** of a vector field is expressed in terms of its three rectangular components A_x, A_y, A_z, so that the values of A_x, A_y, A_z are each given as a function of x, y, and z, the divergence of the vector field **A** (abbreviated $\nabla \cdot \mathbf{A}$) is the sum of the three scalars obtained by taking the derivatives of each component in the direction of its axis, or

$$\text{div}\mathbf{A} \equiv \nabla \cdot \mathbf{A} = \frac{\partial A_x}{\partial x} + \frac{\partial A_y}{\partial y} + \frac{\partial A_z}{\partial z}$$

Examples: If a vector field **A** represents velocity of flow such as the material flow of water or a gas or the imagined flow of heat or electricity, the divergence of **A** at any point is the net outward rate of flow per unit volume and per unit time. It is the time rate of decrease in density of the fluid at that point. Because the density of an incompressible fluid cannot change, the divergence of an incompressible fluid is always zero. The divergence of the flow of heat at a point in a body is equal to the rate of generation of heat per unit volume at the point. The divergence of the electric field strength at a point is proportional to the volume density of charge at the point. 210

divergence loss (acoustic wave). The part of the transmission loss that is due to the divergence or spreading of the sound rays in accordance with the geometry of the system (for example, spherical waves emitted by a point source). *See also:* **loudspeaker.** 176

diversity, dual (data transmission). The term applied to the simultaneous combining of four signals and their detection through the use of space, frequency or pluralization characteristics. 59

diversity factor (system diversity factor). The ratio of the sum of the individual maximum demands of the various subdivisions of a system to the maximum demand of the whole system. *Note:* The diversity factor of a part of the system may be similarly defined as the ratio of the sum of the individual maximum demands of the various subdivisions of the part of the system to the maximum demand of the part of the system under consideration. *See also:* **alternating-current distribution; direct-current distribution.** 64

diversity gain. The reduction in predetection signal-to-interference energy ratio required to achieve a given level of performance, relative to that of a non-diversity radar, resulting from the use of diversity in frequency, polarization, space or other characteristic. 13

diversity reception. That method of radio reception whereby, in order to minimize the effects of fading, a resultant signal is obtained by combination or selection, or both, of two or more sources of received-signal energy that carry the same modulation or intelligence, but that may differ in strength or signal-to-noise ratio at any given instant. *See also:* **radio receiver.** 328

diversity, seasonal (electric power utilization). Load diversity between two (or more) electric systems which occurs when their peak loads are in different seasons of the year. 112

diversity, time zone (electric power utilization). Load diversity between two (or more) electric systems which occurs when their loads are in different time zones. 112

divided code ringing (divided ringing). A method of code ringing that provides partial ringing selectivity by connecting one-half of the ringers from one side of the line to ground and the other half from the other side of the line to the ground. This term is not ordinarily applied to selective and semiselective ringing systems. 328

divided ringing. *See:* **divided code ringing.**

divider. (1) A device capable of dividing one variable by another. (2) A device capable of attenuating a variable by a constant or adjustable amount, as an attenuator. *See also:* **electronic analog computer.** 9, 77

dividing network (crossover network) (loadspeaker dividing network). A frequency selective network that divides the spectrum into two or more frequency bands for distribution to different loads. *See also:* **loudspeaker.**
239, 176

D layer (radio wave propagation). An ionized layer in the D region. *See also:* **radio wave propagation.** 146

DME. *See:* **distance-measuring equipment.**

document (information processing) (computer). (1) A medium and the data recorded on it for human use, for example, a report sheet, a book. (2) By extension, any record that has permanence and that can be read by man or machine. 255, 77

documents (nuclear power generating stations). Drawings and other records significant to the design, construction, testing, maintenance, and operation of Class 1E equipment and systems for nuclear power generating stations. *Note:* Documents include: (1) Drawings such as instrument diagrams, functional control diagrams, one line diagrams, schematic diagrams, equipment arrangements, cable and tray lists, wiring diagrams. (2) Instrument data sheets. (3) Design specifications. (4) Instruction manuals. (5) Test specifications, procedures, and reports. (6) Device lists. Not to be included as documents are: project schedules, financial reports, meeting minutes, correspondence such as letters and memoranda, and equipment procurement documentation covered by quality assurance programs.
31

Doherty amplifier. A particular arrangement of a radio-frequency linear power amplifier wherein the amplifier is divided into two sections whose inputs and outputs are connected by quarter-wave (90-degree) networks and whose operating parameters are so adjusted that, for all values of the input signal voltage up to one-half maximum amplitude, Section No. 2 is inoperative and Section No. 1 delivers all the power to the load, which presents an impedance at the output of Section No. 1 that is twice the optimum for maximum output. At one-half maximum input level, Section No. 1 is operating at peak efficiency, but is beginning to saturate. Above this level, Section No. 2 comes into operation, thereby decreasing the impedance presented to Section No. 1, which causes it to deliver additional power into the load until, at maximum signal input, both sections are operating at peak efficiency and each section is delivering one-half the total output power to the load. *See also:* **amplifier.** 111

Dolph-Chebyshev array antenna. An array antenna with uniform interelement spacing and excitation coefficients chosen such that the array factor can be expressed as a Chebyshev polynominal. 111

domestic induction heater. A cooking device in which the utensil is heated by current, usually of commercial line frequency, induced in it by a primary inductor associated with it. *See also:* **induction heating.**   14, 114

dominant mode (fundamental mode) (waveguide transmission). The mode of propagation with the lowest cutoff frequency. *Note:* Designations for this mode are $TE_{1,0}$ and $TE_{1,1}$ for rectangular and circular waveguides, respectively. *See also:* **waveguide.**   328

dominant wave (uniconductor waveguide). The guided wave having the lowest cutoff frequency. *Note:* It is the only wave that will carry energy when the excitation frequency is between the lowest cutoff frequency and the next higher cutoff frequency. *See also:* **guided waves; waveguide.**
            267, 179

dominant wavelength (colored light, not purple). The wavelength of the spectrum light that, when combined in suitable proportions with the specified achromatic light, yields a match with the light considered. *Note:* When the dominant wavelength cannot be given (this applies to purples) its place is taken by the complementary wavelength. *See also:* **color; complementary wavelength.**
            18, 244, 178

Donnan potential (electrobiology). The potential difference across an inert semipermeable membrane separating mixtures of ions, attributed to differential diffusion. *See also:* **electrobiology.**   192

donor (semiconductor). *See:* **impurity, donor.** *See also:* **semiconductor.**

door contact (burglar-alarm system). An electric contacting device attached to a door frame and operated by opening or closing the door. *See also:* **protective signaling.**   328

door (gate closer). A device that closes a manually opened hoistway door, a car door, or gate by means of a spring or by gravity. *See also:* **hoistway (elevator or dumbwaiter).**
            328

door or gate power operator. A device, or assembly of devices, that opens a hoistway door and/or a car door or gate by power other than by hand, gravity, springs, or the movement of the car; and that closes them by power other than by hand, gravity, or the movement of the car. *See also:* **hoistway (elevator or dumbwaiter).**   328

dopant (1) (acceptor) (semiconductor). An impurity that may induce hole conduction. *See also:* **semiconductor device.** *Syn.* **impurity.**

(2) (donor) (semiconductor). An impurity that may induce electron conduction. *See also:* **semiconductor device.** *Syn.* **impurity.**   66

doping (semiconductor). Addition of impurities to a semiconductor or production of a deviation from stoichiometric composition, to achieve a desired characteristic. *See also:* **semiconductor.**   245, 210, 66

doping compensation (semiconductor). Addition of donor impurities to a *p*-type semiconductor or of acceptor impurities to an *n*-type semiconductor. *See also:* **semiconductor.**   66

Doppler effect (1) (acoustics, speech and signal processing). The phenomenon evidenced by the change in the observed frequency of a wave in a transmission system caused by a time rate of change in the effective length of the path of travel between the source and the point of observation.
            176, 59

(2) (communication satellite). The effective change of frequency of a received signal due to the relative velocity of a transmitter with respect to receiver. In space com-

munications the frequency shifts due to the Doppler effect may be significant when the velocity of the spacecraft relative to earth is high; the frequency shifts are used to determine the velocity of vehicles.   85

Doppler filter (radar). An equipment to separate targets and clutter echoes of different Doppler frequencies.
            13

Doppler-inertial navigation equipment. Hybrid navigation equipment that employs both Doppler navigation radar and inertial sensors. *See also:* **navigation.**   187, 13

Doppler navigator. A self-contained dead-reckoning navigation aid transmitting two or more beams of electromagnetic or acoustic energy outward and downward from the vehicle and utilizing the Doppler effect of the reflected energy, a refeence direction, and the relationship of the beams to the vehicle to determine speed and direction of motion over the reflecting surface. *See also:* **navigation.**   187, 13

Doppler radar. A radar which utilizes the Doppler effect to determine the radial component of relative radar-target velocity or to select targets having particular radial velocities.   13

Doppler shift. The magnitude of the change in the observed frequency of a wave due to the Doppler effect. The unit is the hertz.   176

Doppler system, pulsed (electronic navigation). *See:* **pulsed Doppler radar system.**

Doppler tracking (communication satellite). A method of determining the position of an observer on earth using the known [exact] satellite transmission frequency and the known satellite ephemeris and measuring the Doppler frequency shift of the signal received from the satellite.
            84

Doppler very-high-frequency omnidirectional range. A very-high-frequency radio range, operationally compatible with conventional very-high-frequency omnidirectional ranges, less susceptible to siting difficulties because of its increased aperture; in it the variable signal (the signal producing azimuthal information) is developed by sequentially feeding a radio-frequency signal to a multiplicity of antennas disposed in a ring-shaped array. *Note:* The array usually surrounds the central source of reference signal. *See also:* **navigation.**   187, 13

dose (dosage) (photovoltaic power system). The radiation delivered to a specified area or the whole body. *Note:* Units of dose are rads or roentgens for X or gamma rays and rads for beta rays and protons. *See also:* **photovoltaic power system; solar cells (in a photovoltaic power system).**
            186

dose rate (photovoltaic power system). Radiation dose delivered per unit time. *See also:* **photovoltaic power system; solar cells (in a photovoltaic power system).**   186

dot cycle. One cycle of a periodic alternation between two signaling conditions, each condition having unit duration. *Note:* Thus, in two-condition signaling, it consists of a dot, or marking element, followed by a spacing element. *See also:* **telegraphy.**   59, 194

dot product. *See:* **scalar product.**

dot-product line integral. *See:* **line integral.**

dot-sequential (color television). Sampling of primary colors in sequence with successive picture elements.   18, 178

dot signal (data transmission). A series of binary digits having equal and opposite states, such as a series of alternate "1" and "0" states. The dot cycle per second rate is one half the baud rate. 59

double aperture seal (nuclear power generating stations). Two single aperture seals in series. *See:* **single aperture seal.** 26

double-break switch (circuit). A switch that opens a conductor at two points. 103, 202, 27

double bridge (Thomson bridge). *See:* **Kelvin bridge.**

double-buffered DAC (DAM) (hybrid computer linkage components). A digital-to-analog converter (DAC) or digital-to-analog multiplier (DAM) with two registers in cascade, one a holding register, and the other the dynamic register. *See:* **dynamic register; holding register.** 10

double-circuit system (protective signaling). A system of protective wiring in which both the positive and the negative sides of the battery circuit are employed, and that utilizes either an open or a short circuit in the wiring to initiate an alarm. *See also:* **protective signaling.** 328

double connection (telephone switching systems). A fault condition whereby two separate calls are connected together. 55

double-current generator. A machine that supplies both direct and alternating currents from the same armature winding. *See:* **direct-current commutating machine.** 63

double diode (electron device). An electron tube or valve containing two diode systems. *See:* **multiple tube (valve).** 190

double electric conductor seal (nuclear power generating stations). Two single electric conductor seals in series. *See:* **single electric conductor seal.** 26

double-end control. A control system in which provision is made for operating a vehicle from either end. *See also:* **multiple-unit control.** 328

double-faced tape. Fabric tape finished on both sides with a rubber or synthetic compound. *See also:* **power distribution, underground construction.** 64

double-fed asynchronous machine (rotating machinery). An asynchronous machine of which the stator winding and the rotor winding are fed by supply frequencies each of which may be either constant or variable. *See:* **asynchronous machine.** 63

double-gun cathode-ray tube. A cathode-ray tube containing two separate electron-gun systems. 190

double-integrating gyro. A single-degree-of-freedom gyro having no intentional elastic or viscous restraint of the gimbal about the output axis so that the dynamic behaviour is primarily established by the inertial properties of the gimbal. In this gyro, an output signal is produced by gimbal angular displacement, relative to the case, which is proportional to the double integral of the angular rate of the case about the input axis. 46

double length. Pertaining to twice the normal length of a unit of data or a storage device in a given computing system. *Note:* For example, a double-length register would have the capacity to store twice as much data as a single-length or normal register; a double-length word would have twice the number of characters or digits as a normal or single-length word. *See also:* **double precision.** 235

double-length number (double-precision number). A number having twice as many digits as are ordinarily used in a

given computer. 210

double modulation. The process of modulation in which a carrier wave of one frequency is first modulated by a signal wave and a resultant wave is then made to modulate a second carrier wave of another frequency. 328

double pole-piece magnetic head (electroacoustics). A magnetic head having two separate pole pieces in which pole faces of opposite polarity are on opposite sides of the medium. *Note:* One or both of these pole pieces may be provided with an energizing winding. *See also:* **phonograph pickup.** 176

double-pole relay. *See:* **relay, double-pole.**

double precision (electronic computation). Pertaining to the use of two computer words to represent a number in order to preserve or gain precision. 77

double-precision number. *See:* **double-length number.**

double pulse (pulse terms). Two pulse waveforms of the same polarity which are adjacent in time and which are considered or treated as a single feature. 254

double-secondary current transformer. One which has two secondary coils each on a separate magnetic circuit with both magnetic circuits excited by the same primary winding. 53

double-secondary voltage transformer. One which has two secondary windings on the same magnetic circuit insulated from each other and the primary. 53

double-sideband transmitter. A transmitter that transmits the carrier frequency and both sidebands resulting from the modulation of the carrier by the modulating signal. *See also:* **radio transmitter.** 111

double squirrel cage (rotating machinery). A combination of two squirrel-cage windings mounted on the same induction-motor rotor, one at a smaller diameter than the other. *Note:* It is common but not essential for the two windings to have the same number of slots. In any case, each bar of the lower (smaller-diameter) winding is located at the bottom of a slot containing a bar of the upper winding. A narrow portion of the slot (called the **leakage slot**) is provided in the radial separation between the two bars. *See also:* **asynchronous machine.** 63

double-superheterodyne reception (triple detection). The method of reception in which two frequency converters are employed before final detection. *See also:* **radio receiver.** 328

doublet antenna. *See:* **dipole antenna.**

double-throw (mechanical switching device). A qualifying term indicating that the device can change the circuit connections by utilizing one or the other of its two operating positions. *Note:* A double-throw air switch changes circuit connections by moving the switchblade from one of two sets of contact clips into the other. 103, 202

double-tuned amplifier. An amplifier of one or more stages in which each stage utilizes coupled circuits having two frequencies of resonance, for the purpose of obtaining wider bands than those obtainable with single tuning. *See also:* **amplifier.** 328

double-tuned circuit. A circuit whose response is the same as that of two single-tuned circuits coupled together. 328

double-way rectifier (power semiconductor). A rectifier unit which makes use of a double-way rectifier circuit. 66

double-way rectifier circuit. A rectifier circuit in which the current between each terminal of the alternating-voltage circuit and the rectifier circuit elements conductively connected to it flows in both directions. *Note:* The terms single-way and double-way provide a means for describing the effect of the rectifier circuit on current flow in transformer windings connect to rectifier circuit elements. Most rectifier circuits may be classified into these two general types. Double-way rectifier circuits are also referred to as bridge rectifier circuits. *See also:* **rectification; rectifier circuit element; power rectifier; single-way rectifier circuit; bridge rectifier circuit.** 208

double-winding synchronous generator. A generator that has two similar windings, in phase with one another, mounted on the same magnetic structure but not connected electrically, designed to supply power to two independent external circuits. 63

doughnut (electronic device). *See:* **toroid.**

dovetail projection. A tenon, commonly flared; used for example, to fasten a pole to the spider. *See also:* **stator.** 63

dovetail slot. (1) A recess along the side of a coil slot into which a coil-slot wedge is inserted. (2) A flaring slot into which a dovetail projection is engaged; used for example, to fasten a pole to the spider. *See also:* **stator.** 63

dowel (dowel pin). A pin fitting with close tolerance into a hole in abutting pieces to establish and maintain accurate alignment of parts. Frequently designed to resist a shear load at the interface of the abutting pieces. 63

down conductor. The vertical portion of run of conductor that ends at the ground. *See also:* **lightning protection and equipment.**

down lead (lightning protection). The conductor connecting an overhead ground wire or lightning conductor with the grounding system. *See also:* **direct-stroke protection (lightning).** 64

downlight. A small direct lighting unit which can be recessed, surface mounted, or suspended. *See also:* **luminaire.** 167

down link (communication satellite). A transmission link carrying information from a satellite or spacecraft to earth. Typically down links carry telemetry, data and voice. 83

down time. (1) (general). The period during which a system or device is not operating due to internal failures, scheduled shut down, or servicing. *See:* **electric drive.** 219, 206

(2) (availability). The period of time during which an item is not in a condition to perform its required function. *Notes:* (A) The down time of an item will be made up of active maintenance time and delays due to awaiting labour, awaiting spares, facilities, movement, etc. (B) Unless otherwise stated, down time of an item, due to failure, is considered to commence at the instant the item is determined to have failed. (C) Unless otherwise stated, down time will include any additional time necessary to reach the same stage in the working programme of the item as at the time of failure. 164

downward component (illuminating engineering). That portion of the luminous flux from a luminaire that is emitted at angles below the horizontal. *See also:* **inverse-square law (illuminating engineering).** 167

downward modulation. Modulation in which the instan-taneous amplitude of the modulated wave is never greater than the amplitude of the unmodulated carrier. 339

drag-in (electroplating). The quantity of solution that adheres to cathodes when they are introduced into a bath. *See also:* **electroplating.** 328

drag magnet. *See:* **retarding magnet.**

drag-out (electroplating). The quantity of solution that adheres to cathodes when they are removed from a bath. *See also:* **electroplating.** 328

drain. The current supplied by a cell or battery when in service. *See also:* **battery (primary or secondary).** 328

drainage (corrosion). Conduction of current (positive electricity) from an underground metallic structure by means of a metallic conductor. 221, 205

drain coil (power-system communication). A reactor or choke connected between the carrier-current lead-in terminal of a coupling capacitor and ground, to present a low impedance to the flow of power current and a high impedance to the flow of carrier-frequency current. *Note:* Its purpose is to prevent high voltage at power frequency from being impressed on the carrier-current lead and to limit the loss of carrier-frequency energy to ground. *See also:* **power-line carrier.** 59

drain line (rotating machinery) (bearing oil system). A return pipe line using gravity flow. *See also:* **oil cup (rotating machinery).** 63

drawbar pull (cable plowing). The effective pulling force delivered. 52

drawbridge coupler. *See:* **movable-bridge coupler.**

drawdown. The distance that the water surface of a reservoir is lowered from a given elevation as the result of the withdrawal of water. 112

drawout-mounted device. A device having disconnecting devices and in which the removable portion may be removed from the stationary portion without the necessity of unbolting connections or mounting supports. *Note:* Compare with **stationary-mounted device.** 103, 202

D region (radio wave propagation). The region of the terrestrial ionosphere between about 40 and 90 kilometers altitude responsible for most of the attenuation of radio waves in the range 1 to 100 megahertz. *See also:* **radiation; radio wave propagation.** 180, 146

dribble-up*. *See:* **distortion, pulse.**

* Deprecated

drift (1) (rotating machinery). A long-time change in synchronous-machine regulating system error resulting from causes such as aging of components, self-induced temperature changes, and random phenomena. *Note:* Maximum acceptable drift is normally a specified change for a specified period of time, for specified conditions. 63

(2) (industrial control). An undesired but relatively slow change in output over a period of time, with a fixed reference input. *Note:* Drift is usually expressed in percent of the maximum rated value of the variable being measured. *See:* **control system, feedback.** 219, 206

(3) (sound recording and reproducing). Frequency modulation of the signal in the range below approximately 0.5 Hz resulting in distortion which may be perceived as a slow changing of the average pitch. *Note:* Measurement of drift is not covered by this definition. 145

(4) (analog computer). A slowly varying error in an integrator, caused by the integration of offset errors at the inputs, capacitor leakage, or both. Also, any slowly varying error in a computer component.   9

(5) (electronic navigation). *See:* **G drift; G² drift.**

(6) (power supplies). *See:* **stability, long term.**

(7) (excitation systems) (automatic control). An undesired change in output over a period of time, which change is unrelated to input, environment, or load. *Note:* The change is a plus of minus variation of short periods which may be superimposed on plus or minus variations of a long-time period. On a practical system drift is determined as the change in output over a specified time with fixed command and fixed load, with specified environmental conditions.

105, 56

(8) (oscilloscopes). *See:* **stability.**

drift angle (navigation). The angular difference between the heading and the track. *See also:* **navigation.**

187, 13

drift band of amplification (magnetic amplifier). The maximum change in amplification due to uncontrollable causes for a specified period of time during which all controllable quantities have been held constant. *Note:* The units of this drift band are the amplification units per the time period over which the drift band was determined.

171

drift compensation (industrial control). The effect of a control function, device, or means to decrease overall systems drift by minimizing the drift in one or more of the control elements. *Note:* Drift compensation may apply to feedback elements, reference input, or other portions of a system. *See:* **control system, feedback.**   225, 206

drift correction angle (navigation). The angular difference between the course and the heading. Sometimes called the crab angle. *See also:* **navigation.**   13, 187

drift, direct-current. *See:* **drift.**

drift, G². *See* **G² drift.**

drift, kinematic (radar). *See:* **misalignment drift.**

drift mobility (homogeneous semiconductor). The ensemble average of the drift velocities of the charge-carriers per unit electric field. *Note:* In general, the mobilities of electrons and holes are different. *See also:* **semiconductor device.**   210, 186, 245, 66

drift offset (magnetic amplifier). The change in quiescent operating point due to uncontrollable causes over a specified period of time when all controllable quantities are held constant.   171

drift rate (1) (voltage regulators or reference tubes). The slope at a stated time of the smoothed curve of tube voltage drop with time at constant operating conditions.

190, 125

(2) (gyro). Except for rate gyros, the drift rate is the time rate of angular deviation of the spin axis from the desired orientation in inertial space. It consists of random and systematic components and is expressed as an equivalent input angular displacement per unit time. For laser gyros, drift rate is the indicated time rate of angular deviation of the case from the desired orientation in inertial space.

46

drift space (electron tube). A region substantially free of externally applied alternating fields, in which a relative repositioning of the electrons takes place.   190, 125

drift, stability (electric conversion). Gradual shift or change

in the output over a period of time due to change or aging of circuit components. (All other variables held constant). *See also:* **electric conversion equipment.**   186

drift stabilization (analog computer). Any automatic method used to minimize the drift of a direct-current amplifier. *See also:* **electronic analog computer.**   9, 77

drift tunnel (velocity-modulated tube). A piece of metal tubing, held at a fixed potential, that forms the drift space. *Note:* The drift tunnel may be divided into several parts, which constitute the drift electrodes. *See also:* **velocity-modulated tube.**   244, 190

drift, zero. Drift with zero input. *See also:* **electronic analog computer.**   9, 77

dripproof. So constructed or protected that successful operation is not interfered with when falling drops of liquid or solid particles strike or enter the enclosure at any angle from 0 to 15 degrees from the downward vertical unless otherwise specified. *See also:* **asynchronous machine; direct-current commutating machine.**

3, 232, 63, 103, 202

dripproof enclosure (1) (general). An enclosure, usually for indoor application; so constructed or protected that falling drops of liquid or solid particles which strike the enclosure at any angle within a specified deviation from the vertical shall not interfere with the successful operation of the enclosed equipment.   27, 103, 53

(2) (metal enclosed bus). An enclosure usually for indoor application; so constructed or protected that falling drops of liquid or solid particles which strike the enclosure at any angle not greater than 15° from the vertical shall not interfere with the successful operation of the metal-enclosed bus.   78

dripproof machine. An open machine in which the ventilating openings are so constructed that drops of liquid or solid particles falling on the machine at any angle not greater than 15 degrees from the vertical cannot enter the machine either directly or by striking and running along a horizontal or inwardly inclined surface. *See:* **asynchronous machine; direct-current commutating machine.**

328

driptight (transformer). So constructed or protected as to exclude falling dirt or drops of liquid, under specified test conditions.   53

driptight enclosure. An enclosure so constructed that falling drops of liquid or solid particles striking the enclosure at any angle within a specified variation from the vertical cannot enter the enclosure either directly or by striking and running along a horizontal or inwardly inclined surface.   103, 202, 27, 53

drive (1) (industrial control). The equipment used for converting available power into mechanical power suitable for the operation of a machine. *See:* **electric drive.**

225, 206

(2) (electronic computation and recording). *See:* **tape drive.**

driven element (antenna). A radiating element coupled directly to the feed line. *See also:* **antenna.**   179

drive pattern (facsimile). Density variation caused by periodic errors in the position of the recording spot. When caused by gears this is called gear pattern. *See also:* **recording (facsimile).**   12

drive pin (disc recording). A pin similar to the center pin, but located to one side thereof, that is used to prevent a disc record from slipping on the turntable. *See also:* **phonograph pickup.** 176

drive-pin hole (disc recording). A hole in a disc record that accommodates the turntable drive pin. 176

drive pulse (static magnetic storage). A pulsed magnetomotive force applied to a magnetic cell from one or more sources. 331

driver (communication practice). An electronic circuit that supplies input to another electronic circuit. 328

drive strip (chafing strip) (rotating machinery). An insulating strip located in the coil slots between the wedge and the top of the slot armor or the top coil side, to provide protection during assembly of the wedges. *See also:* **stator.** 63

driving machine. The power unit that applies the energy necessary to raise and lower an elevator or dumbwaiter car or to drive an escalator or a private-residence inclined lift. 328

driving-point admittance (between the jth terminal and the reference terminal of an n-terminal network). The quotient of (1) the complex alternating component I_j of the current flowing to the jth terminal from its external termination by (2) the complex alternating component V_j of the voltage applied to the jth terminal with respect to the reference point when all other terminals have arbitrary external terminations. *Note:* In specifying the driving-point admittance of a given pair of terminals of a network or transducer having two or more pairs of terminals, no two pairs of which contain a common terminal, all other pairs of terminals are connected to arbitrary admittances. *See also:* **electron-tube admittances; linear passive networks; network analysis.** 190, 125

driving-point function (linear passive network). A response function for which the variables are measured at the same port (terminal pair). *See also:* **linear passive network.** 238

driving-point impedance (network). At any pair of terminals the ratio of an applied potential difference to the resultant current at these terminals, all terminals being terminated in any specified manner. *See also:* **linear passive networks; self-impedance.** 328

driving power, grid. *See:* **grid driving power.**

driving signals (television). Signals that time the scanning at the pickup point. *Note:* Two kinds of driving signals are usually available from a central synchronizing generator. One is composed of pulses at line frequency and the other is composed of pulses at field frequency. *See also:* **television.** 178

driving test circuit (telephone equipment). A test circuit used to convert an exciting test voltage into balanced longitudinal voltages on tip and ring leads. *See also:* **metallic transmission port.** 39

droop* (pulse techniques). *See:* **istortion, pulse.**

* Deprecated

droop, frequency (power systems). The absolute change in frequency between steady state no load and steady state full load. 89

drop (drop signal). A visual shutter device consisting of an electromagnet and a visual target either moved or tripped

magnetically to a position indicating the condition supervised. 328

drop-away. The electrical value at which the movable member of an electromagnetic device will move to its de-energized position. 328

dropout (relay) (1) (power switchgear). A term for contact operation (opening or closing) as a relay just departs from pickup. Also identifies the maximum value of an input quantity which will allow the relay to dropout. *Note:* Dropout has been used to denote reset for electromechanical relays that have no intermediate steady-state position between pickup and reset and that reset instantaneously. *See also:* **relay.** 103, 202, 127

(2) (protective relaying of utility-consumer interconnections). Contact operation (opening or closing) as a relay just departs from pickup. The value at which dropout occurs is usually stated as a percentage of pickup. For example, dropout ratio of a typical instantaneous overvoltage relay is 90 percent. 128

dropout fuse. A fuse in which the fuseholder or fuse unit automatically drops into an open position after the fuse has interrupted the circuit. 103, 202

dropout ratio (relay). The ratio of dropout to pickup of an input quantity. *Note:* This term has been used mostly with relays for which reset is not differentiated from dropout. Hence a similar term, reset ratio, the ratio of reset to pickup, is not generally used, though technically correct. 127, 103

dropouts (data transmission). This is a loss of discrete data signals due to noise or attenuation hits. 59

dropout time (relay). The time interval to dropout following a specified change of input conditions. *Note:* When the change of input conditions is not specified it is intended to be a sudden change from pickup value of input to zero input. 103, 202, 60, 127

dropout voltage (or current) (magnetically operated device). The voltage (or current) at which the device will release to its de-energized position. *See also:* **contactor; control switch; extinguishing voltage.** 89

drop, voltage, anode (glow-discharge cold-cathode tube). *See:* **anode voltage drop.**

drop, voltage, starter (glow-discharge cold-cathode tube). *See:* **starter voltage drop.**

drop, voltage, tube (glow-discharge cold-cathode tube). *See:* **tube voltage drop.**

drop wire (drop). Wire suitable for extending an open wire or cable pair from a pole or cable terminal to a building. *See also:* **cable; open wire.** 328

drum. *See:* **magnetic drum.**

drum controller (industrial control). An electric controller that utilizes a drum switch as the main switching element. *Note:* A drum controller usually consists of a rum switch and a resistor. *See also:* **electric controller.** 206, 3

drum factor (facsimile). The ratio of usable drum length to drum diameter. *Note:* Before a picture is transmitted, it is necessary to verify that the ratio of used transmitter drum length to transmitter drum diameter is not greater than the receiver drum factor if the receiver is of the drum type. *See also:* **facsimile (electrical communication).** 194

drum speed (facsimile). The angular speed of the transmitter or recorder drum. *Note:* This speed is measured in revolutions per minute. *See also:* **recording (facsimile); scanning.** 12

drum switch (industrial control). A switch in which the electric contacts are made of segments or surfaces on the periphery of a rotating cylinder or sector, or by the operation of a rotating cam. *See also:* **control switch; switch.** 206

dry-arcing distance (insulator). The shortest distance through the surrounding medium between terminal electrodes, or the sum of the distances between intermediate electrodes, whichever is the shorter, with the insulator mounted for dry flashover test. *See also:* **insulator.** 261

dry cell. A cell in which the electrolyte is immobilized. *See also:* **electrochemistry.** 328

dry contact. One through which no direct current flows. 193, 55

dry friction. *See:* **Coulomb friction.**

dry location (electric system). A location not normally subject to dampness or wetness. *Note:* A location classified as dry may be temporarily subject to dampness or wetness, as in the case of a building under construction. *See also:* **distribution center.** 256

dry reed relay. A reed relay with dry (nonmercury-wetted) contacts. 341

dry-type (1) (current-limiting reactor) (grounding device). Having the coils immersed in an insulating gas. *See also:* **reactor.** 309, 91
(2) (regulator). Having the core and coils not immersed in an insulating liquid. *See also:* **voltage regulator.** 257
(3) (transformer). Having the core and coils neither impregnated with an insulating fluid nor immersed in an insulating oil. *See:* **dry-type transformer.** 203

dry-type forced-air-cooled transformer (class AFA). A dry-type transformer which derives its cooling by the forced circulation of air. 53

dry-type nonventilated self-cooled transformer (class ANV). A dry-type self-cooled transformer which is so constructed as to provide no intentional circulation of external air through the transformer, and operating at zero gauge pressure. *See also:* **reactor; transformer; voltage regulator.** 53, 257

dry-type, self-cooled/forced-air-cooled transformer (class AA/FA). A dry-type transformer which has a self-cooled rating with cooling obtained by the natural circulation of air and a forced-air-cooled rating with cooling obtained by the forced circulation of air. 53

dry-type self-cooled transformer (classAA). A dry-type transformer which is cooled by the natural circulation of air. 53

dry-type transformer. A transformer in which the core and coils are in a gaseous or dry compound insulating medium. *See also:* **ventilated dry-type transformer; nonventilated dry-type transformer; sealed transformer; gas-filled transformer; compound-filled transformer; reactor; transformer; voltage regulator.** 53, 257

dry vault. A ventilated, enclosed area not subject to flooding. 79

***D*-scan (radar).** *See:* ***D*-display.**
***D*-scope (radar).** *See:* ***D*-display.**
dual-beam oscilloscope. A multibeam oscilloscope in which

the cathode-ray tube produces two separate electron beams that may be individually or jointly controlled. *See:* **multibeam oscilloscope.** *See also:* **oscillograph.** 185

dual benchboard. A combination assembly of a benchboard and a vertical-hinged panel switchboard placed back to back (no aisle) and enclosed with a top and ends. *Note:* No primary switching devices are located between the benchboard and panels. 103, 202

dual control. A term applied to signal appliances provided with two authorized methods of operation. 328

dual-element bolometer unit. An assembly consisting of two bolometer elements and a bolometer mount in which they are supported. *Note:* The bolometer elements are effectively in series to the bias power and in parallel to the RF power. 115

dual-element electrothermic unit. An assembly consisting of two thermopile elements and an electrothermic mount in which they are supported. The thermopile elements are effectively in series to the output voltage and in parallel to the RF power. The thermopiles also serve as the power absorber. 47

dual-element fuse (single fuse). A fuse having current-responsive elements of two different fusing characteristics in series. 103, 202

dual-element substitution effect (error). A component of substitution error, peculiar to dual-element bolometer units, that can cause the effective efficiency to vary with RF input power level. *Note:* This component, usually very small, is included in the effective efficiency correction for substitution error only with reference conditions for input RF power level and frequency. It results from a different division of RF and bias powers between the two elements. 115

dual-frequency induction heater or furnace. A heater in which the charge receives energy by induction, simultaneously or successively, from a work coil or coils operating at two different frequencies. *See also:* **induction heating.** 14, 114

dual-headlighting system. A system consists of two double headlighting units, one mounted on each side of the front end of a vehicle. *Note:* Each unit consists of two sealed-beam lamps mounted in a single housing. The upper or outer lamps have two filaments that supply the lower beam and part of the upper beam, respectively. The lower or inner lamps have one filament that provides the primary source of light for the upper beam. *See also:* **headlamp.** 167

dual-mode. *See:* **control system, dual-mode.**

dual modulation (facsimile). The process of modulating a common carrier wave or subcarrier by two different types of modulation. For example, amplitude and frequency modulation, each conveying separate information. *See also:* **facsimile transmission.** 12

dual networks. *See:* **structurally dual networks.**

dual overcurrent trip. *See:* **dual release (dual trip); overcurrent release (overcurrent trip).**

dual release (dual trip). A release that combines the function of a delayed and an instantaneous release. 103, 202

dual service (plural service). Two separate services, usually of different characteristics, supplying one consumer. *Note:* A dual service might consist of an alternating-current and direct-current service, or of 208Y/120 volt 3-phase, 4-wire service for light and some power and a 13.2-kilovolt service for power, etcetera. *See:* **service; duplicate service; emergency service; loop service.**   64

dual switchboard. A control switchboard with front and rear panels separated by a comparatively short distance and enclosed at both ends and top. *Notes:* (1) The panels on at least one side are hinged for access to the panel wiring. (2) No primary switching devices are located between the panels.   103, 202

dual-tone multifrequency pulsing (telephone switching systems). A means of pulsing utilizing a simultaneous combination of one of a lower group of frequencies and one of a higher group of frequencies to represent each digit or character.   55

dual trace (displays). A multitrace operation in which a single beam in a cathode-ray tube is shared by two signal channels. *See:* **alternate display; chopped display; multitrace; oscillograph.**   185, 106

dubbing (electroacoustics). A term used to describe the combining of two or more sources of sound into a complete recording at least one of the sources being a recording. *See also:* **phonograph pickup; re-recording.**   176

duck tape. Tape of heavy cotton fabric, such as duck or drill, which may be impregnated with an asphalt, rubber, or synthetic compound.   64

duct (1) (underground electric systems). A single enclosed runway for conductors or cables.   64
(2) (transmission and distribution). A single enclosed raceway for conductors or cable.   262

duct bank (conduit run). An arrangement of conduit providing one or more continuous ducts between two points. *Note:* An underground runway for conductors or cables, large enough for workmen to pass through, is termed a gallery or tunnel.   64

duct edge fair-lead (cable shield). A collar or thimble, usually flared, inserted at the duct entrance in a manhole for the purpose of protecting the cable sheath or insulation from being worn away by the duct edge.   64, 57

duct entrance. The opening of a duct at a manhole, distributor box, or other accessible space.   64

ducting (radar). Confinement of electromagnetic wave propagation to a restricted atmospheric layer by steep gradients in the index of refraction with altitude.   13

duct rodding (rodding a duct). The threading of a duct by means of a jointed rod of suitable design for the purpose of pulling in the cable-pulling rope, mandrel, or the cable itself.   64

duct sealing. The closing of the duct entrance for the purpose of excluding water, gas, or other undesirable substances.   64

duct spacer (vent finger) (rotating machinery). A spacer between adjacent packets of laminations to provide a radial ventilating duct.   63

duct ventilated (pipe ventilated) (rotating machinery). A term applied to apparatus that is so constructed that a cooling gas can be conveyed to or from it through ducts.   63

dumbwaiter. A hoisting and lowering mechanism equipped with a car that moves in guides in a substantially vertical direction, the floor area of which does not exceed 9 square feet, whose total inside height whether or not provided with fixed or removable shelves does not exceed 4 feet, the capacity of which does not exceed 500 pounds, and which is used exclusively for carrying materials.   328

dummy antenna. A device that has the necessary impedance characteristics of an antenna and the necessary power-handling capabilities, but that does not radiate or receive radio waves. *Note:* In receiver practice, that portion of the impedance not included in the signal generator is often called **dummy antenna.** *See also:* **radio receiver.**   339

dummy-antenna system. An electric network that simulates the impedance characteristics of an antenna system. *See also:* **navigation.**   278, 187, 13

dummy coil (rotating machinery). A coil that is not required electrically in a winding, but that is installed for mechanical reasons and left unconnected. *See also:* **rotor (rotating machinery); stator.**   63

dummy load (radio transmission). A dissipative but essentially nonradiating substitute device having impedance characteristics simulating those of the substituted device. *See also:* **artificial load; radio transmission.**   185

dump (computing systems). (1) To copy the contents of all or part of a storage, usually from an internal storage into an external storage. (2) A process as in (1). (3) The data resulting from the process as in (1). *See:* **dynamic dump; postmortem; selective dump; snapshot dump; static dump.**   255, 77, 54

dump power. Power generated from water, gas, wind, or other source that cannot be stored or conserved and that is beyond the immediate needs of the electric system producing the power. *See also:* **generating station.**   64

duodecimal. (1) Pertaining to a characteristic or property involving a selection, choice, or condition in which there are twelve possibilities. (2) Pertaining to the numeration system with a radix of twelve.   255, 77

duolateral coil. *See:* **honeycomb coil.**

duplex (communications). Pertaining to a simultaneous two-way independent transmission in both directions. *See:* **half duplex; full duplex.**   255, 77

duplex artificial line (balancing network). A balancing network, simulating the impedance of the real line and distant terminal apparatus, that is employed in a duplex circuit for the purpose of making the receiving device unresponsive to outgoing signal currents. *See also:* **telegraphy.**   328

duplex benchboard. A combination assembly of a benchboard and a vertical control switchboard placed back to back and enclosed with a top and ends (not grille). *Notes:* (1) Access space with entry doors is provided between the benchboard and vertical control switchboard. (2) No primary switching devices are located between the benchboard and panels.   103, 202

duplex cable. (1) A cable composed of two insulated stranded conductors twisted together. *Note:* They may or may not have a common insulating covering. *See also:* **power distribution, underground construction.**   345
(2) A cable composed of two insulated single-conductor cables twisted together. *Note:* The assembled conductors may or may not have a common covering of binding or protecting material. *See also:* **power distribution, underground construction.**   64

duplex cavity (radar). *See:* **transmit-receive cavity**

(radar).

duplexer (radar practice). A device that utilizes the finite delay between the transmission of a pulse and the echo thereof so as to permit the connection of the transmitter and receiver to a common antenna. *Note:* A duplexer commonly employs a transmit-receive switch and an antitransmit-receive switch, though the latter is sometimes omitted. *See also:* **radar.**   328

duplexing assembly, radar. *See:* **transmit-receive switch.**

duplex lap winding (rotating machinery). A lap winding in which the number of parallel circuits is equal to twice the number of poles.   63

duplex operation (1) (general). The operation of transmitting and receiving apparatus at one location in conjunction with associated transmitting and receiving equipment at another location, the processes of transmission and reception being concurrent. *See also:* **telegraphy.**
                    111, 59

(2) (radio communication) (two-way radio communication circuit). The operation utilizing two radio-frequency channels, one for each direction of transmission, in such manner that intelligence may be transmitted concurrently in both directions.   181

duplex signaling (telephone switching systems). A form of polar-duplex signaling for a single physical circuit.   55

duplex switchboard. A control switchboard consisting of panels placed back to back and enclosed with a top and ends (not grille). *Notes:* (1) Access space with entry doors is provided between the rows of panels. (2) No primary switching devices are located between the panels.
                    103, 202

duplex system. A telegraph system that affords simultaneous independent operation in opposite directions over the same channel. *See also:* **telegraphy.**   328

duplex wave winding (rotating machinery). A wave winding in which the number of parallel circuits is four, whatever the number of poles.   63

duplicate. *See:* **copy.**

duplicate lines (power transmission). Lines of substantially the same capacity and characteristics, normally operated in parallel, connecting the same supply point with the same distribution point. *See also:* **center of distribution.**   64

duplicate service (power transmission). Two services, usually supplied from separate sources, of substantially the same capacity and characteristics. *Note:* The two services may be operated in parallel on the consumer's premises, but either one alone is of sufficient capacity to carry the entire load. *See:* **service; dual service; emergency service; loop service.**   64

duplication check. A check based on the consistency of two independent performances of the same task.   167, 77

duration (pulse terms). The absolute value of the interval during which a specified waveform or feature exists or continues.   254

dust-ignition proof machine. A totally enclosed machine whose enclosure is designed and constructed in a manner that will exclude ignitable amounts of dusts or amounts that might affect performance or rating, and that, when installation and protection are in conformance with the National Electrical Code (ANSI C1-1975; section 502-1), will not permit arcs, sparks, or heat otherwise generated or liberated inside of the enclosure to cause

ignition of exterior accumulations or atmospheric suspensions of a specific dust on or in the vicinity of the enclosure. *See:* **asynchronous machine; direct-current commutated machine.**   232, 63

dustproof (1) (general). So constructed or protected that the accumulation of dust will not interfere with successful operation.   225, 206, 103, 202, 27

(2) (enclosure). An enclosure so constructed or protected that any accumulation of dust that may occur within the enclosure will not prevent the successful operation of, or cause damage to, the enclosed equipment.   3

(3) (luminaire). Luminaire so constructed or protected that dust will not interfere with its successful operation. *See also:* **luminaire.**   167

dust seal (rotating machinery). A sealing arrangement intended to prevent the entry of a specified dust into a bearing. *See also:* **asynchronous machine; direct-current commutating machine.**   63

dusttight (1) (enclosure). An enclosure so constructed that dust will not enter the enclosing case.   3, 225, 206, 27

(2) (luminaire) (transformer). An enclosure so constructed that dust will not enter the enclosing case under specified conditions. *See also:* **luminaire.**   103, 202, 167, 53

duty (1) (general). A statement of loads including no-load and rest and de-energized periods, to which the machine or apparatus is subjected including their duration and sequence in time.   63

(2) (rating of electric equipment). A statement of the operating conditions to which the machine or apparatus is subjected, their respective durations, and their sequence in time.   310, 233

(3) (industrial control) (transformers). A requirement of service that defines the degree of regularity of the load.
                    203, 53

(4) (excitation systems). Those voltage and current loadings imposed by the synchronous machine upon the excitation system including short circuits and all conditions of loading. *Note:* The duty will include the action of limiting devices to maintain synchronous machine loading at or below that defined by American National Standard Requirements for Cylindrical Rotor Synchronous Generators, C50.13-1965.   105

duty continuous (thyristor converter). A duty where the converter equipment carries a direct current of fixed value for an interval sufficiently long for the components of the converter to reach equilibrium temperatures corresponding to the said value of current.   121

duty cycle (1) (general). The time interval occupied by a device on intermittent duty in starting, running, stopping, and idling.   111

(2) (rotating machinery). A variation of load with time which may or may not be repeated, and in which the cycle time is too short for thermal equilibrium to be attained. *See also:* **asynchronous machine; direct-current commutating machine.**   63

(3) (pulse systems). The ratio of the sum of all pulse durations to the total period, during a specified period of continuous operation. *See also:* **navigation.**   187

(4) (radio transmitter performance). A criterion defining the ratio of average to peak power output from a transmitter, as a function of carrier-on time versus time available.   181

(5) (arc-welding apparatus). The ratio of arc time to total time.   264

(6) (relay). *See: relay duty cycle.*

(7) (automatic control). Of a device operating intermittently, the sequence of operating stages (start, run, stop, idle, etc.).   56

(8) (radar). *See: duty factor.*

(9) (excitation systems). An initial operating condition and a subsequent sequence of events of specified duration to which the excitation system will be exposed. *Note:* The duty cycle usually involves a three-phase fault of specified duration which is located electrically close to the synchronous generator. Its primary purpose is to specify the duty that the excitation system components can withstand without incurring maloperation or damage.   105

duty-cycle rating (rotating machinery). The statement of the loads and conditions assigned to the machine by the manufacturer, at which the machine may be operated on duty cycles. *See also:* **asynchronous machine; direct-current commutating machine.**   63

duty factor (1) (pulse techniques). The ratio of the pulse duration to the pulse period of a periodic pulse train. *See also:* **pulse; pulse carrier.**   185

(2) (electron tubes). The ratio of the ON period to the total period during which an electronic valve or tube is operating. *See also:* **ON period (electron tubes); pulse.**
                                                               244, 190

(3) (radar). In any system with intermittent operation, the ratio of the active or "ON" ("ON DUTY") time within a specified period to the duration of the specified period. *Note:* In systems with a constant repetition cycle, the **duty factor** usually specified (but not always) is the ratio of one ON period to the total period of one cycle.   13

(4) (automatic control). The ratio of working time to the time taken for the complete sequence of a duty cycle.
                                                                     56

duty factor control system. *See:* **control system, duty factor.**

duty, peak load (thyristor converter). A type of duty where the rating of the converter is specified in terms of the magnitude and duration of the peak load together with the time of no-load between peaks.   121

duty ratio (1) (pulse system). The ratio of average to peak pulse power. *See also:* **navigation.**   13, 187

(2) (radar). *See:* **duty factor.**

dwell (numerically controlled machines). A timed delay of programmed or established duration, not cyclic or sequential, that is, not an interlock or hold.   224, 207

dyadic operation. An operation on two operands.
                                                                255, 77

dynamic (industrial control) (excitation control systems). A state in which one or more quantities exhibit appreciable change within an arbitrarily short time interval. *Note:* For excitation control systems, this time interval encompasses up to 15–20 sec., that is, sufficient time to ascertain whether oscillations are decaying or building up with time. *See:* **control system, feedback.**   219, 206, 353

dynamic accuracy. Accuracy determined with a time-varying output. Contrast with **static accuracy.** *See also:* **electronic analog computer.**   9, 77, 16

dynamic braking (1) (general). A system of electric braking in which the traction motors, when used as generators, convert the kinetic energy of the vehicle into electric energy, and in so doing, exert a retarding force on the vehicle.
                                                                    328

(2) (rotating machinery). A system of electric braking in which the excited machine is disconnected from the supply system and connected as a generator, the energy being dissipated in the winding and, if necessary, in a separate resistor.   63

dynamic braking envelope. A curve that defines the dynamic braking limits in terms of speed and tractive force as restricted by such factors as maximum current flow, maximum permissible voltage, minimum field strength, etcetera. *See also:* **dynamic braking.**   328

dynamic braking, resistor (industrial control). A control function which brakes the drive by dissipating its stored energy in a resistor.   70

dynamic characteristic (electron tube) (operating characteristic). *See:* **load (dynamic) characteristic (electron tube).**

dynamic check. *See:* **problem check.** *See also:* **electronic analog computer.**

dynamic computer check. *See:* **problem check.** *See also:* **electronic analog computer.**

dynamic deviation (control) (industrial control). The difference between the ideal value and the actual value of a specified variable when the reference input is changing at a specified constant rate and all other transients have expired. *Note:* Dynamic deviation is expressed either as a percentage of the maximum value of the directly controlled variable or, in absolute terms, as the numerical difference between the ideal and the actual values of the directly controlled variable. *See:* **control system, feedback.**
                                                                219, 206

dynamic dump (computing systems). A dump that is performed during the execution of a program.   255, 77

dynamic dumping (test, measurement and diagnostic equipment). The printing of diagnostic information without stopping the program being tested.   54

dynamic electrode potential. An electrode potential when current is passing between the electrode and the electrolyte. *See also:* **electrochemistry.**   328

dynamic energy sensitivity (photoelectric devices). *See:* **sensitivity dynamic.**

dynamic equilibrium (electromagnetic system). (1) Any two circuits carrying current tend to dispose themselves so that the magnetic flux linking the two will be a maximum. (2) Every electromagnetic system tends to change its configuration so that the magnetic flux will be a maximum. *See also:* **network analysis.**   210

dynamic error. An error in a time-varying signal resulting from inadequate dynamic response of a transducer. Contrast with **static error.** *See also:* **electronic analog computer.**   9, 77

dynamic holding brake. A braking system designed for the purpose of exerting maximum braking force at a fixed speed only and used primarily to assist in maintaining this fixed speed when a train is descending a grade, but not to effect a deceleration. *See also:* **dynamic braking.**   328

dynamic load line (electron device). The locus of all simultaneous values of total instantaneous output electrode current and voltage for a fixed value of load impedance.
                                                                    190

dynamic loudspeaker. *See:* **moving-coil loudspeaker.**

dynamic microphone. *See:* **moving-coil microphone.**

dynamic problem check. *See:* **problem check.** *See also:* **electronic analog computer.**

dynamic range (1) (general). The difference, in decibels, between the overload level and the minimum acceptable signal level in a system or transducer. *Note:* The minimum acceptable signal level of a system or transducer is ordinarily fixed by one or more of the following: noise level, low-level distortion, interference, or resolution level. *See also:* **electronic analog computer; signal.**

239, 188, 9, 77

(2) (control system or element). The ratio of two instantaneous signal magnitudes, one being the maximum value consistent with specified criteria of performance, the other the maximum value of noise.   56

(3) (gyro; accelerometer). The ratio of the input range to the threshold. *See:* **input-output characteristics.**   46

(4) (non-real time spectrum analyzer). The maximum ratio of two signals simultaneously present at the input which can be measured to a specified accuracy. (A) Harmonic Dynamic Range: The maximum ratio of two harmonically related sinusoidal signals simultaneously present at the input which can be measured with a specified accuracy. (B) Non-harmonic Dynamic Range: The maximum ratio of two non-harmonically related sinusoidal signals simultaneously present at the input which can be measured with a specified accuracy. (C) Display Dynamic Range: The maximum ratio of two non-harmonically related sinusoids each of which can be simultaneously measured on the screen to a specified accuracy.   68

(5) (analog computer). The ratio of the specified maximum signal level capability of a system or component to its noise or resolution level, usually expressed in decibels. Also, the ratio of the maximum to minimum amplitudes of a variable during a computer solution.   9

dynamic range, reading (storage tubes). The range of output levels that can be read, from saturation level to the level of the minimum discernible output signal. *See also:* **storage tube.**   174, 190

dynamic range, writing (storage tubes). The range of input levels that can be written under any stated condition of scanning, from the input that will write the minimum usable signal. *See also:* **storage tube.**   174, 190

dynamic register (hybrid computer linkage components). The register that produces the analog equivalent voltage or coefficient.   10

dynamic regulation. Expresses the maximum or minimum output variations occurring during transient conditions, as a percentage of the final value. *Note:* Typical transient conditions are instantaneous or permanent input or load changes.

$$\text{Dynamic Regulation} = \frac{E_{max} - E_{final}}{E_{final}}(100)\%$$

$$= \frac{E_{final} - E_{min}}{E_{final}}(100\%) \qquad 176$$

dynamic regulator. A transmission regulator in which the adjusting mechanism is in self-equilibrium at only one or a few settings and requires control power to maintain it at any other setting. *See also:* **transmission regulator.**

328

dynamic response. *See:* **time response; control system, feedback.**

dynamic signal to noise ratio (digital delay line). The ratio of the minimum peak output signal to the maximum peak noise output when operated with a random bit sequence at a specified clock frequency.   81

dynamic slowdown (industrial control). Dynamic braking applied for slowing down, rather than stopping, a drive. *See:* **electric drive.**   219, 206

dynamic test (test, measurement and diagnostic equipment). A test of one or more of the signal properties or characteristics of the equipment or of any of its constituent items performed while the equipment is energized.   54

dynamic transmission (acoustically tunable optical filter). The ratio of the intensity of the light transmitted by the device at the wavelength to be filtered to the light intensity at this wavelength incident on the device, namely: $T(\lambda) = [I(\lambda)/I_o(\lambda)]$. It includes all static losses as well as the diffraction efficiency of the interaction. For a given design, the dynamic transmission is a function of acoustic drive power.   72

dynamic variable brake. A dynamic braking system designed to allow the operator to select (withing the limits of the electric equipment) the braking force best suited to the operation of a train descending a grade and to increase or decrease this braking force for the purpose of reducing or increasing train speed. *See also:* **dynamic braking.**

328

dynamic vertical. *See:* **apparent vertical.**

dynamometer (conductor stringing equipment). A device designed to measure loads or tension on conductors. Various models of these devices are used to tension guys or sag conductors. *Syn.* **clock; load cell.**   45

dynamometer, electric (rotating machinery). An electric generator, motor or eddy-current load absorber equipped with means for indicating torque. *Note:* When used for determining power input or output of a coupled machine, means for indicating speed are also provided.   63

dynamometer test (rotating machinery). A braking or motoring test in which a dynamometer is used. *See also:* **braking test; asynchronous machine; direct-current commutator machine.**   63

dynamotor (rotating machinery). A converter that combines both motor and generator action, with one magnetic field and with two armatures, or with one armature having separate windings. *See:* **converter.**   63, 3

dynatron effect (electron tubes) (dynatron characteristic). An effect equivalent to a negative resistance, which results when the electrode characteristic (or transfer characteristic) has a negative slope. *Example:* Anode characteristic of a tetrode, or tetrode-connected valve or tube. *See also:* **electronic tube.**   244, 190

dynatron oscillation. Oscillation produced by negative resistance due to secondary emission. *See also:* **oscillatory circuit.**   111

dynatron oscillator. A negative-resistance oscillator in which negative resistance is derived between plate and cathode of a screen-grid tube operating so that secondary electrons produced at the plate are attracted to the higher potential screen grid. *See:* **oscillatory circuit.**   111

dyne. The unit of force in the cgs (centimeter-gram-second) systems. The dyne is 10^{-5} newton.   210
dynode (electron tubes). An electrode that performs a useful function, such as current amplification, by means of secondary emission. *See also:* **electrode (electron tube);**

electron tube.   190 117, 125
dynode spots (image orthicons). A spurious signal caused by variations in the secondary-emission ratio across the surface of a dynode that is scanned by the electron beam.
                                                      190, 125

E

E & M signaling (telephone switching systems). A technique for transferring information between a trunk circuit and a separate signaling circuit over leads designated "E" and "M." The "M" lead transmits to the signaling circuit and the "E" lead transmits to the trunk circuit.   55
early failure period (reliability). That possible early period, beginning at a stated time and during which the failure rate decreases rapidly in comparison with that of subsequent period. *See also:* **constant failure rate period.**
                                                      164, 182

earth station (communication satellite). A ground station designed to transmit to and receive transmission from communication satellites.   83
earth terminal. *See:* **ground terminal.**
E bend (E-plane bend) (waveguide technique). A smooth change in the direction of the axis of a waveguide, throughout which the axis remains in a plane parallel to the direction of polarization. *See also:* **waveguide.**   328
eccentric groove (eccentric circle) (disc recording). A locked groove whose center is other than that of the disc record

Failure periods.

early warning radar. Radar employed to search for distant enemy aircraft. *See also:* **radar.**   328
earphone (receiver). An electroacoustic transducer intended to be closely coupled acoustically to the ear. *Note:* The term receiver should be avoided when there is risk of ambiguity. *See also:* **loudspeaker.**   176
earphone coupler. A cavity of predetermined size and shape that is used for the testing of earphones. The coupler is provided with a microphone for the measurement of pressures developed in the cavity. *Note:* Couplers generally have a volume of 6 cubic centimeters for testing regular earphones and a volume of 2 cubic centimeters for testing insert earphones. *See also:* **loudspeaker.**   176
earth, effective radius (radio wave propagation). A value for the radius of the earth that is used in place of the geometrical radius to correct approximately for atmospheric refraction when the index of refraction in the atmosphere changes linearly with height. *Note:* Under conditions of standard refraction the effective radius of the earth is 8.5 $\times 10^6$ meters, or 4/3 the geometrical radius. *See also:* **radiation; radio wave propagation.**   180, 59
earth-fault protection. *See:* **ground protection.**
earth inductor. *See:* **generating magnetometer.**
earth's rate correction (gyro). A command rate applied to a gyro to compensate for the rotation of the earth with respect to the gyro input axis.   46

(generally used in connection with mechanical control of phonographs). *See also:* **phonograph pickup.**   176
eccentricity (power distribution, underground cables) (1) (general). The ratio of the difference between the minimum and average thickness to the average thickness of an annular element, expressed in percent. *See also:* **power distribution, underground construction.**   57
(2) (disc recording). The displacement of the center of the recording groove spiral, with respect to the record center hole. *See also:* **phonograph pickup.**   176
Eccles-Jordan circuit. A flip-flop circuit consisting of a two-stage resistance-coupled electron-tube amplifier with its output similarly coupled back to its input, the two conditions of permanent stability being provided by the alternate biasing of the two stages beyond cutoff. *See also:* **trigger circuit.**   328
ECCM. *See:* **electronic counter-countermeasures.**
ECCM (electronic counter-countermeasures) improvement factor (radar). The power ratio of the ECM (electronic counter measures) signal level required to produce a given output signal from a receiver using an ECCM technique to the ECM signal level producing the same output from the same receiver without the ECCM technique.   13
echelon (calibration). A specific level of accuracy of calibration in a series of levels, the highest of which is represented by an accepted national standard. *Note:* There may

be one or more auxiliary levels between two successive echelons. *See also:* **measurement system.**

293, 183, 155

echo (1) (general). A wave that has been reflected or otherwise returned with sufficient magnitude and delay to be perceived in some manner as a wave distinct from that directly transmitted. *Note:* Echoes are frequently measured in decibels relative to the directly transmitted wave. *See also:* **recording (facsimile).** 239, 59
(2) (radar). The portion of energy of the transmitted pulse that is reflected to a receiver. *See also:* **navigation.** 13
(3) (facsimile). A wave which has been reflected at one or more points with sufficient magnitude and time difference to be perceived in some manner as a wave distinct from that of the main transmission. 12

echo area, effective (radar). The area of a fictitious perfect electromagnetic reflector that would reflect the same amount of energy back to the radar as the target. *See also:* **navigation.** 187

echo attenuation (data transmission). In a 4-wire or 2-wire circuit in which the two directions of transmission can be separated from each other, the attenuation of the echo currents (which return to the input of the circuit under consideration) is determined by the ratio of the transmitted power to the echo power received expressed in decibels. 59

echo box (radar). A calibrated resonant cavity which stores part of the transmitted pulse power and feeds this exponentially decaying energy into the receiving system after completion of the pulse transmission. 13

echo check. A method of checking the accuracy of transmission of data in which the received data are returned to the sending end for comparison with the original data. 255, 77, 54

echo, second-time-around (radar). *See:* **second-time-around echo.**

echo sounding system (depth finder). A system for determination of the depth of water under a ship's keel, based on the measurement of elapsed time between the propagation and projection through the water of a sonic or supersonic signal, and reception of the echo reflected from the bottom. 328

echo suppressor (1) (telephony). A voice-operated device for connection to a two-way telephone circuit to attenuate echo currents in one direction caused by telephone currents in the other direction. *See also:* **voice-frequency telephony.** 59
(2) (navigation). A circuit component which desensitizes the receiving equipment for a period after the reception of one pulse, for the purpose of rejecting pulses arriving later over indirect reflection paths. 187, 13

eclipsing (pulse Doppler radar). The loss of information on radar echo during intervals when the receiver is blanked because of the occurrence of a transmitter pulse. Numerous such blankings can occur in high PRF (pulse repetition frequency) Doppler systems. 13

ECM. *See:* **electronic countermeasures.**

economic dispatch (electric power systems). The distribution of total generation requirements among alternative sources for optimum system economy with due consideration of both incremental generating costs and incremental transmission losses. 94

economy power. Power produced from a more economical source in one system and substituted for less economical power in another system. *See also:* **generating station.** 64

eddy-current braking (rotating machinery). A form of electric braking in which the energy to be dissipated is converted into heat by eddy currents produced in a metallic mass. *See also:* **asynchronous machine; direct-current commutating machine.** 63

eddy-current loss (1) (parts, hybrids, and packaging). Power dissipated due to eddy currents. *Note:* The eddy-current loss of a magnetic device includes the eddy-current losses in the core, windings, case, and associated hardware. 197
(2) (transformer). The energy loss resulting from the flow of eddy currents in a metallic material. 53

eddy currents (1) (general). Those currents that exist as a result of voltages induced in the body of a conducting mass by a variation of magnetic flux. *Note:* The variation of magnetic flux is the result of a varying magnetic field or of a relative motion of the mass with respect to the magnetic field. *Syn:* **Foucault currents.** 210
(2) (transformers). The currents that are induced in the body of a conducting mass by the time variation of magnetic flux. 197, 53

Edison distribution system. A three-wire direct-current system, usually about 120–240 volts, for combined light and power service from a single set of mains. *See also:* **direct-current distribution.** 64

Edison effect. *See:* **thermionic emission.**

Edison storage battery. An alkaline storage battery in which the positive active material is nickel oxide and the negative an iron alloy. *See also:* **battery (primary or secondary).** 328

E-Display (radar). An intensity modulated rectangular display in which targets appear as blips with distance indicated by the horizontal coordinate and elevation angle by the vertical coordinate. Similar to the RHI (range-height indicator) in which height or altitude is the vertical coordinate. 13

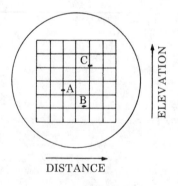

E display.

edit (computing systems). To modify the form or format of data, for example, to insert or delete characters such as page numbers or decimal points. 255, 77, 54

EDR. *See:* **electrodermal reaction.**

effective aperture (EM-radiation collection device) (radar).

Synonymous with **effective area** for an antenna; also, the effective area of other EM-(electromagnetic) radiation collecting devices, such as lenses.   13

effective area (antenna). In a given direction, the ratio of the power available at the terminals of a receiving antenna to the power per unit area of a plane wave incident on the antenna from that direction, polarized coincident with the polarization that the antenna would radiate.   111

effective band (facsimile). The frequency band of a facsimile signal wave equal in width to that between zero frequency and maximum keying frequency. *Note:* The frequency band occupied in the transmission medium will in general be greater than the effective band.   12

effective bunching angle (reflex klystrons). In a given drift space, the transit angle that would be required in a hypothetical drift space in which the potentials vary linearly over the same range as in the given space and in which the bunching action is the same as in the given space.   125

effective center of mass (accelerometer). That point defined by the intersection of the pendulous axis and an axis parallel to the output axis about which angular acceleration results in minimum accelerometer output.   46

effective echo area (radar). *See:* **radar cross section.**

effective efficiency (1) (bolometer units). The ratio of the substitution power to the total RF power dissipated within the bolometer unit.   115

(2) (electrothermic unit). The ratio of the substituted reference power (direct current, audio or radio frequency) in the electrothermic unit to the power dissipated within the electrothermic unit for the same direct current output voltage from the electrothermic unit at a prescribed frequency, power level, and temperature. *Notes:* (A) Calibration factor and effective efficiency are related as follows:

$$\frac{K_b}{\eta_c} = 1 - |\Gamma|^2$$

where K_b, η_c and Γ are the calibration factor, effective efficiency, and reflection coefficient of the electrothermic unit, respectively. (B) The reference frequency is to be supplied with the calibration factor.   47

effective height (antenna). (1) **High-frequency usage.** The height of the antenna center of radiation above the ground level. *Note:* For an antenna with symmetrical current distribution, the center of radiation is the center of distribution. For an antenna with asymmetrical current distribution, the center of radiation is the center of current moments when viewed from directions near the direction of maximum radiation. (2) **(Low-frequency usage).** *See:* **effective length of an antenna.**   111

effective induction area of the control current loop (Hall-effect devices). The effective area of the loop enclosed by the control current leads and the relevant conductive path through the Hall element.   107

effective induction area of the output loop (Hall-effect devices). The effective induction area of the loop enclosed by the leads to the Hall terminals and the relevant conductive path through the Hall plate.   107

effective length of linearly polarized antenna. For a linearly polarized antenna receiving a plane wave from a given direction, the ratio of the magnitude of the open circuit voltage developed at the terminals of the antenna to the magnitude of the electric field strength in the direction of the antenna polarization. *Notes:* (1) Alternatively, the effective length is the length of a thin straight conductor oriented perpendicular to the given direction and parallel to the antenna polarization, having a uniform current equal to that at the antenna terminals and producing the same far-field strength as the antenna in that direction. (2) In low-frequency usage the effective length of a vertically polarized ground based antenna is frequently referred to as effective height. Such usage should not be confused with **effective height of an antenna (high frequency usage).**   111

effectively grounded (1) (transformer). An expression that means grounded through a grounding connection of sufficiently low impedance (inherent or intentionally added or both) that fault grounds that may occur cannot build up voltages in excess of limits established for apparatus, circuits, or systems so grounded. *Note:* An alternating-current system or portion thereof may be said to be effectively grounded when, for all points on the system or specified portion thereof, the ratio of zero-sequence reactance to positive-sequence reactance is not greater than three and the ratio of zero-sequence resistance to positive-sequence reactance is not greater than one for any condition of operation and for any amount of connected generator capacity.   53

(2) (transmission and distribution). Intentionally connected to earth through a ground connection or connections of sufficiently low impedance and having sufficient current-carrying capacity to prevent the buildup of voltages which may result in undue hazard to connected equipment or to persons.   262

effective multiplication factor (k_{eff}) (nuclear power generation). The ratio of the average number of neutrons produced by nuclear fission in each generation to the total number of corresponding neutrons absorbed or leaking out of the system. *Note:* If $k_{eff} = 1$, that is, the number of neutrons produced is equal to the number being absorbed or leaking out of the system, a stable, self-sustaining chain reaction exists and the assembly is said to be critical. If $k_{eff} < 1$, the chain reaction is not self-sustaining and will terminate; such a system is said to be subcritical. If $k_{eff} > 1$, the chain reaction is divergent and the system is supercritical.   112

effective radiated power (ERP). In a given direction, the power gain of a transmitting antenna multiplied by the net power accepted by the antenna from the connected transmitter.   111

effective radius of the earth. An effective value for the radius of the earth, which is used in place of the geometrical radius to correct approximately for atmospheric refraction when the index of refraction in the atmosphere changes linearly with height. *Note:* Under conditions of standard refraction the effective radius of the earth is 8.5 $\times 10^6$ meters, or $\frac{4}{3}$ the geometrical radius.   146

effective temperature (laser-maser). The temperature which must be used in the Boltzmann formula to describe the relative populations of two energy levels, which may or may not be in thermal equilibrium.   363

effective value*. *See:* **root-mean-square value of a periodic**

function. 210

* Deprecated

efficiency (1) (engineering). The ratio of the useful output to the input (energy, power, quantity of electricity, etcetera). *Note:* Unless specifically stated otherwise, the term **efficiency** means efficiency with respect to power. 210
(2) (electric conversion). The ratio of output power to input power expressed in percent; Eff = P_o/P_{in} (100%). *Note:* It is an evaluation of power losses within the conversion equipment and may be also expressed as ratio of the output power to the sum of the output power and the power losses, and is expressed in percent. 186

$$Eff = \frac{P_o}{P_o + P_L} \ (100\%).$$

(3) (by direct calculation) (rotating machinery). The method by which the efficiency is calculated from the input and output, these having been measured directly. *See also:* **asynchronous machine; direct-current commutating machine.** 63
(4) (conventional) (rotating machinery). The efficiency calculated from the summation of the component losses measured separately. *See:* **asynchronous machine; direct-current commutating machine.** 63
(5) (from total loss) (rotating machinery). The method of indirect calculation of efficiency from the measurement of total loss. *See also:* **asynchronous machine; direct-current commutating machine.** 63
(6) (transformer). The ratio of the useful power output to the total power input. 203, 53
(7) (rectifier). The ratio of the power output to the total power input. *Note:* It may also be expressed as the ratio of the power output to the sum of the output and the losses. 291, 208
(8) (rectification). Ratio of the direct-current component of the rectified voltage at the input terminals of the apparatus to the maximum amplitude of the applied sinusoidal voltage in the specified conditions. 224, 190
(9) (arc-welding apparatus) (electric arc-welding power supply). The ratio of the power output at the welding terminals to the total power input. *See also:* **losses, excitation current, and impedance voltage.** 264
(10) (bolometer units). The ratio of the radio-frequency power absorbed by the bolometer element to the total radio-frequency power dissipated within the bolometer unit. *Note:* The bolometer unit efficiency is a measure of the radio-frequency losses in the bolometer unit and is a function of the amount of energy dissipated in the dielectric supports, metal surfaces, etcetera, of the bolometer mount. Bolometer unit efficiency is generally independent of power level. *See also:* **bolometric power meter.** 183
(11) (storage battery). The ratio of the output of the cell or battery to the input required to restore the initial state of charge under specified conditions of temperature, current rate, and final voltage. 328
(12) (station or system). The ratio of the energy delivered from the station or system to the energy received by it under specified conditions. *See also:* **generating station.** 64
(13) (radiation-counter tube). The probability that a tube

count will take place with a specified particle or quantum incident in a specified manner. *See also:* **gas-filled radiation-counter tubes.** 190, 96, 125
(14) (antenna). *See:* **antenna efficiency; aperture illumination efficiency; radiation efficiency.**
efficiency, effective (bolometer units). The ratio of the substitution power to the total radio-frequency power dissipated within the bolometer unit. *Note:* Effective efficiency includes the combined effect of the direct-current-radio-frequency substitution error and bolometer unit efficiency. *See also:* **bolometric power meter.** 183
efficiency, generator (thermoelectric device). *See:* **generator efficiency (thermoelectric couple).**
efficiency, generator, overall (thermoelectric device). *See:* **overall generator efficiency (thermoelectric couple).**
efficiency, generator, reduced (thermoelectric device). *See:* **reduced generator efficiency.**
efficiency, load circuit. *See:* **load circuit efficiency.**
efficiency, overall electrical. *See:* **overall electrical efficiency.**
efficiency, quantum (phototubes). *See:* **quantum efficiency.**
effluent (nuclear power generating stations). The liquid or gaseous waste streams released to the environment. 31
effluve. *See:* **convective discharge.**
E-H tee (waveguide components). A junction composed of E and H plane tee junctions wherein the axes of the arms intersect at a common point in the main guide. *See also:* **hybrid tee.** 166
E-H tuner (waveguide components). An E-H tee having E and H arms terminated in moveable open or short circuit terminations. 166
EI. *See:* **end injection.**
eight-hour rating (magnetic contactor). The rating based on its current-carrying capacity for eight hours, starting with new clean contact surfaces, under conditions of free ventilation, with full-rated voltage on the operating coil, and without causing any of the established limitations to be exceeded. 206
einschleichender stimulus. *See:* **accumulating stimulus.**
Einstein's law (photoelectric device). The law according to which the absorption of a photon frees a photo-electron with a kinetic energy equal to that of the photon less the work function

$$\frac{1}{2} \, mv^2 = h\upsilon - p \ (\text{if } h\upsilon > p).$$

See also: **photoelectric effect.** 244
elapsed time printout (sequential events recording systems). The recording of time interval between first and successive detected events. 48, 58
elastance. The reciprocal of capacitance. 210
elastances (system of conductors) (coefficients of potential—Maxwell). A set of *n* conductors of any shape that are insulated from each other and that are mounted on insulating supports within a conducting shell, or on one side of a conducting sheet of infinite extent or above the surface of the earth constitutes a system of *n* capacitors having mutual elastances and capacitances. *Note:* If the shell (or the earth) is regarded as the electrode common

to all n capacitors and the transfers of charge as taking place between shell and the individual electrodes, the sum of the charges on the conductors will be equal and opposite in sign to the charge on the common electrode. The shell (or the earth) is taken to be at zero potential. Let Q_r represent the value of the charge that has been transferred from the shell to the other electrode of the rth capacitor, and let V_r represent the algebraic value of the potential of this electrode resulting from the charges in all n capacitors. If the charges are known the values of the potentials can be computed from the equations:

$$V_1 = S_{11}Q_1 + S_{12}Q_2 + S_{13}Q_3 + \cdots$$
$$V_2 = S_{21}Q_1 + S_{22}Q_2 + S_{23}Q_3 + \cdots$$
$$V_3 = S_{31}Q_1 + S_{32}Q_3 + S_{33}Q_3 + \cdots$$

.

.

.

$$V_r = \sum_{c=1}^{c=n} S_{r,c}Q_c.$$

The multiplying operators $S_{r,r}$ are the self-elastances and the multipliers $S_{r,c}$ are the mutual elastances of the system. Maxwell termed them the coefficients of potential of the system. Their values can be measured by noting that the defining equation for the mutual elastance $S_{r,c}$ is

$$S_{r,c}(\text{reciprocal farad*}) = \frac{V_r \ (\text{volt})}{Q_c \ (\text{coulomb})}$$

(every Q except Q_c being zero).

It can be shown that $S_{r,c} = S_{c,r}$ and that under the conventions stated all the elastances have positive values.   210

*formerly sometimes called the daraf.

elastic restraint coefficient (gyro). The ratio of gimbal restraining torque about an output axis to the output angle.   46

elastic restraint drift rate (gyro). That component of systematic drift rate which is proportional to the angular displacement of a gyro gimbal about an output axis. The relationship of this component of drift rate to gimbal angle can be stated by means of a coefficient having dimensions of angular displacement per unit time per unit angle. This coefficient is equal to the elastic restraint coefficient divided by angular momentum.   46

elastomer (rotating machinery). Macromolecular material that returns rapidly to approximately the initial dimensions and shape after substantial deformation by a weak stress and release of the stress. *See also:* **asynchronous machine; direct-current commutating machine.**   63

E layer (radio wave propagation). An ionized layer in the E region. *Note:* The principal layer corresponds roughly to what was formerly called the Kennelly-Heaviside layer. In addition, areas of abnormally intense ionization frequently occur, which are called **sporadic E.** *See also:* **radiation; radio wave propagation.**   146, 180

elbow (1) (sharp bend) (interior wiring). (A) A short curved piece of metal raceway of comparatively short radius. *See*

also: **raceway.**   328
(B) A curved section of rigid steel conduit threaded on each end.   101
(2) (waveguide techniques). *See:* **corner.**

electret. A dielectric body possessing separated electric poles of opposite sign of a permanent or semipermanent nature. *Note:* It is the electrostatic analog of a permanent magnet.   210

electric. Containing, producing, arising from, actuated by, or carrying electricity, or designed to carry electricity and capable of so doing. Examples: Electric eel, energy, motor, vehicle, wave. *Note:* Some dictionaries indicate electric and electrical as synonymous but usage in the electrical engineering field has in general been restricted to the meaning given in the definitions above. It is recognized that there are borderline cases wherein the usage determines the selection. *See also:* **electrical.**   103, 202

electric air-compressor governor. A device responsive to variations in air pressure that automatically starts or stops the operation of a compressor for the purpose of maintaining air pressure in a reservoir between predetermined limits.   328

electrical. Related to, pertaining to, or associated with electricity, but not having its properties or characteristics. *Examples:* Electrical engineer, handbook, insulator, rating, school, unit. *Note:* Some dictionaries indicate electric and electrical as synonymous but usage in the electrical engineering field has in general been restricted to the meaning given in the definitions above. It is recognized that there are borderline cases wherein the usage determines the selection. *See also:* **electric.**   103, 202

electrical anesthesia (medical electronics). More or less complete suspension of general or local sensibility produced by electric means.   192

electrical back-to-back test (rotating machinery). *See:* **pump-back test.**

electrical boresight (antenna). *See:* **electric boresight.**

electrical center. *See:* **electric center.**

electrical codes (1) (general). A compilation of rules and regulations covering electric installations.
(2) official electrical code. One issued by a municipality, state, or other political division, and which may be enforced by legal means.
(3) unofficial electrical code. One issued by other than political entities such as engineering societies, and the enforcement of which depends on other than legal means.
(4) National Electrical Code (N.E.C.). The code of rules and regulations as recommended by the National Fire Protection Association and approved by the American National Standards Institute. *Note:* This code is the accepted minimum standard for electric installations and has been accepted by many political entities as their official code, or has been incorporated in whole or in part in their official codes.
(5) National Electrical Safety Code. A set of rules, prepared by the National Electrical Safety Code committee (secretariat held by the Institute of Electrical and Electronics Engineers) and approved by the American National Standards Institute governing: (A) Methods of grounding. (B) Installation and maintenance of electric-supply stations and equipment. (C) Installation and

maintenance of overhead supply and communication lines. (D) Installation and maintenance of underground and electric-supply and communications lines. (E) Operation of electric-supply and communication lines and equipment (Work Rules).

electrical conductor seal, single (nuclear power generating stations). A mechanical assembly providing a single pressure barrier between the electrical conductors and the electrical penetration assembly.   31

electrical conductor seal, double (nuclear power generating stations). An assembly of two single electrical conductor seals in series and arranged in such a.way that there is a double pressure barrier seal between the inside and the outside of the containment structure along the axis of the conductors.   31

electrical degree (rotating machinery). The 360th part of the angle subtended, at the axis of a machine, by two consecutive field poles of like polarity. One mechanical degree is thus equal to as many electrical degrees as there are pairs of poles in the machine. *See:* **direct-current commutating machine.**   63

electrical distance (electronic navigation). The distance between two points expressed in terms of the duration of travel of an electromagnetic wave in free space between the two points. *Note:* A convenient unit of electrical distance is the light-microsecond, approximately 300 meters (983 feet). This unit is widely used in radar technology. *See also:* **electrical range; navigation; signal wave.**   13, 187

electrical interchangeability (fuse links or fuse units). The characteristic that permits the designs of various manufacturers to be used interchangeably so as to provide a uniform degree of overcurrent protection and fuse coordination.   103, 202

electrical length (1) (general). The physical length expressed in wavelengths, radians, or degrees. *Note:* When expressed in angular units, it is distance in wavelengths multiplied by 2π to give radians or by 360 to give degrees. *See also:* **radio wave propagation; signal wave; wave guide.**   146 **(2) (two-port network at a specified frequency).** The length of an equivalent lossless reference waveguide or reference air line (which in the ideal case would be evacuated) introducing the same total phase shift as the two-port when each is terminated in a reflectionless termination. *Note:* It is usually expressed in fractions or multiples of waveguide wavelength. When expressed in radians or degrees it is equal to the phase angle of the transmission coefficient $+ 2n\pi$. *See also:* **waveguide.**   267, 179, 185, 59

electrically connected. Connected by means of a conducting path or through a capacitor, as distinguished from connection merely through electromagnetic induction. *See also:* **inductive coordination.**   328

electrically heated airspeed tube. A Pitot-static or Pitot-Venturi tube utilizing a heating element for deicing purposes.   328

electrically heated flying suit. A garment that utilizes sewn-in heating elements energized by electric means designed to cover the torso and all or part of the limbs. *Note:* It may be a one-piece garment or consist of a coat, trousers, and the like. The lower portion of the one-piece suit is in trouser form.   328

electrically interlocked manual release of brakes (control)

(industrial control). A manual release provided with a limit switch that is operated when the braking surfaces are disengaged manually. *Note:* The limit switch may operate a signal, open the control circuit, or perform other safety functions. *See:* **switch.**   225, 206

electrically release-free (electrically operated switching device) (electrically trip-free). A term indicating that the release can open the device even though the closing control circuit is energized. *Note:* Electrically release-free switching devices are usually arranged so that they are also anti-pump. With such an arrangement the closing mechanism will not reclose the switching device after opening until the closing control circuit is opened and again closed.   103, 202

electrically reset relay. A relay that is so constructed that it remains in the picked-up condition even after the input quantity is removed; an independent electrical input is required to reset the relay.   127, 103

electrically trip-free. *See:* **electrically release-free.**

electrical metallic tubing. A thin-walled metal raceway of circular cross section constructed for the purpose of the pulling in or the withdrawing of wires or cables after it is installed in place. *See also:* **raceways.**   328

electrical noise (control systems). Unwanted electrical signals, which produce undesirable effects in the circuits of the control systems in which they occur.   43

electrical null (gyro; accelerometer). The minimum electrical output. It may be specified in terms of root-mean-square, peak-to-peak, in-phase component, or other electrical measurements.   46

electrical null position (gyro; accelerometer). The angular or linear position of a pickoff corresponding to electrical null.   46

electrical operation. Power operation by electric energy.   103

electrical penetration assembly (nuclear power generating stations). An electrical penetration assembly provides the means to allow passage of one or more electrical circuits through a single aperture (nozzle or other opening) in the containment pressure barrier, while maintaining the integrity of the pressure barrier.   31

electrical penetration assembly current capacity (nuclear power generating stations). The maximum current that each conductor in the assembly is specified to carry for its duty cycle in the design service environment without causing stabilized temperatures of the conductors or the penetration nozzle-concrete interface (if applicable) to exceed their design limits.   31

electrical penetration assembly short-time overload rating (nuclear power generating stations). The limiting overload current that any one third of the conductors (but in no case less than three of the conductors) in the assembly can carry, for a specified time, in the design service environment, while all remaining conductors carry rated continuous current, without causing the conductor temperatures to exceed those values recommended by the insulated conductor manufacturer as the short-time overload conductor temperature and without causing the stabilized temperature of the penetration nozzle-concrete interface (if applicable) to exceed its design limit.   31

electrical zero*. *See:* **electrical null position.**   46

* Deprecated

electrical range. The range expressed in equivalent electrical units. *See also:* **electrical distance; instrument.**
295

electrical arc (gas). A discharge characterized by a cathode drop that is small compared with that in a glow discharge. *Note:* The electron emission of a cathode is due to various causes (thermionic emission, high-field emission, etcetera) acting simultaneously or separately, but secondary emission plays only a small part. *See also:* **discharge (gas).**
244, 190

electric back-to-back test. *See:* **pump-back test.**

electric bell. An audible signal device consisting of one or more gongs and an electromagnetically actuated striking mechanism. *Note:* The gong is the resonant metallic member that produces an audible sound when struck. However, the term gong is frequently applied to the complete electric bell.
328

electric bias, relay. *See:* **relay electric bias.**

electric blasting cap. A device for detonating charges of explosives electrically. *See also:* **blasting unit.**

electric boresight (antenna). The tracking axis as determined by an electric indication, such as the null direction of a conical-scanning or monopulse antenna system, or the beam-maximum direction of a highly directive antenna. *See also:* **reference boresight.**
111

electric braking. A system of braking wherein electric energy, either converted from the kinetic energy of vehicle movement or obtained from a separate source, is one of the principal agents for the braking of the vehicle or train. *See:* **electromagnetic braking (magnetic braking); electropneumatic brake; magnetic track braking; regenerative braking.**
328

electric bus. A passenger vehicle operating without track rails, the propulsion of which is effected by electric motors mounted on the vehicle. *Note:* A prefix diesel-electric, gas-electric, etcetera, may replace the word electric. *See:* **trolley coach (trolley bus) (trackless trolley coach).** 328

electric-cable-reel mine locomotive. An electric mine locomotive equipped with a reel for carrying an electric conductor cable that is used to conduct power to the locomotive when operating beyond the trolley wire. *See also:* **electric mine locomotive.**
328

electric capacitance altimeter. An altimeter, the indications of which depend on the variation of an electric capacitance with distance from the earth's surface. *See also:* **air transportation, electric equipment.**
328

electric center (power system out of synchronism). A point at which the voltage is zero when a machine is 180 degrees out of phase with the rest of the system. *Note:* There may be one or more electric centers depending on the number of machines and the interconnections among them.
103, 202, 60

electric charge (charge) (quantity of electricity). Electric charge, like mass, length, and time, is accepted as a fundamentally assumed concept, required by the existence of forces measurable experimentally; other definitions of electromagnetic quantities are developed on the basis of these four concepts. *Note:* The electric charge on (or in) a body or within a closed surface is the excess of one kind of electricity over the other kind. A plus sign indicates that the positive electricity is in excess, a minus sign indicates that the negative is in excess.
210

electric charge time constant (detector). The time required, after the instantaneous application of a sinusoidal input voltage of constant amplitude, for the output voltage across the load capacitor of a detector circuit to reach 63 percent of its steady-state value. *See also:* **electromagnetic compatibility.**
220, 199

electric coal drill. An electric motor-driven drill designed for drilling holes in coal for placing blasting charges.
328

electric components. The electric equipment, assemblies, and conductors that together form the electric power systems.
58

electric conduction and convection current density. At any point at which there is a motion of electric charge, a vector quantity whose direction is that of the flow of positive charge at this point, and whose magnitude is the limit of the time rate of flow of net (positive) charge across a small plane area perpendicular to the motion, divided by this area, as the area taken approaches zero in a macroscopic sense, so as to always include this point. *Note:* The flow of charge may result from the movement of free electrons or ions but is not, in general, except in microscopic studies, taken to include motions of charges resulting from the polarization of the dielectric.
210

electric console lift. An electrically driven mechanism for raising and lowering an organ console and the organist. *See also:* **elevators.**
328

electric constant (permittivity or capacitivity of free space) (pertinent to any system of units). The scalar ϵ_0 that in that system relates the electric flux density D, in empty space, to the electric field strength E ($D = \epsilon_0 E$). *Note:* It also relates the mechanical force between two charges in empty space to their magnitudes and separation. Thus in the equation

$$F = Q_1 Q_2 / (n \epsilon_0 r^2)$$

for the force F between charges Q_1 and Q_2 separated by a distance r, ϵ_0 is the electric constant, and n is a dimensionless factor that is unity in unrationalized systems and 4π in a rationalized system. *Note:* In the International System of Units (SI) the manitude of ϵ_0 is that of $10^7/(4\pi c^2)$ and the dimension is $[L^{-3}M^{-1}T^4I^2]$. Here c is the speed of light expressed in the appropriate system of units.
210

electric contact. The junction of conducting parts permitting current to flow.
328

electric controller. A device or a group of devices, that serves to govern, in some predetermined manner, the electric power delivered to the apparatus to which it is connected.
206, 3

electric-control trail car. A trail car used in a multiple-unit train, provided at one or both ends with a master controller and other apparatus necessary for controlling the train. *See also:* **electric motor car; electric trail car.** 328

electric coupler. A group of devices (plugs, receptacles, cable, etcetera) that provides for readily connecting or disconnecting electric circuits.
328

electric coupler plug. The removable portion of an electric coupler.
328

electric coupler receptacle (electric coupler socket). The fixed portion of an electric coupler.
328

electric coupler socket. *See:* **electric coupler receptacle.**

electric coupling. *See:* **coupling, electric.**

electric course recorder. A device that operates, under control of signals from a master compass, to make a continuous record of a ship's heading with respect to time. 328

electric crab-reel mine locomotive. An electric mine locomotive equipped with an electrically driven winch, or crab reel, for the purpose of hauling cars by means of a wire rope from places beyond the trolley wire. *See also:* **electric mine locomotive.** 328

electric depth recorder. A device for continuously recording, with respect to time, the depth of water determined by an echo sounding system. 328

electric dipole (1) (general). An elementary radiator consisting of a pair of equal and opposite oscillating electric charges an infinitesimal distance apart. *Note:* It is equivalent to a linear current element. 246, 179, 111
(2) (antenna). The limit of an electric doublet as the separation approaches zero while the moment remains constant. 210

electric dipole moment (two point charges, *q* and -*q*, a distance *a* apart). A vector at the midpoint between them, whose magnitude is the product *qa* and whose direction is along the line between the charges from the negative toward the positive charge. 210

electric-discharge lamp. A lamp in which light (or radiant energy near the visible spectrum) is produced by the passage of an electric current through a vapor or a gas. *Note:* Electric-discharge lamps may be named after the filling gas or vapor that is responsible for the major portion of the radiation; that is, mercury lamps, sodium lamps, neon lamps, argon lamps, etcetera. A second method of designating electric-discharge lamps is by physical dimensions or operating parameters; that is short-arc lamps, high-pressure lamps, low-pressure lamps, etcetera. A third method of designating electric-discharge lamps is by their application; in addition to lamps for illumination there are photochemical lamps, bactericidal lamps, black-light lamps, sun lamps, etcetera. *See also:* **lamp.** 167

electric-discharge time constant (detector). The time required, after the instantaneous removal of a sinusoidal input voltage of constant amplitude, for the output voltage across the load capacitor of the detector circuit to fall to 37 percent of its initial value. *See also:* **electromagnetic compatibility.** 220, 199

electric displacement. *See:* **electric flux.**

electric displacement density. *See:* **electric flux density.**

electric doublet. A separated pair of equal and opposite charges. 210

electric drive (industrial control). A system consisting of one or several electric motors and of the entire electric control equipment designed to govern the performance of these motors. The control equipment may or may not include various rotating electric machines. 206

electric driving machine. A machine where the energy is applied by an electric motor. *Note:* It includes the motor and brake and the driving sheave or drum together with its connecting gearing, belt, or chain, if any. *See also:* **driving machine (elevators).** 328

electric elevator. A power elevator where the energy is applied by means of an electric motor. *See also:* **elevators.** 328

electric energy (energy). The electric energy delivered by an electric circuit during a time interval is the integral with respect to time of the instantaneous power at the terminals of entry of the circuit to a delimited region. *Note:* If the reference direction for energy flow is selected as into the region, the net delivery of energy will be into the region when the sign of the energy is positive and out of the region when the sign is negative. If the reference direction is selected as out of the region, the reverse will apply. Mathematically where

$$W = \int_{t_0}^{t+t_0} p \, dt$$

W = electric energy
p = instantaneous power
t = time during which energy is determined.
When the voltages and currents are periodic, the electric energy is the product of the active power and the time interval, provided the time interval is one or more complete periods or is quite long in comparison with the time of one period. The energy is expressed by

$$W = Pt$$

where
P = active power
t = time interval.
If the instantaneous power is constant, as is true when the voltages and currents form polyphase symmetrical sets, there is no restriction regarding the relation of the time interval to the period. If the voltages and currents are quasi-periodic and the amplitudes of the voltages and currents are slowly varying, the electric energy is the integral with respect to time of the active power, provided the integration is for a time that is one or more complete periods or that is quite long in comparison with the time of one period. Mathematically

$$W = \int_{t_0}^{t_0+t} P \, dt$$

where P = active power determined for the condition of voltages and currents having slowly varying amplitudes. Electric energy is expressed in joules (watt-seconds) or watthours when the voltages are in volts and the currents in amperes, and the time interval is in seconds or hours, respectively. 210

electric explosion-tested mine locomotive. An electric mine locomotive equipped with explosion-tested equipment. *See also:* **electric mine locomotive.** 328

electric field (1) (general). A vector field of electric field strength or of electric flux density. *Note:* The term is also used to denote a region in which such vector fields have a significant magnitude. *See:* **vector field.** 210
(2) (static). A state of the region in which stationary charged bodies are subject to forces by virtue of their charges. 180, 146
(3) (signal-transmission system). A state of a medium characterized by spatial potential gradients (electric field vectors) caused by conductors at different potentials, that is, the field between conductors at different potentials that have capacitance between them. *See also:* **signal.** 188

electric field intensity*. *See:* electric field strength.
210

* Deprecated

electric field strength (electric field) (electric field intensity*) (electric force*) (1) (general). At a given point, the vector limit **E** of the quotient of the force that a small stationary charge at that point will experience, by virtue of its charge, to the charge as the charge approaches zero in a macroscopic sense. *Note:* The concept of a tunnel-like (Kelvin) cavity, properly oriented along **E,** is frequently employed to visualize and compute the **E** vector in material media.
210

* Deprecated

(2) (signal-transmission system). The magnitude of the potential gradient in an electric field expressed in units of potential difference per unit length in the direction of the gradient. *See also:* signal.   60
(3) (radio wave propagation). The magnitude of the electric field vector. *Note:* This term has sometimes been called the **electric field intensity,** but such use of the word intensity is deprecated in favor of field strength. *See also:* radio wave propagation.   146, 199, 180
electric field vector (at a point in an electric field). The force on a stationary positive charge per unit charge. *Note:* This may be measured either in newtons per coulomb or in volts per meter. This term is sometimes called the **electric field intensity,** but such use of the word intensity is deprecated since intensity connotes power in optics and radiation. *See also:* radio wave propagation; waveguide.
267, 180, 146
electric filter (electric wave filter). A filter designed to separate electric waves of different frequencies. *Note:* An electric wave filter may be classified in terms of the reactive elements that it includes, for example, inductors and capacitors, piezoelectric crystal units, coaxial lines, resonant cavities, etcetera. *See also:* filter.   59
electric flux (through a surface) (electric displacement). The surface integral of the normal component of the electric flux density over the surface.   210
electric flux density (electric displacement density) (electric induction*). A quantity related to the charge displaced within the dielectric by application of an electric field. *Notes:* (1) Electric flux density at any point in an isotropic dielectric is a vector that has the same direction as the electric field strength and a magnitude equal to the product of the electric field strength and the absolute capacitivity. The electric flux density is that vector point function whose divergence is the charge density, and that is proportional to the electric field in region free of polarized matter. The electric flux density is given by

$$\mathbf{D} = \epsilon_0 \epsilon \mathbf{E}$$

where **D** is the electric flux density, $\epsilon_0 \epsilon$ is the absolute capacitivity, and **E** is the electric field strength. (2) In a nonisotropic medium, ϵ becomes a tensor represented by a matrix and **D** is not necessarily parallel to **E.** (3) The concept of a disk-like (Kelvin) cavity, properly oriented normal to **D,** is frequently employed to visualize and compute the **D** vector in material media. (4) The electric flux density at a point is equal to the charge per unit area that would appear on one face of a small thin metal plate introduced in the electric field at the point and so oriented that this charge is a maximum. (5) The symbol Γ_ϵ is often used in modern practice in place of ϵ_0; the symbol ϵ_v has occasionally been used.   210, 180, 146

* Deprecated

electric focusing (microwave tubes). The combination of electric fields that acts upon the electron beam in addition to the forces derived from momentum and space charge. *See also:* microwave tube or valve.   190
electric force*. *See:* electric field strength.   210

* Deprecated

electric freight locomotive. an electric locomotive, commonly used for hauling freight trains and generally designed to operate at higher tractive force values and lower speeds than a passenger locomotive of equal horsepower capacity. *Note:* A prefix diesel-electric, gas-electric, turbine-electric, etcetera, may replace the word electric. *See also:* electric locomotive.   328
electric gathering mine locomotive. An electric mine locomotive, the chief function of which is to move empty cars into, and remove loaded cars from, the working places. *See also:* electric mine locomotive.   328
electric generator. A machine that transforms mechanical power into electric power. *See also:* direct-current commutating machine.   3
electric gun heater. An electrically heated element attached to the gun breech to prevent the oil from congealing or the gun mechanism from freezing.   328
electric haulage mine locomotive. An electric mine locomotive used for hauling trains of cars, that have been gathered from the working faces of the mine, to the point of delivery of the cars. *See also:* electric mine locomotive.
328
electric horn. A horn having a diaphragm that is vibrated electrically. *See also:* protective signaling.   328
electric-hydraulic governor (speed governing of hydraulic turbines). A governor in which the control signal is proportional to speed error and the stabilizing and auxiliary signals are developed electrically, summed by appropriate electrical networks, and are then hydraulically amplified. Electrical signals may be analog or digitally derived.
8, 58
electric hygrometer. An instrument for indicating by electric means the humidity of the ambient atmosphere. *Note:* Electric hygrometers usually depend for their operation on the relation between the electric conductance of a film of hygroscopic material and its moisture content. *See also:* instrument.   328
electric incline railway. A railway consisting of an electric hoist operating a single car with or without counterweights, or two cars in balance, which car or cars travel on inclined tracks. *See also:* elevators.   328
electric indication lock. An electric lock connected to a lever of an interlocking machine to prevent the release of the lever or latch until the signals, switches, or other units operated, or directly affected by such lever, are in the proper position. *See also:* interlocking (interlocking plant).
328

electric indication locking. Electric locking adapted to prevent manipulation of levers that would bring about an unsafe condition for a train movement in case a signal, switch, or other operated unit fails to make a movement corresponding with that of its controlling lever; or adapted directly to prevent the operation of one unit in case another unit to be operated first, fails to make the required movement. *See also:* **interlocking (interlocking plant).**
328

electric induction*. *See:* **electric flux density.**

* Deprecated

electric interlocking machine. An interlocking machine designed for the control of electrically operated functions. *See also:* **interlocking (interlocking plant).** 328

electricity meter (meter). A device that measures and registers the integral of an electrical quantity with respect to time. *Note:* The term **meter** is also used in a general sense to designate any type of measuring device, including all types of electric measuring instruments. Such use as a suffix or as part of a compound word (for example, voltmeter, frequency meter) is universally accepted. Meter may be used alone with this wider meaning when the context is such as to prevent confusion with the narrower meaning here defined. 328

electric larry car. A burden-bearing car for operation on track rails used for short movements of materials, the propulsion of which is effected by electric motors mounted on the vehicle. *Note:* A prefix diesel-electric, gas-electric, etcetera, may replace the word electric. *See also:* **electric motor car.** 328

electric loading (rotating machinery). The average ampere-conductors of the primary winding per unit length of the air-gap periphery. *See also:* **rotor (rotating machinery); stator.** 63

electric lock. A device to prevent or restrict the movement of a lever, a switch, or a movable bridge unless the locking member is withdrawn by an electric device such as an electromagnet, solenoid, or motor. *See also:* **interlocking (interlocking plant).** 328

electric locking. The combination of one or more electric locks and controlling circuits by means of which levers of an interlocking machine, or switches, or other units operated in connection with signaling and interlocking, are secured against operation under certain conditions, as follows: (1) approach locking, (2) indication locking, (3) switch-lever locking, (4) time locking, (5) traffic locking. *See also:* **interlocking (interlocking plant).** 328

electric locomotive. A vehicle on wheels, designed to operate on a railway for haulage purposes only, the propulsion of which is effected by electric motors mounted on the vehicle. *Note:* While this is a generic term covering any type of locomotive driven by electric motors, it is usually applied to locomotives receiving electric power from a source external to the locomotive. The prefix electric may also be applied to cars, buses, etcetera, driven by electric motors. A prefix diesel-electric, gas-electric, turbine-electric, etcetera, may replace the word electric. 328

electric-machine regulator (rotating machinery). A specified element or a group of elements that is used within an electric-machine regulating system to perform a regulating function by acting to maintain a designated variable (or variables) at a predetermined value, or to vary it according to a predetermined plan. 63

electric-machine regulating system (rotating machinery). A feedback control system that includes one or more electric machines and the associated control. 63

electric mechanism (demand meter). That portion, the action of which, in response to the electric quantity to be measured, gives a measurement of that quantity. *Note:* For example, the electric mechanism of certain demand meters is similar to the ordinary ammeter or wattmeter of the deflection type; in others it is a watt-hour meter or other integrating meter; and in still others it comprises an electric circuit that heats temperature-responsive elements, such as bimetallic spirals, that deflect to move the indicating means. The electrical quantity may be measured in kilowatts, kilowatt-hours, kilovolt-amperes, kilovolt-ampere-hours, amperes, ampere-hours, kilovars, kilovar-hours, or other suitable units. *See also:* **demand meter.**
328

electric mine locomotive (1) (general). An electric locomotive designed for use underground; for example, in such places as coal, metal, gypsum, and salt mines, tunnels, and in subway construction.

(2) (storage-battery type). An electric locomotive that receives its power supply from a storage battery mounted on the chassis of the locomotive.

(3) (trolley type). An electric locomotive that receives its power supply from a trolley-wire distribution system.

(4) (combination type). An electric locomotive that receives power either from a trolley-wire distribution system or from a storage battery carried on the locomotive.

(5) (separate tandem). An electric mine locomotive consisting of two locomotive units that can be coupled together or operated from one controller as a single unit, or else separated and operated as two independent units.

(6) (permanent tandem). A locomotive consisting of two locomotive units permanently connected together and provided with one set of controls so that both units can be operated by a single operator. 328

electric moment (electric doublet or dipole). The product of the magnitude of either charge by the distance between them. 210

electric motive power unit. A self-contained electric traction unit, comprising wheels and a superstructure capable of independent propulsion from a power supply system, but not necessarily equipped with an independent control system. *Note:* While this is a generic term covering any type of motive power driven by electric motors, it is usually applied to locomotives receiving electric power from an external source. A prefix diesel-electric, gas-electric, turbine-electric, etcetera, may replace the word electric. *See also:* **electric locomotive.** 328

electric motor. A machine that transforms electric power into mechanical power. *See:* **asynchronous machine; direct-current commutating machine.** 328

electric motor car. A vehicle for operating on track rails, used for the transport of passengers or materials, the propulsion of which is effected by electric motors mounted on the vehicle. *Note:* A prefix diesel-electric, gas-electric, etcetera, may replace the word electric. 328

electric motor controller. A device or group of devices that serve to govern, in some predetermined manner, the electric power delivered to the motor. *Note:* An electric motor controller is distinct functionally from a simple disconnecting means whose principal purpose in a motor circuit is to disconnect the circuit, together with the motor and its controller, from the source of power. *See:* **electric controller.**   225, 206

electric movable-bridge (drawbridge) lock. A device used to prevent the operation of a movable bridge until the device is released. *See also:* **interlocking (interlocking plant).**   328

electric network. *See:* **network.**

electric noise (1) (general). Unwanted electrical energy other than crosstalk present in a transmission system. **(2) (interference terminology).** A form of interference introduced into a signal system by natural sources that constitutes for that system an irreducible limit on its signal-resolving capability. *Note:* Noise is characterized by randomness of amplitude and frequency distribution and therefore cannot be eliminated by band-rejection filters tuned to preselected frequencies. *See also:* **distortion; interference.**   111, 63

electric operation. Power operation by electric energy.   103, 202

electric orchestra lift. An electrically driven mechanism for raising and lowering the musicians' platform and the musicians. *See also:* **elevators.**   328

electric parachute-flare-launching tube. A tube mounted on an aircraft through which a metal container carrying a parachute flare is launched, the tube being so designed that as the parachute-flare container passes through the tube, an electric circuit is completed that ignites a slow-burning fuse in the container, the fuse being so designed as to permit the container to clear the aircraft before it ignites the parachute flare. *See also:* **air transportation, electric equipment.**   328

electric passenger locomotive. An electric locomotive, commonly used for hauling passenger trains and generally designed to operate at higher speeds and lower tractive-force values than a freight locomotive of equal horsepower capacity. *Note:* A prefix diesel-electric, gas-electric, turbine-electric, etcetera, may replace the word electric. *See also:* **electric locomotive.**

electric penetration assembly (nuclear power generating stations). An assembly of insulated electric conductors, conductor seals, and aperture seals that provides the passage of the electric conductors through a single aperture in the nuclear containment structure, while providing a pressure barrier between the inside and the outside of the containment structure.   26

electric permissible mine locomotive. An electric locomotive carrying the official approval plate of the United States Bureau of Mines. *See also:* **electric mine locomotive.**   328

electric pin-and-socket coupler (connector). A readily disconnective assembly used to connect electric circuits between components of an aircraft electric system by means of mating pins and sockets. *See also:* **air transportation wiring and associated equipment.**   328

electric polarizability. Of an isotropic medium for which the direction of electric polarization and electric field strength are the same at any point in the medium, the magnitude P of the electric polarization at that point divided by the electric field strength there, E. *Note:* In a rationalized system, the electric polarizability $P_e = P/E = \epsilon_0(\epsilon - 1)$.   210

electric polarization (electric field). At any point, the vector difference between the electric flux density at that point and the electric flux density that would exist at that point for the same electric field strength there, if the medium were a vacuum there. *Note:* Electric polarization is the vector limit of the quotient of the vector sum of electric dipole moments in a small volume surrounding a given point, and this volume, as the volume approaches zero in a microscopic sense.   210

electric port (optoelectronic device). A port where the energy is electric. *Note:* A designated pair of terminals may serve as one or more electric ports. *See also:* **optoelectronic device.**   191

electric power distribution panel. A metallic or nonmetallic, open or enclosed, unit of an electric system. The operable and the indicating components of an electric system, such as switches, circuit breakers, fuses, indicators, etcetera, usually are mounted on the face of the panel. Other components, such as terminal strips, relays, capacitors, etcetera, usually are mounted behind the panel. *See also:* **air transportation wiring and associated equipment.**   328

electric propulsion apparatus. Electric apparatus (generators, motors, control apparatus, etcetera) provided primarily for ship's propulsion. *Note:* For certain applications, and under certain conditions, auxiliary power may be supplied by propulsion apparatus. *See also:* **electric propulsion system.**   328

electric propulsion system. A system providing transmission of power by electric means from a prime mover to a propeller shaft with provision for control, partly or wholly by electric means, of speed and direction. *Note:* An electric coupling (which see) does not provide electric propulsion.   328

electric reset relay. A relay that is so constructed that it remains in the picked-up condition even after the input quantity is removed; an independent electric input is required to reset the relay.   103, 202, 60

electric resistance-type temperature indicator. A device that indicates temperature by means of a resistance bridge circuit.   328

electric road locomotive. An electric locomotive designed primarily for hauling dispatched trains over the main or secondary lines of a railroad. *Note:* A prefix diesel-electric, gas-electric, turbine-electric, etcetera, may replace the word electric. *See also:* **electric locomotive.**   328

electric road-transfer locomotive. An electric locomotive designed primarily so that it may be used either for hauling dispatched trains over the main or secondary lines of a railroad or for transferring relatively heavy cuts of cars for short distances within a switching area. *Note:* A prefix diesel-electric, gas-electric, turbine-electric, etcetera, may replace the word electric. *See also:* **electric locomotive.**   328

electric sign. A fixed or portable self-contained electrically illuminated appliance with words or symbols designed to convey information or attract attention. *See also:* **appliance.**   328

electric-signal storage tube. A storage tube into which the information is introduced as an electric signal and read at

a later time as an electric signal. *See also:* **storage tube.**
174, 190

electric sounding machine. A motor-driven reel with wire line and weight for determination of depth of water by mechanical sounding. 328

electric squib. A device similar to an electric blasting cap but containing a gunpowder composition that simply ignites but does not detonate an explosive charge. *See also:* **blasting unit.** 328

electric stage lift. An electrically driven mechanism for raising and lowering various sections of a stage. 328

electric strength (dielectric strength). The maximum potential gradient that the material can withstand without rupture. *See:* **dielectric strength.** *See also:* **insulation testing (large alternating-current rotating machinery).**
6

electric stroboscope. An instrument for observing rotating or vibrating objects or for measuring rotational speed or vibration frequency, or similar periodic quantities, by electrically produced periodic changes in illumination. *See also:* **instrument.** 328

electric-supply equipment (1) (general). Equipment that produces, modifies, regulates, controls, or safeguards a supply of electric energy. *Note:* Similar equipment, however, is not included where used in connection with signaling systems under the following conditions: (1) where the voltage does not exceed 150, and (2) where the voltage is between 150 and 400 and the power transmitted does not exceed 3 kilowatts. *See also:* **distribution center.**
178

(2) (transmission and distribution) (supply equipment). Equipment which produces, modifies, regulates, controls, or safeguards a supply of electric energy. 262

electric-supply lines (1) (general). Those conductors used to transmit electric energy and their necessary supporting or containing structures. Signal lines of more than 400 volts to ground are always supply lines within the meaning of these rules, and those of less than 400 volts may be considered as supply lines, if so run and operated throughout. *See also:* **center of distribution.** 178, 64

(2) (transmission and distribution) (supply lines). Those conductors used to transmit electric energy and their necessary supporting or containing structures. Signal lines of more than 400 volts are always supply lines within the meaning of the rules, and those of less than 400 volts may be considered as supply lines, if so run and operated throughout. 262

electric-supply station (1) (general). Any building, room, or separate space within which electric-supply equipment is located and the interior of which is accessible, as a rule, only to properly qualified persones. *Note:* This includes generating stations and substations and generator, storage-battery, and transformer rooms, but excludes manholes and isolated transformer vaults on private premises. *See:* **transformer vault.** *See also:* **generating station.** 178

(2) (transmission and distribution) (supply station). Any building, room, or separate space within which electric-supply equipment is located and the interior of which is accessible, as a rule, only to properly qualified persons. *Note:* This includes generating stations and substations and generator, storage battery, and transformer rooms.
262

electric surges (nuclear power generating station). Any spurious voltage or current pulses conducted into the module from external sources. 355

electric susceptibility. Of an isotropic medium, for which the direction of electric polarization and electric field strength are the same, at any point in the medium, the magnitude of the electric polarization at that point of the medium, divided by the electric flux density that would exist at that point for the same electric field strength, if the medium there were a vacuum. *Note:* In a rationalized system the electric susceptibility $\chi_\epsilon = P/\mathrm{D}(\epsilon - 1)$. 210

electric switching locomotive. An electric locomotive designed for yard movements of freight or passenger cars, its speed and continuous electrical capacity usually being relatively low. *Note:* A prefix diesel-electric, gas-electric, turbine-electric, etcetera, may replace the word electric. *See also:* **electric locomotive.** 328

electric switch-lever lock. An electric lock used to prevent the movement of a switch lever or latch in an interlocking machine until the lock is released. *See also:* **interlocking (interlocking plant).** 328

electric switch-lever locking. A general term for route or section locking. *See also:* **interlocking (interlocking plant).**
328

electric switch lock. An electric lock used to prevent the operation of a switch or a switch movement until the lock is released. *See also:* **interlocking (interlocking plant).**
328

electric tachometer (marine usage). An instrument for measuring rotational speed by electric means. *See also:* **instrument.** 328

electric telegraph. A telegraph having the relationship of the moving parts of the transmitter and receiver maintained by the use of self-synchronous motors or equivalent devices. 328

electric telemeter. The measuring, transmitting, and receiving apparatus, including the primary detector, intermediate means (excluding the channel), and end devices for electric telemetering. *Note:* A telemeter that measures current is called a teleammeter; voltage, a televoltmeter, power, a telewattmeter; one that measures angular or linear position, a position telemeter. The names of the various component parts making up the telemeter are, in general, self-defining; for example, the transmitter, receiver, indicator, etcetera. *See also:* **telemetering.**
103, 202

electric telemetering (electric telemetry*). Telemetering performed by an electric translating means separate from the measurand. *See also:* **telemetering.** 103, 202

* Deprecated

electric thermometer (temperature meter). An instrument that utilizes electric means to measure temperature. *Note:* An electric thermometer may employ any electrical or magnetic property that is dependent on temperature. The most commonly used properties are the thermoelectric effects in a circuit of two or more metals and the change of electric resistance of a metal. 328

electric tower car. A rail vehicle, the propulsion of which is effected by electric means and that is provided with an elevated platform, generally arranged to be raised and lowered, for the installation, inspection, and repair of a contact wire system. *Note:* A prefix diesel-electric, gas-electric, etcetera, may replace the word electric. *See also:* **electric motor car.**   328

electric trail car (electric trailer). A car not provided with motive power that is used in a train with one or more electric motor cars. *Note:* A prefix diesel-electric, gas-electric, etcetera, may replace the word electric to identify the motor cars. *See also:* **electric-control trail car, electric motor car.**   328

electric trainline (land transportation vehicles). A continuous electric conductor extended between cars by means of jumper cables or couplers so that power or control signals can be transmitted to and/or from each car to permit simultaneous control of traction motors and other vehicle carried equipment. An electric trainline may include circuits for auxiliary electric brakes, communication, engine controls and other devices.   1

electric transducer. A transducer in which all of the waves concerned are electric. *See also:* **transducer.**   252, 210

electric transfer locomotive. An electric locomotive designed primarily for transferring relatively heavy cuts of cars for short distances within a switching area. *Note:* A prefix diesel-electric, gas-electric, turbine-electric, etcetera, may replace the word electric. *See also:* **electric locomotive.**   328

electric-tuned oscillator. An oscillator whose frequency is determined by the value of a voltage, current, or power. Electric tuning includes electronic tuning, electrically activated thermal tuning, electromechanical tuning, and tuning methods in which the properties of the medium in a resonant cavity are changed by an external electric means. An example is the tuning of a ferrite-filled cavity by changing an external magnetic field. *See also:* **tunable microwave oscillators.**   174, 190

electric turn-and-bank indicator. A device that utilizes an electrically driven gyro for turn determination and a gravity-actuated inclinometer for bank determination.   328

electric valve operator (nuclear power generating stations). An electric powered mechanism for opening and closing a valve, including all electric and mechanical components that are integral to the mechanism and are required to operate and control valve action.   31, 142

electric vector (radio wave propagation). Electric field vector. *See also:* **radio wave propagation.**   180

electric wave filter. *See:* **electric filter.**

electric wind. *See:* **convective discharge.**

electrification by friction. *See:* **triboelectrification.**

electrified track. A railroad track suitably equipped in association with a contact conductor or conductors for the operation of electrically propelled vehicles that receive electric power from a source external to the vehicle. *See also:* **electric locomotive.**   328

electroacoustical reciprocity theorem. For an electroacoustic transducer satisfying the reciprocity principle, the quotient of the magnitude of the ratio of the open-circuit voltage at output terminals (or the short-circuit current) of the transducer, when used as a sound receiver, to the free-field sound pressure referred to an arbitrarily selected reference point on or near the transducer, divided by the magnitude of the ratio of the sound pressure apparent at a distance δ from the reference point to the current flowing at the transducer input terminals (or the voltage applied at the input terminals), when used as a sound emitter, is a constant, called the reciprocity constant, independent of the type or constructional details of the transducer. *Note:* The reciprocity constant is given by

$$\left|\frac{M_o}{S_s}\right| = \left|\frac{M_s}{S_s}\right| = \frac{2\delta}{\rho f}$$

where

M_o = open free-field voltage response, as a sound receiver, in open-circuit volts per newton per square meter, referred to the arbitrary reference point on or near the transducer.

M_s = free-field current response in short-circuit amperes per newton per square meter, referred to the arbitrary reference point on or near the transducer

S_o = sound pressure in newtons per square meter per ampere of input current produced at a distance δ meters from the arbitrary reference point

S_s = sound pressure in newtons per square meter per volt applied at the input terminals produced at a distance δ meters from the arbitrary reference point

f = frequency in hertz

ρ = density of the medium in kilograms per cubic meter

δ = distance in meters from the arbitrary reference point on or near the transducer to the point in which the sound pressure established by the transducer when emitting is evaluated.

See also: **loudspeaker.**   176

electroacoustic transducer (electric system). A transducer for receiving waves and delivering waves to an acoustic system, or vice versa. *See also:* **loudspeaker; transducer.**   176

electroanalysis. The electrodeposition of an element or compound for the purpose of determining its quantity in the solution electrolyzed. *See also:* **electrodeposition.** 328

electrobiology. The study of electrical phenomena in relation to biological systems.   192

electrocapillary phenomena. The phenomena depending on the variation in surface tension, at the boundary of two liquids, with the potential difference established between these two liquids.   210

electrocardiogram. The graphic record of the variation with time of the voltage associated with cardiac activity. *See:* **electrocorticogram (electrobiology); electrodermogram (electrobiology); Galvani's experiment (electrobiology); spindle wave (electrobiology); vector electrocardiogram (electrobiology).**   192

electrocautery (electrotherapy). An instrument for cauterizing the tissues by means of a conductor brought to a high temperature by an electric current. *See also:* **electrotherapy.**   192

electrochemical cell. A system consisting of an anode, cathode, and an electrolyte plus such connections (electric and mechanical) as may be needed to allow the cell to deliver or receive electric energy.   223, 186

electrochemical equivalent (element, compound, radical, or ion) (1) (general). The weight of that substance involved in a specified electrochemical reaction during the passage of a specified quantity of electricity, such as a faraday, ampere-hour, or coulomb.   328

(2) (oxidation). The weight of an element or group of elements oxidized or reduced at 100-percent efficiency by a unit quantity of electricity. *See also:* **electrochemistry.**   221, 205

electrochemical recording (facsimile). Recording by means of a chemical reaction brought about by the passage of signal-controlled current through the sensitized portion of the record sheet. *See also:* **recording (facsimile).**   12

electrochemical series. *See:* **electromotive series.**

electrochemical valve. An electric valve consisting of a metal in contact with a solution or compound across the boundary of which current flows more readily in one direction than in the other direction and in which the valve action is accompanied by chemical changes.   328

electrochemical valve metal. A metal or alloy having properties suitable for use in an electrochemical valve. *See also:* **electrochemical valve.**   328

electrochemistry. That branch of science and technology that deals with interrelated transformations of chemical and electric energy.   328

electrocoagulation (medical electronics). The clotting of tissue by heat generated within the tissue by impressed electric currents.   192

electrocorticogram (medical electronics). A graphic record of the variation with time of voltage taken from exposed cortex cerebra.   192

electroculture (medical electronics). The stimulation of growth, flowering, or seeding by electric means.   192

electrocution. The destruction of life by means of electric current.   192

electrode (1) (general). A conductor through which an electric current enters or leaves an electrolyte, gas, or vacuum.   328

(2) (electrochemistry). An electric conductor for the transfer of charge between the external circuit and the electroactive species in the electrolyte. *Note:* Specifically, in an electrolytic cell, an electrode is a conductor at the surface of which a change occurs from conduction by electrons to conduction by ions or colloidal ions. *See also:* **electrolytic cell; electrochemical cell.**   223, 186

(3) (electron tube). A conducting element that performs one or more of the functions of emitting, collecting, or controlling by an electric field the movements of electrons or ions.   190, 125

(4) (semiconductor device) (light emitting diodes). An element that performs one or more of the functions of emitting or collecting electrons or holes, or of controlling their movements by an electric field. *See also:* **semiconductor.**   66,162

(5) (biological electronics) (reference, inactive, diffuse, dispersive, indifferent electrode). (A) A pickup electrode that, because of averaging, shunting, or other aspects of the tissue-current pattern to which it connects, shows potentials not characteristic of the region near the active electrode. (B) Any electrode, in a system of stimulating electrodes, at which due to its dispersive action, excitation is not produced. (C) An electrode of relatively large area applied to some inexcitable or distant tissue in order to complete the circuit with the active electrode that is used for stimulation.   192

electrode, accelerating (electron-beam tube). *See:* **accelerating electrode.**

electrode admittance (*j*th electrode of an *n*-electrode electron tube). The short-circuit driving-point admittance between the *j*th electrode and the reference point measured directly at the *j*th electrode. *Note:* To be able to determine the intrinsic electronic merit of an electron tube, the driving-point and transfer admittances must be defined as if measured directly at the electrodes inside the tube. The definitions of electrode admittance and electrode impedance are included for this reason. *See also:* **electron-tube admittances.**   190, 125

electrode alternating-current resistance (electron device). The real component of the electrode impedance. *See:* **self-impedance.**   190

electrode bias (electron tubes). The voltage at which an electrode is stabilized under operating conditions with no incoming signal, but taking into account the voltage drops in the connected circuits. *See:* **electrode voltage.**   244, 190

electrode capacitance (*n*-terminal electron tube). The capacitance determined from the short-circuit driving-point admittance at that electrode. *See also:* **electron-tube admittances.**   190, 125

electrode characteristic. A relation, usually shown by a graph, between the electrode voltage and the current of an electrode, all other electrode voltages being maintained constant.   190, 125

electrode conductance. The real part of the electrode admittance.   190, 125

electrode, control. *See:* **control electrode.**

electrode current (electron tube). The current passing to or from an electrode through the interelectrode space. *Note:* The terms cathode current, grid current, anode current, plate current, etcetera, are used to designate electrode currents for these specific electrodes. Unless otherwise stated, an electrode current is measured at the available terminal.   190

electrode current, average. *See:* **average electrode current.**

electrode-current averaging time (electron tubes). The time interval over which the current is averaged in defining the operating capabilities of the electrode. *See also:* **electrode current (electron tube).**   190

electrode dark current (1) (phototubes). The component of electrode current remaining when ionizing radiation and optical photons are absent. *Notes:* (A) Optical photons are photons with energies corresponding to wavelengths between 2000 and 1500 angstroms. (B) Since the dark current may change considerably with temperature, the temperature should be specified. *See also:* **phototube.**

(2) (camera tubes). The current from an electrode in a photoelectric tube under stated conditions of radiation shielding. *See also:* **camera tube.**   178, 190

electrode dissipation. The power dissipated in the form of heat by an electrode as a result of electron or ion bombardment, or both, and radiation from other electrodes. *See:* **grid driving power; modes, degenerate.**   190

electrode drop (arc-welding apparatus). The voltage drop in the electrode due to its resistance (or impedance).   264

electrode economizer. A collar that makes a seal between an electrode and the roof of a covered electric furnace with substantial exclusion of air, thereby preventing serious oxidation of the part of the electrode within the furnace. *See also:* **electrothermics.**   328

electrode impedance. The reciprocal of the electrode admittance. *See also:* **electron-tube admittances.**   190

electrode impedance, biological. The ratio between two vectors, the numerator being the vector that represents the potential difference between the electrode and biological material, and the denominator being the vector that represents the current between the electrode and the biological material. *See:* **loss angle (biological); polarization capacitance (biological); polarization reactance (biological); polarization resistance (biological).**   192

electrodeposition. The process of depositing a substance upon an electrode by electrolysis or electrophoresis. Electrodeposition includes electroplating, electroforming, electrorefining, and electrowinning.   328

electrode, pad. *See:* **pad electrode.**

electrode potential. The difference in potential between an electrode and the immediately adjacent electrolyte, referred to an arbitrary zero of potential. *See also:* **electrolytic cell.**   328

electrode potential, biological. The potential between an electrode and biological material.   192

electrode radiator (cooling fin) (electron tubes). A metallic piece, often of large area, extending the electrode to facilitate the dissipation of the heat generated in the electrode. *See also:* **electron tube.**   244, 190

electrode reactance (electron device). The imaginary component of the electrode impedance. *See:* **self-impedance.**   190

electrode resistance (1) (general). The reciprocal of the electrode conductance. *Note:* This is the effective parallel resistance and is not the real component of the electrode impedance.   190
(2) (at a stated operating point) (electron device). The quotient of the direct electrode voltage by the direct electrode current. *See:* **self-impedance.**   190

electrodermal reaction (EDR) (medical electronics). The change in electric resistance of the skin during emotional stress.   192

electrodermogram (electromyogram) (electroretinogram) (electrobiology). A graphic record of the variation with time of voltage taken from the given anatomical structure (skin, muscle, and retina, respectively). *See also:* **electrocardiogram.**   192

electrodesiccation (fulguration). The superficial destruction of tissue by electric sparks from a movable electrode. *See also:* **electrotherapy.**   192

electrode, signal (camera tube). *See:* **signal electrode.**

electrodes or applicators. *See:* **applicators or electrodes.**

electrode susceptance (electron device). The imaginary component of the electrode admittance. *See:* **self-impedance.**   190

electrode voltage. The voltage between an electrode and the cathode or a specified point of a filamentary cathode. *Note:* The terms grid voltage, anode voltage, plate voltage, etcetera, are used to designate the voltage between these specific electrodes and the cathode. Unless otherwise stated, electrode voltages are understood to be measured at the available terminals.   190, 125

electrodiagnosis. The study of functional states of parts of the body either by studying their responses to electric stimulation or by studying the electric potentials (or currents) that they spontaneously produce.   192

electrodissolution. The process of dissolving a substance from an electrode by electrolysis. *See also:* **electrodeposition.**   328

electrodynamic instrument. An instrument that depends for its operation on the reaction between the current in one or more movable coils and the current in one or more fixed coils. *Note:* Electrodynamic instruments may or may not have iron cores or shields. They may or may not have control or restoring torque springs. An example of the latter is the crossed coil design used in ratio meters, power-factor meters, etcetera, when two sets of moving coils are set at an angle to each other to produce the necessary operating torque. *See also:* **instrument.**   328

electroencephalogram (medical electronics). A graphic record of the changes with time of the voltage obtained by means of electrodes applied to the skin over the cerebrum.   192

electroextraction. The extraction by electrochemical processes of metals or compounds from ores and intermediate compounds. *See also:* **electrowinning.**   328

electroforming. The production or reproduction of articles by electrodeposition.   328

electrographic recording (electrostatography). The branch of electrostatic electrography that employs a charge transfer between two or more electrodes to form directly electrostatic-charge patterns on an insulating medium for producing a viewable record. *See also:* **electrostatography.**   191, 236

electrographitic brush (rotating machinery). A brush composed of selected amorphous carbon that, in the process of manufacture, is carried to a temperature high enough to convert the carbon to the graphitized form. *Note:* This type of brush is exceedingly versatile in that it can be made soft or very hard, also nonabrasive or slightly abrasive. Grades of brushes of this type have a high current-carrying capacity, but differ greatly in operating speed from low to high. *See:* **brush (rotating machinery).**   63, 279

electrohydraulic elevator. A direct-plunger elevator where liquid is pumped under pressure directly into the cylinder by a pump driven by an electric motor. *See also:* **elevators.**   328

electrokinetic potential (zeta potential) (medical electronics). A set of four electric or velocity potentials that accompany relative motion between solids and liquids.   192

electroluminescence (1) (general). The emission of light from a phosphor excited by an electromagnetic field. *See also:* **lamp.**   167
(2) (light emitting diodes). The emission of light from a material (phosphor or semiconductor) where the exciting mechanism is the application of an electromagnetic field.   162

electroluminescent display panel. A thin, usually flat, electroluminescent display device. *See also:* **optoelectronic device.**   191

electroluminescent display device. An optoelectronic device with a multiplicity of electric ports, each capable of independently producing an optic output from an associated

electroluminator element. *See also:* **optoelectronic device.**
191

electrolysis (1) (general). The production of chemical changes by the passage of current from an electrode to an electrolyte, or vice versa. *See also:* **electrolytic cell.**
328

(2) (underground structures). The destructive chemical action caused by stray or local electric currents to pipes, cables, and other metalwork. *See also:* **corrosion; power distribution, underground construction.** 64

electrolyte. A conducting medium in which the flow of electric current takes place by migration of ions. *Note:* Many physical chemists define electrolyte as a substance that when dissolved in a specified solvent, usually water, produces an ionically conducting solution. *See also:* **electrolytic cell.** 210

electrolyte cells. *See:* **cascade.**

electrolytic. *See:* **cathode.**

electrolytic capacitor. A combination of two capacitors at least one of which is a valve metal, separated by an electrolyte, and between which a dielectric film is formed adjacent to the surface of one or both of the capacitors. *See:* **asymmetrical cell; blocking capacitor; electrochemical valve; nonpolarized electrolytic capacitor; polarized electrolytic capacitor; unit-area capacitance.** 328

electrolytic cell. A cell in which electrochemical reactions are produced by applying electric energy, or conversely, that supplies electric energy as a result of electrochemical action. The latter cell may be called a galvanic cell. Each electrolytic cell comprises two or more electrodes and one or more electrolytes contained in a suitable vessel. The cell is a unit which may be used singly or electrically connected to other like units. 328

electrolytic cleaning. The process of degreasing or descaling a metal by making it an electrode in a suitable bath. 221

electrolytic dissociation. In a solution, a process whereby some fraction of the molecules of a solute is ionized to form ions. *See also:* **electrochemistry.** 328

electrolytic oxidation. An anodic process by which electrons are removed from or positive charges added to atoms or ions, or by which when applied to organic molecules, the oxygen or acidic proportion is increased. *See also:* **electrochemistry; electrolytic reduction.** 328

electrolytic parting (electrorefining). The electrolysis of a silver-gold alloy anode in which the silver is deposited in a pure form at the cathode, and the gold remains in the anode slime. *See also:* **electrorefining.** 328

electrolytic recording (facsimile). That type of electrochemical recording in which the chemical change is made possible by the presence of an electrolyte. *See also:* **recording (facsimile).** 12

electrolytic rectifier. A rectifier in which rectification of an alternating current is accompanied by electrolytic action. *See also:* **electrochemical valve; rectification.** 328

electrolytic reduction. A cathode process by which electrons are added to or positive charges removed from atoms or ions, or by which when applied to organic molecules, the hydrogen or basic proportion is increased. *See also:* **electrochemistry; electrolytic oxidation.** 328

electrolytic tank. A vessel containing a poorly conducting liquid, in which are inserted conductors that are scale

models of an electrode system. *Note:* It is used to obtain potential diagrams. *See also:* **electron optics.** 244, 190

electrolyzer. An electrolytic cell for the production of chemical products.

electromagnet. A device consisting of a ferromagnetic core and a coil, that produces appreciable magnetic effects only when an electric current exists in the coil. 210

electromagnetic braking (magnetic braking). A system of electric braking in which a mechanical braking system is actuated by electromagnetic means. *See also:* **electric braking.** 328

electromagnetic compatibility. The capability of electronic equipments or systems to be operated in the intended operational electromagnetic environment at designed levels of efficiency. 199

electromagnetic delay line. A delay line whose operation is based on the time of propagation of electromagnetic waves through distributed or lumped capacitance and inductance. 255, 77

electromagnetic disturbance (electromagnetic compatibility). An electromagnetic phenomenon that may be superimposed on a wanted signal. *See also:* **electromagnetic compatibility.** 199

electromagnetic environment. The electromagnetic field(s) and or signals existing in a transmission medium. *See also:* **electromagnetic compatibility.** 199

electromagnetic induction. The production of an electromotive force in a circuit by a change in the magnetic flux linking with that circuit. 222, 197

electromagnetic interference (electromagnetic compatibility). Impairment of a wanted electromagnetic signal by an electromagnetic disturbance. *See also:* **electromagnetic compatibility.** 199

electromagnetic lens. An electronic lens in which the result is obtained by a magnetic field. *See also:* **electron optics.** 190, 244

electromagnetic noise (electromagnetic compatibility). An unwanted electromagnetic disturbance that is not of a sinusoidal character. *See also:* **electromagnetic compatibility.** 220, 199

electromagnetic oscillograph. An oscillograph in which the record is produced by means of a mechanical motion derived from electromagnetic forces. *See also:* **oscillograph.** 328

electromagnetic radiation (laser-maser). The flow of energy consisting of orthogonally vibrating electric and magnetic fields lying transverse to the direction of propagation. X-rays, ultraviolet, visible, infrared, and radio waves occupy various portions of the electromagnetic spectrum and differ only in frequency and wavelength. 363

electromagnetic relay (relay systems) (power switchgear). An electromechanical relay that operates principally by action of an electromagnetic element which is energized by the input quantity. 127, 103

electromagnetic wave. A wave characterized by variations of electric and magnetic fields. *Note:* Electromagnetic waves are known as radio waves, heat rays, light rays, etcetera, depending on the frequency. *See also:* **radio wave propagation; waveguide.** 267, 146, 179, 180

electromechanical bell. A bell having a prewound spring-driven mechanism, operation of which is initiated by actuation of an electric tripping mechanism. *See also:* **pro-**

tective signaling. 328

electromechanical device (control equipment). A device that is electrically operated and has mechanical motion such as relays, servos, etcetera. 94

electromechanical interlocking machine. An interlocking machine designed for the control of both mechanically and power-operated functions. *See also:* **interlocking (inter-locking plant).** 328

electromechanical recorder. An equipment for transforming electric signals into mechanical motion of approximately like form and inscribing such motion in an appropriate medium by cutting or embossing. 328

electromechanical recording (facsimile). Recording by means of a signal-actuated mechanical device. *See also:* **recording (facsimile).** 12

electromechanical relay. A relay that operates by physical movement of parts resulting from electromagnetic, electrostatic, or electrothermic forces created by the input quantities. 103, 202, 60, 127

electromechanical switching system (telephone switching systems). An automatic switching system in which the control functions are performed principally by electromechanical devices. 55

electromechanical transducer. A transducer for receiving waves from an electric system and delivering waves to a mechanical system, or vice versa. *See also:* **transducer.** 52, 210, 176

electrometallurgy. The branch of science and technology that deals with the application of electric energy to the extraction or treatment of metals. 328

electrometer. An electrostatic instrument for measuring a potential difference or an electric charge by means of the mechanical forces exerted between electrically charged surfaces. *See also:* **instrument.** 328

electrometer tube. A vacuum tube having a very low control-electrode conductance to facilitate the measurement of extremely small direct current or voltage. 190, 125

electromotive force. *See:* **voltage.**

electromotive force series. A list of elements arranged according to their standard electrode potentials. 221, 205

electromotive series (electrochemical series). A table that lists in order the standard potentials of specified electrochemical reactions. *See also:* **electrochemistry.** 328

electromyograph (medical electronics). An instrument for recording action potentials or physical movements of muscles. 192

electron (1) (noun). An elementary particle containing the smallest negative electric charge. *Note:* The mass of the electron is approximately equal to 1/1837 of the mass of the hydrogen atom. 244
(2) (adjective). Operated by, containing, or producing electrons. *Examples:* Electron tube, electron emission, and electron gun. *See:* **electronic; electronics.** 328

electron accelerator, linear. *See:* **linear electron accelerator.**

electronarcosis. The production of transient insensibility by means of electric current applied to the cranium at intensities insufficient to cause generalized convulsions. *See also:* **electrotherapy.** 192

electron (ion) beam. A beam of electrons (ions) emitted from a single source and moving in neighboring paths that are confined to a desired region. 244, 190

electron-beam instrument (cathode-ray instrument). An instrument that depends for its operation on the deflection of a beam of electrons by an electric or magnetic field or by both. *Note:* The deflection is usually observed by causing the beam to strike a fluorescent screen. *See also:* **instrument.** 328

electron-beam magnetometer. A magnetometer that depends for its operation upon the change of the intensity or direction of an electron beam immersed in the field to be measured. *See also:* **magnetometer.** 328

electron-beam tube. An electron tube, the performance of which depends upon the formation and control of one or more electron beams. 125

electron collector (microwave tube). The electrode that receives the electron beam at the end of its path. *Note:* The power of the beam is used to produce some desired effect before it reaches the collector. *See also:* **velocity-modulated tube.** 244, 190

electron-coupled oscillator. An oscillator employing a multigrid tube with the cathode and two grids operating as an oscillator in any conventional manner, and in which the plate circuit load is coupled to the oscillator through the electron system. *See also:* **oscillatory circuit.** 111, 240

electron (proton) damage coefficient. The change in a stated quantity (such as minority carrier inverse squared diffusion length) of a given material per unit particle fluence of a stated energy spectrum. 113

electron device. A device in which conduction is principally by electrons moving through a vacuum, gas, or semiconductor. 190, 94, 125

electron-device transducer. *See:* **admittance, short-circuit forward.**

electron emission. The liberation of electrons from an electrode into the surrounding space. *Note:* Quantitatively, it is the rate at which electrons are emitted from an electrode. 190, 125

electron gun (electron tubes). An electrode structure that produces and may control, focus, deflect, and converge one or more electron beams. *See also:* **electrode (electron tube.)** 190, 125

electron-gun density multiplication (electron tubes). The ratio of the average current density at any specified aperture through which the stream passes to the average current density at the cathode surface. 190, 125

electronic. Of, or pertaining to, devices, circuits, or systems utilizing electron devices. *Examples:* Electronic control, electronic equipment, electronic instrument, and electronic circuit. *See also:* **electron device; electronics.** 190, 125

electronically de-spun antenna (communication satellite). A directional antenna, mounted to a rotating object (namely spin stabilized communication satellite), with beam switching and phasing such that the antenna beam points into the same direction in space regardless of its mechanical rotation. 83

electronic analog computer. An automatic computing device that operates in terms of continuous variation of some physical quantities, such as electric voltages and currents, mechanical shaft rotations, or displacements, and that is used primarily to solve differential equations. *Note:* The equations governing the variation of the physical quantities have the same or very nearly the same form as the mathematical equations under investigation and therefore yield a solution analogous to the desired solution of the problem. Results are measured on meters, dials, oscillograph recorders, or oscilloscopes.   9

electronic contactor (industrial control). A contactor whose function is performed by electron tubes. *See also:* **contactor.**   206

electronic controller (industrial control). An electric controller in which the major portion or all of the basic functions are performed by electron tubes.   206

electronic counter-countermeasures (ECCM) (radar). Any electronic technique designed to make a radar less vulnerable to ECM (electronic countermeasures).   13

electronic countermeasures (ECM) (radar). Any electronic technique designed to deny accurate radar information to radar operators. Screening with noise, confusion with false targets, and deception by affecting tracking circuits are typical types.   13

electronic direct-current motor controller (industrial control). A phase-controlled rectifying system using tubes of the vapor- or gas-filled variety for power conversion to supply the armature circuit or the armature and shunt-field circuits of a direct-current motor, to provide adjustable-speed, adjustable- and compensated-speed, or adjustable- and regulated-speed characteristics. *See:* **electronic controller.**   225, 206

electronic direct-current motor drive (industrial control). The combination of an electronic direct-current motor controller with its associated motor or motors. *See:* **electronic controller.**   225, 206

electronic efficiency (electron tubes). The ratio of (1) the power at the desired frequency delivered by the electron stream to the circuit in an oscillator or amplifier to (2) the average power supplied to the stream.   190, 125

electronic frequency changer. An electric power converter for changing the frequency of electric power. *See also:* **rectification.**   328

electronic instrument. An instrument that utilizes for its operation the action of one or more electron tubes.   328

electronic keying. A method of keying whereby the control is accomplished solely by electronic means. *See also:* **telegraphy.**   111

electronic line scanning (facsimile). That method of scanning that provides motion of the scanning spot along the scanning line by electronic means. *See also:* **scanning.**   12, 178

electronic microphone. A microphone that depends for its operation on a change in the terminal electrical characteristic of an active device when a force is applied to some part of the device. *See also:* **microphone.**   176

electronic navigation. *See:* **navigation.**

electronic power converter. Electronic devices for transforming electric power. *See also:* **rectification.**   328

electronic raster scanning (facsimile). That method of scanning in which motion of the scanning spot in both dimensions is accomplished by electronic means. *See also:* **scanning.**   12, 178

electronic rectifier (industrial control). A rectifier in which electron tubes are used as rectifying elements. *See also:* **electronic controller; rectification of an alternating current.**   206

electronics (1) (adjective). Of, or pertaining to, the field of electronics. *Examples:* Electronics engineer, electronics course, electronics laboratory, and electronics committee. *See:* **electron; electronic.**

(2) (noun). That field of science and engineering that deals with electron devices and their utilization. *See also:* **electron device.**   190

(3) (electric installations on shipboard). That branch of science and technology which relates to the conduction of electricity through gases or in vacuo.   3

electronic scanning (antenna). Scanning an antenna beam by electronic or electric means without moving parts. *See also:* **radiation.** *Syn.* **inertialess scanning.**   179, 111

electronic switching system (telephone switching systems). An automatic switching system in which the control functions are performed principally by electronic devices.   55

electronic transformer. Any transformer intended for use in a circuit of system utilizing electron or solid-state devices. *Note:* Mercury-arc rectifier transformers and luminous-tube transformers are normally excluded from this classification.   197, 53

electronic trigger circuit (industrial control). A network containing electron tubes in which the output changes abruptly with an infinitesimal change in input at one or more points in the operating range.   206

electronic voltmeter (vacuum-tube voltmeter). A voltmeter that utilizes the rectifying and amplifying properties of electron tubes and their associated circuits to secure desired characteristics, such as high input impedance, wide frequency range, crest indications, etcetera. *See also:* **instrument.**   328

electronic warfare support measures (ESM) (radar). Actions taken to search for, intercept, locate, record, and analyze radiated electromagnetic energy for the purpose of exploiting such radiations in support of military operations.   13

electron injector. The electron gun of a betatron.   190

electron lens. A device for the purpose of focusing an electron (ion) beam. *See also:* **electron optics.**   244, 190

electron microscope. An electron-optical device that produces a magnified image of an object. *Note:* Detail may be revealed by virtue of selective transmission, reflection, or emission of electrons by the object. *See also:* **electron devices, miscellaneous.**   190

electron mirror. An electronic device causing the total reflection of an electron beam. *See also:* **electron optics.**   244, 190

electron multiplier. A structure, within an electron tube, that employs secondary electron emission from solids to produce current amplification. *See:* **amplifier; electron emission.**   190, 117, 125

electron multiplier section (electron tube). A section of an electron tube in which an electron current is amplified by one or more successive dynode stages. *See also:* **electrode (electron tube).**   328

electron multiplier transit time. That portion of photomultiplier transit time corresponding to the time delay between an electron packet leaving the first dynode and the multiplied packet striking the anode.   117

electron optics. The branch of electronics that deals with the operation of certain electronic devices, based on the analogy between the path of electron (ion) beams in magnetic or electric fields and that of light rays in refractive media.   244, 190

electron-ray indicator tube. An elementary form of cathode-ray tube used to indicate a change of voltage. *Note:* Such a tube used to indicate the tuning of a circuit is sometimes called a **magic eye.** *See also:* **cathode-ray tubes.**   244, 190

electron resolution. The ability of the electron multiplier section of the photomultiplier to resolve inputs consisting of n and $n + 1$ electrons. This may be expressed as a fractional full width at half maximum of the nth peak, or as the peak to valley ratio of the nth peak to the valley between the nth and $n \times$ 1th peaks.   117

electrons, conduction (semiconductor). The electrons in the conduction band of a solid that are free to move under the influence of an electric field. *See also:* **semiconductor.**   210, 186

electron sheath (gas) (ion sheath). A film of electrons (or of ions) that has formed on or near a surface that is held at a potential different from that of the discharge. *See also:* **discharge (in a gas).**   244, 190

electron-stream potential (electron tubes) (any point in an electron stream). The time average of the potential difference between that point and the electron-emitting surface. *See:* **electron emission.**   190, 125

electron-stream transmission efficiency (electron tubes) (electrode through which the electron stream passes). The ratio of (1) the average stream current through the electrode to (2) the average stream current approaching the electrode. *Note:* In connection with multitransit tubes, the term electron stream should be taken to include only electrons approaching the electrode for the first time. *See:* **electron emission.**   190, 125

electron telescope. An optical instrument for astronomy including an electronic image transformer associated with an optical telescope. *See also:* **electron optics.**   244, 190

electron tube. An electron device in which conduction by electrons takes place through a vacuum or gaseous medium within a gastight envelope. *Note:* The envelope may be either pumped during operation or sealed off.   125

electron-tube admittances. The cross-referenced terms generalize the familiar electron-tube coefficients so that they apply to all types of electron devices operated at any frequency as linear transducers. *Note:* The generalizations include the familiar low-frequency tube concepts. In the case of a triode, for example, at relatively low frequencies the short-circuit input admittance reduces to substantially the grid admittance, the short-circuit output admittance reduces to substantially the plate admittance, the short-circuit forward admittance reduces to substantially the grid-plate transconductance, and the short-circuit feedback admittance reduces to substantially the admittance of the grid-plate capacitance. When reference is made to alternating-voltage or -current components, the components are understood to be small enough so that linear relations hold between the various alternating voltages and currents. Consider a generalized network or transducer having n available terminals to each of which is flowing a complex alternating component I_j of the current and

between each of which and a reference point (which may or may not be one of the n network terminals) is applied a complex alternating voltage V_j. This network represents an n-terminal electron device in which each one of the terminals is connected to an electrode.   328

electron-tube amplifier. An amplifier that obtains its amplifying properties by means of electron tubes.   206

electronvolt (eV). The kinetic energy acquired by an electron in passing through a potential difference of 1 V in vacuum; 1 eV = 1.602 19 $\times$ 10^{-19} J approximately.   21

electron-wave tube. An electron tube in which mutually interacting streams of electrons having different velocities cause a signal modulation to change progressively along their length.   190, 125

electrooptical effect (dielectrics) (Kerr electrostatic effect). Certain transparent dielectrics when placed in an electric field become doubly refracting. *Note:* The strength of the electrooptical effect for unit thickness of the dielectric varies directly as the square of the electric field strength.   210

electroosmosis. The movement of fluids through diaphragms that is as a result of the application of an electric current.   328

electroosmotic potential (electrobiology). The electrokinetic potential gradient producing unit velocity of liquid flow through a porous structure. *See also:* **electrobiology.**   192

electrophonic effect. The sensation of hearing produced when an alternating current of suitable frequency and magnitude from an external source is passed through an animal.   176

electrophoresis. A movement of colloidal ions as a result of the application of an electric potential. *See also:* **ion.**   328

electrophoretic potential (electrobiology). The electrokinetic potential gradient required to produce unit velocity of a colloidal or suspended material through a liquid electrolyte. *See also:* **electrobiology.**   192

electroplating. The electrodeposition of an adherent coating upon an object for such purposes as surface protection or decoration.   328

electropneumatic brake. An air brake that is provided with electrically controlled valves for control of the application and release of the brakes. *Note:* The electric control is usually in addition to a complete air brake equipment to provide a more prompt and synchronized operation of the brakes on two or more vehicles. *See also:* **electric braking.**   328

electropneumatic contactor (1) (industrial control). A contactor actuated by air pressure. *See also:* **contactor.**   206

(2) (electropneumatic unit switch). A contactor or switch controlled electrically and actuated by air pressure. *See also:* **contactor; control switch.**   1

electropneumatic controller. An electrically supervised controller having some or all of its basic functions performed by air pressure. *See:* **electric controller.** *See also:* **multiple-unit control.**   206

electropneumatic interlocking machine. An interlocking machine designed for electric control of electropneumatically operated functions. *See also:* **centralized traffic control system.**   328

electropneumatic valve. An electrically operated valve that controls the passage of air. 328

electropolishing (electroplating). The smoothing or brightening of a metal surface by making it anodic in an appropriate solution. *See also:* **electroplating.** 328

electrorefining. The process of electrodissolving a metal from an impure anode and depositing it in a more pure state. 328

electroretinogram. *See:* **electrodermogram.**

electroscope. An electrostatic device for indicating a potential difference or an electric charge. *See also:* **instrument.** 328

electroshock therapy. The production of a reaction in the central nervous system by means of electric current applied to the cranium. *See also:* **electrotherapy.** 192

electrostatic actuator. An apparatus constituting an auxiliary external electrode that permits the application of known electrostatic forces to the diaphragm of a microphone for the purpose of obtaining a primary calibration. *See also:* **microphone.** 176

electrostatic coupling (interference terminology). *See:* **coupling, capacitive:** *See also:* **signal.**

electrostatic deflection (cathode-ray tube). Deflecting an electron beam by the action of an electric field. *See also:* **cathode-ray tubes.** 244, 190

electrostatic electrography. The branch of electrostatography that employs an insulating medium to form, without the aid of electromagnetic radiation, latent electrostatic-charge patterns for producing a viewable record. *See also:* **electrostatography.** 236, 191

electrostatic electron microscope. An electron microscope with electrostatic lenses. *See also:* **electron optics.** 244, 190

electrostatic electrophotography. The branch of electrostatography that employs a photoresponsive medium to form, with the aid of electromagnetic radiation, latent electrostatic-charge patterns for producing a viewable record. *See also:* **electrostatography.** 236, 191

electrostatic focusing (electron beam). *See:* **focusing, electrostatic.**

electrostatic instrument. An instrument that depends for its operation on the forces of attraction and repulsion between bodies charged with electricity. *See also:* **instrument.** 328

electrostatic lens. An electron lens in which the result is obtained by an electrostatic field. *See also:* **electron optics.**

electrostatic loudspeaker (capacitor loudspeaker) (condenser loudspeaker*). A loudspeaker in which the mechanical forces are produced by the action of electrostatic fields. *See also:* **loudspeaker.** 176

*Deprecated

electrostatic microphone (capacitor microphone) (condenser microphone*). A microphone that depends for its operation upon variations of its electrostatic capacitance. *See also:* **microphone.** 176

*Deprecated

electrostatic potential (any point). The potential difference between that point and an agreed-upon reference point, usually the point at infinity. 210

electrostatic potential difference (between two points). The scalar-product line integral of the electric field strength along any path from one point to the other in an electric field resulting from a static distribution of electric charge. 210

electrostatic recording (facsimile). Recording by means of a signal-controlled electrostatic field. *See also:* **recording (facsimile).** 12

electrostatic relay. A relay in which operation depends upon the application or removal of electrostatic charge. 341

electrostatics. The branch of science that treats of the electric phenomena associated with electric charges at rest in the frame of reference. 210

electrostatic storage. A storage device that stores data as electrostatically charged areas on a dielectric surface. 255, 77

electrostatic voltmeter. A voltmeter depending for its action upon electric forces. An electrostatic voltmeter is provided with a scale, usually graduated in volts or kilovolts. *See also:* **instrument.** 328

electrostatography. The formation and utilization of latent electrostatic-charge patterns for the purpose of recording and reproducing patterns in viewable form. 236, 191

electrostenolysis. The discharge of ions or colloidal ions in capillaries through the application of an electric potential. *See also:* **ion.** 328

electrostriction (electric field). The variation of the dimensions of a dielectric under the influence of an electric field. 210

electrotaxis (electrobiology) (galvanotaxis). The act of a living organism in arranging itself in a medium in such a way that its axis bears a certain relation to the direction of the electric current in the medium. *See also:* **electrobiology.** 192

electrotherapy. The use of electric energy in the treatment of disease. 192

electrothermal efficiency. The ratio of energy usefully employed in a furnace to the total energy supplied. *See also:* **electrothermics.** 328

electrothermal recording (facsimile). That type of recording that is produced principally by signal-controlled thermal action. *See also:* **recording (facsimile).** 12

electrothermic-coupler unit (electrothermic power meter). A three-port directional coupler with an electrothermic unit attached to either the side arm or the main arm which is normally used as a feed-through power measuring system. Typically, an electrothermic unit is attached to the side arm of the coupler so that the power at the output port of the main arm can be determined from a measurement of the power in the side arm. This system also can be used as a terminating powermeter by terminating the output port of the directional coupler. 47

electrothermic element (electrothermic power meter). A power absorber and a thermocouple (or thermopile) which are either two separate units or where the thermocouple (or thermopile) is also the power absorber. 47

electrothermic instrument. An instrument that depends for its operation on the heating effect of a current or currents. *Note:* Among the several possible types are (1) the expansion type, including the hot-wire and hot-strip instruments; (2) the thermocouple type; and (3) the bolometric type. *See also:* **instrument.** 328

electrothermic mount (electrothermic power meter). A waveguide or transmission line structure which is designed to accept the electrothermic element. It normally contains internal matching devices or other reactive elements to obtain specified impedance conditions at its input terminal when an electrothermic element is installed. It usually contains a means of protecting the electrothermic element and the immediate environment from thermal gradients which would cause an undesirable thermoelectric output. 47

electrothermic power indicator (electrothermic power meter). An instrument which may or may not amplify the low level dc output voltage from the electrothermic unit and provides a display, usually in the form of the D'Aronval type indication or a digital readout. 47

electrothermic power indicator error (electrothermic power meter). Ability of the metering circuitry to indicate exactly the substituted power within an electrothermic unit. Included are such factors as meter calibration, open loop gain, meter linearity, tracking errors, range switching errors, line voltage errors, and temperature compensation errors. 47

electrothermic power meter. This consists of an electrothermic unit and an electrothermic power indicator. 47

electrothermics. The branch of science and technology that deals with the direct transformations of electric energy and heat. 328

electrothermic substitution power (electrothermic power meter). The power at a reference frequency which, when dissipated in the electrothermic element, produces the same dc electrothermic output voltage that the element produces when subjected to radio frequency power. 47

electrothermic technique of power measurement (electrothermic power meter). A technique wherein the heating effect of power dissipated in an electrothermic element (which consists of an energy absorber and a thermocouple or thermopile) is used to generate a dc voltage. The power is dissipated either in a separate absorber or in the resistance of the electrothermic element. The resultant heat causes a temperature rise in a portion of the element. This temperature rise is sensed by the thermocouple which generates a dc output voltage proportional to the power. 47

electrothermic unit (electrothermic power meter). An assembly consisting of the electrothermic element installed in the electrothermic mount. 47

electrotonic wave (electrobiology). A brief nonpropagated change of potential on an excitable membrane in the vicinity of an applied stimulus; it is often accompanied by a propagated response and always by electrotonus. *See also:* **excitability (electrobiology).** 192

electrotonus (1) (physical). The change in distribution of membrane potentials in nerve and muscle during or after the passage of an electric current. *See also:* **excitability (electrobiology).** 192

(2) (physiological). The change in the excitability of a nerve or muscle during the passage of an electric current. *See also:* **excitability (electrobiology).** 192

electrotyping. The production or reproduction of printing

plates by electroforming. *See also:* **electroforming.** 328

electrowinning. The electrodeposition of metals or compounds from solutions derived from ores or other materials using insoluble anodes. 328

element (1) (electrical engineering). Any electric device (such as inductor, resistor, capacitor, generator, line, active device tube) with terminals at which it may be directly connected to other electric devices. *See also:* **network analysis.** 210, 67

(2) (electron tubes). A constituent part of the tube that contributes directly to its electrical operation. 125

(3) (semiconductor device). Any integral part that contributes to its operation. *See also:* **semiconductor device.** 245, 210, 66

(4) (integrated circuit). A constituent part of the integrated circuit that contributes directly to its operation. *See also:* **integrated circuit.** 312, 191

(5) (computing systems). *See:* **combinational logic element; logic element; sequential logic element; threshold element.**

(6) (storage cell). Consists of the positive and negative groups with separators, or separators and retainers, assembled for one cell. *See also:* **battery (primary or secondary).** 328

(7) (data transmission). Synonymous with bit as the minimum subdivision within a code grouping representing a character. 59

(8) (primary detecting). That portion of the feedback elements which first either utilizes or transforms energy from the controlled medium to produce a signal that is a function of the value of the directly controlled variable. 105

(9) (telephone equipment). Any electric device (such as inductor, resistor, capacitor, generator, line) with terminals at which it may be directly connected to other electric devices, elements, or apparatus. *See:* **device; network analysis.** 39

(10) (antenna). *See:* **array element; director element; driven element; parasitic element; radiating element; reflector element.** 111

elemental area (facsimile). Any segment of the scanning line of the subject copy the dimension of which along the line is exactly equal to the nominal line width. *Note:* Elemental area is not necessarily the same as the scanning spot. *See also:* **scanning.** 12

elementary diagram. *See:* **schematic diagram.** 210

element cell. In an array having a regular arrangement of elements, any area surrounding an array element which can be identically repeated for all elements without gaps or overlap between cells. *Note:* There are infinitely many possible choices for such a cell. Some may be more convenient than others for analysis purposes. 111

element linear. *See:* **linear system or element.**

element, measuring (automatic control). That portion of the feedback elements which converts the signal from the primary detecting element to a form compatible with the reference input. 56

element, primary detecting (automatic control). That portion of the feedback elements which first either utilizes or transforms energy from the controlled medium to produce a signal which is a function of the value of the directly controlled variable.   56

elements, feedback (control system). The elements in the controlling system that change the feedback signal in response to the directly controlled variable. See Figure B. *See:* **control system, feedback.**   56, 219, 206, 105

elements, forward (automatic control). Those elements situated between the actuating signal and the controlled variable in the closed loop being considered. See Figure B. *See also:* **control system, feedback.**   56, 105

elements, loop (control system, feedback) (a closed loop). All elements in the signal path that begins with the loop error signal and ends with the loop return signal. See Figures A and B. *See also:* **control system, feedback.**   56

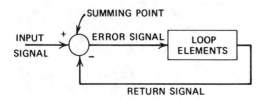

Figure A. Block diagram of a closed loop.

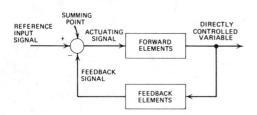

Figure B. Simplified block diagram indicating essential elements of an automatic control system.

elements of a fix (electronic navigation). The specific values of the navigation coordinates necessary to define a position. *See also:* **navigation.**   13, 187

elements, reference-input (automatic control). The portion of the controlling system that changes the reference input signal in response to the command. See Figure C. *See:* **control system, feedback.**   219, 206, 56

elevated-zero range. A range where the zero value of the measured variable, measured signal, etcetera, is greater than the lower range value. *Note:* The zero may be between the lower and upper range values, at the upper range value, or above the upper range value. For example: (a) −20 to 100, (b) −40 to 0, and (c) −50 to −10. *See also:* **instrument.**   295

elevation (illuminating engineering). The angle between the axis of a searchlight drum and the horizontal. *Note:* For angles above the horizontal, elevation is positive, and below the horizontal negative. *See also:* **searchlight.**   167

elevation rod (lightning protection). The vertical portion of conductor in an air terminal by means of which it is elevated above the object to be protected. *See also:* **lightning protection and equipment.**   328

elevator. A hoisting and lowering mechanism equipped with a car or platform that moves in guides in a substantially vertical direction, and that serves two or more floors of a building or structure.   328

elevator automatic dispatching device. A device, the principal function of which is to either: (1) operate a signal in the car to indicate when the car should leave a designated landing; or (2) actuate its starting mechanism when the car is at a designated landing. *See also:* **control (elevators).**   328

elevator automatic signal transfer device. A device by means of which a signal registered in a car is automatically transferred to the next car following, in case the first car passes a floor, for which a signal has been registered, without making a stop. *See also:* **control.**   328

elevator car. The load-carrying unit including its platform, car frame, enclosure, and car door or gate. *See also:* **ele-**

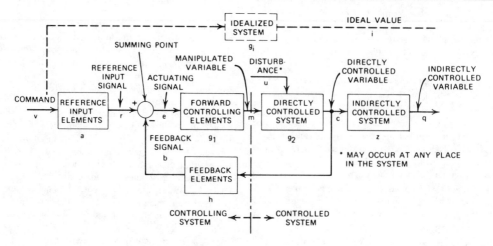

Figure C. Block diagram of an automatic control system illustrating expansion of the simplified block diagram to a more complex system.

vator.   328

elevator car bottom runby (elevator car). The distance between the car buffer striker plate and the striking surface of the car buffer when the car floor is level with the bottom terminal landing. *See also:* **elevator.**   328

elevator-car flash signal device. One providing a signal light, in the car, that is illuminated when the car approaches the landings at which a landing-signal-registering device has been actuated. *See:* **control.**   328

elevator car-leveling device. Any mechanism that will, either automatically or under the control of the operator, move the car within the leveling zone toward the landing only, and automatically stop it at the landing. *Notes:* (1) Where controlled by the operator by means of up-and-down continuous-pressure switches in the car, this device is known as an **inching device.** (2) Where usd with a hydraulic elevator to correct automatically a change in car level caused by leakage in the hydraulic system, this device is known as an **anticreep device.** *See:* **elevator; leveling zone (elevators); one-way automatic leveling device; (elevators); two-way automatic maintaining leveling device (elevators); two-way automatic nonmaintaining leveling device (elevators).**   328

elevator counterweight bottom runby. The distance between the counterweight buffer striker plate and the striking surface of the counterweight buffer when the car floor is level with the top terminal landing. *See also:* **elevator.**   328

elevator landing. That portion of a floor, balcony, or platform used to receive and discharge passengers or freight. *See also:* **bottom terminal landing (elevators); elevators; landing zone; top terminal landing (elevators).**   328

elevator-landing signal registering device. A button or other device, located at the elevator landing that, when actuated by a waiting passenger, causes a stop signal to be registered in the car. *See:* **control.**   328

elevator-landing stopping device. A button or other device, located at an elevator landing that, when actuated, causes the elevator to stop at that floor. *See:* **control.**   328

elevator parking device. An electric or mechanical device, the function of which is to permit the opening, from the landing side, of the hoistway door at any landing when the car is within the landing zone of that landing. The device may also be used to close the door. *See:* **control.**   328

elevator pit. That portion of a hoistway extending from the threshold level of the lowest landing door to the floor at the bottom of the hoistway. *See also:* **elevators.**   328

elevator separate-signal system. A system consisting of buttons or other devices located at the landings that, when actuated by a waiting passenger, illuminate a flash signal or operate an annunciator in the car, indicating floors at which stops are to be made. *See also:* **control (elevators).**   328

elevator signal-transfer switch. A manually operated switch, located in the car, by means of which the operator can transfer a signal to the next car approaching in the same direction, when he desires to pass a floor at which a signal has been registered in the car. *See:* **control.**   328

elevator starter's control panel. An assembly of devices by means of which the starter may control the manner in which an elevator, or group of elevators, functions. *See:*

control.   328

elevator truck zone. The limited distance above an elevator landing within which the truck-zoning device permits movement of the elevator car. *See:* **control.**   328

elevator truck-zoning device. A device that will permit the operator in the car to move a freight elevator, within the truck zone, with the car door or gate and a hoistway door open. *See:* **control.**   328

elliptically polarized field vector. A field vector whose extremity describes an ellipse as a function of time. *Note:* Any single-frequency field vector is elliptically polarized if "elliptical" is understood in the wide sense as including circular and linear. Often, however, the expression is used in the strict sense meaning noncircular and nonlinear.   111

elliptically polarized plane wave (antenna). A plane wave in which the electric field is elliptically polarized.   111

elliptically polarized wave (1) (general). A transverse wave in which the displacement vector at any point describes an ellipse.   210

(2) (radio wave propagation) (given frequency). An electromagnetic wave for which the component of the electric vector in a plane normal to the direction of propagation describes an ellipse. *See also:* **electromagnetic wave; radiation; radio transmitter; waveguide.**   267, 179, 146

elliptical orbit (communication satellite). An orbit of a satellite in which the distance between the centers of mass of the satellite and of the primary body is not constant. The general type of orbit. A circular orbit is a special case.   74

elliptic filter (circuits and systems). A filter having an equiripple pass band and an equiminima stop band.   67

ellipticity. *See note under* **axial ratio.** *See also:* **waveguide.**

embedded coil side (rotating machinery). That part of a coil side which lies in a slot between the ends of the core. 63

embedded temperature detector (rotating machinery). An element, usually a resistance thermometer or thermocouple, built into apparatus for the purpose of measuring temperature. *Note:* (1) This is ordinarily installed in a stator slot between coil sides at a location at which the highest temperature is anticipated.   63, 3

(2) Examination or replacement of an embedded detector after the apparatus is placed in service is usually not feasible.   7

embossing stylus. A recording stylus with a rounded tip that displaces the material in the recording medium to form a groove. *See also:* **phonograph pickup.**   176

embrittlement. Severe loss of ductility of a metal or alloy.   221, 205

emergency announcing system. A system of microphones, amplifier, and loud speakers (similar to a public address system) to permit instructions and orders from a ship's officers to passengers and crew in an emergency and particularly during abandon-ship operations.   328

emergency cells (storage cell). End cells that are held available for use exclusively during emergency discharges. *See also:* **battery (primary or secondary).**   328

emergency electric system (marine). All electric apparatus and circuits the operation of which, independent of ship's service supply, may be required under casualty conditions for preservation of a ship or personel.   328

emergency generator (marine). An internal-combustion-engine-driven generator so located in the upper part of a vessel as to permit operation as long as the ship can remain afloat, and capable of operation, independent of any other apparatus on the ship, for supply of power to the emergency electric system upon failure of a ship's service power. *See also:* **emergency electric system.**   328

emergency lighting. Lighting designed to supply illumination essential to the safety of life and property in the event of failure of the normal supply. *See also:* **general lighting.**   167

emergency lighting storage battery (marine transportation). A storage battery for instant supply of emergency power, upon failure of a ship's service supply, to certain circuits of special urgency principally temporary emergency lighting. *See also:* **emergency electric system.**   328

emergency operations area(s) (nuclear power generating station). Functional area(s) allocated for the displays used to assess the status of safety systems and the controls for manual operations required during emergency situations.   358

emergency power feedback. An arrangement permitting feedback of emergency-generator power to a ship's service system for supply of any apparatus on the ship within the limit of the emergency-generator rating. *See also:* **emergency electric system.**   328

emergency power system. An independent reserve source of electric energy which, upon failure or outage of the normal source, automatically provides reliable electric power within a specified time to critical devices and equipment whose failure to operate satisfactorily would jeopardize the health and safety of personnel or result in damage to property.   89

emergency service. An additional service intended only for use under emergency conditions. *See:* **dual service; duplicate service; loop service.** *See also:* **service.**   64

emergency stop switch (elevators). A device located in the car that, when manually operated, causes the electric power to be removed from the driving-machine motor and brake of an electric elevator or from the electrically operated valves and/or pump motor of a hydraulic elevator. *See:* **control.**   328

emergency switchboard. A switchboard for control of sources of emergency power and for distribution to all emergency circuits. *See also:* **emergency electric system.**   328

emergency-terminal stopping device (elevators). A device that automatically causes the power to be removed from an electric elevator driving-machine motor and brake, or from a hydraulic elevator machine, at a predetermined distance from the terminal landing, and independently of the functioning of the operating device and the normal-terminal stopping device, if the normal-terminal stopping device does not slow down the car as intended. *See:* **control.**   328

emergency transfer capability (electric power supply). The maximum amount of power that can be transmitted following a loss of transmission or generation capacity without causing additional transmission outages. *Syn.* **maximum transmission transfer capability.**   112

EMF. *See:* **voltage (electromotive force).**

emission (laser-maser). The transfer of energy from matter to a radiation field.   363

emission characteristic. A relation, usually shown by a graph, between the emission and a factor controlling the emission (such as temperature, voltage, or current of the filament or heater). *See also:* **electron emission.**   190, 125

emission current. The current resulting from electron emission.   244, 190

emission current, field-free (cathode). *See:* **field-free emission current (cathode).**

emission efficiency (thermionics). The quotient of the saturation current by the heating power absorbed by the cathode. *See also:* **electron emission.**   244, 190

emissivity (photovoltaic power system). The emittance of a specimen of material with an optically smooth, clean surface and sufficient thickness to be opaque. *See also:* **photovoltaic power system; solar cells (photovoltaic power system).**   186

emittance (photovoltaic power system). The ratio of the radiant flux-intensity from a given body to that of a black body at the same temperature. *See also:* **photovoltaic power system; solar cells (photovoltaic power system).**   186

emitter (transistor). A region from which charge carriers that are minority carriers in the base are injected into the base.   210, 245, 66

emitter, majority (transistor). An electrode from which a flow of majority carriers enters the interelectrode region. *See also:* **transistor.**   66

emitter, minority (transistor). An electrode from which a flow of minority carriers enters the interelectrode region. *See also:* **semiconductor; transistor.**   328

emitting sole (microwave tubes). An electron source in crossed-field amplifiers that is extensive and parallel to the slow-wave circuit and that may be a hot or cold electron-emitter. *See also:* **microwave tube or valve.**   190

empirical propagation model (electromagnetic compatibility). A propagation model that is based solely on measured path-loss data. *See also:* **electromagnetic compatibility.**   199

enabling pulse (1) (navigation). A pulse which prepares a circuit for some subsequent action.   13, 187
(2). A pulse that opens an electric gate normally closed, or otherwise permits an operation for which it is a necessary but not a sufficient condition. *See also:* **pulse.**   328

enabling signal. A logic (binary) signal, one of whose states permits a number of other logic (binary) events to occur. *See also:* **digital.**   59

enamel (1) (general). A paint that is characterized by an ability to form an especially smooth film.   215, 63
(2) (wire) (rotating machinery). A smooth film applied to wire usually by a coating process. *See also:* **rotor (rotating machinery); stator.**   63

encapsulated (rotating machinery). A machine in which one or more of the windings is completely encased by molded insulation. *See also:* **asynchronous machine; direct-current commutating machine.**   63

encapsulation (semiconductor radiation detector). The packaging of a detector for protective and/or mounting purposes.   118, 119

enclosed (1) (general). Surrounded by a case that will prevent a person from accidentally contacting live parts. *Syn.*

inclosed.	328
(2) (transmission and distribution). Surrounded by a case, cage or fence, designed to protect the contained equipment and prevent accidental contact with energized parts.	262

enclosed brake (industrial control). A brake that is provided with an enclosure that covers the entire brake, including the brake actuator, the brake shoes, and the brake wheel. *See:* **electric drive.**	225, 206
enclosed cutout. A cutout in which the fuse clips and fuseholder or disconnecting blade are mounted completely within an insulating enclosure.	103, 202
enclosed relay. A relay that has both coil and contacts protected from the surrounding medium. *See also:* **relay.**	259

enclosed self-ventilated machine. A machine having openings for the admission and discharge of the ventilating air that is circulated by means integral with the machine, the machine being otherwise totally enclosed. *Note:* These openings are so arranged that inlet and outlet ducts or pipes may be connected to them. *Note:* Such ducts or pipes, if used, must have ample section and be so arranged as to furnish the specified volume of air to the machine, otherwise the ventilation will not be sufficient.	3
enclosed separately ventilated machine. A machine having openings for the admission and discharge of the ventilating air that is circulated by means external to and not a part of the machine, the machine being otherwise totally enclosed. *Note:* These openings are so arranged that inlet and outlet duct pipes may be connected to them.	3
enclosed switch (industrial control) (safety switch). A switch either with or without fuse holders, meter-testing equipment, or accommodation for meters, having all current-carrying parts completely enclosed in metal, and operable without opening the enclosure. *See also:* **switch.**	206
enclosed switchboard. A dead-front switchboard that has an overall sheet metal enclosure (not grille) covering back and ends of the entire assembly. *Note:* Access to the interior of the enclosure is usually provided by doors or removable covers. The top may or may not be covered.	103, 202, 79
enclosed switches, indoor or outdoor (power switchgear). Switches designed for service within a housing restricting heat transfer to the external medium.	27, 103
enclosed switchgear assembly. A switchgear assembly that is enclosed on all sides and top.	103, 202
enclosed ventilated (rotating machinery). A term applied to an apparatus with a substantially complete enclosure in which openings are provided for ventilation only. *See also:* **asynchronous machine; direct-current commutating machine.**	63
enclosed ventilated apparatus. Apparatus totally enclosed except that openings are provided for the admission and discharge of the cooling air. *Note:* These openings may be so arranged that inlet and outlet ducts or pipes may be connected to them. An enclosed ventilated apparatus or machine may be separately ventilated or self-ventilated.	328

enclosure (1) (general). A surrounding case or housing used to protect the contained conductor or equipment and prevent personnel from accidentally contacting live parts. *Note:* Material and finish shall conform to the standards

for the switchgear enclosed.	103, 202, 78
(2 (transformer). A surrounding case or housing used to protect the contained equipment and prevent personnel from accidentally contacting live parts.	53
(3) (class 1E equipment and circuits). An identifiable housing such as a cubicle, compartment, terminal box, panel, or enclosed raceway used for electrical equipment or cables.	31, 131
encode (1) (general). To express a single character or a message in terms of a code.	235
(2) (electronic control). To produce a unique combination of a group of output signals in response to each of a group of input signals.	235
(3) (computing systems). To apply the rules of a code. *See:* **code.** *See also:* **decode; matrix; translate.**	255, 77
encoder (electronic computation). A network or system in which only one input is excited at a time and each input produces a combination of outputs. *Note:* Sometimes called matrix.	210
end-around carry (computing systems). A carry generated in the most significant place and forwarded directly to the least significant place, for example, when adding two negative numbers, using nines complement. *See:* **carry.**	235, 255, 77

end bell. *See:* **cable terminal.**
end bracket (rotating machinery). A beam or bracket attached to the frame of a machine and intended for supporting a bearing.	63
end cells (storage battery) (storage cell). Cells that may be cut in or cut out of the circuit for the purpose of adjusting the battery voltage. *See also:* **battery (primary or secondary).**	328
end device (telemeter). The final system element that responds quantitatively to the measurand through the translating means and performs the final measurement operation. *Note:* An end device performs the final conversion of measurement energy to an indication, record, or the initiation of control. *See:* **instrument; measurement system.**	103, 202, 295
end distortion (start-stop teletypewriter signals). The shifting of the end of all marking pulses from their proper positions in relation to the beginning of the start pulse. *See also:* **telegraphy.**	328
end finger (outside space-block) (rotating machinery). A radially extending finger piece at the end of a laminated core to transfer pressure from an end clamping plate or flange to a tooth. *See also:* **rotor (rotating machinery); stator.**	63
end-fire array antenna. A linear array antenna whose direction of maximum radiation lies along the line of the array.	111
end injection (Charles or Kino gun (EI)) (microwave tubes). A gun used in the presence of crossed electric and magnetic fields to inject an electron beam into the end of a slow-wave structure. *See also:* **microwave tube.**	190
end-of-block signal (numerically controlled machines). A symbol or indicator that defines the end of one block of data.	224, 207
end-of-copy signal (facsimile). A signal indicating termination of the transmission of a complete subject copy. *See also:* **facsimile signal (picture signal).**	12
end office (telephone switching systems). A local office that

is part of the toll hierarchy of World Zone 1. An end office is classified as a Class 5 office. *See also:* **office class.**
    55

end of program (numerically controlled machines). A miscellaneous function indicating completion of workpiece. *Note:* Stops spindle, coolant, and feed after completion of all commands in the block. Used to reset control and/or machine. Resetting control may include rewinding of tape or progressing a loop tape through the splicing leader. The choice for a particular case must be defined in the format classification sheet.
    224, 207

end of tape (numerically controlled machines). A miscellaneous function that stops spindle, coolant, and feed after completion of all commands in the block. *Note:* Used to reset control and/or machine. Resetting control will include rewinding of tape, progressing a loop tape through the splicing leader, or transferring to a second tape reader. The choice for a particular case must be defined in the format classification sheet.
    224, 207, 54

end-on armature relay. *See:* **armature, end-on.** *See also:* **relay.**

end plate, rotor (rotating machinery). An annular disk (ring) fitted at the outer end of the retaining ring.
    63

end-play washers (rotating machinery). Washers of various thicknesses and materials used to control axial position of the shaft.
    63

end rail (rotating machinery). A rail on which a bearing pedestal can be mounted. *See also:* **bearing.**
    63

end ring, rotor (rotating machinery). The conducting structure of a squirrel-cage or amortisseur (damper) winding that short-circuits all of the rotor bars at one end. *See also:* **rotor (rotating machinery).**
    63

end-scale value (electric instrument). The value of the actuating electrical quantity that corresponds to end-scale indication. *Notes:* (1) When zero is not at the end or at the electrical center of the scale, the higher value is taken. (2) Certain instruments such as power-factor meters, ohmmeters, etcetera, are necessarily excepted from this definition. (3) In the specification of the range of multiple-range instruments, it is preferable to list the ranges in descending order, as 750/300/150. *See also:* **accuracy rating (instrument); instrument.**
    280

end shield (1) (rotating machinery). A solid or skeletal structure, mounted at one end of a machine, for the purpose of providing a specified degree of protection for the winding and rotating parts or to direct the flow of ventilating air. *Note:* Ordinarily a machine has an end shield at each end. For certain types of machine, one of the end shields may be constructed as an integral part of the stator frame. The end shields may be used to align and support the bearings, oil deflectors, and, for a hydrogen-cooled machine, the hydrogen seals.
    63

(2) (magnetrons). A shield for the purpose of confining the space charge to the interaction space. *See also:* **magnetrons.**
    190, 125

end-shift frame (rotating machinery). A stator frame so constructed that it can be moved along the axis of the machine shaft for purposes of inspection. *See also:* **stator.**
    63

endurance limit. The maximum stress a metal can withstand without failure during a specified large number (usually 10 million) cycles of stress.
    221, 205

endurance test (reliability). An experiment carried out to investigate how the properties of an item are affected by the application of stresses and the elapse of time.
    164

end winding (rotating machinery). That portion of a winding extending beyond the slots. *Note:* It is outside the major flux path and its purpose is to provide connections between parts of the winding within the slots of the magnetic circuit. *See:* **asynchronous machine; direct-current commutating machine; rotor (rotating machinery); stator.**
    63

end-winding cover (winding shield) (rotating machinery). A cover to protect an end winding against mechanical damage and/or to prevent inadvertent contact with the end winding.
    63

end-winding support (rotating machinery). The structure by which coil ends are braced against gravity and electromagnetic forces during start-up (for motors), running, and abnormal conditions such as sudden short-circuit, for example, by blocking and lashings between coils and to brackets or rings. *See also:* **stator.**
    63

end-window counter tube (radiation). A counter tube designed for the radiation to enter at one end. *See:* **anticoincidence (radiation counters).**
    190

end-wire insulation (rotating machinery). Insulation members placed between the end wires of individual coils such as between main and auxiliary windings. *See also:* **rotor (rotating machinery); stator.**
    63

end wire, winding (rotating machinery). The portion of a random-wound winding that is not inside the core. *See also:* **rotor (rotating machinery); stator.**
    63

energized (1) (conductor stringing equipment). Electrically connected to a source of potential difference, or electrically charged so as to have a potential different from that of the ground. *See also:* **alive.** *Syn.* **alive; current carrying; hot; live.**
    45

(2) (transmission and distribution) (alive or live). Electrically connected to a source of potential difference, or electrically charged so as to have a potential significantly different from that of earth in the vicinity. The term **live** is sometimes used in place of the term **current-carrying,** where the intent is clear, to avoid repetitions of the longer term.
    262

energy (system). The available energy is the amount of work that the system is capable of doing. *See:* **electric energy.**
    210

energy and torque (International System of Units) (SI). The vector product of force and moment arm is widely designated by the unit newton meter. This unit for bending moment of torque results in confusion with the unit for energy, which is also newton meter. If torque is expressed as newton meter per radian, the relationship to energy is clarified, since the product of torque and angular rotation is energy:

$$(N \cdot m/rad) \cdot rad = N \cdot m.$$

See also: **units and letter symbols.**
    21

energy, byproduct (transmission and distribution). Electric energy produced as a byproduct incidental to some other operation.
    112

energy costs, incremental (electric power supply). The additional cost of producing or transmitting electric energy above some base cost.
    112

energy density (point in a field) (audio and electroacoustics). The energy contained in a given infinitesimal part of the medium divided by the volume of that part of the medium. *Notes:* (1) The term energy density may be used with prefatory modifiers such as instantaneous, maximum, and peak. (2) In speaking of average energy density in general, it is necessary to distinguish between the space average (at a given instant) and the time average (at a given point).
176

energy density spectrum (burst measurements) (finite energy signal). The square of the magnitude of the Fourier transform of a burst. *See also:* **burst; network analysis.**
292, 61

energy distribution (solar cells). The distribution of the flux or fluence of particles with respect to particle energy.
113

energy, dump (electric power supply). Energy generated from water, gas, wind, or other source which cannot be stored and which is beyond the immediate needs of the electric system producing the energy.
112

energy, economy (electric power supply). Energy produced in one system and substituted for less economical energy in another system. *See also:* **generating station.**
112

energy efficiency (specified electrochemical process). The product of the current efficiency and the voltage efficiency. *See also:* **electrochemistry.**
328

energy flux (audio and electroacoustics). The average rate of flow of energy per unit time through any specified area. *Note:* For a sound wave in a medium of density ρ and for a plane or spherical free wave having a velocity of propagation c, the sound-energy flux through the area S corresponding to an effective sound pressure p is

$$\mathbf{J} = \frac{p^2 S}{\rho c} \cos \theta$$

where θ is the angle between the direction of propagation of the sound and the normal to the area S.
176

energy, fuel replacement (transmission and distribution). Energy generated at a hydroelectric plant as a substitute for energy which would otherwise have been generated by a thermal-electric plant.
112

energy gap (semiconductor). The energy range between the bottom of the conduction band and the top of the valence band. *See also:* **semiconductor device.**
245, 66, 210, 186

energy, interchange (transmission and distribution). Energy delivered to or received by one electric system from another.
112

energy-limiting transformer. A transformer that is intended for use on an approximately constant-voltage supply circuit and that has sufficient inherent impedance to limit the output current to a thermally safe maximum value.
53

energy loss (transmission and distribution). The difference between energy input and output as a result of transfer of energy between two points.
112

energy metering point (tie line) (electric power systems). The location of the integrating metering equipment used to measure energy transfer on the tie line. *See also:* **center of distribution.**
94

energy, net system (transmission and distribution). Energy requirements of a system, including losses, defined as (1)

net generation of the system, plus (2) energy received from others, less (3) energy delivered to other systems.
112

energy, nuclear (transmission and distribution). The energy with which nucleons are bound together to form nuclei. When a nucleus is changed or rearranged in a nuclear reaction, (fission, fusion, etc.) or by radioactive decay nuclear energy may be released or absorbed in the form of kinetic energy of the reactants or products.
112

energy, off-peak (transmission and distribution). Energy supplied during designated periods of relatively low system demands.
112

energy, on-peak (transmission and distribution). Energy supplied during designated periods of relatively high system demands.
112

energy, partial discharge (dielectric tests). The energy dissipated by an individual discharge. The partial discharge energy is expressed in joules.
139

energy, potential hydro (electric power supply). The possible aggregate energy obtainable over a specified period by practical use of the available stream flow and river gradient.
112

energy, Q (laser-maser). The capacity for doing work. Energy content is commonly used to characterize the output from pulsed lasers, and is generally expressed in joules.
363

energy ratio (radar). The ratio of signal energy to noise power density in the receiver, which also gives the maximum output signal-to-noise power ratio for a matched-filter system.
13

energy resolution (semiconductor radiation detector). (1) (FWHM) (full width at half maximum). The detector's contribution (including detector leakage current noise), expressed in units of energy, to the FWHM of a pulse-height distribution corresponding to an energy spectrum. (2) (percent). One hundred times the energy resolution divided by the energy for which the resolution is specified.
23, 118, 119

energy straggling. *See:* **straggling, energy.**

engine-driven generator for aircraft. A generator mechanically, hydraulically, or pneumatically coupled to an aircraft propulsion engine to provide power for the electric and electronic systems of an aircraft. It may be classified as follows: (1) Engine-mounted; (2) Remote-driven: (A) Flexible-shaft-driven; (B) Variable-ratio-driven; (C) Air-turbine-driven. *See also:* **air transportation electric equipment.**
328

engineered safety features (nuclear power generating stations). Features of a unit other than reactor trip or those used only for normal operation, that are provided to prevent, limit, or mitigate the release of radioactive material.
31, 102

engine-generator system (electric power supply). A system in which electric power for the requirements of a railway vehicle (other than propulsion) is supplied by an engine-driven generator carried on the vehicle, either as an independent source of electric power or supplemented by a storage battery. *See also:* **axle generator system.**
328

engine-room control. Apparatus and arrangement providing for control in the engine room, on order from the bridge, of the speed and direction of a vessel.
328

engine synchronism indicator. A device that provides a remote indication of the relative speeds between two or more engines. 328

engine-temperature thermocouple-type indicator. A device that indicates temperature of an aircraft engine cylinder by measuring the electromotive force of a thermocouple. 328

engine-torque indicator. A device that indicates engine torque in pound-feet. *Note:* It is usually converted to horsepower with reference to engine revolutions per minute. 328

English language programming (test, measurement and diagnostic equipment). A technique of programming which allows the programmer to write programs and routines in English language statements. 54

enterprise service (telephone switching systems). A service in which calls from certain designated exchanges are completed and billed to a number in another exchange. 55

entrance terminal (distribution oil cutouts). A terminal with an electrical connection to the fuse contact and suitable insulation where the connection passes through the housing. 79

entropy (information theory). *See:* **average information content.** *See also:* **information theory.**

entry point (routine) (1) (computing systems). Any place to which control can be passed. 255, 77

(2) (test, measurement and diagnostic equipment). One of a set of points in an automatic test equipment program where the test conditions are completely stated and are not dependent on previous tests or setups in any way. Such points are the only ones at which it is permissible to begin part of the complete test program. *See:* **re-run point.** 54

envelope (1) (general) (wave). The boundary of the family of curves obtained by varying a parameter of the wave. For the special case

$$y = E(t) \sin (\omega t + \theta),$$

variation of the parameter θ yields $E(t)$ as the envelope. 210

(2) (automatic control) (wave). Another wave composed of the instantaneous peak values of the original wave of an alternating quantity, and which indicates the variation in amplitude undergone by that quantity. 56

envelope delay (1) (television radio wave propagation) (fac-simile). The time of propagation, between two points, of the envelope of a wave. *Note:* It is equal to the rate of change with angular frequency of the difference in phase between these two points. It has significance over the band of frequencies occupied by the wave only if this rate is approximately constant over that band. If the system distorts the envelope, the envelope delay at a specified frequency is defined with reference to a modulated wave that occupies a frequency bandwidth approaching zero. *See:* **television.** *See also:* **facsimile transmission; radio wave propagation.** 12, 146, 178

(2) (circuits and systems). The time that the envelope of a modulated signal takes to pass from one point in a network (or transmission system) to a second point in the network. *Note:* Envelope delay is often defined the same as group delay, that is, as the rate of change, with angular frequency, of the phase shift between two points in a network. *See also:* **group delay time; time delay.** 67

(3) (facsimile). The time of propagation, between two points of the envelope of a wave. *Note:* The envelope delay is measured by the slope of the phase shift in cycles plotted against the frequency in cycles per second. If the system distorts the envelope the envelope delay at a specified frequency is defined with reference to a modulated wave which occupies a frequency bandwidth approaching zero. 12

(4) (non-real time spectrum analyzer). The display produced on a spectrum analyzer when the resolution bandwidth is greater than the spacing of the individual frequency components. 68

envelope delay distortion. *See:* **distortion, envelope delay.**

envelope, vacuum (electron tubes). *See:* **bulb.**

environment. The universe within which the system must operate. All the elements over which the designer has no control and that affect the system or its inputs and outputs. *See also:* **system.** 209

environmental change of amplification (magnetic amplifier). The change in amplification due to a specified change in one environmental quantity while all other environmental quantities are held constant. *Note:* Use of a coefficient implies a reasonable degree of linearity of the considered quantity with respect to the specified environmental quantity. If significant deviations from linearity exist within the environmental range over which the amplifier is expected to operate, particularly if the amplification, for example, is not a monotonic function of the environmental quantity, the existence of such deviations should be noted. 171

environmental coefficient (control systems). (1) (output from a control system or element having a specified input). The ratio of a change of output to the change in the specified environment (temperature, pressure, humidity, vibration, etc.), measured from a specified reference level, which causes it; in a linear system, it includes the "coefficient of sensitivity," and the "coefficient of zero error." **(2) (sensitivity).** The ratio of a change in sensitivity to the change in the specified environment (measured from a specified reference level) which causes it. **(3) (zero error).** The ratio of a change in zero error to the change in the specified environment (measured from a specified reference level) which causes it. 56

environmental coefficient of amplification (magnetic amplifier). The ratio of the change in amplificatin to the change in the specified environmental quantity when all other environmental quantities are held constant. *Note:* The units of this coefficient are the amplification units per unit of environmental quantity. 171

environmental coefficient of offset (magnetic amplifier). The ratio of the change in quiescent operating point to the change in the specified environmental quantity when all other environmental quantities are held constant. *Note:* The units of this coefficient are the output units per unit of environmental quantity. 171

environmental coefficient of trip-point stability (magnetic amplifier). The ratio of the change in trip point to the change in the specified environmental quantity when all other environmental quantities are held constant. *Notes:* (1) The units of this coefficient are the control signal units per unit of environmental quantity. (2) Use of a coefficient implies a reasonable degree of linearity of the considered quantity with respect to the specified environmental quantity. If significant deviations from linearity exist within the environmental range over which the amplifier is expected to operate, particularly if the amplification, for example, is not a monotonic function of the environmental quantity, the existence of such deviations should be noted. 171

environmental conditions (nuclear power generating stations). Physical conditions external to the electric penetration assembly including but not limited to ambient temperature, pressure, radiation, humidity, and chemical spray expected as a result of normal operating requirements, and postulated conditions appropriate for the design basis events of the station applicable to the electric penetration assembly. *See:* **design basis events; electric penetration assembly.** 26

environmental offset (magnetic amplifier). The change in quiescent operating point due to a specified change in one environmental quantity (such as line voltage) while all other environmental quantities are held constant. 171

environmental temperature (separable insulated connectors). The temperature of the surrounding medium, such as air, water, or earth, into which the heat of the connector is dissipated directly, including the effect of heat dissipation from associated cables and apparatus. 134

environmental trip-point stability (magnetic amplifier). The change in the magnitude of the trip point (either trip OFF or trip ON, as specified) control signal due to a specified change in one environmental quantity (such as line voltage) while all other environmental quantities are held constant. 171

ephapse. The electric junction of two parallel or crossing nerve fibers at which there may occur phenomena similar to those occurring at a synapse. 192

ephemeris (communication satellite). The position vector of a satellite or spacecraft in space with respect to time. 74

E-plane bend (corner) (waveguide components). A waveguide bend (corner) in which the longitudinal axis of the guide remains in a plane parallel to the electric field vector throughout the bend (corner). 166

E-plane, principal (linearly polarized antenna). The plane containing the electric field vector and the direction of maximum radiation. *See also:* **antenna; radiation.** 179

E-plane tee junction (series tee) (waveguide components). A waveguide tee junction in which the electric field vector of the dominant mode in each arm is parallel to the plane of the longitudinal axes of the guides. 166

equal-energy source (television). A light source from which the emitted power per unit of wavelength is constant throughout the visible spectrum. 18

equal interval (isophase) (light). A rhythmic light in which the light and dark periods are equal. 167

equalization (1) (data transmission). The process of reducing frequency and/or phase distortion of a circuit by the introduction of networks to compensate for the difference in attenuation and/or time delay at the various frequencies in the transmission band. 59

(2) (telephony). The process of compensating a telephone circuit that ideally causes both transmit and receive responses to be inverse functions of direct current, thus tending to equalize variations in loop loss. 122

(3) (feedback control system). Any form of compensation used to secure a closed-loop gain characteristic which is approximately constant over a desired range of frequencies. *See also:* **compensation.** 56

(4) (electroacoustics). *See:* **frequency response equilization.**

equalizer (1) (general). A device designed to compensate for an undesired characteristic of a system or component. 178

(2))signal circuits). A device designed to compensate for an undesired amplitude-frequency or phase-frequency characteristic, or both, of a system or transducer. 328

(3) (rotating machinery). A connection made between points on a winding to minimize any undesirable potential voltage between these points. *See also:* **asynchronous machine; direct-current commutating machine.** 63

equalizing charge (storage battery) (storage cell). An extended charge to a measured end point that is given to a storage battery to insure the complete restoration of the active materials in all the plates of all the cells. *See also:* **charge.** 328

equalizing pulses (television). Pulse trains in which the pulse-repetition frequency is twice the line frequency and that occur just before and just after a vertical synchronizing pulse. *Note:* The equalizing pulses minimize the effect of line-frequency pulses on the interlace. *See also:* **television.** 254, 178

equally tempered scale. A series of notes selected from a division of the octave (usually) into 12 equal intervals, with a frequency ratio between any two adjacent notes equal to the twelfth root of two. 176

Equally tempered intervals

Name of Interval	Frequency Ratio*	Cents
Unison	1:1	0
Minor second or semitone	1.059463:1	100
Major second or whole tone	1.122462:1	200
Minor third	1.189207:1	300
Major third	1.259921:1	400
Perfect fourth	1.334840:1	500
Augmented fourth } Diminished fifth }	1.414214:1	600
Perfect fifth	1.498307:1	700
Minor sixth	1.587401:1	800
Major sixth	1.681793:1	900
Minor seventh	1.781797:1	1000
Major seventh	1.887749:1	1100
Octave	2:1	1200

* The frequency ratio is $[(2)^{1/12}]^n$ where n equals the number of the interval. (The number of the interval is its value in cents divided by 100.)

equal vectors. Two vectors are equal when they have the same magnitude and the same direction. 210

equation. See: **computer equation.**

equational format (pulse measurement). One or more algebraic equations which specify a waveform wherein, typically, a first equation specifies the waveform from t_0 to t_1, a second equation specifies the waveform from t_1 to t_2, etc. The equational format is typically used to specify hypothetical, ideal, or reference waveforms. 15

equatorial orbit (communication satellite). An inclined orbit with an inclination of zero degrees. The plane of an equatorial orbit contains the equator of the primary body. 74

equilibrium electrode potential. A static electrode potential when the electrode and electrolyte are in equilibrium with respect to a specified electrochemical reaction. *See also:* **electrochemistry.** 328

equilibrium point (control system). A point in state space of a system where the time derivative of the state vector is identically zero. *See also:* **control system.** 56

equilibrium potential. The electrode potential at equilibrium. 221, 205

equilibrium reaction potential. The minimum voltage at which an electrochemical reaction can take place. *Note:* It is equal to the algebraic difference of the equilibrium potentials of the anode and cathode with respect to the specified reaction. It can be computed from the free energy of the reaction. Thus

$$\Delta F = -nFE$$

where ΔF is the free energy of the reaction, n is the number of chemical equivalents involved in the reaction, F is the value of the Faraday expressed in calories per volt gram-equivalent (23 060.5) and E is the equilibrium reaction potential (in volts). *See also:* **electrochemistry.** 328

equilibrium voltage. *See:* **storage-element equilibrium voltage.** *See also:* **storage tube.**

equiphase surface (radio wave propagation). Any surface in a wave over which the field vectors at the same instant are in the same phase or 180 degrees out of phase. *See also:* **radio wave propagation.** 146, 180

equiphase zone (electronic navigation). The region in space within which difference in phase of two radio signals is indistinguishable. *See also:* **radio navigation.** 13, 187

equipment (electrical engineering). A general term including materials, fittings, devices, appliances, fixtures, apparatus, machines, etcetera, used as a part of, or in connection with, an electrical installation. 210, 256, 262, 260, 53

equipment ground (1) (general). A ground connection to noncurrent-carrying metal parts of a wiring installation or of electric equipment, or both. *See:* **ground.** 64

(2) (surge arresters). A grounding system connected to parts of an installation, normally not alive, with which persons may come into contact. 244, 62

equipment noise (sound recording and reproducing system). The noise output that is contributed by the elements of the equipment during recording and reproducing, excluding the recording medium, when the equipment is in normal operation. *Note:* Equipment noise usually comprises hum, rumble, tube noise, and component noise. *See also:* **noise (sound recording and reproducing system).** 350

equipment number (telephone switching systems). A unique, physical or other identification of an input or output termination of a switching network. 193

equipment qualification (nuclear power generating stations). The generation and maintenance of evidence to assure that the equipment will operate on demand, to meet the system performance requirements. 31, 120

equipment signature (test, measurement and diagnostic equipment). The special characteristics of an equipment's response to, or reflection of, impinging impulsive energy, or of its electromagnetic, infrared or acoustical emissions. 54

equipotential (conductor stringing equipment). An identical state of electrical potential for two or more items. 45

equipotential cathode. *See:* **cathode, indirectly heated.**

equipotential line or contour. The locus of points having the same potential at a given time. *See also:* **ground.** 313

equisignal localizer (navigation). A localizer in which the localizer on-course line is established as an equality of the amplitudes of two signals. 13

equisignal zone (radio navigation). The region in space within which the difference in amplitude of two radio signals (usually emitted by a single station) is indistinguishable. *See also:* **radio navigation.** 328

equivalence (computing systems). A logic operator having the property that if P is a statement, Q is a statement, R is a statement, . . . , then the equivalence of $P, Q, R, \ldots$ is true if and only if all statements are true or all statements are false. 255, 77

equivalent binary digits (computing systems). The number of binary places required to count the elements of a given set. 255, 77

equivalent circuit (1) (general). An arrangement of circuit elements that has characteristics, over a range of interest, electrically equivalent to those of a different circuit or device. *Note:* In many useful applications, the equivalent circuit replaces (for convenience of analysis) a more-complicated circuit or device. *See also:* **network analysis.** 210

(2) (piezoelectric crystal unit). An electric circuit that has the same impedance as the unit in the frequency region of resonance. *Note:* It is usually represented by an inductance, capacitance, and resistance in series, shunted by the direct capacitance between the terminals of the crystal unit. *See also:* **crystal.** 328

equivalent concentration (ion type). The concentration equal to the ion concentration divided by the valency of the ion considered. *See also:* **ion.** 328

equivalent conductance (1) (acid, base, or salt). The conductance of the amount of solution that contains one gram equivalent of the solute when measured between parallel electrodes that are one centimeter apart and large enough in area to include the necessary volume of solution. *Note:* Equivalent conductance is numerically equal to the conductivity multiplied by the volume in cubic centimeters containing one gram equivalent of the acid, base, or salt. *See also:* **electrochemistry.** 328

(2) (microwave gas tube). The normalized conductance of the tube in its mount measured at its resonance frequency. *Note:* Normalization is with respect to the characteristic impedance of the transmission line at its junction with the tube mount. *See:* **electron-tube admittances; element (electron tube).** 190, 125

equivalent continuous rating (rotating machinery). The statement of the load and conditions assigned to the machine for test purposes, by the manufacturer, at which the machine may be operated until thermal equilibrium is reached, and which is considered to be equivalent to the duty or duty type.   63

equivalent core-loss resistance. A hypothetical resistance, assumed to be in parallel with the magnetizing inductance, that would dissipate the same power as that dissipated in the core of the transformer winding for a specified value of excitation.   197

equivalent dark-current input (phototubes). The incident luminous (or radiant) flux required to give a signal output current equal to the output electrode dark current. *Note:* Since the dark current may change considerably with temperature, the temperature should be specified. *See:* **phototubes.**   190

equivalent diode (triode or a multielectrode tube). *See:* **diode, equivalent.**

equivalent diode voltage. *See:* **composite controlling voltage.**

equivalent flat plate area (scattering object). Equal to the wavelength times the square root of the ratio of the back-scattering cross section to 4π. *Note:* A perfectly reflecting plate parallel to the incident wavefront and having this area, if it is large compared to the wavelength, will have approximately the same back-scattering cross section as the object.   111

equivalent four-wire (data transmission). In equivalent four wire systems, different frequency bands are used to form a "high group" and "low group" for the two directions of transmission, thereby permitting operation over a single pair of conductors.   59

equivalent network. A network that, under certain conditions of use, may replace another network without substantial effect on electrical performance. *Note:* If one network can replace another network in any system whatsoever without altering in any way the electrical operation of that portion of the system external to the networks, the networks are said to be **networks of general equivalence.** If one network can replace another network only in some particular system without altering in any way the electrical operation of that portion of the system external to the networks, the networks are said to be **networks of limited equivalence.** Examples of the latter are networks that are equivalent only at a single frequency, over a single band, in one direction only, or only with certain terminal conditions (such as H and T networks). *See also:* **network analysis.**   210

equivalent noise bandwidth (interference terminology) (signal system). The frequency interval, determined by the response-frequency characteristics of the system, that defines the noise power transmitted from a noise source of specified characteristics. *Note:* For Gaussian noise

$$\Delta f = \int_0^\infty y(f)^2 \, df$$

where $y(f) = Y(0)/Y(f)$ is the relative frequency dependent response characteristic. *See also:* **interference.**   188

equivalent noise conductance (interference terminology). A quantitative representation in conductance units of the spectral density of a noise-current generator at a specified frequency. *Notes:* (1) The relation between the equivalent noise conductance G_n and the spectral density W_i of the noise-current generator is

$$G_n = \pi W_i / (k T_0)$$

where k is Boltzmann's constant and T_0 is the standard noise temperature (290 kelvins) and $k T_0 = 4.00 \times 10^{-21}$ watt-seconds. (2) The equivalent noise conductance in terms of the mean-square noise-generator current i^2 within a frequency increment Δf is

$$G_n = i^2 / (4 k T_0 \Delta f).$$

See: **electron-tube admittances; signal-to-noise signal.**   188, 190

equivalent noise current (electron tubes) (interference terminology). A quantitative representation in current units of the spectral density of a noise current generator at a specified frequency. *Notes:* (1) The relation between the equivalent noise current I_n and the spectral density W_i of the noise-current generator is

$$I_n = (2\pi W_i)/e$$

where e is the magnitude of the electron charge. (2) The equivalent noise current in terms of the mean-square noise-generator current i^2 within a frequency increment Δf is

$$I_n = i^2 / (2e \Delta f).$$

See also: **circuit characteristics of electrodes; interference; signal-to-noise ratio.**   188, 190

equivalent noise input (phototube). The value of incident luminous (or radiant) flux that, when modulated in a stated manner, produces a root-mean-square signal output current equal to the root-mean-square dark-current noise both in the same specified bandwidth (usually 1 hertz). *See also:* **phototube.**   174, 190

equivalent noise referred to input (semiconductor radiation detectors) (linear amplifier). The value of noise at the input that would produce the same value of noise at the output as does the actual noise source.   23, 118, 119

equivalent noise resistance. A quantitative representation in resistance units of the spectral density of a noise voltage generator at a specified frequency. *Notes:* (1) The relation between the equivalent noise resistance R_n and the spectral density W_e of the noise-voltage generator is

$$R_n = (\pi W_e)/(k T_0)$$

where k is Boltzmann's constant and T_0 is the standard noise temperature (290 kelvins) and $k T_0 = 4.00 \times 10^{-21}$ watt-seconds. (2) The equivalent noise resistance in terms of the mean-square noise-generator voltage $\overline{e^2}$ within a frequency increment Δf is

$$R_n = \overline{e^2} / (4 k T_0 \Delta f).$$

See also: **interference; signal-to-noise ratio.**   188, 190, 125

equivalent noise resistance referred to input (semiconductor radiation detectors) (linear amplifier). That value of resistor which when applied to the input of a hypothetical noiseless amplifier with the same gain and bandwidth would produce the same output noise.   23, 118, 119

equivalent 1-megaelectronvolt electron flux. The flux of electrons of 1-megaelectronvolt energy that changes a stated physical quantity (such as minority carrier diffusion length) of a given material or device to the same value as would the flux of penetrating particles of another stated energy spectrum. 113

equivalent parallel circuit elements (magnetic core testing). Under stated conditions of excitation and coil configuration, the values of inductance and resistance connected parallel so that they give representation to the real permeability of the core (μ'_p) and the total losses in the core. (μ''_p)

$$L_p , \mu'_p L_0$$

$$R_p = \omega \mu''_p L_0$$

$$\frac{1}{Z} = \frac{1}{j\omega L_p} + \frac{1}{R_p} = \frac{1}{j\omega \overline{\mu} L_0}$$

$\overline{\mu}$ = complex relative permeability
μ'_p = real component of $\overline{\mu}$ parallel representation
μ''_p = imaginary component of $\overline{\mu}$, parallel representation
L_0 = self inductance of coil with a core of unit relative permeability, but with the same flux distribution as with a ferromagnetic core.
L_p = parallel equivalent self inductance of the coil with a core of $\overline{\mu}$ permeability.
R_p = parallel equivalent loss resistance of the core
ω = angular frequency in radians/sec. 165

equivalent periodic line (uniform line). A periodic line having the same electrical behavior, at a given frequency, as the uniform line when measured at its terminals or at corresponding section junctions. *See also:* **transmission line.** 210

equivalent series circuit elements (magnetic core testing). Under stated conditions of excitation and coil configuration, values of a reactance and a resistance connected in series so that they give representation to the real permeability of the core (μ'_s) and to the total losses in the core (μ''_s)

$$L_s = \mu'_s L_0$$

$$R_s = \omega \mu''_s L_0$$

$$Z = R_s + j\omega L_s = j\omega \overline{\mu} L_0$$

L_s = self inductance of oil with a core of $\overline{\mu}$ permeability; series equivalent inductance.
R_s = equivalent series resistance of coil in ohms with a core of $\overline{\mu}$ permeability.
ω = angular frequency in radians/sec. 165

equivalent two-winding kVA rating. The equivalent two-winding rating of multi-winding transformers or autotransformers is one-half the sum of the kVA ratings of all windings. *Note:* It is customary to base this equivalent two-winding kVA rating on the self-cooled rating of the transformer. 53

equivocation (information theory). The conditional information content of an input symbol given an output symbol, averaged over all input-output pairs. *See also:* **information theory.** 160

erase (charge-storage tubes). To reduce by a controlled operation the amount of stored information. *See also:* **storage tube.** 174, 190, 125, 54

erasing head (electroacoustics). A device for obliterating any previous magnetic recordings. *See:* **alternating-current erasing head; direct-current erasing head; permanent-magnet erasing head; phonograph pickup.** 176

erasing rate (charge-storage tubes). The time rate of erasing a storage element line or area, from one specified level to another. Note the distinction between this and erasing speed. *See also:* **storage tube.** 174, 190

erasing, selective (storage tubes). Erasing of selected storage elements without disturbing the information stored on other storage elements. *See also:* **storage tube.** 174, 190

erasing speed (charge-storage tubes). The linear scanning rate of the beam across the storage surface in erasing. Note the distinction between this and erasing rate. *See also:* **electron devices, miscellaneous; storage tube.** 174, 190, 125

erasing time, minimum usable (storage tubes). The time required to erase stored information from one specified level to another under stated conditions of operation and without rewriting. *Note:* The qualifying adjectives **minimum usable** are frequently omitted in general usage when it is clear that the minimum usable erasing time is implied. *See also:* **storage tube.** 174, 190

erection (gyro). The process of aligning, by precession, a reference axis with respect to the vertical. 46

erection cut-out (gyro). The feature wherein the signal supplying the erection torque is disconnected in order to minimize vehicle maneuver effects. 46

erection or slaving rate (gyro). The angular rate at which the spin axis is precessed to a reference position. It is expressed as angular displacement per unit time. 46

E region. The region of the terrestrial ionosphere between about 90 and 160 kilometers above the earth's surface. *See also:* **radiation; radio wave propagation.** 146

erg. The unit of work and of energy in the centimeter-gram-second systems. The erg is 10^{-7} joule. 210

erlang (1) (telephone switching systems). Unit of traffic intensity, measured in number of arrivals per mean service time. For carried traffic measurements, the number of erlangs is the average number of simultaneous connections observed during a measurement period. 55
(2) (data transmission). A term used in message loading of telephone leased facilities. One erlang is equal to the number of call-seconds divided by 3600 and is equal to a fully loaded circuit over a one hour period. 59

erosion (corrosion). Deterioration by the abrasive action of fluids, usually accelerated by the presence of solid particles of matter in suspension. When deterioration is further increased by corrosion, the term erosion-corrosion is often used. 221, 205

error (1) (mathematics). Any discrepancy between a computed, observed, or measured quantity and the true, specified, or theoretically correct value or condition. *See:* **absolute error; inherited error.** *Note:* A positive error denotes that the indication of the instrument is greater than the true value. Error = Indication − True. *See:* **absolute error; correction; inherited error.** *See also:* **accuracy rating of an instrument; low-frequency testing; power-system testing; surge testing.** 255, 77, 105
(2) (computer or data-processing system). Any incorrect step, process, or result. *Note:* In the computer field the

term commonly is used to refer to a machine malfunction as a machine error (or computer error) and to a human mistake as a human error (or operator error). Frequently it is helpful to distinguish between these errors as follows: an **error** results from approximations used in numerical methods; a **mistake** results from incorrect programming, coding, data transcription, manual operation, etcetera, a **malfunction** results from a failure in the operation of a machine component such as a gate, a flip-flop, or an amplifier. *See:* **dynamic error; linearity error; loading error; resolution error; static error.** *See also:* **electronic analog computer.**   210, 77, 54, 59

(3) **(analog computer).** (A) In science, the difference between the true value and a calculated or observed value. A quantity (equal in absolute magnitude to the error) added to a calculated, indicated, or observed value to obtain the true value is called a correction. (B) In a computer or data processing system, any incorrect step, process, or result. In the computer field the following terms are commonly used: a machine malfunction is a "machine error" (or "computer error"); an incorrect program is a "program error"; and a human mistake is a "human error" (or "operator error"). Frequently it is helpful to distinguish among these errors as follows: an error results from approximations used in numerical methods or imperfections in analog components; a mistake results from incorrect programming, coding, data transcription, manual operation, etcetera; a malfunction results from a failure in the operation of a machine component such as a gate, a flip-flop, or an amplifier.

(4) **(automatic control).** An indicated value minus an accepted standard value, or true value. *Note:* ASA C85 deprecates use of the term as the negative of deviation. *See also:* **accuracy, precision.**   56

(5) **(unbalanced transmission-line impedance).** "In any measurement of a particular quantity, the difference between the measurement concerned and the true value of the magnitude of this quantity, taken positive or negative accordingly as the measurement is greater or less than the true value" (Churchill Eisenhart, "Realistic Evaluation of the Precision and Accuracy of Instrument Calibration Systems," *Journal of Research of the National Bureau of Standards,* Vol. 67C, No. 2, April–June 1963).   147

(6) **(measurement).** The algebraic difference between a value that results from measurement and a corresponding true value.   94

error and correction. The difference between the indicated value and the true value of the quantity being measured. *Note:* It is the quantity that algebraically subtracted from the indicated value gives the true value. A positive error denotes that the indicated value of the instrument is greater than the true value. The correction has the same numerical value as the error of the indicated value, but the opposite sign. It is the quantity that algebraically added to the indicated value gives the true value. If T, I, E and C represent, respectively, the true value, the indicated value, the error, and the correction, the following equations hold:

$$E = I - T; C = T - I.$$

Example: a voltmeter reads 112 volts when the voltage applied to its terminals is actually 110 volts.

$$\text{Error} = 112 - 110 = +2 \text{ volts};$$
$$\text{correction} = 110 - 112 = -2 \text{ volts}.$$

See also: **accuracy rating (instrument).**   328

error band (gyro; accelerometer). A band about the specified output function which contains the output data. The error band contains the composite effects of non-linearity, resolution, non-repeatability, hysteresis and other uncertainties in the output data. *See:* **input-output characteristics.**   46

error burst. A group of bits in which two successive erroneous bits are always separated by less than a given number of correct bits.   194, 59

error coefficient (control system, feedback). The real number C_n by which the nth derivative of the reference input signal is multiplied to give the resulting nth component of the actuating signal. *Note:* The error coefficients may be obtained by expanding in a Maclaurin series the error transfer function as follows:

$$\frac{1}{1 + GH(s)} = C_0 + C_1 s + C_2 s^2 + \ldots + C_n s^n.$$

See also: **control system, feedback.**   329

error constant (control system, feedback). The real number K_n by which the nth derivative of the reference input signal is divided to give the resulting nth component of the actuating signal. *Note:* $K_n = 1/C_n$; $K_0 = 1 + K_p$, where K_p is position constant; $K_1 = K_v$ velocity constant; $K_2 = K_a$ acceleration constant; $K_3 = K_j$ jerk constant. In some systems these constants may equal infinity. *See also:* **control system, feedback.**   56

error-correcting code. A code in which each telegraph or data signal conforms to specific rules of construction so that departures from this construction in the received signals can be automatically detected, and permits the automatic correction, at the received terminal, or some or all of the errors. *Note:* Such codes require more signal elements than are necessary to convey the basic information. *See also:* **error-detecting code; error-detecting and feedback system; error-detecting system.**   194

error-detecting code. A code in which each expression conforms to specific rules of construction, so that if certain errors occur in an expression the resulting expression will not conform to the rules of construction and thus the presence of the errors is detected. *Note:* Such codes require more signal elements than are necessary to convey the fundamental information. *See also:* **check, forbidden-combination; error correcting code.**   255, 77, 194

error-detecting and feedback system. A system employing an error-detecting code and so arranged that a character or block detected as being in error automatically initiates a request for retransmission of the signal detected as being in error.   194

error-detecting system (data transmission). A system employing an error-detecting code and so arranged that any signal detected as being in error is either (1) deleted from the data delivered to the receiver, in some cases with an indication that such deletion has taken place, or (2) delivered to the receiver together with an indication that it has been detected as being in error. 194

error, dynamic (analog computer). An error in a time-varying signal resulting from imperfect dynamic response of a transducer. 9

error, fractional (measurement). The magnitude of the ratio of the error to the true value. 147

error, linearity (analog computer). An error which is the deviation of the output quantity, from a specified linear reference curve. 9

error, loading (analog computer). An error due to the effect of a load impedance upon the transducer or signal source driving it. 9

error, matching (analog computer). An error resulting from inaccuracy in matching (two resistors) or mating (a resistor and a capacitor) passive elements. *See also:* **electronic analog computer.** 167

error, random (measurement). A component of error whose magnitude and direction vary in a random manner in a sequence of measurements made under nominally identical conditions. 147

error range. The difference between the highest and lowest error values. 255, 77

error rate (data transmission). Ratio of the number of characters of a message incorrectly received to the number of characters of the message received. 194, 59

error, resolution (analog computer). The error due to the inability of a transducer to manifest changes of a variable smaller than a given increment. 9

error signal. *See:* **signal, error.**

error, static (analog computer). An error independent of the time-varying nature of a variable. *Syn.* **D.C. error.** 9

error, systematic (measurement). The inherent bias (offset) of a measurement process or of one of its components. 147

erythema. The temporary reddening of the skin produced by exposure to ultraviolet energy. *Note:* The degree of erythema is used as a guide to dosages applied in ultraviolet therapy. *See also:* **ultraviolet radiation.** 167

erythemal effectiveness. The capacity of various portions of the ultraviolet spectrum to produce erythema. *See also:* **ultraviolet radiation.** 167

erythemal efficiency (radiant flux for a particular wavelength). The ratio of the erythemal effectiveness of that wavelength to that of wavelength 296.7 nanometers which is rated as unity. *Note:* This term formerly was called **relative erythemal factor.** The erythemal efficiency of various wavelengths of radiant flux for producing a minimum perceptible erythema is given in the table. These values have been accepted for evaluating the erythemal effectiveness of sun lamps. *See also:* **ultraviolet radiation.** 167

erythemal exposure. The product of erythemal flux density on a surface, and time. It usually is measured in erythemal microwatt-minutes per square centimeter. *Note:* For average untanned skin a minimum perceptible erythema requires about 300 microwatt-minutes per square centimeter of radiation at 296.7 nanometers. *See also:* **ultraviolet radiation.** 167

erythemal flux. Radiant flux evaluated according to its capacity to produce erythema of the untanned human skin. *Notes:* (1) It usually is measured in microwatts of ultra-

violet radiation weighted in accordance with its erythemal efficiency. Such quantities of erythemal flux would be in erythemal microwatts. (2) A commonly used practical unit of erythemal flux is the erythemal unit (EU) or E-viton (erythema) that is equal to the amount of radiant flux that will produce the same erythemal effect as 10 microwatts of radiant flux at wavelength 296.7 nanometers. *See also:* **ultraviolet radiation.** 167

erythemal flux density. The erythemal flux per unit area of the surface being irradiated. *Notes:* (1) It is equal to the quotient of the incident erythemal flux divided by the area of the surface when the flux is uniformly distributed. It usually is measured in microwatts per square centimeter erythemally weighted ultraviolet radiation (erythemal microwatts per square centimeter). (2) A suggested practical unit of erythemal flux density is the finsen, which is equal to one E-viton per square centimeter. *See also:* **ultraviolet radiation.** 167

Erythemal and bactericidal efficiency of ultraviolet radiation

Wavelength (nanometers)	Erythemal Efficiency	Tentative Bactericidal Efficiency
*235.3	—	0.35
240.0	0.56	—
*244.6	0.57	0.58
*248.2	0.57	0.70
250.0	0.57	—
*253.7	0.55	0.85
*257.6	0.49	0.94
260.0	0.42	—
265.0	—	1.00
*265.4	0.25	0.99
*267.5	0.20	0.98
*270.0	0.14	0.95
*275.3	0.07	0.81
*280.4	0.06	0.68
285.0	0.09	—
*285.7	0.10	0.55
*289.4	0.25	0.46
290.0	0.31	—
*292.5	0.70	0.38
295.0	0.98	—
*296.7	1.00	0.27
300.0	0.83	—
*302.2	0.55	0.13
305.0	0.33	—
310.0	0.11	—
*313.0	0.03	0.01
315.0	0.01	—
320.0	0.005	—
325.0	0.003	—
330.0	0.000	—

* Emission lines in the mercury spectrum; other values interpolated. 167

escalator. A power-driven, inclined, continuous stairway used for raising or lowering passengers. *See also:* **elevators.** 328

E scan (electronic navigation). *See:* **E display.**

escape character (computing systems). A character used to indicate that the succeeding one or more characters are expressed in a code different from the code currently in use. 255, 77

escape ratio (charge-storage tubes). The average number of secondary and reflected primary electrons leaving the vicinity of a storage element per primary electron entering that vicinity. *Note:* The escape ratio is less than the secondary-emission ratio when, for example, some secondary electrons are returned to the secondary-emitting surface by a retarding field. *See also:* **charge-storage tube.**

174, 190

E scope (radar). *See:* **E display.**

ESM. *See:* **electronic warfare support measures.**

essential performance requirements (nuclear power generating stations). Requirements that must be met if a component, module, or channel is to carry out its part in the implementation of a protective function. 31, 109

EU. *See:* **erythemal flux.**

evacuating equipment. The assembly of vacuum pumps, instruments, and other parts for maintaining and indicating the vacuum. *See also:* **rectification.** 328

evanescent mode (cutoff mode) (waveguide). A field configuration in a waveguide such that the amplitude of the field diminishes along the waveguide, but the phase is unchanged. The frequency of this mode is less than the critical frequency. *See:* **waveguide.** 179

evanescent waveguide. *See:* **cutoff waveguide.**

event (sequential events recording systems). A change in a process or a change in operation of equipment which is detected by bistable sensors. 48, 58

event recognition (sequential events recording systems). The capability to detect and process changes of state of one or more inputs. 48, 58

E-viton. *See:* **erythemal flux.**

evolving fault (switchgear). A change in the condition of the current flow during interruption whereby the magnitude of current increases either to a fault or a change in fault in a single-phase or multi-phase mode. 79

EW. Abbreviation for early warning and electronic warfare. 13

exalted carrier reception. *See:* **reconditioned carrier reception.**

excess meter. An electricity meter that measures and registers the integral, with respect to time, of those portions of the active power in excess of the predetermined value. *See also:* **electricity meter (meter).** 328

excess-three code (electronic computation). Number code in which the decimal digit n is represented by the four-bit binary equivalent of $n + 3$. Specifically:

decimal digit	excess-three code
0	0011
1	0100
2	0101
3	0110
4	0111
5	1000
6	1001
7	1010
8	1011
9	1100

See also: **binary-coded-decimal system.**

235, 210, 255, 77

exchangeable power (per unit bandwidth, at a port). The extreme value of the power flow per unit bandwidth from or to a port under arbitrary variations of its terminating impedance. *Notes:* (1) The exchangeable power p_e at a port with a mean-square open-circuit voltage spectral density e^2 and an internal impedance with a real part R is given by the relation

$$p_e = \frac{\overline{e^2}}{4R}$$

(2) The exchangeable power is equal to the available power when the internal impedance of the port has a positive real part. *See:* **signal-to-noise ratio; waveguide.**

190

exchangeable power gain (two-port linear transducer). At a pair of selected input and output frequencies, the ratio of (1) the exchangeable signal power of the output port of the transducer to (2) the exchangeable signal power of the source connected to the input port. *Note:* The exchangeable power gain is equal to the available power gain when the internal impedances of the source and the output port of the transducer have positive real parts. *See:* **signal-to-noise ratio; waveguide.** 190

exchange area (telephone switching systems). The territory included within the boundaries of a telecommunications exchange. 55

exchange, central office. *See:* **central office exchange.**

exchange service (data transmission). A service permitting interconnection of any two customers' telephones through the use of a switching equipment. 59

excitability (electrobiology) (irritability). The inherent ability of a tissue to start its specific reaction in response to an electric current. 192

excitability curve (medical electronics). A graph of the excitability of a given tissue as a function of time, where excitability is expressed either as the reciprocal of the intensity of an electric current just sufficient at a given instant to start the specific reaction of the tissue, or as the quotient of the initial (or conditioning) threshold intensity for the tissue by subsequent threshold intensities. 192

excitation (1) (drive). A signal voltage applied to the control electrode of an electron tube. *See also:* **composite controlling voltage (electron tube).** 111

(2) (antenna). *See:* **aperture illumination.**

excitation anode (pool-cathode rectifier tube). An electrode that is used to maintain an auxiliary arc in the vacuum tank. *See also:* **electrode (electron tube); rectification.**

328

excitation coefficients (array antenna). The relative values of the excitation currents of the radiating elements. *Syn.* **feeding coefficients.** 179

excitation control system. A feedback control system that includes the synchronous machine and its excitation system. 105

excitation current (1) (general). The current supplied to unloaded transformers or similar equipment. 103

(2) (transformer) (no-load current). The current which flows in any winding used to excite the transformer when all other windings are open-circuited. It is usually expressed in percent of the rated current of the winding in which it is measured. 53

(3) (voltage regulator). The current that maintains the

excitation of the regulator. *Note:* It is usually expressed in per unit or in percent of the rated series-winding current of the regulator. *See also: efficiency; voltage regulator.*
257

excitation equipment (rectifier). The equipment for starting, maintaining, and controlling the arc. *See also: rectification.*
328

excitation losses (1) (transformer or regulator). Losses that are incident to the excitation of the transformer at its alternating-current winding terminals. *Note:* Excitation losses include core loss, dielectric loss, conductor loss in the windings due to excitation current, and conductor loss due to circulating current in parallel windings. Interphase transformer losses are not included in excitation losses. *See also:* **rectifier transformer; voltage regulator; no-load losses.**
203, 303, 258

(2) (series transformer). The losses in the transformer with the secondary winding open-circuited when the primary winding is excited at rated frequency and at a voltage that corresponds to the primary voltage obtained when the transformer is operating at nominal rated load. *Note:* The measurement should be made with a constant voltage source of supply with not more than 3-percent harmonic deviation from sine wave.
274

(3) (instrument transformer). The watts required to supply the energy necessary to excite the transformer which includes the dielectric watts, the core watts, and the watts in the excited winding due to the excitation current. 53

excitation purity (purity) (color). The ratio of: (1) the distance from the reference point to the point representing the sample, to (2) the distance along the same straight line from the reference point to the spectrum locus or to the purple boundary, both distances being measured (in the same direction from the reference point) on the (International Commission on Illumination) CIE chromaticity diagram. *Note:* When giving excitation purity and domiant (or complementary) wavelength as a pair of values to determine the chromaticity coordinates, the reference point must be the same in all cases, and it must represent the reference standard light (specified achromatic light) mentioned in the definition of dominant wavelength.
18

excitation-regulating winding (two-core regulating transformer). In some designs, the main unit will have one winding operating as an autotransformer which performs both functions listed under excitation and regulating windings. Such a winding is called the excitation-regulating winding. 53

excitation response. *See: voltage response, exciter.*

excitation system (rotating machinery). The source of field current for the excitation of a principal electric machine, including means for its control. *See: direct-current commutating machine.*
63, 105

excitation system, high initial response. An excitation system having an excitation system voltage response time of 0.1 second or less.
105

excitation-system stability (rotating machinery). The ability of the excitation system to control the field voltage of the principal electric machine so that transient changes in the regulated voltage are effectively suppressed and sustained oscillations in the regulated voltage are not produced by the excitation system during steady-load

conditions following a change to a new steady-load condition. *See: direct-current commutating machine.*
63

excitation system voltage response (synchronous machines). The rate of increase or decrease of the excitation system output voltage determined from the excitation system voltage-time response curve, that if maintained constant would develop the same voltage-time area as obtained from the curve for a specified period. The starting point for determining the rate of voltage change shall be the initial value of the excitation system voltage-time response curve. *Notes:* (1) Similar definitions can be applied to the excitation system major components such as the exciter and regulator (2) A system having an excitation system voltage response time of 0.1 second or less is defined as a high initial response excitation system.
105

excitation system voltage response ratio (synchronous machine). The numerical value that is obtained when the excitation system voltage response, in volts per second, measured over the first half-second interval, unless otherwise specified, is divided by the rated-load field voltage of the synchronous machine. *Note:* Unless otherwise specified, the excitation system voltage response ratio shall apply only to the increase in excitation system voltage.
105

excitation system voltage response time. The time in seconds for the excitation voltage to attain 95 percent of the difference between ceiling voltage and rated load field voltage under specific conditions.
105

excitation system voitage time response. The excitation system output voltage expressed as a function of time, under specified conditions. *Note:* A similar definition can be applied to the excitation system major components, the exciter and regulator, separately.
105

excitation voltage. The nominal voltage of the excitation circuit.
164

excitation winding (two-core regulating transformer). The winding of the main unit which draws power from the system to operate the two-core transformer. *See also:* **field winding.**
53

excite (rotating machinery). To initiate or develop a magnetic field in (such as in an electric machine). *See also:* **asynchronous machine; direct-current commutating machine.**
63

excited-field loudspeaker. A loudspeaker in which the steady magnetic field is produced by an electromagnet. *See also: loudspeaker.*
328

excited-state maser (laser-maser). A maser in which the terminal level of the amplifying transition is not appreciably populated at thermal equilibrium for the ambient temperature
363

excited winding (two-core regulating transformer). The winding of the series unit which is excited from the regulating winding of the main unit.
53

exciter (1) (rotating machinery). The source of all or part of the field current for the excitation of an electric machine. *Note:* Familiar sources include direct-current commutator machines; alternating-current generators whose output is rectified; and batteries. *See also:* **asynchronous machine; direct-current commutating machine.**
63, 105

(2) (data transmission). In antenna practice, an exciter is the portion of a transmitting array, of the type which includes a reflector or director, which is directly connected with the source of power. 59

exciter, alternator—rectifier (synchronous machines). An exciter whose energy is derived from an alternator and converted to direct current by rectifiers. *Notes:* (1) The exciter includes an alternator and power rectifiers which may be either noncontrolled or controlled, including gate circuitry. (2) It is exclusive of input control elements. (3) The alternator may be driven by a motor, prime mover, or by the shaft of the synchronous machine. (4) The rectifiers may be stationary or rotating with the alternator shaft. 105

exciter ceiling voltage. The maximum voltage that may be attained by an exciter under specified conditions. *See:* **direct-current commutating machine.** 63

exciter ceiling voltage, nominal (rotating machinery). The ceiling voltage of an exciter loaded with a resistor having an ohmic value equal to the resistance of the field winding to be excited and with this field winding at a temperature of: (1) 75 degrees Celsius for field windings designed to operate at rating with a temperature rise of 60 degrees Celsius or less. (2) 100 degrees Celsius for field windings designed to operate at rating with a temperature rise greater than 60 degress Celsius. *See:* **direct-current commutating machine.** 328

exciter, compound-rectifier (synchronous machines). An exciter whose energy is derived from the currents and potentials of the alternating current terminals of the synchronous machine and converted to direct current by rectifiers. *Notes:* (1) The exciter includes the power transformers (current and potential), power reactors, and power rectifiers which may be either noncontrolled or controlled, including gate circuitry. (2) It is exclusive of input control elements. 105

exciter, direct current generator—commutator (synchronous machines). An exciter whose energy is derived from a direct current generator. *Notes:* (1) The exciter includes a direct current generator with its commutator and brushes. It is exclusive of input control elements. (2) The exciter may be driven by a motor, prime mover, or by the shaft of the synchronous machine. 105

exciter dome (rotating machinery). Exciter housing for a vertical machine. 63

exciter losses (synchronous machine). The total of the electric and mechanical losses in the equipment supply excitation. 244, 63

exciter, main (synchronous machines). The source of all or part of the field current for the excitation of an electric machine, exclusive of another exciter. 105

exciter, pilot (synchronous machines). The source of all or part of the field current for the excitation of another exciter. 105

exciter platform (rotating machinery). A deck on which to stand while inspecting the exciter. 63

exciter, potential-source—rectifier (synchronous machines). An exciter whose energy is derived from a stationary alternating current potential source and converted to direct current by rectifiers. *Notes:* (1) The exciter includes the power potential transformers, where used, and power rectifiers which may be either noncontrolled or controlled, including gate circuitry. (2) It is exclusive of input control elements. 105

exciter response. *See:* **voltage response, exciter.**

exciter response ratio, main (synchronous machines). The numerical value obtained when the response, in volts per second, is divided by the rated-load field voltage, which response, if maintained constant, would develop, in one half-second, the same excitation voltage—time area as attained by the actual exciter. *Note:* The response is determined with no load on the exciter with the exciter voltage initially equal to the rated-load field voltage, and then suddenly establishing circuit conditions which would be used to obtain nominal exciter ceiling voltage. For a rotating exciter, response should be determined at rated speed. This definition does not apply to main exciters having one or more series fields, except a light differential series field, or to electronic exciters. 105

exciter voltage response ratio (rotating machinery). *See:* **voltage response ratio.**

exciter voltage-time response (rotating machinery). *See:* **voltage-time response, synchronous-machine excitation system.**

exciting current. (1) The total current applied to a coil that links a ferromagnetic core. 210
(2) The component of the primary current of a transformer that is sufficient by itself to cause the counter electromotive force to be induced in the primary winding. 197

excitron. A single-anode pool tube provided with means for maintaining a continuous cathode spot. 190

exclusive OR. A logic operator having the property that if P is a statement and Q is a statement, when P exclusive OR Q is true if either but not both statements are true, false if both are true or both are false. *Note:* P exclusive OR Q is often represented by $P \oplus Q$, $P \neq Q$. *See:* **OR.** 255, 77

excursion (computing system). *See:* **reference excursion.** 9, 77

executive routine (computing systems). A routine that controls the execution of other routines. *See:* **supervisory routine.** 255, 77, 54

exercise (test, measurement and diagnostic equipment). To operate an equipment in such a manner that it performs all its intended functions to allow observation, testing, measurement and diagnosis of its operational condition. 54

exfoliation (corrosion). A thick layer-like growth of corrosion product. 221, 205

existing installation (elevators). An installation, prior to the effective date of a code: (1) all work of installation was completed, or (2) the plans and specifications were filed with the enforcing authority and work begun not later than three months after the approval of such plans and specifications. *See also:* **elevators.** 328

expanded sweep. A sweep of the electron beam of a cathode-ray tube in which the movement of the beam is speeded up during a part of the sweep. *See also:* **magnified sweep; radar.** 328

expander. A transducer that for a given amplitude range of input voltages produces a larger range of output voltages. *Note:* One important type of expander employs the envelope of speech signals to expand their volume range. *See also:* **attenuation.** 239, 59

expandor (telephone switching systems). A switching entity for connecting a number of inlets to a greater number of outlets. 55

expansion (1)(modulation systems). A process in which the effective gain applied to a signal is varied as a function of the signal magnitude, the effective gain being greater for large than for small signals. 57
(2) (oscillography). A decrease in the deflection factor, usually as the limits of the quality area are exceeded.
242, 185

expansion chamber (oil cutout). A sealed chamber separately attachable to the vent opening to provide additional air space into which the gases developed during circuit interruption can expand and cool. 103, 202

expansion orbit (electronic device). The last part of the electron path that terminates at the target. It is outside the equilibrium orbit. *See also:* **electron device.** 190

expendable cap (expendable-cap cutout). A replacement part or assembly for clamping the button head of a fuse link and closing one end of the fuseholder. *Note:* It includes a pressure-responsive section that opens to relieve the pressure within the fuseholder, when a predetermined value is exceeded during circuit interruption. 103, 202

expendable-cap cutout. An open cutout having a fuse support designed for and equipped with a fuseholder having an expendable cap. 13, 202

exploder. *See:* **blasting unit.**

explosionproof apparatus (1) (general). Designed and constructed to withstand an explosion of a specified gas or vapor that may occur within it and to prevent the ignition of the specified gas or vapor surrounding the enclosure by sparks, flashes, or explosions of the specified gas or vapor that may occur within the enclosure.
(2) National Electrical Code. Apparatus enclosed in a case that is capable of withstanding an explosion of a specified gas or vapor which may occur within it and of preventing the ignition of a specified gas or vapor surrounding the enclosure by sparks, flashes, or explosion of the gas or vapor within, and which operates at such an external temperature that a surrounding flammable atmosphere will not be ignited thereby. *Notes:* (A) Without any significant change in the text of this definition, the word enclosure is frequently replaced by more specific terms, such as luminaire, machine, apparatus, etcetera. In some contexts, flameproof appears as a synonym of explosionproof. (B) Explosionproof apparatus should bear Underwriters' Laboratories approval ratings of the proper class and group consonant with the spaces in which flammable volatile liquids, highly flammable gases, mixtures, or highly flammable substances may be present.
3, 256, 178, 167, 103, 202, 232, 65
(3) (explosionproof) (mine apparatus). Apparatus capable of withstanding explosion tests as established by the United States Bureau of Mines, namely, internal explosions of methane-air mixtures, with or without coal dust present, without ignition of surrounding explosive methane-air mixtures and without damage to the enclosure or discharge of flame. *See;* **hazardous area groups; hazardous area classes.** *See also:* **distribution center; luminaire.**
3, 256, 178, 103, 202, 232, 65

explosionproof fuse. A fuse, so constructed or protected, that for all current interruptions within its rating shall not be damaged nor transmit flame to the outside of the fuse.
103

explosion-tested equipment. Equipment in which the housings for the electric parts are designed to withstand internal explosions of methane-air mixtures without causing ignition of such mixtures surrounding the housings. 328

explosives (conductor stringing equipment). Mixtures of solids, liquids or a combination of the two which, upon detonation, transform almost instantaneously into other products which are mostly gaseous and which occupy much greater volume than the original mixtures. This transformation generates heat which rapidly expands the gases, causing them to exert enormous pressure. Dynamite and Primacord are explosives as manufactured. Aerex, Triex and Quadrex are manufactured in two components and are not true explosives until mixed. Explosives are commonly used to build construction roads, blast hole for anchors, structure footings, etc. *Syn.* **aerex; dynamite; fertilizer; powder; primacord; triex; quadrex.** 45

exponential-cosine (exponential mius cosine) (ex-cos) envelope (transient recovery voltage) (1) (general). A voltage-versus-time curve which represents the maximum at any time of the 1-cosine (1 minus cosine) envelope and the exponential envelope. 103
(2) (transient recovery voltage). The greater at any instant of: (A) The curve traced by the multiple exponential, transient voltage across Z when a switch is closed on the circuit shown below. It reaches its crest E_1 at $t = \infty$. (B) The 1-cosine curve with its initial crest at P.

Representation of transient terminology

E_1 represents the alternating current driving or ceiling voltage which is considered at its peak at the time of a current zero and remains practically constant during that portion of the transient defined by the first curve. Hence, it can be considered as a direct current source during this time. The voltage application is simulated by the closing of the switch. e represents the transient voltage across the circuit breaker pole unit. L represents the equivalent effective inductance on the source side of the circuit breaker. Z represents the equivalent surge impedance of associated transmission lines. C represents the equivalent lumped capacitance on the source side of the breaker and modifies the ex-cos envelope by what may be considered as a slight initial time delay, T_1. R is the transient recovery voltage rate. Besides forming a basis of rating the above definition is also useful in discussing the changes of transient voltage caused by varying the parameters. *Note:* The ex-cos curve is the standard envelope for rating circuit breaker transient recovery voltage performance for circuit breakers rated 121 kV and above.

exponential (ex) envelope (of a transient recovery voltage). A voltage-versus-time curve of the general exponential form $e_1 = E_1 [1 - \text{ex} (t/T)]$ in which e_1 represents the transient voltage across a switching device pole unit, reaching its crest E_1 at infinite time. *Note:* In practice this envelope curve is derived from a circuit in which a voltage E_1 charges, by means of a switch, a circuit with inductance L in series with impedance Z and capacitance C in parallel. The voltage of e_1 is measured across Z. E_1 represents the ac driving or ceiling voltage which is considered at its peak at the time of a current zero and remains practically constant during that portion of the transient defined by the first curve. Hence, it can be considered as dc source during this time. The voltage application is simulated by the closing of the switch. e_1 represents the transient voltage across the circuit breaker pole unit. L represents the equivalent effective inductance on the source side of the circuit breaker. Z represents the equivalent surge impedance of associated transmission lines. C represents the equivalent lumped capacitance on the source side of the breaker and modifies the exponential envelope by what may be considered as a slight initial time delay, T_1. R is the transient recovery voltage rate, corresponding to the initial slope of the exponential envelope.   103

exponential function. One of the form $y = ae^{bx}$, where a and b are constants and may be real or complex. An exponential function has the property that its rate of change with respect to the independent variable is proportional to the function, or $dy/dx = by$.   210

exponential horn. A horn the cross-sectional area of which increases exponentially with axial distance. *Note:* If S is the area of a plane section normal to the axis of the horn at a distance x from the throat of the horn, S_0 is the area of the plane section normal to the axis of the horn at the throat, and m is a constant that determines the rate of taper or flare of the horn, then

$$S = S_0 e^{mx}$$

See also: **loudspeaker.**   176
exponential lag. *See:* **lag (first order).**
exponentially damped sine function. A generalized sine function of the form $Ae^{-bx} \sin (x + \alpha)$ where $b > 0$.   210

exponential reference atmosphere (radar). A model of atmospheric refractivity adopted by the National Bureau of Standards in which the index of refraction n is approximated, as a function of height, by a single exponential expression: $n(h) = 1 + 313 \times 10^{-6} \exp (-0.1438h)$ where h is in kilometers above sea level.   13
exponential transmission line. A tapered transmission line whose characteristic impedance varies exponentially with electrical length along the line. *See:* **transmission line.** *See also:* **waveguide.**   267, 179
exposed (1) (live parts). A live part that can be inadvertently touched or approached nearer than a safe distance by a person. *Note:* It is applied to parts not suitably guarded,

isolated, or insulated. *See:* **accessible; concealed.** *See also:* **distribution center.**   256
(2) (wiring methods). Not concealed.   328
(3) (transmission and distribution). Not isolated or guarded.   262
exposed installation (lightning). An installation in which the apparatus is subject to overvoltages of atmospheric origin. *Note:* Such installations are usually connected to overhead transmission lines either directly or through a short length of cable. *See:* **surge arrester (surge diverter).**   244, 62
exposure (laser-maser). The product of an irradiance and its duration.   363
exposure time (electric power generation). The time during which a component is performing its intended function and is subject to outage.   112
expression. An ordered set of one or more characters.   235, 77
expulsion arrester. A surge arrester that includes an expulsion element.   2
expulsion element (arrester). A chamber in which an arc is confined and brought into contact with gas-evolving material. *See:* **surge arrester (surge diverter).**   2, 299, 62
expulsion-fuse unit (expulsion fuse). A vented fuse unit in which the expulsion effect of gases produced by the arc and lining of the fuseholder, either alone or aided by a spring, extinguishes the arc.   103, 202
expulsion-type surge arrester (expulsion-type arrester). *See:* **arrester, expulsion-type.**
extended capability (electric power supply). The generating capability increment in excess of net dependable capability which can be obtained under emergency operating procedures. *See also:* **generating station.**   200, 112
extended delta connection (transformer). A connection similar to a delta, but with a winding extension at each corner of the delta, each of which is 120 degrees apart in phase relationship. *Note:* The connection may be used as an autotransformer to obtain a voltage change or a phase shift, or a combination of both.   53
extended-service area (telephone switching systems). That part of the local service area that is outside of the boundaries of the exchange area of the calling customer.   55
extended source (laser-maser). An extended source of radiation can be resolved by the eye into a geometrical image, in contrast to a point source of radiation, which cannot be resolved into a geometrical image.   363
extended-time rating (grounding device). A rated time in which the period of time is greater than the time required for the temperature rise to become constant but is limited to a specified average number of days operation per year.   91
extension cord. An assembly of a flexible cord with an attachment plug on one end and a cord connector on the other.   328
extension station. A telephone station associated with a main station through connection to the same subscriber line and having the same call number designation as the associated main station. *See also:* **telephone station.**   328
external connector (aerial lug). A connector that joins the

external conductor to the other current-carrying parts of a cable termination.   4

external field influence (electric instrument). The percentage change (of full-scale value) in indication caused solely by a specified external field. Such a field is produced by a standard method with a current of the same kind and frequency as that which actuates the mechanism. This influence is determined with the most unfavorable phase and position of the field in relation to the instrument. *Note:* The coil used in the standard method shall be approximately 40 inches in diameter not over 5 inches long, and carrying sufficient current to produce the required field. The current to produce a field to an accuracy of ±1 percent in air shall be calculated without the instrument in terms of the specific dimensions and turns of the coil. In this coil, 400 ampere-turns will produce a field of approximately 5 oersteds. The instrument under test shall be placed in the center of the coil. *See also:* **accuracy rating (instrument).**   280, 294

external insulation (1) (apparatus) (surge arresters). The external insulating surfaces and the surrounding air. *Note:* The dielectric strength of external insulation is dependent on atmospheric conditions.   62, 103, 53

(2) (outdoor ac high-voltage circuit breakers). Insulation that is designed for use outside of buildings and for exposure to the weather.   149, 103

externally commutated inverters. An inverter in which the means of commutation is not included within the power inverter. *See:* **self-commutated inverters.**   208

externally heated arc (gas). An electric arc characterized by the fact that the thermionic cathode is heated by an external source. *See also:* **discharge (gas).**   244, 190

externally programmed automatic test equipment (test, measurement and diagnostic equipment). An automatic tester using any programming technique in which the programming instructions are not read directly from within the ATE (automatic test equipment), but from a medium which is added to the equipment such as punched tape, punched cards, and magnetic tape.   54

externally quenched counter tube. A radiation counter tube that requires the use of an external quenching circuit to inhibit reignition. *See also:* **gas-filled radiation-counter tube.**   190

externally ventilated machine (rotating machinery). A machine that is ventilated by means of a separate motor-driven blower. The blower is usually mounted on the machine enclosure but may be separately mounted on the foundation for large machines. *See also:* **open-pipe ventilated machine; separately ventilated machine.**   63

external remanent residual voltage (Hall-effect devices). That portion of the zero field residual voltage which is due to remanent magnetic flux density in the external electromagnetic core.   107

external series gap (expulsion-type arrester). An intentional gap between spaced electrodes, in series with the gap or gaps in the arcing chamber.   308, 62

external storage (test, measurement and diagnostic equipment). Information storage off-line in media such as magnetic tape, punched tape, and punched cards.   54

external temperature influence (1) (electric instrument). The percentage change (of full-scale value) in the indication of an instrument that is caused solely by a difference in

ambient temperature from the reference temperature after equilibrium is established. *See also:* **accuracy rating (instrument).**   280

(2) (electric recording instrument). The change in the recorded value that is caused solely by a difference in ambient temperature from the reference temperature. *Note:* It is to be expressed as a percentage of the full-scale value. *See also:* **accuracy rating (instrument).**   294

external termination (*j*th terminal of an *n*-terminal network). The passive or active two-terminal network that is attached externally between the *j*th terminal and the reference point. *See also:* **electron-tube admittances.**
                                                190, 125

extinction voltage (gas tube). The anode voltage at which the discharge ceases when the supply voltage is decreasing. *See also:* **gas-filled rectifier.**   244, 190

extinguishing voltage (drop-out voltage) (glow lamp). Dependent upon the impedance in series with the lamp, the voltage across the lamp at which an abrupt decrease in current between operating electrodes occurs and is accompanied by the disappearance of the negative glow. *Note:* In recording or specifying extinguishing voltage, the impedance must be specified.   283

extracameral effect (nuclear power generating stations). Apparent response of an instrument caused by radiation on any other portion of the system than the detector.
                                                       31

extract (electronic computation). To form a new word by juxtaposing selected segments of given words.   210

extract instruction (electronic digital computation). An instruction that requests the formation of a new expression from selected parts of given expressions.   235, 255, 77

extraction liquor. The solvent used in hydrometallurgical processes for extraction of the desired constituents from ores or other products. *See also:* **electrowinning.**   328

extra high voltage (EHV) (transmission and distribution). A term applied to voltage levels which are higher than 240 000 volts.   112

extra-high-voltage system (electric power). An electric system having a maximum root-mean-square alternating-current voltage above 242,000 volts to 800,000 volts. *See also:* **low-voltage system; medium voltage system; high voltage system.**   49

extraordinary wave (X wave) (1) (radio wave propagation). The magneto-ionic wave component that, when viewed below the ionosphere in the direction of propagation, has clockwise or counterclockwise elliptical polarization; respectively, according as the earth's magnetic field has a positive or negative component in the same direction. *See also:* **radiation.**

(2) (radio wave propagation). The magneto-ionic wave component in which the electric vector rotates in the opposite sense to that for the ordinary-wave component. *See:* **ordinary-wave component.** *See also:* **radiation.**
                                                  146, 180

extrapolated range for electrons (solar cells). The distance of travel in a material by electrons of a given energy, at which the flux of primary electrons extrapolates to zero.
                                                      113

extreme operating conditions (automatic null-balancing electrical instrument). The range of operating conditions within which a device is designed to operate and under which operating influences are usually stated. *See also:* **measurement system.**   295

extrinsic properties (semiconductor). The properties of a semiconductor as modified by impurities or imperfections within the crystal. *See also:* **semiconductor; semiconductor device.**   245, 210, 66, 186

extrinsic semiconductor. A semiconductor whose charge-carrier concentration is dependent upon impurities. *See also:* **semiconductor.**   237, 66

eye bolt (rotating machinery). A bolt with a looped head used to engage a lifting hook.   63

eyelet (soldered connections). A hollow tube inserted in a printed circuit or terminal board to provide electric connection or mechanical support for component leads. *See also:* **soldered connections (electronic and electric applications).**   284

eye light (television). Illumination on a person to provide a specular reflection from the eyes (and teeth) without adding a significant increase in light on the subject. *See also:* **television lighting.**   167

eye pattern (data transmission). An oscilloscope display of the detector voltage wave form in a data modem. This pattern gives a convenient representation of cross over distortion which is indicated by a closing of the center of the eye.   59

F

FA. *See:* **transformer, oil-immersed.**

fabric (rotating machinery). A planar structure comprising two or more sets of fiber yarns interlaced in such a way that the elements pass each other essentially at right angles and one set of elements is parallel to the fabric axis.
                                                           216, 63

faceplate (cathode-ray tube). The large transparent end of the envelope through which the image is viewed or projected. *See also:* **cathode-ray tubes.**   244, 190

faceplate controller (industrial control). An electric controller consisting of a resistor and a faceplate switch in which the electric contacts are made between flat segments, arranged on a plane surface, and a contact arm. *See also:* **electric controller.**   206

faceplate rheostat (industrial control). A rheostat consisting of a tapped resistor and a panel with fixed contact members connected to the taps, and a lever carrying a contact rider over the fixed members for adjustment of the resistance.   206

facilitation. The brief rise of excitability above normal either after a response or after a series of subthreshold stimuli. *See also:* **biological.**   192

facility (communication). Anything used or available for use in the furnishing of communication service.   59

facing (planar structure) (rotating machinery). A fabric, mat, film, or other material used in intimate conjunction with a prime material and forming a relatively minor part of the composite for the purpose of protection, handling, or processing. *See also:* **asynchronous machine: direct-current commutating machine.**   63

facsimile (electrical communication). The process, or the result of the process, by which fixed graphic material including pictures or images is scanned and the information converted into signal waves which are used either locally or remotely to produce in record form a likeness (facsimile) of the subject copy.   12

facsimile signal (picture signal). A signal resulting from the scanning process.   12

facsimile-signal level. The maximum facsimile signal power or voltage (root-mean-square or direct-current) measured at any point in a facsimile system. *Note:* It may be expressed in decibels with respect to some standard value such as 1 milliwatt. *See also:* **facsimile signal (picture signal).**   12

facsimile system. An integrated assembly of the elements used for facsimile. *See also:* **facsimile (electrical communication).**   12

facsimile transient. A damped oscillatory transient occurring in the output of the system as a result of a sudden change in input. *See also:* **facsimile transmission.**   12

facsimile transmission. The transmission of signal waves produced by the scanning of fixed graphic material, including pictures, for reproduction in record form.
                                                           111, 12

factor of assurance (wire or cable insulation). The ratio of the voltage at which completed lengths are tested to that at which they are used.   64

factory-renewable fuse. A fuse that, after circuit interruption, must be returned to the manufacturer to be restored for service.   103, 202

fade in. To increase signal strength gradually in a sound or television channel.   328

fade out. To decrease signal strength gradually in a sound or television channel.   328

fading (radio wave propagation). The variation of radio field intensity caused by changes in the transmission medium, and transmission path, with time. *See also:* **radiation; radio wave propagation.**   180

fading flat (data transmission). That type of fading in which all frequency components of the received radio signal fluctuate in the same proportions simultaneously.   59

fail safe (reliability). A designed property of an item which prevents its failures being critical failures.   164

failure (1) (reliability). The termination of the ability of an item to perform a required function.   182
(2) (catastrophic). Failure that is both sudden and complete.   182
(3) (complete). Failure resulting from deviations in characteristic(s) beyond specified limits such as to cause complete lack of the required function. *Note:* The limits referred to in this category are special limits specified for this purpose.   182
(4) (critical). Failure which is likely to cause injury to persons or significant damage to material.   164
(5) (degradation). Failure which is both gradual and partial. *Note:* In time such a failure may develop into a complete failure.   164
(6) (dependent) (test, measurement, and diagnostic equip-

ment). A failure which is caused by the failure of an associated item, distinguished from independent failure.
54

(7) (gradual). Failures that could be anticipated by prior examination or monitoring.

(8) (independent) (test, measurement and diagnostic equipment). A failure which occurs without being related to the failure of associated items, distinguished from dependent failure.
54

(9) (inherent weakness). Failure attributable to weakness inherent in the item itself when subjected to stresses within the stated capabilities of the item.
164

(10) (intermittent). Failure of an item for a limited period of time, following which the item recovers its ability to perform its required function without being subjected to any external corrective action. *Note:* Such a failure is often recurrent.
164

(11) (major). Failure, other than a critical failure, which is likely to reduce the ability of a more complex item to perform its required function.
164

(12) (minor). Failure, other than a critical failure, which does not reduce the ability of a more complex item to perform its required function.
164

(13) (misuse). Failure attributable to the application of stresses beyond the stated capabilities of the item.

(14) (nonrelevant). Failure to be excluded in interpreting test results or in calculating the value of a reliability characteristic. *Note:* The criteria for the exclusion should be stated.
164

(15) (partial). Failure resulting from deviations in characteristic(s) beyond specified limits, but not such as to cause complete lack of the required function. *Note:* The limits referred to in this category are special limits specified for this purpose.
164

(16) (primary). Failure of an item, not caused either directly or indirectly by the failure of another item. 164

(17) (random). Any failure whose cause and/or mechanism make its time of occurrence unpredictable.

(18) (relevant). Failure to be included in interpreting test results or in calculating the value of a reliability characteristic. *Note:* The criteria for the inclusion should be stated.
164

(19) (secondary). Failure of an item caused either directly or indirectly by the failure of another item.

(20) (sudden). Failure that could not be anticipated by prior examination or monitoring.

(21) (wear-out). Failure whose probability of occurrence increases with the passage of time and which occurs as a result of processes which are characteristic of the population.
164

(22) (nuclear power generating stations). The termination of the ability of an item to perform its required function. Failures may be unannounced and not detected until the next test (unannounced failure), or they may be announced and detected by any number of methods at the instant of occurrence (announced failure). 29, 31, 356

failure analysis (test, measurement and diagnostic equipment). The logical, systematic examination of an item or its diagram(s) to identify and analyze the probability, causes, and consequences of potential and real failures.
54

failure cause (reliability). The circumstances during design,

manufacture or use which have led to failure. 164

failure commutation (thyristor converter). A failure to commutate the direct current from the conducting arm to the succeeding arm of a thyristor connection. *Note:* In inverter operation, a commutation failure results in a conduction-through.
121

failure criteria (reliability). Rules for failure relevancy such as specified limits for the acceptability of an item. *See also:* **reliability.**
182

failure mechanism (reliability). The physical chemical, or other process which results in a failure. *See also:* **reliability.**
182

failure mode (reliability). The effect by which a failure is observed. *Note:* For example, an open- or short-circuit condition, or a gain change. *See also:* **reliability.**
182, 164

failure rate (any point in the life of an item) (1) (general) (reliability). The incremental change in the number of failures per associated incremental change in time.

(2) (nuclear power generating stations). The expected number of failures of a given type, per item, in a given time interval (for example, capacitor short-circuit failures per million capacitor hours).
29, 31

failure rate acceleration factor (reliability). The ratio of the accelerated testing failure rate to the failure rate under stated reference test conditions. Both failure rates refer to the same time period in the life of the tested items.
164

failure rate, assessed (reliability). The failure rate of an item determined by a limiting value or values of the confidence interval associated with a stated confidence level, based on the same data as the observed failure rate of nominally identical items. *Note:* (1) The source of the data shall be stated. (2) Results can be accumulated (combined) only when all conditions are similar. (3) The assumed underlying distribution of failures against time shall be stated. (4) It should be stated whether a one-sided or a two-sided interval is being used. (5) Where one limiting value is given this is usually the upper limit. 164

failure rate, extrapolated (reliability). Extension by a defined extrapolation or interpolation of the observed or assessed failure rate for durations and/or conditions different from those applying to the observed or assessed failure rate. *Note:* The validity of the extrapolation shall be justified.
164

failure rate, instantaneous. *See:* **instantaneous failure rate.**

failure rate level (reliability). For the assessed failure rate, a value chosen from a specific series of failure rate values and used for stating requirements or for the presentation of test results. *Note:* In a requirement, it denotes the highest permissible assessed failure rate. 164

failure rate, observed (reliability). For a stated period in the life of an item, the ratio of the total number of failures in a sample to the cumulative observed time on that sample. The observed failure rate is to be associated with particular, and stated time intervals (or summation of intervals) in the life of the items, and with stated conditions. *Note:* (1) The criteria for what constitutes a failure shall be stated. (2) Cumulative time is the sum of the times during which each individual item has been performing its required function under stated conditions. 164

failure rate, predicted (reliability). For the stated conditions of use, and taking into account the design of an item, the failure rate computed from the observed, assessed, or extrapolated failure rates of its parts. *Note:* Engineering and statistical assumptions shall be stated, as well as the bases used for the computation (observed or assessed). 164

failure-rate acceleration factor (reliability). The ratio of the accelerated testing failure rate to the failure rate under stated reference test conditions and time period. *See also:* **reliability.** 182

failure to trip (relay or relay system). In the performance the lack of tripping that should have occurred, considering the objectives of the relay system design. *See also:* **relay.** 281, 60, 127

fairlead (aircraft). A tube through which a trailing wire antenna is fed from an aircraft, with particular care in the design as to voltage breakdown and corona characteristics. *Note:* An antenna reel and counter are frequently a part of the assembly. *See also:* **air transportation electronic equipment.** 328

fall time (1) (electric indicating instruments). The time, in seconds, for the pointer to reach 0.1 (plus or minus a specified tolerance) of the end scale from a steady end-scale deflection when the instrument is short-circuited. *See also:* **moving element (instrument).** 280

(2) (high- and low-power pulse transformers) (last transition duration) (T_f). The time interval of the pulse trailing edge between the instants at which the instantaneous value first reaches specified upper and lower limits of 90 percent and 10 percent of A_T. *See also:* **input pulse shape.** 32, 33

fall time of a pulse. The time interval of the trailing edge of a pulse between stated limits. *See also:* **pulse.** 185

false alarm (1) (radar). An erroneous radar target detection decision caused by noise or other interfering signals exceeding the detection threshold. 13

(2) (test, measurement and diagnostic equipment). An indicated fault where no fault exists. 54

false-alarm number (radar). The average number of possible detection decisions during the false alarm time. 13

false alarm probability (radar). The probability that noise or other interfering signals will erroneously cause a target detection decision. 13

false-alarm time (radar). The average time between false alarms, that is, the average time between crossings of the target decision threshold by signals not representing targets. In the early work of Marcum, it is the time in which the probability of one or more false alarms is one-half, but the usage is deprecated. 13

false course (navigation systems normally providing one or more course lines). A spurious additional course line indication due to undesired reflections or to a maladjustment of equipment. 13, 187

false-proceed operation. The creation or continuance of a condition of the vehicle apparatus in an automatic train control or cab signal installation that is less restrictive than is required by the condition of the track of the controlling section, when the vehicle is at a point where the apparatus, is or should be, in operative relation with the controlling track elements. *See also:* **automatic train control.** 328

false-restrictive operation. The creation or continuance of a condition of the automatic train control or cab signal

vehicle apparatus that is more restrictive than is required by the condition of the track of the controlling section when the vehicle apparatus is in operative relation with the controlling track elements, or which is caused by failure or derangement of some part of the apparatus. *See also:* **automatic train control.** 328

false tripping (relay or relay system). In the performance the tripping that should not have occurred, considering the objectives of the relay system design. *See also:* **relay.** 281, 60, 127, 103

fan (blower) (rotating machinery). The part that provides an air stream for ventilating the machine. 63

fan-beam antenna. An antenna producing a major lobe whose transverse cross section has a large ratio of major to minor dimensions. 111

fan cover (rotating machinery). An enclosure for the fan that directs the flow of air. *See:* **fan (rotating machinery).** 63

fan-duty resistor (motors). A resistor that is intended for use in the armature or rotor circuit of a motor in which the current is approximately proportional to the speed of the motor. *See:* **asynchronous machine; direct-current commutating machine; electric drive.** 3, 225, 206

fan housing (rotating machinery). The structure surrounding a fan and which forms the outer boundary of the coolant gas passing through the fan. 63

fan-in network (power-system communication). A logic network whose output is a binary code in parallel form of n bits and having up to 2^n inputs with each input producing one of the output codes. *See also:* **digital.** 59

fan marker (electronic navigation). A marker having a vertically directed fan beam intersecting an airway to provide a position fix. *See also:* **radio navigation.** 13, 187

fan-out network (power-system communication). A logic network taking n input bits in parallel and producing a unique logic output on the one and only one of up to 2^n outputs that corresponds to the input code. *See also:* **digital.** 59

fan shroud (rotating machinery). A structure, either stationary or rotating, that restricts leakage of gas past the blades of a fan. 63

farad. The unit of capacitance in the International System of Electrical Units (SI units). The farad is the capacitance of a capacitor in which a charge of 1 coulomb produces 1 volt potential difference between its terminals. 210

faraday. The number of coulombs (96 485) required for an electrochemical reaction involving one chemical equivalent. *See also:* **electrochemistry.** 328

Faraday cell (laser gyro). A biasing device consisting of an optical material with a Verdet constant, such as quartz, which is placed between two quarter wave plates and surrounded by a magnetic field in such a fashion that a differential phase change is produced for oppositely directed plane polarized waves. 46

Faraday dark space (gas tube). The relatively nonluminous region in a glow-discharge cold-cathode tube between the negative flow and the positive column. *See:* **gas tubes.** 190

Faraday effect. *See:* **magnetic rotation (polarized light).**

Faraday rotation (1) (radio wave propagation). The process of rotation of the polarization ellipse of an electromagnetic wave in a magnetoionic medium. *See:* **elliptically polarized wave.** *See also:* **radio wave propagation.** 180

(2) (communication satellite). The rotation of the plane of polarization of an electromagnetic wave when traveling through a magnetic field. In space communications this effect occurs when signals transverse the ionosphere. 85

Faraday's law (electromagnetic induction; circuit). The electromotive force induced is proportional to the time rate of change of magnetic flux linked with the circuit. 210

faradic current (electrotherapy). An asymmetrical alternating current obtained from or similar to that obtained from the secondary winding of an induction coil operated by repeatedly interrupting a direct current in the primary. *See also:* **electrotherapy.** 192

faradization (faradism) (electrotherapy). The use of a faradic current to stimulate muscles and nerves. *See:* **faradic current.** *See also:* **electrotherapy.** 192

far-end crosstalk. Crosstalk that is propagated in a disturbed channel in the same direction as the direction of propagation of the current in the disturbing channel. The terminal of the disturbed channel at which the far-end crosstalk is present and the energized terminal of the disturbing channel are ordinarily remote from each other. *See also:* **coupling.** 328

far-field region in physical media. That region of the field of an antenna where the angular field distribution is essentially independent of the distance from a specified point in the antenna region. *Note:* If the antenna has a maximum overall dimension D which is large compared to $\pi/|\gamma|$, the far-field region can be taken to begin approximately at a distance equal to $|\gamma| D^2/\pi$ from the antenna, γ being the propagation constant in the medium. 111

fastener (lightning protection). A device used to secure the conductor to the structure that supports it. *See also:* **lighting protection and equipment.** 297

fast groove (disk recording) (fast spiral). An unmodulated spiral groove having a pitch that is much greater than that of the recorded grooves. *See also:* **phonograph pickup.** 176

fast-operate, fast-release relay. A high-speed relay specifically designed for both short operate and short release time. 341

fast-operate relay. A high-speed relay specifically designed for short operate time but not necessarily short release time. 341

fast-operate, slow-release relay. A relay specifically designed for short operate time and long release time. 341

fast spiral. *See:* **fast groove.**

fast-time-constant circuit (radar). A circuit with short time-constant used to emphasize signals of short duration to produce discrimination against extended clutter, long-pulse jamming, or noise. 13

fatigue. The tendency for a metal to fracture in brittle manner under conditions of repeated cyclic stressing at stress levels below its tensile strength. 221, 205, 103

fault (1) (wire or cable). A partial or total local failure in the insulation or continuity of a conductor. *See also:* **center of distribution.** 64

(2) (components). A physical condition that causes a device, a component, or an element to fail to perform in a required manner, for example, a short-circuit, a broken wire, an intermittent connection. *See:* **pattern-sensitive fault; program-sensitive fault.** 255, 77

(3) (surge arresters). A disturbance that impairs normal operation, for example, insulation failure or conductor breakage. 244, 62

(4) (thyristor power converter). A condition existing when the conduction cycles of some semiconductors are abnormal. *Note:* This usually results in fault currents of substantial magnitude. 121

(5) (power switchgear). *See:* **short circuit.**

(6) (test, measurement and diagnostic equipment). A degradation in performance due to detuning, maladjustment, misalignment, failure of parts, and so forth. 54

fault bus (fault ground bus). A bus connected to normally grounded parts of electric equipment, so insulated that all of the ground current passes to ground through fault-detecting means. 103, 202, 60, 127

fault-bus protection (relaying). A method of ground-fault protection which makes use of a fault bus. 60, 79

fault, circulating current (thyristor converter). A circulating current in excess of the design value. *Note:* In a double converter precaution must be taken to control circulating direct current between the forward and reverse sections. 121

fault current (1) (general). A current that flows from one conductor to ground or to another conductor owing to an abnormal connection (including an arc) between the two. *Note:* A fault current flowing to ground may be called a ground fault current. 64, 244, 62

(2) (surge arrester). The current from the connected power system that flows in a short circuit. 2

fault detection (test, measurement and diagnostic equipment). One or more tests performed to determine if any malfunctions or faults are present in a unit. 54

fault-detector relay. A monitoring relay whose function is to limit the operation of associated protective relays to specific system conditions. 103, 202, 60, 127

fault electrode current (1) (electron tubes) (surge electrode current*). The peak current that flows through an electrode under fault conditions, such as arc-backs and load short-circuits. *See also:* **electrode current (electron tube); gas-filled rectifier.** 244, 190, 125

* Deprecated

fault ground bus. *See:* **fault bus.**

fault incidence angle. The phase angle as measured between the instant of fault inception and a selected reference, such as the zero point on a current or voltage wave. 79

fault indicator (test, measurement and diagnostic equipment). A device which presents a visual display, audible alarm, and so forth, when a failure or marginal condition exists. 54

fault-initiating switch. A mechanical switching device used in applied-fault protection to place a short-circuit on an energized circuit and to carry the resulting current until the circuit has been deenergized by protective operation. *Notes:* (1) This switch is operated by a stored-energy mechanism capable of closing the switch within a specified rated closing time at its rated making current. The switch may be opened either manually or by a power-operated mechanism. (2) The applied short-circuit may be intentionally limited to avoid excessive system disturbance.

103, 202, 27

fault interrupter. A device for interrupting an electrical overcurrent. *Note:* in present usage this term refers to a non-reclosing self-controlled device for automatically interrupting an alternating current circuit. When applicable, it includes an assembly of control elements to detect overcurrents and control the fault interrupter. 79

fault isolation (test, measurement and diagnostic equipment). Tests performed to isolate faults within the unit under test.

54

fault resistance (surge arresters). The resistance of that part of the fault path associated with the fault itself.

244, 62

fault symptom (test, measurement and diagnostic equipment). A measurable or visible abnormality in an equipment parameter. 54

fault withstandability. The ability of electrical apparatus to withstand the effects of prescribed electrical fault current conditions without exceeding specified damage criteria. 70

Faure plate (storage cell) (pasted plate). A plate consisting of electroconductive material, which usually consists of lead-antimony alloy covered with oxides or salts of lead, that is subsequently transformed into active material. *See also:* **battery (primary or secondary).** 328

F display (radar). A rectangular display in which a target appears as a centralized blip when the radar antenna is aimed at it. *Note:* Horizontal and vertical aiming errors are respectively indicated by the horizontal and vertical displacement of the blip. *See also:* **navigation.** 13, 187

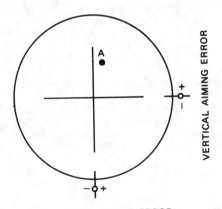

HORIZONTAL AIMING ERROR

F display.

FDNR. *See:* **frequency dependent negative resistor.** 67

feature (pulse terms). A specified portion or segment of, or a specified event in, a waveform. 254

feed (1) (machines). (A) To supply the material to be operated upon to a machine. (B) A device capable of feeding as in (A). 235

(2) (antenna). That portion of an antenna coupled to the·

terminals which functions to produce the aperture illumination. *Note:* A feed may consist of a distribution network or a primary radiator. 111

feedback (transmission system or section thereof). The returning of a fraction of the output to the input.

111, 240

feedback admittance, short-circuit (electron-device transducer). *See:* **admittance, short-circuit feedback.**

feedback control system. *See:* **control system, feedback.**

feedback limiter. *See:* **limiter circuit.** *See also:* **analog computer.** 9, 77

feedback loop (numerically controlled machines). The part of a closed-loop system that provides controlled response information allowing comparison with a referenced command. 224, 207

feedback node (branch) (network analysis). A node (branch) contained in a loop. *See also:* **linear signal flow graphs.**

282

feedback oscillator. An oscillating circuit, including an amplifier, in which the output is coupled in phase with the input, the oscillation being maintained at a frequency determined by the parameters of the amplifier and the feedback circuits such as inductance-capacitance, resistance-capacitance, and other frequency-selective elements. *See also:* **oscillatory circuit.** 111

feedback signal (control) (industrial control). *See:* **signal, feedback.**

feedback winding (saturable reactor). A control winding to which a feedback connection is made. 328

feeder (power distribution). A set of conductors originating at a main distribution center and supplying one or more secondary distribution centers, one or more branch-circuit distribution centers, or any combination of these two types of equipment. *Notes:* (1) Bus tie circuits between generator and distribution switchboards, including those between main and emergency switchboards, are not considered as feeders. (2) Feeders terminate at the overcurrent device that protects the distribution center of lesser order.

3

feeder cable (communication practice). A cable extending from the central office along a primary route (main feeder cable) or from a main feeder cable along a secondary route (branch feeder cable) and providing connections to one or more distribution cables. *See also:* **cable.** 328

feeder distribution center. A distribution center at which feeders or subfeeders are supplied. *See also:* **distribution center.** 328

feeder reactor. A current-limiting reactor for connection in series with an alternating-current feeder circuit for the purpose of limiting and localizing the disturbance due to faults on the feeder. 53

feed function (numerically controlled machines). The relative velocity between the tool or instrument and the work due to motion of the programmed axis (axes).

224, 207

feed groove (rotating machinery). A groove provided to direct the flow of oil in a bearing. *See also:* **bearing.**

63

feeding coefficients. *See:* **excitation coefficients.**

feeding point. The point of junction of a distribution feeder with a distribution main or service connection. *See also:* **center of distribution.** 64

feed line (rotating machinery). A supply pipe line. *See also:* **oil cup (rotating machinery).** 63

feed rate bypass (numerically controlled machines). A function directing the control system to ignore programmed feed rate and substitute a selected operational rate. 224, 207

feed rate override (numerically controlled machines). A manual function directing the control system to modify the programmed feed rate by a selected multiplier. 224, 207

feedthrough power meter (1) (bolometric power meters). A power-measuring system in which the detector structure is inserted or incorporated in a waveguide or coaxial transmission line to provide a means for measuring (monitoring) the power flow through or beyond the system. 115

(2) (electrothermic power meters) (measuring system). A device which is inserted or incorporated in a waveguide or transmission line and provides a means for measuring (monitoring) the power flow through or beyond the system. 47

feedthrough signal (dispersive and nondispersive delay lines). The undelayed signal resulting from direct coupling between the input and the output of the device. 81

feed tube (cable plowing). A tube attached to the blade of a plow which guides and protects the cable as it enters the earth. *See:* **fixed feed tube; hinged removable feed tube; floating removable feed tube.** 52

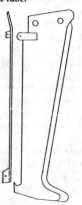

Feed blade with fixed feed tube.

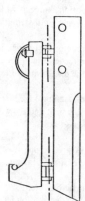

Feed blade and hinged feed tube.

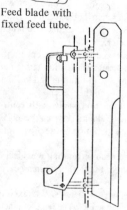

Feed blade with floating feed tube.

fenestra method (lighting calculation). The procedure for predicting the interior illumination received from daylight through windows. *See also:* **inverse-square law (illuminating engineering).** 167

fenestration. Any opening or arrangement of openings (normally filled with media for control) for the admission of daylight. *See also:* **sunlight.** 167

ferreed relay. Coined name (Bell Telephone Laboratories) for a special form of dry reed switch having a return magnetic path of high remanence material that provides a bistable, or latching, transfer contact. 341

ferritic. The body-centered cubic crystal structure of ferrous metals. 221, 205

ferrodynamic instrument. An electrodynamic instrument in which the forces are materially augmented by the presence of ferromagnetic material. *See also:* **instrument.** 328

ferroelectric axis. The crystallograph direction is parallel to the spontaneous polarization vector. *Note:* In some materials the ferroelectric axis may have several possible orientations with respect to the macroscopic crystal. *See also:* **ferroelectric domain.** 247

ferroelectric Curie point. The temperature T_C at which a ferroelectric material undergoes a structural phase transition to a state in which the spontaneous polarization vanishes in the absence of an applied electric field. *Note:* In a normal ferroelectric, the Curie point can be shifted by application of an external electric field, a mechanical stress or by doping with chemical impurities. *See:* **Curie-Weiss temperature; spontaneous polarization.** 80

ferroelectric Curie temperature. The temperature above which ferroelectric materials do not exhibit reversible spontaneous polarization. *Note:* As the temperature is lowered from above the ferroelectric Curie temperature spontaneous polarization is detected by the onset of a hysteresis loop. The ferroelectric Curie temperature should be determined only with unstrained crystals, at atmospheric pressure, and with no externally applied direct-current fields. (In some ferroelectrics multiple hysteresis-loop patterns may be observed at temperatures slightly higher than the ferroelectric Curie temperature under alternating-current fields.) *See also:* **ferroelectric domain.** 247

ferroelectric domain. A region of a crystal exhibiting homogeneous and uniform spontaneous polarization. *Note:* An unpoled ferroelectric material may exhibit a complex domain structure consisting of many domains, each with a different polarization orientation. The direction of the spontaneous polarization within each domain is constrained to a small number of equivalent directions (*see* **polar axis**) dictated by the symmetry of the crystal structure above the ferroelectric Curie point. The transition region between two ferroelectric domains is called a domain wall. Domains can usually be detected by pyroelectric, optical, powder decoration or electrooptic means. *See:* **coercive electric field; ferroelectric; ferroelectric Curie point; polar axis; poling; spontaneous polarization.** 80

ferroelectric material. A crystalline material that exhibits, over some range of temperature, a remanent polarization which can be reversed or reoriented by application of an external electric field. *Note:* The saturation remanent polarization is equal to the spontaneous polarization in a

single domain ferroelectric material. Since the spontaneous polarization in a ferroelectric material is strongly temperature dependent, poled ferroelectric materials exhibit a large pyroelectric effect near the Curie point. Ferroelectric materials also exhibit anomalies in small-signal dielectric permittivity, dielectric loss tangent, piezoelectric coefficients, and electrooptic coefficients near their ferroelectric Curie point. *See:* **antiferroelectric material; coercive electric field; Curie-Weiss temperature; ferroelectric Curie point; ferroelectric domain; paraelectric region; polar axis; polarization; poling; remanent polarization; small-signal permittivity; spontaneous polarization.**
80

ferromagnetic material. Material whose relative permeability is greater than unity and depends upon the magnetizing force. A ferromagnetic material usually has relatively high values of relative permeability and exhibits hysteresis.210

ferroresonance (transformer). A phenomenon usually characterized by overvoltages and very irregular wave shapes and associated with the excitation of one or more saturable inductors through capacitance in series with the inductor.53

ferrule (cartridge fuse). A fuse terminal of cylindrical shape at the end of a cartridge fuse.103, 202

fiber-optic plate (camera tubes). An array of fibers, individually clad with a lower index-of-refraction material, that transfers an optical image from one surface of the plate to the other. *See also:* **camera tube.**190

fibrillation (medical electronics). A continued, uncoordinated activity in the fibers of the heart, diaphragm, or other muscles consisting of rhythmical but asynchronous contraction and relaxation of individual fibers.192

fictitious power (polyphase circuit). At the terminals of entry, a vector equal to the (vector) sum of the fictitious powers for the individual terminals of entry. *Note:* The fictitious power for each terminal of entry is determined by considering each phase conductor and the common reference point as a single-phase circuit, as described for distortion power. The sign given to the distortion power in determining the fictitious power for each single-phase circuit shall be the same as that of the total active power. Fictitious power for a polyphase circuit has as its two rectangular components the reactive power and the distortion power. If the voltages have the same waveform as the corresponding currents, the magnitude of the fictitious power becomes the same as the reactive power. Fictitious power is expressed in volt-amperes when the voltages are in volts and the currents in amperes.210

fictitious power (single-phase two-wire circuit). At the two terminals of entry into a delimited region, a vector quantity having as its rectangular components the reactive power and the distortion power. *Note:* Its magnitude is equal to the square root of the difference of the squares of the apparent power and the amplitude of the active power. Its magnitude is also equal to the square root of the sum of the squares of the amplitudes of reactive power and distortion power. If voltage and current have the same waveform, the magnitude of the fictitious power is equal to the reactive power. The magnitude of the fictitious power is given by the equation

$$
\begin{aligned}
F &= (U^2 - p^2)^{1/2} \\
&= (Q^2 + D^2)^{1/2} \\
&= \left\{ \sum_{r=1}^{r=\infty} \sum_{q=1}^{q=\infty} \left[E_r^2 I_q^2 \right. \right. \\
&\quad \left. \left. - E_r E_q I_r I_q \cos (\alpha_r - \beta_r) \cos (\alpha_q - \beta_q) \right] \right\}^{1/2}
\end{aligned}
$$

where the symbols are those of **power, apparent (single-phase two-wire circuit).** In determining the vector position of the fictitious power, the sign of the distortion power component must be assigned arbitrarily. Fictitious power is expressed in volt-amperes when the voltage is in volts and the current in amperes. *See:* **distortion power (single-phase two-wire circuit).**210

fidelity. The degree with which a system, or a portion of a system, accurately reproduces at its output the essential characteristics of the signal that is impressed upon its input.111, 59

field (1) (television). One of the two (or more) equal parts into which a frame is divided in interlaced scanning. *See also:* **television.**178

(2) (record) (computing systems). A specified area used for a particular category of data, for example, a group of card columns used to represent a wage rate or a set of bit locations in a computer word used to express the address of the operand. *See:* **threshold field.**255, 77

field accelerating relay (industrial control). A relay that functions automatically to maintain the armature current within limits, when accelerating to speeds above base speed, by controlling the excitation of the motor field.
206

field application relay. A relay used to apply the excitation to the field of a synchronous machine. *Note:* This is usually a polarized relay sensitive to the slip frequency induced in the field. It may also remove excitation during an out-of-step condition.127, 103

field bar (waveform test signals) (TV). A composite pulse, nominally of 8 ms duration, of reference-white amplitude. The field bar is composed of line bars as defined.42

field coil (rotating machinery) (1) (direct-current and salient-pole alternating-current machines). A suitably insulated winding to be mounted on a field pole to magnetize it.

(2) (cylindrical-rotor synchronous machines). A group of turns in the field winding, occupying one pair of slots. *See:* **asynchronous machine; direct-current commutating machine.**63

field-coil flange (rotating machinery). Insulation between the field coil and the pole shoe, and between the field coil and the member carrying the pole body, in a salient-pole machine. *See also:* **rotor (rotating machinery); stator.**
63

field contacts (sequential events recording systems). Electrical contacts which define the state of monitored equipment or a process.48, 58

field contact voltage (sequential events recording systems). The voltage applied to field contacts for the purpose of sensing contact status.48, 58

field control (motor) (industrial control). A method of controlling a motor by means of a change in the magnitude of the field current. *See:* **control.**   219, 206

field, critical (magnetrons). *See:* **critical field.**

field, cutoff (magnetrons). *See:* **critical field.**

field data (reliability). Data from observations during field use. *Note:* The time stress conditions, and failure or success criteria should be stated in detail.   164

field decelerating relay (industrial control). A relay that functions automatically to maintain the armature current or voltage within limits, when decelerating from speeds above base speed, by controlling the excitation of the motor field. *See also:* **relay.**   206

field discharge (switching device). A qualifying term indicating that the switching device has main contacts for energizing and deenergizing the field of a generator, motor, synchronous capacitor, or exciter; and has auxiliary contacts for short-circuiting the field through a discharge resistor at the instant preceding the opening of the main contacts. The auxiliary contacts also disconnect the field from the discharge resistor at the instant following the closing of the main contacts. *Note:* For direct-current generator operation the auxiliary contacts may open before the main contacts close.   103, 202

field discharge protection (industrial control). A control function or device to limit the induced voltage in the field when the field circuit is disrupted or when an attempt is made to change the field current suddenly. *See:* **control.**   219, 206

field emission. Electron emission from a surface due directly to high-voltage gradients at the emitting surface. *See also:* **electron emission.**   190, 125

field-enhanced photoelectric emission. The increased photoelectric emission resulting from the action of a strong electric field on the emitter. *See:* **phototubes.**   190, 125

field-enhanced secondary emission. The increased secondary emission resulting from the action of a strong electric field on the emitter. *See:* **electron emission.**   190, 125

field excitation current (Hall-effect devices). The current producing the magnetic flux density in a Hall multiplier.   107

field-failure protection (industrial control). The effect of a device, operative on the loss of field excitation, to cause and maintain the interruption of power in the motor armature circuit.   206

field-failure relay (industrial control). A relay that functions to disconnect the motor armature from the line in the event of loss of field excitation. *See also:* **relay.**   206

field forcing (industrial control). A control function that temporarily overexcites or underexcites the field of a rotating machine to increase the rate of change of flux. *See:* **control.**   225, 206

field forcing relay (industrial control). A relay that functions to increase the rate of change of field flux by underexciting or overexciting the field of a rotating machine. *See also:* **relay.**   206

field frame. *See:* **frame yoke.**

field-free emission current (1) (general). The emission current from an emitter when the electric gradient at the surface is zero.   244, 190

(2) (cathode). The electron current drawn from the cathode when the electric gradient at the surface of the cathode is zero. *See also:* **electron emission.**   190, 125

field frequency (television). The product of frame frequency multiplied by the number of fields contained in one frame. *See also:* **television.**   178

field I^2R loss. The product of the measured resistance, in ohms, of the field winding, corrected to a specified temperature, and the square of the field current in amperes.   244, 63

field-intensity meter* (electric process heating). A calibrated radio receiver for measuring field intensity. *See:* **interference; interference measurement; field-strength meter.**   14

* Deprecated

field-lead insulation (rotating machinery). The dielectric material applied to insulate the enclosed conductor connecting the collector rings to the coil end windings. *Note:* Field leads also include the pole jumpers forming the series connection between the concentric windings on each pole. Where rectangular strap leads are employed, the insulation may consist of either taped mica and glass or moduled mica and glass composites. Where circular rods are used, moulded laminate tubing is frequently employed as the primary insulation. *See also:* **asynchronous machine; collector-ring lead insulator; direct-current commutating machine.**   63

field-limiting adjusting means (industrial control). The effect of a control function or device (such as a resistor) that limits the maximum or minimum field excitation of a motor or generator. *See:* **control.**   225, 206

field loss relay (industrial control). *See:* **motor-field failure relay.**

field molded joint (power cable joints). A joint in which the solid-dielectric joint insulation is fused and curved thermally at the job site.   34

field pole (rotating machinery). A structure of magnetic material on which a field coil may be mounted. *Note:* There are two types of field poles: main and commutating. *See also:* **asynchronous machine; direct-current commutatin machine.**   328

field protection (industrial control). The effect of a control function or device to prevent overheating of the field excitation winding by reducing or interrupting the excitation of the shunt field while the machine is at rest. *See:* **control.**   225, 206

field protective relay (industrial control). A relay that functions to prevent overheating of the field excitation winding by reducing or interrupting the excitation of the shunt field. *See also:* **relay.**   206

field-reliability test. A reliability compliance or determination test made in the field where the operating and environmental conditions are recorded and the degree of control is stated.   164

field-renewable fuse. *See:* **renewable fuse.**

field-reversal permanent-magnet focusing (microwave tubes). Magnetic focusing by a limited series of field reversals, not periodic, whose location is usually related to breaks in the slow-wave circuit. *See:* **magnetrons.**   190

field rheostat. A rheostat designed to control the exciting current of an electric machine.   206

field sequential (color television). Sampling of primary colors in sequence with successive television fields.   18

field shunting control (shunted-field control). A system of regulating the tractive force of an electrically driven vehicle by shunting, and thus weakening, the traction motor series fields by means of a resistor. *See also:* **multiple-unit control.**   328

field splice (nuclear power generating stations). A permanent joining and reinsulating of conductors in the field to meet the service conditions required.   31, 141

field spool (rotating machinery). A structure for the support of a field coil in a salient-pole machine, either constructed of insulating material or carrying field-spool insulation. *See also:* **asynchronous machine; direct-current commutating machine.**   63

field-spool insulation (rotating machinery). Insulation between the field spool and the field coil in a salient-pole machine. *See also:* **asynchronous machine; direct-current commutating machine.**   63

field strength (electromagnetic wave). A general term that usually means the magnitude of the electric field vector, commonly expressed in volts per meter, but that may also mean the magnitude of the magnetic field vector, commonly expressed in amperes (or ampere-turns) per meter. *Note:* At frequencies above about 100 megahertz, and particularly above 1000 megahertz, field strength in the far zone is sometimes identified with power flux density P. For a linearly polarized wave in free space $P = E^2/(\mu_v/\epsilon_v)$, where E is the electric field strength, and μ_v and ϵ_v are the magnetic and electric constants of free space, respectively. When P is expressed in watts per square meter and E in volts per meter, the denominator is often rounded off to 120π. *See:* **electric field strength; magnetic field strength.** *See also:* **measurement system.**
                                                                                213, 193

field-strength meter. A calibrated radio receiver for measuring field strength. *See also:* **induction heater; intensity meter.**   14, 114

field system (rotating machinery). The portion of a direct-current or synchronous machine that produces the excitation flux. *See also:* **asynchronous machine; direct-current commutating machine.**   63

field terminal (rotating machinery). A termination for the field winding. *See also:* **rotor (rotating machinery); stator.**
                                                                                                          63

field tests (1) (cable systems). Tests which may be made on a cable system [including the high-voltage cable termination(s)] by the user after installation, as an acceptance or proof test.   4

(2) (power cable joints). Tests which may be made on the cable and accessories after installation.   34

(3) (metal enclosed bus). Tests made after the assembly has been installed at its place of utilization.   78

(4) (switchgear). Tests made on operating systems usually for the purpose of investigating the performance of switchgear or its component parts under conditions that cannot be duplicated in the factory. *Note:* Field tests are usually supplementary to factory tests and therefore may not provide a complete investigation of capabilities.
                                                                                103, 202

field-time waveform distortion (FD) (video signal transmission measurement). The linear TV waveform distortion of time components from 64 μs to 16 ms, that is, time components of the field-time domain.   42

field-turn insulation (rotating machinery). Insulation in the form of strip or tape separating the individual turns of a

field winding. *See also:* **asynchronous machine; direct-current commutating machine.**   63

field voltage, base. The synchronous machine field voltage required to produce rated voltage on the air-gap line of the synchronous machine at field temperatures of (1) 75°C for field windings designed to operate at rating with a temperature rise of 60°C or less; or (2) 100°C for field windings designed to operate at rating with a temperature rise greater than 60°C. *Note:* This defines one per unit excitation system voltage for use in computer representation of excitation systems.   105

field voltage, no-load. The voltage required across the terminals of the field winding of an electric machine under conditions of no load, rated speed and terminal voltage, and with the field winding at 25°C.   105

field voltage, rated-load. The voltage required across the terminals of the field winding of an electric machine under rated continuous-load conditions with the field winding at (1) 75°C for field windings designed to operate at rating with a temperature rise of 60°C or less; or (2) 100°C for field windings designed to operate at rating with a temperature rise greater than 60°C.   105

field winding (excitation winding) (rotating machinery). A winding on either the stationary or the rotating part of a synchronous machine whose sole purpose is the production of the main electromagnetic field of the machine.   63

fifth voltage range. *See:* **voltage range.**

fifty percent disruptive-discharge voltage (dielectric tests). The voltage that has a 50 percent probability of producing a disruptive discharge. The term applies mostly to impulse tests and has significance only in cases where the loss of dielectric strength resulting from a disruptive discharge is temporary.   150

figure of merit (1) (magnetic amplifier). The ratio of power amplification to time constant in seconds.   171

(2) (thermoelectric couple).

$$\alpha^2[(\rho_1\kappa_1)^{1/2} + (\rho_2\kappa_2)^{1/2}]^{-2}$$

where α is the Seebeck coefficient of the couple and ρ_1, ρ_2, κ_1, and κ_2 are the respective electric resistivities and thermal conductivities of materials 1 and 2. *Note:* This figure of merit applies to materials for the thermoelectric devices whose operation is based on the Seebeck effect or the Peltier effect. *See also:* **thermoelectric device.**

(3) (thermoelectric couple, ideal).

$$\overline{\alpha}^{-2}[(\overline{\rho_1\kappa_1})^{1/2} + (\overline{\rho_2\kappa_2})^{1/2}]^{-2}$$

where $\overline{\alpha}$ is the average value of the Seebeck coefficient of the couple and $\rho_1\kappa_1$ and $\rho_2\kappa_2$ are the average values of the products of the respective electric resistivities and thermal conductivities of materials 1 and 2, where the averages are found by integrating the parameters over the specified temperature range of the couple. *See also:* **thermoelectric device.**

(4) (thermoelectric material). The quotient of (A) the square of the absolute Seebeck coefficient α by (B) the product of the electric resistivity ρ and the thermal conductivity κ.

$$\alpha^2/\rho\kappa.$$

Note: This figure of merit applies to materials for ther-

moelectric devices whose operation is based on the Seebeck effect or the Peltier effect. *See also:* **thermoelectric device.**
                                                            248, 191

filament (electron tube). A hot cathode, usually in the form of a wire or ribbon, to which heat may be supplied by passing current through it. *Note:* This is also known as a filamentary cathode. *See also:* **electrode (electron tube).**
                                                            190, 125

filament current. The current supplied to a filament to heat it. *See also:* **electronic controller; heater current.**   190

filament power supply (electron tube). The means for supplying power to the filament. *See also:* **power pack.**
                                                                    190

filament voltage. The voltage between the terminals of a filament. *See also:* **electrode voltage (electron tube); electronic controller.**   328

file (computing systems). A collection of related records treated as a unit. *Note:* Thus in inventory control, one line of an invoice forms an item, a complete invoice forms a record, and the complete set of such records forms a file.
                                                            255, 77

file gap (computing systems). An area on a storage medium, such as tape, used to indicate the end of a file.   255,777

file maintenance (computing systems). The activity of keeping a file up to date by adding, changing, or deleting data.   255, 77

filiform corrosion. *See:* **underfilm corrosion.**

filled-core annular conductor. A conductor composed of a plurality of conducting elements disposed around a nonconducting supporting material that substantially fills the space enclosed by the conducting elements. *See also:* **conductor.**   64

filled-system thermometer. An all-metal assembly consisting of a bulb, capillary tube, and Bourdon tube (bellows and diaphragms are also used) containing a temperature-responsive fill. A mechanical device associated with the Bourdon is designed to provide an indication or record of temperature. *See:* **Bourdon.**   7

filled tape. Fabric tape that has been thoroughly filled with a rubber or synthetic compound, but not necessarily finished on either side with this compound. *See also:* **conductor.**   64

filler (filler strip) (1) (rotating machinery). Additional insulating material used to insure a tight depth-wise fit in the slot. *See also:* **rotor (rotating machinery); stator.**
                                                                    63

(2) (mechanical recording). The inert material of a record compound as distinguished from the binder. *See also:* **phonograph pickup.**   176

filler strip. *See:* **filler.**

filling compound (power cable joints). A dielectric material poured or otherwise injected into the joint housing. Filling compounds may require heating or mixing prior to filling. Some filling compounds may also serve as the insulation.
                                                                    34

fill light (television). Supplementary illumination to reduce shadow or contrast range. *See also:* **television lighting.**
                                                                    167

film (1) (rotating machinery). Sheeting having a nominal thickness not greater than 0.030 centimeters and being substantially homogeneous in nature. *See also:* **asynchronous machine; direct-current commutating machine;**

electrochemical valve.   63

(2) (electrochemical valve). The layer adjacent to the valve metal and in which is located the high-potential drop when current flows in the direction of high impedance. *See also:* electrochemical valve.   328

film integrated circuit. An integrated circuit whose elements are films formed in situ upon an insulating substrate. *Note:* To further define the nature of a film integrated circuit, additional modifiers may be prefixed. Examples are (1) thin-film integrated circuit, (2) thick-film integrated circuit. *See:* **magnetic thin film; thin film.** *See also:* **electrochemical valve; integrated circuit.**
                                                            312, 191

filter (1) (wave filter). A transducer for separating waves on the basis of their frequency. *Note:* A **filter** introduces relatively small insertion loss to waves in one or more frequency bands and relatively large insertion loss to waves of other frequencies.   239

(2) (computing systems). (A) A device or program that separates data, signals, or material in accordance with specified criteria. (B) A mask.   255, 77

filter, active (circuits and systems). (1) A filter network containing one or more voltage-dependent or current-dependent sources in addition to passive elements. (2) A filter containing energy generating elements.   67

filter, all-pass (circuits and systems). A filter designed to introduce phase shift or delay over a band of frequencies without introducing appreciable attenuation distortion over those frequencies.   67

filter attenuation band (filter stop band) (circuits and systems). A continuous range of frequencies over which the filter introduces an insertion loss whose minimum value is greater than a specified value.   67

filter, band-elimination (1) (signal-transmission system). A filter that has a single attenuation band, neither of the cutoff frequencies being zero or infinite. *See also:* **rejection filter; filter.**   59, 239

(2) (circuits and systems). A network designed to eliminate a band of frequencies. Its frequency response has a single attenuation band bounded by two pass bands.   67

filter, band-pass (1) (general). A filter that has a single transmission band, neither of the cutoff frequencies being zero or infinite. *See also:* **filter.**   239

(2) (circuits and systems). A filter designed to pass a band or frequencies. Its frequency response has a single pass band bounded by two attenuation bands.   67

filter, Butterworth (circuits and systems). A filter whose pass-band frequency response has a maximally flat shape brought about by the use of Butterworth polynomials as the approximating function.   67

filter capacitor. A capacitor used as an element of an electric wave filter. *See also:* **electronic controller.**
                                                                    206

filter, Chebyshev (circuits and systems). A filter whose pass-band frequency response has an equal-ripple shape brought about by the use of Chebyshev cosine polynomials as the approximating function.   67

filter, comb (circuits and systems). A filter whose insertion loss forms a sequence of narrow pass bands or narrow stop bands centered at multiples of some specified frequency.
                                                                    67

filter factor (illuminating engineering). The transmittance of black light by a filter. *Note:* The relationship among these terms is illustrated by the following formula for determining the luminance (photometric brightness) of fluorescent materials exposed to black light:

$$\text{footlamberts} = \frac{\text{fluorens}}{\text{square feet}}$$

$\times$ glow factor $\times$ filter factor.

When integral-filter black-light lamps are used, the filter factor is dropped from the formula because it already has been applied in assigning fluoren ratings to these lamps. *See also:* **ultraviolet radiation.**   167

filter, high-pass. A filter having a single transmission band extending from some cutoff frequency, not zero, up to infinite frequency. *See also:* **filter.**   239

filter impedance compensator. An impedance compensator that is connected across the common terminals of electric wave filters when the latter are used in parallel in order to compensate for the effects of the filters on each other. *See also:* **network analysis; filter.**   210

filter inductor. An inductor used as an element of an electric wave filter. *See also:* **electronic controller.**
                                                                      206

filter, low-pass. A filter having a single transmission band extending from zero to some cutoff frequency, not infinite. *See also:* **filter.**   239

filter pass band (circuits and systems). A frequency band of low attenuation (low relative to other regions termed stop bands). *See also:* **filter transmission band.**   67

filter, passive (circuits and systems). A filter network containing only passive elements, such as inductors, capacitors, resistors and transformers.   67

filter reactor (transformers). A reactor used to reduce harmonic voltage in alternating-current or direct-current circuits.   53

filter, rejection (signal-transmission system). *See:* **rejection filter.** *See also:* **filter.**

filters (power supplies). Resistance-capacitance or inductance-capacitance networks arranged as low-pass devices to attenuate the varying component that remains when alternating-current voltage is rectified. *Note:* In power supplies without subsequent active series regulators, the filters determine the amount of ripple that will remain in the direct-current output. In supplies with active feedback series regulators, the regulator mainly controls the ripple, with output filtering serving chiefly for phase-gain control as a lag element. *See also:* **power supply.**   228, 186

filter, sound effects (electroacoustics). A filter used to adjust the frequency response of a system for the purpose of achieving special aural effects. *See:* **filter.**   239

filter stop band (circuits and systems). A frequency band of high attenuation (high relative to other regions termed pass bands). *See also:* **filter attenuation band.**   67

filter transmission band (circuits and systems). A continuous range of frequencies over which the filter introduces an insertion loss whose maximum value does not exceed a specified value. *See also:* **filter pass band.**   67

final approach path (electronic navigation). *See:* **approach path.**

final contact pressure. *See:* **contact pressure, final.**

final controlling element (electric power systems). The

controlling element that directly changes the value of the manipulated variable.   94

final emergency circuits. All circuits (including temporary emergency circuits) that, after failure of a ship's service supply, may be supplied by the emergency generator. *See also:* **emergency electric system.**   328

final emergency lighting. Temporary emergency lighting plus manually controlled lighting of the boat deck and overside to facilitate lifeboat loading and launching. *See also:* **emergency electric system.**   328

final sag (transmission and distribution). The sag of a conductor under specified conditions of loading and temperature applied after it has been subjected for an appreciable period to the loading prescribed for the loading district in which it is situated, or equivalent loading, and the loading removed. Final sag includes the effect of inelastic deformation (creep).   262

final-terminal stopping device (elevators). A device that automatically causes the power to be removed from an electric elevator or dumbwaiter driving-machine motor and brake or from a hydraulic elevator or dumbwaiter machine independent of the functioning of the normal-terminal stopping device, the operating device, or any emergency terminal stopping device, after the car has passed a terminal landing. *See:* **control.**   328

final trunk (data transmission). A group of trunks to the higher class office which has no alternate route, and in which the number of trunks provided results in a low probability of calls encountering "all trunks busy."   59

final-trunk group (telephone switching systems). A trunk group in the last choice position of a hierarchical network of alternate-route trunk groups.   55

final unloaded conductor tension (electric systems). The longitudinal tension in a conductor after the conductor has been stretched by the application for an appreciable period, and subsequent release, of the loadings of ice and wind, and temperature decrease, assumed for the loading district in which the conductor is strung (or equivalent loading). *See also:* **conductor; initial conductor tension.**
                                                                      178

final unloaded sag (1) (general). The sag of a conductor after it has been subjected for an appreciable period to the loading prescribed for the loading district in which it is situated, or equivalent loading, and the loading removed. *See:* **sag.**   178

(2) (transmission and distribution). The sag of a conductor after it has been subjected for an appreciable period to the loading prescribed for the loading district in which it is situated, or equivalent loading, and the loading removed. Final unloaded sag includes the effect of inelastic deformation (creep).   262

final value (industrial control). The steady-state value of a specified variable. *See:* **control.**   219, 206

final voltage. *See:* **cutoff voltage.**

finder switch. An automatic switch for finding a calling subscriber line or trunk and connecting it to the switching apparatus.   328

finding (telephone switching systems). Locating a circuit requesting service.   55

fine chrominance primary.* (NTSC color television). *See:* I chrominance signal.   18

* Obsolete

fines (cable plowing). Particles of earth or rock smaller than 1/8″ in greatest dimension. 52

finger (rotating machinery). *See:* **end finger.**

finishing (electrotype). The operation of bringing all parts of the printing surface into the same plane, or, more strictly speaking, into positions having equal printing values. *See also:* **electroforming.** 328

finishing rate (storage battery) (storage cell). The rate of charge expressed in amperes to which the charging current for some types of lead batteries is reduced near the end of charge to prevent excessive gassing and temperature rise. *See also:* **charge.** 328

finite energy signal. *See:* **energy density signal.**

finite-time stability. *See:* **stability, finite-time.**

finsen. The recommended practical unit of erythemal flux or intensity of radiation. It is equal to one unit of erythemal flux per square centimeter. 328

fire-alarm system. An alarm system signaling the presence of fire. *See also:* **protective signaling.** 328

fire-control radar. A radar whose prime function is to provide information for the manual or automatic control of artillery or other weapons. 13

fire-door magnet. An electromagnet for holding open a self-closing fire door. 328

fire-door release system. A system providing remotely controlled release of self-closing doors in fire-resisting bulkheads to check the spread of fire. *See also:* **marine electric apparatus.** 328

fired tube (microwave gas tubes). The condition of the tube during which a radio-frequency glow discharge exists at either the resonant gap, resonant window, or both. *See:* **gas tubes.** 125, 190

fireproofing (cables). The application of a fire-resisting covering to protect them from arcs in an adjacent cable or from fires from any cause. *See also:* **power distribution, underground construction.** 64

fire-resistance rating. The measured time, in hours or fractions thereof, that the material or construction will withstand fire exposure as determined by fire tests conducted in conformity to recognize standards. 328

fire-resistant. So constructed or treated that it will not be injured readily by exposure to fire. *Syn:* **fire-resistive.** 328

fire-resistive construction. A method of construction that prevents or retards the passage of hot gases or flames as defined by the fire-resistance rating. 328

firing angle (semiconductor rectifier operating with phase control). *See:* **angle of retard (semiconductor rectifier operating with phase control).** 66

firm capacity purchases or sales (electric power supply). That firm capacity that is purchased, or sold, in transactions with other systems and that is not from designated units, but is from the over-all system of the seller. *Note:* It is understood that the seller provides reserve capacity for this type of transaction. *See also:* **generating station.** 200

firm power. Power intended to be always available even under emergency conditions. *See also:* **generating station.** 64

firm transfer capability (transmission) (electric power supply). The maximum amount of power that can be inter-

changed continuously, over an extended period of time. *See also:* **generating station.** 200

first dial (register). That graduated circle or cyclometer wheel, the reading on which changes most rapidly. The test dial or dials, if any, are not considered. *See also:* **watthour meter.** 328

first Fresnel zone (optics and radio communication). The circular portion of a wave front, transverse to the line between an emitter and a more-distant point where the resultant disturbance is being observed, whose center is the intersection of the front with the direct ray and whose radius is such that the shortest path from the emitter through the periphery to the receiving point is one half-wave longer than the ray. *Note:* A second zone, a third, etcetera, are defined by successive increases of the path by half-wave increments. *See also:* **radiation.** 59

first (last) transition duration (pulse terms). The transition duration of the first (last) transition waveform in a pulse waveform. *See:* The single pulse diagram below the **waveform epoch** entry. 254

first-line release (telephone switching systems). Release under the control of the first line that goes on-hook. 55

first Townsend discharge (gas). A semi-self-maintained discharge in which the additional ions that appear are due solely to the ionization of the gas by electron collisions. *See also:* **discharge (gas).** 244, 190

first transition (pulse terms). The major transition waveform of a pulse waveform between the base and the top. 254

first voltage range. *See:* **voltage range.**

fishbone antenna. An antenna consisting of a series of end fed coplanar dipole antennas closely coupled to a balanced transmission line through lumped circuit elements. 111

fish tape (fishing wire) (snake). A tempered steel wire, usually of rectangular cross section, that is pushed through a run of conduit or through an inaccessible space, such as a partition, and that is used for drawing in the wires. 328

fission. The splitting of a nucleus into parts (which are nuclei of lighter elements), accompanied by the release of a relatively large amount of energy (about 200 million electron volts per fission in the case of ^{235}U fission) and frequently one or more neutrons. 112

fitting (electric system). An accessory such as a locknut, bushing, or other part of a wiring system that is intended primarily to perform a mechanical rather than an electrical function. *See also:* **raceways.** 256

fix (1) (navigation). A position determined without reference to any former position. 13

(2) (interference) (electromagnetic compatibility). A device or equipment modification to prevent interference or to reduce an equipment's susceptibility to interference. *See also:* **electromagnetic compatibility.** 199

fixed block format (numerically controlled machines). A format in which the number and sequence of words and characters appearing in successive blocks is constant. 224, 207

fixed-called-address line (telephone switching systems). A line for originating calls to a fixed called address. 55

fixed cycle (numerically controlled machines). A preset

series of operations that direct machine axis movement and/or cause spindle operation to complete such actions as boring, drilling, tapping, or combinations thereof.

        224, 207

fixed-cycle operation. An operation that is completed in a specified number of regularly timed execution cycles.

        255, 77

fixed feed tube (cable plowing). A feed tube permanently attached to a blade. It may have a removable back plate. *See:* **feed tube.**   52

fixed-frequency transmitter. A transmitter designed for operation on a single carrier frequency. *See:* **radio transmitter.**   111, 240, 59

fixed impedance-type ballast. A reference ballast designed for use with one specific type of lamp and, after adjustment during the original calibration, is expected to hold its established impedance throughout normal use.   271

fixed light. A light having a constant luminous intensity when observed from a fixed point.   167

fixed motor connections. A method of connecting electric traction motors wherein there is no change in the motor interconnections throughout the operating range. *Note:* This term is used to indicate that a transition from series to parallel relation is not provided. *See also:* **traction motor.**   328

fixed point (electronic computation). Pertaining to a numeration system in which the position of the point is fixed with respect to one end of the numerals, according to some convention. *See:* **floating point, variable point.**

        235, 255, 77, 54

fixed-point system (electronic computation). *See:* **point.**

fixed sequential format. A means of identifying a word by its location in the block. *Note:* Words must be presented in a specific order and all possible words preceding the last desired word must be present in the block.   224, 207

fixed signal. A signal of fixed location indicating a condition affecting the movement of a train or engine.

        328

fixed storage (computing systems). A storage device that stores data not alterable by computer instructions, for example, magnetic core storage with a lockout feature or punched paper tape. *See:* **nonerasable storage; permanent storage; read-only storage.**   255, 77

fixed temperature heat detector (fire protection devices). A device which will respond when its operating element becomes heated to a predetermined level.   71

fixed transmitter. A transmitter that is operated in a fixed or permanent location. *See also:* **radio transmitter.**

        111, 240

fixed word length (test, measurement and diagnostic equipment). Property of a storage device in which the capacity for bits in each storage word is fixed.   54

fixing (electrostatography). The act of making a developed image permanent. *See also:* **electrostatography.**

        236, 191

fixture stud (stud). A threaded fitting used to mount a lighting fixture to an outlet box. *See also:* **cabinet.**

        328

flag (computing systems). (1) Any of various types of indicators used for identification, for example, a wordmark. (2) A character that signals the occurrence of some condition, such as the end of a word.   255, 77

flag alarm. An indicator in certain types of navigation instruments used to warn when the readings are unreliable. *See also:* **navigation.**   13, 187

flame detector (fire protection devices). A device which detects the infrared, or ultraviolet, or visible radiation produced by a fire.   71

flame flicker detector (fire protection devices). A photoelectric flame detector including means to prevent response to visible light unless the observed light is modulated at a frequency characteristic of the flicker of a flame.   71

flameproof. *See:* **explosionproof.**

flameproof apparatus. Apparatus so treated that it will not maintain a flame or will not be injured readily when subjected to flame.   328

flameproof terminal box. A terminal box so designed that it may form part of a flameproof enclosure.   63

flame protection of vapor openings. Self-closing gauge hatches, vapor seals, pressure-vacuum breather valves, flame arresters, or other reasonably effective means to minimize the possibility of flame entering the vapor space of a tank. *Note:* Where such a device is used, the tank is said to be flameproofed.   297

flame-resistant cable. A portable cable that will meet the flame test requirements of the United States Bureau of Mines. *See also:* **mine feeder circuit.**   328

flame resisting. *See:* **flame retarding.**

flame-retardant (1) (general). So constructed or treated that it will not support or convey flame.   103, 202

(2) **(nuclear power generating stations).** Capable of preventing the propagation of a fire beyond the area of influence of the energy source that initiated the fire.   31

flame-retarding (flame resisting). Flame-retarding materials and structures should have such fire-resisting properties that they will not convey flame nor continue to burn for longer times than specified in the appropriate flame test. *Note:* Compliance with the requirements of the preceding paragraph should be determined with apparatus and according to the methods described in the Underwriters' Laboratory Standards for the materials and structures.   3

flammable air-vapor mixtures. When flammable vapors are mixed with air in certain proportions, the mixture will burn rapidly when ignited. *Note:* The combustion range for ordinary petroleum products, such as gasoline, is from $1\frac{1}{2}$ to 6 percent of vapor by volume, the remainder being air.   297

flammable vapors. The vapors given off from a flammable liquid at and above its flash point.   297

flange. *See:* **coupling flange.**

flange, choke (waveguide components). A flange designed with auxiliary transmission line elements to form a choke joint when used with a cover flange.   166

flange, contact (waveguide components). A flat flange used in conjunction with another flat flange to provide a contact joint.   166

flange, cover (waveguide components). A flat flange used in conjunction with a choke flange to provide a choke joint.   166

flange, flat. *See:* **flange, cover.**

flange, plane. *See:* **flange, cover.**

flare-out (navigation). The portion of the approach path of an aircraft in which the slope is modified to provide the appropriate rate of descent at touchdown. *See also:* **navigation.** 187, 13

flarescan (electronic navigation). A ground-based navigation system used in conjunction with an instrument approach system to provide flare-out vertical guidance to an aircraft by the use of a pulse-space-coded vertically scanning fan beam that provides elevation angle data. *See also:* **navigation.** 187, 13

flash barrier (rotating machinery). A screen of fire-resistant material to prevent the formation of an arc or to minimize the damage caused thereby. 63

flash current (primary cell). The maximum electric current indicated by an ammeter of the dead-beat type when connected directly to the terminals of the cell or battery by wires that together with the meter have a resistance of 0.01 ohm. *See also:* **electrolytic cell.** 328

flasher. A device for alternately and automatically lighting and extinguishing electric lamps. *See also:* **appliances (including portable).** 328

flasher relay. A relay that is so designed that when energized its contacts open and close at predetermined intervals. *See also:* **appliances (including portable).** 328

flashing light. A rhythmic light in which the periods of light are of equal duration and are clearly shorter than the periods of darkness. 167

flashing-light signal. A railroad-highway crossing signal the indication of which is given by two red lights spaced horizontally and flashed alternately at predetermined intervals to give warning of the approach of trains, or a fixed signal in which the indications are given by color and flashing of one or more of the signal lights. *See also:* **railway signal and equipment.** 328

flashing signal (telephone switching systems). A signal for indicating a change or series of changes of state, such as on-hook/off-hook, used for supervisory purposes. 55

flashlight battery. A battery designed or employed to light a lamp of an electric hand lantern or flashlight. *See also:* **battery (primary or secondary).** 328

flashover (1) (general). A disruptive discharge through air around or over the surface of solid or liquid insulation, between parts of different potential or polarity, produced by the application of voltage wherein the breakdown path becomes sufficiently ionized to maintain an electric arc. *See also:* **test voltage and current.**
210, 64, 299, 62, 307, 201

(2) (surge arresters). A disruptive discharge around or over the surface of a solid or liquid insulator. 2

(3) (high-voltage a-c cable termination). A disruptive discharge around or over the surface of an insulating member, between parts of different potential or polarity, produced by the application of voltage wherein the breakdown path becomes sufficiently ionized to maintain an electric arc. 4

flash plate. A thin electrodeposited coating produced in a short time. 328

flash point. The minimum temperature at which a liquid will give off vapor in sufficient amount to form a flammable air-vapor mixture that can be ignited under specified conditions. 297

flat-compounded. A qualifying term applied to a compound-wound generator to denote that the series winding is so proportioned that the terminal voltage at rated load is the same as at no load. *See:* **direct-current commutating machine.** 328

flat leakage power (microwave gas tubes). The peak radio-frequency power transmitted through the tube after the establishment of the steady-state radio-frequency discharge. *See:* **gas tubes.** 125, 190

flat loss (gain) (circuits and systems). The frequency independent contribution to the total transfer-function loss (or gain) of a four-terminal network. 67

flat-rate call (telephone switching systems). A call for which no billing is required. 55

flat-rate service (telephone switching systems). Service in which a fixed charge is made for all answered local calls during the billing interval. 55

flat-strip conductor. *See:* **strip (-type) transmission line.**

F layer. An ionized layer in the F region. *See also:* **radiation.** 328

F^1 layer (radio wave propagation). The lower of the two ionized layers normally existing in the F region in the day hemisphere. *See also:* **radiation; radio wave propagation.** 146, 180

F^2 layer (radio wave propagation). The single ionized layer normally existing in the F region in the night hemisphere and the higher of the two layers normally existing in the F region in the day hemisphere. *See also:* **radiation; radio wave propagation.** 146, 180

flection-point emission current. That value of current on the diode characteristic for which the second derivative of the current with respect to the voltage has its maximum negative value. *Note:* This current corresponds to the upper flection point of the diode characteristic. *See also:* **electron emission.** 125, 190

flexible connector (rotating machinery). An electric connection that permits expansion, contraction, or relative motion of the connected parts. 63

flexible coupling (rotating machinery). A coupling having relatively high transverse or torsional compliance. *Notes:* (1) May be used to reduce or eliminate transverse loads or deflections of one shaft from being carried, or felt by the other coupled shaft. (2) May be used to reduce the torsional stiffness between two rotating masses in order to change torsional natural frequencies of the shaft system or to limit transient or pulsating torques carried by the shafts. *See also:* **rotor (rotating machinery).** 63

flexible metal conduit. A flexible raceway of circular cross section specially constructed for the purpose of the pulling in or the withdrawing of wires or cables after the conduit and its fittings are in place. *See also:* **raceway.** 328

flexible mounting (rotating machinery). A flexible structure between the core and foundation used to reduce the transmission of vibration. 63

flexible nonmetallic tubing (loom). A mechanical protection for electric conductors that consists of a flexible cylindrical tube having a smooth interior and a single or double wall of nonconducting fibrous material. *See also:* **raceways.** 328

flexible tower (frame). A tower that is dependent on the line conductors for longitudinal stability but is designed to resist transverse and vertical loads. *See also:* **tower.** 64

flexible waveguide (waveguide components). A waveguide constructed to permit limited bending, twisting, stretching or any combination thereof, without appreciable change in its electrical properties. 166

flexure (tuned rotor gyro). An elastic element in a tuned rotor gyro which connects the rotor or gimbal to the motor shaft and permits limited angular freedom about axes perpendicular to the spin axis. 46

flicker (1) (general). Impression of fluctuating brightness or color, occurring when the frequency of the observed variation lies between a few hertz and the fusion frequencies of the images. 244, 178

(2) (television). A repetitive variation in luminance of given area in a monochromatic or color display, the visibility of which is a function of repetition rate, duty cycle, luminance, and the decay characteristic. 18

flicker effect (electron tubes). The random variations of the output-current in a valve or tube with an oxide-coated cathode. *Note:* Its value varies inversely with the frequency. *See also:* **electron tube.** 244, 190

flicker fusion frequency. The frequency of intermittent stimulation of the eye at which flicker disappears. *Note:* It also is called **critical fusion frequency** or **critical flicker frequency.** *See also:* **visual field.** 167

flicker threshold (television). The luminance at which flicker is just perceptible at a given repetition rate with other variables held constant. 18

flight path (navigation). A proposed route in three dimensions. *See also:* **course line; navigation.** 13, 187

flight-path computer (electronic navigation). Equipment providing outputs for the control of the motion of a vehicle along a flight path. *See also:* **navigation.** 187, 13

flight-path deviation (electronic navigation). The amount by which the flight track of a vehicle differs from its flight path expressed in terms of either angular or linear measurement. *See also:* **navigation.** 187, 13

flight-path-deviation indicator (electronic navigation). A device providing a visual display of flight-path deviation. *See also:* **navigation.** 187, 13

flight track (electronic navigation). The path in space actually traced by a vehicle. *See also:* **track.** 13, 187

flip-flop (electronic computation or control). (1) A circuit or device, containing active elements, capable of assuming either one of two stable states at a given time, the particular state being dependent upon (A) the nature of an input signal, for example, its polarity, amplitude, and duration, and (B) which of two input terminals last received the signal. *Note:* The input and output coupling networks, and indicators, may be considered as an integral part of the flip-flop. (2) A device, as in (1) above, that is capable of counting modulo 2, in which case it might have only one input terminal. (3) A sequential logic element having properties similar to (1) or (2) above. *See also:* **control system, feedback; toggle.** 235

flip-flop circuit. A trigger circuit having two conditions of permanent stability, with means for passing from one to the other by an external stimulus. *See also:* **trigger circuit.** 59

float (gyro). An enclosed gimbal assembly housing the spin motor and other components such as the pickoff and torquer. This assembly is immersed in a fluid usually at the condition of neutral buoyancy. 46

floating. A method of operation for storage batteries in which a constant voltage is applied to the battery terminals

sufficient to maintain an approximately constant stage of charge. *See:* **trickle charge.** *See also:* **charge.** 328

floating battery. A storage battery that is kept in operating condition by a continuous charge at a low rate. *See also:* **railway signal and equipment.** 328

floating carrier. *See:* **controlled carrier.**

floating control. *See:* **control system, floating.**

floating grid (electron tubes). An insulated grid, the potential of which is not fixed. *See also:* **electronic tube.** 244, 190

floating network or component (circuits and systems). A network or component having no terminal at ground potential. 67

floating neutral. One whose voltage to ground is free to vary when circuit conditions change. *See also:* **center of distribution.** 59

floating point. Pertaining to a system in which the location of the point does not remain fixed with respect to one end of the numerical expressions, but is regularly recalculated. The location of the point is usually given by expressing a power of the base. *See also:* **fixed point; variable point.** 235

floating-point system (electronic computation). *See:* **point.**

floating removable feed tube (cable plowing). A feed tube removably attached to a blade so relative motion may occur between the feed tube and the blade around axis that are essentially vertical and horizontal (perpendicular to direction of travel). *See:* **feed tube.** 52

floating speed (process control). In single-speed or multiple-speed floating control systems, the rate of change of the manipulated variable. 56

float storage (gyro). The sum of attitude storage and the torque command storage in a rate integrating gyro. *See:* **attitude storage, torque command storage.** 46

floating zero (numerically controlled machines). A characteristic of a numerical machine control permitting the zero reference point on an axis to be established readily at any point in the travel. *Note:* The control retains no information on the location of any previously established zeros. *See:* **zero offset.** 224, 207

float switch (liquid-level switch) (industrial control). A switch in which actuation of the contacts is effected when a float reaches a predetermined level. *See:* **switch.** 244, 206

flood (charge-storage tubes) (verb). To direct a large-area flow of electrons, containing no spatially distributed information, toward a storage assembly. *Note:* A large-area flow of electrons with spatially distributed information is used in image-converter tubes. *See also:* **charge-storage tube.** 174, 190

floodlight. A projector designed for lighting a scene or object to a brightness considerably greater than its surroundings. *Notes:* (1) It usually is capable of being pointed in any direction and is of weatherproof construction. (2) The beam spread of floodlights may range from relatively narrow (10 degrees) to wide (more than 100 degrees). *See also:* **floodlighting.** 167

floodlighting. A system designed for lighting a scene or object to a brightness greater than its surroundings. *Note:* It may be for utility, advertising, or decorative purposes. 167

flood-lubricated bearing (rotating machinery). A bearing in which a continuous flow of lubricant is poured over the top of the bearing or journal at about normal atmospheric pressure. *See also:* **bearing.** 63

flood projection (facsimile). The optical method of scanning in which the subject copy is floodlighted and the scanning spot is defined in the path of the reflected or transmitted light. *See also:* **scanning (facsimile).** 12

floor acceleration (nuclear power generating stations). The acceleration of a particular building floor (or equipment mounting) resulting from a given earthquake's motion. The maximum floor acceleration can be obtained from the floor response spectrum as the acceleration at high frequencies (in excess of 33 Hz) and is sometimes referred to as the ZPA (zero period acceleration). 31

floor bushing. A bushing intended primarily to be operated entirely indoors in a substantially vertical position to carry a circuit through a floor or horizontal grounded barrier. Both ends must be suitable for operating in air. *See also:* **bushing.** 348

floor lamp. A portable luminaire on a high stand suitable for standing on the floor. *See also:* **luminaire.** 167

floor trap (burglar-alarm system). A device designed to indicate an alarm condition in an electric protective circuit whenever an intruder breaks or moves a thread or conductor extending across a floor space. *See also:* **protective signaling.** 328

flowchart (computing systems). A graphical representation for the definition, analysis, or solution of a problem, in which symbols are used to represent operations, data, flow, and equipment. *See also:* **logic diagram.** 255, 77, 54

flow diagram (electronic computers). Graphic representation of a program or a routine. 210

flow relay. A relay that responds to a rate of fluid flow. 103, 202, 60, 127

flow soldering. *See:* **dip soldering.**

fluctuating power (rotating machinery). A phasor quantity of which the vector represents the alternating part of the power, and that rotates at a speed equal to double the angular velocity of the current. *See also:* **asynchronous machine.** 63

fluctuating target. A radar target whose return echo varies up and down in amplitude as a function of time. The fluctuation rate may vary from essentially independent return amplitudes from pulse-to-pulse to significant variation only on a scan-to-scan basis. 13

fluctuation (1) (pulse terms). Dispersion of the pulse amplitude or other magnitude parameter of the pulse waveforms in a pulse train with respect to a reference pulse amplitude or a reference magnitude. Unless otherwise specified by a mathematical adjective, peak-to-peak fluctuation is assumed. *See:* **mathematical adjectives.** 254

(2) (radar). *See:* **target fluctuation.**

fluctuation loss (radar). The apparent loss in radar detectability or measurement accuracy for a target of given average echo return power due to target fluctuation. It may be measured as the increase in required average echo return power of a fluctuating target to achieve the same radar detection performance as for a target of constant echo return. 13

fluctuation noise. *See:* **random noise.**

fluence (solar cells). The total time-integrated number of particles that cross a plane unit area from either side. 113

fluid filled joints (power cable joints). Joints in which the joint housing is filled with an insulating material that is fluid at all operating temperatures. 34

fluid loss (rotating machinery). That part of the mechanical losses in a machine having liquid in its air gap that is caused by fluid friction. *See also:* **asynchronous machine: direct-current commutating machine.** 63

fluidly delayed overcurrent trip. *See:* **fluidly delayed release (fluidly delayed trip) and overcurrent release (overcurrent trip).**

fluidly delayed release (fluidly delayed trip). A release delayed by fluid displacement or adhesion. 103, 202

fluid pressure supply system (speed governing of hydraulic turbines). The pumps, means for driving them, pressure and sump tanks, valves and piping connecting the various parts of the governing system and associated and accessory devices. 8

fluorescence. Emission of light from a substance (a phosphor) as a result of, and only during, excitation by radiant energy. *See:* **lamp; oscillograph.** 167

fluorescent lamp. A low-pressure mercury electric-discharge lamp in which a fluorescing coating (phosphor) transforms some of the ultraviolet energy generated by the discharge into light. 167

fluorescent-mercury lamp. An electric-discharge lamp having a high-pressure mercury arc in an arc tube, and an outer envelope coated with a fluorescing substance (phosphor) that transforms some of the ultraviolet energy generated by the arc into light. *See also:* **lamp.** 167

flush antenna (aircraft). An antenna having no projections outside the streamlined surface of the aircraft. In general, flush antennas may be considered as slot antennas. *See also:* **air transportation electronic equipment.** 328

flush-mounted antenna. An antenna that does not protrude from the exterior side of the surface to which it is attached. 111

flush-mounted device. A device in which the body projects only a small specified distance in front of the mounting surface. 103, 202, 53

flush mounted or recessed (luminaire). A luminaire that is mounted above the ceiling (or behind a wall or other surface) with the opening of the luminaire level with the surface. *See also:* **suspended (pendant).** 167

flush mounting (transformers). So designed as to have a minimal front projection when set into and secured to a flat surface. 53

flutter (sound recording and reproducing equipment). Frequency modulation of the signal in the range of approximately 6 Hz to 100 Hz resulting in distortion which may be perceived as a roughening of the sound quality of a tone or program. 145

flutter echo. A rapid succession of reflected pulses resulting from a single initial pulse. 176

flutter rate (sound recording and reproducing). The number of frequency excursions in hertz, in a tone that is frequency-modulated by flutter. *Notes:* (1) Each cyclical variation is a complete cycle of deviation, for example, from maximum-frequency to minimum-frequency and back to maximum-frequency at the rate indicated. (2) If the over-all flutter is the resultant of several components having different repetition rates, the rates and magnitudes of the individual components are of primary importance. *See also:* **sound recording and reproducing.**   145

flux (1) (photovoltaic power system). The rate of flow of energy through a surface. *See also:* **photovoltaic power system.**   186

(2) (soldering) (connections). A liquid or solid which when heated exercises a cleaning and protective action upon the surfaces to which it is applied. *See also:* **soldered connections (electronic and electric applications).**   281

(3) (solar cells). The number of particles that cross a plane unit area per unit time from either side.   113

flux guide (induction heating usage). Magnetic material used to guide electromagnetic flux in desired channels. *Note:* The guides may be used either to direct flux to preferred locations or to prevent the flux from spreading beyond definite regions. *See also:* **induction heater.**
                                                   14, 114

flux linkages. The sum of the fluxes linking the turns forming the coil, that is, in a coil having N turns the flux linkage is

$$\lambda = \phi_1 + \phi_2 + \phi_3 \cdots \phi_N$$

where ϕ_1 = flux linking turn 1, ϕ_2 = flux linking turn 2, etcetera, and ϕ_N = flux linking the Nth turn.   197

fluxmeter. An instrument for use with a test coil to measure magnetic flux. It usually consists of a moving-coil galvanometer in which the torsional control is either negligible or compensated. *See also:* **magnetometer.**
                                                        328

flyback (television). The rapid return of the beam in a cathode-ray tube in the direction opposite to that used for scanning.   163

flying spot scanner (optical character recognition). A device employing a moving spot of light to scan a sample space, the intensity of the transmitted or reflected light being sensed by a photoelectric transducer.   255, 77

flywheel ring (rotating machinery). A heavy ring mounted on the spider for the purpose of increasing the rotor moment of inertia. *See also:* **rotor (rotating machinery).**
                                                         63

FM. *See:* **frequency modulation.**

FMCW. Abbreviation for **frequency modulated continuous waves.**   13

FM-FM telemetry. *See:* **frequency-modulation—frequency modulation (FM-FM) telemetry.**

FOA. *See:* **transformer, oil-immersed.**

focal length (laser-maser). The distance from the secondary nodal point of a lens to the primary focal point. In a thin lens, the focal length is the distance between the lens and the focal point.   363

focal point (laser-maser). The point toward which radiation converges or from which radiation diverges or appears to diverge.   363

focus (oscilloscopes). Maximum convergence of the electron beam manifested by minimum spot size on the phosphor screen. *See also:* **astigmatism; oscillograph.**
                                                        185

focusing (electron tubes). The process of controlling the convergence of the electron beam.   244, 178, 125

focusing and switching grille (color picture tubes). A

color-selecting-electrode system in the form of an array of wires including at least two mutually-insulated sets of conductors in which the switching function is performed by varying the potential difference between them, and focusing is accomplished by maintaining the proper average potentials on the array and on the phosphor screen.
                                                        125

focusing coil. *See:* **focusing magnet.**

focusing device. An instrument used to locate the filament of an electric lamp at the proper focal point of lens or reflector optical systems.   328

focusing, dynamic (picture tubes). The process of focusing in accordance with a specified signal in synchronism with scanning.   125, 178, 190

focusing electrode (beam tube). An electrode the potential of which is adjusted to focus an electron beam. *See also:* **electrode (electron tube).**   244, 178, 125, 117

focusing, electrostatic (electron beam). A method of focusing an electron beam by the action of an electric field.
                                                        125

focusing grid (pulse techniques). *See:* **focusing electrode.**

focusing magnet. An assembly producing a magnetic field for focusing an electron beam.   190, 125

focusing, magnetic (electron beam). A method of focusing an electron beam by the action of a magnetic field.
                                                     190, 178

fog (adverse-weather) lamps. Lamps that may be used in lieu of headlamps to provide road illumination under conditions of rain, snow, dust, or fog. *See also:* **headlamp.**
                                                        167

fog-bell operator. A device to provide automatically the periodic bell signals required when a ship is anchored in fog.   328

foil (foil tape) (burglar-alarm system). A fragile strip of conducting material suitable for fastening with an adhesive to glass, wood, or other insulating material in order to carry the alarm circuit and to initiate an alarm when severed. *See also:* **protective signaling.**   328

folded-dipole antenna. An antenna composed of two or more parallel, closely spaced dipole antennas connected together at their ends with one of the dipole antennas fed at its center. *See also:* **antenna.**   179, 111

follow (power) (surge arresters). The current from the connected power source that flows through an arrester during and following the passage of discharge current.
                                                          2

follower drive (slave drive) (industrial control). A drive in which the reference input and operation are direct functions of another drive, called the master drive. *See:* **control system, feedback.**   219, 206

font (computing systems). A family or assortment of characters of a given size and style. *See:* **type font.**
                                                     255, 77

foot (rotating machinery). The part of the stator structure, end shield, or base, that provides means for mounting and fastening a machine to its foundation. *See also:* **stator.**
                                                         63

footcandle. *See:* **illumination.**

footings (foundations). Structures set in the ground to support the bases of towers, poles, or other overhead structures. *Note:* Footings are usually skeleton steel pyramids, grilles, or piers of concrete. *See also:* **tower.**   64

footlambert (1) (television color terms). A unit of luminance (photometric brightness) equal to $1/\pi$ candela per square foot ($10.7643/\pi$ candelas per square meter), or to the uniform luminance of a perfectly diffusing surface emitting or reflecting light at the rate of one lumen per square foot (10.7643 lumens per square meter), or to the average luminance of any surface emitting or reflecting light at that rate. *Notes:* (1) A footcandle is a unit of incident light, and a footlambert is a unit of emitted or reflected light. For a perfectly reflecting and perfectly diffusing surface, the number of footcandles is equal to the number of footlamberts. (2) The average luminance of any reflecting surface in footlamberts is, therefore, the product of the illumination in footcandles by the luminous reflectance of the surface. 18

(2) (light emitting diodes) (fL*). A unit of luminance (photometric brightness) equal to $1/\pi$ candela per square foot, or to the uniform luminance of a perfectly diffusing surface emitting or reflecting light at the rate of one lumen per square foot. *See also:* **illumination.** 162

* Deprecated

foot switch (industrial control). A switch that is suitable for operation by an operator's foot. *See:* **switch.**
219, 206

forbidden combination. A code expression that is defined to be nonpermissible and whose occurrence indicates a mistake or malfunction. 125

forbidden-combination check (electronic computation). *See:* **check, forbidden-combination.**

force. Any physical cause that is capable of modifying the motion of a body. The vector sum of the forces acting on a body at rest or in uniform rectilinear motion is zero.
210

forced-air cooling system (rectifier). An air cooling system in which heat is removed from the cooling surfaces of the rectifier by means of a flow of air produced by a fan or blower. *See also:* **rectification.** 291, 208

forced drainage (underground metallic structures). A method of controlling electrolytic corrosion whereby an external source of direct-current potential is employed to force current to flow to the structure through the earth, thereby maintaining it in a cathodic condition. *See also:* **inductive coordination.** 328

forced interruption (electric power systems). An interruption caused by a forced outage. *See also:* **outage.**
200

forced-lubricated bearing (rotating machinery). A bearing in which a continuous flow of lubricant is forced between the bearing and journal. 63

forced oscillation (linear constant-parameter system). The response to an applied driving force. *See also:* **network analysis.** 210

forced outage (emergency and standby power systems). An outage that results from conditions directly associated with a component requiring that it be taken out of service immediately, either automatically or as soon as switching operations can be performed, or an outage caused by improper operation of equipment or human error. *Notes:* (1) This definition derives from transmission and distribution applications and does not necessarily apply to generation outages. (2) The key test to determine if an outage should be classified as forced or scheduled is as follows. If it is possible to defer the outage when such deferment is desirable, the outage is a scheduled outage; otherwise, the outage is a forced outage. Deferring an outage may be desirable, for example, to prevent overload of facilities or an interruption of service to consumers. 89

forced release (telephone switching systems). Release initiated from sources other than the calling or called line.
55

forced response (circuits and systems). The response of a system resulting from the application of an energy source with the system initially free of stored energy. 67

forced-ventilated machine. *See:* **open pipe-ventilated machine.**

force factor (1) (electroacoustic transducer). (A) The complex quotient of the pressure required to block the acoustic system divided by the corresponding current in the electric system; (B) the complex quotient of the resulting open-circuit voltage in the electric system divided by the volume velocity in the acoustic system. *Note:* Force factors (A) and (B) have the same magnitude when consistent units are used and the transducer satisfies the principle of reciprocity. 176

(2) (electromechanical transducer). (A) The complex quotient of the force required to block the mechanical system divided by the corresponding current in the electric system; (B) the complex quotient of the resulting open-circuit voltage in the electric system divided by the velocity in the mechanical system. *Notes:* (1) Force factors (A) and (B) have the same magnitude when consistent units are used and the transducer satisfies the principle of reciprocity. (2) It is sometimes convenient in an electrostatic or piezoelectric transducer to use the ratios between force and charge or electric displacement, or between voltage and mechanical displacement. 328

forcing (industrial control). The application of control impulses to initiate a speed adjustment, the magnitude of which is greater than warranted by the desired controlled speed in order to bring about a greater rate of speed change. *Note:* Forcing may be obtained by directing the control impulse so as to effect a change in the field or armature circuit of the motor, or both. *See also:* **electric drive.** 206

foreign area (telephone switching systems). A numbering plan area other than the one in which the calling customer is located. 55

foreign exchange line (1) (communication switching). A subscriber line by means of which service is furnished to a subscriber at his request from an exchange other than the one from which service would normally be furnished.
193, 59

(2) (telephone switching systems). A loop from an exchange other than the one from which service would normally be furnished. 55

forestalling switch. *See:* **acknowledger; forestaller.**

fork beat (facsimile). *See:* **carrier beat.**

form. Any article such as a printing plate, that is used as a pattern to be reproduced. *See also:* **electroforming.**
328

formal logic. The study of the structure and form of valid argument without regard to the meaning of the terms in the argument. 255, 77

format (1) (computing systems). The general order in which information appears on the input medium.

(2) (data transmission). Arrangement of code characters within a group, such as a block or message.   194

(3) Physical arrangement of possible locations of holes or magnetized areas. *See:* **address format.**   224, 207

format classification (numerically controlled machines). A means, usually in an abbreviated notation, by which the motions, dimensional data, type of control system, number of digits, auxiliary functions, etcetera, for a particular system can be denoted.   224, 207

format detail (numerically controlled machines). Describes specifically which words and of what length are used by a specific system in the format classification.   224, 207

formation light. A navigation light specially provided to facilitate formation flying.   167

formation voltage. The final impressed voltage at which the film is formed on the valve metal in an electrochemical valve. *See also:* **electrochemical valve.**   328

formette (rotating machinery). *See:* **form-wound motorette.**

form factor (1) (periodic function). The ratio of the root-mean-square value to the average absolute value, averaged over a full period of the function. *See also:* **power rectifier; rectification.**   237, 66, 210, 208

(2) (electric process heating). Coil ratio of conductor width to turn to turn space. *See:* **coil shape factor.**   14

forming (1) (electrical) (semiconductor devices). The process of applying electric energy to a semiconductor device in order to modify permanently the electrical characteristics. *See also:* **semiconductor.**   66

(2) (semiconductor rectifier). The electrical or thermal treatment, or both, of a semiconductor rectifier cell for the purpose of increasing the effectiveness of the rectifier junction. *See also:* **rectification.**   237, 66

(3) (electrochemical). The process that results in a change in impedance at the surface of a valve metal to the passage of current from metal to electrolyte, when the voltage is first applied. *See also:* **electrochemical valve.**   328

form-wound (preformed winding) (rotating machinery). Applied to a winding whose coils are formed essentially to their final shape prior to assembly into the machine. *See also:* **rotor (rotating machinery); stator.**   63

form-wound motorette (formette) (rotating machinery). A motorette for form-wound coils. *See also:* **asynchronous machine; direct-current commutating machine.**   63

fortuitous distortion. A random distortion of telegraph signals such as that commonly produced by interference. *See also:* **telegraphy.**   194

fortuitous telegraph distortion. Distortion that includes those effects that cannot be classified as bias or characteristic distortion and is defined as the departure, for one occurrence of a particular signal pulse, from the average combined effects of bias and characteristic distortion. *Note:* Fortuitous distortion varies from one signal to another and is measured by a process of elimination over a long period. It is expressed in percent of unit pulse. *See also:* **distortion.**   111

forward-acting regulator. A transmission regulator in which the adjustment made by the regulator does not affect the quantity that caused the adjustment. *See also:* **transmission regulator.**   328

forward admittance, short-circuit (electron-device transducer). *See:* **admittance, short-circuit forward.**

forward bias (forward voltage) (V_F) (light emitting diodes). The bias voltage which tends to produce current flow in the forward direction.   162

forward current (1) (metallic rectifier). The current that flows through a metallic rectifier cell in the forward direction. *See also:* **rectification.**   328

(2) (semiconductor rectifier device). The current that flows through a semiconductor rectifier device in the forward direction. *See also:* **rectification.**   237, 66

(3) (reverse-blocking or reverse-conducting thyristor). The principal current for a positive anode-to-cathode voltage. *See also:* **principal current.**   243, 66, 208, 191

(4) (light emitting diodes) (I_F). The current that flows through a semiconductor junction in the forward direction.   162

forward current, average, rating. *See:* **average forward current rating.**

forward direction (1) (metallic rectifier). The direction of lesser resistance to current flow through the cell; that is, from the negative electrode to the positive electrode. *See also:* **rectification.**   328

(2) (semiconductor rectifier device). The direction of lesser resistance to steady direct-current flow through the device; for example, from the anode to the cathode. *See also:* **semiconductor; semiconductor rectifier stack.**   237, 208, 162

(3) (semiconductor rectifier diode). The direction of lower resistance to steady-state direct-current; that is, from the anode to the cathode.   66

forward error-correcting system. A system employing an error-correcting code and so arranged that some or all signals detected as being in error are automatically corrected at the receiving terminal before delivery to the data sink or to the telegraph receiver.   194

forward gate current (thyristor). The gate current when the junction between the gate region and the adjacent anode or cathode region is forward biased. *See also:* **principal current.**   243, 66, 208, 190

forward gate voltage (thyristor). The voltage between the gate terminal and the terminal of the adjacent anode or cathode region resulting from forward gate current. *See also:* **principal voltage-current characteristic (principal characteristic).**   243, 66, 208, 190

forward path (signal-transmission system) (feedback-control loop). The transmission path from the loop-error signal to the loop-output signal. *See:* **feedback.**   188

forward period (rectifier circuit element) (rectifier circuit). The part of an alternating-voltage cycle during which forward voltage appears across the rectifier circuit element. *Note:* The forward period is not necessarily the same as the conducting period because of the effect of circuit parameters and semiconductor rectifier cell characteristics. *See also:* **rectifier circuit element.**   237, 66

forward power dissipation (semiconductor). The power dissipation resulting from forward current.   66

forward power loss (semiconductor device). The power loss within a semiconductor rectifier device resulting from the flow of forward current. *See also:* **rectification; semiconductor rectifier stack.**   237, 66, 208

forward recovery time (semiconductor diode). The time

required for the current or voltage to recover to a specified value after instantaneous switching from a stated reverse voltage condition to a stated forward current or voltage condition in a given circuit. *See also:* **rectification.**
237, 66

forward resistance (metallic rectifier). The resistance measured at a specified forward voltage drop or a specified forward current. *See also:* **rectification.** 328

forward voltage (1) (rectifiers). Voltage of the polarity that produces the larger current, hence, the voltage across a semiconductor rectifier diode resulting from forward current. *See also:* **on-state voltage; forward voltage drop.**

(2) (reverse blocking or reverse conducting thyristor). A positive anode-to-cathode voltage. *See also:* **principal characteristic (principal voltage-current characteristic).**
243, 66, 208, 191

forward voltage drop (1) (metallic rectifier). The voltage drop in the metallic rectifying cell resulting from the flow of current through a metallic rectifier cell in the forward direction.

(2) (semiconductor rectifier). *See:* **forward voltage; on-state voltage.**
237, 66, 208

* Deprecated

forward wave (traveling-wave tubes). A wave whose group velocity is in the same direction as the electron steam motion. *See also:* **electron devices, miscellaneous.** 190

forward-wave structure (microwave tubes). A slow-wave structure whose propagation is characterized on a ω/β diagram (ω versus phase shift/section) by a positive slope in the region $0 < \beta < \pi$ (in which the group and phase velocity therefore have the same sign). *See also:* **microwave tube or valve.** 190

Foster's reactance theorem (circuits and systems). States that the driving-point impedance of a network composed of purely capacitive and inductive reactances is an odd rational function of frequency (ω) which has the following characteristics; (1) a positive slope (2) the poles and zeros of the function are on the $j\omega$-axis, they are simple, they occur in complex conjugate pairs and they alternate.
67

Foucault currents. *See:* **eddy currents.**

foul electrolyte. An electrolyte in which the amount of impurities is sufficient to cause an undesirable effect on the operation of the electrolytic cells in which it is employed. *See also:* **electrorefining.** 328

fouling. The accumulation and growth of marine organisms on a submerged metal surface. 221, 205

fouling point (railway practice). The location in a turnout back of a frog at or beyond the clearance point at which insulated joints or details are placed. 328

foundation (rotating machinery). The structure on which the feet or base of a machine rest and are fastened. 63

foundation bolt (rotating machinery). A bolt used to fasten a machine to a foundation. 63

foundation-bolt cone (rotating machinery). A cone placed around a foundation bolt when imbedded in a concrete foundation to provide clearance for adjustment during erection. 63

four-address. Pertaining to an instruction code in which each instruction has four address parts. *Note:* In a typical four-address instruction the addresses specify the location of two operands, the destination of the result, and the location of the next instruction to be interpreted. *See also:* **three-plus-one address.** 125

four-address code (electronic computation). *See:* **instruction code.**

Fourier series. A single-valued periodic function (that fulfills certain mathematical conditions) may be represented by a Fourier series as follows

$$f(x) = 0.5A_0 + \sum_{n=1}^{n=\infty} [A_n \cos nx + B_n \sin nx]$$

$$= 0.5A_0 + \sum_{n=1}^{n=\infty} C_n \sin (nx + \theta_n)$$

where

$$A_n = \frac{1}{\pi} \int_0^{2\pi} f(x) \cos nx \, dx$$

$n = 0,1,2,3, \cdots$

$$B_n = \frac{1}{\pi} \int_0^{2} f(x) \sin nx \, dx$$

$$C_n = +(A_n^2 + B_n^2)^{1/2}$$

and

$$\theta_n = \arctan A_n/B_n$$

Note: $0.5A_0$ is the average of a periodic function $f(x)$ over one primitive period. 210

four-plus-one address (computing systems). Pertaining to an instruction that contains four operand addresses and a control address. 255, 77

four-pole. See: **two-terminal pair network.**

four quadrant DAM (hybrid computer linkage components). A digital-to-analog multiplier (DAM) that accepts both signs of the digital value, giving correct sign output in all four quadrants. 10

four-terminal network. A network with four accessible terminals. *Note:* See **two-terminal-pair network** for an important special case. *See:* **two-terminal-pair network; quadripole.** 210

fourth voltage range (railway signal and interlocking). *See:* **voltage range.**

fourth-wire control (telephone switching systems). The wire (in addition to the tip, ring, and sleeve wires) used for transmission of special signals necessary in the establishment or supervision of a call. 55

four-wire circuit. A two-way circuit using two paths so arranged that the electric waves are transmitted in one direction only by one path and in the other direction only by the other path. *Note:* The transmission paths may or may not employ four wires. *See also:* **transmission line.**
328

four-wire repeater. A telephone repeater for use in a four-wire circuit and in which there are two amplifiers, one serving to amplify the telephone currents in one side of the four-wire circuit and the other serving to amplify the telephone currents in the other side of the four-wire circuit. *See also:* **repeater.** 59

four-wire switching (telephone switching systems). Switching using a separate path, frequency, or time interval for each direction of transmission. 55

four-wire terminating set. A hybrid set for interconnecting a four-wire and two-wire circuit.   59

fovea. A small region at the center of the retina, subtending about two degrees, that contains only cones and forms the site of most distinct vision. *See also:* **retina.**   167

foveal vision. *See:* **central vision.**

FOW. *See:* **transformer, oil-immersed.**

fractional-horsepower brush (rotating machinery). A brush with a cross-sectional area of $\frac{1}{4}$ square inch (thickness × width) or less and not exceeding $1\frac{1}{2}$ inches in length, but larger than a miniature brush and smaller than an industrial brush. *See:* **brush (1) (electric machines).**   63

fractional-horsepower motor. A motor built in a frame smaller than that of a motor of open construction having a continuous rating of 1 horsepower at 1700–1800 revolutions per minute. *See:* **asychronous machine; direct-current commutating machine.**   63

fractional-slot winding (rotating machinery). A distributed winding in which the average number of slots per pole per phase is not integral, for example $3\frac{2}{7}$ slots per pole per phase. *See also:* **asynchronous machine; direct-current commutating machine.**   63

fragility (nuclear power generating stations) (seismic qualification of class 1E equipment). Susceptibility of equipment to malfunction as the result of structural or operational limitations, or both.   28, 31

fragility level (nuclear power generating stations) (seismic qualification of class 1E equipment). The highest level of input excitation, expressed as a function of input frequency, that an equipment can withstand and still perform the required Class 1E functions.   28, 31

fragility response spectrum (FRS) (nuclear power generating stations) (seismic qualification of class 1E equipment). A TRS (test response spectrum) obtained from tests to determine the fragility level of equipment. *See also:* **test response spectrum.**   28, 31

frame (1) (television). The total area, occupied by the picture, that is scanned while the picture signal is not blanked.

(2) (facsimile). A rectangular area, the width of which is the available line and the length of which is determined by the service requirements.   12, 178

(3) (test, measurement and diagnostic equipment). A cross section of tape containing one bit in each channel and possibly a parity bit. *Syn.* **tape line.**   54

(4) (data transmission). A set of consecutive digit time slots in which the position of each digit time slot can be identified by reference to a framing signal.   59

framed plate (storage cell). A plate consisting of a frame supporting active material. *See also:* **battery (primary or secondary).**   328

frame frequency (television). The number of times per second that the frame is scanned. *See also:* **television.**   328

frame, intermediate distributing. *See:* **intermediate distributing frame.**

frame, main distributing. *See:* **main distributing frame.**

frame rate (data transmission). The repetition rate of the frame.   59

frame ring (rotating machinery). A plate or assembly of flat plates forming an annulus in a radial plane and serving as a part of the frame to stiffen it.   63

frame size (low-voltage circuit breaker). A term which

denotes the maximum continuous current rating in amperes for all parts except the coils of the direct-acting trip device.   103

frame split (rotating machinery). A joint at which a frame may be separated into parts.   63

frame synchronization (data transmission). The process whereby a given channel at the receiving end is aligned with the corresponding channel at the transmitting end.   59

framework (rotating machinery). A stationary supporting structure.   63

frame yoke (field frame) (rotating machinery). The annular support for the poles of a direct-current machine. *Note:* It may be laminated or of solid metal and forms part of the magnetic circuit.   63

framing (facsimile). The adjustment of the picture to a desired position in the direction of line progression. *See also:* **recording (facsimile).**   12

framing signal (facsimile). A signal used for adjustment of the picture to a desired position in the direction of line progression. *See also:* **facsimile signal (picture signal).**   12

Fraunhofer pattern (antenna). A radiation pattern obtained in the Fraunhofer region. *See also:* **antenna.**   179, 111

Fraunhofer region (1) (antenna). The region in which the field of an antenna is focused. *See:* **far-field region.** *See also:* **antenna.**   179, 111

(2) (data transmission). That region of the field in which the energy flow from an antenna proceeds essentially as though coming from a point source located in the vicinity of the antenna. *Note:* If the antenna has a well-defined aperture D in a given aspect, the Fraunhofer region in that aspect is commonly taken to exist at distances greater than $2D^2/$ from the aperture, being the wavelength.   59

free capacitance (1) (conductor). The limiting value of its self-capacitance when all other conductors, including isolated ones, are infinitely removed.

(2) (between two conductors). The limiting value of the plenary capacitance as all other, including isolated, conductors are infinitely removed.   210

free-code call (telephone switching systems). A call to a service or office code for which no charge is made.   55

free cyanide (electrodepositing solution) (electroplating). The excess of alkali cyanide above the minimum required to give a clear solution, or above that required to form specified soluble double cyanides. *See also:* **electroplating.**   328

free field. A field (wave or potential) in a homogeneous, isotropic medium free from boundaries. In practice, a field in which the effects of the boundaries are negligible over the region of interest. *Note:* The actual pressure impinging on an object (for example, electroacoustic transducer) placed in an otherwise free sound field will differ from the pressure that would exist at that point with the object removed, unless the acoustic impedance of the object matches the acoustic impedance of the medium.   176

free-field current response (receiving current sensitivity) (electroacoustic transducer used for sound reception). The ratio of the current in the output circuit of the transducer when the output terminals are short-circuited to the free-field sound pressure existing at the transducer location prior to the introduction of the transducer in the sound field. *Notes:* (1) The free-field response is defined for a plane progressive sound wave whose direction of propagation has a specified orientation with respect to the principal axis of the transducer. (2) The free-field current response is usually expressed in decibels, namely, 20 times the logarithm to the base 10 of the quotient of the observed ratio divided by the reference ratio, usually 1 ampere per newton per square meter. *See also:* **loudspeaker.**   176

free-field voltage response (receiving voltage sensitivity) (electroacoustic transducer used for sound reception). The ratio of the voltage appearing at the output terminals of the transducer when the output terminals are open-circuited to the free-field sound pressure existing at the transducer location prior to the introduction of the transducer in the sound field. *Notes:* (1) The free-field response is determined for a plane progressive sound wave whose direction of propagation has a specified orientation with respect to the principal axis of the transducer. (2) The free-field voltage response is usually expressed in decibels, namely, 20 times the logarithm to the base 10 of the quotient of the observed ratio divided by the reference ratio, usually 1 volt per newton per square meter. *See also:* **loudspeaker.**   176

free gyro. A two-degree-of-freedom gyro in which the spin axis may be oriented in any specified attitude. In this gyro, output signals are produced by an angular displacement of the case about an axis other than the spin axis.   46

free impedance (transducer). The impedance at the input of the transducer when the impedance of its load is made zero. *Note:* The approximation is often made that the free electric impedance of an electroacoustic transducer designed for use in water is that measured with the transducer in air. *See also:* **loudspeaker; self-impedance.**   176

free-line call (telephone switching systems). A call to a directory number for which no charge is made.   55

free motion (automatic control). One whose nature is determined only by parameters and initial conditions for the system itself, and not by external stimuli. *Note:* For a linear system, this motion is described by the complementary function of the associated homogeneous differential equation. *Syn.* **free oscillation.**   56

free motional impedance (transducer) (electroacoustics). The complex remainder after the blocked impedance has been subtracted from the free impedance. *See also:* **self-impedance.**   176

free oscillation. The response of a system when no external driving force is applied and energy previously stored in the system produces the response. *Note:* The frequency of such oscillations is determined by the parameters in the system or circuit. The term shock-excited oscillation is commonly used. *See:* **oscillatory circuit.**   111, 210

free progressive wave (free wave) (acoustics). A wave in a medium free from boundary effects. A free wave in a steady state can only be approximated in practice.   176

free-radiation frequencies for industrial, scientific, or medical (ISM) apparatus (electromagnetic compatibility). Center of a band of frequencies assigned to industrial, scientific, or medical equipment either nationally or internationally for which no power limit is specified. *See:* **ISM apparatus.** *See also:* **electromagnetic compatibility.**   220, 167

free-running frequency. The frequency at which a normally synchronized oscillator operates in the absence of a synchronizing signal.   178

free-running sweep (oscilloscopes) (non-real time spectrum analyzer). A sweep that recycles without being triggered and is not synchronized by any applied signal. *See:* **oscillograph.**   185, 68

free-space field intensity. The radio field intensity that would exist at a point in a uniform medium in the absence of waves reflected from the earth or other objects. *See also:* **radiation.**   328

free-space loss. The loss between two isotropic antennas in free space, expressed as a power ratio. *Note:* The free-space loss is usually expressed in decibels and is given by the formula $20 \log (4\pi D/\lambda)$, where D is the separation of the two antennas and λ is the wavelength.   111

free-space transmission (mobile communication). Electromagnetic radiation that propagates unhindered by the presence of obstructions, and whose power or field intensity decreases as a function of distance squared. *See also:* **mobile communication system.**   181

free time (availability). The period of time during which an item is in a condition to perform its required function but is not required to do so.   164

free wave (acoustics). *See:* **free progressive wave.**

freeze-out (telephone circuit). A short-time denial to a subscriber by a speech-interpolation system.   328

***F* region (radio wave propagation).** The region of the terrestrial ionosphere above about 160 kilometers altitude. *See also:* **radiation; radio wave propagation.**   180, 146

freight elevator. An elevator primarily used for carrying freight on which only the operator and the persons necessary for loading and unloading the freight are permitted to ride. *See also:* **elevators.**   328

Frenkel defect (solar cells). A defect consisting of the displacement of a single atom from its place in the atomic lattice of a crystal, the atom then occupying an interstitial position.   113

frequency (periodic function) (wherein time is the independent variable) (1) (general). The number of periods per unit time.   3, 210

(2) (automatic control). The number of periods, or specified fractions of periods, per unit time. *Note:* The frequency may be stated in cycles per second, or in radians per second, where 1 cycle = 2π radians.   56

(3) (transformer). The number of periods occurring per unit time.   53

(4) (pulse terms). The reciprocal of period.   254

frequency-agile radar. A pulse radar in which the transmitter carrier frequency is changed between pulses in a random or pseudo-random way by an amount comparable to the reciprocal of the pulsewidth, or a multiple thereof.   13

frequency allocation (table) (electromagnetic compatibility). (1) The process of designating radio-frequency bands for use by specific radio services; or (2) The resulting table (of frequency allocations). *See also:* **electromagnetic compatibility.**   199

frequency allotment (plan) (electromagnetic compatibility). (1) The process of designating radio frequencies within an allocated band for use within specific geographic areas. (2) The resulting plan (of frequency allotment). *See also:*

electromagnetic compatibility. 199

frequency assignment (list) (electromagnetic compatibility).
(1) The process of designating radio frequency for use by a specific station under specified conditions of operations. (2) The resulting list of frequency assignments. *See also:* electromagnetic compatibility. 199

frequency band. A continuous range of frequencies extending between two limiting frequencies. *Note:* The term frequency band or band is also used in the sense of the term bandwidth. *See also:* **channel; signal; signal wave.**
111, 188, 210, 240, 68

frequency-band number. The number N in the expression 0.3×10^N that defines the range of band N. Frequency band N extends from 0.3×10^N hertz to 3×10^N hertz, the lower limit exclusive, the upper limit inclusive.
210

frequency band of emission (communication band). The band of frequencies effectively occupied by that emission, or the type of transmission and the speed of signaling used. *See also:* **radio transmission.** 111

frequency bands (mobile communication). The frequency allocations that have been made available for land mobile communications by the Federal Communications Commission, including the spectral bands: 25.0 to 50.0 megahertz, 150.8 to 173.4 megahertz, and 450.0 to 470.0 megahertz. *See also:* **mobile communication system.** 181

frequency bias (electric power systems). An offset in the scheduled net interchange power of a control area that varies in proportion to the frequency deviation. *Note:* This offset is in a direction to assist in restoring the frequency to schedule. *See also:* **power system.** 94, 200

frequency bias setting (control area). A factor with negative sign that is multiplied by the frequency deviation to yield the frequency bias. *See also:* **power system.** 200

frequency changer (1) (general). A motor-generator set that changes power of an alternating-current system from one frequency to one or more different frequencies, with or without a change in the number of phases, or in voltage. *See:* **converter.** 328

(2) **(rotating machinery).** A motor-generator set or other equipment which changes power of an alternating-current system from one frequency to another. 63

frequency-changer set (rotating machinery). A motor-generator set that changes the power of an alternating-current system from one frequency to another. 63

frequency-change signaling (telecommunication). A method in which one or more particular frequencies correspond to each desired signaling condition. *Note:* The transition from one set of frequencies to the other may be either a continuous or a discontinuous change in frequency or in phase. *See also:* **frequency modulation.** 194

frequency characteristics (telephone sets). Electrical and acoustical properties as functions of frequency. *Note:* Examples include an amplitude-frequency characteristic and an impedance-frequency characteristic. 122

frequency, chopped. *See:* **rate, chopping.**

frequency control. The regulation of frequency within a narrow range. *See also:* **generating station.** 64

frequency-conversion transducer. *See:* **conversion transducer.**

frequency converter. *See:* **frequency changer.**

frequency converter, commutator type (rotating machinery).

A polyphase machine the rotor of which has one or two windings connected to slip rings and to a commutator. *Note:* By feeding one set of terminals with a voltage of given frequency, a voltage of another frequency may be obtained from the other set of terminals. *See also:* **asynchronous machine.** 63

frequency, corner (asymptotic form of Bode diagram) (control system, feedback). The frequency indicated by a breakpoint, that is, the junction of two confluent straight lines asymptotic to the log gain curve. *Note:* One breakpoint is associated with each distinct real root of the characteristic equation, one with each set of repeated roots, and one with each pair of complex roots. For a single real root, corner frequency (in radians per second) is the reciprocal of the corresponding time constant (in seconds), and the corresponding phase angle is halfway between the phase angles belonging to the asymptotes extended to infinity. *See also:* **control system, feedback.** 56, 329

frequency, cyclotron. *See:* **cyclotron frequency.**

frequency, damped (automatic control). The apparent frequency of a damped oscillatory time response of a system resulting from a nonoscillatory stimulus. *Note:* The value of the frequency in a particular system depends somewhat on the subsidence ratio. *See also:* **control system, feedback.** 56, 329

frequency departure (telecommunication). The amount of variation of a carrier frequency or center frequency from its assigned value. *Note:* The term frequency deviation, which has been used for this meaning, is in conflict with this essential term as applied to phase and frequency modulation and is therefore deprecated for future use in the above sense. *See also:* **radio transmission.** 111

frequency-dependent negative resistor (circuits and systems). An impedance of the form $1/(Ks^2)$, where K is a real positive constant and s is the complex frequency variable.
67

frequency deviation (1) (power system). System frequency minus the scheduled frequency. *See also:* **frequency modulation; frequency departure.** 94

(2) **(telecommunication; frequency modulation).** The peak difference between the instantaneous frequency of the modulated wave and the carrier frequency.
111, 242, 185

(3) **(frequency modulation broadcast receivers).** The difference between the instantaneous frequency of the modulated wave and the carrier frequency. 16

frequency distortion. A term commonly used for that form of distortion in which the relative magnitude of the different frequency components of a complex wave are changed in transmission. *Note:* When referring to the distortion of the phase-versus-frequency characteristic, it is recommended that a more specific term such as phase-frequency distortion or delay distortion be used. *See also:* **amplitude distortion; distortion; distortion, amplitude-frequency.** 111

frequency diversity (telecommunication). *See:* **frequency diversity reception.**

frequency-diversity reception (frequency diversity telecommunication). That form of diversity reception that utilizes transmission at different frequencies. *See also:* **radio receiver.** 328

frequency divider. A device for delivering an output wave whose frequency is a proper fraction, usually a submultiple, of the input frequency. *Note:* Usually the output frequency is an integral submultiple or an integral proper fraction of the input frequency. *See also:* **harmonic conversion transducer.** 111, 59

frequency division multiple access (communication satellite). A method of providing multiple access to a communication satellite in which the transmissions from a particular earth station occupy a particular assigned frequency band. In the satellite the signals are simultaneously amplified and transposed to a different frequency band and retransmitted. The earth station identifies its receiving channel according to its assigned frequency band in the satellite signal. 84

frequency-division multiplex (telecommunication). The process or device in which each modulating wave modulates a separate subcarrier and the subcarriers are spaced in frequency. *Note:* Frequency division permits the transmission of two or more signals over a common path by using different frequency bands for the transmission of the intelligence of each message signal. 111, 242

frequency-division switching (telephone switching systems). A method of switching that provides a common path with a separate frequency band for each of the simultaneous calls. 55

frequency doubler. A device delivering output voltage at a frequency that is twice the input frequency. 111

frequency drift (non-real time spectrum analyzer). Gradual shift or change in displayed frequency over a period of time due to change in components (Hz/sec), (Hz/°C), etcetera. 68

frequency droop (electric power system). The absolute change in frequency between steady state no load and steady state full load. 89

frequency hopping (communication satellite). A modulation technique used for multiple access; frequency-hopping systems employ switching of the transmitted frequencies at a rate equal to or lower than the sampling rate of the information transmitted. Selection of the particular frequency to be transmitted can be made from a fixed sequence or can be selected in a pseudo-random manner from a set of frequencies covering a wide bandwidth. The intended receiver would frequency-hop in the same manner as the transmitter in order to retrieve the desired information. 84

frequency, image (heterodyne frequency converters in which one of the two sidebands produced by beating is selected). An undesired input frequency capable of producing the selected frequency by the same process. *Note:* The word **image** implies the mirrorlike symmetry of signal and image frequencies about the beating-oscillator frequency or the intermediate frequency, whichever is the higher. *See also:* **radio receiver.** 328

frequency influence (electric instrument) (instruments other than frequency meters). The percentage change (of full-scale value) in the indication of an instrument that is caused solely by a frequency departure from a specified reference frequency. *Note:* Because of the dominance of 60 hertz as the common frequency standard in the United States, alternating-current (power-frequency) instruments are always supplied for that frequency unless otherwise specified. *See also:* **accuracy rating (instrument).**
280, 294

frequency, instantaneous. *See:* **instantaneous frequency.**

frequency interlace (color television). The effect of inter-

meshing of the frequency spectrum of a modulated color subcarrier and the harmonics of the horizontal scanning frequency for the purpose of minimizing the visibility of the modulated color subcarrier. 18

frequency linearity (non-real time spectrum analyzer). The linearity of the relationship between the input frequency and the displayed frequency. *See also:* **linearity.** 68

frequency lock (power-system communication). A means of recovering in a single-sideband suppressed-carrier receiver the exact modulating frequency that is applied to a single-sideband transmitter. *See also:* **power-line carrier.**
59

frequency locus (control systems). For a nonlinear system or element whose describing function is both frequency-dependent and amplitude-dependent, a plot of the describing function, in any convenient coordinate system.
56

frequency meter. An instrument for measuring the frequency of an alternating current. *See also:* **instrument.**
328

frequency meter, cavity resonator (waveguide components). A cavity resonator used to determine frequency. *See also:* **cavity resonator.** 166

frequency-modulated cyclotron. A cyclotron in which the frequency of the accelerating electric field is modulated in order to hold the positively charged particles in synchronism with the accelerating field despite their increase in mass at very high energies. *See also:* **electron devices, miscellaneous.** 190

frequency-modulated radar (FM radar). A form of radar in which the radiated wave is frequency modulated and the returning echo beats with the wave being radiated, thus enabling the range to be measured. *See also:* **radar.** 328

frequency-modulated transmitter. A transmitter that transmits a frequency-modulated wave. *See also:* **radio transmitter.** 111, 240

frequency modulation (1) (electrical conversion). The cyclic or random dynamic variation, or both, of instantaneous frequency about a mean frequency during steady-state electric system operation. *See also:* **electric conversion equipment.** 186
(2) (FM) (telecommunication). Angle modulation in which the instantaneous frequency of a sine-wave carrier is caused to depart from the carrier frequency by an amount proportional to the instantaneous value of the modulating wave. *Note:* Combinations of phase and frequency modulation are commonly referred to as **frequency modulation.**
111, 59, 194, 242

frequency modulation—frequency modulation (FM-FM) telemetry (communication satellite). A method of multiplexing many telemetry channels by first frequency modulating subcarriers, combining the modulated subcarriers and finally frequency modulating the radio carrier. This method is widely used for satellite transmissions and follows standards set by Inter Range Instrumentation Group (IRIG). 84

frequency-modulation (friction) noise ("scrape flutter"). Frequency modulation of the signal in the range above approximately 100 Hz resulting in distortion which may be perceived as a noise added to the signal (that is, a noise not present in the absence of a signal). 145

frequency monitor. An instrument for indicating the amount of deviation of a frequency from its assigned value. *See also:* **instrument.** 328

frequency multiplier. A device for delivering an output wave whose frequency is an exact integral multiple of the input frequency. *Note:* Frequency doublers and triplers are common special cases of frequency multipliers. *See also:* **harmonic conversion transducer.**   111, 59

frequency of charging, resonance. The frequency at which resonance occurs in the charging circuit of a pulse-forming network.   137

frequency pulling (oscillator). A change of the generated frequency of an oscillator caused by a change in load impedance. *See also:* **oscillatory circuit; waveguide.**   190, 125

frequency, pulse repetition. The number of pulses per unit time of a periodic pulse train or the reciprocal of the pulse period. *Note:* This term also includes the average number of pulses per unit time of aperiodic pulse trains where the periods are of random duration. *See also:* **pulse.**   185

frequency range (1) (general). A specifically designated part of the frequency spectrum.
(2) (transmission system). The frequency band in which the system is able to transmit power without attenuating or distorting it more than a specified amount.
(3) (device). The range of frequencies over which the device may be considered useful with various circuit and operating conditions. *Note:* Frequency range should be distinguished from bandwidth, which is a measure of useful range with fixed circuits and operating conditions. *See also:* **signal wave.**   210, 190, 125
(4) (acousto-optic deflector). The frequency range, Δf, over which the diffraction efficiency is greater than some specified minimum.   72
(5) (non-real time spectrum analyzer). That range of frequency over which the instrument performance is specified (hertz to hertz).   68

frequency record (electroacoustics). A recording of various known frequencies at known amplitudes, usually for the purpose of testing or measuring. *See also:* **phonograph pickup.**   176

frequency regulation (electric power system). The percentage change in frequency from steady state no load to steady state full load, which is a function of the engine and governing system:

$$\%R = \frac{F_{nl} - F_{fl}}{F_{nl}} \times 100$$

   89

frequency relay. A relay that responds to the frequency of an alternating electrical input quantity.   127, 103

frequency resolution (radar). The ability of a receiver or signal processing system to detect or measure separately two or more signals which differ only in frequency. The classic measure of frequency resolution is the minimum frequency separation of two otherwise identical signals which permits the given system to distinguish that two frequencies are present and to extract the desired information from both of them. When the separation is done by means of a tunable bandpass filter system, the resolution is often specified as the width of the frequency response lobe measured a specific amount (such as 3 decibels) down from the peak response.   13

frequency response (1) (power supplies). The measure of an amplifier or power supply's ability to respond to a sinusoidal program. *Notes:* (1) The frequency response measures the maximum frequency for full-output voltage ex-

cursion. (2) Frequency response connotes amplitude-frequency response, which should be used in full, particularly if phase-frequency response is significant. This frequency is a function of the slewing rate and unity-gain bandwidth. *See also:* **amplitude-frequency response; power supply.**   228, 186
(2) (non-real time spectrum analyzer). The peak-to-peak variation of the displayed amplitude over a specified center frequency range, measured at the center frequency, (decibel) (dB).   68
(3) (speed governing of hydraulic turbines). A characteristic, expressed by formula or graph, which describes the dynamic and steady-state response of a physical system in terms of the magnitude ratio and the phase displacement between a sinusoidally varying input quantity and the fundamental of the corresponding output quantity as a function of the fundamental frequency.   8

frequency-response characteristic (linear system). (1) (signal-transmission system, industrial control). The frequency-dependent relation, in both gain and phase difference, between steady-state sinusoidal outputs. *Notes:* (1) With nonlinearity, as evidenced by distortion of a sinusoidal input of specified amplitudes, the relation is based on that sinusoidal component of the output having the frequency of the input. (2) Mathematically, the frequency-response characteristic is the complex function of $S = j\omega$:

$$A_o(j\omega)/A_i(j\omega) \exp\{j[\theta_o(j\omega) - \theta_i(j\omega)]\}$$

where
A_i = input amplitude
A_o = output amplitude
θ_i = input phase angle
   (relative to fixed reference)
θ_o = output phase angle
   (relative to same reference)
See also: **control system, feedback; signal.**
   56, 188, 56, 219, 206
(2) (excitation control systems). In a linear system, the frequency-dependent relation, in both gain and phase difference between steady-state sinusoidal inputs and the resultant steady-state sinusoidal outputs. *Note:* A plot of gain in logarithmic terms and phase angle in degrees vs logarithmic frequency is commonly called a Bode diagram. See figure below for identification of the principal characterics of interest.   353

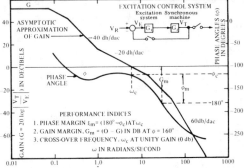

Typical open-loop frequency response of an excitation control system with the synchronous machine open circuited.

frequency-response equalization (1) (acoustics). The effect of all frequency discriminative means employed in a transmission system to obtain a desired over-all frequency response.   176

(2) (circuits and systems). The process of modifying a frequency response of one network by introducing a frequency response of another network so that, within the band of interest, the combined response follows a specified characteristic.   67

frequency-selective ringing (telephone switching systems). Selective ringing that employs currents of several frequencies to activate ringers, each of which is tuned mechanically or electrically, or both, to one of the frequencies so that only the desired ringer responds.   55

frequency-selective voltmeter. A selective radio receiver, with provisions for output indication.   314, 199

frequency selectivity (selectivity). (1) A characteristic of an electric circuit or apparatus in virtue of which electric currents or voltages of different frequencies are transmitted with different attenuation. (2) The degree to which a transducer is capable of differentiating between the desired signal and signals or interference at other frequencies. *See also:* **transducer.**   328

frequency-sensitive relay. A relay that operates when energized with voltage, current, or power within specific frequency bands. *See also:* **relay.**   259

frequency sensitivity (characteristic insertion loss) (attenuator). Peak-to-peak variation in decibels through the specified frequency range.   110

frequency-shift keying (FSK) (telecommunication). The form of frequency modulation in which the modulating wave shifts the output frequency between predetermined values, and the output wave has no phase discontinuity. *Note:* Commonly, the instantaneous frequency is shifted between two discrete values termed the mark and space frequencies. *See also:* **telegraphy.**   242, 111, 59

frequency-shift pulsing (telephone switching systems). A means of transmitting digital information in which a sequence of two frequencies is used.   55

frequency span (non-real time spectrum analyzer). The magnitude of the frequency segment displayed (Hz, Hz/div).   68

frequency stability. *See:* **carrier frequency stability of a transmitter.**

frequency stabilization. The process of controlling the center or carrier frequency so that it differs from that of a reference source by not more than a prescribed amount. *See also:* **frequency modulation.**   111

frequency standard (1) (electric power systems). A device that produces a standard frequency. *See:* **standard frequency.** *See also:* **speed-governing system.**   94

(2) (facsimile). A local precision source supplying a stable frequency which is used, among other things, for control of synchronous scanning and recording devices.   11

frequency swing (frequency modulation). The peak difference between the maximum and the minimum values of the instantaneous frequency. *Note:* The term frequency swing is sometimes used to describe the maximum swing permissible under specified conditions. Such usage should preferably include a specific statement of the conditions. *See also:* **frequency modulation.**   111, 59

frequency time matrix (communication satellite). A modulation technique used for multiple access; frequency-time

matrix systems require the simultaneous presence of energy in more than one time and frequency assignment to produce an output signal. The requirement for presence in several time and/or frequency slots reduces the probability of mutual interference when a number of users are simultaneously transmitting.   84

frequency tolerance (radio transmitter). The extent to which a characteristic frequency of the emission, for example, the carrier frequency itself or a particular frequency in the sideband, may be permitted to depart from a specified reference frequency within the assigned band. *Note:* The frequency tolerance may be expressed in hertz or as a percentage of the reference frequency. *See also:* **radio transmitter.**   328

frequency transformation (circuits and systems). The replacing of the frequency variable s in a function $f(s)$ with a new variable z implicitly defined by $s = g(z)$. This may be done, as examples, to convert a low-pass function into a band-pass function or to make calculations less affected by rounding errors.   67

frequency translation (data transmission). The amount of frequency difference between the received audio signals and the original audio signals after passing through a communication channel.   59

frequency tripler. A device delivering output voltage at a frequency that is three times the input frequency.   111

frequency-type telemeter. A telemeter that employs the frequency of a periodically recurring electric signal as the translating means. *See also:* **telemetering.**   103, 202

frequency, undamped (frequency, natural). (1) Of a second-order linear system without damping, the frequency of free oscillation in radians per unit time or in hertz. (2) Of any system whose transfer function contains the quadratic factor $s^2 + 2\zeta\omega_n s + \omega_n^2$ in the denominator, the value $\omega_n (0 \leq \zeta < 1)$. (3) Of a closed loop control system or controlled system, a frequency at which continuous oscillation (hunting) can occur without periodic stimuli. *Note:* In linear systems, the undamped frequency is the phase crossover frequency. With proportional control action only, the undamped frequency of a linear system may be obtained by raising (in most cases) the proportional gain until hunting occurs. This value of gain has been called the "ultimate gain" and the undamped period the "ultimate period." *Syn.* **natural frequency.**

  56, 329

frequently-repeated overload rating (power converter). The maximum direct current that can be supplied by the converter on a repetitive basis under normal operating conditions. *See:* **power rectifier.**   208

freshening charge (nuclear power generating stations) (lead storage batteries). The charge given to a storage battery following nonuse or storage.   31, 76

Fresnel lens antenna. An antenna consisting of a feed and a planar lens which transmits the radiated power from the feed through the central zone and alternate Fresnel zones of the illuminating field on the lens. *Syn.* **zone-plate lens antenna.**   111

Fresnel pattern (antenna). A radiation pattern obtained in the Fresnel region. *See also:* **antenna.**   179, 111

Fresnel region (1) (general). The region (or regions) adjacent to the region in which the field of an antenna is focused (that is, just outside the Fraunhofer region). *See:*

radiating near-field region. *See also:* **antenna.**
179, 111

(2) (data transmission). The region between the antenna and the Fraunhofer region. *Note:* If the antenna has a well-defined aperture *D* in a given aspect, the Fresnel region in that aspect is commonly taken to extend a distance of $2D^2/\lambda$ in that aspect, λ being the wavelength.   59

fretting (corrosion). Deterioration resulting from repetitive slip at the interface between two surfaces. *Note:* When deterioration is further increased by corrosion, the term fretting-corrosion is used.   221, 205

friction and windage loss (rotating machinery). The power required to drive the unexcited machine at rated speed with the brushes in contact, deducting that portion of the loss that results from: (1) Forcing the gas through any part of the ventilating system that is external to the machine and cooler (if used). (2) The driving of direct-connected flywheels or other direct-connected apparatus. *See also:* **asynchronous machine; direct-current commutating machine.**   244, 63

friction electrification. *See:* **triboelectrification.**

friction tape. A fibrous tape impregnated with a sticky moisture-resistant compound that provides a protective covering for insulation.

fritting, relay. *See:* **relay fritting.**

frogging (telephone equipment). A switching technique whereby the tip and ring leads of the test specimen are reversed relative to the driving and/or terminating test circuits. *See also:* **metallic transmission port.**   39

frog-leg winding (rotating machinery). A composite winding consisting of one lap winding and one wave winding placed on the same armature and connected to the same commutator. *See:* **direct-current commutating machine.**   63

front (motor or generator). The front of a normal motor or generator is the end opposite the largest coupling or driving pulley. *See also:* **asynchronous machine; direct-current commutating machine.**   63

front-and-back connected switch. A switch having provisions for some of the circuit connections to be made in front of, and others in back of, the mounting base.   27

front-connected switch. A switch in which the current-carrying conductors are connected to the fixed terminal blocks in front of the mounting base.   27

front contact (1) (general). A part of a relay against which, when the relay is energized, the current-carrying portion of the movable neutral member is held so as to form a continuous path for current. *See also:* **a contact.**

(2) (relay systems). A contact which is closed when the relay is picked up. *Syn.* **"a" contact.**   127

(3) (utility-consumer interconnections). A contact that is open when the relay is deenergized.   128

front end (communication satellite). The first stage of amplification or frequency conversion immediately following the antenna in a receiving system.   83

front end computer (data communication). A communications computer associated with a host computer. It may perform line control, message handling, code conversion, error control and applications functions such as control and operation of special purpose terminals. *See also:* **communications computer.**   12

front-of-wave impulse sparkover voltage (arrester). The

impulse sparkover voltage with a wavefront that rises at a uniform rate and causes sparkover on the wavefront.   62

front-of-wave lightning impulse test (transformer). A voltage impulse, with a specified rate-of-rise, that is terminated intentionally by sparkover of a gap which occurs on the rising front of the voltage wave with a specified time to sparkover, and a specified minimum crest voltage. Complete front-of-wave tests involve application of the following sequence of impulse waves: (1) one reduced full-wave; (2) two front-of-waves; (3) two chopped waves; (4) one full wave.   53

front pitch (rotating machinery). The coil pitch at the connection end of a winding (usually in reference to a wave winding).   63

front porch (television). The portion of a composite picture signal that lies between the leading edge of the horizontal blanking pulse and the leading edge of the corresponding synchronizing pulse. *See also:* **television.**   178

front-to-back ratio (1) (general). The ratio of the directivity of an antenna to directive gain in a specified direction toward the back.   179, 111

(2) (directional antenna). The ratio of its effectiveness toward the front, to its effectiveness toward the back.
59

fruit. *See:* **fruit pulse.**

fruit pulse (fruit*). A pulse reply received as the result of interrogation of a transponder by interrogators not associated with the responsor in question.   254

* Deprecated

***F* scan.** *See:* ***F* display.**

***F* scope.** *See:* ***F* display.**

FSK (telecommunication). *See:* **frequency-shift keying.**

FS to AM converter (facsimile). *See:* **receiving converter, facsimile.**

FTC. *See:* **fast-time-constant circuit.**

fuel (fuel cells). A chemical element or compound that is capable of being oxidized. *See also:* **electrochemical cell.**
223, 186

fuel adjustment clause (electric power utilization). A clause in a rate schedule that provides for adjustment of the amount of the bill as the cost of fuel varies from a specified base amount per unit.   112

fuel-and-oil quantity electric gauge. A device that measures, by means of bridge circuits and an indicator with separate pointers and scales, the quantity of fuel and oil in the aircraft tanks.   328

fuel battery. An energy-conversion device consisting of more than one fuel cell connected in series, parallel, or both. *See also:* **fuel cell.**   223, 186

fuel-battery power-to-volume ratio. The kilowatt output per envelope volume of the fuel battery (exclusive of the fuel, oxidant, storage, and auxiliaries). *See also:* **fuel cell.**
223, 186

fuel-battery power-to-weight ratio. The kilowatt output per unit weight of the fuel battery (exclusive of the fuel, oxidant, storage, and auxiliaries). *See also:* **fuel cell.**
223, 186

fuel cell. An electrochemical cell that can continuously change the chemical energy of a fuel and oxidant to electric energy by an isothermal process involving an essentially invariant electrode-electrolyte system.   223, 186

fuel-cell Coulomb efficiency.   The ratio of the number of electrons obtained from the consumption of a mole of the fuel to the electrons theoretically available from the stated reaction.

$$\text{Coulomb Efficiency} = \frac{\int_0^{t_m} i\,dt}{nF} \times 100$$

t_m = time required to consume a mole of fuel

i = instantaneous current

n = number of electrons furnished in the stated reaction by the fuel molecule

F = Faraday's constant = 96485.3 ± 10.0 absolute joules per absolute volt gram equivalent.

See also: **fuel cell.**   223, 186

fuel-cell standard voltage (at 25 degrees Celsius).   The voltage associated with the stated reaction and determined from the equation

$$E^0 = \frac{-J\Delta G^0}{nF}$$

E^0 = fuel-cell standard voltage

J = Joule's equivalent = 4.1840 absolute joules per calorie

ΔG^0 = standard free energy changes in kilo-calories/mole of fuel

n = number of electrons furnished in the stated reaction by the fuel molecule

F = Faraday's constant = 96485.3 ± 10.0 absolute joules per absolute volt gram equivalent.

See also: **fuel cell.**   223, 186

fuel-cell system.   An energy conversion device consisting of one or more fuel cells and necessary auxiliaries. *See also:* **fuel cell.**   223, 186

fuel-cell-system energy-to-volume ratio.   The kilowatt-hour output per displaced volume of the fuel-cell system (including the fuel, oxidant, and storage). *See also:* **fuel cell.**   223, 186

fuel-cell-system energy-to-weight ratio.   The kilowatt-hour output per unit weight of the fuel-cell system (including the fuel, oxidant, and storage). *See also:* **fuel cell.**   223, 186

fuel-cell-system power-to-volume ratio.   The kilowatt output per displaced volume of the fuel-cell system (exclusive of the fuel, oxidant, and storage). *See also:* **fuel cell.**   223, 186

fuel-cell-system power-to-weight ratio.   The kilowatt output per unit weight of the fuel-cell system (exclusive of the fuel, oxidant, and storage). *See also:* **fuel cell.**   223, 186

fuel-cell-system standard thermal efficiency.   The efficiency of a system made up of a fuel cell and auxiliary equipment. *Note:* This efficiency is expressed as the ratio of (1) the electric energy delivered to the load circuit to (2) the enthalpy change for the stated cell reaction.

$$\text{Thermal Efficiency} = \frac{\int_0^{t_m} (E_{IL} \times i_L)\,dt}{\Delta H^0}$$

t_m = time required to consume a mole of fuel

E_{IL} = fuel-cell-system working voltage

i_L = instantaneous current into the load

ΔH^0 = enthalpy change for the stated cell reaction at standard conditions.

See also: **fuel cell.**   223, 186

fuel-cell-system working voltage.   The voltage at the load terminals of a fuel-cell system delivering current into the load. *See also:* **fuel cell.**   223, 186

fuel-cell working voltage.   The voltage at the terminals of a single fuel-cell delivering current into system auxiliaries and load. *See also:* **fuel cell.**   223, 186

fuel-control mechanism (gas turbines).   All devices, such as power-amplifying relays, servomotors, and interconnections required between the speed governor and the fuel-control valve. *See also:* **speed-governing system.**   98, 58

fuel-control system (gas turbines).   Devices that include the fuel-control valve and all supplementary fuel-control devices and interconnections necessary for adequate control of the fuel entering the combustion system of the gas turbine. *Note:* The supplementary fuel-control devices may or may not be directly actuated by the fuel-control mechanism. *See also:* **speed-governing system.**   98, 58

fuel-control valve (gas turbines).   A valve or any other device operating as a final fuel-metering element controlling fuel input to the gas turbine. *Notes:* (1) This valve or device may be directly or indirectly controlled by the fuel-control mechanism. (2) Variable-displacement pumps, or other devices that operate as the final fuel-control element in the fuel-control system, and that control fuel entering the combustion system are fuel-control valves. *See also:* **speed-governing system.**   98, 58

fuel economy.   The ratio of the chemical energy input to a generating station to its net electric output. *Note:* Fuel economy is usually expressed in British thermal units per kilowatthour. *See also:* **generating station.**   64

fuel elements, nuclear (nuclear power generating station).   An assembly of rods, tubes, plates, or other geometrical forms into which nuclear fuel is contained for use in a reactor.   112

fuel-pressure electric gauge.   A device that measures the fuel pressure (usually in pounds per square inch) at the carburetor of an aircraft engine. *Note:* It provides remote indication by means of a self-synchronous generator and motor.   328

fuel reprocessing, nuclear (nuclear power generating station).   The processing of irradiated reactor fuel to recover the unused fissionable material, or fission products, or both.   112

fuel stop valve (gas turbines).   A device that, when actuated, shuts off all fuel flow to the combustion system, including that provided by the minimum fuel limiter. *See also:* **speed-governing system.**   98, 58

fulguration.   *See:* **electrodesiccation.**

full automatic plating.   Mechanical plating in which the cathodes are automatically conveyed through successive cleaning and plating tanks.   328

full availability (telephone switching systems).   Availability that is equal to the number of outlets in the desired group.   55

full-direct trunk group (telephone switching systems). A full trunk group between end offices. 55

full duplex (communication circuit) (telecommunication). Method of operation where each end can simultaneously transmit and receive. *Note:* Refers to a communications system or equipment capable of transmission simultaneously in two directions. *See:* **duplex.** 194, 59

full-field relay (industrial control). A relay that functions to maintain full field excitation of a motor while accelerating on reduced armature voltage. *See also:* **relay.** 206

full impulse voltage. An aperiodic transient voltage that rises rapidly to a maximum value and falls, usually less rapidly, to zero. See the following figure. *See also:* **test voltage and current; full-wave voltage impulse.** 307, 201

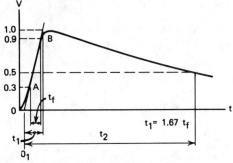

Full impulse voltage.

full-impulse wave (surge arresters). An impulse wave in which there is no sudden collapse. 244, 62

full load (test, measurement and diagnostic equipment). The greatest load that a circuit is designed to carry under specific conditions; any additional load is overload. 54, 59

full-load speed (electric drive) (industrial control). The speed that the output shaft of the drive attains with rated load connected and with the drive adjusted to deliver rated output at rated speed. *Note:* In referring to the speed with full load connected and with the drive adjusted for a specified condition other than for rated output at rated speed, it is customary to speak of the full-load speed under the (stated) conditions. *See also:* **electric drive.** 206

full-magnetic controller (industrial control). An electric controller having all of its basic functions performed by devices that are operated by electromagnets. *See also:* **electric controller.** 206, 3

full-pitch winding (rotating machinery). A winding in which the coil pitch is 100 percent, that is, equal to the pole pitch. *See also:* **asynchronous machine; direct-current commutating machine.** 63

full scale (analog computer). The nominal maximum value of a computer variable or the nominal maximum value at the output of a computing element. *Note:* Also sometimes used to indicate the entire computing voltage range, such as 20V is full scale for a computer whose voltage ranges from +10V to −10V. The latter definition is generally used in manufacturers' specifications, that is, 0.01% of full scale. 9, 77

full-scale value. The largest value of the actuating electrical quantity that can be indicated on the scale or, in the case of instruments having their zero between the ends of the scale, the full-scale value is the arithmetic sum of the values of the actuating electrical quantity corresponding to the two ends of the scale. *Note:* Certain instruments, such as power-factor meters, are necessarily excepted from this definition. *See also:* **accuracy rating (instrument); instrument.** 280, 294

full span—max span (non-real time spectrum analyzer). A mode of operation in which the spectrum analyzer scans an entire band. 68

full speed (data transmission). Referring to transmission of data in teleprinter systems at the full rated speed of the equipment. 59

full-trunk group (telephone switching systems). A trunk group, other than a final trunk group, that does not overflow calls to another trunk group. 55

full-voltage starter (industrial control). A starter that connects the motor to the power supply without reducing the voltage applied to the motor. *Note:* Full-voltage starters are also designated as across-the-line starters. *See also:* **starter.** 206

full-wave lightning impulse test (transformer). Application of the "standard lightning impulse" wave, a full-wave having a front time of 1.2 microseconds and a time to half value of 50 microseconds, described as a 1.2/50 impulse. 53

full-wave rectification (rectifying process) (power supplies). Full-wave rectification inverts the negative half-cycle of the input sinusoid so that the output contains two half-sine pulses for each input cycle. A pair of rectifiers arranged as shown with a center-tapped transformer or a bridge arrangement of four rectifiers and no center tap are both methods of obtaining full-wave rectification. *See also:* **rectification; rectifier.** 208, 228, 186

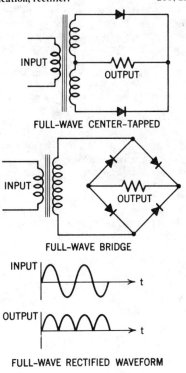

FULL-WAVE CENTER-TAPPED

FULL-WAVE BRIDGE

FULL-WAVE RECTIFIED WAVEFORM

Full-wave rectification.

full-wave rectifier. *See:* **full-wave rectification.**

full-wave rectifier circuit. A circuit that changes single-phase alternating current into pulsating unidirectional current, utilizing both halves of each cycle. *See also:* **rectification.** 237, 66, 118

full-wave voltage impulse (surge arresters). A voltage impulse that is not interrupted by sparkover, flashover, or puncture. *See:* **full-impulse voltage.** 62

full width at half maximum (FWHM) (scintillation counters). The full width of a distribution measured at half the maximum ordinate. For a normal distribution, it is equal to $2(2 \ln 2)^{1/2}$ times the standard deviation (σ).

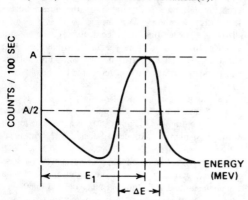

Full width at half maximum (in this case, ΔE).

Note: The expression **full width at half maximum,** given either as an absolute value or as a percentage of the value of the argument at the maximum of the distribution curve, is frequently used in nuclear physics as an approximate description of a distribution curve. Its significance can best be made clear by reference to a typical distribution curve, shown in the figure, of the measurement of the energy of the gamma rays from Cs^{137} with a scintillation counter spectrometer. The measurement is made by determining the number of gamma-ray photons detected in a prescribed interval of time, having measured energies falling within a fixed energy interval (channel width) about the values of energy (channel position) taken as argument of the distribution function. The abscissa of the curve shown is energy in megaelectronvolts (MeV) units and the ordinate is counts per given time interval per megaelectronvolt energy interval. The maximum of the distribution curve shown has an energy E_1 megaelectronvolts. The height of the peak is A_1 counts/100 seconds/megaelectronvolts. The full width at half maximum ΔE is measured at a value of the ordinate equal to $A_1/2$. The percentage full width at half maximum is $100 \, \Delta E/E_1$. It is an indication of the width of the distribution curve, and where (as in the example cited) the gamma-ray photons are monoenergetic, it is a measure of the resolution of the detecting instrument. When the distribution curve is a Gaussian curve, the percentage full width at half maximum is related to the standard deviation σ by

$$100 \, \frac{\Delta E}{E_1} = 100 \times 2(2 \ln 2)^{1/2} \times \sigma.$$

See also: **scintillation counter.** 117, 118

fully connected network (data communication). A network in which each node is directly connected with every other node. 12

fume-resistant (industrial control). So constructed that it will not be injured readily by exposure to the specified fume. 103, 202, 206

function (1) (general). When a mathematical quantity u depends on a variable quantity x so that to each value of x (within the interval of definition) there correspond one or more values of u, then u is a function of x written $u = f(x)$. The variable x is known as the independent variable or the argument of the function. When a quantity u depends on two or more variables $x_1, x_2, \cdots, x_n$ so that for every set of values of $x_1, x_2, \cdots, x_n$ (within given intervals for each of the variables) there correspond one or more values of u, then u is a function of $x_1, x_2, \cdots, x_n$ and is written $u = f(x_1, x_2, \cdots, x_n)$. The variables $x_1, x_2, \cdots, x_n$ are the independent variables or arguments of the function. 210

(2) (vector). When a scalar or vector quantity u depends upon a variable vector V so that if for each value of V (within the region of definition) there correspond one or more values of u, then u is a function of the vector V. 210

(3) (test, measurement and diagnostic equipment). The action or purpose which a specific item is intended to perform or serve. 54

functional adjectives (pulse terms) (1) linear. Pertaining to a feature whose magnitude varies as a function of time in accordance with the following relation or its equivalent:

$$m = a + bt$$

254

(2) exponential. Pertaining to a feature whose magnitude varies as a function of time in accordance with either of the following relations or their equivalents:

$$m = ae^{-bt}$$
$$m = a\left(1 - e^{-bt}\right)$$

254

(3) Gaussian. Pertaining to a waveform or feature whose magnitude varies as a function of time in accordance with the following relation or its equivalent:

$$m = ae^{-b(t-c)^2}, \quad b > 0$$

254

(4) trigonometric. Pertaining to a waveform or feature whose magnitude varies as a function of time in accordance with a specified trigonometric function or by a specified relationship based on trigonometric functions (for example, cosine squared). 254

functional area(s) (nuclear power generating station). Location(s) designated within the control room to which displays and controls relating to specific function(s) are assigned. 358

functional component (power switchgear). A device which performs a necessary function for the proper operation and application of a unit of equipment. 103

functional designation (abbreviation) (1) (general). Letters, numbers, words, or combinations thereof, used to indicate the function of an item or a circuit, or of the position or state of a control of adjustment. Compare with: **letter**

combination, reference designation, symbol for a quantity. *See also:* **abbreviation.**   173

(2) (electric and electronics parts and equipments). Words, abbreviations, or meaningful number or letter combinations, usually derived from the function of an item (for example: slew, yaw), used on drawings, instructional material, and equipment to identify an item in terms of its function. *Note:* A functional designation is not a reference designation nor a substitute for it.   17

functional diagram (test, measurement and diagnostic equipment). A diagram that represents the functional relationships among the parts of a system.   54

functional nomenclature. Words or terms which define the purpose, equipment or system, for which the component is required.   58

functional unit. A system element that performs a task required for the successful operation of the system. *See also:* **system.**   209

function class-A (back-up) current-limiting fuse. A fuse capable of interrupting all currents from the rated maximum interrupting current down to the rated minimum interrupting current. *Note:* The rated minimum interrupting current for such fuses is higher than the minimum melting current that causes melting of the fusible element in one hour.   103

function class-G (general purpose) current-limiting fuse (as applied to a high-voltage current-limiting fuse). A fuse capable of interrupting all currents from the rated maximum interrupting current down to the current that causes melting of the fusible element in one hour.   103

function, coupling (control systems). A mathematical, graphical, or tabular statement of the influence which one element or subsystem has on another element or subsystem, expressed as the effect/cause ratio of related variables or their transforms. *Note:* For a multi-terminal system described by m differential equations and having m input transforms $R_1 \cdots R_m$ and m output transforms $C_1 \cdots C_m$, the coupling functions consist of all effect/cause ratios which can be formed from transforms bearing unlike-numbered subscripts.   56

function, describing (nonlinear element under periodic input) (control system, feedback). A transfer function based solely on the fundamental, ignoring other frequencies. *Note:* This equivalent linearization implies amplitude dependence with or without frequency dependence. *See also:* **control system, feedback.**   56, 329

function, error transfer (closed loop) (control system, feedback). The transfer function obtained by taking the ratio of the Laplace transform of the error signal to the Laplace transform of its corresponding input signal. *See also:* **control system, feedback.**   56, 329

function generator (analog computer). A computing element whose output is a specified nonlinear function of its input or inputs. Normal usage excludes multipliers and resolvers.   9

(2) (electric power systems). A device in which a mathematical function such as $y = f(x)$ can be stored so that for any input equal to x, an output equal to $f(x)$ will be obtained. *See also:* **speed-governing system.**   94

function generator, bivariant. A function generator having two input variables. *See also:* **electronic analog computer.**   9

function generator, card set (analog computer). A diode function generator whose values are stored and set by means of a punched card and mechanical card reading device.   9

function generator, curve-follower (analog computer). A function generator that operates by automatically following a curve $f(x)$ drawn or constructed on a surface, as the input x varies over its range.   9

function generator, digitally controlled (analog computer). A hybrid component using digital-to-analog converters and digital-to-analog multipliers to insert the linear segment approximation values to the desired arbitrary function. The values are stored in a self-contained digital core memory, which is accessed by the digital-to-analog converters and digital-to-analog multipliers at digital computer speeds (microseconds).   9

function generator, diode. A function generator that uses the transfer characteristics of resistive networks containing biased diodes. *Note:* The desired function is approximated by linear segments whose values are manually inserted by means of potentiometers and switches.   9

function generator, map-reader. A variant function generator using a probe to detect the voltage at a point on a conducting surface and having coordinates proportional to the inputs. *See also:* **electronic analog computer.**   9

function generator, servo. A function generator consisting of a position servo driving a function potentiometer. *See also:* **electronic analog computer.**   9

function generator, switch-type. A function generator using a multitap switch rotated in accordance with the input and having its taps connected to suitable voltage sources. *See also:* **electronic analog computer.**   9

function, loop-transfer (closed loop) (control system, feedback). The transfer function obtained by taking the ratio of the Laplace transform of the return signal to the Laplace transform of its corresponding error signal. *See also:* **control system, feedback.**   56, 329

function, output-transfer (closed loop) (control system, feedback). The transfer function obtained by taking the ratio of the Laplace transform of the output signal to the Laplace transform of the input signal. *See also:* **control system, feedback.**   56, 329

function, probability density (control systems). Pertaining to a real random variable x, the derivative with respect to an arbitrary value X of the variable x, of the probability distribution function of X, if a derivative exists. *Note:* The mathematical expression for this function is

$$g(X) = \frac{d}{dX}[f(X)] = \frac{d}{dX}[P\,(x \leqslant X)]$$

56

function, probability distribution (control systems). Pertaining to a real random variable x, the function of an arbitrary value X- X of this variable, whose value is the probability, P, that the random variable is less than or equal to X. *Note:* The mathematical expression for this function is

$$f(X) = P(x \leqslant X)$$

56

function relay (analog computer). A relay used as a computing element, generally driven by a comparator.   9

function, return-transfer (closed loop) (control system, feedback). The transfer function obtained by taking the ratio of the Laplace transform of the return signal to the Laplace transform of its corresponding input signal. *See also:* **control system, feedback.**   56

function switch (analog computer). A manually operated switch used as a computing element, for example to modify a circuit, to add or delete an input function or constant, etcetera.   9

function, system-transfer (automatic control). The transfer function obtained by taking the ratio of the Laplace transform of the signal corresponding to the ultimately controlled variable to the Laplace transform of the signal corresponding to the command. *See also:* **control system, feedback.**   56, 329

function, transfer (1) (control system, feedback). A mathematical, graphic, or tabular statement of the influence that a system or element has on a signal or action compared at input and at output terminals. *Note:* For a linear system, general usage limits the transfer function to mean the ratio of the Laplace transform of the output to the Laplace transform of the input in the absence of all other signals, and with all initial conditions zero. *See also:* **control system, feedback; transfer function.**   56, 329
(2) (antenna). The complex ratio of the output of the device to its input. It is also the combined phase and frequency responses.   151

function, weighting (control system, feedback). A function representing the time response of a linear system, or element to a unit-impulse forcing function; the derivative of the time response to a unit-step forcing function. *Notes:* (1) The Laplace transform of the weighting function is the transfer function of the system or element. (2) The time response of a linear system or element to an arbitrary input is described in terms of the weighting function by means of the convolution integral. *See also:* **control system, feedback.**   56, 329

function, work. *See:* **work function.**

fundamental component. The fundamental frequency component in the harmonic analysis of a wave. *See also:* **signal wave.**   349

fundamental frequency (1) (signal-transmission system). The reciprocal of the period of a wave.
(2) (mathematically). The lowest frequency component in the Fourier representation of a periodic quantity. *See also:* **signal wave.**   188, 176
(3) (data transmission) (periodic quantity). The frequency of a sinusoidal quantity having the same period as the periodic quantity.   59

fundamental mode. *See:* **dominant mode.**

fundamental mode of propagation (laser-maser). The mode in a **beamguide** or **beam resonator** which has a single maximum for the transverse field intensity over the cross-section of the beam.   363

fundamental-type piezoelectric crystal unit. A unit designed to utilize the lowest frequency of resonance for a particular mode of vibration. *See also:* **crystal.**   328

furnace transformer. A transformer that is connected to an electric arc furnace.   53

fuse. An overcurrent protective device with a circuit-opening fusible part that is heated and severed by the passage of overcurrent through it. *Note:* A fuse comprises all the parts that form a unit capable of performing the prescribed functions. It may or may not be the complete device necessary to connect it into an electric circuit. *See:* **control.**   103, 202, 206

fuse blade (cartridge fuse). A cartridge-fuse terminal having a substantially rectangular cross section.   103, 202

fuse carrier (oil cutout). An assembly of a cap that closes the top opening of an oil-cutout housing, an insulating member, and fuse contacts with means for making contact with the conducting element and for insertion into the fixed contacts of the fuse support. *Note:* The fuse carrier does not include the conducting element (fuse link).   103, 202

fuse clips. The contacts on a base, fuse support, or fuse mounting for connecting into the circuit the current-responsive element with its holding means, if such means are used for making a complete device.   103, 202

fuse condenser. A device that, added to a vented fuse, converts it to a nonvented fuse by providing a sealed chamber for condensation of the gases developed during circuit interruption.   103, 202

fuse contact. *See:* **fuse terminal.**

fuse cutout. A cutout having a fuse link or fuse unit. *Note:* A fuse cutout is a fuse disconnecting switch.   103, 202

fused capacitor unit (series capacitor). A capacitor unit in combination with a fuse, either external or internal to the case, intended to isolate a failed unit from the associated units.   86

fused electrolyte (bath) (fused salt) (electrolyte). A molten anhydrous electrolyte.   328

fused-electrolyte cell. A cell for the production of electric energy when the electrolyte is in a molten state. *See also:* **electrochemistry.**   328

fuse-disconnecting switch (disconnecting fuse). A disconnecting switch in which a fuse unit or fuse holder and fuse link forms all or a part of the blade.   103, 202

fused salt (bath). *See:* **fused electrolyte.**

fused trolley tap (mining). A specially designed holder with enclosed fuse for connecting a conductor of a portable cable to the trolley system or other circuit supplying electric power to equipment in mines. *See also:* **mine feeder circuit.**   328

fused-type voltage transformer. One which is provided with the means for mounting a fuse, or fuses, as an integral part of the transformer in series with the primary winding.   53

fuse filler. *See:* **arc-extinguishing medium.**

fuseholder (cutout base*). A device intended to support a fuse mechanically and connect it electrically in a circuit. *See also:* **cabinet.**   328

* Deprecated

fuseholder (1) (high-voltage fuse). An assembly of a fuse tube or tubes together with parts necessary to enclose the conducting element and provide means of making contact with the conducting element and the fuse clips. *Note:* The fuseholder does not include the conducting element (fuse link or refill unit).   103, 202
(2) (low-voltage fuse). An assembly of base, fuse clips, and necessary insulation for mounting and connecting into the

circuit the current-response element, with its holding means if used for making a complete device. *Notes:* (1) For low-voltage fuses the current-responsive element and holding means are called a fuse. (2) For high-voltage fuses the general type of assembly defined above is called a fuse support or fuse mounting. The holding means (fuseholder) and the current-responsive or conducting element are called a fuse unit.   103, 202

fuse hook. A hook provided with an insulating handle for opening and closing fuses or switches and for inserting the fuseholder, fuse unit, or disconnecting blade into and removing it from the fuse support.   103, 202

fuselage lights. Aircraft aeronautical lights, mounted on the top and bottom of the fuselage, used to supplement the navigation lights.   167

fuse link. A replaceable part or assembly, comprised entirely or principally of the conducting element, required to be replaced after each circuit interruption to restore the fuse to operating condition.   103, 202

fuse support (high-voltage fuse) (fuse mounting). An assembly of base, mounting support or oil-cutout housing, fuse clips, and necessary insulation for mounting and connecting into the circuit the current-responsive element with its holding means if such means are used for making a complete device. *Notes:* (1) For high-voltage fuses the holding means is called a fuse carrier or fuseholder and in combination with the current-responsive or conducting element is called a fuse unit. (2) For low-voltage fuses the general type of assembly defined above is called a fuseholder.   103, 202

fuse terminal (fuse contact). The means for connecting the current-responsive element or its holding means, if such means are used for making a complete device, to the fuse clips.   103, 202

fuse time-current characteristic. The correlated values of time and current that designate the performance of all or a stated portion of the functions of the fuse. *Note:* The time-current characteristics of a fuse are usually shown as a curve.   103, 202

fuse time-current tests. Tests that consist of the application of current to determine the relation between the root-mean-square alternating current or direct current and the

time for the fuse to perform the whole or some specified part of its interrupting function.   103, 202

fuse tongs. Tongs provided with an insulating handle and jaws. Fuse tongs are used to insert the fuseholder or fuse unit into the fuse support or to remove it from the support.   103, 202

fuse tube. A tube of insulating material that surrounds the current-responsive element, the conducting element or the fuse link.   103, 202

fuse unit. An assembly comprising a conducting element mounted in a fuseholder with parts and materials in the fuseholder essential to the operation of the fuse.   103, 202

fusible element (fuse). That part, having predetermined current-responsive melting characteristics, which may be all or part of the current-responsive element.   103, 202

fusible enclosed (safety) switch (industrial control). A switch complete with fuse holders and either with or without meter-testing equipment or accommodation for meters, having all current-carrying parts completely enclosed in metal, and operable without opening the enclosure. *See also:* **switch.**   206

fusion (transmission and distribution). The formation of a heavier nucleus from two lighter ones with the attendant release of energy.   112

fusion frequency (television). Frequency of succession of retinal images above which their differences of luminosity or color are no longer perceptible. *Note:* The fusion frequency is a function of the decay characteristic of the display.   18

future point (for supervisory control or indication or telemeter selection). Provision for the future installation of equipment required for a point. *Note:* A future point may be provided with (1) space only (2) drilling or other mounting provisions only, or (3) drilling or other mounting provisions and wiring only.   103, 202

FW. *See:* **forward wave.**

FWHM. *See:* **full-width at half maximum.**

FWTM. Same as **FWHM** except that measurement is made at one-tenth the maximum ordinate rather than at one-half. *See also:* **full width at half maximum.**   118

G

gain (1) (transmission gain). General terms used to denote an increase in signal power in transmission from one point to another. *Note:* Gain is usually expressed in decibels and is widely used to denote transducer gain.   59

(2) (magnitude ratio) (linear system or element). The ratio of the magnitude (amplitude) of a steady-state sinusoidal output relative to the causal input; the length of a phasor from the origin to a point of the transfer locus in a complex plane. *Note:* The quantity may be separated into two factors: (A) a proportional amplification often denoted as K, which is frequency independent and associated with a dimensioned scale factor relating the units of input and output; (B) a dimensionless factor often denoted as $G(j\omega)$, which is frequency dependent. Frequency, conditions of

operation, and conditions of measurement must be specified. A loop gain characteristic is a plot of log gain versus log frequency. In nonlinear systems, gains are often amplitude dependent. *See also:* **signal.**

(3) (closed-loop) (industrial control). The gain of a closed-loop system, expressed as the ratio of the directly controlled output to the reference input. *See:* **control system, feedback.**   239, 178, 59, 56, 219, 207

(4) (antenna). *See:* **directive gain; directivity; power gain; power gain referred to a specified polarization; realized gain; relative power gain; superdirectivity.**   111

gain (photomultipliers). *See:* **current amplification.**

gain, available conversion power. *See:* **available conversion power gain.**

gain, available-power (two-port linear transducer). *See:* available-power gain.

gain, available-power, maximum (two-port linear transducer). The gain of the transducer at a specified frequency obtained when the transducer is conjugately matched to source and load. *Note:* The maximum available-power gain is not defined unless both the input and output impedances of the two-port transducer have positive real parts for arbitrary passive input and output terminations.   125

gain-bandwidth product (circuits and systems). The product of gain of an amplifier at midband and the difference between two frequencies at which the power output is above a specified fraction (usually one-half) of the midband value.   67

gain characteristic, loop (closed loop) (control system, feedback). The magnitude of the loop transfer function for real frequencies. *See also:* **control system, feedback.**   56

gain control. A device for adjusting the gain of a system or transducer.   239, 178

gain control, instantaneous automatic. *See:* **instantaneous automatic gain control.**

gain, conversion voltage (conversion transducer). The ratio of (1) the magnitude of the output-frequency voltage across the output termination, with the transducer inserted between the input-frequency generator and the output termination, to (2) the magnitude of the input-frequency voltage across the input termination of the transducer.   125

gain-crossover frequency (transfer function of an element or system) (1) (speed governing of hydraulic turbines). The frequency at which the gain becomes unity (and its decibel value zero). *See also:* **control system, feedback.**   8, 329
(2) (control systems). Of integral control action, the frequency at which the gain becomes unity. *Note:* The term integral action factor has been used for 2π times the gain cross-over frequency of integral control action.   56

gain, insertion. *See:* **insertion gain.**

gain, insertion voltage (electric transducer). The complex ratio of (1) the alternating component of voltage across the external termination of the output with the transducer inserted between the generator and the output termination, to (2) the voltage across the external termination of the output when the generator is connected directly to the output termination. *See also:* **network analysis; transducer.**   190

gain, loop (gain open loop). *See:* **loop gain.**

gain margin (1) (control system, feedback) (loop transfer function for a stable feedback system) (excitation control system). The reciprocal of the gain at the frequency at which the phase angle reaches minus 180 degrees. *Note:* Gain margin, sometimes expressed in decibels, is a convenient way of estimating relative stability by Nyquist, Bode, or Nichols diagrams, for systems with similar gain and phase characteristics. In a conditionally stable feedback system, gain margin is understood to refer to the highest frequency at which the phase angle is minus 180 degrees. *See also:* **control system, feedback.**   56 353
(2) (speed governing of hydraulic turbines). The reciprocal of the gain at the frequency at which the open-loop phase angle reaches 180 degrees.   8

gain of an antenna. *See:* **directive gain; power gain; relative gain; directivity.** *See also:* **antenna.**

gain, open loop. *See:* **loop gain.**

gain, power. *See:* **power gain.**

gain, proportional. The ratio of the change in output due to proportional control action to the change in input. Illustration: $Y = \pm PX$
where
P = proportional gain
X = input transform
Y = output transform
                                                         56, 105

gain sensitivity (circuits and systems). A measure of the change in gain produced by a change in a system parameter. Several specific definitions exist employing various normalizations.   67

gain sensitivity control. *See:* **differential gain control.**

gain, static (zero-frequency gain). Of gain of an element, or loop gain of a system, the value approached as a limit as frequency approaches zero. *Note:* Its value is the ratio of change of steady state output to a step change in input, provided the output does not saturate.   56

gain time control. *See:* **sensitivity time control.**

gain, transducer. *See:* **transducer gain.**

gain turn-down (transponder). The automatic receiver gain control incorporated for the purpose of protecting the transmitter from overload when the number of interrogations exceeds the design limitations. *See also:* **navigation.**   187

gain, zero frequency. *See:* **gain, static.**

galactic noise (communication satellite). A component of background noise produced by the galaxy.   85

Galvanic cell. A cell in which chemical change is the source of electric energy. *Note:* It usually consists of two dissimilar conductors in contact with each other and with an electrolyte, or two similar conductors in contact with each other and with dissimilar electrolytes. *See also:* **electrochemistry.**   221, 205

Galvanic current (medical electronics). An essentially steady unidirectional current (direct current).   192

Galvanic series. A list of metals and alloys arranged according to their relative potentials in a given environment.   221, 205

Galvani's experiment (electrobiology). Consists of bringing a muscle, or a motor nerve connected with a muscle, into simultaneous contact with two dissimilar metals, such as copper and iron, and producing a muscular contraction by closing the circuit. *See also:* **electrocardiogram.**   192

galvanism (galvanization) (voltaisation) (electrotherapy). The use of a Galvanic current for its biological or medical effects. *See:* **Galvanic current.** *See also:* **electrotherapy.**   192

galvanization (voltaisation) (electrotherapy). *See:* **galvanism.**

galvanometer. An instrument for indicating or measuring a small electric current, or a function of the current, by means of a mechanical motion derived from electromagnetic or electrodynamic forces that are set up as a result of the current. *See also:* **instrument.**   328

galvanometer recorder (photographic recording). A combination of mirror and coil suspended in a magnetic field. The application of a signal current to the coil causes a reflected light beam from the mirror to pass across a slit in front of a moving photographic film, thus providing a photographic record of the signal. 176

gamma (1) (photographic material). The gamma is the slope of the straight-line portion of the Hurter and Driffield curve. It represents the rate of change of photographic density with the logarithm of exposure. *Notes:* (A) Gamma is a measure of the contrast properties of the film. (B) Both gamma and density specifications are commonly used as controls in the processing of photographic film. 176

(2) (television). The exponent of that power law that is used to approximate the curve of output magnitude versus input magnitude over the region of interest. *See also:* **television.** 190, 18, 178, 125

gamma correction (television). The insertion of a nonlinear output-input characteristic for the purpose of changing the system transfer characteristic. 18

gamma ray (photovoltaic power system) (Γ ray). Penetrating short-wavelength electromagnetic radiation of nuclear origin. *See also.* **photovoltaic power system; solar cells (photovoltaic power system).** 186

gangway cable (mining). A cable designed to be installed horizontally (or nearly so) for power circuits in mine gangways and entries. *See also:* **mine feeder circuit.** 328

gap (rotating machinery). *See:* **air gap.**

gap admittance (electronic) (electron tubes). The difference between (1) the gap admittance with the electron stream traversing the gap and (2) the gap admittance with the stream absent. *See:* **admittance gap (electronic); electron-tube admittances.** 190, 125

gap admittance circuit (electron tubes). The admittance of the circuit at a gap in the absence of an electron stream. *See:* **admittance, gap (circuit); electron-tube admittances; gas tubes.** 190, 125

gap capacitance, effective (electron tubes). One half the rate of change with angular frequency of the resonator susceptance, measured at the gap, for frequencies near resonance. *See:* **electron-tube admittances; gas tubes.** 190, 125

gap coding (navigation) (electronic). A process of communicating in which a normally continuous signal (such as a radio-frequency carrier) is interrupted so as to form a telegraph-type message. *See also:* **navigation.** 187, 13

gap, head. *See:* **head gap.**

gap, inter-record. *See:* **inter-record gap.**

gap length (longitudinal magnetic recording). The gap length is the physical distance between the adjacent surfaces of the poles of a magnetic head. *Note:* The value of gap length necessary in certain studies exceeds that defined above and usually is referred to as the effective gap length. *See:* **magnetic head.** *See also:* **phonograph pick-up.**

gap loading, multipactor (electron tubes). The electronic gap admittance, resulting from a sustained secondary-emission discharge existing within a gap as a result of the motion of the secondary electrons in synchronism with the electric field in the gap. *See:* **electron-tube admittances; gas tubes.** 190, 125

gap loading, primary transit-angle (electron tubes). The

electronic gap admittance that results from the traversal of the gap by an initially unmodulated electron stream. *Note:* This is exclusive of secondary emission in the gap. *See:* **electron-tube admittances; gas tubes.** 190, 125

gap loading, secondary electron (electron tubes). The electronic gap admittance that results from the traversal of a gap by secondary electrons originating in the gap. *See:* **electron-tube admittances; gas tubes.** 190, 125

gap, record. *See:* **record gap.**

gas (rotating machinery). Any gas or gas mixture used as a cooling medium. *Note:* The term is usually not applied to air. 63

gas-accumulator relay. A relay that is so constructed that it accumulates all or a fixed proportion of gas released by the protected equipment and operates by measuring the volume of gas so accumulated. 103, 202, 60, 127

gas amplification (gas-filled radiation-counter tube). The ratio of the charge collected to the charge liberated by the initial ionizing event. *See also:* **counter tubes; gas-filled radiation; gas tubes.** 190, 125

gas amplification factor (gas phototube). The ratio of radiant or luminous sensitivies with and without ionization of the contained gas. *See also:* **gas tubes; phototubes.** 190

gas cell. A cell in which the action of the cell depends on the absorption of gases by the electrodes. *See also:* **electrochemistry.** 328

gas counter tube (radiation). A counter tube into which the sample to be measured is introduced in the form of a gas. *See:* **anticoincidence (radiation counters).** 190

gas (ionization) current (vacuum tube). The ionization current flowing to a negatively biased electrode and composed of positive ions that are produced by an electron current flowing between other electrodes. *Note:* These positive ions are a result of collisions between electrons and molecules of the residual gas. *See also:* **electron device.** 190

gas dryer (rotating machinery). Any device used to remove moisture from gas. 63

gas-electric drive (gasoline-electric drive). A self-contained system of power generation and application in which the power generated by a gasoline engine is transmitted electrically by means of a generator and a motor (or multiples of these) for propulsion purposes. *Note:* The prefix gas-electric is applied to ships, locomotives, cars, buses, etcetera, that are equipped with this drive. *See also:* **electric locomotive.** 328

gaseous discharge. The emission of light from gas atoms excited by an electric current. *See also:* **lamp.** 167

gaseous tube generator. A power source comprising a gas-filled electron-tube oscillator, a power supply, and associated control equipment. *See also:* **induction heating; oscillatory circuit.** 14, 114

gas-filled cable. A self-contained pressure cable in which the pressure medium is an inert gas having access to the insulation. *See:* **pressure cable; self-contained pressure cable.** *See also:* **power distribution, underground construction.** 64

gas filled joint (power cable joints). Joints in which the fluid filling the joint housing is in the form of a gas. 34

gas-filled phototube. A phototube into which a quantity of gas has been introduced, usually for the purpose of increasing its sensitivity.   190

gas-filled pipe cable. A pipe cable in which the pressure medium is an inert gas having access to the insulation. *See:* **pressure cable; pipe cable.** *See also:* **power distribution, underground construction.**   64

gas-filled radiation-counter tube. A gas tube, in a radiation counter, used for the detection of radiation by means of gas ionization.   328

gas-filled rectifier. A gas-filled tube whose function is to rectify an alternating current.

gas-filled transformers. A sealed transformer, except that the windings are immersed in a dry gas which is other than air or nitrogen.   53

gas-flow counter tube. A radiation-counter tube in which an appropriate atmosphere is maintained by a flow of gas through the tube. *See also:* **gas-filled radiation-counter tube.**   190

gas flow error (laser gyro). The error resulting from the flow of gas in direct current discharge tubes. It is caused by complex interactions among atoms, ions, electrons and tube walls.   46

gas focusing (electron-beam tubes). A method of concentrating an electron beam by gas ionization within the beam. *See:* **focusing, gas; gas tubes.**   190

gas grooves (electrometallurgy). The hills and valleys in metallic deposits caused by streams of hydrogen or other gas rising continuously along the surface of the deposit while it is forming. *See also:* **electrowinning.**   328

gasket-sealed relay. A relay in an enclosure sealed with a gasket. *See also:* **relay.**   259

gasket, waveguide (waveguide components). A resilient insert usually between flanges intended to serve one or more of the following primary purposes: (1) to reduce gas leakage affecting internal waveguide pressure, (2) to prevent intrusion of foreign material into the waveguide, or (3) to reduce power leakage and arcing.   166

gas multiplication factor (radiation-counter tubes). The ratio of (1) the charge collected from the sensitive volume to (2) the charge produced in this volume by the initial ionizing event.   96

gas-oil-sealed system. A system in which the interior of the tank is sealed from the atmosphere, over the temperature range specified, by means of an auxiliary tank or tanks to form a gas-oil seal operating on the manometer principle.   257, 203, 53

gasoline-electric drive. *See:* **gas-electric drive.**

gas-pressure relay. A relay so constructed that it operates by the gas pressure in the protected equipment.   103, 202, 60

gasproof. So constructed or protected that the specified gas will not interfere with successful operation.   103, 202, 225, 206

gasproof or vaporproof (rotating machinery). So constructed that the entry of a specified gas or vapor under prescribed conditions cannot interfere with satisfactory operating of the machine. *See also:* **asynchronous machine; direct-current commutating machine.**   63

gas ratio. The ratio of the ion current in a tube to the electron current that produces it. *See also:* **electrode current.**   190

gas seal (rotating machinery). A sealing arrangement intended to minimize the leakage of gas to or from a machine along a shaft. *Note:* It may be incorporated into a ball or roller bearing assembly.   63

gassing. The evolution of gases from one or more of the electrodes during electrolysis. *See also:* **electrolytic cell.**   328

gas system (rotating machinery). The combination of parts used to ventilate a machine with any gas other than air, including facilities for charging and purging the gas in the machine.   63

gastight (1) (lightning protection). So constructed that gas or air can neither enter nor leave the structure except through vents or piping provided for the purpose.   297 **(2) (power switchgear).** So constructed that the specified gas will not enter the enclosing case under specified pressure conditions.   103

gas tube. An electron tube in which the pressure of the contained gas or vapor is such as to affect substantially the electrical characteristics of the tube.   190

gas-tube relaxation oscillator (arc-tube relaxation oscillator). A relaxation oscillator in which the abrupt discharge is provided by the breakdown of a gas tube. *See also:* **oscillatory circuit.**   328

gas-turbine-electric drive. A self-contained system of power generation and application in which the power generated by a gas turbine is transmitted electrically by means of a generator and a motor (or multiples of these) for propulsion purposes. *Note:* The prefix gas-turbine-electric is applied to ships, locomotives, cars, buses, etcetera, that are equipped with this drive. *See also:* **electric locomotive.**   328

gate (1) (industrial control). A device or element that, depending upon one or more specified inputs, has the ability to permit or inhibit the passage of a signal. *See also:* **control.**   219, 206 **(2) (electronic computers).** (A) A device having one output channel and one or more input channels, such that the output channel state is completely determined by the contemporaneous input channel states, except during switching transients. (B) A combinational logic element having at least one input channel. (C) An AND gate. (D) An OR gate.   235, 210, 255, 77 **(3) (cryotron).** An output element of a cryotron. *See also:* **superconductivity.**   191 **(4) (thyristor).** An electrode connected to one of the semiconductor regions for introducing control current. *See also:* **anode.**   243, 66, 208, 191 **(5) (navigation systems).** (A) An interval of time during which some portion of the circuit or display is allowed to be operative, or (B) the circuit which provides gating. *See also:* **navigation.**   187, 13

gate-controlled delay time (thyristor). The time interval, between a specified point at the beginning of the gate pulse and the instant when the principal voltage (current) has dropped (risen) to a specified value near its initial value during switching of a thyristor from the OFF state to the ON state by a gate pulse. *See also:* **principal voltage-current characteristic.**   243, 66, 208, 191

gate-controlled rise time (thyristor). The time interval between the instants at which the principal voltage (current) has dropped (risen) from a specified value near its initial value to a specified low (high) value, during switching of a thyristor from the OFF state to the ON state

by a gate pulse. *Note:* This time interval will be equal to the rise time of the ON-state current only for pure resistive loads. *See also:* **principal voltage-current characteristic.**
                                                        243, 66, 208, 191
gate-controlled turn-off time (turn-off thyristor). The time interval, between a specified point at the beginning of the gate pulse and the instant when the principal current has decreased to a specified value, during switching from the ON state to the OFF state by a gate pulse. *See also:* **principal voltage-current characteristic.**   243, 66, 208, 191
gate-controlled turn-on time (thyristor). The time interval, between a specified point at the beginning of the gate pulse and the instant when the principal voltage (current) has dropped (risen) to a specified low (high) value during switching of a thyristor from the OFF state to the ON state by a gate pulse. Turn-on time is the sum of delay time and rise time. *See also:* **principal voltage-current characteristic; delay time; rise time.**   243, 204, 208, 191
gate current (semiconductor). The current that results from the gate voltage. *Notes:* (1) Positive gate current refers to conventional current entering the gate terminal. (2) Negative gate current refers to conventional current leaving the gate terminal.   66
gated sweep (oscilloscopes). A sweep controlled by a gate waveform. Also, a sweep that will operate recurrently (free-running, synchronized, or triggered) during the application of a gating signal. *See:* **oscillograph.**   185
gate limit (speed governing system, hydraulic turbines). A device which acts on the governor system to prevent the turbine-control servomotor from opening beyond the position for which the device is set.   58
gate nontrigger current (thyristor). The maximum gate current that will not cause the thyristor to switch from the OFF state to the ON state. *See also:* **principal current; gate trigger current.**   243, 204, 208, 191
gate nontrigger voltage (thyristor). The maximum gate voltage that will not cause the thyristor to switch from the OFF state to the ON state. *See also:* **principal voltage-current characteristic (principal characteristic); gate trigger voltage.**   243, 204, 208, 191
gate protective action (thyristor converter). Protective action that takes advantage of the switching property in the converter protection network.   121
gate suppression (thyristor power converter). Removal of gating pulses.   121
gate terminal (thyristor). A terminal that is connected to a gate. *See also:* **anode.**   243, 204, 191
gate trigger current (thyristor). The minimum gate current required to switch a thyristor from the OFF state to the ON state. *See also:* **principal current.**   243, 204, 208, 191
gate trigger voltage (thyristor). The gate voltage required to produce the gate-trigger current. *See also:* **principal voltage-current characteristic.**   243, 204, 208, 191
gate turn-off current (gate turn-off thyristor). The minimum gate current required to switch a thyristor from the ON state to the OFF state. *See also:* **principal current.**   243, 204, 208, 191
gate turn-off voltage (gate turn-off thyristor). The gate voltage required to produce the gate turn-off current. *See also:* **principal voltage-current characteristic.**   243, 204, 208, 191
gate voltage (thyristor). The voltage between a gate ter-

minal and a specified main terminal. *See also:* **principal voltage-current characteristic.**   243, 204, 208, 191
gating (1) (antennas). (A) The process of selecting those portions of a wave that exist during one or more selected time intervals or (B) that have magnitudes between selected limits. *See also:* **modulation; wave front.**   111
(2) (radar). The application of selector pulses during part of a cycle of equipment operation when a particular equipment function is desired.   13
gating signal (keying signal). A signal that activates or deactivates a circuit during selected time intervals.
                                                                            328
gauss (centimeter-gram-second electromagnetic system). The gauss is 10^{-4} webers per square meter or 1 maxwell per square centimeter.   210
Gaussian beam (laser-maser). A beam of radiation having an approximately spherical wave front at any point along the beam and having transverse field intensity over any wave front which is a Gaussian function of the distance from the axis of the beam.   363
Gaussian density function (radar). Sometimes referred to as normal probability distribution, the Gaussian probability-density function is given by

$$ f(X) = \frac{1}{\sigma\sqrt{2\pi}} \exp - \left(\frac{x^2}{2\sigma^2} \right). $$

Often used to describe statistical nature of random noise, where σ = standard deviation.
Gaussian filter (circuits and systems). A polynomial filter whose magnitude-frequency response approximates the ideal Gaussian response, the degree of approximation depending on the complexity of the filter. The ideal Gaussian response is given by

$$ |H(j\omega)| = \exp \left[-0.3466 \, (\omega/\omega_c)^2 \right] $$

where ω_c = 3 dB frequency. Gaussian filters, because of their good transient characteristics (small overshoot and ringing), find applications in pulse systems.   67
Gaussian response (amplifiers). A particular frequency-response characteristic following the curve $y(f) = e^{-af^2}$. *Note:* Typically, the frequency response approached by an amplifier having good transient response characteristics. *See:* **amplifier; response, Gaussian.**   185
Gaussian system (units). A system in which centimeter-gram-second electrostatic units are used for electric quantities and centimeter-gram-second electromagnetic units are used for magnetic quantities. *Note:* When this system is used, the factor c (the speed of light) must be inserted at appropriate places in the electromagnetic equations.   210
Gauss law (electrostatics). States that the integral over any closed surface of the normal component of the electric flux density is equal in a rationalized system to the electric charge Q_0 within the surface. Thus

$$ \overset{\text{closed surface}}{\int (\mathbf{D} \cdot \mathbf{n}) \, dA} = \overset{\text{volume enclosed}}{\int \rho_0 \, dV} = Q_0. $$

Here, $\mathbf{D}$ is the electric flux density, $\mathbf{n}$ is a unit normal to the surface, dA the element of area, ρ_0 is the space charge density in the volume V enclosed by the surface. 210

gaussmeter. A magnetometer provided with a scale graduated in gauss or kilogauss. *See also:* **magnetometer.**
328

GCA. *See:* **ground-controlled approach.**

GCI (ground controlled interception). A radar system by means of which a controller on the ground may direct an aircraft to make an interception of another aircraft.
13

G display (radar). A rectangular display in which a target appears as a laterally centralized blip when the radar antenna is aimed at it in azimuth, and wings appear to grow on the blip as the distance to the target is diminished; horizontal and vertical aiming errors are respectively indicated by horizontal and vertical displacement of the blip. *See also:* **navigation.** 13, 187

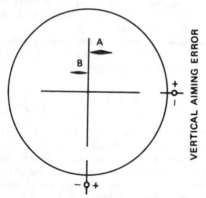

AZIMUTH AIMING ERROR

G display

GDOP. *See:* **geometric dilution of position.**

G drift (electronic navigation). A drift component in gyros (sometimes in accelerometers) proportional to the nongravitational acceleration and caused by torques resulting from mass unbalance. Jargon. *See also:* **navigation.**
13, 187

G^2 drift (electronic navigation). A drift component in gyros (sometimes in accelerometers) proportional to the square of the nongravitational acceleration and caused by anisoelasticity of the rotor supports. Jargon. *See also:* **navigation.** 187, 13

geared-drive machine. A direct-drive machine in which the energy is transmitted from the motor to the driving sheave, drum, or shaft through gearing. *See also:* **driving machine (elevators).**
328

geared traction machine (elevators). A geared-drive traction machine. *See also:* **driving machine (elevators).**
328

gearless motor. A traction motor in which the armature is mounted concentrically on the driving axle, or is carried by a sleeve or quill that surrounds the axle, and drives the axle directly without gearing. *See also:* **traction motor.**
328

gearless traction machine (elevators). A traction machine, without intermediate gearing, that has the traction sheave and the brake drum mounted directly on the motor shaft.

See also: **driving machine (elevators).** 328

gear pattern (facsimile). *See:* **drive pattern.**

gear ratio (watthour meter). The number of revolutions of the rotor for one revolution of the first dial pointer. *Note:* This is commonly denoted by the symbol R_g. *See also:* **electricity meter (meter); moving element of an instrument.**
212

Geiger-Mueller counter tube. A radiation-counter tube designed to operate in the Geiger-Mueller region. 328

Geiger-Mueller region (radiation counter tube). The range of applied voltage in which the charge collected per isolated count is independent of the charge liberated by the initial ionizing event. *See also:* **gas-filled radiation-counter tubes.**
190, 96, 125

Geiger-Mueller threshold (radiation-counter tube). The lowest applied voltage at which the charge collected per isolated tube count is substantially independent of the nature of the initial ionizing event. *See also:* **gas-filled radiation-counter tubes.**
190, 96, 125

Geissler tube. A special form of gas-filled tube for showing the luminous effects of discharges through rarefied gases. *Note:* The density of the gas is roughly one-thousandth of that of the atmosphere. *See:* **gas tubes.** 244, 190

general alarm system. A system of bells energized from a storage battery (the emergency lighting battery, if provided) controlled from the wheelhouse and so located as to be audible to all persons on a ship, to give an alarm in case of a major casualty. 328

general coördinated methods (general application to electric supply or communication systems). Those methods reasonably available that contribute to inductive coordination without specific consideration of the requirements for individual inductive exposures. *See also:* **inductive coordination.**
328

general diffuse lighting. Lighting involving luminaires that distribute 40 to 60 percent of the emitted light downward and the balance upward sometimes with a strong component at 90 degrees (horizontal). *See also:* **general lighting.**
167

generalized entity. *See:* **generalized property.**

generalized impedance converter (circuits and systems). A two-port active network characterized by the conversion factor $f(s)$ of the complex frequency variable s and satisfying the following property: when port B is terminated with impedance $Z(s)$ the impedance at port A is given by $2(s)f(s)$; when port A is terminated with impedance $Z(s)$ the impedance at port B is given by $2(s)/f(s)$. 67

generalized property (generalized quantity) (generalized entity). Any of the physical concepts in terms of examples of which observable physical systems and phenomena are described quantitatively. *Notes:* (1) Examples are the abstract concepts of length, electric current, energy, etcetera. (2) A generalized property is characterized by the qualitative attribute of physical nature, or dimensionality, but not by a quantitative magnitude. 210

generalized quantity. *See:* **generalized property.**

general lighting. Lighting designed to provide a substantially uniform level of illumination throughout an area, exclusive of any provision for special local requirements.
167

general-purpose circuit breaker (alternating current high-voltage circuit breakers). A circuit breaker that is not specifically designed for capacitance current switching. **130**

general-purpose computer. A computer that is designed to solve a wide class of problems. **255, 77**

general-purpose controller (industrial control). Any controller having ratings, characteristics, and mechanical construction for use under usual service conditions. *See:* **electric controller.** **225, 206**

general purpose digital computer. *See:* **digital computer.**

general-purpose enclosure (1) (general). An enclosure used for usual service applications where special types of enclosures are not required. **103, 202**

(2) (rotating machinery). An enclosure that primarily protects against accidental contact and slight indirect splashing but is neither dripproof nor splash-proof. *See also:* **asynchronous machine; direct-current commutating machine.** **3**

general-purpose floodlight. A weatherproof unit so constructed that the housing forms the reflecting surface. *Note:* The assembly is enclosed in a cover glass. *See also:* **floodlighting.** **167**

general-purpose induction motor (rotating machinery). Any open motor having a continuous rating of 50-degrees Celsius rise by resistance for Class A insulation, or of 80-degrees Celsius rise for Class B, a service factor as listed in the following tabulation, and designed, listed, and offered in standard ratings with standard operating characteristics and mechanical construction, for use under usual service conditions without restrictions to a particular application or type of application.

Service Factor

Horse-power	Synchronous Speed, revolutions per minute			
	3600	1800	1200	900
$1/20$	1.4	1.4	1.4	1.4
$1/12$	1.4	1.4	1.4	1.4
$1/8$	1.4	1.4	1.4	1.4
$1/6$	1.35	1.35	1.35	1.35
$1/4$	1.35	1.35	1.35	1.35
$1/3$	1.35	1.35	1.35	1.35
$1/2$	1.25	1.25	1.25	
$3/8$	1.25	1.25		
1	1.25			

See also: **asynchronous machine.** **63**

general-purpose low-voltage dc power circuit breaker (power switchgear). A low-voltage dc power circuit breaker which, during interruption, does not usually prevent the fault current from rising to its E/R value. **103**

general-purpose motor (rotating machinery). Any motor designed, listed and offered in standard ratings with operating characteristics and mechanical construction suitable for use under usual service conditions without restrictions to a particular application or type of application. **63**

general-purpose relay. A relay that is adaptable to a variety of applications. *See also:* **relay.** **259**

general purpose test equipment (GPTE) (test, measurement and diagnostic equipment). Test equipment which is used for the measurement of a range of parameters common to two or more equipments or systems of basically different design. **54**

general-purpose transformers. Step-up or step-down transformers or auto-transformers generally used in secondary distribution circuits of 600 volts or less in connection with power and lighting service. **53**

general support maintenance. *See:* **maintenance, depot.** **54**

general-use snap switch. A form of general-use switch so constructed that it can be installed in flush device boxes, or on outlet box covers, or otherwise used in conjunction with wiring systems. **256**

general-use switch (industrial control). A switch intended for use in general distribution and branch circuits. *Note:* It is rated in amperes, and it is capable of interrupting the rated current at the rated voltage. *See also:* **switch.** **206**

generate (computing systems). To produce a program by selection of subsets from a set of skeletal coding under the control of parameters. **255, 77**

generated voltage (rotating machinery). A voltage produced in a closed path or circuit by the relative motion of the circuit or its parts with respect to magnetic flux. *See:* **Faraday's law; induced voltage; asynchronous machine; direct-current commutating machine.** **63**

generating electric field meter (gradient meter). A device in which a flat conductor is alternately exposed to the electric field to be measured and then shielded from it. *Note:* The resulting current to the conductor is rectified and used as a measure of the potential gradient at the conductor surface. *See also:* **instrument.** **328**

generating magnetometer (earth inductor). A magnetometer that depends for its operation on the electromotive force generated in a coil that is rotated in the field to be measured. *See also:* **magnetometer.** **328**

generating station (electric power supply). A plant wherein electric energy is produced from some other form of energy (for example, chemical, mechanical, or hydraulic) by means of suitable apparatus. **64**

generating-station auxiliaries. Accessory units of equipment necessary for the operation of the plant. *Example:* Pumps, stokers, fans, etcetera. *Note:* Auxiliaries may be classified as essential auxiliaries or those that must not sustain service interruptions of more than 15 seconds to 1 minute, such as boiler feed pumps, forced draft fans, pulverized fuel feeders, etcetera; and nonessential auxiliaries that may, without serious effect, sustain service interruptions of 1 to 3 minutes or more, such as air pumps, clinker grinders, coal crushers, etcetera. *See also:* **generating station.** **64**

generating-station auxiliary power. The power required for operation of the generating station auxiliaries. *See also:* **generating station.** **64**

generating-station efficiency. *See:* **efficiency.**

generating-station reserve. *See:* **reserve equipment.**

generating unit. *See:* **unit.**

generating voltmeter (rotary voltmeter). A device in which a capacitor is connected across the voltage to be measured and its capacitance is varied cyclically. The resulting current in the capacitor is rectified and used as a measure of the voltage. *See also:* **instrument.** **328**

generation rate (semiconductor). The time rate of creation of electron-hole pairs. *See also:* **semiconductor device.**
210, 245, 66

generator (1) (rotating machinery). A machine that converts mechanical power into electric power. *See also:* **asynchronous machine; direct-current commutating machine.**
63

(2) (computing systems). A controlling routine that performs a generate function, for example, report generator, input-output generator. *See also:* **function generator; noise generator.**
255, 77

generator, alternating-current. *See:* **alternating-current generator.**

generator, arc welder (1) (generator, alternating-current arc welder). An alternating-current generator with associated reactors, regulators, control, and indicating devices required to produce alternating current suitable for arc welding.

(2) (generator-rectifier, alternating-current–direct current arc welder). A combination of static rectifiers and the associated alternating-current generator, reactors, regulators, control, and indicating devices required to produce either direct or alternating current suitable for arc welding.

(3) (generator, direct-current arc welder). A direct-current generator with associated reactors, regulators, control, and indicating devices required to produce direct current suitable for arc welding.
264

(4) (generator-rectifier, direct-current arc welder). A combination of static rectifiers and the associated alternating-current generator, reactors, regulators, controls, and indicating devices required to produce direct current suitable for arc welding.
264

generator efficiency (thermoelectric couple). The ratio of (1) the electric power output of a thermoelectric couple to (2) its thermal power input. *Note:* This is an idealized efficiency assuming perfect thermal insulation of the thermoelectric arms. *See also:* **thermoelectric device.**
248, 191

generator-field accelerating relay (industrial control). A relay that functions automatically to maintain the armature current within prescribed limits when a motor supplied by a generator is accelerated to any speed, up to base speed, by controlling the generator field current. *Note:* This definition applies to adjustable-voltage direct-current drives. *See:* **relay.**
225, 206

generator-field control. A system of control that is accomplished by the use of an individual generator for each elevator or dumbwaiter wherein the voltage applied to the driving-machine motor is adjusted by varying the strength and direction of the generator field. *See also:* **control (elevators).**

generator field decelerating relay (industrial control). A relay that functions automatically to maintain the armature current within prescribed limits when a motor, supplied by a generator, is decelerated from base speed, or less, by controlling the generator field-current. *Note:* This definition applies to adjustable-voltage direct-current drives. *See:* **relay.**
225, 206

generator/motor. A machine that may be used as either a generator or a motor usually by changing rotational direction. *Note:* (1) This type of machine has particular application in a pumped-storage operation, in which water is pumped into a reservoir during off-peak periods and released to provide generation for peaking loads. (2) This definition eliminates the confusion of terminology for this type of machine. A slant is used between the terms to indicate their equality, and also the machine serves one function or the other and not both at the same time. The word **generator** is placed first to provide a distinction in speech between this term and the commonly used term **motor-generator,** which has an entirely different meaning. *See also:* **asynchronous machine.**
63

generator set. A unit consisting of one or more generators driven by a prime mover. *See also:* **asynchronous machine; direct-current commutating machine.**
63

generette (rotating machinery). A test jig designed on the principle of a motorette, for endurance tests on sample lengths of coils or bars for large generators. *See also:* **asynchronous machine; direct-current commutating machines.**
63

geocentric latitude (navigation). The acute angle between (1) a line joining a point with the earth's geometric center and (2) the earth's equatorial plane.
187, 13

geocentric vertical. *See:* **geometric vertical.**

geodesic. The shortest line between two points measured on any mathematically derived surface that includes the points. *See also:* **navigation.**
187, 13

geodesic lens antenna. A lens antenna having a two-dimensional path-delay lens with uniform index of refraction in which propagating electromagnetic waves are constrained by the lens boundaries to follow minimal and approximately equal path lengths from the feed to a straight line in the aperture.
111

geodetic latitude (navigation). The angle between the normal to the spheroid and the earth's equatorial plane; the latitude generally used in maps and charts. Also called **geographic latitude.** *See also:* **navigation.**
187, 13

geographic latitude. *See:* **geodetic latitude.**

geographic (map) vertical. The direction of a line normal to the surface of the geoid. *See also:* **navigation.**
187, 13

geoid. The shape of the earth as defined by the hypothetical extension of mean sea level continuously through all land masses. *See also:* **navigation.**
187, 13

geometrical adjectives (pulse terms). (1) trapezoidal. Having or approaching the shape of a trapezoid. (2) rectangular. Having or approaching the shape of a rectangle. (3) triangular. Having or approaching the shape of a triangle. (4) sawtooth. Having or approaching the shape of a right angle. *See:* **composite waveform,** Fig. 2, waveform D. (5) rounded. Having a curved shape characterized by a relatively gradual change in slope.
254

geometrical factor (navigation). The ratio of the change in a navigational coordinate to the change in distance, taken in the direction of maximum navigational coordinate change; the magnitude of the gradient of the navigational coordinate. *See also:* **navigation.**
13, 187

geometric dilution of position (GDOP) (radar). An expression which refers to increased measurement errors in certain regions of coverage of the measurement system. It applies to systems which combine several surface of position measurements such as range only, angle only, or hyperbolic (range difference) to locate the object of interest. When two lines of position cross at a small acute angle, the measurement accuracy is reduced along the axis of the acute angle. 13

geometric distortion (television). The displacement of elements in the reproduced picture from the correct relative positions in the perspective plane projection of the original scene. *See also:* **television.** 163

geometric factor (cable calculations) (power distribution, underground cables). A parameter used and determined solely by the relative dimensions and geometric configuration of the conductors and insulation of a cable. *See also:* **power distribution, underground construction.** 57

geometric inertial navigation equipment. The class of inertial navigation equipment in which the geographic navigational quantities are obtained by computations (generally automatic) based upon the outputs of accelerometers whose vertical axes are maintained parallel to the local vertical, and whose azimuthal orientations are maintained in alignment with a predetermined geographic direction (for example, north). *See also:* **navigation.** 187, 13

geometric rectification error (accelerometer). The error caused by an angular motion of a linear accelerometer input reference axis when this angular motion is coherent with a vibratory cross acceleration input. This is an error that can occur in the application of a linear accelerometer and is not caused by imperfections in the accelerometer. This error is proportional to the square of the cross acceleration and varies with the frequency. 46

geometric (geocentric) vertical (navigation). The direction of the radius vector drawn from the center of the earth through the location of the observer. *See also:* **navigation.** 187

geometry (oscilloscopes). The degree to which a cathode-ray tube can accurately display a rectilinear pattern. *Note:* Generally associated with properties of a cathode-ray tube; the name may be given to a cathode-ray-tube electrode or its associated control. *See:* **oscillograph.** 185

geometry, detector. *See:* **detector geometry.**

getter (electron tubes). A substance introduced into an electron tube to increase the degree of vacuum by chemical or physical action on the residual gases. *See also:* **electrode (of an electron tube); electronic tube.** 244, 190

ghost (television). A spurious image resulting from an echo. *See also:* **television.** 328

ghost pulse. *See:* **ghost signals.**

ghost signals (loran). (1) Identification pulses that appear on the display at less than the desired loran station full pulse repetition frequency. (2) Signals appearing on the display that have a basic repetition frequency other than that desired. *See also:* **navigation.** 187

ghost target (radar). Apparent target signals in a radar which do not correspond in position and/or frequency with any real target, but which result from distortion or misinterpretation by the radar circuitry of other real target signals which are present. For example, they may result from range-Doppler ambiguities in the radar waveform used, from intermodulation distortion due to circuit amplitude non-linearities, or from combining of data from two antenna systems or waveforms. 13

Gibb's phenomenon (circuits and systems). Overshoot phenomenon obtained near a discontinuity point of a signal

when the spectrum of that signal is truncated abruptly. 67

GIC. *See:* **generalized impedance converter.**

gilbert (centimeter-gram-second electromagnetic system). The unit of magnetomotive force. The gilbert is one oersted-centimeter. 210

Gill-Morrell oscillator. An oscillator of the retarding-field type in which the frequency of oscillation is dependent not only on electron transit time within the tube, but also on associated circuit parameters. *See also:* **oscillatory circuits.** 111

gimbal (gyro). A device which permits the spin axis to have one or two angular degrees of freedom. 46

gimbal error (gyro). The error resulting from angular displacements of gimbals from their reference positions such that gimbal pickoffs do not measure the true angular-motion of the case about the input reference axis. 46

gimbal freedom (gyro). The maximum angular displacement of a gimbal about its axis. 46

gimbal lock (gyro). A condition of a two-degree-of-freedom gyro wherein the alignment of the spin axis with an axis of freedom deprives the gyro of a degree-of-freedom and, therefore, of its useful properties. 46

gimbal retardation (gyro). A measure of output axis friction torque when the gimbal is rotated about the output axis. It is expressed as an equivalent input. 46

gimbal unbalance torque (tuned rotor gyro). The acceleration-sensitive torque caused by gimbal unbalance along the spin axis due to non-intersection of the flexure axes. Under constant acceleration, it appears as a second harmonic of the rotor spin frequency because of the single degree of freedom of the gimbal relative to the support shaft. When the gyro is subjected to vibratory acceleration applied normal to the spin axis at twice the rotor spin frequency, this torque results in a rectified unbalance drift rate. 46

gland seal (rotating machinery). A seal used to prevent leakage between a moveable and a fixed part. 63

glare. The sensation produced by brightnesses within the visual field that are sufficiently greater than the luminance to which the eyes are adapted to cause annoyance, discomfort, or loss in visual performance and visibility. *Note:* The magnitude of the sensation of glare depends upon such factors as the size, position, and luminance of a source, the number of sources, and the luminance to which the eyes are adapted. *See also:* **visual field.** 167

glass half cell (glass electrode). A half cell in which the potential measurements are made through a glass membrane. *See also:* **electrolytic cell.** 328

GLC circuit. *See:* **simple parallel circuit.**

glide path (electronic navigation). The path used by an aircraft in approach procedures as defined by an instrument landing facility. *See also:* **navigation.** 13, 187

glide-path receiver. An airborne radio receiver used to detect the transmissions of a ground-installed glide-path transmitter. *Note:* It furnishes a visual, audible, or electric signal for the purpose of vertically guiding an aircraft using an instrument landing system. *See also:* **air transportation; electronic equipment.** 328

glide slope (electronic navigation). An inclined surface generated by the radiation of electromagnetic waves and used with a localizer in an instrument landing system to create a glide path. *See also:* **navigation.**   186, 13

glide-slope angle (electronic navigation). The angle in the vertical plane between the glide slope and the horizontal. *See also:* **navigation.**   186, 13

glide-slope deviation (electronic navigation). The vertical location of an aircraft relative to a glide slope, expressed in terms of the angle measured at the intersection of the glide slope with the runway; or the linear distance above or below the glide slope. *See also:* **navigation.**   186, 13

glide slope facility (navigation). The ground station of an ILS (instrument landing system) which generates the glide slope.   13

glide-slope sector (instrument landing system). A vertical sector containing the glide slope and within which the pilot's indicator gives a quantitative measure of the deviation above and below the glide slope; the sector is bounded above and below by a specified difference in depth of modulation, usually that which gives full-scale deflection of the glide-slope deviation indicator. *See also:* **navigation.**   186, 13

glint (radar). The random component of target location error caused by variations in the phase front of the target signal (as contrasted with **scintillation error**). Glint may affect angle, range or Doppler measurements, and may have peak values corresponding to locations beyond the true target extent in the measured coordinate.   13

glitch. A perturbation of the pulse waveform of relatively short duration and of uncertain origin. *See:* **distortion, pulse.**   185

global stability. *See:* **stability, global.**

globe. A transparent or diffusing enclosure intended to protect a lamp, to diffuse and redirect its light, or to change the color of the light. *See also:* **bare (exposed) lamp.**   167

glossmeter. An instrument for measuring gloss as a function of the regular and diffuse reflection characteristics of a material. *See also:* **photometry.**   167

glow discharge (1) (general). The phenomenon of electric conduction in gases shown by a slight luminosity, without great hissing or noise, and without appreciable heating or volatilization of the electrode, when the voltage gradient exceeds a certain value. *See also:* **abnormal glow discharge.**   210

(2) (electron tubes). A discharge of electricity through gas characterized by (A) a change of space potential, in the immediate vicinity of the cathode, that is much higher than the ionization potential of the gas. (B) A low, approximately constant, current density at the cathode, and a low cathode temperature. (C) The presence of a cathode glow. *See:* **gas tubes.** *See also:* **lamp.**   190, 167, 125

glow-discharge tube. A gas tube that depends for its operation on the properties of a glow discharge.   125

glow factor (illuminating engineering). A measure of the visible light response of a fluorescent material to black light. It is equal to the luminance (photometric brightness) in footlamberts produced on the material divided by the incident black-light flux density in milliwatts per square foot. *Notes:* (1) It may be measured in lumens per milliwatt. (2) See note after **filter factor.** *See also:* **ultraviolet radiation.**   167

glow lamp. An electric-discharge lamp whose mode of operation is that of a glow discharge and in which light is generated in the space close to the electrodes. *See also:* **lamp.**   167

glow, negative. *See:* **negative glow.**

glow-switch. An electron tube containing contacts operated thermally by means of a glow discharge.   190

glow-tube. *See:* **glow discharge tube.**

glue-line heating (dielectric heating usage). An arrangement of electrodes designed to give preferential heating to a thin film of material of relatively high loss factor between alternate layers of relatively low loss factor material. *See also:* **dielectric heating.**   14, 114

go. *See:* **go/no-go.**

goniometer (electronic navigation). A combining device used with a plurality of antennas so that the direction of maximum radiation or of greatest response may be rotated in azimuth without physically moving the antenna array.   187, 13

goniophotometer. A photometer for measuring the directional light distribution characteristics of sources, luminaires, media, and surfaces. *See also:* **photometry.**   167

go/no-go (test, measurement and diagnostic equipment). A set of terms (in colloquial usage) referring to the condition or state of operability of a unit which can only have two parameters: (1) go, functioning properly, or (2) no-go, not functioning properly.   54

governing system (hydraulic turbines). The combination of devices and mechanisms which detect speed deviation and convert it into a change in servomotor position. It includes the speed sensing elements, the governor control actuator, the hydraulic pressure supply system, and the turbine control servomotor.   8, 58

governor actuator rating (speed governing systems, hydraulic turbines). The governor actuator rating is the flow rate in volume per unit time which the governor actuator can deliver at a specified pressure drop. The pressure drop shall be measured across the terminating pipe connections to the turbine control servomotors at the actuator. This pressure drop is measured with the specified minimum normal working pressure of the pressure supply system delivered to the supply port of the actuator distributing valve.   58

governor control actuator (speed governing of hydraulic turbines). The combination of devices and mechanisms which detects a speed error and develops a corresponding hydraulic control output to the turbine control servomotors but does not include the turbine control servomotors. *Notes:* (1) Includes gate, blade, deflector, or needle control, or all equipment as appropriate.   8
(2) Includes gate and blade control equipment in the case of adjustable blade turbines and deflector and/or needle control equipment in the case of impulse turbines.   58

governor control actuator rating (speed governing of hydraulic turbines). The governor actuator rating is the flow rate in volume per unit time which the governor actuator can deliver at a specified pressure drop. The pressure drop shall be measured across the terminating pipe connections to the turbine control servomotors at the actuator. This pressure drop is measured with the specified minimum normal working pressure of the fluid pressure supply system delivered to the supply port of the actuator distributing valve. 8

governor-controlled gates (hydro-turbine). Include the gates that control the energy input to the turbine and that are normally actuated by the speed governor directly or through the medium of the speed-control mechanism. *See also:* **speed-governing system.** 94

governor-controlled valves (steam turbine). Include the valves that control the energy input to the turbine and that are normally actuated by the speed governor directly or through the medium of the speed-control mechanism. *See also:* **speed-governing system.** 94

governor dead band (automatic generation control). The magnitude of the total change in steady-rate speed within which there is no resulting measurable change in the position of the governor-controlled valves. *Note:* Dead band is the measure of the insensitivity of the speed-governing system and is expressed in percent of rated speed. 94

governor dead time (speed governing of hydraulic turbines). Dead time is the time interval between the initiation of a specified change in steady state speed and the first detectable movement of the turbine control servomotor. 8, 58

governor pin. *See:* **centrifugal-mechanism pin.**

governor speed-changer. A device by means of which the speed-governing system may be adjusted to change the speed or power output of the turbine while the turbine is in operation. *See also:* **speed-governing system.** 94

governor speed-changer position. The position of the speed changer indicated by the fraction of its travel from the position corresponding to minimum turbine speed to the position corresponding to maximum speed and energy input. It is usually expressed in percent. *See also:* **speed-governing system.** 94

governor spring. *See:* **centrifugal-mechanism spring.**

governor weights. *See:* **centrifugal-mechanism weights.**

GPI. *See:* **ground-position indicator.**

graded-time step-voltage test (rotating machinery). A controlled overvoltage test in which calculated voltage increments are applied at calculated time intervals. *Note:* Usually, a direct-voltage test with the increments and intervals so calculated that dielectric absorption appears as a constant shunt-conductance; to simplify interpretation. *See also:* **asynchronous machine; direct-current commutating machine.** 63

grade-of-service (telephone switching systems). The proportion of total calls, usually during the busy hour, that cannot be completed immediately or served within a prescribed time. 55

gradient (scalar field). At a point, a vector (denoted by ∇u) equal to, and in the direction of, the maximum space rate of change of the field. It is obtained as a vector field by applying the operator ∇ to a scalar function. Thus if $u = f(x,y,z)$

$$\nabla u = \text{grad } u$$
$$= \mathbf{i}\frac{\partial u}{\partial x} + \mathbf{j}\frac{\partial u}{\partial y} + \mathbf{k}\frac{\partial u}{\partial z}.$$

210

gradient meter. *See:* **generating electric field meter.**

gradient microphone. A microphone the output of which corresponds to a gradient of the sound pressure. *Note:* Gradient microphones may be of any order as, for example, zero, first, second, etcetera. A pressure microphone is a gradient microphone of zero order. A velocity microphone is a gradient microphone of order one. Mathematically, from a directivity standpoint for plane waves the root-mean-square response is proportional to $\cos^n \theta$, where θ is the angle of incidence and n is the order of the microphone. *See also:* **microphone.** 328

grading (telephone switching systems). Partial commoning or multipling of the outlets of connecting networks where there is limited availability to the outgoing group or subgroup of outlets. 55

grading group (telephone switching systems). That part of a grading in which all inlets have access to the same outlets. 55

grading ring (arrester) (control ring). A metal part, usually circular or oval in shape, mounted to modify electrostatically the voltage gradient or distribution. 2, 299, 62

gradual failure. *See:* **failure, gradual.**

graduated (control) (industrial control). Marked to indicate a number of operating positions. 206

grain (photographic material). A small particle of metallic silver remaining in a photographic emulsion after development and fixing. *Note:* In the agglomerate, these grains form the dark area of a photographic image. 176

graininess (photographic material). The visible coarseness under specified conditions due to silver grains in a developed photographic film. 176

granular-filled fuse unit. A fuse unit in which the arc is drawn through powdered, granular, or fibrous material. 103, 202

graph determinant (network analysis). One plus the sum of the loop-set transmittances of all nontouching loop sets contained in the graph. *Notes:* (1) The graph determinant is conveniently expressed in the form

$$\Delta = (1 - \Sigma L_i + \Sigma L_i L_j - \Sigma L_i L_j L_k + \cdots)$$

where L_i is the loop transmittance of the ith loop of the graph, and the first summation is over all of the different loops of the graph, the second is over all of the different pairs of nontouching loops, and the third is over all the different triplets of nontouching loops, etcetera. (2) The graph determinant may be written alternatively as

$$\Delta = [(1 - L_1)(1 - L_2) \cdots (1 - L_n)]\dagger$$

where $L_1, L_2, \cdots, L_n$, are the loop transmittances of the n different loops in the graph, and where the dagger indicates that, after carrying out the multiplications within the brackets, a term will be dropped if it contains the transmittance product of two touching loops. (3) The graph determinant reduces to the return difference for a graph having only one loop. (4) The graph determinant is equal to the determinant of the coefficient equations. *See also:* **linear signal flow graphs.** 282

graphic symbol (1) (abbreviation). A geometric representation used to depict graphically the generic function of an item as it normally is used in a circuit. *See also:* **abbreviation.** 173

(2) (electrical engineering). A shorthand used to show graphically the functioning or interconnections of a circuit.

A graphic symbol represents the functions of a part in the circuit. For example, when a lamp is employed as a nonlinear resistor, the nonlinear resistor symbol is used. Graphic symbols are used on single-line (one-line) diagrams, on schematic or elementary diagrams, or, as applicable, on connection or wiring diagrams. Graphic symbols are correlated with parts lists, descriptions, or instructions by means of designations.　25

graphite brush. A brush composed principally of graphite. *Note:* This type of brush is soft. Grades of brushes of this type differ greatly in current-carrying capacity and in operating speed from low to high. *See also:* **brush (rotating machinery).**　63, 279

graphitic corrosion. Corrosion of gray cast iron in which the metallic constituents are converted to corrosion products leaving the graphite intact.　221, 205

graphitization (corrosion). Decomposition of a metal carbide to matrix-metal and graphite. *See:* **graphitic corrosion.**　221, 205

graph transmittance (network analysis). The ratio of signal at some specified dependent node, to the signal applied at some specified source node. *Note:* The graph transmittance is the weighted sum of the path transmittances of the different open paths from the designated source node to the designated dependent node, where the weight for each path is the path factor divided by the graph determinant. *See also:* **linear signal flow graphs.**　282

grass (radar). A descriptive colloquialism referring to the appearance of noise on certain displays, such as **A-display.**　13

graticule (oscilloscopes). A scale for measurement of quantities displayed on the cathode-ray tube of an oscilloscope. *See:* **oscillograph.**　185

graticule area. The area enclosed by the continuous outer graticule lines. *Note:* Unless otherwise stated the graticule area shall be equal to or less than the viewing area. *See:* **oscillograph.** *See also:* **quality-area; viewing area.**　185

graticule, internal (oscilloscopes). A graticule whose rulings are a permanent part of the inner surface of the cathode-ray tube faceplate.　106

grating. *See:* **ultrasonic space grating.**

grating lobe. A lobe, other than the main lobe, produced by an array antenna when the interelement spacing is sufficiently large to permit the in-phase addition of radiated fields in more than one direction.　111

gravitational acceleration unit (g). The symbol g denotes a unit of acceleration equal in magnitude to the local value of gravity, unless otherwise specified. *Note:* (1) In some applications, a standard value of g may be specified. (2) For an earthbound accelerometer, the attractive force of gravity acting on the proof mass must be treated as an applied upward acceleration of one g.　46

gravity elevator. An elevator utilizing gravity to move the car. *See also:* **elevators.**　328

gravity gradient stabilization (communication satellite). The use of the gravity gradient along a satellite structure for controlling its attitude. This method usually requires long booms to create the necessary mass distribution.　74

gravity vertical (navigation). *See:* **mass-attraction vertical.**

graybody. A temperature radiator whose spectral emissivity is less than unity and the same at all wavelengths.

See also: **radiant energy (illuminating engineering).**　167

Gray code (computing systems). A binary code in which sequential numbers are represented by binary expressions, each of which differs from the preceding expression in one place only. *See:* **reflected binary code.**　255, 77

gray scale (television). An optical pattern in discrete steps between light and dark. *Note:* A gray scale with ten steps is usually included in resolution test charts. *See also:* **television.**　163

greasing truck. An electrically driven service vehicle to transport greases and oil for servicing the underground mine machinery. *Note:* It may include a compressor, air storage tank, and fittings to place lubricant at the proper points in the mining machinery.　328

Gregorian reflector antenna. A double reflector antenna with a subreflector (usually ellipsoidal) located at a distance greater than the focal length from the vertex of the main reflector.　111

grid (1) (electron tube). An electrode having one or more openings for the passage of electrons or ions.　244, 190, 125

(2) (storage cell). A metallic framework employed in a storage cell for supporting the active material and conducting the electric current. For an alkaline battery the grid may be called a frame. *See also:* **primary or secondary cells.**　328

grid bearing (navigation). Bearing relative to grid north. *See also:* **navigation.**　13

grid circuit (industrial control). A circuit that includes the grid-cathode path of an electron tube in series connection with other elements. *See also:* **electronic controller.**　206

grid control (industrial control). Control of anode current of an electron tube by means of proper variation (control) of the control-grid potential with respect to the cathode of the tube. *See also:* **electronic controller.**　206

grid-controlled mercury-arc rectifier. A mercury-arc rectifier in which one or more electrodes are employed exclusively to control the starting of the discharge. *See also:* **rectifier.**　111

grid course (navigation). Course relative to grid north. *See also:* **navigation.**　187, 13

grid current (analog computer). The current flowing between the summing junction and the grid of the first amplifying stage of an operational amplifier. *Note:* Grid current results in an error voltage at the amplifier output. *See also:* **electronic analog computer; electronic controller.**　9

grid-drive characteristic (electron tubes). A relation, usually shown by a graph, between electric or light output and control-electrode voltage measured from cutoff.　190, 125

grid driving power (electron tubes). The average of the product of the instantaneous values of the alternating components of the grid current and the grid voltage over a complete cycle. *Note:* This power comprises the power supplied to the biasing device and to the grid. *See also:* **electrode dissipation.**　190, 125

grid emission. Electron or ion emission from a grid. *See also:* **electron emission.**　190, 125

grid emission, primary (thermionic). Current produced by electrons or ions thermionically emitted from a grid. *See also:* **electron emission.** 190

grid emission, secondary. Electron emission from a grid due directly to bombardment of its surface by electrons or other charged particles. *See also:* **electron emission.** 190

grid-glow tube. A glow-discharge cold-cathode tube in which one or more control electrodes initiate but do not limit the anode current, except under certain operating conditions. *Note:* This term is used chiefly in the industrial field. 190

grid, ground. *See:* **ground grid.**

grid heading (navigation). Heading relative to grid north. *See also:* **navigation.** 187, 13

grid-leak detector. A triode or multielectrode tube in which rectification occurs because of electron current to the grid. *Note:* The voltage associated with this flow through a high resistance in the grid circuit appears in amplified form in the plate circuit. 328

grid modulation (electron tubes). Modulation produced by the application of the modulating voltage to the control grid of any tube in which the carrier is present. *Note:* Modulation in which the grid voltage contains externally generated pulses is called **grid pulse modulation.** 111

grid neutralization (electron tubes). The method of neutralizing an amplifier in which a portion of the grid-cathode alternating-current voltage is shifted 180 degrees and applied to the plate-cathode circuit through a neutralizing capacitor. *See also:* **amplifier; feedback.** 111

grid north (navigation). An arbitrary reference direction used in connection with a system of rectangular coordinates superimposed over a chart. *See also:* **navigation.** 187, 13

grid number *n* (electron tubes). A grid occupying the *n*th position counting from the cathode. *See also:* **electron tube.** 244, 190

grid pitch (electron tubes). The pitch of the helix of a helical grid. *See also:* **electron tube.** 244, 190

grid pulse modulation. Modulation produced in an amplifier or oscillator by application of one or more pulses to a grid circuit. 111

grid (circuit) resistor (industrial control). A resistor used to limit grid current. *See also:* **electronic controller.** 206

grids (high-power rectifier). Electrodes that are placed in the arc stream and to which a control voltage may be applied. *See also:* **rectification.** 328

grid transformer (industrial control). Supplies an alternating voltage to a grid circuit or circuits. 206

grid voltage. *See:* **electrode voltage; electronic controller.**

grid voltage, critical. *See:* **control-grid voltage.**

grid voltage supply (electron tube). The means for supplying to the grid of the tube a potential that is usually negative with respect to the cathode. *See also:* **power pack.** 328

grip, conductor (conductor stringing equipment). A device designed to permit the pulling of conductor without splicing on fittings, eyes, etc. It also permits the pulling of a "continuous" conductor where threading is not possible. The designs of these grips vary considerably. Grips such as the Klein (Chicago) and Crescent utilize an open sided rigid body with opposing jaws and swing latch. In addition

to pulling conductors, this type is commonly used to tension guys and in some cases, pull wire rope. The design of the Comealong (Pocketbook, Suitcase, "4" Bolt, etc.) incorporates a bail attached to the body of a clamp which folds to completely surround and envelop the conductor. "Bolts" are then used to close the clamp and obtain a grip. *Syn.* **buffalo; chicago grip; comealong; crescent; grip; klein; pocketbook; slip-grip; suit case; "4" bolt; "6" bolt; "7" bolt.** 45

grip, woven wire (conductor stringing equipment). A device designed to permit the temporary joining or pulling of conductors without the need of special eyes, links or grips. *Syn.* **basket; chinese finger; kellem; sock; wire mesh grip.** 45

groove (mechanical recording). The track inscribed in the record by the cutting or embossing stylus. *See also:* **phonograph pickup.** 176

groove angle (disk recording). The angle between the two walls of an unmodulated groove in a radial plane perpendicular to the surface of the recording medium. *See also:* **phonograph pickup.** 176

groove diameter. *See:* **tape-wound core.**

groove shape (disk recording). The contour of the groove in a radial plane perpendicular to the surface of the recording medium. *See also:* **phonograph pickup.** 176

groove speed (disk recording). The linear speed of the groove with respect to the stylus. *See also:* **phonograph pickup.** 176

groove width. *See:* **tape-wound core.**

gross demonstrated capacity. The gross steady output that a generating unit or station has produced while demonstrating its maximum performance under stipulated conditions. *See also:* **generating station.** 64

gross generation (electric power systems). The generated output power at the terminals of the generator. 94

gross head (hydroelectric power plant). The difference of elevations between water surfaces of the forebay and tailrace under specified conditions. 112

gross information content. A measure of the total information, redundant or otherwise, contained in a message. *Note:* It is expressed as the number of bits or hartleys required to transmit the message with specified accuracy over a noiseless medium without coding. *See also:* **bit.** 328

gross rated capacity. The gross steady output that a generating unit or station can produce for at least two hours under specified operating conditions. *See also:* **generating station.** 64

ground (earth) (1) (electric system). A conducting connection, whether intentional or accidental, by which an electric circuit or equipment is connected to the earth, or to some conducting body of relatively large extent that serves in place of the earth. *Note:* It is used for establishing and maintaining the potential of the earth (or of the conducting body) or approximately that potential, on conductors connected to it, and for conducting ground current to and from the earth (or the conducting body). 256, 64, 91 **(2) (transmission path).** (A) A direct conducting connection to the earth or body of water that is a part thereof. (B) A conducting connection to a structure that serves a function similar to that of an earth ground (that is, a structure such as a frame of an air, space, or land vehicle that is not conductively connected to earth). 25

groundable parts. Those parts that may be connected to ground without affecting operation of the device.
103, 202

ground absorption. The loss of energy in transmission of radio waves due to dissipation in the ground. *See also:* **radiation.** 59

ground acceleration (nuclear power generating stations) (seismic qualification of class 1E equipment). The acceleration of the ground resulting from a given earthquake's motion. The maximum ground acceleration can be obtained from the ground response spectrum as the acceleration at high frequencies (in excess of 33 hertz).
28, 31

ground-area open floodlight (illumination). A unit providing a weatherproof enclosure for the lamp socket and housing. *Note:* No cover glass is required. *See also:* **floodlighting.** 167

ground-area open floodlight with reflector insert. A weatherproof unit so constructed that the housing forms only part of the reflecting surface. *Note:* An auxiliary reflector is used to modify the distribution of light. No cover glass is required. *See also:* **floodlighting.** 167

ground bar (lightning). A conductor forming a common junction for a number of ground conductors. 244, 62

ground-based navigation aid. An aid that requires facilities located upon land or sea. *See also:* **navigation.** 187, 13

ground bus (electric system). A bus to which the grounds from individual pieces of equipment are connected, and that, in turn, is connected to ground at one or more points. *See:* **ground.** 103, 202, 64

ground cable bond. A cable bond used for grounding the armor or sheaths of cables or both. *See:* **ground.** 64

ground clamp (grounding clamp). A clamp used in connecting a grounding conductor to a grounding electrode or to a thing grounded. *See:* **ground.** 64

ground clutter (radar). Clutter resulting from the ground or objects on the ground. 13

ground conductor (lightning). A conductor providing an electric connection between part of a system, or the frame of a machine or piece of apparatus, and a ground electrode or a ground bar. *See:* **grounded conductor.** 244, 62

ground conduit. A conduit used solely to contain one or more grounding conductors. *See:* **ground.** 64

ground connection. *See:* **grounding connection.**

ground contact (switchgear assembly). A self-coupling separable contact provided to connect and disconnect the ground connection between the removable element and the ground bus of the housing and so constructed that it remains in contact at all times except when the primary disconnecting devices are separated by a safe distance. *Note:* Safe distance, as used here, is a distance at which the equipment will meet its withstand-voltage ratings, both low-frequency and impulse, between line and load terminals with the switching device in the closed position.
103, 202

ground-controlled approach (GCA) (radar). A ground radar system providing information by which aircraft approaches to landing may be directed via radio communications; the system consists of a precision approach radar (PAR) and an airport surveillance radar (ASR). *See also:* **navigation.** 187, 13

ground controlled interception. *See:* **GCI.**

ground current. Current flowing in the earth or in a grounding connection. *See:* **ground.** 64

ground-derived navigation data (air navigation). Data obtained from measurements made on land or sea at locations external to the vehicle. *See also:* **navigation.**
187, 13

ground detection rings (rotating machinery). Collector rings connected to a winding and its core to facilitate the measurement of insulation resistance on a rotor winding. *See also:* **rotor (rotating machinery).** 63

ground detector. An instrument or an equipment used for indicating the presence of a ground on an ungrounded system. *See:* **ground.** 64

grounded (1) (electric systems). Connected to earth or to some extended conducting body that serves instead of the earth, whether the connection is intentional or accidental.
256, 178, 45, 53, 91
(2) (transmission and distribution). Connected to or in contact with earth or connected to some conductive body which serves instead of the earth. 262

grounded capacitance. *See:* **self-capacitance of a conductor; ground.**

grounded-cathode amplifier. An electron-tube amplifier with the cathode at ground potential at the operating frequency, with input applied between the control grid and ground, and the output load connected between plate and ground. *Note:* This is the conventional amplifier circuit. *See also:* **amplifier.** 111

grounded circuit. A circuit in which one conductor or point (usually the neutral conductor or neutral point of transformer or generator windings) is intentionally grounded, either solidly or through a noninterrupting current limiting grounding device. *See:* **ground; grounded conductor; grounded system.** 64, 91

grounded concentric wiring system. A grounded system in which the external (outer) conductor is solidly grounded and completely surrounds the internal (inner) conductor through its length. The external conductor is usually uninsulated. *See:* **ground.** 64, 91

grounded conductor (1) (electric system). A conductor that is intentionally grounded, either solidly or through a current limiting device. *See also:* **ground.** 256, 91
(2) (transmission and distribution). A conductor which is intentionally grounded, either solidly or through a noninterrupting current limiting device. 262

grounded, directly. *See:* **grounded, solidly.**

grounded, effectively. An expression that means grounded through a grounding connection of sufficiently low impedance (inherent or intentionally added or both) that ground fault that may occur cannot build up voltages in excess of limits established for apparatus, circuits, or systems so grounded. *Notes:* (1) An alternating-current system or portion thereof may be said to be effectively grounded when, for all points on the system or specified portion thereof, the ratio of zero-sequence reactance to positive-sequence reactance is not greater than three and the ratio of zero-sequence resistance to positive-sequence reactance is not greater than one for any condition of operation and for any amount of connected generator capacity. (2) This definition is basically used in the application of line-to-neutral surge arresters. Surge arresters

with less than line-to-line voltage ratings are applicable on effectively grounded systems. *See:* **ground; grounded.**
152, 203, 64

grounded-grid amplifier. An electron-tube amplifier circuit in which the control grid is at ground potential at the operating frequency, with input applied between cathode and ground, and output load connected between plate and ground. *Note:* The grid-to-plate impedance of the tube is in parallel with the load instead of acting as a feedback path. *See also:* **amplifier.** 111

grounded, impedance. Grounded through impedance. *Note:* The components of the impedance need not be at the same location. 91

grounded neutral system (surge arresters). A system in which the neutral is connected to ground, either solidly or through a resistance or reactance of low value.
244, 62

grounded-neutral terminal type voltage transformer. A voltage transformer which has the neutral end of the high-voltage winding connected to the case or mounting base. 53

grounded parts (industrial control). Parts that are intentionally connected to ground. 103, 202, 64

grounded-plate amplifier (cathode-follower). An electron-tube amplifier circuit in which the plate is at ground potential at the operating frequency, with input applied between control grid and ground, and the output load connected between cathode and ground. *See also:* **amplifier.** 111

grounded, solidly (directly grounded). Grounded through an adequate ground connection in which no impedance has been inserted intentionally. *Notes:* (1) Adequate as used herein means suitable for the purpose intended. (2) This term, though commonly used, is somewhat confusing since a transformer may have its neutral solidly connected to ground, and yet the connection may be so small in capacity as to furnish only a very-high-impedance ground to the system to which it is connected. In order to define grounding positively and logically as to degree, the term effective grounding has come into use. *See:* **grounded.**
91, 152, 203

grounded system (1) (electric power). A system of conductors in which at least one conductor or point (usually the middle wire or neutral point of transformer or generator windings) is intentionally grounded, either solidly or through a noninterrupting current current-limiting device. *Note:* Various degrees of grounding are used, from solid or effective to the high-impedance grounding obtained from a small grounding transformer used only to secure enough ground current for relaying, to the high-resistance grounding which secures control of transient overvoltages but may not furnish sufficient current for ground-fault relaying. *See also:* **ground; grounded; grounded circuit.**
152, 64, 53

(2) (surge arrester). An electric system in which at least one conductor or point (usually the neutral conductor or neutral point of transformer or generator windings) is intentionally grounded, either solidly or through a grounding device. 2

(3) (transmission and distribution). A system of conductors in which at least one conductor or point is intentionally grounded, either solidly or through a noninterrupting

current-limiting device. 262

ground electrode (surge arresters). A conductor or group of conductors in intimate contact with the ground for the purpose of providing a connection with the ground.
244, 62

ground end (grounding device). The end or terminal of the device that is grounded directly or through another device. *See:* **grounding device.** 64, 91

ground equalizer inductors (antenna). Coils of relatively low inductance, placed in the circuit connected to one or more of the grounding points of an antenna to distribute the current to the various points in any desired manner. *Note:* Broadcast usage only and now in disuse. *See also:* **antenna.** 111, 179

ground fault (surge arresters). An insulation fault between a conductor and ground or frame. 244, 62

ground-fault neutralizer. A grounding device that provides an inductive component of current in a ground fault that is substantially equal to and therefore neutralizes the rated-frequency capacitive component of the ground-fault current, thus rendering the system resonant grounded.
91

ground-fault neutralizer grounded (resonant grounded). Reactance grounded through such values of reactance that, during a fault between one of the conductors and earth, the rated-frequency current flowing in the grounding reactances and the rated-frequency capacitance current flowing between the unfaulted conductors and earth shall be substantially equal. *Notes:* (1) In the fault these two components of current will be substantially 180 degrees out of phase. (2) When a system is ground-fault neutralizer grounded, it is expected that the quadrature component of the rated-frequency single-phase-to-ground fault current will be so small that an arc fault in air will be self-extinguishing. *See:* **ground; grounding device.**
91, 203, 64, 53

ground grid (1) (ground resistance). A system of grounding electrodes consisting of interconnected bare cables buried in the earth to provide a common ground for electric devices and metallic structures. *Note:* It may be connected to auxiliary grounding electrodes to lower its resistance. *See also:* **grounding device.** 313

(2) (conductor stringing equipment). A system of interconnected bare conductors arranged in a pattern over a specified area and buried below the surface of the earth. Normally, it is bonded to ground rods driven around and within its perimeter to increase its grounding capabilities and provide convenient connection points for grounding devices. The primary purpose of the grid is to provide safety for workmen by limiting potential differences within its perimeter to safe levels in case of high currents which could flow if the circuit being worked became energized for any reason, or if an adjacent energized circuit faulted. Metallic surface mats and gratings are sometimes utilized for this same purpose. When used these grids are employed at pull tension and mid-span splice sites. *See also:* **counterpoise; ground mat.** *Syn.* **counterpoise; ground gradient mat; ground mat.** 45

ground indication. An indication of the presence of a ground on one or more of the normally ungrounded conductors of a system. *See:* **ground.** 64

grounding cable. A cable used to make a connection to

ground. *See:* **grounding conductor.**   63

grounding-cable connector. The terminal mounted on the end of a grounding cable.   63

grounding clamp. *See:* **ground clamp.**

grounding, coefficient of. The ratio E_{LG}/E_{LL}, expressed as a percentage, of the highest root-mean-square line-to-ground power-frequency voltage E_{LG} on a sound phase, at a selected location, during a fault to ground affecting one or more phases to the line-to-line power-frequency voltage E_{LL} which would be obtained, at the selected location, with the fault removed. *Notes:* (1) Coefficients of grounding for three-phase systems are calculated from the phase-sequence impedance components as viewed from the selected location. For machines use the subtransient reactance. (2) The coefficient of grounding is useful in the determination of an arrester rating for a selected location. (3) A value not exceeding 80 percent is obtained approximately when for all system conditions the ratio of zero-sequence reactance to positive-sequence reactance is positive and less than 3, and the ratio of zero-sequence resistance to positive-sequence reactance is positive and less than 1.   2

grounding conductor (ground conductor) (1) (general). The conductor that is used to establish a ground and that connects an equipment, device, wiring system, or another conductor (usually the neutral conductor) with the grounding electrode or electrodes. *See:* **conductor; ground; grounding device.**   256, 64

(2) (mining) (safety ground conductor) (safety ground) (frame ground). A metallic conductor used to connect the metal frame or enclosure of an equipment, device, or wiring system with a mine track or other effective grounding medium. *See also:* **mine feeder circuit.**   328

(3) (transmission and distribution). A conductor which is used to connect the equipment or the wiring system with a grounding electrode or electrodes.   262

grounding connection (ground connection). A connection used in establishing a ground and consists of a grounding conductor, a grounding electrode, and the earth (soil) that surrounds the electrode or some conductive body which serves instead of the earth. *See:* **ground.**   64

grounding device (electric power). An impedance device used to connect conductors of an electric system to ground for the purpose of controlling the ground current or voltages to ground or a nonimpedance device used to temporarily ground conductors for the purpose of the safety of workmen. *Note:* The grounding device may consist of a grounding transformer or a neutral grounding device, or a combination of these. Protective devices, such as surge arresters, may also be included as an integral part of the device.   91, 64

grounding electrode (ground electrode). A conductor used to establish a ground. *See:* **ground.**   64

grounding jumper (electric appliances). A strap or wire to connect the frame of the range to the neutral conductor of the supply circuit. *See also:* **appliance outlet.**   263

grounding outlet (safety outlet*). An outlet equipped with a receptacle of the polarity type having, in addition to the current-carrying contacts, one grounded contact that can be used for the connection of an equipment grounding conductor. *Note:* This type of outlet is used for connection

of portable appliances. *See:* **ground.**   64

* Deprecated

grounding pad (rotating machinery). A contact area, usually on the stator frame, provided to permit the connection of a grounding terminal. *See also:* **stator.**   63

grounding switch (general). A mechanical switching device by means of which a circuit or piece of apparatus may be electrically connected to ground. *Note:* A grounding switch is used to ground a piece of equipment to permit maintenance personnel to work with safety.

103, 202, 27

grounding system (1) (general). Consists of all interconnected grounding connections in a specific area. *See also:* **grounding device.**   313

(2) (surge arresters). A complete installation comprising one or more ground electrodes, ground conductors, and ground bars as required.   244, 62

grounding terminal (rotating machinery). A terminal used to make a connection to a ground. *See also:* **stator.**   63

grounding transformer (electric power). A transformer intended primarily to provide a neutral point for grounding purposes. *Note:* It may be provided with a delta winding in which resistors or reactors are connected. *See also:* **grounded; grounding device; transformer; voltage rating (grounding transformer); rated kilovolt ampere (grounding transformer).**   91, 203, 53

ground insulation (rotating machinery). Insulation used to insure the electric isolation of a winding from the core and mechanical parts of a machine. *See also:* **asynchronous machine; coil insulation; direct-current commutating machine.**   63

ground isolation (sequential events recording systems). The disconnection of selected field contact circuits from the contact voltage supply to allow identification of the grounded field contact wires. *See:* **field contacts.**

48, 58

ground level (mobile communication). The elevation of the ground above mean sea level at the antenna site or other point of interest. *See also:* **mobile communication system.**

181

ground light. Visible radiation from the sun and sky reflected by surfaces below the plane of the horizon. *See also:* **sunlight.**   167

ground loop (analog computer). A potentially detrimental loop formed when two or more points in an electrical system that are nominally at ground potential are connected by a conducting path such that either or both points are not at the same ground potential.   9

ground mat. A system of bare conductors, on or below the surface of the earth, connected to a ground or a ground grid to provide protection from dangerous touch voltages. *Note:* Plates and gratings of suitable area are common forms of **ground mats.** *See also:* **grounding device.**   313

ground overcurrent (power switchgear). The net (phasor sum) current flowing in the phase and neutral conductors or the total current flowing in the normal neutral to ground connection which exceeds a predetermined value.   103

ground plane (1) (transmission and distribution). An assumed plane of true ground or zero potential. *See also:* **direct-stroke protection (lightning).**   64

(2) (electromagnetic compatibility). A conducting surface or plate used as a common reference point for circuit returns and electric or signal potentials.   199

(3) (antenna). A conducting or reflecting plane functioning to image a radiating structure. *Syn:* **imaging plane.**
                    111

ground plane, effective (mobile communication). The height of the average terrain above mean sea level as measured for a distance of 100 meters out from the base of the antenna in the desired direction of communication. It may be considered the same as ground level only in open flat country. *See also:* **mobile communication system.**   181

ground plate (grounding plate). A plate of conducting material buried in the earth to serve as a grounding electrode. *See:* **ground.**   64

ground-position indicator (GPI) (electronic navigation). A dead-reckoning tracer or computer similar to an air position indicator (API) with provision for taking account of drift. *See also:* **navigation; radio navigation.**   187, 13

ground protection (ground-fault protection). A method of protection in which faults to ground within the protected equipment are detected irrespective of system phase conditions.   103, 202, 60, 127

ground-referenced navigation data. Data in terms of a coordinate system referenced to the earth or to some specified portion thereof. *See also:* **navigation.**   187, 13

ground-reflected wave. The component of the ground wave that is reflected from the ground. *See also:* **radiation.**
                    59

ground relay. A relay that by its design or application is intended to respond primarily to system ground faults.
                    103, 202, 60, 127

ground resistance (grounding electrode). The ohmic resistance between the grounding electrode and a remote grounding electrode of zero resistance. *Note:* By remote is meant at a distance such that the mutual resistance of the two electrodes is essentially zero. *See also:* **ground.**
                    313

ground return. *See:* **ground clutter.**

ground-return circuit (earth-return circuit) (1) (transmission and distribution). A circuit in which the earth is utilized to complete the circuit. *See also:* **ground; telegraphy; transmission line.**   64

(2) (data transmission). A circuit which has a conductor (or two or more in parallel) between two points and which is completed through the ground or earth.   59

ground-return current (line residual current) (electric supply line). The vector sum of the currents in all conductors on the electric supply line. *Note:* Actually the ground-return current in this sense may include components returning to the source in wires on other pole lines, but from the inductive coordination standpoint these components are substantially equivalent to components in the ground. *See also:* **inductive coordination.**   328

ground-return system. A system in which one of the conductors is replaced by ground.   244, 62

ground rod (ground electrode). A rod that is driven into the ground to serve as a ground terminal, such as a copper-clad rod, solid copper rod, galvanized iron rod, or galvanized iron pipe. *Note:* Copper-clad steel rods are commonly used during conductor stringing operations to provide a means of obtaining an electrical ground using portable grounding devices. *Syn.* **ground electrode.**   45, 297

ground speed (navigation). The speed of a vehicle along its track. *See also:* **navigation.**   13, 187

ground-start signaling (telephone switching systems). A method of signaling using direct current in a ground return path to indicate a service request.   55

ground-state maser (laser-maser). A **maser** in which the terminal level of the amplifying transition is appreciably populated at thermal equilibrium for the ambient temperature.   363

ground support equipment (GSE) (test, measurement and diagnostic equipment). All equipment (implements, tools, test equipment devices [mobile or fixed], and so forth) required on the ground to make an aerospace system (aircraft, missile, and so forth) operational in its intended environment.   54

ground surveillance radar. A radar set operated at a fixed point for observation and control of the position of aircraft or other vehicles in the vicinity. *See also:* **navigation.**
                    187

ground system (antenna). That portion of an antenna consisting of a system of conductors or a conducting surface in or on the ground.   111

ground system of an antenna. That portion of an antenna closely associated with and including an extensive conducting surface which may be the earth itself. *See also:* **antenna.**   179

ground terminal (earth terminal) (1) (industrial control). A terminal intended to ensure, by means of a special connection, the grounding (earthing) of part of an apparatus.
                    244, 206

(2) (lightning protection system). The portion extending into the ground, such as a ground rod, ground plate, or the conductor itself, serving to bring the lightning protection system into electric contact with the ground.
                    297, 244, 62

(3) (surge arrester). The conducting part provided for connecting the arrester to ground.   2

ground transformer. *See:* **grounding transformer.**

ground wave (1) (antennas). A radio wave that is propagated over the earth and is ordinarily affected by the presence of the ground and troposphere. *Notes:* (A) The ground wave includes all components of a radio wave over the earth except ionospheric and tropospheric waves. (B) The ground wave is refracted because of variations in the dielectric constant of the troposphere including the condition known as a surface duct. *See also:* **radiation; radio wave propagation.**   246

(2) (data transmission). Any radio wave that is propagated along the earth except ionospheric and tropospheric waves. *Note:* The ground wave is refracted because of variations of the dielectric constant of the troposphere, including the condition known as a surface duct.   59

(3) (radio wave propagation). From a source in the vicinity of a planetary surface, that wave which would exist in the vicinity of that surface in the absence of an ionosphere.
                    146

ground wire (1) (telecommunication). A conductor leading to an electric connection with the ground. *See also:* **antenna; grounded.**   179, 59

(2) (overhead power line). A conductor having grounding connections at intervals, that is suspended usually above but not necessarily over the line conductor to provide a degree of protection against lightning discharges. *See:* **ground; overhead ground wire.** 64

group (1) (communications). *See:* **channel group.**

(2) (storage cell). An assembly of plates of the same polarity burned to a connecting strap. *See also:* **battery (primary or secondary).** 328

(3) (electric and electronics parts and equipments). A collection of units, assemblies, or subassemblies which is a subdivision of a set or system, but which is not capable of performing a complete operational function. Typical examples: antenna group, indicator group. 17

group alerting (telephone switching systems). A central office feature for simultaneously signaling a group of customers from a control station providing an oral or recorded announcement. 55

group ambient temperature (cable or duct) (power distribution, underground cables). The no-load temperature in a group with all other cables or ducts in the group loaded. *See also:* **power distribution, underground construction.** 57

group-busy tone (telephone switching systems). A tone that indicates to operators that all trunks in a group are busy. 55

group, commutating. A group of thyristor converter circuit elements and the alternating-voltage supply elements conductively connected to them in which the direct current of the group is commutated between individual elements that conduct in succession. 121

group delay (dispersive and nondispersive delay line). The derivative of radian phase with respect to radian frequency, $\partial\phi/\partial\omega$. It is equal to the phase delay for an ideal nondispersive delay device, but may differ greatly in actual devices where there is ripple in the phase vs. frequency characteristic. 81

group delay time. The rate of change, with angular frequency, of the total phase shift through a network. *Notes:* (1) Group delay time is the time interval required for the crest of a group of interfering waves to travel through a 2-port network, where the component wave trains have slightly different individual frequencies. (2) Group delay time is usually very close in value to envelope delay and transmission time delay, and in the case of vanishing spectrum bandwidth of the signal these quantities become identical. *See also:* **measurement system.** 293, 183

group flashing light (navigation). A flashing light in which the flashes are combined in groups, each including the same number of flashes, and in which the groups are repeated at regular intervals. *Note:* The duration of each flash is clearly less than the duration of the dark periods between flashes, and the duration of the dark periods between flashes is clearly less than the duration of the dark periods between groups. 167

grouping (1) (facsimile). Periodic error in the spacing of recorded lines. *See also:* **facsimile signal (picture signal).** 12

(2) (electroacoustics). Nonuniform spacing between the grooves of a disk recording. 176

group operation (switchgear). The operation of all poles of a multipole switching device by one operating mechanism. 103, 202, 27

groups, commutating, set of (thyristor converter). Two or more commutating groups that have simultaneous commutations. 121

group-series loop insulating transformer. An insulating transformer whose secondary is arranged to operate a group of series lamps and/or a series group of individual-lamp transformers. 53

group velocity (traveling wave). The velocity of propagation of the envelope, provided that this moves without significant change of shape. *Notes:* (1) The magnitude of the group velocity is equal to the reciprocal of the change of phase constant with angular frequency. (2) Group velocity differs in magnitude from phase velocity if the phase velocity varies with frequency and differs in direction from phase velocity if the phase velocity varies with direction. *See also:* **radio wave propagation.** 180, 146

grout (rotating machinery). A very rich concrete used to bond the feet, sole plates, bedplate, or rail of a machine to its foundation. 63

grown junction. *See:* **junction, grown.**

G scan (radar). *See:* **G display.**

G scope (radar). *See:* **G display.**

G/T radio (antennas). The ratio of the maximum power gain to the noise temperature of an antenna. 111

guard (interference terminology). A conductor situated between a source of interference and a signal path in such a way that interference currents are conducted to the return terminal of the interference source without entering the signal path. *See also:* **interference.** 60

guard band (data transmission). A frequency band between two channels which gives a margin of safety against mutual interference. 59

guard circle (disk recording). An inner concentric groove inscribed, on disk records, to prevent the pickup from being damaged by being thrown to the center of the record. 176

guarded (1) (electric system). Covered, shielded, fenced, inclosed, or otherwise protected, by means of suitable covers or casings, barrier rails or screens, mats or platforms, to remove the likelihood of dangerous contact or approach by persons or objects to a point of danger. *Note:* Wires that are insulated, but not otherwise protected, are not considered as guarded. *See also:* **grounded.** 256, 178

(2) (transmission and distribution). Covered, fenced, enclosed, or otherwise protected, by means of suitable covers or casings, barrier rails or screens, mats or platforms, designed to prevent dangerous approach or contact by persons or objects. *Note:* Wires which are insulated, but not otherwise protected, are not considered as guarded. 262

guarded input (1) (amplifiers). Means of connecting an input signal so as to prevent any common-mode signal from causing current to flow in the input, thus differences of source impedance do not cause conversion of the common-mode signal into a differential signal. *See:* **amplifier.** 185

(2) (oscilloscopes). A shielded input where the shield is driven by a signal in phase with and equal in amplitude to the input signal. 106

guarded machine (rotating machinery). An open machine in which all openings giving direct access to live or rotating parts (except smooth shafts) are limited in size by the design of the structural parts, or by screens, grilles, expanded metal, etcetera, to prevent accidental contact with such parts. Such openings are of such size as not to permit the passage of a cylindrical rod ½ inch in diameter, except where the distance from the guard to the live or rotating parts is more than 4 inches; they are of such size as not to permit the passage of a cylindrical rod ¾ inch in diameter. *See:* **asynchronous machine; direct-current commutating machine.**   63

guarded release (telephone switching systems). A technique for retaining a busy condition during the restoration of a circuit to its idle state.   55

guard electrode (testing of electric power system components). One or more electrically conducting elements, arranged and connected in an electric instrument or measuring circuit so as to divert unwanted conduction or displacement currents from, or confine wanted currents to, the measurement device.   201

guard-ground system (interference terminology). A combination of guard shields and ground connections that protects all or part of a signal transmission system from common-mode interference by eliminating ground loops in the protected part. *Note:* Ideally the guard shield is connected to the source ground. The source is usually grounded also to the source ground by bonding of the transducer to the test body. The filter, signal receiver, etcetera, are floating with respect to their own grounded cases. This necessitates physically isolating the signal receiver and filter chassis from the cases and using isolation transformers in power supplies, or isolating input circuits from cases and using isolating input transformers. This arrangement in effect places the signal receiver and filter electrically at the source. By means of a similar guard, the load can be placed effectively at the source. See the accompanying figures.   60

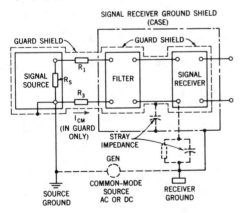

Ideal guard shield. The guard shield consists of the signal source shield (if present), the signal shield, the filter shield, and the signal receiver shield.

See also: **interference.**

guard shield (interference terminology). A guard that is in the form of a shielding enclosure surrounding all or part of the signal path. *Note:* A guard shield is effective against both capacitively coupled and conductively coupled interference whereas a simple guard conductor is usually effective only against conductively coupled interference. *See also:* **guard-ground system; interference.**   60

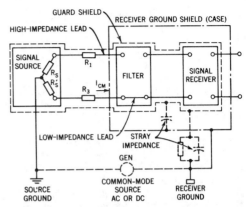

Guard shield connections when connection at source is not convenient. When $(R_3 + R_s') \ll (R_1 + R_s)$, this arrangement causes common-mode current in the low-impedance lead, but protects the more critical high-impedance lead from current flow.

guard signal. A signal sent over a communication channel to make the system secure against false information by preventing or guarding against the relay operation of a circuit breaker or other relay action until the signal is removed and replaced by a releasing (trip) or permissive signal.   103, 202, 60

guard wire. A grounded wire erected near a lower-voltage circuit or public crossing in such a position that a high (or higher) voltage overhead conductor cannot come into accidental contact with the lower-voltage circuit, or with persons or objects on the crossing without first becoming grounded by contact with the guard wire. *See:* **ground.**
                                                        64

guidance (missile). The process of controlling the flight path through space through the agency of a mechanism within the missile. *See also:* **guided missile.**   328

guide bearing (rotating machinery). A bearing arranged to limit the transverse movement of a vertical shaft. *See also:* **bearing.**   63

guided missile. An unmanned device whose flight path through space may be controlled by a self-contained mechanism. *See:* **beam rider guidance; command guidance; guidance; homing guidance; preset guidance.**   328

guided wave. A wave whose energy is concentrated within or near boundaries between materials of different properties and that is propagated along the path so defined. *Notes:* (1) The waves that can be transmitted by cylindrical waveguides may be classified into two types: transverse electric (TE) for which the axial electric field is zero, and transverse magnetic (TM) for which the axial magnetic field is zero. If the line or lines which bound the cross section of the waveguide are lines for which one coordinate remains constant in a curvilinear coordinate system in which it is possible to separate the variables of the wave equation, the type of wave is designated by TE or TM followed by two subscripts indicating the appropriate solution in each of the two coordinates. The interpretation of the subscripts for waveguides of the more usual cross sections is given in their respective definitions. (2) If the cross section of the waveguide does not fulfill the aforementioned requirement, the type of wave is designated by TE or TM with one numerical subscript starting with the wave that has the lowest frequency as 1 and numbering the remaining waves in the order of their cutoff frequencies. *See:* **cutoff wavelength; dominant wave; resonant modes; resonant wavelength; transverse electric wave; transverse magnetic wave.** *See also:* **radio wave propagation; waveguide.**   179, 180

guide flux. *See:* **form factor; shield.**

guide wavelength. The wavelength in a waveguide, measured in the longitudinal direction. *See also:* **waveguide.**
179

gun-control switch. A switch that closes an electric circuit, thereby actuating the gun-trigger-operating mechanism of an aircraft, usually by means of a solenoid. *See also:* **air transportation wiring and associated equipment.**   328

guy. A tension member having one end secured to a fixed object and the other end attached to a pole, crossarm, or other structural part that it supports. *See:* **guy wire.** *See also:* **tower.**   64

guy anchor. The buried element of a guy assembly that provides holding strength or resistance to guy wire pull. *Note:* The anchor may consist of a plate, a screw or expanding device, a log of timber, or a mass of concrete installed at sufficient depth and of such size as to develop strength proportional to weight of earth or rock it tends to move. The anchor is designed to provide attachment for the anchor rod which extends above surface of ground for convenient guy connection. *See:* **guy.** *See also:* **tower.**
64

guy insulator. An insulating element, generally of elongated form with transverse holes or slots for the purpose of insulating two sections of a guy or provide insulation between structure and anchor and also to provide protection in case of broken wires. Porcelain guy insulators are generally designed to stress the porcelain in compression, but wood insulators equipped with suitable hardware are generally used in tension. *See also:* **tower.**   64

guy wire. A stranded cable used for a semiflexible tension support between a pole or structure and the anchor rod, or between structures. *See:* **guy.** *See also:* **tower.**   64

gyration impedance (circuits and systems). A characteristic of a gyrator that may be expressed in terms of the impedance matrix elements as $\sqrt{z_{12} \times z_{21}}$. *See also:* **gyrator.**
67

gyrator (1) (antennas and propagation). (A) A directional phase changer in which the phase changes in opposite directions differ by π radians or 180 degrees. (B)* Any nonreciprocal passive element employing gyromagnetic properties. *See:* **waveguide.**   244, 179

* Deprecated

(2) (circuits and systems). A two port having an immittance matrix containing zero elements on the main diagonal and having remaining elements of opposite sign. For example:

$$\begin{bmatrix} i_1 \\ i_2 \end{bmatrix} = \begin{bmatrix} 0 & y_{12} \\ y_{21} & 0 \end{bmatrix} \begin{bmatrix} v_1 \\ v_2 \end{bmatrix}$$

or

$$\begin{bmatrix} v_1 \\ v_2 \end{bmatrix} = \begin{bmatrix} 0 & z_{12} \\ z_{21} & 0 \end{bmatrix} \begin{bmatrix} i_1 \\ i_2 \end{bmatrix}$$
67

gyro (gyroscope). A device using angular momentum (usually of a spinning rotor) to sense angular motion of its case with respect to inertial space about one or two axes orthogonal to the spin axis. *Notes:* (1) This definition does not include more complex systems such as stable platforms using gyros as components. (2) Certain devices that perform similar functions but do not use angular momentum are sometimes classified as gyros.   46

gyrocompass. A compass consisting of a continuously driven Foucault gyroscope whose supporting ring normally confines the spinning axis to a horizontal plane, so that the earth's rotation causes the spinning axis to assume a position in a plane passing through the earth's axis, and thus to point to true north.   328

gyrocompass alignment (inertial systems). A process of self-alignment in azimuth based upon measurements of misalignment drift about the nominal east-west axis of the system. *See also:* **navigation.**   187

gyrocompassing. *See:* **gyrocompass alignment.**

gyro flux-gate compass (gyro flux-valve compass). A device that uses saturable reactors in conjunction with a vertical gyroscope, to sense the direction of the magnetic north with respect to the aircraft heading.   328

gyro frequency (radio wave propagation). The lowest natural frequency at which charged particles spiral in a fixed magnetic field. *Note:* It is a vector quantity expressed by

$$\mathbf{f}_k = \frac{1}{2\pi} \frac{q\mathbf{B}}{m}$$

where q is the charge of the particles, $\mathbf{B}$ is the magnitude magnetic induction vector, and m is the mass of the particles. *See also:* **radio wave propagation.**   180

gyro gain. The ratio of the output angle of the gimbal to the input angle of a rate integrating gyro at zero frequency. *Note:* It is numerically equal to the ratio of the rotor angular momentum to the damping coefficient.   46

gyro horizon electric indicator. An electrically driven device for use in aircraft to provide the pilot with a fixed artificial horizon. *Note:* It indicates deviation from level flight.   328

H

H (*H* **beacon) (electronic navigation).** A designation applied to two types of facilities: (1) A nondirectional radio beacon for homing by means of an airborne direction finder. (2) A radar air navigation system using an airborne interro-

gator to measure the distances from two ground transponders. *See also:* **navigation.**   187, 13

halation (cathode-ray tube). An annular area surrounding a spot, that is due to the light emanating from the spot being reflected from the front and rear sides of the face plate. *See also:* **cathode-ray tubes.**   244, 190

half-adder. A combinational logic element having two outputs, S and C, and two inputs, A and B, such that the outputs are related to the inputs according to the following equations:

$$S = A \text{ OR } B \text{ (exclusive OR).}$$
$$C = A + B.$$

S denotes sum without carry, C denotes carry. Two half-adders may be used for performing binary addition.
                                                                                            77

half-amplitude recovery time (Geiger-Müller counters). The time interval from the start of a full-amplitude pulse to the instant a succeeding pulse can attain an amplitude of 50 percent of the maximum amplitude of a full-amplitude pulse.   96

half cell. An electrode immersed in a suitable electrolyte. *See also:* **electrolytic cell.**   328

half coil. *See:* **armature bar.**

half duplex (communications). Pertaining to an alternate, one way at a time, independent transmission. *See:* **duplex.** *See also:* **communication.**   255, 77, 194

half-duplex operation (telegraph system). Operation of a duplex system arranged to permit operation in either direction but not in both directions simultaneously. *See also:* **telegraphy.**   328

half-duplex repeater. A duplex telegraph repeater provided with interlocking arrangements that restrict the transmission of signals to one direction at a time. *See also:* **telegraphy.**   328

half-period average-value (symmetrical periodic-quantity). The absolute value of the algebraic average of the values of the quantity taken throughout a half-period, beginning with a zero value. If the quantity has more than two zeros during a cycle, that zero shall be taken that gives the largest half-period average value.   210

half-power beamwidth (plane containing the direction of the maximum value of a beam). The angle between the two directions in which the radiation intensity is one-half the maximum value of the beam. *See also:* **antenna.**   179

half section (circuits and systems). A bisected tee or pi section. A basic L-section building block of image-parameter filters.   67

halftone characteristic (facsimile). A relation between the density of the recorded copy and the density of the subject copy. *Note:* The term may also be used to relate the amplitude of the facsimile signal to the density of the subject copy or the record copy when only a portion of the system is under consideration. In a frequency-modulation system an appropriate parameter is to be used instead of the amplitude. *See also:* **recording (facsimile).**   12

halftones* (storage tubes). *See:* **level.** *See also:* **storage tube.**

* Deprecated

half-wave rectification (power supplies). In the rectifying process, half-wave rectification passes only one-half of each incoming sinusoid, and does not pass the opposite half-cycle. The output contains a single half-sine pulse for each input cycle. A single rectifier provides half-wave rectification. Because of its poorer efficiency and larger alternating-current component, half-wave rectification is usually employed in noncritical low-current circumstances. See the accompanying figure. *See also:* **power**

supply; rectifier circuit element; rectification; rectifier.**
                                                                     228, 186

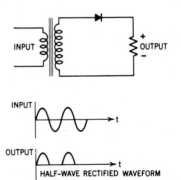

Half-wave rectification

Hall analog multiplier. A Hall multiplier specifically designed for analog multiplication purposes.   107

Hall angle. The angle between the electric field vector and the current density vector.   107

Hall coefficient. The coefficient of proportionality R in the relation

$$E_H = R(J \times B)$$

where

E_H is the resulting transverse electric field,
J is the current density,
B is the magnetic flux density.

Note: The sign of the majority carrier charge can usually be inferred from the sign of the Hall coefficient.   107

Hall effect (1) (in conductors and semiconductors). The change of the electric conduction caused by that component of the magnetic field vector normal to the current density vector, which, instead of being parallel to the electric field, forms an angle with it. *Note:* In conductors and semiconductors of noncubic single crystals, the current density and electric field vectors may not be parallel in the absence of an applied magnetic field. For such crystals the more general definition below should be used.

(2) (in any material, including ferromagnetic and similar materials). The change of the electric conduction caused by that component of the magnetic field vector applied normal to the current density vector, which causes the angle between the current density vector and the electric field to change from the magnitude that existed prior to the introduction of the magnetic field. *Note:* For ferromagnetic and similar materials there are two effects, the "ordinary" Hall effect due to the applied external magnetic flux as described for conductors and semiconductors and the "extraordinary" Hall effect due to the magnetization in the ferromagnetic or similar material. In the absence of the "extraordinary" Hall effect and the effects outlined in the preceding note, the current density vector and the electric field vector will be parallel when there is no external magnetic flux.   107

Hall effect device. A device in which the Hall effect is utilized.   107

Hall generator. A Hall plate, together with leads, and, where used, encapsulation and ferrous or nonferrous backing plate(s).   107

Hall mobility (electric conductor). The quantity μ_H in the relation $\mu_H = R\sigma$, where R = Hall coefficient and σ = conductivity. *See also:* **semiconductor.** 245, 210, 66

Hall modulator. A Hall effect device that is specifically designed for modulation purposes. 107

Hall multiplier. A Hall effect device that contains a Hall generator together with a source of magnetic flux density and that has an output that is a function of the product of the control current and the field excitation current. 107

Hall plate. A three-dimensional configuration of any material in which the Hall effect is utilized. 107

Hall probe. A Hall effect device specifically designed for measurement of magnetic flux density. 107

Hall terminals. The terminals between which the Hall voltage appears. 107

Hall voltage. The voltage generated in a Hall plate due to the Hall effect. 107

halogen-quenched counter tube. A self-quenched counter tube in which the quenching agent is a halogen, usually bromine or chlorine. 190

Hamming distance (computing systems). *See:* **signal distance.**

hand burnishing (electroplating). Burnishing done by a hand tool, usually of steel or agate. *See also:* **electroplating.** 328

hand (head or butt) cable (mining). A flexible cable used principally in making electric connections between a mining machine and a truck carrying a reel of portable cable. *See also:* **mine feeder circuit.** 328

***H* and *D* curve.** *See:* **Hurter and Driffield curve.**

hand elevator. An elevator utilizing manual energy to move the car. *See also:* **elevators.** 328

handhole (transmission and distribution). An opening in an underground system containing cable, equipment, or both into which workmen reach but do not enter. 262

handling device (metal-clad switchgear). The accessory that is used for the removal, replacement, or transportation of the removable element. 103, 202

hand operation (industrial control). Actuation of an apparatus by hand without auxiliary power. *See:* **switch.** 244, 206

hand receiver. An earphone designed to be held to the ear by the hand. *See also:* **loudspeaker.** 328

hand-reset relay (mechanically reset relay). A relay that is so constructed that it remains in the picked-up condition even after the input quantity is removed; specific manual action is required to reset the relay. 103, 202, 60, 123

handset (telephony). An assembly that includes a handle, a telephone transmitter, and receiver. Other components, such as the speech network, also may be located in the handset. 122

handset telephone. *See:* **hand telephone set.**

handshake (test, measurement and diagnostic equipment). A hardware or software sequence of events requiring mutual consent of conditions prior to change. 54

handshake cycle (digital interface for programmable instrumentation). The process whereby digital signals effect the transfer of each data byte across the interface by means of an interlocked sequence of status and control signals. Interlocked denotes a fixed sequence of events in which one event in the sequence must occur before the next event may occur. 40

hand telephone set (handset telephone). A telephone set having a handset and a mounting that serves to support the handset when the latter is not in use. *Note:* The prefix desk, wall, drawer, etcetera, may be applied to the term **hand telephone set** to indicate the type of mounting. *See also:* **telephone station.** 328

handwheel (industrial control). A wheel the rim of which serves as a handle for manual operation of a rotary device. 206

hand winding (rotating machinery). A winding placed in slots or around poles by a human operator. *See also:* **rotor (rotating machinery); stator.** 63

hangover (facsimile). *See:* **tailing.**

hang-up hand telephone set (suspended-type handset telephone) (bracket-type handset telephone). A hand telephone set in which the mounting is arranged for attachment to a vertical surface and is provided with a switch bracket from which the handset is suspended. *See also:* **telephone station.** 328

hang-up signal (telephone switching systems). A signal transmitted over a line or trunk to indicate that the calling party has released. 55

hard limiting. *See:* **limiter circuit.**

hard line (test, measurement and diagnostic equipment). Any direct electrical connection between the unit under test and the testing device. 54

hardware. (1) Physical entities such as computers, circuits, tape readers, etcetera. Contrasted with software. (2) Parts made of metal such as fasteners, hinges, etcetera. *See also:* **software.** 235

hardwire (test, measurement and diagnostic equipment). Circuitry with the absence of electrical elements, such as resistors, inductors, capacitors; circuits containing only wire and terminal connections with no intervening switching inherent. 54

harmful interference (electromagnetic compatibility). Any emission, radiation, or induction that endangers the functioning, or seriously degrades, obstructs, or repeatedly interrupts a radiocommunication service or any other equipment or system operating in accordance with regulations. *See also:* **electromagnetic compatibility.** 199

harmonic. A sinusoidal component of a periodic wave or quantity having a frequency that is an integral multiple of the fundamental frequency. *Note:* For example, a component the frequency of which is twice the fundamental frequency is called the second harmonic. *See also:* **signal wave.** 111, 59, 176

harmonic analyzer. A mechanical device for measuring the amplitude and phase of the various harmonic components of a periodic function from its graph. *See also:* **instrument; signal wave; wave analyzer.** 328

harmonic components (harmonics). The harmonic components of a Fourier Series are the terms $C_n \sin(nx + \theta_n)$. *Note:* For example, the component that has a frequency twice that of the fundamental ($n = 2$) is called the second harmonic. 210

harmonic conjugate. *See:* **Hilbert transform.**

harmonic content (nonsinusoidal periodic wave) (1) (static power converters). The deviation from the sinusoidal form, expressed in terms of the order and magnitude of the Fourier series terms describing the wave. *See also:* **power rectifier; rectification.** 208

(2) (emergency and standby power systems). A measure of the presence of harmonics in a voltage or current waveform expressed as a percentage of the amplitude of the fundamental frequency at each harmonic frequency. The total harmonic content is expressed as the square root of the sum of the squares of each of the harmonic amplitudes (expressed as a percentage of the fundamental).   89

harmonic conversion transducer (frequency multiplier, frequency divider). A conversion transducer in which the output-signal frequency is a multiple or submultiple of the input frequency. *Notes:* (1) In general, the output-signal amplitude is a nonlinear function of the input-signal amplitude. (2) Either a frequency multiplier or a frequency divider is a special case of harmonic conversion transducer. *See also:* **heterodyne conversion transducer (converter); transducer.**   210, 252, 190, 125

harmonic distortion. *See:* **distortion, harmonic.**

harmonic factor. The ratio of the root-mean-square (rms) value of all the harmonics to the root-mean-square (rms) value of the fundamental.   53

$$\text{Harmonic Factor} \atop \text{(for voltage)} = \frac{\sqrt{E_3^2 + E_5^2 + E_7^2 \dots}}{E_1}$$

$$\text{Harmonic Factor} \atop \text{(for current)} = \frac{\sqrt{I_3^2 + I_5^2 + I_7^2 \dots}}{I_1}$$

harmonic leakage power (TR and pre-TR tubes). The total radio-frequency power transmitted through the fired tube in its mount at frequencies other than the fundamental frequencies generated by the transmitter.   125

harmonic-restraint relay. A relay so constructed that its operation is restrained by selecting harmonic components of one or more separate input quantities. *See also:* **relay.**   281, 60, 103, 202, 127

harmonics. *See:* **harmonic components.**

harmonic series. A series in which each component has a frequency that is an integral multiple of a fundamental frequency.   176

harmonic telephone ringer. A telephone ringer that responds only to alternating current within a very narrow frequency band. *Note:* A number of such ringers, each responding to a different frequency, are used in one type of selective ringing. *See also:* **telephone station.**   328

harmonic test (rotating machinery). A test to determine directly the value of one or more harmonics of the waveform of a quantity associated with a machine, relative to the fundamental of that quantity. *See also:* **asynchronous machine.**   63

hartley (information theory). A unit of information content, equal to one decadal decision, or the designation of one of ten possible and equally likely values or states of anything used to store or convey information. *Notes:* (1) A hartley may be conveyed by one decadal code element. One hartley equals (log of 10 to base 2) times one bit. (2) If, in the definition of information content, the logarithm is taken to the base ten, the result will be expressed in hartleys. *Syn.* **dit.** *See also:* **bit.**   59

Hartley oscillator. An electron tube or solid state circuit in which the parallel-tuned tank circuit is connected between grid and plate, the inductive element of the tank having an intermediate tap at cathode potential, and the necessary feedback voltage obtained across the grid-

cathode portion of the inductor. *See:* **radio frequency generator.**   14

hauptnutzzeit. *See:* **utilization time.**

Hay bridge. A 4-arm alternating-current bridge in which the arms adjacent to the unknown impedance are nonreactive resistors and the opposite arm comprises a capacitor in series with a resistor. *Note:* Normally used for the measurement of inductance in terms of capacitance, resistance, and frequency. Usually, the bridge is balanced by adjustment of the resistor that is in series with the capacitor, and of one of the nonreactive arms. The balance depends upon the frequency. It differs from the Maxwell bridge in that in the arm opposite the inductor, the capacitor is in series with the resistor. *See also:* **bridge.**   328

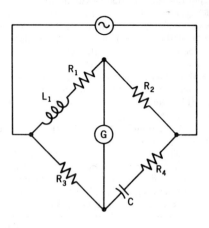

$$L_1 = R_2 R_3 \frac{C}{1 + \omega^2 C^2 R_4^2}$$

$$R_1 = R_2 R_3 \frac{\omega^2 C^2 R_4}{(1 + \omega^2 C^2 R_4^2)}$$

Hay bridge.

hazard or obstruction beacon. An aeronautical beacon used to designate a danger to air navigation. *See also:* **signal lighting.**   167

hazardous area class I. The locations in which flammable gases or vapors are or may be present in the air in quantities sufficient to produce explosive or ignitible mixtures. *See also:* **explosionproof apparatus.**   256, 65

hazardous (classified) area (National Electrical Code) class I locations. Class I locations are those in which flammable gases or vapors are or may be present in the air in quantities sufficient to produce explosive or ignitible mixtures. Class 1 locations shall include those specified in (a) and (b) below.

(a) class I, division 1. A class I, division 1 location is a location: (1) in which hazardous concentrations of flammable gases or vapors exist continuously, intermittently, or periodically under normal operating conditions; or (2) in which hazardous concentrations of such gases or vapors may exist frequently because of repair or maintenance operations or because of leakage or (3) in which breakdown or faulty operation of equipment or processes that might release hazardous concentrations of flammable gases or vapors, and might also cause simultaneous failure of electric equipment.

(b) class I, division 2. A class I, division 2 location is a location: (1) in which volatile flammable liquids or flammable gases are handled, processed, or used, but in which the hazardous liquids, vapors, or gases will normally be confined within closed containers or closed systems from which they can escape only in case of accidental rupture or breakdown of such containers or systems, or in case of abnormal operation of equipment; or (2) in which hazardous concentrations of gases or vapors are normally prevented by positive mechanical ventilation, but which might become hazardous through failure or abnormal operation of the ventilating equipment; or (3) that is adjacent to a class I, division 1 location, and to which hazardous concentrations of gases or vapors might occasionally be communicated unless such communication is prevented by adequate positive-pressure ventilation from a source of clean air, and effective safeguards against ventilation failure are provided.

class II locations. Class II locations are those that are hazardous because of the presence of combustible dust. Class II locations shall include those specified in (a) and (b) below.

(a) Class II, Division 1. A class II, division 1 location is a location: (1) in which combustible dust is or may be in suspension in the air continuously, intermittently or periodically under normal operating conditions, in quantities sufficient to produce explosive or ignitible mixtures, or (2) where mechanical failure or abnormal operation of machinery or equipment might cause such explosive or ignitible mixtures to be produced, and might also provide a source of ignition through simultaneous failure of electric equipment, operation of protection devices, or from other causes; or (3) in which combustible dusts of an electrically conductive nature may be present.

(b) Class II, Division 2. A class II, division 2 location is a location in which combustible dust will not normally be in suspension in the air or will not be likely to be thrown into suspension by the normal operation of equipment or apparatus in quantities sufficient to produce explosive or ignitible mixtures, but: (1) where deposits or accumulations of such combustible dust may be sufficient to interfere with the safe dissipation of heat from electric equipment or apparatus; or (2) where such deposits or accumulations of combustible dust on, in, or in the vicinity of electric equipment might be ignited by arcs, sparks, or burning material from such equipment.

class III locations. Class III locations are those that are hazardous because of the presence of easily ignitible fibers or flyings, but in which such fibers or flyings are not likely to be in suspension in the air in quantities sufficient to produce ignitible mixtures. Class III locations shall include those specified in (a) and (b) below.

(a) class III, division 1. A class III, division 1 location is a location in which easily ignitible fibers or materials producing combustible flyings are handled, manufactured, or used.

(b) class III, division 2. A class III, division 2 location is a location in which easily ignitible fibers are stored or handled. 65

Groups of chemicals.

Group A Atmospheres

Chemical
acetylene

Group B Atmospheres
butadiene[1]
ethylene oxide[2]
hydrogen
manufactured gases containing more than 30% hydrogen
  (by volume)
propylene oxide[2]

Group C Atmospheres
acetaldehyde
cyclopropane
diethyl ether
ethylene
unsymmetrical dimethyl hydrazine
  (UDMH 1, 1-dimethyl hydrazine)

Group D Atmospheres

Chemical
acetone
acrylonitrile
ammonia
benzene
butane
1-butanol (butyl alcohol)
2-butanol (secondary butyl alcohol)
n-butyl acetate
isobutyl acetate
ethane
ethanol (ethyl alcohol)
ethyl acetate
ethylene dichloride
gasoline
heptanes
hexanes
isoprene
methane (natural gas)
methanol (methyl alcohol)
3-methyl-1-butanol (isoamyl alcohol)
methyl ethyl ketone
methyl isobutyl ketone
2-methyl-1-propanol
  (isobutyl alcohol)
2-methyl-2-propanol
  (tertiary butyl alcohol)
petroleum naphtha
octanes
pentanes
1-pentanol (amyl alcohol)
propane
1-propanol (propyl alcohol)
2-propanol (isopropyl alcohol)
propylene
styrene
toluene
vinyl acetate
vinyl chloride
xylenes

Groups of dusts.

Group E: Atmospheres containing metal dust, including aluminum, magnesium, and their commercial alloys, and other metals of similarly hazardous characteristics.

Group F: Atmospheres containing carbon black, charcoal, coal or coke dusts which have more than 8 percent total volatile material (carbon black per ASTM D1620, charcoal, coal and coke dusts per ASTM D271) or atmospheres containing these dusts sensitized by other materials so that they present an explosion hazard.

Group G: Atmospheres containing flour, starch, or grain dust.

Certain chemical atmospheres may have characteristics that require safeguards beyond those required for any of the above groups. Carbon disulfide is one of these chemicals because of its low ignition temperature, 100°C (212°F), and the small joint clearance required to arrest its flame. For a complete list noting properties of flammable liquids, gases, and solids, see Fire-Hazard Properties of Flammable Liquids, Gases, Volatile Solids (NFPA No. 325M-1969).   65

hazardous electrical condition (electrolytic cell line working zone). Exposure to surfaces, contact with which may result in the flow of injurious electrical current.   133

hazardous location. An area where ignitible vapors or dust may cause a fire or explosion created by energy emitted from lighting or other electric equipment or by electrostatic generation. *See also:* **luminaire.**   167

hazardous materials. Those vapors, dusts, fibers or flyings which are explosive under certain conditions.   70

H **bend (waveguide technique).** *See:* **H-plane bend.**

HCL. *See:* **relay, high, common, low.**

H **display (radar).** A *B* display modified to include indication of angle of elevation. *Note:* The target appears as two closely spaced blips which approximate a short bright line, the slope of which is in proportion to the sine of the angle of target elevation.   13, 187

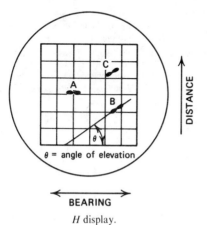

H display.

head (computing system). A device that reads, records, or erases data on a storage medium. *Note:* For example, a small electromagnet used to read, write, or erase data on a magnetic drum or tape, or the set of perforating, reading, or marking devices used for punching, reading, or printing on paper tape.   255, 77

headed brush (rotating machinery). A brush having a top (cylindrical, conical, or rectangular) with a smaller cross section than the cross section of the body of the brush. *Note:* The length of the head shall not exceed 25 percent of the overall length. *See also:* **brush.**   279

head-end system (railways). A system in which the electrical requirements of a train are supplied from a generator or generators, located on the locomotive or in one of the cars, customarily at the forward part of the train. *Note:* The generators may be driven by steam turbine, internal-combustion engine, or, if located in one of the cars, by a mechanical drive from a car axle. *See also:* **axle generator system.**   328

head gap (test, measurement and diagnostic equipment). The space or gap intentionally inserted into the magnetic circuit of the head in order to force or direct the recording flux into or from the recording medium.   54

heading (navigation). The horizontal direction in which a vehicle is pointed, expressed as an angle between a reference line and the line extending in the direction the vehicle is pointed, usually measured clockwise from the reference line. *See also:* **navigation.**   187, 13

heading-effect error (navigation). A manifestation of polarization error causing an error in indicated bearing that is dependent upon the heading of a vehicle with respect to the direction of signal propagation. *Note:* Heading-effect error is a special case of attitude-effect error where the vehicle is in a straight level flight; it is sometimes referred to as course push (or pull). *See also:* **navigation.**   187, 13

headlamp. A major lighting device mounted on a vehicle and used to provide illumination ahead of it. Also called **headlight.**   167

head or butt cable (mining). *See:* **hand cable.**

headquarters system (direct-connected system). A local system to which has been added means of transmitting system signals to and receiving them at an agency maintained by the local government, for example, in a police precinct house, or fire station. *See also:* **protective signaling.**   328

head receiver. An earphone designed to be held to the ear by a headband. *Note:* One or a pair (one for each ear) of head receivers with associated headband and connecting cord is known as a headset. *See also:* **loudspeaker.**   328

head space (test, measurement and diagnostic equipment). The space between the reading or recording head and the recording medium, such as tape, drum or disc.   54

head water benefits (hydroelectric power plant). The benefits brought about by the storage or release of water by a reservoir project upstream. Application of the term is usually in reference to benefits to a downstream hydroelectric power plant.   112

hearing loss (hearing level) (1) (for speech). The difference in decibels between the speech levels at which the average normal ear and the defective ear, respectively, reach the same intelligibility, often arbitrarily set at 50 percent. **(2) (hearing-threshold level) (ear at a specified frequency).** The amount, in decibels, by which the threshold of audibility for that ear exceeds a standard audiometric threshold. *Notes:* (A) See: Current issue of American Standard Specification for Audiometers for General Diagnostic Purposes. (B) This concept was at one time called deafness; such usage is now deprecated. (C) Hearing loss and deafness are both legitimate qualitative terms for the medical condition of a moderate or severe impairment of hearing, respectively. Hearing level, however, should only be used to designate a quantitative measure of the deviation of the hearing threshold from a prescribed standard. *See also:* **loudspeaker.** 176

heat coil. A protective device that grounds or opens a circuit, or does both, by means of a mechanical element that is allowed to move when the fusible substance that holds it in place is heated above a predetermined temperature by current in the circuit. *See also:* **electromagnetic relay.** 59

heat detector (1) (burglar-alarm system). A temperature-sensitive device mounted on the inside surface of a vault to initiate an alarm in the event of an attack by heat or burning. *See also:* **protective signaling.** 328
(2) (fire alarm system). A device which detects abnormally high temperature or rate-of-temperature rise to initiate a fire alarm. 71

heater (electron tube). An electric heating element for supplying heat to an indirectly heated cathode. *See also:* **electrode (electron tube).** 190, 125

heater coil. *See:* **load, work or heater coil (induction heating usage).**

heater connector (heater plug). A cord connector designed to engage the male terminal pins of a heating or cooking appliance. 328

heater current. The current flowing through a heater. *See:* **cathode preheating time; filament current.** *See also:* **electronic controller.** 190

heater transformer (industrial control). Supplies power for electron-tube filaments or heaters of indirectly heated cathodes. *See also:* **electronic controller.** 206

heater voltage. The voltage between the terminals of a heater. *See also:* **electronic controller; electrode voltage (electron tube).** 190

heater warm-up time. *See:* **cathode heating time.**

heat exchanger. *See:* **cooler.**

heat-exchanger cooling system (rectifier). A cooling system in which the coolant, after passing over the cooling surfaces of the rectifier, is cooled in a heat exchanger and recirculated. *Note:* The coolant is generally water of a suitable purity, or water that has been treated by a corrosion-inhibitive chemical. Antifreeze solutions may also be used where there is exposure to low temperatures. The heat exchanger is usually either: (1) water-to-water where the heat is removed by raw water, (2) water to-air where the heat is removed by air supplied by a blower, (3) air-to-water, (4) air-to-air, (5) refrigeration cycle. The liquid in the closed system may be other than water, and the gas in the closed system may be other than air. *See also:* **rectification; rectifier.** 208

heating cycle. One complete operation of the thermostat from ON to ON or from OFF to OFF. 263

heating element. A length of resistance material connected between terminals and used to generate heat electrically. *See also:* **appliance outlet.** 263

heating, glue line. *See:* **glue-line heating.**

heating pattern. The distribution of temperature in a load or charge. *See also:* **induction heating.** 14, 114

heating station (dielectric heating usage). The assembly of components, which includes the work coil or applicator and its associated production equipment. 14

heating time, tube (mercury-vapor tube). *See:* **preheating time.**

heating unit (electrical appliances). An assembly containing one or more heating elements, electric terminals or leads, electrical insulation, and a frame, casing, or other suitable supporting means. *See also:* **appliance outlet; appliances (including portable).** 263

heat loss. The part of the transmission loss due to the conversion of electric energy into heat. *See also:* **waveguide.** 267

heat rate (generating station). A measure of generating station thermal efficiency, generally expressed as BTU per kilowatt hour. *Note:* It is computed by dividing the total BTU content of the fuel burned (or of heat released from a nuclear reactor) by the resulting kilowatt hours generated. 112

heat rejection rate (nuclear power generating station). The rate at which a module emits heat energy to its environment (watts/hr or btu). 355

heat-shield (cathode) (electron tubes). A metallic surface surrounding a hot cathode, in order to reduce the radiation losses. *See also:* **electron tube.** 244, 190

heat sink (1) (general). A part used to absorb heat. 63
(2) (semiconductor rectifier diode). A mass of metal generally having much greater thermal capacity than the diode itself and intimately associated with it. It encompasses that part of the cooling system to which heat flows from the diode by thermal conduction only and from which heat may be removed by the cooling medium. *See:* **semiconductor rectifier stack.** 208
(3) (photovoltaic power system). A material capable of absorbing heat; a device utilizing such material for the thermal protection of components or systems. *See also:* **photovoltaic power system; solar cells (photovoltaic power system).** 186

Heaviside-Campbell mutual-inductance bridge. A mutual-inductance bridge of the Heaviside type in which one of the inductive arms contains a separate inductor that is included in the bridge arm during the first of a pair of measurements and is short-circuited during the second. *Note:* The balance is independent of the frequency. *See:* **Heaviside mutual-inductance bridge.** *See also:* **bridge.** 328

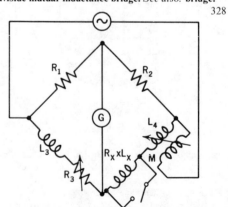

Heaviside-Campbell mutual inductance bridge.

$$R_x = (R_3 - R'_3)\frac{R_2}{R_1}$$

$$L_x = (M - M')\left(1 + \frac{R_2}{R_1}\right)$$

Heaviside-Lorentz system of units. A rationalized system based on the centimeter, gram, and second and is similar to the Gaussian system but differs in that a factor 4π is explicitly inserted to multiply r^2 in each of the formulations of the Coulomb Laws.   210

Heaviside mutual-inductance bridge. An alternating-current bridge in which two adjacent arms contain self-inductance, and one or both of these have mutual inductance to the supply circuit, the other two arms being normally nonreactive resistors. *Note:* Normally used for the comparison of self- and mutual inductances. The balance is independent of the frequency. *See also:* **bridge.**   328

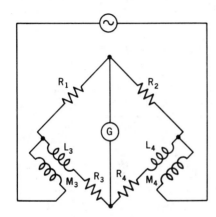

$$R_1 R_4 = R_2 R_3$$

$$L_3 - L_4\left(\frac{R_1}{R_2}\right) = -(M_3 - M_4)\left(1 + \frac{R_1}{R_2}\right)$$

Heaviside mutual-inductance bridge.

heavy-duty floodlight. A weatherproof unit having a substantially constructed metal housing into which is placed a separate and removable reflector. *Note:* A weatherproof hinged door with cover glass encloses the assembly but provides an unobstructed light opening at least equal to the effective diameter of the reflector. *See also:* **floodlighting.**   167

height (antenna). *See:* **effective height.**

height-finding radar. A radar whose function is to measure the range and elevation angle to a target, thus permitting computation of altitude or height; such a radar usually accompanies a surveillance radar which determines other target parameters.   13

height marker (radar). *See:* **calibration marks.**

height marks (radar). *See:* **calibration marks.**

height, pulse*. *See:* **amplitude, pulse.**

* Deprecated

helical antenna. An antenna whose configuration is that of a helix. *Note:* The diameter, pitch, and number of turns in relation to the wavelength provide control of the polarization state and directivity of helical antennas. *See also:* **antenna.**   179

helical plate (storage cell). A plate of large area formed by helically wound ribbed strips of soft lead inserted in supporting pockets or cells of hard lead. *See also:* **battery (primary or secondary).**   328

HEM wave. *See:* **hybrid electromagnetic wave.**

hemispherical reflectance. The ratio of all the flux leaving a surface or medium by reflection to the incident flux. *Note:* If reflectance is not preceded by an adjective descriptive of the angles of view, hemispherical reflectance is implied. *See also:* **lamp.**   167

hemispherical transmittance (illuminating engineering). The ratio of the transmitted flux leaving a surface or medium to the incident flux. *Note:* If transmittance is not preceded by an adjective descriptive of the angles of view, hemispherical transmittance is implied. *See also:* **transmission (illuminating engineering).**   167

henry (circuit). The unit of inductance in the International System of Units (SI units). The inductance for which the induced voltage in volts is numerically equal to the rate of change of current in amperes per second.   210

heptode. A seven-electrode electron tube containing an anode, a cathode, a control electrode, and four additional electrodes that are ordinarily grids.   190, 125

hermetically sealed relay. A relay in a gastight enclosure that has been completely sealed by fusion or other comparable means to insure a low rate of gas leakage over a long period of time. *See also:* **relay.**   259

hermetic motor. A stator and rotor without shaft, end shields, or bearings for installation in refrigeration compressors of the hermetically sealed type.   63

hermitian form (circuits and systems). The $n \times n$ matrix $[A]$ is hermitian if its conjugate transpose is equal to $[A]$ itself. In terms of a set of complex variables; $x_1, x_2, \ldots x_n$; the quantity

$$[\bar{x}_1 \; \bar{x}_2 \ldots \bar{x}_n][A]\begin{bmatrix} x_1 \\ x_2 \\ \vdots \\ x_n \end{bmatrix}$$

is the hermitian form of $[A]$.   67

hertz (1) (general). The unit of frequency, one cycle per second.   3, 59

(2) (transformer). The unit of frequency, (cycles per second).   53

(3) (laser-maser). The unit which expresses the frequency of a periodic oscillation in cycles per second.   363

heterodyne conversion transducer (converter). A conversion transducer in which the useful output frequency is the sum or difference of (1) the input frequency and (2) an integral multiple of the frequency of another wave usually derived from a local oscillator. *Note:* The frequency and voltage or power of the local oscillator are parameters of the conversion transducer. Ordinarily, the output-signal amplitude is a linear function of the input-signal amplitude over its useful operating range.   190, 252, 210, 125

heterodyne frequency. *See:* **beats.**

heterodyne reception (beat reception). The process of reception in which a received high-frequency wave is combined in a nonlinear device with a locally generated wave, with the result that in the output there are frequencies equal to the sum and difference of the combining frequencies. *Note:* If the received waves are continuous waves of constant amplitude, as in telegraphy, it is customary to adjust the locally generated frequency so that the difference frequency is audible. If the received waves are modulated the locally generated frequency is generally such that the difference frequency is superaudible and an additional operation is necessary to reproduce the original signal wave. *See:* **superheterodyne reception.** 328

heteropolar machine (rotating machinery). A machine having an even number of magnetic poles with successive (effective) poles of opposite polarity. *See also:* **asynchronous machine; direct-current commutating machine.** 63

heuristic. Pertaining to exploratory methods of problem solving in which solutions are discovered by evaluation of the progress made toward the final result. *See:* **algorithm.** 255, 77, 54

Hevea rubber. Rubber from the *Hevea brasiliensis* tree. *See also:* **insulation.**

hexadecimal. (1) Pertaining to a characteristic or property involving a selection, choice, or condition in which there are sixteen possibilities. (2) Pertaining to the numeration system with a radix of sixteen. (3) More accurately called sexadecimal. *See also:* **positional notation.** 77

hexode. A six-electrode electron tube containing an anode, a cathode, a control electrode, and three additional electrodes that are ordinarily grids. 190, 125

HF (high-frequency) radar. A radar operating at frequencies between 3 to 30 megahertz. 13

hickey. (1) A fitting used to mount a lighting fixture in an outlet box or on a pipe or stud. *Note:* It has openings through which fixture wires may be brought out of the fixture stem. (2) A pipe-bending tool. 328

high direct voltage (insulation test). A direct voltage above 5000 volts supplied by portable test equipment of limited capacity. *See also:* **insulation testing (large alternating-current rotating machinery).** 6

higher-order mode (waveguide or transmission line). Any mode of propagation characterized by a field configuration other than that of the fundamental or first-order mode with lowest cutoff frequency. *See also:* **waveguide.** 185

higher-order mode of propagation (laser-maser). A mode in a **beamguide** or **beam resonator** which has a plurality of maxima for the transverse field intensity over the cross-section of the beam. 363

high-fidelity signal (speech quality measurements). A signal transmitted over a system comprised of a microphone, amplifier, and loudspeaker or earphones. A tape recorder may be part of the system. All components should be of the best quality the state of the art permits. 126

high-field-emission arc (gas). An electric arc in which the electron emission is due to the effect of a high electric field in the immediate neighborhood of the cathode, the thermionic emission being negligible. *See also:* **discharge (gas).** 244, 190

high-frequency carrier telegraphy. The form of carrier telegraphy in which the carrier currents have their frequencies above the range transmitted over a voice-frequency telephone channel. *See also:* **telegraphy.** 59

high-frequency furnace (coreless-type induction furnace). An induction furnace in which the heat is generated within the charge, or within the walls of the containing crucible, or in both, by currents induced by high-frequency flux from a surrounding solenoid. 328

high-frequency induction heater or furnace. A device for causing electric current flow in a charge to be heated, the frequency of the current being higher than that customarily distributed over commercial networks. *See also:* **induction heating.** 14, 114

high-frequency stabilized arc welder. A constant-current arc-welding power supply including a high-frequency arc stabilizer and suitable controls required to produce welding current primarily intended for tungsten-inert-gas arc welding. *See:* **constant-current arc-welding power supply.** 264

high-gain direct-current amplifier. An amplifier that is capable of amplification substantially greater than required for a specified operation throughout a frequency band extending from zero to some maximum. Also, an operational amplifier without feedback circuit elements. *See:* **operational amplifier.** 9, 77

high-impedance rotor. An induction-motor rotor having a high-impedance squirrel cage, used to limit starting current. *See also:* **rotor (rotating machinery).** 63

high-key lighting (television). A type of lighting that, applied to a scene, results in a picture having graduations falling primarily between gray and white; dark grays or blacks are present, but in very limited areas. *See also:* **television lighting.** 167

high-level firing time (microwave) (switching tubes). The time required to establish a radio-frequency discharge in the tube after the application of radio-frequency power. *See:* **gas tubes.** 190, 125

high level language (test, measurement and diagnostic equipment). A programming language which resembles a common written language, such as English, and thus has the property of being readable and understandable by a person untrained in computer programming. 54

high-level modulation. Modulation produced at a point in a system where the power level approximates that at the output of the system. 111, 240

high-level radio-frequency signal (microwave gas tubes). A radio-frequency signal of sufficient power to cause the tube to become fired. *See:* **gas tubes.** 190, 125

high-level voltage standing-wave ratio (microwave switching tubes). The voltage standing-wave ratio due to a fired tube in its mount located between a generator and matched termination in the waveguide. *See:* **gas tubes.** 190, 125

high lights (any metal article). Those portions that are most exposed to buffing or polishing operations, and hence have the highest luster. 328

high-low signaling (telephone switching systems). A method of loop signaling in which a high-resistance bridge is used to indicate an on-hook condition and a low resistance bridge is used to indicate an off-hook condition. 55

high-pass filter. *See:* **filter, high-pass.**

high peaking. The introduction of an amplitude-frequency characteristic having a higher relative response at the higher frequencies. *See also:* **television.** 178

high pot. *See:* **high-potential test.**

high potential test (overvoltage test) (rotating machinery).
A test that consists of the application of a voltage higher
than the rated voltage for a specified time for the purpose
of determining the adequacy against breakdown of insulating materials and spacings under normal conditions.
Note: The test is used as a proof test of new apparatus, a
maintenance test on older equipment, or as one method
of evaluating developmental insulation systems. *See also:*
**asynchronous machine; direct-current commutating ma-
chine.**   232, 63

high-power-factor mercury-lamp ballast. A multiple-
supply type power-factor-corrected ballast, so designed
that the input current is at a power factor of not less than
90 percent when the ballast is operated with center rated
voltage impressed upon its input terminals and with a
connected load, consisting of the appropriate reference
lamp(s), operated in the position for which the ballast is
designed.   271

high power factor transformer. A high-reactance transformer that has a power-factor-correcting device such as
a capacitor, so that the input current is at a power factor
of not less than 90 percent when the transformer delivers
rated current to its intended load device.   53

**high-pressure contact (as applied to high-voltage discon-
necting switches).** One in which the pressure is such that
the stress in the material of either of the contact surfaces
is near the elastic limit of the material so that conduction
is a function of pressure.   27, 103

high-pressure vacuum pump. A vacuum pump that discharges at atmospheric pressure. *See also:* **rectification.**
                                                    328

high-reactance rotor. An induction-motor rotor having a
high-reactance squirrel cage, used where low starting
current is required and where low locked-rotor and
breakdown torques are acceptable. *See also:* **rotor (rotating machinery).**   63

high-reactance transformer. An energy-limiting transformer that has sufficient inherent reactance to limit the
output current to a maximum value.   53

high-resistance rotor (rotating machinery). An induction
motor rotor having a high-resistance squirrel cage, used
when reduced locked-rotor current and increased
locked-rotor torque are required.   63

high-speed carry (electronic computation). A carry process
such that if the current sum in a digit place is exactly one
less than the base, the carry input is bypassed to the next
place. *Note:* The processing necessary to allow the bypass
occurs before the carry input arrives. Further processing
required in the place as a result of the carry input, occurs
after the carry has passed by. Contrasted with **cascaded
carry.** *See also:* **standing-on-nines carry.**   235

high-speed excitation system. An excitation system capable
of changing its voltage rapidly in response to a change in
the excited generator field circuit. *See also:* **generating
station.**   64

high-speed grounding switch. *See:* **fault-initiating
switch.**

high-speed low-voltage dc power circuit breaker. A low-
voltage dc power circuit breaker which, during interruption, limits the magnitude of the fault current so that its
crest is passed not later than a specified time after the

beginning of the fault current transient, where the system
fault current, determined without the circuit breaker in
the circuit, falls between specified limits of current at a
specified time. *Note:* The specified time in present practice
is 0.01 second.   103

high-speed regulator (power supplies). A power supply
regulator circuit that, by the elimination of its output capacitor, has been made capable of much higher slewing
rates than are normally possible. *Note:* High-speed regulators are used where rapid step-programming is needed;
or as current regulators, for which they are ideally suited.
See: **slewing rate.** *See also:* **power supply.**   228, 186

high-speed relay (electric-power circuits). A relay that
operates in less than a specified time. *Note:* The specified
time in present practice is fifty milliseconds (three cycles
on a 60-hertz basis).   103, 202, 60, 127

high-speed short-circuiting switch. *See:* **fault-initiating
switch.**

high state (programmable instrumentation). The relatively
more positive signal level used to assert a specific message
content associated with one of two binary logic states.
                                                     40

high usage trunk (data transmission). A group of trunks for
which an engineered alternate route is provided, and for
which the number of trunks is determined on the basis of
relative trunk efficiencies and economic considerations.
                                                     59

high-usage trunk group (telephone switching systems). A
trunk group engineered on the basis of relative trunk efficiencies and economic considerations which will overflow
traffic.   55

**high-velocity camera tube (anode-voltage stabilized camera
tube) (electron device).** A camera tube operating with a
beam of electrons having velocities such that the average
target voltage stabilizes at a value approximately equal
to that of the anode.   190

high-voltage and low-voltage windings. *See:* **windings,
high-voltage and low-voltage.**   53

high-voltage cable termination. A device used for terminating alternating current power cables having laminated
or extruded insulation rated 2.5 kV and above, which are
classified according to the following: Class 1 Termination.
Provides electric stress control for the cable insulation
shield terminus; provides complete external leakage insulation between the cable conductor(s) and ground; and
provides a seal to the end of the cable against the entrance
of the external environment and maintains the pressure,
if any, of the cable system. Class 2 Termination. Provides
electric stress control for the cable insulation shield terminus; and provides complete external leakage insulation
between the cable conductor(s) and ground. Class 3 Termination. Provides electric stress control for the cable
insulation shield terminus. *Note:* Some cables do not have
an insulation shield. Terminations for such cables would
not be required to provide electric stress control. In such
cases, this requirement would not be part of the definition.

high-voltage relay. (1) A relay adjusted to sense and
function in a circuit or system at a specific maximum
voltage. (2) A relay designed to handle elevated voltages
on its contacts, coil, or both.   341

high-voltage system (electric power for industrial and com-

mercial systems only). An electric system having a maximum root-mean-square alternating-current voltage above 72,500 volts to 242,000 volts. *See also:* **low-voltage system; medium voltage system; extra-high-voltage system.**
49

high-voltage time test. An accelerated life test on a cable sample in which voltage is the factor increased. *See also:* **power distribution, underground construction.**   64

highway (CAMAC system). An interconnection between CAMAC crate assemblies or between one or more CAMAC crate assemblies and an external controller.
51

highway crossing back light (railway practice). An auxiliary signal light used for indication in a direction opposite to that provided by the main unit of a highway crossing signal.   328

highway crossing bell (railway practice). A bell located at a railroad-highway grade crossing and operated to give a characteristic and arrestive signal to give warning of the approach of trains.   328

highway crossing signal. An electrically operated signal used for the protection of highway traffic at railroad-highway grade crossings.   328

Hilbert transform (harmonic conjugate) (real function $x(t)$ of the real variable t). The real function $x(t)$ that is the Cauchy principal value of

$$\frac{1}{\pi} \int_{-\infty}^{\infty} \frac{x(\tau)\mathrm{d}\tau}{t - \tau}.$$

See: **analytic signal; network analysis.**   61

hinge clip (switching device). The clip to which the blade is movably attached.   103, 202, 27

hinged-iron ammeter. A special form of moving-iron ammeter in which the fixed portion of the magnetic circuit is arranged so that it can be caused to encircle the conductor, the current in which is to be measured. This conductor then constitutes the fixed coil of the instrument. *Note:* The combination of a current transformer of the split-core type with an ammeter is often used similarly to measure alternating current, but should be distinguished from the hinged-iron ammeter. *See also:* **instrument.**
328

hinged removable feed tube (cable plowing). A feed tube removably attached to a blade so relative motion may occur between the feed tube and the blade around an essentially vertical axis.   52

hinged removable feed tube. *See:* **removable feed tube.**

hipot (test, measurement and diagnostic equipment). A colloquialism for high potential test: A testing technique whereby a high voltage source is applied to an insulating material to determine the condition of that material.
54

hiss (electron device). Noise in the audio-frequency range, having subjective characteristics analogous to prolonged sibilant sounds.   239, 190

hit (radar). A target echo from one single pulse in a series of echoes.   13

H network. A network composed of five branches, two connected in series between an input terminal and an output terminal, two connected in series between another input terminal and output terminal, and the fifth con-

nected from the junction point of the first two branches to the junction point of the second two branches. *See also:* **network analysis.**   210·

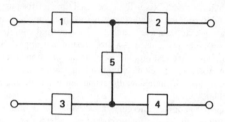

H network. Branches 1 and 2 are the first two branches between an input and an output terminal; branches 3 and 4 are the second two branches; and branch 5 is the branch between the junction points.

hodoscope. An apparatus for tracing the path of a charged particle in a magnetic field. *See also:* **electron optics.**
244, 190

Hoeppner connection. A three-phase transformer connection involving transformation from a wye winding to the combination of a delta winding and a zigzag winding which are connected permanently in parallel. *Note:* This connection is used when a wye-delta connection is needed, with ground connections on both primary and secondary windings.   53

hoist (conductor stringing equipment). An apparatus for moving a load by the application of a pulling force and not including a car or platform running in guides. These devices are normally designed using roller or link chain and built-in leverage to enable heavy loads to be lifted or pulled. They are often used to deadend a conductor during sagging and clipping-in operations and when tensioning guys. *Syn.* **chain hoist; chain tugger; coffing; coffing hoist.**
45

hoist back-out switch (mining). A switch that permits operation of the hoist only in the reverse direction in case of overwind. *See also:* **mine hoist.**   328

hoisting-rope equalizer. A device installed on an elevator car or counterweight to equalize automatically the tensions in the hoisting wire ropes. *See also:* **elevator.**   328

hoist overspeed device (mining). A device that can be set to prevent the operation of a mine hoist at speeds greater than· predetermined values and usually causes an emergency brake application when the predetermined speed is exceeded. *See also:* **mine hoist.**   328

hoist overwind device (mining). A device that can be set to cause an emergency break application when a cage or skip travels beyond a predetermined point into a danger zone. *See also:* **mine hoist.**   328

hoist signal code (mining). Consists of prescribed signals for indicating to the hoist operator the desired direction of travel and whether men or materials are to be hoisted or lowered in mines. *See also:* **mine hoist.**   328

hoist signal system (mining). A system whereby signals can be transmitted to the hoist operator (and in some instances by him to the cager) for control of mine hoisting operations. *See also:* **mine hoist.**   328

hoist slack-brake switch (mining). A device for automatically cutting off the power from the hoist motor and causing the brake to be set in case the links in the brake rigging require tightening or the brakes require relining. *See also:* **mine hoist.**   328

hoist trip recorder (mining). A device that graphically records information such as the time and number of hoists made as well as the delays or idle periods between hoists. *See also:* **mine hoist.**   328

hoistway (elevator or dumbwaiter). A shaftway for the travel of one or more elevators or dumbwaiters. *Note:* It includes the pit and terminates at the underside of the overhead machinery space floor or grating, or at the underside of the roof where the hoistway does not penetrate the roof.   328

hoistway access switch (elevators). A switch, located at a landing, the function of which is to permit operation of the car with the hoistway door at this landing and the car door or gate open, in order to permit access at the top of the car or to the pit. *See also:* **control (elevators).**   328

hoistway-door combination mechanical lock and electric contact (elevators). A combination mechanical and electric device, the two related, but entirely independent, functions of which are: (1) to prevent operation of the driving machine by the normal operating device unless the hoistway door is in the closed position, and (2) to lock the hoistway door in the closed position and prevent it from being opened from the landing side unless the car is within the landing zone. *Note:* As there is no positive mechanical connection between the electric contact and the door-locking mechanism, this device insures only that the door will be closed, but not necessarily locked, when the car leaves the landing. Should the lock mechanism fail to operate as intended when released by a stationary or retiring car-cam device, the door can be opened from the landing side even though the car is not at the landing. If operated by a stationary car-cam device, it does not prevent opening the door from the landing side as the car passes the floor. *See also:* **hoistway (elevator or dumbwaiter).**   328

hoistway-door electric contact (elevators). An electric device, the function of which is to prevent operation of the driving machine by the normal operating device unless the hoistway door is in the closed position. *See also:* **hoistway (elevator or dumbwaiter).**   328

hoistway-door interlock (elevators). A device having two related and interdependent functions that are (1) to prevent the operation of the driving machine by the normal operating device unless the hoistway door is locked in the closed position; and (2) to prevent the opening of the hoistway door from the landing side unless the car is within the landing zone and is either stopped or being stopped. *See also:* **hoistway (elevator or dumbwaiter).**   328

hoistway-door or gate locking device (elevators). A device that secures a hoistway or gate in the closed position and prevents it from being opened from the landing side except under specified conditions. *See also:* **hoistway (elevator or dumbwaiter).**   328

hoistway enclosure. The fixed structure, consisting of vertical walls or partitions, that isolates the hoistway from all other parts of the building or from an adjacent hoistway and in which the hoistway doors and door assemblies are installed. *See also:* **hoistway (elevator or dumbwaiter).**   328

hoistway-gate separate mechanical lock (elevators). A mechanical device, the function of which is to lock a hoistway gate in the closed position after the car leaves a landing and prevent the gate from being opened from the landing side unless the car is within the landing zone. *See also:* **hoistway (elevator or dumbwaiter).**   328

hoistway-unit system (elevators). A series of hoistway-door interlocks, hoistway-door electric contacts, or hoistway-door combination mechanical locks and electric contacts, or a combination thereof, the function of which is to prevent operation of the driving machine by the normal operating device unless all hoistway doors are in the closed position and, where so required, are locked in the closed position. *See also:* **hoistway (elevator or dumbwaiter).**   328

hold (1) (electronic digital computer). An untimed delay in the program, terminated by an operator or interlock action.   224, 207

(2) (analog computer). The computer control state in which the problem solution is stopped and held at its last values usually by automatic disconnect of integrator input signals.   9

(3) (industrial control). A control function that arrests the further speed change of a drive during the acceleration or deceleration portion of the operating cycle. *See also:* **control system, feedback.**   219, 206

(4) (charge-storage tubes) (verb). To maintain storage elements at an equilibrium voltage by electron bombardment. *See also:* **charge-storage tube; data processing.**   174, 190, 125

(5) (test, measurement and diagnostic equipment). (A) The function of retaining information in one storage device after transferring it to another device; and (B) A designed stop in testing.   54

hold-closed mechanism (automatic circuit recloser). A device that holds the contacts in the closed position following the completion of a predetermined sequence of operations as long as current flows in excess of a predetermined value.   103, 202

hold-closed operation (automatic circuit recloser). An opening followed by the number of closing and opening operations that the hold-closed mechanism will permit before holding the contacts in the closed position.   103, 202

holding current (thyristor). The minimum principal current required to maintain the thyristor in the ON-state. *See also:* **principal current.**   243, 66, 208, 191

holding-down bolt. A bolt that fastens a machine to its bedplate, rails, or foundation.   63

holding frequency (take the swings). A condition of operating a generator or station to maintain substantially constant frequency irrespective of variations in load. *Note:* A plant so operated is said to be regulating frequency. *See also:* **generating station.**   64, 94

holding load. A condition of operating a generator or station at substantially constant load irrespective of variations in frequency. *Note:* A plant so operated is said to be operating on base load. *See:* **base load.** *See also:* **generating station.**   64

holding register (hybrid computer linkage components). The register, in a double-buffered digital-to-analog converter (DAC) or a digital-to-analog multiplier (DAM), that holds the next digital value to be transferred into the dynamic register. 10

holding time (1) (data transmission). The length of time a communication channel is in use for each transmission. Includes both message time and operating time. 59

(2) (telephone switching systems). A call duration. 55

holdoff. *See:* intervals; sweep holdoff.

hold-off diode. A diode that is placed in series with the charging inductor and connected to the common junction of the switching element and the pulse-forming network in a pulse generator. *Note:* The use of a hold-off diode in the charging circuit of a pulse-forming network allows the capacitors of the network to charge to full voltage and remain at this voltage until the switch conducts. This permits the use of pulse-repetition frequencies of equal to or less than twice the frequency of resonance charging. 137

holdup-alarm attachment. A general term for the various alarm-initiating devices used with holdup-alarm systems, including holdup buttons, footrails, and others of a secret or unpublished nature. *See also:* protective signaling. 328

holdup-alarm system. An alarm system signaling a robbery or attempted robbery. *See also:* protective signaling. 328

hole (semiconductor). A mobile vacancy in the electronic valence structure of a semiconductor that acts like a positive electron charge with a positive mass. *See also:* semiconductor. 237, 245, 66, 210, 186

hole burning (of an absorption or an emission line) (laser-maser). The frequency dependent saturation of attenuation or gain that occurs in an inhomogeneously broadened transition when the saturating power is confined to a frequency range small compared with the inhomogeneous linewidth. 363

hollow-core annular conductor (hollow-core conductor). A conductor composed of a plurality of conducting elements disposed around a supporting member that does not fill the space enclosed by the elements; alternatively, a plurality of such conducting elements disposed around a central channel and interlocked one with the other or so shaped that they are self-supporting. *See also:* conductor. 64

hollow-core conductor. *See:* hollow-core annular conductor.

home area (telephone switching systems). The numbering plan area in which the calling customer is located. 55

home signal (railway practice). A fixed signal at the entrance of a route or block to govern trains or engines entering or using that route or block. 328

homing (1) (navigation). Following a course directed toward a point by maintaining constant some navigational coordinate (other than altitude). *See also:* radio navigation. 13, 187

(2) (telephone switching systems). Resetting of a sequential switching operation to a fixed starting point. 55

homing guidance. That form of missile guidance wherein the missile steers itself toward a target by means of a mechanism actuated by some distinguishing characteristic of the target. *See also:* guided missile. 328

homing relay. A stepping relay that returns to a specified starting position prior to each operating cycle. *See also:* relay. 259

homochromatic gain (optoelectronic device). The radiant gain or luminous gain for specified identical spectral characteristics of both incident and emitted flux. *See also:* optoelectronic device. 191

homodyne reception (zero-beat reception). A system of reception by the aid of a locally generated voltage of carrier frequency. 328

homogeneous line-broadening (laser-maser). An increase of the width of an absorption or emission line, beyond the natural linewidth, produced by a disturbance (for example, collisions, lattice vibrations, etcetera) which is the same for each of the emitters. 363

homopolar machine* (rotating machinery). A machine in which the magnetic flux passes in the same direction from one member to the other over the whole of a single air-gap area. Preferred term is acyclic machine, which see. 63

* Deprecated

honeycomb coil. A coil in which the turns are wound in crisscross fashion to form a self-supporting structure or to reduce distributed capacitance. *Syn.* duolateral coil. 341

hook, conductor lifting (conductor stringing equipment). A device resembling an open boxing glove designed to permit the lining of conductors from a position above them. Normally used during "clipping-in" operations. Suspension clamps are sometimes used for this purpose. *Syn.* boxing glove; conductor hook; lifting shoe; lip. 45

hook operation. *See:* stick hook operation.

hook ring (air switch). A ring provided on the switch blade for operation of the switch with a switch stick. 27, 79

hook stick. *See:* switch stick (switch hook).

hopper (computing systems). *See:* card hopper.

horizontal amplifier (oscilloscopes). An amplifier for signals intended to produce horizontal deflection. *See:* oscillograph. 185

horizontal hold control (television). A synchronizing control that adjusts the free-running period of the horizontal deflection oscillator. 163

horizontally polarized field vector. A linearly polarized field vector whose direction is horizontal. 111

horizontally polarized wave (1) (general). A linearly polarized wave whose direction of polarization is horizontal. *See also:* radiation. 328

(2) (radio wave propagation). A linearly polarized wave whose electric field vector is horizontal. *See also:* radio wave propagation. 146, 180

horizontal machine. A machine whose axis of rotation is approximately horizontal. 63

horizontal plane (searchlight). The plane that is perpendicular to the elevation axis and in which the train lies. *See also:* searchlight; train. 167

horizontal ring induction furnace. A device for melting metal, comprising an annular horizontal placed open trough or melting channel, a primary inductor winding, and a magnetic core which links the melting channel with the primary winding. 14

horn (acoustic practice). A tube of varying cross-sectional area for radiating or receiving acoustic waves. *Note:* Normally it has different terminal areas that provide a change of acoustic impedance and control of the directional response pattern. *See also:* loudspeaker. 176

horn antenna. A radiating element having the shape of a horn. *See also:* **antenna.** 179, 59

horn-gap switch. A switch provided with arcing horns. 103, 202, 27

horn mouth. Normally the opening, at the end of a horn, with larger cross-sectional area. *See also:* **loudspeaker.** 176

horn reflector antenna (1) (general). An antenna consisting of a section of a paraboloidal reflector fed with an offset horn that intersects the reflector surface. *Note:* The horn is usually pyramidal or conical. *See:* **antenna.** 179
(2) (communication satellite). A form of reflector antenna, where the energy coming from the throat of a horn is reflected by a segment of a paraboloid. This type of antenna has a very low backlobe. *See:* **cassegrainian feed.** 83

horn throat (audio and electroacoustics). Normally the opening, at the end of a horn, with the smaller cross-sectional area. *See also:* **loudspeaker.** 176

horsepower rating, basis for single-phase motor. A system of rating for single-phase motors, whereby horsepower values are determined, for various synchronous speeds, from the minimum value of breakdown torque that the motor design will provide. 63

hose (liquid cooling) (rotating machinery). The flexible insulated or insulating hydraulic connections applied between the conductors and either a central manifold or coolant passage. 63

hoseproof (rotating machinery). *See:* **waterproof machine.**

host computer (data communication). A computer, attached to a network, providing primary services such as computation, data base access or special programs or programming languages. *See:* **communications computer.** 12

hot cathode (thermionic cathode). A cathode that functions primarily by the process of thermionic emission. 190, 125

hot-cathode lamp (fluorescent lamps). An electric discharge lamp in which the electrodes operate at incandescent temperatures and in which the cathode drop is relatively low (10–20 volts). *Note:* The current density at the cathodes is relatively high, and lamps may be designed to carry any desired current up to several amperes. The energy to maintain the cathodes at incandescence may come either from the arc (arc heating), from circuit elements, or from both. 268

hot-cathode tube (thermionic tube). An electron tube containing a hot cathode. 190, 125

hot plate. An appliance fitted with heating elements and arranged to support a flat-bottomed utensil containing the material to be heated. *See also:* **appliances (including portable).** 328

hot reserve. The thermal reserve generating capacity maintained at a temperature and in a condition to permit it to be placed into service promptly. *See also:* **generating station.** 64

hottest spot temperature. The highest temperature inside the transformer winding. It is greater than the measured average temperature (using the resistance change method) of the coil conductors. 53

hottest-spot temperature allowance (equipment rating). A conventional value selected to approximate the degrees of temperature by which the limiting insulation temperature rise exceeds the limiting observable temperature rise. *See also:* **limiting insulation temperature.** 233

hot-wire instrument. An electrothermic instrument that depends for its operation on the expansion by heat of a wire carrying a current. *See also:* **instrument.** 328

hot-wire microphone. A microphone that depends for its operation on the change in resistance of a hot wire produced by the cooling or heating effects of a sound wave. *See also:* **microphone.** 328

hot-wire relay. A relay in which the operating current flows directly through a tension member whose thermal expansion actuates the relay. *See also:* **relay.** 259

house cable (communication practice). A distribution cable within the confines of a single building or a series of related buildings but excluding cable run from the point of entrance to a cross-connecting box, terminal frame, or point of connection to a block cable. *See also:* **cable.** 328

house turbine. A turbine installed to provide a source of auxiliary power. *See also:* **generating station.** 64

housing (1) (rotating machinery). Enclosing structure, used to confine the internal flow of air or to protect a machine from dirt and other harmful material. 63
(2) (body; oil cutout). A part of the fuse support that contains the oil and provides means for mounting the fuse carrier, entrance terminals, and fixed contacts. *Note:* The housing includes the means for mounting the cutout on a supporting structure and openings for attaching accessories such as a vent or an expansion chamber. 103, 202
(3) (power cable joints). A metallic or other enclosure for the insulated splice. 34

howler (telephone switching systems). A circuit which generates a receiver-off-hook tone. 55

H-plane bend (corner) (waveguide components). A waveguide bend (corner) in which the longitudinal axis of the guide remains in a plane parallel to the plane of the magnetic field vectors throughout the bend (corner). 166

H **plane, principal (linearly polarized antenna).** The plane containing the magnetic field vector and the direction of maximum radiation. *See also:* **antenna; radiation.** 179, 111

H-plane tee junction (shunt tee) (waveguide components). A waveguide tee junction in which the magnetic field vectors of the dominant mode in all arms are parallel to the plane containing the longitudinal axes of the arms. 166

H **scan (radar).** *See:* **H display.**

H **scope (radar).** *See:* **H display.**

hub (1) (rotating machinery). (A) The central part of a rotor used for mounting it on a shaft. (B) A small axial extension of an end shield. *Note:* The hub is commonly, but not always, used in conjunction with a resilient or rigid mounting to attach and support the stator on the base. *See also:* **rotor (rotating machinery).** 63
(2) (conductor stringing equipment). A reference point established through a land survey. A hub or P.O.T. (point on tangent) is a reference point for use during construction of a line. The number of such points established will vary with the job requirements. Monuments, however, are usually associated with state or federal surveys and are intended to be permanent reference points. Any of these points may be used as a reference point for transit sagging operations provided all necessary data pertaining to them

is known. It is quite common to establish additional temporary hubs as required for this purpose. *Syn.* **monument; P.O.T.**   45

hue (television). The attribute of visual sensation designated by: blue, green, yellow, red, purple, etc. *Notes:* (1) This attribute is the psychosensorial correlate (or nearly so) of the colorimetric quantity dominant wavelength. (2) This is the IEC definition of **hue**, 45-25-070.   18

hum (interference terminology). A low-pitched droning noise, consisting of several harmonically related frequencies, resulting from an alternating-current power supply, or from ripple from a direct-current power supply, or from induction due to exposure to a power system. *Note:* By extension, the term is applied in visual systems to interference from similar sources. *See also:* **phonograph pickup.**   59

hum and noise (receiver performance) (receiver). The low-pitched composite tone at the receiver output caused by fluctuations in the power supply. *Note:* Noise is any audible undesired signal present at the output. *See also:* **receiver performance.**   181

humidity coefficient. *See:* **environmental coefficient.**

hum modulation. Modulation of a radio-frequency or detected signal by hum.   339

hum noise level (transmitter performance). The ratio of the total average harmonics thereof, audio-frequency noise, Johnson noise, shot noise, and thermal noise generated within the transmitter, to the average carrier-power output delivered by the system into an output termination. *See also:* **audio-frequency distortion.**   181

hump. *See:* **hump yard.**

hump cab signal system (railway practice). A system of signals displayed in the engine-cab and indicating the desired direction and speed of movement of the hump engine.   328

hump repeater signal. A signal that repeats the hump-signal indication.   328

hump signal. A signal located at the hump and used to indicate to the hump engineman the desired direction and speed of movement of his train.   328

hump-signal controller. A unit located at the hump that includes the hump signal lever, the trimmer signal lever, and the signal-repeater lights.   328

hump-signal emergency lever (hump-signal lever). A lever usually located in one of the control towers and used to operate electric contacts that control the hump-signal circuit to set up a stop indication on the hump signal, the trimmer signal, and the repeater signals when in an emergency the operator desires to stop operation.   328

hump-signal lever. *See:* **hump-signal emergency lever.**

hump yard. A railroad classification yard in which the classification of cars is accomplished by pushing them over a summit, known as a hump, beyond which they run by gravity into tracks selected by the operator.   328

hum sidebands (non-real time spectrum analyzer). Undesired responses appearing on the display that are separated from the desired response by the fundamental or harmonics of the power line frequency.   68

hundred call-seconds (CCS) (telephone switching systems). A commonly used unit of traffic intensity, equal to a hundred seconds of call volume or usage in an hour. *Syn.* **unit call.**   55

hunting (1) (control systems). An undesired oscillation generally of low frequency that persists after external stimuli disappear. *Note:* In linear systems, hunting is evidence of operation at or near the stability limit; nonlinearities may cause hunting of well-defined amplitude and frequency. *See also:* **control system, feedback.**
                                                       219, 206

(2) (rotating machinery). A fluctuation of angular velocity about a state of uniform rotation due to electric circuit constants, governor instability, or the like.   63

(3) (telephone switching systems). Searching for an available, idle circuit of a group.   55

Hurter and Driffield curve (*H* and *D* curve) (photographic techniques). A characteristic curve of a photographic emulsion that is a plot of density against the logarithm of exposure. *Note:* It is used for the control of photographic processing and for defining the response characteristics to light of photographic emulsions.   176

Huygens source. An elementary radiator consisting of an infinitesimal area of a given electromagnetic wavefront. *Note:* Its radiation properties are equivalent to those of a suitably oriented and excited combination of electric and magnetic dipoles. *See also:* **antenna.**   179

hybrid balance (1) (power system communications). A measure of the degree of balance between two impedances connected to two conjugate sides of a hybrid set; it is given by the formula for return loss.   59

(2) (circuits and systems). The degree of impedance balance between impedances Z_1 and Z_2 connected to two conjugate sides of a hybrid set. Hybrid balance is measured as a return loss using one of the remaining two sides of the hybrid set as the sending end and the other as the receiving end. With reference return loss set to zero for $Z_1 = 0$ or $Z_2 = 0$ the measure of hybrid balance is given by:

$$20 \log_{10} \left[\frac{Z_2 + Z_1}{Z_2 - Z_1} \right] \qquad 67$$

hybrid coil (1) (bridge transformer). A single transformer, having effectively three windings, that is designed to be connected to four branches of a circuit so as to render these branches conjugate in pairs.   239, 59

(2) (powerline carrier). A hybrid coil used with power-line-carrier equipment to provide electric isolation between two carrier transmitters or transmitters and receivers using common line-tuning equipment. *See also:* **power-line carrier.**   59

hybrid computer (analog computer). A computer which consists of two main computers, one, a direct-current analog computer, and the other a digital computer, with appropriate control and signal interface, such that they may simultaneously operate upon, and/or solve different portions of a single problem. *See also:* **hybrid computer linkage components.**   9

hybrid control (electric power systems). Utilizing both analog and digital computers in combination to control a process. *See also:* **power system.**   200

hybrid coupler (3-decibel coupler) (1) (antennas). A hybrid junction in the form of a directional coupler. The coupling factor is normally 3 decibels. *See:* **hybrid junction; waveguide.**   244, 179

(2) (circuits and systems). A circuit with four branches having the property that when two branches are terminated by reflectionless terminations, energy from any one branch is divided equally between two of the remaining three branches.   67

hybrid electromagnetic wave (HEM wave). An electromagnetic wave having components of both the electric and magnetic field vectors in the direction of propagation. *See also:* **waveguide.**   267, 179

hybrid integrated circuit. An integrated circuit consisting of a combination of two or more integrated-circuit types or one integrated-circuit type and discrete elements. *See also:* **integrated circuit.**   310, 191

hybrid junction (waveguide components). A waveguide or transmission line arrangement with four ports which, when the ports have reflectionless terminations, has the property that energy entering any one port is transferred (usually equally) to two of the remaining three.   166

hybrid navigation equipment. Navigation equipment which employs two or more types of sensors. *See also:* **navigation.**   187, 13

hybrid network (1) (instrumentation and measurement). A network with four pairs of terminals or four ports with such characteristics that when two are properly terminated, energy can be transferred from any one port into only two of the remaining three. Usually this energy is equally divided between the two ports. *Notes:* (A) 3-decibel couplers have hybrid properties when two ports are terminated by reflectionless terminations. (B) An advance form contains matching elements such that it is reflectionless for waves propagating into any port of the hybrid when the other three ports are terminated by reflectionless terminations. *See also:* **waveguide.**   185

(2) (circuits and systems). (A) A network having four ports with the property that when two of the ports are properly terminated, energy can be transferred from any one port to only two of the remaining three ports. Usually this energy is equally divided between the two ports. (B) A communication network which operates with analog and digital signals. (C) An electric network composed of a mixture of integrated and discrete components.   67

hybrid parameter (circuits and systems). In the mathematical description of an n-port, if a subset of the port currents and the port voltages at the remaining ports (a nonempty set) are chosen as dependent variables, then the mathematical description is said to be a hybrid representation. If the relationship is linear then the coefficients of the independent variables are called the hybrid parameters. For example:

$$\begin{bmatrix} v_1 \\ i_2 \end{bmatrix} = \begin{bmatrix} h_{11} & h_{12} \\ h_{21} & h_{22} \end{bmatrix} \begin{bmatrix} i_1 \\ v_2 \end{bmatrix} \qquad 67$$

hybrid power supply. A combination of disparate elements to form a common circuit. In power supplies, the combination of vacuum tubes and transistors in the regulating circuit. *See also:* **power supply.**   228, 186

hybrid ring (rat race). A hybrid junction consisting of a waveguide or transmission line forming a closed ring into which lead four guides or lines appropriately spaced around the circle. *See also:* **waveguide.**   244, 179

hybrid set. Two or more transformers interconnected to form a network having four pairs of accessible terminals

to which may be connected four impedances so that the branches containing them may be made conjugate in pairs when the impedances have the proper values but not otherwise.   239, 59

hybrid system (data processing). A system with the property that at least one of its dependent variables is a function of a continuous variable and at least one of its dependent variables is a function of a discrete variable.   198

hybrid tee (magic tee) (waveguide components). A hybrid junction composed of an E-H tee with interval matching elements, which is reflectionless for a wave propagating into the junction from one port when the other three ports have reflectionless terminations.   166

hybrid transformer (circuits and systems). A transformer or transformer network with four distinct pairs of terminals designed to provide a means of achieving a balanced bridge condition so that transmission can take place between certain pairs of terminals while isolation is maintained between others. In its simplest form the transformer consists of two windings with a center-tapped secondary as shown. When $Z_1 = Z_2$, a balance is achieved and no

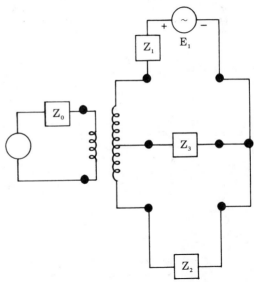

current from the generator E_0 will flow in Z_3 although the generator E_1 will produce a current in Z_3 and the generator E_0 will produce a current in both Z_1 and Z_2.   67

hybrid traveling-wave multiple-beam klystron (microwave tubes). An O-type tube having three or more electron beams, in which the input and output circuits are lateral traveling-wave structures, and the intermediate circuits are cavities uncoupled either laterally or axially. *See also:* **microwave tube or valve.**   190

hybrid wave (radio wave propagation). An electromagnetic wave in which either the electric or magnetic field vector is linearly polarized normal to the plane of propagation and the other vector is elliptically polarized in this plane. *See:* **transverse-electric hybrid wave; transverse-magnetic hybrid wave.** *See also:* **radio wave propagation.**   180, 146

hydraulically release-free (hydraulically operated switching device) (trip-free). A term indicating that by hydraulic control the switching device is free to open at any position in the closing stroke if the release is energized. *Note:* This release-free feature is operative even though the closing control switch is held closed. 103, 202

hydraulic driving machine. A machine in which the energy is applied by means of a liquid under pressure in a cylinder equipped with a plunger or piston. *See also:* **driving machine (elevators).** 328

hydraulic elevator. A power elevator where the energy is applied by means of a liquid under pressure in a cylinder equipped with a plunger or piston. *See also:* **elevators.** 328

hydraulic operation. Power operation by movement of a liquid under pressure. 103, 202

hydraulic pressure supply system (speed governing systems, hydraulic turbines). The pumps, means for driving them, pressure and sump tanks, valves and piping connecting the various parts of the governing system and associated and accessory devices. 58

hydro capability (electric power supply). The capability supplied by hydroelectric sources under specified water conditions. *See also:* **generating station.** 200, 112

hydrogen-cooled machine (rotating machinery). A machine designed to use hydrogen for cooling. It has a strong sealed gastight enclosure, and heat exchangers to remove heat from hydrogen circulated within the enclosure. 63

hydrogen embrittlement. Embrittlement of a metal caused by absorption of hydrogen during a cleaning, pickling, or plating process. 328

hydro-generator. A generator driven by a hydraulic turbine or water wheel. *See:* **asynchronous machine; direct-current commutating machine.** 63

hydrogen overvoltage (electrolytic cell). Overvoltage associated with the liberation of hydrogen gas. 221, 205

hydrometallurgy. The extraction of metals from their ores by the use of aqueous solutions. *See also:* **electrowinning.** 328

hydrophone. An electroacoustic transducer that responds to water-borne sound waves and delivers essentially equivalent electric waves. *Note:* There are many types of hydrophones whose definitions are analogous to those of corresponding microphones, for example, crystal hydrophone, magnetic hydrophone, pressure hydrophone, etcetera. *See also:* **microphone.** 328

hypersonic* (sonic frequency). The preferred term is pretersonic to avoid confusion with hypersonic speed. 352

* Deprecated

hypot test. *See:* **high potential test.**

hysteresigraph. A device for experimentally presenting or recording the hysteresis loop for a magnetic specimen. *See also:* **magnetometer.** 328

hysteresis (1) (control system). The property of an element evidenced by the dependence of the value of the output, for a given excursion of the input, upon the history of prior excursions and the direction of the current traverse. *Note:* Some reversal of output may be expected for any small reversal of input; this distinguishes the hysteretic effect from a dead band. The property may be either useful, as in magnetic amplifiers and computer storage elements, or undesirable, as when it contributes to a dead band in elements expected to show a unique transfer function. Originally used to describe magnetic phenomena, the term

has been extended to inelastic behavior. *See also:* **control system, feedback.** 56, 219, 206

(2) (oscillator). A behavior that may appear in an oscillator wherein multiple values of the output power and/or frequency correspond to given (usually cyclic) values of an operating parameter. *See:* **oscillatory circuit; tuning hysteresis.** 190, 125

(3) (radiation-counter tube). The temporary change in the counting rate versus voltage characteristic caused by previous operation. *See also:* **gas-filled radiation-counter tubes; gas tubes.** 190, 125

(4) (measurement or control). The difference between the increasing input value and the decreasing input value that effect the same output value. *Note:* This term applies only where the output value is a continuous function of the input value. 94

(5) (nuclear power generating station). The maximum difference for the same input between the upscale and downscale output values during a full range traverse in each direction. 355

hysteresis coupling. An electric coupling in which torque is transmitted by forces arising from the resistance to reorientation of established magnetic fields within a ferromagnetic material, usually of high coercitivity. *See:* **electric coupling.** 63

hysteresis error (gyro; accelerometer). The maximum separation due to hysteresis between up-scale-going and down-scale-going indications of the measured variable (during a full range traverse, unless otherwise specified) after transients have decayed. It is generally expressed as an equivalent input. *See also:* **input-output characteristics.** 46, 56

hysteresis heater. An induction device in which a charge or a muffle about the charge is heated principally by hysteresis losses due to a magnetic flux that is produced in it. *Note:* A distinction should be made between hysteresis heating and the enhanced induction heating in a magnetic charge. *See also:* **induction heating.** 14, 114

hysteresis loop. For a magnetic material in a cyclically magnetized condition, a curve (usually with rectangular coordinates) showing, for each value of the magnetizing force, two values of the magnetic flux density—one when the magnetizing force is increasing, the other when it is decreasing. *Note:* The cyclic magnetic force is not necessarily periodic. 210

hysteresis loss (magnetic) (transformer). The energy loss in magnetic material which results from an alternating magnetic field as the elementary magnets within the material seek to align themselves with the reversing magnetic field. 53

hysteresis motor. A synchronous motor without salient poles and without direct-current excitation, that starts by virtue of the hysteresis losses induced in its hardened-steel secondary member of the revolving field of the primary and operates normally at synchronous speed due to the retentivity of the secondary core. 63

hysteretic error (industrial control). The maximum separation due to hysteresis between up-scale-going and down-scale-going indications of the measured variable (during a full-range traverse, unless otherwise specified) after transients have decayed. *Note:* Use of the term to define half the separation is deprecated. *Syn.:* **hysteresis error.** *See:* **control system, feedback.** 219, 206

I

IAGC. *See:* **instantaneous automatic gain control.**

ice-detection light. An inspection light designed to illuminate the leading edge of the wing to check for ice formation. 167

ice proof. So constructed or protected that ice of a specified composition and thickness will not interfere with successful operation. 79

I chrominance signal (NTSC color television). The sidebands resulting from suppressed-carrier modulation of the chrominance subcarrier by the *I* Video Signal. *Note:* The signal is transmitted in vestigal form, the upper sideband being limited to a frequency within the top of the picture transmission channel (approximately 0.6 MHz above the chrominance subcarrier), and the lower sideband extending to approximately 1.5 MHz below the subcarrier. The phase of the signal, for positive *I* Video Signals, is 123° with respect to the $B - Y$ axis. 18

ICI*. *See:* **CIE.**

* Deprecated

iconoscope. A camera tube in which a beam of high-velocity electrons scans a photoemissive mosaic that is capable of storing an electric charge pattern. *See also:* **television.** 178, 190, 125

ICW. *See:* **interrupted continuous wave.**

ideal capacitor (nonlinear capacitor). A capacitor whose transferred charge characteristic is single-valued. *See also:* **nonlinear capacitor.** 191

ideal dielectric. *See:* **dielectric, perfect.**

ideal filter (circuits and systems). (1) (frequency domain) A filter that passes, without attenuation, all frequencies inside specified frequency limits while rejecting all other frequencies. (2) (time domain) A filter with a time domain response identical to the excitation except for a constant delay. 67

ideal noise diode. *See:* **noise diode, ideal.**

ideal paralleling (rotating machinery). Paralleling by adjusting the voltage, and frequency and phase angle for alternating-current machines, such that the conditions of the incoming machine are identical with those of the system with which it is being paralleled. *See also:* **asynchronous machine.** 63

ideal transducer. *See:* **transducer, ideal.**

ideal transformer. *See:* **transformer, ideal.**

ideal value (1) (control systems; general) (control) (industrial control) (automatic control). The value of a selected variable that would result from a perfect system operating from the same command as the actual system under consideration. *See:* **control system, feedback.** 219, 206
(2) (synchronous-machine regulating system). The value of a controlled variable (for example, generator terminal voltage) that results from a desired or agreed-upon relationship between it and the commands (commands such as voltage regulator setting, limits, and reactive compensators). 63

identification (radar). The knowledge that a particular radar return signal is a specific target. This knowledge may be obtained by size, shape, timing, position, maneuvers, rate of change of any of these parameters and by means of coded responses through secondary radar. 13

identification beacon. An aeronautical beacon emitting a coded signal by means of which a particular point of reference can be identified. 167

identifier. A symbol whose purpose is to identify, indicate, or name a body of data. 255

identity friend or foe (IFF). Equipment used for transmitting radio signals between two stations located on ships, aircraft, or ground, for automatic identification. *Notes:* (1) The usual basic parts of equipment are interrogators, transpondors, and respondors. (2) Usually the initial letters of the name (IFF) are used instead of the full name. *See also:* **radio transmission.** 328

I-display (in radar). A display in which a target appears as a complete circle when the radar antenna is pointed at it and in which the radius of the circle is proportional to target distance; incorrect aiming of the antenna changes the circle to a segment whose arc length is inversely proportional to the magnitude of the pointing error, and the position of the segment indicates the reciprocal of the pointing direction of the antenna. 13

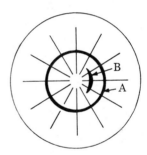

I display.

Two targets (A, B) at different distances. Radar aimed on target A.

idle bar (rotating machinery). An open circuited conductor bar in the rotor of a squirrel-cage motor, used to give low starting current in a moderate torque motor. *See also:* **rotor (rotating machinery).** 63

idle period (gas tube). That part of an alternating-voltage cycle during which a certain arc path is not carrying current. *See also:* **gas-filled rectifier.** 244, 190

idler circuit (parametric device). A portion of a parametric device that chiefly determines the behavior of the device at an idler frequency. *See also:* **parametric device.** 277, 191

idler frequency (parametric device). A sum frequency (or difference frequency) generated within the parametric device other than the input, output, or pump frequencies that requires specific circuit consideration to achieve the desired device performance. *See also:* **parametric device.** 277, 191

idle time (electric drive). The portion of the available time during which a system is believed to be in good operating condition but is not in productive use. *See:* **electric drive.** 219, 206

IF. *See:* **intermediate frequency.**
IFF. *See:* **identity friend or foe.**

ignition control (industrial control). Control of the starting instant of current flow in the anode circuit of a gas tube. *See also:* **electronic controller.** 206

ignition switch (industrial control). A manual or automatic switch for closing or interrupting the electric ignition circuit of an internal-combustion engine at the option of the machine operator, or by an automatic function calling for unattended operation of the engine. *Note:* Provisions for checking individual circuits of the ignition system for relative performance may be incorporated in such switches. *See also:* **switch.** 206

ignition transformer. Step-up transformer generally used for electrically igniting oil, gas or gasoline in domestic, commercial, or industrial heating equipment. 53

ignitor. A stationary electrode that is partly immersed in the cathode pool and has the function of initiating a cathode spot. *See also:* **electrode (electron tube); electronic controller.** 190

ignitor-current temperature drift (microwave gas tubes). The variation in ignitor electrode current caused by a change in ambient temperature of the tube. *See also:* **gas tube.** 190, 125

ignitor discharge (microwave switching tubes). A direct-current glow discharge, between the ignitor electrode and a suitably located electrode, used to facilitate radio-frequency ionization. *See also:* **gas tube.** 190, 125

ignitor electrode (microwave switching tubes). An electrode used to initiate and sustain the ignitor discharge. *See also:* **gas tube.** 190, 125

ignitor firing time (microwave switching tubes). The time interval between the application of a direct voltage to the ignitor electrode and the establishment of the ignitor discharge. *See also:* **gas tube.** 190, 125

ignitor interaction (microwave gas tubes). The difference between the insertion loss measured at a specified ignitor current and that measured at zero ignitor current. *See also:* **gas tube.** 190, 125

ignitor leakage resistance (microwave switching tubes). The insulation resistance, measured in the absence of an ignitor discharge, between the ignitor electrode terminal and the adjacent radio-frequency electrode. *See also:* **gas tube.** 190, 125

ignitor oscillations (microwave gas tubes). Relaxation oscillations in the ignitor circuit. *Note:* If present, these oscillations may limit the characteristics of the tube. *See also:* **gas tube.** 190, 125

ignitor voltage drop (microwave switching tubes). The direct voltage between the cathode and the anode of the ignitor discharge at a specified ignitor current. *See also:* **gas tube.** 190, 125

ignitron. A single-anode pool tube in which an ignitor is employed to initiate the cathode spot before each conducting period. *See also:* **electronic controller.** 190, 125

ignored conductor. *See:* **isolated conductor.**

illuminance. *See:* **illumination.**

illumination (1) (general). The density of the luminous flux incident on a surface, it is the quotient of the luminous flux by the area of the surface when the latter is uniformly illuminated. *Notes:* (A) The term **illumination** also is commonly used in a qualitative or general sense to designate the act of illuminating or the state of being illumi-

nated. Usually the context will indicate which meaning is intended, but occasionally it is desirable to use the expression level of illumination to indicate that the quantitative meaning is intended. The term **illuminance,** which sometimes is used in place of **illumination,** is subject to confusion with luminance and illuminants, especially when not clearly pronounced. (B) The units of measurement are: Footcandle (lumen per square foot, lm/ft^2) Lux (lumen per square meter, lx or $1m/m^2$). This unit of illumination is recommended by the IEC (45-10-075). Phot (lumen per square centimeter, $1m/cm^2$). *See also:* **light.** 18, 167

(2) (light emitting diodes). ($E_v = do_v/dA$). The density of the luminous flux incident on a surface; illumination is the quotient of the luminous flux by the area of the surface. When the area of the surface is uniformly illuminated, the SI unit for illumination is the lux (lx) that is, lumen per square meter. 162

(3) (antenna). *See:* **aperture illumination.**

illumination (footcandle) meter. An instrument for measuring the illumination on a surface. *Note:* Most such instruments consist of barrier-layer cells connected to a meter calibrated in footcandles. *See also:* **photometry.** 167

illuminator (radar). A system designed to impose electromagnetic radiation on a designated target so that the reflections can be used by another sensor typically for purposes of homing. 13

illustrative diagram (industrial control). A diagram whose principal purpose is to show the operating principle of a device or group of devices without necessarily showing actual connections or circuits. Illustrative diagrams may use pictures or symbols to illustrate or represent devices or their elements. Illustrative diagrams may be made of electric, hydraulic, pneumatic, and combination systems. They are applicable chiefly to instruction books, descriptive folders, or other media whose purpose is to explain or instruct. *See also:* **control.** 210, 206

ILS (navigation). An internationally adopted instrument landing system for aircraft, consisting of a vhf localizer, a uhf glide slope and 75 MHz markers. 13

ILS marker beacon. *See:* **outer, middle or boundary marker.**

ILS reference point. A point on the centerline of the ILS runway designated as the optimum point of contact for landing; in International Civil Aviation Organization standards this point is from 150 to 300 meters (500 to 1000 feet) from the approach end of the runway. 13

image (1) (optoelectronic device). A spatial distribution of a physical property, such as radiation, electric charge, conductivity, or reflectivity, mapped from another distribution of either the same or another physical property. *Note:* The mapping process may be carried out by a flux of photons, electric charges, or other means. *See also:* **optoelectronic device.**

(2) (computing systems). *See:* **card image.** 191

image antenna. The imaginary counterpart of an actual antenna, assumed for mathematical purposes to be located below the surface of the ground and symmetrical with the actual antenna above ground. *See also:* **antenna.** 179

image attenuation (circuits and systems). The real part of the image transfer constant. *See also:* **image transfer constant.** 67

image burn. *See:* **retained image.**

image camera tube. *See:* **image tube.**

image converter (solid state). An optoelectronic device capable of changing the spectral characteristics of a radiant image. *Note:* Examples of such changes are infrared to visible and X-ray to visible. *See also:* **optoelectronic device.** 191

image-converter panel. A thin, usually flat, multicell image converter. *See also:* **optoelectronic device.** 191

image-converter tube (camera tubes). An image tube in which an infrared or ultraviolet image input is converted to a visible image output. *See also:* **camera tube.** 190

image dissector (optical character recognition) (computing systems). A mechanical or electronic transducer that sequentially detects the level of light in different areas of a completely illuminated sample space. 255, 77

image dissector tube (dissector tube). A camera tube in which an electron image produced by a photoemitting surface is focused in the plane of a defining aperture and is scanned past that aperture. *See also:* **television.** 178, 190, 125

image element (optoelectronic device). The smallest portion of an image having a specified correlation with the corresponding portion of the original. *Note:* In some imaging systems the size of the image elements is determined by the structure of the image space, in others by the carrier employed for the mapping process. *See also:* **optoelectronic device.** 191

image frequency. *See:* **frequency, image.**

image iconoscope. A camera tube in which an electron image is produced by a photoemitting surface and focused on one side of a separate storage target that is scanned on the same side by an electron beam, usually of high-velocity electrons. *See also:* **television.** 178, 190, 125

image impedances (transducer). The impedances that will simultaneously terminate all of its inputs and outputs in such a way that at each of its inputs and outputs the impedances in both directions are equal. *Note:* The image impedances of a four-terminal transducer are in general not equal to each other, but for any symmetrical transducer the image impedances are equal and are the same as the iterative impedances. *See also:* **self-impedance; transducer.** 239, 185, 210

image intensifier (solid state). An optoelectronic amplifier capable of increasing the intensity of a radiant image. *See also:* **optoelectronic device.** 191

image-intensifier panel. A thin, usually flat, multicell image intensifier. *See also:* **optoelectronic device.** 191

image-intensifier tube. An image tube in which the output radiance is (1) in approximately the same spectral region as, and (2) substantially greater than, the photocathode irradiance. *See also:* **camera tube; image tube.** 190

image orthicon. A camera tube in which an electron image is produced by a photoemitting surface and focused on a separate storage target, which is scanned on its opposite side by a low-velocity electron beam. 190, 125

image parameters (circuits and systems). Fundamental network functions, namely image impedances and the image transfer function, that are used to design or describe a filter. *See also:* **image impedances; image transfer constant.** 67

image phase (circuits and systems). The imaginary part of the image transfer constant. *See also:* **image transfer constant.** 67

image phase constant. The imaginary part of the image transfer constant. *See also:* **transducer; transfer constant.** 210

image ratio (heterodyne receiver). The ratio of (1) the field strength at the image frequency to (2) the field strength at the desired frequency, each field being applied in turn, under specified conditions, to produce equal outputs. *See also:* **radio receiver.** 339

image response. Response of a superheterodyne receiver to the image frequency, as compared to the response to the desired frequency. *See also:* **radio receiver.** 328

image-storage device. An optoelectronic device capable of retaining an image for a selected length of time. *See also:* **optoelectronic device.** 191

image-storage panel (optoelectronic device). A thin, usually flat, multicell image-storage device. *See also:* **optoelectronic device.** 191

image-storage tube. A storage tube into which the information is introduced by means of radiation, usually light, and read at a later time as a visible output. *See also:* **storage tube.** 174, 190

image transfer constant (electric transducer) (transfer constant). The arithmetic mean of the natural logarithm of the ratio of input to output phasor voltages and the natural logarithm of the ratio of the input to output phasor currents when the transducer is terminated in its image impedances. *Note:* For a symmetrical transducer the transfer constant is the same as the propagation constant. *See also:* **transducer.** 210

image tube. An electron tube that reproduces on its fluorescent screen an image of an irradiation pattern incident on its photosensitive surface. *See also:* **camera tube.** 190, 125

imaginary part (circuits and systems). If a complex quantity is represented by 2 components $A + jB$, B is called the imaginary part. 67

imaging plane. *See:* **ground plane.**

imbedded temperature-detector insulation (rotating machinery). The insulation surrounding a temperature detector, taking the place of a coil separator in its area. 63

immediate address (computing systems). Pertaining to an instruction in which an address part contains the value of an operand rather than its address. *See:* **zero-level address.** 255, 77

immediate-nonsynchronized ringing (telephone switching systems). An arrangement whereby a pulse of ringing is sent to the called line when the connection is completed, irrespective of the state of the ringing cycle. 55

immediate-synchronized ringing (telephone switching systems). An arrangement whereby the ringing cycle starts with a complete interval of ringing sent to the called line when the connection is completed. 55

immersed gun (microwave tubes). A gun in which essentially all the flux of the confining magnetic field passes perpendicularly through the emitting surface of the cathode. *See also:* **microwave tube.** 190

immersion plating (dip plating). The deposition, without application of an external electromotive force, of a thin metal coating upon a less noble metal by immersing the latter in a solution containing a compound of the metal to be deposited. *See also:* **electroplating.**   328

immittance (linear passive networks). A response function for which one variable is a voltage and the other a current. *Note:* Immittance is a general term for both impedance and admittance, used where the distinction is irrelevant. *See also:* **linear passive networks.**   238

immittance comparator. An instrument for comparing the impedance or admittance of the two circuits, components, etcetera. *See also:* **auxiliary device to an instrument.**
                                                              185

immittance converter (circuits and systems). A two-port circuit capable of making the input immittance of one port (H_{in}) the product of the immittance connected to the other port (H_1) a positive or negative real constant ($\pm k$) and some internal immittance (H_i) i.e. $H_{in} = \pm k H_1 H_i$. 67

immittance matrix (circuits and systems). A two-dimensional array of immittance quantities that relate currents to voltages at the ports of a network.   67

immunity to interference (electromagnetic compatibility). The property of a receiver or any other equipment or system enabling it to reject a radio disturbance. *See also:* **electromagnetic compatibility.**   199

impedance (1) (linear constant-parameter system). (A) The corresponding impedance function with p replaced by $j\omega$ in which ω is real. (B) The ratio of the phasor equivalent of a steady-state sine-wave voltage or voltagelike quantity (driving force) to the phasor equivalent of a steady-state sine-wave current or currentlike quantity (response). *Note:* Definitions (A) and (B) are equivalent. The *real* part of impedance is the resistance. The *imaginary* part is the reactance. (C) A physical device or combination of devices whose impedance as defined in (A) or (B) can be determined. *Note:* This sentence illustrates the double use of the word impedance, namely for a physical characteristic of a device or system (definitions (A) and (B) and for a device (definition (C)). In the latter case the word impedor may be used to reduce confusion. Definition (C) is a second use of **impedance** and is independent of definitions (A) and (B). *Editor's Note:* The ratio Z is commonly expressed in terms of its orthogonal components, thus:

$$Z = R + jX$$

where Z, R, and X are respectively termed the impedance, resistance, and reactance, all being measured in ohms. In a simple circuit consisting of R, L, and C all in series, Z becomes

$$Z = R + j(\omega L - 1/\omega C),$$

where $\omega = 2\pi f$ and f is the frequency. *See:* **reactance; resistance.** *See also:* **network analysis; input impedance; feedback impedance; impedance function.**   210

(2) (electric machine). Linear operator expressing the relation between voltage (increments) and current (increments). Its inverse is called the admittance of an electric machine. *Notes:* (A) If a matrix has as its elements impedances, it is usually referred to as impedance matrix. Frequently the impedance matrix is called impedance for short. (B) Usually such impedances are defined with the mechanical angular velocity of the machine at steady state. *See:* **asynchronous machine.**   63

(3) (two-conductor transmission line). The ratio of the complex voltage between the conductors to the complex current on one conductor in the same transverse plane.
                                                              179

(4) (circular or rectangular waveguide). A nonuniquely defined complex ratio of the voltage and current at a given transverse plane in the waveguide, which depends on the choice of representation of the characteristic impedance. *See also:* **characteristic impedance; waveguide.**   179

(5) (linear system under sinusoidal stimulus) (automatic control). The complex-number ratio of the force-like variable to the resulting velocity-like steady-state variable; a type of transfer function expressed as voltage per unit current, force per unit velocity, pressure difference per unit volume or mass flux, temperature difference per unit heat flux. *See also:* **function, transfer.**   56

(6) (antenna). *See:* **input impedance; intrinsic impedance; mutual impedance; self-impedance.**

impedance bond (railway practice). An iron-core coil of low resistance and relatively high reactance used on electrified railroads to provide a continuous direct-current path for the return propulsion current around insulated joints and to confine the alternating-current signaling energy to its own track circuit.   328

impedance, characteristic wave. *See:* **wave impedance, characteristic.**

impedance compensator. An electric network designed to be associated with another network or a line with the purpose of giving the impedance of the combination a desired characteristic with frequency over a desired frequency range. *See also:* **network analysis.**   210

impedance, conjugate. An impedance the value of which is the complex conjugate of a given impedance. *Note:* For an impedance associated with an electric network, complex conjugate is an impedance with the same resistance component and a reactance component the negative of the original.   239, 210, 185

impedance drop. The phasor sum of the resistance voltage drop and the reactance voltage drop. *Note:* For transformers, the resistance drop, the reactance drop, and the impedance drop are respectively the sum of the primary and secondary drops reduced to the same terms. They are determined from the load loss measurements and are usually expressed in per unit, or in percent.   53

impedance, effective input (output) (1) (electron valve or tube). The quotient of the sinusoidal component of the control-electrode voltage (output-electrode voltage) by the corresponding component of the current for the given electrical conditions of all the other electrodes. *See also:* **ON period.**   244, 190

(2) (circuits and systems). The quotient of voltage by current at the input port of a device when it is operating normally (usually steady-state).   67

impedance, essentially zero source (transformer electrical tests). Source impedance low enough so that the test currents under consideration would cause less than five (5) percent distortion (instantaneous) in the voltage amplitude or waveshape at the load terminals.   95

impedance feedback (analog computer). A passive network connected between the output terminal of an operational amplifier and its summing junction.   9, 77

impedance function (defined for linear constant-parameter

systems or parts of such systems). That mathematical function of p that is the ratio of a voltage or voltage-like quantity (driving force) to the corresponding current or current-like quantity (response) in the hypothetical case in which the former is e^{pt} (e is the natural log base, p is arbitrary but independent of t, t is an independent variable that physically is usually time) and the latter is a steady-state response of the form $e^{pt}/2(p)$. *Note:* In electric circuits **voltage** is always the driving force and **current** is the response even though as in nodal analysis the current may be the independent variable; in electromagnetic radiation **electric field strength** is always considered the driving force and **magnetic field strength** the response, and in mechanical systems **mechanical force** is always considered as a driving force and **velocity** as a response. In a general sense the dimension (and unit) of **impedance** in a given application may be whatever results from the ratio of the dimensions of the quantity chosen as the driving force to the dimensions of the quantity chosen as the response. However, in the types of systems cited above any deviation from the usual convention should be noted. *See also:* **network analysis.** 210

impedance grounded (electric power). Means grounded through impedance. *Note:* The component of the impedance need not be at the same location as the device to be grounded. *See also:* **ground; grounded; grounded systems.** 91, 203, 64, 53

impedance, image. *See:* **image impedances.**

impedance, input. *See:* **input impedance.**

impedance inverter (circuits and systems). (1) network possessing an input (output) impedance that is proportional to the reciprocal of the load (source) impedance. (2) A symmetrical four-terminal network having the impedance inverting and phase characteristics of a quarter-wave length transmission line at its specified frequency or a chain matrix where $A = D = O$, $B = jK$ and $C = j/K$ (K is a constant relating the input impedance Z to the load impedance Z_L by the relationship $Z = K^2/Z_L$). 67

impedance irregularity (data transmission). A term used to denote impedance mismatch in a transmission medium. For example, a section of cable in an open-wire line constitutes an impedance irregularity. 59

impedance, iterative (transducer or a 2-port network). The impedance that, when connected to one pair of terminals, produces a like impedance at the other pair of terminals. *Notes:* (1) It follows that the iterative impedance of a transducer or a network is the same as the impedance measured at the input terminals when an infinite number of identically similar units are formed into an iterative or recurrent structure by connecting the output terminals of the first unit to the input terminals of the second, the output terminals of the second to the input terminals of the third, etcetera. (2) The iterative impedances of a four-terminal transducer or network are, in general, not equal to each other but for any symmetrical unit the iterative impedances are equal and are the same as the image impedances. The iterative impedance of a uniform line is the same as its characteristic impedance. 239, 185

impedance kilovolt-amperes (1) (regulator). The kilovolt-amperes (kVA) measured in the shunt winding with the series winding short-circuited and with sufficient voltage applied to the shunt winding to cause rated current to flow

in the windings. *See also:* **voltage regulator.** 257

(2) **(transformers).** The kilovolt-amperes (kVA) measured in the excited winding with the other winding short-circuited and with sufficient voltage applied to the excited winding to cause rated current to flow in the winding. 53

impedance, load (1) (general). The impedance presented by the load to a source or network. 185

(2) **(semiconductor radiation detectors).** The impedance shunting the detector, and across which the detector output voltage signal is developed. 23, 119

impedance, loaded applicator. *See:* **loaded applicator impedance.**

impedance matching. The connection across a source impedance of a matching impedance that allows optimum undistorted energy transfer. Types of matching are: (1) An impedance having the same magnitude and phase angle as the source impedance. (2) An impedance having the same magnitude and having a phase angle with the same magnitude but opposite signs as the source impedance. *See also:* **self-impedance.** 59

impedance, matching. *See:* **load matching.**

impedance matrix (multiport network). A matrix operator that interrelates the voltages at the various ports to the currents at the same and other ports. 185

impedance, normalized. The ratio of an impedance to a specified reference impedance. *Note:* For a transmission line or a waveguide, the reference impedance is usually a characteristic impedance. 185

impedance, output (1) (device, transducer, or network). The impedance presented by the output terminals to a load. *Notes:* (A) Output impedance is sometimes incorrectly used to designate load impedance. (B) This is a frequency-dependent function, and is used to help describe the performance of the power supply and the degree of coupling between loads. *See also:* **electrical conversion; self-impedance.** 186

(2) **(electron device).** The output electrode impedance at the output electrodes. *See also:* **self-impedance.** 190

(3) **(power supplies).** The effective dynamic output impedance of a power supply is derived from the ratio of the measured peak-to-peak change in output voltage to a measured peak-to-peak change in load alternating current. Output impedance is usually specified throughout the frequency range from direct current to 100 kilohertz. *See also:* **power supply; self-impedance.** 228, 186

(4) **(analog computer).** The impedance presented by the transducer to a load. 9

(5) **(transformer-rectifier system).** Internal impedance in ohms measured at the direct current terminals when the rectifier is continuously providing direct current to a load. This impedance is perferably expressed as a curve of impedance in ohms versus frequency, over the frequency range of interest to the application. 95

(6) **(Hall generator).** The impedance between the Hall terminals. 107

impedance permeability (magnetic core testing). An ac permeability related to the total rms exciting current, including harmonics.

$$\mu_z = \frac{B_i}{H_z \, \mu_0}$$

where

$H_z = \sqrt{2}\ NI/1$ = equivalent peak field strength, amperes/meters
B_i = maximum intrinsic flux density, tesla
I = rms exciting current, amperes
N = exciting coil turns. 165

impedance ratio (divider). The ratio of the impedance of the two arms connected in series to the impedance of the low-voltage arm. *Note:* In determining the ratio, account should be taken of the impedance of the measuring cable and the instrument. The impedance ratio is usually given for the frequency range within which it is approximately independent of frequency. For resistive dividers the impedance ratio is generally derived from a direct-current measurement such as by means of a Wheatstone bridge. 307, 201

impedance relay. A distance relay in which the threshold value of operation depends only on the magnitude of the impedance and is substantially independent of the phase angle of the impedance. 103, 202, 60

impedance, source. The impedance presented by a source of energy to the input terminals of a device, or network. *See also:* **impedance, input; network analysis; self impedance.** 239, 185

impedance, unloaded applicator. *See:* **transformer, matching.**

impedance voltage (1) (transformer). The voltage required to circulate rated current through one of two specified windings of a transformer when the other winding is short-circuited, with the windings connected as for rated voltage operation. *Note:* It is usually expressed in per unit, or percent, of the rated voltage of the winding in which the voltage is measured. 203, 197, 53
(2) (constant-current transformer). The measured primary voltage required to circulate rated secondary current through the short-circuited secondary coil for a particular coil separation. *Note:* It is usually expressed in per unit, or percent, of the rated primary voltage. *See:* **constant-current transformer.** 303, 203
(3) (current-limiting reactor). The product of its rated ohms impedance and rated current. *See also:* **reactor.** 309
(4) (regulator). The voltage required to circulate rated current through one winding of the regulator when another winding is short circuited, with the respective windings connected as for rated voltage operation. *Note:* It is usually referred to the series winding, and then expressed in per unit, or percent, of the rated voltage of the regulator. *See also:* **voltage regulator.** 257
(5) (neutral grounding device). An effective resistance component corresponding to the impedance losses, and a reactance component corresponding to the flux linkages of the winding. 91

impedance, wave. *See:* **wave impedance.**

impedor. A device, the purpose of which is to introduce impedance into an electric circuit. *See also:* **network analysis.** 210

imperfect dielectric. *See:* **dielectric, imperfect.**

imperfection (crystalline solid). Any deviation in structure from that of an ideal crystal. *Note:* An ideal crystal is perfectly periodic in structure and contains no foreign

atoms. *See also:* **miscellaneous electron devices; semiconductor.** 210, 245, 66

impingement attack (corrosion). Localized erosion-corrosion resulting from turbulent or impinging flow of liquids. 221, 205

implicit computation. Computation using a self-nulling principle in which, for example, the variable sought first is assumed to exist, after which a synthetic variable is produced according to an equation and compared with a corresponding known variable and the difference between the synthetic and the known variable driven to zero by correcting the assumed variable. *Note:* Although the term applies to most analog circuits, even to a single operational amplifier, it is restricted usually to computation performed by (1) circuits in which a function is generated at the output of a single high-gain direct-current amplifier by inserting an element generating the inverse function in the feedback path, (2) circuits in which combinations of computing elements are interconnected in closed loops to satisfy implicit equations, or (3) circuits in which linear to nonlinear differential equations yield the solutions to a system of algebraic or transcendental equations in the steady state. *See also:* **electronic analog computer.** 9, 77

imprecision. *See:* **precision.**

impregnant (rotating machinery). A solid, liquid, or semi-liquid material that, under conditions of application, is sufficiently fluid to penetrate and completely or partially fill or coat interstices and elements of porous or semiporous substances or composites. 63

impregnate (rotating machinery). The act of adding impregnant (bond or binder material) to insulation or a winding. *Note:* The impregnant, if thermosetting, is usually cured in the process. *See also:* **vacuum-pressure impregnation.** 63

impregnated (fibrous insulation). A suitable substance replaces the air between the fibers, even though this substance does not fill completely the spaces between the insulated conductors. *Note:* To be considered suitable, the impregnating substance must have good insulating properties and must cover the fibers and render them adherent to each other and to the conductor. 328

impregnated tape (dispersed magnetic powder tape). *See:* **magnetic powder-impregnated tape.**

impregnation, winding (rotating machinery). The process of applying an insulating varnish to a winding and, when required, baking to cure the varnish. 63

improvement threshold (angle-modulation systems). The condition of unity for the ratio of peak carrier voltage to peak noise voltage after selection and before any nonlinear process such as amplitude limiting and detection. *See also:* **amplitude modulation.** 111

impulse (1) (general). A waveform that, from the observer's frame of reference, approximates a Dirac delta function. 185
(2) (power systems) (surge arresters). A surge of unidirectional polarity. 64, 2
(3) (mathematics). (automatic control). A pulse that begins and ends within a time so short that it may be regarded mathematically as infinitesimal although the area remains finite. *See also:* **pulse.** 210, 56

impulse bandwidth (1) (general). When an inpulse is passed through a network with a restricted passband, the output generally consists of a wave train, the envelope of which builds up to a maximum value and then decays approximately exponentially. The impulse bandwidth of such a network is defined as the ratio of that maximum value (when properly corrected for network sine wave gain at a stated reference frequency) to the spectrum amplitude of the pulse applied at the input. In networks with a single humped response, the reference frequency is taken as that at which the gain is maximum. (Overcoupled or stagger-tuned networks should not be used for measurement of spectrum amplitude of impulses.) *See also:* **impulse strength.**   30

(2) (radio noise from overhead power lines). The peak value of the response envelope divided by the spectrum amplitude of an applied impulse. *See also:* **electromagnetic compatibility.**   199, 36

(3) (non-real time spectrum analyzer). The peak value of the time response envelope divided by the spectrum amplitude (assumed flat within the bandpass) of an applied pulse.   68

impulse current (current testing). Ideally, an aperiodic transient current that rises rapidly to a maximum value and falls usually less rapidly to zero. A rectangular impulse current rises rapidly to a maximum value, remains substantially constant for a specified time and then falls rapidly to zero. *See also:* **test voltage and current.**
     307, 201

impulse, doublet (automatic control). An impulse having equal positive and negative peaks.   56

impulse (shock) excitation. A method of producing oscillator current in a circuit in which the duration of the impressed voltage is relatively short compared with the duration of the current produced. *See also:* **oscillatory cir-**
cuit.   111, 240

impulse flashover voltage (1) (insulator). The crest value of the impulse wave that, under specified conditions, causes flashover through the surrounding medium.   261

(2) (surge arresters). The crest voltage of an impulse causing a complete disruptive discharge through the air between electrodes of a test specimen. *See also:* **critical impulse flashover voltage; insulator;**   64, 62

impulse flashover volt-time characteristic. A curve plotted between flashover voltage for an impulse and time to impulse flashover, or time lag of impulse flashover. *See also:* **insulator.**   64

impulse generator (electromagnetic compatibility). A standard reference source of broadband impulse energy.   314, 199

impulse inertia (surge arresters). The property of insulation whereby more voltage must be applied to produce disruptive discharge, the shorter the time of voltage application.   64, 62

impulse insulation level. An insulation strength expressed in terms of the crest value of an impulse withstand voltage. *See also:* **basic impulse insulation level (BIL) (insulation strength).**   276

impulse margin (time margin) (operation of a relay). The difference in time between the duration of a given input and critical impulse time, with the same input values. See the following figure. *See also:* **relay.**   281, 60

impulse noise. Noise characterized by transient disturbances separated in time by quiescent intervals. *Notes:* (1) The frequency spectrum of these disturbances must be substantially uniform over the useful pass band of the transmission system. (2) The same source may produce impulse noise in one system and random noise in a different system. *See also:* **signal-to-noise ratio.**   111

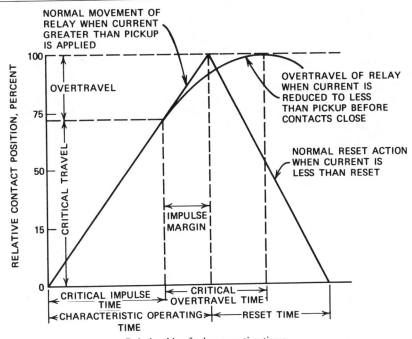

Relationship of relay operating times.

impulse-noise selectivity (receiver) (receiver performance).
A measure of the ability to discriminate against impulse
noise. *See also:* **receiver performance.**   181

impulse protection level (surge arresters). The highest
voltage (crest value) that appears at the terminals of overvoltage protective apparatus under specified impulse-test
conditions. *See also:* **basic impulse insulation level (insulation strength).**   244, 62, 276

impulse ratio (surge arresters). The ratio of the flashover,
sparkover, or breakdown voltage of an impulse to the crest
value of the power-frequency, sparkover, or breakdown
voltage.   64, 62

impulse relay. (1) A relay that follows and repeats current
pulses, as from a telephone dial. (2) A relay that operates
on stored energy of a short pulse after the pulse ends. (3)
A relay that discriminates between length and strength
of pulses, operating on long or strong pulses and not operating on short or weak ones. (4) A relay that alternately
assumes one of two positions as pulsed. (5) Erroneously
used to describe an integrating relay. *See also:* **relay.**
                                                        341

impulse response (linear network) (1) (circuits and systems).
The response, as a function of time, of a network when the
excitation is a unit impulse. Hence, the impulse response
of a network is the inverse Laplace transform of the network function in the frequency domain.   67
(2) (automatic control). *See:* **response, impulse-forced.**

impulse sparkover voltage (surge arrester). The highest
value of voltage attained by an impulse of a designated
wave shape and polarity applied across the terminals of
an arrester prior to the flow of discharge current.   2

impulse sparkover voltage-time curve (arrester). A curve
that relates the impulse sparkover voltage to the time to
sparkover.   62

impulse sparkover volt-time characteristic. (surge arrester).
The sparkover response of the device to impulses of a
designated wave shape and polarity, but of varying magnitudes. *Note:* For an arrester, this characteristic is shown
by a graph of values of crest voltage plotted against time
to sparkover.   2

impulse strength. The area under the amplitude-time
relation for the impulse. *Note:* This definition can be
clarified with the aid of Fig 1. Let $A(t)$ be some function
of time having a value other than zero only between the
times t_1 and $t_1 + \delta$. Then let the area under the curve $A(t)$
be designated by σ:

$$\sigma = \int_{\infty}^{\infty} A(t)dt = \int_{t_1}^{t_1+\delta} + A(t)dt$$

                                                        30

To define the theoretical or ideal impulse, let $A(t)$ vary in
a reciprocal manner with δ such that the value σ remains
constant, so that

$$\sigma = \lim \delta \to 0 \int_{t_1}^{t_1+\delta} A(t)dt$$

In the limit the function $A(t)$ becomes an ideal "impulse"
of "strength" σ. As an example, consider the function
shown in Fig 2. Here a rectangular pulse of finite duration
Δt and height A is shown. Now let $A = \sigma/\Delta t$ where σ is

(for the present argument) an arbitrary constant, and let
$\Delta t \to 0$. In the limit we have an impulse of strength σ.
When $\sigma = 1$, one has a "unit impulse." In many conventional applications the amplitude $A(t)$ has the dimension
volts and σ then has the dimension volt-seconds.

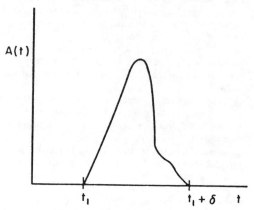

Fig. 1. A pulse of arbitrary shape.

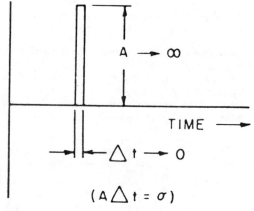

Fig. 2. A rectangular pulse.

impulse test (1) (rotating machinery). A test for applying
to an insulated component an aperiodic transient voltage
having predetermined polarity, amplitude, and wave-form.
See also: **asynchronous machine; direct-current commutating machine.**   91, 63
(2) (surge arresters) (transformers). An insulation test in
which the voltage applied is an impulse voltage of specified
wave shape.   64, 62, 53
(3) (neutral grounding devices). Dielectric test in which
the voltage applied is an impulse voltage of specified wave
shape. The wave shape of an impulse test wave is the graph
of the wave as a function of time or distance. *Note:* It is
customary in practice to express the wave shape by a
combination of two numbers, the first part of which represents the wave front and the second the time between the
beginning of the impulse and the instant at which one-half
crest value is reached on the wave tail, both values being
expressed in microseconds, such as a 1.2 × 50 microsecond
wave.   91

impulse time margin (power switchgear) (in the operation of a relay). The difference between characteristic operating times and critical impulse times.   127, 101

impulse transmission. That form of signaling, used principally to reduce the effects of low-frequency interference, that employs impulses of either or both polarities for transmission to indicate the occurrence of transitions in the signals. *Note:* The impulses are generally formed by suppressing the low-frequency components, including direct current, of the signals. *See also:* **telegraphy.**   328

impulse transmitting relay. A relay that closes a set of contacts briefly while going from the energized to the de-energized position or vice versa. *See also:* **relay.**   259

impulse-type telemeter. A telemeter that employs characteristics of intermittent electric signals, other than their frequency, as the translating means. *See also:* **telemetering.**   328

impulse voltage (current) (surge arresters). Synonymous with voltage of an impulse wave (current of an impulse wave).   244, 62

impulse wave (1) (general). A unidirectional surge generated by the release of electric energy into an impedance network. *See also:* **insulator.**   261

(2) (surge arresters). A unidirectional wave of current or voltage of very short duration containing no appreciable oscillatory components. *See also:* **insulator.**   244, 62

impulse withstand voltage (1) (general). The crest value of an applied impulse voltage that, under specified conditions, does not cause a flashover, puncture, or disruptive discharge on the test specimen. *See also:* **insulator; surge arrester (surge diverter); lightning protection and equipment.**   91, 261, 103, 202, 64, 244, 62

(2) (surge arrester). The crest value of an impulse that, under specified conditions, can be applied without causing a disruptive discharge.   2

impulsive noise (electromagnetic compatibility). Noise, the effect of which is resolvable into a succession of discrete impulses in the normal operation of the particular system concerned. *See also:* **electromagnetic compatibility.**   220, 199

impurity (1) (acceptor) (semiconductor). *See:* **dopant.** *See also:* **semiconductor device.**   66

(2) (donor) (semiconductor). *See:* **dopant.** *See also:* **semiconductor device.**   66

(3) (crystalline solid). An imperfection that is chemically foreign to the perfect crystal. *See also:* **semiconductor.**   245, 66, 210

(4) (chemical) (semiconductor). An atom within a crystal, that is foreign to the crystal. *See also:* **electron devices, miscellaneous.**   66

impurity, stoichiometric. A crystalline imperfection arising from a deviation from stoichiometric composition.   245, 66, 210

inactive region (semiconductor radiation detector). A region of a detector in which charge created by ionizing radiation does not contribute significantly to the signal charge.   23, 118, 119

inadvertent interchange (electric power systems) (control area). The time integral of the net interchange minus the time integral of the scheduled net interchange. *Note:* This includes the intentional interchange energy resulting from the use of frequency and/or other bias as well as the unscheduled interchange energy resulting from human or equipment errors.   94, 200

in-band signaling (telephone switching systems). Analog generated signaling that uses the same path as a message and in which the signaling frequencies are in the same band used for the message.   55

incandescence. The self-emission of radiant energy in the visible spectrum due to the thermal excitation of atoms or molecules. *See also:* **lamp.**   167

incandescent-filament lamp. A lamp in which light is produced by a filament heated to incandescence by an electric current. *See also:* **lamp.**   167

incandescent-filament-lamp transformer (series type). A transformer that receives power from a current-regulated series circuit and that transforms the power to another circuit at the same or different current from that in the primary circuit. *Note:* If of the insulating type, it also provides protection to the secondary circuit, casing, lamp, and associated luminaire from the high voltage of the primary circuit.   271

inching (rotating machinery). Electrically actuated angular movement or slow rotation of a machine, usually for maintenance or inspection. *See also:* **asynchronous machine; direct-current commutating machine.**   63

incidental amplitude-modulation factor (signal generator). That modulation factor resulting unintentionally from the process of frequency modulation and/or phase modulation. *See also:* **signal generator.**   185

incidental and restricted radiation. Radiation in the radio-frequency spectrum from all devices excluding licensed devices. *See also:* **mobile communication system.**   181

incidental frequency modulation (signal generator). The ratio of the peak frequency deviation to the carrier frequency, resulting unintentionally from the process of amplitude modulation. *See also:* **signal generator.**   185

incidental phase modulation (signal generator). The peak phase deviation of the carrier, in radians, resulting unintentionally from the process of amplitude modulation. *See also:* **signal generator.**   185

incidental radiation of conducted power (frequency-modulated mobile communications receivers). Radio-frequency energy generated or amplified by the receiver, which is detectable outside the receiver.   123

incident wave (1) (general). A wave, traveling through a medium, that impinges on a discontinuity or a medium of different propagation characteristics. *See also:* **radiation; radio wave propagation.**   146

(2) (forward wave) (uniform guiding systems). A wave traveling along a waveguide or transmission line in a specified direction toward a discontinuity, terminal plane, or reference plane. *See also:* **reflected wave; waveguide.**   185

(3) (surge arresters). A traveling wave before it reaches a transition point.   244, 62

incipient failure detection (nuclear power generating station). Tests designed to monitor performance characteristics and detect degradation prior to failure(s) which would prevent performance of the Class 1E functions. *Note:* Incipient failure testing requires module test checks, inspection, etcetera, on a sufficient time basis to establish performance trends. At the outset, the test cycle and corresponding limits of deviation of module performance or status must be established. Specific parameter trend patterns, exceeding of performance limits shall require that the module be removed and adjusted, replaced or serviced. As used here "sufficient time basis" would never be less than periodic surveillance test interval. In any event the internal must be justified technically based upon such things as manufacturer recommended, periodic preventive maintenance procedures, past operating experience, etcetera. These module tests require testing on line and/or removal from service. 355

inclined-blade blower (rotating machinery). A fan made with flat blades mounted so that the plane of the blades is parallel to and displaced from the axis of rotation of the rotor. *See:* **fan (rotating machinery).** 63

inclined orbit (communication satellite). An orbit of a satellite which is not equatorial, and not polar. 74

inclosed. *See:* **enclosed.**

inclusive OR (computing systems). *See:* OR.

incoherent scattering (radio wave propagation). When waves encounter matter, a disordered change in the phase of the waves. *See also:* **radio wave propagation.** 180, 146

incoming traffic (telephone switching systems). Traffic received directly from trunks by a switching entity. 55

incomplete diffusion (illuminating engineering). That in which the diffusing medium redirects some of the flux incident upon it in an image-forming state and another portion in a nonimage-forming state. *Note:* Also termed partial diffusion. *See also:* **lamp.** 167

incorrect relaying-system performance (relays or associated equipment). Any operation or lack of operation that, under existing conditions, does not conform to correct relaying-system performance. 103, 202, 60, 127

incorrect relay operation. Any output response or lack of output response by the relay that, for the applied input quantities, is not correct. 103, 202, 60, 127

incremental computer. A special-purpose computer that is specifically designed to process changes in the variables as well as the absolute value of the variables themselves, for example, digital differential analyzer. 225, 77

incremental cost of delivered power (source). The additional per unit cost that would be incurred in supplying another increment of power from that source to the composite system load. 94

incremental cost of reference power (source). The additional per-unit cost that would be incurred in supplying another increment of power from that source to a designated reference point on a transmission system. 94

incremental delivered power (electric power systems). The fraction of an increment in power from a particular source that is delivered to any specified point such as the composite system load, usually expressed in percent. 94

incremental dimension (numerically controlled machines). A dimension expressed with respect to the preceding point in a sequence of points. *See:* **dimension; long dimension; normal dimension; short dimension.** 224, 207

incremental feed (numerically controlled machines). A manual or automatic input of preset motion command for a machine axis. 244, 207

incremental fuel cost of generation (any particular source). The cost, usually expressed in mill/kilowatt-hour, that would be expended for fuel in order to produce an additional increment of generation at any particular source. 94

incremental generating cost (source at any particular value of generation) (electric power systems). The ratio of the additional cost incurred in producing an increment of generation to the magnitude of that increment of generation. *Note:* All variable costs should be taken into account including maintenance. 94

incremental heat rate (steam turbo-generator unit at any particular output). The ratio of a small change in heat input per unit time to the corresponding change in power output. *Note:* Usually, it is expressed in British thermal unit/killowatt-hour. 94

incremental hysteresis loss (magnetic material). The hysteresis loss in a magnetic material when it is subjected simultaneously to a biasing and an incremental magnetizing force. 210

incremental induction. At a point in a material that is subjected simultaneously to a polarizing magnetizing force and a symmetrical cyclically varying magnetizing force, one-half the algebraic difference of the maximum and minimum values of the magnetic induction at that point. 210

incremental loading (electric power systems). The assignment of loads to generators so that the additional cost of producing a small increment of additional generation is identical for all generators in the variable range. 94

incremental magnetizing force (magnetic material). At a point that is subjected simultaneously to a biasing magnetizing force and a symmetrical cyclic magnetizing force, one-half of the algebraic difference of the maximum and minimum values of the magnetizing force at the point. 210

incremental maintenance cost (any particular source) (electric power systems). The additional cost for maintenance that will ultimately be incurred as a result of increasing generation by an additional increment. 94

incremental permeability (magnetic induction) (1) (general). The ratio of the cyclic change in the magnetic induction to the corresponding cyclic change in magnetizing force when the mean induction differs from zero. *Note:* In anisotropic media, incremental permeability becomes a matrix. 210
(2) (magnetic core testing). The permeability with stated alternating magnetic field conditions in the presence of a stated static magnetic field.

$$\mu_\Delta = \frac{1}{\mu_0} \frac{\Delta B}{\Delta H}$$

μ_Δ = relative incremental permeability
ΔH = total cyclic variation of the magnetic field strength
ΔB = corresponding total cyclic change in induction. 165

incremental resistance (forward or reverse of a semiconductor rectifier diode) (semiconductor). The quotient of a small incremental voltage by a small incremental current at a stated point on the static characteristic curve. 66

incremental sweep (oscilloscopes). A sweep that is not a continuous function, but that represents the independent variable in discrete steps. See also: oscillograph; stairstep sweep.   185

incremental transmission loss (electric power systems). The fraction of power loss incurred by transmitting a small increment of power from a point to another designated point. Note: One of these points may be mathematical (rather than physical), such as the composite system load.
                                                        94

incremental worth of power (designated point on a transmission system). The additional per-unit cost that would be incurred in supplying another increment of power from any variable source of a system in economic balance to such designated point. Note: When the designated point is the composite system load, the incremental worth of power is commonly called **lambda** or **Lagrangian multiplier**.   94

increment (network) starter (industrial control). A starter that applies starting current to a motor in a series of increments of predetermined value and at predetermined time intervals in closed-circuit transition for the purpose of minimizing line disturbance. One or more increments may be applied before the motor starts. See: **starter**.
                                                   225, 206

indefinite admittance matrix (network analysis) (circuits and systems). A matrix associated with an n-node network whose elements have the dimension of admittance and, when multiplied into the vector of node voltages, gives the vector of currents entering the n nodes.   67

independence (nuclear power generating stations). No common failure mode for any design basis event.
                                                   31, 102

independent auxiliary. An item capable of performing its function without dependence on a similar item or the component it serves.   58

independent ballast (mercury lamp). A ballast that can be mounted separately outside a lighting fitting or fixture.
                                                       269

independent conformity. See: **conformity**.

independent contact. A contacting member designed to close one circuit only.   328

independent firing (industrial control). The method of initiating conduction of an ignitron by obtaining power for the firing pulse in the ignitor from a circuit independent of the anode circuit of the ignitron. See also: **electronic controller**.   206

independent ground electrode (surge arresters). A ground electrode or system such that its voltage to ground is not appreciably affected by currents flowing to ground in other electrodes or systems.   244, 62

independent linearity. See: **linearity of a signal**.

independent manual operation (switching device). A stored-energy operation where manual energy is stored and released, such that the speed and force of this operation are independent of the action of the attendant.
                                                       103

independent power operation (power switchgear). An operation by means of energy other than manual where the completion of the operation is independent of the continuity of the power supply.   103

independent transformer. A transformer that can be mounted separately outside a luminaire.   274

index (electronic computation). (1) An ordered reference list of the contents of a file or document, together with keys or reference notations for identification or location of those contents. (2) A symbol or a number used to identify a particular quantity in an array of similar quantities. For example, the terms of an array represented by X1, X2, ..., X100 have the indexes 1, 2, ..., 100, respectively. (3) Pertaining to an index register.   255, 77

index of cooperation, international (facsimile in rectilinear scanning). The product of the total length of a scanning or recording line by the number of scanning or recording lines per unit length divided by pi. Notes: (1) For rotating devices the Index of Cooperation is the product of the drum diameter times the number of lines per unit length. (2) The prior IEEE Index of Cooperation was defined for rectilinear scanning or recording as the product of the total line length by the number of lines per unit length. This has been changed to agree with international standards.
                                                        12

index register (computing systems). A register whose content is added to or subtracted from the operand address prior to or during the execution of an instruction.
                                                    255, 54

indicated bearing (direction finding). A bearing from a direction-finder site to a target transmitter obtained by averaging several readings; the indicated bearing is compared to the apparent bearing to determine accuracy of the equipment. See also: **navigation**.   187, 13

indicated value (power meters). The uncorrected value determined by observing the indicating display of the instrument.   47, 115

indicating circuit (control apparatus). The portion of its control circuit that carries intelligence to visual or audible devices that indicate the state of the apparatus controlled. See also: **control**.   3

indicating control switch. A switch that indicates its last control operation.   103, 202

indicating demand meter (register). A demand meter equipped with a scale over which a friction pointer is advanced so as to indicate maximum demand. See also: **electricity meter (meter)**.   212

indicating fuse. A fuse that automatically indicates that the fuse has interrupted the circuit.   103, 202

indicating instrument. An instrument in which only the present value of the quantity measured is visually indicated. See also: **instrument**.   280

indicating or recording mechanism (demand meter). That mechanism that indicates or records the measurement of the electrical quantity as related to the demand interval. Note: This mechanism may be operated directly by and be a component part of the electric mechanism, or may be structurally separate from it. The demand may be indicated or recorded in kilowatts, kilovolt-amperes, amperes, kilovars, or other suitable units. This mechanism may be of an indicating type, indicating by means of a pointer related to its position on a scale or by means of the cumulative reading of a number of dial or cyclometer indicators; or a graphic type, recording on a circular or strip chart; or of a printing type, recording on a tape. It may record the demand for each demand interval or may indicate only the maximum demand. See also: **demand**

meter. 328

indicating scale (recording instrument). A scale attached to the recording instrument for the purpose of affording an easily readable value of the recorded quantity at the time of observation. *Note:* For recording instruments in which the production of the graphic record is the primary function, the chart scale should be considered the primary basis for accuracy ratings. For instruments in which the graphic record is secondary to a control function the indicating scale may be more accurate and more closely related to the control than is the chart scale. *See also:* **moving element (instrument).** 328

indication point (railway practice). The point at which the train control or cab signal impulse is transmitted to the locomotive or vehicle apparatus from the roadway element. 328

indicator. *See:* **display.**

indicator light. A light that indicates whether or not a circuit is energized. *See also:* **appliances outlet.** 263

indicators (nuclear power generating stations). Devices that display information to the operator. 31

indicator symbol (logic diagrams). A symbol that identifies the state or level of an input or output of a logic symbol with respect to the logic symbol definition. 88

indicator travel. The length of the path described by the indicating means or the tip of the pointer in moving from one end of the scale to the other. *Notes:* (1) The path may be an arc or a straight line. (2) In the case of knife-edge pointers and others extending beyond the scale division marks, the pointer shall be considered as ending at the outer end of the shortest scale division marks. *See also:* **moving element (instrument).** 295

indicator tube. An electron-beam tube in which useful information is conveyed by the variation in cross section of the beam at a luminescent target. 190, 125

indicial admittance. The instantaneous response to unit step driving force. *Note:* This is a time function that is not an admittance of the type defined under **admittance.** *See also:* **network analysis.** 210

indirect-acting machine voltage regulator. A machine voltage regulator having a voltage-sensitive element that acts indirectly, through the medium of an interposing device such as contactors or a motor to control the excitation of an electric machine. *Note:* A regulator is called a generator voltage regulator when it acts in the field circuit of a generator and is called an exciter voltage regulator when it acts in the field circuit of the main exciter. 103, 202

indirect-acting recording instrument. A recording instrument in which the level of measurement energy of the primary detector is raised through intermediate means to actuate the marking device. *Note:* The intermediate means are commonly either mechanical, electric, electronic, or photoelectric. *See also:* **instrument.** 328

indirect address (computing systems). An address that specifies a storage location that contains either a direct address or another indirect address. *See:* **multilevel address.** 255, 77

indirect-arc furnace. An arc furnace in which the arc is formed between two or more electrodes. 328

indirect component (illuminating engineering). That portion of the luminous flux from a luminaire that arrives at the work plane after being reflected by room surfaces. *See also:* **inverse-square law (illuminating engineering).** 167

indirect-drive machine (elevators). An electric driving machine, the motor of which is connected indirectly to the driving sheave, drum, or shaft by means of a belt or chain through intermediate gears. *See also:* **driving machine (elevators).** 328

indirect lighting. Lighting involving luminaires that distribute 90 to 100 percent of the emitted light upward. *See also:* **general lighting.** 167

indirectly controlled variable (control) (industrial control) (automatic control). A variable that is not directly measured for control but that is related to, and influenced by, the directly controlled variable. *See also:* **control system, feedback.** 219, 206

indirectly heated cathode (equipotential cathode) (unipotential cathode). A hot cathode to which heat is supplied by an independent heater. *See also:* **electrode (electron tube).** 125

indirect manual operation (switching device). Operation by hand through an operating handle mounted at a distance from, and connected to the switching device by, mechanical linkage. 103, 202

indirect operation (switching device). Operating by means of an operating mechanism connected to the main operating shaft or an extension of it, through offset linkages and bearings. 103, 202, 27

indirect release (switching device) (indirect trip). A release energized by the current in the main circuit through a current transformer, shunt, or other transducing device. 103, 202

indirect stroke (surge arresters). A lightning stroke that does not strike directly any part of a network but that induces an overvoltage in it. 244, 62

indirect-stroke protection (lightning). Lightning protection designed to protect a network or electric installation against indirect strokes. *See also:* **direct-stroke protection (lightning).** 64

individual-lamp autotransformer. A series autotransformer that transforms the primary current to a higher or lower current as required for the operation of an individual street light. 53

individual-lamp insulating transformer. An insulating transformer used to protect the secondary circuit, casing, lamp, and associated luminaire of an individual street light from the high-voltage hazard of the primary circuit. 53

individual line (1) (communication switching). A subscriber line arranged to serve only one main station although additional stations may be connected to the line as extensions. *Note:* An individual line is not arranged for discriminatory ringing with respect to the stations on that line. 193 **(2) (telephone switching systems).** A line arranged to serve one main station. 55

individual trunk (telephone switching systems). A trunk, link, or junctor that serves only one input group of a grading. 55

indoor (prefix) (1) (general). Not suitable for exposure to the weather. *Note:* For example, an indoor capacitor unit is designed for indoor service or for use in a weather-proof housing. *See also:* **outdoor.** 210, 53

(2) (power switchgear). Designed for use only inside buildings, or weatherproof enclosures. 103

indoor arrester. A surge arrester that, because of its construction, must be protected from the weather. 2

indoor termination (cable termination). A termination intended for use where it is protected from direct exposure to both solar radiation and precipitation. Terminations designed for use in sealed enclosures where the external dielectric strength is dependent upon liquid or special gaseous dielectrics are also included in this category. 4

indoor transformer. A transformer which, because of its construction, must be protected from the weather. 53

indoor wall bushing. A wall bushing of which both ends are suitable for operating only where protection from the weather is provided. *See also:* **bushing.** 348

induced charge (ferroelectric device). The charge that flows when the condition of the device is changed from that of zero applied voltage (after having previously been saturated with either a positive or negative voltage) to at least that voltage necessary to saturate in the same sense. *Note:* The induced charge is dependent on the magnitude of the applied voltage, which should be specified in describing this characteristic of ferroelectric devices. *See also:* **ferroelectric domain.** 247

induced control voltage (Hall-effect devices). The electromotive force induced in the loop formed by the control current leads and the current path through the Hall plate by a varying magnetic flux density, when there is no control current. 107

induced current (1) (general). Current in a conductor due to the application of a time-varying electromagnetic field. *See also:* **induction heating.** 14, 114

(2) (interference terminology). The interference current flowing in a signal path as a result of coupling of the signal path with an interference field, that is, a field produced by an interference source. *See:* **interference.** 188

(3) (lightning strokes). The current induced in a network or electric installation by an indirect stroke. *See also:* **direct-stroke protection (lightning).** 64

induced electrification. The separation of charges of opposite sign onto parts of a conductor as a result of the proximity of charges on other objects. *Note:* The charge on a portion of such a conductor is often called an induced charge or a bound charge. 210

induced emission (laser-maser). *See:* **stimulated emission.**

induced field current (synchronous machine). The current that will circulate in the field winding (assuming the circuit is closed) due to transformer action when an alternating voltage is applied to the armature winding, for example, during starting of a synchronous motor. 63

induced-potential tests (electric power). Dielectric tests in which the test voltages are suitable-frequency alternating voltages, applied or induced between the terminals. 91

induced voltage (1) (general). A voltage produced around a closed path or circuit by a change in magnetic flux linking that path. *See also:* **Faraday's law.** *Notes:* (A) Sometimes more narrowly interpreted as a voltage produced around a closed path or circuit by a time rate of change in magnetic flux linking that path when there is no relative motion between the path or circuit and the

magnetic flux. (B) A single-phase stator winding energized with alternating current, produces a pulsating magnetic field which causes a voltage to be induced in a blocked rotor circuit, and the same magnetic field may be interpreted in terms of two magnetic fields of constant amplitude traveling in opposite directions around the air gap, causing two voltages to be generated in a blocked rotor circuit. (C) Whether a voltage is defined as being induced or generated is often simply a matter of point of view. *See also:* **Faraday's law, generated voltage, induction motor.** 63

(2) (lightning strokes). The voltage induced on a network or electric installation by an indirect stroke. *See also:* **direct-stroke protection (lightning).** 64

induced voltage tests. Induced voltage tests are dielectric tests on transformer windings in which the appropriate test voltages are developed in the windings by magnetic induction. *Note:* Power for induced voltage tests is usually supplied at higher-than-rated frequency to avoid core saturation and excessive excitation current. 53

inductance. The property of an electric circuit by virtue of which a varying current induces an electromotive force in that circuit or in a neighboring circuit. 197

inductance coil. *See:* **inductor.**

inductance, effective (1) (general). The imaginary part of an inductive impedance divided by the angular frequency. 185

(2) (winding). The self-inductance at a specified frequency and voltage level, determined in such a manner as to exclude the effects of distributed capacitance and other parasitic elements of the winding but not the parasitic elements of the core. 197

induction coil. (1) A transformer used in a telephone set for interconnecting the transmitter, receiver, and line terminals. (2) A transformer for converting interrupted direct current into high-voltage alternating current. *See also:* **telephone station.** 341

induction compass. A device that indicates an aircraft's heading, in azimuth. Its indications depend on the current generated in a coil revolving in the earth's magnetic field. 328

induction-conduction heater. A heating device in which electric current is conducted through but is restricted by induction to a preferred path in a charge. *See also:* **induction heating.** 14, 114

induction coupling. An electric coupling in which torque is transmitted by the interaction of the magnetic field produced by magnetic poles on one rotating member and induced currents in the other rotating member. *Note:* The magnetic poles may be produced by direct-current excitation, permanent-magnet excitation, or alternating-current excitation. The induced currents may be carried in a wound armature, squirrel cage, or may appear as eddy currents. *See:* **electric coupling.** 328

induction cup (relay). A form of relay armature in the shape of a cylinder with a closed end that develops operating torque by its location within the fields of electromagnets that are excited by the input quantities. 103, 202, 60, 127

induction cylinder (relay). A form of relay armature in the shape of an open-ended cylinder that develops operating torque by its location within the fields of electromagnets that are excited by the input quantities.

103, 202, 60, 127

induction disc (relay) (1) (general). A form of relay armature in the shape of a disc that usually serves the combined function of providing an operating torque by its location within the fields of an electromagnet excited by the input quantities, and a restraining force by motion within the field of a permanent magnet. 103, 202, 60, 127 **(2) (utility consumer interconnections).** A thin circular (or spiraled) disk of nonmagnetic conducting material in which eddy currents are produced to create torque about an axis of rotation. 128

induction factor, A_L (magnetic core testing). Under stated conditions, the self inductance that a coil of specified shape and dimensions placed on the core in a given position should have, if it consisted of one turn.

$$A_L = \frac{L}{N^2}$$

A_L = Induction factor (henrys/turns²)
L = Self inductance of the coil on the core, in henrys
N = Number of turns of the coil. 165

induction frequency converter. A wound-rotor induction machine in which the frequency conversion is obtained by induction between a primary winding and a secondary winding rotating with respect to each other. *Notes:* (1) The secondary winding delivers power at a frequency proportional to the relative speed of the primary magnetic field and the secondary member. (2) In case the machine is separately driven, this relative speed is maintained by an external source of mechanical power. (3) In case the machine is self-driven, this relative speed is maintained by motor action within the machine obtained by means of additional primary and secondary windings with number of poles differing from the number of poles of the frequency-conversion windings. In special cases one secondary winding performs the function of two windings, being short-circuited with respect to the poles of the driving primary winding and open-circuited with respect to the poles of the primary-conversion winding. *See:* **converter.**

63

induction furnace. A transformer of electric energy to heat by electromagnetic induction. 328

induction generator (1) (rotating machinery). An induction machine, when driven above synchronous speed by an external source of mechanical power, used to convert mechanical power to electric power. *See also:* **asynchronous machine.** 63 **(2) (electric installations on shipboard).** An induction machine driven above synchronous speed by an external source of mechanical power. 3

induction heater (interference terminology). A device for causing electric current to flow in a charge of material to be heated. Types of induction heaters can be classified on the basis of frequency of the induced current, for example, a low-frequency induction heater usually induces power-frequency current in the charge; a medium-frequency induction heater induces currents of frequencies between 180 and 540 hertz; a high-frequency induction heater induces currents having frequencies from 1000 hertz upward. 60

induction heater or furnace, core type. *See:* **core type induction heater or furnace.**

induction heating. The heating of a nominally conducting material in a varying electromagnetic field due to its internal losses. 14, 114

induction instrument. An instrument that depends for its operation on the reaction between a magnetic flux (or fluxes) set up by one or more currents in fixed windings and electric currents set up by electromagnetic induction in movable conducting parts. *See also:* **instrument.**

328

induction loop (relay). A form of relay armature consisting of a single turn or loop that develops operating torque by its location within the fields of electromagnets that are excited by the input quantities. 103, 202, 60, 127

induction loudspeaker. A loudspeaker in which the current that reacts with the steady magnetic field is induced in the moving member. *See also:* **loudspeaker.** 176

induction machine. An asynchronous alternating-current machine that comprises a magnetic circuit interlinked with two electric circuits, or sets of circuits, rotating with respect to each other and in which power is transferred from one circuit to another by electromagnetic induction. *Note:* Examples of induction machines are induction motors, induction generators, and certain types of frequency converters and phase converters. 328, 3

induction motor. An alternating-current motor in which a primary winding on one member (usually the stator) is connected to the power source and a polyphase secondary winding or a squirrel-cage secondary winding on the other member (usually the rotor) carries induced current. *See also:* **asynchronous machine.** 328, 3

induction-motor meter. A motor-type meter in which the rotor moves under the reaction between the currents induced in it and a magnetic field. *See also:* **electricity meter (meter).** 212

induction ring heater. A form of core-type induction heater adapted principally for heating electrically conducting charges of ring or loop form, the core being open or separable to facilitate linking the charge. *See also:* **induction heater.** 14, 114

induction vibrator. A device momentarily connected between the airplane direct-current supply and the primary winding of the magneto, thus converting the magneto to an induction coil. *Note:* It provides energy to the sparkplugs of an aircraft engine during its starting period. *See also:* **air transportation electric equipment.** 328

induction voltage regulator. A regulating transformer having a primary winding in shunt and a secondary winding in series with a circuit, for gradually adjusting the voltage or the phase-relation, or both, of the circuit by changing the relative magnetic coupling of the exciting (primary) and series (secondary) windings. 53

inductive coordination (electric supply and communication systems). The location, design, construction, operation, and maintenance in conformity with harmoniously adjusted methods that will prevent inductive interference. 328

inductive coupling (communication circuits). The association of two or more circuits with one another by means of inductance mutual to the circuits or the mutual inductance that associates the circuits. *Note:* This term, when used without modifying words, is commonly used for coupling by means of mutual inductance, whereas coupling by means of self-inductance common to the circuits is called direct inductive coupling. *See also:* **coupling; inductive coordination; network analysis.** 210

inductive exposure. A situation of proximity between electric supply and communication circuits under such conditions that inductive interference must be considered. *See also:* **inductive coordination.** 328

inductive influence (electric supply circuit with its associated apparatus). Those characteristics that determine the character and intensity of the inductive field that it produces. *See also:* **inductive coordination.** 328

inductive interference (electric supply and communication systems). An effect, arising from the characteristics and inductive relations of electric supply and communication systems, of such character and magnitude as would prevent the communication circuits from rendering service satisfactorily and economically if methods of inductive coordination were not applied. *See also:* **inductive coordination.** 328

inductively coupled circuit. A coupled circuit in which the common element is mutual inductance. *See also:* **network analysis.** 210

inductive microphone. *See:* **inductor microphone.**

inductive neutralization (shunt neutralization) (coil neutralization). A method of neutralizing an amplifier whereby the feedback susceptance due to an interelement capacitance is canceled by the equal and opposite susceptance of an inductor. *See also:* **amplifier; feedback.** 111

inductive residual voltage (Hall-effect devices). The electromotive force induced in the loop formed by the Hall voltage leads and the conductive path through the Hall plate by a varying magnetic flux density, when there is no control current. 107

inductive susceptiveness (communication circuit with its associated apparatus). Those characteristics that determine, so far as such characteristics are able to determine, the extent to which the service rendered by the circuit can be adversely affected by a given inductive field. *See also:* **inductive coordination.** 328

inductor (1) (general). A device consisting of one or more associated windings, with or without a magnetic core, for introducing inductance into an electric circuit.
(2) (railway practice). A roadway element consisting of a mass of iron with or without a winding, that acts inductively on the vehicle apparatus of the train control, train stop, or cab signal system. 341

inductor alternator. An inductor machine for use as a generator, the voltage being produced by a variation of flux linking the armature winding without relative displacement of field magnet or winding and armature winding. 63

inductor, charging. An inductive component used in the charging circuit of a pulse-forming network. 137

inductor circuit (railway practice). A circuit including the inductor coil and the two lead wires leading therefrom taken through roadway signal apparatus as required. 328

inductor dynamotor (rotating machinery). A dynamotor inverter having toothed field poles and an associated sta-

tionary secondary winding for conversion of direct current to high-frequency alternating current by inductor-generator action. *See:* **convertor.** 63

inductor frequency-converter (rotating machinery). An inductor machine having a stationary input alternating-current winding, which supplies the excitation, and a stationary output winding of a different number of poles in which the output frequency is induced through change in field reluctance by means of a toothed rotor. *Note:* If the machine is separately driven, the rotor speed is maintained by an external source of mechanical power. If the machine is self-driven, the primary winding and rotor function as in a squirrel-cage induction motor or a reluctance motor. *See also:* **convertor.** 63

inductor machine (rotating machine). A synchronous machine in which one member, usually stationary, carries main and exciting windings effectively disposed relative to each other, and in which the other member, usually rotating, is without windings but carries a number of regular projections. (Permanent magnets may be used instead of the exciting winding). 63

inductor microphone (inductive microphone). A moving-conductor microphone in which the moving element is in the form of a straight-line conductor. *See also:* **microphone.** 328

inductor-type synchronous generator. A generator in which the field coils are fixed in magnetic position relative to the armature conductors, the electromotive forces being produced by the movement of masses of magnetic material. 3

inductor-type synchronous motor. An inductor machine for use as a motor, the torques being produced by forces between armature magnetomotive force and salient rotor teeth. *Note:* Such motors usually have permanent-magnet field excitation, are built in fractional-horsepower ratings, frames and operate at low speeds, 300 revolutions per minute or less. 63

industrial brush (rotating machinery). A brush having a cross-sectional area (width x thickness) of more than $\frac{1}{4}$ square inch or a length of more than $1\frac{1}{2}$ inches, but larger than a fractional-horsepower brush. *See:* **brush (1) (electric machines).** 63

industrial control. Broadly, the methods and means of governing the performance of an electric device, apparatus, equipment, or system used in industry. 206

industrial electric locomotive. An electric locomotive, used for industrial purposes, that does not necessarily conform to government safety regulations as applied to railroads. *Note:* This term is generally applied to locomotives operating in surface transportation and does not include mining locomotives. A prefix diesel-electric, gas-electric, etcetera, may replace the word electric. *See also:* **electric locomotive; industrial electronics.** 328

industrial process supervisory system. A supervisory system that initiates signal transmission automatically upon the occurrence of an abnormal or hazardous condition in the elements supervised, which include heating, air-conditioning, and ventilating systems, and machinery associated with industrial processes. *See also:* **protective signaling.** 328

inertance (automatic control). A property expressible by the quotient of a potential difference (temperature, sound pressure, liquid level) divided by the related rate of change of flow; the thermal or fluid equivalent of electrical inductance or mechanical moment of inertia. *See also:* **control system, feedback.**   56

inert-gas pressure system (regulator). A system in which the interior of the tank is sealed from the atmosphere, over the temperature range specified, by means of a positive pressure of inert gas maintained from a separate inert-gas source and reducing-valve system. *See also:* **transformer, oil-immersed.**   257, 91, 53

inertia compensation (industrial control). The effect of a control function during acceleration or deceleration to cause a change in motor torque to compensate for the driven-load inertia. *See also:* **control system, feedback.**   225, 206

inertia constant (machine). The energy stored in the rotor when operating at rated speed expressed as kilowatt-seconds per kilovolt-ampere rating of the machine. *Note:* The inertia constant is

$$h = \frac{0.231 \times Wk^2 \times n^2 \times 10^{-6}}{kVA}$$

where
h = inertia constant in kilowatt-seconds per kilovolt-ampere
Wk^2 = moment of inertia in pound-feet^2
n = speed in revolutions per minute
kVA = rating of machine in kilovolt-amperes
See: **asynchronous machine; direct-current commutating machine.**   63

inertialess scanning. *See:* **electronic scanning.**

inertial navigation equipment. A type of dead-reckoning navigation equipment whose operation is based upon the measurement of accelerations; accelerations are sensed dynamically by devices stabilized with respect to inertial space, and the navigational quantities (such as vehicle velocity, angular orientation, or positional information) are determined by computers and/or other instrumentation. *See also:* **navigation.**   187, 13

inertial navigator. A self-contained, dead-reckoning navigation aid using inertial sensors, a reference direction, and initial or subsequent fixes to determine direction, distance, and speed; single integration of acceleration provides speed information and a double integration provides distance information. *See also:* **navigation.**   187, 13

inertial space (navigation). A frame of reference defined with respect to the fixed stars. *See also:* **navigation.**   187, 13

inertia relay. A relay with added weights or other modifications that increase its moment of inertia in order either to slow it or to cause it to continue in motion after the energizing force ends. *See also:* **relay.**   259

infinite multiplication factor ($k\infty$) (nuclear power generation). The ratio of the average number of neutrons produced in each generation of nuclear fissions to the average number of corresponding neutrons absorbed. Since neutron leakage out of the system is ignored, $k\infty$ is the effective multiplication factor for an infinitely large assembly.   112

inflection point (tunnel-diode characteristic). The point on the forward current-voltage characteristic at which the slope of the characteristic reaches its most negative value. *See also:* **peak point (tunnel-diode characteristic).**   315, 191

inflection-point current (tunnel-diode characteristic). The current at the inflection point. *See also:* **peak point (tunnel-diode characteristic).**   315, 191

inflection-point emission current (electron tubes). That value of current on the diode characteristic for which the second derivative of the current with respect to the voltage is zero. *Note:* This current corresponds to the inflection point of the diode characteristic and is, under suitable conditions, an approximate measure of the maximum space-charge-limited emission current.   190, 125

inflection-point voltage (tunnel-diode characteristic). The voltage at which the inflection point occurs. *See also:* **peak point (tunnel-diode characteristic).**   315, 191

influence (1) (upon an instrument) (specified variable or condition). The change in the indication of the instrument caused solely by a departure of the specified variable or condition from its reference value, all other variables being held constant. *See also:* **accuracy rating of an instrument.**   280

(2) (upon a recording instrument). The change in the recorded value caused solely by a departure of the specified variable or condition from its reference value, all other variables being held constant. *Note:* If the influences in any direction from reference conditions are not equal, the greater value applies.   294

information (1) (general). The meaning assigned to data by known conventions.   255, 77

(2) (nuclear power generating station). Data describing the status and performance of the plant.   358

information content (message or a symbol from a source). The negative of the logarithm of the probability that this particular message or symbol will be emitted by the source. *Notes:* (1) The choice of logarithmic base determines the unit of information content. (2) The probability of a given message or symbol's being emitted may depend on one or more preceding messages or symbols. (3) The quantity has been called self-information. *See:* **bit; hartley.** *See also:* **information theory.**   160

information content, average. *See:* **average information content.**

information display channel (nuclear power generating station). An arrangement of electrical and mechanical components and/or modules from measured process variable to display device as required to sense and display conditions within the generating station.   361

information display channel failure (nuclear power generating station). A situation where the display disagrees, in a substantive manner, with the conditions or status of the plant. The display may fail to respond to a plant change, may improperly indicate a change when none has occurred, or may fail to provide any meaningful information.   361

information processing. (1) The processing of data that represents information. (2) Loosely, automatic data processing.   255, 77

information rate (1) (from a source, per second). The product of the average information content per symbol and the average number of symbols per second. *See also:* **information theory.**   160

(2) (from a source, per symbol). *See:* **average information content.**

(3) (through a channel, per second). The product of the average transinformation per symbol and the average number of symbols per second. *See also:* **information theory.**   160

information retrieval. The methods and procedures for recovering specific information from stored data.
                                                            255, 77, 54

information theory. (1) In the narrowest sense, is used to describe a body of work, largely about communication problems but not entirely about electrical communication, in which the information measures are central. (2) In a broader sense it is taken to include all statistical aspects of communication problems, including the theory of noise, statistical decision theory as applied to detection problems, and so forth. *Note:* This broader field is sometimes called "statistical communication theory." (3) In a still broader sense its use includes theories of measurement and observation that use other measures of information, or none at all, and indeed work on any problem in which information, in one of its colloquial senses, is important.
                                                            160

information transfer (data transmission). The final result of data transmission from a data source to a data sink. The information transfer rate may or may not be equal to the transmission modulation rate.   59

information writing speed (oscilloscopes). The oscilloscope-recorder characteristic that is a measure of the maximum number of spots of information per second that can be recorded and identified on a single trace. Test conditions must be specified. *See also:* **oscillograph; writing speed (storage tubes).**   185

infrared flame detector (fire protection devices). A device whose sensing element is responsive to radiant energy outside the range of human vision (above approximately 7700 angstroms).   71

infrared radiation (1) (illumination engineering). For practical purposes any radiant energy within the wavelength range 780 to 10^5 nanometers is considered infrared energy. *Note:* In general, unlike ultraviolet energy, infrared energy is not evaluated on a wavelength basis but rather in terms of all of such energy incident upon a surface. Examples of these applications are industrial heating, drying, baking, and photoreproduction. However, some applications, such as infrared viewing devices, involve detectors that are sensitive to a restricted range of wavelengths; in such cases the spectral characteristics of the source and receiver are of importance. *See:* **light; photochemical radiation; units of wavelength.**   167

(2) (laser-maser). Electromagnetic radiation with wavelengths which lie within the range 0.7μm to 1mm.   363

infrared radiation emitting diode (light emitting diodes). A semiconductor device containing a semiconductor junction in which infrared radiant flux is non-thermally produced when a current flows as a result of an applied voltage.
                                                            162

infrared radiation thermometer (temperature measurement). An optical system that accepts electromagnetic radiation in the infrared portion of the spectrum and either concentrates it on a temperature-sensitive element, which in turn activates an indicating device, or transforms it into

radiation in the visible spectrum.   7

infrasonic frequency (subsonic frequency*). A frequency lying below the audio-frequency range. *Notes:* (1) The word infrasonic may be used as a modifier to indicate a device or system intended to operate at infrasonic frequencies. (2) The term subsonic was once used in acoustics synonymously with infrasonic, such usage is now deprecated.   176

* Deprecated

inherent acceleration, by generator field (industrial control). The effect of accelerating the drive by the natural build-up of generator voltage without field forcing or intentional delay. *See:* **electric drive.**   225, 206

inherent harmonics (electrical conversion). Harmonics that are a by-product of the output of an inverter or frequency changer. *Note:* In most cases, these harmonics are the direct result of the methods of inversion using square-wave, quasisquare-wave, or semisinusoidal outputs. Harmonics may also be a by-product of various methods of regulation such as clipping and phase control. *See also:* **electrical conversion.**   186

inherent regulation (1) (rotating machinery). (A) (motor). The change in speed resulting from a load change, the supply voltage and frequency being maintained constant, and due solely to the fundamental characteristics of the motor itself. *See also:* **asynchronous machine; direct-current commutating machine.**   63

(B) (generator). The change in voltage resulting from a load change, the speed being maintained constant, and due solely to the fundamental characteristics of the generator itself.   63

(2) (automatic control). That property of a process or machine which permits attainment of equilibrium after a disturbance, without the intervention of a controller. *Syn.:* **self regulation.**   56

inherent restriking voltage (circuit transient-recovery voltage). The restriking voltage that is associated with a particular circuit and determined by the circuit parameters alone, its form being unmodified by the characteristics of the arrester. *Notes:* (1) It is expressed in terms of amplitude factor and rate-of-rise (or natural frequency). (2) It would be obtained if, in the particular circuit, a sinusoidal symmetrical current were broken by an ideal expulsion-type arrester (no arc voltage and no post-arc conductivity) at a stated recovery voltage.   308, 62

inherent voltage regulation (power rectifier). The voltage regulation without the use of regulating equipment, or other compensating means, and with no phase control. *See also:* **power rectifier; rectification; voltage regulation.**
                                                            328

inherited error. The error in the value of quantities that serve as the initial conditions at the beginning of a step in a step-by-step calculation.   255

inhibit. (1) To prevent an event from taking place. (2) To prevent a device or logic element from producing a specified output.   235

inhibited oil. Mineral transformer oil to which a synthetic oxidation inhibitor has been added. *See:* **oil-immersed transformer.**   203, 91, 53

inhibiting input (electronic computation). A gate that, if in its prescribed state, prevents any output that might otherwise occur.   210

inhibiting signal (test, measurement and diagnostic equipment). A signal derived from the unit under test that delays the test.   54

inhibitor (electroplating). A substance added to a pickle for the purpose of reducing the rate of solution of the metal during the removal from it of oxides or other compounds. *See also:* **electroplating.**   328

inhibit pulse. A drive pulse that tends to prevent flux reversal of a magnetic cell by certain specified drive pulses.   331

inhomogeneous line-broadening (laser-maser). An increase of the width of an **absorption** or **emission** line, beyond the natural linewidth, produced by a disturbance (for example, strain, imperfections, etcetera) which is not the same for all of the source emitters.   363

initial alternating short-circuit current (rotating machinery). The root-mean-square value of the current in the armature winding immediately after the winding has been suddenly short-circuited, the aperiodic component of current, if any, being neglected. *See:* **armature; direct-axis synchronous reactance.**   63

initial condition (analog computer). The value of a variable at the start of computation. A more restricted definition refers solely to the initial value of an integrator. *Note:* Also used as a synonym for the computer control state **"reset."** *See also:* **reset.**   9

initial conditions (inertial navigation). The values of position, velocity, level, azimuth, gyro bias, and accelerometer bias imposed on the system before departure. *See also:* **navigation.**   187, 13

initial conductor tension (power lines). The longitudinal tension in a conductor prior to the application of any external load. *See also:* **conductor.**   262

initial contact pressure. *See:* **contact pressure, initial.**

initial differential capacitance (nonlinear capacitor). Differential capacitance at zero capacitor voltage. *See also:* **nonlinear capacitor.**   191

initial element (sensing element). *See:* **primary detector.**

initial erection (gyro). The mode of operation of a vertical gyro in which the gyro is being erected or slaved initially. The initial erection rate is usually relatively high.   46

initial excitation response (rotating machinery). The initial rate of increase in the excitation voltage when a sudden transition is made from the voltage at the rated conditions of the main machine to the conditions that enable the excitation ceiling voltage to be attained in the shortest possible time. *Note:* This rate can also be expressed by its relative value in relation to the excitation voltage for the rated conditions of the main machine. *See also:* **asynchronous machine; direct-current commutating machine.**   63

initial inverse voltage (rectifier tube). The peak inverse anode voltage immediately following the conducting period. *See also:* **electrode voltage (electron tube); rectification.**   190

initial ionizing event (gas-filled radiation-counter tube). An ionizing event that initiates a tube count. *See also:* **counter tubes; gas-filled radiation.**   190, 96, 125

initialize (1) (computing systems). To set counters, switches, and addresses to zero or other starting values at the beginning of or at prescribed points in a computer routine.   255, 77

(2) (test, measurement and diagnostic equipment). (A) To establish an initial condition or starting state; for example to set logic elements in a digital circuit or the contents of a storage location to a known state so that subsequent application of digital test patterns will drive the logic elements to another known state; and (B) To set counters, switches, and addresses to zero or other starting values at the beginning of, or at prescribed points in, a computer routine. *Syn.* **prestore.**   54

initial luminous exitance. The density of luminous flux leaving a surface within an enclosure before interreflections occur. *Note:* For light sources this is the luminous exitance as defined in luminous flux density (at a surface). For non-selfluminous surfaces it is reflected luminous exitance of the flux received directly from sources within the enclosure or from daylight. *See also:* **inverse-square law (illuminating engineering).**   167

initial output test (storage battery). A service test begun shortly after the manufacture of the battery. (Usually less than one month.) *See also:* **battery (primary or secondary).**   328

initial permeability (magnetic core testing). Under stated conditions, the limiting value of permeability of the core at the origin of the curve of first magnetization.

$$\mu_i = \frac{1}{\mu_0} \operatorname*{Lim}_{H \to 0.} \frac{B}{H}$$

When B and H are in cgs units, $\mu_0 = 1$ μ_i = relative initial permeability   165

initial reverse voltage (semiconductor rectifier) (rectifier circuit). The instantaneous value of the reverse voltage that occurs across a rectifier circuit element immediately following the conducting period and including the first peak of oscillation. *See also:* **rectification.**
                                                        237, 66, 208

initial reversible capacitance (nonlinear capacitor). Reversible capacitance at a constant bias voltage of zero. *See also:* **nonlinear capacitor.**   191

initial test temperature (storage battery). The average temperature of the electrolyte in all cells at the beginning of discharge. *Note:* The standard reference temperature is 25 degree Celsius (77 degrees Fahrenheit) initially, without limit on the final temperature, but the ambient temperature on discharge shall be 5 degrees Celsius to 8 degrees Celsius lower than the temperature of the electrolyte at the beginning of the discharge and shall be kept constant during discharge. *See:* **critical temperature (storage cell); temperature coefficient of electromotive force (storage cell); temperature coefficient of capacity (storage cell).**   328

initial unloaded sag (transmission and distribution). The sag of a conductor prior to the application of any external load. *See:* **sag.**   262

initial value (industrial control). The value of the time response of a system or element at the time a stimulus is applied. *See:* **control system, feedback.**   219, 206

initial voltage (battery). The closed-circuit voltage at the beginning of a discharge. *Note:* It is usually taken after current has been flowing for a sufficient period of time for the rate of change of voltage to become practically constant. *See also:* **battery (primary or secondary).** 328

initiating relay. A programming relay whose function is to constrain the action of dependent relays until after it has operated. 103, 202, 60, 127

injury potential (electrobiology) (demarcation potential) (demarcation current*) (current of injury*). The difference in potential observed between injured and uninjured parts of a living structure such as a muscle or nerve. *See also:* **contact potential.** 192

* Deprecated

ink. *See:* **magnetic ink.**

ink-vapor recording (facsimile). That type of recording in which vaporized ink particles are directly deposited upon the record sheet. *See also:* **recording (facsimile).** 12

inkwell influence. The difference between the recorded values, at full-chart deflection, when the inkwell is full and when it is empty. *Note:* It is to be expressed as a percentage of the full-scale value. *See also:* **accuracy rating of an instrument.** 294

in line (nuclear power generating stations). A system where the detector assembly is adjacent to or immersed in the total effluent stream. 31

inner frame (rotating machinery). The portion of a frame in which the core and stator windings are mounted, which can be inserted and removed from an outer frame as a unit without disturbing the mounting on the foundation. 63

inner marker (electronic navigation). *See:* **boundary marker.**

in-phase rejection. *See:* **common-mode rejection.**

in-phase signal. *See:* **common-mode signal.**

in-phase spring rate (tuned rotor gyro). The residual difference, in a tuned rotor gyro, between the dynamically induced spring rate and the flexure spring rate. 46

input (1) (general). (A) The current, voltage, power, or driving force applied to a circuit or device. (B) The terminals or other places where current, voltage, power, or driving force may be applied to a circuit or device. 59, 197

(2) (electronic digital computer). (A) The data to be processed. (B) The state or sequence of states occurring on a specified input channel. (C) The device or collective set of devices used for bringing data into another device. (D) A channel for impressing a state on a device or logic element. (E) The process of transferring data from an external storage to an internal storage. *See:* **manual input.** 255, 77, 54

(3) (rotating machinery). (A) **(for a generator)** The mechanical power transmitted to its shaft. (B) **(for a motor).** The power (active, reactive, or apparent) supplied to its terminals. *See:* **asynchronous machine.** 63

(4) (relay). A physical quantity or quantities to which the relay is designed to respond. *Note:* A physical quantity that is not directly related to the prescribed response of a relay, though necessary to or in some way affecting the relay operation, is not considered part of input. Time is not considered a relay input, but is a factor in performance. 103, 202, 60, 127

input admittance (circuits and systems). The admittance presented by the input terminals (port) of a transducer or network to a source. When the output terminals are short-circuited this admittance becomes the four-terminal network parameter y_{11} (short-circuit input admittance). *See also:* **admittance short-circuit input.** 67

input angle (gyro). The angular displacement of the case about an input axis. 46

input area. *See:* **block, input.**

input axis (IA) (1) (accelerometer). The axis along or about which an input causes a maximum output.

(2) (gyro). The axis about which a rotation of the case causes a maximum output. For a conventional gyro, the input axis is normal to the spin axis. For a laser gyro, the input axis is perpendicular to the plane established by the laser beams. 46

input axis misalignment (gyro; accelerometer). The angle between an input axis and its associated input reference axis when the device is at electrical null. 46

input capacitance (*n*-terminal electron tubes). *See:* **capacitance, input.** *See also:* **electron-tube admittances.**

input check (sequential events recording systems). A system diagnostic which determines the functional status of the input interface buffers and all system logic through the output. Output may be arranged to indicate either the defective or the operational circuits. *See:* **diagnostic.** 48, 58

input circuit (electronic valve or tube). The external circuit connected to the input electrode and in which the control voltage appears. *See also:* **ON period (electron tubes).** 244, 190

input current circuits (surge withstand capability tests). An input circuit, to which is applied either a voltage or current which is a measure of primary current. 90

input, dark-current, equivalent (phototubes). *See:* **equivalent dark-current input.**

input-dependent overshoot and undershoot (electric conversion). Dynamic regulation for input changes. *See also:* **electric conversion equipment.** 186

input drive power (acousto-optic device). The real portion of the complex input power delivered to the acousto-optic device, thus

$$P = R_e(V^2/Z).$$

 82

input, driving-point or sending-end impedance. The impedance obtained when the response is determined at the point at which the driving force is applied. *Notes:* (1) **Point** is used for generality. In the case of any electric circuit it would be translated **pair of terminals.** (2) The definition does not imply that a driving force has to exist at the point. A hypothetical driving force may be assumed. *See also:* **network analysis.** 210

input electrode (electron tubes). The electrode to which is applied the voltage to be amplified, modulated, detected, etcetera. *See also:* **electron tube.** 244, 190

input, floating (oscilloscopes). Circuits at the input of a differential deflection amplifier that provide rejection, with minimum distortion, of common-mode signals. *Note:* Such signals must be defined for limits of amplitude, frequency content, and impedance balance to ground. 106

input gap. An interaction gap used to initiate a variation in an electron stream. *See:* **gas tubes.** 125

input impedance (1) (antenna). The impedance presented by an antenna at its terminals. *See also:* **antenna.** 246, 179

(2) (transducer device or network). The impedance presented by the transducer device or network to a source. *See also:* **output; self-impedance.** 239, 9, 77, 185, 59

(3) (electron device). The input electrode impedance at the input electrodes. 190

(4) (transmission-line port). The impedance at the transverse plane represented by the port. *Note:* This impedance is independent of the generator impedance. 179

(5) (analog computer). A passive network connected between the input terminal or terminals of an operational amplifier and its summing junction. *See also:* **electronic analog computer.** 9, 77

(6) (transmission line). The impedance between the input terminals with the generator disconnected. *See also:* **self-impedance; waveguide.** 267

(7) (Hall generator). The impedance between the control current terminals of the Hall generator. 107

(8) (acousto-optic device). The complex driving point impedance. It is a function of operating frequency. 82

(9) (transformer-rectifier system). Internal impedance in ohms at rated frequency, when the rectifier is supplying rated direct current into a short circuit. *Note:* It may be measured at the primary terminals by taking the ratio of the primary voltage and the primary current under the conditions stated above. 95

(10) (nuclear power generating station). The internal impedance presented by a module at its input terminals to the source. 355

input limiter. *See:* **limiter circuit.** *See also:* **electronic analog computer.**

input limits (gyro; accelerometer). The extreme values of the input, generally plus or minus, within which performance is of the specified quality. *See:* **input-output characteristics.** 46

input magazine. *See:* **card hopper.**

input noise temperature, effective (signal transmission system) (1) (two-port network or amplifier). The input termination (source) noise temperature that, when the input termination is connected to a noise-free equivalent of the network or amplifier, would result in the same output noise power as that of the actual network or amplifier connected to a noise-free termination (source). *Note:* In terms of the noise factor of the network or amplifier, the effective input noise temperature T_e is

$$T_e = 290 \, (F - 1)$$

where F is the noise factor and 290 kelvins is the standard noise temperature. *See also:* **signal.** 188

(2) (linear multiport transducer). An effective temperature (in kelvins) defined as the noise temperature at a given port, designated as the output port in a linear multiport transducer, divided by the sum of the exchangeable power gains from each of the remaining ports to the designated output port, when these remaining ports, designated as input ports, are connected to specified noise-free impedances having real parts of the same sign. *Notes:* (A) If G_j is the exchangeable power gain from the jth input port to the designated output port, and T_o is the noise temperature at the output port, then the effective input noise temperature T_e is therefore given by the relation

$$T_e = T_o \left/ \sum_j G_j \right.$$

(B) When the real parts of the input termination impedances do not have the same sign, the magnitude of T_e is given by the relation

$$|T_e| = |T_o| \left/ \left| \sum_j G_j \right| \right.$$

(C) If the output place of access has other ports that are coupled either by mode conversion or by frequency conversion to the designated output port, the noise contributed at the selected output frequency by the terminations of these ports is to be included in computing the noise temperature T_o of the output port. The corresponding conversion gains are not to be included in the summation of gains from the input ports to the output port. (D) The effective input noise temperature is a function of the terminating impedances of all ports except the designated output port. *See also:* **transducer; waveguide.** 190

input-output analysis (test, measurement and diagnostic equipment). A quantitative study of the interdependence of system inputs and outputs. 54

input-output characteristic (1) (telephone). (A) (transmit characteristic). The electrical output level of a telephone set as a function of the acoustic input level. The output is measured across a specified impedance connected to the telephone feed circuit, and the input is measured in free field at a specified location relative to the reference point of an artificial mouth.

(B) (receive characteristic). The acoustic output level of a telephone set as a function of the electrical input level. The output is measured in an artificial ear, and the input signal is obtained from an available constant-power source of specified impedance.

(C) (air-to-air characteristic). The acoustic output level of a telephone set as a function of the acoustic input level of another telephone set to which it is connected. The output is measured in an artificial ear, and the input is measured in free field at a specified location relative to the reference point of an artificial mouth. 122

(2) (gyro; accelerometer). The accompanying diagram shows the relationship between the input-output characteristics of an accelerometer or gyro. 46

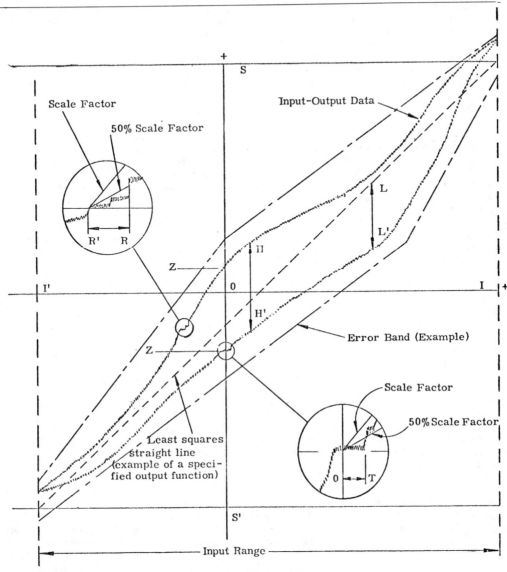

Input-output characteristics.

input-output equipment (data transmission). Any subscriber (user) equipment that introduces data into or extracts data from a data communications system.   194, 54

input-output table. A plotting device used to generate or to record one variable as a function of another variable. *See also:* **electronic analog computer.**   9

input pulse shape (high- and low-power pulse transformers). Current pulse or source voltage pulse applied through associated impedance. The shape of the input pulse is described by a current- or voltage-time relationship and is defined with the aid of Fig. 1. *Note:* (A) designates a general amplitude quantity which may be current (*I*) or voltage (*V*).   32, 33

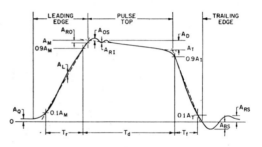

Fig. 1.   Input pulse shape.

input range (gyro; accelerometer). The region between the input limits within which a quantity is measured, expressed by stating the lower-and-upper-range values. For example, an angular displacement input range of −5° to +6°. *See:* **input-output characteristics.** 46

input rate (gyro). The angular displacement per unit time of the case about an input axis. 46

input reference axis (IRA) (gyro; accelerometer). The direction of an axis as defined by the case mounting surfaces and/or external case markings. It is nominally parallel to an input axis. 46

input resonator buncher (electron tubes). A resonant cavity, excited by an external source, that produces velocity modulation of the electron beam. *See also:* **velocity-modulated tube.** 244, 190

input signal (speed governing of hydraulic turbines). A control signal injected at any point into a control system. 8

input span (gyro; accelerometer). The algebraic difference between the upper and lower values of the input range. *See:* **input-output characteristics.** 46

input transient energy (inverters). The product of the average power in the transient and the time the transient exists. *See:* **self-commutated inverters.** 208

input voltage circuit (surge withstand capability tests). An input circuit to which is applied either a voltage or a current which is a measure of primary voltage. 90

input winding(s) (primary winding(s)). The winding(s) to which the input is applied. 197

inrush current (solenoid) (coil) (1) (packaging machinery). The steady-state current taken from the line with the armature blocked in the rated maximum open position. 124

(2) (transformer). The maximum root-mean-square or average current value, determined for a specified interval, resulting from the excitation of the transformer with no connected load, and with essentially zero source impedance, and using the minimum primary turns tap available and its rated voltage. 95

insensitivity. *See:* **dead band.**

insert earphones. Small earphones that fit partially inside the ear. *See also:* **loudspeaker.** 176

insertion (series capacitor). The opening of the capacitor bypass switch to place the series capacitor in service with or without load current flowing. 86

insertion current (series capacitor). The rms current diverted through the series capacitor upon interruption of the bypass current by the opening of the bypass device, except as described in the definition of "reinsertion." 86

insertion gain (1) (general). Resulting from the insertion of a transducer in a transmission system, the ratio of (1) the power delivered to that part of the system following the transducer to (2) the power delivered to that same part before insertion. *Notes:* (A) If the input and/or output power consist of more than one component, such as multifrequency signal or noise, then the particular components used and their weighting must be specified. (B) This gain is usually expressed in decibels. (C) The insertion of a transducer includes bridging of an impedance across the transmission system. 252, 210, 239, 197, 59

(2) (electron tube) (two-port linear transducer). At a specified frequency, the ratio of (A) the actual signal power transferred from the output port of the transducer to its load, to (B) the signal power which the same load would receive if driven directly by the source. 125

insertion loss (1). Resulting from the insertion of a transducer in a transmission system, the ratio of (A) the power delivered to that part of the system following the transducer, before insertion of the transducer, to (B) the power delivered to that same part of the system after insertion of the transducer. *Notes:* (1) If the input and/or output power consist of more than one component, such as multifrequency signal or noise, then the particular components used and their weighting must be specified. (2) This loss is usually expressed in decibels. (3) The insertion of a transducer includes bridging of an impedance across the transmission system. *See also:* **transducer; transmission loss.** 267, 239, 252, 210, 179, 197, 199, 54

(2) (waveguide component). The change in load power, due to the insertion of a waveguide or transmission-line component at some point in a transmission system, where the specified input and output waveguides connected to the component are reflectionless looking in both directions from the component (match-terminated). This change in load power is expressed as a ratio, usually in decibels, of (A) the power received at the load before insertion of the waveguide or transmission-line component to (B) the power received at the load after insertion. *Notes:* (1) A more general definition of insertion loss does not specify match-terminated connecting waveguides, in which case the insertion loss would vary with the load and generator impedances. (2) When the input and output waveguides connected to the component are not alike or do not operate in the same mode, the change in load power is determined relative to an ideal reflectionless and lossless transition between the input and output waveguides. *See also:* **transmission loss; waveguide.** 179

(3) (microwave gas tubes). The decrease in power measured in a matched termination when the unfired tube, at a specified ignitor current, is inserted in the waveguide between a matched generator and the termination. *See also:* **gas tubes.** 190, 125

(4) (A) (pulse delay line). The ratio of the input pulse power to the output power of the main pulse expressed in decibels. (B) (CW delay line). The ratio of input power to total output power, normally expressed in decibels. *Note:* Both the source impedance and the load impedance must be specified. 81

insertion loss ripple (dispersive and nondispersive delay lines). The peak-to-peak variation of the insertion loss, that is the difference between the maximum to minimum insertion loss, over a specified frequency range of the device. 81

insertion phase shift (1) (electric structure). The change in phase caused by the insertion of the structure in a transmission system. *Note:* The usual convention is that shunt capacitance or series inductance produces a positive phase shift. 179

(2) (waveguides). The change in phase of a field quantity or voltage or current, at a specified load port or terminal surface, caused by the insertion of a network at some point in a transmission system. *Note:* The most general definition of insertion phase shift does not specify match-terminated connecting waveguides, in which case the insertion phase shift would vary with the load and generator impedances. *See also:* **measurement system.**   293

insertion/removal life (attenuator). Connect/disconnect cycles due to complete insertion/removal cycle or two-port with complete axial engagement/disengagement without side thrust. *Notes:* (1) All electrical and mechanical specifications must be complied with after specified cycle life. (2) For fixed attenuator repeatability of characteristic insertion loss and for variable attenuator, repeatability of residual characteristic insertion loss must be specified.
110

insertion voltage (series capacitor). The rms voltage appearing across the series capacitor upon the interruption of the bypass current with the opening of the bypass-switch device.   86

insertion voltage gain (circuits and systems). The complex ratio of the alternating component of voltage across the load connected to the output terminals of a system when the transducer (network) is inserted between the source and load, to the output voltage when the source is connected directly to the load.   67

inside air temperature. *See:* **average inside air temperature.**

inside plant (communication practice). That part of the plant within a central office, intermediate station, or subscriber's premises that is on the office or station side of the point of connection with the outside plant. *Note:* The plant in a central office is commonly referred to as central-office plant and that on station premises as station plant. *See also:* **communication.**   328

inside top air temperature. The temperature of the air inside a dry-type transformer enclosure, measured in the space above the core and coils.   53

inspection (watt-hour meters). An observation made to obtain an approximate idea of the condition of the meter. *Note:* In such cases an examination is made of the meter and the conditions surrounding it, for the purpose of discovering defects or conditions that are likely to be detrimental to its accuracy. Such an examination may or may not include an approximate determination of the percentage registration of the meter. *See also:* **service test (field test).**   328

inspection hole (manhole). *See:* **inspection opening.**

inspection opening (inspection hole) (manhole). A port to permit observation by virtue of a transparent or removable cover.   63

installation (1) (industrial control). An assemblage of electric equipment in a given location, designed for coordinated operation, and properly erected and wired.
206

(2) (elevators). A complete elevator, dumbwaiter, or escalator, including its hoistway, hoistway enclosures and related construction, and all machinery and equipment necessary for its operation. *See also:* **elevators.**   328

installation test (electric power devices). A test made upon the consumer's premises within a limited period of time after installation. *See also:* **service test (field test).**   328

installed life (1) (nuclear power generating stations) (electric penetration assemblies). The interval of time from installation to permanent removal from service, during which the electric penetration assembly will meet all design requirements for the specified service conditions.

Note: Components of the assembly may require periodic replacement; thus, the installed life of such components would be less than the installed life of the assembly. *See:* **electric penetration assembly.**   26

(2) (class 1E equipment). The interval from installation to removal, during which the equipment or component thereof may be subject to design service conditions and system or process demands. *Notes:* (A) Equipment may have an installed life of 40 years with certain components changed periodically; thus, the installed life of the components would be less than 40 years. (B) A valve operator may have an installed life of 40 years with certain components (seals and lubricants) changed periodically; thus, the installed life of the components would be less than 40 years.   31, 120, 141

(3) (single-failure criterion to nuclear power generating station protection systems). The time interval for which an equipment or component thereof will be installed.
109

instant (pulse terms). Unless otherwise specified, a time specified with respect to the first datum time, t_0, of a waveform epoch.   254

instantaneous (relay). A qualifying term applied to a relay or other device indicating that no delay is purposely introduced in its action. *See:* **relay.**   103, 202, 60, 127

instantaneous automatic gain control (IAGC). (1) (general). That portion of a system that automatically adjusts the gain of an amplifier for each pulse so as to obtain a substantially constant output pulse peak amplitude with varying input pulse peak amplitudes, the adjustment being sufficiently fast to operate during the time a pulse is passing through the amplifier.

(2) (radar). A quick-acting automatic gain control that responds to variations of mean clutter level, or jamming over different range or angular regions, avoiding receiver saturation.   13

instantaneous burst magnitude (audio and electroacoustics). The absolute value of the instantaneous voltage, current or power for a burst-like excursion measured at a specific time. *Note:* By absolute value is meant the numerical value regardless of sign. See the figure under **burst duration.** *See also:* **burst (audio and electroacoustics).**   253, 176

instantaneous companding (modulation systems). Companding in which the effective gain is determined by the instantaneous value of the signal wave.   242

instantaneous failure rate (hazard) (particular time) (reliability). The rate of change of the number of items that have failed divided by the number of items surviving. *See also:* **reliability.**   182

instantaneous frequency (modulation systems). The time rate of change of the angle of an angle-modulated wave. *Note:* If the angle is measured in radians, the frequency in hertz is the time rate of change of the angle divided by 2π. *See also:* **signal wave.**   242, 111

instantaneous value. The value of a variable quantity at a given instant.   244, 59

instant-starting system (flourescent lamps). The term applied to a system in which an electric discharge lamp is started by the application to the lamp of a voltage sufficiently high to eject electrons from the electrodes by field emission, initiate electron flow through the lamp, ionize the gases, and start a discharge through the lamp without previous heating of the electrodes. 268

instruction (electronic computation). A statement that specifies an operation and the values or locations of its operands. *Notes:* (1) The instruction may specify some operands by the definition of the operation and may use one or more addresses to specify the location in storage of other operands, where the result is to be stored, the next instruction, etcetera. (2) In this context, the term instruction is preferable to the terms command or order, which are sometimes used synonymously; command should be reserved for electronic signals and order should be reserved for sequence, interpolation, and related usage.
235, 255, 77, 54

instruction code (electronic computation). An artificial language for describing or expressing the instructions that can be carried out by a digital computer. *Note:* In automatically sequenced computers, the instruction code is used when describing or expressing sequences of instructions, and each instruction word usually contains a part specifying the operation to be performed and one or more addresses that identify a particular location in storage. Sometimes an address part of an instruction is not intended to specify a location in storage but is used for some other purpose. If more than one address is used, the code is called a multiple-address code. In a typical instruction of a four-address code the addresses specify the location of two operands, the destination of the result, and the location of the next instruction in the sequence. In a typical three-address code, the fourth address specifying the location of the next instruction is dispensed with and the instructions are taken from storage in a preassigned order. In a typical one-address or single-address code, the address may specify either the location of an operand to be taken from storage, the destination of a previously prepared result, or the location of the next instruction. The arithmetic element usually contains at least two storage locations, one of which is an accumulator. For example, operations requiring two operands may obtain one operand from the main storage and the other from a storage location in the arithmetic element that is specified by the operation part. *See also:* **operation code.** 210

instruction counter (computing systems). A counter that indicates the location of the next computer instruction to be interpreted. 255, 77

instruction register (computing systems). A register that stores an instruction for execution. 255, 77

instruction repertory. The set of operations that can be represented in a given operation code. 235, 255, 77

instrument. A device for measuring the value of the quantity under observation. *Note:* An instrument may be an indicating instrument or a recording instrument. The term instrument is used in two different senses: (1) instrument proper consisting of the mechanism and the parts built into the case or made a corporate part thereof; and (2) the instrument proper together with any necessary auxiliary devices, such as shunts, shunt leads, resistors, reactors, capacitors, or instrument transformers. The term meter is also used in a general sense to designate any type of measuring device, including all types of electric measuring instruments. Such use as a suffix or as part of a compound word (for example, voltmeter, frequency meter) is universally accepted. Meter may be used alone with this wider meaning when the context is such as to prevent confusion with the narrower meaning of **electricity meter.** *See:* **meter.** 328

instrumental error (navigation) (radar). Error due to the inaccuracies introduced in any portion of the system by the mechanism that translates path-length differences into navigation coordinate information, including calibration errors and errors resulting from limited sensitivity of the indicating instruments. *See also:* **navigation.** 187, 13

instrument approach (electronic navigation).
The process of making an approach to a landing by the use of navigation instruments without dependence upon direct visual reference to the terrain. *See also:* **navigation.**
187, 13

instrument approach system (electronic navigation). A system furnishing guidance in the vertical and horizontal planes to aircraft during descent from an initial approach altitude to a point near the landing area. *See also:* **navigation.** 187, 13

instrumentation (test, measurement and diagnostic equipment). All those devices (chemical, electrical, hydraulic, magnetic, mechanical, optical, pneumatic) utilized to test, observe, measure, monitor, alter, generate, record, calibrate, manage, or control physical properties, movements or other characteristics. 54

instrument bulb. The sensing portion of certain types of instrument. *See:* **measurement system.** 63

instrument landing system (1) (general). A generic term for a system which provides the necessary lateral, longitudinal and vertical guidance in an aircraft for a low approach or landing. *See also:* **ILS.** 13
(2) (ILS). An internationally adopted instrument landing system for aircraft, consisting of a very-high frequency localizer, an ultra-high-frequency glide slope, and 75-megahertz markers. *See also:* **instrument landing system reference point.** 187

instrument landing system marker beacon (electronic navigation). *See:* **boundary marker.** *See also:* **navigation.**

instrument landing system reference point (electronic navigation). A point on the centerline of the instrument landing system runway designated as the optimum point of contact for landing; in standards of the International Civil Aviation Organization this point is from 500 to 1000 feet from the approach end of the runway. *See also:* **navigation.** 187

instrument multiplier. A particular type of series resistor that is used to extend the voltage range beyond some particular value for which the instrument is already complete. *See also:* **auxiliary device to an instrument; voltage-range multiplier (recording instrument).** 280

instrument relay. A relay whose operation depends upon principles employed in measuring instruments such as the electrodynamometer, iron vane, D'Arsonval galvanometer, and moving magnet. *See also:* **relay.** 259

instrument shunt. A particular type of resistor designed to be connected in parallel with a circuit of an instrument to extend its current range. *Note:* The shunt may be internal or external to the instrument proper. *See also:* **auxiliary device to an instrument; instrument.** 280, 201

instrument switch. A switch used to connect or disconnect an instrument or to transfer it from one circuit or phase to another. 103, 202

instrument terminals of shunts (electric power systems). Those terminals that provide a voltage drop proportional to the current in the shunt and to which the instrument or other measuring device is connected.   201

instrument transformer. A transformer which is intended to reproduce in its secondary circuit, in a definite and known proportion, the current or voltage of its primary circuit, with its phase relations substantially preserved. *See:* **continuous-thermal-current rating factor; transformer correction factor; true ratio; marked ratio; ratio correction factor; percent ratio; percent ratio correction; phase angle; phase angle correction factor; polarity; secondary winding; excitation losses; voltage transformer; cascade-type voltage transformer; grounded-neutral terminal type voltage transformer; insulated-neutral terminal type voltage transformer; double-secondary voltage transformer; fused-type voltage transformer; turn ratio of a voltage transformer; thermal burden rating (voltage transformer); rated voltage (voltage transformer); rated secondary voltage; current transformer; bushing-type current transformer; double-secondary current transformer; multiple-secondary current transformer; multi-ratio current transformer; window type current transformer; wound-type current transformer; three-wire type current transformer; rated current; rated secondary current; turn ratio (current transformer).**   53, 212, 203

instrument-transformer correction factor (watt meter or watthour meter). *See:* **transformer correction factor.**

instrument transformer, dry-type. *See:* **dry-type.**

instrument transformer, liquid-immersed. *See:* **liquid-immersed.**

instrument transformer, low-voltage winding. Winding that is intended to be connected to the measuring or control devices.   203

insulated (transmission and distribution). Separated from other conducting surfaces by a dielectric (including air space) offering a high resistance to the passage of current. *Note:* When any object is said to be insulated, it is understood to be insulated in a suitable manner for the conditions to which it is subjected. Otherwise, it is, within the purpose of these rules, uninsulated.   262

insulated bearing (rotating machinery). A bearing that is insulated to prevent the circulation of stray currents. *See:* **bearing.**   63

insulated bearing housing (rotating machinery). A bearing housing that is electrically insulated from its supporting structure to prevent the circulation of stray currents. *See also:* **bearing.**   63

insulated bearing pedestal (rotating machinery). A bearing pedestal that is electrically insulated from its supporting structure to prevent the circulation of stray currents. *See also:* **bearing.**   63

insulated bolt. A bolt provided with insulation.   328

insulated conductor (transmission and distribution). A conductor covered with a dielectric having a rated insulating strength equal to or greater than the voltage of the circuit in which it is used   262

insulated coupling (rotating machinery). A coupling whose halves are insulated from each other to prevent the circulation of stray current between shafts. *See:* **rotor (rotating machinery).**   63

insulated flange (piping). Element of a flange-type coupling,

insulated to interrupt the electrically conducting path normally provided by metallic piping. *See:* **rotor (rotating machinery).**   63

insulated joint (1) (conduit). A coupling or joint used to insulate adjacent pieces of conduits, pipes, rods, or bars.   1

(2) (cable). A device that mechanically couples and electrically insulates the sheath and armor of contiguous lengths of cable. *See also:* **tower.**   64

insulated rail joint. A joint used to insulate abutting rail ends electrically from one another.   328

insulated splice (power cable joints). A splice with a dielectric medium applied over the connected conductors and adjacent cable insulation. (1) Non-shield Insulated Splice. An insulated splice in which no conducting material is employed over the insulation for electric stress control. (2) Partially Shielded Insulated Splice. An insulated splice in which a conducting material is employed over a portion of the insulation for electric stress control. (3) Shielded Insulated Splice. An insulated splice in which a conducting material is employed over the full length of the insulation for electric stress control.   34

insulated static wire. An insulated conductor on a power transmission line whose primary function is protection of the transmission line from lightning and one of whose secondary function is communications.   59

insulated supply system. *See:* **ungrounded system.**

insulated turnbuckle. An insulated turnbuckle is one so constructed as to constitute an insulator as well as a turnbuckle. *See also:* **tower.**   64

insulating (covering of a conductor, or clothing, guards, rods, and other safety devices). A device that, when interposed between a person and current-carrying parts, protects the person making use of it against electric shock from the current-carrying parts with which the device is intended to be used; the opposite of conducting. *See also:* **insulated; insulation.**   262

insulating cell (rotating machinery). An insulating liner, usually to separate a coil-side from the grounded surface at a slot. *See also:* **rotor (rotating machinery); stator.**   63

insulating (isolating) joint (power cable joints). A cable joint which mechanically couples and electrically separates the sheath, shield and armor of contiguous lengths of cable.   34

insulating material (insulant) (rotating machinery). A substance or body, the conductivity of which is zero or, in practice, very small. *See also:* **asynchronous machine; direct-current commutating machine.**   244, 63

insulating materials. *See:* **insulation, class ratings.**

insulating spacer. Insulating material used to separate parts. *See:* **rotor (rotating machinery); stator.**   63

insulating transformer. A transformer used to insulate one circuit from another.   53

insulation (1) (rotating machinery) (electric system). Material or a combination of suitable nonconducting materials that provide electric isolation of two parts at different voltages.   63

(2) (cable). That part that is relied upon to insulate the conductor from other conductors or conducting parts or from ground. *See also:* **cable; insulation, class ratings.**   64

(3) (power cable joints). A material of suitable dielectric properties, capable of being field-applied, and used to provide and maintain continuity of insulation across the splice. *Note:* The material need not be identical to the cable insulation, but should be electrically and physically compatible, including factory-welded insulating components that are field-installed. 34

insulation breakdown (electrical insulation tests). A rupture of insulation that results in a substantial transient or steady increase in leakage current at the specified test voltage. 116

insulation breakdown current (electrical insulation tests). The current delivered from the test apparatus when a dielectric breakdown occurs. 116

insulation class (1) (outdoor apparatus bushings). The voltage by which the bushing is identified and which designates the level on which the electrical performance requirements are based. 316, 203, 287, 168

(2) (grounding device). A number that defines the insulation levels of the device. 91

(3) (transformer). Deprecated. *See:* **insulation level.** 53

insulation, class ratings (electric-machine-windings and electric cables) (1) (temperature endurance). These temperatures are, and have been in most cases over a long period of time, benchmarks descriptive of the various classes of insulating materials, and various accepted test procedures have been or are being developed for use in their identification. They should not be confused with the actual temperatures at which these same classes of insulating materials may be used in the various specific types of equipment, nor with the temperatures on which specified temperature rise in equipment standards are based. (1) In the following definitions the words **accepted tests** are intended to refer to recognized test procedures established for the thermal evaluation of materials by themselves or in simple combinations. Experience or test data, used in classifying insulating materials, are distinct from the experience or test data derived for the use of materials in complete insulation systems. The thermal endurance of complete systems may be determined by test procedures specified by the responsible technical committees. A material that is classified as suitable for a given temperature may be found suitable for a different temperature, either higher or lower, by an insulation system test procedure. For example, it has been found that some materials suitable for operation at one temperature in air may be suitable for a higher temperature when used in a system operated in an inert gas atmosphere. Likewise some insulating materials when operated in dielectric liquids will have lower or higher thermal endurance than in air. (2) It is important to recognize that other characteristics, in addition to thermal endurance, such as mechanical strength, moisture resistance, and corona endurance, are required in varying degrees in different applications for the successful use of insulating materials. *See also:* **insulation.**

(A) class 90 insulation. Materials or combinations of materials such as cotton, silk, and paper without impregnation. *Note:* Other materials or combinations of materials may be included in this class if by experience or accepted tests they can be shown to have comparable thermal life

at 90 degrees Celsius. 3

(B) class 105 insulation. Materials or combinations of materials such as cotton, silk, and paper when suitably impregnated or coated or when immersed in a dielectric liquid. *Note:* Other materials or combinations may be included in this class if by experience or accepted tests they can be shown to have comparable thermal life at 105 degrees Celsius.

(C) class 130 insulation. Materials or combinations of materials such as mica, glass fiber, asbestos, etcetera, with suitable bonding substances. *Note:* Other materials or combinations of materials may be included in this class if by experience or accepted tests they can be shown to have comparable thermal life at 130 degrees Celsius.

(D) class 155 insulation. Materials or combinations of materials such as mica, glass fiber, asbestos, etcetera, with suitable bonding substances. *Note:* Other materials or combinations of materials may be included in this class if by experience or accepted tests they can be shown to have comparable thermal life at 155 degrees Celsius.

(E) class 180 insulation. Materials or combinations of materials such as silicone elastomer, mica, glass fiber, asbestos, etcetera, with suitable bonding substances such as appropriate silicone resins. *Note:* Other materials or combinations of materials may be included in this class if by experience or accepted tests they can be shown to have comparable thermal life at 180 degrees Celsius.

(F) class 220 insulation. Materials or combinations of materials which by experience or accepted tests can be shown to have the required thermal life at 220 degrees Celsius.

(G) class over-220 insulation. Materials consisting entirely of mica, porcelain, glass, quartz, and similar inorganic materials. *Note:* Other materials or combinations of materials may be included in this class if by experience or accepted tests they can be shown to have the required thermal life at temperatures over 220 degrees Celsius.

(2) (letter symbols).

(A) class O insulation. *See:* **class 90 insulation.**

(B) class A insulation. (1) Cotton, silk, paper, and similar organic materials when either impregnated or immersed in a liquid dielectric. (2) Molded and laminated materials with cellulose filler, phenolic resins, and other resins of similar properties. (3) Films and sheets of cellulose acetate and other cellulose derivatives of similar properties. (4) Varnishes (enamel) as applied to conductors. *Note:* An insulation is considered to be impregnated when a suitable substance replaces the air between its fibers, even if this substance does not completely fill the spaces between the insulated conductors. The impregnating substances, in order to be considered suitable, must have good insulating properties; must entirely cover the fibers and render them adherent to each other and to the conductor; must not produce interstices within itself as a consequence of evaporation of the solvent or through any other cause; must not flow during the operation of the machine at full working load or at the temperature limit specified; and must not unduly deteriorate under prolonged action of heat.

(C) class B insulation. Mica, asbestos, glass fiber, and similar inorganic materials in built-up form with organic binding substances. *Note:* A small proportion of class A materials may be used for structural purposes only. Glass fiber or asbestos magnet-wire insulations are included in this temperature class. These may include supplementary organic materials, such as polyvinylacetal or polyamide films. The electrical and mechanical properties of the insulated winding must not be impaired by application of the temperature permitted for class B material. (The word **impaired** is here used in the sense of causing any change that could disqualify the insulating material for continuous service.) The temperature endurance of different class B insulation assemblies varies over a considerable range in accordance with the percentage of class A materials employed, and the degree of dependence placed on the organic binder for maintaining the structural integrity of the insulation.

(D) class H insulation. Insulation consisting of: (1) mica, asbestos, glass fiber, and similar inorganic materials in built-up form with binding substances composed of silicone compounds or materials with equivalent properties; (2) silicone compounds in rubbery or resinous forms or materials with equivalent properties. *Note:* A minute proportion of class A materials may be used only where essential for structural purposes during manufacture. The electrical and mechanical properties of the insulated winding must not be impaired by the application of the hottest-spot temperature permitted for the specific insulation class. The word **impaired** is here used in the sense of causing any change that could disqualify the insulating materials for continuously performing its intended function, whether creepage spacing, mechanical support, or dielectric barrier action.

(E) class C insulation. Insulation consisting entirely of mica, porcelain, glass, quartz, and similar inorganic materials. 303, 257, 309, 258, 301

(F) class F insulation. Materials or combinations of materials such as mica, glass fiber, asbestos, etc., with suitable bonding substances. Other materials or combinations of materials, not necessarily inorganic, may be included in this class if by experience or accepted tests they can be shown to be capable of operation at 155°C. 3

Editor's note: The detailed definitions above are arranged in two approximately equivalent groups: The first group defines seven classes of insulation in order of increasing ability to withstand stated limiting temperatures, while maintaining a satisfactory expectation of life. Illustrative examples of suitable materials are quoted, but the definitions are primarily temperature based, and are therefore general. The second (partially complete) group is arranged in similar sequence, but the texts include more specific reference to materials and forming processes that have already demonstrated probable ability to satisfy the several limiting-temperature specifications. In this (alphabetical) group, the editor has placed class O ahead of class A to maintain the logical physical sequence. Similarly, class C has been placed toward the end of the sequence. The reader will note the pair-relationship between class 90 and class O; class 105 and class A; class 130 and class B; etcetera. Class over-220 pairs with class C. *Note:* These two overlapping sequences of terms, each sequence descriptive of a graded series of **insulation levels,** remain currently in general use. The sequence based upon letter symbols is particularly favored in reference to the insu-

lation of electric cables. The system based upon numerical values of maximum temperature rating is particularly favored in reference to the insulation of the windings of electric machines.

insulation coordination (insulation strength). *See:* **coordination of insulation (insulation strength).**

insulation fault (surge arresters). Accidental reduction or disappearance of the insulation resistance between conductor and ground or between conductors. 244, 62

insulation, graded. The selective arrangement of the insulation components of a composite insulation system to more nearly equalize the voltage stresses throughout the insulation system. 95

insulation level (1) (insulation strength) (transformer). An insulation strength expressed in terms of a withstand voltage. 53

(2) (surge arresters). A combination of voltage values (both power-frequency and impulse) that characterize the insulation of an equipment with regard to its capability of withstanding dielectric stresses. *See also:* **basic impulse insulation level (BIL) (insulation strength).** 276, 244, 62

(3) (series capacitor). The combination of power frequency and impulse test-voltage values which characterizes the insulation of the capacitor with regard to its capability of withstanding the electric stresses between platform and earth. 86

insulation power factor (1) (dissipation factor). The ratio of the power dissipated in the insulation, in watts, to the product of the effective voltage and current in voltamperes, when tested under a sinusoidal voltage and prescribed conditions. *Note:* If the current is also sinusoidal, the insulation power factor is equal to the cosine of the phase angle between the voltage and the resulting current. 53

(2) (rotating machinery). The ratio of dielectric loss in an insulation system to the applied apparent power, when measured at power frequency under designated conditions of voltage, temperature, and humidity. *Note:* Being the sine of an angle normally small, it is practically equal to loss tangent or dissipation factor, the tangent of the same angle. The angle is the complement of the angle whose cosine is the power factor. *See:* **asynchronous machine; direct-current commutating machine.** *See also:* **loss tangent.** 287, 63

insulation resistance (aircraft, missile, and space equipment). The electrical resistance measured at specified direct-current potentials between any electrically insulated parts, such as a winding and other parts of the machine. 116

insulation resistance, direct current (1) (insulated conductor). The resistance offered by its insulation to the flow of current resulting from an impressed direct voltage. *See also:* **conductor.** 59

(2) (between two electrodes in contact with or embedded in a specimen). The ratio of the direct voltage applied to the electrodes to the total current between them. *Note:* It is dependent upon both the volume and surface resistances of the specimen. 210

(3) (rotating machinery). The quotient of a specified direct voltage maintained on an insulation system divided by the resulting current at a specified time after the application of the voltage under designated conditions of temperature, humidity, and previous charge. *Note:* If steady state has not been reached the apparent resistance will be affected by the rate of absorption by the insulation of electric charge. *See also:* **asynchronous machine; direct-current commutating machine.** 63

insulation-resistance test. A test for measuring the resistance of insulation under specified conditions. 63

insulation-resistance versus voltage test (rotating machinery). A series of insulation-resistance measurements, made at increasing direct voltages applied at successive intervals and maintained for designated periods of time, with the object of detecting insulation system defects by departures of the measured characteristic from a typical form. Usually this is a controlled overvoltage test. *See also:* **asynchronous machine; direct-current commutating machine.** 63

insulation shielding (1) (power distribution). Conducting and/or semiconducting elements applied directly over and in intimate contact with the outer surface of the insulation. Its function is to eliminate ionizable voids at the surface of the insulation and confine the dielectric stress to the underlying insulation. *See also:* **power distribution.** 57
(2) (cable systems). A non-magnetic, metallic material applied over the insulation of the conductor or conductors, to confine the electric field of the cable to the insulation of the conductor or conductors. 35

insulations, laminar (cable-insulation materials). Dielectric materials, either fibrous or film, or composite, comprising two or more layers of insulation arranged in series, and normally impregnated or flooded with an insulating liquid, or both. 97

insulation sleeving (tubing). A varnish-treated or resin-coated flexible braided tube providing insulation when placed over conductors, usually at connections or crossovers. 63

insulations, solid (cable-insulation materials). Firm, essentially homogeneous, dielectric materials comprising virtually complete solid-phase structures and having no liquid phase. 97

insulation system (1) (transformer). An assembly of insulating materials in a particular type, and sometimes size, of equipment. 53
(2) (random-wound ac electric machinery) class A. A system utilizing materials having a preferred temperature index of 105 and operating at such temperature rises above stated ambient temperature as the equipment standard specifies based on experience or accepted test data. This system may alternatively contain materials of any class, provided that experience or a recognized system test procedure for the equipment has demonstrated equivalent life expectancy. The preferred temperature classification for a Class A insulation system is 105°C. 154
class B. A system utilizing materials having a preferred temperature index of 130 and operating at such temperature rises above stated ambient temperature as the equipment standard specifies based on experience or accepted test data. This system may alternatively contain materials of any class, provided that experience or a recognized system test procedure for the equipment has demonstrated equivalent life expectancy. The preferred temperature classification for a Class B insulation system is 130°C. 154

class C. A system utilizing materials having a preferred temperature index of over 240 and operating at such temperatures above stated ambient temperatures as the equipment standard specifies based on experience or accepted test data. This system may alternatively contain materials of any class, provided that experience or a recognized test procedure for the equipment has demonstrated equivalent life expectancy. The preferred temperature classification for a Class C insulation system is over 240°C. 154
class F. A system utilizing materials having a preferred temperature index of 155 and operating at such temperature rises above stated ambient temperatures as the equipment standard specifies based on experience or accepted test data. This system may alternatively contain materials of any class, provided that experience or a recognized test procedure for the equipment has demonstrated equivalent life expectancy. The preferred temperature classification for a Class F insulation system is 155°C. 154
class H. A system utilizing materials having a preferred temperature index of 180 and operating at such temperature rises above stated ambient temperature as the equipment standard specifies based on experience or accepted test data. This system may alternatively contain materials of any class, provided that experience or a recognized test procedure for the equipment has demonstrated equivalent life expectancy. The preferred temperature classification for a Class H insulation system is 180°C. 154
class N. A system utilizing materials having a preferred temperature index of 200 and operating at such temperature rises above stated ambient temperatures as the equipment standard specifies based on experience or accepted test data. This system may alternatively contain materials of any class, provided that experience or a recognized test procedure for the equipment has demonstrated equivalent life expectancy. The preferred temperature classification for a Class N insulation system is 200°C. 154
class R. A system utilizing materials have a preferred temperature index of 220 and operating at such temperatures above stated ambient temperatures as the equipment standard specifies based on experience or accepted test data. This system may alternatively contain materials of any class, provided that experience or a recognized test procedure for the equipment has demonstrated equivalent life expectancy. The preferred temperature classification for a Class R insulation system is 220°C. 154
class S. A system utilizing materials having a preferred temperature index of 240 and operating at such temperatures above stated ambient temperatures as the equipment standard specifies based on experience or accepted test data. This system may alternatively contain materials of any class, provided that experience or a recognized test procedure for the equipment has demonstrated equivalent life expectancy. The preferred temperature classification for a Class S insulation system is 240°C. 154

insulation, temperature class ratings (transformer). These temperatures are and have been, in most cases over a long period of time, benchmarks descriptive of the various classes of insulating materials, and various accepted test procedures have been or are being developed for use in their identification. They should not be confused with the actual temperatures at which these same classes of insulating materials may be used in the various specific types of equipment, nor with the temperatures on which specified temperature rise in equipment standards are based. (1) In the following definitions the words "accepted tests" are intended to refer to recognized test procedures established for the thermal evaluation of materials by themselves or in simple combinations. Experience or test data, used in classifying insulating materials, are distinct from the experience or test data derived for the use of materials in complete insulation systems. The thermal endurance of complete systems may be determined by test procedures specified by the responsible technical committees. A material that is classified as suitable for a given temperature may be found suitable for a different temperature, either higher or lower, by an insulation system test procedure. For example, it has been found that some materials suitable for operation at one temperature in air may be suitable for a higher temperature when used in a system operated in an inert gas atmosphere. Likewise some insulating materials when operated in dielectric liquids will have lower or higher thermal endurance than in air. (2) It is important to recognize that other characteristics, in addition to thermal endurance, such as mechanical strength, moisture resistance, and corona endurance, are required in varying degrees in different applications for the successful use of insulating materials.

class 105 insulation system. Materials or combinations of materials such as cotton, silk, and paper when suitably impregnated or coated or when immersed in a dielectric liquid. *Note:* Other materials or combinations may be included in this class if by experience or accepted tests the insulation system can be shown to have comparable thermal life at 105 degrees Celsius.

class 120 insulation system. Materials or combinations of materials such as cotton, silk, and paper when suitably impregnated or coated or when immersed in a dielectric liquid; and which possess a degree of thermal stability which allows them to be operated at a temperature 15 C higher than Class 105 insulation materials. *Note:* Other materials or combinations may be included in this class if by experience or accepted tests the insulation system can be shown to have comparable thermal life at 120 degrees Celsius.

class 150 insulation system. Materials or combinations of materials such as mica, glass fiber, asbestos, etcetera, with suitable bonding substances. *Note:* Other materials or combinations of materials may be included in this class if by experience or accepted tests the insulation system can be shown to have comparable life at 150 degrees Celsius.

class 185 insulation system. Materials or combinations of materials such as silicone elastomer, mica, glass fiber, asbestos, etcetera, with suitable bonding substances such as appropriate silicone resins. *Note:* Other materials or combinations of materials may be included in this class if by experience or accepted tests the insulation system can be shown to have comparable thermal life at 185 degrees Celsius.

class 220 insulation system. Materials or combinations of materials such as silicone elastomer, mica, glass fiber, asbestos, etcetera, with suitable bonding substances such as appropriate silicone resins. *Note:* Other materials or combinations of materials may be included in this class if by experience or accepted tests, the system can be shown to have comparable thermal life at 220 degrees Celsius.

class over-220 insulation system. Materials consisting entirely of mica, porcelain, glass quartz, and similar inorganic materials. *Note:* Other materials or combinations of materials may be included in this class if by experience or accepted tests the insulation system can be shown to have the required thermal life at temperatures over 220 degrees Celsius.

class O insulation (non-preferred term).

class A insulation (non-preferred term). *See:* **class 105 insulation system.**

class B insulation (non-preferred term).

class C insulation (non-preferred term).

class F insulation (non-preferred term).

class H insulation (non-preferred term).

insulator (1) (power switchgear). A device intended to give flexible or rigid support to electrical conductors or equipment and to insulate these conductors or equipment from ground or from other conductors or equipment. An insulator comprises one or more insulating parts to which connecting devices (metal fittings) are often permanently attached. 70

(2) (transmission and distribution). Insulating material in a form designed to support a conductor physically and electrically separate it from another conductor or object. 262

insulator arcing horn. A metal part, usually shaped like a horn, placed at one or both ends of an insulator or of a string of insulators to establish an arcover path, thereby reducing or eliminating damage by arcover to the insulator or conductor or both. *See also:* **tower.** 64

insulator arcing ring. A metal part, usually circular or oval in shape, placed at one or both ends of an insulator or of a string of insulators to establish an arcover path, thereby reducing or eliminating damage by arcover to the insulator or conductor or both. *See also:* **tower.** 64

insulator arcing shield (insulator grading shield). An arcing ring so shaped and located as to improve the voltage distribution across or along the insulator or insulator string. *See also:* **tower.** 64

insulator arcover. A discharge of power current in the form of an arc, following a surface discharge over an insulator. *See also:* **tower.** 64

insulator grading shield. *See:* **insulator arcing shield.**

insulator string. Two or more suspension insulators connected in series. *See also:* **tower.** 64

insulator unit. An insulator assembled with such metal parts as may be necessary for attaching it to other insulating units or device parts. 103, 202, 118

intake opening (rotating machinery). A port for the entrance of ventilation air. 63

intake port (rotating machinery). An opening provided for the entrance of a fluid. 63

integer adjectives (pulse terms). The ordinal integers (that is, first, second, . . . nth, last) or the cardinal integers (that is, 1, 2, . . . n) may be used as adjectives to identify or distinguish between similar or identical features. The assignment of integer modifiers should be sequential as a function of time within a waveform epoch or within features thereof.   254

integral action rate (process control) (proportional plus integral control action devices). For a step input, the ratio of the initial rate of change of output due to integral control action to the change in steady-state output due to proportional control action. *Note:* Integral action rate is often expressed as the number of repeats per minute because it is equal to the number of times per minute that the proportional response to a step input is repeated by the initial integral response.   56

integral control action (electric power systems). *See:* control action, integral.

integral coupling (rotating machinery). A coupling flange that is a part of a shaft and not a separate piece. *See also:* rotor (rotating machinery).   63

integral-horsepower motor. A motor built in a frame as large as or larger than that of a motor of open construction having a continuous rating of 1 horsepower at 1700–1800 revolutions per minute. *See also:* asynchronous machine; direct-current commutating machine.   328

integral nonlinearity (percent) (semiconductor radiation detectors). The departure from linear response expressed as a percentage of the maximum rated output pulse amplitude.   23

integral-slot winding (rotating machinery). A distributed winding in which the number of slots per pole per phase is an integer and is the same for all poles. *See also:* rotor (rotating machinery); stator.   63

integral test equipment. *See:* built in test equipment (BITE); self-test.

integral time (speed governing of hydraulic turbines). The integral time, T_X, of an integrating element is the time required for the element's percent output to be equal in magnitude to the element's percent input where that input is a step function. The integral gain of an element is the reciprocal of its integral time.   8

INPUT

$\left\{ K\% \right.$ — Integral Element —

t = 0

Time

OUTPUT

$\left. \right\} K\%$

t = 0   t = T_X

Time

integral unit substation. A unit substation in which the incoming, transforming, and outgoing sections are manufactured as a single compact unit.   103, 202, 53

integrated (pulse) (pulse amplifier) (semiconductor radiation detectors). A pulse is "integrated" when it is passed through a lowpass network, such as a single RC (resistance-capacitance) network or a cascaded RC network.   23, 118, 119

integrated antenna system. A radiator with an active or nonlinear circuit element or network incorporated physi-

cally within the structure of the radiator.   111

integrated circuit (solid state). A combination of interconnected circuit elements inseparably associated on or within a continuous substrate. *Note:* To further define the nature of an integrated circuit, additional modifiers may be prefixed. Examples are: (1) dielectric-isolated monolithic integrated circuit, (2) beamlead monolithic integrated circuit, (3) silicon-chip tantalum thin-film hybrid integrated circuit. *See:* **element (integrated circuit); integrated electronics.** *See also:* **monolithic integrated circuit; multichip integrated circuit; film integrated circuit; hybrid integrated circuit.**   312, 191

integrated demand (electric power utilization). The demand averaged over a specified period usually determined by an integrating demand meter or by the integration of a load curve. *See also:* **alternating-current distribution; demand meter.**   200

integrated-demand meter (block-interval demand meter). A meter that indicates or records the demand obtained through integration. *See also:* **demand meter; electricity meter.**   212

integrated electronics. The portion of electronic art and technology in which the interdependence of material, device, circuit, and system-design consideration is especially significant; more specifically, that portion of the art dealing with integrated circuits. *See also:* **integrated circuit.**   312, 191

integrated energy curve (electric power utilization). A curve of demand versus energy showing the amount of energy represented under a load curve, or a load-duration curve, above any point of demand. Also referred to as a **peak percent curve.** *See also:* **generating station.**   200

integrated mica (reconstituted mica). *See:* **mica paper.**

integrated-numbering plan (telephone switching systems). In the world-numbering plan, arrangements for identifying telephone stations within a geographical area identified by a world-zone number which is also used as a country code. *See also:* **world zone number.**   55

integrated numbering-plan area (telephone switching systems). A geographical area of the world that is identified by a world-zone number which is also used as a country code. *See also:* **world zone number.**   55

integrated plow (static or vibratory plows) (cable plowing). A self-contained or integral plow-prime mover unit.

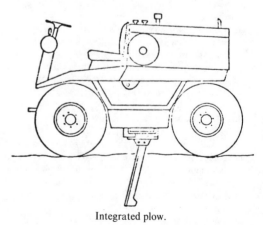

Integrated plow.

integrated radiation, L (laser-maser). The integral of the radiance over the exposure duration. *Syn:* **pulsed radiation** $(W \cdot s \cdot cm^{-2} \cdot sr^{-1})$.   363

integrating accelerometer (accelerometer). A device which produces an output that is proportional to the time integral of an input acceleration.   46

integrating amplifier. An operational amplifier that produces an output signal equal to the time integral of a weighted sum of the input signals. *Note:* In an analog computer, the term integrator is synonymous with integrating amplifier. *See also:* **electronic analog computer.**   9, 77, 10

integrating circuit (integrator) (integrating network). *See:* **integrator.**

integrating network. *See:* **integrator.**

integrating photometer. A photometer that enables total luminous flux to be determined by a single measurement. *Note:* The usual type is the Ulbricht sphere with associated photometric equipment for measuring the luminance (photometric brightness) of the inner surface of the sphere.   167

integrating preamplifier. A pulse preamplifier in which individual pulses are intentionally integrated by passive or active circuits.   118

integrating relay. A relay that operates on the energy stored from a long pulse or a series of pulses of the same or varying magnitude, for example, a thermal relay. *See:* **relay.**

integration loss (radar). The loss incurred by integrating a signal noncoherently instead of coherently.   13

integrator (1) (general). Any device producing an output proportional to the integral of one variable with respect to another, usually time. *See also:* **electronic analog computer; integrating amplifier, watthour meter.**   9

(2) (relaying). A transducer whose output wave form is substantially the time integral of its input wave form.   79

(3) (digital differential analyzer). A device using an accumulator for numerically accomplishing an approximation to the mathematical process of integration.   235

(4) (analog computer). A device producing an output proportional to the integral of one variable or of a sum of variables, with respect to another variable, usually time. *See:* **amplifier, integrating.**   9

integrator, gain (analog computer). For each input, the ratio of the input to the corresponding time rate of change of the output. For fixed input resistors, the "time constant" is determined by the integrating feedback capacitor.   9

intelligence bandwidth. The sum of the audio- (or video-) frequency bandwidths of the one or more channels.   111

intelligibility. *See:* **articulation.**

intensifier (baseline clipper) (non-real time spectrum analyzer). A means of changing the relative brightness between the signal and baseline portion of the display.   68

intensifier electrode. An electrode causing post acceleration. *See:* **post-accelerating electrode.** *See also:* **electrode (electron tube).**   125

intensity (oscilloscopes). A term used to designate brightness or luminance of the spot. *See:* **oscillograph.**   185

intensity amplifier (oscilloscopes). An amplifier for signals controlling the intensity of the spot. *See:* **oscillograph.**   185

intensity level (specific sound-energy flux level) (sound-energy flux density level) (acoustics). In decibels, of a sound is 10 times the logarithm to the base 10 of the ratio of the intensity of this sound to the reference intensity. The reference intensity shall be stated explicitly. *Note:* In discussing sound measurements made with pressure or velocity microphones, especially in enclosures involving normal modes of vibration or in sound fields containing standing waves, caution must be observed in using the terms intensity and intensity level. Under such conditions it is more desirable to use the terms pressure level or velocity level since the relationship between the intensity and the pressure or velocity is generally unknown.   176

intensity modulation (1) (general). The process, or effect, of varying the electron-beam current in a cathode-ray tube resulting in varying brightness or luminance of the trace. *See also:* **oscillograph; television.**   185

(2) (radar). A process used in certain displays whereby the luminance of the signal indication is a function of the received signal strength.   13, 187

intensity of magnetization. *See:* **magnetization.**

interaction (nuclear power generating station). A direct or indirect effect of one device or system upon another.   357

interaction-circuit phase velocity (traveling-wave tubes). The phase velocity of a wave traveling on the circuit in the absence of electron flow. *See also:* **magnetrons; electron devices, miscellaneous.**   190

interaction crosstalk coupling (between a disturbing and a disturbed circuit in any given section). The vector summation of all possible combinations of crosstalk coupling, within one arbitrary short length, between the disturbing circuit and all circuits other than the disturbed circuit (including phantom and ground-return circuits), with crosstalk coupling, within another arbitrary short length, between the disturbed circuit and all circuits other than the disturbing circuit. *See also:* **coupling.**   328

interaction factor (1) (transducer). The factor in the equation for the received current that takes into consideration the effect of multiple reflections at its terminals. *Note:* For a transducer having a transfer constant θ, image impedances Z_{I_1} and Z_{I_2}, and terminating impedances Z_S and Z_R, this factor is

$$\frac{1}{1 - \dfrac{Z_{I_2} - Z_R}{Z_{I_2} + Z_R} \times \dfrac{Z_{I_1} - Z_S}{Z_{I_1} + Z_S} \times e^{-2\theta}}.$$

   210

(2) (electrothermic power meters). The ratio of power incident from an rf source to the power delivered by the source to a nonreflecting load; mathematically, $|1 - \Gamma_g|^2$ where Γ_g is the complex reflection coefficient of the source.   47

(3) (circuits and systems). A factor in the equation for the insertion voltage ratio that takes into account the impedance mismatch variation at one end of the network due to an impedance mismatch at the other end. The factor is written in terms of the source and load impedance, the image impedances and the image transfer function of the four-terminal network. *See:* (**1**) *Note* above. 67

interaction gap. An interaction space between electrodes. *See also:* **electron devices, miscellaneous.** 190, 125

interaction impedance (traveling-wave tubes). A measure of the radio-frequency field strength at the electron stream for a given power in the interaction circuit. It may be expressed by the following equation

$$K = \frac{E^2}{2(\omega/v)^2 P}$$

where E is the peak value of the electric field at the position of electron flow, ω is the angular frequency, v is the interaction-circuit phase velocity, and P is the propagating power. If the field strength is not uniform over the beam, an effective interaction impedance may be defined. *See also:* **electron devices, miscellaneous.** 190, 125

interaction loss (transducer). The interaction loss expressed in decibels is 20 times the logarithm to base 10 of the scalar value of the reciprocal of the interaction factor. *See also:* **attenuation.**

interaction space (traveling-wave tubes). A region of an electron tube in which electrons interact with an alternating electromagnetic field. *See also:* **miscellaneous electron devices.** 190, 125

intercalated tapes (insulation). Two or more tapes, generally of different composition, applied simultaneously in such a manner that a portion of each tape overlies a portion of the other tape. *See also:* **power distribution, underground construction.** 64

intercardinal plane. Any plane bisecting the angle between two adjacent cardinal planes. *Note:* This term is used to relate the regular geometry of antenna array elements to the radiation pattern. 111

intercarrier sound system. A television receiving system in which use of the picture carrier and the associated sound-channel carrier produces an intermediate frequency equal to the difference between the two carrier frequencies. *Note:* This intermediate frequency is frequency modulated in accordance with the sound signal. *See also:* **television.** 328

intercept call (telephone switching systems). A call to a line or an unassigned code that reaches an operator, a recorded announcement, or a vacant-code tone. 55

intercept trunk (telephone switching systems). A central office termination that may be reached by a call to a vacant number, changed number, or line out-of-order. 55

interchangeable bushing (outdoor apparatus bushing). A bushing designed for use in both power transformers and circuit breakers. 168

interchange circuit (data transmission). The length of cable used for signaling between the digital subset and the customer's equipment. 59

interchannel interference (modulation system). In a given channel, the interference resulting from signals in one or more other channels. 242

interclutter visibility (radar). The ability of a moving-target indicator or pulse Doppler radar to detect moving targets which occur in resolution cells between points of strong clutter. The higher the radar resolution, the better the interclutter visibility, since a smaller fraction of the cells contain strong point clutter. 13

intercom (interphone). See: **intercommunicating system.**

intercommunicating system (intercom) (interphone). A privately owned two-way communication system without a central switchboard, usually limited to a single vehicle, building, or plant area. Stations may or may not be equipped to originate a call but can answer any call. 328

interconnected delta connection (transformer). A three-phase connection using six windings (two per phase) connected in a six-sided circuit with six bushings to provide a fixed phase-shift between two three phase circuits without change in voltage magnitude. *Note:* The interconnected delta connection is sometimes described as a "hexagon autotransformer," or a "squashed delta." 53

interconnected system (electric power systems). A system consisting of two or more individual power systems normally operating with connecting tie lines. *See also:* **power system.** 94, 200

interconnected star connection of polyphase circuits. *See:* **zig-zag connection of polyphase circuits.**

interconnecting channel (supervisory system). The transmission link, such as the direct wire, carrier, or microwave channel (including the direct current, tones, etcetera) by which supervisory control or indication signals or selected telemeter readings are transmitted between the master station and the remote station, or stations, in a single supervisory system. 103, 202

interconnection device. *See:* **adapter.**

interconnection diagram (industrial control). A special form of connection diagram that shows only the external connections between controllers and associated machinery, equipment, and extraneous components. 210, 206, 124

interconnection tie. A feeder interconnecting two electric supply systems. *Note:* The normal flow of energy in such a feeder may be in either direction. *See also:* **center of distribution.** 64

interdendritic corrosion. Corrosive attack that progresses preferentially along interdendritic paths. *Note:* This type of attack results from local differences in composition, that is, coring, commonly encountered in alloy castings. 221, 205

interdigital magnetron. A magnetron having axial anode segments around the cathode, alternate segments being connected together at one end, remaining segments connected together at the opposite end. 190, 125

interelectrode capacitance (j-l interelectrode capacitance C_{jl} of an n-terminal electrode tube). The capacitance determined from the short-circuit transfer admittance between the jth and the lth terminals. *Note:* This quantity is often referred to as direct interelectrode capacitance. *See also:* **electron-tube admittance.** 190, 125

interelectrode transadmittance (j-l interelectrode transadmittance of an n-electrode electron tube). The short-circuit transfer admittance from the jth electrode to the lth electrode. *See also:* **electron-tube admittances.** 190, 125

interelectrode transconductance (*j-l* interelectrode transconductance). The real part of the *j-l* interelectrode transadmittance. *See also:* **electron-tube admittances.**
190, 125

interelement influences (polyphase wattmeters). The percentage change in the recorded value that is caused solely by the action of the stray field of one element upon the other element. *Note:* This influence is determined at the specified frequency of calibration with rated current and rated voltage in phase on both elements or such lesser value of equal currents in both elements as gives end-scale deflection. Both current and voltage in one element shall then be reversed, and, for rating purposes, one-half the difference in the readings in percent is the interelement influence. *See also:* **accuracy rating (instrument).** 280, 294

interface (1) (general). A shared boundary. 255, 77

(2) (equipment). Interconnection between two equipments having different functions. 194

(3) (nuclear power generating stations) (class 1E equipment). A junction or junctions between a Class 1E equipment and another equipment or device. (Examples: connection boxes, splices, terminal boards, electrical connections, grommets, gaskets, cables, conduits, enclosures, etcetera.) 120, 31

(4) (programmable instrumentation). A shared boundary between a considered system and another system, or between parts of a system, through which information is conveyed. 40

(5) (test, measurement and diagnostic equipment). A shared boundary involving the specification of the interconnection between two equipments or systems. The specification includes the type, quantity and function of the interconnection circuits and the type and form of signals to be interchanged via those circuits. *See:* **adapter.** 54

(6) (data transmission). (A) A common boundary—for example, physical connection between two systems or two devices. May be mechanical such as the physical surfaces and spacings in mating parts, modules, components, or sub-systems, or electrical, such as matching signal levels, impedances, or power levels of two or more subsystems. **(B)** A concept involving the specification of the interconnection between two equipments or systems. The specification includes the type, quantity, and function of the interconnection circuits and the type and form of signals to be interchanged by these circuits. 59

interface EIA standard RS-232-A or B or C. A standardized method adopted by the Electronic Industries Association to insure uniformity of interface between data communication equipment and data processing terminal equipment. The standard interface has been generally accepted by a great majority of the manufacturers of data transmission and business equipment. 59

interface-mil std 188b. The standard method of interface established by the Department of Defense and is presently mandatory for use by the departments and agencies of the Department of Defense for the installation of all new equipment. This standard provides the interface requirements for interconnection between data communication security devices, data processing equipment, or other special military terminal devices. 59

interface-CCITT. The present European and possible world standard for interface requirements between data processing terminal equipment and data communication equipment. The CCITT standard resembles very closely the American EIA Standard RS-232-A-or B. This standard is considered mandatory in Europe and on the other continents. 59

interface, operating (connector). The surfaces at which a connector is normally separated. 134

interface system (programmable instrumentation). The device-independent mechanical, electrical, and functional elements of an interface necessary to effect communication among a set of devices. Cables, connector, driver and receiver circuits, signal line descriptions, timing and control conventions, and functional logic circuits are typical interface system elements. 40

interfacial connection (soldered connections). A conductor that connects conductive patterns on opposite sides of the base material. *See also:* **soldered connections (electronic and electrical applications).** 284

interference (1) (data transmission). Interference in a signal transmission path is either extraneous power which tends to interfere with the reception of the desired signals or the disturbance of signals which results. 59

(2) (electric-power-system measurements). Any spurious voltage or current appearing in the circuits of the instrument. *Note:* The source of each type of interference may be within the instrument case or external. The instrument design should be such that the effects of interference arising internally are negligible. 295

(3) (induction or dielectric-heating usage). The disturbance of any electric circuit carrying intelligence caused by the transfer of energy from induction- or dielectric-heating equipment. 14, 114

interference, common-mode (signal-transmission system). *See:* **common-mode interference.**

interference coupling ratio (signal-transmission system). The ratio of the interference produced in a signal circuit to the actual strength of the interfering source (in the same units). *See:* **interference.** 188

interference, differential-mode (signal-transmission system). Interference that causes the potential of one side of the signal transmission path to be changed relative to the other side. *Note:* That type of interference in which the interference current path is wholly in the signal transmission path. *See:* **interference.** 188

interference field strength (electromagnetic compatibility). Field strength produced by a radio disturbance. *Note:* Such a field strength has only a precise value when measured under specified conditions. Normally, it should be measured according to publications of the International Special Committee on Radio Interference. *See also:* **electromagnetic compatibility.** 220, 199

interference guard bands. The two bands of frequencies additional to, and on either side of, the communication band and frequency tolerance, which may be provided in order to minimize the possibility of interference. *See also:* **channel.** 111

interference, longitudinal (signal-transmission system). *See:* **interference, common-mode.**

interference measurement (induction or dielectric-heating). A measurement usually of field intensity to evaluate the probability of interference with sensitive receiving apparatus. *See also:* **dielectric heating; induction heating.**

14

interference, normal-mode (signal-transmission system). A form of interference that appears between measuring circuit terminals. *See:* **interference, differential-mode.** *See also:* **accuracy rating (instrument).** 295

interference pattern. The resulting space distribution when progressive waves of the same frequency and kind are superposed. *See also:* **wave front.** 328

interference power (electromagnetic compatibility). Power produced by a radio disturbance. *Note:* Such a power has only a precise value when measured under specified conditions. *See also:* **electromagnetic compatibility.**

220, 199

interference, series-mode (signal-transmission system). *See:* **interference, differential-mode.**

interference susceptibility (mobile communication). A measure of the capability of a system to withstand the effects of spurious signals and noise that tend to interfere with reception of the desired intelligence. *See also:* **mobile communication system.** 181

interference testing (test, measurement and diagnostic equipment). A type of on-line testing that requires disruption of the normal operation of the unit under test. *See:* **non-interference testing.** 54

interference, transverse. *See:* **interference, differential-mode.**

interference voltage (electromagnetic compatibility). Voltage produced by a radio disturbance. *Note:* Such a voltage has a precise value only when measured under specified conditions. Normally, it should be measured according to recommendations of the International Special Committee on Radio Interference. *See also:* **electromagnetic compatibility.** 220, 199

interferometer (radar). A receiving antenna which determines angle of arrival of a signal by phase comparison in the signals at separate points on the antenna. 13

interferometer antenna. An array antenna in which the interelement spacings are large compared to wavelength and element size so as to produce grating lobes. 111

interflectance method (lighting calculation). A lighting design procedure for predetermining the luminances (photometric brightnesses) of walls, ceiling, and floor and the average illumination on the work plane. It takes into account both direct and reflected flux. *See also:* **inverse-square law (in illuminating engineering).** 167

interflected component. That portion of the luminous flux from a luminaire that arrives at the work plane after being reflected one or more times from room surfaces. *See also:* **inverse-square law (illuminating engineering).** 167

interflection. The multiple reflection of light by the various room surfaces before it reaches the work plane or other specified surface of a room. *See also:* **inverse-square law (illuminating engineering).** 167

intergranular corrosion. Corrosion that occurs preferentially at grain boundaries. 221, 205

interior communication systems (marine). Those systems providing audible or visual signals or transmission of information within or on a vessel. 328

interior wiring system ground. A ground connection to one of the current-carrying conductors of an interior wiring system. *See:* **ground.** 64

interlaboratory standards. Those standards that are used for comparing reference standards of one laboratory with those of another, when the reference standards are of such nature that they should not be shipped. *See also:* **measurement system.** 293, 183

interlaced scanning (television). A scanning process in which the distance from center to center of successively scanned lines is two or more times the nominal line width, and in which the adjacent lines belong to different fields. *See also:* **television.** 178

interlace factor (television). A measure of the degree of interlace of nominally interlaced fields. *Note:* In a two-to-one interlaced raster, the interlace factor is the ratio of the smaller of two distances between the centers of adjacent scanned lines to one-half the distance between centers of sequentially scanned lines at a specified point.

163, 178

interlacing impedance voltage of a Scott-connected transformer. The interlacing impedance voltage of Scott-connected transformers is the single-phase voltage applied from the midtap of the main transformer winding to both ends, connected together, which is sufficient to circulate in the supply lines a current equal to the rated three-phase line current. *Notes:* (1) The current in each half of the winding is 50 percent of this value. (2) The per unit or percent interlacing resistance is the measured watts expressed on the base of the rated kilovolt-ampere (kVA) of the teaser winding. (3) The per unit or percent interlacing impedance is the measured voltage expressed on the base of the teaser voltage. *See also:* **impedance voltage of a transformer; resistance drop; reactance drop; impedance drop; impedance kilovolt-ampere (rated); zero-sequence impedance.** 53

interleave. To arrange parts of one sequence of things or events so that they alternate with parts of one or more other sequences of things or events and so that each sequence retains its identity. 255, 77

interleaved windings (transformer). An arrangement of transformer windings where the primary and secondary windings, and the tertiary windings, if any, are subdivided into discs (or pancakes) or layers and interleaved on the same core. 53

interlock (1) (high voltage air switches, insulators, and bus supports) (power switchgear). A device actuated by the operation of some other device with which it is directly associated, to govern succeeding operations of the same or allied devices. *Note:* An interlock system is a series of interlocks applied to associated equipment in such a manner as to prevent or allow operation of the equipment only in a prearranged sequence. Interlocks are classified into three main divisions: mechanical interlocks, electrical interlocks, and key interlocks, based on the type of interconnection between the associated devices. 27, 103

(2) (transformer) (industrial control). A device actuated by the operation of some other device with which it is directly associated, to govern succeeding operations of the same or allied devices. 206, 53

interlock bypass. A command to temporarily circumvent a normally provided interlock. 224, 207

interlocking (interlocking plant) (railway). An arrangement of apparatus in which various devices for controlling track switches, signals, and related appliances are so interconnected that their movements must succeed one another in a predetermined order, and for which interlocking rules are in effect. *Note:* It may be operated manually or automatically. 328

interlocking deactivating means (defeater) (industrial control). A manually actuated provision for temporarily rendering an interlocking device ineffective, thus permitting an operation that would otherwise be prevented. For example, when applied to apparatus such as combination controllers or control centers, it refers to voiding of the mechanical interlocking mechanism between the externally operable disconnect device and the enclosure doors to permit entry into the enclosure while the disconnect device is closed. *See:* **electric controller.** 225, 206

interlocking limits (interlocking territory) (railway). An expression used to designate the trackage between the opposing home signals of an interlocking. *See also:* **interlocking.** 328

interlocking machine (railway). An assemblage of manually operated levers or equivalent devices, for the control of signals, switches, or other units, and including mechanical or circuit locking, or both, to establish proper sequence of movements. *See also:* **interlocking (interlocking plant).** 328

interlocking plant. *See:* **interlocking.**

interlocking relay (railway). A relay that has two independent magnetic circuits with their respective armatures so arranged that the dropping away of either armature prevents the other armature from dropping away to its full stroke. 328

interlocking signals (railway). The fixed signals of an interlocking. 328

interlocking station (railway). A place from which an interlocking is operated. 328

interlocking territory. *See:* **interlocking limits.**

interlock relay. A relay with two or more armatures having a mechanical linkage, or an electric interconnection, or both, whereby the position of one armature permits, prevents, or causes motion of another armature. *See also:* **relay.** 259

intermediate contacts (switching device). Contacts in the main circuit that part after the main contacts and before the arcing contacts have parted. 103, 202

intermediate distributing frame (IDF) (telephone switching systems). A frame where crossconnections are made only between units of central office equipment. 55

intermediate frequency (IF) (1) (general). A frequency to which a signal wave is shifted locally as an intermediate step in transmission or reception. 59

(2) (superheterodyne reception). The frequency resulting from a frequency conversion before demodulation. *See also:* **radio transmission.** 339

intermediate-frequency-harmonic interference (superheterodyne receivers). Interference due to radio-frequency-circuit acceptance of harmonics of an intermediate-frequency signal. 339

intermediate-frequency interference ratio. *See:* **intermediate-frequency response ratio.** *See also:* **radio receiver.**

intermediate-frequency response ratio (superheterodyne receivers). The ratio of (1) the field strength at a specified frequency in the intermediate frequency band to (2) the field strength at the desired frequency, each field being applied in turn, under specified conditions, to produce equal outputs. *See also:* **radio receiver.** 339

intermediate-frequency transformer. A transformer used in the intermediate-frequency portion of a heterodyne system. *Note:* Intermediate-frequency transformers are frequently narrow-band devices. 197

intermediate layer (solar cells). The material on the solar cell surface that provides improved spectral match between the cell and the medium in contact with this surface. 113

intermediate maintenance (test, measurement and diagnostic equipment). Maintenance which is the responsibility of and performed by designated maintenance activities for direct support of using organizations. Its phases normally consist of calibration, repair or replacement of damaged or unserviceable parts, components or assemblies; the emergency manufacture of non-available parts and providing technical assistance to using organizations. 54

intermediate means (measurement sequence). All system elements that are used to perform necessary and distinct operations in the measurement sequence between the primary detector and the end device. *Note:* The intermediate means, where necessary, adapts the operational results of the primary detector to the input requirements of the end device. *See also:* **measurement system.** 295

intermediate office (telephone switching systems). A switching entity where trunks are terminated for purposes of interconnection to other offices. 55

intermediate repeater. A repeater for use in a trunk or line at a point other than an end. *See also:* **repeater.** 59

intermediate subcarrier. A carrier that may be modulated by one or more subcarriers and that is used as a modulating wave to modulate a carrier or another intermediate subcarrier. *See also:* **carrier; subcarrier.** 111

intermittent duty (1) (general). A requirement of service that demands operation for alternate periods of (A) load and no load; or (B) load and rest; or (C) load, no load, and rest, such alternate intervals being definitely specified. 53, 310, 233, 210, 257, 203, 225, 206

(2) (rotating machinery). A duty in which the load changes regularly or irregularly with time. *See also:* **asynchronous machine; direct-current commutating machine; voltage regulator.** 63

intermittent-duty rating. The specified output rating of a device when operated for specified intervals of time other than continuous duty. 111

intermittent failure. *See:* **failure, intermittent.**

intermittent fault (surge arresters). A fault that recurs in the same place and due to the same cause within a short period of time. 244, 62

intermittent inductive train control. Intermittent train control in which the impulses are communicated to the vehicle-carried apparatus inductively. *See also:* **automatic train control.** 328

intermittent rating. *See:* **periodic rating.**

intermittent test (battery). A service test in which the battery is subjected to alternate discharges and periods of recuperation according to a specified program until the cutoff voltage is reached. *See also:* **battery (primary or secondary).** 328

intermittent train control. A system of automatic train

control in which impulses are communicated to the loco-motive or vehicle at fixed points only. *See also:* **automatic train control; intermittent inductive train control.**   328

intermodulation (nonlinear transducer element). The modulation of the components of a complex wave by each other, as a result of which waves are produced that have frequencies equal to the sums and differences of integral multiples of those of the components of the original complex wave. *See also:* **modulation.**   111, 199, 59

intermodulation interference (mobile communication). The modulation products attributable to the components of a complex wave that on injection into a nonlinear circuit produces interference on the desired signal. *See also:* **mobile communication system.**   181

intermodulation rejection (non-real time spectrum analyzer). The ratio of: (1) the level of two equal amplitude input signals which produce any intermodulation product indication at the sensitivity level and, (2) the sensitivity level.   68

intermodulation spurious response (1) (receiver performance). The receiver audio output resulting from the mixing of *n*-th-order frequencies, in the nonlinear elements of the receiver, in which the resultant carrier frequency is equivalent to the assigned frequency. *See:* **spurious response.** *See also:* **receiver performance.**   181

(2) (non-real time spectrum analyzer). The spectrum analyzer response resulting from the mixing of the *n*th order frequencies, in the non-linear elements of the spectrum analyzer, in which the resultant response is equivalent to the tuned frequency. *Syn.* **intermodulation distortion.**   68

(3) (frequency-modulated mobile communications receivers). The response resulting from the mixing of two or more undesired frequencies in the nonlinear elements of the receiver in which a resultant frequency is generated that falls within the receiver passband.   123

internal bias (teletypewriter). Bias, either marking or spacing, that may occur within a start-stop printer receiving mechanism and that will have an effect on the margins of operation.   194

internal blocking (telephone switching systems). The unavailability of paths in a switching network between a given inlet and any suitable idle outlet.   55

internal connector (pothead). A connector that joins the end of the cable to the other current-carrying parts of a pothead. *See also:* **pothead; transformer.**   4, 288, 289

internal correction voltage (electron tubes). The voltage that is added to the composite controlling voltage and is the voltage equivalent of such effects as those produced by initial electron velocity and contact potential. *See also:* **composite controlling voltage (electron tube).**   190, 125

internal heating (electrolysis). The electrolysis of fused electrolytes is the method of maintaining the electrolyte in a molten condition by the heat generated by the passage of current through the electrolyte. *See also:* **fused electrolyte.**   328

internal impedance (rotating machinery). The total self-impedance of the primary winding under steady conditions. *Note:* For a three-phase machine, the primary current is considered to have only a positive-sequence component when evaluating this quantity. *See also:* **asynchronous machine.**   63

internal impedance drop (rotating machinery). The product of the current and the internal impedance. *Note:* This is the phasor difference between the generated internal voltage and the terminal voltage of a machine. *See also:* **asynchronous machine.**   63

internal insulation (apparatus) (surge arresters). The insulation that is not directly exposed to atmospheric conditions. *See:*   308, 62, 53

internally-programmed automatic test equipment (test, measurement and diagnostic equipment). An automated tester using any programming technique in which a substantial amount of programming information is stored within the equipment, although it may originate from external media.   54

internal oxidation. *See:* **subsurface corrosion.**

internal remanent residual voltage (Hall-effect devices). That portion of the zero field residual voltage which is due to the remanent magnetic flux density in the ferromagnetic encapsulation of the Hall generator.   107

internal resistance (battery). The resistance to the flow of an electric current within a cell or battery. *See also:* **battery (primary or secondary).**   328

internal storage (test, measurement and diagnostic equipment). Storage facilities forming an integral part of the machine.   54

internal traffic (telephone switching systems). Traffic originating and terminating within the network being considered.   55

internal triggering (1) (oscilloscopes). The use of a portion of a deflection signal (usually the vertical deflection signal) as a triggering-signal source. *See:* **oscillograph.**   185

(2) (non-real time spectrum analyzer). The use of a deflection signal (usually the vertical deflection signal) as a triggering source.   68

international call (telephone switching systems). A call to a destination outside of the national boundaries of the calling customer.   55

International Commission on Illumination. *See:* **CIE.**

international direct distance dialing (IDDD) (telephone switching systems). The automatic establishing of international calls by means of signals from the calling device of a customer.   55

international distance dialing (IDD) (telephone switching systems). The automatic establishing of international calls by means of signals from the calling device of either a customer or an operator.   55

international interzone call (telephone switching systems). A call to a destination outside of a national or integrated numbering-plan area.   55

international intrazone call (telephone switching systems). A call to a destination within the boundaries of an integrated numbering-plan area, but outside the national boundaries of the calling customer.   55

International Morse code (Continental code). A system of dot and dash signals, differing from the American Morse code only in certain code combinations, used chiefly in international radio and wire telegraphy. *See also:* **telegraphy.**   328

international number (telephone switching systems). The combination of digits representing a country code plus a national number.   55

international operating center (telephone switching systems).

In World Zone 1, a center where telephone operators handle originating and terminating international interzone calls and may also handle international intrazone calls. *See also:* **world zone number.**   55

international originating toll center (telephone switching systems). In World Zone 1, a toll center where telephone operators handle originating international interzone calls. *See also:* **world zone numbers.**   55

international switching center (telephone switching systems). A toll office that normally serves as a point of entry or exit for international interzone calls.   55

International System of Electrical Units. A system that uses the **international ampere** and the **international ohm.** *Notes:* (1) The international ampere was defined as the current that will deposit silver at the rate of 0.00111800 gram per second; and the international ohm was defined as the resistance at 0 degrees Celsius of a column of mercury having a length of 106.300 centimeters and a mass of 14.4521 grams. (2) The International System of Electrical Units was in use between 1893 and 1947 inclusive. By international agreement it was discarded, effective January 1, 1948 in favor of the MKSA system. (3) Experiments have shown that as these units were maintained in the United States of America, 1 international ohm equalled 1.000495 ohm and that 1 international ampere equalled 0.999835 ampere. *See:* **International System of Units.**   210

International System of Units (SI). A universal coherent system of units in which the following six units are considered basic: meter, kilogram, second, ampere, Kelvin degree, and candela. *Notes:* (1) The MKSA system of electrical units (MKSA System of Units) is a constituent part of this system adequate for mechanics and electromagnetism. (2) The electrical units of this system should not be confused with the units of the earlier International System of Electrical Units which was discarded January 1, 1948. (3) The International System of Units (abbreviated SI) was promulgated in 1960 by the Eleventh General Conference on Weights and Measures. *See also:* **units and letter symbols.**   210

interoffice call (telephone switching systems). A call between lines connected to different central offices.   55

interoffice trunk (telephony). A direct trunk between local central offices in the same exchange.   328

interphase rod or shaft (high voltage air switches, insulators, and bus supports). A component of a switch operating mechanism designed to connect two or more poles of a multipole switch for group operation.   27

interphase transformer. An autotransformer, or a set of mutually coupled reactors, used to obtain parallel operation between two or more simple rectifiers that have ripple voltages that are out of phase. *See:* **auto-transformer; rectifier transformer.**   203, 53

interphase-transformer loss (rectifier transformer). The losses in the interphase transformer that are incident to the carrying of rectifier load. *Note:* They include both magnetic core loss and conductor loss. *See also:* **rectifier transformer.**   258

interphase-transformer rating. Consists of the root-mean-square current, root-mean-square voltage, and frequency, at the terminals of each winding, for the rated load of the rectifier unit, and a designated amount of phase

control, as assigned to it by the manufacturer. *See:* **duty; rectifier transformer.**   203

interphone (intercom). *See:* **intercommunicating system.**

interphone equipment (aircraft). Equipment used to provide telephone communications between personnel at various locations in an aircraft. *See also:* **air transportation electronic equipment.**   328

interpolation (signal interpolation) (submarine cable telegraphy). A method of reception characterized by synchronous restoration of unit-length signal elements which are weak or missing in the received signals as a result of one or more of such factors as suppression at the transmitter, attenuation in transmission, or discrimination in the receiving networks. *Note:* This is sometimes referred to as local correction. *See also:* **telegraphy.**   328

interpolation function (burst measurements). A function that may be used to obtain additional values between sampled values. *See also:* **burst.**   292

interpole. *See:* **commutating pole.**

interposing relay (supervisory system). An auxiliary relay at the master or remote station, the contacts of which serve: (1) to energize a circuit (for closing, opening, or other purpose) of an element of remote station equipment when the selection of a desired point has been completed and when suitable operating signals are received through the supervisory equipment from the master station; or (2) to connect in the circuit the telemeter transmitting and receiving equipments, respectively, at the remote and master stations. *Note:* The interposing relays are considered part of a supervisory system.   103, 202

interpreter (computing systems). (1) A program that translates and executes each source language expression before translating and executing the next one. (2) A device that prints on a punched card the data already punched in the card.   255, 77, 54

inter-record gap (test, measurement and diagnostic equipment). An interval of space or time deliberately left between recording portions of data or records. Such spacing is used to prevent errors through loss of data or overwriting and permits tape stop-start operations.   54

interrogation (transponder system) (radar). The signal or combination of signals intended to trigger a response.   187, 13

interrogative supervisory system (power switchgear). A system whereby the master station controls all operations of the system, and whereby all indications are obtained on a master station request or interrogation basis. *Note:* The normal state is usually one of continuous interrogation or polling of the remote stations for changes in status.   103

interrogator (electronic navigation). (1) A radio transmitter and receiver combined to interrogate a transponder and display the resulting replies. (2) The transmitting part of an interrogator-responsor. *See also:* **radio transmission.**   187, 13

interrogator-responsor (IR). *See:* **interrogator.**

interrupt. To stop a process in such a way that it can be resumed. *See:* **pre-emptive control.**   255, 77, 54

interrupted continuous wave (ICW). A continuous wave that is interrupted at a constant audio-frequency rate. *See also:* **radio transmission.**   111, 211, 59

interrupted quick-flashing light. A quick-flashing light in which the rapid alternations are interrupted by periods of darkness at regular intervals.   167

interrupter. An element designed to interrupt specified currents under specified conditions.   103, 202, 27

interrupter blade (of an interrupter switch). A blade used in the interrupter for breaking the circuit.
    103, 202, 27

interrupter switch. An air switch, equipped with an interrupter, for making or breaking specified currents, or both. *Note:* The nature of the current made or broken, or both, may be indicated by suitable prefix; that is, load interrupter switch, fault interrupter switch, capacitor current interrupter switch, etcetera.   103, 202, 27

interruptible loads (electric power utilization). Those loads that by contract can be interrupted in the event of a capacity deficiency on the supplying system. *See also:* **generating station.**   200

interruptible power. Power made available under agreed conditions that permit curtailment or cessation of delivery by the supplier. *See also:* **generating station.**   64

interrupting capacity (packaging machinery). Interrupting capacity is the highest current at rated voltage that the device can interrupt.   124

interrupting current (switching device) (breaking current). The current in a pole at the instant of the initiation of the arc.   103, 202

interrupting tests. Tests that are made to determine or check the interrupting performance of a switching device.
    103, 202

interrupting time (mechanical switching device). The interval between the time when the actuating quantity of the release circuit reaches the operating value, the switching device being in a closed position, and the instant of arc extinction on the primary arcing contacts. *Notes:* (1) Interrupting time is numerically equal to the sum of opening time and arcing time. (2) In multipole devices interrupting time may be measured for each pole or for the device as a whole, in which latter case the interval is measured to the instant of arc extinction in the last pole to clear. *Syn.:* **total break time.**   103 202

interruption (electric power systems). The loss of service to one or more consumers or other facilities. *Note:* It is the result of one or more component outages, depending on system configuration. *See also:* **outage.**   290

interruption duration (electron power systems). The period from the initiation of an interruption to a consumer or other facility until service has been restored to that consumer or facility. *See also:* **outage.**   200

interruption duration index (electric power systems). The average interruption duration for consumers interrupted during a specified time period. It is estimated from operating history by dividing the sum of all consumer interruption durations during the specified period by the number of consumer interruptions during that period. *See also:* **outage.**   200

interruption, forced (electric power systems). An interruption caused by a forced outage.   112

interruption frequency index (electric power systems). The average number of interruptions per consumer served per time unit. *Note:* It is estimated from operating history by dividing the number of consumer interruptions observed in a time unit by the number of consumers served. A

consumer interruption is considered to be one interruption of one consumer. *See also:* **outage.**   200

interruption, momentary (electric power systems). An interruption of duration limited to the period required to restore service by automatic or supervisory-controlled switching operations or by manual switching at locations where an operator is immediately available. *Note:* Such switching operations must be completed in a specified time not to exceed 5 minutes.   112

interruption, scheduled (electric power systems). An interruption caused by a scheduled outage. *See also:* **outages.**   112

interruption, sustained (electric power systems). Any interruption not classified as a momentary interruption.
    112

interruption to service (power switchgear). The isolation of an electrical load from the system supplying that load, resulting from an abnormality in the system.   127, 103

interspersing (rotating machinery). Interchanging the coils at the edges of adjacent phase belts. *Note:* The purpose of interspersing depends on the type of machine in which it is done. In asynchronous motors it is used to reduce harmonics that can cause crawling. *See also:* **asynchronous machine.**   63

intersymbol interference (transmission system) (modulation systems). Extraneous energy from the signal in one or more keying intervals that tends to interfere with the reception of the signal in another keying interval, or the disturbance that results.   242

intertoll dialing (telephony). Dialing over intertoll trunks.
    328

intertoll trunk (telephone switching systems). A trunk between two toll offices.   55

interturn insulation (rotating machinery). The insulation between adjacent turns, often in the form of strips.   63

interturn test. *See:* **turn-to-turn test (rotating machinery).**

interval (1) (acoustics). The spacing between two sounds in pitch or frequency, whichever is indicated by the context. *Note:* The frequency interval is expressed by the ratio of the frequencies or by a logarithm of this ratio.   176
(2) (pulse terms). The algebraic time difference calculated by subtracting the time of a first specified instant from the time of a second specified instant.   254

interval, sweep holdoff (oscilloscopes). The interval between sweeps during which the sweep and/or trigger circuits are inhibited.   106

intrabeam viewing (laser-maser). The viewing condition whereby the eye is exposed to all or part of a laser beam.
    363

intraoffice call (telephone switching systems). A call between lines connected to the same central office.   55

intrinsically safe equipment and wiring (National Electrical Code). Equipment and wiring that are incapable of releasing sufficient electrical or thermal energy under normal or abnormal conditions to cause ignition of a specific hazardous atmospheric mixture in its most easily ignited concentration. This equipment is suitable for use in division 1 locations. Division 2 Equipment and Wiring are equipment and wiring which in normal operation would not ignite a specific hazardous atmosphere in its most easily ignited concentration. The circuits may include sliding or make-and-break contacts releasing insufficient energy to cause ignition. Circuits not containing sliding or make-and-break contacts may have higher energy levels potentially capable of causing ignition under fault conditions.   65

intrinsic coercive force. The magnetizing force at which the intrinsic induction is zero when the material is in a symmetrically cyclically magnetized condition.   210

intrinsic impedance (antenna). The theoretical input impedance for the basic radiating structure when idealized. *Note:* The idealized basic radiating structure usually consists of a uniform-cross-section radiating element, perfectly conducting ground or imaging planes, zero base capacitance (in the case of vertical radiators), and no internal losses.   179

intrinsic induction (magnetic polarization). At a point in a magnetized body, the vector difference between the magnetic induction at that point and the magnetic induction that would exist in a vacuum under the influence of the same magnetizing force. This is expressed by the equation $B_i = B - \mu_0H$. *Note:* In the centimeter-gram-second electromagnetic-unit system, $B_i/4\pi$ is often called magnetic polarization.   210

intrinsic permeability. The ratio of intrinsic normal induction to the corresponding magnetizing force. *Note:* In anisotropic media, intrinsic permeability becomes a matrix.   210

intrinsic properties (semiconductor). The properties of a semiconductor that are characteristic of the pure, ideal crystal. *See also:* **semiconductor.**   245, 66, 210, 186

intrinsic semiconductor. *See:* **semiconductor, intrinsic.**

intrinsic temperature range (semiconductor). The temperature range in which the charge-carrier concentration of a semiconductor is substantially the same as that of an ideal crystal. *See also:* **semiconductor.**   214, 66, 210

introspective testing. *See:* **self-test.**

inverse electrode current. The current flowing through an electrode in the direction opposite to that for which the tube is designed. *See also:* **electrode current (electron tube).**   190, 125

inverse magnitude contours. *See:* **magnitude contours.**

inverse networks. Two two-terminal networks are said to be inverse when the product of their impedances is independent of frequency within the range of interest. *See also:* **network analysis.**   210

inverse neutral telegraph transmission. That form of transmission employing zero current during marking intervals and current during spacing intervals. *See also:* **telegraphy.**   194

inverse Nyquist diagram. *See:* **Nyquist diagram.**

inverse-parallel connection (industrial control). An electric connection of two rectifying elements such that the cathode of the first is connected to the anode of the second, and the anode of the first is connected to the cathode of the second.   206

inverse period (rectifier element). The nonconducting part of an alternating-voltage cycle during which the anode has a negative potential with respect to the cathode. *See also:* **rectification.**   328

inverse-square law. A statement that the strength of the field due to a point source or the irradiance from a point source decreases as the square of the distance from the

source. *Note:* For sources of finite size this gives results that are accurate within one-half percent when distance is at least five times the maximum dimension of the source (or luminaire) as viewed by the observer.   210, 167

inverse time (industrial control). *See:* **inverse-time relay.**

inverse-time delay (power switchgear). A qualifying term indicating that there is purposely introduced a delaying action, the delay decreasing as the operating force increases.   103

inverse-time relay. A relay in which the input quantity and operating time are inversely related throughout at least a substantial portion of the performance range. *Note:* Types of inverse-time relays are frequently identified by such modifying adjectives as definite minimum time, moderately, very, and extremely to identify relative degree of inverseness of the operating characteristics of a given manufacturer's line of such relays.   103, 202, 60, 127

inverse transfer function (control systems). The reciprocal of a transfer function.   56

inverse transfer locus (control systems). The locus of the inverse transfer function.   56

inverse voltage (rectifier). The voltage applied between the two terminals in the direction opposite to the forward direction. This direction is called the backward direction. *See also:* **rectification of an alternating current.**   244, 190

inversion efficiency. The ratio of output fundamental power to input direct power expressed in percent. *See:* **self-commutated inverters.**   208

inversion ratio (laser-maser). In a maser medium, the negative of the ratio of (1) the population difference between two non-degenerate energy states under a condition of population inversion to (2) the population difference at equilibrium.   363

inverted (rotating machinery). Applied to a machine in which the usual functions of the stationary and revolving members are interchanged. *Example:* an induction motor in which the primary winding is on the rotor and is connected to the supply through sliprings, and the secondary is on the stator. *See:* **asynchronous machine.**   63

inverted input (oscilloscopes). An input such that the applied polarity causes a deflection polarity opposite from conventional deflection polarity.   106

inverted-turn transposition (rotating machinery). A form of transposition used on multiturn coils in which one or more turns are given a 180-degree twist in the end winding or at the coil nose or series loop. *See also:* **rotor (rotating machinery); stator.**   63

inverter (electric power). A machine, device, or system that changes direct-current power to alternating-current power. *See:* **inverting amplifier.** *See also:* **electronic analog computer.**   9, 10

inverting amplifier. An operational amplifier that produces an output signal of nominally equal magnitude and opposite algebraic sign to the input signal. *Note:* In an analog computer, the term inverter is synonymous with inverting amplifier. *See also:* **electronic analog computer; power supply.**   9, 228, 77, 186, 10

inverting parametric device. A parametric device whose operation depends essentially upon three frequencies, a harmonic of the pump frequency and two signal frequencies, of which the higher signal-frequency is the difference between the pump harmonic and the lower signal frequency. *Note:* Such a device can exhibit gain at either of the signal frequencies provided power is suitably dissipated at the other signal frequency. It is said to be inverting because if one of the two signals is moved upward in frequency, the other will move downward in frequency. *See also:* **parametric device.**   277, 191

inward-wats service (telephone switching systems). A reverse-charge, flat-rate, or measured-time direct distance dialing service to a specific directory number.   55

I/O (input/output). Input or output or both.   255, 77

ion. An electrically charged atom or radical.   221, 205

ion activity (ion species). The thermodynamic concentration, that is, the ion concentration corrected for the deviation from the law of ideal solutions. *Note:* The activity of a single ion species cannot, however, be measured thermodynamically. *See also:* **ion.**   328

ion burn. *See:* **ion spot.**

ion charging (charge-storage tubes). Dynamic decay caused by ions striking the storage surface. *See also:* **charge-storage tube.**   174, 190

ion concentration (species of ion). The concentration equal to the number of those ions, or of moles or equivalent of those ions, contained in a unit volume of an electrolyte.   328

ion gun. A device similar to an electron gun but in which the charged particles are ions. *Example:* proton gun. *See also:* **electron optics.**   244, 190

ionic-heated cathode (electron tube). A hot cathode that is heated primarily by ionic bombardment of the emitting surface.   125

ionic-heated-cathode tube. An electron tube containing an ionic-heated cathode.   125

ionization (1) (general). The process or the result of any process by which a neutral atom or molecule acquires either a positive or a negative charge. *See also:* **photovoltaic power system; solar cells (in a photovoltaic power system).**   186

(2) (power lines). A breakdown that occurs in parts of a dielectric when the electric stress in those parts exceeds a critical value without initiating a complete breakdown of the insulation system. *Note:* Ionization can occur both on internal and external parts of a device. It is a source of radio noise and can damage insulation.   229, 62

(3) (outdoor apparatus bushings). (A) The formation of limited avalanches of electrons developed in insulation due to an electric field. (B) **Ionization current** is the result of capacitive discharges in an insulating medium due to electron avalanches under the influence of an electric field. *Note:* The occurrence of such currents may cause: (1) radio noise. (2) damage to insulation.   168

ionization current (1) (surge arrester). The electric current resulting from the movement of electric charges in an ionized medium, under the influence of an applied electric field. *See also:* **gas ionization current.**   2

(2) (vacuum tubes). *See:* **gas current.**

ionization extinction voltage (corona level) (cables). The minimum value of falling root-mean-square voltage that sustains electric discharge within the vacuous or gas-filled spaces in the cable construction or insulation. *See also:* **power distribution underground construction.**   57

ionization factor (power distribution, underground cables)

(dielectric). The difference between percent power factors at two specified values of electric stress. The lower of the two stresses is usually so selected that the effect of the ionization on power factor at this stress is negligible. *See also:* **power distribution, underground construction.**   57

ionization-gauge tube. An electron tube designed for the measurement of low gas pressure and utilizing the relationship between gas pressure and ionization current.   190

ionization measurement. The measurement of the electric current resulting from the movement of electric charges in an ionized medium under the influence of the prescribed electric field.   287

ionization or corona detector. *See:* **discharge detector.**

ionization or corona inception voltage. *See:* **discharge inception voltage.**

ionization or corona probe. *See:* **discharge probe.**

ionization smoke detector (fire protection devices). A device which has a small amount of radioactive material which ionizes the air in the sensing chamber, thus rendering it conductive and permitting a current flow through the air between two charged electrodes. This gives the sensing chamber an effective electrical conductance. When smoke particles enter the ionization area, they decrease the conductance of the air by attaching themselves to the ions, causing a reduction in mobility. When the conductance is less than the predetermined level, the detector circuit responds.   71

ionization time (gas tube). The time interval between the initiation of conditions for and the establishment of conduction at some stated value of tube voltage drop. *Note:* To be exact the ionization time of a gas tube should be presented as a family of curves relating such factors as condensed-mercury temperature, anode and grid currents, anode and grid voltages, and regulation of the grid current. *See:* **gas tubes.**   190, 125

ionization vacuum gauge. A vacuum gauge that depends for its operation on the current of positive ions produced in the gas by electrons that are accelerated between a hot cathode and another electrode in the evacuated space. *Note:* It is ordinarily used to cover a pressure range of 10^{-4} to 10^{-10} conventional millimeters of mercury. *See also:* **instrument.**   328

ionization voltage (surge arresters). A high-frequency voltage appearing at the terminals of an arrester, generated by all sources, but particularly by ionization current within the arrester, when a power-frequency voltage is applied across the terminals. *Note:* This voltage provides an indication of internal ionization that might cause deterioration in service. The internal ionization voltage of parts of an arrester may differ from that of a complete arrester.   229, 62, 2

ionizing event (gas-filled radiation-counter tube). Any interaction by which one or more ions are produced. *See also:* **gas-filled radiation-counter tubes.**   190, 125

ionizing radiation (1) (general). Any electromagnetic or particulate radiation capable of producing ions, directly or indirectly, in its passage through matter. *See also:* **photovoltaic power system; solar cells (photovoltaic power system).**   186

(2) (air). (A) Particles or photons of sufficient energy to produce ionization in their passage through air. (B) Par-

ticles that are capable of nuclear interactions with the release of sufficient energy to produce ionization in air.
335

(3) (scintillation counting). Particles or photons of sufficient energy to produce ionization in interactions with matter.
117

ion migration. A movement of ions in an electrolyte as a result of the application of an electric potential. *See also:* **ion.**
328

ionosphere. That part of the earth's outer atmosphere where ions and free electrons are normally present in quantities sufficient to affect the propagation of radio waves. *Notes:* (1) According to present opinion, the lowest level of the ionosphere is approximately 50 kilometers above the earth's surface. (2) Modern usage extends the term to the atmospheres of other planets. *See also:* **radiation; radio wave propagation.**
180, 59, 146

ionosphere disturbance. A variation in the state of ionization of the ionosphere beyond the normally observed random day-to-day variation from average values for the location, date, and time of day under consideration. *Note:* Since it is difficult to draw the line between normal and abnormal variations, this definition must be understood in a qualitative sense. *See also:* **radiation.**
328

ionospheric error (electronic navigation). The total systematic and random error resulting from the reception of the navigational signal via ionospheric reflections; this error may be due to (1) variations in transmission paths, (2) nonuniform height of the ionosphere, and (3) nonuniform propagation within the ionosphere. *See also:* **navigation.**
13, 187

ionosphere-height error (electronic navigation). The systematic component of the total ionospheric error due to the difference in geometrical configuration between ground paths and ionospheric paths. *See also:* **navigation.**
13, 187

ionospheric storm. An ionospheric disturbance characterized by wide variations from normal in the state of the ionosphere, including effects such as turbulence in the *F* region, increases in absorption, and often decreases in ionization density and increases in virtual height. *Note:* The effects are most marked in high magnetic latitudes and are associated with abnormal solar activity. *See also:* **radiation.**
328

ionospheric tilt error (electronic navigation). The component of the ionospheric error due to nonuniform height of the ionosphere. *See also:* **navigation.**
187, 13

ionospheric wave (radio wave propagation). A radio wave propagated by reflection or refraction from an ionosphere. *Note:* This is some times called a sky wave. *See also:* **radiation; radio wave propagation.**
180, 146

ion repeller (charge-storage tubes). An electrode that produces a potential barrier against ions. *See also:* **charge-storage tube.**
174, 190

ion sheath. *See:* **electron sheath.**

ion spot (1) (camera tubes or image tubes). The spurious signal resulting from the bombardment or alteration of the target or photocathode by ions. *See also:* **television.**
178, 190, 125

(2) (cathode-ray-tube screen). An area of localized deterioration of luminescence caused by bombardment with negative ions.
178, 190, 125

ion transfer (electrotherapy) (ionic medication*) (iontophoresis*) (medical ionization*) (ion therapy*) (ionotherapy*). The forcing of ions through biological interfaces by means of an electric field. *See also:* **electrotherapy.**
192

* Deprecated

ion trap (cathode-ray tube). A device to prevent ion burn by removing the ions from the beam. *See also:* **cathode-ray tubes.**
244, 190

IR. *See:* **interrogator-respondor.**

IR drop (electrolytic cell). The drop equal to the product of the current passing through the cell and the resistance of the cell.
328

IR-drop compensation (1) (electric drive) (industrial control). A provision in the system of control by which the voltage drop (and corresponding speed drop) due to armature current and armature-circuit resistance is partially or completely neutralized.
206

(2) (transformer). A provision in the transformer by which the voltage drop due to transformer load current and internal transformer resistance is partially or completely neutralized. Such transformers are suitable only for one-way transformation, that is, not interchangeable for step-up and/or step-down transformations.
53

iris (1) (waveguide technique). A metallic plate, usually of small thickness compared with the wavelength, perpendicular to the axis of a waveguide and partially blocking it. *Notes:* (A) An iris acts like a shunt element in a transmission line; it may be inductive, capacitive, or resonant. (B) When only a single mode can be supported an iris acts substantially as a shunt admittance.
179

(2) (waveguide components). A partial obstruction at a transverse cross-section formed by one or more metal plates of small thickness compared with the wavelength.
166

(3) (laser-maser). The circular pigmented membrane which lies behind the cornea of the human eye. The iris is perforated by the pupil.
363

ironclad plate (storage cell). A plate consisting of an assembly of perforated tubes of insulating material and of a centrally placed conductor. *Note:* "Ironclad" is a registered trademark of ESB Incorporated. *See also:* **battery (primary or secondary).**
328

irradiance (at a point of a surface), E (laser-maser). Quotient of the radiant flux incident on an element of the surface containing the point, by the area of that element. Unit: watt per square centimeter ($W \cdot cm^{-2}$). *See also:* **radiant flux density at a surface.**
363

irreversible process. An electrochemical reaction in which polarization occurs. *See also:* **electrochemistry.**
328

I scan. *See:* **I display**

I scope. *See:* **I display.**

island effect (electron tubes). The restriction of the emission from the cathode to certain small areas of it (islands) when the grid voltage is lower than a certain value. *See also:* **electronic tube.**
244, 190

ISM apparatus (industrial, scientific, and medical apparatus; electromagnetic compatibility). Apparatus intended for generating radio-frequency energy for industrial, scientific or medical purposes. *See also:* **electromagnetic compatibility; industrial electronics.**
220, 199

isochronous speed governing (gas turbines). Governing with steady-state speed regulation of essentially zero magnitude. 98, 58

isocandela line. A line plotted on any appropriate coordinates to show directions in space about a source of light, in which the candlepower is the same. For a complete exploration the line always is a closed curve. A series of such curves, usually for equal increments of candlepower, is called an isocandela diagram. *See also:* **lamp.** 167

isocon mode (camera tube). A low-noise return-beam mode of operation utilizing only back-scattered electrons from the target to derive the signal, with the beam electrons specularly reflected by the electrostatic field near the target being separated and rejected. *See also:* **camera tube.** 190

isoelectric point. A condition of net electric neutrality of a colloid, with respect to its surrounding medium. *See also:* **ion.** 328

isokeraunic level (lightning). The average annual number of thunderstorm days. *See also:* **direct-stroke protection (lightning).** 64

isolated (1) (transmission and distribution). Not readily accessible to persons unless special means for access are used. 262

(2) (conductor stringing equipment). (A) Physically separated electrically and mechanically from all sources of electrical energy. Such separation may not eliminate the effects of electrical induction. (B) An object not readily accessible to persons unless special means for access are used. 45

isolated by elevation. Elevated sufficiently so that persons may walk safely underneath. 328

isolated conductor (ignored conductor). In a multiple-conductor system, a conductor either accessible or inaccessible, the charge of which is not changed and to which no connection is made in the course of the determination of any one of the capacitances of the remaining conductors of the system. 210

isolated impedance of an array element (antennas). The input impedance of a radiating element of an array antenna with all other elements of the array absent. 111

isolated-neutral system. A system that has no intentional connection to ground except through indicating, measuring, or protective devices of very-high impedance. *See:* **grounded system.** 244, 62

isolated-phase bus. A bus in which each phase conductor is enclosed by an individual metal housing separated from adjacent conductor housings by an air space. *Note:* The bus may be self-cooled or may be forced cooled by means of circulating a gas or liquid. 103, 202, 79, 78

isolated plant (electric power). An electric installation deriving energy from its own generator driven by a prime mover and not serving the purpose of a public utility. 328

isolating amplifier (signal-transmission system). *See:* **amplifier, isolating.**

isolating switch. A switch intended for isolating an electric circuit from the source of power. *Note:* It has no interrupting rating, and it is intended to be operated only after the circuit has been opened by some other means. *See also:* **switch.** 206

isolating time (sectionalizer). The time between the ces-sation of a current above the minimum actuating current value that caused the final counting and opening operation and the maximum separation of the contacts. 103, 202

isolating transformer (signal-transmission system). *See:* **transformer, isolating.**

isolation (1) (general). Separation of one section of a system from undesired influences of other sections. 197

(2) (signal transmission). Physical and electrical arrangement of the parts of a signal-transmission system to prevent interference currents within or between the parts. *See:* **signal.** 188

(3) (antennas). A measure of power transfer from one antenna to another. *Note:* The isolation between antennas is the ratio of power input to one antenna to the power received by the other, usually expressed in decibels. *See also:* **radiation.** 179, 111

(4) (multiport device). Between two ports of a nonreciprocal or directional multiport device, the characteristic insertion loss in the nonpreferred direction of propagation, all ports being terminated by reflectionless terminations. *Note:* It is usually expressed in decibels. 185

(5) (industrial control). Connotes the opening of all conductors connected to the source of power. 225, 206

isolation amplifier (buffer). *See:* **amplifier, isolation.**

isolation by elevation. *See:* **isolated by elevation.**

isolation device (nuclear power generating stations) (class 1E equipment and circuits). A device in a circuit which prevents malfunctions in one section of a circuit from causing unacceptable influences in other sections of the circuit or other circuits. 31, 131

isolation transformer*. *See:* **transformer, isolating.**

* Deprecated

isolation voltage (power supplies). A rating for a power supply that specifies the amount of external voltage that can be connected between any output terminal and ground (the chassis). This rating is important when power supplies are connected in series. *See also:* **power supply.** 228, 186

isolator (1) (switchgear). *See:* **disconnecting or isolating switch.**

(2) (waveguide). A passive attenuator in which the loss in one direction is much greater than that in the opposite direction. *See:* **waveguide.** 244, 179

isolux (isocandela) (isofootcandle) line. A line plotted on any appropriate coordinates to show all the points on a surface where the illumination is the same. *Note:* For a complete exploration the line is a closed curve. A series of such lines for various illumination values is called an isolux (isofootcandle) diagram. *See also:* **lamp.** 167

isophase. *See* **equal interval.**

isopreference (speech quality measurements). Two speech signals are isopreferent when the votes averaged over all listeners show an equal preference for the speech test and speech reference signals. 126

isotropic radiator. A hypothetical antenna having equal radiation intensity in all directions. *Note:* An isotropic radiator represents a convenient reference for expressing the directive properties of actual antennas. 111

I·T product (electric supply circuit). The inductive influence usually expressed in terms of the product of its root-mean-square magnitude in amperes I times its telephone influence factor (TIF), abbreviated as $I \cdot T$ product. *See also:* **inductive coordination.**   328

I^2t **characteristic (fuse).** The amount of ampere-squared seconds passed by the fuse during a specified period and under specified conditions. *Notes:* (1) The specified period may be the melting, arcing, or total clearing time. The sum of melting and arcing I^2t is the clearing I^2t. (2) The melting I^2t characteristic is related to a specified current wave shape, and the arcing I^2t to specified voltage and circuit impedance conditions.   103, 202

item (1) (computing systems). A collection of related characters, treated as a unit. *See also:* **file.**   255
(2) (reliability). An all-inclusive term to denote any level of hardware assembly; that is, system, segment of a system, subsystem, equipment, component, part, etcetera. *Note:* Item includes items, population of items, sample, etcetera, where the context of its use so justifies. *See also:* **reliability.**   182

item, nonrepaired (reliability). An item that is not repaired after a failure. *See also:* **reliability.**   182
item, repaired (reliability). An item that is repaired after a failure. *See also:* **reliability.**   182
iterative (test, measurement and diagnostic equipment). Describing a procedure or process which repeatedly exe-

cutes a series of operations until some condition is satisfied. An iterative procedure may be implemented by a loop in a routine.   54

iterative impedance. *See:* **impedance, iterative.** *See also:* **self-impedance.**

iterative operation (analog computer). Similar in many respects to repetitive operation, except that the automatic recycling of the computer is controlled by programmed logic circuits, which generally include a program change for a parameter(s), variable(s), or combinations of these between successive solutions, resulting in an iterative process which tends to converge on desired values of the parameter(s) or variables(s) that have been changed. *See:* **repetitive operation.**   9

I video signal (NTSC color television). One of the two video signals (E_I' and E_Q') controlling the chrominance in the NTSC system. *Note:* It is a linear combination of gamma-corrected primary color signals, E_R', E_G', and E_B' as follows:

$$E_I' = -0.27(E_B' - E_Y') + 0.74(E_R' - E_Y')$$
$$= 0.60E_R' - 0.28E_G' - 0.32E_B' \qquad 18$$

J

jack (electric circuits). A connecting device, ordinarily designed for use in a fixed location, to which a wire or wires of a circuit may be attached and that is arranged for the insertion of a plug.   341
jack bolt (rotating machinery). A bolt used to position or load an object.   63
jacket (1) (primary dry cell). An external covering of insulating material, closed at the bottom. *See also:* **electrolytic cell.**   328
(2) (cable). A thermoplastic or thermosetting covering, sometimes fabric reinforced, applied over the insulation, core, metallic sheath, or armor of a cable. *See also:* **power distribution, underground construction.**   57
jack shaft (rotating machinery). A separate shaft carried on its own bearings and connected to the shaft of a machine. *See also:* **rotor (rotating machinery).**   63
jack system (rotating machinery). A system design to raise the rotor of a machine. *See:* **rotor (rotating machinery).**   63

jamming (radar). A form of electronic countermeasures (ECM) in which noise or noise-like signals are transmitted at frequencies in the receiver bandwidths of a radar to obscure the radar signal.   13
jam transfer (hybrid computer linkage components). The transfer operation, in a double-buffered digital-to-analog converter (DAC) or digital-to-analog multiplier (DAM), in which the digital value is simultaneously loaded into both the holding and dynamic registers. *See:* **dynamic register; holding register.**   10
jar (storage cell). The container for the element and electrolyte of a lead-acid storage cell and unattacked by the electrolyte. *See also:* **battery (primary or secondary).**   328

J display (radar). A modified *A* display in which the time base is a circle and targets appear as radial deflections from the time base. *See also:* **navigation.**   13, 187

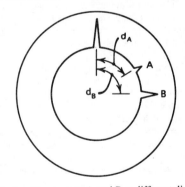

J display. Two targets A and B at different distances.

jitter (1) (repetitive wave). (A) Time-related, abrupt, spurious variations in the duration of any specified, related interval. (B) Amplitude-related, abrupt, spurious variations in the magnitude of successive cycles. (C) Frequency-related, abrupt, spurious variations in the frequency of successive pulses. (D) Phase related, abrupt, spurious variations in the phase of the frequency modulation of successive pulses as referenced to the phase of a continuous oscillator. *Note:* Qualitative use of jitter requires the use of a generic derivation of one of the above categories to identify whether the jitter is time, amplitude, frequency, or phase related and to specify which form within the category, for example, pulse-delay-time jitter,

pulse-duration jitter, pulse-separation jitter, etcetera. Quantitative use of jitter requires that a specified measure of the time- or amplitude-related variation, for example, average, root-mean-square, peak-to-peak, etcetera, be included in addition to the generic term that specifies whether the jitter is time, amplitude, frequency, or phase related. *See also:* **pulse; pulse techniques.** 185
(2) (oscilloscopes, electronic navigation, and television). Small, rapid aberrations in the size or position of a repetitive display indicating spurious deviations of the signal or instability of the display circuit. *Note:* Frequently caused by mechanical and electronic switching systems or faulty components. Generally continuous, but may be random or periodic. *See:* **fortuitous distortion; navigation; oscillograph; television.** 163, 185, 187, 13
(3) (facsimile). Raggedness in the received copy caused by erroneous displacement of recorded spots in the direction of scanning. *See also:* **recording (facsimile).** 12
(4) (radar). (A) Small, rapid and generally continuous variations in the size, shape, or position of observable information, frequently caused by mechanical and electronic switching systems of faulty components. Also refers to zero-mean random errors in successive target positions measurements due to target echo characteristics, propagation or receiver thermal noise. (B) Intentional variation of a radar parameter, for example, pulse interval. 13
(5) (pulse terms). Dispersion of a time parameter of the pulse waveforms in a pulse train with respect to a reference time, interval, or duration. Unless otherwise specified by a mathematical adjective, peak-to-peak jitter is assumed. *See:* **mathematical adjectives.** 254
jog (inch) (control) (industrial control). A control function that provides for the momentary operation of a drive for the purpose of accomplishing a small movement of the driven machine. *See:* **electric drive.** 225, 206
jog control point, supervisory control. *See:* **supervisory control point, jog control.**
jogging (inching). The quickly repeated closure of the circuit to start a motor from rest for the purpose of accomplishing small movements of the driven machine. *See also:* **electric drive.** 328
jogging speed (industrial control). The steady-state speed that would be attained if the jogging pilot device contacts were maintained closed. *Note:* It may be expressed either as an absolute magnitude of speed or a percentage of maximum rated speed. *See:* **control system, feedback.** 219, 206
Johnson noise (interference terminology). The noise caused by thermal agitation (of electron charge) in a dissipative body. *Notes:* (1) The available thermal (Johnson) noise power N from a resistor at temperature T is $N = kT\Delta f$, where k is Boltzmann's constant and Δf is the frequency increment. (2) The noise power distribution is equal throughout the radio frequency spectrum, that is, the noise power is equal in all equal frequency increments. *See:* **signal.** 188
joint (interior wiring). A connection between two or more conductors. 328
joint, compression (conductor stringing equipment). A tubular compression fitting designed and fabricated from aluminum, copper or steel to join conductors or overhead ground wires. It is usually applied through the use of hydraulic or mechanical presses. However, in some cases, automatic, wedge type joints are utilized. *Syn.* **sleeve;**

splice. 45
joint insulation. *See:* **connection insulation.**
joint use (transmission and distribution). Simultaneous use by two or more kinds of utilities. 262
Jordan bearing. A sleeve bearing and thrust bearing combined in a single unit. *See:* **bearing.** 63
joule (1) (general). The unit of work and energy in the International System of Units (SI). The joule is the work done by a force of 1 newton acting through a distance of 1 meter. 210
(2) (laser-maser) (J). A unit of energy; one (1) joule = 1 watt·second. 363
Joule effect. The evolution of thermal energy produced by an electric current in a conductor as a consequence of the electric resistance of the conductor. *See also:* **thermoelectric device; Joule's law.** 248, 191
Joule heat. The thermal energy resulting from the Joule effect. *See also:* **thermoelectric device.** 248, 191
Joule's law (heating effect of a current). The rate at which heat is produced in an electric circuit of constant resistance is equal to the product of the resistance and the square of the current. 210
journal (shaft). A cylindrical section of a shaft that is intended to rotate inside a bearing. *See also:* **bearing; armature.** 63
journal bearing (rotating machinery). A bearing that supports the cylindrical journal of a shaft. *See also:* **bearing.** 63
J scan (electronic navigation). *See:* **J display.**
J scope (class-J oscilloscope). A cathode-ray oscilloscope arranged to present a type-J display. *See:* **J display.** *See also:* **radar.** 328
jump (electronic computation). (1) To (conditionally or unconditionally) cause the next instruction to be obtained from a storage location specified by an address part of the current instruction when otherwise it would be specified by some convention. (2) An instruction that specifies a jump. *Note:* If every instruction in the instruction code specifies the location of the next instruction (for example, in a three-plus-one-address code), then each one is not called a jump instruction unless it has two or more address parts that are conditionally selected for the jump. *See also:* **conditional jump; transfer; unconditional jump.** 235
jumper (1) (general). A short length of cable used to make electric connections within, between, among, and around circuits and their associated equipment. *Note:* It is usually a temporary connection. *See also:* **conductor.** 64, 57
(2) (electronic and electrical applications). A direct electric connection between two or more points on a printed-wiring or terminal board. *See also:* **soldered connections (electronic and electrical applications).** 284
(3) (electric vehicles) (bus line jumper) (train line jumper). A flexible conductor or group of conductors arranged for connecting electric circuits between adjacent vehicles. *Note:* As applied to vehicles, it usually consists of two electric coupler plugs with connecting cable, although sometimes one end of the connecting cable is permanently attached to the vehicle. *See also:* **multiple-control unit.** 328
(4) (telephone switching systems). Crossconnection wire(s). 55
(5) (conductor stringing equipment). (A) The conductor that connects the conductors on opposite sides of a deadend structure. (B) A conductor placed across the clear space

between the ends of two conductors or metal pulling lines which are being spliced together. Its purpose then is to act as a shunt to prevent workmen from accidentally placing themselves in series between the two conductors. *Syn.* **deadend loop.** 45

junction (1). *See:* **summing junction.**

(2) (thermoelectric device). The transition region between two dissimilar conducting materials. *See also:* **thermoelectric device.** 248, 191

(3) A connection between two or more conductors or two or more sections of transmission line.

(4) A contact between two dissimilar metals or materials, as in a rectifier or thermocouple. 328

(5) (semiconductor device) (semiconductor radiation detectors). A region of transition between semiconductor regions of different electrical properties (for example, n-n^+, p-n, p-p^+ semiconductors), or between a metal and a semiconductor. *See also:* **semiconductor; semiconductor device.** 237, 245, 243, 66, 210, 186, 23

junction, alloy (semiconductor). *See:* **alloy or fused junction.**

junction box (1) (interior wiring). A box with a blank cover that serves the purpose of joining different runs of raceway or cable and provides space for the connection and branching of the enclosed conductors. 328

(2). An enclosed distribution panel for connecting or branching one or more corresponding electric circuits without the use of permanent splices. *See also:* **cabinet.** 64

(3) (mine type). A stationary piece of apparatus with enclosure by means of which one or more electric circuits for supplying mining equipment are connected through overcurrent protective devices to an incoming feeder circuit. *See also:* **mine feeder circuit.** 328

junction, collector (semiconductor device). A junction normally biased in the high-resistance direction, the current through which can be controlled by the introduction of minority carriers. *Note:* The polarity of the voltage across the junction reverses when a switching occurs. *See also:* **semiconductor; semiconductor device; transistor.** 210, 245, 66

junction depth (1) (photoelectric converter). The distance from the illuminated surface to the junction in a photoelectric converter. *See also:* **semiconductor.** 186

(2) (p-n semiconductor radiation detector). The distance below the crystal surface at which the conductivity type changes. 118, 119

(3) (solar cells). The distance from the illuminated surface to the center line of the junction in a solar cell. 113

junction, diffused (semiconductor). A junction that has been formed by the diffusion of an impurity within a semiconductor crystal. *See also:* **semiconductor.** 245, 210, 66

junction, doped (semiconductor). A junction produced by the addition of an impurity to the melt during crystal growth. *See also:* **semiconductor.** 245, 210, 66

junction, emitter (semiconductor device). A junction normally biased in the low-resistance direction to inject minority carriers into an interelectrode region. *See also:* **semiconductor; transistor.** 210

junction, fused. *See:* **alloy or fused junction.**

junction, grown (semiconductor). A junction produced during growth of a crystal from a melt. *See also:* **semiconductor device.** 245, 210, 66

junction loss (wire communication). That part of the repetition equivalent assignable to interaction effects arising at trunk terminals. *See also:* **transmission loss.** 328

junction, n-n (semiconductor). A region of transition between two regions having different properties in n-type semiconducting material. *See also:* **semiconductor device.** 328

junction, p-n (semiconductor). A region of transition between p- and n-type semiconducting material. *See also:* **semiconductor device.** 328

junction point. *See:* **node.**

junction pole (wire communication). A pole at the end of a transposition section of an open wire line or the pole common to two adjacent transposition sections. *See also:* **open wire.** 328

junction, p-p (semiconductor). A region of transition between two regions having different properties in p-type semiconducting material. *See also:* **semiconductor device.** 328

junction, rate-grown (semiconductor). A grown junction produced by varying the rate of crystal growth. *See also:* **semiconductor device.** 245, 210, 66

junction resistance (thermoelectric device). The difference between the resistance of two joined materials and the sum of the resistances of the unjoined materials. *See also:* **thermoelectric device.** 248, 191

junction temperature (T_J) (light emitting diodes). The temperature of the semiconductor junction. 162

junction transposition (s-pole transposition) (wire communication). A transposition located at the junction pole (s pole) between two transposition sections of an open wire line. *See also:* **open wire.** 328

junctor (1) (crossbar systems) (wire communication). A circuit extending between frames of a switching unit and terminating in a switching device on each frame. 328

(2) (telephone switching systems). Within a switching system, a connection or circuit between inlets and outlets of the same or different switching networks. 55

just operate value, relay. *See:* **relay just operate value.**

just scale (acoustics). A musical scale formed by taking three consecutive triads each having the ratio 4:5:6, or 10:12:15. *Note:* Consecutive triads are triads such that the highest note of one is the lowest note of the next. 176

K

K-band radar. A radar operating at frequencies between 18 and 27 gigahertz, usually in the International Telecommunications Union assigned band 23 to 24.2 gigahertz. 13

K$_a$-band radar. A radar operating at frequencies between 27 and 40 gigahertz usually in the International Telecommunications Union assigned band 33.4 to 36 gigahertz. 13

K display (radar). A modified *A* display in which a target appears as a pair of vertical deflections; when the radar antenna is correctly pointed at the target the deflections (blips) are of equal height, and when not so pointed, the difference in blip height is an indication of the direction and magnitude of pointing error. *See also:* **navigation.**

187, 13

K display. Two targets A and B at different distances; radar aimed at target A.

keep-alive circuit (transmit-receive or antitransmit-receive switch). A circuit for producing residual ionization for the purpose of reducing the initiation time of the main discharge. *See also:* **navigation.** 13, 187

kelvin. Unit of thermodynamic temperature, is the fraction 1/273.16 of the thermodynamic temperature of the triple point of water. *Notes:* (1) In 1967 the General Conference on Weights and Measures gave the name kelvin to the International Standard Unit (SI) of temperature (which had previously been called degree Kelvin) and assigned the symbol K without the symbol °. (2) The 13th CGPM, 1967, Resolution 3, also decided that the unit kelvin and its symbol K should be used to express an interval or a difference of temperature. (3) In addition to the thermodynamic temperature (symbol *T*), expressed in kelvins, use is also made of Celsius temperature (symbol *t*) defined by the equation

$$t = T - T_0$$

where $T_0 = 273.15$ K by definition. The Celsius temperature is in general expressed in degrees. Celsius (symbol °C). The unit degree Celsius is thus equal to the unit Kelvin and an interval or a difference of Celsius temperature may also be expressed in degrees Celsius. 21

Kelvin bridge (double bridge) (Thomson bridge). A 7-arm bridge intended for comparing the 4-terminal resistances of two 4-terminal resistors or networks, and characterized by the use of a pair of auxiliary resistance arms of known ratio that span the adjacent potential terminals of the two 4-terminal resistors that are connected in series by a conductor joining their adjacent current terminals. *See also:* **bridge.** 328

$$R_x = R_s \frac{R_2}{R_1} - \frac{R_c R_3}{R_3 + R_4 + R_c}(R_4/R_3 - R_2/R_1)$$

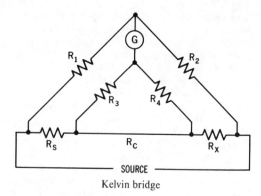

Kelvin bridge

Kendall effect (facsimile). A spurious pattern or other distortion in a facsimile record, caused by unwanted modulation products arising from the transmission of a carrier signal, appearing in the form of a rectified baseband that interferes with the lower sideband of the carrier. *Note:* This occurs principally when the single sideband width is greater than half the facsimile carrier frequency. *See also:* **recording (facsimile).** 12

kenotron (tube or valve). A hot-cathode vacuum diode. *Note:* This term is used primarily in the industrial and X-ray fields. 190

Kerr electrostatic effect. *See:* **electrooptical effect in dielectrics.**

key (1) (telephone switching system). A hand-operated switching device ordinarily comprising concealed spring contacts with an exposed handle or pushbutton, capable of closing or opening one or more parts of a circuit. 193, 55

(2) (rotating machinery). A bar that by being recessed partly in each of two adjacent members serves to transmit a force from one to the other. *See also:* **rotor (rotating machinery).** 63

(3) (computing systems). One or more characters used to identify an item of data. 255, 77

keyboard (test, measurement and diagnostic equipment). A device for the encoding of data by key depression which causes the generation of the selected code element. 54

keyer. A device that changes the output of a transmitter from one value of amplitude or frequency to another in accordance with the intelligence to be transmitted. *Note:* This applies generally to telegraph keying. *See also:* **radio transmission.** 111

keying (1) (modulating systems). Modulation involving a series of selections from a finite set of discrete states. *See also:* **telegraphy.** 242

(2) (telegraph). The forming of signals, such as those employed in telegraph transmission, by an abrupt modulation of the output of a direct-current or an alternating-current source as, for example, by interrupting it or by suddenly changing its amplitude or frequency or some other characteristic. *See also:* **telegraphy.** 111

(3) (television). A signal that enables or disables a network during selected time intervals. *See:* **television.** 178

keying interval (periodically keyed transmission system) (modulation systems). One of the set of intervals starting from a change in state and equal in length to the shortest time between changes of state. 242

keying rate (modulation systems). The reciprocal of the duration of the keying interval.   242

keying wave (telegraphic communication). *See:* **marking wave.**

keyless ringing (telephony). A form of machine ringing on manual switchboards that is started automatically by the insertion of the calling plug into the jack of the called line.   328

key light (television). The apparent principal source of directional illumination falling upon a subject or area. *See also:* **television lighting.**   167

key pulsing (telephone switching systems). A switchboard arrangement using a nonlocking keyset and providing for the transmission of a signal corresponding to each of the keys depressed.   55

key-pulsing signal (telephone switching systems). In multifrequency and key pulsing, a signal used to prepare the equipment for receiving digits.   55

keypunch (electronic computation). A keyboard-actuated device that punches holes in a card to represent data.   255, 77

keyshelf (telephone switching systems). The shelf on which are mounted control keys for use by operators or other personnel.   55

keystone distortion (television). A form of geometric distortion that results in a trapezoidal display of a nominally rectangular raster or picture. *See also:* **television.**   178

keyway (rotating machinery). A recess provided for a key. *See:* **key (rotating machinery).**   63

kick-sorter (British) (pulse techniques). *See:* **pulse-height analyzer.** *See also:* **pulse.**

kilogram. The unit of mass (and not of weight or of force); it is equal to the mass of the international prototype of the kilogram.   21

kilovolt-ampere rating (voltage regulator). The product of the rated load amperes and the rated range of regulation in kilovolts. *Note:* The kilovolt-ampere rating of a three-phase voltage regulator is the product of the rated load amperes and the rated range of regulation in kilovolts multiplied by 1.732. *See also:* **voltage regulator.**   257

kinematic drift (electronic navigation). *See:* **misalignment drift.**

kinescope. *See:* **picture tube.**

kinetic energy. The energy that a mechanical system possesses by virtue of its motion. *Note:* The kinetic energy of a particle at any instant is $(1/2)mv^2$, where m is the mass of the particle and v is its velocity at that instant. The kinetic energy of a body at any instant is the sum of the kinetic energies of its several particles.   210

Kingsbury bearing. *See:* **tilting-pad bearing.**

kino gun. *See:* **end injection.**

Kirchhoff's laws (electric networks). (1) The algebraic sum of the currents toward any point in a network is zero. (2) The algebraic sum of the products of the current and resistance in each of the conductors in any closed path in a network is equal to the algebraic sum of the electromotive forces in that path. *Note:* These laws apply to the instantaneous values of currents and electromotive forces, but may be extended to the phasor equivalents of sinusoidal currents and electromotive forces by replacing algebraic sum by phasor sum and by replacing resistance by impedance. *See also:* **network analysis.**   210

klydonograph (surge voltage recorder). *See:* **Lichtenberg figure camera.**

klystron. A velocity-modulated tube comprising, in principle, an input resonator, a drift space, and an output resonator. *See also:* **reflex klystron; power klystron.**   244

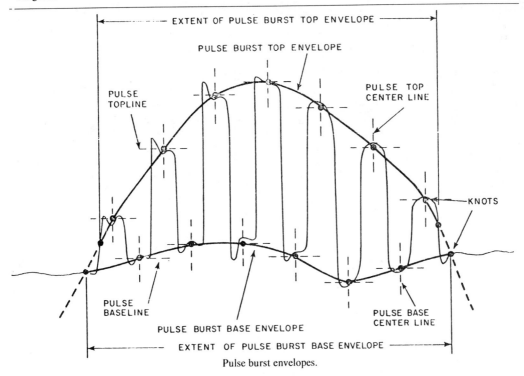

Pulse burst envelopes.

knee of transfer characteristic (image orthicons). The region of maximum curvature in the transfer characteristic. *See also:* **television.** 190

knife switch. A form of switch in which the moving element, usually a hinged blade, enters or embraces the stationary contact clips. *Note:* In some cases, however, the blade is not hinged and is removable. 103, 202

knob (1) (electronic devices). A mechanical element mounted on the end of a shaft which may be held by the fingers or within the hand for the purpose of rotating the shaft and which may include an engraved pointer or marking. 206
(2) (electric power distribution). An insulator, in one or two pieces, having a central hole for a nail or screw and one or more peripheral or chordal grooves for wire, used for supporting conductors at a definite spacing from the surface wired over. 328

knockout. A portion of the wall of a box or cabinet so fashioned that it may be removed readily by a hammer, screwdriver, and pliers, at the time of installation in order to provide a hole for the attachment of a raceway, cable or fitting. *See also:* **cabinet.** 53

knot (pulse terms). A point t_{jk} m_k ($k = 1,2,3, ..., n$) in a sequence of points wherein $t_k \lesssim t_{k+1}$ through which a spline function passes. *See:* **Pulse burst envelopes diagram on page 361.** 254

Kraemer system (rotating machinery). A system of speed control below synchronous speed, for wound-rotor induction motors, giving constant power output over the speed range. *Note:* Slip power is recovered through the medium of an independently mounted rotary converter electrically connected between the secondary windings of the induction motor and a direct-current auxiliary motor that is directly coupled to the induction-motor shaft. *See:* **asynchronous machine.** 63

K scan (electronic navigation). *See:* **K display.**

K scope (electronic navigation). *See:* **k display.**

K$_u$-band radar. A radar operating at frequencies between 12 and 18 gigahertz, usually in one of the International Telecommunications Union assigned bands 13.4 to 14.4 or 15.7 to 17.7 gigahertz. 13

kVA (kilovolt-ampere) or volt-ampere short-circuit input rating (high-reactance transformer). Designates the input kilovolt-amperes or volt-amperes at rated primary voltage with the secondary terminals short circuited. 53

kV·T product (inductive coordination). Inductive influence usually expressed in terms of the product of its root-mean-square magnitude in kilovolts (kV) times its telephone influence factor (TIF) abbreviated as kV·T product. *See also:* **inductive coordination.** 328

L

label. A key attached to the item of data that it identifies. Synonymous with **tag.** 77

laboratory reference standards. The highest ranking order of standards at each laboratory. *Note:* Laboratory reference standards set the values of the standards for that laboratory, and they in turn have their values determined by the standards of another higher-ranking laboratory, or, sometimes, by means of the reference or working standards of another category or echelon. *See also:* **measurement system.** 293, 183

laboratory reliability test. A reliability compliance or determination test made under prescribed and controlled operating and environmental conditions which may or may not simulate field conditions. 164

laboratory test. *See:* **shop test.**

laboratory working standards. Those standards that are used for the ordinary calibration work of the standardizing laboratory. *Note:* Laboratory working standards are calibrated by comparison with the reference standards of that laboratory. *See also:* **measurement system.**
 293, 183, 212

labyrinth seal ring (rotating machinery). Multiple oil catcher ring surrounding a shaft with small clearance. *See also:* **bearing.** 63

lacing, stator-winding end-wire (rotating machinery). Cord or other lacing material used to bind the stator-winding end wire, to hold in place labels and devices placed on or in the end wire, and to position lead cables at their take-off points on the end wire. *See also:* **stator.** 63

lacquer disks (cellulose nitrate disks) (electroacoustics). Mechanical recording disks usually made of metal, glass, or paper and coated with a lacquer compound (often containing cellulose nitrate). *See also:* **phonograph pickup.** 176

lacquer-film capacitor. A capacitor in which the dielectric is primarily a solid lacquer film and the electrodes are thin metallic coatings deposited thereon. 341

lacquer master*. *See:* **lacquer original.**

* Deprecated

lacquer original (electroacoustics) (lacquer master*). An original recording on a lacquer surface for the purpose of making a master. *See also:* **phonograph pickup.** 176

* Deprecated

lacquer recording (electroacoustics). Any recording made on a lacquer recording medium. *See also:* **phonograph pickup.** 176

ladder network (1) (general). A network composed of a sequence of H, L, T, or pi networks connected in tandem. *See also:* **network analysis.** 210
(2) (circuits and systems). A cascade or tandem connection of alternating series and shunt arms. 67

ladder, rope (conductor stringing equipment). A ladder having vertical synthetic or manila suspension members and wood, fiberglass or metal rungs. The ladder is suspended from the arm or bridge of a structure to enable workmen to work at the conductor level, hang travelers, perform clipping-in operations, etcetera. *Syn.* **jacobs ladder.** 45

ladder, tower (conductor stringing equipment). A ladder

complete with hooks and safety chains attached to one end of the side rails. These units are normally fabricated from fiberglass, wood or metal. The ladder is suspended from the "arm" or bridge of a structure to enable workmen to work at the conductor level, to hang travelers, perform clipping-in operations, etc. In some cases, these ladders are also used as linemen's platforms. *Syn.* **hook ladder.**
45

ladder-winding insulation (rotating machinery). An element of winding insulation in the form of a single sheet precut to fit into one or more slots and with a broad area at each end to provide end-wire insulation. *See also:* **rotor (rotating machinery); stator.**
63

lag (1) (general). The delay between two events.
255, 77

(2) (control circuits). Any retardation of an output with respect to the casual input. *See also:* **telegraphy; control system; feedback.**
56, 219, 206

(3) (telegraph system). The time elapsing between the operation of the transmitting device and the response of the receiving device. *See also:* **telegraphy.**
328

(4) (distance/velocity) (automatic control). A delay attributable to the transport of material or the finite rate of propagation of a signal or condition. *Syn.* **lag, transportation; lag transport.** *See also:* **control system; feedback.**
56

(5) (first-order). The change in phase due to a linear element of transfer function. $1/(1 + Ts)$. *Syn.* **linear lag.**

(6) (second order) (automatic control). In a linear system or element, lag which results from changes of energy storage at two separate points in the system, or from effects such as acceleration. *Note:* It is representable by a second-order differential equation, or by a quadratic factor such as $s^2 + 2z\omega_n s + \omega_n^2$ in the denominator of a transfer function. *Syn.* **quadratic lag.**
56

(7) multi-order (automatic control). In a linear system or element, lag of energy storage in two or more separate elements of the system. *Note:* It is evidenced by a differential equation of order higher than one, or by more than one time-constant. It may sometimes be approximated by a delay followed by a first-order or second-order lag.
56

(8) transfer (automatic control). *See:* **lag, first-order; lag, second-order; lag, multi-order.**
56

(9) (camera tubes). A persistence of the electrical-charge image for a small number of frames.
125

lagged-demand meter. A meter in which the indication of the maximum demand is subject to a characteristic time lag by either mechanical or thermal means. *See also:* **electricity meter.**
212

lag networks (power supplies). Resistance-reactance components, arranged to control phase-gain roll-off versus frequency. *Note:* Used to assure the dynamic stability of a power-supply's comparison amplifier. The main effect of a lag network is a reduction of gain at relatively low frequencies so that the slope of the remaining rolloff can be relatively more gentle. *See also:* **power supply.**
228, 186

lambda (electric power systems). The incremental operating cost at the load center, commonly expressed in mils per kilowatt hour.
112

lambert. A unit of luminance (photometric brightness)

equal to $1/\pi$ candela per square centimeter, and, therefore, equal to the uniform luminance of a perfectly diffusing surface emitting or reflecting light at the rate of one lumen per square centimeter. *Note:* The lambert also is the average luminance of any surface emitting or reflecting light at the rate of one lumen per square centimeter. For the general case, the average must take account of variation of luminance with angle of observation; also of its variation from point to point on the surface considered. *See also:* **light.**
18, 178, 167

lambertian surface (laser-maser). An ideal surface whose emitted or reflected radiance is independent of the viewing angle.
363

Lambert's cosine law. A law stating that the flux per solid angle in any direction from a plane surface varies as the cosine of the angle between that direction and the perpendicular to the surface. *See also:* **lamp.**
167

laminated brush (industrial control). A contact part consisting of thin sheets of conducting material fastened together so as to secure individual contact by the edges of the separate sheets.
204

laminated core (rotating machinery). A core made of electrical steel sheets, or pieces cut thereof, usually electrically insulated from each other, stacked in a parallel configuration. *See also:* **core; magnetic core.**
63

laminated frame (rotating machinery). A stator frame formed from laminations clamped, bolted, or riveted together with or without additional strengthening plates. *See also:* **stator.**
63

laminated record (electroacoustics). A mechanical recording medium composed of several layers of material. *Note:* Normally, it is made with a thin face of surface material on each side of a core. *See also:* **phonograph pickup.**
176

lamination. A relatively thin member, usually made of sheet material. A complete structure is made by assembling the laminations in the required number of layers. In a core that carries alternating magnetic flux, the core material is usually laminated to reduce eddy-current losses.
63

lamp. A generic term for a man-made source of light. By extension, the term is also used to denote sources that radiate in regions of the spectrum adjacent to the visible. *Note:* A lighting unit consisting of a lamp with shade, reflector, enclosing globe, housing, or other accessories is also called a lamp. In such cases, in order to distinguish between the assembled unit and the light source within it, the latter is often called a bulb or tube; if it is electrically powered. *See also:* **luminaire.**
167

lampholder (socket*). A device intended to support an electric lamp mechanically and connect it electrically to a circuit.
328

* Deprecated

lampholder adapter. A device that by insertion in a receptacle serves as a lampholder.
328

lampman (mining). A person having responsibility for cleaning, maintaining, and servicing of miners' lamps.
328

lamp post. A standard support provided with the necessary internal attachments for wiring and the external attachments for the bracket and luminaire. *See also:* **street-**

lighting luminaire. 167

lamp regulator. A device for automatically maintaining constant voltage on a lamp circuit from some higher variable voltage power source. 328

lamp shielding angle. The angle between the plane of the baffles or louver grid and the plane most nearly horizontal that is tangent to both the lamps and the louver blades. *Note:* The lamp shielding angle frequently is larger than the lower shielding angle, but never smaller. 167

land (electroacoustics). The record surface between two adjacent grooves of a mechanical recording. *See also:* phonograph pickup. 176

landing direction indicator. A device to indicate visually the direction currently designated for landing and take-off. 167

landing light. An aircraft aeronautical light designed to illuminate a ground area from the aircraft. 167

landing zone (elevator). A zone extending from a point eighteen inches below a landing to a point eighteen inches above the landing. *See also:* elevator landing. 328

landmark beacon. An aeronautical beacon used to indicate the location of a landmark used by pilots as an aid to en-route navigation. 167

lane (navigation system) (electronic navigation). The projection of a corridor of airspace on a navigation chart, the right and left sides of the corridor being defined by the same (ambiguous) values of the navigation coordinate (phase or amplitude), but within which lateral position information is provided (for example, a Decca lane in which there is a 360-degree change of phase). *See also:* navigation. 187, 13

language (1) (general). A set of representations, conventions, and rules used to convey information. 255, 77

(2) (electronic computers). (A) A system consisting of: (1) a well defined, usually finite, set of characters; (2) rules for combining characters with one another to form words or other expressions; (3) a specific assignment of meaning to some of the words or expressions, usually for communicating information or data among a group of people, machines, etcetera. (B) A system similar to the above but without any specific assignment of meanings. Such systems may be distinguished from (A) above, when necessary, by referring to them as formal or uninterpreted languages. Although it is sometimes convenient to study a language independently of any meanings, in all practical cases at least one set of meanings is eventually assigned. *See also:* code; machine language. 235, 210, 54

language code (telephone switching systems). On an international call, an address digit that permits an originating operator to obtain assistance in a desired language. 55

language printout (dedicated-type sequential events recording systems). A word description composed of alphanumeric characters used to further identify inputs and their status. *See:* language. 48, 58

lapel microphone. A microphone adapted to positioning on the clothing of the user. *See also:* microphone. 328

Laplace transform (unilateral) (function f(t)). The quantity obtained by performing the operation

$$F(s) = \int_0^\infty f(t)e^{-st}dt$$

where $s = \sigma + j\omega$. *See also:* control system, feedback. 56

Laplace's equation. The special form taken by Poisson's equation when the volume density of charge is zero throughout the isotropic medium. It is $\nabla^2 v = 0$. 210

lap winding. A winding that completes all its turns under a given pair of main poles before proceeding to the next pair of poles. *Note:* In commutator machines the ends of the individual coils of a simplex lap winding are connected to adjacent commutator bars; those of a duplex lap winding are connected to alternate commutator bars etcetera. *See also:* asynchronous machine; direct-current commutating machine. 328

larry (mine). A motor-driven burden-bearing track-mounted car designed for side or end dumping and used for hauling material such as coal, coke, or mine refuse. 328

laser (laser-maser). A device which produces an intense, coherent, directional beam of light by stimulating electronic, ionic, or molecular transitions to lower energy levels. Also, an acronym for light amplification by stimulated emission of radiation. Also, an optical maser. 363

laser gyro. A device which measures angular rotation by optical heterodyning of counter rotating optical beams. 46

laser safety officer (laser-maser). One who is knowledgeable in the evaluation and control of laser hazards and has authority for supervision of the control of laser hazards. 363

laser system (laser-maser). An assembly of electrical, mechanical, and optical components which includes a laser. 363

lasing medium (laser-maser). A material emitting coherent radiation by virtue of stimulated electronic or molecular transitions to lower energy levels. 363

lasing threshold (laser gyro). The discharge current at which the gain of the laser just exceeds the losses. 46

last-line release (telephone switching systems). Release under control of the last line that goes on-hook. 55

last transition (pulse terms). The major transition waveform of a pulse waveform between the top and the base. 254

latch (high-voltage fuse). An attachment used to hold a fuse in the closed position. 79

latching current (1) (switching device). The making current during a closing operation in which the device latches or the equivalent. 103, 202

(2) (semiconductor). The minimum principal current required to maintain the thyristor in the on-state immediately after switching from the off-state to the on-state has occurred and the triggering signal has been removed. 66

latching relay. A relay that is so constructed that it maintains a given position by means of a mechanical latch until released mechanically or electrically. *See also:* latch-in relay; relay. 103, 202, 60, 259, 127

latch-in relay. A relay that maintains its contacts in the last position assumed without the need of maintaining coil energization. 341

latency (1) (biological electronics). The condition in an excitable tissue during the interval between the application of a stimulus and the first indication of a response. 192

(2) (electronic computation). The time between the completion of the interpretation of an address and the start of

the actual transfer from the addressed location.
255, 77

latent period (electrobiology). The time elapsing between the application of a stimulus and the first indication of a response. *See also:* **excitability (electrobiology).**   192

lateral conductor (1) (pole line work). A wire or cable extending in a general horizontal direction or at an angle to the general direction of the line. *Note:* Service wires either overhead or underground are considered laterals from the street mains. Short branches extended on poles at approximate right angles to lines are also known as laterals. *See also:* **tower.**   178, 64

(2) (transmission and distribution). A wire or cable extending in a general horizontal direction at an angle to the general direction of the line conductors.   262

lateral critical speeds (rotating machinery). The speeds at which the amplitudes of the lateral vibrations of a machine rotor due to shaft rotation reach their maximum values. *See also:* **rotor (rotating machinery).**   63

lateral-cut recording. *See:* **lateral recording.**

lateral insulator (storage cell). An insulator placed between the plates and the side wall of the container in which the element is housed. *See also:* **battery (primary or secondary).**   328

lateral profile (radio noise). The electric field strength at ground level plotted as a function of the horizontal distance from, and at a right angle to, the line conductors.   36

lateral recording (lateral-cut recording). A mechanical recording in which the groove modulation is perpendicular to the motion of the recording medium and parallel to the surface of the recording medium.   176

lateral width (light distribution). The lateral angle between the reference line and the width line, measured in the cone of maximum candlepower. *Note:* This angular width includes the line of maximum candlepower. *See also:* **street-lighting luminaire.**   167

lateral working space (electric power distribution). The space reserved for working between conductor levels outside the climbing space, and to its right and left. *See also:* **conflict (wiring system); tower.**   178, 64

lattice (navigation). A pattern of identifiable intersecting lines of position, which lines are laid down by a navigation aid. *See also:* **navigation.**   187, 13

lattice network. A network composed of four branches connected in series to form a mesh, two nonadjacent junction points serving as input terminals, while the remaining two junction points serve as output terminals. *See also:* **network analysis.**

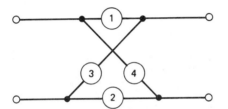

Lattice network. The junction points between branches 4 and 1 and between 3 and 2 are the input terminals; the junction points between branches 1 and 3 and between branches 2 and 4 are the output terminals.   210

launcher (waveguide components). An adapter used to provide a waveguide or transmission line port for a wave propagating structure.   166

lay (1) (cables). The helical arrangement formed by twisting together the individual elements of a cable.
64, 57

(2) (helical element of a cable). The axial length of a turn of the helix of that element. *Notes:* (1) Among the helical elements of a cable may be each strand in a concentric-lay cable, or each insulated conductor in a multiple-conductor cable. (2) Also termed pitch. *See also:* **power distribution, underground construction.**   345

lay-up (nuclear power generating stations). Idle condition of equipment and systems during and after installation, with protective measures applied as appropriate.   31

L-band radar. A radar operating at frequencies between 1 and 2 gigahertz, usually in the International Telecommunications Union assigned band 1.215 to 1.4 gigahertz; may refer also to the 0.89 to 0.94 gigahertz International Telecommunications Union assignment.   13

LC auxiliary switch (railway practice). *See:* **auxiliary switch; contact.**   103, 202

LC contact (railway practice). A latch-checking contact that is closed when the operating mechanism linkage is relatched after an opening operation of the switching device.   103, 202

L display (radar). A display in which a target appears as two horizontal blips, one extending to the right from a central vertical time base and the other to the left; both blips are of equal amplitude when the radar antenna is pointed directly at the target, any inequality representing relative pointing error, and distance upward along the baseline representing target distance. *See also:* **navigation.**
187, 13

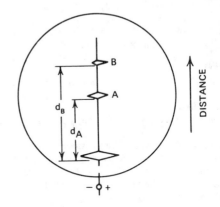

HORIZONTAL POINTING ERROR

L display. Two targets A and B at different distances; radar aimed at target A.

lead (terminal connections) (transformer). A conductor that connects a winding to its termination that is, to a terminal, bushing, terminal board, collector, or connection to another winding. *See:* **rotor (rotating machinery); stator.**
63, 53

lead box (terminal housing) (rotating machinery). A box through which the leads are passed in emerging from the housing.   63

lead cable (rotating machinery). A cable type of conductor connected to the stator winding, used for making connections to the supply line or among circuits of the stator winding. *See:* **stator.**   63

lead clamp (salient-pole construction) (rotating machinery). A device used to retain and support the field leads between the hub and rotor rim along the rotor spider arms. *See: rotor (rotating machinery).*   63

lead collar (salient-pole construction) (rotating machinery). A bushing used to insulate field leads at a point of support between a collector ring and the rotor rim. *See: rotor (rotating machinery).*   63

lead-covered cable (lead-sheathed cable). A cable provided with a sheath of lead for the purpose of excluding moisture and affording mechanical protection. *See: armored cable. See also: power distribution, underground construction.*   64

leader (computing systems). The blank section of tape at the beginning and end of a reel of tape. *Note:* The otherwise blank section may, however, include a parity check.   255, 77, 244, 207

leader cable. A navigational aid consisting of a cable around which a magnetic field is established, marking the path to be followed. *See also: radio navigation.*   328

leader-cable system. A navigational aid in which a path to be followed is defined by the detection and comparison of magnetic fields emanating from a cable system that is installed on the ground or under water. *See also: navigation.*   187, 13

leader cone (conductor stringing equipment). A tapered cone made of rubber, neoprene or polyurethane that is used to lead a conductor splice through the travelers, thus making a smooth transition from the smaller diameter conductor to the larger diameter splice. It is also used at the connection point of the pulling line and running board to assist in a smooth transition of the running board over the travelers, thus significantly reducing the shock loads. *Syn.* tapered hose; nose cone.   45

lead, first-order (control system, feedback). The change in phase due to a factor $(1 + Ts)$ in the numerator of a transfer function. *See also: control system, feedback.*   329

lead-in. That portion of an antenna system that connects the elevated conductor portion to the radio equipment. *See also: antenna.*   179

leading edge (1) (television). The major portion of the rise of a pulse. *See also: television.*   337

(2) (high- and low-power pulse transformers) (first transition). That portion of the pulse occurring between the time the instantaneous value first becomes greater than A_Q to the time of the intersection of straight-line segments used to determine A_M. *See also: input pulse shape.*   32, 33

leading edge (first transition) linearity (A_L) (high- and low-power pulse transformers). The maximum amount by which the instantaneous pulse value deviates during the rise time interval from a straight line intersecting the 10 percent and 90 percent A_M amplitude points used in determining rise time. It is expressed in amplitude units or in percent of $0.8\ A_M$. *See also: input pulse shape.*   32, 33

leading-edge pulse. The first major transition away from the pulse baseline occurring after a reference time. *See also: pulse; television.*   185

leading-edge pulse time. The time at which the instantaneous amplitude first reaches a stated fraction of the peak pulse amplitude. *See also: television.*   254, 210, 162

leading-edge tracking. A radar range tracking technique in which the range error signal is based on the range delay of the leading edge of the received echo. This provides ability to reject delayed interference, chaff and more distant sources.   13

lead-in groove (lead-in spiral) (disk recording) (electroacoustics). A blank spiral groove at the beginning of a recording generally having a pitch that is much greater than that of the recorded grooves. *See also: phonograph pickup.*   334

leading wire (mining). An insulated wire strung separately or as a twisted pair, used for connecting the two free ends of the circuit of the blasting caps to the blasting unit. *See also: blasting unit.*   328

lead-in spiral (disk recording). *See: lead-in groove.*

lead-in wire. A conductor connecting an electrode to an external circuit. *See also: electron tube.*   244, 190

lead networks (power supplies). Resistance-reactance components arranged to control phase-gain roll-off versus frequency. *Note:* Used to assure the dynamic stability of a power-supply's comparison amplifier. The main effect of a lead network is to introduce a phase lead at the higher frequencies, near the unity-gain frequency. *See also: power supply.*   228, 186

lead-out groove (throw-out spiral) (disk recording). A blank spiral groove at the end of a recording generally of a pitch that is much greater than that of the recorded grooves and that terminates in either a locked or an eccentric groove. *See also: phonograph pickup.*   176

lead-over groove (crossover spiral). In disk recording, a groove cut between successive short-duration recordings on the same disk, to enable the pickup stylus to travel from one cut to the next. *See also: phonograph pickup.*   176

lead polarity. A designation of the relative instantaneous direction of the currents in the leads of a transformer. Primary and secondary leads are said to have the same polarity when, at a given instant, the current enters the primary lead in question and leaves the secondary lead in question in the same direction as though the two leads formed a continuous circuit. The lead polarity of a single-phase distribution or power transformer may be either additive or subtractive. If adjacent leads from each of the two windings in question are connected together and voltage applied to one of the windings: (1) the lead polarity is additive if the voltage across the other two leads of the windings in question is greater than that of the higher voltage winding alone; (2) the lead polarity is subtractive if the voltage across the other two leads of the windings in question is less than that of the higher voltage winding alone. The polarity of a three-phase transformer is fixed by the internal connections between phases; it is usually designated by means of a phasor diagram showing the angular displacements of the voltages in the windings and a sketch showing the marking of the leads. The phasors of the phasor diagrams represent induced voltages. The standard rotation of phasors is counterclockwise.

leads, load. *See: load leads or transmission line.*

lead storage battery. A storage battery the electrodes of which are made of lead and the electrolyte consists of a solution of sulfuric acid. *See also: battery (primary or secondary).*   328

leakage (1) (analog computer). (A) Undesirable conductive paths in certain components, specifically, in capacitors, a path through which a slow discharge may take effect; in problem boards, interaction effects between electrical signals through insufficient insulation between patch bay terminals. (B) Current flowing through such paths.
9, 77

(2) (signal-transmission system.) Undesired current flow between parts of a signal-transmission system or between the signal-transmission system and point(s) outside the system. *See:* **signal.** 188

(3) (transmission lines and waveguides). Radiation or conduction of signal power out of or into an imperfectly closed and shielded system. The leakage is usually expressed in decibels below a specified reference power.
185

leakage (conduction) current (1) (rotating machinery). The nonreversible constant current component of measured current which remains after the capacitive current and absorption current have disappeared. *Note:* Leakage current passes through the insulation volume, through any defects in the insulation, and across the insulation surface.
6

(2) (electron tubes). A conduction current that flows between two or more electrodes by any path other than across the vacuous space between the electrodes. *See also:* **electrode current (of an electron tube); electron tube.**
244, 190

leakage current (1) (insulation). The current that flows through or across the surface of insulation and defines the insulation resistance at the specified direct-current potential. 116

(2) (semiconductor radiation detector). The total detector current flowing at the operating bias in the absence of radiation. 23, 118, 119

leakage current, input (amplifiers). A direct current (of either polarity) that would flow in a short circuit connecting the input terminals of an amplifier. 106

leakage distance (insulator). The sum of the shortest distances measured along the insulating surfaces between the conductive parts, as arranged for dry flashover test. *Note:* Surfaces coated with semiconducting glaze shall be considered as effective leakage surfaces, and leakage distance over such surfaces shall be included in the leakage distance. *See also:* **insulator.** 261

leakage distance of external insulation. *See:* **creepage distance.**

leakage flux, relay. *See:* **relay leakage flux.**

leakage inductance (one winding with respect to a second winding). A portion of the inductance of a winding that is related to a difference in flux linkages in the two windings; quantitatively, the leakage inductance of winding 1 with respect to winding 2

$$L_{\text{if}} = \frac{\partial\left(\lambda_{11} - \dfrac{N_1}{N_2}\lambda_{21}\right)}{\partial i_1}$$

where λ_{11} and λ_{21} are the flux linkages of windings 1 and 2, respectively, resulting from current i_1 in winding 1; and N_1 and N_2 are the number of turns of windings 1 and 2, respectively. 197

leakage power (TR and Pre-TR tubes). The radio-frequency power transmitted through a fired tube. *See also:* **flat leakage power; (TR and Pre-TR tubes). harmonic leakage power; (TR and Pre-TR tubes).** 103

leakage radiation (radio transmitting system). Radiation from anything other than the intended radiating system. *See also:* **radio transmission.** 111

leaky-wave antenna. An antenna that couples power in small increments per unit length, either continuously or discretely, from a traveling wave structure to free space.
111

learning system. An adaptive system with memory. *See:* **system science.** 209

leased channel (data transmission). A point-to-point channel reserved for sole use of a single leasing customer.
59

leave-word call (telephone switching systems). A person-to-person call on which the designated called person was not available and the operator left instructions for its later establishment. 55

Lecher wires. Two parallel wires on which standing waves are set up, frequently for the measurement of wavelength. *See also:* **transmission line.** 59

Leduc current* (electrotherapy). A pulsed direct current commonly having a duty cycle of 1:10. *See also:* **electrotherapy.** 192

* Deprecated

left-handed (counterclockwise) polarized wave (radio wave propagation). An elliptically polarized transverse electromagnetic wave in which the rotation of the electric field vector with time is counterclockwise for a stationary observer looking in the direction of the wave normal. *Note:* For an observer looking from a receiver toward the apparent source of the wave, the direction of rotation is reversed. *See also:* **radiation; radio wave propagation.**
180, 146

leg (circuit). Any one of the conductors of an electric supply circuit between which is maintained the maximum supply voltage. *See also:* **center of distribution.** 64

leg, thermoelectric. Alternative term for thermoelectric arm. *See also:* **thermoelectric device.** 248, 191

leg wire (mining). One of the two wires attached to and forming a part of an electric blasting cap or squib. *See also:* **blasting unit.** 328

Lenard tube. An electron beam tube in which the beam can be taken through a section of the wall of the evacuated enclosure. *See also:* **miscellaneous electron devices.**
190

length (1) (electronic computers). (A) A measure of the magnitude of a unit of data, usually expressed as a number of subunits, for example, the length of a record is 32 blocks, the length of a word is 40 binary digits, etcetera. (B) The number of subunits of data, usually digits or characters, that can be simultaneously stored linearly in a given device, for example, the length of the register is 12 decimal digits or the length of the counter is 40 binary digits. (C) A measure of the amount of time that data are delayed when being transmitted from point to point, for example, the length of the delay line is 384 microseconds. *See:* **double length; word length.** *See also:* **storage capacity.** 235

(2) (antenna). *See:* **effective length.**

length of lay (cable) (power distribution underground cables).
The axial length of one turn of the helix of any helical element. *See also:* **power distribution, underground construction.**　　　　　　　　　　　　　　57

lens and aperture (phototube housing) (industrial control).
The cooperating arrangement of a light-refracting member and an opening in an opaque diaphragm through which all light reaching the phototube cathode must pass. *See also:* **electronic controller.**　　　　　　　　206

lens antenna. An antenna consisting of an electromagnetic lens and a radiating feed.　　　　　　　　111

lens distance relay (relaying). A distance relay that has an operating characteristic comprising the common area of two intersecting mho relay characteristics (Figure 1b).　　　　　　　　79

value. *Notes:* (A) Examples of kinds of levels in common use are electric power level, sound-pressure level, voltage level. (B) The level as here defined is measured in two common units; in decibels when the logarithmic base is 10, or in nepers when the logarithmic base is e. The decibel requires that k be 10 for ratios of power, or 20 for quantities proportional to the square root of power. The neper is used to represent ratios of voltage, current, sound pressure, and particle velocity. The neper requires that k be 1. (C) In symbols $L = k \log_r(q/q_0)$ where L = level of kind determined by the kind of quantity under consideration. r = base of the logarithm of the reference ratio. q = the quantity under consideration. q_0 = reference quantity of the same kind. k = a multiplier that depends upon the base of the logarithm and the nature of the ref-

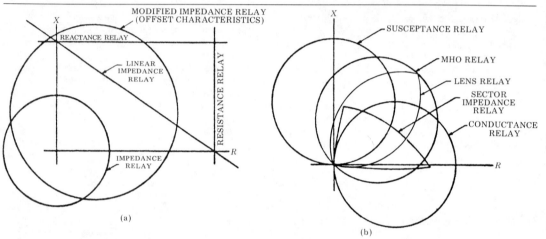

Fig. 1. (a) Operating characteristic of distance relays.　(b) Operating characteristics of distance relays that are inherently directional.

lens, electromagnetic. A three-dimensional structure propagating electromagnetic waves, with an effective index of refraction differing from unity, employed to control the aperture illumination.　　　　111

lens multiplication factor (phototube housing) (industrial control). The maximum ratio of the light flux reaching the phototube cathode with the lens and aperture in place to the light flux with the lens and aperture removed. *See also:* **photoelectric control.**　　　206

Lenz's law (induced current). The current in a conductor as a result of an induced voltage is such that the change in magnetic flux due to it is opposite to the change in flux that caused the induced voltage.　　　210

letter. An alphabetic character used for the representation of sounds in a spoken language.　　　255, 77

letter combination (abbreviation). One or more letters that form a part of the graphic symbol for an item and denote its distinguishing characteristic. *Compare with:* **functional designation; reference designation; symbol for a quantity.** *See also:* **abbreviation.**　　　173

level (1) (general) (quantity). Magnitude, especially when considered in relation to an arbitrary reference value. Level may be stated in the units in which the quantity itself is measured (for example, volts, ohms, etcetera) or in units (for example, decibels) expressing the ratio to a reference

erence quantity. *See:* **blanking level; transmission level.** *See also:* **reference black level; reference white level; signal level (electroacoustics).**　　　59, 176

(2) **(charge-storage tubes).** A charge value which can be stored in a given storage element and distinguished in the output from other charge values. *See also:* **channel.**　　　125

level above threshold (sensation level) (sound) (acoustics). The pressure level of the sound in decibels above its threshold of audibility for the individual observer.　176

level band pressure (for a specified frequency band) (electroacoustics). The sound-pressure level for the sound contained within the restricted band. *Notes:* (1) The reference pressure must be specified. (2) The band may be specified by its lower and upper cutoff frequencies or by its geometric or arithmetic center frequency and bandwidth. The width of the band may be indicated by a prefatory modifier, for example, octave band (sound pressure) level, half-octave band level, third-octave band level, 50-hertz band level. *See also:* **loudspeaker.**　　　176

level compensator (signal transmission). An automatic transmission-regulating feature or device to minimize the effects of variations in amplitude of the received signal. *See also:* **telegraphy.**　　　328

level detector (relaying). A device that produces a change in output at a prescribed input level. 79

leveling (electronic navigation). *See:* **platform erection.**

leveling block (rotating machinery). *See:* **leveling plate.**

leveling plate (leveling block) (rotating machinery). A heavy pad built into the foundation and used to support and align the bed plate or rails using shims for adjustment before grouting. 63

leveling zone (elevator). The limited distance above or below an elevator landing within which the leveling device may cause movement of the car toward the landing. *See also:* **elevator-car leveling device.** 328

level, relay. *See:* **relay level.**

levels, usable (storage tubes). The output levels, each related to a different input, that can be distinguished from one another regardless of location on the storage surface. *Note:* The number of usable levels is normally limited by shading and disturbance. *See also:* **storage tube.**
174, 190

lever blocking device (railway signaling). A device for blocking a lever so that it cannot be operated. 328

lever indication (railway signaling). The information conveyed by means of an indication lock that the movement of an operated unit has been completed. 328

liberator tank (electrorefining). Sometimes known as a depositing-out tank, an electrolytic cell equipped with insoluble anodes for the purpose of either decreasing, or totally removing the metal content of the electrolyte by plating it out on cathodes. *See also:* **electrorefining.**
328

library. A collection of organized information used for study and reference. *See:* **program library.** 255, 77

library routine (computing systems). A commonly used routine that is maintained in a program library.
255, 77

Lichtenberg figure camera (klydonograph) (surge-voltage recorder). A device for indicating the polarity and approximate crest value of the voltage surge by the appearance and dimensions of the Lichtenberg figure produced on a photographic plate or film, the emulsion coating of which is in contact with a small electrode coupled to the circuit in which the surge occurs. *Note:* The film is backed by an extended plane electrode. *See also:* **instrument.**
328

life performance curve (illuminating engineering). A curve that represents the variation of a particular characteristic of a light source (luminous flux, candlepower, etcetera) throughout the life of the source. *Note:* Life performance curves sometimes are called maintenance curves as, for example, lumen maintenance curves. *See also:* **lamp.**
167

life test. *See:* **accelerated test (reliability).**

life test of lamps. A test in which lamps are operated under specified conditions for a specified length of time, for the purpose of obtaining information on lamp life. *Note:* Measurements of photometric and electrical characteristics may be made at specified intervals of time during this test. *See also:* **lamp.** 167

lifetime, volume (semiconductor). The average time interval between the generation and recombination of minority carriers in a homogeneous semiconductor. *See also:* **semiconductor; semiconductor device.** 210, 245, 66

lifter, insulator (conductor stringing equipment). A device

designed to permit insulators to be lifted in a "string" to their intended position on a structure. *Syn.* **insulator saddle; potty seat.** 45

lifting bar. A bar provided as part of an apparatus to permit attachment of a lifting device. 63

lifting eye (fuse holder, fuse unit, or disconnecting blade). An eye provided for receiving a fuse hook or switch hook for inserting the fuse or disconnecting blade into and for removing it from the fuse support. 79

lifting hole. A hole provided to permit attachment of a lifting device. 63

lifting-insulator switch. A switch in which one or more insulators remain attached to the blade, move with it, and lift it to the open position. 103, 202, 27

light (1) (illuminating engineering) (light emitting diodes). Visually evaluated radiant energy. *Note:* Light is psychophysical, neither purely physical nor purely psychological. Light is not synonymous with radiant energy, however restricted, nor is it merely sensation. In a general nonspecialized sense, light is the aspect of radiant energy of which a human observer is aware through the stimulation of the retina of the eye. The present basis for the engineering evaluation of light consists of the color-mixture data $\bar{x}, \bar{y}, \bar{z}$ (*see:* **spectral tristimulus values**) adopted in 1931 by the International Commission on Illumination (CIE). These data include the spectral luminous efficiency data for photopic vision. (*See:* **spectral luminous efficiency for photoptic vision**) adopted in 1924 by the CIE.
18, 167

(2) (color television). Any radiation capable of causing a visual sensation directly. *Note:* The wavelength range of such radiation can be considered, for practical purposes, to lie between 380 and 780 nm. 18

light adaptation. The process by which the retina becomes adapted to a luminance greater than about one footlambert (3 nits). *See also:* **visual field.** 167

light amplifier (optoelectronic device). An amplifier whose signal input and output consist of light. *See also:* **optoelectronic device.** 191

light-beam instrument. An instrument that utilizes the position of a beam of light on a scale as its indicating means. *See also:* **instrument.** 328

light carrier injection (facsimile). The method of introducing the carrier by periodic variation of the scanner light beam, the average amplitude of which is varied by the density changes of the subject copy. *See also:* **scanning (facsimile).**
12

light center (lamp). The center of the smallest sphere that would completely contain the light-emitting element of the lamp. *See also:* **lamp.** 167

light center length (lamp). The distance from the light center to a specified reference point on the lamp. *See also:* **lamp.** 167

light emitting diode (solid state lamp). A semiconductor device containing a semiconductor junction in which visible light is nonthermally produced when a forward current flows as a result of an applied voltage. 162

lighting branch circuit (electric installations on shipboard). A circuit supplying energy to lighting outlets only. *Note:* Lighting branch circuits also may supply portable desk or bracket fans, small heating appliances, motors of ¼ horsepower (186½ W) and less, and other portable apparatus of not over 600 W each. 3

lighting feeder. A feeder supplying principally a lighting load. *See:* **feeder.** 328

lighting outlet. An outlet intended for the direct connection of a lampholder, a lighting fixture, or a pendent cord terminating in a lampholder. *See also:* **outlet.** 328

light-load test. A test on a machine connected to its normal driven or driving member in which the shaft power is restricted to the no-load loss of the member. 63

light microsecond. The distance over which light travels in free space in one microsecond, that is, about 983 feet (300 meters). This distance is employed as a unit for expressing electrical distance. *See:* **electrical distance.** *See also:* **signal wave.** 13

light modulator (audio and electroacoustics). A means for varying a light beam as a function of a controlling signal. *Note:* The resulting variation in the light beam may be used to produce a motion-picture sound track. 176

lightness (perceived object color). The attribute by which an object seems to transmit or reflect a greater or lesser fraction of the incident light. *See also:* **color.** 167

lightning. An electric discharge that occurs in the atmosphere between clouds or between clouds and ground. 299, 62

lightning arrester (surge diverter).* *See:* **surge arrester.** 178, 62

* Deprecated

lightning generator (impulse generator). *See:* **surge generator.**

lightning impulse insulation level. An insulation level expressed in kilovolts of the crest value of a lightning impulse withstand voltage. 53

lightning impulse protection level (protective device). The maximum lightning impulse voltage expected at the terminals of a surge protective device under specified conditions of operation. 53

lightning impulse test (transformer). Application of the following sequence of impulse waves: (1) one reduced full wave; (2) two chopped waves; (3) one full wave. 53

lightning outage. A power outage following a lightning flashover that results in power follow current necessitating the operation of a switching device to clear the fault. *See also:* **direct-stroke protection (lightning).** 64

lightning stroke. A single lightning discharge or series of discharges following the same path between cloud regions or between cloud regions and the earth. *See also:* **direct-stroke protection (lightning).** 64

lightning-stroke component. The high-current discharge of a lightning stroke. *See also:* **direct-stroke protection (lightning).** 64

lightning-stroke current. The crest magnitude of the current in any of the discharges of a lightning stroke. *See also:* **direct-stroke protection (lightning).** 64

lightning-stroke voltage. The crest magnitude of the leader voltage required to produce the stroke current. *See also:* **direct-stroke protection (lightning).** 64

lightning surge. A transient electric disturbance in an electric circuit caused by lightning. *See:* **lightning.** 299, 62, 2

light pattern (optical pattern) (Buchmann-Meyer pattern) (mechanical recording). A pattern that is observed when the surface of the record is illuminated by a light beam of essentially parallel rays. *Note:* The width of the observed pattern is approximately proportional to the signal velocity of the recorded groove. 176

light pipe. An optical transmission element that utilizes unfocused transmission and reflection to reduce photon losses. *Note:* Light pipes have been used to distribute the light more uniformly over a photocathode. *See also:* **phototube.** 335

light source (industrial control). A device to supply radiant energy capable of exciting a phototube or photocell. *See also:* **photoelectric control.** 206

light-source color. The color of the light emitted by the source. *Note:* The color of a point source may be defined by its luminous intensity and chromaticity coordinates; the color of an extended source may be defined by its luminance and chromaticity coordinates. *See also:* **color.** 167

light transition load (rectifier circuit). The transition load that occurs at light load, usually at less than 5 percent of rated load. *Note:* Light transition load is important in multiple rectifier circuits. A similar effect occurs in rectifier units using saturable-reactor control. *See also:* **rectification; rectifier circuit element.** 237, 66

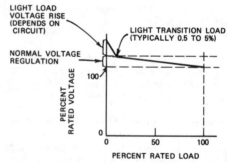

Voltage regulation characteristic showing light transition load.

light valve (electroacoustics). A device whose light transmission can be made to vary in accordance with an externally applied electrical quantity, such as voltage, current, electric field, magnetic field, or an electron beam. *See also:* **phonograph pickup.** 176, 178

lightwatt. *See:* **spectral luminous efficiency.**

limit (1) (mathematical). A boundary of a controlled variable.

(2) (synchronous machine regulating systems). The boundary at or beyond which a limiter functions. 63

(3) (industrial control). The designated quantity is controlled so as not to exceed a prescribed boundary condition. *See also:* **control system, feedback.** 225, 206

(4) (test, measurement and diagnostic equipment). The extreme of the designated range through which the measured value of a characteristic may vary and still be considered acceptable. 54

limit cycle (control systems). A closed curve in the state space of a particular control system, from which state trajectories may recede, or which they may approach, for all initial states sufficiently close to the curve. 56

limit cycle, stable (control systems). One which is approached asymptotically by a state trajectory for all initial states sufficiently close.   56

limit cycle, unstable (control systems). One from which state trajectories recede for all initial states sufficiently close.   56

limited availability (telephone switching systems). Availability that is less than the number of outlets in the desired group.   55

limited proportionality, region of. *See:* **region of limited proportionality.**

limited signal (radar). A signal that is limited in amplitude by the dynamic range of the system.   13, 187

limited stability. A property of a system characterized by stability when the input signal falls within a particular range and by instability when the signal falls outside this range.   349

limiter (1) (general). A device in which some characteristic of the output is automatically prevented from exceeding a predetermined value. More specifically, a circuit in which the output amplitude is substantially linear with regard to the input up to a predetermined value and substantially constant thereafter. *Note:* For waves having both positive and negative values, the predetermined value is usually independent of sign.   59

(2) (rotating machinery). An element or group of elements that acts to limit by modifying or replacing the functioning of a regulator when predetermined conditions have been reached. *Note:* Examples are minimum excitation limiter, maximum excitation limiter, maximum armature-current limiter.   63

(3) (radio receivers). A transducer whose output is constant for all inputs above a critical value. *Note:* A limiter may be used to remove amplitude modulation while transmitting angle modulation. *See also:* **radio receiver; transducer.**   111

(4) (excitation systems). A feedback element of the excitation system which acts to limit a variable by modifying or replacing the function of the primary detector element when predetermined conditions have been reached.   105

limiter circuit (analog computer). A circuit of nonlinear elements that restrict the electric excursion of a variable in accordance with some specified criteria. *Note:* Hard limiting is a limiting action with negligible variation in output in the range where the output is limited. Soft limiting is a limiting action with appreciable variation in output in the range where the output is limited. A bridge limiter is a bridge circuit used as a limiter circuit. In an analog computer, a feedback limiter is a limiter circuit usually employing biased diodes shunting the feedback component of an operational amplifier; an input limiter is a limiter circuit usually employing biased diodes in the amplifier input channel that operates by limiting the current entering the summing junction.   9, 77, 10

limiting (automatic control). The intentional imposition or inherent existence of a boundary on the range of a variable, for example, on the speed of a motor.   56, 105

limiting ambient temperature (equipment rating). An upper or lower limit of a range of ambient temperatures within which equipment is suitable for operation at its rating. Where the term is used without an adjective the upper

limit is meant. *See also:* **limiting insulation system temperature.**   233

limiting angular subtense (α_{min}) (laser-maser). The apparent visual angle which divides intrabeam viewing from extended-source viewing.   363

limiting aperture (laser-maser). The maximum circular area over which radiance and radiant exposure can be averaged.   363

limiting hottest-spot temperature. *See:* **limiting insulation temperature.**

limiting insulation system temperature (limiting hottest-spot temperature). The temperature selected for correlation with a specified test condition of the equipment with the object of attaining a desired service life of the insulation system.   53

limiting insulation temperature rise (equipment rating). The difference between the limiting insulation temperature and the limiting ambient temperature. *Syn.* **limiting hottest-spot temperature.** *See also:* **limiting insulation system temperature.**   233

limiting observable temperature rise (equipment rating). The limit of observable temperature rise specified in equipment standards. *See also:* **limiting insulation system temperature.**   233

limiting polarization (radio wave propagation). The resultant polarization of a wave after it has emerged from a magneto-ionic medium. *See also:* **radio wave propagation.**   180, 146

limiting resolution (television). A measure of overall system resolution usually expressed in terms of the maximum number of lines per picture height discriminated on a test chart. *Note:* For a number of lines N (alternate black and white lines) the width of each line is $1/N$ times the picture height. *See also:* **resolution; television.**   163, 178, 317

limiting temperature. The temperature at which a component or material may be operated continuously with no sacrifice in normal life expectancy.   53

limit, lower. *See:* **lower limit.**

limit of error. *See:* **uncertainty.**

limit of temperature rise (1) (contacts). The temperature rise of contacts, above the temperature of the cooling air, when tested in accordance with the rating shall not exceed the following values. All temperatures shall be measured by the thermometer method. Laminated contacts: 50 degrees Celsius; solid contacts: 75 degrees Celsius.

(2) (resistors). The temperature rise of resistors above the temperature of the cooling air, when test is made in accordance with the rating, shall not exceed the following temperatures for the several classes of resistors: Class A, cast resistors, 450 degrees Celsius; Class B, imbedded resistors, outside of imbedding material, 250 degrees Celsius; Class C, strap or ribbon wound on Class C insulation, 600 degrees Celsius continuous and 800 degrees Celsius intermittent; class D, enameled wire or strap wound resistance, 350 degrees Celsius. Temperatures to be measured by thermocouple method.   1

limits of interference (electromagnetic compatibility). Maximum permissible values of radio interference as specified in International Special Committee on Radio Interference recommendations or by other competent authorities or organizations. *See also:* **electromagnetic compatibility**   220, 199

limit switch. A switch that is operated by some part or motion of a power-driven machine or equipment to alter the electric circuit associated with the machine or equipment. *See also:* **switch.** 206

limit, upper. *See:* **upper limit.**

line (1) (electric power). A component part of a system extending between adjacent stations or from a station to an adjacent interconnection point. A line may consist of one or more circuits. *See:* **system.** 91

(2) (trace) (cathode-ray tube). The path of a moving spot. 190

(3) (electromagnetic theory). *See;* **maxwell.**

(4) (acoustics). *See:* **acoustic delay line; delay line; electromagnetic delay line; magnetic delay line; sonic delay line.**

(5) (data transmission). *See:* **channel.**

line amplifier. *See:* **amplifier, line.**

linear accelerometer. A device which measures translational acceleration along an input axis. In this accelerometer, an output signal is produced by the reaction of the proof mass to a translatory acceleration input. The output is usually an electrical signal proportional to applied translational acceleration. 46

linear amplifier (magnetic). An amplifier in which the output quantity is essentially proportional to the input quantity. *Note:* This may be interpreted as an amplifier that has no intentional discontinuities in the output characteristic over the useful input range of the amplifier. 171

linear antenna. An antenna consisting of one or more segments of straight conducting cylinders. *Note:* This term has restricted usage, and applies to straight cylindrical wire antennas. This term should not be confused with the conventional usage of linear in circuit theory. *See also:* **linear array antenna.** 111

linear array. *See:* **linear array antenna.**

linear array antenna. An array antenna having the centers of the radiating elements lying along a straight line. 111

linear charging (direct current). A special case of resonance charging of the capacitance in an oscillatory series RLC circuit where the capacitor is repetitively discharged at a predetermined voltage at a rate much greater than twice the natural resonance of the RLC circuit. *Note:* The inductance of the charging inductor for linear charging is much greater than that for resonance charging for a given pulse-repetition frequency. Under the above conditions, the current through the charging inductor at the time the capacitance is discharged is not zero and the voltage across the capacitance is still rising. 137

linear cross-field amplifier (microwave tubes). A crossed-field amplifier in which a nonre-entrant beam interacts with a forward wave. *See also:* **microwave tube or valve.** 190

linear detection. That form of detection in which the output voltage is substantially proportional, over the useful range of the detecting device, to the voltage of the input wave. *See also:* **detection.** 328

linear distortion (video signal transmission measurement). That distortion of an electrical signal which is independent of the signal amplitude. *Note:* A small-signal nonuniform frequency response is an example of linear distortion. By contrast, nonlinear distortions of an electrical signal are those distortions that are dependent on the signal amplitude, for example, compression, expansion, and harmonic distortion, etc. 42

linear electrical parameters (uniform line). The series resistance, series inductance, shunt conductance, and shunt capacitance per unit length of line. *Note:* The term **constant** is frequently used instead of **parameter.** *See also:* **transmission line.** 179

linear electron accelerator. An evacuated metal tube in which electrons are accelerated through a series of small gaps (usually in the form of cavity resonators in the high-frequency range) so arranged and spaced that at a specific excitation frequency, the stream of electrons on passing through successive gaps gains additional energy from the electric field in each gap. *See also:* **miscellaneous electron devices.** 244, 190

linear-impedance relay. A distance relay for which the operating characteristic on an R-X diagram is a straight line. *Note:* It may be described by the equation $2 = K/\cos(\theta - \sigma)$ where K and σ are constants and θ is the phase angle by which the input voltage leads the input current. 103, 202, 60, 127

linear interpolation (numerically controlled machines). A mode of contouring control that uses the information contained in a block to produce velocities proportioned to the distance moved in two or more axes simultaneously. 224, 207

linearity (1) (general). The property of being "linear" as defined in various usages for that adjective. 67

(2) (computing systems). A property describing a constant ratio of incremental cause and effect. *Note:* Proportionality is a special case of linearity in which the straight line passes through the origin. Zero-error reference of a linear transducer is a selected straight-line function of the input from which output errors are measured. Zero-based linearity is transducer linearity defined in terms of a zero-error reference where zero input coincides with zero output. *See:* **normal linearity.** *See also:* **electronic analog computer.** 9, 77

(3) (industrial control) (nuclear power generating station). The closeness with which a curve of a function approximates a straight line. *See also:* **control system, feedback.** 219, 206, 355

(4) (test, measurement and diagnostic equipment). The condition wherein the change in the value of one quantity is directly proportional to the change in the value of another quantity. 54

linearity control (television). A control to adjust the variation of scanning speed during the trace interval to minimize geometric distortion. *See also:* **television.** 163, 178

linearity error (1) (computing element). The deviation of the output quantity from a specified zero-error reference. *See also:* **electronic analog computer.** 9, 77

(2) (Hall-effect devices). The deviation of the actual characteristic curve of a Hall generator from the linear approximation to this curve. 107

linearity error, percent of full scale (Hall-effect devices). The maximum deviation, expressed as a percent of full scale, of the actual characteristic curve of a Hall effect device from the straight-line approximation to the characteristic curve derived by minimizing and equalizing the positive and negative deviations of the curve from the straight line. 107

linearity error, percent of reading (Hall-effect devices). The maximum percent deviation of the actual characteristic curve of a Hall effect device from the straight-line approximation to the curve derived by minimizing and equalizing the positive and negative percent deviations of the characteristic curve from the straight line. 107

linearity of a multiplier (analog computer). (1) the ability of an electromechanical or electronic multiplier to generate an output voltage that varies linearly with either one of its two inputs, provided the other input is held constant; (2) the accuracy with which the above requirement is met. 9

linearity of a potentiometer. The accuracy with which a potentiometer yields a linear but not necessarily a proportional relationship between the angle of rotation of its shaft and the voltage appearing at the output arm, in the absence of loading errors. *See also:* **normal linearity.** 9

linearity of a signal (automatic control). The closeness with which its plot against the variable it represents approximates a straight line. *Note:* The property is usually expressed as a "non-linearity," for example, a maximum deviation. The straight line should be specified as drawn to give limited absolute deviation (independent linearity), to give minimum rms deviation (dependent linearity), to pass through the zero point, or to pass through both end points. 56

linearity, programming (power supplies). The linearity of a programming function refers to the correspondence between incremental changes in the input signal (resistance, voltage, or current) and the consequent incremental changes in power-supply output. *Note:* Direct programming functions are inherently linear for the bridge regulator and are accurate to within a percentage equal to the supply's regulating ability. *See also:* **power supply.** 228, 186

linearity region (instrument approach system and similar guidance systems). The region in which the deviation sensitivity remains constant within specified values. *See also:* **navigation.** 187

linearity, sweep (oscilloscopes). Maximum displacement error of the independent variable between specified points on the display area. 106

linearity, vertical (oscilloscopes). The change in deflection factor of an oscilloscope as the display is positioned vertically within the graticule area. *See also:* **compression; expansion.** 106

linear lag. *See:* **lag (5) (first-order).**

linear light. A luminous signal having a perceptible physical length. *See also:* **signal lighting.** 167

linearly polarized field vector (antennas). A field vector for which the polarization ellipse is a line segment. 111

linearly polarized plane wave (antennas). A plane wave in which the electric field is linearly polarized. 111

linearly polarized wave (plane-polarized wave) (1) (general). A transverse wave in which the displacements at all points along a line in the direction of propagation lie in a plane passing through this line. *See also:* **radiation.** 210

(2) (radio wave propagation). An electromagnetic wave whose electric and magnetic field vectors always lie along fixed lines at a given point. *See also:* **radiowave propagation.** 180

linearly rising switching impulse. An impulse in which the impulse voltage rises at an approximately constant rate. Its amplitude is limited by the occurrence of a disruptive discharge that chops the impulse. 108

linear modulator. A modulator in which, for a given magnitude of carrier, the modulated characteristic of the output wave bears a substantially linear relation to the modulating wave. *See also:* **network analysis.** 328

linear network (or system) (circuits and systems). A network (or system) that has both the proportionality and the superposition properties. For example $H(\alpha x_1 + \beta x_2) = \alpha H(x_1) + \beta H(x_2)$. 67

linear power amplifier. A power amplifier in which the signal output voltage is directly proportional to the signal input voltage. *See also:* **amplifier.** 111

linear programming (computing systems). *See:* **programming, linear.**

linear pulse amplifier (pulse techniques). A pulse amplifier in which the peak amplitude of the output pulses is directly proportional to the peak amplitude of the corresponding input pulses, if the input pulses are alike in shape. *See also:* **pulse.** 335

linear rectifier. A rectifier, the output current or voltage of which contains a wave having a form identical with that of the envelope of an impressed signal wave. *See also:* **rectifier.** 111, 240

linear system or element (1) (analog computer) (speed governing of hydraulic turbines). A system or element with the properties that if y_1 is the response to x_1, and y_2 is the response to x_2, then $(y_1 + y_2)$ is the response to $(x_1 + x_2)$, and ky_1 is the response to kx_1. *See also:* **control system, feedback.** 9, 8

(2) (circuits and systems). An electrical element whose cause and effect relationship follows the proportionality rule. 67

(3) (automatic control). One whose time response to several simultaneous inputs is the sum of their independent time responses. *Note:* It is representable by a linear differential equation, and has a transfer function which is constant for any value of input within a specified range. A system or element not meeting these conditions is described as "nonlinear." 56

(4) (excitation control systems). *Note:* The ANSI C.85 standard definition given above does not permit handling of systems with zero input response. Therefore, the following alternate definition is recommended: "Let a system have zero input response $Z(t)$ and response to two independent inputs $R_1(t)$ and $R_2(t)$ is $C_1(t) + Z(t)$ and $C_2(t) + Z(t)$ respectively. Then the system is linear if the input $aR_1(t) + bR_2(t)$ produces output $aC_1(t) + bC_2(t) + Z(t)$. Otherwise, the system is nonlinear. A linear system is both superimposable and homogeneous. It is mathematically modeled by a linear differential equation. If the system is stationary, the transfer function of the system is constant and not a function of time." 353

linear system with one degree of freedom. *See:* **damped harmonic system.**

linear transducer. A transducer for which the pertinent measures of all the waves concerned are linearly related.

Notes: (1) By linearly related is meant any relation of linear character whether by linear algebraic equation, by linear differential equation, by linear integral equation, or by other linear connection. (2) The term **waves concerned** connotes actuating waves and related output waves, the relation of which is of primary interest in the problem at hand. *See also:* **transducer.** 252, 210

linear TV waveform distortion (video signal transmission measurement). The distortion of the shape of a waveform signal where this distortion is independent of the amplitude of the signal. *Notes:* (1) A TV video signal may contain time components with durations from as long as a TV field to as short as a picture element. The shapes of all these time components are subject to distortions. For ease of measurement it is convenient to group these distortions in three separate time domains; short-time waveform distortion, line-time waveform distortion, and field-time waveform distortion. (2) The waveform distortions for times from one field to tens of seconds is not within the scope of this definition. 42

linear-varying parameter (varying parameter). A parameter that varies with time or position or both, but not with any dependent variable. *Note:* Unless otherwise specified, **varying parameter** refers to a linear-varying parameter, not to a nonlinear parameter. 210

linear-varying-parameter network. A linear network in which one or more parameters vary with time. 349

line bar (TV waveform test signals). A pulse, nominally of 20 μs duration, of reference-white amplitude. The rise and fall portions of the line bar are T steps as defined. 42

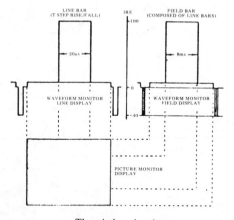

The window signal.

line breaker (line switch) (electrically driven vehicles). A device that combines the functions of a contactor and of a circuit breaker. *Note:* This term is also used for circuit breakers that function to interrupt circuit faults and do not combine the function of a contactor. *See also:* **multiple-control unit.** 328

line, bull (conductor stringing equipment). A high strength line, normally manila or synthetic fiber rope, used for pulling and hoisting large loads. 45

line-busy tone (telephone switching systems). A tone that indicates that a station termination is not available. 55

line-charging capacity (synchronous machine). The reactive

power when the machine is operating synchronously at zero power factor, rated voltage, and with the field current reduced to zero. *Note:* This quantity has no inherent relationship to the thermal capability of the machine. 318, 63

line charging current. The current supplied to an unloaded line or cable. 103, 202

line circuit (1) (railway signaling). A signal circuit on an overhead or underground line. 328

(2) (telephone switching systems). An interface circuit between a line and a switching system. 55

line-closing switching-surge factor (power switchgear). The ratio of the line-closing switching-surge maximum voltage to the crest of the normal-frequency line-to-ground voltage at the source side of the closing switching device immediately prior to closing. 103

line-closing switching-surge maximum voltage (power switchgear). The maximum transient crest voltage to ground measured on a transmission line during a switching surge which results from energizing that line. 103

line concentrator (telephone switching systems). A concentrator in which the inlets are lines. 55

line conductor (1) (electric power). One of the wires or cables carrying electric current, supported by poles, towers, or other structures, but not including vertical or lateral connecting wires. *See also:* **conductor; tower.** 178, 64

(2) (transmission and distribution). A conductor intended to carry electric current, supported by poles, towers, or other structures, but not including vertical or lateral conductors. 262

line coordination (data transmission). The process of insuring that equipment at both ends of a circuit are set up for a specific transmission. 59

line discipline. *See:* **control procedure.**

line display (non-real time spectrum analyzer). The display produced on a spectrum analyzer when the resolution bandwidth is less than the spacing of the individual frequency components. 68

line-drop compensator (voltage regulator). A device which causes the voltage-regulating relay to increase the output voltage by an amount that compensates for the impedance drop in the circuit between the regulator and a predetermined location on the circuit (sometimes referred to as the load center). *See:* **voltage regulator.** 53, 203

line-drop signal (manual switchboard). A drop signal associated with a subscriber line. 328

line-drop voltmeter compensator. A device used in connection with a voltmeter that causes the latter to indicate the voltage at some distant point of the circuit. *See also:* **auxiliary device to an instrument.** 328

line end and ground end (electric power). (1) Line end is that end of a neutral grounding device that is connected to the line circuit directly or through another device. (2) Ground end is that end that is connected to ground directly or through another device. *See also:* **grounding device.** 91, 64

line fill. The ratio of the number of connected main telephone stations on a line to the nominal main-station capacity of that line. *See also:* **cable.** 328

line, finger (conductor stringing equipment). A lightweight line, normally sisal, manila or synthetic fiber rope, which is placed over the traveler when it is hung. It usually ex-

tends from the ground, passes through the traveler and back to the ground. It is used to thread the end of the pilot line or pulling line over the traveler and eliminates the need for workmen on the structure. These lines are not required if pilot lines are installed when the travelers are hung.
45

line-focus tube (X-ray tubes). An X-ray tube in which the focal spot is roughly a line. *See also:* **electron devices, miscellaneous.**   190

line frequency (television). The number of times per second that a fixed vertical line in the picture is crossed in one direction by the scanning spot. *Note:* Scanning during vertical return intervals is counted. *See also:* **television.**
178

line group (telephone switching systems). A multiplicity of lines served by a common set of links.   55

line guy. Tensional support for poles or structures by attachment to adjacent poles or structures. *See also:* **tower.**
64

line hit (data transmission). An electrical interference causing the introduction of spurious signals on a circuit.
59

line impedance stabilization network. *See:* **artificial mains network.**

line-insulation resistance (telephone switching systems). The resistance between the loop conductors and ground or between each other.   55

line insulator (pin, post). An assembly of one or more shells, having means for semirigidly supporting line conductors. *See also:* **insulator.**   261

line integral (dot-product line integral). The line integral between two points on a given path in the region occupied by a vector field is the definite integral of the dot product of a path element and the vector. Thus

$$I = \int_a^b \mathbf{V}\cos\theta\,\mathrm{d}s$$

$$= \int_a^b \mathbf{V}\cdot\mathrm{d}s$$

$$= \int_a^b (V_x\mathrm{d}x + V_y\mathrm{d}y + V_z\mathrm{d}z)$$

where $\mathbf{V}$ is the vector having a magnitude V, ds the vector element of the path, θ the angle between $\mathbf{V}$ and ds. *Example:* The magnetomotive force between two points on a line connecting two points in a magnetic field is the line integral of the magnetic field strength, that is, the definite integral between the two points of the dot product of a vector element of the length of the line and the magnetic strength at the element.   210

line lamp (wire communication). A switchboard lamp for indicating an incoming line signal.   328

line lightning performance. The performance of a line expressed as the annual number of lightning flashovers on a circuit mile or tower-line mile basis. *See also:* **direct-stroke protection (lightning).**   64

line-load control (telephone switching systems). A means of selectively restricting call attempts during emergencies so as to permit the handling of essential traffic.   55

line lockout (telephone switching systems). A means for the handling of permanent line signals to prevent further

recognition as a call attempt.   55

line loss (electric power systems). Energy loss on a transmission or distribution line.   112

line microphone. A directional microphone consisting of a single straight line element, or an array of contiguous or spaced electroacoustic transducing elements disposed on a straight line, or the acoustical equivalent of such an array. *See also:* **microphone; line transducer.**   328

line noise. Noise originating in a transmission line.   59

line number, television (measuring resolution). *See:* **television line number.**

line of nodes (communication satellite). The line which is common to both the orbital plane and the equatorial plane and which passes through the ascending and descending nodes.   74

line of position (LOP) (navigation). The intersection of two surfaces of position, normally plotted as lines on the earth's surface, each line representing the locus of constant indication of the navigational information. *See also:* **navigation.**   187, 13

line or input regulation (electrical conversion). Static regulation caused by a change in input. *See also:* **electrical conversion.**   186

line or trace (cathode-ray tubes). The path of the moving spot on the screen or target.   178, 125

line parameters. A sufficient set of parameters to specify the transmission characteristics of the line.   210

line, pilot (conductor stringing equipment). A lightweight line, normally synthetic fiber rope, used to pull heavier pulling lines which in turn are used to pull the conductor. Pilot lines may be installed by helicopters, when the insulators and travelers are hung or with the aid of finger lines. *Syn.* **lead line; leader; "p" line; straw line.**   45

line printing. The printing of an entire line of characters as a unit.   225, 77

line, pulling (conductor stringing equipment). A high strength line, normally synthetic fiber rope or wire rope, used to pull the conductor. However, on reconstruction jobs where a conductor is being replaced, the old conductor often serves as the pulling line for the new conductor. In such cases, the old conductor must be closely examined for any damage prior to the pulling operations. *Syn.* **bull line; hard line; light line; sock line.**   45

liner (1) (rotating machinery). (A) A separate insulating member that is placed against a grounded surface. (B) A layer of insulating material that is deposited on a grounded surface. *See also:* **slot liner; pole cell insulation.**   63

(2) (dry cell) (primary cell). Usually a paper or pulpboard sheet covering the inner surface of the negative electrode and serving to separate it from the depolarizing mix. *See also:* **electrolytic cell.**   328

line regulation (power supplies). The maximum steady-state amount that the output voltage or current will change as the result of a specified change in input line voltage (usually for a step change between 105–125 or 210–250 volts, unless otherwise specified). Regulation is given either as a percentage of the output voltage or current, or as an absolute change ΔE or ΔI. *See also:* **power supply.**
228

line relay (railway practice). A relay that receives its operating energy over a circuit that does not include the track rails.   328

line replaceable unit (LRU) (test, measurement and diagnostic equipment). A unit which is designated by the plan for maintenance to be removed upon failure from a larger entity (equipment, system) in the latter's operational environment.   54

line residual current. *See:* **ground-return current.**

lines. *See:* **communication lines; electric-supply lines.**

line, safety life (conductor stringing equipment). A safety device normally constructed from manila or synthetic fiber rope and designed to be connected between a fixed object and the body belt of a workman working in an elevated position when his regular safety strap cannot be utilized. *Syn.* **life line; safety line; scare rope.**   45

line side. Data terminal connections to a communications circuit between two data terminals.   194, 59

line source (antenna). A continuous distribution of current lying along a line segment.   179, 111

line source corrector. A linear array antenna feed with radiating element locations and excitations chosen to correct for aberrations present in the focal region fields of a reflector.   111

line spectrum (non-real time spectrum analyzer). A spectrum composed of discrete frequency components.   68

line speed (data transmission). The maximum rate at which signals may be transmitted over a given channel; usually in baud or bits/sec. *See:* **baud.**   59

line stretcher (waveguide components). A section of waveguide or transmission line having an adjustable physical length.   166

line switch (wire communication). A switch attached to a subscriber line, that connects an originating call to an idle part of the switching apparatus.   328

line switching (data transmission). The switching technique of temporarily connecting two lines together so that the stations directly exchange information.   59

line, tag (conductor stringing equipment). A control line, normally manila or synthetic fiber rope, attached to a suspended load to enable a workman to control its horizontal movements.   45

line tap (1) (general). A radial branch connection to a main line. *See also:* **tower.**   64
(2) (system protection). A connection to a line with equipment that does not feed energy into a fault on the line in sufficient magnitude to require consideration in the relay plan.   103, 202, 60, 127

line terminal (1) (surge arrester). The conducting part provided for connecting the arrester to the circuit conductor. *Note:* When a line terminal is not supplied as an integral part of the arrester and the series gap is obtained by providing a specified air clearance between the line end of the arrester and a conductor or arcing electrode etcetera, the words **line terminal** used in the definition refer to the conducting part that is at line potential and that is used as the line electrode of the series gap.   2, 299, 62
(2) (rotating machinery). A termination of the primary winding for connection to a line (not neutral or ground) of the power supply or load. *See:* **asynchronous machine; rotor (rotating machine); stator.**   63
(3) (power system protection). A connection to a line with equipment that can feed energy into a fault on the line in

sufficient magnitude to require consideration in the relay plan and that has means for automatic disconnection.
103, 202, 60, 127

line-time waveform distortion (LD) (video signal transmission measurement). The linear TV waveform distortion of time components from 1 μs to 64 μs, that is, time components of the line-time domain.   42

line-to-line voltage. *See:* **voltage sets.**

line-to-neutral voltage. *See:* **voltage sets.**

line transducer. A directional transducer consisting of a single straight-line element, or an array of contiguous or spaced electroacoustic transducing elements disposed on a straight line, or the acoustical equivalent of such an array. *See also:* **line microphone.**   176

line transformer. A transformer for interconnecting a transmission line and terminal equipment for such purposes as isolation, line balance, impedance matching, or for additional circuit connections.   239

line trap. *See:* **carrier-current line trap.**

line triggering. Triggering from the power line frequency. *See:* **oscillograph.**   185, 68

line-type fire detector (fire protection devices). A device in which detection is continuous along a path.   71

linewidth (laser-maser). The interval in frequency or wavelength units between the points at which the absorbed power (or emitted power) of an absorption (or emission) line is half of its maximum value when measured under specified conditions.   363

link (1) (general). A channel or circuit designed to be connected in tandem with other channels or circuits. *Note:* In automatic switching, a link is a path between two units of switching apparatus within a central office. *See also:* **channel.**   328
(2) (communication satellite). A complete facility over which a certain type of information is transmitted; including all elements from source transducer to output transducer.   83
(3) (telephone switching systems). A connection between switching stages within the same switching system.   55

linkage (programming) (computing systems). Coding that connects two separately coded routines.   255, 77

linkage voltage test, direct-current test (rotating machinery). A series of current measurements, made at increasing direct voltages, applied at successive intervals, and maintained for designated periods of time. *Note:* This may be a controlled overvoltage test. *See:* **asynchronous machine; direct-current commutating machine.**   63

link-break cutout. A load-break fuse cutout that is operated by breaking the fuse link to interrupt the load current
103, 202

link, connector (conductor stringing equipment). A rigid link designed to connect pulling lines and conductors together in series. It will not spin and relieve torsional forces. *Syn.* **link; connector; slug.**   45

link, swivel (conductor stringing equipment). A swivel device designed to connect pulling lines and conductors together in series or connect one pulling line to the drawbar of a pulling vehicle. The device will spin and help relieve the torsional forces which build up in the line or conductor under tension. *Syn.* **swivel.**   45

lin-log receiver (radar). A receiver having a linear amplitude response for small-amplitude signals and a logarithmic response for large-amplitude signals. *See also:* **navigation.** 13, 187

lip microphone. A microphone adapted for use in contact with the lip. *See also:* **microphone.** 328

liquid controller (industrial control). An electric controller in which the resistor is a liquid. *See:* **electric controller.** 206

liquid cooling (rotating machinery). *See:* **manifold insulation.**

liquid counter tube (radiation counters). A counter tube suitable for the assay of liquid samples. It often consists of a thin glass-walled Geiger-Mueller tube sealed into a test tube providing an annular space for the sample. *See:* **anticoincidence (radiation counters).** 190

liquid development (electrostatography). Development in which the image-forming material is carried to the field of the electrostatic image by means of a liquid. *See also:* **electrostatography.** 236, 191

liquid-filled fuse unit. A fuse unit in which the arc is drawn through a liquid. 103, 202

liquid-flow counter tube (radiation counters). A counter tube specially constructed for measuring the radioactivity of a flowing liquid. *See:* **anticoincidence (radiation counters).** 190

liquid-function potential. The contact potential between two electrolytes. It is not susceptible of direct measurement. 328

liquid-immersed transformer. Having the core and coils immersed in an insulating liquid. 53

liquid-in-glass thermometer. A thin-walled glass bulb attached to a glass capillary stem closed at the opposite end, with the bulb and a portion of the stem filled with an expansive liquid, the remaining part of the stem being filled with the vapor of the liquid or a mixture of this vapor and an inert gas. Associated with the stem is a scale in temperature degrees so arranged that when calibrated the reading corresponding to the end of the liquid column indicates the temperature of the bulb. 7

liquid-level switch. *See:* **float switch.**

liquid resistor (industrial control). A resistor comprising electrodes immersed in a liquid. 244, 206

Lissajous figure (oscilloscopes). A special case of an $X-Y$ plot in which the signals applied to both axes are sinusoidal functions. For a stable display the signals must be harmonics. Lissajous figures are useful for determining phase and harmonic relationships. 106

list (computing systems). *See:* **push-down list; pushup list.**

listener echo. Echo that reaches the ear of the listener. 328

listening group (speech quality measurements). A group of persons assembled for the purpose of speech quality testing. Number, selection, characteristics, and training of the listeners depend upon the purpose of the test. 126

lithium-drifted detector (semiconductor radiation detectors). A detector made by the lithium compensation process. 23, 118, 119

lithium drifting (semiconductor radiation detectors). A technique for compensating p-type material by causing lithium ions to move through a crystal under an applied electric field in such a way as to compensate the charge of the bound acceptor impurities. 23, 118, 119

litz wire (litz). A conductor composed of a number of fine, separately insulated strands, usually fabricated in such a way that each strand assumes, to substantially the same extent, the different possible positions in the cross section of the conductor. Litz is an abbreviation for litzendraht. 328

live (electric system). *See:* **alive.**

live cable test cap. A protective structure at the end of a cable that insulates the conductors and seals the cable sheath. 64

live-front (industrial control) (transformer). So constructed that there are exposed live parts on the front of the assembly. *See:* **live-front switchboard.** 206, 53

live-front switchboard. A switchboard having exposed live parts on the front. 103, 202

live-metal part. A part, consisting of electrically conductive material which can be energized under conditions of normal use of the equipment. *See also:* **live-front.** 53

live parts. Those parts that are designed to operate at voltage different from that of the earth. 103, 202, 27

live room (audio and electroacoustics). A room that has an unusually small amount of sound absorption. *See also:* **loudspeaker.** 176

live tank switching device. A switching device in which the vessel(s) housing the interrupter(s) is at a potential above ground. 79

live zone (industrial control). The period(s) in the operating cycle of a machine during which corrective functions can be initiated. 206

L network (1) (general). A network composed of two branches in series, the free ends being connected to one pair of terminals and the junction point and one free end being connected to another pair of terminals. *See also:* **network analysis.** 210

(2) (circuits and systems). An unbalanced ladder network composed of a series arm and a shunt arm. 67

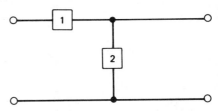

L network. The free ends are the left-hand terminal pair; the junction point and one free end are the right-hand terminal pair.

load (1) (general). (A) A device that receives power, or (B) the power or apparent power delivered to such a device. 328

(2) (signal-transmission system). The device that receives signal power from a signal-transmission system. *See:* **signal.** 188

(3) (induction and dielectric heating usage) (charge). The material to be heated. *See also:* **dielectric heating; induction heating.** 14, 114

(4) (transformer) (output). The apparent power in megavolt-amperes, kilovolt-amperes, or volt-amperes that may be transferred by the transformer. 53

(5) (rotating machinery). All the numerical values of the

electrical and mechanical quantities that signify the demand to be made on a rotating machine by an electric circuit or a mechanism at a given instant. *See also:* **direct-current commutating machine; asynchronous machine.**
63

(6) (programming). To place data into internal storage.
255, 77

(7) (electric) (electric utilization). The electric power used by devices connected to an electrical generating system. *See also:* **generating station.** 200

(8) (automatic control). (A) An energy-absorbing device. (B) The material, force, torque, energy, or power applied to or removed from a system or element. 56

(9) (data transmission). A power-consuming device connected to a circuit. One use of the word load is to denote a resistor or impedance which replaces some circuit element temporarily or permanently removed. For example, if a filter is disconnected from a line, the line may be artificially terminated in an impedance which simulates the filter that was removed. The artificial termination is then called a load, or a dummy load. 59

(10) (test, measurement and diagnostic equipment). (A) To read information from cards or tape into memory; (B) Building block or adapter providing a simulation of the normal termination characteristics of a unit under test; and (C) The effect that the test equipment has on the unit under test or vice versa. 54

(11) (power switchgear). The true or apparent power consumed by power utilization equipment performing its normal function. 103

load-and-go (computing system). An operating technique in which there are no stops between the loading and execution phases of a program, and which may include assembling or compiling. 255, 77

load angle (synchronous machine). The angular displacement, at a specified load, of the center line of a field pole from the axis of the armature magnetomotive force pattern. 63

load-angle curve (load-angle characteristic) (synchronous machine). A characteristic curve giving the relationship between the rotor displacement angle and the load, for constant values of armature voltage, field current, and power factor. 63

load-band of regulated voltage (rotating machinery). The band or zone, expressed in percent of the rated value of the regulated voltage, within which the synchronous-machine regulating system will hold the regulated voltage of a synchronous machine during steady or gradually changing conditions over a specified range of load. 63

load, base (electric power utilization). The minimum load over a given period of time. *See also:* **generating station.**
112

load binder (conductor stringing equipment). A toggle device designed to secure loads in a desired position. Normally used to secure loads on mobile equipment. 45

load-break connector (transformer). A connector designed to close and interrupt current on energized circuits.
53, 134

load-break cutout. A cutout with means for interrupting load currents. 103, 202

load center (electric power utilization). A point at which the load of a given area is assumed to be concentrated. *See*

also: **generating station.** 200

load (dynamic) characteristic (electron tube). For an electron tube connected in a specified operating circuit at a specified frequency, a relation, usually represented by a graph, between the instantaneous values of a pair of variables such as an electrode voltage and an electrode current, when all direct electrode supply voltages are maintained constant. 190, 125

load circuit (1) (industrial control). The complete circuit required to transfer power from a source to a load. *See:* **control.** *See also:* **dielectric heating; induction heating.**
111, 210

(2) (induction and dielectric heating usage). The network including leads connected to the output terminals of the generator. *Note:* The load circuit consists of the coupling network and the load material at the proper position for heating. 14

load-circuit efficiency (induction and dielectric heating usage). The ratio of the power absorbed by the load to the power delivered at the generator output terminals. *See also:* **dielectric heating; induction heating; network analysis; overall electrical efficiency.** 14, 114

load-circuit power input. The power delivered to the load circuit. *Note:* It is the product of the alternating component of the voltage across the load circuit, the alternating component of the current passing through it (both root-mean-square values), and the power factor associated with these two quantities. *See also:* **network analysis.** 111

load coil (induction heating usage). An electric conductor that, when energized with alternating current, is adapted to deliver energy by induction to a charge to be heated. *See also:* **induction heating load, work, or heater coil.** 114

load current (tube) (electron tubes). The current output utilized in an external load circuit. *See also:* **electron tube.**
244, 190

load current output (arc-welding apparatus). The current in the welding circuit under load conditions. 264

load curve. A curve of power versus time showing the value of a specific load for each unit of the period covered. *See also:* **generating station.** 64

load curves, daily (electric power supply). Curves of net 60-minute integrated demand for each clock hour of a 24-hour day. *See also:* **generating station.** 200

load diversity (electric power utilization). The difference between the sum of the maxima of two or more individual loads and the coincident or combined maximum load, usually measured in kilowatts over a specified period of time. *See also:* **generating station.** 64, 200

load diversity power. The rate of transfer of energy necessary for the realization of a saving of system capacity brought about by load diversity. *See also:* **generating station.** 64

load division (load balance) (industrial control). A control function that divides the load in a prescribed manner between two or more power devices connected to the same load. *See also:* **control system, feedback.** 219, 206

load, dummy (artificial load). See: **dummy load.**

load duration curve. A curve showing the total time, within a specified period, during which the load equaled or exceeded the power values shown. *See also:* **generating station.** 64

load duty repetitive (thyristor converter). A type of load duty where overloads appear intermittent but cyclic and so frequent that thermal equilibrium is not obtained between all overloads. 121

loaded applicator impedance (dielectric heating). The complex impedance measured at the point of application, with the load material at the proper position for heating, at a specified frequency. *See also:* **dielectric heating.** 14, 114

loaded impedance (transducer). The impedance at the input of the transducer when the output is connected to its normal load. *See also:* **self-impedance.** 176

loaded Q (1) (working Q) (electric impedance). The value of Q of such impedance when coupled or connected under working conditions. 328

(2) (switching tubes). The unloaded Q of the tube modified by the coupled impedances. *Note:* As here used, Q is equal to 2π times the energy stored at the resonance frequency divided by the energy dissipated per cycle in the tube, or for cell-type tubes, in the tube and its external resonant circuit. *See also:* **gas tubes.** 190, 125

load factor. The ratio of the average load over a designated period of time to the peak load occurring in that period. *See also:* **generating station.** 64, 200

load-frequency control. The regulation of the power output of electric generators within a prescribed area in response to changes in system frequency, tie line loading, or the relation of these to each other, so as to maintain the scheduled system frequency or the established interchange with other areas within predetermined limits or both. *See also:* **generating station.** 64

load ground (signal-transmission system). The potential reference plane at the physical location of the load. *See:* **signal.** 188

load group (nuclear power generating stations). An arrangement of buses, transformers, switching equipment, and loads fed from a common power supply. 31, 102

load impedance. *See:* **impedance, load.**

load-impedance diagram (oscillators). A chart showing performance of the oscillator with respect to variations in the load impedance. *Note:* Ordinarily, contours of constant power and of constant frequency are drawn on a chart whose coordinates are the components of either the complex load impedance or of the reflection coefficient. *See:* **Rieke diagram.** *See also:* **oscillatory circuit.** 190, 125

load-indicating automatic reclosing equipment (power distribution). An automatic reclosing equipment that provides for reclosing the circuit interrupter automatically in response to sensing of predetermined conditions of the load circuit. *Note:* This type of automatic reclosing equipment is generally used for direct-current load circuits. 103, 202

load-indicating resistor. A resistor used, in conjunction with suitable relays or instruments, in an electric circuit for the purpose of determining or indicating the magnitude of the connected load. 103, 202

loading (1) (communication practice). Insertion of reactance in a circuit for the purpose of improving its transmission characteristics in a given frequency band. *Note:* The term is commonly applied, in wire communication practice, to the insertion of loading coils in series in a transmission line at uniform intervals, and in radio practice, to the insertion of one or more loading coils anywhere in a transmission

circuit. *See also:* **storm loading; transmission line.** 328

(2) (antenna). The modification of a basic **antenna,** such as a **dipole** or monopole, by adding conductors or circuit elements that change the current distribution or input impedance. *See also:* **antenna.** 179, 111

(3) (data transmission) (H88). A commonly used type of reactive load used on leased lines to produce a specific bandwidth characteristic. 59

(4) (automatic control). Act of transferring energy into or out of a system. 56

loading coil. An inductor inserted in a circuit to increase its inductance for the purpose of improving its transmission characteristics in a given frequency band. *See also:* **loading.** 341

loading-coil spacing. The line distance between the successive loading coils of a coil-loaded line. *See also:* **loading.** 328

loading error (1) (analog computers). An error due to the effect of a load upon the transducer or signal source driving it. 9, 77

(2) (test, measurement and diagnostic equipment). The error introduced when data are incorrectly transferred from one medium to another. 54

loading machine (mining). A machine for loading materials such as coal, ore, or rock into cars or other means of conveyance for transportation to the surface of the mine. 328

load, internal. *See:* **load, system.**

load-interrupter switch (power distribution). An interrupter switch designed to interrupt currents not in excess of the continuous-current rating of the switch. *Notes:* (1) It may be designed to close and carry abnormal or short-circuit currents as specified. (2) In international practice (International Electrotechnical Commission), a device with such performance characteristics is called a switch. 103, 202

load, interruptible (electric power utilization). A load which can be interrupted as defined by contract. *See also:* **generating station.** 112

load leads or transmission lines (induction and dielectric heating usage). The connections or transmission line between the power source or generator and load, load coil or applicator. *See also:* **dielectric heating; induction heating.** 14, 114

load-limit changer (speed-governing system). A device that acts on the speed-governing system to prevent the governor-controlled fuel valves from opening beyond the position for which the device is set. *See also:* **speed-governing system.** 94, 98, 58

load losses (1) (transformer). Those losses which are incident to the carrying of a specified load. Load losses include I^2R loss in the winding due to load and eddy currents, stray loss due to leakage fluxes in the windings, core clamps, and other parts; and the loss due to circulating currents (if any) in parallel windings, or in parallel winding strands. 53

(2) (copper losses) (series transformer). The load losses of a series transformer are the I^2R losses, computed from the rated currents for the windings and the measured direct-current resistances of the windings corrected to 75 degrees Celsius. 274

load matching (1) (induction and dielectric heating). The process of adjustment of the load-circuit impedance to produce the desired energy transfer from the power source to the load.   14, 114

(2) (circuits and systems). The technique of either adjusting the load-circuit impedance or inserting a network between two parts of a system to produce the desired power transfer or signal transmission.   67

load-matching network. An electric network for accomplishing load matching. *See also:* **induction heating; dielectric heating.**   14, 114

load-matching switch (induction and dielectric heating). A switch in the load-matching network to alter its characteristics to compensate for some sudden change in the load characteristics, such as passing through the Curie point. *See also:* **dielectric heating; induction heating.**   14, 114

load range (watthour meter). Denotes the range in amperes over which the meter is designed to operate continuously. *See also:* **watthour meter.**   212

load regulation (control) (industrial control) (automatic control). A steady-state decrease of the value of the specified variable resulting from a specified increase in load, generally from no-load to full-load unless otherwise specified. *Notes:* (1) In cases where the specified variable increases with increasing load, load regulation would be expressed as a negative quantity. (2) Regulation is given as a percentage of the output voltage or current or as an absolute change ΔV or ΔI. *See* **control system, feedback; power supply.**   219, 206, 186

load resistance (semiconductor radiation detector). The resistive component of the load impedance.

23, 118, 119

load restoration. The process of scheduled load restoration when the abnormality causing load shedding has been corrected.   127, 103

load rheostat. A rheostat whose sole purpose is to dissipate electric energy. *Note:* Frequently used for load tests of generators. *See:* **control.**   244, 206

load saturation curve (load characteristic) (synchronous machine). The saturation curve of a machine on a specified constant load current.   63

load shedding (power systems). The process of deliberately removing preselected loads from a power system in response to an abnormal condition in order to maintain the integrity of the system.   127, 103

load-shifting resistor (electric circuit). A resistor used to shift load from one circuit to another.   103, 202

load, sliding. A load sliding inside or along a fixed length of waveguide or transmission line. *See also:* **waveguide.**

185

load switch or contactor (induction heating). The switch or contactor in an induction heating circuit which connects the high-frequency generator or power source to the heater coil or load circuit. *See also:* **induction heating.**

14, 114

load, system (electric power systems). Total loads within the system including transmission and distribution losses.

112

load, system maximum hourly (electric power systems). The maximum hourly integrated system load. This is an energy quantity usually expressed in kilowatt hours per hour.

112

load-tap-changer (LTC) (transformer). A selector switch

device, which may include current interrupting contactors, used to change transformer taps with the transformer energized and carrying full load.   53

load-tap-changing transformer. A transformer used to vary the voltage, or phase angle, or both, of a regulated circuit in steps by means of a device that connects different taps of tapped winding(s) without interrupting the load. *See:* **primary circuits; regulated circuit; series unit; main unit; series winding; excited winding; regulating winding; excitation winding; excitation-regulating winding; voltage-regulating relay; line-drop compensator; voltage winding (or transformer) for regulating equipment.**   53

load time (hybrid computer linkage components). The time required to read in the digital value to a register of the digital-to-analog converter, measured from the instant that the digital computer commands a digital-to-analog converter or a digital-to-analog multiplier "load."   10

load transfer switch. A switch used to connect a generator or power source optionally to one or another load circuit. *See also:* **dielectric heating; induction heating.**   14, 114

load, transition. The load at which a thyristor converter changes from one mode of operation to another. *Note:* The load current corresponding to a transition load is determined by the intersection of extensions of successive portions of the direct-voltage regulation curve where the curve changes shape or slope.   121

load variation within the hour (electric power utilization). The short-time (three minutes) net peak demand minus the net 60-minute clock-hour integrated peak demand of a supplying system. *See also:* **generating station.**   200

load voltage (arc-welding apparatus). The voltage between the output terminals of the welding power supply when current is flowing in the welding circuit.   264

load, work, or heater coil (induction heating). An electric conductor which when energized with alternating current is adapted to deliver energy by induction to a charge to be heated. *See also:* **induction heating.**   14

lobe (antenna). A portion of the directional pattern bounded by one or two cones of nulls. *See also:* **beam of an antenna; major lobe; minor lobe; radiation lobe; side lobe.**   179, 59

lobe switching (radar). A means of direction finding in which a directive radiation pattern is periodically shifted in position so as to produce a variation of the signal at the target; the amount of signal variation is related to the amount of displacement of the target from the pattern mean position. *See also:* **antenna.** *Syn.* **sequential lobing.**

187, 111, 13

local action (self discharge) (battery). The loss of otherwise usable chemical energy by spontaneous currents within the cell or battery regardless of its connections to an external circuit. *Note:* Corrosion may result from the action of local cells, that is, galvanic cells resulting from inhomogeneities between adjacent areas on a metal surface exposed to an electrolyte. *See also:* **battery (primary or secondary).**   221, 205

local automatic message accounting (LAMA) (telephone switching systems). An arrangement at a local office for collecting automatic message accounting information.

55

local backup (power systems). A form of backup protection in which the relays and circuit breakers are at the same station as the primary protective relays and circuit breakers. *See also:* **relay.**   103, 202, 60

local-battery-talking—common-battery-signaling telephone set. A local-battery telephone set in which current for signaling by the telephone station is supplied from a centralized direct-current power source. *See also:* **telephone station.**   328

local call (telephone switching systems). A call for a destination within the local-service area of the calling customer.   55

local cell. A galvanic cell resulting from inhomogeneities between areas on a metal surface in an electrolyte. *Note:* The inhomogeneities may be of physical or chemical nature in either the metal or its environment. *See:* **electrolytic cell.**   221, 205

local central office. A central office arranged for terminating subscriber lines and provided with trunks for establishing connections to and from other central offices.   193

local channel (data transmission). A channel connecting a communications subscriber to a central office.   59

local control (1) (radio transmitters). A system or method of radio-transmitter control whereby the control functions are performed directly at the transmitter. *See also:* **radio transmitter.**   111

(2) (programmable instrumentation). A method whereby a device is programmable by means of its local (front or rear panel) controls in order to enable the device to perform different tasks. *Syn.* **manual control.**   40

local correction (submarine cable telegraphy). A method of signal shaping used to restore deficient direct-current and low-frequency signal components that have been suppressed at the receiver principally for the purpose of minimizing susceptibility to low-frequency disturbances. *Note:* Local correction is also used to compensate for low-frequency signal distortion, commonly called **zero wander,** inherent in some types of receivers. The local correction network, comprising low-pass and delay elements, is energized under control of the received signals, and its output is applied to the receiver in feedback relationship. *See:* **interpolation.** *See also:* **telegraphy.**   328

localized general lighting. Lighting utilizing luminaires above the visual task and contributing also to the illumination of the surround. *See also:* **general lighting.**   167

localizer (electronic navigation). A radio transmitting facility that provides signals for lateral guidance of aircraft with respect to a runway center line. *See also:* **navigation; radio navigation.**   13, 187

localizer on-course line. A line in a vertical plane passing through a localizer and on either side of which indications of opposite sense are received. *See also:* **radio navigation.**   328

localizer receiver. An airborne radio receiver used to detect the transmissions of a ground-installed localizer transmitter. *Note:* It furnishes a visual, audible, or electric signal for the purpose of laterally guiding an aircraft using an instrument landing system. *See also:* **air-transportation electronic equipment.**   328

localizer sector (equisignal localizer). The sector included between two radial lines from the localizer, the lines being defined by specified equal differences in depth of modu-

lation (usually full-scale right and left, respectively, of the flight-path deviation indicator). *See also:* **navigation.**   187, 13

local level (navigation). The plane normal to the local vertical. *See also:* **navigation.**   187

local lighting. Lighting that provides illumination over a relatively small area or confined space without providing any significant general surrounding lighting. *See also:* **general lighting.**   167

local loop (data transmission). That part of a communication circuit between the subscriber's equipment and the line terminating equipment in the exchange (either two-wire or four-wire).   59

local manual fire-alarm system (1) (general alarm type). A local fire-alarm system in which the alarm signal is sounded on all sounding devices installed.

(2) (presignal type). A local fire-alarm system in which the initial alarm signal is sounded by selected sounding devices, with provision for the subsequent sounding of a general alarm at the option of those responsible for system operation. *See also:* **protective signaling.**   328

local office (telephone switching systems). A central office arranged for terminating lines and that may be provided with trunks to other switching entities.   55

local oscillator (1) (general). An oscillator whose output is mixed with a wave for frequency conversion.   339

(2) (superheterodyne circuit). An oscillator whose output is mixed with the received signal to produce a sum or difference frequency equal to the intermediate frequency of the receiver. *See also:* **oscillatory circuit.**   59

local-oscillator tube. An electron tube in a heterodyne conversion transducer to provide the local heterodyning frequency for a mixer tube.   190, 125

local service area. The area within which are located the stations that a customer may call at rates in accordance with the local tariff.   55

local side (data transmission). Data terminal connections to input-output devices.   194

local system (protective signaling). A system in which the alarm or supervisory signal is sounded, as by a bell, horn, or whistle, locally at the protected premises. *See also:* **protective signaling.**   328

local vertical (navigation). The vertical at the location of the observer. *See also:* **navigation.**   187

locating bearing (rotating machinery). A bearing arranged to limit the axial movement of a horizontal shaft but that is not intended to carry any continuous thrust load. It may be combined with the load-carrying bearing. *See also:* **bearing.**   63

location (computing systems). Loosely, any place in which data may be stored. *See:* **protected location.**   255, 77, 54

locator (electronic navigation). *See:* **nondirectional beacon.**

locator joint (rotating machinery). Any joint used to position a part when assembled.   63

locked groove (concentric groove) (disk recording). A blank and continuous groove at the end of modulated grooves whose function is to prevent further travel of the pickup.   176

locked rotor (rotating machinery). The condition existing when the circuits of a motor are energized, but the rotor is not turning. *See:* **rotor (rotating machinery).**   63

locked-rotor current (1)(motor)(general). The steady-state current taken from the line with the rotor locked and with rated voltage (and rated frequency in the case of alternating-current motors) applied to the motor. *See:* **asynchronous machine; direct-current commutating machine.**   124

(2)(motor and starter). The current taken from the line with the rotor locked, with the starting device in the starting position, and with rated voltage and frequency applied.   328

(3)(rotating machinery). The maximum measured steady-state current taken from the line with the motor at rest, for all angular positions of its rotor, with rated voltage and frequency applied.   63

locked-rotor impedance characteristic (asynchronous machine) (rotating machinery). The relationship between the primary winding current and the primary winding voltage with the rotor held stationary and with the secondary winding short-circuited.   63

locked-rotor temperature-rise rate (rotor end ring or of a winding). The average rate of temperature rise of the rotor end rings, or a winding, at locked rotor with rated voltage applied at rated frequency to the primary winding, in increasing from an initial (usually ambient) temperature to a specified ultimate temperature, expressed in degrees Celsius per second or degrees Fahrenheit per second. *See also:* **rotor (rotating machinery).**   63

locked-rotor test (rotating machinery). A test on an energized machine with its rotor held stationary. *See also:* **asynchronous machine; direct-current commutating machine.**   63

locked-rotor torque. The minimum torque that a motor will provide with locked rotor, at any angular position of the rotor, at a winding temperature of 25 degrees Celsius plus or minus 5 degrees Celsius, with rated voltage applied at rated frequency. *See:* **asynchronous machine; direct-current commutating machine.**   63, 3

lock-in (laser gyro). The phenomenon characterized by the oscillation at the same frequency of the clockwise and counter-clockwise beams of the laser gyro. It is caused by the coupling of energy between the laser beams and results in a threshold error near zero input angular rate unless a corrective means such as biasing is used.   46

locking-in. The shifting and automatic holding of one or both of the frequencies of two oscillating systems that are coupled together, so that the two frequencies have the ratio of two integral numbers.   59

locking plate (rotating machinery). *See:* **lockplate.**

locking ring (rotating machinery). A ring used to prevent motion of a second part.   63

lock-in rate (laser gyro). One half of the absolute value of the algebraic difference between the two rates defining the region over which lock-in occurs.   46

lockout (telephone circuit controlled by two voice-operated devices). The inability of one or both subscribers to get through, either because of excessive local circuit noise or continuous speech from either or both subscribers. *See:* **sweep lockout.**   328, 185

lockout-free (recloser or sectionalizer). A general term denoting that the lockout mechanism can operate even though the manual operating lever is held in the closed

position. *Note:* When used as an adjective modifying a device, the device has this operating capability.   103, 202

lockout mechanism (automatic circuit recloser). A device that locks the contacts in the open position following the completion of a predetermined sequence of operations.   103, 202

lockout operation (recloser). An opening operation followed by the number of closing and opening operations that the mechanism will permit before locking the contacts in the open position.   103, 202

lockout protection device (series capacitor). A device to block the automatic opening of the bypass switch and insertion of the capacitor segment following its closing from a cause which warrants inspection or maintenance.   86

lockout relay. An electrically reset or hand-reset auxiliary relay whose function is to hold associated devices inoperative until it is reset.   103, 202, 60, 127

lockplate (locking plate) (rotating machinery). A plate used to prevent motion of a second part (for example, to prevent a bolt or nut from turning).   63

lock-up relay. (1) A relay that locks in the energized position by means of permanent magnetic bias (requiring a reverse pulse for releasing) or by means of a set of auxiliary contacts that keep its coil energized until the circuit is interrupted. *Note:* Differs from a latching relay in that locking is accomplished magnetically or electrically rather than mechanically. (2) Sometimes used for latching relay. *See also:* **relay.**   259

logarithmic decrement (1) (underdamped harmonic system). The natural logarithm of the ratio of a maximum of the free oscillation to the next following maximum. The logarithmic decrement of an under-damped harmonic system is

$$\ln\left(\frac{X_1}{X_2}\right) = \frac{2\pi F}{(4MS - F^2)^{1/2}}$$

where X_1 and X_2 are the two maxima.   210

(2) (automatic control). A measure of damping of a second-order linear system, expressed as the Napierian logarithm (with negative sign) of the ratio of the greater to the lesser of a pair of consecutive excursions of the variable (in opposite directions) about an ultimate steady-state value. *Note:* For the system characterized by a quadratic factor

$$1/(s^2 + 2z\omega_n s + \omega_n{}^2) \text{ its value is}$$
$$-\pi z/(1 - z^2)^{1/2}$$

Twice this value defines the envelopes of the damping, but C85 prefers the above definition for reasons of convenience noted under damping factor. *See also:* **damping, relative; damping coefficient, viscous; subsidence ratio.**   56

logic. (1) The result of planning a data-processing system or of synthesizing a network of logic elements to perform a specified function. (2) Pertaining to the type or physical realization of logic elements used, for example, diode logic, AND logic. *See also:* **formal logic; logic design; symbolic logic.**   235, 54

logical operations with pulses (pulse terms). This section considers the pulse as a logical operator. Some operations defined in **operations on a pulse, operations by a pulse,** and operations involving the interaction of pulses, frequently are logical operations in the sense of this section. (1) general. AND, NAND, OR, NOR, EXCLUSIVE OR, INVERSION, inhibiting, enabling, disabling, counting, or other logical operations may be performed. (2) slivering. A process in which a (typically, unwanted) pulse of relatively short duration is produced by a logical operation. Typically, slivering is a result of partial pulse coincidence. (3) gating. A process in which a first pulse enables or disables portions of a second pulse or other event for the duration of the first pulse. (4) shifting. A process in which logical states in a specified sequence are transferred without alteration of the sequence from one storage element to another by the action of a pulse.   254

logic board (power-system communication). An assembly of decision-making circuits on a printed-circuit mounting board. *See also:* **digital.**   59'

logic design (electronic computation). (1) The planning of a computer or data-processing system prior to its detailed engineering design. (2) The synthesizing of a network of logic elements to perform a specified function. (3) The result of (1) and (2) above, frequently called the logic of the system, machine, or network.   235, 210

logic diagram (1) (digital computers). A diagram representing the logic elements and their interconnections without necessarily expressing construction or engineering details.   235, 210, 54

(2) (graphic symbols for logic diagrams). A diagram that depicts the two-state device implementation of logic functions with logic symbols and supplementary notations, showing details of signal flow and control, but not necessarily the point-to-point wiring.   88

logic element (1) (general). A combinational logic element or sequential logic element.   235

(2) (computer or data-processing system). The smallest building blocks that can be represented by operators in an appropriate system of symbolic logic. Typical logic elements are the AND gate and the flip-flop, which can be represented as operators in a suitable symbolic logic.   210

(3) (graphic symbols for logic diagrams). An element whose input and output signals represent logic variables, and whose output or outputs are defined functions of the inputs and of the variables (including time dependencies) internal to the element.   88

logic function (graphic symbols for logic diagrams). A combinational, delay, or sequential logic element defined in terms of the relationships which hold between input and output logic variables.   88

logic instruction (computing systems). An instruction that executes an operation that is defined in symbolic logic, such as AND, OR, NOR.   255, 77

logic, 0–1. The representation of information by two states termed 0 and 1. *See also:* **bit; dot cycle; mark or space; digital.**   59

logic operation (electronic computation) (1) (general). Any nonarithmetical operation. *Note:* Examples are: extract, logical (bit-wise) multiplication, jump, data transfer, shift, compare, etcetera.

(2) (sometimes). Only those nonarithmetical operations that are expressible bit-wise in terms of the propositional calculus or two-valued Boolean algebra.   235, 210

logic operator. *See:* AND; exclusive OR; NAND; NOT; OR.

logic shift (computing systems). A shift that affects all positions.   255, 77

logic symbol (1) (electronics computation). (A) A symbol used to represent a logic element graphically. (B) A symbol used to represent a logic connective.   235, 255, 77

(2) (graphic symbols for logic diagrams). The graphic representation in diagrammatic form of a logic function.   88

logic variable (graphic symbols for logic diagrams). A variable which may assume one of two discrete states, the 1-state or the 0-state.   88

log normal density function (radar). A probability density function describing some random variable, given by

$$f(X) = \frac{1}{X\sigma\sqrt{2\pi}} \exp - \left(\frac{(\ln X)^2}{2\sigma^2} \right)$$

Often used to describe the statistical nature of clutter, particularly sea clutter.   13

log periodic antenna. Any one of a class of antennas having a structural geometry such that its impedance and radiation characteristics repeat periodically as the logarithm of frequency.   111

long dimension (numerically controlled machines). Incremental dimensions whose number of digits is one more to the left of the decimal point than for a normal dimension, and the last digit shall be zero, that is, XX.XXX0 for the example under normal dimension.   224, 207

long-distance navigation. Navigation utilizing self-contained or external reference aids or methods usable at comparatively great distances. *Note:* Examples of long-distance aids are loran, Doppler, inertial, and celestial navigation. *See:* **short-distance navigation; approach navigation.** *See also:* **navigation.**   187, 13

long distance (LD) trunk (data transmission). That type of trunk, which permits trunk-to-trunk connection and which interconnects local, secondary, primary and zone centers.   59

longitudinal balance (data transmission). A measure of the similarity of impedance to ground (or common) for the two or more conductors of a balanced circuit. This term is used to express the degree of susceptibility to common mode interference.   59

longitudinal balance, degree of (telephone equipment). The ratio of the disturbing longitudinal voltage (Vs) and the resulting metallic voltage (Vm) of the network under test, expressed in decibels as follows: Longitudinal balance = 20 log |Vs/Vm| (in dB) where Vs and Vm are at the same frequency. *Note:* "log" is assumed to mean "log to the base 10."   39

longitudinal circuit (telephone equipment). A circuit formed by one communication conductor (or by two or more communication conductors in parallel) with a return through ground or through any other conductors except those which are taken with the original conductor or conductors to form a metallic circuit. *See also:* **communication conductor; metallic circuit.**   39

longitudinal circuit port (telephone equipment). A place of access in the longitudinal transmission path of a device or network where energy may be supplied or withdrawn, or where the device or network variables may be measured. *See also:* **device; network.** 39

longitudinal impedance (telephone equipment). Impedance presented by a longitudinal circuit at any given single frequency. *See also:* **impedance; longitudinal circuit.**
39

longitudinal interference. *See:* **common-mode interference.** *See also:* **accuracy rating (instrument); signal.**

longitudinal magnetization (magnetic recording). Magnetization of the recording medium in a direction essentially parallel to the line of travel. 176

longitudinal mode (1) (laser-maser). The term refers to modes which have the same field distributions transverse to the beam but a different number of half period field variations along the axis of the beam. *See also:* **longitudinal resonances.** 363

(2) (interference terminology). (*See:* **interference, commonmode.**

longitudinal redundancy check (LRC) (data transmission). A system of error control based on the formation of a block check following preset rules. *Note:* The check formation rule is applied in the same manner to each character.
194

longitudinal resonances (in a beam resonator) (laser-maser). Resonances corresponding to modes having the same field distribution transverse to the beam, but differing in the number of half period field variations along the axis of the beam. *Note:* Such resonances are separated in frequency by approximately v/2L where v is the speed of light in the resonator and 2L is the round trip length of the beam in the resonator. 363

longitudinal wave. A wave in which the direction of displacement at each point of the medium is the same as the direction of the propagation. 210

long-line adapter (telephone switching systems). Equipment inserted between a line circuit and the associated station(s) to allow conductor loop resistances greater than the maximum for which a system is designed. 55

long-line current (corrosion). Current (positive electricity) flowing through the earth from an anodic to a cathodic area that returns along an underground metallic structure. *Note:* Usually used only where the areas are separated by considerable distance and where the current results from concentration cell action. *See also:* **stray-current corrosion.** 221, 205

long-pitch winding (rotating machinery). A winding in which the coil pitch is greater than the pole pitch. *See:* **direct-current commutating machine.** 63

long-time-delay phase trip element (power switchgear). A direct-acting trip device element that functions with a purposely delayed action (seconds). 103

longwall machine (mining). A power-driven machine used for undercutting coal on relatively long faces. *See also:* **mining.** 328

long-wire antenna. A wire antenna that, by virtue of its considerable length in comparison with the operating wavelength, provides a directional pattern. *See also:* **antenna.** 179, 111

look (radar). A colloquial expression associated with a single attempt at detection of a target. 13

look-up (computing systems). *See:* **table look-up.**

loom. *See:* **flexible nonmetallic tubing.**

loop (1) (signal-transmission system and network analysis). A set of branches forming a closed current path, provided that the omission of any branch eliminates the closed path. *See:* **signal; ground loop; mesh.** *See also:* **linear signal-flow graphs.** 282, 188

(2) (standing wave). A point at which the amplitude is a maximum. 210

(3) (computing systems). A sequence of instructions that is executed repeatedly until a terminal condition prevails.
255, 77

(4) (data transmission) (telephone circuit). In communications signifies a type of facility, normally the circuit between the subscriber and central office. *Note:* Usually a metallic circuit. 59

loop antenna. An antenna whose configuration is that of a loop. *Note:* If the current in the loop, or in multiple parallel turns of the loop, is essentially uniform and the loop circumference is small compared with the wavelength, the radiation pattern approximates that of a magnetic dipole. 111

loop circuit (railway signaling). A circuit that includes a source of electric energy, a line wire that conducts current in one direction, and connections to the track rails at both ends of the line to complete the circuit through the two rails in parallel in the other direction. 328

loop control (industrial control). The effect of a control function or a device to maintain a specified loop of material between two machine sections by automatic speed adjustment of at least one of the driven sections. *See:* **control system, feedback.** 225, 206

loop converter (data transmission). A device used for conversion of direct-current loop current pulses to relay contact closures thereby providing circuit isolation. 59

loop current. The electric current in a loop circuit. *See:* **mesh current.** 328

loop (leakage) current (power supplies). A direct current flowing in the feedback loop (voltage control) independent of the control current generated by the reference Zener diode source and reference resistor. *Note:* The loop (leakage) current remains when the reference current is made zero. It may be compensated for, or nulled, in special applications to achieve a very-high impedance (zero current) at the feedback (voltage control) terminals. *See also:* **power supply.** 228, 186

loop equations. *See:* **mesh or loop equations.**

loop factor (network analysis). *See:* **path factor.**

loop feeder (power distribution). A number of tie feeders in series, forming a closed loop. *Note:* There are two routes by which any point on a loop feeder can receive electric energy, so that the flow can be in either direction. *See also:* **center of distribution.** 64

loop gain (1). The ratio, under specified conditions, of the steady-state sinusoidal magnitude of the feedback signal to that of the actuating signal when the loop is open at the summing point. *Syn.* **open-loop gain.** *See:* **control system, feedback.** 219, 206, 67

(2) (power supplies). A measure of the feedback in a closed-loop system, being equal to the ratio of the open-loop to the closed-loop gains, in decibels. *Note:* The magnitude of the loop gain determines the error attenuation and, therefore, the performance of an amplifier used as a voltage regulator. *See:* **open-loop and closed-loop gain.** *See also:* **power supply.**   228, 186

(3) (automatic control). The absolute magnitude of the loop gain characteristic at a specified frequency. *Note:* The gain of the loop elements is often measured by opening the loop, with appropriate termination. The gain so measured is often called the open loop gain.   56

(4) (data transmission). The sum of the gains which are given to a signal of a particular frequency in passing around a closed loop. The loop may be a repeater, carrier terminal, or a complete system. The loop gain may be less than the sum of the individual amplifier gains because singing may occur if full amplification is used. The maximum usable gain is determined by, and may not exceed, the losses in the closed path.   59

loop gain characteristic. *See:* **gain characteristic, loop.**

loop graph (network analysis). A signal flow graph each of whose branches is contained in at least one loop. *Note:* Any loop graph embedded in a general graph can be found by removing the cascade branches. *See also:* **linear signal-flow graphs.**   282

looping-in (interior wiring). A method of avoiding splices by carrying the conductor or cable to and from the outlet to be supplied.   328

loop noise bandwidth (communication satellite). One of the fundamental parameters of a phase lock loop. It is the equivalent bandwidth of a square cut-off lowpass filter, which, when multiplied by a flat input noise spectral density, produces the loop noise variance.   85

loop phase angle. *See:* **phase angle, loop.**

loop pulsing (telephone switching systems). Dial pulsing using loop signaling.   55

loop service (power distribution). Two services of substantially the same capacity and characteristics supplied from adjacent sections of a loop feeder. *Note:* The two sections of the loop feeder are normally tied together on the consumer's bus through switching devices. *See also:* **service.**   64

loop-service (ring) feeder (power distribution). A feeder that supplies a number of separate loads distributed along its length and that terminates at the same bus from which it originated.   103, 202

loop-set transmittance (network analysis). The product of the negatives of the loop transmittances of the loops in a set. *See also:* **linear signal-flow graphs.**   282

loop signaling (telephone switching systems). A method of signaling over direct current circuit paths that utilize the metallic loop formed by the trunk conductors and terminating bridges.   55

loop stick antenna. A loop receiving antenna with a ferrite rod core used for increasing its radiation efficiency.   111

loop test. A method of testing employed to locate a fault in the insulation of a conductor when the conductor can be arranged to form part of a closed circuit or loop.   328

loop transmittance (1) (network analysis). The product of the branch transmittances in a loop.

(2) (branch). The loop transmittance of an interior node inserted in that branch. *Note:* A branch may always be replaced by an equivalent sequence of branches, thereby creating interior nodes.

(3) (node). The graph transmittance from the source node to the sink node created by splitting the designated node. *See also:* **linear signal-flow graphs.**   282

loose coupling. Any degree of coupling less than the critical coupling. *See also:* **coupling.**   328

loose leads (rotating machinery). A form of termination in which the machine terminals are loose cable leads.   63

LOP (electronic navigation). *See:* **line of position.**

loran. A long-distance radio-navigation system in which hyperbolic lines of position are determined by measuring arrival-time differences of pulses transmitted in fixed time relationship from two fixed-base transmitters. *Note:* **Loran A,** generally useful to distances of 500 to 1500 nautical miles (900 to 2800 kilometers) over water, depending upon the availability of sky wave, uses a baseline of about 300 nautical miles (550 kilometers), operated at approximately 2 megahertz and gives time-difference measurement by the matching of the leading edges of the pulses, usually with the aid of an oscilloscope. **Loran C,** generally useful to distances of 1000 to 1500 nautical miles (1850 to 2800 kilometers) over water, uses a baseline of about 500 nautical miles, operates at approximately 100 kilohertz; it provides a coarse measurement of time-difference through the matching of pulse envelopes, and a fine measurement by the comparing of phase between the carrier waves. **Loran D** is a shorter-baseline and lower-power adaptation of loran C for tactical applications. *See also:* **radio direction-finder (radio compass); radio navigation.**   13, 187, 3

loran repetition rate (electronic navigation). *See:* **pulse repetition frequency.**

Lorenz number. The quotient of (1) the electronic thermal conductivity by (2) the product of the absolute temperature and the component of the electric conductivity due to electrons and holes. *See also:* **thermoelectric device.**   248, 191

lorhumb line (navigation system chart, such as a loran chart with its overlapping families of hyperbolic lines). A line drawn so that it represents a path along which the change in values of one of the families of lines retains a constant relation to the change in values of another of the families of lines. *See also:* **navigation.**   187, 13

loss (1) (power). (A) Power expended without accomplishing useful work. Such loss is usually expressed in watts. **(2) (communications).** The ratio of the signal power that could be delivered to the load under specified reference conditions to the signal power delivered to the load under actual operating conditions. Such loss is usually expressed in decibels. *Note:* Loss is generally due to dissipation or reflection due to an impedance mismatch or both. *See:* **transmission loss.**   197

(2) (network analysis). *See:* **related transmission terms.**

loss angle (1) (biological). The complement Φ of the phase angle θ (between the electrode potential vector and the current vector).

$$\Phi = \frac{\pi}{2} - \theta$$

See also: **electrode impedance (biological).**   192

(2) (magnetic core testing). (A) Dissipation factor. The angle by which the fundamental component of the magnetizing current lags the fundamental component of the exciting current in a coil with a ferromagnetic core. The tangent of this angle is defined as the ratio of the in-phase and quadrature components of the impedance of the coil.

$$\tan \delta_n = \frac{R_s}{\omega L_s} = \frac{\omega L_p}{R_p} = \frac{\mu''_s}{\mu'_s} = \frac{\mu'_p}{\mu''_p}$$

$$\delta_n = \text{loss angle}$$

(B) Relative dissipation factor.
Defined as:

$$\frac{\tan \delta_n}{\mu_i} = \frac{\mu''_s}{(\mu'_s)^2} = \frac{R_s}{\mu_i \omega L_s} = \frac{\omega L_p}{\mu_i R_p} = \frac{\omega L_0}{R_p}$$

Quality factor Q. See: General definition.
For inductive devices it is defined as the inverse of the tangent of the loss angle.

$$Q = 1/(\tan \delta_n)$$

Q = quality factor

δ_n = loss angle. 165

loss compensator. See: transformer-loss compensator.
loss, electric system. Total electric energy loss in the electric system. It consists of transmission, transformation, and distribution losses between sources of supply and points of delivery. 112
losses (grounding device) (electric power). I^2R loss in the windings, core loss, dielectric loss (for capacitors), losses due to stray magnetic fluxes in the windings and other metallic parts of the device, and in cases involving parallel windings, losses due to circulating currents. *Note:* The losses as here defined do not include any losses produced by the grounding device in adjacent apparatus or materials not part of the device. Losses will normally be considered at the maximum rated neutral current but may in some cases be required at other current ratings, if more than one rating is specified, or at no load, as for grounding transformers. *Note:* The losses may be given at 25 degrees Celsius or at 75 degrees Celsius. See also: **grounding device.** 91
losses of a current-limiting reactor. Losses that are incident to the carrying of current. They include: (1) The resistance and eddy-current in the winding due to load current. (2) Losses caused by circulating current in parallel windings. (3) Stray losses caused by magnetic flux in other metallic parts of the reactor and in the reactor enclosure when the enclosure is supplied as an integral part of the reactor installation. (4) The losses produced by magnetic flux in adjacent apparatus or material not an integral part of the reactor or its enclosure are not included. See also: **reactor.** 309
loss factor (1) (electric power generation). The ratio of the average power loss to the peak-load power loss during a specified period of time. See also: **generating station.** 64
(2) (dielectric heater material). The product of its dielectric constant and the tangent of its dielectric loss angle. See also: **dielectric heating; depth of current penetration; electric constant.** 14
loss function (control system). An instantaneous measure

of the cost of being in state x and of using control u at time t. See also: **performance index.** 329
loss, insertion. See: insertion loss.
loss of control-power protection (series capacitor). A means to initiate the closing of the bypass switch upon the loss of normal control power. 86
loss of excitation relay (power switchgear). A relay that produces an output when the input to a synchronous machine indicates that the machine has substantially lost its field excitation. 127, 103
loss of forming (semiconductor rectifier). A partial loss in the effectiveness of the rectifier junction. See also: **rectification.** 237, 66
loss, return. See: return loss.
loss tangent (1) (general). The ratio of the imaginary part of the complex dielectric constant of a material to its real part. 185
(2) (tan δ) (rotating machinery). The ratio of dielectric loss in an insulation system, to the apparent power required to establish an alternating voltage across it of a specified amplitude and frequency, the insulation being at a specified temperature. *Note:* It is the cotangent of the power-factor angle. See also: **asynchronous machine; direct-current commutating machine.** 63
loss-tangent test (dissipation-factor test) (rotating machinery). A test for measuring the dielectric loss of insulation at predetermined values of temperature, frequency, and voltage or dielectric stress, in which the dielectric loss is expressed in terms of the tangent of the complement of the insulation power-factor angle. See also: **asynchronous machine; direct-current commutating machines.** 63
loss, total (1) (rotating machinery). The difference between the active electrical power (mechanical power) input and the active electrical power (mechanical power) output. 63
(2) (transformer or voltage regulator). The sum of the excitation losses and the load losses. See also: **efficiency; voltage regulator.** 303, 203
loss, transmission. See: transmission loss.
lossy medium (laser-maser). A medium which absorbs or scatters radiation passing through it. 363
lost call (telephone switching systems). A call that cannot be completed due to blocking. 55
loudness. The intensive attribute of an auditory sensation in terms of which sound may be ordered on a scale extending from soft to loud. *Note:* Loudness depends primarily upon the sound pressure of the stimulus, but it also depends upon the frequency and waveform of the stimulus. See also: **loudspeaker.** 176
loudness contour. A curve that shows the related values of sound pressure level and frequency required to produce a given loudness sensation for the typical listener. See also: **loudspeaker.** 176
loudness level (sound). A value, expressed in phons, that is numerically equal to the median sound pressure level, in decibels, relative to 2×10^{-5} newton per square meter, of a free progressive wave of frequency 1000 hertz presented to listeners facing the source, which in a number of trials is judged by the listeners to be equally loud. *Note:* The manner of listening to the unknown sound, which must be

stated, may be considered to be one of the characteristics of that sound. *See also:* **loudspeaker.** 176

loudspeaker (loudspeaker measurements). Understood, unless otherwise stated, to include both loudspeaker drivers along with their mountings or enclosures (which together comprise a loudspeaker system) and also loudspeaker systems which comprise one or more loudspeaker drivers and such devices as integral crossover filters, transformers, and any other integral passive or active elements. 19

loudspeaker system. A combination of one or more loudspeakers and all associated baffles, horns, and dividing networks arranged to work together as a coupling means between the driving electric circuit and the acoustic medium. *See also:* **loudspeaker.** 176

loudspeaker voice-coil. The moving coil of a dynamic loudspeaker. *See also:* **loudspeaker.** 176

louver (louver grid) (illuminating engineering). A series of baffles used to shield a source from view at certain angles or to absorb unwanted light. *Note:* The baffles usually are arranged in a geometric pattern. *See also:* **bare (exposed) lamp.** 167

louvered ceiling (illuminating engineering). A ceiling area-lighting system comprising a wall-to-wall installation of multicell louvers shielding the light sources mounted above it. *See also:* **luminaire.** 167

louver shielding angle (illuminating engineering). The angle between the horizontal plane of the baffles or louver grid and the plane at which the louver conceals all objects above. *Note:* The planes usually are so chosen that their intersection is parallel with the louvered blade. *See also:* **bare (exposed) lamp.** 167

low-clearance area (instrument approach system). Any area containing only low-clearance points. *See also:* **navigation.** 13, 187

low-clearance points (instrument landing systems). Locations in space outside the course sector at which course deviation indicator current is below some arbitrary value, usually the full-scale deflection value. *See also:* **navigation.** 187, 13

low-cycle fatigue (nuclear power generating stations) (seismic qualification of class 1E equipment). A progressive fracture or cumulative fatigue of the material which may occur for less than 1000 cycles because of localized stress concentration at high strains under fluctuating loads. *Note:* As the strain is reduced, the fatigue life increases exponentially. Ultimately, the number of cycles can be increased indefinitely at a reduced stress designated as the endurance limit of the material. 28, 31

low-energy power circuit (interior wiring). A circuit that is not a remote-control or signal circuit but that has the power supply limited in accordance with the requirements of Class 2 remote-control circuits. *Note:* Such circuits include electric door openers and circuits used in the operation of coin-operated phonographs. *See also:* **appliances.** 256

lower (passing) beams (motor vehicle). One or more beams directed low enough on the left to avoid glare in the eyes of oncoming drivers, and intended for use in congested areas and on highways when meeting other vehicles within a distance of 1000 feet. Formally **traffic beam.** *See also:* **headlamp.** 167

lower bracket (rotating machinery). A bearing bracket

mounted below the level of the core of a vertical machine. 63

lower burst reference (audio and electroacoustics). A selected multiple of the long-time average magnitude of a quantity, smaller than the upper burst reference. See the figure under **burst duration.** *See also:* **burst (audio and electroacoustics).** 253, 176

lower coil support (rotating machinery). A support to restrain field-coil motion in the direction away from the air gap. *See:* **rotor (rotating machinery); stator.** 63

lower guide bearing (rotating machinery). A guide bearing mounted below the level of the core of a vertical machine. 63

lower-half bearing bracket (rotating machinery). The bottom half of a bracket that can be separated into halves for mounting or removal without access to a shaft end. 63

lower limit (test, measurement and diagnostic equipment). The minimum acceptable value of the characteristic being measured. 54

lower-range value. The lowest quantity that a device is adjusted to measure. *Note:* The following compound terms are used with suitable modifications in the units: **measured-variable lower-range value, measured signal lower-range value,** etcetera. *See also:* **instrument.** 295

lower-sideband parametric down-converter.* An inverting parametric device used as a parametric down-converter. *See also:* **parametric device.** 277, 191

* Deprecated

lower-sideband parametric up-converter.* An inverting parametric device used as a parametric up-converter. *See also:* **parametric device.** 277, 191

* Deprecated

lowest useful high frequency (radio wave propagation). The lowest high frequency effective for ionospheric propagation of radio waves between two specified points, under specified ionospheric conditions, and under specified factors such as absorption, transmitter power, antenna gain, receiver characteristics, type of service, and noise conditions. *See also:* **radiation; radio wave propagation.** 180, 146

low-frequency dry-flashover voltage. The root-mean-square voltage causing a sustained disruptive discharge through the air between electrodes of a clean dry test specimen under specified conditions. 64

low-frequency flashover voltage (insulator). The root-mean-square value of the low-frequency voltage that, under specified conditions, causes a sustained disruptive discharge through the surrounding medium. *See also:* **insulator.** 261

low-frequency furnace (core-type induction furnace). An induction furnace that includes a primary winding, a core of magnetic material, and a secondary winding of one short-circuited turn of the material to be heated. 328

low-frequency high-potential test (rotating machinery). A high-potential test which applies a low-frequency voltage, between 0.1 hertz and 1.0 hertz, to a winding. 63

low-frequency impedance corrector. An electric network designed to be connected to a basic network, or to a basic network and a building-out network, so that the combination will simulate at low frequencies the sending-end impedance, including dissipation, of a line. *See also:* **network analysis.**   328

low-frequency induction heater or furnace. A device for inducing current flow of commercial power-line frequency in a charge to be heated. *See also:* **dielectric heating, induction heating.**   14, 114

low-frequency puncture voltage (insulator). The root-mean-square value of the low-frequency voltage that, under specified conditions, causes disruptive discharge through any part of the insulator. *See also:* **insulator.**
        261

low-frequency wet-flashover voltage. The root-mean-square voltage causing a sustained disruptive discharge through the air between electrodes of a clean test specimen on which water of specified resistivity is being sprayed at a specified rate.   64

low-frequency withstand voltage (insulator). The root-mean-square value of the low-frequency voltage that, under specified conditions, can be applied without causing flashover or puncture. *See also:* **insulator.**   261

low-key lighting (television). A type of lighting that, applied to a scene, results in a picture having graduations falling primarily between middle gray and black, with comparatively limited areas of light grays and whites. *See also:* **television lighting.**   167

low-level analog signal cable (cable systems). Used for transmitting variable current or voltage signals for the control and/or instrumentation of plant equipment and systems.   35

low-level digital signal circuit cable (cable systems). Used for transmitting coded information signals, such as those derived from the output of an analog to digital converter, or the coded output from a digital computer or other digital transmission terminals.   35

low-level modulation (communication). Modulation produced at a point in a system where the power level is low compared with the power level at the output of the system.   111, 240

low-level radio-frequency signal (TR, ATR, and Pre-TR tubes). A radio-frequency signal with insufficient power to cause the tube to become fired.   125

low-pass filter. *See:* **filter, low-pass.**

low (normal) power-factor mercury lamp ballast. A ballast of the multiple-supply type that does not have a means for power-factor correction.   271

low power-factor transformer. A high-reactance transformer that does not have means for power-factor correction.   53

low-pressure contact (area contact) (power switchgear). One in which the pressure is such that stress in the material is well below the elastic limit of both contact surface materials, such that conduction is a function of area.   27, 103

low-pressure vacuum pump. A vacuum pump that compresses the gases received directly from the evacuated system. *See also:* **rectification.**   328

low state (programmable instrumentation). The relatively less positive signal level used to assert a specific message content associated with one of two binary logic states.   40

low-velocity camera tube (electron device) (cathode-volt-

age-stabilized camera tube). A camera tube operating with a beam of electrons having velocities such that the average target voltage stabilizes at a value approximately equal to that of the electron-gun cathode.   190

low-voltage ac power circuit breaker. *See:* **circuit breaker.**

low-voltage integrally fused power circuit breaker (power switchgear). An assembly of a general-purpose ac low-voltage power circuit breaker and integrally mounted current-limiting fuses which together function as a coordinated protective device.   103

low voltage power cable (cable systems). Designed to supply power to utilization devices of the plant auxiliary system, rated 600 volts or less.   35

low-voltage release. The effect of a device, operative on the reduction or failure of voltage, to cause the interruption of power supply to the equipment, but not preventing the reestablishment of the power supply on return of voltage. *See also:* **generating system.**   262

low-voltage system (electric power). An electric system having a maximum root-mean-square alternating-current voltage of 1000 volts or less. *See also:* **voltage classes.**
        49

LRC (data transmission). *See:* **longitudinal redundancy check.**

LRU. *See:* **line replaceable unit.**

L scan (electronic navigation). *See:* **L display.**

L scope (electronic navigation). *See:* **L display.**

LS dividing network. *See:* **dividing network.**

LTS (long-term stability). *See:* **stability, long-term.**

lug (electric installations on shipboard). A wire connector device to which the electrical conductor is attached by mechanical pressure or solder.   3

lug, stator mounting (rotating machinery). A part attached to the outer surface of stator core or a stator shell to provide a means for the bolting or equivalent attachment to the appliance, machine, or other foundation. *See:* **stator.**
        63

lumen (1) (general). The unit of luminous flux. It is equal to the flux through a unit solid angle (steradian), from a uniform point source of one candela (candle), or to the flux on a unit surface all points of which are at unit distance from a uniform point source of one candela. *Note:* For some purposes, the kilolumen, equal to 1000 lumens, is a convenient unit. *See also:* **light.**   167, 244, 178, 162
(2) (color terms). The unit of luminous flux. The luminous flux emitted within unit solid angle (one steradian) by a point source having a uniform intensity of one candela.
        18

lumen hour. A unit of quantity of light (luminous energy). It is the quantity of light delivered in one hour by a flux of one lumen. *See also:* **light.**   167

lumen (or flux) method (lighting calculation). A lighting design procedure used for predetermining the number and types of lamps or luminaires that will provide a desired average level of illumination on the work plane. *Note:* It takes into account both direct and reflected flux. *See also:* **inverse-square law (illuminating engineering).**   167

lumen-second (lm.s.). A unit of quantity of light (luminous energy). It is the quantity of light delivered in one second by a flux of one lumen.   162

luminaire. A complete lighting unit consisting of a lamp or lamps together with the parts designed to distribute the light, to position and protect the lamps, and to connect the lamps to the power supply. 167

luminaire efficiency. The ratio of luminous flux (lumens) emitted by a luminaire to that emitted by the lamp or lamps used therein. *See also:* **inverse-square law (in illuminating engineering).** 167

luminance (1) (at a point of a surface and in a given direction). The quotient of the luminous intensity in the given direction of an infinitesimal element of the surface containing the point under consideration, by the orthogonally projected area of the element on a plane perpendicular to the given direction. *Notes:* (A) Typical units in this system are the candela per square meter and the candela per square foot. (B) The term **luminance** is recommended for the photometry quality, which has been called **brightness.** Use of this term permits brightness to be used entirely with reference to sensory response. The photometric quantity has been confused often with the sensation merely because of the use of one name for two distinct ideas. Brightness will continue to be used properly in nonquantitative statements, especially with reference to sensations and perceptions of light. *See also:* **light.** 18, 185, 244, 178 (2) **(light emitting diodes) (photometric brightness)** $L = d^2\Phi/(d\omega\,dA\,\cos\theta)$. In a direction, at a point of the surface of a source, or of a receiver, or of any other real or virtual surface is the quotient of the luminous flux leaving, passing through, or arriving at an element of the surface surrounding the point, and propagated in direction defined by an elementary cone containing the given direction, by the product of the solid angle of the cone and the area of the orthogonal projection of the element of the surface on a plane perpendicular to the given direction per unit of projected area of the surface as viewed from that direction. *Note:* In the defining equation θ is the angle between the direction of observation and the normal to the surface. In common usage the term brightness usually refers to the intensity of sensation which results from viewing surfaces or spaces from which light comes to the eye. This sensation is determined in part by the definitely measurable luminance (photometric brightness) defined above and in part by conditions of observation such as the state of adaptation of the eye. In much of the literature the term brightness, used alone, refers to both luminance and sensation. The context usually indicates which meaning is intended. 162

(3) **(average luminance) (average photometric brightness).** The total lumens actually leaving the surface per unit area. *Notes:* (A) Average luminance specified in this way is identical in magnitude with **luminous exitance,** which is the preferred term. (B) In general, the concept of average luminance is useful only when the luminance is reasonably uniform throughout a very wide angle of observation and over a large area of the surface considered. It has the advantage that it can be computed readily for reflecting surfaces by multiplying the incident luminous flux density (illumination) by the luminous reflectance of the surface. For a transmitting body it can be computed by multiplying the incident luminous flux density by the luminous transmittance of the body. 167

luminance channel (color-television system). Any path that is intended to carry the luminance signal. *See also:* **television.** 163, 178

luminance channel bandwidth (television). The bandwidth of the path intended to carry the luminance signal. *See also:* **television.** 163, 178

luminance (photometric brightness) contrast. The relationship between the luminances (photometric brightnesses) of an object and its immediate background. It is equal to $(L_1 - L_2)/L_1$ or $(L_2 - L_1)/L_1$, where L_1 and L_2 are the luminances of the background and object, respectively. *Note:* The form of the equation must be specified. Because of the relationship among luminance, illumination, and reflectance, contrast often is expressed in terms of reflectance when only reflecting surfaces are involved. Thus contrast is equal to $(\rho_1 - \rho_2)/\rho_1$ or $(\rho_2 - \rho_1)/\rho_1$ where ρ_1 and ρ_2 are the reflectances of the background and object, respectively. This method of computing contrast holds only for perfectly diffusing surfaces; for other surfaces it is only an approximation unless the angles of incidence and view are taken into consideration. *See also:* **reflectance; visual field.** 167

luminance (photometric brightness) difference. The difference in luminance (photometric brightness) between two areas. *Note:* It usually is applied to contiguous areas, such as the detail of a visual task and its immediate background, in which case it is quantitatively equal to the numerator in the formula for luminance contrast. See the note under **luminance.** *See also:* **visual field.** 167

luminance factor. The ratio of the luminance (photometric brightness) of a surface or medium under specified conditions of incidence, observation, and light source, to the luminance (photometric brightness) of a perfectly reflecting or transmitting, perfectly diffusing surface or medium under the same conditions. *Note:* Reflectance or transmittance cannot exceed unity, but luminance factor may have any value from zero to values approaching infinity. *See also:* **lamp.** 167

luminance (photometric brightness) factor of room surfaces. Factor by which the average work-plane illumination is multiplied to obtain the average luminances of walls, ceilings, and floors. *See also:* **inverse-square law (illuminating engineering).** 167

luminance flicker. The flicker that results from fluctuation of luminance only. 18, 178

luminance primary (color television). One of a set of three transmission primaries whose amount determines the luminance of a color. *Note:* This is an obsolete term because it is useful only in a linear system. 18, 178

luminance (photometric brightness) ratio. The ratio between the luminances (photometric brightnesses) of any two areas in the visual field. See the note under **luminance.** *See also:* **visual field.** 167

luminance signal (NTSC color television). A signal that has major control of the luminance. *Note:* It is a linear combination of gamma-corrected primary color signals, E'_R, E'_G, and E'_B as follows:

$$E'_Y = 0.30E'_R + 0.59E'_G + 0.11E'_B$$

The proportions expressed are strictly true only for television systems using the NTSC original standard receiver primaries having the CIE color points listed below, when they are mixed to produce white light having the same appearance as standard illuminant *C*.

Color	*x*	*y*
Red (R)	0.67	0.33
Green (G)	0.21	0.71
Blue (B)	0.14	0.08

18

luminance (photometric brightness) threshold. The minimum perceptible difference in luminance for a given state of adaptation of the eye. See the note under **luminance.** *See also:* **visual field.** 167

luminescence (light emitting diodes). Any emission of light not ascribable directly to incandescence. *See also:* **lamp.** 162

luminescent-screen tube. A cathode-ray tube in which the image on the screen is more luminous than the background. *See also:* **cathode-ray tubes.** 244, 190

luminosity. Ratio of luminous flux to the corresponding radiant flux at a particular wavelength. It is expressed in lumens per watt. 18, 178

luminosity coefficients. The constant multipliers for the respective tristimulus values of any color, such that the sum of the three products is the luminance of the color. 18, 178

luminous ceiling. A ceiling area-lighting system comprising a continuous surface of transmitting material of a diffusing or light-controlling character with light sources mounted above it. *See also:* **luminaire.** 167

luminous density (w = dQ_v/dV) (light emitting diodes). Quantity of light (luminous energy) per unit volume. 162

luminous efficacy (1) (radiant flux). The quotient of the total luminous flux by the total radiant flux. It is expressed in lumens per watt. *See also:* **light.** 167

(2) (source of light) (light emitting diodes). (ϵ_v). The quotient of the luminous flux emitted by the power dissipation (electrical). *Note:* The term **luminous efficiency** has in the past been extensively used for this concept. 162

luminous efficiency (1) (television). The ratio of the luminous flux to the radiant flux. *Note:* Luminous efficiency is usually expressed in lumens per watt of radiant flux. It should not be confused with the term efficiency as applied to a practical source of light, since the latter is based upon the power supplied to the source instead of the radiant flux from the source. For energy radiated at a single wavelength, luminous efficiency is synonymous with luminosity. 18, 178

(2) (light emitting diodes) (K), (K_λ), (K_m), (radiant flux). The quotient of luminous flux by the corresponding radiant flux. The symbol *K* represents the luminous efficiency of any radiant flux, which may include contributions of any or all wavelengths. The symbol K_λ represents luminous efficiency of monochromatic radiant flux of wavelength λ. The symbol K_m represents the maximum luminous efficiency of monochromatic radiant flux which will be obtained at the wavelength $\lambda = \lambda_m$ at which $V\lambda = 1$; K_m is equal approximately to 680 lumens per watt. 162

luminous flux (1) (illuminating engineering). The time rate of flow of light. *See also:* **light.** 18, 167, 178

(2) (light emitting diodes). (ϕ_v) (A) The time rate of flow of light. (B) The quantity derived from radiant flux by

evaluating the radiant energy according to its action upon a selective receptor, the spectral sensitivity of which is defined by a standard spectral luminous efficiency function. Unless otherwise indicated, the luminous flux relates to photopic vision, and is connected with the radiant flux by the following formula adopted by the CIE (International Commission on Illumination).

$$\phi_v = K_m \int \phi_\lambda V_\lambda d\lambda.$$

Here $\phi_\lambda d\lambda$ is the radiant flux emitted in the wavelength interval $d\lambda$ containing the wavelength λ, and V_λ is the photopic relative luminous efficiency function. The factor K_m is the maximum luminous efficiency corresponding to the wavelength for which $V_\lambda = 1$. 162

luminous flux density (surface) ($M_v = d\varphi_{v/dA}$, $E_v = d\varphi_{v/dA}$) (light emitting diodes). Luminous flux per unit area of the surface. *Note:* When referring to luminous flux emitted from a surface, this has been called **luminous emittance*** (symbol M_v). The preferred term for **luminous flux** leaving a surface is **luminous exitance** (symbol M_v). When referring to flux incident on a surface, it is identical with illumination (symbol E_v). 162

*Deprecated

luminous gain (optoelectronic device). The ratio of the emitted luminous flux to the incident luminous flux. *Note:* The emitted and incident luminous flux are both determined at specified ports. *See also:* **optoelectronic device.** 191

luminous intensity ($I_v = d\phi_v/d\omega$) (light emitting diodes). Luminous flux per unit solid angle in the direction in question. Hence, it is the luminous flux on a small surface normal to that direction, divided by the solid angle (in steradians) which the surface subtends at the source. *Note:* Mathematically a solid angle must have a point as its apex; the definition of luminous intensity, therefore, applies strictly only to a point source. In practice, however, light emanating from a source whose dimensions are negligible in comparison with the distance from which it is observed may be considered as coming from a point. The "apparent luminous intensity" of an extended source at a specified distance is the luminous intensity of a point source which would produce the same illumination at that distance. 162

luminous intensity of angular distribution (light emitting diodes). A curve, generally polar, which represents the variation of luminous intensity of a lamp in a plane through the light center. *Note:* A vertical luminous intensity distribution curve is obtained by taking measurements at various angles of elevation in a vertical plane through the light center; unless the plane is specified, the vertical curve is assumed to represent an average such as would be obtained by rotating the lamp or luminaire about its vertical axis. A horizontal distribution curve represents measurements made at various angles of azimuth in a horizontal plane through the light center. 162

luminous sensitivity (phototube). *See:* **sensitivity (camera tube or phototube).**

luminous-tube transformers. Transformers, autotransformers, or reactors (having a secondary open-circuit root-mean-square of 1000 volts or more) for operation of cold cathode and hot cathode luminous tubing generally used for signs, illumination, and decoration purposes.   53

lumped. Effectively concentrated at a single point.   328

lumped capacitive load (power switchgear). A lumped capacitance which is switched as a unit.   103

lumped element circuit (microwave tubes). A circuit consisting of discrete inductors and capacitors. *See also:* **microwave tube or valve.**   190

Luneburg lens antenna. A lens antenna with a circular cross section having an index of refraction varying only in the radial direction such that a feed located on or near a surface or edge of the lens produces a major lobe diametrically opposite the feed.   111

lux. In the International System of Units (SI) the unit of illumination on a surface one square meter in area on which there is a uniformly distributed flux of one lumen, or the illumination produced at a surface all points of which are at a distance of one meter from a uniform point source of one candela. *See also:* **light.**   167

Luxemburg effect. A nonlinear effect in the ionosphere as a result of which the modulation on a strong carrier wave is transferred to another carrier passing through the same region. *See also:* **radiation.**   328

Lyapunov function (control system) (equilibrium point x_e of a system). A scalar differentiable function $V(\mathbf{x})$ defined in some open region including $\mathbf{x}_e$ such that in that region

$$V(\mathbf{x}) > 0 \text{ for } \mathbf{x} \pm \mathbf{x}_e \qquad (1)$$

$$V(\mathbf{x}_e) = 0 \qquad (2)$$

$$V'(\mathbf{x}) \leq 0 \qquad (3)$$

Notes: (1) The open region may be defined by (norm of $\mathbf{x} - \mathbf{x}_e$) < constant. (2) For the system $\mathbf{x}' = \mathbf{f}(\mathbf{x})$, $V'(\mathbf{x}) \equiv$ [grad $V(\mathbf{x}) \cdot \mathbf{f}(\mathbf{x})$]. *See also:* **control system, feedback.**
329

M

machine (1) (general). An article of equipment consisting of two or more resistant, relatively constrained parts that, by a certain predetermined intermotion, may serve to transmit and modify force, motion, or electricity so as to produce some given effect or transformation or to do some desired kind of work.   210

(2) (computing systems). *See:* **Turing machine; universal Turing machine.**

(3) (sound measurements). A machine is any rotating electrical device of which the acoustical characteristics are to be measured. (A) A **small machine** has a maximum linear dimension of 250 mm. This dimension is over major surfaces, excluding minor surface protuberances as well as shaft extension, and is measured either parallel to the shaft, or at right angles to it, according to which dimension gives the greater measurement. (B) A **medium machine** has a maximum linear dimension from 250 mm to 1 m as measured for **small machine.** (C) A **large machine** has a maximum linear dimension in excess of 1 m as measured for **small machine.**

machine address (computing systems). *See:* **absolute address.**

machine, aircraft electric. An electric machine designed for operation aboard aircraft. *Note:* Minimum weight and size and extreme reliability for a specified (usually short) life are required while operating under specified conditions of coolant temperature and flow, and for air-cooled machines, pressure and humidity.   63

machine check. (1) An automatic check. (2) A programmed check of machine functions. *See also:* **check; automatic.**   235

machine code (computing systems). An operation code that a machine is designed to recognize.   255, 77, 54

machine, electric. An electric apparatus depending on electromagnetic induction for its operation and having one or more component members capable of rotary and/or linear movement. *See also:* **asynchronous machine.**   63

machine equation. *See:* **computer equation.**

machine final-terminal stopping device (stop-motion switch) (elevators). A final-terminal stopping device operated directly by the driving machine. *See:* **control.**   328

machine instruction (computing systems). An instruction that a machine can recognize and execute.   255, 77, 54

machine language (electronic digital computers). (1) A language, occurring within a machine, ordinarily not perceptible or intelligible to persons without special equipment or training. (2) A translation or transliteration of (1) above into more conventional characters but frequently still not intelligible to persons without special training.   235, 210

machine oriented language (test, measurement and diagnostic equipment). (1) A language designed for interpretation and use by a machine without translation; (2) a system for expressing information which is intelligible to a specific machine; for example, a computer or class of computers. Such a language may include instructions which define and direct machine operations, and information to be recorded by or acted upon by these machine operations; and (3) the set of instructions expressed in the number system basic to a computer, together with symbolic operation codes with absolute addresses, relative addresses, or symbolic addresses. *Syn.* **machine language** and **assembly language.**   54

machine positioning accuracy, precision, or reproducibility (numerically controlled machines). Accuracy, precision, or reproducibility of position sensor or transducer and interpeting system, the machine elements, and the machine positioning servo. *Note:* Cutter, spindle, and work deflection, and cutter wear are not included. (May be the same as control positioning accuracy, precision, or reproducibility in some systems.)   224, 207

machine ringing (telephone switching systems). Ringing that once started continues automatically, rhythmically until the call is answered or abandoned.   55

machine time. *See:* **time.**

machine tool control transformers. Step-down transformers which may be equipped with fuse or other over-current protection device, generally used for the operation of solenoids, contactors, relays, portable tools, and localized lighting. 53

machine winding (rotating machinery). A winding placed in slots or around poles directly by a machine. *See:* **rotor (rotating machinery); stator.** 63

machine word (computing systems). *See:* **computer word.**

machining accuracy, precision, or reproducibility. Accuracy, precision, or reproducibility obtainable on completed parts under normal operating conditions. 224, 207

macro instruction (computing systems). An instruction in a source language that is equivalent to a specified sequence of machine instructions. 255, 77

magazine (computing systems). *See:* **input magazine.**

magic tee (waveguides). *See:* **hybrid tee.**

magner. *See:* **reactive power.**

magnesium cell. A primary cell with the negative electrode made of magnesium or its alloy. *See also:* **electrochemistry.** 328

magnet. A body that produces a magnetic field external to itself. 210

magnet, focusing. *See:* **focusing magnet.**

magnetic (switching device). A term indicating that interruption of the circuit takes place between contacts separable in an intense magnetic field. *Note:* With respect to contactors, this term indicates the means of operation. 103, 202

magnetically shielded type instrument. An instrument in which the effect of external magnetic fields is limited to a stated value. The protection against this influence may be obtained either through the use of a physical magnetic shield or through the instrument's inherent construction. *See also:* **instrument.** 280

magnetic amplifier. A device using one or more saturable reactors, either alone or in combination with other circuit elements, to secure power gain. Frequency conversion may or may not be included. 341

magnetic area moment. *See:* **magnetic moment.**

magnetic-armature loudspeaker. A magnetic loudspeaker whose operation involves the vibration of a ferromagnetic armature. *See also:* **loudspeaker.** 328

magnetic axis (coil or winding) (rotating machinery). The line of symmetry of the magnetic-flux density produced by current in a coil or winding, this being the location of approximately maximum flux density, with the air gap assumed to be uniform. *See also:* **rotor (rotating machinery); stator.** 63

magnetic bearing (navigation). Bearing relative to magnetic north. *See also:* **navigation.** 187, 13

magnetic biasing (magnetic recording). The simultaneous conditioning of the magnetic recording medium during recording by the superposing of an additional magnetic field upon the signal magnetic field. *Note:* In general, magnetic biasing is used to obtain a substantially linear relationship between the amplitude of the signal and the remanent flux density in the recording medium. *See:* **alternating-current magnetic biasing; direct-current magnetic biasing.** *See also:* **phonograph pickup.** 176

magnetic bias, relay. *See:* **relay magnetic bias.**

magnetic blowout (industrial control). A magnet, often

electrically excited, whose field is used to aid the interruption of an arc drawn between contacts. *See also:* **contactor.** 206

magnetic brake (industrial control). A friction brake controlled by electromagnetic means. 206

magnetic-brush development (electrostatography). Development in which the image-forming material is carried to the field of the electrostatic image by means of ferromagnetic particles acting as carriers under the influence of a magnetic field. *See also:* **electrostatography.**
236, 191

magnetic card (computing systems). A card with a magnetic surface on which data can be stored by selective magnetization of portions of the flat surface. 255, 77

magnetic circuit. A region at whose surface the magnetic induction is tangential. *Note:* The term is also applied to the minimal region containing essentially all the flux, such as the core of a transformer. 210

magnetic compass. A device for indicating the direction of the horizontal component of a magnetic field. *See also:* **magnetometer.** 328

magnetic-compass repeater indicator. A device that repeats the reading of a master direction indicator, through a self-synchronous coupling means. 328

magnetic constant (pertinent to any system of units) (permeability of free space). The magnetic constant is the scalar dimensional factor that in that system relates the mechanical force between two currents to their magnitudes and geometrical configurations. More specifically, μ_0 is the magnetic constant when the element of force $d\mathbf{F}$ of a current element $I_1 d\mathbf{I}_1$ on another current element $I_2 d\mathbf{I}_2$ at a distance r is given by

$$d\mathbf{F} = \mu_0 I_1 I_2 d\mathbf{I}_1 \times (d\mathbf{I}_2 \times \mathbf{r}_1)/nr^2$$

where $\mathbf{r}_1$ is a unit vector in the direction from $d\mathbf{I}_1$ to $d\mathbf{I}_2$, and n is a dimensionless factor which is unity in unrationalized systems and 4π in a rationalized system. *Note:* In the centimeter-gram-second (cgs) electromagnetic system μ_0 is assigned the magnitude unity and the dimension numeric. In the centimeter-gram-second (cgs) electrostatic system the magnitude of μ_0 is that of $1/c^2$ and the dimension is $[L^{-2}T^2]$. In the International System of Units (SI) μ_0 is assigned the magnitude $4\pi \cdot 10^{-7}$ and has the dimension $[LMT^{-2}I^{-2}]$. 210

magnetic contactor (industrial control). A contactor actuated by electromagnetic means. *See also:* **contactor.**
206

magnetic control relay. A relay that is actuated by electromagnetic means. *Note:* When not otherwise qualified, the term refers to a relay intended to be operated by the opening and closing of its coil circuit and having contacts designed for energizing and/or de-energizing the coils of magnetic contactors or other magnetically operated device. *See:* **relay.** 225, 206

magnetic core. A configuration of magnetic material that is, or is intended to be, placed in a rigid spatial relationship to current-carrying conductors and whose magnetic properties are essential to its use. *Note:* For example, it may be used: (1) to concentrate an induced magnetic field as in a transformer, induction coil, or armature; (2) to retain a magnetic polarization for the purpose of storing

data; or (3) for its nonlinear properties as in a logic element. It may be made of iron wires, iron oxide, coils of magnetic tape, ferrite, thin film, etcetera.
235, 255, 77, 54

magnetic course (navigation). Course relative to magnetic north. *See also:* **navigation.**   187, 13

magnetic deflection (cathode-ray tube). Deflecting an electron beam by the action of a magnetic field. *See also:* **cathode-ray tubes.**   244, 190

magnetic delay line (computing systems). A delay line whose operation is based on the time or propagation of magnetic waves.   255, 77

magnetic deviation. Angular difference between compass north and magnetic north caused by magnetic effects in the vehicle. *See also:* **navigation.**   187

magnetic device (packaging machinery). A device actuated by electromagnetic means.   124

magnetic dipole. An elementary radiator consisting of an infinitesimally small current loop. *See also:* **antenna.**
246, 179, 111

magnetic dipole moment* (centimeter-gram-second electromagnetic-unit system). The volume integral of **magnetic polarization** is often called magnetic dipole moment.
210

*Deprecated

magnetic direction indicator (MDI). An instrument providing compass indication obtained electrically from a remote gyro-stabilized magnetic compass or equivalent. *See also:* **radio navigation.**   328

magnetic disc. A flat circular plate with a magnetic surface on which data can be stored by selective polarization of portions of the flat surface.   235, 255, 77, 54

magnetic dissipation factor (magnetic material). The cotangent of its loss angle or the tangent of its hysteretic angle.   210

magnetic drum. A right circular cylinder with a magnetic surface on which data can be stored by selective polarization of portions of the curved surface.
235, 255, 77, 54

magnetic field (1) (generator action). A state produced in a medium, either by current flow in a conductor or by a permanent magnet, that can induce voltage in a second conductor in the medium when the state changes or when the second conductor moves in prescribed ways relative to the medium. *See:* **signal.**   188
(2) (motor action). A state of a region such that a moving charged body in the region would be accelerated in proportion to its charge and to its velocity.   146
(3) (general). The vector function (**B**) of position describing the magnetic state of a region.   210

magnetic field intensity*. *See:* **magnetic field strength.**

*Deprecated

magnetic field interference. A form of interference induced in the circuits of a device due to the presence of a magnetic field. *Note:* It may appear as common-mode or normal-mode interference in the measuring circuit. *See also:* **accuracy rating (instrument).**   294

magnetic field strength (1) (general). The magnitude of the magnetic field vector.   146, 199, 180
(2) (magnetizing force). That vector point function whose

curl is the current density, and that is proportional to magnetic flux density in regions free of magnetized matter. *Note:* A consequence of this definition is that the familiar formula

$$ \mathbf{H} = \frac{1}{4\pi} \int \mathbf{J} \times \nabla(1/r) \, dv $$
$$ - \frac{1}{4\pi} \nabla \int \mathbf{M} \cdot \nabla(1/r) \, dv $$

(where **H** is the magnetizing force, **J** is current density, and **M** is magnetization) is a mathematical identity.   210

magnetic field strength produced by an electric current (Biot-Savart law) (Ampere's law). The magnetic field strength, at any point in the neighborhood of a circuit in which there is an electric current i, can be computed on the assumption that every infinitesimal length of circuit produces at the point an infinitesimal magnetizing force and the resulting magnetizing force at the point is the vector sum of the contributions of all the elements of the circuit. The contribution, $d\mathbf{H}$, to the magnetizing force at a point P caused by the current i in an element ds of a circuit that is at a distance r from P, has a direction that is perpendicular to both ds and r and a magnitude equal to

$$ \frac{i \, ds \sin \theta}{r^2} $$

where θ is the angle between the element ds and the line r. In vector notation

$$ dH = \frac{i[\mathbf{r} \times \mathbf{ds}]}{r^2} . $$

This law is sometimes attributed to Biot and Savart, sometimes to Ampere, and sometimes to Laplace, but no one of them gave it in its differential form.   210

magnetic field vector (any point in a magnetic field) (radio wave propagation). The magnetic induction divided by the permeability of the medium. *See also:* **radio wave propagation.**   180, 146

magnetic figure of merit. The ratio of the real part of complex apparent permeability to magnetic dissipation factor. *Note:* The magnetic figure of merit is a useful index of the magnetic efficiency of a material in various electromagnetic devices.   210

magnetic flux (through an area). The surface integral of the normal component of the magnetic induction over the area. Thus

$$ \phi_A = \int_A (\mathbf{B} \cdot dA) $$

where ϕ_A is the flux through the area A, and **B** is the magnetic induction at the element dA of this area. *Note:* The net magnetic flux through any closed surface is zero.
210

magnetic flux density (magnetic induction). That vector quantity **B** producing a torque on a plane current loop in accordance with the relation $\mathbf{T} = I A \mathbf{n} \times \mathbf{B}$ where **n** is the positive normal to the loop carrying current I and A is its area. *Note:* The concept of flux density is extended to a point inside a solid body by defining the flux density at such a point as that which would be measured in a thin disk-shaped cavity in the body centered at that point, the axis of the cavity being in the direction of the flux density. 210

magnetic focusing. *See:* **focusing, magnetic.**

magnetic freezing. *See:* **relay magnetic freezing.**

magnetic friction clutch. A friction clutch in which the pressure between the friction surfaces is produced by magnetic attraction. *See:* **electric coupling.** 328

magnetic head (magnetic recording). A transducer for converting electric variations into magnetic variations for storage on magnetic media, or for reconverting energy so stored into electric energy, or for erasing such stored energy. 176

magnetic heading (navigation). Heading relative to magnetic north. *See also:* **navigation.** 187, 13

magnetic hysteresis. The property of a ferromagnetic material exhibited by the lack of correspondence between the changes in induction resulting from increasing magnetizing force and from decreasing magnetizing force. 210

magnetic hysteresis loss (magnetic material). (1) The power expended as a result of magnetic hysteresis when the magnetic induction is periodic. (2) The energy loss per cycle in a magnetic material as a result of magnetic hysteresis when the induction is cyclic (not necessarily periodic). *Note:* Definitions (1) and (2) are not equivalent; both are in common use. 210

magnetic hysteretic angle. The mean angle by which the exciting current leads the magnetizing current. *Note:* Because of hysteresis, the instantaneous value of the hysteretic angle will vary during the cycle; the hysteretic angle is taken to be the mean value. 210

magnetic induction (signal-transmission system). The process of generating currents or voltages in a conductor by means of a magnetic field. *See:* **magnetic flux density; signal.** 188

magnetic ink. An ink that contains particles of a magnetic substance whose presence can be detected by magnetic sensors. 255, 77

magnetic-ink character recognition. The machine recognition of characters printed with magnetic ink. *See:* **optical character recognition.** 255, 77

magnetic latching relay. (1) A relay that remains operated from remanent magnetism until reset electrically. (2) A bistable polarized (magnetically latched) relay. 341

magnetic loading (rotating machinery). The average flux per unit area of the air-gap surface. *See also:* **asynchronous machine; direct-current commutating machine.** 63

magnetic loss angle (core). The angle by which the fundamental component of the core-loss current leads the fundamental component of the exciting current in an inductor having a ferromagnetic core. *Note:* The loss angle is the complement of the hysteretic angle. 210

magnetic loss factor, initial (material). The product of the real component of its complex permeability and the tangent of its magnetic loss angle, both measured when the magnetizing force and the induction are vanishingly small. *Note:* In anisotropic media, magnetic loss factor becomes a matrix. 210

magnetic loudspeaker. A loudspeaker in which acoustic waves are produced by mechanical forces resulting from magnetic reactions. *See also:* **loudspeaker.** 328

magnetic microphone. *See:* **variable-reluctance microphone.**

magnetic (electron) microscope. An electron microscope with magnetic lenses. *See also:* **electron optics.** 244, 190

magnetic mine. A submersible explosive device with a detonator actuated by the distortion of the earth's magnetic field caused by the approach of a mass of magnetic material such as the hull of a ship. *See also:* **degauss.** 328

magnetic moment (1) (magnetized body). The volume integral of the magnetization

$$\mathbf{m} = \int \mathbf{M} \, dv \qquad 210$$

(2) (current loop).

$$\mathbf{m} = I \int \mathbf{n} \, da$$
$$= (I/2) \int \mathbf{r} \times d\mathbf{r}$$

where **n** is the positive normal to a surface spanning the loop, and **r** is the radius vector from an arbitrary origin to a point on the loop. *Notes:* (A) The numerical value of the moment of a plane current loop is IA, where A is the area of the loop. (B) The reference direction for the current in the loop indicates a clockwise rotation, when the observer is looking through the loop in the direction of the positive normal. 210

magnetic north. The direction of the horizontal component of the earth's magnetic field toward the north magnetic pole. *See also:* **navigation.** 187, 13

magnetic overload relay. An overcurrent relay the electric contacts of which are actuated by the electromagnetic force produced by the load current or a measure of it. *See:* **relay.** 225, 206

magnetic-particle coupling. An electric coupling that transmits torque through the medium of magnetic particles in a magnetic field between coupling members. *See:* **electric coupling.** 328

magnetic pickup. *See:* **variable-reluctance pickup.**

magnetic-plated wire. A magnetic wire having a core of nonmagnetic material and a plated surface of ferromagnetic material. 176

magnetic-platform influence (electric instrument). The change in indication caused solely by the presence of a magnetic platform on which the instrument is placed. *See also:* **accuracy rating (instrument).** 280

magnetic polarization*. In the centimeter-gram-second electromagnetic-unit system, the intrinsic induction divided by 4π is sometimes called **magnetic polarization** or **magnetic dipole moment per unit volume.** *See:* **intrinsic induction.** 210

*Deprecated

magnetic poles (magnet). Those portions of the magnet toward which or from which the external magnetic induction appears to converge or diverge, respectively. *Notes:* (1) By convention, the north-seeking pole is marked with N, or plus, or is colored red. (2) The term is also sometimes applied to a fictitious magnetic charge. 210

magnetic pole strength (magnet). The magnetic moment

divided by the distance between its poles. *Note:* Many authors use the above quantity multiplied by the magnetic constant; the two choices are numerically equal in the centimeter-gram-second electromagnetic-unit system.
210

magnetic-powder-impregnated tape (impregnated tape) (dispersed-magnetic-powder tape). A magnetic tape that consists of magnetic particles uniformly dispersed in a nonmagnetic material. 176

magnetic power factor. The cosine of the magnetic hysteretic angle (the sine of the magnetic loss angle). 210

magnetic recorder. Equipment incorporating an electromagnetic transducer and means for moving a magnetic recording medium relative to the transducer for recording electric signals as magnetic variations in the medium. *Note:* The generic term **magnetic recorder** can also be applied to an instrument that has not only facilities for recording electric signals as magnetic variations, but also for converting such magnetic variations back into electric variations. *See also:* **phonograph pickup.** 176

magnetic recording (facsimile). Recording by means of a signal-controlled magnetic field. *See also:* **recording (facsimile).** 12

magnetic recording head. In magnetic recording, a transducer for converting electric currents into magnetic fields, in order to store the electric signal as a magnetic polarization of the magnetic medium. 176

magnetic recording medium. A material usually in the form of a wire, tape, cylinder, disk, etcetera, on which a magnetic signal may be recorded in the form of a pattern of magnetic polarization. 176

magnetic reproducer. Equipment incorporating an electromagnetic transducer and means for moving a magnetic recording medium relative to the transducer, for reproducing magnetic signals as electric signals. 176

magnetic reproducing head. In magnetic recording, a transducer for collecting the flux due to stored magnetic polarization (the recorded signal) and converting it into an electric voltage. 176

magnetic rotation (polarized light) (Faraday effect). When a plane polarized beam of light passes through certain transparent substances along the lines of a strong magnetic field, the plane of polarization of the emergent light is different from that of the incident light. 210

magnetic sensitivity (Hall-effect devices). The ratio of the voltage across the Hall terminals to the magnetic flux density for a given magnitude of control current. 107

magnetic spectrograph. An electronic device based on the action of a constant magnetic field on the paths of electrons, and used to separate electrons with different velocities. *See also:* **electron device.** 244, 191

magnetic starter (packaging machinery). A starter actuated by electromagnetic means. 124

magnetic storage. A method of storage that uses the magnetic properties of matter to store data by magnetization of materials such as cores, films, or plates, or of material located on the surfaces of tapes, discs, or drums, etcetera. *See also:* **magnetic-core; magnetic drum; magnetic tape.** 235

magnetic storm. A disturbance in the earth's magnetic field, associated with abnormal solar activity, and capable

of seriously affecting both radio and wire transmission. *See also:* **radio transmitter.** 328

magnetic susceptibility (isotropic medium). In rationalized systems, the relative permeability minus unity.

$$k = \mu_r - 1 = B_i/\mu_0 H$$

Notes: (1) In unrationalized systems, $k = (\mu_r - 1)4\pi$. (2) The susceptibility divided by the density of a body is called the susceptibility per unit mass, or simply the mass susceptibility. The symbol is χ. Thus

$$\chi = k/\rho$$

where ρ is the density. χ multiplied by the atomic weight is called the atomic susceptibility. The symbol is χ_A. (3) In anisotropic media, susceptibility becomes a matrix.
210

magnetic tape (homogeneous or coated). (1) A tape with a magnetic surface on which data can be stored by selective polarization of portions of the surface. (2) A tape of magnetic material used as the constituent in some forms of magnetic cores. *See also:* **coated magnetic tape.**
235, 255, 77, 54

magnetic tape handler (test, measurement and diagnostic equipment). A device which handles magnetic tape and usually consists of a tape transport and magnetic tape reader with associated electrical and electronic equipments. Most units provide for tape to be wound and stored on reels; however, some units provide for the tape to be stored loosely in closed bins. 54

magnetic tape reader (test, measurement and diagnostic equipment). A device capable of converting information from magnetic tape where it has been stored as variations in magnetizations into a series of electrical impulses.
54

magnetic test coil (search coil) (exploring coil). A coil that, when connected to a suitable device, can be used to measure a change in the value of magnetic flux linked with it. *Note:* The change in the flux linkage may be produced by a movement of the coil or by a variation in the magnitude of the flux. Test coils used to measure magnetic induction **B** are often called **B** coils; those used to determine magnetizing force **H** may be called **H** coils. A coil arranged to rotate through an angle of 180 degrees about an axis of symmetry perpendicular to its magnetic axis is sometimes called a flip coil. *See also:* **magnetometer.** 328

magnetic thin film. A layer of magnetic material, usually less than 10 000 angstroms thick. *Note:* In electronic computers, magnetic thin films may be used for logic or storage elements. *See also:* **coated magnetic tape; magnetic core; magnetic tape.** 235, 77

magnetic track braking. A system of braking in which a shoe or slipper is applied to the running rails by magnetic means. *See also:* **electric braking.** 328

magnetic variometer. An instrument for measuring differences in a magnetic field with respect to space or time. *Note:* The use of variometer to designate a continuously adjustable inductor is deprecated. *See also:* **magnetometer.**
328

magnetic vector (radio wave propagation). Magnetic field vector. *See also:* **radio wave propagation.** 180

magnetic vector potential. An auxiliary solenoidal vector point function characterized by the relation that its curl is equal to the magnetic induction and its divergence vanishes.

$$\text{Curl } \mathbf{A} = \mathbf{B} \qquad \text{Divergence } \mathbf{A} = \mathbf{0}$$

Note: These relations are satisfied identically by

$$A = (\mu_0/4\pi)[\int \mathbf{M} \times \nabla(1/r)\mathrm{d}v + \int (\mathbf{J}/r)\mathrm{d}v]$$

where v is the volume. 240

magnetization (intensity of magnetization) (at a point of a body). The intrinsic induction at that point divided by the magnetic constant of the system of units employed:

$$M = \mathbf{B}_i/\mu_0 = (\mathbf{B} - \mu_0\mathbf{H})/\mu_0.$$

Note: The magnetization can be interpreted as the volume density of magnetic moment. 210

magnetizing current (1) (transformers). A hypothetical current assumed to flow through the magnetizing inductance of a transformer. 197

(2) (rotating machinery). The quadrature (leading) component (with respect to the induced voltage) of the exciting current supplied to a coil. 210, 63

magnetizing force. *See:* **magnetic field strength.**

magnetizing inductance. A hypothetical inductance, assumed to be in parallel with the core-loss resistance, that would store the same amount of energy as that stored in the core for a specified value of excitation. 197

magnet meter (magnet tester). An instrument for measuring the magnetic flux produced by a permanent magnet under specified conditions of use. It usually comprises a torque-coil or a moving-magnet magnetometer with a particular arrangement of pole-pieces. *See also:* **magnetometer.** 328

magneto. *See:* **magnetoelectric generator.**

magneto central office. A telephone central office for serving magneto telephone sets. 193

magnetoelectric generator (magneto). An electric generator in which the magnetic flux is provided by one or more permanent magnets. *See:* **direct-current commutating machine.** 3

magneto-ionic medium (radio wave propagation). An ionized gas that is permeated by a fixed magnetic field. *See also:* **radio wave propagation.** 46, 180

magneto-ionic mode (radio wave propagation). Magneto-ionic wave component. *See also:* **radio wave propagation.** 180

magneto-ionic wave component (radio wave propagation) (1) (incidence of a linearly polarized wave upon a magneto-ionic medium). Either of the two elliptically polarized wave components into which a linearly polarized wave incident on the ionosphere is separated because of the earth's magnetic field. 328

(2) (at a given frequency) (wave-propagation within a magneto-ionic medium). Either of the two plane electromagnetic waves that can travel in a homogeneous magneto-ionic medium without change of polarization. *See also:* **radiation; radio wave propagation.** 146, 180

magnetometer. An instrument for measuring the intensity or direction (or both) of a magnetic field or of a component of a magnetic field in a particular direction. *Note:* The term is more usually applied to instruments that measure

the intensity of a component of a magnetic field, such as horizontal-intensity magnetometers, vertical-intensity magnetometers, and total-intensity magnetometers. 328

magnetomotive force (acting in any closed path in a magnetic field). The line integral of the magnetizing force around the path. 210

magnetoresistive coefficient (Hall generator). The ratio at a specified magnetic flux density B of the rate of change of resistance with magnetic flux density to the resistance R_B at the specified magnetic flux density B: defined by the equation

$$\alpha_B = \frac{1}{R_B}\frac{dR_B}{dB}.$$

107

magnetoresistive effect. The change in the resistance of a current-carrying Hall plate when acted upon by a magnetic field. *Notes: (1)* An increase in magnetic field may cause either an increase or a decrease in ferromagnetic and similar Hall plates, whereas there is usually an increase with Hall plates made of other material. *(2)* There are two factors affecting the changes in resistance: first, a bulk effect due to the characteristics of the Hall plate, and second, a geometric effect due to the shape of the Hall plate and to the presence or absence of shorting bars made of conducting material deliberately, as in the shorting bars plated on some magnetoresistors or the microconductors dispersed in other magnetoresistors or inadvertently, as in the case of the control current electrodes in a Hall generator, added to the current-carrying Hall plate. 107

magnetoresistive ratio (Hall generator). The ratio of the resistance R_B, at a magnetic flux density B, to the resistance R_0, at zero magnetic flux density: defined by the equation

$$\alpha_M = \frac{R_B}{R_0}.$$

107

magnetostriction. The phenomenon of elastic deformation that accompanies magnetization. 210

magnetostriction loudspeaker. A loudspeaker in which the mechanical displacement is derived from the deformation of a material having magnetostrictive properties. *See also:* **loudspeaker.** 328

magnetostriction microphone. A microphone that depends for its operation on the generation of an electromotive force by the deformation of a material having magnetostrictive properties. *See also:* **microphone.** 328

magnetostriction oscillator. An oscillator with the plate circuit inductively coupled to the grid circuit through a magnetostrictive element, the frequency of oscillation being determined by the magnetomechanical characteristics of the coupling element. *See also:* **oscillatory circuit.** 111

magnetostrictive relay. A relay in which operation depends upon dimensional changes of a magnetic material in a magnetic field. *See also:* **relay.** 259

magneto switchboard (telephone switching systems). A telecommunication switchboard for serving magneto telephone sets. 55

magneto telephone set. A local-battery telephone set in which current for signaling by the telephone station is supplied from a local hand generator, usually called a magneto. *See also:* **telephone station.** 328

magneto-telluric (M-T). An adjective denoting natural magnetic and electric fields, and effects produced by them. 132

magnetron (induction and dielectric heating). An electron tube characterized by the interaction of electrons with the electric field of a circuit element in crossed steady electric and magnetic fields to produce alternating-current power output. 14

magnetron injection gun (microwave tubes). A gun that produces a hollow beam of high total permeance that flows parallel to the axis of a magnetic field. *See:* **magnetrons; microwave tube (or valve).** 190

magnetron oscillator. An electron tube in which electrons are accelerated by a radial electric field between the cathode and one or more anodes and by an axial magnetic field that provides a high-energy electron stream to excite the tank circuits. *See also:* **magnetron.** 111

magnet valve (electric controller). A valve controlling a fluid, usually air, operated by an electromagnet. *See also:* **multiple-unit control.** 1

magnet wire (rotating machinery). Single-strand wire with a thin flexible insulation, suitable for winding coils. *See:* **rotor (rotating machinery); stator.** 63

magnified sweep (oscilloscopes). A sweep whose time per division has been decreased by amplification of the sweep waveform rather than by changing the time constants used to generate it. *See:* **oscillograph.** 185

magnitude. The quantitative attribute of size, intensity, extent, etcetera, that allows a particular entity to be placed in order with other entities having the same attribute. *Notes:* (1) The magnitude of the length of a given bar is the same whether the length is measured in feet or in centimeters. (2) The word magnitude is used in other senses. The definition given here is the basic one needed for the logical buildup of later definitions. 210

magnitude characteristic (linear passive networks). The absolute value of a response function evaluated on the imaginary axis of the complex-frequency plane. *See also:* **linear passive networks.** 238

magnitude contours (control system, feedback). Loci of selected constant values of the magnitude of the return transfer function drawn on a plot of the loop' transfer function for real frequencies. *Note:* Such loci may be drawn on the Nyquist or inverse Nyquist diagrams, or Nichols chart. *See also:* **control system, feedback.** 56

magnitude origin line (pulse terms). A line of specified magnitude which, unless otherwise specified, has a magnitude equal to zero and extends through the waveform epoch. *See:* The single pulse diagram below the **waveform epoch** entry. 254

magnitude parameters and references (pulse terms). (Unless otherwise specified, derived from data within the waveform epoch.) *See also:* **base magnitude; top magnitude; pulse amplitude; magnitude reference lines; magnitude reference points.** 254

magnitude ratio (speed governing of hydraulic turbines). The ratio of the peak magnitude of the output signal to the peak magnitude of a constant-frequency, constant-amplitude, sinusoidal input signal. *See:* **gain.** 8

magnitude referenced point (pulse terms). A point at the intersection of a magnitude reference line and a waveform. 254

magnitude reference line (pulse terms). A line parallel to the magnitude origin line at a specified magnitude. 254

magnitude reference lines (pulse terms). (1) **baseline (topline).** The magnitude reference line at the base (top) magnitude. *See:* The single pulse diagram below the **waveform epoch** entry. 254
(2) **percent reference magnitude.** A reference magnitude specified by:

$$(x)\%M_r = M_b + \frac{x}{100}(M_t - M_b)$$

where
$0 < x < 100$
$(x)\%M_r$ = percent reference magnitude
M_b = base magnitude
M_t = top magnitude
M_b, M_t and $(x)\%M_r$ are all in the same unit of measurement
(3) **proximal (distal) line.** A magnitude reference line at a specified magnitude in the proximal (distal) region of a pulse waveform. Unless otherwise specified, the proximal (distal) line is at the 10 (90) percent reference magnitude. *See:* The single pulse diagram below the **waveform epoch** entry.
(4) **mesial line.** A magnitude reference line at a specified magnitude in the mesial region of a pulse waveform. Unless otherwise specified, the mesial line is at the 50 percent reference magnitude. *See:* The single pulse diagram below the **waveform epoch** entry.

magnitude reference points (pulse terms). *See:* proximal (distal) point; mesial point. 254

magnitude-related adjectives (pulse terms). (1) **proximal (distal).** Of or pertaining to a region near to (remote from) a first state or region of origin.
(2) **mesial.** Of or pertaining to the region between the proximal and distal regions. 254

main (interior wiring). A feeder extending from the service switch, generator bus, or converter bus to the main distribution center. 328

main anode (pool-cathode tube). An anode that conducts load current. *Note:* The word main is used only when it is desired to distinguish the anode to which it is applied from an auxiliary electrode such as an excitation anode. It is used only in connection with pool-tube terms. *See also:* **electrode (electron tube).** 190

main bang (radar). A transmitted pulse. 254

main capacitance (capacitance potential device). The capacitance between the network connection and line. *See also:* **outdoor coupling capacitor.** 351

main contacts (switching device). Contacts that carry all or most of the current of the main circuit. 103, 202, 27

main distributing frame, (MDF) (telephone switching systems). A frame where crossconnections are made between the outside plant and central office equipment.   55

main distribution center. A distribution center supplied directly by mains. *See also:* **distribution center.**   328

main exciter (rotating machinery). An exciter that supplies all or part of the power required for the excitation of the principal electric machine or machines. *See also:* **direct-current commutating machine; asynchronous machine.**
   63

main exciter response ratio (nominal exciter response). The numerical value obtained when the response, in volts per second, is divided by the rated-load field voltage; which response, if maintained constant, would develop, in one-half second, the same excitation voltage-time area as attained by the actual exciter. *Note:* The response is determined with no load on the exciter, with the exciter voltage initially equal to the rate-load field voltage, and then suddenly establishing circuit conditions that would be used to obtain nominal exciter ceiling voltage. For a rotating exciter, the response should be determined at the rated speed. This definition does not apply to main exciters having one or more series fields, except a light differential series field, or to electronic exciters. *See:* **direct-current commutating machine.**   328

main gap (glow-discharge tubes). The conduction path between a principal cathode and a principal anode.
   190, 125

main lead (rotating machinery). A conductor joining a main terminal to the primary winding. *See:* **asynchronous machine.**   63

main lobe. *See* **major lobe.**

main protection. *See:* **primary protection.**

mains. *See:* **primary distribution mains; secondary distribution mains.** *See also:* **center of distribution.**

mains coupling coefficient (electromagnetic compatibility). *See:* **mains decoupling factor.**

mains decoupling factor (mains coupling coefficient) (electromagnetic compatibility). The ratio of the radio-frequency voltage at the mains terminal to the interfering apparatus to the radio-frequency voltage at the aerial terminals of the receiver. *Note:* Generally expressed in logarithmic units. *See also:* **electromagnetic compatibility.**
   220, 199

main secondary terminals. The main secondary terminals provide the connections to the main secondary winding. *See:* **main secondary winding.**   351

main secondary winding (capacitance potential device). Provides the secondary voltage or voltages on which the potential device ratings are based. *See:* **main secondary terminals.**   351

mains-interference immunity (mains-interference ratio). The degree of protection against interference conducted by its supply mains as measured under specified conditions. *Note:* see International Special Committee on Radio Interference recommendation 25/1 and International Electrotechnical Commission publication 69 or subsequent publications where the term mains-interference ratio is used. *See also:* **electromagnetic compatibility.**
   220, 199

main station. A telephone station with a distinct call number designation, directly connected to a central office.

See also: **telephone station.**   328

main-station code, (telephone switching systems). The digits designating a main station; these usually follow an office code.   55

main switchgear connections (primary switchgear connection). Those that electrically connect together devices in the main circuit, or connect them to the bus, or both.
   103, 202

maintainability. Ability of an item, under stated conditions of use, to be retained in or restored to a state in which it can perform its required functions, when maintenance is performed under stated conditions and using prescribed procedures and resources. *Notes* (1): Maintainability can, depending on the particular analysis situation, be stated by one or several maintainability characteristics, such as discrete probability distribution, mean active maintenance time, etcetera. (2) The value of the maintainability characteristic may differ for different maintenance situations. (3) When the term maintainability is used as a maintainability characteristic, it always denotes the probability that the active maintenance is carried out within a given period of time. (4) The required function may be defined as a stated condition.   164

maintaining voltage (glow lamp) (operating voltage). The voltage measured across the lamp electrodes when the lamp is operating.   283

maintenance (1) (computing systems). Any activity intended to keep equipment, programs or a data base in satisfactory working condition, including tests, measurements, replacements, adjustments, and repairs. *See:* **file maintenance.**   255, 77

(2) (test, measurement and diagnostic equipment). Activity intended to keep equipment (hardware) or programs (software) in satisfactory working condition, including tests, measurements, replacements, adjustments, repairs, program copying, and program improvement. Maintenance is either preventive or corrective.   54

(3) (reliability). The combination of all technical and corresponding administrative actions intended to retain an item in, or restore it to, a state in which it can perform its required function. *Note:* The required function may be defined as a stated condition.   164

maintenance concept (test, measurement and diagnostic equipment). A description of the general scheme for maintenance and support of an item in the operational environment.   54

maintenance, depot (test, measurement and diagnostic equipment). Maintenance performed on material requiring major overhaul or a complete rebuild of parts, subassemblies, and end items, including the manufacture of parts, modification, testing, and reclamation as required. Depot maintenance serves to support lower categories of maintenance by providing technical assistance and performing that maintenance beyond their responsibility. Depot maintenance provides stocks of serviceable equipment by using more extensive facilities for repair than are available in lower level maintenance activities.
   54

maintenance engineering analysis (test, measurement and diagnostic equipment). A process performed during the development stage to derive the required maintenance resources such as personnel, technical data, support equipment, repair parts, and facilities.   54

maintenance factor (illuminating engineering). The ratio of the illumination on a given area after a period of time to the initial illumination on the same area. *Note:* The maintenance factor is used in lighting calculations as an allowance for the depreciation of lamps, light control elements, and room surfaces to values below the initial or design conditions, so that a minimum desired level of illumination may be maintained in service. The maintenance factor had formerly been widely interpreted as the ratio of average to initial illumination. *See also:* **inverse-square law (illuminating engineering).**   167

maintenance, intermediate. *See:* **intermediate maintenance.**

maintenance level (test, measurement and diagnostic equipment). The level at which maintenance is to be accomplished, that is, organizational, intermediate, and depot.   54

maintenance operation device (power switchgear). A removable device for use with power operated circuit breakers which is used for manual operation of a de-energized circuit breaker during maintenance only. *Note:* This device is not to be used for closing the circuit breaker on an energized circuit.   103

maintenance, organizational. *See:* **organizational maintenance.**

maintenance proof test (rotating machinery). A test applied to an armature winding after being in service to determine that it is suitable for continued service. It is usually made at a lower voltage than the acceptance proof test. *See also:* **insulation testing (large alternating-current rotating machinery).**   6

maintenance, scheduled (generation). Capability which has been scheduled to be out of service for maintenance.
                    112

main terminal (rotating machinery). A termination for the primary winding. *See:* **asynchronous machine.**   63

main-terminal 1 (bidirectional thyristor). The main terminal that is named 1 by the device manufacturer. *See also:* **anode.**   243, 66, 191

main-terminal 2 (bidirectional thyristor). The main terminal that is named 2 by the device manufacturer. *See also:* **anode.**   243, 66, 191

main terminals (thyristor). The terminals through which the principal current flows. *See also:* **anode.**
                    243, 66, 191

main transformer (polyphase power). The term, as applied to two single-phase Scott-connected units for three-phase to two-phase, or two-phase to three-phase operation, designates the transformer that is connected directly between two of the phase wires of the three-phase lines. *Note:* A tap is provided at the midpoint for connection to the teaser transformer. *See:* **autotransformer.**   203, 53

main unit (two-core voltage-regulating transformer). The core and coil unit that furnishes excitation to the series unit.   203, 53

main winding, single-phase induction motor. A system of coils acting together, connected to the supply line, that determines the poles of the primary winding, and that serves as the principal winding for transfer of energy from the primary to the secondary of the motor. *Note:* In some multispeed motors, the same main winding will not be used

for both starting operation and running operation. *See:* **asynchronous machine.**   63

major alarm, (telephone switching systems). An alarm indicating trouble or the presence of hazardous conditions needing immediate attention in order to restore or maintain the system capability.   55

major cycle (electronic computation). In a storage device that provides serial access to storage positions, the time interval between successive appearances of a given storage position.   235, 210

major failure. *See:* **failure, major.**

major insulation (outdoor apparatus bushings). Insulating material internal to the bushing between the line potential conductor and ground.   168

majority (computing systems). A logic operator having the property that if P is a statement, Q is a statement, R is a statement, . . . , then the majority of $P, Q, R, . . .$, is true if more than half the statements are true, false if half or less are true.   255, 77

majority carrier (semiconductor). The type of charge carrier constituting more than one half the total charge-carrier concentration. *See also:* **semiconductor; semiconductor device.**   245, 210, 186, 66

major lobe. The radiation lobe containing the direction of maximum radiation. *Note:* In certain antennas, such as multilobed or split-beam antennas, there may exist more than one major lobe. *Syn.:* **main lobe.** *See also:* **antenna.**
                    179

major loop (control). A continuous network consisting of all of the forward elements and the primary feedback elements of the feedback control system. *See also:* **control system, feedback.**   219, 206

major pulse waveform features (pulse terms). *See:* **base; top; first transition; last transition.**   254

major scheduled generation station shutdown. Periodic shutdowns of the generating station for an extended time scheduled for major reconditioning of the station, for example, fuel reloading.   31

make-break operation (pulse operation) (data transmission). Used to describe a method of data transmission by means of opening and closing a circuit to produce a series of current pulses.   59

make busy, (telephone switching systems). Conditioning a circuit to be unavailable for service.   55

make-busy signal (telephone switching systems). A signal transmitted from the terminating end of a trunk to prevent the seizure of the originating end.   55

making capacity (industrial control). The maximum current or power that a contact is able to make under specified conditions. *See:* **contactor.**   244, 206

making current (switching device). The value of the available current at the time the device closes. *Notes:* (1) Its root-mean-square value is measured from the envelope of the current wave at the time of the first major current peak. (2) The making current may also be expressed in terms of instantaneous value of current in which case it is measured at the first major peak of the current wave. This is designated peak making current.   27, 103

making current, rated (switching device). The maximum root-mean-square current against which the recloser is required to close under specified conditions. *Note:* The root-mean-square value is measured from the envelope of the current wave at the time of the first major current peak. [See ANSI C37.05-1964 (R1969), Methods for Determining the Values of a Sinusoidal Current Wave and Normal-Frequency Recovery Voltage for AC High-Voltage Circuit Breakers.] 92

malfunction (1) (electronic digital computer). An error that results from failure in the hardware. *See also:* **error; mistake; fault.** 235, 54, 9

(2) (test, measurement, and diagnostic equipment). *See:* **fault.** 54

(3) (seismic qualification of class 1E equipment). The loss of capability of Class 1E equipment to initiate or sustain a required function, or the initiation of undesired spurious action which might result in consequences adverse to safety. 28, 31

(4) (analog computer). *See:* **error.** 10

malicious call, (telephone switching systems). A call of an harassing, abusive, obscene, or threatening nature. 55

manhole (electric systems) (1) (More accurately termed **splicing chamber** or **cable vault**). A subsurface chamber, large enough for a man to enter, in the route of one or more conduit runs, and affording facilities for placing and maintaining in the runs, conductors, cables, and any associated apparatus. *See also:* **cable vault; splicing chamber.** 64

(2) An opening in an underground system that workmen or others may enter for the purpose of installing cables, transformers, junction boxes, and other devices, and for making connections and tests. *See:* **cable vault; distribution center; splicing chamber.** 178

(3) (transmission and distribution). A subsurface enclosure which personnel may enter and which is used for the purpose of installing, operating, and maintaining submersible equipment, cable, or both. 262

manhole chimney. A vertical passageway for workmen and equipment between the roof of the manhole and the street level. 64

manhole cover frame. The structure that caps the manhole chimney at ground level and supports the cover. *See also:* **power distribution, underground construction.** 64

manifold insulation (liquid cooling) (rotating machinery). The insulation applied between ground and a manifold connecting several parallel liquid-cooling paths in a winding. *See:* **stator.** 63

manifold-pressure electric gauge. A device that measures the pressure of fuel vapors entering the cylinders of an aircraft engine. *Note:* The gauge is provided with a scale, usually graduated in inches of mercury, absolute. It provides remote indication by means of a self-synchronous generator and motor. 328

manipulated variable (control) (industrial control). A quantity or condition that is varied as a function of the actuating signal so as to change the value of the directly controlled variable. *Note:* In any practical control system, there may be more than one manipulated variable. Accordingly, when using the term it is necessary to state which manipulated variable is being discussed. In process control work, the one immediately preceding the directly controlled system is usually intended. *See:* **control system, feedback.** 56, 219, 206

man-made noise (electromagnetic compatibility). Noise generated in machines or other technical devices. *See also:*

electromagnetic compatibility. 199

manned space flight network (MSFN) (communication satellite). A network of ground communication and tracking facilities maintained for the support of manned space flight programs. 83

manual (1) (electric systems). Operated by mechanical force, applied directly by personal intervention. *See also:* **distribution center.** 206

(2) (transmission and distribution). Capable of being operated by personal intervention.

manual block-signal system. A block or a series of consecutive blocks governed by block signals operated manually upon information by telegraph, telephone, or other means of communication. *See also:* **block signal system.** 328

manual central office. A central office of a manual telephone system. 193

manual checkout (test, measurement and diagnostic equipment). A checkout system which relies completely on manual operation, operator decision and evaluation of results. 54

manual control. Control in which the main devices, whether manually or power operated, are controlled by an attendant. *See:* **control.** 103, 202

manual controller (industrial control). An electric controller having all of its basic functions performed by devices that are operated by hand. *See:* **electric controller.** 206, 3

manual data input (numerically controlled machines). A means for the manual insertion of numerical control commands. 224, 207

manual fire-alarm system. A fire-alarm system in which the signal transmission is initiated by manipulation of a device provided for the purpose. *See also:* **protective signaling.** 328

manual holdup-alarm system. An alarm system in which the signal transmission is initiated by the direct action of the person attacked or of an observer of the attack. *See also:* **protective signaling.** 328

manual input (computing systems). (1) The entry of data by hand into a device at the time of processing. (2) The data entered as in (1). 255, 77

manual load (armature current) division (industrial control). The effect of a manually operated device to adjust the division of armature currents between two or more motors or two or more generators connected to the same load. *See also:* **control system, feedback.** 225, 206

manual lockout device. A device that holds the associated device inoperative unless a predetermined manual function is performed to release the locking feature. 103, 202

manually operated door or gate. A door or gate that is opened and closed by hand. *See also:* **hoistway (elevator or dumbwaiter).** 328

manually release-free (manually trip-free). *See:* **mechanically release-free.**

manually trip-free. *See:* **mechanically release-free.**

manual mobile telephone system. A mobile communication system manually interconnected with any telephone network, or a mobile communication system manually interconnected with a telephone network. 181

manual operation. Operation by hand without using any other source of power. 103, 202

manual release (electromagnetic brake) (industrial control).

A device by which the braking surfaces may be manually disengaged without disturbing the torque adjustment. *See:* **electric drive.** 225, 206

manual-reset, manual-release (control-brakes). A manual release that requires an additional manual action to re-engage the braking surfaces. 225, 206

manual-reset relay. *See:* **relay, manual-reset.**

manual-reset thermal protector (rotating machinery). A thermal protector designed to perform the function by opening the circuit to or within the protected machine, but requiring manual resetting to close the circuit. *See also:* **starting-switch assembly.** 63

manual ringing (telephone switching systems). Ringing that is started by the manual operation of a key and continues only while the key is held operated. 55

manual speed adjustment (industrial control). A speed adjustment accomplished manually. *See also:* **electric drive.** 206

manual switchboard (telephone switching systems). A telecommunication switchboard for making intercon-nections manually by plugs and jacks or keys. 55

manual telecommunications exchange (telephone switching systems). A telecommunications exchange in which connections between stations are manually set by means of plugs and jacks or keys. 55

manual telecommunication system (telephone switching systems). A telecommunications system in which con-nections between customers are ordinarily established manually by operators in accordance with orders given orally by the calling parties. 55

manual test equipment (test, measurement and diagnostic equipment). Test equipment that requires separate ma-nipulations for each task (for example, connection to signal to be measured, selection of suitable range, and insertion of stimuli). 54

manual trip device (power switchgear). A device which is connected to the tripping linkage and which can be oper-ated manually to trip a switching device. 103

manuscript (numerically controlled machines). An ordered list of numerical control instructions. *See:* **programming.** 224, 207

map. To establish a correspondence between the elements of one set and the elements of another set. 225, 77

map vertical (navigation). *See:* **geographic map vertical.**

margin (orientation margin) (1) (printing telegraphy) (tele-typewriter). That fraction of a perfect signal element through which the time of selection may be varied in one direction from the normal time of selection, without causing errors while signals are being received. *Note:* There are two distinct margins, determined by varying the time of selection in either direction from normal. *See also:* **telegraphy.** 194

(2) (nuclear power generating stations). (A) The difference between the most severe specified service conditions of the plant and the conditions used in type testing to account for normal variations in commercial production of equipment and reasonable errors in defining satisfactory performance. 26

(B) The difference between the most severe specified service conditions and the conditions used in design to account for uncertainties in defining satisfactory perfor-mance requirements under accident and post accident

conditions. 361

(3) (data transmission). (A) Digital: Of a receiving equipment, the maximum degree of distortion of the re-ceived signal which is compatible with the correct trans-lation of all of the signals which it may possibly receive. *Note:* This maximum degree of distortion applies without reference to the form of distortion effecting the signals. In other words, it is the maximum degree of the most un-favorable distortion acceptable, beyond which incorrect translation occurs, which determines the value of the margin. The condition of the measurements of the margin are to be specified in accordance with the requirements of the system; (B) Analog: The excess of receive level beyond that needed for proper operation. 59

marginal check (electronic computation). A preventive maintenance procedure in which certain operating con-ditions (for example, supply voltage or frequency) are varied about their nominal values in order to detect and locate incipient defective parts. *See also:* **check.** 235, 210, 255, 77

marginal checking (test, measurement and diagnostic equipment). A system or method of determining circuit weaknesses and incipient malfunctions by varying the operating conditions of the circuitry. 54

marginal relay. A relay that functions in response to pre-determined changes in the value of the coil current or voltage. *See also:* **relay.** 259

marginal testing (test, measurement and diagnostic equip-ment). Testing that presents results on an indicator that has tolerance bands for evaluating the signal or charac-teristic being tested. (For example: a green band might indicate an acceptable tolerance range; a yellow band, a tolerance range representing marginal operation; and a red band, a tolerance that is unsatisfactory for operation of the item). 54

margin of commutation γ (margin angle). The time, ex-pressed in degrees (1 cycle of the ac waveform = 360°) from the termination of commutation in inverter operation to the next point of intersection between the two halfwaves of the voltage phases which have just commutated. *Note:* At this point of intersection, the converter circuit element which has just terminated conduction changes from re-verse blocking state to OFF state. 121

marine distribution panel. A panel receiving energy from a distribution or subdistribution switchboard and dis-tributing energy to energy-consuming devices or other distribution panels or panelboards of a ship. *See also:* **marine electric apparatus.** 328

marine electric apparatus. Electric apparatus designed especially for use on shipboard to withstand the conditions peculiar to such application. 328

marine generator and distribution switchboard. Receives energy from the generating plant and distributes directly or indirectly to all equipment of a ship supplied by the generating plant. *See also:* **marine electric apparatus.** 328

marine panelboard. A single panel or a group of panel units assembled as a single panel, usually with automatic overcurrent circuit breakers or fused switches, in a cabinet for flush or surface mounting in or on a bulkhead and ac-cessible only from the front, serving lighting branch cir-cuits or small power branch circuits of a ship. *See also:*

marine electric apparatus. 328

mariner's compass. A magnetic compass used in navigation consisting of two or more parallel polarized needles secured to a circular compass card that is delicately pivoted and enclosed in a glass-covered bowl filled with alcohol to support by flotation the weight of the moving parts. *Note:* The compass bowl is supported in gimbals mounted in the binnacle. The compass card is graduated to show the 32 points of the compass in addition to degrees. 328

marine subdistribution switchboard. Essentially a section of the marine generator and distribution switchboard (connected thereto by a bus feeder and remotely located) that distributes energy in a certain section of a vessel. *See also:* **marine electric apparatus.** 328

mark (computing systems). *See:* **flag.**

marked ratio (transformer). The ratio of the rated primary value to the rated secondary value as stated on the nameplate. 53

marker (1) (air navigation). A radio transmitting facility whose signals are geographically confined so as to serve as a position fix. *See:* **boundary marker; fan marker; middle marker; outer marker; Z marker.** *See also:* **radio navigation.** 187, 13
(2) (telephone switching systems). A wired-logic control circuit that, among other functions, tests, selects, and establishes paths through a switching stage or stages. 55

marker-beacon receiver. A receiver used in aircraft to receive marker-beacon signals that identify the position of the aircraft when over the marker-beacon station. *See also:* **air-transportation electronic equipment.** 328

marker lamp (railway practice). A signal lamp placed at the side of the rear end of a train or vehicle, displaying light of a particular color to indicate the rear end and to serve for identification purposes. 328

marker light (railway practice). A light that by its color or position, or both, is used to qualify the signal aspect. 328

marker signal (oscilloscopes). A signal introduced into the presentation for the purpose of identification, calibration, or comparison. 185

marking and spacing intervals (telegraph communication). Intervals that correspond, according to convention, to one condition or position of the originating transmitting contacts, usually a closed condition; spacing intervals are the intervals that correspond to another condition of the originating transmitting contacts, usually an open condition. *Note:* The terms **mark** and **space** are frequently used for the corresponding conditions. The waves corresponding to the marking and spacing intervals are frequently designated as marking and spacing waves, respectively. *See also:* **telegraphy.** 328

marking pulse (teletypewriter). The signal pulse that, in direct current, neutral, operation, corresponds to a circuit-closed or current-on condition. 194

marking wave (keying wave) (telegraph communication). The emission that takes place while the active portions of the code characters are being transmitted. *See also:* **radio transmitter.** 111

M-array glide slope (instrument landing systems). A modified null-reference glide-slope antenna system in which the modification is primarily an additional antenna used to obtain a high degree of energy cancellation at the low elevation angles. *Note:* Called *M* because it was 13th in a series of designs. This system is used at locations where higher terrain exists in front of the approach end of the runway, in order to reduce unwanted reflections of energy into the glide-slope sector. *See also:* **navigation.** 187, 13

maser (1) (laser-maser). A device for amplifying or generating radiation by induced transitions of electrons, atoms, molecules, or ions between two energy levels having a **population inversion.** Also, an acronym for **microwave amplification by stimulated emission of radiation.** 363
(2) (data transmission). The general class of microwave amplifiers based on molecular interaction with electromagnetic radiation. The non-electronic nature of the maser principle results in very low noise. 59

mask (computing systems). (1) A pattern of characters that is used to control the retention or elimination of portions of another pattern of characters. (2) A filter. 255, 77

masking (1) (acoustics). (A) The process by which the threshold of audibility for one sound is raised by the presence of another (masking) sound. (B) The amount by which the threshold of audibility of a sound is raised by the presence of another (masking) sound. The unit customarily used is the decibel. *See also:* **loudspeaker.**
(2) (color television). A process to alter color rendition in which the appropriate color signals are used to modify each other. *Note:* The modification is usually accomplished by suitable cross coupling between primary color-signal channels. *See:* **television.** 176, 178

masking audiogram. A graphic presentation of the masking due to a stated noise. *Note:* This is plotted in decibels as a function of the frequency of the masked tone. *See also:* **loudspeaker.** 176

mass (body) (1) (general). The property that determines the acceleration the body will have when acted upon by a given force. 210
(2) (International System of Units) (SI). The SI unit of mass is the kilogram. This unit, or one of the multiples formed by attaching an SI prefix to gram, is preferred for all applications. Among the base and derived units of SI, the unit of mass is the only one whose name, for historical reasons, contains a prefix. Names of decimal multiples and submultiples of the unit of mass are formed by attaching prefixes to the word gram. The megagram (Mg) is the appropriate unit for measuring large masses such as have been expressed in tons. However, the name ton has been given to several large mass units that are widely used in commerce and technology—the long ton of 2240 lb, the short ton of 2000 pounds, and metric ton of 1000 kilograms (also called the tonne). None of these terms are SI. The term metric ton should be restricted to commercial usage, and no prefixes should be used with it. Use of the term tonne is deprecated. *See also:* **units and letter symbols.** 21

mass-attraction vertical. The normal to any surface of constant geopotential; it is the direction that would be indicated by a plumb bob if the earth were not rotating. *See also:* **navigation.** 13, 187

mass spectrograph. An electronic device based on the action of a constant magnetic field on the paths of ions, used to separate ions of different masses. *See also:* **electron device.** 244, 190

mass unbalance (gyro). The characteristic of the gyro resulting from lack of coincidence of the center of supporting forces and the center of mass. It gives rise to torques caused by linear accelerations which lead to acceleration-sensitive drift rates.   46

mast (power transmission and distribution). A column or narrow-base structure of wood, steel, or other material, supporting overhead conductors, usually by means of arms or brackets, span wires, or bridges. *Note:* Broad-base lattice steel supports are often known as towers; narrow-base steel supports are often known as masts. *See:* **pole.** *See also:* **tower.**   64

mast arm. *See:* **bracket.**

master clock. *See:* **clock.**

master compass. A magnetic or gyro compass arranged to actuate repeaters, course recorders, automatic pilots, or other devices.   328

master controller (load-frequency control system) (1) (electric power generators). The central device that develops corrective action, in response to the area control error, for execution at one or more generating units.   94
(2) (car retarders). A controller that governs the operation of one or more magnetic or electropneumatic controllers. *Note:* It is designed to coordinate the movement or the pressure of the retarder with the movement of the retarder level. *See also:* **car retarder; multiple-unit control.**   1
(3) (land transportation vehicles). A device which generates local and trainlike control signals to the propulsion and/or brake systems.   1

master direction indicator. A device that provides a remote reading of magnetic heading. It receives a signal from a magnetic sensing element.   328

master drive (industrial control). A drive that sets the reference input for one or more follower drives. *See also:* **control system, feedback.**   219, 206

master form. An original form from which, directly or indirectly, other forms may be prepared. *See also:* **electroforming.**   328

master oscillator. An oscillator so arranged as to establish the carrier frequency of the output of an amplifier. *See also:* **oscillatory circuit.**   240, 59

master reference system for telephone transmission. Adopted by the International Advisory Committee for Long Distance Telephony (CCIF), a primary reference telephone system for determining, by comparison, the performance of other telephone systems and components with respect to the loudness, articulation, or other transmission qualities of received speech. *Note:* The determination is made by adjusting the loss of a distortionless trunk in the master reference system for equal performance with respect to the quality under consideration.   328

master routine (electronic computation). *See:* **subroutine.**

master/slave operation (power supplies). A system of interconnection of two regulated power supplies in which one (the master) operates to control the other (the slave). *Note:* Specialized forms of the master/slave configuration are used in (1) complementary tracking (plus and minus tracking around a common point), (2) parallel operation to obtain increased current output for voltage regulation, (3) compliance extension to obtain increased voltage

output for current regulation. *See also:* **power supply.**   228, 186

master station (1) (supervisory system). The station from which remotely located units of switchgear or other equipment are controlled by supervisory control or that receives supervisory indications or selected telemeter readings.   103, 202
(2) (electronic navigation). One station of a group of stations, as in loran, that is used to control or synchronize the emission of the other stations. *See also:* **radio navigation.**   187, 13

master-station supervisory equipment. That part of a (single) supervisory system that includes all necessary supervisory control relays, keys, lamps, and associated devices located at the master station for selection, control, indication, and other functions to be performed.   103, 202

master switch (industrial control). A switch that dominates the operation of contactors, relays, or other remotely operated devices. *See also:* **switch.**   206, 3

mast-type antenna for aircraft. A rigid antenna of streamlined cross section consisting essentially of a formed conductor or conductor and supporting body. *See also:* **air-transportation electronic equipment.**   328

mat (rotating machinery). A randomly distributed unwoven felt of fibers in a sheetlike configuration having relatively uniform density and thickness. *See:* **rotor (rotating machinery); stator.**   63

matched condition. *See:* **termination, matched.**

matched filter (radar). A filter which maximizes the output peak signal-to-mean-noise power. For white Gaussian noise, it has a frequency response that is the complex conjugate of the transmitted spectrum, or equivalently, has an impulse response that is the time inverse of the transmitted waveform.   13

matched impedances. Two impedances are matched when they are equal. *Note:* Two impedances associated with an electric network are matched when their resistance components are equal and when their reactance components are equal. *See also:* **network analysis.**   210

matched termination, waveguide. *See:* **termination, matched.**

matched transmission line. A line is matched at any transverse section if there is no wave reflection at that section. *See also:* **matched waveguide; transmission line; waveguide.**   59

matched waveguide. A waveguide having no reflected wave at any transverse section. *See also:* **waveguide.**   267, 319, 179

matching, impedance. *See:* **load, matching.**

matching, load. *See:* **load matching.**

matching loss (radar). The loss in S/N (signal-to-noise) output relative to a matched filter, caused by using a filter of other than matched response to the transmitted signal. *Syn.* **mismatch loss.**   13

matching section (transforming section) (waveguide transformer) (waveguide). A length of waveguide of modified cross section, or with a metal or dielectric insert, used for impedance transformation. *See:* **waveguide.**   179

matching transformer (induction heater). A transformer for matching the impedance of the load to the optimum output characteristic of the power source.   14

mathematical adjectives (pulse terms). All definitions in this section are stated in terms of time (the independent variable) and magnitude (the dependent variable). Unless otherwise specified, the following terms apply only to waveform data within a waveform epoch. These adjectives may be used to describe the relation(s) between other specified variable pairs (for example, time and power, time and voltage, etcetera). (1) instantaneous. Pertaining to the magnitude at a specified time. (2) positive (negative) peak. Pertaining to the maximum (minimum) magnitude. (3) peak-to-peak. Pertaining to the absolute value of the algebraic difference between the positive peak magnitude and the negative peak magnitude. (4) root-mean-square (rms). Pertaining to the square root of the average of the square of the magnitude. If the magnitude takes on n discrete values m_j, the root-mean-square magnitude is

$$M_{rms} = \left[\left(\frac{1}{n} \right) \sum_{j=1}^{j=n} m_j^2 \right]^{1/2}$$

If the magnitude is a continuous function of time $m(t)$,

$$M_{rms} = \left[\left(\frac{1}{t_2 - t_1} \right) \int_{t_1}^{t_2} m^2(t)dt \right]^{1/2}$$

The summation or the integral extends over the interval of time for which the rms magnitude is desired or, if the function is periodic, over any integral number of periodic repetitions of the function. (5) average. Pertaining to the mean of the magnitude. If the magnitude takes on n discrete values m_j, the average magnitude is

$$\overline{M} = \left(\frac{1}{n} \right) \sum_{j=1}^{j=n} m_j$$

If the magnitude is a continuous function of time $m(t)$

$$\overline{M} = \left(\frac{1}{t_2 - t_1} \right) \int_{t_1}^{t_2} m(t)dt$$

The summation or the integral extends over the interval of time for which the average magnitude is desired or, if the function is periodic, over any integral number of periodic repetitions of the function. (6) average absolute. Pertaining to the mean of the absolute magnitude. If the magnitude takes on n discrete values m_j, the average absolute magnitude is

$$|\overline{M}| = \left(\frac{1}{n} \right) \sum_{j=1}^{j=n} |m_j|$$

If the magnitude is a continuous function of time $m(t)$

$$|\overline{M}| = \left(\frac{1}{t_2 - t_1} \right) \int_{t_1}^{t_2} |m(t)|dt$$

The summation or the integral extends over the interval of time for which the average absolute magnitude is desired or, if the function is periodic, over any integral number of periodic repetitions of the function. (7) root sum of squares (rss). Pertaining to the square root of the arithmetic sum of the squares of the magnitude. If the magnitude takes on n discrete values m_j, the root sum of squares magnitude is

$$M_{rss} = \left[\sum_{j=1}^{j=n} m_j^2 \right]^{1/2}$$

If the magnitude is a continuous function of time $m(t)$,

$$M_{rss} = \left[\int_{t_1}^{t_2} m^2(t)dt \right]^{1/2}$$

The summation or the integral extends over the interval of time for which the root sum of squares magnitude is desired r, if the function is periodic, over any integral number of periodic repetitions of the function. 254

mathematical check. A programmed check of a sequence of operations that makes use of the mathematical properties of the sequence. Sometimes called a **control.** *See also:* **programmed check.** 235

mathematical model (analog computer). A set of equations used to represent a physical system. *See also:* **electronic analog computer; model.** 9

mathematical quantity. *See:* **mathematico-physical quantity.**

mathematical symbol (abbreviation). A graphic sign, a letter or letters (which may have letters or numbers, or both, as subscripts or superscripts, or both), used to denote the performance of a specific mathematical operation, or the result of such operation, or to indicate a mathematical relationship. *Compare with:* **symbol for a quantity, symbol for a unit.** *See also:* **abbreviation.** 173

mathematico-physical quantity (symbolic quantity) (mathematical quantity) (abstract quantity). A concept, amenable to the operations of mathematics, that is directly related on one (or more) physical quantity and is represented by a letter symbol in equations that are statements about that quantity. *Note:* Each mathematical quantity used in physics is related to a corresponding physical quantity in a way that depends on its defining equation. It is characterized by both a qualitative and a quantitative attribute (that is, dimensionality and magnitude). 210

matrix (1) (mathematics). (A) A two-dimensional rectangular array of quantities. Matrices are manipulated in accordance with the rules of matrix algebra. (B) By extension, an array of any number of dimensions. 255, 77

(2) (electronic computers). A logic network whose configuration is an array of intersections of its input-output leads, with elements connected at some of these intersections. The network usually functions as an encoder or decoder. *Note:* A translating matrix develops several output signals in response to several input signals; a decoder develops a single output signal in response to several input signals (therefore sometimes called an AND matrix); an encoder develops several output signals in response to a single input signal and a given output signal may be generated by a number of different input signals (therefore sometimes called an OR matrix). *See also:* **decode; encode; translate.** 235, 210

(3) Loosely, any encoder, decoder, or translator. 210

(4) (noun) (color television). An array of coefficients symbolic of a color coordinate transformation. *Note:* This definition is consistent with mathematical usage. 18, 178

(5)(verb)(color television). To perform a color coordinate transformation by computation or by electrical, optical, or other means. 18, 178

(6)(electrochemistry). A form used as a cathode in electroforming. *See also:* **electroforming.** 328

matrix circuit (color television). *See:* **matrix unit.**

matrix, fundamental. *See:* **matrix, transition.**

matrix, system (control systems). A matrix of transfer functions which relate the Laplace transforms of the system outputs and of the system inputs. 56

matrix, transition (control systems). A matrix which maps the state of a linear system at one instant of time into another state at a later instant of time, provided that the system inputs are zero over the closed time interval between the two instants of time. *Note:* This is also the matrix of solutions of the homogeneous equations. *Syn:* **matrix, fundamental.** 56

matrix unit (matrix circuit)(color television). A device that performs a color coordinate transformation by electrical, optical, or other means. 18

matte dip (electroplating). A dip used to produce a matte surface on a metal. *See also:* **electroplating.** 328

matte surface. A surface from which the reflection is predominantly diffuse, with or without a negligible specular component. *See:* **diffuse reflection.** *See also:* **bare (exposed) lamp.** 167

maximum asymmetric short-circuit current (rotating machinery). The instantaneous peak value reached by the current in the armature winding within a half of a cycle after the winding has been suddenly short-circuited, when conditions are such that the initial value of any aperiodic component of current is the maximum possible. 63

maximum average power (attenuator). That maximum specified input power applied for a minimum of one hour (unless specified for a longer period) at the maximum operating temperature with output terminated in the characteristic impedance which will not permanently change the specified properties of the attenuator after return to ambient temperature at a power level 20 dB below maximum specified input power. 110

maximum average power output (television). The maximum radio-frequency output power that can occur under any combination of signals transmitted, averaged over the longest repetitive modulation cycle. *See also:* **television.** 328

maximum common-mode signal. *See:* **common-mode signal maximum.**

maximum continuous rating (rotating machinery). The maximum values of electric and mechanical loads at which a machine will operate successfully and continuously. *Note:* An overload may be implied, along with temperature rises higher than normal standards for the machine. *See also:* **asynchronous machine; direct-current commutating machine.** 63

maximum control current (magnetic amplifier). The maximum current permissible in each control winding either continuously or for designated operating intervals as specified by the manufacturer and shall be specified as either root-mean-square or average. 171

maximum current (instrument)(wattmeter or power-factor meter). A stated current that, if applied continuously at maximum stated operating temperature and with any

other circuits in the instrument energized at rated values, will not cause electric breakdown or any observable physical degradation. *See also:* **instrument.** 280

maximum demand (installation or system). The greatest of all the demands that have occurred during the specified period of time. *Note:* The maximum demand is determined by measurement, according to specification, over a definitely prescribed time interval. *See also:* **alternating-current distribution; demand meter; direct-current distribution.** 212, 64

maximum-demand pointer (demand meter)(friction pointer of a demand meter*). A means used to indicate the maximum demand that has occurred since its previous resetting. The maximum-demand pointer is advanced up the scale of an indicating demand meter by the pointer pusher. When not being advanced, it is held stationary, usually by friction, and it is reset manually when the meter is read for billing purposes. *See also:* **demand meter.** 328

* Deprecated

maximum design voltage (1)(device). The highest voltage at which the device is designed to operate. *Note:* When expressed as a rating this voltage is termed rated maximum voltage. 103, 202, 127

(2)(to ground)(outdoor electric apparatus). The maximum voltage at which the bushing is designed to operate continuously. 287

(3)(transformer). The highest root-mean-square phase-to-phase voltage that equipment components are designed to withstand continuously, and to operate in a satisfactory manner without derating of any kind. 53

maximum-deviation sensitivity (in frequency-modulation receivers). Under maximum system deviation, the least signal input for which the output distortion does not exceed a specified limit. *See also:* **frequency modulation.** 339

maximum excursion (electric conversion). The maximum positive or negative deviation from the initial or steady value caused by a transient condition. *See also:* **electric conversion equipment.** 186

maximum instantaneous fuel change (gas turbines). The fuel change allowable for an instantaneous or sudden increased or decreased load or speed demand. *Note:* It is expressed in terms of equivalent load change in percent of rated load. 98, 58

maximum keying frequency (fundamental scanning frequency)(facsimile). The frequency in hertz numerically equal to the spot speed divided by twice the scanning spot X dimension. *See also:* **scanning (facsimile).** 12

maximum modulating frequency (facsimile). The highest picture frequency required for the facsimile transmission system. *Note:* The maximum modulating frequency and the maximum keying frequency are not necessarily equal. *See also:* **facsimile transmission.** 12

maximum momentary speed variation (speed governing hydraulic turbines). The maximum momentary change of speed when the load is suddenly changed a specified amount. 8, 58

maximum OFF voltage (magnetic amplifier). The maximum output voltage existing before trip ON control signal is reached as the control signal is varied from trip OFF to trip ON. 171

maximum operating voltage (household electric ranges).

The maximum voltage to which the electric parts of the range may be subjected in normal operation. *See also:* **appliances outlet.** 263

maximum output (receivers). The greatest average output power into the rated load regardless of distortion. *See also:* **radio receiver.** 339

maximum output voltage (magnetic amplifier). The voltage across the rated load impedance with maximum control current flowing through each winding simultaneously in a direction that increases the output voltage. *Notes:* (1) Maximum output voltage shall be specified either as root-mean-square or average. (2) While specification may be either root-mean-square or average, it remains fixed for a given amplifier. 171

maximum peak power (attenuator). That maximum peak power at the maximum specified pulse-length and average power which, when applied for a minimum of one hour (unless specified for a longer period) at the maximum operating temperature, while the output is terminated in the characteristic impedance, will not permanently change the specified properties of the attenuator when returned to ambient temperature at a power level 20 dB below the maximum specified input power or lower. 110

maximum power output (hydraulic turbines). The maximum output which the turbine-generator unit is capable of developing at rated speed with maximum head and maximum gate. 8, 58

maximum pulse repetition rate (digital delay line). The maximum pulse repetition rate shall be equal to $1/2 \Delta t$, where Δt is the time spacing between the peaks of the output doublet. 81

maximum rate of fuel change (gas turbines). The rate of fuel change that is allowable after the maximum instantaneous fuel change, when an instantaneous speed or load demand upon the turbine is greater than that corresponding to the maximum instantaneous fuel change. *Note:* It is expressed in percent of equivalent load change per second. 98, 58

maximum safe input power (1) (non-real time spectrum analyzer). The power applied at the input which will not cause degradation of the instrument characteristics. 68

(2) (electrothermic unit). The maximum peak pulse or cw input power which will cause no permanent change in the calibration or characteristics of the electrothermic unit. Specify in watt-microseconds the maximum (safe) input energy per pulse and the applicable pulse repetition frequency in hertz or in kilohertz. Specify in watts or milliwatts the maximum (safe) input peak pulse power. 47

maximum sensitivity (frequency-modulation systems). The least signal input that produces a specified output power. 339

maximum sine-current differential permeability (toroidal magnetic amplifier cores). The maximum value of sine-current differential permeability obtained with a specified sine-current magnetizing force. 170

maximum sound pressure (for any given cycle of a periodic wave). The maximum absolute value of the instantaneous sound pressure occurring during that cycle. *Note:* In the case of a sinusoidal sound wave this maximum sound pressure is also called the pressure amplitude. 176

maximum speed (industrial control). The highest speed within the operating speed range of the drive. *See also:*

electric drive. 206

maximum surge current rating (rectifier circuit) (nonrepetitive) (semiconductor rectifier). The maximum forward current having a specified waveform and short specified time interval permitted by the manufacturer under stated conditions. *See also:* **average forward current rating (rectifier circuit element).** 237, 66, 208

maximum system deviation (frequency-modulation systems). The greatest frequency deviation specified in the operation of the system. *Note:* Maximum system deviation is expressed in kilohertz. In the case of FCC authorized frequency modulation broadcast systems in the range from 88 to 108 megahertz, the maximum system deviation is ±75 kHz. 16

maximum (highest) system voltage ((1) surge arrester). The highest voltage at which a system is operated. *Note:* This is generally considered to be the maximum system voltage as prescribed in American National Standard Voltage Ratings for Electric Power Systems and Equipment (60 Hz). C84.1-1970. 2

(2) (transformer). The highest root-mean-square phase-to-phase voltage which occurs on the system under normal operating conditions, and the highest root-mean-square phase-to-phase voltage for which equipment and other system components are designed for satisfactory continuous operation without derating of any kind. (This voltage excludes voltage transients and temporary overvoltages caused by abnormal system conditions such as faults, load rejection, etcetera.) 53

maximum test output voltage (magnetic amplifier) (1) (nonreversible output). The output voltage equivalent to the summation of the minimum output voltage plus 66⅔ percent of the difference between the rated and minimum output voltages. 171

(2) (reversible output). (A) Positive maximum test output voltage is the output voltage equivalent to 66⅔ percent of the rated output voltage in the positive direction. (B) Negative maximum test output voltage is the output voltage equivalent to 66⅔ percent of the rated output voltage in the negative direction. 171

maximum total sag (electric systems) (transmission and distribution). The sag at the midpoint of the straight line joining the two points of support of the conductor. *See:* **sag.** 178, 262

maximum undistorted output (maximum useful output) (1) (radio receivers). For sinusoidal input, the greatest average output power into the rated load with distortion not exceeding a specified limit. *See:* **radio receiver.** 339

(2) (electroacoustics). The maximum power delivered under specified conditions with a total harmonic not exceeding a specified percentage. *See also:* **level.** 328

maximum usable frequency (radio transmission by ionospheric reflection) (MUF). The upper limit of frequencies that can be used at a particular time for transmission between two specified points by reflection from regular ionized layers. *Note:* Higher frequencies may be transmitted by sporadic and scattered reflections. The maximum usable frequency is usually but not always controlled by electron limitation. *See also:* **radiation; radio wave propagation.** 146

maximum useful output. *See:* **maximum undistorted output.**

maximum voltage (instrument) (wattmeter, power-factor

meter or frequency meter). A stated voltage that, if applied continuously at the maximum stated operating temperature and with any other circuits in the instrument energized at rated values, will not cause electric breakdown or any observable physical degradation. *See also:* **accuracy rating (instrument); instrument.**   280

maxwell (line). The unit of magnetic flux in the centimeter-gram-second electromagnetic system. *Note:* The maxwell is 10^{-8} weber.   210

Maxwell bridge (general). A 4-arm alternating-current bridge characterized by having in one arm an inductor in series with a resistor and in the opposite arm a capacitor in parallel with a resistor, the other two arms being normally nonreactive resistors. *Note:* Normally used for the measurement of inductance (or capacitance) in terms of resistance and capacitance (or inductance). The balance

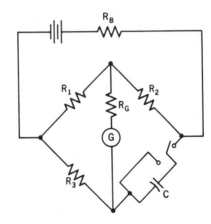

$$C = \frac{R_1}{nR_2R_3} \times \frac{\left[1 - \dfrac{R_1{}^2}{(R_1 + R_2 + R_B)(R_1 + R_3 + R_G)}\right]}{\left[1 + \dfrac{R_1R_B}{R_3(R_1 + R_2 + R_B)}\right]\left[1 + \dfrac{R_1R_G}{R_2(R_1 + R_3R_G)}\right]}$$

is independent of the frequency, and at balance the ratio of the inductance to the capacitance is equal to the product of the resistances of either pair of opposite arms. It differs from the Hay bridge in that in the arm opposite the inductor, the capacitor is shunted by the resistor. *See also:* **bridge.**   328

Maxwell direct-current commutator bridge

Maxwell inductance bridge. A 4-arm alternating-current bridge characterized by having inductors in two adjacent arms and usually, nonreactive resistors in the other two arms. *Note:* Normally used for the comparison of inductances. The balance is independent of the frequency. *See also:* **bridge.**   328

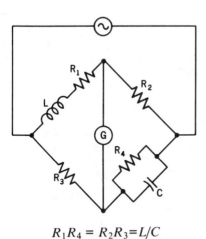

$$R_1R_4 = R_2R_3 = L/C$$

Maxwell bridge

Maxwell direct-current commutator bridge. A 4-arm bridge characterized by the presence in one arm of a commutator, or 2-way contactor, that, with a known periodicity, alternately connects the unknown capacitor in series with the bridge arm and then opens the bridge arm while short-circuiting the capacitor, the other three arms being nonreactive resistors. *Note:* Normally used for the measurement of capacitance in terms of resistance and time. The bridge is normally supplied from a battery and the detector is a direct-current galvanometer. *See also:* **bridge.**   328

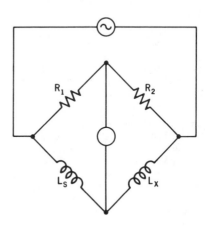

$$R_x = R_s \frac{R_2}{R_1} \qquad L_x = L_s \frac{R_2}{R_1}$$

Maxwell inductance bridge

Maxwell mutual-inductance bridge. An alternating-current bridge characterized by the presence of mutual inductance between the supply circuit and that arm of the network that includes one coil of the mutual inductor, the other three arms being normally non-reactive resistors. *Note:*

Normally used for the measurement of mutual inductance in terms of self-inductance. The balance is independent of the frequency. *See also:* **bridge.**   328

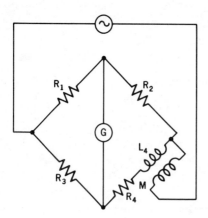

$$R_1 R_4 = R_2 R_3 \qquad L_4 = -M\left(1 + \frac{R_2}{R_1}\right)$$

Maxwell mutual-inductance bridge

Maxwell's equations (Maxwell's laws). The fundamental equations of macroscopic electrmagnetic field theory. All real (physical) electric and magnetic fields satisfy Maxwell's equations, namely

$$\nabla \times \mathbf{E} = \frac{\partial \mathbf{B}}{\partial t} \qquad \nabla \cdot \mathbf{B} = 0$$

$$\nabla \times \mathbf{H} = \frac{\partial \mathbf{D}}{\partial t} + \mathbf{J} \qquad \nabla \cdot \mathbf{D} = q_r$$

where $\mathbf{E}$ is electric field strength, $\mathbf{D}$ is electric flux density, $\mathbf{H}$ is magnetic field strength, $\mathbf{B}$ is magnetic flux density, $\mathbf{J}$ is current density, and q_r is volume charge density.   210

May Day. *See:* **radio distress signal.**

MBWO (reentrant field type microwave backward-wave oscillator) (microwave tubes). *See:* **carcinotron.**

McCulloh circuit. A supervised, metallic loop circuit having manually or automatically operated switching equipment at the receiving end, that, in the event of a break, a ground, or a combination of a break and a ground at any point in the metallic circuit, conditions the circuit, by utilizing a ground return, for the receipt of signals from suitable signal transmitters on both sides of the point of trouble. *See also:* **protective signalling.**   328

MCW. *See:* **modulated continuous wave.**

MDI. *See:* **magnetic direction indicator.**

M display (radar). A type-*A* display in which the target distance is determined by moving an adjustable pedestal signal along the baseline until it coincides with the horizontal position of the target-signal deflections; the control which moves the pedestal is calibrated in distance. *See also:* **navigation.**   13, 187

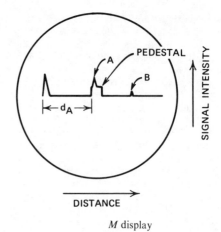

M display

MDS. *See:* **minimum detectable signal; minimum discernible signal.**

MEA (electronic navigation). *See:* **minimum en-route altitude.**

mean charge (nonlinear capacitor). The arithmetic mean of the transferred charges corresponding to a particular capacitor voltage, as determined from a specified alternating charge characteristic. *See also:* **nonlinear capacitor.**   191

mean-charge characteristic (nonlinear capacitor). The function relating mean charge to capacitor voltage. *Note:* Mean-charge characteristic is always single-valued. *See also:* **nonlinear capacitor.**   191

mean first slip time (communication satellite). The mean time for a phase-lock loop starting to lock to a slip one or more cycles.   85

mean free path (acoustics). For sound waves in an enclosure, the average distance sound travels between successive reflections in the enclosure.   191, 176

mean Hall plate temperature (Hall effect devices). The value of the temperature averaged over the volume of the Hall plate.   107

mean life (reliability). The arithmetic mean of the times to failure of a group of nominally identical items. *See also:* **reliability.**   182

mean life, assessed (non-repaired items) (reliability). The mean life of an item determined by a limiting value or values of the confidence interval associated with a stated confidence level, based on the same data as the observed mean life of nominally identical items. *Notes:* (1) The source of the data shall be stated. (2) Results can be accumulated (combined) only when all conditions are similar. (3) The assumed underlying distribution of failures against time shall be stated. (4) It should be stated whether a one-sided or a two-sided interval is being used. (5) Where one limiting value is given this is usually the lower limit.   164

mean life, extrapolated (non-repaired items) (reliability). Extension by a defined extrapolation or interpolation of the observed or assessed mean life for stress conditions different from those applying in the observed or assessed mean life. *Note:* The validity of the extrapolation shall be justified.   164

mean life, observed (non-repaired items) (reliability). The mean value of the lengths of observed times to failure of all items in a sample under stated conditions. *Note:* The criteria for what constitutes a failure shall be stated.
    164

mean life, predicted (non-repaired items) (reliability). For the stated conditions of use, and taking into account the design of an item, the mean life computed from the observed, assessed or extrapolated mean life of its parts. *Note:* Engineering and statistical assumptions shall be stated, as well as the bases used for the computation (observed or assessed).
    164

mean of reversed direct-current values (alternating-current instruments). The simple average of the indications when direct current is applied in one direction and then reversed and applied in the other direction. *See also:* **accuracy rating (instrument).**
    280

mean pulse time. The arithmetic mean of the leading-edge pulse time and the trailing-edge pulse time. *Note:* For some purposes, the importance of a pulse is that it exists (or is large enough) at a particular instant of time. For such applications the important quantity is the mean pulse time. The leading-edge pulse time and trailing-edge pulse time are significant primarily in that they may allow a certain tolerance in timing.
    254, 210, 162

mean temperature coefficient of output voltage (Hall effect devices). The arithmetic average of the percentage changes in output voltage per degree Celsius taken over a given temperature range for a given control current magnitude and a given magnetic flux density.
    107

mean time between errors (MTE). The average time between the generation of a single false code of a given set in use in a code transmission system, assuming a continuous source of random inputs and the bandwidth limited in a manner identical to the normally expected signal.
    59

mean time between failures (1) (power system communications). (MTF). The average time (preferably expressed in hours) between failures of a continuously operating device, circuit, or system.
    59

(2) (nuclear power generating stations) (MTBF). The arithmetic average of operating times between failures of an item.
    29, 31

mean time between (or before) failures (MTBF) (power supplies). A measure of reliability giving either the time before first failure or, for repairable equipment, the average time between repairs. Mean time between (before) failures may be approximated or predicted by summing the reciprocal failure rates of individual components in an assembly. *See also:* **power supply.**
    228, 186

mean time between failures (repairable items) (reliability). The product of the number of items and their operating time divided by the total number of failures. *See also:* **reliability.**
    182

mean time between failures, assessed (repaired items) (reliability). The mean time between failures of an item determined by a limiting value or values of the confidence interval associated with a stated confidence level, based on the same data as the observed mean time between failures of nominally identical items. *Notes:* (1) The source of the data shall be stated. (2) Results can be accumulated (combined) only when all conditions are similar. (3) The

assumed underlying distribution of failures against time shall be stated. (4) It should be stated whether a one-sided or a two-sided interval is being used. (5) Where one limiting value is given this is usually the lower limit.
    164

mean time between failures, extrapolated (repaired items) (reliability). Extension by a defined extrapolation or interpolation of the observed or assessed mean time between failures for duration and/or conditions different from those applying to the observed or assessed mean time between failures. *Note:* The validity of the extrapolation shall be justified.
    164

mean time between failures, observed (repaired items) (reliability). For a stated period in the life of an item, the mean value of the length of time between consecutive failures, computed as the ratio of the cumulative observed time to the number of failures under stated conditions. *Notes:* (1) The criteria for what constitutes a failure shall be stated. (2) Cumulative time is the sum of the times during which each individual item has been performing its required function under stated conditions. (3) This is the reciprocal of the observed failure rate during the period.
    164

mean time between failures, predicted (repaired items) (reliability). For the stated conditions of use, and taking into account the design of an item, the mean time between failures computed from the observed, assessed, or extrapolated failure rates of its parts. *Note:* Engineering and statistical assumptions shall be stated, as well as the bases used for the computation (observed or assessed).
    164

mean time to failure (nonrepaired items) (reliability). The total operating time of a number of items divided by the total number of failures. *See also:* **reliability.**
    182

mean time to failure, assessed (non-repaired items) (reliability). The mean time to failure of an item determined by a limiting value or values of the confidence interval associated with a stated confidence level, based on the same data as the observed mean time to failure of nominally identical items. *Notes:* (1) The source of the data shall be stated. (2) Results can be accumulated (combined) only when all conditions are similar. (3) The assumed underlying distribution of failures against time shall be stated. (4) It should be stated whether a one-sided or a two-sided interval is being used. (5) Where one limiting value is given this is usually the lower limit.
    164

mean time to failure, extrapolated (non-repaired items) (reliability). Extension by a defined extrapolation or interpolation of the observed or assessed mean time to failure for durations and/or conditions different from those applying to the observed or assessed mean time to failure. *Note:* The validity of the extrapolation shall be justified.
    164

mean time to failure, observed (non-repaired items) (reliability). For a stated period in the life of an item, the ratio of the cumulative time for a sample to the total number of failures in the sample during the period, under stated conditions. *Notes:* (1) The criteria for what constitutes a failure shall be stated. (2) Cumulative time is the sum of the times during which each individual item has been performing its required function under stated conditions. (3) This is the reciprocal of the observed failure rate during the period.
    164

mean time to failure, predicted (non-repaired items) (reliability). For the stated conditions of use, and taking into account the design of an item the mean time to failure computed from the observed, assessed or extrapolated mean times to failure of its parts. *Note:* Engineering and statistical assumptions shall be stated, as well as the bases used for the computation (observed or assessed).   164

mean time to repair (MTTR) (nuclear power generating stations). The arithmetic average of time required to complete a repair activity.   31, 29

measurand. A physical or electrical quantity, property, or condition that is to be measured. *See:* **measurement system.**   103, 202, 54

measure. The number (real, complex, vector, etcetera) that expresses the ratio of the quantity to the unit used in measuring it.   210

measured current (insulation test). The total direct current resulting from the application of direct voltage to insulation and including the leakage current, the absorption current, and theoretically the capacitive current. *Note:* The measured current is the value read on the microammeter during a direct high-voltage test of insulation. *See also:* **insulation testing (large alternating-current rotating machinery).**   6

measured service (telephone switching systems). Service in which charges are assessed in terms of the number of message units during the billing interval.   55

measured signal (automatic null-balancing electric instrument). The electrical quantity applied to the measuring-circuit terminals of the instrument. *Note:* It is the electrical analog of the measured variable. *See also:* **measurement system.**   295

measured value (power meters). An estimate of the value of a quantity obtained as a result of a measurement. *Note:* Indicated values may be corrected to give measured values.   115, 47

measured variable (measurand) (automatic null-balancing electric instrument). The physical quantity, property, or condition that is to be measured. *Note:* Common measured variables are temperature, pressure, thickness, speed, etcetera. *See also:* **measurement system.**   295

measure equations. Equations in which the quantity symbols represent pure numbers, the measures of the physical quantities corresponding to the symbols.   210

measurement. The determination of the magnitude or amount of a quantity by comparison (direct or indirect) with the prototype standards of the system of units employed. *See:* **absolute measurement.**   210

measurement component. A general term applied to parts or subassemblies that are primarily used for the construction of measurement apparatus. *Note:* It is used to denote those parts made or selected specifically for measurement purposes and does not include standard screws, nuts, insulated wire, or other standard materials. *See:* **measurement system.**   328

measurement device. An assembly of one or more basic elements with other components and necessary parts to form a separate self-contained unit for performing one or more measurement operations. *Note:* It includes the protecting, supporting, and connecting, as well as the functioning, parts, all of which are necessary to fulfill the application requirements of the device. It should be noted that end devices (which see) are frequently but not always complete measurement devices in themselves, since they often are built-in with all or part of the intermediate means or primary detectors to form separate self-contained units. *See:* **measurement system.**   328

measurement energy. The energy required to operate a measurement device or system. *Note:* Measurement energy is normally obtained from the measurand or from the primary detector. *See:* **measurement system.**   328

measurement equipment. A general term applied to any assemblage of measurement components, devices, apparatus, or systems. *See:* **measurement system.**   328

measurement inverter (chopper). *See:* **measuring modulator.**

measurement mechanism. An assembly of basic elements and intermediate supporting parts for performing a mechanical operation in the sequence of measurement. *Note:* For example, it may be a group of components required to effect the proper motion of an indicating or recording means and does not include such parts as bases, covers, scales, and accessories. It may also be applied to a specific group of elements by substituting a suitable qualifying term; such as time-switch mechanism or chart-drive mechanism. *See:* **measurement system.**   328

measurement range (instrument). That part of the total range within which the requirements for accuracy are to be met. *See also:* **instrument.**   328

measurement system. One or more measurement devices and any other necessary system elements interconnected to perform a complete measurement from the first operation to the end result. *Note:* A measurement system can be divided into general functional groupings, each of which consists of one or more specific functional steps or basic elements.   328

measurement uncertainty (test, measurement and diagnostic equipment). The limits of error about a measured value between which the true value will lie with the confidence stated.   54

measurement voltage divider (voltage ratio box) (volt box*). A combination of two or more resistors, capacitors, or other circuit elements so arranged in series that the voltage across one of them is a definite and known fraction of the voltage applied to the combination, provided the current drain at the tap point is negligible or taken into account. *Note:* The term volt box is usually limited to resistance voltage dividers intended to extend the range of direct-current potentiometers. *See also:* **auxiliary device to an instrument.**   328

* Deprecated

measuring accuracy, precision, or reproducibility (numerically controlled machines). Accuracy, precision, or reproducibility of position sensor or transducer and interpreting system.   224, 207

measuring and test equipment (M&TE) (1) (nuclear power generating stations). Devices or systems used to calibrate, measure, gauge, test, inspect, or control in order to acquire research, development, test, or operational data; to determine compliance with design, specifications, or other technical requirements, M&TE does not include permanently installed operating equipment or test equipment used for preliminary checks where accuracy is not required, for example, circuit checking multimeters.   31, 41

(2) (test, measurement and diagnostic equipment). *See: test, measurement and diagnostic equipment.*   54

measuring mechanism (recording instrument). The parts that produce and control the motion of the marking device. *See also:* **moving element (instrument).**   294

measuring modulator (measurement inverter) (chopper). An intermediate means in a measurement system by which a direct-current or low-frequency alternating-current input is modulated to give a quantitatively related alternating-current output usually as a preliminary to amplification. *Note:* The modulator may be of any suitable type such as mechanical, magnetic, or varistor. The mechanical types, which are actuated by vibrating or rotating members, may be classified as (1) contacting, (2) microphonic, or (3) generating (capacitive or inductive). *See also:* **auxiliary device to an instrument.**   328

measuring units and relay logic (surge withstand capability tests). Analog or digital devices which analyze the input currents and voltages to determine the immediate status of that part of the power system that they were installed to protect and to provide the control signal to trip circuit breakers.   90

mechanical back-to-back test (rotating machinery). A test in which two identical machines are mechanically coupled together and the total losses of both machines are calculated from the difference between the electrical input to one machine and the electrical output of the other machine. *See:* **efficiency.**   63

mechanical bias. *See:* **relay mechanical bias.**

mechanical braking (industrial control). The kinetic energy of the drive motor and the driven machinery is dissipated by the friction of a mechanical brake. *See also:* **electric drive.**   206

mechanical current rating (neutral grounding device) (electric power). The symmetrical root-mean-square alternating-current component of the completely offset current wave that the device can withstand without mechanical failure. *Note:* The mechanical forces depend upon the maximum crest value of the current wave. However, for convenience, the mechanical current rating is expressed in terms of the root-mean-square value of the alternating-current component only of a completely displaced current wave. Specifically, the crest value of the completely offset wave will then be 2.82 times the mechanical current rating. *See also:* **grounding device.**   91, 64

mechanical cutter (mechanical recorder). An equipment for transforming electric or acoustic signals into mechanical motion of approximately like form and inscribing such motion in an appropriate medium by cutting or embossing.   176

mechanical fatigue test (rotating machinery). A test designed to determine the effect of a specific repeated mechanical load on the life of a component. *See also:* **asynchronous machine; direct-current commutating machine.**   63

mechanical filter. *See:* **mechanical wave filter.**

mechanical freedom (accelerometer). The maximum linear or angular displacement of the accelerometer's proof mass relative to its case.   46

mechanical-hydraulic governor (speed governing of hydraulic turbines). A governor in which the control signal(s) proportional to speed error and necessary stabilizing and auxiliary signals are developed mechanically, summed by a mechanical system, and are then hydraulically amplified.   8, 58

mechanical-impact strength (insulator). The impact that, under specified conditions, the insulator can withstand without damage. *See also:* **insulator.**   261

mechanical impedance. The complex quotient of the effective alternating force applied to the system, by the resulting effective velocity.   176

mechanical inertia time (speed governing of hydraulic turbines). A characteristic of the machine due to the inertia of the rotating components of the machine defined as:

$$T_M = \frac{(Wk^2)(N)^2(10^{-6})}{(1.61)(HP)}$$

where: W = weight of machine rotating parts, pounds; k = radius of gyration, feet; N = rated speed, rev/min; HP = rated output of turbine, horsepower. *Notes:* (1) T_M is also approximately equal to $2H$ where H is the inertia constant. (2) To calculate T_M using International SI units:

$T_M = J\omega_0^2/P_0$ where
J = polar moment of inertia in kgm² calculated by dividing Wk^2 in Newton-meters by acceleration of gravity 9.81 m/second².

$$= Mk^2 = \frac{GD}{4g}$$

ω_0 = $\pi N/60$, rad/second.
P_0 =rated output of turbine, watts.   8

mechanical interchangeability (fuse links). The characteristic that permits the designs of various manufacturers to be interchanged physically so they fit into and withstand the tensile stresses imposed by various types of prescribed cutouts made by different manufacturers.   103, 202

mechanical interlocking machine. An interlocking machine designed to operate the units mechanically. *See also:* **interlocking (interlocking plant).**   328

mechanical latching relay. A relay in which the armature or contacts may be latched mechanically in the operated or unoperated position until reset manually or electrically.   341

mechanical limit (neutral grounding devices). The rated maximum instantaneous value of current, in amperes, that the device will withstand without mechanical failure.   91

mechanically delayed overcurrent trip. *See:* **mechanically delayed release (mechanically delayed trip); overcurrent release (overcurrent trip).**

mechanically delayed release (mechanically delayed trip). A release delayed by a mechanical device.   103, 202

mechanically de-spun antenna (communication satellite). A rotating directional antenna, mounted to a rotating object (namely spin stabilized communication satellite); the rotation of the antenna is counter to the rotation of the body it is mounted to, such that the antenna beam points into the same direction of space.   83

mechanically release-free (mechanically trip-free) (switching device). A term indicating that the release can open the device even though (1) in a manually operated switching device the operating lever is being moved toward the closed position; or (2) in a power-operated switching device, such as solenoid- or spring-actuated types, the operating mechanism is being moved toward the closed position either by continued application of closing power or by means of a maintenance closing lever.   103, 202

mechanically reset relay. *See:* **hand-reset relay.**

mechanically timed relays. Relays that are timed mechanically by such features as clockwork, escapement, bellows, or dashpot. *See also:* **relay.**   259

mechanical modulator (electronic navigation). (1) A device that varies some characteristic of a carrier wave so as to transmit information, the variation being accomplished by physically moving or changing a circuit element. (2) In instrument landing systems, a particular arrangement of radio-frequency transmission lines and bridges with resonant sections coupled to the lines and motor-driven capacitor plates that alter the resonance so as to produce 90- and 150-hertz modulations. *See also:* **navigation.**
                          187, 13

mechanical operation (switch) (power switchgear). Operation by means of an operating mechanism connected to the switch by mechanical linkage. *Note:* Mechanically operated switches may be actuated either by manual, electrical, or other suitable means.   27, 103

mechanical part (electric machine). Any part having no electric or magnetic function.   63

mechanical plating. Any plating operation in which the cathodes are moved mechanically during the deposition. *See also:* **electroplating.**   328

mechanical recorder. *See:* **mechanical cutter.**

mechanical rectifier. A rectifier in which rectification is accomplished by mechanical action. *See also:* **rectification.**   328

mechanical reproducer. *See:* **phonograph pickup.**

mechanical shock. A significant change in the position of a system in a nonperiodic manner in a relatively short time. *Note:* It is characterized by suddenness and large displacements and develops significant internal forces in the system. *See also:* **shock motion.**   176

mechanical short-time current rating (current transformer). The root-mean-square value of the alternating-current component of a completely displaced primary current wave that the transformer is capable of withstanding, with secondary short-circuited. *Note:* Capable of withstanding means that after a test the current transformer shows no visible sign of distortion and is capable of meeting the other specified applicable requirements. *See:* **instrument transformer.**   203

mechanical switching device. A switching device designed to close and open one or more electric circuits by means of guided separable contacts. *Note:* The medium in which the contacts separate may be designated by suitable prefix, that is, air, gas, oil, etcetera.   103, 202

mechanical time constant (critically damped indicating instrument). The period of free oscillation divided by 2π. *See also:* electromagnetic compatibility.   220, 199

mechanical transmission system. An assembly of elements adapted for the transmission of mechanical power. *See also:* **phonograph pickup.**   176

mechanical trip (trip arm) (railway practice). A roadway element consisting in part of a movable arm that in operative position engages apparatus on the vehicle to effect an application of the brakes by the train-control system.
                          328

mechanical wave filter (mechanical filter). A filter designed to separate mechanical waves of different frequencies. *Note:* Through electromechanical transducers, such a filter may be associated with electric circuits. *See also:* **filter.**   328

mechanical wrap or connection (soldered connections). The securing of a wire or lead prior to soldering. *See also:* **soldered connections (electronic and electrical applications).**
                          284

mechanism (1) (indicating instrument). The arrangement of parts for producing and controlling the motion of the indicating means. *Note:* It includes all the essential parts necessary to produce these results but does not include the base, cover, dial, or any parts, such as series resistors or shunts, whose function is to adapt the instrument to the quantity to be measured. *See also:* **instrument; moving element (instrument).**   280

(2) (recording instrument). Includes (A) the arrangement for producing and controlling the motion of the marking device; (B) the marking device; (C) the device (clockwork, constant-speed motor, or equivalent) for driving the chart; (D) the parts necessary to carry the chart. *Note:* It includes all the essential parts necessary to produce these results but does not include the base, cover, indicating scale, chart, or any parts, such as series resistors or shunts, whose function is to make the recorded value of the measured quantity correspond to the actual value. *See also:* **moving element (instrument).**   328

(3) (switching device). The complete assembly of levers and other parts that actuates the moving contacts of a switching device.   103, 202

medium (computing systems). The material, or configuration thereof, on which data are recorded, for example, paper tape, cards, magnetic tape.   255, 77

medium noise (sound recording and reproducing system). The noise that can be specifically ascribed to the medium. *See also:* **noise (sound recording and reproducing system).**   350

medium voltage (cable systems). 601 to 15,000 volts.
                          35

medium voltage power cable (cable systems). Designed to supply power to utilization devices of the plant auxiliary system, rated 601 to 15,000 volts.   35

medium-voltage system (electric power for industrial and commercial systems only). An electric system having a maximum root-mean-square alternating-current voltage above 1000 volts to 72,500 volts. *See also:* **voltage classes.**
                          49

Meissner oscillator. An oscillator that includes an isolated tank circuit inductively coupled to the input and output circuits of an amplifying device to obtain the proper feedback and frequency. *See also:* **oscillatory circuit.**
                          111

mel. A unit of pitch. By definition, a simple tone of frequency 1000 hertz, 40 decibels above a listener's threshold, produces a pitch of 1000 mels. *Note:* The pitch of any sound that is judged by the listener to be n times that of the 1-mel tone is n mels.   176

melting channel. The restricted portion of the charge in a submerged resistor or horizontal-ring induction furnace in which the induced currents are concentrated to effect high energy absorption and melting of the charge. *See* *also:* **induction heating.** 14, 114

melting-speed ratio (fuse). A ratio of the current magnitudes required to melt the current-responsive element at two specified melting times. *Notes:* (1) Specification of the current wave shape is required for time less than one-tenth of a second. (2) The lower melting time in present use is 0.1 second, and the higher minimum melting current times are 100 seconds for low-voltage fuses and 300 or 600 seconds, whichever specified, for high-voltage fuses.
103

melting time (fuse). The time required for overcurrent to sever the current-responsive element. 103, 202

membrane potential. The potential difference, of whatever origin, between the two sides of a membrane. *See also:* **electrobiology.** 192

memory (electronic computation). *See:* **storage; storage medium.**

memory action (relay). A method of retaining an effect of an input after the input ceases or is greatly reduced, so that this input can still be used in producing the typical response of the relay. *Note:* For example, memory action in a high-speed directional relay permits correct response for a brief period after the source of voltage input necessary to such response is short-circuited. 103, 202, 60, 127

memory capacity (electronic computation). *See:* **storage capacity.**

memory cycle (test, measurement and diagnostic equipment). The time required to read information from memory and replace it. 54

memory relay. (1) A relay having two or more coils, each of which may operate independent sets of contacts, and another set of contacts that remain in a position determined by the coil last energized. (2) Sometimes erroneously used for polarized relay. *See also:* **relay.** 259

mercury-arc converter, pool-cathode. *See:* **pool-cathode mercury-arc converter.** *See also:* **oscillatory circuit.**

mercury-arc rectifier. A gas-filled rectifier tube in which the gas is mercury vapor. *See also:* **gas-filled rectifier; rectification.** 244, 190

mercury cells. Electrolytic cells having mercury cathodes with which deposited metals form amalgams. 328

mercury-contact relays. **(1) mercury plunger relay:** A relay in which the magnetic attraction of a floating plunger by a field surrounding a sealed capsule displaces mercury in a pool to effect contacting between fixed electrodes. **(2) mercury-wetted-contact relay:** A form of reed relay in which the reeds and contacts are glass enclosed and are wetted by a film of mercury obtained by capillary action from a mercury pool in the base of a glass capsule vertically mounted. **(3) mercury-contact relay:** A relay mechanism in which mercury establishes contact between electrodes in a sealed capsule as a result of the capsule's being tilted by an electromagnetically actuated armature, either on pick-up or dropout or both. *See also:* **mercury relay.** 341

mercury-hydrogen spark-gap converter (dielectric heater usage). A spark-gap generator or power source which utilizes the oscillatory discharge of a capacitor through an inductor and a spark gap as a source of radio-frequency power. The spark gap comprises a solid electrode and a pool of mercury in a hydrogen atmosphere. *See also:* **in-**

duction heating. 14, 114

mercury-lamp ballast. *See:* **ballast.**

mercury-lamp transformer (series type). *See:* **constant-current (series) mercury-lamp transformer.**

mercury motor meter. A motor-type meter in which a portion of the rotor is immersed in mercury, which serves to direct the current through conducting portions of the rotor. *See also:* **electricity meter.** 341

mercury oxide cell. A primary cell in which depolarization is accomplished by oxide of mercury. *See also:* **electrochemistry.** 341

mercury-pool cathode (gas tube). A pool cathode consisting of mercury. *See also:* **gas-filled rectifier.** 244, 190

mercury relay. A relay in which the movement of mercury opens and closes contacts. *See also:* **mercury-contact relay.** 259

mercury vapor lamp transformers (multiple-supply type). Transformers, autotransformers or reactors for operating mercury or metallic iodide vapor lamps for all types of lighting applications, including indoor, outdoor area, roadway, uviarc, and other process and specialized lighting. 53

mercury-vapor tube. A gas tube in which the active gas is mercury vapor. 190, 125

merge (computing systems). To combine two or more sets of items into one, usually in a specified sequence. 255, 77

mesh. A set of branches forming a closed path in a network, provided that if any one branch is omitted from the set, the remaining branches of the set do not form a closed path. *Note:* The term *loop* is sometimes used in the sense of mesh. *See also:* **network analysis.** 210

mesh-connected circuit. A polyphase circuit in which all the current paths of the circuit extend directly from the terminal of entry of one phase conductor to the terminal of entry of another phase conductor, without any intermediate interconnections among such paths and without any connection to the neutral conductor, if one exists. *Note:* In a three-phase system this is called the delta (or Δ) connection. *See also:* **network analysis.** 210

mesh current. A current assumed to exist over all cross sections of a given closed path in a network. *Note:* A mesh current may be the total current in a branch included in the path, or it may be a partial current such that when combined with others the total current is obtained. *See also:* **network analysis.** 210

mesh or loop equations. Any set of equations (of minimum number) such that the independent mesh or loop currents of a specified network may be determined from the impressed voltages. *Notes:* (1) For a given network, different sets of equations, equivalent to one another, may be obtained by different choices of mesh or loop currents. (2) The equations may be differential equations, or algebraic equations when impedances and phasor equivalents of steady-state single-frequency sine-wave quantities are used. *See also:* **network analysis.** 210

mesial point (pulse terms). A magnitude referenced point at the intersection of a waveform and a mesial line. *See:* The single pulse diagram below the **waveform epoch** entry.
254

mesopic vision. Vision with luminance conditions between those of photopic and scotopic vision, that is, between about 1.0 footlambiert (3.0 nits) and 0.01 footlambert (0.03 nit). *See also:* **visual field.** 167

message (1) (general). An arbitrary amount of information whose beginning and end are defined or implied. 255, 77, 194

(2) (telephone switching systems). An answered call or the information content thereof. 55

message source. That part of a communication system where messages are assumed to originate. *See also:* **information theory.** 160

message switching (data communications). A method of handling messages over communications networks. The entire message is transmitted to an intermediate point (that is, a switching computer), stored for a period of time, perhaps very short, and then transmitted again towards its destination. The destination of each message is indicated by an address integral to the message. *See:* **circuit switching.** 12

message telecommunication network (telephone switching systems). An arrangement of switching and transmission facilities to provide telecommunication services to the public. 55

message-timed release (telephone switching systems). Release effected automatically after a measured interval of communication. 55

message unit (telephone switching systems). A basic chargeable unit based on the duration and destination of a call. 55

message-unit call (telephone switching systems). A call for which billing is in terms of accumulated message units. 55

metal clad. The conducting parts are entirely enclosed in a metal casing. 328

metal-clad switchgear. Metal-enclosed power switchgear characterized by the following necessary features. (1) The main switching and interrupting device is of the removable type arranged with a mechanism for moving it physically between connected and disconnected positions and equipped with self-aligning and self-coupling primary and secondary disconnecting devices. (2) Major parts of the primary circuit, that is, the circuit switching ... power transformers are completely enclosed by grounded metal barriers, which have no intentional openings between compartments. Specifically included is a metal barrier in front of or a part of the circuit interrupting device to insure that when, in the connected position, no primary current components are exposed by the opening of a door. (3) All live parts are completely enclosed within grounded metal compartments. *Note:* Automatic shutters prevent exposures of primary-circuit elements when the removable element is in the disconnected, test, or removed position. (4) Primary bus conductors and connections are covered with insulating material throughout. (5) Mechanical interlocks are provided to insure a proper and safe operating sequence. (6) Instruments, meters, relays, secondary control devices, and their wiring are isolated by grounded metal barriers from all primary-circuit elements with the exception of short lengths of wire such as at instrument-transformer terminals. (7) The door through which the circuit-interrupting device is inserted into the housing may serve as an instrument or relay panel and may also provide access to a secondary or control compartment within the housing. *Note:* Auxiliary vertical sections may be required for mounting devices or for use as bus transition. 79

metal distribution ratio (electroplating). The ratio of the thicknesses (weights per unit areas) of metal upon two specified parts of a cathode. *See also:* **electroplating.** 328

metal enclosed (switchgear assembly or components) (metal enclosed bus). Surrounded by a metal case or housing, usually grounded or with provisions for grounding. 103, 202, 78

metal-enclosed bus (1) (power switchgear). An assembly of rigid conductors with associated connections, joints, and insulating supports within a grounded metal enclosure. *Note:* In general, three baisc types of construction are used: nonsegregated phase, segregated phase, and isolated phase. 103, 202, 79

(2) (isolated phase bus). An assembly of conductors with associated connections, joints, and insulating supports within a grounded metal enclosure. The conductors may be either rigid or flexible. 78

metal-enclosed interrupter switchgear. Metal-enclosed power switchgear including the following equipment as required: (1) interrupter switches; (2) power fuses; (3) bare bus and connections; (4) instrument transformers; and (5) control wiring and accessory devices. *Note:* The interrupter switches and power fuses may be stationary or removable type. When removable type, mechanical interlocks are provided to insure a proper and safe operating sequence. 103, 202

metal-enclosed low-voltage power circuit-breaker switchgear. Metal-enclosed power switchgear of multiple or individual enclosure including the following equipment as required: (1) low-voltage power circuit breaker (fused or unfused); (2) bare bus and connections; (3) instrument and control power transformers; (4) instruments, meters, and relays; and (5) control wiring and accessory devices. *Note:* The low-voltage power circuit breakers are contained in individual grounded metal compartments and controlled either remotely or from the front of the panels. The circuit breakers may be stationary or removable type. When removable type, mechanical interlocks are provided to insure a proper and safe operating sequence. 103, 202, 79

metal-enclosed power switchgear. A switchgear assembly completely enclosed on all sides and top with sheet metal (except for ventilating openings and inspection windows) containing primary power circuit switching or interrupting devices, or both, with buses and connections and may include control and auxiliary devices. *Notes:* (1) Access to the interior of the enclosure is provided by doors or removable covers. (2) Metal-clad switchgear, station-type cubicle switchgear, metal-enclosed interrupter switchgear, and low-voltage power circuit-breaker switchgear are specific types of metal-enclosed power switchgear. 103, 202, 79, 93

metal fog (metal mist) (electrolysis). A fine dispersion of metal in a fused electrolyte. *See also:* **fused electrolyte.** 328

metal-graphite brush (rotating machinery). A brush composed of varying percentages of metal and graphite, copper or silver being the metal generally used. *Note:* This type of brush is soft. Grades of brushes of this type have extremely high current-carrying capacities, but differ greatly in operating speed from low to high. *See also:* **brush (rotating machinery).**   63, 279

metallic circuit (data transmission). A circuit of which the ground or earth forms no part. *See also:* **transmission line.**   59, 39

metallic impedance (telephone equipment). Impedance presented by a metallic circuit at any given single frequency, at or across the terminals of one of its transmission ports.   39

metallic longitudinal induction ratio. (M-L ratio). The ratio of the metallic-circuit current or noise-metallic arising in an exposed section of open wire telephone line, to the longitudinal-circuit current or noise-longitudinal in sigma. It is expressed in microamperes per milliampere or the equivalent. *See:* **noise-metallic.** *See also:* **inductive coordination.**   328

metallic rectifier (electric installations on shipboard). A metallic rectifier cell is a device consisting of a conductor and semiconductor forming a junction. The junction exhibits a difference in resistance to current flow in the two directions through the junction. This results in effective current flow in one direction only. A metallic rectifier stack is a single columnar structure of one or more metallic rectifier cells.   3

metallic rectifier cell. A device consisting of a conductor and a semiconductor forming a junction. *Notes:* (1) Synonymous with **metallic rectifying cells.** (2) Such cells conduct current in each direction but provide a rectifying action because of the large difference in resistance to current flow in the two directions. (3) A metallic rectifier stack is a single columnar structure of one or more metallic rectifier cells. *See also:* **rectification.**   3

metallic rectifier stack assembly. The combination of one or more stacks consisting of all the rectifying elements used in one rectifying circuit. *See also:* **rectification.**   328

metallic rectifier unit. An operative assembly of a metallic rectifier, or rectifiers, together with the rectifier auxiliaries, the rectifier transformers, and the essential switchgear. *See also:* **rectification.**   328

metallic transmission port (telephone equipment). A place of access in the metallic transmission path of a device or network where energy may be supplied or withdrawn, or where the device or network variables may be measured. The terminals of such a port are sometimes referred to as the tip and ring terminals. *Note:* In any particular case, the transmission ports are determined by the way the device is used, and not by its structure alone. *See also:* **device; network.**   39

metallic voltage (telephone equipment). The voltage across a metallic circuit. *See also:* **metallic circuit.**   39

metallized brush. *See:* **metal-graphite brush.**

metallized paper capacitor. A capacitor in which the dielectric is primarily paper and the electrodes are thin metallic coatings deposited thereon.   341

metallized screen (cathode-ray tube). A screen covered on its rear side (with respect to the electron gun) with metallic film, usually aluminized, transparent to electrons and with a high optical reflection factor, which passes on to the viewer a large part of the light emitted by the screen on the electron-gun side. *See also:* **cathode-ray tubes.**   244, 190

metal master (metal negative) (no. 1 master) (disk recording) (electroacoustics). *See:* **original master.**

metal mist (electrolysis). *See:* **metal fog.**

metal negative (metal master) (no. 1 master) (disk recording) (electroacoustics). *See:* **original master.**

metamers. Lights of the same color but of different spectral energy distribution. *See also:* **color.**   167

meter (1) (metric practice). The length equal to 1,650,763.73 wavelengths in vacuum of the radiation corresponding to the transition between the levels $2p_{10}$ and $5d_5$ of the krypton-86 atom.   21

(2) (laser-maser) (m). A unit of length in the international systems of units; currently defined as a fixed number of wavelengths, in vacuum, of the orange-red line of the spectrum of krypton 86. Typically, the meter is sub-divided into the following units:

$$\text{Centimeter} = 10^{-2}\text{m(cm)}$$
$$\text{Millimeter} = 10^{-3}\text{m(mm)}$$
$$\text{Micrometer} = 10^{-6}\text{m}(\mu\text{m})$$
$$\text{Nanometer} = 10^{-9}\text{m(nm)}$$

363

meter. *See:* **electricity meter.**

meter laboratory. A place where working standards are maintained for use in routine testing of electricity meters and auxiliary devices. *Note:* A meter laboratory may, but need not, assume some or all the functions of a standards laboratory. *See also:* **measurement system.**   212

meter relay. Sometimes used for instrument relay. *See also:* **relay.**   259

meter shop. A place where meters are inspected, tested, and repaired, and may be at a fixed location or may be mobile. *See also:* **measurement system.**   212

method of pulse measurement. A method of making a pulse measurement comprises: the complete specification of the functional characteristics of the devices, apparatus, instruments, and auxiliary equipment to be used; the essential adjustments required; the procedure to be used in making essential adjustments; the operations to be performed and their sequence; the corrections that will ordinarily need to be made; the procedures for making such corrections; the conditions under which all operations are to be carried out. *See also:* **pulse measurement.**   15

metrology (test, measurement and diagnostic equipment). The science of measurement for determination of conformance to technical requirements including the development of standards and systems for absolute and relative measurements.   54

MEW. *See:* **microwave early warning.**

mho (siemens). The unit of conductance (and of admittance) in the International System of Units (SI). The mho is the conductance of a conductor such that a constant voltage of 1 volt between its ends produces a current of 1 ampere in it.   210

mho relay. A distance relay for which the inherent operating characteristic on an *R-X* diagram is a circle that passes through the origin. *Note:* The operating characteristic may be described by the equation $Z = K\cos(\theta - \alpha)$ where K and α are constants and θ is the phase angle by which the input voltage leads the input current.   103, 202, 60, 127

MIC (electromagnetic compatibility). *See:* **mutual interference chart.**

mica flake (rotating machinery). Mica lamina in thickness not over approximately 0.0028 centimeter having a surface area parallel to the cleavage plane under 1.0 centimeter square. *See:* **rotor (rotating machinery); stator.** 63

mica folium (rotating machinery). A relatively thin flexible bonded sheet material composed of overlapping mica splittings with or without backing or facing. *See:* **rotor (rotating machinery); stator.** 63

mica paper (integrated mica) (reconstituted mica) (rotating machinery). Mica flakes having an area under approximately 0.200 centimeter square combined laminarly into a substantial sheet-like configuration with or without binder, backing, or facing. *See:* **rotor (rotating machinery); stator.** 63

mica sheet (rotating machinery). A composite of overlapping mica splittings bonded into a planar structure with or without backing or facing. *See:* **rotor (rotating machinery); stator.** 63

mica splitting (rotating machinery). Mica lamina in thickness approximately 0.0015 centimeter to 0.0028 centimeter having a surface area parallel to the basal cleavage plane of at least 1.0 centimeter square. *See:* **rotor (rotating machinery); stator.** 63

mica tape (rotating machinery). A composite tape composed of overlapping mica splittings bonded together with or without backing or facing. *See:* **rotor (rotating machinery); stator.** 63

MICR. *See:* **magnetic ink character recognition.**

microbar. A unit of pressure formerly in common usage in acoustics. One microbar is equal to 1 dyne per square centimeter and equals 0.1 newton per square meter. The newton per square meter is now the preferred unit. *Note:* The term bar properly denotes a pressure of 10^6 dynes per square centimeter. Unfortunately the bar was once used in acoustics to mean 1 dyne per square centimeter, but this is no longer correct. 176

microchannel plate (electron image tube). An array of small aligned channel multipliers usually used for intensification. *See also:* **amplifier; camera tube.** 190

microelectronic device (electric and electronics parts and equipments). An item of inseparable parts and hybrid circuits, usually produced by integrated circuit techniques. Typical examples are microcircuit, integrated circuit package, micromodule. 17

micrometer (μm) (laser-maser). A unit of length equal to 10^{-6} meter. In common parlance a micrometer is a micron. 363

micron (metric system). The millionth part of a meter. *Note:* According to the set of submultiple prefixes now established in the International System of Units, the preferred term would be micrometer. However, use of the same word to denote a small length, and also to denote an instrument for measuring a small length, could occasionally invite confusion. Therefore it seems unwise to deprecate, at this time, the continued use of the word micron.

microphone. An electroacoustic transducer that responds to sound waves and delivers essentially equivalent electric waves. 176

microphonics (1) (general). The noise caused by mechanical shock or vibration of elements in a system. 239

(2) (interference terminology). Electrical interference caused by mechanical vibration of elements in a signal transmission system. *See:* **signal.** 188

(3) (microphonic effect) (microphonism) (electron device) (electron tubes). The undesired modulation of one or more of the electrode currents resulting from the mechanical vibration of one or more of the valve or tube elements. *See also:* **electron tube.** 244, 190, 125

microphonism. *See:* **microphonics (electron tubes).**

microradiometer (radio-micrometer*). A thermosensitive detector of radiant power in which a thermopile is supported on and connected directly to the moving coil of a galvanometer. *Note:* This construction minimizes lead losses and stray electric pickup. *See also:* **electric thermometer (temperature meter).** 328

* Deprecated

microstrip. *See:* **strip-type transmission line; (waveguides).**

microwave early warning (MEW). A particular (World War 2) United States high-power, long-distance radar with a number of separate displays giving high resolution and large traffic-handling capacity in detecting and tracking targets. *See also:* **navigation.** 187, 13

microwave landing system (MLS). An airfield approach radar generating a guideline for target landing. 13

microwave pilot protection. A form of pilot protection in which the communication means between relays is a beamed microwave-radio channel. 103, 202, 60, 127

microwaves. A term used rather loosely to signify radio waves in the frequency range from about 1000 megahertz upwards. *See also:* **radio transmission.** 59

microwave therapy. The therapeutic use of electromagnetic energy to generate heat within the body, the frequency being greater than 100 megahertz. *See also:* **electrotherapy.** 192

microwave tube (or valve). An electron tube or valve whose size is comparable to the wavelength of operation and that has been designed in conjunction with its circuits to provide effective electron interaction. *Note:* In many tubes the circuit is self-contained within the vacuum envelope so that the electrode terminals do not need to conduct radio-frequency signals.

middle marker. A marker facility in an ILS which is installed approximately 1000 meters (3500 feet) from the approach end of the runway on the localizer course line to provide a fix. 13

MIG. *See:* **magnetic injection gun.**

mile of standard cable** (MSC). Two units, both loosely designated as a mile of standard cable, were formerly used as measures of transmission efficiency. One, correctly known as an 800-hertz mile of standard cable, signified an attenuation constant, independent of frequency, of 0.109. The other signified the effect upon speech volume of an actual mile of standard cable, equivalent to an attenuation constant of approximately 0.122. Both units are now obsolete, having been replaced by the decibel. One 800-hertz mile of standard cable is equal to approximately 0.95 decibel. One standard cable mile is equivalent in effect on speech volume to approximately 1.06 decibels. 328

** Obsolete

mill scale (corrosion). The heavy oxide layer formed during hot fabrication or heat treatment of metals. Especially applied to iron and steel.   221, 205

Mills cross antenna system. A multiplicative array antenna system consisting of two linear receiving arrays positioned at right angles to one another and connected together by a phase modulator or switch such that the effective angular response of the output is related to the product of the radiation patterns of the two arrays.   111

mimic bus. A single-line diagram of the main connections of a system constructed on the face of a switchgear or control panel, or assembly.   103, 202

mine-fan signal system. A system that indicates by electric light or electric audible signal, or both, the slowing down or stopping of a mine ventilating fan. *See also:* **dispatching system.**   328

mine feeder circuit. A conductor or group of conductors, including feeder and sectionalizing switches or circuit breakers, installed in mine entries or gangways and extending to the limits set for permanent mine wiring beyond which limits portable cables are used.   328

mine hoist. A device for raising or lowering ore, rock, or coal from a mine and for lowering and raising men and supplies.

mine jeep. a special electrically driven car for underground transportation of officials, inspectors, repair, maintenance, surveying crews, and rescue workers.   328

mine radio telephone system. A means to provide communication between the dispatcher and the operators on the locomotives where the radio impulses pass along the trolley wire and down the trolley pole to the radio telephone set. *See also:* **dispatching system.**   328

miner's electric cap lamp. A lamp for mounting on the miner's cap and receiving electric energy through a cord that connects the lamp with a small battery.   328

miner's hand lamp. A self-contained mine lamp with handle for convenience in carrying.   328

mine tractor. A trackless, self-propelled vehicle used to transport equipment and supplies and for general service work.   328

mine ventilating fan. A motor-driven disk, propeller, or wheel for blowing (or exhausting) air to provide ventilation of a mine.   328

miniature brush (electric machines). A brush having a cross-sectional area of less than $1/64$ square inch with the thickness and width thereof less than $1/8$ inch or, in the case of a cylindrical brush, a diameter less than $1/8$ inch. *See also:* **brush.**   279

minimal perceptible erythema. The erythemal threshold. *See also:* **ultraviolet radiation.**   167

minimum access code (test, measurement and diagnostic equipment). A system of coding which minimizes the effect of delays for transfer of data or instructions between storage and other machine units.   54

minimum clearance between poles (phases) (switchgear). The shortest distance between any live parts of adjacent poles (phases). *Note:* Cautionary differentiation should be made between clearance and spacing or center-to-center distance.   103, 202, 27

minimum clearance to ground (switchgear). The shortest distance between any live part and adjacent grounded parts.   103, 202, 27

minimum conductance function (linear passive networks). *See:* **minimum resistance (conductance) function.** *See also:* **linear passive networks.**

minimum detectable signal (MDS) (radar). The minimum signal level which gives reliable detection in the presence of white gaussian noise. Being a statistical quantity, it must be described in terms of a probability of detection and a probability of false alarm.   13

minimum discernible signal (radar). The minimum detectable signal for a system using an operator and display or aural device for detection.   13

minimum-distance code (computing systems). A binary code in which the signal distance does not fall below a specified minimum value.   255, 77

minimum-driving-point function (linear passive networks). A driving-point function that is a minimum-resistance, minimum-conductance, minimum-reactance, and minimum-susceptance function. *See also:* **linear passive networks.**   238

minimum en-route altitude (MEA) (electronic navigation). The lowest altitude between radio fixes that assures acceptable navigational signal coverage and meets obstruction clearance requirements for instrument flight. *See also:* **navigation.**   187, 13

minimum firing power (microwave switching tubes). The minimum radio-frequency power required to initiate a radio-frequency discharge in the tube at a specified ignitor current. *See:* **gas tube.**   190, 125

minimum flashover voltage (impulse). The crest value of the lowest voltage impulse, of a given wave shape and polarity that causes flashover.   64, 62

minimum fuel limiter (gas turbines). A device by means of which the speed-governing system can be prevented from reducing the fuel flow below the minimum for which the device is set as required to prevent unstable combustion or blowout of the flame.   98, 58

minimum illumination (sensitivity) (industrial control). The minimum level, in footcandles of a photoelectric lighting control, at which it will operate. *See also:* **photoelectric control.**   206

minimum impulse flashover voltage (neutral grounding devices). The crest value of the lowest voltage impulse at a given wave shape and polarity that causes flashover.   91

minimum melting current (fuse) (power switchgear). The smallest current at which a current-responsive fuse element will melt.   103

minimum ON-state voltage (thyristor). The minimum positive principal voltage for which the differential resistance is zero with the gate open-circuited. *See also:* **principal voltage-current characteristic (principal characteristic).**   243, 66, 208, 191

minimum ON voltage (magnetic amplifier). The minimum output voltage existing before the trip OFF control signal is reached as the control signal is varied from trip ON to trip OFF.   171

minimum output voltage (magnetic amplifier). The minimum voltage attained across the rated load impedance as the control ampere-turns are varied between the limits established by (1) positive maximum control currents flowing through all the corresponding control windings simultaneously and (2) negative maximum control currents flowing through all the corresponding control windings simultaneously.   171

minimum-phase function (linear passive networks). A transmittance from which a nontrivial realizable allpass function cannot be factored without leaving a nonrealizable remainder. *Note:* For lumped-parameter networks, this is equivalent to specifying that the function has no zeros in the interior of the right half of the complex-frequency plane. *See also:* **linear passive networks.**   238

minimum-phase network (1). A network for which the phase shift at each frequency equals the minimum value that is determined uniquely by the attenuation-frequency characteristic in accordance with the following equation

$$B_c = \frac{1}{\pi} \int_{-\infty}^{+\infty} \frac{dA}{du} \log \coth \frac{|u|}{2} \, du$$

where B_c is phase shift in radians at a particular frequency f_c, A is attenuation in nepers as a function of frequency f, and u is $\log (f/f_c)$.   59

(2) (excitation control systems). See definition above. *Notes:* (A) A network for which the transfer function expressed as a function of s has neither poles nor zeros in the right-hand s plane. Networks (elements) or systems having either poles or zeros in the right half s-plane do not have minimum phase characteristics, assuming there is no right-hand S-plane pole-zero cancellation. (B) Elements whose response is described by transfer functions having transport lags also exhibit non-minimum phase characteristics. The frequency response characteristics of a typical element with transport lag is given in figure below.   353

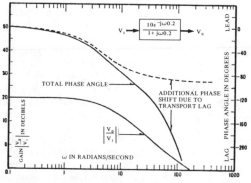

Element with one time constant and a transport lag.

(3) (two specified terminals or two branches). A network for which the transfer admittance expressed as a function of p (*See:* **impedance function**) has neither poles nor zeros in the right-hand p plane. *Note:* A simple T section of real lumped constant parameters without coupling between branches is a minimum-phase network, whereas a bridged-T or lattice section of all-pass type may not be. *See also:* **network analysis; impedance function.**   210

minimum-reactance function (linear passive networks). A driving-point impedance from which a reactance function cannot be subtracted without leaving a nonrealizable remainder. *Notes:* (1) For lumped-parameter networks, this is equivalent to specifying that the impedance function has no poles on the imaginary axis of the complex-frequency plane, including the point at infinity. (2) A driving-point impedance (admittance) having neither poles nor zeros on the imaginary axis is both a minimum-reactance and a minimum-susceptance function. *See also:* **linear passive networks.**   238

minimum reception altitude (MRA) (electronic navigation). The lowest en-route altitude at which adequate signals can be received to determine specific radio-navigation fixes. *See also:* **navigation.**   187, 13

minimum-resistance (conductance) function (linear passive networks). A driving-point impedance (admittance) from which a positive constant cannot be subtracted without leaving a nonrealizable remainder. *See also:* **linear passive networks.**   238

minimum speed (adjustable-speed drive) (industrial control). The lowest speed within the operating speed range of the drive. *See also:* **electric drive.**   206

minimum-susceptance function (linear passive networks). A driving-point admittance from which a susceptance function cannot be subtracted without leaving a nonrealizable remainder. *Notes:* (1) For lumped-parameter networks, this is equivalent to specifying that the admittance function has no poles on the imaginary axis of the complex-frequency plane, including the point at infinity. (2) A driving-point immitance having neither poles nor zeros on the imaginary axis is both a minimum-susceptance and a minimum-reactance function. *See also:* **linear passive networks.**   238

minimum test output voltage (nonreversible output) (magnetic amplifier). The output voltage equivalent to the summation of the minimum output voltage plus 33⅓ percent of the difference between the rated and minimum output voltages.   171

minitrack (communication satellite). A ground based tracking system for satellites using interferometers. It requires a minimum satellite instrumentation, hence the name.   83

minor alarm, (telephone switching systems). An alarm indicating trouble which does not seriously impair the system capability.   55

minor cycle (electronic computation). In a storage device that provides serial access to storage positions, the time interval between the appearance of corresponding parts of successive words.   235, 210

minor failure. *See:* **failure, minor.**

minority carrier (semiconductor). The type of charge carrier constituting less than one half the total charge-carrier concentration. *See also:* **semiconductor.**   245, 66, 210

minor lobe (antenna pattern). Any lobe except a major lobe. *See also:* **antenna.**   246, 179, 111

minor loop (industrial control). A continuous network consisting of both forward elements and feedback elements and is only a portion of the feedback control system. *See:* **control system, feedback.**   219, 206

minor railway tracks. Railway tracks included in the following list. (1) Spurs less than 2000 feet long and not exceeding two tracks in the same span. (2) Branches on which no regular service is maintained or which are not operated during the winter season. (3) Narrow-gauge tracks or other tracks on which standard rolling stock cannot, for physical reasons, be operated. (4) Tracks used only temporarily for a period not exceeding 1 year. (5) Tracks not operated as a common carrier, such as industrial railways used in logging, mining, etcetera.   262

minus input. *See:* **inverted input.**

misalignment drift (gyros). The part of the total apparent drift component due to uncertainty of orientation of the gyro input axis with respect to the coordinate system in which the gyro is being used. *See also:* **navigation.**
187, 13

miscellaneous function (numerically controlled machines). An on-off function of a machine such as spindle stop, coolant on, clamp.   224, 207

misfire (1) (gas tube). A failure to establish an arc between the main anode and cathode during a scheduled conducting period. *See also:* **gas tubes; rectification.**
190, 190

(2) (mining). The failure of a blasting charge to explode when expected. *Note:* In electric firing, this usually is the result of a broken blasting circuit or insufficient current through the electric blasting cap. *See also:* **blasting unit.**
328

mismatch. The condition in which the impedance of a load does not match the impedance of the source to which it is connected. *See also:* **self-impedance.**   179

mismatch factors (power meters). Resulting from a combination of interaction factor and reflection factor resulting from reflective source and load impedances which relate incident, absorbed, and delivered power to a nonreflection load.   47

mismatch loss. *See:* **matching loss.**

mismatch uncertainty (power meters). Uncertainty in an assigned value that is caused by uncorrected or uncertain values for one or both of the mismatch factors.   115, 47

missing (misgating) (thyristor converter). A condition where the onset of conduction of an arm is substantially delayed from its correct instant of time. *Note:* If an arm fails to turn on during inverter service, there is a commutation failure resulting in a conduction-through.   121

mission (1) (systems, man, and cybernetics). The operating objective for which the system was intended. *See also:* **system.**   209

(2) (nuclear power generating stations). The singular objective, task, or purpose of an item or system.   31, 29

mission time (nuclear power generating stations). The time during which the mission must be performed without interruption.   29, 31

mistake (1) (electronic computation). A human action that produces an unintended result. *Note:* Common mistakes include incorrect programming, coding, manual operation, etcetera.   255, 77, 235

(2) (analog computer). *See:* **error.**   9

mixed highs (color television). Those high-frequency components of the picture signal that are intended to be reproduced achromatically in a color picture.   18, 178

mixed logic (logic diagram). The defining of the 1-state of the variables as the more positive or less positive of the two possible levels, depending upon the absence or presence

of the polarity indicator symbol. *Syn:* **direct polarity indication.**   88

mixed-loop series street-lighting system. A street-lighting system that comprises both open loops and closed loops. *See also:* **alternate-current distribution; direct-current distribution.**   64

mixed radix. Pertaining to a numeration system that uses more than one radix, such as the biquinary system.
255, 77

mixed sweep (oscilloscopes). In a system having both a delaying sweep and a delayed sweep, a means of displaying the delaying sweep to the point of delay pickoff and displaying the delayed sweep beyond that point. *See:* **oscillograph.**   185

mixer (1) (general). A circuit having two or more input signals and an output signal that is some desired function of the input.   59

(2) (sound transmission, recording, or reproducing system.) A device having two or more inputs, usually adjustable, and a common output, which operates to combine linearly in a desired proportion the separate input signals to produce an output signal. *Note:* The stage in a heterodyne receiver in which the incoming signal is modulated with the signal from the local oscillator to produce the intermediate-frequency signal.   59

mixer tube. An electron tube that performs only the frequency-conversion function of a heterodyne conversion transducer when it is supplied with voltage or power from an external oscillator.   190, 125

MKSA system of units. A system in which the basic units are the meter, kilogram, and second, and the ampere is a derived unit defined by assigning the magnitude $4\pi \times 10^{-7}$ to the rationalized magnetic constant (sometimes called the permeability of space). *Notes:* (1) At its meeting in 1950 the International Electrotechnical Commission recommended that the MKSA system be used only in the rationalized form. (2) The electrical units of this system were formerly called the *practical* electrical units. (3) If the MKSA system is used in the unrationalized form the magnetic constant is 10^{-7} henry/meter and the electric constant is $10^7/c^2$ farads/meter. Here c, the speed of light, is approximately 3×10^8 meters/second. (4) In this system, dimensional analysis is customarily used with the four independent (basic) dimensions: mass, length, time, current.   210

M-L ratio. *See:* **metallic-longitudinal induction ratio.**

mobile communication system. Combinations of interrelated devices capable of transmitting intelligence between two or more spatially separated radio stations, one or more of which shall be mobile.   181

MLS. See **microwave landing system.**

mnemonic (test, measurement and diagnostic equipment). Assisting or intending to assist a human memory and understanding. Thus a mnemonic term is usually an abbreviation, that is easy to remember; for example, mpy for multiply and acc for accumulator.   54

mnemonic code (test, measurement and diagnostic equipment). A pseudo code in which information, usually instructions, is represented by symbols or characters which are readily identified with the information.   54

mobile radio service. Radio service between a radio station at a fixed location and one or more mobile stations, or between mobile stations. *See also:* **radio transmission.**

mobile station (communication). A radio station designed for installation in a vehicle and normally operated when in motion. *See also:* **mobile communication system.**
  181

mobile telemetering. Electric telemetering between points that may have relative motion, where the use of interconnecting wires is precluded. *Note:* Space radio is usually employed as an intermediate means for mobile telemetering, but radio may also be used for telemetering between fixed points. *See also:* **telemetering.** 328

mobile telephone system (automatic channel access). A mobile telephone system capable of operation on a plurality of frequency channels with automatic selection at either the base station or any mobile station of an idle channel when communication is desired. *See also:* **mobile communication system.** 181

mobile transmitter. A radio transmitter designed for installation in a vessel, vehicle, or aircraft, and normally operated while in motion. *See also:* **radio transmitter.**
  111

mobile unit substation. A unit substation mounted and readily movable as a unit on a transportable device.
  103, 202

mobility (semiconductor). *See:* **drift mobility.**

mobility, Hall (electric conductor). *See:* **Hall mobility.**

modal analysis (power-system communication). A method of computing the propagation of a wave on a multiconductor power line. 59

modal distance (telephony). The distance between the center of the grid of a telephone-handset transmitter cap and the center of the lips of a human talker (or the reference point of an artificial mouth), when the handset is in the modal position. 122

modal position (telephony). The position a telephone handset assumes when the receiver of the handset is held in close contact with the ear of a person with head dimensions that are modal for a population. For this standard, the modal position is defined, by the modal head adopted by the CCITT (*Comité Consultatif International Télégraphique et Téléphonique*) Laboratory for the measurement of AEN. The point of reference for the handset and the head is the center of the circular plane of contact of the handset earcap and the ear. If the handset earcap is not circular or has no external plane of contact, an effective center and an effective plane of contact must be determined. The modal point is the position of the center of the lips with respect to the center and plane of the earcap point of reference. 122

mode. A state of a vibrating system to which corresponds one of the possible resonance frequencies (or propagation constants). *Note:* Not all dissipative systems have modes. *See:* **modes, degenerate; oscillatory circuit.** 190, 125

mode coupler (waveguides). A coupler that provides preferential coupling to a specific wave mode. *See also:* **waveguide.** 185

mode filter (waveguide components). A device designed to pass energy along a waveguide in one or more selected modes of propagation, and substantially to reject energy carried in other modes. 166

mode, higher-order (waveguide or transmission line). Any mode of propagation characterized by a field configuration

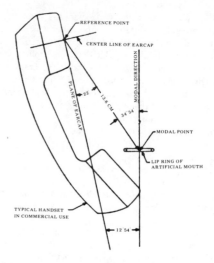

REFERENCE POINT
CENTER LINE OF EARCAP
MODAL DIRECTION
PLANE OF EARCAP
13.6 CM
34 54
22
MODAL POINT
LIP RING OF ARTIFICIAL MOUTH
TYPICAL HANDSET IN COMMERCIAL USE
12 54

other than that of the fundamental or first-order mode with lowest cutoff frequency. 185

model. A mathematical or physical representation of the system relationships. *See also:* **mathematical model; system.** 209

modeling. Technique of system analysis and design using mathematical or physical idealizations of all or a portion of the system. Completeness and reality of the model are dependent on the questions to be answered, the state of knowledge of the system, and its environment. *See also:* **system.** 209

modem. A contraction of modulator-demodulator, an equipment that connects data terminal equipment to a communication line. 194

mode of operation (rectifier circuit). The characteristic pattern of operation determined by the sequence and duration of commutation and conduction. *Note:* Most thyristor converters and rectifier circuits have several modes of operation, which may be identified by the shape of the current wave. The particular mode obtained at a given load depends upon the circuit constants. *See also:* **rectification; rectifier circuit element.** 237, 66, 121

mode of propagation (waveguides). A form of propagation of guided waves that is characterized by a particular field pattern in a plane transverse to the direction of propagation, which field pattern is independent of position along the axis of the waveguide. *Note:* In the case of uniconductor waveguides the field pattern of a particular mode of propagation is also independent of frequency. *See also:* **waveguide.** 267, 319, 179

mode of resonance. A form of natural electromagnetic oscillation in a resonator characterized by a particular field pattern that is invariant with time. 267, 179

mode of vibration (vibratory body, such as a piezoelectric crystal unit). A pattern of motion of the individual particles due to (1) stresses applied to the body; (2) its properties; and (3) the boundary conditions. Three common modes of vibration are (A) flexural, (B) extensional, and (C) shear. *See also:* **crystal.** 328

mode, π. *See:* **pi (π) mode.**

mode purity (waveguides). The ratio of power present in the forward traveling wave of a desired mode to the total power present in the forward traveling waves of all modes.

319, 179, 125

modes, degenerate. A set of modes having the same resonance frequency (or propagation constant). *Note:* The members of a set of degenerate modes are not unique. *See:* **electrode dissipation; oscillatory circuits.**   190, 125

mode separation (oscillators). The frequency difference between resonator modes of oscillation. *See:* **oscillatory circuits.**   190, 125

mode switch (nuclear power generating stations). A switch used to set up protection systems for different modes of reactor operation (for example, startup, heatup, power operation, refueling). *Note:* A mode switch usually supplies signals to redundant protection channels to change set points or bypass the function of certain channels.   109, 31

mode transducer (waveguide components). A device for transforming an electromagnetic wave from one mode of propagation to another.   166
Note: Also termed **mode changer** or **mode transformer.** *See also:* **transducer; waveguide.**

244, 179, 267, 252, 210

mode transformer (transmission lines and waveguides). *See:* **mode transducer (waveguides).**

mode voltage. That voltage that occurs most frequently. It is of primary significance in evaluating certain critical factors affecting the design of equipment.   260

modified autocorrelation function (burst measurements). The time integral of a function $x(t)$ multiplied by itself shifted in time by τ

$$\Gamma_s(\tau) = \int_{-\infty}^{\infty} x(t) \cdot x(t + \tau)\,dt$$

See also: **burst.**   292

modified circuit transient recovery voltage. The circuit transient recovery voltage modified in accordance with the normal-frequency recovery voltage and the asymmetry of the current wave obtained on a particular interruption. *Note:* This voltage indicates the severity of the particular interruption with respect to recovery-voltage phenomena.   103, 202

modified constant-voltage charge (storage battery) (storage cell). A charge in which the bus voltage of the charging circuit is held substantially constant, but a fixed resistance is present in the battery circuit that results in an increase in voltage at the battery terminals as the charge progresses. *See also:* **charge.**   328

modified impedance relay. An impedance form of distance relay for which the operating characteristic of the distance unit on an R-X diagram is a circle having its center displaced from the origin. *Note:* It may be described by the equation

$$Z^2 - 2K_1 Z \cos(\theta - \alpha) = K_2^2 - K_1^2$$

where K_1, K_2, and α are constants and θ is the phase angle by which the input voltage leads the input current.

103, 202, 60, 127

modified index of refraction (modified refractive index) (troposphere). The index of refraction at any height increased by h/a, where h is the height above sea level and a is the mean geometrical radius of the earth. *Note:* When the index of refraction in the troposphere is horizontally stratified, propagation over a hypothetical flat earth through an atmosphere with the modified index of refraction is substantially equivalent to propagation over a curved earth through the real atmosphere. *See also:* **radiation.**   180, 146

modified Kraemer system. A system of speed control below synchronous speed for wound-rotor induction motors. Slip power is recovered through the medium of a converting device electrically connected between the secondary of the induction motor and a direct-current auxiliary motor. The auxiliary motor may be directly coupled to the induction-motor shaft or may form part of a motor-generator set that can return slip power to the supply system. *Note:* This system is not necessarily associated with constant power output over the speed range.   63

modified refractive index (troposphere). *See:* **modified index of refraction.**

modular (power supplies). The term modular is used to describe a type of power supply designed to be built into other equipment either chassis or rack mount. *Note:* It is usually distinguished from laboratory bench equipment by a large choice of mounting configurations and by a lack of meters and controls. *See also:* **power supply.**

228, 186

modulator amplifier. An amplifier stage in which the carrier is modulated by introduction of a modulating signal. *See also:* **amplifier; modulating systems; radio transmitter; signal.**   211, 188

modulated continuous wave (MCW). A wave in which the carrier is modulated by a constant audio-frequency tone. *Note:* In telegraphic service, it is understood that the carrier is keyed. *See also:* **modulation; wave front.**

111, 59

modulated photoelectric system (protective signaling). A photoelectric system in which the light beam is interrupted or modulated in a predetermined manner and in which the receiving equipment is designed to accept only the modulated light. *See also:* **protective signaling.**   328

modulated 12.5 T pulse (MOD 12.5 T) (waveform test signals) (TV). A burst of color subcarrier frequency of nominally 3.58 MHz. The envelope of the burst is $\sin^2$ shaped with a half-amplitude duration (HAD) of nominally 1.56 μs. The MOD 12.5 T pulse consists of a luminance and a chrominance component. The envelope of the frequency spectrum consists of two parts, namely signal energy concentrated in the luminance region below 0.6 MHz and

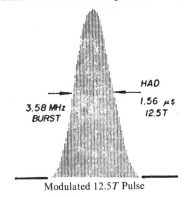

Modulated 12.5 T Pulse

Envelope of Frequency Spectrum of Modulated 12.5 T Pulse

in the chrominance region from roughly 3 MHz to 4.2 MHz. 42

modulated wave. A wave, some characteristic of which varies in accordance with the value of a modulating function. *See also:* **modulation.** 242

modulating function. A function, the value of which is intended to determine a variation of some characteristic of a carrier. 242

modulating signal (modulating wave). A wave which causes a variation of some characteristic of a carrier. *See also:* **modulation.** 111, 339

modulation (1) (carrier). (A) The process by which some characteristic of a carrier is varied in accordance with a modulating wave. (B) The variation of some characteristic of a carrier. 111
(2) (signal transmission system). (A) A process whereby certain characteristics of a wave, often called the carrier, are varied or selected in accordance with a modulating function; (B) the result of such a process.
242, 59, 56, 188, 219, 206, 194
(3) (automatic control). The process, or the result of the process, whereby some characteristic of one wave is varied in accordance with some characteristic of another wave. 56

modulation acceptance bandwidth (receiver performance). The selectivity characteristic of a receiver that limits the maximum permissible modulation deviation of the radio-frequency input signal that a receiver can accept, without degradation of the 12-decibel SINAD, when the radio-frequency input signal is 6 decibels greater than the reference sensitivity level. 123

modulation and noise spectrum (transmitter performance). The composite signal output from a transmitter as measured on either a power or spectrum distribution basis. *See also:* **audio-frequency distortion.** 181

modulation angle (data transmission). The process of causing the angle of the carrier wave to vary in accordance with the signal wave. Phase and frequency modulation are two particular types of angle modulation. 59

modulation bandwidth (Δf) (amplitude, frequency, and pulse modulation). The frequency range between direct-current and the frequency where the depth of modulation is 50 percent (−3 decibel bandwidth). This is sometimes called the information bandwidth. 72

modulation capability (of a transmitter). The maximum percentage modulation that can be obtained without exceeding a given distortion figure. 111

modulation, effective percentage. *See:* **percentage modulation, effective.**

modulation eliminator (electronic navigation). (1) A limiting device designed to remove the modulation components from a carrier. (2) In very-high-frequency omnidirectional radio range, a specific unit used to proportion the transmitter output between the reference and variable signal functions and to remove the modulation from that portion of carrier delivered to the goniometer for use as the variable signal. *See also:* **navigation.** 187, 13

modulation factor (amplitude-modulated wave). The ratio (usually expressed in percent) of the peak variation of the envelope from its reference value, to the reference value. *Notes:* (1) For modulating functions having unequal positive and negative peaks, both positive and negative modulation factors are defined, respectively, in terms of the positive or negative peak variations. (2) The reference value is usually taken to be the amplitude of the unmodulated wave. (3) In conventional amplitude modulation the maximum design variation is considered that for which the instantaneous amplitude of the modulated wave reaches zero. *See also:* **amplitude modulation.**
111, 242, 211

modulation index (angle modulation with a sinusoidal modulating function). The ratio of the frequency deviation of the modulated wave to the frequency of the modulating function. *Note:* The modulation index is numerically equal to the phase deviation expressed in radians. *See also:* **angle modulation.** 111, 242, 240

modulation, intensity. *See:* **intensity modulation.**

modulation limiting (transmitter performance). That action, performed within the modulator, that intentionally contains the signal within the spectral and power limitations of the system. *See also:* **audio-frequency distortion.** 181

modulation meter (modulation monitor). An instrument for measuring the degree of modulation (modulation factor) of a modulated wave train, usually expressed in percent. *See also:* **instrument.** 328

modulation noise (1) (sound recording and reproducing system). Noise that exists only in the presence of a signal and is a function of the instantaneous amplitude of the recorded signal. (The signal is not to be included as part of the noise). *Note:* This type of noise results from such characteristics of a recording medium as light transmission in a film, slope of a groove, magnetism in a wire or tape, etcetera. *See also:* **noise (sound recording and reproducing system).** 350
(2) (noise behind the signal). The noise arising when the recorded signal randomly differs from the desired recording due to imperfections of the recording process. *Note:* Imperfections causing modulation noise include those of the recording medium, the recording and/or reproducing transducer, and the interaction of these elements. 176

modulation pulse (pulse train). The modulation of one or more of the parameters with the intent to transmit information. *See also:* **pulse.** 185

modulation rate. Reciprocal of the unit interval measured in seconds. This rate is expressed in bauds. *See also:* **modulation.** 194

modulation stability (transmitter performance). The measure of the ability of a radio transmitter to maintain a constant modulation factor as determined on both a short-term and long-term basis. *See also:* **audio-frequency distortion.** 181

modulation transfer (camera tubes). *See:* **amplitude response.**

modulation transfer function (MTF) (amplitude, frequency and pulse modulation). The depth of modulation versus modulation frequency. The MTF depends on optical beam diameter in the interaction region and optical wavelength. 72

modulator (1) general. A device to effect the process of modulation.
(2) (radar). A device for generating a succession of short

pulses of energy that cause a transmitter tube to oscillate during each pulse. *See also:* **modulation; radar; amplitude modulation; frequency modulation; pulse modulator.**
59

modulator, asymmetrical (signal-transmission system). A modulator in which the signal attentuation is different for the alternate half-cycles of the carrier. *See:* **signal.**
188

modulator, balanced (signal-transmission system). *See:* **modulator, symmetrical.**

modulator electrode. A control electrode in the case of an electron gun.   244, 190

modulator, full-wave (signal-transmission system). *See:* **modulator, symmetrical.**

modulator, half-wave (signal-transmission system). Extreme case of asymmetrical modulator in which the signal transmission is zero during one half-cycle. *See:* **signal.**
188

modulator, symmetrical (signal-transmission system). A modulator in which the signal attenuation is the same for alternate half-cycles of the carrier. (This implies also equal leakage impedances to the reference plane for the two half-cycles). *See:* **signal.**   188

modulator, unbalanced (signal-transmission system). *See:* **modulator, asymmetrical; signal.**

module (nuclear power generating stations). Any assembly of interconnected components which constitutes an identifiable device, instrument, or piece of equipment. A module can be disconnected, removed as a unit, and replaced with a spare. It has definable performance characteristics which permit it to be tested as a unit. A module could be a card or other subassembly of a larger device, provided it meets the requirements of this definition.
20, 31, 355, 361

module (CAMAC). *See:* **CAMAC module.**

module accelerated aging (advanced life conditioning) (nuclear power generating station). The acceleration process designed to achieve an advanced life condition in a short period of time. It is the process of subjecting a module or component to stress conditions in accordance with known measurable physical or chemical laws of degradation in order to render its physical and electrical properties similar to those it would have at an advanced age operating under expected service conditions. In addition, when operations of a device are cyclical, acceleration is achieved by subjecting the device to the number of cycles anticipated during its qualified life.   355

module accuracy (nuclear power generating station). Conformity of a measurement value to an accepted standard value or true value. *Note:* For further information, see Process Measurement and Control Terminology SAMA PMC-20.1-1973.   355

module aging (natural) (nuclear power generating station). The change with passage of time of physical chemical, or electrical properties of a component or module under design range operating conditions which may result in degradation of significant performance characteristics.
355

module calibration (nuclear power generating station). Adjustment of a device, to bring the module's output to a desired value or series of values, within a specified tolerance, for a particular value or series of values of the input

or measurements used to establish the input-output function of the module.   355

module common mode rejection (nuclear power generating station). The ability of a module with a differential input stage to cancel or reject a signal applied equally to both inputs.   355

module components (nuclear power generating station). Items from which the module is assembled (for example, resistors, capacitors, wires, connectors, transistors, springs, etcetera.)   355

module conformity (nuclear power generating station). The closeness with which the curve of a function approximates a specified curve.   355

module contact rating (nuclear power generating station). The electrical power-handling capability of relay or switch contacts. This should be specified as continuous or interrupting, resistive or inductive, ac or dc.   355

module design range operating conditions (nuclear power generating station). The range or environmental and energy supply operating conditions within which a module is designed to operate.   355

module drift (nuclear power generating station). A change in output-input relationship over a period of time, normally determined as the change in output over a specified period of time for one or more input values which are held constant under specified reference operating conditions.
355

module electromagnetic interference (nuclear power generating station). Any unwanted electromagnetically transmitted energy appearing in the circuitry of a module.
355

module energy supply (nuclear power generating station). Electrical energy, compressed fluid, manual force or other such input to the module which will establish the power for its operation.   355

module failure trending (nuclear power generating station). Systematic documentation and analysis of the frequency of a particular failure mode.   355

module frequency response (nuclear power generating station). The frequency-dependent relation, in both amplitude and phase, between steady-state sinusoidal inputs and the resulting fundamental sinusoidal outputs.   355

module input overrange constraints (nuclear power generating station). The upper and/or lower values of the input signal which may be applied to a module without causing damage or otherwise altering permanent characteristics of the module or causing undesired saturation effects.
355

module input signal range (nuclear power generating station). The region between the limits within which a quantity is measured or received, expressed by stating the lower and upper values of the input signal.   355

module isolation characteristics (nuclear power generating station). Provisions for electrical isolation of particular sections of a module from each other; such as input and output circuitry, control and protection circuitry, and redundant protection circuitry.   355

module load capability (nuclear power generating station). The range of load values within which a module will perform to its specified performance characteristics.   355

module output impedance (nuclear power generating station). The internal impedance presented by a module at its output terminals to a load.   355

module output ripple (nuclear power generating station). The ac component of a dc output signal harmonically related in frequency to either the supply voltage or a voltage generated within the module (for example, carrier demodulation).   355

module output signal range (nuclear power generating station). The region between the limits within which a quantity is transmitted, expressed by stating the lower and upper values of the output signal.   355

module pulse characteristics (nuclear power generating station). Information such as pulse duration, amplitude, rise time, decay time, separation and shape.   355

module qualified life (nuclear power generating station). The life expectancy in years (or cycles of operation, if applicable) over which the module has been demonstrated to be qualified for use, as established by type tests, analysis or other qualification method.   355

module range and characteristics of adjustments (nuclear power generating station). Such information as upper and lower range-limits of calibration capability and where applicable, their relationship to the calibrated range of the module.   355

module reference operating conditions (nuclear power generating station). The range of environmental operating conditions of a module within which environmental influences are negligible.   355

module reproducibility (nuclear power generating station). The closeness of agreement among repeated measurements of the output for the same value of input made under the same operating conditions over a period of time, approaching from both directions.   355

module response time (nuclear power generating station). The time required for an output change from an initial value to a specified percentage of the final steady-state value, resulting from the application of a specified input change under specified conditions. For digital equipment (that is, relays, solid state logic, delay networks, etcetera). Response time is the time required for a change from an initial state to a specified final state resulting from application of specified input under specified conditions.   355

module signal to noise rato (nuclear power generating station). The output signal with input signal applied minus the output signal with no input signal applied divided by the output signal with no input signal applied.   355

module type tests (nuclear power generating station). Tests made on one or more production units to demonstrate that the performance characteristics of the module(s) conform to the module's specifications.   355

modulo N check (1) (data transmission). A form of check digits, such that the number of ones in each number A operated upon is compared with a check number B, carried along with A and equal to the remainder of A when divided by N; for example, in a modulo 4 check, the check number will be 0, 1, 2, or 3 and the remainder of A when divided by 4 must equal the reported check number B, or else an error or malfunction has occurred; a method of verification by congruences; for example, casting out nines. *See also:* **residue check.**   194

(2) (computing systems). *See:* **residue check.**

modulus (phasor). Its absolute value. The modulus of a

phasor is sometimes called its amplitude.   210

Moho (Mohorovičić discontinuity). Seismic discontinuity situated about 35 kilometers below the continents and about 10 kilometers below the oceans. Crudely speaking, it separates the earth's crust and mantle.   132

moiré (1) (television). The spurious pattern in the reproduced picture resulting from interference beats between two sets of periodic structures in the image. *Notes:* (A) The most common cause of moiré is interference between scanning lines and some other periodic structure such as a line pattern or dot pattern in the original scene, a mesh or dot pattern in the camera sensor (for example the target mesh in an image orthicon), or the phosphor dots or other structure in a shadow-mask picture tube. (B) Moiré may result from the interference between the subcarrier elements of the chrominance signal and another periodic structure. (C) In systems using an FM carrier, such as magnetic or video-disc record-playback systems, moiré may also be caused by interference between the upper sidebands of the FM carrier and lower sidebands of harmonics of the FM carrier. (D) In general, moiré may be caused by interference beats between any two periodic structures that are not perfectly aligned and not of the same frequency.   18

(2) (storage tubes). A spurious pattern resulting from interference beats between two sets of periodic structures or scan patterns, or between a periodic structure and a scan pattern. *Note:* In a storage tube, moiré may be produced by writing a resolved parallel-line scan pattern and reading using another parallel-line scan pattern. Since mesh elements usually have a periodic structure, moiré may also be produced between the mesh pattern and a scan pattern, or between two mesh patterns. *See also:* **storage tube.**   174, 190

moisture-repellent. So constructed or treated that moisture will not penetrate.   328

moisture-resistant (industrial control). So constructed or treated that it will not be injured readily by exposure to a moist atmosphere.   103, 225, 202, 206, 27, 124

molar conductance (of an acid, base, or salt). The conductance of a solution containing one gram mole of the solute when measured in a like manner to equivalent conductance. *See also:* **electrochemistry.**   328

mold (1) (disk recording). A mold is a metal part derived from a master by electroforming that results in a positive of the recording, that is, it has grooves similar to a recording and thus can be played in a manner similar to a record. *See also:* **phonograph pickup.**   176

(2) (electrotyping). A matrix made by taking an impression, for example, of a printing plate in wax, lead, or plastic. *See also:* **electroforming.**   328

molded-case circuit breaker. A breaker assembled as an integral unit in a supporting and enclosing housing of insulating material; the overcurrent and tripping means being of the thermal type, the magnetic type, or a combination of both. *See also:* **switch.**   3, 103, 202

molded glass flare (electron tubes and electric lamps). A pressed glass article of the general shape shown in the figure below. *Note:* It is intended for use in the assembly of an electron tube or an electric lamp for the purpose of sealing electric terminals, or an exhaust tube, or both.

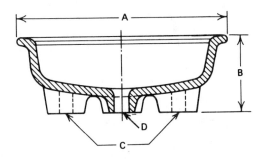

A—Major diameter
B—Over-all height
C—Holes for electrical terminals
D—Hole for exhaust tube
Molded glass flare

mole. That amount of substance of a system which contains as many elementary entities as there are atoms in 0.012 kilogram of carbon 12. *Note:* When the mole is used, the elementary entities must be specified and may be atoms, molecules, ions, electrons, other particles, or specified groups of such particles.   21

momentary current. The current flowing in a device, an assembly, or a bus at the major peak of the maximum cycle as determined from the envelope of the current wave. *Note:* The current is expressed as the root-mean-square value, including the direct-current component, and may be determined by the method shown in American National Standard Methods for Determining the Values of a Sinusoidal Current Wave and Normal-Frequency Recovery Voltage for Alternating-Current High-Voltage Circuit Breakers, C37.05-1964 (R 1969) or revision thereof.
103, 202, 27

momentary event time (sequential events recording systems). The minimum time in which a change of state and return to original state of one input is detected and processed. The time interval is usually expressed in milliseconds.
48, 58

momentary interruption. *See:* **interruption, momentary.**

monadic operation (computing systems). An operation on one operand, for example, negation. *See:* **unary operation.**
255, 77

monitor (1) (computing systems). Software or hardware that observes, supervises, controls, or verifies the operations of a system.   77
(2) (test, measurement and diagnostic equipment). To check the operation and performance of an equipment or system by sampling the result of the operation.   54

monitoring (communication). Observation of the characteristics of transmitted signals.   194, 59

monitoring key. A key that when operated makes it possible for an attendant or operator to listen on a telephone circuit without appreciably impairing transmission on the circuit.   328

monitoring radio receiver. A radio receiver arranged to permit a check to be made on the operation of a transmitting station. *See also:* **radio receiver.**   328

monitoring relay. A relay that has as its function to verify that system or control-circuit conditions conform to prescribed limits.   103, 202, 60, 127

monochromatic (color). Having spectral emission over an extremely small region of the visible spectrum.   18
monochrome. Having only one chromaticity, usually achromatic.   18, 178
monochrome channel (television). Any path that is intended to carry the monochrome signal.   163
monochrome channel bandwidth (television). The bandwidth of the path intended to carry the monochrome signal.   163, 178
monochrome signal (1) (monochrome television). A signal wave for controlling the luminance values in the picture.
(2) (color television). * *See:* **luminance signal.**   18

* Deprecated

monochrome transmission (television). The transmission of a signal wave for controlling the luminance values in the picture, but not the chromaticity values. *Note:* Also termed black-and-white transmission. *See also:* **television.**
18, 178

monocular visual field. The visual field of a single eye.
167

monoenergetic source (ionizing radiation) (semiconductor radiation detectors). A radiation source emitting ionizing radiation of essentially a single energy.   23

monolithic conduit. A monolithic concrete structure built to the desired duct formation by an automatic conduit-forming machine or flexible tubular rubber forms.   64

monolithic integrated circuit. An integrated circuit whose elements are formed in situ upon or within a semiconductor substrate with at least one of the elements formed within the substrate. *Note:* To further define the nature of a monolithic integrated circuit, additional modifiers may be prefixed. Examples are: (1) *p-n*-junction-isolated monolithic integrated circuit, (2) dielectric-isolated monolithic integrated circuit, (3) beam-lead monolithic integrated circuit. *See also:* **integrated circuit.**   312, 191

monophonic (frequency modulation). Audio information carried by a single channel. *Note:* A monophonic receiver responds only to main channel signals, lacking the capability of detecting subcarrier information. The main channel signals may be either monophonic or the left-plus-right (L+R) information from the left and right channels of a composite stereophonic signal. Monophonic receivers have no output from L = −R stereophonic signals. Stereophonic receivers may be manually or automatically switched to the monophonic mode, responding only to monophonic or main channel L+R signals.   16

monopolar electrode system. *See:* **unipolar electrode system.**

monopole. Any one of a class of antennas constructed above an imaging plane to produce a radiation pattern approximating that of an electric dipole in the half-space above the imaging plane.   111

monopulse (radar). A radar technique in which information concerning the angular location of a source or target is derivable from each pulse or signal detection by comparison of signals received simultaneously in two or more antenna beams, as distinguished from techniques such as lobe switching or conical scanning, in which angle information requires multiple pulses. The comparison of signals may be done directly or indirectly, for example by taking

ratios of their differences to their sum. *See also:* **amplitude-comparison monopulse; phase-comparison monopulse.**
13

monoscope. A signal-generating electron-beam tube in which a picture signal is produced by scanning an electrode, parts of which have different secondary-emission characteristics. *See also:* **television.** 178, 190, 125

monostable. Pertaining to a device that has one stable state.
255, 77

monostatic cross section. *See:* **back-scattering cross section.**

monotonic decreasing function. A single-valued function that satisfies the relation $f(x_1) > f(x_2)$ if $x_1 < x_2$. 210

monotonic increasing function. A single-valued function that satisfies the relation $f(x_1) < f(x_2)$ if $x_1 < x_2$. 210

monotonic nondecreasing function. A single-valued function that satisfies the relation $f(x_1) \leq f(x_2)$ if $x_1 < x_2$.
210

monotonic nonincreasing function. A single-valued function that satisfies the relation $f(x_1) \geq f(x_2)$ if $x_1 < x_2$.
210

Monte Carlo analysis (test, measurement and diagnostic equipment). The analysis of system behavior characteristics by evaluating the variation of randomly selected samples of the characteristics with changes in discrete parameters. 54

monthly peak duration curves (electric power utilization). Curves showing the total number of normal weekdays within the month during which a given net 60-minute clock-hour integrated peak demand equals or exceeds the percent of monthly peak values shown. Day one is 100 percent. *See also:* **generating station.** 200

MOPA. Abbreviation for a master oscillator/power amplifier type of coherent transmitter. 13

Morse code (American Morse code). A system of dot and dash signals, invented by Samuel F. B. Morse, now used to a limited extent for wire telegraphy in North America. *See also:* **international Morse code; telegraphy.** 328

Morse signal light (blinker signal). A fixture, installed usually on a standard above the wheelhouse for best visibility all around, controlled by a telegraph key for visual communication by Morse or other code. 328

Morse telegraphy. That form of telegraphy in which the signals are formed in accordance with one of the so-called Morse codes, particularly the International (also called Continental) or American. *See also:* **telegraphy.** 328

Morton wave current* (electrotherapy). An interrupted current obtained from a static machine by applying to the part to be treated a flexible metal electrode connected to the positive terminal of the machine, the negative terminal being grounded and a suitable spark gap being employed between the terminals. (Deprecated as representing one of several types of current, ill-defined as to waveform, frequency, voltage, and tissue-intensity, formerly used in electrotherapy). *See also:* **electrotherapy.** 192

* Deprecated

most-probable position (navigation). A position obtained by using all available position information, weighted statistically in accordance with individual estimated errors. *See also:* **navigation.** 187, 13

motional impedance (loaded motional impedance) (trans-
ducer). The complex remainder after the blocked impedance has been substracted from the loaded impedance. *See:* Note (2) under **acoustic impedance.** *See also:* **self-impedance.** 178

motor. A machine for the purpose of producing mechanical power or mechanical torque or force by means of a rotating shaft or through linear motion. *See also:* **asynchronous machine; direct-current commutating machine.** 63

motorboating. An undesired oscillation in an amplifying system or transducer, usually of a pulse type, occurring at a subaudio or low audio frequency. 239, 178

motor branch circuit. A branch circuit supplying energy only to one or more motors and associated motor controllers. *See also:* **branch circuit.** 3

motor-circuit switch (industrial control). A switch intended for use in a motor branch circuit. *Note:* It is rated in horsepower, and it is capable of interrupting the maximum operating overload current of a motor of the same rating at the rated voltage. *See also:* **operating overload; switch.** 256, 206, 124

motor conduit box (packaging machinery). An enclosure on a motor for the purpose of terminating a conduit run and joining motor to power conductors. 124

motor controller. A device or group of devices that serves to govern, in some predetermined manner, the electric power delivered to the motor or group of motors to which it is connected. *See also:* **multiple-unit control.** 1

motor-converter. A combination of an induction motor with a synchronous converter on a common shaft, the current produced in the rotor of the motor flowing through the armature of the synchronous converter. *See:* **converter.** 63

motor cut-out switch (land transportation vehicles). An isolating switch that isolates one or more of the traction motors from the supply and electric braking circuits. 1

motor-driven relay. A relay in which contact actuation is controlled through an electric motor, cams, and systems of gears. *Note:* An electromagnetic clutch may be used to engage and disengage the gear system. *See also:* **relay.** 259

motor effect (1) (general). The repulsion force exerted between adjacent conductors carrying currents in opposite directions. 114

(2) (induction heating). The force of repulsion or attraction between adjacent current-carrying conductors one of which may be the charge. *See also:* **induction heating.** 14

motor, electric. *See:* **electric motor.**

motor element (electroacoustic receiver). That portion that receives energy from the electric system and converts it into mechanical energy. *See also:* **loudspeaker.** 328

motorette (rotating machinery). A model for endurance tests on samples of rotating machine insulation. *Note:* It consists of a mounting plate, carrying two or more metal channels in which one or more test coils are wound as in the slots of a machine, and provided with suitable insulated terminals. It is designed for exposure to high temperature, moisture, vibration, or other chosen or specified insulation-aging influences, and for electric tests at designated stages of the aging processes. *See:* **asynchronous machine; direct-current commutating machines.** 63

motor-field accelerating relay (industrial control). A relay that functions automatically to maintain the armature current within limits, when accelerating to speeds above base speed, by controlling the excitation of the motor field. *See:* **relay.** 225, 206

motor-field control. A method of controlling the speed of a motor by means of a change in the magnitude of the field current. *See:* **control.** 206

motor-field failure relay (field loss relay) (industrial control). A relay that functions to disconnect the motor armature from the line in the event of loss of field excitation. *See:* **relay.** 225, 206

motor-field induction heater. An induction heater in which the inducing winding typifies that of an induction motor of rotary or linear design. *See also:* **induction heating.** 14, 114

motor-field protective relay (industrial control). A relay that functions to prevent overheating of the field excitation winding by reducing the excitation of the shunt field. *See:* **relay.** 225, 206

motor-generator electric locomotive. An electric locomotive in which the main power supply for the traction motors is changed from one electrical characteristic to another by means of a motor-generator set carried on the locomotive. *See also:* **electric locomotive.** 328

motor-generator set. A machine that consists of one or more motors mechanically coupled to one or more generators. *See:* **converter, generator-motor.** 63, 3

motor meter. A meter comprising a rotor, one or more stators, and a retarding element by which the resultant speed of the rotor is made proportional to the quantity being integrated (for example, power or current) and a register connected to the rotor by suitable gearing so as to count the revolutions of the rotor in terms of the accumulated integral (for example, energy or charge). *See also:* **electricity meter (meter).** 212

motor parts (electric). A term applied to a set of parts of an electric motor. Rotor shaft, conventional stator-frame (or shell), end shields, or bearings may not be included, depending on the requirements of the end product into which the motor parts are to be assembled. 63

motor reduction unit (electric installations on shipboard). A motor, with an integral mechanical means of obtaining a speed differing from the speed of the motor. *Note:* Motor reduction units are usually designed to obtain a speed lower than that of the motor, but may also be built to obtain a speed higher than that of the motor. 3

motor synchronizing. Synchronizing by means of applying excitation to a machine running at slightly below synchronous speed. *See also:* **asynchronous machine.** 63

mount (switching tubes). The flange or other means by which the tube, or tube and cavity, are connected to a waveguide. *See:* **gas tubes.** 190, 125

mounting lug, stator. *See:* **lug, stator mounting.**

mounted plow (static or vibratory plows) (cable plowing). A unit which, to be operable, is semipermanently attached to and dependent upon a prime mover.

mounting position (switch or fuse support). The position determined by and corresponding to the position of the base of the device. *Note:* The usual positions are (1) horizontal upright; (2) horizontal underhung; (3) vertical, and (4) angle. 103, 202, 27

mounting ring (rotating machinery). A ring of resilient or

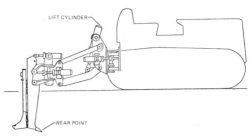

Mounted plow.

nonresilient material used for mounting an electric machine into a base at the end shield hub. 63

movable bridge coupler (drawbridge coupler). A device for engaging and disengaging signal or interlocking connections between the shore and a movable bridge span. 328

movable bridge (drawbridge) rail lock. A mechanical device used to insure that the movable bridge rails are in proper position for the movement of trains. *See also:* **interlocking.** 328

moving-base-derived navigation data. Data obtained from measurements made at moving cooperative facilities located external to the navigated vehicle. *See also:* **navigation.** 187, 13

moving-base navigation aid. An aid that requires cooperative facilities located upon a moving vehicle other than the one being navigated. *Notes:* (1) The cooperative facilities may move along a predictable path that is referenced to a specified coordinate system such as in the case of a nongeostationary navigation satellite. (2) Such an aid may also be designed solely to permit one moving vehicle to home upon another. *See also:* **navigation.** 187, 13

moving-base-referenced navigation data. Data in terms of a coordinate system referenced to a moving vehicle other than the one being navigated. *See also:* **navigation.** 187, 13

moving-coil loudspeaker (dynamic loudspeaker). A moving-conductor loudspeaker in which the moving conductor is in the form of a coil conductively connected to the source of electric energy. *See also:* **loudspeaker.** 328

moving-coil microphone (dynamic microphone). A moving-conductor microphone in which the movable conductor is in the form of a coil. *See also:* **microphone.** 328

moving-conductor loudspeaker (moving conductor). A loudspeaker in which the mechanical forces result from magnetic reactions between the field of the current and a steady magnetic field. *See also:* **loudspeaker.** 328

moving-conductor microphone. A microphone the electric output of which results from the motion of a conductor in a magnetic field. *See also:* **microphone.** 328

moving contact (power switchgear). A conducting part which bears a contact surface arranged for movement to and from the stationary contact. 27

moving-contact assembly (rotating machinery). That part of the starting switch assembly that is actuated by the centrifugal mechanism. *See:* **centrifugal starting switch.** 328

moving element (instrument). Those parts that move as a direct result of a variation in the quantity that the instrument is measuring. *Notes:* (1) The weight of the moving element includes one-half the weight of the springs, if any. (2) The use of the term movement is deprecated.

280, 294

moving-iron instrument. An instrument that depends for its operation on the reactions resulting from the current in one or more fixed coils acting upon one or more pieces of soft iron or magnetically similar material at least one of which is movable. *Note:* Various forms of this instrument (plunger, vane, repulsion, attraction, repulsion-attraction) are distinguished chiefly by mechanical features of construction. *See also:* **instrument.**   328

moving-magnet instrument. An instrument that depends for its operation on the action of a movable permanent magnet in aligning itself in the resultant field produced either by another permanent magnet and by an adjacent coil or coils carrying current, or by two or more current-carrying coils, the axes of which are displaced by a fixed angle. *See also:* **instrument.**   328

moving-magnet magnetometer. A magnetometer that depends for its operation on the torques acting on a system of one or more permanent magnets that can turn in the field to be measured. *Note:* Some types involve the use of auxiliary magnets (Gaussian magnetometer), others electric coils (sine or tangent galvanometer). *See also:* **magnetometer.**   328

moving target indication (MTI) (radar). A technique that enhances the detection and display of moving radar targets. Doppler processing is one method for implementation.   13

moving-target indicator improvement factor. *See:* **MTI improvement factor.**

MPE (laser-maser). maximum permissible exposure.

M peak (closed loop) (1) (control system, feedback). The maximum value of the magnitude of the return transfer function for real frequencies, the value at zero frequency being normalized to unity. *See also:* **control system, feedback.**   56

(2) **(excitation control systems) (Mp).** See definition above. *Note:* See Figure under **bandwidth.** Mp is the maximum value of the closed-loop amplitude response.

353

m-phase circuit. A polyphase circuit consisting of *m* distinct phase conductors, with or without the addition of a neutral conductor. *Note:* In this definition it is understood that *m* may be assigned the integral value of three or more. For a two-phase circuit see: **two-phase circuit; two-phase, three-wire circuit; two-phase, four-wire circuit; two-phase, five-wire circuit.** *See also:* **network analysis.**   210

MRA. *See:* **minimum reception altitude.**

MSC. *See:* **mile of standard cable.**

M scan (radar). *See:* **M display.**

M scope (radar). *See:* **M display.**

MSFN. *See:* **manned space flight network.**

M-T. *See:* **magneto-telluric.**

MTBF (reliability). *See:* **mean time between failures.**

MTE. *See:* **mean time between errors.**

MTF. *See:* **mean time between failures.**

MTI. *See:* **moving-target indication.**

MTI improvement factor (radar MTI). The signal-to-clutter ratio at the output of the system divided by the signal-to-clutter ratio at the input of the system, averaged

uniformly over all target radial velocities of interest. *See also:* **clutter attenuation; moving-target indication.**   13

mu (μ) circuit (feedback amplifier). That part that amplifies the vector sum of the input signal and the fed-back portion of the output signal in order to generate the output signal. *See also:* **feedback.**   328

MUF. *See:* **maximum usable frequency.**

mu factor (μ factor) (n-terminal electron tubes). The ratio of the magnitude of infinitesimal change in the voltage at the *j*th electrode to the magnitude of an infinitesimal change in the voltage at the *l*th electrode under the conditions that the current to the *m*th electrode remain unchanged and the voltages of all other electrodes be maintained constant. *See also:* **electron-tube admittances.**

190, 125

muffler (fuse). An attachment for the vent of a fuse, or a vented fuse, that confines the arc and substantially reduces the venting from the fuse.   103, 202

multiaddress (computers). *See:* **multiple-address.**

multianode tank (multianode tube). An electron tube having two or more main anodes and a single cathode. *Note:* This term is used chiefly for pool-cathode tubes.   190

multibeam oscilloscopes. An oscilloscope in which the cathode-ray tube produces two or more separate electron beams that may be individually, or jointly, controlled. *See:* **dual-beam oscilloscope; oscillograph.**   185

multicable penetrator (marine transportation). A device consisting of multiple preformed nonmetallic cable seals assembled in a surrounding metal frame, for insertion in openings in decks, bulkheads or equipment enclosures and through which cables may be passed to penetrate decks or bulkheads or to enter equipment without impairing their original fire or watertight integrity.   3

multicavity magnetron. A magnetron in which the circuit includes a plurality of cavities. *See also:* **magnetron.**

190,125

multicellular horn (electroacoustics). A cluster of horns with juxtaposed mouths that lie in a common surface. *Note:* The purpose of the cluster is to control the directional pattern of the radiated energy. *See also:* **loudspeaker.**   176

multichannel radio transmitter. A radio transmitter having two or more complete radio-frequency portions capable of operating on different frequencies, either individually or simultaneously. *See also:* **radio transmitter.**   111

multichip integrated circuit. An integrated circuit whose elements are formed on or within two or more semiconductor chips that are separately attached to a substrate. *See also:* **integrated circuit.**   312, 191

multi-constant speed motor (rotating machinery). A multi-speed motor whose two or more definite speeds are constant or substantially constant over its normal range of loads: for example A synchronous or an induction motor with windings capable of various pole groupings.   63

multidimensional system (control system). A system whose state vector has more than one element. *See also.* **control system.**   56

multielectrode tube. An electron tube containing more than three electrodes associated with a single electron stream.

190, 125

multifrequency transmitter. A radio transmitter capable of operating on two or more selectable frequencies, one at a time, using present adjustments of a single radio-frequency portion. *See also:* **radio transmitter.** 111, 240

multigrounded neutral system (transformer). A distribution system of the four-wire type where all transformer neutrals are grounded, and neutral conductors are directly grounded at frequent points along the circuit. 53

multilateration (radar). The location of an object by means of two or more range measurements from different reference points. It is a useful technique with radar because of the inherent accuracy of radar range measurement. The use of three reference points, **trilateration,** is common practice. 13

multilevel address (computing systems). *See:* **indirect address.**

multimeter. *See:* **circuit analyzer.**

multimode waveguide. A waveguide used to propagate more than one mode at the frequency of interest. *See:* **waveguide.** 179

multioffice exchange (telephone switching systems). A telecommunications exchange served by more than one local central office. 55

multioutlet assembly (interior wiring). A type of surface or flush raceway designed to hold conductors and attachment plug receptacles, assembled in the field or at the factory. *See also:* **raceway.** 256

multiparty ringing (telephone switching systems). By custom, any arrangement that provides for the individual ringing of more than four parties. 55

multipath (facsimile). *See:* **multipath transmission.**

multipath error (radar). The error caused by the reception of a composite radio signal which arrives via two or more different paths. *See also:* **navigation.** 187, 13

multipath transmission (multipath). The propagation phenomenon which results in signals reaching the radio receiving antenna by two or more paths. *Note:* In facsimile, multipath causes jitter. 12

multiple (1) (noun). A group of terminals arranged to make a circuit or group of circuits accessible at a number of points at any one of which connection can be made.
(2) (verb). To connect in parallel, or to render a circuit accessible at a number of points at any one of which connection can be made. 328
(3) (analog computer). A junction into which patch cords may be plugged to form a common connection. 10

multiple access (communication satellite). The capability of having simultaneous access to one communication satellite from a number of ground stations. 84

multiple-address (multiaddress) (computers). Pertaining to an instruction that has more than one address part. 235, 255, 77

multiple-address code (electronic computation). *See:* **instruction code:** *See also:* **electronic computation.**

multiple-beam headlamp. A headlamp so designed as to permit the driver of a vehicle to use any one of two or more distributions of light on the road. *See also:* **headlamp.** 167

multiple-beam klystron (microwave tubes). An *O*-type tube having more than one electron beam, and resonators coupled laterally but not axially. *See also:* **microwave tube or valve.** 190

multiple circuit. Two or more circuits connected in parallel. *See also:* **center of distribution.** 64

multiple-conductor cable. A combination of two or more conductors cabled together and insulated from one another and from sheath or armor where used. *Note:* Specific cables are referred to as 3-conductor cable, 7-conductor cable, 50-conductor cable, etcetera. *See also:* **power distribution, underground construction.** 64

multiple-conductor concentric cable. A cable composed of an insulated central conductor with one or more tubular stranded conductors laid over it concentrically and insulated from one another. *Note:* This cable usually has only two or three conductors. Specific cables are referred to as 2-conductor concentric cable, 3-conductor concentric cable, etcetera. *See also:* **power distribution, underground construction.** 64

multiple-current generator. A generator capable of producing simultaneously currents or voltages of different values, either alternating-current or direct-current. *See:* **direct-current commutating machine.** 63

multiple feeder. (1) Two or more feeders connected in parallel. *See also:* **center of distribution.** 64
(2) One that is connected to a common load in multiple with one or more feeders from independent sources. 103, 202

multiple-gun cathode-ray tube. A cathode-ray tube containing two or more separate electron-gun systems. 190

multiple hoistway (elevators). A hoistway for more than one elevator or dumbwaiter. *See:* **hoistway (elevator or dumbwaiter).** 328

multiple lampholder (current tap*). A device that by insertion in a lampholder, serves as more than one lampholder. 328

* Deprecated

multiple lightning stroke. A lightning stroke having two or more components. *See also:* **direct-stroke protection (lightning).** 64

multiple metallic rectifying cell. An elementary metallic rectifier having one common electrode and two or more separate electrodes of the opposite polarity. *See also:* **rectification.** 328

multiple modulation. A succession of processes of modulation in which the modulated wave from one process becomes the modulating wave for the next. *Note:* In designating multiple-modulation systems by their letter symbols, the processes are listed in the order in which the modulating function encounters them. For example, PPM-AM means a system in which one or more signals are used to position-modulate their respective pulse subcarriers which are spaced in time and are used to amplitude-modulate a carrier. 111, 242

multiple plug (cube tap) (plural tap*). A device that, by insertion in a receptacle, serves as more than one receptacle. 328

*Deprecated

multiple rectifier cirucit. A rectifier circuit in which two or more simple rectifier circuits are connected in such a way that their direct currents add, but their commutations do not coincide. *See also:* **rectification; rectifier circuit element.** 237, 66, 208

multiple rho (electronic navigation). A generic term referring to navigation systems based on two or more distance measurements for determination of position. *See also:* **navigation.** 187, 13

multiple-secondary current transformer. One which has three of more secondary coils each on a separate magnetic circuit with all magnetic circuits excited by the same primary winding. 53

multiple-shot blasting unit. A unit designed for firing a number of explosive charges simultaneously in mines, quarries, and tunnels. *See also:* **blasting unit.** 328

multiple sound tract. Consists of a group of sound tracks, printed adjacently on a common base, independent in character but in a common time relationship, for example, two or more have been used for stereophonic sound recording. *See also:* **multitrack recording system; phonograph pickup.** 176

multiple speed floating. *See:* **control system, multiple-speed floating.**

multiple spot scanning (facsimile). The method in which scanning is carried on simultaneously by two or more scanning spots, each one analyzing its fraction of the total scanned area of the subject copy. *See also:* **scanning (facsimile).** 12

multiple street-lighting system. A street-lighting system in which street lights, connected in multiple, are supplied from a low-voltage distribution system. *See also:* **alternating-current distribution; direct-current distribution.** 64

multiple-supply-type ballast. A ballast designed specifically to receive its power from an approximately constant-voltage supply circuit and that may be operated in multiple (parallel) with other loads supplied from the same source. *Note:* The deviation in source voltage ordinarily does not exceed plus or minus 5 percent, but in the case of ballasts designed for a stated input voltage range, the deviation may by greater as long as it is within the stated range. 271

multiple switchboard (telephone switching systems). A telecommunications switchboard having each line connected to two or more jacks so that the line is within the reach of several operators. 55

multiple system (electrochemistry). The arrangement in a multielectrode electrolytic cell whereby in each cell all of the anodes are connected to the positive bus bar and all of the cathodes to the negative bus bar. *See also:* **electrorefining.** 328

multiple transit signals (dispersive and nondispersive delay lines). Spurious signals having delay time related to the main signal delay by small odd integers. *Notes:* (1) Specific multiple transit signals may be labeled the third transit (triple transit), fifth transit, etc. (2) There is often a tradeoff available between multiple transit signal levels and bandwidth, delay time, insertion loss, and VSWR (voltage standing-wave ratio). 81

multiple tube (or valve). A space-charge-controlled tube or valve containing within one envelope two or more units or groups of electrodes associated with independent electron streams, through sometimes with one or more common electrodes. Examples: Double diode, double triode, triode-heptode, etcetera. *See also:* **multiple-unit tube.** 190, 125

multiple tube counts (radiation counter tubes). Spurious

counts induced by previous tube counts. *See also:* **gas-filled radiation-counter tubes.** 190, 125, 96

multiple-tuned antenna. A low-frequency antenna having an elevated horizontal section and a multiplicity of vertical wires at successive intervals along the horizontal section, each of which is series resonated with inductances to the ground system. 111

multiple twin quad (telephony). A quad in which the four conductors are arranged in two twisted pairs, and the two pairs twisted together. *See also:* **cable.** 328

multiple-unit control (electric traction). A control system in which each motive-power unit is provided with its own controlling apparatus and arranged so that all such units operating together may be controlled from any one of a number of points on the units by means of a master controller. 328

multiple-unit electric car. An electric car arranged either for independent operation or for simultaneous operation with other similar cars (when connected to form a train of such cars) from a single control station. *Note:* A prefix diesel-electric, gas-electric, etcetera, may replace the word electric. *See also:* **electric motor car.** 328

multiple-unit electric locomotive. A locomotive composed of two or more multiple-unit electric motive-power units connected for simultaneous operation of all such units from a single control station. *Note:* A prefix diesel-electric, turbine-electric, etcetera, may replace the word electric. *See also:* **electric locomotive.** 328

multiple-unit electric motive-power unit. An electric motive-power unit arranged either for independent operation or for simultaneous operation with other similar units (when connected to form a single locomotive) from a single control station. *Note:* A prefix diesel-electric, gas-electric, turbine-electric, etcetera, may replace the word electric. *See also:* **electric locomotive.** 328

multiple-unit electric train. A train composed of multiple-unit electric cars. *See also:* **electric motor car.** 328

multiple-unit tube. *See:* **multiple tube (or valve).**

multiplex (communication). To interleave or simultaneously transmit two or more messages on a single channel. 255, 77, 54

multiplexing (modulation systems). The combining of two or more signals into a single wave (the multiplex wave) from which the signals can be individually recovered. 242

multiplex lap winding (rotating machinery). A lap winding in which the number of parallel circuits is equal to a multiple of the number of poles. 63

multiplex operation (communication). Simultaneous transmission of two or more messages in either or both directions over the same transmission path. *See also:* **telegraphy.** 59

multiplexor (analog-to-digital converter). An electronic multi-position switch under the control of a digital computer, generally used in conjunction with an analog-to-digital converter, that allows for the selection of any one of a number of analog signals (up to the maximum capacity of the multiplexor), as the input to the analog-to-digital converter. A device that allows the interleaving of two or more signals to a single line or terminus. 10

multiplex printing telegraphy. That form of printing telegraphy in which a line circuit is employed to transmit in turn one character (or one or more pulses of a character) for each of two or more independent channels. *See:* **frequency-division multiplexing; time-division multiplexing.** *See also:* **telegraphy.** 328

multiplex radio transmission. The simultaneous transmission of two or more signals using a common carrier wave. *See also:* **radio transmission.** 111

multiplex wave winding (rotating machinery). A wave winding in which the number of parallel circuits is equal to a multiple of two, whatever the number of poles. 63

multiplication factor (1) (multiplier type of valve or tube) (thermionics). The ratio of the output current to the primary emission current. *See also:* **electron emission.** 244, 190

(2) (κ) (nuclear power generating station). A measure of the change in the neutron population in a reactor core from one generation to the subsequent generation. 112

multiplicative array antenna system. A signal-processing antenna system consisting of two or more receiving antennas and circuitry in which the effective angular response of the output of the system is related to the product of the radiation patterns of the separate antennas. 111

multiplier (1) (general). A device that has two or more inputs and whose output is a representation of the product of the quantities represented by the input signals. 210

(2) (analog computer). A device capable of multiplying one variable by another. *See also:* **electronic analog computer.** 9, 10

(3) (linearity). *See:* **constant multiplier; normal linearity; servo multiplier.**

multiplier, constant (computing systems). A computing element that multiplies a variable by a constant factor. *See also:* **electronic analog computer; multiplier (linearity).** 9, 77, 10

multiplier, electronic. An all-electronic device capable of forming the product of two variables. *Note:* Examples are a time-division multiplier, a square-law multiplier, an amplitude-modulation-frequency-modulation (AM-FM) multiplier, and a triangular-wave multiplier. *See also:* **electronic analog computer.** 9

multiplier, four-quadrant (analog computer). A multiplier in which operation is unrestricted as to the sign of both of the input variables. 10

multiplier, one-quadrant. A multiplier in which operation is restricted to a single sign of both input variables. *See also:* **electronic analog computer.** 9

multiplier phototube. A phototube with one or more dynodes between its photocathode and output electrode. *See also:* **amplifier; photocathode.** 190, 125, 117

multiplier section, electron (electron tubes). *See:* **electron multiplier.**

multiplier servo. An electromechanical multiplier in which one variable is used to position one or more ganged potentiometers across which the other variable voltages are applied. *See also:* **electronic analog computer; multiplier (linearity).** 9, 77, 10

multiplier, two-quadrant. A multiplier in which operation is restricted to a single sign of one input variable only. *See also:* **electronic analog computer.** 9, 10

multiplying-digital-to-analog converter (MDAC). *See:* **DAM.**

multipoint circuit (data transmission). A circuit interconnecting several locations where information transmitted over the circuit is available at all locations simultaneously. 59

multipoint connection (data communication). A configuration in which more than two stations are connected to a shared communications channel. 12

multipole fuse. *See second note under:* **pole (pole unit) (switching device or fuse).**

multiposition relay. A relay that has more than one operate or nonoperate position, for example, a stepping relay. *See also:* **relay.** 259

multiposition switches (industrial control). (1) **self-returning switch.** A switch that returns to a stated position when it is released from any one of a stated set of other positions. (2) **spring return switch.** A switch in which the self-returning function is effected by the action of a spring. (3) **gravity-return switch.** A switch in which the self-returning function is effected by the action of weight. (4) **self-positioning switch.** A switch that assumes a certain operating position when it is placed in the neighborhood of the position. *See:* **switch.** 225, 206

multipressure zone pothead (electric power distribution). A pressure-type pothead intended to be operated with two or more separate pressure zones that may be at different pressures. *See:* **pressure-type pothead; single-pressure zone potheads.** 4

multiprocessing (computing systems). Pertaining to the simultaneous execution of two or more programs or sequences of instructions by a computer network consisting of two or more processors. *See:* **parallel processing; multiprogramming.** 77

multiprocessor (computing systems). A computer capable of multiprocessing. 255, 77

multiprogramming (computing systems). Pertaining to the interleaved execution of two or more programs by a computer. *See:* **parallel processing.** 255, 77

multi-radio-frequency-channel transmitter. *See:* **multichannel radio transmitter.**

multirate meter. A meter that registers at different rates or on different dials at different hours of the day. *See also.* **electricity meter.** 328

multi-ratio current transformer. One from which more than one ratio can be obtained by the use of taps on the secondary winding. 53

multirestraint relay. A restraint relay that is so constructed that its operation is restrained by more than one input quantity. 103, 202, 60, 127

multisection coil (rotating machinery). A coil consisting of two or more coil sections or a group of turns, each section or group being individually insulated. 63

multisegment magnetron. A magnetron with an anode divided into more than two segments, usually by slots parallel to its axis. *See also:* **magnetrons.** 190, 125

multispeed motor. One that can be operated at any one of two or more definite speeds, each being practically independent of the load. *Note:* For example, a direct-current motor with two armature windings or an induction motor with windings capable of various pole groupings. *See:* **asynchronous machine; direct-current commutating machine.** 63

multistage tube (X-ray tubes). An X-ray tube in which the cathode rays are accelerated by multiple ring-shaped anodes at progressively higher potential. *See also:* **electron devices, miscellaneous.** 190

multi-step control. *See:* **control system, step.**

multitrace (oscilloscopes). A mode of operation in which a single beam in a cathode-ray tube is shared by two or more signal channels: *See:* **alternate display; chopped display; dual trace; oscillograph.** 185

multitrack recording system. A system that provides two or more recording tracks on a medium, resulting in either related or unrelated recordings in common time relationship. *See also:* **multiple sound track; phonograph pickup.** 176

multivalent function. If to any value of u there corresponds more than one value of x (or more than one set of values of $x_1, x_2, \cdots, x_n$) then u is a multivalent function. Thus $u = \sin x, u = x^2$ are multivalent. 210

multivalued function. If to any value of x (or any set of values of $x_1, x_2, \cdots, x_n$) there corresponds more than one value of u, then u is a multivalued function. Thus $u = \cos^{-1} x$ is multivalued. 210

multivariable system (control system). A system whose input vector and/or output vector has more than one element. 56

multivibrator. A relaxation oscillator employing two electron tubes to obtain the in-phase feedback voltage by coupling the output of each to the input of the other through, typically, resistance-capacitance elements. *Notes:* (1) The fundamental frequency is determined by the time constants of the coupling elements and may be further controlled by an external voltage. (2) A multivibrator is termed free-running or driven, according to whether its frequency is determined by its own circuit constants or by an external synchronizing voltage. The name multivibrator was originally given to the free-running multivibrator, having been suggested by the large number of harmonics produced. (3) When such circuits are normally in a nonoscillating state and a trigger signal is required to start a single cycle of operation, the circuit is commonly called a one-shot, a flip-flop, or a start-stop multivibrator. *See also:* **oscillatory circuit.** 111, 59

multivoltage control (elevators). A system of control that is accomplished by impressing successively on the armature of the driving-machine motor a number of substantially fixed voltages such as may be obtained from multicommutator generators common to a group of elevators. *See also:* **control (elevators).** 328

municipal fire-alarm system. A manual fire-alarm system in which the stations are accessibly located for operation by the public, and the signals of which register at a central station maintained and operated by the municipality. *See also:* **protective signaling.** 328

municipal police report system. A system of strategically located stations from any one of which a patrolling policeman may report his presence to a supervisor in a central office maintained and operated by the municipality. *See also:* **protective signaling.** 328

Munsell chroma. (1) The dimension of the Munsell system of color that corresponds most closely to saturation. *Note:* Chroma is frequently used, particularly in English works, as the equivalent of saturation. *See also:* **saturation.** 18, 178

(2) The index of saturation of the perceived object color

defined in terms of the Y value and chromaticity coordinates (x,y) of the color of light reflected or transmitted by the object. *See also:* **Munsell color system.** 167

Munsell color system. A system of surface-color specification based on perceptually uniform color scales for the three variables: Munsell hue, Munsell value, and Munsell chroma. *Notes:* (1) For an observer of normal color vision, adapted to daylight and viewing the specimen when illuminated by daylight and surrounded with a middle gray to white background, the Munsell hue, value, and chroma of the color correlate well with the hue, lightness, and saturation of the perceived color. (2) A number of other color specification systems have been developed, usually for specific commercial purposes. 167

Munsell hue. The index of the hue of the perceived object color defined in terms of the Y value and chromaticity coordinates (x,y) of the color of the light reflected or transmitted by the object. 167

Munsell value. The index of the lightness of the perceived object color defined in terms of the Y value. *Note:* Munsell value is approximately equal to the square root of the reflectance expressed in percent. 167

musa antenna* (multiple-unit steerable antenna). *See:* **electronically scanned antenna.** *See also:* **antenna.**

* Deprecated

must operate value. *See:* **relay must operate value.**

mutual capacitance between two conductors*. *See:* **balanced capacitance between two conductors.**

* Deprecated

mutual conductance. The control-grid-to-anode transconductance. *See also:* ON **period (electron tubes).** 244, 190

mutual impedance (1) (between any two pairs of terminals of a network). The ratio of the open-circuit potential difference between either pair of terminals, to the current applied at the other pair of terminals, all other terminals being open. *Note:* Mutual impedance may have either of two signs depending upon the assumed direction of input current and output voltage; the negative of the above ratio is usually employed. Mutual impedance is ordinarily additive if two coils of a transformer are connected in series or in parallel adding, and is subtractive if two coils of a transformer are in series or in parallel opposing. *See also:* **self-impedance.** 328

(2) (between two meshes). The factor by which the phasor equivalent of the steady-state sine-wave current in one mesh must be multiplied to give the phasor equivalent of the steady-state sine-wave voltage in the other mesh caused by the current in the first mesh. *See also:* **network analysis.** 210

(3) (antenna). The mutual impedance between any two terminal pairs in a multielement array antenna is equal to the open-circuit voltage produced at the first terminal pair divided by the current supplied to the second when all other terminal pairs are open circuited. *See also:* **antenna.** 246, 179

mutual inductance. The common property of two electric circuits whereby an electromotive force is induced in one circuit by a change of current in the other circuit. *Notes:* (1) The coefficient of mutual inductance M between two windings is given by the following equation

$$M = \frac{\partial i}{\partial \lambda}$$

where λ is the total flux linkage of one winding and i is the current in the other winding. (2) The voltage e induced in one winding by a current i in the other winding is given by the following equation

$$e = - \left[M \frac{di}{dt} + i \frac{dM}{dt} \right]$$

If M is constant　　$e = -M \dfrac{di}{dt}$

197

mutual inductor. An inductor for changing the mutual inductance between two circuits. 210

mutual information. *See:* **transinformation.**

mutual interference chart (MIC) (electromagnetic compa- tibility). A plot or matrix, with ordinate and abscissa representing the tuned frequencies of a single transmitter-receiver combination, that indicates potential interference to normal receiver operation by reason of interaction of the two equipments under consideration at any combination of tuned transmit/receive frequencies. *Note:* This interaction includes transmitter harmonics and other spurious emissions, and receiver spurious responses and images. *See also:* **electromagnetic compatibility.** 199

mutually exclusive events (nuclear power generating stations). Events that cannot exist simultaneously. 29, 31

mutual resistance of grounding electrodes. Equal to the voltage change in one of them produced by one ampere of direct current in the other, and is expressed in ohms. *See also:* **grounding device.** 313

mutual surge impedance (surge arresters). The apparent mutual impedance between two lines, both of infinite length. *Note:* It determines the relationship between the surge voltage induced into one line by a surge current of short duration in the other. 244, 62

N

nameplate (rating plate) (rotating machinery). A plaque giving the manufacturer's name and the rating of the machine. 63

NAND. A logic operator having the property that if P is a statement, Q is a statement, R is a statement, ... then the NAND of P, Q, R, ... is true if at least one statement is false, false if all statements are true. 255, 77

narrow-angle diffusion (illuminating engineering). Diffusion in which flux is scattered at angles near the direction that the flux would take by regular reflection or transmission. *See also:* **lamp.** 167

narrow-angle luminaire. Luminaire that concentrates the light within a cone of comparatively small solid angle. *See also:* **luminaire.** 167

narrow-band axis (color television) (phasor representation of the chrominance signal). The direction of the phasor representing the coarse chrominance primary. 18, 178

narrow-band interference (electromagnetic compatibility). For purposes of measurement, a disturbance of spectral energy lying within the bandpass of the measuring receiver in use. *See also:* **electromagnetic compatibility.** 199

narrow-band radio noise. Radio noise having a spectrum exhibiting one or more sharp peaks, narrow in width compared to the nominal bandwidth of the radio noise meter, and far enough apart to be resolvable by the receiver. *See also:* **radio-noise field strength.** 285

narrow band response spectrum (nuclear power generating stations) (seismic qualification of class 1E equipment). A response spectrum that describes the motion indicating that a single frequency excitation predominates. 28, 31

N-ary code (information theory). A code employing N distinguishable types of code elements. *See also:* **information theory.** 160

national call (telephone switching systems). A toll call to a destination outside of the local service area of the calling customer but within the boundaries of the country in which he is located. 55

national distance dialing (telephone switching systems). The automatic establishing of a national call by means of signals from the calling device of either a customer or an operator. 55

national number (telephone switching systems). The combination of digits representing an area code and a directory number that, for the purpose of distance dialing, uniquely identifies each main station within each of the world's geographical areas that is identified by a country code. 55

national-numbering plan (telephone switching systems). Any plan for identifying telephone stations within a geographical area identified by a unique country code. 55

national numbering-plan area (telephone switching systems). A geographical area of the world where a country code and the national boundaries are related uniquely. 55

nationwide toll dialing (telephony). A system of automatic switching whereby an outward toll operator can complete calls to any basic-numbering-plan area in the country covered by the system. 328

native system demand (electric power utilization). The monthly net 60-minute clock-hour integrated peak demands of the system. *See also:* **alternating-current distribution.** 200

natural air cooling system (rectifier). An air cooling system in which heat is removed from the cooling surfaces of the rectifier only by the natural action of the ambient air. *See also:* **rectification.** 291, 208

natural frequency (1) (antenna). Its lowest resonance frequency without added inductance or capacitance. *See also:* **antenna.** 179

(2) **(surge arresters).** The frequency or frequencies at which the circuit will oscillate if it is free to do so. *See:* **frequency, undamped.** 308, 62

(3) **(gyro; accelerometer).** That frequency at which the output lags the input by ninety degrees. It is generally

applied only to inertial sensors with approximate second order response. 46

(4) (nuclear power generating stations) (seismic qualification of class 1E equipment). The frequency or frequencies at which a body vibrates due to its own physical characteristics (mass, shape) and elastic restoring forces brought into play when the body is distorted in a specific direction and then released, while restrained or supported at specified points. 28, 31

(5) (automatic control). *See:* **frequency, undamped.**

natural language. A language whose rules reflect and describe current usage rather than prescribe usage. *See:* **artificial language.** 255, 77

natural linewidth (laser-maser). The **linewidth** of an **absorption** or **emission** line when **spontaneous emission** is the dominant process determining spectral distribution. 363

natural noise (electromagnetic compatibility). Noise having its source in natural phenomena and not generated in machines or other technical devices. *See also:* **electromagnetic compatibility.** 220, 199

natural period. The period of the periodic part of a free oscillation of the body or system. *Notes:* (1) When the period varies with amplitude, the natural period is the period as the amplitude approaches zero. (2) A body or system may have several modes of free oscillation, and the period may be different for each. 210

navigation. The process of directing a vehicle so as to reach the intended destination. 13, 187

navigational radar (surface search radar). A high-frequency radio transmitter-receiver for the detection, by means of transmitted and reflected signals, of any object (within range) projecting above the surface of the water and for visual indication of its bearing and distance. *See also:* **radio direction-finder (radio compass).**

navigation coordinate. Any one of a set of quantities, the set serving to define a position. *See also:* **navigation.** 187, 13

navigation light system. A set of aircraft aeronautical lights provided to indicate the position and direction of motion of an aircraft to pilots of other aircraft or to ground observers. 167

navigation parameter. A measurable characteristic of motion or position used in the process of navigation. *See also:* **navigation.** 187, 13

navigation quantity. A measured value of a navigation parameter. *See also:* **navigation.** 187, 13

n-conductor cable (electric power distribution). *See:* **multiple-conductor cable.**

n-conductor concentric cable (electric power distribution). *See:* **multiple-conductor concentric cable.**

N-contours. *See:* **phase contours.**

NDB (electronic navigation). *See:* **nondirectional beacon.**

N-display (radar). A *K* display having an adjustable pedestal signal, as in the *M* display, for the measurement of distance. *See also:* **navigation.** 187, 13

near-end crosstalk. Crosstalk that is propagated in a disturbed channel in the direction opposite to the direction of propagation of the current in the disturbing channel. *Note:* The terminal of the disturbed channel at which the near-end crosstalk is present is ordinarily near to or coin-

cides with the energized terminal of the disturbing channel. *See also:* **coupling.** 328

near field (radar). The region in the vicinity of an antenna or radar target in which amplitude components of the field having range (R) dependence other than $1/R$ cannot be neglected. 13

near-field region in physical media. That part of space between the antenna and the far-field region. *Note:* In lossless media, the near-field region can be further subdivided into reactive and radiating near-field regions. 111

near-field region, radiating. The region of the field of an antenna between the reactive near-field region and the far-field region wherein radiation fields predominate and wherein the angular field distribution is dependent upon distance from the antenna. *Notes:* (1) If the antenna has a maximum over-all dimension which is not large compared to the wavelength, this field region may not exist. (2) For an antenna focused at infinity, the radiating near-field region is sometimes referred to as the Fresnel region on the basis of analogy to optical terminology. *See also:* **radiation.** 179, 111

near-field region, reactive. The region of the field immediately surrounding the antenna wherein the reactive field predominates. *Note:* For most antennas the outer boundary of the region is commonly taken to exist at a distance $\lambda/2\pi$ from the antenna surface, where λ is the wavelength. *See also:* **radiation.** 179, 111

neck (cathode-ray tube). The small tubular part of the envelope near the base. *See also:* **cathode-ray tube.** 244, 190

negate. To perform the logic operation NOT. 255, 77

negative after-potential (electrobiology). Relatively prolonged negativity that follows the action spike in a homogeneous fiber group. *See also:* **contact potential.** 192

negative conductance (circuits and systems). The conductance of a negative-resistance device. 67

negative conductor. A conductor connected to the negative terminal of a source of supply. *Note:* A negative conductor is frequently used as an auxiliary return circuit in a system of electric traction. *See also:* **center of distribution.** 64

negative-differential-resistance region (thyristor). Any portion of the principal voltage-current characteristic in the switching quadrant(s) within which the differential resistance is negative. *See also:* **principal voltage-current characteristic (principal characteristic).** 243, 66, 191

negative electrode (1) (primary cell). The anode when the cell is discharging. *Note:* The negative terminal is connected to the negative electrode. *See also:* **electrolytic cell.**

(2) (metallic rectifier). The electrode from which the forward current flows within the cell. *See also:* **rectification.** 328

negative feedback (1) (circuits and systems). The process by which part of the signal in the output circuit of an amplifying device reacts upon the input circuit in such a manner as to counteract the initial power, thereby decreasing the amplification. 67

(2) (control) (industrial control). A feedback signal in a direction to reduce the variable that the feedback represents. *See also:* **control system, feedback; feedback.**

219, 206

negative glow (gas tube). The luminous glow in a glow-discharge cold-cathode tube between the cathode dark space and the Faraday dark space. *See:* **gas tubes.** 190

negative matrix (negative). A matrix the surface of which is the reverse of the surface to be ultimately produced by electroforming. *See also:* **electroforming.** 328

negative modulation (amplitude-modulation television system). That form of modulation in which an increase in brightness corresponds to a decrease in transmitted power. *See also:* **television.** 328

negative-phase-sequence impedance (rotating machinery). The quotient of the negative-sequence rated-frequency component of the voltage, assumed to be sinusoidal, at the terminals of a machine rotating at synchronous speed, and the negative-sequence component of the current at the same frequency. *Note:* It is equal to the asynchronous impedance for a slip equal to 2. *See also:* **asynchronous machine.** 63

negative-phase-sequence reactance (rotating machinery). The quotient of the reactive fundamental component of negative-sequence primary voltage due to sinusoidal negative-sequence primary current of rated frequency, and the value of this current, the machine running at rated speed. *See also:* **asynchronous machine.** 58

negative-phase-sequence (phase-reversal) relay. A relay that responds to the negative-phase-sequence component of a polyphase input quantity. *Note:* Frequently employed in three-phase systems. *See also:* **relay.**

103, 202, 60, 127

negative phase-sequence resistance (rotating machinery). The quotient of the in-phase fundamental component of negative-sequence primary voltage, due to sinusoidal negative-sequence primary current of rated frequency, and the value of this current, the machine running at rated speed. *See also:* **asynchronous machine.** 63

negative-phase-sequence symmetrical components. Of an unsymmetrical set of polyphase voltages or currents of M phases, that set of symmetrical components that have the $(m - 1)$st phase sequence. That is, the angular phase lag from the first member of the set to the second, from every other member of the set to the succeeding one, and from the last member to the first, is equal to $(m - 1)$ times the characteristic angular phase difference, or $(m - 1)2\pi/m$ radians. The members of this set will reach their positive maxima uniformly but in the reverse order of their designations. *Note:* The negative-phase-sequence symmetrical components for a three-phase set of unbalanced sinusoidal voltages ($m = 3$), having the primitive period, are represented by the equations

$$e_{a2} = (2)^{1/2}E_{a2} \cos (\omega t + \alpha_{a2})$$

$$e_{b2} = (2)^{1/2}E_{a2} \cos \left(\omega t + \alpha_{a2} - \frac{4\pi}{3} \right)$$

$$e_{c2} = (2)^{1/2}E_{a2} \cos \left(\omega t + \alpha_{a2} - \frac{2\pi}{3} \right)$$

derived from the equation of symmetrical components of a set of polyphase (alternating) voltages. Since in this case $r = 1$ for every component (of first harmonic), the third subscript is omitted. Then k is 2 for $(m - 1)$st sequence,

and s takes on the algebraic values 1, 2, and 3 corresponding to phases a, b, and c. The sequence of maxima occurs in the order, a, c, b, which is the reverse or negative of the order for $k = 1$. *See:* **polyphase alternating currents; symmetrical components.** 210

negative plate (storage cell). The grid and active material to which current flows from the external circuit when the battery is discharging. *See also:* **battery (primary or secondary).**

negative-polarity lightning stroke. A stroke resulting from a negatively charged cloud that lowers negative charge to the earth. *See also:* **direct-stroke protection (lightning).** 64

negative-resistance device. A resistance in which an increase in current is accompanied by a decrease in voltage over the working range. 210

negative-resistance oscillator. An oscillator produced by connecting a parallel-tuned resonant circuit to a two-terminal negative-resistance device. (One in which an increase in voltage results in a decrease in current.) *Note:* Dynatron and transitron oscillators, arc converters, and oscillators of the semiconductor type are examples. *See also:* **oscillatory circuit.** 111, 211

negative-resistance repeater. A repeater in which gain is provided by a series or a shunt negative resistance or both. *See also:* **repeater.** 59

negative-sequence reactance. The ratio of the fundamental reactive component of negative-sequence armature voltage, resulting from the presence of fundamental negative-sequence armature current of rated frequency, to this current, the machine being operated at rated speed. *Notes:* (1) The rated current value of negative-sequence reactance is the value obtained from a test with a fundamental negative-sequence current equal to rated armature current. The rated voltage value of negative-sequence reactance is the value obtained from a line-to-line short-circuit test at two terminals of the machine at rated speed, applied from no load at rated voltage, the resulting value being corrected when necessary for the effect of harmonic components in the current. (2) For any unbalanced short-circuits, certain harmonic components of current, if present, may produce fundamental reactive components of negative-sequence voltage that modify the ratio of the total fundamental reactive component of negative-sequence voltage to the fundamental component of negative-sequence current. This effect can be included by multiplying the negative-sequence reactance, before it is used for short-circuit calculations, by a wave distortion factor, equal to or less than unity, that depends primarily upon the type of short-circuit, and upon the characteristics of the machine and the external circuit, if any, between the machine and the point of short-circuit. 328

negative-sequence resistance. The ratio of the fundamental component of in-phase armature voltage, due to the fundamental negative-sequence component of armature current to this component of current at rated frequency. *Note:* This resistance, which forms a part of the negative-sequence impedance for use in circuit calculations to establish relationships between voltages and currents, is not directly applicable in the calculations of the total loss in the machine caused by negative-sequence currents. This loss is the product of the square of the fundamental component of the negative-sequence current and the difference between twice the negative-sequence resistance and the positive-sequence resistance, that is, $I_2{}^2 (2R_2 - R_1)$.

328

negative temperature (laser-maser) An **effective temperature** used in the Boltzmann factor to describe a **population inversion.** *Note:* If n_2 particles populate the higher of two states and n_1 particles populate the lower state, their ratio is conventionally expressed by the Boltzmann factor

$$\frac{n_2}{n_1} = \exp(-kT)$$

where k is the Boltzmann constant and T the (effective) absolute temperature. 363

negative terminal (of a battery). The terminal toward which positive electric charge flows in the external circuit from the positive terminal. *Note:* The flow of electrons in the external circuit is to the positive terminal and from the negative terminal. *See also:* **battery (primary or secondary).** 328

negative-transconductance oscillator. Oscillator in which the output of the device is coupled back to the input without phase shift, the condition for oscillation being satisfied by the negative conductance of the device. *See also:* **oscillatory circuit.** 111

negative vectors. Two vectors are mutually negative if their magnitudes are the same and their directions opposite.
210

negentropy (information theory). *See:* **average information content.**

neighborhood. Of any point u_0, in a three-dimensional space, the volume enclosed by a sphere drawn with u_0 as center. Of any point u_0, in a two-dimensional space, the area enclosed by a circle drawn with u_0 as center. 210

neon indicator (tube). A cold-cathode gas-filled tube containing neon, used as a visual indicator of a potential difference or a field. *See:* **gas tubes.** 244, 190

neper. A division of the logarithmic scale such that the number of nepers is equal to the natural logarithm of the scalar ratio of two currents or two voltages. *Notes:* (1) With I_1 and I_2 designating the scalar value of two currents, and n the number of nepers denoting their scalar ratio: $n = \log_e(I_1/I_2)$. (2) 1 neper equals 0.8686 bel. (3) The neper is a dimensionless unit. 111, 210, 59

nerve-block (electrobiology). The application of a current to a nerve so as to prevent the passage of a propagated potential. *See also:* **excitability (electrobiology).** 192

nest or section (multiple system). A group of electrolytic cells placed close together and electrically connected in series for convenience and economy of operation. *See also:* **electrorefining.** 328

net assured capability (electric power supply). The net dependable capability of all power sources available to a system, including firm power contracts and applicable emergency interchange agreements, less that reserve assigned to provide for scheduled maintenance outages, equipment and operating limitations, and forced outages of power sources. 112

net capability (electric power supply). The maximum generation expressed in kilowatt hours per hour which a generating unit, station, power source, or system can be

expected to supply under optimum operating conditions.
112

net dependable capability (1) (electric power supply). The maximum generation, expressed in kilowatt-hours per hour which a generating unit, station, power source, or system can be depended upon to supply on the basis of average operating conditions. 112
(2) (power system engineering). The maximum system load, expressed in kilowatt-hours per hour that a generating unit, station, or power source can be depended upon to supply on the basis of average operating conditions. *Note:* This capability takes into account average conditions of weather, quality of fuel, degree of maintenance and other operating factors. It does not include provision for maintenance outages. *See also:* **generating station.**
64, 200

net generation (electric power systems). Gross generation less station or unit power requirements. 94

net head (hydroelectric power plant). The gross head less all hydraulic losses except those chargeable to the turbine.
112

net information content. A measure of the essential information contained in a message. *Note:* It is expressed as the minimum number of bits or hartleys required to transmit the message with specified accuracy over a noiseless medium. *See also:* **bit.** 328

net interchange (power and/or energy) (control area). The algebraic sum of the power and/or energy on the area tie lines. *Note:* Positive net interchange is due to excess generation out of the area. 94

net interchange deviation (control area) (electric power systems). The net interchange minus the scheduled net interchange. 94

net interchange schedule programmer (speed-governing system). A means of automatically changing the net interchange schedule from one level to another at a predetermined time and during a predetermined period or at a predetermined rate. *See also:* **speed-governing system.**
94

net load capability. The maximum system load expressed in kilowatt-hours per hour that a generating unit, station, or power source can be expected to supply under good operating conditions. *Notes:* (1) This capability provides for variations of load within the hour that it is assumed are to be spread among all of the power sources. (2) This is sometimes called net rated capability. *See also:* **generating station.** 64, 200

net loss (circuit equivalent*) (circuit). The net loss is the sum of all the transmission losses occurring between the two ends of the circuit, minus the sum of all the transmission gains. *See also:* **transmission loss.** 328
* Obsolescent

network (1) (distribution of electric energy). An aggregation of interconnected conductors consisting of feeders, mains, and services. *See also:* **alternating-current distribution; network analysis.** 64
(2) (computing systems). *See:* **computer network.** *See also:* **stabilization network.**
(3) (data transmission). (A) A series of points interconnected by communication channels. (B) The switched telephone network is the network of telephone lines normally used for dialed telephone calls. (C) A private net-

work is a network of communications channels confined to the use of one customer. 59

(4) (circuits and systems). Any interconnection of electrical elements, devices, apparatus, sources, transducers, switches, etcetera. 67

(5) (telephone equipment). A combination of elements or devices. *See also:* **element; device.** 39

network analysis (network). The derivation of the electrical properties, given its configuration and element values. 210

network analyzer. An aggregation of electric circuit elements that can readily be connected so as to form models of electric networks. *Note:* Each model thus formed can be used to infer the electrical quantities at various points on the prototype system from corresponding measurements on the model. *See also:* **electronic analog computer; oscillograph.** 328

network capacitance (charging inductor). The effective capacitance of the pulse-forming circuit. 137

network control (telephone switching systems). The means of determining and establishing the required connections in response to information received from the system control. 55

network feeder. A feeder that supplies energy to a network. *See also:* **center of distribution.** 64

network function. Any impedance function, admittance function, or other function of p that can be expressed in terms of or derived from the determinant of a network and its cofactors. *Notes:* (1) This includes not only impedance and admittance functions as previously defined, but also voltage ratios, current ratios, and numerous other quantities. (2) Certain network functions are sometimes classified together for a given purpose (for example, those with common zeros or common poles). These represent subgroups that should be specifically defined in each special case. (3) In the case of distributed-parameter networks, the determinant may be an infinite one; the definition still holds. *See also:* **network analysis.** 210

network leased line or private wire (data transmission). A series of points interconnected by telegraph or telephone channels, and reserved for the exclusive use of one customer. 59

network, load matching. *See:* **load matching network.**

network limiter. An enclosed fuse for disconnecting a faulted cable from a low-voltage network distribution system and for protecting the unfaulted portions of that cable against serious thermal damage. 103, 202

network master relay. A relay that functions as a protective relay by opening a network protector when power is back-fed into the supply system and as a programming relay by closing the protector in conjunction with the network phasing relay when polyphase voltage phasors are within prescribed limits. 103, 202, 60, 127

network phasing relay. A monitoring relay that has as its function to limit the operation of a network master relay so that the network protector may close only when the voltages on the two sides of the protector are in a predetermined phasor relationship. 103, 202, 60, 127

network primary distribution system. A system of alternating-current distribution in which the primaries of the distribution transformers are connected to a common network supplied from the generating station or substation

distribution buses. *See also:* **alternating-current distribution.** 64

network, private line telegraph. *See:* **private line telegraph network.**

network, private line telephone. *See:* **private line telephone network.**

network protector (transformer). An assembly comprising a circuit breaker and its complete control equipment for automatically disconnecting a transformer from a secondary network in response to predetermined electric conditions on the primary feeder or transformer, and for connecting a transformer to a secondary network either through manual control or automatic control responsive to predetermined electrical conditions on the feeder and the secondary network. *Note:* The network protector is usually arranged to connect automatically its associated transformer to the network when conditions are such that the transformer, when connected, will supply power to the network and to automatically disconnect the transformer from the network when power flows from the network to the transformer. *See also:* **network tripping and reclosing equipment.** 103, 202, 53

network restraint mechanism. A device that prevents opening of a network protector on transient power reversals that do not either exceed a predetermined value or persist for a predetermined time. 103, 202

network secondary distribution system. A system of alternating-current distribution in which the secondaries of the distribution transformers are connected to a common network for supplying light and power directly to consumers' services. *See also:* **alternating-current distribution.** 64

network synthesis (network). The derivation of the configuration and element values, with given electrical properties. *See also:* **network analysis.** 210

network transformer. A transformer designed for use in a vault to feed a variable capacity system of interconnected secondaries. *Note:* A network transformer may be of the submersible or of the vault type. It usually, but not always, has provision for attaching a network protector. 53

network tripping and reclosing equipment (networking equipment). An equipment that automatically connects its associated power transformer to an alternating-current network when conditions are such that the transformer, when connected, will supply power to the network and that automatically disconnects the transformer from the network when power flows from the network to the transformer. *See also:* **network protector.** 103, 202

neuroelectricity. Any electric potential maintained or current produced in the nervous system. *See also:* **electrobiology.** 192

neutral (rotating machinery). The point along an insulated winding where the voltage is the instantaneous average of the line terminal voltages during normal operation. *See:* **asynchronous machine.** 63

neutral conductor (when one exists) (circuit consisting of three or more conductors). The conductor that is intended to be so energized, that, in the normal steady state, the voltages from every other conductor to the neutral conductor, at the terminals of entry of the circuit into a delimited region, are definitely related and usually equal in amplitude. *Note:* If the circuit is an alternating-current circuit, it is intended also that the voltages have the same period and the phase difference between any two successive voltages, from each of the other conductors to the neutral conductor, selected in a prescribed order, have a predetermined value usually equal to 2π radians divided by the number of phase conductors m. *See also:* **center of distribution; network analysis.** 210

neutral direct-current telegraph system (single-current system) (single Morse system). A telegraph system employing current during marking intervals and zero current during spacing intervals for transmission of signals over the line. *See also:* **telegraphy.** 328

neutral ground. An intentional ground applied to the neutral conductor or neutral point of a circuit, transformer, machine, apparatus, or system. *See:* **ground.**
203, 64, 53

neutral grounding capacitor (electric power). A neutral grounding device the principal element of which is capacitance. *Note:* A neutral grounding capacitor is normally used in combination with other elements, such as reactors and resistors. *See also:* **grounding device.**
91, 64

neutral grounding device (electric power). A grounding device used to connect the neutral point of a system of electric conductors to earth. *Note:* The device may consist of a resistance, inductance, or capacitance element, or a combination of them. *See also:* **grounding device; neutral grounding impedor.** 91, 64

neutral grounding impedor (electric power). A neutral grounding device comprising an assembly of at least two of the elements resistance, inductance, or capacitance. *See also:* **grounding device; neutral grounding device.**
91, 64

neutral grounding reactor (1) (transformer). A current-limiting inductive reactor for connection in the neutral for the purpose of limiting and neutralizing disturbances due to ground faults. 53

(2) (neutral grounding devices). One in which the principal element of which is inductive reactance. 91

neutral grounding resistor (electric power). A neutral grounding device, the principal element of which is resistance. *See also:* **grounding device.** 91, 64

neutral grounding wave trap. A neutral grounding device comprising a combination of inductance and capacitance designed to offer a very high impedance to a specified frequency or frequencies. *Note:* The inductances used in neutral grounding wave traps should meet the same requirements as neutral grounding reactors. 91

neutralization. A method of nullifying the voltage feedback from the output to the input circuits of an amplifier through the tube interelectrode impedances. *Note:* Its principal use is in preventing oscillation in an amplifier by introducing a voltage into the input equal in magnitude but opposite in phase to the feedback through the interelectrode capacitance. *See also:* **amplifier; feedback.**
111, 211

neutralizing indicator. An auxiliary device for indicating the degree of neutralization of an amplifier. (For example, a lamp or detector coupled to the plate tank circuit of an amplifier.) *See also:* **amplifier.** 111

neutralizing voltage. The alternating-current voltage specifically fed from the grid circuit to the plate circuit (or

vice versa), deliberately made 180 degrees out of phase with, and equal in amplitude to, the alternating-current voltage similarly transferred through undesired paths, usually the grid-to-plate tube capacitance. *See also:* **amplifier.** 111

neutral keying (data transmission). Form of telegraph signal which has current either on or off in the circuit with ON as mark, OFF as space. 59

neutral lead (rotating machinery). A main lead connected to the common point of a star-connected winding. *See:* **asynchronous machine.** 63

neutral point (1) (system). The point that has the same potential as the point of junction of a group of equal nonreactive resistances if connected at their free ends to the appropriate main terminals or lines of the system. *Note:* The number of such resistances is 2 for direct-current or single-phase alternating-current; 4 for two-phase (applicable to 4-wire systems only) and 3 for three-, six- or twelve-phase systems. *See also:* **alternating-current distribution; center of distribution; direct-current distribution.**
64

(2) (transformer). (A) The common point of a Y connection in a polyphase system. (B) The point of a symmetrical system which is normally at zero voltage. 53

neutral relay (1) (sometimes called nonpolarized relay). A relay in which the movement of the armature does not depend upon the direction of the current in the circuit controlling the armature. 259

(2) (power circuits). A relay that responds to quantities in the neutral of a power circuit. *See also:* **electromagnetic relay; relay.** 103, 202, 60, 127

neutral terminal. The terminal connected to the neutral of a machine or apparatus. *See:* **asynchronous machine.**
63

neutral wave trap (electric power). A neutral grounding device comprising a combination of inductance and capacitance designed to offer a very high impedance to a specified frequency or frequencies. *See also:* **grounding device.** 91, 64

neutral zone (control element). The range of values of input for which no change in output occurs. *Note:* The neutral zone is an adjustable parameter in many two-step and floating control systems. *See also:* **dead band.** 56

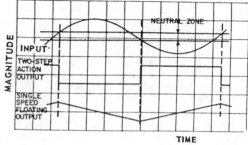

Neutral zone

new installation (elevators). Any installation not classified as an existing installation by definition, or an existing elevator, dumbwaiter, or escalator moved to a new location subsequent to the effective date of a code. *See also:* **elevators.** 328

new sync (data transmission). Allows for a rapid transition from one transmitter to another on multipoint private line data networks.   59

newton. The unit of force in the International System of Units (SI). The newton is the force that will impart an acceleration of 1 meter per second per second to a mass of 1 kilogram. One newton equals 10^5 dynes.   210

n-gate thyristor. A thyristor in which the gate terminal is connected to the n region adjacent to the region to which the anode terminal is connected and which is normally switched to the ON state by applying a negative signal between gate and anode terminals. See also: thyristor.
243, 66, 208, 191

Nichols chart (Nichols diagram) (control system, feedback). A plot showing magnitude contours and phase contours of the return transfer function referred to ordinates of logarithmic loop gain and to abscissae of loop phase angle. See also: control system, feedback.   56

nickel-cadmium storage battery. An alkaline storage battery in which the positive active material is nickel oxide and the negative contains cadmium. See also: battery (primary or secondary).   328

night. The hours between the end of evening civil twilight and the beginning of morning civil twilight. Note: Civil twilight ends in the evening when the center of the sun's disk is six degrees below the horizon and begins in the morning when the center of the sun's disk is six degrees below the horizon. See also: sunlight.   167

night alarm. An electric bell or buzzer for attracting the attention of an operator to a signal when the switchboard is partially attended.   328

night effect (radio navigation systems). A special case of error occurring predominantly at night when skywave propagation is at the maximum. See also: navigation.
187, 13

911 call (telephone switching systems). A call to an emergency service bureau.   55

nines complement. The radix-minus-one complement of a numeral whose radix is ten.   255, 77

ninety-percent response time of a thermal converter (electric instrument). The time required for 90 percent of the change in output electromotive force to occur after an abrupt change in the input quantity to a new constant value. See note 1 of response time of a thermal converter. See also: thermal converter.   280

nipple (rigid metal conduit). A straight piece of rigid metal conduit not more than two feet in length and threaded on each end. See also: raceways.   101

nit (nt) (1) (light emitting diodes). The unit of luminance (photometric brightness) equal to one candela per square meter. Note: Candela per square meter is the International Standard (SI) unit of luminance. The nit is the name recommended by the International Commission on Illumination. (CIE). (nit (nt) × .2919 = footlamberts*).
162

(2) (information theory). The unit in which information is measured when natural logarithms are used.   61

* Deprecated

n-level address (computing systems). A multilevel address that specifies n levels of addressing.   255, 77

noble potential. A potential substantially cathodic to the standard hydrogen potential. See also: stray current corrosion.   221, 205

no-busy test call (telephone switching systems). A call in which busy testing is inhibited.   55

node (network analysis). One of the set of discrete points in a flow graph. See also: linear signal flow graphs.
282

node (1) (network) (junction point) (branch point) (vertex). A terminal of any branch of a network or a terminal common to two or more branches of a network. See also: network analysis.   210

(2) (standing wave). A point at which the amplitude is zero. See also: radio transmission; transmission line.   210

node absorption (network analysis). A flow-graph transformation whereby one or more dependent nodes disappear and the resulting graph is equivalent with respect to the remaining node signals. Note: For example, a circuit analog of node absorption is the star-delta transformation. See also: linear signal flow graphs.   282

node equations (network). Any set of equations (of minimum number) such that the independent node voltages of a specified network may be determined from the impressed currents. Notes: (1) The number of node equations for a given network is not necessarily the same as the number of mesh or loop equations for that network. (2) Notes for mesh or loop equations, with appropriate changes, apply here. See also: network analysis.   210

node signal (network analysis). A variable X_k associated with node k. See also: linear signal flow graphs.   282

node voltage (network). The voltage from a reference point to any junction point (node) in a network. Note: The assumptions of lumped-network theory are such that the path of integration is immaterial.   210

no-go. See: go/no-go.

noise (1) (general). Unwanted disturbances superposed upon a useful signal that tend to obscure its information content. See also: electric noise; electromagnetic noise; electronic analog computer; signal-to-noise ratio; random noise.   9, 56

(2) An undesired disturbance within the useful frequency band. Note: Undesired disturbances within the useful frequency band produced by other services may be called interference.   111

(3) (sound recording and reproducing system). Any output power that tends to interfere with the utilization of the applied signals except for output signals that consist of harmonics and subharmonics of the input signals, intermodulation products, and flutter or wow.   350

(4) (phototubes). The random output that limits the minimum observable signal from the phototube. See also: phototube.   335, 117

(5) (facsimile). Any extraneous electric disturbance tending to interfere with the normal reception of a transmitted signal. See also: facsimile transmission.   12

(6) (hybrid computer linkage components). Unwanted disturbances superimposed upon a useful signal that tends to obscure its information content expressed in millivolts peak and referred to the input voltage.   10

(7) (oscilloscopes). Any extraneous electric disturbance tending to interfere with the normal display.   106

noise, audio-frequency. Any unwanted disturbance in the audio-frequency range.   239

noise-current generator. A current generator, the output of which is described by a random function of time. *Note:* At a specified frequency, a noise-current generator can often be adequately characterized by its mean-square current within the frequency increment Δf or by its spectral density. If the circuit contains more than one noise-voltage generator or noise-current generator, the correlation coefficients among the generators must also be specified. *See also:* **network analysis; signal; signal-to-noise ratio.**   188, 190

noise definitions (data transmission). dBa. For F1A weighted noise measurement; usually obtained with a WECO 2B noise meter. 0 dBa equivalent to 1000 hertz tone with a power of −85 dBm. Or, a 3 kilohertz white noise band of −82 dBm. Filter produces a 3 decibel loss over flat indication.

dBrnC. For C message weighted noise; 0 dBrnC is equivalent to a 1000 hertz tone with a power of −90 dBm (10^{-12} watt or 90 decibel below 1 milliwatt into R). Or, 3 kilohertz white noise band of a power of 88 dBm (actually 88.5 dBm). Filter produces a 2 decibel loss compared to no weighting.

pwp. For psophometrically weighted noise; 1 pWp is equivalent to a 1000 hertz tone with a power of −91 dBm (or an 800 hertz tone of −90 dBm). Or, a 3 kilohertz band of white noise with a power of 88 dBm.

$$1pWp = \frac{1}{1.78}\,pW = 0.56\,pW$$

dBmp. Filter produces a 2.5 decibel loss compared to no weighting. For psophometrically weighted noise according to CCITT; 0 dBmp is equivalent to a 1000 hertz tone with a power of −1 dBm (or an 800 hertz tone with a power of 0 dBm). Or, a 3 kilohertz band of white noise with a power of +2.5 dBm. Filter produces a 2.5 decibel loss compared to no weighting.

dBrn 144. Obsolete unit for Type 144 weighted noise measurement.

dBnrCo, dBaO, pWpO, dBmOp. Noise units as above measured at (or referred to) a 0 transmission on level point.

S/N psoph. S/N measured with psophometrically weighted receiver; expressed in decibels (or dBmOp).

NPR. (noise-power ratio) (expressed in decibels). Term commonly used in noise loading technique. Usually an uncalibrated receiver is used to measure the noise power in a channel of a system loaded with noise, first with full noise in the channel and the noise source. The ratio of these readings is the NPR. An NPR reading is independent of the noise bandwidth of the receiver; provided the same bandwidth is used in both noise measurements and the band stop filters are wide enough.   59

noise diode, ideal. A diode that has an infinite internal impedance and in which the current exhibits full shot noise fluctuations. *See also:* **signal-to-noise ratio.**   190

noise, electrical. *See:* **electrical noise.**

noise factor (two-port transducer). At a specified input frequency the ratio of (1) the total noise power per unit bandwidth at a corresponding output frequency available at the output port when the noise temperature of its input termination is standard (290°K) at all frequencies (Ref-

erence: Definition for average noise factor to (2) that portion of (1) engendered at the input frequency by the input termination at the standard noise temperature 290°K). *Notes:* (A) For heterodyne systems there will be, in principle, more than one output frequency corresponding to a single input frequency, and vice versa; for each pair of corresponding frequencies a noise factor is defined. (2) includes only that noise from the input termination which appears in the output via the principal-frequency transformation of the system, that is, via the signal-frequency transformation(s), and does not include spurious contributions such as those from an unused image-frequency or an unused idler-frequency transformation. (B) The phrase "available at the output port" may be replaced by "delivered by system into an output termination." (C) To characterize a system by a noise factor is meaningful only when the admittance (or impedance) of the input termination is specified. *Syn:* **noise figure.**   125

noise factor, average (two-port transducer). The ratio of (1) the total noise power delivered by the transducer into its output termination when the noise temperature of its input termination is standard (290°K) at all frequencies, to (2) that portion of (1) engendered by the input termination. *Notes:* (A) For heterodyne systems, (2) includes only that noise from the input termination which appears in the output via the principal-frequency transformation of the system, that is, via the signal-frequency transformation(s), and does not include spurious contributions such as those from an unused image-frequency or an unused idler-frequency transformation. (B) A quantitative relation between the average noise factor $\overline{F}$ and the spot noise factor $F(f)$ is

$$\overline{F} = \frac{\displaystyle\int_0^\infty F(f)G(f)df}{\displaystyle\int_0^\infty G(f)df}$$

where f is the input frequency, and $G(f)$ is the transducer gain, that is, the ratio of (1) the signal power delivered by the transducer into its output termination, to (2) the corresponding signal power available from the input termination at the input frequency. For heterodyne systems, (1) comprises only power appearing in the output via the principal-frequency transformation that is, via the signal-frequency transformation(s) of the system; for example, power via unused image-frequency or unused idler-frequency transformation is excluded. (C) To characterize a system by an average noise factor is meaningful only when the admittance (or impedance) of the input termination is specified. *Syn:* **noise figure.**   125

noise figure, average (communication satellite). Of a two-port transducer the ratio of the total noise power to the input noise power, when the input termination is at the standard temperature of 290° kelvin. *See also:* **noise factor.**   85

noise-free equivalent amplifier (signal-transmission system). An ideal amplifier having no internally generated noise that has the same gain and input/output characteristics as the actual amplifier. *See:* **signal.**   188

noise generator (analog computer). A computing element used purposely to introduce noise of specified amplitude distribution, spectral density, and/or root-mean-square value into other computing elements. *See also:* **electronic analog computer.**   9, 77

noise generator diode (electron device). A diode in which the noise is generated by shot effect and the noise power of which is a definite function of the direct current.
                                                                    190

noise killer (telegraph circuit). An electric network inserted usually at the sending end, for the purpose of reducing interference with other communication circuits. *See also:* **telegraphy.**   328

noise level (1) (audio and electroacoustics). (A) The noise power density spectrum in the frequency range of interest, (B) the average noise power in the frequency range of interest, or (C) the indication on a specified instrument. *Notes:* (1) In (C), the characteristics of the instrument are determined by the type of noise to be measured and the application of the results thereof. (2) Noise level is usually expressed in decibels relative to a reference value. *See also:* **signal-to-noise ratio.**   239

(2) (speech quality measurements). The *A*-weighted sound level of the noise.   126

noise-level test (rotating machinery). A test taken to determine the noise level produced by a machine under specified conditions of operation and measurement. *See also:* **asynchronous machine; direct-current commutating machine.**   63

noise linewidth (semiconductor radiation detectors). The contribution of noise to the width of a spectral peak.
                                                          23, 118, 119

noise-longitudinal (telephone practice). The 1/1000th part of the total longitudinal-circuit noise current at any given point in one or more telephone wires. *See also:* **inductive coordination.**   59

noise measurement units (data transmission). The following units are used to express weighted and unweighted circuit noise. They include terms used in American and International practice. (A) DBA-DBRN adjusted. Weighted circuit noise power, in db, referred to 3.16 picowatts (−85 dbm) which is 0 dba. Use of F1A line or HA 1 receiver weighting shall be indicated in parenthesis, as required. (B) DBRN-(Describes above reference noise.) Weighted circuit noise power in db referred to 1.0 picowatt (−90 dbm) which is 0 dbrn. Use of 144 lines, 144 receiver, or C message weighting, as required. With C message weighting, as 1 milliwatt, 1000 hertz tone will read +90 dbrn, but the power as white noise randomly distributed over a 3kc band (nominally 300 to 3300 hertz) will read approximately +88.5 dbrn (rounded off to +88 dbrn, due to frequency weighting). With 144 weighting, as 1 milliwatt, 1000 hertz tone will also read +90 dbrn, but the same 3kc white noise power would read only +82 dbrn, due to the different frequency weighting.

noise measuring set. *See:* **circuit noise meter.**

noise-metallic (metallic circuit). The weighted noise current at a given point when the circuit is terminated at that point in the nominal characteristic impedance of the circuit. *See also:* **inductive coordination.**   59

noise power (Hall effect devices). The power generated by a random electromagnetic process.   107

noise pressure equivalent (electroacoustic transducer or system used for sound reception). The root-mean-square sound pressure of a sinusoidal plane progressive wave that, if propagated parallel to the principal axis of the transducer, would produce an open-circuit signal voltage equal to the root-mean-square of the inherent open-circuit noise voltage of the transducer in a transmission band having a bandwidth of 1 hertz and centered on the frequency of the plane sound wave. *Note:* If the equivalent noise pressure of the transducer is a function of secondary variables, such as ambient temperature or pressure, the applicable values of these quantities should be stated explicitly. *See also:* **phonograph pickup.**   176

noise quieting (receiver) (receiver performance). A measure of the quantity of radio-frequency energy, at a specified deviation from the receiver center frequency, required to reduce the noise output by a specified amount. *See also:* **receiver performance.**   181

noise reduction (photographic recording and reproducing). A process whereby the average transmission of the sound track of the print (averaged across the track) is decreased for signals of low level and increased for signals of high level. *Note:* Since the noise introduced by the sound track is less at low transmission, this process reduces film noise during soft passages. The effect is normally accomplished automatically. *See also:* **phonograph pickup.**   176

noise sidebands (non-real time spectrum analyzer). Undesired response caused by noise internal to the spectrum analyzer appearing on the display around a desired response.   68

noise temperature (1) (general) (at a pair of terminals and at a specific frequency). The temperature of a passive system having an available noise power per unit bandwidth equal to that of the actual terminals. *Note:* Thus, the noise temperature of a simple resistor is the actual temperature of the resistor, while the noise temperature of a diode may be many times the observed absolute temperature. *See also:* **signal; signal-to-noise ratio.**   188

(2) (standard). The standard reference temperature T_0 for noise measurements is 290°K. *Note:* $kT_0/e = 0.0250$ volt, where e is the magnitude of the electronic charge and k is Boltzmann's constant.   125

(3) (antenna). The temperature of a resistor having an available thermal noise power per unit bandwidth equal to that at the antenna output at a specified frequency. *Note:* Noise temperature of an antenna depends on its coupling to all noise sources in its environment as well as noise generated within the antenna. *See also:* **antenna.**
                                                          179, 111

(4) (at a port and at a selected frequency). A temperature given by the exchangeable noise-power density divided by Boltzmann's constant, at a given port and at a stated frequency. *Notes:* (A) When expressed in units of kelvins, the noise temperature T is given by the relation

$$T = N/k$$

where N is the exchangeable noise-power density in watts per hertz at the port at the stated frequency and k is Boltzmann's constant expressed as joules per kelvin ($k \simeq 1.38 \times 10^{-23}$ joules per kelvin). (B) Both N and T are negative for a port with an internal impedance having a negative real part. *See also:* **signal-to-noise ratio; wave-guide.**   190

(5)(at a port). The temperature of a passive system having an available noise power per unit bandwidth equal to that of the actual port, at a specified frequency. A uniform temperature throughout the passive system is implied. *See also:* **thermal noise.**   155, 125

noise temperature, average operating (communication satellite). Equivalent temperature of passive system having an available noise power equal to that of the operating system. In space communication systems the noise temperature is generally composed of contributions from the background, atmospheric and receiver front end noise.
                                                                85

noise-to-ground (telephone practice). The weighted noise current through the 100 000-ohm circuit of a circuit noise meter, connected between one or more telephone wires and ground. *See also:* **inductive coordination.**   59

noise transmission impairment (NTI). The noise transmission impairment that corresponds to a given amount of noise is the increase in distortionless transmission loss that would impair the telephone transmission over a substantially noisefree circuit by an amount equal to the impairment caused by the noise. Equal impairments are usually determined by judgment tests or intelligibility tests.   328

noise unit.* An amount of noise judged to be equal in interfering effect to the one-millionth part of the current output of a particular type of standard generator of artificial noise, used under specified conditions. *Note:* This term was formerly used in connection with ear balance measurements, but has been largely superseded by dBa employed with indicating noise meter. Approximately seven noise units of noise on a telephone line are frequently taken as equivalent to reference noise. *See also:* **signal-to-noise ratio.**   328

* Obsolescent

noise-voltage generator (interference terminology). A voltage generator the output of which is described by a random function of time. *Note:* At a specified frequency, a noise-voltage generator can often be adequately characterized by its mean-square voltage with the frequency increment Δf or by its spectral density. If the circuit contains more than one noise-current generator or noise-voltage generator, the correlation coefficients among the generators must be specified. *See:* **network analysis; signal; signal-to-noise ratio.**   188, 190, 125

noise weighting (data transmission). In measurement of circuit noise, a specific amplitude frequency characteristic of a noise measuring set. It is designed to give numerical readings which approximate the amount of transmission impairment due to the noise, to an average listener, using a particular class of telephone subset. The noise weightings generally used were established by the agencies concerned with public telephone service and are based on characteristics of specific commercial telephone subsets, representing successive stages of technological development. The coding of commercial apparatus appears in the nomenclature of certain weightings. The same weighting nomenclature and units are used in the military versions as well as the commercial noise measuring sets. (1) 144

Line Weighting. A noise weighting used in a noise measuring set to measure noise on a line that would be terminated by a subset with a number 144 receiver or a similar subset. Meter scale readings are in dbrn (144 line). (2) F1A Line Weighting. A noise weighting used in a noise measuring set to measure noise on a line that would be terminated by a 302 type, or similar, subset. The meter scale readings are in dba (F1A). (3) HA1 Receiver Weighting. A noise weighting used in a noise measuring set to measure noise across the HA1 receiver of a 302 type, or similar, subset. The meter scale readings are in the dba (HA1).   59

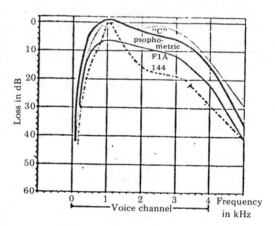

Comparison of weighting curves

(4) C-Message Weighting. A noise weighting used in a noise measuring set to measure noise on a line that would be terminated by a 500 type, or similar, subset. The meter scale readings are in dbrn (C-Message). (5) Flat Weighting. A noise measuring set measuring amplitude frequency characteristics which are flat over a specified frequency range. The frequency range must be stated. Flat noise power may be expressed in dbrn (F1–F2) or in dbm (F1–F2). The terms "3kc flat weighting" and "15kc flat weighting" refer to a range from 30 hertz to the upper frequency indicated.   59

no-load (adjective) (rotating machinery). The state of a machine rotating at normal speed under rated conditions, but when no output is required of it. *See also:* **asynchronous machine; direct-current commutating machine.**
                                                                63

no-load field voltage. The voltage required across the terminals of the field winding of an electric machine under conditions of no load, rated speed and terminal voltage, and with the field winding at 25 degrees Celsius. *See:* **direct-current commutating machine.**   63

no-load (excitation) losses. Those losses which are incident to the excitation of the transformer. No-load (excitation) losses include core loss, dielectric loss, conductor loss in the winding due to exciting current, and conductor loss due to circulating current in parallel windings. These losses change with the excitation voltage.   53

no-load saturation curve (no-load characteristic) (of a synchronous machine). The saturation curve of a machine on no-load.   63

no-load speed (industrial control) (of an electric drive). The speed that the output shaft of the drive attains with no external load connected and with the drive adjusted to deliver rated output at rated speed. *Note:* In referring to the speed with no external load connected and with the drive adjusted for a specified condition other than for rated output at rated speed, it is customary to speak of the no-load speed under the (stated) conditions. *See also:* **electric drive.**   206

no-load test (synchronous machine). A test in which the machine is run as a motor providing no useful mechanical output from the shaft.   63

nomenclature (electric power system). The words and terms used to identify electric power systems.   58

nominal band of regulated voltage. The band of regulated voltage for a load range between any load requiring no-load field voltage and any load requiring rated-load field voltage with any compensating means used to produce a deliberate change in regulated voltage inoperative. *See:* **direct-current commutating machine.**   328

nominal collector ring voltage (rotating machinery). *See:* **rated-load field voltage.**

nominal control current (Hall generator). That value of control current that, if exceeded, will cause the linearity error of the device to exceed a rated magnitude.   107

nominal discharge current (arrester). The discharge current having a designated peak value and waveshape, that is used to classify an arrester with respect to protective characteristics. *Note:* It is also the discharge current that is used to initiate follow current in the operating duty test.   62

nominal input power (loudspeaker measurements). Nominal input power is equal to the square of the true rms voltage at the input terminals of the loudspeaker, divided by the rated impedance. *See also:* **rated impedance.**   19

nominal line pitch (television). The average separation between centers of adjacent lines forming a raster. *See also:* **television.**   163, 178

nominal line width (1) (television). The reciprocal of the number of lines per unit length in the direction of line progression. *See also:* **television.**   328

(2) (facsimile). The average separation between centers of adjacent scanning or recording lines. *See also:* **recording (facsimile); scanning (facsimile).**   12

nominal metallic impedance (telephone equipment). Impedance based on lumped constants of a metallic circuit at a given single frequency. *See also:* **impedance; metallic circuit.**   39

nominal pull-in torque (synchronous motor). The torque it develops as an induction motor when operating at 95 percent of synchronous speed with rated voltage applied at rated frequency. *Note:* This quantity is useful for comparative purposes when the inertia of the load is not known.   63

nominal rate of rise (impulse) (electric power) (transformer). The slope of the line that determines the virtual zero. *Note:* It is usually expressed in volts or amperes per microsecond.   2, 299, 62, 53

nominal synchronous-machine excitation-system ceiling voltage. The ceiling voltage of the excitation system with: (1) The exciter and all rotating elements at rated speed. (2) The auxiliary supply voltages and frequencies at rated values. (3) The excitation system loaded with a resistor having a value equal to the resistance of the field winding to be excited at a temperature of: (A) 75 degrees Celsius for field windings designed to operate at rating with a temperature rise of 60 degrees Celsius or less, and (B) 100 degrees Celsius for field windings designed to operate at rating with a temperature rise greater than 60 degrees Celsius. (4) The manual control means adjusted as it would be to produce the rated voltage of the excitation system, if this manual control means is not under the control of the voltage regulator when the regulator is in service, unless otherwise specified. (The means used for controlling the exciter voltage with the voltage regulator out of service is normally called the manual control means.) (5) The voltage sensed by the synchronous machine voltage regulator reduced to give the maximum output from the regulator. Note that the regulator action may be simulated in test by applying to the field of the exciter under regulator control the maximum output developed by the regulator.   63

nominal system voltage (1) (surge arrester). A nominal value assigned to designate a system of a given voltage class. *Notes:* (A) See American National Standard Guide for Preferred Voltage Ratings for Alternating Current Systems and Equipment, C84.1-1970. (B) The actual voltage may be at variance with the nominal voltage. In fact, depending on the location on the circuit and the conditions of loading under which it is measured, the voltage usually will be other than the nominal value. (C) The term nominal voltage designates the line-to-line voltage, as distinguished from the line-to-neutral voltage. It applies to all parts of the system or circuit. (D) The term "nominal system voltage" designates not a single voltage, but a range of voltages over which the actual voltage at any point on the system may vary and still provide satisfactory operation of equipment connected to the system.   2, 103, 299, 62, 202

(2) (transformer). The system voltage by which the system is designated and to which certain operating characteristics of the system are related. (The nominal voltage of a system is near the voltage level at which the system normally operates and provides a per unit base voltage for system study purposes. To allow for operating contingencies, systems generally operate at voltage levels about 5 to 10 percent below the maximum system voltage for which system components are designed.)   53

nominal thickness (cable element) (power distribution, underground cables). The specified, indicated, or named thickness. *Note:* In general, measured thicknesses will approximate but will not necessarily be identical with nominal thicknesses. *See also:* **power distribution, underground construction.**   57

nonatmospheric paths (atmospheric correction factors to dielectric tests). Paths, such as through a gas or vacuum sealed from the atmosphere, through a liquid such as oil, or through a solid, or a combination thereof.   50

nonautomatic. The implied action that requires personal intervention for its control. *Note:* As applied to an electric controller, nonautomatic control does not necessarily imply a manual controller, but only that personal intervention is required. *See also:* **electric controller.**   206

nonautomatic opening (nonautomatic tripping). The opening of a switching device only in response to an act of an attendant. 103, 202

nonautomatic operation. Operation controlled by an attendant. 103, 202

nonautomatic tripping. *See:* **nonautomatic opening.**

nonblocking switching network (telephone switching systems). A switching network in which any idle outlet can always be reached from any given inlet under all traffic conditions. 55

noncode fire-alarm system. A local fire-alarm system in which the alarm signal is continuous and is usually sounded by vibrating bells. *See also:* **protective signaling.** 328

noncoherent MTI (moving-target indication). An MTI in which a moving target is detected without use of an internal reference phase signal. 13

noncoincident demand (electric power utilization). The sum of the individual maximum demands regardless of time of occurrence within a specified period. *See also:* **alternating-current distribution.** 200

noncomposite color-picture signal (NTSC color television). The electric signal that represents complete color-picture information, but excludes line and field synchronizing signals. 18

nonconforming load (electric power systems). A customer load, the characteristics of which are such as to require special treatment in deriving incremental transmission losses. 99

noncontact plunger. *See:* **choke piston.**

noncontinuous electrode. A furnace electrode the residual length of which is discarded when too short for further effective use. *See also:* **electrothermics.** 328

noncritical failure (test, measurement and diagnostic equipment). Any failure which degrades performance or results in degraded operation requiring special operating techniques or alternative modes of operation which could be tolerated throughout a mission but should be corrected immediately upon completion of mission. 54

nondestructive read (computing systems) (1) (general). A read process that does not erase the data in the source. 255, 77

(2) (magnetic cores). A method of reading the magnetic state of a core without changing its state. 331

nondestructive reading (charge-storage tubes). Reading that does not erase the stored information. *See also:* **charge-storage tube.** 174, 190, 54

nondestructive testing (test, measurement and diagnostic equipment). Testing of a nature which does not impair the usability of the item. 54

nondirectional beacon (NDB) (air navigation). A radio facility that can be used with an airborne direction-finder to provide a line of position; also known as a compass locator, *H, H* beacon. *See also:* **navigation.** 187, 13

nondirectional microphone. *See:* **omnidirectional microphone.**

nondisconnecting fuse. An assembly consisting of a fuse unit or fuseholder and a fuse support having clips for directly receiving the associated fuse unit or fuseholder, and that has no provision for guided operation as a disconnecting switch. 103, 202

nonenclosed (transformer). Not surrounded by a medium which will prevent a person accidentally contacting live parts. 53

nonenclosed switches, indoor or outdoor (power switchgear).

Switches designed for service without a housing restricting heat transfer to the external medium. 27, 103

nonenergy-limiting transformer. A constant-potential transformer that does not have sufficient inherent impedance to limit the output to a thermally safe maximum value. 53

nonerasable storage (computing systems). *See:* **fixed storage.**

nonexposed installation (lightning). An installation in which the apparatus is not subject to overvoltages of atmospheric origin. *Note:* Such installations are usually connected to cable networks. 244, 62

nonhoming (telephone switching systems). Resumption of a sequential switching operation from its last setting. 55

noninjecting contact (semiconductor radiation detector). A contact at which the carrier density in the adjacent semiconductor material is not changed from its equilibrium value. 23, 118, 119

noninterference testing (test, measurement and diagnostic equipment). A type of on-line testing that may be carried out during normal operation of the unit under test without affecting the operation. *See:* **interference testing.** 54

noninverting parametric device. A parametric device whose operation depends essentially upon three frequencies, a harmonic of the pump frequency and two signal frequencies, of which one is the sum of the other plus the pump harmonic. *Note:* Such a device can never provide gain at either of the signal frequencies. It is said to be noninverting because if either of the two signals is moved upward in frequency, the other will move upward in frequency. *See also:* **parametric device.** 277, 191

nonlinear capacitor. A capacitor having a mean-charge characteristic or a peak-charge characteristic that is not linear, or a reversible capacitance that varies with bias voltage. 191

nonlinear circuit. *See:* **nonlinear network.**

nonlinear distortion. Distortion caused by a deviation from a desired linear relationship between specified measures of the output and input of a system. *Note:* The related measures need not be output and input values of the same quantity; for example, in a linear detector the desired relation is between the output signal voltage and the input modulation envelope; or the modulation of the input carrier and the resultant detected signal. *See also:* **close-talking pressure-type microphone; distortion.** 239

nonlinear element capacitance (varactor measurements). The capacitance of the high frequency equivalent series RC circuit. 136

nonlinear ideal capacitor. An ideal capacitor whose transferred-charge characteristic is not linear. *See also:* **nonlinear capacitor.** 191

nonlinearity (gyro; accelerometer). The systematic deviations from the least squares straight line for input-output relationships which nominally can be represented by a linear equation. 46

nonlinear network (signal-transmission system). (1) A network (circuit) not specifiable by linear differential equations with time and/or position coordinates as the independent variable. *Note:* It will not operate in accordance with the superposition theorem. (2) A network (circuit) in which the signal transmission characteristics depend on the input signal magnitude. *See also:* **network analysis; signal.** 210, 188

nonlinear parameter. A parameter dependent on the magnitude of one or more of the dependent variables or driving forces of the system. *Note:* Examples of dependent variables are current, voltage, and analogous quantities. 210

nonlinear resistor-type arrester (valve type). An arrester having a single or a multiple spark gap connected in series with nonlinear resistance. *Note:* If the arrester has no series gap, the characteristic element limits the follow current to a magnitude that does not interfere with the operation of the system. *See:* **arrester, valve type.** 62

nonlinear series resistor (arrester). The part of the lightning arrester that, by its nonlinear voltage-current characteristics, acts as a low resistance to the flow of high discharge currents thus limiting the voltage across the arrester terminals, and as a high resistance at normal power-frequency voltage thus limiting the magnitude of follow current. 62

nonlinear system or element. *See:* **linear system or element.**

nonlined construction (primary cell) (dry cell). A type of construction in which a layer of paste forms the only medium between the depolarizing mix and the negative electrode. *See also:* **electrolytic cell.** 328

nonload-break connector. A connector designed to be separated and engaged on deenergized circuits. 134

nonloaded Q (basic Q) (of an electric impedance). The value of Q of such impedance without external coupling or connection. 328

nonmagnetic relay armature stop. *See:* **relay armature stop, nonmagnetic.**

nonmagnetic ship. A ship constructed with an amount of magnetic material so small that it causes negligible distortion of the earth's magnetic field. *See also:* **degauss.** 328

nonmechanical switching device. A switching device designed to close or open, or both, one or more electric circuits by means other than by separable mechanical contacts. 103, 202

nonmetallic sheathed cable. An assembly of two or more insulated conductors having an outer sheath of moisture-resistant, flame-retardant, nonmetallic material. *See also:* **armored cable (in interior wiring).** 328

nonminimum phase function (linear networks) (circuits and systems). A network function that is not minimum-phase. *See also:* **minimum phase function.** 67

nonmultiple switchboard (telephone switching systems). A telecommunications switchboard having each line connected to only one jack. 55

nonmultiple transit spurious signals (dispersive and nondispersive delay lines). Signals not related to the main signal delay by a simple integer may be labeled by the delay time of that signal. 81

nonnumerical action (switch). That action that does not depend on the called number (such as hunting an idle trunk). 328

nonoperate value. *See:* **relay nonoperate value.**

nonphantomed circuit. A two-wire or four-wire circuit that is not arranged to form part of a phantom circuit. *See also:* **transmission line.** 328

nonphysical primary (color). A primary represented by a point outside the area of the chromaticity diagram enclosed by the spectrum locus and the purple boundary. *Note:* Nonphysical primaries cannot be produced because they require negative power at some wavelengths. However, they have properties that facilitate colorimetric calculation. 18, 178

nonplanar network. A network that cannot be drawn on a plane without crossing of branches. *See also:* **network analysis.** 210

nonpolarized electrolytic capacitor. An electrolytic capacitor in which the dielectric film is formed adjacent to both metal electrodes and in which the impedance to the flow of current is substantially the same in both directions. *See also:* **electrolytic capacitor.** 328

n-on-p solar cells (photovoltaic power system). Photovoltaic energy-conversion cells in which a base of p-type silicon (having fixed electrons in a silicon lattice and positive holes that are free to move) is overlaid with a surface layer of n-type silicon (having fixed positive holes in a silicon lattice with electrons that are free to move). *See also:* **photovoltaic power system; solar cells (photovoltaic power system).** 186

nonquadded cable. *See:* **paired cable.**

nonreciprocal (circuits and systems). A device or network that does not have the property of reciprocity. 67

nonrelevant failure. *See:* **failure, nonrelevant.**

nonrenewable fuse (one-time fuse). A fuse not intended to be restored for service after circuit interruption. 103, 202

nonrepaired item (reliability). *See:* **item, nonrepaired.**

nonrepetitive peak OFF-state voltage (thyristor). The maximum instantaneous value of any nonrepetitive transient OFF-state voltage that occurs across the thyristor. *See also:* **principal voltage-current characteristic (principal characteristic).** 243, 66, 208, 191

non-repetitive transient reverse voltage (1) (reverse-blocking thyristor). The maximum instantaneous value of any unrepetitive transient reverse voltage that occurs across a thyristor. *See also:* **principal voltage-current characteristic (principal characteristic).** 243, 66, 208, 191 **(2) (semiconductor rectifier).** The maximum instantaneous value of the reverse voltage, including all nonrepetitive transient voltages but excluding all repetitive transient voltages, that occurs across a semiconductor rectifier cell, rectifier diode, or rectifier stack. *See also:* **rectification; semiconductor rectifier stack.** 237, 208

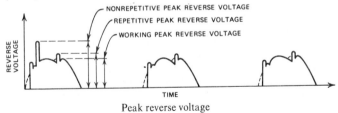

Peak reverse voltage

non-repetitive peak reverse voltage rating (rectifier circuit) (semiconductor rectifier). The maximum value of non-repetitive peak reverse voltage permitted by the manufacturer under stated conditions. *See also:* **average forward-current rating (of a rectifier circuit).**

237, 66, 208

nonrequired time (availability). The period of time during which the user does not require the item to be in a condition to perform its required function. 164

nonrestorable fire detector (fire protection devices). A device whose sensing element is designed to be destroyed by the process of detecting a fire. 71

nonreturn to zero (NRZ) code (power-system communication). A code form having two states termed zero and one, and no neutral or rest condition. *See also:* **digital.** 59

nonreversible output (magnetic amplifier). *See:* **maximum test output voltage.**

nonreversible power converter (semiconverter). An equipment containing assemblies of mixed power thyristor and diode devices that is capable of transferring energy in only one direction (that is, from the alternating-current side to the direct-current side). *See:* **power rectifier.** 208

nonreversing (industrial control). A control function that provides for operation in one direction only. *See also:* **control system, feedback.** 225, 206

nonsalient pole (rotating machinery). The part of a core, usually circular, that by virtue of direct-current excitation of a winding embedded in slots and distributed over the interpolar (and possibly over some or all of the polar) space, acts as a pole. *See also:* **asynchronous machine.**

63

nonsalient-pole machine. *See:* **cylindrical-rotor machine.**

nonsegregated-phase bus (metal enclosed bus). One in which all phase conductors are in a common metal enclosure without barriers between the phases. *Note:* When associated with metal-clad switchgear, the primary bus conductors and connections are or may be covered with insulating material throughout. 79, 78, 103, 202

nonselective collective automatic operation (elevators). Automatic operation by means of one button in the car for each landing level served and one button at each landing, wherein all stops registered by the momentary actuation of landing or car buttons are made irrespective of the number of buttons actuated or of the sequence in which the buttons are actuated. *Note:* With this type of operation the car stops at all landings for which buttons have been actuated, making the stops in the order in which the landings are reached after the buttons have been actuated, but irrespective of its direction of travel. *See also:* **control (elevators).** 328

nonself-maintained discharge (gas). A discharge characterized by the fact that the charged particles are produced solely by the action of an external ionizing agent. *See also:* **discharge (gas).** 244, 190

nonself-restoring insulation. An insulation which loses its insulating properties or does not recover them completely, after a disruptive discharge caused by the application of a test voltage; insulation of this kind is generally, but not necessarily, internal insulation. 53

nonstop switch (elevators). A switch that, when operated, will prevent the elevator from making registered landing

stops. *See:* **control.** 328

nonstorage display (display storage tubes). Display of nonstored information in the storage tube without appreciably affecting the stored information. *See also:* **storage tube.** 174, 190

nonsynchronous transmission (data transmission). A transmission process such that between any two significant instants in the same group, there is always an integral number of unit intervals. Between two significant instants located in different groups, there is not always an integral number of unit intervals. *Note:* In data transmission this group is a block or a character. In telegraphy this group is a character. 194

nontouching loop set (network analysis). A set of loops no two of which have a common node. *See also:* **linear signal flow graphs.** 282

nonuniformity (transmission lines and waveguides). The degree with which a characteristic quantity, for example, impedance, deviates from a constant value along a given path. *Note:* It may be defined as the maximum amount of deviation from a selected nominal value. For example, the nonuniformity of the characteristic impedance of a slotted coaxial line may be 0.05 ohm due to dimensional variations. 185

nonvented fuse. A fuse without intentional provision for the escape of arc gases, liquids, or solid particles to the atmosphere during circuit interruption. 103, 202

nonventilated (transformer). So constructed as to provide no intentional circulation of external air through the enclosure. 53

nonventilated dry-type transformer. A dry-type transformer which is so constructed as to provide no intentional circulation of external air through the transformer. 53

nonventilated enclosure (1) (general). An enclosure designed to provide limited protection against the entrance of dust, dirt, or other foreign objects. *Note:* Doors or removable covers are usually gasketed and humidity control may be provided by filtered breathers. 103, 202
(2) (metal enclosed bus). An enclosure so constructed as to provide no intentional circulation of external air through the enclosure. 78

nonvolatile storage (test, measurement and diagnostic equipment). A storage device which can retain information in the absence of power. Contrast to volatile storage. 54

no op (computing systems). An instruction that specifically instructs the computer to do nothing, except to proceed to the next instruction in sequence. 255, 77

NOR. A logic operator having the property that if P is a statement, Q is a statement, R is a statement, . . . then the NOR of $P, Q, R, . . .$ is true if all statements are false, false if at least one statement is true. 255, 77

norator (circuits and systems). A two-terminal ideal element the current through which and the voltage across which can each be arbitrary. 67

normal (1) (state of a superconductor). The state of a superconductor in which it does not exhibit superconductivity. *Example:* Lead is normal at temperatures above a critical temperature. *See:* **superconducting.** *See also:* **superconductivity.** 191
(2) (power generation). *See:* **preferred.**

normal clear. A term used to express the normal indication of the signals in an automatic block system in which an indication to proceed is displayed except when the block is occupied. 328

normal clear system. A term describing the normal indication of the signals in an automatic block signal system in which an indication to proceed is displayed except when the block is occupied. *See also:* **centralized traffic control system.** 328

normal contact. A contact that is closed when the operating unit is in the normal position. 328

normal (through) dielectric heating applications. The metallic electrodes are arranged on opposite sides of the material so that the electric field is established through it. *Note:* The electrodes may be classified as plate electrodes, roller electrodes, or concentric electrodes. (1) Plate electrodes may have plane surfaces or surfaces of any desired shape to meet a particular condition and the spacing between them may be uniform or varied. (2) Roller electrodes are rollers separated by the material which moves between them. (3) Concentric electrodes consist of an enclosed and a surrounding electrode, with the material placed between the two. *See also:* **dielectric heating.** 14

normal dimension. Incremental dimensions whose number of digits is specified in the format classification. For example, the format classification would be plus 14 for a normal dimension: X.XXXX. *See:* **dimension; incremental dimension; long dimension; short dimension.** 224, 207

normal frequency. The frequency at which a device or system is designed to operate. 103, 202

normal-frequency (low-frequency) dew withstand voltage. The normal-frequency withstand voltage applied to insulation completely covered with condensed moisture. *See:* **normal-frequency withstand voltage.** 103, 202

normal-frequency (low-frequency) dry withstand voltage. The normal-frequency withstand voltage applied to dry insulation. *See:* **normal-frequency (low-frequency) withstand voltage.** 103, 202

normal-frequency line-to-line recovery voltage. The normal-frequency recovery voltage, stated on a line-to-line basis, that occurs on the source side of a three-phase circuit-interrupting device after interruption is complete in all three poles. 103, 202

normal-frequency pole-unit recovery voltage. The normal-frequency recovery voltage that occurs across a pole unit of a circuit-interrupting device upon circuit interruption. 103, 202

normal-frequency recovery voltage. The normal-frequency root-mean-square voltage that occurs across the terminals of an alternating-current circuit-interrupting device after the interruption of the current and after the high-frequency transients have subsided. *Note:* For determination of the normal-frequency recovery voltage, see American National Standard C37.05-1964 (R 1969). 103, 202

normal-frequency (low-frequency) wet withstand voltage. The normal-frequency withstand voltage applied to wetted insulation. *See:* **normal-frequency withstand voltage.** 103, 202

normal-frequency (low-frequency) withstand voltage. The normal-frequency voltage that can be applied to insulation under specified conditions for a specified time without causing flashover or puncture. *Note:* This value is usually

expressed as a root-mean-square value. *See also:* **normal-frequency dew withstand voltage; normal-frequency dry withstand voltage; normal-frequency wet withstand voltage.** 103, 202

normal-glow discharge (gas). The glow discharge characterized by the fact that the working voltage decreases or remains constant as the current increases. *See also:* **discharge (gas).** 244, 190

normal induction (magnetic material). The maximum induction in a magnetic material that is in a symmetrically cyclically magnetized condition. 210

normalize (1) (computing systems). To adjust the representation of a quantity so that the representation lies in a prescribed range. 255, 77

(2) (test, measurement and diagnostic equipment). (A) To adjust the characteristic and fraction of a floating decimal point number thus eliminating leading zeros in the fraction; and (B) To adjust a measured parameter to a value acceptable to an instrument or measurement technique. 54

(3) (circuits and systems). To divide an impedance or frequency by a reference quantity thereby making the result dimensionless. 67

normalized admittance. The reciprocal of the normalized impedance. *See also:* **waveguide.** 267, 179

normalized frequency (circuits and systems). A frequency $\hat{\omega} = \omega/\omega_0$, where ω is the unnormalized frequency variable in a network or system function and ω_0 is chosen such that $\hat{\omega} = 1$ has some desired significance, for example, $\hat{\omega} = 1$ could be the center frequency of a band-pass filter or the cutoff frequency of a low-pass filter. 67

normalized impedance (waveguide). The ratio of an impedance and the corresponding characteristic impedance. *Note:* The normalized impedance is independent of the convention used to define the characteristic impedance, provided that the same convention is also taken for the impedance to be normalized. *See:* **waveguide.** 267, 179

normalized response (automatic control). One obtained by dividing a measured value and dimension by some convenient reference value and dimension; usually the quotient is nondimensional. *See also:* **indicial response; gain, static.** 56

normalized transimpedance (magnetic amplifier). The ratio of differential output voltage to the product of differential control current and control winding turns. 171

normal joint (power cable joints). A joint which is designed not to restrict movement of dielectric fluid between cables being joined. 34

normal linearity (computing systems). Transducer linearity defined in terms of a zero-error reference, that is chosen so as to minimize the linearity error. *Note:* In this case, the zero input does not have to yield zero output and full-scale input does not have to yield fullscale output. The specification of normal linearity, therefore, is less stringent than zero-based linearity. *See also:* **electronic analog computer; linearity of a multiplier (linearity).** 9, 77

normally closed. *See:* **normally open and normally closed.**

normally closed contact. A contact, the current-carrying members of which are in engagement when the operating unit is in its normal position. 328

normally open contact (open contact). A contact, the current-carrying members of which are not in engagement when the operating unit is in the normal position. 328

normally open and normally closed (industrial control). When applied to a magnetically operated switching device, such as a contactor or relay, or to the contacts thereof, these terms signify the position taken when the operating magnet is deenergized. Applicable only to nonlatching types of devices. *See also:* **contacter.** 206

normal-mode interference. *See:* **interference, normal-mode.**

normal mode noise (transverse or differential) (cable systems). The noise voltage appearing differentially between two signal wires and which acts on the signal sensing circuit in the same manner as the desired signal. Normal mode noise may be caused by one or more of the following: (1) Electrostatic induction and differences in distributed capacitance between the signal wires and the surroundings. (2) Electromagnetic induction and magnetic fields linking unequally with the signal wires. (3) Junction or thermal potentials due to the use of dissimilar metals in the connection system. (4) Common mode to normal mode noise conversion. 35

normal operations area (nuclear power generating station). A functional area allocated for those displays and controls necessary for the tasks routinely performed during plant startup, shutdown and power operation modes. 358

normal permeability. The ratio of normal induction to the corresponding maximum magnetizing force. *Note:* In anisotropic media, normal permeability becomes a matrix. 210

normal position (of a device). A predetermined position that serves as a starting point for all operations. 328

normal stop system. A term used to describe the normal indication of the signals in an automatic block signal system in which the indication to proceed is given only upon the approach of a train to an unoccupied block. *See also:* **centralized traffic-control system.** 328

normal-terminal stopping device (elevators). A device, or devices, to slow down and stop an elevator or dumbwaiter car automatically at or near a terminal landing independently of the functioning of the operating device. *See:* **control.** 328

normal transfer capability, (transmission) (electric power supply). The maximum amount of power that can be transmitted continuously. 112

normal weather (electric power systems). All weather not designated as adverse. *See also:* **outage.** 200

normal weather persistent-cause forced-outage rate (for a particular type of component) (electric power systems). The mean number of outages per unit of normal weather time per component. *See also:* **outage.** 200

north-stabilized plan-position indicator (radar). A special case of azimuth-stabilized plan-position indicator in which the reference direction is magnetic north. *See also:* **navigation.** 187, 13

Norton's theorem (circuits and systems). States that a linear time-invariant one-port is equivalent to a circuit which consists of the driving-point admittance of the one-port shunted by the short-circuit current of the one-port. 67

Norton transformation (circuits and systems). A four-terminal network transformation of a series (shunt) ladder element to an equivalent pi (tee) network in cascade with an ideal transformer. The pi (tee) network arms are identical to the series (shunt) arm except for multiplying factors related to the ideal transformer turns-ratio. 67

nose suspension. A method of mounting an axle-hung motor or a generator to give three points of support, consisting of two axle bearings (or equivalent) and a lug or nose projecting from the opposite side of the motor frame, the latter supported by a truck or vehicle frame. *See also:* **traction motor.** 328

NOT. A logic operator having the property that if P is a statement, then the NOT of P is true if P is false, false if P is true. The NOT of P is often represented by $\overline{P}$, $\sim P$, $\neg P, P'$. 255, 77

notation. *See:* **positional notation.**

notch filter (circuits and systems). A band-elimination filter, sometimes used to eliminate a single frequency for example, 60 Hz). 67

notching (relays). A qualifying term applied to a relay indicating that a predetermined number of separate impulses is required to complete operation. *See:* **relay.** 328

notching relay. A programming relay in which the response is dependent upon successive impulses of the input quantity. 103, 202, 60, 127

note. A conventional sign used to indicate the pitch, or the duration, or both, of a tone. It is also the tone sensation itself or the oscillation causing the sensation. *Note:* The word serves when no distinction is desired among the symbol, the sensation, and the physical stimulus. 176

N scan (electronic navigation). *See:* **N display.**

N scope. *See:* **N display.**

n-terminal electron tubes. *See:* **capacitance, input.**

N-terminal network. A network with N accessible terminals. *See also:* **network analysis.** 210

N-terminal pair network. A network with $2N$ accessible terminals grouped in pairs. *Note:* In such a network one terminal of each pair may coincide with a network node. *See also:* **network analysis.** 210

Nth harmonic. The harmonic of frequency N times that of the fundamental component. 349

NTI. *See:* **noise transmission impairment.**

n-type crystal rectifier. A crystal rectifier in which forward current flows when the semiconductor is negative with respect to the metal. *See also:* **rectifier.** 328

n-type semiconductor. *See:* **semiconductor, n-type.**

nuclear generating station (station). A plant wherein electric energy is produced from nuclear energy by means of suitable apparatus. The station may consist of one or more units which may or may not share some common auxiliaries. 102, 120, 31

nuclear power generating stations, class ratings. (A) class 1 electric equipment. The electric equipment that is essential to the safe shutdown and isolation of the reactor or whose failure or damage could result in significant release of radioactive material.
(B) class-1 structures and equipment. Structures and equipment that are essential to the safe shutdown and isolation of the reactor or whose failure or damage could result in significant release of radioactive material.
(C) class 1E. The safety classification of the electric equipment and systems that are essential to emergency

reactor shutdown, containment isolation, reactor core cooling, and containment and reactor heat removal, or are otherwise essential in preventing significant release of radioactive material to the environment.

(D) class 1E control switchboard. A rack panel, switchboard, or similar type structure fitted with any Class 1E equipment.

(E) class 1E electric systems. The systems that provide the electric power used to shut down the reactor and limit the release of radioactive material following a design basis event.

(F) class II structures and equipment. Structures and equipment that are important to reactor operation but are not essential to the safe shutdown and isolation of the reactor and whose failure cannot result in a significant release of radioactive material.

(G) class III structures and equipment. Structures and equipment that are not essential to the operation, safe shutdown, or isolation of the reactor and whose failure cannot result in the release of radioactive material.

28, 143, 31, 102, 142, 140, 120, 141

nuclear reactor. An apparatus by means of which a fission chain reaction can be initiated, maintained, and controlled.
112

nuclear safety related (nuclear power generating stations). That term used to call attention to safety classifications incorporated in the body of the document so marked. *Note:* As used in IEEE Std 494-1974, the term calls attention to the safety classification Class 1E. 31, 156

null (1) (signal-transmission system). The condition of zero error-signal achieved by equality at a summing junction between an input signal and an automatically or manually adjusted balancing signal of phase or polarity opposite to the input signal. *See also:* **signal.** 188

(2) (direction-finding systems). The condition of minimum output as a function of the direction of arrival of the signal, or of the rotation of the response pattern of the direction-finder antenna system. *See also:* **navigation.**
187, 13

nullator (circuits and systems). An idealized one-port that is simultaneously an open and short circuit, that is, $V = I = O$. The nullator is a bilateral, lossless one-port. 67

null balance (automatic null-balancing electric instrument) (instruments). The condition that exists in the circuits of an instrument when the difference between an opposing electrical quantity within the instrument and the measured signal does not exceed the dead band. *Note:* The value of the opposing electrical quantity produced within the instrument is related to the position of the end device. *See also:* **control system, feedback; measurement system.**
295

null-balance system. A system in which the input is measured by producing a null with a calibrated balancing voltage or current. *See also:* **signal.** 188

nullity (degrees of freedom on mesh basis) (network). The number of independent meshes that can be selected in a network. The nullity N is equal to the number of branches B minus the number of nodes V plus the number of separate parts P. $N = B - V + P$. *See also:* **network analysis.**
210

null junction (power supplies). The point on the Kepco bridge at which the reference resistor, the voltage-control resistance, and one side of the comparison amplifier coincide. *Note:* The null junction is maintained at almost zero potential and is a virtual ground. *See:* **summing point.** *See also:* **power supply.** 228, 186

null-reference glide slope. A glide-slope system using a two-element array and in which the slope angle is defined by the first null above the horizontal in the field pattern of the upper antenna. *See also:* **navigation.** 187, 13

number (electronic computation). (1) Formally, an abstract mathematical entity that is a generalization of a concept used to indicate quantity, direction, etcetera. In this sense a number is independent of the manner of its representation. (2) Commonly: A representation of a number as defined above (for example, the binary number 10110, the decimal number 3695, or a sequence of pulses). (3) An expression, composed wholly or partly of digits, that does not necessarily represent the abstract entity mentioned in the first meaning. *Note:* Whenever there is a possibility of confusion between meaning (1) and meaning (2) or (3), it is usually possible to make an unambiguous statement by using **number** for meaning (1) and **numerical expression** for meaning (2) or (3). 235, 210

number group (telephone switching systems). An arrangement for associating equipment numbers with mainstation codes. 55

numbering plan (telephone switching systems). A plan employing codes and directory numbers for identifying main stations and other terminations within a telecommunication system. 55

numbering-plan area (NPA) (telephone switching systems). A geographical subdivision of the territory covered by a national or integrated numbering plan. An NPA is identified by a distinctive area code (NPA code). 55

numbering-plan area code, (NPA) (telephone switching systems). A one, two-, or three-digit number that, for the purpose of distance dialing, designates one of the geographical areas within a country (and in some instances neighboring territories) that is covered by a separate numbering plan. 55

number of loops (magnetically focused electron beam). The number of maxima in the beam diameter between the electron gun and the target, or between a point on the photocathode and the target. 125

number of rectifier phases (rectifier circuit). The total number of successive, nonsimultaneous commutations occuring within that rectifier circuit during each cycle when operating without phase control. *Note:* It is also equal to the order of the principal harmonic in the direct-current potential wave shape. The number of rectifier phases influences both alternating-current and direct-current waveforms. In a simple single-way rectifier the number of rectifier phases is equal to the number of rectifying elements. *See also:* **rectification; rectifier circuit element.** 237, 66

number of scanning lines (television) (numerically). The total number of lines, both active and blanked, in a frame. *Note:* In any specified scanning system, this number is inherently the ratio of the line frequency to the frame frequency, and is always a whole number. In a two-to-one off-line interlaced system, it is always an odd whole number. 163

number 1 master (metal master) (metal negative) (disk recording) (electroacoustics). *See:* **original master.**

number 1 mold (mother) (metal positive). A mold derived by electroforming from the original master. *See also:* **phonograph pickup.** 176

number system (electronic computation). Loosely, a numeration system. *See also:* **positional notation.** 255, 77

number 2, number 3, etcetera master. A master produced by electroforming from a number 1, number 2, etcetera mold. *See also:* **phonograph pickup.** 176

number 2, number 3, etcetera mold. A mold derived by electroforming from a number 2, number 3, etcetera master. *See also:* **phonograph pickup.** 176

numeral. A representation of a number. *See:* **binary numeral.** 255, 77

numeration system (numeral system). A system for the representation of numbers, for example, the decimal system, the Roman numeral system, the binary system. 255, 77

numerical action (switch). That action that depends on at least part of the called number. 328

numerical analysis. The study of methods of obtaining useful quantitative solutions to problems that have been expressed mathematically, including the study of the errors and bounds on errors in obtaining such solutions. 255, 77

numerical aperture (fiber-optic plate) (nominal) (camera tubes). The square root of the difference of (1) the square of the index of refraction of the fiber core, and (2) the square of the index of refraction of the cladding material. *Note:* Numerical aperture of a fiber-optic plate is a measure of the range of angles of incidence of entering radiation that can be conducted through the plate. *See:* **camera tube.** 190

numerical control. Pertaining to the automatic control of processes by the proper interpretation of numerical data. 255, 77

numerical control system (numerically controlled machines). A system in which actions are controlled by the direct insertion of numerical data at some point. *Note:* The system must automatically interpret at least some portion of these data. 224, 207

numerical data. Data in which information is expressed by a set of numbers or symbols that can only assume discrete values or configurations. 224, 207

numeric printout (dedicated-type sequential events recording systems). A brief coded method of identifying inputs using numeric characters only. 48, 58

nutating feed (electronic navigation). *See:* **conical scan.**

nutation (gyro). The oscillation of the spin axis of a two-degree-of-freedom gyro about two orthogonal axes normal to the mean position of the spin axis. 46

nutation field (radar). *See:* **conical scan.**

Nyquist diagram (1) (general). A polar plot of the loop transfer function. *Note:* The inverse Nyquist diagram is a polar plot of the reciprocal function. The generalized Nyquist diagram comprises plots of the loop transfer function of the complex variable s, where $s = \sigma + j\omega$ and σ and ω are arbitrary constants including zero. 56, 59

(2) (feedback amplifier). A plot, in rectangular coordinates, of the real and imaginary parts of the factor μ/β for frequencies from zero to infinity, where μ is the amplification in the absence of feedback, and β is the fraction of the output voltage that is superimposed on the amplifier input. *Note:* The criterion for stability of a feedback amplifier is that the curve of the Nyquist diagram shall not inclose the point $X = 1$, $Y = 0$, where μ/β equals $X + jY$. *See also:* **feedback.** 328

Nyquist interval (channel) (1) (general). The reciprocal of the Nyquist rate. 242

(2) The maximum separation in time that can be given to regularly spaced instantaneous samples of a wave of bandwidth W for complete determination of the waveform of the signal. Numerically, it is equal to $\frac{1}{2} W$ seconds. 59

Nyquist rate (channel). The maximum rate at which independent signal values can be transmitted over the specified channel without exceeding a specified amount of mutual interference. *Note:* The Nyquist rate of a channel is directly related to its bandwidth. For example, an ideal (physically unrealizable) low-pass filter of bandwidth W (constant amplitude and linear phase response within band W and zero transmission elsewhere) would permit a Nyquist rate of $2W$ without interference. For a physical channel, the Nyquist rate, as defined, depends not only on the specified amount of mutual interference, but also on the amplitude and phase response of the channel. 242

O

OA. *See:* **transformer, oil-immersed.**

OBI (omnibearing indicator). An instrument which presents an automatic and continuous indication of an omnibearing. 13

object color. The color of the light reflected or transmitted by the object when illuminated by a standard light source, such as source *A, B,* or *C* of the Commission Internationale de l'Eclairage (CIE). *See:* **standard source.** *See also:* **color.** 167

object language. *See:* **target language.**

object program. *See:* **target program.**

observable (control system). A property of a component of a state whereby its value at a given time can be computed from measurements on the output over a finite past interval. *See also:* **control system.** 329

observable, completely (control system). The property of a plant whereby all components of the state are observable. *See also:* **control system; observable; plant.** 329

observable temperature (equipment rating). The temperature of equipment obtained on test or in operation. *See also:* **limiting insulation temperature.** 320

observable temperature rise (equipment rating). The difference between the observable temperature and the ambient temperature at the time of the test. *See also:* **limiting insulation temperature.** 320

observation time (radar). The time interval over which a radar echo signal may be integrated for detection or measurement.   13

observed instantaneous availability. At a stated instant of time the proportion of occasions when an item can perform a required function. *Notes:* (1) Occasions can refer to either a number of items at a single instant of time or to one or more items at instants repeated in time. (2) The run-up time is counted in down-time after repair and is counted in the up-time when the equipment is brought into use for the first time. (3) The observed instantaneous availability is to be associated with a period of time and with stated conditions of use and maintenance.   164

observed mean active maintenance time. The ratio of the sum of the active maintenance times to the total number of maintenance actions. *Note:* The maintenance conditions applied shall be stated.   164

observed mean availability. The ratio of the cumulative time for which an item can perform a required function to the cumulative time under observation, or at instants of time (chosen by a sampling technique), the mean of the proportion of a number of nominally identical items which can perform their required function. *Notes:* (1) When one limiting value is given, this is usually the lower limit. (2) The observed mean availability is to be associated with a stated period of time and with stated conditions of use and maintenance.   164

observed reliability (reliability) (1) non-repaired items. For a stated period of time, the ratio of the number of items which performed their functions satisfactorily at the end of the period to the total number of items in the sample at the beginning of the period.   164
(2) repaired item or items. The ratio of the number of occasions on which an item or items performed their functions satisfactorily for a stated period of time to the total number of occasions the item or items were required to perform for the same period. *Note:* The criteria for what constitutes satisfactory function shall be stated.   164

obstruction lights. Aeronautical ground lights provided to indicate obstructions.   167

occulting light. A rhythmic light in which the periods of light are clearly longer than the periods of darkness.   167

occupied bandwidth (mobile communication). The frequency bandwidth such that below its lower and above its upper frequency limits, the mean powers radiated are each equal to 0.5 percent of the total power radiated. *See also:* **mobile communication system.**   181

OCR. *See:* **optical character recognition.**

octal. (1) Pertaining to a characteristic or property involving a selection, choice, or condition in which there are eight possibilities. (2) Pertaining to the numeration system with a radix of eight. *See also:* **positional notation.**   255, 77

octantal error (electronic navigation). An error in measured bearing caused by the finite spacing of the antenna elements in systems using spaced antennas to provide bearing information (such as very-high-frequency omnidirectional radio ranges); this error varies in a sinusoidal manner throughout the 360 degrees and has four positive and four negative maximums. *See also:* **navigation.**   187, 13

octave (communication) (1) (general). The interval between

two frequencies having a ratio of 2:1. *See also:* **signal wave.**   239, 59
(2) The interval between two sounds having a basic frequency ratio of two.
(3) The pitch interval between two tones such that one tone may be regarded as duplicating the musical import of the other tone at the nearest possible higher pitch. *Notes:* (A) The interval, in octaves between any two frequencies, is the logarithm to the base 2 (or 3.322 times the logarithm to the base 10) of the frequency ratio. (B) The frequency ratio corresponding to an octave pitch interval is approximately, but not always exactly, 2:1. *See also:* **signal wave.**   176
(4) (nuclear power generating stations) (seismic qualification of class 1E equipment). The interval between two frequencies which have a frequency ratio of two.   28, 31

octave-band pressure level (acoustics) (octave pressure level) (sound). The band pressure level for a frequency band corresponding to a specified octave. *Note:* The location of an octave-band pressure level on a frequency scale is usually specified as the geometric mean of the upper and lower frequencies of the octave.   176

octode. An eight-electrode electron tube containing an anode, a cathode, a control electrode, and five additional electrodes that are ordinarily grids.   190, 125

octonary (electronic computation). *See:* **octal.**

odd-even check (computing systems). *See:* **parity check.**

O **display (radar).** An *A* display modified by the inclusion of an adjustable notch for measuring distance. *See also:* **navigation.**   187, 13

oersted. The unit of magnetic field strength in the unrationalized centimeter-gram-second (cgs) electromagnetic system. The oersted is the magnetic field strength in the interior of an elongated uniformly wound solenoid that is excited with a linear current density in its winding of one abampere per 4π centimeters of axial length.   210

off-axis mode (laser-maser). *See:* **higher order mode of propagation.** An off-axis mode will incorporate one or more of the maxima which lie off the axis of a beam.   363

off-center display. A plan-position-indicator display, the center of which does not correspond to the position of the radar antenna. *See also:* **radar.**   328

off-center PPI (radar). A PPI (plan position indicator) which has the zero position of the time base at a position other than at the center of the display, thus providing the equivalent of a larger display for a selected portion of the service area. *Syn.* **offset PPI.**   13

offered traffic (telephone switching systems). A measure of the calls requesting service during a given period of time.   55

off-hook (telephone switching systems). A closed station line or any supervisory or pulsing condition is indicative of this state.   55

office class (telephone switching systems). A designation (Class 1, 2, 3, 4, 5) given to each office in World Zone 1 involved in the completion of toll calls. The class is determined according to the office's switching function, its interrelation with other switching offices, and its transmission requirements. The class designation given to the switching points in the network determines the routing pattern for all calls. Class 1 is higher in rank than Class 2; Class 2 is higher than Class 3; and so on. *See also:* **world zone number.**   55

office code (telephone switching systems). The digits that designate a block of main-station codes within a numbering-plan area.   55

office test (meter). A test made at the request or suggestion of some department of the company to determine the cause of seemingly abnormal registration. *See also:* **service test (field test).**   328

OFF-impedance (thyristor). The differential impedance between the terminals through which the principal current flows, when the thyristor is in the OFF state at a stated operating point. *See also:* **principal voltage-current characteristic (principal characteristic).**   243, 66, 191

off line (1) (computing systems). Pertaining to equipment or devices not under direct control of the central processing unit.   255, 77
(2) (nuclear power generating stations). A system where an aliquot is withdrawn from the effluent stream and conveyed to the detector assembly.   31
(3) (test, measurement and diagnostic equipment). (A) Operation of input/output and other devices not under direct control of a device; and (B) Peripheral equipment operated outside of, and not under control of the system, for example, the off-line printer.   54

off-line testing (test, measurement and diagnostic equipment). Testing of the unit under test removed from its operational environment or its operational equipment. Shop testing.   54

off-net call (telephone switching systems). A call from a switched-service network to a station outside that network.   55

off-peak power. Power supplied during designated periods of relatively low system demands. *See also:* **generating station.**   64

OFF period: (1) (electron tubes). The time during an operating cycle in which the electronic tube is nonconducting. *See also:* ON **period (electron tubes).**   244, 190
(2) (circuit switching element) (inverters). The part of an operating cycle during which essentially no current flows in the circuit switching element. *See also:* **self-commutated inverters.**   208

off road vehicle (conductor stringing equipment). A vehicle specifically designed and equipped to traverse sand, swamps, muddy tundra or rough mountainous terrain. Vehicles falling into this category are usually "all wheel drive" or tracked units. In some cases, units equipped with special air bag rollers having a soft footprint are utilized. *Syn.* **all terrain vehicle; swamp buggy.**   45

offset (1) (transducer) (analog computer). The component of error that is constant and independent of the inputs, often used to denote bias.   9, 77, 10
(2) (control system, feedback). The steady-state deviation when the command is fixed. *Note:* The offset resulting from a no-load to full-load change (or other specified limits) is often called (speed) droop or load regulation. *See also:* **control system, feedback.**   56
(3) (distance relay). The displacement of the operating characteristic on an *R-X* diagram from the position inherent to the basic performance class of the relay. *Note:*

A relay with this characteristic is called an offset relay.   103, 202, 127
(4) (course computer) (electronic navigation). An automatic computer that translates reference navigational coordinates into those required for a predetermined course. *See also:* **navigation.**   187, 13
(5) (pulse terms). The algebraic difference between two specified magnitude reference lines. Unless otherwise specified, the two magnitude reference lines are the waveform baseline and the magnitude origin line. *See:* The single pulse diagram below the **waveform epoch** entry.   254

offset angle (electroacoustics) (lateral disk reproduction). The offset angle is the smaller of the two angles between the projections into the plane of the disk of the vibration axis of the pickup stylus and the line connecting the vertical pivot (assuming a horizontal disk) of the pickup arm with the stylus point. *See also:* **phonograph pickup.**   176

offset bearing (air switch). A component of a switch operating mechanism designed to provide support for a torsional operating member and a crank which provides reciprocating motion for switch operation. *Syn.* **outboard bearing.**   27, 79

offset, clipping. *See:* **clipping offset.**

offset paraboloidal reflector (antennas). A paraboloidal reflector which is not symmetrical with respect to its focal axis.   111

offset voltage (power supplies). A direct-current potential remaining across the comparison amplifier's input terminals (from the null junction to the common terminal) when the output voltage is zero. The polarity of the offset voltage is such as to allow the output to pass through zero and the polarity to be reversed. It is often deliberately introduced into the design of power supplies to reach and even pass zero-output volts. *See also:* **power supply.**   228, 186

offset waveform (pulse terms). A waveform whose baseline is offset from, unless otherwise specified, the magnitude origin line.   254

OFF state (thyristor). The condition of the thyristor corresponding to the high-resistance low-current portion of the principal voltage-current characteristic between the origin and the breakover point(s) in the switching quadrant(s). *See also:* **principal voltage-current characteristic (principal characteristic).**   243, 66, 208, 191

OFF-state current (thyristor). The principal current when the thyristor is in the OFF state. *See also:* **principal current.**   243, 66, 208, 191

OFF-state power dissipation (thyristor). The power dissipation resulting from OFF-state current.   66

OFF-state voltage (thyristor). The principal voltage when the thyristor is in the OFF state. *See also:* **principal voltage-current characteristic (principal characteristic).**   243, 66, 208, 191

ohm. The unit of resistance (and of impedance) in the International System of Units (SI). The ohm is the resistance of a conductor such that a constant current of one ampere in it produces a voltage of one volt between its ends.   210

ohmic contact (1) (general) (semiconductor radiation detectors). A purely resistive contact; that is, one that has a linear voltage-current characteristic throughout its entire operating range. *See also:* **semiconductor.**

245, 66, 210, 118, 23

(2) (semiconductor). A contact between two materials, possessing the property that the potential difference across it is proportional to the current passing through. *See also:* **semiconductor.** 186

ohmic resistance test (rotating machinery). A test to measure the ohmic resistance of a winding, using direct current. *See also:* **asynchronous machine.** 63

ohmmeter. A direct-reading instrument for measuring electric resistance. It is provided with a scale, usually graduated in either ohms, megohms, or both. If the scale is graduated in megohms, the instrument is usually called a megohmmeter. *See also:* **instrument.** 328

Ohm's law. The current in an electric circuit is inversely proportional to the resistance of the circuit and is directly proportional to the electromotive force in the circuit. *Note:* Ohm's law applies strictly only to linear constant-current circuits. 210

OHR. *See:* **over-the-horizon radar.**

oil (1) (prefix). The prefix oil applied to a device that interrupts an electric circuit indicates that the interruption occurs in oil. 225, 206, 124

(2) (transformers). Includes the following insulating and cooling liquids: type I mineral oil (unhibited oil), type II mineral oil (inhibited oil), and askarel. 53

(3) (outdoor apparatus bushings). Mineral transformer oil. 168

oil buffer (elevators). A buffer using oil as a medium that absorbs and dissipates the kinetic energy of the descending car or counterweight. *See also:* **elevators.** 328

oil-buffer stroke (oil buffer) (elevators). The oil-displacing movement of the buffer plunger or piston, excluding the travel of the buffer-plunger accelerating device. *See also:* **elevators.** 328

oil catcher (rotating machinery). A recess to carry off oil. *See also:* **oil cup (rotating machinery).** 63

oil cup (rotating machinery). An attachment to the oil reservoir for adding oil and controlling its upper level. 63

oil cutout (oil-filled cutout). A cutout in which all or part of the fuse support and its fuse link or disconnecting blade are mounted in oil with complete immersion of the contacts and the fusible portion of the conducting element (fuse link), so that arc interruption by severing of the fuse link or by opening of the contacts will occur under oil. 103, 202

oil feeding reservoirs. Oil storage tanks situated at intervals along the route of an oil-filled cable or at oil-filled joints of solid cable for the purpose of keeping the cable constantly filled with oil under pressure. *See also:* **power distribution, underground construction.** 64

oil-filled (designated liquid-filled) (prefix). The prefix oil-filled or designated liquid-filled as applied to equipment indicates that oil or the designated liquid is the surrounding medium. 328

oil-filled bushing (outdoor electric apparatus). A bushing in which the space between the inside surface of the weather casing and the major insulation (or conductor where no major insulation is used) is filled with oil. 168

oil-filled cable. A self-contained pressure cable in which the pressure medium is low-viscosity oil having access to the insulation. *See:* **pressure cable; self-contained pressure cable.** *See also:* **power distribution, underground construction.** 64

oil-filled pipe cable. A pipe cable in which the pressure medium is oil having access to the insulation. *See:* **pressure cable; pipe cable.** *See also:* **power distribution, underground construction.** 64

oil-fill stand pipe (rotating machinery). *See:* **oil overflow plug.**

oil groove (rotating machinery). A groove cut in the surface of the bearing lining or sometimes in the journal to help to distribute the oil over the bearing surface. *See:* **oil cup (rotating machinery).** 63

oil-immersed (1) (transformers, reactors, regulators, and similar components). Having the coils immersed in an insulating liquid. *Note:* The insulating liquid is usually (though not necessarily) oil. *See:* **transformer, oil-immersed.** 257

(2) (grounding device). Means that the windings are immersed in an insulating oil. 91

oil-immersed forced-oil-cooled transformer with forced-air cooler (class FOA). A transformer having its core and coils immersed in oil and cooled by the forced circulation of this oil through external oil-to-air heat-exchanger equipment utilizing forced circulation of air over its cooling surface. 53

oil-immersed forced-oil-cooled transformer with forced-water cooler (Class FOW). A transformer having its core and coils immersed in oil and cooled by the forced circulation of this oil through external oil-to-water heat-exchanger equipment utilizing forced circulation of water over its cooling surface. 53

oil-immersed self-cooled/forced-air-cooled/forced-air-cooled transformer (Class OA/FA/FA). A transformer having its core and coils immersed in oil and having a self-cooled rating obtained by the natural circulation of air over the cooling surface, a forced-air-cooled rating obtained by the forced circulation of air over a portion of the cooling surface, and an increased forced-air-cooled rating obtained by the increased forced circulation of air over a portion of the cooling surface. 53

oil-immersed self-cooled/forced-air-cooled/forced-oil-cooled transformer (Class OA/FA/FOA). A transformer having its core and coils immersed in oil and having a self-cooled rating with cooling obtained by the natural circulation of air over the cooling surface, a forced-air-cooled rating with cooling obtained by the forced circulation of air over this same cooling surface, and a forced-oil-cooled rating with cooling obtained by the forced circulation of oil over the core and coils and adjacent to this same cooling surface over which the air is being forced-circulated. 53

oil-immersed self-cooled/forced-air, forced-oil-cooled/forced-air, forced-oil-cooled transformer (Class OA/FOA/FOA). A transformer similar to Class OA/FA/FOA transformer except that its auxiliary cooling controls are arranged to start a portion of the oil pumps and a portion of the fans for the first auxiliary rating and the remainder of the pumps and fans for the second auxiliary rating. 53

oil-immersed self-cooled/forced-air-cooled transformer (Class OA/FA). A transformer having its core and coils immersed in oil and having a self-cooled rating with cooling obtained by the natural circulation of air over the cooling surface, and a forced-air-cooled rating with cooling obtained by the forced circulation of air over this same cooling surface.   53

oil-immersed self-cooled transformer (Class OA). A transformer having its core and coils immersed in oil, the cooling being effected by the natural circulation of air over the cooling surface.   53

oil-immersed transformer. A transformer in which the core and coils are immersed in an insulating oil.   53

oil-immersed water-cooled/self-cooled transformer (Class OW/A). A transformer having its core and coils immersed in oil and having a water-cooled rating with cooling obtained by the natural circulation of oil over the water-cooled surface, and a self-cooled rating with cooling obtained by the natural circulation of air over the cooling surface.   53

oil-immersed water-cooled transformer (Class OW). A transformer having its core and coils immersed in oil, the cooling being effected by the natural circulation of oil over the water-cooled surface.   53

oil-impregnated paper-insulated bushing. A bushing in which the major insulation is provided by paper impregnated with oil.   168

oilless circuit breaker. *See:* note under **circuit breaker.**

oil level gauge (rotating machinery). An indicating device showing oil level in the oil reservoir.   63

oil-lift bearing (rotating machinery). A journal bearing in which high-pressure oil is forced under the shaft journal or thrust runner to establish a lubricating film. *See also:* **bearing.**   63

oil-lift system (rotating machinery). A system that lubricates a bearing before starting by forcing oil between the journal or thrust runner and bearing surfaces. *See also:* **oil cup (rotating machinery).**   63

oil-overflow plug (oil-fill stand-pipe) (rotating machinery). An attachment to the oil reservoir that can be opened to allow excess oil to escape, to inspect the oil level, or to add oil.   63

oil pot (oil reservoir) (rotating machinery). A bearing reservoir for a vertical-shaft bearing. *See also:* **oil cup (rotating machinery).**   63

oil-pressure electric gauge. A device that measures the pressure of oil in the line between the oil pump and the bearings of an aircraft engine. The gauge is provided with a scale, usually graduated in pounds per square inch. It provides remote indication by means of self-synchronous generator and motor.   328

oilproof enclosure. An enclosure so constructed that oil vapors, or free oil not under pressure, which may accumulate within the enclosure, will not prevent successful operation of, or cause damage to, the enclosed equipment. *See also:* **asynchronous machine; direct-current commutating machine.**   3

oil reservoir (rotating machinery). *See:* **oil pot.**

oil-resistant gaskets. Those made of material which is resistant to oil or oil fumes.   53

oil ring (rotating machinery). A ring encircling the shaft in such a manner as to bring oil from the oil reservoir to the sleeve bearing and shaft. *See also:* **bearing.**   63

oil-ring guide (rotating machinery). A part whose main purpose is the restriction of the motion of the oil ring. *See also:* **oil cup (rotating machinery).**   63

oil-ring lubricated bearing. A bearing in which a ring, encircling the journal, and rotated by it, raises oil to lubricate the bearing from a reservoir into which the ring dips.   63

oil-ring retainer (rotating machinery). A guard to keep the oil ring in position on the shaft.   63

oil seal (bearing seal) (bearing oil seal) (rotating machinery). A part or combination of parts in a bearing assembly intended to prevent leakage of oil from the bearing.   63

oil slinger. A rotating member mounted on the shaft, or integral with it, that expels oil by centrifugal force and tends to prevent it from creeping along adjacent surfaces. *See also:* **oil cup (rotating machinery).**   328

oil switch. *See:* note under **mechanical switching device.**

oil thrower (oil slinger) (rotating machinery). A peripheral ring or ridge on a shaft adjacent to the journal and which is intended to prevent any flow of oil along the shaft. *See also:* **oil cup (rotating machinery).**   63

oiltight (adjective). So constructed as to exclude oils, coolants and similar liquids under specified test conditions.   53

oiltight enclosure. An enclosure so constructed that oil vapors, or free oil, not under pressure, which may be present in the surrounding atmosphere, cannot enter the enclosure.   3

oiltight pilot devices (industrial control). Devices such as push-button switches, pilot lights, and selector switches that are so designed that, when properly installed, they will prevent oil and coolant from entering around the operating or mounting means. *See also:* **switch.**   225, 206

oil, uninhibited. Mineral transformer oil to which no synthetic oxidation inhibitor has been added. *See:* **oil-immersed transformer.**   203

oil-well cover (rotating machinery). A cover for an oil reservoir. *See:* **oil cup (rotating machinery).**   63

oil wick (rotating machinery). Wool, cotton, or similar material used to bring oil to the journal surface by capillary action. *See:* **oil cup (rotating machinery).**   63

omega. A very-long-distance navigation system operating at approximately 10 kilohertz in which hyperbolic lines of position are determined by measurement of the differences in travel time of continuous-wave signals from two transmitters separated by 5000 to 6000 nautical miles (9000 to 11 000 kilometers), or in which changes in distances from the transmitters are measured by counting radio-frequency wavelengths in space or lanes as the vehicle moves from a known position, the lanes being counted by phase comparison with a stable oscillator aboard the vehicle. *See also:* **navigation.**   187, 13

omnibearing (electronic navigation). A magnetic bearing indicated by a navigational receiver on transmissions from an omnidirectional radio range. *See also:* **navigation; radio navigation.**   87, 13

omnibearing converter (electronic navigation). An electromechanical device that combines the omnibearings signal with vehicle heading information to furnish electric signals for the operation of the pointer of a radio-magnetic indicator. *See also:* **navigation.**   187, 13

omnibearing-distance facility (electronic navigation). A combination of an omnidirectional radio range and a distance-measuring facility, so that both bearing and distance information may be obtained; tactical air navigation (tacan) and combined very-high-frequency omnidirectional radio range and distance-measuring equipment (VOR/DME) are omnibearing distance facilities. *See also:* **navigation.** 187, 13

omnibearing-distance navigation. Radio navigation utilizing a polar-coordinate system as a reference, making use of omnibearing-distance facilities; often called rho-theta navigation. *See also:* **navigation.** 13, 187

omnibearing indicator. *See:* **OBI.**

omnibearing selector (electronic navigation). A control used with an omnidirectional-radio-range receiver so that any desired omnibearing may be selected; deviations from on-course for any selected bearing is displayed on the course-line-deviations indicator. *See also:* **navigation.** 187, 13

omnidirectional antenna. An antenna having an essentially nondirectional pattern in azimuth and a directional pattern in elevation. *See:* **isotropic radiator.** *See also:* **antenna.** 179, 111

omnidirectional microphone (nondirectional microphone). A microphone the response of which is essentially independent of the direction of sound incidence. *See also:* **microphone.** 328

omnidirectional range (omnirange) (electronic navigation). A radio range that provides bearing information to vehicles at all azimuths within its service area. *See also:* **navigation; radio navigation.** 13, 187

omnirange (electronic navigation). *See:* **omnidirectional range.**

on-course curvature (navigation). The rate of change of the indicated course with respect to distance along the course line or path. *See also:* **navigation.** 13, 187

one-address. Pertaining to an instruction code in which each instruction has one address part. Also called single address. In a typical one-address instruction the address may specify either the location of an operand to be taken from storage, the destination of a previously prepared result, or the location of the next instruction to be interpreted. In a one-address machine, the arithmetic unit usually contains at least two storage locations, one of which is an accumulator. For example, operations requiring two operands may obtain one operand from the main storage and the other from the storage location in the arithmetic unit that is specified by the operation part. *See also:* **single-address.** 235

one-address code (electronic computation). *See:* **instruction code.**

one-fluid cell. A cell having the same electrolyte in contact with both electrodes. *See also:* **electrochemistry.** 328

one hundred percent disruptive-discharge voltage (dielectric tests). A specified minimum voltage that is to be applied to a test object in a 100 percent disruptive-discharge test under specified conditions. The term applies mostly to impulse tests and has significance only in cases where the loss of dielectric strength resulting from a disruptive discharge is temporary. 150

one-level address (computing systems). *See:* **direct address.** 255, 77

one-line diagram (single-line) (industrial control). A diagram which shows, by means of single lines and graphic symbols, the course of an electric circuit or system of circuits and the component devices or parts used therein. 210, 206, 25

one minus cosine (1-cosine) envelope (transient recovery voltage) (1) (power switchgear). A voltage-versus-time curve of the general form $e_2 = E_2 (1 - \cos Kt)$ in which e_2 represents the transient voltage across a switching device pole unit, reaching its crest E_2 at a time T_2. 103 **(2) (high voltage circuit breakers).** The 1-cosine curve starting at zero and reaching a peak of E_2 at time T_2. The crest is denoted by P. *Note:* The 1-cosine curve is the standard envelope for rating circuit breaker transient recovery voltage performance for circuit breakers rated 72.5 kV and below. 148

ONE output (magnetic cell). (1) The voltage response obtained from a magnetic cell in a ONE state by a reading or resetting process. (2) The integrated voltage response obtained from a magnetic cell in a ONE state by a reading or resetting process. *See also:* **distributed-ONE output; ONE state.** 331

one-plus (1+) call (telephone switching systems). A type of station-to-station call in which the digit one is dialed as an access code. 55

one-plus-one address. Pertaining to an instruction that contains one operand address and a control address. 255, 77

ones complement. The radix-minus-one complement of a numeral whose radix is two. *Note:* In this context the terms **base** and **radix** are interchangeable. 255, 77

one-sided switching array (telephone switching systems). A switching array where each crosspoint interconnects multiples within one group. 55

ONE state. A state of a magnetic cell wherein the magnetic flux through a specified cross-sectional area has a positive value, when determined from an arbitrarily specified direction of positive normal to that area. A state wherein the magnetic flux has a negative value, when similarly determined, is a ZERO state. A ONE output is (1) the voltage response obtained from a magnetic cell in a ONE state by a reading or resetting process, or (2) the integrated voltage response obtained from a magnetic cell in a ONE state by a reading or resetting process. A ZERO output is (1) the voltage response obtained from a magnetic cell in a ZERO state by a reading or resetting process, or (2) the integrated voltage response obtained from a magnetic cell in a ZERO state by a reading or resetting process. A ratio of a ONE output to a ZERO output is a ONE-to-ZERO ratio. A pulse, for example a drive pulse, is a **write pulse** if it causes information to be introduced into a magnetic cell or cells, or is a **read pulse** if it causes information to be acquired from a magnetic cell or cells. 331

one-time fuse. *See:* **nonrenewable fuse.**

ONE-to-partial-select ratio. The ratio of a ONE output to a partial-select output. *See also:* **coincident-current selection.** 331

ONE-to-ZERO ratio. A ratio of a ONE output to a ZERO output. *See:* **ONE state.** 331

***O* network.** A network composed of four impedance branches connected in series to form a closed circuit, two adjacent junction points serving as input terminals while the remaining two junction points serve as output terminals. *See also:* **network analysis.** 328

one-unit call (telephone switching systems). A call for which there is a single-unit charge for an initial minimum interval. 55

one-way automatic leveling device. A device that corrects the car level only in case of under-run of the car, but will not maintain the level during loading and unloading. *See also:* **elevator-car leveling device.** 328

one-way correction (industrial control). A method of register control that effects a correction in register in one direction only. 206

one-way trunk (telephone switching systems). A trunk between two switching entities accessible by calls from one end only. At the originating end, the one-way trunk is known as an outgoing trunk; at the terminating end, it is known as an incoming trunk. 55

one-wire circuit. *See:* **direct-wire circuit.**

one-wire line (open-wire lead). *See:* **open-wire pole line.**

on-hook (telephone switching systems). An open station line or any supervisory or pulsing condition is indicative of this state. 55

ON impedance (thyristor). The differential impedance between the terminals through which the principal current flows, when the thyristor is in the ON state at a stated operating point. *See also:* **principal voltage-current characteristic (principal characteristic).** 243, 66, 191

online. (1) Pertaining to equipment or devices under direct control of the central processing unit. (2) Pertaining to a user's ability to interact with a computer. 77, 54

on-line testing (test, measurement and diagnostic equipment). Testing of the unit under test in its operational environment. *See:* **interference testing; non-interference testing.** 54

on-net call (telephone switching systems). A call within a switched-service network. 55

ON-OFF keying (modulation systems). A binary form of amplitude modulation in which one of the states of the modulated wave is the absence of energy in the keying interval. *Note:* The terms mark and space are often used to designate, respectively, the presence and absence of energy in the keying interval. *See also:* **telegraphy.** 242

ON-OFF test (test, measurement and diagnostic equipment). A test conducted by repeatedly switching on and off either the signal, power, or load connected to the unit under test while observing the reaction or performance of some parameter of that unit under test. A test frequently used to isolate offending equipment while conducting compatibility, interference, or system performance evaluations. 54

on-peak power. Power supplied during designated periods of relatively high system demands. *See also:* **generating station.** 64

ON period (electron tube or valve). The time during an operating cycle in which the electron tube or valve is conducting. 244, 190

on site (nuclear power generating stations). Location within a facility that is controlled with respect to access by the general public. 31

ON state (thyristor). The condition of the thyristor corresponding to the low-resistance low-voltage portion of the principal voltage-current characteristic in the switching quadrant(s). *Note:* In the case of reverse-conducting thyristors, this definition is applicable only for a positive anode-to-cathode voltage. *See also:* **principal voltage-current characteristic (principal characteristic).** 242, 66, 208, 191

ON-state current (thyristor). The principal current when the thyristor is in the ON state. 242, 66, 208, 191

ON-state voltage (thyristor). The principal voltage when the thyristor is in the ON state. *See also:* **principal voltage-current characteristic (principal characteristic).** 242, 66, 208, 191

opacity (electroacoustics) (optical path). The reciprocal of transmission. *See also:* **transmission (transmittance).** 176

open amortisseur. An amortisseur that has no connections between poles. 328

open area (electromagnetic compatibility). *See:* **test site.**

open-center display. A plan-position-indicator display on which zero range corresponds to a ring around the center of the display. *See also:* **radar.** 328

open center PPI (plan-position indicator) (radar). A PPI in which the display of the initiation of the time base precedes that of the transmitted pulse. 13

open-circuit characteristic (synchronous machine). *See:* **open-circuit saturation curve.**

open-circuit control. A method of controlling motors employing the open-circuit method of transition from series to parallel connections of the motors. *See also:* **multiple-unit control.** 328

open circuit cooling (rotating machinery). A method of cooling in which the coolant is drawn from the medium surrounding the machine, passes through the machine and then returns to the surrounding medium. 63

open-circuit impedance (1) (general). An impedance of a network that has a specified pair or group of terminals open circuited. **(2) (four-terminal network or line).** The input-output- or transfer-impedance parameters z_{11}, z_{22}, z_{12} and z_{21} of a four-terminal network when the far-end is open circuited. 67

open-circuit inductance (transformer). The apparent inductance of a winding with all other windings open-circuited. 197

open-circuit potential. The measured potential of a cell from which no current flows in the external circuit. 221, 205

open-circuit saturation curve (open-circuit characteristic) (synchronous machine). The saturation curve of a machine with an open-circuited armature winding. 63

open-circuit signaling. That type of signaling in which no current flows while the circuit is in the idle condition. 59

open-circuit test (synchronous machine). A test in which the machine is run as a generator with its terminals open-circuited. 63

open-circuit transition (1) (multiple-unit control). A method of changing the connection of motors from series to parallel in which the circuits of all motors are open during the transfer. *See also:* **multiple-unit control.** 328
(2) (industrial control) (reduced-voltage controllers, including star-delta controllers). A method of starting in which the power to the motor is interrupted during normal

starting sequence. *See also:* **electric controller.**

225, 206

open circuit transition auto-transformer starting (rotating machinery). The process of auto-transformer starting whereby the motor is disconnected from the supply during the transition from reduced to rated voltage.   63

open-circuit voltage (1) (battery). The voltage at its terminals when no appreciable current is flowing. *See also:* **battery (primary or secondary).**   328

(2) (arc-welding apparatus). The voltage between the output terminals of the welding power supply when no current is flowing in the welding circuit.   264

open conductor (open wire) (transmission and distribution). A type of electric supply or communication line construction in which the conductors are bare, covered or insulated and without grounded shielding, individually supported at the structure either directly or with insulators.

262

open cutout. A cutout in which the fuse clips and fuseholder, fuse unit, or disconnecting blade are exposed.

103, 202

open-delta connection. A connection similar to a delta-delta connection utilizing three single-phase transformers, but with one single-phase transformer removed. *Note:* The two remaining transformers of an open-delta bank will carry 57.7 percent of the load carried by the bank using three identical transformers connected delta-delta.   53

open ended. Pertaining to a process or system that can be augmented.   255, 77

open-ended coil (rotating machinery). A partly preformed coil the turns of which are left open at one end to facilitate their winding into the machine. *See also:* **asynchronous machine; direct-current commutating machine.**   63

opening operating time (switch). The interval during which a switch is being operated to move from the fully closed to the fully open position.   27, 103

opening operation (switching device). *See:* **open operation (switching device).**

opening time (mechanical switching device) (sectionalizer). The time interval between the time when the actuating quantity of the release circuit reaches the operating value and the instant when the primary arcing contacts have parted. Any time-delay device forming an integral part of the switching device is adjusted to its minimum setting or, if possible, is cut out entirely for the determination of opening time. *Note:* The opening time includes the operating time of an auxiliary relay in the release circuit when such a relay is required and supplied as part of the switching device. *See:* **isolating time (sectionalizer).**

103, 202

open line wire charging current. Current supplied to an unloaded open-wire line. *Note:* Current is expressed in rms amperes.   79, 103

open-link cutout. A cutout that does not employ a fuseholder and in which the fuse support directly receives an open-link fuse link or a disconnecting blade.   103, 202

open-link fuse link. A replaceable part or assembly comprised of the conducting element and fuse tube, together with the parts necessary to confine and aid in extinguishing the arc connected directly into the fuse clip of the open-link fuse support.   103, 202

open-link fuse support. An assembly of base or mounting support, insulators or insulator unit, and fuse clips for directly mounting an open-link fuse link and for connecting it into the circuit.   103, 202

open loop (automatic control). A signal path without feedback. *See also:* **control system, feedback.**   56

open-loop control system (1) (general). A system in which the controlled quantity is permitted to vary in accordance with the inherent characteristics of the control system and the controlled power apparatus for any given adjustment of the controller. *Note:* No function of the controlled variable is used for automatic control of the system. It is not a feedback control system. *See also:* **control; control system; network analysis.**   151, 321, 225, 206

(2) (speed governing of hydraulic turbines). A control system that has no means for comparing the output with the input for control purposes.   8

open-loop gain (power supplies). The gain, measured without feedback, is the ratio of the voltage appearing across the output terminal pair to the causative voltage required at the (input) null junction. The open-loop gain is denoted by the symbol A in diagrams and equations. *See:* **closed loop; loop gain.** *See also:* **power supply.**

228, 186

open loop measurement (data transmission). Measurement made in which a circuit has at least one of two hybrid sets disconnected thereby opening the loop.   59

open-loop phase angle.* *See:* **loop phase angle.**

* Deprecated

open-loop series street-lighting system. A street-lighting system in which the circuits each consist of a single line wire that is connected from lamp to lamp and returned by a separate route to the source of supply. *See also:* **alternating-current distribution; direct-current distribution.**

64

open-loop system. A control system that has no means for comparing the output with input for control purposes.

224

open machine (1) (rotating machinery). A machine in which no mechanical protection as such is embodied and there is no restriction to ventilation other than that necessitated by good mechanical construction. *See:* **asynchronous machine; direct-current commutating machine.**   63

(2) (electric installations on shipboard). A machine having ventilating openings which permit passage of external cooling air over and around the windings.   3

open-numbering plan (telephone switching systems). A numbering plan in which the number of digits dialed varies according to the requirements of the telecommunications message network.   55

open operation (switching device). The movement of the contacts from the normally closed to the normally open position. *Note:* The letter O signifies this operation: OPEN.

103, 202

open path (network analysis). A path along which no node appears more than once. *See also:* **linear signal flow graphs.**   282

open-phase protection. A form of protection that operates to disconnect the protected equipment on the loss of current in one phase conductor of a polyphase circuit, or to prevent the application of power to the protected equipment on the absence of one or more phase voltages of a polyphase system.   103, 202, 60, 127

open-phase relay. A polyphase relay designed to operate when one or more input phases of a polyphase circuit are open. *See also:* **relay.**   103, 202, 60, 127

open pipe-ventilated machine. An open machine except that openings for the admission of the ventilating air are so arranged that inlet ducts or pipes can be connected to them. This air may be circulated by means integral with the machine or by means external to and not a part of the machine. In the latter case, this machine is sometimes known as a **separately ventilated machine** or a **forced-ventilated machine.** Mechanical protection may be defined as under **dripproof machine, splashproof machine, guarded machine,** or **semiguarded machine.** *See also:* **asynchronous machine; direct-current commutating machine.**   328

open region (1) (three-dimensional space). A volume that satisfies the following conditions: (A) any point of the region has a neighborhood that lies within the region; (B) any two points of the region may be connected by a continuous space curve that lies entirely in the region.
(2) (two-dimensional space). An area that satisfies the following conditions: (A) any point of the region has a neighborhood that lies within the region; (B) any two points of the region may be connected by a continuous curve that lies entirely in the region.   210

open relay. An unenclosed relay. *See also:* **relay.**   259

open resonator (laser-maser). An open resonator and a **beam resonator** are identical.   363

open-shut control system.* *See:* **control system, ON-OFF.**

* Deprecated

open subroutine (computing systems). A subroutine that must be relocated and inserted into a routine at each place it is used. *See:* **closed subroutine.**   255, 77

open switchgear assembly. One that does not have enclosures as part of the structure.   79

open terminal box (rotating machinery). A terminal box that is, normally, open only to the interior of the machine.   63

open-wire circuit. A circuit made up of conductors separately supported on insulators. *Note:* The conductors are usually bare wire, but they may be insulated by some form of continuous insulation. The insulators are usually supported by crossarms or brackets on poles. *See also:* **open wire.**   328

open-wire lead (open-wire line). *See:* **open-wire pole line.**

open-wire line charging current (high voltage circuit breakers). Current supplied to an unloaded open wire line. *Note:* Current is expressed in root-mean-square amperes.   130

open-wire pole line (open-wire line) (open-wire lead). A pole line whose conductors are principally in the form of open wire. *See also:* **open wire.**   328

open wiring. A wiring method using cleats, knobs, tubes, and flexible tubing for the protection and support of insulated conductors run in or on buildings, and is not concealed by the building structure.   328

operable equipment (test, measurement and diagnostic equipment). An equipment which, from its most recent performance history and a cursory electrical and mechanical examination, displays an indication of operational performance for all required functions.   54

operand. That which is, or is to be, operated upon. *Note:* An operand is usually identified by an address part of an instruction.   235, 255, 77, 54

operate (analog computer). The computer control state in which input signals are connected to all appropriate computing elements, including integrators, for the generation of the solution. *See also:* **electronic analog computer.**   9, 77, 10

operated unit. A switch, signal, lock, or other device that is operated by a lever or other operating means.   328

operating basis earthquake (OBE) (nuclear power generating stations) (seismic qualification of class 1E equipment). That earthquake which could reasonably be expected to affect the plant site during the operating life of the plant; it is that earthquake which produces the vibratory ground motion for which those features of the nuclear power plant necessary for continued operation without undue risk to the health and safety of the public are designed to remain functional.   28, 31

operating bypass (nuclear power generating stations). Normal and permissive removal of the capability to accomplish a protective function that could otherwise occur in response to a particular set of generating station conditions. *Note:* Typically, operating bypasses are used to permit a change to a different mode of generating station operation (for example, prevention of initiation of safety injection during cold shutdown conditions).   20

operating characteristic (relay). The response of the relay to the input quantities that result in relay operation.
     103, 202, 60, 127

operating conditions (1) (general). The whole of the electrical and mechanical quantities that characterize the work of a machine, apparatus, or supply network, at a given time.   210, 310
(2) (measurement systems). Conditions (such as ambient temperature, ambient pressure, vibration, etcetera) to which a device is subjected, but not including the variable measured by the device. *See also:* **measurement system.**
     295

operating cycle (nuclear power generating stations). The complete sequence of operations that occur during a response to a demand function.   142, 31

operating device (elevators). The car switch, pushbutton, lever, or other manual device used to actuate the control. *See:* **control.**   328

operating duty (switching device). A specified number and kind of operations at stated intervals.   103, 202

operating duty cycle (surge arrester). One or more unit operations, as specified.   2

operating-duty test (surge arresters). A test in which working conditions are simulated by the application to the arrester of a specified number of impulses while it is connected to a power supply of rated frequency and specified voltage.   308, 62

operating experience (nuclear power generating stations). Accumulation of verifiable service data for conditions equivalent to those for the equipment to be qualified or for which particular equipment is to be qualified.   120, 31

operating influence. The change in a designated performance characteristic caused solely by a prescribed change in a specified operating variable from its reference operating condition to its extreme operating condition, all other operating variables being held within the limits of reference operating conditions. *Notes:* (1) It is usually expressed as a percentage of span. (2) If the magnitude of the influence is affected by direction, polarity, or phase, the greater value shall be taken.   295

operating life (gyro; accelerometer). The accumulated time of operation throughout which a gyro or accelerometer exhibits specified performance when maintained and calibrated in accordance with a specified schedule.   46

operating line; operating curve (electron device). The locus of all simultaneous values of total instantaneous electrode voltage and current for given external circuit conditions.   190

operating mechanism (switching device). That part of the mechanism that actuates all the main-circuit contacts of the switching device either directly or by the use of pole-unit mechanisms.   27, 103

operating modes (nuclear power generating station). The nuclear power plant modes as defined by the technical specifications for the plant.   358

operating noise temperature. The temperature in kelvins given by

$$T_{op} = \frac{N_0}{kG_s}$$

where N_0 is the output noise power per unit bandwidth at a specified output frequency flowing into the output circuit (under operating conditions), k is Boltzmann's constant, and G_s is the ratio of (1) the signal power delivered at the specified output frequency into the output circuit (under operating conditions) to (2) the signal power available at the corresponding input frequency or frequencies to the system (under operating conditions) at its accessible input terminations. *Notes:* (A) In a nonlinear system T_{op} may be a function of the signal level. (B) In a linear two-port transducer with a single input and a single output frequency, if the noise power originating in the output termination and reflected at the output port can be neglected, T_{op} is related to the noise temperature of the input termination T_i and the effective input noise temperature T_e by the equation

$$T_{op} = T_i + T_e.$$

See also: **transducer.**   125, 190

operating overload (industrial control). The overcurrent to which electric apparatus is subjected in the course of the normal operating conditions that it may encounter. *Notes:* (1) The maximum operating overload is to be considered six times normal full-load current for alternating-current industrial motors and control apparatus; four times normal full-load current for direct-current industrial motors and control apparatus used for reduced-voltage starting; and ten times normal full-load current for direct-current industrial motors and control apparatus used for full-voltage starting. (2) It should be understood that these overloads are currents that may persist for a very short time only, usually a matter of seconds.   206

operating point (working point) (electron device). The point on the family of characteristic curves corresponding to the average voltages or currents of the electrodes in the ab-

sence of a signal. *See also:* **quiescent point.**   190

operating speed range. The range between the lowest and highest rated speeds at which the drive may perform at full load. *See also:* **electric drive.**   206

operating system (computing systems). An organized collection of techniques and procedures for operating a computer.   255, 77

operating tap voltage (capacitance potential devices). Indicates the root-mean-square voltage to ground at the point of connection (potential tap) of the device network to the coupling capacitor or bushing. This is the voltage on which certain insulation tests are based. *See also:* **outdoor coupling capacitor.**   351

operating temperature (1) (power supplies). The range of environmental temperatures in which a power supply can be safely operated (typically, 20 to 50 degrees Celsius). *See:* **ambient operating temperature (range).** *See also:* **power supply.**   228, 186

(2) (gyro; accelerometer). The temperature at one or more gyro or accelerometer elements when the device is in the specified operating environment. These elements may include the spin motor winding, the flange, the pickoff, the torquer, the temperature sensor, etcetera.   46

operating temperature limits (attenuator). Maximum temperature in degrees Celsius at which attenuator will operate with full input power. *Note:* Derating function for maximum power versus temperature must be specified to show maximum temperature in degrees Celsius at which attenuator will operate 10 dB below full input power.   110

operating temperature, maximum, (electrical insulation tests). The stabilized temperature obtained from operation of the equipment at rated load and duty cycle in the maximum ambient temperature specified for the device under test.   116

operating temperature range (Hall-effect devices). The range of ambient temperature over which the Hall effect device may be operated with nominal control current and a specified maximum magnetic flux density.   107

operating temperature, room (electrical insulation tests). The temperature of the equipment expected at rated load and duty cycle in an ambient temperature of 20°C ± 5° (68°F ± 9°). An equipment item that has been operated through its normal duty cycle or has stabilized to the approximate normal running temperature may be assumed to be at room ambient operating temperature.   116

operating time (1) (relay). The time interval from occurrence of specified input conditions to a specified operation. See the figure attached to **impulse margin.**
       103, 202, 60, 127

(2) (availability). The period of time during which an item performs its intended function.   164

operating voltage (glow lamp). The voltage of the system on which a device is operated. *Note:* This voltage, if alternating, is usually expressed as a root-mean-square value. *See also:* **maintaining voltage.**   103, 202

operating voltage range (analog-to-digital converter with multiplexor with sample and hold). The minimum and maximum values of the analog input voltage which can be represented by the output to within a given accuracy.
       10

operation (1) (switching device). Action of the parts of the device to perform its normal function.

235, 103, 202, 27

(2) (train control). The functioning of the automatic train-control or cab-signaling system that results from the movement of an equipped vehicle over a track element or elements for a block with the automatic train-control apparatus in service, or which results from the failure of some part of the apparatus. *See also:* **automatic train control.**

328

(3) (elevators). The method of actuating the control. *See:* **control.**

(4) (electronic digital computers). (A) A defined action, namely, the act of obtaining a result from one or more operands in accordance with a rule that completely specifies the result for any permissible combination of operands. (B) The set of such acts specified by such a rule, or the rule itself. (C) The act specified by a single computer instruction. (D) A program step undertaken or executed by a computer, for example, addition, multiplication, extraction, comparison, shift, transfer. The operation is usually specified by the operator part of an instruction. (E) The event of specific action performed by a logic element. (F) Loosely: command. 235, 255, 77

operational amplifier (analog computer). (1) An amplifier, usually a high-gain direct-current amplifier, designed to be used with external circuit elements to perform a specified computing operation or to provide a specified transfer function. (2) An amplifier, usually a high-gain direct-current amplifier, with external circuit elements, used for performing a specified computing operation. *See:* **integrating amplifier, summing amplifier, and inverting amplifier.** *Notes:* (A) The gain and phase characteristics are generally designed to permit large variations in the feedback circuit without instability. (B) The input terminal of an operational amplifier (a) is the summing junction of an operational amplifier (b) and is generally designed to draw current that is negligibly small relative to signal currents in the feedback impedance. 9

operational availability (nuclear power generating stations). The measured characteristic of an item expressed by the probability that it will be operable when needed as determined by periodic test and resultant analyses. 75

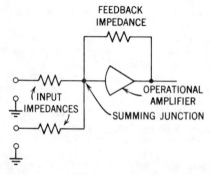

Operational amplifier, typical arrangement

operational gain. *See:* **closed-loop gain.**
operational maintenance. *See:* **non-interference testing.**
operational maintenance influence (instruments). The effect

of routine operations that involve opening the case, such as to inspect or mark records, change charts, add ink, alter control settings, etcetera. *See also:* **accuracy rating (instrument).** 295

operational power supply. A power supply whose control amplifier has been optimized for signal-processing applications rather than the supply of steady-state power to a load. A self-contained combination of operational amplifier, power amplifier, and power supplies for higher-level operations. *See also:* **power supply.** 228, 186

operational programming. The process of controlling the output voltage of a regulated power supply by means of signals (which may be voltage, current, resistance, or conductance) that are operated on by the power supply in a predetermined fashion. Operations may include algebraic manipulations, multiplication, summing, integration, scaling, and differentiation. *See also:* **power supply.**

228, 186

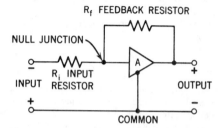

Operational programming

operational relay. A relay that may be driven from one position or state to another by an operational amplifier or a relay amplifier. *See also:* **function relay.** 9, 77, 10

operational reliability. *See:* **reliability, operational.**

operation code. (1) The operations that a computing system is capable of executing, each correlated with its equivalent in another language; for example, the binary or alphanumeric codes in machine language along with their English equivalents; the English description of operations along with statements in a programming language such as **Cobol, Algol,** or **Fortran.** (2) The code that represents or describes a specific operation. The operation code is usually the operation part of the instruction.

235, 210

operation, coordinated (electric power supply). Operation of generation and transmission facilities of two or more interconnected electrical systems to achieve greater reliability and economy. 112

operation, coordinated, of hydro plants. Operation of a group of hydro plants and storage reservoirs so as to obtain optimum power benefits with due consideration to all other uses. 112

operation factor. The ratio of the duration of actual service of a machine or equipment to the total duration of the period of time considered. *See also:* **generating station.**

64

operation influence (electrical influence). The maximum variation in the reading of an instrument from the initial reading, when continuously energized at a prescribed point on the scale under reference conditions over a stated interval of time, expressed as a percentage of full-scale value. *See also:* **accuracy rating (instrument).**   280

operation indicator (relay). *See:* **target (relay).**

operation part (instruction) (electronic computation). The part that usually specifies the kind of operation to be performed, but not the location of the operands. *See also:* **instruction code.**   235, 210

operation, quantizing. *See:* **quantizing operation.**

operations by a pulse (pulse terms) (general). Activation, blanking, clearing, deactivation, deflection, reading, resetting, selection, sequencing, setting, starting, stopping, storing, switching, and writing may occur or be performed.   254

operations involving the interaction of pulses (pulse terms). Addition, chopping, coding, comparison, decoding, encoding, mixing, modulation, subtraction, summation, and superposition may occur or be performed. *See:* **complex waveforms.**   254

operations on a pulse (pulse terms) (general). Amplification, attenuation, conditioning, conversion, coupling demodulation, detection, discrimination, filtering, inversion, reception, reflection, and transmission may occur or be performed.   254

operation, synchronized (power operations). An operation wherein power facilities are electrically connected and controlled to operate at the same frequency.   112

operation time (electron tubes). The time after simultaneous application of all electrode voltages for a current to reach a stated fraction of its final value. Conventionally the final value is taken as that reached after a specified length of time. *Note:* All electrode voltages are to remain constant during measurement. The tube elements must all be at room temperature at the start of the test.   190, 125

operator (1) (general) (in the description of a process). (A) That which indicates the action to be performed on operands. (B) A person who operates a machine.   255, 77
(2) (telephone switching systems). A person who handles switching and signaling operations needed to establish connections between stations or who performs various auxiliary functions associated therewith.   55
(3) (nuclear power generating station). A person licensed to operate the plant.   358

operator code (telephone switching systems). The digits dialed by operators to reach other operators.   55

operator-handled call (telephone switching systems). A call in which information necessary for its completion, other than the number of the calling station, is verbally given by or to an operator.   55

operator loss (radar). A loss in effective signal-to-noise ratio (measured by reduced detection probability or increased false alarm rate) due to the radar operator's inability to report accurately all displayed target reports.   13

operator number identification (ONI) (telephone switching systems). An arrangement in which the operator requests the identity of the calling station and enters it into the system for automatic message accounting.   55

operator's telephone set (operator's set). A set consisting of a telephone transmitter, a head receiver, and associated cord and plug, arranged to be worn so as to leave the operator's hands free. *See also:* **telephone station.**   328

opposition (electrical engineering). The relation between two periodic functions when the phase difference between them is one-half of a period.   210

optical ammeter. An electrothermic instrument in which the current in the filament of a suitable incandescent lamp is measured by comparing the resulting illumination with that produced when a current of known magnitude is used in the same filament. The comparison is commonly made by using a photoelectric cell and indicating instrument. *See also:* **instrument.**   328

optical bandwidth (acoustically tunable optical filter). The width at the 50 percent (-3 decibel) points of the optical intensity versus optical wavelength response curve of the device, measured under the conditions of white light input and fixed acoustic frequency.   72

optical character recognition (OCR). Machine identification of printed characters through use of light-sensitive devices. *See:* **magnetic-ink character recognition.**   255, 77

optical (optoelectronic device) coupling coefficient (between two designated ports). The fraction of the radiant or luminous flux leaving one port that enters the other port. *See also:* **optoelectronic device.**   191

optical density, D_λ (laser-maser). Logarithm to the base ten of the reciprocal of the transmittance: $D_\lambda = -\log_{10\tau\lambda}$, where τ is transmittance. *See also:* **photographic transmission density.**   363

optically pumped laser (laser-maser). A laser in which the electrons are excited into an upper energy state by the absorption of light from an auxiliary light source.   363

optical pattern (mechanical recording). *See:* **light pattern.**

optical photons (scintillation counting). Photons with energies corresponding to wavelengths between approximately 120 = 1800 m.   117

optical pyrometer. A temperature-measuring device comprising a standardized comparison source of illumination and source convenient arrangement for matching this source, either in brightness or in color, against the source whose temperature is to be measured. The comparison is usually made by the eye. *See also:* **electric thermometer (temperature meter).**   328

optical scanner (character recognition). (1) A device that scans optically and usually generates an analog or digital signal. (2) A device that optically scans printed or written data and generates their digital representations. *See:* **visual scanner.** *See also:* **electronic analog computer.**   255, 77

optical sound recorder. *See:* **photographic sound recorder.**

optical sound reproducer. *See:* **photographic sound reproducer.**

optic amplifier. An optoelectronic amplifier whose signal input and output ports are electric. *Note:* This is in accord with the accepted terminologies of other electric-signal input and output amplifiers such as dielectric, magnetic, and thermionic amplifiers. *See also:* **optoelectronic device.**   191

optic port. A port where the energy is electromagnetic radiation, that is, photons. *See also:* **optoelectronic device.**   191

optimal control. An admissible control law that gives a performance index an extremal value. *See also:* **control system.**   329

optimization. The procedure used in the design of a system to maximize or minimize some performance index. May entail the selection of a component, a principle of operation, or a technique. *See also:* **system.** 209

optimum bunching (electron tubes) (traveling-wave tube). The bunching condition that produces maximum power at the desired frequency in an output gap. *See also:* **electron devices, miscellaneous.** 190, 125

optimum linearizing load resistance (Hall generator). The load resistance that produces the least linearity error. 107

optimum working frequency (radio transmission by ionospheric reflection). The frequency at which transmission between two specified points can be expected to be most effectively maintained regularly at a certain time of day. *Note:* A frequency 15 percent below the monthly median value of the $F2$ maximum usable frequency for a specified time of day and path is often taken as the optimum working frequency for propagation by way of the $F2$ layer. *See also:* **radiation.** 146

optional stop (numerically controlled machines). A miscellaneous function command similar to a program stop except that the control ignores the command unless the operator has previously pushed a button to validate the command. 224, 207

optoelectronic amplifier. An optoelectronic device capable of power gain, in which the signal ports are either all electric ports or all optic ports. *See also:* **optoelectronic device.** 191

optoelectronic cell. The smallest portion of an optoelectronic device capable of independently performing all the specified input and output functions. *Note:* An optoelectronic cell may consist of one or more optoelectronic elements. *See also:* **optoelectronic device.** 191

optoelectronic device. An electronic device combining optic and electric ports. 191

optoelectronic element. A distinct constituent of an optoelectronic cell, such as an electroluminor, photoconductor, diode, optical filter, etcetera. *See also:* **optoelectronic device.** 191

OR. A logic operator having the property that if P is a statement, Q is a statement, R is a statement, . . . , then the OR of $P, Q, R,$ . . . is true if at least one statement is true, false if all statements are false. P OR Q is often represented by $P + Q, P \vee Q.$ *See:* **exclusive OR.** 255, 77

orbit (communication satellite). The path described by the center of mass of a satellite or other object in space, relative to a specified frame of reference. 74

orbital inclination (communication satellite). The angle between the plane of the orbit and the plane of the equator measured at the ascending node. 74

orbital plane (communication satellite). The plane containing the radius vector and the velocity vector of a satellite, the system of reference being that specified for defining the orbital elements. In the idealized case of the unperturbed orbit, the orbital plane is fixed relative to the equatorial plane of the primary body. 74

orbital stability. *See:* **stability of a limit cycle.**

OR circuit (electronic computation). *See:* **OR gate.**

order (1) (general). To put items in a given sequence. 255, 77

(2) (in electronic computation). (A) Synonym for instruction. (B) Synonym for command. (C) Loosely, synonym for **operation part.** *Note:* The use of **order** in the computer field as a synonym for terms similar to the above is losing favor owing to the ambiguity between these meanings and the more-common meanings in mathematics and business. *See also:* **instruction.** 235, 210, 54

order tone (telephone switching systems). A tone that indicates to an operator that verbal information can be transferred to another operator. 55

order wire (communication practice). An auxiliary circuit for use in the line-up and maintenance of communication facilities. 193, 101

ordinary wave (O wave). The magnetoionic wave component that, when viewed below the ionosphere in the direction of propagation, has counterclockwise or clockwise elliptical polarization, respectively, according as the earth's magnetic field has a positive or negative component in the same direction. *See also:* **radiation.** 328

ordinary-wave component (radio wave propagation). The magnetoionic wave component deviating the least, in most of its propagation characteristics, relative to those expected for a wave in the absence of a fixed magnetic field. More exactly, if at fixed electron density, the direction of the fixed magnetic field were rotated until its direction were transverse to the direction of phase propagation, the wave component whose propagation is then independent of the magnitude of the fixed magnetic field. *See also:* **radio wave propagation.** 180, 146

organizational maintenance (test, measurement and diagnostic equipment). Maintenance which is the responsibility of and performed by using organizations on its assigned equipment. Its phases normally consist of inspecting, servicing, lubricating, adjusting and the replacing of parts, minor assemblies and subassemblies. 54

organizing. *See:* **self-organizing.**

OR gate (OR circuit) (electronic computation). A gate whose output is energized when any one or more of the inputs is in its prescribed state. An OR gate performs the function of the logical OR. 210

orifice. An opening or window in a side or end wall of a waveguide or cavity resonator, through which energy is transmitted. *See also:* **waveguide.** 328

orifice plate (rotating machinery). A restrictive opening in a passage to limit flow. 63

original master (electroacoustics) (metal master) (metal negative) (number 1 master) (disk recording). The master produced by electroforming from the face of a wax or lacquer recording. 176

originating traffic (telephone switching systems). Traffic received from lines. 55

orthicon. A camera tube in which a beam of low-velocity electrons scans a photoemissive mosaic capable of storing an electric-charge pattern. *See also:* **television.** 190, 125

orthogonality (oscilloscopes). The extent to which traces parallel to the vertical axis of a cathode-ray-tube display are at right angles to the horizontal axis. *See:* **oscillograph.** 185

orthogonal polarization. *See:* **cross polarization.**

O scan (radar). *See:* **O display.**

oscillating current. A current that alternately increases and decreases but is not necessarily periodic.   210

oscillation (1) (general). The variation, usually with time, of the magnitude of a quantity with respect to a specified reference when the magnitude is alternately greater and smaller than the reference. *See:* **vibration.**   176
(2) (vibration). A generic term referring to any type of a response that may appear in a system or in part of a system. *Note:* Vibration is sometimes used synonymously with oscillation, but it is more properly applied to the motion of a mechanical system in which the motion is in part determined by the elastic properties of the body.   210

(3) (gas turbines). The periodic variation of a function between limits above or below a mean value, for example, the periodic increase and decrease of position, speed, power output, temperature, rate of fuel input, etcetera within finite limits.   98, 58
oscillator (1) (general). Apparatus intended to produce or capable of maintaining electric or mechanical oscillations.   244, 210
(2) (electronics). A nonrotating device for producing alternating current, the output frequency of which is determined by the characteristics of the device. *See also:* **industrial electronics.**   14, 111, 114
oscillator starting time, pulsed. *See:* **pulsed oscillator starting time.**   254
oscillator tube, positive grid. *See:* **positive-grid oscillator tube.**
oscillatory circuit. A circuit containing inductance and capacitance so arranged that when shock excited it will produce a current or a voltage that reverses at least once. If the losses exceed a critical value, the oscillating properties will be lost.   328
oscillatory surge (surge arresters). A surge that includes both positive and negative polarity values.   299, 62, 64, 2
oscillogram. A record of the display presented by an oscillograph or an oscilloscope. *See also:* **oscillograph.**   185
oscillograph. An instrument primarily for producing a record of the instantaneous values of one or more rapidly varying electrical quantities as a function of time or of another electrical or mechanical quantity. *Note:* (1) Incidental to the recording of instantaneous values of electrical quantities, these values may become visible, in which case the oscillograph performs the function of an oscilloscope. (2) An oscilloscope does not have inherently associated means for producing records. (3) The term includes mechanical recorders.
oscillograph tube (oscilloscope tube). A cathode-ray tube used to produce a visible pattern that is the graphic representation of electric signals, by variations of the position of the focused spot or spots in accordance with these signals.   125
oscillography. The art and practice of utilizing the oscillograph. *See also:* **oscillograph.**   185
oscilloscope. An instrument primarily for making visible the instantaneous value of one or more rapidly varying electrical quantities as a function of time or of another electrical or mechanical quantity. *See also:* **cathode-ray oscilloscope; oscillograph.**   328
oscilloscope, dual-beam. An oscilloscope in which the cathode-ray tube produces two separate electron beams that may be individually or jointly controlled. *See also:* **oscilloscope, multibeam; oscillograph.**   106
oscilloscope, multibeam. An oscilloscope in which the cathode-ray tube produces two or more separate electron beams that may be individually or jointly controlled. *See also:* **oscilloscope, dual-beam.**   106
O scope (radar). *See: O* **display.**
other tests. Tests so identified in individual product standards which may be specified by the purchaser in addition to routine tests. (Examples: impulse, insulation power factor, audible sound). *Note:* Transformer "General Requirements" Standards (such as C57.12.00) classify various tests as "routine," "design," "other" depending on the size, voltage, and type of transformer involved.   53
OTH radar. *See:* **over-the-horizon radar.**
O-type tube or valve (microwave tubes). A microwave tube in which the beam, the circuit, and the focusing field have symmetry about a common axis. The interaction between the beam and the circuit is dependent upon velocity modulation, the suitably focused beam being launched by a gun structure outside one end of the microwave circuit and principally collected outside the other end of the microwave structure. *See also:* **microwave tube.**   190
Oudin current (desiccating current) (resonator current) (medical electronics). A brush discharge produced by a high-frequency generator that has an output range of 2 to 10 kilovolts and a current sufficient to evaporate tissue water without charring. It is usually applied through a small needlelike electrode with the reference or ground electrode being relatively large and diffuse.   192
Oudin resonator. A coil of wire often with an adjustable number of turns, designed to be connected to a source of high-frequency current, such as a spark gap and induction coil, for the purpose of applying an effluve (convective discharge) to a patient.   192
outage (electric power systems). The state of a component when it is not available to perform its intended function due to some event directly associated with that component. *Notes:* (1) An outage may or may not cause an interruption of service to consumers; depending on system configuration. (2) This definition derives from transmission and distribution applications and does not necessarily apply to generation outages.   200, 112
outage duration (electric power systems). The period from the initiation of an outage until the affected component once again becomes available to perform its intended function. *Note:* Outage durations may be defined for specific types of outages, for example, permanent forced outage duration, transient forced outage duration, and scheduled outage duration. *See also:* **outage.**   112, 200
outage duration, permanent forced (electric power systems). The period from the initiation of the outage until the component is replaced or repaired.   112
outage duration, scheduled (electric power systems). The period from the initiation of the outage until construction, preventive maintenance, or repair work is completed.   112
outage duration, transient forced (electric power systems). The period from the initiation of the outage until the component is restored to service by switching or fuse re-

placement. *Note:* Thus transient forced outage duration is really switching time. 112

outage, equipment (relay systems). The electrical isolation of equipment from the electric system such that it can no longer perform usefully for the duration of such isolation. *Note:* Since the term "outage" can also refer to service as well as equipment, it should always carry the appropriate modifier. 127

outage, forced (electric power systems). An outage that results from conditions directly associated with a component requiring that it be taken out of service immediately, either automatically or as soon as switching operations can be performed, or an outage caused by improper operation of equipment or human error. *Notes:* (1) This definition derives from transmission and distribution applications and does not necessarily apply to generation outages. (2) The key test to determine if an outage should be classified as forced or scheduled is as follows. If it is possible to defer the outage when such deferment is desirable, the outage is a scheduled outage; otherwise, the outage is a forced outage. Deferring an outage may be desirable, for example, to prevent overload of facilities or an interruption of service to consumers. 89, 112

outage, permanent forced (electric power systems). An outage whose cause is not immediately self-clearing, but must be corrected by eliminating the hazard or by repairing or replacing the component before it can be returned to service. An example of a permanent forced outage is a lightning flashover which shatters an insulator thereby disabling the component until repair or replacement can be made. *Note:* This definition derives from transmission and distribution applications and does not necessarily apply to generation outages. 112

outage, power. Complete absence of power at the point of use. 89

outage rate (electric power systems). For a particular classification of outage and type of component, the mean number of outages per unit exposure time per component. *Note:* Outage rates may be defined for specific weather conditions and types of outages. For example, permanent forced outage rates may be separated into adverse weather permanent forced outage rate and normal weather permanent forced outage rate. 112

outage rate, adverse weather permanent forced (electric power systems). For a particular type of component, the mean number of outages per unit of adverse weather exposure time per component. 112

outage rate, normal weather permanent forced (electric power supplies). For a particular type of component, the mean number of outages per unit of normal weather exposure time per component. 112

outage, scheduled (electric power systems). A loss of electric power that results when a component is deliberately taken out of service at a selected time, usually for purposes of construction, preventive maintenance, or repair. *Notes:* (1) This derives from transmission and distribution applications and does not necessarily apply to generation outages. (2) The key test to determine if an outage should be classified as forced or scheduled is as follows. If it is possible to defer the outage when such deferment is desirable, the outage is a scheduled outage; otherwise, the outage is a forced outage. Deferring an outage may be

desirable, for example, to prevent overload of facilities or an interruption of service to consumers. 112

outage, transient forced (electric power systems). An outage whose cause is immediately self-clearing so that the affected component can be restored to service either automatically or as soon as a switch or circuit breaker can be reclosed or a fuse replaced. *Notes:* (1) An example of a transient forced outage is a lightning flashover which does not permanently disable the flashed component. (2) This definition derives from transmission and distribution applications and does not necessarily apply to generation outages. 112

outdoor (1) (transformer). Suitable for installation where exposed to the weather. 53

(2) (power switchgear). Designed for use outside buildings and weather proof enclosures. 103

outdoor arrester. A surge arrester that is designed for outdoor use. 2

outdoor coupling capacitor. A capacitor designed for outdoor service that provides, as its primary function, capacitance coupling to a high-voltage line. *Note:* It is used in this manner to provide a circuit for carrier-frequency energy to and from a high-voltage line and to provide a circuit for power-frequency energy from a high-voltage line to a capacitance potential device or other voltage-responsive device. 351

outdoor termination (cable termination). A termination intended for use where it is not protected from direct exposure to either solar radiation or precipitation. 4

outdoor transformer. A transformer of weather-resistant construction suitable for service without additional protection from the weather. 53

outdoor wall bushing. A wall bushing on which one or both ends (as specified) are suitable for operating continuously outdoors. *See also:* **bushing.** 348

outdoor weatherproof enclosure (series capacitor). An enclosure so constructed or protected that exposure to the weather will not interfere with the successful operation of the equipment contained therein. 86

outer frame (rotating machinery). The portion of a frame into which the inner frame with its assembled core and winding is installed. 63

outer marker (electronic navigation). A marker facility in an instrument landing system that is installed at approximately 5 nautical miles (9 kilometers) from the approach end of the runway on the localizer course line to provide height, distance, and equipment functioning checks to aircraft on intermediate and final approach. *See also:* **navigation.** 13, 187

outgoing traffic (telephone switching systems). Traffic delivered directly to trunks from a switching entity. 55

outlet. A point on the wiring system at which current is taken to supply utilization equipment. *See:* **lighting outlet; receptacle outlet.** 328

outlet box. A box used on a wiring system at an outlet. *See also:* **cabinet.** 328

out-of-band signaling (telephone switching systems). Analog generated signaling that uses the same path as a message and in which the signaling frequencies are lower or higher than those used for the message. 55

out-of-phase (power switchgear) (as prefix to a characteristic

quantity). A qualifying term indicating that the characteristic quantity applies to operation of the circuit breaker in out-of-phase conditions. *See also:* **out-of-step.** 157

out-of-phase conditions (power switchgear). Abnormal circuit conditions of loss or lack of synchronism between the parts of an electrical system on either side of a circuit breaker in which, at the instant of operation of the circuit breaker, the phase angle between rotating phasors representing the generated voltages on either side exceeds the normal value and may be as much as 180° (phase opposition). 157

out-of-roundness (conductor). The difference between the major and minor diameters at any one cross section. *See also:* **waveguide.** 265

out of step. A system condition in which two or more synchronous machines have lost synchronism with respect to one another and are operating at different average frequencies. 103, 202, 66, 127

out-of-step protection (power system). A form of protection that separates the appropriate parts, or prevents separation that might otherwise occur, in the event of loss of synchronism. 103, 202, 66, 127

outpulsing (telephone switching systems). Pulsing from a sender. 55

output (1) (general). (A) The current, voltage, power, or driving force delivered by a circuit or device. (B) The terminals or other places where current, voltage, power or driving force may be delivered by a circuit or device. *See:* **signal, output; variable, output.** 59, 197, 54
(2) (rotating machinery). (A) (generator). The power (active, reactive, or apparent) supplied from its terminals. (B) (motor). The power supplied by its shaft. *See:* **asynchronous machine; direct-current commutating machine.** 63
(3) (electronic digital computer). (A) Data that have been processed. (B) The state or sequence of states occurring on a specified output channel. (C) The device or collective set of devices used for taking data out of a device. (D) A channel for expressing a state of a device or logic element. (E) The process of transferring data from an internal storage to an external storage. 235, 255, 77, 59

output, acoustic (telephony). The sound pressure level developed in an artificial ear, measured in dB referred to an rms sound pressure of 2×10^{-5} newtons per square meter (N/m^2). 122

output angle (gyro). The angular displacement of a gimbal about its output axis with respect to its support. 46

output attenuation (signal generator). The ratio, expressed in decibels (dB), of any selected output, relative to the output obtained when the generator is set to its calibration level. *Note:* It may be necessary to eliminate the effect of carrier distortion and/or modulation feedthrough by the use of suitable filters. *See also:* **signal generator.** 185

output axis (OA) (gyro; accelerometer). An axis of freedom provided with a pickoff which generates an output signal as a function of the output angle. 46

output axis angular acceleration drift rate (gyro). A drift rate that is proportional to the angular acceleration with respect to inertial space of the gyro case about the output axis. The relationship of this component of drift rate to angular acceleration can be stated by means of a coefficient having dimensions of angular displacement per unit time divided by angular displacement per unit time-

squared. 46

output capacitance (*n*-terminal electron tube). The short-circuit transfer capacitance between the output terminal and all other terminals, except the input terminal, connected together. *See also:* **electron-tube admittances.** 190, 125

output-capacitor discharge time (power supply). The interval between the time at which the input power is disconnected and the time when the output voltage of the unloaded regulated power supply has decreased to a specified safe value. *See also:* **regulated power supply.** 347

output circuit (1) (electron tube or valve). The external circuit connected to the output electrode to provide the load impedance. *See also:* ON **period (electron tube or valve).** 244, 190
(2) (protective relay system). An output from a relay system which exercises direct or indirect control of a power circuit breaker, such as trip or close. 90

output-dependent overshoot and undershoot. Dynamic regulation for load changes. *See also:* **electric conversion equipment.** 186

output electrode (electron tubes). The electrode from which is received the amplified, modulated, detected, etcetera, voltage. *See also:* **electron tube.** 244, 190

output factor. The ratio of the actual energy output, in the period of time considered, to the energy output that would have occurred if the machine or equipment had been operating at its full rating throughout its actual hours of service during the period. *See also:* **generating station.** 64

output frequency stability (inverters). The deviation of the output frequency from a given set value. *See:* **self-commutated inverters.** 208

output gap (electron tubes) (traveling-wave tubes). An interaction gap by means of which usable power can be abstracted from an electron stream. *See:* **electron devices, miscellaneous.** 190

output impedance. *See:* **impedance, output.**

output phase displacement (power inverters that have polyphase output) (inverters). The angular displacement between fundamental phasors. *See:* **self-commutated inverters.** 208

output power (1) (general). The power delivered by a system or transducer to its load. 239
(2) (electron tube or valve). The power supplied to the load by the electron tube or valve at the output electrode. *See also:* ON **period (electron tube or valve).** 244, 190

output pulse amplitude (digital delay line). Peak amplitude of the output doublet which is obtained across the specified output load for a given amplitude of input step. 81

output pulse duration (digital delay line). Time spacing between the 10 percent amplitude point of the rise of the first peak to the 10 percent amplitude point of the fall of the second peak. 81

output pulse shape (low and high power pulse transformers). Load current pulse flowing in a winding or voltage pulse developed across a winding in response to application of an input pulse. *Notes:* (1) The shape of the output pulse is described by a current- or voltage-time relationship. (2) Typically, a prominent feature of the output pulse is an accentuated backswing (last transition overshoot) (A_{BS}).

See also: **input pulse shape; pulse amplitude; rise time; pulse duration; fall time; trailing edge; tilt; overshoot; backswing; return swing; roll off; ringing; leading edge; quiescent value; leading edge; pulse top.** 32, 33

output range (gyro; accelerometer). The product of input range and scale factor. *See:* **input-output characteristics.** 46

output reference axis (ORA) (gyro; accelerometer). The direction of an axis defined by the case mounting surfaces and/or external case markings. It is nominally parallel to the output axis. 46

output resonator (catcher) (electron tubes). A resonant cavity, excited by density modulation of the electron beam, that supplies useful energy to an external circuit. *See also:* **velocity-modulated tube.** 244, 190

output ripple voltage (regulated power supply). The portion of the output voltage harmonically related in frequency to the input voltage and arising solely from the input voltage. *Note:* Unless otherwise specified, percent ripple is the ratio of root-mean-square value of the ripple voltage to the average value of the total voltage expressed in percent. In television, ripple voltage is usually expressed explicitly in peak-to-peak volts to avoid ambiguity. *See also:* **regulated power supply.** 347

output signal (speed governing of hydraulic turbines). The physical reaction of any element of a control system to an input signal. 8

output span (gyro; accelerometer). The algebraic difference between the upper and lower values of the output range. *See:* **input-output characteristics.** 46

output-structure transit time. That portion of the photomultiplier transit time occurring with the output structure. 117

output voltage regulation (power supply). The change in output voltage, at a specified constant input voltage, resulting from a change of load current between two specified values. *See also:* **regulated power supply.** 347

output voltage stabilization (power supply). The change in output voltage, at a specified constant load current, resulting from a change of input voltage between two specified values. *See also:* **regulated power supply.** 347

output winding (1) (saturable reactor). A winding, other than a feedback winding, associated with the load and through which power is delivered to the load. *See also:* **magnetic amplifier.** 328
(2) (secondary winding(s)). The winding(s) from which the output is obtained. *See also:* **magnetic amplifier.** 61

outrigger (switching-device terminal). An attachment that is fastened to or adjacent to the terminal pad of a switching device to maintain electrical clearance between the conductor and other parts or, when fastened adjacent, to relieve mechanical strain on the terminal, or both. 103, 202, 27

outside plant (communication practice). That part of the plant extending from the line side of the main distributing frame to the line side of the station or private-branch-exchange protector or connecting block, or to the line side of the main distributing frame in another central office building. 328

outside space block. *See:* **end finger.**

outward-wats service (telephone switching systems). A flat-rate or measured-time direct distance dialing service

for defined geographical groups of numbering plan areas. 55

oven. An enclosure and associated sensors and heaters for maintaining components at a controlled and usually constant temperature. 9, 10

oven, wall-mounted. A domestic oven for cooking purposes designed for mounting in or on a wall or other surface. *See also:* **appliances.** 256

over-all electrical efficiency (dielectric and induction heater). The ratio of the power absorbed by the load material to the total power drawn from the supply lines. *See:* **load circuit efficiency.** *See also:* **dielectric heating; induction heating.** 14,114

over-all generator efficiency (thermoelectric device). The ratio of (1) electric power output to (2) thermal power input. *See also:* **thermoelectric device.** 248, 191

overall power efficiency (laser-maser). The ratio of the useful power output of the device to the total input power. 363

overbunching (electron tubes). The bunching condition produced by the continuation of the bunching process beyond the optimum condition. *See also:* **electron devices, miscellaneous.** 190, 125

overburden (earth conductivity). The surface layers or regions of the earth that are water bearing and are subject to weathering. They comprise predominantly sand, gravel, clays, and poorly consolidated rocks. 132

overcompounded. A qualifying term applied to a compound-wound generator to denote that the series winding is so proportioned that the terminal voltage at rated load is greater than at no load. *See:* **direct-current commutating machine.** 328

overcurrent (rotating machinery). An abnormal current greater than the full load value. *See also:* **asynchronous machine; direct-current commutating machine.** 244, 63, 124

overcurrent protection. A form of protection that operates when current exceeds a predetermined value. *See also:* **circuit breaker; fuse; overcurrent relay.** 60, 127, 53

overcurrent protection (electric installations on shipboard). The effect of a device operative on excessive current (but not necessarily on short circuit) to cause and maintain the interruption of current flow to the device governed. *Syn.* **overload protection.** 3

overcurrent protective device (packaging machinery). A device operative on excessive current that causes and maintains the interruption of power in the circuit. 124

overcurrent relay (power switchgear). A relay that operates when its input current exceeds a predetermined value. 127, 103

overcurrent release (overcurrent trip). A release that operates when the current in the main circuit is equal to or exceeds the release setting. 103, 202

overcutting (disk recording). The effect of excessive level characterized by one groove cutting through into an adjacent one. *See also:* **phonograph pickup.** 176

overdamped (industrial control). A degree of damping that is more than sufficient to prevent the oscillation of the output following an abrupt stimulus. *Note:* For a linear second order system the roots of the characteristic equation must then be real and unequal. *See also:* **control system, feedback.** 67, 219, 206

overdamping (aperiodic damping*). The special case of damping in which the free oscillation does not change sign. A damped harmonic system is overdamped if $F^2 > 4MS$. *See:* **damped harmonic system** for equation, definitions of letter symbols, and referenced terms.   210

* Deprecated

overflow (electronic computation). (1) The condition that arises when the result of an arithmetic operation exceeds the capacity of the number representation in a digital computer. (2) The carry digit arising from this condition.
235, 54

overflow traffic (telephone switching systems). That part of the offered traffic that cannot be carried by a group of servers.   55

overhang packing (rotating machinery). Insulation inserted in the end region of the winding to provide spacing and bracing. *See also:* **rotor (rotating machinery); stator.**
63

overhead electric hoist. A motor-driven hoist having one or more drums or sheaves for rope or chain, and supported overhead. It may be fixed or traveling. *See also:* **hoist.**
328

overhead ground wire (1) (lightning protection). Grounded wire or wires placed above phase conductors for the purpose of intercepting direct strokes in order to protect the phase conductors from the direct strokes. They may be grounded directly or indirectly through short gaps. *See also:* **direct-stroke protection (lightning).**   45
(2) (conductor stringing equipment) (lightning protection). Multiple grounded wire or wires placed above phase conductors for the purpose of intercepting direct strokes in order to protect the phase conductors from the direct strokes. *Syn.* **earth wire; shield wire; skywire; static wire.**
45

overhead line charging current. *See:* **open line wire charging current.**

overhead structure (elevators). All of the structural members, platforms, etcetera, supporting the elevator machinery, sheaves, and equipment at the top of the hoistway. *See also:* **elevators.**   328

overinterrogation control (electronic navigation). *See:* **gain turn-down.**

overlap. The distance the control of one signal extends into the territory that is governed by another signal or other signals. *See also:* **neutral zone.**   328

overlap angle (1) (gas tube). The time interval, in angular measure, during which two consecutive arc paths carry current simultaneously. *See also:* **gas-filled rectifier.**
244, 190

(2) (rectifier circuits). *See:* **commutating angle.**

overlap control. *See:* **control system, two-step.**

overlap testing (nuclear power generating stations). Overlap testing consists of channel, train or load group verification by performing individual tests on the various components and subsystems of the channel, train, or load group. The individual component and subsystem tests shall check parts of adjacent subsystems, such that the entire channel, train, or load group will be verified by testing of individual components or subsystems.   31, 75, 366

overlap X (facsimile). The amount by which the recorded spot X dimension exceeds that necessary to form a most

nearly constant density line. *Note:* This effect arises in that type of equipment which responds to a constant density in the subject copy by a succession of discrete recorded spots. *See also:* **recording (facsimile).**   12
overlap Y (facsimile). The amount by which the recorded spot Y dimension exceeds the nominal line width. *See also:* **recording (facsimile).**   12
overlay (computing systems). The technique of reusing the same blocks of internal storage during different stages of a problem. When one routine is no longer needed in storage, another routine can replace all or part of it.
255, 77
overload (1) (general). Output of current, power, or torque, by a device, in excess of the rated output of the device on a specified rating basis.   53
(2) (analog computer). A condition existing within or at the output of a computing element that causes a substantial computing error because of the voltage or current saturation of one or more of the parts of the computing element. Similar to an overflow of an accumulator in a digital computer.   9, 77
(3) (test, measurement and diagnostic equipment). To exceed the rated capacity of.   54
(4) (thyristor power computer). A condition existing when the load current exceeds the continuous rating of the converter unit in magnitude or time, or both, but the conduction cycles and waveforms remain essentially normal.   121
(5) (electric power systems). Loading in excess of normal rating of equipment.   112
overload capacity (1) (antennas) (industrial control). The current, voltage, or power level beyond which permanent damage occurs to the device considered. This is usually higher than the rated load capacity. *Note:* To carry load greater than the continuous rating, may be acceptable for limited use.   111, 206
(2) (accelerometer). The maximum acceleration to which an accelerometer may be subjected beyond the normal operating range without causing a permanent change in the specified performance characteristics.   46
overload factor (electromagnetic compatibility). The ratio of the maximum value of a signal for which the operation of the predetector circuits of the receiver does not depart from linearity by more than one decibel, to the value corresponding to full-scale deflection of the indicating instrument. *See also:* **electromagnetic compatibility.**
220, 199
overload level (system or component). That level above which operation ceases to be satisfactory as a result of signal distortion, overheating, or damage. *See also:* **level.**
239, 59
overload ON-state current (thyristor). An ON-state current of substantially the same wave shape as the normal ON-state current and having a greater value than the normal ON-state current. *See also:* **principal current.**
66, 208, 191
overload point, signal (electronic navigation). The maximum input signal amplitude at which the ratio of output to input is observed to remain within a prescribed linear operating range. *See also:* **navigation.**   187, 13
overload protection (industrial control). The effect of a device operative on excessive current, but not necessarily

on short-circuit, to cause and maintain the interruption of current flow to the device governed. *See:* **overcurrent protection.**   206

overload relay (1) (general). A relay that responds to electric load and operates at a preset value of overload. *Note:* Overload relays are usually current relays but they may be power, temperature, or other relays.   103, 202
(2) (overcurrent). An overcurrent relay that functions at a predetermined value of overcurrent to cause disconnection of the load from the power supply. *Note:* An overload relay is intended to protect the load (for example, motor armature) or its controller, and does not necessarily protect itself. *See:* **overcurrent relay; undercurrent relay.**   206

overmoded waveguide. A waveguide used to propagate a single mode, but capable of propagating more than one mode at the frequency of interest. *See:* **waveguide.**   179
overpotential. *See:* **overvoltage.**

overrange (noun) (1) (system or element). Any excess value of the response above its nominal full-scale value, or deficiency below the nominal minimum value.   56
(2) (test, measurement and diagnostic equipment). An input to a measuring device which exceeds in magnitude the capability of a given range.   54

overreach (relay). The extension of the zone of protection beyond that indicated by the relay setting.
                                          103, 202, 60, 127

overreaching protection. A form of protection in which the relays at one terminal operate for faults beyond the next terminal. They may be constrained from tripping until an incoming signal from a remote terminal has indicated whether the fault is beyond the protected line section.
                                          103, 202, 60, 127

overshoot (1) (instrument). The amount of the overtravel of the indicator beyond its final steady deflection when a new constant value of the measured quantity is suddenly applied to the instrument. The overtravel and deflection are determined in angular measure and the overshoot is expressed as a percentage of the change in steady deflection. *Notes:* (1) Since in some instruments the percentage depends on the magnitude of the deflection, a value corresponding to an initial swing from zero to end scale is used in determining the overshoot for rating purposes. (2) Overshoot and damping factor have a reciprocal relationship. The percentage overshoot may be obtained by dividing 100 by the damping factor. *See also:* **accuracy rating (instrument); moving element (instrument).**   280
(2) (power supplies). A transient rise beyond regulated output limits, occurring when the alternating-current power input is turned on or off, and for line or load step changes. See the accompanying figure. *See also:* **recovery time (voltage regulation); power supply.**   228, 186
(3) (pulse terms). A distortion which follows a major transition. *See:* Note in **preshoot** entry.   254
(4) (television). That part of a distorted wave front characterized by a rise above (or a fall below) the final value followed by a decaying return to the final value. *Note:* Generally overshoots are produced in transfer devices having excessive transient response.   163
(5) (circuits and systems). The amount by which a pulse amplitude exceeds a desired reference value after it has changed its state from some other value. *See also:* **dis-**

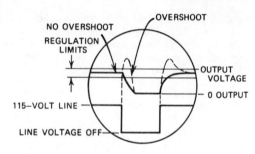

Scope view of turn-off/turn-on effects on a power supply, showing overshoot.

tortion (pulse).   67
(6) (oscilloscopes). In the display of a step function (usually of time), that portion of the waveform which, immediately following the step, exceeds its nominal or final amplitude.   106
(7) (high- and low-power pulse transformers). (first transition overshoot) (A_{OS}). The amount by which the first maximum occurring in the pulse top region exceeds the straightline segment fitted to the top of the pulse in determining A_M. It is expressed in amplitude units or in percent of A_M. *See also:* **input pulse shape.**   32, 33

overshoot switch-off (transformer-rectifier system). The transient output voltage pulse occurring as the result of deenergization of the core on switch-off.   95

overshoot switch-on (transformer–rectifier system) is the transient voltage on the output direct voltage following completion of capacitor charging in the direct current circuit. It may be expressed as a percentage of excess over the steady-state direct voltage.   95

overshoot, system*. *See:* **deviation, system.**   56

* Deprecated

overshoot transient*. *See:* **deviation transient.**   56

* Deprecated

oversized waveguide. A waveguide operated in its dominant mode, but far above cutoff; sometimes termed **quasi-optical waveguide.** *See:* **waveguide.**   179
overslung car frame. A car frame to which the hoisting-rope fastenings or hoisting-rope sheaves are attached to the crosshead or top member of the car frame. *See also:* **hoistway (elevator or dumbwaiter).**   328
overspeed (speed governing of hydraulic turbines). Any speed in excess of rated speed expressed as a percent of rated speed.   8
overspeed and overtemperature protection system (gas turbines). The overspeed governor, overtemperature detector, fuel stop valve(s), blow-off valve, other protective devices and their interconnections to the fuel stop valve, and to the blow-off valve, if used, that are required to shut off all fuel flow and shut down the gas turbine.   98, 58
overspeed governor (gas turbines). A control element that is directly responsive to speed and that actuates the overspeed and overtemperature protection system when the turbine reaches the speed for which the device is set.
                                          98, 58
overspeed protection (1) (industrial control). The effect of a device operative whenever the speed rises above a preset

value to cause and maintain an interruption of power to the protected equipment or a reduction of its speed. *See:* **relay.** 225, 206

(2) (relay systems). A form of protection that operates when the speed of rotation exceeds a predetermined value. 127

overspeed test (rotating machinery). A test on a machine rotor to demonstrate that it complies with specified overspeed requirements. *See also:* **rotor (rotating machinery).** 63

overtemperature (rotating machinery). Unusually high temperature from causes such as overload, high ambient temperature, restricted ventilation, etcetera. 37

overtemperature detector (gas turbines). The primary sensing element that is directly responsive to temperature and that actuates the overspeed and overtemperature protection system when the turbine temperature reaches the value for which the device is set. 98, 58

overtemperature protection (power supplies). A thermal relay circuit that turns off the power automatically should an overtemperature condition occur. *See also:* **power supply.** 228, 186

over-the-horizon radar. Radar using sufficiently low carrier frequencies, usually high-frequency so that ground wave or ionospherically refracted sky wave propagation can allow detection far beyond the ranges allowed by microwave line-of-sight propagation. 13

overtone. *See:* **harmonic.**

overtone-type piezoelectric-crystal unit (circuits and systems). (1) An overtone driven by the action of the piezoelectric effect; (2) **(crystal unit)** A resonator constructed from a piezoelectric crystal material and designed to operate in the vicinity of an overtone of that device. 67

overtravel (relay). The amount of continued movement of the responsive element after the input is changed to a value below pickup. 103, 127

overvoltage (overpotential) (1) (general). A voltage above the normal rated voltage or the maximum operating voltage of a device or circuit. A direct test overvoltage is a voltage above the peak of the line alternating voltage. *See also:* **insulation testing (large alternating-current rotating machinery).** 6

(2) (rotating machinery). An abnormal voltage higher than the normal service voltage, such as might be caused from switching or lightning surges. 37

(3) (surge arresters) (system voltage). Abnormal voltage between two points of a system that is greater than the highest value appearing between the same two points under normal service conditions. *Note:* Overvoltages may be low frequency, temporary, and transient—meaning a lightning or switching surge overvoltage. 53, 244, 62

(4) (radiation-counter tubes). The amount by which the applied voltage exceeds the Geiger-Mueller threshold. *See also:* **gas-filled radiation-counter tube.** 190

(5) (electrochemistry). The displacement of an electrode potential from its equilibrium (reversible) value because of flow of current. *Note:* This is the irreversible excess of potential required for an electrochemical reaction to proceed actively at a specified electrode, over and above the reversible potential characteristic of that reaction. *See also:* **electrochemistry.** 221, 205

overvoltage due to resonance (surge arresters). Overvoltage

at the fundamental frequency of the installation, or of a harmonic frequency, resulting from oscillation of circuits. 244, 62

overvoltage protection. The effect of a device operative on excessive voltage to cause and maintain the interruption of power in the circuit or reduction of voltage to the equipment governed. 206

overvoltage relay (power switchgear). A relay that operates when its input voltage exceeds a predetermined value. 103, 127

overvoltage release (overvoltage trip). A release that operates when the voltage of the main circuit is equal to or exceeds the release setting. 103, 202

overvoltage test (rotating machinery). A test at voltages above the rated operating voltage. 63

overwriting (charge-storage tubes). Writing in excess of that which produces write saturation. *See also:* **charge-storage tube.** 174, 190

OW. *See:* **oil-immersed water-cooled transformer.**

O **wave (radio wave propagation).** *See:* **ordinary-wave component.** *See also:* **radio wave propagation.**

Owen bridge. A 4-arm alternating-current bridge in which one arm, adjacent to the unknown inductor, comprises a capacitor and resistor in series; the arm opposite the unknown consists of a second capacitor,

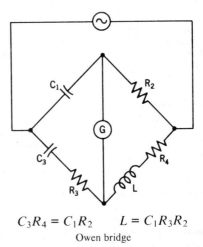

$$C_3 R_4 = C_1 R_2 \qquad L = C_1 R_3 R_2$$

Owen bridge

and the fourth arm of a resistor. *Note:* Normally used for the measurement of self-inductance in terms of capacitance and resistance. Usually, the bridge is balanced by adjustment of the resistor that is in series with the first capacitor and of another resistor that is inserted in series with the unknown inductor. The balance is independent of frequency. *See also:* **bridge.** 328

oxidant. A chemical element or compound that is capable of being reduced. *See also:* **electrochemical cell.** 223

oxidation (electrochemical cells and corrosion). Loss of electrons by a constituent of a chemical reaction. *See:* **electrochemical cell.** 223, 221, 205, 185

oxide-cathode (thermionics). *See:* **oxide-coated cathode.**

oxide-coated cathode (oxide-cathode) (thermionics). A cathode whose active surface is a coating of oxides of alkaline earths on a metal. *See also:* **electron emission.** 244, 190

oxidizing (electrotyping). The treatment of a graphited wax surface with copper sulfate and iron filings to produce a conducting copper coating. *See also:* **electroforming.**
328

oxygen-concentration cell. A galvanic cell resulting primarily from differences in oxygen concentration. *See:* **electrolytic cell.**
221, 205

ozone-producing radiation. Ultraviolet energy shorter than about 220 nanometers that decomposes oxygen O_2 thereby producing ozone O_3. Some ultraviolet sources generate energy at 184.9 nanometers, which is particularly effective in producing ozone. *See also:* **ultraviolet radiation.**
167

P

PABX. *See:* **private automatic branch exchange.**

pace voltage (surge arresters). A voltage generated by ground current between two points on the surface of the ground at a distance apart corresponding to the conventional length of an ordinary pace.
244, 62

pack. To compress several items of data in a storage medium in such a way that the individual items can later be recovered.
255, 77

package, core (rotating machinery). The portion of core lying between two adjacent vent ducts or between an end plate and the nearest vent duct.
63

packaged magnetron. An integral structure comprising a magnetron, its magnetic circuit, and its output matching device. *See also:* **magnetron.**
190, 125

packet (data communication). A group of binary digits including data and control elements which is switched and transmitted as a composite whole. The data and control elements and possibly error control information are arranged in a specified format.
12

packet switching (data communication). A data transmission process, utilizing addressed packets, whereby a channel is occupied only for the duration of transmission of the packet. *See also:* **circuit switching; message switching; store-and-forward switching.**
12

packing density (computing systems). The number of useful storage cells per unit of dimension, for example, the number of bits per inch stored on a magnetic tape or drum track.
255, 77, 54

packing gland. An explosionproof entrance for conductors through the wall of an explosionproof enclosure, to provide compressed packing completely surrounding the wire or cable, for not less than $\frac{1}{2}$ inch measured along the length of the cable. *See also:* **mine feeder circuit.**
328

pad (attenuating pad). A nonadjustable passive network that reduces the power level of a signal without introducing appreciable distortion. *Note:* A pad may also provide impedance matching. *See also:* **transducer.**
239, 178, 59

pad electrode (dielectric heater usage). One of a pair of electrode plates between which a load is placed for dielectric heating. *See also:* **dielectric heating.**
14, 114

pad-mounted. A general term describing equipment positioned on a surface mounted pad located outdoors. *Note:* The equipment is usually enclosed with all exposed surfaces at ground potential.
79

pad-mounted transformer. A transformer utilized as part of an underground distribution system, with enclosed compartment(s) for high voltage and low voltage cables

entering from below, and mounted on a foundation pad.
53

pad-type bearing (rotating machinery). A journal or thrust-type bearing in which the bearing surface is not continuous but consists of separate pads. *see also:* **bearing.**
63

pair. A term applied in electric transmission to two like conductors employed to form an electric circuit. *See also:* **cable.**
328

paired brushes (pair of brushes) (rotating machinery). Two individual brushes that are joined together by a common shunt or terminal. *Note:* They are not to be confused with a split brush. *See also:* **brush (rotating machinery).**
279, 63

paired cable (nonquadded cable). A cable in which all of the conductors are arranged in the form of twisted pairs, none of which is arranged with others to form quads. *See also:* **cable.**
328

pairing (scanning) (television). The condition in which lines appear in groups of two instead of being equally spaced. *See also:* **television.**
163, 178

pair of brushes (rotating machinery). *See:* **paired brushes.**

PAM. *See:* **pulse-amplitude modulation.**

pancake coil. A coil having the shape of a pancake, usually with the turns arranged in the form of a flat spiral.
328

pancake motor. A motor that is specially designed to have an axial length that is shorter than normal.
63

panel (1) (general). A unit of one or more sections of flat material suitable for mounting electric devices.
103, 202

(2) (industrial control). An element of an electric controller consisting of a slab or plate on which various component parts of the controller are mounted and wired.
206, 124

(3) (photovoltaic converter). Combination of shingles or subpanels as a mechanical and electric unit required to meet performance specifications. *See also:* **semiconductor.**

(4) (computing systems). *See:* **control panel; problem board.**

(5) (solar cells). The largest unit combination of solar cells or subpanels that is mechanically designed to facilitate manufacture and handling and that will establish a basis for electrical performance by test.
113

panelboard (electric system). A single panel or a group of panel units designed for assembly in the form of a single panel, including buses and with or without switches and/or automatic overcurrent-protective devices for the control of light, heat, or power circuits of small individual as well as aggregate capacity; designed to be placed in a cabinet or cutout box placed in or against a wall or partition, and accessible only from the front. *See:* **switchboard.** *See also:* **distribution center.** 256

panel efficiency (1) (photoelectric converter). The ratio of available power output to incident radiant power intercepted by a panel composed of photoelectric converters. *Note:* This is less than the efficiency of the individual photoelectric converters because of area not covered by photoelectric converters, Joule heating, and photoelectric-converter mismatches. *See also:* **semiconductor.** 186

(2) (solar cells). The ratio of available electric power output to total incident radiant power intercepted by the area of a panel composed of solar cells. *Note:* This depends on the spectral distribution of the radiant power source and junction temperature(s), requires uniform normal illumination on the intercepting area, and results in an efficiency less than the efficiency of the individual solar cells because of area not covered by solar cells, incident energy heating, solar cell mismatch, optical losses, and wiring losses. 113

panel-frame mounting (switching device). Mounting on a panel frame in the rear of a panel with the operating mechanism on the front of the panel. 103, 202

panel system. An automatic telephone switching system that is generally characterized by the following features: (1) The contacts of the multiple banks over which selection occurs are mounted vertically in flat rectangular panels. (2) The brushes of the selecting mechanism are raised and lowered by a motor that is common to a number of these selecting mechanisms. (3) The switching pulses are received and stored by controlling mechanisms that govern the subsequent operations necessary in establishing a telephone connection. 328

paper-lined construction (dry cell) (primary cell). A type of construction in which a paper liner, wet with electrolyte, forms the principal medium between the negative electrode, usually zinc, and the depolarizing mix. (A layer of paste may lie between the paper liner and the negative electrode.) *See also:* **electrolytic cell.** 328

PAR. *See:* **precision approach radar.**
parabolic torus reflector (antennas). A reflector which is a portion of a surface generated by rotating a parabola about a line located in the plane of and perpendicular to the axis of the parabola. 111

paraboloidal reflector. A reflector that is a portion of a paraboloid of revolution. *See also:* **antenna.** 111

paraelectric Curie temperature (of a ferroelectric material). The intercept of the linear portion of the plot of $1/\epsilon$ versus T, where ϵ is the small signal dielectric permittivity measured at zero bias field and T is the absolute temperature in the region above the ferroelectric Curie temperature where ϵ generally follows the Curie-Weiss relation. *See also:* **ferroelectric domain.** 247

paraelectric region (ferroelectric material). The region above the Curie point where the small signal permittivity increases with decreasing temperature. *See also:* **ferroe-**

lectric Curie point; small signal permittivity. 80
parallel (parallel elements) (1) (network). (A) Two-terminal elements are connected in parallel when they are connected between the same pair of nodes. (B) Two-terminal elements are connected in parallel when any cut-set including one must include the others. *See also:* **network analysis.** 210

(2) (electronic computers). (A) Pertaining to the simultaneity of two or more processes. (B) Pertaining to the simultaneity of two or more similar or identical processes. (C) Pertaining to simultaneous processing of the individual parts of a whole, such as the bits of a character and the characters of a word, using separate facilities for the various parts. *See:* **serial-parallel.** 235, 210, 255, 77

parallel-connected capacitance (interrupter switches). Capacitances are defined to be parallel-connected when the crest value of inrush current to the capacitance being switched exceeds the switch inrush current capability for single capacitance. 27, 103

parallel connection. The arrangement of cells in battery made by connecting all positive terminals together and all negative terminals together, the voltage of the group being only that of one cell and the current drain through the battery being divided among the several cells. *See also:* **battery (primary or secondary).** 328.

parallel digital computer. One in which the digits are handled in parallel. Mixed serial and parallel machines are frequently called serial or parallel according to the way arithmetic processes are performed. An example of a parallel digital computer is one that handles decimal digits in parallel although it might handle the bits that comprise a digit either serially or in parallel. *See also:* **serial digital computer.** 210

parallel feeder. One that operates in parallel with one or more feeders of the same type from the same source. *Note:* These feeders may be of the stub-, multiple-, or tie-feeder type. *See:* **multiple feeder.** *See also:* **center of distribution.** 103, 202, 64

paralleling (rotating machinery). The process by which a generator is adjusted and connected to run in parallel with another generator or system. *See also:* **asynchronous machine; direct-current commutating machine.** 63

paralleling reactor. A current-limiting reactor for correcting the division of load between parallel-connected transformers which have unequal impedance voltages. 53

parallel-mode interference (signal-transmission system). *See:* **interference, common-mode.**
parallel operation (power supplies). Voltage regulators, connected together so that their individual output currents are added and flow in a common load. Several methods for parallel connection are used: spoiler resistors, master/slave connection, parallel programming, and parallel padding. Current regulators can be paralleled without special precaution. *See also:* **power supply.** 228, 186

parallel padding (power supplies). A method of parallel operation for two or more power supplies in which their current limiting or automatic crossover output characteristic is employed so that each supply regulates a portion of the total current, each parallel supply adding to the total and padding the output only when the load current demand exceeds the capability—or limit setting—of the first supply. *See also:* **power supply.** 228, 186

parallel (or perpendicular) polarization. A linear polarization for which the field vector is parallel (or perpendicular) to some reference plane. *Note:* These terms are applied mainly to uniform plane waves incident upon a plane of discontinuity (surface of the earth, surface of a dielectric or a conductor). Then the convention is to take as reference the plane of incidence, that is, the plane containing the direction of propagation and the normal to the surface of discontinuity. If these two directions coincide, the reference plane must be specified by some other convention. 11

parallel processing (computing systems). Pertaining to the simultaneous execution of two or more sequences of instructions or one sequence of instructions operating on two or more sets of data, by a computer having multiple arithmetic and/or logic units. *See:* **multiprocessing; multiprogramming.** 255, 77

parallel programming (power supplies). A method of parallel operation for two or more power supplies in which their feedback terminals (voltage-control terminals) are also paralleled. These terminals are often connected to a separate programming source. *See also:* **power supply.** 228, 186

parallel rectifier. A rectifier in which two or more similar rectifiers are connected in such a way that their direct currents add and their commutations coincide. *See:* **power rectifier.** 208

parallel rectifier circuit. A rectifier circuit in which two or more simple rectifier circuits are connected in such a way that their direct currents add and their commutations coincide. *See also:* **rectification; rectifier circuit element.** 237, 66

parallel resonance (circuits and systems). The sinusoidal steady state condition that exists in a circuit composed of an inductor and a capacitor connected in parallel when the applied frequency is such that (1) the driving-point impedance is a maximum, or (2) the susceptance of the two parallel arms are equal in magnitude, or (3) the phase-angle of the driving-point impedance is zero. Sometimes defined as above for more general RLC (resistance-inductance-capacitance) networks. 67

parallel search storage (computing systems). *See:* **associative storage.** 255, 77

parallel storage (computing systems). A storage device in which characters, words, or digits are dealt with simultaneously. 255, 77

parallel-T network (twin-T network). A network composed of separate T networks with their terminals connected in parallel. *See also:* **network analysis.** 328

parallel transmission (data transmission). Simultaneous transmission of the bits making up a character, either over separate channels or on different carrier frequencies on one channel. 194

parallel two-terminal pair networks. Two-terminal pair networks are connected in parallel at the input or at the output terminals when their respective input or output terminals are in parallel. *See also:* **network analysis.** 332

paramagnetic material. Material whose relative permeability is slightly greater than unity and practically independent of the magnetizing force. 210

parameter (1) (mathematical). A variable that is given a constant value for a specific purpose or process. 255, 77

(2) (physical). One of the constants entering into a functional equation and corresponding to some characteristic property, dimension, or degree of freedom.

(3) (electrical). One of the resistance, inductance, mutual inductance, capacitance, or other element values included in a circuit or network. Also called **network constant.** 210

(4) (control systems). A quantity of property treated as a constant but which may sometimes vary or be adjusted. 56

(5) (test, measurement and diagnostic equipment). (A) Any specific quantity or value affecting or describing the theoretical or measurable characteristics of a unit being considered which behaves as an independent variable or which depends upon some functional interaction of other quantities in a theoretically determinable manner; and (B) In programming, a variable that is given a constant value for a specific purpose or process. 54

parametric amplifier. An inverting parametric device used to amplify a signal without frequency translation from input to output. *Note:* In common usage, this term is a synonym for reactance amplifer. *See also:* **parametric device.** 277, 191

parametric converter. An inverting parametric device or noninverting parametric device used to convert an input signal at one frequency into an output signal at a different frequency. *See also:* **parametric device.** 277, 191

parametric device. A device whose operation depends essentially upon the time variation of a characteristic parameter usually understood to be a reactance. 277, 191

parametric down-converter. A parametric converter in which the output signal is at a lower frequency than the input signal. *See also:* **parametric device.** 277, 191

parametric up-converter. A parametric converter in which the output signal is at a higher frequency than the input signal. *See also:* **parametric device.** 277, 191

parametric variation (automatic control). A change in those system properties generally regarded as constants which affect the dependent variables describing system operation. 56

parasitic element (1) (antennas). A radiating element that is not coupled directly to the feed lines of an antenna and that materially affects the radiation pattern and/or impedance of an antenna. *Note:* Compare with driven element. *See also.* **antenna.** 179, 111

(2) (circuits and systems). An unwanted circuit element that is an unavoidable adjunct of a wanted circuit element. 67

parasitic oscillations. Unintended self-sustaining oscillations, or transient impulses. *See also:* **oscillatory circuit.** 111, 211, 59

parcel plating. Electroplating upon only a part of the surface of a cathode. *See also:* **electroplating.** 328

parity. Pertaining to the use of a self-checking code employing binary digits in which the total number of ONEs (or ZEROs) in each permissible code expression is always even or always odd. A check may be made for either even parity or odd parity. 235

parity bit (computing systems). A binary digit appended to an array of bits to make the sum of all the bits always odd or always even.   255, 77, 54

parity check (electronic computation). A summation check in which the bits in a character or block are added (modulo 2) and the sum checked against a single, previously computed parity digit; that is, a check that tests whether the number of ones is odd or even.   255, 77, 194, 54

parity code (power-system communication). A binary code so chosen that the count of ONES is even (even parity) or odd (odd parity) and used for error detection. *See also:* **analog.**   59

parking lamp. A lamp placed on a vehicle to indicate its presence when parked. *See also:* **headlamp.**   167

parsec (pc). The distance at which 1 astronomical unit subtends an angle of 1 second of arc; approximately, 1 pc $= 206\ 265$ AU $= 3057 \times 10^{12}$ m.   21

partial (audio and electroacoustics). (1) A physical component of a complex tone. (2) A component of a sound sensation that may be distinguished as a simple tone that cannot be further analyzed by the ear and that contributes to the timbre of the complex sound. *Notes:* (A) The frequency of a partial may be either higher or lower than the basic frequency and may or may not be an integral multiple or submultiple of the basic frequency. If the frequency is not a multiple or submultiple, the partial is inharmonic. (B) When a system is maintained in steady forced vibration at a basic frequency equal to one of the frequencies of the normal modes of vibration of the system, the partials in the resulting complex tone are not necessarily identical in frequency with those of the other normal modes of vibration.   176

partial-automatic station. A station that includes protection against the usual operating emergencies, but in which some or all of the steps in the normal starting or stopping sequence, or in the maintenance of the required character of service, must be performed by a station attendant or by supervisory control.   103, 202

partial-automatic transfer equipment (partial-automatic throwover equipment). An equipment that automatically transfers load to another (emergency) source of power when the original (preferred) source to which it has been connected fails, but that will not automatically retransfer the load to the original source under any conditions. *Note:* The restoration of the load to the preferred source from the emergency source upon the reenergization of the preferred source after an outage may be of the continuous-circuit restoration type or the interrupted-circuit restoration type.   103, 202

partial body irradiation (electrobiology). Pertains to the case in which part of the body is exposed to the incident electromagnetic energy. *See also:* **electrobiology.**   322

partial carry (parallel addition). A technique in which some or all of the carries are stored temporarily instead of being allowed to propagate immediately. *See:* **carry.**
                                                          255, 77

partial discharge (PD) (1) (transformer). An electric discharge which only partially bridges the insulation between conductors, and which may or may not occur adjacent to a conductor. *Notes:* (A) Partial discharges occur when the local electric field intensity exceeds the dielectric strength of the dielectric involved, resulting in local ionization and breakdown. Depending on intensity, partial discharges are often accompanied by emission of light, heat, sound, and radio influence voltage (with a wide frequency range). (B) The relative intensity of partial discharge can be observed at the transformer terminals by measurement of the apparent charge (coulombs). However, the apparent charge (terminal charge) should not be confused with the actual charge transferred across the discharging element in the dielectric which in most cases can not be ascertained.

Partial discharge tests using the Radio Influence Voltage techniques which are responsive to the apparent terminal charges are generally used for measurement of relative discharge intensity. (C) Partial discharges can also be detected and located using sonic techniques. (D) "Corona" has also been used to describe partial discharges. This is a non-preferred term since it has other unrelated meanings.   53

(2) (dielectric tests). An electrical discharge that only partially bridges the insulation between conductors. *Note:* The term corona has also frequently been used with this connotation. It is recommended that such usage be discontinued in favor of the term partial discharge.   139

partial discharge (corona) extinction voltage (cable termination). The voltage at which partial discharge (corona) is no longer detectable on instrumentation adjusted to a specified sensitivity, following the application of a specified higher voltage.   4

partial failure. *See:* **failure, partial.**

partial-fraction expansion (circuits and systems). A sum of fractions that is used to represent a function which is a ratio of polynomials. The denominators of the fractions are the poles of the function.   67

partial-read pulse. Any one of the currents applied that cause selection of a cell for reading. *See:* **coincident-current selection.**   331

partial reference designation (electric and electronics parts and equipments). A reference designation that consists of a basic reference designation and which may include, as prefixes, some but not all of the reference designations that apply to the subassemblies or assemblies within which the item is located.   17

partial-select output. (1) The voltage response of an unselected magnetic cell produced by the application of partial-read pulses or partial-write pulses. (2) The integrated voltage response of an unselected magnetic cell produced by the application of partial-read pulses or partial-write pulses. *See:* **coincident-current selection.**
                                                          331

partial system test. *See:* **stimulation, physical.** *See also:* **electronic analog computer.**

partial-write pulse. Any one of the currents applied that cause selection of a cell for writing. *See:* **coincident-current selection.**   331

particle accelerator. Any device for accelerating charged particles to high energies, for example, cyclotron, betatron, Van der Graaff generator, linear accelerator, etcetera. *See also:* **electron devices, miscellaneous.**   190

particle velocity (sound field). The velocity of a given infinitesimal part of the medium, with reference to the medium as a whole, due to the sound wave. *Note:* The terms **instantaneous particle velocity, effective particle velocity, maximum particle velocity,** and **peak particle velocity** have meanings that correspond with those of the related terms used for sound pressure. 176

parting (corrosion). The selective corrosion of one or more components of a solid solution alloy. 221, 205

parting limit (corrosion). The maximum concentration of a more-noble component in an alloy, above which parting does not occur within a specific environment. 221, 205

partition noise (electron device). Noise caused by random fluctuations in the distribution of current between the various electrodes. 190

part programming, computer (numerically controlled machines). The preparation of a manuscript in computer language and format required to accomplish a given task. The necessary calculations are to be performed by the computer. 224, 207

part programming, manual (numerically controlled machines). The preparation of a manuscript in machine control language and format required to accomplish a given task. The necessary calculations are to be performed manually. 224, 207

part-winding starter (industrial control). A starter that applies voltage successively to the partial sections of the primary winding of an alternating-current motor. *See:* **starter.** 225, 206

part-winding starting (rotating machinery). A method of starting a polyphase induction or synchronous motor, by which certain specially designed circuits of each phase of the primary winding are initially connected to the supply line. The remaining circuit or circuits of each phase are connected to the supply in parallel with initially connected circuits, at a predetermined point in the starting operation. 63

party line (telephone switching systems). A line arranged to serve more than one main station, with distinctive ringing for each station. 55

Paschen's law (gas). The law stating that, at a constant temperature, the breakdown voltage is a function only of the product of the gas pressure by the distance between parallel plane electrodes. *See also:* **discharge (gas).** 244, 190

pass band (1) (circuits and systems). A band of frequencies that pass through a filter with little loss (relative to other frequency bands such as a stop band). 67

(2) (data transmission). Range of frequency spectrum which can be passed at low attenuation. *See:* **bandpass filter; band pass.** 59

pass-band ripple (circuits and systems). The difference between maxima and minima of loss in a filter passband. If the differences are of constant amplitude then the filter is said to be equiripple. 67

pass element (power supplies). A controlled variable-resistance device, either a vacuum tube or power transistor, in series with the source of direct-current power. The pass element is driven by the amplified error signal to increase its resistance when the output needs to be lowered or to decrease its resistance when the output must be raised. *See:* **series regulator.** *See also:* **power supply.** 228, 186

passenger elevator. An elevator used primarily to carry persons other than the operator and persons necessary for loading and unloading. *See also:* **elevators.** 328

passivation (corrosion). The process or processes (physical or chemical) by means of which a metal becomes passive. 221, 205

passivator (corrosion). An inhibitor that changes the potential of a metal appreciably to a more cathodic or noble value. 221, 205

passive-active cell (corrosion). A cell composed of passive and active areas. *See:* **electrolytic cell.** 221, 205

passive electric network. An electric network containing no source of energy. *See also:* **network analysis.** 328

passive satellite (communication satellite). A communication satellite which is a reflector and performs no active signal processing. 83

passive station (data transmission). All stations on a multipoint network, other than the master and slave (s), which during the information message transfer state, monitor the line for supervisory sequences, ending characters, etcetera. 59

passive test (test, measurement and diagnostic equipment). A test conducted upon an equipment or any part thereof when the equipment is not energized. *Syn.* **cold test.** 54

passive transducer. *See:* **transducer, passive.**

passivity (1) (chemical). The condition of a surface that retards a specified chemical reaction at that surface. *See also:* **electrochemistry.** 328

(2) (electrolytic or anodic). Such a condition of an anode that the normal anodic reaction is retarded. *See also:* **electrochemistry.** 328

paste (dry cell) (primary cell). A gelatinized layer containing electrolyte that lies adjacent to the negative electrode. *See also:* **electrolytic cell.** 328

pasted sintered plate (alkaline storage battery). A plate consisting of fritted metal powder in which the active material is impregnated. *See also:* **battery (primary or secondary).** 328

patch (1) (in general). To connect circuits together temporarily by means of a cord, known as a patch cord. 328

(2) (computing systems). To modify a routine in a rough or expedient way. 255, 77

patch bay (analog computer). A concentrated assembly of the inputs and outputs of computing elements, control elements, tie points, reference voltages and ground points that offers a means of electrical connection. 9

patch board (analog computer). *See:* **problem board.**

patchcord (test, measurement and diagnostic equipment). An interconnecting cable for plugging or patching between terminals; commonly employed on patchboard, plugboard, and in maintenance operations. 54

patch panel. *See:* **problem board.** *See also:* **electronic analog computer.**

path (1) (navigation). A line connecting a series of points in space and constituting a proposed or traveled route. *See:* **flight path; flight track.** *See also:* **navigation.** 13, 187

(2) (network analysis). Any continuous succession of branches, traversed in the indicated branch directions. *See also:* **linear signal flow graphs.** 282

(3) (data transmission). *See:* **channel.**

(4) (telephone switching systems). The set of links and junctors joined in series to establish a connection. 55

path (loop) factor (network analysis). The graph determinant of that part of the graph not touching the specified path (loop). *Notes:* (1) A path (loop) factor is obtainable from the graph determinant by striking out all terms containing transmittance products of loops that touch that path (loop). (2) For loop L_k, the loop factor is

$$-\partial\Delta/\partial L_k$$

See also: **linear signal flow graphs.** 282

path length (1) (general). The length of a magnetic flux line in a core. *Note:* In a toroidal core with nearly equal inside and outside diameters, the value

$$l_m = \frac{\pi}{2}(O.D. + I.D.)$$

where O.D. and I.D. are the outside and inside diameters of the core, is commonly used. 331

(2) (laser gyro). The length of the optical path traversed in a single pass by the laser beams. 46

path transmittance (network analysis). The product of the branch transmittances in that path. *See also:* **linear signal flow graphs.** 282

patina (corrosion). A green coating consisting principally of basic sulfate and occasionally containing small amounts of carbonate or chloride, that forms on the surface of copper or copper alloys exposed to the atmosphere a long time. 221, 205

pattern. *See:* **radiation pattern.**

pattern, heating. *See:* **heating station.**

pattern-propagation factor (radar). Ratio of the field strength that is actually present at a point in space to that which would have been present if free space propagation had occurred along the axis of maximum antenna gain. The factor is used in the radar equation to modify the strength of the transmitted and/or received signal. 13

pattern recognition. The identification of shapes, forms, or configurations by automatic means. 255, 77

pattern-sensitive fault. A fault that appears in response to some particular pattern of data. 255, 77

PAX (telephony). *See:* **private automatic exchange.**

pay station. *See:* **public telephone station.**

P-band radar. Occasionally used to denote a radar operating in the 420 to 450-megahertz International Telecommunications Union assigned band, more generally described as ultrahigh frequency radar. 13

PBX. *See:* **private branch exchange.**

PBX trunk. *See:* **private-branch-exchange trunk.**

P.D. *See:* **control action, proportional plus derivative.**

P display (electronic navigation). *See:* **plan-position indicator.**

PDM. *See:* **pulse-duration modulation.**

PEAK. Channel number corresponding to the peak of a distribution. *See:* **crest.** 117

peak alternating gap voltage (electron tube) (traveling-wave tubes). The negative of the line integral of the peak alternating electric field taken along a specified path across the gap. *Note:* The path of integration must be stated. *See:* **electron devices, miscellaneous.** 190, 125

peak anode current. The maximum instantaneous value of the anode current. *See also:* **electronic controller.** 206

peak burst magnitude (audio and electroacoustics). The maximum absolute peak value of voltage, current, or power for a burstlike excursion. *See:* The figure attached to the definition of **burst duration.** *See also:* **burst (audio and electroacoustics).** 253, 176

peak cathode current (steady-state). The maximum instantaneous value of a periodically recurring cathode current. 125

peak-charge characteristic (nonlinear capacitor). The function relating one-half the peak-to-peak value of transferred charge in the steady state to one-half the peak-to-peak value of a specified applied symmetrical alternating capacitor voltage. *Note:* Peak-charge characteristic is always single-valued. *See also.* **nonlinear capacitor.** 191

peak detector. A detector, the output voltage of which approximates to the true peak value of an applied signal. *See also:* **electromagnetic compatibility.** 220, 199

peak distortion. The largest total distortion of telegraph signals noted during a period of observation. *See also:* **telegraphy.** 194, 59

peak electrode current (electron tube). The maximum instantaneous current that flows through an electrode. *See also:* electrode current (electron tube). 190

peak flux density. The maximum flux density in a magnetic material in a specified cyclically magnetized condition. 331

peak forward anode voltage (electron tube). The maximum instantaneous anode voltage in the direction in which the tube is designed to pass current. *See also:* **electrode voltage (electron tube); electronic controller.** 190

peak forward current rating (repetitive) (rectifier circuit element). The maximum repetitive instantaneous forward current permitted by the manufacturer under stated conditions. *See also:* **average forward current rating (rectifier circuit element).** 237, 66, 208

peak forward voltage (of a rectifying element). The maximum instantaneous voltage between the anode and cathode during the positive nonconducting period. *See also:* **rectification.** 328

peak induction (of toroidal magnetic amplifier cores). The magnetic induction corresponding to the peak applied magnetizing force specified. *Note:* It will usually be slightly less than the true saturation induction. *Syn.* **peak flux density.** 170

peaking circuit. A circuit capable of converting an input wave into a peaked waveform. 328

peaking network. A type of interstage coupling network in which an inductance is effectively in series (series peaking network) or in shunt (shunt peaking network) with the parasitic capacitance to increase the amplification at the upper end of the frequency range. *See also:* **network analysis.** 328

peaking time (semiconductor radiation detectors). The time elapsed from the first zero crossing of the defined zero level to the departure from peak amplitude of a pulse equal to the maximum rated amplifier output. 23

peak inrush current (transformer). The peak instantaneous current value resulting from the excitation of the transformer with no connected load, and with essentially zero source impedance, and using the minimum turns primary tap and rated voltage. 95

peak inverse anode voltage (electron tube). The maximum instantaneous anode voltage in the direction opposite to

that in which the tube is designed to pass current. *See also:* **electrode voltage (electron tube); electronic controller.**
190

peak inverse voltage (PIV) (1) (semiconductor diode). The maximum instantaneous anode-to-cathode voltage in the reverse direction that is actually applied to the diode in an operating circuit. *Notes:* (A) This is an applications term not to be confused with **breakdown voltage,** which is a property of the device. (B) In semiconductor work the preferred term is **peak reverse voltage.** *See also:* **semiconductor; peak reverse voltage.** 245, 210, 66
(2) (rectifying element). The maximum instantaneous voltage between the anode and cathode during the inverse period. *See also:* **rectification.** 328
peak inverse voltage, maximum rated (semiconductor diode). The recommended maximum instantaneous anode-to-cathode voltage that may be applied in the reverse direction. *See also:* **semiconductor.** 245, 210, 66
peak jitter. *See:* **jitter.**
peak let-through characteristic curve (current-limiting fuse). *See:* **current-limiting characteristic curve (current-limiting fuse).**
peak let-through current (current-limiting fuse) (peak cutoff current). The highest instantaneous current passed by the fuse during the interruption of the circuit. 103, 202
peak limiter. A device that automatically limits the magnitude of a signal to a predetermined maximum value by changing its amplification. *Notes:* (1) The term is frequently applied to a device whose gain is quickly reduced and slowly restored when the instantaneous magnitude of the input exceeds a predetermined value. (2) In this context, the terms **instantaneous magnitude** and **instantaneous peak power** are used interchangeably. 178, 59
peak load (1) (general). The maximum load consumed or produced by a unit or group of units in a stated period of time. It may be the maximum instantaneous load or the maximum average load over a designated interval of time. *Note:* Maximum average load is ordinarily used. In commercial transactions involving peak load (peak power) it is taken as the average load (power) during a time interval of specified duration occurring within a given period of time, that time interval being selected during which the average power is greatest. *See also:* **generating station.**
64
(2) (motor) (rotating machinery). The largest momentary or short-time load expected to be delivered by a motor. It is expressed in percent of normal power or normal torque. *See also.* **asynchronous machine; direct-current commutating machine.** 63
peak load station (electric power supply). A generating station that is normally operated to provide power during maximum load periods. *See also:* **generating station.**
200
peak magnetizing force (1) (toroidal magnetic amplifier cores). The maximum value of applied magnetomotive force per mean length of path of the core. 170
(2) (peak field strength). The upper or lower limiting value of magnetizing force associated with a cyclically magnetized condition. 331
peak overvoltages (current-limiting fuses). The peak value of the voltage that can exist across a current limiting fuse during its arcing interval. 103

peak point (tunnel-diode characteristic). The point on the forward current-voltage characteristic corresponding to the lowest positive (forward) voltage at which $dI/dV = 0$. 315,191
peak-point current (tunnel-diode characteristic). The current at the peak point. *See also:* **peak point (tunnel-diode characteristic).** 315, 191
peak-point voltage (tunnel-diode characteristic). The voltage at which the peak point occurs. *See also:* **peak point (tunnel-diode characteristic).** 315, 191
peak power output (modulated carrier system). The output power, averaged over a carrier cycle, at the maximum amplitude that can occur with any combination of signals to be transmitted. *See also:* **radio transmitter; television.**
111
peak pulse amplitude (television). The maximum absolute peak value of the pulse excluding those portions considered to be unwanted, such as spikes. *Note:* Where such exclusions are made, it is desirable that the amplitude chosen be illustrated pictorially. *See also:* **pulse.** 59
peak pulse power, carrier-frequency. The power averaged over that carrier-frequency cycle that occurs at the maximum of the pulse of power (usually one half the maximum instantaneous power). 254
peak repetitive ON-state current (thyristor). The peak value of the ON-state current including all repetitive transient currents. *See also:* **principal current.**
243, 66, 208, 191
peak responsibility. The load of a customer, a group of customers, or a part of the system at the time of occurrence of the system peak load. *See also:* **generating station.**
64
peak (crest) restriking voltage (surge arresters). The maximum instantaneous voltage that is attained by the restriking voltage. 308, 62
peak reverse voltage (semiconductor rectifier). The maximum instantaneous value of the reverse voltage that occurs across a semiconductor rectifier device, or rectifier stack. *See also:* **rectification.** 237, 66
peak sound pressure (for any specified time interval). The maximum absolute value of the instantaneous sound pressure in that interval. *Note:* In the case of a periodic wave, if the time interval considered is a complete period, the peak sound pressure becomes identical with the maximum sound pressure. 176
peak switching current (rotating machinery). The maximum peak transient current attained following a switching operation on a machine. *See also:* **asynchronous machine; direct-current commutating machine.** 63
peak value (1) (general). The largest instantaneous value of a time function during a specified time period. *See:* **crest value.** 59
(2) (alternating voltage). The maximum value excluding small high-frequency oscillations arising, for instance, from partial discharges in the circuit. *See also:* **test voltage and current.** 307, 201
(3) (crest value) (voltage or current) (surge arresters). The maximum value of an impulse. If there are small oscillations superimposed at the peak, the peak value is defined by the maximum value and not the mean curve drawn through the oscillations. 308, 62

(4) (impulse current) (virtual peak value). Normally the maximum value. With some test circuits, overshoot or oscillations may be present on the current. The maximum value of the smooth curve drawn through the oscillations is defined as the virtual peak value. It will depend on the type of test whether the value of the test current shall be defined by the actual peak or a virtual peak value. *Note:* The term **peak value** is to be understood as including the term **virtual peak value** unless otherwise stated. *See also:* **test voltage and current.**   307, 201

(5) (impulse voltage) (virtual peak value). Normally the maximum value. With some test circuits, oscillations or overshoot may be present on the voltage. If the amplitude of the oscillations is not greater than 5 percent of the peak value and the frequency is at least 0.5 megahertz or, alternatively, if the amplitude of the

pedestal bearing (rotating machinery). A complete assembly of a bearing with its supporting pedestal.   63

pedestal bearing insulation (rotating machinery). The insulation applied either below the bearing liner shell and the adjacent pedestal support or between the base of the pedestal and the machine bed plate, to break the current path that may be formed through the shaft to the outboard bearing to the frame to the drive-end bearing and thence back to the shaft. *Note:* The voltage is usually very low. However, very destructive bearing currents can flow in this path if some insulating break is not provided. High-pressure moulded laminates are usually employed for this type of insulation.   63

pedestal delay time (amplitude, frequency, and pulse modulation). The time elapsed between the application of an electronic command signal to the electronic driver and the

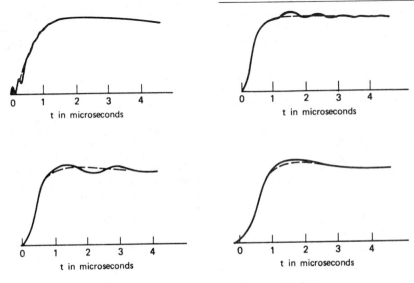

Construction for derivation of virtual peak values

overshoot is not greater than 5 percent of the peak value and the duration not longer than 1 microsecond, when for the purpose of measurement a mean curve may be drawn, the maximum amplitude of which is defined as the virtual peak value. (See the accompanying figure.) If the frequency is less than or if the duration is greater than described above, the peak of the oscillation may be used as the peak value. *Note:* The term **peak value** is to be understood as including the term **virtual peak value** unless otherwise stated. *See also:* **test voltage and current.**
     307, 301

peak wavelength, (λp) (light emitting diodes). The wavelength at which the spectral radiant intensity is a maximum.   162

peak working voltage (high- and low power pulse transformers). The maximum instantaneous voltage stress that may appear under operation across the insulation being considered, including abnormal and transient conditions.   32, 33

pedestal. A substantially flat-topped pulse that elevates the base level for another wave. *See also:* **pulse.**   328

time the diffracted light reaches the 10 percent intensity point.   72

peeling. The unwanted detachment of a plated metal coating from the base metal. *See also:* **electroplating.**
     328

peg count (telephone switching systems). The notation of the number of occurrences of an event.   55

Peltier coefficient, absolute. The product of the absolute temperature and the absolute Seebeck coefficient of the material; the sign of the Peltier coefficient is the same as that of the Seebeck coefficient. *Note:* The opposite sign convention has also been used in the technical literature. *See also.* **thermoelectric device.**   248, 191

Peltier coefficient of a couple. The quotient of (1) the rate of Peltier heat absorption by the junction of the two conductors by (2) the electric current through the junction; the Peltier coefficient is positive if Peltier heat is absorbed by the junction when the electric current flows from the second-named conductor to the first conductor. *Notes:* (A) The opposite sign convention has also been used in the technical literature. (B) The Peltier coefficient of a couple is the algebraic difference of either the relative or absolute Peltier coefficients of the two conductors. *See also:* thermoelectric device.   248, 191

Peltier coefficient, relative. The Peltier coefficient of a couple composed of the given material as the first-named conductor and a specified standard conductor. *Note:* Common standard conductors are platinum, lead, and copper. *See also:* **thermoelectric device.**   248, 191

Peltier effect. The absorption or evolution of thermal energy, in addition to the Joule heat, at a junction through which an electric current flows; and in a nonhomogeneous, isothermal conductor, the absorption or evolution of thermal energy, in addition to the Joule heat, produced by an electric current. *Notes:* (1) For the case of a nonhomogeneous, nonisothermal conductor, the Peltier effect cannot be separated from the Thomson effect. (2) A current through the junction of two dissimilar materials causes either an absorption or liberation of heat, depending on the sense of the current, and at a rate directly proportional to it to a first approximation. *See also:* **thermoelectric device.**   210, 248, 191

Peltier heat. The thermal energy absorbed or evolved as a result of the Peltier effect. *See also:* **thermoelectric device.**   248, 191

penalty factor (electric power system). A factor that, when multiplied by the incremental cost of power at a particular source, produces the incremental cost of delivered power from that source. Mathematically, it is

$$\frac{1}{(1 - \text{Incremental Transmission Loss})^*}$$

* Expressed as a decimal   94

pencil beam (radar) (antenna). A unidirectional antenna having a narrow major lobe with approximately circular contours of equal radiation intensity in the region of the major lobe.   13, 179, 111

pendant. A device or equipment that is suspended from overhead either by means of the flexible cord carrying the current or otherwise.   328

pendulosity (gyro; accelerometer). The product of the mass and the distance from the center of mass to the center of supporting forces or pivot measured along the pendulous axis.   46

pendulous accelerometer. A device that employs a proof mass which is suspended to permit a rotation about an axis perpendicular to an input axis.   46

pendulous axis (accelerometer). A line through the mass center of the proof mass, perpendicular to and intersecting the output axis in pendulous devices. The positive direction is defined from the output axis to the proof mass.   46

pendulous integrating gyro accelerometer. A device using a single-degree-of-freedom gyro having an intentional pendulosity along the spin axis which is servo driven about the input axis at a rate which balances the torque induced by acceleration along the input axis. The angle through which the servoed axis rotates is proportional to the integral of applied acceleration.   46

pendulous reference axis (PRA) (accelerometer). The direction of an axis as defined by the case mounting surfaces and/or external case markings. It is nominally parallel to the pendulous axis.   46

penetration, depth of. *See:* **depth of penetration.**

penetration frequency (radio wave propagation). Critical frequency. *See also:* **radio wave propagation.**   180

pentode. A five-electrode electron tube containing an anode, a cathode, a control electrode, and two additional electrodes that are ordinarily grids.   190, 125

pen travel. The length of the path described by the pen in moving from one end of the chart scale to the other. The path may be an arc or a straight line. *See also:* **moving element (instrument).**   295

perceived light-source color. The color perceived to belong to a light source. *See also:* **color.**   167

perceived object color. The color preceived to belong to an object, resulting from characteristics of the object, of the incident light, and of the surround, the viewing direction, and observer adaptation. *See also:* **color.**   167

percentage differential relay. A differential relay in which the designed response to the phasor difference between incoming and outgoing electrical quantities is modified by a restraining action of one or more of the input quantities. *Note:* The relay operates when the magnitude of the phasor difference exceeds the specified percentage of one or more of the input quantities.   103, 127

percentage error (watthour meter). The difference between its percentage registration and 100 percent. A meter whose percentage registration is 95 percent is said to be 5 percent slow, or its error is −5 percent. A meter whose percentage registration is 105 percent is 5 percent fast or its error is +5 percent. *See also:* **electricity meter (meter).**   212

percentage immediate appreciation (telephone transmission system). The percentage of the total number of spoken sentences that are immediately understood without conscious deductive effort when each sentence conveys a simple and easily understandable idea. *See also:* **volume equivalent.**   328

percentage modulation. (1) In angle modulation, the fraction of a specified reference modulation, expressed in percent. (2) In amplitude modulation, the modulation factor expressed in percent. *Note:* It is sometimes convenient to express percentage modulation in decibels below 100-percent modulation. *See also:* **radio transmission.**   211

percentage modulation, effective (single, sinusoidal input component). The ratio of the peak value of the fundamental component of the envelope to the direct-current component in the modulated conditions, expressed in percent. *Note:* It is sometimes convenient to express percentage modulation in decibels below 100-percent modulation.   111

percentage registration (watthour meter) (accuracy*) (percentage accuracy*). The ratio of the actual registration of the meter to the true value of the quantity measured in a given time, expressed as a percentage. *See also:* **electricity meter (meter).**   212

* Deprecated

percent articulation. *See:* **articulation.**

percent energy resolution. *See:* **energy resolution, percent.**

percent flutter (reproduced tone) (sound recording and reproducing). The root-mean-square deviation from the average frequency, expressed as a percentage of average

frequency. *See also:* **sound recording and reproducing.**
145

percent harmonic distortion. *See:* **distortion, percent harmonic.**

percent impairment of hearing (percent hearing loss). An estimate of a person's ability to hear correctly. It is usually based, by means of an arbitrary rule, on the pure-tone audiogram. The specific rule for calculating this quantity from the audiogram now varies from state to state according to a rule or law. *Note:* The term disability of hearing is sometimes used for impairment of hearing. Impairment refers specifically to a person's illness or injury that affects his personal efficiency in the activities of daily living. Disability has the additional medicolegal connotation that an impairment reduces a person's ability to engage in gainful activity. Impairment is only a contributing factor to the disability.
176

percent impedance (rectifier transformer). The percent of rated alternating-current winding voltage required to circulate current equivalent to rated line kilovolt-amperes in the alternating-current winding with all direct-current winding terminals short-circuited. *See also.* **rectifier transformer.**
258

percent intelligibility. *See:* **articulation.**

percent-make-and-break (telephone switching systems). The proportions of a dial pulse cycle during which the circuit is closed (make) and opened (break) respectively.
55

percent pulse waveform distortion (pulse terms). Pulse waveform distortion expressed as a percentage of, unless otherwise specified, the pulse amplitude of the reference pulse waveform.
254

percent pulse waveform feature distortion (pulse terms). Pulse waveform feature distortion expressed as a percentage of, unless otherwise specified, the pulse amplitude of the reference pulse waveform.
254

percent ratio. The true ratio expressed in percent of the marked ratio.
53

percent ratio correction (instrument transformer). The difference between the ratio correction factor and unity expressed in percent. *Note:* The percent ratio correction is positive if the ratio correction factor is greater than unity. If the percent ratio correction is positive, the measured secondary current or voltage will be less than the primary value divided by the marked ratio.
53

percent ripple (1) (general). The ratio of the effective (root-mean-square) value of the ripple voltage or current to the average value of the total voltage or current, expressed in percent. *See:* **interference.**
111, 188

(2) (electrical conversion). The percent ripple voltage is

Note: In most applications the definition has been revised to simplify the calculations by defining percent ripple as the ratio of the root-mean-square (RMS) value of the voltage pulsations to the nominal no-load output voltage of the converter $E_{nominal}$

$$\text{Percent Ripple} = \frac{\text{RMS Ripple}}{E_{nominal}}(100\%)$$

See also: **electrical conversion.**
186

percent ripple voltage or current. *See:* **percent ripple.**

percent steady-state deviation (control). The difference between the ideal value and the final value, expressed as a percentage of the maximum rated value of the directly controlled variable (or another variable if specified). *See:* **control system, feedback.**
219, 206

percent syllable articulation. *See:* **syllable articulation.**

percent system deviation (control). At any given point on the time response, the difference between the ideal value and the instantaneous value, expressed as a percentage of the maximum rated value of the directly controlled variable (or another variable if specified). *See:* **deviation (control).** *See also:* **control system, feedback.** 219, 206

percent total flutter (sound recording and reproducing). The value of flutter indicated by an instrument that responds uniformly to flutter of all rates from 0.5 up to 200 hertz. *Note:* Except for the most critical tests, instruments that respond uniformly to flutter of all rates up to 120 hertz are adequate, and their indications may be accepted as showing percent total flutter. *See also:* **sound recording and reproducing.**
145

percent transformer correction-factor error. Difference between the transformer correction factor and unity expressed in percent. *Note:* The percent transformer correction-factor error is positive if the transformer correction factor is greater than unity. If the percent transformer correction-factor error is positive, the measured watts or watthours will be less than the true value. 305, 203

percent transient deviation (control). The difference between the instantaneous value and the final value, expressed as a percentage of the maximum rated value of the directly controlled variable (or another variable if specified). *See:* **control system, feedback.**
219, 206

percent unbalance of phase voltages (electrical conversion). The ratio of the maximum deviation of a phase voltage from the average of the total phases to the average of the phase voltages, expressed in percent.

$$\% \text{ Unbalance} = \frac{\text{RMS Phase Voltage} - \text{RMS Average Phase Voltages}}{\text{RMS Average Phase Voltage}} \times 100\%$$

defined as the ratio of the root-mean-square(RMS) value of the voltage pulsations (E_{max} to E_{min}) to the average value of the total voltage.

$$\text{Percent Ripple} = \frac{\text{RMS Ripple}}{E_{av}}(100\%)$$

See also: **electrical conversion.**
186

perfect dielectric (ideal dielectric). A dielectric in which all of the energy required to establish an electric field in the dielectric is recoverable when the field or impressed voltage is removed. Therefore, a perfect dielectric has zero

conductivity and all absorption phenomena are absent. A complete vacuum is the only known perfect dielectric.
210

perfect diffusion. The diffusion in which flux is scattered in accord with Lambert's cosine law. *See also:* **lamp.**
163

perfect transformer. *See:* **ideal transformer.**

perforated tape. Tape in which a code hole(s) and a tape-feed hole have been punched in a row. 224, 207

perforator (1) (telegraph practice). A device for punching code signals in paper tape for application to a tape transmitter. *Note:* A perforating device that is automatically controlled by incoming signals is called a reperforator. *See also:* **telegraphy.** 255, 77, 194
(2) (test, measurement and diagnostic equipment). A device for punching digital information into tape for application to a tape transmitter or tape reader. Sometimes called tape punch.
54

performance characteristic (device) (1) (power switchgear). An operating characteristic, the limit or limits of which are given in the design test specifications. 103, 202
(2) (transformer). Those characteristics (such as impedance, losses, dielectric test levels, temperature rise, sound level, etcetera) which describe the performance of the equipment under specified conditions of operation. 53

performance chart (magnetron oscillators). A plot on coordinates of applied anode voltage and current showing contours of constant magnetic field, power output, and over-all efficiency. *See:* **magnetron.** 190, 125

performance index (1) (system). The system function or objective to be optimized. *See also:* **system.** 209
(2) (control system). A scalar measure of the quality of system behavior. The performance index is usually a function of the plant output and control input over some time interval. *See also:* **control system.** 329
(3) (excitation control systems). A scalar measure of the quality of system behavior. It is frequently a function of system output and control input over some specified time interval and/or frequency range. A quadratic performance index is a quadratic function of system states and this form finds wide applications to linear systems. 353

performance monitor (test, measurement and diagnostic equipment). A device which continuously or periodically scans a selected number of test points to determine if the unit is operating within specified limits. The device may include provisions for insertion of stimuli. 54

performance test (lead storage batteries) (nuclear power generating stations). A capacity test made on a battery (as found) after being in service to detect any change in the capacity determined by the acceptance test. 31, 38

performance tests (rotating machinery). The tests required to determine the characteristics of a machine and to determine whether the machine complies with its specified performance. *See also:* **asynchronous machine; direct-current commutating machine.** 63

performance verification (test, measurement and diagnostic equipment). A short, precise check to verify that the unit under test is operational and performing its intended function. 54

periapsis (communication satellite). The least distant point from the center of a primary body (or planet) to an orbit around it. 74

perigee (communication satellite). The least distant point from the center of the earth to an orbit around it. 74

perimeter lights. Aeronautical ground lights provided to indicate the perimeter of a landing pad for helicopters.
167

period (1) (primitive period) (function). The smallest number $k_1 \neq 0$ for which the following identity holds: $f(x) = f(x + k_1)$. *Note:* For example, the primitive period of $\sin x$ is 2π. Not all periodic functions have primitive periods. For example, $f(x) = A$ (a constant) has no primitive period. It is common engineering practice to exclude from the class of periodic functions, any function that has no primitive period. 210
(2) (periodic time) (electric instrument). The time between two consecutive transits of the pointer or indicating means in the same direction through the rest position. *See also:* **moving element (instrument).** 280
(3) (electric power systems). The minimum interval of the independent variable after which the same characteristics of a periodic phenomenon recur. 244, 59
(4) (pulse terms). The absolute value of the minimum interval after which the same characteristics of a periodic waveform or a periodic feature recur. 254

period, critical hydro (electric power supply). Period when the limitations of hydroelectric energy supply due to water conditions are most critical with respect to system load requirements. 112

periodically sampled equivalent time format (pulse measurement). A format which is identical to the periodically sampled real time format, below, except that the time coordinate is equivalent to and convertible to real time. Typically, each datum point is derived from a different measurement on a different wave in a sequence of waves. *See also:* **sampled format.** 15

periodically sampled real time format (pulse measurement). A finite sequence of magnitudes $m_0, m_1, m_2, \cdots, m_n$ each of which represents the magnitude of the wave at times t_0, $t_0 + \Delta t, t_0 + 2\Delta t, \cdots, t_0 + n\Delta t$, respectively, wherein $\cdots$ the data may exist in a pictorial format or as a list of numbers. *See also:* **sampled format.** 15

periodic-automatic reclosing equipment. An equipment that provides for automatically reclosing a circuit-switching device a specified number of times at specified intervals between reclosures. *Note:* This type of automatic reclosing equipment is generally used for alternating-current circuits. 103, 202

periodic check (test, measurement and diagnostic equipment). A test or series of tests performed at designated intervals to determine if all elements of the unit under test are operating within their designated limits. 54

periodic damping*. *See:* **underdamping.**

* Deprecated

periodic duty. A type of intermittent duty in which the load conditions are regularly recurrent. *See:* **duty.** *See also:* **voltage regulator.** 210, 257, 203, 206, 53, 124
periodic electromagnetic wave (radio wave propagation). A wave in which the electric field vector is repeated in detail in either of two ways: (1) At a fixed point, after the lapse of a time known as the period. (2) At a fixed time, after the addition of a distance known as the wavelength. *See also:* **radio wave propagation.** 146, 180

periodic frequency modulation (inverters). The periodic variation of the output frequency from the fundamental. *See:* **self-commutated inverters.** 208

periodic function. A function that satisfies $f(x) = f(x + nk)$ for all x and for all integers n, k being a constant. For example,

$$\sin (x + a) = \sin (x + a + 2n\pi).$$

210

periodic line (transmission lines). A line consisting of successive identically similar sections, similarly oriented, the electrical properties of each section not being uniform throughout. *Note:* The periodicity is in space and not in time. An example of a periodic line is the loaded line with loading coils uniformly spaced. *See also:* **transmission line.** 210

periodic output voltage modulation (inverters). The periodic variation of output voltage amplitude at frequencies less than the fundamental output frequency. *See:* **self-commutated inverters.** 208

periodic permanent-magnet focusing (PPM) (microwave tubes). Magnetic focusing derived from a periodic array of permanent magnets. *See:* **magnetron.** 190

periodic pulse train (automatic control). A pulse train made up of identical groups of pulses, the groups repeating at regular intervals. 56

periodic rating (1) (electric power sources). The load that can be carried for the alternate periods of load and rest specified in the rating, the apparatus starting at approximately room temperature, and for the total time specified in the rating, without causing any of the specified limitations to be exceeded. *See also:* **asynchronous machine; direct-current commutating machine; relay.** 206
(2) (relay). A rating that defines the current or voltage that may be sustained by the relay during intermittent periods of energization as specified, starting cold and operating for the total time specified without causing any of the prescribed limitations to be exceeded. 103, 202, 60

periodic slow-wave circuit (microwave tubes). A circuit whose structure is periodically recurring in the direction of propagation. *See also:* **microwave (tube or valve).** 190

periodic tests (periodic testing). (nuclear power generating stations). Tests performed at scheduled intervals to detect failures and verify operability. 109, 31, 355, 356

periodic wave. A wave in which the displacement at each point of the medium is a periodic function of the time. Periodic waves are classified in the same manner as periodic quantities. 210

periodic waveguide. A waveguide in which propagation is obtained by periodically arranged discontinuities or periodic modulations of the material boundaries. *See:* **waveguide.** 179

peripheral air-gap leakage flux (rotating machinery). The component of air-gap magnetic flux emanating from the rotor or stator, that flows from pole to pole without entering the radially opposite surface of the air gap. *See also:* **rotor (rotating machinery); stator.** 63

peripheral equipment (test, measurement and diagnostic equipment). Equipment external to a basic unit. A tape unit, for example, is peripheral equipment to a computer. 54

peripheral vision. The seeing of objects displaced from the primary line of sight and outside the central visual field. *See also:* **visual field.** 167

peripheral visual field. That portion of the visual field that falls outside the region corresponding to the foveal portion of the retina. *See also:* **visual field.** 167

permanent echo (primary radar system). A signal reflected from an object fixed with respect to the radar site. *See also:* **radar.** 13, 187

permanent fault (surge arresters). A fault that can be cleared only by action taken at the point of fault. 244, 62

permanent-field synchronous motor. A synchronous motor similar in construction to an induction motor in which the member carrying the secondary laminations and windings carries also permanent-magnet field poles that are shielded from the alternating flux by the laminations. It starts as an induction motor but operates normally at synchronous speed. *See:* **permanent-magnet synchronous motor.** 63

permanently grounded device. A grounding device designed to be permanently connected to ground, either solidly or through current transformers and/or another grounding device. 91

permanent magnet (PM). A ferromagnetic body that maintains a magnetic field without the aid of external electric current. 244, 210

permanent-magnet erasing head (electroacoustics). A head that uses the fields of one or more permanent magnets for erasing. *See also:* **phonograph pickup.** 176

permanent-magnet focusing (microwave tubes). Magnetic focusing derived from the use of a permanent magnet. *See:* **magnetrons.** 190

permanent-magnet generator (magneto). A generator in which the open-circuit magnetic flux field is provided by one or more permanent magnets. 63

permanent-magnet loudspeaker. A moving-conductor loudspeaker in which the steady field is produced by means of a permanent magnet. *See also:* **loudspeaker.** 328

permanent-magnet moving-coil instrument (d'Arsonval instrument). An instrument that depends for its operation on the reaction between the current in a movable coil or coils and the field of a fixed permanent magnet. *See also:* **instrument.** 328

permanent-magnet moving-iron instrument (polarized-vane instrument). An instrument that depends for its operation on the action of an iron vane in aligning itself in the resultant magnetic field produced by a permanent magnet and the current in an adjacent coil of the instrument. *See also:* **instrument.** 328

permanent-magnet, second-harmonic, self-synchronous system. A remote-indicating arrangement consisting of a transmitter unit and one or more receiver units. All units have permanent-magnet rotors and toroidal stators using saturable ferromagnetic cores and excited with alternating current from a common external source. The coils are tapped at three or more equally spaced intervals, and the corresponding taps are connected together to transmit voltages that consist principally of the second harmonic of the excitation voltage. The rotors of the receiver units will assume the same angular position as that of the transmitter rotor. *See:* **synchro system.** 328

permanent-magnet synchronous motor (rotating ma-

chinery). A synchronous motor in which the field system consists of one or more permanent magnets. *See:* **permanent-field synchronous motor.**   62

permanent signal (telephone switching systems). A sustained off-hook supervisory signal originating outside a switching system.   55

permanent-signal alarm, (telephone switching systems). An alarm resulting from the simultaneous accumulation of a predetermined number of permanent signals.   55

permanent-signal tone, (telephone switching systems). A tone that indicates to an operator or other employee that a line is in a permanent-signal state.   55

permanent-split capacitor motor. A capacitor motor with the same value of effective capacitance for both starting and running operations. *See:* **asynchronous machine.**
                                    232, 63

permanent storage (computing systems). *See:* **fixed storage.**

permeability. A general term used to express various relationships between magnetic induction and magnetizing force. These relationships are either (1) absolute permeability, that in general is the quotient of a change in magnetic induction divided by the corresponding change in magnetizing force; or (2) specific (relative) permeability, which is the ratio of the absolute permeability to the magnetic constant. *Notes:* (1) Relative permeability is a pure number that is the same in all unit systems; the value and dimension of absolute permeability depend upon the system of units employed. (2) In anisotropic media, permeability becomes a matrix.   210

permeability of free space. *See:* **magnetic constant.**

permeameter. An apparatus for determining corresponding values of magnetizing force and flux density in a test specimen. From such values of magnetizing force and flux density, normal induction curves or hysteresis loops can be plotted and magnetic permeability can be computed. *See also:* **magnetometer.**   328

permeance. The reciprocal of reluctance.   210

permissible mine equipment. Equipment that complies with the requirements of and is formally approved by the United States Bureau of Mines after having passed the inspections and the explosion and/or other tests specified by that Bureau. *Note:* All equipment so approved must carry the official approval plate required as identification for permissible equipment.   328

permissible response rate (steam generating unit). The maximum assigned rate of change in generation for load-control purposes based on estimated and known limitations in the turbine, boiler, combustion control, or auxiliary equipment. The permissible response rate for a hydro-generating unit is the maximum assigned rate of change in generation for load-control purposes based on estimated and known limitations of the water column, associated piping, turbine, or auxiliary equipment. *See also:* **speed-governing system.**   94

permissive (relay system). A general term indicating that functional cooperation of two or more relays is required before control action can become effective.
                                    103, 202, 60, 127

permissive block. A block in manual or controlled manual territory, governed by the principle that a train other than a passenger train may be permitted to follow a train other

than a passenger train in the block. *See:* **block signal system; controlled manual block signal system.**   328

permissive control (electric power systems). A control mode in which generating units are allowed to be controlled only when the change will be in the direction to reduce area-control error. *See also:* **speed-governing system.**   94

permittivity. *See* **absolute capacitivity.**

permittivity in physical media (antennas). The real part of the complex permittivity. *See also:* **complex dielectric constant.**   111

permittivity of free space. *See:* **electric constant.**

perpendicular magnetization (magnetic recording). Magnetization of the recording medium in a direction perpendicular to the line of travel and parallel to the smallest cross-sectional dimension of the medium. *Note:* In this type of magnetization, either single pole-piece or double pole-piece magnetic heads may be used. *See also:* **phonograph pickup.**   176

persistence (oscilloscopes). The decaying luminosity of the luminescent screen [phosphor/screen] after the stimulus has been reduced or removed. *See also:* **phosphor decay.**
                                    106

persistence characteristic (1) (camera tubes). The temporal step response of a camera tube to illumination. *See:* **methods of measurement.** *See also* **television.**   190, 125
(2) (decay characteristic) (luminescent screen). A relation, usually shown by a graph, between luminance (or emitted radiant power) and time after excitation is removed.
                                    190, 178, 125

persistent-cause forced outage (electric power systems). A component outage whose cause is not immediately self-clearing but must be corrected by eliminating the hazard or by repairing or replacing the affected component before it can be returned to service. *Note:* An example of a persistent-cause forced outage is a lightning flashover that shatters an insulator thereby disabling the component until repair or replacement can be made. *See also:* **outage.**
                                    200

persistent-cause forced-outage duration (electric power systems). The period from the initiation of a persistent-cause forced outage until the affected component is replaced or repaired and made available to perform its intended function. *See also:* **outage.**   200

persistent current (superconducting material). A magnetically induced current that flows undiminished in a superconducting material or circuit. *See also:* **superconductivity.**   191

persistent-image device. An optoelectronic amplifier capable of retaining a radiation image for a length of time determined by the characteristics of the device. *See also:* **optoelectronic device.**   191

persistent-image panel (optoelectronic device). A thin, usually flat, multicell persistent-image device. *See also:* **optoelectronic device.**   191

person-to-person call, (telephone switching systems). A call intended for a designated person.   55

per-unit quantity (rotating machinery). The ratio of the actual value of a quantity to the base value of the same quantity. The base value is always a magnitude, or in mathematical terms, a positive, real number. The actual value of the quantity in question (current, voltage, power, torque, frequency, etcetera) can be of any kind: root-mean-square, instantaneous, phasor, complex, vector, envelope, etcetera. *Note:* The base values, though arbitrary, are usually related to characteristic values, for example, in case of a machine, the base power is usually chosen to be the rated power (active or apparent), the base voltage to be the rated root-mean-square voltage, the base frequency, the rated frequency. Despite the fact that the choice of base values is rather arbitrary, it is of advantage to choose base values in a consistent manner. The use of a consistent per-unit system becomes a practical necessity when a complicated system is analyzed. *See:* **asynchronous machine; direct-current commutating machine.** 63

per-unit resistance. The measured watts expressed in per-unit on the base of the rated kilovolt-amperes of the teaser winding. *See:* **efficiency.** 328

per-unit system (rotating machinery). The system of base values chosen in a consistent manner to facilitate analysis of a device or system, when per-unit quantities are used. Its importance becomes paramount when analog facilities (network analyzer, analog and hybrid computers) are utilized. *Note:* In electric network analysis and electromechanical system studies, usually four independent fundamental base values are chosen. The rest of the base values are derived from the fundamental ones. In most cases power, voltage, frequency, and time are chosen as fundamental base values. The base power must be the same for all types; apparent, active, reactive, instantaneous. The base time is usually 1 second. From the above, all other base values can be found, for example, base power times base time equals base energy, etcetera. The per-unit system can cover extensive networks because the base voltages of network sections connected by transformers can differ, in which case an **ideal per-unit transformer** is usually introduced having a turns ratio equal to the quotient of the effective turns ratio of the actual transformer and the ratio of base voltage values. By keeping the power, frequency, and time bases the same, only those base quantities will differ for different network sections that are directly or indirectly related to voltage (for example, current, impedance, reactance, inductance, capacitance, etcetera) but those related to power, frequency, and time only (for example, energy, torque, etcetera) will remain unchanged. *See:* **asynchronous machine; direct-current commutating machine.** 63

per-unit value (ac rotating machinery). The actual value divided by the value of the base quantity when both actual and base values are expressed in the same unit. 5

perveance. The quotient of the space-charge-limited cathode current by the three-halves power of the anode voltage in a diode. *Note:* Perveance is the constant G appearing in the Child-Langmuir-Schottky equation

$$i_k = Ge_b{}^{3/2}.$$

When the term perveance is applied to triode or multigrid tube, the anode voltage e_b is replaced by the composite controlling voltage e' of the equivalent diode. 190

PFM telemetry. *See:* pulse frequency modulation (PFM) telemetry.

p gate thyristor. A thyristor in which the gate terminal is connected to the p region adjacent to the region to which the cathode terminal is connected and that is normally switched to the ON state by applying a positive signal between gate and cathode terminals. *See also:* **thyristor.** 243, 66, 208, 191

pH (of a solution A). The pH is obtained from the measurements of the potentials E of a galvanic cell of the form H_2; solution A; saturated potassium chloride (KCl); reference electrode with the aid of the equation

$$pH = \frac{E - E_0}{(RT/F) \ln 10} = \frac{E - E_0}{2.303 \, RT/F}$$

in which E_0 is a constant depending upon the nature of the reference electrode, R is the gas constant in joules per mole per degree, T is the absolute temperature in kelvins, and F is the Faraday constant in coulombs per gram equivalent. Historically pH was defined by

$$pH = \log \frac{1}{[H^+]}$$

in which $[H^+]$ is the hydrogen ion concentration. According to present knowledge there is no simple relation between hydrogen ion concentration or activity and pH. *See also:* **ion activity.** Values of pH may be regarded as a convenient scale of acidities. *See also:* **ion.** 328

phanotron. A hot-cathode gas diode. *Note:* This term is used primarily in the industrial field. 190

phantom circuit (data transmission). A third circuit derived from two physical circuits by means of repeating coils installed at the terminals of the physical (side) circuits. A phantom circuit is a superposed circuit derived from two suitably arranged pairs of wires, called side circuits, the two wires of each pair being effectively in parallel. 59

phantom-circuit loading coil. A loading coil for introducing a desired amount of inductance in a phantom circuit and a minimum amount of inductance in the constituent side circuits. *See also:* **loading.** 328

phantom-circuit repeating coil (phantom-circuit repeat coil). A repeating coil used at a terminal of a phantom circuit, in the terminal circuit extending from the midpoints of the associated side-circuit repeating coils. 328

phantom group. A group of four open wire conductors suitable for the derivation of a phantom circuit. *See also:* **open wire.** 328

phantom target (radar). (1) An echo box, or other reflection device, that produces a particular blip on the radar indicator. (2) A condition, maladjustment, or phenomenon (such as a temperature inversion) that produces a blip on the radar indicator resembling blips of targets for which the system is being operated. *See:* **echo box.** *See also:* **navigation.** 13

phase (of a periodic phenomenon $f(t)$, for a particular value of t). The fractional part t/P of the period P through which t has advanced relative to an arbitrary origin. *Note:* The origin is usually taken at the last previous passage through zero from the negative to the positive direction. *See:* **simple sine-wave quantity.** *See also:* **control system, feedback.** 329

phase advance. *See:* phase lead.

phase advancer. A phase modifier that supplies leading reactive volt-amperes to the system to which it is connected. Phase advancers may be either synchronous or asynchronous. *See:* **converter.** 328

phase angle (1) (general). The measure of the progression of a periodic wave in time or space from a chosen instant or position. *Notes:* (A) The phase angle of a field quantity, or of voltage or current, at a given instant of time at any given plane in a waveguide is $(\omega t - \beta z + \theta)$, when the wave has a sinusoidal time variation. The term **waveguide** is used here in its most general sense and includes all transmission lines; for example, **rectangular waveguide, coaxial line, strip line,** etcetera. The symbol β is the imaginary part of the propagation constant for that waveguide, propagation is in the $+z$ direction, and θ is the phase angle when $z = t = 0$. At a reference time $t = 0$ and at the plane z, the phase angle $(- \beta + \theta)$ will be represented by Φ. (B) Phase angle is obtained by multiplying the phase by 360 degrees or by 2π radians. *See also:* **sinewave quantity.** 293, 56, 183
(2) (current transformer). The angle between the current leaving the identified secondary terminal and the current entering the identified primary terminal. *Note:* This angle is conveniently designated by the Greek letter beta (β) and is considered positive when the secondary current leads the primary current. *See also:* **instrument transformer.** 212
(3) (potential (voltage) transformer). The angle between the secondary voltage from the identified to the unidentified terminal and the corresponding primary voltage. *Note:* This angle is conveniently designated by the Greek letter gamma (γ) and is considered positive when the secondary voltage leads the primary voltage. *See:* **instrument transformer.** 212
(4) (instrument transformer). Phase displacement, in minutes, between the primary and secondary values. 305, 203
Note: The phase angle of a current transformer is designated by the Greek letter beta (B) and is positive when the current leaving the identified secondary terminal leads the current entering the identified primary terminal. The phase angle of a voltage transformer is designated by the Greek letter gamma (γ) and is positive when the secondary voltage from the identified to the unidentified terminal leads the corresponding primary voltage. 53
(5) (speed governing of hydraulic turbines). Referring to a simultaneous phasor diagram of the input and output, the angle by which the output signal lags or leads the input signal. 8
phase angle correction factor. The ratio of the true power factor to the measured power factor. It is a function of both the phase angles of the instrument transformer and the power factor of the primary circuit being measured. *Note:* The phase angle correction factor is the factor which corrects for the phase displacement of the secondary current or voltage, or both, due to the instrument transformer phase angles. The measured watts or watthours in the secondary circuits of instrument transformers must be multiplied by the phase angle correction factor and the true ratio to obtain the true primary watts or watthours. 53
phase angle, dielectric. *See:* **dielectric phase angle.**
phase angle, loop (automatic control) (closed loop). The value of the loop phase characteristic at a specified frequency. *See:* **phase characteristic.** *See also:* **control system, feedback.** 56

phase-balance relay. A relay that responds to differences between quantities of the same nature associated with different phases of a normally balanced polyphase circuit. 103, 202, 60, 127
phase belt (coil group). A group of adjacent coils in a distributed polyphase winding of an alternating-current machine that are ordinarily connected in series to form one section of a phase winding of the machine. Usually, there are as many such phase belts per phase as there are poles in the machine. *Note:* The adjacent coils of a phase belt do not necessarily occupy adjacent slots; the intervening slots may be occupied by coils of another winding on the same core. Such may be the case in a two-speed machine. *See also:* **rotor (rotating machinery); stator.** 63
phase center (antenna). In a given direction and for a specified polarization, the center of curvature of the wavefront of the radiation from an antenna in a given plane. 111
phase characteristic (1). The variation with frequency of the phase angle of a phasor quantity. 210
(2) (linear passive networks). The angle of a response function evaluated on the imaginary axis of the complex-frequency plane. *See also:* linear passive networks. 238
(3) synchronous machine) (V curve). A characteristic curve between armature current and field current, with constant mechanical load (or no load) and with constant terminal voltage maintained. 328
phase characteristic, loop (automatic control) (closed loop). The phase angle of the loop transfer function for real frequencies. *See also:* **control system, feedback.** 56
phase-coil insulation (rotating machinery). Additional insulation between adjacent coils that are in different phases. *See also:* **asynchronous machine** 63
phase-comparison monopulse. A variant of monopulse employing receiving beams with different phase centers, as obtained, for example, from side-by-side antennas or separate portions of an array. *Note:* If the amplitude-versus-angle patterns of the beams are identical, the monopulse is pure phase-comparison; otherwise, it is a combination of phase-comparison and amplitude-comparison. *See also:* **monopulse** and **amplitude-comparison monopulse.** 13
phase-comparison protection. A form of pilot protection that compares the relative phase-angle position of specified currents at the terminals of a circuit. 103, 202, 60, 127
phase conductor (alternating-current circuit). The conductors other than the neutral conductor. *Note:* If an alternating-current circuit does not have a neutral conductor, all the conductors are phase conductors. 210
phase connections (rotating machinery). The insulated conductors (usually arranged in peripheral rings) that make the necessary connections between appropriate phase belts in an alternating-current winding. *See:* **rotor (rotating machinery); stator.** 63
phase constant (traveling plane wave at a given frequency) (wavelength constant). The imaginary component of the propagation constant. *Note:* This is the space rate of decrease of phase of a field component (or of the voltage or current), in the direction of propagation, in radians per unit length. *See also:* **waveguide.**
267, 210, 179, 180, 185, 146

phase contours. Loci of the return transfer function at constant values of the phase angle. *Note:* Such loci may be drawn on the Nyquist, inverse Nyquist, or Nichols diagrams for estimating performance of the closed loop with unity feedback. In the complex plane plot of $KG(j\omega)$, these loci are circles with centers at $-1/2, j/2\,N$ and radiuses such that each circle passes through the origin and the point $-1, j0$. In the inverse Nyquist diagram they are straight lines $\gamma = -N(x + 1)$ radiating from the point -1, 0. *See:* **Nichols chart; Nyquist diagram; inverse Nyquist diagram.** *See also:* **control system, feedback.** 329

phase control (rectifier circuits). The process of varying the point within the cycle at which forward conduction is permitted to begin through the rectifier circuit element. *Note:* The amount of phase control may be expressed in two ways: (1) the reduction in direct-current voltage obtained by phase control or (2) the angle of retard or advance. 237, 66, 208

phase converter (rotating machinery). A converter that changes alternating-current power of one or more phases to alternating-current power of a different number of phases but of the same frequency. *See:* **converter.** 63

phase-corrected horn. A horn designed to make the emergent electromagnetic wave front substantially plane at the mouth. *Note:* Usually this is achieved by means of a lens at the mouth. *See:* **circular scanning; waveguide.** 244, 179

phase correction (telegraph transmission). The process of keeping synchronous telegraph mechanisms in substantially correct phase relationship. *See also:* **telegraphy.** 328

phase corrector. A network that is designed to correct for phase distortion. *See also:* **network analysis.** 210

phase-crossover frequency (1) (loop transfer function) (automatic control). The frequency at which the phase angle reaches plus or minus 180 degrees. *See also:* **control system, feedback.** 56

(2) (speed governing of hydraulic turbines). The frequency at which the phase angle reaches 180 degrees. 8

phased-array antenna. An array antenna whose beam direction or radiation pattern is controlled primarily by the relative phases of the excitation coefficients of the radiating elements. *See also:* **antenna.** 179, 111

phase delay (1) (facsimile) (in the transfer of a single-frequency wave from one point to another in a system). The time delay of a part of the wave identifying its phase. *Note:* The phase delay is measured by the ratio of the total phase shift in cycles to the frequency in hertz. *See also:* **facsimile transmission.** 12

(2) (relaying). An equal delay of both the leading and trailing edges of a locally generated block. 79

(3) (dispersive and nondispersive delay lines). The ratio of total radian phase shift, ϕ, to the specified radian frequency, ω. Phase delay is nominally constant over the frequency band of operation for nondispersive delay devices. *See also:* **phase lag.** 81

phase delay time. In the transfer of a single-frequency wave from one point to another in a system, the time delay of a part of the wave identifying its phase. *Note:* The phase delay time is measured by the ratio of the total phase delay through the network, in cycles, to the frequency, in hertz. *See also:* **measurement system.** 293, 183

phase deviation (angle modulation) (phase modulation). The peak difference between the instantaneous angle of the modulated wave and the angle of the carrier. *Note:* In the case of a sinusoidal modulating function, the value of the phase deviation, expressed in radians, is equal to the modulation index. *See also:* **phase modulation.** 111, 242

phase difference (1) (general). The difference in phase between two sinusoidal functions having the same periods. 210

(2) (automatic control). (A) Between sinusoidal input and output of the same frequency, phase angle of the output minus phase angle of the input; it is called "phase lead" if the input angle is the smaller, "phase lag" if the larger. (B) Of two periodic phenomena (for example, in nonlinear systems) the difference between the phase angles of their two fundamental waveforms. *Note:* Regarded as part of the transfer function which relates output to input at a specified frequency, phase difference is simply the phase angle $\theta(j\omega)$ in $A(j\omega)\,\exp j\theta(j\omega)$. Measurement of phase difference in the complex case is sometimes made in terms of the angular interval between respective crossings of a mean reference line, but values so measured will generally differ from those made in terms of the fundamental waveforms. *See also:* **phase shift.** 56

phase distortion (1) (power system communication). (A) Lack of direct proportionality of phase shift to frequency over the frequency range required for transmission, or (B) the effect of such departure on a transmitted signal. *See also:* **distortion; distortion, phase delay.** 59

(2) (facsimile). *See:* **phase-frequency distortion.**

phased satellite (communication satellite). A satellite, the center of mass of which is maintained in a desired relation relative to other satellites, to a point on earth or to some other point of reference such as the sub-solar point. *Note:* If it is necessary to identify those satellites that are not phased satellites, the term "unphased satellites" may be used. 74

phase-failure protection. *See:* **open-phase protection and phase-undervoltage protection.**

phase-frequency distortion (facsimile). Distortion due to lack of direct proportionality of phase shift to frequency over the frequency range required for transmission. *Notes:* (1) **delay distortion** is a special case. (2) This definition includes the case of a linear phase-frequency relation with the zero frequency intercept differing from an integral multiple of π. *See:* **distortion; distortion, phase delay; facsimile transmission; phase distortion.** 12

phase-insulated terminal box (rotating machinery). A terminal box so designed that the protection of phase conductors against electric failure within the terminal box is by insulation only. 63

phase inverter. A stage whose chief function is to change the phase of a signal by 180 degrees, usually for feeding one side of a following push-pull amplifier. *See also:* **amplifier.** 59

phase jitter (data transmission). *See:* **jitter.**

phase lag (phase delay) (2-port network). The phase angle of the input wave relative to the output wave ($\Phi_{in} - \Phi_{out}$), or the initial phase angle of the output wave relative to the final phase angle of the output wave ($\Phi_i - \Phi_f$). *Note:* Under matched conditions, **phase lag** is the negative of the angle of the transmission coefficient of the scattering matrix for a 2-port network. *See:* **phase difference.** *See also:* **measurement system.** 293, 183

phase lead (2-port network). The phase angle of the output wave relative to the input wave ($\Phi_{out} - \Phi_{in}$), or the final phase angle of the output wave relative to the initial phase angle of the output wave ($\Phi_f - \Phi_i$). *Note:* Under matched conditions, **phase lead** is the angle of the transmission coefficient of the scattering matrix for a 2-port network. *Syn.* **phase advance.** *See:* **phase difference.** *See also:* **measurement system.** 293, 183

phase localizer (electronic navigation). A localizer in which the on-course line is defined by the phase reversal of energy radiated by the sideband antenna system, a reference carrier signal being radiated and used for the detection of phase. *See also:* **radio navigation.** 187, 13

phase locking. The control of an oscillator or periodic generator so as to operate at a constant phase angle relative to a reference signal source. *See also:* **oscillatory circuit.** 59

phase lock loop (communication satellite). A circuit for synchronizing a variable local oscillator with the phase of a transmitted signal. Widely used in space communication for coherent carrier tracking, and threshold extension, bit synchronization and symbol synchronization. 83

phase locus (for a loop transfer function, say $G(s) H(s)$). A plot in the s plane of those points for which the phase angle, ang GH, has some specified constant value. *Note:* The phase loci for 180 degrees plus or minus n 360 degrees are also root loci. *See also:* **control system, feedback.** 56

phase margin (1) (loop transfer function for a stable feedback control system) (excitation control systems). 180 degrees minus the absolute value of the loop phase angle at a frequency where the loop gain is unity. *Note:* Phase margin is a convenient way of expressing relative stability of a linear system under parameter changes, in Nyquist, Bode, or Nichols diagrams. In a conditionally stable feedback control system where the loop gain becomes unity at several frequencies, the term is understood to apply to the value of phase margin at the highest of these frequencies. *See also:* **control system, feedback.** 56, 353 **(2) (speed governing of hydraulic turbines).** 180 degrees minus the absolute value of the open-loop phase angle at a frequency where the open-loop gain is unity. 8

phase meter (phase-angle meter). An instrument for measuring the difference in phase between two alternating quantities of the same frequency. *See also:* **instrument.** 328

phase modifier (rotating machinery). An electric machine, the chief purpose of which is to supply leading or lagging reactive power to the system to which it is connected. Phase modifiers may be either synchronous or asynchronous. *See:* **converter.** 63, 3

phase-modulated transmitter. A transmitter that transmits a phase-modulated wave. 111, 240

phase modulation (PM). Angle modulation in which the angle of a carrier is caused to depart from its reference value by an amount proportional to the instantaneous value of the modulating function. *Notes:* (1) A wave phase modulated by a given function can be regarded as a wave frequency modulated by the time derivative of that func-

tion. (2) Combinations of phase and frequency modulation are commonly referred to as frequency modulation. *See also:* **phase deviation; pulse duration; reactance modulator.** 242, 111, 194

phase-modulation telemetering (electric power systems). A type of telemetering in which the phase difference between the transmitted voltage and a reference voltage varies as a function of the magnitude of the measured quantity. *See also:* **telemetering.** 94

phase overcurrent (power switchgear). The current flowing in a phase conductor which exceeds a predetermined value. 103

phase recovery time (microwave gas tubes). The time required for a fired tube to deionize to such a level that a specified phase shift is produced in the low-level radio-frequency signal transmitted through the tube. *See:* **gas tubes.** 190, 125

phase relay. A relay that by its design or application is intended to respond primarily to phase conditions of the power system. 103, 202, 60, 127

phase resolution. The minimum change of phase that can be distinguished by a system. *See also:* **measurement system.** 293, 183

phase-reversal protection. *See:* **phase-sequence reversal protection.**

phase-reversal relay. *See:* **negative-phase-sequence relay.**

phase-segregated terminal box. A terminal box so designed that the protection of phase conductors against electric failure within the terminal box is by insulation, and additionally by grounded metallic barriers forming completely isolated individual phase compartments so as to restrict any electric breakdown to a ground fault. 63

phase-selector relay. A programming relay whose function is to select the faulted phase or phases thereby controlling the operation of other relays or control devices. 103, 202, 60, 127

phase-separated terminal box. *See:* **phase-segregated terminal box.**

phase separator (rotating machinery). Additional insulation between adjacent coils that are in different phases. *See:* **rotor (rotating machinery); stator.** 63

phase sequence (1) (set of polyphase voltages or currents). The order in which the successive members of the set reach their positive maximum values. *Note:* The phase sequence may be designated in several ways. If the set of polyphase voltages or currents is a symmetrical set, one method is to designate the phase sequence by specifying the integer that denotes the number of times that the angular phase lag between successive members of the set contains the characteristic angular phase difference for the number of phases m. If the integer is zero, the set is of zero phase sequence; if the integer is one, the set is of first phase sequence; and so on. Since angles of lag greater than 2π produce the same phase position for alternating quantities as the same angle decreased by the largest integral multiple of 2π contained in the angle of lag, it may be shown that there are only m distinct symmetrical sets normally designated from 0 to $m - 1$ phase sequence. It can be shown that only for the first phase sequence do all the members of the set reach their positive maximum in the

order of identification at uniform intervals of time.
210, 63

(2) (transformer). The order in which the voltages successively reach their positive maximum values. *See also:* **direction of rotation of phasors.** 53

phase-sequence indicator. A device designed to indicate the sequence in which the fundamental components of a polyphase set of potential differences, or currents, successively reach some particular value, such as their maximum positive value. *See also:* **instrument.** 328

phase-sequence relay. A relay that responds to the order in which the phase voltages or currents successively reach their maximum positive values. *See also:* **relay.**
103, 202, 60, 127

phase-sequence reversal (industrial control). A reversal of the normal phase sequence of the power supply. For example, the interchange of two lines on a three-phase system will give a phase reversal. 206

phase-sequence reversal protection. A form of protection that prevents energization of the protected equipment on the reversal of the phase sequence in a polyphase circuit.
225, 103, 202, 60, 206, 127

phase-sequence test (rotating machinery). A test to determine the phase sequence of the generated voltage of a three-phase generator when rotating in its normal direction. *See:* **asynchronous machine.** 63

phase shift (1) (general). The absolute magnitude of the difference between two phase angles. *Notes:* (A) The phase shift between two planes of a 2-port network is the absolute magnitude of the difference between the phase angles at those planes. The total phase shift, or absolute phase shift, is expressed as the total number of cycles, including any fractional number, between the two planes, where one complete cycle is 2π radians or 360 degrees. Relative phase shift is the total or absolute phase shift less the largest integral number of 2π radians or 360 degrees. The unit of phase shift is, therefore, the radian or the electrical degree. The term **2-port network** is used in its most general sense to include structures of passive or active elements. This includes the case of a given length of waveguide but may also refer to any two ports of a multiport device, where it is understood that a signal is incident only at one port. (B) A phase shift can be either a phase lead (advance) or a phase lag (delay). *See also:* **measurement system.** 293, 183

(2) (electrical conversion). The displacement between corresponding points in similar wave shapes and is expressed in degrees lead or lag. *See also:* **electrical conversion.** 186

(3) (transfer function). A change of phase angle with frequency, as between points on a loop phase characteristic. *See also:* **control system, feedback.**

(4) (signal). A change of phase angle with transmission.
329

(5) (dispersive and nondispersive delay lines). The total number of degrees or radians that a CW (continuous wave) signal experiences as it is transmitted through the delay device at a given frequency within the band of operation. The phase shift is nominally linearly proportional to frequency within the frequency band of operation for a nondispersive delay device. 81

phase-shift circuit (industrial control). A network that provides a voltage component shifted in phase with respect to a reference voltage. *See also:* **electronic controller.**
206

phase shifter. A device in which the output voltage (or current) may be adjusted, in use or in its design, to have some desired phase relation with voltage (or current). *See also:* **auxiliary device to an instrument; phase-shifting transformer; waveguide.** 328

phase shifter, waveguide (waveguide components). An essentially lossless device for adjusting the phase of a forward traveling electromagnetic wave at the output of the device relative to the phase at the input. 166

phase-shifting transformer (1) (general). A transformer that advances or retards the phase-angle relationship of one circuit with rspect to another. *Notes:* (A) The terms "advance" and "retard" describe the electrical angular position of the load voltage with respect to the source voltage. (B) If the load voltage reaches its positive maximum sooner than the source voltage, this is an "advance" position. (C) Conversely, if the load voltage reaches its positive maximum later than the source voltage, this is a "retard" position. *See:* **primary circuits; regulated circuit; series unit; main unit; series winding; excited winding; regulating winding; excitation winding; excitation-regulating winding; voltage-regulating relay; line-drop compensator; voltage winding (or transformer) for regulating equipment.** 53

(2) (measurement equipment). An assembly of one or more transformers intended to be connected across the phases of a polyphase circuit so as to provide voltages in the proper phase relations for energizing varmeters, varhour meters, or other measurement equipment. *See also:* **auxiliary device to an instrument; transformer.** 212

(3) (rectifier circuits). An autotransformer used to shift the phase position of the voltage applied to the alternating-current winding of a rectifier transformer for the purpose of decreasing the magnitude of the harmonic content in both the alternating-current and direct-current systems of a multiple-unit rectifier station. 258

phase-shift keying (PSK) (modulation systems). The form of phase modulation in which the modulating function shifts the instantaneous phase of the modulated wave between predetermined discrete values. 242

phase-shift oscillator. An oscillator produced by connecting any network having a phase shift of an odd multiple of 180 degrees (per stage) at the frequency of oscillation, between the output and the input of an amplifier. When the phase shift is obtained by resistance-capacitance elements, the circuit is an *R-C* phase-shift oscillator. *See also:* **oscillatory circuit.** 111

phase space (control system). (1) The state space augmented by the independent time variable. (2) One used synonymously with the state space, usually with the state variables being successive time derivatives of each other. *See also:* **control system.** 329

phase spacing (switching device). The distance between center-lines of the current-carrying parts of the adjacent poles of the switching device. 27, 103

phase splitter (phase-splitting circuit). A device that produces, from a single input wave, two or more output waves that differ in phase from one another. 59

phase-splitting circuit. *See:* **phase splitter.**

phase swinging (rotating machinery). Periodic variations in the speed of a synchronous machine above or below the normal speed due to power pulsations in the prime mover or driven load, possibly recurring every revolution.   63

phase-to-ground per unit overvoltage. The ratio of a phase-to-ground overvoltage to the phase-to-ground voltage corresponding to the maximum system voltage.   53

phase-to-phase per unit overvoltage. The ratio of a phase-to-phase overvoltage to the phase-to-phase voltage corresponding to the maximum system voltage.   53

phase-tuned tube (microwave gas tubes). A fixed-tuned broad-band transmit-receive tube, wherein the phase angle through and the reflection introduced by the tube are controlled within limits. *See:* **gas tubes.**   189, 125

phase-undervoltage protection. A form of protection that disconnects or inhibits connection of the protected equipment on deficient voltage in one or more phases of a polyphase circuit.   103, 202, 60, 127

phase-undervoltage relay (power switchgear). A relay that operates when one or more phase voltages in a normally balanced polyphase circuit is less than a predetermined value.   127, 103

phase vector (of a wave). The vector in the direction of the wave normal, whose magnitude is the phase constant.   180, 146

phase vector in physical media (antenna). The imaginary part of the propagation vector.   111

phase velocity (of a traveling plane wave at a single frequency). The velocity of an equiphase surface along the wave normal. *See also:* **radio wave propagation; waveguide.**   267, 179, 180, 146

phase-versus-frequency response characteristic. A graph or tabulation of the phase shifts occurring in an electric transducer at several frequencies within a band. *See also:* **transducer.**   111

phase voltage of a winding (machine or apparatus). The potential difference across one phase of the machine or apparatus. *See also:* **asynchronous machine.**   244, 63

phasing. The adjustment of picture position along the scanning line. *See also:* **scanning (facsimile).**   12

phasing signal. A signal used for adjustment of the picture position along the scanning line. *See also:* **facsimile signal (picture signal).**   12

phasing time (facsimile). The time interval during which the start positions of the scanning and recording strokes are aligned so as to ensure against a split image at the recorder.   11

phasing voltage (network protector). The voltage across the open contacts of a selected phase. *Note:* This voltage is equal to the phasor difference between the transformer voltage and the corresponding network voltage.   103, 202

phasor (vector*). A phasor is a complex number. Unless otherwise specified the term **phasor** is to be assumed to be used only in connection with quantities related to the steady alternating state in a linear network or system. *Notes:* (1) The term **phasor** is used instead of **vector** to avoid confusion with space vectors. (2) In polar form any phasor can be written $Ae^{j\theta_A}$ or $A\angle\theta_A$, in which A, real, is the modulus, absolute value, or amplitude of the phasor

and θ_A its phase angle (which may be abbreviated phase when no ambiguity will arise).   210

*Deprecated

phasor diagram (synchronous machine). A diagram showing the relationships of as many of the following phasor quantities as are necessary: armature current, armature voltages, the direct and quadrature axes, armature flux linkages due to armature and field winding currents, magnetomotive forces due to armature and field-winding currents, and the various components of air-gap flux. Figure 1 shows a "complete" phasor diagram for an over-excited generator, using the "generator convention" of current and voltages, shown for a nonsalient-pole machine in Figure 2. Figure 3 shows a phasor diagram of the basic quantities for an under-excited motor, using the "motor convention" of Figure 4.

phasor difference. *See:* **phasor sum (difference).**

phasor function. A functional relationship that results in a phasor.   210

phasor power (rotating machinery). The phasor representing the complex power. *See:* **asynchronous machine.**   63

phasor power factor. The ratio of the active power to the amplitude of the phasor power. The phasor power factor is expressed by the equation

$$F_{pp} = \frac{P}{S}$$

where F_{pp} is the phasor power factor, P is the active power, S is the amplitude of phasor power. If the voltages and currents are sinusoidal and, for polyphase circuits, form symmetrical sets,

$$F_{pp} = \cos(\alpha - \beta).$$

See also: **displacement power factor (rectifier).**   210

phasor product (quotient). A phasor whose amplitude is the product (quotient) of the amplitudes of the two phasors and whose phase angle is the sum (difference) of the phase angles of the two phasors. If two phasors are

$$\mathbf{A} = |A|e^{j\theta_A}$$
$$\mathbf{B} = |B|e^{j\theta_B}$$

the phasor product is

$$\mathbf{AB} = |AB|e^{j(\theta_A + \theta_B)}$$

and the quotient is

$$\frac{\mathbf{A}}{\mathbf{B}} = \left|\frac{A}{B}\right| e^{j(\theta_A - \theta_B)}$$

   210

phasor quantity. (1) A complex equivalent of a simple sine-wave quantity such that the modulus of the former is the amplitude A of the latter, and the phase angle (in polar form) of the former is the phase angle of the latter. (2) Any quantity (such as impedance) that is expressed in complex form. *Note:* In case (1), sinusoidal variation with t enters; in case (2), no time variation (in constant-parameter circuits) enters. The term **phasor quantity** covers both cases.   210

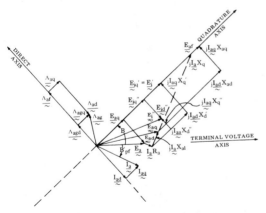

Fig. 1. Phasor diagram of overexcited generator (generator convention).

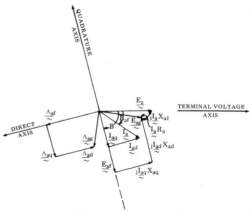

Fig. 3. Phasor diagram of motor operating at underexcited power factor (motor convention).

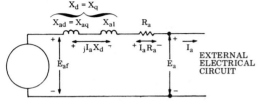

Fig. 2. Steady-state equivalent circuit for nonsalient-pole synchronous machine (generator convention).

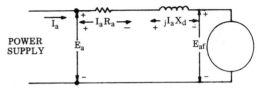

Fig. 4. Steady-state equivalent circuit for nonsalient-pole synchronous machine (motor convention).

Armature Voltages

E_a,	Terminal Voltage.
$\widetilde{E}_{ad}$	Direct-axis component of terminal voltage.
$\widetilde{E}_{aq}$	Quadrature-axis component of terminal voltage.
$\widetilde{E}_{ag}$,	Voltage behind leakage reactance; this is the voltage due to net air-gap flux, also called "virtual voltage".
E_{af},	Voltage due to flux produced only by the field-winding current.
$\widetilde{I}_a R_a$,	The voltage across armature resistance.
$j\widetilde{I}_a X_{al}$	The voltage across armature leakage reactance.
$j\widetilde{I}_{ad} X_{ad}$,	The voltage across direct-axis armature magnetizing reactance.
$j\widetilde{I}_{aq} X_{aq}$,	The voltage across quadrature-axis armature magnetizing reactance.
$j\widetilde{I}_a X_q$	A voltage based on quadrature-axis synchronous reactance, frequently used to locate the quadrature axis.
$j\widetilde{I}_{ad} X_d'$	The voltage across direct-axis transient reactance.
$j\widetilde{I}_{aq} X_q'$	The voltage across quadrature axis transient reactance.
$\widetilde{E}_i'$	Voltage behind transient reactance; this is the transient internal voltage.
$j\widetilde{I}_{ad} X_d''$	The voltage across direct-axis subtransient reactance.
$j\widetilde{I}_{aq} X_q''$	The voltage across quadrature-axis subtransient reactance.
$\widetilde{E}_i''$	Voltage behind subtransient reactance; this is the subtransient internal voltage.
$\widetilde{E}_{id}''$	Direct-axis component of subtransient internal voltage.
$\widetilde{E}_{iq}''$	Quadrature-axis component of subtransient internal voltage.

Armature Current

$\underset{\sim}{I_a}$,   Armature Current.

$\underset{\sim}{I_{ad}}$,   Direct-axis component of armature current. The sense of phasor I_{ad} in Figure 1, indicates that flux produced by armature current is opposed to that produced by the field current.

$\underset{\sim}{I_{aq}}$ Quadrature-axis component of armature current.

Flux Linkages with the Armature Winding

$\underset{\sim}{\wedge_{ag}}$,   Flux linkage due to net air-gap flux.

$\underset{\sim}{\wedge_{agd}}$,   Direct-axis component of flux linkage due to net air-gap flux.

$\underset{\sim}{\wedge_{agq}}$,   Quadrature-axis component of flux linkage due to net air-gap flux.

$\underset{\sim}{\wedge_{ad}}$,   Flux linkage due to the direct-axis component of armature current. The sense of $\wedge_{ad}$ in Fig. 1, indicates that the flux produced is in opposition to that produced by the field winding.

$\underset{\sim}{\wedge_{aq}}$,   Flux linkage due to the quadrature-axis component of armature current.

$\underset{\sim}{\wedge_{af}}$,   Flux linkages due to field-winding current.

Circuit Elements

R_a Positive-sequence resistance of armature winding.

Reactances

X_{al} Armature leakage reactance.

X_{ad} Direct-axis armature magnetizing reactance.

X_{aq} Quadrature-axis armature magnetizing reactance.

X_q Quadrature-axis synchronous reactance.

X_d' Direct-axis transient reactance.

X_q' Quadrature-axis transient reactance.

X_d'' Direct-axis subtransient reactance.

X_q'' Quadrature-axis subtransient reactance.

General

$\underset{\sim}{}$ Symbol for a phasor.

phasor quotient. *See:* **phasor product (quotient).**

phasor reactive factor. The ratio of the reactive power to the amplitude of the phasor power. The phasor reactive factor is expressed by the equation

$$F_{qp} = \frac{Q}{S}$$

where F_{qp} is the phasor reactive factor, Q is the reactive power, S is the amplitude of the phasor power. If the voltages and currents are sinusoidal and, for polyphase circuits, form symmetrical sets,

$$F_{pp} = \sin(\alpha - \beta).$$

210

phasor sum (difference). A phasor of which the real component is the sum (difference) of the real components of two phasors and the imaginary component is the sum (difference) of the imaginary components of two phasors. If two phasors are

$$A = a_1 + ja_2$$
$$B = b_1 + jb_2$$

the phasor sum (difference) is

$$A \pm B = (a_1 \pm b_1) + j(a_2 \pm b_2).$$

210

Philips gauge. A vacuum gauge in which the gas pressure is determined by measuring the current in a glow discharge. *See also:* **instrument.**   328

phi (Φ) polarization. The state of the wave in which the E vector is tangential to the lines of latitude of a given spherical frame of reference. *Note:* The usual frame of reference has the polar axis vertical and the origin at or near the antenna. Under these conditions, a vertical dipole will radiate only theta (θ) polarization, and a horizontal loop will radiate only phi (Φ) polarization. *See also:* **antenna.**   111, 246

phon. The unit of loudness level as specified in the definition of loudness level. *See also:* **loudspeaker.**   176

phonograph pickup (mechanical reproducer). A mechanoelectrical transducer that is actuated by modulations present in the groove of the recording medium and that transforms this mechanical input into an electric output. *Note:* (1) Where no confusion is likely the term **phonograph pickup** may be shortened to **pickup.** (2) A phonograph pickup generally includes a pivoted mounting arm and the transducer itself (the pickup cartridge).   176

phosphene (electrical) (electrotherapy). A visual sensation experienced by a human subject during the passage of current through the eye. *See also:* **electrotherapy.**   192

phosphor. A substance capable of luminescence. *See also:* **cathode-ray tube; fluorescent lamp; radio navigation; television.**   328

phosphor decay. A phosphorescence curve describing energy emitted versus time. *See also:* **oscillograph.**   185

phosphorescence. The emission of light as the result of the absorption of radiation, and continuing for a noticeable length of time after excitation, has been removed. *See also:* **lamp; oscillograph.**   167, 185

phosphor screen. All the visible area of the phosphor on the cathode-ray tube faceplate. *See also:* **oscillograph.**   185

phot. The unit of illumination when the centimeter is taken as the unit of length; it is equal to one lumen per square centimeter. *See also:* **light.**   167

photocathode. An electrode used for obtaining a photoelectric emission when irradiated. *See also:* **electrode (electron tube); phototube.**   178, 190, 117

photocathode blue response. The photoemission current produced by a specified luminous flux from a tungsten filament lamp at 2854 kelvins color temperature when the flux is filtered by a CS 5-58 blue filter of half stock thickness (1.75–2.25 mm). This parameter is useful in characterizing response to scintillation counting sources.   117

photocathode luminous sensitivity. *See:* **sensitivity, cathode luminous.**

pathocathode radiant sensitivity. *See:* **sensitivity, cathode radiant.**

photocathode, semitransparent. *See:* **semitransparent photocathode.**

photocathode spectral-sensitivity characteristic. *See:* **spectral-sensitivity characteristic photocathode.**

photocathode transit time. That portion of the photomultiplier transit time corresponding to the time for photoelectrons to travel from the photocathode to the first dynode.   117

photocathode transit-time difference. The difference in transit time between electrons leaving the center of the photocathode and electrons leaving the photocathode at some specified point on a designated diameter.   117

photocell (photoelectric cell). (1) A solid-state photosensitive electron device in which use is made of the variation of the current-voltage characteristic as a function of incident radiation. *See also:* **phototube.**   117
(2) A device exhibiting photovoltaic or photoconductive effects. *See:* **phototube.**   244, 190

photochemical radiation. Energy in the ultraviolet, visible, and infrared regions to produce chemical changes in materials. *Note:* Examples of photochemical processes are accelerated fading tests, photography, photoreproduction, and chemical manufacturing. In many such applications a specific spectral region is of importance. *See also:* **infrared radiation; light; ultraviolet radiation.**   167

photoconductive cell. A photocell in which the photoconductive effect is utilized. *See:* **phototube.**   244, 190

photoconductive effect (photoconductivity). A photoelectric effect manifested as a change in the electric conductivity of a solid or a liquid and in which the charge carriers are not in thermal equilibrium with the lattice. *Note:* Many semiconducting metals and their compounds (notably selenium, selenides, and tellurides) show a marked increase in electric conductance when electromagnetic radiation is incident on them. *See:* **photoelectric effect; phototube; photovoltaic effect; photoemissive effect.**   210, 190

photoelectric beam-type smoke detector (fire protection devices). A device which consists of a light source which is projected across the area to be protected into a photosensing cell. smoke between the light source and the receiving photosensing cell reduces the light reaching the cell, causing actuation.   71

photoelectric cathode. *See:* **photocathode.**

photoelectric color-register controller. A photoelectric control system used as a longitudinal position regulator for a moving material or web to maintain a preset register relationship between repetitive register marks in the first color and reference positions of the printing cylinders of successive colors. *See also:* **photoelectric control.**   206

photoelectric control (industrial control). Control by means of which a change in incident light effects a control function.   206

photoelectric counter (industrial control). A photoelectrically actuated device used to record the number of times a given light path is intercepted by an object. *See also:* **photoelectric control.**   206

photoelectric current. The current due to a photoelectric effect. *See also:* **photoelectric effect.**   244, 206

photoelectric cutoff register controller (industrial control). A photoelectric control system used as a longitudinal position regulator that maintains the position of the point of cutoff with respect to a repetitively referenced pattern on a moving material. *See also:* **photoelectric control.**   206

photoelectric directional counter. A photoelectrically actuated device used to record the number of times a given light path is intercepted by an object moving in a given direction. *See also:* **photoelectric control.**   204

photoelectric door opener. A photoelectric control system used to effect the opening and closing of a power-operated door. *See also:* **photoelectric control.**   204

photoelectric effect. Interaction between radiation and matter resulting in the absorption of photons and the consequent liberation of electrons. *Note:* This is a general term covering three separate phenomena in which an electric effect is influenced by incident electromagnetic radiation: the photoconductive effect, the photoemissive effect, and the photovoltaic effect.   210, 244, 190

photoelectric emission (electron tube). The ejection of electrons from a solid or liquid by electromagnetic radiation. *See also:* **field-enhanced photoelectric emission.**   125

photoelectric flame detector (fire protection devices). A device whose sensing element is a photocell which either changes its electrical conductivity or produces an electrical potential when exposed to radiant energy.   71

photoelectric lighting controller. A photoelectric relay actuated by a change in illumination to control the illumination in a given area or at a given point. *See also:*

photoelectric control. 206

photoelectric loop control (industrial control). A photoelectric control system used as a position regulator for a strip processing line that matches the average linear speed in one section to the speed in an adjacent section to maintain the position of the loop located between the two sections. *See also:* **photoelectric control.** 206

photoelectric pinhole detector. A photoelectric control system that detects the presence of minute holes in an opaque material. *See also:* **photoelectric control.** 204

photoelectric power system. *See:* **photovoltaic power system.**

photoelectric pyrometer (industrial control). An instrument that measures the temperature of a hot object by means of the intensity of radiant energy exciting a phototube. *See also:* **electronic control.** 206

photoelectric relay. A relay that functions at predetermined values of incident light. *See also:* **photoelectric control.** 206

photoelectric scanner (industrial control). A single-unit combination of a light source and one or more phototubes with a suitable optical system. *See also:* **photoelectric control.** 206

photoelectric side-register controller (industrial control). A photoelectric control system used as a lateral position regulator that maintains the edge of, or a line on, a moving material or web at a fixed position. *See also:* **photoelectric control.** 206

photoelectric smoke-density control. A photoelectric control system used to measure, indicate, and control the density of smoke in a flue or stack. *See also:* **photoelectric control.** 206

photoelectric smoke detector (industrial control). A photoelectric relay and light source arranged to detect the presence of more than a predetermined amount of smoke in air. *See also:* **photoelectric control.** 206

photoelectric spot-type smoke detector (fire protection devices). A device which contains a chamber with either overlapping or porous covers which prevent the entrance of outside sources of light but which allow the entry of smoke. The unit contains a light source and a special photosensitive cell in the darkened chamber. The cell is either placed in the darkened area of the chamber at an angle different from the light path or has the light blocked from it by a light stop or shield placed between the light source and the cell. With the admission of smoke particles, light strikes the particles and is scattered and reflected into the photosensitive cell. This causes the photosensing circuit to respond to the presence of smoke particles in the smoke chamber. 71

photoelectric system (protective signaling). An assemblage of apparatus designed to project a beam of invisible light onto a photoelectric cell and to produce an alarm condition in the protection circuit when the beam is interrupted. *See also:* **protective signaling.** 328

photoelectric tube. An electron tube, the functioning of which is determined by the photoelectric effect. *See also:* **phototube.** 190

photo-electron. An electron liberated by the photoemissive effect. *See also:* **photoelectric effect.** 244, 190

photoemission spectrum (scintillator material). The relative numbers of optical photons emitted per unit wavelength as a function of wavelength interval. The emission spectrum may also be given in alternative units such as wave number, photon energies, frequency, etcetera. *Note:* Optical photons are photons with energies corresponding to wavelengths between 2000 and 15 000 angstroms. 335

photoemissive effect. Electromagnetic radiation (light, ultraviolet, X-rays, etcetera may cause the emission of electrons from matter, if the wavelength of the radiation is less than a critical maximum value that depends on the material. For many metals, the critical wavelength is in the visible spectrum. *See:* **photoelectric effect; photovoltaic effect; photoconductive effect (photoconductivity).** 210

photoflash lamp. A lamp in which combustible metal or other solid material is burned in an oxidizing atmosphere to produce light of high intensity and short duration for photographic purposes. *See also:* **lamp.** 167

photoformer. A function generator that operates by means of a cathode-ray beam optically tracking the edge of a mask placed on a screen. *See also:* **electronic analog computer.** 9

photographic emulsion. The light-sensitive coating on photographic film consisting usually of a gelatin containing silver halide. 176

photographic sound recorder (optical sound recorder). Equipment incorporating means for producing a modulated light beam and means for moving a light-sensitive medium relative to the beam for recording signals derived from sound signals. 176

photographic sound reproducer (optical sound reproducer). A combination of light source, optical system, photoelectric cell, or other light-sensitive device such as a photoconductive cell, and a mechanism for moving a medium carrying an optical sound record (usually film), by means of which the recorded variations may be converted into electric signals of approximately like form. 176

photographic transmission density (optical density). The common logarithm of opacity. Hence, film transmitting 100 percent of the light has a density of zero, transmitting 10 percent a density of 1, and so forth. Density may be diffuse, specular, or intermediate. Conditions must be specified. 176

photometer. An instrument for measuring photometric quantities such as luminance (photometric brightness), luminous intensity, luminous flux, and illumination. *See also:* **photometry.** 167

photometric brightness. *See:* **luminance.**

photometry (1) (general). The measurement of quantities referring to radiation evaluated according to the visual effect which it produces, as based on certain conventions.

(2) (visual). That branch of photometry in which the eye is used to make comparison.

(3) (physical). That branch of photometry in which the measurement is made by means of physical receptors. 18

photomultiplier. *See:* **multiplier phototube.**

photomultiplier transit time (scintillation counting). The time difference between the incidence of a delta-function light pulse on the photocathode of the photomultiplier and the occurrence of the half-amplitude point on the output-pulse leading edge. 117

photomultiplier tube. *See:* **multiplier phototube.**

photon. An elementary quantity of radiant energy (quantum) whose value is equal to the product of Planck's constant and the frequency of the electromagnetic radiation: *hν*. *See also:* **photoelectric effect; photovoltaic power system; solar cells (in a photovoltaic power system).**
                                                        244, 190

photon emission spectrum, scintillator material (scintillation counting). The relative numbers of optical photons emitted per unit wavelength as a function of wavelength interval. The emission spectrum may also be given in alternative units such as wavenumber, photon energies, frequency, and so on.   117

photon emitting diode (light emitting diodes). A semiconductor device containing a semiconductor junction in which radiant flux is non-thermally produced when a current flows as a result of an applied voltage.   162

photopic spectral luminous efficiency function (V_λ) (photometric standard observer for photopic vision). The photopic spectral luminous efficiency function gives the ratio of the radiant flux at wavelength λ_m to that at wavelength λ, when the two fluxes produce the same photopic luminous sensations under specified photometric conditions, λ_m being chosen so that the maximum value of this ratio is unity.

Unless otherwise indicated, the values used for the spectral luminous efficiency function relate to photopic vision by the photometric standard observer having the characteristics laid down by the International Commission on Illumination (CIE).   162

photopic vision. Vision mediated essentially or exclusively by the cones. It is generally associated with adaptation to a luminance of at least 1.0 footlambert (3.0 nits). *See also:* **visual field.**   167

photosensitive recording (facsimile). Recording by the exposure of a photosensitive surface to a signal-controlled light beam or spot. *See also:* **recording (facsimile).**   12

photosensitive tube. *See:* **photoelectric tube.**

photosensitizers (laser-maser). Substances which increase the sensitivity of a material to irradiation by electromagnetic radiation.   363

phototube (photoelectric tube). An electron tube that contains a photocathode and has an output depending at every instant on the total photoelectric emission from the irradiated area of the photocathode. *See also:* **field-enhanced photoelectric emission.**   190, 117

phototube housing (industrial control). An enclosure containing a phototube and an optical system. *See also:* **photoelectric control.**   206

phototube, gas. *See:* **gas phototube.**

phototube, multiplier. *See:* **multiplier phototube.**

phototube, vacuum. *See:* **vacuum phototube.**

photovaristor. A varistor in which the current-voltage relation may be modified by illumination, for example, cadmium sulphide or lead telluride. *See also:* **semiconductor device.**   342

photovoltaic effect. For certain combinations of transparent conducting films separated by thin layers of semiconducting materials, electromagnetic radiation incident on one of the films can create no-load potential differences. *See:* **photoconductive effect (photoconductivity); photoemissive effect; photoelectric effect.**   210

physical circuit (data transmission). A two-wire metallic circuit not arranged for phantom use.   59

physical concept. Anything that has existence or being in the ideas of man pertaining to the physical world. Examples are magnetic fields, electric currents, electrons.
                                                        210

physical damage (rotating machinery). This contributes to electrical insulation failure by opening leakage paths through the insulation. Included here are: physical shock, vibration, overspeed, short-circuit forces, erosion by foreign matter, damage by foreign objects, and thermal cycling.   37

physical entity. *See:* **physical quantity.**

physical photometer. An instrument containing a physical receptor (photoemissive cell, barrier-layer cell, thermopile, etcetera) and associated filters, that is calibrated so as to read photometric quantities directly.   167

physical property. Any one of the generally recognized characteristics of a physical system by which it can be described.   210

physical quantity (physical entity) (concrete quantity). A particular example of a measurable physical property of a physical system. It is characterized by both a qualitative and a quantitative attribute (that is, kind and magnitude). It is independent of the system of units and equations by which it and its relation to other physical quantities are described quantitatively.   210

physical system. A part of the real physical world that is directly or indirectly observed or employed by mankind.
                                                        210

physical unit. *See:* **unit.**

P.I. *See:* **control action, proportional plus integral.**

pickle (corrosion) (electroplating). A solution or process used to loosen or remove corrosion products such as oxides, scale, and tarnish from a metal. *See also:* **electroplating.**
                                                    221, 205

pickling (electroplating) (1) (chemical). The removal of oxides or other compounds from a metal surface by means of a solution that acts chemically upon the compounds.
(2) (electrolytic). Pickling during which a current is passed through the pickling solution to the metal (cathodic pickling) or from the metal (anodic pickling). *See also:* **electroplating.**   328

pickoff (1) (gyro; accelerometer). A device which produces a signal output, generally a voltage, as a function of the relative linear or angular displacement between two elements.   46
(2) (test, measurement and diagnostic equipment). A sensing device that responds to movement to create a signal or to effect some type of control.   54

pickup (1) (electronics). A device that converts a sound, scene, or other form of intelligence into corresponding electric signals (for example, a microphone, a television camera, or a phonograph pickup). *See also:* **microphone; phonograph pickup; television.**   178, 54
(2) (interference terminology)*. Interference arising from sources external to the signal path. *See also:* **interference.**
                                                        188

---

* Deprecated

(3) (relay). (A) The action of a relay as it makes designated response to increase of input. (B) As a qualifying term, the state of a relay when all response to increase of input has been completed. (C) Also used to identify the minimum current, voltage, power, or other value of an input quantity reached by progressive increases that will cause the relay to reach the pickup state from reset. *Note:* In describing the performance of relays having multiple inputs, pickup has been used to denote contact closing, in which case pickup value of any input is meaningful only when related to all other inputs. *See also:* **microphone; phonograph pickup; relay; television.**   60, 127, 103

(4) (protective relaying of utility-consumer interconnections). The minimum input that will cause a device to complete contact operation or similar designated action. For example, an overcurrent relay set for 4 A will operate only if the current through the relay is equal to or greater than the pickup level of 4 A.   128

pickup and seal voltage (magnetically operated device) (industrial control). The minimum voltage at which the device moves from its deenergized into its fully energized position. *See also:* **initial contact pressure.**
    302, 225, 206

pickup current. *See:* **pickup value.**

pickup factor, direction-finder antenna system. An index of merit expressed as the voltage across the receiver input impedance divided by the signal field strength to which the antenna system is exposed, the direction of arrival and polarization of the wave being such as to give maximum response. *See also:* **navigation.**   278, 187, 13

pickup spectral characteristic (color television). The set of spectral responses of the device, including the optical parts, that converts radiation to electric signals, as measured at the output terminals of the pickup tubes. *Note:* Because of nonlinearity, the spectral characteristics of some kinds of pickup tubes depend upon the magnitude of radiance used in the measurement.   18, 178

pickup tube*. *See:* **camera tube.**

* Deprecated

pickup value. The minimum input that will cause a device to complete contact operation or similar designated action. *Note:* In describing the performance of devices having multiple inputs, the pickup value of an input is meaningful only when related to all other inputs.   103, 202

pickup voltage (or current) (magnetically operated device). The voltage (or current) at which the device starts to operate when its operating coil is energized under conditions of normal operating temperature. *See also:* **contactor.**
    1, 206

pictorial format (pulse measurement). A graph, plot, or display in which a waveform is presented for observation or analysis. Any of the waveform formats defined in the following subsections may be presented in the pictorial format.   15

picture element. The smallest area of a television picture capable of being delineated by an electric signal passed through the system or part thereof. *Note:* It has three important properties, namely P_v, the vertical height of the picture element; P_h, the horizontal length of the picture element, and P_a, the aspect ratio of the picture element. In addition, N_p, the total number of picture elements in a complete picture, is of interest since this number provides a convenient way of comparing systems. For convenience

P_v and P_h are normalized for V, the vertical height of the picture; that is, P_v or P_h must be multiplied by V to obtain the actual dimension in a particular picture. P_v is defined as $P_v = 1/N$, where N is the number of active scanning lines in the raster. P_h is defined as $P_h = t_r A/t_e$, where t_r is the average value of the rise and delay times (10 percent to 90 percent) of the most rapid transition that can pass through the system or part thereof, t_e is the duration of the part of a scanning line that carries picture information, and A is the aspect ratio of the picture. (At present all broadcast television systems have a horizontal to vertical aspect ratio of 4/3.) P_a is defined as $P_a = P_h/P_v = t_r AN/t_e$ and N_p is defined as $N_p = (1/P_v) \times (A/P_h) = Nt_e/T_r$. *See:* **television.**   177

picture frequencies (facsimile). The frequencies which result solely from scanning subject copy. *Note:* This does not include frequencies that are part of a modulated carrier signal. *See also:* **scanning (facsimile).**   12

picture inversion (facsimile). A process that causes reversal of the black and white shades of the recorded copy. *See also:* **facsimile transmission.**   12

picture signal (television or facsimile). The signal resulting from the scanning process. *See also:* **television.**   178

picture transmission (telephotography). The electric transmission of a picture having a gradation of shade values.   328

picture tube (kinescope) (television). A cathode-ray tube used to produce an image by variation of the beam current as the beam scans a raster. *See also:* **television.**   163

P.I.D. *See:* **control action, proportional plus integral plus derivative.**

Pierce gun (microwave tubes). A gun that delivers an initially convergent electron beam. If a magnetic focusing scheme is used, the beam is made to enter the field at the minimum beam diameter or else, if the magnetic field threads through the cathode, the magnetic field must have a shape that is consistent with the desired beam that imparts certain flow characteristics to the electron beam. In Brillouin flow, angular electron velocity about the axis is imparted to the beam on entry into the magnetic field and the resulting inwardly directed force balances both the space charge and centrifugal forces. In practice, values of field up to twice the theoretical equilibrium value may be found necessary. In confined flow there is no overall angular velocity of the beam about the beam axis. Individual electron trajectories are tight helices (of radius small compared to beam radius) whose axis is along a magnetic-field line. The required magnetic field is several times greater than the Brillouin value and the flux must intersect the cathode surface. *See also:* **microwave tube or valve.**
    190

Pierce oscillator. An oscillator that includes a piezoelectric crystal connected between the input and the output of a three-terminal amplifying element, the feedback being determined by the internal capacitances of the amplifying elements. *Note:* This is basically a Colpitts oscillator. *See also:* **Colpitts oscillator; oscillatory circuit.**   111, 211

piezoelectric accelerometer. A device that employs a piezoelectric material as the principal restraint and pickoff. It is generally used as a vibration sensor.   46

piezoelectric crystal cut, type. *See:* **type of piezoelectric crystal cut.**

piezoelectric-crystal element. A piece of piezoelectric material cut and finished to a specified geometrical shape and orientation with respect to the crystallographic axes of the material. *See also: crystal.* 341

piezoelectric-crystal plate. A piece of piezoelectric material cut and finished to specified dimensions and orientation with respect to the crystallographic axes of the material, and having two major surfaces that are essentially parallel. *See also: crystal.* 341

piezoelectric-crystal unit. A complete assembly, comprising a piezoelectric-crystal element mounted, housed, and adjusted to the desired frequency, with means provided for connecting it in an electric circuit. Such a device is commonly employed for purposes of frequency control, frequency measurement, electric wave filtering, or interconversion of electric waves and elastic waves. *Note:* Sometimes a piezoelectric-crystal unit may be an assembly having in it more than one piezoelectric-crystal plate. Such an assembly is called a mutliple-crystal unit. *See also: crystal.* 341

piezoelectric effect. Some materials become electrically polarized when they are mechanically strained. The direction and magnitude of the polarization depend upon the nature and amount of the strain, and upon the direction of the strain. In such materials the converse effect is observed, namely, that a strain results from the application of an electric field. 210

piezoelectric loudspeaker. *See: crystal loudspeaker.*

piezoelectric microphone. *See: crystal microphone.*

piezoelectric pickup. *See: crystal pickup.*

piezoelectric transducer. A transducer that depends for its operation on the interaction between electric charge and the deformation of certain materials having piezoelectric properties. *Note:* Some crystals and specially processed ceramics have piezoelectric properties. 176

pigtail. A flexible metallic conductor, frequently stranded, attached to a terminal of a circuit component and used for connection into the circuit. 328

pileup. *See: relay pileup.*

pillbox antenna. A reflector antenna having a cylindrical reflector enclosed by two parallel conducting plates perpendicular to the cylinder, spaced less than one wavelength apart. *Syn: cheese antenna.* 111

pilot (transmission system). A signal wave, usually a single frequency, transmitted over the system to indicate or control its characteristics. 328

pilotage. The process of directing a vehicle by reference to recognizable landmarks or soundings, or to electronic or other aids to navigation. Observations may be by any means including optical, aural, mechanical, or electronic. *See also: navigation.* 187, 13

pilot cell (storage battery). A selected cell whose condition is assumed to indicate the condition of the entire battery. *See also: battery (primary or secondary).* 328

pilot channel. A channel over which a pilot is transmitted. 328

pilot circuit (industrial control). The portion of a control apparatus or system that carries the controlling signal from the master switch to the controller. *See: control.* 206

pilot director indicator. A device that indicates to the pilot information as to whether or not the aircraft has departed from the target track during a bombing run. 328

pilot exciter (rotating machinery). The source of all or part of the field current for the excitation of another exciter. *See: direct-current commutating machine.* 63, 3

pilot fit (spigot fit) (rotating machinery). A clearance hole and mating projection used to guide parts during assembly. 63

pilot-house control (illuminating engineering). A mechanical means for controlling the elevation and train of a searchlight from a position on the other side of the bulkhead or deck on which it is mounted. *See also: searchlight.* 167

pilot lamp. A lamp that indicates the condition of an associated circuit. In telephone switching, a pilot lamp is a switchboard lamp that indicates a group of line lamps, one of which is or should be lit. 328

pilot light. A light, associated with a control, that by means of position or color indicates the functioning of the control. 328

pilot protection (power switchgear) (relay systems). A form of line protection that uses a communication channel as a means to compare electrical conditions at the terminals of a line. 103, 127

pilot streamer (lightning). The initial low-current discharge that begins when the voltage gradient exceeds the breakdown voltage of air. *See also: direct-stroke protection (lightning).* 64

pilot wire. An auxiliary conductor used in connection with remote measuring devices or for operating apparatus at a distant point. *See also: center of distribution.* 64

pilot-wire-controlled network. A network whose switching devices are controlled by means of pilot wires. *See also: alternating-current distribution.* 64

pilot-wire protection. *See: wire-pilot protection.*

pilot-wire regulator. An automatic device for controlling adjustable gains or losses associated with transmission circuits to compensate for transmission changes caused by temperature variations, the control usually depending upon the resistance of a conductor or pilot wire having substantially the same temperature conditions as the conductors of the circuits being regulated. *See also: transmission regulator.* 328

pi (π) mode (magnetrons). The mode of operation for which the phases of the fields of successive anode openings facing the interaction space differ by π radians. *See also: magnetrons.* 190, 125

pinboard. A perforated board that accepts manually inserted pins to control the operation of equipment. 255, 77, 54

pinch (electron tubes). The part of the envelope of an electron tube or valve carrying the electrodes and through which pass the connections to the electrodes. *See also: electron tube.* 244, 190

pinch effect (1) (rheostriction). The phenomenon of transverse contraction and sometimes momentary rupture of a fluid conductor due to the mutual attraction of the different parts carrying currents. *See also: electrothermics; induction heating.* 14, 114, 210
(2) (disk recording). A pinching of the reproducing stylus tip twice each cycle in the reproduction of lateral recordings due to a decrease of the groove angle cut by the recording stylus when it is moving across the record as it

swings from a negative to a positive peak.   176
(3) **(induction heating).** The result of an electromechanical force that constricts, and sometimes momentarily ruptures, a molten conductor carrying current at high density. *See:* **motor effect; skin effect.**   14
p-i-n detector (semiconductor radiation detectors). A detector consisting of an intrinsic or nearly intrinsic region between a **p** and **n** region.   23, 118, 119
pi (π) network. A network composed of three branches connected in series with each other to form a mesh, the three junction points forming an input terminal, an output terminal, and a common input and output terminal, respectively. See accompanying figure. *See also:* **network analysis.**   210

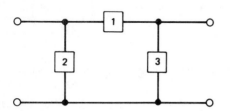

Pi network. The junction point between branches 1 and 2 forms an input terminal, that between branches 1 and 3 forms an output terminal, and that between branches 2 and 3 forms a common input and output terminal.

pin insulator. A complete insulator, consisting of one insulating member or an assembly of such members without tie wires, clamps, thimbles, or other accessories, the whole being of such construction that when mounted on an insulator pin it will afford insulation and mechanical support to a conductor that has been properly attached with suitable accessories. *See also:* **insulator; tower.**   64
pin jack. A single-conductor jack having an opening for the insertion of a plug of very small diameter.   341
pink noise (speech quality measurements). A random noise whose spectrum level has a negative slope of 10 decibels per decade.   126
pins (electron tube or valve). Metal pins connected to the electrodes that plug into the holder. They ensure the electric connection between the electrodes and the external circuit and also mechanically fix the tube in its holder. *See also:* **electron tube.**   244, 190
pip. A popular term for a sharp deflection in a visible trace. *See also:* **radar.**   328
pipe cable. A pressure cable in which the container for the pressure medium is a loose-fitting rigid metal pipe. *See:* **pressure cable; oil-filled pipe cable; gas-filled pipe cable.** *See also:* **power distribution, underground construction.**   64
pipe guide. A component of a switch operating mechanism designed to maintain alignment of a vertical rod or shaft.   27
pipe-ventilated (rotating machinery). *See:* **duct-ventilated.**
pip-matching display (electronic navigation). A display in which the received signal appears as a pair of blips, the comparison of the characteristics of which provides a

measure of the desired quantity. *See:* **K, L, or N display.** *See also:* **navigation.**   13, 187
pi (π) point. A frequency at which the insertion phase shift of an electric structure is 180 degrees or an integral multiple thereof.   328
Pirani gauge. A bolometric vacuum gauge that depends for its operation on the thermal conduction of the gas present; pressure being measured as a function of the resistance of a heated filament ordinarily over a pressure range of 10^{-1} to 10^{-4} conventional millimeter of mercury. *See also:* **instrument.**   328
piston (high-frequency communication practice) (plunger). A conducting plate movable along the inside of an enclosed transmission path and acting as a short-circuit for high-frequency currents. *See also:* **waveguide.**   328
piston attenuator (waveguide). A variable cutoff attenuator in which one of the coupling devices is carried on a sliding member like a piston. *See:* **waveguide.**   244, 179
pistonphone. A small chamber equipped with a reciprocating piston of measurable displacement that permits the establishment of a known sound pressure in the chamber. *See also:* **loudspeaker.**   176
pit (rotating machinery). A depressed area in a foundation under a machine.   63
pitch (acoustics) (audio and electroacoustics). The attribute of auditory sensation in terms of which sounds may be ordered on a scale extending from low to high, such as a musical scale. *Notes:* (1) Pitch depends primarily upon the frequency of the sound stimulus, but it also depends upon the sound pressure and wave form of the stimulus. (2) The pitch of a sound may be described by the frequency of that simple tone, having a specified sound pressure or loudness level, that seems to the average normal ear to produce the same pitch. (3) The unit of pitch is the mel.   176
pitch angle (electronic navigation). *See:* **pitch attitude.**
pitch attitude (electronic navigation). The angle between the longitudinal axis of the vehicle and the horizontal. *See also:* **navigation.**   187, 13
pitch factor (rotating machinery). The ratio of the resultant voltage induced in a coil to the arithmetic sum of the magnitudes of the voltages induced in the two coil sides. *See:* **armature.**   63
pits. Depressions produced in metal surfaces by nonuniform electrodeposition or from electrodissolution; for example, corrosion. *See also:* **electrodeposition.**   328
pitting (corrosion). Localized corrosion taking the form of cavities at the surface.   221, 205
pitting factor (corrosion). The depth of the deepest pit resulting from corrosion divided by the average penetration as calculated from weight loss.   221, 205
PIV. *See:* **peak inverse voltage.** *See also:* **peak reverse voltage (semiconductor rectifier).**
pivot-friction error. Error caused by friction between the pivots and the jewels; it is greatest when the instrument is mounted with the pivot axis horizontal. *Note:* This error is included with other errors into a combined error defined in **repeatability.** *See also:* **moving element (instrument).**   280

place. In positional notation, a position corresponding to a given power of the base, a given cumulated product, or a digit cycle of a given length. It can usually be specified as the nth character from one end of the numerical expression.   235

plain conductor. A conductor consisting of one metal only. *See also:* **conductors.**   64

plain flange (plane flange) (plain connector) (waveguide). A coupling flange with a flat face. *See:* **waveguide.**   179

planar array antenna. An array antenna having the centers of the radiating elements lying in a plane. *See also:* **antenna.**   179, 111

planar network. A network that can be drawn on a plane without crossing of branches. *See also:* **network analysis.**
                                                    210

Planckian locus. The locus of chromaticities of Planckian (blackbody) radiators having various temperatures. *See:* **chromaticity diagram.**   18, 178

Planck radiation law. An expression representing the spectral radiance of a blackbody as a function of the wavelength and temperature. This law commonly is expressed by the formula

$$L_\lambda = I_\lambda/A' = c_{1L}\lambda^{-5}[e^{(c_2/\lambda T)} - 1]^{-1}$$

in which L_λ is the spectral radiance, I_λ is the spectral radiant intensity, A' is the projected area ($A \cos \theta$) of the aperture of the blackbody, e is the base of natural logarithms (2.718+), T is absolute temperature, c_{1L} and c_2 are constants designated as the first and second radiation constants. *Note:* The designation c_{1L} is used to indicate that the equation in the form given here refers to the radiance L, or to the intensity I per unit projected area A', of the source. Numerical values are commonly given not for c_{1L} but for c_1 which applies to the total flux radiated from a blackbody aperture, that is, in a hemisphere (2π steradians), so that, with the Lambert cosine law taken into account, $c_1 = \pi c_{1L}$. The currently recommended value of c_1 is 3.7415×10^{-16} W·m² or 3.7415×10^{-12} W · cm². Then c_{1L} is 1.1910×10^{-16} W · m² · sr⁻¹ or 1.1910×10^{-12} W · cm² · sr⁻¹. If, as is more convenient, wavelengths are expressed in micrometers and area in square centimeters, $c_{1L} = 1.1910 \times 10^1$ W · μm⁴ · cm⁻² · sr − 1, L_λ being given in W · cm⁻² · sr⁻¹ · μm⁻¹. The presently recommended value of c_2 is 1.43879 cm kelvin. The Planck law in the following form gives the energy radiated from the blackbody in a given wavelength interval ($\lambda_1 - \lambda_2$):

$$Q = \int_{\lambda_1}^{\lambda_2} Q_\lambda d\lambda$$
$$= Atc_1 \int_{\lambda_1}^{\lambda_2} \lambda^{-5} (e^{(c_2/\lambda T)} - 1)^{-1} d\lambda$$

If A is the area of the radiation aperture or surface in square centimeters, t is time in seconds, λ is wavelength in micrometers, and $c_1 = 3.7415 \times 10^4$ W · μm⁴ · cm⁻², then Q is the total energy in watt-seconds emitted from this area (that is, in the solid angle 2π), in time t, within the wavelength interval ($\lambda_1 - \lambda_2$). *Note:* It often is convenient, as is done here, to use different units of length in specifying wavelengths and areas, respectively. If both quantities are expressed in centimeters and the corresponding value for c_1 (3.7415×10^{-5} erg · cm · sec⁻¹) is used, this equation

gives the total emission of energy in ergs from area A (that is in the solid angle 2π), for time t, and for the interval $\lambda_1 - \lambda_2$ in centimeters. *See also:* **radiant energy (in illuminating engineering).**   167

plane angle (International System of Units) (SI). The SI unit for plane angle is the radian. Use of the degree and its decimal submultiples is permissible when the radian is not a convenient unit. Use of the minute and second is discouraged except for special fields such as cartography. *See also:* **units and letter symbols.**   21

plane-earth factor (radio wave propagation). The ratio of the electric field strength that would result from propagation over an imperfectly conducting plane earth to that which would result from propagation over a perfectly conducting plane. *See also:* **radio wave propagation.**
                                                146, 180

plane flange. *See:* **plain flange.**

plane of polarization (1) (plane-polarized wave). The plane containing the electric intensity and the direction of propagation. *See also:* **radiation; radio wave propagation.**
                                                 59, 146
(2) (antennas). A plane containing the polarization ellipse. *Notes:* (A) When the ellipse degenerates into a line segment, the plane of polarization is not uniquely defined. In general, any plane containing the segment is acceptable; however, for a plane wave in an isotropic medium, the plane of polarization is taken to be normal to the direction of propagation. (B) In optics the expression plane of polarization is associated with a linearly polarized plane wave (sometimes called a plane polarized wave) and is defined as a plane containing the field vector of interest and the direction of propagation. This usage would contradict the above one and is deprecated.   111

plane of propagation (electromagnetic wave). The plane containing the attenuation vector and the wave normal; in the common degenerate case where these vectors have the same direction, the plane containing the electric vector and the phase vector and the wave normal. *See also:* **radio wave propagation.**   180, 146

plane-parallel resonator (laser-maser). A beam resonator comprising a pair of plane mirrors oriented perpendicular to the axis of the beam.   363

plane-polarized wave (homogeneous isotropic medium). A wave whose electric intensity at all times lies in a fixed plane that contains the direction of propagation. *See also:* **linearly polarized wave; radiation.**   146

plane wave (radio wave propagation). A wave whose equiphase surfaces form a family of parallel planes. *See also:* **radiation; radio wave propagation.**   146, 180

planned stop. *See:* **optional stop.**

plan position indicator (PPI) (radar). A display on which blips produced by signals from reflecting objects and transponders are shown in plan position, thus forming a map-like display.   13

plan-position-indicator scope. A cathode-ray oscilloscope arranged to present a plan-position-indicator display. *See also:* **radar.**   13, 187

plant (control system). For a given system, that part which is to be controlled and whose parameters are unalterable. *See also:* **process equipment.**   56

plant-capacity factor. *See:* **plant factor.**

plant dynamics (control system). Equations which describe

the behavior of the plant. *See also:* **control system.**
329

Planté plate (storage cell). A formed lead plate of large area, the active material of which is formed in thin layers at the expense of the lead itself. *See also:* **battery (primary or secondary).**
328

plant factor (plant-capacity factor). The ratio of the average load on the plant for the period of time considered to the aggregate rating of all the generating equipment installed in the plant. *See also:* **generating station.** 64

plasma. A gas made up of charged particles. *Note;* Usually plasmas are neutral, but not necessarily so, as, for example, the space charge in an electron tube.
210

plasma frequency (radio wave propagation). A natural frequency of oscillation of charged particles in a plasma given by

$$f_N = \frac{1}{2\pi} \left(\frac{Nq^2}{\epsilon_0 m} \right)^{1/2}$$

where q is the charge per particle, m is the particle mass, N is the charge density, and ϵ_0 is the permittivity of free space, all quantities being expressed in meter-kilogram-second (MKS) units. Note: For electrons, $f_n = 8.979\ N^{1/2}$ in the International System of Units (SI). *See also:* **radio wave propagation.**
180, 146

plasma sheath. A layer of charged particles of substantially one sign that accumulates around a body in a plasma. *See also:* **radio wave propagation.**
180, 146

plastic (rotating machinery). A material that contains as an essential ingredient an organic substance of large molecular weight, is solid in its finished state, and, at some stage in its manufacture or in its processing into finished articles, can be shaped by flow.
217, 63

plate (electron tubes). A common name for an anode in an electron tube. *See also:* **electrode (electron tube).** 125

plateau (radiation-counter tubes). The portion of the counting-rate-versus-voltage characteristic in which the counting rate is substantially independent of the applied voltage. *See also:* **gas-filled radiation counter tubes.**
190, 125, 96

plateau length (radiation-counter tubes). The range of applied voltage over which the plateau of a radiation-counter tube extends. *See also:* **gas-filled radiation-counter tubes.**
190, 96, 125

plateau slope, normalized (radiation counter tubes). The slope of the substantially straight portion of the counting rate versus voltage characteristic divided by the quotient of the counting rate by the voltage at the Geiger-Mueller threshold. *See also:* **gas-filled radiation counter tubes.**
190, 125

plateau slope, relative (radiation-counter tubes). The average percentage change in the counting rate near the midpoint of the plateau per increment of applied voltage. *Note:* Relative plateau slope is usually expressed as the percentage change in counting rate per 100-volt change in applied voltage (see accompanying figure). *See also:* **gas-filled radiation-counter tubes.**
190, 96,125

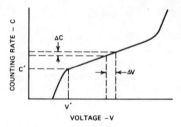

Counting rate–voltage characteristic in which

$$\text{Relative plateau slope} = 100\ \frac{\Delta C/C}{\Delta V}$$

$$\text{Normalized plateau slope} = \frac{\Delta C/\Delta V}{C'/V'} = \frac{\Delta C/C'}{\Delta V/V'}$$

plate-circuit detector. A detector functioning by virtue of a nonlinearity in its plate-circuit characteristic. 328

plated-through hole (electronic and electrical applications). Deposition of metal on the side of a hole and on both sides of a base to provide electric connection, and an enlarged portion of conductor material surrounding the hole on both sides of the base.
284

plate (anode) efficiency. The ratio of load circuit power (alternating current) to the plate power input (direct current). *See also:* **network analysis.** 111

plate keying. Keying effected by interrupting the plate-supply circuit. *See also:* **telegraphy.** 111

plate (anode) load impedance. The total impedance between anode and cathode exclusive of the electron stream. *See also:* **network analysis.** 111, 59

plate (anode) modulation. Modulation produced by introducing the modulating signal into the plate circuit of any tube in which the carrier is present. *See also:* **modulator.**
111, 211

plate (anode) neutralization. A method of neutralizing an amplifier in which a portion of the plate-to-cathode alternating voltage is shifted 180 degrees and applied to the grid-cathode circuit through a neutralizing capacitor. *See also:* **amplifier, feedback.** 111, 211

plate out (nuclear power generating stations). A thermal, electrical, chemical, or mechanical action that results in a loss of material by deposition on surfaces between sampling point and detector.
31

plate (anode) power input. The power delivered to the plate (anode) of an electron tube by the source of supply. *Note:* The direct-current power delivered to the plate of an electron tube is the product of the mean plate voltage and the mean plate current.
111, 59

plate (anode) pulse modulation. Modulation produced in an amplifier or oscillator by application of externally generated pulses to the plate circuit.
111

platform, aerial (conductor stringing equipment). A device designed to be attached to the boom tip of a crane or aerial lift and support a workman in an elevated working position. Platforms are constructed with surrounding railings, fabricated from aluminum, steel or fiber reinforced plastic, and utilized with larger aerial lifts such as the Truco or Condor. Occasionally, a platform is suspended from the load line of a large crane. *Syn.* **cage; platform.** 45

platform control power (series capacitor). Energy sources available at platform potential for performing operational and control functions.   86

platform erection (alignment of inertial systems). The process of bringing the vertical axis of a stable platform system into agreement with the local vertical. *See also:* **navigation.**   187, 13

platform fault protection device (series capacitor). A device to detect insulation failure, which results in current flowing to the platform.   86

platform, lineman's (conductor stringing equipment). A device designed to be attached to a wood pole and/or metal structure to serve as a supporting surface for workmen engaged in deadending operations, clipping-in, insulator work, etcetera. The designs of these devices vary considerably. Some resemble short cantilever beams, others resemble swimming pool diving boards, and still others as long as 40 feet are truss structures resembling bridges. Materials commonly used for fabrication are wood, fiberglass, and metal. *Syn.* **baker board; "D" board; deadend board; deadend platform; diving board.**   45

platform-to-ground signaling devices (series capacitor). Devices to transmit operating, control, alarm, and indication functions to and from the platform insulated from ground.   86

plating rack (electroplating). Any frame used for suspending one or more electrodes and conducting current to them during electrodeposition. *See also:* **electroplating.**   328

playback. A term used to denote reproduction of a recording.   328

playback loss. *See:* **translation loss.**

plenary capacitance (between two conductors). The capacitance between two conductors when the changes in the charges on the two are equal in magnitude but opposite in sign and the other $n-2$ conductors are isolated conductors. *See:* **direct capacitances of a system of conductors.**   210

pliotron. A hot-cathode vacuum tube having one or more grids. *Note:* This term is used primarily in the industrial field.   190

plotting board. *See:* **recorder, X-Y (plotting board).**

plow (cable plowing). Equipment capable of laying cable, flexible conduit, etcetera, underground.   52

plow blade (cable plowing). A soil-cutting tool.   52

plow blade amplitude (cable plowing). Maximum displacement of plow blade tip from mean position induced by the vibrator (half the stroke).   52

plow blade frequency (cable plowing). Rate of blade tip vibration in hertz.   52

plowing (cable plowing). A process for installing cable, flexible conduit, etc., by cutting or separating the earth, permitting the cable or flexible conduit to be placed or pulled in behind the blade.   52

plug. A device, usually associated with a cord, that by insertion in a jack or receptacle establishes connection between a conductor or conductors associated with the plug and a conductor or conductors connected to the jack or receptacle.   341

plug adapter (plug body*). A device that by insertion in a lampholder serves as a receptacle.   328
* Deprecated

plug adapter lampholder (current tap*). A device that by

insertion in a lampholder serves as one or more receptacles and a lampholder.   328
* Deprecated

plugboard (1)(general). A perforated board that accepts manually inserted plugs to control the operation of equipment. *See:* **control panel.**   255, 77

(2) (test, measurement and diagnostic equipment). Patchboard the use of which is restricted to punched card machines. *See:* **patchboard.**   54

plug braking (rotating machinery). A form of electric braking of an induction motor obtained by reversing the phase sequence of its supply. *See:* **asynchronous machine.**   63

plugging (industrial control). A control function that provides braking by reversing the motor line voltage polarity or phase sequence so that the motor develops a countertorque that exerts a retarding force. *See also:* **electric drive.**   225, 206

plug-in. A communication device when it is so designed that connections to the device may be completed through pins, plugs, jacks, sockets, receptacles, or other forms of ready connectors.   328

plug-in-type bearing (rotating machinery). A complete journal bearing assembly, consisting of a bearing liner and bearing housing and any supporting structure that is intended to be inserted into a machine end-shield. *See also:* **bearing.**   63

plug-in unit (CAMAC). *See:* **CAMAC plug-in unit.**

plumb-bob vertical (electronic navigation). The direction indicated by a simple, ideal, frictionless pendulum that is motionless with respect to the earth; it indicates the direction of the vector sum of the gravitational and centrifugal accelerations of the earth at the location of the observer. *See also:* **navigation.**   187, 13

plumbing (1) (communication practice). A colloquial expression employed to designate coaxial lines or waveguides and accessory equipment for radio-frequency transmission. *See also:* **waveguide.**   59, 187, 179

(2) (radar). A colloquial expression for pipelike waveguide circuit elements and transmission lines.   13

plunger relay. A relay operated by a movable core or plunger through solenoid action. *See also:* **relay.**   259

plunger, waveguide. *See:* **short circuit, adjustable.**

plural service. *See:* **dual service.**

plural tap (cube tap). *See:* **multiple plug.**

plus input (oscilloscopes). An input such that the applied polarity causes a deflection polarity in agreement with conventional deflection polarity.   106

PM. *See:* **permanent magnet.**

PM. *See:* **phase modulation.**

PM (microwave tubes). *See:* **permanent-magnet focusing.**

pneumatically release-free (pneumatically operated switching device)(pneumatically trip-free). A term indicating that by pneumatic control the switching device is free to open at any position in the closing stroke if the release is energized. *Note:* This release-free feature is operative even though the closing control switch is held closed.   103, 202

pneumatic bellows, relay. *See:* **relay pneumatic bellows.**

pneumatic controller. A pneumatically supervised device or group of devices operating electric contacts in a predetermined sequence. *See also:* **multiple-unit control.**

1

pneumatic loudspeaker. A loudspeaker in which the acoustic output results from controlled variation of an air stream. *See also:* **loudspeaker.** 328

pneumatic operation. Power operation by means of compressed gas. 103, 202

pneumatic switch. A pneumatically supervised device opening or closing electric contacts, and differs from a pneumatic controller in being purely an ON and OFF type device. *See also:* **control switch.** 1

pneumatic transducer. A unilateral transducer in which the sound output results from a controlled variation of an air stream. 176

pneumatic tubing system (protective signaling). An automatic fire-alarm system in which the rise in pressure of air in a continuous closed tube, upon the application of heat, effects signal transmission. *Note:* Most pneumatic tubing systems contain means for venting slow pressure changes resulting from temperature fluctuations and therefore operate on the so-called rate-of-rise principle. *See also:* **protective signaling.** 328

PN sequence. *See:* **pseudonoise sequence.**

pocket-type plate (of a storage cell). A plate of an alkaline storage battery consisting of an assembly of perforated oblong metal pockets containing active material. *See also:* **battery (primary or secondary).** 328

poid. The curve traced by the center of a sphere when it rolls or slides over a surface having a sinusoidal profile.

176

point (1) (positional notation). (A) The character, or the location of an implied symbol, that separates the integral part of a numerical expression from its fractional part. For example, it is called the binary point in binary notation and the decimal point in decimal notation. If the location of the point is assumed to remain fixed with respect to one end of the numerical expressions, a fixed-point system is being used. If the location of the point does not remain fixed with respect to one end of the numerical expression, but is regularly recalculated, then a floating-point system is being used. *Note:* A fixed-point system usually locates the point by some convention, while a floating-point system usually locates the point by expressing a power of the base. *See:* **branchpoint; breakpoint; checkpoint; entry point; fixed point; floating point; rerun point; variable point.**

210, 255, 77

(B) The character, or implied location of such a character, that separates the integral part of a numerical expression from the fractional part. Since the place to the left of the point has unit weight in the most commonly used systems, the point is sometimes called the units point, although it is frequently called the binary point in binary notation and the decimal point in decimal notation. *See:* **breakpoint; fixed point; floating point.** 235

(2) (lightning protection). The pointed piece of metal used at the upper end of the elevation rod to receive a lightning discharge. *See also:* **lightning protection and equipment.**

328

(3) (supervisory control or indication or telemeter selection). All of the supervisory control or indication devices, in a system, exclusive of the common devices, in the master station and in the remote station that are necessary for (A)

energizing the closing, opening, or other positions of a unit, or set of units of switchgear or other equiment being controlled; (B) automatic indication of the closed or open or other positions of a unit, or set of units of switchgear or other equipment for which indications are being obtained; or (C) connecting a telemeter transmitting equipment into the circuit to be measured and to transmit the telemeter reading over a channel to a telemeter receiving equipment. *Note:* A point may serve for any two or all three of the purposes described above; for example, when a supervisory system is used for the combined control and indication of remotely operated equipment, point (for supervisory control) and point (for supervisory indication) are combined into a single control and indication point.

103, 202

point-by-point method (lighting calculation). A lighting design procedure for predetermining the illumination at various locations in lighting installations, by use of luminaire photometric data. *Note:* Since interreflections are not taken into account, the point-by-point method may indicate lower levels of illumination than are actually realized. *See also:* **inverse-square law (illuminating engineering).** 167

point contact (semiconductors). A pressure contact between a semiconductor body and a metallic point. *See also:* **semiconductor; semiconductor device.** 210, 245, 66

point detector. A device that is a part of a switch-operating mechanism and is operated by a rod connected to a switch, derail, or movable-point frog to indicate that the point is within a specified distance of the stock rail. 328

pointer pusher (demand meter). The element that advances the maximum demand pointer in accordance with the demand and in integrated-demand meters is reset automatically at the end of each demand interval. *See also:* **demand meter.** 328

pointer shift due to tapping. The displacement in the position of a moving element of an instrument that occurs when the instrument is tapped lightly. The displacement is observed by a change in the indication of the instrument. *See also:* **moving element (instrument).** 280

point ID printout (sequential events recording systems). A brief coded method of identifying inputs using alphanumeric characters, usually used in computer based systems.

48, 58

pointing accuracy (communication satellite). The angular difference between the direction in which the main beam of an antenna points and the required pointing direction.

85

point of fixation. A point or object in the visual field at which the eyes look and upon which they are focused. *See also:* **visual field.** 167

point of observation. The midpoint of the base line connecting the centers of rotation of the two eyes. For practical purposes, the center of the pupil of the eye often is taken as the point of observation. *See also:* **visual field.**

167

point source (laser-maser). A source of radiation whose dimensions are small enough compared with the distance between source and receptor for them to be neglected in calculations. 363

point-to-point control system. *See:* **positioning control system.**

point-to-point radio communication. Radio communication between two fixed stations. *See also:* **radio transmission.** 328

point transposition. A transposition, usually in an open wire line, that is executed within a distance comparable to the wire separation, without material distortion of the normal wire configuration outside this distance. *See also:* **open wire.** 328

Poisson's equation. In rationalized form:

$$\nabla^2 V = -\frac{\rho}{\epsilon_0 \epsilon}$$

where $\epsilon_0 \epsilon$ is the absolute capacitivity of the medium, V the potential, and ρ the charge density at any point. 210

polar axis (ferroelectric material). A direction that is parallel to the spontaneous polarization vector. *Note:* In all single crystal ferroelectrics except those of triclinic and monoclinic symmetry, the polar axis has a small number of possible orientations with respect to the macroscopic crystal. *See also:* **ferroelectric domain; spontaneous polarization.** 80

polar contact. A part of a relay against which the current-carrying portion of the movable polar member is held so as to form a continuous path for current. 328

polar direct-current telegraph system. A system that employs positive and negative currents for transmission of signals over the line. *See also:* **telegraphy.** 328

polar-duplex signaling (telephone switching systems). Any method of bidirectional signaling over a line using ground potential compensation and polarity sensing. 55

polarential telegraph system. A direct-current telegraph system employing polar transmission in one direction and a form of differential duplex transmission in the other direction. *Note:* Two kinds of polarential systems, known as types A and B, are in use. In half-duplex operation of a type-A polarential system the direct-current balance is independent of line resistance. In half-duplex operation of a type-B polarential system the direct-current balance is substantially independent of the line leakage. *See also:* **telegraphy.** 194

polarity (1) (battery). An electrical condition determining the direction in which current tends to flow on discharge. By common usage the discharge current is said to flow from the positive electrode through the external circuit. *See also:* **battery (primary or secondary).** 328

(2) (television) (picture signal). The sense of the potential of a portion of the signal representing a dark area of a scene relative to the potential of a portion of the signal representing a light area. Polarity is stated as black negative or black positive. *See also:* **television.** 178

(3) (instrument transformer). The designation of the relative instantaneous directions of the currents entering the primary terminals and leaving the secondary terminals during most of each half cycle. *Note:* Primary and secondary terminals are said to have the same polarity, when, at a given instant during most of each half cycle, the current enters the identified, similarly marked primary lead and leaves the identified, similarly marked secondary terminal in the same direction as though the two terminals formed a continuous circuit. 53

polarity and angular displacement (regulator). Relative lead polarity of a regulator or a transformer is a designation of the relative instantaneous direction of current in its leads. In addition to its main transformer windings, a regulator commonly has auxiliary transformers or auxiliary windings as an integral part of the regulator. The same principles apply to the polarity of all transformer windings. *Notes:* (1) Primary and secondary leads are said to have the same polarity when at a given instant the current enters an identified secondary lead in the same direction as though the two leads formed a continuous circuit. (2) The relative lead polarity of a single-phase transformer may be either additive or subtractive. If one pair of adjacent leads from the two windings is connected together and voltage applied to one of the windings, then: (A) The relative lead polarity is additive if the voltage across the other two leads of the windings is greater than that of the higher-voltage winding alone. (B) The relative lead polarity is subtractive if the voltage across the other two leads of the winding is less than that of the higher-voltage winding alone. (3) The polarity of a polyphase transformer is fixed by the internal connections between phases as well as by the relative locations of leads; it is usually designated by means of a vector line diagram showing the angular displacement of windings and a sketch showing the marking of leads. The vector lines of the diagram represent induced voltages and the recognized counterclockwise direction of rotation is used. The vector line representing any phase voltage of a given winding is drawn parallel to that representing the corresponding phase voltage of any other winding under consideration. *See also:* **voltage regulator.** 257

polarity marks (instrument transformer). The identifications used to indicate the relative instantaneous polarities of the primary and secondary current and voltages. *Notes:* (1) On voltage transformers during most of each half cycle in which the identified primary terminal is positive with respect to the unidentified primary terminal, the identified secondary terminal is also positive with respect to the unidentified secondary terminal. (2) The polarity marks are so placed on current transformers that during most of each half-cycle, when the direction of the instantaneous current is into the identified primary terminal, the direction of the instantaneous secondary current is out of the correspondingly identified secondary terminal. (3) This convention is in accord with that by which standard terminal markings H_1, X_1, etcetera, are correlated. *See:* **instrument transformer.** 203

polarity-related adjectives (pulse terms). (1) **unipolar.** Of, having, or pertaining to a single polarity.

(2) **bipolar.** Of, having, or pertaining to both polarities. 254

polarity test (rotating machinery). A test taken on a machine to demonstrate that the relative polarities of the windings are correct. *See also:* **asynchronous machine; direct-current commutating machine.** 63

polarizability. The average electric dipole moment produced per molecule per unit of electric field strength. 210

polarization (1) (ferroelectric material). The electric moment per unit volume. *Note:* The polarization $\bar{P}$ may be expressed as the bound surface charge per unit area, which in macroscopic electric theory is equal to the component of the polarization vector normal to the surface. The polarization is related to the electric displacement through the expression

$$\vec{D} = \vec{P} + \epsilon_0 \vec{E}$$

where the constant ϵ_0 (usually called the permittivity of free space) equals 8.854×10^{-12} coulomb per volt-meter. In ferroelectric materials both D and P are nonlinear functions of E, that may depend on previous history of the material. If the electric field is applied along the polar axis of the crystal, this expression may then be regarded as a scalar equation, since D, E and P all point along the same direction. When the term $\epsilon_0 E$ in the above expression is negligible compared to P (as is the case for most ferroelectric materials), D is nearly equal to P; therefore, the D−versus −E and P− versus −E plots of the hysteresis loop become, in practice, equivalent. *See also:* **ferroelectric domain; coercive electric field; remanent polarization; spontaneous polarization.**

(2) (radiated wave). That property of a radiated electromagnetic wave describing the time-varying direction and amplitude of the electric field vector; specifically, the figure traced as a function of time by the extremity of the vector at a fixed location in space, as observed along the direction of propagation. *Note:* In general the figure is elliptical and it is traced in a clockwise or counterclockwise sense. The commonly referenced circular and linear polarizations are obtained when the ellipse becomes a circle or a straight line, respectively. Clockwise sense rotation of the electric vector is designated **right-hand polarization** and counterclockwise sense rotation is designated **left-hand polarization.** *See also:* **radiation.**   179, 111, 146

(3) (desired) (electronic navigation). The polarization of the radio wave for which an antenna system is designed. *See also:* **navigation.**   278, 187, 13

(4) (antenna). In a given direction, the polarization of the wave radiated by the antenna. Alternatively, the polarization of a plane wave incident from the given direction which results in maximum available power at the antenna terminals. *Notes:* (A) The polarization of these two waves is the same in the following sense: In the plane perpendicular to the direction considered, their electric fields describe similar ellipses. The sense of rotation on these ellipses is the same if each one is referred to the corresponding direction of propagation, outgoing for the radiated field, incoming for the incident plane wave. (B) When the direction is not stated, the polarization is taken to be the polarization in the direction of maximum gain.   111

(5) (electrolytic). A change in the potential of an electrode produced during electrolysis, such that the potential of an anode always becomes more positive (more noble), or that of a cathode becomes more negative (less noble), than their respective static electrode potentials. The polarization is equal to the difference between the static electrode potential for the specified electrode reaction and the dynamic potential (that is, the potential when current is flowing) at a specified current density. *See also:* **electrochemistry.**   328

(6) (battery). The change in voltage at the terminals of the cell or battery when a specified current is flowing, and is

equal to the difference between the actual and the equilibrium (constant open-circuit condition) potentials of the plates, exclusive of the IR drop. *See:* **polarization (electrolytic).** *See also:* **battery (primary or secondary).**   328

(7) (relay). A term identifying the input that provides a reference for establishing the direction of system phenomena such as direction of power or reactive flow, or direction to a fault or other disturbance on a power system.   103, 202, 127

polarization capacitance (biological). The reciprocal of the product of electrode capacitive reactance and 2π times the frequency.

$$C_p = \frac{1}{2\pi f X_p}$$

See also: **electrode impedance (biological).**   192

polarization current. *See:* **current, polarization.**

polarization, desired (radar). The polarization of the radio wave, for which an antenna system is designed.   13

polarization diversity reception. That form of diversity reception that utilizes separate vertically and horizontally polarized receiving antennas. *See also:* **radio receiver.**   328

polarization ellipse (field vector). The locus of positions for variable time of the terminus of an instantaneous field vector of one frequency at a point in space. *See also:* **waveguide.**   267, 179

polarization error (radar and navigation). The error arising from the transmission or reception of an electromagnetic wave having a polarization other than that intended for the system. *See also:* **navigation.**   13, 187

polarization index (rotating machinery). The ratio of the insulation resistance of a machine winding measured at 1 minute after voltage has been applied divided into the measurement at 10 min.   6

polarization index test (rotating machinery). A test for measuring the ohmic resistance of insulation at specified time intervals for the purpose of determining the polarization index. *See also:* **asynchronous machine; direct-current commutating machine.**   63

polarization of a field vector. For a field vector at a single frequency at a fixed point in space, the polarization is that property which describes the shape and orientation of the locus of the extremity of the field vector and the sense in which this locus is traversed. *Notes:* (1) For a time harmonic (or single-frequency) vector, the locus is an ellipse with center at the origin. In some cases, this ellipse becomes a circle or a segment of a straight line. The polarization is then called, respectively, circular and linear. (2) The orientation of the ellipse is defined by its plane, called the plane of polarization, and by the direction of its axes. (For a linearly polarized field, any plane containing the segment locus of the field vector is a plane of polarization.) (3) The shape of the ellipse is defined by the axial ratio (major axis)/(minor axis). This ratio varies between infinity and 1 as the polarization changes from linear to circular. Sometimes the ratio is defined as (minor axis)/(major axis). (4) The sense of polarization is indicated by an arrow placed on the ellipse. Alternatively, if the observation is made from a particular side of the plane of polarization the sense can be qualified as clockwise or counterclockwise. It can also be called right hand (or left hand) if, when placing the thumb of the right hand (or left hand) in a specified reference direction normal to the plane of polarization, the sense of travel on the ellipse is indicated by the fingers of the hand. (5) The field vector considered may be the electric field, the magnetic field, or any other field vector, for example, the velocity field in a warm plasma. 111

polarization potential (biological). The boundary potential over an interface. *See also:* **electrobiology.** 192

polarization reactance (biological). The impedance multiplied by the sine of the angle between the potential vector and the current vector.

$$X_p = Z_p \sin \theta$$

See also: **electrode impedance (biological).** 192

polarization receiving factor. The ratio of the power received by an antenna from a given plane wave of arbitrary polarization to the power received by the same antenna from a plane wave of the same power density and direction of propagation, whose state of polarization has been adjusted for the maximum received power. *Note:* It is equal to the square of the absolute value of the scalar product of the polarization unit vector of the given plane wave with that of the radiation field of the antenna along the direction opposite to the direction of propagation of the plane wave. *See also:* **waveguide.** 267

polarization resistance (biological). The impedance multiplied by the cosine of the phase angle between the potential vector and the current vector.

$$R_p = Z_p \cos \theta$$

See also: **electrode impedance (biological).** 192

polarization unit vector (field vector) (at a point). A complex field vector divided by its magnitude. *Notes:* (1) For a field vector of one frequency at a point, the polarization unit vector completely describes the state of polarization, that is, the axial ratio and orientation of the polarization ellipse and the sense of rotation on the ellipse. (2) A complex vector is one each of whose components is a complex number. The magnitude is the positive square root of the scalar product of the vector and its complex conjugate. *See also:* **waveguide.** 267

polarized electrolytic capacitor. An electrolytic capacitor in which the dielectric film is formed adjacent to only one metal electrode and in which the impedance to the flow of current in one direction is greater than in the other direction. *See also:* **electrolytic capacitor.** 328

polarized plug (packaging machinery). A plug so arranged that it may be inserted in its counterpart only in a predetermined position. 124

polarized relay. A relay that consists of two elements, one of which operates as a neutral relay and the other of which operates as a polar relay. *See:* **neutral relay; polar relay.** 328

polarizer. A substance that when added to an electrolyte increases the polarization. *See also:* **electrochemistry.** 328

polar mode (analog computer). *See:* **resolver.**

polar operation (data transmission). Circuit operation in which mark and space transitions are represented by a current reversal. 59

polar orbit (communication satellite). An inclined orbit with an inclination of 90°. The plane of a polar orbit contains the polar axis of the primary body. 74

polar relay. A relay in which the direction of movement of the armature depends upon the direction of the current in the circuit controlling the armature. *See also:* **electromagnetic relay; neutral relay; polarized relay.** 328

pole (pole unit) (1) (switching device or fuse). That portion of the device associated exclusively with one electrically separated conducting path of the main circuit of the device. *Notes:* (A) Those portions that provide a means for mounting and operating all poles together are excluded from the definition of a pole. (B) A switching device or fuse is called single-pole if it has only one pole. If it has more than one pole, it may be called multipole (two-pole, three-pole, etcetera) and provided, in the case of a switching device, that the poles are or can be coupled in such a manner as to operate together. 103, 202, 27 **(2) (electric power or communication).** A column of wood or steel, or some other material, supporting overhead conductors, usually by means of arms or brackets. *See:* **field pole; pole shoe; tower.** 64 **(3) (network function).** (A) Any value p_j of p, real or complex, for which the network function is infinite, provided that there exists some positive integer m such that, if the network function is multiplied by $(p - p_j)^m$, the resulting function is not infinite or zero when $p = p_j$. (B) The corresponding point in the P plane. *See also:* **control system, feedback; Laplace transform; network analysis; zero.** 210, 56

pole body (rotating machinery). The part of a field pole around which the field winding is fitted. *See also:* **asynchronous machine; direct-current commutating machine.** 63

pole-body insulation (rotating machinery). Insulation between the pole body and the field coil. *See also:* **asynchronous machine; direct-current commutating machine.** 63

pole bolt (rotating machinery). A bolt used to fasten a pole to the spider. 63

pole-cell insulation (salient pole) (rotating machinery). Insulation that constitutes the liner between the field pole coil and the salient pole body. *See:* **rotor (rotating machinery).** 63

pole-changing winding (rotating machinery). A winding so designed that the number of poles can be changed by simple changes in the coil connections at the winding terminals. *See also:* **rotor (rotating machinery); stator.** 63

pole end plate (rotating machinery). A plate or structure at each end of a laminated pole to maintain axial pressure on the laminations. *See also:* **asynchronous machine; direct-current commutating machine.** 63

pole face (rotating machinery). The surface of the pole shoe or nonsalient pole forming one boundary of the air gap. *See also:* **asynchronous machine; direct-current commutating machine.** 63

pole-face bevel (rotating machinery). The portion of the pole shoe that is beveled so as to increase the length of the radial air gap. *See also:* **asynchronous machine; direct-current commutating machine.** 63

pole face, relay. *See:* **relay pole face.**

pole-face shaping (rotating machinery). The contour of the pole shoe that is shaped other than by being beveled, so as to produce nonuniform radial length of the air gap. *See:* **rotor (rotating machinery); stator.** 63

pole fixture. A structure installed in lieu of a single pole to increase the strength of a pole line or to provide better support for attachments than would be provided by a single pole. Examples are *A* fixtures, *H* fixtures, etcetera. *See also:* **open wire.** 328

pole guy. A tension member having one end securely anchored and the other end attached to a pole or other structure that it supports against overturning. *See also:* **tower.** 64

pole line. A series of poles arranged to support conductors above the surface of the ground; and the structures and conductors supported thereon. *See also:* **open wire.**
328

pole, offset marker (conductor stringing equipment). A small diameter, lightweight pole with a marking device attached to one end, having sufficient length to enable a workman to mark the conductor directly below him from a position on the bridge or arm of the structure. This device is normally utilized when it is necessary to mark the conductor immediately after completion of initial sag for "offset clipping" required to balance the horizontal forces on the structure. *Syn.* **offset marker; marker.** 45

pole piece. A piece or an assembly of pieces of ferromagnetic material forming one end of a magnet and so shaped as to appreciably control the distribution of the magnetic flux in the adjacent medium. 210

pole pitch (rotating machinery). The peripheral distance between corresponding points on two consecutive poles; also expressed as a number of slot positions. *See:* **armature; rotor (rotating machinery); stator.** 63

pole shoe. The portion of a field pole facing the armature that serves to shape the air gap and control its reluctance. *Note:* For round-rotor fields, the effective pole shoe includes the teeth that hold the field coils and wedges in place. *See:* **field pole; rotor (rotating machinery); stator.**
63

pole slipping (rotating machinery). The process of the secondary member of a synchronous machine slipping one pole pitch with respect to the primary magnetic flux.
63

pole steps. Devices attached to the side of a pole, conveniently spaced to provide a means for climbing the pole. *See also:* **tower.** 64

pole tip (rotating machinery). The leading or trailing extremity of the pole shoe. *See:* **rotor (rotating machinery); stator.** 63

pole-type transformer. A transformer which is suitable for mounting on a pole or similar structure. 53

pole-unit mechanism (switching device). That part of the mechanism that actuates the moving contacts of one pole.
103, 202

policy (control system). *See:* **control law.**

poling (1) (general). The adjustment of polarity. Specifically, in wire line practice, it signifies the use of transpositions between transposition sections of open wire or be-

tween lengths of cable to cause the residual crosstalk couplings in individual sections or lengths to oppose one another. *See also:* **open wire.** 328

(2) (ferroelectric material). The process by which a direct-current electric field exceeding the coercive field is applied to a multidomain ferroelectric to produce a net remanent polarization. *See also:* **coercive electric field; ferroelectric domain; polarization; remanent polarization.**
80

polishing (electroplating). The smoothing of a metal surface by means of abrasive particles attached by adhesive to the surface of wheels or belts. *See also:* **electroplating.**
328

poll (data transmission). A flexible, systematic method, centrally controlled for permitting stations on a multipoint circuit to transmit without contending for the line. 59

polyphase (relay). A descriptive term indicating that the relay is responsive to polyphase alternating electric input quantities. *Note:* A multiple-unit relay with individual units responsive to single-phase electric inputs is not a polyphase relay even though the several single-phase units constitute a polyphase set. 103, 202, 60, 127

polyphase circuit. An alternating-current circuit consisting of more than two intentionally interrelated conductors that enter (or leave) a delimited region at more than two terminals of entry and that are intended to be so energized that in the steady state the alternating voltages between successive pairs of terminals of entry of the phase conductors, selected in a systematic chosen sequence, have: (1) the same period, (2) definitely related and usually equal amplitudes, and (3) definite and usually equal phase differences. If a neutral conductor exists, it is intended also that the voltages from the successive phase conductors to the neutral conductor be equal in amplitude and equally displaced in phase. *Note:* For all polyphase circuits in common use except the two-phase three-wire circuit, it is intended that the voltage amplitudes and the phase differences of the systematically chosen voltages between phase conductors be equal. For a two-phase three-wire circuit it is intended that voltages between two successive pairs of terminals be equal and have a phase difference of $\pi/2$ radians, but that the voltage between the third pair of terminals have an amplitude $(2)^{1/2}$ times as great as the other two, and a phase difference from each of the other two of $3\pi/4$ radians. *See:* **mesh connection of polyphase circuit; star connection of polyphase circuit; zig-zag connection of polyphase circuits.** *See also:* **center of distribution; network analysis.** 210

polyphase machine (rotating machinery). A machine that generates or utilizes polyphase alternating-current power. These are usually three-phase machines with three voltages displaced 120 electrical degrees with respect to each other. *See also:* **asynchronous machine.** 63

polyphase symmetrical set (1) (polyphase voltages). A symmetrical set of polyphase voltages in which the angular phase difference between successive members of the set is not zero, π radians, or a multiple thereof. The equations of **symmetrical set of polyphase voltages** represent a polyphase symmetrical set of polyphase voltages if k/m is not zero, $\frac{1}{2}$, or a multiple thereof. (The symmetrical set of voltages represented by the equations of **symmetrical set of polyphase voltages** may be said to have polyphase symmetry if k/m is not zero, $\frac{1}{2}$, or a multiple of $\frac{1}{2}$.) *Note:* This definition may be applied to a two-phase four-wire or five-wire circuit if m is considered to be 4 instead of 2. It is not applicable to a two-phase three-wire circuit. *See: **symmetrical set of polyphase voltages.** See also:* **network analysis.**   210

(2) (polyphase currents). This definition is obtained from the corresponding definitions for voltage by substituting the word current for voltage, the symbol I for E, and β for α wherever they appear. The subscripts are unaltered. *See also:* **network analysis.**   210

polyphase synchronous generator (electric installations on shipboard). A generator whose alternating-current circuits are so arranged that two or more symmetrical alternating electromotive forces with definite phase relationships are produced at its terminals. Polyphase synchronous generators are usually two-phase, producing two electromotive forces displaced 90 electrical degrees with respect to one another, or three-phase, with three electromotive forces displaced 120 electrical degrees with respect to each other. *Note:* Polyphase generators as used for marine service are generally three-phase; for special cases they may be two-phase.   3

polyplexer (radar). Equipment combining the functions of duplexing and lobe switching. *See also:* **navigation.**   13, 187

pondage (hydroelectric generating station). Hydro reserve and limited storage capacity that provides only daily or weekly regulation of streamflow.   112

pondage station. A hydroelectric generating station with storage sufficient only for daily or weekend regulation of flow. *See also:* **generating station.**   64

p-on-n solar cells (photovoltaic power system). Photovoltaic energy-conversion cells in which a base of n-type silicon (having fixed positive holes in a silicon lattice and electrons that are free to move) is overlaid with a surface layer of p-type silicon (having fixed electrons in a silicon lattice and positive holes that are free to move). *See also:* **photovoltaic power system; solar cells (photovoltaic power system).**   186

pool-cathode mercury-arc converter. A frequency converter using a mercury-arc pool-type discharge device. *See also:* **industrial electronics.**   14, 114

pool rectifier. A gas-filled rectifier with a pool cathode, usually mercury. *See also:* **gas-filled rectifier.**   244, 190

pool tube. A gas tube with a pool cathode. *See also:* **electronic controller.**   190

population, conceptual (results from a measurement process). The set of measurements that would result from infinite repetition of a measurement process in a state of statistical control.   115

population inversion (laser-maser). A nonequilibrium condition of a system of weakly interacting particles (electronics, atoms, molecules, or ions) which exists when more than one-half of the particles occupy the higher of two energy states.   363

pores (electroplating). Micro discontinuities in a metal coating that extend through to the base metal or underlying coating. *See also:* **electroplating.**   328

port (1) (electronic devices or networks). A place of access to a device or network where energy may be supplied or withdrawn or where the device or network variables may be observed or measured. *Notes:* (A) In any particular case, the ports are determined by the way the device is used and not by its structure alone. (B) The terminal pair is a special case of a port. (C) In the case of a waveguide or transmission line, a port is characterized by a specified mode of propagation and a specified reference plane. (D) At each place of access, a separate port is assigned to each significant independent mode of propagation. (E) In frequency changing systems, a separate port is also assigned to each significant independent frequency response. *See also:* **network analysis; optoelectronic device; waveguide.**   191, 190, 185

(2) (rotating machinery). An opening for the intake or discharge of ventilating air.   63

portable battery. A storage battery designed for convenient transportation. *See also:* **battery (primary or secondary).**   328

portable concentric mine cable. A double-conductor cable with one conductor located at the center and with the other conductor strands located concentric to the center conductor with rubber or synthetic insulation between conductors and over the outer conductor. *See also:* **mine feeder circuit.**   328

portable lighting. Lighting involving lighting equipment designed for manual portability. *See also:* **general lighting.**   167

portable luminaire. A lighting unit that is not permanently fixed in place. *See also:* **luminaire.**   167

portable mine blower. A motor-driven blower to provide secondary ventilation into spaces inadequately ventilated by the main ventilating system and with the air directed to such spaces through a duct.   328

portable mine cable. An extra-flexible cable, used for connecting mobile or stationary equipment in mines to a source of electric energy when permanent wiring is prohibited or impracticable. *See also:* **mine feeder circuit.**   328

portable mining-type rectifier transformer. A rectifier transformer that is suitable for transporting on skids or wheels in the restrictive areas of mines. *See also:* **rectifier transformer.**   258

portable parallel duplex mine cable. A double or triple-conductor cable with conductors laid side by side without twisting, with rubber or synthetic insulation between conductors and around the whole. The third conductor, when present, is a safety ground wire. *See also:* **mine feeder circuit.**   328

portable shunts (electric power systems). Instrument shunts with insulating bases that may be laid on or fastened to any flat surface. *Note:* They may be used also for switchboard applications where the current is relatively low and connection bars are not used.   205

portable standard meter. A portable form of meter principally used as a standard for testing other meters. It is usually provided with several current and voltage ranges and with dials indicating revolutions and fractions of a revolution of the rotor. *See also:* **electricity meter (meter).**   212

portable station (mobile communication). A mobile station designed to be carried by or on a person. **Personal** or **pocket** stations are special classes of portable stations. *See also:*

mobile communication system. 181

portable traffic-control light. A signaling light producing a controllable distinctive signal for purposes of directing aircraft operations in the vicinity of an aerodrome.
167

portable transmitter. A transmitter that can be carried on a person and may or may not be operated while in motion. *Notes:* (1) This has been called a transportable transmitter, but the designation portable is preferred. (2) This includes the class of so-called **walkie-talkies, handy-talkies,** and **personal** transmitters. *See also:* **radio transmission; radio transmitter; transportable transmitter.** 211, 111

port difference (hybrid). A port that yields an output proportional to the difference of the electric field quantities existing at two other ports of the hybrid. *See also:* **waveguide.** 185

port signal (data transmission). The signal used to telemeter the real time occurrence of polarity reversals of a power voltage or current. The signal may be a pulse train, a square voltage wave, an FSK (frequency-shift keying) tone or an FSK carrier wave. Use is generally for frequency or phase angle telemetering. 59

port sum (hybrid). A port that yields an output proportional to the sum of the electric-field quantities existing at two other ports of the hybrid. *See also:* **waveguide.** 185

position. *See:* **punch position, sign position.**

position (navigation). The location of a point with respect to a specific or implied coordinate system. *See also:* **navigation.** 187, 13

positional crosstalk (multibeam cathode-ray tubes). The variation in the path followed by an one electron beam as the result of a change impressed on any other beam in the tube. 190, 125

positional notation (1) (general). A number representation that makes use of an ordered set of digits, such that the value contributed by each digit depends on its position as well as on the digit value. *Note:* The Roman numeral system for example, does not use positional notation. *See:* **binary system; binary-coded-decimal system; biquinary system; decimal system; Gray code.**
(2) One of the schemes for representing numbers, characterized by the arrangement of digits in sequence, with the understanding that successive digits are to be interpreted as coefficients of successive powers of an integer called the base (or radix) of the number system. *Notes:* (A) In the binary number system the successive digits are interpreted as coefficients of the successive powers of the base two, just as in the decimal number system they relate to successive powers of the base ten. (B) In the ordinary number systems each digit is a character that stands for zero or for a positive integer smaller than the base. (C) The names of the number systems with bases from 2 to 20 are: binary, ternary, quaternary, quinary, senary, septenary, octonary (also octal), novenary, decimal, unidecimal, duodecimal, terdenary, quaterdenary, quindenary, sexadecimal (also hexadecimal), septendecimal, octodenary, novendenary, and vicenary. The sexagenary number system has the base 60. The commonly used alternative of saying **base-3, base-4,** etcetera, in place of ternary, quaternary, etcetera, has the advantage of uniformity and clarity. (D) In the most common form of positional nota-

tion the expression $\pm\, a_n a_{n-1} \cdots a_2 a_1 a_0 \cdot a_{-1} a_{-2} \cdots a_{-m}$ is an abbreviation for the sum

$$\pm \sum_{i=-m}^{n} a_i r^i$$

where the point separates the positive powers from the negative powers, the a_i are integers ($0 \le a_i < r$) called digits, and r is an integer, greater than one, called the base (or radix). *See also:* **base; radix.**
(3) A number-representation system having the property that each number is represented by a sequence of characters such that successive characters of the sequence represent integral coefficients of accumulated products of a sequence of integers (or reciprocals of integers) and such that the sum of these products, each multiplied by its coefficient, equals the number. Each occurrence of a given character represents the same coefficient value. *Note:* The biquinary system is an example of (3).
(4) A number-representation system such that if the representations are arranged vertically in order of magnitude with digits of like significance in the same column, then each column of digits consists of recurring identical cycles (for numbers sufficiently large in absolute value) whose length is an integral multiple of the cycle length in the column containing the next-less-significant digits. *Note:* (2), (3), and (4) are not mutually exclusive. The biquinary system is an example of (3) and (4); whereas the Gray code system is an example of (4) only. The binary and decimal systems are examples of (2), (3), and (4).
235, 210, 77

positional response (close-talking pressure microphone). The response-frequency measurements conducted with the principal axis of a microphone collinear with the axis of the artificial voice and the combination of microphone and artificial voice placed at various angles to the horizontal plane. *Note:* Variations in positional response of carbon microphones may be due to gravitational forces. *See also:* **close-talking pressure-type microphone.** 249

position-control system (industrial control). A control system that attempts to establish and/or maintain an exact correspondence between the reference input and the directly controlled variable, namely physical position. *See also:* **control system, feedback.** 219, 206

position index. The position index is a factor that represents the relative average luminance (photometric brightness), for a sensation at the borderline between comfort and discomfort for a source located anywhere within the visual field. *See also:* **inverse-square law (illuminating engineering).** 167

position indicator (elevators). A device that indicates the position of the elevator car in the hoistway. It is called a hall position indicator when placed at a landing or a car position indicator when placed in the car. *See:* **control.**
328

position influence (electric instrument). The change in the indication of an instrument that is caused solely by a position departure from the normal operating position. *Note:* Unless otherwise specified, the maximum change in the recorded value caused solely by an inclination in the most unfavorable direction from the normal operating position. *See also:* **accuracy rating (instrument).** 280, 294, 295

positioning-control system (numerically controlled ma-

chines). A system in which the controlled motion is required only to reach a given end point, with no path control during the transition from one end point to the next.
224, 207

position lights. Aircraft aeronautical lights that form the basic or internationally recognized navigation light system. *Note:* The system is composed of a red light showing from dead ahead to 110 degrees to the left, a green light showing from dead ahead to 110 degrees to the right, and a white light showing to the rear through 140 degrees. Position lights are also called **navigation lights.** 167

position light signal. A fixed signal in which the indications are given by the position of two or more lights. 328

position of the effective short (microwave switching tube). The distance between a specified reference plane and the apparent short-circuit of the fired tube in its mount. *See:* **gas tube.** 190, 125

position sensor or position transducer (numerically controlled machines). A device for measuring a position and converting this measurement into a form convenient for transmission. 224, 207

position stopping (industrial control). A control function that provides for stopping the driven equipment at a preselected position. *See:* **electric controller.** 225, 206

position-type telemeter. *See:* **ratio-type telemeter.**

positive after-potential (electrobiology). Relatively prolonged positivity that follows the negative after-potential. *See also:* **contact potential.** 192

positive column (gas tube). The luminous glow, often striated, in a glow-discharge cold-cathode tube between the Faraday dark space and the anode. *See:* **gas tube.** 190

positive conductor. A conductor connected to the positive terminal of a source of supply. *See also:* **center of distribution.** 64

positive creep effect (semiconductor rectifier). The gradual increase in reverse current with time, that may occur when a direct-current reverse voltage is applied to a semiconductor rectifier cell. *See also:* **rectification.** 237, 66

positive electrode (1) (primary cell). The cathode when the cell is discharging. The positive terminal is connected to the positive electrode. *See also:* **electrolytic cell.** 328 **(2) (metallic rectifier).** The electrode to which the forward current flows within the metallic rectifying cell. *See also:* **rectification.** 328

positive feedback. The process by which a part of the power in the output circuit of an amplifying device reacts upon the input circuit in such a manner as to reinforce the initial power, thereby increasing the amplification. *See also:* **control system, feedback; feedback.** 328

positive grid. *See:* **retarding field (positive-grid) oscillator.**

positive-grid oscillator tube (Barkhausen tube). A triode operating under oscillating conditions such that the quiescent voltage of the grid is more positive than that of either of the other electrodes. 190

positive logic (logic diagram). When the 1-state of the variables is defined as the more positive of the two possible values, positive logic is used in the diagram. 88

positive matrix (positive). A matrix with a surface like that which is to be ultimately produced by electroforming. *See also:* **electroforming.** 328

positive modulation (in an amplitude-modulation television

system). That form of modulation in which an increase in brightness corresponds to an increase in transmitted power. *See also:* **television.** 328

positive nonconducting period (rectifier element). The nonconducting part of an alternating-voltage cycle during which the anode has a positive potential with respect to the cathode. *See also:* **power rectifier; rectification.** 208

positive noninterfering and successive fire-alarm system. A manual fire-alarm system employing stations and circuits such that, in the event of simultaneous operation of several stations, one of the operated stations will take control of the circuit, transmit its full signal, and then release the circuit for successive transmission by other stations that are held inoperative until they gain circuit control. *See also:* **protective signaling.** 328

positive-phase-sequence reactance (rotating machinery). The quotient of the reactive fundamental component of the positive-sequence primary voltage due to the sinusoidal positive-sequence primary current of rated frequency, and the value of this current, the machine running at rated speed. *See also:* **asynchronous machine.** 63

positive-phase-sequence relay. A relay that responds to the positive-phase-sequence component of a polyphase input quantity. *See also:* **relay.** 60, 127

positive-phase-sequence resistance (rotating machinery). The quotient of the in-phase component of positive-sequence primary voltage corresponding to direct load losses in the primary winding and stray load losses due to sinusoidal positive-sequence primary current, and the value of this current, the machine running at rated speed. 63

positive-phase-sequence symmetrical components (of an unsymmetrical set of polyphase voltages or currents of m phases). The set of symmetrical components that have the first phase sequence. That is, the angular phase lag from the first member of the set to the second, from every other member of the set to the succeeding one, and from the last member to the first, is equal to the characteristic angular phase difference, or $2\pi/m$ radians. The members of this set will reach their positive maxima uniformly in their designated order. The positive-phase-sequence symmetrical components for a three-phase set of unbalanced sinusoidal voltages ($m = 3$), having the primitive period, are represented by the equations

$$e_{a1} = (2)^{1/2} E_{a1} \cos(\omega t + \alpha_{a1})$$

$$e_{b1} = (2)^{1/2} E_{a1} \cos\left(\omega t + \alpha_{a1} - \frac{2\pi}{3}\right)$$

$$e_{c1} = (2)^{1/2} E_{a1} \cos\left(\omega t + \alpha_{a1} - \frac{4\pi}{3}\right)$$

derived from the equation of **symmetrical components of a set of polyphase (alternating) voltages.** Since in this case $r = 1$ for every component (of 1st harmonic) the third subscript is omitted. Then k is 1 for 1st sequence and s takes on the algebraic values 1, 2, and 3 corresponding to phases a, b, and c. The sequence of maxima occurs in the order a, b, c. *See also:* **network analysis.** 210

positive plate (storage cell). The grid and active material from which current flows to the external circuit when the battery is discharging. *See also:* **battery (primary or secondary).** 328

positive-polarity lightning stroke. A stroke resulting from a positively charged cloud that lowers positive charge to the earth. *See also:* **direct-stroke protection (lightning).**
          64

positive-sequence impedance. The quotient of that component of positive-sequence rated-frequency voltage, assumed to be sinusoidal, that is due to the positive-sequence component of current, divided by the positive-sequence component of current. *See also:* **asynchronous machine.**
          63

positive-sequence resistance. That value of resistance that, when multiplied by the square of the fundamental positive-sequence rated-frequency component of armature current and by the number of phases, is equal to the sum of the copper loss in the armature and the load loss resulting from that current, when the machine is operating at rated speed. Positive-sequence resistance is normally that corresponding to rated armature current. *Note:* Inasmuch as the load loss may not vary as the square of the current, the positive-sequence resistance applies accurately only near the current for which it was determined.   328

positive terminal (battery). The terminal from which the positive electric charge flows through the external circuit to the negative terminal when the cell discharges. *Note:* The flow of electrons in the external circuit is to the positive terminal and from the negative terminal. *See also:* **battery (primary or secondary).**   328

post (waveguide). A cylindrical rod placed in a transverse plane of the waveguide and behaving substantially as a shunt susceptance. *See also:* **waveguide.**   179

post-accelerating (deflection) electrode (intensifier electrode). An electrode to which a potential is applied to produce post-acceleration. *See also:* **electrode (electron tube).**
          90

post acceleration (electron-beam tube). Acceleration of the beam electrons after deflection.   190, 125

post-deflection acceleration. *See:* **post-acceleration.**

post-dialing delay (telephone switching systems). In an automatic telecommunication system that time interval between the receipt of the last called address digit from the calling station and the application of ringing to the called station.   55

post emphasis. *See:* **deemphasis.**

post equalization. *See:* **deemphasis.**

post insulator. An insulator of generally columnar shape, having means for direct and rigid mounting. *See also:* **insulator.**   261

post-mortem (computing systems). Pertaining to the analysis of an operation after its completion.
          255, 77, 54

post-mortem dump (computing systems). A static dump used for debugging purposes that is performed at the end of a machine run.   255, 77

post processor (numerically controlled machines). A set of computer instructions that transform tool centerline data into machine motion commands using the proper tape code and format required by a specific machine/control system. Instructions such as feedrate calculations, spindle-speed calculations, and auxiliary-function commands may be included.   224, 207

post puller. An electric vehicle having a powered drum handling wire rope used to pull mine props, after coal has

been removed, for the recovery of the timber.   328

post, waveguide (waveguide components). A cylindrical rod placed in a transverse plane of the waveguide and behaving substantially as a shunt susceptance.   166

potential diagram (electrode-optical system). A diagram showing the equipotential curves in a plane of symmetry of an electron-optical system. *See also;* **electron optics.**
          244, 190

potential difference, electrostatic. *See:* **electrostatic potential difference.**

potential, electrostatic. *See:* **electrostatic potential.**

potential energy (of a body or of a system of bodies, in a given configuration with respect to an arbitrarily chosen reference configuration). The work required to bring the system from an arbitrarily chosen reference configuration to the given configuration without change in other energy of the system.   210

potential false-proceed operation. The existence of a condition of vehicle or roadway apparatus in an automatic train control or cab-signal installation under which a false-proceed operation would have occurred had a vehicle approached or entered a section where normally a restrictive operation would occur.   328

potential gradient. A vector of which the direction is normal to the equipotential surface, in the direction of decreasing potential, and of which the magnitude gives the rate of variation of the potential.   244, 59

potential profile. A plot of potential as a function of distance along a specified path. *See also:* **ground.**   313

potential transformer (voltage transformer). An instrument transformer that is intended to have its primary winding connected in shunt with a power-supply circuit, the voltage of which is to be measured or controlled. *See also:* **instrument transformer.**   212

potential transformer, cascade-type. A single high-voltage line-terminal potential transformer with the primary winding distributed on several cores with the cores electromagnetically coupled by coupling windings and the secondary winding on the core at the neutral end of the high-voltage winding. Each core is insulated from the other cores and is maintained at a fixed potential with respect to ground and the line-to-ground voltage. *See also:* **instrument transformer.**   305, 203

potential transformer, double-secondary. One that has two secondary windings on the same magnetic circuit insulated from each other and the primary. Either or both of the secondary windings may be used for measurement or control. *See also:* **instrument transformer.**   305, 203

potential transformer, fused-type. One that is provided with the means for mounting a fuse, or fuses, as an integral part of the transformer in series with the primary winding. *See also:* **instrument transformer.**   305, 203

potential transformer, grounded-neutral terminal type. One that has the neutral end of the high-voltage winding connected to the case or mounting base. *See also:* **instrument transformer.**   305, 203

potential transformer, insulated-neutral terminal type. One that has the neutral end of the high-voltage winding insulated from the case or base and connected to a terminal that provides insulation for a lower-voltage insulation class than required for the rated insulation class of the transformer. *See also:* **instrument transformer.**   305, 203

potential transformer, single-high-voltage line terminal. One that has the line end of the primary winding connected to a terminal insulated from ground for the rated insulation class. The neutral end of the winding may be (1) insulated from ground but for a lower insulation class than the line end (insulated neutral) or (2) connected to the case or base (grounded neutral). *See also:* **instrument transformer.**   305, 203

potential transformer, two-high-voltage line terminals. One that has both ends of the high-voltage winding connected to separate terminals that are insulated from each other, and from other parts of the transformer, for the rated insulation class of the transformer. *See also:* **instrument transformer.**   305, 203

potentiometer (1) (general). A three-terminal rheostat, or a resistor with one or more adjustable sliding contacts, that functions as an adjustable voltage divider. *See:* **potentiometer, function; normal linearity; potentiometer, multiplier; potentiometer, parameter.** *See also:* **attenuator; electronic analog computer.**   9, 206, 77
(2) (measurement techniques). An instrument for measuring an unknown electromotive force or potential difference by balancing it, wholly or in part, by a known potential difference produced by the flow of known currents in a network of circuits of known electrical constants. *See:* **instrument.**   328
(3) (analog computer). A resistive element with two end terminals and a movable contact. *See also:* **attenuator.**   9

potentiometer, digital coefficient (analog computer). *See:* **digital coefficient potentiometer (hybrid computer linkage component).**

potentiometer, follow-up. A servo potentiometer that generates the signal for comparison with the input signal. *See also:* **electronic analog computer.**   9

potentiometer, function. A multiplier potentiometer in which the voltage at the movable contact follows a prescribed functional relationship to the displacement of the contact. *See:* **linearity.** *See also:* **electronic analog computer.**   9, 77

potentiometer granularity. The physical inability of a potentiometer to produce an output voltage that varies in other than discrete steps, due either to contacting individual turns of wire in a wire-wound potentiometer or to discrete irregularities of the resistance element of composition or film potentiometers. *See also:* **electronic analog computer.**   9, 10

potentiometer, grounded. A potentiometer with one end terminal attached directly to ground. *See also:* **electronic analog computer.**   9, 10

potentiometer, linear. A potentiometer in which the voltage at a movable contact is a linear function of the displacement of the contact. *See:* **linearity.** *See also:* **electronic analog computer.**   9

potentiometer, manual (analog computer). A potentiometer which is set by hand. *Syn.* **hand set potentiometer.**   9

potentiometer, multiplier. Any of the ganged potentiometers of a servo multiplier that permit the multiplication of one variable by a second variable. *See also:* **electronic analog computer.**   9, 77, 10

potentiometer, parameter (scale-factor potentiometer) (coefficient potentiometer). A potentiometer used in an analog computer to represent a problem parameter such as a coefficient or a scale factor. *See also:* **electronic analog computer.**   9, 77, 10

potentiometer, servo. A potentiometer driven by a positional servomechanism. *See also:* **electronic analog computer.**   9, 10

potentiometer set. A computer control state that supplies the same operating potentiometer loading as under computing conditions and thus allows correct potentiometer adjustment. *See also:* **problem check.**   9, 10

potentiometer, sine-cosine. A function potentiometer with movable contacts attached to a rotating shaft so that the voltages appearing at the contacts are proportional to the sine and cosine of the angle of rotation of the shaft, the angle being measured from a fixed referenced position. *See also:* **electronic analog computer.**   9

potentiometer, tapered. A function potentiometer that achieves a prescribed functional relationship by means of a nonuniform winding. *See also:* **electronic analog computer.**   9

potentiometer, tapped. A potentiometer, usually a servo potentiometer, that has a number of fixed contacts (or taps) to the resistance element in addition to the end and movable contacts. *See also:* **electronic analog computer.**   9

potentiometer, ungrounded. A potentiometer with neither end terminal attached directly to ground. *See also:* **electronic analog computer.**   9, 10

pothead. A device that seals the end of a cable and provides insulated egress for the conductor or conductors.   288, 289, 323

pothead body. The part of a pothead that joins the entrance fitting to the insulator or to the insulator lid. *See also:* **pothead; transformer.**   288, 289, 323

pothead bracket or mounting plate. The part of the pothead used to attach the pothead to the supporting structure. *See also:* **pothead; transformer.**   323

pothead bracket or mounting-plate insulator. An insulator used to insulate the pothead from the supporting structure for the purpose of controlling cable sheath currents. *See also:* **pothead; transformer.**   323

pothead entrance fitting. A fitting used to seal or attach the cable sheath, armor, or other coverings to the pothead. *See also:* **pothead; transformer.**   323

pothead insulator. An insulator used to insulate and protect each conductor passing through the pothead. *See also:* **pothead; transformer.**   288, 289, 323

pothead insulator lid. The part of a multi-conductor pothead used to join two or more insulators to the body. *See also:* **pothead; transformer.**   288, 289, 323

pothead mounting plate. The part of the pothead used to attach the pothead to the supporting structure. *See also:* **transformer.**   288, 289

pothead mounting-plate insulator. An insulator used to insulate the pothead from the supporting structure for the purpose of controlling cable sheath currents. *See also:* **transformer.**   288, 289

pothead sheath insulator. An insulator used to insulate an electrically conductive cable sheath or armor from the metallic parts of the pothead in contact with the supporting structure for the purpose of controlling cable sheath currents. *See also:* **pothead.**   323

Potier reactance (rotating machinery). An equivalent reactance used in place of the primary leakage reactance to
calculate the excitation on load by means of the Potier
method. *Note:* It takes into account the additional leakage
of the excitation winding on load and in the overexcited
region; it is greater than the real value of the primary
leakage reactance. It is useful for the calculation of excitation of the machine at other loads and power factors.[†]
The height of a Potier reactance triangle determines the
reactance drop, and the reactance X_P is equal to the reactance drop divided by the current. The value of Potier
reactance is that obtained from the no-load normal-frequency saturation curve; and normally with the excitation
for rated voltage and current at zero power factor (overexcited), and at rated frequency. Approximate values of
Potier reactance may be obtained from test load excitations at loads differing from rated load, and at power
factors other than zero.

[†] The excitation results in the range from zero power factor overexcited to unity power factor are close enough to the test values for
most practical applications   63

poured joint (power cable joints). A joint insulated by the
means of a hot or cold poured insulating medium which
solidifies.   34

power (1) (noun). The time rate of transferring or transforming energy.   210
(2) (adjective). A general term used, by reason of specific
physical or electrical characteristics, to denote application
or restriction or both, to generating stations, switching
stations, or substations. The term may also denote use or
application to energy purposes as contrasted with use for
control purposes.   103, 202
(3) (generating station). The rate (in kilowatts) of generating, transferring, or using energy.   112
(4) (laser-maser) (ϕ). The time rate at which energy is
emitted, transferred, or received; usually expressed in
watts (or in joules per second).   363
power, active (polyphase circuit) (power[†]). At the terminals
of entry of a polyphase circuit into a delimited region, the
algebraic sum of the active powers for the individual terminals of entry when the voltages are all determined with
respect to the same arbitrarily selected common reference
point in the boundary surface (which may be the neutral
terminal of entry). *Notes:* (1) The active power for each
terminal of entry is determined by considering each conductor and the common reference point as a single-phase
two-wire circuit and finding the active power for each in
accordance with the definition of **power, active (single-
phase two-wire circuit).** If the voltages and currents are
sinusoidal and of the same period, the active power P for
a three-phase circuit is given by

$$P = E_a I_a \cos(\alpha_a - \beta_a) + E_b I_b \cos(\alpha_b - \beta_b) + E_c I_c \cos(\alpha_c - \beta_c)$$

where the symbols have the same meaning as in **power,
instantaneous (polyphase circuit).** (2) If there is no neutral
conductor and the common point for voltage measurement
is selected as one of the phase terminals of entry, the expression will be changed in the same way as that for **power,
instantaneous (polyphase circuit).** (3) If both the voltages

and the currents in the preceding equations constitute
symmetrical sets of the same phase sequences

$$P = 3E_a I_a \cos(\alpha_a - \beta_a).$$

(4) In general the active power P at the $(m + 1)$ terminals
of entry of a polyphase circuit of m phases to a delimited
region, when one of the terminals is the neutral terminal
of entry, is expressed by the equation

$$P = \sum_{s=1}^{s=m} \sum_{r=1}^{r=\infty} E_{sr} I_{sr} \cos(\alpha_{sr} - \beta_{sr})$$

where E_{sr} is the root-mean-square amplitude of the rth
harmonic of the voltage e_s, from phase conductor to neutral. I_{sr} is the root-mean-square amplitude of the rth
harmonic of the current i_s through terminal s. α_{sr} is the
phase angle of the rth harmonic of e_s with respect to a
common reference. β_{sr} is the phase angle of the rth harmonic of i_s with respect to the same reference as the
voltages. The indexes s and r have the same meaning as
in **power, instantaneous (polyphase circuit).** (5) The active
power can also be stated in terms of the root-mean-square
amplitudes of the symmetrical components of the voltages
and currents as

$$P = m \sum_{k=0}^{k=m-1} \sum_{r=1}^{r=\infty} E_{kr} I_{kr} \cos(\alpha_{kr} - \beta_{kr})$$

where m is the number of phase conductors, k denotes the
number of the symmetrical component, and r denotes the
number of the harmonic component. (6) When the voltages and currents are quasi-periodic and the amplitudes
of the voltages and currents are slowly varying, the active
power for the circuit of each conductor may be determined
for this condition as in **power, active (single-phase two-wire
circuit).** The active power for the polyphase circuit is the
sum of the active power values for the individual conductors. The active power is also the time average of the instantaneous power for the polyphase circuit. (7) Mathematically the active power at any time t_0 is

$$P = \frac{1}{T} \int_{t_0 - T/2}^{t_0 + T/2} p \, dt$$

where p is the instantaneous power and T is the period.
This formulation may be used when the voltage and current are periodic or quasi-periodic so that the period is
defined. The active power is expressed in watts when the
voltages are in volts and the currents in amperes.

[†] When it is clear that average power and not instantaneous power
is meant, **power** is often used for **active power**   210

power, active (single-phase two-wire circuit) (power)[†]. At
the terminals of entry of a single-phase, two-wire circuit
into a delimited region, when the voltage and current are
periodic or quasi-periodic, the time average of the values
of the instantaneous power, the average being taken over
one period. *Notes:* (1) Mathematically, the active power
P at a time t_0 is given by the equation

$$P = \frac{1}{T} \int_{t_0 - T/2}^{t_0 + T/2} p \, dt$$

where T is the period, and p is the instantaneous power.

(2) If both the voltage and current are sinusoidal and of the same period the active power P is given by

$$P = EI \cos(\alpha - \beta)$$

in which the symbols have the same meaning as in **power, instantaneous (two-wire circuit)**. (3) If both the voltage and current are sinusoidal, the active power P is also equal to the real part of the product of the phasor voltage and the conjugate of the phasor current, or to the real part of the product of the conjugate of the phasor voltage and the phasor current. Thus

$$P = \text{Re } \mathbf{EI}^*$$
$$= \text{Re } \mathbf{E}^*\mathbf{I}$$
$$= \frac{1}{2}[\mathbf{EI}^* + \mathbf{E}^*\mathbf{I}]$$

in which $\mathbf{E}$ and $\mathbf{I}$ are the root-mean-square phasor voltage and root-mean-square phasor current, respectively (see **phasor quantity**), and the * denotes the conjugate of the phasor to which it is applied. (4) If the voltage is an alternating voltage and the current is an alternating current (see **alternating voltage** and **alternating current**), the active power is given by the equations

$$P = E_1 I_1 \cos(\alpha_1 - \beta_1)$$
$$+ E_2 I_2 \cos(\alpha_2 - \beta_2) + \cdots$$
$$= \sum_{r=1}^{r=\infty} \mathbf{E}_r \mathbf{I}_r \cos(\alpha_r - \beta_r)$$
$$= \text{Re} \sum_{r=1}^{r=\infty} \mathbf{E}_r I_r$$
$$= \frac{1}{2} \sum_{r=1}^{r=\infty} [\mathbf{E}_r I_r + \mathbf{E}_r I_r]$$

in which r is the order of the harmonic component of the voltage (see **harmonic components (harmonics)**) and r is also the order of the harmonic component of the current. $\mathbf{E}_r$ and $\mathbf{I}_r$ are the phasors corresponding to the r th harmonic of the voltage and current, respectively. (5) If the voltage and current are quasi-periodic functions of the form given in **power, instantaneous (two-wire circuit)**, the integral over the period T will not result in the simple expressions that are obtained when E_r and I_r are constant. However, if the relative rates of change of the quantities are so small that each may be considered to be constant during any one period, but to have slightly different values in successive periods, the active power at any time t is very closely approximated by

$$P = \sum_{r=1}^{r=\infty} E_r(t) I_r(t) \cos(\alpha_r - \beta_r)$$

which is analogous to the preceding expression. When the amplitudes of voltage and current are slowly changing, the active power may be represented by this expression. (6) Active power is expressed in watts when the voltage is in volts and the current in amperes.

† When it is clear that average power and not instantaneous power is meant, **power** is often used for **active power**.   210

power amplification (1) (general). The ratio of the power level at the output terminals of an amplifier to that at the input terminals. Also called **power gain**. *See also:* **amplifier; power gain.**   111
(2) (magnetic amplifier). The product of the voltage amplification and the current amplification.   171
power amplifier. *See:* **amplifier, power.**
power, apparent (1) (rotating machinery). The product of the root-mean-square current and the root-mean-square voltage. *Note:* It is a scalar quantity equal to the magnitude of the phasor power. *See also:* **asynchronous machine.**   63
(2) (polyphase circuit). At the terminals of entry of a polyphase circuit, a scalar quantity equal to the magnitude of the vector power. *Notes:* (A) In determining the apparent power, the reference terminal for voltage measurement shall be taken as the neutral terminal of entry, if one exists, otherwise as the true neutral point. (B) If the ratios of the components of the vector power, for each of the terminals of entry, to the corresponding apparent power are the same for every terminal of entry, the total apparent power is equal to the arithmetic apparent power for the polyphase circuit; otherwise the apparent power is less than the arithmetic apparent power. (C) If the voltages have the same wave form as the corresponding currents, the apparent power is equal to the amplitude of the phasor power. (D) Apparent power is expressed in volt-amperes when the voltages are in volts and the currents in amperes.   210
(3) (single-phase two-wire circuit). At the two terminals of entry of a single-phase two-wire circuit into a delimited region, a scalar equal to the product of the root-mean-square voltage between one terminal of entry and the second terminal of entry, considered as the reference terminal, and the root-mean-square value of the current through the first terminal. *Notes:* (A) Mathematically the apparent power U is given by the equation

$$U = EI$$
$$= (\pm)(E_1^2 + E_2^2 + \cdots + E_r^2 + \cdots)^{1/2}$$
$$\times (I_1^2 + I_r^2 + \cdots + I_q^2 + \cdots)^{1/2}$$

in which E and I are the root-mean-square amplitudes of the voltage and current, respectively. E_r and I_q are the root-mean-square amplitudes of the rth harmonic of voltage and the qth harmonic of current, respectively. (B) If both the voltage and current are sinusoidal and of the same period, so that the distortion power is zero, the apparent power becomes

$$U = EI = E_1 I_1$$

in which E_1 and I_1 are the root-mean-square amplitudes of voltage and current of the primitive period. The apparent power is equal to the amplitude of the phasor power. (C) If the voltage and current are quasiperiodic and the amplitude of the voltage and current components are slowly varying, the apparent power at any instant may be taken as the value derived from the amplitudes of the components at that instant. (D) Apparent power is expressed in volt-amperes when the voltage is in volts and the current in amperes. Because apparent power has the property of magnitude only and its sign is ambiguous, it does not have a definite direction of flow. For convenience it is usually treated as positive.   210

power, auxiliary (thyristor power converter). Input power used by the thyristor converter to perform its various auxiliary functions as opposed to the power that may be flowing between the ac supply and the load. 121

power, auxiliary, general (thyristor power converter). The power required for fans or blowers, relays, breaker control, phase loss detection, etcetera. 121

power, available (1) (audio and electroacoustics). The maximum power obtainable from a given source by suitable adjustment of the load. *Note:* For a source this is equivalent to a constant sinusoidal electromotive force in series with an impedance independent of amplitude, the available power is the mean-square value of the electromotive force divided by four times the resistive part of the impedance of the source. 239, 252, 210

(2) (at a port). The maximum power that can be transferred from the port to a load. *Note:* At a specified frequency, maximum power transfer will take place when the impedance of the load is the conjugate of that of the source. The source impedance must have a positive real part. *See also:* **network analysis.** 190

(3) (of a sound field with a given object placed in it). The power that would be extracted from the acoustic medium by an ideal transducer having the same dimensions and the same orientation as the given object. The dimensions and the orientation with respect to the sound field must be specified. *Note:* The acoustic power available to an electroacoustic transducer, in a plane-wave sound field of given frequency, is the product of the free-field intensity and the effective area of the transducer. For this purpose the effective area of an electroacoustic transducer, for which the surface velocity distribution is independent of the manner of excitation of the transducer, is $\frac{1}{4}\pi$ times the product of the receiving directivity factor and the square of the wavelength of a free progressive wave in the medium. If the physical dimensions of the transducer are small in comparison with the wavelength, the directivity factor is near unity, and the effective area varies inversely as the square of the frequency. If the physical dimensions are large in comparison with the wavelength, the directivity factor is nearly proportional to the square of the frequency, and the effective area approaches the actual area of the active face of the transducer. 176

(4) (signal generators). The power at the output port supplied by the generator into a specified load impedance. *See also:* **signal generator.** 185

power, average phasor (single-phase two-wire, or polyphase circuit). A phasor of which the real component is the average active power and the imaginary component is the average reactive power. The amplitude of the phasor power is

$$S_{av} = [(P_{av})^2 + (Q_{av})^2]^{1/2}$$

in which P_{av} and Q_{av} are the active and the reactive power, respectively. 210

power, carrier-frequency, peak pulse. *See:* **peak pulse power, carrier-frequency.**

power-circuit limit switch (industrial control). A limit switch the contacts of which are connected into the power circuit. *See also:* **switch.** 206

power circuit protector. An assembly consisting of a modified low-voltage power circuit breaker, which has no direct-acting tripping devices, with a current-limiting fuse in series with the load terminals of each pole. 79, 158

power-closed car door or gate (elevators). A door or gate that is closed by a car-door or gate power closer or by a door or gate power operator. *See also:* **elevators.** 328

power coefficient (characteristic insertion loss) (attenuator). Temporary and reversible variation in decibels when input power is varied from 20 dB below full rated power or lower to full rated power after steady-state condition has been reached. 110

power, commercial. Power furnished by an electric power utility company; when available, it is usually the prime power source. 89

power control center. The location where the area-control error of a control area is computed for the purpose of controlling area generation. 94

power density. Value of the Poynting vector at a point in space. *See also:* **electromagnetic compatibility.** 199, 146

power-density spectrum (1) (electromagnetic compatibility). A plot of power density per unit frequency as a function of frequency. 199

(2) (automatic control). A measure of the power content of the periodic components of a signal, with respect to frequency, the Fourier transform of the power autocorrelation function. 56

power detection. That form of detection in which the power output of the detecting device is used to supply a substantial amount of power directly to a device such as a loudspeaker or recorder. *See also:* **detection.** 328

power dissipation, (P) (light emitting diodes). The time average product of current times voltage of the device. 162

power distribution, underground cables. *See:* **cable bedding; cable separator; base ambient temperature; aluminum-covered steel wire.**

power divider (waveguide). A device for producing a desired distribution of power at a branch point. *See also:* **waveguide.** 244, 179

power, effective radiated (mobile communication). The product in a given direction of the effective gain of the antenna in that direction over a half-wave dipole antenna, and the antenna power input. *See also:* **mobile communication system.** 181

power elevator. An elevator utilizing energy other than gravitational or manual to move the car. *See also:* **elevators.** 328

power, emergency (electric power systems). Power required by a system to make up a deficiency between the current firm power demand and the immediately available generating capability. 112

power factor (1) (general). The ratio of total watts to the total root-mean-square (RMS) volt-amperes.

$$F_P = \frac{\Sigma \text{ Watts per Phase}}{\Sigma \text{ RMS Volt-Amperes per Phase}} = \frac{\text{Active Power}}{\text{Apparent Power}}$$

Note: If the voltages have the same waveform as the corresponding currents, power factor becomes the same as phasor power factor. If the voltages and currents are sinusoidal and, for polyphase circuits, form symmetrical sets, $F_P = \cos(\alpha - \beta)$. *See also:* **asynchronous machine.**

210, 244, 186, 63

(2) (rectifier or rectifier unit) (thyristor converter). The ratio of the total watts input (total power input in watts) to the total volt-ampere input to the rectifier, rectifier unit or converter. *Notes:* (A) This definition includes the effect of harmonic components of current and voltage, the effect of phase displacement between the current and voltage, and the exciting current of the transformer. Volt-ampere is the product of root-mean square volts and root-mean-square amperes. (B) It is determined at the alternating-current line terminals of the thyristor converter or rectifier unit. *See also:* **power rectifier; rectification.** 208, 121

(3) (dielectric) (rotating machinery). The cosine of the dielectric phase angle or the sine of the dielectric loss angle.

63

(4) (outdoor apparatus bushings) (insulation). The ratio of the power dissipated in the insulation, in watts, to the product of the effective voltage and current in voltamperes, when tested under a sinusoidal voltage and prescribed conditions. *Note:* The insulation power factor is equal to the cosine of the phase angle between the voltage and the resulting current when both the voltage and current are sinusoidal. 168

power factor adjustment clause (electric power utilization). A clause in a rate schedule that provides for an adjustment in the billing if the customer's power factor varies from a specified reference. 112

power-factor angle. The angle whose cosine is the power factor. *See also:* **asynchronous machine.** 63

power factor, arithmetic. The ratio of the active power to the arithmetic apparent power. The arithmetic power factor is expressed by the equation

$$ F_{pa} = \frac{P}{U_a} $$

where F_{pa} = arithmetic power factor
P = active power
U_a = arithmetic apparent power.

Note: Normally power factor, rather than arithmetic power factor, will be specified, but in particular cases, especially when the determination of the apparent power for a polyphase circuit is impracticable with the available instruments, arithmetic power factor may be used. When arithmetic power factor and power factor differ, arithmetic power factor is the smaller. 210

power factor, coil Q. *See:* **coil Q.**

power-factor-corrected mercury-lamp ballast. A ballast of the multiple-supply type that has a power-factor-correcting device, such as a capacitor, so that the input current is at a power factor in excess of that of an otherwise comparable low-power-factor ballast design, but less than 90 percent, when the ballast is operated with center rated voltage impressed upon its input terminals and with a connected load, consisting of the appropriate reference lamp(s), operated in the position for which the ballast is designed. The minimum input power factor of such a ballast should be specifically stated. 271

power factor, dielectric. *See:* **dielectric power factor.**

power-factor influence (electric instruments). The change in the recorded value that is caused solely by a power-factor departure from a specified reference power factor maintaining constant power (or vars) at rated voltage, and not exceeding 120 percent of rated current. It is to be expressed as a percentage of the full-scale value. *See also:* **accuracy rating (instrument).** 280, 294

power-factor meter. A direct-reading instrument for measuring power factor. It is provided with a scale graduated in power factor. *See also:* **instrument.** 328

power-factor tip-up (rotating-machinery stator-coil insulation). The difference between the power-factors measured at two different designated voltages applied to an insulation system, other conditions being constant. *Notes:* (1) Used mainly as a measure of discharges, and hence of voids, within the system at the higher voltage. (2) The incremental change in power factor divided by incremental change in voltage applied to an insulation system. *See also:* **asynchronous machine.** (3) Tip-up tests may be made using dissipation factor (tan δ) instead of power factor. In this case the tip-up is often identified as Δ tan δ or delta tan delta. 22, 63

power-factor tip-up test (rotating machinery). A test applied to insulation to determine the power-factor tip-up. *See also:* **asynchronous machine.** 63

power-factor—voltage characteristic (rotating machinery stator-coil insulation). The relation between the magnitude of the applied test voltage and the measured power factor of the insulation. *Note:* The characteristic is usually shown as a curve of power factor plotted against test voltage. *See also:* **asynchronous machine.** 22, 63

power failure (electric power system). Any variation in electric power supply which causes unacceptable performance of the user's equipment. 89

power feeder. A feeder supplying principally a power or heating load. *See also:* **feeder.**

power flux density (electromagnetic). *See:* **power density.**

power-frequency current-interrupting rating (expulsion arrester). A designation of the range of the symmetrical root-mean-square fault currents of the system for which the arrester is designed to operate. An expulsion arrester is given a maximum current-interrupting rating and may also have a minimum current-interrupting rating. 2

power-frequency sparkover voltage (surge arrester). The root-mean-square value of the lowest power-frequency sinusoidal voltage that will cause sparkover when applied across the terminals of an arrester. 2

power-frequency withstand voltage (surge arrester). A specified root-mean-square test voltage at power frequency that will not cause a disruptive discharge. 2

power fuse. A fuse consisting of an assembly of a fuse support and a fuse unit or fuseholder which may or may not include the refill unit or fuse link. *Note:* The power fuse is identified by the following characteristics: (1) Dielectric withstand (basic impulse insulation level) strengths at power levels; (2) Application primarily in stations and substations; (3) Mechanical construction basically adapted to station and substation mounting. 103, 202

power gain (1) (general). The ratio of the signal power that a transducer delivers to its load to the signal power absorbed by its input circuit. *Notes:* (A) Power gain is usually expressed in decibels. (B) If more than one component is involved in the input or output, the particular components used must be specified. (C) If the output signal power is at a frequency other than the input signal power, the gain is a conversion gain.   239, 59

(2) (antenna) (A) **(referred to a specified polarization).** The power gain of an antenna, reduced by the ratio of that portion of the radiation intensity corresponding to the specified polarization to the radiation intensity.

(B) (in physical media). In a given direction and at a given point in the far field, the ratio of the power flux per unit area from an antenna to the power flux per unit area from an isotropic radiator at a specified location with the same power input as the subject antenna. *Note:* The isotropic radiator must be within the smallest sphere containing the antenna. Suggested locations are antenna terminals and points of symmetry if such exist.   111

(3) (two-port linear transducer). At a specified frequency, the ratio of (1) the signal power that the transducer delivers to a specified load, to (2) the signal power delivered to its input Port. *Note:* The power gain is not defined unless the input impedance of the transducer has a positive real part.   125, 179, 111

power, instantaneous (1) (circuit). At the terminals of entry into a delimited region the rate at which electric energy is being transmitted by the circuit into or out of the region. *Note:* Whether power into the region or out of the region is positive is a matter of convention and depends upon the selected reference direction of energy flow. *See:* **sign of power or energy.**   210

(2) (polyphase circuit). At the terminals of entry into a delimited region, the algebraic sum of the products obtained by multiplying the voltage between each terminal of entry and some arbitrarily selected common point in the boundary surface (which may be the neutral terminal of entry) by the current through the corresponding terminal of entry. *Notes:* (A) The reference direction of each current must be the same, either into or out of the delimited region. The reference polarity for each voltage must be consistently chosen, either with all the positive terminals at the terminals of entry and all negative terminals at the common reference point, or vice-versa. If the reference direction for currents is into the delimited region and the positive reference terminals for voltage are at the phase terminals of entry, the power will be positive when the energy flow is into the delimited region and negative when the flow is out of the delimited region. Reversal of either the reference direction or the reference polarity will reverse the relation between the sign of the power and the direction of energy flow. (B) When the circuit has a neutral terminal of entry, it is usual to select the neutral terminal as the common point for voltage measurement, because one of the voltages is then always zero, and, when both the currents and voltages form symmetrical polyphase sets of the same phase sequence, the average power for each single-phase circuit consisting of one phase conductor and the neutral conductor, will be the same. When the voltages and currents are sinusoidal and the voltages are measured to the neutral terminal of entry as the common point, the instantaneous power at the four points of entry of a three-phase circuit with neutral is given by

$$
\begin{aligned}
p = \ & E_a I_a \left[\cos(\alpha_a - \beta_a) + \cos(2\omega t + \alpha_a + \beta_a) \right] \\
+ \ & E_b I_b \left[\cos(\alpha_b - \beta_b) + \cos(2\omega t + \alpha_b + \beta_b) \right] \\
+ \ & E_c I_c \left[\cos(\alpha_c - \beta_c) + \cos(2\omega t + \alpha_c + \beta_c) \right]
\end{aligned}
$$

where E_a, E_b, E_c are the root-mean-square amplitudes of the voltages from the phase conductors a, b, and c, respectively, to the neutral conductor at the terminals of entry. I_a, I_b, I_c are the root-mean-square amplitudes of the currents in the phase conductors a, b, and c. α_a, α_b, α_c are the phase angles of the voltages E_a, E_b, E_c with respect to a common reference. β_a, β_b, β_c are the phase angles of the currents I_a, I_b, I_c with respect to the same reference as the voltages. (C) If there is no neutral conductor, so that there are only three terminals of entry, the point of entry of one of the phase conductors may be chosen as the common voltage point, and the voltage from that conductor to the common point becomes zero. If, in the preceding, the terminal of entry of phase conductor b is chosen as the common point E_a is replaced by E_{ab} in the first line, E_c is replaced by E_{cb} in the third line, and the second line, being zero, is omitted. (D) If both the voltages and currents in the preceding equations constitute symmetrical polyphase sets of the same phase sequence, then $p = 3 E_a I_a \cos(\alpha_a - \beta_a)$. Because this expression and similar expressions for m phases are independent of time, it follows that the instantaneous power is constant when the voltages and currents constitute polyphase symmetrical sets of the same phase sequence. (E) However, if the polyphase sets have single-phase symmetry or zero-phase symmetry rather than poly-phase symmetry, the higher frequency terms do not cancel, and the instantaneous power is not a constant. (F) In general, the instantaneous power p at the $(m + 1)$ terminals of entry of a polyphase circuit of m phases to a delimited area, when one of the terminals is that of the neutral conductor, is expressed by the equation

$$
\begin{aligned}
p = \ & \sum_{s=1}^{s=m} e_s i_s \\
= \ & \sum_{s=1}^{s=m} \sum_{r=1}^{r=\infty} \sum_{q=1}^{q=\infty} E_{sr} I_{sq} \\
& \left[\cos[(r - q)\omega t + \alpha_{sr} - \beta_{sq}] \right. \\
& \left. + \cos[(r + q)\omega t + \alpha_{sr} + \beta_{sq}] \right]
\end{aligned}
$$

where e_s is the instantaneous alternating voltage between the sth terminal of entry and the terminal of voltage reference, which may be the true neutral point, the neutral conductor, or another point in the boundary surface. i_s is the instantaneous alternating current through the sth terminal of entry. E_{sr} is the root-mean-square amplitude of the rth harmonic of voltage e_s. I_{sq} is the root-mean-square amplitude of the qth harmonic of current i_s. α_{sr} is the phase angle of the rth harmonic of e_s with respect to a common reference. β_{sq} is the phase angle of the qth harmonic of i_s with respect to the same reference as the voltages. The index s runs through the phase letters identifying the m-phase conductor of an m-phase system, a, b, c, etcetera, and then concludes with the neutral conductor n if one exists. The indexes r and q identify the order of the harmonic term in each e_s and i_s, respectively, and run through all the harmonics present in the Fourier series representation of each alternating voltage and current. If the terminal voltage reference is that of the neutral conductor, the terms for $s = n$ will vanish. If the voltages and current are quasi-periodic, of the form given in **power, instantaneous (two-wire circuit)**, this expression is still valid but E_{sr} and I_{sq} become aperiodic functions of time. (G) Instantaneous power is expressed in watts when the voltages are in volts and the currents in amperes. *See:* **single-phase symmetrical set of polyphase voltages; zero-phase symmetrical set of polyphase voltage; single-phase symmetrical set of polyphase currents; zero-phase symmetrical set of polyphase currents.** 210

(3) **(two-wire circuit)**. At the two terminals of entry into a delimited region, the product of the instantaneous voltage between one terminal of entry and the second terminal of entry, considered as the reference terminal, and the current through the first terminal. *Notes:* (A) The entire path selected for the determination of each voltage must lie in the boundary surface of the delimited region or be so selected that the voltage is the same as that along such a path. (B) Mathematically the instantaneous power p is given by $p = ei$ in which e is the voltage between the first terminal of entry and the second (reference) terminal of entry and i is the current through the first terminal of entry in the reference direction. (C) If both the voltage and current are sinusoidal and of the same period, the instantaneous power at any instant t is given by the equation

$$
\begin{aligned}
p = ei &= [(2)^{1/2} E \cos(\omega t + \alpha)] \\
&\times [(2)^{1/2} I \cos(\omega t + \beta)] \\
&= 2EI \cos(\omega t + \alpha)\cos(\omega t + \beta) \\
&= EI[\cos(\alpha - \beta) + \cos(2\omega t + \alpha + \beta)]
\end{aligned}
$$

in which E and I are the root-mean-square amplitudes of voltage and current, respectively, and α and β are the phase angles of the voltage and current, respectively, from the same reference. (D) If the voltage is an alternating voltage and the current is an alternating current of the same primitive period (see **alternating voltage, alternating current**, and **(period (primitive period) of a function)**), the instantaneous power is given by the equation

$$
\begin{aligned}
p = ei \\
&= E_1 I_1 [\cos(\alpha_1 - \beta_1) \\
&+ \cos(2\omega t + \alpha_1 + \beta_1)] \\
&+ E_2 I_2 [\cos(\omega t + \alpha_2 - \beta_1) \\
&+ \cos(3\omega t + \alpha_2 + \beta_1)] \\
&+ E_1 I_2 [\cos(\omega t - \alpha_1 + \beta_2) \\
&+ \cos(3\omega t + \alpha_1 + \beta_2)] \\
&+ E_2 I_2 [\cos(\alpha_2 - \beta_2) \\
&+ \cos(4\omega t + \alpha_2 + \beta_2)] \\
&+ \cdots
\end{aligned}
$$

This equation can be written conveniently as a double summation

$$
\begin{aligned}
p = \sum_{r=1}^{r=\infty} \sum_{q=1}^{q=\infty} E_r I_q \{&\cos[(r - q)\omega t + \alpha_r - \beta_q] \\
&+ \cos[(r + q)\omega t + \alpha_r - \beta_q]\}
\end{aligned}
$$

in which r is the order of the harmonic component of the voltage and q is the order of the harmonic component of the current (see **harmonic components (harmonics)**), and E, I, α, and β apply to the harmonic denoted by the subscript. (E) If the voltage and current are quasi-periodic functions of the form

$$
e = (2)^{1/2} \sum_{r=1}^{r=\infty} E_r(t)\cos(r\omega t + \alpha_r)
$$

$$
i = (2)^{1/2} \sum_{q=1}^{q=\infty} I_q(t)\cos(q\omega + \beta_r)
$$

where $E_r(t), I_q(t)$ are aperiodic functions of t, the instantaneous power is given by the equation

$$
\begin{aligned}
p = ei \\
&= E_1(t)I_1(t))[\cos(\alpha_1 - \beta_1) \\
&+ \cos(2\omega t + \alpha_1 + \beta_1)] \\
&+ E_2(t)I_1(t)[\cos(\omega t + \alpha_2 - \beta_1) \\
&+ \cos(3\omega t + \alpha_2 + \beta_1)] \\
&+ E_1(t)I_2(t)[\cos(\omega t - \alpha_1 + \beta_2) \\
&+ \cos(3\omega t + \alpha_1 + \beta_2)] \\
&+ E_2(t)I_2(t)[\cos(\alpha_2 - \beta_2) \\
&+ \cos(4\omega t + \alpha_2 + \beta_2)] \\
&+ \cdots
\end{aligned}
$$

(F) Instantaneous power is expressed in watts when the voltage is in volts and the current in amperes. *Note:* See note 3 of **reference direction of energy.** The sign of the energy will be positive if the flow of power is in the reference direction and negative if the flow is in the opposite direction. 210

power inverter. A rectifier unit in which the direction of average energy flow is from the direct-current circuit to the alternating-current circuit. *See also:* **power rectifier; rectification.** 208, 3

power klystron (microwave tubes). A klystron, usually an amplifier, with two or more cavities uncoupled except by the beam, designed primarily for power amplification or generation. *See also:* **microwave tube or valve.** 190

power level. The magnitude of power averaged over a specified interval of time. *Note:* Power level may be expressed in units in which the power itself is measured or in decibels indicating the ratio to a reference power. This ratio is usually expressed either in decibels referred to one milliwatt, abbreviated dBm, or in decibels referred to one watt, abbreviated dBW. *See also:* **level.** 239

power-line carrier (1) (power system communications). The use of radio-frequency energy, generally below 600 kilohertz, to transmit information over transmission lines whose primary purpose is the transmission of power. 59

(2) **(protective relaying of utility-consumer interconnections).** A high-frequency signal superimposed on the normal voltage on a power circuit. It is customarily coupled to the power line by means of a coupling capacitor.

A tuning device provides series resonance at the carrier frequency. Prevention of shorting of the carrier signal by a fault external to the protected line is ordinarily provided by a line trap.   128

power-line carrier receiver. A receiver for power-line carrier signals. *See also:* **power-line carrier.**   59

power-line carrier relaying. The use of power-line carrier for protective relaying. *See also:* **power-line carrier**   59

power-line carrier transmitter. A device for producing radio-frequency power for purposes of transmission on power lines. *See also:* **power-line carrier.**   59

power loss (1) (from a circuit, in the sense that it is converted to another form of power not useful for the purpose at hand (for example, RI^2 loss). A physical quantity measured in watts in the International System of Units (SI) and having the dimensions of power. For a given R, it will vary with the current in R.   252

(2) (defined as the ratio of two powers). If P_o is the output power and P_i the input power of a transducer or network under specified conditions, P_i/P_o is a dimensionless quantity that would be unity if $P_o = P_i$.   57, 59

(3) (logarithmic). Loss may also be defined as the logarithm, or a quantity directly proportional to the logarithm of a power ratio, such as P_o/P_i. Thus if loss = 10 $\log_{10}(P_o/P_i)$ the loss is zero when $P_o = P_i$. This is the standard for measuring loss in decibels. *Notes:* (A) It should be noted that in cases (2) and (3) the loss for a given linear system is the same whatever may be the power levels. Thus (2) and (3) give characteristics of the system, and do not depend, as (1) does, on the value of the current or other dependent quantity. (B) If more than one component is involved in the input or output, the particular components used must be specified. This ratio is usually expressed in decibels. (C) If the output signal power is at a frequency other than the input signal power, the loss is a conversion loss. *See also:* **transducer; transmission; transmission loss.**   328

(4) (electric instrument) (watt loss). In the circuit of a current- or voltage-measuring instrument, the active power at its terminals for end-scale indication. *Note:* For other than current- or voltage-measuring instruments, for example, wattmeters, the power loss of any circuit is expressed at a stated value of current or of voltage. *See also:* **accuracy rating (instrument).**   280

power, nonfirm (electric power systems). Power supplied or available under an arrangement which does not have the availability feature of firm power.   112

power, nonreactive (polyphase circuit). At the terminals of entry of a polyphase circuit, a vector equal to the (vector) sum of the nonreactive powers for the individual terminals of entry. *Note:* The nonreactive power for each terminal of entry is determined by considering each phase conductor and the common reference point as a single-phase circuit, as described for distortion power. The sign given to the distortion power in determining the nonreactive power for each single-phase circuit shall be the same as that of the total active power. Nonreactive power for a polyphase circuit has as its two rectangular components the active power and the distortion power. If the voltages have the same waveform as the corresponding currents, the magnitude of the nonreactive power becomes the same as the active power. Nonreactive power is ex-

pressed in volt-amperes when the voltages are in volts and the currents in amperes.   210

power, nonreactive (single-phase two-wire circuit). At the two terminals of entry of a single-phase two-wire circuit into a delimited region, a vector quantity having as its rectangular components the active power and the distortion power. Its magnitude is equal to the square root of the difference of the squares of the apparent power and the amplitude of the reactive power. Its magnitude is also equal to the square root of the sum of the squares of the amplitudes of the active power and the distortion power. If voltage and current have the same waveform, the magnitude of the nonreactive power is equal to the active power. The amplitude of the nonreactive power is given by the equation

$$N = (U^2 - Q^2)^{1/2} = (P^2 + D^2)^{1/2}$$

$$= \left\{ \sum_{r=1}^{r=\infty} \sum_{q=1}^{q=\infty} [E_r{}^2 I_q{}^2 - E_r E_q I_r I_q \right.$$

$$\left. \times \sin(\alpha_r - \beta_r)\sin(\alpha_q - \beta_q)] \right\}^{1/2}$$

where the symbols are those in **power, apparent (single-phase two-wire circuit).** In determining the vector position of the nonreactive power, the sign of the distortion power component must be assigned arbitrarily. Nonreactive power is expressed in volt-amperes when the voltage is in volts and the current in amperes. *See:* **distortion power (single-phase two-wire circuit).**   210

power-operated door or gate (elevators). A hoistway door and/or a car door or gate that is opened and closed by a door or gate power operator. *See also:* **hoistway (elevator or dumbwaiter).**   328

power operation. Operation by other than hand power.   103, 202, 27

power output (generating unit) (speed governing of hydraulic turbines). The electrical output of the turbine generator unit as measured at the generator terminals.   8, 58

power output, instantaneous. The rate at which energy is delivered to a load at a particular instant. *See also:* **radio transmitter.**   111

power pack. A unit for converting power from an alternating-current or direct-current supply into alternating-current or direct-current power at voltages suitable for supplying an electronic device.   328

power, partial discharge (dielectric tests). The power fed into the terminals of the test object due to partial discharges. The average discharge power is expressed in watts.   139

power, peak pulse. *See:* **peak pulse power.**

power, phase control (thyristor converter). The power used to synchronize the phase control of the thyristor converter to the ac supply input phases.   121

power, phasor (polyphase circuit). At the terminals of entry of a polyphase circuit into a delimited region, a phasor (or plane vector) that is equal to the (phasor) sum of the phasor powers for the individual terminals of entry when the voltages are all determined with respect to the same arbitrarily selected common reference point in the boundary surface (which may be the neutral terminal of entry). The reference direction for the currents and the reference polarity for the voltages must be the same as for instantaneous power, active power, and reactive power. The phasor power for each terminal of entry is determined by considering each conductor and the common reference point as a single-phase, two-wire circuit and finding the phasor power for each in accordance with the definition of **power, phasor (single-phase two-wire circuit).** The phasor power S is given by $S = P + jQ$ where P is the active power for the polyphase circuit and Q is the reactive power for the same terminals of entry. If the voltages and currents are sinusoidal and of the same period, the phasor power S for a three-phase circuit is given by

$$S = E_a I_a^* + E_b I_b^* + E_c I_c^*$$

where E_a, E_b, and E_c are the phasor voltages from the phase conductors a, b, and c, respectively, to the neutral conductor at the terminals of entry. I_a, I_b, and I_c are the conjugate of the phasor currents in the phase conductors at the terminals of entry. If there is no neutral conductor, so that there are only three terminals of entry, the point of entry of one of the phase conductors may be chosen as the common voltage point, and the phasor from that conductor to the common voltage point becomes zero. If the terminal of entry of phase conductor b is chosen as the common point, the phasor power of a three-phase, three-wire circuit becomes

$$S = E_{ab} I_a^* + E_{cb} I_c^*$$

where E_{ab}, E_{cb} are the phasor voltages from phase conductor a to b and from c to b, respectively. If both the voltages and currents in the preceding equations constitute symmetrical sets of the same phase sequence $S = 3E_a I_a^*$. In general the phasor power at the $(m + 1)$ terminals of entry of a polyphase circuit of m phases to a delimited region, when one of the terminals is the neutral terminal of entry, is expressed by the equation

$$S = \sum_{s=1}^{s=m} \sum_{r=1}^{r=\infty} E_{sr} I_{sr}^*$$

where E_{sr} is the phasor representing the rth harmonic of the voltage from phase conductor s to neutral at the terminals of entry. I_{sr}^* is the conjugate of the phasor representing the rth harmonic of the current through the sth terminal of entry. The phasor power can also be stated in terms of the symmetrical components of the voltages and currents as

$$S = m \sum_{k=0}^{k=m-1} \sum_{r=1}^{r=\infty} E_{kr} I_{kr}^*$$

where E_{kr} is the phasor representing the symmetrical component of kth sequence of the rth harmonic of the line-to-neutral set of polyphase voltages at the terminals of entry. I_{kr}^* is the conjugate of the phasor representing the symmetrical component of the kth sequence of the rth harmonic of the polyphase set of currents through the terminals of entry. Phasor power is expressed in voltamperes when the voltages are in volts and the currents in amperes. *Note:* This term was once defined as **vector power.** With the introduction of the term **phasor quantity,** the name of this term has been altered to correspond. The definition has also been altered to agree with the change in the sign of reactive power. *See:* **power, reactive (magner)**

(single-phase two-wire circuit); power, reactive (magner) (polyphase circuit). 210

power, phasor (single-phase two-wire circuit). At the two terminals of entry of a single-phase two-wire circuit into a delimited region, a phasor (or plane vector) of which the real component is the active power and the imaginary component is the reactive power at the same two terminals of entry. When either component of phasor power is positive, the direction of that component is in the reference direction. The phasor power S is given by $S = P + jQ$ where P and Q are the active and reactive power, respectively. If both the voltage and current are sinusoidal, the phasor power is equal to the product of the phasor voltage and the conjugate of the phasor current.

$$E = Ee^{j\alpha}; \qquad I = Ie^{j\beta};$$

the phasor power is

$$S = P + jQ = EI^* = EIe^{j(\alpha - \beta)}$$
$$= EI[\cos(\alpha - \beta) + j\sin(\alpha - \beta)].$$

If the voltage is an alternating voltage and the current is an alternating current, the phasor power for each harmonic component is defined in the same way as for the sinusoidal voltage and sinusoidal current. Mathematically the phasor power of the rth harmonic component S_r is given by

$$S_r = P_r + jQ_r = E_r I_r^* = E_r I_r e^{j(\alpha_r - \beta_r)}$$
$$= E_r I_r[\cos(\alpha_r - \beta_r) + j\sin(\alpha_r - \beta_r)].$$

The phasor power at the two terminals of entry of a single-phase two-wire circuit into a delimited region, for an alternating voltage and current, is equal to the (phasor) sum of the values of the phasor power for every harmonic. Mathematically, this relation may be expressed

$$S = S_1 + S_2 + S_3 + \cdots = \Sigma S_r$$
$$= E_1 I_1^* + E_2 I_2^* + E_3 I_3^* \cdots = \Sigma E_r I_r^*$$
$$= (P_1 + P_2 + P_3 + \cdots)$$
$$+ j(Q_1 + Q_2 + Q_3 + \cdots) = \Sigma(P_r + jQ_r).$$

The amplitude of the phasor power is equal to the square root of the sum of the squares of the active power and the reactive power. Mathematically, if S is the amplitude of the phasor power and θ is the angle between the phasor power and the real-power axis,

$$S = Se^{j\theta}$$

$$S = (P^2 + Q^2)^{1/2} = [(P_1 + P_2 + P_3 + \cdots)^2 + (Q_1 + Q_2 + Q_3 + \cdots)^2]^{1/2}$$

$$\theta = \tan^{-1}\frac{Q}{P} = \tan^{-1}\frac{Q_1 + Q_2 + Q_3 + \cdots}{P_1 + P_2 + P_3 + \cdots}.$$

If the voltage and current are quasi-periodic and the amplitude of the voltage and current components are slowly varying, the phasor power may still be taken as the phasor having P and Q as its components, the values of P and Q being determined for these conditions, as specified in **power, active (single-phase two-wire circuit) (average power)**

(power) and **power, reactive (magner) (single-phase two-wire circuit)**, respectively. For this condition the phasor power will be a function of time. If the voltage and current have the same waveform, the amplitude of the phasor power is equal to the apparent power, but they are not the same for all other cases. The phasor power is expressed in volt-amperes when the voltage is in volts and the current in amperes. *Note:* This term was once defined as **vector power**. With the introduction of the term **phasor quantity**, the name of this term has been altered to correspond. The definition has also been altered to agree with the change in the sign of reactive power. *See:* **power, reactive (magner) (single-phase two-wire circuit)**. *See also:* **alternating current**. 210

power pool. Term referring to a group of power systems operating as an interconnected system and pooling their resources. 112

power primary detector (electric power systems). A power-measuring device for producing an output proportional to power input. *See also:* **speed-governing system**. 94

power quantities (single-phase three-wire circuit) and (two-phase circuit). The definitions of the power quantities for a single-phase circuit of more than two wires and of a two-phase circuit are essentially the same as those given for a polyphase circuit. *Note:* Where mathematical expressions involve m, the number of phases or phase conductors, the numeral 2 should be used for single-phase, three-wire systems, and the numeral 4 for two-phase, four-wire and five-wire systems. *See:* **polyphase symmetrical sets (polyphase voltages)**. 210

power rating (waveguide attenuator). The maximum power that, if applied under specified conditions of environment and duration, will not produce a permanent change that causes any performance characteristics to be outside of specifications. This includes characteristic insertion loss and standing-wave ratio. *See also:* **waveguide**. 185

power rating or voltage rating (line and connectors) (coaxial transmission line). That value of transmitted power or voltage that permits satisfactory operation of the line assembly and provides an adequate safety factor below the point where injury or appreciably shortened life will occur. *See also:* **transmission line**. 265

power, reactive. The product of voltage and the out-of-phase component of alternating current. In a passive network, reactive power represents the alternating exchange of stored energy (inductive or capacitive) between two areas. *See also:* **power, reactive (magner) (polyphase circuit)**; **power, reactive (magner) (single-phase two-wire circuit)**. 63

power, reactive (magner) (1) (polyphase circuit). At the terminals of entry into a delimited region, the algebraic sum of the reactive powers for the individual terminals of entry when the voltages are all determined with respect to the same arbitrarily selected common reference point in the boundary surface (which may be the neutral terminal of entry). The reference direction for the currents and the reference polarity for the voltages must be the same as for the instantaneous power and the active power. The reactive power for each terminal of entry is determined by considering each conductor and the common reference point as a single-phase two-wire circuit and

finding the reactive power for each in accordance with the definition of **power, reactive (magner) (single-phase two-wire circuit)**. If the voltages and currents are sinusoidal and of the same period, the reactive power Q for a three-phase circuit is given by

$$Q = E_a I_a \sin(\alpha_a - \beta_a) + E_b I_b \sin(\alpha_b - \beta_b) + E_c I_c \sin(\alpha_c - \beta_c)$$

where the symbols have the same meaning as in **power, instantaneous (polyphase circuit)**. If there is no neutral conductor and the common point for voltage measurement is selected as one of the phase terminals of entry, the expression will be changed in the same way as that for **power, instantaneous (polyphase circuit)**. If both the voltages and currents in the preceding equations constitute symmetrical polyphase set of the same phase sequence

$$Q = 3 E_a I_a \sin(\alpha_a - \beta_a).$$

In general the reactive power Q at the $(m + 1)$ terminals of entry of a polyphase circuit of m phases to a delimited region, when one of the terminals is the neutral terminal of entry, is expressed by the equation

$$Q = \sum_{s=1}^{s=m} \sum_{r=1}^{r=\infty} E_{sr} I_{sr} \sin(\alpha_{sr} - \beta_{sr})$$

where the symbols have the same meaning as in **power, active (polyphase circuit)**. The reactive power can also be stated in terms of the root-mean-square amplitudes of the symmetrical components of the voltages and currents as

$$Q = m \sum_{k=0}^{k=m-1} \sum_{r=1}^{r=\infty} E_{kr} I_{kr} \sin(\alpha_{kr} - \beta_{kr})$$

where the symbols have the same meaning as in **power, active (polyphase circuit)**. When the voltages and currents are quasi-periodic and the amplitudes of the voltages and currents are slowly varying, the reactive power for the circuit of each conductor may be determined for this condition as in **power, reactive (magner) (single-phase two-wire circuit)**. The reactive power for the polyphase circuit is the sum of the reactive power values for the individual conductors. Reactive power is expressed in vars when the voltages are in volts and the currents in amperes. *Note:* The sign of reactive power resulting from the above definition is the opposite of that given by the definition in the 1941 edition of the American Standard Definitions of Electrical Terms. The change has been made in accordance with a recommendation approved by the Standards Committee of the Institute of Electrical and Electronics Engineers, by the American National Standards Institute, and by the International Electrotechnical Commission. 210

(2) (single-phase two-wire circuit). At the two terminals of entry of a single-phase two-wire circuit into a delimited region, for the special case of a sinusoidal voltage and a sinusoidal current of the same period, is equal to the product obtained by multiplying the root-mean-square value of the voltage between one terminal of entry and the second terminal of entry, considered as the reference terminal, by the root-mean-square value of the current through the first terminal and by the sine of the angular phase difference by which the voltage leads the current.

The reference direction for the current and the reference polarity for the voltage must be the same as for active power at the same two terminals. Mathematically the reactive power Q, for the case of sinusoidal voltage and current, is given by

$$Q = EI \sin(\alpha - \beta)$$

in which the symbols have the same meaning as in **power, instantaneous (two-wire circuit)**. For the same conditions, the reactive power Q is also equal to the imaginary part of the product of the phasor voltage and the conjugate of the phasor current, or to the negative of the imaginary part of the product of the conjugate of the phasor voltage and the phasor current. Thus

$$Q = \mathrm{Im}EI^*$$
$$= -\mathrm{Im}E^*I$$
$$= \frac{1}{2j}[EI^* - E^*I]$$

in which E and I are the phasor voltage and phasor current, respectively, and $*$ denotes the conjugate of the phasor to which it is applied. If the voltage is an alternating voltage and the current is an alternating current, the reactive power for each harmonic component is equal to the product obtained by multiplying the root-mean-square amplitude of that harmonic component of the voltage by the root-mean-square amplitude of the same harmonic component of the current and by the sine of the angular phase difference by which that harmonic component of the voltage leads the same harmonic component of the current. Mathematically the reactive power of the rth harmonic component of Q_r is given by

$$Q_r = E_r I_r \sin(\alpha_r - \beta_r)$$
$$= \mathrm{Im}E_r I_r^*$$
$$= \frac{1}{2j}[E_r I_r^* - E_r^* I_r]$$
$$= -\mathrm{Im}E_r^* I_r$$

in which the symbols have the same meaning as in **power, instantaneous (two-wire circuit) and power, active (single-phase two-wire circuit) (average power) (power)**. The reactive power at the two terminals of entry of a single-phase two-wire circuit into a delimited region, for an alternating voltage and current, is equal to the sum of the values of reactive power for every harmonic component. Mathematically the reactive power Q for an alternating voltage and current, is given by

$$Q = Q_1 + Q_2 + Q_3 + Q_4 + \cdots + Q_r + \cdots$$
$$= E_1 I_1 \sin(\alpha_1 - \beta_1)$$
$$+ E_2 I_2 \sin(\alpha_2 - \beta_2) + \cdots$$
$$= \sum_{r=1}^{r=\infty} Q_r$$
$$= \sum_{r=1}^{r=\infty} E_r I_r \sin(\alpha_r - \beta_r)$$

in which the symbols have the same meaning as in **power,**

instantaneous (two-wire circuit). If the voltage and current are quasi-periodic functions of the form given in **power, instantaneous (two-wire circuit)**, and the amplitudes are slowly varying, so that each may be considered to be constant during any one period, but to have slightly different values in successive periods, the reactive power at any time t may be taken as

$$Q = \sum_{r=1}^{r=\infty} E_r(t) I_r(t) \sin(\alpha_r - \beta_r)$$

by analogy with the expression for active power. When the reactive power is positive, the direction of flow of quadergy is in the reference direction of energy flow. Because the reactive power for each harmonic may have either sign, the direction of the reactive power for a harmonic component may be the same as or opposite to the direction of the total reactive power. The value of reactive power is expressed in vars when the voltage is in volts and the current in amperes. *Notes:* (1) The sign of reactive power resulting from the above definition is the opposite of that given by the definition in the 1941 edition of the American Standard Definitions of Electrical Terms. The change has been made in accordance with a recommendation approved by the Standards Committee of the Institute of Electrical and Electronics Engineers, by the American National Standards Institute, and by the Electrotechnical Commission. (2) Any designation of positive reactive power as inductive reactive power and negative reactive power as capacitive reactive power is deprecated. If the reference direction is from the generator toward the load, reactive power is positive if the load is predominantly inductive and negative if the load is predominantly capacitive. Thus a capacitor is a source of quadergy and an inductor is a consumer of quadergy. Designations of two kinds of reactive power are unnecessary and undesirable.
210

(3) (circuits and systems). The product of voltage and the component of alternating current that is 90° out of phase with it. In a passive network reactive power represents the energy that is exchanged alternatively between a capacitive and an inductive storage medium.   67

power rectifier. A rectifier unit in which the direction of average energy flow is from the alternating-current circuit to the direct-current circuit.   208

power rectifier transformer. A rectifier transformer connected to mercury-arc or semiconductor rectifiers for electro-chemical service steel processing applications, electric furnace applications, mining applications, transportation applications, and direct-current transmissions.
53

power relay. (1) A relay that responds to power flow in an electric circuit.   103, 202
(2) A relay that responds to a suitable product of voltage and current in an electric circuit. *See also:* **relay.**
60, 127, 103

power response (close-talking pressure-type microphone). The ratio of the power delivered by a microphone to its load, to the applied sound pressure as measured by a Laboratory Standard Microphone placed at a stated distance from the plane of the opening of the artificial voice. *Note:* The power response is usually measured as a function of frequency in decibels (dB) above 1 milliwatt per newton per square centimeter [mW/(N/m²)] or 1 milliwatt per 10 microbars [mW/10μbar]. *See also:* **close-talking pressure-type microphone.** 249

power selsyn (synchros or selsyns). An inductive type of positioning system having two or more similar mechanically independent slip-ring machines with corresponding slip rings of all machines connected together and the stators fed from a common power source. *See also:* **synchro-system.** 63

power sensitivity error. The maximum deviation from linearity over each power range of either the electrothermic unit or the electrothermic power meter. Expressed in percent. 47

power service protector. An assembly consisting of a modified low-voltage power circuit breaker, which has no direct-acting tripping devices, with a current-limiting fuse connected in series with the load terminals of each pole. 103

power, signal electronics (thyristor converter). The power used for the analog or digital system power supplies, or both, required for the thyristor converter control and protection systems. 121

powers of units (International System of Units) (SI). An exponent attached to a symbol containing a prefix indicates that the multiple or submultiple of the unit (the unit with its prefix) is raised to the power expressed by the exponent. For example:

$$1 \text{ cm}^3 = (10^{-2} \text{ m})^3 = 10^{-6} \text{ m}^3$$
$$1 \text{ ns}^{-1} = (10^{-9} \text{ s})^{-1} = 10^9 \text{ s}^{-1}$$
$$1 \text{ mm}^2/\text{s} = (10^{-3} \text{ m})^2/\text{s} = 10^{-6} \text{ m}^2/\text{s}$$

See also: **units and letter symbols; prefixes and symbols.** 21

power source isolation. Absence of a direct-current circuit (path) between the power source and the system power supply outputs. 48, 58

power spectral density (PSD) (nuclear power generating stations) (seismic qualification of class 1E equipment). The mean squared acceleration density of a random waveform. PSD is usually expressed in square gauss per hertz. 28, 31

power storage. That portion of the water stored in a reservoir available for generating electric power. *See also:* **generating station.** 64

power supply circuit (relay system). An input circuit to a relay system which supplies auxiliary power for the proper functioning of the relay system. 90

power supply, direct-current (power-system communication) (1) (direct-current) (general). A device for converting available electric service energy into direct-current energy at a voltage suitable for electronic components. *See:* **power pack; power supply, direct-current regulated (power-system communication).** 59

(2) (alternating-current to direct-current). Generally, a device consisting of a transformer, rectifier, and filter for converting alternating current to a prescribed direct voltage or current. 228, 186

power supply, direct-current regulated. A direct-current power supply whose output voltage is automatically controlled to remain within specified limits for specified

variations in supply voltage and load current. *See also:* **power supply, direct-current (power-system communication).** 59

power supply, uninterruptible (UPS). A system designed to provide power, without delay or transients, during any period when the normal power supply is incapable of performing acceptably. 89

power-supply voltage range (transmitter performance). The range of voltages over which there is not significant degradation in the transmitter or receiver performance. *See also:* **audio-frequency distortion.** 181

power switchboard. A type of switchboard including primary power circuit switching and interrupting devices together with their interconnections. *Note:* Knife switches, fuses, and air circuit breakers are the commonly used switching and interrupting devices. 103, 202, 79

power system (1) (electric). A group of one or more generating sources and/or connecting transmission lines operated under common management or supervision to supply load. 94, 200

(2) (power generation). The electric power sources, conductors and equipment required to supply electric power. 58

power system, emergency. An independent reserve source of electric energy which, upon failure or outage of the normal source, automatically provides reliable electric power within a specified time to critical devices and equipment whose failure to operate satisfactorily would jeopardize the health and safety of personnel or result in damage to property. 89

power system, standby. An independent reserve source of electric energy which, upon failure or outage of the normal source, provides electric power of acceptable quality and quantity so that the user's facilities may continue in satisfactory operation. 89

power—temperature coefficient. The change in power required to hold the bolometer element at the desired operating resistance per unit change in ambient temperature. *Note:* This quantity is expressed in microwatts per degree Celsius. 115

power transfer relay. A relay so connected to the normal power supply that the failure of such power supply causes the load to be transferred to another power supply. 328

power transformer. A transformer which transfers electric energy in any part of the circuit between the generator and the distribution primary circuits. 53

power type relay. A term for a relay designed to have heavy-duty contacts usually rated 15 amperes or higher. Sometimes called a **contactor.** 341

power, utility. *See:* **power, commercial.**

power, vector (polyphase circuit). At the terminals of entry of a polyphase circuit, a vector of which the three rectangular components are, respectively, the active power, the reactive power, and the distortion power at the same terminals of entry. In determining the components, the reference terminals for voltage measurement shall be taken as the neutral terminal of entry, if one exists, otherwise as the true neutral point. The vector power is also the (vector) sum of the vector powers for the individual terminals of entry. The vector power for each terminal of entry is determined by considering each phase conductor and the common reference point as a single-phase circuit, as described for distortion power. The sign given to the distortion power in determining the vector power for each single-phase circuit shall be the same as that of the total active power. The magnitude of the vector power is the apparent power. If the voltages have the same waveform as the corresponding currents, the magnitude of the vector power is equal to the amplitude of the phasor power. Vector power is expressed in volt-amperes when the voltages are in volts and the currents in amperes. *See also:* **network analysis.**   210

power vector (single-phase two-wire circuit). At the two terminals of entry of a single-phase two-wire circuit into a delimited region, a vector whose magnitude is equal to the apparent power, and the three rectangular components of which are, respectively, the active power, the reactive power, and the distortion power at the same two terminals of entry. Mathematically the vector power U is given by

$$U = iP + jQ + kD$$

where **i, j,** and **k** are unit vectors along the three perpendicular axes, respectively. *P, Q,* and *D* are the active power, reactive power, and distortion power, respectively. The direction cosines of the angles between the vector power U and the three rectangular axes are

$$\cos \phi = \frac{P}{U}$$

$$\cos \Psi = \frac{Q}{U}$$

$$\cos \theta = \frac{D}{U}$$

The magnitude of the vector power is the apparent power, or

$$U = (P^2 + Q^2 + D^2)^{1/2}$$
$$= \left(\sum_{r=1}^{r=\infty} \sum_{q=1}^{q=\infty} E_r{}^2 I_q{}^2 \right)^{1/2}$$

where the symbols are those of the preceding definitions. The geometric power diagram shows the relationships among the different types of power. Active power, reactive power, and distortion power are represented in the directions of the three rectangular axes. The accompanying diagram corresponds to a case in which all three are positive.

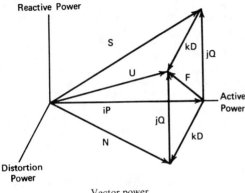

Vector power

Since the sign of D is not definitely determined, kD may be drawn in either direction along the **k** axis. The position of **U** is thus also ambiguous, as it may occupy either of two positions, for D positive or negative. When the sign of D has been assumed, the vector positions of the fictitious power **F** and the nonreactive power **N** are determined. They have been shown in the figure, with the assumption that D has the same sign as P. Vector power is expressed in volt-amperes when the voltage is in volts and the current in amperes. *Notes:* (1) The vector power becomes a plane vector having the same magnitude as the phasor power if the voltage and the current have the same wave form. This condition is fulfilled as a special case when the voltage and current are sinusoidal and of the same period. (2) The term **vector power** as defined in the 1941 edition of the American Standard Definitions of Electrical Terms has now been called **phasor power** (*see* **power, phasor (single-phase two-wire circuit)**) and the present definition of vector power is new. *See also:* **network analysis.**   210

power winding (saturable reactor). A winding to which is supplied the power to be controlled. Commonly the functions of the output and power windings are accomplished by the same winding, which is then termed the output winding. *See also:* **magnetic amplifier.**   328

Poynting's vector. If there is a flow of electromagnetic energy into or out of a closed region, the rate of flow of this energy is, at any instant, proportional to the surface integral of the vector product of the electric field strength and the magnetizing force. This vector product is called Poynting's vector. If the electric field strength is **E** and the magnetizing force is **H,** then Poynting's vector is given by

$$U = E \times H \text{ and } U = E \times H/4\pi$$

in rationalized and unrationalized systems, respectively. Poynting's vector is often assumed to be the local surface density of energy flow per unit time.   210

PPI (radar). *See:* **plan-position indicator.**

ppm (circuits and systems). Parts per million, that is, $x/10^6$ is x parts per million.   67

PPM. *See:* **periodic permanent-magnet focusing.**

PPM. *See:* **pulse-position modulation.**

practical reference pulse waveform (pulse measurement). A reference pulse waveform which is derived from a pulse which is produced by a device or apparatus.   15

preamplifier. An amplifier connected to a low-level signal source to present suitable input and output impedances and provide gain so that the signal may be further processed without appreciable degradation in the signal-to-noise ratio. *Notes:* (1) A preamplifier may include provision for equalizing and/or mixing. (2) Further processing frequently includes further amplification in a main amplifier. *See also:* **amplifier.**   239, 59

preassigned multiple access (communication satellite). A method of providing multiple access in which the satellite channels are preassigned at both ends of the path.   84

precedence call (telephone switching systems). A call on which the calling party has elected to use one of several levels of priority available to him.   55

precession (gyro). A rotation of the spin axis produced by a torque, T, applied about an axis mutually perpendicular to the spin axis and the axis of the resulting rotation. A constant precession rate, ω, is related to rotor angular momentum, H, and applied torque, T, by the equation $T = H\omega$.   46

precision (1) (general). The quality of being exactly or sharply defined or stated. A measure of the precision of a representation is the number of distinguishable alternatives from which it was selected, which is sometimes indicated by the number of significant digits it contains. *See also:* **accuracy; double precision**
    235, 188, 9, 210, 206

(2) (measurement process). The quality of coherence or repeatability of measurement data, customarily expressed in terms of the standard deviation of the extended set of measurement results from a well-defined (adequately specified) measurement process in a state of statistical control. The standard deviation of the conceptual population is approximated by the standard deviation of an extended set of actual measurements. *See also:* **accuracy; reproducibility.**   13, 187, 184, 54

(3) (measurement) (transmission lines and waveguides). The degree of combined repeatability and resolution. *See also:* **measurement system.**   185

(4) (automatic control). Closeness of agreement among repeated measurements of the same characteristic by the same method under the same conditions. *Note:* Random differences among such measurements arise from unassigned causes, and by their magnitudes indicate the "imprecision," which is only one element of inaccuracy. Quantitatively, the precision of a measurement process is inversely proportional to its "imprecision," measured by the RMS deviation from the process average. For a single measurement, precision can be estimated only from prior data obtained by the same method or derived from assumed characteristics of the process. *See also:* **value, rms.**   56

(5) (nuclear power generating stations). The quality of coherence or repeatibility of measurement data.
    31, 41

(6) (pulse measurement). The degree of mutual agreement between the results of independent measurements of a pulse characteristic, property, or attribute yielded by repeated application of a pulse measurement process.   15

precision approach radar (PAR). A radar system located on an airfield for observation of the position of an aircraft with respect to an approach path and specifically intended to provide guidance to the aircraft in the approach. *See also:* **navigation.**   13, 187

precision wound (rotating machinery). A coil wound so that maximum nesting of the conductors occurs, usually with all crossovers at one end, with conductor aligned and positioned with respect to each adjacent conductor. *See also:* **rotor (rotating machinery); stator.**   63

preconditioning (industrial control). A control-function that provides for manually or automatically establishing a desired condition prior to normal operation of the system. *See also:* **control system, feedback.**   225, 206

precursor. *See:* **undershoot.**

predicted mean active maintenance time. The mean active maintenance time of an item calculated by taking into account the reliability characteristics and the mean active maintenance time of all of its parts and other relevant factors according to the stated conditions. *Notes:* (1) Maintenance policy, statistical assumptions and computing methods shall be stated. (2) The source of the data shall be stated.   164

predicted reliability. *See:* **reliability, predicted.**

predissociation. A process by which a molecule that has absorbed energy dissociates before it has had an opportunity to lose energy by radiation. *See also:* **gas-filled radiation-counter tubes.**   190, 125

predistortion (pre-emphasis) (system) (transmitter performance). A process that is designed to emphasize or de-emphasize the magnitude of some frequency components with respect to the magnitude of others. *See:* **pre-emphasis.**
    181

pre-emphasis (pre-equalization) (1) (general). A process in a system designed to emphasize the magnitude of some frequency components with respect to the magnitude of others, to reduce adverse effects, such as noise, in subsequent parts of the system. *Note:* After transmitting the pre-emphasized signal through the noisy part of the system, de-emphasis may be applied to restore the original signal with a minimum loss of signal-to-noise ratio.
    328

(2) (recording). An arbitrary change in the frequency response of a recording system from its basic response (such as constant velocity or amplitude) for the purpose of improvement in signal-to-noise ratio, or the reduction of distortion.   176

pre-emphasis network. A network inserted in a system in order to emphasize one range of frequencies with respect to another. *See also:* **network analysis.**   111, 78

preemption (telephone switching systems). On a precedence call, the disconnection and subsequent reuse of part of an established connection of lower priority if all the relevant circuits are busy.   55

preemptive control (test, measurement and diagnostic equipment). An action or function which, by reason of preestablished priority, is able to seize or interrupt the process in progress and cause to be performed a process of higher priority.   54

pre-envelope. *See:* **analytic signal.**

preference (power-system communication) (1) (channel supervisory control). An assembly of devices arranged to prevent the transmission of any signals over a channel other than supervisory control signals when supervisory control signals are being transmitted. *See also:* **supervisory control system.**   59

(2) (protective relaying). An assembly of devices arranged to prevent the transmission of any signals other than protective relaying signals over a channel when protective relaying signals are being transmitted. *See also:* **relay.**
    59

(3) (speech quality measurements). The proportion of a listening group expressed in percent that prefers the speech test signal to the speech reference signal as a source of information.   126

preference level (speech quality measurements). The signal-to-noise ratio (S/N) of the speech reference signal when it is isopreferent to the speech test signal.   126

preferred (electric power system). That equipment and system configuration selected to supply the power system loads under normal conditions.   58

preferred basic impulse insulation level (insulation strength). A basic impulse insulation level that has been adopted as a preferred American National Standard voltage value. *See:* **basic impulse insulation level (BIL) (insulation strength).** 276

preferred power supply (nuclear power generating stations). That power supply which is preferred to furnish electric energy under accident or post-accident conditions. 102, 31

prefix code (telephone switching systems). One or more digits preceding the national or international number to implement direct distance dialing. 55

prefixes and symbols (International System of Units) (SI). Used to form names and symbols of the decimal mutiples and submultiples of the SI units:

Multiplication Factor	Prefix	Symbol
1 000 000 000 000 000 000 = 10^{18}	exa^a	E
1 000 000 000 000 000 = 10^{15}	peta^a	P
1 000 000 000 000 = 10^{12}	tera	T
1 000 000 000 = 10^{9}	giga	G
1 000 000 = 10^{6}	mega	M
1 000 = 10^{3}	kilo	k
100 = 10^{2}	hecto^b	h
10 = 10^{1}	deka^b	da
0.1 = 10^{-1}	deci^b	d
0.01 = 10^{-2}	centi^b	c
0.001 = 10^{-3}	milli	m
0.000 001 = 10^{-6}	micro	μ
0.000 000 001 = 10^{-9}	nano	n
0.000 000 000 001 = 10^{-12}	pico	p
0.000 000 000 000 001 = 10^{-15}	femto	f
0.000 000 000 000 000 001 = 10^{-18}	atto	a

^{a} Adopted by the CGPM in 1975.
^{b} To be avoided where possible.

These prefixes or their symbols are directly attached to names or symbols of units, forming multiples and submultiples of the units. In strict terms these must be called "multiples and submultiples of SI units," particularly in discussing the coherence of the system. In common parlance, the base units and derived units, along with their multiples and submultiples, are all called SI units. *See also:* **units and letter symbols.**

prefix multipliers. The prefixes listed in the following table, when applied to the name of a unit, serve to form the designation of a unit greater or smaller than the original by the factor indicated. 210

Prefix	Abbreviation	Factor
tera-(megamega-*)	T(MM*)	10^{12}
giga-(kilomega-*)	G(kM*)	10^{9}
mega-	M	10^{6}
myria-		10^{4}
kilo-	k	10^{3}
hecto-	h	10^{2}
deka-		10
deci-	d	10^{-1}
centi-	c	10^{-2}
milli-	m	10^{-3}
decimilli-	dm	10^{-4}
micro-	μ	10^{-6}
nano-(millimicro-*	n (mμ*)	0^{-9}
pico-(micromicro-*)	p($\mu\mu$*)	10^{-12}

* Deprecated

preform (biscuit*) (disk recording) (electroacoustics). A small slab of record stock material as it is prepared for use in the record presses. *See also:* **phonograph pickup.** 176

* Deprecated

preformed coil or coil side (rotating machinery). An element of a preformed winding, composed of conductor strands, usually insulated and sometimes transposed, cooling ducts in some designs, turn insulation where number of turns exceeds one, and coil insulation. *See also:* **rotor (rotating machinery); stator.** 63

preformed winding. A winding consisting of coils which are given their shape before being assembled in the machine. 63

preheating time (mercury-arc valve). The time required for all parts of the valve to attain operating temperature. *See also:* **gas-filled rectifier.** 244, 190

preheat-starting (fluorescent lamps) (switch-starting systems). The designation given to those systems in which hot-cathode electric discharge lamps are started from preheated cathodes through the use of a starting switch, either manual or automatic in its operation. *Note:* The starting switch, when closed, connects the two cathodes in series in the ballast circuit so that current flows to heat the cathodes to emission temperature. When the switch is opened, a voltage surge is produced that initiates the discharge. Only the arc current flows through the cathodes after the lamp is in operation. *See also:* **fluorescent lamp.** 268, 167

pre-molded joint (power cable joints). A joint made of pre-molded components assembled in the field. 34

pre-operational system tests (nuclear power generating station). A pre-operational test is defined as a test to confirm that all individual component parts of a system function as a system and the system functions as designed. The above tests shall be performed following both the initial construction of the facility and subsequent modifications or additions made to the facility at later dates. 354

preparatory function (numerically controlled machines). A command changing the mode of operation of the control such as from positioning to contouring or calling for a fixed cycle of the machine. 224, 207

prepatch panel. *See:* **problem board.**

preregister operation (elevators). Operation in which signals to stop are registered in advance by buttons in the car and at the landings. At the proper point in the car travel, the operator in the car is notified by a signal, visual, audible, or otherwise, to initiate the stop, after which the landing stop is automatic. *See also:* **control.** 328

pre-rip (cable plowing). A process using a plow blade to loosen the earth prior to plowing and installing the cable, flexible tube, etcetera. 52

preselector. (1) A device placed ahead of a frequency converter or other device, that passes signals of desired frequencies and reduces others. (2) In automatic switching, a device that performs its selecting operation before seizing an idle trunk. 328

presence tests (test, measurement and diagnostic equipment). Actions which verify the presence or absence of signals or characteristics. Such signals or characteristics are those which are not tolerance critical to operation of the item. 54

preset. To establish an initial condition, such as the control values of a loop. 255, 77

preset guidance. That form of missile guidance wherein the control mechanism is set, prior to launching, for a predetermined path, with no provision for subsequent adjustment. *See also:* **guided missile.** 328

preset speed (industrial control). A control function that establishes the desired operating speed of a drive before initiating the speed change. *See also:* **electric drive.** 225, 206

preshoot (pulse terms). A distortion which precedes a major transition. *Note:* Colloquial term which qualitatively describes a type of distortion. 254

pressing (disk recording). A pressing is a record produced in a record-molding press from a master or stamper. *See also:* **phonograph pickup.** 176

pressure barrier seal (nuclear power generating stations). Consists of an aperture seal and an electrical conductor seal. 31

pressure cable. An oil-impregnated paper-insulated cable in which positive gauge pressure is maintained on the insulation under all operating conditions. 64

pressure coefficient. *See:* **environmental coefficient.**

pressure connector. A conductor terminal applied with pressure so as to make the connection mechanically and electrically secure. 124

pressure-containing terminal box (rotating machinery). A terminal box so designed that the products of an electric breakdown within the box are completely contained inside the box. 63

pressure-lubricated bearing (rotating machinery). A bearing in which a continuous flow of lubricant is forced into the space between the journal and the bearing. *See also:* **bearing.** 63

pressure microphone. A microphone in which the electric output substantially corresponds to the instantaneous sound pressure of the impressed sound waves. *Note:* A pressure microphone is a gradient microphone of zero order and is nondirectional when its dimensions are small compared to a wavelength. *See also:* **microphone.** 328

pressure relay. A relay that responds to liquid or gas pressure. 103, 202, 60, 127

pressure-relief device (arrester). A means for relieving internal pressure in an arrester and preventing explosive shattering of the housing, following prolonged passage of follow current or internal flashover of the arrester. 62

pressure-relief terminal box (rotating machinery). A terminal box so designed that the products of an electric breakdown within the box are relieved through a pressure-relief diaphragm. 63

pressure-relief test (arresters). A test made to ascertain that an arrester failure will not cause explosive shattering of the housing. 62

pressure retaining boundary (nuclear power generating stations). The pressure retaining boundary includes those surfaces of the aperture seal, the conductor feed-through plate, the conductor seal (or seals), and the conductor (or conductors) which are exposed to the containment environment. 31

pressure switch (industrial control). A switch in which actuation of the contacts is effected at a predetermined liquid or gas pressure. 308, 206

pressure system (protective signaling). A system for protecting a vault by maintaining a predetermined differential in air pressure between the inside and outside of the vault. Equalization of pressure resulting from opening the vault or cutting through the structure initiates an alarm condition in the protection circuit. *See also:* **protective signaling.** 328

pressure-type pothead. A pressure-type pothead is a pothead intended for use on positive-pressure cable systems. *See:* **multipressure zone pothead; single pressure zone potheads.** 323

pressure-type termination. (cable termination). A Class 1 termination intended for use on positive pressure cable systems. (1) single-pressure zone termination: a pressure type termination intended to operate with one pressure zone; (2) multipressure zone termination: a pressure type termination intended to be operated with two or more pressure zones. 4

pressure wire connector. A device that establishes the connection between two or more conductors or between one or more conductors and a terminal by means of mechanical pressure and without the use of solder. 328

pressurized (rotating machinery). Applied to a sealed machine in which the internal coolant is kept at a higher pressure than the surrounding medium. 63

prestressed concrete structures (transmission and distribution). Concrete structures which include metal tendons that are tensioned and anchored either before or after curing of the concrete. 262

prestrike current (lightning). The current that flows in a lightning stroke prior to the return stroke current. *See also:* **direct-stroke protection (lightning).** 64

pretersonic. Ultrasonic and with frequency higher than 500 megahertz. 352

pretransmit-receive tube. A gas-filled radio-frequency switching tube used to protect the transmit-receive tube from excessively high power and the receiver from frequencies other than the fundamental. *See also:* **gas tubes.** 190, 125

preventative autotransformer. An autotransformer (or center-tapped reactor) used in load-tap-changing and regulating transformers, or step-voltage regulators to limit the circulating current when operating on a position in which two adjacent taps are bridged, or during the change of taps between adjacent positions. 53

preventive maintenance (1) (test, measurement and diagnostic equipment). Tests, measurements, replacements, adjustments, repairs and similar activities, carried out with the intention of preventing faults or malfunctions from occurring during subsequent operation. Preventive maintenance is designed to keep equipment and programs in proper operating condition and is performed on a scheduled basis. 54

(2) (reliability). The maintenance carried out at predetermined intervals or corresponding to prescribed criteria, and intended to reduce the probability of failure or the performance degradation of an item. 164

prf (laser-maser). Abbreviation for pulse-repetition frequency. High prf = more than 1 Hz. *See also:* **pulse-repetition frequency.** 363

primaries (color). The colors of constant chromaticity and variable amount that, when mixed in proper proportions, are used to produce or specify other colors. *Note:* Primaries need not be physically realizable.   18, 178

primary (1) (electric machines and devices). The part of a machine having windings that are connected to the power supply line (for a motor or transformer) or to the load (for a generator).   63

(2) (color technology). Any one of three lights in terms of which a color is specified by giving the amounts required to duplicate it by additive combination. *See also:* **color.**   167

(3) (adjective). (A) First to operate; for example, primary arcing contacts, primary detector; (B) First in preference; for example, primary protection; (C) Referring to the main circuit as contrasted to auxiliary or control circuits, for example, primary disconnecting devices; (D) Referring to the energy input side of transformers, or the conditions (voltages) usually encountered at this location; for example, primary unit substation.   103, 202

primary arcing contacts (switching device). The contacts on which the initial arc is drawn and the final current, except for the arc-shunting-resistor current, is interrupted after the main contacts have parted.   103, 202

primary battery. *See also:* **battery (primary or secondary); electrochemistry; primary cell.**

primary calibration (nuclear power generating stations). The determination of the electronic system accuracy when the detector is exposed in a known geometry to radiation from sources of known energies and activity levels traceable to the NBS (National Bureau of Standards).   31

primary cell. A cell that produces electric current by electrochemical reactions without regard to the reversibility of those reactions. Some primary cells are reversible to a limited extent. *See also:* **electrochemistry.**   328

primary center (telephone switching systems). A toll office to which toll centers and toll points may be connected. Primary centers are classified as Class 3 offices. *See also:* **office class.**   55

primary circuit (voltage regulator). The circuit on the input side at the regulator. *See also:* **voltage regulator.**
     257, 53

primary-color unit (television). The area within a color cell occupied by one primary color. *See also:* **television.**
     177, 190, 125

primary current ratio (electroplating). The ratio of the current densities produced on two specified parts of an electrode in the absence of polarization. It is equal to the reciprocal of the ratio of the effective resistances from the anode to the two specified parts of the cathode. *See also:* **electroplating.**   328

primary detector (1) (switchgear). The first system element or group of elements that responds quantitatively to the measurand and performs the initial measurement operation. A primary detector performs the initial conversion or control of measurement energy and does not include transformers, amplifiers, shunts, resistors, etcetera, when these are used as auxiliary means. *Syn:* **sensing element; initial element.** *See also:* **measurement system.**
     103, 202

(2) (power systems). That portion of the measurement device which either utilizes or transforms energy from the controlled medium to produce a measurable effect which is a function of change in the value of the controlled variable.   94

primary disconnecting devices (switchgear assembly). Self-coupling separable contacts provided to connect and disconnect the main circuits between the removable element and the housing.   103, 202

primary distribution feeder. A feeder operating at primary voltage supplying a distribution circuit. *Note:* A primary feeder is usually considered as that portion of the primary conductors between the substation or point of supply and the center of distribution.   204

primary distribution mains. The conductors that feed from the center of distribution to direct primary loads or to transformers that feed secondary circuits. *See also:* **center of distribution.**   204

primary distribution network. A network consisting of primary distribution mains. *See also:* **center of distribution.**   204

primary distribution system. A system of alternating-current distribution for supplying the primaries of distribution transformers from the generating station or substation distribution buses. *See also:* **alternating-current distribution; center of distribution.**   204

primary distribution trunk line. A line acting as a main source of supply to a distribution system. *See also:* **center of distribution.**   204

primary electron (thermionics). An electron in a primary emission. *See also:* **electron emission.**   308, 190

primary emission. Electron emission due directly to the temperature of a surface, irradiation of a surface, or the application of an electric field to a surface.   308, 190

primary fault. The initial breakdown of the insulation of a conductor, usually followed by a flow of power current. *See also:* **center of distribution.**   204

primary flow (carriers). A current flow that is responsible for the major properties of the device.   342

primary line of sight. The line connecting the point of observation and the point of fixation. *See also:* **visual field.**   167

primary line-to-ground voltage (coupling capacitors and capacitance potential devices). Refers to the high-tension root-mean-square line-to-ground voltage of the phase to which the coupling capacitors or potential device, in combination with its coupling capacitor or bushing, is connected. *See:* **rated primary line-to-ground voltage.** *See also:* **outdoor coupling capacitor.**   351

primary network. A network supplying the primaries of transformers whose secondaries may be independent or connected to a secondary network. *See:* **primary distribution network.** *See also:* **center of distribution.**   204

primary overcurrent protective device or apparatus (nuclear power generating stations). A device or apparatus which normally performs the function of circuit interruption.   26

primary protection (relay system). First-choice relay protection in contrast with backup relay protection.
     103, 202, 60, 127

primary radar. A radar in which the reply signal is a portion of the transmitted energy reflected by the target. *See also:* **secondary radar.**   187, 13

primary radiator (antennas). A feed which illuminates a secondary radiator.   111

primary reactor starter (industrial control). A starter that includes a reactor connected in series with the primary winding of an induction motor to furnish reduced voltage for starting. It includes the necessary switching mechanism for cutting out the reactor and connecting the motor to the line. *See also:* **starter.** 225, 206

primary resistor starter (industrial control). A starter that includes a resistor connected in series with the primary winding of an induction motor to furnish reduced voltage for starting. It includes the necessary switching mechanism for cutting out the resistor and connecting the motor to the line. *See also:* **starter.** 206

primary service area (radio broadcast transmitter). The area within which reception is not normally subject to objectionable interference or fading. *See also:* **radio transmitter.** 328

primary standards (illuminating engineering). A light source by which the unit of light is established and from which the values of other standards are derived. This order of standard also is designated as the national standard. *Note:* A satisfactory primary (national) standard must be reproducible from specifications. Primary (national) standards usually are found in national physical laboratories.

primary supply voltage (mobile communication). The voltage range over which a radio transmitter, a radio receiver, or selective signaling equipment is designed to operate without degradation in performance. *See also:* **mobile communication system.** 181

primary switchgear connections. *See:* **main switchgear connections.**

primary transmission feeder. A feeder connected to a primary transmission circuit. *See also:* **center of distribution.** 64

primary unit substation. A unit substation in which the low-voltage section is rated above 1000 volts. 53

primary voltage rating (general purpose specialty transformer). The input circuit voltage for which the primary winding is designed, and to which operating and performance characteristics are referred. 53

primary winding (1) (general). The winding on the energy input side. 203, 53
(2) (voltage (regulator). The shunt winding. *See also:* **voltage regulator.** 257
(3) (rotating machinery) (motor or generator). The winding carrying the current and voltage of incoming power (for a motor) or power output (for a generator). The choice of what constitutes a primary circuit is arbitrary for certain machines having bilateral power flow. In a synchronous or direct-current machine, this is more commonly called the armature winding. See also: **armature.** 63

prime (charge-storage tubes). To charge storage elements to a potential suitable for writing. *Note:* This is a form of erasing. *See also:* **charge-storage tube; television.**
174, 190, 125

prime power (1) (transmission and distribution). The maximum potential power (chemical, mechanical, or hydraulic) constantly available for transformation into electric power. *See also:* **generating station.** 64
(2) (emergency and standby power systems). That source of supply of electric energy utilized by the user which is normally available continuously day and night, usually supplied by an electrical utility company but sometimes

by the owner generation. 89

priming rate (charge-storage tubes). The time rate of priming a storage element, line, or area from one specified level to another. Note the distinction between this and **priming speed.** *See also:* **charge-storage tube.**
174, 190

priming speed (charge-storage tubes). The lineal scanning rate of the beam across the storage surface in priming. Note the distinction between this and **priming rate.** *See also:* **charge-storage tube.** 174, 190, 125

primitive period (function). *See:* **period (function).**

principal axis (1) (close-talking pressure-type microphone). The axis of a microphone normal to the plane of the principal acoustic entrance of a microphone, and that passes through the center of the entrance. *See also:* **close-talking pressure-type microphone.** 249
(2) (transducer used for sound emission or reception). A reference direction used in describing the directional characteristics of the transducer. It is usually an axis of structural symmetry, or the direction of maximum response, but if one of these does not coincide with the reference direction, it must be described explicitly. 176

principal axis of compliance (gyro; accelerometer). An axis along which an applied force results in a displacement along that axis only. The acceleration squared error due to anisoelasticity is zero when acceleration is along a principal axis of compliance. 46

principal characteristics. *See:* **principal voltage-current characteristic.**

principal-city office (telephone switching systems). An intermediate office that has the screening and routing capabilities to accept traffic to all end offices within one or more numbering-plan areas. 55

principal current (thyristor). A generic term for the current through the collector junction. *Note:* It is the current through the main terminals. 243, 66, 208, 191

principal voltage (thyristor). The voltage between the main terminals. *Note:* In the case of reverse-blocking and reverse-conducting thyristors, the principal voltage is called positive when the anode potential is higher than the cathode potential and called negative when the anode potential is lower than the cathode potential. *See also:* **principal voltage-current characteristic.**
243, 66, 208, 191

principal restraint (accelerometer). The means by which a measurable force or torque is generated to oppose the force or torque produced by an acceleration along or about an input axis. 46

principal voltage-current characteristic (thyristor). The function, usually represented graphically, relating the principal voltage to the principal current with gate current, where applicable, as a parameter. *Syn.:* **principal characteristic.** 243, 66, 191

printed circuit (soldered connections). A pattern comprising printed wiring formed in a predetermined design in, or attached to, the surface or surfaces of a common base. 284

printed-circuit assembly. A printed-circuit board on which separately manufactured component parts have been added. 284

printed-circuit board (1) (general). A board for mounting of components on which most connections are made by printed circuitry. 59

(2) (double-sided). A board having printed circuits on both sides. 59

(3) (single-sided) A board having printed circuits on one side only. 59

printed wiring (soldered connections). A portion of a printed circuit comprising a conductor pattern for the purpose of providing point-to-point electric connections only.
284

printer (teleprinter) (teletypewriter). A printing telegraph instrument having a signal-actuated mechanism for automatically printing received messages. It may have a keyboard similar to that of a typewriter for sending messages. The term receiving-only is applied to a printer having no keyboard. *See also:* **telegraphy.** 194

printing. *See:* **line printing.**

printing demand meter. An integrated demand meter that prints on a paper tape the demand for each demand interval and indicates the time during which the demand occurred. *See also:* **electricity meter (meter).** 328

printing recorder (protective signaling). An electromechanical recording device that accepts electric signal impulses from transmitting circuits and converts them to a printed record of the signal received. *See also:* **protective signaling.** 328

printing telegraphy. That method of telegraph operation in which the received signals are automatically recorded in printed characters. *See also:* **telegraphy.** 194

printout (test, measurement and diagnostic equipment). The output of a device which is printed on some type of printer.
54

print-through. The undesirable transfer of a recorded signal from a section of a magnetic recording medium to another section of the same medium when these sections are brought into proximity. *Note:* The resulting copy usually is distorted. 176

priority string (power-system communication). A series connection of logic circuits such that inputs are accommodated in accordance with their position in the string, one end of the string corresponding to the highest priority. *See also:* **digital.** 59

privacy system (radio transmission). A system designed to make unauthorized reception difficult. *See also:* **radio transmission.** 328

private automatic branch exchange (PABX) (telephone switching systems). A private branch exchange that is automatic. 55

private automatic exchange (PAX) (telephone switching systems). A private non-branch exchange that is automatic. 55

private branch exchange (PBX) (telephone switching systems). A private telecommunications exchange that includes access to a public telecommunications exchange. 55

private-branch-exchange (PBX) central office trunk circuit (telephone switching systems). A trunk circuit located at the PBX end of a PBX trunk. 55

private-branch-exchange (PBX) extension number (telephone switching systems). A main station code for a PBX station. 55

private branch exchange hunting (telephone switching systems). An arrangement for searching over a group of trunks at the central office, any one of which would provide a connection to the desired private branch exchange.
55

private-branch-exchange trunk (PBX) (telephone switching systems). A line used as a trunk between a private branch exchange and the central office that serves it. 55

private exchange. A telephone exchange serving a single organization and having no means for connection with a public telephone exchange. 193, 101

private line (data transmission). A channel or circuit furnished a subscriber for his exclusive use. *Syn.:* **private wire.**
59

private line telegraph network (data transmission). A system of points interconnected by leased telegraph channels and providing hard-copy and/or five-track punched paper tape at both sending and receiving points. 59

private line telephone network (data transmission). A series of points interconnected by leased voice-grade telephone lines, with switching facilities or exchange operated by the customer. 59

private non-branch exchange (telephone switching systems). A private automatic telecommunications exchange that has no access to a public telecommunications exchange.
55

private residence. A separate dwelling or a separate apartment in a multiple dwelling that is occupied only by the members of a single family unit. 328

private residence elevator. A power passenger electric elevator, installed in a private residence, and that has a rated load not in excess of 700 pounds, a rated speed not in excess of 50 feet per minute, a net inside platform area not in excess of 12 square feet, and a rise not in excess of 50 feet. *See also:* **elevators.** 328

private-residence inclined lift. A power passenger lift, installed on a stairway in a private residence, for raising and lowering persons from one floor to another. *See also:* **elevator.** 328

private telecommunications exchange (telephone switching systems). A telecommunications exchange for a single organization. 55

probability distribution (nuclear power generating stations). The mathematical function that relates the probability of an event to an elapsed time or to a number of trials.
29, 31

probe (potential) (gas). An auxiliary electrode of small dimensions compared with the gas volume, that is placed in a gas tube to determine the space potential. *See also:* **discharge (gas).** 244, 190

probe loading. The effect of a probe on a network, for example, on a slotted line, the loading represented by a shunt admittance or a discontinuity described by a reflection coefficient. *See also:* **measurement system.** 185

probe pickup, residual. *See:* **residual probe pickup.**

problem. *See:* **benchmark problem.**

problem board (analog computer). In an analog computer, a removable frame of receptacles for patch cords and plugs that, through a patch bay, offers a means for interconnecting the inputs and outputs of computing elements, etcetera, according to the computer diagram. *Syn.:* **patch board, patch panel, prepatch panel.** 9

problem check. One or more tests used to assist in obtaining the correct machine solution to a problem.

Static check consists of one or more tests of computing elements, their interconnections, or both, performed under static conditions. **Dynamic check** consists of one or more tests of computing elements, their interconnections, or both, performed under dynamic conditions. **Rate test** is a test that verifies that the time constants of the integrators are those selected. This term also refers to the computer control state that implements the rate test described above. **Dynamic problem check** is any dynamic check used to ascertain that the computer solution satisfies the given system of equations. **Dynamic computer check** is any dynamic check used to ascertain the correct performance of some or all of the computer components. *See also:* **computer control state.** 9, 77

problem oriented language (computing systems). A programming language designed for the convenient expression of a given class of problems. 255, 77, 54

problem variable (analog computer). *See:* **scale factor.**

procedure (computing systems). The course of action taken for the solution of a problem. 255, 77

procedure-oriented language (computing systems). A programming language designed for the convenient expression of procedures used in the solution of a wide class of problems. 255, 77, 54

process (automatic control). The collective functions performed in and by the equipment in which a variable is to be controlled. *Syn.* **controlled system.** 56

process control. Control imposed upon physical or chemical changes in a material. *See also:* **control system, feedback.** 56

process equipment (automatic control). Apparatus with which physical or chemical changes in a material are produced. *Syn:* **plant.** *See also:* **directly controlled system; indirectly controlled system.** 56

processing. *See:* **data processing; information processing; multiprocessing; parallel processing.**

processor (computing systems) (1) (hardware). A data processor.
(2) (software). A computer program that includes the compiling, assembling, translating, and related functions for a specific programming language, for example, **Cobol** processor, **Fortran** processor. *See:* **data processor; multiprocessor.** 255, 77

production tests (1) (switchgear). Those tests made to check the quality and uniformity of the workmanship and materials used in the manufacture of switchgear or its components. 103, 202
(2) (power cable joints) (routine). Tests made on joint components or sub-assemblies during production for the purpose of quality control. 34

product modulator. A modulator whose modulated output is substantially equal to the product of the carrier and the modulating wave. *Note:* The term implies a device in which intermodulation between components of the modulating wave does not occur. *See also:* **modulation.** 328

product relay. A relay that operates in response to a suitable product of two alternating electrical input quantities. *See also:* **relay.** 103, 202

product sensitivity. The ratio of Hall voltage to the product of control current and magnetic flux density at any point on the product sensitivity characteristic curve of a Hall

generator. 107

prognosis (test, measurement and diagnostic equipment). The use of test data in the evaluation of a system or equipment for potential or impending malfunctions. 54

program (1) (general). A sequence of signals transmitted for entertainment or information. *See also:* **communication.** 239
(2) (electronic computation). (A) A plan for solving a problem. (B) Loosely, a routine. (C) To devise a plan for solving a problem. (D) Loosely, to write a routine. *See also:* **acceleration, programmed; communication; computer program; object program; source program; target program.** 235, 210, 255, 77
(3) (telephone switching systems). A set of instructions arranged in a predetermined sequence to direct the performance of a planned action or actions. 55

program amplifier. *See:* **amplifier, line.**

program level. The magnitude of program in an audio system expressed in volume units. 239

program library (computing systems). A collection of available computer programs and routines. 255, 77

programmable. That characteristic of a device that makes it capable of accepting data to alter the state of its internal circuitry to perform two or more specific tasks. 40

programmable measuring apparatus (programmable instrumentation). A measuring apparatus that performs specified operations on command from the system and, if it is a measuring apparatus proper, may transmit the results of the measurement(s) to the system. 40

programmable stimuli (test, measurement and diagnostic equipment). Stimuli that can be controlled in accordance with instructions from a programming device. 54

programmed check. A check procedure designed by the programmer and implemented specifically as a part of his program. *See:* **check, automatic; check problem; mathematical check.** 255, 77

programmed control (industrial control). A control system in which the operations are determined by a predetermined input program from cards, tape, plug boards, cams, etcetera. *See also:* **control system, feedback.** 219, 206

programmer (1) (power switchgear). An arrangement of operating elements or devices that initiates, and often controls, one or a series of operations in a given sequence. 103, 202
(2) (test, measurement and diagnostic equipment). (A) A device having the function of controlling the timing and sequencing operations; and (B) a person who prepares sequences of instructions for a programmable machine. 54

programmer—comparator (test, measurement and diagnostic equipment). (1) A device which reads commands and data from a sequential program usually on tape or cards; (2) sets up delays, switching, and stimuli, and performs measurements as directed by the program; and (3) compares the results of each measurement with fixed programmed tolerance limits to arrive at a decision. Often numerous other operations, such as branching on no-go or other conditions, are included. 54

programming (1) (electronic computation). The ordered listing of a sequence of events designed to accomplish a given task. *See:* **linear programming; multiprogramming;**

automatic programming.   244, 207, 54
(2) (power supplies). The control of any power-supply functions, such as output voltage or current, by means of an external or remotely located variable control element. Control elements may be variable resistances, conductances, or variable voltage or current sources. *See also:* **power supply.**   228, 186

programming language (computing systems). A language used to prepare computer programs.   255, 77, 54

programming, linear (1) general. Optimization problem characterization in which a set of parameter values are to be determined, subject to given linear constraints, optimizing a cost function that is linear in the parameter. *See also:* **system.**   209

(2) (computing systems). The analysis or solution of problems in which linear function of a number of variables is to be maximized or minimized when those variables are subject to a number of constraints in the form of linear inequalities.   255, 77

the same as the recovery-time specification for current-regulated operation; it is not related to the recovery-time specification for voltage-regulated operation. *See also:* **power supply.**   228, 186

program-sensitive fault (computing systems). A fault that appears in response to some particular sequence of program steps.   255, 77

program stop (numerically controlled machines). A miscellaneous function command to stop the spindle, coolant, and feed after completion of other commands in the block. It is necessary for the operator to push a button in order to continue with the remainder of the program.   244, 207

progressive grading (telephone switching systems). A grading in which the outlets of different grading groups are connected together in such a way that the number of grading groups connected to each outlet is larger for later choice outlets.   55

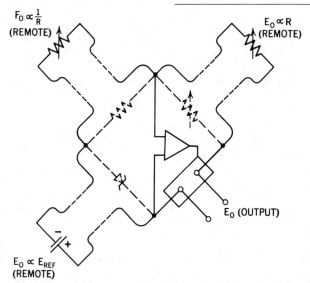

Remote programming connection showing programming of power supplies.

programming, nonlinear. Optimization problem in which any or all of the following are nonlinear in the variables: (1) The objective functions. (2) The defining interrelationships among the variables, the plant description. (3) The constraints. *See also:* **system.**   209

programming, quadratic. Optimization problem in which: (1) The objective function is a quadratic function of the variable. (2) The plant description is linear. *See also:* **system.**   209

programming relay. A relay whose function is to establish or detect electrical sequences.   103, 202, 60, 127

programming speed (power supplies). Describes the time requires to change the output voltage of a power supply from one value to another. The output voltage must change across the load and because the supply's filter capacitor forms a resistance-capacitance network with the load and internal source resistance, programming speed can only be described as a function of load. Programming speed is

progressive scanning* (television). *See:* **sequential scanning.**   163

* Deprecated

program tracking (communication satellite). A technique for tracking a satellite by pointing a high gain antenna towards the satellite which employs a computer program for antenna pointing; known orbital parameters are used as an input to the computer program.   84

projected peak point (tunnel-diode characteristic). The point on the forward current-voltage characteristic where the current is equal to the peak-point current and where the voltage is greater than the valley-point voltage. *See also:* **peak point (tunnel-diode characteristic).**   315, 191

projected peak-point voltage (tunnel-diode characteristic). The voltage at which the projected peak point occurs. *See also:* **peak point (tunnel-diode characteristic).**   315, 191

projection tube (electron device). A cathode-ray tube specifically designed for use with an optical system to produce a projected image. 190

projector. A lighting unit that, by means of mirrors and lenses, concentrates the light to a limited solid angle so as to obtain a high value of luminous intensity. *See also:* **luminaire.** 167

proof (suffix). Apparatus is designated as splashproof, dustproof, etcetera, when so constructed, protected, or treated that its successful operation is not interfered with when subjected to the specified material or condition. 53

proof mass (accelerometer). The effective mass whose inertia transforms an acceleration along or about an input axis into a force or torque. The effective mass takes into consideration flotation and contributing parts of the suspension. 46

proof test (1) (general). A test made to demonstrate that the item of equipment is in satisfactory condition in one or more respects. 210
(2) (rotating machinery) (withstand test). A "fail" or "no-fail" test of the insulation system of a rotating machine made to demonstrate whether the electrical strength of the insulation is above a predetermined minimum value. *See also:* **asynchronous machine; direct-current commutating machine; insulation testing (large alternating-current rotating machinery).** 6, 63

propagated potential (biological). A change of potential involving depolarization progressing along excitable tissue. 192

propagation (electrical practice). The travel of waves through or along a medium. *See also:* **signal wave.** 328

propagation constant (1) (traveling electromagnetic wave in a homogeneous medium). The negative of the natural logarithmic partial derivative, with respect to distance in the direction of the wave normal, of the phasor quantity describing the wave. *Notes:* (A) In the case of cylindrical or spherical traveling waves, the amplitude factors $1/(r)^{1/2}$ and $1/r$, respectively, are not to be included in the phasor quantity. (B) This is a complex quantity whose real part is the attenuation constant in nepers per unit length and whose imaginary part is the phase constant in radians per unit length. *See also:* **radio transmission; radio wave propagation.** 180, 146
(2) (transmission lines and transducers). (A) (per unit length of a uniform line). The natural logarithm of the ratio of the phasor current at a point of the line, to the phasor current at a second point, at unit distance from the first point along the line in the direction of transmission, when the line is infinite in length or is terminated in its characteristic impedance. **(B) (per section of a periodic line).** The natural logarithm of the ratio of the phasor current entering a section, to the phasor current leaving the same section, when the periodic line is infinite in length or is terminated in its iterative impedance. **(C) (of an electric transducer).** The natural logarithm of the ratio of the phasor current entering the transducer, to the phasor current leaving the transducer, when the transducer is terminated in its iterative impedance. 210
(3) (circuits and systems). The image transfer constant for a symmetrical transducer. 67

propagation delay (sequential events recording systems) (power generation). The time interval between the appearance of a signal at any circuit input and the appearance of the associated signal at that circuit output. 48, 58

propagation factor (electromagnetic wave propagating from one point to another). The ratio of the complex electric field strength at the second point to that value that would exist at the second point if propagation took place in a vacuum. 180, 146

propagation loss. The total reduction in radiant power surface density. The propagation loss for any path traversed by a point on a wave front is the sum of the spreading loss and the attenuation loss for that path. *See also:* **radio transmission.** 210

propagation mode (in a periodic beamguide) (laser-maser). A form of propagation characterized by identical field distributions over cross-sections of the beam at positions separated by one period of the guide. 363

propagation model. An empirical or mathematical expression used to compute propagation path loss. *See also:* **electromagnetic compatibility.** 199

propagation ratio. For a wave propagating from one point to another, the ratio of the complex field strength at the second point to that at the first point. 146

propagation vector (traveling electromagnetic wave at a given frequency). The complex vector whose real part is the attenuation vector and whose imaginary part is the phase vector. *See also:* **radio wave propagation.** 180, 146

propagation vector in physical media (antennas). The complex vector $\bar{\gamma}$ in plane wave solutions of the form $e^{-\bar{\gamma} \cdot \bar{r}}$ for an e^{jwt} time variation and $\bar{r}$ the position vector. *See also:* **attenuation vector in physical media; phase vector in physical media; wave vector in physical media; propagation constant in physical media.** 111

propeller-type blower (rotating machinery). An axial-flow fan with air-foil-shaped blades. *See also:* **fan (rotating machinery).** 63

proper operation. The functioning of the train control or cab signaling system to create or continue a condition of the vehicle apparatus that corresponds with the condition of the track of the controlling section when the vehicle apparatus is in operative relation with the track elements of the system. 328

proportional amplifier (industrial control). An amplifier in which the output is a single value and an approximately linear function of the input over its operating range. *See also:* **control system, feedback.** 219, 206

proportional control action (electric power systems). *See:* **control action, proportional.**

proportional counter tube. A radiation-counter tube designed to operate in the proportional region. *See also:* **gas-filled radiation-counter tubes.** 9

proportional gain (speed governing of hydraulic turbines). The proportional gain, G_p, of a proportional element is the ratio of the element's percent output to its percent input. A linear relationship is assumed. 8

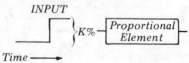

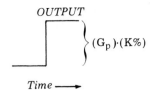

$$(G_p) \cdot (K\%)$$

Time ⟶

proportionality. *See:* **linearity.**

proportional plus derivative control. *See:* **control action, proportional plus derivative.**

proportional plus integral control. *See:* **control action, proportional plus integral.**

proportional plus integral plus derivative control. *See:* **control action proportional plus integral plus derivative.**

proportional region (radiation-counter tubes). The range of operating voltage for a counter tube in which the gas amplification is greater than unity and is independent of the amount of primary ionization. *Notes:* (1) In this region the pulse size from a counter tube is proportional to the number of ions produced as a result of the initial ionizing event. (2) The proportional region depends on the type and energy of the radiation. *See:* **gas-filled radiation-counter tubes.** 190, 125

proprietary system (protective signaling). A local system sounding and/or recording alarm and supervisory signals at a control center located within the protected premises, the control center being under the supervision of employees of the proprietor of the protected premises. *Note:* According to the United States Underwriters' rules, a proprietary system must be a recording system. *See also:* **protective signaling.** 328

propulsion-control transfer switch. Apparatus in the engine room for transfer of control from engine room to bridge and vice versa. *Note:* Engine-room control is provided on all ships. Bridge control with a transfer switch is optional and is used principally on small vessels such as tugs or ferries, usually with a direct-current propulsion system. 328

propulsion set-up switch. Apparatus providing ready means to set up for operation under varying conditions where practicable; for example, cutout of one or more generators when multiple units are provided. *See also:* **electric propulsion system.** 328

prorated section (surge arrester). A complete, suitably housed part of an arrester, comprising all necessary components, including gaseous medium, in such a proportion as to accurately represent, for a particular test, the characteristics of a complete arrester. 2

prorated unit (arrester). A completely housed prorated section of an arrester that may be connected in series with other prorated units to construct an arrester of higher voltage rating. 62

prospective current (available current) (1) (surge arresters). The root-mean-square symmetrical short-circuit current that would flow at a given point in a circuit if the arrester(s) at that point were replaced by links of zero impedance. 308, 62

(2) (circuit) (switching device situated therein). *See:* **available current (circuit) (switching device situated therein) (prospective current).** 103

prospective peak (crest) value (of a chopped impulse) (surge

arresters). The peak (crest) value of the full-wave impulse voltage from which a chopped impulse voltage is derived. 62

prospective peak value of test voltage (switching impulse testing). The voltage that would be obtained if no disruptive discharge occurred before the crest. 108

prospective short-circuit current (at a given point in a circuit). *See:* **available short-circuit current (at a given point in a circuit) (prospective short-circuit current).** 103

prospective short-circuit test current (at the point of test). *See:* **available short-circuit test current (at the point of test) (prospective short-circuit test current).**

protected enclosure. An enclosure in which all openings are protected with wire screen, expanded metal, or perforated covers. *Note:* A common form of specification for protected enclosure is: "The openings should not exceed ½ square inch (323 square millimeters) in area and should be of such shape as not to permit the passage of a rod larger than ½ inch (12.7 millimeters) in diameter, except where the distance of exposed live parts from the guard is more than 4 inches (101.7 millimeters) the openings may be ¾ square inch (484 square millimeters) in area and must be of such shape as not to permit the passage of a rod larger than ¾ inch (19 millimeters) in diameter." 3

protected location (computing systems). A storage location reserved for special purposes in which data cannot be stored without undergoing a screening procedure to establish suitability for storage therein. 255, 77

protected machine. *See:* **guarded machine.**

protected zone. *See:* **cone of protection.**

protection (computing systems). *See:* **storage protection.**

protection system (1) (nuclear power generating stations) (general). The electric and mechanical devices and circuitry (from sensors of the process variable to the actuation device input terminals) involved in generating those signals associated with the protective function. *Note:* These signals include those that actuate reactor trip and that, in the event of a serious reactor accident, actuate engineered safeguards such as containment isolation, core spray, safety injection, pressure reduction, and air cleaning. 31

(2) (protection systems). The electrical and mechanical devices (from measured process variables to protective action system input terminals) involved in generating those signals associated with the protective functions. These signals include those that actuate reactor trip and actuate engineered safety features (for example, containment isolation, core spray, safety injection, pressure reduction, and air cleaning). 20

protective action (1) (nuclear power generating stations) (general). (A) At the channel level, the initiation of a signal by a single channel when the variable sensed exceeds a limit. (B) At the system level, initiation of the operation of a sufficient number of actuators to effect a protective function. 20, 109, 159, 31

(2) (protection systems). The initiation of a signal or operation of equipment within the safety system for the purpose of accomplishing a protective function in response to a generating station condition having reached a limit specified in the design basis. *Notes:* (A) Protective action at the channel level is the initiation of a signal by a single channel when the sensed variable(s) reaches a specified limit. (B) Protective action at the system level is the operation of sufficient actuated equipment including the appropriate auxiliary supporting features to accomplish a protective function. Examples of protective actions at the system level are: rapid insertion of control rods closing of containment isolation valves, and operation of safety injection and core spray.   20, 356

protective action set point (nuclear power generating stations). The reference value to which the measured variable is compared for the initiation of protective action.   31

protective action system (nuclear power generating stations). An arrangement of equipment that performs a protective action when it receives a signal from the protection system.   20

protective covering (power cable joints). A field-applied material to provide environment protection over the joint and/or housing.   34

protective function (nuclear power generating stations) (general). (A) The sensing of one or more variables associated with a particular generating station condition, the signal processing and the initiation and completion of the protective action at values of the variables established in the design bases.   109, 20, 31
(B) Any one of the functions necessary to mitigate the consequences of a design basis event (for example, reduce power, isolate containment, or cool the core). *Note:* A protective function is design basis objective that must be accomplished; a sucessfully completed protective action at the system level, including the sensing of one or more variables, will accomplish the protective function. However, the design may be such that a given protective function may be accomplished by any one of several protective actions at the system level.   356

protective gap. A gap placed between live parts and ground to limit the maximum overvoltage that may occur.   103, 202, 27

protective lighting. A system intended to facilitate the nighttime policing of industrial and other properties. *See also:* **floodlighting.**   167

protective margin (surge arresters). The value of the protective ratio minus one expressed in percent $((PR - 1) \times 100)$.   62

protective power gap (series capacitor). A bypass gap which limits the voltage on the capacitor segment to a predetermined level when system fault occurs on the line, and which is capable of carrying capacitor discharge, system fault, and load currents for specified durations.   86

protective ratio (surge arresters). The ratio of the insulation withstand characteristics of the protected equipment to the arrester protective level, expressed as a multiple of the latter figure.   62

protective relay. A relay whose function is to detect defective lines or apparatus or other power-system conditions of an abnormal or dangerous nature and to initiate appropriate control circuit action. *Note:* A protective relay may be classified according to its input quantities, operating principle, or performance characteristics.   103, 202, 60, 127

protective screen (burglar-alarm system). A lightweight barrier of either solid strip or lattice construction, carrying

electric protection circuits, and barring access through a normal opening to protected premises. *See also:* **protective signaling.**   328

protective signaling. Protective signaling comprises the initiation, transmission, and reception of signals involved in the detection and prevention of property loss or damage due to fire, burglary, robbery, and other destructive conditions, and in the supervision of persons and of equipment concerned with such detection and prevention.   328

protector tube (1) (surge arresters). An expulsion arrester used primarily for the protection of line ard switch insulation.   2, 299, 62
(2) (electron-tube type). A glow-discharge cold-cathode tube that employs a low-voltage breakdown between two or more electrodes to protect circuits against overvoltage.   190

protocol (data communication). A formal set of conventions governing the format and relative timing of message exchange between two communications terminals. *See also:* **control procedure.**   12

proton microscope. A device similar to the electron microscope but in which the charged particles are protons. *See also:* **electron optics.**   244, 190

proton range (solar cells). The maximum distance traversed through a material by a proton of a given energy.   113

prototype standard. A concrete embodiment of a physical quantity having arbitrarily assigned magnitude, or a replica of such embodiment. *Note:* As an illustration of the distinction between prototype standard and unit, the length of the United States Prototype Meter Bar is not exactly one meter.   210

proximal (distal) point (pulse terms). A magnitude referenced point at the intersection of a waveform and a proximal (distal) line. *See:* The single pulse diagram below the **waveform epoch** entry.   254

proximity-coupled dipole array antenna. An array antenna consisting of a series of coplanar dipoles, loosely coupled to the electromagnetic field of a balanced transmission line, the coupling being a function of the proximity and orientation of the dipole to the transmission line.   111

proximity effect (electric circuits and lines). The phenomenon of non-uniform current distribution over the cross section of a conductor caused by the time variation of the current in a neighboring conductor. *See also:* **induction heating.**   14, 210

proximity-effect error (electronic navigation systems). An error in determination of system performance caused by improper use of measurements made in the near field of the antenna system. *See also:* **navigation.**   187, 13

proximity-effect ratio (power distribution, underground cables). The quotient obtained by dividing the alternating-current resistance of a cable conductor subject to proximity effect, by the alternating-current resistance of an identical conductor free of proximity effect. *See also:* **power distribution underground construction.**   57

proximity influence. The percentage change in indication caused solely by the fields produced from two edgewise instruments mounted in the closest possible proximity, one on each side (or above and below for horizontal-scale instruments). *Note:* Proximity influence of alternating-current instruments on either alternating-current or direct-current types is determined by energizing two in-

struments, one on each side of the test instrument (or above and below) at 90 percent of end-scale value (in phase with the current in the instrument under test, if the latter is alternating current). The current in the two outside instruments only shall be reversed. For rating purposes, the proximity influence shall be taken as one-half the difference in the readings in percentage of full scale. In direct-current permanent-magnet moving-coil instruments the field produced by the current in the instrument is small compared with the field from the permanent magnet. The proximity influence on either an alternating-current or direct-current test instrument will be the difference in reading, expressed as a percentage of full-scale value, of the instrument under test mounted alone on the panel, compared with the reading when two direct-current instruments are mounted in closest possible proximity, each with current applied to give 90-percent end-scale deflection. All three instruments shall be of the same manufacture and size. *See also:* **accuracy rating (instrument).**
     280

proximity switch (industrial control). A device that reacts to the proximity of an actuating means without physical contact or connection therewith. *See also:* **switch.**
     225, 206

P **scan (electronic navigation).** *See:* **plan-position indicator.**

P **scope (electronic navigation).** *See:* **plan-position indicator.**

pseudo-code (test, measurement and diagnostic equipment). An arbitrary code, independent of the hardware of a computer, which has the same general form as actual computer code but which must be translated into actual computer code if it is to direct the computer.   54

pseudo-instruction (test, measurement and diagnostic equipment). An instruction which resembles the instructions acceptable to the computer but which must be translated into actual computer instructions in order to control the computer.   54

pseudolatitude (navigation). A latitude in a coordinate system that has been arbitrarily displaced from the earth's conventional latitude system so as to move the meridian convergence zone (polar region) away from the place of intended operation. *See also:* **navigation.**   187, 13

pseudolongitude (navigation). A longitude in a coordinate system that has been arbitrarily displaced from the earth's conventional longitude system so as to move the meridian convergence zone (polar region) away from the place of intended operation. *See also:* **navigation.**   187, 13

pseudonoise (PN) sequence (communication satellite). A binary sequence with a very desirable transorthogonal auto-correlation property. In space communications commonly used for synchronization and ranging.   84

pseudorandom number sequence. A sequence of numbers, determined by some defined arithmetic process, that is satisfactorily random for a given purpose, such as by satisfying one or more of the standard statistical tests for randomness. Such a sequence may approximate any one of several statistical distributions, such as uniform distribution or normal Gaussian distribution.   255, 77

PSK. *See:* **phase-shift keying.**

PTM. *See:* **pulse-time modulation.**

p-**type crystal rectifier.** A crystal rectifier in which forward

current flows when the semiconductor is positive with respective to the metal. *See also:* **rectifier.**   328

p-**type semiconductor.** *See:* **semiconductor,** *p*-**type.**

public-address system. A system designed to pick up and amplify sounds for an assembly of people.   328

public telecommunications exchange (telephone switching systems). A telecommunications exchange that serves the public.   55

public telephone station (pay station). A station available for use by the public, generally on the payment of a fee that is deposited in a coin collector or is paid to an attendant. *See also:* **telephone station.**   328

pull blade (cable plowing). A plow blade used to pull direct burial conductors into position by means of a suitable pulling grip attachment at the heel of the blade.   52

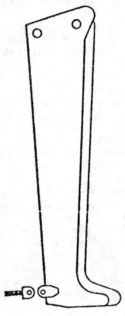

Pull blade.

pull box. A box with a blank cover that is inserted in one or more runs of raceway to facilitate pulling in the conductors, and may also serve the purpose of distributing the conductors. *See also:* **cabinet.**   328

puller (two drum; three drum) (conductor stringing equipment). Drum puller equipped with three drums and thus can pull one, two or three conductors individually or simultaneously. *See also:* **drum puller.** *Syn.* **two drum winch; double drum hoist; double drum winch; three drum winch; triple drum hoist; triple drum winch; tugger.**   45

puller, bullwheel (conductor stringing equipment). A device designed to pull pulling lines and conductors during stringing operations. It normally incorporates one or more pairs of urethane or neoprene lined, power driven, single or multiple groove bullwheels where each pair is arranged in tandem. Pulling is accomplished by friction generated against the pulling line or conductor which is reeved around the grooves of a pair of the "bullwheel." The puller is usually equipped with its own engine which drives the bullwheels mechanically, hydraulically, or through a combination of both. Some of these devices function as either a puller or tensioner. *Syn.* **puller.**   45

puller, drum (conductor stringing equipment). A device designed to pull a conductor during stringing operations. It is normally equipped with its own engine which drives the drum mechanically, hydraulically, or through a combination of both. It may be equipped with synthetic fiber rope or wire rope to be used as the pulling line. The pulling line is payed out from the unit, pulled through the travelers in the sag section, reeved through the tensioner (if one is used), and attached to the conductor on the payout reel. The conductor is then pulled in by winding the pulling line back onto the drum. This unit is sometimes used with the synthetic fiber rope acting as a pilot line to pull heavier pulling lines across canyons, rivers, etcetera. *Syn.* **hoist; single drum hoist; single drum winch; tugger.**   45

puller, reel (conductor stringing equipment). A device designed to pull a conductor during stringing operations. It is normally equipped with its own engine which drives the supporting shaft for the reel mechanically, hydraulically or through a combination of both. The shaft, in turn, drives the reel. The application of this unit is essentially the same as that for the drum puller previously described. Some of these devices function as either a puller or tensioner.   45

pulley (sheave) (rotating machinery). A shaft-mounted wheel used to transmit power by means of a belt, chain, band, etcetera. *See also:* **rotor (rotating machinery).**   63

pulling eye. A device that may be fastened to the conductor or conductors of a cable or formed by or fastened to the wire armor and to which a hook or rope may be directly attached in order to pull the cable into or from a duct. *Note:* Pulling eyes are sometimes equipped, like test caps, with facilities for oil feed or vacuum treatment. *See also:* **power distribution, underground construction.**   64

pulling figure (oscillator). The difference between the maximum and minimum values of the oscillator frequency when the phase angle of the load-impedance reflection coefficient varies through 360 degrees, while the absolute value of this coefficient is constant and equal to a specified value, usually 0.20. (Voltage standing-wave ratio 1.5.) *See also:* **oscillatory circuit; waveguide.**   190, 125

pulling into synchronism (rotating machinery). The process of synchronizing by changing from asynchronous speed to synchronous.   63

pulling out of synchronism (rotating machines). The process of losing synchronism by changing from synchronous speed to a lower asynchronous speed (for a motor) or higher asynchronous speed (for a generator).   63

pulling vehicle (conductor stringing equipment). Any piece of mobile ground equipment capable of pulling pilot lines, pulling lines or conductors. However, helicopters may be considered as a "pulling vehicle" when utilized for the same purpose.   45

pull-in test (synchronous machine). A test taken on a machine that is pulling into synchronism from a specified slip.   63

pull in time (acquisition time) (communication satellite). The time required for achieving synchronization in a phase-lock loop.   85

pull-in torque (synchronous motor). The maximum constant torque under which the motor will pull its connected inertia load into synchronism, at rated voltage and frequency, when its field excitation is applied. *Note:* The speed to which a motor will bring its load depends on the power required to drive it and whether the motor can pull the load into step from this speed depends on the inertia of the revolving parts, so that the pull-in torque cannot be determined without having the Wk^2 as well as the torque of the load.   63

pull or transfer box. A box without a distribution panel, within which one or more corresponding electric circuits are connected or branched.   64

pull-out test (rotating machinery). A test to determine the conditions under which an alternating-current machine develops maximum torque while running at specified voltage and frequency. *See also:* **asynchronous machine.**   63

pull-out torque (synchronous machine). The maximum sustained torque which the machine will develop at synchronous speed with rated voltage applied at rated frequency and with normal excitation.   63, 3

pull-up torque (alternating-current motor). The minimum torque developed by the motor during the period of acceleration from rest to the speed at which breakdown torque occurs with rated voltage applied at rated frequency.   63

pulsating function. A periodic function whose average value over a period is not zero. For example, $f(t) = A + B \sin \omega t$ is a pulsating function where neither A nor B is zero.   210

pulse (1) (general) (relaying). A brief excursion of a quantity from normal (its initial level).   59, 79

(2) (modulation). A single disturbance characterized by the rise and decay in time or space or both of a quantity whose value is normally constant. *Notes:* (A) In these modulation definitions, a radio-frequency carrier, amplitude-modulated by a pulse, is not considered to be a pulse. (B) This definition is broad so that it covers almost any transient phenomenon. The only features common to all pulses are rise, finite duration, and decay. It is necessary that the rise, duration, and decay be of a quantity that is constant (not necessarily zero) for some time before the pulse and has the same constant value for some time afterwards. The quantity has a normally constant value and is perturbed during the pulse. No relative time scale can be assigned.   111

(3) (industrial control). A signal of relatively short duration.   219, 206

(4) (pulse) (impulse*). A sudden change of brief duration produced in the current or voltage of a circuit in order to actuate or control an instrument, meter, or relay.   212

*Deprecated

(5) (automatic control). A variation of a signal whose value is normally constant; this variation is characterized by a rise and a decay, and has a finite duration.   56

(6) (pulse terms). A wave which departs from a first nominal state, attains a second nominal state, and ultimately returns to the first nominal state. Throughout the remainder of this document the term pulse is included in the term wave.   254

pulse accumulator (register) (telemeter system). A device that accepts and stores pulses and makes them available for readout on demand.   103, 202

pulse advance (delay) (pulse terms). The occurrence in time of one pulse waveform before (after) another pulse waveform.   254

pulse advance (delay) interval (pulse terms). The interval by which, unless otherwise specified, the pulse start time of one pulse waveform precedes (follows), unless otherwise specified, the pulse start time of another pulse waveform.
254

pulse amplifier (1) (pulse techniques). An amplifier designed specifically for the amplification of electric pulses. *See also:* **pulse.**   335

(2) (watthour meter). A device used to change the amplitude or waveform of a pulse for retransmission to another pulse device. *See also:* **auxiliary device to an instrument.**
212

pulse amplitude (1) (general). A general term indicating the magnitude of a pulse. *Notes:* (A) For specific designation, adjectives such as average, instantaneous, peak, root-mean-square (effective), etcetera, should be used to indicate the particular meaning intended. (B) Pulse amplitude is measured with respect to the normally constant value unless otherwise stated. *See also:* **average absolute pulse-amplitude; average pulse amplitude; signal.**
254, 178, 188

(2) (light emitting diodes). A general term indicating the magnitude of a pulse measured with respect to the normally constant value unless otherwise stated.   162

(3) (high- and low-power pulse transformers) (A_M). That quantity determined by the intersection of a line passing through the points of the leading edge where the instantaneous value reaches 10 percent and 90 percent of (A_M) and a straight line which is the best least-squares fit to the pulse in the pulse-top region. (Usually this is fitted visually rather than numerically.) For pulses deviating greatly from the ideal trapezoidal pulse shape, a number of successive approximations may be necessary to determine A_M. *Note:* The pulse amplitude (A_M) may be arrived at by applying the following procedure. Step 1: Visually or numerically determine the best straight line fit to the pulse in the pulse-top region and extend this straight line into the leading-edge region. Step 2: An initial estimate of A_M is the first intersection of the pulse (in the late leading-edge or early pulse-top regions) with the straight line fitted to the pulse top. Step 3: Using the estimate of A_M calculate 0.1 A_M and 0.9 A_M and draw a straight line through these two points of the pulse-leading edge. Step 4: The intersection of the leading-edge straight line and the pulse-top straight line gives an improved estimate of A_M. Step 5: Repeat steps 3 and 4, until the estimate of A_M does not change. The converged estimate is the pulse amplitude A_M. *See also:* **input pulse shape.**   32, 33

(4) (pulse terms). The algebraic difference between the top magnitude and the base magnitude. *See:* The single pulse diagram below the **waveform epoch** entry.   254

pulse-amplitude modulation (PAM). Modulation in which the modulating wave is caused to amplitude modulate a pulse carrier.   111, 254, 194

pulse amplitude, peak. *See:* **peak pulse amplitude.**

pulse amplitude, root-mean-square (effective). *See:* **root-mean-square (effective) pulse amplitude.**

pulse average time (t_w) **(light emitting diodes).** The time interval between the instants at which the instantaneous pulse amplitude first and last reaches a specified fraction of the peak pulse amplitude, namely, 50 percent.   162

pulse bandwidth. The smallest continuous frequency interval outside of which the amplitude (of the spectrum) does not exceed a prescribed fraction of the amplitude at a specified frequency. Caution: This definition permits the spectrum amplitude to be less than the prescribed amplitude within the interval. *Notes:* (1) Unless otherwise stated, the specified frequency is that at which the spectrum has its maximum amplitude. (2) This term should really be pulse spectrum bandwidth because it is the spectrum and not the pulse itself that has a bandwidth. However, usage has caused the contraction and for that reason the term has been accepted. *See also:* **signal.**
254, 178, 188, 162

pulse base (pulse waveform) (pulse techniques). That major segment having the lesser displacement in amplitude from the baseline, excluding major transitions. *See also:* **pulse.**
185

pulse, bidirectional. *See:* **bidirectional pulse.**

pulse burst (1) (radar). A sequence of pulses, usually generated coherently and batch-processed for Doppler resolution, and often having a total duration less than the radar echo time.   13

(2) (pulse terms). A finite sequence of pulse waveforms.
254

pulse burst base envelope (pulse terms). Unless otherwise specified, the waveform defined by a cubic natural spline with knots at (1) that point of intersection of the pulse burst top envelope and the pulse burst waveform which precedes the first pulse waveform in a pulse burst, (2) each point of intersection of the base center line and the baseline between adjacent pulse waveforms in a pulse burst, and (3) that point of intersection of the pulse burst top envelope and the pulse burst waveform which follows the last pulse waveform in a pulse burst. *See:* Pulse burst envelopes diagram, below the **knot** entry.   254

pulse burst time-related definitions (pulse terms). (1) pulse burst duration. The interval between the pulse start time of the first pulse waveform and the pulse stop time of the last pulse waveform in a pulse burst. (2) pulse burst separation. The interval between the pulse stop time of the last pulse waveform in a pulse burst and the pulse start time of the first pulse waveform of the immediately following pulse burst. (3) pulse burst repetition period. The interval between the pulse start time of the first pulse waveform in a pulse burst and the pulse start time of the first pulse waveform in the immediately following pulse burst in a sequence of periodic pulse bursts. (4) pulse burst reception frequency. The reciprocal of pulse burst repetition period.
254

pulse burst top envelope (pulse terms). Unless otherwise specified, the waveform defined by a cubic natural spline with knots at (1) the first transition mesial point of the first pulse waveform in a pulse burst, (2) each point of intersection of the top centerline and the topline of each pulse waveform in a pulse burst, and (3) the last transition mesial point of the last pulse waveform in a pulse burst. *See:*

Pulse burst envelopes diagram below the **knot** entry.

254

pulse carrier. A carrier consisting of a series of pulses. *Note:* Usually, pulse carriers are employed as subcarriers. *See also:* **carrier.** 111, 254

pulse, carrier-frequency. *See:* **carrier-frequency pulse.**

pulse code. (1) A pulse train modulated so as to represent information. (2) Loosely, a code consisting of pulses, such as Morse code, Baudot code, binary code. *See also:* **pulse.** 254, 185

pulse-code modulation (1) (general). A modulation process involving the conversion of a waveform from analog to digital form by means of coding. *Notes:* (A) This is a generic term, and additional specification is required for a specific purpose. (B) The term is commonly used to signify that form of pulse modulation in which a code is used to represent quantized values of instantaneous samples of the signal wave. 242, 254

(2) (data transmission) (PCM). A type of pulse modulation where the magnitude of the signal is sampled and each sample is approximated to a nearest reference level (this process is called quantizing). Then a code, which represents the reference level, is transmitted to the distant location. See below for an example of one form of PCM which has eight reference levels. It can be seen that a straight binary code would require a group of three pulses to be transmitted for each sample. The main advantage of PCM is the fact that at the receiving end only the presence or absence of a pulse must be detected. 59

coincidence (noncoincidence) exists. 254

pulse compression (radar). The coding and processing of a signal pulse of long time duration, to one of short time duration and high range resolution, while maintaining the benefits of high pulse energy. 13

pulse corner (pulse waveform feature). A continuous pulse waveform feature of specified extent which includes a region of maximum curvature or a point of discontinuity in the waveform slope. Unless otherwise specified, the extent of the corners in a rectangular or trapezoidal pulse waveform are as specified in the following table:

Corner	First Point	Last Point
First	first base point	first transition proximal point
Second	first transition distal point	top center point
Third	top center point	last transition distal point
Fourth	last transition proximal point	last base point

pulse corrector (telephone switching systems). Equipment to reestablish, within predetermined limits, the make/break ratio of dial pulses. 55

pulse counter (pulse techniques). A device that indicates or records the total number of pulses that it has received during a time interval. *See also:* **pulse.** 117

pulsed Doppler radar. A pulsed radar system which utilizes

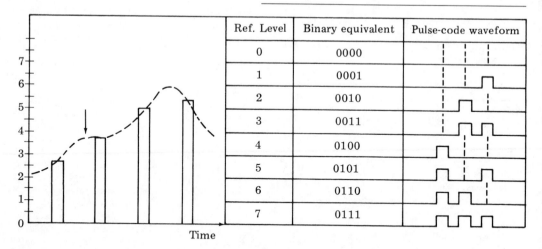

Ref. Level	Binary equivalent	Pulse-code waveform
0	0000	
1	0001	
2	0010	
3	0011	
4	0100	
5	0101	
6	0110	
7	0111	

A form of pulse code modulation

pulse coder (radar). A device for varying one or more of the characteristics of a pulse or of a pulse train so as to transmit information. *See also:* **navigation.** 187, 13

pulse coincidence (noncoincidence) (pulse terms). The occurrence (lack of occurrence) of two or more pulse waveforms in different waveforms either essentially simultaneously or for a specified interval. 254

pulse coincidence (noncoincidence) duration (pulse terms). The interval between specified points on two or more pulse waveforms in different waveforms during which pulse

the Doppler effect for obtaining velocity information relative to the target. *See also:* **navigation.** 187, 13

pulse decay time (1) (light emitting diodes) (general) (pulse fall time) (t_f). The interval between the instants at which the instantaneous amplitude last reaches specified upper and lower limits, namely, 90 percent and 10 percent of the peak pulse amplitude unless otherwise stated. *See also:* **pulse.** 178, 59, 119, 162, 118, 23

(2) (data transmission). The interval of time required for the trailing edge of a pulse to decay from 90 percent to 10 percent of the peak pulse amplitude.   59

(3) (semiconductor radiation detectors) (last transition time) (t_d). The interval between the instants at which the instantaneous value last reaches specified upper and lower limits, namely, 90 and 10 percent of the peak pulse value unless otherwise stated. (In the case of a step function applied to an amplifier that has simple $CR—RC$ shaping, the decay time is given by $t_d = 3.36\ CR$.)   23

pulse decoder (radar). A device for extracting information from a pulse-coded signal. *See also:* **navigation.**   187

pulse delay time (t_d) (light emitting diodes). The interval between the instants at which the instantaneous amplitudes of the input pulse and output pulses first reach a specified fraction of their peak pulse amplitudes, namely, 10 percent.   162

pulse delay, transducer. *See:* **delay, pulse.**

pulse delay, transmitter. *See:* **delay, pulse.**

pulse devices (watthour meter). The functional units for initiating, transmitting, retransmitting, and receiving electric pulses, representing finite quantities of energy normally transmitted from some form of watthour meter to a receiver unit for billing purposes. *See also:* **auxiliary device to an instrument.**   212

pulsed laser (laser-maser). A laser which delivers its energy in the form of a single pulse or train of pulses. The duration of a pulse ≤ 0.25 s.   363

pulsed oscillator. An oscillator that is made to operate during recurrent intervals by self-generated or externally applied pulses. *See also:* **oscillatory circuit.**   111

pulsed-oscillator starting time. The interval between the leading-edge pulse time of the pulse at the oscillator control terminals and the leading-edge pulse time of the related output pulse.   254

pulse droop (television). A distortion of an otherwise essentially flat-topped rectangular pulse characterized by a decline of the pulse top. *See also:* **television.**
                                                     254, 178

pulse duration (1) (loosely). The duration of a rectangular pulse whose energy and peak power equal those of the pulse in question. *Note:* When determining the peak power, any transients of relatively short duration are frequently ignored. *See also:* **phase modulation; pulse.**
                                                          111

(2) (television) (radiation counters) (telecommunications). The time interval between the first and last instants at which the instantaneous amplitude reaches a stated fraction of the peak pulse amplitude. *See also:* **pulse; signal.**   254, 178, 188, 59, 194, 117

(3) (high- and low-power pulse transformers) (90 percent) (T_d). The time interval between the instants at which the instantaneous value reaches 90 percent of A_M on the leading edge and 90 percent of A_T on the trailing edge. *Notes:* (A) Often the input pulse tilt (droop) is only a few percent, and in those cases pulse duration may be considered as the time interval between the first and last instants at which the instantaneous value reaches 90 percent of A_M. (B) Pulse duration may be specified at a value other than 90 percent of A_M and A_T in special cases. *See also:* **input pulse shape.**   32, 33

(4) (pulse terms). The duration between pulse start time and pulse stop time. *See:* The single pulse diagram below the **waveform epoch** entry.   254

(5) (laser-maser). The time duration of a laser pulse; usually measured as the time interval between the half-power points on the leading and trailing edges of the pulse.
                                                          363

pulse-duration discriminator (radar). A circuit in which the output is a function of the deviation of the input pulse duration from a reference. *See also:* **navigation.**
                                                        13, 187

pulse-duration modulation (PDM). Pulse-time modulation in which the value of each instantaneous sample of the modulating wave is caused to modulate the duration of a pulse. *Notes:* (1) The deprecated terms **pulsewidth modulation** and **pulse-length modulation** also have been used to designate this system of modulation. (2) In **pulse-duration modulation,** the modulating wave may vary the time of occurrence of the leading edge, the trailing edge, or both edges of the pulse.   111, 242, 254

pulse-duration telemetering (pulse-width modulation) (electric power systems). A type of telemetering in which the duration of each transmitted pulse is varied as a function of the magnitude of the measured quantity. *See also:* **telemetering.**   94

pulse duration time (t_p) (light emitting diodes). The time interval between the first and last instants at which the instantaneous amplitude reaches a stated fraction of the peak pulse amplitude, namely, 90 percent.   162

pulse duty factor (light emitting diodes). The ratio of the average pulse duration to the average pulse spacing. *Note:* This is equivalent to the product of the average pulse duration and the pulse-repetition rate.   254, 162

pulse energy (pulse terms). The energy transferred or transformed by a pulse(s). Unless otherwise specified by a mathematical adjective, the total energy over a specified interval is assumed. *See:* **mathematical adjectives.**   254

pulse fall time (photomultipliers for scintillation counting). The interval between the instants at which the instantaneous amplitude last reaches specified upper and lower limits, namely, 90 percent and 10 percent of the peak pulse amplitude unless otherwise stated.   117

pulse-forming line (radar modulators). An electric network whose parameters are selected to give a specified shape to the modulation pulse. *See also:* **navigation.**   187, 13

pulse-frequency modulation (PFM). A form of pulse-time modulation in which the pulse repetition rate is the characteristic varied. *Note:* A more precise term for pulse-frequency modulation would be pulse-repetition-rate modulation.   242, 254, 194

pulse frequency modulation (PFM) telemetry (communication satellite). A telemetry system where the information is coded according to subcarrier frequency, pulse duration and pulse repetition rate. Often used for satellite telemetry.
                                                           83

pulse, Gaussian. A pulse shape tending to follow the Gaussian curve corresponding to $A(t) = e^{-a(b-t)^2}$. *See also:* **pulse.**   185

pulse-height analyzer (radiation counters). An instrument capable of indicating the number or rate of occurrence of pulses falling within each of one or more specified amplitude ranges. *See also:* **pulse.**   117

pulse-height resolution constant, electron (photomultipliers). The product of the square of the electron (photomultiplier) pulse-height resolution expressed as the fractional full width at half maximum (FWHM/A_1), and the mean number of electrons per pulse from the photocathode. *See also:* **phototube.**   335

pulse-height resolution, electron (photomultiplier). A measure of the smallest change in the number of electrons in a pulse from the photocathode that can be discerned as a change in height of the output pulse. Quanitatively, it is the fractional standard deviation (σ/A_1) of the pulse-height distribution curve for output pulses resulting from a specified number of electrons per pulse from the photocathode. *Note:* The fractional full width at half maximum of the pulse-height distribution curve (FWHM/A_1) is frequently used as a measure of this resolution, where A_1 is the pulse height corresponding to the maximum of the distribution curve. *See also:* **pulse.**   117

pulse-height selector (pulse techniques). A circuit that produces a specified output pulse when and only when it receives an input pulse whose amplitude lies between two assigned values. *See:* **pulse.**   335

pulse initiator. Any device, mechanical or electrical, used with a meter to initiate pulses, the number of which are proportional to the quantity being measured. *Note:* The complete pulse initiator may include an external amplifier or auxiliary relay assembly or both. *See also:* **auxiliary device to an instrument.**   212

pulse initiator gear ratio (watthour meter). The ratio of disk shaft revolutions to revolutions of pulse initiating shaft. *Note:* It is commonly denoted by the symbol P_g. *See also:* **auxiliary device to an instrument.**   212

pulse-initator ratio (watthour meter). The ratio of revolutions of first gear of pulse device to revolutions of pulse initiating shaft. *Note:* It is commonly denoted by the symbol P_r. *See also:* **auxiliary device to an instrument.**   212

pulse-initiator shaft reduction (watthour meter). The ratio of meter disk shaft revolutions to the revolutions of the first gear of pulse initiator. *Note:* It is commonly denoted by the symbol P_s. *See also:* **auxiliary device to an instrument.**   212

pulse interleaving. A process in which pulses from two or more sources are combined in time-division multiplex for transmission over a common path.   254

pulse interrogation. The triggering of a transponder by a pulse or pulse mode. *Note:* Interrogations by means of pulse modes may be employed to trigger a particular transponder or group of transponders.   254

pulse interval.* *See:* **pulse spacing.**

* Deprecated

pulse-interval modulation. A form of pulse-time modulation in which the pulse spacing is varied.   242, 254

pulse jitter. A relatively small variation of the pulse spacing in a pulse train. *Note:* The jitter may be random or systematic, depending on its origin, and is generally not coherent with any pulse modulation imposed.   254

pulse-length modulation. *See:* **pulse-duration modulation.**

pulse measurement. The assignment of a number and a unit of measurement to a characteristic, property, or attribute of a pulse wherein the number and unit assigned indicate the magnitude of the characteristic which is associated with the pulse. Typically, this assignment is accomplished by comparison of a transform of the pulse, its pulse waveform, with a scale or reference which is calibrated in the unit of measurement. *See:* **method of pulse measurement.**   15

pulse measurement process. A realization of a method of pulse measurement in terms of specific devices, apparatus, instruments, auxiliary equipment, conditions, operators, and observers. *See also:* **method of pulse measurement.**   15

pulse mode. (1) A finite sequence of pulses in a prearranged pattern used for selecting and isolating a communication channel. (2) The prearranged pattern.   254

pulse-mode multiplex. A process or device for selecting channels by means of pulse modes. *Note:* This process permits two or more channels to use the same carrier frequency.   254

pulse mode, spurious. *See:* **spurious pulse mode.**

pulse-modulated radar. A form of radar in which the radiation consists of a series of discrete pulses. *See also:* **radar.**   328

pulse modulation (1) (continuous-wave). (A) Modulation of a carrier by a pulse train. *Note:* In this sense, the term is used to describe the process of generating carrier-frequency pulses. (B) Modulation of one or more characteristics of a pulse carrier. *Note:* In this sense, the term is used to describe methods of transmitting information on a pulse carrier.   242

(2) (data transmission). The use of a series of pulses to convey the information contained in the modulating function. The characteristics of a train of pulses may be modified in one of several ways to convey information including amplitude (PAM) position (PPM) and duration (PDM)   59

pulse modulation, width. *See:* **pulse duration modulation.**

pulse modulator. A device that applies pulses to the element in which modulation takes place. *See also:* **modulation.**   111

pulse multiplex*. *See:* **pulse-mode multiplex.**

* Deprecated

pulse operation. The method of operation in which the energy is delivered in pulses. *Note:* Pulse operation is usually described in terms of the pulse shape, the pulse duration, and the pulse-recurrence frequency. *See also:* **pulse.**   59

pulse packet (pulse-radar system). The volume of space occupied by a single pulse. The dimensions of this volume are determined by the angular width of the beam, the duration of the pulse, and the distance from the antenna. *Syn.* **radar resolution cell.** *See also:* **radio navigation.**   13, 187

pulse-pair resolution (photomultiplier). The time interval between two equal-amplitude delta-function optical pulses such that the valley between the two corresponding anode pulses falls to fifty percent of the peak amplitude.   117

pulse permeability (magnetic core testing). The value of amplitude permeability when the rate of change of induction (that is, the exciting voltage) is held substantially constant over a period of time during each cycle. The frequency, amplitude, duration of the exciting voltage, and the time interval for which the permeability is measured must be stated.

$$\mu_\pi = \frac{1}{\mu_0} \frac{\Delta B}{\Delta H}$$

μ_π = pulse permeability, relative

ΔB = change in induction during the stated time interval

ΔH = associated change in magnetic field strength

Note: When pulse permeability is to be related to a specific circuit condition, a second subscript may be used, for example, $\mu_{\pi a}$ would represent the relative amplitude permeability determined under pulsed excitation.   165

pulse-position modulation (pulse-phase modulation) (PPM). Pulse-time modulation in which the value of each instantaneous sample of a modulating wave is caused to modulate the position in time of a pulse.   111

pulse power (pulse terms). The power transferred or transformed by a pulse(s). Unless otherwise specified by a mathematical adjective average power over a specified interval is assumed. *See:* **mathematical adjectives.**   254

pulse power, carrier-frequency peak. *See:* **peak pulse power, carrier-frequency.**

pulse power, peak. *See:* **peak pulse power.**

pulse quadrant (pulse waveform feature. One of the four continuous and contiguous waveform features of specified extent which include a region of maximum curvature or a point of discontinuity in the waveform slope. Unless otherwise specified, the extent of the quadrants in a rectangular or trapezoidal pulse waveform are as specified in the following table:

Quadrant	First Point	Last Point
First	first base point	first transition mesial point
Second	first transition mesial point	top center point
Third	top center point	last transition mesial point
Fourth	last transition mesial point	last base point

pulse, radio-frequency. *See:* **radio-frequency pulse.**

pulse rate (1) (electronic navigation). *See:* **pulse-repetition frequency.**

(2) (watthour meter). The number of pulses per demand interval at which a pulse device is nominally rated. *See also:* **auxiliary device to an instrument.**   212

pulse rate telemetering (electric power systems). A type of telemetering in which the number of unidirectional pulses per unit time is varied as a function of the magnitude of the measured quantity. *See also:* **telemetering.**   94

pulse receiver (watthour meter). The unit that receives and registers the pulses. It may include a periodic resetting mechanism, so that a reading proportional to demand may be obtained. *See also:* **auxiliary device of an instrument.**   212

pulse regeneration. The process of restoring a series of pulses to their original timing, form, and relative magnitude. *See also:* **pulse.**   194

pulse repeater (transponder). A device used for receiving pulses from one circuit and transmitting corresponding

pulses into another circuit. It may also change the frequency and waveforms of the pulses and perform other functions. *See also:* **pulse; repeater.**   111

pulse-repetition frequency (PRF) (system using recurrent pulses). The number of pulses per unit of time (per second, per minute).   187, 194, 162

pulse repetition interval (radar). The time duration between pulses in a pulse train.   13

pulse-repetition rate (radar). *See:* **pulse-repetition frequency.**

pulse reply. The transmission of a pulse or pulse mode by a transponder as the result of an interrogation.   254

pulse response characteristics (pulse response curve) (electromagnetic compatibility). The relationship between the indication of a quasi-peak voltmeter and the repetition rate of regularly repeated pulses of constant amplitude. *See also:* **electromagnetic compatibility.**   220, 199

pulse rise time (t_r) (1) (general). The interval between the instants at which the instantaneous amplitude first reaches specified lower and upper limits, namely 10 percent and 90 percent, of the peak pulse amplitude unless otherwise stated. *See also:* **pulse.**   185, 194, 59, 162

(2) (semiconductor radiation detectors) (first transition time) (t_r). The interval between the instants at which the instantaneous value first reaches specified lower and upper limits, namely, 10 and 90 percent of the peak pulse value unless otherwise specified. *Notes:* (A) In the case of a step function applied to an RC low-pass filter, the rise time is given by $t_r = 2.2\ RC$. (B) In the case of a step function applied to an amplifier that has simple $CR—RC$ shaping, that is, one high-pass and one low-pass RC filter of equal time constants, the rise time is given by $t_r = 0.57\ RC$.   23

pulse separation. The interval between the trailing-edge pulse time of one pulse and the leading-edge pulse time of the succeeding pulse. *See also:* **leading-edge pulse time; pulse; trailing-edge pulse time.**   254, 178, 185, 162

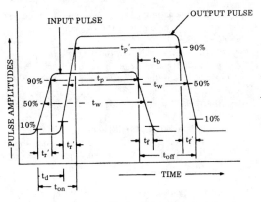

Diagram illustrating pulse time symbology

pulses, equalizing. *See:* **equalizing pulses.**

pulse shape (1) (instrumentation and measurement). The identifying geometrical and/or mathematical characteristics of a pulse waveform used for descriptive purposes. *Note:* For heuristic or tutorial purposes (A) the practical shape of a pulse may be imprecisely described with the following adjectives: exponential, Gaussian, irregular,

rectangular, rounded, trapezoidal, triangular, half sine, etcetera. These or similar adjectives describe the general shape of a pulse and no precise distinctions are defined. (B) A hypothetical pulse may be precisely described by the addition of the adjective ideal; for example, ideal rectangular pulse. *See also:* **pulse.**   185
(2) (**pulse terms**). (A) For descriptive purposes a pulse waveform may be imprecisely described by any of the adjectives, or combinations thereof, in **descriptive adjectives, major (minor); polarity related adjectives; geometrical adjectives; and functional adjectives, exponential.**
(B) For tutorial purposes a hypothetical pulse waveform may be precisely defined by the further addition of the adjective ideal. *See:* **descriptive adjectives, ideal.**
(C) For measurement or comparison purposes a pulse waveform may be precisely defined by the further addition of the adjective reference. *See:* **descriptive adjectives, reference.**   254

pulse shaper (pulse techniques). Any transducer used for changing one or more characteristics of a pulse. *Note:* This term includes pulse regenerators. *See also:* **pulse.**
     254, 178
pulse shaping. Intentionally changing the shape of a pulse.
     254, 178
pulse, single-polarity. *See:* **unidirectional pulse.**
pulse spacing (pulse interval*). The interval between the corresponding pulse times of two consecutive pulses. *Note:* The term pulse interval is deprecated because it may be taken to mean the duration of the pulse instead of the space or interval from one pulse to the next. Neither term means the space between pulses.   254, 178, 162

* Deprecated

pulse spectrum (signal-transmission system). The frequency distribution of the sinusoidal components of the pulse in relative amplitude and in relative phase. *See also:* **signal.**
     254, 188
pulse spike (automatic control). An unwanted pulse of relatively short duration, superimposed on the main pulse. *See also:* **spike.**   56
pulse start (stop) time (pulse terms). The instant specified by a magnitude referenced point on the first (last) transition of a pulse waveform. Unless otherwise specified, the pulse start (stop) time is at the mesial point on the first (last transition). *See:* The single pulse diagram below the **waveform epoch** entry.   254
pulse storage time (t$_s$) **(light emitting diodes).** The interval between the instants at which the instantaneous amplitudes of the input and output pulses last reach a specified fraction of their peak pulse amplitudes, namely, 90 percent.   162
pulse stretcher (pulse techniques) (non-real time spectrum analyzer). A pulse shaper that produces an output pulse whose duration is greater than that of the input pulse and whose amplitude is proportional to that of the peak amplitude of the input pulse. *See also:* **pulse.**   68, 117
pulse techniques. *See:* **burst.**
pulse tilt. A distortion in an otherwise essentially flat-topped rectangular pulse characterized by either a decline or a rise of the pulse top. *See also:* **television.**   254, 178
pulse time, leading edge. *See:* **leading-edge pulse time.**

pulse time, mean. *See:* **mean pulse time.**
pulse-time modulation (PTM). Modulation in which the values of instantaneous samples of the modulating wave are caused to modulate the time of occurrence of some characteristic of a pulse carrier. *Note:* Pulse-duration modulation, pulse-position modulation, and pulse-interval modulation are particular forms of pulse-time modulation.
     111, 254
pulse time reference points (pulse terms). (1) top center point. A specified time referenced point or magnitude referenced point on a pulse waveform top. If no point is specified, the top center point is the time referenced point at the intersection of a pulse waveform and the top center line. *See:* The single pulse diagram below the **waveform epoch** entry. (2) first (last) base point. Unless otherwise specified, the first (last) datum point in a pulse epoch. *See:* The single pulse diagram below the **waveform epoch** entry. *See also:* **Pulse train time-related definitions, base center point.**   254
pulse time, trailing-edge. *See:* **trailing-edge pulse time.**
pulse timing of video pulses (television). The determination of an occurrence of a pulse or a specified portion thereof at a particular time. *See:* **pulse width of video pulses (television); television; time of rise (decay) of video pulses (television).**   336
pulse top (1) (instrumentation and measurement). That major segment of a pulse waveform having the greater displacement in amplitude from the baseline. *See also:* **pulse.**   185
(2) (**high- and low-power pulse transformers**). That portion of the pulse occurring between the time of intersection of straight-line segments used to determine A_M and A_T. *See also:* **input pulse shape.**   32, 33
pulse train (1) (signal-transmission system) (industrial control). A sequence of pulses. *See also:* **pulse; signal.**
     254, 210, 56, 178, 188, 206
(2) (**radar**). A sequence of pulses used to accomplish a function such as MTI or increased effective signal-to-noise ratio. Pulse trains of finite duration which is less than the radar echo time are usually referred to as pulse bursts.
     13
pulse train (pulse terms). A continuous repetitive sequence of pulse waveforms.   254
pulse train, bidirectional. *See:* **bidirectional pulse train.**
pulse train, periodic. *See:* **periodic pulse train.**
pulse-train spectrum (pulse-train frequency-spectrum). The frequency distribution of the sinusoidal components of the pulse train in amplitude and in phase angle.   254
pulse train time-related definitions (pulse terms). (1) pulse repetition period. The interval between the pulse start time of a first pulse waveform and the pulse start time of the immediately following pulse waveform in a periodic pulse train. (2) pulse repetition frequency. The reciprocal of pulse repetition period. (3) pulse separation. The interval between the pulse stop time of a first pulse waveform and the pulse start time of the immediately following pulse waveform in a pulse train. (4) duty factor. Unless otherwise specified, the ratio of the pulse waveform duration to the pulse repetition period of a periodic pulse train. (5) on-off ratio. Unless otherwise specified, the ratio of the pulse waveform duration to the pulse separation of a periodic pulse train. (6) base center line. The time reference

line at the average of the pulse stop time of a first pulse waveform and the pulse start time of the immediately following pulse waveform in a pulse train. (7) base center point. A specified time referenced point or magnitude referenced point on a pulse train waveform base. If no point is specified, the base center point is the time referenced point at the intersection of a pulse train waveform base and a base center line. *See:* **pulse time reference points, first (last) base point.** (8) pulse train epoch. The span of time in a pulse train for which waveform data are known or knowable and which extends from a first base center point to the immediately following base center point.

pulse train top (base) envelope (pulse terms). Unless otherwise specified, the waveform defined by a cubic natural spline with knots at each point of intersection of the top center line and topline (the base center line and the baseline) of each (between adjacent) pulse waveforms(s) in a pulse train. 254

pulse train, unidirectional. *See:* **unidirectional pulse train.**

pulse transmitter. A pulse-modulated transmitter whose peak power-output capabilities are usually large with respect to average power-output rating. *See also:* **pulse; radio transmitter.** 111

pulse turn-off time (t_{off}) (light emitting diodes). The arithmetic sum of the pulse storage time, and the pulse decay time of the output pulse. 162

pulse turn-on time (t_{on}) (light emitting diodes). The arithmetic sum of the pulse delay time, and the pulse rise time of the output pulse. 162

pulse-type telemeter. A telemeter that employs characteristics of intermittent electric signals other than their frequency, as the translating means. *Note:* These pulses may be utilized in any desired manner to obtain the final indication, such as periodically counting the total number of pulses; or measuring their ON time, their OFF time, or both. 103, 202

pulse unidirectional. *See:* **unidirectional pulse.**

pulse waveform distortion (pulse terms). The algebraic difference in magnitude between all corresponding points in time of a pulse waveform and a reference pulse waveform. Unless otherwise specified by a mathematical adjective, peak-to-peak pulse waveform distortion is assumed. *See:* **mathematical adjectives.** *See also:* pulse waveform feature distortion diagram below. 254

pulse waveform feature distortion (pulse terms). The algebraic difference in magnitude between all corresponding points in time of a pulse waveform and a reference pulse waveform feature. Unless otherwise specified by a mathematical adjective, peak-to-peak pulse waveform feature distortion is assumed. *See:* **mathematical adjectives.** *See also:* pulse waveform distortion and pulse waveform feature distortion diagram below the **pulse waveform distortion** entry. 254

pulse width (1) (television)*. *See:* **pulse duration.**

* Deprecated

(2) (data transmission). The time interval between the points on the leading and trailing edges at which the instantaneous value bears a specified relation to the maximum instantaneous value of the pulse. 194

pulsewidth (radar). The time interval between the points on the leading and trailing edges at which the instantaneous value bears a specified relation to the maximum instantaneous value of the pulse. 13

pulse-width modulation. *See:* **pulse-duration modulation.**

pulse width of video pulses (television). The duration of a pulse measured at a specified level. *See also:* **pulse timing of video pulses (television).** 336

pulsing (telephone switching systems). The signaling over the communication path of signals representing one or more digits required to set up a call. 55

pulsing circuit (peaking circuit) (industrial control). A circuit designed to provide abrupt changes in voltage or current of some characteristic pattern. *See also:* **electronic controller.** 206

pulsing transformer (industrial control). Supplies pulses

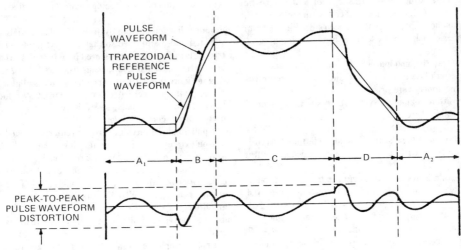

EXTENT OF DATA INCLUDED IN PULSEWAVEFORM FEATURE DISTORTION:
A_1 AND A_2-PULSE BASE DISTORTION   C-PULSE TOP DISTORTION
B-FIRST TRANSITION DISTORTION   D-LAST TRANSITION DISTORTION

Pulse waveform distortion and pulse waveform feature distortion.

of voltage or current. *See also:* **electronic controller.**
                                                                                    206
pump (parametric device). The source of alternating-current power that causes the nonlinear reactor to behave as a time-varying reactance. *See also:* **parametric device.**
                                                                              277, 191
pump-back test (electrical back-to-back test) (rotating machinery). A test in which two identical machines are mechanically coupled together, and they are both connected electrically to a power system. The total losses of both machines are taken as the power input drawn from the system. *See also:* **asynchronous machine; direct-current commutating machine.**
                                                                                     63
pumped-storage hydro capability (electric power supply). The capability supplied by hydro-electric sources under specified water conditions using a reservoir that is alternately filled by pumping and depleted by generating.
                                                                                    112
pumped tube. An electron tube that is continuously connected to evacuating equipment during operation. *Note:* This term is used chiefly for pool-cathode tubes.   190
pump efficiency (laser-maser). The ratio of the power or energy absorbed from the pump to the power or energy available from the pump source.   363
pump-free control. *See:* **antipump (pump-free device) device.**
punch (computing systems). *See:* **keypunch, zone punch.**
punched card. A card on which a pattern of holes or cuts is used to represent data.   77, 54
punched tape (computing systems). A tape on which a pattern of holes or cuts is used to represent data.
                                                                              255, 77, 54
punched tape handler (test, measurement and diagnostic equipment). A device which handles punched tape and usually consists of a tape transport and punched tape reader with associated electrical and electronic equipments. Most units provide for tape to be wound and stored in reels; however, some units provide for the tape to be stored loosely in closed bins.   54
punched tape reader (test, measurement and diagnostic equipment). A device capable of converting information from punched tape, where it has been stored in the form of a series of holes, into a series of electrical impulses.
                                                                                     54
punching (rotating machinery). A lamination made from sheet material using a punch and die. *See also:* **rotor (rotating machinery); stator.**   63
punch position (computing systems). A site on a punched tape or card where holes are to be punched.
                                                                              255, 77, 54
punch-through voltage (semiconductor radiation detector). The voltage at which a junction detector becomes fully depleted.   118, 119
puncture (voltage testing). A disruptive discharge through the body of a solid dielectric. *See also:* **test voltage and current.**   307, 201
puncture voltage (surge arresters). The voltage at which the test specimen is electrically punctured.   64, 62
pupil (laser-maser). The variable aperture in the iris through which light travels toward the interior of the eye.   363
pure tone. *See:* **simple tone.**
purification of electrolyte. The treatment of a suitable

volume of the electrolyte by which the dissolved impurities are removed in order to keep their content in the electrolyte within desired limits. *See also:* **electrorefining.**   328
purity (excitation purity) (color). *See:* **excitation purity (purity) (color).**
Purkinje phenomenon. The reduction in subjective brightness of a red light relative to that of a blue light when the luminances (photometric brightnesses) are reduced in the same proportion without changing the respective spectral distributions. In passing from photopic to scotopic vision, the curve of spectral luminous efficiency changes, the wavelength of maximum efficiency being displaced toward the shorter wavelengths. *See also:* **visual field.**
                                                                                    167
purple boundary. The straight line drawn between the ends of the spectrum locus on a chromaticity diagram. *See also:* **color.**   167, 18, 178
push brace. A supporting member, usually of timber, placed between a pole or other structural part of a line and the ground or a fixed object. *See also:* **tower.**   64
pushbutton (industrial control). Part of an electric device, consisting of a button that must be pressed to effect an operation. *See also:* **switch.**   244, 206
pushbutton dial (telephone switching systems). A type of calling device used in automatic switching that has an activator per digit that generates distinctive pulsing.
                                                                                     55
pushbutton station (industrial control). A unit assembly of one or more externally operable pushbutton switches, sometimes including other pilot devices such as indicating lights or selector switches, in a suitable enclosure. *See also:* **switch.**   206
pushbutton switch (pushbutton). A master switch, usually mounted behind an opening in a cover or panel, and having an operating plunger or button extending forward in the opening. Operation of the switch is normally obtained by pressure of the finger against the end of the button. *See also:* **switch.**   328
pushbutton switching. A reperforator switching system in which selection of the outgoing channel is initiated by an operator.   194
pushdown list (computing systems). A list that is constructed and maintained so that the next item to be retrieved is the most recently stored item in the list, that is, last in, first out. *See also:* **pushup list.**   255, 77
pushing figure (oscillator). The change of oscillator frequency with a specified change in current, excluding thermal effects. *See:* **tuning sensitivity, electronic.** *See also:* **oscillatory circuit; television.**   190, 125
push-pull amplifier circuit. *See:* **amplifier, balanced.**
push-pull circuit. A circuit containing two like elements that operate in 180-degree phase relationship to produce additive output components of the desired wave, with cancellation of certain unwanted products. *Note:* Push-pull amplifiers and push-pull oscillators are examples. *See also:* **amplifier**   59
push-pull currents. Balanced currents. *See also:* **waveguide.**   267
push-pull microphone. A microphone that makes use of two like microphone elements actuated by the same sound waves and operating 180 degrees out of phase. *See also:* **microphone.**   328

push-pull operation (electron device). The operation of two similar electron devices or of an equivalent double-unit device, in a circuit such that equal quantities in phase opposition are applied to the input electrodes, and the two outputs are combined in phase.   190

push-pull oscillator. A balanced oscillator employing two similar tubes in phase opposition. *See also:* **balanced oscillator; oscillatory circuit.**   111

push-pull voltages. Balanced voltages. *See also:* **waveguide.**   267

push-push circuit. A circuit employing two similar tubes with grids connected in phase opposition and plates in parallel to a common load, and usually used as a frequency multiplier to emphasize even-order harmonics.   111

push-push currents. Currents flowing in the two conductors of a balanced line that, at every point along the line, are equal in magnitude and in the same direction. *See also:* **waveguide.**   267

push-push voltages. Voltages (relative to ground) on the two conductors of a balanced line that, at every point along the line, are equal in magnitude and have the same polarity. *See also:* **waveguide.**   267

push-to-talk circuit. A method of communication over a speech channel in which transmission occurs in only one direction at a time, the talker being required to keep a switch operated while he is talking.   59

push-to-type operation. That form of telegraph operation, employing a one-way reversible circuit, in which the operator must keep a switch operated in order to send from his station. It is generally used in radio transmission where the same frequency is employed for transmission and reception. *See also:* **telegraphy.**   328

pushup list (computing systems). A list that is constructed and maintained so that the next item to be retrieved and removed is the oldest item still in the list, that is, first in, first out. *See also:* **pushdown list.**   255, 77

pyramidal horn antenna. A horn antenna the sides of which form a pyramid. *See also:* **antenna.**   179, 111

pyroconductivity. Electric conductivity that develops with rising temperature, and notably upon fusion, in solids that are practically nonconductive at atmospheric temperatures.   328

pyrometer. A thermometer of any kind usable at relatively high temperatures (above 500 degrees Celsius). *See also:* **electric thermometer (temperature meter).**   328

Q

Q chrominance signal (NTSC color television). The sidebands resulting from suppressed-carrier modulation of the chrominance subcarrier by the Q Video Signal. *Note:* The signal is transmitted in double-sideband form, the sidebands extending approximately 0.6 MHz above and below the chrominance subcarrier. The phase of the signal, for positive Q Video Signals, is $33°$ with respect to the $B - Y$ axis.   18

Q coil. *See:* **coil Q.**

Q **of an electrically small tuned antenna.** An inverse measure of the bandwidth or an antenna as determined by its impedance. It is numerically equal to one half the magnitude of the ratio of the incremental change in impedance to the incremental change in frequency at resonance, divided by the ratio of the antenna resistance to the resonance frequency. *Note:* The Q of an antenna also is a measure of the energy stored to the energy radiated or dissipated per cycle.   111

*Q***-meter (quality-factor meter).** An instrument for measuring the quality factor Q of a circuit or circuit element. *See also:* **instrument.**   328

Q-percentile life (1) (assessed) (non-repaired items) (reliability). The Q-percentile life determined as a limiting value or values of the confidence interval with a stated confidence level, based on the same data as the observed Q-percentile life of nominally identical items. *Notes:* (A) The source of the data should be stated. (B) Results can be accumulated (combined) only when all conditions are similar. (C) The assumed underlying distribution of failures against time should be stated. (D) It should be stated whether a one-sided or two-sided interval is being used. (E) Where one limiting value is given this is usually the lower limit.

(2) extrapolated. Extension by a defined extrapolation or interpolation of the observed or assessed Q-percentile life for stress conditions different from those applying to the assessed Q-percentile life and for different percentages. *Note:* The validity of the extrapolation should be justified.

(3) observed. The length of observed time at which a stated proportion (Q%) of a sample of items has failed. *Notes:* (A) The criteria for what constitutes a failure should be stated. (B) The Q-percentile life is also that life at which (100-Q)% reliability is observed.

(4) predicted. For the stated conditions of use, and taking into account the design of an item, the computed Q-percentile life based on the observed, assessed or extrapolated Q-percentile lives of its parts. *Note:* Engineering and statistical assumptions should be stated, as well as the bases used for the computation (observed or assessed).   164

Q-switch (laser-maser). A device for producing very short ($≈30$ns), intense laser pulses by enhancing the storage and dumping of electronic energy in and out of the lasing medium, respectively. *See also:* **static optical transmission; diffraction efficiency.**   363

Q-switched laser (laser-maser). A laser which emits short ($≈30$ns), high-power pulses by utilizing a Q switch.   363

Q-switching (laser-maser). Producing very short ($≈30$ns), intense pulses by enhancing the storage and dumping of electronic energy in and out of the laser-maser medium, respectively.   363

quad. A structural unit employed in cable, consisting of four separately insulated conductors twisted together. *See also:* **cable.**   328

quadded cable. A cable in which at least some of the conductors are arranged in the form of quads. *See also:* **cables.** 328

quadergy. Delivered by an electric circuit during a time interval when the voltages and currents are periodic, the product of the reactive power and the time interval, provided the time interval is one or more complete periods or is quite long in comparison with the time of one period. If the reference direction for energy flow is selected as into the region the net delivery of quadergy will be into the region when the sign of the quadergy is positive and out of the region when the sign is negative. If the reference direction is selected as out of the region, the reverse will apply. The quadergy is expressed by

$$K = Qt$$

where Q is the reactive power and t is the time interval. If the voltages and currents form polyphase symmetrical sets, there is no restriction regarding the relation of the time interval to the period. If the voltages and currents are quasi-periodic and the amplitudes of the voltages and currents are slowly varying, the quadergy is the integral with respect to time of the reactive power, provided the integration is for a time that is one or more complete periods or that is quite long in comparison with the time of one period. Mathematically

$$K = \int_{t_0}^{t_0+t} Q \, dt$$

where Q is the reactive power determined for the condition of voltages and currents having slowly varying amplitudes. Quadergy is expressed in var-seconds or var-hours when the voltages are in volts and the currents are in amperes, and the time is in seconds or hours, respectively. *See also:* **network analysis.** 210

quadrantal error (electronic navigation). An angular error in measured bearing caused by characteristics of the vehicle or station that adversely affect the direction of signal propagation; the error varies in a sinusoidal manner throughout the 360 degrees and has two positive and two negative maximums. *See also:* **navigation.** 187, 13

quadratic lag. *See:* **lag (6) (second-order).**

quadrature. The relation between two periodic functions when the phase difference between them is one-fourth of a period. *See also:* **network analysis.** 210

quadrature acceleration drift rate (tuned rotor gyro). A drift rate about an axis normal to both the spin axis and the axis along which an acceleration is applied. This drift rate results from a torque about the axis of applied acceleration and is in quadrature with that due to mass unbalance. 46

quadrature axis (synchronous machine). The axis that represents the direction of the radial plane along which the main field winding produces no magnetization, normally coinciding with the radial plane midway between adjacent poles. *Notes:* (1) The positive direction of the quadrature axis is 90 degrees ahead of the positive direction of the direct axis, in the direction of rotation of the field relative to the armature. (2) The definitions of currents and voltages given in the terms listed below are applicable to balanced load conditions and for sinusoidal currents and

voltages. They may also be applied under other conditions to the positive-sequence fundamental-frequency components of currents and voltages. More generalized definitions, applicable under all conditions, have not been agreed upon. 63

quadrature-axis component (1) (armature voltage). That component of the armature voltage of any phase that is in time phase with the quadrature-axis component of current in the same phase. *Note:* A quadrature-axis component of voltage may be reproduced by: (A) rotation of the direct-axis component magnetic flux; (B) variation (if any) of the quadrature-axis component of magnetic flux; (C) resistance drop caused by flow of the quadrature-axis component of armature current. The quadrature-axis component of terminal voltage is related to the synchronous internal voltage by

$$\mathbf{E}_{aq} = \mathbf{E}_i - R\mathbf{I}_{aq} - jX_d\mathbf{I}_{ad}.$$

See also: **phasor diagram (synchronous machine).** 328

(2) (armature current). That component of the armature current that produces a magnetomotive-force distribution that is symmetrical about the quadrature axis. *See also:* **quadrature-axis component of armature voltage; quadrature-axis component of magnetomotive force.** 328

(3) (magnetomotive force) (rotating machinery). The component of a magnetomotive force that is directed along an axis in quadrature with the axis of the poles. *See also:* **asynchronous machine; direct-axis synchronous impedance.** 63

quadrature-axis current (rotating machinery). The current that produces quadrature-axis magnetomotive force. *See also:* **direct-axis synchronous impedance.** 63

quadrature-axis magnetic flux (rotating machinery). The magnetic-flux component directed along the quadrature axis. *See also:* **direct-axis synchronous impedance.** 63

quadrature-axis subtransient impedance (rotating machinery). The operator expressing the relation between the initial change in armature voltage and a sudden change in quadrature-axis armature current, with only the fundamental-frequency components considered for both voltage and current, with no change in the voltage applied to the field winding, and with the rotor running at steady speed. In terms of network theory it corresponds to the quadrature-axis impedance the machine displays against disturbances (modulations) with infinite frequency. *Note:* If no rotor winding is along the quadrature axis and/or the rotor is not made out of solid steel, this impedance equals the quadrature-axis synchronous impedance. *See also:* **asynchronous machine; direct-axis synchronous impedance.** 63

quadrature-axis subtransient open-circuit time constant. The time in seconds required for the rapidly decreasing component (negative) present during the first few cycles in the quadrature-axis component of symmetrical armature voltage under suddenly removed symmetrical short-circuit conditions with the machine running at rated speed, to decrease to $1/e \approx 0.368$ of its initial value. 328

quadrature-axis subtransient reactance. The ratio of the fundamental component of reactive armature voltage due to the initial value of the fundamental quadrature-axis component of the alternating-current component of the armature current, to this component of current under suddenly applied balanced load conditions and at rated frequency. Unless otherwise specified, the quadrature-axis subtransient reactance will be that corresponding to rated armature current.   328

quadrature-axis subtransient short-circuit time constant. The time in seconds required for the rapidly decreasing component present during the first few cycles in the quadrature-axis component of the alternating-current component of the armature current under suddenly applied symmetrical short-circuit conditions, with the machine running at rated speed to decrease to $1/e \approx 0.368$ of its initial value.   328

quadrature-axis subtransient voltage (rotating machinery). The quadrature-axis component of the terminal voltage which appears immediately after the sudden opening of the external circuit when the machine is running at a specified load, before any flux variation in the excitation and damping circuits has taken place.   63

quadrature-axis synchronous impedance (synchronous machine) (rotating machinery). The impedance of the armature winding under steady-state conditions where the axis of the armature current and magnetomotive force coincides with the quadrature axis. In large machines where the armature resistance is negligibly small, the quadrature-axis synchronous impedance is equal to the quadrature-axis synchronous reactance.   63

quadrature-axis synchronous reactance. The ratio of the fundamental component of reactive armature voltage, due to the fundamental quadrature-axis component of armature current, to this component of current under steady-state conditions and at rated frequency. Unless otherwise specified, the value of quadrature-axis synchronous reactance will be that corresponding to rated armature current.   328

quadrature-axis transient impedance (rotating machinery). The operator expressing the relation between the initial change in armature voltage and a sudden change in quadrature-axis armature current component, with only the fundamental frequency components considered for both voltage and current, with no change in the voltage applied to the field winding, with the rotor running at steady speed, and by considering only the slowest decaying component and the steady-state component of the voltage drop. In terms of network theory it corresponds to the quadrature-axis impedance the machine displays against disturbances (modulation) with infinite frequency by considering only two pole pairs* (or poles*), namely those with smallest (including zero) real parts, of the impedance function. *Note:* If no rotor winding is along the quadrature axis and/or the rotor is not made out of solid steel, this impedance equals the quadrature-axis synchronous impedance. *See also:* **asynchronous machine; direct-axis synchronous impedance.**

* The term pole refers here to the roots of the denominator of the impedance function.   63

quadrature-axis transient open-circuit time constant. The time in seconds required for the root-mean-square alternating-current value of the slowly decreasing component present in the quadrature-axis component of symmetrical armature voltage on open-circuit to decrease to $1/e \approx$

0.368 of its initial value when the quadrature field winding (if any) is suddenly short-circuited with the machine running at rated speed. *Note:* This time constant is important only in turbine generators.   328

quadrature-axis transient reactance. The ratio of the fundamental component of reactive armature voltage, due to the fundamental quadrature-axis component of the alternating-current component of the armature current, to this component of current under suddenly applied load conditions and at rated frequency, the value of current to be determined by the extrapolation of the envelope of the alternating-current component of the current wave to the instant of the sudden application of load, neglecting the high-decrement currents during the first few cycles. *Note:* The quadrature-axis transient reactance usually equals the quadrature-axis synchronous reactance except in solid-rotor machines, since in general there is no really effective field current in the quadrature axis.   328

quadrature-axis transient short-circuit time constant. The time in seconds required for the root-mean-square alternating-current value of the slowly decreasing component present in the quadrature-axis component of the alternating-current component of the armature current under suddenly applied short-circuit conditions with the machine running at rated speed to decrease to $1/e \approx 0.368$ of its initial value.   328

quadrature-axis transient voltage (rotating machinery). The quadrature-axis component of the terminal voltage that appears immediately after the sudden opening of the external circuit when running at a specified load, neglecting the components with very rapid decay that may exist during the first few cycles. *See also:* **asynchronous machine; direct-axis synchronous impedance.**   63

quadrature-axis voltage (rotating machinery). The component of voltage that would produce quadrature-axis current when resistance limited.   63

quadrature hybrid (waveguide components). A hybrid junction which has the property that a wave leaving one output port is in phase quadrature with the wave leaving the other output port.   66

quadrature modulation. Modulation of two carrier components 90 degrees apart in phase by separate modulating functions.   242

quadrature spring rate (tuned rotor gyro). When the case of a tuned rotor gyro is displaced with respect to the gyro rotor through an angle about an axis perpendicular to the spin axis, a torque proportional to and 90 degrees away from the displacement acts in a direction to reduce this angle and to align the rotor with the case. The torque is usually due to windage, a squeeze-film force, or flexure hysteresis. This spring rate results in a drift rate coefficient having dimensions of angular displacement per unit time per unit angle of displacement about an input axis.   46

quadri pole. *See:* **two-terminal pair network.**

quadrupole parametric amplifier. A beam parametric amplifier having transverse input and ouput couplers for the signal, separated by a quadrupole structure that is excited by a pump to obtain parametric amplification of a cyclotron wave. *See also:* **parametric device.**
    277, 191

qualification (nuclear power generating stations). Dem-

onstration that equipment meets design requirements.
31

qualified (transmission and distribution). Having adequate knowledge of the installation, construction or operation of apparatus and the hazards involved. 262

qualified life (1) (nuclear power generating stations). The period of time that can be verified for which the electric penetration assembly will meet all design requirements for the specified service conditions. *See:* **electric penetration assembly.** 26

(2) (class 1E equipment). The period of time for which satisfactory performance can be demonstrated for a specific set of service conditions. *Note:* The qualified life of a particular equipment item may be changed during its installed life where justified. 31, 120, 141

qualified module (nuclear power generating station). Module which exhibits performance characteristics which are acceptable for Class 1E service in a nuclear power generating station and which satisfy the aging criteria and other requirements of this document. 355

qualifying symbol (logic diagrams). That portion of a rectangular-shape logic symbol that denotes its logic function. 88

qualitative adjectives (pulse terms). *See:* **descriptive adjectives; time related adjectives; magnitude related adjectives; polarity related adjectives; geometrical adjectives.**

qualitative distortion terms (pulse terms). *See:* **preshoot; overshoot; rounding; spike; ringing; tilt; valley.**

quality area. The area of the cathode-ray-tube phosphor screen that is limited by the cathode-ray tube and instrument specification. *Note:* If the quality area and the graticule area are not equal, this must be specified. *See also:* **graticule area; viewing area; oscillograph.** 185

quality assurance (nuclear power generating stations). All those planned and systematic actions necessary to provide adequate confidence that a system or component will perform satisfactorily in service. 31

quality factor (Q) (1) (network, structure, or material). Two pi times the ratio of the maximum stored energy to the energy dissipated per cycle at a given frequency. *Notes:* (A) The Q of an inductor at any frequency is the magnitude of the ratio of its reactance to its effective series resistance at that frequency. (B) The Q of a capacitor at any frequency is the magnitude of the ratio of its susceptance to its effective shunt conductance at that frequency. (C) The Q of a simple resonant circuit comprising an inductor and a capacitor is given by

$$Q = \frac{Q_L Q_C}{Q_L + Q_C}$$

where Q_L and Q_C are the Q's of the inductor and capacitor, respectively, at the resonance frequency. If the resonant circuit comprises an inductance L and a capacitance C in series with an effective resistance R, the value of Q is

$$Q = \frac{1}{R}\left(\frac{L}{C}\right)^{1/2}$$

(D) An approximate equivalent definition, which can be applied to other types of resonant structures, is that the Q is the ratio of the resonance frequency to the bandwidth

between those frequencies on opposite sides of the resonance frequency (known as half-power points) where the response of the resonant structure differs by 3 decibels from that at resonance. (E) The Q of a magnetic or dielectric material at any frequency is equal to 2π times the ratio of the maximum stored energy to the energy dissipated in the material per cycle. (F) For networks that contain several elements, and distributed parameter systems, the Q is generally evaluated at a frequency of resonance. (G) The nonloaded Q of a system is the value of Q obtained when only the incidental dissipation of the system elements is present. The loaded Q of a system is the value of Q obtained when the system is coupled to a device that dissipates energy. (H) The **period** in the expression for Q is that of the driving force, not that of energy storage, which is usually half that of the driving force. *See also:* **network analysis.** 210

(2) (circuits and systems). In active filters the transfer functions are generally broken down into second-order sections expressed by biquadratic functions as follows:

$$T(s) = \frac{n_2 s^2 + n_1 s + n_0}{d_2 s^2 + d_1 s + d_0}$$

Such transfer functions are generally re-arranged in the following form where the zero Q-factor, Q_Z, and the pole Q-factor, Q_p, may be identified.

$$T(s) = K \frac{s^2 + s\,\dfrac{\omega_{0z}}{Q_z} + \omega_{0z}{}^2}{s^2 + s\,\dfrac{\omega_{0p}}{Q_p} + \omega_{0p}{}^2}$$

67

quality of lighting. The distribution of brightness in a visual environment. The term is used in a positive sense and implies that all brightnesses contribute favorably to visual performance, visual comfort, ease of seeing, safety, and esthetics for the specific visual tasks involved. *See also:* **visual field.** 167

quantitative adjectives (pulse terms). *See:* **integer adjectives; mathematical adjectives; functional adjectives.** 254

quantitative testing (test, measurement and diagnostic equipment). Testing that monitors or measures the specific quantity, level or amplitude of a characteristic to evaluate the operation of an item. The outputs of such tests are presented as finite or quantitative values of the associated characteristics. 54

quantity equations. Equations in which the quantity symbols represent mathematico-physical quantities possessing both numerical values and dimensions. 210

quantity of electricity. *See:* **electric charge.**

quantity of light (luminous energy,

$$Q_v = \int \varphi_v dt)$$

(light emitting diodes). The product of the luminous flux by the time it is maintained. It is the time integral of luminous flux. 162

quantization (1) (telecommunication). A process in which the continuous range of values of an input signal is divided into nonoverlapping subranges, and to each subrange a discrete value of the output is uniquely assigned. Whenever the signal value falls within a given subrange, the output has the corresponding discrete value. *Note:* Quantized may be used as an adjective modifying various forms of modulation, for example, quantized pulse-amplitude modulation. *See:* **quantization distortion (quantization noise); quantization level).**   242, 255, 77, 194

(2) (gyro; accelerometer). The analog-to-digital conversion of a gyro or accelerometer output signal which gives an output that changes in discrete steps as the input varies continuously.   46

quantization distortion (quantization noise). The inherent distortion introduced in the process of quantization. *See also:* **quantization.**   59

quantization level (telecommunication). The discrete value of the output designating a particular subrange of the input. *See also:* **quantization.**   242, 59

quantization noise. *See:* **quantization distortion.**

quantize. To subdivide the range of values of a variable into a finite number of nonoverlapping subranges or intervals, each of which is represented by an assigned value within the subrange, for example, to represent a person's age as a number of whole years.   255, 77

quantized pulse modulation. Pulse modulation that involves quantization.   328

quantized system. One in which at least one quantizing operation is present.   56

quantizing error (radar). An error caused by conversion of a variable having a continuous range of values to a quantized form having only discrete values, as in analog-to-digital conversion. The error is the difference between the original (analog) value and its quantized representation. *See also:* **quantization and quantize.**   13

quantizing loss (radar). (1) In phased arrays, a loss that occurs when the beam is phase-steered by digitally controlled phase shifters, due to the quantizing errors in the phase shifts applied to the various elements. (2) In signal processing, a loss that occurs when elements of a composite signal (for example, complex amplitudes of pulses in a pulse train) are quantized (digitized) before being combined. *See also:* **quantizing error.**   13

quantizing operation. One which converts one signal into another having a finite number of predetermined magnitude values.   56

quantum efficiency (1) (photocathodes). The average number of electrons photoelectrically emitted from the photocathode per incident photon of a given wavelength. *Note:* The quantum efficiency varies with the wavelength, angle of incidence, and polarization of the incident radiation. *See also:* **photocathodes; photoelectric converter; phototubes; semiconductors.**   190, 186, 125, 117

(2) (light emitting diodes) (η). The ratio of the number of quanta of radiant energy (photons) emitted per second to the number of electrons flowing per second; that is, photons per electron.   162

(3) (laser, maser, laser material, or maser material) The ratio of the number of photons or electrons emitted by a material at a given transition to the number of absorbed particles.   363

quarter-phase or two-phase circuit. A combination of circuits energized by alternating electromotive forces that differ in phase by a quarter of a cycle, that is, 90 degrees. *Note:* In practice the phases may vary several degrees from

the specified angle. *See also:* **center of distribution.**   64

quarter-thermal-burden ambient-temperature rating. The maximum ambient temperature at which the transformer can be safely operated when the transformer is energized at rated voltage and frequency and is carrying 25 percent of its thermal-burden rating without exceeding the specified temperature limitations.   328

quasi-analog signal (data transmission). A quasi-analog signal is a digital signal after conversion to a form suitable for transmission over a specified analog channel. The specifications of an analog channel would include the frequency of range, frequency of bandwidth, S-N ratio (signal-to-noise ratio), and envelope delay distortion. When this form of signaling is used to convey message traffic over dialed up telephone systems, it is often referred to as voice data.   59

quasi-impulsive noise (electromagnetic compatibility). A superposition of impulsive and continuous noise. *See also:* **electromagnetic compatibility.**   220, 199

quasi-peak detector (electromagnetic compatibility). A detector having specified electrical time constants that, when regularly repeated pulses of constant amplitude are applied to it, delivers an output voltage that is a fraction of the peak value of the pulses, the fraction increasing towards unity as the pulse repetition rate is increased. *See also:* **electromagnetic compatibility.**   220, 199

quasi-peak voltmeter (electromagnetic compatibility). A quasi-peak detector coupled to an indicating instrument having a specific mechanical time-constant. *See also:* **electromagnetic compatibility.**   220, 199

quaternary code (information theory). A code having four distinct code elements. *See also:* **ternary code; information theory; positional notation.**   61

quenched spark gap converter (dielectric heater usage). A spark-gap generator or power source that utilizes the oscillatory discharge of a capacitor through an inductor and a spark gap as a source of radio-frequency power. The spark gap comprises one or more closely spaced gaps operating in series. *See also:* **induction heating; industrial electronics.**   14, 114

quenching (gas-filled radiation-counter tube). The process of terminating a discharge in a Geiger-Mueller radiation-counter tube by inhibiting a reignition. *Note:* This may be effected internally (internal quenching or self-quenching) by use of an appropriate gas or vapor filling, or externally (external quenching) by momentary reduction of the applied potential difference. *See:* **gas-filled radiation-counter tubes.**   190, 125, 96

quenching circuit (radiation counters). A circuit that reduces the voltage applied to a Geiger-Mueller tube after an ionizing event, thus preventing the occurrence of subsequent multiple discharges. Usually the original voltage level is restored after a period that is longer than the natural recovery time of the Geiger-Mueller tube. *See:* **anti-coincidence (radiation counters).**   190

query (data transmission). The process by which a master station asks a slave station to identify itself and to give its status. *See:* **poll.**   59

quick-break. A term used to describe a device that has a high contact-opening speed independent of the operator.   103, 202, 27

quick charge (storage battery). *See:* **boost charge.**

quick-flashing light. A rhythmic light exhibiting very rapid regular alternations of light and darkness. There is no restriction on the ratio of the durations of the light to the dark periods. 167

quick-make. A term used to describe a device that has a high contact-closing speed independent of the operator. 103, 202, 27

quick release (control brakes). The provision for effecting more rapid release than would inherently be obtained. *See also:* **electric drive.** 225, 206

quick set (control brakes). The provision for effecting more rapid setting than would inherently be obtained. *See also:* **electric drive.** 225, 206

quiescent-carrier telephony. That form of carrier telephony in which the carrier is suppressed whenever there are no modulating signals to be transmitted. 328

quiescent current (electron tubes). The electrode current corresponding to the electrode bias voltage. *See also:* **electronic tube.** 244, 190

quiescent operating point (magnetic amplifier). The output obtained under any specified external conditions when the signal is non-time-varying and zero. 171

quiescent point (amplifier). That point on its characteristic that represents the conditions existing when the signal input is zero. *Note:* The quiescent values of the parameters are not in general equal to the average values existing in the presence of the signal unless the characteristic is linear and the signal has no direct-current component. *See:* **operating point.** 328

quiescent supervisory system (power switchgear). A supervisory system that is normally alert but inactive, and transmits information or control signals only when a change in status occurs at the remote station or when a demand operation is initiated at the master station. 103

quiescent value (base magnitude) (A_Q) (high- and low-power pulse transformers). The maximum value existing between pulses. *See also:* **input pulse shape.** 32, 33

quiet automatic volume control. Automatic volume control that is arranged to be operative only for signal strengths exceeding a certain value, so that noise or other weak signals encountered when tuning between strong signals are suppressed. *See also:* **radio receiver.** 328

quieting sensitivity (frequency-modulation receivers). The minimum unmodulated signal input for which the output signal-noise ratio does not exceed a specified limit, under specified conditions. *See also:* **radio receiver; receiver performance.** 181

quiet tuning. A circuit arrangement for silencing the output of a radio receiver except when the receiver is accurately tuned to an incoming carrier wave. *See also:* **radio receiver.** 328

quill drive. A form of drive in which a motor or generator is geared to a hollow cylindrical sleeve, or quill, or the armature is directly mounted on a quill, in either case, the quill being mounted substantially concentrically with the driving axle and flexibly connected to the driving wheels. *See also:* **traction motor.** 328

quinary. *See:* **biquinary.**

quinhydrone electrode. *See:* **quinhydrone half cell.**

quinhydrone half cell. A half cell with an electrode of an inert metal (such as platinum or gold) in contact with a solution saturated with quinhydrone. *Syn.* **quinhydrone electrode.** *See also:* **electrochemistry.** 328

quotation board. A manually or automatically operated panel equipped to display visually the price quotations received by a ticker circuit. Such boards may provide displays of large size or may consist of small automatic units that ordinarily display one item at a time under control of the user. *See also:* **telegraphy.** 328

quotient. *See:* **phasor product.**

quotient relay. A relay that operates in response to a suitable quotient of two alternating electrical input quantities. *See also:* **relay.** 103, 202, 60

Q video signal (NTSC color television). One of the two video signals (E'_I and E'_Q) controlling the chrominance in the NTSC system. *Note:* It is a linear combination of gamma-corrected primary color signals, E'_R, E'_G, and E'_B, as follows:

$$E'_Q = 0.41(E'_B - E'_Y) + 0.48(E'_R - E'_Y)$$
$$= 0.21E'_R - 0.52E'_G + 0.31E'_B$$

18

R

rabbet, mounting (rotating machinery). A male or female pilot on a face or flange type of end shield of a machine, used for mounting the machine with a mating rabbet. The rabbet may be circular, or of other configuration and need not be continuous. 63

raceway (1) (electric system). Any channel for enclosing, and for loosely holding wires, cables, or busbars in interior work that is designed expressly for, and used solely for, this purpose. *Note:* Raceways may be of metal or insulating material and the term includes rigid metal conduit, rigid nonmetallic conduit, flexible metal conduit, electrical metallic tubing, underfloor raceways, cellular concrete-floor raceways, cellular metal-floor raceways, surface metal raceways, structural raceways, wireways and busways, and auxiliary gutters or moldings. 256, 124

(2) (nuclear power generating stations). Any channel that is designed and used expressly for supporting wires, cable, or busbars. Raceways consist primarily of, but are not restricted to, cable trays, conduits, and interlocked armor enclosing cable. 31

(3) (transmission and distribution). Any channel for holding conductors which is designed expressly and used solely for this purpose. 262

rack mounting (1) (general). A method of mounting electric equipment in which metal panels supporting the equipment are attached to predrilled tapped vertical steel channel rails or racks. The dimensions of panels, the spacing and height of rails, and the spacing and size of mounting screws are standardized.   59

(2) (cabinet). A method of rack mounting in which the vertical rail or rack is enclosed in a metal cabinet that may have hinged doors on the front and back.   59

(3) (enclosed). A method of rack mounting in which boxlike covers are attached to the panels to protect and shield the electric equipment.   59

(4) (fixed). A method of cabinet rack mounting in which the rack or rails are fixed to the inside of the cabinet. Equipment is accessible through front and rear doors of the cabinet.   59

(5) (swinging). A method of cabinet rack mounting in which a frame containing the rails or rack is hinged on one side to allow the rack and the equipment on it to be swung out of the front door of the cabinet. This makes both sides of the rack accessible when the back of the cabinet is not accessible.   59

racks, traveler (conductor stringing equipment). A device designed to protect, store and transport travelers. It is normally designed to permit stacking during storage to utilize space, efficient use of transporting vehicles, and spotting by helicopters on the line. The exact design of each rack is dependent upon the specific travelers to be stored. *Syn.* **dollie cars.**   45

racon. *See:* **radar beacon.**

rad (photovoltaic power system). An absorbed radiation unit equivalent to 100 ergs/gram of absorber. *See also:* **photovoltaic power system.**   186

radar (1) (navigation aids). An electromagnetic device for detecting the presence and location of objects. The presence of objects and their distance (range) are determined by the transmission and return of electromagnetic energy; direction is usually obtained also, through the use of a movable or rotating directive antenna pattern. Original name **radar** is derived from **radio detection and ranging.**   13

(2) (electric installations on shipboard). A device which radiates electromagnetic waves and utilizes the reflection of such waves from distant objects to determine their existence or position.   3

radar altimeter (radar). *See:* **radio altimeter.**

radar beacon. A transponder used for replying to interrogations from primary radars.   13

radar camouflage. The art, means, or result of concealing the presence or the nature of an object from radar detection by the use of coverings or surfaces that considerably reduce the radio energy reflected toward a radar. *See also:* **radar.**   328

radar cross section (1) (electromagnetic scatterer or radar target in a particular orientation) (RCS). Usually represented by the symbol σ and sometimes abbreviated RCS, defined as 4π times the ratio of the power per unit solid angle scattered in a specified direction to the power per unit area in a plane wave incident on the scatterer from a specified direction. More precisely, it is the limit of that ratio as the distance from the scatterer to the point where the scattered power is measured approaches infinity. Three cases are distinguished: (A) **monostatic or backscattering RCS** when the incident and pertinent scattering directions are coincident but opposite in sense; (B) **forward-scattering**

RCS when the two directions and senses are the same; and (C) **bistatic RCS** when the two directions are different. If not identified, RCS is usually understood to refer to case (A). In all cases, if the scatterer has appreciable polarization sensitivity, the incident polarization and the polarization component of the scattered wave that is being received should be specified, or else the RCS is understood to be averaged over all polarizations. Equivalent terms: **effective echoing area, equivalent echoing area, scattering (or backscattering, forward-scattering, bistatic scattering) cross section.**   13

(2) (antennas). For a given polarization of the incident wave, that portion of the scattering cross section of a target associated with a specified polarization component of the scattered wave. *See also:* **scattering cross section; equivalent flat plate area of a scattering object; circular scanning; effective echo area.**   111

radar equation. A mathematical expression for primary radar which, in its basic form, relates radar parameters such as transmitter power, antenna gain, wavelength, effective echo area of the target, distance to the target and receiver input power. The equation may be modified to take into account other factors, such as attenuation of energy caused by a radome, attenuation due to unusual atmospheric losses or precipitation, and scanning losses which might affect performance of the radar. Sometimes called radar range equation or range equation.   13

radar letter designation. These designations are consistent with the recommended nomenclature of the International Telecommunications Union (ITU), as shown in Table 2. Note that the HF and VHF definitions are identical in the two systems. The essence of the radar nomenclature is to subdivide the existing ITU bands, in accordance with radar practice, without conflict or ambiguity.

The letter band designations should not be construed as being a substitute for the specific frequency limits of the frequency bands. The specific frequency limits should be used when appropriate, but when a letter designation of a radar frequency band is called for, those of Table 1 shall be used.

The letter designations described in the standard are designed for radar usage and follow current practice. They are not meant to be used for other radio or telecommunication purposes, unless they pertain to radar.

The letter designations for Electronic Countermeasure operations as described in Air Force Regulations No 55-44, Army Regulation No 105-86, and Navy OPNAV Instruction 3430.9B are not consistent with radar practice and shall not be used to describe radar frequency bands.   44

radar performance figure. The ratio of the pulse power of the radar transmitter to the power of the minimum signal detectable by the receiver. *See also:* **navigation.**
                    13, 187

radar range equation. *See:* **radar equation.**

radar relay (radar). An equipment for relaying the radar video and appropriate synchronizing signals to a remote location. *See also:* **navigation.**   13, 187

radar shadow (radar). Absence of radar illumination because of an intervening reflection or absorbing medium; the shadow is manifested on the display by the absence of blips from targets in the shadow area.   13

Table 1
Standard radar-frequency letter band nomenclature

Band Designation	Nominal Frequency Range	Specific Radiolocation (Radar) Bands Based on ITU Assignments for Region 2
HF	3–30 MHz	(Note 1)
VHF	30–300 MHz	138–144 MHz
		216–225
UHF	300–1000 MHz (Note 2)	420–450 MHz (Note 3)
		890–942 (Note 4)
L (Note 5)	1000–2000 MHz	1215–1400 MHz
S	2000–4000 MHz	2300–2500 MHz
		2700–3700
C	4000–8000 MHz	5250–5925 MHz
X	8000–12000 MHz	8500–10 680 MHz
K_u	12.0–18 GHz	13.4–14.0 GHz
		15.7–17.7
K	18–27 GHz	24.05–24.25 GHz
K_a	27–40 GHz	33.4–36.0 GHz
mm	40–300 GHz	(Note 6)

Notes:

(1) There are no official ITU radiolocation bands at HF. So-called HF radars might operate anywhere from just above the Broadcast band (1.605 MHz) to 40 MHz or higher.

(2) The official ITU designation for the Ultra High Frequency band extends to 3000 MHz. In radar practice, however, the upper limit is usually taken as 1000 MHZ, L and S bands being used to describe the higher UHF region.

(3) Sometimes called P band, but use is rare.

(4) Sometimes included in L band.

(5) The Aeronautical Radionavigation band from 960–1215 MHz is sometimes called L_x.

(6) No ITU radiolocation assignments. Experimental operation in this region has generally centered around the 94 GHz, 140 GHz, and 230 GHz windows and at 70 GHz. The region from 300–3000 GHz is called the Submillimeter band.

Table 2
Comparison of radar-frequency letter band nomenclature with ITU nomenclature

Radar Nomenclature		International Telecommunications Union Nomenclature			
Radar Letter Designation	Frequency Range	Frequency Range	Band No.	Adjectival Band Designation	Corresponding Metric Designation
HF	3–30 MHz	3–30 MHz	7	High Frequency (HF)	Decametric Waves
VHF	30–300 MHz	30–300 MHz	8	Very High Frequency (VHF)	Metric Waves
UHF	300–1000 MHz	0.3–3.0 GHz	9	Ultra High Frequency (UHF)	Decimetric Waves
L	1.0–2.0 GHz				
S	2.0–4.0 GHz				
C	4.0–8.0 GHz				
X	8.0–12.0 GHz	3.0–30 GHz	10	Super High Frequency (SHF)	Centimetric Waves
K_u	12.0–18.0 GHz				
K	18.0–27.0 GHz				
K_a	27.0–40 GHz	30–300 GHz	11	Extremely High Frequency (EHF)	Millmetric Waves
mm	40–300 GHz				

radar transmitter. The transmitter portion of a radio detecting and ranging system. 111, 181

radial (navigation). One of a number of lines of position defined by an azimuthal navigational facility; the radial is identified by its bearing (usually the magnetic bearing) from the facility. *See also:* **navigation.** 187, 13

radial air gap (rotating machinery). *See:* **air gap (gap) (ro-** tating machinery).

radial-blade blower (rotating machinery). A fan made with flat blades mounted so that the plane of the blades passes through the axis of rotation of the rotor. *See also:* **fan (blower) (rotating machinery).** 63

radial distribution feeder. *See:* **radial feeder.**

radial feeder. A feeder supplying electric energy to a

substation or a feeding point that receives energy by no other means. *Note:* The normal flow of energy in such a feeder is in one direction only. *See also:* **center of distribution.** 64

radially outer coil side. *See:* **bottom coil slot.**

radial magnetic pull (rotating machinery). The radial force acting between rotor and stator resulting from the radial displacement of the rotor from magnetic center. *Note:* Unless other conditions are specified, the value of radial magnetic pull will be for no load and rated voltage, and for rated no load field current and rated frequency as applicable. 63

radial probable error (RPE) (electronic navigation). *See:* **circular probable error.**

radial system. A system in which independent feeders branch out radially from a common source of supply. *See also:* **alternating-current distribution; direct-current distribution.** 64

radial-time-base display. *See:* **plan-position indicator.**

radial transmission feeder. *See:* **radial feeder.**

radial transmission line. Basically a pair of parallel conducting planes used for propagating circularly cylindrical waves having their axes normal to the planes, sometimes applied to tapered versions, such as biconical lines. *See also:* **waveguide.** 267, 179

radial type. A unit substation which has a single stepdown transformer and which has an outgoing section for the connection of one or more outgoing radial (stub end) feeders. 53

radial unbalance torque (tuned rotor gyro). The acceleration-sensitive torque caused by radial unbalance due to noncoincidence of the flexure axis and the center of mass of the rotor. Under constant acceleration, it appears as a rotating torque at the rotor spin frequency. When the gyro is subjected to vibratory acceleration along the spin axis at the spin frequency, this torque results in a rectified unbalance drift rate. 46

radian. (1) (metric practice). The plane angle between two radii of a circle which cut off on the circumference an arc equal in length to the radius. 21

(2) (laser maser). A unit of angular measure equal to the angle subtended at the center of a circle by an arc whose length is equal to the radius of the circle. One (1) radian ≈ 57.3 degrees; 2π radians = 360 degrees. 363

radiance (1) (radiant intensity per unit area) (at a point on a surface and in a given direction). The quotient of the radiant intensity in the given direction of an infinitesimal element of the surface containing the point under consideration, by the area of the orthogonally projected area of the element on a plane perpendicular to the given direction. *Note:* The usual unit is the watt per steradian per square meter. This is the radiant analog of luminance. 18

(2) (light emitting diodes). $(L_e = d^2\phi_e/d_\omega(dA \cos \theta) = dI_e/(dA \cos \theta))$. At a point of the surface of a source, of a receiver, or of any other real or virtual surface, the quotient of the radiant flux leaving, passing through or arriving at an element of the surface surrounding the point, and propagated in the directions defined by an elementary cone containing the given direction, by the product of the solid angle of the cone and the area of the orthogonal projection of the element of the surface on a plane per-

pendicular to the given direction. *Note:* In the defining equation θ is the angle between the normal to the element of the surface and the direction of observation. 162

(3) (laser-maser) (L). Radiant flux or power output per unit solid angle per unit area (W·sr^{-1}·cm^{-2}). 363

radian frequency (circuits and systems). The number of radians per unit time. The unit of time is generally the second and the radian frequency ω is therefore $2\pi f$ where f is the frequency in hertz. 67

radiant density $(w_e = dQ_e/dV)$ **(light emitting diodes).** Radiant energy per unit volume; joules per m^3. 162

radiant efficiency of a source of radiant flux (ξ_e) **(light emitting diodes).** The ratio of the total radiant flux to the forward power dissipation (total electrical lamp power input). 162

radiant emittance. *See:* **radiant flux density at a surface.**

radiant energy (1) (light emitting diodes) (Q_e). Energy traveling in the form of electromagnetic waves. It is measured in units of energy such as joules, ergs or kilowatt-hours. 162

(2) (laser maser). Energy emitted, transferred, or received in the form of radiation. Unit: joule (J). 363

radiant exitance. *See:* **radiant flux density (at an element of surface).**

radiant exposure, H (laser-maser). Surface density of the radiant energy received. Unit: joules per centimeter squared (J·cm^{-2}). 363

radiant flux (1) (light emitting diodes) $(\phi_e = dQ_e/dt)$. The time rate of flow of radiant energy. *Note:* It is expressed preferably in watts, or in ergs per second. 18, 178, 162, 167

(2) (television). Power emitted, transferred, or received in the form of radiation. *Note:* It is expressed preferably in watts. 18

(3) (laser-maser) (radiant power). Power emitted, transferred, or received in the form of radiation. Unit: watt (W). 363

radiant flux density at a surface $(M_e = d\phi_e/dA, E_e = d\phi_e/dA)$. The quotient of radiant flux at that element of surface of the area of that element: that is, watts per cm^2. When referring to radiant flux emitted from a surface, this has been called **radiant emittance*** (symbol: M). The preferred term for radiant flux leaving a surface is **radiant exitance** (symbol: M). When referring to radiant flux incident on a surface, it is called **irradiance** (symbol: E). 162

*** Deprecated**

radiant gain (optoelectronic device). The ratio of the emitted radiant flux to the incident radiant flux. *Note:* The emitted and incident radiant flux are both determined at specified ports. *See also:* **optoelectronic device.** 191

radiant heater. A heater that dissipates an appreciable part of its heat by radiation rather than by conduction or convection. *See also:* **appliances (including portable).**

radiant intensity (of a source, in a given direction) (1) (television). The quotient of the radiant power emitted by a source, or by an element of source, in an infinitesimal cone containing the given direction, by the solid angle of that cone. *Note:* It is expressed preferably in watts per steradian. 18

(2) (light emitting diodes). $(I_e = d\phi_e/d\omega)$. The radiant flux proceeding from the source per unit solid angle in the direction considered; that is watts per steradian. 162
(3) (laser-maser). Quotient of the radiant flux leaving the source, propagated in an element of solid angle containing the given direction, by the element of solid angle. Unit: watt per steradian $(W \cdot sr^{-1})$. 363
radiant sensitivity (phototube). *See:* **sensitivity, radiant.**
radiated interference (electromagnetic compatibility). Radio interference resulting from radiated noise or unwanted signals. *See also:* **electromagnetic compatibility.** 199
radiated noise (electromagnetic compatibility). Radio noise energy in the form of an electromagnetic field, including both the radiation and induction components of the field. *See also:* **electromagnetic compatibility.** 199
radiated power output (transmitter performance). The average power output available at the antenna terminals, less the losses of the antenna, for any combination of signals transmitted when averaged over the longest repetitive modulation cycle. *See also:* **audiofrequency distortion.** 181
radiating element. A basic subdivision of an antenna that in itself is capable of effectively radiating or receiving radio waves. *Note:* Typical examples of a radiating element are a slot, horn, or dipole antenna. *See also:* **antenna.** 179, 111
radiation (1) (radio communication). The emission of energy in the form of electromagnetic waves. The term is also used to describe the radiated energy. 59
(2) (nuclear) (nuclear work). The usual meaning of radiation is extended to include moving nuclear particles, charged or uncharged. 190, 125, 96
radiation counter. An instrument used for detecting or measuring radiation by counting action. *See also:* **gas-filled radiation-counter tubes.** 190, 125
radiation coupling (interference terminology). *See:* **coupling, radiation.**
radiation detector. Any device whereby radiation produces some physical effect suitable for observation and/or measurement. *See:* **anticoincidence (radiation counters).** 190
radiation efficiency (antenna). The ratio of (1) the total power radiated by an antenna to (2) the net power accepted by the antenna from the connected transmitter. *See also:* **antenna.** 179, 111
radiation, electromagnetic (antenna). The emission of energy in the form of electromagnetic waves. 111
radiation intensity (in a given direction). The power radiated from an antenna per unit solid angle in that direction. *See also:* **antenna.** 246, 179, 111, 59
radiation lobe (antenna pattern). A portion of the radiation pattern bounded by regions of relatively weak radiation intensity. *See also:* **antenna.** 246, 179, 111
radiation loss (transmission system). That part of the transmission loss due to radiation of radio-frequency power. *See also:* **waveguide.** 267, 179
radiation pattern (antenna pattern). A graphical representation of the radiation properties of the antenna as a function of space coordinates. *Notes:* (1) In the usual case the radiation pattern is determined in the far-field region and is represented as a function of directional coordinates.

(2) Radiation properties include power flux density, field strength, phase, and polarization. 111
radiation protection guide (electrobiology). Radiation level that should not be exceeded without careful consideration of the reasons for doing so. *See also:* **electrobiology.** 322
radiation pyrometer (radiation thermometer). A pyrometer in which the radiant power from the object or source to be measured is utilized in the measurement of its temperature. The radiant power within wide or narrow wavelength bands filling a definite solid angle impinges upon a suitable detector. The detector is usually a thermocouple or thermopile or a bolometer responsive to the heating effect of the radiant power, or a photosensitive device connected to a sensitive electric instrument. *See also:* **electric thermometer (temperature meter).** 328
radiation resistance (antenna). The ratio of the power radiated by an antenna to the square of the rms antenna current referred to a specified point. *Note:* This term is of limited utility in lossy media. 111
radiation thermometer. *See:* **radiation pyrometer.**
radiation trapping (laser-maser). The suppression or delay of fluorescence in an optically thick absorbing medium resulting from **absorption** and re-emission. 363
radiative relaxation time (laser-maser). The **relaxation time** which would be observed if only processes involving the radiation of electromagnetic energy were effective in producing **relaxation.** 363
radiator (1) (telecommunication). Any antenna or radiating element that is a discrete physical and functional entity. 111
(2) (illumination) (light emitting diodes). An emitter of radiant energy. *See also:* **radiant energy (illuminating engineering).** 167, 162
radio altimeter. An altimeter using radar principles for height measurement; height is determined by measurement of propagation time of a radio signal transmitted from the vehicle and reflected back to the vehicle from the terrain below. *See also:* **navigation.** 187, 13
radio-autopilot coupler (electronic navigation). Equipment providing means by which electric signals from navigation receivers control the vehicle autopilot. *See also:* **navigation.** 187, 13
radio beacon (electronic navigation). A radio transmitter, usually nondirectional, that emits identifiable signals intended for radio direction finding. *Note:* In air operations a radio beacon is called an aerophare. *See also:* **radio direction-finder; radio navigation.** 13, 187
radio broadcasting. Radio transmission intended for general reception. *See also:* **radio transmission.** 111, 240
radio channel. A band of frequencies of a width sufficient to permit its use for radio communication. *Note:* The width of a channel depends upon the type of transmission and the tolerance for the frequency of emission. Normally allocated for radio transmission in a specified type of service or by a specified transmitter. *See also:* **radio transmission.** 111
radio circuit. A means for carrying out one radio communication at a time in either or both directions between two points. *See also:* **radio channel; radio transmission.** 328

radio compass. A direction-finder used for navigational purposes. *See also:* **radio navigation.** 328

radio compass indicator. A device that, by means of a radio receiver and rotatable loop antenna, provides a remote indication of the relationship between a radio bearing and the heading of the aircraft. 328

radio compass magnetic indicator. A device that provides a remote indication of the relationship between a magnetic bearing, radio bearing, and the aircraft's heading. 328

radio control. The control of mechanism or other apparatus by radio waves. *See also:* **radio transmission.** 328

radio detection (radio warning). The detection of the presence of an object by radiolocation without precise determination of its position. *See also:* **radio transmission.** 328

radio direction-finder. A device used to determine the direction of arrival of radio signals *Note:* In practice, the term is usually applied to a device for ascertaining the apparent direction of an object by means of its own independent emissions (that is, excluding reflected or re-transmitted waves). *See also:* **loran; navigational radar (surface search radar); radio beacon; radio receiver.** 187, 13

radio direction finding. A procedure for determining the bearing, at a receiving point, of the source of a radio signal by observing the direction of arrival and other properties of the signal. *See also:* **navigation.** 187, 13

radio distress signal (SOS). Radiotelegraph distress signal consists of the group · · · · · · · · · in Morse code, transmitted on prescribed frequencies. The radiotelephone distress signal consists of the spoken words **May Day.** *Note:* By international agreement, the effect of the distress signal is to silence all radio traffic that may interfere with distress calls. 328

radio disturbance (electromagnetic compatibility). An electromagnetic disturbance in the radio-frequency range. *See:* **radio interference; radio noise.** *See also:* **electromagnetic compatibility.** 199

radio Doppler. The direct determination of the radial component of the relative velocity of an object by an observed frequency change due to such velocity. *See also:* **radio transmission.** 328

radio fadeout (Dellinger effect). A phenomenon in radio propagation during which substantially all radio waves that are normally reflected by ionospheric layers in or above the *E* region suffer partial or complete absorption. *See also:* **radiation.** 328

radio field strength (1). The electric or magnetic field strength at a given location associated with the passage of radio waves. It is commonly expressed in terms of the electric field strength in microvolts, millivolts, or volts per meter. In the case of a sinusoidal wave, the root-mean-square value is commonly stated. Unless otherwise stated, it is taken in the direction of maximum field strength. *See also:* **radiation.** 328

(2) (radio wave propagation). The electric or magnetic field strength at radio frequency. *See also:* **radio wave propagation.** 186, 146

radio frequency (1) (loosely). The frequency in the portion of the electromagnetic spectrum that is between the audio-frequency portion and the infrared portion. 188, 146, 59

(2) A frequency useful for radio transmission. *Note:* The present practicable limits of radio frequency are roughly 10 kilohertz to 100 000 megahertz. Within this frequency range, electromagnetic radiation may be detected and amplified as an electric current at the wave frequency. *See also:* **radio transmission; radio wave propagation; signal.** 186

radio-frequency alternator. A rotating-type generator for producing radio-frequency power. 111

radio-frequency attenuator (signal-transmission system). A low-pass filter that substantially reduces the radio-frequency power at its output relative to that at its input, but transmits lower-frequency signals with little or no power loss. *See also:* **signal.** 188

radio-frequency converter (industrial electronics). A power source for producing electric power at a frequency of 10 kilohertz and above. *See also:* **industrial electronics.** 114

radio-frequency generator (1) (signal-transmission system). A source of radio-frequency energy. 188

(2) (induction heating). A power source for producing electric power at a frequency of 10 kilohertz and above. 14

radio-frequency generator, electron tube type (induction and dielectric usage). A power source comprising an electron-tube oscillator, an amplifier if used, a power supply and associated control equipment. *See also:* **Colpitts oscillator, Hartley oscillator, magnetron, tuned grid-tuned plate oscillator.** 14

radio-frequency pulse. A radio-frequency carrier amplitude modulated by a pulse. The amplitude of the modulated carrier is zero before and after the pulse. *Note:* Coherence of the carrier (with itself) is not implied. 254

radio-frequency radiation. The propagation of electromagnetic energy in space. *See:* **applicators or electrodes.** 14

radio-frequency switching relay. A relay designed to switch frequencies that are higher than commercial power frequencies with low loss. 341

radio-frequency system loss (mobile communication). The ratio expressed in decibels of (1) the power delivered by the transmitter to its transmission line to (2) the power required at the receiver-input terminals that is just sufficient to provide a specified signal-to-noise ratio at the audio output of the receiver. *See also:* **mobile communication system.** 181

radio-frequency transformer. A transformer for use with radio-frequency currents. *Note:* Radio-frequency transformers used in broadcast receivers are generally shunt-tuned devices that are tunable over a relatively broad range of frequencies. *See also:* **radio transmission.** 197

radio gain (radio system). The reciprocal of the system loss. *See also:* **radio wave propagation.** 180, 146

radio horizon (antenna). The locus of the farthest points at which direct rays from the antenna become tangential to a planetary surface. *Note:* On a spherical surface the horizon is a circle. The distance to the horizon is affected by atmospheric refraction. *See also:* **radiation; radio wave propagation.** 180, 146

radio-influence field (RIF) (electromagnetic compatibility).

Radio-influence field is the radio noise field emanating from an equipment or circuit, as measured using a radio noise meter in accordance with specified methods. *See also:* **electromagnetic compatibility.** 197

radio-influence tests. Tests that consist of the application of voltage and the measurement of the corresponding radio-influence voltage produced by the device being tested. 103, 202

radio influence voltage (RIV) (1) (general) (electromagnetic compatibility). The radio noise voltage appearing on conductors of electric equipment or circuits, as measured using a radio noise meter as a two-terminal voltmeter in accordance with specified methods. *See also:* **electromagnetic compatibility.** 197

(2) (insulator). The radio-frequency voltage produced, under specified conditions, by the application of an alternating voltage of 60 hertz $\pm$ 5 percent. *Note:* A radio influence voltage is a high-frequency voltage generated as a result of ionization and may be propagated by conduction, induction, radiation, or a combined effect of all three. *See also:* **insulator.** 261, 287, 229, 62

(3) (outdoor apparatus bushings). A high-frequency voltage generated as a result of ionization, which may be a propagated by conduction, induction, radiation or a combined effect of all three. 168

(4) (surge arrester). A high-frequency voltage, generated by all sources of ionization current, that appears at the terminals of electric power apparatus or on power circuits. 2

(5) (transformer). A radio frequency voltage generally produced by partial discharge and measured at the equipment terminals for the purpose of determining the electromagnetic interference effect of the discharges. *Notes:* (A) RIV can be measured with a coupled radio interference measuring instrument; and is commonly measured at approximately 1 MHz, athough a wide frequency range is involved. (B) RIV values are often used as an index of partial discharge intensity. (C) The RIV of equipment was historically measured to determine the influence of energized equipment on radio broadcasting, hence—RIV.

(6) (high voltage ac cable terminations). The radio noise appearing on conductors of electric equipment or circuits, as measured using a radio-noise meter as a two-terminal voltmeter in accordance with specified methods. 4

radio interference (electromagnetic compatibility). Impairment of the reception of a wanted radio signal caused by an unwanted radio signal or a radio disturbance. *Note:* For example, the words interference and disturbance are also commonly applied to a radio disturbance. *See:* **radio disturbance; radio noise.** *See also:* **electromagnetic compatibility.** 197

radiolocation (radar). Position determination by means of radio aids for purposes other than those of navigation. Usually synonomous for radar. 13

radio magnetic indicator (RMI). A combined indicating instrument that converts omnibearing indications to a display resembling an automatic-direction-finder display, one in which the indicator points toward the omnirange station; it combines omnibearing, vehicle heading, and relative bearing. *See also:* **radio navigation.** 187, 13

radiometric sextant (electronic navigation). An instrument that measures the direction to a celestial body by detecting and tracking the nonvisible natural radiation of the body; such radiation includes radio, infrared, and ultraviolet. *See also:* **navigation.** 187, 13

radiometry. The measurement of quantities associated with radiant energy. *See also:* **primary standards (illuminating engineering).** 167

radio navigation (radar). Navigation based upon the reception of radio signals. 13

radio noise (1) (electromagnetic compatibility). An electromagnetic noise in the radio-frequency range. *See:* **radio disturbance; radio interference.** 199

(2) (overhead power lines). Any unwanted disturbance within the radio-frequency band, such as undesired electric waves in any transmission channel or device. 36

radio-noise field strength. A measure of the field strength, at a point (as a radio receiving station), of electromagnetic waves of an interfering character. *Notes:* (1) In practice the quantity measured is not the field strength of the interfering waves, but some quantity that is proportional to, or bears a known relation to the field strength. (2) It is commonly measured in average microvolts, quasi-peak microvolts, peak microvolts, or peak microvolts in a unit band per meter, according to which detector function of a radio-noise meter is used. *See also:* **broad-band radio noise; narrow-band radio noise; radiation.** 285

radiophare (electronic navigation). Term often used in international terminology, meaning radio beacon. *See also:* **navigation.** 187, 13

radio propagation path (mobile communication). For a radio wave propagating from one point to another, the great-circle distance between the transmitter and receiver antenna sites. *See also:* **mobile communication system.** 181

radio proximity fuse. A radio device contained in a missile to detonate it within predetermined limits of distance from a target by means of electromagnetic interaction with the target. *See also:* **radio transmission.** 111

radio range (electronic navigation). A radio facility that provides radial lines of position by having characteristics in its emission that are convertible to bearing information and useful in lateral guidance of aircraft. *See also:* **navigation; radio navigation.** 187, 13

radio range-finding. The determination of the range of an object by means of radio waves. *See also:* **radio transmission.** 328

radio receiver. A device for converting radio-frequency power into perceptible signals. 181

radio relay system (radio relay). A point-to-point radio transmission system in which the signals are received and retransmitted by one or more intermediate radio stations. *See also:* **radio transmission.** 328

radio shielding. A metallic covering in the form of conduit and electrically continuous housings for airplane electric accessories, components, and wiring, to eliminate radio interference from aircraft electronic equipment. 328

radiosonde. An automatic radio transmitter in the meteorological-aids service, usually carried on an aircraft, free balloon, kite, or parachute, that transmits meteorological data. *See also:* **radio transmitter.** 111

radio star (communication satellite). A discrete source in the celestial sphere emitting electrical random noise *See also:* **background noise.**   85

radio station. A complete assemblage of equipment for radio transmission or reception, or both. *See also:* **radio transmission.**   328

radio transmission. The transmission of signals by means of radiated electromagnetic waves other than light or heat waves.   328

radio transmitter. A device for producing radio-frequency power, for purposes of radio transmission.   111

radio warning. *See:* **radio detection.**

radio wave. An electromagnetic wave of radio frequency. *See also:* **radio frequency; radio wave propagation.**
                                                                   180, 146

radio wave propagation. The transfer of energy by electromagnetic radiation at radio frequencies.   180, 146

radix. A quantity whose successive integral powers are the implicit multipliers of the sequence of digits that represent a number of some positional-notation systems. For example, if the radix is 5, then 143.2 means 1 times 5 to the second power, plus 4 times 5 to the first power, plus 3 times 5 to the zero power, plus 2 times 5 to the minus one power. Synonymous with base. *See:* **mixed radix.**   77

radix complement. A numeral in radix notation that can be derived from another by subtracting each digit from one less than the radix and then adding one to the least-significant digit of the difference, executing all carries required, for example, tens complement in decimal notation, twos complement in binary notation. *See:* **true complement.**   255, 77

radix-minus-one complement. A numeral in radix notation that can be derived from another by subtracting each digit from one less than the radix, for example, nines complement in decimal notation, ones complement in binary notation.   255, 77

radome (antenna). An enclosure for protecting an antenna from the harmful effects of its physical environment, generally intended to leave the electric performance of the antenna unaffected. *See also:* **antenna, circular scanning.**
                                                              246, 179, 111

rail clamp (mining). A device for connecting a conductor or a portable cable to the track rails that serve as the return power circuit in mines. *See also:* **mine feeder circuit.**
                                                                        328

rain clutter (radar). Return from rain which impairs or obscures return from targets. *See also:* **navigation.**
                                                                   187, 13

rainproof (transformer). So constructed, protected or treated as to prevent rain under specified test conditions from interfering with successful operation of the apparatus.   53

raintight. So constructed or protected as to exclude rain under specified test conditions.   53

raise-lower control point. *See:* **supervisory control point, raise-lower.**

Raman-Nath region (acousto-optic device). The region that occurs when the Bragg Region inequality is reversed, that is $L < n\Lambda^2/\lambda_0$. The angle of incidence is generally zero degrees, and light is diffracted into many diffraction orders.   82

ramp (1) (electronic circuits). A voltage or current that varies at a constant rate for example, that portion of the output waveform of a time-linear-sweep generator used as a time base for an oscilloscope display. *See:* **signal, unit-ramp.** *See also:* **oscillograph.**   185

(2) (railway control). A roadway element consisting of a metal bar of limited length, with sloping ends, fixed on the roadway, designed to make contact with and raise vertically a member supported on the vehicle.   328

(3) (automatic control). *See:* **signal, unit-ramp.**

(4) (pulse terms) (single transition). A linear feature.
                                                                        254

ramp response (1) (null-balancing electric instrument). A criterion of the dynamic response of an instrument when subjected to a measured signal that varies at a constant rate. *See also:* **accuracy rating (instrument).**   295

(2) (automatic control).* *See:* **response, ramp-forced.**
                                                                         56

* Deprecated

ramp response time (null-balancing electric instrument). The time lag, expressed in seconds, between the measured signal and the equivalent positioning of the end device when the measured signal is varying at constant rate. *See also:* **accuracy rating (instrument).**   295

ramp response-time rating (null-balancing electric instrument). The maximum ramp response time for all rates of change of measured signal not exceeding the average velocity corresponding to the span step-response-time-rating of the instrument when the instrument is used under rated operating conditions. *Example:* If the span step-response-time-rating is 4 seconds, the ramp response-time rating shall apply to any rate of change of measured signal not exceeding 25 percent of span per second. *See also:* **accuracy rating (instrument).**   295

ramp shoe. *See:* **shoe.**

random (1) (data transmission). A condition not localized in time nor frequency.   59

(2) (automatic control). Describing a variable whose value at a particular future instant cannot be predicted exactly, but can only be estimated by a probability distribution function.   56

random access (computing systems). (1) Pertaining to the process of obtaining data from, or placing data into storage where the time required for such access is independent of the location of the data most recently obtained or placed in storage. (2) Pertaining to a storage device in which the access time is effectively independent of the location of the data.   255, 77, 54

random access programming (test, measurement and diagnostic equipment). Programming without regard for the sequence required for access to the storage position called for in the program.   54

random drift rate (gyro). The non-systematic time varying component of drift rate under specified operating conditions. It is expressed as an rms value or standard deviation of angular displacement per unit time.   46

random errors (radar). Those errors which cannot be predicted except on a statistical basis. *See also:* **navigation.**
                                                               187, 115, 13, 47

random failure (reliability). *See:* **failure, random.**

random noise (fluctuation noise). Noise that comprises

transient disturbances occurring at random. *Notes:* (1) (The part of the noise that is unpredictable except in a statistical sense.) The term is most frequently applied to the limiting case where the number of transient disturbances per unit time is large, so that the spectral characteristics are the same as those of thermal noise. (Thermal noise and shot noise are special cases of random noise.) *See also:* **drift; signal-to-noise ratio.** 9, 59, 24

random number. *See:* **pseudorandom number sequence.**

random paralleling (rotating machinery). Paralleling of an alternating-current machine by adjusting its voltage to be equal to that of the system, but without adjusting the frequency and phase angle of the incoming machine to be sensibly equal to those of the system. *See also:* **asynchronous machine.** 63

random winding (rotating machinery). A winding in which the individual conductors of a coil side occupy random position in a slot. *Syn.* **mush-wound; pancake-wound.** *See also:* **rotor (rotating machinery); stator.** 63

random-wound motorette. A motorette for random-wound coils. *See also:* **direct-current commutating machine.** 63

range (1) (test, measurement and diagnostic equipment) (instrument). The capability span of an instrument, the region between the lower and upper limits of a measured or generated function. 54

(2) (orientation range) (printing telegraphy). That fraction of a perfect signal element through which the time of selection may be varied so as to occur earlier or later than the normal time of selection, without causing errors while signals are being received. The range of a printing-telegraph receiving device is commonly measured in percent of a perfect signal element by adjusting the indicator. *See also:* **instrument; telegraphy.** 194

(3) (computing systems). (A) The set of values that a quantity or function may assume. (B) The difference between the highest and lowest value that a quantity or function may assume. *See:* **error range.** *See also:* **dynamic range; instrument.** 255, 77

(4) (electronic navigation). An ambiguous term meaning either: (A) a distance, as in artillery techniques and radar measurements or (B) a line of position, located with respect to ground references, such as a very-high-frequency omnidirectional radio range (VOR) station, or a pair of lighthouses, or an aural radio range (A-N) radio beacon. *Note:* In electronic navigation, the reader must be particularly wary, since the two meanings of the word **range** often occur in close proximity. *See:* **radio range.**

(5) (radar techniques and artillery-fire control). Deprecated, because the term invites confusion with **radio range.** Use the common synonym, **distance.**

range and elevation guidance for approach and landing (regal). A ground-based navigation system used in conjunction with a localizer to compute vertical guidance for proper glide-slope and flare-out during an instrument approach and landing; it uses a digitally coded vertically scanning fan beam that provides data for both elevation angle and distance. *See also:* **navigation.** 187, 13

range extender (telephone switching systems). Equipment inserted in a switched connection to allow an increased loop resistance. 55

range lights. Groups of color-coded boundary lights provided to indicate the direction and limits of a preferred landing path normally on an aerodrome without runways but exceptionally on an aerodrome with runways. 167

range mark (radar). *See:* **distance mark.**

range noise (radar). The noise-like variation in the apparent distance of a target, caused by changes in phase and amplitude of the target scattering sources, and including radial components of **glint** and **scintillation error.** 13

range resolution (radar). *See:* **distance resolution.**

ranging (communication satellite). The measurement of distance between two points and a precisely known reference point. A multiplicity of tones or a PN (pseudonoise) sequence ranging code is often used. 84

rank (network) (degrees of freedom on a node basis). The number of independent cut-sets that can be selected in a network. The rank R is equal to the number of nodes V minus the number of separate parts P. Thus $R = V - P$. *See also:* **network analysis.** 210

rapid access loop (test, measurement and diagnostic equipment). In internal memory machines, a small section of memory which has much faster accessibility than the remainder of the memory. 54

rapid-start fluorescent lamp. A fluorescent lamp designed for operation with a ballast that provides a low-voltage winding for preheating the electrodes and initiating the arc without a starting switch or the application of high voltage. *See also:* **fluorescent lamp.** 167

rapid-starting systems (fluorescent lamps). The designation given to those systems in which hot-cathode electric discharge lamps are operated with cathodes continuously heated through low-voltage heater windings built as part of the ballast, or through separate low-voltage secondary transformers. Sufficient voltage is applied across the lamp and between the lamp and fixture to initiate the discharge when the cathodes reach a temperature high enough for adequate emission. The cathode-heating current is maintained even after the lamp is in full operation. *Note:* In Europe this system is sometimes referred to as an instant-start system. 268

raster (cathode-ray tubes). A predetermined pattern of scanning lines that provides substantially uniform coverage of an area. *See also:* **oscillograph; television.**
177, 190, 163, 185, 125

raster burn (camera tubes). A change in the characteristics of that area of the target that has been scanned, resulting in a spurious signal corresponding to that area when a larger or tilted raster is scanned. 177, 190, 125

raster scan (radar). A method of sweeping a cathode ray tube screen or an antenna beam characterized by a network of parallel sweeps either from side to side or from top to bottom. 13

ratchet demand (electric power utilization). The maximum past or present demands that are taken into account to establish billings for previous or subsequent periods. *See also:* **alternating-current distribution.** 200

ratchet demand clause (electric power utilization). A clause in a rate schedule that provides that maximum past or present demands be taken into account to establish billings for previous or subsequent periods. *See also:* **alternating-current distribution.** 200

ratchet relay. A stepping relay actuated by an armature-driven ratchet. *See also:* **relay.**   259

rate action (process control). That component of proportional plus rate control action or of proportional plus reset plus rate control action for which there is a continuous linear relation between the rate of change of the directly controlled variable and the position of a final control element. *See also:* **control action; derivative.**   56

rate base (transmission and distribution). The net plant investment or valuation base specified by a regulatory authority, upon which a utility is permitted to earn a specified rate of return.   112

rate, chopping (cathode-ray oscilloscopes). The rate at which channel switching occurs in chopped display operation.   106

rate compensation heat detector (fire protection devices). A device which will respond when the temperature of the air surrounding the device reaches a predetermined level, regardless of the rate of temperature rise.   71

rate control action (electric power systems). Action in which the output of the controller is proportional to the input signal and the first derivative of the input signal. Rate time is the time interval by which the rate action advances the effect of the proportional control action. *Note:* Applies only to a controller with proportional control action plus derivative control action. *See also:* **speed-governing system.**   94

rated. A qualifying term that, applied to an operating characteristic, indicates the designated limit or limits of the characteristic for application under specified conditions. *Note:* The specific limit or limits applicable to a given device is specified in the standard for that device and included in the title of the rated characteristic, that is, rated maximum voltage, rated frequency range, etcetera.
                                                       103, 202

rated accuracy (automatic null-balancing electric instrument). The limit that errors will not exceed when the instrument is used under any combination of rated operating conditions. *Notes:* (1) It is usually expressed as a percent of the span. It is preferred that a + sign or − sign or both precede the number or quantity. The absence of a sign infers a ± sign. (2) Rated accuracy does not include accuracy of sensing elements or intermediate means external to the instrument. *See also:* **accuracy rating (instrument).**
                                                       295

rated accuracy of instrument shunts (electric power systems). The limit of error, expressed as a percentage of rated voltage drop, with two-thirds rated current applied for one-half hour to allow for self-heating. *Note:* Practically, it represents the expected accuracy of the shunt obtainable over normal operating current ranges. *See also:* **accuracy rating (instrument).**   201

rated alternating voltage (rectifier unit) (rated alternating-current winding voltages) (rectifier). The root-mean-square voltages between the alternating-current line terminals that are specified as the basis for rating. *Note:* When the alternating-current winding of the rectifier transformer is provided with taps, the rated voltage shall refer to a specified tap that is designated as the rated-voltage tap. *See also:* **rectification; rectifier transformer.**
                                                       291, 208, 258

rated average tube current. The current capacity of a tube, in average amperes, as assigned to it by the manufacturer

for specified circuit conditions. *See also:* **rectification.**
                                                       291

rated burden (capacitance potential device). The maximum unity-power-factor burden, specified in watts at rated secondary voltage, that can be carried for an unlimited period when energized at rated primary line-to-ground voltage, without causing the established limitations to be exceeded. *See also:* **outdoor coupling capacitor.**   351

rated circuit voltage. Used to designate the rated, root-mean-square, line-to-line, voltage of the circuit on which coupling capacitors or the capacitance potential device in combination with its coupling capacitor or bushing is designed to operate. *See also:* **outdoor coupling capacitor.**
                                                       351

rated continuous current (1) (electric installations on shipboard). The design limit in rms or dc amperes which a switch or circuit breaker will carry continuously without exceeding the limit of observable temperature rise.
                                                       3, 233

(2) (neutral grounding devices). The current expressed in amperes, root-mean-square, that the device can carry continuously under specified service conditions without exceeding the allowable temperature rise.   91

rated current (1) (machine or apparatus). The value of the current that is used in the specification and from which the conditions of temperature rise, as well as the operation of the equipment, are calculated.

(2) For equipment where the rated output is mechanical power or some related quantity such as air flow per minute, it is determined from the current at rated output under standard test conditions.   210

(3) (instrument) (wattmeter or power-factor meter). The normal value of current used for design purposes.   280

(4) (equipment or of a winding). The current for which the equipment, or winding, is designed, and to which certain operating and performance characteristics are referred.
                                                       203

(5) (neutral grounding device) (electric power). The thermal current rating. The rated current of resistors whose rating is based on constant voltage is the initial root-mean-square symmetrical value of the current that will flow when rated voltage is applied. *See also:* **grounding device.**   91, 64

(6) (current transformer). The primary current selected for the basis of performance specifications of a current transformer.   53

rated direct current. *See:* **direct current, rated.**

rated direct-current winding voltage (rectifier). The root-mean-square voltage of the direct-current winding obtained by turns ratio from the rated alternating-current winding voltage of the rectifier transformer. *See also:* **rectifier transformer.**   258

rated direct voltage (power inverter). The nominal direct input voltage. *See also:* **self-commutated inverters.**
                                                       208

rated duty. That duty that the particular machine or apparatus has been designed to comply with.   210

rated excitation-system voltage (rotating machinery). The main exciter rated voltage.   63

rated frequency (1) (general). The frequency used in the specification of the apparatus and from which the test conditions and the frequency limits for the use of the equipment are calculated.   210

(2) (power system or interconnected system). The normal frequency in hertz for which alternating-current generating equipment operating on such system is designed.
94

(3) (arrester). The frequency, or range of frequencies, of the power systems on which the arrester is designed to be used.   308, 62

(4) (grounding device). The frequency of the alternating current for which it is designed. *Note:* Some devices, such as neutral wave traps, may have two or more rated frequencies: the rated frequency of the circuit and the frequencies of the harmonic or harmonics the devices are designed to control. *See also:* **grounding devices.**
91, 64

rated head (speed governing of hydraulic turbines). The value stated on the turbine nameplate.   8

rated impedance (loudspeaker measurements). The rated impedance of a loudspeaker driver or system is that value of a pure resistance which is to be substituted for the driver or system when measuring the electric power delivered from the source. This should be specified by the manufacturer.   19

rated impulse protective level (arrester). The impulse protective level with the residual voltage referred to the nominal discharge current.   62

rated impulse withstand voltage (apparatus). An assigned crest value of a specified impulse voltage wave that the apparatus must withstand without flashover, disruptive discharge, or other electric failure.   210, 64, 62

rated insulation class (neutral grounding device) (electric power). An insulation class expressed in root-mean-square kilovolts, that determines the dielectric tests that the device shall be capable of withstanding. *See also:* **grounding device; outdoor coupling capacitor.**   92

rated kilovolt-ampere (kVA) (transformer or reactor) (1) (transformer). The output that can be delivered for the time specified at rated secondary voltage and rated frequency without exceeding the specified temperature-rise limitations under prescribed conditions.   203, 53

(2) (current-limiting reactor). The kilovolt-amperes that can be carried for the time specified at rated frequency without exceeding the specified temperature limitations, and within the limitations of established standards. *See also:* **reactor.**   309

(3) (grounding transformer). The short-time kilovolt-ampere rating is the product of the rated line-to-neutral voltage at rated frequency, and the maximum constant current that can flow in the neutral for the specified time without causing specified temperature-rise limitations to be exceeded, and within the limitations of established standards for such equipment. *See also:* **duty.**   203, 53

rated kilovolt-ampere (kVA) tap (transformer). A tap through which the transformer can deliver its rated kilovolt-ampere output without exceeding the specified temperature rise. *Note:* The term **rated kilowatt output** is deprecated, unless power factor is specified. *See:* **windings, high-voltage and low-voltage.**   203, 53

rated kilowatts (constant-current transformer). The kilowatt output at the secondary terminals with rated primary voltage and frequency, and with rated secondary current and power factor, and within the limitations of established standards. *See also:* **duty.**   203, 303, 53

rated life (glow lamp). The length of operating time, expressed in hours, that produces specified changes in characteristics. *Note:* In lamps for indicator use the characteristic usually is light output; the end of useful life is considered to be when light output reaches 50 percent of initial, or when the lamp becomes inoperative at line voltage. In lamps used as circuit components, the characteristic is usually voltage; life is determined as the length of time for a specified change from initial.   283

rated line kilovolt-ampere (kVA) rating (rectifier transformer). The kilovolt-ampere rating assigned to it by the manufacturer corresponding to the kilovolt-ampere drawn from the alternating-current system at rated voltage and kilowatt load on the rectifier under the normal mode of operation. *See also:* **rectifier transformer.**   258

rated load (1) (elevator, dumbwaiter, escalator, or private residence inclined lift). The load which the device is designed and installed to lift at the rated speed. *See also:* **elevator.**   328

(2) (rectifier unit). The kilowatt power output that can be delivered continuously at the rated output voltage. It may also be designated as the one-hundred-percent-load or full-load rating of the unit. *Note:* Where the rating of a rectifier unit does not designate a continuous load it is considered special. *See;* **continuous rating.** *See also:* **rectification.**   208

rated-load field voltage (nominal collector ring voltage) (rotating machinery). The voltage required across the terminals of the field winding of an electric machine under rated continuous-load conditions with the field winding at: (1) 75 degrees Celsius for field windings designed to operate at rating with a temperature rise of 60 degrees Celsius or less. (2) 100 degrees Celsius for field windings designed to operate at rating with a temperature rise greater than 60 degrees Celsius.   63

rated-load impedance (magnetic amplifier). The load impedance specified by the manufacturer that is determined by the specifications to which the amplifier is designed.
171

rated-load torque (rated torque) (rotating machinery). The shaft torque necessary to produce rated power output at rated-load speed. *See also:* **asynchronous machine; direct-current commutating machine.**   63

rated minimum tripping current (automatic circuit reclosers). The minimum rms current which causes a device to operate.   92

rated OFF voltage (magnetic amplifier). The output voltage existing with trip OFF control signal applied.   171

rated ON voltage (magnetic amplifier). The output voltage existing with trip ON control signal applied. Rated ON voltage shall be specified either as root-mean-square or average. *Note:* While specification may be either root-mean-square or average it remains fixed for a given amplifier.   171

rated operating conditions (automatic null-balancing electric instrument). The limits of specified variables or conditions within which the performance ratings apply. *See also:* **measurement system.**   295

rated output capacity (inverters). The kilovolt-ampere output at specified load power-factor conditions. *See also:* **self-commutated inverters.**   208

rated output current (1) (general). The output current at the

continuous rating for a specified circuit. 237, 66
(2) (rectifier unit). The current derived from the rated load and the rated output voltage. *See also:* **average forward-current rating (rectifier circuit); rectification.** 208
(3) (magnetic amplifier). Rated output current that the amplifier is capable of supplying to the rated load impedance, either continuously or for designated operating intervals, under nominal conditions of supply voltage, supply frequency, and ambient temperature such that the intended life of the amplifier is not reduced or a specified temperature rise is not exceeded. Rated output current shall be specified either as root-mean-square or average. *Notes:* (1) When other than rated load impedance is used, the root-mean-square value of the rated output current should not be exceeded. (2) While specification may be either root-mean-square or average, it remains fixed for a given amplifier. *See also:* **rating and testing magnetic amplifiers.** 171
(4) (power inverter). The nominal effective (total root-mean-square) current that can be obtained at rated output voltage. *See also:* **self-commutated inverters.** 208
rated output frequency (inverters). The fundamental frequency or the frequency range over which the output fundamental frequency may be adjusted. *See also:* **self-commutated inverters.** 208
rated output voltage (1) (general). The output voltage at the continuous rating for a specified circuit. 237, 66
(2) (rectifier unit). The voltage specified as the basis of rating. *See also:* **average forward-current rating (rectifier circuit); rectification.** 208
(3) (magnetic amplifier). The voltage across the rated load impedance when rated output current flows. Rated output voltage shall be specified by the same measure as rated output current (that is, both shall be stated as root-mean-square or average). *Note:* While specification may be either root-mean-square or average, it remains fixed for a given amplifer. *See also:* **rating and testing magnetic amplifiers.** 171
(4) (power inverter). The nominal effective (total root-mean-square) voltage that is used to establish the power-inverter rating. *See also:* **self-commutated inverters.** 208
rated output volt-amperes (magnetic amplifier). The product of the rated output voltage and the rated output current. 171
rated performance (automatic null-balancing electric instrument). The limits of the values of certain operating characteristics of the instrument that will not be exceeded under any combination of rated operating conditions. *See also:* **to test an instrument or meter.** 295
rated power output (gas turbines) (1) (normal rated power). The rated or guaranteed power output of the gas turbine when it is operated with an inlet temperature of 80 degrees Fahrenheit (26.67 degrees Celsius) and with inlet and exhaust absolute pressures of 14.17 pounds-force per square inch (9.770 newtons per square centimeter). *See also:* **direct-current commutating machine; asynchronous machine.** 98, 58
(2) (site rated power). The rated or guaranteed power output of the gas turbine when it is operated under specified conditions of compressor inlet temperature, compressor inlet pressure, and gas-turbine exhaust pressure.

It is measured at, or is referred to, the output of the generator terminals, the gas-turbine generator unit, or the specified generator if separately purchased. 98, 58
(3) (speed governing of hydraulic turbines). The value stated on the generator nameplate. 8
rated primary current (current transformer). Current selected for the basis of performance specifications. *See also:* **instrument transformer.** 305, 203
rated primary line-to-ground voltage. The root-mean-square line-to-ground voltage for which the potential device, in combination with its coupling capacitor or bushing, is designed to deliver rated burden at rated secondary voltage. The rated primary line-to-ground voltage is equal to the rated circuit voltage (line-to-line) divided by $(3)^{1/2}$. *See:* **primary line-to-ground voltage.** 351
rated primary voltage (1) (constant-current transformer). The primary voltage for which the transformer is designed, and to which operation and performance characteristics are referred. *See also:* **constant-current transformer.** 53
(2) (constant-voltage transformer). The voltage calculated from the rated secondary voltage by turn ratio. *Notes:* (A) See turn ratio of a transformer and its note, for the definition of the turn ratio to be used. (B) In the case of a multiwinding transformer, the rated voltage of any other winding is obtained in a similar manner. 53
(3) (instrument transformer). (A) The rated primary voltage (of a potential (voltage) transformer) is the voltage selected for the basis of performance guarantees. (B) The rated primary voltage (of a current transformer) designates the insulation class of the primary winding. *Note:* A current transformer can be applied on a circuit having a nominal system voltage corresponding to or less than the rated primary voltage of the current transformer. *See also:* **instrument transformer.** 305, 203
rated range of regulation (voltage regulator). The amount that the regulator will raise or lower its rated voltage. *Note:* The rated range may be expressed in per unit, or in percent, of rated voltage; or it may be expressed in kilovolts. *See also:* **voltage regulator.** 257
rated secondary current (constant-current transformer). The secondary current for which the transformer is designed and to which operation and performance characteristics are referred. *See also:* **constant-current transformer.** 203, 53
rated secondary voltage (1) (constant-voltage transformer). The voltage at which the transformer is designed to deliver rated kilovolt-amperes and to which operating and performance characteristics are referred. *See also:* **duty.** 203, 53
(2) (capacitance potential device). This is the root-mean-square secondary voltage for which the potential device, in combination with its coupling capacitor or bushing, is designed to deliver its rated burden when energized at rated primary line-to-ground voltage. *See:* **secondary voltage.** *See also:* **outdoor coupling capacitor.** 351
(3) (voltage transformer). The rated voltage divided by the marked ratio. 53
rated speed (1) (gas turbines). The speed of the power output shaft corresponding to rated generator speed at which rated power output is developed. 98, 58
(2) (elevators). The speed at which the elevator, dumb-

waiter, escalator, or inclined lift is designed to operate under the following conditions: (A) **Elevator or dumbwaiter:** The speed in the up direction with rated load in the car. (B) **Escalator or private-residence inclined lift:** The rate of travel of the steps or carriage, measured along the angle of inclination, with rated load on the steps or carriage. In the case of a reversible escalator the rated speed shall be the rate of travel of the steps in the up direction, measured along the angle of inclination, with rated load on the steps. *See also:* **elevator.**   328

(3) (hydraulic turbines). The value stated on the unit nameplate.   8

rated supply current (magnetic amplifier). The root-mean-square current drawn from the supply when the amplifier delivers rated output current.   171

rated symmetrical interrupting current (accelerometer). The root-mean-square value of the symmetrical component of the highest current which a device is required to interrupt under the operating duty, rated maximum voltage, and circuit constants specified.   92

rated system deviation. The specified maximum permissible carrier frequency deviation. Nominal values for mobile communications systems are ± 15 kilohertz or ± 5 kilohertz.   123

rated system voltage (current-limiting reactor). The voltage to which operations and performance characteristics are referred. It corresponds to the nominal system voltage of the circuit on which the reactor is intended to be used. *See also:* **reactor.**   309

rated thermal current (neutral grounding devices). The root-mean-square neutral current in amperes which the device is rated to carry under standard operating conditions for rated time without exceeding temperature limits.   91

rated time (grounding device) (electric power). The time during which it will carry its rated current, or withstand its rated voltage, or both, under standard conditions without exceeding standard limitations, unless otherwise specified. *See also:* **grounding device.**   91, 64

rated-time temperature rise (grounding device). The maximum temperature rise above ambient attained by the winding of a device as the result of the flow of rated thermal current (or, for certain resistors, the maintenance of rated voltage across the terminals) under standard operating conditions, for rated time and with a starting temperature equal to the steady-state temperature. It may be expressed as an average or a hot-spot winding rise.   91

rated torque (rotating machinery). *See:* **rated-load torque.**

rated voltage (1) (general) (transformer winding). The voltage to which operating and performance characteristics of apparatus and equipment are referred. *Note:* Deviation from rated voltage may not impair operation of equipment, but specified performance characteristics are based on operation under rated conditions. However, in many cases apparatus standards specify a range of voltage within which successful performance may be expected.   324, 210, 203, 257, 53

(2) (step-voltage or induction-voltage regulator). The voltage between terminals of the series winding, with rated voltage applied to the regulator, when the regulator is in the maximum raise position and is delivering rated kilo-

volt-ampere output at 80-percent power factor. *See also:* **voltage regulator.**   257

(3) (electric instrument). Of an instrument such as a wattmeter, power-factor meter, or frequency meter, the rated voltage is the normal continuous operating voltage of the voltage circuit and the normal value of applied voltage for test purposes. *Note:* This quantity should not be confused with the voltage rating of any auxiliary motors, amplifiers, or other devices. *See also:* **instrument.**   280

(4) (arrester). The designated maximum permissible root-mean-square value of power-frequency voltage between its line and earth terminals at which it is designed to operate correctly.   62

(5) (grounding device) (electric power). The root-mean-square voltage, at rated frequency, that may be impressed between its terminals under standard conditions for its rated time without exceeding standard limitations, unless otherwise specified. *See also:* **grounding device.**   91, 64

rated watts input (household electric ranges). The power input in watts (or kilowatts) that is marked on the range nameplate, heating units, etcetera. *See also:* **appliance outlet.**   263

rated withstand current (surge current) (surge arresters). The crest value of a surge, of given wave shape and polarity, to be applied under specified conditions without causing disruptive discharge on the test specimen.

64, 62

rated withstand voltage (insulation strength). The voltage that electric equipment is required to withstand without failure or disruptive discharge when tested under specified conditions and within the limitations of established standards. *See also:* **basic impulse insulation level (insulation strength).**   276

rate gyro. Generally a single-degree-of-freedom gyro having primarily elastic restraint of the spin axis about the output axis. In this gyro, an output signal is produced by precession of the gimbal, the precession angle being proportional to the angular rate of the case about the input axis.   46

rate-integrating gyro. A single-degree-of-freedom gyro having primarily viscous restraint of the spin axis about the output axis. In this gyro, an output signal is produced by precession of the gimbal, the precession angle being proportional to the integral of the angular rate of the case about the input axis.   46

rate-of-change protection. A form of protection in which an abnormal condition causes disconnection or inhibits connection of the protected equipment in accordance with the rate of change of current, voltage, power, frequency, pressure, etcetera.   103, 202, 60, 127

rate-of-change relay. A relay that responds to the rate of change of current, voltage, power, frequency, pressure, etcetera.   103, 202, 60, 127

rate of decay (audio and electroacoustics). The time rate at which the sound pressure level (or other stated characteristic) decreases at a given point and at a given time. *Note:* Rate of decay is frequently expressed in decibels per second.   176

rate of punching (test, measurement and diagnostic equipment). Number of characters, blocks, words, or frames of information placed in the form of holes distributed on cards or tape per unit time. The number of cards punched per unit time.   54

rate of reading (test, measurement and diagnostic equipment). Number of cards, characters, blocks, words, or frames sensed by a sensing device per unit time.   54

rate-of-rise current tripping. *See:* **rate-of-rise release (rate-of-rise trip).**

rate-of-rise heat detector (fire protection devices). A device which will respond when the temperature rises at a rate exceeding a predetermined amount.   71

rate of rise of restriking voltage (transient recovery voltage rate) (usually abbreviated to rrrv) (surge arresters). The rate, expressed in volts per microsecond, that is representative of the increase of the restriking voltage.
                                                          308, 62

rate-of-rise release (rate-of-rise trip). A release that operates when the rate of rise of the actuating quantity in the main circuit exceeds the release setting.   103, 202

rate-of-rise suppressors (semiconductor rectifiers). Devices used to control the rate of rise of current and/or voltage to the semiconductor devices in a semiconductor power converter. *See also:* **semiconductor rectifier stack.**
                                                             208

rate-of-rise trip. *See:* **rate-of-rise release.**

rate signal (industrial control). A signal that is the time derivative of a specified variable. *See also:* **control system, feedback.**   219, 206

rate, sweep. *See:* **sweep time/division.**

rate test. *See:* **problem check.**

rating (rating of electric equipment) (1) (general). The whole of the electrical and mechanical quantities assigned to the machine, apparatus, etcetera, by the designer, to define its working in specified conditions indicated on the rating plate. *Note:* The rating of electric apparatus in general is expressed in volt-amperes, kilowatts, or other appropriate units. Resistors are generally rated in ohms, amperes, and class of service.   310, 210, 233

(2) (rotating machinery). The numerical values of electrical quantities (frequency, voltage, current, apparent and active power, power factor) and mechanical quantities (power, torque), with their duration and sequences, that express the capability and limitations of a machine. The rated values are usually associated with a limiting temperature rise of insulation and metallic parts. *See also:* **asynchronous machine; direct-current commutating machine.**   63

(3) (device). The designated limit(s) of the rated operating characteristic(s). *Note:* Such operating characteristics as current, voltage, frequency, etcetera, may be given in the rating.   103, 202, 60, 127

(4) (controller). An arbitrary designation of an operating limit. It is based on power governed, the duty and service required. *Note:* A rating is arbitrary in the sense that it must necessarily be established by definite field standards and cannot, therefore, indicate the safe operating limit under all conditions that may occur. *See also:* **electric controller.**   206

(5) (current-limiting reactor). The volt-amperes that it can carry, together with any other characteristics, such as system voltage, current and frequency assigned to it by the manufacturer. *Note:* It is regarded as a test rating that defines an output that can be carried under prescribed

conditions of test, and within the limitations of established standards. *See also:* **reactor.**   309

(6) (arc-welding apparatus). A designated limit of output capacity in applicable terms, such as load, voltage, current, and duty cycle, together with any other characteristics necessary to define its performance. It shall be regarded as a test rating obtained within the limitations of established test standards.   264

(7) (mechanical or thermal short-circuit) (instrument transformer). Short-time emergency rating for fault conditions. Short-circuit rating of a current transformer defines the ability of the transformer to withstand an overload current resulting from a short-circuit or other fault condition in the line in which primary winding is connected. Short-circuit rating of a potential transformer defines the ability of the transformer to withstand a short-circuit directly across the secondary terminals.
                                                          305, 203

(8) (interphase transformer). The root-mean-square current, root-mean-square voltage, and frequency at the terminals of each winding, when the rectifier unit is operating at rated load and with a designated amount of phase control. *See also:* **duty; rectifier transformer.**
                                                           258, 53

(9) (rectifier transformer). The kilovolt-ampere output, voltage, current, frequency, and number of phases at the terminals of the alternating-current winding; the voltage (based on turn ratio of the transformer), root-mean-square current, and number of phases at the terminals of the direct-current winding, to correspond to the rated load of the rectifier unit. *Notes:* (1) Because of the current wave shapes in the alternating- and direct-current windings of the rectifier transformer, these windings may have individual ratings different from each other and from those of power transformers in other types of service. The ratings are regarded as test ratings that define the output that can be taken from the transformer under prescribed conditions of test without exceeding any of the limitations of the standards. (2) For rectifier transformers covered by established standards, the root-mean-square current ratings and kilovolt-ampere ratings of the windings are based on values derived from rectangular rectifier circuit element currents without overlap. *See also:* **rectifier transformer.**
                                                       258, 203, 53

(10) (surge arrester). The designation of an operating limit for a device. *See also:* **current rating, 60-hertz (arrester).**
                                                             2

(11) (power-inverter unit). The kilovolt-amperes, power output, voltages, currents, number of phases, frequency, etcetera, assigned to it by the manufacturer. *See also:* **self-commutated inverters.**   208

(12) (rating of storage batteries) (storage cell). The number of ampere-hours that the batteries are capable of delivering when fully charged and under specified conditions as to temperature, rate of discharge, and final voltage. For particular classes of service different time rates are frequently specified. *See also:* **charge; storage battery; time rate.**   328

(13) (relay). *See:* **relay rating.**

(14) (transformer). A volt-ampere output together with any other characteristics, such as voltage, current, frequency, power factor, and temperature rise, assigned to it by the manufacturer. 53

(15) (semiconductor device). A value which establishes either a limiting capability or a limiting condition for an electronic device. It is determined for specific values of environment and operation, and may be stated in any suitable terms. *Note:* Limiting conditions may either be maxima or minima. 66

rating, emergency (generating station). Capability of installed equipment for a short time interval. 112

rating, normal. Capacity of installed equipment. 112

rating plate (rotating machinery). *See:* **nameplate.**

ratio (magnetic storage). *See:* **squareness ratio.**

ratio correction factor (RCF) (instrument transformer). The ratio of the true ratio to the marked ratio. The primary current or voltage is equal to the secondary current or voltage multiplied by the marked ratio times the ratio correction factor. 53

ratio meter. An instrument that measures electrically the quotient of two quantities. A ratio meter generally has no mechanical control means, such as springs, but operates by the balancing of electromagnetic forces that are a function of the position of the moving element. *See also:* **instrument.** 328

rationalized system of equations. A rationalized system of electrical equations is one in which the proportionality factors in the equations that relate (1) the surface integral of electric flux density to the enclosed charge, and (2) the line integral of magnetizing force to the linked current, are each unity. *Notes:* (A) By these choices, some formulas applicable to configurations having spherical or circular symmetry contain an explicit factor of 4π or 2π; for example, Coulomb's law is $f = q_1 q_2/(4\pi \epsilon_0 r^2)$. (B) The differences between the equations of a rationalized system and those of an unrationalized system may be considered to result from either (a) the use of a different set of units to measure the same quantities or (b) the use of the same set of units to measure quantities that are quantitatively different (though of the same physical nature) in the two systems. The latter consideration, which represents a changed relation between certain mathematicophysical quantities and the associated physical quantities, is sometimes called **total rationalization.** 210

ratio-type (position-type) telemeter. A telemeter that employs the relative phase position between, or the magnitude relation between, two or more electrical quantities as the translating means. *Note:* Examples of ratio-type telemeters include alternating-current or direct-current position-matching systems. *See also:* **telemetering.** 103, 202

ratproof electric installation. Apparatus and wiring designed and arranged to eliminate harborage and runways for rats. *See also:* **marine electric apparatus.** 328

rat race*. *See:* **hybrid ring.**

* Deprecated

Rayleigh density function (radar). A probability density function describing the behavior of some variable, given by

$$f(X) = \frac{1}{\sigma_{AVG}} \exp - \left(\frac{X}{\sigma_{AVG}}\right)$$

Often used to describe the signal statistics after envelope detection. 13

Rayleigh disk. A special form of acoustic radiometer that is used for the fundamental measurement of particle velocity. 176

Rayleigh scattering (laser-maser). Scattering of radiation in the course of its passage through a medium containing particles, the sizes of which are small compared with the wavelength of the radiation. 363

RDF (electronic navigation). *See:* **radio direction finding.** *Note:* At one time this term was used by the British to mean radio distance finding, that is, radar. *See also:* **navigation.** 187, 13

R display (radar). An A display with a segment of the time base expanded near the blip for greater accuracy in distance measurement. *See also:* **navigation.** 187, 13

reach (1) (relay). The extent of the protection afforded by a relay in terms of the impedance or circuit length as measured from the relay location. *Note:* The measurement is usually to a point of fault, but excessive loading or system swings may also come within reach or operating range of the relay. 103, 202, 60, 127

(2) (protective relaying). The maximum distance from the relay location to a fault for which a particular relay will operate. The reach may be stated in terms of miles, primary ohms, or secondary ohms. 128

reactance (1) (general). The imaginary part of impedance. *See also:* **reactor.** 210, 185

(2) (portion of a circuit for a sinusoidal current and potential difference of the same frequency). The product of the sine of the angular phase difference between the current and potential difference times the ratio of the effective potential difference to the effective current, there being no source of power in the portion of the circuit under consideration. The reactance of a circuit is different for each component of an alternating current. If

$$e = E_{1m} \sin(\omega t + \alpha_1) \\ + E_{2m} \sin(2\omega t + \alpha_2) + \ldots$$

and

$$i = I_{1m} \sin(\omega t + \beta_1) \\ + I_{2m} \sin(2\omega t + \beta_2) + \ldots$$

then the reactances, X_1, X_2, etcetera, for the different components are

$$X_1 = \frac{E_{1m} \sin(\alpha_1 - \beta_1)}{I_{1m}} = \frac{E_1 \sin(\alpha_1 - \beta_1)}{I_1}$$

$$X_2 = \frac{E_{2m} \sin(\alpha_2 - \beta_2)}{I_{2m}} = \frac{E_2 \sin(\alpha_2 - \beta_2)}{I_2}$$

etcetera. *Note:* The reactance for the entire periodic current is not the sum of the reactance of the components. A definition of reactance for a nonsinusoidal periodic current has not been agreed upon. 3

reactance amplifier. *See:* **parametric amplifier.**

reactance drop (1) (general). The voltage drop in quadrature with the current. 203, 257

(2) (transformer). The component of the impedance voltage drop in quadrature with the current. 53

reactance, effective synchronous. An assumed value of

synchronous reactance used to represent a machine in a system study calculation for a particular operating condition. 63

reactance frequency divider. A frequency divider whose essential element is a nonlinear reactor. *Note:* The nonlinearity of the reactor is utilized to generate subharmonics of a sinusoidal source. *See also:* **parametric device.**
277, 191

reactance frequency multiplier. A frequency multiplier whose essential element is a nonlinear reactor. *Note:* The nonlinearity of the reactor is utilized to generate harmonics of a sinusoidal source. *See also:* **parametric device.**
277, 191

reactance function (1) (linear passive networks). The driving-point impedance of a lossless network. *Note:* This is an odd function of the complex frequency. *See also:* **linear passive networks.** 238

(2) (circuits and systems). A function that is realizable as a driving-point impedance with ideal inductors and capacitors. It must meet the conditions described in Foster's reactance theorem. *See also:* **linear passive networks.** 67

reactance grounded (electric power) (transformer). Grounded through impedance, the principal element of which is reactance. *Note:* The reactance may be inserted either directly, in the connection to ground, or indirectly by increasing the reactance of the ground return circuit. The latter may be done by intentionally increasing the zero-sequence reactance of apparatus connected to ground or by omitting some of the possible connections from apparatus neutrals to ground. *See also:* **ground.**
91, 203, 64, 53

reactance modulator. A device, used for the purpose of modulation, whose reactance may be varied in accordance with the instantaneous amplitude of the modulating electromotive force applied thereto. *Note:* Such a device is normally an electron-tube circuit and is commonly used to effect phase or frequency modulation. *See also:* **frequency modulation (telecommunication); modulation; phase modulation.** 111, 211

reactance relay. A linear-impedance form of distance relay for which the operating characteristic of the distance unit on an *R-X* diagram is a straight line of constant reactance. *Note:* The operating characteristic may be described by either equation $X = K$, or $Z \sin\theta = K$, where K is a constant and θ is the angle by which the input voltage leads the input current. 103, 202, 60, 127

reaction curve (process control). The plot of a time response. 56

reaction frequency meter. *See:* **absorption frequency meter.**

reaction time. The interval between the beginning of a stimulus and the beginning of the response of an observer. *See also:* **visual field.** 167

reaction torque (or force) (gyro; accelerometer). A torque (or force) exerted on a gimbal, gyro rotor or accelerometer proof mass, usually as a result of applied electrical excitations exclusive of torquer (or forcer) command signals. 46

reaction wheels (communication satellite). A set of gyro wheels used for controlling the attitude of a satellite. 74

reactive attenuator (waveguide). An attenuator that absorbs no energy. *See also:* **waveguide.** 244, 179

reactive current (rotating machinery). The component of a current in quadrature with the voltage. *See also:* **asynchronous machine.** 244, 63

reactive-current compensator (rotating machinery). A compensator that acts to modify the functioning of a voltage regulator in accordance with reactive current. 63

reactive factor. The ratio of the reactive power to the apparent power. The reactive factor is expressed by the equation

$$F_q = \frac{Q}{U}$$

where F_q = reactive factor
Q = reactive power
U = apparent power.

If the voltages have the same waveform as the corresponding currents, reactive factor becomes the same as phasor reactive factor. If the voltages and currents are sinusoidal and for polyphase circuits form symmetrical sets

$$F_q = \sin(\alpha - \beta).$$

See also: **network analysis.** 210

reactive-factor meter. An instrument for measuring reactive factor. It is provided with a scale graduated in reactive factor. *See also:* **instrument.** 328

reactive field (antenna). Electric and magnetic fields surrounding an antenna and resulting in storage of electromagnetic energy. *See also:* **radiation.** 179, 111

reactive ignition cable (electromagnetic compatibility). High-tension ignition cable, the core of which is so constructed to give a high reactive impedance at radio frequencies. *See also:* **electromagnetic compatibility.**
220, 199

reactive near-field region (antenna). That region of the field of an antenna immediately surrounding the antenna wherein the reactive field predominates. *Note:* For most antennas the outer boundary of the region is commonly taken to exist at $\lambda/2\pi$ where λ is the wavelength. *See also:* **antenna.** 246

reactive-power relay. A power relay that responds to reactive power. *See also:* **relay.** 60, 103, 127

reactive volt-ampere-hour meter. *See:* **varhour meter.**

reactive volt-ampere meter. *See:* **varmeter.**

reactivity (nuclear power generating station). A measure of the departure of a nuclear reactor from criticality. Mathematically $\rho = (k_{\text{eff}} - 1)/k_{\text{eff}}$. If ρ is positive (excess reactivity) the reactor is supercritical and its power level is increasing. If ρ is negative, (negative reactivity) the power level of the reactor decreases. For a reactor at criticality, for instance constant power level, the reactivity is zero. *Note:* Other measures are also used to express reactivity. 112

reactivity, excess (nuclear power generating station). More reactivity than that needed to achieve criticality. In order to avoid frequent reactor shutdowns to replace fuel that has been consumed and to compensate for the accumulation of fission products which have high neutron absorption cross sections and negative temperature coefficients, excess reactivity is provided in a reactor by including additional fuel in the core at start-up.   112

reactor (1) (general). A device, the primary purpose of which is to introduce reactance into a circuit. *Notes:* (1) A reactor is a device used for introducing reactance into a circuit for purposes such as motor starting, paralleling transformers, and control of current. (2) **Inductive reactance** is frequently abbreviated **inductor.**   3, 197 **(2) (transformer).** An electromagnetic device, the primary purpose of which is to introduce inductive reactance into a circuit. *See:* **alternating-current saturable reactor; filter reactor; shunt reactor; current-limiting reactor; bus reactor; feeder reactor; neutral grounding reactor; starting reactor; synchronizing reactor; paralleling reactor.**   53

reactor, ac (thyristor converter). An inductive reactor that is inserted between the transformer and the thyristor converter for the purpose of controlling the rate of rise of current in the thyristor and possibly the magnitude of fault current.   121

reactor, amplistat. A reactor conductively connected between the direct-current winding of a rectifier transformer and rectifier circuit elements that when operating in conjunction with other similar reactors, provides a relatively small controlled direct-current voltage range at the rectifier output terminals. *See also:* **reactor.**   203

reactor, bus. A current-limiting reactor for connection between two different buses or two sections of the same bus for the purpose of limiting and localizing the disturbance due to a fault in either bus. *See also:* **reactor.**
                                                         203

reactor, current-balancing. A reactor used in semiconductor rectifiers to achieve satisfactory division of current among parallel-connected semiconductor diodes. *See also:* **reactor.**   203

reactor, current-limiting. A reactor intended for limiting the current that can flow in a circuit under short-circuit conditions, or under other operating conditions such as starting, synchronizing, etcetera. *See also:* **reactor.**
                                                         203

reactor, dc (thyristor converter). An inductive reactor between the dc output of the thyristor converter and the load in order to limit the magnitude of fault current and also, in some cases, to limit the magnitude of ripple current in the load. In this latter case, it is called a ripple reactor.
                                                         121

reactor, diode-current-balancing. A reactor with a set of mutually coupled windings that, operating in conjunction with other similar reactors, forces substantially equal division of current among the parallel paths of a rectifier circuit element. *See also:* **reactor.**   203

reactor, feeder. A current-limiting reactor for connection in series with an alternating-current feeder circuit for the purpose of limiting and localizing the disturbance due to faults on the feeder. *See also;* **reactor.**   203

reactor, filter. A reactor used to reduce harmonic voltage in alternating-current or direct-current circuits. *See also:* **reactor.**   203

reactor, neutral grounding. *See:* **neutral grounding reactor.**
                                                          62

reactor, paralleling. A current-limiting reactor for correcting the division of load between parallel-connected

transformers that have unequal impedance voltages. *See also:* **reactor.**   203

reactor, shunt. *See:* **shunt reactor.**

reactor starting (rotating machinery). The process of starting a motor at reduced voltage by connecting it initially in series with a reactor (inductor) which is short-circuited for the running condition.   63

reactor, starting (transformers). A current-limiting reactor for decreasing the starting current of a machine or device. *See also:* **reactor.**   203

reactor-start motor. A single-phase induction motor of the split-phase type with a main winding connected in series with a reactor for starting operation and an auxiliary winding with no added impedance external to it. For running operation, the reactor is short-circuited or otherwise made ineffective, and the auxiliary winding circuit is opened. *See also:* **asynchronous machine.**   63

reactor, synchronizing. A current-limiting reactor for connecting momentarily across the open contacts of a circuit-interrupting device for synchronizing purposes. *See also:* **reactor.**   203

read (electronic computation). To acquire information, usually from some form of storage. *See also:* **destructive read; nondestructive read; write.**
                                   235, 210, 255, 77, 54, 125

read-around number (storage tubes). The number of times reading operations are performed on storage elements adjacent to any given storage element without more than a specified loss of information from that element. *Note:* The sequence of operations (including priming, writing, or erasing), and the storage elements on which the operations are performed, should be specified. *See also:* **storage tube.**   174, 190, 125

read-around ratio* (storage tubes). *See:* **read-around number.**

* Deprecated

read head (test, measurement and diagnostic equipment). A sensor that converts information stored on punched tape, magnetic tape, magnetic drum, and so forth, into electrical signals.   54

readily accessible. Capable of being reached quickly, for operation, renewal, or inspection, without requiring those to whom ready access is requisite to climb over or remove obstacles or to resort to portable ladders, chairs, etcetera.
                                                         124

readiness test (test, measurement and diagnostic equipment). A test specifically designed to determine whether an equipment or system is operationally suitable for a mission.
                                                          54

reading (recording instrument). The value indicated by the position of the index that moves over the indicating scale. *See also:* **accuracy rating (instrument).**   328

reading rate (storage tubes). The rate of reading successive storage elements. *See also:* **storage tube.**   174, 190

reading speed (storage tubes). The lineal scanning rate of the beam across the storage surface in reading. Note the distinction between this and **reading rate.** *See also:* **data processing; storage tube.**   174, 190, 125

reading speed, minimum usable (storage tubes). The slowest scanning rate under stated operating conditions before a specified degree of decay occurs. *Note:* The qualifying

adjectives **minimum usable** are frequently omitted in general usage when it is clear that the minimum usable reading speed is implied. *See also:* **storage tube.**
174, 190

reading time (storage tubes). The time during which stored information is being read. *See also:* **storage tube.**
174, 190

reading time, maximum usable (storage tubes). The length of time a storage element, line, or area can be read before a specified degree of decay occurs. *Notes:* (1) This time may be limited by static decay, dynamic decay, or a combination of the two. (2) It is assumed that rewriting is not done. (3) The qualifying adjectives **maximum usable** are frequently omitted in general usage when it is clear that the maximum usable reading time is implied. *See also:* **storage tube.**
174, 190

read number (storage tubes). The number of times a storage element, line, or area is read without rewriting. *See also:* **storage tube.**
174, 190, 125

read number, maximum usable (storage tubes). The number of times a storage element, line, or area can be read without rewriting before a specified degree of decay results. *Note:* The qualifying adjectives **maximum usable** are frequently omitted in general usage when it is clear that the maximum usable read number is implied. *See also:* **storage tube.**
174, 190

read-only storage (computing systems). *See:* **fixed storage.**

readout (test, measurement and diagnostic equipment). (1) The device used to present output information to the operator, either in real time or as an output of a storage medium; and (2) the act or reading, transmitting, displaying information either in real time or from an internal storage medium of an operator or an external storage medium or peripheral equipment.
54

readout, command (numerically controlled machines). Display of absolute position as derived from position command. *Note:* In many systems the readout information may be taken directly from the dimension command storage. In others it may result from the summation of command departures.
224, 207

readout, position (numerically controlled machines). Display of absolute position as derived from a position transducer.
224, 207

read pulse. A pulse that causes information to be acquired from a magnetic cell or cells. *See:* **ONE state.**
331

ready-to-receive signal (facsimile). A signal sent back to the facsimile transmitter indicating that a facsimile receiver is ready to accept the transmission. *See also:* **facsimile signal (picture signal).**
12

realizable function (linear passive networks). A response function that can be realized by a network containing only positive resistance, inductance, capacitance, and ideal transformers. *Note:* This is the sense of realizability in the theory of linear, passive, reciprocal, time-invariant networks. *See also:* **linear passive networks.**
238

realized gain. The power gain of an antenna in its environment, reduced by the losses due to the mismatch of the antenna input impedance to a specified impedance.
111

real time (1) (general). (A) Pertaining to the actual time during which a physical process transpires. (B) Pertaining

to the performance of a computation during the actual time that the related physical process transpires in order that results of the computation can be used in guiding the physical process.
255, 77

(2) (analog computer). Using an ordinary clock as a time standard, the number of seconds measured between two events occurring in a physical system. **Computer time:** The number of seconds measured, with the same clock, between corresponding events in the simulated system, computer time is equal to the product of real time and the time scale. **Time scale:** The ratio of the time interval between two events in a simulated system to the time interval between the corresponding events in the physical system. **Real-time computation:** Computer operation in which the time scale is unity. **Machine time** is synonymous with **computer time.** *See:* **scale factor.**
9

real time printout (sequential events recording systems). The recording of actual time that an input signal was received as correlated to a time standard.
48, 58

real time testing (test, measurement and diagnostic equipment). The testing of a system or its components at its normal operating frequency or timing.
54

recalescent point (metal). The temperature at which there is a sudden liberation of heat when metals are lowered in temperature. *See also:* **dielectric heating; induction heating; coupling; Curie point.**
14, 114

receive characteristic (telephony). The acoustic output level of a telephone set as a function of the electrical input level. The output is measured in an **artificial ear,** and the input signal is obtained from an available constant-power source of specified impedance.
114, 160

received power (mobile communication). The root-mean-square value of radio-frequency power that is delivered to a load that correctly terminates an isotropic reference antenna. The reference antenna most commonly used is the half-wave dipole. *See also:* **mobile communication system.**
181

receive-only equipment. Data communication equipment capable of receiving signals, but not arranged to transmit signals.
194

receiver (1). A device on a vehicle so placed that it is in position to be influenced inductively or actuated by the train-stop, train-control, or cab-signal roadway element.
328

(2) (facsimile). The apparatus employed to translate the signal from the communications channel into a facsimile record of the subject copy. *See also:* **facsimile (in electrical communication).**
12

(3) (telephone switching systems). A part of an automatic switching system that receives signals from a calling device or other source for interpretation and action.
55

receiver gating (radar). The application of selector pulses to one or more stages of a receiver only during that part of a cycle of operation when reception is desired. *See also:* **gating.**
187, 13

receiver ground (signal-transmission system). The potential reference at the physical location of the signal receiver. *See also:* **signal.**
188

receiver-off-hook tone (telephone switching systems). A tone on a line to indicate an abnormal off-hook condition.
55

receiver, power-line carrier. *See:* **power-line carrier receiver.**

receiver primaries (color television). *See:* **display primaries.**

receiver pulse delay. *See:* **transducer pulse delay.**

receiver relay. An auxiliary relay whose function is to respond to the output of a communications set such as power-line carrier, wire-line audio or carrier, radio, or microwave receiver. 103, 202, 60, 127

receiver, telephone. An earphone for use in a telephone system. 176

receiving converter, facsimile (frequency-shift to amplitude-modulation converter). A device which changes the type of modulation from frequency shift to amplitude. *See also:* **facsimile transmission.** 12

receiving-end crossfire (telegraph channel). The crossfire from one or more adjacent telegraph channels at the end remote from the transmitting end. *See also:* **telegraphy.** 328

receiving loop loss. That part of the repetition equivalent assignable to the station set, subscriber line, and battery supply circuit that are on the receiving end. *See also:* **transmission loss.** 328

receiving voltage sensitivity. *See:* **free-field voltage response.**

receptacle (1) (electric distribution) (convenience outlet). A contact device installed at an outlet for the connection of an attachment plug and flexible cord to supply portable equipment. 256
(2) (electric installations on shipboard). A device installed in a receptacle outlet to accommodate an attachment plug. 3

receptacle circuit. A branch circuit to which only receptacle outlets are connected. *See also:* **branch circuit.** 328

receptacle outlet. An outlet where one or more receptacles are installed. 3

receptive field (medical electronics). The region in which activity is observed by means of the pickup electrode. 192

reciprocal bearing (navigation). The opposite direction to a bearing. *See also:* **navigation.** 187, 13

reciprocal transducer. A transducer in which the principle of reciprocity is satisfied. *Note:* The use of the term **reversible transducer** as a synonym for reciprocal transducer is deprecated. *See also:* **transducer.** 252, 210

reciprocating mechanism (high voltage air switches, insulators, and bus supports). An operating mechanism which produces longitudinal motion of the operating means to open or close the switching device. 27

reciprocity (multiport network). The property described by symmetry of the impedance matrix. In the case of a network with identical ports, it is also described by symmetry of the scattering matrix. 185

reciprocity theorem. States that if an electromotive force E at one point in a network produces a current I at a second point in the network, then the same voltage E acting at the second point will produce the same current I at the first point. 328

reclosing device (transmission and distribution). A control device which initiates the reclosing of a circuit after it has been opened by a protective relay. 112

reclosing fuse. A combination of two or more fuseholders, fuse units, or fuse links mounted on a fuse support or supports, mechanically or electrically interlocked, so that one fuse can be connected into the circuit at a time and the functioning of that fuse automatically connects the next fuse into the circuit, with or without intentionally added time delay, thereby permitting one or more service restorations without replacement of fuse links, refill units, or fuse units. 103, 202

reclosing interval (automatic circuit recloser). The open-circuit time between an automatic opening and the succeeding automatic reclosure. 103, 202

reclosing relay. A programming relay whose function is to initiate the automatic reclosing of a circuit breaker. 103, 202, 60, 127

reclosing time (circuit breaker). The interval between the time when the actuating quantity of the release (trip) circuit reaches the operating value (the breaker being in the closed position) and the reestablishment of the circuit on the primary arcing contacts on the reclosing stroke. 103, 202

reclosure (relay). The automatic closing of a circuit-interrupting device following automatic tripping. Reclosing may be programmed for any combination of instantaneous, time-delay, single-shot, multiple-shot, synchronism-check, dead-line–live-bus, or dead-bus–live-line operation. 128

recognition. *See:* **character recognition; magnetic-ink character recognition; optical character recognition; pattern recognition.**

recombination center (solar cells). A defect having electrical properties so as to facilitate the recombination of mobile charge carriers (electrons or holes) with one each of the opposite polarity. 137

recombination rate (semiconductor) (1) (surface). The time rate at which free electrons and holes recombine at the surface of a semiconductor. *See also:* **semiconductor device.** 66
(2) (volume). The time rate at which free electrons and holes recombine within the volume of a semiconductor. *See also:* **semiconductor device.** 186, 113

recombination velocity (semiconductor surface). The quotient of the normal component of the electron (hole) current density at the surface by the excess electron (hole) charge density at the surface. *See also:* **semiconductor device.** 245, 66, 210, 186

reconditioned carrier reception (exalted-carrier reception). The method of reception in which the carrier is separated from the sidebands for the purpose of eliminating amplitude variations and noise, and then added at increased level to the sideband for the purpose of obtaining a relatively undistorted output. This method is frequently employed, for example, when a reduced-carrier single-sideband transmitter is used. *See also:* **radio receiver.** 328

reconstituted mica (integrated mica) (rotating machinery). *See:* **mica paper.**

reconstruction (transmission and distribution). Replacement of any portion of an existing installation by new equipment or construction. Does not include ordinary maintenance replacements. 262

record. A collection of related items of data, treated as a unit. *See:* **file.** 255, 77

recorded announcement (telephone switching systems). A prerecorded oral message received on a call.   55

recorded spot (facsimile). The image of the recording spot on the record sheet. *See also:* **recording (facsimile).**   12

recorded spot, X dimension (facsimile). The effective recorded-spot dimension measured in the direction of the recorded line. *Notes:* (1) By effective dimension is meant the largest center-to-center spacing between recorded spots which gives minimum peak-to-peak variation of density of the recorded line. (2) This term applies to that type of equipment which responds to a constant density in the subject copy by a succession of discrete recorded spots. *See also:* **recording (facsimile).**   12

recorded spot, Y dimension (facsimile). The effective recorded-spot dimension measured perpendicularly to the recorded line. *Note:* By effective dimension is meant the largest center-to-center distance between recorded lines which gives minimum peak-to-peak variation of density across the recorded lines. *See also:* **recording (facsimile).**   12

recorded value. The value recorded by the marking device on the chart, with reference to the division lines marked on the chart. *See also:* **accuracy rating (instrument).**   328

recorder (1) (analog computer). A device that makes a permanent record, usually graphic, of varying signals.   9

(2) (facsimile). That part of the facsimile receiver which performs the final conversion of electric picture signal to an image of the subject copy on the record medium. *See also:* **facsimile (electrical communication); recording (facsimile).**   12

recorder, strip-chart. A recorder in which one or more records are made simultaneously as a function of time. *See also:* **electronic analog computer.**   9, 10

recorder-warning tone (telephone switching systems). A tone that indicates periodically that the conversation is being electrically recorded.   55

recorder, X-Y (plotting board). A recorder that makes a record of any one voltage with respect to another. *See also:* **electronic analog computer.**   9, 10

record gap (1) (computing systems) (storage medium). An area used to indicate the end of a record.   255, 77

(2) (test, measurement and diagnostic equipment). An interval of space or time associated with a record to indicate or signal the end of the record.   54

recording (facsimile). The process of converting the electrical signal to an image on the record medium. *See also:* **direct recording; electrochemical recording; electrolytic recording; electromechanical recording; electrostatic recording; electrothermal recording; ink vapor recording; magnetic recording; photosensitive recording.**   12

recording channel (electroacoustics). The term refers to one of a number of independent recorders in a recording system or to independent recording tracks on a recording medium. *Note:* One or more channels may be used at the same time for covering different ranges of the transmitted frequency band, for multichannel recording, or for control purposes. *See also:* **phonograph pickup.**   176

recording-completing trunk (telephone switching systems). A one-way trunk for operator recording, extending, and automatic completing of toll calls.   55

recording demand meter. A demand meter that records on a chart the demand for each demand interval. *See also:* **electricity meter (meter).**   328

recording, instantaneous. A phonograph recording that is intended for direct reproduction without further processing. *See also:* **phonograph pickup.**   176

recording instrument (recorder) (graphic instrument). An instrument that makes a graphic record of the value of one or more quantities as a function of another variable, usually time. *See also:* **instrument.**   328

recording loss (mechanical recording). The loss in recorded level whereby the amplitude of the wave in the recording medium differs from the amplitude executed by the recording stylus. *See also:* **phonograph pickup.**   176

recording spot (facsimile). The image area formed at the record medium by the facsimile recorder. *See also:* **recording (facsimile).**   12

recording stylus. A total that inscribes the groove into the recording medium. *See also:* **phonograph pickup.**   176

recording trunk. A trunk extending from a local central office or private branch exchange to a toll office, that is used only for communication with toll operators and not for completing toll connections.

record medium (facsimile). A physical medium on which the facsimile recorder forms an image of the subject copy. *See also:* **recording (facsimile).**   12

record sheet (facsimile). The medium which is used to produce a visible image of the subject copy in record form. The record medium and the record sheet may be identical. *See also:* **recording (facsimile).**   12

recovered charge (semiconductor). The charge recovered from a semiconductor device after switching from a forward current condition to a reverse condition.   66

recovery current (semiconductor rectifier). The transient component of reverse current associated with a change from forward conduction to reverse voltage. *See also:* **rectification.**   237, 66

recovery cycle (electrobiology). The sequence of states of varying excitability following a conditioning stimulus. The sequence may include periods such as absolute refractoriness, relative refractoriness, supernormality, and subnormality. *See also:* **excitability (electrobiology).**   192

recovery time (1) (automatic gain control). The time interval required, after a sudden decrease in input signal amplitude to a system or transducer, to attain a stated percentage (usually 63 percent) of the ultimate change in amplification or attenuation due to this decrease. *See also:* **radar.**   239

(2) (power supplies). Specifies the time needed for the output voltage or current to return to a value within the regulation specification after a step load or line change. *Notes:* (A) Recovery time, rather than response time, is the more meaningful and therefore preferred way of specifying power-supply performance, since it relates to the regulation specification. (B) For load change, current will recover at a rate governed by the rate-of-change of the compliance voltage across the load. This is governed by the resistance-capacitance time constant of the output filter capacitance, internal source resistance, and load resistance. *See:* **programming speed.** *See also:* **radar.**   228, 186

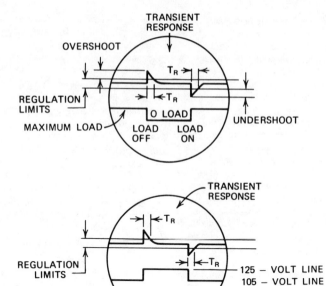

Recovery time. Oscilloscope views showing (top) the effects
of a step load change, and (bottom) the effects of a step
line change. T_R = recovery time.

(3) (radar or component thereof). The time required, after
the end of the transmitted pulse, for recovery to a specified
relation between receiving sensitivity or received signal
and the normal value. *Note:* The time for recovery to 6
decibels below normal sensitivity is frequently specified.
See also: **radar.**   328
(4) (Geiger-Mueller counters). The minimum time from
the start of a counted pulse to the instant a succeeding
pulse can attain a specific percentage of the maximum
value of the counted pulse. *See also:* **gas-filled radiation-
counter tubes; radar.**   190, 125, 96
(5) (antitransmit-receive tubes). The time required for a
fired tube to deionize to such a level that the normalized
conductance and susceptance of the tube in its mount are
within specified ranges. *Note:* Normalization is with respect to the characteristic admittance of the transmission
line at its junction with the tube mount. *See also:* **gas
tubes; radar.**   190, 125
(6) (microwave gas tubes). The time required for a fired
tube to deionize to such a level that the attenuation of a
low-level radio-frequency signal transmitted through the
tube is decreased to a specified value. *See also:* **gas tubes;
radar.**   190
(7) (gas tubes). The time required for the control electrode
to regain control after anode current interruption. *Note:*
To be exact, the deionization and recovery time of a gas
tube should be presented as families of curves relating such
factors as condensed-mercury temperature, anode current,
anode and control electrode voltages, and control-circuit
impedance. *See also:* **radar.**   190, 125
(8) (regulators). The elapsed time from the initiation of
a transient until the output returns to within the regulation
limits. See the figures attached to the definition of **response**

time. *See also:* **electrical conversion; radar; response time.**
                                                               186
(9) (gas turbines). The interval between two conditions of
speed occurring with a specified sudden change in the
steady-state electric load on the gas-turbine-generator
unit. It is the time in seconds from the instant of change
from the initial load condition to the instant when the
decreasing oscillation of speed finally enters a specified
speed band. *Note:* The specified speed band is taken with
respect to the midspeed of the steady-state speed band
occurring at the subsequent steady-state load condition.
The recovery time for a specified load increase and the
same specified load decrease may not be identical and will
vary with the magnitude of the load change.   98, 58
(10) (delay). *See:* **relay recovery time.**
(11) (TR and Pre-TR tubes). The time required for a fired
tube to deionize to such a level that the attenuation of a
low-level radio-frequency signal transmitted through the
tube is decreased to a specified value.   125
(12) (reverse-blocking thyristor or semiconductor diode).
See: **reverse recovery time (thyristor or semiconductor
diode).**
recovery voltage (surge arresters). The voltage that occurs
across the terminals of a pole of a circuit-interrupting
device upon the interruption of the current. *Note:* For an
arrester, this occurs as a result of interruption of the follow
current.   2
rectangular array (antenna). The regular arrangement of
array elements such that lines connecting corresponding
points of adjacent elements form rectangles.   111
rectangular impulse (surge arresters). An impulse that rises
rapidly to a maximum value, remains substantially constant for a specified period, and then falls rapidly to zero.

The parameters that define a rectangular impulse wave are polarity, peak value, duration of the peak, total duration. 62

rectangular mode (analog computer). *See:* **resolver.**

rectangular-shape logic symbol. A logic symbol in which the logic function is indicated by a qualifying symbol in its interior. 88

rectangular wave. A periodic wave that alternately assumes one of two fixed values, the time of transition being negligible in comparison with the duration of each fixed value. *See also:* **network analysis; wavefront.** 210, 59

rectification. The term used to designate the process by which electric energy is transferred from an alternating-current circuit to a direct-current circuit. 237, 66

rectification error (accelerometer). A steady state error in the output while vibratory disturbances are acting on an accelerometer. Anisoelasticity is one source of rectification error. 46

rectification factor. The quotient of the change in average current of an electrode by the change in amplitude of the alternating sinusoidal voltage applied to the same electrode, the direct voltages of this and other electrodes being maintained constant. *See also:* **conductance for rectification; transrectification factor.** 190, 125

rectification of an alternating current. Process of converting an alternating current to a unidirectional current. *See also:* **electronic rectifier; inverse voltage (rectifier); semiconductor device.** 244, 190

rectified unbalance (tuned rotor gyro). *See:* **gimbal unbalance torque.**

rectified value (alternating quantity). The average of all the positive values of the quantity during an integral number of periods. Since the positive values of a quantity y are represented by the expression

$$\frac{1}{2}\Big[y + |y| \Big],$$

$$y_r = \frac{1}{T} \int_0^T \frac{1}{2}\Big[y + |y| \Big] \, dt.$$

Note: The word positive and the sign + may be replaced by the word negative and the sign −. *See also:* **network analysis.** 210

rectifier (power generation). A device for converting alternating-current to direct-current. 58

rectifier anode. An electrode of the rectifier from which the current flows into the arc. *Note:* The direction of current flow is considered in the conventional sense from positive to negative. The cathode is the positive direct-current terminal of the apparatus and is usually a pool of mercury. The neutral of the transformer secondary system is the negative direct-current terminal of the rectifier unit. *See also:* **rectification.** 328

rectifier assembly. A complete unit containing rectifying components, wiring, and mounting structure capable of converting alternating-current power to direct-current power. *See also:* **converter.** 63

rectifier cathode. The electrode of the rectifier into which the current flows from the arc. *Note:* The direction of current flow is considered in the conventional sense from positive to negative. The cathode is the positive direct-

current terminal of the rectifier unit and is usually a pool of mercury. The neutral of the transformer secondary system is the negative direct-current terminal of the rectifier unit. *See also:* **rectification.** 328

rectifier circuit element. A circuit element bounded by two circuit terminals that has the characteristic of conducting current substantially in one direction only. *Note:* The rectifier circuit element may consist of more than one semiconductor rectifier cell, rectifier diode, or rectifier stack connected in series or parallel or both, to operate as a unit. 237, 66

rectifier electric locomotive. An electric locomotive that collects propulsion power from an alternating-current distribution system and converts this to direct current for application to direct-current traction motors by means of rectifying equipment carried by the locomotive. *Note:* A rectifier electric locomotive may be defined by the type of rectifier used on the locomotive, such as **ignitron electric locomotive.** *See also:* **electric locomotive.** 328

rectifier electric motor car. An electric motor car that collects propulsion power from an alternating-current distribution system and converts this to direct current for application to direct-current traction motors by means of rectifying equipment carried by the motor car. *Note:* A rectifier electric motor car may be defined by the type of rectifier used on the motor car, such as **ignitron electric motor car.** *See also:* **electric motor car.** 328

rectifier instrument. The combination of an instrument sensitive to direct current and a rectifying device whereby alternating currents or voltages may be measured. *See also:* **instrument.** 328

rectifier junction (semiconductor rectifier cell or diode). The junction in a semiconductor rectifier cell that exhibits asymmetrical conductivity. *See also:* **semiconductor; semiconductor rectifier stack.** 237, 66, 208

rectifier stack (semiconductor). An integral assembly of one or more rectifier diodes, including its associated mounting and cooling attachments if integral with it. *See also:* **semiconductor rectifier stack.** 66

rectifier transformer. A transformer that operates at the fundamental frequency of an alternating-current system and designated to have one or more output windings conductively connected to the main electrodes of a rectifier. *See also:* **power rectifier transformer; alternating-current winding; direct-current winding; rating (rectifier transformer; interphase transformer); interphase transformer.** 53

rectifier tube (valve). An electronic tube or valve designed to rectify alternating current. 190

rectifier unit. An operative assembly consisting of the rectifier, or rectifiers, together with the rectifier auxiliaries, the rectifier transformer equipment, and the essential switchgear. *See also:* **rectification; rectifier transformer.** 208

rectifying device. An elementary device, consisting of one anode and its cathode, that has the characteristic of conducting current effectively in only one direction. *See also:* **rectification.** 328

rectifying element. A circuit element that has the property of conducting current effectively in only one direction. *Note:* When a group of rectifying devices is connected, either in parallel or series arrangement, to operate as one circuit element, the group of rectifying devices should be considered as a rectifying element. *See also:* **rectification; rectifying device; rectifier circuit element; rectifying junction; metallic rectifying cell.** 328

rectifying junction (barrier layer) (blocking layer). The region in a metallic rectifier cell that exhibits the asymmetrical conductivity. *See also:* **rectification.** 328

rectilinear scanning (television). The process of scanning an area in a predetermined sequence of straight parallel scanning lines. *See also:* **television.** 163, 178

recurrence rate (pulse techniques). *See:* **pulse repetition frequency.**

recurrent sweep. A sweep that repeats or recurs regularly. It may be free-running or synchronized. *See also:* **oscillograph.** 185

redirecting surfaces and media (illuminating engineering). Those that change the direction of the flux without scattering the redirected flux. *See also:* **lamp.** 167

redistribution (storage or camera tubes). The alteration of the charge pattern on an area of a storage surface by secondary electrons from any other part of the storage surface. *See also:* **charge-storage tube.** 174, 190, 125

reduced full-wave test. A wave similar in shape and duration to that involved in a **full-wave lightning impulse test,** but reduced in magnitude. *Note:* The reduced full wave normally has a crest value between 50 and 70 percent of the full-wave value involved, and is used for comparison of oscillograms in failure detection. 53

reduced generator efficiency (thermoelectric device). The ratio of (1) a specified generator efficiency to (2) the corresponding Carnot efficiency. *See also:* **thermoelectric device.** 248, 191

reduced kilovolt-ampere tap (transformer). A tap through which the transformer can deliver only an output less than rated kilovolt-ampere without exceeding the specified temperature rise. The current is usually that of the rated kilovolt-ampere tap. *See also:* **windings, high-voltage and low-voltage.** 53

reduced-voltage starter (industrial control). A starter, the operation of which is based on the application of a reduced voltage to the motor. *See also:* **starter.** 244, 206

reducing joint. A joint between two lengths of cable the conductors of which are not the same size. *See:* **branch joint; cable joint; straight joint.** 64

reduction (1) (electrochemical cells) (corrosion). The gain of electrons by a constituent of a chemical reaction. *See also:* **electrochemical cell.** 223, 186, 221, 205

(2) (data processing). *See:* **data reduction.**

redundancy (1) (transmission of information). The fraction of the gross information content of a message that can be eliminated without loss of essential information. *Note:* Numerically, it is one minus the ratio of the net information content to the gross information content, expressed in percent. 59

(2) (source) (information theory). The amount by which the logarithm of the number of symbols available at the source exceeds the average information content per symbol of the source. *Note:* The term redundancy has been used loosely in other senses. For example, a source whose output is normally transmitted over a given channel has been called redundant if the channel utilization index is less than unity. *See also:* **information theory.** 160

(3) (reliability). In an item, the existence of more than one means for performing a given function. 164

(4) (emergency and standby power systems). Duplication of elements in a system or installation for the purpose of enhancing the reliability or continuity of operation of the system or installation. 89

redundancy, active (reliability). That redundancy wherein all means for performing a given function are operating simultaneously. 164

redundancy check (1) (data transmission). A check based on the insertion of extra bits for the purpose of error control. 194

(2) (electronic computation). *See:* **check, forbidden-combination.**

(3) (power generation). *See:* **check, redundant.**

redundancy, standby (reliability). That redundancy wherein the alternative means for performing a given function are inoperative until needed. 164

redundant. Pertaining to characters that do not contribute to the information content. Redundant characters are often used for checking purposes or to improve reliability. *See also:* **parity; self-checking code; error-detecting code; check, forbidden-combination; check digit.** 235

redundant code (data transmission). A code using more signal elements than necessary to represent the intrinsic information. 194

redundant equipment or systems (nuclear power generating stations). A piece of equipment or a system that duplicates the essential function of another piece of equipment or a system to the extent that either may perform the required function regardless of the state of operation or failure of the other. 31

redundant systems (cable systems). Two or more systems serving the same objective, where they are also either systems where personnel or public safety is involved, such as fire pumps, or systems provided with redundancy because of the severity of economic consequences of equipment damage. *Note:* Turbine-generator alternating-current and direct-current bearing oil pumps are examples of redundant equipment under this definition. 35

reed relay. A relay using glass-enclosed, magnetically closed reeds as the contact members. Some forms are mercury wetted. 341

reel stand (conductor stringing equipment). A device designed to support one or more reels and having the possibility of being skid, trailer or truck mounted. These devices may accommodate rope or conductor reels of varying sizes, and are usually equipped with reel brakes to prevent the reels from turning when pulling is stopped. They are used for either "slack" or "tension" stringing. The designation of "reel trailer" or "reel truck" implies that the trailer or truck has been equipped with a reel stand (jacks) and may serve as a reel transport and/or "payout" unit for stringing operations. Depending upon the sizes of the reels to be carried, the transporting vehicles may range from single axle trailers to semitrucks with trailers having multiple axles. *Syn.* **reel trailer; reel transporter; reel truck.** 45

reentrant beam (microwave tubes). An unterminated recirculating electron beam. *See also:* **microwave tube (or valve).** 190

reentrant-beam crossed-field amplifier (amplitron) (microwave tubes). A crossed-field amplifier in which the beam is reentrant and interacts with either a forward or a backward wave. *See also:* **microwave tube (or valve).**

190

reentrant circuit (microwave tubes). A slow-wave structure that closes upon itself. *See also:* **microwave tube (or valve).**

190

reentrant switching network (telephone switching systems). A switching network in which outlets (usually last choice) from a given connecting stage are connected to inlets of the same or previous stage.

55

re-entry communication (communication satellite). Communications during re-entry of a space vehicle into the atmosphere. Usually the ionization requires a special system of modulation to overcome the communication blackout.

84

referee test. A test made by or in the presence of a representative of a regulatory body or other disinterested agency. *Note:* Such tests usually are made under specific provisions relating thereto, promulgated by the regulator body. *See also:* **demand meter.**

212

reference (computing system). *See:* **linearity.**

reference accuracy (automatic null-balancing electric instrument). A number or quantity that defines the limit of error under reference operating conditions. *Notes:* (1) It is usually expressed as a percent of the span. It is preferred that a + sign or − sign or both precede the number or quantity. The absence of a sign infers a ± sign. (2) Reference accuracy does not include accuracy of sensing elements or intermediate means external to the instrument. *See:* **error and correction.** *See also:* **accuracy rating (instrument).**

295

reference air line. A uniform section of air-dielectric transmission line of accurately calculable characteristic impedance used as a standard of immittance. *See also:* **transmission line.**

185

reference audio noise power output (mobile communications receivers). The average audio noise power present at the output of an unsquelched receiver having no radio-frequency signal input in which the audio gain has been adjusted for the reference audio power output.

123

reference audio power output (mobile communications receivers). The manufacturer's rated audio-frequency power available at the output of a properly terminated receiver, when responding to a standard test modulated radio-frequency input signal at a −80 dBW level.

123

reference ballasts. Specially constructed series ballasts having certain prescribed characteristics. *Note:* They serve as comparison standards for use in testing ballasts or lamps and are used also in selecting the reference lamps that are necessary for the testing of ballasts. Reference ballasts are characterized by a constant impedance over a wide range of operating current. They also have constant characteristics that are relatively uninfluenced by time and temperature. *See also:* **fixed-impedance type (reference ballast); primary standards (illuminating engineering); variable-impedance type (reference ballast).**

167, 270

reference black level (television). The picture-signal level corresponding to a specified maximum limit for black peaks.

178

reference block (numerically controlled machines). A block within the program identified by an *o* (letter *o*) in place of the word address *n* and containing sufficient data to

enable resumption of the program following an interruption. This block should be located at a convenient point in the program that enables the operator to reset and resume operation.

224, 207

reference boresight (antenna). A direction defined by an optical, mechanical, or electrical axis of an antenna, established as a reference for the purposes of beam-direction or tracking-axis alignment. *See also:* **antenna.**

246, 179, 111

reference conditions. The values assigned for the different influence quantities at which or within which the instrument complies with the requirements concerning errors in indication. *See also:* **accuracy rating (instrument).**

280

reference current (fluorescent lamp). The value of current specified in a specific lamp standard. *Note:* It is normally the same as the value of current for which the corresponding lamp is rated. Since the reference ballast is a standard that is representative of the impedance of lamp power sources installed, it is not necessary to change this current value unless major changes in lamp standards require modification of the ballast impedance. For this reason, reference ballast characteristics are specified in terms of, and with reference to, reference current. 270

reference designation (1) (symbols). (abbreviation). Numbers, or letters and numbers, used to identify and locate units, portions thereof, and basic parts of a specific set. *Compare with:* **functional designation** and **symbol for a quantity.** *See also:* **abbreviation.** 173

(2) (electric and electronics parts and equipments). Letters or numbers, or both, used to identify and locate discrete units, portions thereof, and basic parts of a specific set. *Note:* A reference designation is not a letter symbol, abbreviation, or functional designation for an item. 17

reference direction (1) (radar). A direction from which other directions are reckoned; for example, true north, grid north, and so on. *See also:* **navigation.** 187, 13

(2) (voltage). The assigned direction of the positive line element in the integral for voltage. *See also:* **network analysis; radio transmission.** 210

(3) (current). In a conductor, the assigned direction of the positive normal to the cross section of the conductor. *Note:* The reference direction for a current is usually denoted on a circuit diagram by an arrow placed near the line that represents the conductor. *See also:* **network analysis.**

210

(4) (energy) (specified circuit). With reference to the boundary of a delimited region, the arbitrarily selected direction in which electric energy is assumed to be transmitted past the boundary, into or out of the region. *Notes:* (A) When the actual direction of energy flow is the same as the reference direction, the sign of power is positive, and when the actual flow is in the opposite direction, the sign is negative. (B) Unless specifically stated to the contrary, it shall be assumed that the reference direction for all power, energy, and quadergy quantities associated with the circuit is the same as the reference direction of the energy flow. (C) In these definitions it will be assumed that the reference direction of the current in each conductor of the circuit is the same as the reference direction of energy flow.

210

reference distance (sound measurement). A standard 1 m

distance from the major machine surfaces at which mean sound level data shall be reported. 129

reference excursion (analog computer). The range from zero voltage to nominal full-scale operating voltage. *See also:* **electronic analog computer.** 9, 77

reference frequency. A selected frequency from which frequency departure is measured. *Note:* Reference frequency is not necessarily synonymous with carrier frequency. Carrier frequency is the frequency of a periodic wave upon which modulation is imposed. *See also:* **network analysis.** 61

reference frequency, upper and lower. *See:* **bandwidth.**

reference input signal (industrial control). *See:* **signal, reference input.**

reference lamp (1) (mercury). A seasoned lamp that under stable burning conditions, in the specified operating position (usually vertical, base up), and in conjunction with the reference ballast specified for the lamp size and rating, and at reference ballast rated input voltage, operates at values of lamp volts, watts, and amperes, each within ± 2 percent of the nominal values. 274, 271 **(2) (fluorescent).** Seasoned lamps that under stable burning conditions, in conjunction with the reference ballast specified for the lamp size and rating, and at the rated reference ballast supply voltage, operate at values of lamp volts, watts, and amperes each within ± 2½ percent of the values, and under conditions established by present standards. *See:* **reference ballasts.** 268

reference line (1) (navigation). A line from which angular or linear measurements are reckoned. *See also:* **navigation.** 13 **(2) (illuminating engineering).** Either of two radial lines where the surface of the cone of maximum candlepower is intersected by a vertical plane parallel to the curb line and passing through the lightcenter of the luminaire. *See also:* **street lighting luminaire.** 187

reference lines and points (pulse terms). Constructs which are (either actually or figuratively) superimposed on waveforms for descriptive or analytical purposes. Unless otherwise specified, all defined lines and points lie within a waveform epoch. *See also:* **time origin line; magnitude origin line; time reference line; time referenced point; magnitude reference line; magnitude referenced point; knot; cubic natural spline.** 254

reference modulation (very-high-frequency omnidirectional radio range). The modulation of the ground station radiation that produces a signal in the airborne receiver whose phase is independent of the bearing of the receiver; the reference signal derived from this modulation is used for comparison with the variable signal. *See also:* **navigation.** 187, 13

reference noise. The magnitude of circuit noise that will produce a circuit noise meter reading equal to that produced by 10^{-12} watt of electric power at 1000 hertz. *See also:* **signal-to-noise ratio.** 59

reference operating conditions (automatic null-balancing electric instrument). The conditions under which reference performance is stated and the base from which the values of operating influences are determined. *See also:* **measurement system.** 295

reference performance (1) (automatic null-balancing electric instrument). The limits of the values of certain operating

characteristics of the instrument that will not be exceeded under any combination of reference operating conditions. *See also:* **electricity meter (meter); test (instrument or meter).** 295 **(2) (meter).** Its percentage registration under standard conditions with which its percentage registration under a particular condition is to be compared to determine the influence on the meter of the particular condition. *See also:* **electricity meter (meter); service test; test (instrument or meter).** 212

reference plane. A plane perpendicular to the direction of propagation in a waveguide or transmission line, to which measurement of immittance, electrical length, reflection coefficients, scattering coefficients, and other parameters may be referred. *See also:* **waveguide.** 185

reference plane, electrical (standard connector). A transverse plane of the waveguide or transmission line on the drawing standardizing the critical mating dimensions shown in relation to the mechanical reference plane. *Notes:* (1) The electrical reference planes of two mating standard connectors forming a mated standard connector pair nearly coincide. (2) The electrical and mechanical reference planes of standard connectors do not necessarily coincide except for precision coaxial connectors complying with the IEEE Std 287-1968. Precision Coaxial Connectors and many connectors for uniconductor waveguides. 110

reference plane, mechanical (standard connector). A transverse plane of the waveguide or transmission line to which all critical, longitudinal dimensions are referenced to assure nondestructive mating; it is the only plane where a mated standard connector pair butt against one another. *Note:* Usually a stable, rugged metal surface. 110

reference power supply. A regulated, electronic power supply furnishing the reference voltage. *See also:* **electronic analog computer.** 9

reference radius (sound measurement). The sum of the reference distance and one half the maximum linear dimension as defined for **small-, medium-,** or **large machines.** *See:* **machine (3).** 129

reference sensitivity (mobile communications receivers). The level of a radio-frequency signal with standard test modulation which provides a 12-decibel sinad with at least 50 percent reference audio power output. 123

reference standards (nuclear power generating stations). Standards (that is, primary, secondary, and working standards, where appropriate) used in a calibration program. These standards establish the basic accuracy limits for that program. 31, 41

reference test field (direction-finder testing). The field strength, in microvolts per meter, numerically equal to the direction-finder sensitivity. *See also:* **navigation.** 278, 13

reference threshold squelch adjustment (mobile communications receivers). The minimum adjustment position of the squelch control required to reduce the reference audio noise power output by at least 40 decibels. 123

reference time (magnetic storage). An instant near the beginning of switching chosen as an origin for time measurements. It is variously taken as the first instant at which the instantaneous value of the drive pulse, the voltage response of the magnetic cell, or the integrated voltage response reaches a specified fraction of its peak pulse amplitude.   77

reference voltage (analog computer). A voltage used as a standard of reference, usually the nominal full scale of the computer.   9

reference volume (acoustics). The volume that gives a reading of 0 (volume units) on a standard volume indicator.   239

reference waveguide. A uniform section of waveguide with accurately fabricated internal cross-sectional dimensions used as a standard of immittance. *See also:* **waveguide.**   185

reference white (1) (original scene). The light from a non-selective diffuse reflector that is lighted by the normal illumination of the scene. *Notes:* (A) Normal illumination is not intended to include lighting for special effects. (B) In the reproduction of recorded material, the word **scene** refers to the original scene.   18, 178
(2) (color television display). That white with which the display device simulates reference white of the original scene. *Note:* In general, the reference whites of the original scene and of the display device are not colorimetrically identical.   18

reference white level (television). The picture-signal level corresponding to a specified maximum limit for white peaks. *See also:* **television.**   178

refill unit (high-voltage fuse unit). An assembly comprised of a conducting element, the complete arc-extinguishing medium, and parts normally required to be replaced after each circuit interruption to restore the fuse unit to its original operating condition.   103, 202

reflectance (1) (surface or medium). The ratio of the reflected flux to the incident flux. *Note:* Measured values of reflectance depend upon the angles of incidence and view and on the spectral character of the incident flux. Because of this dependence, the angles of incidence and view and the spectral characteristics of the source should be specified. *See also:* **lamp.**   167
(2) (photovoltaic power system). The fraction of radiation incident on an object that is reflected. *See also:* **photovoltaic power system.**   186
(3) (laser-maser) (reflectivity, ρ). The ratio of total reflected radiant power to total incident power.   363

reflected binary code. *See:* **Gray code.**

reflected glare. Glare resulting from specular reflections of high brightnesses in polished or glossy surfaces in the field of view. It usually is associated with reflections from within a visual task or areas in close proximity to the region being viewed. *See:* **veiling reflection; visual field.**   167

reflected harmonics (electrical conversion). Harmonics produced in the prime source by operation of the conversion equipment. These harmonics are produced by current-impedance (IZ) drop due to nonsinusoidal load currents, and by switching or commutating voltages produced in the conversion equipment. *See also:* **electrical conversion.**   186

reflected-light scanning (industrial control). The scanning of changes in the magnitude of reflected light from the surface of an illuminated web.   206

reflected wave (1) (wave propagation). When a wave in one medium is incident upon a discontinuity or a different medium, the reflected wave is the wave component that

results in the first medium in addition to the incident wave. *Note:* The reflected wave includes both the reflected rays of geometrical optics and the diffracted wave. *See also:* **radiation; waveguide.**   267, 179, 59
(2) (surge arresters). A wave, produced by an incident wave, that returns in the opposite direction to the incident wave after reflection at the point of transition.   244, 62
(3) (reverse wave). A wave traveling along a waveguide or transmission line from a reflective termination, discontinuity, or reference plane in the direction usually opposite to the incident wave. *See:* **incident wave.** *See also:* **waveguide.**   185

reflection (1) (illumination engineering). A general term for the process by which the incident flux leaves a surface or medium from the incident side. *Note:* **Reflection** is usually a combination of **regular** and **diffuse reflection.** *See also:* **lamp.**   167
(2) (laser-maser). Deviation of radiation following incidence on a surface.   363

reflection coefficient (at any given point on a transmission line or in a medium). The ratio of the phasor magnitude of the reflected wave to the phasor magnitude of the incident wave under conditions that are specified. *Notes:* (1) Specification of conditions includes defining incident wave and reflected wave; definition applies to one type of physical quantity such as current, voltage, pressure, electric field strength, etcetera. (2) The voltage reflection coefficient is most commonly used and is defined as the ratio of the complex electric field strength (or voltage) of the reflected wave to that of the incident wave. (3) At any specified plane in a uniform transmission line between a source of power and an absorber of power, the voltage reflection coefficient is the vector ratio of the electric field associated with the reflected wave, to that associated with the incident wave. It is given by the formula $(Z_2 - Z_1)/(Z_2 + Z_1)$, where Z_1 is the impedance of the source and Z_2 is the impedance of the load. *See also:* **transmission line; waveguide.**   267, 210

reflection color tube. A color-picture tube that produces an image by means of electron reflection techniques in the screen region. *See also:* **television.**   190, 125

reflection error (radar and navigation) (electronic navigation). The error due to the fact that some of the total received signal arrives from a reflection rather than all via the direct path. *See:* **multipath error (radar).**   187, 13

reflection factor (1) (electric circuits) (mismatch factor). The reflection factor between two impedances Z_1 and Z_2 is

$$\frac{(4Z_1Z_2)^{1/2}}{Z_1 + Z_2}$$

Notes: (A) Physically the reflection factor is the ratio of the current delivered to a load, whose impedance is not matched to the source, to the current that would be delivered to a load of matched impedance. (B) In this context, the synonym **mismatch factor** may also be used. *See also:* **network analysis.**   59, 115
(2) (reflex klystron). The ratio of the number of electrons of the reflected beam to the total number of electrons that enter the reflector space in a given time. *See also:* **velocity-modulated tube.**   244, 190
(3) (electrothermic power meters). The ratio of the power

absorbed in, to the power incident upon, a load; mathematically, $1 - |\Gamma_t|^2$, where $|\Gamma_t|$ is the magnitude of the reflection coefficient of the load.   47

reflectionless termination. *See:* **termination, reflectionless.**

reflectionless transmission line. A transmission line having no reflected wave at any transverse section. *See also:* **transmission line.**   185

reflectionless waveguide. A waveguide having no reflected wave at any transverse section. *See also:* **waveguide.**   185

reflection loss (given frequency). The reflection loss at the junction of a source of power and a load is given by the formula

$$20 \log_{10} \frac{Z_1 + Z_2}{(4Z_1 Z_2)^{1/2}} \text{ decibels}$$

where Z_1 is the impedance of the source of power and Z_2 is the impedance of the load. Physically, the reflection loss is the ratio, expressed in decibels, of the scalar values of the volt-amperes delivered to the load to the volt-amperes that would be delivered to a load of the same impedance as the source. The reflection loss is equal to the number of decibels that corresponds to the scalar value of the reciprocal of the reflection factor. *Note:* When the two impedances have opposite phases and appropriate magnitudes, a reflection gain may be obtained. *See also:* **network analysis; transmission loss.**   59

reflection mode photocathode (photomultipliers for scintillation counting). A photocathode wherein photoelectrons are emitted from the same surface as that on which the photons are incident.   117

reflection modulation (storage tubes). A change in character of the reflected reading beam as a result of the electrostatic fields associated with the stored signal. A suitable system for collecting electrons is used to extract the information from the reflected beam. *Note:* Typically the beam approaches the target closely at low velocity and is then selectively reflected toward the collection system. *See also:* **charge-storage tube.**   174, 190

reflective array antenna. An antenna consisting of a radiating feed and an array of reflecting elements arranged on a surface. *Note:* The reflecting elements are usually waveguides containing electrical phase shifters and are terminated by short circuits.   111

reflectivity (photovoltaic power system). The reflectance of an opaque, optically smooth, clean portion of material. *See also:* **photovoltaic power system.**   186

reflectometer (1). A photometer for measuring reflectance. *Note:* Reflectometers may be visual or physical instruments. *See also:* **photometry.**   167
(2) An instrument for the measurement of the ratio of reflected-wave to incident-wave amplitudes in a transmission system. *Note:* Many instruments yield only the magnitude of this ratio. *See also:* **instrument.**   185

reflectometer, time-domain. An instrument designed to indicate and to measure reflection characteristics of a transmission system connected to the instrument by monitoring the step-formed signals entering the test object and the superimposed reflected transient signals on an oscilloscope equipped with a suitable time-base sweep. The measuring system, basically, consists of a fast-rise step-

function generator, a tee coupler, and an oscilloscope connected to the probing branch of the coupler. *See also:* **instrument.**   185

reflector (1) (wave propagation). A reflector comprises one or more conductors or conducting surfaces for reflecting radiant energy. *See also:* **antenna; reflector element.**   313
(2) (illuminating engineering). A device used to redirect the luminous flux from a source by the process of reflection. *See also:* **bare (exposed) lamp.**   167

reflector antenna. An antenna consisting of a reflecting surface and a feed. *See also:* **antenna.**   179, 111

reflector element (antenna). A parasitic element located in a direction other than forward of the driven element of an antenna intended to increase the directive gain of the antenna in the forward direction. *See also:* **antenna.**   179, 111

reflector space (reflex klystron). The part of the tube following the buncher space, and terminated by the reflector. *See also:* **velocity-modulated tube.**   244, 190

reflex baffle (audio and electroacoustics). A loudspeaker enclosure in which a portion of the radiation from the rear of the diaphragm is propagated outward after controlled shift of phase or other modification, the purpose being to increase the useful radiation in some portion of the frequency spectrum. *See also:* **loudspeaker.**   176

reflex bunching. The bunching that occurs in an electron stream that has been made to reverse its direction in the drift space. *See also:* **electron devices, miscellaneous.**   190, 125

reflex circuit. A circuit through which the signal passes for amplification both before and after a change in its frequency.   328

reflex klystron (microwave tubes). A single-resonator oscillator klystron in which the electron beam is reversed by a negative electrode so that it passes twice through the resonator, thus providing feedback. *See also:* **microwave tube.**   190

reforming (semiconductor rectifier). The operation of restoring by an electrical or thermal treatment, or both, the effectiveness of the rectifier junction after loss of forming. *See also:* **rectification.**   237, 66

refracted wave. That part of an incident wave that travels from one medium into a second medium. *Note:* This is also called the **transmitted wave.** *See also:* **radiation; radio wave propagation; waveguide.**   267, 146, 180, 59

refraction error (radar). Error due to the bending of one or more wave paths by undesired refraction. *See also:* **navigation.**   13, 187

refraction loss. That part of the transmission loss due to refraction resulting from nonuniformity of the medium.   176

refractive index (1) (wave transmission medium). The ratio of the phase velocity in free space to that in the medium. *See also:* **radiation; radio wave propagation.**   146, 180
(2) (dielectric for an electromagnetic wave) (refractivity). The ratio of the sine of the angle of incidence to the sine of the angle of refraction as the wave passes from a vacuum into the dielectric. The angle of incidence θ_i is the angle between the direction of travel of the wave in vacuum and the normal to the surface of the dielectric. The angle of refraction θ_r is the angle between the direction of travel

of the wave after it has entered the dielectric and the normal to the surface. Refractive index is related to the dielectric constant through the following relation

$$n = \frac{\sin \theta_i}{\sin \theta_r} = (\epsilon')^{1/2}$$

where ϵ' is the real dielectric constant. Since ϵ' and n vary with frequency, the above relation is strictly correct only if all quantities are measured at the same frequency. The refractive index is also equal to the ratio of the velocity of the wave in the vacuum to the velocity in the dielectric medium. *See also: radiation.*   210

refractive modulus (radio wave propagation) (troposphere) (excess modified index of refraction). The excess over unity of the modified index of refraction, expressed in millionths. It is represented by M and is given by the equation

$$M = (n + h/a - 1)10^6$$

where a is the mean geometrical radius of the surface and n is the index of refraction at a height h above the local surface. *See also: radiation; radio wave propagation.*
                          180, 59, 146

refractivity. *See: refractive index.*

refractor (illuminating engineering). A device used to redirect the luminous flux from a source, primarily by the process of refraction. *See also: bare (exposed) lamp.*
                                              167

refractory. A nonmetallic material highly resistant to fusion and suitable for furnace roofs and linings.   328

regal. *See: range and elevation guidance for approach and landing.*

regenerate (electronic storage devices) (1). To bring something into existence again after decay of its own accord or after intentional destruction.

(2) (storage tubes). To replace charge to overcome decay effects, including loss of charge by reading.

(3) (storage devices in which physical states used to represent data deteriorate). To restore the device to its latest undeteriorated state. *See also: rewrite.*   235

regenerated leach liquor (electrometallurgy). The solution that has regained its ability to dissolve desired constituents from the ore by the removal of those constituents in the process of electrowinning. *See also: electrowinning.*
                                              328

regeneration (1) (electronic computers). In a storage device whose information storing state may deteriorate, the process of restoring the device to its latest undeteriorated state. *See also: rewrite.*   210

(2) (storage tubes). The replacing of stored information lost through static decay and dynamic decay. *See also: storage tube.*   174, 190, 125

(3) *See: positive feedback.*

regeneration of electrolyte. The treatment of a depleted electrolyte to make it again fit for use in an electrolytic cell. *See also: electrorefining.*   328

regenerative braking (industrial control). A form of dynamic braking in which the kinetic energy of the motor and driven machinery is returned to the power supply system. *See also: asynchronous machine; dynamic breaking; electric drive.*   206, 63

regenerative divider (regenerative modulator). A frequency divider that employs modulation, amplification, and selective feedback to produce the output wave.   328

regenerative fuel-cell system. A system in which the reactance may be regenerated using an external energy source. *See also: fuel cell.*   223, 186

regenerative repeater. A repeater that performs pulse regeneration. *Note:* the retransmitted signals are practically free from distortion. *See also: repeater.*   194

regional center (telephone switching systems). A toll office to which a number of sectional centers are connected. Regional centers are classified as Class 1 offices. *See also: office class.*   55

region, Geiger-Mueller (radiation-counter tubes). *See: Geiger-Mueller region.*

region of limited proportionality (radiation-counter tubes). The range of applied voltage below the Geiger-Mueller threshold, in which the gas amplification depends upon the charge liberated by the initial ionizing event. *See also: gas-filled radiation-counter tubes.*   125

region, proportional (radiation-counter tubes). *See: proportional region.*

regions of electromagnetic spectrum (light emitting diodes). For convenience of reference the electromagnetic spectrum near the visible spectrum is divided as follows.

Spectrum	Wavelength in nanometers
far ultraviolet	10–280
middle ultraviolet	280–315
near ultraviolet	315–380
visible	380–780
infrared	790–10^5

Note: The spectral limits indicated above should not be construed to represent sharp delineations between the various regions. There is a gradual transition from region to region. The above ranges have been established for practical purposes. *See also: radiant energy (illuminating engineering).*   167

register (1) (electronic computation). A device capable of retaining information, often that contained in a small subset (for example, one word), of the aggregate information in a digital computer. *See: address register; circulating register; index register; instruction register; shift register. See also: storage.*   235, 210, 255, 77, 54

(2) (telephone switching systems). A part of an automatic switching system that receives and stores signals from a calling device or other source for interpretation and action.
                                              55

(3) (protective signaling) (signal recorder). An electromechanical recording device that marks a paper tape in response to electric signal impulses received from transmitting circuits. A register may be driven by a prewound spring mechanism, by an electric motor, or by a combination thereof. *Note:* An inking register marks the tape with ink; a punch register marks the tape by cutting holes therein; a slashing register marks the tape by cutting V-shaped slashes therein. *See also: protective signaling.*
                                              328

(4) (meter). That part of the meter that registers the revolutions of the rotor or the number of impulses received from or transmitted to a meter in terms of units of electric energy or other quantity measured. *See also:* **moving element (instrument); watthour meter.** 212

register constant. The factor by which the register reading must be multiplied in order to provide proper consideration of the register, or gear, ratio and of the instrument transformer ratios to obtain the registration in the desired units. *Note:* It is commonly denoted by the symbol K_r. *See also:* **electricity meter (meter); moving element (instrument).** 212

register length (electronic computation). The number of characters that a register can store. 210

register marks or lines (industrial control). Any mark or line printed or otherwise impressed on a web of material, and which is used as a reference to maintain register. *See also:* **photoelectric control.** 206

register, mechanical (pulse techniques). An electromechanical indicating pulse counter. *See also:* **pulse.** 335

register ratio (watthour meters). The number of revolutions of the first gear of the register, for one revolution of the first dial pointer. *Note:* This is commonly denoted by the symbol Rr. *See also:* **electricity meter (meter); moving element (instrument).** 212

register reading. The numerical value indicated by the register. Neither the register constant nor the test dial (or dials), if any exist, is considered. *See also:* **electricity meter (meter).** 328

registration (1) (general). Accurate positioning relative to a reference. 255, 77

(2) (display device). The condition in which corresponding elements of the primary-color images are in geometric coincidence. *See:* **registration (camera device).** 328

(3) (camera device). The condition in which corresponding elements of the primary-color images are scanned in time sequence. *See:* **registration (display device).** 328

(4) (meter). The apparent amount of electric energy (or other quantity being measured) that has passed through the meter, as shown by the register reading. It is equal to the product of the register reading and the register constant. The registration during a given period is equal to the product of the register constant and the difference between the register readings at the beginning and the end of the period. 328

regressed luminaire. A luminaire that is mounted above the ceiling with the opening of the luminaire above the ceiling line. *See also:* **suspended (pendant) luminaire.**
167

regular (specular) reflectance. The ratio of the flux leaving a surface or medium by regular (specular) reflection to the incident flux. *See also:* **lamp.** 167

regular (specular) reflection. Regular (specular) reflection is that process by which incident flux is redirected at the specular angle. *See:* **specular angle.** *See also:* **lamp.**
167

regular transmittance (illuminating engineering). The ratio of the regularly transmitted flux leaving a surface or medium to the incident flux. *See also:* **transmission (illuminating engineering).** 167

regular transmission (illuminating engineering). That process by which incident flux passes through a surface or medium without scattering. *See also:* **transmission (illuminating engineering).** 167

regulated circuit (voltage regulator). The circuit on the output side of the regulator, and in which it is desired to control the voltage, or the phase relation, or both. *Note:* The voltage may be held constant at any selected point on the regulated circuit. *See also:* **voltage regulator.**
257, 53

regulated frequency. Frequency so adjusted that the average value does not differ from a predetermined value by an appreciable amount. *See also:* **generating station.**
64

regulated power supply. A power supply that maintains a constant output voltage (or current) for changes in the line voltage, output load, ambient temperature, or time.
228, 186

regulated-power-supply efficiency. The ratio of the regulated output power to the input power. *See also:* **regulated power supply.** 347

regulated voltage, band of (synchronous machines). The band or zone, expressed in percent of the rated value of the regulated voltage, within which the excitation system will hold the regulated voltage of an electric machine during steady or gradually changing conditions over a specified range of load. 105

regulated voltage, nominal band of (synchronous machines). The band of regulated voltage for a load range between any load requiring no-load field voltage and any load requiring rated-load field voltage with any compensating means used to produce a deliberate change in regulated voltage inoperative. 105

regulating autotransformer (rectifier). A transformer used to vary the voltage applied to the alternating-current winding of the rectifier transformer by means of de-energized autotransformer taps, and with load-tap-changing equipment to vary the voltage over a specified range on any of the autotransformer taps. *See also:* **rectifier transformer.** 258

regulating limit setter (speed-governing system). A device in the load-frequency-control system for limiting the regulating range on a station or unit. *See also:* **speed-governing system.** 94

regulating range (load-frequency control). A range of power output within which a generating unit is permitted to operate. *See also:* **speed-governing system.** 94

regulating relay. A relay whose function is to detect a departure from specified system operating conditions and to restore normal conditions by acting through supplementary equipment. 103, 202, 60

regulating system, synchronous-machine. An electric-machine regulating system consisting of one or more principal synchronous electric machines and the associated excitation system. 63

regulating transformer. A transformer used to vary the voltage, or the phase angle, or both, of an output circuit (referred to as the **regulated circuit**) controlling the output within specified limits, and compensating for fluctuations of load and input voltage (and phase angle, when involved) within specified limits. *See also:* **primary circuits; regulated circuit; series unit; main unit; series winding; excited winding; regulating winding; excitation winding; excitation-regulating winding; voltage-regulating relay; line-drop compensator; voltage winding (or transformer) for regulating equipment.** 53

regulating winding (two-core regulating transformer). The winding of the main unit in which taps are changed to control the voltage or phase angle of the regulated circuit through the series unit. 53

regulation (1) (rotating machinery). The amount of change in voltage or speed resulting from a load change. *See also:* **asynchronous machine; direct-current commutating machine.** 63

(2) (overall) (power supplies). The maximum amount that the output will change as a result of the specified change in line voltage, output load, temperature, or time. *Note:* Line regulation, load regulation, stability, and temperature coefficient are defined and usually specified separately. *See also:* **line regulation; load regulation; power supply; stability; temperature coefficient.** 228, 186

(3) (gas tubes). The difference between the maximum and minimum anode voltage drop over a range of anode current. *See also:* **gas tubes.** 190

(4) (electrical conversion). The change of one of the controlled or regulated output parameters resulting from a change of one or more of the unit's variables within specification limits. *See also:* **electrical conversion.** 186

(5) (transformer-rectifier system). The change in output voltage as the load current is varied. It is usually expressed as a percentage of the rated load voltage when the load current is changed by its rated value.

$$\text{Percent regulation} = 100 \frac{(E_1 - E_2)}{E_2}$$

where E_1 is the no-load voltage and E_2 is the voltage at rated load current and the line voltage is held constant at rated value. 95

(6) (automatic control). *See:* **load regulation.**

regulation changer (speed governing systems, hydraulic turbines). A device by means of which the speed regulation may be adjusted while the turbine is operating. 58

regulation curve (generator). A characteristic curve between voltage and load. The speed is either held constant, or varied according to the speed characteristics of the prime mover. The excitation is held constant for separately excited fields, and the rheostat setting is held constant for self-excited machines. *See also:* **direct-current commutating machine.** 328

regulation, frequency (power systems). The percentage change in frequency from steady state no load to steady state full load, which is a function of the engine and governing system:

$$\%R = \frac{F_{nl} - F_{fl}}{F_{nl}} \times 100$$ 89

regulation, load (synchronous machines). The steady-state decrease of the value of the specified variable resulting from a specified increase in load, generally from no-load to full-load unless otherwise specified. 105

regulation pull-out (regulation drop-out) (power supply). The load currents at which the power supply fails to regulate when the load current is gradually increased or decreased. *See also:* **regulated power supply.** 347

regulator, continuously acting (synchronous machines). A regulator that initiates a corrective action for a sustained infinitesimal change in the controlled variable. 105

regulator, noncontinuously acting (synchronous machines). A regulator that requires a sustained finite change in the controlled variable to initiate corrective action. 105

regulator, rheostatic-type (synchronous machines). A regulator that accomplishes the regulating function by mechanically varying a resistance. *Note:* Historically, rheostatic-type regulators have been further defined as direct acting and indirect acting. An indirect-acting-type regulator is a rheostatic type that controls the excitation of the exciter by acting on an intermediate device which is not considered part of the regulator or exciter. A direct-acting-type regulator is a rheostatic type that directly controls the excitation of an exciter by varying the input to the exciter field circuit. 105

regulator, synchronous-machine (1) (rotating machinery). An electric-machine regulator that controls the excitation of a synchronous machine. 63

(2) (excitation system). One that couples the output variables of the synchronous machine to the input of the exciter through feedback and forward controlling elements for the purpose of regulating the synchronous machine output variables. *Note:* In general, the regulator is assumed to consist of an error detector, preamplifier, power amplifier, stabilizers, auxiliary inputs, and limiters. 105

reguline. A word descriptive of electrodeposits that are firm and coherent. *See also:* **electrodeposition.** 328

reignition (1) (relay). A resumption of current between the contacts of a switching device during an opening operation after an interval of zero current of less than 1/4 cycle at normal frequency. 103, 202, 27

(2) (radiation-counter tubes). A process by which multiple counts are generated within a counter tube by atoms or molecules excited or ionized in the discharge accompanying a tube count. *See also:* **gas-filled radiation-counter tubes.** 190, 96, 125

reinforced plastic (rotating machinery). A plastic with some strength properties greatly superior to those of the base resin, resulting from the presence of high-strength fillers imbedded in the composition. *Note:* The reinforcing fillers are usually fibers, fabrics, or mats made of fibers. The plastic laminates are the most common and strongest. 218, 63

reinsertion (series capacitor). The interruption of the steady-state rms bypass current with the bypass device, with either the protective power gap or capacitor bypass switch to return the series capacitor to service following their operation and the removal of system fault current. 86

reinsertion current (series capacitor). The subsequent steady-state rms load current flowing through the series capacitor upon interruption of bypass current following protective-power-gap sparkover, capacitor discharge, and removal of system fault by other system protective devices. 86

reinsertion voltage (series capacitor). The subsequent steady-state rms voltage appearing across the series capacitor upon the interruption of bypass current following protective-power-gap sparkover, capacitor discharge, and removal of system fault current by other protective devices. 86

rejection band (1) (rejection filter, in a signal-transmission system). A band of frequencies between which the transmission is less than a specified fraction of that at a

reference frequency. The reference frequency is usually that of maximum transmission. *See also:* **signal.** 188
(2) (uniconductor waveguide). The frequency range below the cutoff frequency. *See also:* **waveguide.** 267, 179
(3) (circuits and systems). *See:* **stop band.**

rejection filter (signal-transmission system). A filter that attenuates alternating currents between given upper and lower cutoff frequencies and transmits substantially all others. Also, a filter placed in the signal transmission path to attenuate interference. *See also:* **signal.** 188

related transmission terms (loss and gain). The term **loss** used with different modifiers has different meanings, even when applied to one physical quantity such as power. In view of definitions containing the word **loss** (as well as others containing the word **gain**), the following brief explanation is presented. (1) Power loss from a circuit, in the sense that it is converted to another form of power not useful for the purpose at hand (for example, i^2R loss) is a physical quantity measured in watts in the International System of Units (SI) and having the dimensions of power. For a given R, it will vary with the square of the current in R. (2) Loss may be defined as the ratio of two powers, for example: if P_o is the output power and P_i the input power of a network under specified conditions, P_i/P_o is a dimensionless quantity that would be unity if $P_o = P_i$. Thus, no power loss in the sense of (1) means a loss, defined as the ratio P_i/P_o, of unity. The concept is closely allied to that of efficiency. (3) Loss may also be defined as the logarithm, or as directly proportional to the logarithm, of a power ratio such as P_i/P_o. Thus if loss = 10 $\log_{10} P_i/P_o$ the loss is zero when $P_o = P_i$. This is the standard for measuring loss in decibels. It should be noted that in cases (2) and (3) the **loss** (for a given linear system) is the same whatever may be the power levels. Thus (2) and (3) give characteristics of the system and do not depend (as (1) does) on the value of the current or other dependent quantity. **Power** refers to average power, not instantaneous power. *See also:* **network analysis.** 210

relative address (computing systems). The number that specifies the difference between the absolute address and the base address. 255, 77, 54

relative bearing (navigation). Bearing relative to heading. *See also:* **navigation.** 187

relative capacitivity. *See:* **relative dielectric constant.**

relative coding (computing systems). Coding that uses machine instructions with relative addresses. 255, 77

relative complex dielectric constant (homogeneous isotropic material) (complex capacitivity) (complex permittivity). The ratio of the admittance between two electrodes of a given configuration of electrodes with the material as a dielectric to the admittance between the same two electrodes of the configuration with vacuum as dielectric or

$$\epsilon^* \equiv \epsilon' - j\epsilon'' = Y/(j\omega C_v)$$

where Y is the admittance with the material and $j\omega C_v$ is the admittance with vacuum. Experimentally, vacuum must be replaced by the material at all points where it makes a significant change in the admittance. *Note:* The word **relative** is frequently dropped. *See:* **dielectric loss index; relative dielectric constant.** 210

relative damping (1) (specific damping) (instrument). Under given conditions, the ratio of the damping torque at a given angular velocity of the moving element to the damping torque that, if present at this angular velocity, would produce the condition of critical damping. *See also:* **accuracy rating (instrument).** 328

(2) (automatic control) (under damped system). A number expressing the quotient of the actual damping of a second-order linear system or element by its critical damping. *Note:* For any system whose transfer function includes a quadratic factor $s^2 + 2z\omega_n s + \omega^2_n$, relative damping is the value of z, since z = 1 for critical damping. Such a factor has a root $-\sigma + j\omega$ in the complex s-plane, from which z $= \sigma/\omega_n = \sigma(\sigma^2 + \omega^2)^{1/2}$. 56

relative dielectric constant (homogeneous isotropic material) (relative capacitivity) (relative permittivity). The ratio of the capacitance of a given configuration of electrodes with the material as a dielectric to the capacitance of the same electrode configuration with a vacuum (or air for most practical purposes) as the dielectric or

$$\epsilon' = C_x/C_v$$

where C_x is the capacitance with the material and C_v is the capacitance with vacuum. Experimentally, vacuum must be replaced by the material at all points where it makes a significant change in the capacitance. *See:* **electric flux density; relative capacitivity.** 210

relative directive gain (physical media). In a given direction and at a given point in the far field, the ratio of the power flux per unit area from an antenna to the power flux per unit area from a reference antenna at a specified location and delivering the same power from the antenna to the medium. *Note:* All or part of the reference antenna must be within the smallest sphere containing the subject antenna. 111

relative gain (antenna). *See:* **relative power gain (antenna).**

relative interfering effect (single-frequency electric wave in an electroacoustic system). The ratio, usually expressed in decibels, of the amplitude of a wave of specified reference frequency to that of the wave in question when the two waves are equal in interfering effect. The frequency of maximum interfering effect is usually taken as the reference frequency. Equal interfering effects are usually determined by judgment tests or intelligibility tests. *Note:* When applied to complex waves, the relative interfering effect is the ratio, usually expressed in decibels, of the power of the reference wave to the power of the wave in question when the two waves are equal in interfering effect. 328

relative lead polarity (transformer). A designation of the relative instantaneous directions of current in its leads. *Notes:* (1) Primary and secondary leads are said to have the same polarity when at a given instant during most of each half cycle, the current enters an identified, or marked, primary lead and leaves the similarly identified, or marked, secondary lead in the same direction as though the two leads formed a continuous circuit. (2) The relative lead polarity of a single-phase transformer may be either additive or subtractive. If one pair of adjacent leads from the two windings is connected together and voltage applied to one of the windings, then: (A) The relative lead polarity is additive if the voltage across the other two leads of the windings is greater than that of the higher-voltage winding alone. (B) The relative lead polarity is subtractive if the voltage across the other two leads of the windings is less than that of the higher-voltage winding alone. (3) The relative lead polarity is indicated by identification marks on primary and secondary leads of like polarity, or by other appropriate identification. *See:* **routine test.** *See also:* **constant-current transformer.**   203

relative luminosity. The ratio of the value of the luminosity at a particular wavelength to the value at the wavelength of maximum luminosity.   18, 178

relatively refractory state (electrobiology). The portion of the electric recovery cycle during which the excitability is less than normal. *See also:* **excitability (electrobiology).**   192

relative permittivity in physical media (antenna). The ratio of the complex permittivity to the permittivity of free space. *See:* **relative dielectric constant.**   111

relative permeability. The ratio of normal permeability to the magnetic constant. *Note:* In anisotropic media, relative permeability becomes a matrix.   210

relative power gain (antenna). The ratio of the power gain in a given direction to the power gain of a reference antenna in its reference direction. *Note:* Common reference antennas are half-wave dipoles, electric dipoles, magnetic dipoles, monopoles, and calibrated horn antennas.   111

relative power gain in physical media (antenna). In a given direction and at a given point in the far field, the ratio of the power flux per unit area from an antenna to the power flux per unit area from a reference antenna at a specified location with the same power input to both antennas. *Note:* All or part of the reference antenna must be within the smallest sphere containing the subject antenna.   111

relative redundancy (of a source) (information theory). The ratio of the redundancy of the source to the logarithm of the number of symbols available at the source. *See also:* **information theory.**   160

relative refractive index (two media). The ratio of their refractive indexes. *See also:* **radiation; radio wave propagation.**   146, 180

relative response (audio and electroacoustics). The ratio, usually expressed in decibels, of the response under some particular conditions to the response under reference conditions. Both conditions should be stated explicitly.   176

relaxation (laser-maser). The spontaneous return of a system towards its equilibrium condition.   363

relaxation oscillator. Any oscillator whose fundamental frequency is determined by the time of charging or discharging of a capacitor or inductor through a resistor, producing waveforms that may be rectangular or sawtooth. *Note:* The frequency of a relaxation oscillator may be self-determined or determined by a synchronizing voltage derived from an external source. *See also:* **electronic controller; oscillatory circuit.**   111

relaxation time (laser-maser). The time required for the deviation from equilibrium of some system parameter to diminish to l/e of its initial value.   363

relay (1) (general). An electric device that is designed to interpret input conditions in a prescribed manner and after specified conditions are met to respond to cause contact operation or similar abrupt change in associated electric control circuits. *Notes:* (A) Inputs are usually electric, but may be mechanical, thermal, or other quantities. Limit switches and similar simple devices are not relays. (B) A relay may consist of several units, each responsive to specified inputs, the combination providing the desired performance characteristic.
                    103, 110, 60, 202, 206, 127

(2) (electric and electronics parts and equipments). An electrically controlled, usually two-state, device that opens and closes electrical contacts to effect the operation of other devices in the same or another electric circuit. *Notes:* (A) A relay is a device in which a portion of one or more sets of electrical contacts is moved by an armature and its associated operating coil. (B) This concept is extended to include assembled reed relays in which the armature may act as a contact. For individual magnetic reed switches, see **switch.**   17

relay actuation time. The time at which a specified contact functions.   259

relay actuation time, effective. The sum of the initial actuation time and the contact chatter intervals following such actuation.   259

relay actuation time, initial. The time of the first closing of a previously open contact or the first opening of a previously closed contact.   259

relay actuation time, final. The time of termination of chatter following contact actuation.   259

relay actuator. The part of the relay that converts electric energy into mechanical work.   259

relay adjustment. The modification of the shape or position of relay parts to affect one or more of the operating characteristics, that is, armature gap, restoring spring, contact gap.   259

relay air gap. Air space between the armature and the pole piece. This is used in some relays instead of a nonmagnetic separator to provide a break in the magnetic circuit.   259

relay, alternating-current. A relay designed for operation from an alternating-current source. *See also:* **relay.**   259

relay amplifier. An amplifier that drives an electromechanical relay. *See also:* **electronic analog computer.**   9, 77, 10

relay antifreeze pin. Sometimes used for **relay armature stop, nonmagnetic.**   259

relay armature (electromechanical relay). The moving element that contributes to the designed response of the relay and that usually has associated with it a part of the relay contact assembly.   103, 202

relay armature, balanced. An armature that is approximately in equilibrium with respect to both static and dynamic forces.   259

relay armature bounce. *See:* **relay armature rebound.**   259

relay armature card. An insulating member used to link the movable springs to the armature.   259

relay armature contact. (1) A contact mounted directly on the armature. (2) Sometimes used for **relay contact, movable.**   259

relay armature, end-on. An armature whose motion is in

the direction of the core axis, with the pole face at the end of the core and perpendicular to this axis. 259

relay armature, flat-type. An armature that rotates about an axis perpendicular to that of the core, with the pole face on a side surface of the core. 259

relay armature gap. The distance between armature and pole face. 259

relay armature hesitation. Delay or momentary reversal of armature motion in either the operate or release stroke. 259

relay armature lifter. *See:* **relay armature stud.**

relay armature, long-lever. An armature with an armature ratio greater than 1:1. 259

relay armature overtravel. The portion of the available stroke occurring after the contacts have touched. 259

relay armature ratio. The ratio of the distance through which the armature stud or card moves to the armature travel. 259

relay armature rebound. Return motion of the armature following impact on the backstop. 259

relay armature, short-lever. An armature with an armature ratio of 1:1 or less. 259

relay armature, side. An armature that rotates about an axis parallel to that of the core, with the pole face on a side surface of the core. 259

relay armature stop. Sometimes used for **relay backstop.** 259

relay armature stop, nonmagnetic. A nonmagnetic member separating the pole faces of core and armature in the operated position, used to reduce and stabilize the pull from residual magnetism in release. 259

relay armature stud. An insulating member that transmits the motion of the armature to an adjacent contact member. 259

relay armature travel. The distance traveled during operation by a specified point on the armature. 259

relay, automatic reset. (1) A stepping relay that returns to its home position either when it reaches a predetermined contact position, or when a pulsing circuit fails to energize the driving coil within a given time. May either pulse forward or be spring reset to the home position. (2) An overload relay that restores the circuit as soon as an overcurrent situation is corrected. 259

relay back contacts. Sometimes used for **relay contacts, normally closed.** 259

relay backstop. The part of the relay that limits the movement of the armature away from the pole face or core. In some relays a normally closed contact may serve as backstop. 259

relay backup. That part of the backup protection that operates in the event of failure of the primary relays. 103, 202

relay bank. *See:* **relay level.**

relay bias winding. An auxiliary winding used to produce an electric bias. 259

relay blades. Sometimes used for **relay contact springs.** 259

relay bracer spring. A supporting member used in conjunction with a contact spring. 259

relay bridging. (1) A result of contact erosion, wherein a metallic protrusion or bridge is built up between opposite contact faces to cause an electric path between them. (2)

A form of contact erosion occurring on the break of a low-voltage, low-inductance circuit, at the instant of separation, that results in melting and resolidifying of contact metal in the form of a metallic protrusion or bridge. (3) Make-before-break contact action, as when a wiper touches two successive contacts simultaneously while moving from one to the other. 259

relay brush. *See:* **relay wiper.**

relay bunching time. The time during which all three contacts of a bridging contact combination are electrically connected during the armature stroke. 259

relay bushing. Sometimes used for **relay spring stud.** 259

relay chatter time. The time interval from initial actuation of a contact to the end of chatter. 259

relay coil. One or more windings on a common form. 259

relay coil, concentric-wound. A coil with two or more insulated windings, wound one over the other. 259

relay-coil dissipation. The amount of electric power consumed by a winding. For the most practical purposes, this equals the I^2R loss. 259

relay-coil resistance. The total terminal-to-terminal resistance of a coil at a specified temperature. 259

relay-coil serving. A covering, such as thread or tape, that protects the winding from mechanical damage. 259

relay-coil temperature rise. The increase in temperature of a winding above the ambient temperature when energized under specified conditions for a given period of time, usually the time required to reach a stable temperature. 259

relay-coil terminal. A device, such as a solder lug, binding post, or similar fitting, to which the coil power supply is connected. 259

relay-coil tube. An insulated tube upon which a coil is wound. 259

relay comb. An insulating member used to position a group of contact springs. 259

relay contact actuation time. The time required for any specified contact on the relay to function according to the following subdivisions. When not otherwise specified contact actuation time is **relay initial actuation time.** For some purposes, it is preferable to state the actuation time in terms of final actuation time or effective actuation time. 259

relay contact arrangement. The combination of contact forms that make up the entire relay switching structure. 259

relay contact bounce. Sometimes used for **relay contact chatter,** when internally caused. 259

relay contact chatter. The undesired intermittent closure of open contacts or opening of closed contacts. It may occur either when the relay is operated or released or when the relay is subjected to external shock or vibration. 259

relay contact chatter, armature hesitation. Chatter ascribed to delay or momentary reversal in direction of the armature motion during either the operate or the release stroke. 259

relay contact chatter, armature impact. Chatter ascribed to vibration of the relay structure caused by impact of the armature on the pole piece in operation, or on the backstop

in release.   259
relay contact chatter, armature rebound. Chatter ascribed to the partial return of the armature to its operated position as a result of rebound from the backstop in release.
259
relay contact chatter, externally caused. Chatter resulting from shock or vibration imposed on the relay by external action.   259
relay contact chatter, external shock. Chatter ascribed to impact experienced by the relay or by the apparatus of which it forms a part.   259
relay contact chatter, initial. Chatter ascribed to vibration produced by opening or closing the contacts themselves, as by contact impact in closure.   259
relay contact chatter, internally caused. Chatter resulting from the operation or release of the relay.   259
relay contact chatter, transmitted vibration. Chatter ascribed to vibration originating outside the relay and transmitted to it through its mounting.   259
relay contact combination. (1) The total assembly of contacts on a relay. (2) Sometimes used for contact form.
259
relay contact, fixed. *See:* **relay contact, stationary.**
relay contact follow. The displacement of a stated point on the contact-actuating member following initial closure of a contact.   259
relay contact follow, stiffness. The rate of change of contact force per unit contact follow.   259
relay contact form. A single-pole contact assembly.
259
relay contact functioning. The establishment of the specified electrical state of the contacts as a continuous condition.   259
relay contact gap. *See:* **relay contact separation.**
relay contact, movable. The member of a contact pair that is moved directly by the actuating system.   259
relay contact pole. Sometimes used for **relay contact, movable.**   259
relay contact rating. A statement of the conditions under which a contact will perform satisfactorily.   259
relay contacts. The current-carrying parts of a relay that engage or disengage to open or close electric circuits.
259
relay contacts, auxiliary. Contacts of lower current capacity than the main contacts; used to keep the coil energized when the original operating circuit is open, to operate an audible or visual signal indicating the position of the main contacts, or to establish interlocking circuits, etcetera.   259
relay contacts, back. Sometimes used for **relay contacts, normally closed.**   259
relay contacts, break. *See:* **relay contacts, normally closed.**
relay contacts, break-make. A contact form in which one contact opens its connection to another contact and then closes its connection to a third contact.   259
relay contacts, bridging. A contact form in which the moving contact touches two stationary contacts simultaneously during transfer.   259
relay contacts, continuity transfer. Sometimes used for **relay contacts, make-break.**   259
relay contacts, double break. A contact form in which one

contact is normally closed in simultaneous connection with two other contacts.   259
relay contacts, double make. A contact form in which one contact, which is normally open, makes simultaneous connection when closed with two other independent contacts.   259
relay contacts, dry. (1) Contacts which neither break nor make current. (2) Erroneously used for **relay contacts, low level.**   259
relay contacts, early. Sometimes used for **relay contacts, preliminary.**   259
relay contacts, front. Sometimes used for **relay contacts, normally open.**   259
relay contacts, interrupter. An additional set of contacts on a stepping relay, operated directly by the armature.
259
relay contacts, late. Contacts that open or close after other contacts when the relay is operated.   259
relay contacts, low-capacitance. A type of contact construction providing low intercontact capacitance.   259
relay contacts, low-level. Contacts that control only the flow of relatively small currents in relatively low-voltage circuits; for example, alternating currents and voltages encountered in voice or tone circuits, direct currents and voltages of the order of microamperes and microvolts, etcetera.   259
relay contacts, make. *See:* **relay contacts, normally open.**
relay contacts, make-break. A contact form in which one contact closes connection to another contact and then opens its prior connection to a third contact.   259
relay contacts, multiple-break. Contacts that open a circuit in two or more places.   259
relay contacts, nonbridging. A contact arrangement in which the opening contact opens before the closing contact closes.   259
relay contacts, normally closed. A contact pair that is closed when the coil is not energized.   259
relay contacts, normally open. A contact pair that is open when the coil is not energized.   259
relay contacts, off-normal. Contacts on a multiple switch that are in one condition when the relay is in its normal position and in the reverse condition for any other position of the relay.   259
relay contacts, preliminary. Contacts that open or close in advance of other contacts when the relay is operating.
259
relay contacts, sealed. A contact assembly that is sealed in a compartment separate from the rest of the relay.
259
relay contact separation. The distance between mating contacts when the contacts are open.   259
relay contacts, snap-action. A contact assembly having two or more equilibrium positions, in one of which the contacts remain with substantially constant contact pressure during the initial motion of the actuating member, until a condition is reached at which stored energy snaps the contacts to a new position of equilibrium.   259
relay contact spring. (1) A current-carrying spring to which the contacts are fastened. (2) A non-current-carrying spring that positions and tensions a contact-carrying member   259

relay contact, stationary. The member of a contact pair that is not moved directly by the actuating system.
259

relay contact wipe. The sliding or tangential motion between two contact surfaces when they are touching.
259

relay core. The magnetic member about which the coil is wound.
259

relay critical voltage (current). That voltage (current) that will just maintain thermal relay contacts operated.
259

relay cycle timer. A controlling mechanism that opens or closes contacts according to a preset cycle.
259

relay damping ring, mechanical. A loose member mounted on a contact spring to reduce contact chatter.
259

relay, direct-current. A relay designed for operation from a direct-current source. *See also:* **relay.**
259

relay, double-pole. A term applied to a contact arrangement to denote that it includes two separate contact forms; that is, two single-pole contact assemblies.
259

relay, double-throw. A term applied to a contact arrangement to denote that each contact form included is a break-make.
259

relay driving spring. The spring that drives the wipers of a stepping relay.
259

relay drop-out. *See:* **relay release.**

relay, dry circuit. Erroneously used for a relay with either dry or low-level contacts. *See:* **relay contacts, low-level.**
259

relay duty cycle. A statement of energized and deenergized time in repetitive operation, as: 2 seconds on, 6 seconds off.
259

relay electric bias. An electrically produced force tending to move the armature towards a given position.
259

relay, electric reset. A relay that may be reset electrically after an operation.
259

relay, electromagnetic. A relay, controlled by electromagnetic means, that opens and closes electric contacts. *See also:* **relay.**
259

relay, electrostatic. A relay in which the actuator element consists of nonconducting media separating two or more conductors that change their relative positions because of the mutual attraction or repulsion by electric charges applied to the conductors. *See also:* **relay.**
259

relay, electrostrictive. A relay in which an electrostrictive dielectric serves as the actuator. *See also:* **relay.**
259

relay electrothermal expansion element. An actuating element in the form of a wire strip or other shape having a high coefficient of thermal expansion.
259

relay element. A subassembly of parts. *Note:* The combination of several relay elements constitutes a relay unit.
103, 202, 60, 127

relay finish lead. The outer termination of the coil.
259

relay, flat-type. *See:* **relay armature, flat-type.**

relay frame. The main supporting portion of a relay. This may include parts of the magnetic structure.
259

relay freezing, magnetic. Sticking of the relay armature to the core as a result of residual magnetism.
259

relay fritting. Contact erosion in which the electrical discharge makes a hole through the film and produces molten matter that is drawn into the hole by electrostatic forces

and solidifies there to form a conducting bridge.
259

relay front contacts. Sometimes used for **relay contacts, normally open.**
259

relay functioning time. The time between energization and operation or between de-energization and release.
259

relay functioning value. The value of applied voltage, current, or power at which the relay operates or releases.
259

relay header. The subassembly that provides support and insulation to the leads passing through the walls of a sealed relay.
259

relay heater. A resistor that converts electric energy into heat for operating a thermal relay.
259

relay heel piece. The portion of a magnetic circuit of a relay that is attached to the end of the core remote from the armature.
259

relay, high, common, low (HCL). A type of relay control used in such devices as thermostats and in relays operated by them, in which a momentary contact between the common lead and another lead operates the relay, that then remains operated until a momentary contact between the common lead and a third lead causes the relay to return to its original position.
259

relay hinge. The joint that permits movement of the armature relative to the stationary parts of the relay structure.
259

relay hold. A specified functioning value at which no relay meeting the specification may release.
259

relay housing. An enclosure for one or more relays, with or without accessories, usually providing access to the terminals.
259

relay hum. The sound emitted by relays when their coils are energized by alternating current or in some cases by unfiltered rectified current.
259

relay inside lead. *See:* **relay start lead.**

relay inverse time. A qualifying term applied to a relay indicating that its time of operation decreases as the magnitude of the operating quantity increases.
259

relay just-operate value. The measured functioning value at which a particular relay operates.
259

relay just-release value. The measured functioning value for the release of a particular relay.
259

relay leakage flux. The portion of the magnetic flux that does not cross the armature-to-pole-face gap.
259

relay level. A series of contacts served by one wiper in a stepping relay.
259

relay load curves. The static force displacement characteristic of the total load of the relay.
259

relay magnetic bias. A steady magnetic field applied to the magnetic circuit of a relay.
259

relay magnetic gap. Nonmagnetic portion of a magnetic circuit.
259

relay, manual reset. A relay that may be reset manually after an operation.
259

relay mechanical bias. A mechanical force tending to move the armature towards a given position.
259

relay mounting plane. The plane to which the relay mounting surface is fastened.
259

relay must-operate value. A specified functioning value at which all relays meeting the specification must operate.
259

relay must-release value. A specified functioning value, at

which all relays meeting the specification must release.
259

relay nonfreeze pin. Sometimes used for **relay armature stop, nonmagnetic.** 259

relay nonoperate value. A specified functioning value at which no relay meeting the specification may operate.
259

relay normal condition. The de-energized condition of the relay. 259

relay operate. The condition attained by a relay when all contacts have functioned. *See also:* **relay contact actuation time.** 259

relay operate time. The time interval from coil energization to the functioning time of the last contact to function. Where not otherwise stated the functioning time of the contact in question is taken as its initial functioning time.
259

relay operate time characteristic. The relation between the operate time of an electromagnetic relay and the operate power. 259

relay operating frequency. The rated alternating-current frequency of the supply voltage at which the relay is designed to operate. 259

relay outside lead. *See:* **relay finish lead.**

relay overtravel. Amount of contact wipe. *See:* **relay armature overtravel; relay contact wipe.** 259

relay pickup value. Sometimes used for **relay must-operate value.** 259

relay pileup. A set of contact arms, assemblies, or springs, fastened one on top of the other with insulation between them. 259

relay pneumatic bellows. Gas-filled bellows, sometimes used with plunger-type relays to obtain time delay.
259

relay pole face. The part of the magnetic structure at the end of the core nearest the armature. 259

relay pole piece. The end of an electromagnet, sometimes separable from the main section, and usually shaped so as to distribute the magnetic field in a pattern best suited to the application. 259

relay pull curves. The force-displacement characteristics of the actuating system of the relay. 259

relay pull-in value. Sometimes used for **relay must-operate value.** 259

relay pusher. Sometimes used for **relay armature stud.** *See also:* **relay.** 259

relay rating. A statement of the conditions under which a relay will perform satisfactorily. 259

relay recovery time. A cooling time required from heater de-energization of a thermal time-delay relay to subsequent re-energization that will result in a new operate time equal to 85 percent of that exhibited from a cold start.
259

relay recovery time, instantaneous. Recovery time of a thermal relay measured when the heater is de-energized at the instant of contact operation. 259

relay recovery time, saturated. Recovery time of a thermal relay measured after temperature saturation has been reached. 259

relay release. The condition attained by a relay when all contacts have functioned and the armature (where applicable) has reached a fully opened position. 259

relay release time. The time interval from coil deenergization to the functioning time of the last contact to function. Where not otherwise stated the functioning time of the contact in question is taken as its initial functioning time. 259

relay reoperate time. Release time of a thermal relay.
259

relay reoperate time, instantaneous. Reoperate time of a thermal relay measured when the heater is deenergized at the instant of contact operation. 259

relay reoperate time, saturated. Reoperate time of a thermal relay measured when the relay is de-energized after temperature saturation (equilibrium) has been reached. 259

relay residual gap. Sometimes used for **relay armature stop, nonmagnetic.** 259

relay restoring spring. A spring that moves the armature to the normal position and holds it there when the relay is de-energized. 259

relay retractile spring. Sometimes used for **relay restoring spring.** 259

relay return spring. Sometimes used for **relay restoring spring.** 259

relay saturation. The condition attained in a magnetic material when an increase in field intensity produces no further increase in flux density. 259

relay sealing. Sometimes used for **relay seating.** 259

relay seating. The magnetic positioning of an armature in its final desired location. 259

relay seating time. The elapsed time after the coil has been energized to the time required to seat the armature of the relay. 259

relay shading coil. Sometimes used for **relay shading ring.**
259

relay shading ring. A shorted turn surrounding a portion of the pole of an alternating-current magnet, producing a delay of the change of the magnetic field in that part, thereby tending to prevent chatter and reduce hum.
259

relay shields, electrostatic spring. Grounded conducting members located between two relay springs to minimize electrostatic coupling. 259

relay shim, nonmagnetic. Sometimes used for **relay armature stop, nonmagnetic.** 259

relay, single-pole. A relay in which all contacts connect, in one position or another, to a common contact. 259

relay, single-throw. A relay in which each contact form included is a single contact pair. 259

relay sleeve. A conducting tube placed around the full length of the core as a short-circuited winding to retard the establishment or decay of flux within the magnetic path. 259

relay slow-release time characteristic. The relation between the release time of an electromagnetic relay and the conductance of the winding circuit or of the conductor (sleeve or slug) used to delay release. The conductance in this definition is the quantity N^2/R, where N is the number of turns and R is the resistance of the closed winding circuit. (For a sleeve or slug $N = 1$). 259

relay slug. A conducting tube placed around a portion of the core to retard the establishment or decay of flux within the magnetic path. 259

relay soak. The condition of an electromagnetic relay when its core is approximately saturated. 259

relay soak value. The voltage, current, or power applied to the relay coil to insure a condition approximating magnetic saturation. 259

relay spool. A flanged form upon which a coil is wound. 259

relay spring buffer. Sometimes used for **relay spring stud.** 259

relay spring curve. A plot of spring force on the armature versus armature travel. 259

relay spring stop. A member that controls the position of a pretensioned spring. 259

relay spring stud. An insulating member that transmits the motion of the armature from one movable contact to another in the same pileup. 259

relay stack. Sometimes used for **relay pileup.** 259

relay stagger time. The time interval between the actuation of any two contact sets. 259

relay starting switch (rotating machinery). A relay, actuated by current, voltage, or the combined effect of current and voltage, used to perform a circuit-changing function in the primary winding of a single-phase induction motor within a predetermined range of speed as the rotor accelerates; and to perform the reverse circuit-changing operation when the motor is disconnected from the supply line. One of the circuit changes that is usually performed is to open or disconnect the auxiliary-winding circuit. *See also:* **starting-switch assembly.** 63

relay start lead. The inner termination of the coil. 259

relay static characteristic. The static force-displacement characteristic of the spring system or of the actuating system. 259

relay station (mobile communication). A radio station used for the reception and retransmission of the signals from another radio station. *See also:* **mobile communication system.** 181

relay system (surge withstand capability). An assembly usually consisting of measuring units, relay logic, communications interfaces, and necessary power supplies. The communications link is not considered as a part of a relay system. 90

relay thermal. A relay that is actuated by the heating effect of an electric current. *See also:* **relay** 259

relay, three-position. Sometimes used for a center-stable polar relay. *See also:* **relay.** 259

relay transfer contacts. Sometimes used for **relay contacts, break-make.** 259

relay transfer time. The time interval between opening the closed contact and closing the open contact of a break-make contact form. 259

relay unit. An assembly of relay elements that in itself can perform a relay function. *Note:* One or more relay units constitutes a relay. 103, 202, 60, 127

relay winding. Sometimes used for **relay coil.** 257

relay wiper. The moving contact on a rotary stepping switch or relay. 259

release (telephone switching systems). Disengaging the apparatus used in a connection and restoring it to its idle condition upon recognizing a disconnect signal. 55

release (trip) coil (mechanical switching device). A coil used in the electromagnet that initiates the action of a release (trip). 103, 202

release delay (mechanical switching device). Intentional time delay introduced into contact-parting time in addition to opening time. *Note:* In devices employing a shunt release, release delay includes the operating time of protective and auxiliary relays external to the device. In devices employing direct or indirect release, release delay consists of intentional delay introduced into the function of the release. *Syn.* **tripping delay.** 103, 202

release-delay (trip-delay) setting. A calibrated setting of the time interval between the time when the actuating value reaches the release setting and the time when the release operates. 103, 202

release-free (1) (mechanical switching device). A descriptive term that indicates that the opening operation can prevail over the closing operation during specified parts of the closing operation. *Syn.* **trip-free.** *Note:* When used as an adjective modifying a device, the device has this operating capability. 92, 103, 202

(2) (trip-free in any position). A descriptive term indicating that a switching device is release-free at any part of the closing operation. *Note:* If the release circuit is completed through an auxiliary switch, electric release will not take place until such auxiliary switch is closed. 103, 202

release-free (trip-free) relay. An auxiliary relay whose function is to open the closing circuit of an electrically operated switching device so that the opening operation can prevail over the closing operation. 103, 202, 60

release mechanism (mechanical switching device). A device, mechanically connected to the mechanical switching device, that releases the holding means and permits the opening or closing of the switching device. *Syn.* **tripping mechanism.** 103, 202

release (trip) setting. A calibrated point at which the release is set to operate. 103, 202

release signal (telephone switching systems). A signal transmitted from one end of a line or trunk to indicate that the called party has disconnected. 55

release time, relay. *See:* **relay release time.**

relevant failure. *See:* **failure, relevant.**

reliability (1) (general). (A) The ability of an item to perform a required function under stated conditions for a stated period of time. *Note:* The term reliability is also used as a reliability characteristic denoting a probability of success, or a success ratio. *See also:* **wear-out failure period; observed reliability; assessed reliability; extrapolated reliability.** 182, 164
(B) The probability that a device will function without failure over a specified time period or amount of usage. *Notes:* (1) Definition (B) is most commonly used in engineering applications. In any case where confusion may arise, specify the definition being used. (2) The probability that the system will perform its function over the specified time should be equal to or greater than the reliability. 255, 77, 209

(2) (relay or relay system). A measure of the degree of certainty that the relay, or relay system, will perform correctly. *Note:* Reliability denotes certainty of correct operation together with assurance against incorrect operation from all extraneous causes. *See also:* **dependability; security.** 103, 202, 60, 127

(3) (nuclear power generating stations). The characteristic

of an item expressed by the probability that it will perform a required (function) (mission) under stated conditions for a stated (mission) (period of) time.   29, 31, 357
(4) (automatic control). The mathematical probability that a device will perform its objective adequately, for the period of time intended, under the operating conditions specified.   56

reliability allocation (nuclear power generating stations). The assignment of reliability subgoals to subsystems and elements thereof within a system for the purpose of meeting the overall reliability goal for the system, if each of these subgoals is attained.   31, 75

reliability, assessed. The reliability of an item determined within stated confidence limits from tests or failure data on nominally identical items. The source of the data shall be stated. Results can only be accumulated (combined) when all the conditions are similar. *Note:* Alternatively, point estimates may be used, the basis of which shall be defined. *See also:* **reliability.**   182

reliability compliance test (reliability). An experiment used to show whether or not the value of a reliability characteristic of an item complies with its stated reliability requirements.   164

reliability determination test (reliability). An experiment used to determine the value of a reliability characteristic of an item. *Note:* Analysis of available data may also be used for reliability determination.   164

reliability, extrapolated (reliability). Extension by a defined extrapolation or interpolation of the observed or assessed reliability for durations and/or conditions different from those applying to the observed or assessed reliability. *Note:* The validity of the extrapolation shall be justified.
164

reliability goal (nuclear power generating stations). A design objective, stated numerically, applied to reliability or availability.   31

reliability, inherent. The potential reliability of an item present in its design. *See also:* **reliability.**   182

reliability modeling (nuclear power generating stations). A logical display in a block diagram format and a mathematical representation of component functions as they occur in sequence which is required to produce system success.   31, 75

reliability, operational. The assessed reliability of an item based on field data. *See also:* **reliability.**   182

reliability, predicted. For the stated conditions of use, and taking into account the design of an item, the reliability computed from the observed, assessed, or extrapolated reliabilities of its parts. *Note:* Engineering and statistical assumptions shall be stated, as well as the bases used for the computation (observed or assessed).   164

reliability, test. The assessed reliability of an item based on a particular test with stated stress and stated failure criteria. *See also:* **reliability.**   182

relief door (rotating machinery). A pressure-operated door to prevent excessive gas pressure within a housing.   63

relieving (electroplating). The removal of compounds from portions of colored metal surfaces by mechanical means. *See also:* **electroplating.**   328

relieving anode (pool-cathode tube). An auxiliary anode that provides an alternative conducting path for reducing the current to another electrode. *See also:* **electrode**

(electron tube).   328

relocate (computing systems) (programming). To move a routine from one portion of storage to another and to adjust the necessary address references so that the routine, in its new location, can be executed.   255, 77

reluctance (magnetic circuit). The ratio of the magnetomotive force to the magnetic flux through any cross section of the magnetic circuit.   210

reluctance motor. A synchronous motor similar in construction to an induction motor, in which the member carrying the secondary circuit has salient poles, without permanent magnets or direct-current excitation. It starts as an induction motor, is normally provided with a squirrel-cage winding, but operates normally at synchronous speed.   63

reluctance synchronizing (rotating machinery). Synchronizing by bringing the speed of a salient pole synchronous machine to near-synchronous speed, but without applying excitation to it.   63

reluctance torque (synchronous motor). The torque developed by the motor due to the flux produced in the field poles by action of the armature-reaction magnetomotive force.   63

reluctivity. The reciprocal of permeability. *Note:* In anisotropic media, reluctivity becomes a matrix.   210

remanence. The magnetic flux density that remains in a magnetic circuit after the removal of an applied magnetomotive force. *Note:* This should not be confused with **residual flux density.** If the magnetic circuit has an air gap, the remanence will be less than the residual flux density.
331

remanent charge (ferroelectric device). The charge remaining when the applied voltage is removed. *Note:* The remanent charge is essentially independent of the previously applied voltage, provided this voltage was sufficient to cause saturation. If the device was not or cannot be saturated, the value of the previously applied voltage should be stated when measurements of remanent charge are reported. See the figure under **total charge.** *See also:* **ferroelectric domain.**   247

remanent induction (magnetic material). The induction when the magnetomotive force around the complete magnetic circuit is zero. *Note:* If there are no air gaps or other inhomogeneities in the magnetic circuit, the remanent induction will equal the residual induction; if there are air gaps or other inhomogeneities, the remanent induction will be less than the residual induction.   210

remanent polarization (ferroelectric material). The value of the polarization P_r which remains after an applied electric field is removed. Remanent polarization can be measured by integrating the bound surface charge released on heating a poled ferroelectric to a temperature above its Curie point. *Notes:* (1) When the magnitude of this electric field is sufficient to saturate the polarization (usually $3E_c$, that is three times the coercive electric field), the polarization remaining after the field is removed is termed the saturation remanent polarization P_R. (2) The saturation remanent polarization is equal to the spontaneous polarization in a single domain ferroelectric material. *See also:* **coercive electric field (illustration); spontaneous polarization; domain.**   80

remote access (test, measurement and diagnostic equipment).

Pertaining to communication with a data processing facility by one or more stations that are distant from that facility. 54

remote backup. A form of backup protection in which the protection is at a station or stations other than that which has the primary protection. 103, 202, 60, 127

remote concentrator (telephone switching systems). A concentrator located away from a serving system control. 55

remote control (1) (general). Control of an operation from a distance: this involves a link, usually electrical, between the control device and the apparatus to be operated. *Note:* Remote control may be over (A) direct wire, (B) other types of interconnecting channels such as carrier-current or microwave, (C) supervisory control, or (4) mechanical means. *See also:* **control.** 103, 244, 202, 200
(2) (digital interface for programmable instrumentation). A method whereby a device is programmable via its electrical interface connection in order to enable the device to perform different tasks. 40

remote-control circuits. Any electric circuit that controls any other circuit through a relay or an equivalent device. 256

remote-cutoff tube. *See:* **variable μ tube.**

remote data logging (power switchgear). An arrangement for the numerical representation of selected telemetered quantities on log sheets or paper or magnetic tape, or the like, by means of an electric typewriter, teletype, or other suitable devices. 103

remote error-sensing (power supplies). A means by which the regulator circuit senses the voltage directly at the load. This connection is used to compensate for voltage drops in the connecting wires. *See also:* **power supply.** 228, 186

remote indication. Indication of the position or condition of remotely located devices. *Note:* Remote indication may be over (1) direct wire, (2) other types of interconnecting channels such as carrier-current or microwave, (3) supervisory indication, or (4) mechanical means. 103, 202

remote job entry (data communication). Submission of jobs through an input device that has access to a computer through a communications link. 12

remote line (electroacoustics). A program transmission line between a remote pickup point and the studio or transmitter site. *See also:* **transmission line.** 239

remotely controlled operation (power switchgear). Operation of a device by remote control. 27, 103

remote manual operation. *See:* **indirect manual operation.**

remote metering. *See:* **telemetering.**

remote operation. *See:* **remotely controlled operation.**

remote release. *See* **remote trip.**

remote station (supervisory system). A remotely located station wherein units of switchgear or other equipment are controlled by supervisory control or from which supervisory indications or selected telemeter readings are obtained. 103

remote-station supervisory equipment (single supervisory system). That part that includes all supervisory control relays and associated devices located at the remote station for selection, control, indication, telemetering, and other functions to be performed. 103, 202

remote switching entity (telephone switching systems). An entity for switching inlets to outlets located away from a serving system control. 55

remote trip (power switchgear). A general term applied to a relay installation to indicate that the switching device is located physically at a point remote from the initiating protective relay, device, or source of release power or all these. *Note:* This installation is commonly called transfer trip when a communication channel is used to transmit the signal for remote tripping. *Syn:* **remote release.** 103, 127

removable element (switchgear assembly). That portion that normally carries the circuit-switching and circuit-interrupting devices and the removable part of the primary and secondary disconnecting devices. 103, 202

renewable (field-renewable) fuse. A fuse that, after circuit interruption, may be restored readily for service by the replacement of the renewal element, fuse link, or refill unit. 103, 202

renewal element (low-voltage fuse). That part of a renewable fuse that is replaced after each interruption to restore the fuse to operating condition. 103, 202

renewal parts. Those parts that it is necessary to replace during maintenance as a result of wear. 103, 202

reoperate time, relay. *See:* **relay reoperate time.**

repair (test, measurement and diagnostic equipment). The restoration or replacement of parts or components of material as necessitated by wear and tear, damage, failure of parts or the like in order to maintain the specific item of material in efficient operating condition. 54

repairable item (reliability). *See:* **item, repaired.**

repair rate (nuclear power generating stations). The expected number of repair actions of a given type completed per unit of time. 29, 31

repeatability (measurement) (control equipment) (1) (general). The closeness of agreement among repeated measurements of the same variable under the same conditions. *See also:* **measurement system.** 94
(2) (zero-center instruments). The ability of an instrument to repeat its readings taken when the pointer is deflected upscale, compared to the readings taken when the pointer is deflected downscale, expressed as a percentage of the full-scale value. *See also:* **measurement system; moving element (instrument).** 280
(3) (voltage regulator, or voltage-reference tubes). The ability of a tube to attain the same voltage drop at a stated time after the beginning of any conducting period. *Note:* The lack of repeatability is measured by the change in this voltage from one conducting period to any other, the operating conditions remaining unchanged. *See also:* **transmission regulator.** 190
(4) (analog computer). A quantitative measure of the agreement among repeated operations. 9
(5) attenuator, variable in fixed steps. Maximum difference in decibels of residual or incremental characteristic insertion loss for a selected position between the extreme values of a first and second set of ten measurements before and after the specified stepping life. 110
(6) two-port, due to insertion/removal cycle. The maximum difference in decibels between the extreme value of ten measurements before and ten measurements after the

number of complete insertion/removal cycles specified in insertion/removal life. 110

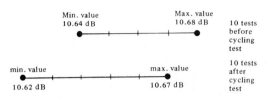

Example of repeatability of a two-port due to insertion/removal cycles

Repeatability = 10.68 dB − 10.62 dB = 0.06 dB

(7) continuously variable attenuator, due to cycling. Maximum difference in decibels between the extreme values of a first set of ten measurements for a selected calibration point, five of which are approached from one direction and five of which are approached from the opposite direction, and the extreme values of a similar second set after the specified cycling life. 110

(8) (nuclear power generating station). The closeness of agreement among a number of consecutive measurements of the output for the same value of the input under the same operating conditions, approaching from the same direction. 355

repeater (1) (power systems). A combination of apparatus for receiving either one-way or two-way communication signals and delivering corresponding signals that are either amplified or reshaped or both. A repeater for one-way communication signals is termed a **one-way repeater** and one for two-way communication signals a **two-way repeater.** 59

(2) (communication satellite). A receiver-transmitter combination, often aboard a satellite or spacecraft, which receives a signal, performs signal processing (amplification, frequency translation, etcetera) and retransmits it. Used in active communication satellite to relay signals between earth stations. *Syn.* **transponder.** 83

repeater station (data transmission). An intermediate point in a transmission system where line signals are received amplified or reshaped, and retransmitted. 59

repeater, 21-type. A two-wire telephone repeater in which there is one amplifier serving to amplify the telephone currents in both directions, the circuit being arranged so that the input and output terminals of the amplifier are in one pair of conjugate branches, while the lines in the two directions are in another pair of conjugate branches. *See also:* **repeater.** 59

repeater, 22-type. A two-wire telephone repeater in which there are two amplifiers, one serving to amplify the telephone currents being transmitted in one direction and the other serving to amplify the telephone currents being transmitted in the other direction. *See also:* **repeater.** 59

repeating coil* (repeat coil*). A term sometimes used in telephone practice to designate a transformer. 59

* Obsolescent

repeller (reflector) (electron tubes). An electrode whose primary function is to reverse the direction of an electron stream. *See also:* **electrode (electron tube).** 190, 125

reperforator. *See:* **perforator.**

reperforator switching center. A message-relaying center at which incoming messages are received on a reperforator that perforates a storage tape from which the message is retransmitted into the proper outgoing circuit. The reperforator may be of the type that also prints the message on the same tape, and the selection of the outgoing circuit may be manual or under control of selection characters at the head of the message. *See also:* **telegraphy.** 194

repertory. *See:* **instruction repertory.**

repetition equivalent (of a complete telephone connection, including the terminating telephone set). A measure of the grade of transmission experienced by the subscribers using the connection. It includes the combined effects of volume, distortion, noise, and all other subscriber reactions and usages. The repetition equivalent of a complete telephone connection is expressed numerically in terms of the trunk loss of a working reference system when the latter is adjusted to given an equal repetition rate. 328

repetition instruction. An instruction that causes one or more instructions to be executed an indicated number of times. 255, 77

repetition rate. Repetition rate signifies broadly the number of repetitions per unit time. Specifically, in appraising the effectiveness of transmission over a telephone connection, repetition rate is the number of repetitions per unit time requested by users of the connection. *See also:* **volume equivalent.** 328

repetitively pulsed laser (laser-maser). A laser with multiple pulses of radiant energy occurring in a sequence. 363

repetitive operation (analog computer). A condition in which the computer operates as a repetitive device; the solution time may be a small fraction of a second or as long as desired, after which the problem is automatically and repetitively cycled through reset, hold, and operate. *See also:* **electronic analog computer.** 9, 77

repetitive peak forward current (semiconductor). The peak value of the forward current including all repetitive transient currents. 66

repetitive peak OFF-state current (semiconductor). The maximum instantaneous value of the OFF-state current that results from the application of repetitive peak-OFF-state voltage. 66

repetitive peak OFF-state voltage. The maximum instantaneous value of the OFF-state voltage that occurs across a thyristor, including all repetitive transient voltages, but excluding all nonrepetitive transient voltages. 243, 66, 208, 191

repetitive peak ON-state current (semiconductor). The peak value of the ON-state current including all repetitive transient currents. 66

repetitive peak reverse current (semiconductor). The maximum instantaneous value of the reverse current that results from the application of repetitive peak reverse voltage. 66

repetitive peak reverse voltage (1) (semiconductor rectifier). The maximum instantaneous value of the reverse voltage, including all repetitive transient voltages but excluding all nonrepetitive transient voltages, that occurs across a semiconductor rectifier cell, rectifier diode, or rectifier stack. *See also:* **principal voltage-current characteristic; rectification; semiconductor rectifier stack.**
237, 243, 66, 208, 191

(2) (reverse blocking thyristor). The maximum instantaneous value of the reverse voltage which occurs across the thyristor, including all repetitive transient voltages, but excluding all non-repetitive transient voltages. 66

repetitive peak reverse-voltage rating (rectifier circuit element). The maximum value of repetitive peak reverse voltage permitted by the manufacturer under stated conditions. *See also:* **average forward-current rating (rectifier circuit).** 237, 66, 208

replacement part. A part for use in place of an existing component of switching equipment. 27

replica temperature relay (power switchgear). A thermal relay whose internal temperature rise is proportional to that of the protected apparatus or conductor, over a range of values and durations of overloads. 103, 127

reply (transponder operation). A radio-frequency signal or combination of signals transmitted as a result of interrogation. *See also:* **navigation.** 13, 187

reproducibility (1) (general). The ability of a system or element to maintain its output/input precision over a relatively long period of time. *See also:* **accuracy; precision.** 56, 207

(2) (transmission lines and waveguides). The degree to which a given set of conditions or observations, using different components or instruments each time, can be reproduced. *See also:* **measurement system.** 185

(3) (automatic null-balancing electric instrument). The closeness of agreement among repeated measurements by the instrument for the same value of input made under the same operating conditions, over a long period of time, approaching from either direction. *Notes:* (1) It is expressed as a maximum nonreproducibility in percent of span for a specified time. (2) Reproducibility includes drift, repeatability, and dead band. *See also:* **measurement system.** 295

reproducing stylus. A mechanical element adapted to following the modulations of a record groove and transmitting the mechanical motion thus derived to the pickup mechanism. *See also:* **phonograph pickup.** 176

reproduction speed (facsimile). The area of copy recorded per unit time. *See also:* **recording (facsimile).** 12

repulsion-induction motor. A motor with repulsion-motor windings and short-circuited brushes, without an additional device for short-circuiting the commutator segments, and with a squirrel-cage winding in the rotor in addition to the repulsion motor winding. 63

repulsion motor. A single-phase motor that has a stator winding arranged for connection to a source of power and a rotor winding connected to a commutator. Brushes on the commutator are short-circuited and are so placed that the magnetic axis of the rotor winding is inclined to the magnetic axis of the stator winding. This type of motor has a varying-speed characteristic. *See also:* **asynchronous machine.** 63

repulsion-start induction motor. A single-phase motor with repulsion-motor windings and brushes, having a commutator-short-circuiting device that operates at a predetermined speed of rotation to convert the motor into the equivalent of a squirrel-cage motor for running operation. For starting operation, this motor performs as a repulsion motor. *See also:* **asynchronous machine.** 63

request test. A test made at the request of a customer. *See*

also: **service test (field test).** 212

required response spectrum (RRS) (nuclear power generating stations) (seismic qualification of class 1E equipment). The response spectrum issued by the user or his agent as part of his specifications for proof testing, or artificially created to cover future applications. The RRS constitutes a requirement to be met. 28, 31

required time (availability). The period of time during which the user requires the item to be in a condition to perform its required function. 164

reradiation. (1) The scattering of incident radiation, or (2) the radiation of signals amplified in a radio receiver. *See also:* **radio receiver.** 328

rerecording (electroacoustics). The process of making a recording by reproducing a recorded sound source and recording this reproduction. *See also:* **dubbing.** 176

rerecording system (electroacoustics). An association of reproducers, mixers, amplifiers, and recorders capable of being used for combining or modifying various sound recordings to provide a final sound record. *Note:* Recording of speech, music, and sound effects may be so combined. *See also:* **dubbing; phonograph pickup.** 176

rering signal (telephone switching systems). A signal initiated by an operator at the calling end of an established connection to recall the operator at the called end or the customer at either end. 55

rerun point (computing systems). The location in the sequence of instructions in a computer program at which all information pertinent to the rerunning of the program is available. 255, 77, 54

reserve (1) (electric power system). A qualifying term to identify equipment and capability that is available and is in excess of that required for the load. *Note:* The reserve may be connected to the system and partially loaded or may be made available by closing switches, contactors or circuit breakers. Reserve not in operation and requiring switching is sometimes called **standby equipment.** 58

(2) (test, measurement and diagnostic equipment). The setting aside of a specific portion of memory for a storage area. 54

reserve cell. A cell that is activated by shock or other means immediately prior to use. *See also:* **electrochemistry.** 328

reserve, electrical (electric power supply). The capability in excess of that required to carry the system load. 112

reserve equipment. The installed equipment in excess of that required to carry peak load. *Note:* Reserve equipment not in operation is sometimes referred to as **standby equipment.** *See also:* **generating station.** 64

reserve, installed (electric power supply). The reserve capability installed on a system. 112

reserve, nonspinning (electric power supply). That operating reserve capable of being connected to the bus and loaded within a specified time. 112

reserve, operating (electric power supply). That reserve above firm system load required to provide for: (1) Regulation within the hour to cover minute to minute variations (2) Load forecasting error (3) Loss of equipment (4) Local area protection. The operating reserve consists of spinning or nonspinning reserve, or both. 112

reserve, required (electric power supply). The system planned reserve capability needed to insure a specified standard of service. 112

reset (1) (electronic digital computation) (industrial control).
(A) To restore a storage device to a prescribed state, not
necessarily that denoting zero. (B) To place a binary cell
in the initial or zero state. *See also:* **set.**

235, 210, 255, 77, 206, 54

(2) (analog computer). The computer control state in
which integrators are held constant and the proper initial
condition voltages or charges are applied or reapplied. *See:*
initial condition. 9

(3) (relay). The action of a relay as it makes designated
response to decreases in input. Reset as a qualifying term
denotes the state of a relay when all response to decrease
of input has been completed. Reset is also used to identify
the maximum value of an input quantity reached by progressive decreases that will permit the relay to reach the
state of complete reset from pickup. *Note:* In defining the
designated performance of relays having multiple inputs,
reset describes the state when all inputs are zero and also
when some input circuits are energized, if the resulting
state is not altered from the zero-input condition.

103, 202, 60, 127

(4) (industrial control). To restore a mechanism, electrical
circuit or device to a prescribed state. 70

reset action (process control). A component of control
action in which the final control element is moved at a
speed proportional to the extent of proportional-position
control action. *Note:* This term applies only to a multiple
control action including proportional-position control
action. *See also:* **control action, proportional plus integral;**
control system, positioning. 56

reset, automatic. A function which operates to automatically re-establish specific conditions. 70

reset control action (electric power systems). Action in
which the controller output is proportional to the input
signal and the time integral of the input signal. The
number of times per minute that the integral control action
repeats the proportional control action is called the reset
rate. *Note:* Applies only to a controller with proportional
control action plus integral control action. *See also:*
speed-governing system. 94

reset device. A device whereby the brakes may be released
after an automatic train-control brake application.

328

reset dwell time. The time spent in reset. In cycling the
computer from reset, to operate, to hold, and back to reset,
this time must be long enough to permit the computer to
recover from any overload and to charge or discharge all
integrating capacitors to appropriate initial voltages. *See
also:* **electronic analog computer.** 9

**reset interval (1) (automatic circuit recloser or automatic line
sectionalizer).** The time required, after a counting operation, for the counting mechanism to return to the starting
position of that counting operation. 103

(2) (automatic circuit recloser). The time required for the
counting mechanism to return to the starting position.

92

reset, manual. A function which requires a manual operation to re-establish specific conditions. 70

reset pulse. A drive pulse that tends to reset a magnetic cell.

331

reset rate (proportional plus reset control action or propor-
tional plus reset plus rate control action) (process control).
The number of times per minute that the effect of proportional-position control action is repeated. *See also:*
integral action rate. 56

reset switch (industrial control). A machine-operated device
that restores normal operation to the control system after
a corrective action. *See also:* **photoelectric control.**

206

resettability (oscillator). The ability of the tuning element
to retune the oscillator to the same operating frequency
for the same set of input conditions. *See also:* **tunable**
microwave oscillators. 174, 190

reset time (1) (relay). The time interval from occurrence
of specified conditions to reset. *Note:* When the conditions
are not specified it is intended to apply to a picked-up relay
and to be a sudden change from pickup value of input to
zero input. 103, 202, 60, 127

(2) (automatic circuit recloser or automatic line section-
alizer). The time required, after one or more counting
operations, for the counting mechanism to return to the
starting position. 103

residual-component telephone-influence factor (three-phase
synchronous machine). The ratio of the square root of the
sum of the squares of the weighted residual harmonic
voltages to three times the root-mean-square no-load
phase-to-neutral voltage. 63

residual current (1) (electric supply circuit). The vector sum
of the currents in the several wires. *Note:* When the vector
sum is of the currents in the phase conductors only it is
sometimes termed the **circuit residual current,** to distinguish it from the vector sum of the currents in all conductors on a line termed the **line residual current** or more
commonly, **ground return current.** *See also:* **inductive**
coordination. 328

(2) (thermionics). The value of the current in the residual-current state. *See also:* **electron emission; inductive**
coordination. 244, 190

(3) (protective relaying). The sum of the three-phase
currents on a three-phase circuit. The current that flows
in the neutral return circuit of three wye-connected current
transformers is residual current. 128

residual-current state (thermionics). The state of working
of an electronic valve or tube in the absence of an accelerating field from the anode of a diode or equivalent diode,
in which the cathode current is due to the nonzero velocity
of emission of electrons. *See also:* **electron emission; in-**
ductive coordination. 244, 190

residual error (electronic navigation). The sum of the
random errors and the uncorrected systematic errors. *See
also:* **navigation.** 13, 187

residual-error rate. *See:* **undetected-error rate.**

residual flux density. The magnetic flux density at which
the magnetizing force is zero when the material is in a
symmetrically cyclically magnetized condition. *See also:*
remanence. 331

residual frequency-modulation (non-real time spectrum
analyzer). Short term displayed frequency instability or
jitter of the spectrum analyzer local oscillators. Given in
terms of peak-to-peak frequency deviation, hertz (Hz).
Any influencing factors (for example, phase lock on or off)
must be given. 68

residual induction (1) (magnetic material). The magnetic induction corresponding to zero magnetizing force in a material that is in a symmetrically cyclically magnetized condition. 210

(2) (residual flux density) (toroidal magnetic amplifier cores). The magnetic induction at which the magnetizing force is zero when the magnetic core is cyclically magnetized with a half-wave sinusoidal magnetizing force of a specified peak magnitude. *Note:* This use of the term re-sidual induction differs from the standard definition that requires symmetrically cyclically magnetized conditions. 170

(3) (magnetic core testing) (B_r). The value of induction corresponding to zero magnetizing force in a ferromagnetic material which is symmetrically cyclically magnetized with a specific maximum magnetic field strength. Remanent induction is the induction in the ferromagnetic core when the magnetizing force around the complete magnetic circuit is zero. *Notes:* (A) If there are no airgaps or other discontinuities in the magnetic circuit, the re-manent induction is equal to the residual induction, otherwise the remanent induction will be less than the residual induction. (B) Squareness ratio is the ratio of remanent induction to peak induction with the magnetic material in a symmetrical cyclically magnetized condition.

residual magnetism (ferromagnetic bodies). A property by which they retain a certain magnetization (induction) after the magnetizing force has been removed. 244, 210

residual modulation. *See:* **carrier noise; carrier noise level.**

residual probe pickup (constancy of probe coupling) (slotted line). The noncyclical variation of the amplitude of the probe output over its complete range of travel when re-flected waves are eliminated on the slotted section by proper matching at the output and the input, discounting attenuation along the slotted section. It is defined by the ratio of one-half of the total variation to the average value of the probe output, assuming linear amplitude response of the probe, at a specified frequency(ies) within the range of usage. *Note:* This quantity consists of two parts of which one is reproducible and the other is not. The repeatable part can be eliminated by subtraction in repeated mea-surements, while the nonrepeatable part must cause an error. The residual probe pickup depends to some extent on the insertion depth of the probe. *See also:* **measurement system; residual standing-wave ratio.** 185

residual reflected coefficient (reflectometer). The erroneous reflection coefficient indicated when the reflectometer is terminated in reflectionless terminations. *See also:* **mea-surement system.** 185

residual relay. A relay that is so applied that its input, derived from external connections of instrument trans-formers, is proportional to the zero phase sequence com-ponent of a polyphase quantity. *See also:* **relay.** 103, 202, 60

residual response (non-real time spectrum analyzer). A spurious response in the absence of an input. 68

residual standing-wave ratio (SWR) (slotted line). The standing-wave ratio measured when the slotted line is terminated by a reflectionless termination and fed by a signal source that provides a nonreflecting termination for waves reflected toward the generator. *Note:* Residual standing-wave ratio does not include the residual noncy-

clical probe pickup or the attenuation encountered as the probe is moved along the line. *See:* **residual probe pickup.** *See also:* **waveguide.** 185

residual voltage (1) (electric supply circuit). The vector sum of the voltages to ground of the several phase wires of the circuit. *See also:* **inductive coordination.** 328

(2) (arrester) (discharge voltage). The voltage that appears between the line and ground terminals of an arrester during the passage of discharge current. *See also:* **inductive coordination.** 308, 62

(3) (protective relaying). The sum of the three line-to-neutral voltages on a three-phase circuit. 128

residue check (computing systems). A check in which each operand is accompanied by the remainder obtained by dividing this number by n, the remainder then being used as a check digit or digits. *See also:* **modulo n check.** 255, 77

resin (rotating machinery). Any of various hard brittle solid-to-soft semisolid amorphous fusible flammable substances of either natural or synthetic origin; generally of high molecular weight, may be either thermoplastic or thermosetting. 63

resin-bonded paper-insulated bushing (outdoor electric ap-paratus). A bushing in which the major insulation is provided by paper bonded with resin. 287, 168

resist (electroplating). Any material applied to part of a cathode or plating rack to render the surface noncon-ducting. *See also:* **electroplating.** 328

resistance (1) (network analysis). (A) That physical prop-erty of an element, device, branch, network, or system that is the factor by which the mean-square conduction current must be multiplied to give the corresponding power lost by dissipation as heat or as other permanent radiation or loss of electromagnetic energy from the circuit. (B) The real part of impedance. *Note:* Definitions (A) and (B) are not equivalent but are supplementary. In any case where confusion may arise, specify definition being used. *See:* **resistor.** 210, 185, 206

(2) (shunt). The quotient of the voltage developed across the instrument terminals to the current passing between the current terminals. In determining the value, account should be taken of the resistance of the instrument and the measuring cable. The resistance value is generally derived from a direct-current measurement such as by means of a double Kelvin bridge. *See also:* **test voltage and current.** 307, 201

(3) (automatic control). A property opposing movement of material, or flow of energy, and involving loss of po-tential (voltage, temperature, pressure, level). 56

(4) (antenna). *See:* **antenna resistance; radiation resis-tance.**

resistance, apparent (insulation testing). Ratio of the voltage across the electrodes in contact with the specimen to the current between them as measured under the specified test conditions and specified electrification time. 97

resistance box. A rheostat consisting of an assembly of resistors of definite values so arranged that the resistance of the circuit in which it is connected may be changed by known amounts. 210

resistance braking. A system of dynamic braking in which electric energy generated by the traction motors is dissi-

pated by means of a resistor. *See also:* **dynamic braking.**
328

resistance bridge smoke detector (fire protection devices).
A device which responds to an increase of smoke particles
and moisture, present in products of combustion, which
fall on an electrical bridge grid. As these conductive substances fall on the grid they reduce the resistance of the
grid and cause the detector to respond.
71

resistance-capacitance characteristic, input (oscilloscopes).
The direct-current resistance and parallel capacitance to
ground present at the input of an oscilloscope.
106

resistance-capacitance (RC) coupling. Coupling between
two or more circuits, usually amplifier stages, by means
of a combination of resistance and capacitance elements.
See also: **coupling.**
328

resistance-capacitance oscillator (RC oscillator). Any oscillator in which the frequency is determined principally
by resistance-capacitance elements. *See also:* **oscillatory
circuit.**
328

resistance drop. The component of the impedance voltage
drop in phase with the current.
53

resistance furnace. An electrothermic apparatus, the heat
energy for which is generated by the flow of electric current against ohmic resistance internal to the furnace.
328

resistance grading (cr corona shielding). A form of corona
shielding embodying high resistance material on the surface of the coil. *Syn.* **corona shielding.**
63

resistance grounded (electric power). Grounded through
impedance, the principal element of which is resistance.
Note: The resistance may be inserted either directly in the
connection to the ground or indirectly, as for example, in
the secondary of a transformer, the primary of which is
connected between neutral and ground, or in series with
the delta-connected secondary of a wye-delta grounding
transformer. *See also:* **ground; grounded.**
91, 203, 64, 53

resistance lamp. An electric lamp used to prevent the
current in a circuit from exceeding a desired limit.
328

resistance magnetometer. A magnetometer that depends
for its operation upon the variation of electrical resistance
of a material immersed in the field to be measured. *See
also:* **magnetometer.**
328

**resistance method of temperature determination (electric
power).** The determination of the temperature by comparison of the resistance of a winding at the temperature
to be determined, with the resistance at a known temperature. *See also:* **routine test.**
257, 91, 203, 53

resistance modulation (bolometric power meters). A change
in resistance of the bolometer resulting from a change in
power (RF, ac, or dc) dissipated in the element. *Note:* The
resistance modulation sensitivity is the (dc) change in
resistance per unit (dc) change in power at normal bias and
at a constant ambient temperature. Resistance modulation
frequency response is the frequency of repetitive (sinusoidal) power change for which the peak-to-peak resistance
change is 3 dB lower than the asymptotic, maximum value
at zero frequency.
115

resistance modulation effect (bolometric power meters). A
component of substitution error (for dc power substitution)
in bolometer units in which both ac and dc bias is used.

Note: This component is dependent upon the frequency
of the ac bias and the frequency response of the element;
it is usually very small, and usually not included in the
effective efficiency correction for substitution error. It is
caused by resistance modulation of the element, and is
more pronounced in barretters than in thermistors.
115

resistance relay. A linear-impedance form of distance relay
for which the operating characteristic on an *R-X* diagram
is a straight line of constant resistance. *Note:* The operating characteristic may be described by the equation $R
= K$, or $Z \cos \theta = K$, where K is a constant, and θ is the
angle by which the input voltage leads the input current.
103, 202, 60, 127

resistance-start motor. A form of split-phase motor having
a resistance connected in series with the auxiliary winding.
The auxiliary circuit is opened when the motor has attained a predetermined speed. *See also:* **asynchronous
machine.**
328

resistance starting (industrial control). A form of reduced-voltage starting employing resistances that are
short-circuited in one or more steps to complete the
starting cycle. *See:* **resistance starting, generator-field;
resistance starting, motor-armature.**
225, 206

resistance starting, generator-field. Field resistance
starting provided by one or more resistance steps in series
with the shunt field of a generator, the output of which is
connected to a motor armature. *See:* **resistance starting;
resistance starting, motor-armature.**
225, 206

resistance starting, motor-armature. Motor resistance
starting provided by one or more resistance steps connected in series with the motor armature. *See:* **resistance
starting; resistance starting, generator-field.**
225, 206

**resistance temperature detector (resistance thermometer
detector) (resistance thermometer resistor).** A resistor
made of some material for which the electrical resistivity
is a known function of the temperature and that is intended
for use with a resistance thermometer. It is usually in such
a form that it can be placed in the region where the temperature is to be determined. *Note:* A resistance temperature detector with its support and enclosing envelope, is
often called a resistance thermometer bulb. *See also:*
**electric thermometer (temperature meter); embedded
temperature detector.**
328

resistance thermometer (resistance temperature meter). An
electric thermometer that operates by measuring the
electric resistance of a resistor, the resistance of which is
a known function of its temperature. The temperature-
responsive element is usually called a **resistance temperature detector.** *Note:* The resistance thermometer is also
frequently used to designate the sensor and its enclosing
bulb alone, for example, as in platinum thermometer,
copper-constantan thermometer, etcetera. *See also:*
electric thermometer (temperature meter); instrument.
7, 328

resistance to ground (surge arresters). The ratio, at a point
in a grounding system, of the component of the voltage to
ground that is in phase with the ground current, to the
ground current that produces it.
244, 62

resistant (suffix) (1) (rotating machinery). Material or apparatus so constructed, protected or treated, that it will
not be injured readily when subjected to the specified

material or condition, for example, fire-resistant, moisture-resistant. *See also:* **asynchronous machine; direct-current commutating machine**

103, 225, 60, 63, 202, 206, 127, 27

(2) (transformers). So constructed, protected, or treated that the apparatus will not be damaged when subjected to the specified material or conditions for a specified time.

53

resistive attenuator (waveguide). A length of waveguide designed to introduce a transmission loss by the use of some dissipative material. *See also:* **absorptive attenuator (waveguide); waveguide.** 244, 179

resistive conductor. A conductor used primarily because it possesses the property of high electric resistance.

225, 206, 64

resistive coupling. The association of two or more circuits with one another by means of resistance mutual to the circuits. *See also:* **coupling.** 328

resistive distributor brush (electromagnetic compatibility). Resistive pickup brush in an ignition distributor cap. *See also:* **electromagnetic compatibility.** 220, 199

resistive ignition cable (electromagnetic compatibility). High-tension ignition cable, the core of which is made of resistive material. *See also:* **electromagnetic compatibility.**

220, 199

resistivity (material). A factor such that the conduction-current density is equal to the electric field in the material divided by the resistivity. 210

resistivity, volume. The reciprocal of volume conductivity, measured in siemens per centimeter, which is a steady-state parameter. 97

resistor. A device the primary purpose of which is to introduce resistance into an electric circuit. *Note:* A resistor as used in electric circuits for purposes of operation, protection, or control, commonly consists of an aggregation of units. Resistors as commonly supplied consist of wire, metal ribbon, cast metal, or carbon compounds supported by or imbedded in an insulating medium. The insulating medium may enclose and support the resistance material as in the case of the porcelain-tube type, or the insulation may be provided only at the points of support as in the case of heavy-duty ribbon or cast iron grids mounted in metal frames. 3, 210, 206

resistor, bias (semiconductor radiation detector). The resistor through which bias voltage is applied to a detector.

23, 118, 119

resistor furnace. A resistance furnace in which the heat is developed in a resistor that is not a part of the charge.

328

resistor-start motor. A single-phase induction motor with a main winding and an auxiliary winding connected in series with a resistor, with the auxiliary winding circuit opened for running operation. 63

resolution (1) (general communication). (A) The act of deriving from a sound, scene, or other form of intelligence, a series of discrete elements wherefrom the original may subsequently be synthesized. (B) The degree to which nearly equal values of a quantity can be discriminated. (C) The fineness of detail in a reproduced spatial pattern. (D) The degree to which a system or a device distinguishes fineness of detail in a spatial pattern. 328

(2) (storage tube). A measure of the quantity of information that may be written into and read out of a storage tube. *Notes:* (A) Resolution can be specified in terms of number of bits, spots, lines, or cycles. (B) Since the relative amplitude of the output may vary with the quantity of information, the true representation of the resolution of a tube is a curve of relative amplitude versus quantity. *See also:* **storage tube.** 174, 190

(3) (television). A measure of ability to delineate picture detail. *Note:* Resolution is usually expressed in terms of a number of lines discriminated on a televised test chart. For a number of lines N (normally alternate black and white lines) the width of each line is $1/N$ times the picture height. In television practice, where the raster has a $4/3$ aspect ratio, resolution, measured in either the horizontal or the vertical direction, is the number of test chart lines observable in a distance equal to the vertical dimension of the raster. 163

(4) (oscilloscopes). A measure of the total number of trace lines discernible along the coordinate axes, bounded by the extremities of the graticule or other specific limits. *See also:* **oscillograph.** 185

(5) (power supplies). The minimum voltage (or current) increment within which the power supply's output can be set using the panel controls. For continuous controls, the minimum increment is taken to be the voltage (or current) change caused by one degree of shaft rotation. *See also:* **power supply.** 228, 186

(6) (numerically controlled machines). The least interval between two adjacent discrete details that can be distinguished one from the other. 56, 207

(7) (transmission lines and waveguides). The degree to which nearly equal values of a quantity can be discriminated. 185

(8) (industrial control). The smallest distinguishable increment into which a quantity is divided in a device or system. *See also:* **control system, feedback.** 219, 206

(9) (digital delay line). The time spacing between peaks of the doublet. 81

(10) (acousto-optic deflector). The ratio of the angular swing to the minimum resolvable angular spread of one spot. The minimum spot size depends on the optical beam amplitude and phase distribution, as well as the criteria used to define minimum spot size. When the Rayleigh criteria is used for minimum spot size, resolution, N, is given by $N = 1/\alpha \ \tau\Delta f$, with $\alpha = 1$ rectangular beam, constant amplitude; 1.22 circular beam, constant amplitude; 1.34 circular beam, Gaussian amplitude. For operation in the scanning mode, the resolution will be reduced as the scan time approaches the access time. 72

(11) (non-real time spectrum analyzer). The ability to display adjacent responses discretely (hertz, hertz decibel down). The measure of resolution is the frequency separation of two responses which merge with a 3 decibel notch. (A) Equal Amplitude Signals. As a minimum instruments will be specified and controls labeled on the basis of two equal amplitude responses under the best operation conditions. (B) Unequal Amplitude Signals. The frequency difference between two signals of specified unequal amplitude when the notch formed between them is 3 decibel down from the smaller signal shall be termed Skirt Resolution. (C) Optimum Resolution. For every combination of frequency span and sweep time there exists a minimum obtainable value of resolution (R). This is the Optimum Resolution (Ro), which is defined theoretically as:

$$Ro = K \sqrt{\frac{\text{Frequency Span}}{\text{Sweep Time}}}$$

The factor "K" shall be unity unless otherwise specified. 68

(12)(pulse measurement). The smallest change in the pulse characteristic, property, or attribute being measured which can unambiguously be discerned or detected in a pulse measurement process. 15

(13)(electrothermic power meters). The smallest discrete or discernible change in power that can be measured. In this standard, resolution includes the estimated uncertainty with which the power changes can be determined on the readout scale. 47

(14)(gyro; accelerometer). The largest value of the minimum change in input, for inputs greater than the threshold, which produces a change in output equal to some specified percentage (at least 50 percent) of the change in output expected using the nominal scale factor. *See:* **input-output characteristics.** 46

re-solution (electrodeposition). The passing back into solution of metal already deposited on the cathode. *See also:* **electrodeposition.** 328

resolution bandwidth (non-real time spectrum analyzer). The width in hertz, of the spectrum analyzer's response to a (continuous wave) signal. This width is usually defined as the frequency difference at specified points on the response curve, such as the 3 or 6 decibel down points. The manufacturer will specify the decibel down points to be used. 68

resolution element (radar). A spatial and velocity region contributing echo energy that can be separated from that of adjacent regions by action of the antenna or receiving system; in conventional radar its dimensions are given by the beamwidths of the antenna, the transmitter pulse width, and the receiver bandwidth; its dimensions may be increased by the presence of spurious response regions (sidelobes), or decreased by use of specially coded transmissions and appropriate processing techniques (such as moving-target indicator). *See also:* **navigation.** 187, 13

resolution, energy (percent). *See:* **energy resolution (percent).** 23

resolution error. The error due to the inability of a transducer to respond to changes of a variable smaller than a given increment. *See also:* **electronic analog computer.** 9, 77

resolution of output adjustment (of any output parameter, voltage, frequency, etcetera) (inverters). The minimum increment of change in setting. *See also:* **self-commutated inverters.** 208

resolution response (square-wave) (television) (1) (camera device). The ratio of (A) the peak-to-peak signal amplitude given by a test pattern consisting of alternate black and white vertical* bars of equal widths corresponding to a specified line number to (B) the peak-to-peak signal amplitude given by large-area blacks and large-area whites having the same luminances as the black and white bars in the test pattern. *Note:* Horizontal scanning lines are

assumed. *See also:* **television.**

(2) (display device). The ratio of (A) the peak-to-peak luminance given by a square-wave test signal whose half-period corresponds to a specified line number to (B) the peak-to-peak luminances given by a test signal producing large-area blacks and large-area whites having the same amplitudes as the signal of (A). *Note:* In a display device, resolution response, at relatively high line numbers, is sometimes called **detail contrast.** 317, 178

resolution, structural (color-picture tubes). The resolution as limited by the size and shape of the screen elements. See also: **television.** 178, 190

resolution time (1) (counter tube or counting system) (radiation counters). The minimum time interval between two distinct events that will permit both to be counted. *See also:* **anticoincidence (radiation counters).** 190

(2) (sequential events recording systems). The minimum time interval between any two distinct events that will permit both to be recorded in sequence of occurrence. *See also:* **event.** 48, 58

resolution time correction (radiation counters). Correction to the observed counting rate to allow for the probability of the occurrence of events within the resolution time. *See also:* **anticoincidence (radiation counters).** 190

resolution wedge. A narrow-angle wedge-shaped pattern calibrated for the measurement of resolution and composed of alternate contrasting strips that gradually converge and taper individually to preserve equal widths along any given line at right angles to the axis of the wedge. *Note:* Alternate strips may be black and white of maximum contrast or strips of different colors. *See also:* **television.** 163, 178

resolver (analog computer). A device or computing element used for vector resolution or composition. *Notes:* (1) The **rectangular mode** is the mode of operation that produces a transformation from polar to rectangular coordinates or a rotation of rectangular coordinates. (2) The **polar mode** is the mode of operation that produces a transformation from rectangular to polar coordinates. 9

resolving time (1) (electronic navigation). The minimum time interval by which two events must be separated to be distinguishable in a navigation system by the time measurement alone. *See also:* **navigation.** 13, 187

(2) (radiation counters). The time from the start of a counted pulse to the instant a succeeding pulse can assume the minimum strength to be detected by the counting circuit. *Note:* This quantity pertains to the combination of tube and recording circuit. *See also:* **gas-filled radiation-counter tubes; ionizing radiation.** 190, 117, 96, 125

resonance (1) (circuits and systems). The enhancement of the response of a physical system to a periodic excitation when the excitation frequency is equal to a natural frequency of the system. 67

(2) (automatic control). Of a system or element, a condition evidenced by large oscillatory amplitude, which results when a small amplitude of a periodic input has a frequency approaching one of the natural frequencies of the driven system. *Note:* In a feedback control system, this occurs near the stability limit. 56

(3) (data transmission). A condition in a circuit containing inductance and capacitance in which the capacitive reactance is equal to the inductive reactance. This condition occurs at only one frequency in a circuit with fixed constants, and the circuit is said to be "tuned" to this frequency. The resonance frequency can be changed by varying the value of the capacitance or inductance of the circuit.   59

resonance bridge. A 4-arm alternating-current bridge in which both an inductor and a capacitor are present in one arm, the other three arms being (usually) nonreactive resistors, and the adjustment for balance includes the establishment of resonance for the applied frequency. *Note:* Normally used for the measurement of inductance, capacitance, or frequency. Two general types can be distinguished according as the inductor and capacitor are effectively in series or in parallel. *See also:* **bridge.**
                                                                328

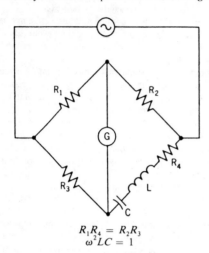

$$R_1R_4 = R_2R_3$$
$$\omega^2 LC = 1$$

Series resonance bridge

resonance charging (charging inductors) (1) (alternating current). The charging of the capacitance or a pulse-forming network to the peak value of voltage selected, in an oscillatory series RLC circuit, when supplied by an alternating voltage. *Note:* In order to provide a pulse train, the network capacitance is repetitively discharged at a time in the charging cycle when the current through the charging inductor is zero. At these times the voltage may be essentially:

$$\frac{\pi E_p}{2}, \ \pi E_p, \ \frac{3\pi E_p}{2}, \ \text{etc.}$$

The value chosen depends upon the pulse-repetition rate and the frequency of the alternating voltage. (E_p = peak value alternating-voltage supply.)   137

(2) (direct current). The charging of the capacitance of a pulse-forming network to the initial peak value of voltage in an oscillatory series RLC circuit, when supplied by a direct voltage. *Note:* In order to provide a pulse train, the network capacitance is repetitively discharged by a synchronous switch at the time when the current through the charging inductor is zero and the peak voltage to which the network capacitance is charged approaches two times the power-supply direct voltage.   137

resonance curve, carrier-current line trap (power-system communication). A graphical plot of the ohmic impedance of a carrier current line trap with respect to frequency at frequencies near resonance. *See also:* **power-line carrier.**
                                                                59

resonance frequency (resonant frequency*) (1) network. Any frequency at which resonance occurs. *Note:* For a given network, resonance frequencies may differ for different quantities, and almost always differ from natural frequencies. For example, in a simple series resistance-inductance-capacitance circuit there is a resonance frequency for current, a different resonance frequency for capacitor voltage, and a natural frequency differing from each of these. *See also:* **network analysis.**   210, 59

* Deprecated

(2) (crystal unit). The frequency for a particular mode of vibration to which, discounting dissipation, the effective impedance of the crystal unit is zero. *See also:* **crystal.**
                                                                328

resonance mode (laser-maser). A natural oscillation in a resonator characterized by a distribution of fields which have the same harmonic time dependence throughout the resonator.   363

resonant frequency (1) (antenna). A frequency at which the input impedance of an antenna is nonreactive.   111
(2) (network). *See:* **resonance frequency.**

resonant gap (microwave gas tubes). The small region in a resonant structure interior to the tube, where the electric field is concentrated. *See also:* **gas tubes.**   190, 125

resonant grounded. *See:* **ground-fault neutralizer grounded.**

resonant grounded system (arc-suppression coil) (surge arresters). A system grounded through a reactor, the reactance being of such value that during a single line-to-ground fault, the power-frequency inductive current passed by this reactor essentially neutralizes the power-frequency capacitive component of the ground-fault current. *Note:* With resonant grounding of a system, the net current in the fault is limited to such an extent that an arc fault in air would be self-extinguishing. *See also:* **ground.**
                                                            91, 244, 62

resonant iris (waveguide components). An iris designed to have equal capacitive and inductive susceptances at the resonant frequency.   166

resonant line oscillator. An oscillator in which the principal frequency-determining elements are one or more resonant transmission lines. *See also:* **oscillatory circuit.**
                                                            111, 211

resonant mode (1) (general). A component of the response of a linear device that is characterized by a certain field pattern, and that when not coupled to other modes is representable as a single-tuned circuit. *Note:* When modes are coupled together, the combined behavior is similar to that of the corresponding single-tuned circuits correspondingly coupled. *See also:* **waveguide.**   328

(2) (cylindrical cavities). When a metal cylinder is closed by two metal surfaces perpendicular to its axis a cylindrical cavity is formed. The resonant modes in this cavity are designated by adding a third subscript to indicate the number of half-waves along the axis of the cavity. When the cavity is a rectangular parallelepiped the axis of the cylinder from which the cavity is assumed to be made should be designated since there are three possible cylinders out of which the parallelepiped may be made. *See also:* **guides wave.**   343

resonant wavelengths (cylindrical cavities). Those given by $\lambda_r = 1[(1/\lambda_c)^2 + (l/2c)^2]^{1/2}$ where λ_c is the cutoff wavelength for the transmission mode along the axis, l is the number of half-period variations of the field along the axis, and c is the axial length of the cavity. *See also:* **guided wave.**   343

resonating (steady-state quantity or phasor). The maximizing or minimizing of the amplitude or other characteristic provided the maximum or minimum is of interest. *Notes:* (1) Unless otherwise specified, the quantity varied to obtain the maximum or minimum is to be assumed to be frequency. (2) Phase angle is an example of a quantity in which there is usually no interest in a maximum or a minimum. (3) In the case of amplification, transfer ratios, etcetera, the amplitude of the phasor is maximized or minimized; in the case of currents, voltages, charges, etcetera, it is customary to think of the amplitude of the steady-state simple sine-wave quantity as being maximized or minimized. *See also:* **network analysis.**   210

resonator (1) (general). A device, the primary purpose of which is to introduce resonance into a system. *See also:* **network analysis.**   210

(2) (circuits and systems). (A) A resonating system. (B) A device designed to operate in the vicinity of a natural frequency of that device. (C) (electrical circuit) An electrical network designed to present a given natural frequency at its terminal.   67

resonator grid (electron tubes). An electrode, connected to a resonator, that is traversed by an electron beam and that provides the coupling between the beam and the resonator. *See also:* **velocity-modulated tube.**   244, 190

resonator mode (oscillator). A condition of operation corresponding to a particular field configuration for which the electron stream introduces a negative conductance into the coupled circuit. *See also:* **oscillatory circuit.**   190, 125

resonator, waveguide (waveguide components). A waveguide or transmission line structure which can store oscillating electromagnetic energy for time periods that are long compared with the period of the resonant frequency, at or near the resonant frequency.   166

responder beacon (electronic navigation). *See:* **transponder.**

response (device or system). A quantitative expression of the output as a function of the input under conditions that must be explicitly stated. *Note:* The response characteristic, often presented graphically, gives the response as a function of some independent variable such as frequency or direction.   197, 79

response, acceleration-forced (automatic control). The total (transient plus steady-state) time response resulting from a sudden increase in the rate of the rate of change of input from zero to some finite value.   56

response, forced (automatic control). A time response which is produced by a stimulus external to the system or element under consideration. *Note:* The response may be described in terms of the causal variable. *See:* **response, acceleration-forced; response, impulse-forced.**   56

response function (linear passive networks). The ratio of

response to excitation, both expressed as functions of the complex frequency, $s = \sigma + j\omega$. *Note:* The response function is the Laplace transform of the response due to unit impulse excitation. *See also:* **linear passive networks.**   238

response, Gaussian (amplifiers) (oscilloscopes). A particular frequency response characteristic following the curve $y(f) = e^{-af^2}$. Typically, the frequency response approached by an amplifier having good transient response characteristics.   106

response, impulse-forced (automatic control). The total (transient plus steady-state) time response resulting from an impulse at the input. *Syn:* **impulse response.**   56

response, indicial (process control). The output of a system or element, expressed as a function of time, when forced from initial equilibrium by a unit-step input. *Note:* In the time domain, it is the graphic statement of the characteristic of a system or element analogous to the frequency-response characteristic of the transfer function.   56

response, instrument (1) (dynamic). The behavior of the instrument output as a function of the measured signal, both with respect to time. *See:* **damping characteristic; frequency response; ramp response; step response.** *See also:* **accuracy rating (instrument).**   295

(2) (forced). The total steady-state plus transient time response resulting from an external input.   329

response, ramp-forced automatic control. The total (transient plus steady-state) time response resulting from a sudden increase in the rate of change of input from zero to some finite value. *Syn.* **ramp response.**   56

response, sinusoidal (sine-force). The forced response due to a sinusoidal stimulus. *Note:* A set of steady-state sinusoidal responses for sinusoidal inputs at different frequencies is called the frequency-response characteristic. *See also:* **control system, feedback.**   329

response spectrum (nuclear power generating stations) (seismic qualification of class 1E equipment). A plot of the maximum response of single-degree-of-freedom bodies, at a damping value expressed as a percent of critical damping of different natural frequencies, when these bodies are rigidly mounted on the surface of interest (that is, on the ground for the ground response spectrum or on the floor for the floor response spectrum) when that surface is subjected to a given earthquake's motion as modified by any intervening structures.   28, 31

response, steady-state (system or element). The part of the time response remaining after transients have expired. *Note:* The term **steady-state** may also be applied to any of the forced-response terms; for example **steady-state sinusoidal response.** *See also:* **control system, feedback; response, sinusoidal.**   329

response, step-forced (automatic control). The total (transient plus steady-state) time response resulting from a sudden change from one constant level of input to another.   56

response time (1) (magnetic amplifier). The time (preferably in seconds; may also be in cycles of supply frequency) required for the output quantity to change by some agreed-upon percentage of the differential output quantity in response to a step change in control signal equal to the differential control signal. *Note:* The initial and final output quantities shall correspond to the test output quantities. The response time shall be the maximum obtained including differences arising from increasing or decreasing output quantity or time phase of signal application. 171

(2) (turn ON response time) (control devices). The time required for the output voltage to change from rated OFF voltage to rated ON voltage in response to a step change in control signal equal to 120 percent of the differential trip signal. *Note:* The absolute magnitude of the initial signal condition shall be the absolute magnitude of the trip OFF control signal plus 10 percent of the differential trip signal. 171

(3) (turn OFF response time) (control devices). The time required for the output voltage to change from rated ON voltage to rated OFF voltage in response to a step change in control signal equal to 120 percent of the differential trip signal. *Note:* The absolute magnitude of the initial signal condition shall be the absolute magnitude of the trip ON control signal minus 10 percent of the differential trip signal. 171

(4) (electrically tuned oscillator). The time following a change in the input to the tuning element required for a characteristic to reach a predetermined range of values within which it remains. *See also:* **tunable microwave oscillators.** 174, 190

(5) (instrument). The time required after an abrupt change has occurred in the measured quantity to a new constant value until the pointer, or indicating means, has first come to apparent rest in its new position. *Note:* (A) Since, in some instruments, the response time depends on the magnitude of the deflection, a value corresponding to an initial deflection from zero to end scale is used in determining response time for rating purposes. (B) The pointer is at apparent rest when it remains within a range on either side of its final position equal to one-half the accuracy rating, when determined as specified above. *See also:* **accuracy rating (instrument); moving element (instrument).** 294, 280

(6) (bolometric power meter). The time required for the bolometric or electrothermic power indication to reach 90 percent of its final value after a fixed amount of radio-frequency power is applied to the bolometer or electrothermic unit. 185, 47

(7) (thermal converter). The time required for the output electromotive force to come to its new value after an abrupt change has occurred in the input quantity (current, voltage, or power) to a new constant value. *Notes:* (A) Since, in some thermal converters, the response time depends upon the magnitude and direction of the change, the value obtained for an abrupt change from zero to rated input quantity is used for rating purposes. (B) The output electromotive force shall be considered to have come to its new value when all but 1 percent of the change in electromotive force has been indicated. *See also:* **thermal converter.** 280

(8) (industrial control). The time required, following the initiation of a specified stimulus to a system, for an output going in the direction of necessary corrective action to first reach a specified value. *Note:* The response time is expressed in seconds. *See also:* **control system, feedback.** 219, 206

(9) (electrical conversion). The elapsed time from the initiation of a transient until the output has recovered to 63 percent of its maximum excursion. See accompanying figures. *See also:* **electrical conversion equipment; recovery time.** 186

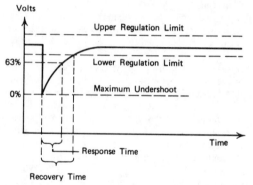

Response and recovery time for a critically damped circuit.

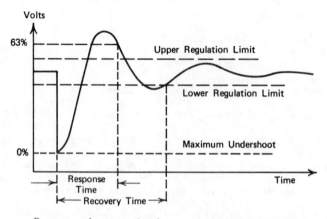

Response and recovery time for an underdamped circuit

(10) (arc-welding apparatus). The time required to attain conditions within a specified amount of their final value in an automatically regulated welding circuit after a definitely specified disturbance has been initiated. *See also:* **photoelectric control.**   264

(11) (photoelectric lighting control) (industrial control). The time required for operation following an abrupt change in illumination from 50 percent above to 50 percent below the minimum illumination sensitivity. *See also:* **photoelectric control.**   219, 200

(12) (control system or element) (time of response) (control system, feedback). The time required for an output to make the change from an initial value to a large specified percentage of the steady state, either before overshoot or in the absence of overshoot. *Note:* If the term is unqualified, time of response of a first-order system to a unit-step stimulus is generally understood; otherwise the pattern and magnitude of the stimulus should be specified. Usual percentages are 90, 95, or 99.   56

(13) (sequential events recording systems). The time interval between receiving a finite input status change and the recognition by the system of the status change. The time interval is usually expressed in milliseconds.   48, 58

(14) (nuclear power generating stations). The time interval from a step change in the input concentration at the instrument inlet to a reading of 90 percent (nominally equivalent to 2.2 time constants) of the ultimate recorded output.   31

(15) (data transmission). The amount of time elapsed between generation of an inquiry at a data communications terminal and receipt of a response at that same terminal. Response time, thus defined includes: (A) transmission time to the computer; (B) processing time at the computer, including access time to obtain any file records needed to answer the inquiry; and (C) transmission time back to the terminal.   59

(16) (temperature measurement). The time required for the indication of a thermometer, which has been subjected to an essentially instantaneous change in temperature, to traverse 63 percent of the temperature interval involved. Following such a temperature change the indication of the thermometer may be expected to traverse 99 percent of the temperature interval in a period ranging from 5 to 8 time constants so defined, depending on the details of its construction.   7

response time, ramp-forced (automatic control). The time interval by which an output lags an input, when both are varying at a constant rate.   56

responsor (electronic navigation). The receiving component of an interrogator. *Note:* It is, therefore, a receiver intended to receive and interpret the signals from a transponder. *See also:* **radio transmission.**   13, 187

rest and de-energized (rotating machinery). The complete absence of all movement and of all electric or mechanical supply. *See also:* **asynchronous machine; direct-current commutating machine.**   63

restart (computing systems). To reestablish the execution of a routine, using the data recorded at a checkpoint.   255, 77, 54

resting potential (biological). The voltage existing between the two sides of a living membrane or interface in the ab-

sence of stimulation.   192

restorable fire detector (fire protection devices). A device whose sensing element is not ordinarily destroyed by the process of detecting a fire. Restoration may be manual or automatic.   71

restoring force gradient (direct-acting recording instrument). The rate of change, with respect to the displacement, of the resultant of the electric, or of the electric and mechanical, forces tending to restore the marking device to any position of equilibrium when displaced from that position. *Note:* The force gradient may be constant throughout the entire travel of the marking device or it may vary greatly over this travel, depending upon the operating principles and the details of construction. *See also:* **accuracy rating (instrument).**   328

restoring torque gradient (instrument). The rate of change, with respect to the deflection, of the resultant of the electric, or electric and mechanical, torques tending to restore the moving element to any position of equilibrium when displaced from that position. *See also:* **accuracy rating (instrument).**   328

restraint relay (power switchgear). A relay so constructed that its operation in response to one input is restrained by a second input.   127, 103

restricted radiation frequencies for industrial, scientific, and medical equipment. Center of a band of frequencies assigned to industrial, scientific, and medical equipment either nationally or internationally and for which a power limit is specified. *See also:* **electromagnetic compatibility.**   220, 199

restricted-service tone (telephone switching systems). A class-of-service tone that indicates to an operator that certain services are denied the caller.   55

restrike (switching device). A resumption of current between the contacts during an opening operation after an interval of zero current of ¼ cycle at normal frequency or longer.   103, 202, 27

restriking voltage (1) (gas tube). The anode voltage at which the discharge recommences when the supply voltage is increasing before substantial deionization has occurred. *See also:* **gas-filled rectifier.**   244, 190

(2) (industrial control). The voltage that appears across the terminals of a switching device immediately after the breaking of the circuit. *Note:* This voltage may be considered as composed of two components. One, which subsists in steady-state conditions, is direct current or alternating current at service frequency, according to the system. The other is a transient component that may be oscillatory (single or multifrequency) or nonoscillatory (for example, exponential) or a combination of these depending on the characteristics of the circuit and the switching device. *See also:* **switch.**   244, 206

retained image (image burn). A change produced in or on the target that remains for a large number of frames after the removal of a previously stationary light image and that yields a spurious electric signal corresponding to that light image. See also: **camera tube.**   178, 190, 125

retainer. *See:* **separator (storage cell).**

retaining ring (rotating machinery) (1) (steel). A mechanical structure surrounding parts of a rotor to restrain radial movement due to centrifugal action. *See also:* **rotor (rotating machinery).**   63

(2) (insulation). The insulation forming a dielectric and mechanical barrier between the rotor end windings and the high-strength steel retaining ring. *See also:* **rotor (rotating machinery).** 63

retaining ring liner (rotating machinery). Insulating ring between the end winding and the metallic ring which secures the coil ends against centrifical force. 63

retardation (deceleration) (industrial control). The operation of reducing the motor speed from a high level to a lower level or zero. *See also:* **electric drive.** 206

retardation coil. *See:* **inductor.**

retardation test (rotating machinery). A test in which the losses in a machine are deduced from the rate of deceleration of the machine when only these losses are present. *See also:* **asynchronous machine; direct-current commutating machine.** 63

retard coil*. *See:* **inductor.**

* Deprecated

retarding-field (positive-grid) oscillator. An oscillator employing an electron tube in which the electrons oscillate back and forth through a grid maintained positive with respect to the cathode and the plate. The frequency depends on the electron-transit time and may also be a function of the associated circuit parameters. The field in the region of the grid exerts a retarding effect that draws electrons back after passing through it in either direction. Barkhausen-Kurz and Gill-Morell oscillators are examples. *See also:* **oscillatory circuit.** 111

retarding magnet. A magnet used for the purpose of limiting the speed of the rotor of a motor-type meter to a value proportional to the quantity being integrated. *See also:* **watthour meter; braking magnet; drag magnet.** 328

retard transmitter. A transmitter in which a delay period is introduced between the time of actuation and the time of transmission. *See also:* **protective signaling.** 328

retention time, maximum (storage tubes). The maximum time between writing into a storage tube and obtaining an acceptable output by reading. *See also:* **storage tube.** 174, 190, 125

retentivity (magnetic material). That property that is measured by its maximum residual induction. *Note:* The maximum residual induction is usually associated with a hysteresis loop that reaches saturation, but in special cases this is not so. 210

retina (laser-maser). That sensory membrane which receives the incident image formed by the cornea and lens of the human eye. The retina lines the inside portion of the eye. 363

retrace (oscillography). Return of the spot on the cathode-ray tube to its starting point after a sweep; also that portion of the sweep waveform that returns the spot to its starting point. *See also:* **oscillograph.** 185

retrace blanking. *See:* **blanking.**

retrace interval (television) (return interval¢). The interval corresponding to the direction of sweep not used for delineation. 163

* Deprecated

retrace line. The line traced by the electron beam in a cathode-ray tube in going from the end of one line or field to the start of the next line or field. 339

retrieval. *See:* **information retrieval.**

retrodirective antenna. An antenna whose back-scattering cross section is comparable to the product of its directivity and its area projected in the direction toward the source, and is relatively independent of the source direction. 111

retrograde orbit (communication satellite). An inclined orbit with an inclination between 90° and 180°. 74

retro-reflector (reflex reflector). A device designed to reflect light in a direction close to that at which it is incident, whatever the angle of incidence. *See also:* **headlamp.** 167

return. *See:* **carriage return.**

return-beam mode (camera tube). A mode of operation in which the output current is derived, usually through an electron multiplier, from that portion of the scanning beam not accepted by the target. *See also:* **camera tubes.** 190

return difference (network analysis). One minus the loop transmittance. *See also:* **linear signal flow graphs.** 282

return interval* (television). *See:* **retrace interval.**

* Deprecated

return loss. (1) At a discontinuity in a transmission system, the difference between the power incident upon the discontinuity and the power reflected from the discontinuity. (2) The ratio in decibels of the power incident upon the discontinuity to the power reflected from the discontinuity. *Note:* This ratio is also the square of the reciprocal of the magnitude of the reflection coefficient. Return loss = $20 \log_{10}(1/\Gamma)$. *See also:* **transmission loss; waveguide.** (3) More broadly, the return loss is a measure of the dissimilarity between two impedances, being equal to the number of decibels that corresponds to the scalar value of the reciprocal of the reflection coefficient, and hence being expressed by the formula:

$$20 \log_{10} \left| \frac{Z_1 + Z_2}{Z_1 - Z_2} \right| \text{ decibel}$$

where Z_1 and Z_2 are the two impedances. *See also:* **transmission loss.** 267, 179, 185, 59

return stroke (lightning). The luminescent, high-current discharge that is initiated after the stepped leader and pilot streamer have established a highly ionized path between charge centers. *See also:* **direct-stroke protection (lightning).** 64

return swing (last transition ringing) (A_{RS}) (high- and low-power pulse transformers). The maximum amount by which the instantaneous pulse value is above the zero axis in the region following the backswing. It is expressed in amplitude units or in percent of A_M. *See also:* **input pulse shape.** 32, 33

return-to-zero (RZ) code (power-system communication). A code form having two information states termed ZERO and ONE and having a third state or an at-rest condition to which the signal returns during each period. *See also:* **digital.** 59

return trace (television). The path of the scanning spot during the return interval. *See also:* **oscillograph; television.** 163, 178, 185

reverberant sound. Sound that has arrived at a given location by a multiplicity of indirect paths as opposed to a single direct path. *Notes:* (1) Reverberation results from multiple reflections of sound energy contained within an enclosed space. (2) Reverberation results from scattering from a large number of inhomogeneities in the medium or reflection from bounding surfaces. (3) Reverberant sound can be produced by a device that introduces time delays that approximate a multiplicity of reflections. *See:* **echo.**   176

reverbation. The presence of reverberant sound.   176

reverberation chamber. An enclosure especially designed to have a long reverberation time and to produce a sound field as diffuse as possible. *See also:* **anechoic chamber.**
                                                                           176

reverberation room. *See:* **reverberation chamber.**

reverberation time. The time required for the mean-square sound pressure level, or electric equivalent, originally in a steady state, to decrease 60 decibels after the source output is stopped.   176

reverberation-time meter. An instrument for measuring the reverberation time of an enclosure. *See also:* **instrument.**   328

reversal (storage battery) (storage cell). A change in normal polarity of the cell or battery. *See also:* **charge.**   328

reverse-battery signaling (telephone switching systems). A method of loop signaling in which the direction of current in the loop is changed to convey on-hook and off-hook signals.   55

reverse-battery supervision (telephone switching systems). A form of supervision employing reverse-battery signaling.
                                                                            55

reverse bias (reverse voltage) (V_R) (light emitting diodes). The bias voltage that is applied to an LED (light emitting diode) in the reverse direction.   162

reverse-blocking current (reverse-blocking thyristor). The reverse current when the thyristor is in the reverse-blocking state. *See also:* **principal current.**
                                                          243, 66, 208, 191

reverse-blocking diode-thyristor. A two-terminal thyristor that switches only for positive anode-to-cathode voltages and exhibits a reverse-blocking state for negative anode-to-cathode voltages, *See also:* **thyristor.**
                                                          243, 66, 204, 191

reverse-blocking impedance (reverse-blocking thyristor). The differential impedance between the two terminals through which the principal current flows, when the thyristor is in the reverse-blocking state at a stated operating point. *See also:* **principal voltage-current characteristic (principal characteristic).**   243, 66, 191

reverse-blocking state (reverse-blocking thyristor). The condition of a reverse-blocking thyristor corresponding to the portion of the anode-to-cathode voltage-current characteristic for reverse currents of lower magnitude than the reverse-breakdown current. *See also:* **principal voltage-current characteristic (principal characteristic).**
                                                          243, 66, 204, 191

reverse-blocking triode-thyristor. A three-terminal thyristor that switches only for positive anode-to-cathode voltages and exhibits a reverse-blocking state for negative anode-to-cathode voltages. *See also:* **thyristor.**
                                                            243, 66, 24, 191

reverse-breakdown current (reverse-blocking thyristor). The principal current at the reverse-breakdown voltage. *See also:* **principal current.**   243, 66, 24, 191

reverse-breakdown voltage (reverse-blocking thyristor). The value of negative anode-to-cathode voltage at which the differential resistance between the anode and cathode terminals changes from a high value to a substantially lower value. *See also:* **principal voltage-current characteristic (principal characteristic).**   243, 66, 24, 191

reverse-conducting diode-thyristor. A two-terminal thyristor that switches only for positive anode-to-cathode voltages and conducts large currents at negative anode-to-cathode voltages comparable in magnitude to the ON-state voltages. *See also:* **thyristor.**   243, 66, 24, 191

reverse-conducting triode-thyristor. A three-terminal thyristor that switches only for positive anode-to-cathode voltages and conducts large currents at negative anode-to-cathode voltages comparable in magnitude to the ON-state voltages. *See also:* **thyristor.**   243, 66, 24, 191

reverse contact. A contact that is closed when the operating unit is in the reverse position.   328

reverse current (1) (general). Current that flows upon application of reverse voltage.   328

(2) (reverse-blocking or reverse-conducting thyristor). The principal current for negative anode-to-cathode voltage. *See also:* **principal current.**   243, 66, 24, 191

(3) (metallic rectifier). The current that flows through a metallic rectifier cell in the reverse direction. *See also:* **rectification.**   328

(4) (semiconductor rectifier). The total current that flows through a semiconductor rectifier device in the reverse direction. *See also:* **rectification.**   237, 245, 210, 66

(5) (capacitive). The current that flows through a semiconductor rectifier cell during the blocking period due to the capacitance of the cell.   237, 66

(6) (resistive) (semiconductor rectifier). The inphase current that flows through a semiconductor rectifier cell during the blocking period exclusive of the reverse recovery current. *See also:* **rectification.**   237, 66

(7) (light emitting diodes) (I_R). The current that flows through a semiconductor junction in the reverse direction.
                                                                            162

reverse-current cleaning. *See:* **anode cleaning.**

reverse-current cutout. A magnetically operated direct-current device that operates to close an electric circuit when a predetermined voltage condition exists and operates to open an electric circuit when more than a predetermined current flows through it in the reverse direction. (1) **Fixed-voltage type:** A reverse-current cutout that closes an electric circuit whenever the voltage at the cutout terminal exceeds a predetermined value and is of the correct polarity. It opens the circuit when more than a predetermined current flows through it in the reverse direction. (2) **Differential-voltage type:** A reverse-current cutout that closes an electric circuit when a predetermined differential voltage appears at the cutout terminal, provided this voltage is of the correct polarity and exceeds a predetermined value. It opens the circuit when more than a predetermined current flows through it in the reverse direction.   328

reverse-current relay (power switchgear). A relay that operates on a current flow in a direct-current circuit in a direction opposite to a predetermined reference direction.
                                                                      103, 127

reverse-current release (reverse-current trip) (direct current).

A release that operates upon reversal in the main circuit from a predetermined direction.   103, 202

reverse-current tripping; reverse-power tripping. *See:* reverse-current release (reverse-current trip).

reverse direction (1) (semiconductor rectifier diode). The direction of higher resistance to steady-state direct-current; that is, from the cathode to the anode.   66
(2) (light emitting diodes). The direction of higher resistance to steady direct-current flow through the device; that is, from the cathode to the anode.   162

reverse emission (back emission) (vacuum tubes). The inverse electrode current from an anode during that part of a cycle in which the anode is negative with respect to the cathode. *See also:* **electron emission.**   190

reverse gate current (thyristor). The gate current when the junction between the gate region and the adjacent anode or cathode region is reverse biased. *See also:* **principal current.**   243, 66, 208, 191

reverse gate voltage (thyristor). The voltage between the gate terminal and the terminal of an adjacent region resulting from reverse gate current. *See also:* **principal voltage-current characteristic (principal characteristic).**   243, 66, 208, 191

reverse period (rectifier circuit) (rectifier circuit element). The part of an alternating-voltage cycle during which the current flows in the reverse direction. See note under **blocking period.** *See also:* **rectifier circuit element.**   237, 66, 208

reverse position (device). The opposite of the normal position.   328

reverse power dissipation (semiconductor). The power dissipation resulting from reverse current.   66

reverse power loss (semiconductor rectifier). The power loss resulting from the flow of reverse current. *See also:* **rectification; semiconductor rectifier stack.**   237, 66, 208

reverser. A switching device for interchanging electric circuits to reverse the direction of motor rotation. *See also:* **multiple-unit control.**   328

reverse recovery current (semiconductor rectifier). The transient component of reverse current associated with a change from ON state conduction to reserve voltage. *See also:* **rectification.**   237, 66

reverse recovery time (reverse-blocking thyristor or semiconductor diode). The time required for the principal current or voltage to recover to a specified value after instantaneous switching from an ON state to a reverse-voltage or current. *See also:* **principal voltage-current characteristic (principal characteristic); rectification.**   237, 243, 66, 208, 191

reverse resistance (metallic rectifier). The resistance measured at a specified reverse voltage or a specified reverse current. *See also:* **rectification.**   328

reverse voltage (1) (rectifier). Voltage of that polarity that produces the smaller current. *See also:* **principal voltage-current characteristic; rectification.**   328
(2) (semiconductor) (reverse blocking or reverse conducting thyristor). A negative anode-to-cathode voltage.   66

reverse voltage dividers (rectifier). Devices employed to assure satisfactory division of reverse voltage among series-connected semiconductor rectifier diodes. Transformers, bleeder resistors, capacitors, or combinations of these may be employed. *See also:* **power rectifier.**   208

reverse wave. *See:* **reflected wave.**

reversibility (Hall generator). The ratio of the change in absolute magnitude of the Hall voltage to the mean absolute magnitude of the Hall voltage, when the control current is kept constant and the magnetic field is changed from a given magnitude of one polarity to the same magnitude of the opposite polarity.   107

reversible capacitance (nonlinear capacitor). The limit, as the amplitude of an applied sinusoidal capacitor voltage approaches zero, of the ratio of the amplitude of the resulting in-phase fundamental-frequency component of transferred charge to the amplitude of the applied voltage, for a given constant bias voltage superimposed on the sinusoidal voltage. *See also:* **nonlinear capacitor.**   191

reversible-capacitance characteristic (nonlinear capacitor). The function relating the reversible capacitance to the bias voltage. *See also:* **nonlinear capacitor.**   191

reversible motor. A motor whose direction of rotation can be selected by change in electric connections or by mechanical means but the motor will run in the selected direction only if it is at a standstill or rotating below a particular speed when the change is initiated. *See also:* **asynchronous machine; direct-current commutating machine.**   328

reversible permeability. The limit of the incremental permeability as the incremental change in magnetizing force approaches zero. *Note:* In anisotropic media, reversible permeability becomes a matrix.   210

reversible permittivity (ferroelectric material). The change in displacement per unit field when a very small relatively high-frequency alternating signal is applied to a ferroelectric at any point of a hysteresis loop. *See also:* **ferroelectric domain.**   247

reversible potential. *See:* **equilibrium potential.**

reversible power converter (1) (static power converters). An equipment containing thyristor converter assemblies connected in such a way that energy transfer is possible from the alternating-current side to the direct-current side and from the direct-current side to the alternating-current side with or without reversing the current in the direct-current circuit. *See also:* **power rectifier.**   208
(2) (thyristor converter). A converter in which the transfer of energy is possible both from the ac side to the dc side and vice versa.   121

reversible process. An electrochemical reaction that takes place reversibly at the equilibrium electrode potential. *See also:* **electrochemistry.**   328

reversing (industrial control). The control function of changing motor rotation from one direction to the opposite direction. *See also:* **electric drive.**   206

reversing motor. One the torque and hence direction of rotation of which can be reversed by change in electric connections or by other means. These means may be initiated while the motor is running at full speed, upon which the motor will come to a stop, reverse, and attain full speed in the opposite direction. *See also:* **asynchronous machine; direct-current commutating machine.**   328

reversing starter (industrial control). An electric controller for accelerating a motor from rest to normal speed in either direction of rotation. *See also:* **electric controller; starter.**   206

reversing switch. A switch intended to reverse the connections of one part of a circuit. *See also:* **switch.** 206

reverting call (telephone switching systems). A call between two stations on the same party line. 55

reverting-call tone (telephone switching systems). A tone that indicates to a calling customer that the called party is on the same line. 55

revertive pulsing (telephone switching systems). A means of pulsing for controlling distant selections whereby the near end receives signals from the far end. 55

rewind (test, measurement and diagnostic equipment). To return a tape to its beginning or a passed location. 54

rewrite. (1) To write again. (2) In a destructive-read storage device, to return the data to the state it had prior to reading. *See also:* **regenerate.** 235, 210

RF. *See:* **radio frequency.**

RF (radio frequency) link (test, measurement and diagnostic equipment). A radio frequency channel or channels used to connect the unit under test with the testing device. 54

rheobase (medical electronics). The intensity of the steady cathodal current of adequate duration that when suddenly applied just suffices to excite a tissue. 192

rheostat. An adjustable resistor so constructed that its resistance may be changed without opening the circuit in which it may be connected. 210, 206

rheostatic braking (industrial control). A form of dynamic braking in which electric energy generated by the traction motors is controlled and dissipated by means of a resistor whose value of resistance may be varied. *See also:* **dynamic braking; electric drive.** 206

rheostatic control (elevators). A system of control that is accomplished by varying resistance and/or reactance in the armature and/or field circuit of the driving-machine motor. *See also:* **control (elevators).** 328

rheostatic-type voltage regulator (rotating machinery). A regulator that accomplishes the regulating function by mechanically varying a resistance. 63

rheostat loss (synchronous machine). The I^2R loss in the rheostat controlling the field current. 298, 63

rheostriction. *See:* **pinch effect.**

(RHI) display. Range-height indicator. *See:* **E-Display.** 13

rho rho (electronic navigation). A generic term referring to navigation systems based on the measurement of only two distances for determination of position. *See also:* **navigation.** 187, 13

rho theta (electronic navigation). A generic term referring to polar-coordinate navigation systems for determination of position of a vehicle through measurement of distance and direction. *See also:* **navigation.** 187, 13

rhombic antenna. An antenna composed of long-wire radiators comprising the sides of a rhombus. The antenna usually is terminated in a resistance. *Note:* The length of the sides of the rhombus, the angle between the sides, the elevation above ground, and the value of the termination resistance are proportioned to give the desired radiation properties. 111

rhumbatron (electron tube) (microwave tube). A resonator, usually in the form of a torus. *See also:* **velocity-modulated tube.** 244, 190

rhythmic light. A light that when observed from a fixed point has a luminous intensity that changes periodically. 167

ribbon microphone. A moving-conductor microphone in which the moving conductor is in the form of a ribbon that is directly driven by the sound waves. *See also:* **microphone.** 328

ribbon transducer. A moving-conductor transducer in which the movable conductor is in the form of a thin ribbon. 176

Richardson-Dushmann equation (thermionics). An equation representing the saturation current of a metallic thermionic cathode in the saturation-current state:

$$ J = A_0 \left(1 - r\right) T^2 \exp\left(-\frac{b}{T}\right) $$

where

J = density of the saturation current

T = absolute temperature

A_0 = universal constant equal to 120 amperes per centimeter^2 kelvin^2

b = absolute temperature equivalent to the work function

r = reflection coefficient, which allows for the irregularities of the surface.

See: **work function.** *See also:* **electron emission.** 244, 190

Richardson effect. *See:* **thermionic emission.**

ridge waveguide. A waveguide with interior projections extending along the length and in contact with the boundary wall. *See also:* **waveguide.** 179

Rieke diagram (oscillator performance). A chart showing contours of constant power output and constant frequency drawn on a polar diagram whose coordinates represent the components of the complex reflection coefficient at the oscillator load. *See:* **load impedance diagram.** *See also:* **oscillatory circuit.** 190, 125

RIF (electromagnetic compatibility). *See:* **radio-influence field.**

right-handed (clockwise) polarized wave (radio wave propagation). An elliptically polarized electromagnetic wave in which the rotation of the electric field vector with time is clockwise for a stationary observer looking in the direction of the normal wave. *Note:* For an observer looking from a receiver toward the apparent source of the wave, the direction of rotation is reversed. *See also:* **radiation; radio wave propagation.** 180, 146

right-hand (or left-hand) polarization (1) (plane wave). The polarization of a plane wave when the electric field vector is right-hand (or left-hand) polarized, taking as the reference the direction of propagation. 111

(2) field vector. A polarization such that the sense of rotation of the extremity of the field vector with time is in the direction of the fingers of the right hand (or left hand) when the thumb of that hand is in some reference direction perpendicular to the plane of polarization. *Note:* For a linearly polarized field vector the sense of polarization is not defined. 111

rigid metal conduit. A raceway specially constructed for the purpose of the pulling in or the withdrawing of wires or cables after the conduit is in place and made of metal pipe of standard weight and thickness permitting the cutting of standard threads. *See also:* **raceway.** 328

rigid tower. A tower that depends only upon its own structural members to withstand the load that may be placed upon it. *See:* **angle tower; dead-end-tower; flexible tower.** *See also:* **tower.** 64

rim (spider rim) (rotating machinery). The outermost part of a spider. A rotating yoke. *See also:* **rotor (rotating machinery).** 63

ring (plug). A ring-shaped contacting part, usually placed in back of the tip but insulated therefrom. 328

ring around (secondary radar). (1) The undesired triggering of a transponder by its own transmitter. (2) The triggering of a transponder at all bearings causing a ring-type presentation on a PPI (plan-position indicator). *See also:* **navigation.** 13, 187

ring array (antennas). An arrangement of elements with centers on a circle. 111

ringback signal (telephone switching systems). A signal initiated by an operator at the called end of an established connection to recall the originating operator. 55

ringback tone (telephone switching systems). A tone that indicates to a caller that a ringing signal is being applied to a destination outlet. 55

ring circuit (waveguide practice). A hybrid T having the physical configuration of a ring with radial branches. *See also:* **waveguide.** 328

ring counter. A re-entrant multistable circuit consisting of any number of stages arranged in a circle so that a unique condition is present in one stage, and each input pulse causes this condition to transfer one unit around the circle. *See also:* **trigger circuit.** 328

ringdown signaling (telephone switching systems). A method of alerting an operator in which ringing is sent over the line to operate a device or circuit to produce a steady indication (normally a visual signal). 55

ringer **(station ringer).** *See:* **telephone ringer.**

ringer box. *See:* **bell box.**

ring feeder. *See:* **loop-service feeder.**

ring head (electroacoustics). A magnetic head in which the magnetic material forms an enclosure with one or more air gaps. The magnetic recording medium bridges one of these gaps and is in contact with or in close proximity to the pole pieces on one side only. 176

ring heater, induction. *See:* **induction ring heater.**

ringing (1) (telephone switching systems). An alternating or pulsing current primarily intended to produce a signal at a station or switchboard. 55

(2) (high- and low-power pulse transformers) (first transition ringing) (A_{RI}). The maximum amount by which the instantaneous pulse value deviates from the straight-line segment fitted to the top of the pulse in determining A_M in the pulse top region following rolloff, or overshoot, or both. It is expressed in amplitude units or in percent of A_M. *See also:* **input pulse shape.** 32, 33

(3) (facsimile). *See:* **facsimile transient.**

(4) (pulse terms). A distortion in the form of a superimposed damped oscillatory waveform which, when present, usually follows a major transition. *See:* Note in **preshoot** entry. 254

ringing cycle (telephone switching systems). A recurring sequence made up of ringing signals and the intervals between them. 55

ringing key. A key whose operation sends ringing current

over the circuit to which the key is connected. 328

ring oscillator. An arrangement of two or more pairs of tubes operating as push-pull oscillators around a ring, usually with alternate successive pairs of grids and plates connected to tank circuits. Adjacent tubes around the ring operate in phase opposition. The load is supplied by coupling to the plate circuits. *See also:* **oscillatory circuit.** 111, 240

ring time (radar). The time during which the indicated output of an echo box remains above a specified signal-to-noise level. The ring time is used in measuring the performance of radar equipment. *See also:* **navigation.** 13, 187

ripple (1) (general). The alternating-current component from a direct-current power supply arising from sources within the power supply. *Notes:* (A) Unless specified separately, ripple includes unclassified noise. (B) In In electrical-conversion technology, ripple is expressed in peak, peak-to-peak, root-mean-square volts, or as percent root-mean-square. (C) Unless otherwise specified, **percent ripple** is the ratio of the root-mean-square value of the ripple voltage to the absolute value of the total voltage, expressed in percent. *See:* **percent ripple.** 210, 228, 186

(2) (dielectric tests). The periodic deviation from the arithmetic-mean value. The magnitude of the ripple is defined as half the difference between the maximum and minimum values. *See also:* **ripple factor.** 150

(3) (circuits and systems). *See:* **pass-band ripple; stop-band ripple.** 67

ripple amplitude (1) (power converters). The maximum value of the instantaneous difference between the average and instantaneous value of a pulsating unidirectional wave. *Note:* The amplitude is a useful measure of ripple magnitude when a single harmonic is dominant. Ripple amplitude is expressed in percent or per unit referred to the average value of the wave. *See also:* **power rectifier; rectification.** 204, 208

(2) (circuits and systems). The fine variations on a frequency plot of an impedance function or of a transfer function are called ripple. The ripple amplitude is the difference between the maximum and the minimum value of the function. 67

ripple current. *See:* **ripple voltage or current.**

ripple factor. The ratio of the ripple magnitude to the arithmetic mean value of the voltage. *See also:* **electrical conversion; interference; power pack; power supply; radio receiver.** 307, 201

ripple filter. A low-pass filter designed to reduce the ripple current, while freely passing the direct current, from a rectifier or generator. *See also:* **filter.** 59

ripple voltage (rectifier or generator). The alternating-voltage component of the unidirectional voltage from a direct-current power supply arising from sources within the power supply. *See:* **interference.** *See also:* **rectifier.** 111, 211, 188

ripple voltage or current. The alternating component whose instantaneous values are the difference between the average and instantaneous values of a pulsating unidirectional voltage or current. *See also:* **rectification.** 291, 237, 66, 208

rise. *See:* **travel.**

rise-and-fall pendant. A pendant the height of which can be regulated by means of a cord adjuster.   328

riser cable (communication practice). The vertical portion of a house cable extending from one floor to another. In addition, the term is sometimes applied to other vertical sections of cable. *See also:* **cable.**   328

rise time (1) (industrial control). The time required for the output of a system (other than first-order) to make the change from a small specified percentage (often 5 or 10) of the steady-state increment to a large specified percentage (often 90 or 95), either before overshoot or in the absence of overshoot. *Note:* If the term is unqualified, response to a step change is understood; otherwise the pattern and magnitude of the stimulus should be specified. *See also:* **control system, feedback.**   219, 206

(2) (instrument). The time, in seconds, for the pointer to reach 0.9 plus or minus a specified tolerance of the end scale when constant electric power is suddenly applied from a source of sufficiently high impedance so as not to influence damping (100 times the impedance of the instrument). *See also:* **moving element of an instrument.**   280

(3) (high- and low-power pulse transformers). (**First transition duration) (T_r).** The time interval of the leading edge between the instants at which the instantaneous value first reaches the specified lower and upper limits of 10 percent and 90 percent of A_M. Limits other than 10 percent and 90 percent may be specified in special cases. *See also:* **input pulse shape.**   32, 33

(4) (thyristor). *See:* **gate controlled rise time (thyristor).**   66

(5) (radiation-counter tubes). The interval between the instants at which the instantaneous value first reaches specified lower and upper limits, namely, 10 and 90 percent of the peak pulse value.   96

rise time, fall time (amplitude, frequency, and pulse modulation). The time for the light intensity to increase from the 10 to 90 percent intensity points. The fall time is the time for the light intensity to fall from the 90 to 10 percent intensity points.   72

rise time; pulse. *See:* **pulse rise time.**

rising-sun magnetron. A multicavity magnetron in which resonators of two different resonance frequencies are arranged alternately for the purpose of mode separation. *See also:* **magnetrons.**   190, 125

risk (nuclear power generating stations). The expected detriment per unit time to a person or a population from a given cause.   29, 31, 357

RIV. *See:* **radio influence voltage.**

RLC circuit. *See:* **simple series circuit.**

RMI (electronic navigation). *See:* **radio magnetic indicator.**

RMS. *See:* **root-mean-square value (of a periodic quantity).**

roadway (transmission and distribution). The portion of highway, including shoulders, for vehicular use. A divided highway has two or more roadways. *Notes:* (1) the shoulder is the portion of the roadway contiguous with the traveled way for accomodation of stopped vehicles for emergency use and for lateral support of base and surface courses. (2) The traveled way is the portion of the roadway for the movement of vehicles, exclusive of shoulders and full-time parking lanes.   262

roadway element (track element). That portion of the roadway apparatus associated with automatic train stop, train control, or cab signal systems, such as a ramp, trip arm, magnet, inductor, or electric circuit, to which the locomotive apparatus is directly responsive. *See also:* **automatic train control.**   328

ROC curves (radar). Abbreviation for receiver operating characteristic curves which are plots of probability of detection versus probability of false alarm for various input signal-to-noise ratios and detection threshold settings.   13

rock duster (rock-dust distributor). A machine that distributes rock dust over the interior surfaces of a coal mine by means of air from a blower or pipe line or by means of a mechanical contrivance, to prevent coal dust explosions.   328

rodding a duct. *See:* **duct rodding.**

rod, ground. *See:* **ground rod.**

Roebel transposition (rotating machinery). An arrangement of strands occupying two heightwise tiers in a bar (half coil), wherein at regular intervals through the core length, one top strand and one bottom strand cross over to the other tier in such a way that each strand occupies every vertical position in each tier so as to equalize the voltage induced in each of the strands, thereby eliminating current that would otherwise circulate among the strands. Looking from one end of the slot, the strands are seen to progress in a clockwise direction through the core length through what may be interpreted as an angle of 360 degrees so that the strands occupy the same position at both ends of the core. There are several variations of the Roebel transposition in use. In a bar having four tiers of copper, the two pairs of tiers would each have a Roebel transposition. The uninsulated bar, then, would be assembled as two Roebel-transposed bars, side-by-side. In order to transpose against voltages induced by end-winding flux, various modifications of the transposition in the slot, and extension of the Roebel transposition into the end winding have been used. *See also:* **rotor (rotating machinery; stator.**   63

roller bearing (rotating machinery). A bearing incorporating a peripheral assembly of rollers. *See also:* **bearing.**   63

roller, uplift. *See:* **uplift roller.**

rolling contacts (industrial control). A contact arrangement in which one cooperating member rolls on the other. *See also:* **contactor.**   244, 206

rolling transposition. A transposition in which the conductors of an open wire circuit are physically rotated in a substantially helical manner. With two wires a complete transposition is usually executed in two consecutive spans. *See also:* **open wire.**   328

roll-in-jewel error. Error caused by the pivot rolling up the side of the jewel and then falling to a lower position when tapped. This effect is not present when instruments are mounted with the axis of the moving element in a vertical position. (Roll-in-jewel error includes pivot-friction error that is small compared to the roll-in-jewel error.) *See also:* **moving element of an instrument.**   280

roll-off (1) (electroacoustics). A gradually increasing loss or attenuation with increase or decrease of frequency beyond the substantially flat portion of the amplitude-frequency response characteristic of a system or transducer. 328

(2) (high- and low-power pulse transformers). (rounding after first transition) (A_{RO}). The amount by which the instantaneous pulse value is less than A_M at the point in time of the intersection of straight-line segments used to determine A_M. It is expressed in amplitude units or in percent of A_M. *See also:* **input pulse shape.** 32, 33

roof bushing. A bushing intended primarily to carry a circuit through the roof, or other grounded barriers of a building, in a substantially vertical position. Both ends must be suitable for operating in air. At least one end must be suitable for outdoor operation. *See also:* **bushing.** 348

roof conductor. The portion of the conductor above the eaves running along the ridge, parapet, or other portion of the roof. *See also:* **lightning protection and equipment.** 328

room ambient temperature (electrical insulation tests). 20°C ± 5° (68°F ± 9°). 116

room coefficient. A number computed from wall and floor areas. It is used to indicate room proportions in tables of luminance factors of room surfaces. *Note:* The room coefficient is computed from

$$K_r = \frac{\text{height} \times (\text{length} + \text{width})}{2 \times \text{length} \times \text{width}}$$

See also: **inverse-square (illuminating engineering).** 167

room index. The room index is a letter designation for a range of room ratios. *See also:* **inverse-square law (illuminating engineering).** 167

room ratio. A number indicating room proportions, calculated from the length, width, and ceiling height (or luminaire mounting height) above the work plane. It is used to simplify lighting design tables by expressing the equivalence of room shapes with respect to the utilization of direct or interreflected light. *Note:* The room ratio depends upon the light distribution characteristics of luminaires. The following formulas are used: (1) For direct, semidirect, and general diffuse distributions,

$$\text{numerical room ratio} = wl/[h_m(w + l)]$$

(2) For semi-indirect and indirect distributions,

$$\text{numerical room ratio} = 3wl/[2h_c(w + l)]$$

where w is the width, l is the length, h_m is the mounting height above the work plane, and h_c is the ceiling height above the work plane. *See also:* **inverse-square law (illuminating engineering).** 167

room utilization factor. The ratio of the luminous flux (lumens) received on the work plane to that emitted by the luminaire. *Note:* This ratio sometimes is called **interflectance.** Room utilization factor is based on the flux emitted by a complete luminaire, whereas coefficient of utilization is based on the rated flux generated by the lamps in a luminaire. *Syn.* **utilance.** *See also:* **inverse-square law (illuminating engineering).** 167

root locus (1) (control system, feedback) (for a closed loop whose characteristic equation is $KG(s)H(s) + 1 = 0$). A plot in the s plane of all those values of s that make

$G(s)H(s)$ a negative real number; those points that make the loop transfer function $KG(s)H(s) = -1$ are roots. *Note:* The locus is conveniently sketched from the factored form of $KG(s)H(s)$; each branch starts at a pole of that function with $K = 0$. With increasing K, the locus proceeds along its several branches toward a zero of that function and, often asymptotic to one of several equiangular radial lines, toward infinity. Roots lie at points on the locus for which (1) the sum of the phase angles of component $G(s)H(s)$ vectors totals 180 degrees, and for which (2) $1/K = |G(s)H(s)|$. Critical damping of the closed loop occurs when the locus breaks away from the real axis; instability when it crosses the imaginary axis. *See also:* **control system, feedback.** 56

(2) (excitation control systems). Consider a linear, stationary, system with closed loop transfer function $C(S)/R(S)$ where $R(S)$ is the Laplace Transform of the excitation (input) driving function of the closed loop system and $C(S)$ is the Laplace Transform of the response (output) function of the closed loop system. When $C(S)/R(S)$ is a function of the gain, K, of one element in either the forward or reverse signal path, the poles of $C(S)/R(S)$ in the S-plane will in general be a function of K. A plot in the S-plane of the loci of poles of the closed loop transfer function as K varies is known as a root locus. 353

root-mean-square detector. A detector, the output voltage of which approximates the root-mean-square values of an applied signal. *See also:* **electromagnetic compatibility.** 220, 199

root-mean-square (effective) burst magnitude (audio and electroacoustics). The square root of the average square of the instantaneous magnitude of the voltage or current taken over the burst duration. See the figure attached to the definition of **burst duration.** *See also:* **burst (audio and electroacoustics).** 253, 176

root-mean-square (effective) pulse amplitude. The square root of the average of the square of the instantaneous amplitude taken over the pulse duration. 254

root-mean-square reverse-voltage rating (rectifier device). The maximum sinusoidal root-mean-square reverse voltage permitted by the manufacturer under stated conditions. *See also:* **average forward-current rating (rectifier circuit).** 237, 66

root-mean-square ripple. The effective value of the instantaneous difference between the average and instantaneous values of a pulsating unidirectional wave integrated over a complete cycle. *Note:* The root-mean-square ripple is expressed in percent or per unit referred to the average value of the wave. *See also:* **rectification.** 291, 208

root-mean-square sound pressure. *See:* **effective sound pressure.**

root-mean-square value (1) (periodic function) (effective value*). The square root of the average of the square of the value of the function taken throughout one period. Thus, if y is a periodic function of t

$$Y_{\text{rms}} = \left[\frac{1}{T} \int_a^{a+T} y^2 \, dt \right]^{1/2}$$

where Y_{rms} is the root-mean-square value of y, a is any value of time, and T is the period. If a periodic function is represented by a Fourier series, then:

$$Y_{\text{rms}} = \frac{1}{(2)^{1/2}} \left(\frac{1}{2} A_0{}^2 + A_1{}^2 + A_2{}^2 \cdots + B_1{}^2 \right.$$

$$\left. + B_2{}^2 + \cdots \right)^{1/2}$$

$$= \frac{1}{(2)^{1/2}} \left(\frac{1}{2} A_0{}^2 + C_1{}^2 + C_2{}^2 + \cdots + C_n{}^2 \right)^{1/2}$$

210

* Deprecated

(2) (alternating voltage or current). The square root of the mean of the square of the voltage, or current, during a complete cycle. 308, 62

root-sum-square. The square root of the sum of the squares. *Note:* Commonly used to express the total harmonic distortion. *See also:* **radio receiver.** 339

rope-lay conductor or cable. A cable composed of a central core surrounded by one or more layers of helically laid groups of wires. *Note:* This kind of cable differs from a concentric-lay conductor in that the main strands are themselves stranded. In the most common type of rope-lay conductor or cable, all wires are of the same size and the central core is a concentric-lay conductor. *See also:* **conductor.** 64

roped-hydraulic driving machine (elevators). A machine in which the energy is applied by a piston, connected to the car with wire ropes, that operates in a cylinder under hydraulic pressure. It includes the cylinder, the piston, and the multiplying sheaves if any and their guides. *See:* **roped-hydraulic elevator.** *See also:* **driving machine (elevators).** 328

roped-hydraulic elevator. A hydraulic elevator having its piston connected to the car with wire ropes. *See:* **roped-hydraulic driving machine (elevators).** *See also:* **elevators.** 328

rosette. An enclosure of procelain or other insulating material, fitted with terminals and intended for connecting the flexible cord carrying apendant to the permanent wiring. *See also:* **cabinet.** 328

rotary attenuator (waveguide). A variable attenuator in circular waveguide having absorbing vanes fixed diametrically across one section; the attenuation is varied by rotation of this section about the common axis. *See also:* **waveguide.** 179

rotary converter. A machine that combines both motor and generator action in one armature winding connected to both a commutator and slip rings, and is excited by one magnetic field. It is normally used to change alternating-current power to direct-current power. 63

rotary dial (telephone switching systems). A type of calling device used in automatic switching that generates pulses by manual rotation and release of a dial, the number of pulses being determined by how far the dial is rotated before being released. 55

rotary generator (induction heating). An alternating-current generator adapted to be rotated by a motor or prime mover. *See also:* **dielectric heating; industrial electronics.** 14, 114

rotary inverter. A machine that combines both motor and generator action in one armature winding. It is excited by one magnetic field and changes direct-current power to alternating-current power. (Usually it has no amortisseur winding.) 63

rotary joint (waveguide components). A coupling for efficient transmission of electromagnetic energy between two waveguide or transmission line structures designed to permit unlimited mechanical rotation of one structure. 166

rotary phase changer (rotary phase shifter) (waveguide). A phase changer that alters the phase of a transmitted wave in proportion to the rotation of one of its waveguide sections. 244, 179

rotary relay. (1) A relay whose armature moves in rotation to close the gap between two or more pole faces (usually with a balanced armature). (2) Sometimes used for stepping relay. *See also:* **relay.** 259

rotary solenoid relay. A relay in which the linear motion of the plunger is converted mechanically into rotary motion. *See also:* **relay.** 259

rotary switch. A bank-and-wiper switch whose wipers or brushes move only on the arc of a circle. *See also:* **switch.** 328

rotary system. An automatic telephone switching system that is generally characterized by the following features: (1) The selecting mechanisms are rotary switches. (2) The switching pulses are received and stored by controlling mechanisms that govern the subsequent operations necessary in establishing a telephone connection. 328

rotary voltmeter. *See:* **generating voltmeter.**

rotatable frame (rotating machinery). A stator frame that can be rotated by a limited amount about the axis of the machine shaft. *See also:* **stator.** 63

rotatable phase-adjusting transformer (phase-shifting transformer). A transformer in which the secondary voltage may be adjusted to have any desired phase relation with the primary voltage by mechanically orienting the secondary winding with respect to the primary. The primary winding of such a transformer usually consists of a distributed symmetrical polyphase winding and is energized from a polyphase circuit. *See also:* **auxiliary device to an instrument.** 328

rotating amplifier. An electric machine in which a small energy change in the field is amplified to a large energy change at the armature terminals. *See also:* **asynchronous machine.** 63

rotating-anode tube (X-ray). An X-ray tube in which the anode rotates. *Note:* The rotation continually brings a fresh area of its surface into the beam of electrons, allowing greater output without melting the target. *See also:* **electron devices, miscellaneous.** 190

rotating control assembly (rotating machinery). The complete control circuits for a brushless exciter mounted to permit rotation. *See also:* **rotor (rotating machinery).** 63

rotating field. A variable vector field that appears to rotate with time. 210

rotating-insulator switch. A switch in which the opening and closing travel of the blade is accomplished by the rotation of one or more of the insulators supporting the conducting parts of the switch. 103, 202, 27

rotating joint (waveguides). A coupling for transmission of electromagnetic energy between two waveguide structures designed to permit mechanical rotation of one structure. *See also:* **waveguide.**   330

rotating machinery. *See:* **machine, electric.**

rotation plate (rotating machinery). A plaque showing the proper direction of rotor rotation. *See also:* **rotor (rotating machinery).**   63

rotation test (rotating machinery). A test to determine that the rotor rotates in the specified direction when the voltage applied agrees with the terminal markings. *See:* **asynchronous machine; direct-current commutating machine.**   63

rotor (1) (rotating machinery). The rotating member of a machine, with shaft. *Note:* In a direct-current machine with stationary field poles, universal, alternating-current series, and repulsion-type motors, it is commonly called the armature.   63

(2) (meter) (rotating element). That part of the meter that is directly driven by electromagnetic action. *See also:* **watthour meter.**   212

rotor angular momentum (gyro). the product of spin angular velocity and rotor moment of inertia about the spin axis.   46

rotor bar (rotating machinery). A solid conductor that constitutes an element of the slot section of a squirrel-cage winding. *See also:* **rotor (rotating machinery).**   63

rotor bushing (rotating machinery). A ventilated or non-ventilated piece or assembly used for mounting onto a shaft, an assembled rotor core whose inside opening is larger than the shaft. *See also:* **rotor (rotating machinery).**   63

rotor coil (rotating machinery). A unit of a rotor winding of a machine. *See also:* **rotor (rotating machinery).**   63

rotor core (rotating machinery). That part of the magnetic circuit that is integral with, or mounted on, the rotor shaft. It frequently consists of an assembly of laminations.   63

rotor-core assembly (rotating machinery). The rotor core with a squirrel-cage or insulated-conductor winding, put together as an assembly. *See also:* **rotor (rotating machinery).**   63

rotor core lamination (rotating machinery). A sheet of magnetic material, containing teeth, slots, or other perforations dictated by design, which forms the rotor core when assembled with other identical or similar laminations.   63

rotor displacement angle (load angle) (rotating machinery). The displacement caused by load between the terminal voltage and the armature voltage generated by that component of flux produced by the field current. *See also:* **rotor (rotating machinery).**   63

rotor end ring (rotating machinery). The conducting structure of a squirrel-cage winding that short-circuits all of the rotor bars at one end. *See also:* **rotor (rotating machinery).**   63

rotor moment of inertia (gyro). The moment of inertia of a gyro rotor about its spin axis.   46

rotor-resistance starting (rotating machinery). The process of starting a wound-rotor induction motor by connecting the rotor initially in series with starting resistors that are short-circuited for the running operation. *See also:* **asynchronous machine.**   63

rotor rotation detector (gyro). A device which produces a signal output as a function of the speed of the rotor.   46

rotor slot armor (cylindrical-rotor synchronous machine) (rotating machinery). Main ground insulation surrounding the slot or core portions of a field coil assembled on a slotted rotor. *See also:* **rotor (rotating machinery).**   63

rotor speed sensitivity (tuned rotor gyro). The change in in-phase spring rate due to a change in gyro rotor speed.   46

rotor spider. *See:* **spider.**

rotor winding (rotating machinery). A winding on the rotor of a machine. *See also:* **rotor (rotating machinery).**   63

roughness (navigational-system display). Irregularities resembling scalloping, but distinguished by their random, noncyclic nature; sometimes called course roughness. *See also:* **navigation.**   187, 13

round conductor. Either a solid or stranded conductor of which the cross section is substantially circular. *See also:* **conductor.**   64

rounding (pulse terms). A distortion in the form of a rounded feature which occurs where a relatively abrupt change in slope is desired or expected. *See:* Note in **preshoot** entry.   254

rounding error (test, measurement and diagnostic equipment). The error resulting from deleting the less significant digits of a quantity and applying some rule of correction to the part retained. A common round-off rule is to take the quantity to the nearest digit. Thus the value of Pi = 3.14159265 . . . , rounded to four decimals is 3.1416.   54

round off. To delete the least-significant digit or digits of a numeral and to adjust the part retained in accordance with some rule.   255, 77, 54

round rotor (cyclindrical rotor) (rotating machinery). A rotor of cylindrical shape in which the coil sides of the winding are contained in axial slots. *See also:* **rotor (rotating machinery).**   63

route (telephone switching systems). A particular order of a set of switching entities through which call connections may be established.   55

route locking. Locking effective when a train passes a signal and adapted to prevent manipulation of levers that would endanger the train while it is within the limits of the route entered. It may be so arranged that a train in clearing each section of the route releases the locking affecting that section. *See also:* **interlocking (interlocking plant).**   328

routine (electronic computation). A set of instructions arranged in proper sequences to cause a computer to perform a desired operation, such as the solution of a mathematical problem.   235, 210, 255, 77

routine tests (1) (general). Tests made for quality control by the manufacturer on every device or representative samples, or on parts or materials as required to verify during production that the product meets the design specifications.   4, 210, 203, 53

(2) (rotating machinery). The tests applied to a machine to show that it has been constructed and assembled correctly, is able to withstand the appropriate high-voltage tests, is in sound working order both electrically and mechanically, and has the proper electrical characteristics. *See also:* **asynchronous machine; direct-current commutating machine; limiting insulation temperature (limiting**

hottest-spot temperature).   63

(3) (switchgear). *See:* **production tests (switchgear).**

(4) (cable termination). Tests made on each high-voltage cable termination or upon a representative number of devices, or parts thereof, during production for the purpose of quality control.   4

(5) (surge arrester). Tests made by the manufacturer on every device or representative samples, or on parts or materials, as required, to verify that the product meets the design specifications.   2

routing code (telephone switching systems). A digit or combination of digits used to direct a call towards its destination.   55

routing pattern (telephone switching systems). The implementation of a routing plan with reference to an individual automatic exchange.   55

routing plan (telephone switching systems). A plan for directing calls through a configuration of switching entities.   55

roving (rotating machinery). A loose assemblage of fibers drawn or rubbed into a single strand with very little twist. In spun yarn systems, the product of the stage or stages just prior to spinning.   216, 63

row. A path perpendicular to the edge of a tape along which information may be stored by presence or absence of holes or magnetized areas.   224, 207

row binary. Pertaining to the binary representation of data on punched cards in which adjacent positions in a row correspond to adjacent bits of data, for example, each row in an 80-column card may be used to represent 80 consecutive bits of two 40-bit words.   255, 77

RPE. (radial probable error). *See:* **circular probable error.**

RRRV. *See:* **rate of rise of restriking voltage.**

R scan. *See:* **R display.**

R scope. *See:* R **display.**

RT box* (radar). *See:* **antitransmit-receive switch.**

* Deprecated

rubber tape. A tape composed of rubber or rubberlike compounds that provides insulation for joints.   328

rudder-angle-indicator system. A system consisting of an indicator (usually in the wheel house) so controlled by a transmitter connected to the rudder stock as to show continually the angle of the rudder relative to the center line of the ship.   328

ruling span (conductor stringing equipment). A calculated deadend span length which will have the same changes in conductor tension due to changes of temperature and conductor loading as will be found in a series of spans of varying lengths between deadends.   45

rumble (electroacoustics). Low-frequency vibration of the recording or reproducing drive mechanism superimposed on the reproduced signal. *See also:* **phonograph pickup.**   176

rumble, turntable. *See:* **turntable rumble.**

run (computing systems). A single, continuous performance of a computer routine.   255, 77, 54

run-down time (gyro). The time interval required for the gyro rotor to reach a specified speed, or during which the gyro exhibits specified performance, after removal of rotor excitation at a specified speed.   46

running board (conductor stringing equipment). A pulling device designed to permit stringing more than one conductor simultaneously with a single pulling line. For distribution stringing, it is usually made of lightweight tubing with the forward end curved gently upward to provide smooth transition over pole crossarm rollers. For transmission stringing, the device is often made of sections hinged transversely to the direction of pull with a flexible pendulum tail suspended from the rear. This configuration stops the conductors from twisting together and permits smooth transition over two or more travelers which have been grouped together ("bundled") on a common shaft. *Syn.* **alligator; bird; birdie; monkey tail; sled.**   45

running-light-indicator panel (telltale). A panel in the wheelhouse providing audible and visible indication of the failure of any running light connected thereto.   328

running lights (navigation lights). Lanterns constructed and located as required by navigation laws, to permit the heading and approximate course of a vessel to be determined by an observer on a nearby vessel. *Note:* Usual running lights are port side, starboard side, mast-head, range, and stern lights.   328

running open-phase protection (industrial control). The effect of a device operative on the loss of current in one phase of a polyphase circuit to cause and maintain the interruption of power in the circuit.   206

running operation (single-phase motor). (1) For a motor employing a starting switch or relay: operation at speeds above that corresponding to the switching operation. (2) For a motor not employing a starting switch or relay: operation in the range of speed that includes breakdown-torque speed and above. *See also:* **asynchronous machine.**   63

running tension control (industrial control). A control function that maintains tension in the material at operating speeds. *See also:* **control system, feedback.**   225, 206

run-of-river station. A hydroelectric generating station that utilizes all or a part of the stream flow without storage. *See also:* **generating station.**   64

runout rate (industrial control). The velocity at which the error in register accumulates.   204

run time variable (RTV) (test, measurement and diagnostic equipment). An application program condition in which the stimuli is varied under system control based on a measurement result.   54

run-up time (gyro). The time interval required for the gyro rotor to reach a specified speed from standstill.   46

runway alignment indicator. An alignment indicator consisting of a group of aeronautical ground lights arranged and located to provide early direction and roll guidance on the approach to a runway.   167

runway centerline lights. Runway lights installed in the surface of the runway along the centerline indicating the location and direction of the runway centerline and are of particular value in conditions of very poor visibility.   167

runway edge lights. Lights installed along the edges of a runway marking its lateral limits and indicating its direction.   167

runway end identification light. A pair of flashing aeronautical ground lights symmetrically disposed on each side of the runway at the threshold to provide additional threshold conspicuity. 167

runway exit lights. Lights placed on the surface of a runway to indicate a path to the taxiway centerline. 167

runway lights. Aeronautical ground lights arranged along or on a runway. 167

runway visibility. The meteorological visibility along an identified runway. Where a transmissometer is used for measurement, the instrument is calibrated in terms of a human observer; that is, the sighting of dark objects against the horizon sky during daylight and the sighting of moderately intense unfocused lights of the order of 25 candelas at night. 167

runway visual range (RVR) (navigation). (1) The forward distance a human pilot can see along the runway during an approach to landing; this distance is derived from electro-optical instruments operated on the ground and it is improved (increased) by the use of lights (such as high-intensity runway lights). *See also:* **navigation.**
187

(2) (in the United States). An instrumentally derived value, based on standard calibrations, that represents the horizontal distance a pilot will see down the runway from the approach end; it is based on the sighting of either high-intensity runway lights or on the visual contrast of other targets, whichever yields the greater visual range. 167

rural districts: (transmission and distribution). All places not urban. This may include thinly settled areas within city limits. 262

rural line. A line serving one or more subscribers in a rural area. 328

rust (corrosion). A corrosion product consisting primarily of hydrated iron oxide. *Note:* This term is properly applied only to iron and ferrous alloys. 221, 205

rust-resistant. So constructed, protected or treated that rust will not exceed a specified limit when subjected to a specified rust resistance test. 53

RVR. *See:* **runway visual range.**

R-X diagram (power switchgear). A graphic presentation of the characteristics of a relay unit in terms of the ratio of voltage to current and the phase angle between them. *Note:* For example, if a relay just operates with ten volts and ten amperes in phase, one point on the operating curve of the relay would be plotted as one ohm on the R axis (that is, $R = 1$, $X = 0$ where R is the abscissa and X is the ordinate). 103, 127

R–X plot (protective relaying). A graphical method of showing the characteristics of a relay element in terms of the ratio of voltage to current and the angle between them. For example, if a relay barely operates with 10 V and 10 A in phase, one point on the operating curve of the relay would be plotted as 1 Ω on the R axis (that is, $R = 1$, $X = 0$). 128

S

sabin (audio and electroacoustics). A unit of absorption having the dimensions of area. *Notes:* (1) The metric sabin has dimensions of square meters. (2) When used without a modifier, the sabin is the equivalent of one square foot of a perfectly absorptive surface. 176

sacrificial protection (corrosion). Reduction or prevention of corrosion of a metal in an environment acting as an electrolyte by coupling it to another metal that is electrochemically more active in that particular electrolyte. *See also:* **stray-current corrosion.** 221, 205

safeguard (electrolytic cell line working zones). Precautionary measure or stipulation, or a technical contrivance to prevent accidents. 133

safe shutdown earthquake (SSE) (nuclear power generating stations) (seismic qualification of class 1E equipment). That earthquake which produces the maximum vibratory ground motion for which certain structures, systems, and components are designed to remain functional. These structures, systems, and components are those necessary to assure: (1) the integrity of the reactor coolant pressure boundary, (2) the capability to shutdown the reactor and maintain it in a safe shutdown condition, or (3) the capability to prevent or mitigate the consequences of accidents which could result in potential offsite exposures comparable to the guideline exposures of Code of Federal Regulations, Title 10, Part 100 (December 5, 1973).
28, 31

safe working voltage to ground (electric recording instru-ment). The highest safe voltage in terms of maximum peak value that should exist between any circuit of the instrument and its case. *See also:* **test (instrument or meter).** 294

safety class structures (nuclear power generating stations) (Class 1E systems and equipment). Structures designed to protect Class IE equipment against the effects of the design basis events. *Note:* For the purpose of this document, separate safety class structures can be separate rooms in the same building. The rooms can share a common wall. 31, 131

safety, conductor (conductor stringing equipment). A sling arranged in a vertical basket configuration, with both ends attached to the supporting structure and passed under the clipped-in conductor(s). These devices are normally utilized with bundled conductors to act as a safety device in case of insulator failure while workmen in conductor cars are installing spacers between the subconductors, or as an added safety measure when crossing above energized circuits. These devices may be fabricated from synthetic-fiber rope or wire rope. 45

safety control feature (deadman's feature). That feature of a control system that acts to reduce or cut off the current to the traction motors or to apply the brakes, or both, if the operator relinquishes personal control of the vehicle. *See also:* **multiple-unit control.** 328

safety control handle (deadman's handle). A safety attachment to the handle of a controller, or to a brake valve, causing the current to the traction motors to be reduced or cut off, or the brakes to be applied, or both, if the pressure of the operator's hand on the handle is released. *Note:* This function may be applied alternatively to a foot-operated pedal or in combination with attachments to the controller or the brake valve handles, or both. *See also:* **multiple-unit control.** 328

safety outlet.* *See:* **grounding outlet.**

* Deprecated

safety-related (nuclear power generating station). Any Class IE power or protection system device included in the scope of IEEE-279-1971 or IEEE-308-1974. 357

safety system (nuclear power generating station). The collection of systems required to mitigate the consequences of design basis events. 356

safe working space (electrolytic cell line working zone). The space required to safeguard personnel from hazarous electrical conditions during the conduct of their work in operating and maintaining cells and their attachments. This shall include space allowance for tools and equipment that may be involved. 133

safe work practices (electrolytic cell line working zones). Those operating and maintenance procedures that are effective in preventing accidents. 133

SAFI. *See:* **semiautomatic flight inspection.**

sag (1) (transmission and distribution). The distance measured vertically from a conductor to the straight line joining its two points of support. Unless otherwise stated, the sag referred to is the sag at the midpoint of the span. 64

See also: **initial unloaded sag; final sag; final unloaded sag; total sag; maximum total sag; apparent sag of a span; sag of a conductor at any point in a span; apparent sag at any point in the span.** 262

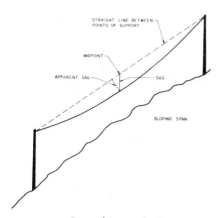

Sag and apparent sag.

(2) (apparent sag at any point). The departure of the wire at the particular point in the span from the straight line between the two points of support of the span, at 60 degrees Fahrenheit, with no wind loading. 178

sag of a conductor at any point in a span (transmission and distribution). The distance measured vertically from the particular point in the conductor to a straight line between its two points of support. 262

sag section (conductor stringing equipment). The section of line between snub structures. More than one sag section may be required in order to properly sag the actual length of conductor which has been strung. *Syn.* **pull; setting; stringing section.** 45

sag span (conductor stringing equipment). A span selected within a sag section and used as a control to determine the proper sag of the conductor, thus establishing the proper conductor level and tension. A minimum of two, but normally three, sag spans are required within a sag section in order to sag properly. In mountainous terrain or where span lengths vary radically, more than three sag spans could be required within a sag section. *Syn.* **control span.** 45

sal ammoniac cell. A cell in which the electrolyte consists primarily of a solution of ammonium chloride. *See also:* **electrochemistry.** 328

salient pole (rotating machinery). A field pole that projects from the yoke or hub towards the primary winding core. *See also:* **rotor (rotating maxhinery).** 63

salient-pole machine. An alternating-current machine in which the field poles project from the yoke toward the armature and/or the armature winding self-inductance undergoes a significant single cyclic variation for a rotor displacement through one pole pitch. *See also:* **asynchronous machine.** 63

salient pole synchronous induction motor (rotating machinery). A salient pole synchronous motor having a coil winding for starting purposes embedded in the pole shoes. The terminal leads of this coil winding are connected to collector rings. 63

salinity indicator system. A system, based on measurement of varying electric resistance of the solution, to indicate the amount of salt in boiler feed water, the output of an evaporator plant, or other fresh water. *Note:* Indication is usually in grains per gallon. 328

sampled data. Data in which the information content can be, or is, ascertained only at discrete intervals of time. *Note:* Sampled data can be analog or digital. *See also:* **control system, feedback.** 224, 207

sampled-data control system (industrial control). A system that operates with sampled data. *See also:* **control system, feedback.** 219, 206

sampled format (pulse measurement). A waveform which is a series of sample magnitudes taken sequentially or nonsequentially as a function of time. It is assumed that nonsequential samples may be rearranged in time sequence to yield the following samples formats. *See also:* **periodically sampled real time format; periodically sampled equivalent time format; aperiodically sampled real time format; aperiodically sampled equivalent time format.** 15

sampled signal. The sequence of values of a signal taken at discrete instants. *See also:* **control system.** 56

sample equipment (nuclear power generating stations). Production equipment tested to obtain data that are valid over a range of ratings and for specific services. 31, 120

sample valve operator (nuclear power generating stations). A production valve operator type tested to obtain data that are valid over a range of sizes and for the specific services. *Note:* All salient factors must be shown to be common to the sample valve operator and to the intended service valve operator. Commonality of factors such as materials of construction, lubrication, mechanical stresses and clearances, manufacturing processes, and dielectric properties may be established by specification, test, or analyses.

31, 142

sampling (1) (modulation systems). The process of obtaining a sequence of instantaneous values of a wave; at regular or intermittent intervals.   242, 255, 77, 54
(2) (pulse terms). A process in which strobing pulses yield signals which are proportional to the magnitude (typically, as a function of time) of a second pulse or other event.

254

sampling circuit (sampler). A circuit whose output is a series of discrete values representative of the values of the input at a series of points in time.   328
sampling control. *See:* **control system, sampling.**
sampling gate (radar). A device that extracts information from the input wave only when activated by a selector pulse. *See also:* **navigation.**   13, 187
sampling, instantaneous. The process for obtaining a sequence of instantaneous values of a wave. *Note:* These values are called instantaneous samples.   111
sampling interval (automatic control). The time between samples in a sampling control system. *See also:* **control system, feedback.**   329
sampling period (automatic control). The time interval between samples in a periodic sampling control system. *See also:* **control system, feedback.**   329
sampling smoke detector (fire protection devices). A device which consists of tubing distributed from the detector unit to the area(s) to be protected. An air pump draws air from the protected area back to the detector through the air sampling ports and piping. At the detector, the air is analyzed for smoke particles.   71
sampling tests. Tests carried out on a few samples taken at random out of one consignment. *See also:* **asynchronous machine; direct-current commutating machine.**   63
satellite (communication satellite). A body which revolves around another body and which has a motion primarily and permanently determined by the force of attraction of this body. *Note:* A body so defined which revolves round the sun is called a planet or planetoid. By extension, a natural satellite of a planet may itself have a satellite.

74

satellite phasing (communication satellite). Maintaining the center of mass of a satellite by propulsion within a prescribed small tolerance in a desired relation with respect to other satellites or a point on the earth or some other point of reference, such as the subsolar point. 74
saturable-core magnetometer. A magnetometer that depends for its operation on the changes in permeability of a ferromagnetic core as a function of the field to be measured. *See also:* **magnetometer.**   328
saturable-core reactor. *See:* **saturable reactor.**
saturable reactor (saturable-core reactor). (1) A magnetic-core reactor whose reactance is controlled by changing the saturation of the core through variation of a superimposed unidirectional flux. (2) A magnetic-core reactor operating in the region of saturation without independent control means. *Note:* Thus, a reactor whose impedance varies cyclically with the alternating current

(or voltage). *See also:* **magnetic amplifier.**

210, 206, 203, 53

saturated signal.* *See:* **saturating signal.**

* Deprecated

saturated sleeving. A flexible tubular product made from braided cotton, rayon, nylon, glass, or other fibers, and coated or impregnated with varnish, lacquer, a combination of varnish and lacquer, or other electrical insulating materials. The impregnant or coating need not form a continuous film.   328
saturating reactor. A magnetic-core reactor operating in the region of saturation without independent control means. *See also:* **magnetic amplifier.**   328
saturating signal (electronic navigation). A signal of an amplitude greater than can be accommodated by the dynamic range of a circuit. *See also:* **navigation.**   187, 13
saturation (1) (signal-transmission system). A natural phenomenon or condition in which any further change of input no longer results in appreciable change of output.

219, 206

(2) (automatic control) A condition caused by the presence of a signal or interference large enough to produce the maximum limit of response, resulting in loss of incremental response. *See also:* **control system, feedback; signal.**

56, 188

(3) (visual). The attribute of a visual sensation which permits a judgment to be made of the proportion of pure chromatic color in the total sensation. *Note:* This attribute is the psychosensorial correlate (or nearly so) of the colorimetric quantity "purity."
(4) (color television). In a tristimulus reproducer, the degree to which the color lies on the triangle as defined by the three reproducing primaries. *Note:* Full saturation is achieved when one or two of the reproduced primary colors have zero intensity.   18
(5) (perceived light-source color). The attribute used to describe its departure from a light-source color of the same brightness perceived to have no hue. *See also:* **color.**

167

(6) (maser, laser, maser material, or laser material). A condition in which the **attenuation** or gain of a material or a device remains at a fixed level or decreases as the input signal is increased.   363
saturation characteristics (nuclear power generating station). A description of the steady state or dynamic conditions or limitations under which a further change in input produces an output response which no longer conforms to the specified steady-state or dynamic input-output relationship.   355
saturation current (1) (thermionics). The value of the current in the saturation state. *See also:* **electron emission.**

244, 190

(2) (semiconductor diode). That portion of the steady-state reverse current that flows as a result of the transport across the junction of minority carriers thermally generated within the regions adjacent to the junction. *See also:* **semiconductor device.**   210, 245, 66
saturation curve (machine or other apparatus). A characteristic curve that expresses the degree of magnetic saturation as a function of some property of the magnetic excitation. *Note:* For a direct-current or synchronous ma-

chine the curve usually expresses the relation between armature voltage and field current for no load or some specified load current, and for specified speed. *See also:* **direct-current commutating machine.** 63

saturation factor (1) (direct-current or synchronous machine). The ratio of a small percentage increase in field excitation to the corresponding percentage increase in voltage thereby produced. *Note:* Unless otherwise specified, the saturation factor of a machine refers to the no-load excitation required at rated speed and voltage. *See also:* **asynchronous machine; direct-current commutating machine.** 63

(2) (rotating machinery). The ratio of the unsaturated value of a quantity to its saturated value. The reciprocal of this definition is also used. 63

saturation flux density. *See:* **saturation induction.**

saturation induction, (B_s) (magnetic core testing). The maximum intrinsic value of induction possible in a material. *Notes:* (1) This term is often used for the maximum value of induction at a stated high value of field strength where further increase in intrinsic magnetization with increasing field strength is negligible. (2) S.I. unit: Tesla; cgs unit: Gauss (1 Tesla) = 10^4 Gauss. (3) Peak induction (B_m) is the magnetic induction corresponding to the peak applied magnetizing force specified in a test. 165

saturation level (storage tubes). The output level beyond which no further increase in output is produced by further writing (then called write saturation) or reading (then called read saturation). *Note:* The word saturation is frequently used alone to denote saturation level. *See also:* **storage tube.** 174, 190

saturation state (thermionics). The state of working of an electron tube or valve in which the current is limited by the emission from the cathode. *See also:* **electron emission.** 244, 190

sawtooth. *See:* **sawtooth waveform.**

sawtooth sweep. A sweep generated by the ramp portion of a sawtooth waveform. *See also:* **oscillograph.** 185

sawtooth wave. A periodic wave whose instantaneous value varies substantially linearly with time between two values, the interval required for one direction of progress being longer than that for the other. *Note:* In television practice, the waveform during the retrace interval is not necessarily linear, since only the trace interval is used for active scanning. *See also:* **television.** 163, 178

sawtooth waveform. A waveform containing a ramp and a return to initial value, the two portions usually of unequal duration. *See also:* **oscillograph.** 185

S-band radar. A radar operating at frequencies between 2 and 4 gigahertz usually in one of the International Telecommunications Union assigned bands 2.3 to 2.5 or 2.7 to 3.7 gigahertz. 13

scalar. A quantity that is completely specified by a single number. 210

scalar field. The totality of scalars in a given region represented by a scalar function $S(x,y,z)$ of the space coordinates x,y,z. 210

scalar function. A functional relationship that results in a scalar. 210

scalar product (of two vectors) (dot product). The scalar obtained by multiplying the product of the magnitudes of the two vectors by the cosine of the angle between them.

The scalar product of the two vectors **A** and **B** may be indicated by means of a dot **A** · **B**. If the two vectors are given in terms of their rectangular components, then

$$\mathbf{A} \cdot \mathbf{B} = A_x B_x + A_y B_y + A_z B_z.$$

Example: Work is the scalar product of force and displacement. 210

scale (1) (acoustics). A musical scale is a series of notes (symbols, sensations, or stimuli) arranged from low to high by a specified scheme of intervals, suitable for musical purposes. 176

(2) (computing systems). To change a quantity by a factor in order to bring its range within prescribed limits. 255, 77

(3) (instrument scale). *See:* **full scale.**

scale factor (1) (instrument or device). The factor by which the number of scale divisions indicated or recorded by an instrument should be multiplied to compute the value of the measurand. *Note:* Deflection factor is a more general term than scale factor in that the instrument response may be expressed alternatively in units other than scale divisions. *See also:* **accuracy rating (instrument) test voltage and current.** 307, 201

(2) (computing systems). A number used as a multiplier, so chosen that it will cause a set of quantities to fall within a given range of values. *Note:* To scale the values 856, 432, −95, and −182 between −1 and +1, a scale factor of 1/1000 would be suitable. 255, 77, 54

(3) (analog computer). The multiplication factor necessary to transform problem variables to computer variables. *Notes:* (A) problem variable is a variable appearing in the mathematical model of the problem. (B) A computer variable is a dependent variable as represented on the computer. *See:* **time.** 9

(4) (gyro; accelerometer). The ratio of a change in output to a change in the input intended to be measured. Scale factor is generally evaluated as the slope of the straight line that can be fitted by the method of least squares to input-output data obtained by varying the input cyclically over the input range. *See:* **input-output characteristics.** 46

scale-factor potentiometer. *See:* **parameter potentiometer.**

scale length (electric instrument). The length of the path described by the indicating means or the tip of the pointer in moving from one end of the scale to the other. *Notes:* (1) In the case of knife-edge pointers and others extending beyond the scale division marks, the pointer shall be considered as ending at the outer end of the shortest scale division marks. In multiscale instruments the longest scale shall be used to determine the scale length. (2) In the case of antiparallax instruments of the step-scale type with graduations on a raised step in the plane of and adjacent to the pointer tip, the scale length shall be determined by the end of the scale divisions adjacent to the pointer tip. *See also:* **accuracy rating (instrument); instrument.** 280

scale-of-two counter. A flip-flop circuit in which successive similar pulses, applied at a common point, cause the circuit to alternate between its two conditions of permanent stability. *See also:* **trigger circuit.** 328

scaler (radiation counters). An instrument incorporating one or more scaling circuits and used for registering the number of counts received. *See also:* **anticoincidence (radiation counters).**   190

scaler, pulse (pulse techniques). A device that produces an output signal whenever a prescribed number of input pulses has been received. It frequently includes indicating devices for interpolation. *See also:* **pulse.**   335

scale span (instrument). The algebraic difference between the values of the actuating electrical quantity corresponding to the two ends of the scale. *See also:* **instrument.**
                                                                        328

scaling (corrosion). (1) The formation at high temperatures of thick corrosion product layer(s) on a metal surface. (2) The deposition of water-insoluble constituents on a metal surface (as on the interior of water boilers).   221, 205

scaling circuit (radiation counters). A device that produces an output pulse whenever a prescribed number of input pulses has been received. *See also:* **anticoincidence (radiation counters).**   190

scalloping (in a navigation system such as localizer, glideslope, very-high-frequency omnidirectional radio range, Tacan). The irregularities in the field pattern of the ground facility due to unwanted reflections from obstructions or terrain features, exhibited in flight as cyclical variations in bearing error. Also called **course scalloping.** *See also:* **navigation.**   187, 13

scan (1) (general). To examine sequentially part by part.
                                                                    255, 77

(2) (oscillography). The process of deflecting the electron beam. *See:* **graticule area; uniform luminance area; phosphor screen.** *See also:* **oscillograph.**   185

scan angle (antenna). The angle between the maximum of the major lobe and a reference direction. *Note:* Reference boresight is usually chosen as the reference direction. *Syn:* **beam angle.**   111

scanner (1) (facsimile). That part of the facsimile transmitter which systematically translates the densities of the subject copy into signal waveform. *See also:* **scanning (facsimile).**   12

(2) (industrial control). (A) A multiplexing arrangement that sequentially connects one channel to a number of channels. (B) An arrangement that progressively examines a surface for information. *See also:* **control system, feedback.**   219, 206

(3) (test, measurement and diagnostic equipment). A device that sequentially samples a number of data points. *See:* **flying-spot scanner; optical scanner; visual scanner.**   54

scanning (1) (radar). A programmed motion given to the major lobe of an antenna for the purpose of searching a larger volume than can be covered with a single direction of the beam, or for measuring angular location of a target; also, the analogous process using range gates or frequency-domain filters.   13

(2) (television). The process of analyzing or synthesizing successively, according to a predetermined method, the light values or equivalent characteristics of elements constituting a picture area. *See also:* **television.**   178

(3) (navigation aids). A periodic motion given to the major lobe of an antenna. *See also:* **antenna; navigation.**
                                                                    179, 187

(4) (facsimile). The process of analyzing successively the densities of the subject copy according to the elements of

a predetermined pattern. *Note:* The normal scanning is from left to right and top to bottom of the subject copy as when reading a page of print. Reverse direction is from right to left and top to bottom of the subject copy.   12

(5) (telephone switching systems). The periodic examination of circuit states under common control.   55

(6) (antenna beam). A repetitive motion given to the major lobe of an antenna.   111

scanning, high-velocity (electron tube). The scanning of a target with electrons of such velocity that the secondary-emission rate is greater than unity. *See also:* **television.**
                                                              178, 190, 125

scanning line (television). A single continuous narrow strip that is determined by the process of scanning. *Note:* In most television systems, the scanning lines that occur during the retrace intervals are blanked. The total number of scanning lines is numerically equal to the ratio of line frequency to frame frequency. *See also:* **television.**
                                                                        163

scanning linearity (television). A measure of the uniformity of scanning speed during the unblanked trace interval. *See also:* **television.**   163

scanning line frequency (facsimile). *See:* **stroke speed (scanning or recording line frequency).**

scanning line length (facsimile). The total length of scanning line is equal to the spot speed divided by the scanning line frequency. *Note:* This is generally greater than the length of the available line. *See also:* **scanning (facsimile).**
                                                                         12

scanning loss (1) (radar system employing a scanning antenna). The reduction in sensitivity, usually expressed in decibels, due to scanning across a target, compared with that obtained when the beam is directed constantly at the target. *See also:* **antenna.**   111

(2) (navigation). The reduction in sensitivity of a scanning radar due to motion of the beam between transmission and reception of the signal. *See also:* **beam-shape loss.**   13

scanning, low-velocity (electron tube). The scanning of a target with electrons of velocity less than the minimum velocity to give a secondary-emission ratio of unity. *See also:* **television.**   178, 190, 125

scanning speed (television). The time rate of linear displacement of the scanning spot. *See also:* **television.**
                                                                        163

scanning spot (1) (television). The area with which the scanned area is being explored at any instant in the scanning process. *See also:* **television.**   328

(2) (facsimile). The area on the subject copy viewed instantaneously by the pickup system of the scanner. *See also:* **scanning (facsimile).**   12

scanning spot, X dimension (facsimile). The effective scanning-spot dimension measured in the direction of the scanning line on the subject copy. *Note:* The numerical value of this will depend upon the type of system used. *See also:* **scanning (facsimile).**   12

scanning spot, Y dimension (facsimile). The effective scanning-spot dimension measured perpendicularly to the scanning line on the subject copy. *Note:* The numerical value of this will depend upon the type of system used. *See also:* **scanning (facsimile).**   12

scanning velocity (non-real time spectrum analyzer). Frequency span divided by sweep time.   68

scan pitch (facsimile). The number of scanning lines per unit length measured perpendicular to the direction of scanning. 11

scan sector (antenna). The angular interval over which the major lobe of an antenna is scanned. 111

scan time (1) (sequential events recording systems). The time required to examine the state of all inputs. 48, 58
(2) (acousto-optic deflector). The time for the light beam to be scanned over the angular swing of the deflector. 72

scatterband (interrogation systems). The total bandwidth occupied by the various received signals from interrogators operating with carriers on the same nominal radio frequency; the scatter results from the individual deviations from the nominal frequency. *See also:* **navigation.** 187

scattering (1) (general). The production of waves of changed direction, frequency, or polarization when radio waves encounter matter. *Note:* The term is frequently used in a narrower sense, implying a disordered change in the incident energy. *See also:* **radiation.** 328
(2) (laser-maser). The angular dispersal of power from a beam of radiation (or the perturbation of the field distribution of a resonance mode) either with or without a change in frequency, caused for example by inhomogeneities or nonlinearities of the medium or by irregularities in the surfaces encountered by the beam. 363

scattering coefficient. Element of the scattering matrix. *See:* **scattering matrix.** 185

scattering cross section (of an object in a given orientation). 4π times the ratio of the radiation intensity of the scattered wave in a specified direction to the power per unit area in an incident plane wave of a specified polarization. *Note:* The term **bistatic cross section** denotes the scattering cross section in any specified direction other than back towards the source. *See also:* **antenna; radiation.** 179

scattering loss (1) (acoustics). That part of the transmission loss that is due to scattering within the medium or due to roughness of the reflecting surface. 176
(2) (laser-maser). That portion of the loss in received power which is due to **scattering.** 363

scattering matrix (1) (waveguide components). A square array of complex numbers consisting of the transmission and reflection coefficients of a waveguide component. As most commonly used, each of these coefficients relates the complex electric field strength (or voltage) of a reflected or transmitted wave to that of an incident wave. The subscripts of a typical coefficient S_{ij} refer to the output and input ports related by the coefficient. These coefficients, which may vary with frequency, apply at a specified set of input and output reference planes. 319, 179
(2) (circuits and systems). An nxn (square) matrix used to relate incident waves and reflected waves for an n-port network. If the incident wave quantities for the ports are denoted by the vector A and the reflected wave quantities by the vector B then the scattering matrix S is defined such that $B = SA$. where:

$$a_i = \frac{1}{\sqrt{R_e Z_i}}(V_i + Z_i I_i)$$

$$b_i = \frac{1}{\sqrt{R_e Z_i}}(V_i - Z_i I_i).$$

Z_i is the port normalization impedance with $R_e Z_i > 0$. One formula for the scattering matrix is $S = [Z + R]^{-1}[Z - R]$ where Z is the open circuit impedance matrix that describes the network and R is a diagonal matrix representing the source or load resistances at each port. It should be noted that the scattering matrix is defined with respect to a specific set of port terminations. Physical interpretations can be given to the scattering coefficients for example, $|S_{ij}|^2$ is the fraction of available power that is delivered to the port termination at port i due to a source at port j. 67

scheduled frequency (electric power systems). The frequency that a power system or an interconnected system attempts to maintain. 94

scheduled frequency offset (electric power systems). The amount, usually expressed in hundredths of a hertz, by which the frequency schedule is changed from rated frequency in order to correct a previously accumulated time deviation. 94

scheduled interruption (electric power systems). An interruption caused by a scheduled outage. *See also:* **outage.** 200

scheduled maintenance. *See:* **preventive maintenance.**

scheduled net interchange (electric power systems) (control area). The mutually prearranged intended net power and/or energy on the area tie lines. 94

scheduled outage (electric power systems). An outage that results when a component is deliberately taken out of service at a selected time, usually for purposes of construction, preventive maintenance, or repair. *See also:* **outage.** 200

scheduled outage duration (electric power systems). The period from the initiation of the outage until construction, preventive maintenance, or repair work is completed and the affected component is made available to perform its intended function. *See also:* **outage.** 200

schedule, electric rate (electric power supply). A statement of an electric rate and the terms and conditions governing its application. 112

schedule setter or set-point device (speed-governing system). A device for establishing or setting the desired value of a controlled variable. *See also:* **speed-governing system.** 94

schematic diagram (elementary diagram). A diagram which shows, by means of graphic symbols, the electrical connections and functions of a specific circuit arrangement. The schematic diagram facilitates tracing the circuit and its functions without regard to the actual physical size, shape, or location of the component device or parts. 25

Scherbius machine (rotating machinery). A polyphase alternating-current commutator machine capable of generator or motor action, intended for connection in the secondary circuit of a wound-rotor induction motor supplied from a fixed-frequency polyphase power system, and used for speed and/or power-factor control. The magnetic circuit components are laminated and may be of the salient-pole type or of the cylindrical-rotor uniformly slotted type, either type having a series-connected armature reaction compensating winding as part of the field system. The control field winding may be separately or shunt-excited with or without an additional series-excited field winding. *See also:* **asynchronous machine.** 63

Schering bridge. A 4-arm alternating-current bridge in which the unknown capacitor and a standard loss-free capacitor form two adjacent arms, while the arm adjacent to the standard capacitor consists of a resistor and a capacitor in parallel, and the fourth arm is a nonreactive resistor. *Note:* Normally used for the measurement of capacitance and dissipation factor. Usually, one terminal of the source is connected to the junction of the unknown capacitor with the standard capacitor. With this connection, if the impedances of the capacitance arms are large compared to those of the resistance arms, most of the applied voltage appears across the former, the maximum test voltage being limited by the rating of the standard capacitor. If the detector and the source of electromotive force are interchanged the resulting circuit is called a **conjugate Schering bridge.** The balance is independent of frequency. *See also:* **bridge.** 328

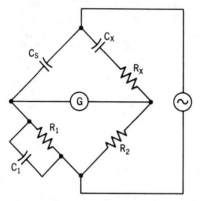

$$C_x R_2 = C_S R_1$$
$$C_x R_x = C_1 R_1$$

Schering bridge.

Schlieren method (acoustics). The technique by which light refracted by the density variations resulting from acoustic waves is used to produce a visible image of a sound field. 176

Schmitt trigger. A solid state element that produces an output when the input exceeds a specified turn-on level, and whose output continues until the input falls below a specified turn-off level. 70

Schottky effect. *See:* **Schottky emission.**

Schottky emission (electron tubes). The increased thermionic emission resulting from an electric field at the surface of the cathode. *See also:* **electron emission.** 190, 125

Schottky noise (electron tubes). The variation of the output current resulting from the random emission from the cathode. *See also:* **electronic tube.** 244, 190

Schuler tuning (design of inertial navigation equipment). The application of parameter values such that accelerations do not deflect the platform system from any vertical to which it has been set; a Schuler-tuned system, if fixed to the mean surface of a nonrotating earth, exhibits a natural period of 84.4 minutes. *See also:* **navigation.** 187, 13

scintillation (1) (radio propagation). A random fluctuation of the received field about its mean value, the deviations

usually being relatively small. *Note:* This use of the term scintillation is an extension of the astronomical term for the twinkling of stars, and the underlying explanation may be similar. *See also:* **radiation.** 328

(2) (radar). Variations in the signal received from a complex target due to changes in aspect angle, etcetera. *Note:* Because this term has been applied variously to **target fluctuation** and **scintillation error,** use of one of the more specific terms is recommended to avoid ambiguity. 13, 187

(3) (scintillators). The optical photons emitted as a result of the incidence of a particle or photon of ionizing radiation on a scintillator. *Note:* Optical photons unless otherwise specified are photons with energies corresponding to wavelengths between 2000 and 15 000 angstroms. *See also:* **ionizing radiation; radiation.** 117

(4) (laser-maser). This term is used to describe the rapid changes in irradiance levels in a cross section of a laser beam. 363

scintillation counter. The combination of scintillation-counter heads and associated circuitry for detection and measurement of ionizing radiation. 117

scintillation-counter cesium resolution. The scintillation-counter energy resolution for the gamma ray or conversion electron from cesium-137. *See also:* **scintillation counter.** 117

scintillation-counter energy resolution. A measure of the smallest difference in energy between two particles or photons of ionizing radiation that can be discerned by the scintillation counter. Quantitatively it is the fractional standard deviation (σ/E_1) of the energy distribution curve. *Note:* The fractional full width at half maximum of the energy distribution curve (FWHM/E_1) is frequently used as a measure of the scintillation-counter energy resolution where E_1 is the mode of the distribution curve. See the accompanying figure. *See also:* **scintillation counter.** 117

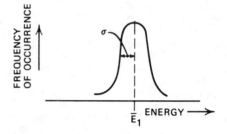

Scintillation-counter energy resolution

scintillation-counter energy-resolution constant. The product of the square of the scintillation-counter energy resolution, expressed as the fractional full width at half maximum (FWHM/E_1), and the specified energy. *See also:* **scintillation counter.** 335

scintillation counter head. The combination of scintillators and phototubes or photocells that produces electric pulses or other electric signals in response to ionizing radiation. *See also:* **phototube; scintillation counter.** 335

scintillation-counter time discrimination. A measure of the smallest interval of time between two individually discernible events. Quantitatively it is the standard deviation

σ of the time-interval curve. See the accompanying figure. *Note:* The full width at half maximum of the time-interval curve is frequently used as a measure of the time discrimination. *See also:* **scintillation counter.**   333

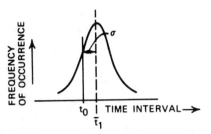

Scintillation-counter time discrimination

scintillation decay time. The time required for the rate of emission of optical photons of a scintillation to decrease from 90 percent to 10 percent of its maximum value. *Note:* Optical photons, for the purpose of this Standard, are photons with energies corresponding to wavelengths between 2000 and 15 000 angstroms. *See also:* **scintillation counter.**   117

scintillation duration. The time interval from the emission of the first optical photon of a scintillation until 90 percent of the optical photons of the scintillation have been emitted. *Note:* Optical photons are photons with energies corresponding to wavelengths between 2000 and 15 000 angstroms. *See also:* **scintillation counter.**   117

scintillation error (radar). An error in target location or Doppler produced in the radar measuring circuits by target fluctuation (as contrasted with glint), and subject to reduction by processes such as monopulse angle tracking, split-gate range tracking, and fast AGC (automatic gain control).   13

scintillation rise time. The time required for the rate of emission of optical photons of a scintillation to increase from 10 percent to 90 percent of its maximum value. *Note:* Optical photons are photons with energies corresponding to wavelengths between 2000 and 15 000 angstroms. *See also:* **scintillation counter.**   117

scintillator. The body of scintillator material together with its container. *See also:* **scintillation counter.**   117

scintillator conversion efficiency. The ratio of the optical photon energy emitted by a scintillator to the incident energy of a particle or photon of ionizing radiation. *Note:* The efficiency is generally a function of the type and energy of ionizing radiation. Optical photons are photons with energies corresponding to wavelengths between 2000 and 15 000 angstroms. *See also:* **scintillation counter.**   335

scintillator material. A material that emits optical photons in response to ionizing radiation. *Notes:* (1) There are five major classes of scintillator materials, namely: (A) inorganic crystals such as NaI(Tl) single crystals, ZnS(Ag) screens, (B) organic crystals (such as, anthracene, *trans*-stilbene), (C) solution scintillators: (1) liquid, (2) plastic, (3) glass, (D) gaseous scintillators, (E) Cerenkov scintillators. (2) Optical photons are photons with energies corresponding to wavelengths between 2000 and 15 000 angstroms. *See also:* **scintillation counter.**   117

scintillator-material total conversion efficiency. The ratio of the optical photon energy produced to the energy of a particle or photon of ionizing radiation that is totally absorbed in the scintillator material. *Note:* The efficiency is generally a function of the type and energy of the ionizing radiation. Optical photons are photons with energies corresponding to wavelengths between 2000 and 15 000 angstroms. *See also:* **scintillation counter.**   117

scintillator photon distribution (in number). The statistical distribution of the number of optical photons produced in the scintillator by total absorption of monoenergetic particles. *Note:* Optical photons are photons with energies corresponding to wavelengths between 2000 and 15 000 angstroms. *See also:* **scintillation counter.**   117

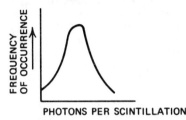

PHOTONS PER SCINTILLATION

Scintillator photon distribution

scope (radar). The face of a cathode-ray tube or a display of similar appearance. Colloquial abbreviation of a cathode-ray oscilloscope.   13

scoring system (electroacoustics) (motion-picture production). A recording system used for recording music to be reproduced in timed relationship with a motion picture.   176

scotopic spectral luminous efficiency function, (V_λ'), (photometric standard observer for scotopic vision) (light emitting diodes). The ratio of the radiant flux at wavelength λ_m, to that at wavelength λ, when the two fluxes produce the same scotopic luminous sensations under specified photometric conditions, λ_m, being chosen so that the maximum value of this ratio is unity. Unless otherwise indicated, the values used for the spectral luminous efficiency function relate to scotopic vision by the photometric standard observer having the characteristics laid down by the International Commission on Illumination.   162

scotopic vision. Vision mediated essentially or exclusively by the rods. It is generally associated with adaptation to a luminance below about 0.01 footlambert (0.03 nit). *See also:* **visual field.**   167

Scott connected transformer (T-connected transformer). An assembly used to transfer energy from a three-phase circuit to a two-phase circuit, or vice versa; or from a three-phase circuit to another three-phase circuit. The assembly consists of a main transformer with a tap at its midpoint connected directly between two of the phase wires of a three-phase circuit, and of a teaser transformer connected between the mid-tap of the main transformer and a third phase wire of the three-phase circuit. The other windings of the transformers may be connected to provide either two-phase or three-phase output. Alternatively, this may be accomplished with an assembly utilizing a three-legged core with main and teaser coil assemblies located on the two outer legs, and with a center leg which has no coil assembly and provides a common magnetic circuit for the two outer legs. *See also:* **main transformer; teaser transformer; interlacing impedance voltage of a Scott-connected transformer.** 53

Scott-connected transformer, interlacing impedance voltage. The single-phase voltage applied from the midtap of the main transformer winding to both ends, connected together, that is sufficient to circulate in the supply lines a current equal to the three-phase line current. The current in each half of the winding is 50 percent of this value. *See:* **efficiency.** 207

Scott-connected transformer per-unit resistance. The measured watts expressed in per-unit on the base of the rated kilovolt-ampere of the teaser winding. 207

SCR (1). *See:* **semiconductor controlled rectifier.**
(2) silicon controlled rectifier. *See:* Note under **semiconductor controlled rectifier.**
(3) selenium controlled rectifier. *See:* Note under **semiconductor controlled rectifier.**

scram (nuclear power generating station). The rapid shutdown of a nuclear reactor. **Notes:** (1) Usually, a scram is accomplished by rapid insertion of safety or control rods, or both. (2) Emergencies or deviations from normal operation may require scramming the reactor by manual or automatic means. 112

scratchpad memory. *See:* **temporary storage.** 54

scraper hoist. A power-driven hoist operating a scraper to move material (generally ore or coal) to a loading point. 328

screen (1) (rotating machinery). A port cover with multiple openings used to limit the entry of foreign objects. 63
(2) (cathode-ray tubes). The surface of the tube upon which the visible pattern is produced. *See also:* **electrode (electron tube).** 190, 125

screened conductor cable (insulated conductors). A cable in which the insulated conductor or conductors is/are enclosed in a conducting envelope or envelopes. 57

screen factor (electron-tube grid). The ratio of the actual area of the grid structure to the total area of the surface containing the grid. *See also:* **electron tube.** 244, 190

screen grid. A grid placed between a control grid and an anode, and usually maintained at a fixed positive potential, for the purpose of reducing the electrostatic influence of the anode in the space between the screen grid and the cathode. *See also:* **electrode (of an electron tube); grid.** 190, 125

screen-grid modulation. Modulation produced by application of a modulating voltage between the screen grid and the cathode of any multigrid tube in which the carrier is present. 211

screening (telephone switching systems). The ability to accept or reject calls by using trunk or line class or trunk or line number information 55

screening test (reliability). A test, or combination of tests, intended to remove unsatisfactory items or those likely to exhibit early failures. *See also:* **reliability.** 164

screen protected. *See:* **guarded.**

screen, viewing. *See:* **viewing area.**

screw machine (elevators). An electric driving machine, the motor of which raises and lowers a vertical screw through a nut with or without suitable gearing, and in which the upper end of the screw is connected directly to the car

frame or platform. The machine may be of direct or indirect drive type. 328

seal (window) (in a waveguide). A gastight or watertight membrane or cover designed to present no obstruction to radio-frequency energy. *See also:* **waveguide.** 179

seal, aperture. *See:* **aperture seal.**

seal, double electric conductor (nuclear power generating stations). An assembly of two single electric conductor seals in series and arranged in such a way that there is a double pressure barrier seal between the inside and the outside of the containment structure along the axis of the conductors. 26

sealable equipment (electric system). Equipment so arranged or enclosed that it may be sealed or locked to prevent operation or access to live parts. *Note:* Enclosed equipment may or may not be operable without opening the enclosure. *See also:* **cabinet.** 256

sealed (1) (rotating machinery). Provided with special seals to minimize either the leakage of the internal coolant out of the enclosure or the leakage of medium surrounding the enclosure into the machine. *See also:* **asynchronous machine; direct-current commutating machine.** 63
(2) (transformers). So constructed that the enclosure will remain hermetically sealed within specified limits of temperature and pressure. 53

sealed-beam headlamp. An integral optical assembly designed for headlighting purposes, identified by the name **Sealed Beam** branded on the lens. *See also:* **headlamp.** 167

sealed cell (nuclear power generating stations) (lead storage batteries). A cell in which the only passage for the escape of gases from the interior of the cell is provided by a vent of effective spray-trap design adapted to trap and return to the cell particles of liquid entrained in the escaping gases. 31, 76

sealed end (cable) (shipping seal). The end fitted with a cap for protection against the loss of compound or the entrance of moisture. 64

sealed dry-type transformer, self-cooled (Class GA). A dry-type self-cooled transformer with an hermetically sealed tank. *Note:* The insulating gas may be air, nitrogen, or other gases (such as fluorocarbons) with high dielectric strength. 53

sealed refrigeration compressor (hermetic type). A mechanical compressor consisting of a compressor and a motor, both of which are enclosed in the same sealed housing, with no external shaft or shaft seals, the motor operating in the refrigerant atmosphere. *See also:* **appliances.** 256

sealed-tank system. A method of oil preservation in which the interior of the tank is sealed from the atmosphere and in which the gas plus the oil volume remains constant over the temperature range. 53

sealed transformer. A dry-type transformer with a hermetically sealed tank. 53

sealed tube. An electron tube that is hermetically sealed. *Note:* This term is used chiefly for pool-cathode tubes. 190

sealing gap (industrial control). The distance between the armature and the center of the core of a magnetic circuit-closing device when the contacts first touch each

other. *See also:* **electric controller; initial contact pressure.**
302, 225, 206

sealing voltage (or current) (contactors). The voltage (or current) necessary to complete the movement of the armature of a magnetic circuit-closing device from the position at which the contacts first touch each other. *See also:* **contactor; control switch.** 1, 206

seal-in relay. An auxiliary relay that remains picked up through one of its own contacts which bypasses the initiating circuit until de-energized by some other device.
103, 202, 60, 127

seal, pressure barrier (nuclear power generating stations). A seal that consists of an aperture seal and an electric conductor seal. 26

seal, single electric conductor (nuclear power generating stations). A mechanical assembly providing a single pressure barrier between the electric conductors and the electric penetration. 26

search (1) (information processing). To examine a set of items for those that have a desired property. *See:* **binary search; dichotomizing search.** 255, 77

(2) (test, measurement and diagnostic equipment). The scanning of information contained on a storage medium by comparing the information of each field with a predetermined standard until an identity is obtained. 54

searchlight. A projector designed to produce an approximately parallel beam of light and having an optical system with an aperture of eight inches or more. 167

searchlighting (radar). The process of projecting a radar beam continuously at a particular object or in a particular direction, as contrasted to scanning. 13

search radar. A radar used primarily for the detection of targets in a particular volume of interest. 13

sea return (radar). Clutter resulting from irregularities of the sea surface. 13

season cracking (corrosion). Cracking resulting from the combined effect of corrosion and internal stress. A term usually applied to stress-corrosion cracking of brass.
221, 205

SEC. *See:* **secondary-electron conduction.**

second. The duration of 9 192 631 770 periods of the radiation corresponding to the transition between the two hyperfine levels of the ground state of the cesium-133 atom. 233

secondary (adjective). (1) Operates after the primary device; for example, secondary arcing contacts. (2) Second in preference. (3) Referring to auxiliary or control circuits as contrasted with the main circuit; for example, secondary disconnecting devices, secondary and control wiring. (4) Referring to the energy output side of transformers or the conditions (voltages) usually encountered at this location; for example, secondary fuse, secondary unit substation.
103, 202

secondary and control wiring (switchgear assemblies) (small wiring). Wire used for control circuits and for connections between instrument transformer secondaries, instruments, meters, relays, or other equipment. 103, 202

secondary arcing contacts (switching device). The contacts on which the arc of the arc-shunting resistor current is drawn and interrupted. 103, 202

secondary calibration (nuclear power generating stations). The determination of the response of a system with an applicable source whose effect on the system was established at the time of a primary calibration. 31

secondary current rating (transformer). The secondary current existing when the transformer is delivering rated kilovolt-amperes at rated secondary voltage. *See also:* **transformer.** 203

secondary disconnecting devices (switchgear assembly). Self-coupling separable contacts provided to connect and disconnect the auxiliary and control circuits between the removable element and the housing. 103, 202

secondary distribution feeder. A feeder operating at secondary voltage supplying a distribution circuit. 64

secondary distribution mains. The conductors connected to the secondaries of distribution transformers from which consumers' services are supplied. *See also:* **center of distribution.** 64

secondary distribution network. A network consisting of secondary distribution mains. *See also:* **center of distribution.** 64

secondary distribution system. A low-voltage alternating-current system that connects the secondaries of distribution transformers to the consumers' services. *See:* **alternating-current distribution.** *See also:* **center of distribution.** 64

secondary distribution trunk line. A line acting as a main source of supply to a secondary distribution system. *See also:* **center of distribution.** 64

secondary electron (thermionics). An electron detached from a surface during secondary emission by an incident electron. *See also:* **electron emission.** 244, 190

secondary-electron conduction (SEC). The transport of charge under the influence of an externally applied field in low-density structured materials by free secondary electrons traveling in the interparticle spaces (as opposed to solid-state conduction). *See also:* **camera tube.** 190

secondary-electron conduction (SEC) camera tube. A camera tube in which an electron image is generated by a photocathode and focused on a target composed of (1) a backplate and (2) a secondary-electron-conduction layer that provides charge amplification and storage. *See also:* **camera tube.** 190

secondary emission. Electron emission from solids or liquids due directly to bombardment of their surfaces by electrons or ions. *See also:* **electron emission.** 190, 125

secondary-emission characteristic (surface) (thermionics). The relation, generally shown by a graph, between the secondary-emission rate of a surface and the voltage between the source of the primary emission and the surface. *See also:* **electron emission.** 244, 190

secondary-emission ratio (electrons). The average number of electrons emitted from a surface per incident primary electron. *Note:* The result of a sufficiently large number of events should be averaged to ensure that statistical fluctuations are negligible. 190, 125

secondary failure (reliability). *See:* **failure, secondary.**

secondary fault. An insulation breakdown occurring as a result of a primary fault. *See also:* **center of distribution.**
64

secondary fuse (transformers). A fuse used on the secondary-side circuits. *Note:* In high-voltage fuse parlance such a fuse is restricted for use on a low-voltage secondary distribution system that connects the secondaries of distribution transformers to consumers' services.   103, 202

secondary grid emission. Electron emission from a grid resulting directly from bombardment of its surface by electrons or other charged particles.   125

secondary neutral grid. A network of neutral conductors, usually grounded, formed by connecting together within a given area all the neutral conductors of individual transformer secondaries of the supply system. *See also:* **center of distribution.**   64

secondary power. The excess above firm power to be furnished when, as, and if available. *See also:* **generating station.**   64

secondary radar. A radar designed to operate only with targets carrying transponders.   13, 187

secondary radiator. That portion of an antenna having the largest radiating aperture, consisting of a reflecting surface or a lens, as distinguished from its feed.   111

secondary-selective type (low-voltage-selective type). A unit substation which has two stepdown transformers each connected to an incoming high-voltage circuit. The outgoing side of each transformer is connected to a separate bus through a suitable switching and protective device. The two sections of bus are connected by a normally open switching and protective device. Each bus has one or more outgoing radial (stub end) feeders.   53

secondary service area (radio broadcast station). The area within which satisfactory reception can be obtained only under favorable conditions. *See also:* **radio transmitter.**   328

secondary short-circuit current rating (high-reactance transformer). Designates the current in the secondary winding when the primary winding is connected to a circuit of rated primary voltage and frequency and when the secondary terminals are short circuited.   53

secondary, single-phase induction motor. The rotor or stator member that does not have windings that are connected to the supply line. *See also:* **asynchronous machine; induction motor.**   63

secondary standard (illuminating engineering). A constant and reproducible light source calibrated directly or indirectly by comparison with a primary standard. This order of standard also is designated as a reference standard. *Note:* National secondary (reference) standards are maintained at national physical laboratories; laboratory secondary (reference) standards are maintained at other photometric laboratories. *See also:* **primary standard (illuminating engineering).**   167

secondary unit substation (transformers). A unit substation in which the low-voltage section is rated 1000 volts and below.   53

secondary voltage (1) (capacitance potential device). The root-mean-square voltage obtained from the main secondary winding, and when provided, from the auxiliary secondary winding. *See:* **rated secondary voltage.** *See also:* **outdoor coupling capacitor.**   351

(2) (wound-rotor motor). The open-circuit voltage at standstill, measured across the collector rings with rated voltage applied to the primary winding. *See:* **asynchronous machine.**   328

secondary voltage rating (transformer). The load-circuit voltage for which the secondary winding is designed. *See*

also: **duty; transformer, specialty; transformer.**   301, 53

secondary winding (1) (general). The winding on the energy output side.   203, 53

(2) (voltage regulator). The series winding. *See also:* **voltage regulator.**   257

(3) (instrument transformer). The winding that is intended to be connected to the measuring or control devices.   53

secondary winding (rotating machinery). Any winding that is not a primary winding. *See:* **asynchronous machine; voltage regulator.**   63

second-channel attenuation. *See:* **selectance.**

second-channel interference. Interference in which the extraneous power originates from a signal of assigned (authorized) type in a channel two channels removed from the desired channel. *See:* **interference.** *See also:* **radio receiver.**   339

second order non-linearity coefficient (accelerometer). The proportionality constant that relates a variation of the output to the square of the input applied parallel to an input reference axis.   46

second-time-around echo (radar). An echo received after an interval exceeding one pulse repetition interval but less than two pulse repetition intervals. Third-time-around (etcetera) echoes are defined in a corresponding manner. The generic term **multiple-time-around** is sometimes used.   13

second Townsend discharge (gas). A semi-self-maintained discharge in which the additional ionization is due to the secondary electrons emitted by the cathode under the action of the bombardment by the positive ions present in the gas. *See also:* **discharge (gas).**   244, 190

second voltage range (railway signal). *See:* **voltage range.**

section (1) (rectifier unit). A part of a rectifier unit with its auxiliaries that may be operated independently. *See also:* **rectification.**   208

(2) (thyristor converter). Those parts of a thyristor converter unit containing the power thyristors (and when also used, the power diodes) together with their auxiliaries (including individual transformers or cell windings of double converters and circulating current reactors, if any), in which the main direct current when viewed from the converter unit dc terminals always flows in the same direction. A thyristor converter section is supposed to be operated independently. *Note:* A converter equipment may have either only one section or one forward and one reverse section.   121

sectional center, (telephone switching systems). A toll office to which may be connected a number of primary centers, toll centers, or toll points. Sectional centers are classified as Class 2 offices. *See also:* **office class.**   55

sectionalized linear antenna. A linear antenna in which reactances are inserted at one or more points along the length of the antenna.   111

section locking. Locking effective while a train occupies a given section of a route and adapted to prevent manipulation of levers that would endanger the train while it is within that section. *See also:* **interlocking (interlocking plant).**   328

section, sag. *See:* **sag section.**   45

sectoral horn antenna. A horn antenna with two opposite sides of the horn parallel and the two remaining sides diverging. 111

sector cable. A multiple-conductor cable in which the cross section of each conductor is substantially a sector of a circle, an ellipse, or a figure intermediate between them. *Note:* Sector cables are used in order to obtain decreased overall diameter and thus permit the use of larger conductors in a cable of given diameter. *See also:* **power distribution; underground construction.** 64

sector display (1) (radar). A limited display in which only a sector of the total service area of the radar system is shown; usually the sector to be displayed is selectable. 13

(2) (continuously rotating radar-antenna system). A range-amplitude display used with a radar set, the antenna system of which is continuously rotating. The screen, which is of the long-persistence type, is excited only while the beam of the antenna is within a narrow sector centered on the object. *See also:* **radar.** 187

sector impedance relay. A form of distance relay that by application and design has its operating characteristic limited to a sector of its operating circle on the *R-X* diagram. 103, 202, 60, 127

sector scanning. A modification of circular scanning in which the direction of the antenna beam generates a portion of a cone or a plane. 111

security (relay or relay system). That facet of reliability that relates to the degree of certainty that a relay or relay system will not operate incorrectly. 127, 103

sedimentation potential (electrobiology). The electrokinetic potential gradient resulting from unity velocity of a colloidal or suspended material forced to move by gravitational or centrifugal forces through a liquid electrolyte. *See also:* **electrobiology.** 192

sediment separator (rotating machinery). Any device, used to collect foreign material in the lubricating oil. *See:* **oil cup (rotating machinery).** 63

Seebeck coefficient (of a couple) (for homogeneous conductors). The limit of the quotient of (1) the Seebeck electromotive force by (2) the temperature difference between the junctions as the temperature difference approaches zero; by convention, the Seebeck coefficient of a couple is positive if the first-named conductor has a positive potential with respect to the second conductor at the cold junction. *Note:* The Seebeck coefficient of a couple is the algebraic difference of either the relative or absolute Seebeck coefficients of the two conductors. *See also:* **thermoelectric device.** 248, 191

Seebeck coefficient, absolute. The integral, from absolute zero to the given temperature, of the quotient of (1) the Thomson coefficient of the material by (2) the absolute temperature. *See also:* **thermoelectric device.** 248,191

Seebeck coefficient, relative. The Seebeck coefficient of a couple composed of the given material as the first-named conductor and a specified standard conductor. *Note:* Common standards are platinum, lead, and copper. *See also:* **thermoelectric device.** 248, 191

Seebeck effect. The generation of an electromotive force by a temperature difference between the junctions in a circuit composed of two homogeneous electric conductors of dissimilar composition; or, in a nonhomogeneous conductor, the electromotive force produced by a temperature gradient in a nonhomogeneous region. *See:* **thermoelectric effect.** *See also:* **thermoelectric device.** 248, 191

Seebeck electromotive force. The electromotive force resulting from the Seebeck effect. *See also:* **thermoelectric device.** 248, 191

segmental conductor. A stranded conductor consisting of three or more stranded conducting elements, each element having approximately the shape of the sector of a circle, assembled to give a substantially circular cross section. The sectors are usually lightly insulated from each other and, in service, are connected in parallel. *Note:* This type of conductor is known as type-*M* conductor in Canada. *See also:* **conductor.** 64

segmental-rim rotor (rotating machinery). A rotor in which the rim is composed of interleaved segmental plates bolted together. *See also:* **rotor (rotating machinery).** 63

segment shoe (bearing shoe) (rotating machinery). A pad that is part of the bearing surface of a pad-type bearing. *See also:* **bearing.** 63

segregated-phase bus. A bus in which all phase conductors are in a common metal enclosure, but are segregated by metal barriers between phases. 103, 202, 78

seizure signal (telephone switching systems). A signal transmitted from the sending end of a trunk to the far end to indicate that its sending end has been selected. 55

selectance (1) (general). A measure of the falling off in the response of a resonant device with departure from resonance. It is expressed as the ratio of the amplitude of response at the resonance frequency, to the response at some frequency differing from it by a specified amount.

(2) (radio receivers). The reciprocal of the ratio of the sensitivity of a receiver tuned to a specified channel to its sensitivity at another channel separated by a specified number of channels from the one to which the receiver is tuned. *Notes:* (A) Unless otherwise specified, selectance should be expressed as a voltage or field-strength ratio. (B) Selectance is often expressed as adjacent-channel attenuation (ACA) or second-channel attenuation (2 ACA). *See also:* **cutoff frequency.** 328

select before operate, supervisory control. *See:* **supervisory control system, select before operate.**

selecting (telephone switching systems). Choosing a particular group of one or more servers in the establishment of a call connection. 55

selection (computing systems). *See:* **amplitude selection; coincident-current selection.**

selection check (electronic computation). A check (usually an automatic check) to verify that the correct register, or other device, is selected in the interpretation of an instruction. 235, 255, 77

selection ratio. The least ratio or a magnetomotive force used to select a cell to the maximum magnetomotive force used that is not intended to select a cell. *See:* **coincident-current selection.** 331

selective collective automatic operation (elevators). Automatic operation by means of one button in the car for each landing level served and by UP and DOWN buttons at the landings, wherein all stops registered by the momentary actuation of the car buttons are made as defined under **nonselective collective automatic operation,** but wherein the stops registered by the momentary actuation of the

landing buttons are made in the order in which the landings are reached in each direction of travel after the buttons have been actuated. With this type of operation, all UP landing calls are answered when the car is traveling in the up direction and all DOWN landing calls are answered when the car is traveling in the down direction, except in the case of the uppermost or lowermost calls, which are answered as soon as they are reached irrespective of the direction of travel of the car. *See also:* **control (elevators).**
    328

selective dump (computing systems). A dump of a selected area of storage.   255, 77

selective fading (radio wave propagation). Fading that is different at different frequencies in a frequency band occupied by a modulated wave. *See also:* **radiation; radio wave propagation.**   146, 180

selective opening (selective tripping) (devices carrying fault current). The application of switching devices in series such that only the device nearest the fault will open and the devices closer to the source will remain closed and carry the remaining load.   103, 202

selective overcurrent trip. *See:* **selective release (selective trip); overcurrent release (overcurrent trip).**

selective overcurrent tripping. *See:* **overcurrent release (overcurrent trip); selective opening (selective tripping).**

selective release (selective trip). A delayed release with selective settings that will automatically reset if the actuating quantity falls and remains below the release setting for a specified time.   103, 202

selective ringing (telephone switching systems). Ringing in which only the ringer at the desired main station on a party line responds.   55

selective signaling equipment (mobile communication). Arrangements for signaling, selective from a base station, of any one of a plurality of mobile stations associated with the base station for communication purposes. *See also:* **mobile communication system.**   181

selectivity (1) (receiver performance). A measure of the extent to which a receiver is capable of differentiating between the desired signal and disturbances at other frequencies. *See also:* **radio receiver.**   339
(2) (circuits and systems). The characteristic of a filter that determines the extent to which the filter is capable of altering the frequency spectrum of a signal. A highly selective filter has an abrupt transition between a pass-band region and a stop-band region.   67
(3) (protective system). A general term describing the interrelated performance of relays and breakers, and other protective devices; complete selectivity being obtained when a minimum amount of equipment is removed from service for isolation of a fault or other abnormality.
    103, 202, 60, 127

selector, amplitude (pulse techniques). *See:* **selector, pulse-height.**

selector pulse (electronic navigation). A pulse that is used to identify, for selection, one event in a series of events. *See also:* **navigation.**   13, 187

selector, pulse-height (pulse techniques). *See:* **pulse-height selector.**

selector switch (1) (general). A switch arranged to permit connecting a conductor to any one of a number of other conductors.   103, 202, 27

(2) (automatic telephone switching). A remotely-controlled switch for selecting a group of trunk lines fixed by part of the call number and connecting to an idle trunk in that group. *See also:* **switch.**   328
(3) (industrial control). A manually operated multiposition switch for selecting alternative control circuits.   206
(4) (control systems). A machine-operated device that establishes definite zones in the operating cycle within which corrective functions can be initiated.   328

self-adapting. Pertaining to the ability of a system to change its performance characteristics in response to its environment.   255, 77

self-aligning bearing (rotating machinery). A sleeve bearing designed so that it can move in the end shield to align itself with the journal of the shaft. *See also:* **bearing.**   63

self-capacitance (conductor) (grounded capacitance) (total capacitance). In a multiple-conductor system, the capacitance between this conductor and the other $(n - 1)$ conductors connected together. *Note:* The self-capacitance of a conductor equals the sum of its $(n - 1)$ direct capacitances to the other $(n - 1)$ conductors.   210

self-checking code (electronic computation). A code that uses expressions such that one (or more) error(s) in a code expression produces a forbidden combination. Also called an error-detecting code. *See also:* **check, forbidden combination; error-detecting code; parity.**   235

self-closing door or gate (elevators). A manually opened hoistway door and/or a car door or gate that closes when released. *See also:* **hoistway (elevator or dumbwaiter).**
    328

self-commutated inverters. An inverter in which the commutation elements are included within the power inverter.   208

self-contained instrument. An instrument that has all the necessary equipment built into the case or made a corporate part thereof. *See also:* **instrument.**   280

self-contained navigation aid. An aid that consists only of facilities carried by the vehicle. *See also:* **navigation.**
    187

self-contained pressure cable. A pressure cable in which the container for the pressure medium is an impervious flexible metal sheath, reinforced if necessary, that is factory assembled with the cable core, *See:* **gas-filled cable; oil-filled cable; pressure cable.**   64

self-coupling separable contacts (switchgear assembly disconnecting device). Contacts, mounted on the stationary and removable elements of a switchgear assembly, that align and engage or disengage automatically when the two elements are brought into engagement or disengagement.
    202

self-excited. A qualifying term applied to a machine to denote that the excitation is supplied by the machine itself. *See also:* **direct-current commutating machine.**   63

self-field (Hall generator). The magnetic field caused by the flow of control current through the loop formed by the control current leads and the relevant conductive path through the Hall plate.   107

self-impedance (1) (mesh). The impedance of a passive mesh or loop with all other meshes of the network open-circuited.   210
(2) (array element). The input impedance of a radiating element of an array antenna with all other elements in the

array open-circuited. *Note:* In general, the self-impedance of a radiating element in an array is not equal to its isolated impedance. 111

(3) (network). At any pair of terminals, the ratio of an applied potential difference to the resultant current at these terminals, all other terminals being open. 328

self-inductance. The property of an electric circuit whereby an electromotive force is induced in that circuit by a change of current in the circuit. *Notes:* (1) The coefficient of self-inductance L of a winding is given by the following expression:

$$L = \frac{\partial \lambda}{\partial i}$$

where λ is the total flux-linkage of the winding and i is the current in the winding. (2) The voltage e induced in the winding is given by the following equation:

$$e = -\left[L\frac{di}{dt} + i\frac{dL}{dt} \right]$$

If L is constant

$$e = -L\frac{di}{dt}.$$

(3) The definition of self-inductance L is restricted to relatively slow changes in i, that is, to low frequencies, but by analogy with the definitions, equivalent inductances may often be evolved in high-frequency applications such as resonators, waveguide equivalent circuits, etcetera. Such inductances, when used, must be specified. The definition of self-inductance L is also restricted to cases in which the branches are small in physical size compared with a wavelength, whatever the frequency. Thus in the case of a uniform 2-wire transmission line it may be necessary even at low frequencies to consider the parameters as distributed rather than to have one inductance for the entire line. 210, 177

self-information. *See:* **information content.**

self-lubricating bearing (rotating machinery). A bearing lined with a material containing its own lubricant such that little or no additional lubricating fluid need be added subsequently to ensure satisfactory lubrication of the bearing. *See also:* **bearing.** 63

self-maintained discharge (gas). A discharge characterized by the fact that it maintains itself after the external ionizing agent is removed. *See also:* **discharge (gas).**
244, 190

self-organizing. Pertaining to the ability of a system to arrange its internal structure. 255, 77

self-phasing array antenna system. A receiving antenna system which introduces a phase distribution among the array elements so as to maximize the received signal regardless of the direction of incidence. *See also:* **retrodirective antenna.** 111

self-propelled electric car. An electric car requiring no external source of electric power for its operation. *Note:* Diesel-electric, gas-electric, and storage-battery-electric cars are examples of self-propelled cars. The prefix self-propelled is also applied to buses. *See also:* **electric motor car.** 328

self-propelled electric locomotive. An electric locomotive

requiring no external source of electric power for its operation. *Note:* Storage-battery, diesel-electric, gas-electric and turbine-electric locomotives are examples of self-propelled electric locomotives. *See also:* **electric locomotive.** 328

self-pulse modulation. Modulation effected by means of an internally generated pulse. *See:* **blocking oscillator.** *See also:* **oscillatory circuit.** 111

self-quenched counter tube. A radiation counter tube in which reignition of the discharge is inhibited by internal processes. *See also:* **gas-filled radiation-counter tubes.**
328

self-rectifying X-ray tube. An X-ray tube operating on alternating anode potential. *See also:* **electron devices, miscellaneous.** 190

self regulation. *See:* **inherent regulation.**

self-reset manual release (control) (industrial control). A manual release that is operative only while it is held manually in the release position. *See:* **electric controller.**
225, 206

self-reset relay (automatically reset relay). A relay that is so constructed that it returns to its reset position following an operation after the input quantity is removed.
103, 202, 60, 127

self-restoring fire detector (fire protection devices). A restorable fire detector whose sensing element is designed to be returned to normal automatically. 71

self-restoring insulation (transformer). Insulation which completely recovers its insulating properties after a disruptive discharge caused by the application of a test voltage; insulation of this kind is generally, but not necessarily, external insulation. 53

self-saturation (magnetic amplifier). The saturation obtained by rectifying the output current of a saturable reactor. 328

self-supporting aerial cable. A cable consisting of one or more insulated conductors factory assembled with a messenger that supports the assemblage, and that may or may not form a part of the electric circuit. *See also:* **conductor.** 64

self-surge impedance. *See:* **surge impedance.**

self-test (test, measurement and diagnostic equipment). A test or series of tests, performed by a device upon itself, which shows whether or not it is operating within designed limits. This includes test programs on computers and automatic test equipment which check out their performance status and readiness. 54

self-test capability (test, measurement and diagnostic equipment). The ability of a device to check its own circuitry and operation. The degree of self-test is dependent on the ability to fault detect and isolate. 54

self-ventilated (rotating machinery). Applied to a machine which has its ventilating air circulated by means integral with the machine. *See also:* **asynchronous machine; direct-current commutating machine.** 3, 325, 63

semantics. The relationships between symbols and their meanings. 255, 77

semaphore signal. A signal in which the day indications are given by the vertical angular position of a blade known as the semaphore arm. 328

semiactive guidance (radar). A bistatic-radar homing system in which a receiver in the guided vehicle derives guidance information from electromagnetic signals scattered from the target, which is illuminated by a transmitter at a third location.   13

semianalytic inertial navigation equipment. The same as geometric inertial navigation equipment except that the horizontal measuring axes are not maintained in alignment with a geographic direction. *Note:* The azimuthal orientations are automatically computed. *See also:* **navigation.**   187, 13

semiautomatic. Combining manual and automatic features so that a manual operation is required to supply to the automatic feature the actuating influence that causes the automatic feature to function.   328

semiautomatic controller. An electric controller in which the influence directing the performance of some of its basic functions is automatic. *See also:* **electric controller.**   206

semiautomatic flight inspection (SAFI) (electronic navigation). A specialized and largely automatic system for evaluating the quality of information in signals from ground-based navigational aids; data from navigational aids along and adjacent to any selected air route are simultaneously received by a specially equipped semiautomatic-flight-inspection aircraft as it proceeds under automatic control along the route, evaluated at once for gross errors and recorded for subsequent processing and detailed analysis at a computer-equipped central ground facility. *Note:* Flight inspection means the evaluation of performance of navigational aids by means of in-flight measurements. *See also:* **navigation.**   187, 13

semiautomatic gate (elevators). A gate that is opened manually and that closes automatically as the car leaves the landing. *See also:* **hoistway (elevator or dumbwaiter).**   328

semiautomatic holdup-alarm system. An alarm system in which the signal transmission is initiated by the indirect and secret action of the person attacked or of an observer of the attack. *See also:* **protective signalling.**   328

semiautomatic plating. Mechanical plating in which the cathodes are conveyed automatically through only one plating tank. *See also:* **electroplating.**   328

semiautomatic signal. A signal that automatically assumes a stop position in accordance with traffic conditions, and that can be cleared only by cooperation between automatic and manual controls.   328

semiautomatic telephone systems. A telephone system in which operators receive orders orally from the calling parties and establish connections by means of automatic apparatus.   328

semiautomatic test equipment (test, measurement and diagnostic equipment). Any automatic testing device which requires human participation in the decision-making, control, or evaluative functions.   54

semiconducting jacket. A jacket of such resistance that its outer surface can be maintained at substantially ground potential by contact at frequent intervals with a grounded metallic conductor, or when buried directly in the earth.   64, 57

semiconducting material. A conducting medium in which the conduction is by electrons, and holes, and whose temperature coefficient of resistivity is negative over some temperature range below the melting point. *See:* **semi-**

conductor. *See also:* **semiconductor device.**   210

semiconducting paint (rotating machinery). A paint in which the pigment or portion of pigment is a conductor of electricity and the composition is such that when converted into a solid film, the electrical conductivity of the film is in the range between metallic substances and electrical insulators.   63

semiconducting tape (power distribution, underground cables). A tape of such resistance that when applied between two elements of a cable the adjacent surfaces of the two elements will maintain substantially the same potential. Such tapes are commonly used for conductor shielding and in conjunction with metallic shielding over the insulation.   57

semiconductive ignition cable (electromagnetic compatibility). High-tension ignition cable, the core of which is made of semiconductive material. *Note:* Semiconductive is understood here as referring to conductivity and no other physical properties. *See also:* **electromagnetic compatibility.**   220, 199

semiconductor. An electronic conductor, with resistivity in the range between metals and insulators, in which the electric-charge-carrier concentration increases with increasing temperature over some temperature range. *Note:* Certain semiconductors possess two types of carriers, namely, negative electrons and positive holes.   237, 245, 210, 186, 66

semiconductor, compensated. A semiconductor in which one type of impurity or imperfection (for example, donor) paritally cancels the electric effects of the other type of impurity or imperfection (for example, acceptor). *See also:* **semiconductor.**   245, 210, 66, 23, 119, 118

semiconductor controlled rectifier (SCR). An alternative name used for the reverse-blocking triode-thyristor. *Note:* The name of the actual semiconductor material (selenium, silicon, etcetera) may be substituted in place of the word **semiconductor** in the name of the components. *See also:* **thyristor.**   245, 210, 66

semiconductor converters, classification. The following designations are intended to describe the functional characteristics of converters, but not necessarily the circuits or components used.
Note: Forms A through D refer only to the converters. Rotational direction of motors may be changed by field or armature reversal.
(1) form A converter. A single converter unit in which the direct current can flow in one direction only and which is not capable of inverting energy from the load to the ac supply. Operates in quadrant I only (semiconverter).
(2) form B converter. A double converter unit in which the direct current can flow in either direction but which is not capable of inverting energy from the load to the ac supply. Operates in quadrants I and III only.
(3) form C converter. A single converter unit in which the direct current can flow in one direction only and which is capable of inverting energy from the load to the ac supply. Operates in quadrants I and IV.
(4) form D converter. A double converter unit in which the direct current can flow in either direction and which is capable of inverting energy from the load to the ac supply. Operates in quadrants I, II, III, and IV.   121

semiconductor device (1) (general). An electron device

in which the characteristic distinguishing electronic conduction takes place within a semiconductor. *See also:* **semiconductor.** 237, 210, 66

(2) (light emitting diodes). A device whose essential characteristics are due to the flow of charge carriers within a semiconductor. 162

semiconductor device, multiple unit. A semiconductor device having two or more sets of electrodes associated with independent carrier streams. *Note:* It is implied that the device has two or more output functions that are independently derived from separate inputs, for example, a duo-triode transistor. *See also:* **semiconductor.** 66

semiconductor device, single unit. A semiconductor device having one set of electrodes associated with a single carrier stream. *Note:* It is implied that the device has a single output function related to a single input. *See also:* **semiconductor.** 66

semiconductor diode (1) (light emitting diodes). A semiconductor device having two terminals and exhibiting a nonlinear voltage-current characteristic. *See also:* **semiconductor; semiconductor rectifier cell.** 237, 66, 162

(2) (circuits and systems). A two-terminal device formed of a semiconductor junction having a nonlinear characteristic which will conduct electric current more in one direction than in the other. 67

semiconductor-diode parametric amplifier. A parametric amplifier using one or more varactors. *See also:* **parametric device.** 277, 191

semiconductor, extrinsic (1) (general). A semiconductor with charge-carrier concentration dependent upon impurities. *See also:* **semiconductor.** 245, 210

(2) (power semiconductor). A semiconductor in which the concentrations of holes and electrons are unbalanced by the introduction of impurities. 66

semiconductor frequency changer. A complete equipment employing semiconductor devices for changing from one alternating-current frequency to another. *See also:* **semiconductor rectifier stack.** 208

semiconductor, intrinsic (1) (general). A semiconductor whose charge-carrier concentration is substantially the same as that of the ideal crystal. *See also:* **semiconductor.** 245, 210, 66

(2) (power semiconductor). A semiconductor in which holes and electrons are created solely by thermal excitation across the energy gap. In an intrinsic semiconductor the concentration of holes and electrons must always be the same. 66

(3) (semiconductor radiation detectors). A semiconductor containing an equal number of free holes and electrons throughout its volume. 23

semiconductor junction (light emitting diodes). A region of transition between semiconductor regions of different electrical properties. 162

semiconductor, n-type. An extrinsic semiconductor in which the conduction electron concentration exceeds the mobile hole concentration. *Note:* It is implied that the net ionized impurity concentration is donor type. *See also:* **semiconductor.** 245, 210, 66

semiconductor, n^+-type. An n-type semiconductor in which the excess conduction electron concentration is very large. *See also:* **semiconductor.** 245, 210, 66

semiconductor, p-type. An extrinsic semiconductor in which the mobile hole concentration exceeds the conduction electron concentration. *Note:* It is implied that the net ionized impurity concentration is acceptor type. *See also:* **semiconductor.** 245, 210, 66

semiconductor, p^+-type. A p-type semiconductor in which the excess mobile hole concentration is very large. *See also:* **semiconductor.** 245, 210, 66

semiconductor power converter. A complete equipment employing semiconductor devices for the transformation of electric power. *See also:* **semiconductor rectifier stack.** 208

semiconductor radiation detector. A semiconductor device that utilizes the production and motion of excess free charge carriers for the detection and measurement of incident radiation. 23, 118, 119

semiconductor rectifier. An integral assembly of semiconductor rectifier diodes or stacks including all necessary auxiliaries such as cooling equipment, current balancing, voltage divider, surge suppression equipment, etcetera, and housing, if any. *See also:* **semiconductor rectifier stack.** 208

semiconductor rectifier cell. A semiconductor device consisting of one cathode, one anode, and one rectifier junction. *See also:* **semiconductor; semiconductor rectifier stack.** 237, 66

semiconductor rectifier cell combination. The arrangement of semiconductor rectifier cells in one rectifier circuit, rectifier diode, or rectifier stack. The semiconductor rectifier cell combination is described by a sequence of four symbols written in the order 1-2-3-4 with the following significances: (1) Number of rectifier circuit elements. (2) Number of semiconductor rectifier cells in series in each rectifier circuit element. (3) Number of semiconductor rectifier cells in parallel in each rectifier circuit element. (4) Symbol designating circuit. If a semiconductor rectifier stack consists of sections of semiconductor rectifier cells insulated from each other, the total semiconductor rectifier cell combination becomes the sum of the semiconductor rectifier cell combinations of the individual insulated sections. If the insulated sections have the same semiconductor rectifier cell combination, the total semiconductor rectifier cell combination may be indicated by the semiconductor rectifier cell combination of one section preceded by a figure showing the number of insulated sections. Example: 4(4-1-1-B) indicates four single-phase full-wave bridges insulated from each other assembled as one semiconductor rectifier stack. *Notes:* (A) The total number of semiconductor rectifier cells in each semiconductor rectifier cell combination is the product of the numbers in the combination. (B) This arrangement can also be applied by analogy to give a semiconductor rectifier diode combination.

Symbol	Circuit	Example
H	half wave	1-1-1-H
C	center tap	2-1-1-C
B	bridge	4-1-1-B
		6-1-1-B
Y	wye	3-1-1-Y
S	star	6-1-1-S
D	voltage doubler	2-1-1-D

See also: **semiconductor rectifier cell.** 237, 66

semiconductor rectifier diode. A semiconductor diode having an asymmetrical voltage-current characteristic, used for the purpose of rectification, and including its associated housing, mounting, and cooling attachments if integral with it. *See also:* **semiconductor; semiconductor rectifier cell.** 237, 66, 208

semiconductor rectifier stack. An integral assembly, with terminal connections, of one or more semiconductor rectifier diodes, and includes its associated mounting and cooling attachments if integral with it. *Note:* It is a subassembly of, but not a complete semiconductor rectifier. 237, 66, 208

semiconverter, bridge. A bridge in which one commutating group uses thyristors and the other uses diodes. 121

semidirect lighting. Lighting that involves luminaires that distribute 60 to 90 percent of the emitted light downward and the balance upward. *See also:* **general lighting.** 167

semienclosed. (1) Having the ventilating openings in the case protected with wire screen, expanded metal, or perforated covers or (2) having a solid enclosure except for a slot for an operating handle or small openings for ventilation, or both. 328

semienclosed brake (industrial control). A brake that is provided with an enclosure that covers the brake shoes and the brake wheel but not the brake actuator. *See also:* **control.** 225, 206

semiflush-mounted device. One in which the body of the device projects in front of the mounting surface a specified distance between the distances specified for flush-mounted and surface-mounted devices. 103, 202

semiguarded machine (rotating machinery). One in which part of the ventilating openings, usually in the top half, are guarded as in the case of a guarded machine but the others are left open. *See also:* **direct-current commutating machine.** 63

semi-high-speed low-voltage dc power circuit breaker. A low-voltage dc power circuit breaker which, during interruption, limits the magnitude of the fault current so that its crest is passed not later than a specified time after the beginning of the fault current transient, where the system fault current, determined without the circuit breaker in the circuit, falls between specified limits of current at a specified time. *Note:* The specified time in present practice is 0.03 second. 103

semi-indirect lighting. Lighting involving luminaires that distribute 60 to 90 percent of the emitted light upward and the balance downward. *See also:* **general lighting.** 167

semimagnetic controller. An electric controller having only part of its basic functions performed by devices that are operated by electromagnets. *See also:* **electric controller.** 206, 3

semioutdoor reactor. A reactor suitable for outdoor use provided that certain precautions in installation (specified by the manufacturer) are observed. For example, protection against rain. 309

semiprotected enclosure. An enclosure in which all other openings, usually in the top half, are protected as in the case of a protected enclosure, but the others are left open. *See:* **protected enclosure.** *See also:* **asynchronous machine; direct-current commutating machine.** 3

semiremote control. A system or method of radio-transmitter control whereby the control functions are performed near the transmitter by means of devices connected to but not an integral part of the transmitter. *See also:* **radio transmitter.** 111

semiselective ringing (telephone switching systems). Ringing wherein the ringers at two or more of the main stations on a party line respond simultaneously, differentiation being by the number of rings. 55

semi-stop joint (power cable joints). A joint which is designed to restrict movement of the dielectric fluid between cables being joined. 34

semistrain insulator (semitension assembly). Two insulator strings at right angles, each making an angle of about 45 degrees with the line conductor. *Note:* These assemblies are used at intermediate points where it may be desirable to partially anchor the conductor to prevent too great movement in case of a broken wire. *See also:* **tower.** 64

semit (half-step). *See:* **semitone.**

semitone (semit) (half-step). The interval between two sounds having a basic frequency ratio approximately the twelfth root of two. *Note:* In equally tempered semitones, the interval between any two frequencies is 12 times the logarithm to the base 2 (or 39.86 times the logarithm to the base 10) of the frequency ratio. 176

semitransparent photocathode (camera tube or phototube). A photocathode in which radiant flux incident on one side produces photoelectric emission from the opposite side. *See also:* **electrode (electron tube); phototubes.** 190, 125

sender (telephone switching systems). Equipment that generates and transmits signals in response to information received from another part of the system. 55

sending-end crossfire. The crossfire in a telegraph channel from one or more adjacent telegraph channels transmitting from the end at which the crossfire is measured. *See also:* **telegraphy.** 328

sending-end impedance (line). The ratio of an applied potential difference to the resultant current at the point where the potential difference is applied. The sending-end impedance of a line is synonymous with the driving-point impedance of the line. *Note:* For an infinite uniform line the sending-end impedance and the characteristic impedance are the same; and for an infinite periodic line the sending-end impedance and the iterative impedance are the same. *See also:* **self-impedance; waveguide.** 328

send-only equipment. Data communication channel equipment capable of transmitting signals, but not arranged to receive signals. 194

sensation level (sound) (acoustics). *See:* **level above threshold.**

sense (navigation). The pointing direction of a vector representing some navigation parameter. *See also:* **navigation.** 187, 13

sense finder. That portion of a direction-finder that permits determination of direction without 180-degree ambiguity. *See also:* **radio receiver.** 328

sensibility, deflection (oscilloscopes). The number of trace widths per volt of input signal that can be simultaneously resolved anywhere within the quality area. 106

sensing (electronic navigation). The process of finding the sense, as, for example, in direction finding, the resolution of the 180-degree ambiguity in bearing indication; and, as in phase- or amplitude-comparison systems like instrument landing systems and omnidirectional radio range, the establishment of a relation between course displacement signal and the proper response in the control of the vehicle. *See also:* **navigation.**   187, 13

sensing element (initial element). *See:* **primary detector.**

sensitive volume (radiation-counter tubes). That portion of the tube responding to specific radiation. *See also:* **gas-filled radiation-counter tubes.**   190, 96, 125

sensitive relay. A relay that operates on comparatively low input power, commonly defined as 100 milliwatts or less. *See also:* **relay.**   259

sensitivity (1) (general comment). Definitions of sensitivity fall into two contrasting categories. In some fields, sensitivity is the ratio of response to cause. Hence increasing sensitivity is denoted by a progressively larger number. In other fields, sensitivity is the ratio of cause to response. Hence increasing sensitivity is denoted by a progressively smaller number. *See:* **sensitivity coefficient.**

(2) (measuring device). The ratio of the magnitude of its response to the magnitude of the quantity measured. *Notes:* (A) It may be expressed directly in divisions per volt, millimeters per volt, milliradians per microampere, etcetera, or indirectly by stating a property from which sensitivity can be computed (for example, ohm per volt for a stated deflection. (B) In the case of mirror galvanometers it is customary to express sensitivity on the basis of a scale distance of 1 meter. *See also:* **accuracy rating (instrument).**   54

(3) (radio receiver or similar device). Taken as the minimum input signal required to produce a specified output signal having a specified signal-to-noise ratio. *Note:* This signal input may be expressed as power or as voltage, with input network impedance stipulated.   59

(4) (transmission lines, waveguides, and nuclear techniques). The least signal input capable of causing an output signal having desired characteristics. *See:* **ionizing radiation.**   185

(5) (bolometric detection). The sensitivity is given by the change of resistance in ohms per milliwatt of energy dissipated in the bolometer.   210, 185

(6) (close-talking pressure-type microphone). The voltage or power response of a microphone measured at a stated frequency. *See also:* **close-talking pressure-type microphone.**   249

(7) (camera tube or phototube). The quotient of output current by incident luminous flux at constant electrode voltages. *Note:* (A) The term output current as here used does not include the dark current. (B) Since luminous sensitivity is not an absolute characteristic but depends on the special distribution of the incident flux, the term is commonly used to designate the sensitivity to light from a tungsten-filament lamp operating at a color temperature of 2870 kelvins. *See:* **sensitivity, cathode luminous.** *See also:* **phototubes.**   328

(8) (gyro; accelerometer). The ratio of a change in a parameter to a change in an undesirable or secondary input. For example; a scale factor temperature sensitivity of a gyro or accelerometer is the ratio of change in scale factor to a change in temperature.   46

(9) (electrothermic unit) (A) dissipated power: The ratio of the dc output voltage of the electrothermic unit to the microwave power dissipated within the electrothermic unit at a prescribed frequency, power level, and temperature. **(B) incident power:** The ratio of the dc output voltage of the electrothermic unit to the microwave power incident upon the electrothermic unit at a prescribed frequency, power level, and temperature.   47

(10) (electrothermic-coupler unit). The ratio of the dc output voltage of the electrothermic unit on the side arm of the directional coupler to the power incident upon a nonreflecting load connected to the output port of the main arm of the directional coupler at a prescribed frequency, power level, and temperature. If the electrothermic unit is attached to the main arm of the directional coupler, the sensitivity is the ratio of the dc output voltage of the electrothermic unit attached to the main arm of the directional coupler to the microwave power incident upon a nonreflecting load connected to the output port of the side arm of the directional coupler at a prescribed frequency, power level, and temperature.   47

(11) (non-real time spectrum analyzer) (volts, decibels above or below one milliwatt). Measure of a spectrum analyzer's ability to display minimum level signals. IF (intermediate frequency) bandwidth, display mode, and any other influencing factors must be given. *Notes:* (A) equivalent input noise. The average level of a spectrum analyzer's internally generated noise referenced to the input. (B) input signal level. The input signal level that produces an output equal to twice the value of the average noise alone. This may be power or voltage relationship, but must be so stated.   68

(12) (nuclear power generating stations). (A) The minimum amount of contaminant that can repeatedly be detected by an instrument.   31

(B) The ratio of a change in output magnitude to the change in input which causes it, after the steady-state has been reached.   355

(13) (automatic control). Of a control system or element, or combination, the ratio of a change in output magnitude to the change of input which causes it, after the steady state has been reached. *Note:* ASA C85 deprecates use of "sensitivity" to describe smallness of a dead-band. *See also:* **amplification; gain.**   56

sensitivity analysis (nuclear power generating stations). An analysis which determines the variation of a given function caused by changes in one or more parameters about a selected reference value.   29, 31

sensitivity, cathode luminous (photocathodes). The quotient of photoelectric emission current from the photocathode by the incident luminous flux under specified conditions of illumination. *Notes:* (1) Since cathode luminous sensitivity is not an absolute characteristic but depends on the spectral distribution of the incident flux, the term is commonly used to designate the sensitivity to radiation from a tungsten filament lamp operating at a color temperature of 2870 kelvins. (2) Cathode luminous sensitivity is usually measured with a collimated beam at normal incidence. *See also:* **phototube.**   190, 117, 125

sensitivity, cathode radiant (photocathodes). The quotient of the photoelectric emission current from the photocathode by the incident radiant flux at a given wavelength under specified conditions of irradiation. *Note:* Cathode radiant sensitivity is usually measured with a collimated beam at normal incidence. 190, 117, 125

sensitivity coefficient (1) (automatic control) (control system). The partial derivative of a system signal with respect to a system parameter. *See also:* **control system.** 56

(2) (circuits and systems). A coefficient used to relate the change of a system function F due to the variation of one of its parameters x. In some applications (for example control theory) absolute changes are important and the sensitivity coefficient is defined as the ∂F/∂x. In other applications (for example, filter theory) relative changes are important and then sensitivity is defined as

$$\partial(\mathrm{Ln}\ F)/\partial(\mathrm{Lnx}) = (\partial F/\partial x)/(F/x).$$

67

sensitivity, deflection. (1) (magnetic-deflection cathode ray tube and yoke assembly). The quotient of the spot displacement by the change in deflecting-coil current. 190

(2) (oscilloscopes). The reciprocal of the deflection factor (for example, divisions/volt). 106

sensitivity, dynamic (phototubes). The quotient of the modulated component of the output current by the modulated component of the incident radiation at a stated frequency of modulation. *Note:* Unless otherwise stated the modulation wave shape is sinusoidal. *See also:* **phototubes.** 174, 190

sensitivity, illumination (camera tubes or phototubes). The quotient of signal output current by the incident illumination, under specified conditions of illumination. *Notes:* (1) Since illumination sensitivity is not an absolute characteristic but depends on the spectral distribution of the incident flux, the term is commonly used to designate the sensitivity to radiation from a tungsten-filament lamp operating at a color temperature of 2870°K. (2) Illumination sensitivity is usually measured with a collimated beam at normal incidence. (3) *See also:* **transfer characteristic; (camera tubes).** 125

sensitivity, incremental (instrument) (nuclear techniques). A measure of the smallest change in stimulus that produces a statistically significant change in response. Quantitatively it is usually expressed as the change in the stimulus that produces a change in response equal to the standard deviation of the response. *See also:* **ionizing radiation.** 335

sensitivity level (response level) (sensitivity) (response) (in electroacoustics) (of a transducer) (in decibels). 20 times the logarithm to the base 10 of the ratio of the amplitude sensitivity S_A to the reference sensitivity S_0, where the amplitude is a quantity proportional to the square root of power. The kind of sensitivity and the reference sensitivity must be indicated. *Note:* For a microphone, the free-field voltage/pressure sensitivity is the kind often used and a common reference sensitivity is $S_0 = 1$ volt per newton per square meter. The square of the sensitivity is proportional to a power ratio. The free-field voltage sensitivity-squared level, in decibels, is therefore $S_A = 10 \log (S_A{}^2/S_0{}^2) = 20 \log (S_A/S_0)$. Often, **sensitivity-squared level** in decibels can be shortened, without ambiguity, to **sensitivity level** in decibels, or simply **sensitivity** in decibels. 176

sensitivity, luminous (camera tubes or phototubes). The quotient of signal output current by incident luminous flux, under specified conditions of illumination. *Notes:* (1) Since luminous sensitivity is not an absolute characteristic but depends on the spectral distribution of the incident flux, the term is commonly used to designate the sensitivity to radiation from a tungsten-filament lamp operating at a color temperature of 2870°K. (2) Luminous sensitivity is usually measured with a collimated beam at normal incidence. 125

sensitivity, quieting (test, measurement and diagnostic equipment). The level of a continuous wave (CW) input signal which will reduce the noise output level of a frequency-modulation (FM) receiver by a specified amount, usually 20 decibels (dB). 54

sensitivity, radiant (camera tube or phototube). The quotient of signal output current by incident radiant flux at a given wavelength, under specified conditions of irradiation. *Note:* Radiant sensitivity is usually measured with a collimated beam at normal incidence. *See also:* **luminous flux; phototubes; radiant flux.** 190, 125

sensitivity, threshold (test, measurement and diagnostic equipment). The smallest quantity that can be detected by a measuring instrument or automatic control system. 54

sensitivity time control (STC). Programmed variation of the gain (sensitivity) of a radar receiver as a function of time within each pulse repetition interval or observation time in order to prevent overloading of the receiver by strong echoes from targets or clutter at close ranges. 13

sensitizing (electrostatography). The act of establishing an electrostatic surface charge of uniform density on an insulating medium. *See also:* **electrostatography.** 236, 191

sensitometry. The measurement of the light response characteristics of photographic film under specified conditions of exposure and development. 176

sensor (1) (nuclear power generating stations). (A) That portion of a channel which first responds to changes in, and performs the primary measurement of, a plant variable or condition. 20

(B) A device directly responsive to the value of the measured quantity. 355

(2) (temperature measurement). That portion of a temperature-measuring system that responds to the temperature being measured. 7

(3) (test, measurement and diagnostic equipment). A transducer which converts a parameter at a test point to a form suitable for measurement by the test equipment. *See:* **pick-up; pick-off.** 54

sensor, active (test, measurement and diagnostic equipment). A sensor requiring a source of power other than the signal being measured. 54

sensor, passive (test, measurement and diagnostic equipment). A sensor requiring no source of power other than the signal being measured. 54

sensory saturation (nuclear power generating station). The impairment of effective operator response to an event due to excessive amount of display information which must be evaluated prior to taking action. 358

sentinel (computing systems). *See:* **flag.**

separable insulated connector (transformer) (power distri-bution systems). A system for terminating and electrically connecting an insulated power cable to electrical apparatus, other power cables, or both, so designed that the electrical connection can be readily established or broken by engaging or separating mating parts of the connector at the operating interface.   53, 134

separate parts of a network. The parts that are not connected. *See also:* **network analysis.**   210

separate terminal enclosure (rotating machinery). A form of termination in which the ends of the machine winding are connected to the incoming supply leads inside a chamber that need not be fully enclosed and may be formed by the foundations beneath the machine.   63

separately excited (rotating machinery). A qualifying term applied to a machine to denote that the excitation is obtained from a source other than the machine itself. *See also:* **direct-current commutating machine.**   63

separately ventilated machine (rotating machinery). A machine that has its ventilating air supplied by an independent fan or blower external to the machine. *See also:* **asynchronous machine; direct-current commutating machine; externally ventilated machine; open-pipe ventilated machine.**   3, 325, 63

separation (frequency modulation). The process of deriving individual channel signals (for example, for stereophonic systems) from a composite transmitted signal. *Note:* Separation describes the ability of a receiver to produce left and right stereophonic channel signals at its output terminals and is a measured parameter for stereo receivers only. Left-channel signal separation is defined as the ratio in decibels of the output voltage of the left output of the receiver to that of the right output when an "L"-only signal is received. Right-channel separation is similarly defined.   16

separation criteria (electromagnetic compatibility). Curves that relate the frequency displacement to the minimum distance between a receiver and an undesired transmitter to insure that the signal-to-interference ratio does not fall below a specified value. *See also:* **electromagnetic compatibility.**   199

separation distance (nuclear power generating stations). Space without interposing structures, equipment, or materials that could aid in the propagation of fire or that could disable the Class 1E system.   31, 131

separator (1) (storage cell). A spacer employed to prevent metallic contact between plates of opposite polarity within the cell. (Perforated sheets are usually called retainers.) *See also:* **battery (primary or secondary).**   328
(2) (computing systems). *See:* **delimiter.**

separator, insulation slot (rotating machinery). Insulation member placed in a slot between individual coils, such as between main and auxiliary windings. *See also:* **rotor (rotating machinery); stator.**   63

sequence. *See:* **calling sequence; collating sequence; pseudorandom number sequence.**

sequence network (relaying). An electrical circuit that produces an output proportional to one or more of the sequence components of a polyphase system of voltages or currents, that is positive-sequence network, negative sequence network, or zero-sequence network.   79

sequence number. A number identifying the relative location of blocks or groups of blocks on a tape.   224, 207

sequence-number readout. Display of the sequence number punched on the tape. *See:* **block-count readout.**   224, 207

sequence switch. A remotely controlled power-operated switching device used as a secondary master controller. *See also:* **multiple-unit control.**   328

sequence table (electric controller). A table indicating the sequence of operation of contactors, switches, or other control apparatus for each step of the periodic duty. *See also:* **multiple-unit control.**   1

sequential (formatted system) (telecommunication). If the signal elements are transmitted successively in time over a channel, the transmission is said to be **sequential.** If the signal elements are transmitted at the same time over a multiwire circuit, the transmission is said to be **coincident.** *See also:* **bit.**   194

sequential access (test, measurement and diagnostic equipment). A system in which the information becomes available in a one after the other sequence only, whether all of it is desired or not.   54

sequential control (computing systems). A mode of computer operation in which instructions are executed consecutively unless specified otherwise by a jump.   255, 77, 54

sequential events recording system (SERS). A system which monitors bistable equipment operations and process status and records changes of state in the order of detected occurrences. This monitoring may be accomplished using a device dedicated solely to this function, or using a multifunction system such as a data acquisition computer system.   48, 58

sequential lobing (electronic navigation). A direction-determining technique utilizing the signals of partially overlapped lobes occurring in sequence. *See also:* **antenna.**   246, 179, 111

sequential logic element. A device having at least one output channel and one or more input channels, all characterized by discrete states, such that the state of each output channel is determined by the previous states of the input channels.   255, 77

sequential logic function. A logic function wherein for at least one combination of states of the input or inputs, there exists more than one state of the output or outputs. The outputs are functions of variables in addition to the inputs, such as time, previous internal states of the element, etcetera.   88

sequential memory (sequential events recording systems). The memory which stores events in the same order in which they were received by the system. The memory capacity can be expressed as the number of events or levels. *See also:* **event; level.**   48, 58

sequential operation. Pertaining to the performance of operations one after the other.   255, 77

sequential programming (test, measurement and diagnostic equipment). The programming of a device by which only one arithmetical or logical operation can be executed at one time.   54

sequential relay. A relay that controls two or more sets of

contacts in a predetermined sequence. *See also:* **relay.**
259

sequential scanning (television). A rectilinear scanning process in which the distance from center to center of successively scanned lines is equal to the nominal line width. *See also:* **television.** 328

serial. (1) Pertaining to the time sequencing of two or more processes. (2) Pertaining to the time sequencing of two or more similar or identical processes, using the same facilities for the successive processes. (3) Pertaining to the time-sequential processing of the individual parts of a whole, such as the bits of a character, the characters of a word, etcetera, using the same facilities for successive parts. *See also:* **serial-parallel.** 235, 255, 77

serial access (computing systems). Pertaining to the process of obtaining data from, or placing data into, storage when there is a sequential relation governing the access time to successive storage locations. 255, 77

serial by bit. *See:* **serial transmission (data transmission) (telecommunications).**

serial digital computer. A digital computer in which the digits are handled serially. Mixed serial and parallel machines are frequently called serial or parallel according to the way arithmetic processes are performed. An example of a serial digital computer is one that handles decimal digits serially although it might handle the bits that comprise a digit either serially or in parallel. *See also:* **parallel digital computer.** 210

serial operation (telecommunication). The flow of information in time sequence, using only one digit, word, line, or channel at a time. *See also:* **bit.** 194

serial-parallel. Pertaining to processing that includes both serial and parallel processing, such as one that handles decimal digits serially but handles the bits that comprise a digit in parallel. 235

serial transmission (data transmission) (telecommunication). Used to identify a system wherein the bits of a character occur serially in time. Implies only a single transmission channel. Also called **serial by bit.** *See also:* **bit.** 194

series capacitor. A device that has the primary purpose of introducing capacitive reactance into an electric circuit. 86

series capacitor bank. An assembly at one location consisting of capacitors with fuses and associated auxiliaries such as structures, support insulators, and protective devices, with control equipment required for a complete operating installation. 86

series circuit. A circuit supplying energy to a number of devices connected in series, that is, the same current passes through each device in completing its path to the source of supply. *See also:* **center of distribution.** 64

series circuit lighting transformer. Dry-type individual lamp insulating transformer, auto-transformer, and group series loop transformers for operation of incandescent or mercury lamps on series lighting circuits such as for street and airport lighting. 53

series coil sectionalizer (power switchgear). A sectionalizer in which main circuit current impulses above a specified value, flowing through a solenoid or operating coil, provide the energy required to operate the counting mechanism. 103

series connected starting-motor starting (rotating ma-

chinery). The process of starting a motor by connecting its primary winding to the supply in series with the primary windings of a starting motor, this latter being short-circuited for the running condition. 63

series connection. The arrangement of cells in a battery made by connecting the positive terminal of each successive cell to the negative terminal of the next adjacent cell so that their voltages are additive. *See also:* **battery (primary or secondary).** 328

series distribution system. A distribution system for supplying energy to units of equipment connected in series. *See also:* **alternating-current distribution; direct-current distribution.** 64

series elements (network). (1) Two-terminal elements are connected in series when they form a path between two nodes of a network such that only elements of this path, and no other elements, terminate at intermediate nodes along the path. (2) Two-terminal elements are connected in series when any mesh including one must include the others. *See also:* **network analysis.** 210

series-fed vertical antenna. A vertical antenna that is insulated from ground and energized at the base. *See also:* **antenna.** 111, 179

series gap (surge arrester). An intentional gap(s) between spaced electrodes; it is in series with the valve or expulsion element of the arrester, substantially isolating the element from line or ground or both under normal line voltage conditions. 2

series loading. Loading in which reactances are inserted in series with the conductors of a transmission circuit. *See also:* **loading.** 328

series-mode interference (signal-transmission system). *See:* **interference, differential-mode.**

series modulation. Modulation in which the plate circuits of a modulating tube and a modulated amplifier tube are in series with the same plate voltage supply. 328

series operation (power supplies). The output of two or more power supplies connected together to obtain a total output voltage equal to the sum of their individual voltages. Load current is equal and common through each supply. The extent of series connection is limited by the maximum specified potential rating between any output terminal and ground. For series connection of current regulators, master/slave (compliance extension) or automatic crossover is used. *See:* **isolation voltage.** *See also:* **power supply.** 228, 186

series overcurrent tripping. *See:* **direct release (series trip); overcurrent release (overcurrent trip).**

series-parallel connection. The arrangement of cells in a battery made by connecting two or more series-connected groups, each having the same number of cells so that the positive terminals of each group are connected together and the negative terminals are connected together in a corresponding manner. *See also:* **battery (primary or secondary).** 328

series-parallel control. A method of controlling motors wherein the motors, or groups of them, may be connected successively in series and in parallel. *See also:* **multiple-unit control.** 328

series-parallel network. Any network, containing only two-terminal elements, that can be constructed by successively connecting branches in series and/or in parallel.

Note: An elementary example is the parallel combination of two branches, one containing resistance and inductance in series, the other containing capacitance. This network is sometimes called a **simple parallel circuit***. *See also:* **network analysis.** 210

* Deprecated

series-parallel starting (rotating machinery). The process of starting a motor by connecting it to the supply with the primary winding phase circuits initially in series, and changing them over to a parallel connection for running operation. *See also:* **asynchronous machine.** 63

series rectifier circuit. A rectifier circuit in which two or more simple rectifier circuits are connected in such a way that their direct voltages add and their commutations coincide. *See also:* **rectification; rectifier circuit element.** 237, 66

series regulator (power supplies). A device placed in series with a source of power that is capable of controlling the voltage or current output by automatically varying its series resistance. *See:* **passive element.** *See also:* **power supply.** 228, 186

series relay. *See:* **current relay.** *See also:* **relay.** 259

series resistor (electric instrument). A resistor that forms an essential part of the voltage circuit of an instrument and generally is used to adapt the instrument to operate on some designated voltage or voltages. The series resistor may be internal or external to the instrument. *Note:* Inductors, capacitors, or combinations thereof are also used for this purpose. *See also:* **auxiliary device to an instrument.** 280

series resonance. The steady-state condition that exists in a circuit comprising inductance and capacitance connected in series, when the current in the circuit is in phase with the voltage across the circuit. 59

series street-lighting transformer. A series transformer that receives energy from a current-regulating series circuit and that transforms the energy to another winding at the same or different current from that in the primary. 53

series system. The arrangement in a multielectrode electrolytic cell whereby in each cell an anode connected to the positive bus bar is placed at one end and a cathode connected to the negative bus bar is placed at the other end, with the intervening unconnected electrodes acting as bipolar electrodes. *See also:* **electrorefining.** 328

series tee junction. *See:* **E-plane tee.**

series transformer. A transformer with a series winding and an exciting winding, in which the series winding is placed in a series relationship in a circuit to change voltage and/or phase in that circuit as a result of input received from the exciting winding. *Note:* Applications of series transformers include: (1) use in a transformer such as a load-tap-changing or regulating transformer to change the voltage or current duty of the tap-changing mechanism; (2) inclusion in a circuit for power factor correction to indirectly insert series capacitance in a circuit by connecting capacitors to the exciting winding. 53

series transformer rating. The lumen rating of the series lamp, or the wattage rating of the multiple lamps, that the transformer is designed to operate. 53

series-trip recloser. A recloser in which main-circuit current above a specified value, flowing through a solenoid

or operating coil, provides the energy necessary to open the main contacts. 103, 202

series two-terminal pair networks. Two-terminal pair networks are connected in series at the input or at the output terminals when their respective input or output terminals are in series. *See also:* **network analysis.** 332

series undercurrent tripping. *See:* **direct release (series trip); undercurrent release (undercurrent trip).**

series unit (two-core regulating transformer). The core and coil unit which has one winding connected in series in the line circuit. 53

series winding (1) (autotransformer). That portion of the autotransformer winding which is not common to both the primary and the secondary circuits, but is connected in series between the input and output circuits. 53

(2) (two-core regulating transformer). The winding of the series unit which is connected in series in the line circuit. *Note:* If the main unit of a two-core transformer is an autotransformer, both units will have a series winding. In such cases, one is referred to as the series winding of the autotransformer and the other, the series winding of the series unit. 53

series-wound (rotating machinery). A qualifying term applied to a machine to denote that the excitation is supplied by a winding or windings connected in series with or carrying a current proportional to that in the armature winding. *See:* **asynchronous machine; direct-current commutating machine.** 63

series-wound motor (1) (rotating machinery). A commutator motor in which the field circuit and armature circuit are connected in series. *See:* **asynchronous machine; direct-current commutating machine.** 63

(2) (electric installations on shipboard). A commutator motor in which the field circuit and armature circuit are connected in series. *Note:* It operates at a much higher speed at light load than at full load. 3

service (1) (electric systems). The conductors and equipment for delivering electric energy from the secondary distribution or street main, or other distribution feeder, or from the transformer, to the wiring system of the premises served. *Note:* For overhead circuits, it includes the conductors from the last line pole to the service switch or fuse. The portion of an overhead service between the pole and building is designated as service drop. 256, 262

(2) (controller) (industrial control). The specific application in which the controller is to be used, for example: (A) general purpose, (B) definite purpose, for example, crane and hoist, elevator, machine tool, etcetera. *See:* **electric controller.** 225, 206

service area (1) (navigation). The area within which a navigational aid provides either generally satisfactory service or a specific quality of service. *See also:* **navigation.** 187, 13

(2) (electric power utilization). Territory in which a utility system is required or has the right to supply or make available electric service to ultimate consumers. 112

service band. A band of frequencies allocated to a given class of radio service. *See also:* **radio transmission.** 111, 240

service bits (telecommunication). Those bits that are neither check nor information bits. *See also:* **bit.** 194

service cable. Service conductors made up in the form of a cable. *See:* **service conductors.** *See also:* **service.** 64

service capacity (cell or battery). The electric output (expressed in ampere-hours, watthours, or similar units) on a service test before its working voltage falls to a specified cutoff voltage. *See also:* **battery (primary or secondary).** 328

service circuit (telephone switching systems). A circuit used for signaling purposes connected to and disconnected from a communication path during the progress of a call. 55

service code (telephone switching systems). Any of the destination codes for use by customers to obtain directory assistance or repair service, or to reach the business office of the telecommunications company. 55

service conditions (nuclear power generating stations). Environmental, power, and signal conditions expected as a result of normal operating requirements, expected extremes in operating requirements, and postulated conditions appropriate for the design basis events of the station (applicable to the electric penetration assembly). *See:* **design basis events; electric penetration assembly.** 26, 31, 120, 355

service conductors. That portion of the supply conductors that extends from the street mains or feeder or transformer to the service equipment of the premises served. For an overhead system it includes the conductors from the last line pole to the service equipment. *See also:* **service.** 64

service corrosion (dry cell). The consumption of the negative electrode as a result of useful current delivered by the cell. *See also:* **electrolytic cell.** 328

service current, continuous (1) (thyristor converter). The value of direct current which a converter unit or section can supply to its load for unlimited time periods under specified conditions. **(2) long-time.** The rms value and duration (minutes) of direct current which may be applied to the converter unit or section within the service current profile. *Note:* This value establishes point B on the service current profile and it may be identical to the long-time test current. **(3) profile.** The time-current profile that defines the allowable rms currents the converter section can sustain. *Note:* The profile is defined for times from zero to infinity, and the rms current derived from any current-time diagram must not exceed this profile. **(4) short time.** The peak rms value and duration (seconds) of direct current which may be applied to the converter unit or section within the service current profile. *Note:* This value establishes point C on the service current profile. 121

service drop (transmission and distribution). The overhead conductors between the electric supply or communication line and the building or structure being served. 262

service entrance conductors (electric system) (1) (overhead system). The service conductors between the terminals of the service equipment and a point usually outside the building, clear of building walls, where joined by tap or splice to the service drop. 256
(2) (underground system). The service conductors between the terminals of the service equipment and the point of connection to the service lateral. *Note:* Where service equipment is located outside the building walls, there may

be no service-entrance conductors, or they may be entirely outside the building. 256, 64

service environment (nuclear power generating stations). The aggregate of conditions surrounding the equipment while providing the intended service in the normal, accident, and post-accident conditions of the nuclear-fueled power generating system. Each service environment must have three separate sets of conditions defined; one each for the normal, accident, and post-accident conditions. 31

service equipment (electric system). The necessary equipment, usually consisting of circuit breaker or switch and fuses, and their accessories, located near point of entrance of supply conductors to a building and intended to constitute the main control and means of cutoff for the supply to that building. *See also:* **distribution center; service.** 256, 64

service evaluation (telephone switching systems). Determination of the quality of service received by the customer. 55

service factor (general-purpose alternating-current motor). A multiplier that, when applied to the rated power, indicates a permissible power loading that may be carried under the conditions specified for the service factor. *See also:* **asynchronous machine; direct-current commutating machine.** 232, 63

service ground. A ground connection to a service equipment or a service conductor or both. *See:* **ground.** 64

service lateral (electric system). The underground service conductors between the street main, including any risers at a pole or other structure or from transformers, and the first point of connection to the service entrance conductors in a terminal box inside or outside the building wall. Where there is no terminal box, the point of connection shall be considered to be the point of entrance of the service conductors into the building. *See also:* **service.** 256

service life (1) (primary cell or battery). The period of useful service before its working voltage falls to a specified cutoff voltage.
(2) (storage cell or battery). The period of useful service under specified conditions, usually expressed as the period elapsed before the ampere-hour capacity has fallen to a specified percentage of the rated capacity. *See also:* **battery (primary or secondary); charge.** 328

service period (illuminating engineering). The number of hours per day for which the day lighting provides a specified illumination level. It often is stated as a monthly average. *See also:* **sunlight.** 167

service pipe. The pipe or conduit that contains underground service conductors and extends from the junction with outside supply wires into the customer's premises. *See:* **distributor duct.** *See also:* **service.** 64

service raceway (electric system). The rigid metal conduit, electrical metallic tubing, or other raceway, that encloses the service entrance conductors. *See also:* **service.** 256

service rating (rectifier transformer). The maximum constant load that, after a transformer has carried its continuous rated load until there is no further measurable increase in temperature rise, may be applied for a specified time without injury. *See also:* **rectifier transformer.** 258

service routine (computing systems). A routine in general support of the operation of a computer, for example, an input-output, diagnostic, tracing, or monitoring routine. *See:* **utility routine.**   255, 77, 54

service, standby (electric power utilization). Service through a permanent connection not normally used but available in lieu of, or as a supplement to, the usual source of supply.
112

service, station. Facilities which provide energy for station use in a generating, switching, converting, or transforming station.   112

service test (1) (primary battery). A test designed to measure the capacity of a cell or battery under specified conditions comparable with some particular service for which such cells are used.   328

(2) (field test) (meter). A test made during the period that the meter is in service. *Note:* A service test may be made on the consumer's premises without removing the meter from its support, or by removing the meter for test, either on the premises or in a laboratory or meter shop.   212

(3) (nuclear power generating stations) (lead storage batteries). A special capacity test made to demonstrate the capability of the battery to meet the design requirements of the system to which it is connected.   31, 38

servicing time (electric drive). The portion of down time that is necessary for servicing due to breakdowns or for preventive servicing measures. *See:* **electric drive.**
1, 206

serving (cable). A wrapping applied over the core of a cable before the cable is leaded, or over the lead if the cable is armored. *Note:* Materials commonly used for serving are jute or cotton. The serving is for mechanical protection and not for insulating purposes.   64

servo. *See:* **servomechanism.**

servo amplifier. An amplifier, used as part of a servomechanism, that supplies power to the electric input terminals of a mechanical actuator. *See also:* **electronic analog computer.**   9, 77, 10

servomechanism. (1) A feedback control system in which at least one of the systems signals represents mechanical motion. (2) Any feedback control system. (3) An automatic feedback control system in which the controlled variable is mechanical position or any of its time derivatives. *See also:* **control system, feedback.**
56, 255, 77, 207

servomechanism, positional. A servomechanism in which a mechanical shaft is positioned, usually in the angle of rotation, in accordance with one or more input signals. *Note:* Frequently, the shaft is positioned (excluding transient motion) in a manner linearly related to the value of the input signal. However, the term also applies to any servomechanism in which a loop input signal generated by a transmitting transducer can be compared to a loop feedback signal generated by a compatible or identical receiving transducer to produce a loop error signal that, when reduced to zero by movement of the receiving transducer, results in a shaft position related in a prescribed and repeatable manner to the position of the transmitting transducer. *See also:* **electronic analog computer; servomechanism, repeater.**   9

servomechanism, rate. A servomechanism in which a mechanical shaft is translated or rotated at a rate proportional to an input signal amplitude. *See also:* **electronic**

analog computer.   9

servomechanism, repeater. A positional servomechanism in which loop input signals from a transmitting transducer are compared with loop feedback signals from a compatible or identical receiving transducer mechanically coupled to the servomechanism to produce a mechanical shaft motion or position linearly related to motion or position of the transmitting transducer. *See also:* **electronic analog computer.**   9

servomechanism type number. In control systems in which the loop transfer function is

$$\frac{K(1 + a_1 s + a_2 s^2 + \ldots + a_i s^i)}{s^n(1 + b_1 s + b_2 s^2 + \ldots + b_k s^k)}$$

where K, a_1, b_1, b_2, etcetera, are constant coefficients, the value of the integer n. *Note:* The value of n determines the low-frequency characteristic of the transfer function. The log-gain—log-frequency curve (Bode diagram) has a zero-frequency slope of zero for $n = 0$, slope -1 for $n = 1$, etcetera. *See also:* **control system, feedback.**   56

servomotor limit (speed governing of hydraulic turbines). A device which acts on the governor system to prevent the turbine-control servomotor from opening beyond the position for which the device is set.   8

servomotor position (speed governing systems, hydraulic systems). The instantaneous position of the turbine control servomotor expressed as a percent of the servomotor stroke. This is commonly referred to as gate position, needle position, blade position or deflector position, although the relationship between servomotor stroke and the position of the controlled device may not always be linear.   8, 58

servomotor stroke (speed governing systems, hydraulic turbines). Travel of the turbine control servomotor from zero to maximum without overtravel at the maximum position or "squeeze" at the minimum position. *Notes:* (1) For a gate servomotor this shall be established as the change in gate position from no discharge to maximum discharge. (2) For a blade servomotor this shall be established as the change in blade position from "flat" to "steep". (3) For a deflector servomotor this shall be established as the change in deflector position from "no deflection" position to "full flow deflected" position with maximum discharge under maximum specified head including overpressure due to water hammer. *Syn:* **effective servomotor stroke.**   8, 58

servomotor time (speed governing of hydraulic turbines). The equivalent elapsed time for one servomotor stroke (either opening or closing) corresponding to maximum servomotor velocity. Servomotor time can be qualified as: (1) gate; (2) blade; (3) deflector; (4) needle.   8

servomotor velocity limiter (speed governing of hydraulic turbines). A device which functions to limit the servomotor velocity in either the opening, closing or both directions exclusive of the operation of the slow closure device. *See also:* **slow closure device.**   8, 58

set (1) (electronic computation). (A) To place a storage device into a specified state, usually other than that denoting ZERO or BLANK (B) To place a binary cell into the state denoting ONE. *See:* **preset; reset.**
235, 210, 255, 77

(2) (instruments). To position the various adjusting devices so as to secure the desired operating characteristic. *Note:* Typical adjustment devices are taps, dials, levers, and scales suitably marked, rheostats that may be adjusted during tests, and switches with numbered positions that refer to recorded operating characteristics.

103, 202, 60, 127

(3) (polyphase currents) (of *m* phases). A group of *m* interrelated alternating currents, each in a separate phase conductor, that have the same primitive period but normally differ in phase. They may or may not differ in amplitude and waveform. The equations for a set of *m*-phase currents, when each is sinusoidal, and has the primitive period, are

$$i_a = (2)^{1/2} I_a \cos(\omega t + \beta_{a1})$$
$$i_b = (2)^{1/2} I_b \cos(\omega t + \beta_{b1})$$
$$i_c = (2)^{1/2} I_c \cos(\omega t + \beta_{c1})$$
$$\vdots$$
$$i_m = (2)^{1/2} I_m \cos(\omega t + \beta_{m1})$$

where the symbols have the same meaning as for the general case given later. The general equations for a set of *m*-phase alternating currents are

$$i_a = (2)^{1/2}[I_{a1}\cos(\omega t + \beta_{a1}) + I_{a2}\cos(2\omega t + \beta_{a2}) + \ldots + I_{aq}\cos(q\omega t + \beta_{aq}) + \ldots]$$

$$i_b = (2)^{1/2}[I_{b1}\cos(\omega t + \beta_{b1}) = I_{b2}\cos(2\omega t + \beta_{b2}) + \ldots + I_{bq}\cos(q\omega t + \beta_{bq}) + \ldots]$$

$$\vdots$$

$$i_m = (2)^{1/2}[I_{m1}\cos(\omega t + \beta_{m1}) + I_{m2}\cos(2\omega t + \beta_{m2}) + \ldots + I_{mq}\cos(q\omega t + \beta_{mq}) + \ldots]$$

where $i_a, i_b, \ldots, i_m$ are the instantaneous values of the currents, and $I_{a1}, I_{a2}, \ldots, I_{aq}$ are the root-mean-square amplitudes of the harmonic components of the individual currents. The first subscript designates the individual current and the second subscript denotes the number of the harmonic component. If there is no second subscript, the quantity is assumed to be sinusoidal. $\beta_{a1} \beta_{a2}, \ldots, \beta_q$ are the phase angles of the components of the same subscript determined with relation to a common reference. *Notes:* (1) If the circuit has a neutral conductor, the current in the neutral conductor is generally not considered as a separate current of the set, but as the negative of the sum of all the other currents (with respect to the same reference direction). (2) See Note (3) of **voltage sets (polyphase circuit).** *See also:* **network analysis.**   210

(4) (polyphase voltages) (*m* phases). A group of *m* interrelated alternating voltages that have the same primitive period but normally differ in phase. They may or may not differ in amplitude and wave form. The equations for a set of *m*-phase voltages, when each is sinusoidal and has the primitive period, are

$$e_a = (2)^{1/2} E_a \cos(\omega t + \alpha_{a1})$$
$$e_b = (2)^{1/2} E_b \cos(\omega t + \alpha_{b1})$$
$$e_c = (2)^{1/2} E_c \cos(\omega t + \alpha_{c1})$$
$$\vdots$$
$$e_m = (2)^{1/2} E_m \cos(\omega t + \alpha_{m1})$$

where the symbols have the same meaning as for the general case given below. The general equations for a set of *m*-phase alternating voltages are

$$e_a = (2)^{1/2}[E_{a1}\cos(\omega t + \alpha_{a1}) + E_{a2}\cos(2\omega t + \alpha_{a2}) + \ldots + E_{ar}\cos(r\omega t + \alpha_{ar}) + \ldots]$$

$$e_b = (2)^{1/2}[E_{b1}\cos(\omega t + \alpha_{b1}) + E_{b2}\cos(2\omega t + \alpha_{b2}) + \ldots + E_{br}\cos(r\omega t + \alpha_{br}) + \ldots]$$

$$\vdots$$

$$e_m = (2)^{1/2}[E_{m1}\cos(\omega t + \alpha_{m1}) + E_{m2}\cos(2\omega t + \alpha_{m2}) + \ldots + E_{mr}\cos(r\omega t + \alpha_{mr}) + \ldots]$$

where $e_a, e_b, \ldots, e_m$ are the instantaneous values of the voltages, and $E_{a1}, E_{a2}, \ldots, E_{ar}$ the root-mean-square amplitudes of the harmonic components of the individual voltages. The first subscript designates the individual voltage and the second subscript denotes the number of the harmonic component. If there is no second subscript, the quantity is assumed to be sinusoidal. $\alpha_{a1}, \alpha_{a79\ 2}, \ldots, \alpha_{ar79}$ are the phase angles of the components with the same subscript determined with relation to a common reference. *Note:* This definition may be applied to a two-phase four-wire or five-wire circuit if *m* is considered to be 4 instead of 2. A two-phase three-wire circuit should be treated as a special case. *See also:* **network analysis.**

210

(5) (test, measurement and diagnostic equipment). (A) A collection; (B) To place a storage device into a specified state, usually other than that denoting zero or blank; and (C) To place a binary cell into the one state.   54

(6) (electric and electronics parts and equipments). A unit or units and necessary assemblies, subassemblies, and basic parts connected or associated together to perform an operational function. Typical examples: search radar set, radio transmitting set, sound measuring set; these include such parts, assemblies, and units as cables, microphone, and measuring instruments.   17

set light (television). The separate illumination of the background or set, other than that provided for principal subjects or areas. *See also:* **television lighting.**   167

set of commutating groups (rectifier). Two or more commutating groups that have simultaneous commutations. *See also:* **rectification; rectifier circuit element.**

237, 66, 208

set point (1) (process control systems). A fixed or constant (for relatively long time periods) command. *See also:* **control system, feedback.**   56

(2) (nuclear power generating stations). A predetermined level at which a bistable device changes state to indicate that the quantity under surveillance has reached the selected value.   31, 143

set pulse. A drive pulse that tends to set a magnetic cell.   331

setting (noun) (1) (general). The desired characteristic, obtained as a result of having set a device, stated in terms of calibration markings or of actual performance bench marks such as pickup current and operating time at a given value of input. *Note:* When the setting is made by adjusting the device to operate as desired in terms of a measured input quantity, the procedure may be the same as in calibration. However, since it is for the purpose of finding one particular position of an adjusting device, which in the general case may have several marked positions that are not being calibrated, the word setting is to be preferred over the word calibration. 103, 202

(2) (circuit breaker). The value of the current at which it is set to trip. *See also:* **contactor.** 256

setting error. The departure of the actual performance from the desired performance resulting from errors in adjustment or from limitations in testing or measuring techniques. 103, 202, 60

setting limitation. The departure of the actual performance from the desired performance resulting from limitations of adjusting devices. 103, 202, 60

settling time (1) (automatic control). The time required, following the initiation of a specified stimulus to a linear system, for the output to enter and remain within a specified narrow band centered on its steady-state value. *Note:* The stimulus may be a step, impulse, ramp, parabola, or sinusoid. For a step or impulse, the band is often specified as ± 2 per cent. For nonlinear behavior, both magnitude and pattern of the stimulus should be specified. *Syn:* **correction time.** 56

(2) (digital-to-analog converter). The time required from the instant after the "load" has been completed until the digital-to-analog converter (DAC) or digital-to-analog multiplier (DAM) output voltage is available within a given accuracy, (under the condition of a jam transfer for a double-buffered DAC). 10

setup (television). The ratio between reference black level and reference white level, both measured from blanking level. It is usually expressed in percent. *See also:* **television.** 178

sexadecimal. (1) Pertaining to a characteristic or property involving a selection, choice, or condition in which there are sixteen possibilities. (2) Pertaining to the numeration system with a radix of sixteen. *Note:* More commonly called **hexadecimal.** *See:* **positional notation.** 77

shade. A screen made of opaque or diffusing material that is designed to prevent a light source from being directly visible at normal angles of view. *See also:* **bare (exposed) lamp.** 163

shaded-pole motor (rotating machinery). A single-phase induction motor with a main winding and one or more short-circuited windings (or shading coils) disposed about the air gap. The effect of the winding combination is to produce a rotating magnetic field which in turn induces the desired motor action. 63

shading (1) (storage tubes). The type of spurious signal, generated within a tube, that appears as a gradual variation or a small number of gradual variations in the amplitude of the output signal. These variations are spatially fixed with reference to the target area. Note the distinction between this and **disturbance.** *See also:* **storage tube; television.** 174, 190

(2) (television). The process of compensating for the spurious signal generated in a camera tube during the trace intervals. *See also:* **television.** 328

(3) (audio and electroacoustics). A method of controlling the directional response pattern of a transducer through control of the distribution of phase and amplitude of the

transducer action over the active face. *See also:* **television.** 176

(4) (camera tubes). A brightness gradient in the reproduced picture, not present in the original scene, but caused by the tube. 125

shading coil (1) (rotating machinery). The short-circuited winding used in a shaded-pole motor, for the purpose of producing a rotating component of magnetic flux.

(2) (direct-current motors and generators). A short-circuited winding used on a main (excitation) pole to delay the shift in flux caused by transient armature current. Transient commutation is aided by the use of this coil. *See also:* **rotor (rotating machinery); stator.** 63

shading wedge (rotating machinery). A strip of magnetic material placed between adjacent pole tips of a shaded-pole motor to reduce the effective separation between the pole tips. The shading wedge usually has a slot running most of its length to provide some separation effect. *See also:* **rotor (rotating machinery); stator.** 63

shadow factor (radio wave propagation). The ratio of the electric field strength that would result from propagation over a sphere to that which would result from propagation over a plane, other factors being the same. *See also:* **radio wave propagation.** 146, 180

shadowing (shielding). The interference of any part of an anode, cathode, rack, or tank with uniform current distribution upon a cathode. 328

shadow loss (mobile communication). The attenuation to a signal caused by obstructions in the radio propagation path. *See also:* **mobile communication system.** 181

shadow mask (color-picture tubes). A color-selecting-electrode system in the form of an electrically conductive sheet containing a plurality of holes that uses masking to effect color selection. *See also:* **television.** 190, 125

shaft (rotating machinery). That part of a rotor that carries other rotating members and that is supported by bearings in which it can rotate. *See:* **rotor (rotating machinery).** 63

shaft current (rotating machinery). Electric current that flows from one end of the shaft of a machine through bearings, bearing supports, and machine framework to the other end of the shaft, driven by a voltage between the shaft ends that results from flux linking the shaft caused by irregularities in the magnetic circuit. *See:* **rotor (rotating machinery).** 63

shaft encoder. A transducer whose input is the turning of a shaft and whose output is a measure of the position of the shaft. *See also:* **transducer.** 59

shaft extension (rotating machinery). The portion of a shaft that projects beyond the bearing housing and away from the core. *See:* **armature.** 63

shaft revolution indicator. A system consisting of a transmitter driven by a propeller shaft and one or more remote indicators to show the speed of the shaft in revolutions per minute, the direction of rotation and (usually) the total number of revolutions made by the shaft. *See also:* **electric propulsion system.** 328

shaft voltage test (rotating machinery). A test taken on an energized machine to detect the induced voltage that is capable of producing shaft currents. *See also:* **rotor (rotating machinery).** 63

shank (cable plowing). A portion of the plow blade to which

a removable wear point is fastened. *See also:* **wear point.**
52

shaped-beam antenna. An antenna which is designed to have a prescribed pattern shape differing substantially from that obtained from a uniform-phase aperture of the same size. 111

shape factor (skirt selectivity) (1) (non-real time spectrum analyzer). A measure of the asymptotic shape of the resolution bandwidth response curve of a spectrum analyzer. Shape factor is defined as the ratio between the frequency difference between two widely spaced points on the response curve, such as the 3 decibels and 60 decibels down points. 68

(2) (induction and dielectric heating equipment). *See:* **coil shape factor.**

shaping (operations on a pulse) (pulse terms). A process in which the shape of a pulse is modified to one which is ideal or more suitable for the intended application wherein time magnitude parameters may be changed. Typically, some property(ies) of the original pulse is preserved. (1) regeneration. A shaping process in which a pulse with desired reference characteristics is developed from a pulse which lacks certain desired characteristics. (2) stretching. A shaping process in which pulse duration is increased. (3) clipping. A shaping process in which the magnitude of a pulse is constrained at one or more predetermined magnitudes. (4) limiting. A clipping process in which the pulse shape is preserved for all magnitudes between predetermined clipping magnitudes. (5) slicing. A clipping process in which the pulse shape is preserved for all magnitudes less (greater) than a predetermined clipping magnitude. (6) differentiation. A shaping process in which a pulse is converted to a wave whose shape is or approximates the time derivative of the pulse. (7) integration. A shaping process in which a pulse is converted to a wave whose shape is or approximates the time integral of the pulse. 254

shaping pulse. The intentional processing of a pulse waveform to cause deviation from a reference waveform. *See also:* **pulse.** 185

shaping time constant (semiconductor radiation detectors). The time constants of the bandwidth defining CR (capacitance-resistance) differentiators and RC (resistance-capacitance) integrators used in pulse amplifiers. 23

sharing. *See:* **time sharing.**

shear pin (rotating machinery). A dowel designed to shear at a predetermined load and thereby prevent damage to other parts. *See also:* **rotor (rotating machinery); shear section shaft.** 63

shear section shaft (rotating machinery). A section of shaft machined to a controlled diameter, or area, designed to shear at a predetermined load and thereby prevent damage to connected machinery. *See also:* **rotor (rotating machinery); shear pin.** 63

shear wave (acoustics) (rotational wave). A wave in an elastic medium that causes an element of the medium to change its shape without a change of volume. *Notes:* (1) Mathematically, a shear wave is one whose velocity field has zero divergence. (2) A shear plane wave in an isotropic medium is a transverse wave. (3) When shear waves combine to produce standing waves, linear displacements may result. 176

shearing machine. An electrically driven machine for making vertical cuts in coal. 328

sheath (jacket) (cable systems). The overall protective covering for the insulated cable. 35

sheave (1) (conductor stringing equipment). (A) The grooved wheel of a traveler or rigging block. Travelers are frequently referred to as sheaves. (B) A shaft-mounted wheel used to transmit power by means of a belt chain, band, etcetera. *Syn:* **pulley; roller; wheel.** 45

(2) (rotating machinery). *See:* **pulley.**

shelf corrosion (dry cell). The consumption of the negative electrode as a result of local action. *See also:* **electrolytic cell.** 328

shelf depreciation. The depreciation in service capacity of a primary cell as measured by a shelf test. *See also:* **battery (primary or secondary).** 328

shelf test. A storage test designed to measure retention of service ability under specified conditions of temperature and cutoff voltage. *See also:* **battery (primary or secondary).** 328

shell (1) (insulators). A single insulating member, having a skirt or skirts without cement or other connecting devices intended to form a part of an insulator or an insulator assembly. *See also:* **insulator.** 261, 64

(2) (electrolysis). The external container in which the electrolysis of fused electrolyte is conducted. *See also:* **fused electrolyte.** 328

(3) (electrotyping). A layer of metal (usually copper or nickel) deposited upon, and separated from, a mold. *See also:* **electroforming.** 328

shell-form transformer. A transformer in which the laminations constituting the iron core surround the windings and usually enclose the greater part of them. 53

shell, stator (rotating machinery). A cylinder in tight assembly around the wound stator core, all or a portion of which is machined or otherwise made to a specific outer dimension so that the stator may be mounted into an appliance, machine, or other end product. *See also:* **stator.** 63

shell-type motor. A stator and rotor without shaft, end shields, bearings or conventional frame. *Note:* A shell-type motor is normally supplied by a motor manufacturer to an equipment manufacturer for incorporation as a built-in part of the end product. Separate fans or fans larger than the rotor are not included. *See:* **asynchronous machine; direct-current commutating machine.** 63

sheltered equipment (test, measurement and diagnostic equipment). Equipment so housed or otherwise protected that the extreme of natural and induced environments are partially or completely excluded or controlled. Examples are laboratory and shop equipment, equipment shielded from sun by a canopy or roof, and so forth. 54

shield (1) (electromagnetic). A housing, screen, or other object, usually conducting, that substantially reduces the effect of electric or magnetic fields on one side thereof, upon devices or circuits on the other side. *See:* **dielectric heating; induction heating; industrial electronics; signal.** 341

(2) (mechanical protection) (rotating machinery). An internal part used to protect rotating parts or parts of the electric circuit. In general, the word **shield** will be preceded by the name of the part that is being protected. 63

(3) (magnetrons). *See:* **end shield.**

(4) (induction heating). A material used to suppress the effect of an electric or magnetic field within or beyond definite regions. 14

(5) (transformer). A conductive protective member placed in relationship to apparatus or test components to control the shape and/or magnitude of electric or magnetic fields, thereby improving performance of apparatus or test equipment by reducing losses, voltage gradients, or interference. 53

(6) (metallic conductors). A housing or other object that substantially reduces the effect of electric or magnetic fields on one side thereof upon devices or circuits on the other side. 57

(7) (cable systems) (instrumentation cables). Metallic sheath (usually copper or aluminum), applied over the insulation of a conductor or conductors for the purpose of providing means for reducing electrostatic coupling between the conductors so shielded and others which may be susceptible to or which may be generating unwanted (noise) electrostatic fields. *Note:* When electromagnetic shielding is intended, the term **electromagnetic** is usually included to indicate the difference in shielding requirement as well as material. To be effective at power system frequencies, electromagnetic shields would have to be made of high permeability steel. Such shielding material is expensive and is not normally applied. Other less expensive means for reducing low frequency electromagnetic coupling, as described herein, are preferred. 35

shielded conductor cable. A cable in which the insulated conductor or conductors is/are enclosed in a conducting envelope or envelopes. 345

shielded ignition harness. A metallic covering for the ignition system of an aircraft engine, that acts as a shield to eliminate radio interference with aircraft electronic equipment. The term includes such items as ignition wiring and distributors when they are manufactured integral with an ignition shielding assembly. 328

shielded joint. A cable joint having its insulation so enveloped by a conducting shield that substantially every point on the surface of the insulation is at ground potential or at some predetermined potential with respect to ground. 64

shielded pair (signal-transmission system). A two-wire transmission line surrounded by a sheath of conducting material to protect it from the effects of external fields, or to confine fields produced by the transmission line. *See:* **signal.** *See also:* **waveguide.** 267, 188

shielded strip transmission line. A strip conductor between two ground planes. Some common designations are: Stripline (trade mark); Tri-plate (trade mark); slab line (round conductor); balanced strip line.* *See:* **strip (-type) transmission line; unshielded strip transmission line.** 179

* Deprecated

shielded transmission line (signal-transmission system). A transmission line surrounded by a sheath of conducting material to protect it from the effects of external fields, or to confine fields produced by the transmission line. *See:* **signal.** *See also:* **waveguide.** 188

shielded-type cable. A cable in which each insulated conductor is enclosed in a conducting envelope so constructed that substantially every point on the surface of the insulation is at ground potential or at some predetermined potential with respect to ground under normal operating conditions. 64

shield factor (telephone circuit). The ratio of noise, induced current, or voltage when a source of shielding is present, to the corresponding quantity when the shielding is absent. *See also:* **induction coordination.** 328

shield grid (gas tubes). A grid that shields the control electrode in a gas tube from the anode or the cathode, or both, with respect to the radiation of heat and the deposition of thermionic activating material and also reduces the electrostatic influence of the anode. It may be used as a control electrode. *See also:* **electrode (electron tube); grid.** 190, 125

shielding (screening) (power cable joints). A conducting layer, applied to control the dielectric stresses within tolerable limits and minimize voids. It may be applied over the entire joint insulation, on the tapered insulation ends only, or over irregular conductor or connector surfaces. 34

shielding angle (1) (lightning protection). The angle between the vertical line through the overhead ground wire and a line connecting the overhead ground wire with the shielded conductor. *See also:* **direct stroke protection (lightning).** 64

(2) (luminaire). The angle between a horizontal line through the light center and the line of sight at which the bare source first becomes visible. *See also:* **bare (exposed) lamp.** 167

shielding effectiveness (electromagnetic compatibility). For a given external source, the ratio of electric or magnetic field strength at a point before and after the placement of the shield in question. *See also:* **electromagnetic compatibility.** 199

shielding failure (lightning protection). The occurrence of a lightning stroke that bypasses the overhead ground wire and terminates on the phase conductor. *See also:* **direct-stroke protection (lightning).** 64

shield wire (electromagnetic fields). A wire employed for the purpose of reducing the effects on electric supply or communication circuits from extraneous sources. *See also:* **inductive coordination.** 328

shift (electronic computation). A displacement of an ordered set of characters one or more places to the left or right. If the characters are the digits of a numerical expression, a shift may be equivalent to multiplying by a power of the base. *See also:* **arithmetic shift; cyclic shift; logic shift.** 235, 210

shift, direct-current (oscilloscopes). A deviation of the displayed response to an input step, occurring over a period of several seconds after the input has reached its final value. 106

shift pulse. A drive pulse that initiates shifting of characters in a register. 331

shift register (computing systems). (1) A logic network consisting of a series of memory cells such that a binary code can be caused to shift into the register by serial input to only the first cell. *See also:* **digital.** 59

(2) A register in which the stored data can be moved to the right or left. 255, 77

shim (rotating machinery) (1) *mechanical.* A lamination usually machined to close-tolerance thickness, for assembly between two parts to control spacing. (2) *magnetic.* A lamination added to adjust or change the effective air gap in a magnetic circuit. *See also:* **rotor (rotating machinery); stator.** 63

shingle (photoelectric converter). Combination of photoelectric converters in series in a shingle-type structure. *See also:* **semiconductor.** 186

ship control telephone system. A system of sound-powered telephones (requiring no external power supply for talking) with call bells, exclusively for communication among officers responsible for control and operation of a ship. *Note:* Call bells are usually energized by hand-cranked magneto generators. 328

shipping brace (rotating machinery). Any structure provided to reduce motion or stress during shipment, that must be removed before operation. 63

shipping seal (cable). *See:* **sealed end.**

ship's service electric system. On any vessel, all electric apparatus and circuits for power and lighting, except apparatus provided primarily either for ship propulsion or for the emergency system. *Note:* Emergency and interior communication circuits are normally supplied with power from the ship's service system, upon failure of which they are switched to an independent emergency generator or other sources of supply. *See also:* **marine electric apparatus.** 328

shock excitation (1) (oscillatory systems). The excitation of natural oscillations in an oscillatory system due to a sudden acquisition of energy from an external source or a sudden release of energy stored with the oscillatory system. *See also:* **oscillatory circuit.** 328 **(2) (signal-transmission system).** The type of excitation supplied by a voltage, current, temperature, etcetera, variation of relatively short duration. *See:* **signal.** 188

shock motion (mechanical system). Transient motion that is characterized by suddenness, by significant relative displacements, and by the development of substantial internal forces in the system. *See:* **mechanical shock.** 176

shockproof electric apparatus. Electric apparatus designed to withstand, to a specified degree, shock of specified severity. *Note:* The severity is stated in footpounds impact on a special test stand equivalent to shock of gunfire, explosion of mine or torpedo, etcetera. *See also:* **marine electric apparatus.** 328

shoe (ramp shoe). Part of a vehicle-carried apparatus that makes contact with a ramp. 328

shop instruments. Instruments and meters that are used in regular routine shop or field operations. 212

shop test (laboratory test). A test made upon the receipt of a meter from a manufacturer, or prior to reinstallation. Such tests are made in a shop or a laboratory of a meter department. *See also:* **service test (field test).** 328

shoran. A precision position-fixing system using a pulse transmitter and receiver in connection with two transponder beacons at fixed points. *See also:* **radio navigation.** 328

shore feeder. Permanently installed conductors from a distribution switchboard to a connection box (or boxes) conveniently located for the attachment of portable leads for supply of power to a ship from a source on shore. *See also:* **marine electric apparatus.** 328

short circuit (power switchgear). An abnormal connection (including an arc) of relatively low impedance, whether made accidentally or intentionally, between two points of different potential. *Note:* The term **fault** or **short-circuit fault** is used to describe a short circuit. 103

short circuit, adjustable, waveguide (waveguide components). A longitudinally movable obstacle which reflects essentially all the incident energy. 166

short-circuit driving-point admittance. *See:* **admittance, short-circuit driving-point.**

short-circuit duration rating (magnetic amplifier). The length of time that a short circuit may be applied to the load terminals nonrecurrently without reducing the intended life of the amplifier or exceeding the specified temperature rise. 171

short-circuiter. A device designed to short circuit the commutator bars when the motor has attained a predetermined speed in some forms of single-phase commutator-type motors. *See:* **asynchronous machine.** 328

short circuit; external dc (thyristor power converter). A short circuit on the dc side outside the converter. *Note:* External short circuits may require different protecting means, depending on the character of the short circuit. Complete dc short circuit occurs when the short-circuit impedance is negligible compared to internal impedance of the converter. Limited dc short circuit occurs when the short-circuit impedance is large enough to limit the fault current. Feeder dc short circuit is a short circuit in a feeder with a separate protective device with much lower rating than the feeding converter (multimotor drives). 121

short-circuit feedback admittance. *See:* **admittance, short-circuit feedback.**

short-circuit flux (magnetic sound records). That flux from a magnetic record which flows across a plane normal to the recorded medium, through a magnetic short circuit placed in intimate contact with the record. 161

short-circuit flux per unit width (magnetic sound records). The measured short-circuit flux divided by the measured width of the recorded track. *Note:* The term **fluxivity** has been proposed for the quantity short-circuit flux per unit width. 161

short-circuit forward admittance. *See:* **admittance, short-circuit forward.**

short-circuit impedance (1) (general). A qualifying adjective indicating that the impedance under consideration is for the network with a specified pair or group of terminals short-circuited. *See also:* **network analysis; self-impedance.** 210 **(2) (line or four-terminal network).** The driving-point impedance when the far-end is short-circuited. *See also:* **self-impedance.** 328

short-circuit inductance. The apparent inductance of a winding of a transformer with one or more specified windings short circuited often taken as a means of determining the leakage inductance of a winding. 197, 67

short-circuit input admittance. *See:* **admittance, short-circuit input.**

short circuit, internal (thyristor power converter). A short circuit caused by converter faults. *Note:* An internal short circuit may be fed from both ac and dc circuits: for example, in the cases of (1) converters with battery or motor loads, (2) converters in a double converter, (3) converters operating as inverters.   121

short-circuit loss (rotating machinery). The difference in power required to drive a machine at normal speed, when excited to produce a specified balanced short-circuit armature current, and the power required to drive the unexcited machine at the same speed. *See also:* **asynchronous machine; direct-current commutating machine.**   63

short-circuit output admittance. *See:* **admittance, short-circuit output.**

short-circuit output capacitance. *See:* **capacitance, short-circuit output.**

short-circuit protection (power supplies) (automatic). Any automatic current-limiting system that enables a power supply to continue operating at a limited current, and without damage, into any output overload including short circuits. The output voltage must be restored to normal when the overload is removed, as distinguished from a fuse or circuit-breaker system that opens at overload and must be closed to restore power. *See:* **current limiting.** *See also:* **power supply.**   228, 186

short-circuit ratio (synchronous machine). The ratio of the field current for rated open-circuit armature voltage and rated frequency to the field current for rated armature current on sustained symmetrical short-circuit at rated frequency.   63

short-circuit saturation curve (synchronous machine). The relationship between the current in the short-circuited armature winding and the field current.   63

short-circuit time constant (1) (armature winding). The time in seconds for the asymmetrical (direct-current) component of armature current under suddenly applied short-circuit conditions, with the machine running at rated speed, to decrease to $1/e \approx 0.368$ of its initial value. The rated current value of the short-circuit time constant of the armature winding will be that obtained from the test specified for the rated current value of direct-axis transient reactance. The rated voltage value of the short-circuit time constant of the armature winding will be that obtained from a short-circuit test at the terminals of the machine at no load and rated speed and at rated armature voltage.   328

(2) (primary winding) (rotating machinery). The time required for the direct-current component present in the short-circuit primary-winding current following a sudden change in operating conditions to decrease to $1/e \approx 0.368$ of its initial value, the machine running at rated speed. *See also:* **asynchronous machine.**   63

short-circuit transfer admittance. *See:* **admittance, short-circuit transfer.**

short-circuit transfer capacitance. *See:* **capacitance, short-circuit transfer.**

short dimension. Incremental dimensions whose number of digits is the same as normal dimensions except the first digit shall be zero, that is 0.XXXX for the example under normal dimension. *See:* **dimension; incremental dimension; long dimension; normal dimension.**   224, 207

short-distance navigation. Navigation utilizing aids usable only at comparatively short distances; this term covers navigation between approach navigation and long-distance

navigation, there being no distinct, universally accepted demarcation between them. *See also:* **navigation.**   187, 13

short field (tapped field*). Where two field strengths are required for a series machine, short field is the minimum-strength field connection. *See:* **asynchronous machine; direct-current commutating machine.**   328

* Deprecated

short-line-fault transient recovery voltage (power switchgear). The transient recovery voltage obtained when a circuit switching device interrupts a nearby fault on the line. *Note:* It differs from terminal fault conditions in that the length of line adds a high-frequency sawtooth component to the transient recovery voltage. As the distance to the fault becomes greater, the amplitude of the sawtooth component increases, the rate of rise of the sawtooth component decreases and the fault current decreases.

The increased amplitude adversely affects the interrupting capability of the circuit switching device while the decrease in rate of rise and the decrease in current makes interruption easier. The effects are not proportional and a distance is reached where interruption is most severe even though the current is less than for a terminal fault. The critical value varies considerably with the type of circuit switching device (oil, air-blast, gas-blast, etc), and with the particular design. The critical distance may be in the order of a mile at the higher voltages. The critical distance is less as lower voltages are considered.   103

short-pitch winding (rotating machinery). A winding in which the coil pitch is less than the pole pitch. *See also:* **rotor (rotating machinery); stator.**   63

short time capability (industrial control). The ability to carry a specified electrical overload under prescribed conditions within specified performance criteria.   70

short-time current. The current carried by a device, an assembly, or a bus for a specified short time interval.   103, 202, 27

short-time current rating (insulated connectors). The designated rms current which a connector can carry for a specified time under specified conditions.   134

short-time current rating tests (high voltage air switches). Tests to determine the maximum current, including the direct-current component, that equipment will carry without injury for specified short-time intervals. This rating recognizes the limitations imposed by both thermal and electromagnetic effects.   144

short-time delay phase or ground trip element (power switchgear). A direct-acting trip device element that functions with a purposely delayed action (milliseconds)   103

short-time duty (rating of electric equipment). A duty that demands operation at a substantially constant load for a short and definitely specified time.   53

Note: The specified conditions normally require duty at constant load during a given time less than that required to obtain constant temperature in continuous duty at the same load, followed by a rest of sufficient duration to re-establish equality of temperature with the cooling medium. *See also:* **asynchronous machine; direct-current commutating machine; industrial control; voltage regulator.**   210, 264, 310, 233, 203, 244, 206

short-time operation influence (electric instrument). The operation influence arising from continuous operation over a period of 15 minutes. *See also:* **accuracy rating (of an instrument).**   280

short-time overload rating (electric penetration assembly). The limiting overload current that any one third of the conductors (but in no case less than three of the conductors) in the assembly can carry, for a specified time, in the design service environment, while all remaining conductors carry rated continuous current, without causing the conductor temperatures to exceed those values recommended by the insulated conductor manufacturer as the short-time overload conductor temperature and without causing the stabilized temperature of the penetration nozzle—concrete interface (if applicable) to exceed its design limit.   26

short-time rating (1) (general). Defines the load that can be carried for a short and definitely specified time, the machine, apparatus, or device being at approximately room temperature at the time the load is applied. *See also:* **duty.**   310, 210, 233, 225, 206, 124

(2) (rotating machinery). The statement of the load, duration, and conditions assigned to a machine by the manufacturer at which the machine may be operated for a limited period, starting at ambient temperature. *See also:* **asynchronous machine; direct-current commutating machine.**   63

(3) (transformer). Defines the maximum constant load that can be carried for a specified short time without exceeding established temperature-rise limitations, under prescribed conditions of test.   53

(4) (voltage regulator). Defines the maximum constant load that can be carried for a specified short time, the regulator being approximately at room temperature at the time the load is applied without exceeding the specified temperature limitation, and within the limitations of established standards.   257

(5) (reactor). Defines the maximum constant load that can be carried for a specified short time, the reactor being approximately at room temperature at the time the load is applied, without exceeding the specified temperature limitation, and within the limitations of established standards.   309

(6) (arc-welding apparatus). Defines the load that can be carried for a short and definitely specified time, the apparatus being approximately at room temperature at the time the temperature test is started.   264

(7) (power inverter unit). Defines the maximum load that can be carried for a specified short time, without exceeding the specified limitations under prescribed conditions of test, and within the limitation of established standards. *See also:* **self-commutated inverters.**   208

(8) (rectifier unit). The maximum load that can be carried for a specified short time, without exceeding the specified temperature-rise limitations under prescribed conditions of test, and within the limitations of established standards. *Note:* The short-time rating includes loads of two hours duration. *See also:* **rectification.**   208

(9) (grounding device). A rated time of ten minutes or less.   91

(10) (diesel-generator unit, nuclear power generating station). The electric power output capability which the diesel-generator unit can maintain in the service environment for two hours in any twenty-four hour period,

without exceeding the manufacturer's design limits and without reducing the maintenance interval established for the continuous rating. *Note:* Operation at this rate does not limit the use of the diesel-generator unit at its continuous rating.   31

(11) (relay). The highest value of current or voltage or their product that the relay can stand, without injury, for specified short-time intervals (for alternating-current circuits, root-mean-square total value including the direct-current component shall be used). The rating shall recognize the limitations imposed by both thermal and electromagnetic effects.   127

short-time waveform distortion (SD) (video signal transmission measurement). The linear TV waveform distortion of time components from 125 ns to 1 μs, that is, time components of the short-time domain.   42

shot effect (electron tubes). The variations in the output of an electron valve or tube due to: (1) Random variation in the emission of electrons from the cathode. (2) Instantaneous variations in the distribution of the electrons among the electrodes. *See also:* **electron tube.**
                    244, 190

shot-firing (blasting) cord. A two-conductor cable used for completing the circuit between the electric blasting cap (or caps) and the blasting unit or other source of electric energy. *See also:* **blasting unit.**   328

shot-firing unit. *See:* **blasting unit.**

shot noise, full (electron tubes) (interference terminology). The fluctuation in the current of charge carriers passing through a surface at statistically independent times. *Notes:* (1) **Shot noise** has a uniform spectral density W_i given by

$$W_i = \frac{eI_0}{2\pi}$$

where e is the charge of the carrier and I_0 is the average current. (2) The mean-square noise current $\bar{i}^2$ of full shot noise within a frequency increment Δf is

$$\bar{i}^2 = 2eI_0\Delta f.$$

(3) The mean-square noise current $\bar{i}^2$ within a frequency increment Δf associated with an average current I_0 is often expressed in terms of full shot noise through a shot noise reduction factor Γ^2, in general a function of frequency, by the formula:

$$\bar{i}^2 = \Gamma^2 2eI_0\Delta f.$$

When $\Gamma^2 < 1$, $\bar{i}^2$ is called **reduced shot noise.** *See also:* **interference; signal-to-noise ratio.**   88, 190

shot noise, reduced. *See:* **shot noise, full.**

shrink link (rotating machinery). A bar with an enlarged head on each end for use like a rivet but slipped into place after expansion by heat. It tightens on cooling by shrinkage only.   63

shunt (1) (general). A device having appreciable resistance or impedance connected in parallel across other devices or apparatus, and diverting some (but not all) of the current from it. Appreciable voltage exists across the shunted device or apparatus and an appreciable current may exist in it.   210

(2) (air switches). A flexible electrical conductor comprised of braid, cable, or flat laminations designed to conduct current around the mechanical joint between two conductors. 27, 79

shunt capacitor bank current. Current, including harmonics, supplied to a shunt capacitor bank. *Note:* Current is expressed in rms amperes. 130, 103

shunt control. A method of controlling motors employing the shunt method of transition from series to parallel connections of the motors. *See also:* **multiple-unit control.** 328

shunt-fed vertical antenna. A vertical antenna connected to ground at the base and energized at a point suitably positioned above the grounding point. *See also:* **antenna.** 111, 179

shunting transition. *See:* **shunt transition.**

shunt leads (instrument). Those leads that connect a circuit of an instrument to an external shunt. The resistance of these leads is taken into account in the adjustment of the instrument. *See also:* **auxiliary device to an instrument; instrument.** 280

shunt loading. Loading in which reactances are applied in shunt across the conductors of a transmission circuit. *See also:* **loading.** 328

shunt noninterfering fire-alarm system. A manual fire-alarm system employing stations and circuits such that, in case two or more stations in the same premises are operated simultaneously, the signal from the operated box electrically closest to the control equipment is transmitted and other signals are shunted out. *See also:* **protective signaling.** 328

shunt reactor. A reactor intended for connection in shunt to an electric system for the purpose of drawing inductive current. *Note:* The normal use for shunt reactors is to compensate for capacitive currents from transmission lines, cable, or shunt capacitors. The need for shunt reactors is most apparent at light load. 53, 203

shunt regulator (power supplies). A device placed across the output that controls the current through a series dropping resistance to maintain a constant voltage or current output. *See also:* **power supply.** 228, 186

shunt release (shunt trip). A release energized by a source of voltage. *Note:* The voltage may be derived either from the main circuit or from an independent source. 103, 202

shunt tee junction. *See:* **H-plane tee.**

shunt transition (shunting transition). A method of changing the connection of motors from series to parallel in which one motor, or group of motors, is first shunted or short circuited, then open circuited, and finally connected in parallel with the other motor or motors. *See also:* **multiple-unit control.** 328

shunt trip. *See:* **shunt release.**

shunt-trip recloser. A recloser in which the tripping mechanism, by releasing the holding means, permits the main contacts to open, with both the tripping mechanism and the contact-opening mechanism deriving operating energy from other than the main circuit. 103, 202

shunt-wound. A qualifying term applied to a direct-current machine to denote that the excitation is supplied by a winding connected in parallel with the armature in the case of a motor, with the load in the case of a generator, or is connected to a separate source of voltage. 328

shunt-wound generator. A direct-current generator in which ordinarily the entire field excitation is derived from one winding consisting of many turns with a relatively high resistance. This one winding is connected in parallel with the armature circuit for a self-excited generator and to the load side of another generator or other source of direct current for a separately excited generator. *See also:* **direct-current commutating machine.** 3

shunt-wound motor. A direct-current motor in which the field circuit and armature circuit are connected in parallel. *See also:* **direct-current commutating machine.** 3

shutter (1) (electric machine). A protective covering used to close, or to close partially, an opening in a stator frame or end shield. In general, the word **shutter** will be preceded by the name of the part to which it is attached. As used for an electric machine, a shutter is rigid and hence not adjustable. *See also:* **stator.** 63

(2) (switchgear assembly). A device that is automatically operated to completely cover the stationary portion of the primary disconnecting devices when the removable element is either in the disconnected position, test position, or has been removed. 103, 202

shuttle car. A vehicle on rubber tires or caterpillar treads and usually propelled by electric motors, electric energy for which is supplied by a diesel-driven generator, by storage batteries, or by a power distribution system through a portable cable. Its chief function is the transfer of raw materials, such as coal and ore, from loading machines in trackless areas of a mine to the main transportation system. 328

shuttle car, explosion-tested. A shuttle car equipped with explosion-tested equipment. 328

SI (International System of Units). *See:* **units and letter symbols.**

side back light (television). Illumination from behind the subject in a direction not parallel to a vertical plane through the optical axis of the camera. *See also:* **television lighting.** 167

sideband attenuation. That form of attenuation in which the transmitted relative amplitude of some component(s) of a modulated signal (excluding the carrier) is smaller than that produced by the modulation process. *See also:* **wave front.** 111

sideband null (rectilinear navigation system). The surface of position along which the resultant energy from a particular pair of sideband antennas is zero. *See also:* **navigation.** 187, 13

sideband-reference glide slope (instrument landing systems). A modified null-reference glide-slope antenna system in which the upper (sideband) antenna is replaced with two antennas, both at lower heights, and fed out of phase, so that a null is produced at the desired glide-slope angle. *Note:* This system is used to reduce unwanted reflections of energy into the glide-slope sector at locations where rough terrain exists in front of the approach end of the runway, by producing partial cancellation of energy at low elevation angles. *See also:* **navigation.** 187, 13

sidebands. (1) The frequency bands on both sides of the carrier frequency within which fall the frequencies of the wave produced by the process of modulation. (2) The wave components lying within such bands. *Note:* In the process of amplitude modulation with a sine-wave carrier, the upper sideband includes the sum (carrier plus modulating) frequencies; the lower sideband includes the difference (carrier minus modulating) frequencies. *See also:* **amplitude modulation; radio receiver.** 111, 59, 188

sideband suppression (power-system communication). A process that removes the energy of one of the sidebands from the modulated carrier spectrum. 179

side-break switch. A switch in which the travel of the blade is in a plane parallel to the base of the switch.

103, 202, 27

side circuit. A circuit arranged for deriving a phantom circuit. *Note:* In the case of two-wire side circuits, the conductors of each side circuit are placed in parallel to form a side of the phantom circuit. In the case of four-wire side circuits, the lines of the two side circuits that are arranged for transmission in the same direction provide a one-way phantom channel for transmission in that same direction, the two conductors of each line being placed in parallel to provide a side for that phantom channel. Similarly the conductors of the other two lines provide a phantom channel for transmission in the opposite direction. *See also:* **transmission line.** 59

side-circuit loading coil. A loading coil for introducing a desired amount of inductance in a side circuit and a minimum amount of inductance in the associated phantom circuit. *See also:* **loading.** 328

side-circuit repeating coil (side-circuit repeat coil). A repeating coil that functions simultaneously as a transformer at a terminal of a side circuit and as a device for superposing one side of a phantom circuit on that side circuit.

328

sideflash (lightning). A spark occurring between nearby metallic objects or from such objects to the lightning protection system or to ground. *See also:* **direct-stroke protection (lightning).** 297, 64

side flashover (lightning). A flashover of insulation resulting from a direct lightning stroke that bypasses the overhead ground wire and terminates on a phase conductor of a transmission line. *See also:* **direct-stroke protection (lightning).** 64

side frequency. One of the frequencies of a sideband. *See also:* **amplitude modulation.** 111, 242

side lobe (antenna pattern). A radiation lobe in any direction other than that of the intended lobe. *Note:* When the intended lobe is not specified it shall be taken to be the major lobe. 111

sidelobe blanker (radar). A device which employs an auxiliary wide-angle antenna and receiver to sense whether a received pulse originates in the sidelobe region of the main antenna and to gate it from the output signal if it does. 13

sidelobe canceller (radar). A device which employs one or more auxiliary antennas and receivers to allow linear subtraction of interfering signals from the desired output if they are sensed to originate in the sidelobe of the main antenna. 13

side-lobe, level, maximum relative (antenna). The relative level of the highest side lobe. *See also:* **radiation.** 179

side lobe, relative level of (antenna). The ratio of the radiation intensity of a side lobe in the direction of its maximum value to that of the intended lobe, usually expressed in decibels. *Note:* When the intended lobe is not specified,

side lobe level will be measured from the maximum value of the major lobe. *See also:* **radiation.** 111

side-lock. Spurious synchronization in an automatic frequency synchronizing system by a frequency component of the applied signal other than the intended component. *See:* **television.** 178

sidelooking radar. A ground mapping radar used aboard aircraft involving the use of a fixed antenna beam pointing out the side of an aircraft either abeam or squinted with respect to the aircraft axis. The beam is usually a vertically oriented fan beam having a narrow azimuth width. The narrow azimuth resolution can either be obtained with a long aperture mounted along the axis of the aircraft or by the use of synthetic aperture radar processing. 13

side marker lamp. A light indicating the presence of a vehicle when seen from the front and sometimes serving to indicate its width. *See also:* **headlamp.** 167

side panel (rotating machinery). A structure enclosing or partly enclosing one side of a machine. 63

sidereal period (communication satellite). The time duration of one orbit measured relative to the stars. 74

side thrust (skating force) (disk recording). The radial component of force on a pickup arm caused by the stylus drag. *See also:* **phonograph pickup.** 176

sidetone (telephone set). The acoustic signal resulting from a portion of the transmitted signal being coupled to the receiver of the same handset. 122

sidetone path loss (telephony). The difference in dB of the acoustic output level of the receiver of a given telephone set to the acoustic input level of the transmitter of the same telephone set. 122

sidetone telephone set. A telephone set that does not include a balancing network for the purpose of reducing sidetone. *See also:* **telephone station.** 328

sidewalk elevator. A freight elevator that operates between a sidewalk or other area exterior to the building and floor levels inside the building below such area, that has no landing opening into the building at its upper limit of travel, and that is not used to carry automobiles. *See also:* **elevators.** 328

siemens. *See:* **mho.**

sigma (σ). The term **sigma** designates a group of telephone wires, usually the majority or all wires of a line, that is treated as a unit in the computation of noise or in arranging connections to ground for the measurement of noise or current balance ratio. *See also:* **induction coordination.** 328

sign (1) (power or energy). Positive, if the actual direction of energy flow agrees with the stated or implied reference direction; negative, if the actual direction is opposite to the reference direction. *See also:* **network analysis.** 210

(2) (test, measurement and diagnostic equipment). The symbol which distinguishes positive from negative numbers. 54

signal (1). A visual, audible, or other indication used to convey information.

(2) The intelligence, message, or effect to be conveyed over a communication system.

(3) A signal wave; the physical embodiment of a message. *See also:* **information theory.** 239, 199, 242

(4) (computing systems). The event or phenomenon that conveys data from one point to another. 255, 77

(5) (control) (industrial control). Information about a variable that can be transmitted in a system. 219, 206

(6) (programmable instrumentation). The physical representation which conveys data from one point to another. *Note:* For the purpose of this standard, this is a restricted definition of what is often called "signal" in more general sense, and is hereinafter referred to digital electrical signals only. 40

(7) (telephone switching systems). An audible, visual or other indication of information. 55

(8) (circuits and systems). A phenomenon (visual, audible, or otherwise) used to convey information. The signal is often coded, such as a modulated waveform, so that it requires decoding to be intelligible. 67

signal, actuating (control system, feedback). *See:* actuating signal.

signal aspect. The appearance of a fixed signal conveying an indication as viewed from the direction of an approaching train; the appearance of a cab signal conveying an indication as viewed by an observer in the cab. 328

signal back light. A light showing through a small opening in the back of an electrically lighted signal, used for checking the operation of the signal lamp. 328

signal charge (ferroelectric device). The charge that flows when the condition of the device is changed from that of zero applied voltage (after having previously been saturated with either a positive or negative voltage) to at least that voltage necessary to saturate in the reverse sense. *Note:* The signal charge Q_s equals the sum of Q_r and Q_t, as illustrated in the accompanying figure. It is dependent on the magnitude of the applied voltage, which should be specified in describing this characteristic of ferroelectric devices. *See also:* ferroelectric domain. 247

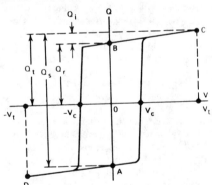

Hysteresis loop for a ferroelectric device

signal circuit (1). Any electric circuit that supplies energy to an appliance that gives a recognizable signal. Such circuits include circuits for door bells, buzzers, code-calling systems, signal lights, and the like. *See also:* appliances. 256

(2) (protective relay system). Any circuit other than input voltage circuits, input current circuits, power supply circuits, or those circuits which directly or indirectly control power circuit breaker operation. 90

signal contrast (facsimile). The ratio expressed in decibels between white signal and black signal. *See also:* facsimile signal (picture signal). 12

signal converter (test, measurement and diagnostic equipment). A device for changing a signal from one form or value to another form or value. 54

signal delay. The transmission time of a signal through a network. The time is always finite, may be undesired, or may be purposely introduced. *See:* delay line. *See also:* oscillograph. 185

signal, difference. *See:* differential signal.

signal distance (computing systems). The number of digit positions in which the corresponding digits of two binary words of the same length are different. *See:* Hamming distance. 255, 77

signal distributing (telephone switching systems). Delivering of signals from a common control to other circuits. 55

signal electrode (camera tube). An electrode from which the signal output is obtained. *See also:* electrode (electron tube). 178

signal element (unit interval). The part of a signal that occupies the shortest interval of the signaling code. It is considered to be of unit duration in building up signal combinations. *See also:* telegraphy. 59, 194

signal, error (1) (automatic control device) (general). A signal whose magnitude and sign are used to correct the alignment between the controlling and the controlled elements. 328

(2) (power supplies). The difference between the output voltage and a fixed reference voltage compared in ratio by the two resistors at the null junction of the comparison bridge. The error signal is amplified to drive the pass elements and correct the output. *See also:* power supply. 228, 186

(3) (closed loop) (control system, feedback). The signal resulting from subtracting a particular return signal from its corresponding input signal. See the accompanying figure. *See also:* control system, feedback. 56, 105

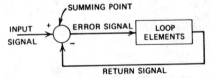

Block diagram of a closed loop

signal, feedback (1) (general). A function of the directly controlled variable in such form as to be used at the summing point. *See also:* control system, feedback. 219, 206

(2) (control system, feedback). The return signal that results from the reference input signal. See the accompanying figure. *See also:* control system, feedback. 56, 105

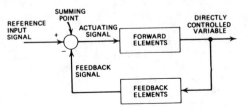

Simplified block diagram indicating essential elements of an automatic control system

signal flow graph (network analysis). A network of directed branches in which each dependent node signal is the algebraic sum of the incoming branch signals at that node. *Note:* Thus,

$$x_1 t_{1k} + x_2 t_{2k} + \ldots + x_n t_{nk} = x_k$$

at each dependent node k, where t_{jk} is the branch transmittance of branch jk. *See also:* **linear signal flow graphs.**
282

signal frequency shift (frequency-shift facsimile system). The numerical difference between the frequencies corresponding to white signal and black signal at any point in the system. *See also:* **facsimile signal (picture signal).**
12

signal generator. A shielded source of voltage or power, the output level and frequency of which are calibrated, and usually variable over a range. *Note:* The output of known waveform is normally subject to one or more forms of calibrated modulation.
185

signal identifier (non-real time spectrum analyzer). A means to identify the frequency of the input when spurious responses are possible.
68

signal indication. The information conveyed by the aspect of a signal.
328

signaling (1) (telephone switching systems). The transmission of address and other switching information between stations and central offices and between switching entities.
55

(2) (data transmission). In a telephone system, any of several methods used to alert subscribers or operators. *See:* **ringing.**
59

signaling and doorbell transformers. Stepdown transformers, (having a secondary of 30 volts or less) generally used for the operation of signals, chimes, and doorbells.
53

signaling common battery (data transmission). A method of actuating a line or supervisory signal at the distant end of a telephone line by the closure of a direct-current circuit.
59

signaling in band (data transmission). Signaling which utilizes frequencies within the voice or intelligence band of a channel.
59

signaling light. A projector used for directing light signals toward a designated target zone.
167

signaling out of band (data transmission). Signaling which utilizes frequencies within the guard band between channels. This term is also used to indicate the use of a portion of a channel bandwidth provided by the medium such as a carrier channel, but denied to the speech or intelligence path by filters. It results in a reduction of the effective available bandwidth.
59

signaling ringdown (data transmission). The application of a signal to the line for the purpose of bringing in a line signal or supervisory signal at a switchboard or ringing a users instrument. (Historically, this was a low frequency signal of about 20 hertz from the user on the line for calling the operator or for disconnect).
59

signal, input (control system, feedback). A signal applied to a system or element. See the figure attached to the

definition (3) of **signal, error.** *See also:* **control system, feedback.**
56, 105

signal integration. The summation of a succession of signals by writing them at the same location on the storage surface. *See also:* **storage tube.**
174, 190

signal level (1) (electroacoustics). The magnitude of a signal, especially when considered in relation to an arbitrary reference magnitude. *Note:* Signal level may be expressed in the units in which the quantity itself is measured (for example, volts or watts) or in units expressing a logarithmic function of the ratio of the two magnitudes. *See:* **level.**
239

(2) (programmable instrumentation). The magnitude of a signal when considered in relation to an arbitrary reference magnitude (voltage in the case of this standard).
40

signal line (programmable instrumentation). One of a set of signal conductors in an interface system used to transfer messages among interconnected devices.
40

signal lines. The passive transmission lines through which the signal passes from one to another of the elements of the signal transmission system. *See:* **signal.**
188

signal operation (elevators). Operation by means of single buttons or switches (or both) in the car, and up-or-down direction buttons (or both) at the landings, by which predetermined landing stops may be set up or registered for an elevator or for a group of elevators. The stops set up by the momentary actuation of the car buttons are made automatically in succession as the car reaches those landings, irrespective of its direction of travel or the sequence in which the buttons are actuated. The stops set up by the momentary actuation of the up-and-down buttons at the landing are made automatically by the first available car in the group approaching the landing in the corresponding direction, irrespective of the sequence in which the buttons are actuated. With this type of operation, the car can be started only by means of a starting switch or button in the car. *See also:* **control (elevators).**
328

signal, output (control system, feedback). A signal delivered by a system or element. *See also:* **control system, feedback.**
56, 105

signal output current (camera tubes or phototubes). The absolute value of the difference between output current and dark current. *See:* **phototubes.**
190, 178, 125

signal parameter (programmable instrumentation). That parameter of an electrical quantity whose values or sequence of values convey information.
40

signal-processing antenna system. An antenna system having circuit elements associated with its radiating elements that perform functions such as multiplication, storage, correlation, and time modulation of the input signals. *See also:* **antenna.**
179, 111

signal, rate. A signal that is responsive to the rate of change of an input signal.
105

signal, reference input (1) (general). The command expressed in a form directly usable by the system. The reference input signal is in the terms appropriate to the form in which the signal is used, that is, voltage, current, ampere-turns, etcetera. *See also:* **control system feedback.**
219, 206

(2) (control system, feedback). A signal external to a

control loop that serves as the standard of comparison for
the directly controlled variable. See the figure attached
to the definition of **signal, feedback.** *See also:* **control
system, feedback.** 56, 105

signal relay. *See:* **alarm relay.**

signal repeater lights. A group of lights indicating the
signal displayed for humping and trimming. 328

signal, return (control system, feedback) (closed loop). The
signal resulting from a particular input signal, and
transmitted by the loop and to be subtracted from that
input signal. See the figure attached to the definition of
signal, error. *See also:* **control system, feedback.**
56, 105

signal, sampled. *See:* **sampled signal.**

signal-shaping amplifier (telegraph practice). An amplifier
and associated electric networks inserted in the circuit,
usually at the receiving end of an ocean cable, for amplifying and improving the waveshape of the signals. *See also:*
telegraphy. 328

signal-shaping network (wave-shaping set). An electric
network inserted (in a telegraph circuit) for improving the
waveshape of the received signals. *See also:* **telegraphy.**
328

signal shutter (illuminating engineering). A device that
modulates a beam of light by mechanical means for the
purpose of transmitting intelligence. *See also:* **searchlight.**
167

signal-to-clutter ratio (radar). The ratio of target echo
power to the power received from clutter sources lying
within the same resolution element prior to detection.
13

signal-to-interference ratio. The ratio of the magnitude
of the signal to that of the interference or noise. *Note:* The
ratio may be in terms of peak values or root-mean-square
values and is often expressed in decibels. The ratio may
be a function of the bandwidth of the system. *See:* **signal.**
188

signal-to-noise ratio (1) (general). The ratio of the value of
the signal to that of the noise. *Notes:* (A) This ratio is
usually in terms of peak values in the case of impulse noise
and in terms of the root-mean-square values in the case of
the random noise. (B) Where there is a possibility of ambiguity, suitable definitions of the signal and noise should
be associated with the term; as, for example: peak-signal
to peak-noise ratio; root-mean-square signal to root-
mean-square noise ratio; peak-to-peak signal to peak-
to-peak noise ratio, etcetera. (C) This ratio may be often
expressed in decibels. (D) This ratio may be a function of
the bandwidth of the transmission system. *See also:* **signal;
signal-to-interference ratio; television.** 111, 178

(2) (camera tubes). The ratio of peak-to-peak signal output
current to root-mean-square noise in the output current.
Note: Magnitude is usually not measured in tubes where
the signal output is taken from target. *See also:* **camera
tube; television.**

(3) (television transmission). The signal-to-noise ratio at
any point is the ratio in decibels of the maximum peak-
to-peak voltage of the video television signal, including
synchronizing pulse, to the root-mean-square voltage of
the noise. *Note:* The signal-to-noise ratio is defined in this
way because of the difficulty of defining the root-mean-
square value of the video signal or the peak-to-peak value

of random noise. *See also:* **television.** 328

(4) (mobile communication). The ratio of a specified
speech-energy spectrum to the energy of the noise in the
same spectrum. *See also:* **television.** 181

(5) (sound recording and reproducing system). The ratio
of the signal power output to the noise power in the entire
pass band. 350

(6) (digital delay line). Ratio of the peak amplitude of the
output doublet to the maximum peak of any noise response
(or signal) outside of the doublet interval. (Includes overshoot.) 81

(7) (speech quality measurements) (S/N). In decibels of
a speech signal, the difference between its speech level and
the noise level. 126

signal-transfer point (telephone switching systems). A
switching entity where common channel signaling facilities
are interconnected. 55

signal transmission system. *See:* **carrier.**

signal, TV waveform. An electrical signal whose amplitude
varies with time in a generally nonsinusoidal manner,
whose shape (that is, duration and amplitude) carries the
TV signal information. 42

signal, unit-impulse (automatic control). A signal that is
an impulse having unity area. *See also:* **control system,
feedback.** 210, 56

signal, unit-ramp (automatic control). A signal that is zero
for all values of time prior to a certain instant and equal
to the time measured from that instant. *Note:* The unit-
ramp signal is the integral of the unit-step signal. *See also:*
control system-feedback. 210, 56

signal, unit-step (automatic control). A signal that is zero
for all values of time prior to a certain instant and unity
for all values of time following. *Note:* The unit-step signal
is the integral of the unit-impulse signal. *See also:* **control
system, feedback.** 210, 56

signal wave. A wave whose shape conveys some intelligence, message, or effect. 199

signal winding (input winding) (saturable reactor). A control
winding to which the independent variable (signal wave)
is applied. 328

**signature diagnosis (test, measurement and diagnostic
eqiupment).** The examination of signature of an equipment for deviation from known or expected characteristics
and consequent determination of the nature and location
of malfunctions. 54

sign digit (1) (electronic computation). A character used
to designate the algebraic sign of a number. 235, 210

(2). The digit in the sign position. 255, 77

significance (test, measurement and diagnostic equipment).
The value or weight given to a position, or to a digit in a
position, in a positional numeration system. In most positional numeration systems positions are grouped in sequence of significance, usually more significant towards
the left. 54

significant (nuclear power generating stations). Demonstrated to be important by the safety analysis of (for) the
station. 31, 102, 355

significant digit. A digit that contributes to the accuracy
or precision of a numeral. The number of significant digits
is counted beginning with the digit contributing the most
value, called the most-significant digit, and ending with
the one contributing the least value, called the least-sig-

nificant digit. 255, 77

sign position. The position at which the sign of a number is located. 255, 77

silent lobing (radar). A method for scanning an antenna beam to achieve angle tracking without revealing the scanning pattern on the transmitted signal. 13

silvering (electrotyping). The application of a thin conducting film of silver by chemical reduction upon a plastic or wax matrix. *See also:* **electroforming.** 328

silver oxide cell. A cell in which depolarization is accomplished by oxide of silver. *See also:* **electrochemistry.** 328

silver storage battery. An alkaline storage battery in which the positive active material is silver oxide and the negative contains zinc. *See also:* **battery (primary or secondary).** 328

silver-surfaced or equivalent. The term indicates metallic materials having satisfactory long-term performance and that operate within the temperature rise limits established for silver-surfaced electric contact parts and conducting mechanical joints. 103, 202

simple GCL circuit. *See:* **simple parallel circuit.**

simple parallel circuit (simple GCL circuit). A linear, constant-parameter circuit consisting of resistance, inductance, and capacitance in parallel. *See also:* **network analysis.** 210

simple rectifier. A rectifier consisting of one commutating group if single-way or two commutating groups if double-way. *See also:* **rectification.** 291, 208

simple rectifier circuit. A rectifier circuit consisting of one commutating group if single-way, or two commutating groups if double-way. *See also:* **rectification; rectifier circuit element.** 237, 66

simple RLC circuit. *See:* **simple series circuit.**

simple scanning (facsimile). Scanning of only one scanning spot at a time during the scanning process. 12

simple series circuit. A resistance, inductance, and capacitance in series. *Syn:* **simple RLC circuit.** *See also:* **network analysis.** 210

simple sine-wave quantity. A physical quantity that is varying with time t as either $A \sin(\omega t + \theta_A)$ or $A \cos(\omega + \theta_B)$ where $A, \omega, \theta_A, \theta_B$ are constants. (**Simple** denotes that $A, \omega, \theta_A, \theta_B$ are constants.) *Notes:* (1) It is immaterial whether the sin or cos form is used, so long as no ambiguity or inconsistency is introduced (2) A is the amplitude or maximum value, $\omega t + \theta_A$ (or $\omega t + \theta_B$) the phase, θ_A (or θB) the phase angle. However, when no ambiguity may arise, **phase angle** may be abbreviated **phase.** (3) In certain special applications, for example, modulation, $\omega t + \theta$ is called the angle (of a sine wave), (not phase angle) in order to clarify particular uses of the word "phase." Another permissible term for $(\omega t + \theta)$ is argument (sine wave). 210

simple sound source. A source that radiates sound uniformly in all directions under free-field conditions. 176

simple target (radar). A target having a reflecting surface such that the amplitude of the reflected signal does not vary with the aspect of the target; for example, a metal sphere. *See also:* **navigation.** 13, 187

simple tone (pure tone). (1) A sound wave, the instantaneous sound pressure of which is a simple sinusoidal function of the time. (2) A sound sensation characterized by its singleness of pitch. *Note:* Whether or not a listener hears a tone as simple or complex is dependent upon the ability, experience, and listening attitude. *See:* **complex tone.** 176

simplex circuit. A circuit derived from a pair of wires by using the wires in parallel with ground return. *See also:* **transmission line.** 328

simplexed circuit. A two-wire metallic circuit from which a simplex circuit is derived, the metallic and simplex circuits being capable of simultaneous use. *See also:* **transmission line.** 328

simplex lap winding (rotating machinery). A lap winding in which the number of parallel circuits is equal to the number of poles. 63

simplex operation. A method of operation in which communication between two stations takes place in one direction at a time. *Note:* This includes ordinary transmit-receive operation, press-to-talk operation, voice-operated carrier and other forms of manual or automatic switching from transmit to receive. *See also:* **radio transmission; telegraphy.** 59, 111, 240

simplex signaling (telephone switching systems). A method of signaling over a pair of conductors by producing current flow in the same direction through both of the conductors. 55

simplex supervision (telephone switching systems). A form of supervision employing simplex signaling. 55

simplex wave winding (rotating machinery). A wave winding in which the number of parallel circuits is two, whatever the number of poles. 63

simply connected region (two-dimensional space). A region, such that any closed curve in the region encloses points all of which belong to the region. 210

simply mesh-connected circuit. A circuit in which two, and only two, current paths extend from the terminal of entry of each phase conductor, one to the terminal of entry that precedes and the other to the terminal of entry that follows the first terminal in the normal sequence, and from which the amplitude of the voltages to the first terminal is normally the smallest (when the number of phases is greater than three). *See also:* **network analysis.** 210

simulate (computing systems). To represent the functioning of one system by another, for example, to represent one computer by another, to represent a physical system by the execution of a computer program, to represent a biological system by a mathematical model. *See also:* **electronic analog computer.** 255, 77

simulation (1) (analog computer). The representation of an actual or proposed system by the analogous characteristics of some device easier to construct, modify, or understand. 9

(2) (physical). The use of a model of a physical system in which computing elements are used to represent some but not all of the subsystems. *See also:* **electronic analog computer.**

(3) (mathematical). The use of a model of mathematical equations generally solved by computers to represent an actual or proposed system. 9

simulator (1) (test, measurement and diagnostic equipment). A device or program used for test purposes which simulates a desired system or condition providing proper inputs and terminations for the equipment under test. 54

(2) (analog computer). A device, used to represent the behavior of a physical system by virtue of its analogous characteristics. In this general sense, all computers are, or can, be simulators. However, in a more restricted definition, a simulator is a device used to interact with or to train a human operator in the performance of a given task or tasks. 9

simultaneous lobing (1) (electronic navigation). A direction-determining technique utilizing the received energy of two concurrent and partially overlapped signal lobes; the relative phase, or the relative power, of the two signals received from a target is a measure of the angular displacement of the target from the equiphase or equisignal direction. Compare with lobe switching. 187, 246, 179

(2) (radar) (antennas). A direction-determining technique utilizing the signals of overlapping lobes existing at the same time. *Syn:* **(radar) monopulse.** 13, 111

sinad ratio (mobile communication). A measure expressed in decibels of the ratio of (1) the signal plus noise plus distortion to (2) noise plus distortion produced at the output of a receiver that is the result of a modulated-signal input. *See also:* **mobile communication system; receiver performance.** 181, 123

sinad sensitivity (receiver performance). The minimum standard modulated carrier-signal input required to produce a specified sinad ratio at the receiver output. *See also:* **receiver performance.** 181

sine beats (nuclear power generating stations) (seismic qualification of class 1E equipment). A continuous sinusoid of one frequency, amplitude modulated by a sinusoid of a lower frequency. *Notes:* (1) The amplitudes of the sinusoids represent acceleration and the modulated frequency represents the frequency of the applied seismic stimulus. (2) Beats are usually considered to be the result of the summation of two sinusoids of slightly different frequencies with the frequencies within the beats as the average of the two, and the beat frequency as one-half the difference between the two. However, the sine beats may be an amplitude-modulated sinusoid with pauses between the beats. 31, 28

sine-square (sin^2) pulse (video signal transmission measurement). One cycle of a sine wave, starting and finishing at its negative peaks with an added constant amplitude component of half the peak-to-peak value, thus raising the negative peaks to zero. *Note:* A sin^2 pulse is obtained by squaring a half-cycle of a sine wave. 42

sine-square (sin^2) step (video signal transmission measurement). A step function whose transition from zero to the final value is the sum of a ramp and a negative sinusoid of equal durations, with zero slope at both the zero and the final value of the step. *Notes:* (1) A sin^2 step is obtained by integrating a sin^2 pulse. (2) The attractiveness of both the sin^2 pulse and the sin^2 step lies in the fact that their frequency spectra are limited; that is, they are effectively at zero amplitude beyond a given frequency. For the sin^2, pulse this frequency is a function of the half-amplitude duration (HAD) of the pulse; for the sin^2 step the frequency is a function of the 10 percent to 90 percent rise time. 42

sine-current coercive force (toroidal magnetic amplifier cores). The instantaneous value of sine-current mag-

netizing force at which the dynamic hysteresis loop passes through zero induction. 170

sine-current differential permeability (toroidal magnetic amplifier cores). The slope of the sides of the dynamic hysteresis loop obtained with a sine-current magnetizing force. 170

sine-current magnetizing force (toroidal magnetic amplifier cores). The applied magnetomotive force per unit length for a core symmetrically cyclically magnetized with sinusoidal current. 170

sine wave. A wave that can be expressed as the sine of a linear function of time, or space, or both. 111

sine-wave generator. An alternating-current generator whose output voltage waveform contains a single main frequency with low harmonic content of prescribed maximum level. *See also:* **asynchronous machine.** 63

sine-wave response (camera tubes). *See:* **amplitude response.**

sine-wave sweep. A sweep generated by a sine function. *See:* **oscillograph.** 185

singing. An undesired self-sustained oscillation existing in a transmission system or transducer. *Note:* Very-low-frequency oscillation is sometimes called motor-boating. *See also:* 111, 239, 178, 59

singing margin (gain margin). The excess of loss over gain around a possible singing path at any frequency, or the minimum value of such excess over a range of frequencies. *Note:* Singing margin is usually expressed in decibels. 239, 59

singing point (circuit coupled back to itself). The point at which the gain is just sufficient to make the circuit break into oscillation. *See also:* **oscillatory circuit.** 59

singing point margin (singing margin) (gain margin) (data transmission). The amount of additional gain (decibels) which can be inserted into a loop without sustained oscillations developing. 59

single-address. Pertaining to an instruction that has one address part. In a typical single-address instruction the address may specify either the location of an operand to be taken from storage, the destination of a previously prepared result, the location of the next instruction to be interpreted, or an immediate address operand. Synonymous with one-address. 77

single-address code (electronic computation). *See:* **instruction code.**

single-anode tank (single-anode tube). An electron tube having a single main anode. *Note:* This term is used chiefly for pool-cathode tubes. 190

single aperture seal (nuclear power generating stations). A single seal between the containment aperture and the electric penetration assembly. *See:* **electric penetration assembly.** 26

single automatic operation (elevators). Automatic operation by means of one button in the car for each landing level served and one button at each landing so arranged that if any car or landing button has been actuated, the actuation of any other car or landing operating button will have no effect on the operation of the car until the response to the first button has been completed. *See also:* **control (elevators).** 328

single-break switch (circuit). A switch that opens each conductor at one point only. 103, 202

single-buffered DAC (DAM) (hybrid computer linkage components). A digital-to-analog converter (DAC) or digital-to-analog multiplier (DAM) with one dynamic register, which also serves as the holding register for the digital value. *See:* **dynamic register; holding register.**
                                                                10

single capacitance (interrupter switches). A capacitance is defined to be a single capacitance when the crest of its inrush current does not exceed the switch inrush current capability for single capacitance.   27, 103

single-circuit system (protective signaling). A system of protective wiring that employs only the nongrounded side of the battery circuit, and consequently depends primarily on an open circuit in the wiring to initiate an alarm. *See also:* **protective signaling.**   328

single-degree-freedom gyro. A gyro in which the rotor is free to precess (relative to the case) about only the axis orthogonal to the rotor spin axis. *See also:* **navigation.**
                                                            187,13

single electric conductor seal (nuclear power generating stations). A mechanical assembly arranged in such a way that there is a single pressure barrier seal between the inside and the outside of the containment structure along the axis of the electric conductor.   26

single electron distribution (scintillation counting). The pulse-height distribution associated with single electrons originating at the photocathode.   117

single-electron PHR (pulse-height resolution) (scintillation counting). The fractional FWHM (full width at half maximum) of the single-electron distribution of a photo-multiplier.   117

single-electron rise time (scintillation counting). The anode-pulse rise time associated with single electrons originating at the photocathode.   117

single-electron transit-time spread (scintillation counting). Transit-time spread measured with single-electron events.
                                                                117

single-element fuse. A fuse having a current-responsive element comprising one or more parts with single fusing characteristic.   103, 202

single-element relay. An alternating-current relay having a set of coils energized by a single circuit.   328

single-end control (single-station control). A control system in which provision is made for operating a vehicle from one end or one location only. *See also:* **multiple-unit control.**
                                                                328

single-ended amplifier. An amplifier in which each stage normally employs only one active element (tube, transistor, etcetera) or, if more than one active element is used, in which they are connected in parallel so that operation is asymmetric with respect to ground. *See also:* **amplifier.**
                                                            111, 239

single-ended push-pull amplifier circuit (electroacoustics). An amplifier circuit having two transmission paths designed to operate in a complementary manner and connected so as to provide a single unbalanced output without the use of an output transformer. *See:* **amplifier.**   239

single-faced tape. Fabric tape finished on one side with rubber or synthetic compound.   64

single feeder. A feeder that forms the only connection between two points along the route considered.   64

single-frequency pulsing (telephone switching systems). Dial pulsing using the presence or absence of a single frequency to represent break or make intervals, respectively or vice versa.   55

single-frequency signaling (telephone switching systems). A method for conveying dial pulse and supervisory signals from one end of a trunk to the other using the presence or absence of a single specified frequency.   55

single-frequency signal-to-noise ratio (sound recording and reproducing system). The ratio of the single-frequency signal power output to the noise power in the entire pass band. *See also:* **noise (sound recording and reproducing system).**   350

single-frequency simplex operation (radio communication). The operation of a two-way radio-communication circuit on the same assigned radio-frequency channel, which necessitates that intelligence can be transmitted in only one direction at a time. *See also:* **channel spacing.** 181

single hoistway (elevators). A hoistway for a single elevator or dumbwaiter. *See also:* **hoistway (elevator or dumbwaiter).**   328

single-layer winding (rotating machinery). A winding in which there is only one actual coil side in the depth of the slot. (Also known as one-coil-side-per-slot winding). *See also:* **asynchronous machine; direct-current commutating machine.**   63

single-line diagram. *See:* **one-line diagram.**

single-office exchange (telephone switching systems). A telecommunications exchange served by one central office.
                                                                55

single-operator arc welder. An arc-welding power supply designed to deliver current to only one welding arc.
                                                                264

single-phase circuit (1) (transformer). An alternating-current circuit consisting of two or three intentionally interrelated conductors that enter (or leave) a delimited region at two or three terminals of entry. If the circuit consists of two conductors, it is intended to be so energized that, in the steady state, the voltage between the two terminals of entry is an alternating voltage. If the circuit consists of three conductors, it is intended to be so energized that, in steady state, the alternating voltages between any two terminals of entry have the same period and are in phase or in phase opposition. *See also:* **center of distribution; network analysis.**   53
(2) (electric installations on shipboard). A circuit energized by a single alternating electromotive force. *Note:* A single-phase circuit is usually supplied through two wires. The currents in these two wires, counted outward from the source, differ in phase by 180 degrees or a half cycle.   3

single-phase electric locomotive. An electric locomotive that collects propulsion power from a single phase of an alternating-current distribution system. *See also:* **electric locomotive.**   328

single-phase machine. A machine that generates or utilizes single-phase alternating-current power. *See also:* **asynchronous machine.**   63

single-phase motor (rotating machinery). A machine that converts single-phase alternating-current electric power into mechanical power, or that provides mechanical force or torque.   63

single-phase symmetrical set (1) (polyphase voltages). A symmetrical set of polyphase voltages in which the angular

phase difference between successive members of the set is π radians or odd multiples thereof. The equations of **symmetrical set (polyphase voltages)** represent a single-phase symmetrical set of polyphase voltages if k/m is $\frac{1}{2}$ or an odd multiple thereof. (The symmetrical set of voltages represented by the equations of **symmetrical set (polyphase voltages)** may be said to have single-phase symmetry if k/m is an odd (positive or negative) multiple of $\frac{1}{2}$.) *Notes:* (1) A set of polyphase voltages may have single-phase symmetry only if m, the number of members of the set, is an even number. (2) This definition may be applied to a two-phase four-wire or five-wire circuit if m is considered to be 4 instead of 2. It is not applicable to a two-phase three-wire circuit. *See also:* **network analysis.**
210

(2) (polyphase currents). This definition is obtained from the corresponding definitions for voltage by substituting the word **current** for **voltage,** and the symbol I for E and β for α wherever they appear in the equations of **symmetrical set (polyphase voltages).** The subscripts are unaltered. *See also:* **network analysis.** 210

single-phase synchronous generator. A generator that produces a single alternating electromotive force at its terminals. It delivers electric power that pulsates at double frequency. 3

single-phase three-wire circuit. A single-phase circuit consisting of three conductors, one of which is identified as the neutral conductor. *See also:* **network analysis.**
210

single-phase two-wire circuit. A single-phase circuit consisting of only two conductors. *See also:* **network analysis.**
210

single-phasing (rotating machinery). An abnormal operation of a polyphase machine when its supply is effectively single-phase. *See also:* **asynchronous machine.** 63

single-polarity pulse. A pulse in which the sense of the departure from normal is in one direction only. *See:* **unidirectional pulse.** *See also:* **pulse.** 328

single-pole relay. *See:* **relay, single-pole.**

single-pressure-zone potheads. A pressure-type pothead intended to operate with one pressure zone. *See:* **multipressure-zone pothead; pressure-type pothead.** 323

single service. One service only supplying a consumer. *Note:* Either or both lighting and power load may be connected to the service. *See also:* **service.** 64

single-shot blasting unit. A unit designed for firing only one explosive charge at a time. *See also:* **blasting unit.**
328

single-shot blocking oscillator. A blocking oscillator modified to operate as a single-shot trigger circuit. *See also:* **trigger circuit.** 328

single-shot multivibrator (single-trip multivibrator). A multivibrator modified to operate as a single-shot trigger circuit. *See also:* **trigger circuit.** 328

single-shot trigger circuit (single-trip trigger circuit). A trigger circuit in which a triggering pulse intiates one complete cycle of conditions ending with a stable condition. *See also:* **trigger circuit.** 328

single-sideband modulation (SSB). Modulation whereby the spectrum of the modulating function is translated in frequency by a specified amount either with or without inversion. *See also:* **modulation.** 111, 242, 59

single-sideband transmission. The method of operation in which one sideband is transmitted and the other sideband is suppressed. The carrier wave may be either transmitted or suppressed. 111, 59

single-sideband transmitter. A transmitter in which one sideband is transmitted and the other is effectively eliminated. 111

single-sided printed-circuit board. *See:* **printed-circuit board.**

single-station control. *See:* **single-end control.**

single step (computing systems). Pertaining to a method of operating a computer in which each step is performed in response to a single manual operation. 255, 77

single-stroke bell. An electric bell that produces a single stroke on its gong each time its mechanism is actuated. *See also:* **protective signaling.** 328

single sweep (non-real time spectrum analyzer). Operating mode for a triggered sweep instrument in which the sweep must be reset for each operation, thus preventing unwanted multiple displays. This mode is useful for trace photography. In the interval after the sweep is reset and before it is triggered it is said to be an armed sweep.
68

single throw (switching device). A qualifying term used to indicate that the device has an open and a closed circuit position only. 79

single-tone keying (modulation systems). That form of keying in which the modulating function causes the carrier to be modulated by a single tone for one condition, which may be either a mark or a space, the carrier being unmodulated for the other condition. *See also:* **telegraphy.**
111, 242, 194

single-track (standard track) (electroacoustics). A variable-density or variable-area sound track in which both positive and negative halves of the signal are linearly recorded. *See also:* **phonograph pickup.** 176

single-trip multivibrator. *See:* **single-shot multivibrator.**

single-trip trigger circuit. *See:* **single-shot trigger circuit.**

single-tuned amplifier (circuits and systems). An amplifier characterized by a resonance at a single frequency as indicated by the s-plane representation of its gain which is $A(s) = A_0 s / (s^2 + \omega_0 \xi s + \omega_0^2)$. It rejects low and high frequencies while having a peak gain at a center frequency $s = j\omega_0$. *See also:* **amplifier.** 67

single-tuned circuit. A circuit that may be represented by a single inductance and a single capacitance, together with associated resistances. 328

single-valued function. A function u is single valued when to every value of x (or set of values of $x_1, x_2, \ldots, x_n$) there corresponds one and only one value of u. Thus $u = ax$ is single valued if a is an arbitrary constant. 210

single-way rectifier. A rectifier unit which makes use of a single-way rectifier circuit. 66

single way rectifier circuit. A rectifier circuit in which the current between each terminal of the alternating voltage circuit and the rectifier circuit element or elements conductively connected to it flows in only one direction.
66

single-winding multispeed motor. A type of multispeed motor having a single winding capable of reconnection in two or more pole groupings. *See:* **asynchronous machine;**

direct-current commutating machine. 328

single-wire line (waveguides). A surface-wave transmission line consisting of a single conductor so treated as to confine the propagated energy to the neighborhood of the wire. The treatment may consist of a coating of dielectric. *See:* **waveguides.** 179

singular point (control system). Synonymous with equilibrium point. *See also:* **control system.** 329

sink (1) (oscillator). The region of a Rieke diagram where the rate of change of frequency with respect to phase of the reflection coefficient is maximum. Operation in this region may lead to unsatisfactory performance by reason of cessation or instability of oscillations. *See:* **oscillatory circuit.** 190, 125

(2) (communication practice). (A) A device that drains off energy from a system. (B) A place where energy from several sources is collected or drained away. 328

sink node (network analysis). A node having only incoming branches. *See also:* **linear signal flow graphs.** 282

sin² pulse. *See:* **sine-square pulse.**

sin² step. *See:* **sine-square step.**

sinusoidal electromagnetic wave (homogeneous medium). A wave whose electric field vector proportional to the sine (or cosine) or an angle that is a linear function of time, or a distance, or of both. *See also:* **radio propagation.** 180, 146

sinusoidal field. A field in which the field quantities vary as a sinusoidal function of an independent variable, such as space or time. 210

sinusoidal function. A function of the form $A \sin (x + a)$. A is the amplitude, x is the independent variable, and a the phase angle. Note that $\cos (x)$ may be expressed as $\sin [x + (\pi/2)]$. *See also:* **simple sinewave quantity.** 210

siphon recorder. A telegraph recorder comprising a sensitive moving-coil galvanometer with a siphon pen that is directed by the moving coil across a traveling strip of paper. *See also:* **telegraphy.** 328

site error (electronic navigation). Error due to the distortion in the electromagnetic field by objects in the vicinity of the navigational equipment. *See also:* **navigation.** 187, 13

site, pull (conductor stringing equipment). The location of the line where the puller, reel winder and anchors (snubs) are located. This site may also serve as the pull or tension site for the next sag section. *Syn:* **reel set-up; tugger set-up.** 45

site, tension (conductor stringing equipment). The location on the line where the tensioner, reel stands, and anchor (snubs) are located. This site may also serve as the pull or tension site for the next sag section. *Syn:* **conductor payout station; payout site; reel set-up.** 45

six-phase circuit. A combination of circuits energized by alternating electromotive forces which differ in phase by one-sixth of a cycle, that is, 60 degrees. *Note:* In practice the phases may vary several degrees from the specified angle. *See also:* **center of distribution.** 64, 53

size threshold. The minimum perceptible size of an object. It also is defined as the size that can be detected some specific fraction of the times, it is presented to an observer, usually 50 percent. It usually is measured in minutes of arc. *See also:* **visual field.** 167

skate machine. A mechanism, electrically controlled, for placing on, or removing from, the rails a skate that, if al-

lowed to engage with the wheels of a car, provides continuous braking until the car is stopped and that may be electrically or pneumatically operated. 328

skating force. *See:* **side thrust.**

skeleton frame (rotating machinery). A stator frame consisting of a simple structure that clamps the core but does not enclose it. 63

skew (1) (facsimile). The deviation of the received frame from rectangularity due to asynchronism between scanner and recorder. Skew is expressed numerically as the tangent of the angle of the deviation. *See also:* **recording (facsimile).** 12

(2) (magnetic storage). The angular displacement of an individual printed character, group of characters, or other data, from the intended or ideal placement.

255, 77, 54

skewed slot (rotating machinery). A slot of a rotor or stator of an electric machine, placed at an angle to the shaft so that the angular location of the slot at one end of the core is displaced from that at the other end. Slots are commonly skewed in many types of machines to provide more uniform torque, less noise, and better voltage waveform. *See also:* **rotor (rotating machinery); stator.** 63

skiatron (electronic navigation). (1) A dark-trace storage-type cathode-ray tube. (2) A display employing an optical system with a dark-trace tube. *See:* **dark-trace tube.** 13, 187

skid wire (pipe-type cable) (power distribution, underground cables). Wire or wires, usually D shaped, applied open spiral with curved side outward with a suitable spacing between turns over the outside surface of the cable. Its purpose is to facilitate cable pulling and to provide mechanical protection during installation. 57

skim tape. Filled tape coated on one or both sides with a thin film of uncured rubber or synthetic compound to produce a coating suitable for vulcanization. 64

skin depth. For a conductor carrying currents at a given frequency as a result of the electromagnetic waves acting upon its surface, the depth below the surface at which the current density has decreased one neper below the current density at the surface. *Note:* Usually the skin depth is sufficiently small so that for ordinary configurations of good conductors, the value obtained for a plane wave falling on a plane surface is a good approximation.

267, 179

skin effect (1) (conductor). The phenomenon of nonuniform current distribution over the cross section caused by the time variation of the current in the conductor itself. *See also:* **transmission line.** 210

(2) (induction heating). Tendency of an alternating current to concentrate in the areas of lowest impedance. 14

skip (computing systems). To ignore one or more instructions in a sequence of instructions. 255, 77

skip distance. The minimum separation for which radio waves of a specified frequency can be transmitted at a specified time between two points on the earth by reflection from the regular ionized layers of the ionosphere. *See also:* **radiation.** 328

sky factor. The ratio of the illumination on a horizontal plane at a given point inside a building due to the light received directly from the sky, to the illumination due to an unobstructed hemisphere of sky of uniform luminance (photometric brightness) equal to that of the visible sky. *See also:* **sunlight.** 167

sky light. Visible radiation from the sun redirected by the atmosphere. *See also:* **sunlight.** 167

sky noise (communication satellite). Noise contribution of the sky (often the galaxies). *See also:* **background noise.** 85

sky wave. Ionospheric wave. 180

sky-wave contamination (electronic navigation). Degradation of the received ground-wave signal, or of the desired sky-wave signal, by the presence of delayed ionospheric-wave components of the same transmitted signal. *See also:* **navigation.** 187, 13

sky-wave correction (electronic navigation). A correction for sky-wave propagation errors applied to measured position data; the amount of the correction is established on the basis of an assumed position and an assumed ionosphere height. *See also:* **navigation.** 13, 187

sky-wave station-error (sky-wave synchronized loran). The error of station synchronization due to the effect of variations of the ionosphere on the time of transmission of the synchronizing signal from one station to the other. *See also:* **navigation.** 13, 187

slabbing or arcwall machine. A power-driven mobile-cutting machine that is a single-purpose cutter in that it cuts only a horizontal kerf at variable heights. 328

slack-rope switch (elevators). A device that automatically causes the electric power to be removed from the elevator driving-machine motor and brake when the hoisting ropes of a winding-drum machine become slack. *See:* **control.** 328

slant distance (radar). The distance between two points not at the same elevation. Used in contrast to ground distance. *See also:* **navigation.** 13, 187

slant range (radar). Slant distance when one point is a radar site. 13

slave (test, measurement and diagnostic equipment). Device that follows an order given by a master remote control. 54

slave drive. *See:* **electric drive; follower drive.** 13, 187

slaved tracking (power supplies). A system of interconnection of two or more regulated supplies in which one (the master) operates to control the others (the slaves). The output voltage of the slave units may be equal or proportional to the output voltage of the master unit. (The slave output voltages track the master output voltage in a constant ratio.) *See:* **complementary tracking, master/slave.** *See also:* **power supply.** 228, 186

slave relay. *See:* **auxiliary relay.** *See also:* **relay.**

slave station (electronic navigation). A station of a synchronized group whose emissions are controlled by a master station. *See also:* **radio navigation.** 13, 187

slaving (gyro). The use of a torquer to maintain the orientation of the spin axis relative to an external reference such as a pendulum or magnetic compass. 46

sleet hood (switch). A cover for the contacts to prevent sleet from interfering with successful operation of the switch. 27, 103, 202

sleetproof (1) (general). So constructed or protected that the accumulation of sleet will not interfere with successful operation. 225, 206, 103, 202, 27

(2) (transformer). So constructed or protected that the accumulation of sleet (ice) under specified test conditions will not interfere with the successful operation of the apparatus. 53

sleeve (1) (plug) (three-wire telephone-switchboard plug). A cylindrically shaped contacting part, usually placed in back of the tip or ring but insulated therefrom. 328

(2) (rotating machinery). A tubular part designed to fit around another part. *Note:* In a sleeve bearing, the sleeve is that component that includes the cylindrical inner surface within which the shaft journal rotates. 63

sleeve bearing (rotating machinery). A bearing with a cylindrical inner surface in which the journal of a rotor (or armature) shaft rotates. *See:* **rotor (rotating machinery).** 63

sleeve conductor. *See:* **sleeve wire.**

sleeve-dipole antenna. A dipole antenna surrounded in its central portion by a coaxial conducting sleeve. *See also:* **antenna.** 179, 111

sleeve-monopole antenna. An antenna consisting of half of a sleeve-dipole antenna projecting from a ground plane. *Syn:* **sleeve-stub antenna.** 111

sleeve supervision. The use of the sleeve circuit for transmitting supervisory signals. 328

sleeve-type suppressor (electromagnetic compatibility). A suppressor designed for insertion in a high-tension ignition cable. *See also:* **electromagnetic compatibility.** 220, 199

sleeve wire (telephone switching systems). That conductor, usually accompanying the tip and ring leads of a switched connection, which provides for miscellaneous functions necessary to the control and supervision of the connection. In cord-type switchboards, the sleeve wire is that conductor associated with the sleeve contacts of the jacks and plugs. 55, 183

slewing (gyro). The rotation of the spin axis about an axis parallel to that of the applied torque causing the rotation. 46

slewing rate (power supplies). A measure of the programming speed or current-regulator-response timing. The slewing rate measures the maximum rate-of-change of voltage across the output terminals of a power supply. Slewing rate is normally expressed in volts per second ($\Delta E/\Delta T$) and can be converted to a sinusoidal frequency-amplitude product by the equation $f(E_{pp}) = $ slewing rate$/\pi$, where E_{pp} is the peak-to-peak sinusoidal volts. Slewing rate $= \pi f(E_{pp})$. *See:* **high-speed regulator.** *See also:* **power supply.** 228, 186

slewing speed (test, measurement and diagnostic equipment). A continuous speed, usually the maximum at which a tape reader or other rotating device can search for information. 54

slicer (amplitude gate) (clipper-limiter*). A transducer that transmits only portions of an input wave lying between two amplitude boundaries. *Note:* The term is used especially when the two amplitude boundaries are close to each other as compared with the amplitude range of the input. 328

* Deprecated

slide rail (rotating machinery). A special form of soleplate which is long in the direction of the machine axis to permit sliding the stator frame in the axial direction. 63

slide-screw tuner (1) (transmission lines and waveguides). An impedance or matching transformer that consists of a slotted waveguide or coaxial-line section and an adjustable screw or post that penetrates into the guide or line and can be moved axially along the slot. *See also:* **waveguide.** 185

(2) (waveguide components). A waveguide or transmission line tuner employing a post of adjustable penetration, adjustable in position along the longitudinal axis of the waveguide. 166

sliding contact. An electric contact in which one conducting member is maintained in sliding motion over the other conducting member. *See also:* **contactor.** 328

sliding load. *See:* **load, sliding.**

sliding short circuit. A short-circuit termination that consists of a section of waveguide or transmission line fitted with a sliding short-circuiting piston (contacting or noncontacting) that ideally reflects all the energy back toward the source. *See also:* **waveguide.** 185

slime, anode. Finely divided insoluble metal or compound forming on the surface of an anode or in the solution during electrolysis. *See also:* **electrodeposition.** 328

slinging wire. A wire used to suspend and carry current to one or more cathodes in a plating tank. *See also:* **electroplating.** 328

sling, traveler. ·*See:* **traveler sling.**

slip (1) (rotating machinery). The quotient of (A) the difference between the synchronous speed and the actual speed of a rotor, to (B) the synchronous speed, expressed as a ratio, or as a percentage.

(2). The difference between the speed of a rotating magnetic field and that of a rotor, expressed in revolutions per minute.

(3) (electric couplings). The difference between the speeds of the two rotating members. *See:* **asynchronous machine.** 63

slip regulator (rotating machinery). A device arranged to produce a reduction in speed below synchronous speed greater than would be obtained inherently. Such a device is usually in the form of a variable impedance connected in the secondary circuit of a slip ring induction motor. 63

slip relay. A relay arranged to act when one or more pairs of driving wheels increase or decrease in rotational speed with respect to other driving wheels of the same motive power unit. *See also:* **multiple-unit control.** 328

slip ring. *See:* **collector ring.**

slip-ring induction motor. *See:* **wound-rotor induction motor.**

slope angle (electronic navigation). *See:* **glide-slope angle.** *See also:* **navigation.**

slot (rotating machinery). A channel or tunnel opening onto or near the air gap and passing essentially in an axial direction through the rotor or stator core. A slot usually contains the conductors of a winding, but may be used exclusively for ventilation. *See also:* **rotor (rotating machinery); stator.** 63

slot antenna. A radiating element formed by a slot in a conducting surface. *See also:* **antenna.** 111, 246, 179

slot array. An antenna array formed of slot radiators. *See also:* **antenna.** 244, 179

slot cell (rotating machinery). A sheet of insulation material used to line a slot before the winding is placed in it. *See also:* **rotor (rotating machinery); stator.** 63

slot coupling factor (slot-antenna array). The ratio of the desired slot current to the available slot current, controlled by changing the depth of penetration of the slot probe into the waveguide. *See also:* **navigation.** 187, 13

slot current ratio (slot-antenna array). The relative slot currents in the slots of the waveguide reading from its center to its end, with the maximum taken as 1; this ratio is dependent upon the slot spacing factor and the slot coupling factor. *See also:* **navigation.** 187, 13

slot discharge (rotating machine). Sparking between the outer surface of coil insulation and the grounded slot surface, caused by capacitive current between conductors and iron. The resulting current pulses have a fundamental frequency of a few kilohertz. *See also:* **asynchronous machine; direct-current commutating machine.** 63

slot-discharge analyzer (rotating machinery). An instrument designed for connection to an energized winding of a rotating machine, to detect pulses caused by slot discharge, and to discriminate between them and pulses otherwise caused. *See also:* **asynchronous machine; direct-current commutating machine.** 63

slot insulation (rotating machinery). A sheet or deposit of insulation material used to line a slot before the winding is placed in it. *See also:* **asynchronous machine; direct-current commutating machine.** 63

slot liner (rotating machinery). Separate insulation between an embedded coil side and the slot which can provide mechanical and electrical protection. 63

slot packing (filler) (rotating machinery). Additional insulation used to pack embedded coil sides to ensure a tight fit in the slots. *See also:* **rotor (rotating machinery); stator.** 63

slot pitch (tooth pitch) (rotating machinery). The peripheral distance between fixed points in corresponding positions in two consecutive slots. 63

slot separator insulation. *See:* **separator insulation, slot.**

slot space factor (rotating machinery). The ratio of the cross-sectional area of the conductor metal in a slot to the total cross-sectional area of the slot. *See also:* **asynchronous machine; direct-current commutating machine.** 63

slot spacing factor (slot-antenna array). A value proportional to the size of the angle between the slot location and the null of the internal standing wave; this factor is dependent upon frequency. 187

slotted armature (rotating machine). An armature with the winding placed in slots. *See also:* **armature.** 244, 63

slotted section (slotted line) (slotted waveguide). A section of a waveguide or shielded transmission line the shield of which is slotted to permit the use of a carriage and travelling probe for examination of standing waves. *See also:* **auxiliary device to an instrument.** 244, 179, 185

slot-type antenna (aircraft). A slot in the normal streamlined metallic surface of an aircraft, excited electromagnetically by a structure within the aircraft. Radiation is thus obtained without projections that would disturb the aerodynamic characteristics of the aircraft. Radiation from a slot is essentially directive. 328

slot wedge (rotating machinery). The element placed above the turns or coil sides in a stator or rotor slot, and held in place by engagement of wedge (slots) grooves along the sides of the coil slot, or by projections from the sides of the slot tending to close the top of the slot. *Note:* A wedge may be a thin strip of material provided solely as insulation or to provide temporary retention of the coils during the manufacturing process. It may be a piece of structural insulating material or high-strength metal to hold the coils in the slot. Slots in laminated cores are normally wedged with insulating material. *See also:* **rotor (rotating machinery); stator.** 63

slow-closure device (speed governing of hydraulic turbines). A cushioning device which retards the closing velocity of servomotor travel from a predetermined servomotor position to zero servomotor position. 8, 58

slow-operate relay. A slugged relay that has been specifically designed for long operate time but not for long release time. *Caution:* The usual slow-operate relay has a copper slug close to the armature, making it also at least partially slow to release. 341

slow-operating relay. A relay that has an intentional delay between energizing and operation. *See also:* **electromagnetic relay.** 328

slow-release relay. A relay that has an intentional delay between de-energizing and release. *Note:* The reverse motion need not have any intentional delay. *See also:* **electromagnetic relay.** 341

slow release time characteristic, relay. *See:* **relay slow release time characteristic.**

slow-speed starting (industrial control). A control function that provides for starting an electric drive only at the minimum-speed setting. *See:* **starter.** 225, 206

slow-wave circuit (microwave tubes). A circuit whose phase velocity is much slower than the velocity of light. For example, for suitably chosen helixes the wave can be considered to travel on the wire at the velocity of light but the phase velocity is less than the velocity of light by the factor that the pitch is less than the circumference. *See also:* **microwave tube or valve.** 190

slug, relay. *See:* **relay slug.**

slug tuner (waveguide). A waveguide or transmission-line tuner containing one or more longitudinally adjustable pieces of metal or dielectric. *See:* **waveguide.** 179

slug tuning. A means for varying the frequency of a resonant circuit by introducing a slug of material into either the electric or magnetic fields or both. *See also:* **network analysis; radio transmission.** 111

slush compound (corrosion). A non-drying oil, grease, or similar organic compound that, when coated over a metal, affords at least temporary protection against corrosion. 221, 205

small-signal (light emitting diodes). A signal which when doubled in magnitude does not produce a change in the parameter being measured that is greater than the required accuracy of the measurement. 162

small-signal forward transadmittance. The value of the forward transadmittance obtained when the input voltage is small compared to the beam voltage. *See:* **electron-tube admittances.** 190, 125

small-signal permittivity (ferroelectric material). The incremental change in electric displacement per unit electric field when the magnitude of the measuring field is very small compared to the coercive electric field. (Measure-

ments are usually made at a frequency of 1 kilohertz or higher). The small signal relative permittivity κ is equal to the ratio of the absolute permittivity ϵ to the permittivity of free space ϵ_0, that is, $\kappa = \epsilon/\epsilon_0$. *Note:* The value of the small-signal permittivity may depend on the remanent polarization, electric field, mechanical stress, sample history, or frequency of the measuring field. *See also:* **Curie-Weiss temperature; paraelectric; remanent polarization; coercive electric field.** 80

small-signal resistance (semiconductor rectifier). The resistive part of the quotient of incremental voltage by incremental current under stated operating conditions. *See also:* **rectification.** 237, 66

small wiring. *See:* **secondary and control wiring.**

smashboard signal. A signal so designed that the arm will be broken when passed in the stop position. 328

smoke detector (fire protection devices). A device which detects the visible or invisible particles of combustion. 71

smooth. To apply procedures that decrease or eliminate rapid fluctuations in data. 255, 77

smothered-arc furnace. A furnace in which the arc or arcs is covered by a portion of the charge. 328

snake. *See:* **fish tape.**

snapover. When used in connection with alternating-current testing, a quasi-flashover or quasi-sparkover, characterized by failure of the alternating-current power source to maintain the discharge, thus permitting the dielectric strength of the specimen to recover with the test voltage still applied. *See also:* **test voltage and current.** 307, 201

snapshot dump (computing systems). A selective dynamic dump performed at various points in a machine run. 255, 77

snow (intensity-modulated display). A varying speckled background caused by noise. *See also:* **radar; television.** 178

snub structure (conductor stringing equipmemt). A structure located at one end of a sag section, and considered as a "zero" point for sagging and clipping offset calculations. The section of line between two such structures is the sag section, but more than one sag section may be required in order to properly sag the actual length of conductor which has been strung. *Syn:* **"0" structure (zero structure).** 45

soak, relay. *See:* **relay soak.**

socket*. *See:* **lampholder.**

* Deprecated

sodium vapor lamp transformers (multiple supply type). Transformers, auto-transformers or reactors for operating sodium vapor lamps for all types of lighting applications, including indoor, outdoor area, roadway, and other process and specialized lighting. 53

soft limiting. *See:* **limiter circuit.** *See also:* **electronic analog computer.** 9, 10

software (electronic computers). (1) Computer programs, routines, programming languages and systems. (2) The collection of related utility, assembly, and other programs that are desirable for properly presenting a given machine to a user. (3) Detailed procedures to be followed, whether expressed as programs for a computer or as procedures for an operator or other person. (4) Documents, including hardware manuals and drawings, computer-program listings and diagrams, etcetera. (5) Items such as those in (1), (2), (3), and (4) as contrasted with hardware.

            235, 54

solar array (photovoltaic power system). A group of electrically interconnected solar cells assembled in a configuration suitable for oriented exposure to solar flux. *See also:* **photovoltaic power system; solar cells (photovoltaic power system).** 186

solar noise (communication satellite). Electrical noise generated by the sun. Exceeds other background noise sources by several orders of magnitude. 85

solar panel (photovoltaic power system). *See:* **solar array.**

solar wind (communication satellite). Energetic particles emitted by the sun and travelling through space. 74

solderability. That property of a metal surface to be readily wetted by molten solder. 284

soldered joints. The connection of similar or dissimilar metals by applying molten solder, with no fusion of the base metals. 284

solder projections. Icicles, nubs, and spikes are undesirable protrusions from a solder joint. 284

solder splatter. Unwanted fragments of solder. 284

solenoid. An electric conductor wound as a helix with a small pitch, or as two or more coaxial helixes. *See:* **solenoid magnet.** 210

solenoid magnet (solenoid) (industrial control). An electromagnet having an energizing coil approximately cylindrical in form, and an armature whose motion is reciprocating within and along the axis of the coil.

            210, 206

solenoid relay. *See:* **plunger relay.** *See also:* **relay.**

soleplate (rotating machinery). A support fastened to a foundation on which a stator frame foot or a bracket arm can be mounted. *See also:* **slide rail.** 63

solid angle (1) (illuminating engineering). A ratio of the area on the surface of a sphere to the square of the radius of the sphere. It is expressed in steradians. *Note:* Solid angle is a convenient way of expressing the area of light sources and luminaires for computation of discomfort glare factors. It combines into a single number the projected area A_p of the luminaire and the distance D between the luminaire and the eye. It usually is computed by means of the approximate formula

$$\omega = \frac{A_p}{D^2}$$

in which A_p and D^2 are expressed in the same units. This formula is satisfactory when the distance D is greater than about three times the maximum linear dimension of the projected area of the source. Larger projected areas should be subdivided into several elements. *See also:* **inverse-square law (illuminating engineering).** 167

(2) (laser-maser) (ω). The ratio of the area on the surface of a sphere to the square of the radius of that sphere. It is expressed in **steradians.** 363

solid bushing (outdoor electric apparatus). A bushing in which the major insulation is provided by a ceramic or analogous material. 287, 168

solid conductor. A conductor consisting of a single wire.

See also: **conductor.** 64

solid contact. A contact having relatively little inherent flexibility and whose contact pressure is supplied by another member. 103, 202

solid coupling (rotating machinery). A coupling that makes a rigid connection between two shafts. *See also:* **rotor (rotating machinery).** 63

solid electrolytic capacitor. A capacitor in which the dielectric is primarily an anodized coating on one electrode, with the remaining space between the electrodes filled with a solid semiconductor. 341

solid enclosure. An enclosure that will neither admit accumulations of flyings or dust nor transmit sparks or flying particles to the accumulations outside. 328

solid-iron cylindrical-rotor generator. *See:* **cylindrical-rotor generator.**

solidly grounded. Grounded through all adequate ground connection in which no impedance has been inserted intentionally. *Note:* Adequate as used herein means suitable for the purpose intended. 53

solid-material fuse unit. A fuse unit in which the arc is drawn through a hole in solid material. 103, 202

solid-pole synchronous motor. A salient-pole synchronous motor having solid steel pole shoes, and either laminated or solid pole bodies. 63

solid rotor (rotating machinery). (1) A rotor, usually constructed of a high-strength forging, in which slots may be machined to accommodate the rotor winding. (2) A spider-type rotor in which spider hub is not split. *See also:* **rotor (rotating machinery).** 63

solid-state. An adjective used to describe a device, circuit, or system whose operation is dependent upon any combination of optical, electrical, or magnetic phenomena within a solid. Specifically excluded are devices, circuits, or systems dependent upon the macroscopic physical movement, rotation, contact, or noncontact of any combination of solids, liquids, gases, or plasmas. 59

solid-state component. A component whose operation depends on the control of electric or magnetic phenomena in solids, for example, a transistor, crystal diode, ferrite core. 255, 77

solid-state device (control equipment). A device that may contain electronic components that do not depend on electronic conduction in a vacuum or gas. The electrical function is performed by semiconductors or the use of otherwise completely static components such as resistors, capacitors, etcetera. 94

solid-state relay (or relay unit). A static relay or relay unit constructed exclusively of solid-state components.

            127, 103

solid-state scanning (facsimile). A method in which all or part of the scanning process is due to electronic commutation of a solid-state array of thin-film photosensitive elements. *See also:* **facsimile (electrical communication).**

            194

solid-type paper-insulated cable. Oil-impregnated, paper-insulated cable, usually lead covered, in which no provision is made for control of internal pressure variations. 64

solution. *See:* **check solution.**

solvent cleaning (electroplating). Cleaning by means of organic solvents. *See also:* **electroplating.** 328

solventless (rotating machinery). A term applied to liquid or semiliquid varnishes, paints, impregnants, resins, and similar compounds that have essentially no change in weight or volume when converted into a solid or semisolid. 63

sonar. A general name for sonic and ultrasonic underwater ranging, sounding and communication systems. *Note:* The name is derived from the initial letters of the phrase **sound navigation and ranging.** A device that radiates underwater acoustic energy and utilizes reflection of this energy is termed an **active sonar.** A device that merely receives underwater acoustic energy generated at a distant source is termed a **passive sonar.** *See also:* **loudspeaker.** 13

sonic delay line. *See:* **acoustic delay line.**

sonne. A radio navigation aid that provides a number of characteristic signal zones that rotate in a time sequence; a bearing may be determined by observation (by interpolation) of the instant at which transition occurs from one zone to the following zone. *See:* **consol.** *See also:* **navigation.** 13, 187

sort. To arrange data or items in an ordered sequence by applying specific rules. 255, 77, 54

sorter. A person, device, or computer routine that sorts. 255, 77

SOS. *See:* **radio distress signal.**

sound. (1) An oscillation in pressure, stress, particle displacement, particle velocity, etcetera, in a medium with internal forces (for example, elastic, viscous), or the superposition of such propagated oscillations. (2) An auditory sensation evoked by the oscillation described above. *Notes:* (A) In case of possible confusion, the term sound wave or elastic wave may be used for concept (1) and the term sound sensation for concept (2). Not all sound waves can evoke an auditory sensation, for example, an ultrasonic wave. (B) The medium in which the sound exists is often indicated by an appropriate adjective, for example, airborne, water-borne, structure-borne. 176

sound absorption. (1) The change of sound energy into some other form, usually heat, in passing through a medium or on striking a surface. (2) The property possessed by material and objects, including air, of absorbing sound energy. 176

sound-absorption coefficient (surface). The ratio of sound energy absorbed or otherwise not reflected by the surface, to the sound energy incident upon the surface. Unless otherwise specified, a diffuse sound field is assumed. 176

sound analyzer. A device for measuring the band pressure level, or pressure spectrum level, of a sound at various frequencies. *Notes:* (1) A sound analyzer usually consists of a microphone, an amplifier and wave analyzer, and is used to measure amplitude and frequency of the components of a complex sound. (2) The band pressure level of a sound for a specified frequency band is the effective root-mean-square sound pressure level of the sound energy contained within the bands. *See also:* **instrument.** 334

sound articulation (percent sound articulation). The percent articulation obtained when the speech units considered are fundamental sounds (usually combined into meaningless syllables). *See also:* **volume equivalent.** 328

sound-detection system (protective signaling). A system for the protection of vaults by the use of sound-detecting devices and relay equipment to pick up and convert noise, caused by burglarious attack on the structure, to electric impulses in a protection circuit. *See also:* **protective signaling.** 328

sound-effects filter. *See:* **filter, sound-effects.**

sound energy. Of a given part of a medium, the total energy in this part of the medium minus the energy that would exist in the same part of the medium with no sound waves present. 176

sound field. A region containing sound waves. 176

sound intensity (sound-energy flux density) (sound power density) (in a specified direction at a point). The average rate of sound energy transmitted in the specified direction through a unit area normal to this direction at the point considered. *Notes:* (1) The sound intensity in any specified direction a of a sound field is the sound-energy flux through a unit area normal to that direction. This is given by the expression

$$I_a = \frac{1}{T} \int_0^T p v_a \, dt$$

where
T = an integral number of periods or a time long compared to a period
p = the instantaneous sound pressure
v_a = the component of the instantaneous particle velocity in the direction a
t = time.
(2) In the case of a free plane or spherical wave having an effective sound pressure, p, the velocity of propagation c, in a medium of density ρ, the intensity in the direction of propagation is given by

$$I = \frac{p^2}{\rho c}$$

176

sound level. A weighted sound pressure level obtained by the use of metering characteristics and the weightings A, B, or C specified in American National Standard Specification for Sound-Level Meters, S1.4-1971 (or latest revision thereof). The weighting employed must always be stated. The reference pressure is 2×10^{-5} newton per meter^2. *Notes:* (1) The meter reading (in decibels) corresponds to a value of the sound pressure integrated over the audible frequency range with a specified frequency weighting and integration time. (2) A suitable method of stating the weighting is, for example, "The sound level (A) was 43 decibels." *See also:* **level.** 176

sound level, A-weighted (airborne sound measurements on rotating electric machinery). The A-weighted sound level is the weighted sound pressure level, obtained by use of metering characteristics and A-weighting specified in ANSI S1.4-1971, Specification for Sound Level Meters. 129

sound-level meter. An instrument including a microphone, an amplifier, an output meter, and frequency-weighting networks for the measurement of noise and sound levels in a specified manner. *Notes:* (1) The measurements are intended to approximate the loudness level of pure tones that would be obtained by the more-elaborate ear balance method. (2) Loudness level in phons of a sound is numerically equal to the sound pressure level in decibels relative to 0.0002 microbar of a simple tone of frequency 1000 hertz that is judged by the listeners to be equivalent in loudness. (3) Specifications for sound-level meters are given in American National Standard Specification for Sound-Level Meters, S1.4-1971 (or latest revision thereof). *See also:* **instrument.** 176

sound power (source). The total sound energy radiated by the source per unit of time. 176

sound-powered telephone set. A telephone set in which the transmitter and receiver are passive transducers. *See also:* **telephone station.** 328

sound power level (airborne sound measurements on rotating electric machinery). The sound power level, in decibels, is equal to 10 times the logarithm to the base 10 of the ratio of a given power to the reference power, 10^{-12} W (see ANSI S1.8-1969).

$$L_W = 10 \log_{10} \left(\frac{W}{W_0} \right)$$

where
L_W = sound power level
W = measured sound power in watts
W_0 = reference power 129

sound power level, A-weighted (airborne sound measurements on rotating electric machinery). The *A*-weighted sound power level, in decibels, is equal to the sound power level determined by weighting each of the frequency bands. 129

sound pressure (at a point). The total instantaneous pressure at that point, in the presence of a sound wave, minus the static pressure at that point. 122, 196

sound pressure, effective (root-mean-square sound pressure). At a point over a time interval, the root-mean-square value of the instantaneous sound pressure at the point under consideration. In the case of periodic sound pressures, the interval must be an integral number of periods or an interval long compared to a period. In the case of nonperiodic sound pressures, the interval should be long enough to make the value obtained essentially independent of small changes in the length of the interval. *Note:* The term **effective sound pressure** is frequently shortened to **sound pressure.** 176

sound pressure, instantaneous (at a point). The total instantaneous pressure at that point minus the static pressure at that point. *Note:* The commonly used unit is the newton per square meter. 176

sound pressure level. The sound pressure level, in decibels, is 20 times the logarithm to the base 10 of the ratio of the pressure of this sound to the reference pressure. The reference pressure must be explicitly stated. *Notes:* (1) Unless otherwise explicitly stated, it is to be understood that the sound pressure is the effective root-mean-square sound pressure. (2) The following reference pressures are in common use: (A) 2×10^{-4} dyne per square centimeter. (B) 1 dyne per square centimeter. Reference pressure (A) has been in general use for measurements dealing with hearing and sound-level measurements in air and liquids, while (B) has gained widespread use for calibrations and many types of sound-level measurements in liquids. It is to be noted that in many sound fields the sound pressure ratios are not proportional to the square root of corre-

sponding power ratios and hence cannot be expressed in decibels in the strict sense; however, it is common practice to extend the use of the decibel to these cases. *See also:* **level.** 176, 122, 196

sound probe. A device that responds to some characteristic of an acoustic wave (for example, sound pressure, particle velocity) and that can be used to explore and determine this characteristic in a sound field without appreciably altering the field. *Note:* A sound probe may take the form of a small microphone or a small tubular attachment added to a conventional microphone. *See also:* instrument. 176

sound recording system. A combination of transducing devices and associated equipment suitable for storing sound in a form capable of subsequent reproduction. *See also:* **phonograph pickup.** 176

sound reflection coefficient (surface). The ratio of the sound reflected by the surface to the sound incident upon the surface. Unless otherwise specified, reflection of sound energy in a diffuse sound field is assumed. 176

sound reproducing system. A combination of transducing devices and associated equipment for reproducing recorded sound. *See also:* **loudspeaker.** 176

sound spectrum analyzer (sound analyzer). A device or system for measuring the band pressure level of a sound as a function of frequency. 176

sound tract (electroacoustics). A band that carries the sound record. In some cases, a plurality of such bands may be used. In sound film recording, the band is usually along the margin of the film. *See also:* **phonograph pickup.** 176

sound transmission coefficient (interface or partition). The ratio of the transmitted to incident sound energy. Unless otherwise specified, transmission of sound energy between two diffuse sound fields is assumed. 176

source (1) (electroacoustics). That which supplies signal power to a transducer or system. *Note:* Modifiers are usually used to differentiate between sources of different kinds as signal source, interference source, common-mode source, etcetera. *See:* **signal.** 239, 188

(2) (laser-maser). The term source is taken to mean either laser or laser-illuminated reflecting surface. 363

source ground (signal-transmission system). Potential reference at the physical location of a source, usually the signal source. *See:* **signal.** 188

source impedance. *See:* **impedance, source.** *See also:* **self-impedance.**

source language. A language that is an input to a given translation process. 255, 77

source node (network analysis). A node having only outgoing branches. *See also:* **linear signal flow graphs.** 282

source program (computing systems). A program written in a source language. 255, 77

source resistance. The resistance presented to the input of a device by the source. *See also:* **measurement system.** 295

source resistance rating. The value of source resistance that, when injected in an external circuit having essentially zero resistance, will either (1) double the dead band, or (2) shift the dead band by one-half its width. *See also:* **measurement system.** 295

space (1) (computing devices). (A) A site intended for the storage of data, for example, a site on a printed page or a location in a storage medium. (B) A basic unit of area, usually the size of a single character. (C) One or more blank characters. (D) To advance the reading or display position according to a prescribed format, for example, to advance the printing or display position horizontally to the right or vertically down.   77
(2) (data transmission). One of the two possible conditions of an element (bit); an open line in a neutral circuit.
                                                                 59
space charge (1) (general). A net excess of charge of one sign distributed throughout a specified volume.
(2) (thermionics). Electric charge in a region of space due to the presence of electrons and/or ions. *See also:* **electron emission.**   210, 244, 190, 125
space-charge-control tube. *See:* **density-modulated tube.**
space-charge debunching. Any process in which the mutual interactions between electrons in the stream disperse the electrons of a bunch. *See also:* **electron devices, miscellaneous.**   190, 125
space-charge density (thermionics). The space charge per unit volume. *See also:* **electron emission.**
                                                            244, 190, 125
space-charge generation (semiconductor radiation detector). The thermal generation of free charge carriers in the space-charge region.   23, 118, 119
space-charge grid. A grid, usually positive, that controls the position, area, and magnitude of a potential minimum or of a virtual cathode in region adjacent to the grid. *See also:* **electrode (electron tube); grid.**   190, 125
space-charge-limited current (electron vacuum tubes). The current passing through an interelectrode space when a virtual cathode exists therein. *See also:* **electrode current (electron tube).**   190, 125
space-charge region (semiconductor device). A region in which the net charge density is significantly different from zero. *See also:* **depletion layer; semiconductor device; depletion region.**   245, 210, 66, 118, 119, 23
space correction (industrial control). A method of register control that takes the form of a sudden change in the relative position of the web.   206
spacecraft (communication satellite). Any type of space vehicle, including an earth satellite or deep-space probe, whether manned or unmanned, and also rockets and high-altitude balloons which penetrate the earth's outer atmosphere.   74
space current (electron tubes). Synonym in a diode or equivalent diode of cathode current. *See:* **electrode current (electron tube); leakage current (electron tubes); load current (electron tubes); quiescent current (electron tubes).**
                                                            244, 190
space diversity. *See:* **space diversity reception.**
space-diversity reception (space diversity). That form of diversity reception that utilizes receiving antennas placed in different locations. *See also:* **radio receiver.**   328
space-division digital switching (telephone switching systems). Digital switching with separate paths for each call.
                                                                 55
space-division switching (telephone switching systems). A method of switching that provides a separate path for each of the simultaneous calls.   55

space factor (rotating machinery). The ratio of (1) the sum of the cross-sectional areas of the active or specified material to (2) the cross-sectional area within the confining limits specified. *See:* **asynchronous machine; direct-current commutating machine; slot space factor.**   63
space, head. *See:* **head space.**
space heater (1) (general). A heater that warms occupied spaces.
(2) (rotating machinery). A device that warms the ventilating air within a machine and prevents condensation of moisture during shut-down periods. *See also:* **appliances (including portable).**   63
space pattern (television). A geometrical pattern appearing on a test chart designed for the measurement of geometric distortion. *See also:* **television.**   337
space probe (communication satellite). A spacecraft with a trajectory extending into deep space.   74
spacer cable (transmission and distribution). A type of electric supply line construction consisting of an assembly of one or more covered conductors, separated from each other and supported from a messenger by insulating spacers.   262
space-referenced navigation data. Data in terms of a coordinate system referenced to inertial space. *See also:* **navigation.**   187, 13
spacer shaft (rotating machinery). A separate shaft connecting the shaft ends of two machines. *See:* **armature.**
                                                                 63

space, state. *See:* **state space.**
space-tapered array antenna. An array antenna whose radiation pattern is shaped by varying the density of driven radiating elements over the array surface. *Syn:* **density-tapered array antenna.**   111
spacing (data transmission). A term which originated with telegraph indicating an open key condition. Present usage implies the absence of current or carrier on a circuit. It also indicates the binary digit '0' in computer language.   59
spacing pulse (data transmission). A spacing pulse or **space** is the signal pulse that, in direct-current neutral operation, corresponds to a **circuit open** or **no current** condition. *See also:* **pulse.**   194
spacing wave (back wave) (telegraph communication). The emission that takes place between the active portions of the code characters or while no code characters are being transmitted. *See also:* **radio transmitter.**   111
spalling (corrosion). Spontaneous separation of a surface layer from a metal.   221, 205
span (1) (measuring devices). The algebraic difference between the upper and lower values of a range. *Notes:* (A) For example: (a) Range 0 to 150, span 150; (b) Range -20 to 200, span 220; (c) Range 20 to 150, span 130; (d) Range -100 to -20, span 80. (B) The following compound terms are used with suitable modifications in the units: measured variable span, measured signal span, etcetera. (C) For multirange devices, this definition applies to the particular range that the device is set to measure. *See also:* **instrument.**   295
(2) (overhead conductors). (A) The horizontal distance between two adjacent supporting points of a conductor. (B) That part of any conductor, cable, suspension strand, or pole line between two consecutive points of support. *See also:* **cable; open wire.**   64

span frequency-response rating. The maximum frequency in cycles per minute of sinusoidal variation of measured signal for which the difference in amplitude between output and input represents an error no greater than five times the accuracy rating when the instrument is used under rated operating conditions. The peak-to-peak amplitude of the sinusoidal variation of measured signal shall be equivalent to full span of the instrument. It must be recognized that the span frequency-response rating is a measure of dynamic behavior under the most adverse conditions of measured signal (that is, the maximum sinusoidal excursion of the measured signal). The frequency response for an amplitude of measured signal less than full span is not proportional to the frequency response for full span. The relationship between the frequency response of different instruments at any particular amplitude of measured signal is not indicative of the relationship that will exist at any other amplitude. *See also:* **accuracy rating of an instrument.** 295

span length. The horizontal distance between two adjacent supporting points of a conductor. 262

span, ruling. *See:* **ruling span.**

span, sag. *See:* **sag span.**

span step-response-time rating. The time that the step-response time will not exceed for a change in measured signal essentially equivalent to full span when the instrument is used under rated operating conditions. The actual span step-response time shall not be less than $\frac{2}{3}$ of the span step-response-time rating. (For example, for an instrument of 3-second span step-response-time rating, the span step-response time, under rated operating conditions, will be between 3 and 2 seconds.) It must be recognized that the step-response time for smaller steps is not proportional to the step-response time for full span. *Note:* The end device shall be considered to be at rest when it remains within a band of plus and minus the accuracy rating from its final position. *See also:* **accuracy rating (instrument).** 295

span wire (transmission and distribution). An auxiliary suspension wire which serves to support one or more trolley contact conductors or a light fixture and the conductors which connect it to a supply system.

spare equipment. Equipment complete or in parts, on hand for repair or replacement. *See:* **reserve equipment.** 64

spare point (for supervisory control or indication or telemeter selection). A point that is not being utilized but is fully equipped with all of the necessary devices for a point. 103, 202

spark. A brilliantly luminous phenomenon of short duration that characterizes a disruptive discharge. *Note:* A disruptive discharge is the sudden and large increase in current through an insulating medium due to the complete failure of the medium under electric stress. *See:* **disruptive discharge.** *See also:* **discharge (gas).** 210

spark capacitor (spark condenser*). A capacitor connected across a pair of contact points, or across the inductance that causes the spark, for the purpose of diminishing sparking at these points. 341

* Deprecated

spark gap. Any short air space between two conductors electrically insulated from or remotely electrically connected to each other. 297

spark-gap converter, mercury-hydrogen. *See:* **mercury-hydrogen spark-gap converter.**

spark-gap converter, quenched. *See:* **quenched spark-gap converter.**

spark-gap modulation. A modulation process that produces one or more pulses or energy by means of a controlled spark-gap breakdown for application to the element in which modulation takes place. *See also:* **oscillatory circuit.** 111

spark killer. An electric network, usually consisting of a capacitor and resistor in series, connected across a pair of contact points, or across the inductance that causes the spark, for the purpose of diminishing sparking at these points. *See also:* **network analysis.** 328

sparkover (1) (general). A disruptive discharge between preset electrodes in either a gaseous or a liquid dielectric. *See also:* **spark gap; test voltage and current.** 307, 201 **(2) (surge arrester).** A disruptive discharge between electrodes of a measuring gap, voltage control gap, or protective device. *See also:* **lightning protection and equipment.** 2

spark-plug suppressor (electromagnetic compatibility). A suppressor designed for direct connection to a spark plug. *See also:* **electromagnetic compatibility.** 199

spark transmitter. A radio transmitter that utilizes the oscillatory discharge of a capacitor through an inductor and a spark gap as the source of its radio-frequency power. *See also:* **radio transmitter.** 111, 240

spatial coherence (electromagnetic) (laser-maser). The correlation between electromagnetic fields at points separated in space. *See also:* **coherence area.**

speaker. *See:* **loudspeaker.**

special-billing call (telephone switching systems). A call charged to a special number. 55

special character (character set). A character that is neither a numeral, a letter, nor a blank, for example, virgule, asterisk, dollar sign, equals sign, comma, period. 255, 77

special-dial tone (telephone switching systems). A tone for certain features that indicates that a customer can use his calling device. 55

specialized common carrier (data communication). A company that provides private line communications services, for example, voice, teleprinter, data, facsimile transmission. *See also:* **common carrier, value added service.** 12

special-purpose computer. A computer that is designed to solve a restricted class of problems. 255, 77

special purpose electronic test equipment. *See:* **special purpose test equipment.**

special-purpose motor. A motor with special operating characteristics or special mechanical construction, or both designed for a particular application and not falling within the definition of a general-purpose or definite-purpose motor. *See:* **asynchronous machine; direct-current commutating machine.** 232, 63

special purpose test equipment (test, measurement and diagnostic equipment). Equipment used for test, repair and maintenance of a specified system, subsystem or module, having application to only one or a very limited number of systems. 54

specialty transformer. A transformer generally intended

to supply electric power for control, machine tool, Class 2, signaling, ignition, luminous-tube, cold-cathode lighting, series street-lighting, low-voltage general purpose, and similar applications. *See:* **individual-lamp transformer; series street-lighting transformer; energy-limiting transformer; high-reactance transformer; non-energy-limiting transformer; high power factor transformer; low power factor transformer; insulating transformer; individual-lamp insulating transformer; group-series loop insulating transformer; luminous-tube transformers; ignition transformer; series circuit lighting transformer; signalling and doorbell transformers; control transformers; machine tool control transformers; general purpose transformers; mercury vapor lamp transformers; saturable reactor; electronic transformer; series transformer rating; primary voltage rating (general purpose transformer); secondary voltage rating; IR-drop compensation transformer; kilovolt or volt-ampere short-circuit input rating (high-reactance transformer); secondary short-circuit current rating (high reactance transformer).**   53

specific acoustic impedance (unit area acoustic impedance) (at a point in the medium). The complex ratio of sound pressure to particle velocity. *See:* Note 2 under **acoustic impedance.**   176

specific acoustic reactance. The imaginary component of the specific acoustic impedance.   176

specific acoustic resistance. The real component of the specific acoustic impedance.   176

specific coordinated methods. Those additional methods applicable to specific situations where general coordinated methods are inadequate. *See also:* **inductive coordination.**   328

specific emission. The rate of emission per unit area.   244

specific inductive capacitance. *See:* **relative capacitivity.**

specific repetition frequency (loran). One of a set of closely spaced pulse repetition frequencies derived from the basic repetition frequency and associated with a specific set of synchronized stations. *See also:* **navigation.**   187, 13

specific repetition rate (electronic navigation). *See:* **specific repetition frequency.** *See also:* **navigation.**

special unit capacity purchases (electric power supply). That capacity that is purchased or sold in transactions with other utilities and that is from a designated unit on the system of the seller. It is understood that the seller does not provide reserve capacity for this type of capacity transaction. *See also:* **generating station.**   200

specified achromatic lights. (1) Light of the same chromaticity as that having an equi-energy spectrum. (2) The standard illuminants of colorimetry *A, B,* and *C,* the spectral energy distributions of which were specified by the International Commission on Illumination (CIE) in 1931, with various scientific applications in view. Standard *A:* incandescent electric lamp of color temperature 2854 kelvins. Standard *B:* Standard *A* combined with a specified liquid filter to give a light of color temperature approximately 4800 kelvins. Standard *C:* Standard *A* combined with a specified liquid filter to give a light of color temperature approximately 6500 kelvins. (3) Any other specified white light. *See also:* **color.**   244, 178

specified breakaway torque (rotating machinery). The torque which a motor is required to develop to break away

its load from rest to rotation.   63

spectral bandwidth (λ_{BW}) (light emitting diodes). The difference between the wavelengths at which the spectral radiant intensity is 50 percent (unless otherwise stated) of the maximum value. The term spectral linewidth is sometimes used.   162

spectral characteristic (1) (color television). The set of spectral responses of the color separation channels with respect to wavelength. *Notes:* (A) The channel terminals at which the characteristics apply must be specified and an appropriate modifier may be added to the term, such as pickup spectral characteristic or studio spectral characteristic. (B) Because of nonlinearity, some spectral characteristics depend upon the magnitude of radiance used in the measurement. (C) Nonlinearizing and matrixing operations may be performed within the channels. (D) The spectral taking characteristics are uniquely related to the chromaticities of the display primaries.   18

(2) (camera tube). A relation, usually shown by a graph, between wavelength and sensitivity per unit wavelength interval *See:* **spectral sensitivity characteristic.**   328

(3) (luminescent screen). The relation, usually shown by a graph, between wavelength and emitted radiant power per unit wavelength interval. *Note:* The radiant power is commonly expressed in arbitrary units.   190, 178, 125

(4) (phototube). A relation, usually shown by a graph, between the radiant sensitivity and the wavelength of the incident radiant flux. *See:* **spectral sensitivity characteristic.** *See also:* **phototubes.**   328

spectral-conversion luminous gain (optoelectronic device). The luminous gain for specified wavelength-intervals of both incident and emitted luminous flux. *See also:* **optoelectronic device.**   191

spectral-conversion radiant gain (optoelectronic device). The radiant gain for specified wavelength intervals of both incident and emitted radiant flux. *See also:* **optoelectronic device.**   191

spectral emissivity (element of surface of a temperature radiator at any wavelength). The ratio of its radiant flux density per unit wavelength interval (spectral radiant exitance) at that wavelength to that of a blackbody at the same temperature. *See also:* **radiant energy (illuminating engineering).**   167

spectral line (semiconductor radiation detectors). A sharply peaked portion of the spectrum that represents a specific feature of the incident radiation, usually the full energy of a monoenergetic radiation.   23, 118, 119

spectral luminous efficacy (radiant flux). The quotient of the luminous flux at a given wavelength by the radiant flux at that wavelength. It is expressed in lumens per watt. *Note:* This term formerly was called **luminosity factor.** The reciprocal of the maximum luminous efficacy of radiant flux is sometimes called **mechanical equivalent of light;** that is, the watts per lumen at the wavelength of maximum luminous efficacy. The most probable value is 0.00147 watt per lumen, corresponding to 680 lumens per watt as the maximum possible luminous efficacy. These values are based on the candela, on the 1948 International Temperature Scale by using Planck's equation with $c_2 = 1.438$ centimeter kelvin, $c_1 = 3.741 \times 10^{-12}$ watt centimeter^2 and 2042 kelvins as the temperature of freezing platinum, and on the standard CIE photopic spectral luminous efficiency values given in the accompanying table. If the standard CIE scotopic spectral luminous efficiency values are used, the maximum luminous efficacy is 1746 "scotopic" lumens per watt. *See:* **radiant flux; spectral radiant flux.** *See also:* **light.** 167

spectral luminous efficiency (radiant flux). The ratio of the luminous efficacy for a given wavelength to the value of the wavelength of maximum luminous efficacy. It is dimensionless. *Notes:* (1) The term **spectral luminous efficiency** replaces the previously used terms **relative luminosity** and **relative luminosity factor.** (2) Values of spectral luminous efficiency for photopic vision at 10-nanometer intervals were provisionally adopted by the International Commission on Illumination in 1924 and were adopted in 1933 by the International Committee on Weights and Measures as a basis for the establishment of photometric standards of types of sources differing from the primary standard in spectral distribution of radiant flux. These standard values of spectral luminous efficiency were determined by observations with a two-degree photometric field having a moderately high luminance (photometric brightness), and photometric evaluations based upon them consequently do not apply exactly to other conditions of observation. Watts weighted in accord with these standard values are often referred to as **lightwatts.** See the accompanying table.
(3) Values of spectral luminous efficiency for scotopic vision at 10-nanometer intervals were provisionally adopted by the International Commission on Illumination in 1951. These values of spectral luminous efficiency were determined by observation by young dark-adapted observers using extra-foveal vision at near-threshold luminance. *See also:* **light.** 167

spectral luminous flux ($\phi v \lambda$) (light emitting diodes). The luminous flux per unit wavelength interval at wavelength λ; that is, lumens per nanometer. 162

spectral luminous gain (optoelectronic device). Luminous gain for a specified wavelength interval of either the incident or the emitted flux. *See also:* **optoelectronic device.** 191

spectral luminous intensity ($I_{v\lambda}$) (light emitting diodes). The luminous intensity per unit wavelength (at wavelength λ), that is, candela per nanometer. 162

spectral-noise density (sound recording and reproducing system). The limit of the ratio of the noise output within a specified frequency interval to the frequency interval, as that interval approaches zero. *Note:* This is approximately the total noise within a narrow frequency band divided by that bandwidth in hertz. *See also:* **noise (sound recording and reproducing system).** 350

spectral power density (radio wave propagation). The power density per unit bandwidth. *See also:* **radio wave propagation.** 180

spectral power flux density (radio wave propagation). *See:* **spectral power density.** *See also:* **radio wave propagation.** 180

spectral quantum yield (photocathode). The average number of electrons photoelectrically emitted from the photocathode per incident photon of a given wavelength. *Note:* The spectral quantum yield may be a function of the angle of incidence and of the direction of polarization of the incident radiation. *See also:* **phototube.** 335

spectral radiance (laser-maser). The power transmitted in a radiation field per unit frequency (or wavelength) interval, per unit solid angle, per unit area normal to a given direction. *See also:* **Planck radiation law; radiant energy (illuminating engineering); spectral radiant intensity.** 363, 167

spectral radiant energy ($Q_\lambda = dQ_e/d\lambda$) (light emitting diodes). Radiant energy per unit wavelength interval at wavelength λ; that is, joules per nanometer. 162

spectral radiant flux ($\phi_\lambda = d\phi_e/d\lambda$) (light emitting diodes). Radiant flux per unit wavelength interval at wavelength λ; that is watts per nanometer. 162

spectral radiant gain (optoelectronic device). Radiant gain for a specified wavelength interval of either the incident or the emitted radiant flux. *See also:* **optoelectronic device.** 190

spectral radiant intensity ($I\lambda = dI_e/d\lambda$) (light emitting diodes). The radiant intensity per unit wavelength interval; for example watts per (steradian-nanometer). 162

spectral range (acoustically tunable optical filter). The wavelength region over which the dynamic transmission is greater than some specified minimum value. 72

spectral reflectance (surface or medium). The ratio of the reflected flux to the incident flux at a particular wavelength λ or within a small band of wavelengths $\Delta\lambda$ about λ. *Note:* The terms **hemispherical, regular,** or **diffuse reflectance** may each be considered restricted to a specific region of the spectrum and may be so designated by the addition of the adjective **spectral.** *See also:* **lamp.** 167

spectral response characteristic (photoelectric devices). *See:* **spectral sensitivity characteristic.**

spectral selectivity (photoelectric device). The change of photoelectric current with the wavelength of the irradiation. *See also:* **photoelectric effect.** 244, 190

spectral sensitivity characteristic (camera tubes or phototubes). The relation between the radiant sensitivity and the wavelength of the incident radiation, under specified conditions of irradiation. *Note:* Spectral sensitivity characteristic is usually measured with a collimated beam at normal incidence. *See:* **phototubes.** 190, 178

spectral temperature (radiation field) (laser-maser). The temperature of a black body which produces the same **spectral radiance** as the radiation field at a given frequency and in a given direction. 363

spectral transmittance (medium). The ratio of the transmitted flux to the incident flux at a particular wavelength λ or within a small band of wavelengths $\Delta\lambda$ about λ. *Note:* The terms **hemispherical, regular,** or **diffuse transmittance** may each be considered restricted to a specific region of the spectrum and may be so designated by the addition of the adjective **spectral.** *See also:* **transmission (illuminating engineering).** 167

spectral tristimulus values. Values per unit wavelength interval and unit spectral radiant flux. *Note:* Spectral tristimulus values have been adopted by the International Commission on Illumination (CIE). They are tabulated as functions of wavelength throughout the spectrum and are the basis for the evaluation of radiant energy as light. 167

spectrophotometer. An instrument for measuring the transmittance and reflectance of surfaces and media as

Photopic spectral luminous efficiency $V(\lambda)$.
(Unity at wavelength of maximum luminous efficacy.)

λ in nanometers	Standard values	Standard values interpolated at intervals of 1 nanometer								
		1	2	3	4	5	6	7	8	9
380	0.00004	0.000045	0.000049	0.000054	0.000059	0.000064	0.000071	0.000080	0.000090	0.000104
390	0.00012	0.000138	0.000155	0.000173	0.000193	0.000215	0.000241	0.000272	0.000308	0.000350
400	0.0004	0.00045	0.00049	0.00054	0.00059	0.00064	0.00071	0.00080	0.00090	0.00104
410	0.0012	0.00138	0.00156	0.00174	0.00195	0.00218	0.00244	0.00274	0.00310	0.00352
420	0.0040	0.00455	0.00515	0.00581	0.00651	0.00726	0.00806	0.00889	0.00976	0.01066
430	0.0116	0.01257	0.01358	0.01463	0.01571	0.01684	0.01800	0.01920	0.02043	0.02170
440	0.023	0.0243	0.0257	0.0270	0.0284	0.0298	0.0313	0.0329	0.0345	0.0362
450	0.038	0.0399	0.0418	0.0438	0.0459	0.0480	0.0502	0.0525	0.0549	0.0574
460	0.060	0.0627	0.0654	0.0681	0.0709	0.0739	0.0769	0.0802	0.0836	0.0872
470	0.091	0.0950	0.0992	0.1035	0.1080	0.1126	0.1175	0.1225	0.1278	0.1333
480	0.139	0.1448	0.1507	0.1567	0.1629	0.1693	0.1761	0.1833	0.1909	0.1991
490	0.208	0.2173	0.2270	0.2371	0.2476	0.2586	0.2701	0.2823	0.2951	0.3087
500	0.323	0.3382	0.3544	0.3714	0.3890	0.4073	0.4259	0.4450	0.4642	0.4836
510	0.503	0.5229	0.5436	0.5648	0.5865	0.6082	0.6299	0.6511	0.6717	0.6914
520	0.710	0.7277	0.7449	0.7615	0.7776	0.7932	0.8082	0.8225	0.8363	0.8495
530	0.862	0.8739	0.8851	0.8956	0.9056	0.9149	0.9238	0.9320	0.9398	0.9471
540	0.954	0.9604	0.9661	0.9713	0.9760	0.9803	0.9840	0.9873	0.9902	0.9928
550	0.995	0.9969	0.9983	0.9994	1.0000	1.0002	1.0001	0.9995	0.9984	0.9969
560	0.995	0.9926	0.9898	0.9865	0.9828	0.9786	0.9741	0.9691	0.9638	0.9581
570	0.952	0.9455	0.9386	0.9312	0.9235	0.9154	0.9069	0.8981	0.8890	0.8796
580	0.870	0.8600	0.8496	0.8388	0.8277	0.8163	0.8046	0.7928	0.7809	0.7690
590	0.757	0.7449	0.7327	0.7202	0.7076	0.6949	0.6822	0.6694	0.6565	0.6437
600	0.631	0.6182	0.6054	0.5926	0.5797	0.5668	0.5539	0.5410	0.5282	0.5156
610	0.503	0.4905	0.4781	0.4658	0.4535	0.4412	0.4291	0.4170	0.4049	0.3929
620	0.381	0.3690	0.3570	0.3449	0.3329	0.3210	0.3092	0.2977	0.2864	0.2755
630	0.265	0.2548	0.2450	0.2354	0.2261	0.2170	0.2082	0.1996	0.1912	0.1830
640	0.175	0.1672	0.1596	0.1523	0.1452	0.1382	0.1316	0.1251	0.1188	0.1128
650	0.107	0.1014	0.0961	0.0910	0.0862	0.0816	0.0771	0.0729	0.0688	0.0648
660	0.061	0.0574	0.0539	0.0506	0.0475	0.0446	0.0418	0.0391	0.0366	0.0343
670	0.032	0.0299	0.0280	0.0263	0.0247	0.0232	0.0219	0.0206	0.0194	0.0182
680	0.017	0.01585	0.01477	0.01376	0.01281	0.01192	0.01108	0.01030	0.00956	0.00886
690	0.0082	0.00759	0.00705	0.00656	0.00612	0.00572	0.00536	0.00503	0.00471	0.00440
700	0.0041	0.00381	0.00355	0.00332	0.00310	0.00291	0.00273	0.00256	0.00241	0.00225
710	0.0021	0.001954	0.001821	0.001699	0.001587	0.001483	0.001387	0.001297	0.001212	0.001130
720	0.00105	0.000975	0.000907	0.000845	0.000788	0.000736	0.000688	0.000644	0.000601	0.000560
730	0.00052	0.000482	0.000447	0.000415	0.000387	0.000360	0.000335	0.000313	0.000291	0.000270
740	0.00025	0.000231	0.000214	0.000198	0.000185	0.000172	0.000160	0.000149	0.000139	0.000130
750	0.00012	0.000111	0.000103	0.000096	0.000090	0.000084	0.000078	0.000074	0.000069	0.000064
760	0.00006	0.000056	0.000052	0.000048	0.000045	0.000042	0.000039	0.000037	0.000035	0.000032

a function of wavelength. *See also:* **photometry.** 167

spectroradiometer. An instrument for measuring radiant flux as a function of wavelength. *See also:* **primary standards (illuminating engineering).** 167

spectrum (1) (general). (A) The distribution of the amplitude (and sometimes phase) of the components of the wave as a function of frequency. (B) A continuous range of components, usually wide in extent, within which waves have some specified common characteristic; for example, audio-frequency spectrum. *Note:* The term **spectrum** can also be applied to functions of variables other than time, such as distance. *See also:* **signal wave.** 59, 176

(2) (radiation). A distribution of the intensity of radiation as a function of energy or its equivalent electric analog (such as charge or voltage) at the output of a radiation detector. 118, 119, 23

spectrum amplitude (impulse strength and impulse bandwidth). The voltage spectrum of a pulse can be expressed as

$$V(\omega) = R(\omega) + jX(\omega) = \int_{-\infty}^{+\infty} v(t)e^{-jwt}\,dt$$

where

$$R(\omega) = \int_{-\infty}^{+\infty} v(t)\cos\omega t\,dt$$

$$X(\omega) = \int_{-\infty}^{+\infty} v(t)\sin\omega t\,dt$$

and $\omega = 2\pi f$.

and the phase characteristic

$$\varphi(\omega) = \tan^{-1}\frac{X(\omega)}{R(\omega)}$$

The inverse transform can be written

$$v(t) = \frac{1}{\pi}\int_0^\infty A(\omega)\cos[\omega t + \varphi(\omega)]dw,$$

$$\text{for real } v(t)$$

The spectrum amplitude is also expressible in volts per hertz (volt-seconds) as follows:

$$S(f) = 2A(\omega) \qquad \text{(eq 1)}$$

It is this form that is used as the basis for calibration of commercially available impulse generators. A practical impulse is a function of time duration short compared with the reciprocals of all frequencies of interest. *Note:* For a rectangular pulse, the spectrum is flat within about 1 dB up to a frequency for which the pulse duration is equal to $\frac{1}{4}$ cycle. See Fig 3. Its spectrum amplitude $S(f)$ is substantially uniform in this frequency range and is equal to twice the area under the impulse time function or 2σ. At frequencies higher than this it is still of interest to define the spectrum amplitude which will usually be less than 2σ.

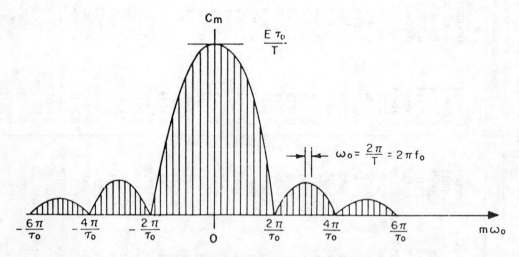

Frequency spectrum of the rectangular pulse train.

Note: See IEEE Std 263-1965, Measurement of Radio Noise Generated by Motor Vehicles and Affecting Mobile Communications Receivers in the Frequency Range 25 to 1000 megahertz.

The spectrum then has the amplitude characteristic

$$A(\omega) = \sqrt{R^2(\omega) + X^2(\omega)} \quad (V/\text{rad})/s$$

In most broadband impulse generators a dc voltage is used to charge a calibrated coaxial transmission line. The pulses are produced when the line is discharged into its terminating impedance through mechanically activated contacts. These mechanical contacts may be parts of either a vibrating diaphragm or mercury wetted relay switches. By proper choice of transmission line length and resistive termination, it is possible to produce impulses having a predictable uniform spectrum amplitude range.

The advent of solid-state switches has made it possible to switch on a sine wave for a precisely measurable time interval (τ), producing in the frequency band in the vicinity of the sine wave a spectrum simulating that produced by an impulse. The spectrum amplitude at that particular frequency can be measured in terms of a measurement of the amplitude of the sine wave when not switched, and a measurement of the on time (τ_0) for the switch. 30

spectrum analyzer. An instrument generally used to display the power distribution of an incoming signal as a function of frequency. *Notes:* (1) Spectrum analyzers are useful in analyzing the characteristics of electrical waveforms in general since, by repetitively sweeping through the frequency range of interest, they display all components of the signal. (2) The display format may be a cathode ray tube or chart recorder. 68

spectrum intensity (impulse strength and impulse bandwidth). (For spectra which have a continuous distribution of components—components are not discrete—over the frequency range of interest). The spectrum intensity is the ratio of the power contained in a given frequency range to the frequency range as the frequency range approaches zero. It has the dimensions watt-seconds or joules and is usually stated quantitatively in terms of watts per hertz. 30

spectrum level (spectrum density level) (acoustics) (specified signal at a particular frequency). The level of that part of the signal contained within a band 1 hertz wide, centered at the particular frequency. Ordinarily this has significance only for a signal having a continuous distribution of components within the frequency range under consideration. The words **spectrum level** cannot be used alone but must appear in combination with a prefatory modifier; for example, pressure, velocity, voltage. *Note:* For illustration, if L_{ps} be a desired pressure spectrum level, p the effective pressure measured through the filter system, p_0 reference sound pressure, Δf the effective bandwidth of the filter system, and $\Delta_0 f$ the reference bandwidth (1 hertz), then

$$L_{ps} = \log_{10} \frac{p^2/\Delta f}{p_0{}^2/\Delta_0 f}$$

For computational purposes, if L_{ps} is the band pressure level observed through a filter of bandwidth Δf, the above relation reduces to

$$L_{ps} = L_p - 10 \log_{10} \frac{\Delta f}{\Delta_0 f}$$

176

spectrum locus (color). The locus of points representing the chromaticities of spectrally pure stimuli in a chromaticity diagram. *See also:* **color.** 18, 178

Chromaticity diagram.

specular angle. The angle between the perpendicular to the surface and the reflected ray that is numerically equal to the angle of incidence and that lies in the same plane as the incident ray and the perpendicular but on the opposite side of the perpendicular to the surface. *See also:* **lamp.** 167

specular reflection (laser-maser). A mirrorlike reflection. 363

specular surface. One from which the reflection is predominantly regular. *See:* **regular (specular) reflection.** *See also:* **bare (exposed) lamp.** 167

speech interpolation. The method of obtaining more than one voice channel per voice circuit by giving each subscriber a speech path in the proper direction only at times when his speech requires it. 328

speech level (speech quality measurements). The speech level defined and measured subjectively by comparison of the speech signal with a signal obtained by passing pink noise through a filter with A-weighting characteristics that has been judged to be equal to it in loudness. *Note:* The value of the speech level is defined to be the A-weighted sound pressure level of this noise [dB(A)]. 126

speech network (telephony). An electrical circuit that connects the transmitter and the receiver to a telephone line or telephone test loop and to each other. 122

speech quality (speech quality measurements). A characteristic of a speech signal that can be described in terms of subjective and objective parameters. Speech quality is evaluated only in terms of the subjective parameter of preference. 126

speech reference signal (speech quality measurements). Used as a standard of reference for the purpose of preference testing, a speech signal which is artificially degraded in a measurable and reproducible way. 126

speech signals. Utterances in their acoustical form or electrical equivalent. 126

speech test signal (speech quality measurements). A speech signal whose speech quality is to be evaluated. 126

speed (speed governing of hydraulic turbines). The instantaneous speed of rotation of the turbine expressed as a percent of rated speed.   8

speed adjustment (control). A speed change of a motor accomplished intentionally through action of a control element in the apparatus or system governing the performance of the motor. *Note:* For an adjustable-speed direct-current motor, the speed adjustment is expressed in percent (or per unit) of base speed. Speed adjustment of all other motors is expressed in percent (or per unit) of rated full-load speed. *See also:* **adjustable-speed motor; base speed of an adjustable-speed motor; electric drive.**
                    206

speed changer (1) (gas turbines). A device by means of which the speed-governing system is adjusted to change the speed or power output of the turbine during operation. *See also:* **asynchronous machine; direct-current commutating machine.**   98, 58
(2) (speed governing of hydraulic turbines). A device by means of which the governor system may be adjusted to change the speed or power output of the turbine while the turbine is operating.   8

speed-changer high-speed stop (gas turbines). A device that prevents the speed changer from moving in the direction to increase speed or power output beyond the position for which the device is set. *See also:* **asynchronous machine; direct-current commutating machine.**   98, 58

speed-control mechanism (electric power systems). Includes all equipment such as relays, servomotors, pressure or power-amplifying devices, levers, and linkages between the speed governor and the governor-controlled valves. *See also:* **speed-governing system.**   94

speed deadband (speed governing of hydraulic turbines). The total magnitude of the change in steady state speed, expressed in percent of rated speed, required to reverse the direction of travel of the turbine control servomotor.

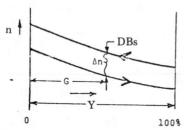

One half of the governor speed deadband is termed the governor speed insensitivity.

$$DB_s = \Delta n$$
                    8, 58

speed deviation (turbines). The instantaneous difference between the actual speed and a reference speed.   8, 58
speed droop (speed governing of hydraulic turbines). The speed droop graph may indicate a non-linear relationship between the two measured variables depending on the adjustment of the governor speed changer and the quantity (servomotor position or generator power output) used to develop the feedback signal used in the governor system. Speed droop is considered positive when speed increases with a decrease in gate position or power output.   8

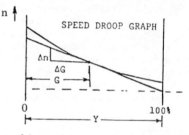

The slope of the speed droop graph at a specified point of operation G. The change in steady state speed expressed in percent of rated speed corresponding to the 100 percent turbine servomotor stroke with no change in setting of any governor adjustments and with the turbine supplying power to a load independently of any other power source.

$$D_S = \left(\frac{-\Delta n}{\Delta G}\right) \cdot (100)$$

Speed droop is classified as either permanent or temporary. (1) Permanent speed droop: The speed droop which remains in steady state after the decay action of the damping device has been completed. (2) Temporary speed droop: The speed droop in steady state which would occur if the decay action of the damping device were blocked and the permanent speed droop were made inactive.   8, 58

speed-droop changer (speed governing of hydraulic turbines). A device by means of which the speed droop may be adjusted while the turbine is operating.   8, 58

speed-governing system. Control elements and devices for the control of the speed or power output of a gas turbine. This includes a speed governor, speed changer, fuel-control mechanism, and other devices and control elements.
                    94, 98, 57

speed governor (electric power system). Includes only those elements that are directly responsive to speed and that position or influence the action of other elements of the speed-governing system. *See also:* **asynchronous machine; direct-current commutating machine; gas turbines; speed-governing system.**   94, 98, 58

speed limit (industrial control). A control function that prevents a speed from exceeding prescribed limits. Speed-limit values are expressed as percent of maximum rated speed. If the speed-limit circuit permits the limit value to change somewhat instead of being a single value, it is desirable to provide either a curve of the limit value of speed as a function of some variable, such as load, or to give limit values at two or more conditions of operation. *See also:* **control system, feedback.**   219, 206

speed-limit indicator. A series of lights controlled by a relay to indicate the speeds permitted corresponding to the track conditions.   328

speed of transmission (telecommunication). The instantaneous rate at which information is transferred over a transmission facility. This quantity is usually expressed in characters per unit time or bits per unit time. **Rate of transmission** is more common usage.   194, 59

speed of transmission, effective. Speed, less than rated, of information transfer that can be averaged over a significant period of time and that reflects effects of control codes, timing codes, error detection, retransmission, tabbing, hand keying, etcetera.   194

speed of vision. The reciprocal of the duration of the exposure time required for something to be seen. *See also: visual field.*   167

speed range (industrial control). All the speeds that can be obtained in a stable manner by action of part (or parts) of the control equipment governing the performance of the motor. The speed range is generally expressed as the ratio of the maximum to the minimum operating speed. *See also: electric drive.*   206

speed ratio (fuse). *See: melting-speed ratio (fuse).*

speed ratio control (industrial control). A control function that provides for operation of two drives at a preset ratio of speed. *See also; control system, feedback.*   225, 206

speed-regulating rheostat (industrial control). A rheostat for the regulation of the speed of a motor. *See: control.*   244, 206

speed regulation (speed governing of hydraulic turbines). The speed regulation graph may indicate a nonlinear relationship between the two measured variables depending on the adjustment of the governor speed changer and the quantity (servomotor position or generator power output) used to develop the feedback signal used in the governor system. Speed regulation is considered positive when speed increases with a decrease in gate position or power output.

The slope of the speed regulation graph at a specified point, P, of operation. The change in steady state speed expressed in percent of rated speed when the power output of the unit is reduced from rated power output to zero power output under rated head with no change in setting of any governor adjustments and with the unit supplying power to a load independently of any other power source.

$$R_S = \left(\frac{-\Delta n/100}{\Delta P/P_r} \right) \cdot (100)$$

8

speed regulation changer (speed governing of hydraulic turbines). A device by means of which the speed regulation may be adjusted while the turbine is operating.   8

speed regulation characteristic (rotating machinery). The relationship between speed and the load of a motor under specified conditions. *See also: asynchronous machine; direct-current commutating machine.*   63

speed regulation of a constant-speed direct-current motor. The change in speed when the load is reduced gradually from the rated value to zero with constant applied voltage

and field-rheostat setting, expressed as a percent of speed at rated load. *See: direct-current commutating machine.*   328

speed-sensing elements (speed governing systems). The speed responsive elements which determine speed (error) and influence the action of other elements of the governing system. Included are the means used to transmit a signal proportional to the speed of the turbine to the governor.   8, 58

speed variation (industrial control). Any change in speed of a motor resulting from causes independent of the control-system adjustment, such as line-voltage changes, temperature changes, or load changes. *See also: electric drive.*   206

spherical hyperbola (electronic navigation). The locus of the points on the surface of a sphere having a specified constant difference in great-circle distances from two fixed points on the sphere.   13, 187

spherical reflector. A reflector that is a portion of a spherical surface. *See also: antenna.*   179, 111

spherical-seated bearing (self-aligning bearing) (rotating machinery). A journal bearing in which the bearing liner is supported in such a manner as to permit the axis of the journal to be moved through an appreciable angle in any direction. *See also: bearing.*   63

spherical support seat (rotating machinery). A support for a journal bearing in which the inner surface that mates with the bearing shell is spherical in shape, the center of the sphere coinciding approximately with the shaft centerline, permitting the axis of the bearing to be aligned with that of the shaft. *See also: bearing.*   63

spherical wave. A wave whose equiphase surfaces form a family of concentric spheres. *See also: radiation; radio wave propagation.*   146, 180

spider (rotor spider) (rotating machinery). A structure supporting the core or poles of a rotor from the shaft, and typically consisting of a hub, spokes, and rim, or some modified arrangement of these.   63

spider rim (rotating machinery). *See: rim.*

spider web (rotating machinery). The component of a rotor that provides radial separation between the hub or shaft and the rim or core. *See also: rotor (rotating machinery).*   63

spike (1) (pulse terms). A distortion in the form of a pulse waveform of relatively short duration superimposed on an otherwise regular or desired pulse waveform. *See:* Note in **preshoot** entry.   254

(2) (relaying). An output signal of short duration and limited crest derived from an alternating input of specified polarity.   79

spike leakage energy (microwave gas tubes). The radio-frequency energy per pulse transmitted through the tube before and during the establishment of the steady-state radio-frequency discharge. *See: gas tubes.*   190, 125

spike train (electrotherapy) (courant iteratif). A regular succession of pulses of unspecified shape, frequency, duration, and polarity. *See also: electrotherapy.*   192

spill (charge-storage tubes). The loss of information from a storage element by redistribution.   125

spillover (antennas). That part of the power radiated by a feed not intercepted by the secondary radiator.   111

spillover loss (radar). In a transmitting antenna having a focusing device such as a reflector or lens illuminated by a feed, spillover loss is the reduction in gain due to the portion of the power radiated by the feed in directions that do not intersect the focusing device. By reciprocity, the same loss occurs when the same antenna is used for reception. 13

spin axis (SA) (gyro). The axis of rotation of the rotor. 46

spindle speed (numerically controlled machines). The rate of rotation of the machine spindle usually expressed in terms of revolutions per minute. 224, 207

spindle wave (electrobiology). A sharp, rather large wave considered of diagnostic importance in the electroencephalogram. *See also:* **electrocardiogram.** 328

spin-input rectification drift rate (gyro). The drift rate in a single-degree-of-freedom gyro resulting from coherent oscillatory rates about the spin reference axis (SRA) and input reference axis (IRA). It occurs only when gyro and loop dynamics allow the gimbal to move away from null in response to the rate about the input reference axis, resulting in a cross-coupling of the spin reference axis rate. This drift rate is a function of the input rate amplitudes and the phase angle between them. 46

spinner (radar). Rotating part of a radar antenna, together with directly associated equipment, used to impart any subsidiary motion, such as conical scanning, in addition to the primary slewing of the beam. *See also:* **radar.** 13, 187

spinning reserve. That reserve generating capacity connected to the bus and ready to take load. *See also:* **generating station.** 64

spin-output rectification drift rate)gyro). The drift rate in a single-degree-of-freedom gyro resulting from coherent oscillatory rates about the spin reference axis (SRA) and output reference axis (ORA). It occurs only when gyro and loop dynamics allow the float motion to lag case motion when subjected to a rate about the output reference axis, resulting in a cross-coupling of the spin reference axis rate. This drift rate is a function of the input rate amplitudes and the phase angle between them. 46

spin reference axis (SRA) (gyro). An axis normal to the input reference axes and nominally parallel to the spin axis when the gyro outputs have specified values, usually null. 46

spiral antenna. An antenna consisting of one or more conducting wires arranged as a spiral. *Note:* Spiral antennas are usually classified according to the surface to which they conform, that is, conical spiral or planar spiral, and according to mathematical form such as equiangular and Archimedean. 111

spiral four (star quad). A quad in which the four conductors are twisted about a common axis, the two sets of opposite conductors being used as pairs. *See also:* **cable.** 328

spiral scanning (electronic navigation). Scanning in which the direction of maximum response describes a portion of a spiral. *See also:* **antenna.** 179

SPL. *See:* **sound pressure level.**

s-plane (circuits and systems). In the Laplace transform, the notation $s = \sigma + j\omega$ is introduced. The s plane is a coordinate system with σ as the abscissa and ω as the ordinate. The letter "p" is sometimes used instead of "s". 67

splashproof (industrial control). So constructed and protected that external splashing will not interfere with successful operation. *See also:* **traction motor.** 1, 103, 202, 225, 206, 27, 124

splashproof enclosure. An enclosure in which the openings are so constructed that drops of liquid or solid particles falling on the enclosure or coming towards it in a straight line at any angle not greater than 100 degrees from the vertical cannot enter the enclosure either directly or by striking and running along a surface. 3

splashproof machine. An open machine in which the ventilating openings are so constructed that drops of liquid or solid particles falling on the machine or coming towards it in a straight line at any angle not greater than 100 degrees downward from the vertical cannot enter the machine either directly or by striking and running along a surface. *See:* **asynchronous machine; direct-current commutating machine.** 63

splice (1) (straight-through joint). A joint used for connecting in series two lengths of conductor or cable. **(2) (power cable joints).** The physical connection of two or more conductors to provide electrical continuity. 34

splice box (mine type). An enclosed connector permitting short sections of cable to be connected together to obtain a portable cable of the required length. *See also:* **mine feeder circuit.**

splice, wire rope (conductor stringing equipment). The point at which two wire ropes are joined together. The various methods of joining (splicing) wire ropes together include "hand tucked" woven splices, compression splices which utilize compression fittings but do not incorporate loops (eyes) in the ends of the ropes and mechanical splices which are made through the use of loops (eyes) in the ends of the ropes held in place by either compression fittings or wire rope clips. The latter are joined together with connector links or steel bobs, and in some cases, rigged "eye to eye." Woven splices are often classified as "short" or "long." A "short" splice varies in length from 7 to 17 feet for $\frac{1}{4}$ to $1\frac{1}{2}$ inch diameter ropes respectively, while a "long" splice varies from 15 to 45 feet for the same size ropes. In one instance, however, splices classified as "long" splices, but only 6 feet in length, have been made in $\frac{5}{8}$ inch diameter "Roepac" wire rope pulling lines using compression fittings at each strand butt joint with restraining compression collar on either side of each fitting to obtain the ultimate breaking strength of new rope. 45

splicing cart, (conductor stringing equipment). A unit which is equipped with a hydraulic compressor (press) and all other necessary equipment for performing splicing operations on conductor. *Syn:* **sleeving trailer; splicing trailer; splicing truck.** 45

splicing chamber. *See:* **cable vault; manhole.**

split-anode magnetron. A magnetron with an anode divided into two segments; usually by slots parallel to its axis. *See also:* **magnetrons.** 190, 125

split-beam cathode-ray tube (double-beam cathode-ray tube). A cathode-ray tube containing one electron gun producing a beam that is split to produce two traces on the screen. 190

split brush (electric machines). Either an industrial or fractional-horsepower brush consisting of two pieces that are used in place of one brush. The adjacent sides of the split brush are parallel to the commutator bars. *Note:* A split brush is normally mounted so that the plane formed by the adjacent contacting brush sides is parallel to or passes through the rotating axis of the rotor. *See:* **asynchronous machine.** *See also:* **brush; brush (rotating machinery); direct-current commutating machine.** 279

split collector ring (rotating machinery). A collector ring that can be separated into parts for mounting or removal without access to a shaft end. *See also:* **rotor (rotating machinery).** 63

split-conductor cable. A cable in which each conductor is composed of two or more insulated conductors normally connected in parallel. *See:* **segmental conductor.** 64

split-core-type current transformer. *See:* **current transformer.**

split fitting. A conduit split longitudinally so that it can be placed in position after the wires have been drawn into the conduit, the two parts being held together by screws or other means. *See also:* **raceway.** 328

split-gate tracker (radar). A form of range tracker using a pair of time gates called an early gate and a late gate, contiguous or partly overlapping in time. When tracking is established, the pair of gates straddles the received pulse that is being tracked. The position of the pair of gates then gives a measure of the time of arrival of the pulse that is, the range of the target from which the echo is received. Deviation of the pair of gates from the proper tracking position increases the signal energy in one gate and decreases it in the other, producing an error signal which moves the pair of gates so as to establish equilibrium. 13

split hub (rotating machinery). A hub that can be separated into parts for ease of mounting on removal from a shaft. *See:* **rotor (rotating machinery).** 63

split hydrophone (audio and electroacoustics). *See:* **split transducer.**

split node (network analysis). A node that has been separated into a source node and a sink node. *Notes:* (1) Splitting a node interrupts all signal transmission through that node. (2) In splitting a node, all incoming branches are associated with the resulting sink node, and all outgoing branches with the resulting source node. *See also:* **linear signal flow graphs.** 282

split-phase electric locomotive. A single-phase electric locomotive equipped with electric devices to change the single-phase power to polyphase power without complete conversion of the power supply. *See also:* **electric locomotive.** 328

split-phase motor. A single-phase induction motor having a main winding and an auxiliary winding, designed to operate with no external impedance in either winding. The auxiliary winding is energized only during the starting operation of the auxiliary-winding circuits and is open-circuited during running operation. *See:* **asynchronous machine.** 63

split projector (audio and electroacoustics). *See:* **split transducer.**

split rotor (rotating machinery). A rotor that can be separated into parts for mounting or removal without access to a shaft end. *See also:* **rotor (rotating machinery).** 63

split-sleeve bearing (rotating machinery). A journal bearing having a bearing sleeve that is split for assembly. *See also:* **bearing.** 63

split-throw winding (rotating machinery). A winding wherein the conductors that constitute one complete coil side in one slot do not all appear together in another slot. *See:* **asynchronous machine; direct-current commutating machine.** 63

split transducer (audio and electroacoustics). A directional transducer in which electroacoustic transducing elements are so divided and arranged that each division is electrically separate. 176

split-winding protection (power switchgear). A form of differential protection in which the current in all or part of the winding is compared to the normally proportional current in another part of the winding. 127, 103

spoiler resistors (power supplies). Resistors used to spoil the load regulation of regulated power supplies to permit parallel operation when not otherwise provided for. *See also:* **power supply.** 228, 186

s pole. *See:* **junction pole.**

sponge (electrodeposition). A loose cathode deposit that is fluffy and of the nature of a sponge, contrasted with a reguline metal. *See also:* **electrodeposition.** 328

spontaneous emission (laser-maser). The emission of radiation from a single electron, atom, molecule, or ion in an excited state at a rate independent of the presence of applied external fields. 363

spontaneous polarization (ferroelectric material). Magnitude of the polarization within a single ferroelectric domain in absence of an external electric field. *See also:* **polarization; remanent polarization.** 80

spool insulator. An insulating element of generally cylindrical form having an axial mounting hole and a circumferential groove or grooves for the attachment of a conductor. *See also:* **insulator; tower.** 64, 261

spot (oscilloscopes) (cathode-ray tube). The illuminated area that appears where the primary electron beam strikes the phosphor screen of a cathode-ray tube. *Note:* The effect of the impact on this small area of the screen is practically instantaneous. *See:* **cathode-ray tubes; oscillograph.** 244, 190, 185

spot-network type. A unit substation which has two step-down transformers, each connected to an incoming high-voltage circuit. The outgoing side of each transformer is connected to a common bus through circuit breakers equipped with relays which are arranged to trip the circuit breaker on reverse power flow to the transformer and to reclose the circuit breaker upon the restoration of the correct voltage, phase angle and phase sequence at the transformer secondary. The bus has one or more outgoing radial (stub end) feeders. 53

spot noise figure (transducer at a selected frequency) (spot noise factor). The ratio of the output noise power per unit bandwidth to the portion thereof attributable to the thermal noise in the input termination per unit-bandwidth, the noise temperature of the input termination being standard (290 kelvins). The spot noise figure is a point function of input frequency. *See:* **noise figure.** *See also:* **signal-to-noise ratio.** 328

spot projection (facsimile). The optical method of scanning or recording in which the scanning or recording spot is defined in the path of the reflected or transmitted light. *See also:* **scanning (facsimile); recording (facsimile).** 12

spot size. *See:* **trace width.**

spot speed (facsimile). The speed of the scanning or recording spot within the available line. *Note:* This is generally measured on the subject copy or on the record sheet. *See also:* **recording (facsimile); scanning (facsimile).** 12

spotting (electroplating). The appearance of spots on plated or finished metals. 328

spot-type fire detector (fire protection devices). A device whose detecting element is concentrated at a particular location. 71

spot wobble (television). A process wherein a scanning spot is given a small periodic motion transverse to the scanning lines at a frequency above the picture signal spectrum. *See also:* **television.** 163, 178

spreading loss (wave propagation). The reduction in radiant-power surface density due to spreading. 210

spread spectrum (communication satellite). A modulation technique for multiple access, or for increasing immunity to noise and interference. Spread spectrum systems makes use of a sequential noise-like signal structure, for example P.N. (pseudonoise) codes, to spread the normally narrowband information signal over a relatively wide band of frequencies. The receiver correlates these signals to retrieve the original information signal. 84

spring (relay). *See:* **relay spring.**

spring attachment (burglar-alarm system) (spring contact) (trap). A device designed for attachment to a movable section of the protected premises, such as a door, window, or transom, so as to carry the electric protective circuit in or out of such section, and to indicate an open- or short-circuit alarm signal upon opening of the movable section. *See also:* **protective signaling.** 328

spring barrel. The part that retains and locates the short-circuiter. *See:* **rotor (rotating machinery).** 328

spring buffer. A buffer that stores in a spring the kinetic energy of the descending car or counterweight. *See also:* **elevators.** 328

spring-buffer load rating (spring buffer) (elevators). The load required to compress the spring an amount equal to its stroke. *See also:* **elevators.** 328

spring-buffer stroke (elevators). The distance the contact end of the spring can move under a compressive load until all coils are essentially in contact. *See also:* **elevators.** 328

spring contact. An electric contact that is actuated by a spring. 328

spring-loaded bearing (rotating machinery). A ball bearing provided with a spring to ensure complete angular contact between the balls and inner and outer races, thereby removing the effect of diametral clearance in both bearings of a machine provided with ball bearing at each end. *See also:* **bearing.** 328

spring operation. Stored-energy operation by means of spring-stored energy. 103, 202

sprinkler supervisory system. A supervisory system attached to an automatic sprinkler system that initiates signal transmission automatically upon the occurrence of abnormal conditions in valve positions, air or water pressure, water temperature or level, the operability of power sources necessary to the proper functioning of the automatic sprinkler, etcetera. *See also:* **protective signaling.** 328

sprocket hole (test, measurement and diagnostic equipment).

The hole in a tape that is used for electrical timing or mechanically driving the tape. 54

spurious count (1) (nuclear techniques). A count from a scintillation counter other than (A) one purposely generated or (B) one due directly to ionizing radiation. *See also:* **scintillation counter.** 117

(2) (radiation-counter tubes). A count caused by any event other than the passage into or through the counter tube of the ionizing radiation to which it is sensitive. 96

spurious emissions (transmitter performance). Any part of the radio-frequency output that is not a component of the theoretical output, as determined by the type of modulation and specified bandwidth limitations. *See also:* **audio-frequency distortion.** 181

spurious output (nonharmonic) (signal generator). Those signals in the output of a source that have a defined amplitude and frequency and are not harmonically related to the fundamental frequency. This definition excludes sidebands due to residual and intentional modulation. *See also:* **signal generator.** 185

spurious pulse (nuclear techniques). A pulse in a scintillation counter other than (1) one purposely generated or (2) one due directly to ionizing radiation. *See also:* **scintillation counter.** 117

spurious pulse mode. An unwanted pulse mode, formed by the chance combination of two or more pulse modes, that is indistinguishable from a pulse interrogation or pulse reply. 254

spurious radiation. Any emission from a radio transmitter at frequencies outside its communication band. *See also:* **radio transmission.** 111, 64

spurious response (1) (general). Any response, other than the desired response, of an electric transducer or device. 59

(2) (mobile communication or electromagnetic compatibility). Output, from a receiver, due to a signal or signals having frequencies other than that to which the receiver is tuned. *See also:* **electromagnetic compatibility.** 181

(3) (non-real time spectrum analyzer). A characteristic of a spectrum analyzer wherein the displayed frequency does not conform to the input frequency. *Syn:* **spurii, spur.** 68

(4) (frequency-modulated mobile communications receivers). Any receiver response that occurs because of frequency conversions other than the desired frequency translations in the receiver. 123

spurious-response ratio (radio receiver). The ratio of (1) the field strength at the frequency that produces the spurious response to (2) the field strength at the desired frequency, each field being applied in turn, under specified conditions, to produce equal outputs. *Note:* Image ratio and intermediate-frequency-response ratio are special forms of spurious response ratio. *See also:* **radio receiver.** 339

spurious transmitter output. Any part of the radio-frequency output that is not implied by the type of modulation (amplitude modulation, frequency modulation, etcetera) and specified bandwidth. *See also:* **radio transmitter.** 240

spurious transmitter output, conducted. Any spurious output of a radio transmitter conducted over a tangible transmission path. *Note:* Power lines, control leads,

radio-frequency transmission lines and waveguides are all considered as tangible paths in the foregoing definition. Radiation is not considered as tangible path in the foregoing definition. Radiation is not considered a tangible path in this definition. *See also:* **radio transmitter.**
240

spurious transmitter output, extraband. Spurious output of a transmitted outside of its specified band of transmission. *See also:* **radio transmitter.** 240

spurious transmitter output, inband. Spurious output of a transmitter within its specified band of transmission. *See also:* **radio transmitter.** 240

spurious transmitter output, radiated. Any spurious output radiated from a radio transmitter. *Note:* The radio transmitter does not include the associated antenna and transmission lines. *See also:* **radio transmitter.** 240

spurious tube counts (radiation-counter tubes). Counts in radiation-counter tubes, other than background counts and those caused by the source measured. *Note:* Spurious counts are caused by failure of the quenching process, electric leakage, and the like. Spurious counts may seriously affect measurement of background counts. *See also:* **gas-filled radiation-counter tubes.** 190, 125

sputtering (electroacoustics) (cathode sputtering). A process sometimes used in the production of the metal master wherein the original is coated with an electric conducting layer by means of an electric discharge in a vacuum. *Note:* This is done prior to electroplating a heavier deposit. *See also:* **phonograph pickup.** 176

square-law detection. That form of detection in which the output voltage is substantially proportional, over the useful range of the detecting device, to the square of the voltage of the input wave. *See also:* **detection.** 328

squareness ratio (material in a symmetrically cyclically magnetized condition) (magnetic storage). The ratio of (1) the flux density at zero magnetizing force to the maximum flux density. (2) The ratio of the flux density when the magnetizing force has changed halfway from zero toward its negative limiting value, to the maximum flux density. *Note:* Both these ratios are functions of the maximum magnetizing force. 77

square wave (1) (regarded as a wave). A periodic wave that alternately for equal lengths of time assumes one of two fixed values, the time of transition being negligible in comparison. *See also:* **wavefront.** 210

(2) (regarded as a pulse train) (pulse techniques). A pulse train that has a normally rectangular shape and a nominal duty factor of 0.5. *See also:* **pulse.** 184

(3) (pulse terms). A periodic rectangular pulse train with a duty factor of 0.5 or an on-off ratio of 1.0. 254

square-wave response (camera tubes). The ratio of (1) the peak-to-peak signal amplitude given by a test pattern consisting of alternate black and white bars of equal widths to (2) the difference in signal between large-area blacks and large-area whites having the same illuminations as the black and white bars in the test pattern. *Note:* Horizontal square-wave response is measured if the bars run perpendicular to the direction of horizontal scan. Vertical square-wave response is measured if the bars run parallel to the direction of horizontal scan. *See:* **amplitude response.** 190, 125

square-wave response characteristic (camera tubes). The

relation between square-wave response and the ratio of (1) a raster dimension to (2) the bar width in the square-wave response test pattern. *Note:* Unless otherwise specified, the raster dimension is the vertical height. *See:* **amplitude response characteristic; television.** 190, 125

squaring amplifier (relaying). A circuit which produces a block. 79

squeezable waveguide (radar). A variable-width waveguide for shifting the phase of the radio-frequency wave traveling through it. *See also:* **navigation.** 13, 187

squeeze section (transmission lines and waveguides). A length of rectangular waveguide so constructed as to permit alteration of the broad dimension with a corresponding alteration in electrical length. *See also:* **waveguide.** 185

squeeze trace (electroacoustics). A variable-density sound track wherein, by means of adjustable masking of the recording light beam and simultaneous increase of the electric signal applied to the light modulator, a track having variable width with greater signal-to-noise ratio is obtained. *See also:* **phonograph pickup.** 176

squelch (radio receivers) (noun). A circuit function that acts to suppress the audio output of a receiver when noise power that exceeds a predetermined level is present. 123

squelch circuit (1) (power system communications). A circuit for preventing a radio receiver from producing audio-frequency output in the absence of a signal having predetermined characteristics. A squelch circuit may be operated by signal energy in the receiver pass band, by noise quieting, or by a combination of the two (ratio squelch). It may also be operated by a signal having special modulation characteristics (selective squelch). *See also:* **radio receiver.** 59

(2) (power switchgear). A circuit for preventing production of an unwanted output in the absence of a signal having predetermined characteristics. 79

(3) (relaying). A circuit for preventing a transducer from producing a spurious output in the absence of a signal having predetermined characteristics. 79

squelch clamping (frequency-modulated mobile communications receivers). The characteristic of the receiver, when receiving a normal signal, in which the squelch circuit under certain conditions of modulation will cause suppression of the audio output. 123

squelch selectivity (frequency-modulated mobile communications receivers). The characteristic that permits the receiver to remain squelched when a radio-frequency signal not on the receiver's tuned frequency is present at the input. 123

squelch sensitivity (1) (receiver performance). A measure of the ability of the squelch circuit to distinguish between a signal and a no-signal condition. *See also:* **receiver performance.** 181

(2) (frequency-modulated mobile communications receivers). The minimum radio-frequency signal input level, with standard test modulation required to increase the audio power output from the reference threshold squelch adjustment condition to within 6 decibels of the reference audio power output. 123

squint (radar). (1) The angle between the major lobe axis of each lobe and the central axis in a lobe-switching or simultaneous-lobing antenna. (2) The angular difference

between the axis of antenna radiation and a selected geometric axis, such as the axis of the reflector or center of the cone formed by movement of the radiation axis. (3) (verb) To position the major lobe of an antenna at an offset angle with respect to a selected geometric reference axis.
13

squint angle. A small difference in pointing angle between a reference beam direction and the direction of maximum radiation. *See also:* **antenna.**   179, 111

squirrel-cage induction motor. An induction motor in which a primary winding on one member (usually the stator) is connected to an alternating-current power source and a secondary cage winding on the other member (usually the rotor) carries alternating current produced by electromagnetic induction. *Note:* The cage winding is suitably disposed in slots in the secondary core. *See also:* **asynchronous machine.**   63

squirrel-cage rotor (rotating machinery). A rotor core assembly having a squirrel-cage winding. *See also:* **rotor (rotating machinery).**   63

squirrel-cage winding (1) (rotating machinery). A winding, usually on the rotor of a machine, consisting of a number of conducting bars having their extremities connected by metal rings or plates at each end.   63

(2) (electric installations on shipboard). A permanently short-circuited winding, usually uninsulated (chiefly used in induction machines) having its conductors uniformly distributed around the periphery of the machine and joined by continuous end rings.   3

squitter (radar). Random output pulses from a transponder caused by ambient noise, or by an intentional random triggering system, but not by the interrogation pulses. *See also:* **navigation.**   187, 13

SSB. *See:* **single-sideband modulation.**

stability (1) (general). The property of a system or element by virtue of which its output will ultimately attain a steady state. *See also:* **control system, feedback.**   219, 206

(2) (circuits and systems). An aspect of system behavior associated with systems having the general property that bounded input perturbations result in bounded output perturbations. *Notes:* (A) A stable system will ultimately attain a steady state. (B) Deviations from this steady state due to component aging or environmental changes do not indicate instability, but a change in the system. *See also:* **steady-state stability; transient stability.**   67

(3) (perturbations). For convenience in defining various stability concepts, only those parameters or signals that are perturbed are explicitly exhibited, or mentioned, that is, for perturbations in initial states, a perturbed solution is denoted

$$\varphi(\mathbf{x}(t_0) + \Delta\mathbf{x}(t_0);t),$$

where $\Delta\mathbf{x}(t_0)$ represents the perturbation in initial state. Finally, the perturbed-state solution is denoted

$$\Delta\varphi = \varphi(\mathbf{x}(t_0) + \Delta\mathbf{x}(t_0);t) - \varphi(\mathbf{x}(t_0);t).$$
329

(4) (power system stability). In a system of two or more synchronous machines connected through an electric network, the condition in which the difference of the angular positions of the rotors of the machines either remains constant while not subjected to a disturbance, or becomes constant following an aperiodic disturbance. *Note:* If automatic devices are used to aid stability, their use will modify the steady-state and transient stability terms to: **steady-state stability with automatic devices; transient stability with automatic devices.** Automatic devices as defined for this purpose are those devices that are operating to increase stability during the period preceding and following a disturbance as well as during the disturbance. Thus relays and circuit breakers are excluded from this classification and all forms of voltage regulators included. Devices for inserting and removing shunt or series impedance may or may not come within this classification depending upon whether or not they are operating during the periods preceding and following the disturbance. *See:* **steady-state stability; transient stability.** *See also:* **alternating-current distribution.**   64

(5) (electric machines). The change in output voltage or current as a function of time, at constant line voltage, load, and ambient temperature (sometimes referred to as drift).
186

(6) (oscilloscopes). The property of retaining defined electrical characteristics for a prescribed time and environment. *Notes:* (A) Deviations from a stable state may be called drift if it is slow, or jitter or noise if it is fast. In triggered-sweep systems, triggering stability may refer to the ability of the trigger and sweep systems to maintain jitter-free displays of high-frequency waveforms for long (seconds to hours) periods of time. (B) Also, the name of the control used on some oscilloscopes to adjust the sweep for triggered, free-running, or synchronized operation. *See also:* **sweep mode control.**   106

(7) (speed governing of hydraulic turbines). Characteristics of the governing system pertaining to limitation of oscillations of speed or power under sustained conditions, to damping of oscillations of speed following rejection of load, and to damping of speed oscillations under isolated load conditions following sudden load changes.   8, 58

(8) (electrothermic power meters). For a constant input rf power, constant ambient temperature and constant power line voltage, the variation in rf power indication over stated time intervals. *Note:* Long term stability or drift is the maximum acceptable change in 1 hour. Short term stability or fluctuation is the maximum (peak) change in 1 minute.   47

(9) (gyro; accelerometer). A measure of the ability of a specific mechanism or performance coefficient to remain invariant when continuously exposed to a fixed operating condition. (This definition does not refer to dynamic or servo stability).   46

(10) (nuclear power generating station). The ability of a module to attain and maintain a steady state.   355

stability, absolute (control system). Global asymptotic stability maintained for all nonlinearities within a given class. *Note:* A typical problem to which the concept of absolute stability has been applied consists of a system with dynamics described by the vector differential equation

$$\dot{\mathbf{x}} = A\mathbf{x} + \mathbf{b}f(\sigma),$$

$$\sigma = \mathbf{c}^\tau\mathbf{x},$$

with a nonlinearity class defined by the conditions

$$f(0) = 0,$$

$$k_1 \leqslant f(\sigma)/\sigma \leqslant k_2.$$

The solution $\mathbf{x}(t) = \mathbf{0}$ is said to be absolutely stable if it is globally asymptotically stable for all nonlinear functions $f(\sigma)$ in the above class. *See also:* **control system.**   329

stability, asymptotic (1) (control system) (of a solution $\phi(\mathbf{x}(t_0);t))$. The solution is (1) Lyapunov stable, (2) such that

$$\lim_{t \to \infty} \|\Delta\phi\| = 0,$$

where $\Delta\phi$ is a change in the solution due to an initial state perturbation. See **stability** for explanation of symbols. *Notes:* (A) The solution $\mathbf{x} = 0$ of the system $\dot{\mathbf{x}} = \mathbf{a}\mathbf{x}$ is asymptotically stable for $\mathbf{a} < 0$, but not for $\mathbf{a} = 0$. In this case

$$\varphi(\mathbf{x}(t_0);t) = \mathbf{x}(t_0) \exp(-\mathbf{a}(t - t_0)).$$

(B) In some cases the rate of convergence to zero depends on both the initial state $\mathbf{x}(t_0)$ and the initial time t_0. See stability, equiasymptotic for stability concepts where this rate of convergence is independent of either $\mathbf{x}(t_0)$ or t_0. *See also:* **control system.**   329

(2) (excitation control systems). A system exhibits asymptotic stability if the output approaches a constant value within an arbitrarily small tolerance after a sufficiently long period of time. *Note:* Asymptotic stability generally refers to systems which are excited by a constant signal. The output approaches a fixed steady state level.   353

stability, bounded-input-bounded-output (1) (control system). Driven stability when the solution of interest is the output solution. *See also:* **control system.**   329

(2) (excitation control systems). A system exhibits bounded input-bounded output (BIBO) stability if the output is bounded for every bounded input. *Note:* BIBO stability is also known as stability in the sense of Lyapunov and it refers to forced systems. In non-linear systems, a bounded limit cycle appearing in the output signal is an example of BIBO stability.   353

stability, conditional (linear feedback control system). A property such that the system is stable for prescribed operating values of the frequency-invariant factor of the loop gain and becomes unstable not only for higher values, but also for some lower values. *See also:* **control system, feedback.**   56, 353

stability, driven (control system) (solution $\phi(\mathbf{u};t))$. For each bounded system input perturbation $\Delta\mathbf{u}(t)$ the output perturbation $\Delta\phi$ is also bounded for $t \geq t_0$. *Note:* A necessary and sufficient condition for a solution of a linear system to be driven-stable is that the solution be uniformly asymptotically stable. See **stability** for explanation of symbols. *See also:* **control system.**   329

stability, equiasymptotic (control system). Asymptotic stability where the rate of convergence to zero of the perturbed-state solution is independent of all initial states in some region $\|\Delta\mathbf{x}(t_0)\| \leq v$. *See also:* **control system.**   329

stability, excitation-system (synchronous machines). The

ability of the excitation system to control the field voltage of the principal electric machine so that transient changes in the regulated voltage are effectively suppressed and sustained oscillations in the regulated voltage are not produced by the excitation system during steady-load conditions or following a change to a new steady-load condition. *Note:* It should be recognized that under some system conditions it may be necessary to use power system stabilizing signals as additional inputs to excitation control systems to achieve stability of the power system including the excitation system.   105

stability factor. The ratio of a stability limit (power limit) to the nominal power flow at the point of the system to which the stability limit is referred. *Note:* In determining stability factors it is essential that the nominal power flow be specified in accordance with the one of several bases of computation, such as rating or capacity of, or average or maximum load carried by, the equipment or the circuits. *See also:* **alternating-current distribution.**   64

stability, finite-time (control system) (solutions). For all initial states that originate in a specified region R at time t_0, the resulting solutions remain in another specified region R_ϵ over the given time interval $t_0 \leq t \leq T$. *Notes:* (1) In the definition of finite-time stability the quantities R_π, R_ϵ, and T are prespecified. Obviously, R must be included in R_ϵ. (2) A system may be Lyapunov unstable and still be finite-time stable. For example, a system with dynamics $\mathbf{x} = \mathbf{a}\mathbf{x}$, $\mathbf{a} > 0$, is Lyapunov unstable, but if

$$R_\delta: |\mathbf{x}| \leq \delta,$$

$$R_\epsilon: |\mathbf{x}| \leq \epsilon,$$

and $T < \mathbf{a}^{-1} \ln(\epsilon/\delta)$, the system is finite-time stable (relative to the given values of δ, ϵ, and T). *See also:* **control system.**   329

stability, global (control system) (solution $\phi(\mathbf{x}(t_0);t))$. Stable for all initial perturbations, no matter how large they may be. *See also:* **control system.**   329

stability in-the-whole (control system). Synonymous with **global stability.** *See:* **stability, global.** *See also:* **control system.**   329

stability, Lagrange (system) (control system). Every solution that is generated by a finite initial state is bounded. *Note:* An example of a system that is Lagrange stable is a second-order system with a single stable limit cycle. Although this system must contain a point inside the limit cycle that is Lyapunov unstable, the system is still Lagrange stable because every solution remains bounded. *See also:* **control system.**   329

stability limit (1) (general). A condition of a linear system or one of its parameters that places the system on the verge of instability.   56

(2) (power limit) (electric systems). The maximum power flow possible through some particular point in the system when the entire system or the part of the system to which the stability limit refers is operating with stability. *See also:* **alternating-current distribution.**   64

stability, long-term (LTS) (power supplies). The change in output voltage or current as a function of time, at constant line voltage, load, and ambient temperature (sometimes referred to as drift). *See also:* **power supply.**   228, 186

stability, Lyapunov (control system) (of a solution $\phi(\mathbf{x}(t_0);t))$.

For every given $\epsilon > 0$ there exists a $\delta > 0$ (which, in general, may depend on ϵ and on t_0) such that $\|\Delta x(t_0)\| \leq \delta$ implies $\|\Delta\phi\| \leq \epsilon$ for $t \geq t_0$. *Notes:* (1) The solution $\mathbf{x} = 0$ of the system $\dot{\mathbf{x}} = a\mathbf{x}$ is Lyapunov stable if $\mathbf{a} < 0$ and is Lyapunov unstable if $\mathbf{a} > 0$. (2) For a linear system with an irreducible transfer function $T(s)$, Lyapunov stability implies that all the poles of $T(s)$ are in the left-half s plane and that those on the $j\omega$ axis are simple. *See also:* **control system.** 329

stability of a limit cycle (control system). Synonymous with orbital stability. *See also:* **control system.** 329

stability of the speed-governing system (gas turbines). A characteristic of the system that indicates that the speed-governing system is capable of actuating the turbine fuel-control valve so that sustained oscillations in turbine speed, or rate of energy input to the turbine, are limited to acceptable values by the speed-governing system. 98, 58

stability of the temperature-control system (gas turbines). A characteristic of the system that indicates that the temperature-control system is capable of actuating the turbine fuel-control valve so that sustained oscillations in rate of energy input to the turbine are limited to acceptable values by the temperature-control system during operation under constant system frequency. 98, 58

stability, orbital (control system) (closed solution curve denoted Γ). Implies that for every given $\epsilon > 0$ there exists a $\delta > 0$ (which, in general, may depend on ϵ and on t_0) such that $\rho(\Gamma, \mathbf{x}(t_0)) \leq \delta$ implies $\rho(\Gamma, \phi(\mathbf{x}(t_0); t)) \leq \epsilon$ for $t \geq t_0$, where $\rho(\Gamma, \mathbf{a})$ denotes the minimum distance between the curve Γ and the point $\mathbf{a}$. Here the point $\mathbf{x}(t_0)$ is assumed to be off the curve Γ. *Notes:* (1) Orbital stability does not imply Lyapunov stability of a closed solution curve, since a point on the closed curve may not travel at the same speed as a neighboring point off the curve. (2) Only non-linear systems can produce the type of solutions for which the concept of orbital stability is applicable. *See also:* **control system.** 329

stability, practical (control system). Synonymous with **finite-time stability.** *See:* **stability, finite-time.** *See also:* **control system.** 329

stability, quasi-asymptotic (control system) (solution $\phi\mathbf{x}(\mathbf{x}(t_0); t)$). Implies

$$\lim_{t \to \infty} \|\Delta\phi\| = 0.$$

Notes: (1) Quasi-asymptotic stability is condition (2) in the definition of asymptotic stability and, hence, need not imply Lyapunov stability. (2) An example of a solution that is quasi-asymptotically stable but not asymptotically stable is the solution $\mathbf{x}(t) = 0$ of the system $\dot{\mathbf{x}} = \mathbf{x}^2$. The solution of the above system for a perturbation $\Delta x(t_0)$ from 0 is

$$\rho(\Delta\mathbf{x}(t_0); t) = \Delta\mathbf{x}(t_0)/[1 - (t - t_0)\Delta\mathbf{x}(t_0)].$$

Obviously, $\phi(\Delta\mathbf{x}(t_0); t)$ approaches zero as t approaches ∞, yet is not Lyapunov stable since it is unbounded at $t = t_0 + (1/\Delta\mathbf{x}(t_0))$. *See also:* **control system.** 329

stability, relative (automatic control) (stable underdamped system). The property measured by the relative setting times when parameters are changed. *See also:* **control**

system, feedback. 329

stability, short-time (control system). Synonymous with **finite-time stability.** *See:* **stability, finite-time.** *See also:* **control system.** 329

stability, synchronous-machine regulating-system. The property of a synchronous-machine-regulating system in which a change in the controlled variable, resulting from a stimulus, decays with time if the stimulus is removed. 63

stability, total (control system) (solution $\phi = \phi(\mathbf{x}(t_0); t)$ of the system $\dot{\mathbf{x}} = \mathbf{f}(\mathbf{x}, t)$). Implies that for every given $\epsilon > 0$ there exist a $\delta_1 > 0$ and a $\delta_2 > 0$ (both of which, in general, may depend on ϵ and t_0) such that $\|\Delta\mathbf{x}(t_0)\| \leq \delta_1$ and $\|\mathbf{g}(\mathbf{x}, t)\| \leq \delta_2$ imply $\|\phi - \Psi\| \leq \epsilon$ for $t \geq t_0$, where $\Psi = \Psi(\mathbf{x}(t_0) + \Delta\mathbf{x}(t_0); t)$ is a solution of the system $\dot{\mathbf{x}} = \mathbf{f}(\mathbf{x}, t) + \mathbf{g}(\mathbf{x}, t)$. *See also:* **control system.** 329

stability, trajectory (control system). Orbital stability where the solution curve is not closed. *See also:* **control system.** 329

stability, uniform-asymptotic (control system). Asymptotic stability where the rate of convergence to zero of the perturbed-state solution is independent of the initial time t_0. *Note:* An example of a solution that is asymptotically stable but not uniformly asymptotically stable is the solution

$$\varphi(\mathbf{x}(t_0); t) = \mathbf{x}(t_0)t_0/t$$

of the system $\dot{\mathbf{x}} = -\mathbf{x}/t$, $t_0 > 0$. Note that the initial rate of decay,

$$\dot{\mathbf{x}}(t_0)/\mathbf{x}(t_0) = -1/t_0,$$

is clearly a function of t_0. Compare with the time-invariant system $\dot{\mathbf{x}} = a\mathbf{x}$ where $\dot{\mathbf{x}}(t_0)/\mathbf{x}(t_0) = \mathbf{a}$ is independent of t_0. The concept of uniformity with respect to the initial time t_0 applies only to time-varying systems. All stable time-invariant systems are uniformly stable. *See also:* **control system.** 329

stabilization (1) (control system, feedback). Act of attaining stability or of improving relative stability. *See also:* **control system, feedback.** 56

(2) (navigation). Maintenance of a desired orientation of a vehicle or device with respect to one or more reference directions. *See also:* **navigation.** 187, 13

(3) (direct-current amplifier). *See:* **drift stabilization.**

stabilization, drift. *See:* **drift stabilization.**

stabilization network (analog computer). As applied to operational amplifiers, a network used to shape the transfer characteristics to eliminate or minimize oscillations when feedback is provided. 9

stabilized feedback. Feedback employed in such a manner as to stabilize the gain of a transmission system or section thereof with respect to time or frequency or to reduce noise or distortion arising therein. *Note:* The section of the transmission system may include amplifiers only, or it may include modulators. *See also:* **feedback.** 111

stabilized flight. The type of flight that obtains control information from devices that sense orientation with respect to external references. *See also:* **navigation.** 187, 113

stabilized shunt-wound generator. A type of compound-wound generator with a series field winding of such proportion and polarity as to minimize voltage regulation or provide sufficient droop so that machines may be operated in parallel without equalizers. *Note:* The voltage regulation of such a generator is always drooping; that is, the voltage at rated load is less than at no load. This definition is not intended to apply to stability of machines operating through a voltage range on a straight-line portion of the saturation curve. *See also:* **direct-current commutating machine.** 3

stabilized shunt-wound motor. A shunt-wound motor having a light series winding added to prevent a rise in speed or to obtain a slight reduction in speed, with increase of load. *See also:* **direct-current commutating machine.** 3

stabilizer, excitation control system (synchronous machines). An element or group of elements that modify the forward signal by either series or feedback compensation to improve the dynamic performance of the excitation control system. 105

stabilizer, power system (synchronous machines). An element or group of elements that provide an additional input to the regulator to improve power system dynamic performance. *Note:* A number of different quantities may be used as input to the power system stabilizer, such as shaft speed, frequency, synchronous machine electrical power, and others. 105

stabilizing winding. A delta connected auxiliary winding used particularly in Y-connected transformers for such purposes as the following: (1) to stabilize the neutral point of the fundamental frequency voltages; (2) to minimize third harmonic voltage and the resultant effects on the system; (3) to mitigate telephone influence due to third harmonic currents and voltages; (4) to minimize the residual direct-current magnetomotive force on the core; (5) to decrease the zero-sequence impedance of transformers with Y-connected windings. *Note:* A winding is regarded as a stabilizing winding if its terminals are not brought out for connection to an external circuit. However, one or two points of the winding which are intended to form the same corner point of the delta may be brought out for grounding, or grounded internally to the tank. For a three-phase transformer, if other points of the winding are brought out, the winding should be regarded as a normal winding as otherwise defined. 53

stable (1) (control system, feedback). Possessing stability. *See also:* **control system, feedback.** 56
(2) (excitation control systems). Possessing stability, where, for a feedback control system or element, stability is the property such that its output is asymptotic, that is, will ultimately attain a steady-state, within the linear range and without continuing external stimuli. For certain nonlinear systems or elements, the property such that the output remains bounded, that is, in a limit cycle of continued oscillation, when the input is bounded. *See also:* **stability, asymptotic; stability, bounded input—bounded output; stability conditional.** 353

stable element (electronic navigation). An instrument or device that maintains a desired orientation independently of the motion of the vehicle. *See also:* **navigation.** 13, 187

stable oscillation. A response that does not increase indefinitely with increasing time; an unstable oscillation is

the converse. *Note:* The response must be specified or understood; a steady current in a pure resistance network would be stable, although the total charge passing any cross section of a network conductor would be increasing continuously. 210

stable platform (electronic navigation). A gimbal-mounted platform, usually containing gyros and accelerometers, whose purpose is to maintain a desired orientation in inertial space independent of the motion of the vehicle. *See also:* **navigation.** 187, 13

stack, insulator. A rigid assembly of two or more switch and bus insulating units. *See also:* **tower.** 261, 64

stacked-beam radar. A radar that forms two or more simultaneous beams at the same azimuth but at different elevation angles. The beams are usually contiguous or partly overlapping. Each stacked beam feeds an independent receiver. 13

stacker (computing systems). *See:* **card stacker.**

stage (1) (communication practice). One step, especially if part of a multistep process, or the apparatus employed in such a step. The term is usually applied to an amplifier. *See also:* **amplifier.** 328
(2) (thermoelectric device). One thermoelectric couple or two or more similar thermoelectric couples arranged thermally in parallel and electrically connected. *See also:* **thermoelectric device.** 248, 191

stage efficiency. The ratio of useful power delivered to the load (alternating current) and the plate power input (direct current). *See also:* **network analysis.** 111

stagger (facsimile). Periodic error in the position of the recorded spot along the recorded line. *See also:* **recording (facsimile).** 12

staggered-repetition-interval moving-target indicator (radar). A moving-target indicator with multiple interpulse intervals. The interval may vary either from pulse-to-pulse or from scan-to-scan. *See:* **moving-target indicator.** 13

staggering. The offsetting of two channels of different carrier systems from exact sideband frequency coincidence in order to avoid mutual interference. 328

staggering advantage. The effective reduction, in decibels, of interference between carrier channels, due to staggering. 328

stagger time, relay. *See:* **relay stagger time.**

stagger-tuned amplifier. An amplifier consisting of two or more single-tuned stages that are tuned to different frequencies. *See also:* **amplifier.** 328

stain spots (electroplating). Spots produced by exudation, from pores in the metal, of compounds absorbed from cleaning, pickling plating solutions. The appearance of stain spots is called spotting out. 328

staircase (pulse terms). Unless otherwise specified, a periodic and finite sequence of steps of equal magnitude and of the same polarity. 254

staircase signal (television). A waveform consisting of a series of discrete steps resembling a staircase. 163

stairstep sweep (oscilloscopes). An incremental sweep in which each step is equal. The electric deflection waveform producing a stairstep sweep is usually called a staircase or stairstep waveform. *See:* **incremental sweep.** *See also:* **oscillograph.** 185

stalled tension control (industrial control). A control function that maintains tension in the material at zero speed. *See also:* **electric drive.** 225, 206

stalled torque control (industrial control). A control function that provides for the control of the drive torque at zero speed. *See also:* **electric drive.** 225, 206

stalo (radar MTI). Stable local oscillator, a highly stable local RF (radio-frequency) oscillator used for heterodyning signals to produce an intermediate frequency. *See also:* **moving target indicator.** 13

stamper (electroacoustics). A negative (generally made of metal by electroforming) from which finished pressings are molded. *See also:* **phonograph pickup.** 176

standard (1) (transmission lines and waveguides). A device having stable, precisely defined characteristics that may be used as a reference. 185
(2) (test, measurement and diagnostic equipment). A laboratory type device which is used to maintain continuity of value in the units of measurement by periodic comparison with higher echelon or national standards. They may be used to calibrate a standard of lesser accuracy or to calibrate test and measurement equipment directly. 54

standard cable*. The standard cable formerly used for specifying transmission losses had, in American practice, a linear series resistance and linear shunt capacitance of 88 ohms and 0.054 microfarad, respectively, per loop mile, with no inductance or shunt conductance. 328

* Obsolete

standard cell. A cell that serves as a standard of electromotive force. *See:* **Weston normal cell; unsaturated standard cell.** *See also:* **auxiliary device to an instrument; electrochemistry.** 328

standard code. The operating, block signal, and interlocking rules of the Association of American Railroads. 328

standard compass. A magnetic compass so located that the effect of the magnetic mass of the vessel and other factors that may influence compass indication is the least practicable. 328

standard connector (fixed and variable attenuators). A connector, the critical mating dimensions of which have been standardized to assure nondestructive mating. *Notes:* (1) It butts against its mating standard connector only in the mechanical reference plane. (2) It joins to its waveguide or transmission line with a minimum discontinuity. (3) All its discontinuities are to the maximum extent possible, self-compensated, not within the mating connector. 110

standard connector pair (fixed and variable attenuators). Two standard connectors designed to mate with each other. 110

standard de-emphasis characteristic (frequency modulation). A falling response with modulation frequency, complementary to the standard pre-emphasis characteristic and equivalent to an RC circuit with a time constant of 75 μs. *Note:* The de-emphasis characteristic is usually incorporated in the audio circuits of the receiver. 16

standard-dimensioned motor. A general or definite-application motor so dimensioned that it is mechanically interchangeable as a whole with any other motor of the same frame size and complying with the same standard specification. *See also:* **asynchronous machine; direct-**

current commutating machine. 63

standard electrode potential. An equilibrium potential for an electrode in contact with an electrolyte, in which all of the components of a specified electrochemical reaction are in their standard states. The standard state for a gas is the pressure of one atmosphere, for an ionic constituent it is unit ion activity, and it is a constant for a solid. *See also:* **electrochemistry.** 328

standard frequency (electric power systems). A precise frequency intended to be used for a frequency reference. *Note:* In the U.S. a frequency of 60 Hz is recognized as a standard for all ac lighting and power systems. 3, 94

standard full impulse voltage wave (1) (insulation strength). An impulse that rises to crest value of voltage in 1.2 microseconds (virtual time) and drops to 0.5 crest value of voltage in 50 microseconds (virtual time), both times being measured from the same origin and in accordance with established standards of impulse testing techniques. *Note:* The virtual value for the duration of the wavefront is 1.67 times the time taken by the voltage to increase from 30 percent to 90 percent of its crest value. The origin from which time is measured is the intersection with the zero axis of a straight line drawn through points on the front of the voltage wave at 30-percent and 90-percent crest value. 276
(2) (mercury lamp transformers). An impulse that rises to crest value of voltage in 1.5 microseconds (nominal time) and drops to 0.5 crest value of voltage in 40 microseconds (nominal time), both times being measured from the same time origin and in accordance with established standards of impulse testing techniques. *See also:* **basic impulse insulation level (BIL) (insulation strength).** 274

standard insulation class (instrument transformer). Denotes the maximum voltage in kilovolts that the insulation of the primary winding is designed to withstand continuously. *See also:* **instrument transformer.** 212

standardization. *See:* **echelon; interlaboratory working standards; laboratory reference standards.**

standardize. *See:* **check (instrument or meter).**

standard lightning impulse. An impulse that rises to crest value of voltage in 1.2 microseconds (virtual time) and drops to 0.5 crest value of voltage in 50 microseconds (virtual time), both times being measured from the same origin and in accordance with established standards of impulse testing techniques. It is described as a 1.2/50 impulse. *See:* **C68.1.** *Note:* The virtual value for the duration of the wavefront is 1.67 times the time taken by the voltage to increase from 30 percent to 90 percent of its crest value. The origin from which time is measured is the intersection with the zero axis of a straight line drawn through points on the front of the voltage wave at 30 percent and 90 percent crest value. 53

standard microphone. A microphone the response of which is accurately known for the condition under which it is to be used. *See also:* **instrument.** 328

standard noise temperature (interference terminology). The temperature used in evaluating signal transmission systems for noise factor 290 kelvins (27 degrees Celsius). *See also:* **interference.** 188

standard observer (color) (CIE 1931). Receptor of radiation whose colorimetric characteristics correspond to the distribution coefficients $\bar{x}_\lambda$, $\bar{y}_\lambda$, $\bar{z}_\lambda$ adopted by the International Commission on Illumination in 1931. 18

standard operating duty. *See:* **operating duty.**

standard pitch. *See:* **standard tuning frequency.**

standard potential (standard electrode potential). The reversible potential for an electrode process when all products and reactants are at unit activity on a scale in which the potential for the standard hydrogen half-cell is zero.
221, 205

standard propagation (radio waves). The propagation of radio waves over a smooth spherical earth of specific uniform dielectric constant and conductivity, under conditions of standard refraction in the atmosphere. *See also:* **radiation; radio wave propagation.** 146, 180, 59

standard reference position (contact). The nonoperated or de-energized position of the associated main device to which the contact position is referred. *Note:* Standard reference positions of typical devices are listed in 9.44 of American National Standard Manual and Automatic Station Control Supervisory and Associated Telemetering Equipments, C37.2-1970, for example: 103

Device	Standard Reference Position
Circuit breaker	Main contacts open
Disconnecting switch	Main contacts open
Relay	De-energized position
Contactor	De-energized position
Valve	Closed position

103, 202

standard refraction (radio waves). The refraction that would occur in an idealized atmosphere in which the refractive index decreases uniformly with height above the earth at the rate of 39×10^{-6} per meter. *Note:* Standard refraction may be included in ground wave calculations by use of an effective earth radius of 8.5×10^{6} meters, or 4/3 the geometrical radius of the earth. Refraction exceeding standard refraction is called superrefraction, and refraction less than standard refraction is called subrefraction. *See also:* **radiation.** 180, 59, 146

standard register (motor meter) (dial register). A four- or five-dial register, each dial of which is divided into ten equal parts, the division marks being numbered from zero to nine, and the gearing between the dial pointers being such that the relative movements of the adjacent dial pointers are in opposite directions and in a 10-to-1 ratio. *See also:* **watthour meter.** 328

standard resistor (resistance standard). A resistor that is adjusted with high accuracy to a specified value, is but slightly affected by variations in temperature, and is substantially constant over long periods of time. *See also:* **auxiliary device to an instrument.** 328

standard rod gap. A gap between the ends of the two one-half-inch square rods cut off squarely and mounted on supports so that a length of rod equal to or greater than one-half the gap spacing overhangs the inner edge of each support. It is intended to be used for the approximate measurement of crest voltages. *See also:* **instrument.** 328

standard source (illuminating engineering). One that has a specified spectral distribution and is used as a standard for colorimetry. *Note:* In 1931 the International Commission on Illumination (CIE) specified the spectral energy distributions for three standard sources, *A*, *B* and *C*. *See:* **color; standard source *A*; standard source *B*; standard source *C*.** 167

standard source *A*. A tungsten filament lamp operated at a color temperature of 2854 degrees Celsius, and approximates a blackbody operating at that temperature. *See also:* **standard source (illuminating engineering).** 167

standard source *B*. An approximation of noon sunlight having a correlated color temperature of approximately 4870 degrees Celsius. It is obtained by a combination of Source A and a special filter. *See also:* **standard source (illuminating engineering).** 167

standard source *C*. An approximation of daylight provided by a combination of direct sunlight and clear sky having a correlated color temperature of approximately 6770 degrees Celsius. It is obtained by a combination of Source *A* and a special filter. *See also:* **standard source (illuminating engineering).** 167

standard sphere gap. A gap between two metal spheres of standard dimensions, mounted and used in a specified manner. It is intended to be used for the measurement of the crest value of a voltage by observing the maximum gap spacing at which sparkover occurs when the voltage is applied under known atmospheric conditions. *See also:* **instrument.** 328

standard switching impulse. A full impulse having a front time of 250 microseconds and a time to half value of 2500 microseconds. It is described as a 250/2500 impulse. *Note:* It is recognized that some apparatus standards may have to use a modified wave shape where practical test considerations or particular dielectric strength characteristics make some modification imperative. Transformers, for example, use a modified switching impulse wave with the following characteristics: (1) Time to crest greater than 100 microseconds. (2) Exceeds 90 percent of crest value for at least 200 microseconds. (3) Time to first voltage zero on tail not less than 1000 microseconds, except where core saturation causes the tail to become shorter. (See C57.12.90) 53

standard systems (electric installations on shipboard). The following systems of distribution are recognized as standard: (1) Two-wire with single-phase alternating current, or direct current. (2) Three-wire with single-phase alternating current, or direct current. (3) Three-phase three-wire, alternating current. (4) Three-phase, four-wire, alternating current. 3

standard television signal. A signal that conforms to certain accepted specifications. *See also:* **television.** 328

standard test modulation (frequency-modulated mobile communications receivers). Sixty percent of the rated system deviation at a frequency of 1 kilohertz. 123

standard test problem (test, measurement and diagnostic equipment). An evaluation of the performance of a system, or any part of it, conducted by setting parameters into the system; the parameters are operated on and the results obtained from system readouts. 54

standard test tone (data transmission). A 1 mw (0 dbm) 1000 hertz signal applied to the 600 ohm audio portion of a circuit at a zero transmission level reference point. If referred to a point with a relative level other than 0, the absolute power of the tone shall be adjusted to suit the relative level at the point of application. 59

standard track (electroacoustics). *See:* **single track (electroacoustics).**

standard tuning frequency (standard musical pitch). The frequency for the note A_4, namely, 440 hertz. See the accompanying table. 176

standard voltages (electric installations on shipboard). The following voltages are recognized as standard:

	Alternating Current (volts)	Direct Current (volts)
Lighting	115	115
Power	115-200-220-440	115 and 230
Generators	120-208-230-450	120 and 240

Note: Satisfactory to use 120 V lamps. 3

standard volume indicator (electroacoustics). *See:* **volume indicator, standard.** *See also:* **instruments.**

standard-wave error (direction-finder measurements). The bearing error produced by a wave whose vertically and horizontally polarized electric fields are equal and phased so as to give maximum error in the direction finder, and whose incidence direction is arranged to be 45 degrees. *See also:* **navigation.** 187, 13

standby. *See:* **reserve equipment.**

standby power supply (nuclear power generating stations) (Class 1E power systems). The power supply that is selected to furnish electric energy when the preferred power supply is not available. 102, 31

standby redundancy (reliability). *See:* **redundancy, standby.**

standing-on-nines carry (parallel addition of decimal numbers). A high-speed carry in which a carry input to a given digit place is bypassed to the next digit place if the current sum in the given place is nine. *See also:* **carry.** 255, 77

standing wave. A wave in which, for any component of the field, the ratio of its instantaneous value at one point to that at any other point does not vary with time. *Notes:* (1) Commonly a periodic wave in which the amplitude of the displacement in the medium is a periodic function of the distance in the direction of any line of propagation of the wave. (2) A standing wave results when the electric variation in a circuit takes the form of periodic exchange of energy between current and potential forms without translation of energy. (3) Also termed stationary wave. *See also:* **radio; radio wave propagation; transmission line; waveguide.** 267, 179, 185, 146, 210, 64

standing-wave detector. *See:* **standing-wave meter.**

standing-wave indicator. *See:* **standing-wave meter.**

standing-wave loss factor. The ratio of the transmission loss in an unmatched waveguide to that in the same waveguide when matched. *See also:* **waveguide.** 267

standing-wave machine. *See:* **standing-wave meter.**

standing-wave meter (standing-wave indicator) (standing-wave detector) (standing-wave machine). An instrument for measuring the standing-wave ratio in a transmission line. In addition a standing-wave meter may include means for finding the location of maximum and minimum amplitudes. See table on facing page. *See also:* **instrument.** 328

standing-wave ratio (1) (general). The ratio of the amplitude of a standing wave at an antinode to the amplitude at a

node. *Note:* The standing-wave ratio in a uniform transmission line is

$$\frac{1+p}{1-p}$$

where p is the reflection coefficient. *See also:* **wave front.** 265

(2) (voltage-standing-wave ratio*) (at a given frequency in a uniform waveguide or transmission line). The ratio of the maximum to the minimum amplitudes of corresponding components of the field (or the voltage or current) appearing along the guide or line in the direction of propagation. *Notation:* SWR (VSWR*). *Note:* Alternatively, the standing-wave ratio may be expressed as the reciprocal of the ratio defined above. 267, 319, 179, 185

* Deprecated

(3) (at a port of a multiport network). The standing-wave ratio (SWR) when reflectionless terminations are connected to all other ports.

standing-wave-ratio indicator (standing-wave-ratio meter). A device or part thereof used to indicate the standing-wave ratio. *Note:* In common terminology, it is the combination of amplifier and meter as a supplement to the slotted line or bridge, etcetera, when performing impedance or reflection measurements. 185

stand, reel. *See:* **reel stand.**

standstill locking (rotating machinery). The occurrence of zero or unusably small torque in an energized polyphase induction motor, at standstill, for certain rotor positions. *See also:* **asynchronous machine.** 63

star chain (electronic navigation). A group of navigational radio transmitting stations comprising a master station about which three or more slave stations are symmetrically located. *See also:* **radio navigation.** 13, 187

star-connected circuit. A polyphase circuit in which all the current paths of the circuit extend from a terminal of entry to a common terminal or conductor (which may be the neutral conductor). *Note:* In a three-phase system this is sometimes called a Y (or wye) connection. *See:* **star network.** *See also:* **network analysis.** 210

star connection. *See:* **Y connection.**

star-delta starter (industrial control). A switch for starting a three-phase motor by connecting its windings first in star and then in delta. *See also:* **starter.** 244, 206

star-delta starting. The process of starting a three-phase motor by connecting it to the supply with the primary winding initially connected in star, then reconnected in delta for running operation. 63

star network. A set of three or more branches with one-terminal of each connected at a common node. *See also:* **network analysis.** 210

star quad. *See:* **spiral four.**

star rectifier circuit. A circuit that employs six or more rectifying elements with a conducting period of 60 electrical degrees plus the commutating angle. *See also:* **rectification.** 328

start-dialing signal (semiautomatic or automatic working) (telecommunication). A signal transmitted from the incoming end of a circuit, following the receipt of a seizing signal, to indicate that the necessary circuit conditions

Frequencies and Frequency Levels of the Usual Equally Tempered Scale; Based on $A_4 = 440$ Hz

Note Name	Frequency Level (semits)	Frequency (hertz)	Note Name	Frequency Level (semits)	Frequency (hertz)	Note Name	Frequency Level (semits)	Frequency (hertz)
C_0	0	16.352	E_3	40	164.81	A_6	81	1760.0
	1	17.324	F_3	41	174.61		82	1864.7
D_0	2	18.354		42	185.00	B_6	83	1975.5
	3	19.445	G_3	43	196.00			
E_0	4	20.602		44	207.65			
F_0	5	21.827	A_3	45	220.00	C_7	84	2093.0
	6	23.125		46	233.08		85	2217.5
G_0	7	24.500	B_3	47	246.94	D_7	86	2349.3
	8	25.957					87	2489.0
A_0	9	27.500				E_7	88	2637.0
	10	29.135	C_4	48	261.63	F_7	89	2793.8
B_0	11	30.868		49	277.18		90	2960.0
			D_4	50	293.66	G_7	91	3136.0
				51	311.13		92	3322.4
C_1	12	32.703	E_4	52	329.63	A_7	93	3520.0
	13	34.648	F_4	53	349.23		94	3729.3
D_1	14	36.708		54	369.99	B_7	95	3951.1
	15	38.891	G_4	55	392.00			
E_1	16	41.203		56	415.30			
F_1	17	43.654	A_4	57	440.00	C_8	96	4186.0
	18	46.249		58	466.16		97	4434.9
G_1	19	48.999	B_4	59	493.88	D_8	98	4698.6
	20	51.913					99	4978.0
A_1	21	55.000				E_8	100	5274.0
	22	58.270	C_5	60	523.25	F_8	101	5587.7
B_1	23	61.735		61	554.37		102	5919.9
			D_5	62	587.33	G_8	103	6271.9
				63	622.25		104	6644.9
C_2	24	65.406	E_5	64	659.26	A_8	105	7040.0
	25	69.296	F_5	65	698.46		106	7458.6
D_2	26	73.416		66	739.99	B_8	107	7902.1
	27	77.782	G_5	67	783.99			
E_2	28	82.407		68	830.61	C_9	108	8372.0
F_2	29	87.307	A_5	69	880.00		109	8869.8
	30	92.499		70	932.33	D_9	110	9397.3
G_2	31	97.999	B_5	71	987.77		111	9956.1
	32	103.83				E_9	112	10548.
A_2	33	110.00				F_9	113	11175.
	34	116.54	C_6	72	1046.5		114	11840.
B_2	35	123.47		73	1108.7	G_9	115	12544.
			D_6	74	1174.7		116	13290.
				75	1244.5	A_9	117	14080.
C_3	36	130.81	E_6	76	1318.5		118	14917.
	37	138.59	F_6	77	1396.9	B_9	119	15084.
D_3	38	146.83		78	1480.0			
	39	155.56	G_6	79	1568.0			
				80	1661.2	C_{10}	120	16744.

have been established for receiving the numerical routing information. 194

start diesel signal (nuclear power generating stations). That input signal to the diesel-generator unit start logic which initiates a diesel-generator unit start sequence. 31

start element (data transmission). In a character transmitted in a start-stop system, the first element in each character, which serves to prepare the receiving equipment for the reception and registration of the character. The start element is a spacing signal. 59

starter (1) (industrial control). An electric controller for accelerating a motor from rest to normal speed and to stop the motor. 206, 3, 124

(2) (illuminating engineering). A device used in conjunction with a ballast for the purpose of starting an electric-discharge lamp. *See also:* **fluorescent lamp.** 167

(3) (glow-discharge cold-cathode tube). An auxiliary electrode used to initiate conduction. *See also:* **electrode (of an electron tube).** 328

(4) (gas tubes). A control electrode, the principal function of which is to establish sufficient ionization to reduce the anode breakdown voltage. *Note:* This has sometimes been referred to as a "trigger electrode." 125

starter gap (gas tube). The conduction path between a starter and the other electrode to which starting voltage is applied. *Note:* Commonly used in the glow-discharge cold-cathode tube. *See also:* **gas-filled rectifier.**

244, 190, 125

starting reactor. A current-limiting reactor for decreasing the starting current of a machine or device. 53

starters (fluorescent lamp). Devices that first connect a fluorescent or similar discharge lamp in a circuit to provide for cathode preheating and then open the circuit so that the starting voltage is applied across the lamp to establish an arc. Starters also include a capacitor for the purpose of assisting the starting operation and for the suppression of radio interference during lamp starting and lamp operation. They may also include a circuit-opening device arranged to disconnect the preheat circuit if the lamp fails to light normally. 326

starter voltage drop (glow-discharge cold-cathode tube). The starter-gap voltage drop after conduction is established in the starter gap. *See also:* **electrode voltage (electron tube); gas tubes.** 190, 125

starting (rotating machinery). The process of bringing a motor up to speed from rest. *Note:* This includes breaking away, accelerating and if necessary, synchronizing with the supply. *See also:* **asynchronous machine; direct-current commutating machine.** 63

starting amortisseur. An amortisseur the primary function of which is the starting of the synchronous machine and its connected load. 328

starting anode. An electrode that is used in establishing the initial arc. *See also:* **rectification.** 328

starting capacitance (capacitor motor). The total effective capacitance in series with the auxiliary winding for starting operation. *See also:* **asynchronous machine.**

63

starting current (1) (rotating machinery). The current drawn by the motor during the starting period. (A function of speed or slip). *See also:* **asynchronous machine; direct-current commutating machine.** 63

(2) (oscillator). The value of electron-stream current through an oscillator at which selfsustaining oscillations will start under specified conditions of loading. *See:* **magnetrons.** 190, 125

starting electrode (1) (gas tube). An auxiliary electrode used to initiate conduction. *See also:* **gas-filled rectifier.**

244, 190

(2) (pool-cathode tube). An electrode used to establish a cathode spot. *See also:* **electrode (electron tube).** 190

starting motor. An auxiliary motor used to facilitate the starting and accelerating of a main machine to which it is mechanically connected. *See also:* **asynchronous machine; direct-current commutating machine.** 63

starting open-phase protection. The effect of a device operative to prevent connecting the load to the supply unless all conductors of a polyphase system are energized.

206

starting operation (single-phase motor). (1) The range of operation between locked rotor and switching for a motor employing a starting-switch or relay. (2) The range of operation between locked rotor and a point just below but not including breakdown-torque speed for a motor not

employing a starting switch or relay. *See also:* **asynchronous machine.** 63

starting resistor (rotating machinery). A resistor connected in a secondary or field circuit to modify starting performance of an electric machine. *See also:* **rotor (rotating machinery); stator.** 63

starting rheostat (industrial control). A rheostat that controls the current taken by a motor during the period of starting and acceleration, but does not control the speed when the motor is running normally. 206

starting sheet (electrorefining). A thin sheet of refined metal introduced into an electrolytic cell to serve as a cathode surface for the deposition of the same refined metal. *See also:* **electrorefining.** 328

starting-sheet blank (electrorefining). A rigid sheet of conducting material designed for introduction into an electrolytic cell as a cathode for the deposition of a thin temporarily adherent deposit to be stripped off as a starting sheet. *See also:* **electrorefining.** 328

starting-switch assembly. The make-and-break contacts, mechanical linkage, and mounting parts necessary for starting or running, or both starting and running, split-phase and capacitor motors. *Note:* The starting-switch assembly may consist of a stationary-contact assembly and a contact that moves with the rotor. 328

starting switch, centrifugal (rotating machinery). *See:* **centrifugal starting switch.**

starting switch, relay. *See:* **relay starting switch.**

starting temperature (grounding device). The winding temperature at the start of the flow of thermal current.

91

starting test (rotating machinery). A test taken on a machine while it is accelerating from standstill under specified conditions. *See also:* **asynchronous machine; direct-current commutating machine.** 63

starting torque (synchronous motor). The torque exerted by the motor during the starting period. (A function of speed or slip). 63

starting voltage (radiation counters). The voltage applied to a Geiger-Mueller tube at which pulses of 1 volt amplitude appear across the tube when irradiated. *See:* **anti-coincidence (radiation counters).** 190

starting winding (rotating machinery). A winding, the sole or main purpose of which is to set up or aid in setting up a magnetic field for producing the torque to start and accelerate a rotating electric machine. *See also:* **asynchronous machine; direct-current commutating machine.**

63

start-pulsing signal (telephone switching systems). A signal transmitted from the receiving end to the sending end of a trunk to indicate that the receiving end is in a condition to receive pulsing. 55

start-record signal. A signal used for starting the process of converting the electric signal to an image on the record sheet. *See also:* **facsimile signal (picture signal).** 12

start signal (1) (start-stop system). Signal serving to prepare the receiving mechanism for the reception and registration of a character, or for the control of a function. 194

(2) (facsimile). A signal that initiates the transfer of a facsimile equipment condition from standby to active. *See also:* **facsimile signal (picture signal).** 12

(3) (telephone switching systems). In multifrequency and key pulsing, a signal used to indicate that all digits have been transmitted. 55

start-stop printing telegraphy. That form of printing telegraphy in which the signal-receiving mechanisms are started in operation at the beginning and stopped at the end of each character transmitted over the channel. *See also:* **telegraphy.** 328

start-stop transmission (data transmission). Asynchronous transmission in which a group of code elements corresponding to a character signal is preceded by a start signal which serves to prepare the receiving mechanism for the reception and registration of a character and is followed by a stop signal which serves to bring the receiving mechanism to rest in preparation for the reception of the next character. 59

start transition (data transmission). In a character transmitted in a start-stop system, the mark-to-space transition at the beginning of the start element. 59

startup (relay). The action of a relay as it just departs from complete reset. Startup as a qualifying term is also used to identify the minimum value of the input quantity that will permit this condition. 103, 202, 60, 127

statcoulomb. The unit of charge in the centimeter-gram-second electrostatic system. It is that amount of charge that repels an equal charge with a force of one dyne when they are in a vacuum, stationary, and one centimeter apart. One statcoulomb is approximately 3.335×10^{-10} coulomb. 210

state (automatic control). The values of a minimal set of functions, which values contain information about the past history of a system sufficient to determine the future behavior, given knowledge of future inputs. 56

statement (computer programming). A meaningful expression or generalized instruction in a source language. 255, 77

state of statistical control (pulse measurement process). That state wherein a degree of consistency among repeated measurements of a characteristic, property, or attribute is attained. 15

state space (automatic control). A space which contains the state vectors of a system. *Note:* The number of state variables in the system determines the dimension of the state space. *See also:* **control system.** 56

state trajectory (automatic control). The vector function describing the dependence of the state on time and initial state. Note: If ϕ is the state trajectory, then

$$1) \ \phi(t_0, x_0) = x_0$$
$$2) \ \phi(t_2, x_0) = \phi[t_2, \phi(t_1, x_0)]$$

56

state variable formulation, (eigenvalue, eigenvector, characteristic equation) (excitation control systems). A system may be mathematically modeled by assigning variables $x_1, x_2, \ldots, x_n$ to system parameters; when these x's comprise the minimum number of parameters which completely specify the system, they are termed "states" or "state variables." System states arranged in a n-vector form a state vector. The mathematical model of the system may be manipulated into the form

$$dx/dt = \dot{X} = AX + bu$$
$$Y = CX + du$$

where X is the system state vector, u is the input vector, Y is the output vector, and A, b, C, d are matrices of appropriate dimension which specify the system. Such a model is known as a state variable or modern control formulation.

The nth order polynominal in λ,

$$\det (A - \lambda I) = 0$$

is called the characteristic equation and has n roots which are called eigenvalues (det $(\cdot)$ denotes determinant). When eigenvalues are real, they are the negative inverses of closed loop system time constants. Eigenvalues are also the pole locations of the closed loop transfer function. Any vector e_i such that

$$(A - \lambda_i I)e_i = 0$$
$$\|e_i\| \neq 0$$

is called an eigenvector of the eigenvalue λ_i ($\| \cdot \|$) denotes the square root of the sum of the squares of all entries of a vector. All n eigenvectors of a system form a modal matrix of matrix A when arranged side-by-side in a square matrix. The modal matrix is used in certain analytic procedures in modern control theory whereby large, complex systems are decoupled into many first order systems. 353

state variables (automatic control). Those whose values determine the state. 56

state vector (automatic control). One whose components are the state variables. 56

static (1) (atmospherics). Interference caused by natural electric disturbances in the atmosphere, or the electromagnetic phenomena capable of causing such interference. *See also:* **radio transmitter.** 328

(2) (adjective) (automatic control). Referring to a state in which a quantity exhibits no appreciable change within an arbitrarily long time interval. 56

static accuracy. Accuracy determined with a constant output. Contrast with dynamic accuracy. *See also:* **electronic analog computer.** 9, 77, 10

static breeze. *See:* **convective discharge.**

static characteristic (electron tubes). A relation, usually represented by a graph, between a pair of variables such as electrode voltage and electrode current, with all other voltages maintained constant. 190

static characteristic, relay. *See:* **relay static characteristic.**

static check. *See:* **problem check.** *See also:* **electronic analog computer.**

static converter. A unit that employs static switching devices such as metallic controlled rectifiers, transistors, electron tubes, or magnetic amplifiers. *See also:* **rectifier.** 3

static dump (computing systems). A dump that is performed at a particular point in time with respect to a machine run, frequently at the end of a run. 255, 77

static electrode potential. The electrode potential that exists when no current is flowing between the electrode and the electrolyte. *See also:* **electrolytic cell.** 328

static error. An error independent of the time-varying nature of a variable. *See:* **dynamic error.** *See also:* **electronic analog computer.** 9, 77

static friction. *See:* **stiction.**

static induced current*. The charging and discharging current of a pair of Leyden jars or other capacitors, which current is passed through a patient. *See also:* **electrotherapy.** 192

* Deprecated

staticize (electronic digital computation). (1) To convert serial or time-dependent parallel data into static form. (2) Occasionally, to retrieve an instruction and its operands from storage prior to its execution. 235, 255, 77

staticizer (electronic computation). A storage device for converting time-sequential information into static parallel information. 210

static Kraemer system (rotating machinery). A system of speed control below synchronous speed for wound-rotor induction motors. Slip power is recovered through the medium of a static converter equipment electrically connected between the secondary winding of the induction motor and a power system. *See also:* **asynchronous machine.** 63

static load line (electron device). The locus of all simultaneous average values of output electrode current and voltage, for a fixed value of direct-current load resistance. 190

static noise (atmospherics) (telephone practice). Interference caused by natural electric disturbances in the atmosphere, or the electromagnetic phenomena capable of causing such interference. *See:* **static.** 24

static optical transmission (acousto-optic device). The ratio of the transmitted zero order intensity, I_0, to the incident light intensity, I_{in}, when the acoustic drive power is off; thus $T = I_0/I_n$. 82

static overvoltage (surge arresters). An overvoltage due to an electric charge on an isolated conductor or installation. 327, 62

static plow (cable plowing). A plowing unit that depends upon drawbar pull only for its movement through the soil. 52

static pressure (acoustics) (audio and electroacoustics) (at a point in a medium). The pressure that would exist at that point in the absence of sound waves. 176

static regulation. Expresses the change from one steady-state condition to another as a percentage of the final steady-state condition.

$$\text{Static Regulation} = \frac{E_{\text{initial}} - E_{\text{final}}}{E_{\text{final}}} (100\%).$$

186

static regulator. A transmission regulator in which the adjusting mechanism is in self-equilibrium at any setting and requires control power to change the setting. *See also:* **transmission regulator.** 328

static relay. A relay or relay unit in which there is no armature or other moving element, the designed response being developed by electronic, solid-state, magnetic, or other components without mechanical motion. *Note:* A relay that is composed of both static and electromechanical

units in which the designed response is accomplished by static units may be referred to as a static relay. 103, 202, 60, 127

static resistance (forward or reverse) (semiconductor rectifier device). The quotient of the voltage by the current at a stated point on the static characteristic curve. *See also:* **rectification.** 237, 66

static short-circuit ratio (arc-welding apparatus). The ratio of the steady-state output short-circuit current of a welding power supply at any setting to the output current at rated load voltage for the same setting. 264

static, solid state converter (induction and dielectric heating equipment). A solid state generator or power source which utilizes semiconductor devices to control the switching of currents through inductive and capacitive circuit elements and thus generate a useable alternating current at a desired output frequency. 14

static test (1) (analog computer). The computer control state which applies a predetermined set of voltages and conditions to the analog computer, which allows the static check to be executed. 9

(2) (test, measurement and diagnostic equipment). (A) A test of a non-signal property, such as voltage and current, of an equipment or of any of its constituent units, performed while the equipment is energized; and (B) A test of a device in a stationary of helddown position as a means of testing and measuring its dynamic reactions. 54

static value (light emitting diodes). A non-varying value or quantity of measurement at a specified fixed point, or the slope of the line from the origin to the operating point on the appropriate characteristic curve. 162

static volt-ampere characteristic (arc-welding apparatus). The curve or family of curves that gives the terminal voltage of a welding power supply as ordinate, plotted against output load current as abscissa, is the static volt-ampere characteristic of the power supply. 264

static wave current* (electrotherapy). The current resulting from the sudden periodic discharging of a patient who has been raised to a high potential by means of an electrostatic generator. *See also:* **electrotherapy.** 192

* Deprecated

station. *See:* **tape unit.**

stationary battery. A storage battery designed for service in a permanent location. *See also:* **battery (primary or secondary).** 328

stationary-contact assembly. The fixed part of the starting-switch assembly. *See also:* **starting-switch assembly.** 328

stationary contact member (power switchgear). A conducting part having a contact surface which remains substantially stationary. 27, 103

stationary-mounted device. A device that cannot be removed except by the unbolting of connections and mounting supports. *Note:* Compare with **drawout-mounted device.** 103, 202

stationary satellite (communication satellite). A synchronous satellite with an equatorial, circular and direct orbit. A stationary satellite remains fixed in relation to the surface of the primary body. *Note:* A geo-stationary satellite is a stationary earth satellite. 74

stationary (time invarient) system (excitation control systems). Let a system have zero input response Z(t), then the system is stationary (time invarient) if the response to input R(t) is C(t) + Z(t) and the response to input R(t + T) is C(t + T) + Z(t). Otherwise the system is nonstationary. *Note:* A stationary system is modelled mathematically by a stationary differential equation the coefficients of which are not functions of time. 353

stationary wave. *See:* **standing wave.**

station auxiliary (generating station). Any auxiliary at a generating station not assigned to a specific unit. 58

station changing (communication satellite). The changeover of service from one earth station to another, especially in a system using satellites that are not stationary. 84

station-control error (electric power systems). The station generation minus the assigned station generation. *Note:* Refer to note on polarity under **area control error.** 94

station equipment (data transmission). Station equipment is a broad term used to denote equipment located at the customers' premises. The equipment may be owned by the telephone company or the customer. If the equipment is owned by the customer it is referred to as the customers' equipment. 59

station identification (telephone switching systems). A means of determining which of the customers served by a party line is making a call. 55

station line (telephone switching systems). Conductors carrying direct current between a central office and a main station, private branch exchange, or other end equipment. 55

station-loop resistance (telephone switching systems). The series resistance of the loop conductors, including the resistance of an off-hook station. 55

station, peaking. A generating station which is normally operated to provide power only during maximum load periods. 112

station, pumped storage. A hydroelectric generating station at which electric energy is normally generated during periods of relatively high system demand by utilizing water which has been pumped into a storage reservoir usually during periods of relatively low system demand. 112

station ringer (ringer). *See:* **telephone ringer.**

station, run-of-river. A hydroelectric generation station utilizing limited pondage or the flow of the stream as it occurs. 112

station service transformer. A transformer that supplies power from a station high-voltage bus to the station auxiliaries and also to the unit auxiliaries during unit startup and shutdown and/or when the unit auxiliaries transformer is not available. 58

station, steam-electric. An electric generating station utilizing steam for the motive force of its prime movers. 112

station, storage. A hydroelectric generating station associated with a water storage reservoir. 112

station-to-station call (telephone switching systems). A call intended for a designated main station. 55

station-type transformer. A transformer designed for installation in a station or substation. 53

station-type cubicle switchgear. Metal-enclosed power switchgear composed of the following equipment: (1) primary power equipment for each phase segregated and enclosed by metal; (2) stationary-mounted power circuit breakers; (3) group-operated switches, interlocked with the circuit breakers, for isolating the circuit breakers; (4) bare bus and connections; (5) instrument transformers; (6) control wiring and accessory devices. 103, 202

statistical delay (gas tubes) (electron device). The time lag from (1) the application of the specified voltage to initiate the discharge to (2) the beginning of breakdown. *See:* **gas tubes.** 190

stator (1) (rotating machinery). The portion that includes and supports the stationary active parts. The stator includes the stationary portions of the magnetic circuit and the associated winding and leads. It may, depending on the design, include a frame or shell, winding supports, ventilation circuits, coolers, and temperature detectors. A base, if provided, is not ordinarily considered to be part of the stator. 63
(2) (induction watthour meter). A voltage circuit, one or more current circuits, and a combined magnetic circuit so arranged that their joint effect when energized is to exert a driving torque on the rotor by the reaction with currents induced in an individual, or a common, conducting disk. *Note:* In a 2-wire stator, only one current circuit is present. In a 3-wire stator, two current circuits of the same number of turns are wound on a common core. The term **stator** may be similarly applied to the corresponding parts of certain motor-type meters designed to measure other quantities. *See also:* **watthour meter.** 212

stator coil (rotating machinery). A unit of a winding on the stator of a machine. *See also:* **stator.** 63

stator coil pin (rotating machinery). A rod through an opening in the stator core, extending beyond the faces of the core, for the purpose of holding coils of the stator winding to a desired position. *See also:* **stator.** 63

stator core (rotating machinery). The stationary magnetic-circuit of an electric machine. It is commonly an assembly of laminations of magnetic steel, ready for winding. *See also:* **stator.** 63

stator-core lamination (rotating machinery). A sheet of material usually of magnetic steel, containing teeth and winding slots, or containing pole structures, that forms the stator core when assembled with other identical or similar laminations. *See also:* **stator.** 63

stator frame (rotating machinery). The supporting structure holding the stator core or core assembly. *Note:* In certain types of machines, the stator frame may be made integral with one end shield. *See also:* **stator.** 63

stator iron (rotating machinery). A term commonly used for the magnetic steel material or core of the stator of a machine. *See also:* **stator.** 63

stator resistance starting (rotating machinery). The process of starting a motor at reduced voltage by connecting the primary winding initially in series with starting resistors which are short-circuited for the running condition. 63

stator shell. *See:* **shell, stator.**

stator winding (rotating machinery). A winding on the stator of a machine. *See also:* **stator.** 63

stator winding copper (rotating machinery). A term commonly used for the material or conductors of a stator winding. *See also:* **stator.** 63

status memory (sequential events recording systems). The memory which contains the most recently scanned status of all inputs. 48, 58

status point, supervisory control (power-system communication). *See:* **supervisory control point, status.** *See also:* **supervisory control system.**

STC (electronic navigation). *See:* **sensitivity time control.**

steady current. A current that does not change with time. 210

steady state (1) (signal-transmission system). That in which some specified characteristic of a condition, such as value, rate, periodicity, or amplitude, exhibits only negligible change over an arbitrarily long period of time. *See also:* **signal.** 105
(2) (industrial control). The condition of a specified variable at a time when no transients are present. *Note:* For the purpose of this definition, drift is not considered to be a transient. *See also:* **control system, feedback.** 219, 206
(3) (cable insulation materials). Conditions of current in the material attained when the difference between the maximum and minimum current observed during four consecutive hourly readings is less than 5 percent of the minimum current. 97
(4) (excitation control systems). That in which some specified characteristic of a condition, such as value, rate, periodicity, or amplitude, exhibits only negligible change over an arbitrarily long interval of time. *Note:* It may describe a condition in which some characteristics are static, others dynamic. 353

steady-state condition (gas turbines). A condition or value with limited deviations having a constant mean value over an arbitrarily long interval of time. 98, 58

steady-state deviation (control). *See:* **deviation, steady-state.**

steady state governing load band (speed governing systems). The magnitude of the envelope of cyclic load variations caused by the governing system, expressed as a percent of rated power output, when the generating unit is operating in parallel with other generators and under steady state load demand. *Syn:* **power stability index.** 8, 58

steady state governing speed band. The magnitude of the envelope of the cyclic speed variations caused by the governing system, expressed as a percent of rated speed when the generating unit is operating independently and under steady-state load demand. *Syn:* **speed stability index.** 8, 58

steady-state incremental speed regulation (excluding the effects of deadband) (gas turbines). At a given steady-state speed and power, the rate of change of the steady-state speed with respect to the power output. It is the slope of the tangent to the steady-state speed versus power curve at the point of power output under consideration. It is the difference in steady-state speed, expressed in percent of rated speed, for any two points on the tangent, divided by the corresponding difference in power output, expressed as a fraction of the rated power output. For the basis of comparison, the several points of power output at which the values of steady-state incremental speed regulation are derived are based upon rated speed being obtained at each point of power output. 94, 98, 58

steady-state oscillation. A condition in which some aspect of the oscillation is a continuing periodic function. 176

steady-state short-circuit current (synchronous machine). The steady-state current in the armature winding when short-circuited. 63

steady-state speed regulation (straight condensing and noncondensing steam turbines, nonautomatic extraction turbines, hydro-turbines, and gas turbines). The change in steady-state speed, expressed in percent of rated speed, when the power output of the turbine operating isolated is gradually reduced from rated power output to zero power output with unchanged settings of all adjustments of the speed-governing system. *Note:* Speed regulation is considered positive when the speed increases with a decrease in power output. *See also:* **asynchronous machine; direct-current commutating machine; speed-governing system.** 94, 98, 58

steady-state stability. A condition that exists in a power system if it operates with stability when not subjected to an aperiodic disturbance. *Note:* In practical systems, a variety of relatively small aperiodic disturbances may be present without any appreciable effect upon the stability, as long as the resultant rate of change in load is relatively slow in comparison with the natural frequency of oscillation of the major parts of the system or with the rate of change in field flux of the rotating machines. *See also:* **alternating-current distribution.** 64

steady-state stability factor (system or part of a system). The ratio of the steady-state stability limit to the nominal power flow at the point of the system to which the stability limit is referred. *See:* **stability factor.** *See also:* **alternating-current distribution.** 64

steady-state stability limit (steady-state power limit). The maximum power flow possible through some particular point in the system when the entire system or the part of the system to which the stability limit refers is operating with steady-state stability. *See also:* **alternating-current distribution.** 64

steady-state temperature rise (grounding device). The maximum temperature rise above ambient which will be attained by the winding of a device as the result of the flow of rated continuous current under standard operating conditions. It may be expressed as an average or a hot-spot winding rise. 91

steady-state value. The value of a current or voltage after all transients have decayed to a negligible value. For an alternating quantity, the root-mean-square value in the steady state does not vary with time. *See also:* **asynchro-**

nous machine; direct-current commutating machine.
63

steady voltage. *See:* **steady-current.**

steam capability (electric power supply). The maximum net capability of steam generating units which can be obtained under normal operating practices for a given period of time as demonstrated by total plant tests. Units can be rated at overpressure and with top heaters out, providing one is willing to operate under these conditions and has developed normal procedures for such operation. The effects of any seasonal changes in cooling water temperature are to be taken into account. 112

steam turbine-electric drive. A self-contained system of power generation and application in which the power generated by a steam turbine is transmitted electrically by means of a generator and a motor (or multiples of these) for propulsion purposes. *Note:* The prefix steam turbine-electric is applied to ships, locomotives, cars, buses, etcetera, that are equipped with this drive. *See also:* **electric locomotive; electric propulsion system.** 328

steel container (storage cell). The container for the element and electrolyte of a nickel-alkaline storage cell. This steel container is sometimes called a can. *See also:* **battery (primary or secondary).** 328

steerable antenna. A directional antenna whose major lobe can be readily shifted in direction. *See also:* **antenna.** 111, 179

steering compass. A compass located within view of a steering stand, by reference to which the helmsman holds a ship on the set course.

Stefan-Boltzmann law. The statement that the radiant exitance of a blackbody is proportional to the fourth power of its absolute temperature; that is,

$$M = \sigma T^4.$$

See: National Bureau of Standards Technical News Bulletin, October 1963 (corrected) for constants. *See also:* **radiant energy (illuminating engineering).** 167

step (1) (pulse techniques). A waveform that, from the observer's frame of reference, approximates a Heaviside (unit step) function. *See:* **signal, unit step.** *See also:* **pulse.** 185

(2) (computing systems). (A) One operation in a computer routine. (B) To cause a computer to execute one operation. *See also:* **single step.** 255, 77

(3) (pulse terms) (single transition). A transition waveform which has a transition duration which is negligible relative to the duration of the waveform epoch or to the duration of its adjacent first and second nominal states. 254

step-back relay (motors and industrial control). A relay that functions to limit the current peaks of a motor when the armature or line current increases. A stepback relay may, in addition, remove this limitation when the cause of the high current has been removed. *Note:* When used with a motor having a flywheel, the relay causes the momentary transfer of stored energy from the flywheel to the load but does not limit the current peaks if the load is sustained for a considerable interval after the relay operates. *See also:* **relay.** 3, 225, 206

step-by-step switch. A bank-and-wiper switch in which the wipers are moved by electromagnet ratchet mechanisms

individual to each switch. *Note:* This type of switch may have either one or two types of motion. 328

step-by-step system. An automatic telephone switching system that is generally characterized by the following features. (1) The selecting mechanisms are step-by-step switches. (2) The switching pulses may either actuate the successive selecting mechanisms directly or may be received and stored by controlling mechanisms that, in turn, actuate the selecting mechanisms. 328

step change (control) (industrial control) (step function). An essentially instantaneous change of an input variable from one value to another. *See also:* **control system, feedback.** 219, 206

step compensation (correction) (industrial control). The effect of a control function or a device that will cause a step change in an other function when a predetermined operating condition is reached. 225, 206

step-down transformer. A transformer in which the energy transfer is from a higher voltage circuit to a lower voltage circuit. 53

step line-voltage change (power supplies). An instantaneous change in line voltage (for example, 105-125 volts alternating current); for measuring line regulation and recovery time. *See also:* **power supply.** 228, 186

step load change (power supplies). An instantaneous change in load current (for example, zero to full load); for measuring the load regulation and recovery time. *See also:* **power supply.** 228, 186

stepped antenna. *See:* **zoned antenna.**

stepped leader (lightning). A series of discharges emanating from a region of charge concentration at short time intervals. Each discharge proceeds with a luminescent tip over a greater distance than the previous one. *See also:* **direct-stroke protection (lightning).** 64

stepping life (for attenuator variable in fixed steps). Number of times to switch from any selected position to any other selected positions, after which the residual and incremental characteristic insertion loss remain within the specified repeatability. 110

stepping relay (1) (general). A multiposition relay in which moving wiper contacts mate with successive sets of fixed contacts in a series of steps, moving from one step to the next in successive operations of the relay. *See also:* **relay.** 259

(2) (rotary type). A relay having many rotary positions, ratchet actuated, moving from one step to the next in successive operations, and usually operating its contacts by means of cams. There are two forms: (A) **directly driven,** where the forward motion occurs on energization, and (B) **indirectly (spring) driven,** where a spring produces the forward motion on pulse cessation. *Note:* The term is also incorrectly used for **stepping switch.** 341

stepping relay, spring-actuated. A stepping relay that is cocked electrically and operated by spring action. *See also:* **relay.** 259

step response. A criterion of the dynamic response of an instrument when subjected to an instantaneous change in measured quantity from one value to another. *See also:* **accuracy rating of an instrument.** 295

step-response time. The time required for the end device to come to rest in its new position after an abrupt change to a new constant value has occurred in the measured signal. *See also:* **accuracy rating of an instrument.**   295

step speed adjustment (industrial control). The speed drive can be adjusted in rather large and definite steps between minimum and maximum speed. *See also:* **electric drive.**
  206

step-stress test (reliability). A test consisting of several stress levels applied sequentially, for periods of equal duration, to a (one) sample. During each period a stated stress-level is applied and the stress level is increased from one step to the next. *See also:* **reliability.**   182, 164

step twist, waveguide (waveguide components). A waveguide twist formed by abruptly rotating about the waveguide longitudinal axis, one or more waveguide sections each nominally a quarter wavelength long.   166

step-up transformer. A transformer in which the energy transfer is from a lower voltage circuit to a higher voltage circuit.   53

step voltage (safety) (conductor stringing equipment). The potential difference between two points on the earth's surface, separated by a distance of one pace, that will be assumed to be one meter, in the direction of maximum potential gradient *Note:* This potential difference could be dangerous when current flows through the earth or material upon which a workman is standing, particularly under fault conditions. *Syn:* **step potential.** *See also:* **ground.**   313, 45

step-voltage regulator. A small regulating transformer in which the voltage of the regulated circuit is controlled in steps by means of taps and without interrupting the load. *Note:* Such units are generally 833 kVA (output) and below, single-phase, or 2500 kVA (output) and below, three-phase.   53

step-voltage test (rotating machinery). A controlled overvoltage test in which designated voltage increments are applied at designated times. Time increments may be constant or graded. *See:* **graded-time step-voltage test.** *See also:* **asynchronous machine; direct-current commutating machine.**   63

step wedge* (television). *See:* **gray scale.**   163

* Deprecated

steradian (1) (metric practice). The solid angle which, having its vertex in the center of a sphere, cuts off an area of the surface of the sphere equal to that of a square with sides of length equal to the radius of the sphere (ISO Recommendation R 31, part 1, second edition, December 1965).   21
(2) (laser-maser). (sr). The unit of measure for a solid angle. There are 4π steradians in a sphere.   363

stereophonic (adjective) (frequency modulation). Pertains to audio information carried by a plurality of channels arranged to afford the listener a sense of the spatial distribution of the sound sources. *Note:* A stereophonic receiver responds to both the L + R main channel and the L − R subcarrier channel of a composite stereophonic signal, so that the one output contains substantially only L information, and the other only R. In addition to the main channel, stereophonic program modulation requires transmission of a 19 kHz pilot signal and the sidebands of a suppressed 38 kHz subcarrier carrying L − R information. This combination is called the composite signal, and it may be used alone or with other subcarrier (SCA) sig-

nals to frequency modulate the RF carrier. After pre-emphasis, the left and right channels are added for main channel information. The right-channel program material is subtracted from the left to derive a difference signal that then amplitude modulates a 38 kHz subcarrier. The subcarrier is suppressed, divided by two, and transmitted as a 19 kHz pilot signal to facilitate demodulation of the suppressed carrier information at the receiver.   16

stick circuit. A circuit used to maintain a relay or similar unit energized through its own contact.   328

stickiness. The condition caused by physical interference with the rotation of the moving element. *See also:* **moving element of an instrument.**   280

sticking voltage (luminescent screen). The voltage applied to the electron beam below which the rate of secondary emission from the screen is less than unity. The screen then has a negative charge that repels the primary electrons. *See also:* **cathode-ray tubes.**   244, 190

stick operation (switching device) (hook operation). Manual operation by means of a switch stick.   103, 202, 27

stiction (1). The force in excess of the coulomb friction required to start relative motion between two surfaces in contact.   329
(2) (static friction) (industrial control). The total friction that opposes the start of relative motion between elements in contact. *See also:* **control system, feedback.**
  219, 206

stiffness (industrial control). The ability of a system or element to resist deviations resulting from loading at the output. *See also:* **control system, feedback.**
  56, 219, 206

stiffness coefficient. The factor K (also called spring constant) in the differential equation for oscillatory motion $M\ddot{x} + B\dot{x} + Kx = 0$. *See also:* **control system, feedback.**
  56

stilb. The unit of luminance (photometric brightness) equal to one candela per square centimeter. *Note:* The name **stilb** has been adopted by the International Commission on Illumination (CIE) and is commonly used in European publications. In the United States and Canada the preferred practice is to use self-explanatory terms such as candela per square inch and candela per square centimeter. *See also:* **light.**   167

Stiles-Crawford effect. The reduced luminous efficiency of rays entering the peripheral portion of the pupil of the eye. *See also:* **visual field.**   167

stimulated emission (laser-maser). The **emission** of radiation at a given frequency caused by an applied external radiation field of the same frequency.   363

stimulus (industrial control). Any change in signal that affects the controlled variable; for example, a disturbance or a change in reference input. *See also:* **control system, feedback.**   219, 206, 54

stirring effect (induction heater usage). The circulation in a molten charge due to the combined forces of motor and pinch effects. *See also:* **induction heating; motor effect; pinch effect.**   14, 114

stop (limit stop). A mechanical or electric device used to limit the excursion of electromechanical equipment. *See also:* **limiter circuit.** *See also:* **electronic analog computer.**
  9

stop band (circuits and systems). A band of frequencies that

pass through a filter with a substantial amount of loss (relative to other frequency bands such as a pass band).
67

stop-band ripple (circuits and systems). The difference between maxima and minima of loss in a filter stop band.
67

stop dowel (rotating machinery). A pin fitted into a hole to limit motion of a second part.
63

stop element (data transmission). In a character transmitted in a start-stop system, the last element in each character, to which is assigned a minimum duration, during which the receiving equipment is returned to its rest condition in preparation for the reception of the next character. The stop element is a marking signal.
59

stop-go pulsing (telephone switching systems). A method of pulsing control wherein the pulsing operation may take place in stages, and the sending end is arranged to pulse the digits continuously unless or until the stop-pulsing signal is received. *Note:* When this occurs, the pulsing of the remaining digits is suspended until the sending end receives a start-pulsing signal.
55

stop joint (power cable joints). A joint which is designed to prevent any transfer of dielectric fluid between the cables being joined.
34

stop lamp. A lamp giving a steady warning light to the rear of a vehicle or train of vehicles, to indicate the intention of the operator to diminish speed or to stop. *See also:* **headlamp.**
167

stop-motion switch (elevators). *See:* **machine final-terminal stopping device.**

stopping off. The application of a resist to any part of a cathode or plating rack. *See also:* **electroplating.**
328

stop-pulsing signal (telephone switching systems). A signal transmitted from the receiving end to the sending end of a trunk to indicate that the receiving end is not in a condition to receive pulsing.
55

stop-record signal (facsimile). A signal used for stopping the process of converting the electrical signal to an image on the record sheet. *See also:* **facsimile signal (picture signal).**
12

stop signal (data transmission) (1) (start-stop system). Signal serving to bring the receiving mechanism to rest in preparation for the reception of the next telegraph signal.
194

(2) (facsimile). A signal which initiates the transfer of a facsimile equipment condition from active to standby. *See also:* **facsimile signal (picture signal).**
12

storage (electronic computation). (1) The act of storing information. (2) Any device in which information can be stored, sometimes called a memory device. (3) In a computer, a section used primarily for storing information. Such a section is sometimes called a memory or store (British). *Notes:* (A) The physical means of storing information may be electrostatic, ferroelectric, magnetic, acoustic, optical, chemical, electronic, electric, mechanical, etcetera, in nature. (B) Pertaining to a device in which data can be entered, in which it can be held, and from which it can be retrieved at a later time. *See also:* **store.**
210, 255, 77, 54

storage allocation (computing systems). The assignment of sequences of data or instructions to specified blocks of storage.
255, 77

storage assembly (storage tubes). An assembly of electrodes (including meshes) that contains the target together with electrodes used for control of the storage process, those that receive an output signal, and other members used for structural support. *See also:* **storage tube.**
174, 190

storage battery (secondary battery or accumulator). A battery consisting of two or more storage cells electrically connected for producing electric energy. Common usage permits this designation to be applied also to a single storage cell used independently. *See also:* **battery (primary or secondary).**
328

storage capacity. The amount of data that can be contained in a storage device. *Notes:* (1) The units of capacity are bits, characters, words, etcetera. For example, capacity might be "32 bits," "10 000 decimal digits," "16 384 words with 10 alphanumeric characters each." (2) When comparisons are made among devices using different character sets and word lengths, it may be convenient to express the capacity in equivalent bits, which is the number obtained by taking the logarithm to the base 2 of the number of usable distinguishable states in which the storage can exist. (3) The storage (or memory) capacity of a computer usually refers only to the internal storage section.
235, 210, 255, 77

storage cell (secondary cell or accumulator) (1) (electric energy). A galvanic cell for the generation of electric energy in which the cell, after being discharged, may be restored to a fully charged condition by an electric current flowing in a direction opposite to the flow of current when the cell discharges. *See also:* **electrochemistry.**
328

(2) (computing systems) (information). An elementary unit of storage, for example, a binary cell, a decimal cell. *See also:* **electrochemistry.**
255, 77

storage device. A device in which data can be stored and from which it can be copied at a later time. The means of storing data may be chemical, electrical, mechanical, etcetera. *See also:* **storage.**
235

storage element (storage tubes). An area of a storage surface that retains information distinguishable from that of adjacent areas. *Note:* The storage element may be a portion of a continuous storage surface or a discrete area such as a dielectric island. *See also:* **storage tube.**
174, 190, 125

storage-element equilibrium voltage (storage tubes). A limiting voltage toward which a storage element charges under the action of primary electron bombardment and secondary emission. At equilibrium voltage the escape ratio is unity. *Note:* **Cathode equilibrium voltage, second-crossover equilibrium voltage,** and **gradient-established equilibrium voltage** are typical examples. *See also:* **charge-storage tube.**
174, 190

storage-element equilibrium voltage, cathode (storage tubes). The storage element equilibrium voltage near cathode voltage and below first-crossover voltage. *See also:* **charge-storage tube.**
174, 190

storage-element equilibrium voltage, collector (storage tubes). *See:* **charge storage tube.**

storage-element equilibrium voltage, gradient established (storage tubes). The storage-element equilibrium voltage, between first- and second-crossover voltages, at which the escape ratio is unity. *See also:* **charge-storage tube.**
174, 190

storage-element equilibrium voltage, second-crossover (storage tubes). The storage-element equilibrium voltage at the second-crossover voltage. *See also:* **charge-storage tube.**   174, 190

storage integrator (analog computer). An integrator used to store a voltage in the hold condition for future use while the rest of the computer assumes another computer control state. *See also:* **electronic analog computer.**   9

storage life (gyro; accelerometer). The non-operating time interval under specified conditions, after which a device will still exhibit a specified operating life. *See:* **operating life.**   46

storage light-amplifier (optoelectronic device). *See:* **image-storage panel.**

storage medium. Any device or recording medium into which data can be stored and held until some later time, and from which the entire original data can be obtained.   224, 207

storage protection (computing systems). An arrangement for preventing access to storage for either reading or writing, or both.   77

storage, reservoir (electric power systems). The volume of water in a reservoir at a given time.   112

storage station. A hydroelectric generating station having storage sufficient for seasonal or hold-over operation. *See also:* **generating station.**   64

storage surface (storage tubes). The surface upon which information is stored. *See also:* **storage tube.**   174, 190

storage temperature (1) (power supply). The range of environmental temperatures in which a power supply can be safely stored (for example, −40 to +85 degrees Celsius). *See also:* **power supply.**   186
(2) (semiconductor device). The range of environmental temperatures in which a semiconductor device can be safely stored.   66
(3) (light emitting diodes) (T_{stg}). The temperature at which the device, without any power applied, is stored.   162

storage temperature range. The range of temperatures over which the Hall generators may be stored without any voltage applied, or without exceeding a specified change in performance.   107

storage time* (storage tubes). *See:* **retention time, maximum; decay time.** *See also:* **storage time.**

* Deprecated

storage tube. An electron tube into which information can be introduced and read at a later time. *Note:* The output may be an electric signal and/or a visible image corresponding to the stored information.   174, 190, 125

store. (1) To retain data in a device from which it can be copied at a later time. (2) To put data into a storage device. (3) British synonym for storage. *See also:* **storage.**   235, 210, 255, 77

store-and-forward switching (data communication). A method of switching whereby messages are transferred directly or with interim storage, each in accordance with its own address. *See also:* **packet switching.**   12

store-and-forward switching system (telephone switching systems). A switching system for the transfer of messages, each with its own address or addresses, in which the message can be stored for subsequent transmission.   55

stored energy indicator (power switchgear). An indicator which visibly shows that the stored energy mechanism is in the charged or discharged position.   103

stored-energy operation. Operation by means of energy stored in the mechanism itself prior to the completion of the operation and sufficient to complete it under predetermined conditions. *Note:* This kind of operation may be subdivided according to: (1) How the energy is stored (spring, weight, etcetera); (2) How the energy originates (manual, electric, etcetera); (3) How the energy is released (manual, electric, etcetera).   103, 202

stored logic (telephone switching systems). Instructions in memory arranged to direct the performance of predetermined functions in response to readout.   55

stored program (telephone switching systems). A program in memory that a processor can execute.   55

stored-program computer. A digital computer that, under control of internally stored instructions, can synthesize, alter, and store instructions as though they were data and can subsequently execute these new instructions.   255, 77

stored program control (telephone switching systems). A system control using stored logic.   55

stored-program switching system (telephone switching systems). An automatic switching system having stored program control.   55

storm guys. Anchor guys, usually placed at right angles to direction of line, to provide strength to withstand transverse loading due to wind. *See also:* **tower.**   64

storm loading. The mechanical loading imposed upon the components of a pole line by the elements, that is, wind and/or ice, combined with the weight of the components of the line. *Note:* The United States has been divided into three loading districts, light, medium, and heavy, for which the amounts of wind and/or ice have been arbitrarily defined. *See also:* **cable; open wire.**   328

straggling, energy (semiconductor radiation detectors). The random fluctuations in energy loss whereby those particles having the same initial energy lose different amounts of energy when traversing a given thickness of matter. (This process leads to the broadening of spectral lines.)   23, 118, 119

straight-cut control system (numerically controlled machines). A system in which the controlled cutting action occurs only along a path parallel to linear, circular, or other machine ways.   224, 207

straightforward trunking (manual telephone switchboard system). That method of operation in which the *A* operator gives the order to the *B* operator over the trunk on which talking later takes place.   328

straight joint (1) (transmission and distribution). A joint used for connecting two lengths of cable in approximately the same straight line in series. *Note:* A straight joint is made between two like cables, for example, between two single-conductor cables, between two concentric cables, or between two triplex cables. *See:* **branch joint; cable joint; reducing joint.**   64
(2) (power cable joints). A cable joint used for connecting two lengths of cable, each of which consists of one or more conductors.   34

straight-line coding (computing systems). Coding in which loops are avoided by the repetition of parts of the coding when required.   255, 77

straight-seated bearing (rotating machinery) (cylindrical bearing). A journal bearing in which the bearing liner is constrained about a fixed axis determined by the supporting structure. *See also:* **bearing.**   63

straight storage system (electric power supply). A system in which the electrical requirements of a car are supplied solely from a storage battery carried on the car. *See also:* **axle generator system.**   328

strain element (fuse) (strain wire). That part of the current-responsive element, connected in parallel with the fusible element in order to relieve it of tensile strain. *Note:* The fusible element melts and severs first and then the strain element melts during circuit interruption.
103, 202

strain insulator (1). An insulator generally of elongated shape, with two transverse holes or slots.   261
(2). A single insulator, an insulator string, or two or more strings in parallel, designed to transmit to the tower or other support the entire pull of the conductor and to insulate it therefrom. *See also:* **insulator.**   64

strain wire. *See:* **strain element (fuse).**

strand. (1) One of the wires, or groups of wires, of any stranded conductor. (2) One of a number of paralleled uninsulated conducting elements of a conductor which is stranded to provide flexibility in assembly or in operation. (3) One of a number of paralleled insulated conducting elements which constitute one turn of a coil in rotating machinery. The strands are usually separated electrically through all the turns of a multi-turn coil. Various types of transposition are commonly employed to reduce the circulation of current among the strands. A strand has a solid cross section, or it may be hollow to permit the flow of cooling fluid in intimate contact with the conductor (one form of "conductor cooling"). *See also:* **conductor; rotor (rotating machinery); stator.**   64, 63

stranded conductor. A conductor composed of a group of wires or of any combination of groups of wires. *Note:* The wires in a stranded conductor are usually twisted or braided together. *See also:* **conductor.**   64

stranded wire. *See:* **stranded conductor.**

strand insulation (rotating machinery). The insulation on a strand or lamination or between adjacent strands or laminations which comprise a conductor.   63

strand-to-strand test (rotating machinery). A test that is designed to apply a voltage of specified amplitude and waveform between the strands of a coil for the purpose of determining the integrity of the strand insulation.   63

strap, anode (magnetron). *See:* **anode strap.**

strap key. A pushbutton circuit controller that is biased by a spring metal strip and is used for opening or closing a circuit momentarily.   328

strapped-down (gimbal-less) inertial navigation equipment. Inertial navigation equipment wherein the inertial devices (gyros and accelerometers) are attached directly to the carrier, eliminating the stable platform and gimbal system. *Note:* In this equipment a computer utilizes gyro information to resolve the accelerations that are sensed along the carrier axes and to refer these accelerations to an inertial frame of reference. Navigation is then accomplished in the same manner as in systems using a stable platform. *See also:* **navigation.**   187, 13

strapping (multiple-cavity magnetrons). *See:* **anode strap; jumper.**

stray (circuits and systems). An element or occurrence usually not desired in a theoretical design, but unavoidable

in a practical realization. For example, the relative proximity of wires can cause stray capacitance. *See also:* **parasitic element.**   67

stray-current corrosion. Corrosion caused by current through paths other than the intended circuit or by an extraneous current in the earth. *See:* **cathodic corrosion; long-line current (corrosion); noble potential; sacrificial protection.**   221, 205

stray load loss (synchronous machine). The losses due to eddy currents in copper and additional core losses in the iron, produced by distortion of the magnetic flux by the load current, not including that portion of the core loss associated with the resistance drop.   244, 63

strays*. Electromagnetic disturbances in radio reception other than those produced by radio transmitting systems. *See also:* **radio transmitter.**   328

* Obsolete

streamer (voltage testing). When used in connection with high-voltage testing, an incomplete disruptive discharge in a gaseous or liquid dielectric that does not completely bridge the test piece or gap. *See also:* **test voltage and current.**   307, 57

stream flow. The quantity rate of water passing a given point. *See also:* **generating station.**   64

streaming (audio and electroacoustics). Unidirectional flow currents in a fluid that are due to the presence of acoustic waves.   176

streaming potential (electrobiology). The electrokinetic potential gradient resulting from unit velocity of liquid forced to flow through a porous structure or past an interface. *See also:* **electrobiology.**   192

street-lighting luminaire. A complete lighting device consisting of a light source together with its direct appurtenances such as globe, reflector, refractor, housing, and such support as is integral with the housing. The pole, post, or bracket is not considered part of the luminaire. *See also:* **luminaire.**   167

street-lighting unit. The assembly of a pole or lamp post with a bracket and a luminaire. *See also:* **street lighting luminaire.**   167

strength-duration (time-intensity) curve (medical electronics). A graph of the intensity curve of applied electrical stimuli as a function of the duration just needed to elicit response in an excitable tissue.   192

strength of a sound source (strength of a simple source). The maximum instantaneous rate of volume displacement produced by the source when emitting a wave with sinusoidal time variation. *Note:* The term is properly applicable only to sources of dimension small with respect to the wavelength.   176

stress-accelerated corrosion. Corrosion that is accelerated by stress.   221, 205

stress corrosion cracking. Spontaneous cracking produced by the combined action of corrosion and static stress (residual or applied).   221, 205

stress relief. A predetermined amount of slack to relieve tension in component or lead wires.   284

strike deposit (1) (electroplating). A thin film of deposited metal to be followed by other coatings. *See also:* **electroplating.**   328
(2) (bath) (electroplating). An electrolyte used to deposit a thin initial film of metal. *See also:* **electroplating.** 328

striking (1) (arc) (spark) (gas). The process of establishing an arc or a spark. *See also:* **discharge (gas).** 244, 190
(2) (electroplating). The electrodeposition of a thin initial film of metal, usually at a high current density. *See also:* **electroplating.** 328
striking current (gas tube). The starter-gap current required to initiate conduction across the main gap for a specified anode voltage. *See also:* **gas-filled rectifier.** 244, 190
striking distance (1) (power switchgear). The shortest distance, measured through air, between parts of different polarities. 103, 202
(2) (transformers). The shortest unobstructed distance measured through a dielectric medium such as liquid, gas, or vacuum; between parts of different electric potential. 53
(3) (outdoor apparatus bushings). The shortest tight string distance measured externally over the weather casing between the metal parts which have the operating line to ground voltage between them. 168
string. A connected sequence of entities such as characters or physical elements. 255, 77
stringing (conductor stringing equipment). The pulling of pilot lines, pulling lines and conductors over travelers supported on structures of overhead transmission lines. Quite often, the entire job of stringing conductors is referred to as "stringing operations," beginning with the planning phase and terminating after the conductors have been installed in the suspension clamps. 45
stringing, slack (conductor stringing equipment). The method of stringing conductor slack without the use of a tensioner. The conductor is pulled off the reel by a pulling vehicle and dragged along the ground. As the conductor is dragged to, or past, each supporting structure, the conductor is placed in the travelers, normally with the aid of "finger lines." *Syn:* **conventional stringing.** 45
stringing, tension (conductor stringing equipment). The use of "pullers" and "tensioners" to keep the conductor under tension and positive control during the stringing phase, thus keeping it clear of the earth and other obstacles which could cause damage. 45
string-shadow instrument. An instrument in which the indicating means is the shadow (projected or viewed through an optical system) of a filamentary conductor, the position of which in a magnetic or an electric field depends upon the measured quantity. *See also:* **instrument.** 328
strip (electroplating. A solution used for the removal of a metal coating from the base metal. *See also:* **electroplating.** 328
stripper tank (electrorefining). An electrolytic cell in which the cathode deposit, for the production of starting sheets, is plated on starting-sheet blanks. *See also:* **electrorefining.** 328
stripping (1) (electroplating) (mechanical). The removal of a metal coating by mechanical means. 328
(2) (chemical). The removal of a metal coating by dissolving it. 328
(3) (electrolytic). The removal of a metal coating by dissolving it or an underlying coating anodically with the aid of a current. *See also:* **electroplating.** 328
stripping compound (electrometallurgy). Any suitable material for coating a cathode surface so that the metal

electrodeposited on the surface can be conveniently stripped off in sheets. *See also:* **electrowinning.** 328
strip terminals (rotating machinery). A form of terminal in which the ends of the machine winding are brought out to terminal strips mounted integral with the machine frame or assembly. 63
strip-type transmission line (waveguides). A transmission line consisting of a conductor above or between extended conducting surfaces. *See:* **shielded strip transmission line; unshielded strip transmission line.** *See also:* **waveguides.** 179
strobing (pulse terms). A process in which a first pulse of relatively short duration interacts with a second pulse or other event of relatively longer duration to yield a signal which is indicative (typically, proportional to) the magnitude of the second pulse during the first pulse. 254
stroboscopic tube. A gas tube designed for the periodic production of short light flashes. *See also:* **gas tubes.** 255, 190
stroke (cable plowing). Peak to peak displacement of the plow blade tip. 52
stroke speed (scanning or recording line frequency) (facsimile). The number of times per minute, unless otherwise stated, that fixed line perpendicular to the direction of scanning is crossed in one direction by a scanning or recording spot. *Note:* In most conventional mechanical systems this is equivalent to drum speed. In systems in which the picture signal is used while scanning in both directions, the stroke speed is twice the above figure. *See also:* **recording (facsimile); scanning (facsimile).** 12
structurally dual networks. A pair of networks such that their branches can be marked in one-to-one correspondence so that any mesh of one corresponds to a cut-set of the other. Each network of such a pair is said to be the dual of the other. *See also:* **network analysis.** 210

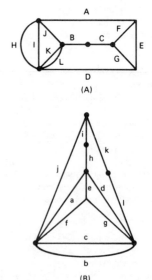

Structurally dual networks. For example, the mesh EFG in (A) corresponds to the cut-set efg in (B), the mesh bc in (B) to the cut-set BC in (A), and the mesh JA-EGCB in (A) to the cut-set jaegcb in (B).

structurally symmetrical network. A network that can be arranged so that a cut through the network produces two parts that are mirror images of each other. *See also:* **network analysis.**   210

structure conflict (transmission and distribution). A line is so situated with respect to a second line that the overturning of the first line will result in contact between its supporting structures or conductors and the conductors of the second line, assuming that no conductors are broken in either line. *Exceptions:* Lines are not considered as conflicting under the following conditions: (1) Where one line crosses another. (2) Where two lines are on opposite sides of a highway, street, or alley and are separated by a distance not less than 60% of the height of the taller pole line and not less than 20 feet. *See also:* **conductor clearance.**   262

structure, crossing. *See:* **crossing structure.**

structure, snub. *See:* **snub structure.**

structures, safety class (Class 1E power systems). Structures designed to protect Class 1E equipment against the effects of design basis events.   102

stub (communication practice). A short length of transmission line or cable that is joined as a branch to another transmission line or cable. *See also:* **cable.**   328

stub antenna. A short thick monopole.   111

stub feeder (radial feeder). A feeder that connects a load to its only source of power.   103, 202

stub-multiple feeder. A feeder that operates as either a stub or a multiple feeder.   103, 202

stub shaft (rotating machinery). A separate shaft not carried in its own bearings and connected to the shaft of a machine. *See also:* **rotor (rotating machinery).**   63

stub-supported coaxial. A coaxial whose inner conductor is supported by means of short-circuited coaxial stubs. *See also:* **waveguide.**   328

stub tuner. A stub that is terminated by movable short-circuiting means and used for matching impedance in the line to which it is joined as a branch. *See also:* **waveguide.**   179

stub, waveguide (waveguide components). A section of waveguide or transmission line joined to the main guide or transmission line and containing an essentially nondissipative termination.   166

stud (switching device). A rigid conductor between a terminal and a contact.   103, 202, 27, 79

stuffing box (watertight gland). A device for use where a cable passes into a junction box or other piece of apparatus and is so designed as to render the joint watertight.   64

stylus (electroacoustics). A mechanical element that provides the coupling between the recording or the reproducing transducer and the groove of a recording medium. *See:* **phonograph pickup.**   176

stylus drag (needle drag) (electroacoustics). An expression used to denote the force resulting from friction between the surface of the recording medium and the reproducing stylus. *See also:* **phonograph pickup.**   176

stylus force (electroacoustics). The vertical force exerted on a stationary recording medium by the stylus when in its operating position. *See also:* **phonograph pickup.**   176

subassembly (electric and electronics parts and equipments). Two or more basic parts which form a portion of an assembly or a unit, replaceable as a whole, but having a part or parts which are individually replaceable. *Notes:* (1) The application, size, and construction of an item may be factors in determining whether an item is regarded as a unit, an assembly, a subassembly, or a basic part. A small electric motor might be considered as a part if it is not normally subject to disassembly. (2) The distinction between an assembly and a subassembly is not always exact; an assembly in one instance may be a subassembly in another where it forms a portion of an assembly. Typical examples: filter network, terminal board with mounted parts.   17

subcarrier (facsimile). A carrier which is applied as a modulating wave to modulate another carrier.   12

subclutter visibility (radar). The ratio by which the target echo power may be weaker than the coincident clutter echo power and still be detected with specified detection and false alarm probabilities. Target and clutter powers are measured on a single pulse return and all target radial velocities are assumed equally likely.   13

subdivided capacitor (condenser box*). A capacitor in which several capacitors known as sections are so mounted that they may be used individually or in combination.   210

* Deprecated

subfeeder. A feeder originating at a distribution center other than the main distribution center and supplying one or more branch-circuit distribution centers. *See:* **feeder.**   328

subharmonic. A sinusoidal quantity having a frequency that is an integral submultiple of the frequency of some other sinusoidal quantity to which it is referred. For example, a wave, the frequency of which is half the fundamental frequency of another wave, is called the second subharmonic of that wave. *See also:* **signal wave.**   239, 111, 188, 59

subharmonic protector (series capacitor). A device to detect subharmonic current of a specified frequency and duration to initiate closing of the capacitor bypass switches.   86

subject copy (facsimile). The material in graphic form which is to be transmitted for facsimile reproduction. *See also:* **facsimile (electrical communication).**   12

subjective brightness (light emitting diodes). The subjective attribute of any light sensation giving rise to the percept of luminous intensity, including the whole scale of qualities of being bright, light, brilliant, dim, or dark. *Note:* The term brightness often is used when referring to the measurable **photometric brightness.** While the context usually makes it clear as to which meaning is intended, the preferable term for the photometric quantity is **luminance,** thus reserving **brightness** for the subjective sensation. *See:* **luminance.** *See also:* **visual field.**   167, 162

submarine cable. A cable designed for service under water. *Note:* Submarine cable is usually a lead-covered cable with a steel armor applied between layers of jute.   64

submerged-resistor induction furnace. A device for melting metal comprising a melting hearth, a depending melting channel closed through the hearth, a primary induction winding, and a magnetic core which links the melting channel and the primary winding. *See also:* **induction heating.**   14, 114

submersible (rotating machinery) (industrial control). So constructed as to be successfully operable when submerged in water under specified conditions of pressure and time. *See also:* **constant current transformer; direct-current commutating machine.** 103, 202, 63, 225, 206, 3, 53

submersible entrance terminals (distribution oil cutouts) (cableheads). A hermetically sealable entrance terminal for the connection of cable having a submersible sheathing or jacket. 103, 202

submersible fuse (subway oil cutout). *See:* **submersible; fuse.**

submersible transformer. A transformer so constructed as to be successfully operable when submerged in water under predetermined conditions of pressure and time. 53

subnormality (electrical depression) (electrobiology). The state of reduced electrical sensitivity after a response or succession of responses. *See also:* **excitability (electrobiology).** 192

subpanel (1) (photoelectric converter). Combination of photoelectric converters in parallel mounted on a flat supporting structure. *See also:* **semiconductor.** 186
(2) (solar cells). A combination of solar cells in series/parallel matrix to provide current at array (bus) voltage. 113

subpost car frame (elevators). A car frame all of whose members are located below the car platform. *See also:* **hoistway (elevator or dumbwaiter).** 328

subprogram (test, measurement and diagnostic equipment). A part of a larger program which can be converted into machine language independently. 54

subreflector (antennas). A reflector which redirects the power radiated from the feed to the main reflector. 111

subroutine (electronic computation). (1) In a routine, a portion that causes a computer to carry out a well-defined mathematical or logic operation. (2) A routine that is arranged so that control may be transferred to it from a master routine and so that at the conclusion of the subroutine, control reverts to the master routine. *Note:* Such a subroutine is usually called a closed subroutine. A single routine may simultaneously be both a subroutine with respect to another routine and a master routine with respect to a third. Usually control is transferred to a single subroutine from more than one place in the master routine and the reason for using the subroutine is to avoid having to repeat the same sequence of instructions in different places in the master routine. 210, 54

subroutine, open. A subroutine that must be relocated and inserted into a routine at each place it is used. Contrast with closed subroutine. *See:* **subroutine.** 54

subscriber equipment (protective signaling). That portion of a system installed in the protected premises or otherwise supervised. *See also:* **protective signaling.** 328

subscriber line (subscriber loop) (data transmission). A telephone line between a central office and a telephone station, private branch exchange, or other end equipment. 193, 59

subscriber multiple. A bank of jacks in a manual switchboard providing outgoing access to subscriber lines, and usually having more than one appearance across the face of the switchboard. 328

subscriber set (customer set). An assembly of apparatus for use in originating or receiving calls on the premises of a subscriber to a communication or signaling service. *See also:* **voice-frequency telephony.** 328

subsidence. *See:* **attenuation; damping.**

subsidence ratio (automatic control). A measure of the damping of a second-order linear oscillation, resulting from step or ramp forcing, expressed as the greater divided by the lesser of two successive excursions in the same direction from an ultimate steady-state value. *Note:* The term is also used loosely to describe the ratio of the first two consecutive peaks of any damped oscillation. 56

subsidiary communications authorization (SCA) subcarrier modulation. The FCC permits broadcasters to transmit privileged information and control signals on subcarriers as specified under the SCA but only when transmitted in conjunction with broadcast programming. *Notes:* (1) With monophonic broadcasting, the SCA service may use from 20 to 75 kHz with no restriction on the number of subcarriers, but the total SCA modulation of the RF (radio frequency) carrier must not exceed 30 percent and the crosstalk into the main channel must be at least 60 dB down. (2) With stereophonic broadcasting, the SCA service is limited to 53–75 kHz, 10 percent modulation of the carrier, and must still comply with the 60 dB crosstalk ratio. A 67 kHz subcarrier with ±6 kHz modulation is often used. 16

subsidiary conduit (lateral). A terminating branch of an underground conduit run, extending from a manhole or handhole to a nearby building, handhole, or pole. *See also:* **cable.** 328

subsonic frequency*. *See:* **infrasonic frequency.**

* Deprecated

substantial (transmission and distribution). So constructed and arranged as to be of adequate strength and durability for the service to be performed under the prevailing conditions. 262

substation (1) (power generation). An area or group of equipment containing switches, circuit breakers, buses, and transformers for switching power circuits and to transform power from one voltage to another or from one system to another. 58
(2) (transmission and distribution). An assemblage of equipment for purposes other than generation or utilization, through which electric energy in bulk is passed for the purpose of switching or modifying its characteristics. Service equipment, distribution transformer installations, or other minor distribution or transmission equipment are not classified as substations. *Note:* A substation is of such size or complexity that it incorporates one or more buses, a multiplicity of circuit breakers, and usually is either the sole receiving point of commonly more than one supply circuit, or it sectionalizes the transmission circuits passing through it by means of circuit breakers. *See also:* **alternating-current distribution; direct-current distribution.** 64

substitution error, direct-current-radio-frequency (bolometers). The error arising in the bolometric measurement technique when a quantity of direct-current or audio-frequency power is replaced by a quantity of radio-frequency power with the result that the different current distributions generate different temperature fields that give the bolometer element different values of resistance for the same amounts of power. This error is expressed as

$$\epsilon_s = \frac{\eta_e - \eta}{\eta}$$

where η_e is the effective efficiency of the bolometer units and η is the efficiency of the bolometer unit. *See also:* **bolometric power meter.** 185

substitution error, dual-element. A substitution error peculiar to dual-element bolometer units that results from a different division of direct-current (or audio-frequency) and radio-frequency powers between the two elements. 185

substitution power (bolometers). The difference in bias power required to maintain the resistance of a bolometer at the same value before and after radio-frequency power is applied. Commonly, a bolometer is placed in one arm of a Wheatstone bridge that is balanced when the bias current (direct current and/or audio frequency) holds the bolometer at its nominal operating resistance. Following the application of the radio-frequency signal, the reduction in bias power is taken as a measure of the radio-frequency power. This reduction in the bias power is the substitution power and is given by

$$P = I_1{}^2R - I_2{}^2R$$

where I_1 and I_2 are the bias currents before and after radio-frequency power is applied and R is the nominal operating resistance of the bolometer. *See also:* **bolometric power meter.** 185, 115

substrate (1) (integrated circuit). The supporting material upon or within which an integrated circuit is fabricated or to which an integrated circuit is attached. 312, 191, 197

(2) (photovoltaic power system). Supporting material or structure for solar cells in a panel assembly. Solar cells are attached to the substrate. *See also:* **photovoltaic power system.** 186

subsurface corrosion. Formation of isolated particles of corrosion product(s) beneath the metal surface. This results from the preferential reaction of certain alloy constituents by inward diffusion of oxygen, nitrogen, sulfur, etcetera (internal oxidation). 221, 205

sub-surface transformer. A transformer utilized as part of an underground distribution system, connected below ground to high voltage and low voltage cables, and located below the surface of the ground. 53

subsynchronous reluctance motor. A form of reluctance motor that has the number of salient poles greater than the number of electrical poles of the primary winding, thus causing the motor to operate at a constant average speed that is a submultiple of its apparent synchronous speed. *See:* **asynchronous machine.** 63

sub-synchronous satellite (communication satellite). A satellite, for which the sidereal period of rotation of the primary body about its own axis is an integral multiple of the mean sidereal period of revolution of the satellite about the primary body. 74

subsystem (1) (power system communications). A division of a system that in itself has the characteristics of a system. 59

(2) (nuclear power generating station protective systems). That part of the system which effects a particular protective function. These subsystems may include, but are not limited to those actuating: reactor shutdown; safety injection; containment isolation; emergency core cooling; containment pressure and temperature reduction; containment air cleaning. 31

subtransient current (rotating machinery). The initial alternating component of armature current following a sudden short circuit. *See also:* **armature.** 63

subtransient internal voltage (synchronous machine) (specified operating condition). The fundamental-frequency component of the voltage of each armature phase that would appear at the terminals immediately following the sudden removal of the load. *Note:* The subtransient internal voltage, as shown in the phasor diagram, is related to the terminal-voltage and phase-current phasors by the equation:

$$\mathbf{E}''_1 = \mathbf{E}_a + R\mathbf{I}_a + jX''_d\mathbf{I}_{ad} + jX''_q\mathbf{I}_{aq}$$

For a machine subject to saturation, the reactances should be determined for the degree of saturation applicable to the specified operating conditions. 63

subway transformer. A submersible-type distribution transformer suitable for installation in an underground vault. 53

sudden failure. *See:* **failure, sudden.**

sudden-pressure relay (power switchgear). A relay that operates by the rate of rise in pressure of a liquid or gas. 103, 127

sudden short-circuit test (synchronous machine). A test in which a short-circuit is suddenly applied to the armature winding of the machine under specified operating conditions. 63

Suez Canal searchlight. A searchlight constructed to the specifications of the Canal Administration that by regulation of the Administration, must be carried by every ship traversing the canal, so located as to illuminate the banks. 328

suicide control (adjustable-speed drive). A control function that reduces and automatically maintains the generator voltage at approximately zero by negative feedback. *See also:* **control system, feedback.** 219, 206

sum frequency (parametric device). The sum of a harmonic (nf_p) of the pump frequency (f_p) and the signal frequency (f_s), where n is a positive integer. *Note:* Usually n is equal to one. *See also:* **parametric device.** 277, 191

sum-frequency parametric amplifier*. *See:* **noninverting parametric device.** *See also:* **parametric device.**

* Deprecated

summary punch (test, measurement and diagnostic equipment). A tape or card punch operating in conjunction with another machine to punch data which have been summarized or calculated by the other machine. 54

summation check (computing systems). A check based on the formation of the sum of the digits of a numeral. The sum of the individual digits is usually compared with a previously computed value. 255, 77

summer (computing systems). *See:* **summing amplifier.**

summing amplifier. An operational amplifier that produces an output signal equal to a weighted sum of the input signals. *Note:* In an analog computer, the term **summer** is synonymous with **summing amplifier.** *See also:* **electronic analog computer.** 9, 77, 10

summing junction. The junction common to the input and feedback impedances used with an operational amplifier. *See also:* **electronic analog computer.** 9, 77, 10

summing point (1). Any point at which signals are added algebraically. *Note:* For example the null junction of a power supply is a summing point because, as the input to a high-gain direct-current amplifier, operational summing can be performed at this point. As a virtual ground, the summing point decouples all inputs so that they add linearly in the output, without other interaction. *See:* **operational programming.** *See also:* **null junction; power supply.** 56, 228, 186

(2) (industrial control). The point in a feedback control system at which the algebraic sum of two or more signals is obtained. *See also:* **control system, feedback.** 219, 206

sun bearing. The angle measured in the plane of the horizon between a vertical plane at a right angle to the window wall and the position of this plane after it has been rotated to contain the sun. *See also:* **sunlight.** 167

sunlight. Direct visible radiation from the sun.

superconducting. The state of a superconductor in which it exhibits superconductivity. *Example:* Lead is superconducting below a critical temperature and at sufficiently low operating frequencies. *See:* **normal (state of a superconductor).** *See also:* **superconductivity.** 191

superconductive. Pertaining to a material or device that is capable of exhibiting superconductivity. *Example:* Lead is a superconductive metal regardless of temperature. The cryotron is a superconductive computer component. *See also:* **superconductivity.** 191

superconductivity. A property of a material that is characterized by zero electric resistivity and, ideally, zero permeability. 191

superconductor. Any material that is capable of exhibiting superconductivity. *Example:* Lead is a superconductor. *See also:* **superconductivity.** 191

superdirectivity (antenna). The directivity of an antenna when its value exceeds the value which could be expected from the antenna on the basis of its dimensions and the excitation that would have yielded in-phase addition in the direction of maximum radiation intensity. *Note:* Superdirectivity is obtained only at the cost of a large increase in the ratio of average stored energy to energy radiated per cycle. *See also:* **radiation.** 179, 111

supergroup. *See:* **channel supergroup.**

superheterodyne reception. A method of receiving radio waves in which the process of heterodyne reception is used to convert the voltage of the received wave into a voltage of an intermediate, but usually superaudible, frequency, that is then detected. *See also:* **radio receiver.** 328

superimposed ringing (telephone switching systems). Selective ringing that utilizes direct current polarity to obtain selectivity. 55

superposed circuit. An additional channel obtained from one or more circuits, normally provided for other channels, in such a manner that all the channels can be used simultaneously without mutual interference. *See also:* **trans-**

mission line. 328

superposition theorem. States that the current that flows in a linear network, or the potential difference that exists between any two points in such a network, resulting from the simultaneous application of a number of voltages distributed in any manner whatsoever throughout the network is the sum of the component currents at the first point, or the component potential differences between the two points, that would be caused by the individual voltages acting separately. 328

superregeneration. A form of regenerative amplification, frequently used in radio receiver detecting circuits, in which oscillations are alternately allowed to build up and are quenched at a superaudible rate. *See also:* **radio receiver.** 328

supersonic frequency*. *See:* **ultrasonic frequency.**

* Obsolescent

supersynchronous satellite (communication satellite). A satellite with mean sidereal period of revolution about the primary body which is an integral multiple of the sidereal period of rotation of the primary body about its axis. 74

supervised circuit (protective signaling). A closed circuit having a current-responsive device to indicate a break in the circuit, and, in some cases, to indicate an accidental ground. *See also:* **protective signaling.** 328

supervision (telephone switching systems). The function of indicating and controlling the status of a call. 55

supervisory control (1) (general). A form of remote control comprising an arrangement for the selective control of remotely located units by electrical means over one or more common interconnecting channels. 103, 202

(2) (power-system communication). A discrete portion of a supervisory control system associated with the control and/or indication of the status of a remote device. *See also:* **supervisory control system.** 59

supervisory control point, control and indication (C & I) (power-system communication). A supervisory control point permitting control of an end device with report back and display of its status. *See also:* **supervisory control system.** 59

supervisory control point, indication (power-system communication). A point on a supervisory control system providing indication only of the status of a remote device. *See also:* **supervisory control system.** 59

supervisory control point, jog control (power-system communication). A point on a supervisory control system that permits changing a controlled device one increment each time a raise or a lower command is executed. *See also:* **supervisory control system.** 59

supervisory control point, raise-lower (power-system communication). A supervisory control point capable of performing a raise or lower control for as long as a manual raise or lower control switch is operated. *See also:* **supervisory control system.** 59

supervisory control point, status (power-system communication). A supervisory control point displaying the one of two or three possible conditions of a remote device. If the device has four or more possible conditions, the remote indication of the condition will be termed telemetering. *See also:* **supervisory control system.** 59

supervisory control receiver (power-system communication). A supervisory control subsystem used to receive codes. *See also:* **supervisory control system.** 59

supervisory control system (1). A remote-control system exchanging coded signals over communications channels so as to effect control and display the status of remote equipment with far fewer than one channel per control point. 59

(2) (power-system communication) (continuous scan). A supervisory control system that continually reports the position of controlled devices except when interrupted for control, telemeter, or other discrete operations. *See also:* **supervisory control system.** 59

supervisory control system, quiescent (power system communication). A supervisory control system that does not transmit codes in an at-rest condition. *See also:* **supervisory control system.** 59

supervisory control system, select before operate (power-system communication). A supervisory control system in which a point is first selected and a return displayed proving the selection after which one of several operations can be performed on that point. *See also:* **supervisory control system.** 59

supervisory control transmitter (power-system communication). A supervisory control subsystem used to transmit codes. *See also:* **supervisory control system.** 59

supervisory indication. A form of remote indication comprising an arrangement for the automatic indication of the position or condition of remotely located units by electrical means over one or more common interconnecting channels. 103, 202

supervisory relay. A relay that, during a call, is generally controlled by the transmitter current supplied to a subscriber line in order to receive, from the associated station, directing signals that control the actions of operators or switching mechanisms with regard to the connection. 328

supervisory routine (computing systems). *See:* **executive routine.**

supervisory signal (telephone switching systems). Any signal used to indicate or control the states of the circuits involved in a particular connection. 55

supervisory station check. The automatic selection in a definite order, by means of a single initiation at the master station, of all of the supervisory points associated with one remote station of a system; and the transmission to the master station of indications of positions or conditions of the individual equipment or device associated with each point. 103, 202

supervisory system. All supervisory control, indicating, and telemeter selection devices in the master station and all of the complementary devices in the remote station, or stations, that utilize a single common interconnecting channel for the transmission of the control or indication signals between these stations. *Note:* The supervisory system may be designed to detect and signal the deviation of supervised persons or devices from an established norm. *See also:* **protective signaling.** 103, 202

supervisory system check. The automatic selection in a definite order, by means of a single initiation at the master station, of all supervisory points associated with all of the remote stations in a system; and the transmission to the master station of indications of positions or conditions of the individual equipment or device associated with each point. *See also:* **protective signaling.** 103, 202

supervisory telemeter selection. A form of remote telemeter selection comprising an arrangement for the selective connection of telemeter transmitting equipment to an appropriate telemeter receiving equipment over one or more common interconnecting channels. 103, 202

supervisory tone (telephone switching systems). A tone that indicates to equipment, an operator or a customer that a particular state in the call has been reached, and which may signify the need for action to be taken. The terms used for the various supervisory tones are usually self-explanatory. 55

supplementary lighting. Lighting used to provide an additional quantity and quality of illumination that cannot readily be obtained by a general lighting system and that supplements the general lighting level, usually for specific work requirements. *See also:* **general lighting.** 167

supply circuit (household electric ranges). The circuit that is the immediate source of the electric energy used by the range. *See also:* **appliance outlet.** 263

supply impedance (inverters). The impedance appearing across the input lines to the power inverter with the power inverter disconnected. *See:* **self-commutated inverters.** 208

supply line, motor (rotating machinery). The source of electric power to which the windings of a motor are connected. *See also:* **asynchronous machine; direct-current commutating machine.** 63

supply transient voltage (inverters). The peak instantaneous voltage appearing across the input lines to the power inverter with the inverter disconnected. *See:* **self-commutated inverters.** 208

supply voltage (electrode) (electron tubes). The voltage, usually direct, applied by an external source to the circuit of an electrode. *See:* **electrode voltage (electron tube).** 244, 190

support equipment (test, measurement and diagnostic equipment). Equipment required to make an item, system or facility operational in its environment. This includes all equipment required to maintain and operate the item, system or facility and the computer programs related thereto. 54

supporting operations area(s) (nuclear power generating station). Functional area(s) allocated for controls and displays which support plant operation. 358

supporting structure (transmission and distribution). The main supporting unit (usually a pole or tower). 262

support ring (rotating machinery). A structure for the support of a winding overhang; either constructed of insulating material, carrying support-ring insulation, or separately insulated before assembly. *See also:* **stator.** 63

support-ring insulation (rotating machinery). Insulation between the winding overhang or end winding and the winding support rings. *See also:* **rotor (rotating machinery); stator.** 63

support test system (test, measurement and diagnostic equipment). A measurement system used to assess the quality of operational equipments and may include (1) test equipment; (2) ancillary equipment; (3) supporting documentation; (4) operating personnel.   54

suppressed-carrier modulation. Modulation in which the carrier is suppressed. *Note:* By **carrier** is meant that part of the modulated wave that corresponds in a specified manner to the unmodulated wave.   242

suppressed-carrier operation. That form of amplitude-modulation carrier transmission in which the carrier wave is suppressed. *See also:* **amplitude modulation.**   328

suppressed time delay (electronic navigation). A deliberate displacement of the zero of the time scale with respect to time of emission of a pulse. *See also:* **navigation.**
     13, 187

suppressed-zero instrument. An indicating or recording instrument in which the zero position is below the end of the scale markings. *See also:* **instrument.**   328

suppressed-zero range. A range where the zero value of the measured variable, measured signal, etcetera, is less than the lower range value. Zero does not appear on the scale. *Note:* For example: 20 to 100.   295

suppression (computing systems). *See:* **zero suppression.**

suppression distributor rotor. Rotor of an ignition distributor with a built-in suppressor. *See also:* **electromagnetic compatibility.**   220, 199

suppression ratio (suppressed-zero range). The ratio of the lower range-value to the span. *Note:* For example: Range 20 to 100

$$\text{Suppression Ratio} = \frac{20}{80} = 0.25$$

     295

suppressive wiring techniques (coupling in control systems). Those wiring techniques which result in the reduction of electric or magnetic fields in the vicinity of the wires which carry current without altering the value of the current. Wires which are candidates for suppressive techniques are generally connected to a noise source, may couple noise into a susceptible circuit by induction. Example: twisting or transposing of alternating-current power lines to reduce the intensity of magnetic field produced by current in these lines. *See also:* **barrier wiring techniques; compensatory wiring techniques.**   43

suppressor grid. A grid that is interposed between two positive electrodes (usually the screen grid and the plate), primarily to reduce the flow of secondary electrons from one electrode to the other. *See also:* **electrode (electron tube); grid.**   190, 125

suppressor spark plug. A spark plug with a built-in interference suppressor. *See also:* **electromagnetic compatibility.**   199

surface active agent. *See:* **wetting agent (electroplating).**

surface barrier contact (semiconductor radiation detectors). A rectifying contact that is characterized by a potential barrier associated with an inversion or accumulation layer; said inversion or accumulation layer being caused by surface charge resulting from the presence of surface states or work function differences, or both.   23, 118, 119

surface barrier radiation detector (semiconductor radiation detectors). A radiation detector for which the principal rectifying junction is a surface barrier contact.
     23, 118, 119

surface duct (radio wave propagation). An atmospheric

duct for which the lower boundary is the surface of the earth. *See also:* **radiation; radio wave propagation.**
     180, 146

surface leakage. The passage of current over the surface of a material rather than through its volume.   210

surface metal raceway (metal molding). A raceway consisting of an assembly of backing and capping. *See also:* **raceways.**   328

surface-mounted device. A device, the entire body of which projects in front of the mounting surface.   103, 202

surface-mounted luminaire. A luminaire that is mounted directly on the ceiling. *See also:* **suspended (pendant) luminaire.**   167

surface noise (mechanical recording). The noise component in the electric output of a pickup due to irregularities in the contact surface of the groove. *See also:* **phonograph pickup.**   176

surface of position (electronic navigation). Any surface defined by a constant value of some navigation quantity. *See also:* **navigation.**   13, 187

surface-potential gradient. The slope of a potential profile, the path of which intersects equipotential lines at right angles. *See also:* **ground.**   313

surface, prescribed (sound measurements). A prescribed surface is a hypothetical surface surrounding the machine on which sound measurements are made.   129

surface search radar. *See:* **navigational radar.**

surface wave antenna. An antenna which radiates power from discontinuities in the structure that interrupt a bound wave on the antenna surface.   111

surface-wave transmission line (waveguides). A transmission line in which propagation in other than a *TEM* mode is constrained to follow the external face of a guiding structure. *See:* **waveguide.**   179

surge (electric power). A transient wave of current, potential, or power in the electric circuit. *Note:* A transient has a high rate of change of voltage (current) in the system. It will be propagated along the length of the circuit. *See also:* **lightning.**   2, 64, 244, 62

surge arrester. A protective device for limiting surge voltages on equipment by discharging or bypassing surge current; it prevents continued flow of follow current to ground, and is capable of repeating these functions as specified. *Note:* Hereafter, the term "arrester" as used in this standard shall be understood to mean "surge arrester." *Syn:* **lightning arrester*.**   2

* Deprecated.

surge-crest ammeter. A special form of magnetometer intended to be used with magnetizable links to measure the crest value of transient electric currents. *See also:* **instrument.**   328

surge diverter. *See:* **surge arrester.**

surge electrode current. *See:* **fault electrode current.**

surge (non-repetitive) forward current (semiconductor). A forward current current of short time duration and specified waveshape.   66

surge generator (impulse generator) (lightning generator*). An electric apparatus suitable for the production of surges. *Note:* Surge generator types common in the art are: transformer-capacitor; transformer-rectifier; transformer-rectifier-capacitor, parallel charging, series discharging. 62, 64

* Deprecated

surge impedance (self-surge impedance). The ratio between voltage and current of a wave that travels on a line of infinite length and of the same characteristics as the relevant line. *See:* **characteristic impedance.** 244, 62

surge (nonrepetitive) ON-state current (thyristor). An ON-state current of short-time duration and specified wave shape. *See also:* **principal current.** 243, 57, 208, 191

surge protection. *See:* **rate-of-change protection.**

surge suppressor (industrial control). A device operative in conformance with the rate of change of current, voltage, power, etcetera, to prevent the rise of such quantity above a predetermined value. 206

surge, switching. *See:* **switching surge.**

surge voltage recorder (klydonograph). *See:* **Lichtenberg figure camera.**

surveillance (nuclear power generating stations). The determination of the state or condition of a system or subsystem. 31

surveillance radar. A radar used to maintain cognizance of selected traffic within a selected area, such as an airport terminal area or air route. 187, 13

susceptance. The imaginary part of admittance. 210, 185

susceptance function (1) (linear passive networks). The driving-point admittance of a lossless network. *Note:* This is an odd function of the complex frequency. *See also:* **linear passive networks.** 238

(2) (circuits and systems). A function that is realizable as a driving-point admittance with ideal inductors and capacitors. *Note:* It must meet the conditions described in Foster's reactance theorem. *See also:* **linear passive networks.** 67

susceptance relay. A mho-type distance relay for which the center of the operating characteristic on the R-X diagram is on the X axis. *Note:* The equation that describes such a characteristic is $Z = K \sin \theta$ where K is a constant and θ is the phase angle by which the input voltage leads the input current. 103, 202, 60, 127

susceptibility meter. A device for measuring low values of magnetic susceptibility. One type is the Curie balance. *See also:* **magnetometer.** 328

susceptiveness (transmission and distribution). The characteristics of a communications circuit including its connected apparatus which determine the extent to which it is adversely affected by inductive fields. 262

susceptor. Energy absorbing device generally used to transfer heat to another load. 14

suspended (pendant) luminaire. A luminaire that is hung from a ceiling by supports. *See:* **flush mounted or recessed; regressed; surface mounted.** 167

suspended-type handset telephone (bracket-type handset telephone). *See:* **hang-up hand telephone.**

suspension insulator. One or a string of suspension-type insulators assembled with the necessary attaching members and designed to support in a generally vertical direction the weight of the conductor and to afford adequate insulation from tower or other structure. *See also:* **insulator; tower.** 64

suspension-insulator unit. An assembly of a shell and hardware, having means for nonrigid coupling to other units or terminal hardware. 261

suspension-insulator weights. Devices, usually cast iron, hung below the conductor on a special spindle supported by the conductor clamp. *Note:* Suspension insulator weights will limit the swing of the insulator string, thus maintaining adequate clearances. In practice, weights of several hundreds of pounds are sometimes used. *See also:* **tower.** 64

suspension strand (messenger). A stranded group of wires supported above the ground at intervals by poles or other structures and employed to furnish within these intervals frequent points of support for conductors or cables. *See also:* **cable; open wire.** 328

sustained gap-arc protection (series capacitor). A means to detect prolonged arcing of the protective power gap or arcing of the backup gap if included to initiate closing of the capacitor bypass switch. 86

sustained interruption (electric power systems). Any interruption not classified as a momentary interruption. *See also:* **outage.** 200

sustained-operation influence. The change in the recorded value, including zero shift, caused solely by energizing the instrument over extended periods of time, as compared to the indication obtained at the end of the first 15 minutes of the application of energy. It is to be expressed as a percentage of the full-scale value. *Note:* The coil used in the standard method shall be approximately 80 inches in diameter, not over 5 inches long, and shall carry sufficient current to produce the required field. The current to produce a field to an accuracy of ±1 percent in air shall be calculated without the instrument in terms of the specific dimensions and turns of the coil. In this coil, 800 ampere-turns will produce a field of approximately 5 oersteds. The instrument under test shall be placed in the center of the coil. *See also:* **accuracy rating (instrument).** 294

sustained oscillation (sustained vibration) (1) (system). The oscillation when forces controlled by the system maintain a periodic oscillation of the system. Example: Pendulum actuated by a clock mechanism. 210

(2) (gas turbines). Those oscillations in which the amplitude does not decrease to zero, or to a negligibly small, final value. 98, 58

sustained overvoltage protection device (series capacitor). A device to detect capacitor voltage that is above rating or predetermined value but is below the sparkover of the protective power gaps, and to initiate the closing of the capacitor bypass switch according to a predetermined voltage-time characteristic. 86

sustained short-circuit test (synchronous machine). A test in which the machine is run as a generator with its terminals short-circuited. 63

sweep. A traversing of a range of values of a quantity for the purpose of delineating, sampling, or controlling another quantity. *Notes:* (1) Examples of swept quantities are (A) the displacement of a scanning spot on the screen of a cathode-ray tube, and (B) the frequency of a wave. (2) Unless otherwise specified, a linear time function is implied; but the sweep may also vary in some other controlled and desirable manner. 178, 185

sweep accuracy (oscilloscopes). Accuracy of the horizontal (vertical) displacement of the trace compared with the reference independent variable, usually expressed in terms of average rate error as a percent of full scale. *See also:* **oscillograph.** 185, 106

sweep-delay accuracy (oscilloscopes). Accuracy of indicated sweep delay, usually specified in error terms. 106

sweep, delayed. *See:* **delayed sweep.**

sweep duration (sawtooth sweep). The time required for the sweep ramp. *See also:* **oscillograph.** 185

sweep duty factor. For repetitive sweeps, the ratio of the sweep duration to the interval between the start of one sweep and the start of the next. *See also:* **oscillograph.** 185

sweep, expanded. *See:* **magnified sweep.**

sweep, external (oscilloscopes). A sweep generated external to the instrument. 106

sweep, free-running. *See:* **free-running sweep.**

sweep frequency (oscilloscopes). The sweep repetition rate. *See also:* **oscillograph.** 185

sweep gate (oscilloscopes). Rectangular waveform used to control the duration of the sweep; usually also used to unblank the cathode-ray tube for the duration of the sweep. *See also:* **oscillograph.** 185

sweep, gated. *See:* **gated sweep.**

sweep generator (oscilloscopes). A circuit that generates a signal used as an independent variable; the signal is usually a ramp, changing value at a constant rate. 106

sweep holdoff interval. The interval between sweeps during which the sweep and/or trigger circuits are inhibited. 185

sweep, incremental. *See:* **incremental sweep.**

sweep lockout (oscilloscopes). Means for preventing multiple sweeps when operating in a single-sweep mode. *See also:* **oscillograph.** 185

sweep magnifier (oscilloscopes). Circuit or control for expanding part of the sweep display. Sometimes known as **sweep expander.** *See also:* **oscillograph.** 185

sweep mode control (oscilloscopes). The control used on some oscilloscopes to set the sweep for triggered, free-running, or synchronized operation. 106

sweep oscillator. An oscillator in which the output frequency varies continuously and periodically between two frequency limits. *See also:* **telephone station.** 122, 196

sweep-out time, charge (semiconductor radiation detector). *See:* **charge collection time.** 23

sweep range (oscilloscopes). The set of sweep-time/division settings provided. *See also:* **oscillograph.** 185

sweep recovery time (oscilloscopes). The minimum possible time between the completion of one sweep and the initiation of the next, usually the sweep holdoff interval. *See also:* **oscillograph.** 185

sweep, recurrent. *See:* **recurrent sweep.**

sweep reset (oscilloscopes). In oscilloscopes with single-sweep operation, the arming of the sweep generator to allow it to cycle once. *See also:* **oscillograph.** 185

sweep, sine-wave. A sweep generated by a sinusoidal function. *See also:* **oscillograph.** 106

sweep, stairstep (oscilloscopes). An incremental sweep in which each step is equal. The electrical deflection waveform producing a stairstep sweep is usually called a staircase or stairstep waveform. *See also:* **sweep, incre-**

mental; oscillograph. 106

sweep switching (automatic). Alternate display of two or more time bases or other sweeps using a single-beam cathode-ray tube; comparable to dual- or multiple-trace operation of the deflection amplifier. 106

sweep time (acoustically tunable optical filter). The time to continuously tune the filter over its spectral range. 72

sweep time/division (oscilloscopes) (non-real time spectrum analyzer). The nominal time required for the spot in the reference coordinate to move from one graticule division to the next. Also the name of the control used to select this time. 68, 106

switch (1) (electrical systems). A device for opening and closing or for changing the connection of a circuit. *Note:* A switch is understood to be manually operated, unless otherwise stated. 206

(2) (computing systems). A device or programming technique for making a selection, for example, a toggle, a conditional jump. 255, 77

(3) (electric and electronics parts and equipments). A device for making, breaking, or changing the connections in an electric circuit. *Note:* a switch may be operated by manual, mechanical, hydraulic, thermal, barometric, or gravitational means, or by electromechanical means not falling within the definition of "relay." 17

(4) (transmission and distribution). A device for opening and closing or for changing the connection of a circuit. A switch is understood to be manually operable unless otherwise stated. 262

switch-and-lock movement. A device, the complete operation of which performs the three functions of unlocking, moving, and locking a switch, movable-point frog, or derail. 328

switch base. The main members to which the insulator units are attached. 27

switchboard (1) (electric power systems). A large single panel, frame, or assembly of panels, on which are mounted, on the face or back or both, switches, overcurrent and other protective devices, buses, and usually instruments. *Note:* Switchboards are generally accessible from the rear as well as from the front and are not intended to be installed in cabinets. *See also:* **center of distribution; distribution center; panelboard.** 256

(2) (power switchgear). A type of switchgear assembly that consists of one or more panels with electric devices mounted thereon, and associated framework. *Note:* Switchboards may be classified by function; that is, power switchboards or control switchboards. Both power and control switchboards, may be further classified by construction as defined. 103

(3) (transmission and distribution). When referred to in connection with supply of electricity, a large single panel, frame, or assembly of panels, on which are mounted (on the face, or back, or both) switches, fuses, buses, and usually instruments. 262

switchboard cord. A cord that is used in conjunction with switchboard apparatus to complete or build up a telephone connection. 328

switchboard lamp (switchboard). A small electric lamp associated with the wiring in such a way as to give a visual indication of the status of a call or to give information concerning the condition of trunks, subscriber lines, and apparatus. 328

switchboard position (telephone switching systems). That portion of a manual switchboard normally provided for the use of one operator. 55

switchboards and panels (marine transportation). A **generator and distribution switchboard** receives energy from the generating plant and distributes directly or indirectly to all equipment supplied by the generating plant. A **subdistribution switchboard** is essentially a section of the generator and distribution switchboard (connected thereto by a bus feeder and remotely located for reasons of convenience or economy) that distributes energy for lighting, heating, and power circuits in a certain section of the vessel. A **distribution panel** receives energy from a distribution or subdistribution switchboard and distributes energy to energy-consuming devices or other distribution panels or panelboards. A **panelboard** is a distribution panel enclosed in a metal cabinet. *See also:* **generating station.** 3

switchboard section (telephone switching systems). A structural unit providing for one or more operator positions. A complete switchboard may consist of one or more sections. 55

switchboard supervisory lamp (cord circuit or trunk circuit). A lamp that is controlled by one or other of the users to attract the attention of the operator. 328

switchboard supervisory relay. A relay that controls a switchboard supervisory lamp. 328

switched-service (SSN) network (telephone switching systems). An arrangement of dedicated switching facilities to provide telecommunications services for a specific customer. 55

switchgear. A general term covering switching and interrupting devices and their combination with associated control, instrumentation, metering, protective and regulating devices, also assemblies of these devices with associated interconnections, accessories and supporting structures used primarily in connection with the generation, transmission, distribution, and conversion of electric power. 93

switchgear assembly. An assembled equipment (indoor or outdoor) including, but not limited to, one or more of the following: switching, interrupting, control, instrumentation, metering, protective and regulating devices, together with their supporting structures, enclosures, conductors, electric interconnections, and accessories. 93

switchgear pothead. A pothead intended for use in a switchgear where the inside ambient air temperature may exceed 40 degrees Celsius. It may be an indoor or outdoor pothead that has been suitably modified by silver surfacing (or the equivalent) the current-carrying parts and incorporates sealing materials suitable for the higher operating temperatures. *See also:* **pothead.** 4

switchgear, protective (thyristor converter). The ac circuit devices and the dc circuit devices that may be used in the thyristor converter unit to clear fault conditions. 121

switchhook (hookswitch). A switch on a telephone set, associated with the structure supporting the receiver or handset. It is operated by the removal or replacement of the receiver or handset on the support. *See also:* **telephone station.** *See:* **switch stick.** 328

switch indicator. A device used at a noninterlocked switch to indicate the presence of a train in a block. 328

switching (1) (single-phase motor). The point in the starting operation at which the stator-winding circuits are switched from one connection arrangement to another. *See also:* **asynchronous machine.** 63

(2) (test, measurement and diagnostic equipment). The act of manually, mechanically or electrically actuating a device for opening or closing an electrical circuit. 54

switching amplifier (industrial control). An amplifier which is designed to be applied so that its output is sustained at one of two specified states dependent upon the presence of specified inputs. *See also:* **control system, feedback.** 219, 206

switching array (telephone switching systems). An assemblage of multipled crosspoints. 55

switching card (test, measurement and diagnostic equipment). A plug-in device which provides the necessary interconnection to the unit under test. 54

switching circuit (data transmission). Term applied to the method of handling traffic through a switching center, either from a local user or from other switching centers, whereby additional electrical connection is established between the calling and the called station. 59

switching coefficient. The derivative of applied magnetizing force with respect to the reciprocal of the resultant switching time. It is usually determined as the reciprocal of the slope of a curve of reciprocals of switching times versus values of applied magnetizing forces. The magnetizing forces are applied as step functions. 77

switching control center (telephone switching systems). A place where maintenance analysis and control activities are centralized for switching entities situated in different locations. 55

switching current (power circuit breaker). The value of current expressed in rms symmetrical amperes which the power circuit breaker element of the circuit protector interrupts at the rated maximum voltage and rated frequency under the prescribed test conditions. 79, 158

switching current rating (power distribution systems). The designated rms current which a load-break connector can connect and disconnect for a specified number of times under specified conditions. 134

switching device (switch). A device designed to close or open, or both, one or more electric circuits. *Note:* The term **switch** in International Electrotechnical Commission (IEC) practice refers to a mechanical switching device capable of opening and closing rated continuous load current. *See:* **mechanical switching device; nonmechanical switching device.** 103, 202, 27

switching entity (telephone switching systems). A switching network and its control. 55

switching impulse. Ideally, an aperiodic transient impulse voltage that rises rapidly to a maximum value and falls, usually less rapidly, to zero. Switching impulses generally have front times of the order of tens to thousands of microseconds, in contrast to lightning impulses, which have front times from fractions of a microsecond to tens of microseconds. 108

switching impulse insulation level. An insulation level expressed in kilovolts of the crest value of a switching impulse withstand voltage. 53

switching impulse protection level (protective device). The maximum switching impulse expected at the terminals of a surge protective device under specified conditions of operation. 53

switching-impulse sparkover voltage (arrester). The impulse sparkover voltage with an impulse having a virtual duration of wavefront greater than 30 microseconds. 62

switching impulse test (transformer). Application of the "standard switching impulse," a full wave having a front time of 250 microseconds and a time to half value of 2500 microseconds, described as a 250/2500 impulse. *Note:* It is recognized that some apparatus standards may have to use a modified wave shape where practical test considerations or particular dielectric strength characteristics make some modification imperative. Transformers for example, use a modified switching impulse wave with the following characteristics: (1) Time to crest greater than 100 microseconds. (2) Exceeds 90 percent of crest value for at least 200 microseconds. (3) Time to first voltage zero on tail not less than 1000 microseconds, except where core saturation causes the tail to become shorter. 53

switching network (telephone switching systems). Switching stages and their interconnections. Within a switching system there may be more than one switching network. 55

switching-network plan (telephone switching systems). The switching stages and their interconnections within a specific switching system. 55

switching plan (telephone switching systems). A plan for the interconnection of switching entities. 55

switching, slave-sweep (oscilloscopes). A combination of sweep switching and multiple-trace operation in which a specific channel is displayed with a specific sweep. 106

switching stage (telephone switching systems). An assemblage of switching arrays within each inlet which can be connected through a single crosspoint to its associated outlet. 55

switching structure. An open framework supporting the main switching and associated equipment, such as instrument transformers, buses, fuses, and connections. It may be designed for indoor or outdoor use and may be assembled with or without switchboard panels carrying the control equipment. 103, 202

switching surge (conductor stringing equipment). A transient wave of over-voltage in an electrical circuit caused by a switching operation. When this occurs, a momentary voltage surge could be induced in a circuit adjacent and parallel to the switched circuit in excess of the voltage induced normally during steady state conditions. If the adjacent circuit is under construction, switching operations should be minimized to reduce the possibility of hazards to the workmen. 45

switching-surge protective level (arrester). The highest value of switching-surge voltage that may appear across the terminals under the prescribed conditions. *Note:* The switching-surge protective levels are given numerically by the maximums of the following quantities: (1) discharge voltage at a given discharge current, and (2) switching-impulse sparkover voltage. 62

switching system (telephone switching systems). A system in which connections are established between inlets and outlets either directly or with intermediate storage. 55

switching-system processor (telephone switching systems). Circuitry to perform a series of switching system operations under control of a program. 55

switching time (1) (electric power circuits). The period from the time a switching operation is required due to a forced outage until that switching operation is performed. Switching operations include reclosing a circuit breaker after a trip-out, opening or closing a sectionalizing switch or circuit breaker, or replacing a fuse link. *See also:* **outage.** 200

(2) (magnetic storage cells). (A) T_s, the time interval between the reference time and the last instant at which the instantaneous voltage response of a magnetic cell reaches a stated fraction of its peak value. (B) T_x, the time interval between the reference time and the first instant at which the instantaneous integrated voltage response reaches a stated fraction of its peak value. 77

(3) (analog-to-digital converter with multiplexor with sample and hold). That time required from the time at which a channel is addressed until the selected analog signal is available at the output within a given accuracy. *Syn:* **settling time.** *See also:* **analog-to-digital converter with sample and hold.** 10

switching torque (motor having an automatic connection change during the starting period). The minimum external torque developed by the motor as it accelerates through switch operating speed. *Note:* It should be noted that if the torque on the starting connection is never less than the switching torque, the pull-up torque is identical with the switching torque; however, if the torque on the starting connection falls below the switching torque at some speed below switch operating speed, the pull-up and switching torques are not identical. *See also:* **asynchronous machine.** 328

switching torque, single-phase motor. The minimum torque which a motor will provide at switching at normal operating temperature, with rated voltage applied at rated frequency. *See also:* **asynchronous machine.** 63

switch inrush current capability for single capacitance (as applied to interrupter switches). This capability is a function of the rated switching current, for single capacitance, the rated differential capacitance voltage (minimum) and the maximum design voltage of the switch. *Note:* This can be calculated from the equation:

$$\text{Capability, in Peak Amperes} = \sqrt{2}\, I_C \sqrt{1 + \frac{0.816 E_m}{\Delta V_{\min}}}$$

where

I_C = rated switching current for single capacitance
$\Delta V_{\min}$ = rated differential capacitance voltage, minimum
E_m = switch rated maximum voltage, in volts, rms. 27, 103

switch, load matching. *See:* **load-matching switch.**
switch, load transfer. *See:* **load transfer switch.**
switch machine. A quick-acting mechanism, electrically controlled, for positioning track switch points, and so arranged that the accidental trailing of the switch points does

not cause damage. A switch machine may be electrically or pneumatically operated. *See also:* **car retarder.**   328

switch machine lever lights. A group of lights indicating the position of the switch machine.   328

switch-machine point detector. *See:* **point detector.**

switch or contactor, load. *See:* **load switch or contactor.**

switchroom (telephone switching systems). That part of a building that houses an assemblage of switching equipment.   55

switch signal. A low two-indication horizontal color light signal with electric lamps for indicating position of switch or derail.   328

switch sleeve. A component of the linkage between the centrifugal mechanism and the starting-switch assembly. *See also:* **starting-switch assembly.**   328

switch starting. *See:* **preheat starting.**

switch stick (switchhook). A device with an insulated handle and a hook or other means for performing stick operation of a switching device.   103, 202

switch train. A series of switches in tandem.   328

SWR. *See:* **standing-wave-ratio indicator.**

syllabic companding (modulation systems). Companding in which the gain variations occur at a rate comparable to the syllabic rate of speech; but do not respond to individual cycles of the audio-frequency signal wave.   242

syllable articulation (percent syllable articulation). The percent articulation obtained when the speech units considered are syllables (usually meaningless and usually of the consonant-vowel-consonant type). *See:* **articulation (percent articulation).** *See also:* **volume equivalent.**   328

symbol. A representation of something by reason of relationship, association, or convention. *See:* **logic symbol.**   255, 77

symbol for a quantity (quantity symbol) (abbreviation). A letter (which may have letters or numbers, or both, as subscripts or superscripts, or both), used to represent a physical quantity or a relationship between quantities. *Compare with:* **abbreviation, functional designation, mathematical symbol, reference designation, and symbol for a unit.** *See also:* **abbreviation.**   173

symbol for a unit (unit symbol) (abbreviation). A letter, a character, or combinations thereof, that may be used in place of the name of the unit. With few exceptions, the letter is taken from the name of the unit. *Compare with:* **abbreviation, mathematical symbol, symbol for a quantity.** *See also:* **abbreviation.**   173

symbolic address (computing systems). An address expressed in symbols convenient to the programmer.   255, 77

symbolic coding (computing systems). Coding that uses machine instructions with symbolic addresses.   255, 77

symbolic logic. The discipline that treats formal logic by means of a formalized artificial language or symbolic calculus whose purpose is to avoid the ambiguities and logical inadequacies of natural languages.   255, 77

symbolic quantity. *See:* **mathematico-physical quantity.**

symmetrical alternating current. A periodic alternating current in which points one-half a period apart are equal and have opposite signs. *See also:* **alternating function; network analysis.**   210

symmetrical component (total current) (alternating-current component). That portion of the total current that constitutes the symmetry.   103, 202

symmetrical components (1) (set of polyphase alternating voltages). The symmetrical components of an unsymmetrical set of sinusoidal polyphase alternating voltages of m phases are the m symmetrical sets of polyphase voltages into which the unsymmetrical set can be uniquely resolved, each component set having an angular phase lag between successive members of the set that is a different integral multiple of the characteristic angular phase difference for the number of phases. The successive component sets will have phase differences that increase from zero for the first set to $(m - 1)$ times the characteristic angular phase difference for the last set. The phase sequence of each component set is identified by the integer that denotes the number of times the angle of lag between successive members of the component set contains the characteristic angular phase difference. If the members of an unsymmetrical set of alternating polyphase voltages are not sinusoidal, each voltage is first resolved into its harmonic components, then the harmonic components of the same period are grouped to form unsymmetrical sets of sinusoidal voltages, and finally each harmonic set of sinusoidal voltages is uniquely resolved into its symmetrical components. Because the resolution of a set of polyphase voltages into its harmonic components is also unique, it follows that the resolution of an unsymmetrical set of polyphase voltages into its symmetrical components is unique. There may be a symmetrical-component set of voltages for each of the possible phase sequences from zero to $(m - 1)$ and for each of the harmonics present from 1 to r, where r may approach infinity in particular cases. Each member of a set of symmetrical component voltages of kth phase sequence and rth harmonic may be denoted by

$$e_{ski} = (2)^{1/2} E_{akr} \cos \left(r\omega t + \alpha_{akr} - (s - 1) K \frac{2\pi}{m} \right)$$

where e_{skr} is the instantaneous voltage component of phase sequence k and harmonic r in phase s. E_{akr} is the root-mean-square amplitude of the voltage component of phase sequence k and harmonic r, using phase a as reference, a_{akr} is the phase angle of the first member of the set, selected as phase a, with respect to a common reference. The letter s as the first subscript denotes the phase identification of the individual member, a, b, c, etcetera for successive members, and a denotes that the first phase, a, has been used as a reference from which other members are specified. The second subscript k denotes the phase sequence of the component, and may run from 0 to $m - 1$. The third subscript denotes the order of the harmonic, and may run from 1 to ∞. The letter s as an algebraic quantity denotes the member of the set and runs from 1 for phase a to m for the last phase. Of the m symmetrical component sets for each harmonic, one will be of zero phase sequence, one of positive phase sequence, and one of negative phase sequence. If the number of phases m ($m > 2$) is even, one of the symmetrical component sets for $k = m/2$ will be a single-phase symmetrical set (polyphase voltages). The zero-phase-sequence component set will constitute a zero-phase symmetrical set (polyphase voltages), and the remaining sequence components will constitute polyphase symmetrical sets (polyphase voltages). *See also:* **network analysis.** 210

(2) (set of polyphase alternating currents). Obtained from the corresponding definition for **symmetrical components (set of polyphase alternating voltages)** by substituting the word **current** for **voltage** wherever it appears. *See also:* **network analysis.** 210

symmetrical fractional-slot winding (rotating machinery). A distributed winding in which the average number of slots per pole per phase is not integral, but in which the winding pattern repeats after every pair of poles, for example, 3½ slots per pole per phase. *See also:* **rotor (rotating machinery); stator.** 63

symmetrically cyclically magnetized condition. A condition of a magnetic material when it is in a cyclically magnetized condition and the limits of the applied magnetizing forces are equal and of opposite sign, so that the limits of flux density are equal and of opposite sign. 210

symmetrical network. *See:* **structurally symmetrical network.**

symmetrical periodic function. A function having the period 2π is symmetrical if it satisfies one or more of the following identities.

$$(1)\, f(x) = -f(-x) \qquad (4)\, f(x) = f(-x)$$
$$(2)\, f(x) = -f(\pi + x) \qquad (5)\, f(x) = f(\pi + x)$$
$$(3)\, f(x) = -f(\pi - x) \qquad (6)\, f(x) = f(\pi - x)$$

See also: **network analysis.** 210

symmetrical set (1) (polyphase voltages). A symmetrical set of polyphase voltages of m phases is a set of polyphase voltages in which each voltage is sinusoidal and has the same amplitude, and the set is arranged in such a sequence that the angular phase difference between each member of the set and the one following it, and between the last member and the first, can be expressed as the same multiple of the characteristic angular phase difference $2\pi/m$ radians. A symmetrical set of polyphase voltages may be expressed by the equations

$$e_a = (2)^{1/2} E_{ar} \cos{(r\omega t + \alpha_{ar})}$$

$$e_b = (2)^{1/2} E_{ar} \cos{\left(r\omega t + \alpha_{ar} - k\,\frac{2\pi}{m}\right)}$$

$$e_c = (2)^{1/2} E_{ar} \cos{\left(r\omega t + \alpha_{ar} - 2k\,\frac{2\pi}{m}\right)}$$

$$e_m = (2)^{1/2} E_{ar} \cos{\left(r\omega t + \alpha_{ar}\right.}$$
$$\left. - (m-1)k\,\frac{2\pi}{m}\right)$$

where E_{ar} is the root-mean-square amplitude of each member of the set, r is the order of the harmonic of each member, with respect to a specified period. a_{ar} is the phase angle of the first member of the set with respect to a selected reference. k is an integer that denotes the phase sequence. *Notes:* (1) Although sets of polyphase voltages

that have the same amplitude and waveform but that are not sinusoidal possess some of the characteristics of a symmetrical set, only in special cases do the several harmonics have the same phase sequence. Since phase sequence is an important feature in the use of symmetrical sets, the definition is limited to sinusoidal quantities. This represents a change from the corresponding definition in the 1941 edition of the American Standard Definitions of Electrical Terms. (2) This definition may be applied to a two-phase four-wire or five-wire circuit if m is considered to be 4 instead of 2. The concept of symmetrical sets is not directly applicable to a two-phase three-wire circuit. 210

(2) (polyphase currents). This definition is obtained from the corresponding definitions for voltage by substituting the word **current** for **voltage,** and the symbol I for E and β for α wherever they appear. The subscripts are unaltered. *See also:* **network analysis.** 210

symmetrical terminal voltage (electromagnetic compatibility). Terminal voltage measured in a delta network across the mains lead. *See also:* **electromagnetic compatibility.** 199

symmetrical transducer (1) (specified pair of terminations). A transducer in which the interchange of that pair of terminations will not affect the transmission. *See also:* **transducer.** 328

(2) (specified terminations in general). A transducer in which all possible pairs of specified terminations may be interchanged without affecting transmission. *See also:* **transducer.** 252, 210

synapse. The junction between two neural elements, which has the property of one-way propagation. *See also:* **biological.** 192

synchro control transformer (synchro or selsyn devices). A transformer with relatively rotatable primary and secondary windings. The primary inputs is a set of two or more voltages from a synchro transmitter that define an angular position relative to that of the transmitter. The secondary output voltage varies with the relative angular alignment of primary and secondary windings, of the control transformer and the position of the transmitter. The output voltage is substantially zero in value at a position known as correspondence. *See also:* **synchro system.** 63

synchro differential receiver (motor) (synchro or selsyn devices). A transformer identical in construction to a synchro differential transmitter but used to develop a torque increasing with the difference in the relative angular displacement (up to about 90 electrical degrees) between the two sets of voltage input signals to its primary and secondary windings, the torque being in a direction to reduce this difference to zero. *See also:* **synchro system.** 63

synchro differential transmitter (generator) (rotating machinery). A transformer with relatively rotatable primary and secondary windings. The primary input is a set of two or more voltages that define an angular position. The secondary output is a set of two or more voltages that represent the sum or difference, depending upon connections, of the position defined by the primary input and the relative angular displacement between primary and secondary windings. *See also:* **synchro system.** 63

synchronism (rotating machinery). The state where connected alternating-current systems, machines, or a combination operate at the same frequency and where the phase-angle displacements between voltages in them are constant, or vary about a steady and stable average value. *See:* **asynchronous machine.** 63

synchronism-check relay. A verification relay whose function is to operate when two input voltages are within predetermined phasor limits. 103, 202, 60, 127

synchronization (data transmission). A means of insuring that both transmitting and receiving stations are operating together (in phase). 59

synchronization error (electronic navigation). The error due to imperfect timing of two operations; this may or may not include signal transmission time. *See also:* **navigation.** 13, 187

synchronized sweep (oscilloscopes) (non-real time spectrum analyzer). A sweep that would free run in the absence of an applied signal but in the presence of the signal is synchronized by it. *See also:* **oscillograph.** 185, 68

synchronizing (1) (rotating machinery). The process whereby a synchronous machine, with its voltage and phase suitably adjusted, is paralleled with another synchronous machine or system. *See also:* **asynchronous machine.** 63

(2) (facsimile). The maintenance of predetermined speed relations between the scanning spot and the recording spot within each scanning line. *See also:* **facsimile (electrical communication).** 12

(3) (television). Maintaining two or more scanning processes in phase. 178

(4) (pulse terms). The process of rendering a first pulse train or other sequence of events synchronous with a second pulse train. 254

synchronizing coefficient (rotating machinery). The quotient of the shaft power and the angular displacement of the rotor. *Note:* It is expressed in kilowatts per electrical radian. Unless otherwise stated, the value will be for rated voltage, load, power-factor, and frequency. *See also:* **asynchronous machine.** 63

synchronizing reactor. A current-limiting reactor for connecting momentarily across the open contacts of a circuit-interrupting device for synchronizing purposes. 53

synchronizing relay. A programming relay whose function is to initiate the closing of a circuit breaker between two alternating-current sources when the voltages of these two sources have a predetermined relationship of magnitude, phase angle, and frequency. 103, 202, 60

synchronizing signal (1) (television). The signal employed for the synchronizing of scanning. *Note:* In television, this signal is composed of pulses at rates related to the line and field frequencies. The signal usually originates in a central synchronizing generator and is added to the combination of picture signal and blanking signal, comprising the output signal from the pickup equipment, to form the composite picture signal. In a television receiver, this signal is normally separated from the picture signal and is used to synchronize the deflection generators. 178

(2) (facsimile). A signal used for maintenance of predetermined speed relations between the scanning spot and recording spot within each scanning line. *See also:* **facsimile signal (picture signal).** 12

(3) (oscillograph). A signal used to synchronize repetitive

functions. *See:* **oscillograph.** 185

(4) (telecommunication). A special signal which may be sent to establish or maintain a fixed relationship in synchronous systems. 194

synchronizing signal compression (television). The reduction in gain applied to the synchronizing signal over any part of its amplitude range with respect to the gain at a specified reference level. *Notes:* (1) The gain referred to in the definition is for a signal amplitude small in comparison with the total peak-to-peak composite picture signal involved. A quantitative evaluation of this effect can be obtained by a measurement of differential gain. (2) Frequently the gain at the level of the peaks of synchronizing pulses is reduced with respect to the gain at the levels near the bases of the synchronizing pulses. Under some conditions, the gain over the entire synchronizing signal region of the composite picture signal may be reduced with respect to the gain in the region of the picture signal. *See also:* **television.** 178

synchronizing signal level (television). The level of the peaks of the synchronizing signal. *See also:* **television.** 178

synchronizing torque (synchronous machine). The torque produced, primarily through interaction between the armature currents and the flux produced by the field winding, tending to pull the machine into synchronism with a connected power system or with another synchronous machine. 63

synchronous (data transmission). A device whose speed of operation is related to the rest of the system to which the device is connected. 59

synchronous booster converter. A synchronous converter having a mechanically connected alternating-current reversible booster connected in series with the alternating-current supply circuit for the purpose of adjusting the output voltage. *See also:* **converter.** 328

synchronous booster inverter. An inverter having a mechanically connected reversible synchronous booster connected in series for the purpose of adjusting the output voltage. *See also:* **converter.** 328

synchronous capacitor (synchronous condenser)* (rotating machinery). A synchronous machine running without mechanical load and supplying or absorbing reactive power to or from a power system. *See also:* **converter.** 63

---

* Deprecated

synchronous computer. A computer in which each event, or the performance of each operation, starts as a result of a signal generated by a clock. 255, 77

synchronous condenser (1) (rotating machinery). A synchronous machine running without mechanical load and supplying or absorbing reactive power. *Syn:* **synchronous compensator.** 63

(2) (electric installations on shipboard). A synchronous phase modifier running without mechanical load, the field excitation of which may be varied so as to modify the power factor of the system; or through such modification, to influence the load voltage. 3

synchronous converter. A converter that combines both motor and generator action in one armature winding and is excited by one magnetic field. It is normally used to change alternating-current power to direct-current power. *See:* **converter.** 328

synchronous gate. A time gate wherein the output intervals are synchronized with an incoming signal. 111

synchronous generator. A synchronous alternating-current machine that transforms mechanical power into electric power. *Notes:* (1) A synchronous machine is one in which the average speed of normal operation is exactly proportional to the frequency of the system to which it is connected. (2) Unless otherwise stated, it is generally understood that a synchronous generator (or motor) has field poles excited with direct current. (3) As a synonym, use of the term **alternator** is now deprecated. 6, 63

synchronous impedance (1) (per unit direct-axis). The ratio of the field current at rated armature current on sustained symmetrical short-circuit to the field current at normal open-circuit voltage on the air-gap line. *Note:* This definition of synchronous impedance is used to a great extent in electrical literature and corresponds to the definition of direct-axis synchronous reactance as determined from open-circuit and sustained short-circuit tests. *See also:* **positive phase-sequence reactance (rotating machinery).** 328

(2) (rotating machinery). The ratio of the value of the phasor difference between the synchronous internal voltage and the terminal voltage of a synchronous machine to the armature current under a balanced steady-state condition. *Note:* This definition is of rigorous application to turbine type machines only, but it gives a good degree of approximation for salient pole machines. *See also:* **synchronous reactance (effective).** 63

synchronous induction motor (rotating machinery). A cylindrical rotor synchronous motor having a secondary coil winding similar to that of a wound rotor induction motor. *Note:* This winding is used for both starting and excitation purposes. 63

synchronous internal voltage (synchronous machine for any specified operating conditions). The fundamental-frequency component of the voltage of each armature phase that would be produced by the steady (or very slowly varying) component of the current in the main field winding (or field windings) acting alone provided the permeance of all parts of the magnetic circuit remained the same as for the specified operating condition. *Note:* The synchronous internal voltage, as shown in the phasor diagram, is related to the terminal-voltage and phase-current phasors by the equation

$$\mathbf{E}_i = \mathbf{E}_a + R\mathbf{I}_a + jX_d\mathbf{I}_{ad} + jX_q\mathbf{I}_{aq}$$

For a machine subject to saturation, the reactances should be determined for the degree of saturation applicable to the specified operating condition. 63

synchronous inverter. An inverter that combines both motor and generator action in one armature winding. It is excited by one magnetic field and changes direct-current power to alternating-current power. *Note:* Usually it has no amortisseur winding. *See also:* **converter.** 328

synchronous machine (rotating machinery). A machine in which the average speed of normal operation is exactly proportional to the frequency of the system to which it is connected. 3

synchronous machine, ideal. A hypothetical synchronous machine that has certain idealized characteristics that facilitate analysis. *Note:* The results of the analysis of ideal machines may be applied to similar actual machines by making, when necessary, approximate corrections for the deviations of the actual machine from the ideal machine. The ideal machine has, in general, the following properties: (1) the resistance of each winding is constant throughout the analysis, independent of current magnitude or its rate of change; (2) the permeance of each portion of the magnetic circuit is constant throughout the analysis, regardless of the flux density; (3) the armature circuits are symmetrical with respect to each other; (4) the electric and magnetic circuits of the field structure are symmetrical about the direct axis or the quadrature axis; (5) the self-inductance of the field, and every circuit on the field structure, is constant; (6) the self-inductance of each armature circuit is a constant or a constant plus a second-harmonic sinusoidal function of the angular position of the rotor relative to the stator; (7) the mutual inductance between any circuit on the field structure and any armature circuit is a fundamental sinusoidal function of the angular position of the rotor relative to the stator; (8) the mutual inductance between any two armature circuits is a constant or a constant plus a second-harmonic sinusoidal function of the angular position of the rotor relative to the stator; (9) the amplitude of the second-harmonic component of variation of the self-inductance of the armature circuits and of the mutual inductances between any two armature circuits is the same; (10) effects of hysteresis are negligible; (11) effects of eddy currents are negligible or, in the case of solid-rotor machines, may be represented by hypothetical circuits on the field structure symmetrical about the direct axis and the quadrature axis. 63

synchronous motor. A synchronous machine that transforms electric power into mechanical power. Unless otherwise stated, it is generally understood that it has field poles excited by direct current. *See:* **asynchronous machine.** 63

synchronous operation (of a machine). Operation where the speed of the rotor is equal to that of the rotating magnetic flux and where there is a stable phase relationship between the voltage generated in the primary winding and the voltage of a connected power system or synchronous machine. 63

synchronous reactance (effective) (rotating machinery). An assumed value of synchronous reactance used to represent a machine in a system study calculation for a particular operating condition. *Note:* The synchronous internal voltage, as shown in the phasor diagram, is related to the terminal-voltage and phase-current phasors by the equation

$$\mathbf{E}'_i = \mathbf{E}_a + R\mathbf{I}_a + jX_{\text{eff}}\mathbf{I}_a$$

See also: **synchronous internal voltage, synchronous impedance.** 63

synchronous satellite (communication satellite). A satellite with a mean sidereal period of revolution about the primary body equal to the sidereal period of rotation of the primary body about its own axis. *Note:* A synchronous earth satellite must be synchronized to the sidereal period of the earth or length of the mean sidereal day, which is about 23 hours and 56 minutes. 74

synchronous speed (rotating machinery). The speed of

rotation of the magnetic flux, produced by or linking the primary winding. 63

synchronous system (telecommunication). A system in which the sending and receiving instruments are operating continuously at substantially the same rate and are maintained by means of correction if necessary, in a fixed relationship. 194

synchronous voltage (traveling-wave tubes). The voltage required to accelerate electrons from rest to a velocity equal to the phase velocity of a wave in the absence of electron flow. *See: magnetron. See also: electron devices, miscellaneous.* 190

synchro receiver (or motor) (rotating machinery). A transformer electrically similar to a synchro transmitter and that, when the secondary windings of the two devices are interconnected, develops a torque increasing with the difference in angular alignment of the transmitter and receiver rotors and in a direction to reduce the difference toward zero. *See also: synchro system.* 63

synchroscope. An instrument for indicating whether two periodic quantities are synchronous. It usually embodies a continuously rotatable element the position of which at any time is a measure of the instantaneous phase difference between the quantities; while its speed of rotation indicates the frequency difference between the quantities; and its direction of rotation indicates which of the quantities is of higher frequency. *Note:* This term is also used to designate a cathode-ray oscilloscope providing either (1) a rotating pattern giving indications similar to that of the conventional synchroscope, or (2) a triggered sweep, giving an indication of synchronism. *See also: instrument.* 328

synchro system (alternating current). An electric system for transmitting angular position or motion. It consists of one or more synchro transmitters, one or more synchro receivers or synchro control transformers and may include differential synchro machines. 63

synchro transmitter (or generator) (rotating machinery). A transformer with relatively rotatable primary and secondary windings, the output of the secondary winding being two or more voltages that vary with and completely define the relative angular position of the primary and secondary windings. *See also: synchro system.* 63

synchrotron. A device for accelerating charged particles (for example, electrons) to high energies in a vacuum. The particles are guided by a changing magnetic field while they are accelerated many times in a closed path by a radio-frequency electric field. *See also: electron devices, miscellaneous.* 190

sync signal. *See: synchronizing signal.*

syntax. (1) The structure of expressions in a language. (2) The rules governing the structure of a language. 255, 77

synthetic aperture radar. A radar system that generates the effect of a long antenna by signal processing means rather than by the actual use of a long physical antenna. The long antenna is synthesized through the motion of a small antenna relative to the target with either the antenna or the target or both moving. The signal received by the antenna is processed coherently over an integration time. The synthesized antenna length is given by the trajectory traversed by the small antenna relative to the target during the coherent integration time. Because of the two-way phase shift in forming the effective radiation pattern, the effective half-power beamwidth must be computed considering twice the synthesized antenna length. 13

system (1) (general). An integrated whole even though composed of diverse, interacting, specialized structures and subjunctions. *Notes:* (1) Any system has a number of objectives and the weights placed on them may differ widely from system to system. (2) A system performs a function not possible with any of the individual parts. Complexity of the combination is implied. 59, 209

(2) An organized collection of men, machines, and methods required to accomplish a specific objective. *See: number system; numeral system; numeration system; operation system.* 255, 77

(3) (control). A controller and a plant combined to perform specific functions. *See also: control system.* 329

(4) (continuous control system). A system that is not discrete and that does not contain sampling or quantization. 329

(5) (electric power). Designates a combination of lines, and associated apparatus connected therewith, all connected together without intervening transforming apparatus.

(6) *See: adjoint.*

(7) (controlling). *See: controlling system.* 91

(8) (nuclear power generating stations). The entire assembled equipment excluding only the sample collecting pipe. 31

(9) (automatic control). A collection of interconnected physical units or mathematical equations or operations. 56

(10) (programmable instrumentation). A set of interconnected elements constituted to achieve a given objective by performing a specified function. 40

(11) (electric and electronics parts and equipments). A combination of two or more sets, generally physically separated when in operation, and such other units, assemblies, and basic parts necessary to perform an operational function or functions. Typical examples: telephone carrier system, ground-controlled approach (GCA) electronic system, telemetering system, facsimile transmission system. 17

systematic drift rate (gyro). That component of drift rate that is correlated with specific operating conditions. It is composed of acceleration-sensitive drift rate and acceleration-insensitive drift rate. It is expressed as angular displacement per unit time. 46

systematic error (1) (general) (electrothermic power meters). The inherent bias (off-set) of a measurement process or (of) one of its components. 184, 47

(2) (electronic navigation). Error capable of identification due to its orderly character. *See also: navigation.* 13, 187

system control (telephone switching systems). The means for collecting and processing pulsing and supervisory signals in a switching system. 55

system delay time (mobile communication). The time required for the transmitter associated with the system to provide rated radio-frequency output after activation of the local control (push to talk) plus the time required for the system receiver to provide useful output. *See also: mobile communication system.* 181

system demand factor. *See:* **demand factor.**

system deviation (control). *See:* **deviation, system.**

system, directly controlled. That portion of the controlled system that is directly guided or restrained by the final controlling element to achieve a prescribed value of the directly controlled variable. *See also:* **control system, feedback.** 219, 329, 206

system, discrete (control system). A system whose signals are inherently discrete. *See also:* **control system.** 329

system, discrete-state (control system) (system, finite-state). A system whose state is defined only for discrete values of time and amplitude. *See also:* **control system.** 329

system diversity factor. *See:* **diversity factor.**

system element. One or more basic elements with other components and necessary parts to form all or a significant part of one of the general functional groups into which a measurement system can be classified. While a system element must be functionally distinct from other such elements it is not necessarily a separate measurement device. Typical examples of system elements are: a thermocouple, a measurement amplifier, a millivoltmeter. *See also:* **measurement system.** 328

system, finite-state (control system). *See:* **system, discrete state.**

system frequency (electric power system). Frequency in hertz of the power system alternating voltage. 94

system frequency stability (radio system) (mobile communication). The measure of the ability of all stations, including all transmitters and receivers, to remain on an assigned frequency-channel as determined on both a short-term and long-term basis. *See also:* **mobile communication system.** 181

system ground (surge arresters). The connection between a grounding system and a point of an electric circuit (for example, a neutral point). 244, 62

system grounding conductor. An auxiliary solidly grounded conductor that connects together the individual grounding conductors in a given area. *Note:* This conductor is not normally a part of any current-carrying circuit including the system neutral. *See also:* **ground.** 64

system, idealized (automatic control). An imaginary system whose ultimately controlled variable has a stipulated relationship to specified commands. *Note:* It is a basis for performance standards. *See also:* **control system, feedback.** 56, 329

system, indirectly controlled (industrial control). The portion of the controlled system in which the indirectly controlled variable is changed in response to changes in the directly controlled variable. *See also:* **control system, feedback.** 219, 56, 329, 206

system interconnection. The connecting together of two or more power systems. *See also:* **alternating-current distribution; direct-current distribution.** 64

system load (electric power utilization). The summation of load served by a given system. *See also:* **generating station.** 200

system logic (nuclear power generating stations). That equipment that monitors the output of two or more channels and supplies output signals in accordance with a prescribed combination rule (for example, 2 of 3, 2 of 4, etcetera). 31, 109, 356

system loss (radio system). The transmission loss plus the losses in the transmitting and receiving antennas. *See also:* **radio wave propagation.** 180

system matrix (control system). A matrix of transfer functions that relate the Laplace transforms of the system outputs and of the system inputs. *See also:* **control system.** 329

system maximum hourly load. The maximum hourly integrated load served by a system. *Note:* An energy quantity usually expressed in kilowatthours per hour. *See also:* **generating station.** 200

system, multidimensional. *See:* **multidimensional system.**

system, multivariable. *See:* **multivariable system.**

system noise (sound recording and reproducing system). The noise output that arises within or is generated by the system or any of its components, including the medium. *See also:* **noise (sound recording and reproducing system).**

system of units. A set of interrelated units for expressing the magnitudes of a number of different quantities. 210

system overshoot (control) (industrial control). The largest value of system deviation following the first dynamic crossing of the ideal value in the direction of correction, after the application of a specified stimulus. *See also:* **control system, feedback.** 219, 206

system performance testing (nuclear power generating stations). Tests performed on complete systems, including all their electric, instrumentation, controls, fluid, and mechanical subsystems under normal or simulated normal process conditions of temperature, flow, level, pressure, etcetera. 31

system, quantized. *See:* **quantized system.**

system recovery time (mobile communication). The elapsed time from deactivation of the local transmitter control until the local receiver is capable of producing useful output. *See also:* **mobile communication system.** 181

system reserve. The capacity, in equipment and conductors, installed on the system in excess of that required to carry the peak load. *See also:* **generating station.** 64

system routing code (telephone switching systems). In World Zone 1, a three-digit code consisting of a country code and two additional numerals that uniquely identifies an international switching center. *See also:* **world zone number.** 55

system, sampled-data (control system). One in which at least one sampled signal is present. *See also:* **control system.** 329, 56

system science. The branch of organized knowledge dealing with systems and their properties, the systematized knowledge of systems. *See also:* **adaptive system; cybernetics; learning system; system; systems engineering; tradeoff.** 209

systems engineering. The application of the mathematical and physical sciences to develop systems that utilize economically the materials and forces of nature for the benefit of mankind. *See:* **system science.** 209

system utilization factor. *See:* **utilization factor.**

system (circuit) voltage (1) (surge arrester). The root-mean-square power-frequency voltage from line to line as distinguished from the voltage from line to neutral. *See also:* **surge arresters.** 2

(2) (transformer). A root-mean-square (RMS) phase-to-phase power frequency voltage on a three-phase alternating current electrical system. 53

T

T **(video signal transmission measurement).** Letter symbol for the duration of a half-period of the nominal upper cut-off frequency of a transmission system. Therefore

$$T = \frac{1}{2f_c}$$

Note: For the TV system M

$$T = \frac{1}{2 \times 4 \, (\text{MHz})} = 125 \, (\text{ns})$$

The duration T is commonly referred to as the Nyquist interval. The concept of T is employed not only when the frequency cut-off is a physical property of a given system but also when the system is flat and there is no interest in the performance of the system beyond a given frequency. 42

table. A collection of data, each item being uniquely identified either by some label or by its relative position. *See:* **decision table; truth table.** 255, 77

table lamp. A portable luminaire with a short stand suitable for standing on furniture. *See also:* **luminaire.** 167

table look-up. A procedure for obtaining the function value corresponding to an argument from a table of function values. 255, 77

tab sequential format (numerically controlled machines). A means of identifying a word by the number of tab characters preceding the word in the block. The first character in each word is a tab character. Words must be presented in a specific order but all characters in a word, except the tab character, may be omitted when the command represented by that word is not desired. 224, 207

tabulate. (1) To form data into a table. (2) To print totals. 255, 77

tacan (tactical air navigation). A complete ultra-high-frequency polar coordinate (rho-theta) navigation system using pulse techniques; the distance (rho) function operates as distance-measuring equipment (DME) and the bearing function is derived by rotating the ground transponder antenna so as to obtain a rotating multilobe pattern for coarse and fine bearing information. *See also:* **navigation.** 187, 13

tachometer. A device to measure speed or rotation. *See also:* **rotor (rotating machinery).** 63

tachometer electric indicator. A device that provides an indication of the speed of an aircraft engine, of a helicopter rotor, of a jet engine, and of similar rotating apparatus used in aircraft. Such tachometer indicators may be calibrated directly in revolutions per minute or in percent of some particular speed in revolutions per minute. 328

tachometer generator (rotating machinery). A generator, mechanically coupled to an engine, whose main function is to generate a voltage, the magnitude or frequency of which is used either to determine the speed of rotation of the common shaft or to supply a signal to a control circuit to provide speed regulation. 63

tachometric relay (industrial control). A relay in which actuation of the contacts is effected at a predetermined speed of a moving part. *See also:* **relay.** 244, 206

taffrail log. A device that indicates distance traveled based on the rotation of a screw-type rotor towed behind a ship which drives, through the towing line, a counter mounted on the taffrail. *Note:* An electric contact made (usually) each tenth of a mile causes an audible signal to permit ready calculation of speed. 328

tag (1) (computing system). (A) Same as flag. (B) Same as label. 77, 54

(2) (transmission and distribution). "Men at work" tags of distinctive appearance, indicating that the equipment or lines so marked are being worked on. 262

tag address. *See:* **symbolic address.**

tags. Men-at-work tags of distinctive appearance, indicating that the equipment or lines so marked are being worked on. 262

tailing (hangover) (1) (facsimile). The excessive prolongation of the decay of the signal. *See also:* **facsimile signal (picture signal).** 12

(2) (electrometallurgy) (hydrometallurgy and ore concentration). The discarded residue after treatment of an ore to remove desirable minerals. *See also:* **electrowinning.** 328

tail lamp. A lamp used to designate the rear of a vehicle by a warning light. *See also:* **headlamp.** 167

tail-of-wave (chopped wave) impulse test voltage (insulation strength). The crest voltage of a standard impulse wave that is chopped by flashover at or after crest. 287

talker echo. Echo that reaches the ear of the talker. 328

talking key. A key whose operation permits conversation over the circuit to which the key is connected. 328

talk-ringing key (listening and ringing key). A combined talking key and ringing key operated by one handle. 328

tamper-resistant (enclosure). A metal-enclosure for a power switchgear assembly which is designed to resist damage to or improper operation of the switchgear from willful acts of destruction and which is designed to provide reasonably safe protection against tampering by unauthorized persons who may attempt to gain entry by forcible means, to insert foreign substances into, or otherwise tamper with the assembly. 79, 93

tandem (cascade) (network) (circuits and systems). Networks are in tandem when the output terminals of one network are directly connected to the input terminals of the other network. *See also:* **network analysis.** 67

tandem central office (tandem office). A central office used primarily as a switching point for traffic between other central offices. 93

tandem-completing trunk. A trunk, extending from a tandem office to a central office, used as part of a telephone connection between stations. 328

tandem control (electric power systems). A means of control whereby the area control error of an area or areas A, connected to the interconnected system B only through the facilities of another area C, is included in control of area C generation. 94

tandem drive (industrial control). Two or more drives that are mechanically coupled together. *See also:* **electric drive.** 219, 206

tandem office (telephone switching systems). An intermediate office used primarily for interconnecting end offices with each other and with toll connecting trunks.   55

tandem trunk (telephone switching systems). A trunk to a tandem office.   55

tangential wave path (radio wave propagation over the earth). A path of propagation of a direct wave that is tangential to the surface of the earth. The tangential wave path is curved by atmospheric refraction. *See also:* **radiation.**
          328

tank (storage cell). A lead container, supported by wood, for the element and electrolyte of a storage cell. *Note:* This is restricted to some relatively large types of lead-acid cells. *See also:* **battery (primary or secondary).**   328

tank circuit (1) (signal-transmission system). A circuit consisting of inductance and capacitance, capable of storing electric energy over a band of frequencies continuously distributed about a single frequency at which the circuit is said to be resonant, or tuned. *Note:* The selectivity of the circuit is proportional to the ratio of the energy stored in the circuit to the energy dissipated. The ratio is often called the Q of the circuit. *See also:* **dielectric heating; oscillatory circuit; Q quality factor; signal.**
          111, 188

(2) (induction heating). A circuit consisting of inductance and capacitance, capable of storing electric energy.   14

tank, single-anode. *See:* **single-anode tank.**

tank voltage. The total potential drop between the anode and cathode bus bars during electrodeposition. *See also:* **electroplating.**   328

tap (1) (general). An available connection that permits changing the active portion of the device in the circuit. *See also:* **grounding device.**   91

(2) (transformer). a connection brought out of a winding at some point between its extremities to permit changing the voltage, or current, ratio. *See:* **windings, high-voltage and low-voltage.**   257, 203, 53

(3) (reactor). A connection brought out of a winding at some point between its extremities, to permit changing the impedance. *See also:* **reactor.**   309

(4) (rotating machinery). A connection made at some intermediate point in a winding. *See also:* **rotor (rotating machinery); stator; voltagé regulator.**   63

tap-changer (deenergized operation). A selector switch device used to change transformer taps with the transformer deenergized.   53

tape (1) (rotating machinery). A relatively narrow, long, thin, flexible fabric, mat, or film, or a combination of them with or without binder, not over 20 centimeters in width. *See also:* **rotor (rotating machinery); stator.**

(2) (electronic computation). *See:* **magnetic tape.**   63

tape block (test, measurement and diagnostic equipment). A group of frames or tape lines.   54

taped insulation. Insulation of helically wound tapes applied over a conductor or over an assembled group of insulated conductors. (1) When successive convolutions of a tape overlie each other for a fraction of the tape width, the taped insulation is lap wound. This is also called positive lap wound. (2) When a tape is applied so that there is an open space between successive convolutions, this construction is known as open butt or negative lap wound. (3) When a tape is applied so that the space between successive convolutions is too small to measure with the unaided eye, it is a closed butt taping.   64

taped joint (power cable joints). A joint with hand applied tape insulation.   34

tape drive. A device that moves tape past a head.
          255, 77

tape line. *See:* **frame.**

tape preparation. The act of translating command information into punched or magnetic tape.   224, 207

tape punch. *See:* **perforator.**

taper (communication practice). A continuous or gradual change in electrical properties with length, as obtained, for example, by a continuous change of cross-section of a waveguide. *See also:* **transmission line.**   210

tape recorder. *See:* **magnetic recorder.**

tapered key (rotating machinery). A wedge-shaped key to be driven into place, in a matching hole or recess.   63

tapered transmission line. *See:* **tapered waveguide.**

tapered waveguide (waveguide components). A waveguide or transmission line in which a physical or electrical characteristic changes progressively with distance along the axis of the guide.   166

taper, waveguide. A section of tapered waveguide. *See also:* **waveguide.**   330, 166

tape station. *See:* **tape unit.**

tape thickness. The lesser of the cross-sectional dimensions of a length of ferromagnetic tape. *See:* **tapewound core.**
          331

tape to card. Pertaining to equipment or methods that transmit data from either magnetic tape or punched tape to punched cards.   255, 77

tape transmitter (telegraphy). A machine for keying telegraph code signals previously recorded on tape. *See also:* **telegraphy.**   194

tape transport (test, measurement and diagnostic equipment). A device which moves magnetic or punched tape past the tape reader. Reels for storage of the tape are usually provided. *See:* **tape drive.**   54

tape unit. A device containing a tape drive, together with reading and writing heads and associated controls.
          255, 77

tape width. The greater of the cross-sectional dimensions of a length of ferromagnetic tape. *See:* **tape-wound core.**
          331

tape-wound core. A length of ferromagnetic tape coiled about an axis in such a way that one convolution falls directly upon the preceding convolution. *See:* **wrap thickness.**
          331

tapped field. *See:* **short field.**

tapped field control. A system of regulating the tractive force of an electrically driven vehicle by changing the number of effective turns of the traction motor series-field windings by means of an intermediate tap or taps in those windings. *See also:* **multiple-unit control.**   328

tapper bell. A single-stroke bell having a gong designed to produce a sound of low intensity and relatively high pitch. *See also:* **protective signaling.**   328

target (1) (radar). (A) Specifically, an object of radar search or tracking. (B) Broadly, any discrete object which reflects energy back to the radar equipment. *See also:* **navigation; radar.**   13, 187

(2) (camera tube). A structure employing a storage surface that is scanned by an electron beam to generate a signal output current corresponding to a charge-density pattern stored thereon. *Note:* The structure may include the storage surface that is scanned by an electron beam, the backplate, and the intervening dielectric. *See also:* **radar; television.**   190, 178, 125

(3) (relay) (operation indicator). A supplementary device, operated either mechanically or electrically, to visibly indicate that the relay has operated or completed its function. *Notes:* (A) A mechanically operated target indicates the physical operation of the relay. (B) An electrically operated target, when not further described, is actuated by the current in the control circuit associated with the relay and hence indicates that the relay has not only operated, but also that it has completed its function by causing current to flow in the associated control circuit. (C) A shunt-energized target only indicates operation of the relay contacts and does not necessarily show that current has actually flowed in the associated control circuit.   103, 202, 60, 127

(4) (storage tubes). The storage surface and its immediate supporting electrodes. *See also:* **radar; storage tube.**
                    174, 190

(5) (anode) (anticathode*) (X-ray tube). An electrode, or part of an electrode, on which a beam of electrons is focused and from which X-rays are emitted. *See also:* **electrode (electron tube); radar.**   190

* Deprecated

target capacitance (camera tubes). The capacitance between the scanned area of the target and the backplate. *See also:* **television.**   178, 190, 125

target cutoff voltage (camera tubes). The lowest target voltage at which any detectable electric signal corresponding to a light image on the sensitive surface of the tube can be obtained. *See also:* **television.**   190, 125

target fluctuation (radar). Variation in the amplitude of a target signal, caused by changes in target aspect angle, rotation or vibration of target scattering sources, or changes in radar wavelength (the amplitude component of **target noise**). The similar terms **scintillation** and **amplitude noise** have been used also to denote location errors caused by fluctuation, and should be avoided because of this ambiguity.   13

target glint (radar). *See:* **glint.**

target language. A language that is an output from a given translation process. *See:* **object language.**   255, 77

target noise (radar). Random variations in observed amplitude, location, and Doppler of a target, caused by changes in target aspect angle, rotation or vibration of target scattering sources, or changes in radar wavelength. *See also:* **target fluctuation; scintillation error; glint.**
                    13

target program. A program written in a target language. *See:* **object program.**   255, 77

target, sag (conductor stringing equipment). A device used as a reference point to sag conductors. It is placed on one structure of the sag span. The sagger, on the other structure of the sag span, can use it as his reference to determine the proper conductor sag. *Syn:* **sag board; target.**   45

target transmitter (electronic navigation). A source of radio-frequency energy suitable for providing test signals at a test site. *See also:* **navigation.**   278, 187, 13

target voltage (camera tube with low-velocity scanning). The potential difference between the thermionic cathode and the backplate. *See also:* **television.**   178, 190, 125

tariff (1) (data transmission). The published rate for a particular approved commercial service of a common carrier.   59

(2) (electric power utilization). A published volume of rate schedules and general terms and conditions.   112

tarnish (corrosion). Surface discoloration of a metal caused by formation of a thin film of corrosion product.
                    221, 205

taut-band suspension (electric instrument). A mechanical arrangement whereby the moving element of an instrument is suspended by means of ligaments, usually in the form of a thin flat conducting ribbon, at each of its ends. The ligaments normally are in tension sufficient to restrict the lateral motion of the moving element to within limits that permit freedom of useful motion when the instrument is mounted in any position. A restoring torque is produced within the ligaments with rotation of the moving element. *See also:* **moving element (instrument).**   280

taxi light. An aircraft aeronautical light designed to provide necessary illumination for taxiing.   167

taxi-channel light. Aeronautical ground lights arranged along a taxi-channel of a water aerodrome to indicate the route to be followed by taxiing aircraft.   167

taxiway centerline lights. Taxiway lights placed along the centerline of a taxiway except that on curves or corners having fillets, these lights are placed a distance equal to half the normal width of the taxiway from the outside edge of the curve or corner.   167

taxiway-edge lights. Taxiway lights placed along or near the edges of a taxiway.   167

taxiway holding-post light. A light or group of lights installed at the edge of a taxiway near an entrance to a runway, or to another taxiway, to indicate the position at which the aircraft should stop and obtain clearance to proceed.   167

taxiway lights. Aeronautical ground lights provided to indicate the route to be followed by taxiing aircraft.
                    167

TB cell. *See:* **transmitter-blocker cell.**

T-connected (or tee-connected) transformer. A three-phase to three-phase transformer, similar to a Scott-connected transformer. *See also:* **Scott or T-connected transformer.**
                    53

TE. *See:* **transverse electric.**

tearing (television). An erratic lateral displacement of some scanning lines of a raster caused by disturbance of synchronization. *See also:* **television.**   178

teaser transformer. The term as applied to two single-phase Scott-connected units for three-phase to two-phase or two-phase to three-phase operation, designates the transformer that is connected between the midpoint of the main transformer and the third-phase wire of the three-phase system.   53

teed feeder. A feeder that supplies two or more feeding points. *See also:* **center of distribution.**   203

tee junction (waveguide components). A junction of waveguides or transmission lines in which the longitudinal guide axes form a T.   166

telautograph. A system in which writing movement at the transmitting end causes corresponding movement of a writing instrument at the receiving end. *See also:* **telegraphy.** 328

telecommunication (electrical practice). The transmission of information from one point to another. 194

telecommunication loop (telephone switching systems). A channel between a telecommunications station and a switching entity. 55

telecommunications customer (telephone switching systems). One for whom telecommunications service is provided (formerly referred to as a "subscriber"). 55

telecommunications exchange (telephone switching systems). A means of providing telecommunications services to a group of users within a specified geographical area. 55

telecommunications switchboard (telephone switching systems). A manual means of interconnecting telecommunications lines, trunks, and associated circuits, and including signaling facilities. 55

telecommunications switching (telephone switching systems). The function of selectively establishing and releasing connections among telecommunication transmission paths. 55

telecommunications system (telephone switching systems). An assemblage of telecommunications stations, lines, and channels, and switching arrangements for their interconnection, together with all the accessories for providing telecommunications services. 55

telegraph (marine transportation). A mechanized or electric device for the transmission of stereotyped orders or information from one fixed point to another. *Note:* The usual form of telegraph is a transmitter and a receiver, each having a circular dial in sectors upon which are printed standard orders. When the index of the transmitter is placed at any order, the pointer of the receiver designates that order. Dual mechanism is generally provided to permit repeat back or acknowledgment of orders. 328

telegraph channel. A channel suitable for the transmission of telegraph signals. *Note:* Three basically different kinds of telegraph channels used in multichannel telegraph transmissions are (1) one of a number of paths for simultaneous transmission in the same frequency range as in bridge duplex, differential duplex and quadruplex telegraphy; (2) one of a number of paths for simultaneous transmission in different frequency ranges as in carrier telegraphy; (3) one of a number of paths for successive transmission as in multiplex printing telegraphy. Combinations of these three types may be used on the same circuit. *See also:* **channel.** 328, 59

telegraph concentrator. A switching arrangement by means of which a number of branch or subscriber lines or station sets may be connected to a lesser number of trunk lines or operating positions or instruments through the medium of manual or automatic switching devices in order to obtain more efficient use of facilities. *See also:* **telegraphy.** 194

telegraph distortion (telecommunication). The condition in which the significant intervals have not all exactly their theoretical durations. The reference point used when measuring telegraph distortion is the initial space-to-mark transition of each character which occurs at the beginning of each "start" element. The slicing level for all measurements is at the 50% point on the rising or falling current waveforms. Percent distortion is expressed by:

$$\text{Percent Distortion} = \frac{\Delta t}{t_e} \times 100$$

where Δt is the time difference between the actual slicing point and the ideal cross over point and t_e is the time interval of one signal element. *See also:* **telegraphy.** 194

telegraph distributor. A device that effectively associates one direct-current or carrier telegraph channel in rapid succession with the elements of one or more signal sending or receiving devices. *See also:* **telegraphy.** 328

telegraph key. A hand-operated telegraph transmitter used primarily in Morse telegraphy. *See also:* **telegraphy.** 328

telegraph repeater. An arrangement of apparatus and circuits for receiving telegraph signals from one line and retransmitting corresponding signals into another line. *See also:* **telegraphy.** 328

telegraph selector. A device that performs a switching operation in response to a definite signal or group of successive signals received over a controlling circuit. *See also:* **telegraphy.** 328

telegraph sender. A transmitting device for forming telegraph signals. Examples are a manually operated telegraph key and a printer keyboard. *See also:* **telegraphy.** 328

telegraph signal (telecommunication). The set of conventional elements established by the code to enable the transmission of a written character (letter, figure, punctuation sign, arithmetic sign, etcetera) or the control of a particular function (spacing, shift, line-feed, carriage return, phase correction, etcetera); this set of elements being characterized by the variety, the duration and the relative position of the component elements or by some of these features. 194

telegraph signal distortion. Time displacement of transitions between conditions, such as marking and spacing, with respect to their proper relative positions in perfectly timed signals. *Note:* The total distortion is the algebraic sum of the bias and the characteristic and fortuitous distortions. *See also:* **telegraphy.** 328

telegraph sounder. A telegraph receiving instrument by means of which Morse signals are interpreted aurally (or read) by noting the intervals of time between two diverse sounds. *See also:* **telegraphy.** 328

telegraph speed (telecommunication). *See:* **modulation rate.**

telegraph transmission speed. The rate at which signals are transmitted, and may be measured by the equivalent number of dot cycles per second or by the average number of letters or words transmitted, and received per minute. *Note:* A given speed in dot cycles per second (often abbreviated to dots per second) may be converted to **bauds** by multiplying by 2. The baud is the unit of signaling transmission speed recommended by the International Consultative Committee on Telegraph Communication. Where words per minute are used as a measure of transmission speed, five letters and a space per word are assumed. *See also:* **telegraphy.** 328

telegraph transmitter. A device for controlling a source of electric power so as to form telegraph signals. *See also:* **telegraphy.** 328

telegraph word (conventional). A word comprising five letters together with one letter-space, used in computing telegraph speed in words per minute or traffic capacity. *See also:* **telegraphy.** 194

telegraphy. A system of telecommunication for the transmission of graphic symbols, usually letters or numerals, by the use of a signal code. It is used primarily for record communication. The term may be extended to include any system of telecommunication for the transmission of graphic symbols or images for reception in record form, usually without gradation of shade values. 328

telemeter service. Metered telegraph transmission between paired telegraph instruments over an intervening circuit adapted to serve a number of such pairs on a shared-time basis. *See:* **electric metering.** *See also:* **telegraphy.** 328

telemetering (remote metering*). Measurement with the aid of intermediate means that permit the measurement to be interpreted at a distance from the primary detector. *Note:* The distinctive feature of telemetering is the nature of the translating means, which includes provision for converting the measurand into a representative quantity of another kind that can be transmitted conveniently for measurement at a distance. The actual distance is irrelevant. 103, 202

* Deprecated

telephone air-to-air input-output characteristic. The acoustical output level of a telephone set as a function of the acoustical input level of another telephone set to which it is connected. The output is measured in an artificial ear, and the input is measured free-field at a specified location relative to the reference point of an artificial mouth. *See also:* **telephone station.** 122, 196

telephone booth. A booth, closet, or stall for housing a telephone station. *See also:* **telephone station.** 328

telephone central office. A telephone switching unit, installed in a telephone system providing service to the general public, having the necessary equipment and operating arrangements for terminating and interconnecting lines and trunks. *Note:* There may be more than one central office in the same building. 193, 59

telephone channel. A channel suitable for the transmission of telephone signals. *See also:* **channel.** 328, 59

telephone connection. A two-way telephone channel completed between two points by means of suitable switching apparatus and arranged for the transmission of telephone currents, together with the associated arrangements for its functioning with the other parts of a telephone system in switching and signaling operations. *Note:* The term is also sometimes used to mean a two-way telephone channel permanently established between two telephone stations. 328

telephone electrical impedance. The complex ratio of the voltage to the current at the line terminals at any given single frequency. *See also:* **telephone station.** 122, 196

telephone equalization. A property of a telephone circuit that ideally causes both transmit and receive responses to be inverse functions of current, thus tending to equalize variations in loop loss. 122, 196

telephone exchange. A unit of a telephone communication system for the provision of communication service in a specified area that usually embraces a city, town, or village, and its environs. 193, 59

telephone feed circuit. An arrangement for supplying direct-current power to a telephone set and an alternating-current path between the telephone set and a terminating circuit. 122, 196

telephone frequency characteristics. Electrical and acoustical properties as functions of frequency. 122, 196

telephone handset. A telephone transmitter and receiver combined in a unit with a handle. *See also:* **telephone station.** 122, 196

telephone influence factor (TIF). Of a voltage or current wave in an electric supply circuit, the ratio of the square root of the sum of the squares of the weighted root-mean-square values of all the sine-wave components (including in alternating-current waves both fundamental and harmonics) to the root-mean-square value (unweighted) of the entire wave. *Note:* This factor was formerly known as **telephone interference factor,** which term is still used occasionally when referring to values based on the original (1919) weighting curve. *See also:* **inductive coordination.** 53

telephone line. (1) The conductors and circuit apparatus associated with a particular communication channel. 194

(2) A multiplicity of conductors between telephone stations and a central office or between central offices. 193

telephone modal distance. The distance between the center of the grid of a telephone handset transmitter and the center of the lips of a human talker (or the reference point of an artificial mouth) when the handset is in the modal position. *See also:* **telephone station.** 122, 196

telephone modal position. The position a telephone handset assumes when the receiver of the handset is held in close contact with the ear of a person with head dimensions that are modal for a population. *See also:* **telephone station.** 122, 196

telephone operator. A person who handles switching and signaling operations needed to establish telephone connections between stations or who performs various auxiliary functions associated therewith. 193, 101

telephone receive input-output characteristic. The acoustical output level of a telephone set as a function of the electric input level. The output is measured in an artificial ear, and the input is measured across a specified termination connected to the telephone feed circuit. *See also:* **telephone station.** 122, 196

telephone receiver. *See:* **receiver, telephone.**

telephone repeater. A repeater for use in a telephone circuit. *See also:* **repeater.** 328

telephone ringer (station ringer) (ringer). An electric bell designed to operate on low-frequency alternating or pulsating current and associated with a telephone station for indicating a telephone call to the station. *See also:* **telephone station.** 328

telephone set (general) (telephone). An assemblage of apparatus including a telephone transmitter, a telephone receiver, and usually a switch, and the immediately associated wiring and signaling arrangements. *See also:* **telephone station.**   328

telephone set, common battery. A telephone set for which both transmitter current and the current for signaling by the telephone station are supplied from a centralized direct-current power source.   59

telephone set, local battery. A telephone set for which the transmitter current is supplied from a battery or other direct-current supply circuit, individual to the telephone set. Current for signaling by the telephone station may be supplied from a local hand-operated generator or from a centralized direct-current power source. *See also:* **telephone station.**   59

telephone sidetone. The ratio of the acoustical output of the receiver of a given telephone set to the acoustical input of the transmitter of the same telephone set. *See:* **telephone air-to-air input-output characteristic.** *See also:* **telephone station.**   122, 196

telephone speech network. An electric circuit that connects the transmitter and the receiver to a telephone line or telephone test loop and to each other. *See also:* **telephone station.**   122, 196

telephone station. An installed telephone set and associated wiring and apparatus, in service for telephone communication. *Note:* As generally applied, this term does not include the telephone sets employed by central-office operators and by certain other personnel in the operation and maintenance of a telephone system.   328

telephone subscriber. A customer of a telephone system who is served by the system under a specific agreement or contract.   193

telephone switchboard. A switchboard for interconnecting telephone lines and associated circuits.   193

telephone system. An assemblage of telephone stations, lines, channels, and switching arrangements for their interconnection, together with all the accessories for providing telephone communication.   193

telephone test connection. Two telephone sets connected together by means of telephone test loops and a telephone feed circuit.   122, 196

telephone test loop. A circuit that is interposed between a telephone set and a telephone feed circuit to simulate a real telephone line.   122, 196

telephone transmit input-output characteristic. The electric output level of a telephone set as a function of the acoustical input level. The output is measured across a specified impedance connected to the telephone feed circuit, and the input is measured free-field at a specified location relative to the reference point of an artificial mouth. *See also:* **telephone station.**   122, 196

telephone transmitter. A microphone for use in a telephone system. *See also:* **telephone station.**   176

telephone-type relay. A type of electromechanical relay in which the significant structural feature is a hinged armature mechanically separate from the contact assembly. This assembly usually consists of a multiplicity of stacked leaf-spring contacts.   103, 202, 60, 127

telephony. *See:* **sleeve conductor; sleeve wire.**

telephotography. *See:* **picture transmission.**

teleprinter (teletypewriter). *See:* **printer.**

teletypewriter (teleprinter). *See:* **printer.**

television (TV). The electric transmission and reception of

transient visual images.   328

television broadcast station. A radio station for transmitting visual signals, and usually simultaneous aural signals, for general reception. *See also:* **television.**   328

television camera. A pickup unit used in a television system to convert into electric signals the optical image formed by a lens. *See also:* **television.**   328

television channel. A channel suitable for the transmission of television signals. The channel for associated sound signals may or may not be considered a part of the television channel. *See also:* **channel.**   328

television line number. The ratio of the raster height to the half period of a periodic test pattern. *Example:* In a test pattern composed of alternate equal-width black and white bars, the television line number is the ratio of the raster height to the width of each bar. *Note:* Both quantities are measured at the camera-tube sensitive surface. *See also:* **television.**   174, 190

television picture tube. *See:* **picture tube.**

television receiver. A radio receiver for converting incoming electric signals into television pictures and customarily associated sound. *See also:* **television.**   328

television repeater. A repeater for use in a television circuit. *See also:* **repeater; television.**   328

television transmitter. The aggregate of such radio-frequency and modulating equipment as is necessary to supply to an antenna system modulated radio-frequency power by means of which all the component parts of a complete television signal (including audio, video, and synchronizing signals) are concurrently transmitted. *See also:* **television.**   111

telltale. *See:* **running light indicator panel.**

telluric. An adjective denoting the electric field effects in the earth due to M-T (magneto-telluric) fields.   132

TEM. *See:* **transverse electromagnetic.**

temperature, ambient air. The temperature of the surrounding air which comes in contact with equipment. *Note:* Ambient air temperature, as applied to enclosed switchgear assemblies, is the average temperature of the surrounding air that comes in contact with the enclosure.   93

temperature coefficient (1) (rotating machinery). The variation of the quantity considered, divided by the difference in temperature producing it. Temperature coefficient may be defined as an average over a temperature range or an incremental value applying to a specified temperature. *See:* **asynchronous machine; direct-current commutating machine.**   63

(2) (power supplies). The percent change in the output voltage or current as a result of a 1 degree-Celsius change in the ambient operating temperature (percent per degree Celsius). *See also:* **power supply.**   228, 186

(3) (automatic control). *See:* **environmental coefficient.**

(4) (variable or fixed attenuator). Maximum temporary and reversible change of insertion loss in decibels per degree Celsius over operating temperature range.   110

temperature coefficient of capacity (storage cell or battery). The change in delivered capacity (ampere-hour or watt-hour capacity) per degree Celsius relative to the capacity of the cell or battery at a specified temperature. *See also:* **initial test temperature.**   328

temperature coefficient of electromotive force (storage cell

or battery). The change in open-circuit voltage per degree Celsius relative to the electromotive force of the cell or battery at a specified temperature. *See also:* **initial test temperature.**　　　　328

temperature coefficient of resistance (rotating machinery). The temperature coefficient relating a change in electric resistance to the difference in temperature producing it. *See also:* **asynchronous machine; direct-current commutating machine.**　　　　63

temperature coefficient of sensitivity (electrothermic power meters). The change in rf sensitivity (microvolts/milliwatts) resulting from a specified temperature change of the electrothermic unit at a specified power level. Expressed in percent per degree celsius.　　　　47

temperature coefficient of voltage drop (glow-discharge tubes). The quotient of the change of tube voltage drop (excluding any voltage jumps) by the change of ambient (or envelope) temperature. *Note:* It must be indicated whether the quotient is taken with respect to ambient or envelope temperature. *See also:* **gas tube.**　　190, 125

temperature-compensated overload relay. A device that functions at any current in excess of a predetermined value essentially independent of the ambient temperature. *See also:* **relay.**　　　　3

temperature control system (gas turbines). The devices and elements, including the necessary temperature detectors, relays, or other signal-amplifying devices and control elements, required to actuate directly or indirectly the fuel-control valve, speed of the air compressor, or stator blades of the compressor so as to limit or control the rate of fuel input or air flow inlet to the gas turbine. By this means the temperature in the combustion system or the temperatures in the turbine stages or turbine exhaust may be limited or controlled.　　98, 58

temperature conversion (tolerance requirements) (International System of Units). Standard practice for converting tolerances from degrees Fahrenheit to kelvins or degrees Celsius is:

Conversion of temperature tolerance requirements

Tolerance °F	Tolerance K or °C
± 1	± 0.5
± 2	± 1
± 5	± 3
±10	± 5.5
±15	± 8.5
±20	±11
±25	±14

Normally, temperatures expressed in a whole number of degrees Fahrenheit should be converted to the nearest 0.5 kelvin (or degree Celsius). As with other quantities, the number of significant digits to retain will depend upon implied accuracy of the original dimension, for example:

100 ± 5°F implied accuracy estimated total 2°F.

37.7777 ± 2.7777°C rounds to 38 ± 3°C

1000 ± 50°F implied accuracy estimated total 20°F.

537.7777 ± 27.7777°C rounds to 540 ± 30°C

See also: **units and letter symbols.**　　　　21

temperature derating (semiconductor device). The reduction in reverse-voltage or forward-current rating, or both, assigned by the manufacturer under stated conditions of higher ambient temperatures. *See also:* **average forward-current rating (rectifier circuit); semiconductor rectifier stack.**　　237, 66, 208

temperature detectors (gas turbines and rotating electric machinery). The primary temperature-sensing elements that are directly responsive to temperature. *See also:* **asynchronous machine; direct-current commutating machines; electric thermometer.**　　98, 58, 63

temperature, equilibrium (thyristor power converter). The steady-state temperature reached by a component of a thyristor converter under specified conditions of load and cooling. *Note:* The steady-state temperatures are, in general, different for different components. The times necessary to establish steady-state temperatures are also different and proportional to the thermal time constants.　　　　121

temperature index. An index that allows relative comparisons of the temperature capability of insulating materials or insulation systems based on specified, controlled test conditions. Preferred values of temperature index numbers are:

Number Range	Preferred Temperature Index
90–104	90
105–129	105
130–154	130
155–179	155
180–199	180
200–219	200

220 and above no preferred indices established.

temperature meter. *See:* **electric thermometer.**

temperature, operating (power supply). *See:* **operating temperature.**

temperature radiator. A radiator whose radiant flux density (radiant exitance) is determined by its temperature and the material and character of its surface, and is independent of its previous history.　　　　167

temperature-regulating equipment (rectifier). Any equipment used for heating and cooling a rectifier, together with the devices for controlling and indicating its temperature. *See also:* **rectification.**　　　　208

temperature relays (1) (gas turbines). Devices by means of which the output signals of the temperature detectors are enabled to control directly or indirectly the rate of fuel energy input, the air flow input, or both, to the combustion system. *Note:* Operation of a temperature relay is caused by a specified external temperature; whereas operation of a thermal relay is caused by the heating of a part of the relay. *See:* **thermal relay.**　98, 58, 103, 202, 60, 127
(2) (power switchgear). A relay whose operation is caused by specified external temperature.　　103, 127

temperature rise (transformer). The difference between the temperature of the part under consideration (commonly the "average winding rise" or the "hottest-spot winding rise") and the ambient temperature.　　　　53

temperature-rise rate, locked rotor, winding. *See:* **locked-rotor temperature-rise rate, winding.**

temperature-rise test (1) (rotating machinery). A test undertaken to determine the temperature rise above ambient of one or more parts of a machine under specified operating conditions. *Note:* The specified conditions may refer to current, load, etcetera. *See also:* **asynchronous machine; direct-current commutating machine.**   202, 63
(2) (power switchgear). A test to determine the temperature rise, above ambient, of various parts of the tested device when subjected to specified test quantities. *Note:* The test quantities may be current, load, etcetera.   103
(3) (high voltage air switches). A test in which rated current at rated frequency is applied to equipment to determine its temperature rise.   144
temperature stability (electrical conversion). Static regulation caused by a shift or change in output that was caused by temperature variation. This effect may be produced by a change in the ambient or by self-heating. *See also:* **electrical conversion.**   186
temporal coherence (electromagnetic) (laser-maser). The correlation in time of electromagnetic fields at a point in space.   363
temporary emergency circuits (marine transportation). Circuits arranged for instantaneous automatic transfer to a storage-battery supply upon failure of a ship's service supply. *See also:* **emergency electric system.**   328
temporary emergency lighting (marine transportation). The lighting of exits and passages to permit passengers and crew, upon failure of a ship's service lighting, readily to find their way to the lifeboat embarkation deck. *See also:* **emergency electric system.**   328
temporary ground. A connection between a grounding system and parts of an installation that are normally alive, applied temporarily so that work may be safely carried out in them.   244, 62
temporary overvoltage. An oscillatory phase-to-ground or phase-to-phase overvoltage at a given location of relatively long duration and which is undamped or only weakly damped. Temporary overvoltages usually originate from switching operations or faults (for example, load rejection, single-phase faults) and/or from non-linearities (ferroresonance effects, harmonics). They may be characterized by their amplitude, their oscillation frequencies, their total duration or their decrement.   53
temporary storage (programming). Storage locations reserved for intermediate results. *See:* **working storage.**   255, 77, 54
tens complement. The radix complement of a numeral whose radix is ten. *Note:* Using synonyms, the definition may also be expressed as "a true complement with a base of ten." *See also:* **base; radix.**   235, 255, 77
tension. *See:* **final unloaded conductor tension; initial conductor tension.** *See also:* **conductor.**
tensioner, bullwheel (conductor stringing equipment). A device designed to hold tension against a pulling line or conductor during the stringing phase. Normally, it consists of one or more pairs of urethane or neoprene lined, power braked, single or multiple groove "bullwheels" where each pair is arranged in tandem. Tension is accomplished by friction generated against the pulling line or conductor which is reeved around the grooves of a pair of the "bullwheels." The tensioner is usually equipped with its own engine which retards the "bullwheels" mechanically, hy-

draulically, or through a combination of both. Some of these devices function as either a puller or tensioner. *Syn:* **retarder; tensioner.**   45
tensioner, reel (conductor stringing equipment). A device designed to generate tension against a pulline line or conductor during the stringing phase. It is normally equipped with its own engine which retards the supporting shaft for the reel mechanically, hydraulically, or through a combination of both. The shaft, in turn, retards the reel. Some of these devices function as either a puller or tensioner. *Syn:* **retarder; tensioner.**   45
tension, unloaded (transmission and distribution). (1) Initial: The longitudinal tension in a conductor prior to the application of any external load. (2) Final: The longitudinal tension in a conductor after it has been subjected for an appreciable period to the loading prescribed for the loading district in which it is situated, or equivalent loading, and the loading removed. Final unloaded tension shall include the effect of inelastic deformation (creep).   262
tenth-power width (in a plane containing the direction of the maximum of a lobe). The full angle between the two directions in that plane about the maximum in which the radiation intensity is one-tenth the maximum value of the lobe. *See also:* **antenna.**   111, 179
terminal (1) (general) (transformer). (A) A conducting element of an equipment or a circuit intended for connection to an external conductor. (B) A device attached to a conductor to facilitate connection with another conductor.   210, 53
(2) (network). A point at which any element may be directly connected to one or more other elements. *See also:* **network analysis.**   332
(3) (semiconductor device) (industrial control) (light emitting diodes). An externally available point of connection to one or more electrodes or elements within the device. *See also:* **anode; semiconductor; semiconductor rectifier cell.**   237, 245, 243, 210, 206, 66, 208, 191, 162
(4) (communication channels) (A) (general). A point in a system or communication network at which data can either enter or leave.   255, 77
(B) (telegraph circuits). A general term referring to the equipment at the end of a telegraph circuit, modems, input-output and associated equipment. *See also:* **telegraph.**   194
(5) (rotating machinery). A conducting element of a winding intended for connection to an external electrical conductor. *See also:* **stator.**   63
terminal block (power switchgear). An insulating base equipped with terminals for connecting secondary and control wiring. *Syn:* **terminal board.**   103
terminal board (1) (rotating machinery). A plate of insulating material that is used to support terminations of winding leads. *Notes:* (1) The terminations, which may be mounted studs or blade connectors, are used for making connections to the supply line, the lead, other external circuits, or among the windings of the machine. (2) Small terminal boards may also be termed **terminal blocks,** or **terminal strips.**   63, 225, 206
(2) (transformer). An assembly including an insulating board supporting terminal board studs, nuts and links arranged to change transformer winding connections from

Δ to Y, series to multiple, and other connection arrangements. 53

terminal board cover (rotating machinery). A closure for the opening which permits access to the terminal board and prevents accidental contact with the terminals. 62

terminal box (conduit box) (rotating machinery). A form of termination in which the ends of the machine winding are connected to the incoming supply leads inside a box that virtually encloses the connections, and is of minimum size consistent with adequate access and with clearance and creepage-distance requirements. The box is provided with a removable cover plate for access. *See also:* **stator.** 63

terminal chamber (metal enclosed bus). A metal-enclosed container which includes all necessary mechanical and electrical items to complete the connections to other equipment. 78

terminal conformity. *See:* **conformity.**

terminal connection detail (lead storage batteries). Connections made between rows of cells or at the positive and negative terminals of the battery, which may include lead-plated terminal plates, cables with lead-plated lugs and lead-plated rigid copper connectors. 31, 38

terminal connector (industrial control). A connector for attaching a conductor to a lead, terminal block, or stud of electric apparatus. *See:* **terminal.** 225, 206

terminal interference voltage (electromagnetic compatibility). *See:* **terminal voltage.**

terminal linearity. *See:* **conformity.**

terminal of entry (for a conductor entering a delimited region). That cross section of the conductor that coincides with the boundary surface of the region and that is perpendicular to the direction of the electric field intensity at its every point within the conductor. In a conventional circuit, in which the conductors have a cross section that is uniform and small by comparison of the largest dimension with the length, the terminal of entry is a cross section perpendicular to the axis of the conductor. If the cross section of the conductor is infinitesimal, the terminal of entry becomes the point at which the conductor cuts the surface. *Notes:* (1) It follows from this definition and **delimited region** that the algebraic sum of the currents directed into a delimited region through all the terminals of entry is zero at every instant. (2) The term **terminal of entry** has been introduced because of the need in precise definitions of indicating definitely the terminations of the paths along which voltages are determined. The terms **phase conductor** and **neutral conductor** refer to a portion of a conductor rather than to a particular cross section although they may be considered by a practical engineer as representing a portion along which the integral of the electric intensity is negligibly small. Hence he may treat these terms as synonymous with **terminal of entry** in particular cases. *See also:* **network analysis.** 210

terminal pad. A usually flat conducting part of a device to which a terminal connector is fastened. 103, 202, 27

terminal pair (network). An associated pair of accessible terminals, such as input pair, output pair, and the like. *See also:* **network analysis.** 153, 210

terminal-per-line system (telephone switching systems). A switching entity having an outlet corresponding to each line. 55

terminal-per-station system (telephone switching systems). A switching entity having an outlet corresponding to each main-station code. 55

terminal room (telephone switching systems). That part of a building that contains distributing frames, relays and similar apparatus associated with switching equipment. 55

terminal repeater. A repeater for use at the end of a trunk or line. *See also:* **repeater.** 328, 59

terminal room. A room, associated with a central office, private branch exchange, or private exchange, that contains distributing frames, relays, and similar apparatus except switchboard equipment. 192

terminals (1) (surge arrester). The conducting parts provided for connecting the arrester across the insulation to be protected. 2

(2) (storage battery) (storage cell). The parts to which the external circuit is connected. *See also:* **battery (primary or secondary).** 328

terminal screw. *See:* **binding screw.**

terminal, stator winding (rotating machinery). The end of a lead cable or a stud or blade of a terminal board to which connections are normally made during installation. *See also:* **stator.** 63

terminal strip. *See:* **terminal board.**

terminal trunk (data transmission). A trunk circuit connecting switching centers used in conjunction with local switching only in these centers. 59

terminal unit (programmable instrumentation). An apparatus by means of which a connection (and translation, if required) is made between the considered interface system and another external interface system. 40

terminal voltage (terminal interference voltage) (electromagnetic compatibility). Interference voltage measured between two terminals of an artificial mains network. *See also:* **electromagnetic compatibility.** 220, 199

terminating (line or transducer). The closing of the circuit at eitherend by the connection of some device thereto. Terminating does not imply any special condition, such as the elimination of reflection. 328

terminating power meter or measuring system. A device which terminates a waveguide or transmission line in a prescribed manner and contains provisions for measuring the incident of absorbed power. 47, 115

terminating test circuit (telephone equipment). A network connected to a transmission port of a circuit to terminate it in a suitable balanced termination for longitudinal balance testing. This circuit is used when a driving test circuit is connected to one such port and the test specimen has additional transmission ports. *See also:* **driving test circuit; longitudinal balance, degree of; network.** 39

terminating toll center code (TTC) (telephone switching systems). In operator distance dialing, the three digits used for identifying the toll center within the area to which a call is routed. 55

terminating traffic (telephone switching systems). traffic delivered directly to lines. 55

termination (1) (general). A one-port load that terminates a section of a transmission system in a specified manner. *See also:* **transmission line.** 185

(2) (rotating machinery). The arrangement for making the connections between the machine terminals and the external conductors. *See also: stator.* 63

(3) (waveguide components). A one port load in a waveguide or transmission line. 166

termination, conjugate. A termination whose input impedance is the complex conjugate of the output impedance of the source or network to which it is connected. *See also: transmission line.* 185

termination insulator (cable termination). An insulator used to protect each cable conductor passing through the device and provide complete external leakage insulation between the cable conductor(s) and ground. 4

termination, matched (waveguide components). A termination matched with regard to the impedance in a prescribed way; for example, (A) a reflectionless termination, or (B) a conjugate termination. *See: termination, reflectionless. See also: transmission line.* 185, 166

termination, reflectionless. A termination that terminates a waveguide or transmission line without causing a reflected wave at any transverse section. *See also: transmission line; waveguide.* 267, 179, 185

ternary. (1) Pertaining to a characteristic or property involving a selection, choice, or condition in which there are three possibilities. (2) Pertaining to the numeration system with a radix of three. *See also: base; positional notation; radix.* 235, 255, 77

ternary code (information theory). A code employing three distinguishable types of code elements. *See also: information theory; positional notation.* 160

terrain avoidance radar. A radar which provides assistance for a pilot for flight around obstacles by displaying the obstacles at or above his altitude. 13

terrain-clearance indicator. An absolute altimeter using the measurement of height above the sea or ground to alert the pilot of danger. *See also: radio navigation.* 187, 13

terrain echoes. *See: ground clutter.*

terrain error (in navigation). The error resulting from the use of a wave that has become distorted by the terrain over which it has propagated. 187, 13

terrain following radar. A radar which works with the aircraft flight control system to provide low level flying following the contour of the earth's surface at some given altitude. 13

terrestrial-reference flight. The type of stabilized flight that obtains control information from terrestrial phenomena, such as earth's magnetic field, atmospheric pressure, etcetera. 13, 187, 13

tertiary winding. An additional winding in a transformer which can be connected to a synchronous condenser, a reactor, an auxiliary circuit, etcetera. For transformers with Y-connected primary and secondary windings, it may also help: (1) to stabilize voltages to the neutral; (2) to reduce the magnitude of third harmonics when delta connected; (3) to control the value of the zero-sequence impedance; and (4) to serve load. 53

tesla. The unit of magnetic induction in the International System of Units (SI). The tesla is a unit of magnetic induction equal to 1 weber per square meter. 210

Tesla current* (electrotherapy) (coagulating current*). A spark discharge having a drop of 5 to 10 kilovolts in air, from monopolar or bipolar electrodes, generated by a special arrangement of transformers, spark gaps, and capacitors, delivered to a tissue surface, and dense enough to precipitate and oxidize (char) tissue proteins. *Note:* The term **Tesla current** is appropriate if the emphasis is on the method of generation; a **coagulating current,** if the emphasis is on the physiological effects. *See also: electrotherapy.* 192

* Deprecated

test (1) (item of equipment). To determine its performance characteristics while functioning under controlled conditions. 210

(2) (electronic digital computation). (A) To ascertain the state or condition of an element, device, program, etcetera. (B) Sometimes used as a general term to include both check and diagnostic procedures. (C) Loosely, same as check. *See also: check; check problem.* 235

(3) (instrument or meter). To ascertain its performance characteristics while functioning under controlled conditions. 328, 54

test analysis (test, measurement and diagnostic equipment). The examination of the test results to determine whether the device is in a go or no-go state or to determine the reasons for or location of a malfunction. 54

test bench (test, measurement and diagnostic equipment). An equipment specifically designed to provide a suitable work surface for testing a unit in a particular test setup under controlled conditions. 54

test block. *See: test switch.*

test board. A switchboard equipped with testing apparatus so arranged that connections can be made from it to telephone lines or central-office equipment for testing purposes. 193

test cabinet (switchgear assembly). An assembly of a cabinet containing permanent electric connections, with cable connections to a contact box arranged to make connection to the secondary contacts on an electrically operated removable element, permitting operation and testing of the removable element when removed from the housing. It includes the necessary control switch and closing relay, if required. 103, 202

test call (telephone switching systems). A call made to determine if circuits or equipment are performing properly. 55

test cap. A protective structure that is placed over the exposed end of the cable to seal the sheath or other covering completely against the entrance of dirt, moisture, air, or other foreign substances. *Note:* Test caps are often provided with facilities for vacuum treatment, oil filling, or other special field operations. *See: live cable test cap.* 64

test connection (telephony). Two telephone sets connected together by means of test loops and a feed circuit. 122

test current (watthour meter). The current marked on the nameplate by the manufacturer (identified as *TA* on meters manufactured since 1960) and is the current in amperes that is used as the base for adjusting and determining the percent registration of a watthour meter at heavy and light loads. *See also: watthour meter.* 212

test current, continuous (thyristor converter). The value of direct current that a converter unit or section can supply to its load for unlimited time periods under specified conditions. 121

test current in alternating-current circuits (insulation tests).

The normal current flowing in the test circuit as the result of insulation leakage and, in alternating-current circuits, is the vector sum of the inphase leakage currents and quadrature capacitive currents.   116

test current, long-time (thyristor converter). The specified value of direct current that a converter unit or section shall be capable of carrying for a sustained period (minutes or hours) following continuous operation at a specified lower dc value under specific conditions.   121

test current, short-time (thyristor converter). The value of direct current that may be applied to a unit or section for a short period (seconds) following continuous operation at a specified lower dc value under specific conditions.   121

test data (reliability). Data from observations during tests. *Note:* All conditions should be stated in detail, for example, time, stress conditions and failure or success criteria.   164

test desk (telephone switching systems). A position equipped with testing apparatus so arranged that connections can be made from it to telephone lines or central office equipment for testing purposes.   55

test duration (nuclear power generating stations). The elapsed time between the test initiation and the test termination.   366

test enclosure (for low-voltage ac power circuit breakers). A single unit enclosure used for test purposes for a specific frame size circuit breaker and which conforms to the manufacturer's recommendation for minimum volume, minimum electrical clearances, effective areas and locations of ventilation openings, and configuration of connections to terminals.   103

test frequency (reliability analysis) (nuclear power generating station). The number of tests of the same type per unit time interval; the inverse of the test interval.   29, 31

test handset (hand test telephone). A handset used for test purposes in a central office or in the outside plant. It may contain in the handle other components in addition to the transducer, as for example a dial, keys, capacitors, and resistors. *See also:* **telephone station.**   328

test initiation (nuclear power generating stations). The application of a test input or removal of equipment tram service to perform a test.   366

test input (nuclear power generating stations). A real or simulated, but deliberate action which is imposed upon a sensor, channel, train, load group, or other system or device for the purpose of testing.   366

test interval (nuclear power generating stations). The elapsed time between the initiation of identical tests on the same sensor, channel, train, load group, or other specified system or device.   29, 31, 75, 366

test logic (test, measurement and diagnostic equipment). The logical, systematic examination of circuits and their diagrams to identify and analyze the probability and consequence of potential malfunctions for determining related maintenance or maintainability design requirements.   54

test loop (telephony). A circuit which is interposed between a telephone set and a telephone feed circuit to simulate a real telephone line.   122

test, measurement and diagnostic equipment (TMDE). Any system or device used to evaluate the operational condition of a system or equipment to identify and isolate or both any actual or potential malfunction.   54

test oriented language (test, measurement and diagnostic equipment). A computer language utilizing English mnemonics that are commonly used in testing. Examples are measure, apply, connect, disconnect, and so forth.   54

test point (1) (test, measurement and diagnostic equipment). A convenient, safe access to a circuit or system so that a significant quantity can be measured or introduced to facilitate maintenance, repair, calibration, alignment, and checkout.   54

(2) (separable insulated connectors). A capacitively coupled terminal for use with sensing devices.   134

test point selector (test, measurement and diagnostic equipment). A device capable of selecting test points on an item being tested in accordance with instructions from the programmer.   54

test position (switchgear assembly removable element). That position in which the primary disconnecting devices of the removable element are separated by a safe distance from those in the housing and some or all of the secondary disconnecting devices are in operating contact. *Notes:* (1) A set of test jumpers or mechanical movement of secondary disconnecting devices may be used to complete all secondary connections for test in the test position. This may correspond with the disconnected position. (2) Safe distance, as used here, is a distance at which the equipment will meet its withstand-voltage ratings, both low-frequency and impulse, between line and load terminals with the switching device in the closed position.   103, 202

test procedure (test, measurement and diagnostic equipment). A document that describes step by step the operation required to test a specific unit with a specific test system.   54

test program (test, measurement and diagnostic equipment). A program specifically intended for the testing of a unit under test (UUT). *See:* **program; unit under test.**   54

test program documentation. *See:* **test programming procedures.**

test programming procedures (test, measurement and diagnostic equipment). Documents which explain in detail the composition of test programs including definitions and logic used to compose the program. Provides instructions to implement changes in the program.   54

test provisions (test, measurement and diagnostic equipment). The capability included in the design for conveniently evaluating the performance of a prime equipment, module, assembly, or part.   54

test requirement analysis (test, measurement and diagnostic equipment). The examination of documents such as schematics, assembly drawings and specifications for the purpose of deriving test requirements for a unit.   54

test requirement document (TRD) (test, measurement and diagnostic equipment). The document that specifies the tests and test conditions required to test and fault isolate a unit under test.   54

test response spectrum (TRS) (nuclear power generating stations) (seismic qualification of class 1E equipment). The response spectrum that is constructed using analysis or derived using spectrum analysis equipment based on the actual motion of the shake table. *Note:* When qualifying equipment by utilizing the response spectrum, the TRS is to be compared to the RRS, using the methods described in IEEE Std 344-1975. 28, 31

test routine. (1) Usually a synonym for check routine. (2) Sometimes used as a general term to include both check routine and diagnostic routine. 210

test schedule (nuclear power generating stations). The pattern of testing applied to parts of a system. In general, there are two patterns of interest: (1) simultaneous: Redundant items are tested at the beginning of each test interval, one immediately following the other. (2) perfectly staggered: Redundant items are tested such that the test interval is divided into equal subintervals. 29, 31

test sequence (test, measurement and diagnostic equipment). (1) A unique setup of measurements; and (2) A specific order of related tests. 54

test sequence number (test, measurement and diagnostic equipment). Identification of a test sequence. 54

test signal, telephone channel (power-system communication). A signal of specified power and frequency that is applied to the input terminals of a telephone channel for measurement purposes. At any other point in the system, the test signal may be specified in terms applicable to the specific system. 59

test site (electromagnetic compatibility). A site meeting specified requirements suitable for measuring radio interference fields radiated by an appliance under test. *See also:* **electromagnetic compatibility.** 220, 199

test software (test, measurement and diagnostic equipment). Maintenance instructions which control the testing operations and procedures of the automatic test equipment. This software is used to control the unique stimuli and measurement parameters used in testing the unit under test. 54

test specimen (insulator). An insulator that is representative of the product being tested; it is a specimen that is undamaged in any way that would influence the result of the test. 261

test spectrum (test, measurement and diagnostic equipment). A range of test stimuli and measurements based on analysis of prime equipment test requirements. 54

test stand (test, measurement and diagnostic equipment). An equipment specifically designed to provide suitable mountings, connections, and controls for testing electrical, mechanical, or hydraulic equipment as an entire system. 54

test stimulus (electrical). A single shock or succession of shocks, used to characterize or determine the state of excitability or the threshold of a tissue. 192

test support software (test, measurement and diagnostic equipment). Computer programs used to prepare, analyze, and maintain test software. Test software includes automatic test equipment (ATE) compilers, translation/analysis programs and punch/print programs. 54

test switch (test block). A combination of connection studs, jacks, plugs, or switch parts arranged conveniently to connect the necessary devices for testing instruments, meters, relays, etcetera. 103, 202

test termination. The removal of a test input with results of the test being known, or the committal of the equipment for repair based on the results of the test. 31

test testboard (telephone switching systems). A position equipped with testing apparatus so arranged that connections can be made from it to trunks for testing purposes. 55

test voltage (electrical insulation tests). The voltage applied across the specimen during a test. 116

test voltage, partial discharge-free. A specified voltage applied in a specified test procedure, at which the test object is free from partial discharges exceeding a specified level. This voltage is expressed as a peak value divided by the square root of two. *Note:* The term **corona-free test voltage** has frequently been used with this connotation. It is recommended that such usage be discontinued in favor of the term **partial discharge-free test voltage.** 139

tetanizing current (electrotherapy). The current that, when applied to a muscle or to a motor nerve connected with a muscle stimulates the muscle with sufficient intensity and frequency to produce a smoothly sustained contraction as distinguished from a succession of twitches. *See also:* **electrotherapy.** 192

tetrode. A four-electrode electron tube containing an anode, a cathode, a control electrode, and one additional electrode that is ordinarily a grid. 190

theoretical cutoff frequency (theoretical cutoff) (electric structure). A frequency at which, disregarding the effects of dissipation, the attenuation constant changes from zero to a positive value or vice versa. *See also:* **cutoff frequency.** 328

theory. *See:* **information theory.**

thermal aging (rotating machinery). Normal load/temperature deteriorating influence on insulation. 37

thermal burden rating (1) (potential transformer). The volt-amperes that the potential transformer will carry continuously at rated voltage and frequency without causing the specified temperature limitations to be exceeded. *See:* **instrument transformer.** 212, 207 **(2) (voltage transformer).** The volt-ampere output that the transformer will supply continuously at rated secondary voltage without causing the specified temperature limitations to be exceeded. 53

thermal cell. A reserve cell that is activated by the application of heat. *See also:* **electrochemistry.** 328

thermal conduction. The transport of thermal energy by processes having rates proportional to the temperature gradient and excluding those processes involving a net mass flow. *See also:* **thermoelectric device.** 248, 191

thermal conductivity. The quotient of (1) the conducted heat through unit area per unit time by (2) the component of the temperature gradient normal to that area. *See also:* **thermoelectric device.** 248, 191

thermal conductivity, electronic. The part of the thermal conductivity resulting from the transport of thermal energy by electrons and holes. *See also:* **thermoelectric device.** 248, 191

thermal converter (electric instrument) (thermocouple converter) (thermoelement*). A device that consists of one or more thermojunctions in thermal contact with an electric heater or integral therewith, so that the electromotive force developed at its output terminals by thermoelectric action gives a measure of the input current in its heater. *Note:* The combination of two or more thermal converters when connected with appropriate auxiliary equipment so that its combined direct-current output gives a measure of the active power in the circuit is called a thermal watt converter.   328

*Deprecated

thermal current converter (electric instrument). A type of thermal converter in which the electromotive force developed at the output terminals gives a measure of the current through the input terminals. *See also:* **thermal converter.**   280

thermal current rating (1) (neutral grounding device) (electric power). The root-mean-square neutral current in amperes that it will carry under standard conditions for its rated time without exceeding standard temperature limitations, unless otherwise specified. *See also:* **grounding device.**
                                                           91, 64

(2) (resistor). The initial root-mean-square symmetrical value of the current that will flow when rated voltage is applied.   91

thermal cutout (industrial control). An overcurrent protective device that contains a heater element in addition to and affecting a renewable fusible member that opens the circuit. *Note:* It is not designed to interrupt short-circuit currents. *See also:* **relay.**   256, 206

thermal electromotive force. Alternative term for **Seebeck electromotive force.** *See also:* **thermoelectric device.**
                                                           248, 191

thermal endurance (electric insulation) (rotating machinery). The relationship, between temperature and time spent at that temperature, required to produce such degradation of an electrical insulation that it fails under specified conditions of stress, electric or mechanical, in service or under test. For most of the chemical reactions encountered, this relationship is a straight line when plotted with ordinates of logarithm of time against abscissae of reciprocal of absolute temperature (Arrhenius plot). *See also:* **asynchronous machine; direct-current commutating machine.**   63

thermal equilibrium (rotating machinery). The state reached when the observed temperature rise of the several parts of the machine does not vary by more than 2 degrees Celsius over a period of one hour. *See:* **asynchronous machine; direct-current commutating machine.**   63

thermal flow switch. *See:* **flow relay.**

thermally delayed overcurrent trip. *See* **thermally delayed release** and **overcurrent release.**

thermal-mechanical cycling (rotating machinery). The experience undergone by rotating-machine windings, and particularly their insulation, as a result of differential movement between copper and iron on heating and cooling. Also denotes a test in which such actions are simulated for study of the resulting behavior of an insulation system, particularly for machines having a long core length. *See also:* **asynchronous machine; direct-current commutating machine.**   63

thermal noise (circuit) (resistance noise) (1) (power system communications). Random noise associated with the thermodynamic interchange of energy necessary to maintain thermal equilibrium between the circuit and its surroundings. *Note:* The average square of the open-circuit voltage across the terminals of a passive two-terminal network of uniform temperature, due to thermal agitation, is given by

$$V_T^2 = 4kT \int R(f)\, df$$

where T is the absolute temperature in degrees Celsius, R is the resistance component in ohms of the network impedance at the frequency f measured in hertz, and k is the Boltzmann constant, 1.38×10^{-23}. *See also:* **Johnson noise; signal-to-noise ratio.**   59

(2) (telephone practice). Noise occurring in electric conductors and resistors and resulting from the random movement of free electrons contained in the conducting material. The name derives from the fact that such random motion depends on the temperature of the material. Thermal noise has a flat power spectrum out to extremely high frequencies.   24

(3) (electron tube). The noise caused by thermal agitation in a dissipative body. *Note:* The available thermal noise power N, from a resistor at temperature T, is $N = kT\Delta f$, where k is Boltzmann's constant and Δf is the frequency increment.   125

thermal-overload protection (series capacitor). A means to detect excessive heating of capacitor units as a result of a combination of current, ambient temperature, and solar radiations, and to initiate an alarm signal or the closing of the associated capacitor bypass switch, or both.
                                                           86

thermal power converter (thermal watt converter*) (electric instrument). A complex type of thermal converter having both potential and current input terminals. It usually contains both current and potential transformers or other isolating elements, resistors, and a multiplicity of thermoelements. The electromotive force developed at the output terminals gives a measure of the power at the input terminals. *See also:* **thermal converter.**   280

*Deprecated

thermal protection (motor). The words **thermal protection** appearing on the nameplate of a motor indicate that the motor is provided with a thermal protector. *See also:* **contactor.**   256

thermal protector (rotating machinery). A protective device, for assembly as an integral part of a machine, that protects the machine against dangerous overheating due to overload or any other reason. *Notes:* (1) It may consist of one or more temperature-sensing elements integral with the machine and a control device external to the machine. (2) When a thermal protector is designed to perform its function by opening the circuit to the machine and then automatically closing the circuit after the machine cools to a satisfactory operating temperature, it is an automatic-reset thermal protector. (3) When a thermal protector is designed to perform its function by opening the circuit to the machine but must be reset manually to close the circuit, it is a manual-reset thermal protector. *See also:* **contactor.**   256, 63

thermal relay (1) (industrial control). A relay in which the displacement of the moving contact member is produced by the heating of a part of the relay under the action of electric currents. *Note:* Compare with **temperature relay.** *See also:* **relay.**   244, 206, 60, 202

(2) (power switchgear). A relay whose operation is caused by heat developed within the relay as a result of specified external conditions. *Note:* Compare with **temperature relay.** 103, 127

thermal residual voltage (Hall effect devices). That component of the zero field residual voltage caused by a temperature gradient in the Hall plate. 107

thermal resistance (1) (cable). The resistance offered by the insulation and other coverings to the flow of heat from the conductor or conductors to the outer surface. *Note:* The thermal resistance of the cable is equal to the difference of temperature between the conductor or conductors and the outside surface of the cable divided by rate of flow of heat produced thereby. It is preferably expressed by the number of degrees Celsius per watt per foot of cable.
64

(2) (semiconductor device) (light emitting diodes) (formerly θ) R_{θ}. The temperature difference between two specified points or regions divided by the power dissipation under conditions of thermal equilibrium. *See also:* **principal voltage-current characteristic (principal characteristic).**
243, 66, 208, 191, 162

(3) (Hall generator). The difference between the mean Hall plate temperature and the temperature of an external reference point, divided by the power dissipation in the Hall plate. 107

thermal resistance case-to-ambient, ($R_{\theta CA}$) (light emitting diodes). The thermal resistance (steady-state) from the device case to the ambient. 162

thermal resistance, effective (semiconductor rectifier) (semiconductor device). The effective temperature rise per unit power dissipation of a designated junction, above the temperature of a stated external reference point under conditions of thermal equilibrium. *Note:* Thermal impedance is the temperature rise of the junction above a designated point on the case, in degrees Celsius per watt of heat dissipation. *See also:* **rectification; semiconductor; semiconductor rectifier stack.** 237, 245, 210, 66, 208

thermal resistance junction-to-ambient (formerly θ_{J-A}) ($R_{\theta JA}$) (light emitting diodes). The thermal resistance (steady-state) from the semiconductor junction (s) to the ambient.
162

thermal resistance junction-to-case, (formerly θ_{J-C}) ($R_{\theta JC}$) (light emitting diodes). The thermal resistance (steady-state) from the semiconductor junction (s) to a stated location on the case. 162

thermal short-time current rating (current transformer). The root-mean-square symmetrical primary current that may be carried for a stated period (five seconds or less) with the secondary winding short-circuited, without exceeding a specified maximum temperature in any winding. *See also:* **instrument transformer.** 203

thermal telephone receiver (thermophone). A telephone receiver in which the temperature of a conductor is caused to vary in response to the current input, thereby producing sound waves as a result of the expansion and contraction of the adjacent air. *See also:* **loudspeaker.** 328

thermal tuning. *See:* **tuning, thermal.**

thermal voltage converter (electric instrument). A thermoelement of low-current input rating with an associated series impedance or transformer, such that the electromotive force developed at the output terminals gives a measure of the voltage applied to the input terminals. *See also:* **thermal converter.** 280

thermal watt converter. *See:* **thermal power converter.** *See also:* **thermal converter.**

thermally delayed overcurrent trip. *See:* **thermally delayed release.**

thermally delayed release (trip). A release delayed by a thermal device. 103, 202

thermionic arc (gas). An electric arc characterized by the fact that the thermionic cathode is heated by the arc current itself. *See also:* **discharge (gas).** 244, 190

thermionic emission (Edison effect) (Richardson effect). The liberation of electrons or ions from a solid or liquid as a result of its thermal energy. *See also:* **electron emission.**
333, 210, 190, 125

thermionic generator. A thermoelectric generator in which a part of the circuit, across which a temperature difference is maintained, is a vacuum or a gas. *See also:* **thermoelectric device.** 248, 191

thermionic grid emission. Current produced by electrons thermionically emitted from a grid. *See:* **electron emission.**
333, 190, 125

thermionic tube. An electron tube in which the heating of one or more of the electrodes is for the purpose of causing electron or ion emission. *See also:* **hot cathodetube.**
125

thermistor (1) (general). An electron device that makes use of the change of resistivity of semiconductor with change in temperature. *See also:* **bolometric detector; electron devices, miscellaneous; semiconductor.**
245, 210, 185, 7

(2) (power semiconductor). A semiconductor device whose electric resistance is dependent upon temperature. 66

(3) (waveguide components). A form of bolometer element having a negative temperature coefficient of resistivity which typically employs a semiconductor bead. 166

thermistor mount (bolometer mount) (waveguide). A waveguide termination in which a thermistor (bolometer) can be incorporated for the purpose of measuring electromagnetic power. *See also:* **waveguide.** 244, 179

thermocouple. A pair of dissimilar conductors so joined at two points that an electromotive force is developed by the thermoelectric effects when the junctions are at different temperatures. *See also:* **electric thermometer (temperature meter); thermoelectric effect.** 188, 47, 7

thermocouple converter (thermoelement). *See:* **thermal converter.**

thermocouple extension wire. A pair of wires having such electromotive-force-temperature characteristics relative to the thermocouple with which the wires are intended to be used that, when properly connected to the thermocouple, the reference junction is in effect transferred to the other end of the wires. 7

thermocouple instrument. An electrothermic instrument in which one or more thermojunctions are heated directly or indirectly by an electric current or currents and supply a direct current that flows through the coil of a suitable direct-current mechanism, such as one of the permanent-magnet moving-coil type. *See also:* **instrument.**
328

thermocouple leads. A pair of electrical conductors that connect the thermocouple to the electromotive force measuring device. One or both leads may be simply extensions of the thermoelements themselves or both may be of copper, dependent on the thermoelements in use and upon the physical location of the reference junction or junctions relative to the measuring device. 7

thermocouple thermometer. A temperature-measuring instrument comprising a device for measuring electromotive force, a sensing element called a thermocouple that produces an electromotive force of magnitude directly related to the temperature difference between its junctions, and electrical conductors for operatively connecting the two. 7

thermocouple vacuum gauge. A vacuum gauge that depends for its operation on the thermal conduction of the gas present, pressure being measured as a function of the electromotive force of a thermocouple the measuring junction of which is in thermal contact with a heater that carries a constant current. It is ordinarily used over a pressure range of 10^{-1} to 10^{-3} conventional millimeter of mercury. *See also:* **instrument.** 328

thermoelectric cooling device. A thermoelectric heat pump that is used to remove thermal energy from a body. *See also:* **thermoelectric device.** 248, 191

thermoelectric device. A generic term for thermoelectric heat pumps and thermoelectric generators. 248, 191

thermoelectric effect. *See:* **Seebeck effect.**

thermoelectric effect error (bolometric power meter). An error arising in bolometric power meters that employ thermistor elements in which the majority of the bias power is alternating current and the remainder direct current. The error is caused by thermocouples at the contacts of the thermistor leads to the metal oxides of the thermistors. *See also:* **bolometric power meter.** 185

thermoelectric generator. A device that converts thermal energy into electric energy by direct interaction of a heat flow and the charge carriers in an electric circuit, and that requires for this process the existence of a temperature difference in the electric circuit. *See also:* **thermoelectric device.** 248, 191

thermoelectric heating device. A thermoelectric heat pump that is used to add thermal energy to a body. *See also:* **thermoelectric device.** 248, 191

thermoelectric heat pump. A device that transfers thermal energy from one body to another by the direct interaction of an electric current and the heat flow. *See also:* **thermoelectric device.** 248, 191

thermoelectric power*. *See:* **Seebeck coefficient.** *See also:* **thermoelectric device.**

* Deprecated

thermoelement (electric instrument). The simplest type of thermal converter. It consists of a thermocouple, the measuring junction of which is in thermal contact with an electric heater or integral therewith. *See also:* **thermal converter; thermoelectric arm; thermoelectric couple.** 280

thermoelectric thermometer (thermocouple thermometer). An electric thermometer that employs one or more thermocouples of which the set of measuring junctions is in thermal contact with the body, the temperature of which is to be measured, while the temperature of the reference junctions is either known or otherwise taken into account. *See also:* **electric thermometer (temperature meter).** 328

thermogalvanic corrosion. Corrosion resulting from a galvanic cell caused primarily by a thermal gradient. *See also:* **electrolytic cell.** 221, 205

thermojunction. One of the surfaces of contact between the two conductors of a thermocouple. The thermojunction that is in thermal contact with the body under measurement is called the measuring junction, and the other thermojunction is called the reference junction. *See also:* **electric thermometer (temperature meter).** 328

thermometer. An instrument for determining the temperature of a body or space. 7

thermometer method of temperature determination. The determination of the temperature by mercury, alcohol, resistance, or thermocouple thermometer, any of these instruments being applied to the hottest accessible part of the device. 53

thermophone. An electroacoustic transducer in which sound waves of calculable magnitude result from the expansion and contraction of the air adjacent to a conductor whose temperature varies in response to a current input. *Note:* When used for the calibration of pressure microphones, a thermophone is generally used in a cavity the dimensions of which are small compared to a wavelength. *See also:* **microphone.** 176

thermopile. A group of thermocouples connected in series aiding. This term is usually applied to a device used either to measure radiant power or energy or as a source of electric energy. *See also:* **electric thermometer (temperature meter).** 328, 47, 7

thermoplastic (1) (noun). A plastic that is thermoplastic in behavior.
(2) (adjective). Having the quality of softening when heated above a certain temperature range and of returning to its original state when cooling below that range. 63

thermoplastic insulating tape. A tape composed of a thermoplastic compound that provides insulation for joints. 328

thermoplastic insulations and jackets (power distribution, underground cables). Insulations and jackets made of materials that are softened by heat for application to the cable and then become firm, tough and resilient upon cooling. Subsequent heating and cooling will reproduce similar changes in the physical properties of the material. 57

thermoplastic protective tape. A tape composed of a thermoplastic compound that provides a protective covering for insulation. 328

thermoset. A plastic that, when cured by the application of heat or chemical means, changes into a substantially infusible and insoluble product. 63

thermosetting insulations and jackets (power distribution, underground cables). Insulations and jackets made of materials that may be applied to the cable in a relatively soft state and then crosslinked or vulcanized by suitable means to develop firm, tough, and resilient properties. 57

thermostat. A temperature-sensitive device that automatically opens and closes an electric circuit to regulate the temperature of the space with which it is associated. *See also:* **appliance outlet.** 263

thermostatic switch (thermostat). A form of temperature-operated switch that receives its operating energy by

theta (θ) polarization
718
three-phase seven-wire system

thermal conduction or convection from the device being controlled or operated. *See also:* **switch.** 206

theta (θ) polarization. The state of the wave in which the *E* vector is tangential to the meridian lines of a given spherical frame of reference. *Note:* The usual frame of reference has the polar axis vertical and the origin at or near the antenna. Under these conditions, a vertical dipole will radiate only theta (θ) polarization and the horizontal loop will radiate only phi (ϕ) polarization. *See also:* **antenna.** 246, 244, 179

Thévenin's theorem. States that the current that will flow through an impedance Z', when connected to any two terminals of a linear network between which there previously existed a voltage *E* and an impedance *Z*, is equal to the voltage *E* divided by the sum of *Z* and Z'. 328

thickener (hydrometallurgy) (electrometallurgy). A tank in which suspension of solid material can settle so that the solid material emerges from a suitable opening with only a portion of the liquid while the remainder of the liquid overflows in clear condition at another part of the thickener. *See also:* **electrowinning.** 328

thick film technology (circuits and systems). A technology in which a thick film (about 1 mil) is screen-printed onto an insulating substrate and then fused to the substrate by firing. *Note:* Resistors, capacitors, and conductors are commonly made by this technology. 67

thin film. Loosely, magnetic thin film. 255, 77

thin film technology (circuits and systems). A technology in which a thin film (a few hundred to a few thousand angstroms in thickness) is applied by vacuum deposition to an insulating substrate. Resistors, capacitors, and conductors are commonly made by this technology. 67

thinned array antenna. An array antenna which contains substantially fewer driven radiating elements than a conventional uniformly spaced array with the same beamwidth having identical elements. *Note:* Interelement spacings in the thinned array are chosen such that no large grating lobes are formed and side lobes are minimized. 111

thin-wall counter (radiation counters). A counter tube in which part of the envelope is made thin enough to permit the entry of radiation of low penetrating power. *See:* **anticoincidence (radiation counters).** 190

third order non-linearity coefficient (accelerometer). The proportionality constant that relates a variation of the output to the cube of the input applied parallel to an input reference axis. 46

third-rail clearance line (railroad). The contour that embraces all cross sections of third rail and its insulators, supports, and guards located at an elevation higher than the top of the running rail. *See also:* **electric locomotive.** 328

third-rail electric car. An electric car that collects propulsion power through a third-rail system. *See also:* **electric motor car.** 328

third-rail electric locomotive. An electric locomotive that collects propulsion power from a third-rail system. *See also:* **electric locomotive.** 328

third voltage range. *See:* **voltage range.**

Thomson bridge (double bridge). *See:* **Kelvin bridge.**

Thomson coefficient (Thomson heat coefficient). The quotient of (1) the rate of Thomson heat absorption per unit volume of conductor by (2) the scalar product of the electric current density and the temperature gradient. The Thomson coefficient is positive if Thomson heat is absorbed by the conductor when the component of the electric current density in the direction of the temperature gradient is positive. *See also:* **thermoelectric device.** 248, 191

Thomson effect. The absorption or evolution of thermal energy produced by the interaction of an electric current and a temperature gradient in a homogeneous electric conductor. *Notes:* (1) An electromotive force exists between two points in a single conductor that are at different temperatures. The magnitude and direction of the electromotive force depend on the material of the conductor. A consequence of this effect is that if a current exists in a conductor between two points at different temperatures, heat will be absorbed or liberated depending on the material and on the sense of the current. (2) In a nonhomogeneous conductor, the Peltier effect and the Thomson effect cannot be separated. *See also:* **thermoelectric device.** 248, 210, 191

Thomson heat. The thermal energy absorbed or evolved as a result of the Thomson effect. *See also:* **thermoelectric device.** 248, 191

thread (control) (industrial control). A control function that provides for maintained operation of a drive at a preset reduced speed such as for setup purposes. *See:* **electric drive.** 225, 206

threaded coupling (rigid steel conduit). An internally threaded steel cylinder for connecting two sections of rigid steel conduit. 101

three-address. Pertaining to an instruction code in which each instruction has three address parts. Also called triple-address. In a typical three-address instruction the addresses specify the location of two operands and the destination of the result, and the instructions are taken from storage in a preassigned order. *See also:* **two-plus-one-address.** 235

three-address code. *See:* **instruction code.**

three-dimensional (3D) radar. A radar capable of producing three-dimensional position data on a multiplicity of targets. *See also:* **radar.** 187, 13

three-phase circuit. A combination of circuits energized by alternating electromotive forces which differ in phase by one-third of a cycle, that is, 120 degrees. *Note:* In practice the phases may vary several degrees from the specified angle. *See also:* **center of distribution.** 3, 64, 53

three-phase electric locomotive. An electric locomotive that collects propulsion power from three phases of an alternating-current distribution system. *See also:* **electric locomotive.** 328

three-phase four-wire system. A system of alternating-current supply comprising four conductors, three of which are connected as in a three-phase three-wire system, the fourth being connected to the neutral point of the supply, which may be grounded. *See also:* **alternating-current distribution.** 64

three-phase seven-wire system. A system of alternating-current supply from groups of three single-phase transformers connected in Y so as to obtain a three-phase

four-wire grounded-neutral system for lighting and a three-phase three-wire grounded-neutral system of a higher voltage for power, the neutral wire being common to both systems. *See also:* **alternating-current distribution.**
64

three-phase three-wire system. A system of alternating-current supply comprising three conductors between successive pairs of which are maintained alternating differences of potential successively displaced in phase by one-third of a period. *See also:* **alternating-current distribution.**
64

three-plus-one-address. Pertaining to a four-address code in which one address part always specifies the location of the next instruction to be interpreted.
235

three-position relay. A relay that may be operated to three distinct positions.
328

three-terminal capacitor. Two conductors (the active electrodes) insulated from each other and from a surrounding third conductor that constitutes the shield. When the capacitor is provided with properly designed terminals and used with shielded leads, the direct capacitance between the active electrodes is independent of the presence of other conductors. (Specialized usage.)
210

three-terminal network. A four-terminal network having one input and one output terminal in common. *See also:* **network analysis.**
59

three-wire control (industrial control). A control function that utilizes a momentary-contact pilot device and a holding-circuit contact to provide undervoltage protection. *See:* **undervoltage protection.** *See also:* **relay.**
225, 206

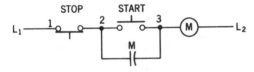

Three-wire control

three-wire system (direct current or single-phase alternating current). A system of electric supply comprising three conductors, one of which (known as the **neutral wire**) is maintained at a potential midway between the potential of the other two (referred to as the outer conductors). *Note:* Part of the load may be connected directly between the outer conductors, the remainder being divided as evenly as possible into two parts each of which is connected between the neutral and one outer conductor. There are thus two distinct voltages of supply, the one being twice the other. *See also:* **alternating-current distribution; direct-current distribution.**
64

three-wire type current transformer. One which has two primary windings each completely insulated for the rated insulation level of the transformer. This type of current transformer is for use on a three-wire, single-phase service. *Note:* The primary windings and secondary windings are permanently assembled on the core as an integral structure. The secondary current is proportional to the phasor sum of the primary currents.
53

threshold (1) (general). The value of a physical stimulus (such as size, luminance, contrast, or time) that permits an object to be seen a specific percentage of the time or at a specific accuracy level. *Note:* In many psychophysical experiments, thresholds are presented in terms of 50-percent accuracy or accurately 50 percent of the time. However, the threshold also is expressed as the value of the physical variable that permits the object to be just barely seen. The threshold may be determined by merely detecting the presence of an object or it may be determined by discriminating certain details of the object.
167

(2) (ultimate sensitivity*). The least change in the measurand that can be detected with a specified degree of certainty by use of the instrument under specified conditions in a specified time. The limiting factor that fixes the threshold may be any of a number of effects such as the uncertainty of reading, friction, spurious external disturbances or internal thermal agitation, depending upon the type of instrument and the conditions of use. *See also:* **accuracy rating (instrument).**
328

* Deprecated

(3) (audiology) (medical electronics). The lowest amplitude of sound that may be heard by an individual.
192

(4) (modulation systems). The smallest value of carrier-to-noise ratio at the input to the demodulator for all values above which a small percentage change in the input carrier-to-noise ratio produces a substantially equal or a smaller percentage change in the output signal-to-noise ratio. *Note:* Where precision is required, the method of determining the value of the threshold must be specified.
242

(5) (electrobiology) (limen). (A) The least current or voltage needed to elicit a minimal response. (B) That needed to elicit a minimal response in 50 percent of the trials. *See also:* **biological; excitability (electrobiology).**
192

(6) (logic operations). (A) A logic operator having the property that if P is a statement, Q is a statement, R is a statement, . . . , then the threshold of $P, Q, R, . . . ,$ is true if at least N statements are true, false if less than N statements are true, where N is a specified nonnegative integer called the threshold condition. (B) The threshold condition as in (A).
255, 77

(7) (gyro; accelerometer). The largest absolute value of the minimum input that produces an output equal to at least 50 percent of the output expected using the nominal scale factor. *See:* **input-output characteristics.**
46

(8) (communications) (radar). A value of voltage or current equal in value to the smallest signal to be detected by the receiver.
13

(9) (laser-maser). The condition of a **maser** or **laser** wherein the gain of its medium is just sufficient to permit the start of oscillation.
363

threshold audiogram. *See:* **audiogram.**

threshold current (current-limiting fuse). A current magnitude of specified wave shape at which the melting of the current-response element occurs at the first instantaneous peak current for that wave shape. *Note:* The current magnitude is usually expressed in root-mean-square amperes.
103, 202

threshold element. (1) A combinational logic element such that the output channel is in its ONE state if and only if at least n input channels are in their ONE states, where n is a specified fixed nonnegative integer, called the threshold of the element. (2) By extension, a similar element whose output channel is in its ONE state if and only if at least n input channels are in states specified for them, not necessarily the ONE state but a fixed state for each input channel. 235

(3) A device that performs the logic threshold operation but in which the truth of each input statement contributes to the output determination a weight associated with that statement. 255, 77

threshold field. The least magnetizing force in a direction that tends to decrease the remanence, that, when applied either as a steady field of long duration or as a pulsed field appearing many times, will cause a stated fractional change of remanence. 77

threshold frequency (photoelectric device) (photoelectric tubes). The frequency of incident radiant energy below which there is no photoemissive effect.

$$v_0 = p/h.$$

See also: **photoelectric effect.** 244, 190

threshold lights. Runway lights so placed as to indicate the longitudinal limits of that portion of a runway, channel or landing path usable for landing. 167

threshold of audibility (threshold of detectability) (specified signal). The minimum effective sound pressure level of the signal that is capable of evoking an auditory sensation in a specified fraction of the trials. The characteristics of the signal, the manner in which it is presented to the listener, and the point at which the sound pressure is measured must be specified. *Notes:* (1) Unless otherwise specified, the ambient noise reaching the ears is assumed to be negligible. (2) The threshold is usually given as a sound pressure level in decibels relative to 20 micronewtons per square meter. (3) Instead of the method of constant stimuli, which is implied by the phrase in a specified fraction of the trials, another psychophysical method (which should be specified) may be employed. 176

threshold of discomfort (for a specified signal) (audio and electroacoustics). The minimum effective sound pressure level at the entrance to the external auditory canal that, in a specified fraction of the trials, will stimulate the ear to a point at which the sensation of feeling becomes uncomfortable. 176

threshold of feeling (tickle) (for a specified signal) (audio and electroacoustics). The minimum effective sound pressure level at the entrance to the external auditory canal that, in a specified fraction of the trials, will stimulate the ear to a point at which there is a sensation of feeling that is different from the sensation of hearing. 176

threshold of pain (for a specified signal) (audio and electroacoustics). The minimum effective sound pressure level at the entrance to the external auditory canal that, in a specified fraction of the trials, will stimulate the ear to a point at which the discomfort gives way to definite pain that is distinct from the mere nonnoxious feeling of discomfort. 176

threshold ratio (current-limiting fuse). The ratio of the threshold current to the fuse current rating. 103, 202

threshold signal (1) (signal-transmission system). The minimum increment in signal magnitude that can be dis-

tinguished from the interference variations. *See:* **signal.** 188

(2) (navigation). The smallest signal capable of effecting a recognizable change in navigational information. *See also:* **navigation.** 13, 187

threshold signal-to-interference ratio (TSI) (electromagnetic compatibility). The minimum signal to interference power, described in a prescribed way, required to provide a specified performance level. *See also:* **electromagnetic compatibility.** 199

threshold voltage (semiconductor rectifiers). The zero-current–voltage intercept of a straight-line approximation of the forward current-voltage characteristic over the normal operating range. *See:* **semiconductor rectifier stack.** 208

threshold wavelength (photoelectric tubes) (photoelectric device). The wavelength of the incident radiant energy above which there is no photoemission effect. *See also:* **photoelectric effect.** 244, 190

throat microphone. A microphone normally actuated by mechanical contact with the throat. *See also:* **microphone.** 328

through bolt (rotating machinery). A bolt passing axially through a laminated core, that is used to apply pressure to the end plates. 63

throughput (1) (data transmission). The total capability of equipment to process or transmit data during a specified time period. 59

(2) (automatic control). *See:* **capacity.**

through supervision (1) (communication switching). The automatic transfer of supervisory signals through one or more trunks in a manual telephone switchboard. 193

(2) (telephone switching systems). The capability of apparatus within a switched connection to pass or repeat signaling. 55

throwing power (of a solution) (electroplating). A measure of its adaptability to deposit metal uniformly upon a cathode of irregular shape. In a given solution under specified conditions it is equal to the improvement (in percent) of the metal distribution ratio above the primary-current distribution ratio. *See also:* **electroplating.** 328

thrust bearing (rotating machinery). A bearing designed to carry an axial load so as to prevent or to limit axial movement of a shaft, or to carry the weight of a vertical rotor system. *See also:* **bearing.** 63

thrust block (rotating machinery). A support for a thrust-bearing runner. *See also:* **bearing.** 63

thrust collar (rotating machinery). The part of a shaft or rotor that contacts the thrust bearing and transmits the axial load. *See:* **rotor (rotating machinery).** 63

thump. A low-frequency transient disturbance in a system or transducer characterized audibly by the onomatopoeic connotation of the word. *Note:* In telephony, thump is the noise in a receiver connected to a telephone circuit on which a direct-current telegraph channel is superposed caused by the telegraph currents. *See also:* **signal-to-noise ratio.** 239

Thury transmission system. A system of direct-current transmission with constant current and a variable high voltage. *Note:* High voltage used on this system is obtained by connecting series direct-current generators in series at the generating station; and is utilized by connecting series direct-current motors in series at the substations. *See also:* **direct-current distribution.** 64

thyratron. A hot-cathode gas tube in which one or more control electrodes initiate but do not limit the anode current except under certain operating conditions. 190

thyristor. A bistable semiconductor device comprising three or more junctions that can be switched from the OFF state to the ON state or vice versa, such switching occurring within at least one quadrant of the principal voltage-current characteristic. 243, 191, 66, 208

thyristor converter (thyristor converter unit, thyristor converter equipment). An operative unit comprising one or more thyristor sections together with converter transformers, essential switching devices, and other auxiliaries, if any of these items exist. System control equipments are optionally included. 121

thyristor converter, bridge (double-way). A bridge thyristor converter in which the current between each terminal of the alternating-voltage circuit and the thyristor converter circuit elements conductively connected to it flows in both directions. *Note:* The terms single-way and double-way (bridge) provide a means for describing the effect of the thyristor converter circuit on current flow in the transformer windings connected to the converter. Most thyristor converters may be classified into these two general types. The term bridge relates back to the single-phase "bridge" which resembles the Wheatstone bridge. 121

thyristor converter, cascade. A thyristor converter in which two or more simple converters are connected in such a way that their direct voltages add, but their commutations do not coincide. 121

thyristor converter circuit element. A group of one or more thyristors, connected in series or parallel or any combination of both, bounded by no more than two circuit terminals and conducting forward current in the same direction between these terminals. *Note:* A circuit element is also referred to as a leg or arm, and in the case of paralleled thyristors each path is referred to as a branch. 121

thyristor converter, multiple. A thyristor converter in which two or more simple thyristor converters are connected in such a way that their direct currents add, but their commutations do not coincide. 121

thyristor converter, parallel. A thyristor converter in which two or more simple converters are connected in such a way that their direct currents add and their commutations coincide. 121

thyristor converter, series. A thyristor converter in which two or more simple converters are connected in such a way that their direct voltages add and their commutations coincide. 121

thyristor converter, simple. A thyristor converter that consists of one commutating group. 121

thyristor converter, single-way. A thyristor converter in which the current between each terminal of the alternating-voltage circuit and the thyristor converter circuit element or elements conductively connected to it flows only in one direction. 121

thyristor converter transformer. A transformer that operates at the fundamental frequency of the ac system and is designed to have one or more output windings conductively connected to the thyristor converter elements. 121

thyristor converter unit, double. Two converters connected to a common dc circuit such that this circuit can accept or give up energy with direct current in both directions. *Note:* The converters may be supplied from separate cell windings on a common transformer, from common cell windings, or from separate transformers, The converter connections may be single way or symmetrical double way. Where two converters are involved, the designated forward converter arbitrarily operates in quadrant I. Quadrant I implies motoring torque in the agreed-upon forward direction. 121

thyristor converter unit, single. A thyristor converter unit connected to a dc circuit such that the direct current supplied by the converter is flowing in only one direction. The single converter section is referred to as a forward converter section. *Note:* When used without a reversing switch a single converter can be used in a reversible power sense only in those cases where single-way thyristor connections or symmetrical double-way thyristor connections are used and where the dc circuit can change from accepting energy to giving up energy without the need for current reversal, for example, a heavily inductive load. When used with a reversing switch, a single converter can be used in a reversible power sense in all cases where single-way thyristor connections or uniform double-way thyristor connections are used. 121

thyristor fuses. Fuses of special characteristics connected in series with one or more thyristors to protect the thyristor or other circuit components, or both. 121

thyristor, reverse-blocking triode. A monocrystalline reverse-blocking semiconductor device with bistable character in the forward direction normally having three pn junctions and a gate electrode at which a suitable electrical signal will cause switching from the off state to the on state within the first quadrant of the anode to cathode voltage—current characteristics. If cooling means are integrated, they are included. *Note:* In this document the word thyristor means a reverse-blocking triode thyristor. *See also:* **junction, P-N.** 121

ticker. A form of receiving-only printer used in the dissemination of information such as stock quotations and news. *See also:* **telegraphy.** 194

tie (rotating machinery). A binding of the end turns used to hold a winding in place or to hold leads to windings for purpose of anchoring. *See also:* **rotor (rotating machinery); stator.** 63

tie feeder. A feeder that connects together two or more independent sources of power and has no tapped load between the terminals. *Notes:* (1) If a feeder has any tapped load between the two sources, it is designated as a multiple feeder. (2) The source of power may be a generating system, substation, or feeding point. The normal flow of energy in such a feeder may be in either direction. *See also:* **center of distribution.** 103, 202, 64

tie line (electric power systems). A transmission line connecting two or more power systems. *See also:* **transmission line.** 94

tie-line bias control (control area). A mode of operation under load-frequency control in which the area control error is determined by the net interchange minus the biased scheduled net interchange. *See also:* **speed-governing system.** 94

tie point (tie line) (electric power systems). The location of the switching facilities, when closed, permit power to flow between the two power systems. *See also:* **center of distribution.** 94

tier (rotating machinery). A concentric winding is said to have one, two or more tiers according to whether the periphal extremities of the end windings of groups of coils at each end of the machine form one, two or more solids of revolution around the axis of the machine. 63

tie trunk. A telephone line or channel directly connecting two private branch exchanges. 193, 59, 101, 55

tie wire. A short piece of wire used to bind an overhead conductor to an insulator or other support. *See also:* **conductor; tower.** 64

TIF. *See:* **telephone influence factor.**

tight (suffix) (1) (general). So constructed that the specified material is excluded under specified conditions. 27, 127, 103, 124

(2) (transformers). Apparatus is designated as watertight, dusttight, etcetera, when so constructed that the enclosing case will exclude the specified material under specified conditions. 225, 53

tight coupling. *See:* close coupling.

tilt (1) (in a directive antenna). The angle that the antenna axis forms with the horizontal. *See also:* **navigation.** 187, 13

(2) (pulse terms). A distortion of a pulse top or pulse base wherein the overall slope over the extent of the pulse top or the pulse base is essentially constant and other than zero. Tilt may be of either polarity. *See:* Note in **preshoot** entry. 254

(3) (high- and low-power pulse transformers). (droop) (A_D). The difference between A_M and A_T. It is expressed in amplitude units or in percent of A_M. *See also:* **input pulse shape.** 32, 33

tilt angle (radar). The vertical angle between the axis of maximum radiation and a reference axis; the reference is normally horizontal. *See also:* **navigation.** 13, 187

tilt error (electronic navigation). *See:* **ionospheric tilt error.**

tilting-insulator switch. A switch in which the opening and closing travel of the blade is accomplished by a tilting movement of one or more of the insulators supporting the conducting parts of the switch. 103, 202, 27

tilting-pad bearing (Kingsbury bearing) (rotating machinery). A pad-type bearing in which the pads are capable of moving in such a manner as to improve the flow of lubricating fluid between the bearing and the shaft journal or collar (runner). *See:* **bearing.** 63

timbering machine. An electrically driven machine to raise and hold timbers in place while supporting posts are being set after being cut to length by the machine's power-driven saw. 328

timbre. The attribute of auditory sensation in terms of which a listener can judge that two sounds similarly presented and having the same loudness and pitch are dissimilar. *Note:* Timbre depends primarily upon the spectrum of the stimulus, but it also depends upon the waveform, the sound pressure, the frequency location of the spectrum, and the temporal characteristics of the stimulus.

See also: **loudspeaker.** 176

time (1) (reliability). Any duration of observations of the considered items either in actual operation or in storage, readiness, etcetera, but excluding down time due to a failure. *Note:* In definitions where time is used, this parameter may be replaced by distance, cycles, or other measures of life as may be appropriate. This refers to terms such as acceleration factor, wear-out failure, failure rate, mean life, mean time between failures, mean time to failure, reliability, and useful life. *See also:* **reliability.** 182

(2) (electronic computation). *See:* **access time; downtime; real time; reference time; switching time; word time.**

(3) (International System of Unts) (SI). The SI unit of time is the second. This unit is preferred and should be used if practical, particularly when technical calculations are involved. In cases where time relates to life customs or calendar cycles, the minute, hour, day, and other calendar units may be necessary. For example, vehicle velocity will normally be expressed in kilometers per hour. *See also:* **units and letter symbols.** 21

time above 90 percent (switching impulse testing). The time interval T_d during which the switching impulse exceeds 90 percent of its crest value. 108

time-and-charge-request call, (T&C) (telephone switching systems). A call for which a request is made to be informed of its duration and cost upon its completion. 55

time base (oscilloscopes). The sweep generator in an oscilloscope. *See:* **oscillograph.** 185

time bias (electric power systems). An offset in the scheduled net interchange power of a control area that varies in proportion to the time deviation. This offset is in a direction to assist in restoring the time deviation to zero. *See also:* **power system.** 200, 94

time bias setting (electric power systems). For a control area, a factor with negative sign that is multiplied by the time deviation to yield the time bias. 200, 94

time, build-up (automatic control). In a continuous step-forced response, the fictitious time interval, which would be required for the output to rise from its initial to its ultimate value, assuming that the entire rise were to take place at the maximum rate. *Note:* It can be evaluated as π/ω_o, where ω_o is the cut-off frequency of an ideal low-pass filter. 56

time constant (1) (automatic control). The value T in an exponential response term $A \exp(-1/T)$, or in one of the transform factors. *Notes:* (A) For the output of a first-order system forced by a step or an impulse, T is the time required to complete 63.2 percent of the total rise or decay. In higher-order systems, there is a time constant for each of the first-order factors of the process. In a Bode diagram, breakpoints occur at $\omega = 1/T$. *See also:* **control system, feedback.** (B) In terms of Laplace transforms it is the absolute value of the reciprocal of the real part of pole's position. The term **pole** refers here to the roots of the denominator of the Laplace transform. 329

(2) (electrothermic unit). The time required for the dc electrothermic output voltage to reach $1 - (1/e)$, or 63 percent of its final value after a fixed amount of power is applied to the electrothermic unit. 47

(3) (excitation control systems). The value T in an exponential response term A exp $(-t/T)$ or in one of the transform factors $1 + sT, 1 + jwT, 1/(1 + sT), 1/(1 + jwT)$. *Note:* For the output of a first-order (lag or lead) system forced by a step or an impulse, T is the time required to complete 63.2 percent of the total rise or decay; at any instant during the process, T is the quotient of the instantaneous rate of change divided into the change still to be completed. In higher order systems, there is a time constant for each of the first-order components of the process. In a Bode diagram, breakpoints occur at $w = 1/T$.   353

time constant, derivative action (automatic control). A parameter whose value is equal to $1/2\pi f_d$ where f_d is the frequency (cycles/unit time) on a Bode diagram of the lowest frequency gain corner resulting from derivative control action.   56

time constant of an exponential function (ae^{-bt}). $1/b$, if t represents time and b is real.   210

time constant of fall (pulse). The time required for the pulse to fall from 70.7 percent to 26.0 percent of its maximum amplitude excluding spike. *See also:* **pulse.**   328, 59

time constant of integrator (for each input). The ratio of the input to the corresponding time rate of change of the output. *See also:* **electronic analog computer.**   9, 77

time constant of rise (pulse). The time required for the pulse to rise from 26.0 percent to 70.7 percent of its maximum amplitude excluding spike. *See also:* **pulse.**   328, 59

time constant of the damping device (speed governing of hydraulic turbines). A time constant which describes the decay of the output signal from the damping device.   8

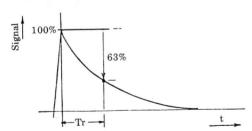

time-current characteristic (fuse). *See:* **fuse time-current characteristic.**

time-current tests (fuse). *See:* **fuse time-current tests.**

time delay (1) (general). The time interval between the manifestation of a signal at one point and the manifestation or detection of the same signal at another point. *Notes:* (1) Generally, the term **time delay** is used to describe a process whereby an output signal has the same form as an input signal causing it but is delayed in time; that is, the amplification of all frequency components of the output are related by a single constant to those of corresponding input frequency components but each output component lags behind the corresponding input component by a phase angle proportional to the frequency of the component. (2) Transport delay is synonymous with time delay but usually is reserved for applications that involve the flow of material. *See also:* **electronic analog computer.**   9, 10

(2) (industrial control). A time interval is purposely introduced in the performance of a function. *See also:* **control system, feedback.**   225, 206

time-delay relay. *See:* **delay relay.** *See also:* **relay.**

time deviation (power system). The integrated or accumulated difference in cycles between system frequency and rated frequency. This is usually expressed in seconds by dividing the deviation in cycles by the rated frequency.   94

time dial (relay) (time lever). An adjustable, graduated element by which, under fixed input conditions, the prescribed relay operating time can be varied.   103, 202, 60

time discriminator (electronic navigation). A circuit in which the sense and magnitude of the output is a function of the time difference of the occurrence, and relative time sequence, of two pulses. *See also:* **navigation.**   13, 187

time distribution analyzer (nuclear techniques). An instrument capable of indicating the number or rate of occurrence of time intervals falling within one or more specified time interval ranges. The time interval is delineated by the separation between pulses of a pulse pair. *See also:* **ionizing radiation.**   335

time-division analog switching (telephone switching systems). Analog switching with common time-divided paths for simultaneous calls.   55

time-division digital switching (telephone switching systems). Digital switching with common time-divided paths for simultaneous calls.   55

time division multiple access (TDMA) (communication satellite). A technique whereby earth stations communicate with each other on the basis of non-overlapping time sequenced bursts of transmissions through a common satellite repeater.   84

time-division multiplex. The process or device in which each modulating wave modulates a separate pulse subcarrier, the pulse subcarriers being spaced in time so that no two pulses occupy the same time interval. *Note:* Time division permits the transmission of two or more signals over a common path by using different time intervals for the transmission of the intelligence of each message signal.   111, 242

time-division switching (telephone switching systems). A method of switching that provides a common path with separate time intervals assigned to each of the simultaneous calls.   55

time-domain reflectometer. *See:* **reflectometer, time domain.**

timed release (telephone switching systems). Release accomplished after a specified delay.   55

time, electrification (cable-insulation material). Time during which a steady direct voltage is applied to electrical insulating materials before the current is measured.   97

time gain control (electronic navigation). *See:* **differential gain-control circuit.**

time gate. A transducer that gives output only during chosen time intervals.   111

time-interval selector (nuclear techniques). A circuit that produces a specified output pulse when and only when the time interval between two pulses lies between specified limits. *See also:* **scintillation counter.**   335

time lag. *See:* **lag.**

time lag of impulse flashover (surge arresters). The time between the instant when the voltage of the impulse wave first exceeds the power-frequency flashover crest voltage and the instant when the impulse flashover causes the abrupt drop in the testing wave. 62, 64

time-load withstand strength (of an insulator). The mechanical load that, under specified conditions, can be continuously applied without mechanical or electrical failure. *See also:* **insulator.** 261

time locking. A method of locking, either mechanical or electric, that, after a signal has been caused to display an aspect to proceed, prevents, until after the expiration of a predetermined time interval after such signal has been caused to display its most restrictive aspect, the operation of any interlocked or electrically locked switch, movable-point frog, or derail in the route governed by that signal, and that prevents an aspect to proceed from being displayed for any conflicting route. *See also:* **interlocking (interlocking plant).** 328

time-of-arrival location (radar). A process whereby the position of a radiating transmitter can be located by means of the relative time delay between its signals as received in multiple receivers of known relative position. The process is very similar to **loran** operated in reverse, that is with one transmitting element and three receiving elements. *See also:* **loran.** 13

time of rise (decay) of video pulses (television). The duration of the rising (decaying) portion of a pulse measured between specified levels. *See also:* **pulse timing of video pulses (television).** 336

time origin line (pulse terms). A line of constant and specified time which, unless otherwise specified, has a time equal to zero and passes through the first datum time, t_0, of a waveform epoch. *See:* The single pulse diagram below the **waveform epoch** entry. 254

time-overcurrent relay. An overcurrent relay in which the input current and operating time are inversely related throughout a substantial portion of the performance range. 103, 202, 60

time parameters and references (pulse terms). *See:* **pulse start (stop) time; pulse duration; time reference lines; pulse time reference points; transition duration; first (last) transition duration.**

time pattern (television). A picture tube presentation of horizontal and vertical lines or dot rows generated by two stable frequency sources operating at multiples of the line and field frequencies. *See also:* **television.** 337

time per point (multiple-point recorders). The time interval between successive points on printed records. *Note:* For some instruments this interval is variable and depends on the magnitude of change in measured signal. For such instruments, time per point is specified as the minimum and maximum time intervals. 295

time rate (storage cell). The current in amperes at which a storage battery will be discharged in a specified time, under specified conditions of temperature and final voltage. *See also:* **battery (primary or secondary).** 328

time referenced point (pulse terms). A point at the intersection of a time reference line and a waveform. 254

time reference line (pulse terms). A line parallel to the time origin line at a specified instant. 254

time reference lines (pulse terms). (1) pulse start (stop) line. The time reference line at pulse start (stop) time. *See:* The single pulse diagram below the **waveform epoch** entry. (2) top center line. The time reference line at the average of

pulse start time and pulse stop time. *See:* The single pulse diagram below the **waveform epoch** entry. 254

time-related adjectives (pulse terms). (1) periodic (aperiodic). Of or pertaining to a series of specified waveforms or features which repeat or recur regularly (irregularly in time. (2) coherent (incoherent). Of or pertaining to two or more repetitive waveforms whose constituent features have (lack) time correlation. (3) synchronous (asynchronous). Of or pertaining to two or more repetitive waveforms whose sequential constituent features have (lack) time correlation. 254

time-related definitions (pulse terms). *See:* **instant; interval; duration; period; frequency; cycle.** 254

time release. A device used to prevent the operation of an operative unit until after the expiration of a predetermined time interval after the device has been actuated. 328

time, response. *See:* **response time.**

time response (1) (control system, feedback). An output, expressed as a function of time, resulting from the application of a specified input under specified operating conditions. *Syn:* **dynamic response.** *Note:* It consists of a transient component that depends on the initial conditions of the system, and a steady-state component that depends on the time pattern of the input. 219, 56, 329, 206
(2) (synchronous-machine regulator). The output of the synchronous-machine regulator (that is, voltage, current, impedance, or position) expressed as a function of time following the application of prescribed inputs under specfied conditions. 63
(3) (excitation control systems). An output expressed as a function of time, resulting from the application of a specified input under specified operating conditions. See figure below. For a typical time response of a system to step increase of input and for identification of the principal characteristics of interest.

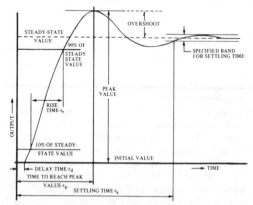

Typical time response of a feedback control system to a step change in input.

time share. To use a device or other resource for two or more interleaved purposes. 255, 77

time sharing. Pertaining to the interleaved use of the time of a device. 255, 77, 54

time skew (analog-to-digital converter). In an analog to digital conversion process, the time difference between the conversion of one analog channel and any other analog channel, such that the converted (digital) representations of the analog signals do not correspond to values of the analog variables that existed at the same instant of time. Time skew is eliminated, where necessary, by the use of a multiplexor with a sample/hold feature, allowing all input channels to be simultaneously sampled and stored for later conversion. *See also:* **switching time; converters, analog to digital.**   9, 10

time sorter (nuclear techniques). *See:* **time distribution analyzer.**

time-to-amplitude converter (scintillation counting). An instrument producing an output pulse whose amplitude is proportional to the time difference between start and stop pulses.   117

time to chopping (switching impulse testing). The time interval T_c between actual zero and the instant when the chopping occurs.   108

time to crest. The time interval T_{cr} between actual zero and the instant when the voltage has reached its crest value.   108

time to half value. The time interval T_h between actual zero and the instant on the tail when the impulse has decreased to half its crest value.   108

time to half-value on the wavetail (virtual duration of an impulse) (surge arresters). *See:* **virtual time to half-value (on the wavetail).**

time to impulse flashover. The time between the initial point of the voltage impulse causing flashover and the point at which the abrupt drop in the voltage impulse takes place.   64

time to impulse-sparkover (surge arresters). The time between virtual zero of the voltage impulse causing sparkover and the point on the voltage wave at which sparkover occurs. *Note:* The time is expressed in microseconds. *See also:* **lightning protection and equipment.**   2, 299, 62

time, turn-around (test, measurement and diagnostic equipment). The time needed to service or check out an item for recommitment.   54

time-undervoltage protection. A form of undervoltage protection that disconnects the protected equipment upon a deficiency of voltage after a predetermined time interval.   103, 202, 60, 127

time update (sequential events recording systems). The correction or resetting of a real time clock to match a time standard. *See also:* **real time.**   48, 58

timing deviation (demand meter). The difference between the elapsed time indicated by the timing element and the true elapsed time, expressed as a percentage of the true elapsed time. *See also:* **demand meter.**   212

timing mechanism (1) (demand meter). That mechanism through which the time factor is introduced into the result. The principal function of the timing mechanism of a demand meter is to measure the demand interval, but it has a subsidiary function, in the case of certain types of demand meters, to provide also a record of the time of day at which any demand has occurred. A timing mechanism consists either of a clock or its equivalent, or of a lagging device that delays the indications of the electric mechanism. In thermally lagged meters the time factor is introduced by the thermal time lag of the temperature responsive elements. In the case of curve-drawing meters, the timing element merely provides a continuous record

of time on a chart or graph. *See also:* **demand meter.** 328

(2) (recording instrument). The time-regulating device usually includes the motive power unit necessary to propel the chart at a controlled rate (linear or angular). *See also:* **moving element (instrument).**   294

timing relay (or relay unit). An auxiliary relay or relay unit whose function is to introduce one or more definite time delays in the completion of an associated function. *See also:* **relay.**   103, 202, 60

timing table. That portion of central-station equipment at which means are provided for operators' supervision of signal reception. *See also:* **protective signaling.**   328

tinning (electrotyping). The melting of lead-tin foil or tin plating upon the back of shells. *See also:* **electroforming.** 328

tinsel cord. A flexible cord in which the conducting elements are thin metal ribbons wound helically around a thread core. *See also:* **transmission line.**

tip (plug) (1). The contacting part at the end of the plug. 328

(2) (electron tubes) (pip). A small protuberance on the envelope resulting from the sealing of the envelope after evacuation. *See also:* **electronic tube.**   244, 190

tip and ring wires (telephone switching systems). A pair of conductors associated with the transmission portions of circuits and apparatus. Tip or ring designation of the individual conductors is arbitrary except when applied to cord-type switchboard wiring in which case the conductors are designated according to their association with tip or ring contacts of the jacks and plugs.   55

T junction (waveguide). A junction of waveguides in which the longitudinal guide axes form a T. *Note:* The guide that continues through the junction is the main guide; the guide that terminates at a junction is the branch guide. *See also:* **waveguide.**   330, 179

T network. A network composed of three branches with one end of each branch connected to a common junction point, and with the three remaining ends connected to an input terminal, an output terminal, and a common input and output terminal, respectively. *See also:* **network analysis.**   332, 210

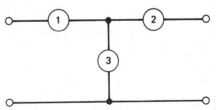

T network. One end of each of the branches 1, 2, and 3 is connected to a common point. The other ends of branches 1 and 2 form, respectively, an input and an output terminal, and the other end of branch 3 forms a common input and output terminal.

to-from indicator (omnirange receiver). A supplementary device used with an omnibearing selector to resolve the ambiguity of measured omnibearings.   13

toe and shoulder (of a Hurter and Driffield (H and D) curve) (photographic techniques). The terms applied to the nonlinear portions of the H and D curve that lie, respec-

tively, below and above the straight portion of this curve.
176

toggle. Pertaining to any device having two stable states. *See:* **flip-flop.** 255, 77

tolerable out-of-service time (nuclear power generating station). The time an information display channel is allowed to be unavailable for use as a post accident monitoring display. 361

tolerance (1) (test, measurement and diagnostic equipment). The total permissible variation of a quantity from a designated value. 54
(2) (nuclear power generating stations). The allowable deviation from a specified or true value. 31, 41

toll board. A switchboard used primarily for establishing connections over toll lines. 328

toll call (telephone switching systems). A call for a destination outside the local-service area of the calling station. 55

toll center (telephone switching systems). A toll office where trunks from end offices are connected to intertoll trunks and where operator's assistance is provided in completing incoming calls and where other traffic operating functions are performed. Toll centers are classified as Class 4C offices. *See also:* **office class.** 55

toll connecting trunk (telephone switching systems). A trunk between a local office and a toll office or switchboard. 55

toll line. A telephone line or channel between two central offices in different telephone exchanges. 328

toll office (telephone switching systems). An intermediate office serving toll calls. 55

toll point (telephone switching systems). A toll office where trunks from end offices are connected to the distance dialing network and where operators handle only outward calls or where there are no operators present. Toll points are classified as Class 4P offices. 55

toll restriction (telephone switching systems). A method that prevents private automatic branch exchange stations from completing certain or any toll calls or reaching a toll operator, except through the attendant. 55

toll station. A public telephone station connected directly to a toll telephone switchboard. *See also:* **telephone station.** 328

toll switching trunk (telephone switching systems). A trunk for completing calls from a toll office or switchboard to a local office. 55

toll switch train (toll train). A switch train that carries a connection from a toll board to a subscriber line. *See also:* **switching system.** 328

toll terminal loss (toll connection). That part of the over-all transmission loss that is attributable to the facilities from the toll center through the tributary office to and including the subscriber's equipment. *Note:* The toll terminal loss at each end of the circuit is ordinarily taken as the average of the transmitting loss and the receiving loss between the subscriber and the toll center. *See also:* **transmission loss.** 328

toll train. See: **toll switch train.**

toll transmission selector. A selector in a toll switch train that furnishes toll-grade transmission to the subscriber and controls the ringing. 328

tone (1) (general). (A) A sound wave capable of exciting an auditory sensation having pitch. (B) A sound sensation having pitch. 176
(2) (telephone switching systems). An audible signal transmitted over the telecommunications network. 55

tone, call (telephone switching systems). A tone that indicates to an operator or attendant that a call has reached the position or console. 55

tone control. A means for altering the frequency response at the audio-frequency output of a circuit, particularly of a radio receiver or hearing aid, for the purpose of obtaining a quality more pleasing to the listener. *See also:* **amplifier; radio receiver.** 328

tone localizer. *See:* **equisignal localizer.**

tone-modulated waves. Waves obtained from continuous waves by amplitude modulating them at audio frequency in a substantially periodic manner. *See also:* **telegraphy.** 328

tone-operated net-loss adjuster. *See:* **tonlar.**

toner (electrostatography). The image-forming material in a developer that, deposited by the field of an electrostatic-charge pattern, becomes the visible record. *See also:* **electrostatography.** 236, 191

tone receiver (power-system communication). A device for receiving a specific voice-frequency carrier telegraph signal and converting it into direct current. *See also:* **telegraphy.** 59

tone transmitter (power-system communication). A device for transmitting a voice-frequency telegraph signal. *See also:* **telegraphy.** 59

tonlar. A system for stabilizing the net loss of a telephone circuit by means of a tone transmitted between conversations. The name is derived from the initial letters of the expression **tone-operated net-loss adjuster.** *See also:* **tone-frequency telephony.** 328

tool function (numerically controlled machines). A command identifying a tool and calling for its selection either automatically or manually. The actual changing of the tool may be initiated by a separate tool-change command. 224, 207

tool offset (numerically controlled machines). A correction for tool position parallel to a controlled axis. 224, 207

tools, access (switchgear assembly). Keys or other special accessories with unique characteristics that make them suitable for gaining access to the tamper-resistant switchgear assembly. 93

tooth (1) (rotating machinery). A projection from a core, separating two adjacent slots, the tip of which forms part of one surface of the air gap. *See also:* **rotor (rotating machinery); stator.** 63
(2) (cable plowing).* *See:* **wear point.** 52

* Deprecated

tooth pitch (rotating machinery). *See:* **slot pitch.**

tooth tip (rotating machinery). That portion of a tooth that forms part of the inner or outer periphery of the air gap. It is frequently considered to be the section of a tooth between the radial location of the wedge and the air gap. *See also:* **rotor (rotating machinery); stator.** 63

top (pulse terms). The portion of a pulse waveform which represents the second nominal state of a pulse. 254

top cap (side contact) (electron tubes). A small metal shell on the envelope of an electron tube or valve used to connect

one electrode to an external circuit. *See also:* **electron tube.**
244, 190

top car clearance (elevators). The shortest vertical distance between the top of the car crosshead, or between the top of the car where no car crosshead is provided, and the nearest part of the overhead structure or any other obstruction when the car floor is level with the top terminal landing. *See also:* **hoistway (elevator or dumbwaiter).**
328

top coil side (radially inner coil side) (rotating machinery). The coil side of a stator slot nearest the bore of the stator or nearest the slot wedge. *See also:* **stator.** 63

top counterweight clearance (elevator counterweight) (elevator). The shortest vertical distance between any part of the counterweight structure and the nearest part of the overhead structure or any other obstruction when the car floor is level with the bottom terminal landing. *See also:* **hoistway (elevator or dumbwaiter).** 328

top half bearing (rotating machinery). The upper half of a split sleeve bearing. *See also:* **bearing.** 63

top-loaded vertical antenna. A vertical monopole with additional metallic structure at the top intended to increase the effective height of the antenna. 111

top magnitude (pulse terms). The magnitude of the top as obtained by a specified procedure or algorithm. *See:* The single pulse diagram below the **waveform epoch** entry. *See also:* IEEE Std 181-1975, Section 4.3, for suitable algorithms.) 254

top side sounding (communication satellite). Ionospheric sounding from medium altitude satellites for measuring ionospheric densities at high altitudes. 85

top terminal landing (elevators). The highest landing served by the elevator that is equipped with a hoistway door and hoisting-door locking device that permits egress from the hoistway side. *See also:* **elevator landing.** 328

torchere. An indirect floor lamp that sends all or nearly all of its light upward. *See also:* **luminaire.** 167

toroid (doughnut) (electron device). A toroidal-shaped vacuum envelope in which electrons are accelerated. *See also:* **electron device.** 244, 190

toroidal coil. A coil wound in the form of a toroidal helix.
329

torque (1) (instrument). The turning moment on the moving element produced by the quantity to be measured or some quantity dependent thereon acting through the mechanism. This is also termed the deflecting torque and in many instruments is opposed by the controlling torque, which is the turning moment produced by the mechanism of the instrument tending to return it to a fixed position. *Note:* Full-scale torque is the particular value of the torque for the condition of full-scale deflection and as an index of performance should be accompanied by a statement of the angle corresponding to this deflection. *See also:* **accuracy rating (instrument).** 328

(2) (International System of Units) (SI). *See:* **energy and torque.**

torque (force) balance accelerometer. A device that measures acceleration by applying a principal restraint through capturing. 46

torque-coil magnetometer. A magnetometer that depends for its operation on the torque developed by a known current in a coil that can turn in the field to be measured.

See also: **magnetometer.** 328

torque command storage (gyro). The transient deviation of the output of a rate integrating gyro from that of an ideal integrator when the gyro is subjected to a torquer command signal. It is a function of the gyro's characteristic time and the torquer time constant. *See also:* **attitude storage; float storage.** 46

torque control (1) (general) (relay). A method of constraining the pickup of a relay by preventing the torque-producing element from developing operating torque until another associated relay unit operates.
103, 202, 60, 127

(2) (protective relaying of utility-consumer interconnections). A means of supervising the operation of one relay element with another. For example, an overcurrent relay cannot operate unless the lag coil circuit is closed. It may be closed by the contact of an undervoltage element.
128

torque-generator reaction torque (gyro; accelerometer). *See also:* **torquer reaction torque.** 46

torque margin. The increase in torque, under any steady-state operating condition, that a synchronous propulsion motor will deliver without pulling out of step. *Note:* Rated torque margin is the additional torque available when the propulsion system is operating at its designed rating. *See also:* **converter; electric propulsion system.** 328

torque motor. A motor designed primarily to exert torque through a limited travel or in a stalled position. *Note:* Such a motor may be capable of being stalled continuously or only for a limited time. *See:* **asynchronous machine; direct-current commutating machine.** 63

torquer (gyro; accelerometer). A device which exerts a torque (or force) on a gimbal, a gyro rotor or a proof mass in response to a command signal. *Syn:* **forcer.** 46

torquer reaction torque (gyro; accelerometer). The reaction torque which is a function of the frequency and amplitude of the command torque signal. 46

torquing (gyro; accelerometer). The application of torque to a gimbal or a gyro rotor about an axis-of-freedom for the purpose of precessing, capturing, slaving or slewing.
46

torquing rate (inertial navigation equipment). The angular rate at which the orientation of a gyro, with respect to inertial space, is changed in response to a command. *See also:* **navigation.** 187, 13

torsional critical speed (rotating machinery). The speed at which the amplitudes of the angular vibrations of a machine rotor due to shaft torsional vibration reach a maximum. *See:* **rotor (rotating machinery).** 63

torsional mechanism. An operating mechanism which transfers rotary motion by torsion through a pipe or shaft from the operating means to open or close the switching device. 27

torsionmeter. A device to indicate the torque transmitted by a propeller shaft based on measurement of the twist of a calibrated length of the shaft. *See also:* **electric propulsion system.** 328

total average power dissipation (semiconductor). The sum of the full cycle average forward and full cycle average reverse power dissipation. 66

total break time (mechanical switching device). *See:* **interrupting time.**

total capability for load (electric power supply). The capability available to a system from all sources including purchases. *See also:* **generating station.** 200

total capacitance. *See:* **self-capacitance (conductor).**

total charge (ferroelectric device). One-half of the charge that flows as the condition of the device is changed from that of full applied positive voltage to that of full negative voltage (or vice versa). *Note:* Total charge is dependent on the amplitude of the applied voltage which should be stated when measurements of total charge are reported. *See also:* **ferroelectric domain.** 247

total clearing time (fuse). *See:* **clearing time (fuse).**

total current (asymmetrical current). The combination of the symmetrical component and the direct-current component of the current. 103, 202

total-current regulation (axle generator). That type of automatic regulation in which the generator regulator controls the total current output of the generator. *See also:* **axle generator system.**

total cyanide (in a solution for metal deposition) (electroplating). The total content of the cyanide radical (CN), whether present as the simple or complex cyanide of an alkali or other metal. *See also:* **electroplating.** 328

total electric current density. At any point, the vector sum of the conduction-current density vector, the convection-current density vector, and the displacement-current density vector at that point. 210

total electrode capacitance (electron tubes). The capacitance of one electrode to all other electrodes connected together. 244, 190

total emissivity (element of surface of a temperature radiator). The ratio of its radiant-flux density (radiant exitance) to that of a blackbody at the same temperature. 167

total for load capability (electric power supply). The net dependable capability available to a system from all sources including purchases. 112

totalizing relay. A device used to receive and totalize pulses from two or more sources for proportional transmission to another totalizing relay or to a receiver. *See also:* **auxiliary device to an instrument.** 212

total losses (transformer or regulator). The sum of the no-load and load losses, excluding losses due to accessories. 53

totally depleted detector (semiconductor radiation detectors). A detector in which the thickness of the depletion region is essentially equal to the thickness of the semiconductor material. 23, 118, 119

totally enclosed (rotating machinery). A term applied to apparatus with an integral enclosure that is constructed so that while it is not necessarily airtight, the enclosed air has no deliberate connection with the external air except for the provision for draining and breathing. 63

totally enclosed fan-cooled (totally enclosed fan-ventilated). A term applied to a totally enclosed apparatus equipped for exterior cooling by means of a fan or fans, integral with the apparatus but external to the enclosing parts. *See:* **asynchronous machine; direct-current commutating machine.** 63, 3

totally-enclosed fan-ventilated air-cooled (rotating machinery). Applied to a totally-enclosed machine having an air-to-air heat exchanger in the internal air circuit, the external air being blown through the heat exchanger by a fan mechanically driven by the machine shaft. 63

totally enclosed machine. A machine so enclosed as to prevent the free exchange of air between the inside and the outside of the case but not sufficiently enclosed to be termed airtight. *See:* **asynchronous machine; direct-current commutating machine.** 328, 3

totally enclosed nonventilated (rotating machinery). A term applied to a totally enclosed apparatus that is not equipped for cooling by means external to the enclosing parts. *See:* **asynchronous machine; direct-current commutating machine.** 63

totally enclosed pipe-ventilated machine. A totally enclosed machine except for openings so arranged that inlet and outlet ducts or pipes may be connected to them for the admission and discharge of the ventilating air. This air may be circulated by means integral with the machine or by means external to and not a part of the machine. In the latter case, these machines shall be known as separately ventilated or forced ventilated machines. *See:* **asynchronous machine; closed air-circuit; direct-current commutating machine.** 328

totally enclosed ventilated apparatus. Apparatus totally enclosed in which the cooling air is carried through the case and apparatus by means of ventilating tubes and the air does not come in direct contact with the windings of the apparatus. 328

totally unbalanced currents (balanced line). Push-push currents. *See also:* **waveguide.** 267

total power loss (semiconductor rectifier). The sum of the forward and reverse power losses. *See also:* **rectification.** 237, 66, 208

total range (instrument). The region between the limits within which the quantity measured is to be indicated or recorded and is expressed by stating the two end-scale values. *Notes:* (1) If the span passes through zero, the range is stated by inserting zero or 0 between the end-scale values. (2) In specifying the range of multiple-range instruments, it is preferable to list the ranges in descending order, for example, 750/300/150. *See also:* **instrument.** 294

total sag. The distance measured vertically from the conductor to the straight line joining its two points of support, under conditions of ice loading equivalent to the total resultant loading for the district in which it is located. 262

total start-stop telegraph distortion. Refers to the time displacement of selecting-pulse transitions from the beginning of the start pulse expressed in percent of unit pulse. 111

total switching time (ferroelectric device). The time required to reverse the signal charge. *Note:* Total switching time is measured from the time of application of the voltage pulse, which must have a rise time much less than and a duration greater than the total switching time. The magnitude of the applied voltage pulse should be specified as part of the description of this characteristic *See also:* **ferroelectric domain.** 247

total telegraph distortion. Telegraph transmisson impairment, expressed in terms of time displacement of mark-space and space-mark transitions from their proper positions relative to one another, in percent of the shortest perfect pulses called the unit pulse. (Time lag affecting all transitions alike does not cause distortion). Telegraph distortion is specified in terms of its effect on code and terminal equipment. Total Morse telegraph distortion for a particular mark or space pulse is expressed as the algebraic sum of time displacements of space-mark and mark-space transitions determining the beginning and end of the pulses, measured in percent of unit pulse. Lengthening of mark is positive, and shortening, negative. *See also:* **distortion.**   111

total varactor capacitance. The capacitance between the varactor terminals under specified conditions.   136

total voltage regulation (rectifier). The change in output voltage, expressed in volts, that occurs when the load current is reduced from its rated value to zero or light transition load with rated sinusoidal alternating voltage applied to the alternating-current line terminals, but including the effect of the specified alternating-current system impedance as if it were inserted between the line terminals and the transformer, with the rectifier transformer on the rated tap. *Note:* The measurement shall be made with zero phase control and shall exclude the corrective action of any automatic voltage-regulating means, but not impedance. *See also:* **power rectifier; rectification.**   291, 208

touchdown tone lights. Barettes of runway lights installed in the surface of the runway between the runway edge lights and the runway centerline lights to provide additional guidance during the touchdown phase of a landing in conditions of very poor visibility.   167

touch voltage (1) (safety). The potential difference between a grounded metallic structure and a point on the earth's surface separated by a distance equal to the normal maximum horizontal reach, approximately one meter. *See also:* **ground.**   313, 45

(2) (conductor stringing equipment). This potential difference could be dangerous, and could result from induction and/or fault conditions. *Syn:* **touch potential.**   45

tower. A broad-base latticed steel support for line conductors.   64

tower footing resistance (lightning protection). The resistance between the tower grounding system and true ground. *See also:* **direct-stroke protection (lightning).**   64

tower loading. The load placed on a tower by its own weight, the weight of the wires with or without ice covering, the insulators, the wind pressure normal to the line acting both on the tower and the wires and the pull from the wires in the direction of the line. *See also:* **tower.**   64

towing light. A lantern or lanterns fixed to the mast or hung in the rigging to indicate that a ship is towing another vessel or other objects.   328

Townsend coefficient (gas). The number of ionizing collisions per centimeter of path in the direction of the applied electric field. *See also:* **discharge (gas).**   244, 190

T pulse (TV waveform test signals). A $\sin^2$ pulse with a half-amplitude duration (HAD) of 125 ns. The amplitude of the envelope of the frequency spectrum at 4 MHz is 0.5 of the amplitude at zero frequency and effectively zero at and beyond 8 MHz.   42

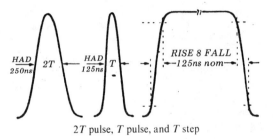

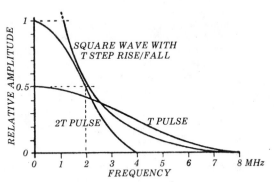

2T pulse, T pulse, and T step

Envelope of frequency spectrum of 2T pulse, T pulse and square wave with T step rise and fall

TR. *See:* **transmit-receive.**

trace. The cathode-ray-tube display produced by a moving spot. *See:* **spot.** *See also:* **oscillograph.**   185

traceability (test, measurement and diagnostic equipment). Process by which the assigned value of a measurement is compared, directly or indirectly, through a series of calibrations to the value established by the U. S. national standard.   54

trace finder. *See:* **beam finder.**

trace interval (television). The interval corresponding to the direction of sweep used for delineation. *See also:* **television.**   163, 178

trace, return (oscillography) (television). The path of the scanning spot during the retrace. *See also:* **oscillograph; television.**   106

trace width (oscilloscope). The distance between two points on opposite sides of a trace perpendicular to the direction of motion of the spot, at which luminance is 50 percent of maximum. With one setting of the beam controls, the width of both horizontally and vertically going traces within the quality area should be stated. *See:* **oscillograph.**   185

tracing distortion. The nonlinear distortion introduced in the reproduction of mechanical recording because the curve traced by the motion of the reproducing stylus is not an exact replica of the modulated groove. For example, in the case of a sine-wave modulation in vertical recording the curve traced by the center of the tip of a stylus is a poid. *See also:* **phonograph pickup.**   176

tracing routine (computing systems). A routine that provides a historical record of specified events in the execution of a program.   255, 77

track (1) (navigation). (A) The resultant direction of actual travel projected in the horizontal plane and expressed as a bearing. (B) The component of motion that is in the horizontal plane and represents the history of accomplished travel. *See also:* **navigation.** 187, 13

(2) (in electronic computers). The portion of a moving-type storage medium that is accessible to a given reading station; for example, as on film, drum, tapes, or discs. *See also:* **band.** 338, 210, 255, 224, 77, 207

(3) (test, measurement and diagnostic equipment). *See:* **channel.**

track circuit. An electric circuit that includes the rails of a track relay as essential parts. 328

track element. *See:* **roadway element.**

track homing (navigation). The process of following a line of position known to pass through an objective. *See also:* **radio navigation.** 13, 187

track indicator chart. A maplike reproduction of railway tracks controlled by track circuits so arranged as to indicate automatically for defined sections of track whether such sections are or are not occupied. 328

tracking (1) (radar). The process of following a moving object or a variable input quantity, using a servomechanism. *Note:* In radar, tracking is carried out by keeping a narrow beam or angle cursor centered on the target angle, a range mark or gate on the delayed echo, or a narrowband filter on the signal frequency. 13

(2) (electric). The maintenance of proper frequency relations in circuits designed to be simultaneously varied by gang operation.

(3) (phonographic technique). The accuracy with which the stylus of a phonograph pickup follows a prescribed path. 111, 187

(4) (instrument). The ability of an instrument to indicate at the division line being checked, when energized by corresponding proportional value of actual end-scale excitation, expressed as a percentage of actual end-scale value. 280

(5) (communication satellite). (A) The determination of the orbit and the ephemeris of a satellite or spacecraft. (B) Maintaining the pointing of a high gain antenna at a moving spacecraft. 84

(6) (antenna). A motion given to the major lobe of an antenna such that a selected moving target is contained within the major lobe. *Syn:* **angle tracking.** 111

tracking error (1) (general). The deviation of a dependent variable with respect to a reference function. *Note:* As applied to power inverters, tracking error may be the deviation of the output volts per hertz from a prescribed profile or the deviation of the output frequency from a given input synchronizing signal or others. *See:* **self-commutated inverters.** 208

(2) (phonographic techniques) (lateral mechanical recording). The angle between the vibration axis of the mechanical system of the pickup and a plane containing the tangent to the unmodulated record groove that is perpendicular to the surface of the recording medium at the point of needle contact. *See also:* **phonograph pickup.** 176

tracking radar. A radar (or mode of a radar) in which the target is tracked by a closed-loop servo which controls the antenna (for angle tracking) or receiver gates or filters (for range or Doppler tracking). 13

track instrument. A device in which the vertical movement of the rail or the blow of a passing wheel operates a contact to open or close an electric circuit. 328

trackless trolley coach. *See:* **trolley coach (trolley bus).**

track relay. A relay receiving all or part of its operating energy through conductors of which the track rails are an essential part and that responds to the presence of a train on the track. 328

track/store (analog computer). In an analog computer, a component, controlled by digital logic signals, whose output equals the input, when in the "track" mode, and whose output becomes constant and is held (stored) at the value it possessed at the instant its mode was switched to the "store" mode. 9

track-while-scan (radar). A target tracking process in which the radar antenna and receiver are not part of the tracking loop, but provide periodical video data from the search scan, as inputs to computer channels which follow individual targets. 13

traction machine (elevators). A direct-drive machine in which the motion of a car is obtained through friction between the suspension ropes and a traction sheave. *See also:* **driving machine (elevators).** 328

traction motor. An electric propulsion motor used for exerting tractive force through the wheels of a vehicle. 328

tractive force (tractive effort) (electrically propelled vehicle). The total propelling force measured at the rims of the driving wheels, or at the pitch line of the gear rack in the case of a rack vehicle. *Note:* Tractive force of an electrically propelled vehicle is commonly qualified by such terms as: maximum starting tractive force; short-time-rating tractive force; continuous-rating tractive force. *See also:* **electric locomotive.** 328

tractor, crawler (conductor stringing equipment). A tracked unit employed to pull pulling lines, sag conductor, level or clear pull and tension sites, and miscellaneous other work. It is also frequently used as a temporary anchor. Sagging winches on this unit are usually arranged in a vertical configuration. *Syn:* **cat; crawler; tractor.** 45

tractor, wheel (conductor stringing equipment). A wheeled unit employed to pull pulling lines, sag conductor, and miscellaneous other work. Sagging winches on this unit are usually arranged in horizontal configuration. It has some advantages over crawler tractors in that it has a softer footprint, travels faster, and is more manuverable. *Syn:* **logger; sagger; skidder; tractor.** 45

tradeoff. Parametric analysis of concepts or components for the purpose of optimizing the system or some trait of the system. *See:* **system science.** 209

traffic-control system. A block signal system under which train movements are authorized by block signals whose indications supersede the superiority of trains for both opposing and following movements on the same track. *See also:* **centralized traffic-control system.** 328

traffic locking. Electric locking adapted to prevent the manipulation of levers or other devices for changing the direction of traffic on a section of track while that section is occupied or while a signal is displayed for a train to proceed into that section. *See also:* **interlocking (interlocking plant).** 328

traffic service (telephone switching systems). The services rendered to customers by telephone company operators.   55

traffic usage recorder (telephone switching systems). A device or system for sampling and recording the occupancy of equipment.   55

trailer plow (static or vibratory plows) (cable plowing). A unit which is self-contained except for drawbar pull that is furnished by a prime mover.   52

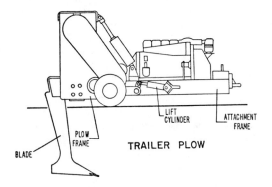

TRAILER PLOW

trailing edge (1) (television). The major portion of the decay of a pulse. *See also:* **television.**   178

(2) (high- and low-power pulse transformers) (last transition). That portion of the pulse occurring between the time of intersection of straight-line segments used to determine A_T and the time at which the instantaneous value reduces to zero. *See also:* **input pulse shape.**   32, 33

trailing edge (last transition) amplitude (A_T) (high- and low-power pulse transformers). That quantity determined by the intersection of a line passing through the points on the trailing edge where the instantaneous value reaches 90 percent and 10 percent of A_T, and the straight-line segment fitted to the top of the pulse in determining A_M. *See also:* **input pulse shape.**   32, 33

trailing edge, pulse. The major transition towards the pulse baseline occuring before a reference time. *See also:* **pulse.**   185

trailing-edge pulse time. The time at which the instantaneous amplitude last reaches a stated fraction of the peak pulse amplitude.   254, 178, 162

trailing-type antenna (aircraft). A flexible conductor usually wound on a reel within the aircraft passing through a fairlead to the outside of the aircraft, terminated in a streamlined weight or wind sock and fed out to the proper length for the desired radio frequency of operation. It has taken other forms such as a capsule that when exploded releases the antenna.   328

train (illuminating engineering). The angle between the vertical plane through the axis of the searchlight drum and the plane in which this plane lies when the searchlight is in a position designated as having zero train. *See also:* **searchlight.**   167

train-control territory. That portion of a division or district equipped with an automatic train-control system. *See also:* **automatic train control.**   328

train describer. An instrument used to give information regarding the origin, destination, class, or character of trains, engines, or cars moving or to be moved between given points. *See also:* **railway signal and equipment.**   328

trained listening group (speech-quality measurements). Six to ten listeners who understand thoroughly the purpose of the speech quality test and respond properly throughout the test. All persons of the group shall meet the requirements on auditory acuity as described by USAS S3.2-1960 (Monosyllabic Word Intelligibility). The training of the listeners will depend on the special type of tests to be conducted.   126

trainline coupler (land transportation vehicles). A group of devices which connects the electric training circuits of two adjacent vehicles in the train consist through one or both of the following: (1) A **jumper,** which is the removable member of a circuit connecting two adjacent cars or locomotives together, although sometimes one end of the connecting cable is permanently fixed. *Note:* This term is usually preceded by a designating name, such as control jumper, bus line jumper, train line jumper, etcetera. (2) A **coupler** plug, which is that portion of a jumper that serves to directly connect the jumper wiring to the coupler socket or receptacle. (3) A **coupler socket** or **receptacle,** which is that portion of the jumper attached to the locomotive or car and serves to directly connect the car or locomotive wiring to the jumper coupler plug. *See also:* **car wiring apparatus.**

(4) Electric coupler—(A) Electric contact head. An electric contact head is that portion of the electrical coupler which is attached to, and rotates with, the mechanical coupler and contains electrical contacts which mate with corresponding contacts on the adjacent vehicle. (B) Coupler Cable. A coupler cable is the cable which carries the trainline from the carbody to the electric contact head.   1

trajectory, state. *See:* **state trajectory.**

transactor (relaying). A magnetic device with an air-gapped core having an input winding which is energized with an alternating current and having an output winding which produces a voltage that is a function of the input current. *Note:* The term transactor is a contraction of the words transformer and reactor.   79

transadmittance. For harmonically varying quantities at a given frequency, the ratio of the complex amplitude of the current at one pair of terminals of a network to the complex amplitude of the voltage across a different pair of terminals. *See:* **interelectrode transadmittance ($j - 1$ interelectrode transadmittance of an n-electrode electron tube).**   185

transadmittance compression ratio (electron tubes). The ratio of the magnitude of the small-signal forward transadmittance of the tube to the magnitude of the forward transadmittance at a given input signal level.   125

transadmittance, forward (electron tubes). The complex quotient of (1) the fundamental component of the short-circuit current induced in the second of any two gaps and (2) the fundamental component of the voltage across the first.   125

transceiver. The combination of radio transmitting and receiving equipment in a common housing, usually for portable or mobile use, and employing common circuit components for both transmitting and receiving. *See also:* **radio transmission.**   328

transconductance. The real part of the transadmittance. *Note:* Transconductance is, as most commonly used, the interelectrode transconductance between the control grid and the plate. At low frequencies, transconductance is the slope of the control-grid-to-plate transfer characteristic. *See also:* **electron-tube admittances; interelectrode transconductance.** 185, 190, 125

transconductance meter (mutual-conductance meter). An instrument for indicating the transconductance of a grid-controlled electron tube. *See also:* **instrument.**
328

transcribe (electronic computation). To convert data recorded in a given medium to the medium used by a digital computing machine or vice versa. 235

transcriber (electronic computation). Equipment associated with a computing machine for the purpose of transferring input (or output) data from a record of information in a given language to the medium and the language used by a digital computing machine (or from a computing machine to a record of information). 210

transducer (communication and power transmission). A device by means of which energy can flow from one or more transmission systems or media to one or more other transmission systems or media. *Note:* The energy transmitted by these systems or media may be of any form (for example, it may be electric, mechanical, or acoustical), and it may be of the same form or different forms in the various input and output systems or media.
111, 239, 252, 210, 56, 255, 188, 77, 329, 59, 206, 54

transducer, active. A transducer whose output waves are dependent upon sources of power, apart from that supplied by any of the actuating waves, which power is controlled by one or more of the waves. *Note:* The definition of active transducer is a restriction of the more general **active network;** that is, one in which there is an impressed driving force. *See also:* **transducer.** 239, 252, 210

transducer gain (1) (general). The ratio of the power that the transducer delivers to the specified load under specified operating conditions to the available power of the specified source. *Notes:* (A) If the input and/or output power consist of more than one component, such as multifrequency signals or noise, then the particular components used and their weighting must be specified. (B) This gain is usually expressed in decibels. *See also:* **transducer.**
239, 252, 210

(2) (two-port linear transducer). At a specified frequency, the ratio of (A) the actual signal power transferred from the output port of the transducer to its load, to (B) the available signal power from the source driving the transducer. 125

transducer, ideal (for connecting a specified source to a specified load). A hypothetical passive transducer that transfers the maximum available power from the source to the load. *Note:* In linear transducers having only one input and one output, and for which the impedance concept applies, this is equivalent to a transducer that (1) dissipates no energy and (2) when connected to the specified source and load presents to each its conjugate impedance. *See also:* **transducer.** 239, 252, 210

transducer, line. *See:* **line transducer.**

transducer loss. The ratio of the available power of the specified source to the power that the transducer delivers to the specified load under specified operating conditions. *Notes:* (1) If the input and/or output power consist of more than one component, such as multifrequency signals or noise, then the particular components used and their weighting must be specified. (2) This loss is usually expressed in decibels. *See also:* **transducer.**
239, 252, 210, 197

transducer, passive. A transducer that has no source of power other than the input signal(s), and whose output signal-power cannot exceed that of the input. *Note:* The definition of a passive transducer is a restriction of the more general **passive network,** that is, one containing no impressed driving forces. *See also:* **transducer.**
239, 252, 210

transfer (1) (electronic computation). (A) To transmit, or copy, information from one device to another. (B) To jump. (C) The act of transferring. *See also:* **jump; transmit.** 235, 210

(2) (electrostatography). The act of moving a developed image, or a portion thereof, from one surface to another, as by electrostatic or adhesive forces, without altering the geometric configuration of the image. *See also:* **electrostatography.** 236, 191

transfer admittance (1) (linear passive networks, general). A transmittance for which the excitation is a voltage and the response is a current. *See also:* **linear passive networks.**
238

(2) (from the *i*th terminal to the *j*th terminal of an *n*-terminal network). The (complex) current flowing to the ith terminal divided by the (complex) voltage applied between the jth terminal with respect to the reference point when all other terminals have arbitrary terminations. For example, for a 3-terminal network terminated in short circuits

$$y_{12} = \left.\frac{I}{v_2}\right|_{v_1} = 0 \qquad 67$$

transfer characteristic (1) (electron tubes). A relation, usually shown by a graph, between the voltage of one electrode and the current to another electrode, all other electrode voltages being maintained constant. *See also:* **electrode (electron tube).** 328, 125

(2) (camera tubes). A relation between the illumination on the tube and the corresponding signal output current, under specified conditions of illumination. *Note:* The relation is usually shown by a graph of the logarithm of the signal output current as a function of the logarithm of the illumination. *See also:* **illumination; sensitivity; television.** 178, 190, 125

transfer check (electronic computation). A check (usually an automatic check) on the accuracy of a data transfer. *Note:* In particular, a check on the accuracy of the transfer of a word. 235, 210, 255, 77

transfer constant (electric transducer). *See:* **image transfer constant.**

transfer control (electronic computation). *See:* **jump.**

transfer current (1) (glow-discharge cold-cathode tube). The starter-gap current required to cause conduction across the main gap. *Note:* The transfer current is a function of the anode voltage. *See:* **gas tubes.** 190

(2) (gas tubes). The current to one electrode required to initiate breakdown to another electrode. *Note:* The transfer current is a function of the voltage of the second electrode. 125

transfer-current ratio (linear passive network). A transmittance for which the variables are currents. *Note:* The word **transfer** is frequently dropped in present usage. *See also:* **linear passive networks.** 238

transfer function (1) (circuits and systems). A response function for which the variables are measured at different ports (terminal pairs). The variables are usually considered to represent an input signal and a response to that excitation. *See also:* **function, transfer.** 67

(2) (low-power wide-band transformers). The complex ratio of the output of the device to its input. It is also the combined phase and frequency responses. 151

(3) (nuclear power generating station). A mathematical, graphical, or tabular statement of the influence which a module has on a signal or action compared at input and at output terminals. This should be specified as to whether it is transient or steady state. 355

(4) (excitation control systems). A mathematical, graphical, or tabular statement of the influence which a system or element has on a signal or action compared at input and output terminals. *Note:* For a linear system, general usage limits the transfer function to mean the ratio of the Laplace transform of the output to the Laplace transform of the input in the absence of all other signals, and with all initial conditions zero. 353

transfer immittance. *See:* **transmittance.**

transfer impedance (1) (linear passive networks). A transmittance for which the excitation is a current and the response is a voltage. *Note:* It is therefore the impedance obtained when the response is determined at a point other than that at which the driving force is applied, all terminals being terminated in any specified manner. In the case of an electric circuit, the response would be determined in any branch except that in which the driving force is. *See also:* **self-impedance; linear passive networks; network analysis.** 238, 210

(2) (circuits and systems). (A) **(linear passive networks) (general).** A transmittance for which the excitation is a current and the response is a voltage. *See also:* **linear passive networks.** (B) (from the ith terminal to the jth terminal of an n-terminal network). The (complex) voltage measured between the ith terminal and the reference point divided by the (complex) current applied to the jth terminal when all other terminals have arbitrary terminations. For example, for a 3-terminal network terminated in open circuits

$$Z_{12} = \frac{v_1}{I_2}\bigg|_{I_1} = 0 \qquad 6$$

transfer instruction. *See:* **branch instruction.**

transfer locus (linear system or element). A plot of the transfer function as a function of frequency in any convenient coordinate system. *Note:* A plot of the reciprocal of the transfer function is called the inverse transfer locus. *See:* **amplitude frequency locus; phase locus.** *See also:* **control system, feedback.** 329

transfer of control. Same as jump.

transfer ratio. A dimensionless transfer function. 210

transfer ratio correction (correction to setting). The de-

viation of the output phasor from nominal, in proportional parts of the input phasor.

$$\frac{\text{Output}}{\text{Input}} = A + \alpha + j\beta \qquad \text{INPUT} \boxed{} \begin{array}{l} A \\ \text{OUTPUT} \end{array}$$

A = setting
α = in-phase transfer ratio correction
β = quadrature transfer ratio correction. 184

transferred charge (capacitor). The net electric charge transferred from one terminal of a capacitor to another via an external circuit. *See also:* **nonlinear capacitor.** 191

transferred-charge characteristic (nonlinear capacitor). The function relating transferred charge to capacitor voltage. *See also:* **nonlinear capacitor.** 191

transferred information. *See:* **transinformation.**

transfer standards, alternating-current—direct-current. Devices used to establish the equality of a root-mean-square current or voltage (or the average value of alternating power) with the corresponding steady-state direct-current quantity that can be referred to the basic standards through potentiometric techniques. *See also:* **auxiliary device to an instrument.** 212

transfer switch (1) (high-voltage switch). A device for transferring one or more load conductor connections from one power source to another. *Note:* High-voltage transfer switches are arranged to permit transferring a conductor connection from one circuit to another without interrupting the current. (A) A tandem transfer switch is a switch with two blades, each of which can be moved into or out of only one contact. (B) A double-blade double-throw transfer switch is a switch with two blades, each of which can be moved into or out of either of two contacts. (C) In contrast to high-voltage switches, many low-voltage control and instrument transfer switches interrupt current during transfer. Compare with **selector switch.** Also compare with an **automatic transfer switch,** that connects a load to an alternate source after failure of an original source. 70

(2) (emergency and standby power systems). A device for transferring one or more load conductor connections from one power source to another. 73

transfer switch, load. *See:* **load transfer switch.**

transfer time, relay. *See:* **relay transfer time.**

transfer trip. A form of remote release in which a communication channel is used to transmit the signal for release from the relay location to a remote location. 103, 60, 202, 127

transfer voltage ratio (linear passive networks). A transmittance for which the variables are voltages. *Note:* The word **transfer** is frequently dropped in present usage. *See also:* **linear passive networks.** 238

transfer winding (rotating machinery). A winding for which coils are form-wound to a suitable shape and then inserted into slots or around poles by a mechanical means. *See also:* **rotor (rotating machinery); stator.** 63

transform (computing systems). To change the form of data according to specific rules. 255, 77

transformation (impedance or admittance) (rotating ma-

chinery). The result of one of many mathematical processes that transforms the original impedance or admittance of a machine or system into a more manageable form. During the process all currents and voltages undergo consistent transformations. *Note:* Following are some of the better-known transformations: (1) Symmetrical component transformation, also known as Fortescue transformation, sequence component transformation, and phase-sequence transformation. (2) Synchronously rotating reference frame transformation, also known as Park transformation and *d-q* transformation. (3) Clark transformation, also known as *a-b* component transformation, and equivalent two-phase transformations. (4) Complex rotating transformation, also known as *Ku* transformation, *f-b* transformation, and forward-backward component transformation. *See also:* **asynchronous machine; direct-current commutating machine.**   63

transformer. A static electric device consisting of a winding, or two or more coupled windings, with or without a magnetic core, for introducing mutual coupling between electric circuits. *Note:* Transformers are extensively used in electric power systems to transfer power by electromagnetic induction between circuits at the same frequency, usually with changed values of voltage and current.   53

transformer, alternating-current arc welder. A transformer with isolated primary and secondary windings and suitable stabilizing, regulating, and indicating devices required for transforming alternating current from normal supply voltages to an alternating-current output suitable for arc welding.   264

transformer class designations. *See:* **transformer, oil-immersed.**

transformer, constant-voltage (constant-potential transformer). *See:* **constant-voltage transformer.**

transformer correction factor (TCF). The ratio of true watts or watthours to the measured watts or watthours, divided by the marked ratio. *Note:* The transformer correction factor for a current or voltage transformer is the ratio correction factor multiplied by the phase angle correction factor for a specified primary circuit power factor. The true primary watts or watthours are equal to the watts or watthours measured, multiplied by the transformer correction factor and the marked ratio. The true primary watts or watthours, when measured using both current and voltage transformers, are equal to the current transformer correction factor times the voltage transformer correction factor multiplied by the product of the marked ratios of the current and voltage transformers multiplied by the observed watts or watthours.   53

transformer, dry-type. *See:* **dry-type transformer.**

transformer, energy-limiting. A transformer that is intended for use on an approximately constant-voltage supply circuit and that has sufficient inherent impedance to limit the output current to a thermally safe maximum value. *See:* **transformer, specialty.**   203

transformer equipment rating. A volt-ampere output together with any other characteristics, such as voltage, current, frequency, and power factor, assigned to it by the manufacturer. *Note:* It is regarded as a test rating that defines an output that can be taken from the item of transformer equipment without exceeding established

temperature-rise limitations, under prescribed conditions of test and within the limitations of established standards. *See:* **duty.**   203

transformer, grounding. *See:* **grounding transformer.**

transformer grounding switch and gap (capacitance potential device). Consists of a protective gap connected across the capacitance potential device and transformer unit to limit the voltage impressed on the transformer and the auxiliary or shunt capacitor, when used; and a switch that when closed removes voltage from the potential device to permit adjustment of the potential device without interrupting high-voltage line operation and carrier-current operation when used. *See also:* **outdoor coupling capacitor.**   341

transformer, group-series loop insulating. An insulating transformer whose secondary is arranged to operate a group of series lamps and/or a series group of individual-lamp transformers. *See:* **transformer, specialty.**
                                                        203

transformer, high-power-factor. A high-reactance transformer that has a power-factor-correcting device such as a capacitor, so that the input current is at a power factor of not less than 90 percent when the transformer delivers rated current to its intended load device. *See:* **transformer, specialty.**   301

transformer, high-reactance (1) (output limiting). An energy-limiting transformer that has sufficient inherent reactance to limit the output current to a maximum value. *See:* **transformer, specialty.**   203
(2) secondary short-circuit current rating. The current in the secondary winding when the primary winding is connected to a circuit of rated primary voltage and frequency and when the secondary terminals are short-circuited. *See:* **transformer, specialty.**   203
(3) (kilovolt-ampere or voltampere short-circuit input rating). The input kilovolt-amperes or volt-amperes at rated primary voltage with the secondary terminals short-circuited. *See:* **transformer, specialty.**   203

transformer, ideal. A hypothetical transformer that neither stores nor dissipates energy. *Note:* An ideal transformer has the following properties: (1) Its self and mutual impedances are equal and are pure inductances of infinitely great value. (2) Its self-inductances have a finite ratio. (3) Its coefficient of coupling is unity. (4) Its leakage inductance is zero. (5) The ratio of the primary to secondary voltage is equal to the ratio of secondary to primary current.   239, 197

transformer, individual-lamp insulating. An insulating transformer used to protect the secondary circuit, casing, lamp, and associated luminaire of an individual street light from the high-voltage hazard of the primary circuit. *See:* **transformer, specialty.**   203

transformer, insulating. A transformer used to insulate one circuit from another. *See:* **transformer, specialty.**   203

transformer integrally mounted cable terminating box. A weatherproof air-filled compartment suitable for enclosing the sidewall bushings of a transformer and equipped with any one of the following entrance devices: (1) Single or multiple-conductor potheads with couplings or wiping sleeves. (2) Wiping sleeves. (3) Couplings with or without stuffing boxes for conduit-enclosed cable, metallic-sheathed cable, or rubber-covered cable.   289

transformer, interphase (thyristor power converter). An

autotransformer, or a set of mutually coupled inductive reactors, used to obtain multiple operation between two or more simple converters that have ripple voltages that are out of phase.   121

transformer, isolating (1) (signal-transmission system). A transformer inserted in a system to separate one section of the system from undesired influences of the other sections. *Example:* A transformer having electrical insulation and electrostatic shielding between its windings such that it can provide isolation between parts of the system in which it is used. It may be suitable for use in a system that requires a guard for protection against common-mode interference. *See:* **signal.**   188
(2) (electroacoustics). A transformer inserted in a system to separate one section of the system from undesired influences of other sections. *Note:* Isolating transformers are commonly used to isolate system grounds and prevent the transmission of undesired currents.   239

transformer, line. *See:* **line transformer.**

transformer loss (communication). The ratio of the signal power that an ideal transformer would deliver to a load, to the power delivered to the same load by the actual transformer, both transformers having the same impedance ratio. *Note:* Transformer loss is usually expressed in decibels. *See also:* **transmission loss.**   239

transformer-loss compensator. A passive electric network that is connected in series-parallel with a meter to add to or to subtract from the meter registration active or reactive components of registration proportional to predetermined iron and copper losses of transformers and transmission lines. *See also:* **auxiliary device to an instrument.**   212

transformer, low-power-factor. A high-reactance transformer that does not have means for power-factor correction. *See:* **transformer, specialty.**   301

transformer, matching. *See:* **matching transformer.**

transformer, network. *See:* **network transformer.**

transformer, nonenergy-limiting. A constant-potential transformer that does not have sufficient inherent impedance to limit the output current to a thermally safe maximum value. *See:* **transformer, specialty.**   203

transformer, oil-immersed. *See:* **oil-immersed transformer.**

transformer, outdoor. *See:* **outdoor transformer.**

transformer overcurrent tripping. *See:* **indirect release and overcurrent release.**

transformer, protected outdoor. A transformer that is not of weatherproof construction but that is suitable for outdoor use if it is so installed as to be protected from rain or immersion in water. *See also:* **transformer.**   203

transformer, phase-shifting. *See:* **phase-shifting transformer.**

transformer, pole-type. *See:* **pole-type transformer.**

transformer primary voltage rating. *See:* **primary voltage rating.**

transformer-rectifier, alternating-current—direct-current arc welder. A combination of static rectifier and the associated isolating transformer, reactors, regulators, control, and indicating devices required to produce either direct or alternating current suitable for arc-welding purposes.   264

transformer-rectifier, direct-current arc welder. A combination of static rectifiers and the associated isolating transformer, reactors, regulators, control, and indicating devices required to produce direct current suitable for arc welding.   264

transformer relay. A relay in which the coils act as a transformer.   328

transformer removable cable-terminating box. A weatherproof air-filled compartment suitable for enclosing the sidewall bushings of a transformer and equipped with mounting flange(s) (one or two) to accommodate either single-conductor or multiconductor potheads or entrance fittings, depending upon the type of cable termination to be used and the number of three-phase cable circuits (one or two) to be terminated.   288

transformer, assembly, Scott-connected. *See:* **Scott-connected transformer assembly.**

transformer secondary current rating. *See:* **secondary current rating (transformer).**

transformer, series. A transformer in which the primary winding is connected in series with a power-supply circuit, and that transfers energy to another circuit at the same or different current from that in the primary circuit. *See:* **transformer.**   203

transformer, series street-lighting. A series transformer that receives energy from a current-regulating series circuit and that transforms the energy to another winding at the same or different current from that in the primary. *See:* **transformer, specialty.**   203

transformer, series street-lighting, rating. The lumen rating of the series lamp, or the wattage rating of the multiple lamps, that the transformer is designed to operate. *See:* **specialty transformer.**   203

transformer, shunt. A transformer in which the primary winding is connected in shunt with a power-supply circuit, and that transfers energy to another circuit at the same or different voltage from that of the primary circuit. *See:* **transformer.**   203

transformer, specialty. *See:* **specialty transformer.**
                                                          203

transformer, station-type. *See:* **station-type transformer.**

transformer, submersible. *See:* **submersible transformer.**

transformer, subway-type. *See:* **subway-type transformer.**

transformer undercurrent tripping. *See:* **indirect release and undercurrent release.**

transformer, unit-substation. *See:* **unit substation transformer.**

transformer vault (1) (general). An isolated enclosure either above or below ground, with fire-resistant walls, ceiling, and floor, for unattended transformers and their auxiliaries. *See also:* **power distribution, underground construction.**   262, 64
(2) (transmission and distribution). An isolated inclosure either above or below ground with fire-resistant walls, ceiling, and floor, in which transformers and related equipment are installed, and which is not continuously attended during operation.

transformer, vault-type. *See:* **vault-type transformer.**

transformer voltage (network protector). The voltage between phases or between phase and neutral on the transformer side of a network protector.   103, 202

transformer, variable-voltage. *See:* **variable-voltage transformer.**

transformer, waveguide. A structure added to a waveguide or transmission line for the purpose of impedance transformation. *See also:* **waveguide.**   166

transient (1) (industrial power and control). That part of the change in a variable that disappears during transition from one steady-state operating condition to another. *Note:* Using the term to mean the total variation during the transition between two steady states is deprecated.   73
(2) (excitation control systems) (noun). In a variable observed during transition from one steady-state operating condition to another, that part of the variation which ultimately disappears. *Note:* ANSI C85 deprecates using the term to mean the total variable during the transition between two steady-states.   353

transient analyzer. An electronic device for repeatedly producing in a test circuit a succession of equal electric surges of small amplitude and of adjustable waveform, and for presenting this waveform on the screen of an oscilloscope. *See also:* **oscillograph.**   328

transient blocking (relaying). A circuit function which blocks tripping during the interval in which an external fault is being cleared.   79

transient-cause forced outage (electric power systems). A component outage whose cause is immediately self-clearing so that the affected component can be restored to service either automatically or as soon as a switch or circuit breaker can be reclosed or a fuse replaced. *Note:* An example of a transient-cause forced outage is a lightning flashover that does not permanently disable the flashed component. *See also:* **outage.**   200

transient-cause forced outage duration (electric power systems). The period from the initiation of the outage until the affected component is restored to service by switching or fuse replacement. *See also:* **outage.**   200

transient current (rotating machinery). (1) The current under nonsteady conditions. (2) The alternating component of armature current immediately following a sudden short-circuit, neglecting the rapidly decaying component present during the first few cycles.   63

transient-decay current (photoelectric device). The decreasing current flowing in the device after the irradiation has been abruptly cut off. *See also:* **phototubes.**   190

transient deviation (control). *See:* **deviation, transient.**

transient fault (surge arresters). A fault that disappears of its own accord.   244, 62

transient inrush current (power switchgear). Current which results when a circuit interrupting device is closed to energize a capacitance. *Note:* Current is expressed by the highest peak value in amperes and frequently in hertz.   130,103

transient insulation level (TIL) (transformer). An insulation level expressed in kilovolts of the crest value of the withstand voltage for a specified transient wave shape; that is, lightning or switching impulse.   53

transient internal voltage (synchronous machine) (for any specified operating condition). The fundamental-frequency component of the voltage of each armature phase that would be determined by suddenly removing the load, without changing the excitation voltage applied to the field, and extrapolating the envelope of the voltage back to the instant of load removal, neglecting the voltage components of rapid decrement that may be present during the first few cycles after removal of the load. *Note:* The transient internal voltage, as shown in the phasor diagram, is related to the terminal-voltage and phase-current phasors by the equation

$$E_i' = E_a + RI_a + jX_d'I_{ad} + jX_q'I_{aq}.$$

For a machine subject to saturation, the reactances should be determined for the degree of saturation applicable to the specified operating condition. *See:* **direct-axis synchronous reactance; phasor diagram.**   63

transient motion (audio and electroacoustics). Any motion that has not reached or that has ceased to be a steady state.   176

transient overshoot. An excursion beyond the final steady-state value of output as the result of a step-input change. *Note:* It is usually referred to as the first such excursion; expressed as a percent of the steady-state output step. *See also:* **accuracy rating (instrument); control system, feedback).**   295

transient overvoltage. The peak voltage during the transient conditions resulting from the operation of a switching device. *Note:* The location and units of measurement are specified in apparatus standards. Compare with **transient overvoltage ratio (factor).**   103

transient overvoltage ratio (factor). The ratio of the transient overvoltage to the closed-switching-device operating line-to-neutral peak voltage with the load connected. *Note:* The location of measurement is specified in the apparatus standards.   103

transient performance (synchronous-machine regulating system). The performance under a specified stimulus, before the transient expires.   63

transient phenomena (rotating machinery). Phenomena appearing during the transition from one operating condition to another.   244, 63

transient recovery voltage (TRV) (circuit-switching device). The voltage transient that occurs across the terminals of a pole upon interruption of the current. *Notes:* (1) It is the difference between the transient voltages to ground occurring on the terminals. It may be a circuit transient recovery voltage, a modified circuit transient recovery voltage, or an actual transient recovery voltage. (2) In a multipole circuit breaker, the term is usually applied to the voltage across the first pole to interrupt. For circuit breakers having several interrupting units in series, the term may be applied to the voltage across units or groups of units.   103, 202

transient recovery voltage rate (circuit switching device). The rate at which the voltage rises across the terminals of a pole upon interruption of the current. *Note:* It is usually determined by dividing the voltage at one of the crests of the transient recovery voltage by the time from current zero to that crest. In case no definite crest exists, the rate may be taken to some stated value usually arbitrarily selected as a certain percentage of the crest value of the normal-frequency recovery voltage. In case the transient is an exponential function the rate may also be taken at the point of zero voltage. It is the rate of rise of the algebraic difference between the transient voltages occurring on the terminals of the switching device upon interruption of the current. The transient recovery voltage rate may be a circuit transient recovery voltage rate or a modified circuit transient recovery voltage rate, or an actual transient recovery voltage rate according to the type of transient from which it is obtained. When giving actual transient recovery voltage rates, the points between which the rate is measured should be definitely stated. 103, 202

transient response (1) (pulse techniques). The time response of a system or device under test to a stated input stimulus. *Note:* Step functions, impulse functions, and ramps are the commonly used stimuli. *See also:* **pulse.** 185

(2) (relaying). The manner in which a relay, relay unit or relay system responds to a sudden change in the input(s). 79

(3) (oscilloscopes). Time-domain reactions to abruptly varying inputs. 106

transient speed deviation (gas turbines). (1) Load decrease. The maximum instantaneous speed above the steady-state speed occurring after the sudden decrease from one specified steady-state electric load to another specified steady-state electric load having values within limits of the rated output of the gas-turbine-generator unit. It is expressed in percent of rated speed. 98, 58

(2) (Load increase). The minimum instantaneous speed below the steady-state speed occurring after the sudden increase from one specified steady-state electric load to another specified steady-state electric load having values within the limits of rated output of the gas-turbine—generator unit. It is expressed in percent of rated speed. 98, 58

transient stability. A condition that exists in a power system if, after an aperiodic disturbance, the system regains steady-state stability. *See also:* **alternating-current distribution.** 64

transient stability factor (system or part of a system). The ratio of the transient stability limit to the nominal power flow at the point of the system to which the stability limit is referred. *See:* **stability factor.** *See also:* **alternating-current distribution.** 64

transient stability limit (transient power limit). The maximum power flow possible through some particular point in the system when the entire system or the part of the system to which the stability limit refers is operating with transient stability. *See also:* **alternating-current distribution.** 64

transient suppression networks. Capacitors, resistors, or inductors so placed as to control the discharge of stored energy banks. They are commonly used to suppress transients caused by switching. 95

transient thermal impedance (semiconductor device). The change in the difference between the virtual junction temperature and the temperature of a specified reference point or region at the end of a time interval divided by the step function change in power dissipation at the beginning of the same time interval which causes the change of temperature-difference. *Note:* It is the thermal impedance of the junction under conditions of change and is generally given in the form of a curve as a function of the duration of an applied pulse. *See also:* **principal voltage-current characteristic (principal characteristic); semiconductor rectifier stack.** 243, 191, 66, 208

transimpedance (1) (general). For harmonically varying

quantities at a given frequency, the ratio of the complex amplitude of the voltage at one pair of terminals of a network to the complex amplitude of the current across a different pair of terminals. 185

(2) (of a magnetic amplifier). The ratio of differential output voltage to differential control current. 171

transinformation (of an output symbol about an input symbol) (information theory). The difference between the information content of the input symbol and the conditional information content of the input symbol given the output symbol. *Notes:* (1) If x_i is an input symbol and y_j is an output symbol, the transinformation is equal to

$$[-\log p(x_i)] - [-\log p(x_i|y_j)]$$
$$= \log \frac{p(x_i|y_j)}{p(x_i)} = \log \frac{p(x_i,y_j)}{p(x_i)p(y_j)}$$

where $p(x_i|y_j)$ is the conditional probability that x_i was transmitted when y_j is received, and $p(x_i,y_j)$ is the joint probability of x_i and y_j (2) This quantity has been called **transferred information, transmitted information,** and **mutual information.** *See also:* **information theory.** 160

transistor. An active semiconductor device with three or more terminals. It is an analog device. 245, 210, 66

transistor, conductivity-modulation. A transistor in which the active properties are derived from minority-carrier modulation of the bulk resistivity of a semiconductor. *See also:* **semiconductor; transistor.** 245, 210, 66

transistor, filamentary. A conductivity-modulation transistor with a length much greater than its transverse dimensions. *See also:* **semiconductor; transistor.** 245, 210, 66

transistor, junction. A transistor having a base electrode and two or more junction electrodes. *See also:* **transistor.** 245, 210, 66

transistor, point-contact. A transistor having a base electrode and two or more point-contact electrodes. *See also:* **semiconductors; transistor.** 245, 210, 66

transistor, point-junction. A transistor having a base electrode and both point-contact and junction electrodes. *See also:* **transistor.** 328

transistor, unipolar. A transistor that utilizes charge carriers of only one polarity. *See also:* **semiconductor; transistor.** 245, 210, 66

transit (1) (electronic navigation). A radio navigation system using low-orbit satellites to provide world-wide coverage, with transmissions from the satellites at very- and ultra-high frequencies, in which fixes are determined from measurements of the Doppler shift of the continuous-wave signal received from the moving satellite. *See also:* **navigation.** 187, 13

(2) (conductor stringing equipment). An instrument primarily used during construction of a line to survey the route, set "hubs" and P.O.T. (point on tangent) locations, "plumb" structures, determine "downstrain" angles for locations of anchors at the pull and tension sites, and to sag conductors. *Syn:* **level; scope; site marker.** 45

transit angle. The product of angular frequency and the time taken for an electron to traverse a given path. *See also:* **electron emission.** 190, 125

transition (1) (motor control). The procedure of changing from one scheme of motor connections to another scheme of connections, such as from series to parallel. *See also:* **multiple-unit control.**

(2) (signal transmission). The change from one circuit condition to the other, that is, the change from mark to space or from space to mark. *See also:* **multiple-unit control.**

(3) (waveform) (pulse techniques). A change of the instantaneous amplitude from one amplitude level to another amplitude level. *See also:* **pulse.**

(4) *See:* **adapter.** 185, 194

(5) (pulse terms). A portion of a wave or pulse between a first nominal state and a second nominal state. Throughout the remainder of this document the term transition is included in the term pulse and wave. 254

transition duration (pulse terms). The duration between the proximal point and the distal point on a transition waveform. 254

transition frequency (disk recording system) (crossover frequency) (turnover frequency). The frequency corresponding to the point of intersection of the asymptotes to the constant-amplitude and the constant-velocity portions of its frequency response curve. This curve is plotted with output voltage ratio in decibels as the ordinate and the logarithm of the frequency as the abscissa. *See also:* **phonograph pickup.** 176

transition joint (power cable joints). A cable joint which connects two different types of cable. 34

transition load (rectifier circuit). The load at which a rectifier unit changes from one mode of operation to another. *Note:* The load current corresponding to a transition load is determined by the intersection of extensions of successive portions of the direct-current voltage-regulation curve where the curve changes shape or slope. *See also:* **rectification; rectifier circuit element.** 237, 66

transition loss (1) (wave propagation). (A) At a transition or discontinuity between two transmission media, the difference between the power incident upon the discontinuity and the power transmitted beyond the discontinuity that would be observed if the medium beyond the discontinuity were match-terminated. (B) The ratio in decibels of the power incident upon the discontinuity to the power transmitted beyond the discontinuity that would be observed if the medium beyond the discontinuity were match terminated. *See:* **waveguide.** 267

(2) (junction between a source and a load). The ratio of the available power to the power delivered to the load. Transition loss is usually expressed in decibels. *See:* **waveguide.** *See also:* **transmission loss.** 267, 239, 197

transition point (circuit). A point in a transmission system at which there is change in the surge impedance. 244, 62

transition pulse (pulse waveform). That segment comprising a change from one amplitude level to another amplitude level. *See also:* **pulse.** 185

transition region (semiconductor). The region, between two homogeneous semiconductor regions, in which the impurity concentration changes. *See also:* **semiconductor; transistor.** 186, 113

transition shape (pulse terms). (1) For descriptive purposes a transition waveform may be imprecisely described by any of the adjectives, or combinations thereof, in **descriptive adjectives, major (minor); polarity related adjec-** tives; geometrical adjectives, round; and functional adjectives. When so used, these adjectives describe general shape only, and no precise distinctions are defined.

(2) For tutorial purposes a hypothetical transition waveform may be precisely defined by the further addition of the adjective ideal. *See:* **descriptive adjectives, ideal.**

(3) For measurement or comparison purposes a transition waveform may be precisely defined by the further addition of the adjective reference. *See:* **descriptive adjectives, reference.** 254

transitron oscillator. A negative-transconductance oscillator employing a screen-grid tube with negative transconductance produced by a retarding field between the negative screen grid and the control grid that serves as the anode. *See also:* **oscillatory circuit.** 111

transit time (1) (electron tube). The time taken for a charge carrier to traverse a given path. *See also:* **electron emission.** 190

(2) (multiplier-phototube). The time interval between the arrival of a delta-function light pulse at the entrance window of the tube and the time at which the output pulse at the anode terminal reaches peak amplitude. *See also:* **electron emission; phototube.** 174, 190

transit-time mode (electron tubes). A condition of operation of an oscillator corresponding to a limited range of drift-space transit angle for which the electron stream introduces a negative conductance into the coupled circuit. 125

transit-time spread (1) (electron tubes). The time interval between the half-amplitude points of the output pulse at the anode terminal, arising from a delta function of light incident on the entrance window of the tube. *See also:* **phototube.** 174, 190

(2) (scintillation counting). The FWHM (full-width-at-half-maximum) of the time distribution of a set of pulses each of which corresponds to the photomultiplier transit time for that individual event. 117

translate. (1) To convert expressions in one language to synonymous expressions in another language. (2) To encode or decode. *See also:* **matrix; translator.** 235, 255, 77

translation (telecommunication). The process of converting information from one system of representation into equivalent information in another system of representation. 194

translation loss (playback loss) (reproduction of a mechanical recording). The loss whereby the amplitude of motion of the reproducing stylus differs from the recorded amplitude in the medium. *See also:* **phonograph pickup.** 334, 176

translator (1) (general) (telephone switching systems). Equipment capable of interpreting and converting information from one form to another form. 193, 55

(2) (electronic computation). A network or system having a number of inputs and outputs and so connected that signals representing information expressed in a certain code, when applied to the inputs, cause output signals to appear that are a representation of the input information in a different code. Sometimes called **matrix.** *See also:* **matrix.** 210

(3) (test, measurement and diagnostic equipment). An automatic means, usually a program, to translate machine language mnemonic symbols for computer operations into true machine language. Memory locations and input-output lines must be written in numerical code, not symbolically. 54

transliterate. To convert the characters of one alphabet to the corresponding characters of another alphabet. 255, 77

transmission (1) (illuminating engineering). A general term for the process by which incident flux leaves a surface or medium on a side other than the incident side. *Note:* Transmission through a medium is often a combination of regular and diffuse transmission. 167

(2) (laser-maser). Passage of radiation through a medium. 363

transmission band (uniconductor waveguide). The frequency range above the cutoff frequency. *See also:* **waveguide.** 267, 179

transmission coefficient (1) (transition or discontinuity between two transmission media, at a given frequency). The ratio of some quantity associated with the transmitted wave at a specified point in the second medium to the same quantity associated with the incident wave at a specified point in the first medium, the second medium being match terminated. *See also:* **waveguide.** 267

(2) (transmission medium) (given frequency, at a given point, and for a given mode of transmission). The ratio of some quantity associated with the resultant field, which is the sum of the incident and reflected waves, to the corresponding quantity in the incident wave. *Note:* The transmission coefficient may be different for different associated quantities, and the chosen quantity must be specified. The **voltage transmission coefficient** is commonly used and is defined as the complex ratio of the resultant electric field strength (or voltage) to that of the incident wave. *See also:* **waveguide.** 267

(3) (multiport). Ratio of the complex amplitude of the wave emerging from a port of a multiport terminated by reflectionless terminations to the complex amplitude of the wave incident upon another port. *See also:* **reflection coefficient; scattering coefficient.** 185

transmission detector (semiconductor radiation detectors). A totally depleted detector whose thickness including its entrance and exit windows is sufficiently small to permit radiation to pass completely through the detector. 23, 118, 119

transmission facility (data transmission). The transmission medium and all the associated equipment required to transmit a message. 59

transmission feeder. A feeder forming part of a transmission circuit. *See also:* **center of distribution.** 64

transmission frequency meter (waveguide). A cavity frequency meter that, when tuned, couples energy from a waveguide into a detector. *See also:* **waveguide.** 244, 179

transmission gain stability, environmental (power-system communication). The variation in the gain or loss of a transmission medium with change in ambient temperature, ambient humidity, supply-voltage variations, and frequency variation in supply voltage. 59

transmission level (signal power at any point in a transmission system). The ratio of the power at that point to the power at some point in the system chosen as a reference

point. *Note:* This ratio is usually expressed in decibels. The transmission level at the transmitting switchboard is frequently taken as the zero level reference point. *See also:* **level.** 111, 59

transmission line (1) (signal-transmission system). (A) The conductive connections between system elements which carry signal power. (B) A waveguide consisting of two or more conductors. *See also:* **waveguide.** 267, 210, 60

(2) (electric power). A line used for electric power transmission. 64

(3) (electromagnetic wave guidance). A system of material boundaries or structures for guiding electromagnetic waves. Frequently, such a system for guiding electromagnetic waves, in the TEM (transverse electromagnetic) mode. Commonly a two-wire or coaxial system of conductors. *See also:* **waveguide.** 179, 185

(4) (induction heating). *See:* **load leads or transmission line.**

transmission-line capacity (electric power supply). The maximum continuous rating of a transmission line. The rating may be limited by thermal considerations, capacity of associated equipment, voltage regulation, system stability, or other factors. *See also:* **generating station.** 200

transmission line, coaxial. *See:* **coaxial transmission line.**

transmission loss (1) (electric power system). (A) The power lost in transmission between one point and another. It is measured as the difference between the net power passing the first point and the net power passing the second. (B) The ratio in decibels of the net power passing the first point to the net power passing the second. 267

(2) (communication). A general term used to denote a decrease in power in transmission from one point to another. This loss is usually expressed in decibels. *See also:* **waveguide.** 197, 199, 59

(3) (radio system). In a system consisting of a transmitting antenna, receiving antenna, and the intervening propagation medium, the ratio of the power radiated from the transmitting antenna to the resultant power that would be available from an equivalent loss-free receiving antenna. *See also:* **radio wave propagation.** 180, 146

transmission-loss coefficients (electric power systems). Mathematically derived constants to be combined with source powers to provide incremental transmission losses from each source to the composite system load. These coefficients may also be used to calculate total system transmission losses. 94

transmission measuring set. A measuring instrument comprising a signal source and a signal receiver having known impedances, that is designed to measure the insertion loss or gain of a network or transmission path connected between those impedances. *Note:* This name also applies to the signal receiver as a separate unit when it is used at a location remote from the signal source. *See also:* **instrument.** 328, 59

transmission mode. A form of propagation along a transmission line characterized by the presence of any one of the elemental types of TE, (transverse electric), TM (transverse magnetic), or TEM (transverse electromagnetic) waves. *Note:* Waveguide transmission modes are designated by integers (modal numbers) associated with the orthogonal functions used to describe the waveform. These integers are known as waveguide mode subscripts. They may be assigned from observations of the transverse field components of the wave and without reference to mathematics. A waveguide transmission mode is commonly described as a $TE_{m,n}$ or $TM_{m,n}$ mode, $_{m,n}$ being numerics according to the following system. **(1)** **(waves in rectangular waveguides).** If a single wave is transmitted in a rectangular waveguide, the field that is everywhere transverse may be resolved into two components, parallel to the wide and narrow walls respectively. In any transverse section, these components vary periodically with distance along a path parallel to one of the walls. m = the total number of half-period variations of either component of field along a path parallel to the wide wall. n = the total number of half-period variations of either component of field along a path parallel to the narrow walls. **(2)** **(waves in circular waveguides).** If a single wave is transmitted in a circular waveguide, the transverse field may be resolved into two components, radial and angular, respectively. These components vary periodically along a circular path concentric with the wall and vary in a manner related to the Bessel function of order m along a radius, where m = the total number of full-period variations of either component of field along a circular path concentric with the wall. n = one more than the total number of reversals of sign of either component of field along a radial path. This system can be used only if the observed waveform is known to correspond to a single mode. *See also:* **waveguide.**

328

transmission-mode photocathode. A photocathode in which radiant flux incident on one side produces photoelectric emission from the opposite side. 117

transmission modulation (storage tubes). Amplitude modulation of the reading-beam current as it passes through apertures in the storage surface, the degree of modulation being controlled by the charge pattern stored on that surface. *See also:* **storage tube.** 174, 190

transmission network. A group of interconnected transmission lines or feeders. *See also:* **transmission line.**

64

transmission primaries (color television). The set of three colorimetric primaries that, if used in a display and controlled linearly and individually by a corresponding set of three channel signals generated in the color television camera, would result in exact colorimetric rendition (over the gamut defined by the primaries) of the scene viewed by the camera. *Note:* Ideally the primaries used at the receiver display would be identical with the transmission primaries, but this is not usually possible since developments in display phosphors occurring since the setting of transmission standards, for example, may result in the use of receiver display primaries that differ from the transmission primaries. Within a linear part of the overall system, it is always possible to compensate for differences existing between transmission and display primaries by means of matrixing. Because of the capability afforded by matrixing, the transmission primaries need not be real. There exists a unique relationship between the chromaticity coordinates of the transmission primaries and the spectral taking characteristics used at the camera to generate the three respective channel signals. 18

transmission quality (mobile communication). The measure of the minimum usable speech-to-noise ratio, with reference to the number of correctly received words in a specified speech sequence. *See also:* **mobile communication system.** 181

transmission regulator (electric communication). A device that functions to maintain substantially constant transmission over a transmission system. 328

transmission route. The route followed by a transmission circuit. *See also:* **transmission line.** 64

transmission system (1) (communication practice). An assembly of elements capable of functioning together to transmit signal waves. 328, 59 **(2) (electric power supply).** An interconnected group of electric transmission lines and associated equipment for the movement or transfer of electric energy in bulk between points of supply and points for delivery. 112

transmission throughput. *See:* **speed of transmission, effective.**

transmission time (signal). The absolute time interval from transmission to reception. 328, 59

transmissometer. A photometer for measuring transmittance. *Note:* Transmissometers may be visual or physical instruments. *See also:* **photometry.** 167

transmit (computing machines). To move data from one location to another location. *See:* **transfer (2).** 255, 77

transmit characteristic (telephony). The electrical output level of a telephone set as a function of the acoustic input level. The output is measured across a specified impedance connected to the telephone feed circuit, and the input is measured in free field at a specified location relative to the reference point of an **artificial mouth.** 122

transmit-receive cavity (radar). The resonant portion of a transmit-receive switch. *See also:* **navigation.** 13, 187

transmit-receive cell (tube) (waveguide). A gas-filled waveguide cavity that acts as a short circuit when ionized but is transparent to low-power energy when un-ionized. It is used in a transmit-receive switch for protecting the receiver from the high power of the transmitter but is transparent to low-power signals received from the antenna. *See also:* **waveguide.** 179

transmit-receive switch, duplexer. A switch, frequently of the gas discharge type, employed when a common transmitting and receiving antenna is used, that automatically decouples the receiver from the antenna during the transmitting period. *See also:* **navigation.** 13, 187

transmit-receive switch (TR switch) (TR box). An automatic device employed in a radar for substantially preventing the transmitted energy from reaching the receiver but allowing the received energy to reach the receiver without appreciable loss. *See also:* **radar.** 328

transmit-receive (TR) tube. A gas-filled radio-frequency switching tube used to protect the receiver in pulsed radio-frequency systems. *See also:* **gas tube.** 190, 125

transmittance (medium) (1) (illuminating engineering). The ratio of the transmitted flux to the incident flux. *Note:* Measured values of transmittance depend upon the angle of incidence, the method of measurement of the transmitted flux, and the spectral character of the incident flux. Because of this dependence, complete information on the technique and conditions of measurement should be specified. It should be noted that transmittance refers to the ratio of flux emerging to flux incident; therefore, reflections at the surface as well as absorption within the material operate to reduce the transmittance. *See also:* **transmission (illuminating engineering).** 167

(2) (photovoltaic power system). The fraction of radiation incident on an object that is transmitted through the object. *See also:* **photovoltaic power system; solar cells.** 186

(3) (transfer function) (linear passive networks). A response function for which the variables are measured at different ports (terminal pairs). *See also:* **linear passive networks.** 238

(4) (laser-maser) τ. The ratio of total transmitted radiant power to total incident radiant power. 363

transmitted-carrier operation. That form of amplitude-modulation carrier transmission in which the carrier wave is transmitted. *See also:* **amplitude modulation.** 328

transmitted harmonics (induced harmonics) (electrical conversion). Harmonics that are transformed or pass through the conversion device from the input to the output. *See also:* **electrical conversion.** 186

transmitted information. *See:* **transinformation.**

transmitted light scanning (industrial control). The scanning of changes in the magnitude of light transmitted through a web. *See also:* **photoelectric control.** 206

transmitted wave (1) (circuit). A wave (or waves) produced by an incident wave that continue(s) beyond the transition point. 244

(2) (at a discontinuity). When a wave in a medium of certain propagation characteristics is incident upon a discontinuity or a second medium, the forward traveling wave that results in the second medium. *Note:* In a single medium the transmitted wave is that wave which is traveling in the forward direction. *See also:* **waveguide.** 267, 179

(3) (radio wave propagation). Refracted wave. *See:* **reflected wave; refracted wave.** 180

transmitter (protective signaling). A device for transmitting a coded signal when operated by any one of a group of actuating devices. *See also:* **protective signaling.** 328

transmitter-blocker cell (TB cell) (antitransmit-receive tube) (with reference to a waveguide). A gas-filled waveguide cavity that acts as a short circuit when ionized but as an open circuit when un-ionized. It is used in a transmit-receive switch for directing the energy received from the aerial to the receiver, no matter what the transmitter impedance may be. *See also:* **waveguide.** 179

transmitter, facsimile. The apparatus employed to translate the subject copy into signals suitable for delivery to the communication system. *See also:* **facsimile (in electrical communication).** 12

transmitter, pulse delay. *See:* **pulse delay transducer.**

transmitter performance. *See:* **audio input power; audio input signal.**

transmitter, telephone. *See:* **telephone transmitter.**

transmitting converter (facsimile) (amplitude-modulation to frequency-shift-modulation converter). A device which changes the type of modulation from amplitude to frequency shift. *See also:* **facsimile transmission.** 12

transmitting current response (electroacoustic transducer used for sound emission). The ratio of the sound pressure apparent at a distance of 1 meter in a specified direction

from the effective acoustic center of the transducer to the current flowing at the electric input terminals. *Note:* The sound pressure apparent at a distance of 1 meter can be found by multiplying the sound pressure observed at a remote point (where the sound field is spherically divergent) by the number of meters from the effective acoustic center of the transducer to that point. *See also:* **loudspeaker.** 176

transmitting efficiency (electroacoustic transducer) (projector efficiency). The ratio of the total acoustic power output to the electric power input. *Note:* In computing the electric power input, it is customary to omit any electric power supplied for polarization or bias. 176

transmitting loop loss. That part of the repetition equivalent assignable to the station set, subscriber line, and battery supply circuit that are on the transmitting end. *See also:* **transmission loss.** 328

transmitting power response (projector power response) (electroacoustic transducer used for sound emission). The ratio of the mean-square sound pressure apparent at a distance of 1 meter in a specified direction from the effective acoustic center of the transducer to the electric power input. *Note:* The sound pressure apparent at a distance of 1 meter can be found by multiplying the sound pressure observed at a remote point (where the sound field is spherically divergent) by the number of meters from the effective acoustic center of the transducer to that point. *See also:* **loudspeaker.** 176

transmitting voltage response (electroacoustic transducer used for sound emission). The ratio of the sound pressure apparent at a distance of 1 meter in a specified direction from the effective acoustic center of the transducer to the signal voltage applied at the electric input terminals. *Note:* The sound pressure apparent at a distance of 1 meter can be found by multiplying the sound pressure observed at a remote point (where the sound field is spherically divergent) by the number of meters from the effective acoustic center of the transducer to that point. *See also:* **loudspeaker.** 176

trans-μ-factor (multibeam electron tubes). The ratio of (1) the magnitude of an infinitesimal change in the voltage at the control grid of any one beam to (2) the magnitude of an infinitesimal change in the voltage at the control grid of a second beam. The current in the second beam and the voltage of all other electrodes are maintained constant. 190, 125

transparency (data communication). A capability of a communications medium to pass within specified limits a range of signals having one or more defined properties, for example, a channel may be code transparent, or an equipment may be bit pattern transparent. 12

transponder (1) (radar). A receiver-transmitter facility the function of which is to transmit signals automatically when the proper interrogation is received. 13

(2) (communication satellite). A receiver-transmitter combination, often aboard a satellite, or spacecraft, which receives a signal and retransmits it at a different carrier frequency. Transponders are used in communication satellites for reradiating signals to earth stations or in spacecraft for returning ranging signals. *See also:* **repeater.** 83

transponder beacon (electronic navigation). *See:* transponder.

transponder, crossband (electronic navigation). A transponder that replies in a different frequency band from that of the received interrogation. *See also:* navigation.
13, 187

transponder reply efficiency (electronic navigation). The ratio of the number of replies emitted by a transponder to the number of interrogations that the transponder recognizes as valid; the interrogations recognized as valid include those accidentally combined to form recognizable codes, a statistical computation of them normally being made. *See also:* navigation.
13, 187

transport (computing machines). *See:* tape transport.

transportable transmitter. A transmitter designed to be readily carried or transported from place to place, but which is not normally operated while in motion. *Note:* This has been commonly called a portable transmitter, but the term transportable transmitter is preferred. *See also:* radio transmitter; radio transmission.
111, 211

transportation and storage conditions. The conditions to which a device may be subjected between the time of construction and the time of installation. Also included are the conditions that may exist during shutdown. *Note:* No permanent physical damage or impairment of operating characteristics shall take place under these conditions, but minor adjustments may be needed to restore performance to normal.
295

transportation lag.* *See:* lag, distance/velocity.

*Deprecated

transport delay. *See:* time delay.

transport lag.* *See:* lag, distance/velocity.

*Deprecated

transport standards. Standards of the same nominal value as the basic reference standards of a laboratory (and preferably of equal quality), that are regularly intercompared with the basic group but that are reserved for periodic interlaboratory comparison tests that act as checks on the stability of the basic reference group. *See also:* measurement system.
212

transport time (industrial control) (feedback system). The time required to move an object, element or information from one predetermined position to another. *See also:* control system, feedback.
219, 206

transposition (1) (transmission lines). An interchange of positions of the several conductors of a circuit between successive lengths. *Notes:* (A) It is normally used to reduce inductive interference on communication or signal circuits by cancellation. (B) The term is most frequently applied to open wire circuits. *See also:* open wire; signal; tower.
188, 64

(2) (rotating machinery). An arrangement of the strands or laminations of a conductor or of the conductors comprising a turn or coil whereby they take different relative positions in a slot for the purpose of reducing eddy current losses.
63

transposition section. A length of open wire line to which a fundamental transposition design or pattern is applied as a unit. *See also:* open wire.
328

transreactance. The imaginary part of the transimpedance.
185

transrectification factor. The quotient of the change in average current of an electrode by the change in the amplitude of the alternating sinusoidal voltage applied to another electrode, the direct voltages of this and other electrodes being maintained constant. *Note:* Unless otherwise stated, the term refers to cases in which the alternating sinusoidal voltage is of infinitesimal magnitude. *See also:* rectification factor.
190, 125

transrectifier. A device, ordinarily a vacuum tube in which rectification occurs in one electrode circuit when an alternating voltage is applied to another electrode. *See also:* rectifier.
328

transresistance. The real part of the transimpedance.
185

transsusceptance. The imaginary part of the transadmittance.
185

transverse-beam traveling-wave tube. A traveling-wave tube in which the direction of motion of the electron beam is transverse to the average direction in which the signal wave moves. *See also:* miscellaneous electron devices.
190, 125

transverse crosstalk coupling (between a disturbing and a disturbed circuit in any given section). The vector summation of the direct couplings between adjacent short lengths of the two circuits, without dependence on intermediate flow in other nearby circuits. *See also:* coupling.
328

transverse-electric hybrid wave. An electromagnetic wave in which the electric field vector is linearly polarized normal to the plane of propagation and the magnetic field vector is elliptically polarized in this plane. *See also:* radio wave propagation.
180, 146

transverse electric ($TE_{m,n,p}$) resonant mode (cylindrical cavity). In a hollow metal cylinder closed by two plane metal surfaces perpendicular to its axis, the resonant mode whose transverse field pattern is similar to the $TE_{m,n}$ wave in the corresponding cylindrical waveguide and for which p is the number of half-period field variations along the axis. *Note:* When the cavity is a rectangular parallelepiped, the axis of the cylinder from which the cavity is assumed to be made should be designated since there are three such axes possible. *See also:* waveguide.
267, 179

transverse electric wave (TE wave) (1) (general). In a homogeneous isotropic medium, an electromagnetic wave in which the electric field vector is everywhere perpendicular to the direction of propagation. *See also:* waveguide.
267, 179, 146

(2) ($TE_{m,n}$ wave) (rectangular waveguide) (hollow rectangular metal cylinder). The transverse electric wave for which m is the number of half-period variations of the field along the x coordinate, which is assumed to coincide with the larger transverse dimension, and n is the number of half-period variations of the field along the y coordinate, which is assumed to coincide with the smaller transverse dimension. *Note:* The dominant wave in a rectangular waveguide is $TE_{1,0}$; its electric lines are parallel to the shorter side. *See also:* guided waves; waveguide.
267, 179

(3) ($TE_{m,n}$ wave) (circular waveguide) (hollow circular metal cylinder). The transverse electric wave for which m is the number of axial planes along which the normal component

of the electric vector vanishes, and n is the number of coaxial cylinders (including the boundary of the waveguide) along which the tangential component of the electric vector vanishes. *Notes:* (A) $TE_{0,n}$ waves are circular electric waves of order n. The $TE_{0,1}$ wave is the circular electric wave with the lowest cutoff frequency. (B) The $TE_{1,1}$ wave is the dominant wave. Its lines of electric force are approximately parallel to a diameter. *See also:* **waveguide.** 267, 179

transverse electromagnetic (TEM) mode (waveguide). A mode in which the longitudinal components of the electric and magnetic fields are everywhere zero. *See also:* **waveguide.** 244, 179

transverse electromagnetic wave (TEM wave). In a homogeneous isotropic medium, an electromagnetic wave in which both the electric and magnetic field vectors are everywhere perpendicular to the direction of propagation. *See also:* **radio-wave propagation; waveguide.** 267, 179, 180, 146

transverse-field traveling-wave tube. A traveling-wave tube in which the traveling electric fields that interact with electrons are essentially transverse to the average motion of the electrons. *See also:* **electron devices, miscellaneous.** 190, 125

transverse interference (signal-transmission system). *See:* **interference, differential-mode; interference, normal-mode.** *See also:* **accuracy rating (instrument); signal.**

transverse magnetic hybrid wave. An electromagnetic wave in which the magnetic field vector is linearly polarized normal to the plane of propagation and the electric field vector is elliptically polarized in this plane. *See also:* **radio wave propagation.** 180, 146

transverse magnetic ($TM_{m,n,p}$) resonant mode (cylindrical cavity). In a hollow metal cylinder closed by two plane metal surfaces perpendicular to its axis, the resonant mode whose transverse field pattern is similar to the $TM_{m,n}$ wave in the corresponding cylindrical waveguide and for which p is the number of half-period field variations along the axis. *Note:* When the cavity is a rectangular parallelepiped, the axis of the cylinder from which the cavity is assumed to be made should be designated since there are three such axes possible. *See also:* **waveguide.** 267, 179

transverse magnetic wave (TM wave) (1) (general). In a homogeneous isotropic medium, an electromagnetic wave in which the magnetic field vector is everywhere perpendicular to the direction of propagation. *See also:* **waveguide.** 267, 179, 180, 146

(2) ($TM_{m,n}$ wave) (circular waveguide) (hollow circular metal cylinder). The transverse magnetic wave for which m is the number of axial planes along which the normal component of the magnetic vector vanishes, and n is the number of coaxial cylinders to which the electric vector is normal. *Note:* $TM_{0,n}$ waves are circular magnetic waves of order n. The $TM_{0,1}$ wave is the circular magnetic wave with the lowest cutoff frequency. *See also:* **guided wave; circular magnetic wave; waveguide.** 267, 179

(3) ($TM_{m,n}$ wave) (rectangular waveguide) (hollow rectangular metal cylinder). The transverse magnetic wave for which m is the number of half-period variations of the magnetic field along the longer transverse dimension, and n is the number of half-period variations of the magnetic field along the shorter transverse dimension. *See also:* **guided wave; circular magnetic wave; waveguide.** 267, 179

transverse magnetization (magnetic recording). Magnetization of the recording medium in a direction perpendicular to the line of travel and parallel to the greatest cross-sectional dimension. *See also:* **phonograph pickup.** 176

transverse mode (laser-maser). A mode which is detected by measuring one or more maxima in transverse field intensity in the cross-section of a beam. 363

transverse-mode interference (signal-transmission system). *See:* **interference, differential-mode.** *See also:* **signal.**

transverse wave. A wave in which the direction of displacement at each point of the medium is perpendicular to the direction of propagation. *Note:* In those cases where the displacement makes an acute angle with the direction of propagation, the wave is considered to have longitudinal and transverse components. 210

trap (1) (computing machines). An unprogrammed conditional jump to a known location, automatically activated by hardware, with the location from which the jump occurred recorded. 255, 77

(2) (burglar-alarm system). An automatic device applied to a door or window frame for the purpose of producing an alarm condition in the protective circuit whenever a door or window is opened. *See also:* **protective signaling.** 328

trap circuit. A circuit used at locations where it is desirable to protect a section of track on which it is impracticable to maintain a track circuit. It usually consists of an arrangement of one or more stick circuits so connected that when a train enters the trap circuit the stick relay drops and cannot be picked up again until the train has passed through the other end of the trap circuit. 328

trapezium distortion (cathode-ray tube). A fault characterized by a variation of the sensitivity of the deflection parallel to one axis (vertical or horizontal) as a function of the deflection parallel to the other axis and having the effect of transforming an image that is a rectangle into one which is a trapezium. *See also:* **cathode-ray tubes.** 244, 290

trapped flux (superconducting material). Magnetic flux that links with a closed superconducting loop. *See also:* **superconductivity.** 191

travel (1) (relay). The amount of movement in either direction (towards pickup or reset) of the principal responsive element or a contact part of it. *Note:* Travel may be specified in linear, angular, or other measure. 103, 60, 202

(2) (rise) (elevators). Of an elevator, dumbwaiter, escalator, or of a private-residence inclined lift, the vertical distance between the bottom terminal landing and the top terminal landing. *See also:* **elevators.** 328, 127

traveler (conductor stringing equipment). A single groove wheel complete with suspension arm or frame used separately or in groups and suspended from structures to permit the stringing of conductors. These devices are sometimes "bundled" with a center drum, or another traveler, and used to string more than one conductor simultaneously. For protection of conductors that should not be nicked or scratched, the travelers are often lined with nonconductive or semi-conductive neoprene, or with nonconductive urethane. Any one of these materials acts as a "padding" or "cushion" for the conductor as it passes over the traveler. Traveler grounds must be used with "lined" travelers in order to establish an electrical ground. *Syn:* **block; dolly; sheave; stringing block; stringing sheave; stringing traveler.** 45

traveler sling (conductor stringing equipment). A single "vertical" sling of wire rope utilized in place of insulators to support the traveler during stringing operations. Normally, it is used when insulators are not readily available or when adverse stringing conditions might impose severe downstrains and cause damage or complete failure of the insulators. *Syn:* **choker.** 45

traveling cable (elevators). A cable made up of electric conductors that provides electric connection between an elevator or dumbwaiter car and fixed outlet in the hoistway. *See:* **control.** 328

traveling overvoltage (surge arresters). A surge propagated along a conductor. 244, 62

traveling plane wave. A plane wave each of whose frequency components has an exponential variation of amplitude and a linear variation of phase in the direction of propagation. *See also:* **radio wave propagation; waveguide.** 267, 180, 146

traveling wave. The resulting wave when the electric variation in a circuit takes the form of translation of energy along a conductor, such energy being always equally divided between current and potential forms. *See also:* **direction of propagation; traveling plane; waveguide.** 64, 62

traveling-wave magnetron. A traveling-wave tube in which the electrons move in crossed static electric and magnetic fields which are substantially normal to the direction of wave propagation. 125

traveling-wave magnetron oscillations. Oscillations sustained by the interaction between the space-charge cloud of a magnetron and a traveling electromagnetic field whose phase velocity is approximately the same as the mean velocity of the cloud. *See:* **magnetron, slow-wave circuit.** 190, 161

traveling-wave parametric amplifier. A parametric amplifier that has a continuous or iterated structure incorporating nonlinear reactors and in which the signal, pump, and difference-frequency wave are propagated along the structure. *See also:* **parametric device.** 277, 191

traveling-wave tube. An electron tube in which a stream of electrons interacts continuously or repeatedly with a guided electromagnetic wave moving substantially in synchronism with it, and in such a way that there is a net transfer of energy from the stream to the wave. *See also:* **transverse-beam traveling-wave tube; transverse-field traveling-wave tube.** 125

traveling-wave-tube interaction circuit. An extended electrode arrangement in a traveling-wave tube designed to propagate an electromagnetic wave in such a manner that the traveling electromagnetic fields are retarded and extended into the space occupied by the electron stream. *Note:* traveling-wave tubes are often designated by the type of interaction circuit used, as in helix traveling-wave tube. 125

tray (storage cell) (storage battery). A support or container

for one or more storage cells. *See also:* **battery (primary or secondary).** 328

TR box. *See:* **TR switch.**

treated fabric (treated mat) (rotating machinery). A fabric or mat in which the elements have been essentially coated but not filled with an impregnant such as a compound or varnish. *See also:* **rotor (rotating machinery); stator.** 63

treated mat (rotating machinery). *See:* **treated fabric.**

treble boost. An adjustment of the amplitude-frequency response of a system or transducer to accentuate the higher audio frequencies. 239

tree. A set of connected branches including no meshes. *See also:* **network analysis.** 322, 210

trees and nodules. Projections formed on a cathode during electrodeposition. Trees are branched whereas nodules are rounded. *See also:* **electrodeposition.** 328

tree wire. A conductor with an abrasion-resistant outer covering, usually nonmetallic, and intended for use on overhead lines passing through trees. *See also:* **armored cable; conductor; covered wire.** 64

triangular array (antennas). The regular arrangement of array elements such that lines connecting corresponding points of adjacent elements form triangles (usually equilateral triangles). 111

triboelectrification (electrification by friction). The mechanical separation of electric charges of opposite sign by processes such as (1) the separation (as by sliding) of dissimilar solid objects; (2) interaction at a solid-liquid interface; (3) breaking of a liquid-gas interface. 210

tributary office. A telephone central office that passes toll traffic to, and receives toll traffic from, a toll center. 328

tributary trunk (data transmission). A trunk circuit connecting a local exchange with a toll center or other toll office through which access to the LD (long distance) network is achieved. 59

trickle charge (storage battery) (storage cell). A continuous charge at a low rate approximately equal to the internal losses and suitable to maintain the battery in a fully charged condition. *Note:* This term is also applied to very low rates of charge suitable not only for compensating for internal losses but to restore intermittent discharges of small amount delivered from time to time to the load circuit. *See:* **floating.** *See also:* **charge.** 328

trigatron (1) (general). A triggered spark-gap switch on which control is obtained by a voltage applied to a trigger electrode. *Note:* This voltage distorts the field between the two main electrodes converting the sphere-to-sphere gap to a point to sphere gap. *See also:* **electron device.** 244, 190

(2) (radar). An electronic switch in which conduction is initiated by the breakdown of an auxiliary gap. 13, 187

trigger (1) (verb). To start action in another circuit which then functions for a period of time under its own control. 328

(2) (noun). A pulse used to initiate some function, for example, a triggered sweep or delay ramp. *Note:* Trigger may loosely refer to a waveform of any shape used as a signal from which a trigger pulse is derived as in **trigger source, trigger input,** etcetera. *See:* **triggering signal.** *See* *also:* **oscillograph.**　　　　　　　　　185

(3) (thyristor) (verb). The act of causing a thyristor to switch from the off-state to the on-state. *See also:* **gate trigger current.**　　　　　　　　　66

trigger circuit. A circuit that has two conditions of stability, with means for passing from one to the other when certain conditions are satisfied, either spontaneously or through application of an external stimulus.　　328

trigger countdown. A process that reduces the repetition rate of a triggering signal. *See also:* **oscillograph.** 185

triggered sweep. A sweep that can be initiated only by a trigger signal, not free running. *See also:* **oscillograph.**
　　　　　　　　　185

trigger, external (oscilloscopes). A triggering signal introduced into the trigger circuit from an external source.
　　　　　　　　　106

trigger gap (series capacitor). Enclosed electrodes relatively unaffected by barometric pressure and temperature changes that initiate the sparkover of the protective power gap. It is not intended to handle currents large enough to affect its calibration.　　　　　　　　　86

triggering (pulse terms). A process in which a pulse initiates a predetermined event or response.　　　　254

triggering level. The instantaneous level of a triggering signal at which a trigger is to be generated. Also, the name of the control that selects the level. *See also:* **oscillograph.**
　　　　　　　　　185

triggering, line (oscilloscopes). Triggering from the power-line frequency. *See also:* **oscillograph.**　　106

triggering signal. The signal from which a trigger is derived. *See also:* **oscillograph.**　　　　185

triggering slope. The positive-going (+slope) or negative-going (−slope) portion of a triggering signal from which a trigger is to be derived. Also, the control that selects the slope to be employed. *Note:* + and − slopes apply to the slope of the waveform only and not to the absolute polarity. *See also:* **oscillograph.**　　　185

triggering stability. *See:* **stability.**

trigger level (transponder). The minimum input to the receiver that is capable of causing the transmitter to emit a reply. *See also:* **navigation.**　　　　13, 77

trigger lockout. *See:* **sweep lockout.**

trigger pickoff. A process or a circuit for extracting a triggering signal. *See also:* **oscillograph.**　　185

trigger-starting systems (fluorescent lamps). Applied to systems in which hot-cathode electric discharge lamps are started with cathodes heated through low-voltage heater windings built into the ballast. Sufficient voltage is applied across the lamp and between the lamp and fixture to initiate the discharge when the cathodes reach a temperature high enough for adequate emission. The ballast is so designed that the cathode-heating current is greatly reduced as soon as the arc is struck.　　　　　　268

trigger tube (electron device). A cold-cathode gas-filled tube in which one or more electrodes initiate, but do not control, the anode current.　　　　　　　　　190

trilateration. *See:* **multilateration.**

trimmer capacitor (trimming capacitor). A small adjustable capacitor associated with another capacitor and used for fine adjustment of the total capacitance of an element or part of a circuit.　　　　　　　　　329

trimmer signal. A signal that gives indication to the en-

gineman concerning movements to be made from the classification tracks into the switch and retarder area.
　　　　　　　　　328

triode. A three-electrode electron tube containing an anode, a cathode, and a control electrode.　　190, 125

trip (1) (verb). (A) To release in order to initiate either an opening or a closing operation or other specified action. (B) To release in order to initiate an opening operation only. (C) To initiate and complete an opening operation. *Note:* All terms employing **trip, tripping,** or their derivatives are referred to the term that expresses the intent of the usage.　　　　　　　　　127

(2) (noun). (A) A release that initiates either an opening or a closing operation or other specified action. (B) A release that initiates an opening operation only. (C) A complete opening operation.　　　103, 202, 60, 127

(3) (tripping) (adjective). (A) Pertaining to a release that initiates either an opening or a closing operation or other specified action. (B) Pertaining to a release that initiates an opening operation only. (C) Pertaining to a complete opening operation. *Note:* All terms employing **trip, tripping,** or their derivatives are referred to the term that expresses the intent of the usage.　　　103, 202

trip arm. *See:* **mechanical trip.**

trip coil. *See:* **release coil.**

trip delay setting. *See:* **release delay setting.**

trip device (opening release), impulse (low voltage dc power circuit breakers). A trip device designed to operate only by the discharge of a capacitor into its release (trip) coil.
　　　　　　　　　360

trip-free. *See:* **release-free.**

trip-free in any position. *See:* **release-free in any position.**

trip-free relay (power switchgear). An auxiliary relay whose function is to open the closing circuit of an electrically operated switching device so that the opening operation can prevail over the closing operation. *Syn:* **release-free relay.**　　　　　　　103, 127

trip lamp. A removable self-contained mine lamp, designed for marking the rear end of a train (trip) of mine cars.
　　　　　　　　　328

triple-address. Same as three-address.

triple detection. *See:* **double superheterodyne reception.**

triplen (rotating machinery). An order of harmonic that is a multiple of three. *See also:* **asynchronous machine; direct-current commutating machine.**　　　63

triplet (electronic navigation systems). Three radio stations operated as a group for the determination of positions. *See also:* **navigation.**　　　　13, 187

triplex cable. A cable composed of three insulated single-conductor cables twisted together. *Note:* The assembled conductors may or may not have a common covering of binding or protecting material.　　　64

trip OFF control signal (magnetic amplifier). The final value of signal measured when the amplifier has changed from the ON to the OFF state as the signal is varied so slowly that an incremental increase in the speed with which it is varied does not affect the measurement of the trip OFF control signal. That is, the change in trip OFF control signal is below the sensitivity of the measuring instrument.　171

trip ON control signal (magnetic amplifier). The final value of signal measured when the amplifier has changed from the OFF to the ON state as the signal is varied so slowly than an incremental increase in the speed with which it is varied does not affect the measurement of the trip ON control signal. That is, the change in trip ON control signal is below the sensitivity of the measuring instrument.   171

tripping delay. *See:* **release delay.**

tripping mechanism. *See:* **release (tripping) mechanism.**

trip-point repeatability (magnetic amplifier). The change in trip point (either trip OFF or trip ON, as specified) control signal due to uncontrollable causes over a specified period of time when all controllable quantities are held constant.   171

trip point repeatability coefficient (magnetic amplifier). The ratio of (1) the maximum change in trip point control signal due to uncontrollable causes to (2) the specified time period during which all controllable quantities have been held constant. *Note:* The units of this coefficient are the control signal units per the time period over which the coefficient was determined.   171

trip setting. *See:* **release setting.**

tristimulus values (light). The amounts of the three reference or matching stimuli required to give a match with the light considered, in a given trichromatic system. *Notes:* (1) In the standard colorimetric system, CIE (1931), the symbols X, Y, Z are recommended for the tristimulus values. (2) These values may be obtained by multiplying the spectral concentration of the radiation at each wavelength by the distribution coefficients ($\bar{x}_\lambda$, $\bar{y}_\lambda$, $\bar{z}_\lambda$) and integrating these products over the whole spectrum.   18

troffer (illuminating engineering). A long recessed lighting unit usually installed with the opening-flush with the ceiling. The term is derived from **trough** and **coffer.** *See also:* **luminaire.**   167

troland. A unit used for expressing the magnitude of the external light stimulus applied to the eye. When the eye is viewing a surface of uniform luminance, the number of trolands is equal to the product of the area in square millimeters of the limiting pupil, natural or artificial, and the luminance of the surface in candelas per square meter. *See also:* **visual field.**   167

trolley. A current collector, the function of which is to make contact with a contact wire. *See also:* **contact conductor.**   1

trolley bus. *See:* **trolley coach.**

trolley car. An electric motor car that collects propulsion power from a trolley system. *See also:* **electric motor car.**   328

trolley coach (trolley bus) (trackless trolley coach). An electric bus that collects propulsion power from a trolley system. *See also:* **electric bus.**   328

trolley locomotive. An electric locomotive that collects propulsion power from a trolley system. *See also:* **electric locomotive.**   328

trombone line (transmission lines and waveguides). A U-shaped length of waveguide or transmission line of adjustable length. *See also:* **waveguide.**   185

troposphere. That part of the earth's atmosphere in which temperature generally decreases with altitude, clouds form and convection is active. *Note:* Experiments indicate that the troposphere occupies the space above the earth's surface up to a height ranging from about 6 kilometers at the poles to about 18 kilometers at the equator. *See also:* **radiation; radio wave propagation.**   180, 59, 146

tropospheric wave. A radio wave that is propagated by reflection from a place of abrupt change in the dielectric constant or its gradient in the troposphere. *Note:* In some cases the ground wave may be so altered that new components appear to arise from reflections in regions of rapidly changing dielectric constant. When these components are distinguishable from the other components, they are called tropospheric waves. *See also:* **radiation.**   328, 59

trouble recorder (telephone switching systems). A device or system associated with one or more switching systems for automatically recording data on calls encountering trouble.   55

troubleshoot. *See:* **debug; fault isolation.**

troughing. An open channel of earthenware, wood, or other material in which a cable or cables may be laid and protected by a cover.   64

TR switch (radar). An RF switch, employed when a common transmitting and receiving antenna is used, which automatically decouples the receiver from the antenna during the transmitting period.   13

truck generator suspension. A design of support for an axle generator in which the generator is supported by the vehicle truck.   328

true bearing (navigation). Bearing relative to true north. *See also:* **navigation.**   187, 13

true complement. (1) *See:* **radix complement.**
(2) A number representation that can be derived from another by subtracting each digit from one less than the base and then adding one to the least significant digit and executing all carries required. Tens complements and twos complements are true complements.   235

true course (navigation). Course relative to true north. *See also:* **navigation.**   187, 13

true heading (navigation). Heading relative to true north. *See also:* **navigation.**   187, 13

true-motion-display (radar). A display that shows the absolute motion of a vehicle mounted radar relative to a fixed background.   13

true neutral point (at terminals of entry). Any point in the boundary surface that has the same voltage as the point of junction of a group of equal nonreactive resistors placed in the boundary surface of the region and connected at their free ends to the appropriate terminals of entry of the phase conductors of the circuit, provided that the resistance of the resistors is so great that the voltages are not appreciably altered by the introduction of the resistors. *Notes:* (1) The number of resistances required is two for direct-current or single-phase alternating-current circuits, four for two-phase four-wire or five-wire circuits, and is equal to the number of phases when the number of phases is three or more. Under normal symmetrical conditions the number of resistors may be reduced to three for six- or twelve-phase systems when the terminals are properly selected, but the true neutral point may not be obtained by this process under all abnormal conditions. The concept of a true neutral point is not considered applicable to a two-phase, three-wire circuit. (2) Under abnormal conditions the voltage of the true neutral point may not be the same as that of the neutral conductor. *See also:* **network**

analysis. 210

true north (navigation). The direction of the north geographical pole. *See also:* **navigation.** 187, 13

true ratio. The ratio of the root-mean-square (rms) primary value to the rms secondary value under specified conditions. 53

truncate. To terminate a computational process in accordance with some rule, for example, to end the evaluation of a power series at a specified term. 255, 77

trunk (1) (telephone switching systems). A channel provided as a common traffic artery between switching entities. 55

(2) (analog computer). A connecting line between one analog computer and another, or between an analog computer and an external point, allowing the input (or output) of an analog component to communicate directly with the output (or input) at another component which is located outside of the analog computer. 9

(3) (data transmission). A telephone line or channel between two central offices or switching devices, which is used in providing telephone connections between subscribers. *See:* **final trunk; high usage trunk; long distance (LD) trunk; terminal trunk; tie trunk; tributary trunk; tanden trunk.** 59

trunk circuit (telephone switching systems). An interface circuit between a trunk and a switching system. 55

trunk circuit, combined line and recording (CLR). (1) Name given to a class of trunk circuits that provide access to operator positions generally referred to by abbreviation only. (2) Recording-completing trunk circuit for operator recording and completing of toll calls originated by subscribers of central offices. 193

trunk concentrator (telephone switching systems). A concentrator in which all inlets and outlets are trunks. 55

trunk feeder. A feeder connecting two generating stations or a generating station and an important substation. *See also:* **center of distribution.** 64

trunk group (telephone switching systems). A number of trunks that can be used interchangeably between two switching entities. 55

trunk hunting. The operation of a selector or other similar device, to establish connection with an idle circuit of a chosen group. This is usually accomplished by successively testing terminals associated with this group until a terminal is found that has an electrical condition indicating it to be idle. 328

trunk-line conduit. A duct-bank provided for main or trunk-line cables. 64

trunk loss. That part of the repetition equivalent assignable to the trunk used in the telephone connection. *See also:* **transmission loss.** 328

trunk multifrequency pulsing (telephone switching systems). A means of pulsing embodying a simultaneous combination of two out of six frequencies to represent each digit or character. 55

trunk transmission line. A transmission line acting as a source of main supply to a number of other transmission circuits. *See also:* **transmission line.** 64

trussed blade (switching device). A blade that is reinforced by truss construction to provide stiffness. 103, 202, 27

truth table. A table that describes a logic function by listing all possible combinations of input values and indicating, for each combination, the true output values. 255, 77, 54

TSI. *See:* **threshold signal-to-interference ratio.**

T step (waveform test signals) (TV). A $\sin^2$ step with a 10 percent—90 percent rise (fall) time of nominally 125 ns. The amplitude of the envelope of the frequency spectrum is effectively zero at and beyond 8 MHz. *Note:* In practice the T step is part of a square wave, so that there is a T step rise and fall. 42

tube (1) (interior wiring). A hollow cylindrical piece of insulating material having a head or shoulder at one end, through which an electric conductor is threaded where passing through a wall, floor, ceiling, joist, stud, etcetera. *See also:* **raceways.** 328

(2) (primary cell). A cylindrical covering of insulating material, without closure at the bottom. *See also:* **electrolytic cell.** 328

tube count (radiation-counter tubes). A terminated discharge produced by an ionizing event in a radiation-counter tube. *See also:* **gas-filled radiation-counter tubes.** 190, 96, 125

tube current averaging time. The time interval over which the current is averaged in defining the operating capability of the tube. *See also:* **rectification.** 291

tube, display. *See:* **display tube.**

tube, electron. *See:* **electron tube.**

tube fault current. The current that flows through a tube under fault conditions, such as arc-back or short circuit. *See also:* **rectification.** 291

tube, fuse. *See:* **fuse tube.**

tube heating time (mercury-vapor tube). The time required for the coolest portion of the tube to attain operating temperature. *See:* **preheating time.** *See also:* **electronic controller; gas tubes.** 328, 125

tubelet (soldered connections). *See:* **eyelet.**

tuberculation (corrosion). The formation of localized corrosion products scattered over the surface in the form of knoblike mounds. 221, 205

tube scintillation pulses (photomultipliers). Dark pulses caused by scintillations within the photomultiplier structure. Example: cosmic-ray-induced events. 117

tube-type plate (storage cell). A plate of an alkaline storage battery consisting of an assembly of metal tubes filled with active material. *See also:* **battery (primary or secondary).** 328

tube, vacuum. *See:* **vacuum tube.**

tube voltage drop (electron tube). The anode voltage during the conducting period. *See also:* **electrode voltage (electron tube); electronic controller.** 190, 125

tubing (rotating machinery). A tubular flexible insulation, extruded or made of layers of film plastic, into which a conductor is inserted to provide additional insulation. Tubing is frequently used to insulate connections and crossovers. *See also:* **asynchronous machine; direct-current commutating machine.** 63

Tudor plate (storage cell). A lead storage battery plate obtained by molding and having a large area. *See also:* **battery (primary and secondary).** 328

tumbling (gyro). The loss of reference in a two-degree-of-freedom gyro due to gimbal lock or contact between a gimbal and a mechanical stop. This is not to be confused with tumble testing which is a method of evaluating gyro performance. 46

tuned-grid oscillator. An oscillator whose frequency is determined by a parallel-resonance circuit in the grid circuit coupled to the plate to provide the required feedback. *See also:* **oscillatory circuit.** 111

tuned-grid—tuned plate oscillator. An electron tube circuit in which both grid and plate circuits are tuned to resonance where the feedback voltage normally is developed through the inter-electrode capacity of the tube. *See:* **radio frequency generator.** 14

tuned-plate oscillator. An oscillator whose frequency is determined by a parallel-resonance circuit in the plate circuit coupled to the grid to provide the required feedback. *See also:* **oscillatory circuit.** 111

tuned rotor gyro. A two-degree-of-freedom gyro in which a flexure and gimbal mechanism both supports the rotor and provides angular freedom about axes perpendicular to the spin axis. The rotor is tuned by adjusting the gimbal inertia or flexure spring rate so that the dynamically induced (negative) spring rate cancels the spring rate of the flexure suspension at a selected rotor speed. 46

tuned transformer. A transformer, the associated circuit elements of which are adjusted as a whole to be resonant at the frequency of the alternating current supplied to the primary, thereby causing the secondary voltage to build up to higher values than would otherwise be obtained. *See also:* **power pack.** 329

tuner (1) (radio receiver). In the broad sense, a device for tuning. Specifically, in radio receiver practice, it is (A) a packaged unit capable of producing only the first portion of the functions of a receiver and delivering either radio-frequency, intermediate-frequency, or demodulated information to some other equipment, or (B) that portion of a receiver that contains the circuits that are tuned to resonance at the received-signal frequency. *See also:* **radio receiver.** 328

(2) (transmission line) (waveguide). An ideally lossless, fixed or adjustable, network capable of transforming a given impedance into a different impedance. *See also:* **transmission loss; waveguide.** 179, 185

tuner, waveguide (waveguide components). An adjustable waveguide transformer. 166

tuning (of circuits). The adjustment in relation to frequency of a circuit or system to secure optimum performance; commonly the adjustment of a circuit or circuits to resonance. *See also:* **radio transmission.** 210, 59

tuning creep (oscillator). The change of an essential characteristic as a consequence of repeated cycling of the tuning element. 174, 190

tuning, electronic. The process of changing the operating frequency of a system by changing the characteristics of a coupled electron stream. Characteristics involved are, for example: velocity, density, or geometry. *See also:* **oscillatory circuit.** 190, 125

tuning hysteresis (microwave (oscillator). The difference in a characteristic when a tuner position, or input to the tuning element, is approached from opposite directions. 174, 190

tuning indicator (electron device). An electron-beam tube in which the signal supplied to the control electrode varies the area of luminescence of the screen. 190

tuning probe (waveguides). An essentially lossless probe of

adjustable penetration extending through the wall of the waveguide or cavity resonator. *See also:* **waveguide.** 330

tuning range (1) (switching tubes). The frequency range over which the resonance frequency of the tube may be adjusted by the mechanical means provided on the tube or associated cavity. *See:* **gas tubes.** 190, 125

(2) (oscillator). The frequency range of continuous tuning within which the essential characteristics fall within prescribed limits. 174, 190

tuning range, electronic. The frequency range of continuous tuning between two operating points of specified minimum power output for an electronically tuned oscillator. *Note:* The reference points are frequently the half-power points, but should always be specified. *See also:* **oscillatory circuit.** 190, 125

tuning rate, thermal. The initial time rate of change in frequency that occurs when the input power to the tuner is instantaneously changed by a specified amount. *Note:* This rate is a function of the power input to the tuner as well as the sign and magnitude of the power change. *See also:* **oscillatory circuit.** 190, 125

tuning screw (waveguide technique). An impedance-adjusting element in the form of a rod whose depth of penetration through the wall into a waveguide or cavity is adjustable by rotating the screw. *See also:* **waveguide.** 179

tuning sensitivity (oscillator). The rate of change of frequency with the control parameter (for example, the position of a mechanical tuner, electric tuning voltage, etcetera) at a given operating point. *See also:* **tunable microwave oscillators.** 174, 190

tuning sensitivity, electronic. At a given operating point, the rate of change of oscillator frequency with the change of the controlling electron stream. For example, this change may be expressed in terms of an electrode voltage or current. *See:* **pushing figure (oscillator).** *See also:* **oscillatory circuit.** 190, 125

tuning sensitivity, thermal. The rate of change of resonator equilibrium frequency with respect to applied thermal tuner power. 190, 125

tuning susceptance (anti-transmit-receive tube) (ATR tube). The normalized susceptance of the tube in its mount due to the deviation of its resonance frequency from the desired resonance frequency. *Note:* Normalization is with respect to the characteristic admittance of the transmission line at its junction with the tube mount. 125

tuning, thermal. The process of changing the operating frequency of a system by using a controlled thermal expansion to alter the geometry of the system. *See:* **oscillatory circuit.** 190, 125

tuning time constant, thermal. The time required for the frequency to change by a fraction $(1 - 1/e)$ of the change in equilibrium frequency after an incremental change of the applied thermal tuner power. *Notes:* (1) If the behavior is not exponential, the initial conditions must be stated. (2) Here e is the base of natural logarithms. *See:* **oscillatory circuit.** 190, 125

tuning time thermal (1) (cooling). The time required to tune through a specified frequency range when the tuner power is instantaneously changed from the specified maximum to zero. *Note:* The initial condition must be one of equilibrium. *See also:* **electron emission.** 190, 125

(2) (heating). The time required to tune through a specified frequency range when the tuner power is instantaneously changed from zero to the specified maximum. *Note:* The initial condition must be one of equilibrium. *See:* **electron emission.** 190, 125

turbine-control servomotor (speed governing systems, hydraulic turbines). The actuating element which moves the turbine-control mechanism in response to the action of the governor control actuator. Turbine-control servomotors are designated as: (1) gate servomotor; (2) blade servomotor; (3) deflector servomotor; (4) needle servomotor. 8, 58

turbine-driven generator. An electric generator driven by a turbine. *See also:* **direct-current commutating machine.** 328

turbine end (rotating machinery). The driven or power-input end of a turbine-driven generator. 63

turbine-generator. *See:* **cylindrical-rotor generator.**

turbine-generator unit. An electric generator with its driving turbine. *See also:* **direct-current commutating machine.** 328

turbine-nozzle control system (gas turbines). A means by which the turbine diaphragm nozzles are adjusted to vary the nozzle angle or area, thus varying the rate of energy input to the turbine(s). 98, 58

turbine, reversible. A hydraulic turbine, normally installed in a pumped storage station, which can be used alternately as a pump or as a prime mover. 112

turbine-type (rotating machinery). Applied to alternating-current machines designed for high-speed operation and having an excitation winding embedded in slots in a cylindrical steel rotor made from forgings or thick disks. *See:* **asynchronous machine.** 63

turbo-machine (turbo-generator) (rotating machinery). A machine of special design intended for high-speed operation. Turbo-generators usually are directly connected to gas or steam turbines. 63

Turing machine. A mathematical model of a device that changes its internal state and reads from, writes on, and moves a potentially infinite tape, all in accordance with its present state, thereby constituting a model for computerlike behavior. *See:* **universal Turing machine.** 255, 77

turn (rotating machinery). The basic coil element which forms a single conducting loop comprising one insulated conductor. *Note:* The conductor may consist of a number of strands or laminations. Each strand or lamination is in the form of a wire, rod, strip or bar, depending on its cross-section, and may be either uninsulated or insulated for the sole purpose of reducing eddy currents. 63

turnbuckle. A threaded device inserted in a tension member to provide minor adjustment of tension or sag. *See also:* **tower.** 64

turn error (gyro). An error in gyro output due to cross-coupling and acceleration encountered during vehicle turns. 46

turning gear (rotating machinery). A separate drive to rotate a machine at very low speed for the purpose of thermal equilization at a time when it would otherwise be at rest. *See also:* **rotor (rotating machinery).** 63

turn insulation (rotating machinery). Insulation applied to provide electrical separation between turns of a coil. *Note:* In the usual case, the insulation encircles each turn. However, in the case of edgewise-wound field coils for salient pole synchronous machines, the outer edges may be left bare to facilitate cooling. 63

turn-off thyristor. A thyristor that can be switched from the ON state to the OFF state and vice versa by applying control signals of appropriate polarities to the gate terminal, with the ratio of triggering power to triggered power appreciably less than one. *See also:* **thyristor.** 249, 66, 191

turn-off time (thyristor). *See:* **gate controlled turn-off time (thyristor); circuit commutated turn-off time (thyristor).**

turnover frequency. *See:* **transition frequency.**

turn ratio (1) (transformer). The ratio of the number of turns in a higher voltage winding to that in a lower voltage winding. *Note:* In the case of a constant-voltage transformer having taps for changing its voltage ratio, the nominal turn ratio is based on the number of turns corresponding to the normal rated voltage of the respective windings, to which operating and performance characteristics are referred. 53

(2) (constant-current transformer). The ratio of the number of turns in the primary winding to that in the secondary winding. *Note:* In case of a constant-current transformer having taps for changing its voltage ratio, the turn ratio is based on the number of turns corresponding to the normal rated voltage of the respective windings, to which operation and performance characteristics are referred. 303

(3) (potential transformer) (voltage transformer). The ratio of the primary winding turns to the secondary winding turns. 305, 203, 53

(4) (rectifier transformer). The ratio of the number of turns in the alternating-current winding to that in the direct-current winding. *Note:* The turn ratio is based on the number of turns corresponding to the normal rated voltage of the respective windings to which operating and performance characteristics are referred. *See also:* **rectifier transformer.** 258

(5) (current transformer). The ratio of the secondary winding turns to the primary winding turns. 53

turn separator (rotating machinery). An insulation strip between turns; a form of turn insulation. *See also:* **rotor (rotating machinery); stator.** 63

turns factor (magnetic core testing). Under stated conditions the number of turns that a coil of specified shape and dimensions placed on the core in a given position should have to obtain a given unit of self inductance. When measured with a measuring coil of the specified shape and dimensions and placed in the same position, it is defined as:

$$\propto = \frac{N}{\sqrt{L}}$$

$\propto$ = Turns factor
N = Number of turns of the measuring coil
L = Self inductance in henrys of the measuring coil placed on the core. 165

turn-signal operating unit. That part of a signal system by which the operator of a vehicle indicates the direction a turn will be made, usually by a flashing light. *See also:* **headlamp.** 167

turns per phase, effective (rotating machinery). The product of the number of series turns of each coil by the number of coils connected in series per phase and the winding factor. 63

turnstile antenna. An antenna composed of two dipole antennas, perpendicular to each other, with their axes intersecting at their midpoints. Usually, the currents on the two dipole antennas are equal and in phase quadrature. 111

turntable rumble (audio and electroacoustics). Low-frequency vibration mechanically transmitted to the recording or reproducing turntable and superimposed on the reproduction. *See also:* **phonograph pickup; rumble.** 339

turn-to-turn test (interturn test) (rotating machinery). A test for applying or more often introducing between adjacent turns of an insulated component, a voltage of predetermined amplitude, for the purpose of checking the integrity of the interturn insulation. 63

turn-to-turn voltage (rotating machinery). The voltage existing between adjacent turns of a coil. *See also:* **rotor (rotating machinery); stator.** 63

TV. *See:* **television.**

TV waveform signal (video signal transmission measurement). An electrical signal whose amplitude varies with time in a generally nonsinusoidal manner, whose shape (that is, duration and amplitude) carries the TV signal information. 42

twin cable. A cable composed of two insulated conductors laid parallel and either attached to each other by the insulation or bound together with a common covering. 64

twin-T network. *See:* **parallel-T network.**

twin wire. A cable composed of two small insulated conductors laid parallel, having a common covering. *See also:* **conductor.** 345

twisted-lead transposition (rotating machinery). A form of transposition used on a distributed armature winding wherein the strands comprising each turn are kept insulated from each other throughout all the coils in a phase belt, and the last half turn of each coil is given a 180-degree twist prior to connecting it to the first half turn of the next coil in the series. *See also:* **rotor (rotating machinery); stator.** 63

twisted pair. A cable composed of two small insulated conductors, twisted together without a common covering. *Note:* The two conductors of a twisted pair are usually substantially insulated, so that the combination is a special case of a cord. *See also:* **conductors.** 64

twist, waveguide (waveguide components). A waveguide section in which there is progressive rotation of the cross section about the longitudinal axis. *See also:* **step twist.** 166

two-address. Pertaining to an instruction code in which each instruction has two address parts. Some two-address instructions use the addresses to specify the location of one operand and the destination of the result, but more often they are one-plus-one-address instructions. 235

two-degree-freedom gyro. A gyro in which the rotor axis is free to move in any direction. *See also:* **navigation.** 187, 13

two-dimensional scanning. Scanning the beam of a directive antenna using two degrees of freedom to provide solid angle coverage. 111

two-element relay. An alternating-current relay that is controlled by current from two circuits through two co-operating sets of coils. 328

two-fluid cell. A cell having different electrolytes at the two electrodes. *See also:* **electrochemistry.** 328

two-frequency simplex operation (radio communication). The operation of a two-way radio-communication circuit utilizing two radio-frequency channels, one for each direction of transmission, in such manner that intelligence can be transmitted in only one direction at a time. *See also:* **channel spacing.** 181

two-layer winding (two-coil-side-per-slot winding). A winding in which there are two coil sides in the depth of a slot. *See also:* **rotor (rotating machinery); stator.** 63

two-out-of-five code. A code in which each decimal digit is represented by five binary digits of which two are one kind (for example, ones) and three are the other kind (for example, zeros). 255, 77

two-phase circuit (transformer). A polyphase circuit of three, four, or five distinct conductors intended to be so energized that in the steady state the alternating voltages between two selected pair of terminals of entry, other than the neutral terminal when one exists, have the same periods, are equal in amplitude, and have a phase difference of 90 degrees. When the circuit consists of five conductors, but not otherwise, one of them is a neutral conductor. *Note:* A **two-phase circuit** as defined here does not conform to the general pattern of polyphase circuits. Actually a two-phase, four-wire or five-wire circuit could more properly be called a four-phase circuit, but the term **two-phase** is in common usage. A two-phase, three-wire circuit is essentially a special case, as it does not conform to the general pattern of other polyphase circuits. *See also:* **network analysis.** 210, 53

two-phase five-wire system. A system of alternating-current supply comprising five conductors, four of which are connected as in a four-wire two-phase system, the fifth being connected to the neutral points of each phase. *Note:* The neutral is usually grounded. Although this type of system is usually known as the two-phase five-wire system, it is strictly a four-phase five-wire system. *See also:* **alternating-current distribution; network analysis.** 64

two-phase four-wire system. A system of alternating-current supply comprising two pairs of conductors between one pair of which is maintained an alternating difference of potential displaced in phase by one-quarter of a period from an alternating difference of potential of the same frequency maintained between the other pair. *See also:* **alternating-current distribution; network analysis.** 64

two-phase three-wire system. A system of alternating-current supply comprising three conductors between one of which (known as the common return) and each of the other two are maintained alternating differences of potential displaced in phase by one quarter of a period with relation to each other. *See also:* **alternating-current distribution; network analysis.** 64

two-plus-one address (electronic computation). Pertaining to an instruction that contains two operand addresses and a control address. *See:* **control address; instruction; oper-**

and; three-address code. 235, 255, 77

two quandrant dam (hybrid computer linkage components). A digital-to-analog multiplier (DAM) that multiplies with a single sign only for the digital value. 10

two rate meter. A meter having two sets of register dials with a changeover arrangement such that integration of the quantity will be registered on one set of dials for a specified number of hours each day and on the other set of dials for the remaining hours. *Note:* This is a special case of a multirate meter. *See also:* **electricity meter.** 212

twos complement. The radix complement of a numeral whose radix is two. *Note:* Using approved synonyms, the same definition may be expressed in the alternative form: "A true complement with a base of two." 235, 255, 77

two-signal selectivity (frequency-modulated mobile communications receivers). The characteristic that determines the extent to which the receiver is capable of differentiating between the desired signal and disturbances of signals at other frequencies. It is expressed as the amplitude ratio of the modulated desired signal and the unmodulated disturbing signal when the reference sensitivity sinad of the desired signal is degraded 6 decibels. 123

two-source frequency keying. That form of keying in which the modulating wave abruptly shifts the output frequency between predetermined values, where the values of output frequency are derived from independent sources. *Note:* Therefore, the output wave is not coherent and, in general, will have a phase discontinuity. *See also:* **telegraphy.** 111

two-speed alternating-current control. A control for two-speed driving-machine induction motor that is arranged to run near two different synchronous speeds by connecting the motor windings so as to obtain different numbers of poles. *See also:* **control (elevators).** 328

two-terminal capacitor. Two conductors separated by a dielectric. The construction is usually such that one conductor essentially surrounds the other and therefore the effect of the presence of other conductors, except in the immediate vicinity of the terminals, is eliminated. (Specialized usage). 210

two-terminal pair network (quadripole) (four-pole) (circuits and systems). A network with four accessible terminals grouped in pairs, for example, input pair, output pair. 67

two-tone keying. That form of keying in which the modulating wave causes the carrier to be modulated with a single tone for the marking condition and modulated with a different single tone for the spacing condition. *See also:* **telegraphy.** 111, 242, 194

$2T$ pulse (waveform test signals) (TV). A $\sin^2$ pulse with a half-amplitude duration (HAD) of 250 ns. The amplitude of the envelope of the frequency spectrum at 2 MHz is 0.5 of the amplitude at zero frequency and effectively zero at and beyond 4 MHz. *Note:* The $2T$ pulse is mentioned here for the sake of completeness. The short-time domain may be tested by the $2T$ pulse in conjunction with the T pulse and a reference signal. This method is not used in this standard since the T step alone tests the short-time domain in a simpler and more direct manner. 42

two-value capacitor motor. A capacitor motor using different values of effective capacitance for the starting and

running conditions. *See also:* **asynchronous machine.** 63

two-way automatic maintaining leveling device. A device that corrects the car level on both underrun and overrun, and maintains the level during loading and unloading. *See also:* **elevator-car leveling device.** 328

two-way automatic nonmaintaining leveling device. A device that corrects the car level on both underrun and overrun, but will not maintain the level during loading and unloading. *See also:* **elevator-car leveling device.** 328

two-way correction (industrial control). A method of register control that effects a correction in register in either direction. 204

two-way trunk (telephone switching systems). A trunk between two switching entities used for calls that originate from either end. 55

two-wire circuit. A metallic circuit formed by two adjacent conductors insulated from each other. *Note:* Also used in contrast with **four-wire circuit** to indicate a circuit using one line or channel for transmission of electric waves in both directions. *See also:* **center of distribution; transmission line.** 64, 122

two-wire control (industrial control). A control function which utilizes a maintained-contact type of pilot device to provide undervoltage release. *See:* **undervoltage release.** *See also:* **control.** 225, 206

Two-wire control.

two-wire repeater. A telephone repeater which provides for transmission in both directions over a two-wire telephone circuit. *Note:* In practice this may be either a 21-type repeater or a 22-type repeater. *See also:* **repeater.** 59

two-wire switching (telephone switching systems). Switching using the same path, frequency, or time interval for both directions of transmission. 55

two-wire system. *See:* **two-wire circuit.**

type-A display; type-B display etcetera (radar). *See: A* display; B display; etcetera.

type font. A type face of a given size and design, for example, 10-point Bodoni Book Medium; 9-point Gothic. 255, 77

type of emission (mobile communication). A system of designating emission, modulation, and transmission characteristics of radio-frequency transmissions, as defined by the Federal Communications Commission. *See also:* **mobile communication system.** 181

type of piezoelectric crystal cut. The orientation of a piezoelectric crystal plate with respect to the axes of the crystal. It is usually designated by symbols. For example, *GT, AT, BT, CT,* and *DT* identify certain quartz crystal cuts having very low temperature coefficients. *See also:* **crystal.** 328

type of service (industrial control). The specific type of application in which the controller is to be used, for example: (1) general purpose; (2) special purpose, namely, crane and hoist, elevator, steel mill, machine tool, printing press, etcetera. *See also:* **electric controller.** 206

type tests (1) (rotating machinery). The performance tests taken on the first machine of each type of design. *See:* **asynchronous machine; direct-current commutating machine.** 63

(2) (nuclear power generating stations). Tests made on one or more sample equipments to verify adequacy of design and the manufacturing processes. 31, 120

U

ultimate gain. *See:* **frequency, undamped.**

ultimately controlled variable (control) (industrial control). The variable the control of which is the end purpose of the automatic control system. *See also:* **control system, feedback.** 219, 206, 329

ultimate mechanical strength (insulator). The load at which any part of the insulator fails to perform its function of providing a mechanical support without regard to electrical failure. *See also:* **insulator.** 261

ultimate period. *See:* **frequency, undamped.**

ultra-audible frequency (supersonic frequency)*. *See:* **ultrasonic frequency.**

* Obsolescent

ultra-audion oscillator. *See:* **Colpitts oscillator.**

ultra-high-frequency radar (UHF radar). A radar operating at frequencies between 300 to 1000 megahertz, usually in one of the International Telecommunications Union assigned bands 420 to 450 or 890 to 942 megahertz. Radars between 1 gigahertz and 3 gigahertz, although within the ultrahigh-frequency band as defined by International Telecommunications Union, are described as L-band or S-band radars, as appropriate. 13

ultra-high-voltage system (transformer). An electric system having a maximum rms ac voltage above 800,000 volts to 2,000,000 volts. 53

ultrasonic cross grating (grating). A space grating resulting from the crossing of beams of ultrasonic waves having different directions of propagation. *Note:* The grating may be two- or three-dimensional. 176

ultrasonic delay line. A transmission device, in which use is made of the propagation time of sound to obtain a time delay of a signal. 176

ultrasonic frequency (ultra-audible frequency) (supersonic frequency)*. A frequency lying above the audio-frequency range. The term is commonly applied to elastic waves propagated in gases, liquids, or solids. *Note:* The word **ultrasonic** may be used as a modifier to indicate a device or system employing or pertaining to ultrasonic frequencies. The term **supersonic,** while formerly applied to frequency, is now generally considered to pertain to velocities above those of sound waves. Its use as a synonym of ultrasonic is now deprecated. *See also:* **signal wave.** 176

* Obsolescent

ultrasonic generator. A device for the production of sound waves of ultrasonic frequency. 328

ultrasonic grating constant. The distance between diffracting centers of the sound wave that is producing particular light diffraction spectra. 176

ultrasonic light diffraction. Optical diffraction spectra or the process that forms them when a beam of light is passed

through the field of a longitudinal wave. 176

ultrasonic space grating (grating). A periodic spatial variation of the index of refraction caused by the presence of acoustic waves within the medium. 176

ultrasonic stroboscope. A light interrupter whose action is based on the modulation of a light beam by an ultrasonic field. 176

ultraviolet flame detector (fire protection devices). A device whose sensing element is responsive to radiant energy outside the range of human vision (below approximately 4000 Angstroms). 71

ultraviolet radiation (1) (illuminating engineering). For practical purposes, any radiant energy within the wavelength range 10 to 380 nanometers. *Note:* On the basis of practical applications and the effect obtained, the ultraviolet region often is divided into the following bands: ozone-producing, 180-220 nanometers; bactericidal (germicidal), 220-300 nanometers; erythemal, 280-320 nanometers; black light, 320-400 nanometers. There are no sharp demarcations between these bands, the indicated effects usually being produced to a lesser extent by longer and shorter wave-lengths. For engineering purposes, the black-light region extends slightly into the visible portion of the spectrum. 167

(2) (laser-maser). Electromagnetic radiation with wavelengths smaller than those for visible radiation. For the purposes of this standard, 0.2–0.4 μm. 363

umbrella antenna. A type of top-loaded vertical antenna where the top-loading structure has an appearance similar to the ribs on an umbrella. 111

unary operation (computing machines). *See:* **monadic operation.**

unattended automatic exchange (CDO or CAX). A normally unattended telephone exchange, wherein the subscribers, by means of calling devices, set up in the central office the connections to other subscribers or to a distant central office. 328

unavailability (nuclear power generating stations). (1) The numerical complement of availability. Unavailability may occur as a result of the item being repaired, for example, repair unavailability or it may occur as a result of undetected malfunctions, for example, unannounced unavailability. 29, 31, 75, 366

(2) The numerical complement of availability. Unavailability may occur as a result of the item being repaired (repair unavailability), tested (testing unavailability), or it may occur as a result of undetected malfunctions (unannounced unavailability). 357

unavailability margin (nuclear power generating station). The favorable difference between the desired goal and the calculated or observed unavailability. 357

unbalance. A differential mutual impedance or mutual

admittance between two circuits that ideally would have no coupling. *See also:* **coupling.**   328

unbalanced circuit. A circuit, the two sides of which are inherently electrically unlike with respect to a common reference point, usually ground. *Note:* Frequently, unbalanced signifies a circuit, one side of which is grounded.   239, 188

unbalanced modulator (signal-transmission system). *See:* **modulator, asymmetrical.** *See also:* **signal.**

unbalanced strip line. *See:* **strip (strip-type) transmission line.**

unbalanced wire circuit. A wire circuit whose two sides are inherently electrically unlike. *See also:* **transmission line.**   64

unbiased telephone ringer. A telephone ringer whose clapper-driving element is not normally held toward one side or the other, so that the ringer will operate on alternating current. Such a ringer does not operate reliably on pulsating current. *Note:* A ringer that is weakly biased so as to avoid tingling when dial pulses pass over the lines may be referred to as an unbiased ringer. *See also:* **telephone station.**   328

unblanking. Turning on of the cathode-ray-tube beam. *See:* **oscillograph.**   185

uncertainty (1) (general). The estimated amount by which the observed or calculated value of a quantity may depart from the true value. *Note:* The uncertainty is often expressed as the average deviation, the probable error, or the standard deviation. *See also:* **measurement system; measurement uncertainty.**   185, 54
(2) (electrothermic power meters). The assigned allowance for the systematic error, together with the random error attributed to the imprecision of the measurement process.   184, 47, 115

unconditional jump (unconditional transfer of control) (electronic computation). An instruction that interrupts the normal process of obtaining instructions in an ordered sequence and specifies the address from which the next instruction must be taken. *See also:* **jump.**   235, 210

unconditional transfer of control. *See:* **unconditional jump.**

underbunching. A condition representing less than optimum bunching. *See also:* **electron devices, miscellaneous.**   190, 125

undercounter dumbwaiter. A dumbwaiter that has its top terminal landing located underneath a counter and that serves only this landing and the bottom terminal landing.   328

undercurrent relay (1) (general). A relay that operates when the current through the relay is equal to or less than its setting. *See also:* **relay.**   259, 60, 202
(2) (power switchgear). A relay that operates when the current is less than a predetermined value.   103

undercurrent release (undercurrent trip). A release that operates when the current in the main circuit is equal to or less than the release setting.   103, 202

undercurrent trip. *See:* **undercurrent release.**

underdamped. Damped insufficiently to prevent oscillation of the output following an abrupt input stimulus. *Note:* In an underdamped linear second-order system, the roots of the characteristic equation have complex values. *See:* **damped harmonic system.**   219, 206, 329

underdamped period (instrument) (periodic time). The time between two consecutive transits of the pointer or indicating means in the same direction through the rest position, following an abrupt change in the measurand.   328

underdamping (periodic damping*). The special case of damping in which the free oscillation changes sign at least once. A damped harmonic system is underdamped if $F^2 < MS$. See **damped harmonic system** for equation, definitions of letter symbols, and referenced terms.   210

* Deprecated

underdome bell. A bell whose mechanism is mostly concealed within its gong. *See also:* **protective signaling.**   328

underfilm corrosion. Corrosion that occurs under films in the form of randomly distributed hairlines (filiform corrosion).   221, 205

underfloor raceway. A raceway suitable for use in the floor. *See also:* **raceway.**   328

underflow (computing machines). Pertaining to the condition that arises when a machine computation yields a nonzero result that is smaller than the smallest nonzero quantity that the intended unit of storage is capable of storing.   155, 77

underground cable. A cable installed below the surface of the ground. *Note:* This term is usually applied to cables installed in ducts or conduits or under other conditions such that they can readily be removed without disturbing the surrounding ground. *See also:* **cable; tower.**   64

underground collector or plow. A current collector, the function of which is to make contact with an underground contact rail. *See also:* **contact conductor.**   1

underlap, X (facsimile). The amount by which the center-to-center spacing of the recorded spots exceeds the recorded spot X dimension. *Note:* This effect arises in that type of equipment which responds to a constant density in the subject copy by a succession of discrete recorded spots. *See also:* **recording (facsimile).**   12

underlap, Y (facsimile). The amount by which the nominal line width exceeds the recorded spot Y dimension. *See also:* **recording (facsimile).**   12

underreaching protection. A form of protection in which the relays at a given terminal do not operate for faults at remote locations on the protected equipment, the given terminal being cleared either by other relays with different performance characteristics or by a transferred trip signal from a remote terminal similarly equipped with underreaching relays.   103, 60, 202, 127

undershoot (1) (television). That part of a distorted wave front characterized by a decaying approach to the final value. *Syn:* **rounding.** *Note:* Generally undershoots are produced in transfer devices having insufficient transient response.   163
(2) (oscilloscopes). In the display of a step function (usually of time), that portion of the waveform that, following any overshoot or rounding that may be present, falls below its nominal or final value.   106

underslung car frame. A car frame to which the hoisting-rope fastenings or hoisting rope sheaves are attached at or below the car platform. *See also:* **hoistway (elevator or dumbwaiter).**   328

underspeed (speed governing of hydraulic turbines). Any speed below rated speed expressed as a percent of rated speed. 8

undervoltage protection (1) (industrial control). The effect of a device, operative on the reduction or failure of voltage, to cause and maintain the interruption of power to the main circuit. *Note:* The principal objective of this device is to prevent automatic restarting of the equipment. Standard undervoltage or low-voltage protection devices are not designed to become effective at any specific degree of voltage reduction. 60, 202, 206, 3, 124
(2) (power switchgear) (low-voltage protection). A form of protection that operates when voltage is less than a predetermined value. 103, 127

undervoltage relay (1) (general). A relay that operates when the voltage applied to the relay is equal to or less than its setting. *See:* **relay.** 259, 60, 202, 206
(2) (power switchgear). A relay that operates when its voltage is less than a predetermined value. 103, 127

undervoltage release (1) (industrial control). The effect of a device, operative on the reduction or failure of voltage, to cause the interruption of power to the main circuit but not to prevent the re-establishment of the main circuit on return of voltage. *Note:* Standard undervoltage or low-voltage release protection devices are not designed to become effective at any specific degree of voltage reduction. *Syn:* **low voltage release.** 206, 3
(2) (power switchgear) (trip). A release that operates when the voltage of the main circuit is equal to or less than the release setting. 103

underwater log. A device that indicates a ship's speed based on the pressure differential, resulting from the motion of the ship relative to the water, as developed in a Pitot tube system carried by a retractable support extending through the ship's hull. Continuous integration provides indication of total distance travelled. The ship's draft is indicated, based on static pressure. 328

underwater sound projector. A transducer used to produce sound in water. *Notes:* (1) There are many types of underwater sound projectors whose definitions are analogous to those of corresponding loudspeakers, for example, crystal projector, magnetic projector, etcetera. (2) Where no confusion will result, the term underwater sound projector may be shortened to projector. *See also:* **microphone.** 176

undesired conducted power (frequency-modulated mobile communications receivers). Radio-frequency power that is present at the antenna, power terminals, or any other interfacing terminals. 123

undesired radiated power (frequency-modulated mobile communications receivers). Radio-frequency power radiated from the receiver that can be measured outside a specified area. 123

undetected error rate (data transmission). The ratio of the number of bits, unit elements, characters, blocks incorrectly received but undetected or uncorrected by the error-control equipment, to the total number of bits, unit elements, characters, blocks sent. 194

undisturbed-ONE output (magnetic cell). A ONE output to which no partial-read pulses have been applied since that cell was last selected for writing. *See also:* **coincident-current selection.** 331

undisturbed-ZERO output (magnetic cell). A ZERO output to which no partial-write pulses have been applied since that cell was last selected for reading. *See also:* **coincident-current selection.** 331

unfired tube (microwave gas tubes). The condition of the tube during which there is no radio-frequency glow discharge at either the resonant gap or resonant window. *See:* **gas tubes.** 190, 125

ungrounded (electric power). A system, circuit, or apparatus without an intentional connection to ground except through potential-indicating or measuring devices or other very-high-impedance devices. *Note:* Though called ungrounded, this type of system is in reality coupled to ground through the distributed capacitance of its phase windings and conductors. In the absence of a ground fault, the neutral of an ungrounded system under reasonably balanced load conditions will usually be close to ground potential, being held there by the balanced electrostatic capacitance between each phase conductor and ground. 152

unguarded release (telephone switching systems). A condition during the restoration of a circuit to its idle state when it can be prematurely seized. 55

uniconductor waveguide. A waveguide consisting of a cylindrical metallic surface surrounding a uniform dielectric medium. *Note:* Common cross-sectional shapes are rectangular and circular. *See also:* **waveguide.** 267, 319, 179

unidirectional. A connection between telegraph sets, one of which is a transmitter and the other a receiver. 194

unidirectional antenna. An antenna that has a single well-defined direction of maximum gain. *See also:* **antenna.** 111, 179

unidirectional bus (programmable instrumentation). A bus used by any individual device for one-way transmission of messages only, that is, either input only or output only. 40

unidirectional current. A current that has either all positive or all negative values. 210

unidirectional microphone. A microphone that is responsive predominantly to sound incident from a single solid angle of one hemisphere or less. *See also:* **microphone.** 176

unidirectional pulse train (signal-transmission system). Pulses in which pertinent departures from the normally constant value occur in one direction only. *See also:* **pulse.** 254, 188

unidirectional transducer (unilateral transducer). A transducer that cannot be actuated at its output by waves in such a manner as to supply related waves at its input. *See also:* **transducer.** 252, 210

unified atomic mass unit (u). The unit equal to the fraction $1/12$ of the mass of an atom of the nuclide ^{12}C; $1 u = 1.660 53 \times 10^{-27}$ kg approximately. 21

unified s-band system (communication satellite). A communication system using an S-band carrier (2000–2300 megahertz) combining all links into one spectrum. The functions of spacecraft command, data transmission, tracking, ranging, etcetera, are transmitted on separate carrier frequencies for earth-space and space-earth links. 84

uniform current density. A current density that does not change (either in magnitude or direction) with position within a specified region. (A uniform current density may be a function of time.) 210

uniform field. A field (scalar or vector or other) in which the field quantities do not vary with position. 210

uniform line. A line that has substantially identical electrical properties throughout its length. *See also:* **transmission line.** 328

uniform linear array (antenna). A linear array of identically oriented and equally spaced radiating elements having equal current amplitudes and equal phase increments between excitation currents. *See also:* **antenna.**
 179, 111

uniform luminance area. The area in which a display on a cathode-ray tube retains 70 percent or more of its luminance at the center of the viewing area. *Note:* The corners of the rectangle formed by the vertical and horizontal boundaries of this area may be below the 70-percent luminance level. *See:* **oscillograph.** 185

uniform plane wave (radio wave propagation). A plane wave in which the electric and magnetic intensities have constant amplitude over the equiphase surfaces. *Note:* Such a wave can only be found in free space at an infinite distance from the source. *See also:* **radiation; radio wave propagation.** 146, 180

uniform waveguide (waveguide components). A waveguide or transmission line in which the physical and electrical characteristics do not change with distance along the axis of the guide. *See:* **waveguide.** 267, 179, 166

unilateral area track. A sound track in which one edge only of the opaque area is modulated in accordance with the recorded signal. There may, however, be a second edge modulated by a noise-reduction device. *See also:* **phonograph pickup.** 176

unilateral connection (control system, feedback). A connection through which information is transmitted in one direction only. *See also:* **control system, feedback.** 329

unilateral network. A network in which any driving force applied at one pair of terminals produces a nonzero response at a second pair but yields zero response at the first pair when the same driving force is applied at the second pair. *See also:* **network analysis.** 210

unilateral transducer. *See:* **unidirectional transducer.**

uninhibited oil (neutral grounding devices) (transformer). Mineral transformer oil to which no synthetic oxidation inhibitor has been added. 91, 53

unipolar (power supplies). Having but one pole, polarity, or direction. Applied to amplifiers or power supplies, it means that the output can vary in only one polarity from zero and, therefore, must always contain a direct-current component. *See:* **bipolar.** *See also:* **power supply.**
 228, 186

unipolar electrode system (electrobiology) (monopolar electrode system). Either a pickup or a stimulating system, consisting of one active and one dispersive electrode. *See also:* **electrobiology.** 192

unipole*. *See:* **antenna; isotropic antenna.**

* Deprecated

unipotential cathode. *See:* **cathode, indirectly heated.**

unit (1) (physical unit). An amount of a physical quantity arbitrarily assigned magnitude unity and determined by a specified relationship to one or more prototype standards.
 210

(2) (general). A device having a special function.
 255, 77

(3) (electronic computation). A portion of a computer that constitutes the means of accomplishing some inclusive operation or function, as **arithmetic unit.** *See also:* **arithmetic unit; processing unit; control unit; tape unit.**
 235, 210

(4) (relay). *See:* **relay unit.**

(5) (switchgear). That portion of the switchgear assembly which contains one switching device such as a circuit breaker, interrupter switch, power fuse interrupter switch combination, etcetera and the associated primary conductors. 79, 93, 103

(6) (nuclear power-generating stations). A nuclear steam supply, its associated turbine-generator, auxiliaries and engineered safety features. 31, 102

(7) (generating station). The generator(s), associated prime mover(s), auxiliaries and energy supply(ies) that are normally operated together as a single source of electric power. 58

(8) (electric and electronics parts and equipments). A major building block for a set or system, consisting of a combination of basic parts, subassemblies, and assemblies packaged together as a physically independent entity. The application, size, and construction of an item may be factors in determining whether an item is regarded as a unit, an assembly, a subassembly, or a basic part. A small electric motor might be considered as a part if it is not normally subject to disassembly. Typical examples are: radio receiver, radio transmitter, electronic power supply, antenna. 17

unit-area capacitance (electrolytic capacitor). The capacitance of a unit area of the anode surface at a specified frequency after formation at a specified voltage. *See also:* **electrolytic capacitor.** 328

unit auxiliaries transformer. A transformer intended primarily to supply all or a portion of the unit auxiliaries.
 58

unit auxiliary. An auxiliary intended for a specific generating unit. 58

unit cable construction. That method of cable manufacture in which the pairs of the cable are stranded into groups (units) containing a certain number of pairs and these groups are then stranded together to form the core of the cable. *See also:* **cable.** 328

unit-control error (electric power systems). The unit generation minus the assigned unit generation. *Note:* Refer to note on polarity under **area control error.** 94

unit-impulse function. *See:* **signal, unit-impulse.**

unit interval (data transmission). The longest interval of time such that the nominal durations of the signal elements in a synchronous system or the start and information elements in a start-stop system are whole multiples of this interval. *Note:* A unit interval is the shortest nominal signal element. 59

unit operation (1) (surge arrester). Discharging a surge through an arrester while the arrester is energized. *See also:* **current rating, 60-hertz (arrester).** 2, 299, 62

(2) (CO) (circuit breaker). *See:* **close-open operation (switching device).**

(3) (recloser). An interrupting operation followed by a

closing operation. The final interruption is also considered one unit operation.   103, 202

unit-ramp function. *See:* **signal, unit-ramp.**

unit rate-limiting controller (electric power systems). A controller that limits rate of change of generation of a generating unit to an assigned value or values. *Note:* The limiting action is normally based on a measured megawatt-per-minute rate. *See also:* **speed-governing system.**
     94

units and letter symbols (International System of Units) (SI). The three classes of SI units are: (**1**) **base units,** regarded by convention as dimensionally independent:

Quantity	Unit	Symbol
length	meter	m
mass	kilogram	kg
time	second	s
electric current	ampere	A
thermodynamic temperature	kelvin	K
amount of substance	mole	mol
luminous intensity	candela	cd

(**2**) supplementary units, regarded either as base units or as derived units:

Quantity	Unit	Symbol
plane angle	radian	rad
solid angle	steradian	sr

(**3**) **derived units,** formed by combining base units, supplementary units, and other derived units according to the algebraic relations linking the corresponding quantities. The symbols for derived units are obtained by means of the mathematical signs for multiplication, division, and use of exponents. Derived units which have special names and symbols approved by the General Conference on Weights and Measures (CGPM) are:

units of luminance, (photometric brightness) (light emitting diodes). The luminance (photometric brightness) of a surface in a specified direction may be expressed in luminous intensity per unit of projected area of surface. *Note:* Typical units in this system are the candela per square meter.   162

units of wavelength. The distance between two successive points of a periodic wave in the direction of propagation, in which the oscillation has the same phase. The three commonly used units are listed in the following table:

Name	Symbol	Value
micrometer	μm	$1\ \mu m = 10^{-3}$ millimeters
nanometer	nm	$1\ nm = 10^{-6}$ millimeters
angstrom	Å	$1\ Å = 10^{-7}$ millimeters

See also: **radiant energy.**   167

unit-step function. *See:* **signal, unit-step.**

unit substation. A substation consisting primarily of one or more transformers which are mechanically and electrically connected to and coordinated in design with one or more switchgear or motor control assemblies, or combinations thereof. *See also:* **primary unit substation; secondary unit substation; integral unit substation; articulated unit substation; radial type; distributed network type; spot network type; secondary selective type; duplex type.**

unit-substation transformer. A transformer which is mechanically and electrically connected to, and coordinated in design with, one or more switchgear or motor-control assemblies, or combinations thereof. *See:* **unit substation; primary unit substation; secondary unit substation; integral unit substation; articulated unit substation.**   53

unit symbol (abbreviation). *See:* **symbol for a unit.**

unit test (switchgear). A test performed on a single unit or group of units. *Note:* One widespread use of such tests is extrapolation of test results for the purpose of representing overall performance of a device composed of several units.
     79

unit transformer. A power system supply transformer which transforms all or a portion of the unit power from

Quantity	Unit	Symbol	Formula
frequency (of a periodic phenomenon)	hertz	Hz	$1/s$
force	newton	N	$kg \cdot m/s^2$
pressure, stress	pascal	Pa	N/m^2
energy, work, quantity of heat	joule	J	$N \cdot m$
power, radiant flux	watt	W	J/s
quantity of electricity, electric charge	coulomb	C	$A \cdot s$
electric potential, potential difference, electromotive force	volt	V	W/A
capacitance	farad	F	C/V
electric resistance	ohm	Ω	V/A
conductance	siemens	S	A/V
magnetic flux	weber	Wb	$V \cdot s$
magnetic flux density	tesla	T	Wb/m^2
inductance	henry	H	Wb/A
luminous flux	lumen	lm	$cd \cdot sr$
illuminance	lux	lx	lm/m^2
activity (of radionuclides)	becquerel	Bq	$1/s$
absorbed dose	gray	Gy	J/kg

See also: **prefixes and symbols.**

Some Common Derived Units of the International System of Units

Quantity	Unit	Symbol
acceleration	meter per second squared	m/s^2
angular acceleration	radian per second squared	rad/s^2
angular velocity	radian per second	rad/s
area	square meter	m^2
concentration (of amount of substance)	mole per cubic meter	mol/m^3
current density	ampere per square meter	A/m^2
density, mass	kilogram per cubic meter	kg/m^3
electric charge density	coulomb per cubic meter	C/m^3
electric field strength	volt per meter	V/m
electric flux density	coulomb per square meter	C/m^2
energy density	joule per cubic meter	J/m^3
entropy	joule per kelvin	J/K
heat capacity	joule per kelvin	J/K
heat flux density	watt per square meter	W/m^2
irradiance	watt per square meter	W/m^2
luminance	candela per square meter	cd/m^2
magnetic field strength	ampere per meter	A/m
molar energy	joule per mole	J/mol
molar entropy	joule per mole kelvin	$J/(mol \cdot K)$
molar heat capacity	joule per mole kelvin	$J/(mol \cdot K)$
moment of force	newton meter	$N \cdot m$
permeability	henry per meter	H/m
permittivity	farad per meter	F/m
radiance	watt per square meter steradian	$W/(m^2 \cdot sr)$
radiant intensity	watt per steradian	W/sr
specific heat capacity	joule per kilogram kelvin	$J/(kg \cdot K)$
specific energy	joule per kilogram	J/kg
specific entropy	joule per kilogram kelvin	$J/(kg \cdot K)$
specific volume	cubic meter per kilogram	m^3/kg
surface tension	newton per meter	N/m
thermal conductivity	watt per meter kelvin	$W/(m \cdot K)$
velocity	meter per second	m/s
viscosity, dynamic	pascal second	$Pa \cdot s$
viscosity, kinematic	square meter per second	m^2/s
volume	cubic meter	m^3
wavenumber	1 per meter	$1/m$

Units in Use with the International System of Units

Quantity	Unit	Symbol	Definition
time	minute	min	1 min = 60 s
	hour	h	1 h = 60 min = 3600 s
	day	d	1 d = 24 h = 86 400 s
	other calendar	—	
plane angle	degree	°	$1° = (\pi/180)$rad
temperature	degree Celsius	°C	
volume	liter[a]	l	$1 l = 1 dm^3 = 10^{-3} .m^3$
mass	metric ton	t	1 t = 10^3 kg

[a] Because of the similarity between the lower case letter l and the number 1 in many type fonts it is recommended that when confusion might result the word "liter" be written in full.   21

the unit to the power system voltage.   58

unit under test (UUT) (test, measurement and diagnostic equipment). Any system, set, subsystem, assembly, subassembly, and so forth, undergoing testing.   54

unit vector. A vector whose magnitude is unity.   210

unit warmup time (power supply). The interval between the time of application of input power to the unit and the time at which the regulated power supply is supplying regulated power at rated output voltage. *See also:* **regulated power supply.**   347

unity-gain bandwidth (power supplies). A measure of the gain-frequency product of an amplifier. Unity-gain bandwidth is the frequency at which the open-loop gain becomes unity, based on a 6-decibel-per-octave crossing. See the acompanying figure. *See also:* **power supply.**

228, 186

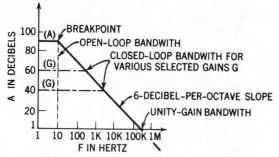

Typical gain-frequency (Bode) plot, showing unity-gain bandwidth.

unity power-factor test (synchronous machine). A test in which the machine is operated as a motor under specified operating conditions with its excitation adjusted to give unity power factor. 63

univalent function. If to every value of u there corresponds one and only one value of x (or one and only one set of values of $x_1, x_2, \cdots, x_n$) then u is a univalent function. Thus $u^2 = ax + b$ is univalent, within the interval of definition. 210

universal fuse links. Fuse links that, for each rating, provide mechanical and electrical interchangeability within prescribed limits over the specified time-current range.

103, 202

universal motor. A series-wound or compensated series-wound motor designed to operate at approximately the same speed and output on either direct current or single-phase alternating current within a specified frequency range and at the same root-mean-square voltage. *See:* **asynchronous machine; direct-current commutating machine.**

63, 3

universal-motor parts (rotating machinery). A term applied to a set of parts of a universal motor. Rotor shaft, conventional stator frame (or shell), end shields, or bearings may not be included, depending on the requirements of the end product into which the universal-motor parts are to be assembled. *See also:* **asynchronous machine; direct-current commutating machine.** 63

universal-numbering plan (telephone switching systems). A numbering plan employing nonconflicting codes so arranged that all main stations can be reached from any point within a telecommunications system. 55

universal or arcshear machine. A power-driven cutter that will not only cut horizontal kerfs, but will also cut vertical kerfs or at any angle, and is designed for operation either on track, caterpillar treads, or rubber tires. 328

universal Turing machine. A Turing machine that can simulate any other Turing machine. 255, 77

unloaded applicator impedance (dielectric heating usage). The complex impedance measured at the point of application, without the load material in position, at a specified frequency. *See also:* **dielectric heating.** 14, 114

unloaded (intrinsic) Q (switching tubes). The Q of a tube

unloaded by either the generator or the termination. *Note:* As here used, Q is equal to 2π times the energy stored at the resonance frequency divided by the energy dissipated per cycle in the tube or, for cell-type tubes, in the tube and its external resonant circuit. *See also:* **gas tubes.**

190, 125

unloaded sag (conductor or any point in a span). The distance measured vertically from the particular point in the conductor to a straight line between its two points of support, without any external load. 263

unloading amplifier (analog computer). An amplifier that is capable of reproducing or amplifying a given voltage signal while drawing negligible current from the voltage source. *Note:* In an analog computer, the term **buffer amplifier** is sometimes used as a synonym for **unloading amplifier,** in an incorrect sense, since a buffer amplifier draws significant current, but at a constant load impedance (seen at the input). *See also:* **electronic analog computer.** 9

unloading circuit (analog computer). A computing element or combination of computing elements capable of reproducing or amplifying a given voltage signal while drawing negligible current from the voltage source, thus decreasing the loading errors. *See also:* **unloading amplifier.** 9

unloading point (electric transmission system used on self-propelled electric locomotives or cars). The speed above or below which the design characteristics of the generators and traction motors or the external control system, or both, limit the loading of the prime mover to less than its full capacity. *Note:* The unloading point is not always a sharply defined point, in which case the unloading point may be taken as the useful point at which essentially full load is provided. *See also:* **traction motor.** 328

unmodulated groove (blank groove) (mechanical recording). A groove made in the medium with no signal applied to the cutter. *See also:* **phonograph pickup.** 176

un-ordered access (communication satellite). A system in which access to a radio frequency channel is gained without determining channel availability. This method is useful in common spectrum or random access discrete address systems. 84

unpack. To separate various sections of packed data.

255, 77

unpropagated potential (electrobiology). An evoked transient localized potential not necessarily associated with changed excitability. *See also:* **excitability (electrobiology).** 328

unregulated voltage (electronically regulated power supply). The voltage at the output of the rectifier filter. *See also:* **regulated power supply.** 347

unsaturated standard cell. A cell in which the electrolyte is a solution of cadmium sulphate at less than saturation at ordinary temperatures. (This is the commercial type of cadmium standard cell commonly used in the United States). *See also:* **electrochemistry.** 328

unshielded strip transmission line. A strip conductor above a single ground plane. Some common designations are. **microstrip (flat-strip conductor); unbalanced strip line.*** *See:* **strip (type) transmission line; shielded strip transmission line; waveguides.** 179

* Deprecated

unstable (control system, feedback). Not possessing stability. *See also:* **control system, feedback.** 56, 329

unusual service conditions. Environmental conditions that may affect the constructional or operational requirements of a machine. This includes the presence of moisture and abrasive, corrosive, or explosive atmosphere. It also includes external structures that limit ventilation, unusual conditions relating to the electrical supply, the mechanical loading, and the position of the machine. 63

uplift roller (conductor stringing equipment). A small single groove wheel designed to fit in or immediately above the throat of the traveler and keep the pulling line in the traveler groove when "uplift" occurs due to stringing tensions. 45

up link (communication satellite). A ground to satellite link, very often the command link. 83

upper (driving) beams. One or more beams intended for distant illumination and for use on the open highway when not meeting other vehicles. Formally **country beam.** *See also:* **headlamp.** 167

upper bracket (rotating machinery). A bearing bracket mounted above the core of a vertical machine. 63

upper burst reference (audio and electroacoustics). A selected multiple of the long-time average magnitude of the quantity mentioned in the definition of **burst.** See the figure attached to the definition of **burst duration.** *See also:* **burst (audio and electroacoustics).** 253, 176

upper coil support (rotating machinery). A coil support to restrain field-coil motion in the direction toward the air gap. *See also:* **rotor (rotating machinery); stator.** 63

upper frequency limit (coaxial transmission line). The limit determined by the cutoff frequency of higher-order waveguide modes of propagation, and the effect that they have on the impedance and transmission characteristics of the normal TEM coaxial-transmission-line mode. The lowest cutoff frequency occurs with the $TE_{1,1}$ mode, and this cutoff frequency in air dielectric line is the upper frequency limit of a practical transmission line. How closely the $TE_{1,1}$ mode cutoff frequency can be approached depends on the application. *See also:* **waveguide.** 265

upper guide bearing (rotating machinery). A guide bearing mounted above the core of a vertical machine. *See also:* **bearing.** 63

upper half bearing bracket (rotating machinery). The top half of a bracket that can be separated into halves for mounting or removal without access to a shaft end. *See also:* **bearing.** 63

upper limit (test, measurement and diagnostic equipment). The maximum acceptable value of the characteristic being measured. 54

upper range-value. The highest quantity that a device is adjusted to measure. *Note:* The following compound terms are used with suitable modifications in the units: **measured variable upper range-value, measured signal upper range-value,** etcetera. *See also:* **instrument.** 295

upper sideband (data transmission). The higher of two frequencies or groups of frequencies produced by a modulation process. 59

upper-sideband parametric down-converter. A noninverting parametric device used as a parametric downconverter. *See also:* **parametric device.** 277, 191

upper-sideband parametric up-converter. A noninverting parametric device used as a parametric up-converter. *See also:* **parametric device.** 277, 191

upset duplex system. A direct-current telegraph system in which a station between any two duplex equipments may transmit signals by opening and closing the line circuit, thereby causing the signals to be received by upsetting the duplex balance. *See also:* **telegraphy.** 328

up time (availability). The period of time during which an item is in a condition to perform its required function. 164

upward component (illuminating engineering). That portion of the luminous flux from a luminaire that is emitted at angles above the horizontal. *See also:* **inverse-square law (illuminating engineering).** 167

urban districts (transmission and distribution). Thickly settled areas (whether in cities or suburbs) or where congested traffic often occurs. A highway, even though in thinly settled areas, on which the traffic is often very heavy, is considered as urban. 262

usable sensitivity. The minimum standard modulated carrier-signal power required to produce usable receiver output. *See also:* **receiver performance.** 181

useful life (1) (reliability). The period from a stated time, during which, under stated conditions, an item has an acceptable failure rate, or until an unrepairable failure occurs." 164, 182

(2) (nuclear power generating stations). The time to failure for a specific service condition. 31

useful line. *See:* **available line.**

useful output power (electron device). That part of the output power that flows into the load proper. 190

usual service conditions. Environmental conditions in which standard machines are designed to operate. The temperature of the cooling medium does not exceed 40 degrees Celsius and the altitude does not exceed 3300 feet. 63

utilance. *See:* **room utilization factor.**

utility (transmission and distribution). An organization responsible for the installation, operation or maintenance of electric supply or communication systems. 262

utility power. *See:* **commercial power.**

utility routine. *See:* **service routine.**

utilization equipment (transmission and distribution). Equipment, devices, and connected wiring which utilize electric energy for mechanical, chemical, heating, lighting, testing, or similar purposes and are not a part of supply equipment, supply lines, or communication lines. *See also:* **distribution center.** 262

utilization factor (system utilization factor). The ratio of the maximum demand of a system to the rated capacity of the system. *Note:* The utilization factor of a part of the system may be similarly defined as the ratio of the maximum demand of the part of the system to the rated capacity of the part of the system under consideration. *See also:* **alternating-current distribution; direct-current distribution.** 64

utilization time (hauptnutzzeit) (medical electronics). (1) The minimum duration that a stimulus of rheobasic strength must have to be just effective. (2) The shortest latent period between stimulus and response obtainable by very strong stimuli.* (3) The latent period following application of a shock of theobasic intensity.* *See also:* **biological.**   192

* Deprecated

UUT. *See:* **unit under test.**

UTT oriented language (test, measurement and diagnostic equipment). A computer language used to program automatic test equipment to test units under test (UUT's), whose characteristics are directed to the test needs of the UUT's and therefore do not imply the use of a specific ATE (automatic test equipment) system or family of ATE systems.   54

V

vacant code (telephone switching systems). A digit or a combination of digits that is unassigned.   55
vacant-code tone (telephone switching systems). A tone that indicates that an unassigned code has been dialed.   55
vacant number (telephone switching systems). An unassigned or unequipped directory number.   55
vacuum envelope (electron tube). The airtight envelope that contains the electrodes. *See also:* **electrode (of an electron tube).**   328
vacuum gauge. A device that indicates the gas pressure in an evacuated system. *Note:* Pressures in vacuum systems are usually expressed in microns absolute, one micron being the pressure that will support a column of mercury 1/1000 of a millimeter high. There are two types of vacuum gauges in common use in rectifier practice; the McLeod type, that operates on the principle that pressure multiplied by volume is constant, and measures the pressure of a gas by compressing a sample of known volume into a calibrated measuring tube; the hot-wire type, that operates on the principle that the heat conductivity of a gas is a function of its pressure. *See also:* **rectification.**   328
vacuum phototube. A phototube that is evacuated to such a degree that its electrical characteristics are essentially unaffected by gaseous ionization.   190
vacuum-pressure impregnation (VPI) (rotating machinery). The filling of voids in a coil or insulation system by withdrawal of air and solvent, if any, from the contained voids by vacuum, admission of a resin or resin solution followed by pressurization, and finally cure, usually with the application of heat. *See also:* **rotor (rotating machinery); stator.**   63
vacuum seal. The airtight junction between component parts of the evacuated system. *See also:* **rectification.**   328
vacuum switch. A switch whose contacts are enclosed in an evacuated bulb, usually to minimize sparking.   328
vacuum tank. The airtight metal chamber that contains the electrodes and in which the rectifying action takes place. *See also:* **rectification.**   328
vacuum tube. An electron tube evacuated to such a degree that its electrical characteristics are essentially unaffected by the presence of residual gas or vapor.   190, 125
vacuum-tube amplifier. An amplifier employing electron tubes to effect the control of power from the local source. *See also:* **amplifier.**   111
vacuum-tube radio frequency generator. *See:* **radio frequency generator, electron tube type.**
vacuum-tube transmitter. A radio transmitter in which

electron tubes are utilized to convert the applied electric power into radio-frequency power. *See also:* **radio transmitter.**   111
vacuum-tube voltmeter. *See:* **electronic voltmeter.**
vacuum valve. A device for sealing and unsealing the passage between two parts of an evacuated system. *See also:* **rectification.**   328
valance. A longitudinal shielding member mounted across the top of a window or along a wall, usually parallel to the wall, to conceal light sources giving both upward and downward distributions. *See also:* **luminaire.**   167
valance lighting. Lighting comprising light sources shielded by a panel parallel to the wall at the top of a window. *See also:* **general lighting.**   167
valence band. The range of energy states in the spectrum of a solid crystal in which lie the energies of the valence electrons that bind the crystal together. *See also:* **electron devices, miscellaneous; semiconductor.**
                                    245, 210, 186, 66
validation (test, measurement and diagnostic equipment). That process in the production of a test program by which the correctness of the program is verified by running it on the automatic test equipment together with the unit under test. The process includes the identification of run-time errors, procedure errors, and other non-compiler errors, not uncovered by pure software methods. The process is generally performed with the customer or designated representative as a witness.   54
valley (pulse terms). A portion of a pulse waveform between two specified peak magnitudes of the same polarity. *See:* Note in **preshoot** entry.   254
valley point (tunnel-diode characteristic). The point on the forward current-voltage characteristic corresponding to the second-lowest positive (forward) voltage at which $di/dV = 0$. *See also:* **peak point (tunnel-diode characteristic).**   315, 191
valley-point current (tunnel-diode characteristic). The current at the valley point. *See also:* **peak point (tunnel-diode characteristic).**   315, 191
valley-point voltage (tunnel-diode characteristic). The voltage at which the valley point occurs. *See also:* **peak point (tunnel-diode characteristic).**   315, 191
value (1) (several) (automatic control). The quantitative measure of a signal or variable. *See also:* **control system, feedback.**   329
(2) (direct-current through test object). The arithmetic mean value. *See also:* **test voltage and current.**   307, 60
(3) (test direct voltages). The arithmetic mean value; that is, the integral of the voltage over a full period of the ripple

divided by the period. *Note:* The maximum value of the test voltage may be taken approximately as the sum of the arithmetic mean value plus the ripple magnitude. *See also:* **test voltage and current.**   307, 201

(4) (alternating test voltage). The peak value divided by $(2)^{1/2}$. *See also:* **test voltage and current.**   307, 201

value added service (data communication). A communications service utilizing communications common carrier networks for transmission and providing added data services with separate additional equipment. Such added service features may be store-and-forward switching, terminal interfacing and host interfacing.   12

value, desired. *See:* **value, ideal.**

value, ideal (automatic control). The value of the ultimately controlled variable of an idealized system under consideration. *Syn:* desired value. *See also:* **ideal value.**   56

value, Munsell. *See:* **Munsell value.**

value, true (measured quantity). The actual value of a precisely defined quantity under the conditions existing during its measurement.   147

valve. *See:* **electron tube.**

valve action (electrochemical). The process involved in the operation of an electrochemical valve. *See also:* **electrochemical valve.**   328

valve arrester. A surge arrester that includes a valve element. *See:* **nonlinear-resistor type arrester; valve-type arrester.**   2, 299, 62

valve element (arrester). A resistor that, because of its nonlinear current-voltage characteristic, limits the voltage across the arrester terminals during the flow of discharge current and contributes to the limitation of follow current at normal power-frequency voltage. *See:* **valve-type arrester.**   299, 62, 2

valve-point loading control (electric power system). A control means for making a unit operate in the more efficient portions of the range of the governor-controlled valves. *See also:* **speed-governing system.**   94

valve ratio (electrochemical valve). The ratio of the impedance to current flowing from the valve metal to the compound or solution, to the impedance in the opposite direction. *See also:* **electrochemical valve.**   328

valve tube. *See:* **kenotron.**

valve-type arrester. *See:* **arrester, valve-type.**

Van Allen belt(s) (communication satellite). Belts of charged particles (electrons and protons) trapped by the earth's (external) magnetic field and which surround the earth at altitudes from 1000 to 6000 kilometers. The paths of the particles are determined by the directions of the (external) lines of force of the earth's magnetic field. The particles migrate from the region above earth's equator toward the North Pole, then toward the South Pole, then return to the region above the equator.   74

vane-type relay. A type of alternating-current relay in which a light metal disc or vane moves in response to a change of the current in the controlling circuit or circuits.   328

V antenna. A V-shaped arrangement of conductors, balanced-fed at the apex and with included angle, length, and elevation proportioned to give the desired directive properties. *See also:* **antenna.**   179, 111

vapor openings. Openings through a tank shell or roof above the surface of the stored liquid. Such openings may be provided for tank breathing, tank gauging, fire fighting, or other operating purposes. *See also:* **lightning protection and equipment.**   297

vapor-safe electric equipment. A unit so constructed that it may be operated without hazard to its surroundings in an atmosphere containing fuel, oil, alcohol, or other vapors that may occur in aircraft; that is, the unit is capable of so confining any sparks, flashes, or explosions of the combustible vapors within itself that ignition of the surrounding atmosphere is prevented. *Note:* This definition closely parallels that given for **explosionproof;** however, it is believed that the new term is needed in order to avoid the connotation of compliance with Underwriter's standards that are now associated with **explosionproof** in the minds of most engineers who are familiar with the use of that term applied to industrial motors and control equipment.   328

vaportight (1) (general). So enclosed that vapor will not enter the enclosure.   103, 202

(2) (luminaire). Designed and approved for installation in damp or wet locations. *Note:* It also is described as **enclosed and gasketed.** *See also:* **luminaire.**   167

var (electric power circuits). The unit of reactive power in the International System of Units (SI). The var is the reactive power at the two points of entry of a single-phase, two-wire circuit when the product of the root-mean-square value in amperes of the sinusoidal current by the root-mean-square value in volts of the sinusoidal voltage and by the sine of the angular phase difference by which the voltage leads the current is equal to one.   210

VAR (visual-aural range). A special type of VHF radio range which provides: (1) two reciprocal radial lines of position displayed to the pilot visually on a course deviation indicator and (2) two reciprocal radial lines of position presented to the pilot as interlocked and alternate A and N aural code signals. The aural lines of position are displaced 90 degrees from the visual and either may be used to resolve the ambiguity of the other.   13

varactor. A two-terminal semiconductor device in which the electrical characteristic of primary interest is a voltage-dependent capacitance.   277, 191

varhour. The unit of a quadrature-energy (quadergy) in the International System of Units (SI). The varhour is the quadrature energy that is considered to have flowed past the points of entry of a reactive circuit when a reactive power of one var has been maintained at the terminals of entry for one hour.   210

varhour meter (reactive volt-ampere-hour meter). An electricity meter that measures and registers the integral, with respect to time, of the reactive power of the circuit in which it is connected. The unit in which this integral is measured is usually the kilovarhour. *See also:* **electricity meter (meter).**   212

variable. (1) A quantity or condition that is subject to change. *See also:* **control system, feedback.**   56, 329

(2) A quantity that can assume any of a given set of values.   255, 77

variable-area track (electroacoustics). A sound track divided laterally into opaque and transparent areas, a sharp line of demarcation between these areas forming an oscillographic trace of the wave shape of the recorded signal. *See also:* **phonograph pickup.**   176

variable-block format. A format that allows the number of words in successive blocks to vary. 224, 207

variable carrier. *See:* **controlled carrier.**

variable, complex. *See:* **complex variable.**

variable-density track (electroacoustics). A sound track of constant width, usually but not necessarily of uniform light transmission on any instantaneous transverse axis, on which the average light transmission varies along the longitudinal axis in proportion to some characteristic of the applied signal. *See also:* **phonograph pickup.** 176

variable, directly controlled. *See:* **directly controlled variable.**

variable field. One that varies with time. 210

variable-frequency telemetering (electric power systems). A type of telemetering in which the frequency of the alternating-voltage signal is varied as a function of the magnitude of the measured quantity. *See also:* **telemetering.** 94

variable, indirectly controlled. *See:* **indirectly controlled variable.**

variable inductor.* *See:* **continuously adjustable inductor.**

* Deprecated

variable, input. A variable applied to a system or element. *See also:* **control system, feedback.** 56, 329

variable, manipulated. *See:* **manipulated variable.**

variable modulation (in very-high-frequency omnidirectional radio ranges). The modulation of the ground station radiation that produces a signal in the airborne receiver whose phase with respect to a radiated reference modulation corresponds to the bearing of the receiver. *See also:* **navigation.** 187, 13

variable-mu tube (variable-μ tube) (remote-cutoff tube). An electron tube in which the amplification factor varies in a predetermined way with control-grid voltage. 190, 125

variable, output. A variable delivered by a system or element. *See also:* **control system, feedback.** 56, 329

variable point. Pertaining to a numeration system in which the position of the point is indicated by a special character at that position. *See:* **fixed point; floating point.** 255, 77

variable-reluctance microphone (magnetic microphone). A microphone that depends for its operation on variations in the reluctance of a magnetic circuit. *See also:* **microphone.** 328

variable-reluctance pickup (magnetic pickup). A phonograph pickup that depends for its operation on the variation in the reluctance of a magnetic circuit. *See also:* **phonograph pickup.** 328

variable-reluctance transducer. An electroacoustic transducer that depends for its operation on the variation in the reluctance of a magnetic circuit. 176

variable-speed axle generator. An axle generator in which the speed of the generator varies directly with the speed of the car. *See also:* **axle generator system.** 328

variable-speed drive (industrial control). An electric drive so designed that the speed varies through a considerable range as a function of load. *See also:* **electric drive.** 206

variable speed motor (rotating machinery). A motor with a positively damped speed-torque characteristic which lends itself to controlled speed applications. 63

variables, state. *See:* **state variables.**

variable-torque motor. (1) A multispeed motor whose rated load torque at each speed is proportional to the speed. Thus the rated power of the motor is proportional to the square of the speed. (2) An adjustable-speed motor in which the specified torque increases with speed. It is common to provide a variable-torque adjustable-speed motor in which the torque varies as the square of the speed and hence the power output varies as the cube of the speed. *See also:* **asynchronous machine.** 63

variable, ultimately controlled. *See:* **ultimately controlled variable.**

variable-voltage transformer. An autotransformer in which the output voltage can be changed (essentially from turn to turn) by means of a movable contact device sliding on the shunt winding turns. 53

varindor. An inductor whose inductance varies markedly with the current in the winding. 328

variocoupler (radio practice). A transformer, the self-impedance of whose windings remains essentially constant while the mutual impedance between the windings is adjustable. 329

variolosser. A device whose loss can be controlled by a voltage or current. 328

variometer. A variable inductor in which the change of inductance is effected by changing the relative position of two or more coils. 341

varioplex. A telegraph switching system that establishes connections on a circuit-sharing basis between a multiplicity of telegraph transmitters in one locality and respective corresponding telegraph receivers in another locality over one or more intervening telegraph channels. Maximum usage of channel capacity is secured by momentarily storing the signals and allocating circuit time in rotation among those transmitters having intelligence in storage. *See also:* **telegraphy.** 328

varistor. (1) A two-terminal resistive element, composed of an electronic semiconductor and suitable contacts, that has a markedly nonlinear volt-ampere characteristic. (2) A two-terminal semiconductor device having a voltage-dependent nonlinear resistance. *Note:* **Varistors** may be divided into two groups, symmetrical and nonsymmetrical, based on the symmetry or lack of symmetry of the volt-ampere curve. *See also:* **semiconductor.** 245, 210, 66

varmeter (reactive volt-ampere meter). An instrument for measuring reactive power. It is provided with a scale usually graduated in either vars, kilovars, or megavars. If the scale is graduated in kilovars or megavars, the instrument is usually designated as a kilovarmeter or megavarmeter. *See also:* **instrument.** 328

varnish (rotating machinery). A liquid composition that is converted to a transparent or translucent solid film after application as a thin layer. 215, 63

varnished fabric (varnished mat) (rotating machinery). A fabric or mat in which the elements and interstices have been essentially coated and filled with an impregnant such as a compound or varnish and that is relatively homogeneous in structure. *See also:* **rotor (rotating machinery); stator.** 63

varnished tubing. A flexible tubular product made from

braided cotton, rayon, nylon, glass, or other fibers, and coated, or impregnated and coated, with a continuous film or varnish, lacquer, a combination of varnish and lacquer, or other electrical insulating materials. 328

varying duty (rating of electric equipment). A requirement of service that demands operation at loads, and for periods of time, both of which may be subject to wide variation. *See also:* **asynchronous machine; direct-current commutating machine; duty; voltage regulator.**
53, 310, 210, 257, 207, 206, 124

varying parameter. *See:* **linear varying parameter.** 210

varying-speed motor. A motor the speed of which varies appreciably with the load, ordinarily decreasing when the load increases, for example, a series or repulsion motor. *See:* **asynchronous machine; direct-current commutating machine.** 63, 3

varying-voltage control. A form of armature-voltage control obtained by impressing on the armature of the motor a voltage that varies considerably with change in load, with a consequent change in speed, such as may be obtained from a differentially compound-wound generator or by means of resistance in the armature circuit. *See also:* **control.** 206

VASIS. *See:* **visual approach slope indicator system.**

vault-type transformer. A transformer that is so constructed as to be suitable for occasional submerged operation in water under specified conditions of time and external pressure. 53

V-beam radar. A ground-based three-dimensional radar system for the determination of distance, bearing and, uniquely, the height or elevation angle of the target. It uses two fan-shaped beams, one vertical and the other inclined, that rotate together in azimuth so as to give two responses from the target; the time difference between these responses, together with distance, being factors used in determining the height of the target. 13

VCO (voltage controlled oscillator). An oscillator whose frequency is a function of the voltage of an input signal. 13

V curve (synchronous machine). The load characteristic giving the relationship between the armature current and the field current for constant values of load, power, and armature voltage. *See also:* **asynchronous machine.** 63

vector. A mathematico-physical quantity that represents a vector quantity. *See:* **mathematico-physical quantity (mathematical quantity) (abstract quantity).** 210

vector electrocardiogram (electrobiology) (vectorcardiogram). The 2-dimensional or 3-dimensional presentation of cardiac electric activity that results from displaying lead pairs against each other rather than against time. More strictly, it is a loop pattern taken from leads placed orthogonally. *See also:* **electrocardiogram.** 192

vector field. The totality of vectors in a given region represented by a vector function $V(x,y,z)$ of the space coordinates x,y,z. 210

vector function. A functional relationship that results in a vector. 210

vector norm (control systems). The measure of the size of a vector, with the usual norm properties. *Note:* (1) Vector norm of x is denoted by $\|x\|$
(2) Norm properties are:

$$\|x\| > 0 \qquad \text{for } x \neq 0$$
$$\|0\| = 0$$
$$\|\alpha x\| = |\alpha| \cdot \|x\|$$
$$\|x_1 + x_2\| \leqslant \|x_1\| + \|x_2\|$$
56

vector operator del ∇. A differential operator defined as follows in terms of Cartesian coordinates:

$$\nabla = \mathbf{i}\frac{\partial}{\partial x} + \mathbf{j}\frac{\partial}{\partial y} + \mathbf{k}\frac{\partial}{\partial z}.$$
210

vector power. *See:* **power, vector.**

vector product (cross product). The vector product of vector $\mathbf{A}$ and a vector $\mathbf{B}$ is a vector $\mathbf{C}$ that has a magnitude obtained by multiplying the product of the magnitudes of $\mathbf{A}$ and $\mathbf{B}$ by the sine of the angle between them; the direction of $\mathbf{C}$ is that traveled by a right-hand screw turning about an axis perpendicular to the plane of $\mathbf{A}$ and $\mathbf{B}$, in the sense in which $\mathbf{A}$ would move into $\mathbf{B}$ by a rotation of less than 180 degrees; it is assumed that $\mathbf{A}$ and $\mathbf{B}$ are drawn from the same point. The vector product of two vectors $\mathbf{A}$ and $\mathbf{B}$ may be indicated by using a small cross: $\mathbf{A} \times \mathbf{B}$. The direction of the vector product depends on the order in which the vectors are multiplied, so that $\mathbf{A} \times \mathbf{B} = -\mathbf{B} \times \mathbf{A}$. If the two vectors are given in terms of their rectangular components, then

$$\mathbf{A} \times \mathbf{B} = \begin{vmatrix} \mathbf{i} \, \mathbf{j} \, \mathbf{k} \\ A_x A_y A_z \\ B_x B_y B_z \end{vmatrix}$$
$$= \mathbf{i}(A_y B_z - A_z B_y) + \mathbf{j}(A_z B_x - A_x B_z) + \mathbf{k}(A_x B_y - A_y B_x).$$

Example: The linear velocity $\mathbf{V}$ of a particle in a rotating body is the vector product of the angular velocity ω and the radius vector $\mathbf{r}$ from any point on the axis to the point in question, or

$$\mathbf{V} = \omega \times \mathbf{r}$$
$$= -\mathbf{r} \times \omega$$
210

vector quantity. Any physical quantity whose specification involves both magnitude and direction and that obeys the parallelogram law of addition. 210

vector, state. *See:* **state vector.**

vehicle. That in or on which a person or thing is being or may be carried. *See also:* **navigation.** 13, 187

vehicle-derived navigation data. Data obtained from measurements made at a vehicle. *See also:* **navigation.** 187

vehicle maneuver effects (gyro). Gyro output errors due to vehicle maneuvers. 46

vehicle, off road. *See:* **off road vehicle.**

vehicle, pulling. *See:* **pulling vehicle.**

veiling brightness. A brightness superimposed on the retinal image that reduces its contrast. It is this veiling effect produced by bright sources or areas in the visual field that results in decreased visual performance and visibility. *See also:* **visual field.** 167

veiling reflection. Regular reflections that are superimposed upon diffuse reflections from an object that partially or totally obscure the details to be seen by reducing the contrast. This sometimes is called **reflected glare.** *See also:* **visual field.** 167

velocity correction (industrial control). A method of register control that takes the form of a gradual change in the relative velocity of the web. 206

velocity level in decibels of a sound (acoustics). Twenty times the logarithm to the base 10 of the ratio of the particle velocity of the sound to the reference particle velocity. The reference particle velocity shall be stated explicitly. *Note:* In many sound fields the particle velocity ratios are not proportional to the square root of corresponding power ratios and hence cannot be expressed in decibels in the strict sense; however, it is common practice to extend the use of the decibel to these cases. 176

velocity microphone. A microphone in which the electric output substantially corresponds to the instantaneous particle velocity in the impressed sound wave. *Note:* A velocity microphone is a gradient microphone of order one, and it is inherently bidirectional. *See:* **gradient microphone.** *See also:* **microphone.** 176

velocity-modulated amplifier (velocity-variation amplifier). An amplifier that employs velocity modulation to amplify radio frequencies. *See also:* **amplifier.** 328

velocity-modulated oscillator. An electron-tube structure in which the velocity of an electron stream is varied (velocity-modulated) in passing through a resonant cavity called a buncher. Energy is extracted from the bunched electron stream at a higher energy level in passing through a second cavity resonator called the catcher. Oscillations are sustained by coupling energy from the catcher cavity back to the buncher cavity. *See also:* **oscillatory circuit.** 111

velocity-modulated tube. An electron-beam tube in which the velocity of the electron stream is alternately increased and decreased with a period comparable with the total transit time. 244, 190

velocity modulation (velocity variation) (of an electron beam). The modification of the velocity of an electron stream by the alternate acceleration and deceleration of the electrons with a period comparable with the transit time in the space concerned. *See also:* **velocity-modulated oscillator; velocity modulated tube.** 125

velocity response factor (radar moving-target indicator). The ratio of voltage gain at a specific target velocity (or Doppler frequency) to the root-mean-square voltage gain evaluated over the entire velocity spectrum; this ratio for the conventional single-delay canceller varies sinusoidally from zero at the blind speeds to 1.414 at the optimum speeds. *See also:* **navigation.** 187, 13

velocity shock. A mechanical shock resulting from a nonoscillatory change in velocity of an entire system. 176

velocity sorting (electronic). Any process of selecting electrons according to their velocities. *See also:* **electron devices, miscellaneous.** 190, 125

velocity storage (accelerometer). The velocity information which is stored in the accelerometer as a result of its dy-namics. 46

velocity variation. *See:* **velocity modulation (electron beam).**

velocity-variation amplifier. *See:* **velocity-modulated amplifier.**

Venn diagram. A diagram in which sets are represented by closed regions. 255, 77

vent (1) (fuse). The means provided for the escape of the gases developed during circuit interruption. *Note:* In distribution oil cutouts, the vent may be an opening in the housing or an accessory attachable to a vent opening in the housing with suitable means to prevent loss of oil. 103, 299, 62, 202

(2) (rotating machinery). An opening that will permit the flow of air. 63

(3) (surge arrester). An intentional opening for the escape of gases to the outside. 2

vented fuse (or fuse unit). A fuse with provision for the escape of arc gases, liquids, or solid particles to the surrounding atmosphere during circuit interruption. 103, 202

vent finger. *See:* **duct spacer.**

ventilated dry-type transformer. A dry-type transformer which is so constructed that the ambient air may circulate through its enclosure to cool the transformer core and windings. 53

ventilated. Provided with a means to permit circulation of the air sufficiently to remove an excess of heat, fumes, or vapors. 53

ventilated enclosure (1) (power switchgear). An enclosure provided with means to permit circulation of sufficient air to remove an excess of heat, fumes, or vapors. *Note:* For outdoor applications ventilating openings or louvers are usually filtered, screened, or restricted to limit the entrance of dust, dirt, or other foreign objects. 103, 202

(2) (metal enclosed bus). An enclosure so constructed as to provide for the circulation of external air through the enclosure to remove heat, fumes, or vapors. 78

ventilating and cooling loss (synchronous machine). Any power required to circulate the cooling medium through the machine and cooler (if used) by fans or pumps that are driven by external means (such as a separate motor) so that their power requirements are not included in the friction and windage loss. It does not include power required to force ventilating gas through any circuit external to the machine and cooler. 244, 63

ventilating duct (cooling duct) (rotating machinery). A passage provided in the interior of a magnetic core in order to facilitate circulation of air or other cooling agent. 244, 63

ventilating passage (rotating machinery). A passage provided for the flow of cooling medium. 63

ventilating slot (rotating machinery). A slot provided for the passage of cooling medium. 63

verification (electronic computation). The process of checking the results of one data transcription against the results of another data transcription. Both transcriptions usually involve manual operations. *See also:* **check.** 210

verification relay. A monitoring relay restricted to functions pertaining to power-system conditions and not involving opening circuit breakers during fault condition.

Note: Such a relay is sometimes referred to as a check or checking relay. 103, 60, 202, 127

verify. (1) To check, usually automatically, one typing or recording of data against another in order to minimize human and machine errors in the punching of tape or cards. 235, 224, 207

(2) To check the results of keypunching. 255, 77

vernier control (industrial control). A method for improving resolution. The amount of vernier control is expressed as either the percent of the total operating range or of the actual operating value, whichever is appropriate to the circuit in use. *See also:* **control system, feedback.** 219, 200

vertex. *See:* **node.**

vertex plate (reflector). A metallic plate placed near the vertex and parallel to the surface of a reflector to reduce the power reflected back to the primary radiator. 111

vertical amplifier (oscilloscope). An amplifier for signals intended to produce vertical deflection. *See also:* **oscillograph.** 185

vertical-break switch. A switch in which the travel of the blade is in a plane perpendicular to the plane of the mounting base. The blade in the closed position is parallel to the mounting base. 103, 202, 79, 27

vertical conductor (pole line work). A wire or cable extending in an approximately vertical direction on the supporting pole or structure. *See also:* **conductor; tower.** 262, 64

vertical, gravity. *See:* **mass-attraction vertical.**

vertical gyro. A two-degree-of-freedom gyro with provision for maintaining the spin axis vertical. In this gyro, output signals are produced by gimbal angular displacements which correspond to angular displacements of the case about two nominally orthogonal, horizontal axes. 46

vertical-hold control (television). A synchronizing control that adjusts the free-running period of the vertical-deflection oscillator. 163

vertically polarized field vector (antennas). A linearly polarized field vector whose direction is vertical. 111

vertically polarized plane wave (antennas). A plane wave in which the electric vector is in the vertical plane containing the direction of propagation. 111

vertically polarized wave. A linearly polarized wave whose electric field vector is vertical. *Note:* The term **vertical polarization** is commonly employed to characterize ground-wave propagation in the medium-frequency broadcast band; these waves, however, have a small component of electric field in the direction of propagation due to finite ground conductivity. *See also:* **radio wave propagation.** 180, 146

vertical machine (rotating machinery). A machine whose axis of rotation is approximately vertical. 63

vertical plane (searchlight). The plane that is perpendicular to the train axis and in which the elevation axis lies. *See also:* **searchlight.** 167

vertical reach switch. One in which the stationary contact is supported by a structure separate from the hinge mounting base. The blade in the closed position is perpendicular to the hinge mounting base. 27, 79

vertical recording. A mechanical recording in which the groove modulaton is in a direction perpendicular to the surface of the recording medium. 328

vertical riser cable. Cable designed for use in long vertical runs, as in tall buildings. 57

vertical rod or shaft. A component of a switch operating mechanism designed to transmit motion from an operating handle or power operator to a switch offset bearing or bell crank. 27

vertical section. That portion of the switchgear assembly between two successive vertical delineations. It may contain one or more units. 79, 93

vertical switchboard. A control switchboard composed only of vertical panels. *Note:* This type of switchboard may be enclosed or have an open rear. An enclosed vertical switchboard has an overall sheet-metal enclosure (not grille) covering back and ends of the entire assembly, access to which is usually provided by doors or removable covers. 103

very-high-frequency omnidirectional radio range. *See:* **VOR.** 187

very-low-frequency high-potential test. An alternating-voltage high-potential test performed at a frequency equal to or less than 1 hertz. *See also:* **asynchronous machine.** 63

very-low-frequency test (VLF test). A test made at a frequency considerably lower than the normal operating frequency. *Note:* In order to facilitate communication and comparison among investigators, this document recommends that the very low frequency used be 0.1 Hz + 25 percent. 135

vestigial sideband (amplitude-modulated transmission). The transmitted portion of the sideband that has been largely suppressed by a transducer having a gradual cutoff in the neighborhood of the carrier frequency, the other sideband being transmitted without much suppression. *See also:* **amplitude modulation; facsimile transmission.** 12

vestigial-sideband modulation. A modulation process involving a prescribed partial suppression of one of the two sidebands. 242

vestigial-sideband transmission (facsimile). That method of signal transmission in which one normal sideband and the corresponding vestigial sideband are utilized. *See also:* **amplitude modulation; facsimile transmission.** 12

vestigial-sideband transmitter. A transmitter in which one sideband and a portion of the other are intentionally transmitted. *See also:* **radio transmitter.** 111

VF. *See:* **voice frequency.**

via net loss (VNL) (data transmission). This term describes the net losses of trunks in the long distance switched telephone network of North America. The trunk is said to be in a via condition when it is an intermediate trunk in a longer switched connection. 59

vibrating bell. A bell having a mechanism designed to strike repeatedly when and as long as actuated. *See also:* **protective signaling.** 328

vibrating circuit (telegraph circuit). An auxiliary local timing circuit associated with the main line receiving relay for the purpose of assisting the operation of the relay when the definition of the incoming signals is indistinct. *See also:* **telegraphy.** 328

vibrating-contact machine regulator (electric machine). A regulator that varies the excitation by changing the average time of engagement of vibrating contacts in the field circuit. 103, 202

vibrating-reed relay. A relay in which the application of an alternating or a self-interrupted voltage to the driving coil produces an alternating or pulsating magnetic field that causes a reed to vibrate and operate contacts. *See also:* **relay.** 259, 341

vibrating string accelerometer. A device that employs one or more vibrating strings whose natural frequencies are affected as a result of acceleration acting on one or more proof masses. 46

vibration. An oscillation wherein the quantity is a parameter that defines the motion of a mechanical system. *See also:* **oscillation.** 176

vibration detection system (protective signaling). A system for the protection of vaults by the use of one or more detector buttons firmly fastened to the inner surface in order to pick up and convert vibration, caused by burglarious attack on the structure, to electric impulses in a protection circuit. *See also:* **protective signaling.** 328

vibration meter. An apparatus including a vibration pickup, calibrated amplifier, and output meter for the measurement of displacement, velocity, and acceleration of vibrations. *See also:* **instrument.** 328

vibration relay. A relay that responds to the magnitude and frequency of a mechanical vibration. 103, 60, 202, 127

vibration test (rotating machinery). A test taken on a machine to measure the vibration of any part of the machine under specified conditions. 63

vibrato. A family of tonal effects in music that depend upon periodic variations in one or more characteristics of the sound wave. *Note:* When the particular characteristics are known, the term **vibrato** should be modified accordingly, for example, **frequency vibrato; amplitude vibrato; phase vibrato** and so forth. 176

vibrator (cable plowing). That device which induces the vibration in a vibratory plow. *See also:* **vibratory plow.** 52

vibratory isolation (cable plowing). Percentage reduction in force transmitted from vibration source to receiver by use of flexible mounting(s) (amount of isolation for a given unit varies with plow blade frequency). 52

vibratory plow (cable plowing). A plow utilizing induced periodic motion(s) of the blade in conjunction with drawbar pull for its movement through the soil. *Note:* Orbital and oscillating plows are types of vibratory plows that are commercially available. 52

vibropendulous error (accelerometer). A cross coupling rectification error caused by angular motion of the pendulum in a pendulous accelerometer in response to a linear vibratory input. The error varies with frequency and is maximum when the vibratory acceleration is applied in a plane normal to a pivot axis and at 45 degrees to the input axis. 46

video (1) (television). A term pertaining to the bandwidth and spectrum position of the signal resulting from television scanning. *Note:* In present usage, video means a bandwidth of the order of several megahertz, and a spectrum position that goes with a direct-current carrier. *See also:* **signal wave.** 328

(2) (radar). Refers to the signal after envelope or phase detection, which in early radar was the displayed signal. Contains the relevant radar information after removal of the carrier frequency. 13

video filter (non-real time spectrum analyzer). A post detection low-pass filter. 68

video-frequency amplifier. A device capable of amplifying such signals as comprise periodic visual presentation. *See also:* **television.** 111

video integration (electronic navigation). A method of utilizing the redundancy of repetitive video signals to improve the output signal-to-noise ratio, by summing successive signals. *See also:* **navigation.** 13, 187

video mapping (electronic navigation). The electronic superposition of geographic or other data on a radar display. *See also:* **navigation.** 187

video stretching (electronic navigation). The increasing of the duration of a video pulse. *See also:* **navigation.** 187

video-telephone call (telephone switching systems). A call between stations equipped to provide video-telephone service. 55

vidicon. A camera tube in which a charge-density pattern is formed by photoconduction and stored on that surface of the photoconductor that is scanned by an electron beam, usually of low-velocity electrons. *See also:* **television.** 178, 190

viewing area (oscilloscope). The area of the phosphor screen of a cathode-ray tube that can be excited to emit light by the electron beam. *See:* **oscillograph; screen, viewing.** 184

viewing time (storage tubes). The time during which the storage tube is presenting a visible output corresponding to the stored information. *See also:* **storage tube.** 174, 190

viewing time, maximum usable (storage tubes). The length of time during which the visible output of a storage tube can be viewed, without rewriting, before a specified decay occurs. *Note:* The qualifying adjectives **maximum usable** are frequently omitted in general usage when it is clear that maximum usable viewing time is implied. *See also:* **storage tube.** 174, 190

virtual cathode (potential-minimum surface) (electron tubes). A region in the space charge where there is a potential minimum that, by reason of the space charge density, behaves as a source of electrons. *See also:* **electronic tube.** 244, 190

virtual duration (of a peak of a rectangular-wave current or voltage impulse) (surge arresters). The time during which the amplitude of the wave is greater than 90 percent of its peak value. 308, 62

virtual duration of wave front impulse (surge arrester). The virtual value for the duration of the wave front is as follows. (1) For voltage waves with wave front durations less than 30 μs, either full or chopped on the front, crest, or tail, 1.67 times the time for the voltage to increase from 30 to 90 percent of its crest value. (2) For voltage waves with wave front durations of 30 or more μs, the time taken by the voltage to increase from actual zero to maximum crest value. (3) For current waves, 1.25 times the time for the current to increase from 10 to 90 percent of crest value. *See also:* **surge arrester.** 2

virtual front time. *See:* **virtual duration of wavefront impulse.**

virtual height (radio wave propagation). The apparent

height of an ionized layer determined from the time interval between the transmitted signal and the ionospheric echo at vertical incidence, assuming that the velocity of propagation is the velocity of light in a vacuum over the entire path. *See also:* **radiation; radio wave propagation.**
146, 180

virtual instant of chopping (voltage testing). The instant preceding point C on the accompanying figures by 0.3 times the (estimated) virtual time of voltage collapse during chopping. *See also:* **test voltage and current.**
307, 201

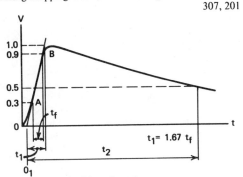

Full impulse voltage.

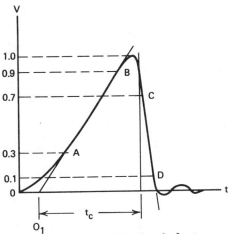

Impulse voltage chopped on the front.

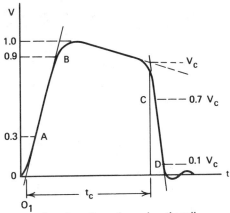

Impulse voltage chopped on the tail.

virtual junction temperature. The temperature of the active semiconductor element of a semiconductor device based on a simplified representation of the thermal and electrical behavior of the device. It is particularly applicable to multi-junction semiconductor devices.
66

virtual origin (impulse current or voltage). *See:* **virtual zero time.**

virtual peak value. *See:* **peak value.**

virtual rate of rise of the front (impulse voltage). The quotient of the peak value and the virtual front time. *Note:* The term **peak value** is to be understood as including the term **virtual peak value** unless otherwise stated. *See also:* **test voltage and current.**
307, 201

virtual steepness of voltage during chopping (surge arresters). The quotient of the estimated voltage at the instant of chopping and the virtual time of voltage collapse. *See also:* **test voltage and current.**
308, 307, 201, 62

virtual steepness of wavefront of an impulse (surge arresters). The slope of the line that determines the virtual-zero time. It is expressed in kilovolts per microsecond or kiloamperes per microsecond.
308, 62

virtual time of voltage collapse during chopping. 1.67 times the time interval between points C and D on the figures attached to the definition of **virtual instant of chopping.** *See also:* **test voltage and current.**
307, 201

virtual time to chopping (impulse voltage). The time interval between the virtual origin and the virtual instant of chopping. *See also:* **test voltage and current.** 307, 201

virtual time to half-value (on the wavetail) (current or voltage impulse) (surge arresters). The time interval between virtual zero and the instant when the voltage or current has decreased to half its peak value. *Note:* This time is expressed in microseconds. The term **peak value** is to be understood as including the term **virtual peak value** unless otherwise stated.
308, 62

virtual total duration (rectangular-wave current or voltage impulse) (surge arresters). The time during which the amplitude of the wave is greater than 10 percent of its peak value. *Note:* If small oscillations are present on the wavefront, a mean curve should be drawn in order to determine the time at which the 10 percent value is reached.
62

virtual zero point (impulse) (surge arrester). The intersection with the zero axis of a straight line drawn through points on the front of the current wave at 10 percent and 90 percent crest value, or through points on the front of the voltage wave at 30 percent and 90 percent crest value. *See also:* **virtual zero time.**
2

virtual zero time (impulse voltage or current in a conductor) (surge arresters) (conventional origin) (virtual origin). The point on a graph of voltage-time or current-time determined by the intersection with the zero voltage or current axis, of a straight line drawn through two points on the front of the wave: (1) for full voltage waves and voltage waves chopped on the front, peak, or tail, the reference points shall be 30 percent and 90 percent of the peak value, and (2) for current waves the reference points shall be 10 percent and 90 percent of the peak value.
308, 62

viscous friction (industrial control). The component of friction that is due to the viscosity of a fluid medium, usually idealized as a force proportional to velocity, and

that opposes motion. *See also:* **control system, feedback.**
219, 329, 206

visibility (1) (general) (light emitting diodes). The quality or state of being perceivable by the eye. In many outdoor applications, visibility is defined in terms of the distance at which an object can be just perceived by the eye. In indoor applications it usually is defined in terms of the contrast or size of a standard test object, observed under standardized viewing conditions, having the same threshold as the given object. *See also:* **visual field.**
167, 162

(2) (meteorological). A term that denotes the greatest distance, expressed in miles, that selected objects (visibility markers) or lights of moderate intensity (25 candelas) can be seen and identified under specified conditions of observation. 167

visibility factor (radar) (1) (pulsed radar). The ratio of single-pulse signal energy to noise power per unit bandwidth that provides stated probabilities of detection and false alarm on a display, measured in the intermediate-frequency portion of the receiver under conditions of optimum bandwidth and viewing environment. *See also:* **navigation.** 187

(2) (continuous-wave radar). The ratio of single-look signal energy to noise power per unit bandwidth using a filter matched to the time on target. The equivalent term for radar using automatic detection is **detectability factor;** for operaton in a clutter environment a **clutter visibility factor** is defined. 187

visible corona. A luminous discharge due to ionization of the air surrounding a device, caused by a voltage gradient exceeding a certain critical value. 27, 79

visible radiation (light) (laser-maser). Electromagnetic radiation which can be detected by the human eye. It is commonly used to describe wavelengths which lie in the range between 0.4μm and 0.7μm. 363

visible radiation emitting diode (light emitting diodes). A semiconductor device containing a semiconductor junction in which visible light is nonthermally produced when a current flows as a result of an applied voltage. 162

visual acuity. A measure of the ability to distinguish fine details. Quantitatively, it is the reciprocal of the angular size in minutes of the critical detail that is just large enough to be seen. *See also:* **visual field.** 167

visual angle. The angle that an object or detail subtends at the point of observation. It usually is measured in minutes of arc. *See also:* **visual field.** 167

visual approach slope indicator system (VASIS). The system of angle-of-approach lights, accepted as a standard by the International Civil Aviation Organization, comprising two bars of lights located at each side of the runway near the threshold and showing red or white or a combination of both (pink) to the approaching pilot depending upon his position with respect to the glide path. 167

visual-aural radio range. *See:* **VAR.**

visual field. The locus of objects or points in space that can be perceived when the head and eyes are kept fixed. The field may be monocular or binocular. 167

visual inspection. Qualitative observation of physical characteristics utilizing the unaided eye or with stipulated levels of magnification. 284

visual perception (light emitting diodes). The interpretation of impressions transmitted from the retina to the brain in terms of information about a physical world displayed before the eye. *Note:* Visual perception involves any one or more of the following: recognition of the presence of something (object, aperture, or medium); identifying it; locating it in space; noting its relation to other things; identifying its movement, color, brightness, or form. *See also:* **visual field.** 167, 162

visual performance. The quantitative assessment of the performance of a task taking into consideration speed and accuracy. *See also:* **visual field.** 167

visual photometer. A visual photometer is one in which the equality of brightness of two surfaces is established visually. *Note:* The two surfaces usually are viewed simultaneously side by side. This method is used in portable visual luminance (photometric brightness) meters. This is satisfactory when the color difference between the test source and comparison source is small. However, when there is a color difference, a flicker photometer provides more precise measurements. In this type of photometer the two surfaces are viewed alternately at such a rate that the color sensations either nearly or completely blend and the flicker due to brightness difference is balanced by adjusting the comparison source. 167

visual radio range. Any radio range (such as very-high-frequency omnidirectional radio range) whose primary function is to provide lines of position to be flown by visual reference to a course-deviation indicator. *See also:* **radio navigation.** 187, 13

visual range (of a light or object). The maximum distance at which that particular light (or object) can be seen and identified. 167

visual scanner (character recognition). *See:* **optical scanner.**

visual signal device (protective signaling). A general term for pilot lights, annunciators, and other devices providing a visual indication of the condition supervised. *See also:* **protective signaling.** 328

visual surround. The visual surround includes all portions of the visual field except the visual task. *See also:* **visual field.** 167

visual task. Those details and objects that must be seen for the performance of a given activity including the immediate background of the details or objects. *See also:* **visual field.** 167

visual transmitter. All parts of a television transmitter that handle picture signals, whether exclusively or not. *See also:* **television.** 111

visual transmitter power. The peak power output during transmission of a standard television signal. *See also:* **television.** 328

vital circuit. Any circuit the function of which affects the safety of train operation. 328

vital services (marine transportation). Normally considered to be those required for the safety of the ship and its passengers and crew. These may include propulsion, steering, navigation, fire fighting, emergency lighting and communications functions. Since the specific identification of vital services is influenced by the type of vessel and its intended service, this matter should be specified by the design agent for the particular vessel under consideration.
3

V-network (electromagnetic compatibility). An artificial mains network of specified disymmetric impedance used for two-wire mains operation and comprising resistors in V formation connected between each conductor and earth. *See also:* **electromagnetic compatibility.** 199

VNL. *See:* **via net loss.**

vodas. A system for preventing the over-all voice-frequency singing of a two-way telephone circuit by disabling one direction of transmission at all times. The name is derived from the initial letters of the expression **voice-operated device anti-sing.** *See also:* **voice-frequency telephony.** 328

vogad. A voice-operated device used to give a substantially constant volume output for a wide range of inputs. The name is derived from the initial letters of the expression **voice-operated gain-adjusted device.** *See also:* **voice-frequency telephony.** 328

voice channel (mobile communication). A transmission facility defined by the constraints of the human voice. For mobile-communication systems, a voice channel may be considered to have a range of approximately 250 to 3000 hertz; since the Rules and Regulations of the Federal Communications Commission do not authorize the use of modulating frequencies higher than 3000 hertz for radiotelephony or tone signaling on radio frequencies below 500 megahertz. *See also:* **channel spacing.** 181

voice frequency (VF). A frequency lying within that part of the audio range that is employed for the transmission of speech. *Note:* Voice frequencies used for commercial transmission of speech usually lie within the range 200 to 3500 hertz. *See also:* **signal wave.** 194, 59

voice-frequency carrier telegraphy. That form of carrier telegraphy in which the carrier currents have frequencies such that the modulated currents may be transmitted over a voice-frequency telephone channel. *See also:* **telegraphy.** 194, 59

voice-frequency telephony. That form of telephony in which the frequencies of the components of the transmitted electric waves are substantially the same as the frequencies of corresponding components of the actuating acoustical waves. 59

voice grade channel (data transmission). A channel suitable for transmission of speech, digital or analog data, or facsimile, generally with a frequency range of about 300 to 3000 hertz. 59

volatile (electronic data processing). Pertaining to a storage device in which data cannot be retained without continuous power dissipation, for example, an acoustic delay line. *Note:* Storage devices or systems employing nonvolatile media may or may not retain data in the event of planned or accidental power removal. 235, 210, 54

volatile flammable liquid. A flammable liquid having a flash point below 100 degrees Fahrenheit or whose temperature is above its flash point. 256

volcas. A voice-operated device that switches loss out of the transmitting branch and inserts loss in the receiving branch under control of the subscriber's speech. The name is derived from the initial letters of the expression **voice-operated loss control and suppressor.** *See also:* **voice-frequency telephony.** 328

volt. The unit of voltage or potential difference in SI units. The volt is the voltage between two points of a conducting wire carrying a constant current of one ampere, when the power dissipated between these points is one watt. 210

volta effect. *See:* **contact potential.**

voltage (electromotive force*) (1) (general) (along a specified path in an electric field). The dot product line integral of the electric field strength along this path. *Notes:* (A) Voltage is a scalar and therefore has no spatial direction. (B) As here defined, voltage is synonymous with potential difference only in an electrostatic field. (C) In cases in which the choice of the specified path may make a significant difference, the path is taken in an equiphase surface unless otherwise noted. (D) It is often convenient to use an adjective with voltage, for example, phase voltage, electrode voltage, line voltage, etcetera. The basic definition of voltage applies and the meaning of adjectives should be understood or defined in each particular case. *See also:* **reference voltage.** 210

* Deprecated

(2) (transmission and distribution). The effective rms (root-mean-square) potential difference between any two conductors or between a conductor and ground. Voltages are expressed in nominal values unless otherwise indicated. The nominal voltage of a system or circuit is the value assigned to a system or circuit of a given voltage class for the purpose of convenient designation. The operating voltage of the system may vary above or below this value. **(A) (circuit not effectively grounded).** The highest nominal voltage available between any two conductors of the circuit. If one circuit is directly connected to and supplied from another circuit of higher voltage (as in the case of an autotransformer), both are considered as of the higher voltage, unless the circuit of the lower voltage is effectively grounded, in which case its voltage is not determined by the circuit of higher voltage. Direct connection implies electric connection as distinguished from connection merely through electromagnetic or electrostatic induction. **(B) (an effectively grounded circuit).** The highest nominal voltage available between any conductor of the circuit and ground unless otherwise indicated. 262

(3) (effectively grounded circuit). The highest effective voltage between any conductor and ground unless otherwise indicated. *See also:* **ground; grounding.** 262

(4) (circuit not effectively grounded). The highest effective voltage between any two conductors unless otherwise indicated. *Notes:* (A) If one circuit is directly connected to another circuit of higher voltage (as in the case of an autotransformer), both are considered as of the higher voltage, unless the circuit of lower voltage is effectively grounded, in which case its voltage is not determined by the circuit of higher voltage. Direct connection implies electric connection as distinguished from connection merely through electromagnetic or electrostatic induction. (B) Where safety considerations are involved, the voltage to ground that may occur in an ungrounded circuit is usually the highest voltage normally existing between the conductors of the circuit, but in special circumstances higher voltages may occur. *See also:* **ground; grounded.** 262

(5) (surge arresters). The voltage between a part of an electric installation connected to a grounding system and points on the ground at an adequate distance (theoretically

at an infinite distance) from any earth electrodes.
244, 62

voltage amplification (1) (general). An increase in signal voltage magnitude in transmission from one point to another or the process thereof. *See also:* **amplifier.**   210

(2) (transducer). The scalar ratio of the signal output voltage to the signal input voltage. *Warning:* By incorrect extension of the term decibel, this ratio is sometimes expressed in decibels by multiplying its common logarithm by 20. It may be correctly expressed in decilogs. *Note:* If the input and/or output power consist of more than one component, such as multifrequency signal or noise, then the particular components used and their weighting must be specified. *See also:* **transducer.**   239, 210

(3) (magnetic amplifier). The ratio of differential output voltage to differential control voltage.   171

voltage attenuation. *See:* **attenuation, voltage.**

voltage buildup (rotating machinery). The inherent establishment of the excitation current and induced voltage of a generator. *See also:* **direct-current commutating machine.**   63

voltage circuit (instrument). That combination of conductors and windings of the instrument to which is applied the voltage of the circuit in which a given electrical quantity is to be measured, or a definite fraction of that voltage, or a voltage or current dependent upon it. *See also:* **instrument; moving element (instrument); watthour meter.**   280

voltage class (power cables). *See:* **medium voltage power cable; low-voltage power cable; control cable; low-level analog signal cable; low-level digital signal circuit cable.**   35

voltage classes. *See:* **voltage classes** chart below.

voltage class, rated nominal (enclosed field discharge circuit breakers for rotating electric machinery). The voltage to which operating and performance characteristics are referred.   359

voltage coefficient of capacitance (nonlinear capacitor). The derivative with respect to voltage of a capacitance characteristic, such as a differential capacitance characteristic or a reversible capacitance characteristic, at a point, divided by the capacitance at that point. *See also:* **nonlinear capacitor.**   191

voltage, common mode (surge withstand capability tests). The voltage common to all conductors of a group as measured between that group at a given location and an arbitrary reference (usually earth).   90

voltage controlled oscillator. *See:* **VCO.**

voltage corrector (power supplies). An active source of regulated power placed in series with an unregulated

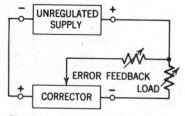

Circuit used to sense output voltage changes.

VOLTAGE CLASSES
NOMINAL SYSTEM VOLTAGE

TWO WIRE	THREE WIRE	FOUR WIRE	MAXIMUM VOLTAGE[3]
	Single-Phase Systems		
(120)	120/240		127 127/254
	Three-Phase Systems		
		208Y/120	220Y/127
	(240)	240/120	245/127
	480	480Y/277	508Y/293
	(600)		635
	(2400)		2540
	4160	4160Y/2400	4400Y/2540
	(4800)		5080
	(6900)		7260
		(8320Y/4800)	8800Y/5080
		(12000Y/6930)	12700Y/7330
		12470Y/7200	13200Y/7620
		13200Y/7620	13970Y/8070
	13800	(13800Y/7970)	14520Y/8380
		(20780Y/12000)	22000Y/12700
		(22860Y/13200)	24200Y/13970
	(23000)		24340
		24940Y/14400	26400Y/15240
	(34500)	34500/19920	36510Y/21080
	(46 kV)		48.3 kV
	69 kV		72.5 kV
	115 kV		121 kV
	138 kV		145 kV
	(161 kV)		169 kV
	230 kV		242 kV
	345 kV [2]		362 kV
	500 kV [2]		550 kV
	765 kV [2]		800 kV
	1100 kV [2]		1200 kV

Vertical bracket labels at left of table: LOW VOLTAGE SYSTEMS — MEDIUM VOLTAGE — HIGH VOLTAGE — EXTRA HIGH VOLTAGE; IEEE Std for Industrial & Commercial Power Systems[1]; ANSI C84.1–1977 no voltage class stated; ANSI C84.1–1977 HIGHER VOLTAGE SYSTEMS; ANSI C92.2–1976.

Preferred nominal voltages as shown without parentheses ( ).

1. Voltage class designations applicable to industrial and commercial power systems, adapted by IEEE Standards Board (LB 100A—April 23, 1975).

2. Typical nominal voltage (for system identification only).

3. A comprehensive list of minimum and maximum voltage ranges is given in ANSI C84.1-1977.

4. ANSI C92.2-1967 designated these as Extra High Voltage Systems.

supply to sense changes in the output voltage (or current); also to correct for the changes by automatically varying its own output in the opposite direction, thereby maintaining the total output voltage (or current) constant. See the accompanying figure.

See also: **power supply.**   228, 186

voltage divider. A network consisting of impedance elements connected in series, to which a voltage is applied, and from which one or more voltages can be obtained across any portion of the network. *Notes:* (1) Dividers may have parasitic impedances affecting the response. These impedances are, in general, the series inductance and the capacitance to ground and to neighboring structures at ground or at other potentials. (2) An adjustable voltage divider of the resistance type is frequently referred to as a potentiometer.   307, 341, 201

voltage doubler. A voltage multiplier that separately rectifies each half cycle of the applied alternating voltage and adds the two rectified voltages to produce a direct voltage whose amplitude is approximately twice the peak amplitude of the applied alternating voltage. *See also:* **rectifier.**   328

voltage drop (1) (general). The difference of voltages at the two terminals of a passive impedance.   63
(2) (supply system). The difference between the voltages at the transmitting and receiving ends of a feeder, main, or service. *Note:* With alternating current the voltages are not necessarily in phase and hence the voltage drop is not necessarily equal to the algebraic sum of the voltage drops along the several conductors. *See also:* **alternating-current distribution.**   64

voltage efficiency (specified electrochemical process). The ratio of the equilibrium reaction potential to the bath voltage.   328

voltage endurance (rotating machinery). A characteristic of an insulation system, obtained by plotting voltage against time to failure, for a number of samples tested to destruction at each of several sustained voltages. Constant conditions of frequency, waveform, temperature, mechanical restraint, and ambient atmosphere are required. Ordinate scales of arithmetical or logarithmic voltage, and abscissa scales of multicycle logarithmic time, normally give approximately linear characteristics. *See also:* **asynchronous machine; direct-current commutating machine.**   63

voltage endurance test (rotating machinery). A test designed to determine the effect of voltage on the useful life of electric equipment. When this test voltage exceeds the normal design voltage for the equipment, the test is voltage accelerated. When the test voltage is alternating and the frequency of alternation exceeds the normal voltage frequency for the equipment, the test is frequency accelerated. *See:* **asynchronous machine; direct-current commutating machine.**   63

voltage, equivalent test alternating (charging inductor). A sinusoidal root-mean-square test voltage equal to 0.707 times the power-supply voltage of the network-charging circuit and having a frequency equal to the resonance frequency of charging. *Note:* This is the alternating component of the voltage that appears across the charging inductor in a resonance-charging circuit of the pulse-forming network.   137

voltage, exciter-ceiling (enclosed field discharge circuit breakers for rotating electric machinery). The maximum voltage that may be attained by an exciter under specified conditions.   359

voltage factor (electron tubes). The magnitude of the ratio of the change in one electrode voltage to the change in another electrode voltage, under the conditions that a specified current remains unchanged and that all other electrode voltages are maintained constant. *See also:* ON

period.   244, 190

voltage generator (network analysis and signal-transmission system). A two-terminal circuit element with a terminal voltage substantially independent of the current through the element. *Note:* An ideal voltage generator has zero internal impedance. *See also:* **network analysis; signal.**   188, 190, 125

voltage impulse. A voltage pulse of sufficiently short duration to exhibit a frequency spectrum of substantially uniform amplitude in the frequency range of interest. As used in electromagnetic compatibility standard measurements, the voltage impulse has a uniform frequency spectrum over the frequency range 25 to 1000 megahertz.   314, 199

voltage influence (electric instrument). In instruments, other than indicating voltmeters, wattmeters, and varmeters, having voltage circuits, the percentage change (of full-scale value) in the indication of an instrument that is caused solely by a voltage departure from a specified reference voltage. *See also:* **accuracy rating (instrument).**   280, 294

voltage jump (glow-discharge tube). An abrupt change or discontinuity in tube voltage drop during operation. *Note:* This may occur either during life under constant operating conditions or as the current or temperature is varied over the operating range. *See also:* **gas-filled radiation-counter tube; gas tube.**   190, 125

voltage level (transmission system). At any point, the ratio of the voltage existing at that point to an arbitrary value of voltage used as a reference. Specifically, in systems such as television systems, where wave shapes are not sinusoidal or symmetrical about a zero axis and where the arithmetical sum of the maximum positive and negative excursions of the wave is important in system performance, the voltage level is the ratio of the peak-to-peak voltage existing at any point in the transmission system to an arbitrary peak-to-peak voltage used as a reference. This ratio is usually expressed in decibels referred to one volt peak-to-peak (dBv). *See also:* **level.**   328, 59

voltage limit (industrial control). A control function that prevents a voltage from exceeding prescribed limits. Voltage limit values are usually expressed as percent of rated voltage. If the voltage-limit circuit permits the limit value to increase somewhat instead of being a single value, it is desirable to provide either a curve of the limit value of voltage as a function of some variable such as current or to give limit values at two or more conditions of operation. *See also:* **control system, feedback.**   219, 206

voltage loss (electric instrument) (current circuits). In a current-measuring instrument, the value of the voltage between the terminals when the applied current corresponds to nominal end-scale deflection. In other instruments the voltage loss is the value of the voltage between the terminals at rated current. *Note:* By convention, when an external shunt is used, the voltage loss is taken at the potential terminals of the shunt. The overall voltage drop resulting may be somewhat higher owing to additional drop in shunt lugs and connections. *See also:* **accuracy rating (instrument).**   280

voltage multiplier. A rectifying circuit that produces a direct voltage whose amplitude is approximately equal to an integral multiple of the peak amplitude of the applied alternating voltage. *See also:* **rectifier.** 328

voltage, nominal (system or circuit). *See:* **nominal system voltage.**

voltage overshoot (arc-welding apparatus). The ratio of transient peak voltage substantially instantaneously following the removal of the short circuit to the normal steady-state voltage value. *See also:* **voltage recovery time.** 264

voltage overshoot, effective (arc-welding apparatus). The area under the transient voltage curve during the time that the transient voltage exceeds the steady-state value. *See also:* **voltage recovery time.** 264

voltage, peak working (charging inductor). The algebraic sum of the maximum alternating crest voltage and the direct voltage of the same polarity appearing between the terminals of the inductor winding or between the inductor winding and the grounded elements. 137

voltage phase-angle method (economic dispatch) (electric power systems). Considers the actual measured phase-angle difference between the station bus and a reference bus in the determination of incremental transmission losses. 94

voltage-phase-balance protection. A form of protection that disconnects or prevents the connection of the protected equipment when the voltage unbalance of the phases of a normally balanced polyphase system exceeds a predetermined amount. 103, 60, 202, 127

voltage range (electrically propelled vehicle). Divided into five voltage ranges, as follows. **first voltage range:** 30 volts or less. **second voltage range:** over 30 volts to and including 175 volts. **third voltage range:** over 175 volts to and including 250 volts. **fourth voltage range:** over 250 volts to and including 660 volts. **fifth voltage range:** over 660 volts. 328

voltage range multiplier (instrument multiplier). A particular type of series resistor or impedor that is used to extend the voltage range beyond some particular value for which the measurement device is already complete. It is a separate component installed external to the measurement device. *See also:* **auxiliary device (instrument).** 294

voltage, rated maximum. The highest rms voltage at which a device is designed to operate. *Note:* This voltage corresponds to the maximum tolerable zone primary voltage at distribution transformers for distribution reclosers and at substations and on transmission systems for power reclosers as given in ANSI C84.1-1977. 92

voltage, rated maximum interrupting of main contacts (en-

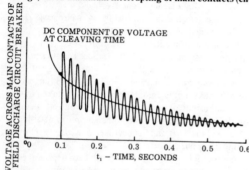

Voltage across field discharge circuit breaker main contacts following interruption of excitation source current.

closed field discharge circuit breakers for rotating electric machinery). The maximum dc voltage including voltage induced in the machine field by current in the machine armature, at which the field discharge circuit breaker main contacts are required to interrupt the excitation source current. The magnitude of the dc component of the total voltage across the main contacts is equal to the displacement of the axis. (See figure above for illustration of the direct component of voltage under three-phase fault conditions.) 359

voltage, rated short-time of main contacts (enclosed field discharge circuit breakers for rotating electric machinery). The highest dc voltage at which the circuit breaker main contacts shall be required to interrupt exciter short-circuit current. 359

voltage rating (1) (surge arrester). The designated maximum permissible operating voltage between its terminals at which an arrester is designed to perform its duty cycle. It is the voltage rating specified on the nameplate. 2
(2) (household electric ranges). The voltage limits within which the range is intended to be used. *See:* **appliance outlet.** 308
(3) (grounding transformer). The maximum "line-to-line" voltage at which it is designed to operate continuously from line to ground without damage to the grounding transformer. 53
(4) (relay). The voltage rating of a relay is the voltage at a specified frequency that may be sustained by the relay for an unlimited period without causing any of the prescribed limitations to be exceeded. 127

voltage rating, maximum (connector). The highest phase-to-ground voltage at which a connector is designed to operate. 134

voltage ratio (1) (transformer). The ratio of the root-mean-square terminal voltage of a higher voltage winding to the root-mean-square terminal voltage of a lower-voltage winding, under specified conditions of the load. *See:* **turn ratio (transformer).** 207, 53
(2) (capacitance potential device, in combination with its coupling capacitor or bushing). The overall ratio between the root-mean-square primary line-to-ground voltage and the root-mean-square secondary voltage. *Note:* It is not the turn ratio of the transformer used in the network. *See also:* **outdoor coupling capacitor.** 341

voltage recovery time (arc-welding apparatus). With a welding power supply delivering current through a short-circuiting resistor whose resistance is equivalent to the normal load at that setting on the power supply, and measurement being made when the short circuit is suddenly removed, the time measured in seconds between the instant the short circuit is removed and the instant when voltage has reached 95 percent of its steady-state value. *See:* **voltage overshoot; voltage overshoot, effective.** 264

voltage reference (power supplies). A separate, highly regulated voltage source used as a standard to which the output of the power supply is continuously referred. *See also:* **power supply.** 186

voltage-reference tube. A gas tube in which the tube voltage drop is approximately constant over the operating range of current and relatively stable with time at fixed values of current and temperature. 190, 125

voltage reflection coefficient. The ratio of the complex number (phasor) representing the phase and magnitude of the electric field of the backward-traveling wave to that representing the forward-traveling wave at a cross section of a waveguide. The term is also used to denote the magnitude of this complex ratio. *See also: waveguide.* 179

voltage-regulating relay. A voltage-sensitive device that is used on an automatically operated voltage regulator to control the voltage of the regulated circuit. *See also: voltage regulator.* 257, 203, 53

voltage-regulating transformer (step-voltage regulator). A voltage regulator in which the voltage and phase angle of the regulated circuit are controlled in steps by means of taps and without interrupting the load. *See also: voltage regulator.* 203

voltage-regulating transformer, two-core. A voltage-regulating transformer consisting of two separate core and coil units in a single tank. *See also: voltage regulator.* 203

voltage-regulating transformer, two-core, excitation-regulating winding. In some designs, the main unit will have one winding operating as an autotransformer that performs both functions listed under regulating winding and excitation winding. Such a winding is called the excitation-regulating winding. *See also: voltage regulator.* 203

voltage-regulating transformer, two-core, excitation winding. The winding of the main unit that draws power from the system to operate the two-core transformer. *See also: voltage regulator.* 203

voltage-regulating transformer, two-core, excited winding. The winding of the series unit that is excited from the regulating winding of the main unit. *See also: voltage regulator.* 203

voltage-regulating transformer, two-core, regulating winding. The winding of the main unit in which taps are changed to control the voltage or phase angle of the regulated circuit through the series unit. *See also: voltage regulator.* 203

voltage-regulating transformer, two-core, series unit. The core and coil unit that has one winding connected in series in the line circuit. *See also: voltage regulator.* 203

voltage-regulating transformer, two-core, series winding. The winding of the series unit that is connected in series in the line circuit. *Note:* If the main unit of a two-core transformer is an autotransformer, both units will have a series winding. In such cases, one is referred to as the series winding of the autotransformer and the other, the series winding of the series unit. *See also: voltage regulator.* 203

voltage regulation (1) (constant-voltage transformer). The change in output (secondary) voltage which occurs when the load (at a specified power factor) is reduced from rated value to zero, with the primary impressed terminal voltage maintained constant. *Note:* In case of multi-winding transformers, the loads on all windings, at specified power factors, are to be reduced from rated kVA to zero simultaneously. The regulation may be expressed in per unit, or percent, on the base of the rated output (secondary) voltage at full load. 53

(2) (rectifier). The change in output voltage that occurs when the load is reduced from rated value to zero, or light transition load, with the values of all other quantities remaining unchanged. *Note:* The regulation may be expressed in volts or percentage of rated output voltage at full load. *See also: rectification; semiconductor rectifier stack.* 208

(3) (outdoor coupling capacitor). The variation in voltage ratio and phase angle of the secondary voltage of the capacitance potential device as a function of primary line-to-ground voltage variation over a specified range, when energizing a constant, linear impedance burden. *See also: outdoor coupling capacitor.* 341

(4) (direct-current generator). The final change in voltage with constant field-rheostat setting when the specified load is reduced gradually to zero, expressed as a percent of rated-load voltage, the speed being kept constant. *Note:* In practice it is often desirable to specify the over-all regulation of the generator and its driving machine thus taking into account the speed regulation of the driving machine. *See also: direct-current commutating machine.* 328

(5) (induction frequency converter). The rise in secondary voltage when the rated load at rated power factor is reduced to zero, expressed in percent of rated secondary voltage, the primary voltage, primary frequency, and the speed being held constant. *See also: asynchronous machine.* 328

(6) (synchronous generator). The rise in voltage with constant field current, when, with the synchronous generator operated at rated voltage and rated speed, the specified load at the specified power factor is reduced to zero, expressed as a percent of rated voltage. 63

(7) (line regulator circuits). *See: Zener diode; pulse-width modulation.*

(8) (thyristor converter). The change in output voltage that occurs when the load current is reduced from its rated value to zero, or light transition load, with rated sinusoidal alternating voltage applied to the thyristor power converter with the transformer on its rated tap, but excluding the corrective action of any voltage regulating means. *Note:* The regulation may be expressed in volts or in percent of rated volts. 121

voltage regulation curve (voltage regulation characteristic) (synchronous generator). The relationship between the armature winding voltage and the load on the generator under specified conditions and constant field current. 63

voltage regulator (transformer type). An induction device having one or more windings in shunt with, and excited from, the primary circuit, and having one or more windings in series between the primary circuit and the regulated circuit, all suitably adapted and arranged for the control of the voltage, or the phase-angle, or of both, of the regulated circuit. *Note:* A voltage regulator is basically an autotransformer. 257, 203, 186

voltage regulator, continuously acting type (rotating machinery). A regulator that initiates a corrective action for a sustained infinitesimal change in the controlled variable. 63

voltage regulator, direct-acting type (rotating machinery). A rheostatic-type regulator that directly controls the excitation of an exciter by varying the input to the exciter field circuits. *See also: direct-current commutating ma-*

chine.   63

voltage regulator, dynamic type (rotating machinery). A continuously acting regulator that does not require mechanical acceleration of parts to perform the regulating function. *Note:* Dynamic-type voltage regulators utilize magnetic amplifiers, rotating amplifiers, electron tubes, semiconductor elements, and/or other static components. *See also:* **direct-current commutating machine.**   63

voltage regulator, indirect-acting type (rotating machinery). A rheostatic-type regulator that controls the excitation of the exciter by acting on an intermediate device not considered part of the voltage regulator or exciter. *See also:* **direct-current commutating machine.**   63

voltage regulator, noncontinuously acting type (rotating machinery). A regulator that requires a sustained finite change in the controlled variable to initiate corrective action.   63

voltage regulator, synchronous-machine (rotating machinery). A synchronous-machine regulator that functions to maintain the voltage of a synchronous machine at a predetermined value, or to vary it according to a predetermined plan.   63

voltage-regulator tube. A glow-discharge cold-cathode tube in which the voltage drop is approximately constant over the operating range of current, and that is designed to provide a regulated direct-voltage output.   190, 125

voltage relay (1) (industrial control). A relay that functions at a predetermined value of voltage. *Note:* It may be an overvoltage relay, an undervoltage relay, or a combination of both. *See also:* **relay.**   206
(2) (power switchgear). A relay that responds to voltage.   103, 127

voltage response (close-talking pressure-type microphone). The ratio of the open-circuit output voltage to the applied sound pressure, measured by a laboratory standard microphone placed at a stated distance from the plane of the opening of the artificial voice. *Note:* The voltage response is usually measured as a function of frequency. *See also:* **close-talking pressure-type microphone.**   249

voltage response, exciter. The rate of increase or decrease of the exciter voltage when a change in this voltage is demanded. It is the rate determined from the exciter voltage response curve that if maintained constant would develop the same exciter voltage-time area as is obtained from the curve for a specified period. The starting point for determining the rate of voltage change shall be the initial value of the exciter voltage-time response curve. *See also:* **asynchronous machine; direct-current commutating machine.**   63

voltage response ratio, excitation-system (rotating machinery). The numerical value that is obtained when the excitation-system voltage response in volts per second, measured over the first $\frac{1}{2}$-second interval unless otherwise specified, is divided by the rated-load field voltage of the synchronous machine. *Note:* This response, if maintained constant, would develop, in $\frac{1}{2}$ second, the same excitation voltage-time area as attained by the actual response.   63

voltage response, synchronous-machine excitation-system. The rate of increase or decrease of the excitation-system output voltage, determined from the synchronous machine excitation-system voltage-time response curve, that if maintained constant would develop the same excitation-system voltage-time areas as are obtained from the curve for a specified period. The starting point for determining the rate of voltage change shall be the initial value of the synchronous-machine excitation-system voltage-time response curve.   63

voltage restraint (relay or control equipment). A method of restraining the operation by means of a voltage input that opposes the typical response of the relay to other inputs.   103, 60, 202, 127, 79

voltage sensing relay. (1) A term correctly used to designate a special-purpose voltage-rated relay that is adjusted by means of a voltmeter across its terminals in order to secure pickup at a specified critical voltage without regard to coil or heater resistance and resulting energizing current at that voltage. (2) A term erroneously used to describe a general-purpose relay for which operational requirements are expressed in voltage.   341

voltage sensitivity (nonlinear capacitor). *See:* **voltage coefficient of capacitance.** *See also:* **nonlinear capacitor.**

voltage sets (polyphase circuit). The voltages at the terminals of entry to a polyphase circuit into a delimited region are usually considered to consist of two sets of voltages: the line-to-line voltages, and the line to-neutral voltages. If the phase conductors are identified in a properly chosen sequence, the voltages between the terminals of entry of successive pairs of phase conductors form the set of line-to-line voltages, equal in number to the number of phase conductors. The voltage from the successive terminals of entry of the phase conductors to the terminal of entry of the neutral conductor, if one exists, or to the true neutral point, form the set of line-to-neutral voltages, also equal in number to the number of phase conductors. In case of doubt, the set intended must be identified. In the absence of other information, stated or implied, the line-to-neutral-conductor set is understood. *Notes:* (1) Under abnormal conditions the voltage of the neutral conductor and of the true neutral point may not be the same. Therefore it may become necessary to designate which is intended when the line-to-neutral voltages are being specified. (2) The set of line-to-line voltages may be determined by taking the differences in pairs of the successive line-to-neutral voltages. The line-to-neutral voltages can be determined from the line-to-line voltages by an inverse process only when the voltage between the neutral conductor and the true neutral point is completely specified, or equivalent additional information is available. If instantaneous voltages are used, algebraic differences are taken, but if root-mean-square voltages are used, information regarding relative phase angles must be available, so that the voltages may be expressed in phasor form and the phasor differences taken. (3) This definition may be applied to a two-phase, four-wire or five-wire circuit. A two-phase, three-wire circuit should be treated as a special case. *See also:* **network analysis.**   210

voltage spread. The difference between maximum and minimum voltages.   260

voltage-stabilizing tube. *See:* **voltage regulator tube.**

voltage standing-wave ratio (VSWR) (mode in a waveguide). The ratio of the magnitude of the transverse electric field in a plane of maximum strength to the magnitude at the equivalent point in an adjacent plane of minimum field strength. *See also:* **waveguide.**   179

voltage surge, internal (thyristor converter). Voltage surge caused by sources within a converter. *Note:* It may originate from blowing fuses, hole storage recovery phenomena, etcetera. Internal voltage surges are substantially under control of the circuit designer.   121

voltage surge suppressor (semiconductor rectifier). A device used in the semiconductor rectifier to attenuate surge voltages of internal or external origin. Capacitors, resistors, nonlinear resistors, or combinations of these may be employed. Nonlinear resistors include electronic and semiconductor devices. *See also:* **semiconductor rectifier stack.**   208

voltage test. *See:* **controlled overvoltage test.**

voltage-time product (high- and low-power pulse transformers). The time integral of a voltage pulse applied to a transformer winding.   32, 33

voltage-time product rating (high- and low-power pulse transformers) (transformer winding). Considered as being a constant and is the maximum voltage-time product of a voltage pulse that can be applied to the winding before a specified level of core saturation-region effects is reached. The level of core saturation-region effects is determined by observing either the shape of the output voltage pulse for a specified degradation (for example, a maximum tilt (droop)), or the shape of the exciting current pulse for a specified departure from linearity (for example, deviation from a linear ramp by a given percentage).   32, 33

voltage-time response, synchronous-machine excitation-system. The output voltage of the excitation system, expressed as a function of time, following the application of prescribed inputs under specified conditions.   63

voltage-time response, synchronous-machine voltage-regulator. The voltage output of the synchronous-machine voltage regulator expressed as a function of time following the application of prescribed inputs under specified conditions.   63

voltage to ground (1) (transformer) (neutral grounding device). The voltage between any live conductor of a circuit and the earth. *Note:* Where safety considerations are involved, the voltage to ground which may occur in an ungrounded circuit is usually the highest voltage normally existing between the conductors of the circuit, but in special circumstances higher voltages may occur. *See also:* **voltage.**   53, 91

(2) (transmission and distribution). (A) (a grounded circuit). The highest nominal voltage available between any conductor of the circuit and that point or conductor of the circuit which is grounded.

(B) (an ungrounded circuit). The highest nominal voltage available between any two conductors of the circuit concerned.   262

voltage to ground of a conductor (transmission and distribution) (1) (a grounded circuit). The nominal voltage between such conductor and that point or conductor of the circuit which is grounded.

(2) (an ungrounded circuit). The highest nominal voltage available between such conductor and any other conductor of the circuit concerned.   262

voltage, touch. *See:* **touch voltage.**

voltage transformer. An instrument transformer intended to have its primary winding connected in shunt with a power supply circuit, the voltage of which is to be measured or controlled.   53

voltage, transverse mode (surge withstand capability tests). The voltage at a given location between two conductors of a group. *Syn.* **differential mode voltage.**   90

voltage-tunable magnetron (microwave tubes). A magnetron in which the resonant circuit is heavily loaded ($Q_L = 1$ to 10) and in which the supply of electrons to the interaction space is restricted whereby the frequency of oscillation becomes proportional to the plate voltage. *See also:* **magnetrons.**   190

voltage-type telemeter. A telemeter that employs the magnitude of a single voltage as the translating means. *See also:* **telemetering.**   103, 202

voltage winding (or transformer) for regulating equipment. The winding (or transformer) which supplies voltage within close limits of accuracy to instruments, such as contact-making voltmeters.   53

voltage-withstand test (1) (insulation materials). The application of a voltage higher than the rated voltage for a specified time for the purpose of determining the adequacy against breakdown of insulation materials and spacing under normal conditions. *See also:* **dielectric tests (voltage-withstand tests).**   328

(2) (rotating machinery). *See:* **overvoltage test.** *See also:* **asynchronous machine; direct-current commutating machine.**

voltaisation (galvanization) (electrotherapy). *See:* **galvanism.**

voltameter (coulometer). *See:* **coulometer; instrument.**

volt-ammeter. An instrument having circuits so designed that the magnitude either of voltage or of current can be measured on a scale calibrated in terms of each of these quantities. *See also:* **instrument.**   328

volt-ampere. The unit of apparent power in the International System of Units (SI). The volt-ampere is the apparent power at the points of entry of a single-phase, two-wire system when the product of the root-mean-square value in amperes of the current by the root-mean-square value in volts of the voltage is equal to one.   210

volt-ampere-hour meter. An electricity meter that measures and registers the integral, with respect to time, of the apparent power in the circuit in which it is connected. The unit in which the integral is measured is usually the kilovolt-ampere-hour. *See also:* **electricity meter (meter).**   328

volt-ampere loss (electric instrument). *See:* **apparent power loss (electric instrument).**

volt-ampere meter. An instrument for measuring the apparent power in an alternating-current circuit. It is provided with a scale graduated in volt-amperes or in kilovolt-amperes. *See also:* **instrument.**   328

volt efficiency (storage battery) (storage cell). The ratio of the average voltage during the discharge to the average voltage during the recharge. *See also:* **electrochemistry; charge.**   328

voltmeter. An instrument for measuring the magnitude of electric potential difference. It is provided with a scale, usually graduated in either volts, millivolts, or kilovolts. If the scale is graduated in millivolts or kilovolts the instrument is usually designated as a millivoltmeter or a kilovoltmeter. *See also:* instrument. 328

voltmeter-ammeter. The combination in a single case, but with separate circuits, of a voltmeter and an ammeter. *See also:* instrument. 328

volts per hertz relay (power switchgear). A relay whose pickup is a function of the ratio of voltage to frequency. 103, 127

volt-time curve (surge arresters) (1) (impulses with fronts rising linearly) (surge arresters). The curve relating the disruptive-discharge voltage of a test object to the virtual time to chopping. The curve is obtained by applying voltages that increase at different rates in approximately linear manner. 308, 62

(2) (standard impulses). A curve relating the peak value of the impulse causing disruptive discharge of a test object to the virtual time to chopping. The curve is obtained by applying standard impulse voltages of different peak values. 308, 62

volume (1) (electric circuit). The magnitude of a complex audio-frequency wave as measured on a standard volume indicator. *Notes:* (1) Volume is expressed in volume units (vu). (2) The term volume is used loosely to signify either the intensity of a sound or the magnitude of an audiofrequency wave. 239

(2) (data transmission). In general, volume is the intensity or loudness of sound. In a telephone or other audio-frequency circuit, a measure of the power corresponding to an audio-frequency wave at that point (expressed in decibels). 59

(3) (International System of Units) (SI). The SI unit of volume is the cubic meter. This unit, or one of the regularly formed multiples such as the cubic centimeter, is preferred for all applications. The special name liter has been approved for the cubic decimeter, but use of this unit is restricted to the measurement of liquids and gases. No prefix other than milli- should be used with liter. *See also:* units and letter symbols. 21

volume control. *See:* gain control.

volume density of magnetic pole strength. At any point of the medium in a magnetic field, the negative of the divergence of the magnetic polarization vector there. 210

volume equivalent (complete telephone connection, including the terminating telephone sets). A measure of the loudness of speech reproduced over it. The volume equivalent of a complete telephone connection is expressed numerically in terms of the trunk loss of a working reference system when the latter is adjusted to give equal loudness. *Note:* For engineering purposes, the volume equivalent is divided into volume losses assignable to (1) the station set, subscriber line, and battery-supply circuit that are on the transmitting end; (2) the station set, subscriber line, and battery supply that are on the receiving end; (3) the trunk; and (4) interaction effects arising at the trunk terminals. 328

volume indicator (standard volume indicator). A standardized instrument having specified electric and dynamic characteristics and read in a prescribed manner, for indicating the volume of a complex electric wave such as that corresponding to speech or music. *Notes:* (1) The reading in volume units is equal to the number of decibels above a reference volume. The sensitivity is adjusted so that the reference volume or zero volume unit is indicated when the instrument is connected across a 600-ohm resistor in which there is dissipated a power of 1 milliwatt at 1000 hertz. (2) Specifications for a volume indicator are given in American National Standard Volume Measurements of Electrical Speech and Program Waves, C16.5. *See also:* instrument; volume unit. 122, 176, 196

volume limiter. A device that automatically limits the output volume of speech or music to a predetermined maximum value. *See also:* peak limiter. 328

volume-limiting amplifier. An amplifier containing an automatic device that functions when the input volume exceeds a predetermined level and so reduces the gain that the output volume is thereafter maintained substantially constant notwithstanding further increase in the input volume. *Note:* The normal gain of the amplifier is restored when the input volume returns below the predetermined limiting level. *See also:* amplifier. 111

volume range (1) (transmission system). The difference, expressed in decibels between the maximum and minimum volumes that can be satisfactorily handled by the system.

(2) (complex audio-frequency signal). The difference, expressed in decibels, beween the maximum and minimum volumes occurring over a specified period of time. 328

volume resistivity. *See:* resistivity, volume.

volume unit (vu). The unit in which the standard volume indicator is calibrated. *Note:* One volume unit equals one decibel for a sine wave but volume units should not be used to express results of measurements of complex waves made with devices having characteristics differing from those of the standard volume indicator. *See:* volume indicator. 239

volumetric radar. *See:* three-dimensional radar.

VOR (very high-frequency omnidirectional range). A navigation aid operating at VHF and providing radial lines of position in any direction as determined by bearing selection within the receiving equipment; it emits a (variable) modulation whose phase relative to a reference modulation is different for each bearing of the receiving point from the station. 13

vortac (electronic navigation). A designation applied to certain navigation stations (primarily in the United States) in which both VOR and tacan are used; the distance function in tacan is used with VOR to provide bearing and distance (VOR/DME) (rho theta) navigation. *See also:* navigation. 187

vowel articulation (percent vowel articulation). The percent articulation obtained when the speech units considered are vowels (usually combined with consonants into meaningless syllables). *See also:* articulation (percent articulation); volume equivalent. 176

VPI. *See:* vacuum-pressure impregnation.

VSWR. *See:* voltage standing-wave ratio.

V-terminal voltage (electromagnetic compatibility). Terminal voltage measured with a V network between each mains conductor and earth. *See also:* electromagnetic compatibility. 220, 199

vu. *See:* volume unit.

W

wait (test, measurement and diagnostic equipment). A programmed instruction which causes an automatic test system to remain in a given state for a predetermined period. 54

waiting, call. A feature providing a signal to a busy called line to indicate that another call is waiting. 103

waiting-passenger indicator (elevators). An indicator that shows at which landings and for which direction elevator-hall stop or signal calls have been registered and are unanswered. *See also:* **control.** 328

walkie-talkie. A two-way radio communication set designed to be carried by one person, usually strapped over the back, and capable of operation while in motion. *See also:* **radio transmission.** 328

walk-off mode (laser-maser). A walk-off mode will be characterized by successive shifts per reflection in the location of a maximum in the transverse field intensity. *See also:* **transverse mode.** 363

wall bushing. A bushing intended primarily to carry a circuit through a wall or other grounded barrier in a substantially horizontal position. Both ends must be suitable for operating in air. *See also:* **bushing.** 348

wall telephone set. A telephone set arranged for wall mounting. *See also:* **telephone station.** 328

wander (radar). *See:* **scintillation.**

warble-tone generator. A voice-frequency oscillator, the frequency of which is varied cyclically at a subaudio rate over a fixed range. It is usually used with an integrating detector to obtain an averaged transmission or crosstalk measurement. *See also:* **auxiliary device to an instrument.** 328

warm-up time (1) (power supplies). The time (after power turn on) required for the output voltage or current to reach an equilibrium value within the stability specification. *See also:* **power supply.** 186
(2) (gyro; accelerometer). The time interval required for a gyro or accelerometer to reach specified performance from the instant that it is energized under specified operating conditions. 46
(3) (nuclear power generating station). The time, following power application to a module, required for the output to stablize within specifications. 355

warning whistle. *See:* **audible cab indicator.**

washer (rotating machinery). *See:* **collar.**

watchman's reporting system. A supervisory system arranged for the transmission of a patrolling watchman's regularly recurrent report signals to a central supervisory agency from stations along his patrol route. *See also:* **protective signaling.** 328

water. *See:* **acoustic properties of water.**

water-air-cooled machine. A machine that is cooled by circulating air that in turn is cooled by circulating water. *Note:* The machine is so enclosed as to prevent the free exchange of air between the inside and outside of the enclosure, but not sufficiently to be termed airtight. It is provided with a water-cooled heat exchanger for cooling the ventilating air and a fan or fans, integral with the rotor shaft or separate, for circulating the ventilating air. *See:* **asynchronous machine; direct-current commutating machine.** 63

water conditions, adverse (hydroelectric generating station). Water conditions which limit electric energy production. 112

water conditions, average (hydroelectric generating station). Precipitation and runoff conditions which provide water for hydroelectric energy production approximating the average amount and distribution available over a long time period, usually the period of record. 112

water conditions, median (hydroelectric generating station). Precipitation and runoff conditions which provide water for hydroelectric energy development approximating the median amount and distribution available over a long-time period, usually the period of record. 112

water-cooled (rotating machinery). (1) A term applied to apparatus cooled by circulating water, the water or water ducts coming in direct contact with major parts of the apparatus. (2) In certain types of machine, it is customary to apply this term to the cooling of the major parts by enclosed air or gas ventilation, where water removes the heat through an air-to-water or gas-to-water heat exchanger. *See:* **asynchronous machine; direct-current commutating machine.** 63

water cooler (rotating machinery). A cooler using water as one of the fluids. 63

waterflow-alarm system (protective signaling). An alarm system in which signal transmission is initiated automatically by devices attached to an automatic sprinkler system and actuated by the flow through the sprinkler system pipes of water in excess of a predetermined maximum. *See also:* **protective signaling.** 328

water inertia time (hydraulic turbines). A characteristic time, usually taken at rated conditions, due to inertia of the water in the water passages from intake to exit defined as:

$$T_w = \frac{Q_r}{gH_r} \int \frac{dL}{A} \approx \frac{Q_r}{gH_r} \sum \frac{L}{A}$$

where:
A = area of each section
L = corresponding length
Q_r = rated discharge
H_r = rated head
g = acceleration due to gravity 8, 58

water load (high-frequency circuits). A matched termination in which the electromagnetic energy is absorbed in a stream of water for the purpose of measuring power by continuous-flow calorimetric methods. *See also:* **waveguide.** 244, 179

water-motor bell. A vibrating bell operated by a flow of water through its water-motor striking mechanism. *See also:* **protective signaling.** 328

waterproof electric blasting cap. A cap specially insulated to secure reliability of firing when used in wet work. *See also:* **blasting unit.** 328

waterproof enclosure. An enclosure so constructed that any moisture or water leakage that may occur into the enclosure will not interfere with its successful operation. In the case of motor or generator enclosures, leakage that may occur around the shaft may be considered permissible provided it is prevented from entering the oil reservoir and provision is made for automatically draining the motor or generator enclosure. 3

waterproof machine (rotating machinery). A machine so constructed that water directed on it under prescribed conditions cannot cause interference with satisfactory operation. *See also:* **asynchronous machine; direct-current commutating machine.** 63

watertight. So constructed that water will not enter the enclosing case under specified conditions. *Note:* A common form of specification for watertight is: "So constructed that there shall be no leakage of water into the enclosure when subjected to a stream from a hose with a 1-inch nozzle and delivering at least 65 gallons per minute, with the water directed at the enclosure from a distance of not less than 10 feet for a period of 5 minutes, during which period the water may be directed in one or more directions as desired." *See also:* **distribution center.**
256, 103, 225, 202, 206, 27, 53

watertight door-control system. A system of control for power-operated watertight doors providing individual local control of each door and, at a remote station in or adjoining the wheelhouse, individual control of any door, collective control of all doors, and individual indication of open or closed condition. *See also:* **marine electric apparatus.**
328

watertight enclosure. An enclosure constructed so that a stream of water from a hose not less than 1 in. (25.4 mm) in diameter under a head of 35 ft. (10.6 m) from a distance of 10 ft. (3.05 m) can be played on the enclosure from any direction for a period of 15 minutes without leakage. The hose nozzle shall have a uniform inside diameter of 1 in. (25.4 mm). 3

water treatment equipment. Any apparatus such as deionizers, electrolytic targets, filters, or other devices employed to control electrolysis, corrosion, scaling, or clogging in water systems. *See also:* **power rectifier.**
208

watt (1) (general). The unit of power in the International System of Units (SI). The watt is the power required to do work at the rate of 1 joule per second. 210
(2) (laser-maser) (W). The unit of power, or radiant flux.
363

watthour. 3600 joules. 210

watthour capacity (storage battery) (storage cell). The number of watthours that can be delivered under specified conditions as to temperature, rate of discharge, and final voltage. *See also:* **battery (primary or secondary).** 328

watthour constant (meter). The registration expressed in watthours corresponding to one revolution of the rotor. *Note:* It is commonly denoted by the symbol K_h. When a meter is used with instrument transformers, the watthour constant is expressed in terms of primary watthours. For a secondary test of such a meter, the constant is the primary watthour constant divided by the product of the nominal ratios of transformation. *See also:* **electricity meter (meter); watthour meter.** 212

watthour-demand meter. A watthour meter and a demand meter combined as a single unit. *See also:* **electricity meter.** 328

watthour efficiency (storage battery) (storage cell). The energy efficiency expressed as the ratio of the watthours

output to the watthours of the recharge. *See also:* **charge.**
328

watthour meter. An electricity meter that measures and registers the integral, with respect to time, of the active power of the circuit in which it is connected. This power integral is the energy delivered to the circuit during the interval over which the integration extends, and the unit in which it is measured is usually the kilowatthour.
212

watt loss (electric instrument). *See:* **power loss (electric instrument).**

wattmeter. An instrument for measuring the magnitude of the active power in an electric circuit. It is provided with a scale usually graduated in either watts, kilowatts, or megawatts. If the scale is graduated in kilowatts or megawatts, the instrument is usually designated as a kilowattmeter or megawattmeter. *See also:* **instrument.**
328

watt-second constant (meter). The registration in watt-seconds corresponding to one revolution of the rotor. *Note:* The wattsecond constant is 3600 times the watthour constant and is commonly denoted by the symbol K_s. *See also:* **electricity meter (meter).** 328

wave (1). A disturbance that is a function of time or space or both. 111
(2). A disturbance propagated in a medium or through space. *Notes:* (A) Any physical quantity that has the same relationship to some independent variable (usually time) that a propagated disturbance has, at a particular instant, with respect to space, may be called a wave. (B) Disturbance, in this definition, is used as a generic term indicating not only mechanical displacement but also voltage, current, electric field strength, temperature, etcetera.
210
(3) (surge arrester). The variation with time of current, potential, or power at any point in an electric circuit. 2
(4) (automatic control). A vibrational disturbance propagated in a medium or energy quanta propagated through space. A wave may be a function of time, or space, or of time and space. *Note:* In control systems, examples of waves are signals and variables. *See also:* **pulse.** 56
(5) (pulse terms). A modification of the physical state of a medium which propagates in the medium as a function of time as a result of one or more disturbances. 254

wave analyzer. An electric instrument for measuring the amplitude and frequency of the various components of a complex current or voltage wave. *See also:* **instrument.**
328

wave antenna. *See:* **Beverage antenna.**

wave clutter. Clutter caused by echoes from waves of the sea. *See also:* **radar.** 328

wave filter (data transmission). A wave filter is a transducer for separating waves on the basis of their frequency. It introduces relatively small insertion loss to waves in one or more frequency bands and relatively large insertion loss to waves of other frequencies. *See:* **filter.** 59

waveform (wave) (1) (pulse techniques). The geometrical shape as obtained by displaying a characteristic of the wave as a function of some variable, usually time, when it is plotted over one primitive period. *Note:* It may be expressed mathematically by dividing the function by the maximum absolute value of the function that occurs during the period. *See also:* **pulse.**   210, 185

(2) (pulse terms) (pulse waveform) (transition waveform). A manifestation or representation (that is, graph, plot, oscilloscope presentation, equation(s), table of coordinate or statistical data, etcetera) or a visualization of a wave, pulse, or transition. (1) The term **pulse waveform** is included in the term waveform. (2) The term **transition waveform** is included in the terms pulse waveform and waveform.   254

waveform-amplitude distortion. Nonlinear distortion in the special case where the desired relationship is direct proportionality between input and output. *Note:* Also sometimes called **amplitude distortion.** *See:* **nonlinear distortion.**   349

waveform distortion (oscilloscopes). A displayed deviation from the representation of the input reference signal. *See:* **oscillograph.**   185

waveform epoch (pulse terms). The span of time for which waveform data are known or knowable. A waveform epoch manifested by equations may extend in time from minus infinity to plus infinity or, like all waveform data, may extend from a first datum time, t_0, to a second datum time, t_1. *See:* The single pulse diagram below.   254

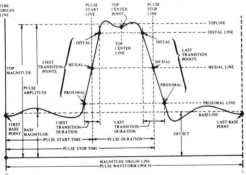

The single pulse.

waveform epoch contraction (pulse measurement). A technique for the determination of the characteristics of individual pulse waveforms (or pulse waveform features) wherein the waveform epoch (or pulse waveform epoch) is contracted in time to a pulse waveform epoch (or transition waveform epoch) for the determination of time or magnitude characteristics. In any waveform epoch contraction procedure two or more sets of time or magnitude reference lines may exist, and the set of reference lines being used in any pulse measurement process shall be specified.   15

waveform epoch expansion (pulse measurement). A technique for the determination of the characteristics of a transition waveform (or pulse waveform) wherein the transition waveform epoch (or pulse waveform epoch) is expanded in time to a pulse waveform epoch (or waveform epoch) for the determination of magnitude or time reference lines. The reference lines determined by analysis of the pulse waveform (or waveform) are transferred to the transition waveform (or pulse waveform) for the determination of characteristics. In any waveform epoch expansion procedure two or more sets of reference lines may exist, and the set of reference lines being used in any pulse

measurement process shall be specified.   15

waveform formats. Waveforms may exiṣt, be recorded, or be stored in a variety of formats. It is assumed that: (1) waveform formats are in terms of Cartesian coordinates, or some transform thereof; (2) conversion from one waveform format to any other is possible; and (3) such waveform format conversions can be made with precision, accuracy, and resolution which is consistent with the accuracy desired in the pulse measurement process.   15

waveform influence of root-mean-square (RMS) responding instruments. The change in indication produced in an RMS responding instrument by the presence of harmonics in the alternating electrical quantity under measurement. In magnitude it is the deviation between an indicated RMS value of an alternating electrical quantity and the indication produced by the measurement of a pure sine-wave form of equal RMS value. *See also:* **instrument.**   280

waveform pulse. A waveform or a portion of a waveform containing one or more pulses or some portion of a pulse. *See also:* **pulse.**   185

waveform reference. A specified waveform, not necessarily ideal, relative to which waveform measurements, derivations, and definitions may be referred.   185

waveforms produced by continuous time superposition of simpler waveforms (pulse terms). *See:* **pulse train; pulse train time-related definitions; square wave.**

waveforms produced by magnitude superposition (pulse terms). *See:* **offset; offset waveform; composite waveform.**

waveforms produced by noncontinuous time superposition of simpler waveforms (pulse terms). *See:* **pulse burst; pulse burst time-related definitions.**

waveforms produced by operations on waveforms (pulse terms). All envelope definitions in this section are based on the cubic natural spline (or its related approximation, the draftsman's spline) with knots at specified points. All burst envelopes extend in time from the first to the last knots specified, the remainder of the waveform being (1) that portion of the waveform which precedes the first knot and (2) that portion of the waveform which follows the last knot. Burst envelopes and their adjacent waveform bases, taken together, comprise a continuous waveform which has a continuous first derivative except at the first and last knots of the envelope. *See also:* **pulse train top (base) envelope; pulse burst top envelope; pulse burst base envelope.**   254

waveform test (rotating machinery). A test in which the waveform of any quantity associated with a machine is recorded. *See also:* **asynchronous machine.**   63

wave front (1) (progressive wave in space). A continuous surface that is a locus of points having the same phase at a given instant.   328

(2) (impulse in a conductor). That part (in time or distance) between the virtual-zero point and the point at which the impulse reaches its crest value.   64

(3) (surge or impulse). That part which occurs prior to the crest value. The rising part of an impulse wave.   244, 62, 2

(4) (signal wave envelope). That part (in time or distance) between the initial point of the envelope and the point at which the envelope reaches its crest.   328

waveguide. (1) Broadly, a system of material boundaries

capable of guiding electromagnetic waves. (2) More specifically, a transmission line comprising a hollow conducting tube within which electromagnetic waves may be propagated or a solid dielectric or dielectric filled conductor for the same purpose. (3) A system of material boundaries or structures for guiding transverse-electromagnetic mode, often and originally a hollow metal pipe for guiding electromagnetic waves.   185

waveguide calorimeter (waveguide components). A waveguide or transmission line structure which uses the temperature rise in a medium as a measure of absorbed power. The medium, typically water or a thermoelectric element, is either the power-absorbing agent or has heat transferred to it from a power-absorbing element.   166

waveguide component. A device designed to be connected at specified ports in a waveguide system.   319, 179

waveguide cutoff frequency (critical frequency). *See:* **cutoff frequency.**

waveguide gasket. *See:* **gasket, waveguide.**

waveguide joint. A connection between two sections of waveguide. *See also:* **waveguide.**   185

waveguide plunger. *See:* **sliding short-circuit.**

waveguide switch (waveguide system). A device for stopping or diverting the flow of high-frequency energy as desired. *See also:* **waveguide.**   244, 179

waveguide termination. *See:* **cavity; unloaded applicator impedance.**

waveguide-to-coaxial transition. A mode changer for converting coaxial line transmission to rectangular waveguide transmission. *See also:* **waveguide.**   244, 179

waveguide wavelength (traveling wave at a given frequency). The distance along a uniform guide between points at which a field component (or the voltage or current) differs in phase by 2π radians. *Note:* It is equal to the quotient of phase velocity divided by frequency. For a waveguide with air dielectric, the waveguide wavelength is given by the formula:

$$\lambda_g = \frac{\lambda}{(1 - (\lambda^2/\lambda_c))^{1/2}}$$

where λ is the free-space wavelength and λ_c is the cut-off wavelength of the guide. *See also:* **waveguide.**

267, 319, 179, 185

wave heater (dielectric heating). A heater in which heating is produced by energy absorption from a traveling electromagnetic wave. *See also:* **dielectric heating.**   14

wave heating. The heating of a material by energy absorption from a traveling electromagnetic wave. *See also:* **dielectric heating; induction heating.**   14, 114

wave impedance (waveguides including transmission lines). The complex factor relating the transverse component of the magnetic field to the transverse component of the electric field at every point in any specified plane, for a given mode. *Note:* Both incident and reflected waves may be present. *See also:* **self-impedance; transmission line; waveguide.**   179

wave impedance, characteristic (traveling wave) (waveguides). (1) The wave impedance with the sign so chosen that the real part is positive. *Note:* In a given mode, in a homogeneously filled waveguide, this is constant for all points and all cross sections. *See also:* **self-impedance; waveguide.**

179

(2) Wave impedance for purely progressing waves. 185

wave interference. The variation of wave amplitude with distance or time, caused by the superposition of two or more waves. *Notes:* (1) As most commonly used, the term refers to the interference of waves of the same or nearly the same frequency. (2) Wave interference is characterized by a spatial or temporal distribution of amplitude of some specified characteristic differing from that of the individual superposed waves. *See also:* **wave front.**

146, 180

wavelength (1) (sinusoidal wave) (light emitting diodes). The distance between points of corresponding phase of two consecutive cycles. *Note:* The wavelength λ is related to the phase velocity v and the frequency f by $\lambda = v/f$. *See also:* **radio wave propagation.**   162

(2) (periodic wave). 2π times the limit of the ratio of the distance between two equiphase surfaces to the phase difference as these quantities go to zero.   210, 180

(3) (laser-maser). The distance between two points in a periodic wave which have the same phase is called a wavelength.   363

wavelength constant. *See:* **phase constant.**

wavelength shifter (scintillator). A photofluorescent compound used with a scintillator material to absorb photons and emit related photons of a longer wavelength. *Note:* The purpose is to cause more efficient use of the photons by the phototube or photocell. *See also:* **phototube.**   335

wavemeter. An instrument for measuring the wavelength of a radio-frequency wave. The following are representative types: resonant-cavity, resonant-circuit, and standing-wave. *Note:* The standing-wave type is exemplified by a Lecher wire. Wavemeters may also be classified as (1) transmission type in which the resonant element and the detector are so arranged as to give a maximum response at resonance; and (2) suppression (absorption or reaction) type, which provide minimum response at resonance. *See also:* **frequency meter; instrument; cavity resonator.**

328

wave normal. A unit vector normal to an equiphase surface with its positive direction taken on the same side of the surface as the direction of propagation. *Note:* In isotropic media, the wave normal is in the direction of propagation. *See also:* **radio wave propagation; waveguide.**   267

waves, electrocardiographic, P, Q, R, S, and T (medical electronics) (in electrocardiograms obtained from electrodes placed on the right arm and left leg). The characteristic tracing consists of five consecutive waves; P, a prolonged, low, positive wave; Q, brief, low, negative; R, brief, high, positive; S, brief, low, negative, and T, prolonged, low, positive.   192

wave shape (impulse test wave). The graph of the wave as a function of time.   2

wave shape designation (impulse) (surge arrester). The wave shape of an impulse (other than rectangular) of a current or voltage is designated by a combination of two numbers. The first, an index of the wave front, is the virtual duration of the wave front in microseconds. The second, an index of the wave tail, is the time in microseconds from virtual zero to the instant at which one-half of the crest value is reached on the wave tail. Examples are 1.2×50 and 8×20 waves. The wave shape of a rectangular impulse of current or voltage is designated by two numbers. The first designates the minimum value of current or voltage which is sustained for the time in microseconds designated by the second number. An example is the 75A × 1000 wave. 2

wave, square (pulse techniques). *See:* **square wave.**

wave tail (impulse in a conductor). That part between the crest value and the end of the impulse. 2

wave tilt (data transmission). The forward inclination of a radio wave due to its proximity to ground. *See also:* **radiation.** 59

wave train. A limited series of wave cycles caused by a periodic disturbance of short duration. *See also:* **pulse train.** 56

wave vector in physical media. The complex vector $\overline{k}$ in plane wave solutions of the form $e^{-j\overline{k}\cdot\overline{r}}$, for an $e^{j\omega t}$ time variation and $\overline{r}$ the position vector. *See also:* **propagation vector in physical media.** 111

wave winding. A winding that progresses around the armature by passing successively under each main pole of the machine before again approaching the starting point. In commutator machines, the ends of individual coils are not connected to adjacent commutator bars. *See also:* **asynchronous machine; direct-current commutating machine.** 328

way point (navigation). A selected point on or near a course line and having significance with respect to navigation or traffic control. *See also:* **navigation.** 187

way station (data transmission). A telegraph term for one of the stations on a multipoint circuit. 59

wear-out failure (reliability). *See:* **failure, wear-out.**

wear-out failure period (reliability). That possible period during which the failure rate increases rapidly in comparison with the preceding period. *See also:* **constant failure rate period** for curve showing the failure rate pattern when the terms of **early failure period** and **constant failure rate period** apply. 164

wearout period (reliability analysis) (nuclear power generating station). The time interval, following the period of constant failure rate, during which failures occur at an increasing rate. 31, 29

wear point (cable plowing). A removable tip on the end of some shanks or plow blades. 52

weather, adverse (generating station). Designates weather conditions which cause an abnormally high rate of forced outages for exposed components during the periods such conditions persist, but do not qualify as major storm disasters. Adverse weather conditions can be defined for a particular system by selecting the proper values and combinations of conditions reported by the Weather Bureau: thunderstorms, tornadoes, wind velocities, precipitation, temperature, etc. *Note:* This definition derives from transmission and distribution applications and does not necessarily apply to generation outages. *See also:* **major storm disaster.** 112

weather, normal (generating station). Includes all weather not designated as adverse or major storm disaster. *Note:* This definition derives from transmission and distribution applications and does not necessarily apply to generation outages. 112

weatherproof (1) (outside exposure). So constructed or protected that exposure to the weather will not interfere with successful operation. *See:* **outdoor.**
103, 225, 202, 206, 27

(2) (conductor covering). Made up of braids of fibrous material that are thoroughly impregnated with a dense moistureproof compound after they have been placed on the conductor, or an equivalent protective covering designed to withstand weather conditions. 328

weatherproof enclosure. An enclosure for outdoor application designed to protect against weather hazards such as rain, snow, or sleet. *Note:* Condensation is minimized by use of space heaters. 103, 202

weather-protected machine. A guarded machine whose ventilating passages are so designed as to minimize the entrance of rain, snow, and airborne particles to the electric parts. *See also:* **asynchronous machine; direct-current commutating machine.** 63

weathertight. *See:* **raintight.**

weber. The unit of magnetic flux in the International System of Units (SI). The weber is the magnetic flux whose decrease to zero when linked with a single turn induces in the turn a voltage whose time integral is one volt-second. 210

wedge (rotating machinery). A tapered shim or key. *See:* **slot wedge.** *See also:* **rotor (rotating machinery); stator.** 63

wedge groove (wedge slot) (rotating machinery). A groove, usually in the side of a coil slot, to permit the insertion of and to retain a slot wedge. 63

wedge, slot. *See:* **wedge groove.**

wedge washer (salient pole) (rotating machinery). Insulation triangular in cross section placed underneath the inner ends of field coils and spanning between field coils. 63

weight coefficient (thermoelectric generator) (thermoelectric generator couple). The quotient of the electric power output by the device weight. *See also:* **thermoelectric device.** 248, 191

weighted peak flutter. Flutter and wow indicated by the weighted peak flutter measuring equipment specified in IEEE Std 193-1971, Method for Measurement of Weighted Peak Flutter of Sound Recording and Reproducing Equipment. *Note:* The meter indicates one-half the peak-to-peak demodulated signal. The frequency-response weighting network is specified in Table I, and the dynamic response of the system is given in Sec. 5.4, of the referenced standard. 145

weighting (1) (antennas). The artificial adjustment of measurements in order to account for factors that in the normal use of the device, would otherwise be different from the conditions during measurement. For example, background noise measurements may be weighted by applying factors or by introducing networks to reduce measured values in inverse ratio to their interfering effects.
111

(2) (sound recording and reproducing equipment). The use of a psychoacoustically determined time response and frequency response in an objective measuring equipment. This is done in order to obtain indications which better predict the subjective values than would wideband measurement with a meter having either an instantaneous time response, or a long-time average or rms response. 145

weight transfer compensation. A system of control wherein the tractive forces of individual traction motors may be adjusted to compensate for the transfer of weight from one axle to another when exerting tractive force. *See also:*

multiple-unit control. 328

weld decay (corrosion). Localized corrosion at or adjacent to a weld. 221, 205

welding arc voltage. The voltage across the welding arc. 264

Weston normal cell. A standard cell of the cadmium type containing a saturated solution of cadmium sulphate as the electrolyte. *Note:* Strictly speaking this cell contains a neutral solution, but acid cells are now in more common use. *See also:* **electrochemistry.** 328

wet cell. A cell whose electrolyte is in liquid form. *See also:* **electrochemistry.** 328

wet contact (telephone switching systems). A contact through which direct current flows. *Note:* The term has significance because of the healing action of direct current flowing through contacts. 55

wet-dry signaling, (telephone switching systems). Two-state signaling achieved by the application and removal of battery at one end of a trunk. 55

wet electrolytic capacitor. A capacitor in which the dielectric is primarily an anodized coating on one electrode, with the remaining space between the electrodes filled with a liquid electrolytic solution. 341

wet location. A location subject to saturation with water or other liquids, such as locations exposed to weather, washrooms in garages, and like locations. Installations underground or in concrete slabs or masonry in direct contact with the earth shall be considered as wet locations. *See also:* **distribution center.** 256

wetting. The free flow of solder alloy, with proper application of heat and flux, on a metallic surface to produce an adherent bond. 284

wetting agent (electroplating) (surface active agent). A substance added to a cleaning, pickling or plating solution to decrease its surface tension. *See also:* **electroplating.** 328

wet-wound (rotating machinery). A coil in which the conductors are coated with wet resin in passage to the winding form, or on to which a bonding or insulating resin is applied on each successive winding layer to produce an impregnated coil. *See also:* **rotor (rotating machinery); stator.** 63

Wheatstone bridge. A 4-arm bridge, all arms of which are predominantly resistive. *See also:* **bridge.** 328

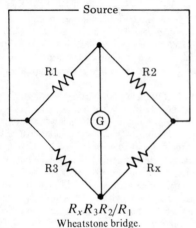

$$R_x R_3 R_2 / R_1$$

Wheatstone bridge.

wheel speed sensitivity (tuned rotor gyro). *See:* **rotor speed sensitivity.**

whip antenna. A thin flexible monopole antenna. 111

whistle operator. A device to provide automatically the timed signals required by navigation laws when underway in fog, and also manual control of electrical operation of a whistle or siren, or both, for at-will signals. 328

white (color television). Used most commonly in the nontechnical sense. More specific usage is covered by the term **achromatic locus,** and this usage is explained in the note under the term **achromatic locus.** *See:* **reference white.** 18

white compression (white saturation) (television). The reduction in gain applied to a picture signal at those levels corresponding to light areas in a picture with respect to the gain at that level corresponding to the midrange light value in the picture. *Notes:* (1) The gain referred to in the definition is for a signal amplitude small in comparison with the total peak-to-peak picture signal involved. A quantitative evaluation of this effect can be obtained by a measurement of differential gain. (2) The overall effect of white compression is to reduce contrast in the highlights of the picture as seen on a monitor. *See also:* **television.** 178

white noise (1) (interference terminology). Noise, either random or impulsive type, that has a flat frequency spectrum at the frequency range of interest. *See:* **Johnson noise; thermal noise.** *See also:* **electromagnetic compatibility.** 199

(2) (telephone practice). Noise, either random or impulsive type, that has a flat frequency spectrum at the frequency range of interest. This type of noise is used in the evaluation of systems on a theoretical basis and is produced for testing purposes by a white-noise generator. The use of the term should be limited and is not good usage in describing message circuit noise. 24

white object (color). An object that reflects all wavelengths of light with substantially equal high efficiencies and with considerable diffusion. 18, 178

white peak (television). A peak excursion of the picture signal in the white direction. *See also:* **television.** 178

white recording (1) (amplitude-modulation facsimile system). That form of recording in which the maximum received power corresponds to the minimum density of the record medium.

(2) (frequency-modulation facsimile system). That form of recording in which the lowest received frequency corresponds to the minimum density of the record medium. *See also:* **recording (facsimile).** 12

white saturation. *See:* **white compression.**

white signal (at any point in a facsimile system). The signal produced by the scanning of a minimum-density area of the subject copy. *See also:* **facsimile signal (picture signal).** 12

white transmission (1) (amplitude-modulation facsimile system). That form of transmission in which the maximum transmitted power corresponds to the minimum density of the subject copy.

(2) (frequency-modulation facsimile system). That form of transmission in which the lowest transmitted frequency corresponds to the minimum density of the subject copy. *See also:* **facsimile transmission.** 12

whole body irradiation (electrobiology). Pertains to the case in which the entire body is exposed to the incident electromagnetic energy or in which the cross section of the body is smaller than the cross section of the incident radiation beam. *See also:* **electrobiology.** 322

wicking. The flow of solder along the strands and under the insulation of stranded lead wires. 284

wick-lubricated bearing (rotating machinery). (1) A sleeve bearing in which a supply of lubricant is provided by the capillary action of a wick that extends into a reservoir of free oil or of oil-saturated packing material. (2) A sleeve bearing in which the reservoir and other cavities in the bearing region are packed with a material that holds the lubricant supply and also serves as a wicking. *See also:* **bearing.** 63

wide-angle diffusion. Diffusion in which flux is scattered at angles far from the direction that the flux would take by regular reflection or transmission. *See also:* **lamp.** 167

wide-angle luminaire. A luminaire that distributes the light through a comparatively wide solid angle. *See also:* **luminaire.** 167

wide-band channel (data transmission). A channel wider in bandwidth than a voice grade channel. 59

wide-band improvement. The ratio of the signal-to-noise ratio of the system in question to the signal-to-noise ratio of a reference system. *Note:* In comparing frequency-modulation and amplitude-modulation systems, the reference system usually is a double-sideband amplitude-modulation system with a carrier power, in the absence of modulation, that is equal to the carrier power of the frequency-modulation system. 111

wide-band ratio. The ratio of the occupied frequency bandwidth to the intelligence bandwidth. 111

width line (illuminating engineering). The radial line (the one that makes the larger angle with the reference line) that passes through the point of one-half maximum candlepower on the lateral candlepower distribution curve plotted on the surface of the cone of maximum candlepower. *See also:* **street-lighting luminaire.** 167

Wiedemann-Franz ratio. The quotient of the thermal conductivity by the electric conductivity. *See also:* **thermoelectric device.** 248, 191

Wien bridge oscillator. An oscillator whose frequency of oscillation is controlled by a Wien bridge. *See also:* **oscillatory circuit.** 328

Wien capacitance bridge. A 4-arm alternating-current bridge characterized by having in two adjacent arms capacitors respectively in series and in parallel with resistors, while the other two arms are normally nonreactive resistors. *Note:* Normally use for the measurement of capacitance in terms of resistance and frequency. The balance depends upon frequency, but from the balance conditions the capacitance of either or both capacitors can be computed from the resistances of all four arms and the frequency. *See also:* **bridge.** 328

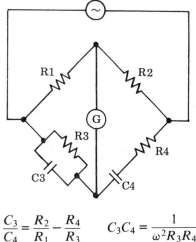

$$\frac{C_3}{C_4} = \frac{R_2}{R_1} - \frac{R_4}{R_3} \qquad C_3C_4 = \frac{1}{\omega^2 R_3 R_4}$$

Wien capacitance bridge.

Wien displacement law. An expression representing, in a functional form, the spectral radiance L_λ of a blackbody as a function of the wavelength λ and the temperature T:

$$L_\lambda = I_\lambda/A'$$

$$= c_1\lambda^{-5}f(\lambda T),$$

where the symbols are those used in the definition of **Planck radiation law.** The two principal corollaries of this law are:

$$\lambda_m T = b$$

$$L_m/T^5 = b'$$

which show how the maximum spectral radiance L_m and the wavelength λ_m at which it occurs are related to the absolute temperature T. *Note:* The currently recommended value of b is 2.8978×10^{-3} m · K or 2.8978×10^{-1} cm · K. From the definition of the **Planck radiation law,** and with the use of the value of b, as given above, b', is found to be 4.10×10^{-12} W · cm^{-3} · K^{-5} · sr^{-1}. 167

Wien inductance bridge. A 4-arm alternating-current bridge characterized by having in two adjacent arms inductors respectively in series and in parallel with resistors, while the other two arms are normally nonreactive resistors. *Note:* Normally used for the measurement of inductance in terms of resistance and frequency. The balance depends upon frequency, but from the balance conditions the inductances of either or both inductors can be computed from the resistances of the four arms and the frequency. *See also:* **bridge.** 328

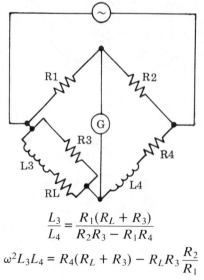

$$\frac{L_3}{L_4} = \frac{R_1(R_L + R_3)}{R_2R_3 - R_1R_4}$$

$$\omega^2 L_3 L_4 = R_4(R_L + R_3) - R_L R_3 \frac{R_2}{R_1}$$

Wien inductance bridge.

Wien radiation law. An expression representing approximately the spectral radiance of a blackbody as a function of its wavelength and temperature. It commonly is expressed by the formula

$$L_\lambda = I_\lambda/A'$$

$$= c_{1L}\lambda^{-5}e^{-(c_2/\lambda T)}$$

where the symbols are those used in the definition of **Planck radiation law.** This formula is accurate to one percent or better for values of λT less than 3000 micrometer kelvins. *See also:* **radiant energy.** 167

wigwag signal. A railroad-highway crossing signal, the indication of which is given by a horizontally swinging disc with or without a red light attached. 328

Williams-tube storage (electronic computation). A type of electrostatic storage. 210

Wilson center (limb center) (V potential) (medical electronics) (electrocardiography). An electric reference contact; the junction of three equal resistors to the limb leads. 192

wind-driven generator for aircraft. A generator used on aircraft that derives its power from the air stream applied on its own air screw or impeller during flight. 328

winder, pilot line (conductor stringing equipment). A device designed to payout and re-wind pilot lines during stringing operations. It is normally equipped with its own engine which drives a drum or a supporting shaft for a reel mechanically, hydraulically, or through a combination of both. These units are usually equipped with multiple drums or reels, depending upon the number of pilot lines required. The pilot line is payed out from the drum or reel, pulled through the travelers in the sag section, and attached to the pulling line on the reel stand or drum puller. It is then re-wound to pull the pulling line through the travelers. 45

winder, reel (conductor stringing equipment). A device designed to serve as a recovery unit for a pulling line. It is normally equipped with its own engine which drives a supporting shaft for a reel mechanically, hydraulically, or through a combination of both. The shaft, in turn, drives the reel. It is normally used to re-wind a pulling line as it leaves the "bullwheel" puller during stringing operations. This unit is not intended to serve as a puller, but sometimes serves this function where low tensions only are involved. *Syn.* **take-up reel.** 45

winding (1) (rotating machinery). An assembly of coils designed to act in consort to produce a magnetic flux field or to link a flux field. 63

(2) (data processing). A conductive path, usually of wire, inductively coupled to a magnetic core or cell. *Note:* When several windings are employed, they may be designated by the functions performed. Examples are: sense, bias, and drive windings. Drive windings include read, write, inhibit, set, reset, input, shift, and advance windings. 331

winding, ac (thyristor converter). The winding of a thyristor converter transformer that is connected to the ac circuit and usually has no conductive connection with the thyristor circuit elements. *Syn:* **primary winding.** 121

winding, autotransformer series. *See:* **series winding.**

winding, control power. The winding (or transformer) that supplies power to motors, relays, and other devices used for control purposes. *See:* **windings, high-voltage and low-voltage.** 203

winding, dc (thyristor converter). The winding of a thyristor converter transformer that is conductively connected to the thyristor converter circuit elements and that conducts the direct current of the thyristor converter. *See:* **secondary winding.** 121

winding-drum machine (elevators). A geared-drive machine in which the hoisting ropes are fastened to and wind on a drum. *See also:* **driving machine (elevators).** 328

winding factor (rotating machinery). The product of the distribution factor and the pitch factor. *See also:* **rotor (rotating machinery); stator.** 63

winding hottest-spot temperature (transformers). The highest temperature inside the transformer winding. It is greater than the measured average temperature (using the resistance change method) of the coil conductors. 53

winding inductance. *See:* **air core inductance.**

winding overhang (rotating machinery). That portion of a winding extending beyond the ends of the core. 63

winding pitch. *See:* **coil pitch.**

winding, primary. *See:* **primary winding.**

winding, secondary. *See:* **secondary winding.**

winding, tertiary. *See:* **tertiary winding.**

winding shield (rotating machinery). A shield secured to the frame to protect the windings but not to support the bearing. 63

windings, high-voltage and low-voltage. The terms high-voltage and low-voltage are used to distinguish the winding having the greater from that having the lesser voltage rating. 53

winding, stabilizing. *See:* **stabilizing winding.**

winding voltage rating. The voltage for which the winding is designed. *See also:* **duty.** 301

window (1) (counter tube) (radiation-counter tubes). That portion of the wall that is made thin enough for radiation of low penetrating power to enter. *See:* **anticoincidence (radiation counters).** 190

(2) (semiconductor radiation detector). *See:* **dead layer thickness.**

window amplifier. *See:* **biased amplifier.**

window-type current transformer. One which has a secondary winding insulated from and permanently assembled on the core, but has no primary winding as an integral part of the structure. Complete insulation is provided for a primary winding in the window through which one turn of the line conductor can be passed to provide the primary winding. 53

window, waveguide (waveguide components). A gas- or liquid-tight barrier or cover designed to be essentially transparent to the transmission of electromagnetic waves. 166

windshield wiper for aircraft. A motor-driven device for removing rain, sleet, or snow from a section of an aircraft windshield, window, navigation dome, or turret. 328

windup. Lost motion in a mechanical system that is proportional to the force or torque applied. 224, 207

wing-clearance lights. A pair of aircraft lights provided at the wing tips to indicate the extent of the wing span when the navigation lights are located an appreciable distance inboard of the wing tips. 167

wink-start pulsing (telephone switching systems). A method of pulsing control and trunk integrity check wherein the sender delays the sending of the address pulses until it receives a momentary off-hook signal from the far end. 55

wiper (brush). That portion of the moving member of a selector or other similar device, that makes contact with the terminals of a bank. 328

wiper relay. *See:* **relay wiper.**

wiping gland (wiping sleeve). A projecting sleeve on a junction box, pothead, or other piece of apparatus serving to make a connection to the lead sheath of a cable by means of a plumber's wiped joint. *See also:* **tower; transformer removable cable-terminating box.** 64

wire. A slender rod or filament of drawn metal. *Note:* The definition restricts the term to what would be ordinarily understood by the term solid wire. In the definition, the word slender is used in the sense that the length is great in comparison with the diameter. If a wire is covered with insulation, it is properly called an insulated wire; while primarily the term wire refers to the metal, nevertheless when the context shows that the wire is insulated, the term wire will be understood to include the insulation. *See also:* **car wiring apparatus; conductors.** 1, 64

wire-band serving (power distribution, underground cables). A short closed helical serving of wire applied tightly over the armor of wire-armored cables spaced at regular intervals, such as on vertical riser cables, to bind the wire armor tightly over the core to prevent slippage. 57

wire broadcasting. The distribution of programs over wire circuits to a large number of receivers, using either voice frequencies or modulated carrier frequencies. 328

wired logic (telephone switching systems). A fixed pattern of interconnections among a group of devices to perform predetermined functions in response to input signals. 55

wired program (telephone switching systems). A program embodied in a pattern of fixed physical interconnections among a group of devices. 55

wired program control (telephone switching systems). A system control using wired logic. 55

wire gages (transmission and distribution). The American Wire Gage (AWG), otherwise known as Brown & Sharpe (B&S), is the standard gage for copper, aluminum, and other conductors, excepting steel, for which the Steel Wire Gage (Stl. WG) is used throughout these rules. 262

wire-grid lens antenna. A lens antenna having a two-dimensional path delay lens consisting of wire grids, in which the effective index of refraction is controlled by the dimensions and the spacing of the wire grids that form the lens boundaries. *See also:* **Luneburg lens antenna; geodesic lens antenna.** 111

wire holder (insulator). An insulator of generally cylindrical or pear shape, having a hole for securing the conductor and a screw or bolt for mounting. *See also:* **insulator.** 261

wire insulation (rotating machinery). The insulation that is applied to a wire before it is made into a coil or inserted in a machine. *See also:* **rotor (rotating machinery); stator.** 63

wireless connection diagram (industrial control). The general physical arrangement of devices in a control equipment and connections between these devices, terminals, and terminal boards for outgoing connections to external apparatus. Connections are shown in tabular form and not by lines. An elementary (or schematic) diagram may be included in the connection diagram. 210, 206

wire, overhead ground. *See:* **overhead ground wire.**

wire-pilot protection. Pilot protection in which an auxiliary metallic circuit is used for the communicating means between relays at the circuit terminals. 103, 60, 202, 127

wire spring relay. A relay design in which the contacts are attached to round wire springs instead of the conventional flat or leaf spring. 341

wireway (packaging machinery). A rigid rectangular raceway provided with a cover. 124

wiring or busing terminal, screw and/or lead. That terminal, screw or lead to which a power supply will be connected in the field. 53

withstand current (surge). The crest value attained by a surge of a given wave shape and polarity that does not cause disruptive discharge on the test specimen. 64

withstand test voltage (cable termination). The voltage that the device must withstand without flashover, disruptive discharge, puncture, or other electric failure when voltage is applied under specified conditions. *Note:* For power frequency voltages, the values specified are rms values and for a specified time. For lightning or switching impulse voltages, the values specified are crest values of a specified wave. For direct voltages the values specified are average values and for a specified time. 4

withstand voltage (1) (general). The specified voltage that, under specified conditions, can be applied to insulation without causing flashover or puncture. 103, 134

(2) (transformer). The voltage that electric equipment is capable of withstanding without failure or disruptive discharge when tested under specified conditions. 53

(3) (impulse) (electric power). The crest value attained by an impulse of any given wave shape, polarity, and amplitude, that does not cause disruptive discharge on the test specimen. 91

(4) (surge arresters). A specified voltage that is to be applied to a test object in a withstand test under specified conditions. During the test, in general no disruptive discharge should occur. *See also:* **basic impulse insulation level (insulation strength); test voltage and current.**

308, 307, 201, 62

withstand voltage test. A high-voltage test that the armature winding must withstand without flashover or other electric failure at a specified voltage for a specified time and under specified conditions. 135

word. An ordered set of characters that is the normal unit in which information may be stored, transmitted, or operated upon within a given computer. *Example:* The set

instantaneous value of the OFF-state voltage that occurs across a thyristor, excluding all repetitive and nonrepetitive transient voltages. *See also:* **principal voltage-current characteristic (principal characteristic).**

243, 191, 66, 208

working peak reverse voltage (semiconductor rectifier) (reverse blocking thyristor). The maximum instantaneous value of the reverse voltage, excluding all repetitive and nonrepetitive transient voltages, that occurs across a semiconductor rectifier cell, rectifier diode, rectifier stack, or reverse-blocking thyristor. *See also:* **principal voltage-current characteristic (principal characteristic); rectification.**

237, 243, 191, 66, 208

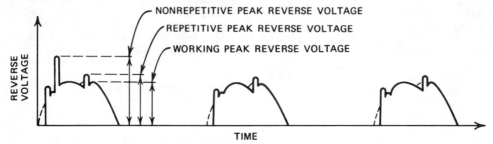

Working peak reverse voltage.

of characters 10692 is a word that may give a command for a machine element to move to a point 10.692 inches from a specified zero. *See also:* **computer word; machine word.** 235, 210, 224, 207

word address format. Addressing each word of a block by one or more characters that identify the meaning of the word. 224, 207

word length (1) (information processing). The number of bits or other characters in a word. 255, 77

(2) (analog-to-digital converter). The number of data bits, including sign, which form the digital representation of the analog input over the voltage range. 10

(3) (digital-to-analog converter). The number of data bits, including sign, in the digital register of a digital-to-analog converter, or a digital-to-analog multiplier. 10

word time (electronic computation). In a storage device that provides serial access to storage locations, the time interval between the appearance of corresponding parts of successive words. *See also:* **minor cycle.** 255, 77

work. The work done by a force is the dot-product line integral of the force. *See:* **line integral.** 210

work coil. *See:* **load, work or heater coil (induction heating usage).**

work function. The minimum energy required to remove an electron from the Fermi level of a material into field-free space. *Note:* Work function is commonly expressed in electron volts. 190, 125

working (electrolysis). The process of stirring additional solid electrolyte or constituents of the electrolyte into the fused electrolyte in order to produce a uniform solution thereof. *See also:* **fused electrolyte.** 328

working optical aperture (acousto-optic device). That aperture which is equal to the size of the acoustic column which the light will encounter. 82

working peak OFF-state voltage (thyristor). The maximum

working peak reverse voltage rating (rectifier circuit). The maximum value of working peak reverse voltage permitted by the manufacturer under stated conditions. *See also:* **average forward current rating (rectifier circuit).**

237, 66, 208

working point. *See:* **operating point.**

working pressure. The pressure, measured at the cylinder of a hydraulic elevator, when lifting the car and its rated load at rated speed. *See also:* **elevators.** 328

working reference system. A secondary reference telephone system consisting of a specified combination of telephone sets, subscriber lines, and battery supply circuits connected through a variable distortionless trunk and used under specified conditions for determining, by comparison, the transmission performance of other telephone systems and components. 328

working standard (illuminating engineering). A standardized light source for regular use in photometry. *See also:* **primary standards (illuminating engineering).** 167

working storage (computing machines). *See:* **temporary storage.**

working value. The electrical value that when applied to an electromagnetic instrument causes the movable member to move to its fully energized position. This value is frequently greater than pick-up. *See:* **pick-up.** 328

working voltage to ground (electric instrument). The highest voltage, in terms of maximum peak value, that should exist between any terminal of the instrument proper on the panel, or other mounting surface, and ground. *See also:* **instrument.** 280

work plane. The plane at which work is usually done, and at which the illumination is specified and measured. Unless otherwise indicated, this is assumed to be a horizontal plane 30 inches above the floor. *See also:* **inverse-square law (illuminating engineering).** 167

world-numbering plan (telephone switching systems). The arrangement whereby, for the purpose of international distance dialing, every telephone main station in the world is identified by a unique number having a maximum of twelve digits representing a country code plus a national number.   55

world-zone number (telephone switching systems). The first digit of a country code. In the world-numbering plan, this number identifies one of the larger geographical areas into which the world is arranged, namely:

Zone 1—North America (includes areas operating with unified regional numbering).
Zone 2—Africa
Zone 3 & 4—Europe
Zone 5—South America, Cuba, Central America including part of Mexico
Zone 6—South Pacific (Austral-Asia)
Zone 7—Union of Soviet Socialist Republics
Zone 8—North Pacific (Eastern Asia)
Zone 9—Far East and Middle East
Zone 0—Spare   55

worm-geared machine (elevators). A direct-drive machine in which the energy from the motor is transmitted to the driving sheave or drum through worm gearing. *See also:* **driving machine (elevators).**   328

wound rotor (rotating machinery). A rotor core assembly having a winding made up of individually insulated wires. *See also:* **asynchronous machine; direct-current commutating machine.**   63

wound-rotor induction motor (1) (rotating machinery). An induction motor in which a primary winding on one member (usually the stator) is connected to the alternating-current power source and a secondary polyphase coil winding on the other member (usually the rotor) carries alternating current produced by electromagnetic induction. *Note:* The terminations of the rotor winding are usually connected to collector rings. The brush terminals may be either short-circuited or closed through suitable adjustable circuits. *See also:* **asynchronous machine.**   63
(2) (electric installations on shipboard). An induction motor in which the secondary circuit consists of polyphase winding or coils whose terminals are either short-circuited or closed through suitable circuits. *Note:* When provided with collector or slip rings, it is also known as a slip-ring induction motor.   3

wound stator core (rotating machinery). A stator core into which the stator winding, with all insulating elements and lacing has been placed, including any components imbedded in or attached to the winding, and including the lead cable when this is used. *See also:* **stator.**   63

wound-type current transformer. One which has a primary winding consisting of one or more turns mechanically encircling the core or cores. The primary and secondary windings are insulated from each other and from the core(s) and are assembled as an integral structure.   53

wow (sound recording and reproducing equipment). Frequency modulation of the signal in the range of approximately 0.5 Hz to 6 Hz resulting in distortion which may be perceived as a fluctuation of pitch of a tone or program. *Note:* Measurement of unweighted wow only is not covered by this standard.   145

wrap. One convolution of a length of ferromagnetic tape about the axis. *See:* **tape-wound core.**   331

wrapper (rotating machinery). (1) A relatively thin flexible sheet material capable of being formed around the slot section of a coil to provide complete enclosure. (2) The outer cylindrical frame component used to contain the ventilating gas. *See also:* **rotor (rotating machinery); stator.**   63

wrap thickness. The distance between corresponding points on two consecutive wraps, measured parallel to the ferromagnetic tape thickness. *See:* **tape-wound core.**   331

wrap width (tape width). *See:* **tape-wound core.**

write (1) (general). To introduce data, usually into some form of storage. *See:* **read.**
  174, 190, 235, 210, 255, 77, 54
(2) (charge-storage tubes). To establish a charge pattern corresponding to the input.   125

write pulse. *See:* ONE **state.**

writing rate (1) (storage tubes). The time rate of writing on a storage element, line, or area to change it from one specified level to another. Note the distinction between this and **writing speed.** *See also:* **storage tube.**
  174, 190
(2) (oscilloscopes). *See:* **writing time division.**

writing speed (storage tubes). Lineal scanning rate of the beam across the storage surface in writing. Note the distinction between this and **writing rate.** *See:* **information writing speed.** *See also:* **storage tube.**
  174, 190, 125

writing speed, maximum usable (storage tubes). The maximum speed at which information can be written under stated conditions of operation. Note the qualifying adjectives **maximum usable** are frequently omitted in general usage when it is clear that the maximum usable writing speed is implied. *See also:* **storage tube.**
  174, 190

writing time/division (oscilloscopes). The minimum time per unit distance required to record a trace. The method of recording must be specified.   106

writing time, minimum usable (storage tubes). The time required to write stored information from one specified level to another under stated conditions of operation. *Note:* The qualifying adjectives **minimum usable** are frequently omitted in general usage when it is clear that the minimum usable writing time is implied. *See also:* **storage tube.**   174, 190

Wullenweber antenna. An antenna consisting of a circular array of radiating elements, each having its maximum directive gain along the outward radial, and a feed system that provides a steerable beam that is narrow in the azimuth plane.   111

wye. *See:* **Y.**

wye junction (waveguide components). A junction of waveguides or transmission lines in which the longitudinal guide axes form a Y.   166

wye rectifier circuit. A circuit that employs three or more rectifying elements with a conducting period of 120 electrical degrees plus the commutating angle. *See also:* **rectification.**   328

X

X-axis amplifier. *See:* horizontal amplifier.
X-band radar. A radar operating at frequencies between 8 and 12 gigahertz, usually in the International Telecommunications Union assigned band 8.5 to 10.68 gigahertz. 13
xerography. The branch of electrostatic electrophotography that employs a photoconductive insulating medium to form, with the aid of infrared, visible, or ultraviolet radiation, latent electrostatic-charge patterns for producing a viewable record. *See also:* electrostatography. 236, 191
xeroprinting. The branch of electrostatic electrography that employs a pattern of insulating material on a conductive medium to form electrostatic-charge patterns for duplicating purposes. *See also:* electrostatography. 236, 191
xeroradiography. The branch of electrostatic electro-

photography that employs a photoconductive insulating medium to form, with the aid of X rays or gamma rays, latent electrostatic-charge patterns for producing a viewable record. *See also:* electrostatography. 236, 191
X-ray tube. A vacuum tube designed for producing X-rays by accelerating electrons to a high velocity by means of an electrostatic field and then suddenly stopping them by collision with a target. 190
X wave (radio wave propagation). Extraordinary-wave component. 180
X-Y display. A rectilinear coordinate plot of two variables. *See also:* oscillograph. 185
XY switch. A remotely controlled bank-and-wiper switch arranged in a flat manner, in which the wipers are moved in a horizontal plane, first in one direction and then in another. 328

Y

Yagi antenna. A linear end-fire array consisting of a driven element, one or more reflector elements, and one or more director elements. 111
Y amplifier. *See:* vertical amplifier.
Y-axis amplifier. *See:* vertical amplifier.
Y-connected circuit. A three-phase circuit that is star connected. *See also:* network analysis. 210
Y connection. So connected that one end of each of the windings of a polyphase transformer (or of each of the windings for the same rated voltage of single phase transformers associated in a polyphase bank) is connected to a common point (the neutral point) and the other end to its appropriate line terminal. *Syn:* wye

connection. 53
Y junction. *See:* Wye junction.
Y network. A star network of three branches. *See also:* network analysis. 210
yoke (magnetic) rotating machinery). The element of ferromagnetic material, not surrounded by windings, used to connect the cores of an electromagnet, or of a transformer, or the poles of a machine, or used to support the teeth of stator or rotor. *Note:* A yoke may be of solid material or it may be an assembly of laminations. *See also:* rotor (rotating machinery); stator. 210, 63
Y-T display. An oscilloscope display in which a time-dependent variable is displayed against time. *See also:* oscillograph. 185

Z

Z-axis amplifier (oscilloscopes). An amplifier for signals controlling a display perpendicular to the *X-Y* plane (commonly intensity of the spot). *See:* intensity amplifier. *See also:* oscillograph. 185, 106
Zeeman effect. If an electric discharge tube, or other light source emitting a bright-line spectrum, is placed between the poles of a magnet, each spectrum line is split by the action of the magnetic field into three or more close-spaced but separate lines. The amount of splitting or the separation of the lines, is directly proportional to the strength of the magnetic field. 210
Zener breakdown (semiconductor device). A breakdown that is caused by the field emission of charge carriers in the depletion layer. *See also:* semiconductor; semiconductor device. 245, 210, 66

Zener diode (semiconductor). A class of silicon diodes that exhibit in the avalanche-breakdown region a large change in reverse current over a very narrow range of reverse voltage. *Note:* This characteristic permits a highly stable reference voltage to be maintained across the diode despite a relatively wide range of current through the diode. *See:* Zener breakdown; avalanche breakdown. 186
Zener diode regulator. A voltage regulator that makes use of the constant-voltage characteristic of the Zener diode to produce a reference voltage that is compared with the voltage to be regulated to initiate correction when the voltage to be regulated varies through changes in either load or input voltage. See the accompanying figure. *See:* Zener diode. *See also:* electrical conversion. 186

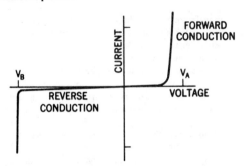

Current and voltage characteristics for a typical Zener diode regulator $|V_A| \gg |V_B|$.

Zener impedance* (semiconductor diode). *See:* **breakdown impedance.** *See also:* **semiconductor.**

* Deprecated

Zener voltage*. *See:* **breakdown voltage.** *See also:* **semiconductor.**

* Deprecated

zero (1) (function) (root of an equation). A zero of a function $f(x)$ is any value of the argument X for which $f(x) = 0$. *Note:* Thus the zeros of sin x are $x_1 = 0$, $x_2 = \pi$, $x_3 = 2\pi$, $x_4 = 3\pi$, ..., $x_n = (n-1)\pi$, ... The roots of the equation $f(x) = 0$ are the zeros of $f(x)$.   210
(2) (transfer function in the complex variable s). (A) A value of s that makes the function zero. (B) The corresponding point in the s plane. *See also:* **control system, feedback; pole (network function).**   56, 329
(3) (network function). Any value of p, real or complex, for which the network function is zero. *See also:* **network analysis.**   210
zero adjuster. A device for bringing the indicator of an electric instrument to a zero or fiducial mark when the electrical quantity is zero. *See also:* **moving element (instrument).**   328
zero-based linearity. *See:* **linearity.** *See also:* **electronic analog computer.**
zero-beat reception. *See:* **homodyne reception.**
zero carryover (1) (bolometric power meters). A characteristic of multirange direct reading bolometer bridges that is a measure of the ability of the meter to maintain a zero setting from range to range without readjustment after initially being set to zero on the most sensitive range.   47
(2) (electrothermic power meter). A characteristic of multirange direct reading electrothermic power indicators which is a measure of the ability of the meter to maintain a zero setting from range to range without readjustment after initially being set to zero on the most sensitive range. Expressed in terms of percentage of full scale.   47
zero control current residual voltage (Hall-effect devices). The voltage across the Hall terminals that is caused by a time-varying magnetic field when there is no control current.   107
zero-error (device operating under the specified conditions of use). The indicated output when the value of the input presented to it is zero. *See also:* **control system, feedback.**   329

zero-error reference. *See:* **linearity.** *See also:* **electronic analog computer.**
zero field residual voltage (Hall-effect devices). The voltage across the Hall terminals that exists when control current flows but there is zero applied magnetic field.   107
zero field residual voltage temperature drift (Hall generator). The maximum change in output voltage per degree Celsius over a given temperature range when operated with zero external field and a given magnitude of control current.   107
zero field resistive residual voltage (Hall-effect devices). That component of the zero field residual voltage which remains proportional to the voltage across the control current terminals of the Hall generator for a specified temperature.   107
zero guy. A line guy installed in a horizontal position between poles to provide clearance and transfer strain to an adjacent pole. *See also:* **tower.**   64
zero lead (medical electronics). *See:* **biolectric null.**
zero-level address (computing machines). *See:* **immediate address.**
zero-minus (0−) call (telephone switching systems). A call for which the digit zero is dialed alone to indicate that operator assistance is desired.   55
zero-modulation medium noise (sound recording and reproducing system). The noise that is developed in the scanning or reproducing device during the reproducing process when a medium is scanned in the zero-modulation state. *Note:* For example, zero-modulation medium noise is produced in magnetic recording by undesired variations of the magnetomotive force in the medium, that are applied across the scanning gap of a demagnetized head, when the medium moves with the desired motion relative to the scanning device. Medium noise can be ascribed to nonuniformities of the magnetic properties and to other physical and dimensional properties of the medium. *See also:* **noise (sound recording and reproducing system).**   350
zero-modulation state (sound recording medium). The state of complete preparation for playback in a particular system except for omission of the recording signal. *Notes:* (1) Magnetic recording media are considered to be in the zero-modulation state when they have been subjected to the normal erase, bias, and duplication printing fields characteristic of the particular system with no recording signal applied. (2) Mechanical recording media are considered to be in the zero-modulation state when they have been recorded upon and processed in the customary specified manner to form the groove with no recording signal applied. (3) Optical recording media are considered to be in the zero-modulation state when all normal processes of recording and processing, including duplication, have been performed in the customary specified manner, but with no modulation input to the light modulator. *See also:* **noise (sound recording and reproducing system).**   350
zero offset (1) (industrial control). A control function for shifting the reference point in a control system. *See also:* **control system, feedback.**   219, 206
(2) (numerically controlled machines). A characteristic of a numerical machine control permitting the zero point

on an axis to be shifted readily over a specified range. The control retains information on the location of the permanent zero. *See:* **floating zero.** 224, 207

(3) (rate gyros). The gyro output when the input rate is zero, generally expressed as an equivalent input rate. It excludes outputs due to hysteresis and acceleration. *See:* **input-output characteristics.** 46

zero output. (1) The voltage response obtained from a magnetic cell in a ZERO state by a reading or resetting process. (2) The integrated voltage response obtained from a magnetic cell in a ZERO state by a reading or resetting process. *See:* **ONE state; ZERO state.** 331

zero-phase-sequence relay. A relay that responds to the zero-phase-sequence component of a input quantity.

103, 60, 202, 127

zero-phase-sequence symmetrical components (unsymmetrical set of polyphase voltages or currents of *m* phases). That set of symmetrical components that have zero phase sequence. That is, the angular phase lag from each member to every other member is 0 radians. The members of this set will all reach their positive maxima simultaneously. The zero-phase-sequence symmetrical components for a three-phase set of unbalanced sinusoidal voltages ($m = 3$) having the primitive period are represented by the equations

$$e_{a0} = e_{b0} = e_{c0} = (2)^{1/2} E_{a0} \cos(\omega t + \alpha_{a0})$$

derived from the equation of **symmetrical components (set of polyphase alternating voltages).** Since in this case $r = 1$ for every component (of first harmonic), the third subscript is omitted. Then k is 0 for the zero sequence, and s takes on the values 1, 2, and 3 corresponding to phases a, b, and c. These voltages have no phase sequence since they all reach their positive maxima simultaneously.

zero-phase symmetrical set (1) (polyphase voltage). A symmetrical set of polyphase voltages in which the angular phase difference between successive members of the set is zero or a multiple of 2π radians. The equations of **symmetrical set (polyphase voltages)** represent a zero-phase symmetrical set of polyphase voltages if k/m is zero or an integer. (The symmetrical set of voltages represented by the equations of **symmetrical set of polyphase voltages** may be said to have zero-phase symmetry if k/m is zero or an integer (positive or negative).) *Note:* This definition may be applied to a two-phase four-wire or five-wire system if m is considered to be 4 instead of 2. 210

(2) (polyphase currents). This definition is obtained from the corresponding definitions for voltage by substituting the word current for voltage, and the symbol I for E and β for α wherever they appear. The subscripts are unaltered. 210

zero pip (non-real time spectrum analyzer). An output indication which corresponds to zero input frequency. 68

zero-plus (0+) (telephone switching systems). A call in which the digit zero is dialed as a prefix where operator intervention is necessary. 55

zero-power-factor saturation curve (zero-power-factor characteristic) (synchronous machine). The saturation curve of a machine supplying constant current with a power-factor of approximately zero, overexcited. 63

zero-power-factor test (synchronous machine). A no-load test in which the machine is overexcited and operates at a power-factor very close to zero. 63

zero-sequence impedance (1) (rotating machinery). The quotient of the zero-sequence component of the voltage, assumed to be sinusoidal, supplied to a synchronous machine, and the zero-sequence component of the current at the same frequency. *See also:* **direct-axis synchronous reactance.** 63

(2) (transformer). An impedance voltage measured between a set of primary terminals and one or more sets of secondary terminals when a single-phase voltage source is applied between the three primary terminals connected together and the primary neutral, with the secondary line terminals shorted together and connected to their neutral (if one exists). *Notes:* (A) For two-winding transformers, the other winding is short-circuited. For multi-winding transformers, several tests are required, and the zero-sequence impedance characteristics are represented by an impedance network. (B) In some transformers, the test must be made at a voltage lower than that required to circulate rated current in order to avoid magnetic core saturation or to avoid excessive current in other windings. (C) Zero-sequence impedances are usually expressed in per unit or percent on a suitable voltage and kVA base. 53

zero-sequence reactance (rotating machinery). The ratio of the fundamental component of reactive armature voltage, due to the fundamental zero-sequence component of armature current, to this component at rated frequency, the machine running at rated speed. *Note:* Unless otherwise specified, the value of zero-sequence reactance will be that corresponding to a zero-sequence current equal to rated armature current. *See also:* **direct-axis synchronous reactance.** 63

zero-sequence resistance. The ratio of the fundamental in-phase component of armature voltage, resulting from fundamental zero-sequence current, to this component of current at rated frequency. 328

ZERO shift error. Error measured by the difference in deflection as between an initial position of the pointer, such as at zero, and the deflection after the instrument has remained deflected upscale for an extended length of time, expressed as a percentage of the end-scale deflection. *See also:* **moving element (instrument).** 280

zero span (non-real time spectrum analyzer). A mode of operation in which the frequency span is reduced to zero. 60

ZERO state. A state wherein the magnetic flux has a negative value, when similarly determined. *See:* **ONE state.** 331

zero-subcarrier chromaticity (color television). The chromaticity that is intended to be displayed when the subcarrier amplitude is zero. *Note:* This chromaticity is also known as reference white for the display. 18

zero suppression. The elimination of nonsignificant zeros in a numeral. 255, 77

zero synchronization (numerically controlled machines). A technique that permits automatic recovery of a precise position after the machine axis has been approximately positioned by manual control. 224, 207

zero vector. A vector whose magnitude is zero. 210

zeta potential. *See:* **electrokinetic potential.**

zigzag connection. A polyphase transformer with Y-connected windings, each one of which is made up of parts in which phase-displaced voltages are induced.
53

zig-zag connection of polyphase circuits (zig-zag or interconnected star). The connection in star of polyphase windings, each branch of which is made up of windings that generate phase-displaced voltage. *See also:* **connections of polyphase circuits; polyphase circuits; polyphase systems.** 244, 63

zig-zag leakage flux. The high-order harmonic air-gap flux attributable to the location of the coil sides in discrete slots. *See also:* **rotor (rotating machinery); stator.** 63

Z marker (zone marker) (electronic navigation). A marker used to define a position above a radio range station.
13

Zobel filters (circuits and systems). A filter designed according to image parameter techniques. 67

zonal constant. A factor by which the mean candlepower emitted by a source of light in a given angular zone is multiplied to obtain the lumens in the zone. 167

zonal factor interflection method (lighting calculation). A procedure for calculating coefficients of utilization that takes into consideration the ultimate disposition of luminous flux from every 10-degree zone from luminaires. *See also:* **inverse-square law (illuminating engineering).**
167

zonal factor method (lighting calculation). A procedure for predetermining, from typical luminaire photometric data in discrete angular zones, the proportion of luminaire output that would be incident initially (without interreflections) on the work plane, ceiling, walls, and

$$F(z,m) = \sum_{n=0}^{\infty} f[nT - (1 - m)T]u[nT - (1 - m)T]z^{-n} \qquad 0 < m < 1.$$

floor of a room. *See also:* **inverse-square law (illuminating engineering).**
167

zone (relay). *See:* **reach (relay).**

zone comparison protection. A form of pilot protection in which the response of fault-detector relays, adjusted to have a zone of response commensurate with the protected line section, is compared at each line terminal to determine whether a fault exists within the protected line section. 103, 60, 202

zoned antenna. A lens or reflector antenna having various portions (called zones or steps) which form a discontinuous surface such that a desired phase distribution of the aperture illumination is achieved. *Syn.* **stepped antenna.**
111

zone leveling (semiconductor processing). The passage of one or more molten zones along a semiconductor body for the purpose of uniformly distributing impurities throughout the material. *See also:* **semiconductor device.**
342

zone of protection (1) (general). The part of an installation guarded by a certain protection. 103, 202
(2) (relays). That segment of the power system in which the occurrence of assigned abnormal conditions should cause the protective relay system to operate. *See also:* **relay.** 60

zone-plate lens antenna. *See:* **Fresnel lens antenna.**

zone punch. A punch in the 0, 11, or 12 row on a Hollerith punched card. 77

zone purification (semiconductor processing). The passage of one or more molten zones along a semiconductor for the purpose of reducing the impurity concentration of part of the ingot. *See also:* **semiconductor device.**
342

zoning (stepping) (lens or reflector). The displacement of various portions (called zones or steps) of the lens or surface of the reflector so that the resulting phase front in the near field remains unchanged. *See also:* **antenna (aerial).** 179

z transform, advanced (data processing). The advanced *z* transform of $f(t)$ is the *z* transform of $f(t + \Delta T)$; that is,

$$\sum_{n=0}^{\infty} f(nT + \Delta T)z^{-n} \qquad 0 < \Delta < 1.$$
198

z transform, delayed (data processing). The delayed *z* transform of $f(t)$, denoted $F(z, \Delta)$, is the *z* transform of $f(t - \Delta T)u(T - \Delta T)$, where $u(t)$ is the unit step function; that is,

$$F(z, \Delta) = \sum_{n=0}^{\infty} f(nT - \Delta T)u(nT$$
$$ - \Delta T)z^{-n} \qquad 0 < \Delta < 1.$$
198

z transform, modified (data processing). The modified *z* transform of $f(t)$, denoted $F(z, m)$, is the delayed *z* transform of $f(t)$ with the substitution $\Delta = 1 - m$; that is,

z transform, one-sided (data processing). Let T be a fixed positive number, and let $f(t)$ be defined for $t \geq 0$. The *z* transform of $f(t)$ is the function

$$[f(t)] = F(z) = \sum_{n=0}^{\infty} f(nT)z^{-n},$$
$$\text{for } |z| > R = 1/\rho$$

where ρ is the radius of convergence of the series and z is a complex variable. If $f(t)$ is discontinuous at some instant $t = kT$, k an integer, the value used for $f(kT)$ in the *z* transform is $f(kT^+)$. The *z* transform for the sequence $\{f_n\}$ is:

$$[\{f_n\}] = F(z) = \sum_{n=0}^{\infty} f_n z^{-n}.$$
198

z transform, two-sided (data processing). The two-sided *z* transform of $f(t)$ is

$$F(z) = \sum_{n=\infty}^{-1} f(nT)z^{-n} + \sum_{n=0}^{\infty} f(nT)z^{-n}$$

where the first summation is for $f(t)$ over all negative time and the second summation is for $f(t)$ over all positive time. 198

Sources

1 IEEE Std 16-1955; ANSI C48.1-1955 (withdrawn 1969).
Electric Control Apparatus for Land Transportation Vehicles.

2 IEEE Std 28-1974; ANSI C62.1-1975
Surge Arresters for AC Power Circuits, Standard for.

3 IEEE Std 45-1971 (reaffirmed 1976); ANSI C110.1-1972
Electric Installations on Shipboard, Recommended Practice for.

4 IEEE Std 48-1975
High Voltage AC Cable Terminations.

5 IEEE Std 86-1961 (reaffirmed 1975)
Basic Per Unit Quantities for AC Rotating Machines, Definitions of.

6 IEEE Std 95-1976
Insulation Testing of Large AC Rotating Machinery with High Direct Voltage, Recommended Practice for.

7 IEEE Std 119-1974
General Principles of Temperature Measurement as Applied to Electrical Apparatus, Recommended Practice for.

8 IEEE Std 125-1977
Preparation of Equipment Specifications for Speed-Governing of Hydraulic Turbines Intended to Drive Electric Generators, Recommended Practice for.

9 IEEE Std 165-1963 (reaffirmed 1972)
Analog Computers, Definitions of Terms for.

10 IEEE Std 166-1972
Hybrid Computer Linkage Components.

11 IEEE Std 167-1966 (reaffirmed 1971)
Facsimile, Test Procedure for.

12 IEEE Std 168-1956 (reaffirmed 1971); ANSI C16.30-1972; Datacom/com 19.4
Facsimile, Definitions of Terms for.

13 IEEE Std 172-1971 (reaffirmed 1975) (as of 7/12/77 to be known as IEEE Std 686; ANSI C16.26-1972
Navigation Aids, Definitions of Terms for. See also code 187.

14 IEEE Std 54-1955
Induction and Dielectric Heating Equipment, Recommended Practice for.

15 IEEE Std 181-1976
Pulse Quantities, Methods of Measurement.

16 IEEE Std 185-1975
Methods of Testing Frequency Modulation Broadcast Receivers.

17 IEEE Std 200-1975; ANSI Y32.16-1975
Reference designations for Electrical and Electronics Parts and Equipment.

18 IEEE Std 201-1955; Video Techniques/Broadcasting 2.2
Television: Color Terms, Definitions of. Video Techniques/Broadcasting 2.2

19 IEEE Std 219-1975
Loudspeaker Measurements, Recommended Practice for.

20 IEEE Std 279-1971; ANSI N42.7-1972
Protection Systems for Nuclear Power Generating Stations, Criteria for. See also code 356.

21 IEEE Std 268-1975
Metric Practice, Standard for

22 IEEE Std 286-1975
Measurement of Power-Factory Tip-Up of Rotating Machinery Stator Coil Insulation, Recommended Practice for.

23 IEEE Std 301-1976
Amplifiers and Preamplifiers for Semiconductor.

24 IEEE Std 305-1976; ANSI C37.93-1976
Protective Relay Applications of Audio Tones over Telephone Channels, Guide for.

25 IEEE Std 315-1975; ANSI Y32.2-1975; CSA Z99-1975
Graphic Symbols for Electrical and Electronics Diagrams.

26 IEEE Std 317-1976
Electric Penetration Assemblies in Containment Structures for Nuclear Power Generating Stations.

27 IEEE Std 324-1971; ANSI C37.30-1971; ANSI C30a-1975
High Voltage Air Switches, Insulators, and Bus Supports, Definitions and Requirements for.

28 IEEE Std 344-1975
Seismic Qualification of Class 1E Equipment for Nuclear Power Generating Stations, Recommended Practice for.

29 IEEE Std 352-1975
Reliability Analysis of Nuclear Power Generating Station Protection Systems, Guide for General Principles of.

30 IEEE Std 376-1975
Measurement of Impulse Strength and Impulse Bandwidth, Standard for the.

31 IEEE Std 380-1975
Definitions of Terms Used in IEEE Nuclear Power Generating Stations Standards.

32 IEEE Std 390-1975
Low-Power Pulse Transformers, Standard for.

33 IEEE Std 391-1976
High Power (300 Watts peak) Pulse Transformers, Standard for.

34 IEEE Std 404-1976; IC/PES 31.1
Power Cable Joints

35 IEEE Std 422-1976
Guide for the Design and Installation of Cable Systems in Power Generating Systems.

36 IEEE Std 430-1976
Measurement of Radio Noise from Overhead Power Lines, Procedures for.

37 IEEE Std 432-1976
Guide for Insulation Maintenance for Rotating Electrical Machinery 5—Less than 10 000 kVA.

38 IEEE Std 450-1975
Maintenance, Testing, and Replacement of Large Lead Storage Batteries for Generating Stations and Substations, Recommended Practice for.

39 IEEE Std 455-1976
Test Procedure for Measuring Longitudinal Balance of Telephone Equipment.

40 IEEE Std 488-1975
Digital Interface for Programmable Instrumentation and Related System Components.

41 IEEE Std 498-1975
Measuring and Test Equipment Used in the Construction and Maintenance of Nuclear Power Generating Stations, IEEE Standard Supplementary Requirements for the Calibration and Control of.

42 IEEE Std 511-1974
Video Signal Transmission Measurement of Linear Waveform Distortion, Trial-Use Standard on.

43 IEEE Std 518-1976
Guide for the Installation of Electrical Equipment to Minimize Electrical Noise Inputs to Controllers from Outside Sources.

44 IEEE Std 521-1976
Standard Letter Designations for Radar-Frequency Bands.

45 IEEE P524 (Draft 4) 1976 (Committee draft).
Definitions for Conductor Stringing Equipment.

46 IEEE 528 (Committee draft).
Standard Gyro and Accelerometer Terminology.

47 IEEE Std 544-1975
Electrothermic Power Meters, Standard for.

48 IEEE Std 549-1975
Sequential Events Recording Systems.

49 IEEE Std 570-1975
System Voltage Nomenclature Table for Use in All Industrial and Commercial Power Systems Standards and Committee Reports.

50 IEEE Std 579-1975
Test Procedures AC HV Circuit Breakers for Load Current Switching Requirements and Test Duties.

51 IEEE Std 583-1975
Modular Instrumentation and Digital Interface System (CAMAC). See also IEEE Std 595 and IEEE Std 596.

52 IEEE Std 590-1977
Cable Plowing Guide.

53 IEEE Std 591-1977; ANSI C57.12.80
Transformer Terminology.

54 Mil. Std. 1309B; Automated Instrumentation 9.8
Terms for Test Measurement, and Diagnostic Equipment, Definitions of.

55 IEEE Std 312-1977
Terms for Communication, Definitions of

56 ANSI C85.1-1963; (a) 1966; (b) 1972
Terminology for Automatic Control. See also code 198.

57 Insulated Conductors 31.1

58 Power Generation/PES 31.2

59 Power System Communication/PES 31.3
Glossary of Current Data Transmission Terms and Parameters.

60 Power System Relaying Committee/PES 31.6

61 Information Theory 12.1. See also codes 242 and 160.

62 Surge Protective Devices/PES 31.7
See also codes 2, 351, and 91.

63 Rotating Machinery/PES 31.8

64 Transmission and Distribution/PES 31.13

65 Petroleum and Chemical Industry/IAS 34.16
Definitions taken from the NFPA.

66 Power Semiconductor/IAS 34.17. See also code 162

67 Circuits and Systems. Network Applications of Circuits and Systems.

68 Non-Real Time Spectrum Analyzer. IM 6.

69 Test, Measurement, and Diagnostic Equipment. IM 8. See also code 54.

70 Switchgear terms derived from C37.100-1975 and C37.23.

71 ANSI SE3.13-1974; NFPA 72E-1974. Standard on Automatic Fire Detectors.

72 Sonics and Ultrasonics.
Definitions for specific (acoustic-optical) devices; delay lines; ferroelectric material terms. See also codes 80, 81, and 82.

73 IAS Subcommittee 2-447-02.
Emergency and Standby Power Systems. See also code 89.

74 Space/Com
Definitions of Communication Satellite Terms; Astrodynamic and Astronomical Terms. See also code 84.

75 See code 159.

76 IEEE Std 484-1975
Installation Design and Installation of Large Lead Storage Batteries for Generating Stations and Substations, Recommended Practice for.

77 Standards/Computer 16.9. Computing Systems.

78 IEEE Std 298-1969; ANSI C37.23-1970
Isolated-Phase Bus, Guide for Calculating Losses in.

79 ANSI C37.100-1975
Power Systems Relaying Committee.

80 Sonics and Ultrasonics.
Definitions of Ferroelectric Material Terms.

81 Sonics and Ultrasonics.
Definitions for Delay Lines, Dispersive and Non-dispersive.

82 Sonics and Ultrasonics.
Definitions for Acousto-Optic Devices.

83 Spacecom/Com
Component parts of communications systems; Communications satellite terms.

84 Spacecom/Com
Communications System Methods and Functions.

85 Spacecom/Com
Transmission and Propagation Terms.

86 ANSI C55.2-1973
Series Capacitor. See also code 138.

87 ANSI C104.2-1968; EIA RS 330-1966
Closed Circuit Television Camera 525/60 Interlaced 2:1, Electrical Performance of.

88 IEEE Std 91-1973; ANSI Y32.14-1973

Included in this volume is a list of abbreviations, symbols, acronyms, functional designations, sigla, and letter combinations. There is nothing official about this list. Nor does it pretend to be comprehensive or definitive. It is intended only as a convenient reference supplement in a time when so many aspects of electrotechnical effort, if rendered fully, have become polysyllabic to the point of tongue-twisting mind-boggle. Thus, an "alphanumeric photocomposer system" is much more easily tagged as an "APS" and discussion can flow smoothly on. Also, if the speaker were required to identify a "floating-gate avalanche-injection metal-oxide semiconductor" in its fully spelled out form each time he had to refer to it, he might forget how his sentence was supposed to end. In this case "FAMOS" makes everything simpler.

The list follows.

A

A-A: air to air.
AA: arithmetic average.
AA: armature accelerator.
AA: Automobile Association.
AA: dry-type self-cooled (transformer)
AAA: American Accounting Association
AAA: antiaircraft artillery.
AAO: antiair output.
AAAS: American Association for the Advancement of Science.
AAB: Aircraft Accident Board.
AABP: aptitude assessment battery programming.
AACE: American Association of Cost Engineers.
AACOM: Army Area Communications.
AACOMS: Army Area Communications System.
AACS: airborne astrographic camera system.
AACSCEDR: Associate and Advisory Committee to the Special Committee on Electronic Data Retrieval.
AADS: Area Air Defense System (SAM-D).
AA/FA: dry-type self-cooled/forced-air-cooled (transformer).
AAFB: Andrews Air Force Base.
AAFSS: Advanced Aerial Fire Support System (IHAS).
AAG: Aeronautical Assignment Group (see IRAC).
AAL: absolute assembly language.
AAL: Arctic Aeromedical Laboratory (USAF; see AAML).
AALC: Advance Airborne Launch Centers (SAC).
AAM: air-to-air missile.
AAMG: antiaircraft machine gun.
AAML: Arctic Aeromedical Laboratory.
AAP: Apollo Applications Program.
AAPL: additional programming language.
AAR: Association of American Railroads.
AAS: advanced antenna system.
AAS: American Astronautical Society.
AAS: arithmetic assignment statement.
AAS: automated accounting system.
AATC: Antiaircraft Training Center.
AATC: automatic air traffic control (system).
AAUP: American Association of University Professors.
AAVD: automatic alternate voice/data.
AAVS: Aerospace Audiovisual Service.
AB: adapter booster.
AB: Aeronautical Board.
AB: afterburner.
AB: air blast.
AB: anchor bolt.
ABA: American Bankers Association.
ABA: American Bar Association.
ABACUS: Air Battle Analysis Center Utility System.
ABAR: advanced battery acquisition radar.
ABAR: alternate battery acquisition radar.
ABC: approach by concept.
ABC: automatic bandwidth control.
ABC: automatic brightness control.
ABCB: air blast circuit breaker.
ABCCTC: Advanced Base Combat Communication Training Center.
ABL: Atlas basic language.

ABLE: activity balance line evaluation.
ABM: antiballistic missile.
ABM: automated batch mixing.
ABMDA: (U.S. Army) Advanced Ballistic Missile Defense Agency.
ABRACE: Brazilian Association of Electronic Computers.
ABRES: advanced ballistic reentry system
ABSW: air-break switch.
ABT: air blast transformer.
ABTF: airborne task force.
ABV: absolute value.
AC: aerodynamic center.
AC: alternating current.
AC: analog computer.
AC: automatic checkout.
AC: automatic computer.
ACA: adjacent channel attenuation.
ACA: American Communications Association
ACA: Automatic Communication Association.
ACAC: automated direct analog computer.
ACAP: automatic circuit analysis program.
ACB: air circuit breaker.
ACBS: Accrediting Commission for Business School.
ACCAP: autocoder to cobol conversion and program (IBM).
ACCEL: automated circuit card etching layout.
ACE: acceptance checkout equipment.
ACE: American Council of Education.
ACE: animated computer education.
ACE: Aviation Construction Engineers.
ACE: automatic checkout equipment.
ACE: automatic computer evaluation.
ACE: automatic computing engine.
ACEAA: Advisory Committee on Electrical Appliances and Accessories.
ACE-S/C: acceptance checkout equipment— spacecraft.
ACCESS: aircraft communication electronic signaling system.
ACCESS: automatic computer controlled electronic scanning system.
ACF: Alternate Communications Facility.
ACF: area computing facilities.
ACFG: automatic continuous function generation.
ACI: Air Combat Intelligence.
ACI: American Concrete Institute.
ACI: automatic card identification.
ACID: automatic classification and interpretation of data.
ACIS: Association for Computing and Information Sciences.
ACL: application control language.
ACL: Association for Computational Linguistics.
ACL: Atlas commercial language.
ACLS: all-weather carrier landing system.
ACLS: American Council of Learned Societies.
ACM: asbestos-covered metal.
ACM: Association for Computing Machinery.
ACME: advanced computer for medical research.

ACME: Association of Consulting Management Engineers.

ACM/GAMM: Association for Computing Machinery/ German Association for Applied Mathematics and Mechanics.

ACMO: Afloat Communications Management Office (NSEC).

ACMR: Air Combat Maneuvering Range.

ACMSC: ACM Standards Committee.

ACNOCOMM: Assistant Chief of Naval Operations for Communications and Cryptology.

ACOM: automatic coding system.

ACOPP: abbreviated cobol preprocessor.

ACP: advanced computational processor (Sylvania).

ACP: aerospace computer program.

ACPA: Association of Computer Programmers and Analysts.

ACR: aircraft control room.

ACR: air-field control radar.

ACR: antenna coupling regulator.

ACR: automatic compression regulator.

ACRE: automatic checkout and readiness.

ACRI: Air Conditioning and Refrigeration Institute.

ACRL: Association of College and Research Libraries.

ACRS: Advisory Committee on Reactor Safeguards.

ACS: accumulator switch.

ACS: Advanced Computer Services (HIS).

ACS: Advanced Computer System (SIBM).

ACS: Alaska Communication System.

ACS: alternating current synchronous.

ACS: American Chemical Society.

ACS: Assembly Control System (IBM).

ACS: attitude command system.

ACS: attitude control system.

ACS: Australian Computer Society.

ACS: automated communications set.

ACS: automatic checkout system.

ACS: auxiliary core storage.

ACSCE: Army Chief on Staff for Communications Electronics (Washington, D. C.).

ACSF: attack carrier striking force.

ACSM: American Congress on Survey and Mapping.

ACSP: Advisory Council on Scientific Policy (GB).

ACSR: aluminum cable steel reinforced.

ACSR: aluminum conductor steel reinforced.

ACSS: Air Combat and Surveillance System.

ACSS: analog computer subsystem.

ACST: access time.

ACT: Air Control Team.

ACT: algebraic compiler and translator.

ACT: automatic code translation.

ACTO: automatic computing transfer oscillator.

ACTRAN: autocoder-to-cobol translating service.

ACTS: acoustic control and telemetry system.

ACTS: automatic computer Telex services.

ACU: address control unit.

ACU: arithmetic and control unit.

ACU: automatic calling unit.

ACUTE: Accountants Computer Users Technical Exchange.

ACV: alarm check valve.

AC&W: air control and warning (Air Force site).

AC&WS: air control and warning stations.

A/D: analog to digital.

AD: ampere demand meter.

AD: advanced design.

ADA: airborne data automation.

ADA: automatic data acquisition.

ADAC: automatic direct analog computer.

ADACC: automatic data acquisition and computer complex.

ADC: air data computing.

ADAM: advanced data management.

ADAM: advanced direct-landing Apollo mission.

ADAM: associometrics data management system.

ADAM: automatic distance and angle measurement.

ADAPS: automatic display and plotting system.

ADAPSO: Association of Data Processing Service Organizations.

ADAPT: adoption of automatically programmed tools.

ADAPTS: air-delivered antipollution transfer system.

ADAR: advanced design array radar.

ADAS: automatic data acquisition system.

ADAT: automatic data accumulator and transfer.

ADC: airborne digital computer.

ADC: air data computer.

ADC: analog-to-digital converter.

ADC: automatic data collection.

ADCI: Air Defense Control Center.

ADCCP: advanced data communications control procedure.

ADCSP: Advanced Defense Communications Satellite Program (DCSP).

ADDAR: automatic digital acquisition and recording.

ADDAS: automatic digital data assembly system.

ADDDS: automatic direct distance dialing system.

ADE: automated design engineering.

ADE: automatic drafting equipment.

ADES: automatic digital encoding system.

ADF: airborne direction finder.

ADF: automatic direction finder.

ADI: American Documentation Institute.

ADI: attitude director indicator.

ADIDAC: Argonne National Laboratory Computer.

ADINTELCEN: advanced intelligence center.

ADIOS: automatic digital input-output system.

ADIS: Air Defense Integrated System.

ADIS: Association for Development of Instructional Systems.

ADIS: automatic data interchange system.

ADIT: analog-digital integration translator.

ADL: automatic data link.

ADM: activity data method.

ADMIRAL: automatic and dynamic monitor with immediate relocation, allocation, and loading.

ADMIRE: automatic diagnostic maintenance information retrieval.

ADMS: automatic digital message switching (autovon/ autodin).

ADMSC: automatic digital message switching centers.

ADOC: Air Defense Operation Center.

ADONIS: automatic digital on-line instrumentation system.

ADP: airborne data processor.

ADP: air defense position.
ADP: automatic data processing.
ADP: airport development program.
ADPC: automatic data processing center.
ADPCM: Association for Data Processing and Computer Management.
ADPE: auxiliary data processing equipment.
ADPE/S: automatic data processing equipment /system.
ADPESO: automatic data processing equipment selection office.
ADPSO: Association of Data Procesing Service Organizations.
ADR: aircraft direction room.
ADRAC: automatic digital recording and control.
ADRS: analog-to-digital data recording system.
ADRT: analog data recorder transcriber.
ADS: accurately defined system.
ADS: activity data sheet.
ADS: advance data system.
ADS: air defense sector.
ADS: Air Development Service (FAA).
ADS: automatic door seal.
ADSAF: Automatic Data System for the Army in the Field.
ADSAS: air-derived separation assurance system.
ADSC: automatic data service center.
ADSC: automatic digital switching center.
ADSCOM: advanced shipboard communication.
ADS-TP: Administrative Data Systems — Teleprocessing.
ADSUP: automated data systems uniform practices.
ADT: automatic data translator.
ADTECH: advanced decoy technology.
ADU: automatic dialing unit.
ADX: automatic data exchange.
AE: aeroelectronic.
AE: air escape.
AE: application engineer.
AE: arithmetic element.
AE: arithmetic expression
AEA: Agriculture Engineers Association.
AEA: Atomic Energy Authority.
AEC: Atomic Energy Commission.
AECB: Atomic Energy Control Board (Canada).
AECL: Atomic Energy of Canada, Ltd. (Canada).
AECT: Association for Educational Communications and Technology.
AED: Algol extended for design.
AED: automated engineering design.
AEDP: Association for Educational Data Processing.
AEDS: Association for Educational Data Systems.
AEDS: Atomic Energy Detection System (EMP, Vela).
AEE: airborne evaluation equipment.
AEE: Atomic Energy Establishment (GB).
AEEC: Airlines Electronic Engineering Committee (Annapolis, Md.).
Ae.Eng.: aeronautical engineer.
AEF: Aviation Engineer Force.
AEG: active element group.
AEI: *Associazione Elettrotecnica Italiana.*
AEI: average efficiency index.

AEL: Aeronautical Engine Laboratory.
AEO: air engineer officer.
AEPG: Army Electronic Proving Ground (Ft. Huachuca, Ariz.).
AERA: American Educational Research Association.
AERDL: Army Electronics Research and Development Laboratory.
AERE: Atomic Energy Research Establishment (GB).
AERNO: Aeronautical Equipment Reference Number (military).
AEROF: aerological officer.
AES: Aerospace and Electronic Systems (IEEE Society).
AES: American Electroplaters Society.
AES: Apollo extension system.
AES: Audio Engineering Society.
AESOP: an evolutionary system for on-line processing.
AEW: airborne early warning.
AEW&C: airborne early warning and control.
AEWTU: airborne early warning training unit.
AF: audio frequency.
AF: automatic following.
AFA: dry-type forced-air-cooled (transformer).
AFACD: Air Force Director of Data Automation.
AFADS: Advanced Forward Air Defense System. (SAM-D)
AFAL: Air Force Avionics Laboratory (WPAFB, Dayton, Ohio).
AFAPL: Air Force Aero Propulsion Laboratory.
AFB: antifriction bearing.
AFC: area frequency coordinator.
AFC: automatic frequency control.
AFCAL: *Association Française de Calcul.*
AFCE: automatic flight control equipment.
AFCEA: Armed Force Communication Electronics Association.
AFCRL: Air Force Cambridge Research Laboratory (Bedford, Mass.).
AFCS: adaptive flight control system.
AFCS: automatic flight control system.
AFDDA: Air Force Director of Data Automation.
AFELIS: Air Force Engineering and Logistics Information System.
AFIPS: American Federation of Information Processing Societies.
AFIT: Air Force Institute of Technology.
AFM: antifriction metal.
AFM: automatic flight management.
AFMR: antiferromagnetic resonance.
AFMTC: Air Force Missile Test Center.
AFNOR: *Association Française des Normes.*
AFOSR: Air Force Office of Scientific Research.
AFPA: automatic flow process analysis.
AF/PC: automatic frequency/phase controlled (loop).
AFRAL: *Association Française de Reglage Automatique.*
AFRDR: Air Force Director of Reconnaissance and Electronic Warfare.
AFRPL: Air Force Rocket Propulsion Laboratory.
AFRRI: Armed Forces Radiobiology Research Institute.
AFRS: Armed Forces Radio Service.

AFRTS: Armed Forces Radio and Television Service.

AFS: audio frequency shift.

AFTN: aeronautical fixed telecommunications network.

AFTRCC: Aerospace Flight Test Radio Coordinating Council.

AFWL: Air Force Weapons Laboratory (Kirkland AFB, N. Mex.).

A-G: air to ground.

AG: air to ground.

AG: armor grating.

AG: arresting gear (and barriers).

AGA: American Gas Association.

AGACS: automatic ground-to-air communications systems.

AGARD: Advisory Group for Aerospace Research and Development (NATO).

AGC: automatic gain control.

AGCA: automatic ground-controlled approach.

AGCL: automatic ground-controlled landing.

AGDS: American Gage Design Standard.

AGDIC: Astro guidance digital computer.

AGE: aerospace ground (support) equipment.

AGED: Advisory Group on Electron Devices.

AGEP: Advisory Group on Electronic Parts.

AGET: Advisory Group on Electron Tubes.

AGI: American Geological Institute.

AGIL: airborne general illumination light (Shed Light).

AGILE: Autonetics general information learning equipment.

AGM: auxiliary general missile.

AGR: advanced gas-cooled reactor.

AGREE: Advisory Group on the Reliability of Electronic Equipment.

AGS: alternating gradient synchrotron.

AGSP: Atlas general survey program.

AGU: American Geophysical Union.

A/H: air over hydraulic.

AH: ampere-hour.

AHAM: Association of Home Appliance Manufacturers.

AHEM: Association of Hydraulic Equipment Manufacturers.

AHM: ampere-hour meter.

AHR: acceptable hazard rate.

AHSR: air-height surveillance radar.

AI: airborne intercept.

AI: automatic input.

AI: Automation Institute.

AIA: American Institute of Architects.

AIA: Aerospace Industries Association.

AIA: American Inventors Association.

AIAA: American Institute of Aeronautics and Astronautics.

AIB: American Institute of Banking.

AICA: *Associazione Italiana per il Calcolo Automatico.*

AIBS: American Institute of Biological Sciences.

AICBM: anti-intercontinental ballistic missile.

AICE: American Institute of Consulting Engineers.

AICHE: American Institute of Chemical Engineers.

AICPA: American Institute of Certified Public Accountants.

AID: algebraic interpretive dialog.

AIDE: automated integrated design engineering.

AIDS: automated integrated debugging system.

AIDS: Aircraft Integrated Data System (AFSC).

AIDSCOM: Army Information Data Systems Command.

AIEE: American Institute of Electrical Engineers (obsolete, see IEEE).

AIENDF: Atomics International Evaluation Nuclear Data File.

AIF: Atomic Industrial Forum.

AIIE: American Institute of Electrical Engineers.

AILAS: Automatic Instrument Landing Approach System.

AILS: Automatic Instrument Landing System.

AIM: air intercept missile.

AIM: air-isolated monolithic.

AIM: American Institute of Management.

AIMACO: air materiel computer.

AIME: American Institute of Mechanical Engineers.

AIME: American Institute of Mining, Metallurgical, and Petroleum Engineers.

AIMES: automated inventory management evaluation system.

AIMS: automated information and management systems.

AIP: American Institute of Physics.

AIPE: American Institute of Plant Engineers.

AIR: aerospace information report.

AIR: air intercept rocket.

AIR: American Institute for Research.

AIRCON: automated information and reservations computer-oriented network.

AIRS: automatic image retrieval system.

AIRS: automatic information retrieval system.

AISC: Association of Independent Software Companies.

AIRSS: ABRES Instrument Range Safety System.

AIS: American Interplanetary Society (obsolete, see ARS).

AISC: American Institute of Steel Construction.

AISE: Association of Iron and Steel Engineers.

AISI: American Iron and Steel Institute.

AIST: automatic informational station.

AIT: American Institute of Technology.

AJ: antijamming.

AJ: assembly jig.

AJD: antijam display.

AK: Alaska.

AL: air lock.

AL: Alabama.

AL: assembly language.

ALA: American Library Association.

ALARM: automatic light aircraft readiness monitor.

ALAS: asynchronous look-ahead simulator.

ALBM: air-launched ballistic missile.

ALC: automatic leveling control.

ALC: automatic load control.

ALCC: airborne launch control center (PACCS).

ALCOM: algebraic compiler.

ALCOR: ARPA Lincoln Coherent Observable Radar (Defender).

ALD: analog line driver.

ALDP: automatic language data processing.

ALERT: automated linguistic extraction and retrieval technique.

ALFTRAN: algol to fortran translator.

ALGM: air-launched guided missile.

ALGOL: algorithmic language.

ALI: automated logic implementation.

ALIS: advanced life information system.

ALMS: analytic language manipulation system.

ALOR: advanced lunar orbital rendezvous.

ALOTS: airborne lightweight optical tracking system.

ALP: assembly language program.

ALP: automated learning process.

ALPS: Advanced Linear Programming System (HIS).

ALPS: associated logic parallel system.

ALPS: automated library processing service.

ALRI: airborne long-range intercept (495L).

ALS: approach-light system.

ALSEP: Apollo lunar surface experiment package.

ALTAC: algebraic transistorized automatic computer translator.

ALTAIR: ARPA long-range tracking and instrumentation radar.

ALTRAN: algebraic translator.

ALU: arithmetic and logic unit.

ALVIN: antenna lobe for variable ionospheric nimbus.

A/m: ampere per meter (unit of magnetic field strength).

AM: amplitude modulation.

AM: *ante meridiem* (before noon).

AM: auxiliary memory.

AMA: American Management Association.

AMA: American Manufacturers Association.

AMA: American Marketing Association.

AMA: American Medical Association.

AMA: automatic memory allocation.

AMA: automatic message accounting.

AMACUS: automated microfilm aperture card updating system.

AMAS: advanced midcourse active system.

AMB: asbestos millboard.

AMBIT: algebraic manipulation by identity translation.

AMC: American Mining Congress.

AMC: Army Missile Command.

AMC; Association of Management Consultants.

AMC: automatic message counting.

AMC; automatic mixture control.

AMCBMC: Air Materiel Command Ballistic Missile Center.

AMCEC: Allied Military Communications Electronics Committee.

AMCS: airborne missile control system.

AMFIS: Automatic Microfilm Information System.

AMFIS: American Microfilm Information Society.

AMI: American Microsystems, Inc.

AMICOM: Army Missile Command.

AMIM: American Journal of Mathematics.

AMM: antimissile missile.

AMNIP: adaptive man-machine nonarithmetical information processing.

AMO: alternant molecular orbit.

AMOS: acoustic meteorological oceanographic survey.

AMOS: automatic meteorological observing system.

AMP: associative memory processor.

AMPLG: amplidyne generator.

AMPLMG: amplidyne motor generator.

AMPS: automatic message processing system.

AMR: Atlantic Missile Range, Air Force (Cocoa Beach, Fla.).

AMR: automated management reports.

AMR; automatic message registering.

AMRAD: Arda Measurements Radar (Defender).

AMS: advanced memory systems.

AMS: Administrative Management Society.

AMS: aeronautical material specification.

AMS: American Mathematical Society.

AMS: American Meteorological Society.

AMST: Association of Maximum Service Telecasters.

AMTB: antimotor torpedo boat.

AMTCL: Association for Machine Translation and Computational Linguistics.

AMTI: airborne moving target indicator.

AMTI: area moving target indicator.

AMTI: automatic moving target indicator.

AMTRAN: automatic mathematical translator.

AMU: astronaut maneuvering unit.

AMU: atomic mass unit.

ANACOM: analog computer.

ANATRAN: analog translator.

ANCS: American Numerical Control Society.

ANDREE: Association for Nuclear Development and Research in Electrical Engineering.

ANF: antinuclear factor.

ANF: Atlantic Nuclear Force (Polaris/Poseidon).

ANI: automatic number identification.

ANL: Argonne National Laboratory (AEC).

ANL: automatic noise limiter.

ANLOR: angle order.

ANMC: American National Metric Council.

ANO: alphanumeric output.

ANOVA: analysis of variance.

ANP: aircraft nuclear propulsion.

ANPA: American Newspaper Publishers Association.

ANPP: Army Nuclear Power Program.

ANPO: Aircraft Nuclear Propulsion Office.

ANPT: aeronautical national taper pipe thread.

ANRAC: Aids Navigation Radio Control.

ANS: American Nuclear Society.

ANSI: American National Standards Institute.

ANTC: antichaff circuit.

ANTIVOX: voice-actuated transmitter keyer inhibitor.

ANTS: airborne night television system (TARS/SEA Niteops).

ANZAAS: Australian and New Zealand Association for the Advancement of Science.

AO: access opening.

AOC: automatic overload control.

AOCS: Alpha Omega computer system.

AOI: and-or invert.

AOL: application-oriented language.

AOML: Atlantic Oceanographic and Meteorological Laboratories (NOAA).

AOPA: Aircraft Owners and Pilots Association.

AOQL: average outgoing quality limit.

AOSO: Advanced Orbiting Solar Observatory.

AOSP: Automatic Operating and Scheduling Program.
AP: access panel.
AP: after peak.
AP: after perpendicular.
AP: Antennas and Propagation (IEEE Society).
AP: application program.
AP: applications processor.
AP: applied physics.
AP: argument programming.
AP: associative processor.
APA: American Psychological Association.
APA: axial pressure angle.
APADS: automatic programmer and data system.
APAR: automatic programming and recording.
APC: aeronautical planning chart.
APC: automatic phase control.
APC: autoplot controller.
APCA: Air Pollution Control Association.
APCHE: automatic programming checkout equipment.
APD: amplitude probability distribution.
APEC: all-purpose electronic computer.
APEL: Aeronautical Photographic Experimental Laboratory.
APF: atomic packing factor.
APFCS: automatic power-factor-control systems.
APGC: Air Proving Ground Center.
APHA: American Public Health Association.
APHI: Association of Public Health Inspectors.
API: air position indicator.
API: American Petroleum Institute.
APIC: Apollo Parts Information Center.
APICS: American Production and Inventory Control Society.
APIN: Association for Programmed Instruction in the Netherlands.
APL: Applied Physics Laboratory (Silver Spring, Md.).
APL: Association of Programmed Learning.
APL: associative programming language.
APL: average picture level.
APLE: Association of Public Lighting Engineers.
APM: analog panel meter.
APMI: area precipitation measurement indicator.
APNIC: Automatic Programming National Information Centre (GB).
APP: associative parallel processor.
APP: auxiliary power plant.
APPA: American Public Power Association.
APPECS: adaptive pattern perceiving electronic computer system.
APPI: Advanced Planning Procurement Information.
APR: airborne profile recorder.
APRS: automatic position reference system.
APS: alphanumeric photocomposer system.
APS: American Physical Society.
APS: American Physiological Society.
APS: assembly programming system.
APS: automatic patching system.
APS: auxiliary power system.
APS: auxiliary program storage.
APT: augmented programming training.
APT: automatic picture taking.
APT: automatic picture transmission.

APT: automatically programmed tools.
APT: automation planning and technology.
APU: auxiliary power unit.
APW: augmented plane wave.
AQ: any quantity.
AQL: acceptable quality level.
AQT: acceptable quality test.
AR: acid resisting.
AR: Arkansas.
AR: assembly and repair.
AR: avionic requirements.
AR: aviation radionavigation, land (FCC services).
ARA: Amateur Rocket Association.
ARAL: automatic record analysis language.
ARC: Aeronautical Research Council.
ARC: Argonne reactor computation.
ARC: automatic relay calculator.
ARC: average response computer.
arccos: inverse cosine.
arccot: inverse cotangent.
arccse: inverse cosecant.
ARCS: Air Resupply and Communication Service.
arcsec: inverse secant.
arcsin: inverse sine.
arctan: inverse tangent.
ARE: automated responsive environment.
ARFA: Allied Radio Frequency Agency.
ARGUS: automatic routine generating and updating system.
ARI: Air-Conditioning and Refrigeration Institute.
ARIA: Apollo Range Instrumented Aircraft.
ARIA: Apollo Range Instrumentation Ships (Satellite Tracking Network, NASA, MSFN).
ARINC: Aeronautical Research, Inc. (Annapolis, Md.).
ARIP: automatic rocket impact predictor.
ARIS: Advanced Range Instrumentation Ship (Satellite Tracking Network, NASA, MSFN).
ARL: acceptable reliability level.
ARL: Association of Research Libraries.
ARM: antiradiation missile.
ARMA: American Records Management Association.
ARMMS: automated reliability and maintainability measurement system.
ARMS: aerial radiological measuring survey.
ARMS: Amateur Radio Mobile Society.
ARO: Air Radio Officer.
AROD: airborne range and orbit determination.
AROU: aviation repair and overhaul unit.
ARP: Advanced Reentry Program (ABRES).
ARP: Aeronautical Recommended Practice (SAE).
ARP: airborne radar platform.
ARPA: Advanced Research Projects Agency (Washington, D. C.).
ARQ: automatic repeat request.
ARR: antirepeat relay.
ARRL: Aeronautical Radio and Radar Laboratory.
ARRL: American Radio Relay League.
ARRS: Aerospace Rescue and Recovery Service.
ARS: Advanced Record System.
ARS: American Radium Society.
ARS: American Rocket Society (merged with Institute of Aerospace Sciences).

ARSP: Aerospace Research Support Program.
ARSR: air route surveillance radar.
ART: Advanced Reactor Technology.
ART: Airborne Radiation Thermometer.
ART: automatic reporting telephone.
ARTCC: Air-Route Traffic Control Center.
ARTE: Admiralty Reactor Test Establishment (GB).
ARTIC: associometrics remote terminal inquiry control system.
ARTRON: artificial neutron.
ARTRAC: advanced range testing, reporting, and control.
ARTRAC: advanced real-time range control.
ARTS: advanced radar terminal system.
ARTS: advanced radar traffic-control system.
ARTU: automatic range tracking unit.
ARU: audio response unit.
AS: ammeter switch.
AS: antisubmarine.
ASA: Acoustical Society of America.
ASA: American Standards Association (later USASI; now ANSI).
ASA: American Statistical Association.
ASA: Army Security Agency (Arlington, Va.).
ASA: Atomic Security Agency.
ASAE: American Society of Agricultural Engineers.
ASAP: antisubmarine attack plotter.
ASAP: Army Scientific Advisory Panel.
ASAP: as soon as possible.
ASBC: American Standard Building Code.
ASBD: advanced sea-based deterrent.
ASBO: Association of School Business Officials.
ASC: Aeronautical Systems Center.
ASC: advanced scientific computer.
ASC: American Society for Cybernetics.
ASC: associative structure computer.
ASC: Atlantic Systems Conference.
ASC: automatic sensitivity control.
ASC: automatic system controller.
ASC: auxiliary switch (breaker) normally closed.
ASCA: automatic science citation alerting.
ASCAP: American Society of Composers, Authors, and Publishers.
ASCATS: Apollo Simulation Checkout and Training System.
ASCC: Air Standardization Coordinating Committee.
ASCC: automatic sequence controlled calculator.
ASCE: American Society of Civil Engineers.
ASCII: American National Standard Code for Information Interchange.
ASCO: automatic sustainer cutoff.
ASCS: area surveillance control system.
ASD: Aeronautical Systems Division, USAF (Dayton, Ohio).
ASDE: airport surface detection equipment.
ASDIC: Antisubmarine Detection and Identification Committee.
ASDL: Automated Ship Data Library.
ASDSRS: automatic spectrum display and signal recognition system.
ASE: airborne search equipment.
ASE: automatic support equipment.
ASEC: American Standard Elevator Code.

ASEE: American Society for Engineering Education.
ASFDO: Antisubmarine Fixed Defenses Officer.
ASFIR: active swept frequency interferometer radar.
ASFTS: airborne systems functional test stand.
ASG: Aeronautical Standards Group.
ASGLS: advanced space ground link subsystem.
ASHRAE: American Society of Heating, Refrigerating, and Air-Conditioning Engineers.
ASI: American Standards Institute.
ASII: American Science Information Institute.
ASIS: abort sensing and instrumentation system.
ASIS: American Society for Information Science.
ASIST: advanced scientific instruments symbolic translator.
ASK: amplitude-shift keying.
ASKA: automatic system for kinematic analysis.
ASKS: automatic station keeping system.
ASL: Association for Symbolic Logic.
ASLBM: air-to-ship launched ballistic missile (Defender).
ASLE: American Society of Lubrication Engineers.
ASLIB: Association of Special Libraries and Information Bureau.
ASLO: American Society of Limnology and Oceanography.
ASLT: advanced solid logic technology.
ASM: advanced surface-to-air missile.
ASM: Association for Systems Management.
ASM: air-to-surface missile.
ASM: American Society for Metals.
ASM: asynchronous state machine.
ASMB: Acoustical Standards Management Board.
ASME: American Society of Mechanical Engineers.
ASMS: advanced surface missile system (SAM-D).
ASN: average sample number.
ASNE: American Society of Naval Engineers.
ASO: auxiliary switch (breaker) normally open.
ASODDS: ASWEPS Submarine Oceanographic Digital Data System.
ASOP: automatic scheduling and operating program.
ASP: association storing processor.
ASP: attached support processor.
ASP: automatic schedule procedures.
ASPEP: Association of Scientists and Professional Engineering Personnel.
ASPER: assembly system peripheral processors.
ASPJ: advanced self-protective jammer.
ASPP: alloy-steel protective plating.
ASQC: American Society for Quality Control.
ASR: accumulators shift right.
ASR: airborne surveillance radar.
ASR: airport surveillance radar.
ASR: automatic send receive.
ASR: available supply rate.
ASRA: automatic stereophonic recording amplifier.
ASRE: American Society of Refrigerating Engineers.
ASS: Aerospace Support Systems (IEEE AESS Technical Committee).
ASSE: American Society of Safety Engineers.
ASSP: Acoustics, Speech, and Signal Processing (IEEE Group).
ASTD: American Society for Training and Development.

ASTEC: Antisubmarine Technical Evaluation Center.
ASTM: American Society for Testing and Materials.
ASTME: American Society of Tool and Manufacturing Engineers.
ASTOR: antiship torpedo.
ASTRAL: analog schematic translator to algebraic language.
ASV: angle stop valve.
ASV: automatic self-verification.
ASVIP: American Standard Vocabulary for Information Processing.
ASW: antisubmarine warfare.
ASW: auxiliary switch.
ASWEPS: antisubmarine warfare environmental prediction system.
ASWG: American Steel Wire Gage.
ASW/SICS: ASW/Ship Command and Control System.
ASWSPO: Antisubmarine Warfare Systems Project Office.
AT: ampere-turn.
AT: automatic ticketing.
ATA: Air Transport Association.
ATAR: antitank aircraft rocket.
ATBM: antitactical ballistic missile (Hawk).
ATC: air traffic control.
ATC: Automation Training Center.
ATCAC: Air-Traffic Control Advisory Committee.
ATCRBS: Air-Traffic Control Radar Beacon System (AIMS).
ATDS: airborne tactical data system.
ATE: automatic test equipment.
ATERM: air-to-air gunnery range.
ATG: air to ground.
ATIS: automatic terminal information service.
ATJS: advanced tactical jamming system (ATEWS, TJS, TEWS).
ATLAS: abbreviated test language for all systems.
ATM: air turbine motor.
ATOLL: acceptance, test or launch language.
ATOMDEF: atomic defense.
ATP: automated test plan.
ATR: air transport radio.
ATR: antitransmit-receive.
ATRAN: automatic terrain recognition and navigation.
ATRC: antitracking control.
ATRT: antitransmit-receive tube.

ATS: astronomical time switch.
AT/W: atomic hydrogen weld.
AU: arithmetic unit.
AUDREY: audio reply.
AUIS: analog input-output unit.
AU: astronomical unit.
AUM: air-to-underwater missile.
AUNT: automatic universal translator.
AUT: advanced user terminals.
AUTODIN: automatic digital network.
AUTONET: automatic network display.
AUTOVON: automatic voice network.
AUTOSTRT: automatic starter.
AUTOSTRTG: automatic starting.
AUTRAN: automatic utility translator.
AUW: antiunderwater warfare.
AV: audiovisual.
AVA: azimuth versus amplitude.
AVC: automatic volume control.
AVE: aerospace vehicle electronics.
AVE: automatic volume expansion.
AVERT: Association of Volunteer Emergency Radio Teams.
AVL: automatic vehicle location.
AVLB: armored vehicle launched bridge.
AVM: automatic vehicle monitoring.
AVNL: automatic video noise limiting.
AVR: automatic volume recognition.
AW: above water.
AW: air warning.
AWCS: Air Weapons Control System.
AWACS: advanced warning and control system.
AWACS: airborne warning and control system.
AWADS: all weather aerial delivery system.
AWAR: area weight average resolution.
AWARS: airborne weather and reconnaissance system.
AWAT: area weighted average T-number.
AWCS: air weapon control system.
AWG: American Wire Gage.
AWN: automated weather network.
AWRNCO: Aircraft Warning Company (Marines).
AWWA: American Water Works Association.
AXD: auxiliary drum.
AXFMR: automatic transformer.
AXP: axial pitch.
AZ: Arizona.
AZS: automatic zero set.

B

B: Broadcasting (IEEE Group).
BA: binary add.
BA: breathing apparatus.
BA: buffer amplifier.
BAAS: British Association for the Advancement of Science.
BABS: blind approach beacon system.
BACAIC: Boeing Airplane Company algebraic interpretive computing system.

BACE: basic automatic checkout equipment.
BACE: British Association of Consulting Engineers.
BAL: basic assembly language.
BALGOL: Burroughs algebraic compiler.
BALLOTS: bibliographic automation of large library operations using time sharing.
BALS: balancing set.
BALUN: balanced to unbalanced.
BAM: basic access method.

BAMS: Bulletin of the American Mathematical Society.

BAMBI: ballistic missile boost intercept.

BAMO: Bureau of Aeronautics Material Officer.

BAP: basic assembly program.

BAR: base address register.

BAR: Bureau of Aeronautics Representative.

BARR: Bureau of Aeronautics Resident Representative.

BARSTUR: Barking Sands Tactical Underwater Range.

BASIC: basic algebraic symbolic interpretive compiler.

BASIC: basic automatic stored instruction computer.

BASIC: beginners' all-purpose symbolic instruction code.

BASICS: Battle Area Surveillance and Integrated Communication.

BASIS: bank automated service information system.

BASIS: Burroughs applied statistical inquiry system.

BASYS: basic system.

BB: base band.

BB: broadband (emission).

BBC: broadband conducted.

BBD: bucket-brigade device.

BBL: basic business language.

BBR: broadband radiated.

BC: back-connected.

BC: between centers.

BC: binary code.

BC: binary counter.

BCAC: British Conference on Automation and Computation.

BCD: binary-coded decimal.

BCDIC: binary coded decimal interchange code.

BCE: Bachelor of Civil Engineering.

BCH: Bose-Chaudhuri-Hocquenguem (cyclic codes).

BCI: binary coded information.

BCI: broadcast interference.

BCIP: Belgium Centre for Information Processing.

BCL: basic contour line.

BCL: Burroughs common language.

BCO: bridge cutoff.

BCO: binary coded octal.

BCOM: Burroughs computer output to microfilm.

BCS: Biomedical Computing Society.

BCS: British Computer Society.

BCT: bushing current transformer.

BCU: binary counting unit.

BCW: buffer control word.

BD: base detonating.

BD: binary decoder.

BDF: base detonating fuse.

BDI: bearing deviation indicator.

BDIA: base diameter.

BDT: binary deck to binary tape.

BDU: basic display unit.

BE: back end.

BE: base injection.

BE: breaker end.

BEA: Business Education Association.

BEAC: Boeing Engineering analog computer.

BEAIRA: British Electrical and Allied Industries Research Association.

BEAM: Burroughs electronic accounting machine.

BEAME: British Electrical and Allied Manufacturers.

BECTO: British Electric Cable Testing Organization.

BED: bridge-element delay.

B.EE.: Bachelor of Electrical Engineering.

BEEC: binary error erasure channel.

BEEF: business and engineering enriched Fortran.

BEFAP: Bell Laboratories Fortran assembly program.

BEMA: Business Equipment Manufacturers Association.

BEPC: British Electrical Power Convention.

BEPO: British Experimental Pile Operation.

BER: bit error rate.

BERT: basic energy reduction technology.

BES: Biological Engineering Society.

BESRL: Behavioral Science Research Laboratory (U.S. Army).

BEST: Business EDP System Technique.

BEST: business electronic systems techniques.

BEST: business equipment software techniques.

beV: billion electronvolts.

BEX: broadband exchange.

BF: backface.

BF: back-feed.

BF: base fuse.

BF: beat frequency.

BF: boldface (type).

BF: bottom face.

BFCO: band filter cutoff.

BFD: back focal distance.

BFG: binary frequency generator.

BFL: back focal length.

BFO: beat frequency oscillator.

BFPDDA: binary floating point digital differential analyzer.

BG: back gear.

BGRR: Brookhaven's Graphite Research Reactor (BNL).

BGRV: boost-glide reentry vehicle.

BH: flux density vs. magnetizing force.

B-H: binary to hexadecimal.

BHA: base helix angle.

BI: base ignition.

BI: blanking input.

BIE: British Institute of Engineers.

BIEE: British Institute of Electrical Engineers.

BIAS: Battlefield Illumination Airborne System (Shed Light).

BIL: basic impulse insulation level.

BIL: basic impulse isolation level.

BIL: block input length.

BILA: Battelle Institute Learning Automation (BMI).

BIMAC: bistable magnetic core.

BiMOS: bipolar metal-oxide semiconductor.

BINAC: binary automatic computer.

BINOMEXP: binomial expansion.

BIOALRT: bioastronautics laboratory research tool.

BIONICS: biological electronics.

BIOR: business input-output rerun.

BIOSIS: bioscience information service of biological abstracts.

BIP: balanced in plane.
BIR: British Institute of Radiology.
BIRE: British Institute of Radio Engineers.
BIRDIE: battery integration and radar display equipment.
BIRS: basic indexing and retrieval system.
BIS: British Interplanetary Society.
BISAM: basic indexed sequential access method (IBM).
BIT: binary digit.
BIT: built-in test.
BITE: built-in test equipment.
BIX: binary information exchange.
BL: base line.
BL: bend line.
BL: bottom layer.
BLADE: basic level automation of data through electronics.
BLADES: Bell Laboratories automatic design system.
BLC: British Lighting Council.
BLD: beam-lead device.
BLEU: blind landing experimental unit (RAE).
BLEVE: boiling liquid expanding vapor explosion.
BLIP: background limited infrared photoconductor.
BLIS: Bell Laboratories interpretive system.
BLL: below lower limit.
BLM: basic language machine.
BLODIB: block diagram compiler B.
BL&T: blind loaded and traced.
BLT: basic language translator.
BLU: basic logic unit.
BM: binary multiply.
BMAR: ballistic missile acquisition radar.
BMB: British Metrication Board.
BMC: bulk-molding compound.
BMEP: brake mean effective pressure.
BMETO: Ballistic Missiles European Task Organization.
BMEWS: Ballistic Missile Early Warning System.
BMI: Battelle Memorial Institute.
BMILS: bottom-mounted impact and location system (ABRES).
BMR: Brookhaven's Medical Reactor (BNL).
BMTS: Ballistic Missile Test System.
BMW: beam width.
BN: binary number.
BNC: baby "N" connector.
BNCS: British Numerical Control Society.
BNCSR: British National Committee on Space Research.
BNF: bomb nose fuse.
BNL: Brookhaven National Laboratory (USAEC).
BNS: binary number system.
B-O: binary to octal.
BO: blocking oscillator.
BOCOL: basic operating consumer oriented language.
BOCS: Bendix Optimum Configuration Satellite.
BOD: biochemical oxygen demand.
BOLD: bibliographic on-line display.
BOM: basic operating monitor.
BoM: U.S. Bureau of Mines.
BOP: binary output program.
BORAM: block-oriented random access memories.

BOS: background operating system.
BOS: basic operating system.
BOT: beginning of tape.
BP: between perpendiculars.
BP: boiling point.
BPA: Bonneville Power Administration (Portland, Ore.).
BPC: back-pressure control.
BPD: bushing potential device.
BPI: bits per inch (packing density).
BPI: bytes per inch.
BPKT: basic programming knowledge test.
BPM: batch processing monitor.
BPPMA: British Power Press Manufacturers Association.
BPR: Bureau of Public Roads.
BPS: basic programming system.
BPS: bits per second.
BPSC: binary phase-shift keying.
BPWR: burnable poison water reactor.
BQL: basic query language.
BR: bend radius.
BRA: British Refrigeration Association.
BRANE: bombing radar navigation equipment.
BREMA: British Radio Equipment Manufacturers Association.
BRH: Bureau of Radiological Health (HEW, Rockville, Md.).
BRL: Ballistics Research Laboratory (U.S. Army).
BRLS: Barrier Ready Light System.
BRT: binary run tape.
BRVMA: British Radio Valve Manufacturers Association.
B&S: Brown & Sharpe Wire Gage.
B.S.: British Standard.
BS: binary subtract.
BSAM: basic sequential access method.
BSC: binary synchronous communication.
BSCA: binary synchronous communications adapter.
BSCN: bit scan.
BSD: Ballistic Systems Division.
BSDC: British Standard Data Code.
BSDL: boresight datum line.
BSI: British Standards Institute.
BSIRA: British Scientific Instrument Research Association.
BSNDT: British Society for Non-Destructive Testing.
BSR: blip-scan ratio.
BSTJ: Bell Systems Technical Journal.
BSWM: Bureau of Solid Waste Management.
BT: bathythermograph.
BTD: binary to decimal.
BTD: bomb testing device.
BTDL: basic transient diode logic.
BTE: battery terminal equipment.
BTF: bomb tail fuse.
BTI: bridged tap isolator.
BTL: beginning tape label.
BTMA: basic telecommunication.
BTR: Broadcast and Television Receivers (IEEE Group).

BTSS: basic time sharing system.
BTTP: British Towing Tank Panel.
Btu: British thermal unit.
BU: binding unit.
BUAER: Bureau of Aeronautics (USN).
BUIC: back-up interceptor control.
BUILD: base for uniform language definition.
BUIS: Barrier Up Indicator System.
BUMED: Bureau of Medicine and Surgery (USN).
BUORD: Bureau of Ordnance (USN).
BUPERS: Bureau of Naval Personnel (USN).
BUSANDA: Bureau of Supply and Accounts (USN).

BUSHIPS: Bureau of Ships (USN, NSSC).
BUWEPS: Bureau of Naval Weapons (USN, NOSC).
BVC: black varnish cambric (insulation).
BW: biological warfare.
BW: bandwidth.
BWA: British Waterworks Association.
BWG: Birmingham Wire Gage.
BWM: backward-wave magnetron.
BWO: backward-wave oscillator.
BWPA: backward-wave power amplifier.
BWR: boiling water reactor.

C

CA: California.
CA: clear aperture.
CA: computers and automation.
CAA: Civil Aeronautics Board (DOT Washington, D.C.).
CAA: computer assisted accounting.
CAARC: Commonwealth Advisory Aeronautical Research Council.
CAAS: Ceylon Association for the Advancement of Science.
CAB: Civil Aeronautics Board.
CACB: compressed-air circuit breaker.
CACDP: California Association of County Data Processors.
CACM: Communications of the Association for Computing Machinery.
CAD: cartridge-activated device.
CAD: computer access device.
CAD: computer-aided design.
CAD: computer-aided detection.
CAD: Computer Applications Digest.
CADAM: computer graphics augmented design and manufacturing system.
CADAR: computer-aided design, analysis and reliability.
CADC: Cambridge automatic digital computer.
CADC: central air data computer.
CADEP: computer-aided design of electronic products.
CADET: computer-aided design experimental translator.
CADETS: classroom-aided dynamic educational time-sharing system.
CADF: cathode-ray tube automatic direction finding.
CADF: commutated antenna direction finder.
CADFISS: computation and data flow integrated subsystems.
CADIC: computer-aided design of integrated circuits.
CADRE: current awareness and document retrieval for engineers.
CADSS: combined analog-digital systems simulator.

CAE: computer-aided education.
CAE: computer-assisted enrollment (IBM).
CAFE: computer-aided film editor.
CAGE: computerized aerospace ground equipment.
CAI: computer-administered instruction.
CAI: computer-aided instruction.
CAI: computer analog input.
CAI/OP: computer analog input/output.
CAI: computer assisted instruction.
CAIS: computer-aided instruction.
CAL: computer animation language.
CAL: computer-assisted learning.
CAL: conversational algebraic language.
CAL: Cornell Aeronautic Laboratory.
CALM: collected algorithms for learning machines.
CALM: computer-assisted library mechanization.
CAM: central address memory.
CAM: computer address matrix.
CAM: computer-aided manufacturing.
CAM: content addressable memory.
CAM: cybernetic anthropomorphous machine.
CAMA: centralized automatic message accounting.
CAMAC: computer automated measurement and control.
CAMAR: common-aperture multifunction array radar.
CAMESA: Canadian Military Electronics Standards Agency.
CAMP: compiler for automatic machine programming.
CAMP: computer-assisted mathematics program.
CAMP: computer-assisted movie production.
CAMP: controls and monitoring processor.
CAMRAS: computer assisted mapping and records activities systems.
CANS: computer-assisted network scheduling system.
CANUNET: Canadian University Computer Network.
CAOS: completely automatic operational system.
CAP: Canadian Association of Physicists.
CAP: Computer Analysts and Programmers of England.

CAP: computer assisted production.

CAP: Council to Advance Programming.

CAP: cryotron associative processor.

CAPAL: Computer and Photographically Assisted Learning (Kodak).

CAPE: coalition of aerospace employees.

CAPE: communications automatic processing equipment.

CAPERTSIM: computer-assisted program evaluation review technique simulation.

CAPM: computer-aided patient management.

CAPRI: computerized advance personnel requirements and inventory.

CAPS: Courtauld's all-purpose simulator.

CAPS: computer assisted problem solving.

CAPT: conversational parts programming language.

CAPST: capacitor start.

CAR: channel address register.

CAR: computer-assisted research.

CARBINE: computer automated real-time betting information network.

CARD: channel allocation and routing data.

CARD: compact automatic retrieval device.

CARDCODER: Card Automatic Code System (IBM).

CARDE: Canadian Armament Research and Development Establishment (DRB).

CARS: computer-aided routing system.

CARS: computerized automotive reporting service.

CART: central automated replenishment technique.

CART: central automatic reliability tester (TI).

CART: computerized automatic rating technique.

CAS: calculated air speed.

CAS: calibrated air speed.

CAS: Circuits and Systems (IEEE Society).

CAS: collision-avoidance system.

CASD: computer-aided system design.

CASE: computer automated support equipment.

CASI: Canadian Aeronautics and Space Institute.

CASPAR: Cambridge Analog Simulator for Predicting Atomic Reactions.

CASS: command active sonobuoy system.

CAST: Clearinghouse Announcements in Science and Technology.

CAST: computerized automatic systems tester.

CAT: carburetor air temperature.

CAT: compile and test.

CAT: computer-aided teaching.

CAT: computer-aided translation.

CAT: computer-assisted testing.

CAT: controlled attenuator timer.

CATO: compiler for automatic teaching operation.

CATS: computer-aided teaching system.

CATS: computer automated test system.

CATV: community antenna television.

CAUML: Computers and Automation Universal Mailing List.

CAW: channel address word.

CAX: community automatic exchange.

CAX: unattended automatic exchange.

CB: circuit breaker.

CB: citizens band (radio).

CB: common base.

CB: common battery.

CB: continuous blowdown.

CBAL: counterbalance.

CB/ATDS: Carrier Based/Airborne Tactical Data System.

CBEMA: Computer Business Manufacturers Association.

CBI: computer-based instruction.

CBIS: computer-based instructional system.

CBMA: Canadian Business Manufacturers Association.

CBMS: Conference Board of the Mathematical Sciences.

CBMU: current bit motor unit.

CBR: chemical, biological, radiological (warfare).

CBS: call box station.

CBX: cam box.

C-C: center to center.

C&C: command and control.

C&CP: Corrosion and Cathodic Protection (IEEE IAS Technical Committee).

CC: central computer.

CC: channel command.

CC: close-coupled.

CC: closing coil.

CC: common collector.

CC: computer community.

CC: condition code.

CC: control center.

CC: control computer.

cc: cubic centimeters.

CCA: carrier-controlled approach.

CCAP: communications control application program.

CCB: circuit concentration bay.

CCC: Canadian Computer Conference.

CCC: computer communication console.

CCD: charge-coupled device.

CCD: computer-controlled display.

CCD: core current driver.

CC&S: central computer and sequencing.

CCFM: cryogenic continuous-film memory.

CCI: combined form of CCIR and CCIT.

CCIA: console computer interface adapter.

CCIF: International Telephone Consultative Committee.

CCIR: International Radio Consultative Committee.

CCIS: Command Control Information System.

CCITT: International Consultative Committee for Telephone and Telegraph.

CCL: control language.

C³L: complementary constant-current logic.

CCM: communications controller multichannel.

CCM: counter countermeasures.

CCMD: continuous current monitoring device.

CCO: current-controlled oscillator.

CCOP: constant-control oil pressure.

CCR: control contactor.

CCROS: card capacitor read-only storage.

CCS: Conversational Compiling System (Xerox).

CCS: continuous commercial service.

CCS: custom computer system.
CCST: Center for Computer Sciences and Technology (NBS).
CCT: communications control team.
CCT: constant current transformer.
CCT: correlated color temperature.
CCTV: close-circuit television.
CCVS: Cobol compiler validation system.
CCW: channel command word.
CCW: counterclockwise.
CD: cable duct.
CD: center distance.
CD: circuit description.
CD: current density.
CDA: command and data acquisition.
CDC: call directing code.
CDC: code directing character.
CDC: configuration data control.
CDCE: central data-conversion equipment.
CDCR: Center for Documentation and Communication Research (Western Reserve University).
CDE: certification in data education.
CDI: collector diffusion isolation.
CDI: Control Data Institute.
CDJM: Canadian Journal of Mathematics.
CDL: common display logic.
CDL: computer description language.
CDL: computer design language.
CDMA: code-division multiple-access.
CDO: unattended automatic exchange.
CDP: central data processor.
CDP: centralized data processing.
CDP: certification in data processing.
CDP: checkout data processor.
CDP: communication data processor.
CDP: compression discharge pressure.
CDPC: central data processing computer.
CDPC: commercial data processing center.
CDPIR: crash data position indicator recorder.
CDR: command destruct receiver.
CDR: current directional relay.
CDS: computer duplex system.
CDSE: computer driven simulation environment.
CDT: control data terminal.
CDU: central display unit.
CDW: computer data word.
CDX: control-differential transmitter.
CE: civil engineering.
CE: common emitter.
CE: communications-electronics.
CE: commutator end.
CE: computer engineer.
CE: conducted emission.
CEA: Cambridge electron accelerator.
CEA: circular error average.
CEARC: Computer Education and Applied Research Center.
CEC: Commonwealth Engineering Conference.
CEDPA: California Educational Data Processing Association.
CEE: International Commission on Rules for the Approval of Electrical Equipment.

CEGB: Central Electricity Generating Board.
CEI: *Commission Electrotechnique Internationale.*
CEI: communication electronics instructions.
CEI: computer extended instruction.
CEI: Council of Engineering Institutions.
CEMF: counterelectromotive force.
CEMON: customer engineering monitor (IBM).
CENEL: Committee for the Coordination of Engineering Standards in the Electrical Field.
CEP: Circle of Equal Probability (see COU).
CEP: circular error probability.
CEP: civil engineering package.
CEP: computer entry punch.
CER: Civil Engineering Report.
CERA: Civil Engineering Research Association.
CERC: Coastal Engineering Research Center (Army).
CERL: Central Electricity Research Laboratories.
CERMET: ceramic metal element.
CERMET: ceramic-to-metal (seal).
CES: constant elasticity of substitution.
CESI: Council for Elementary Science International.
CET: corrected effective temperature.
CET: cumulative elapsed time.
CETS: Conference on European Telecommunications Satellites.
CEV: combat engineer vehicle.
CEV: corona extinction voltage.
CF: cathode follower.
CF: centrifugal force.
CF: concept formulation.
CF: conversion factor.
CF: counterfire.
CFA: Council of Iron Foundry Associations.
CFA: cross-field amplifier.
CFAR: constant false alarm rate.
CFBS: Canadian Federation of Biological Societies.
CFC: capillary filtration coefficient.
CFC: central fire control.
CFL: calibrated focal length.
CFL: context-free language.
CFM: cathode follower mixer.
CFM: cubic foot per minute.
CFR: Code of Federal Regulations.
CFR: Coordinating Fuel Research.
CFS: cubic foot per second.
CFSG: Cometary Feasibility Study Group (ESRO).
CFSSB: Central Flight Status Selection Board.
CFSTI: Clearinghouse for Federal Scientific and Technical Information (Department of Commerce).
CG: center of gravity.
CGS: centimeter-gram-second.
CGSB: Canadian Government Specifications Board.
CH: chain home (radar).
CHABA: Committee on Hearing and Bioacoustics.
CHAD: code to handle angular data.
CHE: channel end.
CHEC: channel evaluation and call.
Ch.E.: chemical engineer.
CHESNAVFAC: Chesapeake Naval Division Facilities Engineering Command.
CHF: critical heat flux.
CHI: computer human interaction.

CHIL: current hogging injection logic.
CHILD: cognitive hybrid intelligent learning and development.
CHILD: computer having intelligent learning and development.
CHORI: Chief of Office of Research and Inventions.
CHPAE: Critical Human Performance and Evaluation.
CHS: Canadian Hydrographic Service.
C&I: control and indication.
CI: card input.
CI: circuit interrupter.
CIA: Central Intelligence Agency.
CIA: Chemical Industries Association.
CIA: communications interrupt analysis.
CIA: Computer Industry Association.
CIA: computer interface adapter.
CIB: Cobol information bulletin.
CIC: Chemical Industries Council.
CIC: Chemical Institute of Canada.
CICA: Chicago Industrial Communications Association.
CID: Centre for Information and Documentation (Euratom).
CID: component identification.
CIDAS: conversational interactive digital/analog simulator (IBM).
CIE: *Commission Internationale de l'Eclairage* (International Commission on Illumination).
CIGRE: *Conférence Internationale des Grandes Reseaux Electriques à Haute Tension* (International Conference on Large High Voltage Electric Systems).
CII: *Compagnie Internationale pour l'Informatique.*
CIM: Communications Improvement Memorandum.
CIM: computer input multiplexer.
CIM: continuous image microfilm.
CIMCO: card image correction.
CIMM: Canadian Institute of Mining and Metallurgy.
CIN: carrier input.
CINS: Cento Institute of Nuclear Science.
CIO: central input output (system).
CIOCS: communication input and output control system.
CIOMS: Council for International Organizations of Medical Sciences.
CIOU: custom input/output unit.
CIP: compatible independent peripherals.
CIP: current injection probe.
CIPASH: Committee on International Programs in Atmospheric Sciences and Hydrology (NAS/NRC).
CIPHONY: cipher and telephony equipment.
CIPM: Council for International Progress in Management.
CIPS: Canadian Information Processing Society.
CIR: Canada India reactor.
CIRC: Centralized Information Reference and Control.
CIRCAL: circuit analysis.
CIRES: Communication Instructions for Reporting Enemy Sightings.
CIRIA: Construction Industry Research and Information Association.
CIS: Central Instructor School.

CIS: *Centre International d'Informations de Sécurité et d'Hygiéne du Travail* (International Occupational Safety and Health Information Centre).
CIS: communication information system.
CIS: cue indexing system.
CIS: current information selection.
CISCO: compass integrated system compiler.
CISIR: Ceylon Institute of Scientific and Industrial Research.
CISPR: International Special Committee on Radio Interference.
CISR: Center for International Systems Research (Department of State).
CIT: California Institute of Technology.
CIT: *Compagnie Industrielle de Télécommunications.*
CIT: Carnegie Institute of Technology.
CIT: compressor inlet temperature.
CITAB: computer instruction and training assistance for the blind.
CITB: Construction Industry Training Board.
CITC: Construction Industry Training Center.
CITE: compression ignition and turbine engine.
CITEC: contractor independent technical effort.
CIU: computer interface unit.
CKMTA: Cape Kennedy Missile Test Annex (now Cape Canaveral) (NASA).
CL: computational linguistics.
CL: control language.
CL: conversion loss.
CLA: center line average.
CLA: clear and add.
CLA: communication line adapter.
CLA: communication link analyzer.
CLAIRA: Chalk Lime and Allied Industries Research Association (now WHRA).
CLAM: chemical low-altitude missile.
CLARA: Cornell Learning and Recognizing Automation.
CLAS: Computer Library Applications Service
CLASP: closed line assembly for single particles.
CLASP: Computer Language for Aeronautics and Space Programming (NASA).
CLASS: closed loop accounting for stores sales.
CLASS: Computer-Based Laboratory for Automated School Systems.
CLASSMATE: computer language to aid and simulate scientific, mathematical, and technical education.
CLC: course line computer.
CLCR: controlled letter contract reduction.
CLCS: current logic, current switching.
CLD: Central Library and Documentation Branch (ILO).
CLDAS: clinical laboratory data acquisition system.
CLEA: Conference on Laser Engineering and Applications (IEEE).
CLEAN: Commonwealth Law Enforcement Assistance Network.
CLEM: composite for the lunar excursion module.
CLEO: clear language for expressing orders.
CLF: capacitive loss factor.

CLIP: compiler language for information processing.
CLR: combined line and recording trunk.
CLR: computer language recorder.
CLR: computer language research.
CLR: Coordinating Lubricants Research.
CLR: Council on Library Resources.
CLR: current limiting resistor.
CLRU: Cambridge Language Research Unit.
CLS: clear and subtract.
CLS: common language system.
CLS: concept learning system.
CLT: communications line terminals.
CLT: computer language translator.
CLU: central logic unit.
CLU: circuit line-up.
CLUT: computer logic unit tester (ADSAF).
C/M: communications multiplexer.
CM: calibrated magnification.
CM: centimeter (10^{-2} meters).
CM: central memory.
CM: construction and machinery.
CM: controlled mine field.
CM: core memory.
CM: countermeasure.
CMA: Canadian Medical Association.
CMA: contact-making (or breaking) ammeter.
CMANY: Communications Managers Association of New York.
CMC: code for magnetic characters.
CMC: communications mode control.
CMCTL: current mode complementary transistor logic.
CMD: core memory driver.
CMERI: Central Mechanical Engineering Research Institute (India).
CMF: coherent memory filter.
CMG: control moment gyro.
CMI: computer managed instruction.
CMIL: circular mil.
CML: current-mode logic.
CMLCENCOM: Chemical Corps Engineering Command.
CMI: Christian Michelsen's Institute (Norway).
CMI: Commonwealth Mycological Institute.
CMM: Commission for Maritime Meteorology (WMO).
CMMP: Commodity Management Master Plan.
CMOS: complementary metal oxide semiconductor.
CMOS/SOS: complementary metal oxide semiconductor/silicon on sapphire.
cmps: centimeter per second.
CMR: common-mode rejection.
CMR: Committee on Manpower Resources.
CMR: Communications Monitoring Report.
CMRR: common-mode rejection ratio.
CMS: compiler monitor system.
CMS: current-mode switching.
CMT: conversational mode terminal (Friden).
C&MS: Consumer and Marketing Service.
CMV: common-mode voltage.
CMVM: contact-making voltmeter.
CMVPCB: California Motor Vehicle Pollution Control Board.

C/N: carrier to noise.
CN: coordination number.
CNA: Canadian Nuclear Association.
CNA: copper nickel alloy.
CNA: cosmic noise absorption.
CNAA: Council for National Academic Awards.
CNAS: Civil Navigation Aids System.
CNC: computerized numerical control.
CNDP: communication network design program (IBM).
CNI: communication navigation identification.
CNL: circuit net loss.
CNLA: Council on National Library Associations.
CNR: carrier-to-noise ratio.
CNR: *Consiglio Nazionale delle Ricerche* (Italian National Research Council).
CNS: Congress of Neurological Surgeons.
CNT: celestial navigation trainer.
CNTP: Committee for a National Trade Policy.
CO: close-open operation.
CO: Colorado.
CO: combined operations.
CO: Communications Officer.
COA: College of Aeronautics.
COBESTCO: computer-based estimating technique for contractors.
COBIS: computer-based instruction system.
COBLIB: Cobol library.
COBLOC: codap language block-oriented compiler.
Cobol: common business-oriented language.
COC: coded optical character.
COCODE: compressed coherency detection (radar technique).
CODAN: carrier-operated device antinoise.
CODAN: coded weather analysis (Navy).
CODAP: control data assembly program.
CODAS: customer-oriented data retrieval and display system.
CODASYL: Conference on Data SystemLanguages.
CODATA: Committee on Data for Science and Technology (ICSU).
CODEL: computer developments limited automatic coding system.
CODES: computer design and evaluation system.
CODIC: color difference computer.
CODIC: computer directed communications.
CODIL: control diagram language.
CODIT: computer direct to telegraph.
CODSIA: Council of Defense and Space Industry Associations.
COED: Char Oil Energy Development.
COED: computer-operated electronic display.
COFIL: core file.
COG: computer operations group.
COGENT: compiler and generalized translator.
COGO: coordinate geometry.
COGS: continuous orbital guidance sensor.
COGS: Continuous Orbital Guidance System.
COHO: coherent oscillator.
COI: communication operation instructions.
COINS: computer and information sciences.
COL: computer oriented language.

COLASL: compiler/Los Alamos Scientific Laboratory
COLINGO: compile on-line and go.
COLT: communication line terminator.
COLT: computerized on-line testing.
COLT: control language translator.
COLT: computer oriented language translator.
COLIDAR: coherent light detection and ranging.
COM: computer output microfilm.
COMAC: continuous multiple-access collator.
COMAR: Committee on Man and Radiation (IEEE).
COMAT: computer-assisted training.
COMCM: communications countermeasures and deception.
COMEINDORS: composite mechanized information and documentation retrieval system.
COMET: computer operated management evaluation technique.
COMIT: compiler/MIT.
COMLOGNET: combat logistics network.
COMLOGNET: communications logistics network.
COMMEN: compiler oriented for multiprogramming and multiprocessing environments.
COMMEND: computer-aided mechanical engineering design system.
COMM: Communications (IEEE Society).
COMP: Computer (IEEE Society).
COMPAC: computer program for automatic control.
COMPACT: compatible algebraic compiler and translator.
COMPACT: computer planning and control technique (HIS).
COMPANDER: compressor expander.
COMPARE: computerized performance and analysis response evaluator.
COMPARE: console for optical measurement and precise analysis of radiation from electronics.
COMPASS: compiler-assembler.
COMPASS: computer-assisted classification and assignment system.
COMPCON: Computer Convention (IEEE).
COMPEL: compute parallel.
COMPOOL: communications pool (Xerox).
COMPRESS: computer research, systems and software.
COMPROG: computer program.
COMPSO: computer software and peripheral show.
COMRADE: computer-aided design environment.
COMSAT: communications satellite.
COMSEC: communications security.
COMSL: communication system simulation language.
COMSOAL: computer method of sequencing operations for assembly lines.
COMTRAN: commercial translator.
COMZ: communications zone.
CONELRAD: control of electromagnetic radiation.
CONFLEX: conditioned reflex (machine).
CONRAD: Committee on Nuclear Radiology.
CONSORT: conversational system with on-line remote terminals.
CONSUL: control subroutine language.
CONTRAN: control translator (HIS).

CONTRANS: conceptual thought, random net simulation.
COOL: checkout oriented language.
COOL: control oriented language.
COP: computer optimization package.
COPE: Committee on Political Education (AFL-CIO).
COPE: communications oriented processing equipment.
COPI: computer-oriented programmed instruction.
CORA: conditioned reflex analog.
CORA: conditioned-response analog (machine).
CORAL: computer on-line real-time applications language.
CORAPRAN: Cobelda radar automatic preflight analyzer.
CORC: Cornell computing language.
CORDIC: coordinate rotation digital computer.
CORDS: coherent on receive doppler system.
CORE: computer-oriented reporting efficiency.
COREP: combined overload repair control.
CORREGATE: correctable gate.
CORS: Canadian Operational Research Society.
CORTS: convert range telemetry systems.
COS: compatible operating system.
COS: concurrent operating system (UNIVAC).
cos: cosine.
COS: cosmic rays and trapped radiation committee (ESRO).
COSATI: Committee on Scientific and Technical Information (FCST).
COSBA: Computer Services Bureau Association (GB).
COSH: hyperbolic cosine.
COSI: Committee on Scientific Information.
COSINE: Committee on Computer Science in Electrical Engineering Education.
COSIP: College Science Improvement Program.
COSMIC: Computer Software Management and Information Center.
COSMOS: Computer Oriented System for Management Order Synthesis (IBM).
COSOS: Conference on Self-Operating Systems.
COSPAR: Committee on Space Research (ICSU).
COSPUP: Committee on Science and Public Policy.
COSRIMS: Committee on Research in the Mathematical Sciences.
cot: cotangent.
coth: hyperbolic cotangent.
COTRAN: Cobol to Cobol translator.
CP: candle power.
CP: card punch.
CP: circuit package.
CP: circular pitch.
CP: circularly polarized.
CP: clock pulse.
CP: coefficient of performance.
CP: command pulse.
CP: communication processor.
CP: control panel.
CP: control processor.
CP: customized processor (IBM).
CP: conference paper.
CP: constant pressure.

CP: current paper.
CPA: Canadian Pharmaceutical Association.
CPA: Canadian Psychological Association.
CPA: closest point of approach.
CPA: critical path analysis.
CPC: card programmed calculator.
CPC: clock-pulse control.
CPC: computer program component.
CPC: computer programming concepts.
CPC: cycle program control.
CPC: cycle program counter.
CPCEI: computer program contract end item.
CPD: consolidated programming document.
CPD: contact potential difference.
CPD: cumulative probability distribution.
CPDS: computer program design specification.
CPE: circular probable error.
CPE: central processing element.
CPE: central programmer and evaluator.
CPEA: College Physical Education Association.
CPEM: Conference on Precision Electromagnetic Measurement.
CPG: College Publishers Group.
CPI: computer prescribed instruction.
CPIA: Chemical Propulsion Information Agency.
CPILS: correlation protection integrated landing system.
CPIP: computer program implementation process.
CPL: combined programming language.
CPL: computer program library.
CPM: critical path method.
CPM: cycles per minute.
CPP: card punching printer.
CPP: Computer Professionals for Peace.
CPPS: critical path planning and scheduling.
CPR: Committee on Polar Research (NAS/NRC).
CPRG: Computer Personnel Research Group.
CPS: characters per second.
CPS: circuit package schematic.
CPS: conversational programming system.
CPS: critical path scheduling.
cps: cycles per second (obsolete); use hertz (Hz).
CPSC: Consumer Product Safety Commission.
CPSE: counterpoise.
CPSS: Common Programming Support System (Xerox).
CPT: control power transformer.
CPT: critical path technique.
CPTA: computer programming and testing activity.
CPU: central processing unit.
CPU: collective protection unit.
CRD: capacitor-resistor-diode.
CPU: computer peripheral unit.
CR: cathode ray.
CR: command register.
CR: conference report.
CR: contract report.
CR: control relay.
CR: crystal rectifier.
CR: current relay.
CRAC: Careers Research and Advisory Center.

CRAD: Committee for Research into Apparatus for the Disabled.
CRAFT: computerized relative allocation of facilities technique.
CRAM: card random access memory.
CRAM: computerized reliability analysis method.
CRAM: conditional relaxation analysis method.
CRC: Coordinating Research Council.
CRC: Copy Research Council.
CRC: cumulative results criterion.
CRC: cyclic redundancy check.
CRCC: cyclic redundancy check character.
CRD: capacitor-resistor-diode.
CRDF: Canadian radio-direction finding or finder.
CRDF: cathode-ray direction finding (radar).
CRDL: Chemical Research and Development Laboratories (Army).
CRDME: Committee for Research into Dental Materials and Equipment.
CRDSD: Current Research and Development in Scientific Documentation.
CRE: corrosion-resistant.
CRES: corrosion-resistant steel.
CRESS: Combined Reentry Effort for Small Systems (ABRES).
CRESS: Computerized Reader Enquiry Service System.
CRESS/AU: Center for Research in Social Systems of the American University.
CRESTS: Courtauld's Rapid Extract, Sort and Tabulate System.
CRF: capital recovery factor.
CRF: Cryptographic Repair Facility (military).
CRI: color rendering index.
CRI: Committee for Reciprocity Information.
CRIME: Censorship Records and Information, Middle East (military).
CRIS: command retrieval information system.
CRIS: current research information system.
CRJE: conversational remote job entry.
CRM: control and reproducibility monitor.
CRM: counter-radar measures.
CRM: counter-radar missile.
CRNL: Chalk River Nuclear Laboratories (Canada).
CRO: cathode ray oscilloscope.
CROM: capacitive read-only memory.
CROM: control read-only memory.
CRPL: Central Radio Propagation Laboratory (ESSA).
CRPM: Communication Registered Publication Memoranda.
CRREL: Cold Regions Research and Engineering Laboratory (Army).
CRT: cathode ray tube.
CRT: circuit requirement table.
CRTU: combined receiving and transmitting unit.
CRWO: Coding Room Watch Officer (Navy).
CRW: community radio watch.
C&S: computers and systems.
CS: commercial standard.
CS: community service.

CS: computer science.
CS: conducted susceptibility.
CS: control switch.
CS: Control Systems (IEEE Society).
C/S: call signal.
c/s: cycle per second.
CSA: Canadian Standards Association.
CSA: Computer Sciences Association.
CSA: Computer Systems Association.
CSA: Community Service Activities (AFL-CIO).
CSC: common signaling channel.
CSC: Computer Society of Canada.
csc: cosecant.
C-SCAN: viewing cathode-ray screen (Air Force).
C-SCOPE: cathode-ray screen (Air Force).
CSD: constant speed drive.
CSD: controlled-slip differentials.
CSE: containment systems experiment.
CSE: control systems engineering.
CSE: core storage element.
CSEE: Canadian Society for Electrical Engineers.
CSI: Construction Specifications Institute.
CSIC: computer system interface circuits.
CSIR: Council for Scientific and Industrial Research (South Africa).
CSIR: Council of Scientific and Industrial Research (India).
CSIRAC: Commonwealth Scientific and Industrial Research Automatic Computer (Australia).
CSIRO: Commonwealth Scientific and Industrial Research Organization (Australia).
CSL: code selection language.
CSL: computer sensitive language.
CSL: control and simulation language.
CSL: current switch logic.
CSM: continuous sheet memory.
CSMA: Chemical Specialties Manufacturers Association.
CSMA: Communications Systems Management Association.
CSMP: Continuous System Modeling Program (IBM).
CSP: Communications Security Publication (military).
CSP: Commercial Subroutine Package (IBM).
CSP: continuous sampling plan.
CSP: Council for Scientific Policy.
CSSA: Computer Society of South Africa, Ltd.
CSSB: compatible single sideband.
CSSE: Conference of State Sanitary Engineers.
CSSL: continuous system simulation language.
CSSS: Canadian Soil Science Society.
CSST: computer system science training.
CST: central standard time.
CSTI: Committee on Scientific and Technical Information.
CSTS: combined system test stand.
CSV: corona start voltage.
CSW: control power switch.
CT: center tap.
CT: commercial translator.
CT: communications technician (military).
CT: computer technology.
CT: Connecticut.

CT: current transformer.
CTA: compatibility test area.
CTDH: command and telemetry data handling.
CTDS: code translation data system
C/TDS: count/time data system.
CTFM: continuous transmission frequency modulated.
CTL: CAGE test language.
CTL: checkout test language.
CTL: complementary transistor logic.
CTL: core transistor logic.
CTMC: Communication Terminal Module Controller (UNIVAC).
CTO: charge transforming operator.
CTP: charge transforming parameter.
CTP: command translator and programmer.
CTS: carrier test switch.
CTS: Computer Telewriter Systems (Broomfield, Colo.).
CTS: communications terminal synchronous.
CTS: conversational terminal system.
CTS: conversational time sharing.
CTS: courier transfer station (military).
CTSS: compatible time shared system.
CTU: central terminal unit.
CU: control unit.
CUAS: computer utilization accounting system.
CUBE: cooperating users of Burroughs equipment.
CUBOL: computer usage business oriented language.
CU: piezoelectric-crystal unit.
CUD: Craft Union Department (AFL-CIO).
CUDOS: continuously updated dynamic optimizing systems.
CUE: computer updating equipment.
CUE: cooperating users exchange.
CUES: computer utility educational system.
CUJT: complementary unijunction transistor.
CULP: computer usage list processor.
CUMM: Council of Underground Machinery Manufacturers.
CUMMFU: complete utter monumental military foul up.
CUP: communications users program.
CURES: computer utilization reporting system.
CURTS: common user radio transmission system.
CURV: cable-controlled underwater research vehicle.
CUSIP: Committee on Uniform Security Identification Procedures.
CUSP: Central Unit for Scientific Photography (RAE).
CUT: Control Unit Tester (UNIVAC).
CUT: circuit under test.
CUTS: computer utilized turning system.
CUW: Committee on Undersea Warfare (DOD).
CV/CC: constant voltage/constant current.
CVD: coupled vibration dissociation.
CVD: current-voltage diagram.
CVDV: coupled vibration dissociation vibration.
CVIC: conditional variable incremental computer.
CW: carrier wave; composite wave; continuous wave.
CW: chemical warfare.
CW: clockwise.
CW: continuous wave.
CWA: Communications Workers of America.

CWAR: continuous wave acquisition radar.
CWAS: contractor weighted average share.

CWMTU: cold-weather material test unit.
CZ: Canal Zone (Panama).

D

D/A: digital to analog.
D-A: digital to analog.
DA: digital to analog.
DA: decimal add.
DA: decimal to analog.
DA: design automation.
DA: differential analyzer.
DA: discrete address.
DA: double amplitude.
DAA: data access arrangement.
DAC: design augmented by computer.
DAC: digital-to-analog converter.
DAC: display analysis console.
DACC: direct access communications channel.
DACON: digital to analog converter.
DACS: data acquisition control system.
DACOR: data correction.
DADEE: dynamic analog differential equation equalizer.
DAFC: digital automatic frequency control.
DAFT: digital analog function table.
DAGC: delayed automatic gain control.
DAFM: discard-at-failure maintenance.
DAIR: direct attitude and identity readout.
DAIR: driver air, information, and routing.
DAIS: defense automatic integrated switching system.
DAISY: double precision automatic interpretive system.
DAM: data addressed memory.
DAM: data association message.
DAM: descriptor attribute matrix.
DAM: digital-to-analog multiplier.
DAM: direct access memory.
DAM: direct access method.
DAMP: down-range antimissile measurement program.
DAMPS: Data Acquisition Multiprogramming System (IBM).
DAP: deformation of aligned phases.
DAP: digital assembly program (EAI).
DAPR: digital automatic pattern recognition.
DAPS: direct access programming system.
DAR: defense acquisition radar.
DARE: document abstract retrieval equipment.
DARE: documentation automated retrieval equipment.
DARE: doppler automatic reduction equipment.
DARES: data analysis and reduction system.
DARLI: digital angular readout by laser interferometry.
DARS: digital adaptive recording system.
DARS: digital attitude and rate system.
DART: daily automatic rescheduling technique.
DART: data analysis recording tape.
DART: director and response tester.
DART: dual axis rate transducer.

DART: dynamic acoustic response trigger.
DAS: data acquisition system.
DAS: data automation system.
DAS: Datatron Assembly System.
DAS: digital analog simulator.
DASD: direct access storage device.
DASS: demand assignment signaling and switching.
DAT: Director of Advanced Technology.
DAT: dynamic address translation.
DATAC: data analog computer.
DATACOL: data collection system.
DATACOM: data communications.
DATAGEM: data file generator.
DATAN: data analysis.
DATAR: digital autotransducer and recorder.
DATICO: digital automatic tape intelligence checkout.
DATRIX: direct access to reference information.
DAU: data acquisition unit.
DAV: data valid.
DAVI: dynamic antiresonant vibration isolator.
DAVC: delayed automatic volume control.
D-B: decimal to binary.
DB: double-biased (relay).
DB: double bottom.
DB: double break.
DB: dry bulb.
dB: decibel.
dBa: decibel adjusted.
DBAO: digital block AND-OR gate.
DBB: detector back bias.
DBB: detector balanced bias.
DBC: diameter bolt circle.
DBF: demodulator band filter.
dBj: relative RF signal levels (j stands for Jerrold Electronics where the term originated).
DBM: data buffer module.
DBMS: data base management system.
dBm: decibels above (or below) 1 milliwatt.
dB/m^2: dB above 1 milliwatt per square meter.
dBm/m^2/MHz: dB above 1 milliwatt per square meter per megahertz.
DBOS: disk-based operating system.
dBμV: dB above 1 microvolt.
dBμV/m/MHz: dB above 1 microvolt per meter per megahertz.
dBμV/MHz: dB above 1 microvolt per megahertz.
dBRAP: decibels above reference acoustical power defined as 10^{-16} watt.
dBRN: decibels above reference noise.
dBRNC: decibels above reference noise, C-message weighted.
DBSP: double-based solid propellant.

DBUT: data base update time.

dBV: increase or decrease in voltage regardless of impedance levels.

dBW: decibels referred to 1 watt.

dBx: decibels above the reference coupling.

DC: data channel.

DC: data check.

DC: data classifier.

DC: data collection.

DC: data communications.

DC: decimal classification.

DC: digital comparator.

DC: digital computer.

dc: direct current.

DC: disc controller.

DC: display console.

DC: District of Columbia.

DC: double contact.

DCA: decade counting assembly.

DCA: Defense Communications Agency.

DCA: digital command assembly.

DCA: Digital Computer Association.

DCAA: Defense Contract Audit Agency.

DCAOC: Defense Communications Agency Operations Center.

DCAS: Deputy Commander Aerospace System.

DCB: data control block.

DCB: Defense Communications Board.

DCBRL: Defense Chemical, Biological, and Radiation Laboratories.

DCC: data communication channel.

DCC: device control character.

DCC: District Communications Center.

DCC: double cotton covered.

DCCS: digital command communications system.

DCCU: data communications control unit.

DCD: dynamic computer display.

DCDS: digital control design system.

DCEO: Defense Communications Engineering Office.

DCIB: data communication input buffer.

DCKP: direct current key pulsing.

DCL: Digital Computer Laboratory.

dcm: dc noise margin.

DCOS: data communication output selector.

DCP: data communication processor.

DCP: design criteria plan (MIL-STD-469).

DCP: digital computer processor.

DCP: digital computer programming.

DCPG: Defense Communications Planning Group.

DCPS: digitally controlled power source.

DCR: data conversion receiver.

DCR: data coordinator and retriever.

DCR: digital conversion receiver.

DCS: data communication system.

DCS: data communications subsystem.

DCS: data control services.

DCS: defense communication systems.

DCS: design control specifications.

DCS: digital command system.

DCS: digital communication system.

DCS: Direct-Couple Operating System (IBM).

DCS: distributed computer system.

DCSP: Defense Communications Satellite Program.

DCTL: direct coupled transistor logic.

DCU: data command unit.

DCU: data control unit.

DCU: digital control unit.

DCWV: direct-current working volts.

DD: data definition.

DD: decimal divide.

DD: Dewey decimal.

DD: digital data.

DD: disconnecting device.

DDA: digital differential analyzer.

DDA: dynamics differential analyzer.

DDAS: digital data acquisition system.

DDC: deck decompression chamber.

DDC: Defense Documentation Center (DOD).

DDC: Dewey decimal classification.

DDC: Digital Development Corporation.

DDC: digital display converter.

DDC: director digital control.

DDCE: digital data conversion equipment.

DDCS: digital data calibration system.

DDD: direct distance dialing.

DDG: digital data generator.

DDG: digital display generator.

DDGE: digital display generator element.

DDE: director design engineering.

DDI: depth deviation indicator.

DDL: dispersive delay line.

DDM: data demand module.

DDM: digital display make-up.

DDM: difference in depth of modulation.

DDOCE: digital data output conversion equipment.

DDP: digital data processor.

DDPU: digital data processing unit.

DDR&E: Directorate of Defense Research and Engineering (DOD).

DDRS: digital data recording system.

DDS: data display system.

DDS: deployable defense system.

DDS: digital dynamics simulator.

DDS: digital display scope.

DDT: deflagration to detonation transition.

DDT: digital data transmission.

DDT: digital debugging tape.

DDT: dynamic debugging technique.

DDTE: digital data terminal equipment.

DDTS: digital data transmission system.

DE: Delaware.

DE: design engineering.

DE: diesel-electric.

DE: digital element.

DE: display electronics.

DEA: display electronics assemblies.

DEAL: decision evaluation and logic.

DECEO: Defense Communication Engineering Office.

DECOM: telemetry decommutators.

DECUS: Digital Equipment Computer Users Society.

DED: Design Engineering Directorate.

DEDUCOM: deductive communicator.

DEE: digital events evaluator.
DEE: digital evaluation equipment.
DEFT: dynamic error-free transmission.
DEI: development engineering inspection.
DEPA: Defense Electric Power Administration.
DEPI: differential equations pseudocode interpreter.
DEPSK: differential-encoded phase shift keying.
DE/Q: design evaluation/qualification.
DER: diesel engine, reduction drive.
DES: Department of Education and Science.
DES: differential equation solver.
DES: digital expansion system.
DESC: Defense Electronics Supply Center (DOD).
DESC: digital equation solving computer.
DESC: Directorate of Engineering Standardization.
DEU: data exchange unit.
DEUA: Diesel Engines and Users Association.
DEUA: Digitronics Equipment Users Association.
DEUCE: digital electronic universal computing engine.
DEW: distant early warning.
DEWIZ: distant early warning identification zone.
DEXAN: digital experimental airborne navigator.
DF: decimal fraction.
DF: decontamination factor.
DF: deflection factor.
DF: direction finder.
DF: disk file.
DF: dissipation factor.
DF: double feeder.
DFB: distribution fuse board.
DFC: diagnostic flow charts.
DFI: developmental flight instrumentation.
DFL: display formatting language.
DFMSR: Directorate of Flight and Missile Safety Research.
DFO: Director Flight Operations.
DFR: decreasing failure rate.
DFR: Dornreay fast reactor.
DFRL: differential relay.
DFS: dynamic flight simulator.
DFSR: Directorate of Flight Safety Research.
DFT: discrete Fourier transform.
DG: differential generator.
DG: diode gate.
DG: double-groove (insulators).
DGB: disk-gap-bond.
DGBC: digital geoballistic computer.
DGDP: double-groove double-petticoat (insulators).
DGS: data ground station.
DGS: degaussing system.
DGS: data gathering system.
D-H: decimal to hexadecimal.
DHC: data handling center.
DHE: data handling equipment.
DI: data input.
DI: digital input.
DI: demand indicator.
DIA: Defense Intelligence Agency.
DIAC: Defense Industry Advisory Council.
DIAL: display Interactive Assembly Language (DEC).
DIAL: Drum Interrogation, Alteration and Loading System (Honeywell).

DIALGOL: dialect of Algol.
DIAN: digital analog.
DIANE: digital integrated attack and navigation equipment.
DIAS: dynamic inventory analysis system.
DIC: data insertion converter.
DIC: Detailed Interrogation Center.
DICON: digital communication through orbiting needles.
DID: Datamation industry directory.
DID: digital information display.
DID: Division of Isotopes Development.
DIDACS: digital data communications system.
DIDAD: digital data display.
DIDAP: digital data processor.
DI/DO: data input/data output.
DIE: Document of Industrial Engineering.
DIFFTR: differential time relay.
DIGCOM: digital computer.
DIGACC: digital guidance and control computer.
DIGICOM: digital communications system.
DIGITAC: digital tactical automatic control.
DIGITAR: digital airborne computer.
DIMATE: depot-installed maintenance automatic test equipment.
DIMES: defense integrated management engineering system.
DIMUS: digital multibeam steering.
DINA: direct noise amplification.
DIOB: digital input/output buffer.
DIP: display information processor.
DIP: dual in-line package.
DIPS: development information processing system.
DIRCOL: direction cosine linkage.
DISAC: digital simulator and computer.
DISCOM: digital selective communications.
DISD: Data and Information Systems Division.
DISTRAM: digital space trajectory measurement system.
DITRAN: diagnostic Fortran.
DIVA: digital input voice answer-back.
DIVOT: digital-to-voice translator.
DL: data link.
DL: dead load.
DL: delay line.
DL: diode logic.
DLA: Defense Logistics Agency.
DLC: direct lift control.
DLC: duplex line control.
DLE: data link escape.
DLI: Defense Language Institute.
DLIEC: Defense Language Institute, East Coast.
DLK: data link.
DLP: data listing programs.
DLSC: Defense Logistics Services Center.
DLT: data line translator.
DLT: data loop transceiver.
DLT: decision logic table.
DLT: depletion-layer transistor.
DM: data management.
DM: decimal multiply.
dm: decimeter.

DM: delta modulation.
DMA: direct memory access.
DMC: digital microcircuit.
DME: distance measuring equipment.
DMED: digital message entrance device.
DMIRR: demand mode integral rocket Ramjet.
DMM: digital multimeter.
DMO: Data Management Office.
DMR: dynamic modular replacement.
DMS: data management service.
DMS: data management system.
DMS: defense missile systems.
DMS: documentation of molecular spectroscopy.
DMSR: Director of Missile Safety Research.
DMT: digital message terminal.
DMTC: digital message terminal computer.
DMTR: Dornreay materials testing reactor.
DMU: digital message unit.
DMU: dual maneuvering unit.
DMZ: demilitarized zone.
DN: decimal number.
DN: delta amplitude.
DNC: direct numerical control.
DNCCC: Defense National Communications Control Center.
DNE: Department of Nuclear Engineering.
DNS: doppler navigation system.
DNSR: Director of Nuclear Safety Research.
D-O: decimal to octal.
DO: data output.
DO: digital output.
DOC: data optimizing computer.
DOC: decimal to octal conversion.
DOC: direct operating costs.
DOCS: disk-oriented computer system.
DOCUS: display-oriented computer usage system.
DOD: Department of Defense.
DOD: Development Operations Division.
DODCI: Department of Defense Computer Institute.
DODGE: Department of Defense Gravity Experiment.
DODIS: distribution of oceanographic data at isentropic levels.
DOF: degree of freedom.
DOF: direction of flight.
DOFIC: domain originated functional integrated circuit.
DOI: differential orbit improvement.
DO/IT: digital output/input translator.
DOL: Director of Laboratories.
DOL: display-oriented language.
DOL: dynamic octal load.
DOM: digital ohmmeter.
DOPIC: documentation of programs in core.
DOPS: digital optical protection system.
DORAN: doppler range and navigation.
DORIS: direct order recording and invoicing system.
DORV: deep ocean research vehicle.
DOS: digital operation system.
DOS: disk operating system.
DOSV: deep ocean survey vehicle.
DOT: deep ocean technology.

DOT: Department of Transportation.
DOVAP: doppler velocity and position.
DOUSER: doppler unbeamed search radar.
DOWB: deep operating work board.
DP: data processing.
DP: deflection plate.
DP: deep penetration.
DP: dew point.
DP: diametral pitch.
DP: disk pack.
DP: distribution point.
DP: double pole.
DP: drum processor.
DP: dynamic programming.
DPA: data processing activities.
DPB: data processing branch.
DPB: Defense Policy Board.
DPBC: double-pole back-connected.
DPC: data processing center.
DPC: direct program control.
DPD: data processing department.
DPD: data processing division.
DPD: digit plane driver.
DPCM: differential pulse code modulation.
DPDT: double pole, double throw (switch).
DPE: data processing equipment.
DPFC: double-pole front-connected.
DPG: digital pattern generator.
DPI: digital pseudorandom inspection.
DPLCS: digital propellant level control system.
DPM: data processing machine.
DPM: digital panel meter.
DPMA: Data Processing Management Association.
DPMS: data project management system.
DPS: data processing standards.
DPS: data processing systems.
DPS: descent power system.
DPS: disk programming system.
DPSA: Data Processing Supplies Association.
DPSK: differential phase-shift keying.
DPSS: data processing system simulator.
DQC: data quality control.
DPST: double pole, single throw (switch).
DR: data receiver.
DR: data recorder.
DR: data reduction.
DR: Division of Research.
DRA: dead reckoning analyzer.
DRADS: degradation of radar defense systems.
DRAI: dead reckoning analog indicator.
DRC: damage risk criterion.
DRC: data reduction compiler.
DRD: Director of Research and Development.
DRDTO: detection radar data take-off.
DRE: dead reckoning equipment.
DRE: Director of Research and Engineering.
DRI: data reduction interpreter.
DRI: dead reckoning indicator.
DRIFT: diversity receiving instrumentation for telemetry.
DRM: digital ratiometer.

DRML: Defense Research Medical Laboratories.
DRMO: District Records Management Officer.
DRO: destructive readout (memory).
DRO: digital readout.
DROD: delayed readout detector.
DROS: direct readout satellite.
DROS: disk resident operating system.
DRP: dead reckoning plotter.
DRS: digital radar simulator.
DRSS: data relay satellite system.
DRT: dead reckoning tracer.
DRTC: Documentation Research and Training Center (India).
DRTE: Defense Research Telecommunications Establishment.
DRTL: diode resistor transistor logic.
DRTR: dead reckoning trainer.
D&S: display and storage.
DS: data scanning.
DS: data set.
DS: decimal subtract.
DS: descent stage.
DS: disconnect switch.
DS: disk storage.
DS: drum storage.
DSA: digital signal analyzer.
DSAP: data systems automation program.
DSAR: data sampling automatic receiver.
DSB: Defense Science Board.
DSB: double-sideband.
DSBAMRC: double-sideband amplitude modulation reduced carrier.
DSC: double silk covered.
DSCB: data set control block.
DSCC: deep space communications complex.
DSCS: defense satellite communications system.
DSCS: desk side computer system.
DSCT: double secondary current transformer.
DSD: digital system design.
DSE: data storage equipment.
DSE: data systems engineering.
DSEA: data storage electronics assembly.
DSEG: Data Systems Engineering Group.
DSIF: deep-space instrumentation facility.
DSIR: Department of Scientific Industrial Research.
DSIS: Directorate of Scientific Information Services.
DSL: data set label.
DSL: data structures language.
DSL: data simulation language.
DSM: digital simulation model.
DSM: dynamic scattering mode.
DSN: Deep Space Network (NASA).
DSR: data scanning and routing.
DSR: discriminating selector repeater.
DSS: deep space station.
DSS: Director of Statistical Services.
DSSB: double single-sideband.
DSSC: double-sideband suppressed carrier (modulation).
DSSCS: defense special secure communications system.
DSTE: data subscriber terminal equipment.

DSU: data storage unit.
DSU: disk storage unit.
DSU: data synchronizer unit.
D/T: disk tape.
DT: data translator.
DT: data transmission.
DT: decay time.
DT: differential time.
DT: digital technique.
DT: double-throw.
DTA: differential thermal analysis.
DTARS: digital transmission and routing system.
DTAS: data transmission and switching.
DTB: decimal to binary.
DTC: data transmission center.
DTCS: digital test command system.
DTE: data transmission equipment.
DTG: display transmission generator.
DTI: Department of Trade and Industry.
DTL: diode transistor logic.
DTM: delay timer multiplier.
DTMS: digital test monitoring system.
DTP: directory tape processor.
DTPL: domain tip propagation logic.
DTR: definite-time relay.
DTR: demand-totalizing relay.
DTS: data transmission system.
DTVM: differential thermocouple voltmeter.
DTS-W: defense telephone service (Washington).
DTU: data transfer unit.
DTU: digital tape unit.
DTU: digital transmission unit.
DTUTF: digital tape unit tape facility.
D/TV: digital to television.
DTVC: digital transmission and verification converter.
DUAL: dynamic universal assembly language.
DUF: diffusion under (epitaxial) film.
DUMS: deep unmanned submersibles.
DUNC: deep underwater measuring device.
DUNC: deep underwater nuclear counter.
DUNS: data universal numbering system.
DUO: Datatron Users Organization.
DUT: device under test.
DV: differential voltage.
DVA: dynamic visual acuity.
DVD: detail velocity display.
DVESO: DOD Value Engineering Services Office.
DVFO: digital variable-frequency oscillator.
DVOM: digital volt-ohm meter.
DVM: digital voltmeter.
DVM: displaced virtual machine.
DVST: direct view storage tube.
DWBA: distorted-wave Born approximation.
DWICA: deep water isotopic current analyzer.
DWL: dominant wavelength.
DWSMC: Defense Weapons Systems Management Center (DOD).
DYNASAR: Dynamic Systems Analyzer (GE).
DYSAC: dynamic storage analog computer.
DYSTAL: dynamic storage allocation language.
DYSTAC: dynamic storage analog computer.

E

EA: effective address.

EAA: Engineer in Aeronautics and Astronautics.

EABRD: electrically activated bank release device.

EACC: error adaptive control computer.

EAES: European Atomic Energy Society.

EAF: electron arc furnace.

EAM: electric accounting machine.

EAM: electronic accounting machinery.

EAM: electronic automatic machine.

EAR: employee attitude research.

EARC: Extraordinary Administrative Radio Conference.

EAROM: electrically alterable read-only memory.

EAS: extensive air shower.

EASCON: Electronics and Aerospace Systems Convention (IEEE).

EASE: electrical automatic support equipment.

EASI: Electrical Accounting for the Security Industry (IBM).

EASL: engineering analysis and simulation language.

EASTT: experimental Army satellite tactical terminals.

EASY: early acquisition system.

EASY: engine analyzer system.

EAUTC: Engineer Aviation Unit Training Center.

EAX: electronic automatic exchange.

EB: electron beam.

EBAM: electron-beam-accessed memories.

EBCDIC: extended binary-coded decimal interchange.

EBMD: electron-beam mode discharge.

EBPA: electron beam parametric amplifier.

EBR: epoxy bridge rectifier.

EBR: experimental breeder reactor.

EBU: European Broadcasting Union.

EBW: exploding bridge wire.

EC: electronic calculators.

EC: enamel covered.

EC: engineering change.

EC: engineering construction.

EC: error correcting.

ECA: Electrical Contractors Association.

ECAC: Electromagnetic Compatibility Analysis Center.

ECAP: electronic circuit analysis program.

ECARS: electronic coordinator graph and readout system.

ECB: event control block.

ECC: Electronic Calibration Center (NBS).

ECC: Electronic Components Conference.

ECC: error correction code.

ECCANE: East Coast Conference on Aerospace and Navigational Electronics.

ECCM: electronic counter-countermeasures.

ECDC: electrochemical diffused collector (transistor).

ECE: engineering capacity exchange.

ECG: electrocardiogram.

ECG: electrochemical grinding.

ECHO: electronic computing hospital oriented.

ECL: emitter-coupled logic.

ECL: equipment component list.

ECM: electric coding machine.

ECM: electrochemical machining.

ECM: electronic countermeasures.

ECMA: Electronic Computer Manufacturers Association.

ECMA: European Computer Manufacturers Association.

ECME: electronic countermeasures.

ECMP: electronic countermeasures program.

ECMSA: Electronics Command Meteorological Support Agency.

ECN: engineering change notice.

ECNE: Electric Council of New England.

ECO: electron-coupled oscillator.

ECO: electronic checkout.

ECO: engineering change order.

ECOM: Electronics Command, Army (Ft. Monmouth, N.J.).

ECP: electromagnetic compatibility program.

ECP: engineering change proposal.

ECPD: Engineers Council for Professional Development.

ECPI: Electronic Computer Programming Institute.

ECR: electronic control relay.

ECR: engineering change request.

ECRC: Electronic Components Reliability Center.

ECS: electronic control switch.

ECS: Emergency Coolant System.

ECS: environmental control system.

ECS: extended core storage.

ECSA: European Communications Security Agency.

ECR: error control receiver.

ECU: electronic conversion unit.

ECU: environmental control unit.

ED: Education (IEEE Group).

ED: Electron Devices (IEEE Group).

ED: engine drive.

EDA: Economic Development Administration.

EDA: Electrical Development Association.

EDA: electronic differential analyzer.

EDA: electronic digital analyzer.

EDAC: error detection and correction.

EDB: educational data bank.

EDC: Economic Development Committee.

EDC: electronic desk calculator.

EDC: electronic digital computer.

EDC: error detection and correction.

EDCOM: editor and compiler.

EDCPF: Environmental Data Collection and Processing Facility.

EDCW: external device control word.

EDD: electronic data display.

EDE: emergency decelerating (relay).

EDE: emitter dip effect.

EDFR: effective date of Federal recognition.

EDG: exploratory development goals.

EDGE: electronic data gathering equipment.

EDICT: engineering document information collection technique.

EDIS: engineering data information system.

EDLCC: Electronic Data Local Communications Central.

EDM: electrodischarge machining.

EDO: effective diameter of objective.

EDO: engineering duties only.

EDOS: extended disk operating system.

EDP: electronic data processing.

EDPC: electronic data processing center.

EDPE: electronic data processing equipment.

EDPI: Electronic Data Processing Institute.

EDPM: electronic data processing machine.

EDPS: electronic data processing system.

EDR: electrodermal reaction.

EDR: equivalent direct radiation.

EDS: electronic data system.

EDS: emergency detection system.

EDS: environmental data service.

EDSAC: electronic delay storage automatic computer.

EDSAC: electronic discrete sequential automatic computer.

EDST: Elastic Diaphragm Switch Technology (IBM).

EDT: electric discharge tube.

EDT: engineer design tests.

EDU: electronic display unit.

EDU: experimental diving unit.

EDVAC: electronic discrete variable automatic computer.

EE: Electrical Engineer.

EE: electrical engineering.

EEA: Electronic Engineering Association.

EECL: emitter-emitter coupled logic.

EED: electroexplosive device.

EEG: electroencephalogram.

EEI: Edison Electric Institute.

EEM: Electronic Engineer's Master.

EEM: electronic equipment monitoring.

EEMJEB: Electrical and Electronic Manufacturers Joint Education Board.

EEMTIC: Electrical and Electronic Measurement and Test Instrument Conference.

EER: explosive echo ranging.

EERL: Electrical Engineering Research Laboratory.

EE&RM: elementary electrical and radio material training school.

EES: Engineering Experiment Station.

EESMB: Electrical and Electronics Standards Management Board.

EF: elevation finder.

EF: emitter follower.

EF: extra fine (thread)

EFG: edge-defined film-fed growth.

EFI: electronic fuel injection.

EFL: effective focal length.

EFL: equivalent focal length.

EFS: electronic frequency selection.

EFT: earliest finish time.

EFTS: electronic funds transfer system.

EFTS: Elementary Flying Training School.

e.g.: *exempli gratia* (for example).

EGD: electrogasdynamics.

EGO: Eccentric (orbit) Geophysical Observatory.

EGPS: Extended General Purpose Simulator (NEC)

EGSMA: Electrical Generating Systems Marketing Association.

EHD: electrohydrodynamic.

EHF: extremely high frequency (30-300 GHz).

EHP: electric horsepower.

EHS: Environmental Health Service.

EI: Electrical Insulation (IEEE Group).

EI: end injection.

EIA: Electronics Industries Association.

EIA: Engineering Industries Association (GB).

EIAC: Electronic Industries Association of Canada.

EIB: Electronics Installation Bulletin.

EIC: Electrical Insulation Conference.

EIC: engineer in charge.

EIC: Engineering Institute of Canada.

EIC: equipment identification code.

EID: Electronic Instrument Digest.

EIL: electron injection laser.

EIM: excitability inducing material.

EIN: education information network.

EIS: electromagnetic intelligence system.

EIT: engineer in training.

EITB: Engineering Industry Training Board.

EJC: Engineers Joint Council.

EJCC: Eastern Joint Computer Conference.

EKG: electrocardiogram.

EL: education level.

EL: electroluminescence.

ELDO: European Space Vehicle Launcher Development Organization.

ELECOM: electronic computer.

ELF: extensible language facility.

ELF: extremely low frequency.

ELFA: Electric Light Fittings Association (GB).

ELG: electrolytic grinding.

ELINT: electromagnetic intelligence.

ELINT: electronic intelligence.

ELIP: electrostatic latent image photography.

ELP: English language programs.

ELPC: electroluminescent-photoconductive.

ELPG: Electric Light and Power Group.

ELPH: elliptical head.

ELR: engineering laboratory report.

ELRO: Electronics Logistics Research Office.

ELSB: edge-lighted status board.

ELSIE: electronic signaling and indicating equipment.

ELV: electrically operated valve.

EM: electromagnetic.

EM: Engineering Management (IEEE Society).

EM: engineering manual.

EM: engineering memorandum.

EM: epitaxial mesa.

EMA: extended mercury autocoder.

EMATS: emergency mission automatic transmission service.

EMB: Engineering in Medicine and Biology (IEEE Group).

EMC: electromagnetic compatibility.

EMC: Electromagnetic Compatibility (IEEE Group).

EMC: Engineering Manpower Commission.

EMC: engineered military circuits.
EMC: European military communication.
EMCAB: Electromagnetic Compatibility Advisory Board.
EMCON: emission control.
EMCP: electromagnetic compatibility program.
EMCS: electromagnetic compatibility standardization (program).
EMCTP: electromagnetic compatibility test plan.
EMD: electric-motor driven.
EMDI: Energy Management Display Indicator.
EME: electromagnetic energy.
EMEC: Electronic Maintenance Engineering Center (Army).
EMETF: Electromagnetic Environmental Test Facility (Ft. Huachuca).
EMF: electromotive force.
EMG: electromyography.
EMI: electromagnetic interference.
EMICE: electromagnetic interference control engineer.
EMINT: electromagnetic intelligence.
EMIS: educational management information system.
EML: Engineering Mechanics Laboratory (NBS).
EML: equipment modification list.
EMM: electromagnetic measurement.
EMMA: electron microscopy and microanalysis.
EMOS: Earth's mean orbital speed.
EMP: electromagnetic pulse.
EMP: electromechanical power.
EMPIRE: early manned planetary-interplanetary round-trip expedition.
EMR: electromagnetic radiation.
EMR: electromechanical research.
EMRA: Electrical Manufacturers Representatives.
EMRIC: Educational Media Research Information Center.
EMS: electromagnetic susceptibility.
EMS: Engineering Management Society.
EMSA: Electron Microscope Society of America.
EMT: electrical metal tubing.
EMT: European Mediterranean Tropo.
EMU: electromagnetic unit.
EMU: extravehicular mobility unit.
EMW: electromagnetic warfare.
ENDOR: electron nuclear double resonance.
ENIAC: electronic numerical integrator and calculator.
ENR: equivalent noise resistance.
ENSI: equivalent-noise-sideband input.
ENT: equivalent noise temperature.
ENTC: engine negative torque control.
EO: engineering order.
EOA: end of address.
EOB: end of block.
EOC: emergency operating center.
EOD: end of data.
EOD: explosive ordnance disposal.
EODD: electrooptic digital deflector.
EOF: end of file.
EOG: electrooculography.
EOJ: end of job.
EOL: end of life.

EOL: expression oriented language.
EOLM: electrooptical light modulator.
EOLT: end of logical tape.
EOM: end of message.
EOP: end of program.
EOP: end output.
EOR: end of record.
EOR: end of reel.
EOR: explosive ordnance reconnaissance.
EOS: electrooptical systems.
EOT: end of tape.
EOT: end of transmission.
EOV: end of volume.
EP: electrically polarized (relay).
EP: engineer personnel.
EP: epitaxial planar.
EPA: Environmental Protection Agency.
EPAM: elementary perceiver and memorizer.
EPC: electronic program control.
EPC: engineering change proposals.
EPCO: Engine Parts Coordinating Office.
EPD: electric power distribution.
EPH: Electric Process Heating (IEEE IAS Technical Committee).
EPIC: earth-pointing instrument carrier.
EPLA: Electronics Precedence List Agency.
EPM: external polarization modulation.
EPMAU: expected present multi-attribute utility.
EPNdB: effective perceived noise level (decibels).
EPR: electron paramagnetic resonance.
EPRI: Electric Power research Institute.
EPROM: electrically programmable read only memory.
EPS: equilibrium problem solver.
EPT: electrostatic printing tube.
EPTE: existed prior to entry.
EPU: electrical power unit.
E&R: Engineering and Repair Department.
ER: echo ranging.
ER: effectiveness report.
ER: electrical resistance.
ER: enhanced radiation (weapons).
ERA: electronic reading automation.
ERA: Electronic Representatives Association.
ERA: Electrical Research Association.
ERB: Equipment Review Board.
ERBM: extended range ballistic missile.
ERC: Electronics Research Center (NASA).
ERC: equatorial ring current.
ERCR: electronic retina computing reader.
ERCS: Emergency Rocket Command System.
ERDA: Energy Research and Development Administration.
ERDL: Engineering Research and Development Laboratory (Ft. Belvoir, Va.).
ERFPI: extended range floating point interpretive system.
ERG: electroretinography.
ERGS: en-route guidance system.
ERIC: Educational Research Information Center.
ERIC: energy rate input controller.
ERL: echo return loss.

ERL: Environmental Research Laboratory.
ERM: earth reentry module.
ERMA: Electronic Recording Machine Accounting (GE).
ERN: engineering release notice.
ERNIE: electronic random numbering and indicating equipment.
ERO: emergency repair overseer.
EROP: extensions and restrictions of operators.
EROS: Earth Resources Observation Satellite (systems).
ERP: effective radiated power.
ERPLD: extended range phased-locked demodulator.
ERS: Economic Research Service.
ERS: Environmental Research Satellite.
ERS: external regulation system.
ERSA: Electronic Research Supply Agency.
ERSR: equipment reliability status report.
ERSS: Earth Resources Satellite System.
ERTS: Earth Resources Technology Satellite.
ERW: electronic resistance welding.
ERX: electronic remote switching.
ES: echo sounding.
ES: electromagnetic switching.
ES: electronic switch.
ES: experimental station.
ESA: Ecological Society of America.
ESAIRA: electronically scanned airborne intercept radar.
ESAR: electronically scanned array radar.
ESAR: electronically steerable array radar.
ESARS: earth surveillance and rendezvous simulator.
ESARS: Employment Service Automated Reporting System.
ESB: electrical simulation of the brain.
ESB: Electrical Standards Board.
ESC: electrostatic compatibility.
ESC: escape character (teletypewriter).
ESCA: electron spectroscopy for chemical analysis.
ESCAPE: expansion symbolic compiling assembly program for engineering.
ESD: Electronics Systems Division, USAF Systems Command.
ESD: electrostatic storage deflection.
ESD: Environmental Services Division.
ESE: electrical support equipment.
ESG: electrically suspended gyroscope.
ESH: equivalent standard hours.
ESHAC: Electric Space Heating and Air Conditioning (IEEE IAS Technical Committee).
ESHU: emergency ship-handling unit.
ESI: engineering and scientific interpreter.
ESLO: European Satellite Launching Organization.
ESM: elastomeric shield material.
ESM: electronic warfare support measures.
ESMA: Electronic Sales and Marketing Association.
ESNE: Engineering Society of New England.
ESONE: European Standards of Nuclear Electronics.
ESP: extrasensory perception.
ESPAR: electronically steerable phased array radar.
ESPOD: electronic systems precision orbit determination.

ESR: electron spin resonance.
ESR: equivalent series resistance.
ESRO: European Space Research Organization.
ESRANGE: European space range.
ESS: electronic switching system.
ESS: emplaced scientific station.
ESSA: Environmental Sciences Services Administration.
ESSG: Engineer Strategic Studies Group (Army).
EST: earliest start time.
EST: eastern standard time.
EST: electrostatic storage tube.
EST: enlistment screening test.
ESTEC: European Space Research Technical Center.
ESTRAC: European Space Satellite Tracking and Telemetry Network (ESRO).
ESW: error status word.
ET: edge-triggered.
ET: electrical time.
ET: Electronic Transformers (IEEE MAGS Technical Committee).
ET: engineering tests.
ET: ephemeris time.
ETAC: Environmental Technical Applications Center (Air Force).
ETCG: elapsed-time code generator.
ETL: Electrotechnical Laboratory (Japan).
ETL: ending tape label.
ETN: equipment table nomenclature.
ETOS: extended tape operating system.
ETP: electrical tough pitch.
ETR: engineering test reactor.
ETS: electronic telegraph system.
ETSQ: electrical time, superquick.
ETX: end of text character.
EUCLID: experimental use computer, London integrated display.
EUFMC: Electric Utilities Fleet Managers Conference.
EURATOM: European Atomic Energy Community.
EUROCAE: European Organization for Civil Aviation Electronics.
EUROCOMP: European Computing Congress.
EUROCON: European Convention (Region 8, IEEE).
EURODOC: Joint Documentation Service of ESRO, EUROSPACE and EODCSVL.
EUROSPACE: European Industrial Space Research Group.
eV: electronvolt.
EVA: electronic velocity analyzer.
EVC: Electric Vehicle Council.
EVM: electronic voltmeter.
EVOM: electronic voltohmmeter.
EW: electronic warfare.
EWC: electric water cooler.
EWCAS: early warning and control aircraft system.
EWCS: European Wideband Communications System.
EWF: Electrical Wholesalers Federation.
EWO: engineering work order.
EWTMI: European Wideband Transmission Media Improvement Program.
EXACT: International Exchange of Authenticated Component Performance Test Data.

EXDAMS: extended debugging and monitoring system.
EXMETNET: experimental meteorological sounding rocket research network.

EXSTA: experimental station.
EXTERRA: Extraterrestial Research Agency.

F

FA: forced-air-cooled (transformer).
F/A: fuel/air.
FA: fully automatic.
FAA: Federal Aviation Administration.
FAAAS: Fellow of the American Association for the Advancement of Science.
FAAB: Frequency Allocation Advisory Board.
FABMDS: Field Army Ballistic Missile Defense System (SAM-D).
FAC: field accelerator.
FAC: Frequency Allocation Committee.
FACE: field-alterable control element.
FACISCOM: Finance and Comptroller Information System Command.
FACS: fine attitude control system.
FACT: flight acceptance composite test.
FACT: flexible automatic circuit tester.
FACT: Foundation for Advanced Computer Technology.
FACT: fully automatic cataloging technique.
FACT: fully automatic compiler translator.
FAD: floating add.
FADAC: field artillery digital automatic computer.
FAE: final approach equipment.
FAETUA: Fleet Airborne Electronics Unit (Atlantic).
FAETUP: Fleet Airborne Electronics Unit (Pacific).
FADP: Finnish Association for Data Processing.
FAHQMT: fully automatic high-quality machine translation.
FAGC: fast automatic gain control.
FAL: frequency allocation list.
FALTRAN: Fortrau to Algol translator.
FAM: fast auxiliary memory.
FAMECE: Family of Military Engineer Construction Equipment.
FAMOS: Fleet Application of Meteorological Observations for Satellites.
FAMOS: floating-gate avalanche-injection metal-oxide semiconductor.
FAP: floating point arithmetic package.
FAP: frequency allocation panel.
FAP: Fortran assembly program.
FAPUS: frequency allocation panel (United States).
FAR: failure analysis report.
FARET: fast reactor test facility.
FARGO: 1401 automatic report generating operation.
FAS: free alongside.
FAS: Frequency Assignment Subcommittee.
FASE: fundamentally analyzable simplified English.
FAST: facility for automatic sorting and testing.
FAST: fast automatic shuttle transfer.

FAST: Fieldata applications, systems, and techniques.
FAST: Flexible Algebraic Scientific Translator (NCR).
FAST: formal auto-indexing of scientific texts.
FAST: formula and statement translator.
FAST: four-address to SOAP translator.
FASTAR: frequency angle scanning, tracking, and ranging.
FASTI: fast access to systems technical information.
FASWC: fleet antisubmarine warfare command.
FAT: formula assembler translator.
FATAL: FADAC automatic test analysis language.
FATE: fusing and arming test experiments.
FATH: fathom (1 fath = 6 ft = 1.8289 m).
FATR: fixed autotransformer.
FB: fuse block.
FBI: Federal Bureau of Investigation.
FBM: Fleet Ballistic Missile (Polaris/Poseidon).
FBMP: Fleet Ballistic Missile Program.
FBMWS: Fleet Ballistic Missile Weapons System.
FBR: feedback resistance.
FC: ferrite core.
FC: file code.
FC: file conversion.
fc: footcandle.
FC: function code.
FC: fuel cell.
FCA: fire control area.
FCA: frequency control analysis.
FCC: Federal Communications Commission.
FCC: flight control computer.
FCFT: fixed cost, fixed time (estimate).
FCI: functional configuration identification.
FCL: feedback control loop.
FCMD: fire command.
FCMV: fuel consuming motor vehicle.
FCOH: Flight Controllers Operations Handbook.
FCP: file control processor.
FCPC: Fleet Computer Programming Center.
FCR: final configuration review.
FCS: flight control system.
FCST: Federal Council for Science and Technology.
FCT: filament center tap.
F/D: focal (length) to diameter ratio.
F&D: facilities and design.
FD: file description.
FD: full duplex.
FD: flange local distance.
FDA: Food and Drug Administration.
FDAS: frequency distribution analysis sheet.
FDB: field dynamic braking.
FDDL: frequency division data link.

FDE: field decelerator.

FDI: field discharge.

FDIC: Federal Deposit Insurance Corporation.

FDM: frequency division multiplex.

FDMA: frequency division multiple access.

FDOS: floppy disk operating system.

FDP: future data processors (DPMA).

FDR: frequency domain reflectometry.

FDS: fluid distribution system.

FE: field engineer.

FEA: failure effect analysis.

FEA: Federal Energy Administration.

FEAT: frequency of every allowable term.

FEB: functional electronic block.

FEC: forward error correction.

FEI: Financial Executives Institute.

FEMF: Floating Electronic Maintenance Facility (FLATTOP).

FEO: field engineering order.

FERPIC: ferroelectric ceramic picture device.

FET: field-effect transistor.

FF: fixed focus.

FF: flip-flop.

FF: full field.

FFAR: folding fin air rocket.

FFC: flip-flop complementary.

FFI: fuel flow indicator.

FFL: first financial language.

FFL: front focal length.

FFSA: field functional system assembly and checkout.

FFT: fast Fourier transform.

FG: filament ground.

FG: function generator.

FGAA: Federal Government Accountants Association.

FGSA: Fellow of the Geological Society of America.

FHA: Federal Housing Administration.

FHA: Federal Highway Administration.

FHD: fixed-head disks.

FHLBB: Federal Home Loan Bank Board.

FHS: forward head shield.

FHT: fully heat treated.

FI: field intensity.

FIA: Federal Insurance Administration.

FIB: Fortran information bulletin.

FIC: Federal Information Centers.

FIC: frequency interference control.

FIDAC: film input to digital automatic computer.

FIFO: first in, first out.

FIIG: Federal Item Identification Guide.

FILS: flare-scan instrument landing system.

FIM: field intensity meter.

FIM: field ion microscope.

FIMATE: factory installed maintenance automatic test equipment.

FINAC: fast interline nonactivate automatic control.

FINAL: financial analysis language.

FIOP: Fortran input-output package.

FIPS: federal information processing standard.

FIR: fuel indicator reading.

FIRETRAC: Firing Error Trajectory Recorder and Computer.

FIRMS: forcasting information retrieval of management system.

FIRST: financial information reporting system.

FIS: field information system.

FIST: fault isolation by semiautomatic techniques.

FJCC: Fall Joint Computer Conference.

FL: Florida.

FL: focal length.

fl: footlambert.

FLAC: Florida automatic computer.

FLAM: forward launched aerodynamic missile.

FLAMR: forward looking advanced multilobe radar.

FLB: Federal Land Bank.

FLDEC: floating point decimal.

FLEA: flux logic element array.

FLF: follow-the-leader feedback (circuit theory).

FLF: fixed length field.

FLINT: floating interpretive language.

FLIP: film library instantaneous presentation.

FLIP: floating indexed point (arithmetic).

FLIP: floating point interpretive program.

FLODAC: Fluid Operated Digital Automatic Computer (Univac).

FLIR: forward looking infrared system.

FLODAC: Fluid Operated Digital Automatic Computer (Univac).

FLOSOST: fluorine one-stage orbital space truck.

FLOP: floating octal point.

FLOTRAN: flowcharting Fortran.

FLOX: fluorine liquid oxygen.

FLPL: Fortran list processing language.

FLR: forward-looking radar.

FLT: fault location technology.

FM: file maintenance.

FM: frequency modulation.

FMAC: Frequency Management Advisory Council.

FMC: Federal Maritime Commission.

FMCW: frequency-modulated continuous wave.

FMEA: failure mode and effect analysis.

FMES: full mission engineering simulator.

FMFB: frequency modulation with feedback.

FMI: frequency modulation intercity relay broadcasting.

FMIC: Frequency Monitoring Interference Control.

FMPS: Fortran mathematical programming system.

FMPS: functional mathematical programming system.

FMX: FM transmitter.

FNH: flashless nonhygroscopic.

FNP: fusion point.

FNS: Foodland Nutrition Service.

FO: fast operate (relay).

FO: oil-immersed forced-oil-cooled (transformer).

FOA: oil-immersed forced-oil-cooled with forced-air cooler (transformer).

FOCAL: formula calculator (DEC).

FOIL: file oriented interpretive language.

FOPT: fiber optic photo transfer.

FORBLOC: Fortran compiled block oriented simulation language.

FORC: formula coder (automatic coding system).

FORDS: Floating Ocean Research and Development Station.
FORESDAT: formerly restricted data.
FORGO: Fortran load and go (system).
FORMAC: formula manipulation compiler.
FORTRAN: formula translation.
FORTRANSIT: Fortran and internal translator system.
FORTRUNCIBLE: Fortran style runcible.
FOSDIC: film optical sensing device for input to computers.
FOT: frequency of optimum operation.
FOW: oil-immersed forced-oil-cooled with forced-water cooler (transformer).
FP: faceplate.
FP: feedback positive.
FP: forward perpendicular.
FP: freezing point.
FPC: Federal Power Commission.
FPCH: Foreign Policy Clearinghouse.
FPIS: forward propagation by ionospheric scatter.
FPL: final protective line.
FPM: feet per minute.
FPM: functional planning matrices.
FPP: floating point processor.
FPS: focus projection and scanning.
FPS: feet per second.
FPS: foot-pound-second.
FPS: frames per second.
FPT: female pipe thread.
FPT: full power trial.
FPTS: forward propagation tropospheric scatter.
FQPR: frequency programmer.
FR: failure rate.
FR: fast release (relay).
FR: Federal Reserve.
FR: field reversing.
FR: flash ranging.
FRB: Federal Reserve Bank.
FRC: File Research Council.
FRC: functional residue capacity.
FRCTF: fast reactor core test facility (AEC).
FRD: functional reference device.
FRED: figure reader electronic device.
FREQMULT: frequency multiplier.
FRESCANNAR: frequency scanning radar.
FRINGE: file and report information processing generator.
FRP: fiberglass-reinforced plastic.
FRS: Fellow of the Royal Society (GB).
FRS: Fire Research Station.
FRS: Federal Reserve System.
FRS: fragility response spectrum.
FS: Federal Standard.
FS: female soldered.
FS: full scale.
fs: functional schematic.

FSC: Federal Supply Classification.
FSD: full-scale deflection.
FSDC: Federal Statistical Data Center.
FSK: frequency shift keying.
FSL: formal semantic language (MIT).
FSLIC: Federal Savings and Loan Insurance Corporation.
FSR: feedback shift register.
FSS: Federal Supply Services.
FSTC: Foreign Science and Technology Center (Army).
FSUC: Federal Statistics Users Conference.
FSVM: frequency selective voltmeter.
F&T: fuel and transportation.
FT: firing tables.
FT: flush threshold.
FT: frequency and time.
FT: Frequency and Time (IEEE IM Technical Committee).
FT: frequency tracker.
FT: functional test.
FTC: fast time constant.
FTC: Federal Trade Commission.
FTF: flared tube fitting.
FTFET: four-terminal field-effect transistor.
FTL: Federal Communications Laboratory.
FTM: flight test missile.
FTP: fuel transfer pump.
FTR: functional test requirement.
FTS: Federal Telecommunications System.
FTS: Free-Time System (GE/PAC).
FTS: frequency and timing subsystem.
FTU: flight test unit.
FUDR: failure and usage data report.
FUIF: fire unit integration facility.
FUNCTLINE: functional line diagram.
fV: femtovolt.
F/V: frequency to voltage.
FV: full voltage.
FVD: front vertex back focal distance.
FVS: Flight Vehicles Systems (IEEE AESS Technical Committee).
FW: face width.
FW: filament wound.
FW: full wave.
FWA: first word address.
FWA: fluorescent whitening agent.
FWA: forward wave amplifier.
FWHM: full width at half maximum.
FWL: fixed word length.
FWPCA: Federal Water Pollution Control Administration.
FWQA: Federal Water Quality Administration.
FWS: filter wedge spectrometer.
FY: fiscal year.
FYDP: five year defense program.

G

g-a: ground to air.
GA: gain of antenna.
ga: gas amplification.
GA: Georgia.
ga: glide angle.
ga: graphic ammeter.
GaAs: gallium arsenide (semiconductor).
GaAsP: gallium arsenide phosphide (semiconductor).
GAFB: Griffiss Air Force Base.
GALCIT: Guggenheim Aeronautical Laboratory of California Institute of Technology (USA).
GAM: graphic access method (IBM).
GAM: guided aircraft missile.
GAMA: Gas Appliance Manufacturers Association.
GAMA: graphics assisted management application.
GAMLOGS: gamma-ray logs.
GAMM: German Association for Applied Mathematics and Mechanics.
GAN: generalized activity network.
GAN: generating and analyzing networks.
GAP: general assembly program.
GAR: growth analysis and review.
gar: guided aircraft rocket.
GARP: global atmospheric research program.
GASP: general activity simulation program.
GASP: generalized academic simulation program.
GASP: graphic applications subroutine package.
GASS: generalized assembly system.
GAT: generalized algebraic translator.
GAT: Georgetown automatic translator.
GATAC: general assessment tridimensional analog computer.
GATB: general aptitude test battery.
GATE: GARP Atlantic Tropical Experiment.
GATE: generalized algebraic translator extended.
GATT: General Agreement on Tariffs and Trade.
gb: gilbert (unit of magnetomotive force).
gc: gigacycles per second (obsolete); use GHz (gigahertz).
GC: guidance computer.
GCA: ground-controlled approach.
GCAP: generalized circuit analysis program.
GCC: ground control center.
GCD: greatest common divisor.
GCMA: Government Contract Management Association of America.
GCFR: gas-cooled fast reactor.
GCFRE: gas-cooled fast reactor experiment.
GCHQ: Government Communications Headquarters.
GCI: ground-controlled interception.
GCN: gage code number.
GCIS: ground control intercept squadron.
GCMS: gas chromatography and mass spectroscopy.
GCS: gate-controlled switch.
GCR: general component reference.
GCR: group coded recording.
GCT: general classification test.
GCT: Greenwich civil time.
GD: ground detector.

GD: grown diffused.
GDE: ground data equipment.
GDG: generation data group.
GDMS: generalized data management system.
GE: Gaussian elimination.
GE: gas ejection.
Ge: germanium (semiconductor).
GEBCO: General Bathymetric Chart of the Oceans.
GECOM: generalized compiler.
GECOS: general comprehensive operating supervisor.
GEEP: General Electric electronic processor.
GEEIA: General Electronics Engineering Installation.
GEESE: General Electric Electronic Systems Evaluator.
GEISHA: geodetic inertial survey and horizontal alignment.
GEK: geomagnetic electrokinetograph.
GEM: general epitaxial monolith.
GEM: ground effect machine.
GE/MAC: General Electric Measurement and Control.
GEMS: general education management system.
GEMS: General Electric Manufacturing Simulator.
GEMSIP: Gemini Stability Improvement Program.
GEO: Geoscience Electronics (IEEE Group).
GEON: gyro erected optical navigation.
GEOS: geodetic earth orbiting satellite.
GE/PAC: General Electric Process Automation Computer.
GEPAC: General Electric Programmable Automatic Comparator.
GERSIS: General Electric Range Safety Instrumentation System.
GERTS: General Electric Remote Terminal System.
GET: ground elapsed time.
GE/TAC: General Electric Telemetering and Control.
GETEL: General Electric Test Engineering Language.
GETOL: General Electric Training Operational Logic.
GETOL: ground effect take-off and landing.
GEVIC: General Electric Variable Increment Computer.
GFAE: Government-furnished avionics equipment.
GFE: Government furnished equipment.
GFRP: glass fiber reinforced plastic.
GFW: ground-fault warning.
G-G: ground to ground.
GGTS: gravity gradient test satellite.
GHCP: Georgia Hospital Computer Group.
GHOST: global horizontal sounding technique.
GHz: gigahertz (1000 MHz or 10⁹ Hz).
GI: geodesic isotensoid.
GI: government and industrial.
GI: government initiated.
GIA: General Industry Applications (IEEE IAS Technical Committee).
GIANT: genealogical information and name tabulating system.
GIC: generalized immittance converter.
GIDEP: government industry data exchange program.
GIGO: garbage in, garbage out.

GIFS: generalized interrelated flow simulation.

GIFT: general internal Fortran translator.

GIM: generalized information management (language).

GIMRADA: Geodesy Intelligence and Mapping Research and Development Agency (U.S. Army).

GIOC: generalized input/output controller.

GIPS: ground information processing system.

GIPSY: generalized information processing system.

GIRL: graph information retrieval language.

GIRLS: Generalized Information Retrieval and Listing System.

GIS: generalized information system.

GIS: Geoscience Information Society.

GIT: graph isomorphism tester.

GJ: grown junction.

GJE: Gauss-Jordan elimination.

GL: gate leads.

GLC: gas-liquid chromatography.

GLEAN: graphic layout and engineering aid method.

GLEEP: graphite low energy experimental pile.

GLINT: global intelligence.

GLOMEX: global meteorological experiment.

GLOPAC: gyroscopic lower power controller.

GLOSS: global ocean surveillance system.

GLOTRAC: global tracking network.

GLV: Gemini launch vehicle.

G-M: Geiger-Mueller (counter).

GM: gaseous mixture.

GM: geometric mean.

GM: guided missile.

GM: metacentric height.

GMCM: guided missile countermeasures.

GMD: Guided Missiles Division.

GMFCS: guided missile fire control system.

GMR: ground mapping radar.

GMSFC: George Marshall Space Flight Center (Huntsville, Ala.).

GMT: Greenwich mean time.

GMV: guaranteed minimum value.

GNP: gross national product.

GOAL: ground operations aerospace language.

GOCI: general operator-computer interaction.

GOE: ground operating equipment.

GOES: geostationary operational environmental satellite.

GOL: general operating language.

GOLD: graphic on-line language.

GOP: general operational plot.

GOR: gained output ratio.

GOR: gas-oil ratio.

GOR: general operational requirement.

GORID: ground optical recorder for intercept determination.

GOSS: ground operational support system.

GP: generalized programming.

GP: ground protective (relay).

GPA: general purpose analysis.

GPA: graphical PERT analog.

GPATS: general purpose automatic test system.

GPC: general purpose computer.

GPCP: generalized process control programming.

GPDC: general purpose digital computer.

GPDS: general purpose display system.

GPGL: general purpose graphic language.

gph: gallons per hour.

GPI: ground position indicator.

GPIS: Gemini problem investigation status.

GPL: general purpose language.

GPL: generalized programming language.

GPLP: general purpose linear programming.

gpm: gallons per minute.

GPM: general purpose macrogenerator.

GPMS: general purpose microprogram simulator.

GPO: General Post Office.

GPO: Government Printing Office (U.S.A.).

gps: gallons per second.

GPS: general problem solver.

GPSS: general purpose system simulation.

GPT: gas power transfer.

GPT: Gemini pad test.

GPX: generalized programming extended.

GR: general reconnaissance.

GR: general reserve.

GRACE: graphic arts composing equipment.

GRAD: general recursive algebra and differentiation.

GRAD: graduate resume accumulation and distribution.

GRADB: generalized remote access data base.

GRADS: generalized remote access data base system.

GRAF: graphic addition to Fortran.

GRAMPA: general analytical model for process analysis.

GRAPE: gamma-ray attenuation porosity evaluator.

GRAPHDEN: graphic data entry.

GRARR: Goddard range and range rate.

graser: gamma-ray amplification by stimulated emission of radiation.

GRASP: generalized retrieval and storage program.

GRASP: graphic service program.

GRATIS: generation, reduction, and training input system.

GRB: Geophysical Research Board (NRC).

GRB: Government Reservation Bureau.

GRCSW: Graduate Research Center of the Southwest.

GRE: Graduate Reliability Engineering.

GREB: galactic radiation experiment background satellite.

GRED: generalized random extract device.

GREMEX: Goddard Research Engineering Management Exercise.

GRID: graphic interactive display.

GRIN: graphical input.

GRINS: general retrieval inquiry negotiation structure.

GRIT: graduated reduction in tensions.

GRM: generalized Reed-Muller (codes).

GRM: global range missile.

GRP: glass-reinforced plastic.

GRR: guidance reference release.

GRS: generalized retrieval system.

GRWT: gross weight.

GS: galvanized steel.

GS: ground speed.

GSA: General Services Administration.

GSA: Genetics Society of America.
GSC: gas-solid chromatography.
GSCU: ground support cooling unit.
GSD: General Supply Depot.
GSD: general systems division.
GSDB: geophysics and space data bulletin.
GSDS: Goldstone Duplicate Standard (standard DSIF equipment).
GSE: ground support equipment.
GSFC: Goddard Space Flight Center.
GSI: Government Source Inspection.
GSI: grand scale integration.
GSL: generation strategy language.
GSL: generalized simulation language.
GSMB: Graphics Standards Management Board.
GSO: Ground Safety Office.
GSOP: Guidance Systems Operation Plan.
GSP: general simulation program.
GSP: guidance signal processor.
GSPR: guidance signal processor repeater.
GSPO: Gemini Spacecraft Project Office.
GSR: galvanic skin response.
GSS: Gamma Scintillation System.
GSS: global surveillance system.
GSSC: ground support simulation computer.
GSTA: ground surveillance and target acquisition.
GSU: general service unit.
GSV: guided space vehicle.

GSWR: galvanized steel wire rope.
GT: game theory.
GT: Gemini Titan.
GT: ground transmit.
GTA: Gemini Titan Agena.
GTC: gain time constant.
GTC: gain time control.
GTC: gas turbine compressor.
GTD: graphic tablet display.
GTE: ground transport equipment.
GTG: gas turbine generator.
GTM: ground test missile.
GTO: gate-turn-off (switches).
GTOW: gross take-off weight.
GTP: general test plan.
GTS: general technical services.
GTV: ground test vehicle.
GU: Guam.
GUHA: general unary hypotheses automation.
GUIDE: guidance for users of integrated data equipment.
GULP: general utility library program.
GUSTO: guidance using stable tuning oscillations.
GVA: graphic kilovolt-ampere meter.
GW: general warning.
GWC: global weather central (Air Force).
GZ: ground zero.

H

ha: high altitude.
HA: Hydraulic Association.
HA: half-adder.
HAD: half-amplitude duration.
HADS: hypersonic air data sensor.
HAF: high abrasion furnace.
HAF: high altitude fluorescence.
HAFB: Holloman Air Force Base (N. Mex.).
HAIC: Hearing Aid Industry Conference.
HAL: highly automated logic.
HALSIM: hardware logic simulator.
HAM: hardware associative memory.
HANE: high altitude nuclear effects.
HAOSS: high altitude orbital space station.
HAP: high-altitude platform.
HAPDAR: hard point demonstration array radar.
HARAC: high altitude resonance absorption calculation.
HARP: Halpern antiradar point.
HARP: high altitude research project.
HARP: Hitachi Arithmetic Processor.
HARTRAN: Hardwell Fortran.
HASL: Health and Safety Laboratory (AEC).
HASP: high-altitude sampling program.
HASP: high-altitude sounding project.
HASP: Houston automatic spooling priority system.
HAZ: heat affected zone.

H-B: hexadecimal to binary.
H/C: hand carry.
HC: heuristic concepts.
HC: high capacity (projectile).
HC: high carbon.
HC: holding coil.
HC: hybrid computer.
HCD: hot-carrier diode.
HCE: hollow-cathode effect.
HCG: horizontal location of center of gravity.
HCL: high, common, low (relay).
HCL: horizontal center line.
HCSS: hospital computer sharing system.
H-D: hexadecimal to decimal.
HD: half duplex.
HD: head diameter.
HD: heavy duty.
HDDR: high density digital recording.
HDDS: high-density data system.
HDL: Harry Diamond Laboratories, Army (Washington, D.C.).
HDLC: high-level data link control.
HDMR: high density moderated reactor.
HDOC: handy dandy orbital computer.
HDST: high density shock tube.
HE: handling duplex.
HE: heat exchange.

HE: high explosive.

HEAP: high explosive armor piercing.

HEAT: high-explosive antitank.

HEC: Hollerith electronic computer.

HECTOR: heated experimental carbon thermal oscillator reactor.

HED: horizontal electrical dipole.

HEEP: Highway Engineering Exchange Program.

HELEN: hydrogenous exponential liquid experiment.

HELP: helicopter electronic landing path.

HELP: highly extendable language processor.

HEM: hybrid electromagnetic (wave).

HEOS: high excentricity orbit satellite.

HEPL: High Energy Physics Laboratory (Stanford University).

HERA: high explosive rocket assisted.

HERF: high energy rate forming.

HERMES: heavy element and radioactive material electromagnetic separator.

HERO: hazards of electromagnetic radiation to ordnance.

HEVAC: Heating, Ventilating, and Air-Conditioning Manufacturers Association.

HEW: Health, Education, and Welfare.

HF: height finder (radar).

HF: high frequency (3 to 30 MHz).

HFBR: high flux beam reactor.

HFE: Human Factors in Engineering Group.

HFG: heavy free gas.

HFIM: high-frequency instruments and measurements.

HFIM: High-Frequency Instruments and Measurement (IEEE IM Technical Committee).

HFIR: high flux isotope reactor.

HFO: high-frequency oscillator.

HFORL: Human Factors Operations Research Laboratory.

HFS: hyperfine structure.

Hg: mercury.

HGT: high group transmit.

HI: Hydraulic Institute.

HIBEX: high impulse boost experiment.

HIC: hybrid integrated circuit.

HID: high intensity discharge (lamps).

HIDM: high information delta modulation.

HIIS: Honeywell Institute for Information Science.

HIL: high-intensity lighting.

HILAC: heavy ion linear accelerator.

HIPAR: high power acquisition radar.

HIPERNAS: high-performance navigation system.

HIPOT: high potential.

HIPOTT: high potential test.

HIR: horizontal impulse reaction.

HIS: Honeywell Information Systems.

HIVOS: high vacuum orbital simulation.

HLL: higher level language.

hm: hectometer.

HMRB: Hazardous Materials Regulation Board.

HMSS: Hospital Management Systems Society.

HOBO: homing official bomb.

HOP: hybrid operating program.

H-P: Hewlett-Packard (Corporation).

HP: high pass.

HP: high pressure.

HP: horsepower.

HPBW: half-power beamwidth.

HPF: highest probable frequency.

HP HR: horsepower-hour.

HPOT: helipotentiometer.

HPS: Health Physics Society.

HPT: high-pressure test.

HPT: horizontal plot table.

H&RP: holding and reconsignment point.

HRC: high rupturing capacity.

HRC: horizontal redundancy check.

HRIR: high resolution infrared radiometer.

HRL: horizontal reference line.

HRRC: Human Resources Research Center.

HS: half subtractor.

HS: high speed.

HSD: high-speed data.

HSD: high-speed displacement.

HSDA: high-speed data acquisition.

HSGT: high-speed ground transport.

HSM: Health Services and Mental Health Administration.

HSM: high-speed memory.

HSP: high-speed printer.

HSR: high-speed reader.

HSRO: high-speed repetitive operation.

HSS: high-speed storage.

HST: hypersonic transport.

HT: high tension.

HT: horizontal tabulation.

HTA: heavier than air.

HTD: hand target designator.

HTGCR: high-temperature gas-cooled reactor.

HTGR: high-temperature gas-cooled reactor.

HTL: high-threshold logic.

HTM: high temperature.

HTM: hypothesis testing model.

HTO: high-temperature oxidation.

HTS: high tensile strength.

HTSS: Honeywell Time-Sharing System.

HTTL: high-power transistor-transistor logic.

HTU: heat transfer unit.

HUD: head-up display.

HUD-FHA: Housing and Urban Development-Federal Housing Administration.

HUMRRO: Human Resources Research Office.

HV: hardware virtualizer.

HV: high voltage.

HVACC: High Voltage Apparatus Coordinating Committee (ANSI).

HVAR: high-velocity aircraft rocket.

HVAT: high-velocity antitank.

HVCA: Heating and Ventilating Contractors Association.

HVHMD: holographic visor helmet mounted display.

HVP: high video pass.

HVR: high-voltage regulator.

HVRA: Heating and Ventilating Research Association.

HW: half wave.

HWCTR: heavy water components test reactor.

HWOCR: heavy water organic cooled reactor.

hxwxl: height by width by length.
HYCOL: hybrid computer link.
HYCOTRAN: hybrid computer translator.
HYDAC: Hybrid Digital Analog Computer (EAI)
HYDAPT: hybrid digital-analog and pulse time.
HYFES: hypersonic flight environmental simulator.

HYPSES: hydrographic precision scanning echo sounder.
HYSTAD: hydrofoil stabilization device.
HYTRESS: high test recorder and simulator system.
Hz: hertz.
HZMP: horizontal impulse.

I

IA: indirect addressing.
IA: Industry Application (IEEE Society).
IA: instrumentation amplifier.
IA: Iowa.
IAA: International Academy of Astronautics.
IAAC: International Agriculture Aviation Center (GB).
IABSE: International Association for Bridge and Structural Engineering.
IAC: Information Analysis Center.
IAC: Industry Advisory Committee.
IAC: international algebraic compiler.
IACP: International Association of Computer Programmers.
IACP: International Association of Chiefs of Police.
IACS: inertial attitude control system.
IACS: integrated armament control system.
IACS: International Annealed Copper Standard.
IAD: integrated automatic documentation.
IADIC: integration analog-to-digital converter.
IADR: International Association for Dental Research.
IAE: integral absolute error.
IAEA: International Atomic Energy Agency.
IAEE: International Association of Earthquake Engineers.
IAESTE: International Association for the Exchange of Students for Technical Experience.
IAF: International Astronautical Federation.
IAG: IFIP Administrative Data Processing Group.
IAG: International Association of Geodesy.
IAGA: International Association of Geomagnetism and Aeronomy.
IAGC: instantaneous automatic gain control.
IAHR: International Association for Hydraulic Research.
IAL: international algorithmic language.
IAL: investment analysis language.
IALA: International Association of Lighthouse Authorities.
IAM: Institute of Aviation Medicine.
IAM: interactive algebraic manipulation.
IAM: International Association of Machinists and Aerospace Workers.
IAMAP: International Association of Meteorology and Atmospheric Physics.
IAMC: Institute for Advancement of Medical Communication.
IAMTCT: Institute of Advance Machine Tool and Control Technology.

IANEC: Inter-American Nuclear Energy Commission.
IAO: internal automation operation.
IAPO: International Association of Physical Oceanography.
IARU: International Amateur Radio Union.
IAS: immediate access storage.
IAS: indicated air speed.
IAS: Industrial Applications Society.
IAS: Institute of Advanced Studies (Army).
IAS: Institute of Aerospace Sciences.
IAS: International Association of Sedimentology.
IASA: Insurance Accounting and Statistical Association.
IASA: International Air Safety Association.
IASF: Instrumentation in Aerospace Simulation Facilities (IEEE AESS Technical Committee).
IASH: International Association of Scientific Hydrology.
IASI: Inter-American Statistical Institute.
IASPEI: International Association of Seismology and Physics of the Earth's Interior.
IAT: International Accountants Society.
IAT: Institute for Advanced Technology.
IAT: Institute for Applied Technology.
IAT: Institute of Automatics and Telemechanics (USSR).
IATA: International Air Transport Association.
IATM: International Association for Testing Materials.
IATUL: International Association of Technical University Libraries.
IAU: International Association of Universities.
IAV: International Association of Volcanology.
IAVC: instantaneous automatic video control.
IAVC: instantaneous automatic volume control.
IBA: Investment Bankers Association.
IBAM: Institute of Business Administration and Management (Japan).
IBCC: International Building Classification Committee.
IBEW: International Brotherhood of Electrical Workers.
IBG: interblock gap.
IBI: Intergovernmental Bureau for Informatics.
IBIS: intense bunched ion source.
IBP: initial boiling point.
IBPA: International Business Press Associates.
IBR: integral boiling reactor.
IBR: integrated bridge rectifier.
IBRL: initial bomb release line.

IBS: Institute for Basic Standards.
IBSAC: Industrialized Building Systems and Components.
IBW: impulse bandwidth.
IBWM: International Bureau of Weights and Measures.
I&C: installation and checkout.
IC: impulse conductor.
IC: input circuit.
IC: instruction counter.
IC: Insulated Conductors (IEEE PES Technical Committee).
IC: integrated circuit.
IC: interface control.
IC: interior communications.
IC: ion chamber.
ICA: Industrial Communications Association.
ICA: International Communications Association.
ICA: International Council on Archives.
ICAD: integrated control and display.
ICARVS: interplanetary craft for advanced research in the vicinity of the sun.
ICAS: intermittent commercial and amateur service.
ICBM: intercontinental ballistic missile.
ICBO: International Conference of Building Officials.
ICC: International Computation Centre (Italy).
ICCC: International Chamber of Commerce.
ICC: Interstate Commerce Commission.
ICCP: Institute for Certification of Computer Professionals.
ICCP: International Conference on Cataloging Principles.
ICE: input checking equipment.
ICE: integrated cooling for electronics.
ICE: intermediate cable equalizers.
ICE: Institution of Civil Engineers.
ICEM: International Council for Educational Media.
ICES: integrated civil engineering system.
ICES: International Conference of Engineering Studies.
ICES: International Council for the Exploration of the Sea.
ICETK: International Committee of Electrochemical Thermodynamics and Kinetics.
ICG: interactive computer graphics.
ICIP: International Conference on Information Processing.
ICL: incoming line.
ICM: advanced IBM.
ICM: improved capability missile.
ICO: Inter-Agency Committee on Oceanography.
ICO: International Commission for Optics.
ICOR: Intergovernmental Conference on Oceanic Research.
ICP: international computer programs.
ICP: International Control Plan.
ICPS: Industrial and Commercial Power Systems (IEEE IAS Technical Committee).
ICPS: International Conference on the Properties of Steam.
ICPS: International Congress of Photographic Science.
ICR: iron-core reactor.
ICRH: Institute for Computer Research in the Humanities.

ICS: inland computer service.
ICS: integrated communication system.
ICS: Institute of Computer Science.
ICS: Intercommunication System.
ICS: interphone control station.
ICSC: Interim Communications Satellite Committee.
ICSI: International Conference on Scientific Information.
ICSU: International Council of Scientific Unions.
ICT: Institute of Computer Technology.
ICtl: Industrial Control (IEEE IAS Technical Committee).
ICW: interrupted continuous wave.
ICWM: International Committee of Weights and Measures.
I/D: instruction/data.
ID: Idaho.
ID: inside diameter.
ID: Interior Department.
ID: intermediate description.
ID: internal diameter.
IDA: Institute for Defense Analysis.
IDA: ionospheric dispersion analysis.
IDAS: information displays automatic drafting system.
IDAST: interpolated data and speech transmission.
IDC: image dissector camera.
IDCSP: Initial Defense Communications Satellite Project.
IDEAS: integrated design and engineering automated system.
IDEEA: information and data exchange experimental activities.
IDEP: interservice data exchange program.
IDEX: initial defense experiment.
IDF: integrated data file.
IDF: intermediate distributing frame.
IDFT: inverse discrete Fourier transform.
IDG: inspector of degaussing.
IDI: improved data interchange.
IDIOT: instrumentation digital on-line transcriber.
IDM: interdiction mission.
IDOC: inner diameter of outer conductor.
IDP: industrial data processing.
IDP: integrated data processing.
IDPC: integrated data processing center.
IDPI: International Data Processing Institute.
IDR: industrial data reduction.
IDS: Instrument Development Section.
IDS: integrated data store.
IDT: isodensitracer.
IDTS: instrumentation data transmission system.
IDU: industrial development unit.
i.e.: *id est* (that is).
I&E: Information and Education.
IE: industrial engineer.
IE: initial equipment.
IEC: integrated electronic component.
IEC: International Electrotechnical Commission.
IECI: Industrial Electronics and Control Instrumentation (IEEE Group).
IECPS: International Electronic Packaging Symposium.

IED: individual effective dose.
IEE: Institution of Electrical Engineers (London).
IEEE: Institute of Electrical and Electronics Engineers.
IEETE: Institution of Electrical and Electronics Technician Engineers.
IEG: information exchange group.
IER: Institutes for Environmental Research.
IERE: Institution of Electronic and Radio Engineers (Australia).
IES: Illuminating Engineering Society.
IES: Institute of Environmental Sciences.
IES: integral error squared.
IES: intrinsic electric strength.
IEV: International Electrotechnical Vocabulary.
I/F: interface.
IF: intermediate frequency.
IFAC: International Federation for Automatic Control.
IFATCA: International Federation of Air Traffic Controllers Association.
IFC: International Finance Corporation.
IFCS: in-flight checkout system.
IFCS: International Federation of Computer Sciences.
IFD: instantaneous frequency discriminator.
IFD: International Federation for Documentation.
IFEMS: International Federation of Electron Microscopes Societies.
IFF: identification friend or foe.
iff: if and only if.
IFIP: International Federation of Information Processing.
IFIPC: International Federation of Information Processing Congress.
IFIPS: International Federation of Information Processing Societies.
IFLA: International Federation of Library Associations.
IFME: International Federation of Medical Electronics.
IFORS: International Federation of Operational Research Societies.
IFR: increasing failure rate.
IFR: inflight refueling.
IFR: instantaneous frequency (indicating) receivers.
IFR: instrument flight rules.
IFRB: International Frequency Registration Board.
IFRF: International Flame Research Foundation.
IFRU: interference frequency rejection unit.
IFS: International Federation of Surveyors.
IFT: Institute of Food Technologists.
IFTC: International Federation of Thermalism and Climatism.
IFTC: International Film and Television Council.
IFVME: Inspectorate of Fighting Vehicles and Mechanical Equipment.
IG: inertial guidance.
IGAAS: Integrated Ground Airborne Avionics System.
IGC: International Geophysical Committee.
IGE: Institution of Gas Engineers.
IGFET: insulated gate field-effect transistor.
IGFET: isolated-gate field-effect transistor.
IGOR: intercept ground optical recorder.
IGPP: Institute of Geophysics and Planetary Physics.

IGS: inertial guidance system.
IGS: integrated graphics system.
IGU: International Gas Union.
IGY: International Geophysical Year.
IHAS: Integrated Helicopter Avionics System.
IHB: International Hydrographic Bureau.
IHD: International Hydrological Decade.
IHE: Institution of Highway Engineers.
IHFM: Institute of High-Fidelity Manufacturers.
IHP: indicated horsepower.
IHPH: indicated horsepower-hour.
IIA: Information Industry Association.
IIAS: International Institute of Administrative Services.
IIC: International Institute of Communications.
IILS: International Institute for Labor Studies.
IIR: International Institute of Refrigeration.
IIRS: Institute for Industrial Research and Standards (Erie).
IIS: Institute of Industrial Supervisors.
IIS: Institute of Information Scientists.
IISO: Institution of Industrial Safety Officers.
IIT: Illinois Institute of Technology.
IITRI: Illinois Institute of Technology Research Institute.
IIW: International Institute of Welding.
IJAJ: International Jitter Antijam.
IJCAI: International Joint Conference on Artificial Intelligence.
IJJU: intentional jitter jamming unit.
I&L: Installation and Logistics.
IL: intermediate language.
ILAAS: integrated light attack avionic system.
ILAR: Institute of Laboratory Animal Resources.
ILAS: Instrument Landing Approach System.
ILAS: interrelated logic accumulating scanner.
ILC: instruction length code.
ILF: inductive loss factor.
ILJM: Illinois Journal of Mathematics.
ILIR: in-house laboratories independent research.
ILLIAC: Illinois Integrator and Automatic Computer.
ILM: information logic machine.
ILO: International Labor Office.
ILO: International Labor Organization.
ILP: intermediate language processor.
ILS: ideal liquidus structures.
ILS: instrument landing system.
ILTS: Institute of Low Temperature Science (Japan).
ILU: Illinois University.
ILW: intermediate-level wastes.
IM: industry (FCC services) motion picture.
IM: installation and maintenance.
IM: Institute of Metrology (Leningrad, USSR).
IM: Instrumentation and Measurement (IEEE Group).
IM: interceptor missile.
IM: intermediate missile.
IMA: Industrial Medical Association.
IMA: International Mineralogical Association.
I Mar E: Institute of Marine Engineers.
IMAS: Industrial Management Assistance Survey.
IMC: instrument meteorological conditions.
IMC: integrated maintenance concept.

IMC: International Micrographic Congress.
IMCO: Intergovernmental Maritime Consultative Organization.
IMD: intermodulation distortion.
IME: Institute of Mining Engineers.
I Mech E: Institute of Mechanical Engineers.
IMEP: indicated mean effective pressure.
IMF: International Monetary Fund.
IMFI: Industrial Mineral Fiber Institute.
IMH: Institute of Materials Handling.
IMI: Irish Management Institute (Eire).
IMITAC: image input to automatic computers.
IMM: Institute of Mathematics Machines (Poland).
IMM: Institute of Mining and Metallurgy.
IMM: integrated maintenance management.
IMMS: International Material Management Society.
IMP: integrated microwave products.
IMP: interface message processor.
IMP: interplanetary monitoring platform.
IMP: Inter-Industry Management Program.
IMP: intrinsic multiprocessing.
IMPACT: inventory management program and control technique.
IMPATT: impact avalanche and transit time (diode).
IMPCM: improved capability missile.
IMPI: International Microwave Power Institute.
IMPRESS: interdisciplinary machine processing for research and education in the social sciences.
IMPS: integrated master programming and scheduling.
IMPTS: improved programmer test station.
IMR: Institute for Materials Research (NBS).
IMRA: International Marine Radio Association.
IMRADS: information management retrieval and dissemination system.
IMS: Industrial Management Society.
IMS: Industrial Mathematics Society.
IMS: information management system.
IMS: Institute of Mathematical Statistics.
IMS: Institute of Marine Science.
IMS: instructional management system.
IMS: Integrated Meteorological system (Army).
IMS: inventory management and simulator.
IMU: inertial measuring unit.
IMU: International Mathematical Union.
IN: Indiana.
IN: interference-to-noise ratio.
INAS: inertial navigation and attack systems.
INCH: integrated chopper.
INCO: International Chamber of Commerce.
INCOMEX: International Computer Exhibition.
INCOR: Intergovernmental Conference on Oceanographic Research.
INCOR: Israeli National Committee for Oceanographic Research.
INCOSPAR: Indian National Committee Space Research.
INCUM: Indiana Computer Users Meeting.
INCR: interrupt control register.
INDEX: Inter-NASA Data Exchange.
INDREG: inductance regulator.
INDTR: indicator-transmitter.
INEL: International Exhibition of Industrial Electronics.

INFANT: Iroquois night fighter and night tracker.
INFO: information network and file organization.
INFOCEN: information center.
INFOL: information oriented language.
INFRAL: information retrieval automatic language.
ING: intense neutron generator.
INGO: International Nongovernmental Organization.
INIS: International Nuclear Information System.
INL: internal noise level.
INLC: initial launch capability.
INP: inert nitrogen protection.
INR: interference-to-noise ratio.
INREQ: information on request.
INS: Institute for Naval Studies.
INS: Institute for Nuclear Study (Tokyo University).
INSPEC: Information Service in Physics, Electrotechnology and Control (IEE).
INSPEX: Engineering Inspection and Quality Control Conference and Exhibition.
INSTARS: information storage and retrieval systems.
INTCO: international code of signals.
INTELSAT: International Telecommunications Satellite Consortium.
INTERCON: International Convention (IEEE).
INTERGALVA: International Galvanizing Conference.
INTERMAG: International Magnetics Conference (IEEE).
INTERMAMA: International Congress for Measurement and Automation.
INTERMAG: International Magnetics Conference.
INTIPS: integrated information processing system.
INTOP: international operations simulation.
INTPHTR: interphase transformer.
INTRAN: input translator.
INTREX: information transfer experiments.
IO: input/output (devices).
IO: interpretive operation.
IOB: input-output buffer.
IOC: initial operational capability.
IOC: input-output controller.
IOC: Intergovernmental Oceanographic Commission (UNESCO).
IOCS: input-output control system.
IOCU: International Organization of Consumers Unions.
IOF: International Oceanographic Foundation.
IOL: instantaneous overload.
IOM: input-output multiplexer.
ION: Ionosphere and Aural Phenomena Advisory Committee (ESRO).
IOP: input-output processor.
IOPS: input-output programming system.
IOR: input-output register.
IOS: input-output selector.
IOS: International Organization for Standardization.
IOTA: information overload testing apparatus.
IOUBC: Institute of Oceanography, University of British Columbia.
IOVST: International Organization for Vacuum Science and Technology.
IP: information processing.
IP: initial phase.
IP: initial point.

IP: Institute of Petroleum.
IP: intermediate pressure.
IP: item processing.
IPA: Information Processing Association (Israel).
IPA: integrated photodetection assemblies.
IPA: intermediate power amplifier.
IPA: International Psychoanalytical Association.
IPC: information processing code.
IPC: Information Processing Center.
IPC: Institute of Printed Circuits.
IPCEA: Insulated Power Cable Engineers Association.
IPE: information processing equipment.
IPFM: integral pulse frequency modulation.
IPL: information processing language.
IPL: initial program load.
IPM: impulses per minute.
IPM: Institute for Practical Mathematics (Germany).
IPM: interference prediction model.
IPM: internal polarization modulation.
IPM: interruptions per minute.
IPOEE: Institution of Post Office Electrical Engineers (GB).
IPP: imaging photopolarimeter.
IPPJ: Institute of Plasma Physics (Japan).
IPPS: Institute of Physics and the Physical Society.
IPS: inch per second.
IPS: International Pipe Standard.
IPS: interruptions per second.
IPSOC: Information Processing Society of Canada.
IPSSB: Information Processing Systems Standards Board.
IPST: International Practical Scale of Temperature.
IPT: internal pipe thread.
IQC: International Quality Center.
IQEC: International Quantum Electronics Conference (IEEE).
IR: information retrieval.
IR: infrared.
IR: infrared radiation.
IR: inside radius.
IR: instruction register.
IR: instrument reading.
IR: insulation resistance.
IR: interrogator-responder.
IRAC: Interdepartment Radio Advisory Committee.
IR&D: Independent Research and Development.
IRAN: inspection and repair as necessary.
IRBM: intermediate range ballistic missile.
IRBO: infrared homing bomb.
IRC: Industrial Reorganization Corporation.
IRCM: infrared countermeasures.
IRDP: information retrieval data bank.
IRE: Institute of Radio Engineers (now IEEE).
IRED: infrared-emitting diode.
IREX: International Research and Exchanges Board.
IRHD: international rubber hardness degrees.
IRI: Industrial Research Institute.
IRI: Institution of the Rubber Industry.
IRIS: instant response information system.
IRG: interrange instrumentation group.
IRIS: infrared interferometer spectrometer.

IRIS: integrated reconnaissance intelligence system.
IRL: information retrieval language.
IRLS: interrogation, recording of location subsystem.
IRM: intermediate range monitor.
IRMA: information revision and manuscript assembly.
IROD: instantaneous readout detector.
IRPA: International Radiation Protection Association.
IRRD: International Road Research Documentation (OECD).
IRRI: International Rice Research Institute.
IRS: isotope removal service.
IRS: information retrieval system.
IRSIA: Institute for the Encouragement of Scientific Research in Industry (Belgium).
IRT: Institute of Reprographic Technology.
IRT: interrogator—responder—transponder.
IRTE: Institute of Road Transport Engineers.
IRU: inertial reference unit.
I&S: Board of Inspection and Survey.
IS: incomplete sequence (relay).
IS: information science.
IS: interference suppressor.
IS: Iowa State University of Science and Technology.
ISA: Instrument Society of America.
ISA: International Society of Acupuncture.
IS&D: integrate sample and dump.
ISALIS: Indian Society for Automation and Information Sciences.
ISAM: indexed sequential access method.
ISAM: integrated switching and multiplexing.
ISAP: information sort and predict.
ISAR: information storage and retrieval.
ISB: independent sideband.
ISB: International Society of Biometeorology.
ISBB: International Society of Bioclimatology and Biometeorology.
ISBN: International Standard Book Number.
ISC: Information Society of Canada.
ISC: Instruction Staticizing Control.
ISCAN: inertialess steerable communication antenna.
ISCAS: International Symposium on Circuits and Systems (IEEE).
IScT: Institute of Science Technology.
ISD: induction system deposit.
ISD: International Subscriber Dialing.
ISDS: integrated ship design system.
I&SE: installation and service engineering.
ISE: Institution of Structural Engineers.
ISFET: ion-sensitive field-effect transistor.
ISI: Indian Standards Institution.
ISI: Institute for Scientific Information.
ISI: International Statistical Institute.
ISI: Iron and Steel Institute.
ISI: Israel Standards Institute.
ISIS: International Satellites for Ionospheric Studies.
ISIS: International Science Information Service.
ISJM: Israeli Journal of Mathematics.
ISK: insert storage key.
I²L: integrated injection logic.
ISL: information search language.
ISL: information system language.

ISL: instructional systems language.
ISL: interactive simulation language.
ISLIC: Israel Society for Special Libraries and Information Centers.
ISM: industrial, scientific, medical (apparatus).
ISM: information systems for management.
ISSMB: Information Systems Standards Management Board.
ISN: International Society for Neurochemistry.
ISO: individual system operation.
ISO: International Standards Organization.
ISP: International Society of Photogrammetry.
ISP: Italian Society of Physics.
ISPEC: insulation specification.
ISPEMA: Industrial Safety (Personal Equipment) Manufacturers Association.
ISPO: International Statistical Programs Office (Department of Commerce).
ISR: information storage and retrieval.
ISR: intersecting storage ring.
ISR: interrupt service routine.
ISS: ideal solidus structures.
ISS: information storage system.
ISS: input subsystem.
ISSCC: International Solid-State Circuits Conference.
ISSN: International Standard Serial Number.
IST: information science and technology.
IST: integrated system test.
ISSS: International Society of Soil Science.
IST: integrated system transformer.
IST: International Standard Thread (metric).
ISTAR: image, storage translation and reproduction.
ISTIM: interchange of scientific and technical information in machine language.
ISTS: International Symposium on Space Technology and Science.
ISU: interface switching unit.
ISU: International Scientific Union.
IT: industry (FCC services) telephone maintenance.
IT: information theory.
IT: Information Theory (IEEE Group).
IT: input translator.
IT: Institute of Technology.
IT: insulating transformer.
IT: interrogator—transponder.
IT: item transfer.
ITA: Independent Television Authority.
ITA: Institute for Telecom and Aeronomy (ESSA).
ITA: International Tape Association.
ITA: International Typographic Association.
ITAE: integrated time and absolute error.
ITB: intermediate block (check).
ITC: International Trade Commission.
ITC: ionic thermoconductivity.
ITCC: International Technical Communications Conference.

ITE: Institute of Telecommunications Engineers.
ITE: Institute of Traffic Engineers.
ITEM: Interference Technology Engineer's Master.
ITEWS: integrated tactical electronic warfare system.
ITF: interactive terminal facility.
ITFS: instructional television fixed services.
ITIRC: IBM Technical Information Retrieval Center.
ITL: integrated-transfer-launch.
ITL: inverse time limit.
ITPS: integrated teleprocessing system.
ITR: inverse time relay.
ITS: insulation test specification.
ITS: integrated tracking system (ARTRAC); obsolete.
ITSA: Institute for Telecommunication Sciences and Aeronomy (ESSA); formerly CPRL of NBS.
ITU: International Telecommunications Union (UNO).
ITV: industrial television.
ITV: instructional television.
IU: interference unit.
IUAPPA: International Union of Air Pollution Prevention Associations.
IUB: International Union of Biochemistry.
IUBS: International Union of Biological Sciences.
IUCAF: Inter-Union Commission on Allocation of Frequencies.
IUCN: International Union for the Conservation of Nature and Natural Resources.
IUCr: International Union of Crystallography.
IUE: International Union of Electrical, Radio, and Machine Workers.
IUGG: International Union of Geodesy and Geophysics.
IUGS: International Union of Geological Sciences.
IUPAC: International Union of Pure and Applied Chemistry.
IUPAP: International Union of Pure and Applied Physics.
IUTAM: International Union of Theoretical and Applied Mechanics.
IUVSTA: International Union for Vacuum Science Techniques and Applications.
IV: intermediate voltage.
IVDS: independent variable depth sonar.
IVMU: inertial velocity measurement unit.
IVR: integrated voltage regulator.
IW: industry (FCC services) power.
IWAHMA: Industrial Warm Air Heater Manufacturers Association.
IWCS: integrated wideband communications system.
IWCS/SEA: integrated wideband communications system/Southeast Asia.
IWS: Institute of Work Study.
IWSA: International Water Supply Association.
IWSc: Institute of Wood Science.
IWSP: Institute of Work Study Practitioners.
IX: industry (FCC services) manufacturers.

J

JACC: Joint Automatic Control Conference.
JACM: Journal of the Association for Computing Machinery.
JAEC: Japan Atomic Energy Commission.
JAERI: Japan Atomic Energy Research Institute.
jaff: electronic and chaff jamming.
JAIF: Japan Atomic Industrial Forum.
JAIEG: Joint Atomic Information Exchange Group.
JAPT: Journal of Approximation Theory.
JASIS: Journal of the American Society for Information Science.
JBMMA: Japanese Business Machine Makers Association.
JCA: Joint Commission on Accreditation of Universities.
JCAE: Joint Committee on Atomic Energy.
JCAM: Joint Commission on Atomic Masses.
JCAR: Joint Commission on Applied Radioactivity.
JCB: Joint Communications Board.
JCC: Joint Communications Center.
JCC: Joint Computer Conference.
JCEC: Joint Communications Electronics Committee.
JCI: Joint Communications Instruction.
JCL: job control language.
JCSS: Journal of Computer and System Sciences.
JDM: Journal of Data Management.
JDS: job data sheet.
JECC: Japan Electronic Computer Center.
JEDEC: Joint Electron Device Engineering Council.
JEIDA: Japan Electronic Industry Development Association.
JEIPAC: JICST Electronic Information Processing Automatic Computer.
JEMC: Joint Engineering Management Conference.
JERC: Joint Electronic Research Committee.
JETEC: Joint Electron Tube Engineering Council.
JFET: junction field-effect transistor.
JFL: Joint Frequency List.
JFP: Joint Frequency Panel.
JHU: Johns Hopkins University.
JIC: Joint Industrial Council.

JICST: Japan Information Center of Science and Technology.
JIE: Junior Institution of Engineers.
JIFTS: Joint In-Flight Transmission System.
JILA: Joint Institute for Laboratory Astrophysics (University of Colorado).
JIMS: Journal of the Indian Mathematical Society.
JIP: joint input processing.
JIS: Japanese Industrial Standards.
JISC: Japanese Industrial Standards Committee.
JLMS: Journal of the London Mathematical Society.
JMA: Japan Meteorological Agency.
JMA: Japan Microphotography Association.
JMC: Joint Meteorological Committee.
JMED: jungle message encoder-decoder.
JMKU: Journal of Mathematics of Kyoto University.
JMSJ: Journal of the Mathematical Society of Japan.
JOLA: Journal of Library Automation.
JPB: Joint Planning Board.
JPDR: Japan Power Demonstration Reactor.
JPL: Jet Propulsion Laboratory (NASA).
JPRS: Joint Publications Research Service.
JQE: Journal of Quantum Electronics (IEEE).
JRATA: Joint Research and Test Activity.
JRDOD: Joint Research and Development Objectives Document.
JRIA: Japan Radioisotope Association.
JS: jam strobe.
JSC: Japan Science Council.
JSC: Joint Security Control.
JSSC: Journal of Solid-State Circuits (IEEE).
JESM: Japan Society of Electrical Discharge Machining.
JSIA: joint service induction area.
JSME: Japan Society of Mechanical Engineers.
JTAC: Joint Technical Advisory Committee.
JTB: Joint Transportation Board.
JTF: Joint Task Force.
JTIDS: joint tactical information distribution system.
JUNE: joint utility notification for excavators.
JUSE: Union of Japanese Scientists and Engineers.

K

kA: kiloampere.
KAEDS: Keystone Association for Educational Data Systems.
KAFB: Kirkland Air Force Base (Albuquerque, NM).
KALDAS: Kidsgrove Algol digital analog simulation.
kc: kilocycle per second; obsolete, use kilohertz (kHz).
KDR: keyboard data recorder.
KE: kinetic energy.
KEAS: knot equivalent air speed.
keV: kiloelectronvolt.

kg: kilogram (10^3 grams; 1 kg = 2.205 pounds).
kg/m^3: kilogram per cubic meter.
kgm: kilogram-meter.
kgps: kilogram per second.
kHz: kilohertz (10^3 hertz).
KIFIS: Kollsman integrated flight instrumentation system.
KIPO: keyboard input printout.
KIS: keyboard input simulation.
KK: 1 000 000.

KL: key length.
km: kilometer (10^3 meters; 1 km = 0.621 miles).
kM: kilomega (obsolete).
kMc: kilomegacycle (obsolete); use gigahertz (GHz).
KMER: Kodak metal etch resist.
kohm: kilohm; (also shown as kΩ).
KORSTIC: Korea Scientific and Technological Information Center.
KP: key punch.
KS: Kansas.
KSC: John F. Kennedy Space Center (NASA); (now Cape Canaveral).
KSR: keyboard send receive (set).

KTS: key telephone system.
KTSA: Kahn test of symbol arrangement.
kV: kilovolt (10^3 volts).
kVA: kilovoltampere.
kVAh: kilovolt-ampere-hour.
kvar: kilovar.
kVAhm: kilovolt-ampere-hour meter.
kVdc: kilovolt direct current.
kW: kilowatt.
kWh: kilowatt-hour.
kWhm: kilowatt-hour meter.
KY: Kentucky.
kΩ: kilohm.

L

LA: lead angle.
LA: lightning arrester.
LA: Louisiana.
LAAV: light airborne ASW vehicle.
LAB: low-altitude bombing.
LABS: low-altitude bombing system.
LACBWR: La Crosse boiling water reactor.
LACES: Los Angeles Council of Engineering Societies.
LAD: location aid device.
LAD: logical aptitude device.
LADAPT: lookup dictionary adaptor program.
ladar: laser detection and ranging.
LADSIRLAC: Liverpool and District Scientific, Industrial and Research Library Advisory Council.
LAE: left arithmetic element.
LAEDP: large area electronic display panel.
LAG: load and go assembler.
LAH: logical analyzer of hypothesis.
LAM: loop adder and modifier.
LANNET: large artificial nerve net.
LANNET: large artificial neuron network.
LAP: lesson assembly program.
LAP: list assembly program.
LAPDOG: low-altitude pursuit dive on ground.
LAPES: low-altitude parachute extraction system.
LAR: low angle reentry.
LARC: Livermore Atomic Research Computer (UNIVAC).
LARIAT: laser radar intelligence acquisition technology (469L spacetrack).
LARS: laser angular rate sensor (Defender).
LAS: large astronomical satellite.
LASA: large-aperture seismic arrays.
LASCR: light-activated silicon-controlled rectifier.
LASCS: light-activated silicon-controlled switch.
laser: light amplification by stimulated emission of radiation.
LASL: Los Alamos Scientific Laboratory.
LASRM: low-altitude supersonic research missile.
LASSO: laser search and secure observer.
LATCC: London air-traffic control centre.

LATINCON: Latin American Convention (Region 9, IEEE).
LATS: long-acting thyroid stimulator.
LAVA: linear amplifier for various applications.
LB: light bombardment.
lb: pound (1 lb = 16 ounces = 453.59 grams).
LBP: length between perpendiculars.
LBT: low bit test.
LC: inductance-capacitance.
LC: lead covered.
LC: Library of Congress.
LC: line carrying.
LC: line connector.
LC: line of communication.
LC: line of contact.
LC: liquid crystals.
LC: load carrier.
LC: load center.
LC: load compensating (relay).
LC: low carbon.
LCAO: linear combination of atomic orbitals.
LCC: landing craft, control.
LCC: launch control center.
LCC: liquid crystal cell.
LCD: liquid crystal display.
LCE: launch complex equipment.
LCF: local cycle fatigue.
LCIE: *Laboratoire Central des Industries Electriques* (France).
LCL: lifting condensation level.
LCL: linkage control language.
LCLV: liquid-crystal light valve.
LCM: large core memory.
LCM: least common multiple.
LCMM: life cycle management model.
LCN: load classification number.
LCNT: link celestial navigation trainer.
LCP: language conversion program.
LCR: inductance-capacitance-resistance.
LCS: large core store.
LCS: loudness-contour selector.

LCVD: least voltage coincidence detection.
LD: lift-drag ratio.
LD: line drawing.
LD: line of departure.
LD: long distance.
LDA: locate drum address.
LDB: light distribution box.
LDC: latitude data computer.
LDC: light direction center.
LDC: linear detonating cord.
LDC: line drop compensator.
LDDS: low density data system.
LDE: laminar defect examination.
LDE: linear differential equation.
LDP: language data processing.
LDR: light dependent resistor.
LDR: low data rate.
LDRI: low data rate input.
LDRS: LEM data reduction system.
LDS: large disk storage.
LDT: logic design translator.
LDT: long distance transmission.
LDX: long distance xerography.
LE: light equipment.
LE: low explosive.
LE: antenna effect length for electric-field antennas.
LE: leading edge.
LEADS: law enforcement automated data system.
LEAP: language for the expression of associative procedures.
LEAR: logistics evaluation and review technique.
LEAS: lease electronic accounting system.
LEC: liquid encapsulated Czochralski.
LED: light-emitting diode.
LEDT: limited-entry decision table.
LEED: laser-energized explosive device.
LEED: low energy electron diffraction.
LEF: light-emitting film.
LEL: lower explosion limit.
LEM: antenna effective length for magnetic-field antennas.
LEM: lunar excursion module.
LES: launch escape system.
LET: linear energy transfer.
LF: leapfrog configuration (circuit theory).
LF: load factor (tabular).
LF: low frequency (30 to 300 kHz).
LFC: laminar flow control.
LFE: Laboratory for Electronics.
LFIM: Low-Frequency Instruments and Measurement (IEEE IM Technical Committee).
LFO: low-frequency oscillator.
LFRD: lot fraction reliability definition.
LG: landing gear.
LG: landing ground.
LGG: light-gun pulse generator.
LGO: Lamont Geological Observatory (Columbia University).
LGT: low group transmit.
LH: left-hand.
LH: liquid hydrogen.
LH: luteinizing hormone.

LHC: left-hand circular (polarization).
LHR: lower hybrid resonance.
LIC: linear integrated circuit.
LID: leadless inverted device.
LID: locked-in device.
LIDAR: laser intensity direction and ranging.
LIDAR: light detection and ranging.
LIFO: last in, first out.
LIFT: logically integrated Fortran translator.
LIL: Lunar International Laboratory.
LIMAC: large integrated monolithic array computer.
LINAC: linear accelerator.
LINC: Laboratory Instruments Computer (DEC).
LINCOMPEX: linked compressor and expander.
LINLOG: linear-logarithmic.
LIPL: linear information processing language.
LIR: line integral refractometer.
LISA: library systems analysis.
LISN: line impedance stabilization network.
LIST: library and information services, tees-side.
LIT: liquid injection technique.
LITE: legal information through electronics.
LITR: low-intensity test reactor.
LITVC: liquid injection thrust vector control.
LIZARDS: library information search and retrieval data system.
LKY: Lefschetz - Kalman - Yakubovich (lemma).
L/L: latitude/longitude.
LL: light line.
LL: low level.
LLL: long-path laser.
LLL: low-level logic.
LLLLLL (or L6): laboratories low-level linked list language. (BTL).
LLLTV: low light-level television.
LLRES: load-limiting resistor.
LLRV: lunar landing research vehicle.
L/M: lines per minute.
LM: lunar module.
LMF: linear matched filter.
LMFBR: liquid metal fast breeder reactor.
LMS: London Mathematical Society.
LMSS: lunar mapping and survey system.
ln: natural logarithm.
LNG: liquefied natural gas.
LO: local oscillator.
LO: lock-on.
LO: longitudinal optical.
LO: lunar orbiter.
LOA: length over all.
LOAMP: logarithmic amplifier.
LOB: line of balance.
LOC: Launch Operations Center.
LOCATE: Library of Congress Automation Techniques Exchange.
loc cit.: *loco citato* (in the place cited).
LOCI: logarithmic computing instrument.
LOCS: librascope operations control system.
LOCS: logic and control simulator.
LODESMP: logistics data element standardization and management process.
LOERO: large orbiting earth resources observatory.

LOF: lowest operating frequency.
LOFAR: low-frequency acquisition and ranging.
LOFT: loss of fluid test.
LOGALGOL: logical algorithmic language.
LOGEL: logic generating language.
LOGIPAC: logical processor and computer.
LOGIT: logical inference tester.
LOGLAN: logical language.
LOGRAM: logical program.
LOGTAB: logic tables.
LOH: light observation helicopter.
LOHAP: light observation helicopter avionics package.
LOL: length of lead (actual).
LOLA: lunar orbit landing approach.
LOLITA: language for the on-line investigation and transformation of abstractions.
LOMUSS: Lockheed multiprocessor simulation system.
LOP: line of position.
LOP: local operational plot.
LOPAD: logarithmic outline processing system for analog data.
LOPAR: low power acquisition radar.
LOR: lunar orbital rendezvous.
LORAN: long range navigation.
LORL: large orbital research laboratory.
LORPGAC: long range proving ground automatic computer.
LOS: line of sight.
LOS: loss of signal.
LOSS: landing observer signal system.
LOSS: large object salvage system.
LOTIS: logic, timing, sequencing (language).
LOX: liquid oxygen.
LP: linear programming.
LP: line printer.
LP: low pass.
LP: low point.
LP: low pressure.
LPAC: Launching Programs Advisory Committee.
LPC: loop-control (relay).
LPD: language processing and debugging.
LPE: liquid phase epitaxial.
LPG: *langage de programmation et de gestion* (C11 French computer language).
LPG: liquefied petroleum gas.
LPGTC: Liquefied Petroleum Gas Industry Technical Committee.
LPL: linear programming language.
LPL: list processing language.
LPM: lines per minute.
LPO: low power output.
LPRINT: lookup dictionary print program.
LPTV: large payload test vehicle.
lpW: lumen per watt.
L/R: locus of radius.
LR: limited recoverable.
LR: load ratio.
LR: load-resistor (relay).
LRC: Langley Research Center (NASA).
LRC: load ratio control.
LRC: Lewis Research Center (NASA, Cleveland).

LRC: longitudinal redundancy check.
LRI: long-range radar input.
LRL: Lawrence Radiation Laboratory (University of California).
LRLTRAN: Lawrence Radiation Laboratory translator.
LRM: limited register machine.
LRP: long range path.
LRPL: Liquid Rocket Propulsion Laboratory (Army).
LRS: linguistic research system.
LRSM: long range seismograph measurements.
LRSS: long range survey system.
LRTF: long range technical forecast.
LRTM: long range training mission.
LRU: line replaceable unit.
LRV: lunar roving vehicle.
LS: language specification.
LS: least significant.
LSA: limited space (charge) accumulation (microwave diodes).
LSB: least significant bit.
LSB: lower sideband.
LSBR: large seed blanket reactor.
LSC: least significant character.
LSCC: line-sequential color composite.
LSD: least significant digit.
LSD: low speed data.
LSECS: life support and environmental control system.
LSFFAR: low-spin folding fin aircraft rocket.
LSHI: large-scale hybrid integration.
LSI: large scale integration.
LSN: line stabilization network.
LSP: low speed printer.
LSPC: Lewis Space Flight Center.
LSR: load shifting resistor.
LSSM: local scientific survey module.
LSU: Louisiana State University.
LT: language translation.
LT: line telecommunications.
LT: link trainer instructor.
LT: low tension.
LTC: long time constant.
LTDS: laser target designation system.
LTP: library technology project.
LTPD: lot tolerance percent defective.
LTS: long-term stability.
LTTL: low-power transistor-transistor logic.
LTV: long tube vertical.
LUCID: language for utility checkout and instrumentation development.
LUF: lowest usable frequency.
LUHF: lowest usable high frequency.
LUME: light utilization more efficient (Shed Light).
LUT: launcher umbilical tower.
LV: low voltage.
LVCD: least voltage coincidence detection.
LVD: low velocity detonation.
LVDA: launch vehicle data adapter.
LVDC: launch vehicle digital computer.
LVDT: linear variable differential transformer.

LVDT: linear velocity displacement transformer.
LVHV: low volume high velocity.
LVP: low voltage protection.
LVRE: low-voltage release effect.
LWBR: light-water breeder reactor.

LWR: light water reactor.
lx: lux (1 lux = 1 lumen/m^2).
LYRIC: language for your remote instruction by computer.

M

MA: Maritime Administration.
MA: Massachusetts.
MA: Mathematical Association (GB).
MA: megampere.
mA: milliampere.
MA: modify address.
MAA: Mathematical Association of America.
MAARC: magnetic annular arc.
MAC: machine-aided cognition.
MAC: maintenance allocation chart.
MAC: maximum allowable concentration.
MAC: mean aerodynamic chord.
MAC: Mineralogical Association of Canada.
MAC: multiple access computer.
MACC: modular alter and compose console.
MACE: management applications in a computer environment.
MACMIS: maintenance and construction management information system.
MACON: matrix connector punched card programmer.
MACSMB: Measurement and Automatic Control Standards Management Board.
MAD: magnetic airborne detector.
MAD: magnetic anomaly detection.
MAD: maintenance, assembly, and dissassembly shop.
MAD: Michigan algorithm decoder.
MAD: multiaperture device.
MAD: multiple access device.
MAD: multiply and add.
MADA: multiple-access discrete address.
MADAEC: Military Application Division of the Atomic Energy Commission.
MADAM: multipurpose automatic data analysis machine.
MADAR: malfunction analysis detection and recording.
MADDAM: macromodule and digital differential analyzer machine.
MADE: microalloy diffused electrode.
MADE: minimum airborne digital equipment.
MADE: multichannel analog-to-digital data encoder.
MADM: Manchester automatic digital machine.
MADP: main air display plot.
MADRE: magnetic drum receiving equipment.
MADREC: malfunction and detection recording.
MADS: machine-aided drafting system.
MADS: missile attitude determination system.
MADT: microalloy diffused-base transistor.
MAECON: Mid-America Electronics Conference.

MAESTRO: machine-assisted educational system for teaching by remote operation.
MAF: major academic field.
MAG: Magnetics (IEEE Society).
MAGAMP: magnetic amplifier.
MAGIC: machine for automatic graphics interface to a computer.
MAGIC: matrix algebra general interpretive coding.
MAGIC: Michigan automatic general integrated computation.
MAGIC: Midac automatic general integrated computation.
MAGLOC: magnetic logic computer.
MAGMOD: magnetic modulator.
MAGPIE: machine automatically generating production inventory evaluation.
MAIDS: multipurpose automatic inspection and diagnostic system.
MAIP: matrix algebra interpretive program.
MAIR: molecular airborne intercept radar.
MAL: macro assembly language.
MAM: multiple access to memory.
MALE: multiaperture logic element.
MALT: mnemonic assembly language translator.
MAMI: machine-aided manufacturing information.
MAMIE: magnetic amplification of microwave integrated emissions.
MAMOS: marine automatic meteorological observing station.
MANIAC: mathematical analyzer numerical integrator and computer.
MANIAC: mechanical and numerical integrator and computer.
MANOVA: multivariate analysis of variance.
MAP: macro arithmetic processor.
MAP: macro assembly program.
MAP: manifold absolute pressure.
MAP: mathematical analysis without programming.
MAP: message acceptable pulse.
MAP: multiple allocation procedure.
MAPCHE: mobile automatic programmed checkout equipment.
MAPID: machine aided program for preparation of instruction data.
MAPORD: Methodology Approach to Planning and Programming Air Force Operational Requirements, Research and Development.
MAPS: multicolor automatic projection system.

MAPS: multivariate analysis and prediction of schedules.

MAR: malfunction array radar.

MAR: memory address register.

MARC: machine-readable cataloging.

MARCAS: maneuvering reentry control and ablation studies.

MARCEP: maintainability and reliability cost-effectiveness program.

MARCIA: mathematical analysis of requirements for career information appraisal.

MARCOM: microwave airborne communications relay.

MARDAN: marine differential analyzer (Polaris/Poseidon).

MARGEN: management report generator.

MARLIS: multiaspect relevance linkage information system.

MARS: machine retrieval system.

MARS: Management Analysis Reporting Service.

MARS: Marconi automatic relay system.

MARS: Martin automatic reporting system.

MARS: Military Affiliated Radio System.

MARS: military amateur radio system.

MARS: multiaperture reluctance switch.

MART: maintenance analysis review technique.

MART: mean active repair time.

MAS: Management Advisory Services.

MAS: Military Agency for Standardization.

MAS: multiaspect signaling.

MASCOT: Motorola automatic sequential computer operated tester.

MASER: microwave amplification by stimulated emission of radiation.

MASIS: management and scientific information system.

MASK: maneuvering and sea-keeping.

MASRT: marine air support radar teams.

MASS: monitor and assembly system.

MASS: Michigan automatic scanning system.

MASS: multiple access sequential selection.

MASSTER: mobile Army sensor system test evaluation and review.

MASTER: matching available student time to educational resources.

MASTER: multiple access shared time executive routine.

MAT: mechanical aptitude test.

MAT: microalloy transistor.

MATA: Michigan Aviation Trades Association.

MATCON: microwave aerospace terminal control.

MATD: mine and torpedo detector.

MATE: multiple-access time-division experiment.

MATE: multisystem automatic test equipment.

MATIC: multiple area technical information center.

MATICO: machine applications to technical information center operations.

MATLAN: matrix language.

MATPS: machine-aided technical processing system.

MAU: multiattribute utility.

MAUDE: morse automatic decoder.

MAVIN: machine-assisted vendor information network.

MATT: missile ASW torpedo target.

MAW: marine aircraft wing.

MB: main battery.

MB: memory buffer.

MB: missile bomber.

MBA: Marine Biological Association.

MBA: Master of Business Administration.

MBD: magnetic-bubble device.

MBL: Marine Biological Laboratory.

MBPS: megabits per second.

MBR: memory buffer register.

MBRV: maneuverable ballistic reentry vehicle.

MBT: metal-base transistor.

MBWO: microwave backward-wave oscillator.

M&C: monitor and control.

MC: magnetic core.

MC: manual control.

MC: master control.

MC: Maritime Commission.

MC: Medical Corps.

mc: megacycle (obsolete); use megahertz (MHz).

MC: military computer.

MC: molded components.

MC: multichip.

MC: multiple contact.

MCA: Manufacturing Chemists Association.

MCA: Material Coordinating Agency.

MCA: Model Cities Administration.

MCAA: Mechanical Contractors Association of America.

MCBF: mean cycles between failures.

MCC: main communications center.

MCC: management control center.

MCC: multiple computer complex.

MCC-H: Mission Control Center-Houston.

MCCU: multiple communication control unit.

MCD: months for cyclical dominance.

MCDP: microprogrammed communication data processor.

MCE: Master of Civil Engineering.

MCEB: Military Communications Electronics Board.

MCF: monolithic crystal filter.

MCG: man-computer graphics.

MCID: multipurpose concealed intrusion detector.

MCIS: maintenance control information system.

MCM: magnetic core memory.

MCM: Monte Carlo method.

MCMJ: Michigan Mathematical Journal.

MCOM: mathematics of computation.

MCP: master control program.

MCP: memory centered processor.

MCP: multichannel communications program.

MCR: magnetic character reader.

MCR: magnetic character recognition.

MCR: master change record.

MCR: master control routine.

MCR: Military Compact Reactor.

MCROA: Marine Corps Reserve Officers Association.

MCS: Marine Corps School.

MCS: master control system.

MCS: medical computer services.

MCS: method of constant stimuli.
MCS: Mobile Communications Systems (IEEE VT Technical Committee).
MCS: modular computer systems.
MCS: multiprogrammed computer system.
MCT: mechanical comprehension test.
MCTI: Metal Cutting Tool Institute.
MCTR: message center.
MCU: microprogram control unit.
MCUG: Military Computers Users Group.
MCW: modulated continuous wave.
MD: magnetic disk.
MD: magnetic drum.
MD: main drum.
MD: manual data.
MD: Maryland.
MD: medical department.
MD: message data.
MD: mine disposal.
MD: monitor displays.
MD: movement directive.
MDA: multidimensional analysis.
MDA: multidocking adapter.
MDAA: Mutual Defense Assistance Act.
MDAC: multiplying digital-to-analog converter.
MDAP: Mutual Defense Assistance Program.
MDC: maintenance data collection.
MDC: maintenance dependency chart.
MDC: Missile Development Center.
MDCS: maintenance data collection system.
MDE: magnetic decision element.
MDF: main distributing frame.
MDF: mild detonating fuse.
MDI: magnetic direction indicator.
MDI: miss distance indicator.
MDIF: manual data input function.
MDL: macro description language.
MDL: Mine Defense Laboratory.
MDM: maximum design meter.
MDM: metal-dielectric-metal (filter).
MD/NC: mechanical drafting/numerical control.
MDR: memory data register.
MDS: memory disk system.
MDS: maintenance data system.
MDS: malfunction detection system.
MDS: microprocessor development system.
MDS: minimum discernible (or detectable) signal.
MDS: modern data systems.
MDSS: meteorological data sounding system.
MDT: mean down time.
MDTA: Man Power Development and Training Act.
MDTS: megabit digital-to-troposcatter subsystem.
MDU: mine disposal unit.
ME: Maine.
ME: Master of Engineering.
ME: mechanical efficiency.
ME: mechanical engineering.
MEA: minimum en-route altitude.
MEAL: master equipment allowance list.
MEAL: master equipment authorization list.
MEAR: maintenance engineering analysis record.
MEAR: maintenance engineering analysis report.

MEC: Metrology Engineering Center, NOSC (Pomona, Calif.).
MECCA: mechanized catalog.
MECL: Motorola emitter-coupled logic.
MECOMSAG: mobility equipment command scientific advisory group.
MED: microelectronic device.
MEDAC: Medical Equipment Display and Conference.
MEDAL: Micromechanized Engineering Data for Automated Logistics.
MEDDA: mechanized defense decision anticipation.
MEDIA: Magnavox electronic data image apparatus.
MEDLARS: medical literature analysis and retrieval system.
MEDS: medical evaluation data system.
MEDSERV: Medical Service Corps.
MEDSMB: Medical Standards Management Board.
MEDSPECC: Medical Specialist Corps.
MEECN: minimum essential emergency communications network.
MEG: message expediting group.
MEI: manual of engineering instructions.
MEIU: mobile explosives investigation unit.
MEL: many-element laser.
MEL: Marine Engineering Laboratory (U.S. Navy).
MEL: Materials Evaluation Laboratory.
MELBA: multipurpose extended lift blanket assembly.
MELEC: microelectronics.
MELEM: microelement.
MEM: Mars excursion module.
MEM: minimum essential medium.
MEMA: Motor and Equipment Manufacturers Association.
M. Eng.: Master of Engineering.
MEO: major engine overhaul.
MEP: mean effective pressure.
MERA: molecular electronics for radar applications.
MERDC: Mobility Equipment Research and Development Center.
MERDL: Medical Equipment Research and Development Laboratory.
MERMUT: mobile electronic robot manipulator and underwater television.
MERS: mobility environmental research study.
MES: manual entry subsystem.
MES: Michigan Engineering Society.
MESA: manned environmental system assessment.
MESA: Mining Enforcement and Safety Administration.
MESG: maximum experimental safe gap.
MESS: monitor event simulation system.
MESUCORA: measurement, control regulation, and automation.
MET: Management Engineering Team.
MET: modified expansion tube.
META: Maintenance Engineering Training Agency.
META: methods of extracting text automatically.
METAPLAN: methods-of-extracting-text-auto-programming language.
METRIC: multiechelon technique for recoverable item control.
MeV: million electronvolts.

MEW: microwave early warning.

MEXE: Military Engineering Experimental Establishment.

MF: medium frequency (300 to 3 MHz).

MF: microfiche.

MF: microfilm.

MFB: mixed functional block.

MFC: microfunctional circuit.

MFCM: multifunction card machine.

MFD: magnetofluid dynamics.

MFI: mobile fuel irradiator.

MFN: most favored nation.

MFS: manned flying system.

MFT: multiprogramming fixed tasks.

MFTG: Manufacturing Technology (IEEE Group).

MFTRS: magnetic flight test recording system.

MG: motor generator.

MG: multigage.

MGC: manual gain control.

MGC: missile guidance computer.

MGL: matrix generator language.

MH: magnetic head.

mH: millihenry.

MHA: Marine Historical Association.

MHA: modified handling authorized.

MHD: magnetohydrodynamics.

MHE: materials handling equipment.

MHEA: Mechanical Handling Engineers Association.

MHEDA: Material Handling Equipment Distributors Association.

MHF: medium high frequency.

MHRST: Medical and Health Related Sciences Thesaurus.

MHT: mild heat treatment.

MI: manual input.

MI: Michigan.

mi: mile.

MIA: metal interface amplifier.

MIAC: minimum automatic computer.

MIB: manual input buffer.

MIB: multilayer interconnection board.

MIC: Michigan instructional computer.

MIC: micrometer.

MIC: microphone.

MIC: microwave integrated circuit.

MIC: minimum ignition current.

MIC: monolithic integrated circuit.

MIC: mutual interference chart.

MICA: macro instruction compiler assembler.

MICAM: microammeter.

MICELEM: microphone element.

MICOM: Missile Command, Army (Huntsville, Ala.).

MICR: magnetic ink character recognition.

MICRO: multiple indexing and console retrieval options.

MICROMIN: microminiature.

MICROPAC: micromodule data processor and computer.

MICS: Management Information and Control System.

MICS: mineral insulated copper sheathed.

MIDAC: Michigan digital automatic computer.

MIDAS: measurement information data analytic system.

MIDAS: missile intercept data acquisition system.

MIDAS: modified integration digital analog simulator.

MIDAS: modulator isolation diagnostic analysis system.

MIDOT: multiple interference determination of trajectory.

MIDS: movement information distribution station.

MIFR: master international frequency register.

MIG: magnetic injection gun.

MIG: metal inert gas.

MIKE: measurement of instantaneous kinetic energy.

MIL: military specification (followed by a single capital letter and numbers).

MILA: Merritt Island launch area.

MIL-I: military specification on interference.

MILADGRU: Military Advisory Group.

MILECON: Military Eletronics Conference.

MILHDBK: military handbook.

MIL-STD: military standard (book).

MIM: metal insulator metal.

MIM: modified index method.

MIMD: multiple instruction multiple data.

MIMO: man in, machine out.

MIMS: multi-item multisource.

MINAC: miniature navigation airborne computer.

MINDAC: marine inertial navigation data assimilation computer.

MINEAC: miniature electronic autocollimator.

MINIRAR: minimum radiation requirements (to accomplish assigned missions).

MINITRACK: satellite tracking network (NASA).

MINPRT: miniature processing time.

MINS: miniature inertial navigation system.

MINSOP: minimum slack time per operation.

MINT: materials identification and new item control technique.

MINTECH: Ministry of Technology.

MIOP: multiplexing input/output processors.

MIP: manual input processing.

MIP: manual input program.

MIP: matrix inversion program.

MIP: missile impact predictor.

MIPE: modular information processing equipment.

MIPIR: missile precision instrumentation radar.

MIPS: million instructions per second.

MIPS: missile impact predictor set.

MIR: memory information register.

MIR: memory input register.

MIRACL: management information report access without computer languages.

MIRACODE: microfilm retrieval access code.

MIRAGE: microelectronic indicator for radar ground equipment.

MIRD: medical internal radiation dose.

MIRF: multiple instantaneous response file.

MIRFAC: mathematics in recognizable form automatically compiled.

MIROS: modulation inducing retrodirective optical system.

MIRR: material inspection and receiving report.
MIRS: man power information retrieval system.
MIRT: molecular infrared track.
MIRV: multiple independent reentry vehicle.
MIS: management information system.
MIS: metal insulator semiconductor.
MISD: multiple instruction single data (stream or pipeline processors).
MISDAS: mechanical impact system design for advanced spacecraft.
MISFET: metal insulator semiconductor field-effect transistor.
MISHAP: missiles high-speed assembly program.
MISP: medical information system programs.
MISS: mechanical interruption statistical summary.
MISS: mobile integrated support system.
MISS: multi-item single source.
MISSIL: management information system symbolic interpretive language.
MISTRAM: missile trajectory measurement system.
MIT: Massachusetts Institute of Technology.
MIT: master instruction tape.
MITE: missile integration terminal equipment.
MITOL: machine-independent telemetry-oriented language.
MITR: Massachusetts Institute of Technology reactor.
MKS: meter-kilogram-second.
ML: machine language.
M-L: metallic-longitudinal.
ML: main lobe.
ML: Materials Laboratory.
ML: methods of limits.
ml: milliliter.
MLCAEC: Military Liaison Committee to the Atomic Energy Commission.
MLD: minimum lethal dose.
MLE: maximum likelihood estimate.
MLG: main landing gear.
MLP: machine language program.
MLPCB: machine language printed circuit boards.
MLPWB: multilayer printed-wiring board.
MLR: main line of resistance.
MLRG: Marine Life Research Group.
MLS: machine literature searching.
MLS: Master of Library Science.
MLS: microwave landing system.
MLS: missile location system.
MLS: multilanguage system.
MLSNPG: Microwave Landing System National Planning Group.
MLT: mean logistical time.
MLT: mean length of turn.
MM: maintenance manual.
MM: materials measurement.
mm: millimeter.
MMAU: master multiattribute utility.
mm-band: 40 to 300 GHz frequency band.
MMC: maximum metal condition.
MMD: Manual of the Medical Department.
MME: Master of Mechanical Engineering.
MMF: magnetomotive force.

MMM: Maintenance and Material Management.
MMOD: micromodule.
MMP: multiplex message processor.
MMPR: missile manufacturers planning reports.
MMPT: man-machine partnership translation.
MMRBM: mobile medium range ballistic missile.
MMS: Man-Machine Systems Group.
MMS: mass memory store.
MMS: missile monitoring system.
MMSE: minimum mean squared error.
MMU: modular maneuvering unit.
MN: Minnesota.
MNOS: metal nitride oxide semiconductor.
MNS: metal nitride semiconductor.
MO: manual output.
MO: master oscillator.
MO: Missouri.
MOA: matrix output amplifier.
MOBIDAC: mobile data acquisition system.
MOBIDIC: mobile digital computer.
MOBL: macro oriented business language.
MOBULA: model building language.
MOC: master operational controller.
MOC: memory operating characteristic.
MOC: mission operation computer.
MODA: motion detector and alarm.
MODAC: mountain systems digital automatic computer.
MODEM: modulator-demodulator.
MODI: modified distribution method.
MODICON: modulator dispersed control.
MODS: major operation data system.
MOERO: medium orbiting earth resources observatory.
MOF: maximum operating frequency.
MOGA: microwave and optical generation and amplification.
MOL: machine-oriented language.
MOL: manned orbiting laboratory.
MOLAB: mobile lunar laboratory.
MOLDS: multiple on-line debugging system.
MOOSE: manned orbital operations safety equipment.
MOP: multiple on-line programming.
MOPA: master oscillator power amplifier.
MOPR: manner of performing rating.
MOPTS: mobile photographic tracking station.
MOS: management operating system.
MOS: metal-oxide semiconductor.
MOSAIC: macro operation symbolic assembler and information compiler.
MOSAIC: Ministry of Supply automatic integrator and computer.
MOSAIC: metal oxide semiconductor advanced integrated circuit.
MOSFET: metallic oxide semiconductor field-effect transistor.
MOSM: metal-oxide semimetal.
MOST: metal oxide semiconductor transistor.
MOTARDES: moving target detection system.
MOTNE: Meteorological Operational Telecommunications Network of Europe.
MOTU: mobile optical tracking unit.

MP: main phase.
MP: minimum phase.
MP: multipole.
MPA: multiple-period average.
MPBE: molten plutonium burn-up experiment.
MPCC: multi-protocol communications controller.
MPD: magnetoplasmadynamic.
MPE: mathematical and physical sciences and engineering.
MPE: maximum permissible exposure.
MPEP: manual of patent examining procedure.
MPG: miles per gallon.
MPG: miniature precision gyrocompass.
MPH: miles per hour.
MPI: magnetic particle inspection.
MPI: mean point of impact.
MPL: maximum permissible level.
MPPL: multipurpose processing language.
MPRE: medium power reactor experiment.
MPS: multiprogramming system.
MPT: male pipe thread.
MPTA: Mechanical Power Transmission Association.
MPU: microprocessing unit.
MPW: modified plane wave.
MR: machine records.
MR: memory register.
mR: milliroentgen.
MR: moisture resistant.
MRA: minimum reception altitude.
MRAD: mass random access disc.
MRADS: mass random access data storage.
MRB: magnetic recording borescope.
MRBM: medium range ballistic missile.
MRC: maintenance requirement card.
MRE: multiple-response enable.
MRE: Microbiological Research Establishment.
MRI: Machine Records Installation.
MRI: Medical Research Institute (USN).
MRI: monopulse resolution improvement.
MRIR: medium resolution infrared radiometer.
MRL: Medical Research Laboratory.
MRL: multiple rocket launcher.
MRM: metabolic rate monitor.
mR/min: milliroentgens per minute.
MRN: meteorological rocket network.
MRN: minimum reject number.
MRO: midrange objectives.
MRR: Mechanical Reliability Report. (FAA).
MRT: mean repair time.
MRT: modified rhyme test.
MRU: machine records unit.
MRU: material recovery unit.
MRU: mobile radio unit.
MRWC: multiple read-write compute.
M&S: maintenance and supply.
M-S: Mitte-Seite (stereo).
MS: machine selection.
MS: macromodular system.
MS: magnetic storage.
MS: mass spectrometry.
MS: Master of Sciences.
MS: material specification.

MS: medical survey.
MS: Medicine and Surgery (Navy).
MS: memory system.
MS: Metallurgical Society.
MS: military standard (sheet).
MS: military standard, prefix to numbered series issued by DOD.
ms: millisecond (10^{-3} seconds).
MS: Mississippi.
MSB: most significant bit.
MSC: macro selection compiler.
MSC: Manned Spacecraft Flight Center (NASA).
MSC: monolithic crystal filter.
MSC: most significant character.
MSCE: main storage control element.
MSD: most significant digit.
MSDT: maintenance strategy diagramming technique.
MSEMPR: Missile Support Equipment Manufacturers Planning Reports.
MSFC: Marshall Space Flight Center (NASA).
MSFN: manned space flight network.
MSI: medium scale integration.
MSIO: mass storage input-output.
MSMB: Mechanical Standards Management Board.
MSOS: mass storage operating system.
MSR: mass storage resident.
MSR: missile site radar.
MSRE: molten salt reactor experiment.
MSRT: missile system readiness test.
MSS: management science systems.
MSS: mass storage system.
MSSR: Mars soil sample return.
MST: monolithic systems technology.
MSTS: multisubscriber time sharing systems.
MT: machine translation.
MT: magnetic tape.
MT: Montana.
MT: maximum torque.
MTA: multiterminal adapter.
MTAC: mathematical tables and other aids to computation.
MTBF: mean time between failures.
MTBM: mean time between maintenance.
MTC: maintenance time constraint.
MTC: master tape control.
MTC: memory test computer.
MTC: Missile Test Center.
MTC: mission and traffic control.
MTCF: mean time to catastrophic failure.
MTCU: magnetic tape control unit.
MTD: minimal toxic dose.
MTDS: manufacturing test data system.
MTDS: marine tactical data system.
MTF: mechanical time fuse.
MTF: Mississippi Test Facility (NASA).
MTF: modulation transfer function.
MTI: Machine Tools Industry (IEEE IAS Technical Committee).
MTI: moving target indicator.
MTIC: moving target indicator coherent.
MTL: merged-transistor logic.
MTM: methods time measurement.

MTNS: metal thick oxide semiconductor.
MTOP: molecular total overlap population.
MTOS: metal thick oxide silicon.
MTP: mechanical thermal pulse.
MTPF: minimal total processing time.
MTPS: magnetic tape programming system.
MTR: magnetic tape recorder.
MTR: materials testing reactor.
MTR: materials testing report.
MTR: missile tracking radar.
MTRE: magnetic tape recorder end.
MTRS: magnetic tape recorder start.
MTS: magnetic tape station.
MTS: magnetic tape system.
MTS: Marine Technology Society.
MTS: missile tracking system.
MTSE: magnetic trap stability experiment.
MT/SC: magnetic tape selective composer.
MTSMB: Material and Testing Standards Management Board.
MTST: magnetic tape selective typewriter.
MTT: magnetic tape terminal.
MTT: magnetic tape transport.
MTT: Microwave Theory and Techniques (IEEE Society).
MTTD: mean time to diagnosis.
MTTF: mean time to failure.
MTTFF: mean time to first failure.
MTTR: mean time to repair.
MTTR: mean time to restore.
MTV: marginal terrain vehicle.
MTU: magnetic tape unit.
MTU: multiplexer and terminal unit.
MU: multiple unit.
μA: microampere.

MUF: maximum usable frequency.
μH: microhenry.
MULTEWS: multiple electronic warfare surveillance.
MULTICS: multiplexed information and computing service.
μmho: micromho ($\mu\Omega$).
MUMS: mobile utility module system.
μohm: microohm ($\mu\Omega$).
μs: microsecond.
MUSA: multiple-unit steerable antenna.
μV: microvolt.
μW: microwatt.
MUX: multiplexer.
MV: mean variation.
MV: megavolt.
mV: millivolt (10^{-3} volts).
MV: multivibrator.
MVA: megavoltampere.
MVB: multivibrator.
MVC: manual volume control.
MVC: multiple variate counter.
MVS: minimum visible signal.
MVS: multiple virtual storage.
MW: megawatt (10^{6} watts).
MW: microwave.
mW: milliwatt (10^{-3} watts).
mW/cm²: milliwatt per square centimeter.
MWd/t: megawatt days per ton.
MW(E): megawatts (electrical).
MW(H): megawatts (heat).
MWI: message waiting indicator.
MWh: megawatt-hour.
MWP: maximum working pressure.
MWR: mean width ratio.
MWV: maximum working voltage.

N

nA: nanoampere.
NAA: National Association of Accountants.
NAATS: National Association of Air Traffic Specialists.
NAB: National Alliance of Businessmen.
NAB: National Association of Broadcasters.
NAB: navigational aid to bombing.
NABE: National Association for Business Education.
NABER: National Association of Business and Educational Radio.
NACA: National Advisory Commission for Aeronautics (now NASA).
NACAA: National Association of Computer Assisted Analysis.
NACATS: North American clear air turbulence tracking system.
NACC: National Automatic Controls Conference.
NACE: National Association of Corrosion Engineers.
NACEIC: National Advisory Council on Education for Industry and Commerce.

NACOA: National Advisory Committee on Oceans and Atmosphere.
NACS: Northern Area Communications System.
NAD: no-acid descaling.
NADWARN: natural disaster warning system.
NAE: National Academy of Engineering.
NAEC: National Aerospace Education Council.
NAECON: National Aerospace Electronics Conference (IEEE).
NAEDS: National Association of Educational Data Systems.
NAET: National Association of Educational Technicians.
NAFEC: National Administrative Facilities Experimental Center.
NAIOP: navigational aid inoperative for parts.
NAK: negative acknowledgment character.
NAM: National Association of Manufacturers.
NAMFI: NATO Missile Firing Installation.
NAMI: Naval Aerospace Medical Institute.

NAMTC: Naval Air Missile Test Center.

NAPCA: National Air Pollution Control Administration.

NAPE: National Association of Power Engineers.

NAPHCC: National Association of Plumbing-Heating-Cooling Contractors.

NAPL: National Association of Photolithographers.

NAPS: Nimbus automatic programming system.

NAPSS: numerical analysis problem solving system.

NAR: net assimilation rate.

NARBA: North American Regional Broadcasting Agreement.

NARDIS: Navy Automated Research and Development Information System.

NAREC: Naval Research Electronic Computer.

NARF: Nuclear Aerospace Research Facility.

NARM: National Association of Relay Manufacturers.

NARS: National Archives and Records Service.

NARTS: National Association of Radio Telephone Systems.

NARTS: Naval Aeronautics Test Station.

NARTS: Naval Air Rocket Test Station.

NARUC: National Association of Regulatory Utility Commissioners.

NAS: National Academy of Sciences.

NAS: National Aircraft Standards.

NAS/NRC: National Academy of Sciences/National Research Council.

NASA: National Aeronautics and Space Administration.

NASARR: North American Search and Range Radar.

NASCAS: NAS/NRC Committee on Atmospheric Sciences.

NASCO: National Academy of Sciences' Committee on Oceanography.

NASCOM: NASA communications (satellite tracking network).

NASDAQ: National Association of Security Dealers' Automated Quotations.

NASIS: National Association for State Information Systems.

NASL: Naval Applied Sciences Laboratory.

NASPA: National Society of Public Accountants.

NASW: National Association of Science Writers.

NAT: normal allowed time.

NATA: National Association of Testing Authorities (Australia).

NATCS: National Air Traffic Control Service.

NATTS: National Association of Trade and Technical Schools.

NAVA: National Audio Visual Association.

NAVAIDS: navigational aids.

NAVAIR: prefix to numbered series issued by Naval Air Systems Command.

NAVAPI: North American voltage and phase indicator.

NAVAR: radar air navigation and control system.

NAVARHO: navigation aid, rho radio navigation system.

NAVASCOPE: airborne radarscope used in navar.

NAVASCREEN: ground screen used in navar.

NAVCM: navigation countermeasures and deception.

NAVCOM: naval communications.

NAVCOMMSTA: naval communications station.

NAVDAC: Navigation Data Assimilation Center.

NAVDOCKS: prefix to numbered series issued by Navy Yards and Docks Bureau.

NAVELEX: Naval Electronics System Command.

NAVFAC: Naval Facilities Engineering Command.

NAVMAT: prefix to numbered series issued by Office of Naval Materiel.

NAVMC: prefix to numbered series issued by the Marine Corps.

NAVMED: prefix to numbered series issued by Naval Aerospace Medical Institute.

NAVORD: prefix to numbered series issued by NOSC.

NAVPERS: prefix to numbered series issued by Bureau of Naval Personnel.

NAVSEC: Naval Ship Engineering Center.

NAVSO: prefix to numbered series issued by Navy Industrial Relations Office.

NAVTRADEVCEN: prefix to numbered series issued by Naval Training Device Center.

NAVWEPS: prefix to numbered series issued by Bureau of Naval Weapons.

NB: narrow band.

NB: Nebraska.

NB: no bias (relay).

NBC: narrow-band conducted.

NBCV: narrow-band coherent video.

NBFM: narrow-band frequency modulation.

NBO: network buildout.

NBR: narrow-band radiated.

NBS: National Bureau of Standards.

NBSD: night bombardment short distance.

NBSFS: NBS frequency standard.

NBSR: NBS Reactor.

NBT: null-balance transmissometer.

NBTDR: narrow-band time-domain reflectometry.

N/C: numerical control.

NC: national coarse (thread).

NC: no coil.

NC: no connection.

NC: noise criteria (value).

NC: normally closed.

NC: North Carolina.

NC: numerical control.

NCA: Naval Communications Annex.

NCA: Northwest Computing Association.

NCAR: National Center for Atmospheric Research.

NCASI: National Council for Air and Stream Improvement.

NCB: National Coal Board.

NCB: Naval Communications Board.

NCC: National Computing Center.

NCC: National Computer Conference.

NCCAT: National Committee for Clear Air Turbulence.

NCEA: North Central Electric Association.

NCET: National Council for Educational Technology.

NCF: nominal characteristics file (see ECAC).

NCFMF: National Committee for Fluid Mechanics Films.

NCGG: National Committee for Geodesy and Geophysics (Pakistan).

NCHEML: National Chemical Laboratory (MIN-TECH).

NCHS: National Center for Health Statistics.

NCHVRFE: National College for Heating, Ventilating, Refrigeration, and Fan Engineering.

NCI: National Computer Institute.

NCI: National Computing Industries.

NCI: Netherlands Centre for Informatics.

NCI: Northeast Computer Institute.

NCL: National Central Library.

NCL: National Chemical Laboratory.

NCOR: National Committee for Oceanographic Research (Pakistan).

NCPTWA: National Clearinghouse for Periodical Title Word Abbreviations.

NCQR: National Council for Quality and Reliability.

NCR: no carbon required.

NCRP: National Committee on Radiation Protection.

NCS: National Communications System.

NCS: National Computer Systems.

NCS: net control station.

NCS: Netherlands Computer Society.

NCS: Numerical Control Society.

NCSC: National Council of Schoolhouse Construction.

NCSE: North Carolina Society of Engineers.

NCSL: National Conference on Standards Laboratories.

NCTA: National Cable Television Association.

NCTM: National Council of Teachers of Mathematics.

NCTSI: National Council of Technical Service Industries.

NCUR: National Committee for Utilities Radio.

ND: North Dakota.

NDB: nondirectional beacon.

NDI: numerical designation index.

NDL: network definition language.

NDP: normal diametral pitch.

NDRO: nondestructive readout.

NDT: nil-ductility transition.

NDT: nondestructive testing.

NDTC: Nondestructive Testing Center.

NEA: National Education Association.

NEAC: Nippon electric automatic computer.

NEADAI: National Education Association Department of Audiovisual Instruction.

NEAT: NCR electronic autocoding technique.

NEBSS: National Examinations Board in Supervisory Studies.

NEC: National Electronics Conference.

NECIES: North East Coast Institution of Engineers and Shipbuilders.

NECPUC: New England Conference of Public Utility Commissioners.

NECS: National Electrical Code Standards.

NECS: nationwide educational computer service.

NECTA: National Electrical Contractors Trade Association.

NEEDS: New England educational data systems.

NEEP: Nuclear Electronic Effect Program.

NEF: national extra fine (thread).

NEHA: National Environmental Health Association.

NEIS: National Engineering Information System.

NEL: National Engineering Laboratory.

NEL: Naval Electronics Laboratory.

NELAT: Naval Electronics Laboratory assembly tester.

NELC: Naval Electronics Laboratory Center.

NELCON NZ: National Electronics Conference, New Zealand (IEEE).

NELEX: Naval Electronics Systems Command Headquarters.

NELIAC: Navy Electronics Laboratory international algebraic compiler.

NELPA: Northwest Electric Light & Power Association.

NEMA: National Electrical Manufacturers Association.

NEMI: National Elevator Manufacturing Industry.

NEMP: nuclear electromagnetic pulse.

NEN: prefix to standards issued by NNI.

NEPTUNE: North-Eastern electronic peak tracing unit and numerical evaluator.

NERC: National Electric Reliability Council.

NERC: Natural Environment Research Council.

NEREM: Northeast Electronics Research and Engineering Meeting.

NERVA: Nuclear Engine for Rocket Vehicle Application.

NES: noise equivalent signal.

NESC: National Electrical Safety Code.

NESC: National Environmental Satellite Center (ESSA).

NEST: Naval Experimental Satellite Terminal.

NESTEF: Naval Electronics System Test and Evaluation.

NESTOR: neutron source thermal reactor.

NET: National Educational Television.

NETSET: network synthesis and evaluation technique.

NEWRADS: Nuclear Explosion Warning and Radiological Data System.

NEWS: naval electronic warfare simulator.

nF: nanofarad.

NF: national fine (thread).

NF: noise figure.

NF: noise frequency.

NF: nose fuse.

NF: prefix to standards issued by AFNOR.

NFAC: Naval Facilities Engineering Command Headquarters.

NFC: not favorably considered.

NFEA: National Federated Electrical Association.

NFETM: National Federation of Engineers' Tools Manufacturers.

NFPA: National Fire Protection Agency.

NFPA: National Fluid Power Association.

NFPA: National Forest Products Association.

NFPEDA: National Farm and Power Equipment Dealers Association.

NFR: no further requirement.

NFSAIS: National Federation of Science Abstracting and Indexing Services.

NFSR: National Foundation for Scientific Research (Belgium).

NG: nitroglycerine.

NGCC: National Guard Computer Center.

NGM: neutron gamma Monte Carlo.
NGRS: Narrow Gauge Railway Society.
nH: nanohenry.
NH: New Hampshire.
NH: nonhygroscopic.
NI: numerical index.
NIAM: Netherlands Institute for Audiovisual Media.
NIAE: National Institute of Agricultural Engineering.
NIB: noninterference basis.
NIC: National Indicational Center.
NIC: National Inspection Council.
NIC: negative impedance converter.
NICAP: National Investigations Committee on Aerial Phenomena.
NICB: National Industrial Conference Board.
NICE: National Institute of Ceramic Engineers.
NICEIC: National Inspection Council for Electrical Installation Contracting.
NID: Nuclear Instruments and Detectors (IEEE NPSS Technical Committee).
NIDA: numerically integrated differential analyzer.
NIER: National Industrial Equipment Reserve.
NIESR: National Institute for Economic and Social Research.
NIF: noise improvement factor.
NIFES: National Industrial Fuel Efficiency Service.
NIFTE: neon indicating functional test equipment.
NIH: National Institutes of Health.
NIH: not invented here.
NIMBUS: Meteorological Observation Satellite.
NINA: National Institute Northern Accelerator.
NIO: National Institute of Oceanography.
NIPCC: National Industrial Pollution Control Council Staff.
NIPO: negative input, positive output.
NIRNS: National Institute for Research in Nuclear Science.
NIS: not in stock.
NISARC: National Information Storage and Retrieval Centers.
NISC: National Industrial Space Committee.
NJ: New Jersey.
NJAC: National Joint Advisory Council.
NJCC: National Joint Computer Committee (now AFIPS).
NJPMB: Navy Jet Propelled Missile Board.
NLG: noise landing gear.
NLI: nonlinear interpolating.
NLL: National Lending Library for Science and Technology.
NLM: National Library of Medicine.
NLO: nonlinear optics.
NM: New Mexico.
NM: noise meter.
NM: nuclear magnetron.
NMA: National Management Association.
NMA: National Microfilm Association.
NMAA: National Machine Accountants Association (now DPMA).
NMAP: National Metric Advisory Panel.
NMC: National Meteorological Center.
NMC: Naval Missile Center.

NMCL: Naval Missile Center Laboratory.
NMCS: Nuclear Materials Control System.
NMEL: Navy Marine Engineering Laboratory.
NMI: NASA Management Instruction.
nmi: nautical mile (1 nmi = 1.151 miles).
NMOS: N-type MOS.
NMR: National Missile Range.
NMR: normal-mode rejection.
NMR: nuclear magnetic resonance.
NMS: Naval Meteorological Service.
NMSS: National Multipurpose Space Station.
NO: normally open.
NOA: National Oceanography Association.
NOAA: National Oceanographic and Atmospheric Administration.
NODAC: Naval Ordnance Data Automation Center.
NODC: National Oceanographic Data Center.
NOF: NCR optical font.
NOMA: National Office Management Association.
NOMAD: Naval Oceanographic Meteorological Automatic Device.
NOMSS: National Operational Meteorological Satellite System.
NONCOHO: noncoherent oscillator.
NORAD: North American Air Defense System.
NORM: not operationally ready maintenance.
NORS: not operationally ready supply.
NOSMO: Norden optics setting, mechanized operation.
NOSS: nimbus operational satellite system.
NOTU: naval operational training unit.
NP: national pipe (thread).
NP: naval publication (numbered series issued by MOD).
NP: neuropsychiatric.
NPA: normal pressure angle.
NPA: numerical production analysis.
NPC: NASA Publication Control.
NPD: National Power Demonstration.
NPDS: nuclear particle detection system.
NPEC: Nuclear Power Engineering Committee (IEEE PES Technical Committee).
NPFO: Nuclear Power Field Office.
NPL: National Physical Laboratory (GB).
NPL: National Physics Laboratory.
NPLG: Navy Program Language Group.
NPM: connector type.
NPN: negative-positive-negative (transistor).
NPR: noise power ratio.
NPRCG: Nuclear Public Relations Contact Group.
NPS: Nuclear and Plasma Sciences (IEEE Society).
NPT: national taper pipe (thread).
NPT: network planning technique.
NPV: nitrogen pressure valve.
NQR: nuclear quadruple resonance.
NR: nonreactive (relay).
NR: nonrecoverable.
NR: nuclear reactor.
NRA: naval radio activity.
NRA: network resolution area.
NRAO: National Radio Astronomy Observatory.
NRC: National Research Council.
NRC: Nuclear Regulatory Commission.

NRCST: National Referral Center for Science and Technology.
NRDC: National Research Development Council.
NRDL: Naval Radiological Defense Laboratory.
NRDS: Nuclear Rocket Development Station.
NRECA: National Rural Electric Cooperative Association.
NRL: National Reference Library for Science and Invention.
NRL: Naval Research Laboratory.
NRM: natural remanent magnetization.
NRMA: National Retail Merchants Association.
NRMEC: North American Rockwell Microelectronics Company.
NRP: normal rated power.
NRRS: Naval Radio Research Station.
NRS: Naval Radio Station.
NRTS: National Reactor Testing Station.
NRTSC: naval reconnaissance and technical support center.
NRZ: nonreturn to zero.
NRZ1: nonreturn to zero change on one.
NRZC: nonreturn to zero change.
NRZI: nonreturn to zero inverted.
NRZL: nonreturn to zero level.
NRZM: nonreturn to zero mark.
NS: national standard.
NSA: National Security Agency.
NSA: Netherlands Society for Automation.
NSB: National Science Board.
NSC: National Safety Council.
NSC: National Security Council.
NSEC: Naval Ships Engineering Center.
ns: nanosecond.
NSEF: Navy Security Engineers Facility.
NSEIP: Norwegian Society for Electronic Information Processing.
NSF: National Science Foundation.
NSI: nonstandard item.
NSIA: National Security Industrial Association.
NSIF: Near Space Instrumentation Facility.
NSL: National Science Library.
NSL: National Science Laboratories.
NSL: Northrop Space Laboratories.
NSM: network status monitor.
NSMB: Nuclear Standards Management Board.
NSP: NASA Support Plan.
NSP: Network Support Plan.
NSPA: National Society of Public Accountants.

NSPE: National Society of Professional Engineers.
NSPI: National Society for Programmed Instruction.
NSPP: Nuclear Safety Pilot Plant.
NSQCRE: National Symposium on Quality Control and Reliability in Electronics.
NSRB: National Security Resources Board.
NSRDS: National Standard Reference Data System.
NSRS: Naval Supply Radio Station.
NSS: Navy Secondary Standards.
NSSL: National Severe Storms Laboratory.
NSSS: National Space Surveillance System.
NST: Network Support Team.
NSTA: National Science Teachers Association.
NSTIC: Naval Scientific and Technical Information Center.
NSTP: Nuffield Science Teaching Project.
NTC: National Telemetering Conference.
NTC: negative temperature coefficient.
NTDS: Navy Tactical Data System.
NTE: Navy Teletypewriter Exchange.
NTG: *Nachrichtentechnische Gesellschaft* (German Communications Society).
NTI: noise transmission impairment.
NTIS: National Technical Information Service.
NTP: normal temperature and pressure.
NTS: negative torque signal.
NTS: Nevada Test Site (NASA).
NTSC: National Television System Committee.
NUDETS: nuclear detection system.
NULACE: nuclear liquid air cycle engine.
NUPAD: nuclear powered active detection.
NUS: Nuclear Utility Service.
NUSL: Naval Underwater Sound Laboratory.
nV: nanovolt.
NV: Nevada.
NVR: no voltage release.
NVSD: Night Vision System Development.
nW: nanowatt.
NWAC: National Weather Analysis Center.
NWG: National Wire Gage.
NWRC: National Weather Records Center.
NWRF: Naval Weather Research Facility.
NWSC: National Weather Satellite Center.
NWSSG: Nuclear Weapons System Satellite Group.
NXDO: Nike X Development Office.
NY: New York.
NYAP: New York assembly program.
NYSE: New York Stock Exchange.

O

OA: oil-immersed self-cooled (transformer).
OA: operations analysis.
OAATM: Office of the Assistant for Automation.
OAM: Office of Aerospace Medicine.

OAME: orbital attitude and maneuvering electronics.
OAMP: optical analog matrix processing.
OAMS: orbital attitude and maneuvering system.
OAO: Orbiting Astronomical Observatory.

OAR: Office of Aerospace Research.
OARAC: Office of Air Research Automatic Computer.
OARC: Ordinary Administrative Radio Conference.
OART: Office of Advanced Research and Technology.
OASF: Orbiting Astronomical Support Facility.
OASIS: Ocean All Source Information System.
OASM: Office of Aerospace Medicine.
OASV: Orbital Assembly Support Vehicle.
O–B: octal to binary.
OB: output buffer.
OBA: oxygen breathing apparatus.
OBGS: Orbital Bombardment Guidance System.
OBI: omnibearing indicator.
OBIFCO: on-board in-flight checkout.
OBS: omnibearing selector.
OC: operating characteristic.
OC: operational computer.
OC: outside circumference.
OC: overcurrent.
OCA: operational control authority.
OCAL: on-line cryptanalytic aid language.
OCAS: on-line cryptanalytic aid system.
OCB: oil circuit breaker.
OCC: operational computer complex.
OCDU: optics coupling display unit.
OCL: operators control language.
OCL: operational control level.
OCO: open-close-open.
OCO: operation capability objectives.
OCP: operating control procedure.
OCP: operational checkout procedure.
OCR: oil circuit recloser.
OCR: optical character reader.
OCR: optical character recognition.
OCR: overcurrent relay.
OCR: overhaul component requirement.
OCRUA: Optical Character Recognition Users Association.
OD: operations directive.
OD: optical density.
OD: output data.
OD: outside diameter.
ODA: operational data analysis.
ODA: operational design and analysis.
ODLRO: off-diagonal long-range order.
ODM: orbital determination module.
ODN: own doppler nullifier.
ODOP: offset doppler.
ODP: original document processing.
ODT: octal debugging technique.
ODT: outside diameter tube.
ODU: output display unit.
oe: oersted (ampere per meter).
OE: Office of Education.
OE: open end.
OEM: original equipment manufacturer.
O/F: orbital flight.
OF: outside face.
OFA: oil-immersed forced-air-cooled (transformer).
OFB: Operational Facilities Branch (NASA).
OFC: operational flight control.
OFO: Office of Flight Operations (NASA).

OFP: operating force plan.
OFR: on-frequency repeater.
OFR: overfrequency relay.
OFSD: operating flight strength diagram.
OG: outer gimbal.
OGA: outer gimbal axis.
OGE: operational ground equipment.
OGMC: Ordnance Guided Missile Center.
OGO: Orbiting Geophysical Laboratory.
OGR: outgoing repeater.
OGT: outgoing trunk.
OGU: outgoing unit.
O–H: octal to hexadecimal.
OH: Ohio.
OH: operational hardware.
OHA: outside helix angle.
OHC: Occupational Health Center.
OHD: over-the-horizon detection.
OHF: occupational health facility.
ohm-cm: ohm-centimeter (also shown as Ω-cm).
OHP: oxygen at high pressure.
OI: oil insulated.
OIB: Operations Integration Branch (NASA).
OIC: Operations Instrumentation Coordinator.
OIC: optical integrated circuit.
OIFC: oil-insulated fan-cooled.
OIG: optically isolated gate.
OII: Office of Invention and Innovation (NBS).
OIRT: International Radio and Television Organization.
OIS: Operational Intercommunication System.
OISA: Office of International Science Activities.
OISC: oil-insulated self-cooled.
OIWC: oil-immersed water-cooled.
OK: Oklahoma.
OL: operating location.
OL: overload.
OL: open loop.
OLC: on-line computer.
OLERT: on-line executive for real time.
OLM: on-line monitor.
OLO: orbital launch operations.
OLPARS: on-line pattern analysis and recognition system.
OLPS: on-line programming system.
OLRT: on-line real-time.
OLSC: on-line scientific computer.
OLSS: on-line software system.
OLVP: Office of Launch Vehicle Programs.
OM: optical master.
OMAT: Office of Manpower Automation and Training.
OMB: outer marker beacon.
OME: Office of Management Engineer.
OMETA: Ordnance Management Engineering Training Agency.
OMIBAC: ordinal memory inspecting binary automatic computer.
OML: Ordnance Missile Laboratories.
OMPR: optical mark page reader.
OMPRA: one-man propulsion research apparatus.
OMPT: observed man point trajectory.
OMR: optical mark reader.

OMSF: Office of Manned Space Flight (NASA).
OMTS: Organizational Maintenance Test Station.
OMU: optical measuring unit.
ONR: Office of Naval Research.
OOK: on-off keying.
OOL: operator oriented language.
OOPS: off-line operating simulator.
OP: operating procedure.
OP: operational priority.
OP: output.
OPA: optoelectronic pulse amplifier.
op. cit.: *opus citatum* (the work quoted from).
OPCOM: operations communications.
OPCON: optimizing control.
OPADEC: optical particle decoy.
OPDAR: optical detection and ranging.
OPE: operations project engineer.
OPEP: orbital plane experimental package.
OP&I: Office of Patents and Inventions.
OPLE: omega position location experiment.
OPM: operations per minute.
OPP: octal print punch.
OPP: oriented polypropylene.
OPPCE: opposite commutator end.
OPPOSIT: Optimization of a Production Process by an Ordered Simulation and Iteration Technique.
OPR: optical page reading.
OPS: on-line process synthesizer.
OPS: operational paging system.
OPSCAN: Optical Scanning Users Group.
OPSCON: operations control (room).
OPSF: orbital propellant storage facility.
OPTA: optimal performance theoretically attainable.
OPW: operating weight.
OPW: orthogonalized plane wave.
OR: operational readiness.
OR: operations requirements.
OR: operations research.
OR: Oregon.
OR: overhaul and repair.
OR: overload relay.
OR: output register.
OR: outside radius.
ORACLE: Oak Ridge automatic computer and logical engine.
ORAN: orbital analysis.
ORATE: ordered random access talking equipment.
ORBIS: orbiting radio beacon ionospheric satellite.
ORBIT: on-line retrieval of bibliographic data.
ORBIT: Oracle binary internal translator.
ORC: Operations Research Center.
ORC: Ordnance Rocket Center.
ORD: optical rotary dispersion.
ORD: operational readiness date.
ORDENG: ordnance engineering.
ORDVAC: ordnance variable automatic computer.
ORG: Operations Research Group.
ORI: operational readiness inspection.
ORIT: operational readiness inspection team.
ORL: Orbital Research Laboratory.
ORLY: overload relay.
ORR: Oak Ridge Reactor.

ORRAS: optical research radiometrical analysis system.
ORSA: Operations Research Society of America.
ORT: operational readiness test.
ORT: overland radar technology.
ORV: orbital rescue vehicle.
OS: operating system.
OS: operational sequence.
OSA: Optical Society of America.
OSCAR: optimum systems covariance analysis results.
OSCAR: Orbiting Satellite Carrying Amateur Radio.
OSD: operational sequence diagram.
OSD: operational systems development.
OSE: operational support equipment.
OSFM: Office of Spacecraft and Flight Missions.
OSFP: Office of Space Flight Programs.
OSHA: Occupational Safety and Health Administration.
OSI: Office of Scientific Intelligence.
OSIC: optimization of subcarrier information capacity.
OSIS: Office of Science Information Service (NSF).
OSMV: one-shot multivibrator.
OSO: Orbital Solar Observatory.
OSO: Orbiting Satellite Observer.
OSO: Orbiting Scientific Observatory.
OSR: Office of Scientific Research.
OSR: output shift register.
OSRD: Office of Standard Reference Data (NBS).
OSS: Ocean Surveillance Satellite.
OSS: Office of Space Sciences.
OSS: Office of Statistical Standards.
OSS: orbital space station.
OSSA: Office of Space Science and Applications (NASA).
OSSS: orbital space station system.
OST: Office of Science and Technology.
OST: operational system test.
OSTAC: Ocean Science and Technology Advisory Committee.
OSTI: Office for Scientific and Technical Information.
OSTS: Office of State Technical Service.
OSURF: Ohio State University Research Foundation.
OSV: orbital support vehicle.
OT: overall test.
OTC: Office of Telecommunications.
OTC: operational test center.
OTCCC: open-type control circuit contacts.
OTDA: Office of Tracking and Data Acquisition.
OTE: operational test equipment.
OTHR: over-the-horizon radar.
OTLP: zero transmission level point.
OTM: Office of Telecommunications Management.
OTP: Office of Telecommunications Policy.
OTP: operational test procedure.
OTR: optical tracking.
OTRAC: oscillogram trace reader.
OTS: Office of Technical Services.
OTS: Office of Technological Service.
OTS: optical technology satellite.
OTS: optical transport systems.
OTT: one-time tape.
OTU: Office of Technology Utilization.
OTU: operational test unit.

OTU: operational training unit.
OTV: operational television.
OUTRAN: output translator.
OV: orbiting vehicle.
OVERS: orbital vehicle reentry simulator.
OVV: overvoltage.
OW: oil-immersed water-cooled (transformer).
OWF: optimum working frequency.

OWM: Office of Weights and Measures.
OWPR: ocean wave profile recorder.
OWRR: Office of Water Resources Research.
OWS: ocean weather station.
OWS: operational weather support.
OWS: orbital workshop.
oz: ounces.

P

PA: pad abort.
PA: pending availability.
PA: pilotless aircraft.
PA: point of aim.
PA: power amplifier.
PA: pressure angle.
PA: probability of acceptance.
PA: product analysis.
PA: program address.
PA: program analysis.
PA: program authorization.
PA: public address (system).
PA: pulse amplifier.
PAAC: program analysis adaptable control.
PABLA: problem analysis by logical approach.
PABX: private automatic branch exchange.
PAC: pedagogic automatic computer.
PAC: personal analog computer.
PACCT: PERT and cost correlation technique.
PACE: precision analog computing equipment.
PACE: prelaunch automatic checkout equipment.
PACE: preflight acceptance checkout equipment.
PACE: programming analysis consulting education.
PACE: projects to advance creativity in education.
PACED: program for advanced concepts in electronic design.
PACER: process assembly case evaluator routine.
PACER: program assisted console evaluation and review.
PACER: programmed automatic communications equipment.
PACE-S/C: preflight acceptance checkout equipment for spacecraft.
PACE/LV: preflight acceptance checkout equipment—launch vehicle.
PACOR: passive correlation and ranging station.
PACT: pay actual computer time.
PACT: production analysis control technique.
PACT: programmed analysis computer transfer.
PACT: programmed automatic circuit tester.
PAD: polyaperture device.
PAD: post-activation diffusion.
PAD: propellant-actuated device.
PADAR: passive detection and ranging.
PADLOC: passive active detection and location.

PADLOCK: passive detection and location of counter-measures.
PADRE: patient automatic data recording equipment.
PADRE: portable automatic data recording equipment.
PADS: passive-active data simulation.
PAEM: program analysis and evaluation model.
PAGE: PERT automated graphical extension.
PAGEOS: passive geodetic earth orbiting satellite.
PAHO: Pan American Health Organization.
PAI: programmer appraisal instrument.
PAIGH: Pan-American Institute of Geography and History.
PAIR: performance and integration retrofit.
PAL: pedagogic algorithmic language.
PAL: permanent artificial lighting.
PAL: phase alternation line.
PAL: process assembly languages.
PAL: Production and Application of Light (IEEE IAS Technical Committee).
PAL: psychoacoustic laboratory.
PAL-D: phase alternation line delay.
PAM: pole amplitude modulation.
PAM: pulse amplitude modulation.
PAMPER: practical application of mid-points for exponential regression.
PAMS: pad abort measuring system.
PAMS: Proceedings of the American Mathematical Society.
PANSDOC: Pakistan National Scientific and Technical Documentation.
PANSMET: Procedures for Air Navigation Services—Meteorology.
PAPA: programmer and probability analyzer.
PAR: performance analysis and review.
PAR: perimeter acquisition radar.
PAR: precision approach radar.
PAR: program appraisal and review.
PARADE: passive-active ranging and determination.
PARAMP: parametric amplifier.
PARASYN: parametric synthesis.
PARC: progressive aircraft reconditioning cycle.
PARD: parts application reliability data.
PARD: periodic and random deviation.
PARD: precision annotated retrieval display.
PARM: program analysis for resource management.

PAROS: passive ranging on submarines.
PARR: procurement authorization and receiving report.
PARSEV: paraglider research vehicle.
PARTNER: proof of analog results through numerically equivalent routine.
PAS: primary alert system.
PAS: Professor of Air Science.
PAS: program address storage.
PASCAL: Philips automatic sequence calculator.
PASE: power assisted storage equipment.
PASS: production automated scheduling system.
PASS: program aid software systems.
PASS: program alternative simulation system.
PAT: parametric artificial talker.
PAT: personalized array translator.
PAT: program attitude test.
PAT: production acceptance test.
PATA: pneumatic all-terrain amphibian.
PATC: professional, administrative, technical, and clerical.
PATE: programmed automatic test equipment.
PATH: performance analysis and test histories.
PATI: passive airborne time-difference intercept.
PATS: precision altimeter techniques study.
PATT: project for the analysis of technology transfer.
PATTERN: planning assistance through technical evaluation of relevance numbers (ORSA).
PAU: pilotless aircraft unit.
PAWOS: portable automatic weather observable station.
PAX: private automatic exchange.
PB: playback.
PB: plot board.
PB: plug board.
PB: push-button.
PBB: polybrominated biphenyl.
PBDG: push-button data generator.
PBHP: pounds per brake horsepower.
PBIT: parity bit.
PBPS: post boost propulsion system.
PBS: Public Broadcasting Service.
PBV: post boost vehicle.
PBW: parts by weight.
PBX: private branch exchange.
pC: picocoulomb.
pC: picocurie.
PC: pitch circle.
PC: pitch control.
PC: point of curve.
PC: printed circuit.
PC: Professional Communication (IEEE Group).
PC: program coordination.
PC: program counter.
PC: pulsating current.
PC: punched card.
PCA: polar cap absorption.
PCAC: partially conserved axial vector current.
PCAM: punched card accounting machine.
PCB: polychlorinated biphenyl.
PCB: power circuit breaker.
PCB: printed-circuit board.
PCBC: partially conserved baryon current.

PCBS: positive control bombardment system.
PCC: partial crystal control.
PCC: point of compound curve.
PCCD: peristaltic charge-coupled device.
PCCS: photographic camera control system.
PCDC: punched card data processing.
PCE: punched card equipment.
PCEA: Pacific Coast Electric Association.
PCEM: process chain evaluation model.
PCF: pounds per cubic foot (use lb/ft^3).
PCG: planning and control guide.
PCI: panel call indicator.
PCI: peripheral command indicator.
PCI: pilot controller integration.
PCI: product configuration identification.
PCM: pitch control motor.
PCM: pulse-code modulation.
PCM: pulse-count modulation.
PCM: punched card machine.
PCME: pulse-code modulation event.
PCMI: photochromic microimage.
PCOS: primary communications oriented system.
PCP: parallel cascade processor.
PCP: parallel circular plate.
PCP: photon-coupled pair.
PCP: primary control program.
PCP: process control processor.
PCP: processor control program.
PCP: program change proposal.
PCP: punched card punch.
PCP: project control plan.
PCR: photoconductive relay.
PCR: procedure change request.
PCR: program change request.
PCR: program control register.
PCR: punched card reader.
PCS: pointing control system.
PCS: Primary Coolant System.
PCS: print contrast scale.
PCS: process control system.
PCSC: power conditioning, switching and control.
PCSIR: Pakistan Council of Scientific and Industrial Research.
PCT: photon-coupled transistor.
PCT: planning and control techniques.
PCTFE: polymonochlorotrifluoroethyle.
PCTM: pulse-count modulation.
PCU: power control unit.
PCU: power conversion unit.
PCU: progress control unit.
PCV: pollution control valve.
PCW: pulsed continuous wave.
PD: peripheral device.
PD: pitch diameter.
PD: preliminary design.
PD: positive displacement.
PD: power distribution.
PD: propellant dispersion.
PD: priority directive.
PDA: post-deflection acceleration.
PDA: precision drive axis.
PDA: probability discrete automata.

PDC: power distribution control.
PDC: premission documentation change.
PDD: Physical Damage Division.
PDD: program design data.
PDF: point detonating fuse.
PDF: probability density function.
PDGDL: Plasma Dynamics and Gaseous Discharge Laboratory.
PDGS: Precision Delivery Glides System.
PDIO: photodiode.
PDIS: Proceedings of the National Symposia.
PDL: procedure definition language.
PDM: pulse delta modulation.
PDM: pulse duration modulation.
PDO: program directive—operations.
PDP: positive displacement pump.
PDP: programmed data processor.
PDP: program definition phase.
PDPS: parts data processing system.
PDQ: programmed data quantizer.
PDR: periscope depth range.
PDR: power directional relay.
PDR: precision depth recorder.
PDR: preliminary design review.
PDR: priority data reduction.
PDR: processed data recorder.
PDR: processing data rate.
PDR: program drum recording.
PDR: program discrepancy report.
PDS: power density spectra.
PDS: power distribution system.
PDS: procedures development simulator.
PDS: program data source.
PDS: propellant dispersion system.
PDSMS: point-defense surface missile system.
PDT: programmable data terminal.
PDU: pressure distribution unit.
PDV: premodulation processor deep-space voice.
PE: peripheral equipment.
PE: permanent echo.
PE: phase encoded.
PE: polyethylene.
PE: probable error.
PE: professional engineer.
PEA: Pennsylvania Electric Association.
PEACU: plastic energy absorption in compression unit.
PEC: photoelectric cell.
PECBI: Professional Engineers Conference Board for Industry.
PECS: portable environmental control system.
PED: personnel equipment data.
PEDN: planned event discrepancy notification.
PEDRO: pneumatic energy detector with remote optics.
PEEP: pilot's electronic eye-level presentation.
PEF: physical electronics facility.
PEI: preliminary engineering inspection.
PEIC: periodic error integrating controller.
PEM: photoelectromagnetic.
PEM: production engineering measure.
PENA: primary emission neuron activation.
PENCIL: pictorial encoding language.
PEOS: propulsion and electrical operating system.

PEP: peak envelope power.
PEP: planar epitaxial passivated (transistor).
PEP: planetary ephemeris program.
PEP: program evaluation procedure.
PEPAG: Physical Electronics and Physical Acoustics Group.
PEPP: planetary entry parachute program.
PEPR: precision encoder and pattern recognition.
PER: preliminary engineering report.
PERA: Production Engineering Research Association.
PERCOS: performance coding system.
PEREF: Propellant Engine Research Environmental Facility.
PERGO: project evaluation and review with graphic output.
PERM: Program Evaluation for Repetitive Manufacture.
PERT: performance evaluation review technique.
PERT: program evaluation and review technique.
PERT: program evaluation research task.
PERTCO: program evaluation review technique with cost.
PERU: production equipment records unit.
PES: photoelectric scanner.
PES: Power Engineering (IEEE Society).
PET: patterned epitaxial technology.
PET: peripheral equipment tester.
PET: polyethylene terephthalate.
PET: production environmental testing.
PETE: pneumatic end to end.
pF: picofarad.
PF: power factor.
PF: pulse frequency.
PFA: pulverized fuel ash.
PFD: primary flash distillate.
PFL: Propulsion Field Laboratory (Rocketdyne).
PFM: power factor meter.
PFM: pulse-frequency modulation.
PFR: parts failure rate.
PFRS: portable field recording system.
PFS: propellant field system.
PG: power gain.
PG: Power Generation (IEEE PES Technical Committee).
PG: pressure gage.
PGEWS: Professional Group on Engineering Writing and Speech (IEEE).
PGNCS: primary guidance and navigation control system.
PGNS: primary guidance and navigation system.
PGR: precision graphic recorder.
PGS: power generator section.
PGS: power generation system.
PGU: pressure gas umbilical.
P&HEP: Plasma and High-Energy Physics (IEEE NPSS Technical Committee).
pH: degree of acidity or alkalinity.
pH: hydrogen-ion concentration.
PH: power house.
PHENO: precise hybrid elements for nonlinear operation.
PHI: position and homing indicator.

PHIN: position and homing inertial navigator.

PHM: phase meter.

PHOENIX: plasma heating obtained by energetic neutral injection experiment.

PHP: Parts, Hybrids, and Packaging (IEEE Group).

PHP: pound per horsepower.

PHR: pound-force per hour.

PI: parallel input.

PI: performance index.

PI: pilotless interceptor.

PI: point initiating.

PI: point insulating.

PI: point of intersection.

PI: priority interrupt.

PI: productivity index.

PI: program indicator.

PI: programmed instruction.

PI: program interrupt.

PI: proportional integral.

PIA: peripheral interface adapter.

PIA: preinstallation acceptance.

PIAPACS: psychophysiological information acquisition, processing and control system.

PIB: polar ionosphere beacon.

PIB: Publishing Information Bulletin.

PIB: Pyrotechnic Installation Building.

PIBMRI: Polytechnic Institute of Brooklyn, Microwave Research Institute.

PIC: particle in cell.

PIC: photographic interpretation center.

PIC: plastic insulated cable.

PIC: polyethylene-insulated conductor.

PIC: program interrupt control.

PICA: power industry computer applications.

PICAC: Power Industry Computer Applications Conference.

PID: proportional integral and differential.

PID: proportional integral derivation.

PIE: pulse interference emitting.

PIF: payload integration facility.

PIGA: pendulous integrating gyroscopic accelerometer.

PII: positive immittance inverter.

PILOT: permutation indexed literature of technology.

PILOT: piloted low-speed test.

PIM: precision indicator of the meridian.

PIM: pulse interval modulation.

PIN: position indicator.

pin: positive-intrinsic-negative (transistor).

PINO: positive input, negative output.

PINS: portable inertial navigation system.

PINT: Purdue interpretive programming and operating system.

PIOCS: physical input-output unit.

PIOSA: Pan Indian Ocean Science Association.

PIOU: parallel input-output unit.

PIP: peripheral interchange program.

PIP: problem input preparation.

PIP: predicted impact point.

PIP: probabilistic information processing.

PIP: programmable integrated processor.

PIP: pulsed integrating pendulum.

PIP: project on information processing.

PIPA: pulse integrating pendulum accelerometer.

PIRD: program instrumentation requirements document.

PIRN: preliminary interface revision notice.

PIRT: precision infrared triangulation.

PISH: program instrumentation summary handbook.

PIT: peripheral input tape.

PIT: processing index terms.

PIT: program instruction tape.

PIU: plug-in unit.

PIV: peak inverse voltage.

PL: phase line.

PL: production language.

PL: programming language.

PL: proportional limit.

PLA: physiological learning aptitude.

PLA: programmable logic array.

PLA: proton linear accelerator.

PLAAR: packaged liquid air-augmented rocket.

PLACE: programming language for automatic checkout equipment.

PLAN: program language analyzer.

PLANIT: programming language for interactive teaching.

PLANS: program logistics and network scheduling system.

PLANT: program for linguistic analysis of natural plants.

PLAT: pilot landing aid television.

PLATO: programmed logic for automatic teaching operations.

PLC: power-line carrier.

PLD: phase-lock demodulator.

PLD: pulse-length discriminator.

PLDTS: propellant loading data transmission system.

PLF: parachute landing fall.

PLIANT: procedural language implementing analog techniques.

PLIM: post-launch information message.

PLL: phase-locked loop.

PLOD: planetary orbit determination.

PLOP: pressure line of position.

PLP: pattern learning parser.

PLRS: position location reporting system.

PLT: program library tape.

PLUS: program library update system.

P/M: parts per million.

pm: permanent magnet.

pm: phase modulation.

PM: photomultiplier.

PM: *post meridiem* (after noon).

PM: pounds per minute.

PM: preventive maintenance.

PM: procedures manual.

PM: pulse modulation.

PMA: Permanent Magnet Association.

PMA: Precision Measurement Association.

PMAF: Pharmaceutical Manufacturers Association Foundation.

PMB: Physical Metallurgy Branch.

PMBX: private manual branch exchange.

PMC: program marginal checking.

PMC: pseudo machine code.
PME: photomagnetoelectric effect.
PME: protective multiple earthing.
PMEE: prime mission electronic equipment.
PMI: preventive maintenance inspection.
PMOS: p-channel (type) metal oxide semiconductor.
PMP: premodulation processor.
PMP: preventive maintenance plan.
PMR: Pacific Missile Range.
PMS: processor, memories and switches.
PMSRP: Physical and Mathematical Sciences Research Paper.
PN: performance number.
PCN: programmed numerical control.
PNDC: parallel network digital computer.
PNEC: Proceedings of the National Electronics Conference.
PndB: perceived noise level expressed in decibels.
PNGCS: Primary Navigation, Guidance, and Control System.
pnp: positive-negative-positive (transistor).
PNTD: personnel neutron threshold detector.
PO: parallel output.
PO: power oscillator.
PO: program objectives.
PO: pulsed carrier without any modulation intended to carry information.
POCP: program objectives change proposal.
POCS: Patent Office classification system.
POD: point of origin device.
PODAF: power density exceeding a specified level over an area within an assigned frequency band.
PODS: post-operative destruct system.
POGO: Polar Orbiting Geophysical Observatory.
POGO: programmer oriented graphics operation.
POI: program of instruction (NASA).
POISE: panel on in-flight scientific experiments.
POL: problem-oriented language.
POL: procedure-oriented language.
POL: process-oriented language.
POLYTRAN: polytranslation analysis and programming.
POMM: Preliminary Operating and Maintenance Manual.
POMS: Panel on Operational Meteorological Satellites.
POMSEE: Performance, Operational and Maintenance Standards for Electronic Equipment.
POP: power on/off protection.
POP: programmed operators and primitives.
POP: program operating plan.
POPI: Post Office Position Indicator.
POPS: pantograph optical projection system.
POR: problem-oriented routine.
POS: point-of-sale (manufacturer).
POS: primary operating system.
POTC: PERT Orientation and Training Center (DOD).
POWS: pyrotechnic outside warning system.
P-P: peak to peak.
P/P: point to point.
P&PI: Pulp and Paper Industry (IEEE IAS Technical Committee).
PP: panel point.

PP: peripheral processor.
PP: preprocessor.
PP: print/punch.
PP: pressureproof.
PP: push-pull.
PPA: photo peak analysis.
PPA: Professional Programmers Association.
PPC: pulsed power circuit.
PPE: polyphenylether.
PPE: premodulation processor equipment.
PPE: problem program efficiency.
PPG: program pulse generator.
PPG: propulsion and power generator.
PPI: plan position indicator.
PPM: periodic permanent-magnet focusing.
PPM: planned preventive maintenance.
PPM: pulse position modulation.
PPMS: program performance measurement systems.
PPS: parallel processing system.
PPS: phosphorous propellant system.
PPS: primary propulsion system.
PPS: pulses per second.
PPT: punched paper tape.
PPU: peripheral processing unit.
PQGS: propellant quantity gauging systems.
PR: pattern recognition.
PR: program register.
PR: program requirements.
PR: pseudorandum.
PR: Puerto Rico.
PR: pulse rate.
PR: pulse ratio.
PRA: precision axis.
PRA: production reader assembly.
PRA: program reader assembly.
PRB: Panel Review Board.
PRBS: pseudorandom binary sequence.
PRC: point of reverse curve.
PRD: program requirements data.
PRD: program requirements document.
PREDICT: prediction of radiation effects by digital computer techniques.
PREF: propulsion research environmental facility.
PREP: programmed educational package.
PRESS: Pacific Range Electromagnetic Signature Studies.
PRESTO: program for rapid earth-to-space trajectory optimization.
PRESTO: program reporting and evaluation system for total operations.
PRESSAR: presentation equipment for slow scan radar.
PRF: pulse repetition frequency.
PRFL: pressure-fed liquid.
PRIDE: programmed reliability in design.
PRIME: precision recovery including maneuverable entry.
PRIME: precision integrator for meteorological echoes.
PRIME: programmed instruction form management education.
PRINCE: programmed reinforced instruction necessary to continuing education.
PRIS: Propeller Revolution Indicator System.

PRISE: program for integrated shipboard electronics.
PRISM: programmed integrated system maintenance.
PRM: power range monitor.
PRM: pulse-rate modulation.
PRN: pseudorandom noise.
PROCOMP: process computer.
PROCOMP: program compiler.
PRODAC: programmed digital automatic control.
PROFAC: propulsive fluid accumulator.
PROFILE: programmed functional indices for laboratory evaluation.
PROFIT: programmed reviewing, ordering, and forecasting inventory technique.
PROGDEV: program device.
PROM: Pockels readout optical memory.
PROM: programmable read-only memory.
PROMPT: program monitoring and planning techniques.
PRONTO: program for numerical tool operation.
PROMPT: program reporting, organization, and management planning technique.
PROP: performance review for operating programs.
PROP: planetary rocket ocean platform.
PROTECT: probabilities recall optimizing the employment of calibration time.
PROXI: projection by reflection optics of xerographic images.
PRP: pulse repetition period.
PRR: pulse repetition rate.
PRS: Pattern Recognition Society.
PRS: pattern recognition system.
PRT: portable remote terminal.
PRT: pulse repetition time.
PRTR: plutonium recycle test reactor.
PRU: programs research unit.
PRV: peak inverse (reverse) voltage.
PRV: pressure-reducing valve.
PS: power source.
PSA: power servo amplifier.
PSAC: President's Science Advisory Committee.
PSALI: permanent supplementary artificial lighting of interiors.
PSAR: programmable synchronous/asynchronous receiver.
PSAT: programmable synchronous/asynchronous transmitter.
PSC: Power System Communications (IEEE PES Technical Committee).
PSCC: Power System Computation Conference.
PSD: phase-sensitive demodulator.
PSD: power spectral density.
PSDF: Propulsion Systems Development Facility.
PSE: Power System Engineering (IEEE PES Technical Committee).
PSEP: passive seismic experiment package.
PSF: pound-force per square foot.
PSI: pound force per square inch.
PSI: preprogrammed self-instruction.
PSI: present serviceability index.
PSIA: pounds per square inch absolute.

PSIEP: Project on Scientific Information Exchange in Psychology.
PSIM: Power System Instrumentation and Measurement (IEEE PES Technical Committee).
PSK: phase-shift keying.
PSL: Photographic Science Laboratory.
PSMR: parts specification management for reliability.
PSMS: Permanent Section of Microbiological Standardization.
PSO: pilot systems operator.
PSP: planet scan platform.
PSP: power system planning.
PSR: Power System Relaying (IEEE PES Technical Committee).
PSRP: physical sciences research paper.
PSS: proprietary software systems.
PSS: propulsion support system.
PST: polished surface technique.
PST: point of spiral tangent.
PSTC: Pressure Sensitive Tape Council.
PSW: program status word.
PT: point of tangency.
PT: potential transformer.
PT: propellant transfer.
PT: pulse time.
PT: punched tape.
PTA: planar turbulence amplifier.
PTA: pulse torquing assembly.
PTC: programmed transmission control.
PTCR: pad terminal connection room.
PTCS: propellant tanking computer system.
PTF: program temporary fix.
PTFE: polytetrafluoroethylene.
PTM: proof test model.
PTM: pulse time modulation.
PTM: pulse time multiplex.
PTML: pnpn transistor magnetic logic.
PTO: power take-off.
PTP: paper tape punch.
PTP: point to point.
PTP: preferred target point.
PTR: paper tape reader.
PTR: pool test reactor.
PTR: Power Transformers (IEEE PES Technical Committee).
PTS: permanent threshold shift.
PTS: pneumatic test set.
PTS: power transient suppressor.
PTS: program test system.
PTS: propellant transfer system.
PTS: pure time sharing.
PTT: program test tape.
PTT: push to talk.
PTV: predetermined time value.
PTVA: propulsion test vehicle assembly.
PU: power unit.
PU: processing unit.
PU: propulsion unit.
PUCK: propellant utilization checkout kit.
PUCS: propellant utilization control system.

PUFFS: passive underwater fire control sonar feasibility study.
PUFFT: Purdue University fast Fortran translator.
PUGS: propellant utilization gauging system.
PUR: Purdue University Research.
PUSAS: proposed US Standard.
PUSS: pilots' universal sighting system.
PUT: programmable unijunction transistors.
PV: positive volume.
PVC: polyvinyl chloride.
PVC: position and velocity computer.
PVOR: precision vhf omnirange.
PVS: performance verification system.
PVST: premate verification system test.
PVT: polyvinyl toluene.
PVT: pressure-volume-temperature.

pW: picowatt.
PW: printed wiring.
PW: pulsewidth.
PWB: printed wiring board.
PWD: power distributor.
PWD: pulsewidth discriminator.
PWF: present worth factor.
PWI: proximity warning indicator.
PWL: piecewise linear.
PWM: pulsewidth modulation.
PWM: pulsewidth multiplier.
PWR: pressurized water reactor.
PWT: propulsion wind tunnel.
PXSTR: phototransistor.
PZT: lead (Pb) zirconate titanate (semiconductor).

Q

QA: query analyzer.
QADS: quality assurance data system.
QB: quick break.
QCE: quality control engineering.
QCR: quality control reliability.
QCW: quadrature phase subcarrier (signal).
QD: quick disconnect.
QDRI: qualitative development requirements information.
QEC: quick engine change.
QF: quick firing.
QFRIC: quick fix interference reduction capability.
QISAM: queued indexed sequential access method.
QIT: quality information and test.
QL: query language.
QLAP: quick look analysis program.
QLDS: quick look data station.

QLIT: quick-look intermediate tape.
QMDO: qualitative material development objective.
QMR: qualitative material requirement.
QOD: quick-opening device.
QPL: qualified products list.
QPPM: quantized pulse position modulation.
QR: quality and reliability.
QRA: quality and reliability assurance.
QRBM: quasi-random band model.
QRC: quick reaction capability.
QRI: qualitative requirements information.
QSAM: queued sequential access method.
QTAM: queued telecommunications access method.
QUAM: quadrature amplitude modulation.
QUEST: quality electrical systems test.
QUIP: query interactive processor.
QVT: quality verification testing.

R

R: Reliability (IEEE Group).
RA: radar altimeter.
RACC: radiation and contamination control.
RACE: random access computer equipment.
RACE: random access control equipment.
RACE: rapid automatic checkout equipment.
RACEP: random access and correlation for extended performance.
RACES: Radio Amateur Civil Emergency Service.
RACON: radar beacon.
RACS: remote access computing system.
RACS: remote access calibration system.
RACT: remote access computer technique.
RAD: radiation absorbed dose.

RAD: random access data.
RAD: random access disc.
RAD: rapid access disk.
RADA: random access discrete address.
RADAC: radar analog digital data and control.
radar: radio detection and ranging.
RADAS: random access discrete address system.
RADATA: radar data transmission and assembly.
RADC: Rome Air Development Center (Air Force).
RADCM: radar countermeasures and deception.
RADCON: radar data converter.
RADEM: random access delta modulation.
RADFAC: radiating facility.
RADHAZ: radiation hazards.

RADIAC: radioactivity detection, identification, and computation.
RADIQUAD: radio quadrangle.
RADIR: random access document indexing and retrieval.
RADIST: radar distance indicator.
RADNOTE: radio note.
RADOME: radar dome.
RADOP: radar operator.
RADOPWEAP: radar optical weapons.
RADPLANBD: radio planning board.
RADPROPCAST: radio propagation forecast.
RADSO: radiological survey officer.
RADTT: radio teletype.
RADVS: radar altimeter and doppler velocity sensor.
RAE: radio astronomy explorer satellite.
RAE: range azimuth elevation.
RAEN: radio amateur emergency network.
RAES: remote access editing system.
RAES: Royal Aeronautical Society.
RAFT: radially adjustable facility tube.
RAI: random access and inquiry.
RAIC: Redstone Arsenal Information Center.
RAID: remote access interactive debugger.
RAIDS: rapid availability of information and data for safety.
RAILS: remote area instrument landing sensor.
RAIR: random access information retrieval.
RAL: Riverbend Acoustical Laboratory.
RALU: register arithmetic and logic unit.
RALW: radioactive liquid waste.
RAM: radar absorbing material.
RAM: radio attenuation measurements.
RAM: random access memory.
RAMA: Rome Air Materiel Area.
RAMAC: random access method of accounting and control.
RAMARK: radar marker.
RAMIS: rapid access management information system.
RAMP: Raytheon airborne microwave platform.
RAMPART: radar advanced measurement program for analysis of reentry techniques.
RAMPS: resource allocation and multiproject scheduling.
RANDAM: random access nondestructive advanced memory.
RAO: Radio Astronomical Observatory.
RAP: redundancy adjustment of probability.
RAP: rocket-assisted projectile.
RAPCON: radar approach control.
RAPEC: rocket-assisted personnel ejection catapult.
RAPID: reactor and plant integrated dynamics.
RAPID: research in automatic photocomposition and information dissemination.
RAPO: Resident Apollo Project Office (NASA).
RAPPI: random access plan position indicator.
RAPS: retrieval analysis and presentation system.
RAPS: risk appraisal of programs system.
RAPT: reusable aerospace passenger transport.
RAR: rapid access recording.
RAREP: radar weather report.
RASC: Rome Air Service Command.

RASPO: Resident Apollo Spacecraft Program Office.
RASSR: reliable advanced solid-state radar.
RASTA: radiation special test apparatus.
RASTAC: random access storage and control.
RASTAD: random access storage and display.
RAT: reliability assurance test.
RATAC: radar target acquisition.
RATAN: radar and television aid to navigation.
RATCC: radar air-traffic control center.
RATE: remote automatic telemetry equipment.
RATER: response analysis tester.
RATIO: radio telescope in orbit.
RATO: rocket-assisted take-off.
RATOG: rocket-assisted take-off gear.
RATSC: Rome Air Technical Service Command.
RATSCAT: radar target scatter.
RATT: radio teletype.
RAVIR: radar video recording.
RAWIN: radar wind sounding.
RAYDAC: Raytheon digital automatic computer.
RB: radar beacon.
RB: return to bias.
RBA: recovery beacon antenna.
RBDE: radar bright display equipment.
RBE: radiation biological effectiveness.
RBE: relative biological effectiveness.
RBI: ripple-blanking input.
RBO: ripple-blanking output.
RBP: Registered Business Programmer.
RBR: radar boresight range.
RBS: radar bomb scoring.
RBS: random barrage system.
RBSS: recoverable booster support system.
R/C: range clearance.
R/C: radio command.
R/C: radio control.
R/C: rate of climb.
RC: range control.
RC: remote control.
RC: research center.
RC: resistance-capacitance.
RCAG: remote-controlled air-ground communication site.
RCAT: radio-code aptitude test.
RCC: recovery control center.
RCC: remote communications central.
RCDC: radar course directing central.
RCEI: range communications electronics instructions.
RCI: radar coverage indicator.
RCL: Radiation Counting Laboratory.
RCM: radar countermeasures.
RCM: radio countermeasures.
RCO: reactor core.
RCO: Remote Control Office.
RCO: remote control oscillator.
RCS: radio command system.
RCS: reaction control subsystem.
RCS: reaction control system.
RCS: reentry control system.
RCS: remote control system.
RCT: resolve control transformer.
RCTL: resistor-capacitor-transistor logic.

RCTSR: radio code test speed on response.
RCVR: receiver.
R&D: research and development.
RD: radiation detection.
RD: readiness date.
RD: register drive.
RD: restricted data.
RD: root diameter.
rd: rutherford.
RDA: reliability design analysis.
RDB: radar decoy balloon.
RDB: Research and Development Board.
RDC: Reliability Data Center.
RDC: remote data collection.
RDE: radial defect examination.
RDF: radio direction finding.
RDL: Rocket Development Laboratory.
RDM: recording demand meter.
RDMU: range drift measuring unit.
RDOS: real-time disc operating system.
RDP: radar data processing.
RDPS: radar data processing system.
RDRINT: radar intermittent.
RDR XMTR: radar transmitter.
RDS: rendezvous and docking simulator.
RDT: remote data transmitter.
RDTE: Research, Development, Test, and Evaluation.
RDTL: resistor diode transistor logic.
R&E: research and engineering.
RE: radiated emission.
RE: rate effect.
RE: reentry.
REA: Rural Electrification Administration.
REAC: Reeves electronic analog computer.
REACT: Radio Emergency Associated Citizens Teams.
REACT: register-enforced automated control technique.
READ: real-time electronic access and display.
READ: remote electronic alphanumeric display.
READI: rocket engine analysis and decision instrumentation.
REALCOM: real-time communications computer.
REAP: remote entry acquisition package.
REAR: Reliability Engineering Analysis Report.
REC: request for engineering change.
RECMFA: Radio and Electronic Component Manufacturers Association.
RECSTA: receiving station.
REDAP: reentrant data processing.
REDSOD: repetitive explosive device for soil displacement.
REEP: regression estimation of event probabilities.
REGAL: range and evaluation guidance for approach and landing.
REI: research-engineering interaction.
REIC: Radiation Effects Information Center.
REINS: requirements electronic input system.
REL: rapidly extensible language.
REM: rapid eye movement.
REM: reliability engineering model.
REM: roentgen equivalent man.
REMOS: real-time event monitor.

REN: remote enable.
RENE: rocket engine/nozzle ejector.
REON: rocket engine operations—nuclear.
REP: range error probable.
REP: rendezvous evaluation pad.
REP: roentgen equivalent physical.
REPLAB: responsive environment programmed laboratory.
RESD: Research and Engineering Support Division.
RESER: reentry system evaluation radar.
RESG: Research Engineering Standing Group.
RESS: radar echo simulation subsystem.
REST: reentry environment and systems technology.
RETAIN: remote technical assistance and information network.
RETMA: Radio Electronics Television Manufacturing Association.
REV: reentry vehicle.
REVS: rotor entry vehicle system.
REWSONIP: reconnaissance electronic warfare special operation and naval intelligence processing.
RF: radio frequency.
RF: range finder.
RF: rating factor.
RFC: radio facility charts.
RFC: radio-frequency choke.
RFCP: radio-frequency compatibility program.
RFD: ready for data.
RFD: reentry flight demonstration.
RFEI: request for engineering information.
RFG: radar field gradient.
RFI: radio-frequency interference.
RFIT: radio-frequency interference tests.
RFM: reactive factor meter.
RFMO: Radio Frequency Management Office.
RFS: radio-frequency shift.
RFSTF: RF Systems Test Facility.
RG: rate gyroscope.
RG: reticulated grating.
RGA: rate gyro assembly.
RGL: report generator language.
RGP: rate gyro package.
RGS: radio guidance system.
RGS: rate gyro system.
RGS: rocket guidance system.
RGT: resonant gate transistor.
R/h: roentgens per hour.
RH: radiological health.
RH: relative humidity.
RHAW: radar homing and warning.
RHC: right-hand circular (polarization).
RHC: rotation hand controller.
RHE: radiation hazard effects.
RHEL: Rutherford High Energy Laboratory.
RHI: range height indicator.
RHOGI: radar homing guidance.
RHP: reduced hard pressure.
RHR: rejectable hazard rate.
RI: reflective insulation.
RI: resistance inductance.
RI: Rhode Island.
RIAA: Record Industry Association of America.

RIAS: Research Institute for Advanced Studies.
RIC: Range Instrumentation Coordination.
RICASIP: Research Information Center and Advisory Service on Information Processing.
RICS: Range Instrumentation Control System.
RIF: radio-influence field.
RIFI: radio interference field intensity.
RIFT: reactor in-flight test.
RIM: radar input mapper.
RIN: regular inertial navigator.
RINAL: radar inertial altimeter.
RINS: rotorace inertial navigation system.
RINT: radar intermittent.
RIOMETER: relative ionospheric opacity meter.
RIOT: real-time input-output transducer.
RIPPLE: radioactive isotope powered pulse light equipment.
RIPPLE: radioisotope powered prolonged life equipment.
RIPS: radio isotope power system.
RIPS: range instrumentation planning study.
RIR: reliability investigation requests.
RIS: range instrumentation ship.
RIS: revolution indicating system.
RISE: research in supersonic environment.
RIST: radar installed system tester.
RIT: radio information test.
RIT: rocket interferometer tracking.
RITE: Rapidata interactive test editor.
RITU: Research Institute of Temple University.
RJE: remote job entry.
RKO: range keeper operator.
RL: radiation laboratory.
RL: research laboratory.
RL: resistance-inductance.
RL: resistor logic.
RLBM: rearward launched ballistic missiles.
RLC: radio launch control system.
RLC: resistance-inductance-capacitance.
RLE: Research Laboratory of Electronics.
RLHTE: Research Laboratory of Heat Transfer in Electronics.
RM: radio monitoring.
RM: Rotating Machinery (IEEE PES Technical Committee).
RMA: Radio Manufacturers Association.
RMC: rod memory computer.
RMI: radio magnetic indicator.
RMI: reliability maturity index.
RML radar microwave link.
RMM: read-mostly memory.
RMO: radio material officer.
RMOS: refractory metal oxide semiconductor.
RMP: reentry measurements program.
RMS: root mean square.
RMU: remote maneuvering unit.
RN: random number.
RO: radar operator.
RO: range operations.
RO: readout.
RO: receive only.

ROAMA: Rome Air Materiel Area.
ROAT: radio operator's aptitude test.
ROB: radar order of battle.
ROBIN: remote on-line business information network.
ROBOMB: robot bomb.
ROC: Range Operations Conference.
ROC: receiver operating characteristics.
ROC: required operation capability.
ROC: reusable orbital carrier.
ROCAPPI: Research on Computer Applications in the Printing and Publishing Industries.
ROCP: radar out of commission for parts.
ROCR: remote optical character recognition.
RODATA: Registered Organization Data Bank.
ROI: range operations instructions.
ROIS: radio operational intercom system.
ROLF: remotely operated longwall face.
ROLS: recoverable orbital launch system.
ROM: read-only memory.
ROM: readout memory.
ROM: rough order of magnitude.
ROMBUS: reusable orbital module booster and utility shuttle.
ROPP: receive-only page printer.
ROS: read-only storage.
ROSE: remotely operated special equipment.
ROSE: retrieval by on-line search.
ROT: reusable orbital transport.
ROTI: recording optical tracking instrument.
ROTR: receive-only tape perforator.
R&PI: Rubber and Plastics Industry (IEEE IAS Technical Committee).
RP: recovery phase.
RP: relative pressure.
RPAO: radium plaque adaptometer operator.
RPC: row parity check.
RPD: radar planning device.
RPD: retarding potential difference.
RPE: radial probable error.
RPE: Registered Professional Engineer.
RPI: radar precipitation integrator.
RPL: radar processing language.
RPL: Radiation Physics Laboratory.
RPL: Rocket Propulsion Laboratory.
RPL: running program language.
RPM: rate per minute.
RPM: reliability performance measure.
RPM: resupply provisions module.
RPM: revolutions per minute.
RPN: reverse Polish notation.
RPS: remote processing service.
RPS: revolutions per second.
RPSM: resources planning and scheduling method.
RPU: radio propagation unit.
RPU: radio phone unit.
R/Q: resolver/quantizer.
R&QA: reliability and quality assurance.
RQA: recursive queue analyzer.
RQL: reference quality level.
RQS: rate quoting system.
R/R: readout and relay.

R/R: record/retransmit.
R&RC: Reactors and Reactor Controls (IEEE NPSS Technical Committee).
RR: register to register instruction.
RR: rendezvous radar.
RR: repetition rate.
RR: return rate.
RR: retro rockets.
RR: round robin.
RRE: Radar Research Establishment.
RRI: range-rate indicator.
RRI: Rocket Research Institute.
RRIS: remote radar integration station.
RRNS: redundant residue number system.
RRRV: rate of rise of restriking voltage.
RRS: Radiation Research Society.
RRS: radio research station.
RRS: Reaction Research Society.
RRS: required response spectrum.
RRS: restraint release system.
RRS: retrograde rocket system.
RRU: radiobiological research unit.
R&S: research and statistics.
RS: radiated susceptibility.
RS: range safety.
RS: remote station.
RSAC: radiological safety analysis computer.
RSAC: Reactor Safety Advisory Committee.
RSCIE: remote station communication interface equipment.
RSDP: remote site data processing.
RSI: Research Studies Institute.
RSIC: Radiation Shielding Information Center.
RSIC: Redstone Scientific Information Center.
RSL: Radio Standards Laboratory.
RSM: resource management system.
RSMPS: Romanian Society for Mathematics and Physical Sciences.
RSN: radiation surveillance network.
RSP: record select program.
RSRI: Regional Science Research Institute.
RSRS: radio and space research station.
RSS: root sum square.
RST: reset-set trigger.
R/T: real time.
RT: radio telephony.
RT: rated time.
RT: receiver transmitter.
RT: remote terminal.
RT: reduction table.
RTB: read tape binary.
RTC: reader tape contact.
RTC: real-time command.
RTC: real-time computer.
RTCA: Radio Technical Commission for Aeronautics.

RTCC: real-time computer complex.
RTCF: real-time computer facility.
RTCM: Radio Technical Commission for Marines.
RTCS: real-time computer system.
RTCU: real-time control unit.
RTD: read tape decimal.
RTD: real-time display.
RTD: resistance-temperature detector (probe).
RTDC: real-time data channel.
RTDHS: real-time data handling system.
RTE: real-time executive.
RTG: radioisotope thermoelectric generator.
RTI: referred to input.
RTI: Research Triangle Institute.
RT/IOC: real-time input/output controller.
RTIRS: real-time information retrieval system.
RTK: range tracker.
RTL: real-time language.
RTL: resistor-transistor logic.
RTM: real-time monitor.
RTM: recording tachometer.
RTMA: Radio-Television Manufacturers Association.
RTMOS: real-time multiprogramming operating system.
RTO: referred to output.
RTOS: real-time operating system.
RTP: real-time peripheral.
RTP: requirement and test procedures.
RTPH: round trips per hour.
RTS: radar tracking station.
RTS: reactive terminal service.
RTSD: Resources and Technical Services Division.
RTSS: real-time scientific system.
RTST: radio technician selection test.
RTT: radioteletypewriter.
RTTDS: real-time telemetry data system.
RTTV: real-time television.
RTTY: radio teletypewriter.
RTU: remote terminal unit.
RTWS: raw tape write submodule.
RU: reproducing unit.
RUSH: remote use of shared hardware.
RVA: reactive volt-ampere meter.
RVA: reliability variation analysis.
RVM: reactive voltmeter.
RVR: runway visual range.
R-W: read-write (head).
RW: resistance welding.
RWG: Roebling Wire Gage.
RWM: rectangular wave modulation.
RWMA: Resistance Welder Manufacturers Association.
RZ: return to zero.
RZL: return to zero level.
RZM: return to zero mark.

S

SA: sense amplifier.
SA: spectrum analyzer.
SA: stress anneal.
SA: successive approximation.
SA: symbolic assembler.
SAB: Scientific Advisory Board.
SAB: System Advisory Board.
SABE: Society for Automating Better Education.
SABE: Society for Automation in Business Education.
SABRE: sales and business reservations done electronically.
SABRE: secure airborne radar bombing equipment.
SABS: South African Bureau of Standards.
SAC: Scientific Advisory Committee.
SAC: semiautomatic coding.
SAC: Society for Analytical Chemistry.
SACMAPS: selective automatic computational matching and positioning system.
SADA: seismic array data analyzer.
SADAP: simplified automatic data plotter.
SADC: sequential analog-digital computer.
SADIC: solid-state analog-to-digital computer.
SADIE: scanning analog-to-digital input equipment.
SADIE: semiautomatic decentralized intercept environment.
SADIE: sterling and decimal invoicing electronically.
SADR: six-hundred-megacycle air-defense radar.
SADSAC: sampled data simulator and computer.
SADSAC: Seiler algol digitally simulated analog computer.
SAE: shaft-angle encoder.
SAE: Society of Automotive Engineers.
SAEH: Society for Automation in English and the Humanities.
SAFA: Society for the Fine Arts.
SAFEA: Space and Flight Equipment Association.
SAFI: semiautomatic flight inspection.
SAFOC: semiautomatic flight operations center.
SAG: standard address generator.
SAGA: Studies Analysis and Gaming Agency.
SAGE: semiautomatic ground environment.
SAHYB: simulation of analog and hybrid computers.
SAILS: simplified aircraft instrument landing system.
SAIMS: selected acquisitions information and management system.
SAINT: satellite interceptor.
SAIS: South African Interplanetary Society.
SAKI: solatron automatic keyboard instructor.
SAL: Supersonic Aerophysics Laboratory.
SAL: symbolic assembly language.
SAL: systems assembly language.
SALE: simple algebraic language for engineers.
SALM: Society for Airline Meteorologists.
SAM: School of Aerospace Medicine.
SAM: selective automonitoring.
SAM: semantic analyzing machine.
SAM: semiautomatic mathematics.
SAM: sequential access memory.
SAM: sequential access method.
SAM: simulation of analog methods.

SAM: sort and merge.
SAM: surface-to-air missile.
SAM: symbolic and algebraic manipulation.
SAM: system activity monitor.
SAM: systems analysis module.
SAM-D: surface-to-air missile development.
SAMA: Scientific Apparatus Makers Association.
SAME: Society of American Military Engineers.
SAMI: socially acceptable monitoring instrument.
SAMIS: structural analysis and matrix interpretive system.
SAMMIE: scheduling analysis model for mission integrated experiments.
SAMOS: satellite and missile observation system.
SAMPE: Society of Aerospace Material and Process Engineers.
SAMS: satellite automonitor system.
SAP: share assembly program.
SAP: systems assurance program.
SAMSO: Space and Missile Systems Organization.
SAMSON: system analysis of manned space operations.
SANOVA: simultaneous analysis of variance.
SAO: Smithsonian Astrophysical Observatory.
SAPE: Society for Professional Education.
SAPIR: system of automatic processing and indexing of reports.
SAR: storage address register.
SAR: successive approximation register.
SAR: synthetic aperture radar.
SARAH: search and rescue and homing.
SARPS: standards and recommended practices.
SARS: single-axis reference system.
SARUC: Southeastern Association of Regulatory Utility Commissioners.
SAS: Society for Applied Spectroscopy.
SAS: surface active substances.
SASM: Society for Science and Mathematics.
SASS: Society for the Social Sciences.
SASTU: signal amplitude sampler and totalizing unit.
SAT: Society of Acoustic Technology.
SAT: stabilization assurance test.
SAT: stepped atomic time.
SATAN: sensor for airborne terrain analysis.
SATANAS: semiautomatic analog setting.
SATCO: signal automatic air-traffic control.
SATCOM: Satellite Communication Agency (DOD).
SATCOM: Scientific and Technical Communication Committee.
SATIF: Scientific and Technical Information Facility.
SATIN: SAGE air-traffic integration.
SATIRE: semiautomatic technical information retrieval.
SATRAC: satellite automatic terminal rendezvous and coupling.
SAVA: Society for Accelerator and Velocity Apparatus.
SAVE: Society of American Value Engineers.
SAVE: system for automatic value exchange.
SAVITAR: Sanders Associates video input/output terminal access resource.
SAVS: status and verification system.

SAW: surface acoustic waves.
SAWE: Society of Aeronautical Weight Engineers.
SB: secondary battery.
SB: serial binary.
SB: sideband.
SB: sleeve bearing.
SB: straight binary.
SB: synchronization bit.
SBA: standard beam approach.
SBD: Schottky-barrier diode.
SBFM: silver-band frequency modulation.
SBK: single-beam klystron.
SBM: system balance measure.
SBP: Society of Biological Psychiatry.
SBS: satellite business system.
SBT: surface-barrier transistor.
SBX: S-band transponder.
SC: saturable core.
SC: search control.
SC: semiconductor.
SC: Simulation Council.
SC: shaped charge.
SC: silvered copper (wire).
SC: single contact.
SC: South Carolina.
SC: superimposed current.
SC: switching cell.
SC: synchrocyclotron.
SCA: sequence control area.
SCADA: supervisory control and data acquisition.
SCADS: scanning celestial attitude determination system.
SCADS: simulation of combined analog digital systems.
SCALE: space checkout and launch equipment.
SCAMA: station conferencing and monitoring arrangement.
SCAMA: switching, conference, and monitoring arrangement.
SCAMPS: small computer analytical and mathematical programming system.
SCAN: selected current aerospace notices (NASA).
SCAN: self-correcting automatic navigation.
SCAN: stock-market computer-answering network.
SCAN: student career automated network.
SCAN: switched circuit automatic network.
SCANS: scheduling control and automation by network systems.
SCAP: silent compact auxiliary power.
SCAR: satellite capture and retrieval.
SCAR: Scientific Committee on Antarctic Research.
SCAR: submarine celestial altitude recorder.
SCARF: Santa Cruz Acoustic Range Facility.
SCAT: share compiler assembler and translator.
SCAT: space communication and tracking.
SCAT· speed command of attitude and thrust.
SCAT: supersonic commercial air transport.
SCATE: Stromberg-Carlson automatic test equipment.
SCATS: simulation checkout and training system.
SCC: simulation control center.
SCC: single-conductor cable.
SCC: single cotton covered (wire).
SCD: space control document.

SCDP: Society of Certified Data Processors.
SCEA: signal conditioning electronic assembly.
SCEL: Signal Corps Engineering Laboratory.
SCEPTRE: systems for circuit evaluation and prediction of transient radiation effects.
SCEPTRON: spectral comparative pattern recognizer.
SCERT: systems and computers evaluation and review technique.
SCF: satellite control facility.
SCF: sequence compatibility firing.
SCI: Society of Chemical Industry.
SCI: Society of Computer Intelligence.
SCIC: semiconductor integrated circuit.
SCIM: speech communication index meter.
SCIP: scanning for information parameters.
SCIP: self-contained instrument package.
SCM: service command module.
SCM: signal conditioning module.
SCM: small core memory.
SCMA: Systems Communications Management Association.
SCN: sensitive command network.
SCOMO: satellite collection of meteorological observations.
SCOOP: scientific computation of optimal programs.
SCOPE: schedule-cost-performance.
SCOPE: sequential customer order processing electronically.
SCOPE: Special Committee on Paperless Entries.
SCOR: Scientific Committee on Oceanographic Research.
SCOR: self-calibration omnirange.
SCORE: satellite computer-operated readiness equipment.
SCORE: selection copy and reporting.
SCORE: signal communications by orbital relay equipment.
SCORPI: subcritical carbon-moderated reactor assembly for plutonium investigations.
SCOST: Special Committee on Space Technology.
SCP: symbolic conversion program.
SCP: system communication pamphlet.
SCR: selective chopper radiometer.
SCR: semiconductor-controlled rectifier.
SCR: short-circuit ratio.
SCR: silicon-controlled rectifier.
SCRAP: super-caliber rocket-assisted projectile.
SCRIPT: scientific and commercial interpreter and program translator.
SCS: silicon-controlled switch.
SCS: simulation control subsystem.
SCS: Society for Computer Simulation.
SCS: Southern Computer Service.
SCS: space cabin simulator.
SCS: standard coordinate system.
SCSE: South Carolina Societies of Engineering.
SCT: scanning telescope.
SCTP: straight channel tape print.
SCTPP: straight channel tape print program.
SCUBA: self-contained underwater breathing apparatus.
SCUP: school computer use plan.

SD: South Dakota.
SDA: shaft drive axis.
SDA: source data acquisition.
SDA: source data automation.
SDAD: satellite digital and analog display.
SDAP: systems development analysis program.
SDAS: scientific data automation system.
SDC: Semiconductor Devices Council of JEDEC.
SDC: stabilization data computer.
SDCC: San Diego Computer Center.
SDCE: Society of Die Casting Engineers.
SDE: Society of Data Educators.
SDE: Students for Data Education.
SDFC: Space Disturbance Forecast Center (ESSA).
SDI: selective dissemination of information.
SDI: source data information.
SDL: system descriptive language.
SDLC: synchronous data link control.
SDM: statistical delta modulation.
SDM: standardization design memoranda.
SDP: site data processor.
SDPL: servomechanisms and data-processing laboratory.
SDR: system design review.
SDS: safety data sheet.
SDS: scientific data systems.
SDS: simulation data subsystems.
SDS: system data synthesizer.
SDW: standing detonation wave.
SE: shielding effectiveness.
SE: systems engineer.
SEA: systems effectiveness analyzer.
SEAC: Standard's eastern automatic computer.
SEAL: standard electronic accounting language.
SEALS: severe environmental air launch study.
SEALS: stored energy actuated lift system.
SEASCO: South East Asia Science Cooperation Office.
SEB: Source Evaluation Board.
SEC: Sanitary Engineering Center.
SEC: secondary electron conduction.
SEC: secondary emission conductivity.
SEC: simple electronic computer.
SECAM: sequential color and memory.
SECAP: system experience correlation and analysis program.
SECAR: secondary radar.
sech: hyperbolic secant.
SECO: sequential coding.
SECOR: sequential collation of range.
SECPS: secondary propulsion system.
SECS: sequential events control system.
SECTAM: Southern Conference on Theoretical and Applied Mechanics.
SED: spectral energy distribution.
SED: Sanitary Engineering Division.
SEDD: Systems Evaluation and Development Division.
SEDIT: sophisticated string editor.
SEDR: Systems Engineering Department Report.
SEDS: space electronics detection system.
SEDS: Society for Educational Data Systems.
SEE: Society of Environmental Engineers.

SEE: Southeastern Electric Exchange.
SEEK: systems evaluation and exchange of knowledge.
SEF: Space Education Foundation.
SEFAR: sonic end fire for azimuth and range.
SEFOR: South-West Experimental Fast Oxide Reactor.
SEG: Society of Exploration Geophysicists.
SEG: Standardization Evaluation Group.
SEG: systems engineering group.
SEI: systems engineering and integration.
SEIP: systems engineering implementation plan.
SEIT: satellite educational and informational television.
SEL: Stanford Electronics Laboratory.
SEL: System Engineering Laboratories.
SELR: Saturn engineering liaison request.
SEM: scanning electron microscope.
SEMA: *Société d'Economie et de Mathematiques Appliques* (France).
SEMLAM: semiconductor laser amplifier.
SEMLAT: semiconductor laser array techniques.
SEMIRAD: secondary electron-mixed radiation dosimeter.
SEMS: severe environmental memory system.
SENL: standard equipment nomenclature list.
SENTOS: sentinel operating system.
SEP: space electronic package.
SEP: standard electronic package.
SEP: star epitaxial planar.
SEPOL: settlement problem-oriented language.
SEPOL: soil-engineering problem-oriented language.
SEPS: service module electrical power system.
SEPS: severe environment power system.
SERAPE: simulator equipment requirements for accelerating procedural evolution.
SERB: study of enhanced radiation belt.
SEREL: *Société d'Exploitation et de Recherches Electioniques*.
SERI: Solar Energy Research Institute.
SERL: Services Electronics Research Laboratory.
SERPS: service propulsion system.
SERT: Society of Electronic and Radio Technicians.
SERT: space electrical rocket test.
SES: Solar Energy Society.
SES: Standards Engineering Society.
SES: Strategic Engineering Survey.
SESA: Society for Environmental Stress Analysis.
SESA: Society for Experimental Stress Analysis.
SESE: secure echo-sounding equipment.
SESL: Space Environmental Simulation Laboratory.
SESOME: service, sort and merge.
SET: self-extending translator.
SET: solar energy thermionic (conversion system).
SETA: simplified electronic tracking apparatus.
SETAB: sets tabular material (photocomposition).
SETAR: serial event time and recorder.
SEURE: systems evaluation code under radiation environment.
SEVAS: secure voice access system.
SEW: sonar early warning.
SF: sampled filter.
S/F: store and forward.
SFA: Scientific Film Association.

SFAR: system failure analysis report.
SFB: semiconductor functional block.
SFC: Solar Forecast Center (Air Force).
SFD: sudden frequency deviation.
SFD: system function description.
SFE: Society of Fire Engineers.
SFF: Solar Forecast Facility (Air Force).
SFIT: Swiss Federal Institute of Technology.
SFPE: Society of Fire Protection Engineers.
SFT: simulated flight test.
SFTS: standard frequency and time signals.
SG: screen grid (electrode).
SG: standing group.
SGHWR: steam generating heavy water reactor.
SGLS: space ground link system.
SGRCA: sodium graphite reactor critical assembly.
SGS: symbol generator and storage.
SGSR: Society for General Systems Research.
S/H: sample and hold.
SHA: sideral hour angle.
SHA: Sofrware Houses Association (GB).
SHEP: solar high-energy particles.
SHF: superhigh frequency.
SHIELD: Sylvania high intelligence electronic defense.
SHIEF: shared information elicitation facility.
SHIRAN: S-band of high-precision short-range electronic navigation.
SHIRTDIF: storage, handling, and retrieval of technical data in image formation.
SHORAN: short-range aid to navigation.
SHORAN: short-range navigation.
SHOT: Society for the History of Technology.
SHTC: short time constant.
SI: Smithsonian Institute.
SI: International System of Units.
SIA: subminiature integrated antenna.
SIA: system integration area.
SIAM: signal information and monitoring.
SIAM: Society for Industrial and Applied Mathematics.
SIAT: single integrated attack team.
SIB: satellite ionospheric beacons.
SIC: Science Information Council.
SIC: semiconductor integrated circuit.
SICEJ: Society of Instrument and Control Engineers of Japan.
SICO: switched in for checkout.
SID: silicon imaging device.
SID: Society for Information Displays.
SID: sudden ionospheric disturbance.
SID: syntax improving device.
SIDASE: significant data selection.
SIDS: stellar inertial doppler system.
SIE: Science Information Exchange.
SIFT: share internal Fortran translator.
SIGGEN: signal generator.
SII: Standards Institute of Israel.
SIMCHE: simulation and checkout equipment.
SIMCOM: simulator compiler.
SIMCON: scientific inventory management and control.
SIMD: single instruction multiple data.

SIMICORE: simultaneous multiple image correlation.
SIMILE: simulator of immediate memory in learning experiments.
SIMM: symbolic integrated maintenance manual.
simp: specific impulse.
SIMPAC: simplified programming for acquisition and control.
SIMS: single-item, multisource.
SIMS: symbolic integrated maintenance system.
sinh: hyperbolic sine.
SINS: ship's inertial marine navigational system.
SIOP: selector input/output processors.
SIOP: single integrated operations plan.
SIOUX: sequential iterative operation unit X.
SIP: symbolic input program.
SIPOP: satellite information processor operational program.
SIPROS: simultaneous processing operating system.
SIPS: simulated input preparation system.
SIR: selective information retrieval.
SIR: semantic information retrieval.
SIR: simultaneous impact rate.
SIR: statistical information retrieval.
SIR: submarine intermediate reactor.
SIR: symbolic input routine.
SIRA: Scientific Instrument Research Association.
SIRA: Safety Investigation Regulations.
SIRS: salary information retrieval system.
SIRS: satellite infrared spectrometer.
SIRSA: Special Industrial Radio Service Association.
SIRU: strapdown inertial reference unit.
SIS: satellite interceptor system.
SIS: simulation interface subsystem.
SIS: shorter interval scheduling.
SISD: single instruction single data (stream or serial processors).
SISS: single item, single source.
SISS: submarine integrated sonar system.
SIT: Society of Instrument Technology.
SIT: software integration test.
SITC: Standard International Trades Classification.
SITE: spacecraft instrumentation test equipment.
SITS: Sage intercept target simulation.
SITVC: secondary injection thrust vector control.
SIXPAC: system for inertial experiment priority and attitude control.
SJ: Simulation Journal.
SJCC: Spring Joint Computer Conference.
SJCM: Standing Joint Committee on Metrication.
SLA: spacecraft LM adapter.
SLA: Special Libraries Association.
SLAC: Stanford Linear Accelerator Center.
SLAET: Society of Licensed Aircraft Engineers and Technicians.
SLAM: space-launched air missile.
SLAM: supersonic low-altitude missile.
SLAMS: simplified language for abstract mathematical structures.
SLANT: simulator landing attachment for night landing training.
SLAR: side-looking airborne radar.

SLASH: Seiler Laboratory Algol simulated hybrid.

SLATE: stimulated learning by automated typewriter environment.

SLB: side-lobe blanking.

SLBM: submarine-launched ballistic missile.

SLBMDWS: submarine-launched ballistic missile detection and warning system.

SLC: side-lobe cancellation.

SLC: simulated linguistic computer.

SLCB: single-line color bar.

SLCC: Saturn launch control computer.

SLD: simulated launch demonstration.

SLE: Society of Logistics Engineers.

SLEAT: Society of Laundry Engineers and Allied Trades.

SLEW: static load error washout system.

SLI: sea level indicator.

SLIP: symmetric list processor.

SLIS: shared laboratory information system.

SLM: statistical learning model.

SLO: swept local oscillator.

SLP: segmented level programming.

SLP: source language processor.

SLR: side-looking radar.

SLS: side-lobe suppression.

SLS: side-looking sonar.

SLT: simulated launch test.

SLT: solid logic technology.

SLT: solid logic technique.

SLWL: straight-line wavelength.

SM: service module.

SM: sequence and monitor.

SM: shared memory.

SM: strategic missile.

SMA: Science Masters Association.

SMAC: special mission attack computer.

SMALGOL: small computer algorithmic language.

SMART: systems management analysis, research, and test.

SMC: Systems, Man, and Cybernetics (IEEE Society).

SMCC: simulation monitor and control console.

SMD: systems measuring device.

SMDC: Superconductive Materials Data Center.

SME: Society of Military Engineers.

SME: Society of Manufacturing Engineers.

SMF: system measurement facility.

SMG: Spacecraft Meteorology Group.

SMGP: strategic missile group.

SMIS: Society for Management Information Systems.

SML: symbolic machine language.

SMLM: simple-minded learning machine.

SMM: standard method of measurement.

SMMP: standard methods of measuring performance.

SMMT: Society of Motor Manufacturers and Traders.

SMOG: special monitor output generator.

SMPS: simplified message processing simulation.

SMPTE: Society of Motion Picture and Television Engineers.

SMRD: spin motor rate detector.

SMRE: Safety in Mines Research Establishment.

SMS: strategic missile squadron.

SMS: surface missile system.

SMS: synchronous-altitude meteorological satellite.

SMSAE: surface missile system availability evaluation.

SMT: service module technician.

SMT: square mesh tracking.

SMTI: selective moving target indicator.

SMWG: strategic missile wing.

SN: semiconductor network.

S/N: signal-to-noise (ratio).

SN: sine of the amplitude.

SNF: system noise figure.

SNAFU: situation normal, all fouled up.

SNAP: simplified numerical automatic programmer.

SNAP: space nuclear auxiliary power.

SNDT: Society of Nondestructive Testing.

SNEMSA: Southern New England Marine Sciences Association.

SNL: standard nomenclature list.

SNM: Society of Nuclear Medicine.

SNPM: standard and nuclear propulsion module.

SNPO: space nuclear propulsion office.

SNR: signal-to-noise ratio.

SNS: simulated network simulations.

SO: slow operate (relay).

SOAP: self-optimizing automatic pilot.

SOC: separated orbit cyclotron.

SOC: set overrides clear.

SOC: simulation operations center.

SOC: specific optimal controller.

SOCO: switched out for checkout.

SOCOM: solar optical communications system.

SOCR: sustained operations control room.

SOCS: spacecraft orientation-control system.

SOD: small object detector.

SODA: source oriented data acquisition.

SODAR: sound detecting and ranging.

SODAS: structure oriented description and simulation.

SOERO: small orbiting earth resources observatory.

SOFAR: sound fixing and ranging.

SOFAR: sound fusing and ranging.

SOFNET: solar observing and forecasting network.

SOFT: simple output format translator.

SOI: specific operating instruction.

SOI: standard operating instruction.

SOL: simulation oriented language.

SOL: systems oriented language.

SOLAR: serialized on-line automatic recording.

SOLID: self-organizing large information dissemination system.

solion: solution ion.

SOLIS: symbionics on-line information system.

SOLO: selective optical lock-on.

SOLRAD: solar radiation.

SOLV: solenoid valve.

SOM: start of message.

SOMADA: self-organizing multiple-access discrete address.

SONAC: sonar nacelle (sonacelle).

SONAR: sound navigation and ranging.

SONCM: sonar countermeasures and deception.

SONCR: sonar control room.

SONIC: system-wide on-line network for informational control.

SOP: simulation operations plan.
SOP: standard operating procedure.
SOP: strategic orbit point.
SOPM: standard orbital parameter message.
SOR: start of record.
SORTI: satellite orbital track and intercept.
SORTIE: supercircular orbital reentry test integrated environment.
SOS: share operating system.
SOS: silicon on sapphire.
SOS: symbolic operating system.
SOSI: shift in, shift out.
SOT: syntax-oriented translator.
SOTIM: sonic observation of the trajectory and impact of missiles.
SOTUS: sequentially operated teletypewriter universal selector.
SOUTHEASTCON: South-Eastern Convention (Region 3, IEEE).
SP: self-propelled.
SP: single pole.
SP: symbol programmer.
SP: system processor.
SPA: servo power assembly.
SPA: Southern Psychological Association.
SPA: Southwestern Psychological Association.
SPAC: spatial computer.
SPACE: self-programming automatic circuit evaluator.
SPACE: sequential position and covariance estimation.
SPACE: sidereal polar axis celestial equipment.
SPACON: space control.
SPAD: satellite position predictor and display.
SPAD: satellite protection for area defense.
SPADATS: space detection and tracking system.
SPADE: Sparta acquisition digital equipment.
SPAM: ship position and attitude measurement.
SPAMS: ship position and attitude measurement system.
SPAN: solar particle alert network.
SPAN: statistical processing and analysis.
SPAN: stored program alphanumerics.
SPANRAD: superposed panoramic radar display.
SPAQUA: sealed package quality assurance.
SPAR: symbolic program assembly routine.
SPARS: space precision attitude reference system.
SPARSA: sferics pulse, azimuth, rate, and spectrum analyzer.
SPARTA: spatial antimissile research tests in Australia.
SPASM: system performance and activity software monitor.
SPAT: silicon precision alloy transistor.
SPC: Static Power Converters (IEEE IAS Technical Committee).
SPC: stored program control.
SPC: stored programmed command.
SPD: Surge Protective Devices (IEEE PES Technical Committee).
SPDP: Society of Professional Data Processors.
SPDS: safe-practice data sheet.
SPDT: single pole double throw (switch).
SP3T: single pole triple throw (switch).
SP4T: single pole quadruple throw (switch).

SPE: Society of Petroleum Engineers.
SPE: Society of Plastics Engineers.
SPE: systems performance effectiveness.
SPEARS: satellite photoelectric analog rectification system.
SPEC: stored program educational computer.
SPECON: system performance effectiveness conference.
SPED: supersonic planetary entry decelerator.
SPEDAC: solid-state parallel expandable differential analyzer computer.
SPEDE: state system for processing educational data electronically.
SPEDTAC: stored program educational transistorized automatic computer.
SPEED: self-programmed electronic equation delineator.
SPEED: subsistence preparation by electronic energy diffusion.
SPERT: schedule performance evaluation and review technique.
SPERT: special power excursion reactor test.
SPES: stored program element system.
SPET: solid propellant electric thruster.
SPFP: single point failure potential.
SPG: single point ground.
SPG: sort program generator.
SPHE: Society of Packaging and Handling Engineers.
SPI: Society of the Plastics Industry.
SPI: specific productivity index.
SPIA: Solid Propellant Information Agency.
SPIC: Society of the Plastics Industry of Canada.
SPIDER: sonic pulse-echo instrument designed for extreme resolution.
SPIE: scavenging—precipitation—ion exchange.
SPIE: self-programmed individualized education.
SPIE: simulated problem, input evaluation.
SPIE: Society of Photooptical Instrumentation Engineers.
SPL: software programming language.
SPL: sound pressure level.
SPL: spaceborne programming language.
SPLIT: Sunstrand processing language internally translated.
SPM: self-propelled mount.
SPM: sequential processing machine.
SPM: source program maintenance.
SPM: symbol processing machine.
SPMS: solar particle monitoring system.
SPOT: satellite positioning and tracking.
SPRA: space probe radar altimeter.
SPRC: self-propelled robot craft.
SPRI: Scott Polar Research Institute.
SPRINT: solid propellant rocket intercept missile.
SPRITE: solid propellant rocket ignition test and evaluation.
SPRT: sequential probability ratio test.
SPS: secondary propulsion system.
SPS: symbolic programming system.
SPSE: Society of Photographic Scientists and Engineers.
SPST: single pole single throw (switch).

SPT: symbolic program tape.
SPT: symbolic program translator.
SPUR: space power unit reactor.
SPURM: special purpose unilateral repetitive modulation.
SPURT: spinning unguided rocket trajectory.
SQR: sequence relay.
SQR: service request.
SQUID: Sperry quick updating of internal documentation.
SQUID: superconducting quantum interference device.
SQW: square wave.
S/R: send and receive.
SR: saturable reactor.
SR: scientific report.
SR: selective ringing.
SR: selenium rectifier.
SR: shift register.
SR: short range.
SR: slip ring.
SR: slow release (relay).
SR: solid rocket.
SR: sound ranging.
SR: sound rating.
SR: speed recorder.
SR: split ring.
SR: storage and retrieval.
S/R: subroutine.
SRAM: short range attack missile.
SRBM: short range ballistic missile.
SRC: Science Research Council.
SRC: sound ranging control.
SRC: standard requirements code.
SRCRA: Shipowners Refrigeration Cargo Research Association.
SRD: secret restricted data.
SRDAS: service recording and data analysis system.
SRDE: Signals Research and Development Establishment.
SRDL: Signals Research and Development Laboratory.
SRDS: standard reference data system.
SRE: sodium reactor experiment.
SRG: Statistical Research Group.
SRI: Southern Research Institute.
SRI: Spalling resistance index.
SRI: Stanford Research Institute.
SRLY: series relay.
SRR: shift register recognizer.
SRR: sound recorder reproducer.
SRS: simulated remote sites.
SRS: subscriber response system (cable television).
SRSA: Scientific Research Society of America.
SRSK: short-range station keeping.
SRSS: simulated remote sites subsystem.
SRT: Society of Radiologic Technologists.
SRT: supporting research and technology.
SRT: systems readiness test.
SS: signal strength.
SS: single shot.
SS: solid state.
SS: spin-stabilized (rockets).
SS: space simulator.

SS: summing selector.
SSAR: spin-stabilized aircraft rocket.
SSB: single sideband.
SSBAM: single-sideband amplitude modulation.
SSBD: single-sideboard.
SSBFM: single-sideband frequency modulation.
SSBN: ship submersible nuclear.
SSBO: single swing blocking oscillator.
SSBSC: single-sideband suppressed carrier.
SSBSCOM: single-sideband suppressed-carrier optical modulator.
SSD: single-station doppler.
SSE: safe shutdown earthquake.
SSEC: selective sequence electronic calculator.
SSEP: system safety engineering plan.
SSESM: spent stage experimental support module.
SSG: small signal gain.
SSGS: standard space guidance system.
SSI: sector scan indicator.
SSI: small scale integration.
SSI: storage-to-storage instruction.
SSIG: single signal.
SSL: solid-state lamp.
SSM: surface-to-surface missile.
SSMTG: solid-state and molecular theory group.
SSP: scientific subroutine package.
SSPM: single-sideband frequency modulation.
SSR: secondary surveillance radar.
SSR: solid state relay.
SSR: synchronous stable relaying.
SSRS: start—stop—restart system.
SSRT: subsystem readiness test.
SSS: scientific subroutine system.
SSS: simulation study series.
SSSA: Soil Science Society of America.
SSSC: surface subsurface surveillance center.
SST: simulated structural test.
SST: subsystems test.
SST: supersonic transport.
SSTC: single-sideband transmitted carrier.
SSTP: subsystems test procedure.
SSW: synchro switch.
ST: sawtooth.
ST: Schmitt trigger.
ST: scientific and technical.
ST: single throw.
STA: shuttle training aircraft.
STAC: Science and Technology Advisory Committee for MSF.
STADAN: satellite tracking and data acquisition network (NASA).
STAF: scientific and technological applications forecast.
STALO: stabilized local oscillator.
STAMO: stabilized master oscillator.
STAMOS: sortie turn-around maintenance operations simulation.
STAMP: systems tape addition and maintenance program.
STAR: Scientific and Technical Aerospace Report.
STAR: self-testing and repairing.
STARFIRE: system to accumulate and retrieve financial information with random extraction.

STARS: satellite telemetry automatic reduction system.
START: selections to activate random testing.
START: systematic tabular analysis of requirements technique.
STATPAC: statistics package.
STB: subsystems test bed.
STC: satellite test center.
STC: Scientific and Technical Committee.
STC: sensitivity time control.
STC: short time constant.
STC: standard transmission code.
STC: system test complex.
STD: salinity temperature depth.
STD: subscriber trunk dialing.
STE: Society of Tractor Engineers.
STEM: stay time extension module.
STEP: scientific and technical exploitation program.
STEP: simple transition to economical processing.
STEP: simple transition electronic processing.
STEP: standard terminal program.
STEPS: solar thermionic electric power system.
STET: specialized technique for efficient typesetting.
STI: scientific technical information.
STIC: Scientific Technical Intelligence Center.
STID: Scientific and Technical Information Division (NASA).
STINFO: scientific and technical information.
STL: Schottky transistor logic.
STL: Space Technology Laboratory.
STL: studio transmitter link.
STLO: Scientific and Technical Liaison Office.
STM: structural test model.
STM: system master tape.
STMIS: system test manufacturing information system.
STMU: special test and maintenance unit.
STN: satellite tracking network (NASA).
STO: system test objectives.
STORET: storage and retrieval.
STORLAB: Space Technology Operations and Research Laboratory.
STORM: statiscally oriented matrix program.
STP: selective tape print.
STP: standard temperature and pressure.
STP: system test plan.
STPO: Science and Technology Policy Office.
STR: synchronous transmitter receiver.
STRAD: signal transmission reception and distribution.
STRESS: structural engineering system solver.
STRIVE: standard techniques for reporting information on value engineering.
STROBES: shared time repair of big electronic systems.
STRUDL: structural design language.
STS: satellite tracking station.
STS: structural transition section.
STU: systems test unit.
STV: surveillance television.
STWP: Society of Technical Writers and Publishers.
SU: service unit.

SU: Sonics and Ultrasonics (IEEE Group).
SU: storage unit.
SUA: State Universities Association.
SUAS: system for upper atmospheric sounding.
SUDT: silicon unilateral diffused transistor.
SUHL: Sylvania ultra-high level logic.
SUI: Stanford University Institute for Plasma Research.
SUM: surface-to-underwater missile.
SUMMIT: supervisor of multiprogramming, multiprocessing, interactive time sharing.
SUMT: sequential unconstrained minimization technique.
SUPARCO: Space and Upper Atmosphere Research Committee.
SUPROX: successive approximation.
SURCAL: surveillance calibration.
SURGE: sorting, updating, report generating.
SURIC: surface ship integrated control.
SUS: single underwater sound.
SUS: silicon unilateral switch.
SUSIE: stock updating sales invoicing electronically.
SV: safety valve.
SVIC: Shock and Vibration Information Center.
SVR: supply-voltage rejection.
SVTL: Services Valve Test Laboratory.
SVTP: sound velocity, temperature and pressure.
SW: short wave.
SW: single weight.
SW: switchband wound (relay).
SWAC: Standards Western automatic computer.
SWAT: sidewinder IC acquisition track.
SWCL: Sea Water Conversion Laboratory.
SWE: Society for Women Engineers.
SWFR: slow write, fast read.
SWG: British Standard Wire Gage.
SWG: Stubs Wire Gage.
SWIFT: software implemented Friden translator.
SWITT: surface wave independent tap transducer.
SWOP: structural weight optimization program.
SWP: safe working pressure.
SWR: standing wave ratio.
SWRI: Southwestern Research Institute.
SWT: supersonic wind tunnel.
SWTL: surface-wave transmission line.
SWW: severe weather warning.
SYCOM: synchronous communications.
SYDAS: system data acquisition system.
SYMPAC: symbolic program for automatic control.
SYNCOM: synchronous-orbiting communications satellite.
SYNSEM: syntax and semantics.
SYNTOL: syntagmatic organization of language.
SYSGEN: systems generation.
SYSIN: system input.
SYSPOP: system programmed operator.
SYSTRAN: systems analysis translator.

T

TA: turbulence amplifier.
TAA: Transportation Association of America.
TAAS: three-axis attitude sensor.
TAB: tabular language.
TAB: technical abstract bulletin.
TAB: Technical Activities Board.
TABS: terminal access to batch service.
TABSIM: tabulator simulator.
TABSOL: tabular systems oriented language.
TAC: trapped air cushion.
TAC: TRANSAC assembler compiler.
TAC: transistorized automatic control.
TAC: translator—assembler—compiler.
TACAN: tactical air navigation.
TACCAR: time-averaged clutter-coherent airborne radar.
TACDEN: tactical data entry device.
TACL: time and cycle log.
TACODA: target coordinate data.
TACOL: thinned aperture computed lens.
TACPOL: tactical procedure oriented language.
TACR: time and cycle record.
TACS: tactical air control system.
TACT: transistor and component tester.
TAD: top assembly drawing.
TADS: teletypewriter automatic dispatch system.
TADSS: tactical automatic digital switching system.
TAE: Technion Aeronautical Engineering (Israel).
TAG: Technical Assistance Group.
TAG: transient analysis generator.
TAI: time to autoignition.
TAM: telephone answering machine.
TAMIS: telemetric automated microbial identification system.
tanh: hyperbolic tangent.
TAO: Technical Assistance Order.
TAP: terminal applications package.
TAP: time-sharing assembly program.
TAPAC: tape automatic positioning and control.
TAPE: tape automatic preparation equipment.
TAPP: two-axis pneumatic pickup.
TAPS: tactical area positioning system.
TAPS: turboalternator power system.
TAR: terrain-avoidance radar.
TAR: trajectory analysis room.
TARAN: tactical attack radar and navigation.
TARE: telemetry automatic reduction equipment.
TARFU: things are really fouled up.
TARGET: thermal advanced reactor gas-cooled exploiting thorium.
TARMAC: terminal area radar moving aircraft.
TARS: terrain and radar simulator.
TARS: three-axis reference system.
TART: twin accelerator ring transfer.
TASC: terminal area sequence and control.
TASCON: television automatic sequence control.
TASI: time-assignment speech interpolation.
TAT: thrust-augmented Thor.
TATC: transatlantic telephone cable.

TATC: terminal air-traffic control.
TAVE: Thor-Agena vibration experiment.
TB: technical bulletin.
TB: terminal board.
TB: time base.
TB: transmitter—blocker (cell).
TBAX: tube axial.
TBC: toss bomb computer.
TBD: target-bearing designator.
TBF: tail bomb fuse.
TBI: target-bearing indicator.
TBL: Terminal Ballistics Laboratory.
TBO: time between overhaul.
TBX: tactical ballistic missile experiment.
TC: tactical computer.
TC: tantalum capacitor.
TC: technical committee.
TC: technical control.
TC: temperature coefficient.
TC: temperature compensating.
TC: terminal computer.
TC: test conductor.
TC: test console.
TC: thermocouple.
TC: thermocurrent.
TC: thrust chamber.
TC: time constant.
TC: timed closing.
TC: tracking camera.
TC: transmission control.
TCAI: tutorial computer-assisted instruction.
TCB: technical coordinator bulletin.
TCBM: transcontinental ballistic missile.
TCC: technical computing center.
TCC: Telecommunications Coordinating Committee.
TCC: temperature coefficient of capacitance.
TCC: test control center.
TCC: test controller console.
TCC: transfer channel control.
TCD: telemetry and command data.
TCD: thyratron core driver.
TCD: transistor-controlled delay.
TCE: telemetry checkout equipment.
TCEA: Training Center for Experimental Aerodynamics.
TCED: thrust control exploratory development.
TCG: tune-controlled gain.
TCI: telemetry components information.
TCI: terrain clearance indicator.
TCI: Theoretical Chemistry Institute.
TCL: time and cycle log.
TCM: terminal-to-computer multiplexer.
TCMF: touch calling multifrequency.
TCP: test checkout procedure.
TCP: thrust chamber pressure.
TCP: traffic control post.
TCPC: tab card punch control.
TCR: temperature coefficient of resistance.
TCS: terminal countdown sequencer.

TCS: terminal count sequence.
TCS: thermal conditioning service.
TCS: traffic control station.
TCS: transportation and communications service.
TCSC: trainer control and simulation computer.
TCWG: telecommunications working group.
TCT: translator and code treatment frame.
TCU: tape control unit.
TCXO: temperature-compensated crystal oscillator.
TCXO: temperature-controlled crystal oscillator.
T&D: Transmission and Distribution (IEEE PES Technical Committee).
TD: tabular data.
TD: terminal distributor.
TD: temperature differential.
TD: testing device.
TD: thoria dispersed.
TD: time delay.
TD: transmitter distributor.
TD: tunnel diodes.
TDA: target docking adapter.
TDA: tracking and data acquisition.
TDA: tunnel diode amplifier.
TDC: time delay closing.
TDCM: transistor driver core memory.
TDCO: torpedo data computer operator.
TDD: target detection device.
TDD: technical data digest.
TDDL: time division data link.
TDDR: technical data department report.
TDEC: Technical Division and Engineering Center.
TDG: test data generator.
TDI: Tool and Die Institute.
TDL: tunnel-diode logic.
TDM: time division multiplex.
TDM: torpedo detection modification.
TDMA: time division multiple access.
TDMS: telegraph distortion measurement set.
TDMS: time-shared data management system.
TDN: target doppler nullifier.
TDO: time delay opening.
TDOS: tape/disk operating system.
TDP: technical development plan.
TDR: target discrimination radar.
TDR: technical data relay.
TDR: time delay relay.
TDR: time domain reflectometry.
TDR: torque differential receiver.
TDS: target designation system.
TDS: technical data system.
TDS: test data sheet.
TDS: tracking and data system.
TDS: translation and docking simulator.
TDT: target designation transmitter.
TDX: thermal demand transmitter.
TE: test equipment.
TE: thermal element.
TE: thermoelectric.
TE: totally enclosed.
TE: transverse electric (field).
TEA: Technical Engineers Association.
TEA: transferred electron amplifier.

TEA: tyrethylaluminum.
TEA: tunnel-emission amplifier.
TEAM: technique for evaluation and analysis of maintainability.
TEAMS: test evaluation and monitoring system.
TEC: tactical electromagnetic coordinator.
TED: Test Engineering Division.
TED: translation error detector.
TEE: Telecommunications Engineering Establishment.
TEIC: tissue equivalent ionization chamber.
Tel: Telephone Group.
TELEDAC: telemetric data converter.
TELEX: teleprinter exchange.
TELSIM: teletypewriter simulator.
TELUS: telemetric universal sensor.
TEM: transmission electron microscope.
TEM: transverse electromagnetic.
TEMP: electrical resistance temperature.
TEO: transferred electron oscillator.
TEOM: transformer environment overcurrent monitor.
TEPG: thermionic electrical power generator.
TEPRSSC: Technical Electronic Product Radiation Safety Standards Committee.
TER: transmission equivalent resistance.
TER: triple ejection rack.
TERAC: Tactical Electromagnetic Readiness Advisory Council.
TEREC: tactical electromagnetic reconnaissance.
TERP: terrain elevation retrieval program.
TERS: tactical electronic reconnaissance system.
TESS: tactical electromagnetic systems study.
TEST: Thesaurus of Engineering and Scientific Terms.
TET: total elapsed time.
TETRA: terminal tracking telescope.
TETROON: tethered meteorological balloon.
TEVROC: tailored exhaust velocity rocket.
TEXTIR: text indexing and retrieval.
TF: test fixture.
TFD: Television Feasibility Demonstration.
TFR: terrain-following radar.
TFT: thin-film technology.
TFT: thin-film transistor.
TFT: threshold failure temperatures.
TG: tuned grid.
TGA: thermogravimetric analysis.
TGC: transmit gain control.
TGS: translator generator system.
TGSE: telemetry ground support equipment.
THC: thermal converter.
THC: thrust hand controller.
THD: total harmonic distortion.
THERP: technique for human error rate prediction.
THI: temperature humidity index.
THIR: temperature humidity infrared radiometer.
THOMIS: total hospital operating and medical information system.
THOPS: tape handling operational system.
THOR: tape handling option routines.
THTR: thorium high temperature reactor.
THTRA: Thorium High Temperature Reactor Association.
THz: terahertz.

THYMOTRO: thyratron motor control.
TI: tape inverter.
TI: time interval.
TI: transfer impedance.
TIARA: target illumination and recovery aid.
TIAS: target identification and acquisition system.
TIB: Technical Information Bureau.
TIBOE: transmittng information by optical electronics.
TIC: tape intersystem connection.
TIC: target intercept computer.
TIC: Technical Information Center.
TIC: Telemetry Instruction Conference.
TIC: temperature indicating controller.
TIC: Transducer Information Center.
TICA: Technical Information Center Administration.
TICE: time integral cost effectiveness.
TICS: teacher interactive computer system.
TID: Technical Information Division.
TID: Test Instrument Division.
TIDAR: time delay array radar.
TIDDAC: time in deadband digital attitude control.
TIDES: time division electronics switching system.
TIE: technical integration and evaluation.
TIES: transmission and information exchange system.
TIF: telephone influence factor.
TIF: telephone interference factor.
TIG: tungsten inert gas.
TIIF: tactical image interpretation facility.
TIIPS: technically improved interference prediction system.
TIL: Technical Information and Library Services.
TIM: time meter.
TIMM: thermionic integrated micromodules.
TIMS: The Institute of Management Sciences.
TIO: time interval optimization.
TIP: technical information processing.
TIP: technical information project.
TIPI: tactical information processing and interpretation.
TIPL: teach information processing language.
TIPP: time-phasing program.
TIPS: technical information processing system.
TIPTOP: tape input, tape output.
TIR: technical information report.
TIR: total indicator reading.
TIROS: television and infrared observation satellite.
TIRP: total internal reflection prism.
TIS: target information sheet.
TIS: technical information service.
TIS: total information system.
TIU: tape identification unit.
TJC: trajectory chart.
TJD: trajectory diagram.
TL: tape library.
TL: target language.
TL: tie line.
TLC: thin layer chromatography.
TLD: thermoluminescent dosimeter.
TLE: tracking light electronics.
TLI: telephone line interface.
TLK: test link.
TLM: telemeter.

TLP: threshold learning process.
TLP: total language processor.
TLS: telecommunication liaison staff.
TLS: terminal landing system.
TLU: table look-up.
TLV: threshold limit value.
TM: tactical missile.
TM: tape mark.
TM: technical manual.
TM: telemetry.
TM: temperature meter.
TM: tone modulation.
TM: transmission matrix.
TM: transverse magnetic (field).
TMAMA: Textile Machinery and Accessory Manufacturers Association.
TMCC: time-multiplexed communication channel.
TMG: thermal meteoroid garment.
TML: tetramethyl lead.
TMN: technical and management note.
TMR: triple modular redundancy.
TMRBM: transportable medium range ballistic missile.
TMS: tactical missile squadron.
TMS: Temperature Measurements Society.
TMS: time-shared monitor system.
TMS: The Metallurgical Society.
TMS: transmission measuring set.
TMX: telemeter transmitter.
TN: technical note.
TN: Tennessee.
TNT: trinitrotoluene.
T&O: test and operation.
TO: transistor outline.
TOCS: technological aides to creative thoughts.
TOD: technical objective directive.
TODS: test-oriented disc system.
tohm: terohmmeter.
TOI: technical operation instruction.
TOJ: track on jamming.
TOL: test oriented language.
TOLIP: trajectory optimization and linearized pitch.
TOMCAT: telemetry on-line monitoring compression and transmission.
TOPP: terminal operated production language.
TOPSY: test operations and planning system.
TOPS: teletype optical projection system.
TOPTS: test-oriented paper-tape system.
TOR: Technical Operations Research.
TOS: terminal-oriented software.
TOS: terminal-oriented system.
TOS: Tiros operational satellite.
TOSBAC: Toshiba scientific and business automatic computer.
TOSS: Tiros operational satellite system.
TOT: time of tape.
TOW: tube-launched optically tracked wire-guided antitank missile.
TOWA: terrain and obstacle warning and avoidance.
TP: technical publication.
TP: teleprocessing.
TP: test procedure.
TPC: Telecommunications Planning Committee.

TPF: terminal phase finalization.

TPG: transmission project group (CEGB).

TPI: tape phase inverter.

TPI: target position indicator.

TPI: terminal phase initiate.

TPL: test parts list.

TPM: tape preventive maintenance.

TPM: telemetry processor module.

TPMA: thermodynamics properties of metals and alloys.

TPP: test point pace.

TPPD: Technical Program Planning Division.

TPRC: Thermophysical Properties Research Center (Purdue University).

TPS: task parameter synthesizer.

TPS: thermal protection system.

TPSI: torque pressure in pounds per square inch.

TPTG: tuned plate, tuned grid.

TPU: tape preparation unit.

TR: tape recorder.

TR: transient response.

TR: transmit-receive.

TR: transmitter receiver.

TRAACS: transit research and attitude control satellite.

TRAC: text reckoning and compiling.

TRAC: transient radiation analysis by computer.

TRACE: tape-controlled reckoning and checkout equipment.

TRACE: teleprocessing recording for analysis by the customer.

TRACE: time-shared routines for analysis, classification, and evaluation.

TRACE: tolls recording and computing equipment.

TRACE: transportable automated control environment.

TRACON: terminal radar control.

TRADIC: transistor digital computer.

TRAIN: telerail automated information network.

TRAMP: time-shared relational associative memory program.

TRANDIR: translation director.

TRANSAC: transistorized automatic computer.

TRAP: terminal radiation airborne program.

TRAP: tracker analysis program.

TRAWL: tape read and write library.

TRC: tape record coordinator.

TRDTO: tracking radar take-off.

TRE: Telecommunications Research Establishment.

TREAT: transient radiation effects automated tabulation.

TREE: transient radiation effects on electronics.

TRF: tuned radio frequency.

TRI: Technical Research Institute.

TRIAL: technique for retrieving information from abstracts of literature.

TRICE: transistorized real-time incremental computer.

TRL: Thermodynamics Research Laboratory.

TRL: transistor-resistor logic.

TRM: thermal remanent magnetization.

TRN: technical research note.

TROS: tape resident operating system.

TRR: target ranging radar.

TRS: test response spectrum.

TRS: time reference system.

TRSB: time reference scanning beam.

TRSSM: tactical range surface-to-surface missile.

TRTL: transistor—resistor—transistor logic.

TRU: transportable radio unit.

TRUMP: target radiation measurement program.

TRUMP: teller register unit monitoring program.

TRV: transient recovery voltage.

TS: tensile strength.

TS: test set.

TS: time sharing.

TS: time switch.

TSA: time series analysis.

TSB: twin sideband.

TSC: technical subcommittee.

TSC: transmitter start code.

TSCLT: transportable satellite communications link terminal.

TSDD: temperature—salinity—density—depth.

TSDM: time-shared data management system.

TSDOS: time-shared disk operating system.

TSF: ten-statement FORTRAN.

TSF: thin solid films.

TSEQ: time sequence.

TSI: threshold signal-to-interference ratio.

TSM: time shared monitor system.

TSO: time sharing option.

TSOS: time sharing operating system.

TSPS: time sharing programming system.

TSS: time-sharing system.

TSU: technical service unit.

TT: terminal timing.

TT: timing and telemetry.

TT: tracking telescope.

TTA: turbine-alternator assembly.

TTBWR: twisted tape boiling water reactor.

TT&C: telemetry tracking and control.

TTE: time to event.

TTG: technical translation group.

TTG: time to go.

TTI: teletype test instruction.

TTI: time temperature indicator.

TTL: transistor-transistor logic.

TTP: tape to print.

TTR: target tracking radar.

TTS: teletypesetting.

TTS: temporary threshold shift.

TTS: transistor-transistor logic Schottky barrier.

TTS: transmission test set.

TTU: terminal time unit.

TTY: teletypewriter.

TUD: Technology Utilization Division (NASA).

TUM: tuning unit member.

TURPS: terrestrial unattended reactor power system.

TUT: transistor under test.

TV: television.

TV: test vehicle.

TV: thermal vacuum.

TVA: Tennessee Valley Authority.

TVA: thrust vector alignment.

TVC: thrust vector control.

TVCS: thrust vector control system.

TVI: television interference.
TVIST: television information storage tube.
TVM: tachometer voltmeter.
TVM: transistor voltmeter.
TVOC: television operations center.
TVOR: terminal very high frequency omnirange.
TVPPA: Tennessee Valley Public Power Association.
TW: thermal wire.
TW: traveling wave.
TWA: traveling-wave amplifier.
TWCRT: traveling-wave cathode-ray tube.
TWM: traveling wave maser.

TWMBK: traveling-wave multiple-beam klystron.
TWMR: tungsten water-moderated reactor.
TWS: track while scan.
TWSB: twin sideband.
TWT: traveling wave tube.
TWTA: traveling-wave-tube amplifier.
TWX: teletypewriter exchange.
TWX: teletypewriter exchange service.
TX: Texas.
TX: transmit.
TXE: telephone exchange electronics.
TYDAC: typical digital automatic computer.

U

UAA: University Aviation Association.
UADPS: uniform automatic data processing system.
UAIDE: users of automatic information display equipment.
UAM: underwater-to-air missile.
UARI: University of Alabama Research Institute.
UART: universal asynchronous receiver transmitter.
UAUM: underwater-to-air-to-underwater missile.
UAW: United Auto Workers.
UBC: universal buffer controller.
UC: unit call.
UC/SSL: University of California/Space Sciences Laboratory.
UCCRS: underwater coded command release system.
UCCS: universal camera control system.
UCL: upper confidence level.
UCORC: University of California/Operations Research Center.
UCRI: Union Carbide Research Institute.
UCRL: University of California Research Laboratory.
UCS: Universal Camera Site.
UCSEL: University of California Structural Engineering Laboratory.
UCW: unit control word.
UCWE: Underwater Countermeasures and Weapons Establishment.
UDAR: universal digital adaptive recognizer.
UDB: up data buffer.
UDC: universal decimal classification.
UDEC: unitized digital electronic calculator.
UDL: up data link.
UDOFFT: universal digital operational flight trainer.
UDOP: uhf doppler.
UDP: United Data Processing.
UET: universal engineer tractor.
UF: ultrasonic frequency.
UFO: unidentified flying object.
UFR: underfrequency relay.
UGLIAC: United Gas Laboratory interally programmed automatic computer.
UH: unit heater.
UHF: ultra high frequency (300 MHz to 3 GHz).

UHMW: ultrahigh molecular weight.
UHT: ultra high temperature.
UHTREX: ultra high temperature reactor experiment.
UIEO: Union of International Engineering Organizations.
UIL: Univac interactive language.
UIM: Union of International Motor Boating.
UJT: unijunction transistor.
UKAEA: United Kingdom Atomic Energy Authority.
UL: Underwriters Laboratories.
ULB: universal logic block.
ULC: universal logic circuit.
ULF: ultra low frequency (300 to 0 Hz).
ULMS: underseas long-range missile system.
ULO: unmanned launch operations.
ULOW: unmanned launch operations western test range.
ULPR: ultra low pressure rocket.
ULSV: unmanned launch space vehicles.
ULTRA-X: universal language for typographic reproduction applications.
UMA: universal measuring amplifier.
UMASS: unlimited machine access from scattered sites.
UNADS: Univac automated documentation system.
UNAMACE: universal automatic map compilation equipment.
UNCAST: United Nations Conference on the Applications of Science and Technology.
UNCOL: universal computer-oriented language.
UNESCO: United Nations Educational, Scientific, and Cultural Organization.
UNICOM: universal integrated communications.
UNICOMP: universal compiler.
UNICON: unidensity coherent light recording.
UNIPOL: universal procedures oriented language.
UNISAP: Univac share assembly program.
UNISIST/ICSU: International Council of Scientific Unions.
UNITAR: United Nations Institute for Training and Research.
UNITRAC: universal trajector compiler.
UNIVAC: universal automatic computer.

UNPS: universal power supply.
UNSCEAR: United Nations Scientific Committee on the Effects of Atomic Radiation.
UOC: ultimate operating capability.
UP: utility path.
UPDATE: unlimited potential data through automation technology in education.
UPL: universal programming language.
UPL: user programming language.
UPOS: utility program operating system.
UPR: ultrasonic parametric resonance.
UPS: uninterruptible power supply.
U/R: up range.
URA: Universities Research Associates.
URBM: ultimate range ballistic missile.
URIPS: undersea radioisotope power supply.
URIR: unified radioactive isodromic regulator.
URPA: University of Rochester, Department of Physics and Astronomy.
URRI: Urban Regional Research Institute (Michigan State University).
URS: unate ringe sum (logic expression).
URS: United Research Service.
URS: universal regulating system.
USA: United States of America.
USAASO: U.S. Army Aeronautical Services Office.
USAAVLABS: U.S. Army Aviation Materiel Laboratories.
USAAVNTA: U.S. Army Aviation Test Activity.
USABAAR: U.S. Army Board for Aviation Accident Research.
USABESRL: U.S. Army Behavioral Science Research Laboratory.
USABRL: U.S. Army Ballistics Research Laboratories.
USACDA: U.S. Arms Control and Disarmament Agency.
USACSC: U.S. Army Computer Systems Command.
USACSSC: U.S. Army Computer Systems Support and Evaluation Command.
USAEC: U.S. Atomic Energy Commission.
USAECOM: U.S. Army Electronics Command.
USAEPG: U.S. Army Electronics Proving Ground.
USAERDL: U.S. Army Engineer Research and Development Laboratories.
USAFETAC: U.S.A.F. Environmental Technical Applications Center.
USAM: Unified Space Applications Mission (NASA).
USANWSG: U.S. Army Nuclear Weapons Surety Group.
USAOMC: U.S. Army Ordnance Missile Command.
USAREPG: U.S. Army Electronic Proving Ground.
USARIEM: U.S. Army Research Institute of Environmental Medicine.
USARP: U.S. Antarctic Research Program.
USARPA: U.S. Army Radio Propagation Agency.
USAS: USA Standard.
USASCII: U.S. American Standard Code for Information Interchange (now ASCII).
USASI: U.S.A. Standards Institute (now ANSI).
USASMSA: U.S. Army Signal Missile Support Agency.

USATEA: U.S. Army Transportation Engineering Agency.
USB: upper sideband.
USBE: unified S-band equipment.
USBS: unified S-band system (tracking).
USCAL: University of Southern California Aeronautical Laboratory.
USCEC: University of Southern California Engineering Center.
USC&GS: U.S. Coast and Geodetic Survey.
USD: ultimate strength design.
USDOD: U.S. Department of Defense.
USE: unit support equipment.
USE: Univac scientific exchange.
USEC: united system of electronic computers.
USFS: U.S. Frequency Standard.
USFSS: U.S. Fleet Sonar School.
USG: United States Gage.
USGPO: U.S. Government Printing Office.
USGR: U.S. Government Report.
USGRDR: U.S. Government Research and Development Report.
USGS: U.S. Geological Survey.
USIA: U.S. Information Agency.
USIC: U. S. Information Center.
USIS: U. S. Information Service.
USJPRS: U. S. Joint Publication Research Service.
USL: Underwater Sound Laboratory.
USM: underwater-to-surface missile.
USNEL: U. S. Navy Electronics Laboratory.
USNRDL: U. S. Naval Radiological Defense Laboratory.
USNUSL: U. S. Navy Undersea Laboratory.
USPO: U. S. Patent Office.
USPO: U. S. Post Office.
USRL: Underwater Sound Reference Laboratory.
USRS: U. S. Rocket Society.
USS: United States Standard.
USW: undersea warfare.
USWB: U. S. Weather Bureau.
UT: Utah.
UTEC: Utah University College of Engineering.
UTIA: University of Toronto Institute of Aerophysics.
UTIAS: University of Toronto Institute for Aerospace Studies.
UTM: universal test message.
UTM: universal transverse mercator.
UTS: ultimate tensile strength.
UTS: unified transfer system.
UTS: universal test station.
UTS: universal time sharing.
UTTC: universal tape-to-tape converter.
UUA: Univac Users Association.
UUM: underwater-to-underwater missile.
UUT: unit under test.
UV: ultraviolet.
UV: undervoltage.
UVASER: ultraviolet amplification by stimulated emission of radiation.
UVD: undervoltage device.
UWAL: University of Washington Aeronautical Laboratory.

V

VA: value analysis.
VA: Virginia.
VA: volt-ampere.
VAB: voice answer back.
VADE: versatile auto data exchange.
VAEP: variable, attributes, error propagation.
VALSAS: variable length word symbolic assembly system.
VAM: vector airborne magnetometer.
VAM: voltammeter.
VAMP: vector arithmetic multiprocessor.
VAMP: visual—acoustic—magnetic pressure.
VAR: visual aural range.
VAR: volt-ampere reactive.
VARAD: varying radiation.
VARHM: var-hour meter.
VARISTOR: variable resistor.
VARITRAN: variable-voltage transformer.
VARR: variable range reflector.
VARR: visual aural radio range.
VARS: vertical azimuth reference system.
VASCA: Electronic Valve and Semiconductor Manufacturers Association.
VASCAR: visual average speed computer and recorder.
VASI: visual approach slope indicator.
VASIS: visual approach slope indicator system.
VAST: versatile automatic specification tester.
VAST: versatile avionics ship test.
VAT: virtual address translator.
VATE: versatile automatic test equipment.
VATLS: visual airborne target locator system.
VB: valve box.
VBD: voice band data.
VBL: Voyager Biological Laboratory.
VC: voice coil.
VC: vector control.
VCC: Visual Communications Congress.
VCC: voice control center.
VCD: variable capacitance diode.
VCF: voltage-controlled frequency.
VCG: vertical location of the center of gravity.
VCI: volatile corrosion inhibitor.
VCL: vertical center line.
VCO: voltage-controlled oscillator.
VCR: video cartridge recorder.
VCS: Victorian Computer Society.
VCS: visually coupled system.
VCT: voltage control transfer.
VCXO: voltage-controlled crystal oscillator.
V/D: voice/data.
VD: voltage drop.
VDAS: vibration data acquisition system.
Vdc: volts dc.
VDDI: Voyager data detailed index.
VDDL: Voyager data distribution list.
VDDS: Voyager data description standards.
VDE: *Verband Deutscher Elektrotechniker.*
VDET: voltage detector.
VDF: video frequency.

VDP: vertical data processing.
VDPI: Voyager data processing instructions.
VDRA: voice and data recording auxiliary.
VDS: variable depth sonar.
VE: value engineering.
VE: vernier engine.
VEA: Value Engineering Association.
VEB: variable elevation beam.
VECI: vehicular equipment complement index.
VECO: vernier engine cutoff.
VECOS: vehicle checkout set.
VECP: value engineering change proposal.
VEDS: vehicle emergency detection system (NASA).
VEFCO: Vertient functional checkout.
VELF: velocity filter.
VEM: vasoexcitor material.
VEPIS: vocational education program information system.
VERA: versatile experimental reactor assembly.
VERNITRAC: vernier tracking by automatic correlation.
VERS: versed sine.
VEST: Volunteer Engineers, Scientists and Technicians.
VEV: voice-excited vocoder.
VEWS: very early warning system.
VF: variable frequency.
VF: video frequency.
VF: voice frequency.
V/F: voltage to frequency.
VFC: voltage-to-frequency converter.
VFO: variable frequency oscillator.
VFU: vertical format unit.
VGPI: visual glide path indicator.
VHAA: very high altitude abort.
VHF: very high frequency.
VHM: virtual hardware monitor.
VI: Virgin Islands.
VIB: vertical integration building.
VIC: variable instruction computer.
VIDAC: visual information display and control.
VIDAMP: video amplifier.
VIDAT: visual data acquisition.
VIDF: video frequency.
VIFI: Voyager information flow instructions.
VINS: velocity inertia navigation system.
VINT: video integration.
VIP: variable information processing.
VIP: visual information projection.
VIPER: video processing and electronic reduction.
VIPP: variable information processing package.
VIPS: voice interruption priority system.
VIR: vertical interval retrace.
VIRNS: velocity inertia radar navigation system.
VITAL: variably initialized translator for algorithmic languages.
VITS: vertical interval test signal.
viz: *videlicet* (namely).
VKIFD: Von Karman Institute for Fluid Dynamics.
VL: video logic.

VLA: very large array.
VLA: very low altitude.
VLCS: voltage-logic, current-switching.
VLED: visible light-emitting diodes.
VLF: variable field length.
VLF: very low frequency (3 to 30 kHz).
VLF: vertical launch facility.
VLR: very long range.
VLVS: voltage-logic, voltage switching.
V/m: volt per meter (unit of picofarad electric field strength).
VM: velocity modulation.
VM: virtual machine.
VM: voltmeter.
VMA: Valve Manufacturers Association.
VMC: visual meteorological conditions.
VMD: vertical magnetic dipole.
VMGSE: vehicle measuring GSE.
V/mil: volt per mil.
VMM: virtual machine monitor.
VMTSS: virtual machine time sharing system.
VOC: voice-operated coder.
VOCOM: voice communications.
VOD: velocity of detonation.
VODACOM: voice data communications.
VODAT: voice-operated device for automatic transmission.
VOGAD: voice-operated gain-adjusting device.
VOIS: visual observation instrumentation subsystem.
VOLTAN: voltage amperage normalizer.
VOM: volt-ohm meter.
VOR: very-high-frequency omnidirectional radio range.
VORDAC: VOR distance measuring equipment for average coverage.
VOR/DMET: very high frequency omnirange/distance measuring equipment compatible with TACAN.
VORTAC: very high frequency omnirange TACAN.
VOX: voice-operated transmitter keyer.
VP: vent pipe.
VP: vertical polarization.
VP: vulnerable point.
VPE: vapor phase epitaxial.
VPI: vacuum pressure impregnation.
VPI: Virginia Polytechnic Institute.
VPLCC: vehicle propellant loading control center.
VPM: vehicles per mile.
VPM: volts per mile.
VPRF: variable pulse repetition frequency.

VR: voltage regulator.
VR: voltage relay.
VRB: VHF recovery beacon.
VRC: vertical redundancy check.
VRC: visual record computer.
VRFWQ: vehicle rapid fire weapons system successor.
VRIC: Variable Resistive Components Institute.
VRL: Vibration Research Laboratory.
VRR: visual radio range.
VRSA: voice-reporting signal assembly.
VS: variable speed.
VS: variable sweep.
VS: virtual storage.
VS: voltmeter switch.
VSAM: virtual storage access method.
VSB: vestigial sideband.
VSC: vibration safety cutoff.
VSCF: variable - speed - constant frequency.
VSI: vertical speed indicator.
VSM: vestigial sideband modulation.
VSMF: visual search microfilm file.
VSS: variable stability system.
VSTOL: vertical and short take-off and landing.
VSWR: voltage standing wave ratio.
VT: vacuum tube.
VT: variable time.
VT: Vehicular Technology (IEEE Group).
VT: Vermont.
VT: vertical tail.
VT: video tape.
VT: voice tube.
VTAM: virtual telecommunications access method.
VTCS: vehicular traffic control system.
VTD: vertical tape display.
VTDC: Vacuum Tube Development Committee.
VTF: vertical test fixture.
VTL: variable threshold logic.
VTM: voltage tunable magnetron.
VTO: voltage tuned oscillator.
VTR: video tape recording.
VTS: vertical test stand.
VTVM: vacuum tube voltmeter.
VU: vehicle unit.
VU: voice unit.
VU: volume unit.
VUTS: verification unit test set.
VWL: variable word length.
VWSS: vertical wire sky screen.

W

WA: Washington.
WAC: World Aeronautical Chart.
WACS: workshop attitude control system.
WADC: Wright Air Development Center.
WADD: Wright Air Development Division.
WADS: wide area data service.

WAF: wiring around frame.
WAGR: wind-scale advanced gas-cooled reactor.
WAM: worth analysis model.
WAML: Wright Aero Medical Laboratory.
WAMOSCOPE: wave-modulated oscilloscope.
WANL: Westinghouse Astronuclear Laboratory.

WAP: work assignment procedure.
WARC: World Administrative Radio Conference.
WARLA: wide aperture radio location array.
WAT: weight, altitude and temperature.
WATS: wide area telephone service.
WB: Weather Bureau.
WB: wide band.
WBD: wire bound.
WB/NWRC: Weather Bureau/National Weather Records Center.
WBAN: Weather Bureau, Air Force—Navy.
WBCO: waveguide below cutoff.
WBCV: wide-band coherent video.
WBD: wide-band data.
WBDL: wide-band data link.
WBFM: wide-band frequency modulation.
WBGT: wet bulb globe thermometer.
WBIF: wide-band intermediate frequency.
WBL: wide-band limiting.
WBP: weather- and boilproof.
WBRS: wide-band remote switch (unit).
WBTS: wide-band transmission system.
WCAP: Westinghouse Commercial Atomic Power.
WCPSC: Western Conference of Public Services Commissioners.
WCS: writable control store.
WD: watt demand meter.
WDB: wide band.
WDC: World Data Center.
WE: write enable.
WEMA: Western Electronic Manufacturers Association.
WES: Women's Engineering Society.
WESCON: Western Electronic Show and Convention (Region 6, IEEE).
WESRAC: Western Research Application Center (USC).
WETAC: Westinghouse electronic tubeless analog computer.
WFCMV: wheeled fuel-consuming motor vehicle.
WF&Eq: Wave Filters and Equalizers (IEEE PHP Technical Committee).
WG: waveguide.
WG: wire gage.
WGBC: waveguide operating below cutoff.
Wh: watt-hour.
WHC: watt-hour meter with contact device.
WHDM: watt-hour demand meter.
WHL: watt-hour meter with loss compensator.
Whm: watt-hour meter.
WHOI: Woods Hole Oceanographic Institution.
WHT: watt-hour demand meter, thermal type.
WI: Wisconsin.
WIDE: wiring integration design.
WIF: water immersion facility.
WINCON: Winter Convention (IEEE).
WIND: weather information network and display.
WINDS: weather information network and display system.

WISTIC: Wisconsin University, Theoretical Chemistry Institute.
WJ: Watkins-Johnson (Corporation).
WJCC: Western Joint Computer Conference.
WKNL: Walter Kidde Nuclear Laboratories.
WL: water line.
WL: wavelength.
WM: wattmeter.
WMEC: Western Military Electronics Center.
WMO: World Meteorological Organization.
WMS: Waste Management System.
WMS: World Magnetic Survey.
WMSI: Western Management Science Institute.
WMSO: Wichita Mountains Seismological Observatory.
WNRC: Washington National Records Center.
WOM: write optional memory.
WORC: Washington Operations Research Council.
WOSAC: world-wide synchronization of atomic clocks.
WP: weatherproof (insulation).
WPAFB: Wright-Patterson Air Force Base (Dayton, Ohio).
Wpc: watts per candle.
WPC: World Power Conference.
WPCF: Water Pollution Control Federation.
WPI: Worcester Polytechnic Institute.
WPM: words per minute.
WPRL: Water Pollution Research Laboratory.
WPS: words per second.
WR: Wilson repeater.
WRA: Water Research Association.
WRAC: Willow Run Aeronautical Center.
WRAIR: Walter Reed Army Institute of Research.
WRC: Water Resources Committee.
WRL: Willow Run Laboratory.
WRQ: Westinghouse resolver/quantizer.
W(RC: Willow Run Research Center.
WRU: Western Reserve University.
WSE: Washington Society of Engineers.
WSE: Western Society of Engineers.
WSIT: Washington State Institute of Technology.
WSL: Warren Spring Laboratory.
WSMR: White Sands Missile Range.
WSP: water supply point.
WSR: weather search radar.
WSTF: White Sands Test Facility.
WTM: wind tunnel model.
WTS: world terminal synchronous.
WU: Western Union.
WUIS: work unit information system.
WV: West Virginia.
WV: working voltage.
WVDC: working voltage, direct current.
WW: wireway.
WW: wire-wound.
WWMCCS: world-wide military command control computer system (trademark of Honeywell, Phoenix, Ariz.).
WWW: World Weather Watch (WMO).
WY: Wyoming.

X

XACT: X automatic code translation.
XBT: expendable bath-thermograph.
XCONN: cross connection.
XDCR: transducer.
XDP: X-ray density probe.
XE: experimental engine.
XECF: experimental engine—cold flow.
XEG: X-ray emission gauge.
XGAM: experimental guided aircraft missile.
XHAIR: cross hair.
XHMO: extended Huckel molecular orbit.
XHV: extreme high vacuum.
XLR: experimental liquid rocket.
XM: experimental missile.
XMAS: extended mission Apollo simulation.
XMFR: transformer.
XMIT: transmit.
XMSN: transmission.
XMTD: transmitted.
XMTG: transmitting.
XMTL: transmittal.

XMTR: transmitter.
XO: crystal oscillator.
XPDR: transponder.
XPL: explosive.
XPT: cross-point.
X ray: energetic high-frequency electromagnetic radiation.
XRM: X-ray microanalyzer.
XRPM: X-ray projection microscope.
XSECT: cross section.
XSM: experimental strategic missile.
XSONAD: experimental sonic azimuth detector.
XSPV: experimental solid propellant vehicle.
XT: cross talk.
XTAL: crystal.
XTASI: exchange of technical Apollo simulation information.
XTEL: cross tell.
XTS: cross-tell simulator.
XUV: extreme ultraviolet.
XWAVE: extraordinary wave.

Y

YAG: yttrium aluminum garnet.
YEA: Yale Engineering Association.
YIG: yttrium iron garnet.

YIL: yellow indicating light.
YSF: yield safety factor.
YSLF: yield strength load factor.

Z

ZA: zero adjusted.
ZA: zero and add.
ZAR: Zeus acquisition radar.
ZD: zero defect.
ZDC: Zeus Defense Center.
ZDCTBS: Zeus defense center tape and buffer system.
ZDP: zero delivery pressure.
ZDR: Zeus discrimination radar.
ZDT: zero-ductility transition.
ZEA: zero energy assembly.
ZEBRA: zero energy Breeder reactor assembly.
ZEEP: zero energy experimental file.
ZENITH: zero energy nitrogen heated thermal reactor.
ZERLINA: zero energy reactor for lattice investigations and study of new assemblies.
ZES: zero energy system.
ZETA: zero energy thermonuclear apparatus.
ZETR: zero energy thermal reactor.
ZEUS: zero energy uranium system.
ZFC: zero failure criteria.

ZFS: zero-field splitting.
ZG: zero gravity.
ZGE: zero gravity effect.
ZGS: zero gradient synchrotron.
ZIF: zero insertion force.
ZIP: zinc impurity photodetector.
zkW: zero kilowatt.
ZMAR: Zeus multifunction array radar.
ZMAR/MAR: Zeus multifunction array radar/multifunction array radar.
ZODIAC: zone defense integrated active capability.
ZPA: Zeus program analysis.
ZPEN: Zeus project engineer network.
ZPO: Zeus Project Office.
ZPPR: zero power plutonium reactor.
ZPR: zero power reactor.
ZPRF: zero power reactor facility.
ZS: zero and subtract.
ZURF: Zeus up-range facility.